D0757638

GLOSSARY OF GEOLOGY—Second Edition

Definition is that which refines
the pure essence of things from
the circumstance.

—Milton

It is not really a mark of distinction
for a geologist's writing to be so
obscure that a glossary is required
for its comprehension.

—Jules Braunstein

Second Edition

GLOSSARY OF GEOLOGY

Robert L. Bates and
Julia A. Jackson, Editors

American Geological Institute
Falls Church, Virginia
1980

Artwork at alphabet-section openings courtesy of the following: **A.** Stromatolites, western Australia (Australian Information Service); **B.** Skyline Arch, southeastern Utah (U.S. Geological Survey); **C.** Concretions weathering out of limestone, Egypt (Tad Nichols, Tucson, Ariz.); **D.** Cooling cracks in diabase sill, Ontario (Wilfried Meyer of Giant Yellowknife Mines Ltd); **E.** Salt polygons, Chile (Guillerma Chong Diaz of Universidad del Norte, Antofagasta); **F.** Permian brachiopods from West Texas (G. Arthur Cooper of the U.S. National Museum); **G.** Glacier in Alaska (from *Glacier Ice,* by Austin Post & Edward R. LaChapelle, University of Washington Press © 1971); **H.** Precambrian sill, Northwest Territories, Canada (Geological Survey of Canada); **I.** Ice at the terminus of Columbia Glacier, Alaska (U.S. Geological Survey); **J.** Chevron folds, northern Chile (Kenneth Segerstrom, U.S. Geological Survey); **K.** Ledge-forming sandstones, southwestern Colorado (sketch by W.H. Holmes, in F.V. Hayden's *Geological and Geographical Atlas of Colorado,* 1877); **L.** Ropy lava in the Galapagos Islands (Eileen Schofield of Ohio State University); **M.** Meanders on the Shenandoah River, near Woodstock, Va. (U.S. Geological Survey); **N.** The San Andreas Fault in Carrizo Plain, California (Robert E. Wallace, U.S. Geological Survey); **O.** Deformed ooids, Maryland (Ernst Cloos, Johns Hopkins University, Baltimore, Md.); **P.** Soil vein over an ice wedge in flood-plain sediments, Siberia (Troy L. Péwé, Arizona State University, Tempe); **Q.** Folded chert from Arkansas (John David McFarland III, Arkansas Geological Commission, Little Rock); **R.** Beach ridges, Florida (U.S. Geological Survey); **S.** Wind-eroded sandstone, northern Libya (L.K. Ratschiller, University of Trieste, Italy); **T.** Striations on stripped thrust surface, Cuzco, Peru (Edith S. Vincent, University of Southern California, Los Angeles); **U.** Synthetic silicate (Edwin Roedder, U.S. Geological Survey); **V.** Volcanic cone, Ecuador (Smithsonian Institution); **W.** Lower Whitewater Falls, South Carolina (Alan-Jon W. Zupan, South Carolina Geological Survey, Columbia); **X.** Square wave troughs, Puerto Rico (Watson H. Monroe, U.S. Geological Survey); **Y.** Microtektite with star-shaped pit (Ralph N. Baker, University of Delaware, Newark); **Z.** Craters on the far side of the Moon (NASA). These details from larger pictures have all, except **W**, appeared in *Geotimes.*

Preface

"If it were not for the occasional appearance of an authoritative glossary," wrote Ian Campbell in his preface to the first edition of this work, "our language . . . would rapidly degenerate into babel." He added that as the science advances and knowledge expands, we modify and improve our concepts, and with this must come modification and enlargement of our vocabulary.

The American Geological Institute had recognized the need for a glossary, and in 1957, in cooperation with the National Academy of Sciences, it published a 14,000-term *Glossary of Geology and Related Sciences.* A revision, with a 4,000-term supplement, appeared in 1960. These gave rise, in 1972, to AGI's one-volume *Glossary of Geology,* containing nearly 33,000 terms.

The present volume, the 36,000-term second edition, incorporates the modifications and growth of the geoscience vocabulary since 1972. Coverage has been expanded and updated, particularly in such active fields as biostratigraphy, caves and karst, igneous petrology, paleomagnetism, remote sensing, plate tectonics, and seismic stratigraphy. Some 450 new mineral names join the 4,000 in the previous edition; more than 100 abbreviations commonly used by geoscientists make their first appearance; and the list of references includes about 400 additional entries.

We hope this *Glossary* will meet the needs of the geoscience community, and will prove to be a bulwark against the babelization of the geological language.

December 1979

Julia A. Jackson
Robert L. Bates

Introduction

The terms listed in this *Glossary* have appeared in English-language publications, and reflect North American usage unless otherwise noted. Foreign-language terms are included if they have been used by writers in English. Many obsolete terms are retained, as they remain valuable for readers using the older literature. Besides giving the current or preferred meaning of a term, some definitions include information on original usage or historical development.

Alphabetization is strictly letter-by-letter. Thus *C wave* appears near the end of the C's, following *cuyamite.*

Italicized terms denote cross-references. An entry followed by a single term in italics means that the terms are synonyms and that the definition will be found under the italicized term. Terms italicized within a definition are themselves defined elsewhere in the *Glossary.*

Citations to the literature are given in many definitions. The works referred to are listed in the References Cited at the end of the book.

Multiple definitions for a term are indicated by (a), (b), (c), and so on. If there is a comment about a term that applies to all its meanings, it is given at the end of the definition after a long dash(—).

Pronunciation is indicated for a few terms, by simple phonetic means.

Brackets enclose tags for terms that have meanings in more than one subject. For example, head [coast], head [hydrogeol], and head [paleont] are each treated as a separate entry. Many tags, for example [coast], are self-explanatory, but many others are abbreviations, as follows:

arch—archaeology
astrogeol—astrogeology
astron—astronomy
biol—biology
bot—botany
cart—cartography
chem—chemistry
clim—climatology
cryoped—cryopedology
cryst—crystallography
drill—drilling
eco geol—economic geology
ecol—ecology
elect—electricity
eng—engineering
eng geol—engineering geology
evol—evolution
exp struc geol—experimental structural geology
geochem—geochemistry
geochron—geochronology
geog—geography
geol—geology
geomorph—geomorphology
geophys—geophysics
glac geol—glacial geology
glaciol—glaciology
grd wat—ground water
hydraul—hydraulics
ign—igneous
ign petrol—igneous petrology
intrus rocks—intrusive rocks
magnet—magnetism
mass move—mass movements
mater—materials
meta—metamorphism
meteorol—meteorology
migma—migmatites
mineral—mineralogy
mtns—mountains
oceanog—oceanography
ore dep—ore deposits
paleoclim—paleoclimatology
paleoecol—paleoecology
paleomag—paleomagnetism
paleont—paleontology
palyn—palynology
part size—particle size
pat grd—patterned ground
periglac—periglacial features
petrog—petrography

philos—philosophy
photo—photography
phys—physics
phys sci—physical science
planet—planetology
pyroclast—pyroclastics
rock mech—rock mechanics
sed—sedimentology
sed struc—sedimentary structures
seis—seismology
speleo—speleology
stat—statistics
stratig—stratigraphy
struc geol—structural geology
struc petrol—structural petrology
surv—surveying
taxon—taxonomy
tect—tectonics
topog—topography
volc—volcanism
wtr res—water resources
weath—weathering

Abbreviations of certain common terms are used in the definitions, as follows:

Abbrev:—abbreviation
adj.—adjective
Ant:—antonym
Cf:—compare (Latin *confer*)
e.g.—for example (Latin *exempli gratia*)
esp.—especially
et al.—and others (Latin *et alii*)
etc.—and so forth (Latin *et cetera*)
Etymol:—etymology
i.e.—that is (Latin *id est*)
n.—noun
Pl:—plural
q.v.—which see (Latin *quod vide*)
Pron:—pronounced
Sing:—singular
specif.—specifically
Syn:—synonym
v.—verb
var.—variant

Acknowledgments

The authority and accuracy of this *Glossary* derive from the specialized knowledge of nearly 150 members of the geoscience community. Each willingly gave the time needed to review definitions, search out and define new terms, recommend deletions, and cite references. Most of these reviewers consulted colleagues, whose help ranged from answering occasional questions to revising an entire subcategory of terms. Thus most of the definitions have been reviewed by at least two qualified persons.

The fields covered, and their reviewers, are listed below. For those entries followed by more than one name, the first name is that of the chief reviewer.

Archaeology Daniel F. Merriam; John A. Gifford, George R. Rapp, Jr.

Astrogeology Paul D. Lowman, Jr.

Cartography, surveying Douglas M. Kinney; William H. Chapman, William J. Jones, Richard D. Kidwell, John B. Rowland, Stanley H. Schroeder, Dale F. Stevens

Climatology, paleoclimatology Rhodes W. Fairbridge; William L. Donn, George J. Kukla

Coal geology Jack A. Simon; Richard D. Harvey

Economic geology, mining geology John M. Guilbert; Robert L. Frantz, Daniel A. Freiberg, Philip M. Giudice, Hans-Friedrich Krausse, George F. Sanders, Jr., Robert W. Schafer, Ty Schuiling, Walter S. White, William V. Yarter

Engineering geology Edward J. Cording; Harold C. Ganow

Environmental geology John C. Frye; Robert E. Bergstrom, Keros Cartwright, Donald O. Doehring, Paul B. DuMontelle, Leon R. Follmer, Robert A. Griffin, John P. Kempton, David E. Lindorff, Ronald W. Tank

Forensic geology Daniel F. Merriam; Raymond C. Murray

General geology Rhodes W. Fairbridge; Hanna Bremer

Geochemistry John Hower; Isidore Adler, Daniel E. Appleman, Alan M. Gaines, Gordon L. Nord, Jr.

Geochronology Francis H. Brown; John R. Bowman

Geomorphology Arthur L. Bloom; Joseph H. Hartshorn, Richard Jarvis, George W. Moore, Ernest H. Muller, Marie Morisawa, Thomas A. Mutch, Sidney E. White, Michael Woldenberg

Geophysics Robert E. Sheriff; Lawrence W. Braile, Michael D. Fuller, William J. Hinze, George V. Keller, Thomas R. LaFehr, Robert P. Lowell, Otto Nuttli, Floyd F. Sabins, Jr., Carl Savit, J. M. Wampler

Glaciers, snow, permafrost Robert F. Black; Jerry Brown, Anthony L. Gow, Joseph H. Hartshorn, Malcolm Mellor, Johannes Weertman, Wilford Weeks

History of geology Daniel F. Merriam; Cecil J. Schneer

Hydrology, hydrogeology Philip E. LaMoreaux; Charles Hains, Charles E. Herdendorf, Doyle B. Knowles, William J. Powell

Igneous and metamorphic petrology Richard V. Dietrich; Daniel S. Barker, Donald M. Burt, Bernard W. Evans, Grant H. Heiken, Richard A. Heimlich, Jean-Claude C. Mercier, Charles P. Thornton

Mathematical geology Daniel F. Merriam

Military geology Daniel F. Merriam; Arthur D. Howard

Mineralogy, crystallography Donald R. Peacor; Charles W. Burnham, Franklin F. Foit, Jr., R. Gaal

Oceanography, marine geology Orrin H. Pilkey, Jr.; Roger W. Baier, William J. Cleary,

Larry J. Doyle, Robert M. Goll, George W. Lynts, William J. Neal

Paleobotany, palynology James M. Schopf; Arthur Cronquist, Robert M. Kosanke

Paleontologic names, taxonomy James M. Schopf; Stig M. Bergström, James W. Collinson, Nicholas Hotton III, Leslie F. Marcus

Paleontologic morphology Donald B. Macurda; Nicholas Hotton III, Roger Batten, Bruce Bell, William Berry, Richard Boardman, Alan Cheetham, Donald Baird, Al Fagerstrom, Carol Faul, R. M. Feldmann, Cris Hughes, Robert Jull, Robert V. Kesling, N. Gary Lane, Matthew Nitecki, John Pojeta, J. Keith Rigby, Albert Rowell, Charles Smith, James Sprinkle, Curt Teichert

Paleontology, paleoecology Roger J. Cuffey; Niles Eldredge

Petroleum geology and technology Jules Braunstein; James A. Hartman, Robert E. Sheriff, Al Singleton

Photogeology Daniel F. Merriam; Alden Colvocoresses, Robert G. Reeves

Sedimentary petrology Walter D. Keller; Harvey Blatt, Frank G. Ethridge, Robert L. Folk, Gerard V. Middleton, Robert F. Schmalz

Soils, weathering M. E. Harward; J. H. Huddleston, G. F. Kling

Stratigraphy Hollis D. Hedberg; William A. Berggren, William A. Cobban, James A. Helwig, Grant Steele

Structural geology, tectonics Winthrop D. Means; Clark B. Burchfiel, Kevin C. Burke, B. E. Hobbs, A. M. Celal Sengör, Edgar W. Spencer, P. F. Williams

Many geoscientists, not formally involved with this revision, have taken the time to send in corrections to the 1972 edition, call attention to new terms and definitions, and provide citations to the literature. Much of this material has filled gaps and has added appreciably to the completeness of the book.

Wendell Cochran, of *Geotimes,* and Robert McAfee, Jr., an editor of the 1972 edition of the *Glossary,* gave advice and counsel on the editorial process. Margery A. Tibbetts, of the Orton Memorial Library of Geology at Ohio State University, tracked down many references to the literature from fragmentary or otherwise baffling citations. Jack Wolfire translated the editors' and designer's demands for the computer. Pauline Pennington, Lesa Read, and especially Kay Yost, expertly keyed the data.

GLOSSARY OF GEOLOGY—Second Edition

aa A Hawaiian term for lava flows typified by a rough, jagged, spinose, clinkery surface. Cf: *pahoehoe; block lava.* Pron: ah-ah. Obs. syn: *aphrolith.*
Aalenian Stage in Great Britain: lowermost Middle Jurassic or uppermost Lower Jurassic (above Yeovilian, below Bajocian).
AAR *accumulation-area ratio.*
a* axis That axis of a reciprocal crystal lattice which is perpendicular to (100). Cf: *b* axis; c* axis.*
***a* axis [cryst]** One of the crystallographic axes used as reference in crystal description. It is the axis that is oriented horizontally, front-to-back. In an orthorhombic or triclinic crystal, it is usually the *brachy-axis.* In monoclinic crystals, it is the *clinoaxis.* The letter *a* usually appears in italics. Cf: *b* axis; *c* axis.
***a* axis [struc petrol]** One of three orthogonal reference axes, *a, b,* and *c,* that are used in two distinct ways. (a) To help describe the geometry of a fabric possessing monoclinic symmetry. The unique symmetry plane is labelled the *a-c* plane, and *a* lies at the intersection of this plane with a prominent fabric surface. (b) In a kinematic sense, to describe a *deformation plan* that possesses monoclinic symmetry, such as progressive simple shear. In this case the *a* axis again lies in the unique plane of symmetry but parallel to the movement plane. It is the direction of maximum displacement and is commonly referred to as the direction of tectonic transport. In a progressive simple shear the *a* axis is the direction of shear. See also: *b* axis; *c* axis. Syn: *a* direction.
abactinal Referring to the *aboral* aspect; e.g. pertaining to the upper side of the test of an echinoid or to the side of a crinoid theca or plate opposite the oral surface. Ant: *actinal.*
abandoned channel (a) A drainage channel along which runoff no longer occurs, as on an alluvial fan. (b) *oxbow.*
abandoned cliff A sea cliff that is no longer undergoing wave attack, as a result of a relative drop of sea level or progradation at the cliff base.
abandoned meander *cutoff meander.*
abapertural Away from the aperture of a gastropod, nautiloid, or tentaculite shell. Ant: *adapertural.*
abapical Away from the apex of a gastropod shell and toward the base, along the axis of spiral or slightly oblique to it.
abathochroal eye A trilobite eye superficially similar to a *schizochroal eye* but possessing no scleral projections.
abaxial Directed or facing away from, or situated on the outside of, the axis or center of the axis, as of an organ, plant, or invertebrate; dorsal or anterior. Also, said of the abaxial side. Ant: *adaxial.*
Abbe refractometer An instrument used for determining the refractive index of liquids, minerals, and gemstones. Its operation is based on measurement of the *critical angle.*
abbreviation Loss of the final ontogenetic stages during the course of evolution.
ABC soil A soil with a distinctly developed profile, including A, B, and C horizons.
ABC system A method of correcting seismic data for the effect of irregular thickness of the surficial *low-velocity layer.* It is based on reciprocal placement of shot holes and seismometers.
abdomen (a) The posterior and often elongated region of the body of an arthropod, behind the thorax or cephalothorax; e.g. the tagma following the thorax of a crustacean, including the telson, and consisting of seven or fewer segments; or the complete, usually unsegmented part of the body of an arachnid or merostome, following the cephalothorax. Cf: *opisthosoma; pygidium.* (b) The third joint of the shell of a nasselline radiolarian.—Pl: *abdomens* or *abdomina.* Adj: *abdominal.*
abelsonite A triclinic mineral: $C_{32}H_{36}N_4Ni$ (nickel porphyrin).
aber The mouth of a river, or the confluence of two rivers. Etymol: Celtic.
abernathyite A mineral: $K(UO_2)(AsO_4)\cdot 4H_2O$.
aberration Any disturbance of the rays of a pencil of light such that they cannot be brought to a sharp focus or form a clear image.
AB interray Right anterior interray in echinoderms situated between A ray and B ray and clockwise of A ray when the echinoderm is viewed from the adoral side; equal to interambulacrum 3 of the Lovenian system.
abiogenesis The development of living organisms from lifeless matter. Cf: *biogenesis.*
abioglyph A hieroglyph of inorganic origin (Vassoevich, 1953, p.38). Cf: *bioglyph.*
ablation [geomorph] Separation and removal of rock material, and formation of residual deposits, esp. by wind action or the washing away of loose and soluble materials. Most writers prefer to restrict the term to wasting of glaciers by melting and evaporation.
ablation [glaciol] (a) All processes by which snow and ice are lost from a glacier, floating ice, or snow cover. These processes include melting, evaporation (sublimation), wind erosion, and calving. Sometimes calving is excluded, or ablation may be restricted to surface phenomena. Cf: *accumulation [glaciol].* (b) The amount of snow or ice removed by the process of ablation.—Syn: *wastage [glaciol].*
ablation [meteorite] Removal of molten surface layers of meteorites and tektites by direct vaporization during flight through the atmosphere.
ablation area The part of a glacier or snowfield in which, over a year's time, ablation exceeds accumulation; the region below the *equilibrium line.* Cf: *accumulation area.* Syn: *zone of ablation.*
ablation breccia *solution breccia.*
ablation cave A *glacier cave,* a few meters in height and width, formed near a glacier terminus by circulating warm air, generally where a meltwater stream flows from beneath the ice.
ablation debris Rock material of all sizes, from blocks to clay, as isolated fragments or discontinuous piles on the glacier surface; not continuous enough to form *ablation moraine.* The term is also used for that material comprising ablation moraine.
ablation form A feature formed on a surface of snow, firn, or ice by melting or evaporation; e.g. *nieve penitente* or *ice pyramid.*
ablation funnel A closed depression, similar to a solution channel, formed by solution processes or by removal of particulate material by circulating ground water.
ablation gradient The change in ablation with altitude on a glacier, usually expressed as millimeters of water equivalent per meter of altitude. Also used incorrectly to specify change of net balance with altitude up to the equilibrium line. Cf: *activity index.*
ablation moraine An uneven pile or continuous layer of *ablation till* or *ablation debris,* either overlying ice in the ablation area or resting on ground moraine derived from the same glacier.
ablation rate The amount of ice or snow loss per unit time from a glacier, floating ice, or snow cover. Usually expressed in millimeters of water equivalent per hour or day.
ablation season *summer season.*
ablation till Loosely consolidated rock debris, formerly in or on a glacier, that accumulated in place as the surface ice was removed by ablation.
ablatograph An instrument that measures the distance through which the surface of snow, ice, or firn changes, because of *ablation* and *regelation* during a given period.
ablykite A clay-mineral material consisting of an aluminosilicate of magnesium, calcium, and potassium. It resembles halloysite in its dehydration characteristics but differs from it in its thermal and X-ray diffraction properties. Syn: *ablikite.*

Abney level A *hand level* consisting of a short telescope, a movable bubble tube, and a graduated vertical arc to which a bubble is attached, and used to measure vertical angles. Named after William de Wiveleslie Abney (1843-1920), English physicist.

abnormal Said of an anticlinorium in which the axial surfaces of the subsidiary folds converge upwards; said of a synclinorium in which the axial surfaces of the subsidiary folds converge downwards. Cf: *normal [fold].*

aboral (a) Located opposite to or directed away from the mouth of an invertebrate; esp. applied to the *abactinal* surface (or to the structures on it) opposite that bearing the mouth and proximal ends of the ambulacral grooves of an echinoderm; or to the part of an echinoderm theca or plate directed away from the mouth (directed downward in an edrioasteroid). Cf: *adoral.* (b) Toward the underside of a conodont element; e.g. "aboral edge", "aboral groove", or "aboral attachment scar".—Ant: *oral.*

aboral margin The trace of the aboral side of a conodont element in lateral view. The term has also been used for the aboral side itself.

aboral pole (a) The end of a flask-shaped chitinozoan that includes the chamber of the body and the base. Cf: *oral pole.* (b) The point of intersection of the oral-aboral axis with the aboral surface of the theca of echinoderms; it marks the center of the aboral surface.

aboral side The underside of a conodont element, to which the basal plate is attached or onto which the basal cavity or attachment scar opens. Cf: *oral side.*

aboral surface Morphologically related unit of an edrioasteroid theca distal to the oral surface plates; commonly forms only part of the lower side of the individual.

aboriginal Said of the original race, fauna, or flora of a particular area, as distinguished from later immigrant or imported forms. Noun: *aborigine.*

abrasion The mechanical wearing, grinding, scraping, or rubbing away (or down) of rock surfaces by friction and impact, in which the solid rock particles transported by wind, ice, waves, running water, or gravity are the tools of abrasion. The term *corrasion* is essentially synonymous. Also, an abraded place or the effect of abrading, such as the abrasion left by glacial action. Verb: *abrade.* Cf: *attrition.*

abrasion pH A term proposed by Stevens & Carron (1948) to designate the characteristic *pH* achieved by a suspension of a pulverized mineral in water, resulting from a complex of hydrolysis and dissolution reactions.

abrasion platform An extensive, gently seaward-sloping intertidal surface produced by long-continued wave erosion. Term introduced by D. W. Johnson (1916, p. 444); see also Bradley & Griggs (1976). Cf: *erosion platform.* See also: *wave-cut platform; plain of marine erosion.*

abrasion shoreline *retrograding shoreline.*

abrasion tableland A broad, elevated region in which the outcrops of various rocks have been reduced to nearly the same level by denuding agents (Stamp, 1961, p. 2).

abrasive [geomorph] n. A rock fragment, mineral particle, or sand grain used by natural agents in abrading rock material or land surfaces.—adj. Possessing the characteristics of a tool for grinding or mechanical wear.

abrasive [mater] Any natural or artificial substance suitable for grinding, polishing, cutting, or scouring. Natural abrasives include diamond, emery, garnet, silica sand, diatomite, and pumice; manufactured abrasives include esp. silicon carbide, fused alumina, and boron nitride.

abrolhos A term used in Brazil for a mushroom-shaped barrier reef spreading widely near the surface. Etymol: Portuguese, "breakers; pointed rocks".

absarokite A basaltic rock, composed of phenocrysts of olivine and clinopyroxene in a groundmass of labradorite with alkali feldspar rims, olivine, and some leucite. Absarokite grades into *shoshonite* with a decrease in the olivine content and with the presence of some dark-colored glass, and into *banakite* with a decrease in the olivine and augite. It was named by Iddings in 1895 from the Absaroka Range, Wyoming.

abscission Separation of plant parts, e.g. of a leaf from a stem, usually by cell-wall dissolution along a certain layer (*abscission layer*).

abscission layer The zone of cells, e.g. at the base of a petiole, along which separation of plant parts occurs. Syn: *separation layer.*

absite A thorian variety of brannerite.

absolute abundance The exact number of individuals of a taxon in a certain area or volume. See also: *abundance [ecol]; relative abundance.*

absolute age The *geologic age* of a fossil organism, rock, or geologic feature or event given in units of time, usually years. Commonly used as a syn. of *isotopic age* or *radiometric age,* but may also refer to ages obtained from tree rings, varves, etc. Term is now in disfavor as it implies a certainty or exactness that may not be possible by present dating methods, i.e. two absolute ages for the same pluton may disagree by hundreds of millions of years. Cf: *relative age.* Syn: *actual age.*

absolute age determination Calculation of *absolute age,* usually but not always on the basis of radioactive isotopes. The ratio of decay products to parent products in the sample is calibrated to a certain number of years as in the *age equation.*

absolute chronology *Geochronology* in which the time-order is based on *absolute age,* usually measured in years by radiometric dating, rather than on superposition and/or fossil content as in *relative chronology.*

absolute date The date of an event usually expressed in years (*absolute age*) and related to a specific time scale.

absolute-gravity instrument A device for measuring the true value of gravity at a point. This type of measurement is much more difficult than relative determinations, because all the physical influences must be evaluated with extreme accuracy. The measurements are accomplished by various forms of reversible pendulums or by timing the motion of a body in free fall. Cf: *relative-gravity instrument.*

absolute humidity The content of water vapor in air, expressed as the mass of water per unit volume of air. Cf: *relative humidity.*

absolute permeability The ability of a rock to conduct a fluid, e.g. gas, at 100% saturation with that fluid. See also: *effective permeability; relative permeability.*

absolute pollen frequency An estimate of the actual amount of pollen deposited per unit area in a given length of time, achieved by correcting the amount of pollen per gram of sediment by factors based on rate of sedimentation. Abbrev: APF.

absolute time *Geologic time* measured in years; specif. time as determined by radioactive decay of elements. Jeletzky (1956, p. 681) proposed that the term be abandoned because its usage, based on criteria peculiar to the Earth and having the present part of geologic history as its starting point, is "incorrect and highly misleading". Cf: *relative time; mineral time; physical time.*

absolute viscosity *viscosity coefficient.*

Absonderung A term, now obsolete, applied by Leonhard in 1823 to the parting in igneous rocks that divides them into more or less regular bodies. The parting results from fractures that developed as a cooling phenomenon (Johannsen, 1939, p.163). Etymol: German, "separation, division".

absorbed water (a) Water retained mechanically within a soil mass and having properties similar to those of ordinary water at the same temperature and pressure. (b) Water entering the lithosphere by any means. Cf: *adsorbed water.*

absorbing well *drainage well.*

absorptance The ratio of the energy absorbed by a material to that incident upon it. Syn: *absorption coefficient.*

absorption Taking up, assimilation, or incorporation; e.g. of liquids in solids or of gases in liquids. Cf: *adsorption.* Syn: *occlusion.*

absorption [grd wat] The entrance of surface water into the lithosphere by any method. Verb: *to absorb.* Cf: *adsorption.*

absorption [optics] The reduction of light intensity in transmission through an absorbing substance or in reflection from a surface. In crystals, the absorption may vary with the wavelength or vibration direction of the transmitted light.

absorption [phys] Any mechanism by which energy, e.g. electromagnetic or seismic, is converted into heat.

absorption band The wavelength interval at which electromagnetic radiation is absorbed by the atmosphere or by other media, e.g. an atmospheric absorption band at 5 to 8 μm, caused by water vapor that absorbs thermal infrared radiation of those wavelengths. Cf: *absorption spectrum; absorption line.*

absorption coefficient *absorptance.*

absorption edge The wavelength at which there is an abrupt change in the intensity of an *absorption spectrum.* The term is usually applied to X-ray spectra.

absorption line Any of the dark lines in the *absorption spectrum* of a substance due to certain wavelengths in the spectrum being selectively absorbed on passing through a medium. Cf: *absorption* *band.*

absorption loss Water lost through absorption by rock and soil during the initial filling of a reservoir.

absorption spectroscopy The observation of an *absorption spectrum* and all processes of recording and measuring that go with it.

absorption spectrum The array of absorption bands or lines seen when a continuous spectrum is transmitted through a selectively absorbing medium. Cf: *atomic absorption spectrum.*

absorptivity The ability of a material to absorb energy incident upon it.

abstraction [streams] The merging of two or more subparallel streams into a single stream course, as a result of competition between adjacent consequent gullies and ravines, as by the deepening and widening of one channel so that it absorbs a shallower and smaller one nearby; the simplest type of *capture.* It usually occurs at the upper end of a drainage line. Syn: *stream abstraction.*

abstraction [water] That part of precipitation that does not become direct runoff (e.g. interception, evaporation, transpiration, depression storage, infiltration). Cf: *precipitation excess; rainfall excess.*

abtragung The part of degradation not resulting directly from stream erosion, i.e. preparation and reduction of rock debris by weathering and transportation of waste (Von Engeln, 1942, p. 265). Etymol: German *Abtragung,* "degradation; denudation".

abukumalite *britholite-(Y).*

Abukuma-type facies series Rocks produced in a type of dynamothermal regional metamorphism named after the Central Abukuma plateau of Japan, and characterized by the index minerals (in order of increasing metamorphic grade) biotite - andalusite - cordierite - sillimanite, representing the greenschist and amphibolite or hornblende-hornfels facies. Pressures are rather low, approaching those in contact metamorphism, i.e. 2500-3500 bars (Hietanen, 1967, p.192). Cf: *Buchan-type facies series.*

abundance [ecol] In ecology, the number of individuals of a particular taxon in a certain area or volume of sediment. See also: *absolute abundance; relative abundance.*

abundance [geochem] The mean concentration of an element in a geochemical reservoir, e.g. the abundance of Ni in meteorites, or the crustal abundance of oxygen. Also used for relative average content, e.g. the order of abundance of elements in the Earth's crust is O, Si, Al, Fe, Ca, etc.; the estimated cosmic abundance of Li in atoms per 10,000 atoms of Si is 1.0 (Suess & Urey, 1956).

abundant In the description of coal constituents, 30-60% of a particular constituent occurring in the coal (ICCP, 1963). Cf: *rare; common; very common; dominant.*

abyss [geomorph] *chasm.*

abyss [oceanog] *deep [oceanog].*

abyssal [intrus rocks] Pertaining to an igneous intrusion that occurs at considerable depth, or to the resulting rock; *plutonic.* Cf: *hypabyssal.*

abyssal [lake] Pertaining to the zones of greatest depth in a lake at which the water is "stagnant" or has a uniform temperature.

abyssal [oceanog] Pertaining to the ocean environment or *depth zone* of 500 fathoms or deeper; also, pertaining to the organisms of that environment.

abyssal benthic Pertaining to the benthos of the abyssal zone of the ocean. Syn: *abyssobenthic.*

abyssal cone A type of *submarine fan.*

abyssal deposit *pelagic deposit.*

abyssal fan *submarine fan.*

abyssal gap A passage that connects two abyssal plains of different levels, through which clastic sediments are transported. Syn: *gap [marine geol].*

abyssal hill A common low-relief feature of the ocean floor, usually found seaward of abyssal plains and in basins isolated by ridges, rises, or trenches. Abyssal hills range up to several hundred meters in height and several kilometers in diameter. About 85% of the Pacific Ocean floor and 50% of the Atlantic Ocean floor are covered by abyssal hills.

abyssal injection The rising of plutonic magma through deep-seated contraction fissures.

abyssal pelagic Pertaining to the open-ocean or pelagic environment at abyssal depths. Syn: *abyssopelagic.*

abyssal plain A flat region of the ocean floor, usually at the base of a continental rise, whose slope is less than 1:1000. It is formed by the deposition of turbidity-current and pelagic sediments that obscure the preexisting topography.

abyssal theory A theory of mineral-deposit formation involving the separation and sinking of minerals below a silicate shell during the cooling of the Earth from a liquid stage, followed by their transport to and deposition in the crust as it was fractured (Shand, 1947, p.204). Modern thought has completely negated such theories.

abyssal tholeiite *oceanic tholeiite.*

abyssobenthic *abyssal benthic.*

abyssolith *batholith.*

abyssopelagic *abyssal pelagic.*

acadialite A flesh-red variety of chabazite, found in Nova Scotia.

Acadian North American provincial series: Middle Cambrian (above Georgian, below Potsdamian). Obsolete syn. of *Albertan.*

Acadian orogeny A Middle Paleozoic deformation, especially in the northern Appalachians; it is named for Acadia, the old French name for the Canadian Maritime Provinces. In Gaspé and adjacent areas the climax of the orogeny can be dated by limiting strata as early in the Late Devonian, but deformational, plutonic, and metamorphic events were prolonged over a more extended period; the last two have been dated radiometrically as between 330 and 360 m.y. ago. The Acadian had best be regarded, not as a single orogenic episode, but as an orogenic era in the sense of Stille. Cf: *Antler orogeny.*

acantharian Any radiolarian belonging to the suborder Acantharina, characterized by a centrogenous skeleton composed of strontium sulfate and a central capsule enclosed by a thin simple membrane.

acanthine septum A corallite septum composed of a vertical or steeply inclined series of *trabeculae* and commonly marked by spinose projections along the axially directed margin of the septum.

acanthite A monoclinic mineral: Ag_2S. It is dimorphous with argentite and constitutes an ore of silver.

Acanthodii A subclass of the Osteichthyes characterized by fixed paired fins supported anteriorly by spines; more than two pairs are usually present. It includes the oldest recorded gnathostomes (Upper Silurian). Stratigraphic range, Upper Silurian to Lower Permian.

acanthopore A small rodlike skeletal structure, originally believed to be hollow, consisting of a solid core surrounded by a sheath of cone-in-cone laminae lying within zooecial walls or extrazooidal skeleton in stenolaemate bryozoans. Acanthopores form spinelike projections at the colony surface.

acanthostyle A monaxonic sponge spicule (*style*) covered with short or tiny spines over most of its surface.

acanthus A secondary deposit in the chamber floor of certain foraminifers (such as *Endothyra*), sharply pointed but not curved toward the anterior (TIP, 1964, pt.C, p.58). Pl: *acanthi.*

acarid Any arachnid belonging to the order Acarida, characterized by the absence of abdominal segmentation but with subdivision of the body into a proterosoma and hysterosoma. Their stratigraphic range is Devonian to present.

acaustobiolith A noncombustible organic rock, or a rock formed by the organic accumulation of purely mineral matter (Grabau, 1924, p. 280). Cf: *caustobiolith.*

acaustophytolith An *acaustobiolith* formed by plant activity; e.g. a pelagic ooze containing diatoms, and a nullipore reef or limestone.

accelerated development The production of a landscape where the rate of uplift is more rapid than the rate of downward erosion or where valley deepening exceeds valley widening, characterized by an increase of the relative relief and the formation of convex slopes. Cf: *declining development; uniform development.* Syn: *waxing development; ascending development.*

accelerated erosion Erosion occurring in a given region at a greater rate than *normal erosion*, usually brought about by the influence of man's activities in disturbing or destroying the natural cover, thus sharply reducing resistance of the land surface and rate of infiltration. It may result from deforestation, improper cultivation of soil, dry-farming, overgrazing of rangelands, burning and clearance of natural vegetation, excavation for buildings and highways, urbanization of drainage areas, strip mining, or copper smelting; and by nonhuman influences, such as lightning or rodent invasion.

acceleration (a) During evolution, the appearance of modifications earlier and earlier in the life cycle of successive generations; adult characters of the ancestor appear earlier in immature stages of the descendants (*tachygenesis*), sometimes to the point that certain steps are omitted (*brachygenesis*). (b) In Paleozoic corals, the addition of more secondary septa in one pair of quadrants than in the other pair.

acceleration due to gravity The acceleration of a freely falling body in a vacuum as a result of gravitational attraction. Although its true value varies with altitude, latitude, and the nature of the underlying rocks, the standard value of 980.665 cm/sec^2 has been adopted by the International Committee on Weights and Measures.

accelerometer A seismometer whose response is linearly proportional to the acceleration of the earth materials with which it is in contact.

accented contour *index contour.*

accessory [mineral] *accessory mineral.*

accessory [paleont] adj. Said of a secondary or minor element of an ammonoid suture; e.g. "accessory lobe" or "accessory saddle". Cf: *auxiliary.*—n. Such a lobe or saddle.

accessory [pyroclast] Said of pyroclastics that are formed from fragments of the volcanic cone or earlier lavas; it is part of a classification of volcanic ejecta based on mode of origin, and is equivalent to *resurgent* ejecta. Cf: *essential; accidental.* See also: *cognate [pyroclast].*

accessory aperture An opening in the test of a planktonic foraminifer that does not lead directly into a primary chamber but extends beneath or through accessory structures (such as bullae and tegilla); e.g. a *labial aperture,* an *infralaminal accessory aperture,* and an *intralaminal accessory aperture.*

accessory archeopyle suture An *archeopyle suture* that consists of a short cleft in the wall adjacent to the principal suture, or that may be more fully developed on the operculum of the dinoflagellate cyst, dividing that structure into two or more separate pieces.

accessory comb The line of large cilia within the preoral cavity in a tintinnid.

accessory element *trace element.*

accessory mineral A mineral whose presence in a rock is not essential to the proper classification of the rock. Accessory minerals generally occur in minor amounts; in sedimentary rocks, they are mostly *heavy minerals.* Cf: *essential mineral.* Syn: *accessory [mineral].*

accessory muscle (a) A convenient noncommittal term for any muscle of a bivalve mollusk (other than an adductor muscle or a muscle withdrawing marginal parts of the mantle) of uncertain origin and having a scar of attachment to the shell. (b) One of a pair of diductor muscles branching posteriorly and ventrally from the main diductor muscles of a brachiopod and inserted in the pedicle valve posterior to the adductor bases (TIP, 1965, pt.H, p.139).

accessory spore A spore present in a rock only in very small quantities. Accessory spores may contain types with a restricted range and they have been used for correlation and for zoning (as of coal measures).

accident (a) A departure from the normal cycle of erosion, caused by events that occur "arbitrarily as to place and time", such as climatic changes and volcanic eruptions (Davis, 1894). Cf: *interruption.* (b) An event, such as drowning, rejuvenation, ponding, or capture, that interferes with, or entirely puts an end to, the normal development of a river system (Scott, 1922, p. 188). (c) An irregular feature in, or an undulation of, a land surface.

accidental Said of pyroclastics that are formed from fragments of nonvolcanic rocks or from volcanic rocks not related to the erupting volcano; it is part of a classification of volcanic ejecta based on mode of origin, and is equivalent to *allothigenous* ejecta. Cf: *cognate; accessory; essential.* Syn: *noncognate.*

accidental error An unpredictable error that occurs without regard to any known mathematical or physical law or pattern and whose occurrence is due to chance only; e.g. an error ascribed to uncontrollable changes of external conditions. Syn: *random error.*

accidental inclusion *xenolith.*

accidented relief Rugged and irregular relief; probably a literal translation of the common French term *relief accidenté* (Stamp, 1961, p. 4).

acclimation *acclimatization.*

acclimatization Physiologic adjustment by an organism to a change in its immediate environment. Syn: *acclimation.*

acclinal A syn. of *cataclinal.* Term used by Powell (1873, p. 463). Not to be confused with *aclinal.*

acclivity A slope that ascends from a point of reference. Ant: *declivity.*

accordance of summit levels *summit concordance.*

accordant Said of topographic features that have the same or nearly the same elevation; e.g. an *accordant* valley whose stream enters the main stream at the same elevation as that of the main stream. Ant: *discordant.*

accordant drainage Drainage that has developed in a systematic relationship with, and consequent upon, the present geologic structure. Ant: *discordant drainage.* Syn: *concordant drainage.*

accordant fold One of several folds having similar orientation.

accordant junction The joining of two streams or two valleys whose surfaces are at the same level at the place of junction. See also: *Playfair's law.* Ant: *discordant junction.* Syn: *concordant junction.*

accordant summit level A hypothetical level or gently sloping surface that regionally intersects hilltops or mountain summits. Accordant summit levels in a region of high topographic relief suggest that the summits are remnants of an erosion plain formed in a previous erosion cycle. See also: *summit concordance; even-crested ridge.* Syn: *concordant summit level.*

accordant summits Hilltops or mountain peaks that regionally reach the same hypothetical level or gently sloping surface. The term cannot be used in the singular.

accordion fold An old term, formerly used with genetic significance; now sometimes used as a syn. of *kink fold.* See also: *zigzag fold; chevron fold.* Syn: *angular fold; concertina fold.*

accreting plate boundary A boundary between two *plates* that are moving apart, with new oceanic-type lithosphere being created at the seam (Dennis & Atwater, 1974, p. 1033). See also: *mid-oceanic ridge.* Syn: *divergent plate boundary.*

accretion [planet] The process whereby small particles and gases in the solar nebula came together to form larger bodies, eventually of planetary size.

accretion [sed] (a) The gradual or imperceptible increase or extension of land by natural forces acting over a long period of time, as on a beach by the washing-up of sand from the sea or on a flood plain by the accumulation of sediment deposited by a stream. Legally, the added land belongs to the owner of the land to which it is added. Cf: *avulsion; reliction.* See also: *lateral accretion; vertical accretion.* Syn: *aggradation; alluvion.* (b) The land so added or resulting from accretion. (c) *continental accretion.*

accretion [sed struc] (a) The process by which an inorganic body increases in size by the external addition of fresh particles, as by adhesion. (b) A *concretion;* specif. one that grows from the center outward in a regular manner by successive additions of material (Todd, 1903). (c) Deposition of eolian sand on a continuous sand surface because of a decrease in wind intensity or an increase in surface roughness (Bagnold, 1941, p.127).

accretion [stream] The filling-up of a stream bed, due to such factors as silting or wave action. Cf: *degradation [stream].*

accretion [struc geol] *continental accretion.*

accretionary Tending to increase by external addition or accumulation; esp. said of a secondary sedimentary structure produced by overgrowth upon a preexisting nucleus, such as a rounded form that originated through rolling, or said of a limestone formed in place by slow accumulation of organic remains.

accretionary lapilli More or less spherical masses, mostly between 1 mm and 1 cm in diameter, of cemented ash; the cementation is often weak. Formed by accretion of particles around wet nuclei, e.g. raindrops falling through a cloud of ash (Macdonald, 1972, p. 133). Syn. for individual mass: *pisolite [volc]; tuff ball.*

accretionary lava ball A rounded mass, ranging in diameter from a few centimeters to several meters, formed on the surface of a lava flow such as *aa,* or on *cinder-cone* slopes, by the molding of viscous lava around a core of already solidified lava.

accretion ridge A beach ridge located inland from the modern beach, representing an ancient beach deposit and showing that the coast has been built out seaward (Fisk, 1959, p. 111). It is often accentuated by the development of dunes.

accretion ripple mark An asymmetric ripple mark having a gentle and curved lee slope, with a maximum angle of dip less than the angle of repose, and composed of cross-strata without conspicuous sorting of particles (Imbrie & Buchanan, 1965, p.151 & 153). Cf: *avalanche ripple mark.*

accretion till *basal till.*

accretion topography A landscape built by accumulation of sediment.

accretion vein A type of vein in which the mineral deposit has been formed by repetition of channelway filling and reopening of the fractures.

accumulated discrepancy The sum of the separate *discrepancies* that occur in the various steps of making a survey or of the computation of a survey.

accumulation (a) All processes that add snow or ice to a glacier, floating ice, or snow cover, including snowfall, condensation, avalanching, snow transport by wind, and freezing of liquid water. Syn: *nourishment [glaciol]; alimentation.* Cf: *ablation [glaciol].* (b) The amount of snow and other solid precipitation added to a glacier or snowfield by the processes of accumulation.

accumulation area The part of a glacier or snowfield in which, over a year's time, accumulation exceeds ablation; the region above the *equilibrium line.* Cf: *ablation area; névé.* Syn: *firn field; accumulation zone; zone of accumulation [snow].*

accumulation-area ratio The ratio of accumulation area to total area of a glacier for any given year, used as a rough guide to the balance between accumulation and ablation. Abbrev: AAR.

accumulation mountain *mountain of accumulation.*

accumulation rate The amount of ice or snow gain per unit time to a glacier, floating ice, or snow cover. Usually expressed in millimeters of water equivalent per hour or day.

accumulation season *winter season.*

accumulation zone (a) *accumulation area.* (b) The area in which the bulk of the snow contributing to an avalanche was originally deposited. Syn: *zone of accumulation.*

accumulative rock *cumulate.*

accumulator plant In geobotanical prospecting, a tree or plant that preferentially concentrates an element.

accuracy The degree of conformity with a standard, or the degree of perfection attained in a measurement. Accuracy relates to the quality of a result, and is distinguished from *precision,* which relates to the quality of the operation by which the result is obtained.

AC demagnetization *alternating-field demagnetization.*

acequia A Spanish word, of Arabic origin, for an irrigation ditch or canal.

acervuline Heaped, or resembling little heaps; e.g. said of some foraminifers (such as *Acervulina*) having chambers in irregular clusters.

acetamide A trigonal mineral: CH_3CONH_2.

acetolysis Any chemical reaction in which acetic acid plays a role similar to that of water in hydrolysis; e.g. a reaction used in maceration in which organic material such as peat is heated in a mixture of nine parts acetic anhydride and one part concentrated sulfuric acid. It breaks down cellulose especially vigorously.

ACF diagram A triangular diagram showing the simplified compositional character of metamorphic rocks and minerals by plotting the molecular quantities of the three components: $A=Al_2O_3 + Fe_2O_3 - (Na_2O + K_2O)$; $C=CaO - 3.3P_2O_5$; and $F=FeO + MgO + MnO$. A+C+F (in mols) are recalculated to 100%; the presence of excess SiO_2 is assumed. Cf: *AFM diagram; A'KF diagram.*

achene A dry one-seeded indehiscent fruit developed from a simple ovary with unfused seed coat and fruit wall. Also spelled: *akene.*

achlamydate Said of a gastropod without a mantle.

achoanitic Said of the condition in a nautiloid in which septal necks are vestigial or absent. Syn: *aneuchoanitic.*

achondrite A *stony meteorite* that lacks chondrules. Achondrites are commonly more coarsely crystallized than *chondrites,* and nickel-iron is almost completely lacking in most of them; they represent meteorites that are most like terrestrial rocks, with sizable fragments of various minerals visible to the naked eye. Adj: *achondritic.* Cf: *chondrite [meteorite].*

achroite A colorless variety of tourmaline, used as a gemstone.

acicular [cryst] Said of a crystal that is needlelike in form. Cf: *fascicular; sagenitic.*

acicular [sed] Said of a sedimentary particle whose length is more than three times its width (Krynine, 1948, p.142). Cf: *platy.*

acicular ice Freshwater ice consisting of numerous long crystals and hollow tubes having variable form, layered arrangement, and a content of air bubbles; it forms at the bottom of an ice layer near its contact with water. Syn: *fibrous ice; satin ice.*

aciculate Needle-shaped, or having a needlelike point; esp. said of a slender gastropod shell that tapers to a sharp point.

acid adj. (a) *silicic.* (b) *acidic.* (c) Said of a plagioclase that is sodic.

acid clay A clay that yields hydrogen ions in a water suspension; e.g. "Japanese acid clay", a variety of fuller's earth occurring in Kambara, Japan.

acidic (a) A descriptive term applied to those igneous rocks that contain more than 60% SiO_2, as contrasted with *intermediate* and *basic.* Sometimes loosely and incorrectly used as equivalent to *felsic* and to *oversaturated,* but these terms include rock types (e.g., nepheline syenite, quartz basalt) that are not generally considered acidic. This is not the chemist's usage; the term is deprecated by some because of its confusing nature. (b) Applied loosely to any igneous rock composed predominantly of light-colored minerals having a relatively low specific gravity. Cf: *felsic.* Syn: *acid; silicic.*

acidity coefficient *oxygen ratio.*

acidity quotient *oxygen ratio.*

acidization The process of forcing acid down a well into a limestone or dolomite, in order to increase permeability and porosity by dissolving a part of the rock constituents. It is also used to remove mud injected during drilling. The general objective of acidization is to increase oil productivity. Syn: *acid treatment.*

acid mine drainage Drainage with a pH of 2.0 to 4.5 from mines and mine wastes. It results from the oxidation of sulfides exposed during mining, which produces sulfuric acid and sulfate salts. The acid dissolves minerals in the rocks, further degrading the quality of the drainage water.

acid plagioclase A variety of plagioclase having a relatively high content of SiO_2; e.g. an Ab-rich member such as albite or oligoclase.

acid soil A soil with a pH of less than 7.0.

acid treatment *acidization.*

ac-joint A *cross joint* in folded sedimentary rock that is parallel to the fold axis.

aclinal A little-used term said of strata that have no inclination; horizontal. Not to be confused with *acclinal.*

acline A syn. of *orthocline* as used to describe the hinge teeth or shell body of a bivalve mollusk.

acline-A twin law *Manebach-Ala twin law.*

acline-B twin law A twin law for parallel twins in feldspar with twin axis *b* and composition plane (100). Cf: *x-Carlsbad twin law.*

aclinic line *magnetic equator.*

acme That point in the phylogeny of a species, genus, or other taxon at which greatest abundance and/or variety occurs. See also: *paracme.*

acme-zone A *biozone* consisting of "a body of strata representing the acme or maximum development — usually maximum abundance or frequency of occurrence — of some species, genus, or other taxon, but not its total range" (ISG, 1976, p. 59-60). It is named for the taxon whose zone of maximum development it delimits, e.g. *Didymograptus* acme-zone. The corresponding geologic-time unit is *hemera.* Cf: *assemblage-zone; range-zone.* Syn: *epibole; flood zone; peak-zone.*

acmite A brown or green mineral of the clinopyroxene group: $NaFe(SiO_3)_2$. It occurs in certain alkali-rich igneous rocks. Symbol: Ac. Syn: *aegirine.*

acmite-augite A mineral intermediate between augite and acmite; a variety of augite rich in sodium and ferric iron. Syn: *aegirine-augite.*

acmolith *akmolith*

acolpate Said of pollen grains without *colpi.* In practice, such pollen grains are sometimes difficult to distinguish from *alete* spores. Cf: *inaperturate.*

acoustic basement Generally the deepest more or less continuous seismic reflector, often an "acoustic unconformity", below which seismic energy returns are extremely poor to absent.

acoustic impedance The product of seismic velocity and density. Syn: *impedance [seis].*

acoustic log Generic term for a *well log* that displays any of several measurements of acoustic waves in rocks exposed in a borehole, e.g. compressional-wave transit time over an interval (*sonic log*) or relative amplitude (*cement-bond log*).

acoustics The study of sound, including its production, transmission, reception, and utilization, especially in fluid media such as air or water. With reference to Earth sciences, it is especially relevant to oceanography. The term is sometimes used to include compressional waves in solids, e.g. seismic waves.

acoustic wave A longitudinal wave. In common usage it is restricted to fluids such as air, but it often includes *P-waves* in the solid Earth. Syn: *sound wave; sonic wave.*

acquired character A character not inherited but acquired by an individual organism during its lifetime as a result of use or disuse according to its mode of life or the conditions under which it lived.

acre A unit of land area used in U.S. and England, equal to 43,560 sq ft, 4840 sq yd, 160 square rods, 10 square chains, 1/640 square mile, or 0.405 hectare. It is based on an old unit thought to be equal to the amount of land that could be plowed by a yoke of oxen in a day.

acre-foot The volume of a one-acre area (43,560 square feet) one foot thick, or 43,560 cubic feet. It is the unit commonly used in measuring volumes of water or reservoir storage space, and in measuring the volume of *reservoir rock* in an oil or gas field. See also: *acre-inch.*

acre-inch The volume of water required to cover one acre to a depth of one inch. See also: *acre-foot.*

acrepid Said of a *desma* (of a sponge) that lacks an axial canal, implying that it was not formed about a *crepis.*

acre-yield The average amount of oil, gas, or water recovered from 1 acre of a reservoir.

acritarch A unicellular, or apparently unicellular, resistant-walled microscopic organic body of unknown or uncertain biologic relationship and characterized by varied sculpture, some being spiny and others smooth. Many if not most acritarchs are of algal affinity, but the group is artificial. They range from Precambrian to Holocene, but are esp. abundant in Precambrian and early Paleozoic. The term was proposed by Evitt (1963, p.300-301) as "an informal, utilitarian, 'catch-all' category without status as a class, order, or other suprageneric unit" consisting of "small microfossils of unknown and probably varied biological affinities consisting of a central cavity enclosed by a wall of single or multiple layers and of chiefly organic composition". See also: *hystrichosphaerid; dinoflagellate.*

acrobatholithic Said of a mineral deposit occurring in or near an exposed batholith dome; also, said of the stage of batholith erosion in which that area is exposed (Emmons, 1933). The term is little used today. Cf: *cryptobatholithic; embatholithic; endobatholithic; epibatholithic; hypobatholithic.*

acrodont adj. Pertaining to vertebrate teeth fused to the occlusal margins of upper and lower jaws.

acrolamella A leaflike extension of the leasurae of megaspores. Cf: *gula; apical prominence.* Pl: *acrolamellae.*

acrolobe The central portion of the cephalon or pygidium of agnostid trilobites in which the axial furrows are effaced.

acron The anteriormost part of the cephalon of a crustacean, carrying the eyes and antennules.

acrotretacean Any inarticulate brachiopod belonging to the superfamily Acrotretacea, characterized by a conical to subconical, rarely convex, pedicle valve.

acrozone *range-zone.*

actinal Referring to the *oral* aspect; e.g. pertaining to the under or mouth side of the test of an echinoid or to the side of a crinoid theca or plate containing the mouth. Ant: *abactinal.*

actine (a) One of the individual branches of the triaene or *triode* in the *ebridian* skeleton. (b) A star-shaped spicule, as of a sponge.

actinium series The radioactive series beginning with uranium-235.

actinodont Said of the dentition of certain bivalve mollusks of early origin having hinge teeth radiating from the beak (the outer teeth being more or less elongate).

actinolite A bright-green or grayish-green monoclinic mineral of the amphibole group: $Ca_2(Mg,Fe)_5Si_8O_{22}(OH)_2$. It may contain manganese. Actinolite is a variety of *asbestos,* occurring in long, slender, needlelike crystals and also in fibrous, radiated, or columnar forms in metamorphic rocks (such as schists) and in altered igneous rocks. Symbol: Ac. Cf: *tremolite.*

actinometer Any device that measures the intensity of radiation capable of effecting photochemical changes, particularly the radiation of the Sun. Actinometers may be classified according to the quantities they measure. See also: *pyrheliometer; pyranometer; pyrgeometer.*

actinomorphic Said of an organism or organ that is radially symmetrical or capable of division into essentially symmetrical halves by any longitudinal plane passing through the axis. Cf: *zygomorphic.*

actinopod Any protozoan belonging to the class Actinopoda and characterized by protoplasmic extensions radiating from the spheroidal main body. Cf: *rhizopod.*

Actinopterygii A subclass of the Osteichthyes characterized by movable paired fins supported by bony rays; ray-finned fish. Range, Lower Devonian to present.

actinosiphonate Said of endosiphuncular structures of a nautiloid, consisting of radially arranged longitudinal lamellae.

actinostele A type of *stele* consisting of alternating or radial groups of xylem and phloem within a pericycle and having a star shape in cross section.

activation [clay] The act or process of treating clay (such as bentonite) with acid so as to improve its adsorptive properties or to enhance its bleaching action, as for use in removing colors from oils.

activation [radioactivity] The process of making a substance radioactive by bombarding it with nuclear particles. The radioactivity so produced is called *induced radioactivity.*

activation analysis A method of identifying stable isotopes of elements in a sample by irradiating the sample with neutrons, charged particles, or gamma rays to render the elements radioactive, after which the elements are identified by their characteristic radiations. Cf: *neutron activation.* Syn: *radioactivation analysis.*

activation energy The extra amount of energy which any particle or group of particles must have in order to go from one energy state into another, such as changes in phase and movement of particles in diffusion. The greater the amount of energy involved, the higher the resistance to the change, or the *potential barrier.*

active Said of a karst feature that contains moving water, or that is still being developed by the action of water.

active cave *live cave.*

active channel A channel, on an alluvial fan, in which runoff flows.

active earth pressure The minimum value of lateral *earth pressure* exerted by soil on a structure, occurring when the soil is allowed to yield sufficiently to cause its internal shearing resistance along a potential failure surface to be completely mobilized. Cf: *passive earth pressure.*

active fault A fault along which there is recurrent movement, which is usually indicated by small, periodic displacements or seismic activity. Cf: *dead fault; capable fault.*

active glacier (a) A glacier that has an *accumulation area,* and in which the ice is flowing. Ant: *dead glacier.* (b) A glacier that moves at a comparatively rapid rate, generally in a maritime environment at a low latitude where accumulation and ablation are both large. Ant: *passive glacier.*

active ice That part of a glacier showing clear evidence of movement, such as crevassing.

active layer [eng geol] The surficial deposit that undergoes seasonal changes of volume, swelling when frozen or wet and shrinking when thawing and drying.

active layer [permafrost] A surface layer of ground, above the permafrost, that is alternately frozen each winter and thawed each summer; it represents *seasonally frozen ground* on permafrost. Its thickness ranges from several centimeters to a few meters. Syn: *annually thawed layer; mollisol.*

active method (a) A seismic method that involves monitoring artificially induced signals. (b) A construction method in permafrost areas by which the frozen ground is thawed and removed or kept unfrozen at and near the structure.—Cf: *passive method.*

active patterned ground Patterned ground that is growing or still developing. Ant: *fossil patterned ground.*

active permafrost Permafrost that is able to revert to a perennially frozen state under present climatic conditions after having been thawed by natural or artificial means. Ant: *fossil permafrost.*

active remote sensing Remote-sensing methods that provide their own source of electromagnetic radiation. *Radar* is an example. Cf: *passive remote sensing.* Syn: *active system.*

active speleothem *live speleothem.*

active system *active remote sensing.*

active volcano A volcano that is erupting; also, one that is not now erupting but is expected to do so. There is no precise distinction between an active and a *dormant volcano.* Cf: *extinct volcano; inactive volcano.*

active water Water with corrosive capabilities.

activity [chem] (a) The ratio between the *fugacity* of a substance in some given state and its standard-state fugacity, f^o. The standard-state fugacity is unity for a gas or vapor; for solids and liquids, it is evaluated at each temperature as the fugacity of the pure substance at 1 atmosphere of pressure. Activity arises in consideration of chemical equilibrium problems involving solids or liquids. (b) The tendency of a substance to react spontaneously and energetically with other substances.

activity [radioactivity] The rate of radioactive decay of a substance, measured as the number of atoms decaying per unit of time. See also: *specific activity.*

activity coefficient The ratio of chemical *activity* to concentration.

activity index The rate of change with altitude of the *net balance* of a glacier, measured in the vicinity of the equilibrium line. High values indicate vigorous transfer of excess accumulation to lower altitudes (e.g. temperate maritime glaciers); low values indicate minimal or sluggish transfer (e.g. polar continental glaciers). Usually measured in millimeters per meter. The term was introduced by Meier in 1961. Syn: *energy of glacierization.* Cf: *ablation gradient.*

activity ratio In a sediment, the ratio of the *plasticity index* to the percentage of clay-sized minerals.

actual age *absolute age.*

actual horizon A great circle on the celestial sphere whose plane is perpendicular to the direction of the plumb line (or the direction of gravity) at the point of observation. It is usually referred to as the *horizon.* Syn: *rational horizon.*

actualism European term for *uniformitarianism,* but allowing the possibility of appreciable differences in the duration and intensity of geologic processes operating in the past.

actuopaleontology The paleontologic study of a present-day area, with the intent of discovering how it will appear later, after eventual burial and fossilization.

acute bisectrix In optically biaxial minerals, the direction bisecting the acute angle between optic axes. Cf: *obtuse bisectrix.*

acyclic In plant morphology, pertaining to attachment of three or more parts, e.g. leaves, in a spiral or helical pattern. Cf: *whorled.*

adamantine luster A brilliant mineral luster, characteristic of minerals with a high *index of refraction,* e.g. diamond and cerussite.

adamantine spar Silky brown *corundum.*

adambulacral Said of the plate adjacent to the ambulacral plate in the arm of an *asteroid.*

adamellite In English-speaking countries and the USSR, a syn. of *quartz monzonite* of U.S. usage (i.e. quartz as 10-50% of the felsic minerals, and a plagioclase/total feldspar ratio of 35-65%). The term was first used in this sense by Brögger in 1895; it was originally used, however, by Cathrein in 1890 for an orthoclase-bearing *tonalite* of Monte Adamello, Italy.

Adamic earth A term used for common clay, in reference to the material of which Adam, the first man, was made; specif. a kind of red clay (Humble, 1843, p. 4).

adamite A colorless, white, or yellow orthorhombic mineral: $Zn_2(AsO_4)(OH)$. It is dimorphous with paradamite. Originally spelled: *adamine.*

adapertural Toward the aperture of a gastropod or cephalopod shell. Ant: *abapertural.* Syn: *adoral; forward.*

adapical (a) Toward the apex of a gastropod or cephalopod shell. (b) Toward the apical system of an echinoid.

adaptation Modification, as the result of natural selection, of an organism or of its parts so that it becomes better fitted to exist under the conditions of its environment.

adaptive grid A changing series of ecologic zones having time as one dimension.

adaptive norm That part of an organic population that can survive and reproduce in the environment usually occupied by the species; the remainder may carry hereditary defects and diseases.

adaptive radiation Subdivision of a group of organisms into diversified groups within a short geologic-time interval (or more or less simultaneously) as a result of evolution; often shown by the occupation of equivalent niches in comparable habitats by ecologically similar but taxonomically distinct organisms. Cf: *convergence [evol].* Syn: *divergence [evol].*

adaptive zone A unit of environment defined in terms of its occupation by a single kind of organism.

adaxial Facing, directed toward, or situated on the same side as, the axis or center of the axis, as of an organ, plant, or invertebrate; ventral or posterior. Also, said of the adaxial side. Ant: *abaxial.*

adcumulate A cumulate formed by *adcumulus growth,* with intercumulus material comprising less than five percent of the rock. Cf: *mesocumulate.*

adcumulus growth Continued growth of cumulus crystals from liquid of the same composition so that the crystals are unzoned. This process reduces the intercumulus liquid by forcing it out of the intercumulus. See also: *adcumulate.*

adder stone A *serpent stone* once believed to be formed by an adder.

addition solid solution The addition of small atoms or ions, at random, in some interstices between atoms of a crystal structure. It may result in an *interstitial defect.* Syn: *interstitial solid solution.*

adductor *adductor muscle.*

adductor muscle (a) A muscle, or one of a pair of muscles, that contracts and thereby closes and/or tends to draw or hold together the valves of a bivalve shell (as in ostracodes, brachiopods, and pelecypods). Two adductor muscles, each dividing dorsally, are commonly present in articulate brachiopods; two pairs of adductor muscles (anterior and posterior), passing almost directly from the dorsal to the ventral side between the valves, are commonly present in inarticulate brachiopods. (b) Any transverse muscle (esp. that of the maxillary segment) for closure of the aperture in a cirripede crustacean. (c) In vertebrates, a muscle that moves an appendage toward a point of reference, usually one on the midsagittal plane. (d) Among *gnathostomes,* a group of muscles that brings the jaws together.—Cf: *diductor muscle.* Syn: *adductor.*

adductor-muscle scar A *muscle scar* showing the final site where an adductor muscle was attached. In articulate brachiopods, a single pair of such scars is located between diductor impressions in the pedicle valve and two pairs (anterior and posterior) in the brachial valve (TIP, 1965, pt.H, p. 139). See also: *cicatrix.* Syn: *adductor scar.*

adductor pit A depression that develops on the interior of a scutum for attachment of an adductor muscle of a cirripede crustacean.

adductor ridge A linear elevation that develops (in association with an adductor pit) on the interior of a scutum of certain cirripede crustaceans (such as Balanomorpha).

adductor testis attachment The place of attachment of the adductor testis muscle, near the lower end of the postcervical groove in some decapods; the position of this point is not clearly marked on the carapace (Holthuis, 1974, p. 735).

adelite A mineral: $CaMg(OH)AsO_4$. It sometimes has appreciable fluorine.

adeoniform Said of a rigid bilamellar lobate erect colony in cheilostome bryozoans, firmly attached by a calcified base.

ader wax *ozocerite.*

adfreezing The process by which two objects adhere to one another due to the binding action of ice; applied in permafrost studies.

adherent In plant morphology, pertaining to the union of parts that are usually separate. Cf: *adnate; connate [bot]; coherent [bot].*

adhesion The molecular attraction between contiguous surfaces. Cf: *cohesion.*

adhesion ripple One of a series of irregularly parallel crests of sand, formed transverse to a wind blowing dry sand over a smooth moist surface. In cross-section the crests are asymmetrical, with the stoss side steeper than the lee side (i.e., the opposite of normal current ripples). The length (chord) of the ripple is generally less than 2 cm and the height less than 2 mm. Originally described by Van Straaten (1953a). Syn: *antiripplet; antiripple.*

adhesive water *pellicular water.*

adiabatic In thermodynamics, pertaining to the relationship of pressure and volume when a gas or fluid is compressed or expanded without either giving or receiving heat. In an adiabatic process, compression causes a rise in temperature, and expansion a drop in temperature (U.S. Naval Oceanographic Office, 1968). See also: *potential density; potential temperature [oceanog].*

adiagnostic Nonrecommended synonym of *cryptocrystalline.* The term was originally used by Zirkel in German as *adiagnostisch.* Ant: *eudiagnostic.*

adinole An argillaceous sediment that has undergone albitization as a result of contact metamorphism along the margins of a sodium-rich mafic intrusion. Cf: *Schalstein; spilosite; spotted slate.* See also: *desmosite.*

adipocere A waxy or unctuous, brownish or light-colored natural substance consisting mainly of free fatty acids, hydroxy acids, and their calcium and magnesium salts, formed on decay of human, animal, or fish remains in damp areas or in fresh or salt water. Its consistency may range from soft and pliable for recent material to hard and brittle for older material. It often replaces and takes the form of the decomposing body.

adipocerite *hatchettine.*

adipocire A syn. of *hatchettine.* Not to be confused with *adipocere.*

a direction *a axis [struc petrol].*

adit A horizontal passage from the surface into a mine. Sometimes called a *tunnel.*

adjusted stream A stream that carves its valley parallel to the strike of the least resistant rocks over which it flows.

adjustment The determination and application of corrections to a series of survey observations for the purpose of reducing errors, removing internal inconsistencies, and coordinating and correlating the derived results within the survey itself or with previously existing basic data; e.g. the determination and application of orthometric corrections in order to make the elevations of all bench marks consistent and independent of the circuit closures, or the positioning of public-land lines on a topographic map to indicate their true, theoretical, or approximate location relative to the adjacent terrain and culture. The term may refer either to mathematical procedures or to corrections applied to instruments used in making observations.

adjustment of cross section The tendency in glaciers and rivers to change the size of every cross section of the channel to accomodate the volume of ice or water that must pass through it.

adjustor muscle One of a pair of two pairs of muscles in many articulate brachiopods, branching from the pedicle, and responsible for moving the position of the shell on the pedicle. A ventral pair is attached posteriorly and laterally from the diductor muscles, and a dorsal pair is on hinge plates or floor of brachial valve behind posterior adductor muscles (TIP, 1965, pt.H, p.139).

admission *admittance [chem].*

admittance [chem] In a crystal structure, substitution of a trace element for a major element of higher valence, e.g. Li^+ for Mg^{++}. Admitted trace elements generally have a lower concentration relative to the major element in the mineral than in the fluid from which the mineral crystallized. Cf: *capture [chem]; camouflage.* Syn: *admission.*

admittance [elect] The reciprocal of *impedance [elect],* or the ratio of complex current to voltage in a linear circuit.

admixture (a) A term applied by Udden (1914) to one of the lesser or subordinate of several particle-size grades of a sediment. See also: *coarse admixture; fine admixture; distant admixture; proximate admixture.* (b) A material that is added to another to produce a desired modification; e.g. a substance (other than aggregate, cement, or water) added during the mixing of concrete.

adnate In plant morphology, pertaining to the union of parts attached throughout their length. Cf: *connate [bot]; coherent [bot]; adherent.*

adobe (a) A fine-grained, usually calcareous, hard-baked clayey deposit mixed with silt, usually forming as sheets in the central or lower parts of desert basins, as in the playas of SW U.S. and in the arid parts of Mexico and Spanish America. It is probably a windblown deposit, although it is often reworked and redeposited by running water. The term was applied originally to a sun-dried brick made of adobe mixed with straw, and later to the clayey material constituting the brick. (b) A heavy-textured clay soil derived from adobe deposits.—Etymol: Spanish. Cf: *loess.*

adobe flat A generally narrow plain formed by sheetflood deposition of fine sandy clay or adobe brought down by an ephemeral stream, and having a smooth, hard surface (when dry) usually unmarked by stream channels.

adolescence A stage following youth and preceding maturity in a developmental sequence such as the cycle of erosion. It is sometimes considered "early maturity". Adolescence is characterized by incipient development of mature features, together with residual features of youth. It may be applied wherever the terms youth and maturity are appropriate, e.g. in the regional erosion cycle, the karst cycle, or development of a valley. Syn: *topographic adolescence.*

adolescent Pertaining to the stage of *adolescence* of the cycle of erosion; esp. said of a valley having a well-cut, smoothly graded stream channel that may reach base level at its mouth, or of a coast marked by low but nearly continuous sea cliffs.

adont Said of a class of ostracode hinges having no teeth, with a ridge or bar in one valve fitting into a groove in the opposed valve.

adoral Located or directed toward or near the mouth of an invertebrate; e.g. an "adoral surface" of an echinoderm theca bearing the mouth or ambulacral grooves, or the "adoral direction" toward the aperture of an ammonoid shell, or an "adoral fiber" of cytoplasm tissue leading from neuromotorium to the edge of peristome in a ciliate protozoan. Cf: *aboral.*

adradial adj. (a) Said of the position corresponding to the boundary between ambulacral and interambulacral areas of an echinoid or edrioasteroid. (b) Directed toward the axis of an asterozoan ray. (c) Pertaining to a radius of the third order in a coelenterate.—n. (a) One of the small plates lining ambulacra in certain edrioasteroids. (b) One of a series of ossicles on the aboral surface of an asterozoan ray.

adradial suture The zone of contact between the oral-ambulacral and interambulacral plates in echinoids and edrioasteroids (Bell, 1976).

adret A mountain slope so oriented as to receive the maximum available amount of light and warmth from the Sun; esp. a southward-facing slope of the Alps. Etymol: French dialect, "good side". Cf: *ubac.*

adsorbed water Water adhering in ionic or molecular layers to the surfaces of soil or mineral particles. See also: *adsorption.* Cf: *absorbed water.*

adsorption Adherence of gas molecules, or of ions or molecules in solution, to the surface of solids with which they are in contact. Cf: *absorption.*

adularescence A floating, billowy, white or bluish light, seen in certain directions as a gemstone (usually adularia) is turned; it is caused by diffused reflection of light from parallel intergrowths of another feldspar of slightly different refractive index from the main mass of adularia. It is often called *schiller.*

adularia A moderate to low-temperature mineral of the alkali feldspar group: $KAlSi_3O_8$. It is weakly triclinic (formerly regarded as apparently monoclinic) and typically occurs in well-developed, usually transparent, and colorless to milky-white (and often opalescent) pseudo-orthorhombic crystals in fissures in crystalline schists, esp. in the region of the Swiss Alps. Adularia displays pearly internal reflections and a fascinating variety of optical behavior between crossed nicols. It typically has a relatively high content of barium.

adularization Introduction of, or replacement by, adularia.

advance [coast] A continuing seaward movement of a shoreline as a result of progradation or emergence. Also, a net seaward movement of the shoreline during a specified period of time. Ant: *recession.* Syn: *progression.*

advance [glaciol] (a) The forward and downslope movement of the terminus of a glacier, generally resulting from a period of positive *net balance,* or an excess of accumulation over ablation. Its rate is usually measured in meters per year. (b) A time interval marked by an advance or general expansion of a glacier.—See also: *readvance.* Syn: *glacial advance; glacier advance.*

advance-cut meander A meander in which the outer bank of the channel is eroded so rapidly that deposition on the inner bank fails to keep pace, thereby widening the channel (Melton, 1936, p. 598-599). Cf: *forced-cut meander.* Syn: *induced meander.*

advanced dune A sand dune formed on the windward side of a larger *attached dune,* remaining separate from it by the eddy motion of the wind.

advection [meteorol] Horizontal transport of air or of an atmospheric property within the Earth's atmosphere. Cf: *convection [meteorol].*

advection [oceanog] The horizontal or vertical flow of seawater as a current.

advection [tect] Lateral mass movements of mantle material. Cf: *convection [tect].*

advection fog A *fog* that results when warm moist air moves across a cold or snow-covered surface. As the air loses heat to the ground, its temperature may drop below the *dew point* and the airborne moisture condenses.

adventitious (a) Said of a plant part that arises from an unusual place, e.g. a root that arises from a leaf or stem rather than from a primary root. (b) Said of plants introduced to areas beyond their customary range.

adventitious avicularium An avicularium of a cheilostome bryozoan occupying a position on a wall of another zooid.

adventitious lobe A secondary lobe of an ammonoid suture, formed by subdivision of the first lateral saddle; also, all later lobes that develop from the first adventitious lobe. Syn: *adventive lobe.*

adventitious stream A stream resulting from accidental variations of conditions, generally in an area that is approaching topographic maturity (Horton, 1945, p. 341-342).

adventive *parasitic [volc].*

adventurine *aventurine.*

advolute Said of a gastropod shell whose whorls barely touch one another but are not distinctly embracing or overlapping, or said of a coiled cephalopod shell in which the outer whorl touches but does not cover any part of the adjacent inner whorls. Cf: *evolute; involute; convolute.*

adyr (a) A term used in Turkmenia for a part of a desert plain devoid of sands and having soft ground. Cf: *kyr.* (b) A term used in Kazakhstan for a small flat top of relict high ground or of a mesa-like hill. (c) A term loosely applied in central Asia to a low mountain, a small hill, an eroded ridge with gentle slopes, etc.—Etymol: Turkish. The definitions above are from Murzaevs & Murzaevs (1959, p. 20).

aegagropile A *lake ball* consisting of radial, outgrowing, hair-like filaments formed by algae. Etymol: Greek *aigagros*, "goat", + *pilus*, "hair". Cf: *algal biscuit.* Syn: *aegagropila; egagropile.*

aegirine A syn. of *acmite.* The term is sometimes applied to acmite containing calcium, magnesium, or aluminum. Syn: *aegirite.*

aegirine-augite A syn. of *acmite-augite.* Also spelled: *aegirite-augite; aegirinaugite.*

aegirite *aegirine.*

Aeneolithic *Eneolithic.*

aenigmatite A mineral: $Na_2Fe_5TiSi_6O_{20}$. Syn: *enigmatite.*

aeolian *eolian.*

aeolianite *eolianite.*

aeolotropic A syn. of *anisotropic.* Also spelled: *eolotropic.*

aeon *eon.*

aeration In a soil, the supplying of air and other gases to the pores.

aeration porosity The volume of interstices that do not hold water at a specified low moisture tension (Jacks et al., 1960). Cf: *noncapillary porosity.*

aerial Pertaining to the air; related to, located in, or consisting of, the Earth's atmosphere. Not to be confused with *areal.*

aerial arch An anticline, the crest of which has been eroded.

aerial magnetometer *airborne magnetometer.*

aerial mapping The taking of aerial photographs for making maps and for geologic interpretation.

aerial mosaic *mosaic [photo].*

aerial photograph Any photograph taken from the air, such as a photograph of a part of the Earth's surface taken by a camera mounted in an aircraft. Syn: *air photograph.*

aerial survey (a) A survey using aerial photographs as part of the surveying operation. (b) The taking of aerial photographs for surveying purposes.

aerial triangulation *aerotriangulation.*

aerobic (a) Said of an organism (esp. a bacterium) that can live only in the presence of free oxygen; also, said of its activities. Noun: *aerobe.* (b) Said of conditions that can exist only in the presence of free oxygen. Cf: *anaerobic.*

aerobic decay Decomposition of organic substances, primarily by microorganisms, in the presence of free oxygen; the ultimate decay products are carbon dioxide and water.

aeroides Pale sky-blue aquamarine.

aerolite *stony meteorite.*

aerolithology An obsolete term for the science that deals with meteorites. Cf: *meteoritics.*

aeromagnetic survey A *magnetic survey* made with an *airborne magnetometer.*

aerosiderite An obsolete syn. of *siderite* meteorite.

aerosiderolite An obsolete syn. of *stony-iron meteorite.*

aerosol A *sol* in which the dispersion medium is a gas (usually air) and the dispersed or colloidal phase consists of solid particles or liquid droplets; e.g. mist, haze, most smoke, and some fog.

aerospace A mnemonic term derived from *aero*nautics + *space* to denote both the Earth's atmosphere and the space beyond as a single unit.

aerotriangulation Process of *phototriangulation* accomplished by means of aerial photographs. Syn: *aerial triangulation.*

aerugite A mineral: $Ni_9As_3O_{16}$.

aeschynite A mineral: $(Ce,Ca,Fe,Th)(Ti,Nb)_2(O,OH)_6$. It is isomorphous with priorite. Syn: *eschynite.*

aeschynite-(Y) An orthorhombic mineral: $(Y,Ca,Fe,Th)(Ti,Nb)_2(O,OH)_6$. It forms a series with aeschynite and tantalaeschynite.

aethoballism A term proposed by Grabau (1904) for local metamorphism resulting from contact with a meteorite. The term is obsolete. Cf: *symphrattism.*

aff. Abbrev. of *affinity.* It implies less certain similarity than does *cf.*

affine adj: Said of a homogeneous deformation, i.e. one in which initially straight lines remain straight after deformation. Affine transformation is a mathematical transformation in which the coordinates of the deformed state are related to the coordinates of the undeformed state in a linear manner. Such a transformation describes a homogeneous deformation.

affinity In biology, the state of being akin to; used to indicate relationship without specifying identity. Abbrev: aff.

affluent adj. Said of a stream flowing toward or into a larger stream or into a lake.—n. An affluent stream; esp. an *influent* flowing into a lake. The term, originally introduced by Jackson (1834, p. 77-78) as distinct from a *confluent,* is obsolescent as the synonym *tributary* is more commonly used.

afflux (a) The upstream rise of water level above the normal surface of the water in a channel, caused by contraction or obstruction of the normal waterway. (b) The difference between high flood levels upstream and downstream of a weir.

afghanite A mineral: $(Na,Ca,K)_{12}(Si,Al)_{16}O_{34}(Cl,SO_4,CO_3)_4 \cdot H_2O$.

AFMAG method The use of natural audio-frequency electromagnetic noise to study lateral changes in Earth resistivity. Etymol: An acronym for *a*udio-*f*requency *mag*netic.

AFM diagram A triangular diagram showing the simplified compositional character of a metamorphosed pelitic rock by plotting molecular quantities of the three components: $A=Al_2O_3$; $F=FeO$; and $M=MgO$. Cf: *ACF diagram; A'KF diagram; AFM projection.*

AFM projection A triangular diagram showing molecular proportions, constructed by projecting the compositions of metamorphic pelitic minerals in the tetrahedron Al_2O_3 (=A), FeO(=F), MgO(=M), K_2O from the composition of muscovite or potassium feldspar onto the plane AFM. Devised by J. B. Thompson, Jr. (1957) and now widely used in the study of pelitic schists and gneisses. Syn: *Thompson diagram.*

A-form The megalospheric form of a foraminifer. Cf: *B-form.*

afterdamp The gas remaining in a coal mine after an explosion of *firedamp* or after a fire. It includes carbon monoxide and carbon dioxide. Cf: *whitedamp; blackdamp.*

aftershock An earthquake that follows a larger earthquake or *main shock* and originates at or near the focus of the larger earthquake. Generally, major earthquakes are followed by many aftershocks, which decrease in frequency and magnitude with time. Such a series of aftershocks may last many days for small earthquakes or many months for large ones. Cf: *foreshock.*

afterworking *creep recovery.*

Aftonian Pertaining to the classical first interglacial stage of the Pleistocene Epoch in North America, following the Nebraskan and preceding the Kansan glacial stages. Etymol: Afton, town in Iowa. See also: *Günz-Mindel.*

afwillite A mineral: $Ca_3Si_2O_4(OH)_6$.

agalite A fine fibrous variety of talc, pseudomorphous after enstatite. Syn: *asbestine.*

agalmatolite A soft, waxy stone, such as pinite, pyrophyllite, or steatite, of a gray, green, yellow, or brown shade, used by the Chinese to simulate jade for carving small images, miniature pagodas, and similar objects. Syn: *figure stone; pagodite; lardite; lard stone.*

agardite A hexagonal mineral: $(Y,Ca)Cu_6(AsO_4)_3(OH)_6 \cdot 3H_2O$.

agaric mineral *moonmilk.*

agate A translucent cryptocrystalline variety of quartz, being a variegated *chalcedony* frequently mixed or alternating with opal, and characterized by colors arranged in alternating stripes or bands, in irregular clouds, or in mosslike forms. Agate is found in virtually all colors, usually of low intensity; it commonly occupies vugs in volcanic rocks and cavities in some other rocks. Cf: *onyx.* See also: *banded agate; clouded agate; moss agate.*

agate jasper An impure variety of agate consisting of jasper with veins of chalcedony. Syn: *jaspagate.*

agate opal Opalized agate. Cf: *opal-agate.*

agathocopalite *kauri.*

agatized wood A syn. of *silicified wood.* See also: *wood agate.*

age [geochron] (a) A geologic-time unit shorter than an *epoch [geochron]* and longer than a *subage,* during which the rocks of the corresponding *stage* were formed. Syn: *chron.* (b) A term used informally to designate a length of geologic time during which the rocks of any stratigraphic unit were formed. (c) A division of time of unspecified duration in the history of the Earth, characterized by a dominant or important type of life form; e.g. the "age of mammals". (d) The time during which a particular geologic event or series of events occurred or was marked by special physical conditions; e.g. the "Ice Age". (e) The position of anything in the geologic time scale; e.g. "the rocks of Miocene age". It is often expressed in years.—See also: *geologic age.*

age [ice] The stage of development of sea ice; the term usually refers to the length of time since the formation of the ice and to its thickness.

aged Said of a topographic feature in the penultimate stage of development, such as a shore that is approaching reduction to base level.

age determination The evaluation of geologic age by faunal or stratigraphic means, or by physical methods involving determination of the relative abundance of a radioactive parent and radiogenic daughter isotope. Some methods of relative age determination are based on the extent of chemical change, like the hydration of obsidian or the fluorine uptake of bone.

age equation The relationship between radioactive decay and geologic time. Expressed mathematically, it is $t = 1/\lambda . ln(1 + D/P)$, where t is the age of a rock or mineral specimen, D is the number of daughter isotopes today, P is the number of parent isotopes today, *ln* is the natural logarithm (logarithm to base e), and λ is the decay constant. Cf: *general age equation.*

age of amphibians An informal designation of the late Paleozoic, i.e., the *Carboniferous* and the *Permian.*

age of coal An informal designation of the *Carboniferous.* Syn: *coal age.*

age of cycads An informal designation of the *Jurassic.*

age of ferns An informal designation of the *Pennsylvanian.*

age of fishes An informal designation of the *Silurian* and the *Devonian.*

age of gymnosperms An informal designation of the *Mesozoic.*

age of mammals An informal designation of the *Cenozoic.*

age of man An informal designation of the *Quaternary.*

age of marine invertebrates An informal designation of the *Cambrian* and the *Ordovician.*

age of reptiles An informal designation of the *Mesozoic.* Syn: *reptilian age.*

age of the Earth The length of time that the Earth has existed essentially as it is now. This is 4.5 b.y. as determined from isotopic investigation of common-lead relationships and uranium-lead and rubidium-strontium dating of meteorites. Oldest dated terrestrial rocks are approximately 3.5 b.y.

age of the Moon The time elapsed in days since the last new Moon.

age of the Universe Usually refers to the Hubble age of the Universe, which, given by the reciprocal of the Hubble constant, is 13.7 b.y. Other estimates include 10 to 15 b.y., the time when nucleosynthesis is thought to have begun, and 4.7 b.y., the age of meteorites.

age of water The length of time since a water mass was last in contact with the atmosphere at the surface of the ocean.

age ratio The ratio of daughter to parent isotope upon which the *age equation* is based. For a valid age determination, the isotope system must have remained closed since solidification, metamorphism, or sedimentation, the decay constant must be known, and the sample must be truly representative of the rock from which it is taken.

age-specific eruption rate In stochastic treatment of volcanologic data based on renewal theory, if the volcano has not erupted in time interval (0,x), the age-specific eruption rate, or *short eruption rate,* ϕ (x), is the limit of the ratio to Δx of the probability of eruption in time interval (x, x + Δx). Its physical dimension is time^{-1}. Approximately ϕ (x) gives the probability of almost immediate eruption, ending repose period of known age x (Wickman, 1966, p. 298). Syn: *eruption rate.*

age spectrum In radiometric dating methods where the radiogenic daughter product is a gas (argon-40/argon-39 dating method, xenon-xenon age spectrum method), it is possible to extract the gas from the sample incrementally, and to compute an apparent age for each fraction of gas released. A plot of apparent age versus temperature or versus cumulative percent of the gas released is termed an age spectrum.

agglomerate A term originally used by Lyell in 1831 for a chaotic assemblage of coarse angular pyroclastic materials. The term has been variously defined since then, and should be defined in context to avoid confusion. Cf: *volcanic breccia.*

agglomerate ice Ice that has formed by a mixture of floating ice fragments congealing, as in a lake.

agglomerating Said of bituminous coal that softens when heated. See also: *caking coal; coke.*

agglomerating value A measure of the binding qualities of a coal, determined by fusing tests in which no inert material is heated with the sample. Cf: *agglutinating value.*

agglomeroplasmic Said of an arrangement in a soil fabric whereby the plasma occurs as loose or incomplete fillings in the intergranular spaces between skeleton grains (Brewer, 1964, p.170). Cf: *porphyroskelic; intertextic.*

agglutinate [lunar] A term for certain particles in the lunar regolith that are held together by and largely composed of glass, probably spatter and melted ejecta from small hypervelocity impacts on the lunar surface, together with mineral grains and small rock fragments.

agglutinate [volc] A welded pyroclastic deposit characterized by vitric material binding the pyroclasts, or "sintered" vitric pyroclasts. Also spelled: *agglutinite.*

agglutinate cone *spatter cone.*

agglutinated Said of the tests of certain protists (some foraminifers, thecamoebians, and tintinnids) composed of foreign particles (sand grains, sponge spicules, mica flakes, etc.) bound together by cement.

agglutinating value A measure of the binding qualities of a coal and an indication of its caking or coking characteristics, determined by the ability of the coal, when fused, to combine with an inert material, such as sand. Cf: *agglomerating value.* Syn: *caking index.*

agglutination A syn. of sedimentary *cementation,* esp. in regard to coarse-grained rocks, such as breccia or conglomerate.

agglutinite *agglutinate [volc].*

aggradation [geomorph] (a) The building-up of the Earth's surface by deposition; specif. the *upbuilding* performed by a stream in order to establish or maintain uniformity of grade or slope. See also: *gradation.* Cf: *degradation.* Syn: *upgrading.* (b) A syn. of *accretion,* as in the development of a beach.

aggradation [permafrost] The spread or growth of permafrost, under present climatic conditions, due to natural or artificial causes. Ant: *degradation [permafrost].*

aggradation [stratig] The vertical component of coastal *onlap* (Mitchum, 1977, p.208). See also: *encroachment [stratig].*

aggradational ice Ice incorporated in the upper part of permafrost during *aggradation,* specif. by a rise of the permafrost table, over a period of years. Commonly ice lenses are formed seasonally in the base of the active layer, and these can be incorporated in the permafrost if the depth of thaw over a period of years does not reach them (Mackay, 1972, p. 10).

aggradational plain A broad, fanlike plain with a nearly straight longitudinal profile, formed in an arid region by stream deposition.

aggradation recrystallization Recrystallization resulting in the enlargement of crystals. Ant: *degradation recrystallization.*

aggraded valley plain An *alluvial plain,* or a stream-built flood plain; a plain formed by a stream aggrading its valley, the infilling with alluvium on the valley floor attaining a thickness greater than that of the stream channel (Cotton, 1958, p. 193).

aggrading neomorphism A kind of *neomorphism* in which the crystal size increases (Folk, 1965, p. 23); e.g. *porphyroid neomorphism* and *coalescive neomorphism.* Ant: *degrading neomorphism.*

aggrading stream (a) A stream that is actively building up its channel or flood plain by being supplied with more load than it is capable of transporting. (b) A stream that is upbuilding approximately at grade.

aggregate (a) A mass or body of rock particles, mineral grains, or a mixture of both. (b) Any of several hard, inert materials, such as sand, gravel, slag, or crushed stone, used for mixing with a cementing or bituminous material to form concrete, mortar, or plaster; or used alone, as in railroad ballast or graded fill. The term sometimes includes rock material used as chemical or metallurgical fluxstone. See also: *coarse aggregate; fine aggregate; lightweight aggregate.*

aggregated Said of a massive sulfide or other ore deposit in which the sulfide or other valuable constituent makes up 20% or more of the total volume. The term is little used.

aggregate polarization In crystal optics, a pattern seen between crossed nicols in a fine-grained crystal aggregate, composed of the different interference colors of variously oriented grains. The presence of aggregate polarization may define a material as crystalline that in ordinary light appears amorphous.

aggregate structure A mass of separate crystals, scales, or grains that extinguish under the polarizing microscope at different intervals during the rotation of the stage.

aggressive intrusion *forcible intrusion.*

aging The process by which a *young lake* becomes an *old lake* as a result of filling and nutrient loading, *eutrophication,* vegetation encroachment, and other actions.
aglet A tiny plate pierced by a single pore in a radiolarian.
agmatite Migmatite with appearance of breccia (Dietrich & Mehnert, 1961).
Agnatha A class of vertebrates, the jawless fishes. A modern example is the lamprey. Range, Ordovician to the present.
agnostid Any trilobite belonging to the order Agnostida, which includes small forms with subequal cephalons and pygidia and only two segments in the thorax. Their stratigraphic range is Lower Cambrian to Upper Ordovician.
Agnotozoic *Proterozoic.*
agonic line An *isogonic line* that connects points of zero magnetic declination. Its position changes according to the secular variation of the Earth's magnetic field.
agouni A term used in Morocco for a broad, generally dry gully carved by a torrent (Termier & Termier, 1963, p. 399). Etymol: Berber. Cf: *asif.*
agpaite A group of feldspathoid-bearing igneous rocks, first described from Ilimaussaq, Greenland, by Ussing in 1912, that includes sodalite-bearing nepheline syenite, *naujaite, lujavrite, kakortokite,* etc., and is distinguished by having $(Na+K)>Al$ on a molecular or atomic basis.
agpaitic (a) Said of crystallization in the presence of an excess of alkali (esp. sodium), so that the amount of aluminum oxide is insufficient for the formation of aluminum silicates (Thrush, 1968, p. 18). (b) Said of rocks with an agpaitic coefficient > 1.
agpaitic coefficient The ratio $Na + K/Al$, where Na, K and Al are amounts of sodium, potassium, and aluminum atoms, respectively, in a rock, esp. an alkalic igneous rock.
agrellite A triclinic mineral: $NaCa_2Si_4O_{10}F$.
agric horizon A diagnostic subsurface soil horizon formed after long-continued cultivation. It has an illuvial accumulation of silt, clay, and humus immediately beneath the plow layer (USDA, 1975).
agricolite *eulytite.*
agricultural geology The application of geology to agricultural needs, e.g. mineral deposits used as fertilizers or the location of ground water. Syn: *agrogeology.*
agricultural pollution Pollution resulting from farming practices and agricultural wastes. Sources include pesticides and fertilizers; animal manure and carcasses; crop residues; sediment from erosion; and dust from plowing.
agrinierite An orthorhombic mineral: $(K_2,Ca,Sr)U_3O_{10} \cdot 4H_2O$.
agrogeology *agricultural geology.*
agrology An old term for *soil science.*
aguada In the karst region of Yucatán, Mexico, a shallow depression that collects water in the rainy season. Etymol: Spanish, "watering place".
aguilarite A mineral: Ag_4SeS.
ahermatypic coral A non-reef-building coral; a coral lacking symbiotic algae; a coral capable of living in cold, deep, dark water. Syn: *ahermatype.* Ant: *hermatypic coral.*
ahlfeldite A mineral: $(Ni,Co)SeO_3 \cdot 2H_2O$.
A horizon The uppermost *soil horizon,* incorporating one or more of the following subdivisions: the A1 horizon, characterized by an accumulation of organic matter; the A2 horizon, characterized by a concentration of quartz or other resistant minerals in the sand- and silt-size fractions due to leaching of clay, iron, or aluminum; and the A3 horizon, transitional to the underlying B horizon. Partial syn: *topsoil.*
Ahren's prism A type of polarizing prism having three cut and recemented segments; the ordinary rays of the light are reflected to the sides, and the extraordinary ray forms the polarized light.
aiguille A prominent sharp rock peak or pinnacle, of the type commonly found in intensely glaciated mountain regions (as near Mont Blanc in the French Alps); it is a remnant of a septum between two cirques, the rest of which has been largely or wholly removed by erosion. Cf: *gendarme.* Pron: ai-*gwee.* Etymol: French, "needle". Syn: *needle [geol].*
aikinite (a) A mineral: $PbCuBiS_3$. Syn: *needle ore.* (b) Wolframite pseudomorphous after scheelite.
ailsyte Fine-grained quartz syenite containing sodic amphibole, named for Ailsa Craig, Scotland, by Heddle in 1897. Not recommended usage.
aimless drainage Drainage without a well-developed system, such as that in an area of glacial drift or karst topography.
ainalite A mixture of cassiterite and columbite.
air [coast] A Scottish term for a beach. Etymol· Old Norse *eyrr,* "gravelly bank".
air [meteorol] *atmosphere.*
air base An imaginary line connecting the points in space at which successive photos in a flight strip were taken; specif., the length of such a line (ASP, 1975, p. 2064). Cf: *photobase.*
airborne magnetometer An instrument used to measure the Earth's magnetic field while being transported by an aircraft. Syn: *aerial magnetometer; flying magnetometer.*
airborne profile recorder An electronic instrument that emits a pulsed-type radar signal from an aircraft to measure vertical distances between the aircraft and the Earth's surface. Abbrev: APR. Syn: *terrain profile recorder.*
airborne scanner A *scanner* designed for use on aircraft or spacecraft, in which the forward motion of the vehicle provides coverage normal to the scan direction.
air dome *dome [beach].*
air drilling *Rotary drilling* using high-velocity air instead of conventional *drilling mud.* It is unsuitable where significant volumes of water may be encountered or where natural gas may create an explosive mixture downhole.
air dry The condition of a substance whose moisture content has come into approximate equilibrium with the humidity of the surrounding air.
air entrainment The incorporation of air into moving water as a result of turbulence or the breaking of small waves.
air-fall deposition Showerlike falling of pyroclastic fragments from an *eruption cloud.* See also: *ash fall; pumice fall.*
air gap *wind gap.*
air gun An energy source extensively used in marine seismic surveys. Air under high pressure is abruptly released into the water to generate a seismic wave. Air guns are also used in boreholes, and in modified form on the land surface.
air heave The process of deformation of plastic sediments by the enlargement of a pocket of air trapped in them, such as by the accretion of small air bubbles rising through sand exposed at low tide on a beach or tidal flat. See also: *gas heave.*
air-heave structure A crumpled sedimentary structure believed to have been produced by air heave, measuring several centimeters across, and characterized by an abrupt upward doming of laminae with a core of unlaminated sandstone (Stewart, 1956, p.159). See also: *gas-heave structure.*
air mass A large, widespread volume of air having particular characteristics of temperature and moisture content that were acquired at its *source region* and are modified as it moves away from its source; for example, a *polar air mass.*
air photograph *aerial photograph.*
air photo lineament "Any linear feature . . . visible primarily on aerial photographs . . . and expressed continuously or semi-continuously for ten to 100 kilometers" (El-Etr, 1976, p. 486).
air photo linear "Any natural linear feature . . . visible primarily on aerial photographs . . . and expressed continuously or semi-continuously for less than ten kilometers" (El-Etr, 1976, p. 486).
air photo lineation "Any natural linear feature of any length . . . visible primarily on aerial photographs" (El-Etr, 1976, p. 486).
air pressure *atmospheric pressure.*
air sac A cavity or *vesicle [palyn]* in a pollen grain of a pine.
air shooting In seismic prospecting, a technique of applying a seismic pulse to the Earth by detonating explosive charges in the air. See also: *Poulter seismic method.*
air shrinkage The volume decrease that a clay undergoes in drying at room temperature.
air-space ratio In a soil, the ratio of the volume of water that can be drained by gravity from a saturated soil to the total volume of the soil's voids. Cf: *air-void ratio.*
air-void ratio In a soil, the ratio of the volume of air-filled voids to the total volume of voids in the soil. Cf: *air-space ratio.*
air volcano A *mud volcano* characterized more by the gas it emits than by the mud and rocks thrown out.
air wave An acoustic wave in the air. Cf: *earthquake sound.*
air well A tower of loose rock, used in some desert countries to collect water, by condensation of moisture from the warm atmosphere on the relatively cooler rock surfaces within the tower, whose temperature fluctuates about the daily and annual mean in a smaller range than does that of the air.
Airy hypothesis A concept of the mechanism of *isostasy,* proposed by George Bedell Airy, that postulates an equilibrium of

crustal blocks of the same density but of different size; thus the topographically higher mountains would be of the same density as other crustal blocks but would have greater mass and deeper *roots.* Cf: *Pratt hypothesis.*

Airy phase (a) The predominant surface-wave group corresponding to a maximum or minimum group velocity. (b) More generally, any seismic *phase [phys]* associated with a stationary group velocity.

Aïstopoda An order of the amphibian subclass Lepospondyli, characterized by wormlike or snakelike habitus and lacking appendicular skeletons. Range, Lower Mississippian to Lower Permian.

ait A British term for a small island in a lake or river; an islet. Syn: *eyot.*

Aitoff projection (a) A map projection derived from the azimuthal equidistant projection by doubling the horizontal distances from the central meridian until the entire spherical surface is within an ellipse whose major axis (equator) is twice the length of its minor axis (central meridian). It resembles the Mollweide projection, but the parallels (except the equator) and meridians (except the central meridian) are represented by curved lines and there is less distortion at the margins. Named after David Aitoff (d.1933), Russian geographer, who introduced the projection in 1889. (b) A name commonly, but incorrectly, given to the *Hammer-Aitoff projection.*

ajkaite A pale-yellow to dark reddish-brown, sulfur-bearing fossil resin found in brown coal. Syn: *ajkite.*

ajoite A mineral: $Cu_6Al_2Si_{10}O_{29} \cdot 5H_2O$.

akaganeite A mineral: beta-FeO(OH,Cl).

akatoreite A mineral: $Mn_9(Si,Al)_{10}O_{23}(OH)_9$.

akdalaite A mineral: $4Al_2O_3 \cdot H_2O$.

akene *achene.*

akenobeite A granodioritic aplite composed of an aggregate of alkali feldspar and oligoclase, the former in excess of the latter, with an aggregate of fine quartz grains in the interstices and with little ferromagnesian material present (Johannsen, 1939, p. 238). Its name is derived from Akenobe district, Japan. Not recommended usage.

akerite A syenite containing alkali feldspar, oligoclase, biotite, sodic clinopyroxene, and quartz. Its name, given by Brögger in 1895, is derived from Aker, Norway. Not recommended usage.

akermanite A mineral of the melilite group: $Ca_2MgSi_2O_7$. It is isomorphous with gehlenite.

A'KF diagram A triangular diagram showing the simplified compositional character of a metamorphic rock by plotting molecular quantities of the three components: $A' = Al_2O_3 + Fe_2O_3 - (Na_2O + K_2O + CaO)$; $K = K_2O$; and $F = FeO + MgO + MnO$. $A' + K + F$ (in mols) are recalculated to 100%; the diagram is used in addition to the *ACF diagram* when K minerals require representation. Cf: *AFM diagram.*

akinetic surface The surface in a sedimentary rock layer that was the outer surface of the lithosphere at the place and time that oscillation of base level at that point passed through its maximum (Goldman, 1921, p. 8-20).

akmolith An igneous intrusion along a zone of décollement, with or without tonguelike extensions into the overlying rock. Also spelled: *acmolith.*

akrochordite A mineral: $Mn_4Mg(AsO_4)_2(OH)_4 \cdot 4H_2O$. It occurs in reddish-brown rounded aggregates.

aksaite A mineral: $MgB_6O_{10} \cdot 5H_2O$.

aktashite A mineral: $Cu_6Hg_3As_5S_{12}$.

aktology Study of nearshore and shallow-water areas, conditions, sediments, life, and environments.

akyrosome A term used by Niggli (1954, p. 191) for a subsidiary mass (such as a vein, nodule, band, lens, or block) of a complex rock; a minor part of a migmatite. This spelling appears in the English translation of Niggli's paper but probably should have been "akyriosome". Not widely used. Cf: *kyriosome.*

ala (a) A wing or winglike process or part of an organism; e.g. a winglike flange in the diatom *Surirella,* or the winglike extension of the ear of a brachiopod shell or of the ventral and lateral part of the carapace of an ostracode, or the triangular lateral part of a compartmental plate of a cirripede crustacean. (b) A smooth, commonly semicircular area flanking the posterior portion of the glabella in harpetid trilobites. Cf: *baccula.* Pl: *alae.*

Ala-A twin law A twin law in plagioclase, in which the twinning axis is the *a* axis and the composition plane is (001). An Ala-A twin is usually multiple and parallel, and often occurs with the Manebach twin law. Cf: *Manebach-Ala twin law; Ala-B twin law.*

alabandite A mineral: MnS. It usually occurs in iron-black massive or granular form. Syn: *manganblende; alabandine.*

alabaster (a) A firm, very fine-grained, massive or compact variety of gypsum, usually snow-white and translucent but sometimes delicately shaded or tinted with light-colored tones of yellow, brown, red, orange, or gray. It is used as an interior decorative stone (esp. for carved ornamental vases and figures) and for statuary. (b) *onyx marble.*

Ala-B twin law A twin law in plagioclase, in which the twinning axis is the *a* axis and the composition plane is (010). It often occurs with the albite twin law. Cf: *Ala-A twin law; albite-Ala twin law.*

aladzha Impure ozokerite containing an admixture of country rocks, found in the region of the Caspian Sea (Tomkeieff, 1954, p.21). Etymol: Tatar.

alaite A dubious mineral: $V_2O_5 \cdot H_2O$.

alalite A light-green variety of diopside.

alamandine *almandine.*

alamosite A mineral: $PbSiO_3$.

alar fossula A *fossula* developed in the position of an alar septum in a rugose coral or adjoining it on the side toward the counter septum.

alar projection A winglike extension of a foraminiferal test.

alar septum One of two protosepta of a rugose corallite located about midway between the *cardinal septum* and the *counter septum,* distinguished from other protosepta by pinnate insertion of newly formed metasepta on the side facing the counter septum (TIP, 1956, pt.F, p.245). Symbol: A.

alas A thermokarst depression with steep sides and a flat grass-covered floor (Washburn, 1973, p. 237). Etymol: Yakutian. Anglicized pl: *alases.* Also spelled: *alass.*

alaskaite A mixture of sulfosalt minerals of lead, silver, copper, and bismuth. Not to be confused with *alaskite.*

Alaskan band A *dirt-band ogive.* Use of the term is to be discouraged because such ogives are not limited to Alaska and are not even the most common forms of ogives on Alaskan glaciers.

alaskite In the U.S., a commonly used term for a granitic rock containing only a few percent of dark minerals. The term was introduced by Spurr in 1900 for a holocrystalline-granular plutonic rock characterized by essential alkali feldspar and quartz, and little or no dark component. Johannsen (1917) proposed two subdivisions of alaskite: *kalialaskite,* which lacks albite, and alaskite proper, which contains albite. In the recommendations of the Terminological Commission of the Petrographic Committee of the USSR of 1969, the term alaskite is used to designate granitoid rocks in which quartz constitutes 20-60% of the felsic minerals and in which the ratio of alkali feldspar to total feldspar is greater than 90%; i.e. the equivalent of alkali granite, or *kaligranite.* Cf: *aplogranite; tarantulite.*

alaskite-quartz *tarantulite.*

alate Having wings or a winged form; e.g. said of an outward lateral extension in the ventral half of an ostracode valve, usually increasing in width backward and terminating abruptly, and tending to have a triangular shape (TIP, 1961, pt.Q, p.47); or said of the form of a brachiopod shell in which the valves are drawn out at the ends of the hinge line to form winglike extensions; or said of a bivalve-mollusk shell having auricles.

A layer The seismic region of the Earth equivalent to the *crust,* extending from the surface to the Mohorovičić discontinuity. It is part of a classification of the Earth's interior made up of layers A to G.

alb A flat or gently inclined narrow shelf separating the nearly vertical side of an alpine glacial trough from the mountain slope above. Cf: *alp.*

albedo The ratio of the amount of electromagnetic energy reflected by a surface to the amount of energy incident upon it.

Albers projection An equal-area projection of the conical type, on which the meridians are straight lines that meet in a common point beyond the limits of the map and the parallels are concentric circles whose center is at the point of intersection of the meridians. Meridians and parallels intersect at right angles and the arcs of longitude along any given parallel are of equal length. The parallels are spaced to retain the condition of equal area. Along two selected parallels, called standard parallels, the scale is held exact: along the other parallels the scale varies with latitude, but is constant along any given parallel. Between the standard parallels the meridional scale is too large; beyond them, too small. When used for the map of the U.S. at 1:2,500,000, the projection has a maximum scale error of 1.25 percent along the northern and southern borders. Named after Heinrich C. Albers (1773-1833), German cartographer, who devised the projection in 1805.

Albertan North American series: Middle Cambrian (above Waucoban, below Croixian).

albert coal An early name for *albertite.*

albertite A dark brown to black asphaltic pyrobitumen with conchoidal fracture occurring as veins 1 to 16 ft wide in the Albert Shale of Albert County, New Brunswick. It is partly soluble in turpentine, but practically insoluble in alcohol. It was earlier called *albert coal.* See also: *libollite; stellarite; impsonite; byerite.*

Albian European stage: uppermost Lower Cretaceous, or Middle Cretaceous of some authors (above Aptian, below Cenomanian). See also: *Gault.*

albic horizon A diagnostic subsurface soil horizon from which clay and free iron oxides have been removed so that its color is determined by the sand and silt particles, and not by their coatings. It is often an A2 horizon and may form immediately beneath a layer of leaf litter (USDA, 1975).

albite (a) A colorless or milky-white triclinic mineral of the feldspar group: $NaAlSi_3O_8$. It is a variety of plagioclase with composition ranging from $Ab_{100}An_0$ to $Ab_{90}An_{10}$; it is also an alkali feldspar, representing the triclinic modification of sodium feldspar. Albite occurs in all groups of rocks, forming a common constituent of granite and of various acid-to-intermediate igneous rocks; it is widely distributed in low-temperature metamorphic rocks (greenschist facies), and is regularly deposited from hydrothermal solutions in cavities and veins. Albite crystals frequently exhibit polysynthetic twinning, predominantly after the *albite twin law.* Cf: *analbite.* Syn: *sodium feldspar; sodaclase; white feldspar; white schorl.* (b) The pure sodium-feldspar end member in the plagioclase series.

albite-Ala twin law A complex twin law in feldspar, in which the twin axis is perpendicular to [100] and the composition plane is (010). Cf: *Ala-B twin law.*

albite-Carlsbad twin law A complex twin law in feldspar, in which the twin axis is perpendicular to [001] and the composition plane is (010).

albite-epidote-amphibolite facies The set of metamorphic mineral assemblages (facies) in which basic rocks are represented by hornblende + albite + epidote. Equivalent to Eskola's *epidote-amphibolite facies,* it is of uncertain status, transitional between the *greenschist facies* and the *amphibolite facies.* It is generally believed to be favored by the higher pressures of regional metamorphism.

albite-epidote-hornfels facies The set of metamorphic mineral assemblages (facies) in which basic rocks are represented by albite + epidote + chlorite + actinolite. It is typical of the outermost parts of contact aureoles. It is not clearly distinct from the *greenschist facies,* and in fact is not given the status of a metamorphic facies by many workers (e.g. Miyashiro, 1973). Cf: *hornblende-hornfels facies.*

albite porphyrite *albitite.*

albite twin law A twin law in triclinic feldspars, in which the twin plane and composition plane are (010). An albite twin is usually multiple and lamellar, and shows fine striations on the (001) cleavage plane.

albitite A porphyritic igneous rock, defined by Turner in 1896, containing phenocrysts of albite in a groundmass chiefly consisting of albite. Muscovite, garnet, apatite, quartz, and opaque oxides are common accessory minerals. Syn: *albitophyre; albite porphyrite.*

albitization Introduction of, or replacement by, albite, usually replacing a more calcic plagioclase.

albitophyre *albitite.*

Alboll In U.S. Dept. of Agriculture soil taxonomy, a suborder of the soil order *Mollisol,* characterized by an albic horizon immediately beneath a mollic epipedon, and by fluctuating ground water. Albolls have either an agrillic or a natric horizon, and mottles and/or iron-manganese concretions indicative of seasonal wetness (USDA, 1975). Cf: *Aquoll; Boroll; Rendoll; Udoll; Ustoll; Xeroll.*

alboranite An olivine-free hypersthene-bearing *basalt* named by Becke in 1899 after Alboran Island of Spain. A *subalkaline basalt* or *tholeiitic basalt.* Not recommended usage.

alcove A large, deep niche formed in a precipitous face of rock; specif. a *spring alcove.*

alcove lands An angular landscape characterized by terraced slopes consisting of resistant beds interbedded with deeply cut softer rocks (Powell, 1875, p. 149-150).

alcyonarian Any coral of the subclass Alcyonaria, colonial forms with eight pinnate tentacles, an endoskeleton, and eight complete septa. Assured range, Permian to present. See also: *octocoral.* See: Pennak (1964).

Aldanian European stage: Lower Cambrian (above Precambrian, below Lenan).

aldanite A variety of thorianite containing lead and uranium.

aldzhanite An orthorhombic mineral: $CaMgB_2O_4Cl \cdot 7H_2O$ (?).

Aleppo stone *eye agate.*

alert The time when a navigation satellite should pass within range so that a location fix can be obtained. See also: *satellite navigation.*

alete Said of a spore without a laesura. In practice, such spores are sometimes difficult to distinguish from *acolpate* pollen. Cf: *inaperturate.*

aleurite An unconsolidated sedimentary deposit intermediate in texture between sand and clay, consisting of particles having diameters in the range of 0.01 to 0.1 mm. The term is common in Russian literature, and is frequently translated as "silt". Etymol: Greek *aleuron,* "flour".

aleurolite A consolidated *aleurite,* intermediate in texture between sandstone and shale; esp. siltstone.

aleutite A porphyritic *belugite* having a fine-grained groundmass. It was named by Spurr in 1900 from the Aleutian Islands. Not recommended usage.

Alexandrian North American provincial series: Lower Silurian (above Cincinnatian of Ordovician, below Niagaran). Obsolete syn: *Medinan.*

alexandrine sapphire An alexandritelike sapphire that is blue in daylight, and violet, purple, or reddish under most artificial light.

alexandrite A transparent variety of chrysoberyl that has a grass-green or emerald-green color in daylight and wine-red to brownish-red color by transmitted or incandescent artificial light.

alexoite A pyrrhotite-bearing *dunite.* The term is used locally, in the Alexo mine, Dundonald Township, Ontario, Canada. Not recommended usage.

Alfisol In U.S. Dept. of Agriculture soil taxonomy, a soil order characterized by an ochric (or rarely an umbric) epipedon, an argillic horizon, moderate to high base saturation, and the retention of water at less than 15 bars tension during a growing season of at least three months. Because these soils have water and bases, they are widely used for agriculture, though tilth problems are common (USDA, 1975). Suborders and great soil groups of this order have the suffix -alf. See also: *Aqualf; Boralf; Udalf; Ustalf; Xeralf.*

alga An individual plant of the taxon *Algae.*

algae Photosynthetic, almost exclusively aquatic plants of a large and diverse division (Algae) of the thallophytes, including seaweeds and their fresh-water allies. They range in size from simple unicellular forms to giant kelps several meters long, and display extremely varied life-cycles and physiological processes, with, for example, different complexes of photosynthetic pigments. Algae range from the Precambrian. An individual plant is called an *alga.*

algae wash Shoreline drift composed mainly of filamentous algae.

algal Of, pertaining to, or composed of algae.

algal ball *algal biscuit.*

algal biscuit Any of various hemispherical or disk-shaped calcareous masses, up to 20 cm in diameter, produced in freshwater as a result of precipitation by various blue-green algae; e.g. a deposit of marl formed around a piece of algal material or other nucleus as a result of photosynthesis and found on the shallow bottoms of hard-water lakes of the temperate region (as in Wisconsin). Cf: *aegagropile.* Syn: *algal ball; water biscuit; lake biscuit; marl biscuit; girvanella; pycnostromid.*

algal coal *boghead coal.*

algal dust (a) Angular to subangular, medium- to dark-colored (usually brown and brownish-gray) grains or crystals of carbonate, commonly 1-5 microns in diameter, derived from breakdown of algal felts, algally precipitated aragonite needles, algal slime, and comminution of phytoplankton (Chilingar et al., 1967, p. 310). Term proposed by Wood (1941). (b) Algal micrite.

algal head A bulbous, dome-shaped or columnar mass, 10-12 cm in diameter, of mechanically transported, laminated sediments collected by algae (esp. blue-green algae) on a tidal flat or in a lake, and bound together by innumerable algal filaments.

algal limestone A limestone composed largely of the remains of calcium-carbonate-producing algae, or one in which such algae serve to bind together the fragments of other calcium-carbonate-producing forms.

algal mound A local thickening of limestone attributed chiefly to the presence of a distinctive suite of rock types (such as massive calcilutite) containing algae.

algal paste A term used in a loose sense by Schlanger (1957) for dark-gray to black, finely divided (micrograined or microcrystalline) flecks forming a rather dense micritic limestone or dolomite, and associated with organic frame-builders such as corals. It is common in, but not restricted to, the reef core, and it may represent compact, dense, diagenetically altered algal dust.

algal pit A small depression containing, or thought to contain, algae, occurring in the ablation area of a glacier or on the surface of sea ice.

algal reef An *organic reef* in which algae are or were the principal organisms secreting calcium carbonate; e.g. off the coast of Bermuda. The reefs may be up to 10 m high and more than 15 m across.

algal ridge A low ridge or elevated margin at the seaward (outer) edge of a reef flat, composed of the calcium-carbonate secretions of actively growing calcareous algae. See also: *lithothamnion ridge.* Cf: *algal rim; coralgal ridge.*

algal rim A low, slight rim built by actively growing calcareous algae on the lagoonal side of a leeward reef or on the windward side of a reef patch in a lagoon; esp. the outer part of the main reef surface or reef top, situated behind and rising gradually from the reef front, frequently culminating in a low *algal ridge,* and varying greatly in width (up to 500 m).

algal stromatolite *stromatolite.*

algal structure A sedimentary structure of definite form and usually of calcareous composition, resulting from secretion and precipitation by colonial algae; it includes crusts, small pseudopisolitic and pseudoconcretionary forms, biscuit- and cabbage-like heads of considerable size, and laminated structures such as stromatolites or bedding modified by blue-green algal mats. Some so-called algal structures may be of inorganic origin.

algarite A bitumen derived from algae.

algarvite A *melteigite* having more biotite and less nepheline than the original melteigite. Not recommended usage.

Algerian onyx A distinctly banded, stalagmitic form of *onyx marble.*

algerite A pinitic pseudomorph after scapolite.

alginite A maceral of coal within the *exinite* group, consisting of algal matter and characteristic of boghead coals. Cf: *algite; cutinite; sporinite; resinite.*

algite A microlithotype of coal within the *liptite* group, consisting of at least 95% alginite. Cf: *boghead coal.*

algodonite A mineral: Cu_6As.

Algoman orogeny Orogeny and accompanying granitic emplacement that affected Precambrian rocks of northern Minnesota and adjacent Ontario about 2400 m.y. ago; it is synonymous with the *Kenoran orogeny* of the Canadian Shield and was the final event of the Archean according to the current Canadian classification.

algon The viscous, organic binding material of *vase,* consisting of finely divided remains of algae (or of land vegetation, as in the upper parts of the estuary) and iron principally in the form of FeS (Bourcart, 1939).

Algonkian *Proterozoic.*

Algophytic *Archeophytic.*

algorithm A fixed step-by-step procedure for accomplishing a given result, such as solving a mathematical problem. It frequently involves repetition of an operation.

algovite *allgovite.*

alias An ambiguity in the frequency represented by sampled seismic data. Where there are fewer than two samples per cycle, an input signal at one frequency yields the same sample values as, and hence appears to be, another frequency at the output of the system. Half of the frequency of sampling is called the "folding" or "Nyquist" frequency, f_N; a frequency larger than this, $f_N + Y$, appears to be the smaller frequency, $f_N - Y$. The two frequencies are "aliases" of each other.

alidade (a) A rule equipped with simple or telescopic sights, used for determining the directions of objects, specif. a part of a surveying instrument consisting of a telescope or other sighting device, with index, and reading or recording accessories. (b) A surveying instrument used with a planetable for mapping; e.g. *peep-sight alidade* and *telescopic alidade.*

aliettite A clay mineral consisting of regularly interstratified layers of talc and saponite.

aligned current structure *directional structure.*

alignment (a) The placing of surveying points along a straight line. Also, the location of such points with reference to a straight line or to a system of straight lines. (b) Representing on a map the correct direction and character of a line or a feature in relation to other lines or features.—Syn: *alinement.*

alimentation The supplying of a glacier with material, such as snow or firn, that turns to ice; the process of *accumulation*.

alimentation facies Facies characteristics that provide evidence of the source of sediments, as revealed mainly by rock composition (such as "sandstone", "clay", and "chert") (Sonder, 1956). Cf: *precipitation facies.*

alinement *alignment.*

alio A French term for an impervious, ferruginous crust formed by the precipitation of iron salts from subsurface water.

aliphatic hydrocarbon A straight or branched open-chain hydrocarbon having the empirical formula C_nH_{2n+2}, such as methane or phytane.

alive In ore deposits, a syn. of *quick.*

alivincular Said of a type of ligament of bivalve mollusks (e.g. *Ostrea*) whose longer axis is transverse to the hinge axis, or that is not elongated in the longitudinal direction or necessarily situated entirely posterior to the beaks, but is located between the cardinal areas (where present) of the respective valves, with the lamellar component both anterior and posterior to the fibrous component.

alkali [chem] n. Any strongly basic substance, such as a hydroxide or carbonate of an alkali metal (e.g. sodium, potassium). Plural: *alkalies.* Adj: *alkaline [chem]; alkalic [chem].*

alkali [mineral] Said of silicate minerals that contain alkali metals but little calcium, e.g. the alkali feldspars.

alkali [petrology] *alkalic [petrology].*

alkali basalt Critically silica-undersaturated *basalt,* containing normative nepheline, diopside, and olivine, with no normative hypersthene. The term was defined by Yoder and Tilley (1962). Cf: *alkali-olivine basalt; alkaline basalt; basalt; olivine basalt.*

alkalic [chem] An adj. of *alkali [chem].*

alkalic [petrology] (a) Said of an igneous rock that contains more alkali metals than is considered average for the group of rocks to which it belongs. (b) Said of an igneous-rock series that contains more sodium and/or potassium than is required to form feldspar with the available silica. (c) Said of an igneous-rock series having an *alkali-lime index* below 51. Cf: *calcic; alkali-calcic; calc-alkalic.* (d) Said of an igneous rock belonging to the *Atlantic suite.*—Syn: *alkali [petrology]; alkaline [petrology].*

alkali-calcic Said of a series of igneous rocks in which the weight percentage of silica is between 51 and 56 when the weight percentages of CaO and of $K_2O + Na_2O$ are equal. See also: *alkali-lime index.*

alkali-calc index *alkali-lime index.*

alkali charnockite According to Tobi (1971), a rock of the charnockite series in which quartz constitutes 20-60% of the felsic constituents and in which the ratio of alkali feldspar to total feldspar is greater than 90%. He uses the term as a replacement for *enderbite.*

alkali feldspar (a) A group of feldspars composed of mixtures, or mixed crystals, of potassium feldspar (Or, or $KAlSi_3O_8$) and sodium feldspar (Ab, or $NaAlSi_3O_8$) in any ratio; a group of feldspars containing alkali metals but little calcium. (b) A mineral of the alkali feldspar group, such as microcline, orthoclase, sanidine, adularia, albite, anorthoclase, and plagioclase in which the proportion of the An molecule is less than 20%.—Cf: *plagioclase.* Syn: *alkalic feldspar.*

alkali-feldspar granite In the *IUGS classification,* a plutonic rock with Q between 20 and 60, and P/(A+P) less than 10.

alkali-feldspar quartz syenite In the *IUGS classification,* a plutonic rock with Q between 5 and 20, and P/(A+P) less than 10.

alkali-feldspar syenite In the *IUGS classification,* a plutonic rock with Q between 0 and 5, and P/(A+P) less than 10.

alkali flat A level area or plain in an arid or semiarid region, encrusted with alkali salts that became concentrated by evaporation and poor drainage; a *salt flat.* See also: *playa.*

alkali gabbro The plutonic equivalent of *alkali basalt.* As defined by Wilkinson (1968), alkali gabbro contains plagioclase at least as calcic as An_{50}, clinopyroxene, and usually olivine; nepheline and/or analcime are present but make up less than 10 per cent of the rock, and alkali feldspar makes up less than 10 per cent of the total feldspar.

alkali lake A *salt lake,* commonly found in an arid region, whose waters contain in solution large amounts of sodium carbonate and potassium carbonate, as well as sodium chloride and other alkaline compounds; e.g. Lake Magadi in the Eastern Rift Valley of Kenya. See also: *potash lake; soda lake.* Syn: *alkaline lake.*

alkali-lime index A means of classifying igneous rocks introduced by Peacock (1931) based on the weight percentage of silica present when the weight percentages of CaO and of K_2O + Na_2O are equal. Four chemical classes of igneous rocks based on this index are recognized: *alkalic* (when the silica percentage is less than 51), *alkali-calcic* (when it is between 51 and 56), *calc-alkalic* (when it is between 56 and 61), *calcic* (when it is over 61). Syn: *alkali-calc index.*

alkaline [chem] An adj. of *alkali [chem].*

alkaline [petrology] *alkalic [petrology].*

alkaline basalt As proposed by Chayes (1964), a replacement for the terms *alkali basalt* and *alkali-olivine basalt.* Basalts with nepheline and/or acmite in the CIPW norm fall in this category.

alkaline soil A soil whose pH value is greater than 7.0.

alkaline-sulfide hypothesis A theory of ore-deposit formation that postulates complex sulfide ions in hydrothermal solutions as a means of ore transportation. The theory accounts satisfactorily for only a few of the common metals (Krauskopf, 1967, p.501-502).

alkalinity [lake] Refers to the quantity and kinds of compounds present in a lake that collectively shift the pH to the alkaline side of neutrality, thereby providing an index to the nature of the rocks within the drainage basin and the degree to which they are weathered. The property of alkalinity is usually imparted by the presence of bicarbonates and hydroxides, and less frequently in inland waters by borates, silicates, and phosphates. Alkalinity is frequently expressed as the total quantity of base, usually in equilibrium with carbonate or bicarbonate, that can be determined by titration with a strong acid.

alkalinity [oceanog] The number of milliequivalents of hydrogen ion that is neutralized by one liter of seawater at 20°C.

alkali-olivine basalt A term proposed by Tilley in 1950 to replace "olivine basalt" as used by Kennedy in 1933. The term was changed to *alkali basalt* by Yoder and Tilley (1962) and to *alkaline basalt* by Chayes (1964).

alkali soil An obsolete syn. of *sodic soil.*

alkalitrophy The quality or state of an arid-region lake exhibiting alkaline characteristics. Adj: *alkalitrophic.*

Alkemade line In a ternary phase diagram, a straight line that connects the composition points of two primary phases whose areas are adjacent and whose interface forms a boundary curve (Levin et al., 1964, p. 5). Cf: *Alkemade theorem.* Incorrect syn: *conjugation line.*

Alkemade theorem The statement in chemical phase studies that, in a ternary phase diagram, the direction of falling temperature on the boundary curve between the areas of two primary phases is always away from the *Alkemade line;* also, that the temperature maximum on the boundary curve is at the point at which the Alkemade line intersects it, or, if it does not intersect it, at the end of the boundary curve which, if prolonged, would intersect the Alkemade line (Levin et al., 1964, p. 5).

allactite A mineral: $Mn_7(AsO_4)_2(OH)_8$.

allalinite An altered gabbro containing saussurite as euhedral pseudomorphs after the original minerals, thus maintaining the original ophitic texture. Its name was derived from Allalin, Switzerland by Rosenbusch in 1896. Not recommended usage.

allanite A monoclinic, cerium-bearing mineral of the epidote group: (Ce,Ca,Y) $(Al,Fe)_3(SiO_4)_3(OH)$. It is typically an accessory mineral in igneous rocks (granite, syenite, diorite, pegmatite) and in their metamorphic equivalents. Syn: *orthite; cerine; bucklandite; treanorite.*

allargentum A mineral: $Ag_{1-x}Sb_x$, with x=0.09-0.16.

allcharite *goethite.*

alleghanyite A mineral: $Mn_5(SiO_4)_2(OH)_2$.

Alleghenian North American provincial stage: Middle Pennsylvanian (above Pottsvillian, below Conemaughian).

Allegheny orogeny A mountain-building event that deformed the rocks of the Valley and Ridge province, and those of the adjacent Allegheny Plateau, in the central and southern Appalachians. It cannot be closely dated, as there are no limiting overlying strata; Pennsylvanian rocks are involved in many places, and Lower Permian (Dunkard) in a few. Most of the orogeny was probably late in the Paleozoic, but phases may have extended into the early Triassic.

allemontite A mineral: a mixture of stibarsen (SbAs) and As or Sb. Syn: *arsenical antimony.*

allenite *pentahydrite.*

Allen's rule In zoology, the statement that warm-blooded animals tend to have longer protruding body parts (i.e. legs, arms, tails) in warmer parts of the environment than in cooler. Named after Joel A. Allen (d.1921), American zoologist. Cf: *Bergmann's rule.*

Allerød n. A term used primarily in Europe for an interval of late-glacial time (centered about 11,000 years ago) following the Older Dryas and preceding the Younger Dryas, during which the climate as inferred from stratigraphic and pollen data (Iversen, 1954) ameliorated, favoring the growth of birch, pine, and willow vegetation.—adj. Pertaining to the late-glacial Allerød interval and to its climate, deposits, biota, and events.

allevardite *rectorite.*

allgovite An obsolete term for a group of igneous rocks containing augite and plagioclase and ranging in composition from *dolerite* to *gabbro,* including porphyritic varieties. Its name is derived from Allgäu Alps. Also spelled: *algovite.*

alliaceous Said of minerals that have an odor of garlic when rubbed, scratched, or heated; e.g. arsenical minerals.

Alling grade scale A metric *grade scale* designed by Harold L. Alling (1888-1960), U.S. geologist, for two-dimensional measurements (such as with thin sections or polished blocks) of sedimentary rocks. It has a constant geometric ratio of 10 for the major divisions (colloid, clay, silt, sand, gravel, cobble, boulder) and one of the fourth root of 10 for the four-fold subdivisions of each major unit (Alling, 1943).

allingite A fossil resin (retinite) containing no succinic acid but considerable sulfur, found at Allinges in Haute-Savoie, France.

allite A rock name for *allitic* material; e.g. bauxite and laterite.

allitic Said of a rock or soil from which silica has been largely removed and which contains a high proportion of aluminum and iron compounds in the clay fraction.

allivalite A *gabbro* containing anorthite and olivine, with pyroxene rare or absent and accessory apatite and opaque iron oxides. Its name was derived from Allival, Isle of Rhum, Scotland by Harker in 1908. Not recommended usage.

allochem A collective term introduced by Folk (1959, p. 4) for one of several varieties of discrete and organized carbonate aggregates that serve as the coarser framework grains in most mechanically deposited limestones, as distinguished from sparry calcite (usually cement) and carbonate-mud matrix (micrite). Important allochems include: silt-, sand-, and gravel-size *intraclasts;* ooliths; pellets; lumps; and fossils or fossil fragments (carbonate skeletons, shells, etc.). Adj: *allochemical.* Cf: *pseudoallochem; orthochem.*

allochemical metamorphism Metamorphism that is accompanied by addition or removal of material so that the bulk chemical composition of the rock is changed (Mason, 1958).

allochetite A porphyritic hypabyssal igneous rock composed of phenocrysts of labradorite, alkali feldspar, titanaugite, nepheline, magnetite, and apatite, in a fine-grained, felty groundmass of augite, biotite, magnetite, hornblende, nepheline, and alkali feldspar. It was named from the Allochet Valley, Italy, by Ippen in 1903. Not recommended usage.

allochromatic Said of a mineral that is colorless in its pure state, but that has color due to submicroscopic inclusions, or to the presence of a closely related element that has become part of the chemical structure of the mineral. Cf: *idiochromatic.*

allochthon [sed] A mass of redeposited sedimentary materials originating from distant sources.

allochthon [tect] A mass of rock that has been moved from its place of origin by tectonic processes, as in a *thrust sheet* or *nappe.* Many allochthonous rocks have been moved so far from their original sites that they differ greatly in facies and structure from those on which they now lie. Ant: *autochthon.* Adj: *allochthonous.* Also spelled: *allochthone.*

allochthone *allochthon.*

allochthonous Formed or produced elsewhere than in its present place; of foreign origin, or introduced. The term is widely applied, e.g. to coal or peat that originated from plant material transported from its place of growth, to an *allochthon* on a low-angle thrust fault, or to a *fallback breccia* at a meteorite impact crater. The term was first used by Naumann (1858, p. 657) to designate rocks of distant origin; it is similar in meaning to *allogenic,* which refers to constituents rather than whole formations. Ant: *autochthonous.* See also: *parautochthonous; exotic; heterochthonous.*

allochthonous mantle Loose debris of rock fragments or soil transported from elsewhere to its present site; not formed in situ. Cf: *autochthonous mantle; block field.*

allochthony Accumulation of plant materials elsewhere than at their site of growth. Ant: *autochthony.* See also: *primary allochthony; secondary allochthony; drift theory [coal]; hypautochthony.*

alloclasite A monoclinic mineral: (Co,Fe)AsS.

alloclastic breccia A breccia that is formed by disruption of nonvolcanic rocks by volcanic processes beneath the Earth's surface; a type of *volcanic breccia.*

allocyclicity The state of cyclothemic deposition that results from changes in the supply of energy or material input into a sedimentary system (Beerbower, 1964, p.32). It involves such mechanisms as uplift, subsidence, climatic variation, eustatic change in sea level, and other changes external to the sedimentary unit. Cf: *autocyclicity.*

allodapic limestone A limestone deposited by turbidity currents, e.g. off reef fronts (Meischner, 1964, p. 156).

allofacial Pertaining to rocks belonging to different metamorphic facies. It is inferred that the present juxtaposition of allofacial rocks has been brought about by tectonic transport. Cf: *isofacial.*

allogene An *allogenic* mineral or rock constituent; e.g. a xenolith in an igneous rock, a pebble in a conglomerate, or a detrital mineral in a placer deposit. Ant: *authigene.* Syn: *allothigene.*

allogenic (a) Formed or generated elsewhere, usually at a distant place; specif. said of rock constituents and minerals that were derived from pre-existing rocks and transported to their present depositional site, or that came into existence before the rock of which they now constitute a part and at some place other than where now found. Ant: *authigenic.* Cf: *chthonic; allochthonous.* Syn: *allothogenic; allothigenic; allothigenous; allothigenetic.* (b) Pertaining to a stream that derives much of its water from a distant terrain or from beyond its surface draining area, such as one originating in a humid or glacial region and flowing across an arid or desert region. (c) Said of an ecologic succession that resulted from factors that arise from outside the natural community and alter its habitat, such as an allogenic drought of prolonged duration. Cf: *autogenic [ecol].*

allokite A clay mineral intermediate in structure between kaolinite and allophane.

allolistostrome A *mélange* of sedimentary origin; an *olistostrome* containing exotic blocks (Raymond, 1978).

allomeric *isomorphous.*

allomerism *isomorphism.*

allometry [biol] (a) The differential growth of a part of an organism in relation to its entirety. Cf: *isometry.* (b) Measurement and study of the growth of a part of an organism in relation to its entirety.

allometry [geomorph] In geomorphology, this term has come to be used in studying the ratios between relative rates of change of variables within geomorphic systems. These systems rarely achieve a steady state (Bull, 1975). In fluvial geomorphology, paired ordered variables form power-function relationships and usually indicate allometric change (Woldenberg, 1966).

allomicrite Allochthonous *orthomicrite.*

allomorph (a) A *polymorph* or a *dimorph.* (b) A *pseudomorph.*

allomorphic *polymorphic.*

allomorphism [cryst] *paramorphism.*

allomorphism [paleont] A term used erroneously for *xenomorphism* in reference to bivalve mollusks.

allomorphite A mineral consisting of barite pseudomorphous after anhydrite.

allomorphosis Biologic evolution characterized by a rapid increase in specialization; evolutionary allometry. Cf: *aromorphosis.*

allomorphous *polymorphic.*

allopatric Said of organisms or biologic events occurring in different geographic areas; e.g. the development of a distinct species from an isolated population. Noun: *allopatry.* Cf: *sympatric.*

allophane An amorphous clay mineral: $Al_2O_3 \cdot SiO_2 \cdot nH_2O$. It consists of a hydrous aluminum silicate gel (of highly variable composition, with minor amounts of bases and accessory acid radicals) that is or appears to be amorphous to X-ray diffraction and that changes, on standing, from a glassy or translucent constituent of clay materials to one with an earthy appearance owing to the loss of water. Allophane has various colors (snow white, blue, green, brown, yellow, or colorless) and often occurs in incrustations, thin seams, or rarely stalactitic masses. Syn: *allophanite.*

allophaneton An obsolescent term used by ceramists (esp. in Europe) for the portion of a clay that is soluble in hydrochloric acid. Cf: *kaolinton.*

allophanite An obsolete syn. of *allophane.*

allophanoid A group name for the clay minerals allophane, halloysite, and montmorillonite.

Allotheria A subclass of Mammalia, characterized by elongate multituberculate molar teeth and peculiarly sectorial lower premolars. It includes a single order of rodentlike habitus, the longest-lived of all mammalian orders; stratigraphic range, Upper Jurassic to Upper Eocene. Syn: *Multituberculata.*

allothigene *allogene.*

allothigenetic *allogenic.*

allothigenic *allogenic.*

allothigenous (a) *allogenic.* (b) In the classification of pyroclastics, the equivalent of *accidental.* Cf: *authigenous.*

allothimorph A constituent of a metamorphic rock which, in the new rock, has not changed its original crystal outlines (Johannsen, 1939).

allothogenic *allogenic.*

allothrausmatic A descriptive term applied to igneous rocks with an orbicular texture in which the nuclei of the orbicules are xenoliths differing in composition from the groundmass (Eskola, 1938, p.476). Cf: *isothrausmatic; crystallothrausmatic; homeothrausmatic; heterothrausmatic.*

allotrio- A prefix meaning "alien", "foreign".

allotrioblast *xenoblast.*

allotriomorphic A syn. of *xenomorphic.* The term *allotriomorphisch* was proposed by Rosenbusch (1887, p.11), but it lacks priority. See also: *anhedral.*

allotriomorphic-granular *xenomorphic.*

allotrope A crystal form of a substance that displays *allotropy.*

allotrophic *heterotrophic.*

allotropy *Polymorphism* in an element, e.g. sulfur as both orthorhombic and monoclinic. See also: *allotrope.*

allowable The amount of oil or gas that a well or leasehold is permitted to produce under *proration* by a regulatory body.

alluaudite A mineral: $(Na,Ca)_{1\text{-}2}(Fe^{+3},Mn^{+2})_3(PO_4)_3$.

alluvia Seldom-used plural of *alluvium.*

Alluvial A name, now obsolete, applied by Jameson (1808) from the teachings of A.G. Werner in the 1790's to the group or series of rocks consisting of unconsolidated or poorly consolidated gravels, sands, clays, and peat that were believed to have been formed after the withdrawal of the ocean from the continents. It constituted the fourth (following the *Floetz*) of the divisions in which Werner placed the rocks of the geologic column. Syn: *Tertiary.*

alluvial [ore dep] Said of a *placer* formed by the action of running water, as in a stream channel or alluvial fan; also, said of the valuable mineral, e.g. gold or diamond, associated with an alluvial placer.

alluvial [sed] adj. Pertaining to or composed of alluvium, or deposited by a stream or running water; e.g. an "alluvial clay" or an "alluvial divide". Syn: *alluvian; alluvious.*—n. *alluvium.*

alluvial apron *bajada.*

alluvial bench A term used by Hobbs (1912, p. 214) for a feature now known as a *bajada.*

alluvial channel A channel whose bed and banks are composed of alluvium.

alluvial cone An *alluvial fan* with very steep slopes; it is generally higher and narrower than a fan, and is composed of coarser and thicker material believed to have been deposited by larger streams. The term is sometimes used synonymously with alluvial fan. Syn: *cone of dejection; cone of detritus; hemicone; debris cone [geomorph]; cone delta; dry delta; wash [sed].*

alluvial dam A deposit of alluvium that is built by an overloaded stream and that obstructs its channel, thereby impounding water behind the deposit; esp. such a dam in a distributary on an alluvial fan.

alluvial-dam lake A lake formed behind an *alluvial dam;* esp. a lake at the apex of an alluvial fan, formed as a result of a *cloudburst flood.*

alluvial deposit *alluvium.*

alluvial fan A low, outspread, relatively flat to gently sloping mass of loose rock material, shaped like an open fan or a segment of a cone, deposited by a stream (esp. in a semiarid region) at the place where it issues from a narrow mountain valley upon a plain or broad valley, or where a tributary stream is near or at its junction with the main stream, or wherever a constriction in a valley abruptly ceases or the gradient of the stream suddenly decreases; it is steepest near the mouth of the valley where its apex points upstream, and it slopes gently and convexly outward with gradually decreasing gradient. Cf: *alluvial cone; bajada.* Syn: *detrital fan; talus fan; dry delta.*

alluvial-fan shoreline A prograding shoreline formed where an alluvial fan is built out into a lake or sea.

alluvial fill A deposit of alluvium occupying a stream valley, conspicuously thicker than the depth of the stream. It represents a single stratigraphic unit.

alluvial flat A small *alluvial plain* bordering a river, on which alluvium is deposited during floods. Syn: *river flat.*

alluvial meander An extremely sinuous bend in an *alluvial river.*

alluvial plain A level or gently sloping tract or a slightly undulating land surface produced by extensive deposition of alluvium, usually adjacent to a river that periodically overflows its banks; it may be situated on a flood plain, a delta, or an alluvial fan. Cf: *alluvial flat.* Syn: *wash plain; waste plain; river plain; aggraded valley plain; bajada.*

alluvial-plain shoreline A prograding shoreline formed where the broad alluvial slope at the base of a mountain range is built out into a lake or sea.

alluvial river A river that occupies a broad flood-plain belt over which the depth of alluvium deposited by the river equals or exceeds the depth to which scour takes place in time of flood.

alluvial slope A surface underlain by alluvium, which slopes down and away from the sides of mountains and merges with a plain or a broad valley floor (Bryan, 1923a, p. 86); an alluvial surface that lacks the distinctive form of an alluvial fan or a bajada. See also: *bajada.*

alluvial-slope spring *boundary spring.*

alluvial soil A young soil on flood plains and deltas that is actively in the process of construction and has primary characteristics of the alluvium itself.

alluvial stone A gemstone that has been transported and deposited by a stream.

alluvial talus Accumulation of rock fragments of any size or shape, derived during and after heavy rains by rainwash, during storm flow, and by melting snow, passing through a gully, chute, or couloir in a cliff face or from a steep rocky slope above. Larger blocks collect at the top, but fine sizes reach the bottom as miniature *debris flows* (White, 1967, p. 237). Cf: *scree; talus cone.*

alluvial terrace A *stream terrace* composed of unconsolidated alluvium (including gravel), produced by renewed downcutting of the flood plain or valley floor by a rejuvenated stream or by the later covering of a terrace with alluvium. Cf: *rock terrace.* Syn: *stream-built terrace; built terrace; fill terrace; drift terrace.*

alluvial tin *stream tin.*

alluviation (a) The subaerial deposition or formation of alluvium or alluvial features (such as cones or fans) at places where stream velocity is decreased or streamflow is checked; the process of aggradation or of building-up of sediments by a stream along its course, or of covering or filling a surface with alluvium. (b) A hydraulic effect on solids suspended in a current of water, whereby the coarsest and heaviest particles are the first to settle out, and the finest muds the last, as gradient or velocity of a stream is decreased (Pryor, 1963).

alluvion [geol] (a) The wash of the sea against the shore, or the flow of a river against its bank. (b) An overflowing; an inundation or flood, esp. when the water is charged with much suspended material. (c) Material deposited by a flood; *alluvium.*—An obsolete term.

alluvion [law] The formation of new land by the gradual or imperceptible action of flowing water or of waves and currents; *accretion.* Also, the land so added. Cf: *diluvion.*

alluvion [volc] An obsolete term for a flood of volcanic-cinder mud and for the resulting consolidated material.

almandine (a) The iron-aluminum end member of the garnet group, characterized by a deep-red to purplish color: $Fe_3Al_2(SiO_4)_3$. It occurs in mica schists and other regionally metamorphosed rocks, and is used as a gemstone. Syn: *almandite; alamandine; almond stone.* (b) A violet or mauve variety of ruby spinel; a reddish-purple to purplish-red spinel. (c) A reddish-purple sapphire.

almandine-amphibolite facies A subdivision of Eskola's *amphibolite facies,* suggested by Fyfe, Turner, and Verhoogen (1958) to distinguish higher-pressure assemblages from those of lower pressure, which they termed *hornblende-hornfels facies.* The term was later abandoned (Fyfe and Turner, 1966) in favor of amphibolite facies, although hornblende-hornfels facies was retained.

almandite *almandine.*

almashite A green or black variety of amber that has a low content of oxygen (2.5-3%).

almeriite *natroalunite.*

almond-shaped bomb *spindle-shaped bomb.*

almond stone *almandine.*

almost atoll An atoll with a minute noncoral island, generally of volcanic origin, in the central lagoon.

alnoite A *lamprophyre* chiefly composed of biotite or phlogopite and melilite as essential minerals, commonly with olivine, calcite, and clinopyroxene. Perovskite, apatite, nepheline, and garnet may be present. Its name (Rosenbusch, 1887) is derived from Alnö, Sweden. Also spelled: *allnöite; alnöite.*

alongshore Along the shore or coast, such as an "alongshore drift" or "alongshore current"; *longshore.*

alp (a) A high, rugged, steep-sided mountain, esp. one that is snow-covered, resembling topographically those in the European Alps. (b) A high pasture or meadowland on a mountain side, between timberline and snowline, like those in the Swiss Alps; also a high *shoulder [glac geol]* or gentle slope, esp. in the Swiss Alps, commonly above a glaciated valley at a marked change of slope. Cf: *alb.*

alpestrine *montane.*

alpha [cryst] (a) In a biaxial crystal, the smallest *index of refraction.* (b) The interaxial angle between the *b* and *c* crystallographic axes.—Cf: *beta [cryst]; gamma [cryst].*

alpha [mineral] adj. Of or relating to one of two or more closely related minerals and specifying a particular physical structure (esp. a polymorphous modification); specif. said of a mineral that is stable at a temperature lower than those of its *beta* and *gamma* polymorphs (e.g. "alpha cristobalite" or "α-cristobalite", the low-temperature tetragonal phase of cristobalite). Some mineralogists reverse this convention, using α for the high-temperature phase (e.g. "alpha carnegieite", the isometric phase of carnegieite stable above 690°C).

alpha* angle The angle of the reciprocal lattice between the *b* axis* and the *c* axis,* which is equal to the interfacial angle between (010) and (001). Cf: *beta* angle; gamma* angle.*

alpha chalcocite *digenite.*

alpha decay Radioactive decay of an atomic nucleus by the emission of an *alpha particle.* Cf: *beta decay.*

alpha level *significance level.*

alpha particle (a) A particle, emitted from an atomic nucleus during one type of radioactive decay, which is positively charged and has two protons and two neutrons. It is physically identical with the nucleus of a ^{4}He atom. Cf: *beta particle; gamma radiation.* (b) By extension, the nucleus of a ^{4}He atom.—Less-preferred syn: *alpha ray.*

alpha-particle recoil tracks The paths of *radiation damage* in a solid substance, caused by the recoil nucleus accompanying the alpha-particle decay of uranium and thorium impurities. The tracks are similar to *fission tracks,* but are much smaller and more numerous. An age determination can be made on the basis of the track density, usually examined with an electron microscope, plus determinations of the thorium and uranium contents of the sample (Huang & Walker, 1967, p. 1103-1106). Syn: *alpha-recoil tracks.*

alpha quartz The polymorph of quartz that is stable below 573°C, that has a vertical axis of three-fold symmetry and three horizontal axes of two-fold symmetry, and that has a higher refractive index and birefringence than those of *beta quartz.* It occurs commonly as a constituent of igneous, metamorphic, and sedimentary rocks, and in veins, geodes, and large pegmatites. Also spelled: *α-quartz.* Syn: *low quartz.*

alpha ray A less-preferred syn. of *alpha particle.*

alpha-recoil tracks *alpha-particle recoil tracks.*

alpha-vredenburgite A homogeneous, metastable mineral: $(Mn,Fe)_3O_4$. It has the same composition as that of *beta-vredenburgite,* and is regarded as an iron-rich hausmannite.

alphitite A term suggested by Salomon (1915, p. 398) for a clay or silt consisting largely of rock flour, such as the fine material produced by a glacier. Twenhofel (1937, p. 84) recommends disuse of the term "because of inability to determine that a clay is a rock flour and not composed of particles of many origins brought together by wind or water".

Alpides A name used by Suess for the great orogenic belt or system of young folded mountains, including the Alps, that extends eastward from Spain into southern Asia. Syn: *Alpine-Himalayan belt.*

alpine [ecol] Characteristic or descriptive of the mountainous regions lying between timberline and snowline; said of the climate, flora, relief, ecology, etc. Less strictly, pertaining to high elevations and cold climates. Cf: *montane.*

alpine [geomorph] Pertaining to, characteristic of, or resembling the European Alps or any lofty mountain or mountain system, esp. one so modified by intense glacial erosion as to contain aiguilles, cirques, horns, etc.; e.g. an *alpine* lake resulting from glacial erosion and situated in or along the border of a high mountain region. Spelled *Alpine* when referring specif. to the European Alps.

alpine [struc geol] A general term for topographical and structural features that resemble in grandeur and complexity those of the European Alps, regardless of the age or location of the mountains and features so described.

alpine glacier Any *glacier* in a mountain range except an *ice cap* or *ice sheet.* It usually originates in a *cirque* and may flow down into a valley previously carved by a stream. Syn: *mountain glacier; valley glacier.*

Alpine-Himalayan belt *Alpides.*

alpine karst Karst formed at high altitude, or in polar regions regardless of altitude. Syn: *glaciokarst; nival karst.*

Alpine Meadow soil A great soil group in the 1938 classification system, an intrazonal, hydromorphic group of dark soils that develop under grasses above the timber line (USDA, 1938). Most of these soils are now classified as *Aquods.*

Alpine orogeny A name for the relatively young orogenic events of southern Europe and Asia, by which the rocks of the Alps and the remainder of the Alpide orogenic belt were strongly deformed. Stille includes in his Alpidic orogenic era all orogenic events from the Jurassic to the end of the Tertiary, but most geologists restrict the era to the Tertiary, with many episodes of varying strength from place to place, ending during the Miocene or Pliocene.

Alpine-type facies series Rocks produced in the highest-pressure type of dynamothermal regional metamorphism at no more than moderate temperature (150° to 400°C), characterized by the presence of the pumpellyite and glaucophane schist facies. It may also involve the zeolite facies in the low-temperature/highest-pressure range and the deep-seated eclogite facies at the highest pressures with moderate temperatures (Hietanen, 1967, p.203).

alpinotype tectonics The tectonics of orogenic belts, regardless of age, produced by *convergent plate boundary* processes (the *orthogeosynclines* of older terminology). They are characterized in their internal parts by deep-seated plastic folding and plutonism, and in their external parts by lateral thrusting, which has produced nappes, thrust sheets, and closely crowded folds. Cf: *germanotype tectonics.* Syn: *orthotectonics.*

alpland An area whose topography resembles that of the Alps.

alquifou A coarse-grained galena, used by potters in preparing a green glaze.

alsbachite A porphyritic *granodiorite* chiefly composed of sodic plagioclase, quartz, and a small amount of alkali feldspar, often with accessory garnet and mica. The quartz and alkali feldspar commonly form the phenocrysts in a granular groundmass. Named for Alsbach, Germany by Chelius in 1892. Not recommended usage.

ALSEP An acronym for Apollo Lunar Surface Experiments Package, a collection of geophysical and other instruments, with auxilary devices for data collection and transmission, powered by radioisotopic thermoelectric generators, and emplaced on the Moon during Apollo 12, 14, 15, 16, and 17 missions.

alstonite A mineral: $BaCa(CO_3)_2$. It is the triclinic, pseudo-orthorhombic dimorph of barytocalcite. Syn: *bromlite.*

alta A miner's term for the black, shaly, highly sheared capping of quicksilver orebodies. Syn: *black alta.*

Altaides A name used by Suess for a late Paleozoic orogenic belt extending across the width of Eurasia, and including also the Appalachian and Ouachita belts of North America. It is named for the Altai Mountains of central Asia, where there was late Paleozoic deformation, but as these mountains are remote and still poorly known, the term is little used now by tectonic geologists; most modern references are to component parts, such as the Variscan belt in Europe.

altaite A tin-white mineral: PbTe.

alteration (a) Any change in the mineralogic composition of a rock brought about by physical or chemical means, esp. by the action of hydrothermal solutions; also, a secondary, i.e. supergene, change in a rock or mineral. Alteration is sometimes considered as a phase of metamorphism, but is usually distinguished from it because of being milder and more localized than metamorphism is generally thought to be. (b) Changes in the chemical or mineralogical composition of a rock produced by weathering.

altered rock A rock that has undergone changes in its chemical and mineralogic composition since its original formation.

alterite A general term for altered, unrecognizable grains of heavy minerals.

alternate In plant morphology, pertaining to the attachment of parts, e.g. leaves, singly at each node; also, said of plant parts in regular occurrence between other organs.

alternate folding Deflection of the surface of a brachiopod shell in which the fold of one valve is opposed by the sulcus of the other valve.

alternate terrace One of several *meander-scar terraces.*

alternating-field demagnetization The process of progressive demagnetization by exposing rock samples to AC fields which are smoothly decreased in the absence of a DC field. This brings about randomization of magnetic remanence carried by material whose magnetization can be switched by fields less than or equal to the maximum peak AC field. Cf: *thermal demagnetization; chemical demagnetization.* Syn: *AC demagnetization.*

alternation of generations The occurrence in the life cycle of a plant or animal of two or more forms having different types of development, usually involving the regular alternation of sexual and asexual generations. Syn: *heterogony; metagenesis [evol].*

althausite An orthorhombic mineral: $Mg(PO_4)(OH,F,O)$.

altimeter An instrument, usually an aneroid barometer, for determining height above ground or above mean sea level, based on the fall of atmospheric pressure accompanying an increase in altitude.

altimetric frequency curve A curve showing the altitudinal distribution of the highest points in a series of small squares that divide the map of a given region.

altiplanation A group of erosion processes, involving solifluction and related mass movement, which tend to produce flat or terrace-like surfaces. Such processes are especially active at high elevations and latitudes where periglacial processes predominate. Cf: *equiplanation; cryoplanation.*

altiplanation terrace A hillside or summit bench that is cut in bedrock, lacks predominant structural control, and is confined to cold climates. These terraces have a veneer of solifluction debris, which may be imprinted with patterned ground. They appear to range in width from about 10 m to 2-3 km and to exceed 10 km in length. Their gradient is commonly 1 to 12 ° (Washburn, 1973, p. 205). Syn: *equiplanation terrace.*

altiplano A high-lying plateau or tableland; specif. the high plateau of western Bolivia, consisting of a string of intermontane basins. Etymol: Spanish. Syn: *altiplanicie.*

Altithermal n. A term proposed by Antevs (1948, p.176) for a dry postglacial interval (from about 7500 to 4000 years ago) following the Anathermal and preceding the Medithermal, during which temperatures were warmer than at present. It corresponds to the *Atlantic* interval or the middle part of the *Hypsithermal.* See also: *thermal maximum; Megathermal; Xerothermic.* —adj. Pertaining to the postglacial Altithermal interval and to its climate, deposits, biota, and events.

altithermal adj. Pertaining or belonging to a climate characterized by rising or high temperatures; e.g. "altithermal soil" of postglacial time.

altitude (a) The vertical distance of a level, a point, or an object considered as a point, above or below the surface of the Earth, measured from a given datum, usually mean sea level. Altitude is positive if the point or object is above the given datum, and negative if it is below it. Cf: *elevation.* (b) The vertical angle between the horizontal plane of the observer and any higher point (such as the summit of a peak).

alto A term used in the SW U.S. for a bluff, height, or hill. Etymol: Spanish, "high ground".

alum (a) A mineral: $KAl(SO_4)_2 \cdot 12H_2O$. It is colorless or white, and has a sweet-sour astringent taste. Cf: *kalinite.* Syn: *potash alum; potassium alum.* (b) A group of minerals containing hydrous aluminum sulfates, including alum, kalinite, soda alum, mendozite, and tschermigite.

alum coal A pyritic, argillaceous brown coal containing alum that formed as a weathering product. Cf: *alum shale.*

alum earth *alum shale.*

aluminite A mineral: $Al_2(SO_4)(OH)_4 \cdot 7H_2O$. Syn: *websterite.*

aluminocopiapite A mineral of the copiapite group: $AlFe_6^{+3}(SO_4)_9(OH)_3 \cdot 30H_2O$.

aluminosilicate A silicate in which aluminum substitutes for the silicon in the SiO_4 tetrahedra.

alumite *alunite.*

alumocalcite A variety of opal containing alumina and lime as impurities.

alumogel An amorphous aluminum hydroxide that is a constituent of bauxite; *cliachite.*

alumohydrocalcite A mineral: $CaAl_2(CO_3)_2(OH)_4 \cdot 3H_2O$. It occurs as white, chalky masses consisting of radially fibrous spherulites.

alumotungstite A trigonal mineral: $(W,Al)_{16}(O,OH)_{48} \cdot H_2O$.

alum rock *alunite.*

alum schist *alum shale.*

alum shale An argillaceous, often carbonaceous, rock impregnated with *alum,* originally containing iron sulfide (pyrite, marcasite) which, when decomposed, formed sulfuric acid that reacted with the aluminous and potassic materials of the rock to produce aluminum sulfates. Syn: *alum earth; alum schist; alum slate.*

alum slate *alum shale.*

alumstone *alunite.*

alunite (a) A mineral: $KAl_3(SO_4)_2(OH)_6$. It is isomorphous with natroalunite, sometimes contains appreciable sodium, generally occurs as a hydrothermal-alteration product in feldspathic igneous rocks, and is used in the manufacture of alum. Syn: *alumstone; alum rock; alumite.* (b) A group of minerals containing hydrous sulfates, including alunite, natroalunite, jarosite, natrojarosite, ammoniojarosite, argentojarosite, and plumbojarosite.

alunitization Introduction of, or replacement by, alunite.

alunogen A mineral: $Al_2(SO_4)_3 \cdot 18H_2O$. It occurs as a white, fibrous incrustation or efflorescence formed by volcanic action or by decomposition of pyrite in alum shales. Syn: *feather alum; hair salt.*

alurgite A manganiferous muscovite.

alushtite A mixture of dickite with clay minerals (such as illite).

alvanite A mineral: $Al_6(VO_4)_2(OH)_{12} \cdot 5H_2O$.

alveolar (a) In invertebrates, having small cavities or pits. (b) In vertebrates, pertaining to a tooth socket.

alveolar weathering *honeycomb weathering.*

alveole A space or cavity, such as a *vacuole* in a foraminiferal test wall; *alveolus.*

alveolinid Any foraminifer belonging to the family Alveolinidae, characterized by an imperforate, porcelaneous, axially elongated test that may be subcylindrical, fusiform, ellipsoidal, or spherical. Range, Lower Cretaceous to present. Although this group resembles the *fusulinids* in shape, the two groups are not genetically related.

alveolitoid Said of a type of reclined corallite having a vaulted upper wall and a nearly plane lower wall parallel to the surface of adherence of the colony (as in the genus *Alveolites*) (TIP, 1956, pt.F, p.245).

alveolus (a) In invertebrates, a small cavity or pit; e.g. a minute blind cavity in the keriotheca of some fusulinids, the conical cavity in the anterior end of the guard of a belemnite, or a pit in the skeleton of a bryozoan colony. (b) In vertebrates, the socket of a tooth. Pl: *alveoli.* Adj: *alveolar.* Syn: *alveole.*

alveozone A trough just beneath the selenizone or periphery in certain gastropods (Batten, 1966, p. 21).

alvikite A hypabyssal rock, the fine-grained equivalent of *sövite* (calcite carbonatite). The name, given by Eckermann in 1928, is for Alvik on Alnö Island, Sweden. Not recommended usage.

amagmatic Said of a structure, region, or process that does not involve magmatic activity.

amakinite A mineral: $(Fe,Mg)(OH)_2$.

amalgam (a) A naturally occurring alloy of silver with mercury; mercurian silver. It is found in the oxidation zone of silver deposits and as scattered grains in cinnabar ores. See also: *gold amalgam; moschellandsbergite.* Syn: *silver amalgam; argental mercury.* (b) A general term for alloys of mercury with one or more of the well-known metals (except iron and platinum); esp. an alloy of mercury with gold, containing 40-60% gold, and obtained from the plates in a mill treating gold ore.

amalgamate Said of a type of wall structure in trepostome bryozoans in which the zooecial boundaries are not visible in tangential section. Cf: *integrate.*

amang A term used in Malaysia for the heavy iron and tungsten minerals (and associated minerals) found with placer cassiterite deposits.

amarantite A dark reddish-purple mineral: $FeSO_4(OH) \cdot 3H_2O$.

amararhysis A *skeletal canal* in dictyonine hexactinellid sponges running longitudinally within the body wall, opening at intervals to the cloaca by slit-like apertures, and opening to the exterior by radial canals terminating in oscula. It is part of the exhalant system. Pl: *amararhyses.*

amargosite A syn. of *bentonite.* Etymol: *Amargosite,* a trade name for a bentonitic clay (montmorillonite) from the Amargosa River, Calif.

amarillite A pale greenish-yellow mineral: $NaFe(SO_4)_2 \cdot 6H_2O$.

amatrice A green gem cut from variscite and its surrounding matrix of gray, reddish, or brownish crystalline quartz or chalcedony.

amausite A finely crystalline rock, e.g. devitrified glass (Thrush, 1968, p. 33). Syn: *petrosilex [petrology].* Obsolete.

amazonite An apple-green, bright green or blue-green variety of microcline, sometimes used as a gemstone. Syn: *amazonstone.*

amazonstone The earlier name for *amazonite.* Also spelled: *Amazon stone.*

amb The contour or outline of a pollen grain (less commonly of a spore) as viewed from directly above one of the poles. See also: *polar view.* Syn: *equatorial limb.*

ambatoarinite A poorly described mineral: $Sr(Ce,La,Nd)O(CO_3)_3$ (?).

amber (a) A hard, brittle fossil resin, usually yellowish to brownish and translucent or transparent, that is derived from coniferous trees, frequently encloses insects and other organisms, takes a fine polish, and is found in alluvial soils, clays, and recent sediments, in beds of lignite, and on some seashores (as of the Baltic Sea). It is used chiefly in making pipe mouthpieces, beads, and other small ornamental objects. Syn: *succinite; bernstein; electrum.* (b) A term applied to a group of fossil resins containing considerable succinic acid and having highly variable C:H:O ratios; e.g. almashite, simetite, delatynite, and ambrosine.

amberite *ambrite.*

amber mica *phlogopite.*

amberoid A gem material consisting of small fragments of genuine amber that have been artificially united or reconstructed by heat and pressure. It may be characterized by an obvious flow structure or by a dull spot left by a drop of ether. Also spelled: *ambroid.* Syn: *pressed amber.*

ambilateral class A topological classification of channel networks of given *magnitude:* two topologically distinct channel networks belong to the same ambilateral class if one can be converted into the other by transposing the right-left arrangement of the tributary links at any of the junctions (Smart, 1969, p. 1761).

ambitus The exterior edge or periphery; e.g. the greatest horizontal circumference of an echinoid test or edrioasteroid theca, or the thecal outline of a dinoflagellate as viewed from the dorsal or ventral side. Pl: *ambitus.*

amblygonite A mineral: $(Li,Na)AlPO_4(F,OH)$. It commonly occurs in white or greenish cleavable masses in pegmatites, and is mined as an ore of lithium. Syn: *hebronite.*

amblyproct Said of a sponge in which the exhalant surface is in the form of an open cup.

amblystegite A dark brownish-green to black variety of hypersthene.

ambonite A group of porphyritic cordierite-bearing hornblende-biotite *andesites* and *dacites* originally described by Verbeck in 1905 from the Indonesian island of Ambon (or Amboina); also, any rock in that group, the three main subdivisions being bronzite-bearing, mica-bearing, and hornblende-bearing andesites and dacites (Johannsen, 1939, p. 239). Not recommended usage.

ambrite A yellowish-gray, subtransparent variety of retinite resembling amber, occurring in large masses in the coalfields of New Zealand. Its approximate formula: $C_{40}H_{66}O_5$. Syn: *amberite.*

ambroid *amberoid.*

ambrosine A yellowish or clove-brown variety of amber containing considerable succinic acid and occurring as rounded masses in phosphate beds.

ambulacral adj. (a) Pertaining to an *ambulacrum* or to ambulacra collectively; e.g. a small needlelike "ambulacral spine" attached to the part of a side plate directed toward the main food groove of the ambulacrum of a blastoid. (b) Corresponding in position to an ambulacrum, or referring to the zone in which an ambulacrum is present; e.g. an "ambulacral ray" representing an area defined by direction of an ambulacrum radiating from the mouth of a crinoid. Cf: *interambulacral.* Syn: *radial [paleont].*—n. A small calcareous plate that covers part of an ambulacrum of an echinoderm; e.g. a side plate or the broader of the floor plates in a cystoid. Pl: *ambulacralia.*

ambulacral bifurcation plate One of two large, unpaired plates in edrioasteroids, one lying at the junction of each lateral pair of ambulacra (I-II and IV-V); the perradial tip of each bifurcation plate lies at the junction of the two ambulacral perradial lines with each other and with one end of the transverse oral midline (Bell, 1976).

ambulacral cover plate *cover plate.*

ambulacral floorplate *floorplate.*

ambulacral groove A passageway, furrow, or depression along the course of an ambulacrum of an echinoderm, commonly covered by ambulacral plates, through which food particles are believed to have been conveyed to the mouth by means of ciliary currents. Syn: *food groove.*

ambulacral pore An opening, in or between ambulacral plates of an echinoderm, for the passage of tube feet or of podia or for the connection of a podium to an ampulla.

ambulacral radius Thecal radius in an edrioasteroid, defined by the midline of an ambulacrum; in curved ambulacra the radius is defined by the midline of the proximal end of the ambulacrum, which is extended directly toward the edge of the theca (Bell, 1976).

ambulacral system An organ system peculiar to echinoderms, consisting of a ring canal encircling the mouth and five ambulacral vessels radiating from the ring canal and lodged in the ambulacral grooves (TIP, 1966, pt.U, p.153). See also: *subvective system; water-vascular system.*

ambulacral tunnel The space enclosed between the ambulacral floor plates and the overlying cover plates of an edrioasteroid; the ambulacral food groove extends along the floor of this space (Bell, 1976).

ambulacrum (a) One of the narrow, usually elongate areas extending radially from the mouth of an echinoderm, along which run the principal nerves, the blood vessels, the radial canals of the water-vascular system, and the ambulacral groove. Commonly it bears numerous tube feet. Most echinoderms have five such areas, as in the blastoids and edrioasteroids. See also: *ray [paleont].* (b) A trough of the coenosteum separating collines on the surface of some meandroid coralla of a scleractinian coral.—Pl: *ambulacra.*

ameghinite A mineral: $NaB_3O_5 \cdot 2H_2O$.

amensalism *antibiosis.*

ament A spike inflorescence of closely spaced, often intricate, apetalous flowers. It may be conelike and is usually deciduous. Syn: *catkin.*

American cut *Tolkowsky theoretical brilliant cut.*

American jade (a) Nephrite from Wyoming. (b) A syn. of *californite.*

American ruby A red garnet (pyrope) found in Arizona and New Mexico.

amerikanite A natural glass from South America (Colombia and Peru), once classed as a tektite but now believed to be of volcanic origin. Syn: *mancusanite.* See also: *columbianite.*

amesite An apple-green mineral: $(Mg,Fe^{+2})_4Al_4Si_2O_{10}(OH)_8$. It may be essentially free of iron. It is structurally related to kaolinite.

amethyst (a) A transparent to translucent, purple to pale violet variety of crystalline quartz, much used as a semiprecious gemstone. The color is due to iron compounds. Syn: *bishop's stone.* (b) A term applied to a deep-purple variety of corundum and to a pale reddish-violet beryl.

Amgan European stage: Middle Cambrian (above Lenan, below Mayan).

amherstite A *syenodiorite* in which the feldspar is andesine antiperthite. It was named by Watson and Taber in 1913, from Amherst County, Virginia. Not recommended usage.

amianthus A syn. of *asbestos,* applied esp. to a fine silky variety such as chrysotile. Syn: *amiantus.*

amictic lake A perennially frozen lake that does not undergo mixing of the water column.

amino acid One of the group of organic compounds, containing both amine ($-NH_2$) and carboxyl ($-COOH$) groups, which are the building blocks of proteins and therefore essential to life processes. All but one (glycine) are optically active and most occur in nature in the L-form.

amino-acid racemization age method *racemization age method.*

aminoffite A tetragonal mineral: $Ca_2(Be,Al)Si_2O_7(OH) \cdot H_2O$.

ammersooite A clay mineral (illite?) occurring in soil.

ammite *ammonite [sed].*

ammonia alum *tschermigite.*

ammonioborite A white mineral: $(NH_4)_2B_{10}O_{16} \cdot 5H_2O$. It is found as aggregates of minute plates in fumarolic deposits.

ammoniojarosite A pale-yellow mineral of the alunite group: $(NH_4)Fe_3(SO_4)_2(OH)_6$.

ammonite [paleont] Any ammonoid belonging to the order Ammonitida, characterized by a thick, strongly ornamented shell with sutures having finely divided lobes and saddles. Range, Ordovician to Cretaceous. Syn: *ammoniticone.*

ammonite [sed] An obsolete term, applied in the 17th and 18th centuries to a sedimentary rock now known as oolite. Syn: *ammite.*

ammonitic suture A type of suture in ammonoids characterized by complex fluting in which all of the smaller secondary and tertiary lobes and saddles (developed on a larger primary set) are denticulate or frilled; specif. a suture in ammonites. Cf: *goniatitic suture; ceratitic suture.*

ammonium alum *tschermigite.*

ammonoid Any extinct cephalopod belonging to the subclass Ammonoidea, characterized by an external shell that is symmetrical and coiled in a plane and has a bulbous protoconch, septa that form angular sutural flexures, and a small marginal siphuncle. Range, Lower Devonian to Upper Cretaceous.

amnicolous Said of an organism living on sand.

amniote adj. Pertaining to a vertebrate egg characterized by a large yolk and covered by a shell which is lined with cellular membranes produced from embryonic tissue, which function to conserve water and for the exchange of gases.—n. Any vertebrate reproducing by means of such an egg; includes all tetrapod classes except Amphibia.

amoebocyte A sponge cell of amoeboid (irregular, changing) form. It includes such cells as archaeocytes, sclerocytes, trophocytes, and collencytes. Syn: *amoeboid cell.*

amoeboid Said of a fold that has no specific shape and a very shallow dip, e.g. a *placanticline.*

amorphous (a) Said of a mineral or other substance that lacks crystalline structure, or whose internal arrangement is so irregular that there is no characteristic external form. The term does not preclude the existence of any degree of order. Ant: *crystalline.* (b) A term formerly used to describe a body of rock occurring in a continuous mass, without division into parts. Cf: *massive.*

amorphous graphite Very fine-grained, generally sooty graphite from metamorphosed coal beds. The word "amorphous" is a misnomer because all graphite is crystalline. The term has also been applied to very fine particles of *flake graphite* that can be sold only for low-value uses (such as foundry facings), and to fine-grained varieties of Ceylon lump graphite.

amorphous peat Peat in which degradation of cellulose matter has destroyed the original plant structures. Cf: *pseudofibrous peat; fibrous peat.*

amosite A commercial term for an iron-rich, asbestiform variety of amphibole occurring in long fibers. It may consist of an orthorhombic amphibole (anthophyllite or gedrite) or of a monoclinic amphibole (cummingtonite or grunerite).

Amosnuria *Lemuria.*

ampangabeite *samarskite.*

ampasimenite An igneous rock, probably equivalent to *ijolite* or *nephelinite,* characterized by the presence of nepheline, titanaugite, hornblende, and magnetite in a fine-grained or glassy groundmass. Named by Lacroix in 1922 for Ampasimena, Malagasy. Not recommended usage.

ampelite An obsolete term for a black carbonaceous or bituminous shale.

amphiaster A siliceous sponge spicule (microsclere) composed of a straight rod with a group of radiating spines at each end.

amphibian Any vertebrate of the class Amphibia: cold-blooded tetrapods that breathe by means of gills in the early stages of life and by means of lungs in the later stages.

amphiblastula A hollow, ovoid, free-swimming sponge larva, composed of small flagellated and large nonflagellated types of cells, one type grouped anteriorly and the other posteriorly. Pl: *amphiblastulae.*

amphibole (a) A group of dark rock-forming ferromagnesian silicate minerals, closely related in crystal form and composition and having the general formula: $A_{2-3}B_5(Si,Al)_8O_{22}(OH)_2$, where A =

Mg, Fe^{+2}, Ca, or Na, and B = Mg, Fe^{+2}, Fe^{+3}, or Al. It is characterized by a cross-linked double chain of tetrahedra with a silicon:oxygen ratio of 4:11, by columnar or fibrous prismatic crystals, and by good prismatic cleavage in two directions parallel to the crystal faces and intersecting at angles of about 56° and 124°; colors range from white to black. Most amphiboles crystallize in the monoclinic system, some in the orthorhombic. They constitute an abundant and widely distributed constituent in igneous and metamorphic rocks (some are wholly metamorphic), and they are analogous in chemical composition to the *pyroxenes.* (b) A mineral of the amphibole group, such as hornblende, anthophyllite, cummingtonite, tremolite, actinolite, riebeckite, glaucophane, or arfvedsonite. (c) A term sometimes used as a syn. of *hornblende.* —Etymol: Greek *amphibolos,* "ambiguous, doubtful", in reference to its many varieties.

amphibolide A general term, for use in the field, to designate any coarse-grained, holocrystalline igneous rock almost entirely composed of amphibole minerals. Syn: *amphibololite.*

amphibolite A crystalloblastic rock consisting mainly of amphibole and plagioclase with little or no quartz. As the content of quartz increases, the rock grades into hornblende plagioclase gneiss. The term was originated by Brongniart. Cf: *feather amphibolite.*

amphibolite facies The set of metamorphic mineral assemblages (facies) in which basic rocks are represented by hornblende + plagioclase, the plagioclase being oligoclase-andesine or some more calcic variety (Eskola, 1939). Epidote and almandine are common in amphibolites. Pelitic assemblages contain micas associated with almandine, staurolite, kyanite, or sillimanite, but not andalusite or cordierite (Turner, 1968). The facies is typical of regional dynamothermal metamorphism under moderate to high pressures (in excess of 3000 bars) with temperatures in the range 450°-700°C. Cf: *hornblende-hornfels facies.*

amphibolitization Introduction of, or replacement by, an amphibole mineral.

amphibololite *amphibolide.*

amphidetic Said of a ligament or ligamental area of certain bivalve mollusks (e.g. *Arca*) extending on both the anterior and posterior sides of the beaks. Cf: *opisthodetic.*

amphidisc A siliceous sponge spicule (microsclere) consisting of a central shaft at each end of which is a transverse, stellate disk or an umbrella-like structure (umbel) containing multiple recurved teeth. Also spelled: *amphidisk.*

amphidont Said of a class of ostracode hinges consisting of three elements, of which the terminal elements are teeth (or sockets in the opposed valve), and a median element is subdivided into an anterior socket and a bar (or a tooth and a groove in the opposed valve).

amphidromic point A geographic position in the ocean where, theoretically, there is no tide range and from which *cotidal lines* radiate in various directions, with the tide amplitude presumably increasing with distance from this point. See also: *amphidromic region.* Syn: *nodal point.*

amphidromic region An oceanic region whose cotidal lines radiate from one *amphidromic point.*

amphidromic system A system of tidal action in which the tide wave rotates around a point or center of little or no tidal movement. Cf: *amphidromic point.*

amphigene *leucite.*

amphineuran A marine mollusk belonging to the class Amphineura, with a flattened body covered by eight articulated dorsal plates. See also: *polyplacophoran.*

amphioxea A slightly curved oxea (sponge spicule).

amphiphloic Pertaining to the siphonostele of certain vascular plants having phloem both internal and external to the xylem. Cf: *ectophloic.*

amphipod Any crustacean belonging to the order Amphipoda, the members of which resemble the *isopods* by the absence of a carapace and by the presence of unstalked sessile eyes, but differ from them in having laterally, rather than dorsoventrally, compressed bodies. Range, Upper Eocene to present.

amphi-sapropel Sapropel containing coarse plant debris (Tomkeieff, 1954, p.23).

amphistylic Pertaining to a jaw suspension in which the cartilaginous or cartilage-replacement skeleton of the upper jaw is directly articulated to the ear region of the neurocranium, and in addition is articulated via elements of the visceral skeleton; both sets of articulation are movable.

amphitheater A concave landform, generally oval or circular in outline, nearly surrounded by steep slopes, and having a relatively flat floor; e.g. a *cirque [glac geol].*

amphitropous Said of a plant ovule whose stalk (funiculus) is curved about it so that the ovule tip and stalk base are near each other (Lawrence, 1951, p.738). Cf: *anatropous.*

amphoteric Having both basic and acidic properties.

amphoterite A chondritic stony meteorite composed essentially of pyroxene (bronzite) and olivine with small amounts of oligoclase and nickel-rich iron.

amplexoid septum A rugose corallite septum characterized by extreme shortness except where it is extended axially on the distal side of a tabula, as in *Amplexus.*

amplitude [ecol] The degree of adaptability exhibited by a particular kind or group of organisms to its surroundings.

amplitude [fold] For a symmetrical, periodic fold system it is analogous to the amplitude of a wave form, i.e. half the orthogonal distance between the antiformal and synformal *enveloping surfaces.* For asymmetrical and nonperiodic systems, various definitions have been proposed (see Fleuty, 1964a).

amplitude [waves] Half the height of the crest of a wave or ripple above the adjacent troughs.

ampulla One of the muscular vesicles of the water-vascular system of an echinoderm (such as an asteroid), being a contractile bulb or saclike structure of a tube foot, either seated externally in a small cup-shaped depression on the surface of an ambulacral plate or internally, and connecting with the podium by a branch canal through the podial pore or passageway. Pl: *ampullae.*

AMT *audio-magneto-telluric method.*

amygdale *amygdule [ign].*

amygdaloid n. An extrusive or intrusive rock containing numerous *amygdules.* —adj. Said of a rock having numerous amygdules. Syn: *amygdaloidal.*

amygdaloidal Said of rocks containing *amygdules* and of the structure of such rocks.

amygdule [ign] A gas cavity or vesicle in an igneous rock, which is filled with such secondary minerals as calcite, quartz, chalcedony, or a zeolite. The term *amygdale* is preferred in British usage. Adj: *amygdaloidal.* See also: *amygdaloid.*

amygdule [sed] An agate pebble.

ana- A prefix meaning "toward" or "up to".

Anabar block A syn. of *Angaraland,* used by Schatsky & Bogdanoff (1957).

anabatic wind A local wind that moves upslope, e.g. a *valley wind.* Ant: *katabatic wind.*

anabohitsite An olivine-bearing *pyroxenite* containing hornblende and hypersthene and a high proportion of magnetite and/or ilmenite, approximately 30 percent according to Johannsen (1939, p. 240). Its name was given by Lacroix in 1914 from Anabohitsy, Malagasy. Not recommended usage.

anaboly Acceleration of the ontogeny. Syn: *hypermorphosis.*

anabranch (a) A diverging branch flowing out of a main stream and later rejoining it farther downstream; esp. one of the several branches composing a braided stream. The term was coined by Jackson (1834, p. 79) and is used mainly in Australia. Etymol: *ana*stomosing + *branch.* Cf: *braid.* Syn: *valley braid; anastomosing branch.* (b) A branch of a stream that loses itself in sandy soil.

anaclinal Said of an *antidip stream* or of a valley that descends in a direction opposite to that of the general dip of the underlying strata it traverses; e.g. applied to an antecedent stream flowing on a surface that has been tilted in a direction opposite to that of the flow of the stream. Term introduced by Powell (1875, p. 160). See also: *obsequent.* Ant: *cataclinal.* Syn: *contraclinal.*

anacline Said of the dorsal and posterior *inclination [paleont]* of the cardinal area in either valve of a brachiopod, lying in the top left or first quadrant moving clockwise from the orthocline position (TIP, 1965, pt.H, p.60, fig.61).

anadiagenesis A term used by Fairbridge (1967) for the compaction, maturation phase of diagenesis, in which clastic sediment grains or chemical sediments become lithified during deep burial (extending to 10,000 m). It is characterized by expulsion and upward migration of connate waters and other nonmagmatic fluids, such as petroleum, and often by reducing conditions; it may pass into metamorphism. See also: *epidiagenesis; syndiagenesis.* Adj: *anadiagenetic.* Syn: *middle diagenesis.*

anaerobic *adj.* 1. Said of an organism (esp. a bacterium) that can live in the absence of free oxygen; also, said of its activities. Noun: *anaerobe.* (b) Said of conditions that exist only in the absence of free oxygen. — Cf: *aerobic.*

anaerobic decay Decomposition of organic substances in the absence or near absence of oxygen; the ultimate decay products are enriched in carbon.

anaerobic sediment A highly organic sediment characteristic of some fjords and basins where restricted circulation of the water results in the absence or near absence of oxygen at the sediment surface, and bottom water is rich in hydrogen sulfide.

anagenesis A progressive, linear evolutionary change.

anagenite Quartz conglomerate in the northern Apennines.

anaglacial A term proposed by Trevisan (1950) for the part of a paleoclimatic cycle transitional from an interglacial to a full glacial or pleniglacial. It is equal to the "early glacial" of Van der Hammen and others (1971). Cf: *kataglacial.*

Anahuac North American (Gulf Coast) stage: Miocene (above Frio, below Napoleonville).

anal adj. Pertaining to or situated near the anus of an animal; e.g. an "anal opening" representing a large thecal orifice marking the position of the anus in the CD interray of a blastoid or crinoid.—n. *anal plate.*

analbite The triclinic polymorph of albite, having a disordered Al-Si distribution, and obtained by heating albite.

analcime A mineral: $NaAlSi_2O_6 \cdot H_2O$. It is an isometric zeolite, commonly found in diabase and in alkali-rich basalts. Syn: *analcite.*

analcimite An extrusive or hypabyssal igneous rock consisting mainly of analcime and pyroxene (usually titanaugite). Feldspathoids, plagioclase, and/or olivine may be present. Apatite, sphene, and opaque oxides may be present as accessories.

analcimization Replacement of feldspars or feldspathoids by analcime, usually in igneous rocks during late magmatic and post-magmatic stages. Syn: *analcitization.*

analcimolith A less preferred synonym of *analcimite.*

analcite *analcime.*

analcitite An olivine-free analcime-bearing *basalt* named by Pirsson in 1896. Not recommended usage.

analcitization *analcimization.*

anal cover plate One of the plates covering the anal opening in the theca of echinoderms, e.g. one of many small polygonal plates that may extend over and conceal the anal opening of a blastoid and that are bordered by anal deltoids or deltoid plates. See also: *cover plate.*

anal deltoid An undivided interradial plate on the posterior (CD) part of a blastoid theca below the circlet of oral plates or the mouth opening.

anal fasciole (a) A *fasciole* generated on gastropod-shell whorls by the indentation of the outer lip (such as a sinus) situated close to the adapical suture. (b) A *fasciole* adoral and lateral to the periproct of an echinoid.

analog adj. Said of the representation of a range of numbers by directly measurable quantities such as voltage or rotation, as in an *analog computer* or analog systems. Cf: *digital.*

analog computer A *computer* that operates with numbers represented by directly measurable quantities (such as length, voltage, or resistance) in a one-to-one correspondence; a measuring device that operates on continuous variables represented by physical or mathematical analogies between the computer variables and the variables of a given problem to be solved. Cf: *digital computer; hybrid computer.*

analogous pole In crystallography, that pole of a crystal which becomes electrically positive when the crystal is heated or expanded by decompression. Cf: *antilogous pole.*

analogy In evolution, likeness in form or function but not in origin. Cf: *homology.*

anal plate Any plate covering the anus or anal opening of an echinoderm; a plate of the CD interray of a crinoid, mostly confined to the dorsal cup but excluding fixed pinnulars incorporated in the theca. Syn: *anal.*

anal pyramid A cone-shaped structure composed of several elongate, more or less triangular plates, commonly imbricate, serving to close the anus in echinoderms such as cystoids, blastoids, and edrioasteroids.

anal sac A variously shaped, generally inflated, and highly elevated part of theca encompassing the anus, as developed among inadunate crinoids.

anal tube A conical to cylindrical structure on a crinoid tegmen, usually of considerable height, bearing the anal opening at its summit. It is typically developed in many camerate, flexible, and articulate crinoids.

anal X A special anal plate in inadunate and flexible crinoids, typically located between posterior (CD) radial plates, adjacent or next adjacent to tegmen.

analysis of variance A statistical technique for simultaneously partitioning the total variance of a set of data into components which can be attributed to different sources of variation, and which can be used to test for differences among several samples. Abbrev: ANOVA.

analytical geomorphology *dynamic geomorphology.*

analytical triangulation Triangulation accomplished by computational routines using measured coordinates and appropriate formulas; e.g. *aerotriangulation* by computing positions and/or elevations of ground stations from measurements made on aerial photographs and known locations of control points.

analytic group A rock-stratigraphic unit formerly classed as a formation but now called a *group* because subdivisions of the unit are considered to be formations (Weller, 1960, p.434). Cf: *synthetic group.*

analyzer That part of a polariscope that receives the light after polarization and exhibits its properties. In a petrographic microscope, it is the polarizing mechanism (*Nicol prism, Polaroid,* etc.) that intersects the light after it has passed through the object under study. See also: *polarizer.*

anamesite A fine-grained basaltic rock intermediate in texture between fine-grained *basalt* and coarse-grained *dolerite.* It was named by Leonhard in 1832. Obsolete.

anamigmatization High-temperature, high-pressure remelting of pre-existing rock to form migma. Cf: *anatexis; metatexis.* Not widely used.

anamniote adj. Pertaining to a vertebrate egg which lacks protective membranes produced from embryonic tissue and is usually small-yolked.—n. Any vertebrate reproducing by means of such an egg; includes all vertebrate classes except Reptilia, Mammalia, and Aves.

anamorphic zone The zone deep in the Earth's crust in which *anamorphism* takes place. The term, originated in 1898 by Van Hise, is now little used. Cf: *katamorphic zone.*

anamorphism Intense metamorphism in the *anamorphic zone* in which rock flowage takes place and simple minerals of low density are changed into more complex ones of greater density by silication, decarbonization, dehydration, and deoxidation. The term was originated by Van Hise in 1904. Cf: *katamorphism.*

anamorphosis In the evolution of a group of animals or plants, the gradual change from one form to another; e.g., certain arthropods acquire an additional body segment after hatching.

anandite A mineral: $(Ba,K)(Fe,Mg)_3(Si,Al,Fe)_4O_{10}(O,OH)_2$.

anapaite A pale-green or greenish-white mineral: $Ca_2Fe(PO_4)_2 \cdot 4H_2O$.

anapeirean Said of rocks of the *Pacific suite.*

anaphoresis *Electrophoresis* in which the movement of suspended negative particles in a fluid is toward the anode. Cf: *cataphoresis.*

anaplasis An evolutionary state characterized by increasing vigor and diversification of organisms; considered to be the first stage in an evolutionary line. Cf: *metaplasis; cataplasis.*

anaprotaspis A small protaspis trilobite in which the protopygidium is poorly defined. Cf: *protaspis; metaprotaspis.* Pl: *anaprotaspides.*

Anapsida A reptilian subclass that includes turtles and tortoises in addition to diverse forms of Late Paleozoic and Early Mesozoic age.

anaptychus A single plate with rounded edges, made of calcite, originally lodged in the body chamber of some ammonoid conchs; it probably formed part of the lower jaw of the animal. It was formerly interpreted as an opercular structure. See also: *aptychus.*

anarakite *zincian paratacamite (?).*

anascan adj. Pertaining to the cheilostome bryozoan suborder Anasca, characterized by parietal muscles attached to the frontal membrane, which is generally exposed.—n. An anascan cheilostome (bryozoan). Cf: *ascophoran.*

anaseism Earth movement away from the focus of an earthquake. Cf: *kataseism.*

Anaspida An order of one-nostril jawless vertebrates characterized by fusiform habitus and unexpanded head shield. Range, Middle Silurian to Upper Devonian.

anastable adj. Stable, with a tendency towards upheaval. Cf: *ca-*

tastable.

anastatic water A syn. of *fringe water* proposed by Meinzer (1939) as one of the three classes of *kremastic water.*

anastomosing Said of a leaf whose veins form a netlike pattern; pertaining to an interveined leaf. Sometimes the vein branches meet only at the margin.

anastomosing branch *anabranch.*

anastomosing stream *braided stream.*

anastomosis [speleo] A net of braided interconnected tubes, commonly confined to a discrete layer of rock. Partial syn: *spongework.*

anastomosis [streams] (a) *braiding.* (b) A product of braiding; esp. an interlacing network of branching and reuniting channels.

anastomosis tube One of many small irregular repeatedly interconnected solution tubes, commonly found along bedding planes.

anastomotic cave A cave consisting of braided interconnected tubular passages. See also: *network cave; spongework cave.*

anatase A brown, dark-blue, or black tetragonal mineral: TiO_2. It is trimorphous with rutile (which has different facial angles) and brookite, and occurs as an alteration product of other titanium minerals. Syn: *octahedrite.*

anatectic magma Magma formed as a result of *anatexis.*

anatectite *anatexite.*

anatexis Melting of pre-existing rock. This term is commonly modified by terms such as intergranular, partial, differential, selective, or complete (Dietrich & Mehnert, 1961). Cf: *metatexis; diatexis; palingenesis [petrology]; syntexis; anamigmatization.*

anatexite Rock formed by anatexis. Also spelled: *anatectite.* See also: *syntectite.* Cf: *arterite.*

Anathermal n. A term proposed by Antevs (1948, p.176) for a postglacial interval (from about 10,000 to 7500 years ago) preceding the Altithermal and representing the period of generally rising temperatures following the last major advance of the continental glaciers. It is equivalent to the Preboreal and Boreal.—adj. Pertaining to the postglacial Anathermal interval and to its climate, deposits, biota, and events.

anathermal n. A period of time during which temperatures are rising. The term was used by Emiliani (1955, p. 547) for a part of a cycle as displayed in a deep-sea sediment core. Ant: *catathermal.*

anatriaene A sponge triaene in which the cladi are bent back toward the rhabdome.

anatropous Said of a plant ovule that is reversed, i.e. one whose opening (micropyle) is closed to the point of funiculous attachment (Lawrence, 1951, p.738). Cf: *amphitropous.*

anauxite A clay consisting of a mixture of kaolinite and amorphous silica. Syn: *ionite.*

anaxial Lacking a distinct axis or axes; e.g. said of an arm or branch of a sponge spicule that has no axial filament or canal.

ancestral river A term applied in Australia to a major, ancient river system, older than a *prior river.*

ancestrula The primary or first-formed zooid of a bryozoan colony, derived by metamorphosis of a free-swimming larva, from which secondary individuals are formed by budding. Pl: *ancestrulae.*

anchi- In petrologic terminology, a prefix signifying "almost".

anchieutectic Said of a rock whose minerals are in almost completely eutectic proportions.

anchimetamorphism A term introduced by Harrassowitz (1927) and still used by German-language authors to indicate changes in mineral content of rocks under temperature and pressure conditions prevailing in the region between the Earth's surface and the zone of true metamorphism, i.e. approximately in the zones of weathering and ground-water circulation. The term has not been accepted by writers in English.

anchimonomineralic Said of an igneous rock (such as anorthosite or dunite) that consists essentially of a single mineral. Term originated by Vogt (1905). Cf: *monomineralic.* Syn: *anchimonomineral.*

anchor A holothurian sclerite in the shape of an anchor, consisting of a *shank,* two or more *flukes,* and usually a *stock.*

anchorate n. (a) A sponge spicule (hexactin) with one long ray and two (coplanar) or four recurved, short rays at one end. (b) A strongly dentate chela of a sponge.—adj. Said of a sponge spicule having one or more processes like the fluke of an anchor.

anchor branch A curved hooklet in the radiolarian suborder Phaeodarina.

anchored dune A sand dune whose movement is arrested or whose form is protected from further wind action, as by a growth of vegetation or by cementation of the sand. Cf: *attached dune; wandering dune.* Syn: *fixed dune; established dune; stabilized dune.*

anchor ice Spongy underwater ice formed on a submerged object or structure, or attached to the bottom of a body of water (as a stream, lake, or very shallow sea) which itself is not frozen; usually forms in cold, clear, and still water. Syn: *bottom ice; ground ice [ice]; depth ice; lappered ice; underwater ice.*

anchor-ice dam An accumulation of anchor ice raising the water level of a river or stream.

anchorite A nodular, veined *diorite,* named by Lapworth in 1898, having isolated patches of mafic minerals and contemporaneous veins of felsic minerals. Not recommended usage.

anchylosis *ankylosis.*

ancient volcano A general term for a volcano of a past geologic age.

ancylite A mineral: $SrCe(CO_3)_2(OH)\cdot H_2O$.

andalusite A brown, yellow, green, red, or gray orthorhombic mineral: Al_2SiO_5. It is trimorphous with kyanite and sillimanite. Andalusite occurs in thick, nearly square prisms in schists, gneisses, and hornfelses; it forms at medium temperatures and pressures of a regionally metamorphosed sequence and is characteristic of contact-metamorphosed argillaceous rocks. In transparent gem quality, andalusite has a very strong pleochroism: brownish green in one direction and brownish red at 90°. See also: *chiastolite.*

Andept In U.S. Dept. of Agriculture soil taxonomy, a suborder of the soil order *Inceptisol,* characterized by low bulk density. They either contain much allophane that has a high exchange capacity, or consist mostly of pyroclastic materials. Most Andepts formed in volcanic ash, have a relatively high organic-carbon content, and contain or have contained some glass (USDA, 1975). Cf: *Aquept; Ochrept; Plaggept; Tropept; Umbrept.*

andersonite A bright yellow-green secondary mineral: $Na_2Ca(UO_2)(CO_3)_3\cdot 6H_2O$.

andesine A mineral of the plagioclase feldspar group with composition ranging from $Ab_{70}An_{30}$ to $Ab_{50}An_{50}$. It occurs as a primary constituent of intermediate igneous rocks, such as andesites and diorites.

andesinite A coarse-grained igneous rock almost entirely composed of andesine. It was named by Turner in 1900. Cf: *anorthosite.* Not recommended usage.

andesite A dark-colored, fine-grained extrusive rock that, when porphyritic, contains phenocrysts composed primarily of zoned sodic plagioclase (esp. andesine) and one or more of the mafic minerals (e.g. biotite, hornblende, pyroxene), with a groundmass composed generally of the same minerals as the phenocrysts, although the plagioclase may be more sodic and quartz is generally present; the extrusive equivalent of *diorite.* Andesite grades into *latite* with increasing alkali feldspar content, and into *dacite* with more alkali feldspar and quartz. It was named by Buch in 1826 from the Andes Mountains, South America.

andesite line The geographic-petrographic boundary between basalts of the *Atlantic suite* and the mainly andesitic rocks of the *Pacific suite.* The boundary on the west is generally drawn from Alaska to the east of New Zealand and Chatham Island, by way of Japan, the Marianas, Palau Islands, Bismarck Archipelago, and the Fiji and Tonga groups. The boundary on the east is less clearly defined but probably runs along the coasts of North and South America; it has not been traced in the South Pacific. Syn: *Marshall line.*

andorite A dark-gray or black mineral: $PbAgSb_3S_6$. It is closely related to ramdohrite and fizelyite. Syn: *sundtite.*

ando soil A black or dark brown soil that is formed from volcanic material and is characterized by an A horizon high in organic matter, low exchangeable bases, and high exchangeable aluminum. Some ando soils have illuvial-clay B horizons, others have AC profiles (Thorp and Smith, 1949). Ando soils are now classified as *Andepts.* Also spelled: *andosol.*

andosol *ando soil.*

andradite The calcium-iron end member of the garnet group: $Ca_3Fe_2(SiO_4)_3$. It has a variety of colors, ranging from yellow, red, and green to brown and black; it often occurs in contact-metamorphosed limestones. Varieties include topazolite, demantoid, melanite, aplome, and bredbergite.

andremeyerite A monoclinic mineral: $BaFe_2Si_2O_7$.

andrewsite A bluish-green mineral: $(Cu,Fe^{+2})Fe_3^{+3}(PO_4)_3(OH)_2$.

anegite An igneous rock characterized by the absence of feldspar and olivine, but containing pyroxene, spinel, pyrope, and hornblende. Obsolete.

anelasticity [exp struc geol] The inelastic relaxation in time of very small deformations, usually studied as forced vibrations.

anelasticity [seis] The effect of absorption of seismic energy. See also: *absorption; absorptance; Q.*

anemochore A plant whose seeds or spores are distributed by the wind. Cf: *anemophily.*

anemoclast A rock fragment that was broken off and more or less rounded by wind action (Grabau, 1904).

anemoclastic rock A rock consisting primarily of anemoclasts.

anemolite An erratically shaped *speleothem,* thought to have been formed by air currents. See also: *helictite.*

anemometer Any type of instrument that measures the velocity of wind.

anemophily Pollination by wind. Adj: *anemophilous.* Cf: *entomophily.*

anemosilicarenite An eolian sand of siliceous composition (Grabau, 1904; 1913, p. 293).

anemousite A silica-deficient variety of albite.

aneroid barometer A type of *barometer* that measures change of atmospheric pressure by its effect on the thin sides of a partially evacuated short hollow cylinder. It is commonly used to measure altitude. The *altimeter* is a barometer of this type. Cf: *mercury barometer.*

aneuchoanitic *achoanitic.*

Angaraland A name used by Suess for a small shield exposing ancient Precambrian rocks in north-central Siberia, supposed to have been the *vertex* or nucleus around which all other structures of Asia were built. Modern Soviet geologists ascribe less significance to the feature. Syn: *Anabar block; Angara shield.*

angaralite A mineral of the chlorite group, occurring in thin black plates. Its formula has been given as: $Mg_2(Al,Fe)_{10}Si_6O_{29}$.

Angara shield *Angaraland.*

angelellite A blackish-brown mineral: $Fe_4As_2O_{11}$.

angiosperm A plant with true flowers, in which the seeds, resulting from double fertilization, are enclosed in an ovary, comprising the fruit. Such plants (*Anthophyta*) range from the Early Cretaceous or possibly before. Examples include grasses, orchids, elms, roses. Cf: *gymnosperm.* Syn: *flowering plant.*

angle The difference in direction between two convergent lines or surfaces; a measure of the amount of rotation required to make either of two intersecting lines coincide with or become parallel to the other, the rotation being in the plane of the lines and about the point of intersection.

angle of departure The acute angle between a structural plane and the vertical plane of a geologic cross section, that relates the structure to the vertical plane as the angle of dip does to the horizontal. It is measured in a plane perpendicular to the trace line of the structural plane in the cross section (Knutson, 1958). Cf: *angle of penetration.*

angle of dip *dip.*

angle of emergence An angle formed between a ray of energy, e.g. a seismic wave, and the horizontal. It is the complement of the *angle of incidence.* Cf: *apparent dip [seis].* Syn: *emergence angle.*

angle of incidence (a) The angle that a ray of energy—optic, acoustic, or electromagnetic—makes with the normal to a boundary surface. It is the complement of the *angle of emergence.* See also: *critical angle.* Syn: *Brewster angle.* (b) In SLAR, the angle between the vertical and a line connecting the antenna and a target. Syn: *aspect angle [remote sensing].*

angle of penetration The minimum angle between a structural plane and the plane of a geologic cross section, comparable to the angle of plunge on a geologic map (Knutson, 1958). Cf: *angle of departure.*

angle of reflection *Bragg angle.*

angle of refraction In optics, the angle of a refracted ray of light, measured from a perpendicular to the surface from which the ray is refracted. Syn: *refraction angle.*

angle of repose The maximum angle of slope (measured from a horizontal plane) at which loose, cohesionless material will come to rest on a pile of similar material. This angle is somewhat less than the slope angle at which sliding will be initiated (angle of slide) and is generally 5° to 10° less than the angle of internal friction of the same material. The angle of repose commonly ranges between 33° and 37° on natural slopes, and is rarely less than 30° or more than 39°. The angle depends on the frictional properties of the material and increases slightly as the size and angularity of the fragments increase. Cf: *angle of slide.* Syn: *angle of rest.*

angle of rest *angle of repose.*

angle of slide The angle (usually measured from a horizontal plane) of minimum slope at which any loose material (such as earth or talus) will start to slide; it is slightly greater than the *angle of repose.*

angle of ultimate stability *critical slope angle.*

anglesite A white orthorhombic mineral: $PbSO_4$. It is a common secondary mineral formed by the oxidation of galena and is a valuable ore of lead. Syn: *lead vitriol; lead spar.*

angrite An achondritic stony meteorite consisting chiefly of purple titaniferous augite (more than 90%) with a little olivine and troilite.

anguclast An angular *phenoclast,* such as a large fragment of a breccia. Cf: *spheroclast.*

angular Having sharp angles or borders; specif. said of a sedimentary particle showing very little or no evidence of abrasion, with all of its edges and corners sharp, such as blocks with numerous (15-30) secondary corners and a roundness value between zero and 0.15 (midpoint at 0.125) (Pettijohn, 1957, p. 58-59). Powers (1953) gives values between 0.17 and 0.25 (midpoint at 0.21). Also, said of the *roundness class* containing angular particles.

angular cross-bedding Cross-bedding in which the inclined beds appear in section as nearly straight lines meeting the underlying surface at high, sharp, or discordant angles; it often implies deposition by water, as in *torrential cross-bedding.* Cf: *tangential cross-bedding.*

angular discordance *angular unconformity.*

angular distance The angle, measured at the Earth's center, that subtends the great-circle path between the earthquake's epicenter and the receiver. Cf: *epicentral distance.*

angular distortion The change in shape of an area on a globe when it is represented on a map projection. See also: *distortion [cart].*

angular drift "Rock debris formed by intensive frost action, derived from underlying or adjacent bedrock" (ADTIC, 1955, p.4).

angular field of view The angle subtended by lines from a remote sensing system to the outer margins of the strip of terrain that is viewed by the system. Cf: *instantaneous field of view.*

angular fold A fold resembling a *kink fold* but with a less angular hinge.

angularity A term often used for the property of a sedimentary particle now commonly known as *roundness,* but used by Lamar (1928, p. 148-151) for the property now referred to as *sphericity.*

angular spreading The lateral extension of ocean waves as they move out of the generating area as *swell.*

angular unconformity An *unconformity* between two groups of rocks whose bedding planes are not parallel or in which the older, underlying rocks dip at a different angle (usually steeper) than the younger, overlying strata; specif. an unconformity in which younger sediments rest upon the eroded surface of tilted or folded older rocks. It is sometimes regarded as a type of *nonconformity.* Cf: *discordance.* Syn: *angular discordance; clinounconformity; structural unconformity; orogenic unconformity.*

angulate drainage pattern A modified *rectangular drainage pattern* developed where streams follow joints or faults that join each other at acute or obtuse angles, rather than at right angles (Zernitz, 1932, p. 517). Examples are found in the Timiskaming and Nipissing areas of Ontario.

anhedral (a) Said of a mineral crystal that has failed to develop its own *rational faces* or that has a rounded or indeterminate form produced by the crowding of adjacent mineral grains during crystallization or recrystallization. (b) Said of a detrital grain that shows no crystal outline. (c) Said of the shape of such a crystal.—The term was originally used in reference to igneous-rock components by Cross et al. (1906, p. 698) in preference to the synonymous terms *xenomorphic* and *allotriomorphic* (as these were originally defined). Cf: *subhedral; euhedral.*

anhedron Geometrical term for a solid figure not limited by plane surfaces. The term was introduced by Pirsson (1896) in reference to an imperfectly defined igneous-rock component (crystal). Pl: *anhedrons; anhedra.*

anhydrite A mineral consisting of anhydrous calcium sulfate: $CaSO_4$. It represents *gypsum* without its water of crystallization, and it alters readily to gypsum, from which it differs in crystal form (anhydrite is orthorhombic) and in being harder and slightly less soluble. Anhydrite usually occurs in white or slightly colored, granular to compact masses, forming large beds or seams in sedimentary rocks or associated with gypsum and halite in evaporites.

Syn: *cube spar.*

anhydrock A sedimentary rock composed chiefly of anhydrite.

anhydrous Said of a substance, e.g. magma or a mineral, that is completely or essentially without water. An anhydrous mineral contains no water in chemical combination.

anhysteretic remanent magnetization Remanent magnetization produced by simultaneous application of a constant magnetic field and an initially larger alternating magnetic field whose amplitude decreases smoothly to zero.

anideltoid Externally visible anal deltoid, which is unaccompanied by any others and lies on the aboral side of the anal opening or the anispiracle (TIP, 1967, pt. S, p. 345).

anidiomorphic *xenomorphic.*

anilite A mineral: Cu_7S_4.

Animikean Var. of *Animikie.*

Animikie A provincial series of the Proterozoic of the Canadian Shield; it is also called the *Animikean.*

animikite A silver ore consisting of a mixture of sulfides, arsenides, and antimonides showing striking intergrowth relations and occurring in white or gray granular masses. It contains nickel and lead. Cf: *macfarlanite.*

anion exchange The displacement of an anion bound to a site on the surface of a solid by an anion in solution. See also: *ion exchange.*

Anisian European stage: lower Middle Triassic (above Scythian, below Ladinian). Syn: *Virglorian.*

anisochela A sponge *chela* having unequal or dissimilar ends. Cf: *isochela.*

anisodesmic Said of a crystal or compound in which the ionic bonds are of unequal strength. Cf: *isodesmic.*

anisomerism (a) Repetition of parts that differ more or less importantly among themselves. (b) Reduction in number and differentiation of similar parts in organisms.

anisometric (a) Said of crystals having unequal dimensions, including those with a significant flattening (see *tabular*), elongation, or both. Ant: *equant; isometric.* (b) An obsolete syn. of *heterogranular.*

anisomyarian adj. Said of a mollusk with anterior adductor muscles much reduced or absent.—n. A mollusk with such muscles. Cf: *heteromyarian.*

anisotropic Having some physical property that varies with direction. All crystals are anisotropic relative to some properties, e.g. propagation of sound waves. Unless otherwise stated, however, the term refers to optical properties. In this sense, all crystals except those of the isometric system are anisotopic. Ant: *isotropic.* Syn: *aeolotropic.*

anisotropy The condition of having different properties in different directions, as in geologic strata that transmit sound waves with different velocities in the vertical and horizontal directions. Adj: *anisotropic.*

anispiracle An enlarged opening in the summit part of the posterior interray of a blastoid, formed by the union of anal opening and posterior spiracle or spiracles.

anitaxis A linear succession of crinoid anal plates. Pl: *anitaxes.*

ankaramite An olivine-bearing *basalt* containing numerous pyroxene and olivine phenocrysts, the former being more abundant than the latter, in a fine-grained groundmass composed of clinopyroxene microlites and calcic plagioclase. It was named by Lacroix in 1916 from Ankaramy, Malagasy.

ankaratrite Olivine *nephelinite* containing biotite. Named by Lacroix in 1916 for Ankaratra, Malagasy. Not recommended usage.

ankerite A white, red, or grayish iron-rich mineral related to dolomite: $Ca(Fe,Mg,Mn)(CO_3)_2$. It is associated with iron ores and commonly forms thin veins of secondary matter in some coal seams. Syn: *ferroan dolomite; cleat spar.*

ankylosis (a) Fusion of columnals or other skeletal elements of an echinoderm, commonly obscuring the sutures. (b) In vertebrates, comparable fusion between adjacent bones, or between base of tooth and supporting bone.—Also spelled: *anchylosis.*

annabergite An apple-green mineral: $(Ni,Co)_3(AsO_4)_2 \cdot 8H_2O$. It is isomorphous with erythrite, and usually occurs in incrustations as an alteration product of nickel arsenides. Syn: *nickel bloom; nickel ocher.*

annealing recrystallization The formation of new grains in a rock after solid-state deformation, while the temperature is still high. This is a recovery process, starting with *nucleation* and evolving through *grain growth,* the larger *neoblasts* consuming the smaller ones at any time. See also: *dynamic recrystallization.*

annelid Any wormlike invertebrate belonging to the phylum Annelida, characterized by a segmented body with a distinct head and appendages. Because the annelids lack skeletal structures (except for chitinous jaws, called *scolecodonts*), they are usually known as fossils only from their burrows and trails.

annerödite A black mineral consisting of samarskite with parallel overgrowths of columbite. Also spelled: *annerodite.*

annotated photograph A photograph on which planimetric, hypsographic, geologic, cultural, hydrographic, or vegetation information has been added to identify, classify, outline, clarify, or describe features that would not otherwise be apparent in examination of an unmarked photograph (ASP, 1975, p. 2065). The term generally does not apply to a photograph marked only with geodetic control or pass points.

annual balance The change in mass of a glacier from the beginning to the end of a hydrologic year (usually October 1 to September 30), or other measurement year defined by fixed calendar dates, determined at a point, as an average for an area, or as a total mass change for the glacier. Millimeters, meters, or cubic meters are the units normally used. Cf: *balance; net balance.*

annual flood (a) The highest discharge of a stream in a given water year. (b) That flood in a given water year which has been equalled or exceeded on the average of once a year.—(ASCE, 1962).

annual growth ring *growth ring.*

annual layer (a) A sedimentary layer deposited or presumed to have been deposited during the course of a year; e.g. a glacial varve. (b) A dark band (in a salt stock) of formerly disseminated anhydrite crystals that accumulated upon being freed by solution of the enclosing salt.

annually thawed layer *active layer.*

annual wave Yearly cyclical heating and cooling of the upper 3 to 5 meters of the Earth, in response to the annual solar cycle. Below this level the temperature is constant. Cf: *diurnal wave.*

annular drainage pattern A drainage pattern in which subsequent streams follow a roughly circular or concentric path along a belt of weak rocks, resembling in plan a ring-like pattern. It is best displayed by streams draining a maturely dissected structural dome or basin where erosion has exposed rimming sedimentary strata of greatly varying degrees of hardness, as in the Red Valley which nearly encircles the domal structure of the Black Hills, S.D.

annular lobe A small secondary dorsal lobe in the center of the main *internal lobe* of a suture of some coiled nautiloid conchs.

annular tracheid In plants, commonly the first *tracheid* to mature, characterized by the deposition of secondary wall material in ringlike form. Cf.: *spiral tracheid; protoxylem.*

annulation A ringlike structure; e.g. a ringlike expansion of an ammonoid conch, either transverse or slightly oblique to the longitudinal axis of the conch.

annulus [bot] (a) In the ferns, a specialized ring of cells on the sporangium that is involved in sporangial bursting. (b) Variously used for other annulate structures in mosses, horsetails, mushrooms, and diatoms.

annulus [drill] The space between the casing in a well and the wall of the hole, or between two concentric strings of casing, or between casing and tubing.

annulus [paleont] (a) A thin, ring-shaped endosiphuncular deposit, semicircular in cross section, on the inner side of a septal neck of a nautiloid (TIP, 1964, pt.K, p.54). (b) The *periphract* of a nautiloid. (c) A ring-shaped plate taking part in construction of the wall of archaeocyathids.

annulus [palyn] A ring bordering a pore of a pollen grain, in which the ektexine is modified (usually thickened). Cf: *margo.* See also: *endannulus.*

anogenic (a) Pertaining to plutonic metamorphism or replacement. (b) Pertaining to eruptive rocks.

anomalous dispersion The extraordinary scattering effects that an atom displays when the wavelength of impinging x-rays is close to a natural absorption edge of the atom. Normal atomic scattering factors may be corrected for anomalous dispersion using correction factors listed in International Tables for X-ray Crystallography, vol. III, p. 213-216.

anomalous lead Lead with isotopic ratios that indicate a model age either older or younger than the accepted age of the lead-bearing mineral or whole-rock sample. Cf: *lead-lead age; J-type lead; B-type lead.*

anomaly (a) A departure from the expected or normal. (b) The difference between an observed value and the corresponding computed value. (c) A geological feature, esp. in the subsurface, distinguished by geological, geophysical, or geochemical means, which is different from the general surroundings and is often of potential economic value; e.g. a magnetic anomaly.

anomite A variety of biotite differing only in optic orientation.

anomoclone A sponge *desma* consisting of a short arm (brachyome) and several longer arms directed at various angles away from the short arm.

Anomodontia A suborder of therapsid synapsid reptiles that includes the infraorder Dicynodontia, highly specialized for herbivorous habit, and a variety of forms, mostly large, of generally primitive aspect. Range, Upper Permian to Upper Triassic.

anomphalous Said of a gastropod shell lacking an umbilicus. Cf: *phaneromphalous.*

anorogenic Not orogenic, lacking in or unrelated to orogenic disturbance; e.g. an anorogenic area, time, or granite.

anorthic Said of crystals having unequal oblique axes; i.e., triclinic crystals.

anorthite (a) A white or grayish triclinic mineral of the plagioclase feldspar group: $CaAl_2Si_2O_8$. It is the most basic member of the plagioclases, its composition ranging from $Ab_{10}An_{90}$ to $Ab_0 An_{100}$. Anorthite occurs in basic and ultrabasic igneous rocks (gabbro, norite, anorthosite), rarely as a well-developed druse mineral, sometimes in tuffs, and very rarely in metamorphic rocks (skarns). Syn: *calcium feldspar; calciclase.* (b) The pure calcium-feldspar end member in the plagioclase series.

anorthitfels *anorthitite.*

anorthitissite A *hornblendite* containing anorthite. Obsolete.

anorthitite An igneous rock almost completely composed of anorthite. Syn: *calciclasite; anorthitfels.* Cf: *anorthosite.* Not recommended usage.

anorthoclase A triclinic mineral of the alkali feldspar group: (Na, K)$AlSi_3O_8$. It is a sodium-rich feldspar ($Or_{40}Ab_{60}$ to $Or_{10}Ab_{90}$) that shows deviations from monoclinic symmetry and that contains very fine-grained intergrowths; it is widespread as a groundmass constituent of slightly alkalic lavas. The term is usually applied to a mixture of several phases that may not even have a stability field of their own at any temperature. Cf: *orthoclase.* Syn: *anorthose; soda microcline.*

anorthoclasite A *trachyte* composed almost entirely of anorthoclase. It was named by Loewinson-Lessing in 1901. Not recommended usage.

anorthose *anorthoclase.*

anorthosite (a) In the *IUGS classification,* a plutonic rock with Q between 0 and 5, P/(A+P) greater than 90, and M less than 10. (b) A group of essentially monomineralic plutonic igneous rocks composed almost entirely of plagioclase feldspar, which is usually labradorite but may be as calcic as bytownite or as sodic as andesine or oligoclase, and little or no dark-colored minerals; also, any rock in that group. Anorthosites occur as large nonstratiform plutonic bodies and as stratiform intrusions; they have been identified in lunar rock samples. Syn: *plagioclasite; plagioclase rock.*

anorthositization Introduction of, or replacement by, anorthosite.

ANOVA *analysis of variance.*

ANT An acronym for a suite of rock types abundant in the lunar highlands, including anorthosite, norite, and troctolite, as well as gradational mixtures of these (Taylor, 1975).

antagonism In ecology, the relationship that exists between two organisms in which one or both are harmed, usually as a result of their trying to occupy the same ecologic niche.

antapical series The series of plates forming the terminal group behind the postcingular series in a dinoflagellate theca. Cf: *apical series.*

antarctic n. The area within the Antarctic Circle; the region of the South Pole.—adj. Pertaining to features, climate, vegetation, and animals characteristic of the antarctic region.

Antarctic Circle The parallel of latitude falling at approx. 66°32′ S; it delineates the frigid zone of the South Pole. Cf: *Arctic Circle.*

Antarctic convergence A natural and distinct oceanographic boundary around the continent of Antarctica, more or less equivalent to the 50°F isotherm for the warmest month. The colder, denser Antarctic waters sink sharply below the warmer, lighter sub-Antarctic waters, with little mixing. The oceanographic aspect of the boundary is reflected in water and air temperatures and in the flora and fauna.

antarcticite A mineral: $CaCl_2 \cdot 6H_2O$.

antecedent Said of a stream, valley, or drainage system that maintains its original course or direction despite subsequent deformation or uplift. The term was first applied by Powell (1875, p. 163) to the valley thus formed.

antecedent moisture The amount of moisture present in a soil mass at the beginning of a runoff period; often expressed in terms of an *antecedent precipitation index.*

antecedent-platform theory A theory of coral-atoll and barrier-reef formation according to which reefs are built upward to the water surface from an extensive submarine platform (perhaps consisting of volcanic debris rapidly leveled by wave erosion), situated 50 m or more below sea level, and formed before its colonization by corals, without the intervention of relative changes in sea level (see Hoffmeister & Ladd, 1944). Cf: *glacial-control theory; subsidence theory.*

antecedent precipitation index The amount of moisture in a drainage basin before a storm. Abbrev: API. See also: *antecedent moisture.*

antecedent stream A stream that was established before local uplift or diastrophic movement was developed across it and that maintained its original course after and in spite of the deformation by incising its channel at approximately the same rate as the land was rising; a stream that existed prior to the present topography.

anteclise A positive or uplifted structure of the continental platform; it is of broad, regional extent (tens to hundreds of thousands of square kilometers) and is produced by slow crustal upwarp during the course of several geologic periods. The term is used mainly in the Russian literature; e.g. the Belorussian anteclises of the Volga-Urals. Also spelled: *anticlise.* Ant: *syneclise.*

anteconsequent adj. Said of a stream, valley, or drainage system that is *consequent* in the earlier stages and *antecedent* in the later stages of an erosion or orogenic sequence. The term is rarely used "doubtless due to the practical difficulties of differentiating the effects of stages in tectonic movements in most areas" (Stamp, 1961, p. 25).—n. An anteconsequent stream.

antediluvian Pertaining to or produced before the flood described in the Bible; antedating the Noachian flood. Syn: *antediluvial; prediluvian.*

antenna (a) One of a pair of anterior sensory appendages of the cephalon of a crustacean, preceded by antennule and followed by mandible. (b) One of a pair of slender multijointed sensory appendages attached to the ventral surface of the cephalon in front of the mouth of a trilobite. (c) An obsolete term for a *chelicera* of an arachnid, used in the past to emphasize its homology with an antenna of a crustacean or of an insect.—Pl: *antennae* or *antennas.*

antennal carina A ridge extending backward from the antennal spine on some decapods (Holthuis, 1974, p. 735).

antennal groove In decapods, a groove that extends forward from the lower end of the cervical groove, and curves around the lower border of the elevation of the carapace behind the antennal spine. Anteriorly it may split into two branches, often indistinct, one going straight forward, the other curving up towards the antennal spine (Holthuis, 1974, p. 732-733).

antennal region The anterior marginal part of the carapace of some decapods, bordering the orbital region laterally and also touching the hepatic, pterygostomial, and, in some forms, the frontal regions (TIP, 1969, pt. R, p. 92).

antennule A small antenna; specif. one of a pair of the most anterior appendages of the cephalon of a crustacean, followed by antennae. Adj: *antennular.* Syn: *first antenna; antennula.*

anter Part of the orifice in ascophoran cheilostome bryozoans that is distal to the condyles or the sinus. Cf: *poster.*

anterior adj. Situated toward the front of an animal, or near or toward the head or head region, as opposed to *posterior*; e.g. in a direction (in the plane of symmetry or parallel to it) away from the pedicle and toward the mantle cavity of a brachiopod, or in a direction (in the plane of bilateral symmetry) parallel to the cardinal axis of a bivalve mollusk and approximating the direction in which the mouth faces, or in a direction toward the aperture of a foraminifer.— n. The forward-moving or head region of an animal.

anterior lateral muscle One of a pair of *retractor muscles* in some lingulid brachiopods, originating on the pedicle valve posteriorly and laterally to the central muscles, and converging dorsally to their insertions anteriorly on the brachial valve (TIP, 1965, pt.H, p.139).

anterior oral midline The perradial line between opposing oral plates in edrioasteroids; it extends anteriorly from the oral pole (Bell, 1976).

anterior side The front end of a conodont; e.g. the convex side of a cusp (the side facing in the direction opposite that toward which the tip of the cusp points) in simple conodont elements, the convex side of the cusp and denticles in compound conodont elements, or the distal end of the free blade in platelike conodont elements. Ant: *posterior side.*

anterior tubercle A polygenetic swelling or small protuberance in the anterior region of the carapace of a phyllocarid crustacean. It includes the "optic tubercle" of some authors.

anterolateral region The lateral part of the carapace of some decapods, bordering the subhepatic or hepatic regions (TIP, 1969, pt. R, p. 92).

antetheca The final septal face in a fusulinid; e.g. the front wall of the last-formed volution of the test of *Triticites.*

ante-turma One of two groupings in which *turmae* are classified: Sporites (spores) and Pollenites (pollen).

anther The pollen-bearing part of a stamen.

antheridium (a) In cryptogamous plants, the male reproductive organ, within which the male sexual cells are organized. (b) In primitive seed plants, a minute structure of only a few cells developed within the pollen grain.

anthill A colloquial term for *termitarium.*

anthoblast The basal portion of the zooid in certain solitary corals, from which the *anthocyathus* is pinched off to form a new zooid; e.g. the stage of ontogeny of *Acrosmilia* derived by transverse division from a solitary *Fungia* individual by extratentacular budding. Cf: *anthocaulus.*

anthocaulus The stalklike basal portion of the zooid in certain solitary corals, from which the *anthocyathus* is pinched off to form a new zooid; e.g. the stage of ontogeny of *Acrosmilia* derived by transverse division from a solitary *Fungia* individual by sexual generation. Pl: *anthocauli.* Cf: *anthoblast.*

anthocyathus The oral disk that is pinched off from the basal portion in some solitary corals and enlarges to become a new zooid; e.g. the neanic stage of *Fungia* after separation from an *anthocaulus* or *anthoblast.* Pl: *anthocyathi.*

anthodite In a cave, *helictites* usually radiating from a common base and composed of needlelike aragonite crystals. See also: *cave flower.*

anthoinite A white mineral: $AlWO_3(OH)_3$.

anthonyite A lavender-colored mineral: $Cu(OH,Cl)_2 \cdot 3H_2O$.

anthophyllite A clove-brown orthorhombic mineral of the amphibole group: $(Mg,Fe)_7Si_8O_{22}(OH)_2$. It contains less iron than *cummingtonite,* and may contain manganese and calcium. Anthophyllite is a variety of *asbestos,* normally occurring in metamorphic rocks as lamellae, radiations, or fibers, or in massive form. See: *gedrite.* Syn: *bidalotite.*

Anthophyta *angiosperms.*

anthozoan Any coelenterate belonging to the class Anthozoa, which includes marine, polypoid, solitary or colonial, mostly sedentary forms and is characterized by the presence of a stomodaeum. Range, Ordovician to present.

anthracite Coal of the highest metamorphic rank, in which fixed-carbon content is between 92% and 98% (on a dry, mineral-matter-free basis). It is hard and black, and has a semimetallic luster and semiconchoidal fracture. Anthracite ignites with difficulty and burns with a short blue flame, without smoke. Syn: *hard coal; stone coal; kilkenny coal; black coal.*

anthracitic Pertaining to *anthracite.*

anthracitization The metamorphic transformation of bituminous coal into anthracite.

anthracology Coal petrography; the analysis of coals by type.

anthraconite A black *bituminous limestone* (or marble) that usually emits a fetid smell on being struck or rubbed; a *stinkstone.* Syn: *swinestone; lacullan.*

Anthracosauria An order of labyrinthodont amphibians characterized by vertebrae with unreduced intercentrum, distinctive pattern of dermal bones of skull, and reptilian phalangeal formula. It includes the most probable candidates for reptilian ancestry. Range, Upper Mississippian to Upper Permian.

anthracoxene A brownish resin which, when treated with ether, dissolves into an insoluble portion, *anthracoxenite,* and a soluble portion, *schlanite.*

anthracoxenite The insoluble resin remaining when *anthracoxene* is treated with ether. See also: *schlanite.*

anthraxolite A hard, black asphaltite with a high fixed-carbon content; it occurs in veins and masses in sedimentary rocks, especially in association with oil shales.

anthraxylon A composite term for the vitreous coal components derived from woody tissues of plants and forming lustrous bands interlayered with dull *attritus* in banded coal. Etymol: Greek *anthrax,* "coal", and *xylon,* "wood". Pron: anthra-*zy*-lon.

anthraxylous-attrital coal A bright coal in which the ratio of anthraxylon to attritus ranges from 3:1 to 1:1. Cf: *attrital-anthraxylous coal; anthraxylous coal; attrital coal.*

anthraxylous coal A bright coal in which the ratio of anthraxylon to attritus is greater than 3:1. Cf: *attrital coal; attrital-anthraxylous coal; anthraxylous-attrital coal.*

anthrinoid Vitrinite that occurs in noncaking anthracites and that has a reflectance higher than 2.0% (ASTM, 1970, p.19). Cf: *xylinoid; vitrinoid.*

anthropic epipedon A diagnostic surface horizon that is similar to a *mollic epipedon* but in which the content of soluble P_2O_5 is greater than 250 ppm. It develops after long periods of use by man, either as a place of residence where bones and shells have been disposed of, or as a site for growing irrigated crops (USDA, 1975).

anthropozoic Said of that span of geologic time since the appearance of man; also, said of the rocks formed during that time. Cf: *Diluvial.*

anthrosol n. A soil that has been materially affected in its physical or chemical properties by human activities. Cf: *anthropic epipedon.*

antibiosis Passive action by one organism which is harmful to another (Ager, 1963, p. 313). Syn: *amensalism.*

anticenter That point on the Earth's surface that is diametrically opposite the *epicenter* of an earthquake.

anticlinal [bot] adj. At right angles to the surface or circumference of a plant organ. Cf: *periclinal [bot].*

anticlinal [struc geol] n. An obsolete form of *anticline.* —adj. Pertaining to an anticline.

anticlinal axis The axis of an anticline; see *axis [fold].*

anticlinal nose *nose.*

anticlinal spring A contact spring occurring along the outcrop of an anticline, from a pervious stratum overlying one that is less pervious (ASCE, 1962).

anticlinal theory The theory that oil and gas tend to accumulate in anticlinal structures. It was well set forth by I. C. White in 1885.

anticlinal valley A valley that follows the axis of an anticline.

anticline A fold, generally convex upward, whose core contains the stratigraphically older rocks. Ant: *syncline.* See also *antiform; synformal anticline.*

anticlinorium A composite anticlinal structure of regional extent composed of lesser folds. Cf: *synclinorium.* Pl: *anticlinoria.*

anticlise *anteclise.*

anticonsequent stream *obsequent stream.*

anticusp The downward projection of the base of a conodont *cusp.* It may or may not bear denticles.

anticyclone An atmospheric high-pressure system with closed isobars, the pressure gradient being directed away from the center so that the wind blows spirally outward in a clockwise direction in the Northern Hemisphere, counterclockwise in the Southern. It was named by Francis Galton in 1861. See also: *cyclone.*

antidip stream A stream flowing in a direction opposite to that of the general dip of the strata; an *anaclinal* stream. It is frequently but not necessarily an *obsequent stream.*

antidune (a) A term used by Gilbert (1914, p.31) for an ephemeral or transient *sand wave* formed on a stream bed (but not believed to be preserved in sediments), similar to a dune but traveling upstream as the individual sand particles move downcurrent, and characterized by erosion on the downstream slope and deposition on the upstream slope. Its profile is more symmetrical than that of a subaqueous *dune;* it is indicated on the water surface by a regular undulating wave. Syn: *regressive sand wave.* (b) Any bed form (whether it moves upstream, downstream, or not at all) that is produced by unidirectional flow and is in phase with surface water waves (Kennedy, 1963). Syn: *sinusbed.* (c) A term used by Lamont (1957) for *flame structure [sed].*

antidune phase The part of stream traction (transitional to the *smooth phase*) whereby a mass of sediment travels in the form of a ridge-like structure having an eroded downcurrent slope and a depositional upcurrent slope (Gilbert, 1914, p. 30-34); it develops when the bed load is large or the current is strong. The antidune form moves upstream as the individual particles move downstream. Cf: *dune phase.*

antiferromagnetism A type of *magnetic order* in which the moments of neighboring magnetic ions are aligned antiparallel, so that there is no macroscopic spontaneous magnetization. Cf: *ferro-*

magnetism; ferrimagnetism. See also: *weak ferromagnetism.*

antiform A fold whose limbs close upward in strata for which the stratigraphic sequence is not known. Cf: *anticline.* Ant: *synform.*

antiformal syncline A *syncline* the limbs of which close upward as in an *antiform* (Turner & Weiss, 1963, p. 106).

antigorite A platy or lamellar brownish-green mineral of the serpentine group: $Mg_3Si_2O_5(OH)_4$. Cf: *chrysotile.* Syn: *picrolite; baltimorite.*

antigravitational gradation A term introduced by Keyes (1913) for wind erosion and deposition operating mainly from a lower to a higher elevation, as on broad intermont valleys of arid regions where "the wind is able to blow sands erodingly and extensively up-hill". See also: *planorasion.*

antilogous pole In crystallography, that pole of a crystal which becomes electrically negative when the crystal is heated or is expanded by decompression. Cf: *analogous pole.*

antimagmatist *transformist.*

antimeridian The meridian that is 180 degrees of longitude from a given meridian. A meridian and its antimeridian constitute a complete great circle.

antimonate A mineral compound characterized by the presence of antimony and oxygen in the radical. An example is swedenborgite, $NaBe_4SbO_7$.

antimonite *stibnite.*

antimonpearceite A mineral: $(Ag,Cu)_{16}(Sb,As)_2S_{11}$. Cf: *arsenpolybasite.*

antimony A hexagonal mineral, the native metallic element Sb. It is brittle and commonly occurs in silvery or tin-white granular, lamellar, or shapeless masses.

antimony blende *kermesite.*

antimony bloom *valentinite.*

antimony glance *stibnite.*

antimony ocher Any of several native antimony oxides such as stibiconite or cervantite.

antinode That point on a standing wave where the vertical motion is greatest and the horizontal velocity is least. Ant: *node.* Syn: *loop.*

antipathetic Said of two or more minerals that are far apart from each other in a crystallization sequence and thus will not be commonly found in association. See also: *antipathies of minerals.*

antipathies of minerals An aspect of the theory of fractional crystallization, which states that minerals that are far apart in a crystallization sequence will not be found in association to any great extent. Such minerals are said to be *antipathetic.*

antiperthite A variety of alkali feldspar consisting of parallel or subparallel intergrowths in which the sodium-rich phase (albite, oligoclase, or andesine) appears to be the host from which the potassium-rich phase (usually orthoclase) exsolved. Cf: *perthite.*

antipodal bulge The tidal effect on the side of the Earth farthest from the Moon, at the point that is antipodal to the *tidal bulge;* lunar attraction is weakest and produces an apparent bulge.

antipodal point *antipode.*

antipode The opposite point with respect to any given point; specif. one of two diametrically opposite parts of the Earth. The term is usually used in the plural and is often extended to include the whole region at the opposite end of a diameter of the Earth, such as Australia and New Zealand which lie roughly opposite to the British Isles. Syn: *antipodal point.*

antiprism An n-gonal form with faces consisting of two parallel regular n-gons connected by $2n$ isosceles triangles.

antiripple *adhesion ripple.*

antiripplet *adhesion ripple.*

antiroot According to the *Pratt hypothesis,* crustal material under the oceans of higher density as isostatic compensation for its lesser mass and lower topographic elevation. Cf: *root.*

antistress mineral A term suggested by Harker (1918) for minerals such as cordierite, the feldspars, the pyroxenes, forsterite, andalusite, etc. whose formation in metamorphosed rocks is believed to be favored by conditions that are not controlled by shearing stress, but by thermal action and by hydrostatic pressure that is probably no more than moderate. Cf: *stress mineral.*

antithetic Pertaining to minor normal faults that are oriented opposite to the major fault with which they are associated. Ant: *synthetic [fault].*

antitrades A layer of westerly winds in the troposphere, above the *trade winds* of the tropics.

antlerite An emerald-green to blackish-green mineral: $Cu_3SO_4(OH)_4$. It occurs in interlaced aggregates of needlelike crystals and constitutes an ore of copper. Syn: *vernadskite.*

Antler orogeny An orogeny that extensively deformed Paleozoic rocks of the Great Basin in Nevada during late Devonian and early Mississippian time; named by R. J. Roberts (1951) for relations in the Antler Peak quadrangle near Battle Mountain, Nevada. Its main expression is the emplacement of eugeosynclinal western rocks over miogeosynclinal eastern rocks along the Roberts Mountains thrust. Minor orogenic pulses followed the main event, extending into the Permian. It is broadly equivalent to the *Acadian orogeny* of eastern North America.

antofagastite *eriochalcite.*

antozonite A dark-violet to black semiopaque variety of fluorite that emits a strong odor when crushed, perhaps due to free fluorine. It is produced by alpha bombardment, as in the inner bands of halos surrounding uraninite and thorite inclusions.

Anura Order of tailless *lissamphibians* that includes frogs and toads. Range, Middle Jurassic to Recent.

anus The posterior or terminal opening of the alimentary canal or digestive tract of an animal.

Apache tear (a) An obsidian nodule weathered from a lava flow. (b) A widely used syn. of *marekanite.*

apachite A *phonolite* characterized by the presence of abundant aenigmatite and sodic amphibole in about the same amount as the pyroxene, but of later crystallization. It was named by Osann in 1896 from the Apache (now Davis) Mountains, Texas. Not recommended usage.

apalhraun An Icelandic term for both *block lava* and *aa.* Cf: *helluhraun.*

apatite (a) A group of variously colored hexagonal minerals consisting of calcium phosphate together with fluorine, chlorine, hydroxyl, or carbonate in varying amounts and having the general formula: $Ca_5(PO_4,CO_3)_3(F,OH,Cl)$. Also, any mineral of the apatite group, such as fluorapatite, chlorapatite, hydroxylapatite, carbonate-apatite, and francolite; when not specified, the term usually refers to *fluorapatite.* The apatite minerals occur as accessory minerals in almost all igneous rocks, in metamorphic rocks, and in veins and other ore deposits; and most commonly as fine-grained and often impure masses as the chief constituent of phosphate rock and of most or all bones and teeth. Syn: *calcium phosphate.* (b) A group of hexagonal minerals having the general formula: $A_5(RO_4)_3(F,OH,Cl)$, where A = Ca, Sr, or Pb, and R = P, As, V, or less commonly Si. Examples include svabite, hedyphane, mimetite, pyromorphite, and vanadinite.—Symbol: Ap.

apatotrophic Pertaining to a lake that is brackish and contains living organisms (Termier & Termier, 1963).

apertural bar A fused pair of costae immediately proximal to the orifice in cribrimorph cheilostome bryozoans.

aperture [paleont] (a) The opening of a univalve shell; e.g. the opening at the last-formed margin of a gastropod shell, through which the head-foot mass is extended or withdrawn. (b) Any of the major openings through the theca or calyx of echinoderms, such as the mouth and anus, and sometimes including hydropores and gonopores. (c) The terminal skeletal openings of zooids in stenolaemate bryozoans. Also used as a syn. of *orifice* in cheilostome bryozoans. (d) An opening in the test or shell of a foraminifer, such as a relatively large opening to the exterior in the last-formed chamber. Also, the large main opening of a radiolarian shell. (e) The opening into the mantle cavity of a cirripede crustacean.

aperture [palyn] Any of the various modifications in the exine of spores and pollen that can be used as a locus for exit of the contents; e.g. laesura, colpus, and pore. See also: *germinal aperture.*

apex [fold] *culmination.*

apex [geomorph] The tip, summit, or highest point of a landform, as of a mountain; specif. the highest point on an alluvial fan, usually the point where the stream that formed the fan emerged from the mountain or from confining canyon walls.

apex [mining] In mining, the highest point of a vein relative to the surface, whether it crops out or not. The concept is used in mining law. See also: *apex law.*

apex [paleont] (a) The first-formed, generally pointed end of an elongate or conical form of an organism, such as the small end of the shell or spire of a gastropod. (b) The first-formed part of a brachiopod valve, around which the shell has grown subsequently. The term is "usually restricted to valves having this point placed centrally or subcentrally" (TIP, 1965, pt.H, p.140). (c) The tip of the basal cavity or of a denticle of a conodont. Also, the juncture of bars, blades, or other processes of conodont elements. (d) The upper (umbonal) angle of a valve or plate of a cirripede crustacean. (e)

The pointed beginning of a straight or slightly curved cephalopod shell.

apex law In U.S. mining law, the individual whose claim contains the *apex* of a vein may follow and exploit the vein indefinitely along its dip, even if it passes downdip under adjoining surface property lines. Syn: *law of extralateral rights.*

APF *absolute pollen frequency.*

aphanic Said of the texture of a carbonate sedimentary rock characterized by individual crystals or clastic grains whose diameters are less than 0.01 mm (Bissell & Chilingar, 1967, p. 150) or 0.005 mm (Chilingar et al., 1967, p. 311). The term was proposed by DeFord (1946) to replace *aphanitic.* See also: *aphanocrystalline.* Cf: *phaneric.*

aphanide An informal term used in the field to designate a wholly or predominantly fine-grained rock.

aphaniphyric *cryptocrystalline.*

aphanite Any fine-grained igneous rock whose components are not distinguishable with the unaided eye; a rock having aphanitic texture. Adj: *aphanitic.* Syn: *kryptomere; felsite; felsitoid.*

aphanitic [ign] Said of the texture of an igneous rock in which the crystalline components are not distinguishable by the unaided eye; both *microcrystalline* and *cryptocrystalline* textures are included. Also, said of a rock or a groundmass exhibiting such texture. The obsolescent syn. *felsitic* has been sometimes restricted to the light-colored rocks with this texture and "aphanitic" to the dark-colored (Johannsen, 1939, p.201). Cf: *phaneritic.* Syn: *fine-grained [geol].*

aphanitic [sed] A term formerly loosely applied to a sedimentary carbonate-rock texture now referred to as *aphanic.*

aphanocrystalline Descriptive of an interlocking texture of a carbonate sedimentary rock having crystals whose diameters are in the range of 0.001-0.004 mm (Folk, 1959). See also: *aphanic.* Syn: *extremely finely crystalline.*

aphanophyre A porphyry with an aphanitic groundmass. Cf: *granophyre; vitrophyre.* Syn: *felsophyre.* Adj: *aphanophyric.*

aphanophyric *microcrystalline.*

Aphebian In a three-part division of the Proterozoic of Canada, the earliest division, before the *Helikian.* Cf: *Hadrynian.*

aphlebia Anomalous pinnae, usually located at the base of the rachis of certain fossil ferns and pteridosperms (Swartz, 1971, p. 36).

aphodus A short canal of uniform diameter in a sponge, leading to an exhalant canal from an apopyle of approximately the same cross-sectional area. Pl: *aphodi.* Cf: *prosodus.*

aphotic Literally, "without light". It refers to a depth in the sea below approximately 200 m.

aphotic zone That part of the ocean in which there isn't enough penetration of light for photosynthesis. Cf: *disphotic zone; euphotic zone.*

aphrite A foliated, lamellar, scaly, or chalky variety of calcite having a white pearly luster. Syn: *earth foam; foaming earth.*

aphrizite A black variety of tourmaline containing iron.

aphrodite *stevensite.*

aphroid Said of a massive corallum similar to *astreoid* type but with septa shortened peripherally and adjacent corallites united by a dissepimental zone.

aphrolith An obsolete syn. of *aa.* Cf: *dermolith.*

aphrosiderite *ripidolite.*

aphthitalite A white rhombohedral mineral: $(K,Na)_3Na(SO_4)_2$. Syn: *glaserite.*

aphylactic projection A map projection that does not possess any one of the three special properties of equivalence, conformality, or equidistance; e.g. a gnomonic projection.

aphyllous Said of a leafless plant.

aphyric Said of the texture of a fine-grained or aphanitic igneous rock that lacks phenocrysts. Also, said of a rock exhibiting such texture.

aphytal zone The plantless part of a lake bottom. Cf: *phytal zone.*

Aphytic A paleobotanic division of geologic time, signifying the time that preceded the first occurrence of plant life. Cf: *Archeophytic; Eophytic; Paleophytic; Mesophytic; Cenophytic.*

API (a) The American Petroleum Institute. (b) *antecedent precipitation index.*

apical (a) Situated at or in the direction or vicinity of the apex of a shell; e.g. "apical horn", a spine at the apex of the shell of a nasselline radiolarian. (b) Located away from the mouth of an echinoderm; aboral.

apical archeopyle An *archeopyle* formed in a dinoflagellate cyst by the loss of the entire apical series of plates. See also: *haplotabular archeopyle; tetratabular archeopyle.*

apical area On an embryophytic spore, the area including the trilete suture.

apical axis The lengthwise axis of a pennate diatom. Cf: *transapical axis; pervalvar axis.*

apical papilla A dotlike thickening of the interradial area of a spore. When present, there is generally one apical papilla per interradial area, hence three per spore.

apical prominence In megaspores, mostly Paleozoic, a variously constructed projection at the intersection of the contact areas. Cf: *acrolamella; gula.*

apical series The series of plates grouped about an open apical pore or forming an apical cluster in the epitheca of a dinoflagellate. Cf: *antapical series.*

apical system A system of primordial plates at the aboral terminus of ambulacra and interambulacra of echinoids, sometimes including an outer circlet of *ocular plates* surrounding an inner circlet of *genital plates,* and one or more supplementary plates. See also: *oculogenital ring.*

apiculate (a) Pertaining to surface ornamentation, as in *palynomorphs,* consisting of short, sharp spines. (b) Said of pollen grains with slightly protuberant poles.

apiculus An open-end process extending from the valve surface in a diatom. Plural: *apiculi.*

API gravity A standard adopted by the American Petroleum Institute for expressing the specific weight of oils. API gravity = (141.5/specific gravity at 60°F) - 131.5. This arbitrary scale simplifies the construction of *hydrometers* because it enables the stems to be calibrated linearly. The lower the specific gravity, the higher the API gravity. Cf: *Baumé gravity.*

API unit An arbitrary unit of *radioactivity log* measurement to which the count rates of *gamma-ray log* or *neutron log* detectors are standardized. Two shallow boreholes at the University of Houston, Texas, serve as the calibration pits. The API gamma unit and the API neutron unit are different.

apjohnite A silky-white, faintly green, or yellow mineral: $MnAl_2(SO_4)_4 \cdot 22H_2O$. It occurs in crusts, fibrous masses, or efflorescences. Syn: *manganese alum.*

aplacophoran A marine mollusk belonging to the class Aplacophora and known only from living forms. See also: *polyplacophoran.*

aplite A light-colored hypabyssal igneous rock characterized by a fine-grained allotriomorphic-granular (i.e. aplitic) texture. Aplites may range in composition from granitic to gabbroic, but the term "aplite" with no modifier is generally understood to mean granitic aplite, consisting essentially of quartz, potassium feldspar, and acid plagioclase. The term, from a Greek word meaning "simple", was in use before 1823. Syn: *haplite.*

aplitic (a) Pertaining to the fine-grained and *saccharoidal* or allotriomorphic-granular texture characteristic of aplites. See also: *sutured.* (b) Said of an igneous rock having such a texture.

aplodiorite A light-colored biotite *granodiorite* with little or no hornblende. It was named by Bailey in 1916. Not recommended usage.

aplogranite A light-colored plutonic rock, named by Bailey in 1916, having granitic texture and essentially composed of alkali feldspar and quartz, with lesser amounts of biotite and with or without muscovite. Cf: *alaskite; two-mica granite.* Not recommended usage.

aplome A dark-brown, yellowish-green, or brownish-green variety of andradite garnet containing manganese. Syn: *haplome.*

aplowite A mineral: $(Co,Mn,Ni)SO_4 \cdot 4H_2O$.

apo- In petrologic terminology, a prefix signifying metasomatic change without destruction of original texture.

apobsidian A devitrified *obsidian.* Obsolete.

apocarpous Said of a plant ovary whose carpels are separate rather than united; also, said of a gynoecium of separate pistils (Lawrence, 1951, p.739). Cf: *syncarpous.*

apochete An *exhalant canal* of a sponge.

apodeme One of the ingrowths from the exoskeleton of many arthropods that provide points of muscle attachment, such as an invagination of the body wall of an arachnid, an inward deflection of a sclerite of a merostome, a downward projection from the dorsal interior of a thoracic segment of a trilobite, or an infold of the exoskeleton of a crustacean. Syn: *apodema.*

apo-epigenesis Post-diagenetic changes that affect sediments while they are remote from the original environment of deposition, as when they are under a relatively thick overburden (Chilingar et al., 1967, p. 311). Cf: *juxta-epigenesis.*

apogean tide A tide of decreased range that occurs monthly, as the Moon is at or near the apogee of its orbit. It is a secondary modification of the tidal cycle. Ant: *perigean tide.*

apogee That point on the orbit of an Earth satellite, natural or man-made, which is farthest from the Earth. Cf: *perigee.*

apogranite Albitized and greisenized granite, located at the peripheral and apical parts of certain intrusives, and commonly mineralized in rare elements (Nb, Ta, Li, Rb, Be, Sn, W, Mo, etc.). The term was originated in 1962 by A. A. Beus and co-workers in the Soviet Union.

apogrit *graywacke.*

Apollonian metamorphic rocks A small proportion of lunar rocks that possess polygonal granular texture and are composed of minerals with constant composition throughout a given rock. These characteristics could only have been developed during prolonged subsolidus annealing in the most deeply sampled parts of the Moon's crust and upper mantle (Stewart, 1975). Etymol: in commemoration of the Apollo program.

Apollonian metamorphism Lunar processes that produced *Apollonian metamorphic rocks.*

apomagmatic Said of a hydrothermal mineral deposit at an intermediate distance from its magmatic source. The term is little used. Cf: *telemagmatic; perimagmatic; cryptomagmatic.*

apomorphy A syn. of *derived character.* Adj: *apomorphic; apomorphous.*

aponeurotic band An anterior or posterior area of attachment of ligaments (mantle and visceral) on the inside of the body chamber of a nautiloid (TIP, 1964, pt.K, p.54).

apophyllite A mineral: $KCa_4Si_8O_{20}(F,OH)\cdot 8H_2O$. It is a secondary mineral related to and occurring with zeolites in geodes in decomposed basalts and other igneous rocks. Syn: *fish-eye stone.*

apophysis [intrus rocks] A syn. of *tongue [intrus rocks].* Cf: *epiphysis [intrus rocks].*

apophysis [paleont] (a) An internal projection from interambulacral basicoronal plates of an echinoid for the attachment of muscles supporting Aristotle's lantern. (b) A lateral transverse process of a radial spine in acantharian radiolarians.—Pl: *apophyses.*

apopore The exit opening of an exhalant canal of a sponge, located either within the sponge (as on the lining of a larger exhalant canal or the spongocoel) or on its surface, in which case it is equivalent to an osculum. Cf: *prosopore.*

apopyle Any opening through which water passes out of a flagellated chamber of a sponge. Cf: *prosopyle.*

aporhyolite A *rhyolite* in which the once glassy groundmass has become devitrified. Not recommended usage.

aporhysis A *skeletal canal* in dictyonine hexactinellid sponges running radially through the body wall, opened at the spongocoel end but closed over at the outer end. Pl: *aporhyses.*

apotaphral Descriptive of a type of tectonics involving lateral outward spreading (under gravity) of the orogenic zone away from the axis of a geosyncline (Carey, 1963, p.A6); it is characterized by nappes, thrusts, and recumbent folds. Cf: *syntaphral; diataphral.*

Appalachia One of the *borderlands* proposed by Schuchert (1923), in this case along the southeast side of North America, seaward from the Appalachian orogenic belt. Most of the evidence for Appalachia, as originally conceived, can now be otherwise interpreted. It is true that during middle and late Paleozoic time much sediment was being shed from lands in the seaward part of the Appalachian belt, but these were probably narrow, ephemeral tectonic lands. No former large extensions of this borderland into the present Atlantic Ocean basin are possible, because of the oceanic crustal structure beyond the edge of the continental shelf.

Appalachian relief Structurally controlled topography developed on folded strata of dissimilar erosional resistance, characterized by homoclinal ridges and subsequent valleys that are adapted to the structure and differential resistance of the rocks. Type example: the relief of the Appalachian Mountains in North America. Cf: *Jurassian relief.*

Appalachian Revolution A concept, widely held in the first part of the 20th Century, that Paleozoic time was closed by a profound crustal disturbance, which especially deformed the rocks in the central and southern Appalachians. The term is misleading, and should be abandoned in favor of the term *Allegheny orogeny.*

apparent ablation *summer balance.*

apparent accumulation *winter balance.*

apparent crater The depression of an explosion or impact crater as it appears after modification of the original shape by postformational processes such as slumping and deposition of material ejected during crater formation; a crater that is visible on the surface and whose dimensions are measured with respect to the original ground level (Nordyke, 1961, p. 3447). The "apparent diameter" and "apparent depth" are measured using the highest points on the rim crest and the deepest part of the observable depression. Cf: *true crater.*

apparent density (a) Rock density calculated from gravity measurements in a borehole. See also: *interval density.* (b) Obsolete syn. of *bulk density.*

apparent dip [seis] In seismology, the angle that an emerging wave front makes with the surface, as related to the dip of the reflector associated with it. The apparent dip is the angle whose tangent is equal to the ratio of the vertical and horizontal components of the displacement produced by the wave front. Cf: *dip [seis]; moveout.* Syn: *emergence angle.*

apparent dip [struc geol] The angle that a structural surface, e.g. a bedding or fault plane, makes with the horizontal, measured in any random, vertical section rather than perpendicular to the strike. It varies from nearly zero to nearly the *true dip,* depending on whether the random section is close to the direction of the strike or of the dip. Syn: *false dip.*

apparent horizon The visible line of demarcation between land or sea and sky. Strictly, it is the circle that bounds the part of the Earth's surface which would be visible from a given point if no irregularities or obstructions were present. The apparent horizon is extended slightly downward because of atmospheric refraction. In popular usage, the term *horizon* usually signifies the "apparent horizon". Cf: *true horizon.* Syn: *visible horizon; local horizon; sensible horizon; geographic horizon; topocentric horizon; natural horizon.*

apparent movement The movement observed in any chance section across a fault. It is a function of several variables: the attitude of the fault, of the disrupted strata, and of the section on which the fault is observed, as well as the net or actual slip of the fault.

apparent onlap *Onlap* observed in a randomly oriented vertical geologic section, which may not be parallel to the depositional dip and thus may give only one component of the true onlap direction (Mitchum, 1977, p.208).

apparent optic angle The *optic angle* as it appears under the conoscope, after being refracted upon emergence from the crystal. It is larger than the actual optic angle within the crystal.

apparent plunge The angle assumed by a normal projection of a geologic structure in the plane of a vertical cross section (Knutson, 1958).

apparent resistivity The resistivity measured at any point on the surface of the real (inhomogeneous) Earth may be equated to the resistivity of some homogeneous Earth, for the same electrode array and the same excitation frequency. The homogeneous Earth would be characterized by the value of apparent resistivity so measured. This quantity varies from point to point over the surface of the real Earth and its values do not relate in any simple way to the *true resistivities* of the homogeneous units making up the inhomogeneous Earth.

apparent stress The product of the seismic efficiency and the average stress of an earthquake.

apparent thickness The *thickness [geol]* of a stratigraphic unit or other tabular body, measured at right angles to the surface of the land. See also: *vertical [stratig].* Cf: *true thickness.*

apparent velocity The velocity with which the phase of a seismic wave train appears to travel along the surface of the Earth. It exceeds the actual velocity if the wave train is not travelling parallel to the surface.

apparent water table *perched water table.*

appendicular skeleton In the vertebrates, the skeleton of the shoulder and hip girdles and of the fore and hind limbs.

appinite A group of dark-colored hornblende-rich plutonic rocks, such as certain syenites, monzonites, and diorites, in which the hornblende occurs as large prismatic phenocrysts and also in the finer-grained groundmass. It was named by Bailey in 1916 for Appin, Loch Linnhe, Scotland. Not recommended usage.

applanation All processes that tend to reduce the relief of an area, causing it to become more and more plainlike. These include lowering of the high parts by erosion and raising of the low parts by addition of material; the latter is usually more effective.

applied geology The application of various fields of geology to economic, engineering, water-supply, or environmental problems; geology related to human activity.

applied geophysics *geophysical exploration.*

applied seismology The use of artificially generated seismic waves in the search for economic deposits such as salt, oil, and gas, or in engineering studies such as determining depth to bedrock or the presence of potentially active faults. Syn: *seismic exploration; prospecting seismology.* Obsolete syn: *explosion seismology.*

apposed glacier A term, not in current usage, for a glacier formed by the union of two glaciers (Swayne, 1956, p.14).

apposition beach One of a series of roughly parallel beaches successively formed on the seaward side of an older beach.

apposition fabric A primary orientation of the elements of a sedimentary rock, developed or formed at the time of deposition of the material by the successive placing of particles upon those already present (Sander, 1970, p. 119). See also: *depositional fabric.* Syn: *primary fabric.*

appressed [paleont] Said of very closely set conodont denticles, each partly or entirely fused to those adjoining. Cf: *discrete.*

appressed [struc geol] Said of a fold whose limbs are almost parallel.

approach The area of the sea extending indefinitely seaward from the shoreline at mean low-water spring tide (Wiegel, 1953, p. 2).

APR *airborne profile recorder.*

apron [geomorph] An extensive blanketlike deposit of alluvial, glacial, eolian, marine, or other unconsolidated material derived from an identifiable source, and deposited at the base of a mountain, in front of a glacier, etc.; e.g. a *bajada* or an *outwash plain.* Syn: *frontal apron.*

apron [glaciol] *ice apron.*

apron [ice] *ram.*

apsacline Said of the ventral and posterior *inclination* of the cardinal area in either valve of a brachiopod, lying in the bottom left or first quadrant moving counterclockwise from the orthocline position (TIP, 1965, pt.H, p.60, fig.61).

Aptian European stage: Lower Cretaceous, or Lower and Middle Cretaceous of some authors (above Barremian, below Albian). See also: *Gault.*

aptychus A pair of symmetrical calcareous or horny plates, joined along straight edges, originally lodged in the body chamber of some ammonoid conchs; it probably formed part of the lower jaw of the animal. It was formerly interpreted as an opercular structure. See also: *anaptychus.*

aquafact An isolated boulder or cobble, commonly on a sandy beach, that has been worn smooth on its seaward face by wave action, so that a sharp ridge parallel to the shore is developed along the exposed surface of the boulder or cobble; a *water-faceted stone* produced by *wet blasting* (Kuenen, 1947).

aquagene tuff *hyaloclastite.*

Aqualf In U.S. Dept. of Agriculture soil taxonomy, a suborder of the soil order *Alfisol,* characterized by an aquic moisture regime (unless the soil has been drained) and by mottles or gray colors indicative of wetness. Cultivation without artificial drainage is difficult, and most Aqualfs have some artificial drainage (USDA, 1975). Cf: *Boralf; Udalf; Ustalf; Xeralf.*

aquamarine (a) A transparent and pale-blue, greenish-blue, or bluish-green gem variety of beryl. (b) A pale-blue, greenish, light greenish-blue, or bluish-green color designation applied to mineral names; e.g. "aquamarine chrysolite" (a greenish-blue beryl), "aquamarine sapphire" (a pale-blue sapphire), "aquamarine topaz" (a greenish topaz), and "aquamarine tourmaline" (a pale-blue or pale greenish-blue tourmaline).

aquatic (a) Living entirely or primarily in or on water. (b) Growing in or on water. (c) Living near or frequenting water.

aquatillite A term proposed by Schermerhorn (1966, p. 832) for a glaciomarine or glaciolacustrine till-like deposit, such as one deposited from a melting iceberg.

aquatolysis A term proposed by Müller (1967, p. 130) for the chemical and physicochemical processes that occur in a freshwater environment during transportation, weathering, and preburial diagenesis of sediments. Cf: *halmyrolysis.*

Aquent In U.S. Dept. of Agriculture soil taxonomy, a suborder of the soil order *Entisol,* characterized by bluish colors where continuously saturated or by gray colors and mottles where periodically saturated. They form from recent sediments in tidal marshes, deltas, lake margins, and flood plains. Cultivation without artificial drainage is difficult (USDA, 1975). Cf: *Arent; Fluvent; Orthent; Psamment.*

aqueoglacial *glacioaqueous.*

aqueo-igneous Pertaining to the formation of a mineral or rock from a magma in which water was influential. Syn: *hydroplutonic; hydatopyrogenic.* Not recommended usage.

aqueo-residual sand A term used by Sherzer (1910, p.627) for a sand in which the particles, produced by various residual agents, were subsequently modified by the action of water. It includes "all water-transported sand, for residual agencies have been present to some extent in the derivation of all sand from the parent material" (Allen, 1936, p.12). Cf: *residuo-aqueous sand.*

aqueous (a) Of, or pertaining to, water. (b) Made from, with, or by means of water; e.g. aqueous solutions. (c) Produced by the action of water; e.g. aqueous sediments.

aqueous fusion Melting in the presence of water, as a magma (Thrush, 1968, p. 48).

aqueous ripple mark A ripple mark made by waves or currents of water, as opposed to one made by air currents (wind).

Aquept In U.S. Dept. of Agriculture soil taxonomy, a suborder of the soil order *Inceptisol,* characterized by poor or very poor natural drainage. Unless they have been artificially drained, ground water stands close to the surface for some part of each year, making cultivation difficult. Most Aquepts have a gray or black surface horizon and a mottled gray subsurface cambic horizon that begins at a depth of less than 50 cm. Some Aquepts also have a fragipan (USDA, 1975). Cf: *Andept; Ochrept; Plaggept; Tropept; Umbrept.*

aquiclude A body of relatively impermeable rock that is capable of absorbing water slowly but does not transmit it rapidly enough to supply a well or spring. Cf: *aquifuge; aquitard; confining bed.*

aquic moisture regime A reducing soil-moisture regime nearly free of dissolved oxygen owing to ground-water saturation, when soil temperature at a depth of 50 cm is more than 5°C (USDA, 1975).

aquifer A body of rock that is sufficiently permeable to conduct ground water and to yield economically significant quantities of water to wells and springs. Syn: *water horizon; ground-water reservoir.*

aquiferous system The entire water-conducting system between the ostia and the oscula of a sponge, including the *inhalant system* and the *exhalant system.* Syn: *canal system.*

aquifer system A heterogeneous body of intercalated permeable and less permeable material that acts as a water-yielding hydraulic unit of regional extent.

aquifer test A test involving the withdrawal of measured quantities of water from, or addition of water to, a well and the measurement of resulting changes in *head [hydraul]* in the aquifer both during and after the period of discharge or addition.

aquifuge An impermeable body of rock; a rock with no interconnected openings and thus lacking the ability to absorb and transmit water. Cf: *aquiclude; aquitard; confining bed.*

Aquilonian Stage in France: uppermost Jurassic. It is equivalent to Purbeckian in Great Britain.

Aquitanian European stage: lowermost Miocene (above Chattian of Oligocene, below Burdigalian). It was formerly regarded by some authors as uppermost Oligocene.

aquitard A *confining bed* that retards but does not prevent the flow of water to or from an adjacent aquifer; a *leaky confining bed.* It does not readily yield water to wells or springs, but may serve as a storage unit for ground water. Cf: *aquifuge; aquiclude.*

Aquod In U.S. Dept. of Agriculture soil taxonomy, a suborder of the soil order *Spodosol,* characterized by formation in an aquic moisture regime due to fluctuating ground water or to an extremely humid climate. Aquods have either a black surface horizon resting on a dark, iron-free spodic horizon, or a thick white albic horizon. Indicators of wetness may include a histic epipedon, mottles in the albic or spodic horizon, or a duripan or placic horizon at relatively shallow depth. In many Aquods the spodic horizon is cemented (USDA, 1975). Cf: *Ferrod; Humod; Orthod.*

Aquoll In U.S. Dept. of Agriculture soil taxonomy, a suborder of the soil order *Mollisol,* characterized by a thick, black epipedon and a gray, distinctly mottled subsoil. Aquolls have an aquic moisture regime unless they have been artificially drained to permit cultivation (USDA, 1975). Cf: *Alboll; Boroll; Rendoll; Udoll; Ustoll; Xeroll.*

Aquox In U.S. Dept. of Agriculture soil taxonomy, a suborder of the soil order *Oxisol,* characterized by formation in periodically flooded depressions or at the base of slopes that receive surface water. Aquox soils in depressions are periodically saturated unless artificially drained and have either a histic epipedon or an oxic horizon with characteristics of wetness, such as mottles. Aquox soils at the base of slopes have continuous plinthite near the surface, formed by recementing of ironstone fragments transported downslope (USDA, 1975). Cf: *Humox; Orthox; Torrox; Ustox.*

Aquult In U.S. Dept. of Agriculture soil taxonomy, a suborder of the soil order *Ultisol,* characterized by formation in wet places where ground water is very close to the surface in winter and spring months. Aquults have an umbric or an ochric epipedon above an argillic horizon that has mottles, manganese-iron concretions, or gray colors indicative of wetness. Aquults are extensive along the Atlantic and Gulf of Mexico coastal plains. Most are forested but some have been artificially drained to allow cultivation (USDA, 1975). Cf: *Humult; Udult; Ustult; Xerult.*

arabesquitic Said of the texture of certain porphyritic rocks in which an apparently homogeneous groundmass breaks up, under crossed nicols, into irregular patches, supposedly resembling arabesques (Johannsen, 1939, p.202).

arachnid Any terrestrial *chelicerate* belonging to the class Arachnida, characterized by the presence of one pair of preoral appendages with two to three joints. Cf: *merostome.*

aragonite (a) A white, yellowish, or gray orthorhombic mineral: $CaCO_3$. It is trimorphous with calcite and vaterite. Aragonite has a greater density and hardness, and a less distinct cleavage, than calcite, and is also less stable and less common. It occurs in fibrous aggregates in beds of gypsum and iron ore; as a deposit from hot springs; and as a major constituent of shallow marine muds and the upper parts of coral reefs. Aragonite is also an important constituent of the pearl, and of some shells. Syn: *Aragon spar.* (b) A group of orthorhombic carbonate minerals, including aragonite, alstonite, witherite, strontianite, and cerussite.

Aragon spar *aragonite.*

arakawaite *veszelyite.*

aramayoite An iron-black mineral: $Ag(Sb,Bi)S_2$.

araneid Any *arachnid* belonging to the order Araneida, characterized by the presence of maxillary lobes and glands and by the similarity of the first pair of legs to the other legs. Range, Carboniferous (possibly Devonian) to present.

arapahite A dark-colored, porous, fine-grained *basalt* that is microscopically holocrystalline, poikilitic, and composed of magnetite (about 50 percent), bytownite, and augite. It was named by Washington and Larsen in 1913 for the Arapaho Indians, Colorado. Not recommended usage.

A ray Anterior ray in echinoderms, located opposite the CD interray; equal to ambulacrum III of the Lovenian system.

arbitrary cutoff An arbitrary boundary separating two laterally intergrading stratigraphic units (Wheeler & Mallory, 1953, p. 2412). See also: *cutoff [stratig].*

arborescent *dendritic.*

arborescent pollen Pollen of trees. Abbrev: AP. Syn: *tree pollen.*

Arbuckle orogeny A name used by Van der Gracht (1931) for the last major deformation in the Wichita orogenic belt of southern Oklahoma (Wichita and Arbuckle mountains, and subsurface). It is placed in the late Pennsylvanian (Virgil) by its relations to limiting fossiliferous strata. The nearby Ouachita Mountains were not supposed to have been materially affected by this orogeny, but to have been deformed later.

arcanite An orthorhombic mineral: K_2SO_4.

arch [geomorph] *natural arch.*

arch [struc geol] A broad, open anticlinal fold on a regional scale; it is usually a basement doming, e.g. the Cincinnati Arch. Cf: *dome.* Less-preferred syn: *swell.*

archaeocyathid Any marine organism belonging to the phylum Archaeocyatha and characterized chiefly by a cone-, goblet-, or vase-shaped skeleton composed of calcium carbonate. The archaeocyathids have been variously classified as corals, sponges, protozoans, and calcareous algae. Range, Lower and Middle Cambrian; worldwide in distribution. Syn: *pleosponge; cyathosponge.*

archaeocyte An amoebocyte of a sponge that has a large nucleus and cytoplasm rich in ribonucleic acid, is capable of ingesting particulate material and serving as the origin of any other type of cell, and is believed to be a persistent undifferentiated embryonic cell. Also spelled: *archeocyte.*

archaeology The study of human cultures through the recovery and analysis of their material relics. Also spelled: *archeology.*

archaeomagnetism The study of natural remanent magnetism of baked clays and recent lavas to determine intensity and direction of the Earth's magnetic field in the archaeologic past.

archaeometry n. The application of methods from the natural sciences to archaeological measurements, e.g. thermoluminescent dating of pottery sherds, or geophysical prospecting for archaeological sites.

Archaeopteryx A genus of birds characterized by teeth, elongate caudal skeleton, and very dinosaurlike details of trunk and limbs. Range, Upper Jurassic.

Archaeozoic *Archeozoic.*

Archaic n. In New World archaeology, a prehistoric cultural stage that follows the Lithic and is characterized in a general way by a foraging pattern of existence and numerous types of stone implements (Jennings, 1968, p.110). It is followed by the Formative. Correlation of relative cultural levels with actual age (and, therefore, with the time-stratigraphic units of geology) varies from region to region.—adj. Pertaining to the Archaic.

arch dam A dam built in the form of a horizontal arch that abuts against the side walls of a gorge and that has its convex side upstream.

Archean Said of the rocks of the *Archeozoic.*

arched iceberg An iceberg eroded in such a manner that a large opening at the water line extends through the ice, forming an arch.

archegonium In plants, the multicellular female gametangium, commonly flask-shaped, produced by gametophytes of all higher cryptogams and gymnosperms (Swartz, 1971, p. 41).

archeocyte *archaeocyte.*

archeology *archaeology.*

archeometry *archaeometry.*

Archeophytic A paleobotanic division of geologic time, signifying the time of initial plant evolution, specif. algae. Cf: *Aphytic; Paleophytic; Mesophytic; Eophytic; Cenophytic.* Syn: *Algophytic; Proterophytic.*

archeopyle An opening in the wall of a dinoflagellate cyst by means of which the cell contents can emerge from the cyst. It is usually more or less polygonal in shape, and operculate. See also: *apical archeopyle; cingular archeopyle; precingular archeopyle; combination archeopyle.*

archeopyle suture A line of dehiscence on the dinoflagellate cyst that more or less completely separates a part of the cyst wall to form an operculum that covers the archeopyle. See also: *accessory archeopyle suture.*

Archeozoic The earlier part of Precambrian time, corresponding to *Archean* rocks. Cf: *Proterozoic.* Also spelled: *Archaeozoic.*

archetype *prototype.*

arch-gravity dam A solid-masonry arch dam that has sufficient mass and breadth of base so that gravity combines with the arch design to provide stability.

archibenthic Pertaining to the benthos of the continental slope. Cf: *bathybenthic.*

archibole An obsolete syn. of *positive element.*

Archie equations Three empirical relationships between *well log* resistivity measurements and the properties of porous, nonshaly sandstone, formulated by G. E. Archie (1942). Equation 1: $R_o = FR_w$, where R_o is the resistivity of fully water-saturated rock, F is the *formation factor,* and R_w is the resistivity of the interstitial water. Equation 2: $F = \phi^{-m}$, where ϕ is the porosity of Gulf Coast and other sandstone core samples and m is the "cementation component". From equation 1, equation 2 may be restated: $R_o = R_w \phi^{-m}$. Equation 3: $S_w = (R_o/R_t)^{1/n}$, where S_w is the percent formation-water saturation of a porous sandstone, R_t is the true resistivity of the sandstone underground, and n is the "saturation exponent". The Archie equations are the foundations of quantitative well-log analysis in hydrocarbon reservoirs. They have been elaborated and their constants refined for specific reservoirs, rock types, and subsurface conditions, e.g. by Winsauer et al. (1952).

Archimedes' principle The statement in fluid mechanics that a fluid buoys up a completely immersed solid so that the apparent weight of the solid is reduced by an amount equal to the weight of the fluid that it displaces.

arching The transfer of stress from a yielding part of a soil or rock mass to adjoining less yielding or restrained parts of the mass (ASCE, 1958, term 22).

archipelagic apron A smooth, fanlike slope or broad cone surrounding a seamount or an island. It is comparable to a continental rise and an abyssal plain in its topography and sedimentary processes.

archipelago A sea or area in a sea that contains numerous islands; also, the island group itself.

architype The type of a genus or species named in a publication prior to the time of the establishment of the current interpretation of types.

Archosauria A subclass of reptiles in which the skull is diapsid and the teeth are set in sockets. It includes the dinosaurs, crocodilians, pterosaurs, and a variety of Triassic forms including the ancestors of these groups.
arc measurement The measurement, in geodesy, that follows a given meridian to determine the shape and size of the Earth along that line.
arcose *arkose.*
arc shooting *fan shooting.*
arc spectrum The spectrum of light emitted by a substance at the temperature of an electric arc when the substance is placed in an arc or applied to one of the poles of the arc as a coating. The spectrum is representative of non-ionized atoms owing to the low potential difference of the arc. Cf: *spark spectrum.*
Arctic [clim] adj. Said of a climate in which the mean temperature of the coldest month is less than 0°C, and the mean temperature for the warmest month is below 10°C.
Arctic [paleoclim] The oldest subunit of the *Blytt-Sernander climatic classification* (Post, 1924), preceding the Preboreal, characterized primarily by tundra vegetation, and recording the cold climate of full and late glacial time (prior to about 10,000 years B.P.). It is subdivided into Oldest Dryas, Bølling, Older Dryas, Allerød, and Younger Dryas. The term is rarely used today.
arctic n. The area within the Arctic Circle; the area of the North Pole.—adj. (a) Pertaining to cold, frigid temperatures. (b) Pertaining to features, climate, vegetation, and animals characteristic of the arctic region.
Arctic Circle The parallel of latitude falling at approx. 66°32′ N; it delineates the frigid zone of the North Pole. Cf: *Antarctic Circle.*
arctic desert *polar desert.*
arctic pack *polar ice.*
Arctic suite A group of basaltic and associated igneous rocks intermediate in composition between rocks of the *Atlantic suite* and the *Pacific suite.*
arc triangulation Triangulation designed to extend in a single general direction, following approximately the arc of a great circle. It is executed in order to connect two distinct control points or two independent and widely separated surveys. Cf: *area triangulation.*
arcuate Said of a fold whose axial trace is curved or bent.
arcuate delta A curved or bowed delta with its convex outer margin facing the sea or lake, such as that formed at the mouth of the Nile River. Syn: *fan-shaped delta; lobate delta.*
arcuate fault A fault that has a curved trace on any given transecting surface. Cf: *peripheral fault; plane fault.*
arculite A textural term for a bow-shaped aggregate of crystallites.
arcus A bandlike thickening in the exine of a pollen grain (as in *Alnus*), running from one pore apparatus to another.
ardealite A white or light-yellow mineral: $Ca_2(HPO_4)(SO_4)\cdot4H_2O$.
Ardennian orogeny One of the 30 or more short-lived orogenies during Phanerozoic time identified by Stille, in this case late in the Silurian, within the Ludlovian Stage.
ardennite A yellow to yellowish-brown mineral: $Mn_4(Al,Mg)_6[(V,As)O_4](SiO_4)_2Si_3O_{10}(OH)_6$.
arduinite *mordenite.*
are A metric unit of area equal to 100 square meters, 0.01 hectare, or 119.60 square yards. Abbrev: a.
area-altitude analysis *hypsometric analysis.*
area contagionis *contact area.*
area curve In hydraulics, a curve that expresses the relation between area and some other variable, as between cross-sectional area of a stream and water-surface elevation, or between surface area of a reservoir and water-surface elevation (ASCE, 1962).
areal geology The geology of an area, esp. the spatial distribution and position of stratigraphic units, structural features, and surface forms. Cf: *regional geology.*
areal map A geologic map showing the horizontal extent and distribution of rock units exposed at the surface.
area of influence The area within which the potentiometric surface of an aquifer is lowered by withdrawal, or raised by injection, of water through a well or other structure designed for the purpose; the outer boundary of the *cone of depression.* Syn: *circle of influence.*
area slope The generalization of slope conditions within a given area.
area triangulation Triangulation designed to extend in every direction from a control point and to cover the region surrounding it. Cf: *arc triangulation.*
areg Arabic plural of *erg.*
arena A term used in Uganda for a large, undulating, relatively low-lying central area more or less completely surrounded by a hilly rim of resistant rock, and representing a dome of softer rock that has been worn away (Wayland, 1920, p. 36-37).
arenaceous (a) Said of a sediment or sedimentary rock consisting wholly or in part of sand-size fragments, or having a sandy texture or the appearance of sand; pertaining to sand or arenite. Also said of the texture of such a sediment or rock. The term implies no special composition, and should not be used as a syn. of "siliceous". Syn: *psammitic; sandy; sabulous; arenarious.* (b) Said of organisms growing in sandy places.
arenarious Composed of sand; *arenaceous.*
arenated Said of a substance that is mixed with sand or that has been reduced to sand.
arendalite [mineral] A dark-green *epidote* from Arendal, southern Norway.
arendalite [rock] A French term for a garnetiferous rock. Not recommended usage.
arenicolite A sand-filled, U-shaped hole in a sedimentary rock (generally a sandstone), interpreted as a burrow of a marine worm and resembling the U-shaped burrow or trail of the modern worm *Arenicola.* It has also been regarded as the trail of a mollusk or crustacean.
Arenigian European stage: Lower Ordovician (above Tremadocian, below Llanvirnian). Syn: *Skiddavian.*
arenilitic Pertaining to, having the quality of, or resembling sandstone.
arenite (a) A general name used for consolidated sedimentary rocks composed of sand-sized fragments irrespective of composition; e.g. sandstone, graywacke, arkose, and calcarenite. The term is equivalent to the Greek-derived term, *psammite,* and was introduced as *arenyte* by Grabau (1904, p.242) who used it with appropriate prefixes in classifying medium-grained rocks (e.g. "autoarenyte", "autocalcarenyte", "hydrarenyte", and "hydrosilicarenyte"). See also: *lutite; rudite.* (b) A "clean" sandstone that is well-sorted, contains little or no matrix material, and has a relatively simple mineralogic composition; specif. a pure or nearly pure, chemically cemented sandstone containing less than 10% argillaceous matrix and inferred to represent a slowly deposited sediment well-washed by currents (Williams, Turner & Gilbert, 1954, p.290). The term is used for a major category of sandstone, as distinguished from *wacke.*—Etymol: Latin *arena,* "sand". Adj: *arenitic.*
arenose Full of grit or fine sand; *gritty.*
Arent In U.S. Dept. of Agriculture soil taxonomy, a suborder of the soil order *Entisol,* characterized by the presence of fragments of former pedogenic horizons scattered through the soil. Arents form by mechanically mixing soil horizons by deep plowing, cut and fill, or other earth-moving operations (USDA, 1975). Cf: *Aquent; Fluvent; Orthent: Psamment.*
arenyte Var. of *arenite.*
areography Description of the surface of the planet Mars.
areola (a) One of the thinner chamberlike or boxlike structures arranged in characteristic pattern within the shell wall of diatoms, being larger and more complex than a *puncta,* lying perpendicular to the valve surface, and permitting the diffusion of gases and nutrients. An areola may be subcircular, elliptical, or hexagonal, and wholly or partly closed on either inner or outer surface. Syn: *areole.* (b) A marginal opening in the frontal shield in some ascophoran cheilostome bryozoans containing an intrazooidal septula. (c) A generally smooth, featureless area of a crinoid columnal articulum, situated between lumen and inner margin of crenularium. It may be granulose or marked by fine vermicular furrows and ridges.—Pl: *areolae.*
areole (a) A *scrobicule* or depression around a boss of an echinoid for the attachment of muscles controlling the movement of spines. (b) An *areola* of a diatom.
areology The scientific study of the planet Mars.
arête A narrow serrate mountain crest or rocky sharp-edged ridge or spur, commonly present above the snowline in rugged mountains (as in the Swiss Alps) sculptured by glaciers, and resulting from the continued backward growth of the walls of adjoining cirques. Etymol: French, "fish bone". See also: *horn [glac geol]; comb ridge; grat.* Syn: *crib [glac geol]; arris.*
aretic *arheic.*
arfvedsonite (a) A black monoclinic mineral of the amphibole

group, approximately: $Na_{2-3}(Fe,Mg,Al)_5Si_8O_{22}(OH)_2$. It may contain some calcium, and it occurs in strongly pleochroic prisms in certain sodium-rich igneous rocks. Syn: *soda hornblende.* (b) An end member of the amphibole group: $Na_3Fe_4{}^{+2}Fe^{+3}Si_8O_{22}(OH)_2$.

argental mercury Naturally occurring *amalgam.*

argentian *argentiferous.*

argentiferous Said of a substance that contains or yields silver, e.g. "argentiferous galena". Syn: *argentian.*

argentine n. A pearly-white variety of calcite with undulating lamellae.—adj. Pertaining to, containing, or resembling silver; silvery.

argentite A dark lead-gray cubic dimorph of acanthite: Ag_2S. Isometric above 179°C, it inverts to acanthite below this temperature. Argentite is a valuable ore of silver. Syn: *silver glance; vitreous silver; argyrite.*

argentojarosite A yellow or brownish mineral of the alunite group: $AgFe_3(SO_4)_2(OH)_6$.

argentopyrite A mineral: $AgFe_2S_3$.

argic water A syn. of *intermediate vadose water* proposed by Meinzer (1939) as one of the three classes of *kremastic water.*

Argid In U.S. Dept. of Agriculture soil taxonomy, a suborder of the soil order *Aridisol,* characterized by the presence of an argillic or natric horizon. Though the soils occur in arid climates, the argillic horizon was probably formed in glacial times under higher precipitation and lower temperatures more effective for leaching (USDA, 1975). Cf: *Orthid.*

argil (a) A clay; esp. a white-colored clay, such as *potter's clay.* (b) *alumina.*

argillaceous (a) Pertaining to, largely composed of, or containing clay-size particles or clay minerals, such as an "argillaceous ore" in which the gangue is mainly clay; esp. said of a sediment (such as marl) or a sedimentary rock (such as shale) containing an appreciable amount of clay. Cf: *shaly; lutaceous; pelitic; argillic.* Syn: *clayey; pelolithic; argillous.* (b) Said of the peculiar odor emitted by an argillaceous rock when breathed upon. (c) Pertaining to argillite.

argillaceous hematite A brown to deep-red variety of natural ferric oxide containing an appreciable portion of clay (or sand).

argillaceous limestone A limestone containing an appreciable amount (but less than 50%) of clay; e.g. *cement rock.*

argillaceous sandstone (a) A term applied loosely to an impure sandstone containing an indefinite amount of fine silt and clay. (b) A relatively weak sandstone, not suitable for building purposes, containing a considerable amount of clay that serves as the cementing material.—Cf: *clayey sandstone.*

argillan A *cutan* composed dominantly of clay minerals (Brewer, 1964, p.212); e.g. a *clay skin.* Syn: *argitan.*

argillation A term used by Keller (1958, p. 233) for the development of kaolinite and other clay minerals by weathering of primary aluminum-silicate minerals.

argille scagliose A thick sheet of chaotic, allochthonous material consisting of highly plastic, churned, and slickensided clays that have been displaced many kilometers by lateral or vertical stresses aided by gravity sliding or by diapiric movement; specif. a stratigraphic unit of Jurassic to Oligocene age exposed along parts of the Apennines, on which great slabs of rock have slid. Pron: ar-*jee*-leh skahl-*yo*-seh. Etymol: Italian, "scaly shale".

argillic Pertaining to clay or clay minerals; e.g. "argillic alteration" in which certain minerals of a rock are converted to minerals of the clay group. Cf: *argillaceous.*

argillic horizon A diagnostic subsurface soil horizon characterized by an accumulation of clay. Its thickness and clay content depend on the thickness and clay content of the overlying eluvial horizon. It usually has clay coatings on the surface of peds or pores, or as bridges between sand grains (USDA, 1975).

argilliferous Abounding in clay; producing clay.

argillite (a) A compact rock, derived either from mudstone (claystone or siltstone) or shale, that has undergone a somewhat higher degree of induration than mudstone or shale but is less clearly laminated than shale and without its fissility, and that lacks the cleavage distinctive of slate. Flawn (1953, p.563-564) regards argillite as a weakly metamorphosed argillaceous rock, intermediate in character between a claystone and a *meta-argillite,* in which less than half of the constituent material (clay minerals and micaceous paste) has been reconstituted to combinations of sericite, chlorite, epidote, or green biotite, the particle size of the reconstituted material ranging from 0.01 to 0.05 mm. Cf: *clay slate.* (b) A term that has been applied to an argillaceous rock cemented by silica (Holmes, 1928, p.35) and to a claystone composed entirely of clay minerals.—Also spelled: *argillyte.*

argillith A term suggested by Grabau (1924, p.298) for claystone. Syn: *argillyte.*

argillization Replacement or alteration of feldspars to form clay minerals, esp. that occurring in wall rocks adjacent to mineral veins.

argillutite A pure lutite. Term introduced as "argillutyte" by Grabau (1904, p.243).

argillyte (a) Var. of *argillite.* (b) *argillith.* (c) Obsolete synonym of orthoclase.

argitan *argillan.*

argon-40/argon-39 age method A variation of the *potassium-argon age method* in which the sample to be dated is first irradiated with neutrons, converting some potassium-39 to argon-39. Argon is then extracted from the sample (either in one step or incrementally), and its isotopic composition is analyzed. The amount of argon-39 is a measure of potassium content, and the ratio of argon-40 to argon-39 is a function of age. It is sometimes possible to detect extraneous argon, and to determine whether or not the dated material has been disturbed by later thermal or chemical events (J. Miller, 1972). Cf: *potassium-argon age method.*

Argovian Substage in Great Britain: Upper Jurassic (lower Lusitanian; above Oxfordian Stage, below Rauracian Substage).

argyrite *argentite.*

argyrodite A steel-gray mineral: Ag_8GeS_6. It is isomorphous with canfieldite.

arheic Said of a drainage basin or region characterized by arheism; without flow. Syn: *areic; aretic; arhetic.*

arheism The condition of a region (such as a desert) in which runoff is nil or surface drainage is almost completely lacking, or where rainfall is so infrequent that the water sinks into the ground or evaporates. Syn: *areism.*

arich A term used in Algeria for a rocky, sand-cloaked butte on which an isolated dune is based (Capot-Rey, 1945, p. 399).

arid Said of a climate characterized by dryness, variously defined as rainfall insufficient for plant life or for crops without irrigation; less than 25 cm of annual rainfall; or a higher evaporation rate than precipitation rate. Syn: *dry.*

arid cycle The cycle of erosion in an arid region. Cf: *normal cycle.*

aridic moisture regime A soil moisture regime characteristic of arid climates. There is little or no leaching, and soluble salts may accumulate. There is no plant-available water for more than half the time the soil is warm enough for plant growth (above 5°C), and the soil never has water continuously available for as long as 90 consecutive days when the soil temperature is above 8°C (USDA, 1975).

Aridisol In U.S. Dept. of Agriculture soil taxonomy, a soil order characterized by an ochric epipedon and other pedogenic horizons, but none of them oxic or sodic. It develops under an aridic moisture regime and has a low percentage of organic carbon (USDA, 1975). Suborders and great soil groups of this soil order have the suffix -id. See also: *Argid; Orthid.*

ariegite A group of pyroxenites chiefly composed of clinopyroxene, orthopyroxene, and spinel, with pyrope and/or hornblende as possible accessories and lacking primary feldspar; a spinel *pyroxenite.* The term is more commonly used by European (esp. French) petrologists. The rock was named by Lacroix in 1901 for Ariège in the Pyrenees. Not recommended usage.

Arikareean North American continental stage: Upper Oligocene (above Whitneyan, below Hemingfordian).

aristarainite A monoclinic mineral: $Na_2MgB_{12}O_{20}\cdot 10H_2O$.

Aristotle's lantern A complex masticatory system of as many as forty calcareous skeletal elements that surround the mouth of an echinoid and function as jaws. Etymol: from a passage in Aristotle where the shape of a sea urchin is said to resemble the frame of a lantern (Webster, 1967, p.118). Syn: *lantern.*

arithmetic mean The sum of the values of n numbers divided by n. It is usually referred to simply as the "mean". Syn: *average.*

arithmetic mean diameter An expression of the average particle size of a sediment or rock, obtained by summing the products of the size-grade midpoints and the frequency of particles in each class, and dividing by the total frequency. Syn: *equivalent grade.*

Arizona ruby A deep-red or ruby-colored pyrope garnet of igneous origin from the SW U.S.

arizonite [mineral] (a) A doubtful mineral: $Fe_2Ti_3O_9$. It is found in irregular metallic steel-gray masses in pegmatite veins near Hackberry, Ariz. Cf: *kalkowskite; pseudorutile.* (b) A mixture of hematite, rutile, ilmenite, and anatase. (c) A type of ore, discovered in Yavapai County, Ariz., whose principal vein material consists of micaceous iron, silver iodide, gold, iron sulfides, and antimony.

arizonite [rock] A light-colored hypabyssal rock composed chiefly of quartz (80 percent) and alkali feldspar (18 percent), with mica and apatite as possible accessories. It was named by Spurr and Washington in 1917 for Arizona, where the rock was first found. Not recommended usage.

Arkansas stone A variety of *novaculite* found in the Ouachita Mountains of western Arkansas. Also, a whetstone made of Arkansas stone.

arkansite A brillant iron-black variety of brookite from Magnet Cove, Arkansas.

arkesine A name, now obsolete, proposed by Jurine (1806, p. 373-374) for the talc- and chlorite-bearing hornblende granite on Mont Blanc in the French Alps. Etymol: Greek *archaios,* "ancient, primitive".

arkite A porphyritic plutonic foidite with the same general composition as *fergusite* but distinguished from it by the presence of melanite. Leucite usually forms the phenocrysts. It was named by Washington in 1901 from Magnet Cove, Arkansas. Syn: *nepheline fergusite; leucite ijolite.* Not recommended usage.

arkose A feldspar-rich sandstone, typically coarse-grained and pink or reddish, that is composed of angular to subangular grains that may be either poorly or moderately well sorted, is usually derived from the rapid disintegration of granite or granitic rocks, and often closely resembles granite; e.g. the Triassic arkoses of eastern U.S. Quartz is usually the dominant mineral, with feldspars (chiefly microcline) constituting at least 25%. Cement (silica or calcite) is commonly rare, and matrix material (usually less than 15%) includes clay minerals (esp. kaolinite), mica, and iron oxide; fine-grained rock fragments are often present. Arkose is commonly a current-deposited sandstone of continental origin, occurring as a thick, wedge-shaped mass of limited geographic extent (as in a fault trough or a rapidly subsiding basin); it may be strongly cross-bedded and associated with coarse granite-bearing conglomerate, and it may denote an environment of high relief and vigorous erosion of strongly uplifted granitic rocks in which the feldspars were not subjected to prolonged weathering or transport before burial. Arkose may also occur at the base of a sedimentary series as a thin blanketlike residuum derived from and resting on granitic rock. Modern definitions of arkose include those by Krynine (1940); Folk (1954); Williams, Turner & Gilbert (1954, p. 294-295); Pettijohn (1957); McBride (1963, p. 667); and Folk (1968, p.124). The term "arkose" was introduced by Brongniart (1823, p.497-498) in an attempt to limit use of "grés" (sandstone) and was defined by him as a rock of granular texture formed principally by mechanical aggregation and composed essentially of large grains of feldspar and glassy quartz mixed together unequally, with mica and clay as fortuitous constituents (see Oriel, 1949, p.825). Roberts (1839, p.11) attributes the term to Bonnard. Etymol: French, probably from Greek *archaios,* "ancient, primitive" (Oriel, 1949, p.826). Adj: *arkosic.* Cf: *graywacke; feldspathic sandstone; subarkose.* Also spelled: *arcose.*

arkose-quartzite *arkosite.*

arkosic arenite A sandstone containing abundant quartz, chert, and quartzite, less than 10% argillaceous matrix, and more than 25% feldspar (chiefly unaltered sodic and potassic varieties), and characterized by an abundance of unstable materials in which the feldspar grains exceed the rock fragments (Williams, Turner & Gilbert, 1954, p.294). It is more feldspathic and less mature than *feldspathic arenite.* See also: *arkosic sandstone.*

arkosic bentonite A term used by Ross & Shannon (1926, p. 79) for a bentonite with 25-75% sandy impurities; a bentonite derived from a volcanic ash whose abundant detrital crystalline grains remained essentially unaltered. Cf: *bentonitic arkose.*

arkosic conglomerate A poorly sorted, lithologically homogeneous *orthoconglomerate* consisting of immature gravels derived from granites in a tectonically active area under conditions of fluvial transport and rapid burial in a subsiding basin; an arkose with scattered pebbles of granite or with lenses of granitic gravel. The sand and silt matrix has the composition of arkose, consisting of quartz and feldspar particles with some finer kaolinitic material. The rock forms thick, wedge-shaped deposits and is commonly interbedded with arkosic sandstones. Syn: *granite-pebble conglomerate.*

arkosic graywacke A graywacke characterized by abundant unstable materials; specif. a sandstone containing more feldspar grains than fine-grained rock fragments, the feldspar content exceeding 25% (Williams, Turner & Gilbert, 1954, p.294). It is more feldspathic than *feldspathic graywacke.*

arkosic limestone An impure limestone containing a relatively high proportion of grains and/or crystals of feldspar, either detrital or formed in place.

arkosic sandstone A sandstone with considerable feldspar, such as one containing minerals derived from coarse-grained quartzofeldspathic rocks (granite, granodiorite, gneiss) or from highly feldspathic sedimentary rocks; specif. a sandstone containing more than 25% feldspar and less than 20% matrix material of clay, sericite, and chlorite (Pettijohn, 1949, p.227). It is more feldspathic than *feldspathic sandstone.* The term is used also as a general term to include *arkosic arenite, arkosic wacke,* and *arkose* (Williams, Turner & Gilbert, 1954, p. 310), or arkose and *subarkose* (Pettijohn, 1954, p.364). See also: *arkosite.*

arkosic wacke A sandstone containing abundant quartz, chert, and quartzite, more than 10% argillaceous matrix, and more than 25% feldspar (chiefly sodic and potassic varieties), and characterized by an abundance of unstable materials in which the feldspar grains exceed the fine-grained rock fragments (Williams, Turner & Gilbert, 1954, p.291-292). It is more feldspathic and less mature than *feldspathic wacke.*

arkosite A quartzite with a notable amount of feldspar; e.g. a well-indurated *arkosic sandstone* (Pettijohn, 1949, p.227) or a well-cemented arkose (Tieje, 1921, p.655). Syn: *arkose-quartzite; quartzitic arkose.*

arkositite A term used by Tieje (1921, p.655) for an arkose so well-cemented that the particles are interlocking.

arm [coast] A long, narrow inlet of water extending inland from another body of water, such as an "arm of the sea". It is usually longer and narrower than a bay.

arm [geomorph] (a) A ridge or spur extending from a mountain. (b) The trailing outer extension of a parabolic dune.

arm [paleont] (a) One of several radially disposed appendages bearing an extension of an ambulacrum and mounted on the oral surface of an echinoderm, such as one of the five radial extensions of the body of an asteroid; also, a free, usually pinnule-bearing extension of a crinoid. It is a major element in the food-gathering structure of many echinoderms, and may or may not be distinct from the disc. Cf: *brachiole.* (b) A raylike structure of a sponge spicule, whether or not it is a true ray or a pseudoactin. (c) A brachium of a brachiopod. (d) A flat extension from the central region of a radiolarian shell.

arm [streams] A tributary or branch of a stream.

armalcolite A mineral of the pseudobrookite group found in Apollo 11 lunar samples: $(Mg,Fe)Ti_2O_5$.

armangite A black rhombohedral mineral: $Mn_3(AsO_3)_2$.

armchair geology Deducing geologic conditions without intensive field work (Shepard, 1967, p.209).

Armenian bole A soft, clayey, bright-red astringent earth found chiefly in Armenia and Tuscany, used in the 14th to 18th centuries for medical purposes and now as a coloring material. Syn: *bole Armoniac.*

armenite A mineral: $BaCa_2Al_6Si_8O_{28}\cdot 2H_2O$.

armored Said of dinoflagellates (such as those of the order Peridiniales) possessing a cellulose envelope or cell wall that is subdivided into or covered by articular plates. Ant: *unarmored.*

armored clay ball *armored mud ball.*

armored mud ball A large subspherical mass of silt or clay, which becomes coated or studded with coarse sand and fine gravel as it rolls along downstream; it is generally 5-10 cm in diameter, although the size ranges between 1 cm and 50 cm. See also: *clay ball; till ball.* Syn: *armored clay ball; mud ball; pudding ball.*

armored relict An *unstable relict* that is prevented from further reaction by a rim of reaction products.

Armorican orogeny A name used by Suess for late Paleozoic deformation in western Europe; it is based on relations in Brittany (Armorica). The term is now little used; modern geologists prefer the names *Hercynian orogeny* or *Variscan orogeny.*

armoring Formation of a reaction rim resulting from, e.g., a loss of equilibrium in a discontinuous reaction series. Cf: *zoning [crystal].*

armstrongite A monoclinic mineral: $CaZrSi_6O_{15}\cdot 2.5H_2O$.

arnimite A hydrous copper-sulfate mineral, perhaps identical with antlerite.

aromatic hydrocarbon A monocyclic or polycyclic relatively stable hydrocarbon having the empirical formula C_nH_{2n-6}, of which the simplest example is benzene. The compounds of higher molecular weight in this series are solid and may fluoresce or be slightly colored. See also: *benzene series.*

aromatite A bituminous stone resembling myrrh in color and odor. Webster (1967, p.120) defines *aromatites* (pl: *aromatitae*) as a precious stone of ancient Arabia and Egypt.

aromorphosis Biologic evolution characterized by an increase in the degree of organization without marked specialization. Cf: *allomorphosis.*

arpent (a) An old French unit of land area, still used in certain French sections of Canada and U.S., equal to about 0.85 acre depending on local custom and usage; e.g. a unit of area in Missouri and Arkansas equal to 0.8507 acre, and in Louisiana, Mississippi, Alabama, and NW Florida equal to 0.84625 acre. (b) A unit of length equal to one side of a square having an area of one arpent (192.50 ft in Missouri and Arkansas, and 191.994 ft in Louisiana, Mississippi, Alabama, and NW Florida).

arquerite A soft, malleable, silver-rich variety of native amalgam, containing about 87% silver and 13% mercury.

array An ordered arrangement of geophysical instruments such as electrodes, magnetometers, or geophones, the data from which feed into a central point or receiver.

arris A term used in the English Lake District for *arête.* Syn: *arridge* (Stamp, 1961, p. 33).

arrival The initial appearance of seismic energy on a seismic record; the buildup of amplitude and the coherent lineup of energy signifying the passage of a wave front. See also: *first arrival.* Syn: *onset; break [seis]; kick [seis].*

arrival time In seismology, the time at which a particular wave phase arrives at a detector, usually measured from the time of generation of the wave.

arrojadite A dark-green monoclinic mineral: nearly $Na_2(Fe,Mn)_5(PO_4)_4$. It is isostructural with dickinsonite.

arrow *pin [surv].*

arroyo (a) A term applied in the arid and semi-arid regions of SW U.S. to the small deep flat-floored channel or gully of an ephemeral stream or of an intermittent stream, usually with vertical or steeply cut banks of unconsolidated material at least 60 cm high; it is usually dry, but may be transformed into a temporary watercourse or short-lived torrent after heavy rainfall. Cf: *dry wash.* (b) The small intermittent stream or rivulet that occupies such a channel.—Etymol: Spanish, "stream, brook; gutter, watercourse of a street". See also: *wadi; nullah.* Syn: *arroya.*

arroyo-running A term applied in the SW U.S. to the phase of local flooding characterized by a temporary mountain torrent debouching from a canyon and spreading out over a great fan (Keyes, 1910, p. 572).

arsenate A mineral compound characterized by pentavalent arsenic and oxygen in the anion. An example is mimetite, $Pb_5(AsO_4)_3Cl$. Cf: *phosphate; vanadate.*

arsenate-belovite A mineral: $Ca_2Mg(AsO_4)_2 \cdot 2H_2O$. Formerly called: *belovite.* Syn: *talmessite.*

arsenic A hexagonal mineral, the native metallic element As. It is brittle and commonly occurs in steel-gray and granular or kidney-shaped masses.

arsenical antimony *allemontite.*

arsenical nickel *nickeline.*

arsenical pyrites *arsenopyrite.*

arsenic bloom (a) *arsenolite.* (b) *pharmacolite.*

arseniopleite A brownish-red mineral consisting of a basic arsenate of manganese, calcium, iron, lead, and magnesium.

arseniosiderite A yellowish-brown mineral: $Ca_3Fe_4(AsO_4)_4(OH)_4 \cdot 4H_2O$.

arsenite A mineral compound characterized by trivalent antimony and oxygen in the anion. An example is trigonite, $Pb_3MnH(AsO_3)_3$.

arsenobismite A yellowish-green mineral: $Bi_2(AsO_4)(OH)_3$.

arsenoclasite A red mineral: $Mn_5(AsO_4)_2(OH)_4$. Also spelled: *arsenoklasite.*

arsenolamprite A lead-gray polymorph of native arsenic. It was formerly regarded as a mixture of arsenic and arsenolite.

arsenolite A cubic mineral: As_2O_3. It usually occurs as a white bloom or crust, and is dimorphous with claudetite. Syn: *arsenic bloom.*

arsenopalladinite A triclinic mineral: $Pd_8(As,Sb)_3$.

arsenopyrite A tin-white or silver-white to steel-gray orthorhombic mineral: FeAsS. It is isomorphous with loellingite. Arsenopyrite occurs chiefly in crystalline rocks and esp. in lead and silver veins, and it constitutes the principal ore of arsenic. Syn: *arsenical pyrites; mispickel; white pyrites; white mundic.*

arsenosulvanite A cubic mineral: $Cu_3(As,V)S_4$. It is isomorphous with sulvanite. Syn: *lazarevicite.*

arsenpolybasite A mineral: $(Ag,Cu)_{16}(As,Sb)_2S_{11}$. Cf: *antimonpearceite.* Syn: *arsenopolybasite.*

arsenuranylite An orange-red mineral: $Ca(UO_2)_4(AsO_4)_2(OH)_4 \cdot 6H_2O$.

arsoite A *trachyte* containing phenocrysts of alkali feldspar, andesine, clinopyroxene, and olivine in a groundmass of alkali feldspar, oligoclase, clinopyroxene, magnetite, and sodalite. Its name is derived from the Arso lava flow of 1302 on Monte Epomeo (Ischia, Italy). Not recommended usage.

arterite Migmatite, the more mobile portion of which was injected magma (Dietrich & Mehnert, 1961). "This is the same as the arteritic gneiss, *injection gneiss,* and lit-par-lit gneiss of some workers" (Dietrich, 1960, p. 36). Originally proposed, along with the term *venite,* to replace *veined gneiss* with terms of genetic connotation (Mehnert, 1968, p. 17). Cf: *phlebite; anatexite; venite; composite gneiss; diadysite.*

artesian An adjective referring to ground water confined under hydrostatic pressure. Etymol: French *artésien,* "of Artois", a region in northern France.

artesian aquifer *confined aquifer.*

artesian basin A terrane, often but not necessarily basin-shaped, including an artesian aquifer whose potentiometric surface typically is above the land surface in the topographically lower portion of the terrane. Examples range in size from areas a few hundred feet across to several hundred miles across. Cf: *ground-water basin.*

artesian discharge Discharge of water from a well, spring, or aquifer by artesian pressure.

artesian head The *hydrostatic head* of an artesian aquifer or of the water in the aquifer.

artesian leakage Slow percolation from a confined aquifer into confining beds.

artesian pressure *Hydrostatic pressure* of artesian water, often expressed in terms of pounds per square inch at the land surface; or height, in feet above the land surface, of a column of water that would be supported by the pressure.

artesian-pressure surface A potentiometric surface that is above the zone of saturation. Cf: *normal-pressure surface; subnormal-pressure surface.*

artesian province A region within which structure, stratigraphy, topography, and climate combine to produce conditions favorable to the existence of one or more artesian aquifers; e.g. the Atlantic and Gulf Coastal Plain.

artesian spring A spring from which the water flows under artesian pressure, usually through a fissure or other opening in the confining bed above the aquifer.

artesian system (a) A structure permitting water confined in a body of rock to rise within a well or along a fissure. (b) Any system incorporating a water source, a body of permeable rock bounded by bodies of distinctly less permeable rock, and a structure enabling water to percolate into and become confined in the permeable rock under pressure distinctly greater than atmospheric.

artesian water *confined ground water.*

artesian weathering Chemical weathering produced by artesian water moving down an aquifer, as in a sandstone overlying an unconformity or in the permeable zone along the unconformable surface.

artesian well A well tapping confined ground water. Water in the well rises above the level of the water table under artesian pressure, but does not necessarily reach the land surface. Sometimes restricted to mean only a *flowing artesian well.* Cf: *water-table well; nonflowing artesian well.*

arthrolite A cylindrical concretion with transverse joints, sometimes found in clays or shales.

arthrophycus A sand-filled rounded furrow, curving and branching, with faint but regularly spaced transverse ridges commonly bearing a median depression, probably representing a feeding burrow but also variously regarded as an inorganic structure or a trail produced by a worm, mollusk, or arthropod crawling over a soft-mud surface. The "branches" of the trace fossil may reach 60 cm in length. It was originally described as a plant fossil (seaweed) and assigned to the genus *Arthrophycus.*

arthropod Any one of a group of solitary marine, freshwater, and aerial invertebrates belonging to the phylum Arthropoda, charac-

terized chiefly by jointed appendages and segmented bodies. Among the typical arthropods are trilobites, crustaceans, chelicerates, and myriapods. Range, Lower Cambrian to present.
arthurite An apple-green mineral: $Cu_2Fe_4[(As,P,S)O_4]_4(O,OH)_4 \cdot 8H_2O$.
article An articulated *segment* of an appendage in an arthropod.
articular adj. Pertaining to a joint or joints, e.g. "articular furrow" and "articular groove" occurring on the margins of the scutum and tergum of a cirripede crustacean and together forming the articulation between these valves.—n. In some vertebrates, e.g. amphibians and reptiles, a cartilage-replacement bone that provides much or all of the lower jaw.
articular facet *articulum.*
articulate [bot] Said of a plant that has nodes or joints, or places where separation may occur.
articulate [paleont] (a) n. Any brachiopod belonging to the class Articulata, characterized by calcareous valves united by hinge teeth and dental sockets.—adj. Said of a brachiopod possessing such valves, or of the valves themselves. Cf: *inarticulate.* (b) n. Any crinoid belonging to the subclass Articulata, characterized by highly differentiated brachial articulations.—adj. Said of a crinoid having highly differentiated brachial articulation.
articulation (a) The action or manner of jointing, or the state of being jointed; e.g. the interlocking of two brachiopod valves by two ventral teeth fitting into sockets of the brachial valve, or the union of adjoined cirrals and columnals of a crinoid stem effected by ligaments attached to articula. (b) Any movable joint between the rigid parts of an invertebrate, as between the segments of an insect appendage, or of a vertebrate, as between the bones of limbs.
articulite *itacolumite.*
articulum A smooth or sculptured surface of a crinoid columnal or cirral serving for articulation with the contiguous stem. Pl: *articula.* Syn: *articular facet.*
articulus The hinge in bivalve mollusks, including the hinge plate, hinge teeth, and ligament. Pl: *articuli.*
artifact An object made or used by man.
artificial brine Brine produced from an underground deposit of salt or other soluble rock material in the process of *solution mining.* Cf: *brine [geol].*
artificial horizon A device for indicating the horizontal, as a bubble, gyroscope, pendulum, or the flat surface of a liquid. It is sometimes simply called a *horizon.* Syn: *false horizon.*
artificial magnetic anomaly A local magnetic field caused by a man-made feature such as a transmission or telegraph line, an electric railway, a steel drill casing, or a pipeline.
artificial radioactivity The radioactivity of a synthetic nuclide. Cf: *induced radioactivity.*
artificial recharge Recharge at a rate greater than natural, resulting from the activities of man.
artinite A snow-white mineral: $Mg_2CO_3(OH)_2 \cdot 3H_2O$. It occurs in orthorhombic crystals and fibrous aggregates.
Artinskian European stage: Lower Permian (above Sakmarian, below Kungurian).
arzrunite A bluish-green mineral consisting of a sulfate and chloride of copper and lead.
ås A Swedish term for *esker.* Pron: *auss.* Pl: *åsar.* See also: *os [glac geol].*
åsar Plural of *ås.*
asbecasite A mineral: $Ca_3(Ti,Sn)(As_6Si_2Be_2O_{20})$.
asbestiform Said of a mineral that is fibrous, i.e. that is like asbestos.
asbestine adj. Pertaining to or having the characteristics of asbestos.—n. A variety of talc; specif. *agalite.*
asbestos (a) A commercial term applied to a group of silicate minerals that readily separate into thin, strong fibers that are flexible, heat resistant, and chemically inert, and therefore are suitable for uses (as in yarn, cloth, paper, paint, brake linings, tiles, insulation, cement, fillers, and filters) where incombustible, nonconducting, or chemically resistant material is required. (b) A mineral of the asbestos group, principally chrysotile (best adapted for spinning) and certain fibrous varieties of amphibole (esp. amosite, anthophyllite, and crocidolite). (c) A term strictly applied to the fibrous variety of actinolite.—Syn: *asbestus; amianthus; earth flax; mountain flax.*
asbolane *asbolite.*
asbolite A soft, black, earthy mineral aggregate, often classed as a variety of *wad,* containing hydrated oxides of manganese and cobalt, the content of cobalt sometimes reaching as much as 32% (or 40% cobalt oxide). Syn: *asbolane; earthy cobalt; black cobalt; cobalt ocher.*
ascending branch Either of two ventral elements of a brachiopod loop, continuous anteriorly with ventrally recurved *descending branches* and joined posteriorly by a transverse band.
ascending development *accelerated development.*
ascension theory A theory of hypogene mineral-deposit formation involving mineralizing solutions rising through fissures from magmatic sources in the Earth's interior. Cf: *descension theory.*
aschaffite An obsolete term originally applied to a *kersantite* that contains large quartz and plagioclase crystals, probably xenocrysts (Johannsen, 1939, p. 241).
ascharite *szaibelyite.*
aschistic Said of the rock of a minor intrusion which has a composition equivalent to that of the parent magma, i.e. in which there has been no significant differentiation. Cf: *diaschistic.*
ascoceroid conch The mature conch of the nautiloid group Ascocerida, typically consisting of an expanded exogastric brevicone having an inflated posterior part with dorsal phragmocone, an anterior cylindric neck, and an apical end formed by a transverse partition of conch comprising specialized thick septum (septum of truncation) (TIP, 1964, pt.K, p.54 & 263). See also: *ascocone.*
ascocone A cephalopod shell (like that of *Ascoceras*) whose early portion is slender, curved, and deciduous; its later portion is the *ascoceroid conch.*
ascon A thin-walled sponge or sponge larva with a single flagellated chamber that is also the spongocoel and lacking either inhalant or exhalant canals. Cf: *leucon; sycon.* Adj: *asconoid.*
ascophoran adj. Pertaining to the cheilostome bryozoan suborder Ascophora, characterized by parietal muscles attached to the floor of the ascus.—n. An ascophoran cheilostome (bryozoan). Cf: *anascan.*
ascopore A median frontal pore that serves as the inlet of the ascus in some ascophoran cheilostomes (bryozoans).
ascus A flexible floored sac beneath the frontal shield in ascophoran cheilostomes (bryozoans), variously developed to serve in protruding the lophophore. Pl: *asci.* Syn: *compensatrix; compensation sac.*
aseismic Said of an area that is not subject to earthquakes.
aseismic ridge A submarine ridge that is a fragment of continental crust; it is so named to distinguish it from the seismically active mid-oceanic ridge. Cf: *microcontinent.*
asexual reproduction Reproduction that does not involve or follow the union of individuals or of germ cells of two different sexes.
ash [coal] The inorganic residue after burning. Ignition generally alters both the weight and the composition of the inorganic matter. See also: *ash content; extraneous ash; inherent ash.*
ash [volc] Fine pyroclastic material (under 2.0mm diameter; under 0.063mm diameter for fine ash). The term usually refers to the unconsolidated material but is sometimes also used for its consolidated counterpart, *tuff.* Syn: *dust; volcanic ash; volcanic dust.*
ashbed diabase An igneous rock that resembles a conglomerate but that is a scoriaceous, amygdaloidal, tabular intrusion that has incorporated sandy material. Not recommended usage.
Ashby North American stage: Middle Ordovician (upper subdivision of older Chazyan, above Marmorian, below Porterfieldian) (Cooper, 1956).
ash cloud *eruption cloud.*
ash content The percentage of incombustible material in coal, determined by burning a sample and measuring the *ash.*
ashcroftine A pink zeolite mineral: $KNaCaY_2Si_6O_{12}(OH)_{10} \cdot 4H_2O$.
ash fall The descent of volcanic ash from an eruption cloud by *airfall deposition.* Cf: *pumice fall.* Syn: *ash shower.*
ash field A more or less well-defined area that is covered by volcanic ash. Cf: *lava field; volcanic field.*
ash flow A density current, generally a highly heated mixture of volcanic gases and ash, traveling down the flanks of a volcano or along the surface of the ground; produced by the explosive disintegration of viscous lava in a volcanic crater, or by the explosive emission of gas-charged ash from a fissure or group of fissures. Ash flows of the type described at Mt. Pelée are considered to represent the feeblest type of the *nuée ardente.* The solid materials contained in a typical ash flow are generally unsorted and ordinarily include volcanic dust, pumice, scoria, and blocks in addition to ash. Cf: *nuée ardente; pumice flow.* Syn: *pyroclastic flow; incandescent tuff flow; glowing avalanche.* Obsolete syn: *sand flow.*
ash-flow tuff A tuff deposited by an ash flow or gaseous cloud; a type of *ignimbrite.* It is a consolidated but not necessarily welded

deposit.

Ashgillian European stage: Upper Ordovician (above upper Caradocian, below Llandoverian of Silurian).

ashlar Rectangular pieces of stone of nonuniform size that are set randomly in a wall.

ash shower *ash fall.*

ashstone An indurated deposit of fine volcanic ash.

ashtonite *mordenite.*

ash tuff Pyroclastic rock made up chiefly of consolidated *ash.*

ashy grit (a) A deposit of sand-size and smaller pyroclastics. (b) A deposit of sand and volcanic ash.

asiderite An obsolete term for a meteorite that lacks metallic iron; a *stony meteorite.*

asif A term used in Morocco for a large, generally dry valley in a mountainous region. Etymol: Berber. Cf: *agouni.*

askeletal Without a skeleton; said esp. of sponges.

asparagus stone A yellow-green variety of apatite. Syn: *asparagolite.*

aspect [slopes] The direction toward which a slope faces with respect to the compass or to the rays of the Sun. Cf: *exposure.*

aspect [stratig] The general appearance of a particular geologic entity or fossil assemblage, considered more or less apart from relations in time and space; e.g. the gross or overall lithologic and/or paleontologic character of a stratigraphic unit as displayed at any single geographic point of observation (as in a borehole or outcrop section), and representing the summation or "flavor" of a *facies.* The aspect of a facies generally has environmental significance.

aspect angle [remote sensing] *angle of incidence.*

aspect angle [slopes] The angle between the *aspect* of a slope and geographic south (if measured in the Northern Hemisphere), usually reckoned positive eastward and negative westward.

asperite Any rough vesicular lava having plagioclase as its chief feldspar (Brown & Runner, 1939, p. 27). Not recommended usage.

asphalt A dark brown to black viscous liquid or low-melting solid bitumen that consists almost entirely of carbon and hydrogen and is soluble in carbon disulfide. Natural asphalt, formed in oil-bearing rocks by the evaporation of the volatiles, occurs in Trinidad, near the Dead Sea, and in the Uinta Basin of Utah. Asphalt can be prepared by the pyrolysis of coals or shales. Var: *asphaltum.* Syn: *pitch.* Obsolete syn: *mineral pitch.* See also: *tabbyite.*

asphalt-base crude Crude oil containing a high percentage of naphthenic and asphaltic hydrocarbons. Cf: *paraffin-base crude; mixed-base crude.* Syn: *naphthene-base crude.*

asphaltene Any of the solid, amorphous, black to dark brown dissolved or dispersed constituents of crude oils and other bitumens that are soluble in carbon disulfide but insoluble in paraffin naphthas. They consist of carbon, hydrogen, and some nitrogen and oxygen, hold most of the inorganic constituents of bitumens, and are isolated by dilution of the bitumen with 10-20 parts of petroleum ether or n-pentane, chilling overnight, and centrifugation or filtration.

asphaltic Pertaining to or containing asphalt; e.g. "asphaltic limestone" or "asphaltic sandstone" impregnated with asphalt, or "asphaltic sand" representing a natural mixture of asphalt with varying proportions of loose sand grains.

asphaltic pyrobitumen A bitumen, usually black and structureless, similar in composition to *asphaltite,* but infusible and insoluble in carbon disulfide; it generally contains less than 5% of oxygen. Examples are albertite, elaterite, impsonite, and wurtzilite.

asphaltic sand A natural mixture of asphalt and sand.

asphaltite Any one of the naturally occurring black solid bitumens that are soluble in carbon disulfide and fuse above 230°F. Examples are uintahite, glance pitch, and grahamite.

asphalt rock A porous rock, such as a sandstone or limestone, that is impregnated naturally with asphalt. Syn: *asphalt stone; rock asphalt.*

asphalt seal The clogging of porosity where an oil-bearing rock crops out, as a result of loss of the lighter fractions of the oil and accumulation of the heavier asphaltic residues.

asphalt stone *asphalt rock.*

asphaltum *asphalt.*

assay v. In economic geology, to analyze the proportions of metals in an ore; to test an ore or mineral for composition, purity, weight, or other properties of commercial interest.—n. The test or analysis itself; its results.

assay foot In determining the *assay value* of an orebody, the multiplication of its *assay grade* by the number of feet along which the sample was taken. Cf: *assay inch.*

assay grade The percentage of valuable constituents in an ore, determined from assay. Cf: *assay value; value.*

assay inch In determining the *assay value* of an orebody, the multiplication of its *assay grade* by the number of inches along which the sample was taken. Cf: *assay foot.*

assay limit The limit of an ore body as determined by assay, rather than by structural, stratigraphic, or other geologic controls. Cf: *cutoff grade.* Syn: *cutoff limit; economic limit.*

assay ton A weight of 29.166+ grams, used in assaying to represent proportionately the *assay value* of an ore. Since it bears the same ratio to one milligram that a ton of 2,000 pounds bears to the troy ounce, the weight in milligrams of precious metal obtained from an assay ton of ore equals the number of ounces to the ton. Abbrev: AT.

assay value (a) The quantity of an ore's valuable constituents, determined by multiplying its *assay grade* or percentage of valuable constituents by its dimensions. Cf: *assay inch; assay foot.* The figure for precious metals is generally given in troy ounces per ton of ore, or *assay ton.* See also: *value.* (b) The monetary value of an orebody, calculated by multiplying the quantity of its valuable constituents by the market price (von Bernewitz, 1931).

assemblage [paleoecol] A group of fossils that occur at the same stratigraphic level; often with a connotation also of localized geographic extent. For its accurate interpretation, an assemblage should be relatively homogeneous or uniformly heterogeneous. Cf: *congregation; association [ecol]; community; faunule.* Syn: *fossil assemblage.*

assemblage [petrology] (a) *mineral assemblage.* (b) *metamorphic assemblage.*

assemblage-zone A *biozone* consisting of "a body of strata whose content of fossils, or of fossils of a certain kind, taken in its entirety, constitutes a natural assemblage or association that distinguishes it in biostratigraphic character from adjacent strata" (ISG, 1976, p. 50). It is named for one or more of the fossils particularly representative of the assemblage. Cf: *faunizone; florizone; range-zone; acme-zone.* Syn: *cenozone.*

assembled stone Any stone constructed of two or more parts of gem materials, whether genuine, synthetic, imitation, or a combination thereof; e.g. a doublet or triplet. Syn: *composite stone.* Cf: *imitation.*

assimilated Said of an ore-forming fluid or mineralizer that is derived from crustal, palingenic magmas (Smirnov, 1968). Cf: *juvenile [ore dep]; filtrational.*

assimilation The incorporation and digestion of solid or fluid foreign material, i.e. wall rock, in magma. The term implies no specific mechanisms or results. Such a magma, or the rock it produces, may be called *hybrid* or *contaminated.* See also: *hybridization; contamination [ign]; cross-assimilation.* Cf: *differentiation.* Syn: *magmatic assimilation; magmatic dissolution.*

assise (a) A term approved by the 2nd International Geological Congress in Bologna in 1881 for a stratigraphic unit next in rank below stage and equivalent to a substage. (b) A term suggested by P. F. Moore (1958) to replace *format.*—Etymol: French, "course, layer, base", from the architectural usage for a continuous row or layer of stones or brick placed side by side between layers of mortar. Pron: a-*seas.*

associated gas Natural gas that occurs in association with oil in a reservoir, either as free gas or in solution. Cf: *nonassociated gas.*

association [ecol] A group of organisms (living or fossil) occurring together because they have similar environmental requirements or tolerances, and usually having one or more dominant species. Cf: *assemblage; community.* Syn: *fossil association.*

association [petrology] *rock association.*

associes A *seral* community in which there are two or more dominant forms. Pron: uh-*so*-sees.

assorted *poorly sorted.*

assortment The inverse of sorting or of the measure of dispersion of particle sizes within a frequency distribution.

assured mineral *developed reserves.*

Assyntian orogeny A name proposed by Stille and widely used in western Europe for orogenies and disturbances at the end of the Precambrian. It is named for the district of Assynt in the northwest highlands of Scotland, and based on the angular truncation of Torridonian strata by Lower Cambrian strata. The name is unfortunate, because the age of the Torridonian is undetermined; it may be very much older than the Cambrian.

assyntite A *nepheline syenite* rich in sphene, containing augite, and with biotite and apatite among the accessories. According to Johannsen (1939, p. 242), the name (from Assynt, Scotland) has been withdrawn. Not recommended usage.

astatic Said of a geophysical instrument that has a negative restoring force which aids a deflecting force, making the instrument more sensitive and/or less stable.

astatic gravimeter An instrument with a mechanical system designed to produce relatively large motions for small changes in gravity. Syn: *unstable gravimeter.*

astatic magnetometer Classic instrument of early paleomagnetic studies. An astatized pair of magnets is used in a torsional experiment to measure fields from samples.

astatic pair A pair of magnets of identical strength and opposed polarity, geometrically positioned so that they experience zero net torque in a homogeneous magnetic field.

astatic pendulum A pendulum having almost no tendency to take a definite position of equilibrium.

aster A sponge spicule (microsclere) that has a starlike appearance, with a relatively large number of rays or pseudoactins radiating from a relatively restricted central area.

asteria Any gemstone that, when cut *en cabochon* in the correct crystallographic direction, displays a rayed figure (a star) by either reflected or transmitted light; e.g. a *star sapphire.* Syn: *star stone.*

asteriated Said of a mineral, crystal, or gemstone that exhibits asterism; e.g. "asteriated beryl". Syn: *star.*

asterism [cryst] Elongation of Laue X-ray diffraction spots produced by stationary single crystals as a result of internal crystalline deformation. The size of the Laue spot is determined by the solid angle formed by the normals to any set of diffracting lattice planes; this angle increases with increasing crystal deformation, producing progressively elongated ("asterated") spots. Measurements of asterism are used as indicators of deformation in crystals subjected to slow stress or to shock waves.

asterism [gem] The optical phenomenon of a rayed or star-shaped figure of light displayed by some crystals when viewed in reflected light, as in a star sapphire or a gemstone cut *en cabochon,* or in transmitted light, as in some mica. It is caused by minute oriented acicular inclusions. See also: *star.*

asteroid [astron] One of the many small celestial bodies in orbit around the Sun. Most asteroid orbits are between those of Mars and Jupiter (NASA, 1966, p. 6). Syn: *planetoid; minor planet.*

asteroid [paleont] Any asterozoan belonging to the subclass Asteroidea, characterized by relatively broad arms usually not separable from the central disc; e.g. a starfish. Range, Lower Ordovician to present.

asterolith A star- or rosette-shaped coccolith with a concave face, formed of a single crystal with the *c*-axis perpendicular to the plane of the disk. See also: *discoaster.*

asterozoan Any free-living echinoderm belonging to the subphylum Asterozoa, having a characteristic depressed star-shaped body composed of a central disc and symmetrical radiating arms. Range, Cambrian to present.

asthenolith A body of magma that was formed by melting in response to heat generated by radioactive disintegration (Willis, 1938). See also: *asthenolith hypothesis.*

asthenolith hypothesis A theory of magmatic activity, both intrusive and extrusive, that postulates local *asthenoliths,* or areas of melting by radioactive heat, that have a repetitive cycle of melting, growth, migration, cooling, solidification, and remelting. Asthenolithic activity is postulated as the cause of uplift and subsidence, orogeny, earthquakes, and metamorphism (Willis, 1938, p.603). Cf: *blister hypothesis.*

asthenosphere The layer or shell of the Earth below the lithosphere, which is weak and in which isostatic adjustments take place, magmas may be generated, and seismic waves are strongly attenuated. It is equivalent to the *upper mantle.* Syn: *zone of mobility.*

Astian European stage: Upper Pliocene (above Plaisancian, below Calabrian). Equivalent to Redonian.

astite A variety of hornfels in which mica and andalusite dominate (Holmes, 1928, p.37). Type locality: Cima d'Asta, Italian Alps. Cf: *aviolite; edolite.*

astogenetic stage A stage in the development of a colony, as applied to bryozoa by Cumings in 1904.

astogeny The life history of a colonial animal, such as a graptolite or a bryozoan.

astrakhanite A syn. of *bloedite.* Also spelled: *astrakanite.*

astreoid Said of a massive rugose corallum in which the septa of each corallite are fully developed but walls between corallites are lacking, and characterized by septa of adjacent corallites generally in alternating position. Cf: *aphroid.*

astringent (a) Said of a mineral (such as alum) having a taste that tends to pucker the tissues of the mouth. (b) Said of a clay containing an astringent salt.

astrobleme An ancient erosional scar on the Earth's surface, produced by the impact of a cosmic body, and usually characterized by a circular outline and highly disturbed rocks showing evidence of intense shock (Dietz, 1961, p. 53); an eroded remnant of a meteoritic or cometary impact crater. The term is generally applied to *cryptoexplosion structures* of great age in which any original extraterrestrial fragments have been destroyed. Term introduced by Dietz (1960). Etymol: Greek *astron,* "star", + *blema,* "wound from a thrown object such as a javelin or stone". Cf. *geobleme.* Syn: *fossil meteorite crater.*

astrogeodetic measurement The direct measurement of the Earth to determine the deflection of the vertical, and hence the separation of the geoid and the ellipsoid. The method contrasts with gravimetric or indirect measurement.

astrogeology A science that applies the principles and techniques of geology, geochemistry, and geophysics to the study of the nature, origin, and history of the condensed matter and gases in the solar system (usually excluding the Earth). It includes: remote-sensing observations and in-situ manned exploration of other planetary bodies (the Moon, Mars); the study of the chemistry, mineralogy, and history of objects that occur on the Earth but are of known or possible extraterrestrial origin (such as meteorites and tektites) or that are returned to the Earth (such as lunar samples); and the study of the effects of extraterrestrial processes (such as meteorite impact, solar energy changes, and tides) on the Earth in the present and past. The term was first used by Lesevich (1877) for a branch of astronomy based primarily on the study of meteorites and secondarily on telescopic spectroscopy (see Milton, 1969). See also: *planetology; planetary geology.* Syn: *extraterrestrial geology; exogeology; space geology; geoastronomy.*

astrolabe A compact optical instrument designed for measuring the altitudes of celestial bodies; e.g. a "prismatic astrolabe" consisting of a telescope in a horizontal position, with a prism (generally of 45 or 60 degrees) in front of the object glass, immediately underneath which is an artificial horizon (pool of mercury). It was formerly used to fix latitude precisely by observing the apparent transit of the Sun across the meridian at midday, but has been superseded by the *sextant.*

astrolithology An obsolete term for the science that deals with meteoritic stones.

astronomical position The latitude and longitude of a point on the Earth as determined from astronomical measurements based on the position of the stars.

astronomical unit A unit of planetary distance equivalent to the mean distance of the Earth from the Sun: $1.496x10^8$ km, or approximately 93 million statute miles. This distance is also equivalent to the length of the semimajor axis of the Earth's orbit.

astronomic azimuth The angle between the astronomic meridian plane of the observer and the plane containing the observed point and the true normal (vertical) of the observer, measured in the plane of the horizon, preferably clockwise from north. Cf: *azimuth; bearing.*

astronomic equator The line on the Earth's surface whose astronomic latitude at every point is zero degrees. When corrected for station error, it becomes the *geodetic equator.*

astronomic horizon A great circle on the celestial sphere formed by the intersection of the celestial sphere and a plane passing through any point (such as the eye of an observer) and perpendicular to the zenith-nadir line; the plane that passes through the observer's eye and is perpendicular to the zenith at that point. It is the projection of a horizontal plane in every direction from the point of orientation. Cf: *celestial horizon.* Syn: *sensible horizon.*

astronomic latitude The *latitude* or angle between the plane of the celestial equator and the plumb line (direction of gravity) at a given point on the Earth's surface; the angle between the plane of the horizon and the Earth's axis of rotation. It represents the latitude resulting directly from observations on celestial bodies; when corrected for station error, it becomes the *geodetic latitude.* Symbol: Φ.

astronomic longitude The *longitude* or angle between the plane of the celestial meridian and the plane of an arbitrarily chosen prime meridian (generally the Greenwich meridian). It represents the longitude resulting directly from observations on celestial bodies; when corrected for station error, it becomes the *geodetic longitude.* Symbol: Λ.

astronomic measurement In geodetic surveying, the determination of latitude, longitude, or azimuth from observations on stars, referred to a plumb line or gravity vector at the point of observation. The sun is sometimes used in lower-order surveying. Cf: *deflection of the vertical.*

astronomic meridian A line on the Earth's surface having the same astronomic longitude at every point. It is an irregular line, not lying in a single plane. Syn: *terrestrial meridian.*

astronomic parallel A line or circle on the Earth's surface having the same astronomic latitude at every point. It is an irregular line, not lying in a single plane.

astronomic tide *equilibrium tide.*

astronomy The study of celestial bodies: their positions, sizes, movements, relative distances, compositions and physical conditions, interrelationships, and history.

astrophyllite A mineral: $(K,Na)_3(Fe,Mn)_7Ti_2Si_8O_{24}(O,OH)_7$.

astrophysics That aspect of astronomy which is concerned with the physics and chemistry of celestial bodies, and with their origins.

astropyle A nipplelike projection from the central capsule of a radiolarian of the suborder Phaeodarina.

astrorhiza A system of regularly spaced, stellate, generally sinuous surficial impressions and internal canals of some living sponges and fossil stromatoporoids. They typically lack walls, may be crossed by thin tabulae, and decrease in diameter and branch distally from their point of divergence, which is marked by one or more vertical small tubes. Pl: *astrorhizae.*

Asturian orogeny One of the 30 or more short-lived orogenies during Phanerozoic time identified by Stille, in this case late in the Carboniferous, between the Westphalian and Stephanian Stages.

asylum *refugium.*

asymmetric Said of crystals of the hemihedral class of the triclinic system, which have no symmetry elements; also, said of any irregular crystal.

asymmetric bedding Bedding characterized by lithologic types or facies that succeed each other vertically in a cyclical arrangement illustrated by the sequence 1-2-3-1-2-3-1-2-3. Cf: *symmetric bedding.*

asymmetric fold (a) A fold whose axial surface is not perpendicular to the *enveloping surface.* (b) A fold whose limbs have different angles of dip relative to the axial surface. Ant: *symmetrical fold.*

asymmetric ripple mark A *ripple mark* having an asymmetric profile in cross section, characterized by a short steep slope facing downcurrent and a long gentle slope facing upcurrent; specif. *current ripple mark.* In plan view, the crest may be relatively straight or markedly curved. Ant: *symmetric ripple mark.*

asymmetric valley A valley with one side steeper than the other.

AT *assay ton.*

atacamite A green orthorhombic mineral: $Cu_2Cl(OH)_3$. It is trimorphous with partacamite and botallackite, and is formed by weathering of copper lodes, esp. under desert conditions. Syn: *remolinite.*

atatschite A porphyritic igneous rock containing microscopic crystals of alkali feldspar, clinopyroxene, and biotite in a glassy groundmass, and characterized by the presence of small amounts of sillimanite and cordierite; a *vitrophyre.* It was named by Morozewicz in 1901 for Atatsch Mountain in the Urals. Not recommended usage.

atavism *reversion.*

ataxic Said of an unstratified mineral deposit. Cf: *eutaxic.*

ataxite [meteorite] An *iron meteorite* containing more than 10% nickel and lacking the structure of either *hexahedrite* or *octahedrite.* Many ataxites show microscopic oriented plates of kamacite in a groundmass of plessite. Symbol: *D.*

ataxite [volc] A *taxite* whose components have mixed in a breccia-like manner. Cf: *eutaxite.*

atelestite A yellow mineral: $Bi_8(AsO_4)_3O_5(OH)_5$.

atexite Basic material that is unchanged during anatexis (Dietrich & Mehnert, 1961). Also spelled: *atectite.* Not widely used.

at grade [eng] On the same level or degree of rise; at design level or slope. The term is applied to highways, walks, culverts, etc., or combinations of these, at the point where they intersect.

at grade [geomorph] *graded [geomorph].*

athabascaite A mineral: Cu_5Se_4.

atheneite A hexagonal mineral: $(Pd,Hg)_3As$.

Atlantic n. A term used primarily in Europe for an interval of Holocene time (from about 7500 to 4500 years ago) following the Boreal and preceding the Subboreal, during which the inferred climate was warmer than at present and generally wet; a subunit of the *Blytt-Sernander climatic classification,* characterized by oak, elm, linden, and ivy vegetation. It corresponds to most of the *Altithermal* and the middle part of the Hypsithermal.—adj. Pertaining to the postglacial Atlantic interval and to its climate, deposits, biota, and events.

Atlantic suite One of two large groups of igneous rocks, characterized by alkalic and alkali-calcic rocks. Harker (1909) divided all Tertiary and Holocene igneous rocks of the world into two main groups, the Atlantic suite and the *Pacific suite,* the former being so named because of the predominance of alkalic and alkali-calcic rocks in the nonorogenic areas of crustal instability around the Atlantic Ocean. Because there is such a wide variety of tectonic environments and associated rock types in the areas of Harker's Atlantic and Pacific suites, the terms are now seldom used to indicate kindred rock types; e.g. Atlantic-type rocks are widespread in the mid-Pacific volcanic islands. Cf: *Mediterranean suite.* See also: *andesite line.*

Atlantic-type coastline A *discordant coastline,* esp. as developed in many areas around the Atlantic Ocean; e.g. the SW coastline of Ireland and the NW coastlines of France and Spain. Ant: *Pacific-type coastline.*

atlantite A nepheline-bearing *basanite* or *tephrite* having a predominance of dark-colored minerals over light. It was defined by Lehmann in 1924. Not recommended usage.

atlas A collection of maps bound into a volume. The use of the term is derived from the figure of Atlas (a Titan of Greek mythology, often represented as supporting the heavens) used as a frontispiece to certain early collections of maps, first appearing on the general title page of Mercator's *Atlas* (1595).

atmidometer *atmometer.*

atmoclast A rock fragment broken off in place by atmospheric weathering, either chemically or mechanically.

atmoclastic rock A clastic rock consisting of atmoclasts that have been recemented without rearrangement by wind or water (Grabau, 1924, p.292).

atmodialeima A term proposed by Sanders (1957, p.295) for an unconformity caused by subaerial processes.

atmogenic Said of a rock, mineral, or deposit derived directly from the atmosphere, as by condensation, wind action, or deposition from volcanic vapors; e.g. snow. See also: *atmolith.*—Little used.

atmolith An *atmogenic* rock (Grabau, 1924, p. 279).

atmometer Any device used to measure the rate of evaporation in the atmosphere. It may be a large evaporation tank, a small evaporation pan, a porous porcelain body, or a porous paper-wick device. Syn: *atmidometer; evaporimeter; evapograph.*

atmophile (a) Said of those elements that are most typical in the Earth's atmosphere: H, C, N, O, I, Hg, and inert gases (Rankama & Sahama, 1950, p.88). (b) Said of those elements that occur in the uncombined state, or that "will concentrate in the gaseous primordial atmosphere" (Goldschmidt, 1954, p.26).

atmosilicarenite A siliceous sand resulting from weathering and disintegration of a parent rock; e.g. *grus.* It results from passive action of the atmosphere rather than from the atmosphere in motion. Originally spelled "atmosilicarenyte" by Grabau (1913).

atmosphere The mixture of gases that surrounds the Earth, being held thereto by gravity. It consists by volume of 78% nitrogen, 21% oxygen, 0.9% argon, 0.03% carbon dioxide, and minute quantities of helium, krypton, neon, and xenon. The atmosphere is so compressed by its own weight that half is below 5.5 km from the Earth's surface. Syn: *air.*

atmospheric argon Argon in the atmosphere and argon absorbed on the surfaces of rocks and minerals that have been exposed to the atmosphere. Cf: *excess argon; inherited argon; radiogenic argon.*

atmospheric pressure The pressure, or force per unit area, exerted by the atmosphere on any surface beneath or within it. Normal pressure at sea level is 1013.25/millibars, or 1,013,250 dynes per cm^2. Equivalent common measures are 76.0 cm, or 29.92 inches, of mercury; 1033.3 cm, or 33.9 feet, of water; and 14.66 lb per in^2. Syn: *air pressure; barometric pressure.*

atmospheric radiation The infrared radiation emitted by the atmosphere in two directions: upward into space and downward toward the Earth. The latter is known as *counterradiation.*

atmospheric tide The rhythmic vertical oscillation of the atmosphere, produced primarily by thermal effects from absorption of radiation at different levels. Syn: *tide.*

atmospheric water Water in the atmosphere, in gaseous, liquid, or solid state.

atmospheric weathering Weathering occurring at the surface of the Earth.

atmospheric window A wavelength interval at which the atmosphere transmits most electromagnetic radiation. Syn: *transmission window.*

Atokan North American provincial series: lower Middle Pennsylvanian (above Morrowan, below Desmoinesian).

atokite A cubic mineral: $(Pd,Pt)_3Sn$.

atoll A coral reef appearing in plan view as roughly circular (though sometimes elliptical or horseshoe-shaped), and surmounted by a chain or ring of closely spaced low coral islets that encircle or nearly encircle a shallow lagoon in which there is no pre-existing land or islands of noncoral origin; the reef is surrounded by deep water of the open sea, either oceanic or continental-shelf. Atolls range in diameter from 1 km to more than 130 km, and are esp. common today in the western and central Pacific Ocean. Several fossil atolls have also been described. Etymol: native name in the Maldive Islands (Indian Ocean) which are typical examples of this structure. Syn: *lagoon island; ring reef; reef ring.*

atoll moor A peat bog that entirely surrounds a lake in the form of a ring and is in turn surrounded by a ditch or ring of open water around the original shoreline of the lake. See also: *moat lake.* Syn: *sphagnum atoll.*

atollon A term used in the Maldive Islands of the Indian Ocean for a large atoll consisting of many smaller ones; the term "atoll" was derived from this name.

atoll structure In a metamorphic rock, porphyroblasts with hollow centers resembling atolls. The ring may be almost complete or consist of a chain of granules (Joplin, 1968).

atoll texture In mineral deposits, the surrounding of one mineral by a ring of one or more other minerals. It commonly results from replacement of pyrite by another mineral, with the outermost pyrite unaffected and constituting the "atoll". Syn: *core texture.* Cf: *tubercle texture.*

atomic absorption spectrometry Chemical analysis performed by vaporizing in a flame a sample, usually in a liquid form, and measuring the absorptance by the unexcited atoms in the vapor of various narrow resonant wavelengths of light that are characteristic of specific elements. The amount of an element present is proportional to the amount of absorption by the vapor.

atomic absorption spectrophotometer An instrument for generating and analyzing an *atomic absorption spectrum.*

atomic absorption spectroscopy The observation of an *atomic absorption spectrum* and all processes of recording and measuring that go with it.

atomic absorption spectrum The *absorption spectrum* seen when the unexcited atoms of a vaporized sample selectively absorb certain wavelengths of light passed through the sample.

atomic clock *radioactive clock.*

atomic plane In a crystal, any plane that contains a regular array of atomic units (atoms, ions, or molecules); it is a potential cleavage face or cleavage plane.

atomic time scale A *geologic time scale* calibrated on the basis of radioactive decay in rocks. Measurements are made in years. Cf: *relative time scale.*

atomous Said of a crinoid arm that does not branch.

atom percent The percentage of an atomic species in a substance, calculated with reference to number of atoms rather than weight, number of molecules, or other criteria.

atopite A yellow or brown variety of *romeite* containing fluorine.

at rest Said of the lateral *earth pressure* when the soil is neither compressed nor allowed to expand and the structure (such as a wall) does not move. See also: *neutral pressure.*

atrium [paleont] The *spongocoel* of a sponge. Pl: *atria.*

atrium [palyn] The space between the external opening (pore) and a much larger internal opening in the endexine of a pollen grain with a complex porate structure. The internal opening is so large that the endexine is missing in the pore area. Pl: *atria.* Cf: *vestibulum.*

atrium oris A preoral cavity in a crustacean, bounded ventrally by the posteriorly directed labrum, dorsally by the ventral surface of the cephalon just behind the mouth, and laterally by metastoma and mandibles.

atrypoid Any articulate brachiopod belonging to the family Atrypidae, characterized by costate or plicate shells that are unequally biconvex or convexo-plane, the brachial valve being the more convex. Range, Middle Ordovician to Upper Devonian.

attached dune A dune that accumulates around a rock or other obstacle in the path of windblown sand, occurring on either the windward or the leeward sides of the obstacle, or on both sides, and varying widely in size and form. Cf: *anchored dune.*

attached ground water Ground water retained on the walls of interstices in the zone of aeration. It is considered equal in amount to the *pellicular water* and is measured by specific retention.

attached operculum The part of a dinoflagellate cyst that is surrounded by archeopyle sutures and hence remains joined to the main part of the theca where the suture is not developed. Cf: *free operculum.* Syn: *attached opercular piece.*

attachment scar (a) An expanded *basal cavity* of a conodont, or one that is larger than a small-sized pit. (b) The part of the aboral side of a conodont element to which the basal plate was attached.

attakolite A mineral: $(Ca,Mn,Sr)_3Al_6(PO_4,SiO_4)_7 \cdot 3H_2O$. Also spelled: *attacolite.*

attapulgite *palygorskite.*

attenuation (a) A reduction in the amplitude or energy of a signal, such as might be produced by passage through a filter. (b) A reduction in the amplitude of seismic waves, as produced by *divergence [seis],* reflection and scattering, and *absorption.* (c) That portion of the decrease in seismic or sonar signal strength with distance that is not dependent on geometrical divergence, but on the physical characteristics of the transmitting medium.

attenuation coefficient If the loss of amplitude A with distance x is expressed as an exponential, $A/A_o = e^{-\alpha x}$, then α is the attenuation coefficient. Cf: *Q.*

attenuation constant *Q.*

attenuation distance *depth of penetration [remote sensing].*

Atterberg grade scale A geometric and decimal *grade scale* devised by Albert Atterberg (1846-1916), Swedish soil scientist; it is based on the unit value 2 mm and involves a fixed ratio of 10 for each successive grade, yielding the diameter limits of 200, 20, 2.0, 0.2, 0.02, and 0.002 (Atterberg, 1905). Subdivisions are the geometric means of the grade limits. The scale has been widely used in Europe and was adopted in 1927 by the International Commission on Soil Science (but not by the U.S. Bureau of Soils).

Atterberg limits In a sediment, the water-content boundaries between the semiliquid and plastic states (known as the *liquid limit*) and between the plastic and semisolid states (known as the *plastic limit*). Syn: *consistency limits.*

attic The very small, uppermost (abaxial) chamberlet in superposed chamberlets in a volution of a foraminiferal shell (as in *Flosculinella* and *Alveolinella*).

Attic orogeny One of the 30 or more short-lived orogenies during Phanerozoic time, identified by Stille; in this case in the Miocene, between the Sarmatian and Pontian Stages.

attitude [photo] (a) The angular orientation of a camera, or of the photograph taken with that camera, with respect to some external reference system. It is usually expressed as "tilt", "swing", and "azimuth", or as "roll", "pitch", and "yaw". (b) The angular orientation of an aerial or space vehicle with respect to a reference system.—(ASP, 1975, p. 2066.)

attitude [struc geol] The position of a structural surface relative to the horizontal, expressed quantitatively by both *strike* and *dip* measurements.

attribute A qualitative *variable,* usually denoted by its presence or absence.

attrital-anthraxylous coal A bright coal in which the ratio of anthraxylon to attritus varies from 1:1 to 1:3. Cf: *anthraxylous coal; anthraxylous-attrital coal; attrital coal.*

attrital coal (a) A coal in which the ratio of anthraxylon to attritus varies from 1:1 to 1:3. Cf: *anthraxylous coal; anthraxylous-attrital coal; attrital-anthraxylous coal.* (b) The groundmass or matrix of *banded coal,* in which vitrain and commonly fusain are embedded.

attrition The act or process of wearing down by friction; specif. the mutual wear and tear that loose rock fragments or particles, moved about by wind, waves, running water, or ice, undergo by rubbing, grinding, knocking, scraping, and bumping against one another, with resulting reduction in size and increase in roundness. Although the term has been used synonymously with *abrasion* and *corrasion,* strictly it is the wearing away and reduction in size undergone by rock fragments serving as tools of abrasion or corrasion.

attritus A composite term for dull grey to nearly black coal components of varying maceral content, unsorted and with fine granular texture, that forms the bulk of some coals or is interlayered with bright bands of *anthraxylon* in others. It is formed of a tightly compacted mixture of altered vegetal materials, especially those that were relatively resistant to complete degradation. Cf: *attrital coal.* Syn: *durain.*

aubrite An achondritic stony meteorite whose essential mineral is enstatite; it contains diopside in minor amounts. It normally has a brecciated structure. See also: *whitleyite.* Syn: *bustite.*

audio-magneto-telluric method A magnetic-telluric survey using frequencies in the natural electromagnetic field above 20 Hz. Abbrev: AMT.

auerlite A variety of thorite containing phosphorus, with a PO_4/SiO_4 ratio of about 0.8:1.

aufeis A syn. of *icing.* Etymol: German.

aufwuchs Aquatic organisms that are attached to but do not penetrate the substrate; e.g. crustaceans. Cf: *periphyton.* Etymol: German, "growth".

auganite An augite-bearing *andesite;* according to Johannsen (1939, p. 242), an olivine-free *basalt.* Not recommended usage.

augelite A colorless, white, or pale-red mineral: $Al_2(OH)_3PO_4$.

augen In foliate metamorphic rocks such as schists and gneisses, large lenticular mineral grains or mineral aggregates having the shape of an eye in cross section, in contrast to the shapes of other minerals in the rock. See also: *augen structure.* Etymol: German, "eyes".

augen-blast An *augen*-forming porphyroblast in dynamically metamorphosed rocks (Bayly, 1968). Cf: *augen-clast.*

augen-clast In dynamically metamorphosed rock, *augen* consisting of clastic fragments in a clastic matrix (Bayly, 1968). Cf: *augen-blast.*

augen gneiss A general term for a gneissic rock containing *augen.*

augen schist A metamorphic rock characterized by the presence of recrystallized minerals as *augen* or lenticles parallel to and alternating with schistose streaks.

augen structure In some gneissic and schistose metamorphic rocks, a structure consisting of minerals like feldspar, quartz, or garnet that have been squeezed into elliptical or lens-shaped forms resembling eyes (*augen*), which are commonly enveloped by essentially parallel layers of contrasting constituents such as mica or chlorite. Cf: *flaser structure.* Syn: *eyed structure; phacoidal structure.*

auger (a) A screwlike boring tool resembling a carpenter's auger bit but much larger, usually motor-driven, designed for use in clay, soil, and other relatively unconsolidated near-surface materials. (b) A rotary drilling device for making seismic shotholes or geophone holes by which the cuttings are mechanically and continuously removed from the bottom of the borehole during the drilling operation without the use of fluids.

Auger spectrum A series of discrete lines representing the energies of emitted electrons arising from a transition between atomic states that involves no emission or absorption of radiation.

augite (a) A common mineral of the clinopyroxene group: $(Ca,Na)(Mg,Fe^{+2},Al)(Si,Al)_2O_6$. It may contain titanium and ferric iron. Augite is usually black, greenish black, or dark green, and occurs as an essential constituent in many basic igneous rocks and in certain metamorphic rocks. Dana (1892) confined the name "augite" to clinopyroxenes containing appreciable $(Al,Fe)_2O_3$, but petrologists have applied it to members of the system $(Mg,Fe,Ca)SiO_3$. Cf: *pigeonite.* (b) A term often used as a syn. of *pyroxene.*— Syn: *basaltine.*

augitite A *tephrite* containing abundant phenocrysts of clinopyroxene with lesser amounts of amphibole, magnetite or ilmenite, apatite, and sometimes nepheline, hauyne, or feldspar in a dark-colored glassy groundmass probably of the composition of analcime. Not recommended usage.

augitophyre A porphyritic basaltic rock in which the phenocrysts are augite. Not recommended usage.

aulacocerid Said of cephalopods belonging to the order Aulacocerida of the subclass Coleoidea.

aulacogen A tectonic trough on a craton, bounded by convergent normal faults. Aulacogens have a radial orientation relative to cratons and are open outward. Cf: *graben; rift.*

aulocalycoid Said of the skeleton of a hexactinellid sponge in which presumed dictyonal strands are diagonally interwoven and connected by synapticulae.

aulos An *axial structure* in a rugose coral, consisting of a tube commonly formed by abrupt sideward deflection of the inner edges of septa and their junction with neighbors.

aureole A zone surrounding an igneous intrusion in which the country rock shows the effects of contact metamorphism. Syn: *contact aureole; contact zone; metamorphic aureole; thermal aureole; exomorphic zone; zone [meta].*

auric Said of a substance that contains gold, esp. gold in its trivalent state. Cf: *auriferous.*

aurichalcite A pale-green or pale-blue mineral: $(Zn,Cu)_5(CO_3)_2(OH)_6$. Syn: *brass ore.*

auricle (a) An ear-like extension of the dorsal region of the shell in certain bivalve mollusks, commonly separated from the body of the shell by a notch or sinus. (b) An internal process arising from basicoronal ambulacral plates of an echinoid and serving for attachment of muscles supporting Aristotle's lantern. (c) In the vertebrates, a blind pocket projecting from a receiving or venous chamber of the heart; or this chamber itself.

auricula One of the thickened "ears" of auriculate spores. Pl: *auriculae.* Cf: *zone [palyn].*

auricular sulcus An external furrow at the junction of an auricle with the body of the shell of a bivalve mollusk.

auriculate Said of spores having exine thickenings (auriculae) in the equatorial region that project like "ears", generally from the area of the ends of the laesura.

auricupride A mineral: Cu_3Au. Syn: *cuproauride.*

auriferous Said of a substance that contains gold, esp. said of gold-bearing mineral deposits. Cf: *auric.*

auriform Said of a mollusk shell (such as that of a gastropod) shaped like the human ear.

aurora A division of geologic time from the appearance of a mutant in a bioseries to the appearance of the following mutant (Glaessner, 1945, p. 225). Pl: *aurorae.*

aurorite A mineral: $(Mn,Ag,Ca)Mn_3O_7 \cdot 3H_2O$.

aurostibite A cubic mineral: $AuSb_2$.

austausch A measure of turbulent mixing, equal to the product of mass and transverse distance traveled in a unit of time by a fluid in turbulent motion as it passes through a unit area parallel to the general direction of flow (Twenhofel, 1939, p. 187). Syn: *eddy conductivity; mixing coefficient; austausch coefficient; eddy coefficient; exchange coefficient.*

austausch coefficient *austausch.*

Austinian North American (Gulf Coast) stage: Upper Cretaceous (above Eaglefordian, below Tayloran).

austinite A colorless or yellowish orthorhombic mineral: $CaZnAsO_4(OH)$.

austral Pertaining to the south, or located in southern regions; southern.

australite A jet-black, usually button-shaped or lensoid, often well-preserved tektite from southern Australia. Syn: *blackfellows' button.*

Austrian orogeny One of the 30 or more short-lived orogenies during Phanerozoic time identified by Stille, in this case at the end of the Early Cretaceous.

autallotriomorphic *sutured.*

autapomorphy In *cladism,* a *derived character* that is unique to a single taxon, usually a single species. Adj: *autapomorphic; autapomorphous.*

autecology The study of the relationships between individual organisms or species (or a particular taxon) and their environment. Also spelled: *autoecology.* Cf: *synecology.*

authalic projection *equal-area projection.*

authigene An *authigenic* mineral or rock constituent, e.g. a mineral of an igneous rock; the cement of a sedimentary rock if deposited directly from solution; or a mineral resulting from metamorphism. The term was introduced by Kalkowsky (1880, p.4). Ant: *allogene.*

authigenesis The process by which new minerals form in place within an enclosing sediment or sedimentary rock during or after deposition, as by replacement or recrystallization, or by secondary enlargement of quartz overgrowths. Cf: *neogenesis.*

authigenetic *authigenic.*

authigenic Formed or generated in place; specif. said of rock constituents and minerals that have not been transported or that crystallized locally at the spot where they are now found, and of minerals that came into existence at the same time as, or subsequently to, the formation of the rock of which they constitute a part. The term, as used, often refers to a mineral (such as quartz or feldspar) formed after deposition of the original sediment. Ant:

allogenic. Cf: *halmeic; autochthonous.* Syn: *authigenous; authigenetic.*

authigenous (a) *authigenic.* (b) In the classification of pyroclastics, the equivalent of *essential.* Cf: *allothigenous.*

authimorph A constituent of a metamorphic rock which, in the formation of the new rock, had its outlines or boundaries altered. The term is obsolete.

autoarenite A sand produced by crushing due to tectonic movements; the sand-size equivalent of an autoclastic breccia.

autoassociation A method for self-comparison of a string or sequence of nonnumeric data (Sackin & Merriam, 1969, p. 7). Cf: *autocorrelation [stat].*

autobreccia A breccia formed by some process that is contemporaneous with the formation or consolidation of the rock unit from which the fragments are derived; specif. a *flow breccia.*

autobrecciation Formation of an autobreccia; e.g. the fragmentation process whereby portions of the first consolidated crust of a lava mass are incorporated into the still-fluid portion.

autochthon [sed] A residual deposit produced in place by decomposition.

autochthon [tect] A body of rocks that remains at its site of origin, where it is rooted to its basement. Although not moved from their original site, autochthonous rocks may be mildly to considerably deformed. Ant: *allochthon.* Cf: *parautochthonous; stationary block.* Also spelled: *autochthone.*

autochthone *autochthon [tect].*

autochthonous Formed or produced in the place where now found. The term is widely applied, e.g. to a coal or peat that originated at the place where its constituent plants grew and decayed, to rocks that have not been displaced by overthrust faulting, or to a breccia at an explosion crater that remains in its original position, with only minor rotation or translation of the fragments. The term was first used by Naumann (1858, p. 657) to designate rock units remaining at the site of their formation; it is similar in meaning to *authigenic,* which refers to constituents rather than whole formations. Ant: *allochthonous.* See also: *parautochthonous.*

autochthonous mantle Loose debris of rock fragments or soil derived from the underlying bedrock and formed in situ. Cf: *allochthonous mantle; block field.*

autochthony Accumulation of plant remains in their original environment or in the place of their growth. Ant: *allochthony.* See also: *euautochthony; hypautochthony; in-situ theory.*

autoclast A rock fragment in an autoclastic rock.

autoclastic rock A term originated by Smyth (1891) for a rock having a broken or brecciated structure, formed in the place where it is found as a result of crushing, shattering, dynamic metamorphism, orogenic forces, or other mechanical processes; e.g. a *fault breccia,* or a brecciated dolomite produced by diagenetic shrinkage followed by recementation. Cf: *cataclastic rock; epiclastic rock.*

autoconsequent Said of a stream whose course is guided by the slopes of material it has deposited; also said of the topographic features (such as waterfalls) developed by such a stream.

autocorrelation [seis] A measure of the statistical dependence of a later portion of a wave form on an earlier portion, or the extent to which future values can be predicted from past values; the *correlation [seis]* of a signal with itself. Cf: *crosscorrelation.*

autocorrelation [stat] A method for self-comparison of a string or sequence of numeric data (Sackin & Merriam, 1969, p. 8). Cf: *autoassociation.*

autocyclicity The state of cyclothemic deposition that requires no change in the total energy and material input into a sedimentary system but involves simply the redistribution of these elements within the system (Beerbower, 1964, p.32). It involves such mechanisms of deposition as channel migration, channel diversion, and bar migration. Cf: *allocyclicity.*

autodermalium A specialized sponge spicule lying within the exopinacoderm. Cf: *hypodermalium.*

autoecology *autecology.*

autogastralium A specialized sponge spicule lying within the endopinacoderm of the cloaca. Cf: *hypogastralium.*

autogenetic (a) Said of landforms that have developed or evolved under strictly local conditions, without interference by orogenic movements; esp. a topography resulting from the action of falling rains and flowing streams upon land surfaces having free drainage to the sea. (b) Said of a type of drainage (and of its constituent streams) that is determined entirely by the conditions of the land surface over which the streams flow, as a drainage system developed solely by headwater erosion. See also: *self-grown stream.*—Syn: *autogenous; autogenic.*

autogenic [ecol] Said of an ecologic succession that resulted from factors originating within the natural community and altering its habitat. Cf: *allogenic (c).*

autogenic [geomorph] *autogenetic.*

autogenous *autogenetic.*

autogeosyncline A *parageosyncline* without an adjoining uplifted area, containing mostly carbonate sediments (Kay, 1947, p. 1289-1293); an *intracratonic basin.* Syn: *residual geosyncline.* Cf: *zeugogeosyncline.*

autoinjection *autointrusion.*

autointrusion [ign] A process wherein the residual liquid of a differentiating magma is injected into rifts formed in the crystallized fraction at a late stage by deformation of unspecified origin. Syn: *autoinjection.*

autointrusion [sed] Sedimentary *intrusion* of rock material from one part of a bed or set of beds in process of deposition into another part.

autolith (a) An inclusion in an igneous rock to which it is genetically related. Cf: *xenolith.* Syn: *cognate inclusion; cognate xenolith; endogenous inclusion.* (b) In a granitoid rock, an accumulation of Fe-Mg minerals of uncertain origin (Balk, 1937, p. 10-12). It may appear as a round, oval, or elongate segregation or clot.

autolysis (a) The process of "self-digestion", as in the albitization of plagioclase in a lava by sodium from the lava itself rather than by newly introduced sodium. (b) Return of a substance to solution, as phosphate removed from seawater by plankton and returned when these organisms die and decay.

autometamorphism (a) A process of recrystallization of an igneous mineral assemblage under conditions of falling temperature, attributed to the action of its own volatiles, e.g. serpentinization of peridotite or spilitization of basalt. (b) The alteration of an igneous rock by its own residual liquors (Tyrrell, 1926). This process should rather be called *deuteric* because it is not considered to be metamorphic.

autometasomatism Alteration of a recently crystallized igneous rock by its own last water-rich liquid fraction, trapped within the rock, generally by an impermeable chilled border. Cf: *deuteric* alteration; *autopneumatolysis; autometamorphism.*

automicrite Autochthonous *orthomicrite.*

automolite A dark-green to nearly black variety of gahnite.

automorphic (a) Said of the holocrystalline texture of an igneous or metamorphic rock, characterized by crystals bounded by their own *rational faces.* Also said of a rock with such a texture. The term (Rohrbach, 1885, p. 17-18) has priority over *idiomorphic* (Rosenbusch, 1887, p. 11), though the latter is more common in American usage. (b) A syn. of *euhedral,* obsolete in American usage, but generally preferred in European usage; the corresponding texture is sometimes referred to as "panautomorphic" (now obsolescent). Syn: *automorphic-granular.*

automorphic-granular *automorphic.*

autopiracy Capture of an upper part of a stream by its lower part, as by the cutting-off of a meander, generally resulting in a shortening of its own course.

autopneumatolysis Autometamorphism involving the crystallization of minerals or the alteration of a rock by gaseous emanations originating in the magma or rock itself.

autopore Tubular *autozooecium* in Paleozoic bryozoans. Obsolete.

autopotamic Said of an aquatic organism adapted to living in flowing fresh water. Cf: *eupotamic; tychopotamic.*

autoradiograph A type of *scan* of radioactivity, e.g. a neutron, X-ray, or gamma-ray photograph. Syn: *radiograph; radioautograph.*

autosite An igneous rock similar in composition to *kersantite* but without feldspar. Not recommended usage.

autoskeleton The endoskeleton of sponges, consisting of spicules or spongin secreted by the cells. Cf: *pseudoskeleton.*

autotheca The largest tube of three regularly produced at each budding in the development of a graptolithine colony. It may have contained a female zooid. Cf: *bitheca; stolotheca; metatheca.*

autotrophic Said of an organism that nourishes itself by utilizing inorganic material to synthesize living matter. Green plants and certain protozoans are autotrophic. Noun: *autotroph.* Cf: *heterotrophic.*

autozooecium The skeleton of a bryozoan autozooid.

autozooid (a) A fully formed octocorallian polyp with eight well developed tentacles and septs. It is the only type of polyp in monomorphic species and it is the major type in dimorphic species. Cf: *siphonozooid.* (b) A feeding bryozoan zooid.

autumn ice Sea ice in an early stage of formation and not yet affected by lateral pressure; it is relatively salty and is crystalline in appearance.

Autunian European stage: Lower Permian (above Stephanian of Carboniferous, below Saxonian).

autunite (a) A lemon-yellow or sulfur-yellow radioactive tetragonal mineral: $Ca(UO_2)_2(PO_4)_2 \cdot 10\text{-}12H_2O$. It is isomorphous with torbernite. Autunite is commonly a secondary mineral and occurs as tabular plates or in micalike scales. Syn: *lime uranite; calcouranite.* (b) A group of isomorphous tetragonal minerals of general formula: $R^{+2}(UO_2)_2(XO_4)_2 \cdot nH_2O$, where R =Ca, Cu, Mg, Ba, Na_2, or other elements, and X = P or As. The group includes minerals such as autunite, torbernite, uranocircite, saléeite, sodium autunite, zeunerite, uranospinite, novacekite, and kahlerite.

Auversian European stage: Eocene (above Lutetian, below Bartonian).

auwai An Hawaiian term for a watercourse or channel, esp. one used for irrigation.

auxiliary adj. Said of an inflection (any lateral lobe or lateral saddle) of the ammonoid suture added later than the first two or three pairs; e.g. "auxiliary lobe" springing from the umbilical lobe and occurring between the second lateral saddle and the umbilicus (TIP, 1959, pt.L, p.18). Cf: *accessory [paleont].*—n. An auxiliary lobe or an auxiliary saddle.

auxiliary contour *supplementary contour.*

auxiliary fault A minor fault abutting against or branching from a major one. Syn: *branch fault.*

auxiliary mineral In the Johannsen classification of igneous rocks, a light-colored, relatively rare or unimportant mineral such as apatite, muscovite, corundum, fluorite, or topaz.

auxiliary plane The plane that is perpendicular to the fault plane and perpendicular to the slip. It is determined from seismic data for earthquakes.

auxotrophic Said of a microscopic organism that requires certain specific nutrients.

available moisture *available water.*

available relief The total relief available for stream dissection in a given area, equal to the vertical distance between the height of the remnants of an original upland surface and the level at which grade is first attained by adjacent streams (Glock, 1932). Cf: *local relief.*

available water Water available to plants; the difference between field capacity and wilting point. Syn: *available moisture.*

avalanche A large mass of snow, ice, soil, or rock, or mixtures of these materials, falling, sliding, or flowing very rapidly under the force of gravity. Velocities may sometimes exceed 500 km/hr.

avalanche bedding Steeply inclined bedding in barchan and related dune forms, produced by an avalanche of sand down the slip face of the dune.

avalanche blast A very destructive *avalanche wind* occurring when an avalanche is stopped abruptly, as when it falls vertically onto a valley floor, or displaces a large volume of air by flowing horizontally for a long distance.

avalanche boulder tongue An *avalanche talus* of a narrow, railroad-embankment shape, controlled by winter snowbanks on both sides of the avalanche track. The snowbanks confine the avalanche, and when they melt the talus stands above the surrounding slopes with steep edges (Rapp, 1959). Cf: *avalanche talus; avalanche track.*

avalanche chute The track or path formed by an avalanche. Cf: *avalanche track.*

avalanche cone The mass of material deposited where an avalanche has fallen, consisting of snow, ice, rock, and all other material torn away and carried along by the avalanche.

avalanche ripple mark Any asymmetric ripple mark having a steep lee-side slope at or near the angle of repose and migrating by a series of small avalanches down the lee slope (Imbrie & Buchanan, 1965, p.151). Cf: *accretion ripple mark.*

avalanche talus An accumulation of rock fragments of any size or shape, derived from snow and ice mixed with soil and rock debris avalanched from a cliff or rocky slope above. It usually occurs on the downwind side of a ridgetop or arête, in mountains where drifting snow builds cornices that collapse and bring down snow, ice, and rocks. Angle of slope is much less than 30° and the slope profile is concave upward (White 1967, p. 237). Cf: *avalanche boulder tongue; scree; avalanche track.*

avalanche track The central channel-like corridor along which an avalanche has moved; it may take the form of an open path in a forest, with bent and broken trees, or an eroded surface marked by pits, scratches, and grooves. Cf: *avalanche chute; avalanche talus; avalanche boulder tongue; devil's slide.*

avalanche wind A high wind or rush of air produced in front of a large landslide or of a fast-moving dry-snow avalanche, and sometimes causing destruction at a considerable distance from the avalanche itself. See also: *avalanche blast.*

avalanching The sudden and rapid downward movement of an avalanche.

Avalonian orogeny An orogenic event near the end of Precambrian time along the southeastern border of North America, especially well shown in the type area where a granite dated at 575 m.y. is separated from the fossiliferous Lower Cambrian by a thick, late Precambrian sequence. Indications of the event, represented by radiometric dates ranging from slightly earlier to later, occur along the southeastern edge of the Appalachian orogenic belt, and in rocks beneath the sediments of the Atlantic Coastal Plain as far southwest as Florida. It was named by Lilly (1966) and Rodgers (1967) for the Avalon Peninsula, southeastern Newfoundland.

avanturine Var. (error) of *aventurine.*

avelinoite *cyrilovite.*

aven (a) A *domepit.* (b) A vertical shaft open to the surface. Partial syn: *vertical cave.*—Etymol: French.

aven (a) A *domepit.* (b) A vertical shaft open to the surface. Partial syn.: *vertical cave.* Etymol: French.

aventurescence In certain translucent minerals, a display of bright or strongly colored reflections from included crystals. Examples are aventurine quartz and aventurine feldspar (sunstone).

aventurine (a) A translucent quartz spangled throughout with tiny inclusions of another mineral; a grayish, greenish, brown, or yellowish *quartzite* that exhibits aventurescence from minute crystals, platelets, flakes, or scales of minerals such as green mica, ilmenite, hematite, and limonite. Syn: *aventurine quartz.* (b) *aventurine feldspar.*—The term is also used as an adjective in referring to the brilliant, spangled appearance of a glass or mineral containing gold-colored or shiny inclusions. Misspelled: *avanturine.* Syn: *adventurine.*

aventurine feldspar A variety of feldspar (oligoclase, albite, andesine, or adularia) characterized by a reddish luster produced by fiery, golden reflections or firelike flashes of color from thin disseminated mineral particles (such as flakes of hematite) oriented parallel to structurally defined planes and probably formed by exsolution; specif. *sunstone.* Syn: *aventurine.*

aventurine glass *goldstone.*

aventurine quartz *aventurine.*

average *arithmetic mean.*

average deviation *mean deviation.*

average discharge As used by the U.S. Geological Survey, the arithmetic average of all complete water years of record of discharge whether consecutive or not.

average igneous rock A theoretical rock whose chemical composition is believed to be similar to the average composition of the outermost layer of the Earth.

average level anomaly Gravity anomaly related to average topographic level in an area, usually of some fixed radius. Syn: *Putnam anomaly.*

average stress One-half the sum of the initial stress and the final stress associated with the occurrence of an earthquake.

average velocity [hydraul] (a) For a stream, discharge divided by the area of a cross section normal to the flow. (b) For ground water, the volume of ground water passing through a given cross-sectional area, divided by the porosity of the material through which it moves. Syn: *mean velocity.*

average velocity [seis] The ratio of the distance traversed along a ray path by a seismic pulse to the time required for that traverse. Cf: *seismic velocity; interval velocity.*

avezacite A dike rock intermediate in composition between *pyroxenite* and *hornblendite,* with amphibole in excess of pyroxene and with ilmenite constituting approximately 20 percent of the rock. Named by Lacroix in 1901 for Avezac in the Pyrenees. Not recommended usage.

avicennite A black cubic mineral: Tl_2O_3.

avicularium A polymorph in cheilostome bryozoans, resembling a bird's head. It is commonly a heterozooid, having the equivalent of the operculum (the *mandible*) and associated muscles relatively larger than those of an autozooid. Pl: *avicularia.*

aviolite A type of hornfels whose main constituents are mica and cordierite. Type locality: Monte Aviolo, Italian Alps. Cf: *astite; edolite.*

avlakogene A syn. of *aulacogen.* The term was introduced by Shatsky (1955). Etymol: Russian.

avogadrite An orthorhombic mineral: $(K,Cs)BF_4$.

avon A river. Etymol: Celtic.

avulsion (a) A sudden cutting off or separation of land by a flood or by an abrupt change in the course of a stream, as by a stream breaking through a meander or by a sudden change in current whereby the stream deserts its old channel for a new one. Legally, the part thus cut off or separated belongs to the original owner. Cf: *accretion.* (b) Rapid erosion of the shore by waves during a storm (Wiegel, 1953, p. 4).

awaruite A mineral consisting of a natural alloy of nickel and iron; *nickel-iron.*

awn *seta [bot].*

axial angle *optic angle.*

axial canal (a) A longitudinal passageway penetrating columnals of an echinoderm and connecting with the body cavity, and generally but not invariably located centrally. (b) An intraspicular cavity left by decay of an axial filament in a sponge.

axial compression In experimental work with cylinders, a compression applied parallel with the cylinder axis.

axial cross The orthogonal crossing of the six axial filaments at the center of a hexactinellid sponge spicule. It is used esp. where some of the axial filaments are reduced to the area about the center, as in spicules with fewer than six rays.

axial diameter The anterior-posterior thecal diameter in edrioasteroids. Cf: *transverse diameter.*

axial dipole field A hypothetical magnetic field, consisting of an ideal *dipole field* centered at the Earth's center and with its axis along the Earth's rotational axis. While the actual geomagnetic field does not have this ideal form, it is hypothesized that it would, after averaging thousands of years of secular variation.

axial elements In crystallography, the ratio of unit distances along crystallographic axes and the angles between these axes.

axial figure An *interference figure* in which one optic axis is centered in the figure.

axial filament An organic fiber about which the mineral substance of a sponge-spicule ray is deposited.

axial filling A deposit of dense calcite developed in the axial region of some fusulinacean foraminifers and formed probably at the same time as excavation of tunnel or foramina and formation of chomata and parachomata (TIP, 1964, pt.C, p.58).

axial furrow (a) One of the two longitudinal grooves bounding the axis of a trilobite. Syn: *dorsal furrow.* (b) A longitudinal groove separating the median lobe or axis of a merostome from the pleural area.

axial increase A type of *increase* (offset formation of corallites) in coralla characterized by the appearance of dividing walls between newly formed corallites approximately in position of the axis of the parent corallite.

axial jet A flow pattern characteristic of *hypopycnal inflow,* in which the inflowing water spreads as a cone with an apical angle of about 20° (Moore, 1966, p. 88). Cf: *plane jet.*

axial lobe The *axis* of a trilobite.

axial plane [cryst] (a) The plane of the optic axes of an optically biaxial crystal. (b) A crystallographic plane that includes two of the crystallographic axes.

axial plane [fold] An *axial surface* that is planar.

axial-plane cleavage Cleavage which is closely related to the axial planes of folds in the rock, either being rigidly parallel to the axes, or diverging slightly on each flank (*fan cleavage*). Most axial-plane cleavage is closely related to the minor folds seen in individual outcrops, but some is merely parallel to the regional fold axes. Most axial-plane cleavage is also *slaty cleavage.* Cf: *bedding-plane cleavage.*

axial-plane separation The distance between axial surfaces of adjacent antiforms and synforms where the folds occur in the same layer or surface.

axial ratio The ratio of the lengths of the crystallographic axes of a crystal, stated in terms of one axis as unity.

axial section A slice bisecting a foraminiferal test in a plane coinciding with the axis of coiling, and intersecting the proloculus.

axial septulum A secondary or tertiary septum located between primary septa of a foraminifer, its plane approximately parallel to the axis of coiling and thus observable in sagittal, parallel, and tangential sections. See also: *primary axial septulum; secondary axial septulum.*

axial skeleton In the vertebrates, the vertebral column, ribs, braincase, and median fins when present. In common usage the axial skeleton also includes the entire skull and visceral skeleton.

axial stream (a) The main stream of an intermontane valley, flowing in the deepest part of the valley and parallel to its longest dimension. (b) A stream that follows the axis of a syncline or anticline.

axial structure A collective term for various longitudinal structures in the axial region of a corallite, whether a solid or spongy rodlike columella or an axial vortex. See also: *clisiophylloid; aulos.*

axial surface A surface that connects the *hinge lines* of the strata in a fold.

axial symmetry In structural petrology, a fabric having a unique axis of symmetry, an infinite number of mirror planes passing through that axis, and a single mirror plane normal to it. Syn: *spheroidal symmetry.*

axial trace The intersection of the *axial surface* of a fold with the surface of the Earth or other given surface.

axial vortex A longitudinal structure in the axial region of a corallite, formed by the twisting-together of the inner edges of major septa associated commonly with the transverse skeletal elements.

axil The distal angle formed between two plant parts, specif. a stalk or petiole and the stem (axis) from which it grows (Swartz, 1971, p. 54).

axil angle Acute angle as shown on a map between two confluent streams, measured upstream from their junction. Symbol: ξ. Syn: *entrance angle; stream-entrance angle.*

axillary A brachial plate supporting two crinoid arm branches.

axinellid Said of a sponge skeleton built of spiculofibers in which the component spicules are all directed obliquely outward from the axes of the fibers.

axinite A brown, violet, blue, green, or gray mineral: $(Ca,Mn,Fe)_3Al_2BSi_4O_{15}(OH)$. It may contain appreciable sodium. Axinite commonly forms glassy wedge-shaped triclinic crystals. Syn: *glass schorl.*

axiolite A spherulitic aggregate in which the needles radiate from a central line or axis, rather than from a point; e.g. a spherulite in a rhyolite composed of welded glass fragments to whose outlines minute acicular crystals or fibers of feldspar are approximately perpendicular and radiate inward; or a subspherical oolith or pisolith in a carbonate sediment, around which acicular needles are axially grouped. Term proposed by Zirkel (1876, p. 167). Syn: *axiolith.*

axiolith *axiolite.*

axiolitic Said of the structure of a rock in which *axiolites* are abundant; also said of a rock containing axiolites.

axiometer A device that permits accurate location and measurement of pebble and cobble axes by means of a clamp and a track-mounted caliper (Schmoll & Bennett, 1961).

axis [cryst] *crystal axis.*

axis [fold] (a) The line which, moved parallel to itself, generates the form of a fold. (b) The trace of the axial surface of a fold on the fold profile plane. This definition is now rarely used.

axis [geomorph] (a) The central or dominant region of a mountain chain. (b) A line that follows the trend of large landforms, such as one following the crest of a ridge or mountain range, or of the bottom or trough of a depression.

axis [paleont] (a) The median lobe of a trilobite, consisting of the longitudinal raised portion of the exoskeleton lying between the pleural regions, particularly of the thorax and the pygidium. Syn: *axial lobe.* (b) The central supporting structure of certain octocorals, such as a spicular structure in a gorgonian, or a horny structure in the order Pennatulacea. (c) A straight line with respect to which an invertebrate is radially or bilaterally symmetrical; e.g. the oral-aboral axis of a corallite, or the axis formed by ambulacral plates in the sheath of radial water vessel in an asterozoan ray (TIP, 1966, pt.U, p.28). (d) An imaginary line through the apex of a gastropod shell, about which the whorls of conispiral and discoid shells are coiled. (e) An imaginary line around which a spiral or cyclical shell of a protist is coiled, transverse to the plane of coiling. (f) In certain algae, the central portion of the thallus, from which branches may originate. (g) In the tetrapods, the second cervical vertebra.—Pl: axes.

axis culmination *culmination.*

axis of divergence The generally vertical or oblique line in a coral septum from which trabeculae incline inward and outward. See also: *fan system.*

axis of symmetry *symmetry axis.*

axis of tilt The line along which a tilted photo intersects the plane of an imaginary vertical photo taken with the same camera from the same point, and along which the tilted photo has the same scale as would the vertical photo (ASP, 1975, p. 2067).

axoblast An individual *scleroblast* that produces the axis of certain octocorals, e.g. *Holaxonia.*

axopodium A semipermanent *pseudopodium,* typically present in radiolarian and heliozoan cells, consisting of an axial rod surrounded by a protoplasmic envelope. Pl: axopodia. Syn: *axopod.*

azimuth [seis] A horizontal angle, measured clockwise, between the north meridian and the arc of the great circle connecting the epicenter of an earthquake and the receiver. When measured at the epicenter, it is called the azimuth from epicenter to receiver; when measured at the receiver, the *back azimuth.*

azimuth [surv] (a) Direction of a horizontal line as measured on an imaginary horizontal circle; the horizontal *direction* reckoned clockwise from the meridian plane of the observer, expressed as the angular distance between the vertical plane passing through the point of observation and the poles of the Earth and the vertical plane passing through the observer and the object under observation. In the basic control surveys of U.S., azimuths are measured clockwise from south, a practice not followed in all countries. Cf: *bearing.* See also: *true azimuth; magnetic azimuth.* (b) *astronomic azimuth.*

azimuthal equal-area projection *Lambert azimuthal equal-area projection.*

azimuthal equidistant projection A map projection (neither equal-area nor conformal) in which all points are placed at their true distances and true directions from the central point of the projection. Any point on the globe may be placed at the center, and a straight line radiating from this point to any other point represents the shortest distance (a great circle in its true azimuth from the center), whose length can be measured to scale. Syn: *zenithal equidistant projection.*

azimuthal projection (a) A map projection in which a portion of the sphere is projected upon a plane tangent to it at the pole or any other point (which becomes the center of the map) and on which the azimuths (directions) of all lines radiating from the central point to all other points are the same as the azimuths of the corresponding lines on the sphere. Distortion at the central point is zero and scale distortions are generated radially from the central point. All great circles through the central point are straight lines intersecting at true angles. (b) A similar projection used in structural petrology.—Syn: *zenithal projection.*

azimuthal survey A resistivity or induced-polarization survey in which an area is traversed by a voltage-measuring electrode pair along azimuths away from a fixed current electrode which may be in a drill hole or in contact with a metallic ore. The second current electrode is placed so far away that its location does not affect the measurements. Syn: *radial array.*

azimuth angle (a) The horizontal angle, less than 180 degrees, between the plane of the celestial meridian and the vertical plane containing the observation point and the observed object (celestial body), reckoned from the direction of the elevated pole. In the astronomic triangle (composed of the pole, the zenith, and the star), it is the spherical angle at the zenith. (b) An angle in triangulation or in a traverse, through which the computation of azimuth is carried.

azimuth circle An instrument for measuring azimuth, consisting of a horizontal graduated circle divided into 360 major divisions; e.g. one attached to a compass to show magnetic azimuths.

azimuth compass A magnetic compass, supplied with vertical sights, for measuring the angle that a line on the Earth's surface, or the vertical circle through a heavenly body, makes with the magnetic meridian; a compass used for taking the magnetic azimuth of a celestial body.

azimuth line A term used in radial triangulation for a radial line from the principal point, isocenter, or nadir point of a photograph, representing the direction to a similar point on an adjacent photograph in the same flight line.

azimuth mark A mark set at a significant distance from a triangulation or traverse station to mark the end of a line for which the azimuth has been determined and to serve as a starting or reference azimuth for later use.

Azoic (a) That earlier part of Precambrian time, represented by rocks in which there is no trace of life. Cf: *Protozoic.* (b) The entire *Precambrian.*

azoic Said of an environment that is devoid of life.

azonal peat *local peat.*

azonal soil In the 1938 soil-classification system, one of the three *soil orders.* Azonal soils lack well-developed horizons and resemble the parent material (USDA, 1938). Cf: *intrazonal soil; zonal soil; Entisol.* Syn: *immature soil.*

azonate Said of spores without a zone or a similar (usually equatorial) extension.

azoproite A mineral of the ludwigite group: $(Mg,Fe^{+2})_2(Fe^{+3},Ti,Mg)BO_5$.

azulite A translucent pale-blue variety of smithsonite often found in large masses (as in Arizona and Greece).

azurchalcedony *azurlite.*

azure quartz A blue variety of quartz; specif. *sapphire quartz.*

azure stone A term applied to lapis lazuli and to blue minerals such as lazulite and azurite.

azurite (a) A deep-blue to violet-blue monoclinic mineral: $Cu_3(CO_3)_2(OH)_2$. It is an ore of copper and is a common secondary mineral associated with malachite in the upper (oxidized) zones of copper veins. Syn: *chessylite; blue copper ore; blue malachite.* (b) A semiprecious stone derived from compact azurite and used chiefly for ornamental objects. (c) A trade name for a sky-blue gem variety of smithsonite.

azurlite A variety of chalcedony colored blue by chrysocolla and used as a gemstone. Syn: *azurchalcedony.*

azurmalachite An intimate mixture or intergrowth of azurite and malachite, usually occurring massive and concentrically banded, and used as an ornamental stone.

azygous basal plate The smallest of the three plates of the basalia of a blastoid, normally located in the AB interray but sometimes in the DE interray. Cf: *zygous basal plate.*

azygous node A special kind of cusp located directly above the basal cavity of certain conodonts (such as *Palmatolepis* and *Panderodella*).

alluvium (a) A general term for clay, silt, sand, gravel, or similar unconsolidated detrital material deposited during comparatively recent geologic time by a stream or other body of running water as a sorted or semisorted sediment in the bed of the stream or on its flood plain or delta, or as a cone or fan at the base of a mountain slope; esp. such a deposit of fine-grained texture (silt or silty clay) deposited during time of flood. The term formerly included (but is not now intended to include) subaqueous deposits in seas, estuaries, lakes, and ponds. Syn: *alluvial; alluvial deposit; alluvion.* (b) A driller's term used "incorrectly" for the broken, earthy rock material directly below the soil layer and above the solid, unbroken bed or ledge rock (Long, 1960). (c) *alluvial soil.* —Etymol: Latin *alluvius,* from *alluere,* "to wash against". Pl: *alluvia; alluvium.* Cf: *eluvium; diluvium.*

B

babefphite A mineral: $BaBe(PO_4)(O,F)$.

Babel quartz A variety of crystalline quartz so named for its fancied resemblance to the successive tiers of the Tower of Babel. Syn: *Babylonian quartz.*

babingtonite A greenish-black triclinic mineral: $Ca_2(Fe^{+2},Mn)Fe^{+3}Si_5O_{14}(OH)$.

bacalite A variety of amber from Baja California, Mexico.

baccula A swollen area, generally sagittally elongate, within the axial furrow and alongside the posterior of the glabella in some trilobites, especially the Agnostidae, Raphiophoridae, Trinucleidae and Telephinidae. It has been commonly referred to as an *ala,* but this term should be restricted to a different structure, in harpetid trilobites. Pl: *bacculae.*

bache A term used in England for the valley of a small stream.

bacillite A rodlike crystallite composed of a group of parallel longulites.

back v. To change wind direction counterclockwise in the northern hemisphere and clockwise in the southern.

back azimuth The horizontal direction from receiver to epicenter of an earthquake. See also: *azimuth [seis].*

back bay A small, shallow bay into which coastal streams drain and which is connected to the sea through a pass between barrier islands, as along the coast of Texas. Cf: *front bay.*

backbeach *backshore.*

back bearing (a) A *bearing* along the reverse direction of a line; the reverse or reciprocal of a given bearing. If the bearing of line AB is N 58° W, the back bearing (bearing of line BA) is S 58° E. Syn: *reverse bearing; reciprocal bearing.* (b) A term used by the U.S. Public Land Survey system for the reverse direction of a line as corrected for the curvature of the line from the forward bearing at the preceding station.

backbone A ridge serving as the main axis of a mountain; the principal mountain ridge, range, or system of an area.

backdeep A syn. of *epieugeosyncline,* so named because of its relative position, away from the craton (Aubouin, 1965, p. 34).

backfill (a) Earth or other material used to replace material removed temporarily during construction or permanently during mining, such as stones and gravel used to fill pipeline trenches or placed behind structures such as bridge abutments, or waste rock used to support the roof after removal of ore from a stope. Also, material such as sand or dirt placed between an old structure and a new lining, as in a shaft or tunnel. Backfill in excavations may or may not be the material originally removed. (b) The process of refilling an excavation, a mine opening, or the space around a foundation.

background [geochem] The abundance of an element, or any chemical property of a naturally occurring material, in an area in which the concentration is not anomalous (Hawkes, 1958, p. 336).

background [radioactivity] *background radiation.*

background radiation The radiation of the environment, e.g. from cosmic rays and from the Earth's naturally radioactive substances. Also, any radiation that is not part of a controlled experiment. Syn: *background.*

backhand drainage Drainage in which the general course of the tributaries on both sides is opposite to that of the main stream (Eakin, 1916, p. 17).

backland [geomorph] The lowland along either side of a river, behind the natural levee; the part of a flood plain extending from the base of a valley slope and separated from the river by the natural levee. Syn: *back lands.*

backland [tect] *hinterland.*

back lead A *lead [eco geol]* or deposit of coastline sands above the high-water mark. Pron: back leed.

backlimb The less steep of the two limbs of an asymmetrical, anticlinal fold. Cf: *forelimb.*

back-limb thrust fault A thrust fault developed on the more gently dipping "back limb" of an asymmetric anticline, in which the direction of tectonic transport is uplimb and the fault dips in the same direction as the limb but at a steeper angle (Douglas, 1950, p. 88-89). See also: *front-limb thrust fault; contraction fault.* Also spelled: *backlimb thrust fault.*

back marsh The low, wet, poorly drained areas of an alluvial flood plain.

back radiation *counterradiation.*

back reef The landward side of a reef, including the area and the contained deposits between the reef and the mainland; the terrestrial deposits connecting the reef with the land; the reef flat. The term is often used adjectivally to refer to the restricted lagoon behind a barrier reef, such as the "back-reef facies" of lagoonal deposits. In some places, as on a platform-edge reef tract, "back reef" refers to the side of the reef away from the open sea, even though no land may be nearby. Cf: *fore reef.* Also spelled: *backreef.*

back-reef moat *boat channel.*

backrush *backwash.*

back scatter The scattering of radiant energy into the hemisphere bounded by a plane normal to the direction of the incident radiation and lying on the same side as the incident ray; the opposite of a *forward scatter.* Atmospheric back scatter depletes 6 to 9 percent of the incident solar energy before it reaches the Earth's surface. In radar usage, back scatter generally refers to the microwave radiation scattered back toward the antenna. Syn: *backward scattering.*

backset bed A cross-bed that dips against the direction of flow of a depositing current; e.g. an inclined layer of sand deposited on the gentle windward slope of a transverse dune, often trapped by tufts of sparse vegetation; or a glacial deposit that formed, as the ice retreated, on the rear slope of an apron or sand plain, or at the front of an esker, consequently dipping toward the retreating ice.

backset eddy A small current revolving in the direction opposite to that of the main ocean current.

backshore (a) The upper or inner, usually dry and narrow, zone of the shore or beach, lying between the high-water line of mean spring tides and the upper limit of shore-zone processes; it is acted upon by waves or covered by water only during exceptionally severe storms or unusually high tides. It is essentially horizontal or slopes landward, and is divided from the *foreshore* by the crest of the most seaward berm. (b) The area lying immediately at the base of a sea cliff. (c) *berm.*—Syn: *backbeach.*

backshore terrace A wave-built terrace on the backshore of a beach; a *berm.*

backsight (a) A sight or bearing on a previously established survey point (other than a closing or check point), taken in a backward direction. (b) A reading taken on a level rod held in its unchanged position on a survey point of previously determined elevation when the leveling instrument has been moved to a new position. It is used to determine the height of instrument prior to making a foresight. Syn: *plus sight.*—Abbrev: BS. Ant: *foresight.*

back slope (a) A syn. of *dip slope;* the term is used where the angle of dip of the underlying rocks is somewhat divergent from the angle of the land surface. (b) The slope at the back of a scarp; e.g. the gentler slope of a cuesta or of a fault block. It may be unrelated to the dip of the underlying rocks.—Also spelled: *backslope.* Cf: *scarp slope.*

backsteinbau Nacreous tablets in the mollusks arranged in a brick-and-mortar pattern (Wise, 1970).

backswamp A swampy or marshy depressed area developed on a flood plain, with poor drainage due to the natural levees of the river.

backswamp deposits Thin layers of silt and clay deposited in the flood basin behind the natural levees of a river.

backswamp depression A low, usually swampy area adjacent to a leveed river. Syn: *levee-flank depression.*

back thrusting Thrust faulting in an orogenic belt, with the direction of displacement towards its interior, or contrary to the general direction of tectonic transport.

backwall *headwall.*

backwash The seaward return of water running down the foreshore of a beach following an *uprush* of waves; also, the seaward-flowing mass of water so moved. Syn: *backrush.*

backwash mark A term used by Johnson (1919, p.517) for a "criss-cross ridge" developed on a beach slope by the return flow of the uprush. Probable syn: *rhomboid ripple mark.*

backwash ripple mark A term used by Kuenen (1950, p. 292) for a broad, flat ripple mark between narrow, shallow troughs, formed on a beach by backrush above the level of maximum wave retreat; its crest is parallel to the shoreline.

backwasting (a) Wasting that causes a slope to retreat without changing its declivity. (b) The recession of the front of a glacier.—Cf: *downwasting.*

backwater (a) Water that is retarded, backed up, or turned back in its course by an obstruction (such as a bridge or dam), an opposing current, or the movement of the tide; e.g. the water in a reservoir or the water obtained at high tide to be discharged at low tide. Also, the resulting *backwater effect.* (b) A body of currentless or relatively stagnant water, parallel to a river and usually fed from it through a single channel at the lower end by the back flow of the river. Loosely, any tranquil body of water joined to a main stream but little affected by its current, such as the water collected in side channels or flood-plain depressions after it overflowed the lowland. (c) A creek, arm of the sea, or series of connected lagoons, usually parallel to the coast, separated from the sea by a narrow strip of land but communicating with it through barred outlets. (d) A backward current of water. Also, the motion of water that is turned back; a *backwash.*

backwater curve (a) The form of the water surface along a longitudinal profile, assumed by a stream above the point where depth is made to exceed the normal depth by a constriction or obstruction in the channel (ASCE, 1962). Cf: *drop-down curve.* (b) A generic term for all *surface profiles* of water; esp. *flow profile.*

backwater effect The upstream increase in height of the water surface of a stream, produced when flow is retarded above a temporary obstruction (such as a bridge or dam) or when the main stream overflows low-lying land and backs up water in its tributaries. The effect is also characterized by an expansion in width of the body of water and by a slackening in the current. Syn: *backwater.*

backwearing Erosion that causes the parallel retreat of an escarpment or of the slope of a hill or mountain, or the sideways recession of a slope without changing its declivity; a process contributing to the development of a pediment or pediplain. Cf: *downwearing.*

backweathering Weathering that contributes to slope retreat.

bacon [sed] A quarrymen's term used in Portland, southern England, for *beef.*

bacon [speleo] *bacon-rind drapery.*

bacon-rind drapery A type of dripstone that projects from the cave walls and ceiling in thin translucent sheets and is characterized by parallel colored bands. Syn: *bacon [speleo].* Partial syn: *drapery.*

bacon stone An old name for a variety of *steatite,* alluding to its greasy appearance. See also: *speckstone.*

bacteriogenic Said of ore deposits formed by the action of anaerobic bacteria, by the reduction of sulfur or the oxidation of metals (Park & MacDiarmid, 1970, p. 105-107). See also: *iron bacteria; sulfur bacteria.*

bacterium A single-celled microorganism that lacks chlorophyll and an evident nucleus. Most bacteria are capable of breaking down extraneous matter; some are pathogens. Range, Precambrian to the present.

bactritoid Any straight cephalopod belonging to the subclass Bactritoidea, characterized by a shell of relatively uniform shape, having a small globular to ovate protoconch and a much larger orthoconic or cyrtoconic conch. Bactritoids have been classified both as nautiloids and as ammonoids. Range, Ordovician to Permian.

baculate Said of sculpture of pollen and spores consisting of bacula. Cf: *baculum.*

baculite A crystallite that appears as a dark rod.

baculum One of the tiny rods, varying widely in size and either isolated or clustered, that make up the ektexine sculpture of pollen or spores. Pl: *bacula.*

baddeleyite A colorless, yellow, brown, or black monoclinic mineral: ZrO_2. It may contain some hafnium, titanium, iron, and thorium.

badenite A steel-gray mineral: $(Co,Ni,Fe)_3(As,Bi)_4$ (?).

badlands Intricately stream-dissected topography, characterized by a very fine drainage network with high *drainage densities* (77 to 747 miles per square mile) and short steep slopes with narrow *interfluves.* Bedlands develop on surfaces with little or no vegetative cover, overlying unconsolidated or poorly cemented clays or silts, sometimes with soluble minerals such as gypsum or halite (Fairbridge, 1968, p. 43). They may also be induced in humid areas by removal of the vegetative cover through overgrazing, or by air pollution from sulfide smelting, as at Ducktown, Tennessee (Strahler, 1956, p. 630). The term was first applied to an area in western South Dakota, which was called "mauvaises terres" by the early French fur traders.

Baer's law *von Baer's law.*

bafertisite An orthorhombic mineral: $Ba(Fe,Mn)_2TiSi_2O_7(O,OH)_2$.

Bagnold dispersive stress A shear stress between two layers in a fluid, caused by the impact between cohesionless particles that are free to collide with each other during current flow in the absence of applied body force. The stress increases as the diameter squared and the large particles, subjected to highest stress, are forced to the bed surface where the stress is zero. The term was used by Leopold et al. (1966, p. 213) and named for Ralph A. Bagnold (b. 1896), British geographer, who quantified the influence of collective mutual collisions on current flow (Bagnold, 1956). The term "Bagnold effect" was proposed by Sanders (1963, p. 174) for the same phenomenon. Syn: *dispersive stress.*

baguette A *step cut* used for small, narrow rectangular gemstones, principally diamonds. Also, the gem so cut. Etymol: French, "rod".

bahada Anglicized var. of *bajada.*

bahamite A term proposed by Beales (1958, p. 1851-1852) for a shallow marine deposit that consists of limestone grains closely resembling the predominant deposits (described by Illing, 1954) now accumulating in the interior of the Bahama Banks. It is very pure, generally fine-grained, massively bedded, widely extensive, and without abundant fossils; the grains are accretionary and commonly composite. The term applies to sediments that accumulate under conditions of coastal shoaling or on offshore banks, and does not imply that the limestone was formed under genetic conditions exactly comparable with those prevailing on present-day Bahama Banks. See also: *grapestone.*

bahiaite A *pyroxenite* containing orthopyroxene, amphibole, olivine, and a small amount of spinel. Named by Washington in 1914 for Bahia, Brazil. Not recommended usage.

bahr A body of water as found in the Saharan region; esp. a deep natural spring, often in the form of a small, crater-shaped lake of great depth, as in some oases of eastern Algeria. Etymol: Arabic, "sea". Pl: *bahar; bahrs.*

Baikalian orogeny A name widely used throughout the U.S.S.R. for orogeny that occurred about the time of the Precambrian-Cambrian transition and was completed in the middle Cambrian. Initial movements are known from the lower Riphean. Different phases are defined by relations within the many late Precambrian stratified sequences of the Soviet Union. Named after Lake Baikal in Siberia.

baikalite A dark-green variety of diopside containing iron and found near Lake Baikal, U.S.S.R.

baikerinite A thick, tarry hydrocarbon that makes up about one third of *baikerite* and from which it may be separated by alcohol.

baikerite A variety of *ozocerite.* See also: *baikerinite.*

bailer A cylindrical steel container with a valve at the bottom for admission of fluid, attached to a wire line and used in *cable-tool drilling* for recovering and removing water, cuttings, and mud from the bottom of a well.

bajada A broad, continuous *alluvial slope* or gently inclined detrital surface extending from the base of mountain ranges out into and around an inland basin, formed by the lateral coalescence of a series of separate but confluent *alluvial fans,* and having an undulating character due to the convexities of the component fans; it occurs most commonly in semiarid and desert regions, as in the

SW U.S. A bajada is a surface of deposition, as contrasted with a *pediment* (a surface of erosion that resembles a bajada in surface form), and its top often merges with a pediment. Originally, the term was used in New Mexico for the gentler of the two slopes of a cuesta. Etymol: Spanish, "descent, slope". Pron: ba-*ha*-da. Syn: *bahada; apron [geomorph]; alluvial apron; mountain apron; fan apron; debris apron; alluvial plain; compound alluvial fan; piedmont alluvial plain; piedmont plain; waste plain; piedmont slope; gravel piedmont; alluvial bench.*

bajada breccia A term used by Norton (1917, p.167) for a wedge-shaped, imperfectly stratified accumulation of coarse, angular, poorly sorted rock fragments mixed with mud, formed in an arid region by an intermittent stream or a mudflow containing considerable water. Cf: *fanglomerate.*

bajir A term applied in the deserts of central Asia to a lake occupying a flat-bottomed basin separating sand hills or dunes.

Bajocian European stage: Middle Jurassic (above Toarcian, below Bathonian).

bakerite A mineral: $Ca_4B_4(BO_4)(SiO_4)_3(OH)_3 \cdot H_2O$. It occurs in white compact nodules resembling marble or unglazed porcelain.

baking The hardening of rock material by heat from magmatic intrusions or lava flows. Prolonged baking leads to contact-metamorphic effects. Cf: *caustic metamorphism.*

balaghat A term used in India for a tableland situated above mountain passes.

balance The change in mass (the difference between total *accumulation* and gross *ablation*) of a glacier over some defined interval of time, determined either as a value at a point, an average over an area, or the total mass change for the glacier. The units normally used are millimeters, meters, or cubic meters of water equivalent, but kilograms per square meter or kilograms are used by some. Syn: *mass balance; mass budget; regimen [glaciol]; regime; economy.* Cf: *net balance; annual balance.*

balanced rock (a) A large rock resting more or less precariously on its base, formed by weathering and erosion in place. See also: *pedestal rock; rocking stone.* (b) *perched block.*

balance of nature "A state of equilibrium in nature due to the constant interaction of the whole biotic and environmental complex, interference with this equilibrium (as by human intervention) being often extremely destructive" (Webster, 1967, p.165).

balance rate The rate of change of mass of a glacier at any time, equal to the difference between accumulation rate and ablation rate.

balance year The period from the time of minimum mass in one year to the time of minimum mass in the succeeding year for a glacier; the period of time between the formation of one *summer surface* and the next. See also: *net balance.* Syn: *budget year.*

balancing The process of systematically distributing corrections through any traverse to eliminate the error of closure and to obtain an adjusted position for each traverse station. Also known as "balancing a survey".

balas ruby A pale rose-red or orange gem variety of spinel, found in Badakhshan (or Balascia) province in northern Afghanistan. See also: *ruby spinel.* Syn: *balas; ballas.*

balavinskite A mineral: $Sr_2B_6O_{11} \cdot 4H_2O$.

bald n. A local term, esp. in southern U.S., for an elevated, grassy area, as a mountain top or high meadow, that is devoid of trees.

bald-headed anticline An anticline whose crest has been eroded prior to deposition of an unconformably overlying sedimentary unit. Commonly used in petroleum geology.

baldite The hypabyssal equivalent of analcime-bearing *basalt,* composed of pyroxene phenocrysts in a groundmass of analcime, augite, and iron oxide. The name (Johannsen, 1938) is derived from Big Baldy Mountain, in the Little Belt Mountains, Montana. Not recommended usage.

balipholite An orthorhombic mineral: $BaMg_2LiAl_3Si_4O_{12}(OH)_8$.

balk A low ridge of earth that marks a boundary line (ASCE, 1954, p. 20).

balkanite An orthorhombic mineral: $Cu_9Ag_5HgS_8$.

ball [coast] A syn. of *longshore bar.* The term is used in the expression *low and ball,* but the elongate character of the bar is "not well indicated by 'ball' which suggests a round object" (Shepard, 1952, p. 1909).

ball [sed] A primary sedimentary structure consisting of a spheroidal mass of material; e.g. an armored mud ball, a slump ball, a lake ball.

ball-and-pillow structure A primary sedimentary structure found in sandstones and some limestones, characterized by hemispherical or kidney-shaped masses resembling balls and pillows, and commonly attributed to foundering; e.g. a *flow roll* or a *pseudonodule.* The term was introduced by Smith in 1916. See also: *pillow structure; ball structure.*

ball-and-socket jointing In basalt columns, cross-jointing surfaces that are concave either upward or downward. Syn: *cup and ball jointing.*

ballas (a) A dense, globular aggregate of minute diamond crystals, having a confused radial or granular structure, whose lack of through-going cleavage planes imparts a toughness that makes it useful as an *industrial diamond.* Cf: *bort; carbonado.* (b) A term incorrectly applied to a rounded, single-crystal form of diamond. (c) *balas ruby.*

ballast (a) Broken stone, gravel, or other heavy material used to provide weight in a ship and therefore improve its stability or control its draft. Jettisoned ballast may be found in samples of marine sediments. Also, similar material used to supply weight in equipment for use in lunar-gravity studies. (b) Gravel, broken stone, expanded slag or similar material used as a foundation for roads, esp. that laid in the roadbed of a railroad to provide a firm bed for the ties, distribute the load, and hold the track in line, as well as to facilitate drainage.

ball clay A highly plastic, sometimes refractory clay, commonly characterized by the presence of organic matter, having unfired colors ranging from light buff to various shades of gray, and used as a bonding constituent of ceramic wares; *pipe clay.* It has high wet and dry strength, long vitrification range, and high firing shrinkage. Ball clay is so named because of the early English practice of rolling the clay into balls weighing 13-22 kg (30-50 lb) and having diameters of about 25 cm (10 in.).

ball coal Coal occurring in spheroidal masses that are probably formed by jointing. Not to be confused with *coal ball.* Syn: *pebble coal.*

balled-up structure A term used by O.T. Jones (1937) for a knot, several centimeters to a few meters in diameter, of highly contorted silty material lying isolated in mud and produced by subaqueous slump. See also: *ball structure.*

ball ice Sea ice, either *frazil ice* or *pancake ice,* consisting of numerous soft, spongy, floating spheres (2.5-5 cm in diameter) shaped by waves, and usually occurring in belts.

ball ironstone A sedimentary rock containing large argillaceous nodules of ironstone.

ballistic magnetometer A type of magnetometer that uses the transient voltage induced in a coil when the magnetized specimen is moved relative to the coil or vice versa.

ball jasper (a) Jasper showing a concentric banding of red and yellow. (b) Jasper occurring in spherical masses.

ballon A rounded, dome-shaped hill, formed either by erosion or by uplift. Etymol: French, "balloon".

ballstone A nodule or large rounded lump of rock in a stratified unit; specif. an ironstone nodule in a coal measure. See also: *iron ball.* Syn: *ball.*

ball structure (a) A ball-shaped primary sedimentary structure. See also: *ball-and-pillow structure; balled-up structure.* (b) The structure of ball coal.

bally A term used in northern California for a mountain. It is thought to be a corruption of "buli", an American Indian term for mountain. Syn: *bolly.*

ballycadder A syn. of *icefoot.* Variant and shortened forms: *bellicatter; catter; cadder.* Rarely used.

balm A concave cliff or precipice, forming a shelter beneath the overhanging rock; a cave. Etymol: Celtic.

balneology The science of the healing qualities of baths, esp. with natural mineral waters.

baltimorite A grayish-green and silky, fibrous, or splintery serpentine mineral found near Baltimore, Md.; *antigorite.*

bambollaite A mineral: $Cu(Se,Te)_2$. Also spelled: *bombollaite.*

banakite A basaltic rock composed of olivine and clinopyroxene phenocrysts in a groundmass of labradorite with alkali feldspar rims, olivine, clinopyroxene, some leucite, and possibly quartz. Banakite grades into *shoshonite* with an increase in the olivine and clinopyroxene and with less alkali feldspar, and into *absarokite* with more olivine and clinopyroxene. It was named by Iddings in 1895 from the Bannock (or Robber) Indians.

banalsite A mineral of the feldspar group: $BaNa_2Al_4Si_4O_{16}$.

banana hole A term used in the Bahamas for a sinkhole, in which it is customary to cultivate bananas and sugar cane. See also: *blue hole.*

banatite A *quartz diorite* containing alkali feldspar. The term has been variously defined since its original usage and is now obsolescent (Johannsen, 1939, p. 242).

banco A term applied in Texas to the part of a stream channel or flood plain cut off and left dry by a change in the stream course; an oxbow lake. Etymol: Spanish, "sandbank, shoal, bench".

band [coal] *dirt band [coal].*

band [glaciol] *glacier band.*

band [phys] A frequency or wavelength interval, e.g. the infrared band, comprising electromagnetic radiation ranging in wavelength from 0.7 μm to 1.0 mm.

band [stratig] (a) A thin stratum with a distinctive lithology or color. (b) A widespread thin stratum that is useful in correlating strata. (c) A deprecated term for any bed or stratum of rock (BSI, 1964, p. 5).

bandaite A *dacite* containing labradorite or bytownite. It was named by Iddings in 1913 for Bandai San, Japan. Not recommended usage.

banded Said of a vein, sediment, or other deposit having alternating layers of ore that differ in color or texture and that may or may not differ in mineral composition, e.g. banded *iron formation.* Cf: *ribbon [ore dep].*

banded agate An *agate* whose various colors (principally different tones of gray, but also white, pale and dark brown, bluish, and other shades) are arranged in delicate parallel alternating bands or stripes of varying thicknesses. The bands are sometimes straight but more often wavy or zigzag and occasionally concentric; they may be sharply demarcated or grade imperceptibly into one another. Banded agate is formed by deposits of silica (from solutions intermittently supplied) in irregular cavities in rocks, and it derives its concentric pattern from the irregularities of the walls of the cavity. Cf: *onyx.*

banded coal Heterogeneous coal, containing bands of varying luster. Banded coal is usually bituminous although banding occurs in all ranks of coal. Cf: *banded ingredients.* See also: *bright-banded coal; dull-banded coal; intermediate coal.*

banded constituents *banded ingredients.*

banded differentiate An igneous rock which, in outcrop, shows bands that reflect layers of differing chemical or mineral composition, usually an alteration of two rock types; a *layered intrusion.* The structure has been attributed to rhythmic crystal settling during convection.

banded gneiss A regularly layered metamorphic or composite rock with alternating layers of different composition and/or texture. Thickness of individual layers is usually not more than a few meters (Dietrich, 1960, p. 36).

banded hematite quartzite A term used in India and Australia for *iron formation.* See also: *banded quartz-hematite.*

banded ingredients Vitrain, clarain, fusain, and durain as they appear as macroscopically visible and separable bands of varying luster in *banded coal.* Cf: *lithotype.* Syn: *banded constituents; primary-type coal; rock type [coal].*

banded iron formation *Iron formation* that shows marked banding, generally of iron-rich minerals and chert or fine-grained quartz. Abbrev: *bif.*

banded ironstone A term used in South Africa for *iron formation* consisting essentially of iron oxides and chert occurring in prominent layers or bands of brown or red and black. This usage of the term *ironstone* is at variance with that applied in the U.S. and elsewhere.

banded peat Peat that consists of alternating bands of plant debris and sapropelic matter. Cf: *mixed peat; marsh peat.*

banded quartz-hematite A syn. of *itabirite.* See also: *banded hematite quartzite.*

banded structure An outcrop feature developed in igneous and metamorphic rocks as a result of alternation of layers, stripes, flat lenses, or streaks differing conspicuously in mineral composition and/or texture. See also: *banding [ign]; banding [meta].*

banding [glaciol] The occurrence of a layered structure in glacier ice, due to alternating layers of coarse-grained and fine-grained ice or bubbly and clear ice. Nonpreferred syn: *foliation [glaciol].*

banding [ign] The appearance of *banded structure* in an outcrop of igneous or metamorphic rock as a result of *layering.* It may be produced by such processes as flow of heterogeneous material (e.g. *flow layering* of rhyolites) or successive deposition of layers of different materials (cf: *phase layering*). Although the term strictly describes the appearance of a two-dimensional feature on a rock exposure, it is often used for any "roughly planar heterogeneity in igneous rocks, whatever its origin" (Wager and Brown, 1967, p. 5).

banding [meta] A *banded structure* of metamorphic rocks consisting of nearly parallel bands having different textures or minerals or both. It may be produced by incomplete segregation of constituents during recrystallization or may be inherited from bedding in sediments or from layering in igneous rocks. Cf: *ribbon [petrology].*

banding [sed] Thin bedding produced by deposition of different materials in alternating layers, and conspicuous in a cross-sectional appearance of laminated sedimentary rocks; e.g. *ribbon banding.*

band spectrum A spectrum that appears to be a number of bands because the array of intensity values occurs over broad ranges of wavelengths of the ordering variable. The term is also applied to lines in a *line spectrum* that are grouped closely and cannot be resolved by available instruments, so that the groups of lines appear as bands. An optical band spectrum arises mainly in molecular transitions.

bandwidth A range of wavelengths or frequencies, such as that to which a detector responds.

bandylite A dark-blue tetragonal mineral: $Cu_2(B_2O_4)Cl_2 \cdot 4H_2O$.

bangar *bhangar.*

bank [coast] (a) An *embankment;* a *sandbank.* (b) A long, narrow island along the Atlantic coast of the U.S., composed of sand, and forming a barrier between the inland lagoon or sound and the ocean. (c) A *shoal,* e.g. Georges Bank. (d) An obsolete term for a *seacoast.* (e) The rising ground bordering a sea.

bank [geomorph] (a) A steep slope or face, as on a hillside, usually of sand, gravel, or other unconsolidated material; rarely of bedrock. (b) A term used in northern England and in Scotland for a hill or a hillside. (c) A term used in South Africa for a moderately high scarp (up to 500 m) consisting of resistant rock layers and forming a high hill or a low mountain, often occurring in groups of two or more with broad longitudinal valleys in between. The term is most often used in the plural: *banke.* Etymol: Afrikaans.

bank [lake] (a) The sharply rising ground, or abrupt slope, bordering a lake. (b) The scarp of a littoral shelf of a lake. (c) Shoal bottom of a lake.

bank [mining] A coal deposit; the surface or face of a coal deposit that is being worked.

bank [oceanog] A relatively flat-topped elevation of the sea floor at shallow depth (generally less than 200 m), typically on the continental shelf or near an island.

bank [sed] A moundlike or ridgelike limestone deposit consisting of skeletal matter, largely unbroken, formed in place by organisms such as crinoids or brachiopods that lack the ecologic potential to erect a rigid, wave-resistant structure (Nelson et al., 1962, p. 242). Banks therefore lack the structural framework of *organic reefs,* and may develop in cold or deep waters far from true reefs. See also: *submarine bank; mudbank; shell bank; lithoherm.* Syn: *organic bank.*

bank [streams] The sloping margin of, or the ground bordering, a stream, and serving to confine the water to the natural channel during the normal course of flow. It is best marked where a distinct channel has been eroded in the valley floor, or where there is a cessation of land vegetation. A bank is designated as right or left as it would appear to an observer facing downstream.

bank atoll *pseudoatoll.*

bank-barrier reef (a) A *coral reef* consisting of a coral-algal frame capping a carbonate rubble-and-sand bar that previously accumulated on a shallow limestone platform. (b) An organic reef intermediate between *fringing reef* and *barrier reef* types; obsolete.

bank caving The slumping or sliding into a stream channel of masses of sand, gravel, silt, or clay, caused by a highly turbulent current undercutting or undermining the channel wall on the outside of a stream bend.

banke Plural of *bank,* an escarpment in South Africa.

banker An Australian term for a stream flowing full to the top of its banks.

banket A general term for a compact, siliceous conglomerate of vein-quartz pebbles of about the size of a pigeon's egg, embedded in a quartzitic matrix. The term was originally applied in the Witwatersrand area of South Africa to the mildly metamorphosed gold-bearing conglomerates containing muffin-shaped quartz pebbles and resembling an almond cake made by the Boers. Etymol: Afrikaans, "a kind of confectionary".

bankfull discharge The discharge at *bankfull stage.*

bankfull stage The elevation of the water surface of a stream flowing at *channel capacity.* Discharge at this stage is called *bankfull discharge.*
bank gravel Gravel found in natural deposits, usually more or less intermixed with sand, silt or clay. Syn: *pit run.*
bank-inset reef A coral reef situated on a submarine flat (such as the shelf of a continent or island, or an offshore bank), well within its locally unrimmed outer margin (Kuenen, 1950, p.426). Cf: *bank reef.*
bank reef Any large reef growth, generally irregular in shape, developed over submerged highs (whether of tectonic or other origin), and more or less completely surrounded by water too deep to support the growth of reef-forming organisms (Henson, 1950, p. 227). See also: *bank-inset reef; shoal reef; shelf-edge reef.*
bank stability The quality of permanence or resistance to change in the slope and contour of the bank of a stream. It can be attained by benching, growth of vegetation, and artificial protections such as retaining walls, drainage systems, and fences. See also: *slope stability.*
bank storage Water absorbed and retained in permeable material adjacent to a stream during periods of high water and returned as effluent seepage or flow during periods of low water. Syn: *lateral storage.*
banner bank *tail [coast].*
bannisterite A mineral: $(Na,K)(Mn,Fe,Al)_5(Si,Al)_6O_{15}(OH)_5 \cdot 2H_2O$.
banquette An embankment at the toe of the land side of a levee, constructed to protect the levee from sliding when saturated with water.
baotite A tetragonal mineral: $Ba_4(Ti,Nb)_8Si_4O_{28}Cl$.
bar [coast] A generic term for any of various elongate offshore ridges, banks, or mounds of sand, gravel, or other unconsolidated material, submerged at least at high tide, and built up by the action of waves or currents on the water bottom, esp. at the mouth of a river or estuary, or at a slight distance from the beach. A bar commonly forms an obstruction to water navigation. Cf: *barrier.*
bar [eco geol] (a) Any band of hard rock, such as a vein or dike, crossing a lode. (b) A hard band of barren rock crossing a stream bed. (c) A mass of inferior rock in a workable deposit of granite. (d) A fault across a coal seam or orebody. (e) A banded ferruginous rock; specif. *jasper bar; jaspilite.*
bar [paleont] (a) The slender shaft of a compound conodont element, commonly bearing denticles, and having an anterior, lateral, or posterior position. Also, any conodont with discrete denticles and with a single large denticle near one end. Cf: *blade.* (b) An elongate, thin skeletal element in archaeocyathids, rectangular in section.
bar [streams] A ridgelike accumulation of sand, gravel, or other alluvial material formed in the channel, along the banks, or at the mouth, of a stream where a decrease in velocity induces deposition; e.g. a *channel bar* or a *meander bar.* See also: *river bar.*
baraboo An ancient monadnock that was buried and later re-exposed by partial erosion of overlying strata. Type locality: Baraboo Ridge, Wisc.
barachois A term used in the Gulf of St. Lawrence region for a lagoon.
bararite A hexagonal low-temperature mineral: $(NH_4)_2SiF_6$. Cf: *cryptohalite.*
baratovite A monoclinic, pseudohexagonal mineral: $KCa_8Li_2(Ti,Zr)_2Si_{12}O_{37}F$.
Barbados earth A fine-grained siliceous deposit of Miocene age occurring in Barbados, West Indies, and containing abundant radiolarian remains. It was formed originally in deep water and later upraised above sea level.
bar beach A syn. of *barrier beach.* The term is not acceptable.
barbed drainage pattern A drainage pattern produced by tributaries that join the main stream in sharp *boathook bends* that point upstream; it is usually the result of stream piracy that has reversed the of flow of the main stream.
barbed tributary A stream that joins the main stream in an upstream direction, forming a sharp bend that points upstream and an acute angle that points downstream at the point of junction.
barbertonite A rose-pink to violet hexagonal mineral: $Mg_6Cr_2(CO_3)(OH)_{16} \cdot 4H_2O$. It is dimorphous with stichtite.
barbierite A name formerly applied to a hypothetical high-temperature monoclinic form of albite, and later changed to *monalbite.* The name was originally applied to a mineral later shown to be finely twinned microcline with about 20% of unmixed albite.
barbosalite A black mineral: $Fe^{+2}Fe_2^{+3}(PO_4)_2(OH)_2$. It is also known as "ferrous ferric lazulite".
barcan *barchan.*
barcenite A mixture of stibiconite and cinnabar.
barchan An isolated crescent-shaped sand dune lying transverse to the direction of the prevailing wind, with a gently sloping convex side facing the wind, wings or horns of the crescent pointing downwind, and a steep concave leeward slope inside the horns; it can grow to heights of greater than 30 m and widths up to 350 m from horn to horn. A barchan forms on a flat, hard surface where the sand supply is limited and the wind is constant with only moderate velocity. It is among the commonest of the dune types, characteristic of very dry, inland desert regions the world over. Etymol: Russian *barkhan,* from Kirghiz; originally a Turkish word meaning a "sand hill" in central Asia. Cf: *parabolic dune.* See also: *snow barchan; ice barchan.* Syn: *barchan dune; barcan; barchane; barkan; barkhan; horseshoe dune; crescentic dune.*
barchan dune *barchan.*
barchane A French variant of *barchan.*
bar diggings A term applied in the western U.S. to diggings for gold or other precious minerals located on a bar or in the shallows of a stream, and worked when the water is low.
bare ice Ice without a snow cover.
bare karst *naked karst.*
bar finger An elongated, lenticular body of sand underlying, but several times wider than, a distributary channel in a bird-foot delta; it is produced and lengthened by the seaward advance of the *lunate bar* at the distributary mouth. Examples in the Mississippi River delta are as much as 30 km long, 8 km wide, and 80 m thick. Syn: *bar-finger sand; finger bar.*
bar-finger sand (a) A deposit of sand in the form of a bar finger. (b) *bar finger.*
bariandite A mineral: $V_2O_4 \cdot 4V_2O_5 \cdot 12H_2O$.
baricite A monoclinic mineral: $(Mg,Fe^{+2},Fe^{+3})_3(PO_4)_2(OH_8 \cdot (8x)H_2O$.
baric type The inferred geothermal gradient corresponding to the different *metamorphic facies series.* Miyashiro (1973) recognizes three baric types: low-, medium-, and high-pressure.
baring *overburden [ore dep].*
barite A white, yellow, or colorless orthorhombic mineral: $BaSO_4$. Strontium and calcium are often present. Barite occurs in tabular crystals, in granular form, or in compact masses resembling marble, and it has a specific gravity of 4.5. It is used in paint, drilling mud, and as a filler for paper and textiles, and is the principal ore of barium. Syn: *barytes; heavy spar; cawk.*
barite dollar A term used esp. in Texas and Oklahoma for a small disk-shaped mass of barite formed in a sandstone or sandy shale.
barite rosette A *rosette* consisting of a cluster or aggregate of tabular sand-filled crystals of barite, usually forming in sandstone. Syn: *barite rose; petrified rose.*
bark Outermost tissues of woody stems, external to the cambium.
barkan *barchan.*
barkevikite A brownish or velvet-black monoclinic mineral of the amphibole group, near arfvedsonite in composition and appearance.
barkhan Var. of *barchan.* Etymol: Russian.
bar lake A lake with a sandbar across its outlet. Examples occur along the east coast of Lake Michigan.
barnesite A mineral: $Na_2V_6O_{16} \cdot 3H_2O$. It is the sodium analogue of hewettite.
Barneveld North American stage: upper Middle Ordovician (above Wilderness, below Edenian).
baroclinic Adj. of *baroclinity.*
baroclinity In oceanography, the condition of stratification of a fluid where surfaces of constant pressure intersect with surfaces of constant density. Adj: *baroclinic.* See also: *barotropy.*
barograph A *barometer* that makes a continuous record of changes in atmospheric pressure. It is usually an aneroid type.
barometer An instrument that is used to measure atmospheric pressure. It may be either a *mercury barometer* or an *aneroid barometer.* See also: *barograph.*
barometric altimeter An instrument that indicates elevation or height above sea level, or above some other reference level, by measuring the weight of air (atmospheric pressure) on the instrument. Also called *aneroid barometer.* Syn: *pressure altimeter.*
barometric efficiency The ratio of the fluctuation of water level in a well to the change in atmospheric pressure causing the fluctuation, expressed in the same units such as feet of water. Symbol:

B. Cf: *tidal efficiency.*

barometric elevation An elevation above mean sea level estimated by the use of a barometer to measure the difference in air pressure between the point in question and a reference base of known value.

barometric leveling A type of indirect leveling in which differences of elevation are determined from differences of atmospheric pressure observed with altimeters or barometers. Cf: *thermometric leveling.*

barometric pressure *atmospheric pressure.*

barometric tendency *pressure tendency.*

barophilic Said of marine organisms that live under conditions of high pressure.

baroque adj. Said of a pearl, or of a tumble-polished gem material, that is irregular in shape.—n. A baroque pearl.

baroque dolomite Dolomite characterized by large crystal size (commonly 4-16 mm), opaque white color (caused by abundant fluid inclusions and excessive cleavage), curving or saddle-shaped crystal faces, and undulose extinction. Common in sulfide ore deposits, veins, and evaporite-related sedimentary strata. Syn: *white sparry dolomite* (Folk & Assereto, 1974). Origin: In allusion to the curved, elaborate form, as in baroque art.

baroseismic storm Strong microseisms caused by barometric changes.

barotropic Adj. of *barotropy.*

barotropy In oceanography, the state of zero *baroclinity;* the coincidence of surfaces of constant density with those of constant pressure. Adj: *barotropic.*

bar plain A term introduced by Melton (1936, p. 594 & 596) for a relatively smooth flood plain with neither a low-water channel nor an alluvial cover, and characterized by a network of elongate and irregularly sized "bars" built from "the tractional and suspended load in the declining stages of the last flood". Cf: *meander plain; covered plain.*

barrage A somewhat archaic syn. of *dam.* Etymol: French, "a barring".

barranca (a) Var. of *barranco.* (b) A large rift in a piedmont glacier or in an ice shelf. Cf: *donga [glaciol].*

barranco [geomorph] (a) A term used in the SW U.S. for a deep, steep-sided, usually rock-walled ravine, gorge, or small canyon, and for a deep cleft, gully, or arroyo made by a heavy rain. (b) A term used in the SW U.S. for a steep bank or a precipice; in New Mexico it is equivalent to *cliff.*—Etymol: Spanish. Cf: *quebrada.* Syn: *barranca.*

barranco [volc] A deep, steep-sided drainage valley on the slope of a volcanic cone, formed by erosion and coalescence of smaller channels. Barrancos form a radiating pattern around a volcanic cone. Etymol: Spanish *barranca,* "ravine".

barrandite A pale-gray orthorhombic mineral: $(Fe,Al)PO_4 \cdot 2H_2O$. It is isomorphous with, and intermediate in composition between, strengite and variscite.

barred basin *restricted basin.*

barrel As used in the petroleum industry, a volumetric unit of measurement equivalent to 42 U.S. gallons (158.76 liters).

barrel copper Pieces of native copper occurring in sizes large enough to be extracted from the gangue, and of sufficient purity to be smelted without mechanical concentration. Syn: *barrel work.*

barrel work A syn. of *barrel copper,* used in the Lake Superior mining region.

Barremian European stage: Lower Cretaceous (above Hauterivian, below Aptian).

barren A word, usually used in the plural, for rugged and unproductive land that is devoid of significant vegetation compared to adjacent areas because of environmental factors such as adverse climate, poor soil, or wind.

barren interzone An interval lacking in fossils between successive *biozones.* It is referred to informally by reference to the adjacent biozones, e.g. *Exus parvus* to *Exus magnus* barren interzone (ISG, 1976, p. 49). Cf: *barren intrazone.*

barren intrazone A barren interval of substantial thickness within a *biozone,* e.g. the barren intrazone near the top of the *Exus albus assemblage-zone* (ISG, 1976, p. 49). Cf: *barren interzone.*

barren zone A quasi-biostratigraphic unit representing a part of the stratigraphic succession devoid of all diagnostic fossils or representatives of the taxonomic categories on which the remainder of the succession is subdivided.

barrerite An orthorhombic mineral of the zeolite group: $(Na,K,Ca)_7(Al_2Si_7)O_{18} \cdot 7H_2O$.

barrier [coast] An elongate offshore ridge or mass, usually of sand, rising above the high-tide level, generally extending parallel to, and at some distance from, the shore, and built up by the action of waves and currents. Examples include *barrier beach* and *barrier island.* Cf: *bar.*

barrier [ecol] A condition, such as a topographic feature or physical quality, that tends to prevent the free movement and mixing of populations or individuals.

barrier [glaciol] *ice barrier [glaciol].*

barrier [grd wat] *ground-water barrier.*

barrier bar A syn. of *longshore bar.* The term is not acceptable.

barrier basin A basin produced by the formation of a natural dam or barrier; it may contain a *barrier lake.*

barrier beach A narrow, elongate sand ridge rising slightly above the high-tide level and extending generally parallel with the shore, but separated from it by a lagoon (Shepard, 1952, p. 1904) or marsh; it is extended by longshore drifting and is rarely more than several kilometers long. See also: *barrier island.* This feature was termed an *offshore bar* by Johnson (1919, p. 259 & 350). Syn: *offshore barrier; offshore beach; bar beach.*

barrier chain A series of barrier islands, barrier spits, and barrier beaches extending along a coast for a considerable distance (Shepard, 1952, p. 1908).

barrier flat A relatively flat area, often occupied by pools of water, separating the exposed or seaward edge of a barrier from the lagoon behind it.

barrier ice (a) A term used by Robert F. Scott in 1902 for the ice constituting the Antarctic ice shelf (which was then called an "ice barrier"). The syn. *shelf ice* seems to be more commonly used for the actual ice itself. (b) A term sometimes used improperly as a syn. of *ice shelf.*

barrier island (a) A long, narrow coastal sandy island, representing a broadened *barrier beach* that is above high tide and parallel to the shore, and that commonly has dunes, vegetated zones, and swampy terranes extending lagoonward from the beach. Also, a long series of barrier beaches. Examples include Long Beach, N.J., and the Lido in Venice. This feature was termed an *offshore bar* by Johnson (1919). (b) A detached portion of a barrier beach between two inlets (Wiegel, 1953, p. 5).

barrier-island marsh A salt or brackish marsh on the low inner margin of a barrier island.

barrier lagoon (a) A *lagoon* that is roughly parallel to the coast and is separated from the open ocean by a strip of land or by a barrier reef. (b) A *lagoon* encircled by coral islands or coral reefs, esp. one enclosed within an atoll.

barrier lake A lake whose waters are impounded by the formation of a natural dam or barrier, such as a landslide, alluvium in a delta, glacial moraine, ice, or lava; also, a freshwater lagoon separated from a lake by a shore dune or a *sandbank.*

barrier reef A long, narrow coral reef roughly parallel to the shore and separated from it by a lagoon of considerable depth and width. It may enclose a volcanic island (either wholly or in part), or it may lie a great distance from a continental coast (such as the Great Barrier Reef off the coast of Queensland, Australia). Generally, barrier reefs follow the coasts for long distances, often with short interruptions, termed *passes* or channels. Cf: *fringing reef.*

barrier spit A *barrier island* or *barrier beach* that is connected at one end to the mainland.

barrier spring A spring resulting from the diversion of a flow of ground water over or underneath an impermeable barrier in the floor of a valley (Schieferdecker, 1959, term 0308). Cf: *contact spring.*

barrier well A type of recharge well, generally one of several along a line, used to inject water of usable quality to build up a ridge of such water between wells used for water supply and a potential source of contamination, as between water-supply wells and the salt-water front in a coastal area.

barringerite A meteorite mineral: $(Fe,Ni)_2P$.

barringtonite A mineral: $MgCO_3 \cdot 2H_2O$ (?).

Barrovian metamorphic zone In complexes of regional dynamothermal metamorphism, one of the belts of progressively increasing metamorphic grade, based on the first appearance of the index minerals chlorite, biotite, almandine, staurolite, kyanite, and sillimanite (Tilley, 1925). This now classical sequence, representing the commonest type of regional metamorphism, is named after George Barrow who in 1893 first described and mapped zones of progressive regional metamorphism in the Grampian Highlands of Scotland.

Barrovian-type facies series Rocks produced in the most common type of dynamothermal regional metamorphism, characterized by the appearance of metamorphic zones of the greenschist and amphibolite facies. Typical index minerals (in order of increasing metamorphic grade) are chlorite - biotite - garnet (almandine) - staurolite - kyanite - sillimanite (but no andalusite). Cf: *Saxonian-type facies series; Idahoan-type facies series.* See also: *Barrovian metamorphic zone.*

barrow pit A term used chiefly in western U.S. for a *borrow pit;* esp. a ditch dug along a roadway to furnish fill for the road.

barsanovite *eudialyte.*

barshawite An *ijolite* containing alkali feldspar and andesine; named by Johannsen (1931) for Barshaw, Renfrewshire, Scotland. Not recommended usage.

Barstovian North American continental stage: upper Lower Miocene (above Hemingfordian, below Clarendon).

bar theory A theory advanced by Ochsenius in 1877 to account for thick deposits of evaporites. It assumes a lagoon separated from the ocean by a bar, in an arid or semiarid climate. As water is lost by evaporation, additional water of normal salinity flows in from the ocean. Because some water in the lagoon is evaporating, the salinity there constantly increases, and it finally reaches a point where gypsum, salt, and other evaporites are deposited.

barthite A variety of austinite containing copper. It was formerly thought to be a variety of veszelyite.

Bartonian European stage: Eocene (above Priabonian, below Rupelian). Essentially equivalent to Biarritzian.

barycenter The location of the center-of-mass of a collection of bodies.

barylite A colorless mineral: $BaBe_2Si_2O_7$.

barysilite A white mineral: $Pb_8Mn(Si_2O_7)_3$.

barysphere The interior of the Earth beneath the lithosphere, including both the mantle and the core. However, it is sometimes used to refer only to the core or only to the mantle. Cf: *pyrosphere.* Syn: *centrosphere.*

barytes A syn. of *barite.* Also spelled: *baryte; barytine; barytite.*

barytocalcite (a) A mineral: $BaCa(CO_3)_2$. It is the monoclinic dimorph of alstonite. (b) A mixture of calcite and barite.

barytolamprophyllite A mineral: $(Na,K)_6(Ba,Ca,Sr)_3(Ti,Fe)_7Si_8O_{32}(O,OH,F,Cl)_4$.

basal adj. Pertaining to, situated at, or forming the base of an animal structure; e.g. referring to the aboral part of the theca of an echinoderm, or pertaining to the under or reverse side of an incrusting or freely growing bryozoan colony.—n. A basal structure of an animal; esp. a *basal plate* of an echinoderm.

basal arkose An arkosic sandstone basal to a sedimentary sequence, resting unconformably on a granitic terrane; the arkosic equivalent of a granitic *basal conglomerate.* It may grade downward into sedentary or residual arkose (Pettijohn, Potter & Siever, 1972, p. 163).

basal cavity A pit or concavity about which a conodont element was built through accretion of lamellae, opening onto the aboral side, and present on all true conodont elements. See also: *cup; attachment scar.* Improper syn: *pulp cavity; escutcheon.*

basal cleavage Mineral cleavage parallel to the basal pinacoid; e.g. in molybdenite.

basal clinker The zone of *autobreccia* that forms the base of an *aa* flow.

basal complex *basement.*

basal conglomerate A well-sorted, lithologically homogeneous conglomerate that forms the bottom stratigraphic unit of a sedimentary series and that rests on a surface of erosion, thereby marking an unconformity; esp. a coarse-grained beach deposit of an encroaching or transgressive sea. It commonly occurs as a relatively thin widespread or patchy sheet, interbedded with quartz sandstone. Cf: *marginal conglomerate.*

basal diaphragm A skeletal partition extending across the zooidal chamber in many stenolaemate bryozoans. It forms the inner end or floor of the living chamber (Boardman, 1971, p. 5).

basal disk An expanded basal part by which certain stalked sessile organisms are attached to the substrate; specif. the aboral fleshy part of a scleractinian coral polyp, typically subcircular in outline, that closes off the lower end of the polyp; or the *proancestrula* in stenolaemate bryozoans. Cf: *oral disk.* Also spelled: *basal disc.*

basal funnel An excavated conelike basal plate whose tip fits into the basal cavity of a conodont.

basal glide In glaciology, translation or slip along basal planes of ice crystals subject to plastic deformation. See also: *crystal gliding.*

basal granule A dotlike body forming part of the neuromotor system in tintinnids.

basal ground water A term that originated in Hawaii and refers to a major body of ground water floating on and in hydrodynamic equilibrium with salt water. Syn: *basal water.*

basalia (a) A circlet of basal plates in the theca of a blastoid, normally consisting of two large zygous plates and one small azygous plate. (b) Prostalia (spicules) protruding from the base of a sponge and serving to stabilize or anchor it to the substrate.

basal ice Ice at the bottom of a glacier or ice sheet.

basal lamina The generally encrusting, calcified colony wall of stenolaemate bryozoans, extending to the growing edge of the colony. See also: *median lamina.*

basal leaf cross Broad wings on radial spines of acantharian radiolarians.

basal lobe One of two lobes set off by furrows in the posterior and lateral parts of the glabella of a trilobite, just in advance of the occipital ring.

basal pinacoid In all crystals except those of the isometric system, the {001} pinacoid. Cf: *front pinacoid; side pinacoid.* Syn: *basal plane.*

basal plane *basal pinacoid.*

basal plate (a) One of a circlet of certain chiefly ventral skeletal plates of an echinoderm; e.g. a plate composing the aboral end of the theca of a blastoid and articulated to the stem aborally and to radial plates on oral borders, or an interambulacral plate just below the arm-bearing radial plates of a crinoid. Syn: *basal.* (b) A thin skeletal plate formed initially as a part of a corallite immediately below the basal disk of the polyp of a scleractinian coral, from which the septa and walls begin to extend upward and outward. (c) A laminated platelike structure of organic material attached to the aboral side of a conodont element along the attachment scar or basal cavity.

basal platform A term used by Linton (1955) as a syn. of *basal surface.*

basal pore One of the pores outlined by connector bars joining the basal ring of a radiolarian skeleton (as in the subfamily Trissocyclinae).

basal ring A ring at or below the base of the sagittal ring of a radiolarian skeleton, commonly with basal spines projecting from it (as in the subfamily Trissocyclinae).

basal sapping The undercutting, or breaking away of rock fragments, along the headwall of a cirque, due to frost action at the bottom of the *bergschrund.* Syn: *sapping [glac geol].*

basal shear stress *Shear stress* that acts at the base of a glacier or ice sheet.

basal sliding (a) The sliding of a glacier on its bed. (b) The velocity or speed of sliding of a glacier on its bed.—Syn: *basal slip.*

basal slip *basal sliding.*

basal slope *wash slope.*

basal surface The generalized boundary between weathered and unweathered rock, or the lower limit to active weathering (Ruxton & Berry, 1959). This contact, which may be regular or irregular, indicates a very rapid or sudden change upward into the base of the mass of weathering debris. Syn: *basal platform; weathering front.*

basalt [ign] A general term for dark-colored mafic igneous rocks, commonly extrusive but locally intrusive (e.g. as dikes), composed chiefly of calcic plagioclase and clinopyroxene; the fine-grained equivalent of *gabbro.* Nepheline, olivine, orthopyroxene, and quartz may be present in the *CIPW norm,* but not all simultaneously: nepheline and olivine can occur together, as can olivine and orthopyroxene, and orthopyroxene and quartz, but nepheline does not coexist with orthopyroxene or quartz, nor quartz with nepheline or olivine. These associations and incompatibilities are discussed by Yoder and Tilley (1962) and by Muir and Tilley (1961). Cf: *basaltic rocks.*

basalt [lunar] An igenous rock from the Moon that is composed chiefly of nearly equal amounts of augite, plagioclase, and ilmenite. The plagioclase is characteristically highly calcic (An_{80}-An_{90}). Lunar basalt contains more titanium dioxide, rare-earth elements, zirconium, and less nickel than terrestrial basalt.

basalt glass *sideromelane.*

basal thrust plane *sole fault.*

basaltic Pertaining to, composed of, containing, or resembling basalt; e.g. "basaltic lava".

basaltic andesite "Rocks of the calc-alkaline kindred that are intermediate, in one or more respects, between typical basalt and typical andesite" (Coats, 1968, p. 689-690). The term has been widely used but never precisely defined; its use is not recommended.

basaltic dome *shield volcano.*

basaltic hornblende A black or brown variety of hornblende rich in ferric iron (ferrous iron having been oxidized) and occurring in *basalt* and other iron-rich basic igneous (volcanic) rocks; a type of *brown hornblende* characterized optically by strong pleochroism and birefringence, high refractive indices, and a small extinction angle. Syn: *lamprobolite; oxyhornblende; basaltine.*

basaltic layer A syn. of *sima*, so named for its supposed petrologic composition. It is also called the *gabbroic layer,* and may be equivalent to the *Conrad layer.* Cf: *granitic layer.* A layer is sometimes termed "basaltic layer" if it possesses the appropriate seismic velocity ($\sim$6.5-7.0 km/s), although nothing may be known about its composition.

basaltic plateau *lava plateau.*

basaltic rocks A general term incorporating fine-grained compact dark-colored extrusive igneous rocks such as *basalt [ign], diabase, dolerite,* and dark-colored *andesite.*

basal till A firm clay-rich till containing many abraded stones dragged along beneath a moving glacier and deposited upon bedrock or other glacial deposits. Cf: *lodgment till.* Syn: *accretion till.*

basaltine n. (a) *basaltic hornblende.* (b) *augite.*—adj. *basaltic.*

basaltite An older term revived by the International Geological Congress in 1900 and applied to olivine-free *basalt.* Not recommended usage.

basalt obsidian *sideromelane.*

basal tunnel A water-supply tunnel excavated along the basal water table in basaltic areas, esp. Hawaii. Cf: *Maui-type well.*

basaluminite A white mineral: $Al_4(SO_4)(OH)_{10}\cdot 5H_2O$. It occurs in veinlets lining crevices in ironstone. Cf: *felsöbanyite.*

basal water *basal ground water.*

basal water table The water table of a body of basal ground water.

basanite [ign] A group of basaltic rocks characterized by calcic plagioclase, clinopyroxene, a feldspathoid (nepheline, leucite), and olivine; also, any rock in that group. Without the olivine, the rock would be called a *tephrite.* The term was coined by Brongniart in 1813.

basanite [sed] (a) A *touchstone* consisting of flinty jasper or finely crystalline quartzite. Syn: *Lydian stone.* (b) A black variety of jasper.

basanitoid n. A term proposed by Bücking in 1881 for a group of rocks intermediate in composition between *basanite* and *basalt* (Johannsen, 1939, p. 243), i.e. having the chemical composition of basanite but without modal feldspathoids and with a glassy groundmass.—adj. Said of a rock resembling basanite, having normative but no modal feldspathoids.

basculating fault *wrench fault.*

base [eng] That part of an engineering structure resting on the subgrade, on supporting soil, or on solid rock; the *base course.*

base [gem] *pavilion.*

base [ign] *mesostasis.*

base [paleont] (a) The aboral end of an echinoderm theca. In cystoids the term is restricted by some to the columnar facet but extended by others to include thecal plates of the basal circlet or aboral circlets. (b) The aboral end of a crinoid calyx, or the part of a crinoid dorsal cup between radial plates and stem, normally composed of basal plates or of basal and infrabasal plates, but in articulate crinoids including the centrale. (c) The area adjacent to the aboral side of a conodont element.

base [petroleum] An informal term for the hydrocarbon series that is dominant in a given crude oil. Cf: *asphalt-base crude; mixed-base crude; paraffin-base crude.*

base [surv] *base line.*

base apparatus Any apparatus (such as wood tubes, metal wires, iron bars, steel rods, or invar tapes) used in geodetic surveying and designed to measure with accuracy and precision the length of a base line in triangulation or the length of a line in a traverse.

base bullion A vein or lode in a gold mine that may or may not contain gold but is recoverable for its silver content. The term is little used.

base-centered lattice A type of *centered lattice* that is centered in one pair of the (001), (010), or (100) faces.

base correction A correction or adjustment of geophysical measurements to express them relative to the values of a *base station.*

base course A bottom layer of coarse gravel or crushed stone, generally of specified character and thickness, constructed on the subgrade or subbase of a highway or structure for the purpose of serving such functions as distributing load, providing drainage, and minimizing frost action. Syn: *base [eng].*

base discharge As used by the U.S. Geological Survey, that discharge above which peak discharge data are published.

base-discordance A term used in seismic stratigraphy to refer to a lack of parallelism between a sequence of strata and its lower boundary, owing to either *onlap* or *downlap* (Mitchum, 1977, p. 206). Cf: *top-discordance.*

base exchange *cation exchange.*

base flow Sustained or fair-weather flow of a stream, whether or not affected by the works of man (Langbein & Iseri, 1960). Cf: *base runoff.*

base-height ratio The ratio between the *air base* and the *flight height* of a stereoscopic pair of aerial photographs.

baselap A term used in seismic stratigraphy to describe the termination of a sequence of strata along its lower boundary where discrimination between *onlap* and *downlap* is difficult or impossible.

base level n. (a) The theoretical limit or lowest level toward which erosion of the Earth's surface constantly progresses but seldom, if ever, reaches; esp. the level below which a stream cannot erode its bed. The general or *ultimate base level* for the land surface is sea level, but *temporary base levels* may exist locally. The base level of eolian erosion may be above or below sea level; that of marine erosion is the lowest level to which marine agents can cut a bottom. Also spelled: *baselevel.* Syn: *base level of erosion.* (b) A curved or planar surface extending inland from sea level, inclined gently upward from the sea and representing the theoretical limit of stream erosion. (c) The surface toward which external forces strive, at which neither erosion nor deposition takes place (Barrell, 1917); a surface of equilibrium.—v. To reduce by erosion to, or toward the condition of a plain at, base level.

base level of deposition The highest level to which a sedimentary deposit can be built (Twenhofel, 1939, p. 8); if built of marine deposits, it coincides with the base level of erosion.

base level of erosion A syn. of *base level.* The term was introduced by Powell (1875, p. 203) for an irregular "imaginary surface, inclining slightly in all its parts toward the lower end of the principal stream", below which the stream and its tributaries were supposed to be unable to erode.

base-level peneplain *peneplain.*

base-level plain A flat surface, area, or lowland produced by the wearing-down of a region to or near its base level; a plain that cannot be materially reduced in elevation by erosion. Cf: *peneplain.*

base line (a) A surveyed line established with more than usual care, which serves as a reference to which surveys are coordinated and correlated. Syn: *base.* (b) The initial measurement in triangulation, being an accurately measured distance constituting one side of one of a series of connected triangles, and used, together with measured angles, in computing the lengths of the other sides. (c) One of a pair of coordinate axes (the other being the *principal meridian*) used in the U.S. Public Land Survey system. It consists of a line extending east and west along the true parallel of latitude passing through the initial point, along which standard township, section, and quarter-section corners are established. (d) An aeromagnetic profile flown at least twice in opposite directions and at the same level, in order to establish a line of reference of magnetic intensities on which to base an aeromagnetic survey. (e) The center line of location of a railway or highway; the reference line for the construction of a bridge or other engineering structure.—Sometimes spelled: *baseline.*

base map (a) A map of any kind showing essential outlines necessary for adequate geographic reference, on which additional or specialized information is plotted for a particular purpose; esp. a topographic map on which geologic information is recorded. (b) *master map.* (c) Obsolete syn. of *outline map.*

basement n. (a) The undifferentiated *complex* of rocks that underlies the rocks of interest in an area. Cf: *basement terrane.* (b) The crust of the Earth below sedimentary deposits, extending downward to the Mohorovičić discontinuity. In many places the rocks of the complex are igneous and metamorphic and of Precambrian age, but in some places they are Paleozoic, Mesozoic, or even Cenozoic. Syn: *basement rock; basal complex; fundamental com-*

plex; basement complex. adj. Said of material, processes, or structures originating or occurring in the basement.

basement complex *basement.*

basement fold A compressional structure (fold or thrust) that formed within a continent and affected the entire thickness of the continental crust. This obsolete term is the English translation of Argand's *plis de fond.*

basement rock *basement.*

basement terrane The lowest mappable mass of rock, generally with complex structure, that underlies the other major rock sequences of a region, specif. of the California Coast Ranges (Berkland et al., 1972, p. 2296). No unconformable upper contact is required. Cf: *basement.*

base metal (a) Any of the more common and more chemically active metals, e.g. lead, copper. (b) The principal metal of an alloy, e.g. the copper in brass.—Cf: *noble metal.*

base net A small *net* of triangles and quadrilaterals, starting from a measured *base line* and connecting with a line of the main scheme of a *triangulation net;* e.g. a triangle formed by sighting a point from both ends of a base line, or two adjacent triangles with the base line common to both. It is the initial figure in a triangulation system.

base of drift A seismic-velocity discontinuity often encountered between glacial drift material and the competent formation beneath. By extension, any similar discontinuity between a layer below the weathering and the topmost competent formation layer.

base of weathering In seismic work, the boundary between a low-velocity surface layer and an underlying, comparatively high-velocity layer. It often corresponds to the water table. It is important in deriving time corrections for seismic records.

base runoff Sustained or fair-weather *runoff [water]* (Langbein & Iseri, 1960). It is primarily composed of effluent ground water, but also of runoff delayed by slow passage through lakes or swamps. The term refers to the natural flow of a stream, unaffected by the works of man. Cf: *base flow; direct runoff.* Syn: *fair-weather runoff; sustained runoff.*

base-saturation percentage The extent to which the adsorption complex of a soil is saturated with exchangeable cations other than hydrogen. It is expressed as a percentage of the total cation-exchange capacity.

base station An observation point used in geophysical surveys as a reference, to which measurements at additional points can be compared. See also: *base correction.*

base surge A ring-shaped cloud of gas and suspended solid debris that moves radially outward at high velocity as a density flow from the base of a vertical explosion column accompanying a volcanic eruption or crater formation by an explosion or hypervelocity impact.

base temperature The temperature in a region of uniform temperature normally found in the lower part of a convecting system.

Bashkirian European stage: Middle Carboniferous (above Namurian, below Moscovian).

basic (a) Said of an igneous rock having a relatively low silica content, sometimes delimited arbitrarily as 44 to 51% to 45 to 52%; e.g. gabbro, basalt. Basic rocks are relatively rich in iron, magnesium, and/or calcium, and thus include most mafic rocks as well as other rocks. "Basic" is one of four subdivisions of a widely used system for classifying igneous rocks based on their silica content: *acidic, intermediate* basic, and *ultrabasic.* Cf: *femic.* (b) Said loosely of any igneous rock composed chiefly of dark-colored minerals.—Cf: *silicic; mafic.* (c) Said of a plagioclase that is calcic.

basic behind In granitization, a zone in which residual mafic components are concentrated.

basic border The marginal area of an igneous intrusion, characterized by a more basic composition than the interior of the rock mass. Cf: *chill zone.* Syn: *mafic margin.*

basic front In granitization, an advancing zone enriched in calcium, magnesium, and iron, which is said to represent those elements in the rock being granitized that are in excess of those required to form granite. During granitization, these elements are believed to be displaced and moved through the rock ahead of the granitization front, to form a zone enriched in minerals such as hornblende and pyroxene. Cf: *basic behind.* Syn: *mafic front; magnesium front.*

basic hydrologic data Data including inventories of land and water features that vary from place to place (e.g. topographic and geologic maps), and records of processes that vary with time and from place to place (e.g. precipitation, streamflow, and ground-water levels) (Langbein & Iseri, 1960). Cf: *basic hydrologic information.*

basic hydrologic information A broader term than *basic hydrologic data,* including surveys and appraisals of the water resources of an area and a study of its physical and related economic processes, interrelations, and mechanisms (Langbein & Iseri, 1960).

basicoronal Pertaining to the corona of an echinoid at the edge of the peristome.

basic plagioclase A variety of plagioclase having relatively low content of SiO_2; e.g. an An-rich member such as bytownite or anorthite.

basic wash A driller's term for material eroded from outcrops of basic igneous rocks (gabbro, basalt) and redeposited to form a rock having approximately the same major mineral constituents as the original rock (Taylor & Reno, 1948, p. 164). Cf: *granite wash.*

basidiospore A *fungal spore* produced by the basidium of a basidiomycete. Such spores with chitinous walls may occur as microfossils in palynologic preparations.

basification Enrichment of a rock in elements such as calcium, magnesium, iron, and manganese.

basimesostasis Said of an augite-rich *mesostasis.*

basin (a) A depressed area with no surface outlet. The term is widely applied, e.g. to a *lake basin,* to a *ground-water basin,* to a shallow depression on the sea floor, to a circular depression on the Moon's surface, or to a *tidal basin.* (b) A *drainage basin* or *river basin.* (c) A low area in the Earth's crust, of tectonic origin, in which sediments have accumulated, e.g. a circular *centrocline* such as the Michigan Basin, a fault-bordered intermontane feature such as the Bighorn Basin of Wyoming, or a linear crustal downwarp such as the Appalachian Basin. Such features were basins at the time of sedimentation but are not necessarily so today. Syn: *structural basin.*

basin-and-range Said of a topography, landscape, or physiographic province characterized by a series of tilted fault blocks forming longitudinal, asymmetric ridges or mountains and broad, intervening basins; specif. the Basin and Range physiographic province in SW U.S. See also: *basin-range structure; basin range.*

basin area For a given stream order *u,* the total area, projected upon a horizontal plane, of a drainage basin bounded by the basin perimeter and contributing overland flow to the stream segment of order *u,* including all tributaries of lower order (Strahler, 1964, 4-48). Symbol: A_u. Cf: *watershed area.* See also: *law of basin areas.*

basin-area ratio Ratio of mean basin area of a given order to the mean basin area of the next lower order within a specified larger drainage basin (Schumm, 1956, p.606). Symbol: R_a.

basin-circularity ratio Ratio of the area of a drainage basin to the area of a circle with the same perimeter as the basin. Symbol: R_c. Syn: *circularity; circularity ratio.*

basin-elongation ratio Ratio of the diameter of a circle having the same area as a drainage basin to the maximum length of that basin. Symbol: R_e. Syn: *elongation ratio.*

basin facies Sediments deposited beyond the outer limits of a land-bordering submarine shelf. See also: *facies* (g).

basining The bending down or settling of part of the Earth's crust in the form of a basin, as by erosion or by solution and transportation of underground deposits of salt or gypsum.

basin length Horizontal distance of a straight line from the mouth of a stream to the farthest point on the drainage divide of its basin, parallel to the principal drainage line (Schumm, 1956, p.612). Symbol: L_b.

basin order The number assigned to an entire drainage basin contributing to the stream segment of a given order and bearing an identical integer designation; e.g. a first-order basin contains all of the drainage area of a first-order stream. See also: *stream order.*

basin peat *local peat.*

basin perimeter Length of the line enclosing the area of a drainage basin. Symbol: P.

basin range A mountain range that owes its elevation and structural form mainly to faulting and tilting of strata and that is surrounded by alluvium-filled basins or valleys. Etymol: from the Great Basin, a region in SW U.S. characterized by fault-block mountains. See also: *basin-and-range.*

basin-range structure Regional geologic structure dominated by generally subparallel fault-block mountains separated by broad alluvium-filled basins; e.g. basin-range faulting characterized by normal fault movements, as in the *basin-and-range* province of SW U.S.

basin relief Difference in elevation between the mouth of a stream and the highest point within, or on the perimeter of, its drainage basin (Strahler, 1952b, p.1119); the maximum relief in the basin. Symbol: H.
basin valley A broad, shallow valley with gently sloping sides.
basiophitic Said of an ophitic rock whose mesostasis is composed of augite. Cf: *oxyophitic; oxybasiophitic.*
basiophthalmite The proximal segment (lowest joint) of the eyestalk of a decapod crustacean, articulating with the *podophthalmite.*
basiphytous Said of a sponge that is attached to the substrate by an encrusting base.
basipinacoderm The pinacoderm delimiting a sponge at the surface of fixation.
basipod The *basis* of a crustacean limb. Syn: *basipodite.*
basis [ign] *mesostasis.*
basis [paleont] The limb segment of a crustacean just distal to the coxa, commonly bearing the exopod and endopod; a membranous or calcareous structure (in nonpedunculate cirripedes) that contacts the substratum. Pl: *bases.* Syn: *basipod.*
basis rami The calcified structure from which a node emanates to start a new branch in a jointed colony of cheilostome bryozoans. It commonly consists of the calcified proximal ends of one or more zooids, the more distal parts of which lie in the node and following internode.
basite A basic igneous rock.
basket-of-eggs topography A landscape characterized by swarms of closely spaced drumlins, distributed more or less en echelon, and commonly separated by small marshy tracts. Syn: *drumlin field.*
bass *batt [coal].*
bassanite A white mineral: $CaSO_4 \cdot 1/2H_2O$. Syn: *vibertite.*
basset An obsolete term for the noun "outcrop" and the verb "to crop out".
basset edge Area of obliquely truncated outcrop, as in cross-bedded strata exposed transverse to their depositional planes.
bassetite A yellow mineral: $Fe(UO_2)_2(PO_4)_2 \cdot 8H_2O$.
bastard adj. (a) Said of an inferior or impure rock or mineral, or of an ore deposit that contains a high proportion of noncommercial material. (b) Said of any metal or ore that gives misleading assays or values. (c) Said of a vein or other deposit close to and more or less parallel to a main vein or deposit, but thinner, less extensive, or of a lower grade.
bastard coal Thin partings of impure coal occurring in the lower part of shale strata immediately overlying a coal seam; any coal with a high ash content. Syn: *batt [coal].*
bastard ganister A silica rock having the superficial appearance of a true *ganister* but characterized by more interstitial matter, a greater variability of texture, and often an incomplete secondary silicification.
bastard quartz (a) A syn of *bull quartz.* (b) A round or spherical boulder of quartz embedded in soft or decomposed rock.
bastard rock A term used in south Wales and in north Staffordshire, England, for an impure sandstone containing thin lenticular layers of shale or coal.
bastard shale *cannel shale.*
bastinite *hureaulite.*
bastion A knob or mass of bedrock projecting into a main glacial valley at the junction with a hanging valley, at or below the level of the hanging valley floor.
bastite An olive-green, blackish-green, or brownish variety of serpentine mineral resulting from the alteration of orthorhombic pyroxene (esp. enstatite), occurring as foliated masses in igneous rocks, and characterized by a schiller (metallic or pearly luster) on the chief cleavage face of the pyroxene. Syn: *schiller spar.*
bastnaesite A greasy, wax-yellow to reddish-brown mineral: $(Ce,La)CO_3(F,OH)$. It occurs in alkaline igneous rocks, esp. *carbonatite,* as at Mountain Pass, Calif. Bastnaesite is the chief U.S. source of rare-earth elements. Also spelled: *bastnäsite.*
bastnaesite-(Y) A hexagonal mineral: $(Y,Ce)(Co_3)F$.
bat *batt [coal].*
batholith A large, generally discordant plutonic mass that has more than 40 sq mi (100 km^2) of surface exposure and no known floor. Its formation is believed by most investigators to involve magmatic processes. Also spelled: *bathylith.* Syn: *abyssolith.*
Bathonian European stage: Middle Jurassic (above Bajocian, below Callovian).
Båth's law A generalization in seismology that the largest aftershock occurring within a few days of a main shock has a magnitude of 1.2 units lower than that of the main shock (Richter, 1958, p.69).
Bath stone A soft, creamy, oolitic limestone, easily quarried and used for building purposes. Type locality: near Bath, England.
bathvillite An amorphous, opaque, very brittle woody resin occurring as fawn-brown porous lumps in torbanite at Bathville, Scotland.
bathy- A prefix meaning "deep".
bathyal Pertaining to the ocean environment or *depth zone* between 200 and 2000 meters; also, pertaining to the organisms of that environment.
bathybenthic Pertaining to the benthos of the bathyal zone of the ocean. Cf: *archibenthic.*
bathydermal Said of deformation or gliding of the lower part of the sialic crust. Cf: *dermal; epidermal.*
bathygenesis Negative or subsident tectonic movement; tectonic lowering of marine basins. It is analogous to *epeirogeny* or positive tectonic movements associated with the continents. Adj: *bathygenic.*
bathygenic Adj. of *bathygenesis.*
bathylimnion The deeper part of a *hypolimnion* characterized by constant rates of heat absorption at different depths. Cf: *clinolimnion.*
bathylith *batholith.*
bathymetric chart A topographic map of the bottom of a body of water (such as the sea floor), with depths indicated by contours (isobaths) drawn at regular intervals.
bathymetric contour *isobath [oceanog].*
bathymetric tint A distinctive shading or coloring of the area between bathymetric contour lines to emphasize the distribution of high and low areas and the range of water depths. Cf: *hypsometric tint.*
bathymetry The measurement of ocean depths and the charting of the topography of the ocean floor.
bathyorographical (a) Pertaining to ocean depths and mountain heights considered together; said of a map that shows both the relief of the land and the depths of the ocean. (b) Pertaining to the description of the relief features on the ocean floor.
bathypelagic Pertaining to the open water of bathyal depth.
bathyscaph A manned, submersible vehicle for deep-sea exploration; it is somewhat navigable, in contrast to a *bathysphere.*
bathyseism A deep-focus earthquake that is instrumentally detected worldwide. The term is little used.
bathysphere (a) A manned submersible sphere that is lowered into the deep ocean by cable for observations; unlike the *bathyscaph,* it is not navigable. (b) A nonrecommended syn. of *barysphere.*
bathythermogram The record (or a photographic print of it) that is made by a *bathythermograph.*
bathythermograph In oceanography, an instrument that records temperature in relation to depth. See also: *bathythermogram.*
batisite A dark-brown orthorhombic mineral: $Na_2BaTi_2(Si_2O_7)_2$.
Batoidea An order, largely marine, of elasmobranch fishes, characterized by flattened body form, expanded pectoral fins, and crushing dentitions; it includes the skates and rays. Range, Upper Jurassic to Recent.
batt [clay] An English term for any hardened clay other than fireclay.
batt [coal] (a) An English term for a compact black carbonaceous shale, which splits into fine laminae and is often interstratified with thin layers of coal or ironstone. Syn: *bass.* (b) *bastard coal.*— Also spelled: *bat.*
battery ore A type of manganese ore, generally a pure crystalline manganese dioxide (pyrolusite or nsutite), that is suitable for use in dry cells.
batture An elevated part of a river bed, formed by gradual accumulation of alluvium; esp. the land between low-water stage and a levee along the banks of the lower Mississippi River. Etymol: Louisiana French, from French *battre,* "to strike upon or against".
batukite A dark-colored extrusive rock composed of phenocrysts of clinopyroxene and minor olivine in a groundmass of clinopyroxene, magnetite, and leucite; a leucite-bearing *basalt.* It was named by Iddings and Morley in 1917 for Batuku, Celebes. Not recommended usage.
baulite *krablite.*
Baumé gravity The specific weight of a liquid, measured on a

scale based on the weight of water; it is used in the petroleum industry for denoting the specific weight of oils. For liquids lighter than water, degrees Baumé=140/(specific gravity of the liquid at 60°F)-130. Cf: *API gravity.*

baumhauerite A lead- to steel-gray monoclinic mineral: $Pb_3As_4S_9$.

baumite A mineral of the kaolinite-serpentine group: $(Mg,Mn,Fe,Zn)_3Si_2O_5(OH)_4$.

baum pot (a) A calcareous concretion in the roof of a coal seam; a *bullion.* (b) A cavity left in the roof of a coal seam due to the dropping downward of a cast of a fossil tree stump after removal of the coal. Cf: *coal pipe; saddle [coal].*

bauranoite A mineral: $BaU_2O_7 \cdot 4\text{-}5H_2O$.

bauxite An off-white, grayish, brown, yellow, or reddish-brown rock composed of a mixture of various amorphous or crystalline hydrous aluminum oxides and aluminum hydroxides (principally gibbsite, some boehmite and diaspore), along with free silica, silt, iron hydroxides, and esp. clay minerals; a highly aluminous *laterite.* It is a common residual or transported constituent of clay deposits in tropical and subtropical regions, and occurs in concretionary, compact, earthy, pisolitic, or oolitic forms. Bauxite is the principal commercial source of aluminum; the term is also used collectively for lateritic aluminous ores. Bauxite was formerly regarded as an amorphous clay mineral consisting essentially of hydrated alumina, $Al_2O_3 \cdot 2H_2O$. Named after Les Baux de Provence, a locality near Arles in southern France.

bauxitic Containing much bauxite; e.g. a "bauxitic clay" containing 47% to 65% alumina on a calcined basis, or a "bauxitic shale" abnormally high in alumina and notably low in silica.

bauxitization Development of bauxite from primary aluminum silicates (such as feldspars) or from secondary clay minerals under aggressive tropical or subtropical weathering conditions of good surface drainage, such as the dissolving (usually above the water table) of silica, iron compounds, and other constituents from alumina-containing material.

bavenite A white fibrous mineral: $Ca_4BeAl_2Si_9O_{24}(OH)_2$. Syn: *duplexite.*

Baveno twin law An uncommon twin law in feldspar, in which the twin plane and composition surface are (021). A Baveno twin usually consists of two individuals.

Baventian British climato-stratigraphic stage: Lower Pleistocene (above Antian, below Pastonian).

***B* axis** Originally used by Sander (1930) to denote an axis of rotation, but has come to be used as a term meaning fold axis. In much recent literature it is replaced by the symbol *F.*

b* axis That axis of a reciprocal crystal lattice which is perpendicular to (010). Cf: *a* axis; c* axis.*

***b* axis [cryst]** One of the crystallographic axes used as reference in crystal description. It is the axis that is oriented horizontally, right-to-left. In an orthorhombic or triclinic crystal, it is usually the *macro-axis.* In a monoclinic crystal, it is the *orthoaxis.* The letter *b* usually appears in italics. Cf: *a* axis; *c* axis.

***b* axis [struc petrol]** One of three orthogonal reference axes, *a, b,* and *c,* that are used in two distinct ways. (a) To help describe the geometry of a fabric possessing monoclinic symmetry. The *b* axis is that axis at right angles to the unique plane of symmetry. In many instances this direction is parallel to a fold axis, but in complicated fabrics this need not necessarily be so. (b) In a kinematic sense, to help describe a *deformation plan* that possesses monoclinic symmetry, such as progressive simple shear. In this case the *b* axis is again the line at right angles to the unique symmetry plane. In a progressive simple shear the *b* axis lies in the shear plane at right angles to the direction of shear. See also: *a* axis; *c* axis. Syn: *b* direction.

bay [coast] (a) A wide, curving open indentation, recess, or inlet of a sea or lake into the land or between two capes or headlands, larger than a cove, and usually smaller than, but of the same general character as, a gulf; the width of the seaward opening is normally greater than the depth of penetration into the land. (b) A large tract of water that penetrates into the land and around which the land forms a broad curve. By international agreement (for purposes of delimiting territorial waters), a bay is a water body having a baymouth less than 24 nautical miles wide and an area that is equal to or greater than the area of a semicircle whose diameter is equal to the width of the baymouth.—See also: *bight [coast]; embayment [coast].*

bay [geog] (a) Any terrestrial formation resembling a bay of the sea, as a recess or extension of lowland along a river valley or within a curve in a range of hills, or an arm of a prairie extending into and partly surrounded by a forest. Also, a piece of low marshy ground producing many bay trees (such as laurel). (b) A *Carolina bay.* (c) A term used in southern Georgia and in Florida for an arm of a swamp extending into the upland as a baylike indentation (Veatch & Humphrys, 1966, p. 23).

bay [ice] (a) *bight [ice].* (b) A part of the sea partly enclosed by ice.

bay [magnet] A transient magnetic disturbance, lasting typically an hour. On a magnetic record it has the appearance of a V or of a bay of the sea.

bay bar A syn. of *baymouth bar* and *bay barrier.* The term is "confusing" because it "fails to indicate whether or not the sand ridge is submerged or stands above the water level" (Shepard, 1952, p. 1908).

bay barrier A term proposed by Shepard (1952, p. 1908) to replace *bay bar,* signifying a spit that has grown "entirely across the mouth of a bay so that the bay is no longer connected to the main body of water". See also: *baymouth bar.*

bay cable A marine seismic cable that is laid in place on the water bottom, as opposed to a drag cable or *streamer,* which is towed into place.

bay delta A delta formed at the mouth of a stream entering, and filling or partially filling, a bay or drowned valley. See also: *bayhead delta.*

bayerite A mineral: $Al(OH)_3$. It is a polymorph of gibbsite. Not to be confused with *beyerite.*

bayhead (a) The part of a bay that lies farthest inland from the larger body of water with which the bay is in contact. (b) A local term in southern U.S. for a swamp at the bayhead.—Also spelled: *bay head.*

bayhead bar A feature similar to a *bayhead barrier,* but smaller and generally submerged.

bayhead barrier A *barrier* formed a short distance from the shore and across a bay near its head. It commonly has a narrow inlet. Cf: *bayhead bar.*

bayhead beach A small, crescentic beach formed at the head of a bay by materials eroded from adjacent headlands and carried to the bayhead by longshore currents and/or storm waves. Syn: *pocket beach; cove beach.*

bayhead delta A delta at the head of a bay or estuary into which a river discharges. See also: *bay delta.*

bay ice (a) Newly formed, relatively smooth sea ice of more than one winter's growth. (b) A term sometimes used in Antarctica for thick ice floes recently broken away from an ice shelf. (c) A term used in Labrador for one-year ice that forms in bays and inlets. (d) An obsolete term for young sea ice sufficiently thick to impede navigation.

bayldonite A grass-green to blackish-green mineral: $PbCu_3(AsO_4)_2(OH)_2$.

bayleyite A yellow monoclinic mineral: $Mg_2(UO_2)(CO_3)_3 \cdot 18H_2O$.

baylissite A monoclinic mineral: $K_2Mg(CO_3)_2 \cdot 4H_2O$.

baymouth The entrance to a bay; the part of a bay that is in contact, and serves as a connection, with the main body of water. Also spelled: *bay mouth.*

baymouth bar A long, narrow bank of sand or gravel, generally submerged, deposited by waves entirely or partly across the mouth or entrance of a bay, so that the bay is no longer connected or is connected only by a narrow inlet with the main body of water; it usually connects two headlands, thus straightening the coast. It can be produced by the convergent growth of two spits from opposite directions, by a single spit extending in one direction, or by a longshore bar being driven shoreward. Syn: *bay barrier; bay bar.*

bayou (a) A term variously applied to many local water features in the lower Mississippi River basin and in the Gulf Coast region of the U.S., esp. Louisiana. Its general meaning is a creek or secondary watercourse that is tributary to another body of water; esp. a sluggish and stagnant stream that follows a winding course through alluvial lowlands, coastal swamps or river deltas. (b) An effluent branch, esp. sluggish or stagnant, of a main river, e.g. a distributary flowing through a delta. Also, the distributary channel that carries floodwater or affords a passage for tidal water through swamps or marshlands. (c) A *bayou lake* or an *oxbow lake.* (d) A *slough* in a salt marsh. (e) An estuarine creek (generally tidal), or an inlet, bay, or open cove on the Gulf Coast. (f) A term used in northern Arkansas and southern Missouri for a clear brook or rivulet that rises in the hills (Webster, 1967, p. 188).—Etymol: American French *boyau,* "gut"; from Choctaw *bayuk,* "small stream". Syn: *girt.*

bayou lake A lake or pool in an abandoned and partly closed channel of a stream, as on the Mississippi River delta. Syn: *bayou.*

bayside beach A beach formed along the side of a bay by materials eroded from adjacent headlands.

bazirite A trigonal mineral: $BaZrSi_3O_9$.

bazzite An azure-blue hexagonal mineral: $Be_3(Sc,Al)_2Si_6O_{18}$. It is the scandium analogue of beryl.

BC interray Right posterior interray in echinoderms, situated between B ray and C ray and clockwise of B ray when the echinoderm is viewed from the adoral side; equal to interambulacrum 4 in the Lovenian system.

bc-joint *longitudinal joint.*

BC soil A soil having only B and C horizons.

***b* direction** *b axis [struc petrol].*

beach (a) The unconsolidated material that covers a gently sloping zone, typically with a concave profile, extending landward from the low-water line to the place where there is a definite change in material or physiographic form (such as a cliff), or to the line of permanent vegetation (usually of the effective limit of the highest storm waves); a *shore* of a body of water, formed and washed by waves or tides, usually covered by sandy or pebbly material, and lacking a bare rocky surface. See also: *strand.* (b) The relatively thick and temporary accumulation of loose water-borne material (usually well-sorted sand and pebbles, accompanied by mud, cobbles, boulders, and smoothed rock and shell fragments) that is in active transit along, or deposited on, the shore zone between the limits of low water and high water. The term was originally used to designate the loose wave-worn shingle or pebbles found on English shores, and is so used in this sense in some parts of England (Johnson, 1919, p. 163). (c) A term used in New Jersey for a low sand island along the coast. (d) A term commonly used for a seashore or lake-shore area, esp. that part of the shore used for recreation.

beach berm *berm [beach].*

beach breccia A breccia formed on a beach where wave action is inefficient and angular blocks are supplied from cliffs, and produced under conditions of rapid submergence (Norton, 1917, p.181).

beachcomber *comber.*

beach concentrate A natural accumulation in beach sand of heavy minerals selectively concentrated (by wave, current, or surf action) from the ordinary beach sands in which they were originally present as accessory minerals; esp. a *beach placer.*

beach crest A temporary ridge or berm marking the landward limit of normal wave activity (Veatch & Humphrys, 1966, p. 24). Cf: *berm crest.*

beach cusp A low seaward projection of sand, pebbles, gravel, or boulders, formed on the foreshore of a beach by wave action; specif. a relatively small *cusp* along a straight beach. Distance between beach cusps is 10-60 m; it generally increases with increase in wave height.

beach cycle The periodic retreat and outbuilding of a beach under the influence of tides and waves: cutting back occurs during periods of spring tides and of high waves produced by winter storms; building out occurs during periods of neap tides and of low waves characteristic of summer.

beach erosion The destruction and removal of beach materials by wave action, tidal currents, littoral currents, or wind.

beach face The section of the beach normally exposed to the action of wave uprush; the *foreshore* of a beach. Not to be confused with *shoreface.*

beach firmness The ability of beach sand to resist pressure; the "strength" of the sand. It is controlled by the degree of packing and sorting, by moisture content, by the quantity of trapped air, and by sand-particle size: the damper and finer-grained the sand, the firmer the beach.

beachline A shoreline characterized by a series of well-developed beaches.

beach mining The extraction and concentration of beach placer ore, usually by dredging.

beach ore *beach placer.*

beach pad A discrete mass of shoreline sand, with the form of an asymmetrical triangle in plan view. The base lies along the beach, the gentle slope faces updrift, and the short steep slope faces downdrift. Pads have a quasi-regular spacing of tens or hundreds of meters and move slowly along the shoreline (Tanner, 1975, p. 175).

beach placer A placer deposit of heavy minerals, e.g. zircon, ilmenite, or rutile, on a contemporary or ancient beach or along a coastline; a *beach concentrate* containing valuable minerals. Syn: *beach ore; seabeach placer; mineral sands.*

beach plain A continuous and level or undulating area formed by closely spaced successive embankments of wave-deposited beach material added more or less uniformly to a prograding shoreline, such as to a growing compound spit or to a cuspate foreland (Johnson, 1919, p. 297 & 319); a *wave-built terrace.*

beach platform *wave-cut bench.*

beach pool (a) A small, usually temporary, body of water between two beaches or two beach ridges, or a lagoon behind a beach ridge. (b) A pool adjoining a lake and resulting from wave action.

beach profile The trace of a beach surface on a vertical plane perpendicular to the shoreline. The beach profile of equilibrium is commonly concave upward as the slope is steeper above high water and gentler seaward.

beach ridge A low, essentially continuous mound of beach or beach-and-dune material (sand, gravel, shingle) heaped up by the action of waves and currents on the backshore of a beach beyond the present limit of storm waves or the reach of ordinary tides, and occurring singly or as one of a series of approximately parallel deposits. The ridges are roughly parallel to the shoreline and represent successive positions of an advancing shoreline. Syn: *full.*

beachrock A friable to well-cemented sedimentary rock, formed in the intertidal zone in a tropical or subtropical region, consisting of sand or gravel (detrital and/or skeletal) cemented with calcium carbonate; e.g. a thin, clearly stratified, seaward-dipping calcarenite found on a sandy coral beach. Also spelled: *beach rock.* Syn: *beach sandstone.*

beach sandstone *beachrock.*

beach scarp An almost vertical slope fronting a berm on a beach, caused by wave erosion. It may range in height from several centimeters to a few meters, depending on the character of the wave action and the nature and composition of the beach.

beach width The horizontal dimension of the beach as measured normal to the shoreline.

beachy Covered with beach materials, esp. pebbles, shingle, or sand.

bead In blowpipe analysis of minerals, a drop of a fused material, such as a *borax bead* used as a solvent in color testing for various metals. The addition of a metallic compound to the bead will cause the bead to assume the color that is characteristic of the metal. See also: *blowpiping.*

beaded drainage A series of small pools connected by short watercourses. The pools result from the thawing of ground ice that commonly represents the intersections of *ice wedges.* Pools may be 1-4 m deep and 100 m across (Washburn, 1973, p. 235).

beaded esker An esker with numerous bulges or swellings (commonly representing fans or deltas) along its length. It may have been formed in lakes or other water bodies during pauses in the retreat of the glacier that nurtured the esker-forming stream.

beaded lake One of a string of lakes, as a *paternoster lake* or a long narrow lake between sand dunes.

beaded stream A stream consisting of a series of small pools or lakes connected by short stream segments; e.g. a stream commonly found in a region of *paternoster lakes* or an area underlain by permafrost.

beak [bot] A long, prominent point of a plant part, e.g. of a fruit or pistil.

beak [coast] *promontory.*

beak [paleont] (a) The generally pointed extremity of a bivalve shell, marking the point of the initial growth of the shell; specif. the noselike projection of the dorsal part of a pelecypod valve, located along or above the hinge line, and typically showing strong curvature and pointing anteriorly, or the tip of the umbo of a brachiopod, located adjacent or posterior to the hinge line and in the midline of the valve. Cf: *umbo.* (b) The prolongation of certain univalve shells, containing the canal (as in a gastropod). The term in this sense is not generally used by paleontologists. (c) The generally calcified tips of the otherwise horny jaws of *Nautilus* and some coleoid cephalopods; also found in fossil cephalopods. Cf: *rhyncholite.* (d) The skeletal rim or *rostrum* around the palate of an avicularium in cheilostome bryozoans. (e) The toothless jaws (of bone sheathed with horn) of a bird, turtle, dicynodont, dinosaur, or pterosaur.

beaked apex The upper angle of tergum of a cirripede crustacean, produced into a long narrow point (TIP, 1969, pt.R, p.91).

beak ridge A more or less angular linear elevation of a brachiopod shell, extending from each side of the umbo so as to delimit all or most of the cardinal area (TIP, 1965, pt.H, p.140).

Beaman stadia arc A specially graduated arc attached to the vertical circle of an alidade or transit to simplify the computation of elevation differences for inclined stadia sights (without the use of vertical angles). The arc is so graduated that each division on the arc is equal to 100(0.5 sin $2A$), where A is the vertical angle. Named after William M. Beaman (1867-1937), U.S. topographic engineer, who designed it in 1904.

beam balance *Westphal balance.*

beam steering A method for emphasizing seismic energy coming from a particular direction by delaying successive channels so that events of a certain dip moveout, or apparent velocity, occur at the same time, and then summing them. This procedure can be repeated for a succession of different dip moveouts to determine direction of approach. See also: *steer.*

beam width In *SLAR,* the angle subtended in the horizontal plane by the radar beam.

bean ore A loose, coarse-grained pisolitic iron ore; limonite occurring in lenticular aggregations. See also: *pea ore.*

bearing (a) The angular *direction* of any place or object at one fixed point in relation to another fixed point; esp. the horizontal direction of a line on the Earth's surface with reference to the cardinal points of the compass, usually expressed as an angle of less than 90 degrees east or west of a reference meridian adjacent to the quadrant in which the line lies and referred to either the north or south point (e.g. a line in the NE quadrant making an angle of 50 degrees with the meridian will have a bearing of N 50° E). Cf: *azimuth [surv].* See also: *true bearing; magnetic bearing; compass bearing; back bearing.* (b) The horizontal direction of one terrestrial point from another (such as an observer on a ship), usually measured clockwise from a reference direction and expressed in degrees from zero to 360; specif. *astronomic azimuth.* (c) Relative position or direction, as in reference to the compass or to surrounding landmarks. Usually used in the plural.

bearing capacity The maximum load per unit of area that the ground can safely support without failing in shear. See also: *ultimate bearing capacity; California bearing ratio.*

bearing tree A tree forming a *corner accessory,* its distance and direction from the corner being recorded. It is identified by prescribed marks cut into its trunk. Syn: *witness tree.*

bearsite A monoclinic mineral: $Be_2(AsO_4)(OH) \cdot 4H_2O$.

Beaufort wind scale A *wind scale* commonly used at sea. It is based on the visible effects of wind on the sea surface or on fixed objects. Code numbers and descriptive terms are assigned to various ranges of wind velocity, e.g. a wind velocity of 8-10 mph (or 7-10 knots) is Beaufort Code Number 3, and is called a "gentle breeze". The scale is a modernized version of that devised by Admiral Beaufort of the British Navy early in the nineteenth century.

beaverite A canary-yellow mineral: $Pb(Cu,Fe,Al)_3(SO_4)_2(OH)_6$.

beaver meadow An area of soft, moist ground resulting from the construction of a beaver dam; a beaver pond that has been changed into a marsh of grass or sedge upon abandonment of the dam by the beavers.

bebedourite A *pyroxenite* that contains biotite, with accessory perovskite, apatite, and titanomagnetite. Not recommended usage.

beck A British term for a small stream or brook, often with a stony bed, a rugged or winding course, and a rapid flow. Etymol: Old Norse.

Becke line In the *Becke test,* a bright line, visible under the microscope, that separates substances of different refractive indices.

beckelite A wax-yellow to brown isometric mineral: $Ca_3(Ce,La,Y)_4(Si,Zr)_3O_{15}$. It may be britholite.

beckerite A brown variety of retinite having a very high oxygen content (20-23%).

Becke test In optical mineralogy, a test under the microscope, at moderate or high magnification, for comparing the indices of refraction of two contiguous minerals, or of a mineral and a mounting medium or immersion liquid, in a thin section or other mount. If these substances differ materially in refractive index, they are separated by a bright line (the *Becke line*), which moves toward the less refractive substance when the tube of the microscope is lowered, and away from that substance when the tube is raised.

beckite Original, but incorrect, spelling of *beekite.*

becquerelite An amber to yellow secondary mineral: $CaU_6O_{19} \cdot 11H_2O$. It occurs in small orthorhombic crystals and crusts on pitchblende.

Becquerel ray A term used, before the terms alpha, beta, and gamma rays were introduced, for the particles emitted during radioactive decay.

bed [geomorph] (a) The ground upon which any body of water rests, or the land covered by the waters of a stream, lake, or ocean; the bottom of a watercourse or of a stream channel. Examples: *stream bed; seabed.* (b) The land surface marking the site of a former body of water, or representing land recently exposed by recession or by drainage.—Syn: *floor; bottom.*

bed [stratig] (a) The smallest formal unit in the hierarchy of *lithostratigraphic units.* In a stratified sequence of rocks it is distinguishable from layers above and below; e.g. the *Baker Coal Bed.* A bed commonly ranges in thickness from a centimeter to a few meters. Customarily, only distinctive *marker beds [stratig],* useful for correlation or reference, are given proper names and considered formal units; intervening beds are left unnamed (ISG, 1976, p. 33). The term is applied primarily to sedimentary strata, but it may be used for other types, e.g. an ash bed. Cf: *lamina.* See also: *beds.* Syn: *layer [stratig]; stratum.* (b) A layer containing a concentration of paleontologic or anthropologic evidence, e.g. a *bone bed.*

bed configuration A group of *bed forms,* produced by flow and making up a particular bed geometry. Cf: *bed phase.*

bedded [ore dep] (a) Said of a vein or other mineral deposit that follows the bedding plane in a sedimentary rock. (b) Said of a layered replacement deposit. Cf: *stratiform [ore dep]; stratabound.*

bedded [stratig] Formed, arranged, or deposited in layers or beds, or made up of or occurring in the form of *beds [stratig];* esp. said of a layered sedimentary rock, deposit, or formation. The term has also been applied to nonsedimentary material that exhibits depositional layering, such as the "bedded deposits" of volcanic tuff alternating with lava in the mantle of a stratovolcano. See also: *stratified; well-bedded.*

bedded chert Brittle, closely jointed, rhythmically layered *chert* occurring in areally extensive deposits, with thicknesses measured in tens of meters, and consisting of distinct, usually even-bedded layers (3-5 cm thick), separated by partings of dark siliceous shale or by layers of siderite. Most bedded cherts are believed to be the result of crystallization of biogenic opaline skeletons such as those of diatoms and radiolaria. Examples include the Monterey (Miocene) and the Franciscan (Jurassic?) cherts of California, and the Mesozoic radiolarian cherts of the Alps and Apennines. See also: *novaculite.*

beddedness index *stratification index.*

bedded volcano A less-preferred syn. of *stratovolcano.*

bedding [mining] (a) A quarrymen's term applied to a structure occurring in granite and other crystalline rocks that tend to split in well-defined planes more or less horizontally or parallel to the land surface. Syn: *sheeting.* (b) The storing and mixing of different ores in thin layers in order to blend them more uniformly in reclamation.

bedding [stratig] (a) The arrangement of a sedimentary rock in beds or layers of varying thickness and character; the general physical and structural character or pattern of the beds and their contacts within a rock mass, such as *cross-bedding* and *graded bedding;* a collective term denoting the existence of beds. Also, the structure so produced. The term may be applied to the layered arrangement and structure of an igneous or metamorphic rock. See also: *stratification [sed].* Syn: *layering [stratig].* (b) *bedding plane.*

bedding cave *bedding-plane cave.*

bedding cleavage *bedding-plane cleavage.*

bedding fault A fault whose surface is parallel to the bedding plane of the constituent rocks. Cf: *bedding glide.* Syn: *bedding-plane fault.*

bedding fissility The property possessed by a sedimentary rock (esp. shale) of tending to split more or less parallel to the bedding; *fissility* along bedding planes. It is a primary foliation that forms in a sedimentary rock while the sediment is being deposited and compacted, and is a result of the parallelism of the platy minerals to the bedding plane.

bedding glide A nearly horizontal overthrust fault produced by bedding-plane slip (Nelson, 1965). Cf: *bedding fault.* Syn: *bedding thrust.*

bedding index *stratification index.*

bedding joint In sedimentary rock, a joint that is parallel to the bedding plane; a joint that follows a bedding plane. Syn: *bed joint.*

bedding plane (a) A planar or nearly planar *bedding surface* that visibly separates each successive layer of stratified rock (of the same or different lithology) from the preceding or following layer; a plane of deposition. It often marks a change in the circumstances of deposition, and may show a parting, a color difference, or both. (b) A term commonly applied to any bedding surface, even when conspicuously bent or deformed by folding. (c) A term commonly applied to a plane of discontinuity (usually the bedding plane) along which a rock tends to split or break readily.—Syn: *bedding [stratig]; bed plane; stratification plane; plane of stratification.*

bedding-plane cave Cave passages, generally broad and flat, that have developed along a bedding plane, usually by dissolution of one of the beds. Syn: *bedding cave.*

bedding-plane cleavage Cleavage that is parallel to the bedding plane. Cf: *axial-plane cleavage.* Syn: *bedding cleavage; parallel cleavage.*

bedding-plane fault *bedding fault.*

bedding-plane parting A *parting* or surface of separation between adjacent beds or along a bedding plane.

bedding-plane sag Depressed and disturbed strata or laminae of tuff or other deposit into which a volcanic *block* or *bomb* has fallen. Cf: *secondary crater.* Syn: *bomb sag.*

bedding-plane slip The slipping of sedimentary strata along bedding planes during folding. It produces *disharmonic folding* and, in extreme form, *décollement.* Syn: *flexural slip.*

bedding surface A surface, usually conspicuous, within a mass of stratified rock, representing an original surface of deposition; the surface of separation or interface between two adjacent beds of sedimentary rock. If the surface is more or less regular or nearly planar, it is called a *bedding plane.*

bedding thrust *bedding glide.*

bedding void An open space between successive lava flows, formed where an overlying flow does not completely conform to the solidified crust of lava beneath it.

Bedford limestone A commercial name for *spergenite,* a uniform gray or buff Mississippian limestone extensively quarried in the vicinity of Bedford, Ind., for building stone. Syn: *Indiana limestone.*

bed form Any deviation from a flat bed, generated by the flow on the bed of an alluvial channel (Middleton, 1965, p.247). See also: *bed configuration.*

bediasite A jet-black to brown tektite from east-central Texas. Named after the town of Bedias, from the Bidai (Bedias) Indians of the Trinity River valley.

bed joint *bedding joint.*

bed load The part of the total *stream load* that is moved on or immediately above the stream bed, such as the larger or heavier particles (boulders, pebbles, gravel) transported by traction or saltation along the bottom; the part of the load that is not continuously in suspension or solution. See also: *bed-material load; contact load; saltation load.* Also spelled: *bedload.* Syn: *bottom load; traction load.*

bed-load function The rate at which various streamflows for a given channel will transport the different particle sizes of the *bed-material load.*

bed material The material of which the bed of a stream is composed; it may originally have been the material of suspended load or of bed load, or may in some cases be partly residual.

bed-material load The part of the total *sediment load* (of a stream) composed of all particle sizes found in appreciable quantities in the bed material; it is the coarser part of the load, or the part that is most difficult to move by flowing water. See also: *bed load.* Cf: *wash load.*

bed moisture *inherent moisture.*

Bedoulian Substage in Switzerland: Lower Cretaceous (lower Aptian; below Gargasian Substage).

bed phase "The aggregate of all bed states that involve a particular kind of bed form or combination of bed forms" (Southard, 1971, p. 904). Thus a bed phase is a general class of *bed configurations* produced over a particular range of hydraulic conditions, by flow acting on an alluvial bed composed of a particular range of bed materials. For example, the ripple bed phase can be defined, for quasi-equilibrium unidirectional flows, by a certain range of conditions of depth, velocity, and grain size. See also: *bed form; bed state.*

bed plane *bedding plane.*

bed rock A general term for the rock, usually solid, that underlies soil or other unconsolidated, superficial material. Adj: *bedrock.* A British syn. of the adjectival form is *solid,* as in *solid geology.*

bedrock valley A valley eroded in bedrock.

beds (a) An informal term for strata that are incompletely known, constitute a lithologically similar succession, or are of local economic significance, e.g. "beds of Permian age", "key beds", "coal beds". (b) A formal unit, consisting of several contiguous strata of similar lithology, e.g. the *Marcus Limestone Beds* (ISG, 1976, p. 48).—See also: *bed [stratig].*

bed separation In mining, the parting of strata along bedding planes caused by differential subsidence above a mine roof. It is an important factor in the engineering structure of mines.

bed state The average of all particular *bed configurations* that can be formed by a given flow over an alluvial bed (Southard, 1971, p. 904). See also: *bed form; bed configuration; bed phase.*

beechleaf marl A term used in Lancashire, England, for brownish, finely laminated marl of glacial origin. Cf: *toadback marl.*

beef A quarrymen's term, used originally in Purbeck, southern England, for thin, flat-lying veins or layers of fibrous calcite, anhydrite, gypsum, halite, or silica, occurring along bedding planes of shale, giving a resemblance to beef. It appears to be due to rapid crystallization in lenticular cavities. Syn: *bacon.*

beegerite A mixture of matildite and schirmerite.

beekite (a) White, opaque silica occurring in the form of subspherical, discoid, rosettelike, doughnut-shaped, or botryoidal accretions, commonly found as bands or layers on silicified fossils or along joint surfaces as a replacement of organic matter; e.g. chalcedony pseudomorphous after coral, shell, or other fossils. See also: *ooloid.* (b) Concretionary calcite commonly occurring in small rings on the surface of a fossil shell that has weathered out of its matrix.—Named after Dr. Beek, dean of Bristol. Originally spelled: *beckite.*

beerbachite A hornfels, originally described as a hypabyssal dike rock, that is typically light-colored, similar to aplite in appearance but chiefly composed of fine-grained labradorite, orthopyroxene, clinopyroxene, and magnetite.

beetle stone (a) A *septarium* of coprolitic ironstone, the enclosed coprolite resembling the body and limbs of a beetle. (b) An old name for *turtle stone.*

before breast A miner's term for that part of the orebody that still lies ahead. See also: *breast.*

beforsite A hypabyssal dolomitic *carbonatite,* named by Eckermann for Bergeforseu, Alnö, Sweden. Cf: *rauhaugite.* Not recommended usage.

beheaded stream The diminished lower part of a stream whose headwaters have been captured by another stream.

beheading (a) The cutting-off of the upper part of a stream and the diversion of its headwaters into another drainage system by *capture.* (b) The removal of the upper part of a stream's drainage area by wave erosion. Cf: *betrunking.*

behierite A mineral: $(Ta,Nb)BO_4$.

behoite A mineral: $Be(OH)_2$.

beidellite (a) A white, reddish, or brownish-gray clay mineral of the montmorillonite group: $(Na,Ca/2)_{0.33}Al_{2.17}(Si_{3.17}Al_{0.83})O_{10}(OH)_2 \cdot nH_2O$. It is an aluminian montmorillonite, characterized by replacement of Si^{+4} by Al^{+3} and by the absence or near absence of magnesium or iron replacing aluminum. Beidellite is a common constituent of soils and certain clay deposits (such as *metabentonite*). Cf: *montmorillonite.* (b) An old name applied to a material that is actually a mixture, esp. an interlayered mixture of illite and montmorillonite or a mixture of clay minerals and hydrated ferric oxide.—"Because of the past use of the term, it is believed desirable to drop it entirely [from clay-mineral terminology] in order to avoid confusion" (Grim, 1968, p.48).

bekinkinite An igneous rock composed chiefly of sodic amphibole, clinopyroxene, nepheline, and olivine, along with plagioclase, biotite, and analcime; essentially a *theralite.* Cf: *fasinite.* Named by Rosenbusch in 1907 for Bekinkina, Malagasy. Not recommended usage.

bel A term applied in India and Pakistan to "sandy islands in the beds of rivers" (Stamp, 1961, p. 60). Etymol: Panjabi. Also spelled: *bhel.*

Belanger's critical velocity *critical velocity* (c).

belemnite Any member of an order of coleoid cephalopods characterized by a well-developed internal shell consisting of a guard, phragmocone, and forward-projecting daggerlike or spadelike proostracum. The body has a ten-armed crown, each arm equipped with a double row of arm hooks. Fossil phragmocenes are cigar-shaped. Range, Carboniferous to Eocene.

belemnoid A broad term applied to any one of the belemnitelike coleoids including, besides the belemnites, those forms having a tripartite proostracum and those with a living chamber and tentacles without arm hooks.

belite A calcium orthosilicate found as a constituent of portland-cement clinkers; specif. *larnite.* Syn: *felite.*

bell A cone-shaped nodule or concretion in the roof of a coal seam, which may fall without warning. Cf: *caldron bottom; pot bottom; camel back.* See also: *bell hole; coal pipe.*

bell hole A cavity in the roof of a coal seam, produced by the falling of a *bell.*

bellicatter Var. of *ballycadder.* Not commonly used.

bellidoite A tetragonal mineral: Cu_2Se.

bellingerite A light-green or bluish-green mineral: $Cu_3(IO_3)_6 \cdot 2H_2O$.

bell-jar intrusion An igneous intrusion similar to a *bysmalith* but differing in that the adjacent strata have become domed and severely fractured.

bell-metal ore A syn. of *stannite,* esp. the bronze-colored variety.

bell-shaped distribution A *frequency distribution* whose plot has the shape of a bell; usually, a *normal distribution.*

belly *pocket [eco geol].*

beloeilite A granular plutonic rock composed of sodalite, less potassium feldspar, and a small amount of mafic minerals. Nepheline may or may not be present. Cf: *tawite.* The rock was named by Johannsen (1931) for Mont Beloeil (now Mont St. Hilaire), Quebec.

belonite An elongated or acicular crystallite having rounded or pointed ends.

belonosphaerite An obsolete term for a spherulite whose minerals are in radial arrangement.

belovite (a) A mineral of the apatite group: $(Sr,Ce,Na,Ca)_5(PO_4)_3(OH)_2$. (b) *arsenate-belovite.*

belt A long area of pack ice, measuring 1 km to more than 100 km in width. Cf: *strip [ice].*

belted coastal plain A broad, maturely dissected coastal plain on which there are a series of roughly parallel cuestas alternating with subsequent lowlands or vales; e.g. the Gulf Coastal Plain through Alabama and Mississippi.

belted metamorphics A geomorphic term proposed by Strahler (1946) to describe mountain areas of folded and faulted metamorphosed sedimentary and igneous rocks that have been differentially eroded in distinct elongate subparallel ridges and valleys.

belteroporic fabric A rock fabric in which the preferred orientation of its mineral constituents was determined solely by the direction of easiest growth.

Beltian An approximate equivalent of *Riphean.*

Beltian orogeny A name proposed by Eardley (1962) for a supposed orogeny at the end of the Precambrian in western North America, based on unconformable relations between the Belt Series and the Cambrian in northwestern Montana. The existence of an orogeny at the time proposed is dubious; strata overlying the Belt Series are Middle Cambrian or younger, and the Belt itself has now been dated radiometrically as much older than 900 m.y.; moreover, it is overlain unconformably farther west by thick sequences of younger Precambrian strata.

belt of no erosion A zone adjacent to a drainage divide where no erosion by overland flow occurs because of insufficient depth and velocity of flow, and a slope that is too gentle to overcome the initial resistance of the soil surface to sheet erosion (Horton, 1945, p. 320); its width is equal to *critical length.*

belt of soil moisture *belt of soil water.*

belt of soil water The upper subdivision of the *zone of aeration,* limited above by the land surface and below by the intermediate belt. This zone contains plant roots and water available for plant growth. Syn: *belt of soil moisture; discrete-film zone; zone of soil water; soil-water zone; soil-water belt.*

belt of wandering The whole breadth of the valley floor that may be worn down by a stream.

belugite A group of intrusive rocks containing andesine and/or labradorite, and thus intermediate in feldspar content between *diorite* and *gabbro;* also, any rock in that group. The term is rarely used and is not recommended. Its name was derived from the Beluga River, Alaska by Spurr in 1900. Cf: *aleutite.*

belyankinite A yellowish-brown mineral: $Ca_{1\text{-}2}(Ti,Zr,Nb)_5O_{12} \cdot 9H_2O(?)$.

bementite A grayish-yellow or grayish-brown mineral: $Mn_8Si_6O_{15}(OH)_{10}$ (?). It may contain small amounts of zinc, magnesium, and iron.

ben A Scottish term for a high hill or mountain, or a mountain peak; it is used only in proper names, esp. in those of the higher summits of mountains, as *Ben* Nevis or *Ben* Lomond. Etymol: Gaelic *beann* or *beinn,* "peak".

bench [coal] A layer of coal; either a coal seam separated from nearby seams by an intervening noncoaly bed, or one of several layers within a coal seam that may be mined separately from the others.

bench [coast] (a) *wave-cut bench.* (b) A nearly horizontal area at about the level of maximum high water on the ocean side of an artificial dike (CERC, 1966, p. A3).

bench [geomorph] A long, narrow, relatively level or gently inclined strip or platform of land, earth, or rock, bounded by steeper slopes above and below, and formed by differential erosion of rocks of varying resistance or by a change of base-level erosion; a small terrace or steplike ledge breaking the continuity of a slope; an eroded bedrock surface between valley walls. The term sometimes denotes a form cut in solid rock as distinguished from one (as a *terrace*) cut in unconsolidated material. See also: *berm [geomorph]; mesa.*

bench gravel A term applied in Alaska and the Yukon Territory to gravel beds on the side of a valley above the present stream bottom, which represent part of the stream bed when it was at a higher level. See also: *bench placer.*

benchland (a) A *bench,* esp. one along a river. Also, the land situated in, or forming, a bench. (b) A land surface composed largely of benches; e.g. a *piedmont benchland.*

bench mark (a) A relatively permanent metal tablet or other mark firmly embedded in a fixed and enduring natural or artificial object, indicating a precisely determined elevation above or below a standard datum (usually sea level) and bearing identifying information, and used as a reference in topographic surveys and tidal observations; e.g. an embossed and stamped disk of bronze or aluminum alloy, about 3.75 in. in diameter, with an attached shank about 3 in. in length, which may be cemented in natural bedrock, in a massive concrete post set flush with the ground, or in the masonry of a substantial building. Abbrev: BM. See also: *permanent bench mark; temporary bench mark.* (b) A well-defined, permanently fixed point in space, used as a reference from which measurements of any sort (such as of elevations) may be made.

bench-mark soil A soil that is representative of many similar soils and has been selected for detailed characterization because of its historical significance, wide geographic extent, or importance to soil classification.

bench placer A *bench gravel* that is mined as a placer. Syn: *river-bar placer; terrace placer.*

bench reef A *coral reef* consisting of a coral-algal frame arising directly from the margin of a pre-existing terrace or shelf generally less than 10 m deep.

bench terrace A shelflike embankment of earth with a flat or gently inclined top and often a steep or vertical downhill side, constructed along the contour of sloping land to control runoff and erosion, or to improve stability, as in a road cut. It is also used in series to convert mountainous slopes to arable land or to building lots.

bend [geomorph] (a) A curve or turn in the course, bed, or channel of a stream, not yet developed into a meander. Also, the land area partly enclosed by a bend or meander. (b) A curved part of a lake, inlet, or coastline.

bend [paleont] A rather sharp angulation in the ventral part of an ostracode valve, usually parallel to the free edge and commonly in the position of a carina, in which case it is called a *carinal* bend (Kesling, 1951, p. 120).

bend [sed] An English miner's term for any hardened clayey substance. See also: *bind.*

bend folding Folding that occurs in response to a (vertical) bending moment. Cf: *buckle folding.*

bend gliding Slip on planes that are being bent about an axis. Usually applied to slip on an atomic scale, but may also refer to slip on bedding or other planes during flexural slip folding.

bendway A term applied along the Mississippi River to one of the deep-water pools occurring at the bends of meanders on alternate sides of the river. See also: *crossing.*

beneficiation Improvement of the grade of ore by milling, flotation, sintering, gravity concentration, or other processes.

Benioff fault plane *Benioff seismic zone.*

Benioff seismic zone A plane beneath the trenches of the circum-

Pacific belt, dipping toward the continents at an angle of about 45°, along which earthquake foci cluster. It is sometimes referred to as the *Benioff fault plane.* According to the theory of plate tectonics and sea-floor spreading, plates of the lithosphere sink into the upper mantle causing earthquakes along the upper boundary of the plate, thus defining this zone.

Benioff zone *Benioff seismic zone.*

benitoite A blue to colorless, transparent, hexagonal mineral: $BaTiSi_3O_9$. It is strongly dichroic, resembles sapphire in appearance, and is sometimes used as a gem.

benjaminite A monoclinic mineral: $(Ag,Cu)_3(Bi,Pb)_7S_{12}$.

benmoreite A silica-saturated to undersaturated igneous rock intermediate between *mugearite* and *trachyte,* with a differentiation index between 65 and 75 and K_2O:Na_2O less than 1:2. It is named for Ben More, Mull, Scotland (Tilley and Muir, 1964).

benstonite A rhombohedral mineral: $(Ca,Mg,Mn)_7(Ba,Sr)_6(CO_3)_{13}$.

benthic Pertaining to the *benthos;* also, said of that environment. Syn: *benthonic; demersal.*

benthogene Said of sediments derived from benthonic plants or animals, or precipitated chemically on the ocean floor (Sander, 1951, p. 6).

benthograph A submersible, spherical container for photographic equipment used in photographic exploration of the deep sea.

benthonic *benthic.*

benthos Those forms of marine life that are bottom-dwelling; also, the ocean bottom itself. Certain fish that are closely associated with the benthos may be included. Adj: *benthic.*

bentonite (a) A soft, plastic, porous, light-colored rock composed essentially of clay minerals of the montmorillonite (smectite) group plus colloidal silica, and produced by devitrification and accompanying chemical alteration of a glassy igneous material, usually a tuff or volcanic ash. It often contains accessory crystal grains that were originally phenocrysts in the parent rock. Its color ranges from white to light green and light blue when fresh, becoming light cream on exposure and gradually changing to yellow, red, or brown. The rock is greasy and soaplike to the touch (without gritty feeling), and commonly has the ability to absorb large quantities of water accompanied by an increase in volume of about 8 times. The term was first used by Knight (1898) to replace *taylorite,* previously proposed for the argillaceous Cretaceous deposits occurring in the Benton Formation (formerly Fort Benton Formation) of the Rock Creek district in eastern Wyoming. Syn: *volcanic clay; soap clay; mineral soap; amargosite.* (b) A commercial term applied to any of numerous variously colored clay deposits (esp. bentonite) containing montmorillonite (smectite) as the essential mineral and presenting a very large total surface area, characterized either by the ability to swell in water or to be slaked and to be activated by acid, and used chiefly (at a concentration of about 3 lb per cu ft of water) to thicken oil-well drilling muds. (c) Any clay composed dominantly of a smectite clay mineral, whose physical properties are dictated by this mineral (Grim & Güven, 1978, p. 1).

bentonite debris flow A *debris flow* associated with the seasonal freezing and thawing and extreme cold of the arctic region, formed where easily hydrated bentonite-rich sediments are exposed to surface water (in moderate quantities for at least several weeks) on slopes of 5-30°, and developed in a smooth-sided, fluted, leveed, and U-shaped mudflow channel. Term proposed by Anderson et al. (1969, p. 173) for such features near Umiat, Alaska.

bentonitic arkose A term used by Ross & Shannon (1926, p. 79) for a sandy volcanic ash containing less than 25% bentonitic clay minerals. Cf: *arkosic bentonite.*

benzene A colorless, volatile, highly inflammable toxic liquid that is the simplest member (formula C_6H_6) of the *aromatic hydrocarbon* series. It is usually produced from coal tar or coke-oven gas or synthesized from open-chain hydrocarbons, and is used chiefly as a solvent, as a motor fuel, as a material in the manufacture of dyes, and in organic synthesis. See also: *benzol.*

benzene series The *aromatic hydrocarbon* series of liquids and solids, empirical formula C_nH_{2n-6}, containing the benzene ring; i.e. it consists of benzene, the simplest member, and the homologues of benzene.

benzol A commercial form of *benzene,* which is at least 80 percent benzene but also contains its homologues toluene and xylene.

beraunite A dark-red or brown mineral: $Fe^{+2}Fe^{+3}_5(PO_4)_4(OH)_5 \cdot 4H_2O$.

berborite A mineral: $Be_2(BO_3)(OH,F) \cdot H_2O$.

Berek compensator In a microscope used for optical analysis of minerals, a type of *compensator [optics]* used for the measurement of the path difference produced by a crystal plate; it is a calcite plate that is cut at right angles to the optic axis and mounted on a rotating axis in the tube slit above the objective. The angle through which it is rotated to reach compensation is the measurement of the path difference.

beresite An aplitic hypabyssal rock altered to a material resembling *greisen,* containing quartz and often pyrite. Beresite was originally described as being predominantly feldspar, was later determined to be free of feldspar, and was still later described as a *quartz porphyry* (Johannsen, 1939, p. 243). Not recommended usage.

berezovite A syn. of *phoenicochroite.* Also spelled: *beresovite.*

berezovskite A variety of chromite with Fe:Mg from 3 to 1. Syn: *beresofskite.*

berg [geomorph] (a) A term used in the Hudson River valley, N.Y., for a mountain or hill. Etymol: Dutch *bergh,* akin to German *Berg,* "mountain". (b) A term used in South Africa for a mountain or mountain range. Etymol: Afrikaans.

berg [glaciol] Shortened form of *iceberg.*

bergalite A *lamprophyre* containing phenocrysts of melilite, hauyne, biotite, and rare clinopyroxene in a fine-grained groundmass of the same minerals and also nepheline, magnetite, perovskite, apatite, and glass. Named by Soellner in 1913. Not recommended usage.

bergenite A yellow secondary mineral: $Ba(UO_2)_4(PO_4)_2(OH)_4 \cdot 8H_2O$.

Bergmann's rule In zoology, the statement that warm-blooded animals tend to be larger in colder parts of the environment than in warmer. Named after Carl Bergmann (d. 1865), German biologist. Cf: *Allen's rule.*

bergmehl *moonmilk.*—Also spelled: *bergmeal.*

bergschrund A deep and often wide gap or crevasse, or a series of closely spaced crevasses, in ice or firn at or near the head of an *alpine glacier* or snowfield, that separates moving ice and snow from the relatively immobile ice and snow (*ice apron*) adhering to the confining headwall of a *cirque.* It may be covered by or filled with snow during the winter, but visible and reopened in the summer. Etymol: German *Bergschrund,* "mountain crack". Cf: *randkluft.* Syn: *schrund.*

bergschrund action Enlargement of a cirque occupied by a glacier through such processes as frost action and abrasion along the bergschrund.

berg till (a) A glacial till deposited intact by grounded icebergs in fresh or saline water bordering an ice sheet. (b) A lacustrine or marine clay containing boulders and stones dropped into it by melting icebergs.—Syn: *floe till; subaqueous till; glacionatant till.*

bergy bit A piece of floating ice, generally less than 5 m above sea level and not more than about 10 m across, larger than a *growler.* It is generally glacier ice but may be a massive piece of sea ice or disrupted hummocked ice.

bergy seltzer A sizzling sound, like that of newly uncorked seltzer water, emitted by an iceberg when melting. It is caused by the release of air bubbles that were retained in the ice under high pressure (Baker et al., 1966, p.21).

beringite A dark-colored sodic-amphibole-bearing *trachyte* containing albite and a smaller amount of potassium feldspar. Its name was derived from Bering Island, U.S.S.R. by Starzynski in 1913. Not recommended usage.

berkeyite A transparent gem *lazulite* from Brazil.

berlinite A colorless to rose-red mineral: $AlPO_4$.

berm [beach] A low, impermanent, nearly horizontal or landward-sloping bench, shelf, ledge, or narrow terrace on the backshore of a beach, formed of material thrown up and deposited by storm waves; it is generally bounded on one side or the other by a beach ridge or beach scarp. Some beaches have no berms, others have one or several. See also: *storm berm.* Syn: *beach berm; backshore; backshore terrace.*

berm [eng] (a) A relatively narrow, horizontal man-made shelf, ledge, or bench built along an embankment, situated part way up and breaking the continuity of a slope. (b) The bank of a canal opposite the towing path. (c) The margin or shoulder of a road, adjacent to and outside the paved portion.—Etymol: Dutch, "strip of ground along a dike". Syn: *berme.*

berm [geomorph] (a) A term introduced by Bascom (1931) for a terracelike or benchlike remnant of a surface developed to middle or late maturity in a former erosion cycle which has since been interrupted by renewed downward cutting following uplift; e.g. the undissected remnant of an earlier valley floor of a rejuvenated stream, or the remnant of an uplifted abrasion platform that underwent wave erosion along the coast. Bascom intended that the term replace *strath,* although "berm" sometimes includes the *valley shoulder* of a new valley together with the remnant of the old valley floor (Von Engeln, 1942, p. 221). See also: *bench [geomorph]; strath terrace.* (b) A horizontal ledge of land bordering either bank of the Nile River and inundated when the river overflows.

bermanite A reddish-brown mineral: $Mn^{+2}Mn_2^{+3}(PO_4)_2(OH)_2 \cdot 4H_2O$.

berm crest The seaward or outer limit or edge, and generally the highest part, of a berm on a beach; a line representing the intersection of two berms or of a berm and the foreshore. The crest of the most seaward berm separates the foreshore from the backshore. Cf: *beach crest.* Syn: *berm edge; crest.*

berm edge *berm crest.*

bermudite A lamprophyric extrusive rock composed of small biotite phenocrysts in a groundmass presumably composed of analcime, nepheline, and alkali feldspar; a biotite *nephelinite.* Named by Pirsson in 1914 for the island of Bermuda. Not recommended usage.

berndtite A mineral: SnS_2.

berndtite-C6 A hexagonal mineral: SnS_2.

Bernoulli effect The observation that, in a stream of fluid, pressure is reduced as velocity of flow increases. Cf: *Bernoulli's theorem.*

Bernoulli's theorem The statement in hydraulics that under conditions of uniform steady flow of water in a conduit or stream channel, the sum of the velocity head, the pressure head, and the head due to elevation at any given point is equal to the sum of these heads at any other point plus or minus the losses in head between the two points due to friction or other causes (plus if the latter point is upstream and minus if downstream). It was developed by the Swiss engineer Daniel Bernoulli in 1738 (ASCE, 1962). Cf: *Bernoulli effect.*

bernstein A syn. of *amber.* Etymol: German *Bernstein,* from Old German *börnen,* "to burn", + *Stein,* "stone".

berondrite A *theralite* similar to *luscladite* but characterized by the presence of amphibole laths and titanaugite as mafic phases, and modal nepheline. Cf: *fasinite; mafraite.* It was named by Lacroix in 1920 for Berondra, Malagasy. Not recommended usage.

Berriasian European stage: lowermost Lower Cretaceous (above Tithonian-Upper Volgian of Jurassic, below Valanginian).

berryite A mineral: $Pb_2(Cu,Ag)_3Bi_5S_{11}$.

berthierine A mineral: $(Fe^{+2},Fe^{+3},Mg)_{2\text{-}3}(Si,Al)_2O_5(OH)_4$, with x about 0.1 to 0.2. Much so-called *chamosite* appears to be identical with berthierine (Hey, 1962, 16.19.16a).

berthierite A dark steel-gray mineral: $FeSb_2S_4$.

berthonite *bournonite.*

Bertillon pattern A pattern of fine raised terrace lines on the dorsal surface of a trilobite exoskeleton, arranged subconcentrically and resembling a fingerprint.

bertossaite A mineral: $(Li,Na)_2(Ca,Fe,Mn)Al_4(PO_4)_4(OH,F)_4$.

bertrandite A colorless to pale-yellow mineral: $Be_4Si_2O_7(OH)_2$.

Bertrand lens A removable lens in the tube of a petrographic microscope that is used in conjunction with convergent light to form an interference figure.

beryl (a) A mineral: $Be_3Al_2Si_6O_{18}$. It usually occurs in green or bluish-green, sometimes yellow or pink, or rarely white, hexagonal prisms in metamorphic rocks and granitic pegmatites and as an accessory mineral in acid igneous rocks. Transparent and colored gem varieties include emerald, aquamarine, heliodor, golden beryl, and vorobievite. Beryl is the principal ore of beryllium. (b) *green beryl.*

beryllite A mineral: $Be_3SiO_4(OH)_2 \cdot H_2O$.

beryllium detector An instrument, commonly portable, that uses the principles of gamma-ray *activation analysis* to detect and analyze for beryllium. An enclosed gamma-ray source, generally ^{124}Sb, transforms ^9Be to ^8Be (2α) plus a neutron (n). Measurement of the neutron production rate allows for quantitative evaluation of beryllium. A popular term for the instrument is *berylometer.*

beryllium-10 age method A method of age determination based on measurement of the activity of beryllium-10 (half life $= 1.5 \times 10^6$ years), a nuclide formed in the upper atmosphere. It has been used in dating deep-sea sediments, and in determining sedimentation rates (Amin et al., 1975, p. 1187-1191).

beryllonite A colorless or yellow mineral: $NaBePO_4$. It occurs in transparent, topazlike orthorhombic crystals.

beryllosodalite *tugtupite.*

berylometer *beryllium detector.*

berzelianite A silver-white mineral: Cu_2Se.

berzeliite A bright-yellow mineral: $(Mg,Mn)_2(Ca,Na)_3(AsO_4)_3$. It is isomorphous with manganberzeliite. Syn: *berzelite.*

beschtauite A sodic *quartz porphyry,* named by Gerassimow in 1910 for Mount Beschtau, Caucasus Mountains. Obsolete.

Bessemer ore An iron ore that contains very little phosphorus (generally less than 0.045%). It is so named because it was suitable for use in the Bessemer process of steelmaking, a process no longer in use.

beta [cryst] (a) In a biaxial crystal, the intermediate *index of refraction.* (b) The interaxial angle between the *a* and *c* crystallographic axes.—Cf: *alpha [cryst], gamma [cryst].*

beta [mineral] adj. Of or relating to one of two or more closely related minerals and specifying a particular physical structure (esp. a polymorphous modification); specif. said of a mineral that is stable at a temperature intermediate between those of its *alpha* and *gamma* polymorphs (e.g. "beta cristobalite" or "β-cristobalite", the high-temperature isometric phase of cristobalite). Some mineralogists reverse this convention, using β for the low-temperature phase (e.g. "beta carnegieite", the triclinic phase of carnegieite produced from alpha carnegieite at temperatures below 690°C).

beta* angle The angle of the reciprocal lattice between the *a* axis* and the *c* axis,* which is equal to the interfacial angle between (100) and (001). Cf: *alpha* angle; gamma* angle.*

beta axis The line of intersection of two or more surfaces. It is usually written as β axis. The β axis is not equivalent to a π-pole. The β axis for two planes lies at right angles to the plane defined by the π-poles of the two planes. See also: *pi pole.*

beta chalcocite *chalcocite.*

beta decay Radioactive decay of an atomic nucleus involving the emission of a *beta particle* or *electron capture.* Cf: *alpha decay.*

beta diagram In structural petrology, a stereographic or equal-area projection upon which the intersections of all surfaces of common origin are plotted, taken two at a time. Usually written as β diagram. Not widely used today, chiefly because from n planes, $n\ (n\text{-}1)/2\ \beta$ intersections are produced from only n lots of data. β diagrams may contain spurious maxima that have no structural significance.

beta-fergusonite A monoclinic mineral: $YNbO_4$. It is dimorphous with fergusonite.

beta-fergusonite-(Ce) A monoclinic mineral: $(Ce,La)NbO_4$. Syn: *brocenite.*

betafite A yellow, brown, greenish, or black mineral of the pyrochlore group: $(Ca,Na,U)_2(Nb,Ta)_2O_6(O,OH)$. It is a uranium-rich variety of pyrochlore found in granitic pegmatites near Betafo, Madagascar. Betafite probably forms a continuous series with pyrochlore; the name is assigned to members of the series with uranium greater than 15%. Syn: *ellsworthite; hatchettolite; blomstrandite.*

beta particle A particle, emitted from an atomic nucleus during a type of radioactive decay, which is physically identical with either the electron or the positron. Cf: *alpha particle; gamma radiation.* Less-preferred syn: *beta ray.*

beta quartz The polymorph of quartz that is stable from 573°C to 870°C, that has a vertical axis of six-fold symmetry and six horizontal axes of two-fold symmetry, and that has a lower refractive index and birefringence than those of *alpha quartz.* It occurs as phenocrysts in quartz porphyries, graphic granite, and granite pegmatites. Also spelled: *β-quartz.* Syn: *high quartz.*

beta ray A less-preferred syn. of *beta particle.*

beta-roselite A triclinic mineral: $Ca_2Co(AsO_4)_2 \cdot 2H_2O$. It is dimorphous with roselite.

beta-uranophane A yellow, monoclinic secondary mineral: $Ca(UO_2)_2Si_2O_7 \cdot 6H_2O$. It is a dimorph of uranophane. Syn: *beta-uranotile.*

beta-uranotile *beta-uranophane.*

beta-vredenburgite An exsolution mixture or oriented intergrowth of jacobsite and hausmannite. Cf: *alpha-vredenburgite.*

betekhtinite An orthorhombic mineral: $Cu_{10}(Fe,Pb)S_6$.

bet lands A term used in India and Pakistan for a flood plain. Etymol: anglicization of Panjabi *bét.*

betpakdalite A lemon-yellow mineral: $CaFe_2H_8(AsO_4)_2(MoO_4)_5 \cdot 10H_2O$.

betrunked river A river that is shorn of its lower course by *be-*

trunking. Cf: *dismembered stream.*

betrunking The removal of the lower part of a stream course by submergence of a valley or by recession of the coast, leaving the several upper branches of the drainage system to enter the sea as independent streams. Cf: *dismembering; beheading.*

betwixt mountains *Zwischengebirge.*

beudantite A green to black rhombohedral mineral: $PbFe_3(AsO_4)(SO_4)(OH)_6$.

beusite A mineral: $(Mn,Fe,Ca,Mg)_3(PO_4)_2$. It is related to graftonite. Syn: *magniophilite.*

bevel (a) Any surface that has been or appears to have been planed off or beveled, such as a flat surface along the crest of a cuesta; esp. an inclined surface that meets another at other than right angles, such as the slope produced by subaerial erosion above the vertical face of a sea cliff. (b) The angle made by a bevel. Also, the slant or inclination of a bevel.

bevel cut Any style of cutting for a gemstone having a very large table and a pavilion that may be step cut, brilliant cut, or any other style; it is used predominantly for less valuable gems. See also: *table cut.*

beveled Said of a geologic structure or landform that is truncated or cut across by an erosion surface.

beveling An act or instance of cutting across a geologic structure or landform; e.g. the planation of an anticline or of the outcropping edges of strata on a mountain summit. Cf: *truncation.* Also spelled: *bevelling.*

beyerite A yellow mineral: $(Ca,Pb)Bi_2(CO_3)_2O_2$. Not to be confused with *bayerite*.

bezel (a) The portion of a brilliant-cut gem above the girdle; the *crown.* (b) More specif. the sloping surface between the girdle and the table, or only a small part (the so-called "setting edge") of that sloping surface just above the girdle.

bezel facet One of the eight large four-sided facets on the crown of a round brilliant-cut gem, the upper points joining the table and the lower points joining the girdle. Cf: *star facet.*

BF *bright field.*

B-form The microspheric form of a foraminifer. Cf: *A-form.*

bhabar A great piedmont composed of gravel and fringing the outer margin of the Siwalik Range in northern India. Etymol: Urdu-Hindi *bhābar,* "porous". Syn: *bhabbar.*

bhangar A term used in India for a high area (as a terrace, scarp, or hill) consisting of an old alluvial plain so situated within a river valley that it is beyond the reach of river floods. Etymol: Urdu-Hindi *bāngar.* Cf: *khadar.* Syn: *bangar.*

bhel *bel.*

BHGM *borehole gravity meter.*

bhil A term applied in the Ganges delta region of India to a brackish and stagnant body of water (such as an oxbow lake or marsh) occupying an *interdistributary bay* that is often below sea level. Etymol: Bengali. Cf: *jheel.* Also spelled: *bil; bheel.*

bhit A term used in West Pakistan for a sand hill or sand ridge. Etymol: Sindhi.

B horizon A *soil horizon* below the A horizon, and characterized by one or more of the following conditions: an illuvial accumulation of humus or clay, iron, or aluminum; a residual accumulation of sesquioxides or clays; darker, stronger, or redder coloring due to the presence of sesquioxides; a blocky or prismatic structure. B1, B2, and B3 horizons may be distinguished; B1 and B3 are transitional upward and downward, respectively. The B horizon is also called the *zone of accumulation [soil]* or the *zone of illuviation.* Partial syn: *subsoil.*

BHP *bottom-hole pressure.*

bhur A term used in India and Pakistan for a hill or patch of windblown sandy soil, frequently capping the high bank of a river. Etymol: Urdu-Hindi *bhūr.*

bianchite A monoclinic mineral occurring as white crystalline crusts: $(Zn,Fe)SO_4 \cdot 6H_2O$.

Biarritzian European stage: upper part of the Middle Eocene, essentially equivalent to Bartonian stage.

bias A purposeful or accidental distortion of observations, data, or calculations in a nonrandom manner.

biaxial Said of a crystal having two optic axes and three indices of refraction, e.g. of an orthorhombic, monoclinic, or triclinic crystal. Cf: *uniaxial.*

biaxial figure An *interference figure* that may display both axes or no axes.

biaxial stress A stress system in which only one of the three principal stresses is zero.

bicchulite A cubic mineral: $Ca_2Al_2SiO_7 \cdot H_2O$.

biconvex Convex on both sides; e.g. said of a brachiopod shell having both valves convex.

bidalotite *anthophyllite.*

bideauxite A mineral: $Pb_2AgCl_3(F,OH)_2$.

bieberite A flesh-red to rose-red mineral occurring esp. in crusts and stalactites: $CoSO_4 \cdot 7H_2O$. Syn: *red vitriol; cobalt vitriol.*

bielenite A *peridotite* that contains olivine and various pyroxenes. It differs from *lherzolite* in containing more pyroxene than olivine. Clinopyroxene, orthopyroxene, chromite, and magnetite are commonly present. The name is from the Biele River, Czechoslovakia. Not recommended usage.

bif *banded iron formation.*

bifacies A term used by Bailey & Childers (1977) in uranium exploration to refer to "varicolored formations characterized by both oxidized and reduced facies." Cf: *monofacies.*

bifoliate Pertaining to a *bilamellar* bryozoan colony.

biforaminate Said of a foraminifer (such as *Discorbis*) having both protoforamen and deuteroforamen.

biform Having forms of two distinct kinds; e.g. said of the rhabdosome of a graptoloid (esp. of a monograptid) with thecae of two conspicuously different shapes, or said of foraminiferal shells having a growth plan that changes during ontogeny. Obsolete var: *biformed.*

biform processes In a spore, exine projections with broad bases that abruptly terminate in sharply pointed tips. The term is most commonly applied to Paleozoic spores.

bifurcating link A link of magnitude μ that is formed at its upstream fork by the confluence of two links, each of a magnitude of $1/2\mu$, and that flows at its downstream fork into a link of a magnitude of less than 2μ (Mock, 1971, p. 1559). Symbol: B. Cf: *link [geomorph]; magnitude [geomorph].*

bifurcation (a) The separation or branching of a stream into two parts. (b) A stream branch produced by bifurcation.

bifurcation ratio Ratio of the number of streams of a given order to the number of streams of the next higher order. According to the law of stream numbers, the ratio tends to be constant for all orders of streams in the basin. It is a measure of the degree of branching within a drainage network. Symbol: R_b.

"big bang" hypothesis The hypothesis that the currently observed expansion of the Universe may be extrapolated back to a primeval cosmic fireball. Depending on the ratio of the initial expansion velocity to the mass of the Universe, which is relatable to currently observable parameters (the deceleration parameter), the Universe may or may not reach a maximum distension and collapse in on itself. Syn: *fireball hypothesis; primeval-fireball hypothesis.*

bight [coast] (a) A long gradual bend or gentle curve, or a slight crescent-shaped indentation, in the shoreline of an open coast or of a bay; it may be larger than a bay, or it may be a segment of or a smaller feature than a bay. (b) A tract of water or a large bay formed by a bight; an *open bay.* Example: the Great Australian Bight. (c) A term sometimes, although rarely, applied to a bend or curve in a river, or in a mountain chain.

bight [ice] An extensive crescent-shaped indentation in the ice edge, formed by either wind or current. Syn: *bay [ice]; ice bay.*

bigwoodite A medium-grained plutonic rock consisting chiefly of microcline, microcline-microperthite, sodic plagioclase (albite), and sodic amphibole, with sodic pyroxene or biotite sometimes substituting for the amphibole; an alkalic *syenite.* It was named by Quirke in 1936 for Bigwood, Ontario, Canada. Not recommended usage.

Biharian European stage: Middle Quaternary (above Villafranchian, below Olderburgian).

bikitaite A white monoclinic mineral: $LiAlSi_2O_6 \cdot H_2O$.

bil *bhil.*

bilamellar (a) Said of a cheilostome bryozoan colony consisting of two layers of zooids growing back to back with separate but touching basal walls. Syn: *bilaminar.* (b) Said of the walls of each chamber (in hyaline calcareous foraminifers) consisting of two primary formed layers.

bilateral symmetry The condition, property, or state of having the individual parts of an organism arranged symmetrically along the two sides of an elongate axis or having a median plane dividing the organism or part into equivalent right and left halves so that they are counterparts one of the other. Cf: *radial symmetry.* Syn: *bilateralism.*

bilinite A white to yellowish mineral: $Fe^{+2}Fe_2^{+3}(SO_4)_4 \cdot 22H_2O$. It occurs in radially fibrous masses.

bill A long, narrow promontory or headland, or a small peninsula, resembling a beak or ending in a prominent spur; e.g. Portland Bill in Dorset, England.

billabong (a) A term applied in Australia to a blind channel leading out from a river and to a usually dry stream bed that may be filled seasonally. (b) An Australian term for an elongated, stagnant backwater or pool produced by a temporary overflow from a stream, or for an *oxbow lake* that may not be permanently filled with water.—Etymol: aboriginal term meaning "dead river".

billietite An amber-yellow secondary mineral: $BaU_6O_{19} \cdot 11H_2O$. It occurs in orthorhombic plates and is closely related to becquerelite.

billingsleyite An orthorhombic mineral: $Ag_7(Sb,As)S_6$.

billitonite An Indonesian tektite from Belitung (Billiton) Island, near Sumatra; a tektite from the East Indies.

bilobite A *trace fossil* consisting of a two-lobed (bilobate) trail; esp. a shallow pocketlike pit, passage, or burrow shoveled or scratched by a trilobite, or a coffee-bean form with a median groove and transverse wrinkles representing a resting trail made by a trilobite.

biloculine Having two chambers; specif. pertaining to or shaped like *Pyrgo* ("Biloculina"), a genus of calcareous imperforate foraminifers having a two-chambered exterior part of the test and found abundantly in the North Sea where their remains form much of the ooze covering the bottom.

bimaceral Said of a coal microlithotype consisting of two macerals. Cf: *monomaceral; trimaceral.*

bimagmatic Nonrecommended generic term for porphyritic rocks in which the minerals occur in two generations. A translation of the German term *bimagmatisch* (Johannsen, 1939, p.203).

bimodal distribution A *frequency distribution* characterized by two localized modes, each having a higher frequency of occurrence than other immediately adjacent individuals or classes. Cf: *polymodal distribution.*

bimodal sediment A sediment whose particle-size distribution shows two maxima; e.g. many coarse alluvial gravels.

binary granite A syn. of *two-mica granite,* used by Keyes in 1895. Not recommended usage.

binary sediment A sediment consisting of a mixture of two components or end members; e.g. a sediment with one clastic component (such as quartz) and one chemical component (such as calcite); or an aggregate containing sand and gravel.

binary system A chemical system containing two components, e.g. the $MgO\text{-}SiO_2$ system.

bind A British coal miner's term for any fine-grained, well-laminated rock (such as shale, clay, or mudstone, but not sandstone) associated with coal. See also: *bend [sed]; blaes.*

binder (a) The material that produces or promotes consolidation in loosely aggregated sediments; e.g. a mineral cement that is precipitated in the pore spaces between grains and that holds them together, or a primary clay matrix that fills the interstices between grains. (b) *soil binder.* (c) A term used in Ireland for a bed of sand in shale, slate, or clay. (d) A coal miner's term used in Pembrokeshire, England, for shale.

bindheimite A secondary mineral: $PB_2Sb_2O_6(O,OH)$.

binding coal *caking coal.*

Bingham substance An idealized material showing linear-viscous behavior above a yield point. Below the yield point the material is presumed to be rigid.

binnacle A term applied in New York and Pennsylvania to a secondary channel of a stream. Etymol: Dutch *binnenkil,* "within channel". Syn: *binnekill.*

binnite A variety of tennantite containing silver.

binocular microscope A microscope adapted to the simultaneous use of both eyes.

binomen Two Latin or Latinized words which, taken together, are the name of a *species.* The first word is its *generic* name and the second its *specific* name. Syn: *binomial.*

binomial n. A syn of *binomen.*

binomial distribution A discrete *frequency distribution* of independent events or occurrences with only two possible outcomes, such as zero and one, or success and failure.

binomial nomenclature A system of naming plants and animals in which the name of each species consists of two words (i.e. a *binomen*), the first designating the genus and the second the particular species; e.g. *Phacops rana.* Syn: *binominal nomenclature.*

bioaccumulated limestone A limestone consisting predominantly of shells and other fossil material that were derived from sedentary but noncolonial organisms and that accumulated essentially in place. It is characterized by an abundance of unsorted and unbroken fossils, diverse organic components, and scarce fine-grained matrix. Cf: *bioconstructed limestone.*

biocalcarenite A calcarenite containing abundant fossils or fossil fragments; e.g. a crinoidal limestone.

biocalcilutite A *calcilutite* containing abundant fossils or fossil fragments.

biocalcilyte A term used by Grabau (1924, p. 297) for a calcareous biogenic clastic rock, such as coral rock, shell rock, or calcareous ooze. The modernized form "biocalcilite" might be useful (Thomas, 1960).

biocalcirudite A *calcirudite* containing abundant fossils or fossil fragments.

biocalcisiltite A *calcisiltite* containing abundant fossils or fossil fragments.

biocenology The branch of ecology concerned with all aspects of natural communities and the relationships between the members of those communities. Also spelled: *biocoenology.* Cf: *biosociology.*

biocenosis *biocoenosis.*

biochemical oxygen demand The amount of oxygen, measured in parts per million, that is removed from aquatic environments rich in organic material by the metabolic requirements of aerobic microorganisms. Abbrev: BOD. Cf: *chemical oxygen demand.* Syn: *biological oxygen demand.*

biochemical rock A sedimentary rock characterized by, or resulting directly or indirectly from, the chemical processes and activities of living organisms; e.g. bacterial iron ores and certain limestones.

biochore (a) A region with a distinctive fauna and/or flora; specif. one or more similar *biotopes [ecol].* (b) The part of the Earth's surface having a life-sustaining climate, characterized by a major type of vegetation. It consists largely of the *dendrochore.*

biochron First defined by Williams (1901, p. 579-580) as the time of duration of a fauna or flora, the term today signifies the "total time represented by a biozone" (ISG, 1976, p. 48).

biochronologic unit (a) A division of time distinguished on the basis of biostratigraphic or objective paleontologic data; a *geologic-time unit* during which sedimentation of a biostratigraphic unit took place (Teichert, 1958a, p. 117). Examples include a moment corresponding to a biostratigraphic zone, and a biochron equivalent to a time-stratigraphic biozone. (b) A term used by Jeletzky (1956, p. 700) to replace biostratigraphic unit and time-stratigraphic unit, being a material rock unit defined in its type locality "by agreement among specialists, elsewhere by criteria of time-correlation found in the contained rocks, which in practice means geochronologically valuable fossils"; a time-stratigraphic unit considered as a biostratigraphic unit.

biochronology *Geochronology* based on the relative dating of geologic events by biostratigraphic or paleontologic methods or evidence; i.e. the study of the relationship between geologic time and organic evolution. See also: *orthochronology; parachronology.*

biochronostratigraphic unit A term used by Henningsmoen (1961) for a time-stratigraphic unit based on fossil evidence.

bioclast (a) A single fossil fragment (Carozzi & Textoris, 1967, p.3). (b) Material derived from "the supporting or protective structures of animals or plants, whether whole or fragmentary" (Sander, 1967, p.327).

bioclastic rock (a) A rock consisting primarily of fragments that are broken from pre-existing rocks, or are pulverized or arranged, by the action of living organisms, such as plant roots or earthworms (Grabau, 1904). The rock need not consist of organic material. The term includes "rocks" (such as concrete) that owe their existence to man's activities. (b) A sedimentary rock consisting of fragmental or broken remains of organisms, such as a limestone composed of shell fragments. Cf: *biogenic rock.*

biocoenology *biocenology.*

biocoenosis (a) A set of fossil remains found in the same place where the organisms lived. Cf: *thanatocoenosis.* Syn: *life assemblage.* (b) A group of organisms that live closely together and form a natural ecologic unit. The term was first defined and introduced by the German zoologist Moebius in 1877. Cf: *community.*—Var: *biocenosis; biocoenose; biocenose.* Plural: *biocoenoses.* Etymol: Greek *bios,* "mode of life" + *koinos,* "general, common".

bioconstructed limestone A limestone consisting predominantly of material resulting from the vital activities of colonial and sediment-binding organisms (such as algae, corals, bryozoans, and stromatoporoids) that erect three-dimensional frameworks in place. Cf: *bioaccumulated limestone.*

biocycle A group of related *biochores* that comprise one of the major divisions of the biosphere; salt water, fresh water, and dry land are biocycles.

biodegradable Subject to decomposition by micro-organisms.

bioecology The branch of ecology concerned with the relationships between plants and animals in their common environment.

bioerosion Removal of consolidated mineral or lithic substrate by the direct action of organisms (Neumann, 1966).

biofacies (a) A subdivision of a stratigraphic unit, distinguished from adjacent subdivisions on the basis of its fossils, without respect to nonbiologic features; esp. such a body of sediment or rock recognized by characters that do not affect lithology, such as the taxonomic identity or environmental implications of fossils (Weller, 1958, p. 634). (b) The biologic aspect or fossil character of a facies of some definite stratigraphic unit, esp. considered as an expression of local biologic conditions; "the total biological characteristics of a sedimentary deposit" (Moore, 1949, p. 17). (c) A distinctive assemblage of organisms formed under one set of environmental conditions, as compared with another assemblage formed at the same time but under different conditions; an ecological association of fossils, or the fossil record of a biocoenosis.—See also: *facies.* Syn: *biologic facies.*

biofacies map A *facies map* based on paleontologic attributes, showing areal variation in the overall paleontologic character of a given stratigraphic unit. It may be based on proportions or population discriminants of the fossil organisms present, or on ratios among them.

biogenesis (a) Formation by the action of organisms; e.g. coral reefs. (b) The doctrine that all life has been derived from previously living organisms. Cf: *abiogenesis.*

biogenetic law The so-called "law" of recapitulation: ontogeny recapitulates phylogeny.

biogenetic rock *biogenic rock.*

biogenic rock An *organic rock* produced directly by the physiological activities of organisms, either plant or animal (Grabau, 1924, p. 280); e.g. coral reefs, shelly limestone, pelagic ooze, coal, and peat. Cf: *bioclastic rock; biolith.* See also: *phytogenic rock; zoogenic rock.* Syn: *biogenous rock; biogenetic rock.*

biogenous rock *biogenic rock.*

biogeochemical cycling The cycling of chemical constituents through a biological system.

biogeochemical prospecting *Geochemical exploration* based on the chemical analysis of systematically sampled plants in a region, in order to detect biological concentrations of elements that might reflect hidden ore bodies. The trace-element content of one or more plant organs is most often measured. Cf: *geobotanical prospecting.*

biogeochemistry A branch of geochemistry that deals with the effects of life processes on the distribution and fixation of chemical elements in the biosphere.

biogeography The science that deals with the geographic distribution of all living organisms. See also: *zoogeography; phytogeography.* Syn: *chorology.*

biogeology The biological aspects of geology, e.g. systematic paleontology, the study of organically influenced sedimentation, or the identification of a concealed rock unit by the type of plant growing at the surface above it. See also: *geobiology; paleobiology; paleontology.*

bioglyph A *hieroglyph* produced by an organism or of biologic origin (Vassoevich, 1953, p.38). Ant: *abioglyph.* Cf: *trace fossil.* Syn: *organic hieroglyph.*

bioherm A moundlike, domelike, lenslike, or reeflike mass of rock built up by sedentary organisms (such as corals, algae, foraminifers, mollusks, gastropods, and stromatoporoids), composed almost exclusively of their calcareous remains, and enclosed or surrounded by rock of different lithology; e.g. an *organic reef* or a nonreef limestone mound. Term proposed by Cumings & Shrock (1928, p. 599), and defined by Cumings (1930, p. 207), as a structural term, although as applied it often stresses calcareous composition. Cf: *biostrome.* Syn: *organic mound.*

biohermal Pertaining to a bioherm, such as a "biohermal limestone" of restricted extent.

biohermite A term used by Folk (1959, p. 13) for a limestone composed of debris broken from a bioherm and forming pocket-fillings or talus slopes associated with reefs; it was formerly used by Folk for a limestone now better known as *biolithite.*

biohorizon A surface of biostratigraphic change or of distinctive biostratigraphic character, esp. valuable for correlation; it is commonly a *biozone* boundary, although biohorizons are sometimes recognized within biozones. In theory, a biohorizon is strictly a surface or interface; in practice, it may be a thin biostratigraphically distinctive bed. Features on which biohorizons are commonly based include "first appearance" of a given fossil form, "last occurrence", change in frequency and abundance, evolutionary change, and change in the character of individual taxons, e.g. in direction of coiling in foraminifers, or in number of septa in corals (ISG, 1976, p. 49). Cf: *chronohorizon; lithohorizon.*

biohydrology The study of the interactions of water, plants, and animals, including both the effects of water on biota and the physical and chemical changes in water or its environment caused by biota. Cf: *hydrobiology.*

biointerval-zone *interval-zone.*

biokinematic Said of sedimentary operations in which "the largest displacement vectors occur between a living organism and the unmodified deposit surrounding the structure produced" (Elliott, 1965, p.196); e.g. the activities shown by trace fossils. Also, said of the sedimentary structures produced by biokinematic operations.

biolite [mineral] A group name for minerals formed by biologic action (Hey, 1963, p. 92).

biolite [sed] (a) *biolith.* (b) An old term for a concretion formed through the action of living organisms.

biolith A rock of organic origin or composed of organic remains; specif. *biogenic rock.* See also: *phytolith; zoolith.* Syn: *biolite.*

biolithite A limestone constructed by organisms (faunal or floral) that grew and remained in place, characterized by a rigid framework of carbonate material that binds allochem grains and skeletal elements. It is typical of reef cores. The major organism should be specified when using the term; e.g. "coral biolithite", "algal-mat biolithite", or "rudist biolithite". See also: *biohermite.*

biological oceanography The study of the plant and animal life of the oceans.

biological oxygen demand *biochemical oxygen demand.*

biologic artifact An organic compound whose chemical structure demonstrates its derivation from living matter.

biologic facies A syn. of *biofacies* as that term is used in stratigraphy and in ecology; e.g. coral reefs and shell banks are "biologic facies" characterized by the organisms themselves.

biologic time scale An uncalibrated *geologic time scale,* based on organic evolution, giving the relative order for a succession of events. Cf: *relative time scale.*

biologic weathering *organic weathering.*

biology The study of all organisms, esp. living ones; includes *neontology* and *paleontology,* but most often is used to imply neontology alone.

biolysis Death and subsequent disintegration of the body.

biomass The amount of living material in a particular area, stated in terms of the weight or volume of organisms per unit area or of the volume of the environment. Syn: *standing crop.*

biome A climax community that characterizes a particular natural region; esp. a particular type of vegetation, climatically bounded, which dominates a large geographic area. Partial syn: *biotic formation.*

biomechanical Said of a rock or deposit formed by detrital accumulation of organic material.

biomere A term proposed by Palmer (1965, p.149-150) for "a regional biostratigraphic unit bounded by abrupt nonevolutionary changes in the dominant elements of a single phylum". The changes are not necessarily related to physical discontinuities, and they may be diachronous.

biometrics Statistics as applied to biologic observations and phenomena.

biomicrite A limestone consisting of a variable proportion of skeletal debris and carbonate mud (micrite); specif. a limestone containing less than 25% intraclasts and less than 25% ooliths, with a volume ratio of fossils and fossil fragments to pellets greater than 3 to 1, and the carbonate-mud matrix more abundant than the sparry-calcite cement (Folk, 1959, p. 14). It is characteristic of environments of relatively low physical energy. The major organism should be specified when using the term; e.g. "crinoid biomicrite" or "brachiopod biomicrite". See also: *sparse biomicrite; packed biomicrite.*

biomicrosparite A biomicrite in which the carbonate-mud matrix has recrystallized to microspar (Folk, 1959, p. 32); a *microsparite* containing fossils or fossil fragments.

biomicrudite A *biomicrite* containing fossils or fossil fragments that are more than one millimeter in diameter.

biomineral A mineral substance "of obviously organic origin", e.g. francolite in a *Lingula* shell or the apatitic mineral of teeth and bones (McConnell, 1973, p. 425-426).

biomineralogy The systematic study of biominerals. See: McConnell, 1973.

biomorphic Pertaining to or incorporating the forms of organisms; e.g. "biomorphic sediments" containing fossil forms.

bionomics *ecology.*

biopelite An organic pelite; specif. a *black shale.*

biopelmicrite A limestone intermediate in content between biomicrite and pelmicrite; specif. a limestone containing less than 25% intraclasts and less than 25% ooliths, with a volume ratio of fossils and fossil fragments to pellets ranging between 3 to 1 and 1 to 3, and the carbonate-mud matrix (micrite) more abundant than the sparry-calcite cement (Folk, 1959, p. 14).

biopelsparite A limestone intermediate in content between biosparite and pelsparite; specif. a limestone containing less than 25% intraclasts and less than 25% ooliths, with a volume ratio of fossils and fossil fragments to pellets ranging between 3 to 1 and 1 to 3, and the sparry-calcite cement more abundant than the carbonate-mud matrix (micrite) (Folk, 1959, p. 14).

biophile (a) Said of those elements that are the most typical in organisms and organic material (Rankama & Sahama, 1950, p.88). (b) Said of those elements that are concentrated in and by living plants and animals (Goldschmidt, 1954, p.26).

biopyribole A mnemonic term coined by Johannsen in 1911 in his classification of igneous rocks to indicate the presence of an otherwise unidentified biotite, pyroxene, or amphibole. Etymol: *bio*tite + *pyr*oxene + amph*ibole.*

biorhexistasy The name given to a theory of sediment production related to variations in the vegetational cover of the land surface, and characterized by long-term stable subtropical weathering conditions resulting in lateritic soils accompanied by removal of calcium, silica, alkalies, and alkaline earths (Erhart, 1956). See also: *rhexistasy; biostasy.*

bioseries "A morphogenetic sequence marked by stages in the progressive structural development of morphological characters of index fossils" (Glaessner, 1945, p. 225).

biosociology The branch of ecology concerned with the social behavior of organisms in communities. Cf: *biocenology.*

biosome (a) A term proposed by Wheeler (1958a, p.648) for an ecologically controlled biostratigraphic unit that is mutually intertongued with one or more biostratigraphic units of differing character; the biostratigraphic equivalent of *lithosome.* Cf: *holosome.* (b) A term used by Sloss (in Weller, 1958, p.625) for a "body of sediment deposited under uniform biological conditions"; the record of a uniform biologic environment or of a biotope; a three-dimensional rock mass of uniform paleontologic content.—Not to be confused with *biostrome.* Cf: *biotope [stratig].*

biospace As used by Valentine (1969, p.686), that part of the *environmental hyperspace lattice* that actually represents conditions existing on the Earth. Syn: *realized ecological hyperspace.*

biosparite A limestone consisting of a variable proportion of skeletal debris and clear calcite (spar); specif. a limestone containing less than 25% intraclasts and less than 25% ooliths, with a volume ratio of fossils and fossil fragments to pellets greater than 3 to 1, and the sparry-calcite cement more abundant than the carbonate-mud matrix (micrite) (Folk, 1959, p. 14). It is generally characteristic of high-energy carbonate environments, with the spar being normally a pore-filling cement. According to Folk (1962), further textural subdivision may be made into "unsorted biosparite", "sorted biosparite", and "rounded biosparite". The major organism should be specified when using the term; e.g. "trilobite biosparite" or "pelecypod biosparite".

biosparrudite A *biosparite* containing fossils or fossil fragments that are more than one millimeter in diameter.

biospecies A *species* defined on the basis of observed interbreeding capability and potential.

biospeleology The scientific study of the organisms that live in caves.

biosphere (a) All the area occupied or favorable for occupation by living organisms. It includes parts of the lithosphere, hydrosphere, and atmosphere. Cf: *ecosphere.* (b) All living organisms of the Earth and its atmosphere.

biostasy Maximum development of organisms at a time of tectonic repose when residual soils form extensively on land and deposition of calcium carbonate is widespread in the sea. See also: *rhexistasy; biorhexistasy.*

biostratic unit *biostratigraphic unit.*

biostratigraphic classification The organization of strata into units based on their fossil content (ISG, 1976, p. 48).

biostratigraphic unit A stratum or body of strata that is "unified by its fossil content or paleontological character and thus differentiated from adjacent strata. A biostratigraphic unit is present only within the limits of observed occurrence of the particular biostratigraphic feature on which it is based" (ISG, 1976, p. 48). The fundamental unit is the *biozone.* If fossil remains are so abundant that in themselves they become lithologically important, a biostratigraphic unit may also be a lithostratigraphic unit. Syn: *biostratic unit.*

biostratigraphy Stratigraphy based on the paleontologic aspects of rocks, or stratigraphy with paleontologic methods; specif. the separation and differentiation of rock units on the basis of the description and study of the fossils they contain. The term was apparently proposed by Louis Dollo, Belgian paleontologist, in 1904 in a wider sense for the entire research field in which paleontology exercises a significant influence upon historical geology. Cf: *stratigraphic paleontology.*

biostratinomy *biostratonomy.*

biostratonomy The branch of paleoecology concerned with the relationships between organisms and their environment after death and before and during burial. Cf: *taphonomy; Fossildiagenese.* Also spelled: *biostratinomy.*

biostromal Pertaining to a biostrome, e.g. "biostromal limestone".

biostrome A distinctly bedded and widely extensive or broadly lenticular, blanketlike mass of rock built by and composed mainly of the remains of sedentary organisms, and not swelling into a moundlike or lenslike form; an "organic layer", such as a bed of shells, crinoids, or corals, or a modern reef in the course of formation, or even a coal seam. Term proposed by Cumings (1932, p. 334). Cf: *bioherm.* Not to be confused with *biosome.*

biota All living organisms of an area; the flora and fauna considered as a unit.

Biot-Fresnel law A statement in crystallography that the directions of extinction in any section of a biaxial crystal are parallel to the traces, on that section, of the planes bisecting the angles between the two planes containing the normal to the section and the optic axes.

biotic Of or pertaining to life or the mode of living of plants and animals collectively.

biotic community *community.*

biotic factor A factor of a biological nature, such as availability of food, competition between species, and predator-prey relationships, that affects the distribution and abundance of species.

biotic formation (a) *biome.* (b) In botany, a broad natural unit consisting of distinctive plants in a climax community.

biotic province A geographic area that supports one or more ecologic associations which are distinct from those of adjacent provinces.

biotic succession An ambiguous term improperly used for either *seral* succession or *faunal succession;* its use is not recommended.

biotite (a) A widely distributed and important rock-forming mineral of the mica group: $K(Mg,Fe^{+2})_3(Al,Fe^{+3})Si_3O_{10}(OH)_2$. It is generally black, dark brown, or dark green, and forms a constituent of crystalline rocks (either as an original crystal in igneous rocks of all kinds or a product of metamorphic origin in gneisses and schists) or a detrital constituent of sandstones and other sedimentary rocks. Biotite is useful in the potassium-argon method of age determination. (b) A general term to designate all ferromagnesian micas.—Syn: *black mica; iron mica; magnesia mica.*

biotitite An ultramafic igneous rock almost entirely composed of biotite. Syn: *glimmerite.*

biotitization Introduction of, or replacement by, biotite.

biotope [ecol] (a) An area of uniform ecology and organic adaptation (Hesse et al., 1937, p.135); the habitat, or physical basis, of a uniform community of animals and plants adapted to its environment; a limited region characterized by certain environmental conditions under which the existence of a given biocoenosis is possible. It is more or less ephemeral and at any moment it is circumscribed by a boundary that is subject to expansion, contraction, or other shift in position. Cf: *biochore.* (b) An association of organisms characteristic of a particular geographic area. See also: *paleobiotope.*

biotope [stratig] The environment under which an assemblage of plants or animals lives or lived (Wells, 1947; ISSC, 1971, p. 32). It is an ecologic term, which should be used in stratigraphy in a sense of paleontologic environment. However, it has been applied to a biostratigraphic surface or area (Wheeler, 1958a, p. 653-654), a faunal or floral unit (Sloss et al., 1949, p. 95-96), and numerous other entities. Cf: *lithotope; biosome.*

bioturbation The churning and stirring of a sediment by organisms.

biotype Any one of a group of organisms having the same genetic constitution.

biozone A general term for any kind of *biostratigraphic unit* (ISG, 1976, p. 49), e.g. *acme-zone, interval-zone, range-zone,* etc.; the basic unit in biostratigraphic classification and generally the smallest biostratigraphic unit on which intercontinental or worldwide correlations can be established. Biozones "vary greatly in thickness and geographic extent. They may range from a local bed to a unit thousands of meters thick" (ISG, 1976, p. 48). See also: *zone [stratig]; subzone; zonule.* Syn: *biostratigraphic zone.*

bipartite oolith An oolith whose central part is divided into two more or less distinct fractions that differ in texture and/or grain size, so that it displays an asymmetric appearance (Choquette, 1955, p. 338).

biphosphammite A tetragonal mineral: $(NH_4,K)H_2PO_4$.

bipocillus A siliceous monaxonic sponge spicule (microsclere), consisting of a shaft, at each end of which is a transverse, bowl-like, and excentrically attached expansion, the concave surfaces facing one another. Pl: *bipocilli.*

bipolarity The similarity or identity of groups of organisms occurring north and south of, but not in, the equatorial zone.

bipole-dipole array A direct-current resistivity array in which the Earth is energized using a fixed bipole source, and the resultant electric field is mapped at numerous locations using orthogonal pairs of dipoles (closely spaced pairs of electrodes) to determine the magnitude and direction of the electric field.

bipyramid *dipyramid.*

biquartz plate A type of *compensator [optics]* in a polarizing microscope, one half of which is *right-handed [cryst]* and the other half *left-handed [cryst],* with superimposed wedges. It is used to detect accurately the position of extinction. Syn: *Wright biquartz wedge.*

biramous Two-branched; said of a crustacean limb in which the basis bears both exopod and endopod, or said of a trilobite appendage consisting of an outer and an inner branch.

Birch discontinuity A seismic-velocity discontinuity in the *C layer* or transition zone of the upper mantle at about 900 km, caused by phase change or chemical change or both.

bird A geophysical measuring device such as a magnetometer, plus the housing in which it is towed behind an aircraft.

bird-foot delta A delta formed by many levee-bordered distributaries extending seaward and resembling in plan the outstretched claws of a bird; e.g. the Mississippi River delta. Syn: *digitate delta; bird's-foot delta.*

bird's-eye A spot, bleb, tube, or irregular patch of sparry calcite commonly found in limestones (such as *dismicrites*) and some dolomites as a precipitate that infills cavities resulting from localized disturbances, such as algal or burrowing activity, escaping gas bubbles, shrinkage cracking, soft-sediment slumping, reworking of sediments, or plant roots. Also applied to the porosity created by the presence of bird's-eyes in a rock. Also spelled: *birdseye.* Syn: *calcite eye.*

bird's-eye coal Anthracite with numerous small fractures that display its semiconchoidal fracture pattern.

bird's-eye limestone A syn. of *dismicrite.* The term was applied in a titular sense in early New York State reports to the Lowville Limestone, a very fine-textured limestone containing spots or tubes of crystalline calcite or having light-colored specks due in part to a characteristic fossil supposed to be a form of coral and now known as *Tetradium cellulosum* (Wilmarth, 1938, p. 192). Cf: *loferite.*

bird's-eye ore A miner's term used in Arkansas for a variety of pisolitic bauxite characteristic of residual deposits. Also spelled: *birdseye ore.*

bird's-foot delta *bird-foot delta.*

bird track A term applied in the mid-19th century to a dinosaur track before its true character was recognized, and now used to denote footprints or tracks made by ancient birds.

bireflectance The ability of a mineral to change color in reflected polarized light with change in crystal orientation. Cf: *pleochroism.* Syn: *reflection pleochroism.*

birefracting *birefringent.*

birefraction *birefringence.*

birefractive *birefringent.*

birefringence The ability of crystals other than those of the isometric system to split a beam of ordinary light into two beams of unequal velocities; the difference between the greatest and the least indices of refraction of a crystal. Cf: *single refraction.* See also: *positive birefringence; refraction.* Adj: *birefringent.* Syn: *double refraction; birefraction.*

birefringent Said of a crystal that displays *birefringence;* such a crystal has more than one *index of refraction.* Syn: *birefractive; birefracting.*

biringuccite A monoclinic mineral: $Na_4B_{10}O_{17} \cdot 4H_2O$.

birkremite A light-colored quartz-bearing *syenite* containing alkali feldspar and some hypersthene; an orthopyroxene-bearing *kalialaskite.* Also spelled: *bjerkreimite.* The name, given by Kolderup in 1896, is for Birkrem, Norway. Not recommended usage.

birne A syn. of *boule.* Etymol: German *Birne,* "pear".

birnessite A mineral: $(Na,Ca)Mn_7O_{14} \cdot 3H_2O$.

birotulate A sponge spicule (microsclere) having two wheel-shaped ends; e.g. an amphidisc, or one of its derivatives such as a hemidisc, staurodisc, or hexadisc. Syn: *birotule.*

birthstone A stone that has been chosen as appropriate to the time (month) of one's birth. The modern list specifies: January (garnet); February (amethyst); March (bloodstone or aquamarine); April (diamond); May (emerald); June (pearl, moonstone, or alexandrite); July (ruby); August (sardonyx or peridot); September (sapphire); October (opal or tourmaline); November (topaz or citrine); December (turquoise or zircon).

bisaccate *bivesiculate.*

bisbeeite A doubtful mineral: $CuSiO_3 \cdot H_2O$ (?). It may be the equivalent of chrysocolla.

bischofite A white to colorless mineral: $MgCl_2 \cdot 6H_2O$.

biscuit *algal biscuit.*

biscuit-board topography A glacial landscape characterized by a rolling upland on the sides of which are cirques that resemble the bites made by a biscuit-cutter in the edge of a slab of dough; e.g. the Wind River Mountains in Wyoming. It may represent an early or partial stage in glaciation.

bisectrix In a biaxial crystal, a line that bisects either of the complementary angles between the two optic axes of a biaxial crystal. See also: *acute bisectrix; obtuse bisectrix.*

biserial Arranged in, characterized by, or consisting of two rows or series; e.g. a "biserial arm" of a crinoid composed of brachial plates arranged in a double row, a "biserial brachiole" of a cystoid consisting of plates arranged in two rows, a "biserial test" of a foraminifer with chambers in a two-row series, or a "biserial rhabdosome" of a scandent graptoloid composed of two rows of thecae in contact either back-to-back (dipleural) or side-by-side (monopleural). Cf: *uniserial.*

bishop's stone *amethyst.*

bismite A monoclinic mineral: Bi_2O_3. It is straw yellow and earthy or powdery, and is polymorphous with sillenite. Syn: *bismuth ocher.*

bismoclite A pale-grayish or creamy-white mineral: BiOCl. It is isomorphous with daubreeite.

bismuth A rhombohedral mineral, the native metallic element Bi. It is brittle and heavy and commonly occurs in silvery-white or grayish-white (with a pinkish or reddish tinge) and arborescent, foliated, or granular forms.

bismuth blende *eulytite.*

bismuth glance *bismuthinite.*

bismuth gold *maldonite.*

bismuthide A mineral compound that is a combination of bismuth with a more positive element.

bismuthine *bismuthinite.*

bismuthinite A lead-gray to tin-white orthorhombic mineral: Bi_2S_3. It has a metallic luster and an iridescent tarnish, and it usually occurs in foliated, fibrous, or shapeless masses associated with copper, lead, and other ore minerals. Syn: *bismuth glance; bismuthine.*

bismuth ocher A group name for earthy oxides and carbonates of bismuth; specif. *bismite.*

bismuth spar *bismutite.*

bismutite A mineral: $(BiO)_2CO_3$. It is earthy and amorphous, and usually dull white, yellowish, or gray. Syn: *bismuth spar.*

bismutoferrite A mineral: $BiFe_2(SiO_4)_2(OH)$.

bismutotantalite A black mineral: $Bi(Ta,Nb)O_4$.
bisphenoid *disphenoid.*
bistatic radar A *radar* with its transmitter and receiver spatially separated.
bit A general term for *drill bit* or *core bit.*
biteplapallidite *merenskyite.*
biteplatinite *moncheite.*
bitheca The smallest of three tubes regularly produced at each budding in the development of a graptolithine colony. It may have contained a male zooid. Cf: *autotheca; stolotheca.*
bitter lake A *salt lake* whose waters contain in solution a high content of sodium sulfate and lesser amounts of the carbonates and chlorides ordinarily found in salt lakes; a lake whose water has a bitter taste. Examples include Carson Lake in Nevada, and the Great Bitter Lake in Egypt.
bittern (a) The bitter liquid remaining after seawater has been concentrated by evaporation until most of the sodium chloride has crystallized out. See also: *bittern salt.* (b) A natural solution, in an evaporite basin, that resembles a saltworks liquor, esp. in its high magnesium content.
bittern salt Any of the salts that may be extracted from the *bittern* of a saltworks or from a comparable natural solution; e.g. magnesium chloride, magnesium sulfate, bromides, iodides, and calcium chloride.
bitter salts *epsomite.*
bitter spar *dolomite [mineral].*
bitumen A generic term applied to natural inflammable substances of variable color, hardness, and volatility, composed principally of a mixture of hydrocarbons substantially free from oxygenated bodies. Bitumens are sometimes associated with mineral matter, the nonmineral constituents being fusible and largely soluble in carbon disulfide, yielding water-insoluble sulfonation products. Petroleums, asphalts, natural mineral waxes, and asphaltites are all considered bitumens.
bitumenite *torbanite.*
bitumenization (a) *coalification.* (b) Enrichment in hydrocarbons.
bitumicarb Low-rank bituminous matter in coal that is derived from resins, waxes, spores, exines, etc. (Tomkeieff, 1954).
bituminous [coal] Pertaining to bituminous coal.
bituminous [mineral] Said of a mineral having an odor like that of bitumen.
bituminous [sed] (a) Said of a sedimentary rock that is naturally impregnated with, contains, or constitutes the source of *bitumen.* (b) Loosely, said of a substance containing much organic or carbonaceous matter; e.g. "bituminous ore" (iron ore whose gangue consists principally of coaly matter).
bituminous brown coal *pitch coal.*
bituminous coal Coal that ranks between subbituminous coal and anthracite and that contains more than 14% volatile matter (on a dry, ash-free basis) and has a calorific value of more than 11,500 BTU/lb (moist, mineral-matter-free) or more than 10,500 BTU/lb if agglomerating (ASTM). It is dark brown to black in color and burns with a smoky flame. Bituminous coal is the most abundant rank of coal; much is Carboniferous in age. Cf: *high-volatile bituminous coal; medium-volatile bituminous coal; low-volatile bituminous coal.* Syn: *soft coal.*
bituminous fermentation Fermentation of vegetable matter under conditions of no air and abundant moisture. Volatiles are retained, resulting in the formation of bitumens, i.e. peat, coal.
bituminous lignite *pitch coal.*
bituminous limestone A dark, dense limestone containing abundant organic matter, believed to have accumulated under stagnant conditions and emitting a fetid odor when freshly broken or vigorously rubbed; e.g. the Bone Spring Limestone of Permian age in west Texas. See also: *stinkstone; anthraconite.*
bituminous wood *woody lignite.*
bityite A mineral of the mica group: $CaLiAl_2(AlBeSi_2)O_{10}(OH)_2$.
biumbilicate Having a central depression (umbilicus) on each side of a foraminiferal test (as in planispiral forms).
biumbonate Said of a foraminifer having two raised umbonal bosses (as in *Lenticulina*).
bivalve adj. Having a shell composed of two distinct and usually movable valves, equal or subequal, that open and shut. Cf: *univalve.* Syn: *bivalved.*—n. A bivalve animal, such as a rostroconch, a brachiopod, or an ostracode; specif. a mollusk of the class Bivalvia (Pelecypoda), including the clams, oysters, scallops, and mussels, generally sessile or burrowing into soft sediment, having no distinct head, and possessing a hatchet-shaped foot and a sheet-like or lamelliform gill on each side of a bilaterally symmetrical body. See also: *pelecypod.*
bivariant *divariant.*
bivariate Pertaining to or involving two mathematical variables; e.g. "bivariate distribution".
bivesiculate Said of pollen with two vesicles. Bivesiculate pollen are usually the pollen of conifers, but can also occur in other gymnosperms, e.g. Caytoniales and other seed ferns. Syn: *bisaccate; disaccate.*
bivium (a) The two posterior ambulacra of a crinoid or an echinoid. (b) The part of an asterozoan containing the madreporite and the rays on each side of it. This usage is not recommended (TIP, 1966, pt.U, p.29).—Pl: *bivia.* Cf: *trivium.*
bixbyite A black isometric mineral: $(Mn,Fe)_2O_3$. Syn: *partridgeite; sitaparite.*
bizardite An *alnoite* that contains nepheline as an essential phase. The name is for the Ile Bizard, Quebec. Not recommended usage.
bjarebyite A monoclinic mineral: $(Ba,Sr)(Mn,Fe,Mg)_2Al_2(PO_4)_3(OH)_3$.
bjerezite A porphyritic igneous rock in which the phenocrysts of nepheline, pyroxene with acmite rims, elongated andesine laths, and alkali feldspar are contained in a fine-grained groundmass of pyroxene, brown mica, andesine, potassium feldspar, nepheline, analcime, and indeterminate zeolites. The rock was named by Erdmannsdoerfer in 1928 for Bjerez, U.S.S.R. Not recommended usage.
bjerkreimite *birkremite.*
black alkali An old term for an *alkali soil* whose sodium tends to disperse organic matter and give a black color. Cf: *white alkali.*
black alta *alta.*
black amber (a) *jet [coal].* (b) *stantienite.*
black-and-white iceberg An iceberg made up of sharply defined alternating layers of dark, opaque ("black") ice containing dirt and stones, and cleaner, light-colored ("white") ice.
blackband (a) A dark, earthy variety of the mineral siderite, occurring mixed with clay, sand, and considerable carbonaceous matter, and frequently associated with coal. Syn: *blackband ore.* (b) A thin layer (up to 10 cm in thickness) of blackband interbedded with clays or shales in blackband ironstone. (c) *blackband ironstone.*
blackband ironstone A dark variety of *clay ironstone* containing sufficient carbonaceous matter (10-20%) to make it self-calcining (without the addition of extra fuel). Syn: *blackband; blackband ore.*
blackband ore (a) *blackband.* (b) *blackband ironstone.*
blackbody An ideal emitter that radiates energy at the maximum possible rate per unit area at each wavelength for any given temperature. A blackbody also absorbs all the radiant energy incident upon it. No actual substance behaves in this way, although some materials, such as lampblack, approach it.
black chalcedony The correct designation for most so-called *black onyx.*
black chalk A bluish-black carbonaceous clay, shale, or slate, used as a pigment or crayon.
black chert (a) Carbonaceous chert, such as that occurring in South Africa. (b) A term used in England for *flint.* Cf: *white chert.*
black cobalt *asbolite.*
black copper *tenorite.*
black cotton soil *Regur.*
blackdamp A coal-mine gas that is nonexplosive and consists of about 15% carbon dioxide and about 85% nitrogen. Cf: *whitedamp; afterdamp; firedamp.* Syn: *chokedamp.*
black diamond [coal] A syn. of *coal.*
black diamond [mineral] (a) *carbonado.* (b) A black gem diamond. (c) Dense black hematite that takes a polish like metal.
black drift *forest bed.*
black earth [coal] Brown coal that is finely ground and used as a pigment. Syn: *Cologne earth; Cologne umber; Cassel brown; Cassel earth; Vandyke brown.*
black earth [soil] (a) *Chernozem.* (b) More generally, any black soil.
blackfellow's button *australite.*
black gold (a) *maldonite.* (b) Placer gold coated with a black or dark-brown substance (such as a film of manganese oxide) so that the yellow color is not visible until the coating is removed.
black granite A *commercial granite* that when polished is dark gray to black. It may be a diabase, diorite, or gabbro.

black hematite A syn. of *romanechite.* The term is a misnomer because romanechite contains no iron. Cf: *red hematite; brown hematite.*

black ice (a) A clear, thin ice layer, formed on the sea, in rivers or lakes, or on land, that appears dark because of its transparency. (b) Dark glacier ice formed by freezing of silt-laden water. Cf: *blue ice; white ice.* (c) A thin sheet or glaze of dark ice formed when a light rain or drizzle falls on a surface whose temperature is below freezing. Cf: *verglas.*

blackjack [coal] (a) A thin stratum of coal interbedded with layers of *slate;* a slaty coal with a high ash content. (b) A British term for a variety of cannel coal.—Also spelled: *black jack.*

blackjack [mineral] A syn. of *sphalerite.* esp. a dark variety. The term was originated by miners who regarded sphalerite as an impish intrusion ("jack") of worthless material in lead ores. Also spelled: *black jack.*

blackjack [sed] A term used in Arkansas for a soft, black carbonaceous clay or earth associated with coal.

blackland A term used in Texas for a *Vertisol.*

black lead *graphite.*

black-lead ore An old name for the black variety of cerussite.

black light (a) A prospector's and miner's term for ultraviolet light, used in exploration and evaluation to detect mineral fluorescence. (b) An instrument, usually portable, that produces ultraviolet light for this purpose.

black lignite *lignite A.*

black manganese A term applied to dark-colored manganese minerals, such as pyrolusite, hausmannite, and psilomelane.

black metal A *black shale* associated with coal measures.

black mica *biotite.*

blackmorite A yellow variety of opal from Mount Blackmore, Mont.

black mud A type of *mud [marine geol]* whose dark color is due to hydrogen sulfide, developed under anaerobic conditions. Syn: *hydrogen-sulfide mud; reduced mud; euxinic mud.*

black ocher *wad [mineral].*

black onyx The popular name for *black chalcedony,* usually artificially colored. Although the word "onyx" is not quite accurate (except for banded material), it has come to be accepted as the usual term for solid-color chalcedony.

black opal A form of precious opal whose *play of color* (usually red or green) is displayed against a dark gray (rarely black) body color; e.g. the fine Australian blue opal with flame-colored flashes. Cf: *white opal.*

black prairie A *prairie* with rich, dark soil.

Blackriverian North American substage: Middle Ordovician (lower Mohawkian Stage), below the Trentonian Substage. See also: *Wilderness.*

blacks (a) Highly carbonaceous black shale; impure cannel coal. (b) A British term for dark coaly shale, clay, or mudstone.

black sand (a) An alluvial or beach sand consisting predominantly of grains of heavy, dark minerals or rocks (e.g. magnetite, rutile, garnet, or basaltic glass), concentrated chiefly by wave, current, or surf action. It may yield valuable minerals. (b) An asphaltic sand.

black-sand beach A beach containing a large quantity of *black sand* concentrated by the action of waves and currents.

black shale (a) A dark, thinly laminated carbonaceous shale, exceptionally rich in organic matter (5% or more carbon content) and sulfide (esp. iron sulfide, usually pyrite), and often containing unusual concentrations of certain trace elements (U, V, Cu, Ni). It is formed by partial anaerobic decay of buried organic matter in a quiet-water, reducing environment (such as in a stagnant marine basin) characterized by restricted circulation and very slow deposition of clastic material. Fossil organisms (principally planktonic and nektonic forms) are preserved as a graphitic or carbonaceous film or as pyrite replacements. Syn: *biopelite.* (b) A finely laminated, sometimes canneloid, carbonaceous shale often found as a roof to a coal seam (Tomkeieff, 1954, p.29). Syn: *black metal.*

black silver *stephanite.*

black tellurium *nagyagite.*

black tin *cassiterite.*

blackwall Originally a quarrymen's term for a tabular body of black or dark-colored mica- or chlorite-rich rock, it is now used in metamorphic petrology for black or dark-colored almost monomineralic biotite, chlorite, and amphibole rocks, formed by contact reaction between ultrabasic rock and a rock of contrasting bulk composition.

bladder *vesicle [palyn].*

blade [bot] The widened portion of a leaf or of a plant structure that resembles a leaf.

blade [mineral] A flattened, elongate mineral crystal.

blade [paleont] A laterally compressed structure of a conodont; e.g. a denticle-bearing posterior or anterior process (based on position with reference to the basal cavity) in a compound conodont, or a generally compressed and denticulate part of the axis anterior to the basal cavity in a platelike conodont. Cf: *bar.*

blade [sed] A bladed or triaxial shape of a sedimentary particle, defined in *Zingg's classification* as having width/length and thickness/width ratios less than 2/3.

blade [speleo] A thin projection from any surface of a cave.

bladed Said of a mineral in the form of aggregates of flattened *blades* or elongate crystals.

blady Like a blade; e.g. "blady calcite" having elongate crystals somewhat wider than those of fibrous calcite.

blaes (a) A Scottish term for a gray-blue carbonaceous shale that weathers to a crumbly mass and eventually to a soft clay. See also: *bind.* (b) A Scottish term for a hard, joint-free sandstone.—Syn: *blaize.*

Blagden's law The statement in chemistry that, for a given salt, the depression of the freezing point is proportional to the concentration of the solution.

blairmorite A porphyritic extrusive rock consisting predominantly of analcime megacrysts in a groundmass of analcime, alkali feldspar, and sodic clinopyroxene, with accessory sphene, melanite, and nepheline; an analcime *phonolite.* The name, given by Knight in 1904, is for Blairmore, Alberta. Not recommended usage.

blaize *blaes.*

blakeite (a) A reddish-brown mineral consisting of a ferric tellurite, found sparingly as crusts from Goldfield, Nev. (b) *zirconolite.*

Blancan North American continental stage: Pleistocene (above Hemphillian, below Irvingtonian).

blanchardite *brochantite.*

blanket A thin, widespread sedimentary body whose width/thickness ratio is greater than 1000 to 1 and may be as great as 50,000 to 1 (Krynine, 1948, p. 146). Cf: *tabular.* Syn: *sheet [sed].*

blanket bog A bog covering a large, fairly horizontal area and depending on high rainfall or high humidity, rather than local water sources, for its supply of moisture. See also: *highmoor bog.*

blanket deposit [ore dep] A miner's term for a horizontal, tabular orebody. The term has no genetic connotation.

blanket deposit [sed] A sedimentary deposit of great lateral or areal extent and of relatively uniform thickness; esp. a *blanket sand* and associated blanket limestones.

blanket moss An accumulation of dead algae, often forming peat. See also: *blanket peat.*

blanket peat Peat that is derived mainly from the algae of *blanket moss.*

blanket sand A *blanket deposit* of sand or sandstone of unusually wide distribution, typically an orthoquartzitic sandstone deposited by a transgressive sea advancing for a considerable distance over a stable shelf area; e.g. the St. Peter Sandstone of the central U.S. Syn: *sheet sand; blanket sandstone.*

blast [geophys] The violent effect produced near an explosion, consisting of a wave of increased atmospheric pressure followed by a wave of decreased atmospheric pressure. Syn: *shock wave.*

blast [meta] (a) A prefix signifying a relict texture. (b) A suffix signifying a texture formed entirely by metamorphism.

blastation A term suggested by Glock (1928) for the destructive action of windblown particles of sand and dust; *blasting.*

blastetrix In an anisotropic medium, any surface to which a direction of greatest ease of growth is perpendicular (Turner, 1948, p.223).

blastic deformation One of the processes of dynamothermal metamorphism that operates by recrystallization according to *Riecke's principle,* in such a way that previously existing minerals are elongated perpendicular to the direction of greatest pressure, and new minerals grow in the same plane. Cf: *clastic deformation; plastic deformation.*

blasting Abrasion or attrition effected by the impact of fine particles moved by wind or water against or past an exposed, stationary surface; esp. *sandblasting.* Syn: *blastation.*

blastogeny In corals, "the phenomena of development of the offset (or asexually developed corallite in the colony) from the parent corallite" (Fedorowski and Jull, 1976, p. 39). Cf: *astogeny; hystero-ontogeny.*

blastogranitic (a) A relict texture in a metamorphic rock in which remnants of the original granitic texture remain. (b) A nonrecommended syn. of *blastogranular.*

blastogranular Said of a *heterogranular* metamorphic texture characterized by volumetrically significant amounts of large *paleoblasts* and smaller *neoblasts,* and by a relatively small strain (>200%) as partly evidenced in hand specimen by a weak foliation. The term is regarded as a syn. of *porphyroclastic* or is applied to a textural subgroup for small strains. Nonrecommended syn: *blastogranitic.*

blastoid Any crinozoan belonging to the class Blastoidea, characterized chiefly by highly developed quinqueradiate symmetry, a dominant meridional growth pattern, uniform arrangement of thecal plates in four cycles, specialized recumbent ambulacral areas, and the presence of hydrospires. Range, Ordovician to Permian.

blastolaminar Said of a *heterogranular* metamorphic texture characterized by volumetrically significant amounts of large *paleoblasts* and smaller *neoblasts,* and by large strains (200% to >1000%) as evidenced in hand specimen by a strong foliation. The term may be regarded as a textural subgroup of *porphyroclastic* for large strains.

blastomylonite A mylonitic rock in which some recrystallization and/or neomineralization has taken place. Cf: *mylonite.*

blastopelitic Said of a texture of a metamorphosed argillaceous rock in which there are relicts of the parent rock.

blastophitic Said of a relict texture in a metamorphic rock in which traces of an original ophitic texture remain.

blastoporphyritic Said of a relict texture in a metamorphic rock in which traces of an original porphyritic texture remain.

blastopsammitic Said of a texture of a metamorphosed sandstone that contains relicts of the parent rock.

blastopsephitic Said of a texture of a metamorphosed conglomerate or breccia that contains relicts of the parent rock.

blast wave A sharply defined wave of increased atmospheric pressure rapidly propagated through a surrounding medium from a center of detonation or similar disturbance. See also: *shock wave.*

blattfuss A platelike mesosomal appendage associated with gill structures of eurypterids (Waterston, 1975, p. 243).

B layer The seismic region of the Earth from the Mohorovičić discontinuity to 410 km. It is part of a classification of the Earth's interior made up of layers A to G. Syn: *low-velocity zone.*

blaze A man-made mark on a tree trunk, usually at about breast height, in which a piece of the bark and a very small amount of the live wood tissue is removed leaving a flat scar that permanently marks the tree. It is made for the purpose of guiding the course of a survey or of a trail in wooded country.

bleached Said of sand and silt that has become pale because of leaching, with consequent loss of iron-oxide coatings.

bleaching clay A clay or earth that, either in its natural state or after chemical activation, has the capacity for adsorbing or removing coloring matter or grease from liquids (esp. oils). Syn: *bleaching earth.*

bleaching earth *bleaching clay.*

bleach spot A greenish or yellowish area in a red rock, developed by the reduction of ferric oxide around an organic particle (Tyrrell, 1926). Syn: *deoxidation sphere.*

bleb In petrology, a small, usually rounded inclusion, e.g. olivine that is poikilitically enclosed in pyroxene.

bleeding n. (a) The process of giving off oil or gas from pore spaces or fractures; it can be observed in drill cores. (b) The exudation of small amounts of water from coal or a stratum of some other rock.

Bleiberg-type lead *B-type lead.*

blende (a) *sphalerite.* (b) Any of several minerals (chiefly metallic sulfides) with bright or resinous but nonmetallic luster, such as zinc blende (sphalerite), antimony blende (kermesite), bismuth blende (eulytite), cadmium blende (greenockite), pitchblende, and hornblende.—Etymol: German *Blende*, "deceiver".

blended unconformity An unconformity having no distinct surface of separation or sharp contact, such as at an erosion surface that was originally covered by a thick residual soil, which graded downward into the underlying rocks and was partly incorporated in the overlying rocks; e.g. a nonconformity between granite and overlying basal arkosic sediments derived as a product of its disintegration. Syn: *graded unconformity.*

blind Said of a mineral deposit that does not crop out. The term is more appropriate for a deposit that terminates below the surface than for one that is simply hidden by unconsolidated surficial debris.

blind apex The near-surface end of a mineral deposit, e.g. the upper end of a seam or vein that is truncated by an unconformity. Syn: *suboutcrop.*

blind coal (a) Anthracite or other coal that burns without a flame. (b) *Natural coke* that resembles anthracite.

blind creek A creek that is dry except during a rainfall. The term is "obsolete or obsolescent and better avoided because of confusion with blind valley" (Stamp, 1961, p. 66).

blind estuary A term used in Australia and South Africa for an *estuarine lagoon.*

blind island A patch of marl or organic matter covered by shallow water in a lake. Cf: *sunken island.*

blind joint In apparently massive rock that is being quarried, a plane of potential fracture along which the rock may break during excavation.

blind lake A colloquial term for a lake that has neither an influent nor an effluent.

blind lead A long narrow passage in pack ice with only one outlet. Syn: *cul-de-sac [ice]; pocket [ice].*

blind shaft *domepit.*

blind valley (a) A valley in karst that ends abruptly downstream at the point at which its stream disappears underground as a *sinking stream.* See also: *half-blind valley.* (b) In older usage, a syn. of *pocket valley.*

blind zone (a) A layer that cannot be detected by seismic (esp. refraction) methods, also called "hidden layer". It may have a velocity lower than that of shallower refractors, which will lead to an overestimate of the depth of deeper refractors; or it may have a velocity intermediate between those of layers above and below but with insufficient velocity difference or thickness to produce first arrivals, thus tending to cause an underestimate of the depth of deeper refractors. (b) A zone from which reflections do not occur; a *shadow zone.*

blink (a) A brightening of the sky near the horizon or the underside of a cloud layer, caused by reflection of light from a snow- or ice-covered surface. See also: *iceblink [meteorol]; snowblink; landblink.* (b) A dark appearance of the sky near the horizon or the underside of a cloud layer, caused by the relative absence of reflected light from a water or land surface. See also: *water sky; land sky.*

blister [coal] In a coal seam, a downward protrusion of roof rock into the seam, probably formed as the filling of a pothole in a stream bed.

blister [volc] A surficial swelling of the crust of a lava flow formed by the puffing-up of gas or vapor beneath the flow. A blister is usually about one m. in diameter, and is hollow. Cf: *shelly pahoehoe; tumulus [volc].*

blister hypothesis A theory of the cause of orogeny, which proposes that, in a zone not more than 80 km deep in the crust, heat from radioactive disintegration created a large convex-upward lens of heated and expanded rock, which produced doming of the overlying crust and in turn formed orogenic structures in the near-surface rocks. The "melting spot" or *asthenolith hypothesis* of B. Willis is similar. Geophysical evidence indicates that the existence of such blisters is unlikely, and the theory is probably obsolete. Cf: *undation theory.*

blixite A mineral: $Pb_2Cl(O,OH)_2$.

block [ice] A fragment of floating sea ice ranging in size from 2 m to 10 m across; the term is being replaced by *ice cake.*

block [part size] (a) A large, angular rock fragment, showing little or no modification by transporting agents, its surfaces resulting from breaking of the parent mass, and having a diameter greater than 256 mm (10 in.); it may be nearly in place or transported by gravity, ice, or other agents. Cf: *boulder.* (b) A term used by Woodford (1925) for a nearly equidimensional, angular rock fragment of any diameter greater than 4 mm. (c) A rock or mineral particle in the soil, having a diameter range of 200-2000 mm (Atterberg, 1905). (d) A layer of sedimentary rock, from 60 cm to 120 cm (2-4 ft) thick, produced by splitting (McKee & Weir, 1953, p. 383).

block [tect] *fault block.*

block [volc] A pyroclast that was ejected in a solid state; it has a diameter greater than 64 mm. It may be essential, accessory, or accidental. Cf: *lapilli; volcanic gravel; cinder.*

block caving A large-production low-cost method of mining, in which the greater part of the bottom area of a block of ore is undercut, the supporting pillars are blasted away, and the ore settles as it caves downward and is removed. As the block caves and settles, the cover follows.

block diagram (a) A plane figure representing an imaginary rectangular block of the Earth's crust (depicting geologic and topographic features) in what appears to be a three-dimensional perspective, showing a surface area on top and including one or more (generally two) vertical cross sections. The top of the block gives a bird's-eye view of the ground surface, and its sides give the underlying geologic structure (Lobeck, 1924). (b) A sketch of a relief model; a representation of a landscape in perspective projection.

block disintegration *joint-block separation.*

blocked-out ore *developed reserves.*

block faulting A type of normal faulting in which the crust is divided into structural or *fault blocks* of different elevations and orientations. It is the process by which *block mountains* are formed.

block field A thin accumulation of usually angular blocks, with no fine sizes in the upper part, over solid or weathered bedrock, colluvium, or alluvium, without a cliff or ledge above as an apparent source. Block fields occur on high mountain slopes above treeline, and in polar regions; they are most extensive along slopes parallel to the contour; and they exist on slopes of less than 5°. Blocks may be subround to subangular suggesting abrasion during transport or in-situ derivation (White, 1976, p. 89). Cf: *allochthonous mantle; autochthonous mantle; block slope; block stream; scree.* Syn: *block sea; felsenmeer; blockmeer; mountain-top detritus; stone field.*

block glide A translational landslide in which the slide mass remains essentially intact, moving outward and downward as a unit, most often along a pre-existing plane of weakness, such as bedding, foliation, joints, or faults. In contrast to rotational slides, the various points within a displaced block-glide slide have predominantly maintained the same mutual difference in elevation in relation to points outside the slide mass.

blocking out In economic geology, delimitation of an orebody on three sides in order to develop it, i.e., to make estimates of its tonnage and quality. The part so prepared is an *ore block.*

blockite *penroseite.*

block lava Lava having a surface of angular blocks; it is similar to *aa* but the fragments are more regular in shape, somewhat smoother, and less vesicular.

blockmeer A syn. of *block field.* Etymol: German *Blockmeer,* "sea of blocks".

block mountain A mountain that is formed by *block faulting.* The term is not applied to mountains that are formed by thrust faulting. Syn: *fault-block mountain.*

block movement [glaciol] A syn. of *Block-Schollen movement* generally used in English-language publications.

block movement [mining] In mining, a general failure of the hanging wall.

Block-Schollen movement A type of glacier flow in which the greater portion of the glacier moves as a solid mass with a nearly uniform velocity; blocks of ice are produced by movement over irregularities in the glacier bed (Finsterwalder, 1950). Etymol: German. Syn: *block movement.*

block sea *block field.*

block slope A thin accumulation of usually angular blocks, on high mountains and in polar regions, on slopes of 5° to 25° (White, 1976, p. 89). Cf: *block field.*

block stream An accumulation of boulders or angular blocks, with no fine sizes in the upper part, over solid or weathered bedrock, colluvium, or alluvium. Block streams usually occur at the heads of ravines, as narrow bodies more extensive downslope than along the slope. They may extend into forests or fill a valley floor; and they may exist on any slope angle, but ordinarily not steeper than 40° (White, 1976, p. 91). Sometimes incorrectly referred to as *block field* or *rock glacier.* Cf: *block field.* Syn: *boulder field; boulder stream; rock stream; stone stream.*

block stripe A short, broad *sorted stripe* containing material that is coarser, and of less uniform size, than that in a *stone stripe.*

block talc A general term for any massive talc or soapstone that can be worked by machines.

blocky iceberg An iceberg with steep, precipitous sides and a horizontal or nearly horizontal upper surface.

bloedite A white or colorless monoclinic mineral: $Na_2Mg(SO_4)_2 \cdot 4H_2O$. Also spelled: *blödite; blodite.* Syn: *astrakhanite.*

blomstrandine *priorite.*

blomstrandite *betafite.*

Blondeau method A seismic method of determining vertical time to a predetermined depth based on first-break data and the assumption that the instantaneous velocity is proportional to a power of the depth.

blood agate (a) Flesh-red, pink, or salmon-colored agate from Utah. (b) *hemachate.*

blood rain Rain with a reddish color caused by dustlike material, containing iron oxide, picked up from the air by raindrops during descent, often leaving a red stain on the ground; e.g. the blood rain of Italy, containing red dust carried north by great storms from the Saharan desert region. Syn: *dust fall.*

bloodstone (a) A semitranslucent leek-green or dark-green variety of chalcedony speckled with red or brownish-red spots of jasper resembling drops of blood. Cf: *plasma [mineral].* Syn: *heliotrope [mineral]; oriental jasper.* (b) *hematite.*

bloom [mineral] *efflorescence.*

bloom [oceanog] *water bloom.*

bloom [ore dep] *blossom.*

blossom The oxidized or decomposed outcrop of a vein or coal bed, more frequently the latter. Syn: *bloom [ore dep].*

blow *blowhole [coast].*

blowhole [coast] A nearly vertical hole, fissure, or natural *chimney* in coastal rocks, leading from the inner end of the roof of a sea cave to the ground surface above, through which incoming waves and the rising tide forcibly compress the air to rush upward or spray water to spout intermittently, often with a noise resembling a geyser outburst. It is probably formed by wave erosion concentrated along planes of weakness, as in a well-jointed rock. Also spelled: *blow-hole.* Syn: *puffing hole; blow; boiler; buller; spouter.*

blowhole [glaciol] An opening that passes through a snowbridge into a crevasse, or system of crevasses, that is otherwise sealed by snowbridges (Armstrong et al., 1966, p. 11). It is commonly characterized by a current of moving air.

blowhole [volc] A minute gas vent on the surface of a lava flow.

blowing Transportation and deposition of sediments effected by the wind acting along the surface of the ground.

blowing cave A cave that has a movement of air through its entrance. See also: *breathing cave; cave breathing.*

blowing well A water well that has a movement of air through its entrance. Syn: *breathing well.* Not to be confused with *blow well.*

blow land Land that is subject to wind erosion.

blown sand Sand that has been transported by the wind; sand consisting of wind-borne particles; eolian sand. See also: *dune sand.*

blowoff The removal of humus and loose topsoil by wind action. Also, the material so moved.

blowout [geomorph] (a) A general term for a small saucer-, cup-, or trough-shaped hollow or depression formed by wind erosion on a pre-existing dune or other sand deposit, esp. in an area of shifting sand or loose soil, or where protective vegetation is disturbed or destroyed; the adjoining accumulation of sand derived from the depression, where recognizable, is commonly included. Some blowouts may be many kilometers in diameter. (b) A butte, the top of which has been blown out by the wind until it resembles a volcanic crater. (c) A shallow basin formed where vegetation has been destroyed by fire or by overgrazing.—See also: *deflation basin.* Also spelled: *blow-out.* Syn: *blowout basin; deflation hollow.*

blowout [grd wat] *sand boil.*

blowout [ore dep] (a) A prospector's term for a weathered exposure considered to be indicative of a mineral deposit. (b) A large mineral-deposit outcrop beneath which the deposit is smaller.

blowout dune A dune consisting of a large accumulation of sand derived from the formation of a *blowout.* An "elongate blowout dune" is characterized by a slight migration of the blowout and its crescent-shaped rim in the direction of the prevailing wind (Stone, 1967, p. 226).

blowout pond A shallow, intermittent pond occupying a *blowout,* as on a dune.

blowover (a) Sand blown by onshore winds over a *barrier* and deposited as a veneer in the lagoon, e.g. along the Gulf Coast of Texas. Cf: *washover.* (b) The process of forming a blowover.

blowpipe A plain brass tube that produces an intense heat by combining a flame from a bunsen burner or other heat source with a stream of air; it is used in simple qualitative analysis of minerals. See also: *blowpiping.*

blowpipe reaction The indicative changes of a mineral specimen as it undergoes *blowpiping;* e.g. color of the flame, odor, nature of the encrustation.

blowpiping In mineralogy, a qualitative test of a mineral made by heating a specimen in the flame of a *blowpipe* and observing its *blowpipe reaction,* such as color of the flame or color of the encrustation, to determine what elements may be present. See also: *bead.*

blow well A syn. of *flowing artesian well.* Not to be confused with *blowing well.*

blue amber A variety of osseous amber with a bluish tinge that is probably due to the presence of calcium carbonate.

blue asbestos *crocidolite.*

blue band [glaciol] (a) A sharply bounded lens or layer of relatively bubble-free glacier ice; a bluish band marking the appearance of such a lens or layer on the surface of a glacier. The bluish tint is due to the low content of air in the ice. Cf: *white band.* (b) The dark-ribbon effect produced on the surface of a glacier by the exposure of blue bands.

blue band [sed] The thin but persistent bed of bluish clay found throughout the Illinois-Indiana coal basin.

blue-black ore *corvusite.*

blue chalcocite *digenite.*

blue copper ore *azurite.*

blue earth *blue ground.*

blue elvan A Cornish term for greenstone occurring in dikes.

blue-green algae A group of algae corresponding to the phylum Cyanophyta, that contains both chlorophyll and the pigment c-phycocyanin. Cf: *brown algae; green algae; red algae; yellow-green algae.*

blue ground Unoxidized slate-blue or blue-green *kimberlite,* usually a breccia (as in the diamond pipes of South Africa) that is found below the surficial oxidized zone of *yellow ground.* Cf: *hardebank.* Syn: *blue earth.*

blue hole (a) A Jamaican term for a *resurgence* that does not fountain. See also: *boiling spring [karst].* (b) A term used on the Bahama Banks for a drowned sinkhole. See also: *banana hole.* Syn: *ocean hole.*

blue ice (a) Nonbubbly, unweathered, coarse-grained glacier ice, often occurring as *blue bands [glaciol];* it is distinguished by a slightly bluish or greenish color. Cf: *black ice; white ice.* (b) An ablation area created by wind erosion on the Antarctic Ice Sheet, characterized by bare glacier ice showing at the surface.

blue iron earth Pale-blue powdery *vivianite.*

blue ironstone A bluish iron-bearing mineral; specif. crocidolite and vivianite.

blue john A massive, fibrous, or columnar and blue or purple variety of fluorite found in Derbyshire, England. It is frequently banded, and is used esp. for the manufacture of vases. Syn: *derbystone.*

blue lead [mineral] A syn. of *galena,* esp. a compact variety with a bluish-gray color. Syn: *blue lead ore.*

blue lead [ore dep] A bluish, gold-bearing *lead [eco geol]* or gravel deposit found in Tertiary river channels of the Sierra Nevada, California. Pron: blue leed.

blue malachite A misnomer for *azurite.*

blue metal A term used in England for a hard bluish-gray shale or mudstone lying at the base of a coal bed and often containing pyrite.

blue mud A hemipelagic type of *mud [marine geol]* whose bluish-gray color is due to iron sulfides and organic matter.

blue ocher *vivianite.*

blue quartz (a) A faintly blue or lavender variety of crystalline quartz, containing needlelike inclusions of rutile. It occurs as grains in metamorphic and igneous rocks. (b) *sapphire quartz.*

blue-rock phosphate A term used for the Ordovician bedded phosphate rock of Tennessee.

blueschist A schistose metamorphic rock with a blue color owing to the presence of sodic amphibole, glaucophane or crossite, and commonly mottled bluish-gray lawsonite. Cf: *glaucophane schist.*

blueschist facies *glaucophane-schist facies.*

blue spar *lazulite.*

bluestone [mineral] *chalcanthite.*

bluestone [rock] (a) A commercial name for a building or paving stone of bluish-gray color; specif. a dense, tough, fine-grained, dark blue-gray or slate-gray feldspathic sandstone that splits easily into thin, smooth slabs and that is extensively quarried near the Hudson River in New York State for use as flagstone. The color is due to the presence of fine black and dark-green minerals, chiefly hornblende and chlorite. The term is applied locally to other rocks, such as dark-blue shale and blue limestone. (b) A highly argillaceous sandstone of even texture and bedding, formed in a lagoon or lake near the mouth of a stream (Grabau, 1920a, p.579). (c) A local term used in Great Britain for a hard shale or clay (as in south Wales), and for a basalt.

blue vitriol *chalcanthite.*

blue-white A misused term, once applied in the gem trade to diamonds without body color, i.e. colorless or slightly bluish stones of the highest quality. Misapplication to yellowish stones has made the term meaningless. Cf: *jager.*

bluff (a) A high bank or bold headland with a broad, precipitous, sometimes rounded cliff face overlooking a plain or a body of water; esp. on the outside of a stream meander; a *river bluff.* (b) Any cliff with a steep broad face.

bluff formation Deposit of coarse *loess,* forming bluffs immediately adjacent to the edges of river flood plains, as in the Mississippi Valley region.

bluff line The side of a valley formed by a river or by ice cutting away the heads of interlocking spurs (Swayne, 1956, p. 25).

blythite A hypothetical member of the garnet group: $Mn_3{}^{+2}Mn_2{}^{+3}(SiO_4)_3$.

Blytt-Sernander climatic classification A classification of late-glacial and Holocene climate inferred originally from bog stratigraphy and megascopic plant remains from Norway and Sweden, and later refined by Post (1924) from pollen evidence. It is the classic system for worldwide research on postglacial climate. It includes six subunits: Arctic, Preboreal, Boreal, Atlantic, Subboreal, and Subatlantic. Named after Axel Gudbrand Blytt (1843-1898), Norwegian botanist, and Johan Rutger Sernander (1866-1944), Swedish botanist.

BM *bench mark.*

board coal *woody lignite.*

boar's back A *horseback [glac geol]* or esker in northern New England, esp. Maine.

boart Var. of *bort.*

boat channel A channel, on or behind a reef flat, separating a *fringing reef* from the shore to which the channel is parallel. It is generally only a few meters in depth and width. Cf: *shipping channel; moat [reef].* Syn: *back-reef moat.*

boathook bend The sharp curvature of a tributary where it joins the main stream in an upstream direction in a *barbed drainage pattern,* resembling in plan a boathook.

bobierrite A mineral: $Mg_3(PO_4)_2 \cdot 8H_2O$. It occurs massive or in crystals in guano.

boca The mouth of a stream, esp. the point where a stream or its channel emerges from a canyon, gorge, or other precipitous valley and flows onto or enters a plain. Etymol: Spanish, "mouth".

bocanne A naturally burning shale bank (Crickmay, 1967, p. 626). Pron: bo-*kahn.* Etymol: French-Canadian, "smoke".

bocca An aperture on any part of a volcano from which magma or gas escapes. Etymol: Italian, "mouth". Pl: *bocche.*

bocche Plural of *bocca.*

BOD *biochemical oxygen demand.*

bodden A broad shallow irregularly shaped inlet or bay along the southern Baltic coast, typically produced by partial submergence of an uneven lowland surface and characterized by seaward islands. Etymol: German *Bodden.* Cf: *förde.*

bodily tide *earth tide.*

body [coal] The fatty, inflammable property that makes a coal combustible; e.g. bituminous coal has more *body* than anthracite.

body [water] A separate entity or mass of water, distinguished from other water masses; e.g. an ocean, sea, stream, lake, pond, pool, and water in an aquifer are distinct "bodies of water".

body cavity A cavity or major space within an animal body, such as a *coelom;* e.g. the principal part of the coelomic space in a brachiopod, situated posteriorly, bounded by the body wall, and containing the alimentary tract, internal organs, etc.

body-centered lattice A type of *centered lattice* in which the unit cell contains two lattice points; the point at the intersection of the four body diagonals is identical with those at the corners. Syn: *I-centered lattice.*

body chamber (a) The undivided anterior space in a cephalopod shell occupied by the living body of the animal, bounded at the back by a septum and open at the front through the aperture. Syn: *living chamber; chamber [paleont].* (b) The interior of the shell containing the soft parts of a cirripede crustacean.

body force Any force acting on a material proportional to the mass of the substance, e.g. gravity, centrifugal force, magnetic force. Cf: *surface force.*

body wall The external surface of the body in animals, enclosing the body cavity; e.g. the part of a sponge between the exterior and a central spongocoel, or the *perisome* of an echinoderm.

body wave A *seismic wave* that travels through the interior of the Earth, with a propagation mode that does not depend on any boundary surface. A body wave may be either longitudinal (a *P wave*) or transverse (an *S wave*).

body-wave magnitude An *earthquake magnitude* determined at teleseismic distances, using the logarithm of the ratio of amplitude to period for *body waves.*

body whorl The outer, last-formed, and typically largest *whorl* of a univalve shell; e.g. the last complete loop in the spiral of a gastropod shell, terminating in the aperture.

boehmite A grayish, brownish, or reddish orthorhombic mineral: AlO(OH). It is a major constituent of some bauxites and it represents the gamma phase dimorphous with diaspore. Also spelled: *böhmite.*

Boehm lamellae Planar structural features produced in deformed mineral grains. They resemble *deformation lamellae* except that they are decorated by trains of fluid inclusions. Cf: *Tuttle lamellae.* Also spelled: *Böhm lamellae.*

bog (a) Waterlogged, spongy ground, consisting primarily of mosses, containing acidic, decaying vegetation that may develop into peat. (b) The vegetation characteristic of this environment, esp. sphagnum, sedges, and heaths.—The term is often used synonymously with *peat bog.* Cf: *fen; marsh; swamp.*

bogan *pokelogan.*

bogaz *solution corridor.*

bog burst The bursting of a bog under the pressure of its swelling, due to water retention by a marginal dam of growing vegetation. The escaping water produces muddy peat flows over the surrounding area.

bog butter A substance, found preserved in Irish peat bogs, that was formerly believed to be a native hydrocarbon but is now known to be "fossil" butter that had been buried for storage and found at a much later date (Tomkeieff, 1954, p. 30). Syn: *butyrellite; butyrite.*

bog coal An earthy type of brown coal.

bog flow The outflow from a *bog burst.* Cf: *peat flow.*

bøggildite A mineral: $Na_2Sr_2Al_2(PO_4)F_9$.

boghead coal A *sapropelic coal* resembling *cannel coal* in its physical properties but consisting dominantly of algal matter rather than spores. Cf: *torbanite; algite.* Syn: *algal coal; gélosic coal; sapromyxite; tomite.*

boghedite An old syn. of *torbanite.*

bog iron ore A general term for a soft, spongy, and porous deposit of impure hydrous iron oxides formed in bogs, marshes, swamps, peat mosses, and shallow lakes by precipitation from iron-bearing waters and by the oxidizing action of algae, iron bacteria, or the atmosphere; a *bog ore* composed principally of limonite that is often impregnated with plant debris, clay, and clastic material. It is a poor-quality iron ore, found in tubular, pisolitic, nodular, concretionary, or thinly layered forms, or in irregular aggregates, in level sandy soils, and is esp. abundant in the glaciated northern regions of North America and Europe (Scandinavia). See also: *murram.* Syn: *limnite; morass ore; meadow ore; marsh ore; lake ore; swamp ore.* (b) A term commonly applied to a loose, porous, earthy form of *limonite* occurring in wet ground.—Syn: *bog iron.*

bog lake A lake or small body of open water surrounded or nearly surrounded by bogs and characterized by a false bottom of organic (peaty) material, high acidity, scarcity of aquatic fauna, and vegetation growing on a firm deposit or on a semifloating mat of peat. See also: *sphagnum bog.*

bog lime An obsolete syn. for the *marl* of freshwater lakes.

bog manganese A *bog ore* of variable composition, but consisting chiefly of hydrous manganese oxide; specif. *wad* formed in bogs or marshes by the action of minute plants.

bog-mine ore A syn. of *bog ore.* Also called: *bog mine.*

bog moat *lagg.*

bog ore A poorly stratified accumulation of earthy metallic-mineral substances, mainly oxides, that are formed in bogs, marshes, swamps, and other low-lying moist places, usually by direct chemical precipitation from surface or near-surface percolating waters; specif. *bog iron ore* and *bog manganese.* Cf: *lake ore.* Syn: *bog-mine ore.*

bog peat *highmoor peat.*

Bog soil A great soil group in the 1938 classification system, an intrazonal, hydromorphic group of soils having a mucky or peaty surface horizon and an underlying peat horizon. These soils developed in swamps or marshes in humid or subhumid climates (USDA, 1938). Bog soils are now classified as *Histosols.* Cf: *Half-Bog soil.*

bogue A term used in Alabama and Mississippi for the mouth or outlet of a stream, or for the stream itself, or for a bayou. Etymol: American French, from Choctaw *bouk,* "stream, creek".

bogusite An intrusive rock of the same general composition as *teschenite* but of lighter color. Named by Johannsen (1931) for Boguschowitz, Czechoslovakia. Not recommended usage.

bohdanowiczite A mineral: $AgBiSe_2$.

Bohemian garnet A yellowish-red to dark, intense-red gem variety of *pyrope* obtained from Bohemia.

Bohemian ruby A red variety of crystalline quartz; specif. *rose quartz* cut as a gem.

Bohemian topaz *citrine.*

böhmite *boehmite.*

Böhm lamellae *Boehm lamellae.*

boil n. A churning agitation of water, esp. at the surface of a water body, such as a river, spring, or the sea.

boiler (a) A small submerged coral reef, esp. one occurring where the sea breaks frequently. Syn: *breaker [reef]; cup reef.* (b) A *blowhole* along the coast.

boiling hole *boiling spring [karst].*

boiling spring [grd wat] (a) A spring, the water from which is agitated by the action of heat. (b) A spring that flows so rapidly that strong vertical eddies develop.

boiling spring [karst] A Jamaican term for a fountaining *resurgence.* See also: *vauclusian spring; blue hole.* Syn: *boiling hole.*

bojite A *gabbro* in which primary hornblende substitutes for most of the pyroxene, although some augite and biotite may be present; a *hornblende gabbro.* The name, given by Weinschenk in 1899, is for the Boii, "a Celtic tribe that settled in Germany" (Johannsen, 1937, p. 227). Not recommended usage.

bokite A black mineral: $KAl_3Fe_6V_6{}^{+4}V_{20}{}^{+5}O_{76}\cdot 30H_2O$.

bold coast A prominent landmass, such as a cliff or promontory, rising or sloping steeply from a body of water, esp. along the seacoast. Syn: *bold.*

bole Any of several varieties of fine, compact, friable, and earthy or unctuous clay (impure halloysite), usually colored red, yellow, or brown due to the presence of iron oxide, and consisting essentially of hydrous silicates of aluminum or less often of magnesium; a waxy decomposition product of basaltic rocks, having the variable composition of lateritic clays. Adj: *bolar.* Syn: *bolus; terra miraculosa.*

boleite An indigo-blue mineral: $Pb_{26}Ag_9Cu_{24}Cl_{62}(OH)_{48}$. Also spelled: *boléite.*

bolide An exploding or exploded meteor or meteorite; a detonating *fireball.*

Boliden gravimeter An electrical *stable gravimeter* with a moving system suspended on a pair of bowed springs. The moving system carries electrical condenser plates at each end, one to measure the position of the moving system, the other to apply a balancing force to bring the system to a fixed position. Syn: *Lindblad-Malmquist gravimeter.*

bolivarite A non-crystalline mineral: $Al_2(PO_4)(OH)_3\cdot 4\text{-}5H_2O$.

Bølling n. A term used primarily in Europe for an interval of late-glacial time (centered about 12,500 years ago) following the Oldest Dryas and preceding the Older Dryas, during which the climate ameliorated favoring birch and park-tundra vegetation. Also spelled: *Bölling.*—adj. Pertaining to the late-glacial Bølling interval and to its climate, deposits, biota, and events.

bolly *bally.*

Bologna stone A nodular, concretionary, or rounded form of barite, composed of radiating fibers; phosphorescent when calcined with charcoal. Syn: *Bolognan stone; Bologna spar.*

bolometer A detector used to measure the radiant temperature or flux by measuring the change in electrical resistance of a metal (e.g. platinum) or of a semiconductor (e.g. a thermistor).

bolson (a) A term applied in the desert regions of SW U.S. to an extensive flat alluvium-floored basin or depression, into which drainage from the surrounding mountains flows centripetally with gentle gradients toward a playa or central depression; an interior basin, or a basin with internal drainage. See also: *semibolson.* Syn: *playa basin.* (b) A temporary lake, usually saline, formed in a bolson.—Etymol: Spanish *bolsón,* "large purse".

bolson plain A broad, intermontane plain in the central part of a bolson or semibolson, composed of deep alluvial accumulations washed into the basin from the surrounding mountains.

boltonite A greenish or yellowish granular variety of forsterite from Bolton, Mass.
boltwoodite A yellow mineral: $K_2(UO_2)_2(SiO_3)_2(OH)_2 \cdot 5H_2O$.
bolus A *bole.* Etymol: Latin, "clod of earth".
bolus alba A syn. of *kaolin.* Etymol: Latin, "white clay".
bomb [geochem] A vessel in which experiments can be conducted at high temperature and pressure. It is used in geochemistry and in experimental petrology. Syn: *pressure vessel.*
bomb [pyroclast] A pyroclast that was ejected while viscous and received its rounded shape while in flight. It is larger than 64 mm in size, and may be vesicular to hollow inside. Actual shape or form varies greatly, and is used in descriptive classification, e.g. rotational bomb; spindle bomb.
bombiccite *hartite.*
bombite A blackish-gray aluminosilicate of ferric iron and calcium from Bombay, India. It resembles Lydian stone and is probably a glassy rock.
bombollaite *bambollaite.*
bomb sag *bedding-plane sag.*
bomby An Australian term for a large, submerged reef clump found in a back-reef area and constituting a hazard for navigation and fishing (Maxwell, 1968, p.133). See also: *bommy.*
bommy A *coral head.* See also: *bomby.*
bonaccordite An orthorhombic mineral: Ni_2FeBO_5.
bonamite A trade name for an apple-green gem variety of smithsonite, resembling the color of chrysoprase.
bonanza A miner's term for a rich body of ore or a rich part of a deposit; a mine is "in bonanza" when it is operating profitably. Also, discontinuous locally rich ore deposits, esp. epithermal ones. Spanish, "prosperity, success". Cf: *borasca.*
bonattite A monoclinic mineral: $CuSO_4 \cdot 3H_2O$.
bonchevite An orthorhombic mineral: $PbBi_4S_7$.
bond clay A clay which, because of its plasticity, serves to bond relatively nonplastic materials in the fabrication of ceramic or other molded products ("green bond"). Also, a clay which, on firing to furnace or vitrification temperature, bonds adjacent ceramic materials that vitrify at a still higher temperature ("fired bond").
bone A tough, very fine-grained, gray, white, or reddish quartz.
bone amber *osseous amber.*
bone bed Any sedimentary stratum (usually a thin bed of sandstone, limestone, or gravel) in which fossil bones or bone fragments are abundant, and often containing other organic remains, such as scales, teeth, and coprolites.
bone breccia An accumulation of bones or bone fragments, often mixed with earth and sand, and cemented with calcium carbonate; esp. such a deposit formed in limestone caves or other animal retreats. Syn: *osseous breccia.*
bone cave A cave that has served as a trap for fossil vertebrates.
bone chert A weathered, residual chert that appears chalky and somewhat porous, and that is usually white but may be stained with red or other colors. When found in insoluble residues, it is an indicator of an unconformity.
bone coal (a) Coal that has a high ash content. It is hard and compact. Syn: *bony coal.* (b) Argillaceous partings in coal, sometimes called *slate.*
bone phosphate of lime Tricalcium phosphate, $Ca_3(PO_4)_2$. The phosphate content of phosphorite may be expressed as percentage of bone phosphate of lime. Abbrev: BPL.
bone turquoise *odontolite.*
boninite A glassy olivine-bronzite *andesite* that contains little or no modal feldspar. Named by Petersen in 1891 for the Bonin Islands, Japan. Not recommended usage.
Bonne projection An equal-area, modified-conic map projection having one standard parallel intersecting the central meridian (a straight line along which the scale is exact) near the center of the map. All parallels are represented by equally spaced arcs of concentric circles (divided to exact scale) and all meridians (except the central meridian) are curved lines connecting corresponding points on the parallels. The projection is commonly used for mapping compactly shaped areas in middle latitudes (such as France) and for mapping continents such as North America and Eurasia. Named after Rigobert Bonne (1727-1795), French cartographer, who is said to have introduced the projection in 1752. See also: *sinusoidal projection.*
bony coal *bone coal.*
bony fish *Osteichthyes.*
book *mica book.*
book clay Clay deposited in thin, leaflike laminae. Syn: *leaf clay.*
bookhouse structure A term introduced by Sloane & Kell (1966, p. 295) for a fabric found in compacted kaolin clays, consisting of parallel and random arrangements of packets of oriented clay particles (flakes). Cf: *cardhouse structure.*
book structure In ore deposits, the alternation of ore with gangue, usually quartz, in parallel sheets. Cf: *ribbon [ore dep].*
boolgoonyakh A syn. of *pingo.* Etymol: Yakutian. Also spelled: *boolyunyakh; bulgunniakh.*
boomer (a) A marine seismic-energy source in which a high-voltage discharge causes two metal plates to separate abruptly in a body of water. (b) A very strong, usually low-frequency event on a seismic recording.
booming dune A term used by Criswell et al. (1975) for a dune of *booming sand.*
booming sand A *sounding sand,* found on a desert, that emits a low-pitched note of considerable magnitude and duration as it slides (either spontaneously or when induced) down the slip face of a dune or drift (Humphries, 1966, p.135). See also: *roaring sand; booming dune.*
boort *bort.*
boothite A blue monoclinic mineral: $CuSO_4 \cdot 7H_2O$. Its blue color is lighter than that of chalcanthite.
bora A *katabatic wind* of the northern Adriatic coast.
boracite A white, yellow, greenish, or bluish orthorhombic mineral: $Mg_3B_7O_{13}Cl$. It is strongly pyroelectric, becomes cubic at high temperatures, and occurs in evaporites and saline deposits. See also: *stassfurtite.*
Boralf In U.S. Dept. of Agriculture soil taxonomy, a suborder of the soil order *Alfisol,* characterized by formation in frigid or cryic temperature regimes and in a udic moisture regime. Most Boralfs have an O horizon and an albic horizon above the argillic horizon. These soils generally formed under coniferous forests and because of the short growing season tend to remain forested (USDA, 1975). Cf: *Aqualf; Udalf; Ustalf; Xeralf.*
borasca A miner's term for an unproductive area of a mine or orebody; a mine is "in borasca" when it is exhausted. Etymol: Mexican Spanish *borrasca,* "exhaustion of a mine". Cf: *bonanza.*
borate A mineral compound characterized by a fundamental structure of BO_3^{-3}. An example of a borate is boracite, $Mg_3B_7O_{13}Cl$. Cf: *carbonate [mineral]; nitrate.*
borax A white, yellowish, blue, green, or gray mineral: $Na_2B_4O_7 \cdot 10H_2O$. It is an ore of boron and occurs as a surface efflorescence or in large monoclinic crystals embedded in muds of alkaline lakes. Borax is used chiefly in glass, ceramics, agricultural chemicals, and pharmaceuticals, and as a flux, cleansing agent, water softener, preservative, and fire retardant. Syn: *tincal.*
borax bead The type of *bead* commonly used in blowpipe analysis of metallic compounds.
borax lake (a) A lake whose shores are encrusted with deposits rich in borax. (b) A dry, borax-rich bed of a lake.
borcarite A mineral: $Ca_4MgH_6(BO_3)_4(CO_3)_2$.
border belt A term used by Chamberlin (1893, p. 263) for superficial glacial deposits now known as a *boulder belt.*
bordered pit A *pit [bot]* in which the margin projects over a chamber separated from the thin pit-closing membrane, as in tracheids of coniferous wood. Cf: *simple pit.*
border facies The marginal portion of an igneous intrusion, which differs in texture and composition from the main body of the intrusion, possibly due to more rapid cooling or to assimilation of material from the country rock.
border fault (a) *boundary fault.* (b) *peripheral fault.*
borderland According to a concept widely held in the first part of the 20th Century, and championed by Schuchert (1923), a crystalline landmass on the seaward borders of the Phanerozoic orogenic belts near the edges of the North American continent. The borderlands were tectonically much more active than the Canadian Shield, and were subsequently lost by foundering into the oceans. The concept is now discredited; continental crust ends near the edges of the continental shelves, and it would be difficult to founder large areas of such crust into the ocean basins beyond. Most of the geological evidence adduced for these lands can be otherwise interpreted. Cf: *hinterland; tectonic land.* See also: *Appalachia; Cascadia; Llanoria.*
bore [marine geol] A submarine sand ridge, in very shallow water, whose crest may rise to intertidal level.
bore [tides] (a) A large, turbulent, wall-like wave of water with a high, abrupt front, caused by the meeting of two tides or by a very rapid rise or rush of the tide up a long, shallow and narrowing

estuary, bay, or tidal river where the tidal range is appreciable; it can be 3-5 m high and moves rapidly (10-15 knots) upstream with and faster than the rising tide. A bore usually occurs after low water of a spring tide. Syn: *tidal bore.*

bore [volc] The outlet of a geyser at the Earth's surface.

bore [water] A syn. of *bored well.* Sometimes applied to any deep well or shaft.

Boreal [clim] n. A climatic zone having a definite winter with snow and a short summer that is generally hot, and characterized by a large annual range of temperature. It includes large parts of North America, central Europe, and Asia, generally between latitudes 60°N and 40°N.

Boreal [paleoclim] n. A term used primarily in Europe for an interval of Holocene time (from about 9000 to 7500 years ago) following the Preboreal and preceding the Atlantic, during which the inferred climate was relatively warm and dry; a subunit of the *Blytt-Sernander climatic classification,* characterized by pine and hazel vegetation.—adj. Pertaining to the postglacial Boreal interval and to its climate, deposits, biota, and events.

boreal (a) Pertaining to the north, or located in northern regions; northern. (b) Pertaining to the northern biotic area (or Boreal region) characterized by *tundra* and *taiga* and by dominant coniferous forests. (c) Pertaining to the Boreal postglacial period, characterized by a cool climate like that of the present Boreal region. Also, said of the climate of such a period. (d) Pertaining to a Boreal climatic zone, or to the climate of such a zone.

bored well A shallow water well, 3 to 30 m deep and 20 to 90 cm in diameter, constructed by hand-operated or power-driven augers. Syn: *bore [water].*

borehole A circular hole made by boring; esp. a deep hole of small diameter, such as an oil well or a water well. Syn: *hole [drill]; well bore.*

borehole gravity meter A *gravity meter* adapted to the borehole environment that measures the gravitational force while hanging stationary at selected depths within a well. Differences in readings at different depths derive from a large volume of the intervening rock; the effects of washed-out zones, mud, filtrate invasion, or casing are not significant. Borehole gravity data may be solved for bulk formation density. Abbrev: BHGM. Syn: *borehole gravimeter.*

borehole log *well log.*

borehole survey A *directional survey.* (b) A *well log.*

borickite A reddish-brown mineral consisting of a hydrous basic phosphate of iron and calcium.

boring (a) A *trace fossil* consisting of an etching, groove, or hollow, produced by plants (fungi, algae) or animals (sponges, worms, bryozoans, barnacles) in shells, bones, or other hard parts of invertebrates and vertebrates. Cf: *burrow [paleont].*

boring porosity *burrow porosity.*

borishanskiite An orthorhombic mineral: $Pd_{1+x}(AsPb)_2$.

bornemanite An orthorhombic mineral: $BaNa_4Ti_2NbSi_4O_{17}F \cdot Na_3PO_4$.

bornhardt A residual peak having the characteristics of an *inselberg;* specif. a large granite-gneiss inselberg associated with the second cycle of erosion in a rejuvenated desert region (King, 1948). Named in honor of F. Wilhelm C.E. Bornhardt (1864-1946), German explorer of Tanganyika, who first described the feature.

bornhardtite A cubic mineral: Co_3Se_4.

bornite A brittle, metallic-looking mineral: Cu_5FeS_4. It has a reddish-brown or coppery-red color on fresh fracture, but tarnishes rapidly to iridescent purple or blue. Bornite is a valuable ore of copper. Syn: *erubescite; variegated copper ore; peacock ore; horseflesh ore; purple copper ore.*

borolanite A plutonic rock composed chiefly of alkali feldspar and melanite, with lesser amounts of nepheline, biotite, and pyroxene; a melanite *nepheline syenite.* The feldspar and nepheline commonly form aggregates that resemble phenocrysts of leucite. The term was originated by Horne & Teall in 1892, for Loch Borolan, Scotland. Not recommended usage.

Boroll In U.S. Dept. of Agriculture soil taxonomy, a suborder of the soil order *Mollisol,* characterized by a frigid or cryic temperature regime and udic or ustic moisture regime. Borolls are common in cool to cold regions that have a continental climate (USDA, 1975). Cf: *Alboll; Aquoll; Rendoll; Udoll; Ustoll; Xeroll.*

boronatrocalcite *ulexite.*

borovskite A cubic mineral: Pd_3SbTe_4.

borrow Earth material (sand, gravel, etc.) taken from one location (such as a borrow pit) to be used for fill at another location; e.g. embankment material obtained from a pit when there is insufficient excavated material nearby to form the embankment. The implication is often present that the borrowed material has suitable or desirable physical properties.

borrow pit An excavated area where borrow has been obtained. See also: *barrow pit.*

bort (a) A granular to very finely crystalline aggregate consisting of imperfectly crystallized diamonds or of fragments produced in cutting diamonds. It often occurs as spherical forms, with no distinct cleavage, and having a radial fibrous structure. (b) A diamond of the lowest quality, so flawed, imperfectly crystallized, or off-color that it is suitable only for crushing into abrasive powders for industrial purposes (as for saws and drill bits); an *industrial diamond.* Originally, any crystalline diamond (and later, any diamond) not usable as a gem. (c) A term formerly used as a syn. of *carbonado.*—Cf: *ballas.* Syn: *boart; boort; bortz; bowr.*

bortz *bort.*

böschung A syn. of *gravity slope.* Etymol: German *Böschung,* term used by Penck (1924) for a rock slope maintaining constant gradient as it retreats.

boss [geomorph] A smooth and rounded mound, hillock, or other mass of resistant bedrock, usually bare of soil or vegetation.

boss [ign] An igneous intrusion that is less than 40 sq mi (100 km^2) in surface exposure and is roughly circular in plan. Cf: *stock [intrus rocks].*

boss [paleont] (a) A rounded and raised knoblike ornamental structure in foraminifera. (b) The part of an echinoid tubercle, below the mamelon, shaped like a truncated cone and supporting the spheroidal summit of the tubercle. (c) A coarse, short nodule occurring on the spire of a gastropod.

bostonite A light-colored hypabyssal rock, characterized by *bostonitic* texture and composed chiefly of alkali feldspar; a fine-grained *trachyte* with few or no mafic components. The name, given by Hunter and Rosenbusch in 1890, is derived from Boston, Massachusetts, for no clear reason. Not recommended usage.

bostonitic Said of the texture of *bostonite,* in which microlites of rough irregular feldspar tend to form clusters of divergent laths within a trachytoid groundmass.

botallackite A bluish-green mineral: $Cu_2(OH)_3Cl \cdot 3H_2O$.

botanical anomaly A local increase above the normal variation in the chemical composition, distribution, ecological assemblage, or morphology of plants, indicating the possible presence of an ore deposit. See also: *geobotanical prospecting.*

botn A Norwegian and Swedish term for the "bottom" of a glacial lake or of a fjord, but used as an equivalent of *cirque [glac geol].*

botryogen A deep-red or deep-yellow mineral, usually botryoidal: $MgFe(SO_4)_2(OH) \cdot 7H_2O$.

botryoid *cave coral.*

botryoidal Having the form of a bunch of grapes. Said of mineral deposits, e.g. hematite, having a surface of spherical shapes; also said of a crystalline aggregate in which the spherical shapes are composed of radiating crystals. Cf: *colloform; reniform.*

botryolite A radiated, columnar variety of datolite with a botryoidal surface.

bottleneck bay A bay with a narrow entrance which is guarded from the waves by features other than barrier islands.

bottle post *drift bottle.*

bottle spring A freshwater spring that issues through the floor of a saline lake or pool. The name is derived from the fact that fresh water can be obtained by submerging a stoppered bottle directly over the spring and then removing the stopper.

bottom [geog] Low-lying, level land, usually highly fertile, esp. in the Mississippi Valley region and farther west where the term signifies a grassy lowland formed by deposition of alluvium along the margin of a watercourse; an alluvial plain or a flood plain; the floor of a valley. The term is usually used in the plural. Syn: *bottomland; flat; interval; lowland.*

bottom [geomorph] (a) The *bed* of any body of water; the *floor* upon which any body of water rests. (b) A term used in England for the former head of a lake in a U-shaped valley, now covered with sediment deposited by inflowing streams. (c) *valley floor.*

bottom [ore dep] (a) Syn. of *gutter [ore dep].* (b) The lower limit of an orebody, either structurally or by economic grade. Syn: *root [ore dep].* See also: *bottoming.*

bottom-hole pressure The pressure produced in a well bore at or near the depth of a reservoir formation. It may be measured as a "flowing bottom-hole pressure" or as *shut-in pressure* to record the rate of pressure build-up during the survey period. Abbrev:

BHP. Syn: *reservoir pressure.*

bottom-hole temperature The temperature of the fluid at or near the bottom of a borehole; significantly lower than the temperature of the formation if borehole fluids have been circulated recently or are being produced with expansion into the well bore.

bottom ice *anchor ice.*

bottoming The downward pinching-out or termination of an orebody, either structurally or by economic grade. See also: *bottom [ore dep].*

bottomland A syn. of *bottom.* Also spelled: *bottom land.*

bottom load *bed load.*

bottom lock A situation where doppler-sonar measurements are based on reflections from the sea bottom. It is the normal operational mode, as opposed to the "water-scatter mode" that occurs in deeper water.

bottom moraine *ground moraine [glac geol].*

bottom peat Peat that is associated with lakes or streams and is derived mainly from mosses such as *Hypnum.*

bottomset *bottomset bed.*

bottomset bed One of the horizontal or gently inclined layers of sediment deposited in front of the advancing *foreset beds* of a delta or bed form. Syn: *bottomset.*

bottom terrace A depositional landform produced by streams having moderate or small *bed loads* coarse sand and gravel, characterized by a broad gently sloping surface, a few meters wide, in the direction of flow and a steep escarpment (about a meter high) facing downstream, and generally trending at right angles to the flow (Russell, 1898b, p. 166-167).

bottom water [oceanog] The deepest and most dense *water mass,* formed by cooling at the surface in high latitudes. Cf: *deep water; intermediate water; surface water [oceanog].*

bottom water [petroleum] The water immediately underlying the oil or gas in an *oil pool* or a *gas pool.* Cf: *edge water.*

boudin (a) One of a series of elongate, sausage-shaped segments occurring in *boudinage* structure, either separate or joined by pinched connections, and having barrel-shaped cross sections. Cf: *tectonic lens.* (b) A term applied loosely, without regard to shape or origin, to any *tectonic inclusion.*—Etymol: French, "bag; blood sausage". Pron: boo-*dan.*

boudinage A structure common in strongly deformed sedimentary and metamorphic rocks, in which an original continuous competent layer or bed between less competent layers has been stretched, thinned, and broken at regular intervals into bodies resembling *boudins* or sausages, elongated parallel to the fold axes. Syn: *sausage structure.*

Bouguer anomaly A gravity anomaly calculated after corrections for latitude, elevation, and terrain. Pron: boo-*gay.* See also: *Bouguer correction.*

Bouguer correction A correction made to gravity data for the attraction of the rock between the station and the datum elevation (commonly sea level); or, if the station is below the datum elevation, for the rock missing between station and datum. The Bouguer correction is 0.01276 ph mgal/ft, or 0.04185 ph mgal/m, where p is the specific gravity of the intervening rock and h is the difference in elevation between station and datum. See also: *double Bouguer correction.*

Bouguer plate An imaginary layer, having infinite length and a thickness equal to the height of the observation point above the reference surface, which is usually the geoid (Mueller & Rockie, 1966, p. 13).

boulangerite A bluish-gray or lead-gray metallic-looking mineral: $Pb_5Sb_4S_{11}$. It occurs in plumose masses.

boulder (a) A detached rock mass larger than a cobble, having a diameter greater than 256 mm (10 in., or -8 phi units, or about the size of a volleyball), being somewhat rounded or otherwise distinctively shaped by abrasion in the course of transport; the largest rock fragment recognized by sedimentologists. In Great Britain, the limiting size of 200 mm (8 in.) has been used. Cf: *block [part size].* See also: *small boulder; medium boulder; large boulder; very large boulder.* (b) *glacial boulder.* (c) *boulder of weathering.* (d) *boulder stone.* (e) A general term for any rock that is too heavy to be lifted readily by hand.—Also spelled: *bowlder.*

boulder barricade An accumulation of many large boulders visible along a coast (such as that of Labrador) between low tide and half tide (Daly, 1902, p. 260).

boulder barrier A shore ridge created by great pressure from floating ice under the influence of strong winds and gentle shore slopes, and measuring over 6 m in height and 800 m in length (Hamelin & Cook, 1967, p. 97).

boulder beach A beach consisting mostly of boulders.

boulder bed (a) A boulder-bearing conglomerate. (b) A glacial deposit, such as a till or tillite, containing a wide range of particle sizes; e.g. the Talchir boulder beds of India.

boulder belt A long, narrow accumulation of glacial boulders derived from distant sources, lying transverse to the direction of movement of the glacier by which it was deposited; also, a zone of such boulders. Cf: *boulder train.* Syn: *border belt.*

boulder clay A term used in Great Britain as an equivalent of *till,* but applied esp. to glacial deposits consisting of striated, subangular boulders of various sizes embedded in stiff, hard, pulverized clay or rock flour. The term "till" is preferable as a general term, applicable not only to material of the character described above but to glacial deposits that contain no boulders or that may be so sandy as to have very little clay. Syn: *drift clay.*

boulder conglomerate A consolidated rock consisting mainly of boulders.

boulder depression A type of *block field* situated in a shallow depression, displaying a flat surface of pure boulder material that gradually decreases in size downward, and found mainly below the timberline. Diameter: a few meters to hundreds of meters.

boulderet A term suggested by Chamberlin (1883, p. 324) for a rounded, coarse fragment of glacial drift, having a diameter range of 6-15 in. (15-38 cm).

boulder facet One of the small plane surfaces on a *faceted boulder.*

boulder fan A fan-shaped assemblage of clasts (mainly boulders) diverging from their bedrock source in the direction of movement of the glacier by which they were transported and deposited.

boulder field *block stream.*

boulder flat A level tract covered with boulders.

boulder gravel An unconsolidated deposit consisting mainly of boulders.

boulder of decomposition A *boulder of weathering* produced by chemical weathering; e.g. a joint block of basalt, modified and rounded by *spheroidal weathering,* leaving a relatively fresh spherical core surrounded by shells of decayed rock.

boulder of disintegration A *boulder of weathering* produced by mechanical weathering; e.g. a boulder fashioned by *exfoliation.*

boulder of weathering A large, detached rock mass whose corners and edges have been rounded in place, at or somewhat below the surface of the ground, by chemical or mechanical weathering; e.g. *boulder of decomposition* and *boulder of disintegration.* Cf: *boulder.* Syn: *residual boulder; weathering boulder.*

boulder opal A miner's term applied in Queensland, Australia, to siliceous ironstone nodules of concretionary origin, containing precious opal and occurring in sandstone or clay.

boulder pavement [geomorph] (a) An accumulation of boulders produced on a terrace by the eroding action of waves or river currents in removing finer material from littoral or fluvial deposits. (b) A slightly inclined surface composed of randomly spaced, flat-surfaced, usually frost-shattered blocks resulting from solifluction or other mass movement. (c) A *desert pavement* consisting of boulders.

boulder pavement [glac geol] (a) An accumulation of glacial boulders once contained in a moraine and remaining nearly in their original positions when the finer material has been removed by waves and currents. (b) A relatively smooth surface strewn with striated and polished boulders, abraded to flatness by the movement of an overriding glacier. Cf: *glacial pavement.*

boulder prospecting The use of boulders and boulder trains from outcrops of mineral deposits as a guide to ore.

boulder quarry A quarry in which weathering has produced so much jointing in the stone that it is not possible to mine large blocks from it.

boulder rampart A *rampart* or narrow ridge of boulders built along the seaward edge of a reef flat, esp. on the side from which the prevailing winds blow. The rampart, which seldom exceeds 1 or 2 m. in height, occurs close behind the *lithothamnion ridge* where present. Syn: *boulder ridge.*

boulder ridge (a) A *beach ridge* composed of boulders. (b) *boulder rampart.*

boulder size A term used in sedimentology for a volume greater than that of a sphere with a diameter of 256 mm (10 in.).

boulder stone An obsolete term for any large rock mass lying on the surface of the ground or embedded in the soil, differing from the country rock of the region, such as an erratic. Syn: *boulder.*

boulderstone A consolidated sedimentary rock consisting of boulder-size particles (Alling, 1943, p.265).

boulder stream *block stream.*

boulder train A line or series of glacial boulders and smaller clasts extending from the same bedrock source, often for many kilometers, in the direction of movement of the glacier by which they were transported and deposited. Cf: *boulder belt; boulder fan; indicator fan.*

boulder wall A boulder-built glacial moraine.

bouldery Characterized by boulders; e.g. a "bouldery soil" containing stones having diameters greater than 60 cm (24 in.) (SSSA, 1965, p. 333).

boule A pear-shaped or carrot-shaped mass, as of sapphire, ruby, spinel, or rutile, that forms during the production of synthetic gem material by the Verneuil process. Etymol: French, "ball". Syn: *birne.*

Bouma cycle A fixed, characteristic succession, of five intervals, that makes up a complete sequence of a turbidite (Bouma, 1962). One or more of the intervals may be missing. The five intervals, from the top: (e) pelitic; (d) upper parallel laminations; (c) current ripple laminations; (b) lower parallel laminations; and (a) graded. Named after Arnold H. Bouma, Dutch sedimentologist.

bounce cast The cast of a *bounce mark,* consisting of a short ridge that fades out gradually at both ends.

bounce mark A shallow *tool mark,* up to 5 cm in length, oriented parallel to the current and produced by an object that struck or grazed against the bottom, rebounded, and was carried upward. The longitudinal profile is symmetrical. The mark is widest and deepest in the middle and fades out gradually in both directions. The term was proposed by Wood & Smith (1958, p. 168). Cf: *prod mark.* See also: *brush mark.*

boundary current A deep ocean current, esp. along the western part of the oceans, characterized by sudden changes in temperature and salinity.

boundary curve *boundary line.*

boundary fault A descriptive term used in coal-mining geology for a fault along which there has been sufficient displacement to truncate the coal-bearing strata and thus bound the coalfield. Syn: *marginal fault.* Partial syn: *border fault.*

boundary lake A lake situated on or crossed by a political boundary line, as between two states or nations.

boundary layer In a fluid, a region of concentrated velocity variation and shear stress close to a solid that is moving relatively to the fluid. It is thin and its flow may be either turbulent or laminar (Middleton, 1965, p. 247).

boundary line [geochem] In a binary system, the line along which any two phase areas adjoin; in a ternary system, the line along which any two liquidus surfaces intersect. In a condensed ternary system, the boundary line represents equilibrium, typically with two solid phases and one liquid phase. See also: *reaction line.* Syn: *boundary curve; phase boundary.*

boundary line [surv] A line along which two areas meet; a line of demarcation between contiguous political or geographic entities.

boundary map A map that delineates a boundary line and the adjacent territory.

boundary monument A *monument* placed on or near a boundary line for the purpose of preserving and identifying its location on the ground.

boundary spring A type of gravity spring whose water issues from the lower slope of an alluvial cone. Syn: *alluvial-slope spring.*

boundary stratotype "A specific point in a specific sequence of rock strata that serves as the standard for definition and recognition of a stratigraphic boundary" (ISG, 1976, p. 24); the upper or lower limit of a *unit stratotype.*. Cf: *stratotype; mutual-boundary stratotype.* Syn: *type-boundary section.*

boundary survey A survey made to establish or re-establish a boundary line on the ground or to obtain data for constructing a map showing a boundary line; esp. such a survey of boundary lines between political territories. Cf: *land survey; cadastral survey.*

boundary-value component *perfectly mobile component.*

boundary-value problem One of three problems of potential theory that have great significance in geodesy. See: *Dirichlet's problem; Neumann's problem.* The third (unnamed) problem is to determine a function that is harmonic outside of a given surface and is such that a certain linear combination of it and its normal derivative assumes prescribed boundary values on the surface.

boundary vista A lane cleared along a boundary line passing through a wooded area.

boundary wave A seismic wave propagated along a free surface or an interface between layers.

bound gravel A hard, lenticular, cemented mass of sand and gravel occurring in the region of the water table; it is often mistaken for bedrock.

boundstone A term used by Dunham (1962) for a sedimentary carbonate rock whose original components were bound together during deposition and remained substantially in the position of growth (as shown by such features as intergrown skeletal matter and lamination contrary to gravity); e.g. most reef rocks and some biohermal and biostromal rocks. Cf: *biolithite.*

bound water Water present in such materials as animal and plant cells and soils, which cannot be removed without changing the structure or composition of the material and cannot react as does *free water* in such ways as dissolving sugar and forming ice crystals.

bourne A small stream or brook; specif. an intermittent stream that flows on the chalk downs and limestone heights of southern England after a heavy rainfall. Syn: *bourn; burn; winterbourne; woebourne; gypsey; chalk stream.*

bournonite A steel-gray to iron-black orthorhombic mineral: $PbCuSbS_3$. It commonly occurs in wheel-shaped twin crystals associated with other copper ores. Syn: *wheel ore; cogwheel ore; endellionite; berthonite.*

bourrelet (a) An externally inflated or elevated part of an interambulacral area of an echinoid, situated adjacent to the peristome. Cf: *phyllode.* (b) Either of two parts of the ligamental area of a bivalve flanking the resilifer on its anterior and posterior sides. Each bourrelet comprises a growth track and a seat of the lamellar ligament.

boussingaultite A mineral: $(NH_4)_2Mg(SO_4)_2 \cdot 6H_2O$.

bowenite A hard, compact, greenish-white to yellowish-green mineral of the serpentine group, representing a translucent, massive, fine-grained variety of antigorite resembling nephrite jade in appearance and composed of a dense feltlike aggregate of colorless fibers, with occasional patches of magnesite, flakes of talc, and grains of chromite. The term has also been applied to a serpentine rock in New Zealand. Syn: *tangiwai.*

Bowen ratio The ratio of sensible to evaporative energy (heat) loss from the surface of a body of water.

Bowen's reaction series A term used interchangeably with *reaction series* for a concept originally proposed by N. L. Bowen.

Bowie effect The indirect effect on gravity due to a warping of the geoid resulting from the application of gravity corrections. Syn: *indirect effect.*

bowlder *boulder.*

bowlingite *saponite.*

bowr *bort.*

bowralite A syenitic *pegmatite* composed chiefly of tabular euhedral alkali feldspar crystals with lesser amounts of sodic amphibole and aegirine, and with quartz, perovskite, zircon, and ilmenite as possible accessories. It was named by Mawson in 1906 for Bowral, New South Wales. Not recommended usage.

box (a) A hollow limonitic concretion. (b) *box-stone.*

box canyon (a) A narrow gorge or canyon containing a stream following a zigzag course, characterized by high, steep rock walls and typically closed upstream with a similar wall, giving the impression as viewed from its bottom of being surrounded or "boxed in" by almost-vertical walls. (b) A steep-walled canyon heading against a cliff; a dead-end canyon.—Syn: *cajon.*

box corer A type of *corer* that retrieves relatively undisturbed and quantitative sediment samples in a block rather than in a cylinder.

box fold A fold with the approximate profile form of three sides of a rectangle.

box level *circular level.*

box-stone A British term applied to a ferruginous concretion (found in Jurassic and Tertiary sands), often of rounded and rectangular or boxlike form, having a hollow interior in which white, powdery sand is sometimes present (P.G.H. Boswell in Wentworth, 1935, 241). Syn: *box.*

box the compass To name or repeat the 32 points of the compass in their contact order, clockwise from north.

boxwork In mineral deposits, a network of intersecting blades or plates of limonite or other iron oxide, deposited in cavities and along fracture planes from which sulfides have been dissolved by processes associated with the oxidation and leaching of sulfide ores, esp. porphyry copper deposits.

BPL *bone phosphate of lime.*

braccianite A melilite-free *cecilite,* named by Lacroix in 1917 for Bracciano, Italy. Not recommended usage.

bracewellite An orthorhombic mineral: CrO(OH). It is trimorphous with grimaldite and guyanaite.

brach (a) *brachial plate.* (b) *brachiopod.*

brachia (a) Plural of *brachium.* (b) A term sometimes used as a syn. of *lophophore.*

brachial adj. Pertaining to an arm or armlike structure of an animal (such as to the rays of a starfish or the brachium of a brachiopod).—n. A brachial part; esp. *brachial plate.*

brachial plate One of the plates that form the arms of a crinoid; any crinoid-ray plate above the radial plates (exclusive of pinnulars). Syn: *brachial; brach.*

brachial process An anteriorly directed bladelike or rodlike projection from the cardinalia of pentameracean brachiopods.

brachial ridge A narrow elevation of the secondary shell of some articulate brachiopods, extending laterally or anteriorly as an open loop from the dorsal adductor-muscle field. The brachial ridges are thought to be the region of attachment of the lophophore.

brachial valve The valve of a brachiopod that invariably contains any skeletal support (brachidium) for the lophophore and never wholly accommodates the pedicle, that is commonly smaller than the *pedicle valve,* and that has a distinctive muscle-scar pattern (TIP, 1965, pt.H, p.141). It typically has a small or indistinguishable beak. Syn: *dorsal valve.*

brachidium The looplike internal calcareous skeletal support structure of the lophophore of certain brachiopods. Pl: *brachidia.*

brachiolar facet An elliptical or subcircular *facet,* indentation, or scarlike area where a brachiole was attached, as in a cystoid or blastoid. Also spelled: brachiole facet.

brachiolar plate One of the biserially arranged plates of a brachiole of a blastoid, semielliptical in cross section and subquadrangular in side view, with a basal pair attached at the brachiolar facet (TIP, 1967, pt.S, p.346).

brachiole A biserial, nonpinnulate exothecal appendage of an echinoderm, springing independently from its surface and containing no extension of the body systems; esp. an erect, food-gathering structure arising from a cystoid thecal plate at the end or along the side of an ambulacrum, and bearing an extension of the ambulacral groove. Cf: *arm [paleont].*

brachiophore One of the short, typically stout, bladelike processes of secondary shell projecting from either side of the notothyrium and forming anterior and median boundaries of sockets in the brachial valves of certain brachiopods.

brachiophore base The basal (dorsal) part of a brachiophore that joins the floor of a brachiopod valve (TIP, 1965, pt.H, p.141).

brachiophore process A distal rodlike extension of a brachiophore that possibly supported the lophophore in some brachiopods.

brachiopod Any solitary marine invertebrate belonging to the phylum Brachiopoda, characterized by a lophophore and by two bilaterally symmetrical valves that may be calcareous or composed of chitinophosphate and that are commonly attached to a substratum but may also be free. Range, Lower Cambrian to present. Syn: *brach; lamp shell.*

brachistochrone *minimum-time path.*

brachitaxis A series of brachial plates in the crinoids. Pl: *brachitaxes.*

brachium (a) Either of the two armlike, coiled, muscular projections from the mouth segment of the lophophore of a brachiopod, variably disposed but symmetrically placed about the mouth. (b) Any process of an invertebrate similar to an arm, such as a tentacle of a cephalopod. (c) In the tetrapods, the humerus and its surrounding soft tissue.—Pl: *brachia.*

brachy- A prefix meaning "short".

brachyanticline A short, broad anticline. Cf: *brachysyncline.*

brachy-axis The shorter lateral axis of an orthorhombic or triclinic crystal; it is usually the *a* axis [cryst]. Cf: *macro-axis.* Also spelled: *brachyaxis.*

brachydome A *first-order prism* in the orthorhombic system; it is rhombic, with four faces parallel to the brachy-axis. Its indices are $\{0kl\}$. Cf: *clinodome.*

brachygenesis The phenomenon in evolution in which part of a presumed recapitulated sequence has evolved out and no longer appears in the course of development. Cf: *acceleration.*

brachygeosyncline A deep, oval depression formed during the later stages of geosynclinal deformation; a type of secondary geosyncline (Peyve & Sinitzyn, 1950).

brachylinear "Any lineation . . . ranging in length from less than two kilometers to the lower limit of visibility of the unaided eye" (El-Etr, 1976, p. 485).

brachyome The short arm of an anomoclone or ennomoclone of a sponge, or the different fourth ray of a *trider* of a sponge.

brachypinacoid *side pinacoid.*

brachysyncline A short, broad syncline. Cf: *brachyanticline.*

brachyuran Any *decapod* belonging to the infraorder Brachyura, characterized by a carapace that becomes progressively shortened and widened, developing a lateral margin; e.g. a crab. Range, Lower Jurassic to present.

brackebuschite A black to reddish mineral: $Pb_2(Mn,Fe)(VO_4)_2 \cdot H_2O$.

brackish water An indefinite term for water with a salinity intermediate between that of normal seawater and that of normal fresh water.

bract (a) A modified leaf associated with a flower or inflorescence, e.g. bearing a flower on its axis or being borne on a floral axis (subtending the flower or inflorescence). (b) A scooplike or spoonlike extension from the lower half of the rim of pores in the wall of an archaeocyathid.

bracteate Said of a plant having *bracts.*

Bradfordian (a) North American provincial stage: uppermost Devonian (above Cassadagan, below Mississippian). Syn: *Conewangoan.* (b) Substage in Great Britain: Middle Jurassic (upper Bathonian Stage).

bradleyite A mineral: $Na_3Mg(PO_4)(CO_3)$.

Bradydonti A group (order?) of marine cartilaginous fishes of uncertain affinity and poorly known general structure, characterized by a dental battery that consists of a few heavy crushing teeth. The teeth are distinctive. Lower taxa are of limited duration but wide distribution and are useful in late Paleozoic biostratigraphy. Range, Upper Devonian to Lower Permian.

bradygenesis *bradytely.*

bradyseism A long-continued, extremely slow vertical instability of the crust, as in the volcanic district west of Naples, Italy, where the Phlegraean bradyseism has involved up-and-down movements between 6 m below sea level and 6 m above over a period of more than 2,000 years (Casertano et al., 1976, p. 162). Etymol: Greek *bradys,* "slow", + *seismos,* "earthquake".

bradytely Retardation in the development of a group of organisms that may gradually cause certain individuals to fall behind the normal rate of progress in some or all of their characteristics. Cf: *horotely; tachytely; lipogenesis.* Syn: *bradygenesis.*

brae A syn. of *glacier* in Scandinavian usage.

Bragg angle In the *Bragg equation,* the angle between the diffracted beam of light and the diffracting crystal planes. It is symbolized by θ (theta). Syn: *reflection angle; angle of reflection.*

Bragg equation A statement in crystallography that the X-ray diffractions from a three-dimensional lattice may be thought of as reflecting from the lattice planes: $n\lambda = 2d \sin\theta$, in which n is any integer, λ is the wavelength of the X-ray, d is the crystal plane separation, also known as *d-spacing,* and θ is the angle between the crystal plane and the diffracted beam, also known as the *Bragg angle.* Syn: *Bragg's law.*

braggite A steel-gray mineral: (Pt,Pd,Ni)S.

Bragg reflection A diffracted beam of X-rays by a crystal plane according to the Bragg equation.

Bragg's law *Bragg equation.*

braid v. To branch and rejoin repeatedly to form an intricate pattern or network of small interlacing stream channels.—n. A reach of a braided stream, characterized by relatively stable branch islands and hence two or more separate channels. Cf: *anabranch.*

braid bar Any exposed sand or gravel bar that divides flow and causes a braided pattern in a stream (Rust, 1972, p. 232). Cf: *channel bar; unit bar.*

braided drainage pattern A drainage pattern consisting of braided streams. Syn: *interlacing drainage pattern.*

braided stream A stream that divides into or follows an interlacing or tangled network of several small branching and reuniting shallow channels separated from each other by branch islands or channel bars, resembling in plan the strands of a complex braid. Such a stream is generally believed to indicate an inability to carry all of its load, such as an overloaded and aggrading stream flowing in a wide channel on a flood plain. Syn: *anastomosing* *stream.*

brait A rough diamond.

braitschite A mineral: $(Ca,Na_2)_7(Ce,La)_2B_{22}O_{43}\cdot 7H_2O$.

brammallite A micaceous clay mineral, representing the sodium analogue of illite. Syn: *sodium illite.*

branch [seis] (a) One of two or more reflecting events that may be observed at a given location from a reflector that is concave upward or discontinuous. See: *buried focus.* (b) A set of values for a multivalued function. (c) A refraction event that may be observed at a given point because of the configuration of the refractor.

branch [streams] (a) A small stream that flows into another, usually larger, stream; a tributary. (b) A term used in the southern U.S. for a creek, or a stream normally smaller than and often tributary to a river. (c) A stream flowing out of the main channel of another stream and not rejoining it, as on a delta or alluvial fan; a distributary. (d) A stream flowing out of another stream and rejoining it, such as an anabranch; a by-channel. (e) A fork of a tidal river; e.g. the fork of the Severn River, Md.

branch fault *auxiliary fault.*

branch gap An interruption in the vascular tissue of a stem at the point at which a *branch trace* occurs. It is most evident in cross section, at the point of branch-trace departure.

branchia (a) A thin-walled, fingerlike or leaflike structure extending outward from a crustacean limb or secondarily from a side of the body, typically occurring in pairs, and functioning for respiration. Syn: *gill [paleont].* (b) A slender, hollow, fingerlike extension of the body wall of an asteroid.—Pl: *branchiae.*

branchial carina A longitudinal ridge extending over the branchial region behind the postcervical groove on some decapods (Holthuis, 1974, p. 735).

branchial chamber The space between the body and the wall of carapace enclosing the branchiae of a crustacean. Syn: *gill chamber.*

branchial region The lateral part of the carapace of some decapods, behind the pterygostomial region and overlying the branchiae; it is divided by some authors into epibranchial, mesobranchial, and metabranchial subregions (TIP, 1969, pt. R, p. 92).

branchial slit A *gill slit* of an echinoid.

branching bay A bay having a dendritic pattern, produced by drowning or flooding of a river valley by the sea. See also: *estuary.*

branching fault A fault that splits into two or more parts or branches.

branching ratio The ratio of the *decay constants* for each of two competing modes of *radioactive decay.*

branchiocardiac groove The groove in some decapods that separates the branchial region from the cardiac region of the carapace. In its posterior part it runs longitudinally, parallel with the median dorsal line of the carapace, and anteriorly it curves down to meet the post-cervical groove (Holthuis, 1974, p. 732).

branchiopod Any crustacean belonging to the class Branchiopoda, characterized by the morphologic similarity of their numerous somites and limbs and by their filter-feeding mode of nourishment. Range, Lower Devonian to present.

branchiostegite The extended portion of carapace covering the branchial chamber of a decapod crustacean.

branch island An island formed by the braiding of the branches of a stream; an island formed between a tributary and the main stream. Term was introduced by Jackson (1834, p. 79).

branchite *hartite.*

branch trace Vascular tissue extending from a stem into a branch. Cf: *leaf trace.* See also: *branch gap.*

branch water Water from a small stream or branch.

branchwork cave A cave in which the passages intersect as tributaries. See also: *network cave; maze cave.*

brandbergite A hypabyssal *granite* having aplitic texture and being composed of potassium feldspar (as whitish Carlsbad twins), quartz grains, arfvedsonite, and aggregates of biotite in a micrographic groundmass. The name, given by Chudoba in 1930, is for Brandberg, S.W. Africa. Not recommended usage.

brandtite A mineral: $Ca_2Mn(AsO_4)_2\cdot 2H_2O$. It is isomorphous with roselite and may contain up to 3% MgO.

brannerite A mineral: $(U,Ca,Ce)(Ti,Fe)_2O_6$.

brannockite A hexagonal mineral of the osumilite group: $KSn_2Li_3Si_{12}O_{30}$.

brash *brash ice.*

brash ice An accumulation of floating fragments not more than 2 m across, and representing the wreckage of other forms of ice; occurs esp. near an ice pack or floe. Syn: *brash; debris ice; mush.*

brass An English term for yellowish iron pyrites (pyrite and marcasite) found in coal or coal seams. Syn: *brasses.*

brassil *brazil.*

brassite An orthorhombic mineral: $MgHAsO_4\cdot 4H_2O$.

brass ore (a) *aurichalcite.* (b) A mixture of sphalerite and chalcopyrite.

braunite A brittle brownish-black or steel-gray tetragonal mineral: $3Mn_2O_3\cdot MnSiO_3$. It sometimes has appreciable ferric iron.

bravaisite A name proposed for a micaceous clay mineral having about half the potassium of muscovite, and later used as a synonym to replace *illite.* Material from the type locality has been shown to be a mixture of montmorillonite and illite, with illite predominating; therefore, bravaisite is not a specific mineral and has no standing as a distinct mineral species (Grim, 1953, p. 36). Cf: *sarospatakite.*

Bravais lattice A syn. of *crystal lattice;* it is named for the nineteenth-century French physicist, Auguste Bravais, who demonstrated that there are only 14 possible unique kinds of crystal lattices.

bravoite A yellow mineral: $(Ni,Fe)S_2$. It is related to pyrite, and has a paler color.

B ray Right anterior ray in echinoderms situated clockwise of A ray when the echinoderm is viewed from the adoral side; equal to ambulacrum IV in the Lovenian system.

brazil An English dialectal term for iron pyrite, esp. associated with coal. Also, by extension, a term applied to a coal seam containing much pyrite. Adj: *brazilly.* Also spelled: *brazzil; brazzle; brassil.*

Brazilian emerald A transparent green variety of tourmaline occurring in Brazil and used as a gemstone.

brazilianite A yellowish-green to greenish-yellow monoclinic mineral: $NaAl_3(PO_4)_2(OH)_4$.

Brazilian ruby A reddish mineral resembling ruby in appearance and occurring in Brazil; e.g. a light rose-red spinel, or a pink to rose-red or deep-red topaz (either natural or artificially heated), or a reddish tourmaline.

Brazilian sapphire A transparent blue variety of tourmaline occurrring in Brazil and used as a gemstone.

Brazilian topaz Topaz mined in Brazil and ranging in color from pure white to blue; esp. yellowish topaz.

brazilite A mixture of baddeleyite, zircon, and altered zircon. The term has also been applied to an oil shale, to a fibrous variety of baddeleyite, and as a syn. of baddeleyite.

Brazil twin law A type of twin law in quartz in which the twin plane is perpendicular to one of the *a* crystallographic axes; an example of *optical twinning.* Cf: *Dauphiné twin law.*

brea A rarely used term for a viscous asphalt formed by the evaporation of volatile components from oil in seepages. In Trinidad, it is used as the name for *maltha.* Etymol: Spanish, "pitch".

breach v. To cut a deep opening in a landform, esp. by erosion.

breached Said of a volcanic cone or crater, the rim of which has been broken through by the outpouring lava.

breached anticline An anticline whose crest has been deeply eroded, so that it is flanked by inward-facing erosional scarps. Cf: *bald-headed anticline.* Syn: *unroofed anticline; scalped anticline.*

bread-crust bomb A type of volcanic bomb characterized by a network of opened cracks on its surface, due to continued expansion of the interior after solidification of the crust. See also: *explosive bomb.*

bread-crust surface A surface, resembling the crust of bread, characterizing certain concretions formed where abundant salts are being precipitated by evaporating water in a semiarid climate (Twenhofel, 1939, p. 40).

break [drill] A change in the penetration rate of a drill; usually said of an increase, i.e. a "fast break". Cf: *shale break.*

break [geomorph] A marked variation of topography, or a tract of land distinct from adjacent land, or an irregular and rough piece of ground; e.g. a deep valley, esp. a ravine or gorge cutting through a ridge or mountain. See also: *breaks.*

break [mining] A general term used in mining geology for any discontinuity in the rock, such as a fault, a fracture, or a small cavity.

break [seis] *arrival.*

break [slopes] A marked or abrupt change or inflection in a slope or profile; a *knickpoint.* Term is used in the expressions "break of slope" and "break of profile".

break [stratig] (a) An abrupt change at a definite horizon in a chronologic sequence of sedimentary rocks, usually indicative of an unconformity (esp. a disconformity) or hiatus; esp. a marked change in lithology, such as one separating a channel sand from an underlying shale. See also: *faunal break.* (b) An interruption of a normal geologic sequence, esp. of stratigraphic continuity; a *discontinuity [stratig].*—Syn: *gap [stratig]; stratigraphic break.*

breakdown *cave breakdown.*

breaker [reef] *boiler.*

breaker [waves] A sea-surface wave that has become so steep (wave steepness of 1/7) that the crest outraces the body of the wave and collapses into a turbulent mass on shore or over a reef or rock. Breaking usually occurs when the water depth is less than 1.28 times the wave height. See also: *plunging breaker; spilling breaker; surging breaker; surf.* Syn: *breaking wave.*

breaker depth The still-water depth at the point where a wave breaks. Syn: *breaking depth.*

breaker line The axis along which a wave breaks as it approaches the shore. Syn: *plunge line.*

breaker zone *surf zone.*

breaking depth *breaker depth.*

breaking strength *fracture strength.*

breaking wave *breaker.*

break of slope *break [slopes].*

breakover A rounded crest that is both structurally and topographically high.

breakpoint bar A longshore bar formed at the breakpoint of waves, where there is a sudden decrease of sand moving landward outside the breakpoint but where sand is moving seaward to this point (King & Williams, 1949, p. 80).

breaks (a) A term used in the western U.S. for a tract of rough or broken land dissected by ravines and gullies, as in a badlands region. (b) Any sudden change in topography, as from a plain to hilly country, or a line of irregular cliffs at the edge of a mesa or at the head of a river; e.g. Cedar Breaks, Utah. See also: *break.*

breakthrough (a) The erosive action of water in wearing or cutting a passage. (b) The channel made by such a breakthrough.

break thrust An overthrust developed during the deformation of an anticline at that point at which folding becomes fracturing, and strata are overthrust along the fault surface.

breakup (a) The melting, loosening, fracturing, or destruction of snow or floating ice during the spring; specif. the destruction of the ice cover on a river during the spring thaw. (b) The period during the spring thaw when a breakup occurs.

breakwater An offshore structure (such as a mole, wall, or jetty) that, by breaking the force of the waves, protects a harbor, anchorage, beach, or shore area. Syn: *water-break.*

breast A miner's term for the *face* of a mine working. See also: *before breast.*

breast wall A wall designed to withstand the force of a natural bank of earth, as of timber used to support the face of a tunnel.

breathing *river breathing.*

breathing cave A cave that has an alternating movement of air through its passages, usually with a period of a few minutes. See also: *blowing cave.*

breathing well A well, generally a water well, that, in response to changes in atmospheric pressure, alternately takes in and emits a strong current of air, often with an alternating sucking and blowing sound. It penetrates, but is uncased in at least part of, a thick zone of aeration that is porous and permeable enough to exchange air freely with the well but otherwise is poorly connected with the atmosphere because of the presence of tight soil or other low-permeability material above the unsaturated material. Syn: *blowing well.*

breccia [geol] A coarse-grained clastic rock, composed of angular broken rock fragments held together by a mineral cement or in a fine-grained matrix; it differs from *conglomerate* in that the fragments have sharp edges and unworn corners. Breccia may originate as a result of talus accumulation (*sedimentary breccia*); igneous processes, esp. explosive (*igneous breccia, volcanic breccia*); disturbance during sedimentation (*intraclastic breccia*); collapse of rock material (*solution breccia, collapse breccia*); or tectonic processes (*fault breccia*). Etymol: Italian, "broken stones, rubble". Pron: *bretsh*-ia. Syn: *rubblerock.*

breccia [lunar] A common lunar rock with clasts produced by meteoroid impact. As impact energy serves to lithify unconsolidated debris, a single breccia specimen may record several successive breaking and annealing events.

breccia-conglomerate A sedimentary rock consisting of both angular and rounded particles (Norton, 1917, p.181); a sedimentary rock that is not clearly referable to either breccia or conglomerate. Syn: *breccio-conglomerate.*

breccia dike A sedimentary dike composed of breccia injected into the country rock.

breccial Pertaining to breccia.

breccia marble Any marble composed of angular fragments. The term was in use before the separate use of "breccia" in geology.

breccia pipe *pipe [volc].*

breccia porosity *Interparticle porosity* in a breccia (Choquette & Pray, 1970, p. 244).

breccia-sandstone A sandstone containing small fragments of breccia.

brecciated [geol] Converted into, characterized by, or resembling a *breccia;* esp. said of a rock structure marked by an accumulation of angular fragments, or of an ore texture showing mineral fragments without notable rounding.

brecciated [meteorite] A term incorrectly applied to a meteorite of intermediate type (between iron and stony-iron), in which the main mass is iron with octahedral or hexahedral structure but there are also relatively large silicate inclusions of rounded or angular form. Also, said of such a texture occurring in a meteorite, including stony meteorites.

brecciation Formation of a breccia, as by crushing a rock into angular fragments.

breccia vein A fissure containing numerous wall-rock fragments, with mineral deposits in the interstices.

brecciform In the form or shape of a breccia, or resembling a breccia.

breccio-conglomerate *breccia-conglomerate.*

breccioid Having the appearance of a breccia.

brecciola A well-graded, intraformational breccia consisting of small, angular limestone fragments in well-defined beds separated by dark shale, such as the breccia occurring in the northern Apennines. Etymol: Italian, diminutive of *breccia.*

bredbergite A variety of andradite garnet containing magnesium.

bredigite A mineral: $Ca_{14}Mg_2(SiO_4)_8$. It is a metastable orthorhombic phase of calcium orthosilicate (but not isomorphous with olivine), stable from about 800° to 1447°C on heating and from 1447° to 670°C on cooling. Cf: *calcio-olivine; larnite.*

breithauptite A copper-red mineral: NiSb.

Breithaupt twin law A rare type of normal twin law in feldspar, in which the twin plane is $(\bar{1}11)$.

Bretonian orogeny One of the 30 or more orogenies during Phanerozoic time identified by Stille. It consisted of several phases, from the Late Devonian to the end of the Devonian; it is considered to be the earliest part of the Variscan orogenic era, which continued to the end of the Paleozoic.

breunnerite A variety of magnesite containing 5-30% iron carbonate.

Brevaxones A group of mid-Cretaceous and younger angiosperm pollen in which the polar axis is shorter than the equatorial diameter, representing an evolutionary advance over *Longaxones,* and including such forms as Normapolles.

brevicone A straight or slightly curved shell characteristic of certain Paleozoic cephalopods, having a short, blunt form; it expands rapidly from the apex to the base of the body chamber, or to a point a short distance before or behind that base. Cf: *longicone.*

Brewster angle *polarizing angle.*

brewsterite A zeolite mineral: $(Sr,Ba,Ca)Al_2Si_6O_{16}\cdot 5H_2O$. It usually contains some calcium.

brewsterlinite Liquid CO_2, found as inclusions in cavities in minerals such as quartz, topaz, and chrysoberyl. It will expand so as to fill cavities under the warmth of the hand.

Brewster's law A statement in optics that when unpolarized light is incident on a surface, it acquires maximum plane polarization at a particular *angle of incidence* whose tangent equals the refractive index of the substance. This angle is called the *polarizing angle,* or the *Brewster angle.*

brezinaite A meteorite mineral: Cr_3S_4.

Brezina's lamellae Lamellae of schreibersite oriented parallel to dodecahedral planes in parent taenite of iron meteorites. Named after M. Aristides S.F. Brezina (1848-1909), Austrian mineralogist.

brianite A mineral: $Na_2CaMg(PO_4)_2$.

briartite A mineral: $Cu_2(Fe,Zn)GeS_4$.

brick clay (a) Any clay suitable for the manufacture of bricks or coarse pottery; a *brick earth.* (b) An impure clay containing iron, calcium, magnesium, and other ingredients.

brick earth Earth, clay, or loam suitable for making bricks; specif. a fine-grained brownish deposit consisting of quartz and flint sand mixed with ferruginous clay and found on river terraces as a result of reworking by water of windblown material, such as that overlying the gravels on certain terraces of the River Thames in England. See also: *brick clay.* Also spelled: *brickearth.*

bridal-veil fall A cataract of great height and such small volume that the falling water is largely dissipated in spray before reaching the lower stream bed, and having a form that suggests a bridal veil. Type example: Bridalveil Fall in Yosemite Valley, Calif.

bridge [drill] n. A rock fragment, *cavings,* or other obstruction that lodges (either accidentally or intentionally) part way down in a drill hole (as in an oil well).—v. To form a bridge in a drill hole.

bridge [geomorph] *natural bridge.*

bridge [speleo] In a cave, a solutional remnant of rock that spans a passage from wall to wall. See also: *partition.*

bridge islet An island that becomes a peninsula during low tide.

bridge plug A mechanical device or a volume of cement deliberately set in a well bore at a selected depth in order to seal off the hole below.

Bridgerian North American continental stage: Middle Miocene (above Wasatchian, below Uintan).

bridging factor A term used by Gruner (1950) for a number that expresses the manner by which SiO_4 tetrahedra are tied together in a mineral. It is equal to 0.8 plus twenty percent of the quotient of the sum of the valence bonds of a silicate divided by the number of cations. A bridging factor of 1.00 is assigned to quartz, whose SiO_4 tetrahedra are all directly tied to other tetrahedra, resulting in the highest number of bridges possible; all other structures have smaller factors. See also: *energy index [mineral].*

bridle v. To connect a group of seismic amplifiers or other devices to a common input.—n. A seismogram produced with the amplifiers bridled. Syn: *parallel shot.*

brigg An English term for a headland formed by "a scarp of hard rock cropping out at or near tide marks" (Stamp, 1961, p. 78). Syn: *brig.*

bright-banded coal *Banded coal* consisting mainly of vitrain and clarain, with some durain and minor fusain. Cf: *dull-banded coal; semisplint coal.*

bright coal A type of banded coal defined microscopically as consisting of more than 5% of anthraxylon and less than 20% of opaque matter; banded coal in which translucent matter predominates. Bright coal corresponds to the microlithotypes vitrite and clarite and in part to duroclarite and vitrinerite (ICCP, 1963). Cf: *dull coal; semibright coal; semidull coal; intermediate coal.* Syn: *brights.*

bright field An image obtained in the transmission electron microscope by deliberately excluding the diffracted beams. This is accomplished by placing an aperture in the back focal plane of the objective lens, permitting only the transmitted beam to form the image. Cf: *dark field.*

brightness temperature (a) The temperature of a blackbody radiating the same amount of energy per unit area at the wavelengths under consideration as the observed body. Cf: *color temperature.* (b) The apparent temperature of a nonblackbody determined by measurement with an optical pyrometer or radiometer.

brights *bright coal.*

bright spot An exceptionally strong signal on a seismic profile, often indicating an accumulation of natural gas (Hammond, 1974, p. 515).

brilliancy The total amount of light reaching the eye after being reflected from both exterior and interior surfaces of a gemstone in the face-up position. Given equal transparency and perfection of cutting, the transparent gem species with the highest refractive index will be the most brilliant. Brilliancy should not be confused with *scintillation* or *dispersion.*

brilliant n. A *brilliant cut* diamond. The term is less correctly applied to any brilliant-cut gemstone.

brilliant cut The most common style of cutting for most gemstones. The standard round brilliant consists of a total of 58 facets: 1 table, 8 bezel facets, 8 star facets, and 16 upper girdle facets on the crown; and 8 pavilion facets, 16 lower girdle facets, and usually a culet on the pavilion, or base. Cf: *Tolkowsky theoretical brilliant cut; step cut; single cut; mixed cut; pear-shaped cut; pendeloque.*

brim The flared or recurved portion of a cyrtochoanitic septal neck of a nautiloid, measured transverse to the longitudinal axis of the siphuncle.

brimstone A common or commercial name for *sulfur,* esp. native sulfur or fine sulfur fused into rolls, sticks, or blocks.

brine [geol] A term used for pore fluids in deep sedimentary basins, stratified hot fluids in restricted basins such as the Red Sea, oil-field waters, and geothermal mineralizing fluids. It denotes warm to hot highly saline waters containing Ca, Na, K, Cl, and minor amounts of other elements. Syn: *natural brine.* Cf: *artificial brine.*

brine [oceanog] (a) Seawater that, owing to evaporation or freezing, contains more than the usual amount (about 35%) of dissolved salts. Cf: *hot brine.* (b) Subsurface water with a high content of dissolved salts.

brine cell A small inclusion, usually in the shape of an elongated tube about 0.05 mm in diameter, containing residual liquid more saline than seawater, formed in sea ice as it develops. Syn: *brine pocket.*

brine content Relative volume of ice composed of brine, expressed as an absolute ratio or in parts per thousand.

brine lake *salt lake.*

brine pit A *salt well,* or an opening at the mouth of a salt spring, from which water is taken to be evaporated for making salt.

brine pocket *brine cell.*

brine slush "A mixture of ice crystals and salt water, which retards or prevents complete freezing, often found between young sea ice and a cover of newly fallen snow" (ADTIC, 1955, p. 14).

brine spring *salt spring.*

Bringewoodian European stage: Upper Silurian (above Eltonian, below Leintwardinian).

brink (a) A bank, edge, or border of a body of water, esp. of a stream. (b) The top of the slip face of a dune. It need not be the same as the crest.

brinkpoint The point on a cross section of a ripple that separates the steeply inclined lee side from the gently inclined stoss side or crestal platform. Term introduced by Allen (1968). Cf: *summitpoint.*

briolette A pear-shaped or drop-shaped gemstone with a circular cross section, having its entire surface cut in triangular, or sometimes rectangular, facets. Cf: *pear-shaped cut; pendeloque.*

britholite A mineral of the apatite group: $(Ca,Ce)_5(SiO_4,PO_4)_3(OH,F)$. Syn: *abukumalite.*

britholite-(Y) A mineral of the apatite group: $(Ca,Y)_5(SiO_4,PO_4)_3(OH,F)$. Syn: *abukmalite.*

brittle Said of a rock that fractures at less than 3-5% deformation or strain. Cf: *ductile.*

brittle mica (a) A group of minerals resembling the true *micas* in crystallographic characters, but having the cleavage flakes less elastic and containing calcium (instead of potassium) as an essential constituent. Syn: *clintonite.* (b) A mineral of the brittle-mica group, such as clintonite, margarite, and ephesite. (c) A micaceous mineral occurring in brittle folia; e.g. chloritoid.

brittle silver ore *stephanite.*

broad A British term for a lake or wide sheet of shallow, reed-fringed fresh water, forming a broadened part of, or joined to, a sluggish river near its estuary; often used in the plural. The feature is typically found in East Anglia (Norfolk and Suffolk), and is believed to have been produced artificially by the cutting and removal of peat in the Middle Ages.

brocenite *beta-fergusonite-(Ce).*

brochantite An emerald-green to dark-green mineral: $Cu_4(SO_4)(OH)_6$. It is common in the oxidation zone of copper-sulfide deposits. Syn: *blanchardite; kamarezite.*

brock An English term for a *brook.*

brockite A red and yellow mineral: $(Ca,Th,Ce)PO_4 \cdot H_2O$.

brockram A term used in Cumberland County, NW England, for breccia whose angular blocks are believed to have accumulated as talus material; e.g. the Brockram of Appleby and Kirkby Stephen, a Permian breccia in which fragments of Carboniferous limestone are held together by cement containing gypsum.

Brodelboden *involution [sed].*

bröggerite A variety of thorian uraninite, $(U,Th)O_2$.

broken belt The transition zone between open water and consolidated pack ice.

broken formation A body of broken strata that contains no exotic elements. Regardless of its broken state, it is a mappable rock-stratigraphic unit (Hsü, 1968, p. 1065).

broken ice An obsolete term for sea-ice concentration of 5/10 to 8/10; now replaced generally by *open pack ice* and *close pack ice.* Syn: *open ice; loose ice; slack ice.*

broken round A term used by Bretz (1929, p. 507) for a roundstone (such as a pebble or cobble) that has undergone breakage and whose spalled corners are believed to indicate exceptionally high-

velocity currents.
broken sand A sandstone containing a mixed sequence of deposits (such as shaly layers).
broken shoreline A shoreline characterized by many closely spaced islands, peninsulas, or jutting headlands.
broken-stick model A model comparing the relative abundances (or *niche* sizes) of several species competing within a habitat to the lengths of segments of a straight stick broken randomly into as many pieces as there are species (MacArthur, 1957). Its principal implication—that species diversity and abundance result from nonoverlapping, randomly sized ecologic niches—has stimulated much controversy among population ecologists.
broken stream A stream that repeatedly disappears and reappears, as in an arid region.
broken water Water whose surface is covered with ripples and eddies.
bromargyrite A yellow isometric mineral: AgBr. Syn: *bromyrite.*
bromellite A white hexagonal mineral: BeO.
bromlite *alstonite.*
bromoform Tribromethane: $CHBr_3$. It is used as a *heavy liquid;* its specific gravity is 2.9. Cf: *methylene iodide; Clerici solution; Sonstadt solution; Klein solution.*
bromyrite *bromargyrite.*
brontolith An obsolete syn. of *stony meteorite.* Also spelled: *brontolite.*
Bronze Age In archaeology, a cultural level that was originally the middle division of the *three-age system,* and is characterized by the technology of bronze. Correlation of relative cultural levels with actual age (and, therefore, with the time-stratigraphic units of geology) varies from region to region. The term is used mainly in Europe, since in Asian archaeology it coincides with written history and in the Americas and Africa bronze was little used (Bray and Trump, 1970, p.43).
bronzite A brown or green variety of *enstatite* containing iron and often having a bronzelike or pearly metallic luster; an orthopyroxene intermediate in composition between enstatite and hypersthene.
bronzitfels *bronzitite.*
bronzitite A *pyroxenite* composed almost entirely of bronzite. Syn: *bronzitfels.*
brood chamber A space within which embryos develop in bryozoans, partly or entirely enclosed by body walls of one or more polymorphs or extrazooidal parts of a colony. It may be part of the body cavity or outside the body cavity of the colony. Cf: *brood pouch.*
brood pouch A sac or cavity of the body of an animal where the eggs or embryos are received and undergo a part of their development; e.g. the gently to strongly swollen part of the heteromorphous (presumed female) carapace of an ostracode, thought to be used for containing the not yet independent young. Partial syn: *brood chamber.*
brook (a) A small *stream* or rivulet, commonly swiftly flowing in rugged terrain, of lesser length and volume than a *creek;* esp. a stream that issues directly from the ground, as from a spring or seep, or that is produced by heavy rainfall or melting snow. Also, one of the smallest branches or ultimate ramifications of a drainage system. (b) A term used in England and New England for any tributary to a small river or to a larger stream. Syn: *brock; bruik.* (c) A general literary term for a *creek.*
brookite A brown, reddish, or sometimes black orthorhombic mineral: TiO_2. It is trimorphous with rutile and anatase, and occurs in druses and cavities. Syn: *pyromelane.*
brooklet A small brook; a rill.
brookside The land adjacent to or bordering on a brook. Syn: *burnside.*
brotocrystal A crystal fragment of a previously consolidated rock that is only partially assimilated in a later magma. Rarely used.
brow (a) The projecting upper part or margin of a steep slope just below the crest; the edge of the top of a hill or mountain, or the place at which a gentle slope becomes abrupt. Syn: *brae.* (b) An English term for a steep slope.
brown algae A group of algae, commonly large seaweeds, corresponding to the phylum Phaeophyta, that owes its greenish-yellow to deep-brown color to the presence of carotenes and xanthophylls in greater amounts than the chlorophylls. Most brown algae are restricted to salt water. Cf: *blue-green algae; green algae; red algae; yellow-green algae.*
brown body A colored mass formed in many bryozoan zooids by the aggregation of the residue of a degenerated polypide.
brown clay *red clay.*
brown coal A brown to brownish black coal, intermediate in rank between peat and lignite A, in which original plant structures may usually be seen. The term is generally used in Europe, Australia, and Great Britain. Cf: *lignite.* Partial syn: *lignite B.* Old syn: *fulvurite.*
Brown earth *Brown Forest soil.*
Brown Forest soil A great soil group in the 1938 classification system, an intrazonal, calcimorphic group of soils that develops in a temperate climate under deciduous forest. It has a calcium-rich parent material, and has a mull horizon but no horizon of clay or sesquioxides (USDA, 1938). Most Brown Forest soils are now classified as *Ochrepts.* Syn: *Brown earth.*
brown hematite A syn. of *limonite.* The term is a misnomer, because true hematite (unlike limonite) is anhydrous. Cf: *red hematite; black hematite.*
brown hornblende A brown variety of hornblende rich in iron; specif. *basaltic hornblende.*
brown iron ore *limonite.*
brown lignite *lignite B.*
brown matter *Humic degradation matter;* cell-wall degradation matter that is brown and translucent in thin section.
Brown Mediterranean soil An obsolete term for a brown soil that forms in a Mediterranean climate. Cf: *Red Mediterranean soil.*
brown mica *phlogopite.*
brownmillerite A mineral: Ca_2AlFeO_5. It is a constituent of portland cement. Syn: *celite.*
brown ocher A *limonite* that is used as a pigment.
brown ore A brown-colored ore; specif. the limonite group of iron ores.
Brown Podzolic soil A great soil group in the 1938 classification, a group of zonal soils that is similar to a Podzol but lacks the leached, light-colored A2 horizon. It is considered by some to be a type of Podzol rather than a separate soil group (USDA, 1938). Most of these soils are now classified as *Andepts* and *Orthods.*
brown rock A term used in Tennessee for dark brown to black phosphorite resulting from the weathering of phosphatic limestone. Cf: *white-bedded phosphate.*
Brown soil A great soil group in the 1938 classification system, a group of zonal soils having a brown surface and a light-colored subsurface over an accumulation of calcium carbonate. It developed in a temperate to cool semi-arid climate (USDA, 1938). Most of these soils are now classified as *Ustolls* and *Xerolls.*
brown spar Any light-colored crystalline carbonate mineral that is colored brown by the presence of iron; e.g. ankerite, dolomite, magnesite, or siderite.
brownstone A brown or reddish-brown sandstone whose grains are generally coated with iron oxide; specif. a dark reddish-brown, ferruginous quartz sandstone of Triassic age, once extensively quarried in the Connecticut River valley for use as building stone.
browser (a) A terrestrial vertebrate that eats the leaves and shoots from branches well above the ground surface. (b) A marine invertebrate that eats tiny algae scraped off the solid substrate.
brucite A hexagonal mineral: $Mg(OH)_2$. It commonly occurs in thin pearly folia and in fibrous form, as in serpentine and impure limestone.
Brückner cycle A climatic cycle of 33-35 years, first noted by Sir Francis Bacon in 1625 and restated by E. Brückner in 1890. It is not seen in many records, but is suggested by tree-ring analyses and solar-cycle indices.
brüggerite A monoclinic mineral: $Ca(IO_3)_2 \cdot H_2O$.
brugnatellite A flesh-red mineral: $Mg_6Fe(OH)_{13}(CO_3) \cdot 4H_2O$.
bruik A Scottish var. of *brook.*
brunckite A colloidal variety of sphalerite.
Brunizem *Prairie soil.*
brunogeierite A mineral of the spinel group: $(Ge,Fe)Fe_2O_4$.
Brunton compass A compact pocket instrument that consists of an ordinary compass, folding open sights, a mirror, and a rectangular spirit-level clinometer, which can be used in the hand or on a staff or light rod for reading horizontal and vertical angles, for leveling, and for reading the magnetic bearing of a line. It is used in sketching mine workings, and in preliminary topographic and geologic surveys on the surface, e.g. in determining elevations, stratigraphic thickness, and strike and dip. Named after its inventor, David W. Brunton (1849-1927), U.S. mining engineer. Usually called a "Brunton". Syn: *pocket transit.*
brush (a) Numerous fine cytoplasmic strands radiating from the

distal end of the caecum and connected to the periostracum of the punctate shells of articulate brachiopods (TIP, 1965, pt.H, p.141). (b) A bunch of fine terminal branches in phaeodarian radiolarians.

brush cast The cast of a *brush mark,* characterized by a crescentic depression around the downcurrent end. Originally defined by Dzulynski and Slaczka in 1959.

brush hook A short, stout, heavy hooked blade with a sharpened iron edge, attached to an axe handle, and used by surveyors for cutting brush.

brushite A nearly colorless mineral: $CaHPO_4 \cdot 2H_2O$.

brush mark A *bounce mark* whose downcurrent end has a small crescentic ridge of mud pushed up by and in front of the impinging object.

brute-force radar *real-aperture radar.*

Bruxellian European stage: lower Middle Eocene (above Ypresian, below Auversian). It includes Cuisian and Lutetian.

bryalgal Said of a rigid, wave-resistant limestone composed largely of materials constructed in place by frame-building bryozoans and algae that often encrust one another. The material so formed is intimately associated with reefs. Term proposed by Bissell (1964, p. 586). Cf: *coralgal.*

bryochore A climatic term for the part of the Earth's surface represented by tundras.

Bryophyta A phylum of nonvascular plants that may have differentiated stems and leaves but have no true roots. It includes liverworts and mosses.

bryophyte A nonvascular plant that may have differentiated stems and leaves, but that has no true roots. Liverworts and mosses are bryophytes. Cf: *thallophyte; pteridophyte.*

bryozoan Any invertebrate belonging to the phylum Bryozoa and characterized chiefly by colonial growth, a calcareous skeleton, or, less commonly, a chitinous membrane, and a U-shaped alimentary canal, with mouth and anus. Range, Ordovician to present, with a possible downward extension into the Upper Cambrian. Syn: *sea mat; moss animal; moss coral; moss polyp; polyzoan.* See also: *ectoproct; entoproct.*

BS *backsight.*

B-tectonite A tectonite whose fabric is dominated by linear elements. Not in common use, having been largely replaced by the term L-tectonite. Cf: *L-tectonite; S-tectonite.*

B-type lead *Anomalous lead* that gives model ages older than the age of the enclosing rock. Cf: *J-type lead.* Syn: *Bleiberg-type lead.*

bubble (a) A small air-filled cavity, or the globule of air or gas, in the glass tube of a spirit level. When the level is adjusted to the horizontal, the center of the bubble comes to rest under a fixed mark or etched line at the highest point possible in the tube. (b) A term sometimes applied to a bubble tube and its contents.

bubble impression A small, shallow depression (2.5 cm in diameter) formed on a beach or sedimentary surface by a bubble of gas after it has been dislodged into the air or water above. It has a smooth surface, is not margined by a raised rim, and may pass downward into a tube. Syn: *bubble mark.*

bubble mark *bubble impression.*

bubble noise Unwanted seismic energy generated by the oscillation of a bubble of high-pressure gas in water. The waste gases from an explosion, or from other seismic-energy sources such as air guns, can give rise to a sequence of bubble oscillations.

bubble-point pressure The pressure at which gas, held in solution in crude oil, breaks out of solution as free gas. Syn: *saturation pressure.*

bubble pulse A pulsation attributable to the bubble produced by a seismic charge fired in deep water. The bubble pulsates several times with a period proportional to the cube root of the charge, each oscillation producing an identical unwanted seismic effect.

bubble trend A planar or linear distribution of bubbles in glacier ice.

bubble tube The circular or slightly curved glass tube containing the liquid and bubble in a spirit level and mounted with the bend convex upward.

bubble-wall texture Texture shown by *phenocrysts* that have a thin coating of vesicular glass.

Bubnoff unit A standard measure of geologic time-distance rates (as for geologic movements and increments), proposed by Fischer (1969) and defined as 1 micron/year (1 mm/thousand years, or 1 m/million years). Named in honor of Serge von Bubnoff (1888-1957), Russian-born German geologist.

bucaramangite A pale-yellow variety of retinite that is insoluble in alcohol, found at Bucaramanga, Colombia.

buccal cavity In crustaceans, a hollow space on the ventral side of the body, containing the mouth parts. In Malacostraca, it is bounded by the epistome in front and the free edges of the carapace on the sides. (b) In Ostreostraci, the hollow space containing the gill bars on the ventral side of the head shield.

buccal frame A structure enclosing the mouth parts of brachyuran decapod crustaceans, bounded laterally by free anterior and lateral edges of the carapace and in front by the epistome and commonly by closed maxillipeds.

buccal mass The mouth parts of molluscs, except bivalves, and the muscles with which they are attached.

buccal plate (a) One of ten large primordial plates of an echinoid, located on the tissue between peristomial margin and mouth, and containing pores for passage of tube feet. (b) *buccal shield.*

buccal shield A large, more or less triangular ossicle in the interradial position, adjoining the mouth in an ophiuroid (TIP, 1966, pt.U, p.29). Syn: *buccal plate.*

Buchan-type facies series Rocks produced in a type of dynamothermal regional metamorphism rather similar to that of the *Abukuma-type facies series* but in a somewhat higher pressure environment, of 3000 to 4000 bars (Hietanen, 1967, p.192).

buchite A vitrified hornfels produced by fusion of an argillaceous rock by intense local thermal metamorphism. Cf: *hyalomylonite.*

buchonite A dark-colored extrusive rock containing hornblende and biotite in addition to plagioclase, nepheline, and augite; an alkali-feldspar-bearing *tephrite.* Named by Sandberger in 1872 for Buchonia, Germany. Not recommended usage.

bucking Pulverization of a representative sample of rock, as for assay or quantitative analysis.

bucklandite (a) A black variety of epidote containing iron and having nearly symmetric crystals. (b) *allanite.*

buckle folding Folding that occurs in response to end-loading of competent layers. Cf: *bend folding.*

buck quartz *bull quartz.*

buckshot *shot [soil].*

bud [bot] The undeveloped or meristematic state of a branch or flower cluster, with or without scales, commonly covered by rudiments of leaves. Buds may be apical, axillary, or adventitious in location on the stem (Swartz, 1971, p. 75).

bud [paleont] (a) An asexual reproductive body (including various types of cells) that is eventually isolated from a parent sponge or polyp. (b) A newly developing, asexually produced zooid in bryozoans.

budding *vegetative reproduction.*

buddingtonite A mineral: $(NH_4)AlSi_3O_8 \cdot nH_2O$, with n about 0.5. It is isostructural with orthoclase.

budget year *balance year.*

bud scale A protective modified leaf, covering or enclosing a bud.

bud scar A scar left on a twig by the falling-away of a bud or a group of bud scales (Fuller & Tippo, 1954, p.952).

buergerite A mineral of the tourmaline group: $NaFe^{+3}_3Al_6Si_6B_3O_{30}F$.

Buerger precession method The recording on film of a single level of the reciprocal lattice of an individual crystal by means of X-ray diffractions, for the purpose of determining unit-cell dimensions and space groups. See also: *precession camera.* Syn: *precession method.*

buetschliite A mineral: $K_2Ca(CO_3)_2$. It is dimorphous with fairchildite. Also spelled: *bütschliite.*

buffalo wallow (a) One of the small undrained shallow depressions that were once common on the Great Plains of the western U.S., usually containing water after a rain (and often remaining as a stagnant water hole for most of the year). It is generally believed to have been deepened or modified, and perhaps initially formed, by the trampling and wallowing of buffalo herds in mud and dust. The diameter ranges from about a meter to 15-20 m, and the depth from several centimeters to a few meters. (b) A term improperly applied to one of the large natural depressions widely distributed throughout the Great Plains of the western U.S. (esp. on the High Plains), often containing an intermittent pond or temporary lake (Veatch & Humphrys, 1966, p. 49).

buffered reaction An equilibrium among metamorphic minerals and pore fluid in which an intensive parameter, such as temperature or the partial pressure of a fluid component, is not free to vary independently, by virtue of the presence in the rock of all the participants in the reaction.

buffer-zone A time-stratigraphic boundary, placed as precisely as possible in a continuous type section but still not truly isochronous in other continuous sections. "It would perhaps be more realistic to think of isochronous 'buffer-zones' instead of isochronous surfaces" (Hornibrook, 1965, p. 1199). Cf: *chronohorizon*.

bug hole A miner's term for *vug*.

bugor A hill or succession of small hills separating creeks or ravines, as on the shore of the Black Sea. Etymol: Russian *bugori*, "hillock".

buhrstone (a) A siliceous rock suitable for use as millstones; e.g. an open-textured, porous but tough, fine-grained sandstone, or a silicified fossiliferous limestone. In some sandstones, the cement is calcareous. Syn: *millstone*. (b) A millstone cut from buhrstone.—Also spelled: *burrstone; burstone*. Syn: *burr*.

building stone A general, nongeneric term for any rock suitable for use in construction. Whether igneous, metamorphic, or sedimentary, a building stone is chosen for its properties of durability, attractiveness, and economy. See also: *dimension stone*.

buildup A nongenetic term used by Merriam (1962, p. 73) for "any extra, stray, or super" limestone bed or beds, in addition to the "normal" sequence, as exemplified in the rhythmic (cyclic) deposits of the northern midcontinent region of U.S.; e.g. a marine bank, a bioherm, and an organic reef.

built platform *wave-built platform*.

built terrace (a) *wave-built terrace*. (b) *alluvial terrace*.

bukovite A tetragonal mineral: $Tl(Cu,Fe)Se_2$.

bukovskyite A mineral: $Fe(AsO_4)(SO_4)(OH)\cdot 7H_2O$.

bulb glacier *expanded-foot glacier*.

bulbous dome *volcanic dome*.

bulge (a) A tumescence of lava. (b) A landmass projecting beyond the general outline of the body of which it is a part; e.g. the "bulge" of Brazil. (c) A diapiric structure with a clay core beneath more competent overlying strata. Cf: *camber*.

Bulitian North American stage: Paleocene (above Ynezian, below Penutian).

bulk density The weight of an object or material divided by its volume, including the volume of its pore spaces; specif. the weight per unit volume of a soil mass that has been oven-dried to a constant weight at 105°C. Syn: *apparent density*.

bulkhead A stone, steel, wood, or concrete wall-like structure primarily designed to resist earth or water pressure, as a retaining wall holding back the ground from sliding into a channel, or a partition preventing water from entering a working area in a mine.

bulk modulus A *modulus of elasticity* which relates a change in volume to the hydrostatic state of stress. It is the reciprocal of *compressibility*. Symbol: k. Syn: *volume elasticity; incompressibility modulus; modulus of incompressibility*.

bulla (a) One of the blisterlike structures that partly or completely cover the apertures in planktonic foraminifers, that are not closely related to primary chambers, and that may be umbilical, sutural, or areal in position and may have one or more marginal accessory apertures. (b) A radially elongated tubercle of an ammonoid. (c) In many mammals, the auditory or tympanic bulla, a cup- or shell-shaped bone that floors the middle ear and supports the ear drum.—Pl: *bullae*. Adj: *bullate*.

Bullard discontinuity The seismic-velocity interface between the *outer core* and the *inner core*.

Bullard's method Computation of the effect of topography for the *Hayford zones*, by first calculating the effect of the *spherical cap* of height equal to the station height, then of the topographic deviations of this cap (Schieferdecker, 1959, term 3489).

bulldust An Australian term for coarse dust or silt.

buller *blowhole [coast]*.

bullet crystal A snow crystal in the shape of a short hexagonal prism with one pointed end, characteristically formed at very cold temperatures.

bullette A siphonal deposit of a nautiloid, similar to an annulus but flatter and more elongated in cross section, in which it appears knoblike or bosslike (TIP, 1964, pt. K, p. 54).

bullion (a) A concretion found in some types of coal. It is composed of carbonate or silica, stained brown by humic derivatives, and may be several centimeters to a meter or more in diameter. Well-preserved plant structures often form the nucleus. Cf: *coal ball*. (b) A nodule of clay, shale, ironstone, or pyrite that generally encloses a fossil.

bull mica Large clusters of diversely oriented and partially intergrown crystals of muscovite with a little interstitial albite and quartz (Skow, 1962, p. 169).

bull pup A miner's term for a worthless claim.

bull quartz A miner's or prospector's term for white massive quartz, essentially free of accessory minerals and valueless as ore. Syn: *bastard quartz; buck quartz*.

bull's-eye level *circular level*.

bultfonteinite A mineral: $Ca_2SiO_2(OH,F)_4$. It is found at Bultfontein, South Africa.

bummock A downward projection from the underside of sea ice; the submariner's counterpart of a *hummock [ice]*.

Bumstead head A lightweight tripod, adapted to use on foot traverses. It carries a 15-inch-square plane-table board, which can be revolved about its axis, but provides no means of leveling. It is used with a peep-sight alidade. Named after Albert H. Bumstead (1875-1940), U.S. cartographer.

bund Any artificial embankment used to control the flow of water in a river or on irrigated land. The term is applied extensively in India to large low dams and dikes and also to the small ridges between rice fields. Also, an embanked causeway or thoroughfare along a river or the sea.

bundle scar In a *leaf scar*, a scar indicating the position of a vascular bundle that had connected the stem and the stalk.

bunsenite A pistachio-green mineral: NiO.

Bunter European stage (esp. in Germany): Lower Triassic (above Permian, below Muschelkalk).

burbankite A pale-yellow hexagonal mineral: $(Na,Ca,Sr,Ba,Ce)_6(CO_3)_5$.

Burdigalian European stage: Miocene (above Aquitanian, below Langhian).

Burgers circuit A loop that encloses a *dislocation* line in a crystal and that fails to close by an amount known as the *Burgers vector*.

Burgers vector The vector required to complete a *Burgers circuit*. It is the same as the vector of, and independent of the position of, the crystal dislocation.

burial Covering up or concealing geologic features by the process of sedimentation.

burial metamorphism A type of low-grade regional metamorphism affecting sediments and interlayered volcanic rocks in a geosyncline without any influence of orogenesis or magmatic intrusion. Original rock fabrics are largely preserved but mineralogical compositions are generally changed (Coombs, 1961). Cf: *dynamothermal metamorphism*.

buried channel An old channel concealed by surficial deposits; esp. a preglacial channel filled with glacial drift.

buried erosion surface An *erosion surface*, such as a peneplain, that has been covered by younger sediments; it may represent a surface of unconformity at depth. Cf: *fossil erosion surface*.

buried focus A situation in seismic prospecting where the concave-upward curvature of a reflector is large enough that the energy focuses before it reaches the recording plane. Several *branches [seis]* (usually three) of a reflection may be observable from the same surface location. Focusing can also be produced by certain velocity distributions that act as a lens. See also: *reverse branch*.

buried ice Any relatively distinct ice mass buried in the ground, esp. surface ice, as that of sea, lake, river, or glacier origin, that has been buried syngenetically by sediments, esp. in a permafrost region.

buried karst *paleokarst*.

buried river A river bed that has been concealed beneath alluvium, lava, pyroclastic rocks, or till.

buried soil *paleosol*.

buried valley A depression in an ancient land surface or in bedrock, now covered by younger deposits; esp. a preglacial valley filled with glacial drift.

burkeite A white, buff, or grayish mineral: $Na_6(CO_3)(SO_4)_2$.

burl An oolith or nodule in fireclay. It may have a high content of alumina or iron oxide.

burley clay A clay containing burls; specif. a diaspore-bearing clay in Missouri, usually averaging 45 to 65% alumina. See also: *diaspore clay*.

burmite A dark-brown, pale-yellow, or reddish variety of retinite, found in Burma, that resembles amber but is tougher and harder. It has also been regarded as a variety of amber low in succinic acid.

burn A term used in Scotland and northern England for a small stream or brook, such as a *bourne*.

burned Said of shale that adheres tightly to and is difficult to remove from the coal with which it is associated. Syn: *frozen [coal]*.

burnie A Scottish term for a little stream or brook.

burnside *brookside*.

burnt stone A gemstone whose color has been altered by heating; e.g. amethyst, which changes from purple to clear, or tiger-eye, which changes from yellowish-brown to reddish-brown. Cf: *heated stone; stained stone.*

Burozem A Russian term for a brown steppe soil.

burr (a) A term used in England for a rough or hard stone, such as a compact siliceous sandstone especially hard to drill. (b) A knob, boss, nodule, or other hard mass of siliceous rock in a softer rock; a hard lump of ore in a softer vein. (c) *buhrstone.* (d) *whetstone.*—Syn: *bur.*

burr ball *lake ball.*

burrow A tubular or cylindrical hole or opening, made in originally soft or loose sediment, by a mud-eating worm, a mollusk, or other invertebrate, extending along a bedding plane or penetrating a rock, and often later filled with clay or sand and preserved as a filling; it may be straight or sinuous, and vertical, horizontal, or inclined. Cf: *boring.*

burrow porosity Porosity resulting from burrowing organisms. Although this type of porosity is uncommon in ancient carbonate rocks due to collapse of burrows, other types such as interparticle porosity can develop within the burrow-filling material, esp. if its permeability is greater than that of the host sediment (Choquette & Pray, 1970, p. 244). Syn: *boring porosity.*

burr rock An aggregate of muscovite books and quartz (Skow, 1962).

burrstone *buhrstone.*

bursa An internal gill pouch in ophiuroids, entered by the gill slit. Pl: *bursae.*

bursaite A mineral: $Pb_5Bi_4S_{11}$.

burst An explosive breaking of brittle rock material; e.g. a *rock burst* in a deep mine or tunnel. In coal mines, a burst may be accompanied by a discharge of methane, carbon dioxide, or coal dust.

burstone *buhrstone.*

bushveld A large, flat grassy area with scattered trees, found in tropical or subtropical regions, esp. in Africa. Etymol: Afrikaans, *bosveld.* Cf: *savanna.*

bustamite A grayish-red mineral: $CaMnSi_2O_6$.

bustite *aubrite.*

butane A gaseous inflammable paraffin hydrocarbon, formula C_4H_{10}, occurring in either of two isomeric forms: n-butane, $CH_3CH_2CH_2CH_3$, or isobutane, $CH_3CH(CH_3)_2$. The butanes occur in petroleum and natural gas.

butlerite A monoclinic mineral: $FeSO_4(OH)\cdot 2H_2O$. Cf: *parabutlerite.*

bütschliite *buetschliite.*

butt cleat The minor cleat system, or jointing, in a coal seam, usually at right angles to the *face cleat.* Syn: *end cleat.*

butte (a) A conspicuous, usually isolated, generally flat-topped hill or small mountain with relatively steep slopes or precipitous cliffs, often capped with a resistant layer of rock and bordered by talus, and representing an erosion remnant carved from flat-lying rocks; the summit is smaller in extent than that of a *mesa,* and many buttes in the arid and semiarid regions of the western U.S. result from the wastage of mesas. Syn: *mesa-butte.* (b) An isolated hill having steep sides and a craggy, rounded, pointed, or otherwise irregular summit; e.g. a volcanic cone (as Mount Shasta, Calif., formerly known as Shasta Butte), or a *volcanic butte.*—Etymol: French, "knoll, hillock, inconspicuous rounded hill; rising ground". Pron: *bewt.*

butterball A clear-yellow, rounded segregation of very pure carnotite found in the soft sandstone of Temple Rock, San Rafael Swell, Utah.

butter rock *halotrichite.*

butte témoin A flat-topped hill representing the former extension of an escarpment edge or plateau, now detached by stream erosion, its surface in broadly the same plane as the main mass; an outlier. In most cases, a *butte* is a "butte témoin". Etymol: French, "witness hill". Syn: *witness butte; zeugenberg.*

buttgenbachite A sky-blue mineral: $Cu_{19}Cl_4(NO_3)_2(OH)_{32}\cdot 2H_2O$. It is isomorphous with connellite.

buttress [geomorph] A protruding rock mass on, or a projecting part of, a mountain or hill, resembling the buttress of a building; a spur running down from a steep slope. Example: a prominent *salient* produced in the wall of a gorge by differential weathering.

buttress [paleont] (a) An internal ridgelike projection from the shell wall of a bivalve mollusk which supports the hinge plate or chondrophore. (b) A ridge of skeletal material extending adapically from an echinoid auricle on the inner surface of the test (TIP, 1966, pt.U, p.253). (c) An aboral tongue-shaped extension of the apical denticle of a conodont, generally on the inner and outer sides.

buttress sand A sandstone that intersects an underlying surface of unconformity, as on the flank of a buried hill or a truncated anticline. It often forms a trap for oil.

buttress unconformity A surface on which onlapping strata abut against a steep topographic scarp of regional extent. A buttress unconformity may be produced on a submarine escarpment along which older rocks crop out, between a deep basin and a platform (Enos, 1974, p. 807).

butyrellite *bog butter.*

butyrite *bog butter.*

Buys Ballot's law The statement in meteorology that describes the relationship between horizontal wind direction and barometric pressure: if an observer stands with his back to the wind in the Northern Hemisphere, the pressure is lower to his left than to his right; the reverse relationship holds true for an observer in the Southern Hemisphere. The law is named after the Dutch meteorologist who formulated it in 1857.

by-channel A stream or branch along one side of the main stream.

byerite A bituminous coal that resembles *albertite.*

bypassing A term applied by Eaton (1929) to sedimentary transport across areas of nondeposition, as in the case where one particle size passes another that is being simultaneously transported, or continues in motion after the other has come to rest; e.g. the normal decrease in average particle size of sediments away from a source area. The term is also applied to transport of coarser particles farther than finer particles ("reverse" bypassing); e.g. gravel bypassing sand along the edge of the continental shelf, probably the result of density-current deposition. Cf: *total passing.* Also spelled: *by-passing.*

bysmalith A roughly vertical cylindrical igneous intrusion, bounded by steep faults. It has been interpreted as a type of *laccolith.* Cf: *bell-jar intrusion.*

by-spine A small accessory spine additional to a radial spine in acantharian and spumellarian radiolarians.

byssal gape An opening between the margins of a bivalve-mollusk shell for the passage of the byssus.

byssal notch The indentation, below the anterior auricle of the right valve in many pectinacean bivalve mollusks, serving for the passage of the byssus or for the protrusion of the foot.

byssiferous Possessing a byssus.

byssolite An olive-green, fibrous variety of amphibole. The term is used in the gem trade for a variety of quartz containing greenish, fibrous inclusions of actinolite or asbestos.

byssus A tuft or bundle of long, tough hairlike strands or filaments, secreted by a gland in a groove of the foot of certain bivalve mollusks and issuing from between the valves, by which a temporary attachment of the bivalve can be made to rocks or other extraneous objects. Pl: *byssi* or *byssuses.*

by-stream Said of the part of a flood plain consisting of a narrow belt of levee deposits immediately adjacent to the stream channel, and composed usually of sandy alluvium.

byströmite (a) A pale blue-gray tetragonal mineral: $MgSb_2O_6$. (b) A monoclinic polymorph of pyrrhotite.—Also spelled: *bystromite.*

by-terrace Said of the part of a flood plain consisting of a narrow belt of clayey deposits adjacent to a bounding terrace relatively distant from the stream channel.

bytownite A bluish to dark-gray mineral of the plagioclase feldspar group with composition ranging from $Ab_{30}An_{70}$ to $Ab_{10}An_{90}$. It occurs in basic and ultrabasic igneous rocks.

bytownitfels *bytownitite.*

bytownitite An igneous rock composed almost entirely of bytownite. Syn: *bytownitfels.* Cf: *anorthosite.* Not recommended usage.

by-wash A channel or spillway designed to carry surplus water from a dam, reservoir, or aqueduct in order to prevent overflow.

by-water A yellow-tinted diamond.

c *carat.*

caballing The mixing of two water masses to produce a blend that sinks because it is denser than its original components. This occurs when the two water masses have the same density but different temperatures and salinities.

cabbage-leaf mark *frondescent cast.*

cable (a) A heavy multiple-strand steel rope used in *cable-tool drilling* as the line between the tools and the *walking beam.* Syn: *drilling cable.* (b) A term used loosely to signify a *wire line.*

cable break An event on a well geophone record caused by elastic energy being transmitted down the supporting cable to the well geophone.

cable tool Any component of a set of bottom-hole tools used in cable-tool drilling, e.g. a *drill bit,* a *drill stem,* a set of *jars,* or a *bailer.*

cable-tool drilling A method of drilling, now largely replaced by *rotary drilling,* in which the rock at the bottom of the hole is broken up by a steel bit with a blunt, chisel-shaped cutting edge. The bit is at the bottom of a heavy string of steel tools suspended on a cable that is activated by a *walking beam,* the bit chipping the rock by regularly repeated blows. The method is adapted to drilling water wells and relatively shallow oil wells.

cabochon (a) An unfaceted cut gemstone of domed or convex form. The top is smoothly polished; the back, or base, is usually flat or slightly convex, sometimes concave, and often unpolished. The girdle outline may be round, oval, square, or of any other shape. (b) The style of cutting such a gem. (c) A polished but uncut gem.—Etymol: French. Pron: cab-o-*shon.* See also: *en cabochon.*

cabocle A compact rolled pebble resembling red jasper, supposed to be a hydrated phosphate of calcium and aluminum, found in the diamond-producing sands of Bahia, Brazil.

cacholong An opaque or feebly translucent and bluish-white, pale-yellowish, or reddish variety of common opal containing a little alumina. Syn: *cachalong; pearl opal.*

cacoxenite A yellow or brownish mineral: $Fe_9^{+3}(PO_4)_4(OH)_{15} \cdot 8H_2O$. Syn: *cacoxene.*

cactolith An irregular intrusive igneous body of obscurely cactus-like form, more or less confined to a horizontal zone and appearing to consist of irregularly related and possibly distorted branching and anastomosing dikes that fed a laccolith. Term introduced by Hunt et al. (1953, p. 151): "a quasi-horizontal chonolith composed of anastomosing ductoliths whose distal ends curl like a harpolith, thin like a sphenolith, or bulge discordantly like an akmolith or ethmolith".

cadacryst *chadacryst.*

cadastral Delineating or recording property boundaries, and sometimes subdivision lines, buildings, and other details. Etymol: French *cadastre,* an official register of the real property of a political subdivision with details of area, ownership, and value, and used in apportioning taxes.

cadastral map A large-scale map showing the boundaries of subdivisions of land, usually with the directions and lengths thereof and the areas of individual tracts, compiled for the purpose of describing and recording ownership. It may also show culture, drainage, and other features relating to use of the land.

cadastral survey A survey relating to land boundaries and subdivisions, made to create units suitable for transfer or to define the limitations of title; esp. a survey of the public lands of the U.S., such as one made to identify or restore property lines. Cf: *land survey; boundary survey.*

cadder Shortened form of *ballycadder,* a syn. of *icefoot.*

cadicone An evolute, coiled cephalopod shell with a strongly depressed whorl section, wide venter, and a deep umbilicus (as in the ammonoid *Cadoceras*).

cadmia (a) *calamine.* (b) A chemical compound: CdO. (c) An impure zinc oxide that forms on the walls of furnaces in the smelting of ores containing zinc.

cadmium blende *greenockite.*

cadmium ocher *greenockite.*

cadmoselite A black hexagonal mineral: CdSe.

caducous Said of the calyx of a flower that is shed before the flower expands (Swartz, 1971, p. 78).

cadwaladerite A mineral: $Al(OH)_2Cl \cdot 4H_2O$.

Caecilia An order of burrowing caudate *lissamphibians* in which limbs have been lost and eyes reduced greatly. Alone among extant amphibians, caecilians retain rudimentary scales and reproduce by means of a large-yolked egg. Syn: *Gymnophiona.*

caecum (a) The sac-shaped apical end of the siphuncle of a nautiloid or ammonoid. Also, a cavity associated with the digestive system in the living *Nautilus.* (b) The evagination of the outer epithelium projecting into the endopuncta of a brachiopod shell.—Pl: *caeca.* Adj: *caecal.*

Caen stone A yellowish or light cream-colored Jurassic limestone, marked with a rippled figure, and largely used for building purposes. Type locality: near Caen, city in Normandy, France.

Caerfaian European stage: Lower Cambrian (above Precambrian, below Solvan).

cafarsite A mineral: $Ca_6(Fe,Ti)_6Mn_2(AsO_4)_{12} \cdot 4H_2O$.

cafemic Said of an igneous rock or magma that contains calcium, iron, and magnesium. Etymol: a mnemonic term derived from *ca*lcium + *fe*rric (or ferrous) + *m*agnesium + *ic.*

cafetite An orthorhombic mineral: $Ca(Fe,Al)_2Ti_4O_{12} \cdot 4H_2O$.

cage A void in a crystal structure that is large enough to trap one or more atoms (such as argon or xenon) foreign to the structure; it is found in tectosilicates, beryl, and organic compounds.

cahemolith *humic coal.*

cahnite A tetragonal mineral occurring in white sphenoidal crystals: $Ca_2B(AsO_4)(OH)_4$.

CAI *computer-aided instruction; computer-assisted instruction.*

Cainophyticum A paleobotanic division of geologic time, corresponding approximately to, and characterized by the plant life of, the Cenozoic. Cf: *Palaeophyticum; Mesophyticum.*

Cainozoic *Cenozoic.*

cairn An artificial mound of rocks, stones, or masonry, usually conical or pyramidal, used in surveying to aid in the identification of a point or boundary.

cairngorm A type of *smoky quartz* from Cairngorm, a mountain southwest of Banff in Scotland. Syn: *cairngorm stone; Scotch topaz.*

cajon (a) *box canyon.* (b) A defile leading up to a mountain pass; also, the pass itself.—Etymol: Spanish *cajón,* "large box". Pron: ca*hone.* The term is used in the SW U.S.

cake [drill] *mud cake [drill].*

cake [ice] *ice cake.*

cake ice An accumulation of *ice cakes.*

caking coal Coal that softens and agglomerates when heated, and on quenching produces a hard gray cellular coke. Not all caking coals are good *coking coals.* Syn: *binding coal.*

caking index *agglutinating value.*

cal A term used in Cornwall, England, for iron tungstate (wolframite).

cala [coast] A short, narrow *ria* formed in a limestone coast; a small semicircular shallow bay along a *cala coast,* as along the coast of Majorca. Etymol: Spanish, "cove, small bay, inlet". See also: *caleta.*

cala [streams] A term applied in SW U.S. to a creek corresponding to a lateral stream of a main drainage. Etymol: Spanish, "creek". See also: *caleta.*

Calabrian European stage: Lower Pleistocene (above Astian-Redonian of Pliocene, below Emilian). It is the marine equivalent, in France and Italy, of the terrestrial Villafranchian.

cala coast A coast formed by the submergence of many small valleys having steep slopes so that *calas,* separated by narrow peninsulas, are formed under the influence of breakers; examples occur along several coasts of the Mediterranean Sea. Syn: *calas coast.*

calaite *turquoise.*

calamine (a) A name used in the U.S. for *hemimorphite.* This name is disapproved by the Commission on New Minerals and Mineral Names of the International Mineralogical Association (Fleischer, 1966, p.1263). (b) A name frequently used in Great Britain for *smithsonite.* (c) *hydrozincite.* (d) A commercial, mining, and metallurgical term for the oxidized ores of zinc (including silicates and carbonates), as distinguished from the sulfide ores of zinc.—Syn: *cadmia.*

calamistrum A spinose comb or row of special bristles on the fourth (hind) metatarsi of certain spiders, used for drawing out a band of special silk from the cribellum. Pl: *calamistra.*

calanque A French term for a cove or inlet, esp. a dry valley excavated in limestone during a wet period and later submerged by the rise of sea level. Examples occur along the Mediterranean coast of France.

calaverite A pale bronze-yellow or tin-white monoclinic mineral: $AuTe_2$. It often contains silver, and is an important source of gold.

calc-alkalic (a) Said of a series of igneous rocks in which the weight percentage of silica is between 56 and 61 when the weight percentages of CaO and of $K_2O + Na_2O$ are equal. See also: *alkali-lime index.* (b) Said of an igneous rock containing plagioclase feldspar.

calc-alkali rock series Those series of comagmatic silica-oversaturated volcanic rocks that are characterized mineralogically by the presence of groundmass augite plus hypersthene (basalts and andesites), hypersthene (dacites), or biotite or hornblende (rhyolites); and chemically by a rather constant rate of enrichment of iron to magnesium during evolution of the series (Kuno, 1959). See also: *tholeiitic rock series.*

calcarenaceous Said of a sandstone containing abundant calcium-carbonate detritus; e.g. "calcarenaceous orthoquartzite" in which the calcareous components constitute up to 50% of the total clastic particles (Pettijohn, 1957, p.404-405).

calcarenite A limestone consisting predominantly (more than 50%) of detrital calcite particles of sand size; a consolidated calcareous sand. The term was introduced by Grabau (1903). Cf: *calcareous sandstone.*

calcarenitic limestone A term used by Powers (1962, p. 145) for a limestone composed dominantly (more than 10%) of original calcareous-mud matrix (particles with diameters less than 0.06 mm) accompanied by more than 10% coarse clastic carbonate grains (sand and gravel).

calcareous Said of a substance that contains calcium carbonate. When applied to a rock name it implies that as much as 50% of the rock is calcium carbonate (Stokes & Varnes, 1955).

calcareous algae A group of algae that remove carbon dioxide from the shallow water in which they live and secrete or deposit it around the thallus as a more or less solid calcareous structure. See also: *coralline algae.*

calcareous clay A clay containing a significant amount of calcium carbonate; specif. a *marl.*

calcareous crust An indurated soil horizon cemented with calcium carbonate; *caliche.* Syn: *calc-crust.*

calcareous dolomite A term used by Leighton & Pendexter (1962, p. 54) for a carbonate rock containing 50-90% dolomite. Cf: *calcitic dolomite.*

calcareous nannoplankton Any of the chromatophore-bearing protists that normally produce coccoliths during some phase in their life cycle; also, in a broader sense, the morphologically diverse group of minute calcareous skeletal elements produced by coccolithophores. Cf: *nannoplankton.*

calcareous ooze A deep-sea pelagic sediment containing at least 30% calcareous skeletal remains, e.g. pteropod ooze. Cf: *siliceous ooze.*

calcareous peat *eutrophic peat.*

calcareous rock A sedimentary rock containing an appreciable amount of calcium carbonate, such as limestone, chalk, tufa, or shelly sandstone. See also: *carbonate rock.*

calcareous sandstone (a) A sandstone cemented with calcite. (b) A sandstone containing appreciable calcium carbonate, but in which clastic quartz is present in excess of 50% (Pettijohn, 1957, p.381).—Cf: *calcarenite.*

calcareous shale A shale containing at least 20% calcium carbonate in the form of finely precipitated materials or small organically-fixed particles (Pettijohn, 1957, p.368-369).

calcareous sinter *travertine.*

calcareous soil A soil whose content of carbonate is sufficient to cause effervescence when tested with hydrochloric acid.

calcareous spar *calcspar.*

calcareous tufa *tufa [sed].*

calcarinate An adjective restricted by Allen (1936, p.21) to designate the calcium-carbonate cement of a sedimentary rock.

calc-crust *calcareous crust.*

calcdolomite *calcitic dolomite.*

calcedony *chalcedony.*

calceoloid Said of a solitary corallite shaped like the tip of a pointed slipper (as in *Calceola*), with angulated edges between flattened and rounded sides.

calc-flinta A fine-grained calc-silicate rock of flinty appearance formed by thermal metamorphism of a calcareous mudstone, possibly with some accompanying pneumatolytic action.

calcian dolomite A dolomite mineral that contains at least 8% calcium in excess of the ideal composition (Ca:Mg = 1:1 molar) of dolomite.

calciborite A white mineral: CaB_2O_4.

calcibreccia A limestone breccia, or a consolidated calcareous rubble; a *calcirudite* whose constituent particles are angular (Carozzi & Textoris, 1967, p. 3).

calcic [geochem] Said of minerals and igneous rocks containing a relatively high proportion of calcium; the proportion required to warrant use of the term depends on circumstances.

calcic [geol] Said of a series of igneous rocks in which the weight percentage of silica is greater than 61 when the weight percentages of CaO and of $K_2O + Na_2O$ are equal. See also: *alkali-lime index.*

calcic horizon A diagnostic subsurface soil horizon, at least 15 cm thick, characterized by enrichment in secondary carbonates (USDA, 1975).

calciclase *anorthite.*

calciclasite *anorthitite.*

calciclastic Pertaining to a clastic carbonate rock (Braunstein, 1961).

calcicole A plant requiring a lime-rich, i.e. alkaline, soil. Cf: *calcifuge; calciphobe.* Syn: *calciphile.*

calcicrete *calcrete.*

calciferous In a stratigraphic sense, pertaining to a series of strata containing limestone, e.g. the Calciferous Sandstone series of Scotland (Challinor, 1978, p. 39).

calcification [paleont] (a) Deposition of calcium salts in living tissue. (b) Replacement of organic material, generally original hard parts, by calcium carbonate ($CaCO_3$) in fossilization.

calcification [soil] A general term used for those processes of soil formation in which the surface soil is kept sufficiently supplied with calcium to keep the soil colloids nearly saturated with exchangeable calcium and thus render them relatively immobile and nearly neutral in reaction. The process is best expressed in *Borolls, Ustolls,* and other soils with calcic horizons.

calcifuge A plant surviving, but not thriving, on a lime-rich soil; it grows better on an acid soil. Cf: *calcicole; calciphobe.*

calcigranite A *granite* in which the plagioclase is labradorite or bytownite, named by Johannsen (1931). Not recommended usage.

calcigravel The unconsolidated equivalent of *calcirudite.*

calcikersantite A *kersantite* that contains labradorite or bytownite. It was named by Johannsen (1931). Not recommended usage.

calcilith (a) A term suggested by Grabau (1924, p. 298) for a limestone. (b) A sedimentary rock composed principally of the calcareous remains of organisms (Pettijohn, 1957, p. 429).

calcilutite A limestone consisting predominantly (more than 50%) of detrital calcite particles of silt and/or clay size; a consolidated calcareous mud. Some authors broaden the term to include calcareous rocks containing chemically precipitated crystalline components (of inorganic or organic origin). The term was introduced by Grabau (1903). See also: *micritic limestone.* Cf: *calcisiltite.* Syn: *calcipelite.*

calcimicrite A term used by Schmidt (1965, p. 127-128) for a limestone whose particles have diameters less than 20 microns and whose micrite component exceeds its allochem component. See also: *micritic limestone.*

calcimixtite A term proposed by Schermerhorn (1966, p. 835) for a *mixtite* that is dominantly calcareous.
calcimorphic Said of an intrazonal soil whose characteristics reflect the influence of the calcification process on its development. Examples of calcimorphic soils are Brown Forest and Rendzina soils.
calcination The heating of a substance to its temperature of dissociation, e.g. of limestone to CaO and CO_2 or of gypsum to lose its water of crystallization.
calciocarnotite *tyuyamunite.*
calciocopiapite A mineral of the copiapite group: $CaFe_4(SO_4)_6(OH)_2 \cdot 19H_2O$.
calcioferrite A yellow or green mineral: $Ca_2Fe_2(PO_4)_3(OH) \cdot 7H_2O$.
calcio-olivine An orthorhombic phase of calcium orthosilicate: γ-Ca_2SiO_4. It is stable below 780° to 830°C, and is isomorphous with olivine. The term has also been applied to a highly calciferous variety of olivine, and to any of the polymorphs of Ca_2SiO_4. Cf: *larnite; bredigite.* Syn: *lime olivine.*
calciouranoite A noncrystalline mineral: $(Ca,Ba,Pb)U_2O_7 \cdot 5H_2O$.
calciovolborthite A green, yellow, or gray mineral: $CaCu(VO_4)(OH)$. Syn: *tangeite.*
calcipelite *calcilutite.*
calciphile *calcicole.*
calciphobe A plant requiring an acid soil; it cannot survive in a lime-rich soil. Cf: *calcicole; calcifuge.*
calciphyre A marble containing conspicuous crystals of calcium silicate and/or magnesium silicate.
calcirhyolite A *rhyolite* that contains labradorite or bytownite as its accessory plagioclase. It was named by Johannsen (1931). Not recommended usage.
calcirudite A limestone consisting predominantly (more than 50%) of detrital calcite particles larger than sand size (larger than 2 mm in diameter), and often also cemented with calcareous material; a consolidated calcareous gravel or rubble, or a limestone conglomerate or breccia. Some authors (e.g. Folk, 1968, p. 162) use one millimeter as the lower limit. The term was introduced by Grabau (1903). Cf: *calcigravel; calcibreccia.*
calcisiltite A limestone consisting predominantly of detrital calcite particles of silt size; a consolidated calcareous silt. Cf: *calcilutite.*
Calcisol A group term for zonal soils developed from alluvium and characterized by a neutral or calcareous A horizon, no B horizon, and a calcareous C horizon (Harper, 1957). Most of these soils are now classified as *Borolls, Orthids, Ustolls,* or *Xerolls,* all with calcic horizons.
calcisponge Any sponge belonging to the class Calcispongea (=Calcarea) and characterized mainly by a skeleton composed of spicules of calcium carbonate. Range, Cambrian to present.
calcisyenite A *syenite* that contains labradorite or bytownite, rather than andesine or oligoclase, as its accessory plagioclase. It was named by Johannsen (1931). Not recommended usage.
calcite A common rock-forming mineral: $CaCO_3$. It is trimorphous with aragonite and vaterite. Calcite is usually white, colorless, or pale shades of gray, yellow, and blue; it has perfect rhombohedral cleavage, a vitreous luster, and a hardness of 3 on the Mohs scale, and it readily effervesces in cold dilute hydrochloric acid. It is the principal constituent of limestone; calcite also occurs crystalline in marble, loose and earthy in chalk, spongy in tufa, and stalactitic in cave deposits. It is commonly found as a gangue mineral in many ore deposits and as a cementing medium in clastic sedimentary rocks; it is also a minor constituent of many igneous rocks and the chief constituent of some *carbonatites.* Calcite crystallizes in a variety of forms, such as nailhead spar, dogtooth spar, and Iceland spar. Symbol: Cc. Cf: *dolomite [mineral].* Syn: *calcspar.*
calcite bubble *cave bubble.*
calcite eye (a) One of the rounded bodies of clear calcite occurring sporadically in the radial zone and central area of foraminifers of the family Orbitolinidae. The term is usually used in the plural. (b) *bird's-eye.*
calcite raft *cave raft.*
calcitic dolomite A dolomite rock in which calcite is conspicuous, but the mineral dolomite is more abundant; specif. a dolomite rock containing 10-50% calcite and 50-90% dolomite and having an approximate magnesium-carbonate equivalent of 22.7-41.0% (Pettijohn, 1957, p. 418), or a dolomite rock whose Ca/Mg ratio ranges from 2.0 to 3.5 (Chilingar, 1957). Cf: *dolomitic limestone; calcareous dolomite.* Syn: *calcdolomite.*
calcitic limestone A limestone that consists essentially of calcite; specif. a limestone whose Ca/Mg ratio exceeds 105 (Chilingar, 1957).
calcitite A term used by Kay (1951) for a rock composed of calcite; e.g. a limestone.
calcitization (a) The act or process of forming calcite, as by alteration of aragonite. (b) The alteration of existing rocks to limestone, due to the replacement of mineral particles by calcite, e.g. of dolomite in dolomite rocks or of feldspar and quartz in sandstones.
calcitostracum An internal layer of various mollusk shells, consisting chiefly of calcite. Cf: *nacre.*
calcium-carbonate compensation depth *carbonate compensation depth.*
calcium-catapleiite A mineral: $CaZrSi_3O_9 \cdot 2H_2O$. See also: *catapleiite.*
calcium feldspar A plagioclase feldspar rich in the An molecule ($CaAl_2Si_2O_8$); specif. *anorthite.* See also: *lime feldspar.* Syn: *Ca-spar.*
calcium-larsenite *esperite.*
calcium mica *margarite [mineral].*
calcium phosphate A mineral name applied to naturally occurring *apatite.*
calcjarlite A monoclinic mineral: $Na(Ca,Sr)_3Al_3(F,OH)_{16}$.
calclacite A mineral: $CaCl_2 \cdot Ca(C_2H_3O_2)_2 \cdot 10H_2O$. It is calcium chloride acetate found as a fibrous efflorescence on calcareous museum specimens.
calclithite (a) A litharenite in which carbonates constitute the most abundant rock fragments (Folk, 1968, p. 124); e.g. the Oakville Sandstone (Miocene) in S. Texas, derived largely from erosion of Cretaceous limestones. Not to be confused with a limestone consisting of intraclasts. Syn: *carbonate-arenite.* (b) A term originally defined by Folk (1959, p. 36) as a limestone that was derived mainly from erosion of older, lithified limestones and that contains more than 50% carbonate rock fragments (extraclasts); e.g. a terrigenous rock formed as an alluvial fan in an area of intense tectonism or very dry climate.
calcolistolith A limestone olistolith.
calcouranite *autunite.*
calcrete (a) A term suggested by Lamplugh (1902) for a conglomerate consisting of surficial sand and gravel cemented into a hard mass by calcium carbonate precipitated from solution and redeposited through the agency of infiltrating waters, or deposited by the escape of carbon dioxide from vadose water. (b) A calcareous *duricrust; caliche.*—Etymol: *cal*careous + con*crete.* Cf: *silcrete; ferricrete.* Syn: *calcicrete.*
calc-sapropel Sapropel containing calcareous algae.
calc-schist A metamorphosed argillaceous limestone with a schistose structure produced by parallelism of platy minerals (Holmes, 1928, p.52).
calc-silicate marble A *marble* in which calcium silicate and/or magnesium silicate minerals are conspicuous.
calc-silicate rock A metamorphic rock consisting mainly of calcium-bearing silicates such as diopside and wollastonite, and formed by metamorphism of impure limestone or dolomite. Syn: *lime-silicate rock.*
calc-sinter *travertine.*
calcspar Crystalline *calcite.* Also spelled: *calc-spar.* Syn: *calcareous spar.*
calcsparite A sparry calcite crystal, as distinguished from *dolosparite.* The term is synonymous with *sparite* when the latter is understood to mean the calcareous variety.
calcspathization Widely distributed crystallization of sparry calcite (Sander, 1951, p. 154).
calc-tufa *tufa [sed].*
calcurmolite A honey-yellow secondary mineral: $Ca(UO_2)_3(MoO_4)_3(OH)_2 \cdot 11H_2O$.
caldera A large, basin-shaped volcanic depression, more or less circular or cirquelike in form, the diameter of which is many times greater than that of the included vent or vents, no matter what the steepness of the walls or form of the floor (Williams, 1941). See also: *collapse caldera; erosion caldera; explosion caldera.*
caldera complex The diverse rock assemblage underlying a caldera, comprising dikes, sills, stocks, and vent breccias; crater fills of lava; talus beds of tuff, cinder, and agglomerate; fault gouge and fault breccias; talus fans along fault escarpments; cinder cones; and other products formed in a *caldera.*
caldera lake A term that may be used for a *crater lake* in a caldera.
calderite The manganese-iron end member of the garnet group:

$Mn_3Fe_2(SiO_4)_3$.

caldron [coal] *caldron bottom.*

caldron [marine geol] A small steep-sided pot-shaped depression in the ocean floor. Also spelled: *cauldron.*

caldron bottom The mud cast of a fossil root or trunk of a tree or fern extending upward into the roof of a coal seam, and resembling the bottom of a caldron or pot; it is commonly surrounded by a film of coal, and may fall without warning. Cf: *pot bottom; bell; coal pipe.* Syn: *kettle bottom; caldron [coal].*

Caledonian orogeny A name commonly used for the early Paleozoic deformation in western Europe that created an orogenic belt, the *Caledonides,* extending from Ireland and Scotland northeastward through Scandinavia. The term should not be used in other regions. The classical Caledonian orogeny has been dated as near the end of the Silurian, but Stille and many others have used the term for an orogenic era that included pulses both earlier and later.

Caledonides The orogenic belt, named by Suess, extending from Ireland and Scotland northeastward through Scandinavia, formed by the early Paleozoic *Caledonian orogeny.*

caledonite A green mineral: $Cu_2Pb_5(SO_4)_3(CO_3)(OH)_6$. Not to be confused with *celadonite.*

caleta [coast] A small *cala;* a cove or inlet. Etymol: Spanish, "cove, small bay".

caleta [streams] The ultimate and smallest headwater ramification of a *cala;* a draw, drain, or coulee. Etymol: Spanish, "creek".

calf A small mass of *calved ice;* specif. a piece of ice that has risen to the surface after breaking loose from the submerged part of an iceberg. Syn: *calf ice.*

calice The oral (upper or distal), generally bowl-shaped surface of a corallite, on which the basal disk of a polyp rests. Pl: *calices.* See also: *calyx.*

caliche [eco geol] (a) Gravel, rock, soil, or alluvium cemented with soluble salts of sodium in the nitrate deposits of the Atacama Desert of northern Chile and Peru; it contains sodium nitrate (14-25%), potassium nitrate (2-3%), sodium iodate (up to 1%), sodium chloride, sodium sulfate, and sodium borate, mixed with brecciated clayey and sandy material in beds up to 2 m thick. It may form by leaching of bird guano, by bacterial fixation of nitrogen, by leaching from volcanic tuffs, or by drying up of former shallow lakes. (b) A term used in various geographic areas for: a thin layer of clayey soil capping a gold vein (Peru); whitish clay in the selvage of veins (Chile); feldspar, white clay, or a compact transition limestone (Mexico); a mineral vein recently discovered, or a bank composed of clay, sand, and gravel in placer mining (Colombia). The term has been extended by some authors to quartzite and kaolinite.

caliche [soil] A term applied broadly in SW U.S. (esp. Arizona) to a reddish-brown to buff or white calcareous material of secondary accumulation, commonly found in layers on or near the surface of stony soils of arid and semiarid regions, but also occurring as a subsoil deposit in subhumid climates. It is composed largely of crusts of soluble calcium salts in addition to such materials as gravel, sand, silt, and clay. It may occur as a thin porous friable horizon within the soil, but more commonly it is several centimeters to a meter or more in thickness, impermeable, and strongly indurated; the cementing material is essentially calcium carbonate, but it may include magnesium carbonate, silica, or gypsum. The term has also been used for the calcium-carbonate cement itself. Caliche appears to form by a variety of processes, e.g. capillary action, in which soil solutions rise to the surface and on evaporation deposit their salt content on or in the surface materials. It is called *hardpan,* calcareous *duricrust,* or *calcrete* in some localities, and *kankar* in parts of India. Syn: *soil caliche; calcareous crust; croute calcaire; nari; sabach; tepetate.*—Etymol: American Spanish, from a Spanish word for almost any porous material (such as gravel) cemented by calcium carbonate. "The Spanish word originally was used for a small stone or pebble accidentally burned with the clay mass when brick or tile was made, and it also was used for a crust of lime or similar material flaking from a wall" (Cottingham, 1951, p. 162).

calichification The production of caliche.

calico rock (a) A term used in South Africa for *iron formation.* (b) A local term used in eastern Pennsylvania for a quarry rock of the Helderberg Limestone.

California bearing ratio A measure of the relative resistance of a soil to penetration under controlled conditions of density and moisture content. It is the ratio of the force per unit area required to penetrate a subgrade soil to that required for corresponding penetration of a standard material (crushed-rock base material) whose resistance under standardized conditions is well established. Abbrev: *CBR.* See also: *bearing capacity.*

California onyx A dark, amber-colored or brown variety of aragonite used in ornamentation.

californite (a) A compact, massive, translucent to opaque variety of vesuvianite, typically dark-green, olive-green, or grass-green, usually mottled with white or gray, closely resembling jade, and used as an ornamental stone. Principal sources are in Fresno, Siskiyou, and Tulare counties in California. Syn: *American jade.* (b) A white variety of grossular garnet from Fresno County, Calif.

caliper log A *well log* that shows the variations with depth in the diameter of an uncased borehole. It is produced by spring-activated arms that measure the varying widths of the hole as the device is drawn upward. Syn: *section-gage log.*

calk (a) *cawk.* (b) *cauk.*—Also spelled: *caulk.*

calkinsite A pale-yellow mineral: $(Ce,La)_2(CO_3)_3 \cdot 4H_2O$.

calkstone *hassock.*

callaghanite A blue mineral: $Cu_2Mg_2(CO_3)(OH)_6 \cdot 2H_2O$.

callainite A massive, waxlike, translucent, apple-green to emerald-green hydrated aluminum phosphate, possibly a mixture of wavellite and turquoise.

callais An ancient name used by Pliny for a precious green or greenish-blue stone, probably turquoise. The name is still sometimes used for turquoise. Pl: callaides. Syn: *callaica; callaina.*

calley stone A term used in Yorkshire, England, for a hard argillaceous sandstone associated with coal.

calliard *galliard.*

callose n. A carbohydrate component of cell walls in certain plants; e.g. the amorphous cell-wall substance that envelops the pollen mother cell during pollen-grain development and acts as a barrier between mother cells but that disappears as the ektexine structure is completed and impregnated with sporopollenin.—adj. Having protuberant hardened spots; e.g. "callose leaves".

Callovian European stage: lowermost Upper Jurassic (above Bathonian, below Oxfordian). The stage has been assigned by some authors to the uppermost Middle Jurassic.

calluna peat Peat that is derived mainly from the common heather *Calluna vulgaris.* Syn: *heath peat.*

callus (a) The thickened inductura on the parietal region of a gastropod shell, or a growth of shelly material extending from the inner lip over the base of the shell and perhaps into the umbilicus in a gastropod. (b) Any excessive thickening of secondary shell (fibrous inner layer) located on the valve floor of a brachiopod and covering interior structures.

calm A Scottish term for a light-colored shale or mudstone, such as a baked shale or clay slate used for making pencils, or a shale that is easily cut with a knife.

calomel A tetragonal mineral: Hg_2Cl_2. It is a colorless, white, or faintly tinted tasteless salt, and is used as a cathartic, fungicide, and insecticide; when fused, it has a horny appearance. Syn: *horn quicksilver; horn mercury; calomelite.*

calorific value For solid fuels and liquid fuels of low volatility, the amount of heat produced by combustion of a specified quantity under specified conditions. See also: *net calorific value; gross calorific value.*

calotte "A French term for the ring of surrounding rock in a tunnel which has become weakened or pressure-relieved by the excavation" (Nelson & Nelson, 1967, p.57).

calthrops A tetraxonic sponge spicule in which the rays are equal or nearly equal in length. Pl: *calthrops.* See also: *candelabrum; cricocalthrops.*

caltonite A dark-colored analcime-bearing *basanite* that contains microphenocrysts of olivine and clinopyroxene in a trachytic groundmass composed of feldspar laths, clinopyroxene, iron oxides, and analcime. It was named by Johannsen (1931) for Calton Hill, Derbyshire, England. Not recommended usage.

calumetite An azure-blue mineral: $Cu(Cl,OH)_2 \cdot 2H_2O$.

calved ice A fragment or fragments of ice, floating in a body of water after *calving* from a larger ice mass. See also: *calf [glaciol].*

calving The breaking away of a mass or block of ice from a glacier, from an ice front (as from an ice shelf), or from an iceberg or floeberg; the process of iceberg formation. See also: *calved ice.*

calymma A frothy layer of cytoplasm in radiolarians.

calyptrolith A basket-shaped coccolith, opening proximally.

calyx (a) The skeletal cover of an echinoderm including the body and internal structures, but excluding the stem and appendages such as free arms or brachioles; e.g. the plated skeletal structure surrounding the viscera of a crinoid and comprising the dorsal cup and tegmen. (b) A small cup-shaped structure or living cavity in which a coral polyp sits. See also: *calice.*

calzirtite A tetragonal mineral: $CaZr_3TiO_9$.

cam A term used in the English Lake District for the crest of a mountain. Etymol: German *Kamm.*

camarophorium A spoon-shaped, adductor-bearing platform in the brachial valve of brachiopods of the superfamily Stenoscismatacea, supported by the median septum, and derived independently of the cardinalia (TIP, 1965, pt.H, p.141).

camarostome A concave space found in certain arachnids, formed by a depression in the common wall of fused pedipalpal coxae and the convex rostrum fitting into it, and serving as a filter of liquefied food before it reaches the mouth (TIP, 1955, pt.P, p.61).

camber A superficial structure resembling an arch or ridge, caused by gravitational sagging toward topographically lower areas. Cf: *bulge.*

cambic horizon A diagnostic subsurface soil horizon that is characterized by alteration or removal of mineral matter as indicated by strong chromas, gray or red colors or mottling. It has a texture finer than loamy sand and is not indurated (USDA, 1975).

cambium In woody plants, a layer of persistent meristematic tissue beneath the bark, from which new bark and wood originate; it gives rise to secondary xylem, secondary phloem, and parenchyma (Swartz, 1971, p. 82). See also: *cork cambium.*

Cambrian The earliest period of the Paleozoic era, thought to have covered the span of time between 570 and 500 million years ago; also, the corresponding system of rocks. It is named after Cambria, the Roman name for Wales, where rocks of this age were first studied. See also: *age of marine invertebrates.* Obsolete syn: *Primordial.*

camel back A coal miner's term for a *bell, potbottom, kettle bottom,* or other rock mass that tends to fall easily from the roof of a coal seam. Syn: *tortoise.*

cameo A carved gem that is actually a miniature bas-relief sculpture, commonly cut from materials of differently colored layers (esp. onyx or a gastropod shell), the upper layer being used for the figure and the lower layer serving as the background. Cf: *intaglio.*

cameo mountain A mountain composed of elevated horizontal strata left behind by two or more subparallel streams carving out deep channels that eventually unite.

camera The space enclosed within a cephalopod shell between two adjacent septa but excluding the siphuncle. It represents a portion of an earlier living space now closed off by a septum. Pl: *camerae.* Syn: *chamber.*

camera axis A line perpendicular to the focal plane of the camera and passing through the interior perspective center or emergent (rear) nodal point of the lens system.

cameral deposit A calcareous deposit secreted against the septa and/or the original walls of camerae during the life of a cephalopod; e.g. *mural deposit.*

camera lucida A simple monocular instrument for hand-copying or tracing a map or diagram onto a sheet of paper. It consists of a half-silvered mirror (or a prism or the optical equivalent) attached to the eyepiece of a microscope, thereby causing a virtual image of an external object to appear as if projected upon a plane. Etymol: Latin, "light chamber". Pl: *camera lucidas.*

camera station The point in the air or on the ground occupied by the camera lens at the moment of exposure. Syn: *exposure station.*

camerate [paleont] n. Any crinoid belonging to the subclass Camerata, characterized by rigidly united calyx plates forming a chamber.—adj. Divided into chambers or forming a chamber.

camerate [palyn] adj. A not widely used syn. of *cavate.*

camouflage In crystal structure, substitution of a trace element for a major element of the same valence, e.g. Ga for Al, and Hf for Zr. The trace element is then said to be camouflaged by the common element. Cf: *admittance [chem]; capture [chem].*

campagiform Said of the loop, or of the growth stage in the development of the loop, of a dallinid brachiopod, marked by a proportionally large hood without lateral lacunae, with the position of attachment of the descending branches to the septum and hood varying in different genera (TIP, 1965, pt.H, p.141). It precedes the *frenuliniform* stage.

campagna An Italian term for a nearly level, open plain; esp. the undulating, uncultivated plain surrounding Rome. Pron: cam-*pahn*-ya.

Campanian European stage: Upper Cretaceous (above Santonian, below Maestrichtian).

campanite An extrusive rock originally described as a *tephrite* containing large leucite crystals, and later defined as a pseudoleucite-bearing extrusive equivalent of *nepheline syenite* (Johannsen, 1939, p. 245); Tröger considered it the extrusive equivalent of *essexite-foidite* (Streckeisen, 1967, p. 208). Not recommended usage.

Campbell's law The general law of migration of drainage divides, which states that the divide tends to migrate toward an axis of uplift or away from an axis of subsidence (Campbell, 1896, p. 580-581). "Where two streams that head opposite to each other are affected by an uneven lengthwise tilting movement, that one whose declivity is increased cuts down vigorously and grows in length headward at the expense of the other" (Cotton, 1948, p. 342). Named in honor of its originator, Marius R. Campbell (1858-1940), American geologist.

camptonite A *lamprophyre,* similar in composition to nepheline *diorite,* being composed essentially of plagioclase (usually labradorite) and brown hornblende (usually barkevikite). Its name, given by Rosenbusch in 1887, is derived from Campton, New Hampshire.

camptospessartite A dark-colored *spessartite* in which the pyroxene is titanaugite. Not recommended usage.

campylite A yellow or brown variety of mimetite sometimes crystallizing in barrel-shaped forms.

campylotropous Said of a plant ovule curved by uneven growth so that its axis is approximately at right angles to its funiculus.

camstone (a) A Scottish term for a compact whitish limestone containing much clay. (b) A Scottish term for a bluish-white pipe clay used for whitening hearths and doorsteps.

canaanite A grayish white to bluish white *pyroxenite* composed chiefly of large white crystals of diopside. It is associated with dolomite at Canaan, Connecticut (Dana, 1892, p. 356). Not recommended usage.

cañada (a) A term used in the western U.S. for a ravine, glen, or narrow valley, smaller and less steep-sided than a *canyon,* such as the V-shaped valley of a dry river bed; a dale or open valley between mountains. (b) A term used in the western U.S. for a small stream; a creek.—Etymol: Spanish caña, "cane, reed". Pron: kan-*yah*-da.

canada balsam A natural cement used in mounting specimens for microscopic analysis; it is exuded as a viscous, yellow-green oleoresin by the balsam fir tree. Syn: *canada turpentine.*

canada turpentine *canada balsam.*

Canadian (a) North American series: Lower Ordovician (above Croixian of Cambrian, below Chazyan). (b) Name of a no-longer-recognized "system" of rocks between the Ozarkian below and the Ordovician above.

canadite A *nepheline syenite* containing albite as the main feldspar, along with calcian and aluminian mafic minerals, so that the normative plagioclase is much more anorthite-rich than the modal plagioclase. Named by Quensel in 1913. Not recommended usage.

canal [astrogeol] Any of various dark faint linear markings on the surface of Mars, generally of low contrast, that have been reported by visual observers. Very few have been confirmed by spacecraft photography; they are now generally considered illusory.

canal [coast] (a) A long, narrow channel or arm of the sea connecting two larger stretches of water, usually extending far inland (sometimes between islands or between an island and the mainland), and approximately uniform in width; e.g. Lynn Canal in Alaska. (b) A term used along the Atlantic coast of the U.S. for a sluggish coastal stream.

canal [paleont] A hollow vessel, tube, passage, channel, or groove of an invertebrate; e.g. a *ring canal* and *stone canal* of an echinoderm, a gutterlike extension of the lower end of a gastropod shell that carries the siphon, a tube running lengthwise along the walls of a foraminifer test, or a tube leading from an external pore of a sponge to the spongocoel and serving for water flow.

canal [speleo] A passage in a cave that is partly filled with water.

canal [streams] An artificial watercourse of relatively uniform dimensions, cut through an inland area, and designed for navigation, drainage, or irrigation by connecting two or more bodies of water; it is larger than a ditch.

canalarium A specialized sponge spicule lining a canal. Pl: *canalaria.*

canaliculate Grooved or channeled longitudinally; e.g. said of the sculpture of pollen and spores consisting of more or less parallel grooves, or said of a foraminifer possessing a series of fine tubular cavities.

canal system A system of passages connecting various cavities of an invertebrate body; e.g. the *aquiferous system* of a sponge.

canary A pale-yellow diamond.

canary stone A yellow variety of carnelian.

canasite A monoclinic mineral: $(Na,K)_6Ca_5Si_{12}O_{30}(OH,F)_4$.

Canastotan Stage in New York State: Upper Silurian (lower Cayugan; below Murderian).

cancellate Having a honeycomblike structure, divided into small spaces by laminae, or marked with numerous crossing plates, bars, lines, threads, etc.; e.g. said of the shell surface of a mollusk marked by subequal, intersecting concentric and radial markings.

cancellus A cylindrical, interzooidal tubelike cavity developed in some cyclostome bryozoans (as in the family Lichenoporidae). Pl: *cancelli.* Syn: *alveolus.*

cancrinite A group of feldspathoid minerals of the general formula (Hey, 1962, p.219): $(Na,K,Ca)_{6\text{-}8}(Al,Si)_{12}O_{24}(SO_4,CO_3,Cl)_{1\text{-}2} \cdot n H_2O$. The nomenclature of the group is obscure, but it includes cancrinite (the carbonate end member), vishnevite (the sulfate end member), and davyne (containing considerable chloride).

cand A term used in Cornwall, England, for fluorite occurring in a vein.

candela A term used in the SW U.S. for a candlelike rocky pinnacle. Etymol: Spanish, "candle".

candelabrum A *calthrops* (sponge spicule) with multiply branched rays, the branches of one ray often differing from those of the others. Pl: *candelabra.*

candite A blue spinel; *ceylonite.*

candle coal A syn. of *cannel coal,* so named because it burns with a steady flame.

candle ice Disintegrating sea or lake ice consisting of ice prisms oriented perpendicular to the ice surface; a form of *rotten ice.* Syn: *candled ice; needle ice.*

caneolith A heterococcolith having a central area with laths, a simple or complex wall, and petaloid upper and lower rims. See also: *complete caneolith; incomplete caneolith.*

canfieldite A black mineral: Ag_8SnS_6. It is isomorphous with argyrodite.

canga A Brazilian term for a tough, well-consolidated, unstratified, iron-rich rock composed of varying amounts of fragments derived from *itabirite,* high-grade hematite, or other ferruginous material, and cemented by limonite (which may range from about 5% to more than 95%). It occurs as a near-surface or surficial deposit and blankets older rocks on or near the present or an ancient erosion surface, and is very resistant to erosion and chemical weathering. Park (1959, p.580) restricts the term to a rock formed by the cementation by hematite of *rubble ore* into "a hard ironstone conglomerate". Some writers use the term to refer to a ferruginous laterite developed from any iron-bearing rock, commonly basalt or gabbro; e.g. as used in Sierra Leone, "canga" is equivalent to "lateritic iron ore".

cannel *cannel coal.*

cannel bass An English term for an inferior or impure carbonaceous shale approaching the character of an oil shale.

cannel coal A compact, tough *sapropelic coal* that consists dominantly of spores and is characterized by dull to waxy luster, conchoidal fracture, and massiveness. It is attrital and high in volatiles; by American standards it must contain less than 5% anthraxylon. Cf: *boghead coal; torbanite.* Syn: *candle coal; kennel coal; cannel; cannelite; parrot coal; curly cannel.* Adj: *canneloid.*

cannelite *cannel coal.*

canneloid Of or pertaining to *cannel coal.*

cannel shale A black shale or oil shale formed by the accumulation of sapropelic sediments, accompanied by an approximately equal amount of silt and clay. Syn: *bastard shale.*

cannizzarite A mineral: lead bismuth sulfide.

cannonball A large, dark concretion, as much as 3 m in diameter, resembling a cannonball, as in the Cannonball Member (Paleocene) of the Fort Union Formation in the Dakotas.

canoe fold A closely folded syncline, the surface expression of which is elongate.

canon An obsolete syn. of *canyon.* Etymol: American Spanish *cañon.* Incorrectly spelled: *canon.*

canopy In a cave, a ledge of flowstone extending from a wall, fringed with stalactites on its outer edge.

cant An inclination from a horizontal, vertical, or other given line; a slope or tilt.

cantalite An old name for a glassy sodic *rhyolite.* The name is from Cantal, France. Not recommended usage.

cantonite A variety of covellite occurring in cubes and probably pseudomorphous after chalcopyrite that had replaced galena.

canyon [geomorph] (a) A long, deep, relatively narrow steep-sided valley confined between lofty and precipitous walls in a plateau or mountainous area, often with a stream at the bottom; similar to, but larger than, a *gorge.* It is characteristic of an arid or semiarid area (such as western U.S.) where stream downcutting greatly exceeds weathering; e.g. Grand Canyon. (b) Any valley in a region where canyons abound.—Etymol: anglicized form of American Spanish *cañón.* Cf: *cañada.* Syn: *cañon.*

canyon [speleo] In a cave, a *passage* that is much higher than wide.

canyon bench One of a series of relatively narrow, flat landforms occurring along a canyon wall and caused by differential erosion of alternating strong and weak horizontal strata. See also: *step [geomorph].*

canyon dune A dune formed in a *box canyon* (Stone, 1967, p. 219).

canyon fill Unconsolidated material filling a canyon, consisting of sediment in transport and sediment permanently or temporarily deposited.

cap [glaciol] Nonpreferred syn. of *ice cap.*

cap [palyn] The thickened proximal surface of the body of a bladdered pollen grain, as found among species of Abietineae or Podocarpineae. Syn: *cappula; cappa.*

capable fault A fault defined by the Nuclear Regulatory Commission as one that is "capable" of "near future" movement; in general, a fault on which there has been movement within the last 35, 000 years. The definition was developed for use in the siting of nuclear power plants. Cf: *active fault.*

capacitance That property of an electrically nonconducting material which permits energy storage, due to electric displacement when opposite surfaces of the nonconductor are maintained at a difference of potential.

capacitive coupling The capacity between two adjacent circuit elements. In induced-polarization surveys, capacitive coupling between a current wire and a potential wire, or between either wire and the ground, can lead to fictitious anomalies.

capacity [grd wat] (a) The ability of a soil to hold water. (b) The yield of a pump, well, or reservoir.

capacity [hydraul] The ability of a current of water or wind to transport detritus, as shown by the amount measured at a given point per unit of time. Capacity may vary according to the detrital grain size. Cf: *competence.* See also: *efficiency.*

capacity curve In hydraulics, a graphic presentation of the relationship between the water-surface elevation of a reservoir and the volume of water below; also, a graphic presentation of the rate of discharge in a pipe or conduit or through porous material (ASCE, 1962). Syn: *storage curve.*

cape [coast] An extensive, somewhat rounded irregularity of land jutting out from the coast into a large body of water, either as a peninsula (e.g. Cape Cod, Mass.) or as a projecting point (e.g. Cape Hatteras, N.C.); a *promontory* or *headland,* generally more prominent than a *point.* Also, the part of the projection extending farthest into the water.

cape [mineral] *Cape diamond.*

Cape blue Crocidolite asbestos from South Africa.

Cape diamond A diamond having a yellowish tinge. Etymol: *Cape* of Good Hope. See also: *silver Cape.* Syn: *cape [mineral].*

cap effect *caprock effect.*

Cape ruby A misleading name applied to a ruby-colored garnet; specif. gem-quality *pyrope* that is perfectly transparent and ruby red in color, such as that obtained in the diamond mines of the Kimberley district of South Africa. Syn: *South African ruby.*

capilla A very fine radial ridge on the external surface of a brachiopod shell; there are usually more than 25 capillae in a width of 10 mm. Cf: *costa; costella.*

capillarity (a) The action by which a fluid, such as water, is drawn up (or depressed) in small interstices or tubes as a result of surface tension. Syn: *capillary action.* (b) The state of being *capillary.*

capillary [mineral] Said of a mineral that forms hairlike or threadlike crystals, e.g. millerite. Syn: *filiform; moss; wire; wiry.*

capillary [water] adj. Said of tubes or interstices with such small openings that they can retain fluids by *capillarity.*

capillary action *capillarity.*

capillary attraction The adhesive force between a liquid and a solid in capillarity.

capillary condensation The formation of rings of *pendular water* around point contacts of grains, and, when the rings around adjacent contacts become large enough to touch, of *funicular water* filling clusters of interstices surrounded by a single closed meniscus.

capillary conductivity The ability of an unsaturated soil or rock to transmit water or another liquid. As the larger interstices are partly occupied by air or other gas, the liquid must move through bodies surrounding point contacts of rock or soil particles. For water, the conductivity increases with the moisture content, from zero in a perfectly dry material to a maximum equal to the hydraulic conductivity, or *permeability coefficient.*

capillary ejecta *Pele's hair.*

capillary flow *capillary migration.*

capillary fringe The lower subdivision of the *zone of aeration*, immediately above the water table, in which the interstices are filled with water under pressure less than that of the atmosphere, being continuous with the water below the water table but held above it by surface tension. Its upper boundary with the intermediate belt is indistinct, but is sometimes defined arbitrarily as the level at which 50 percent of the interstices are filled with water. Syn: *zone of capillarity; capillary-moisture zone.*

capillary head The *capillary potential* expressed as head of water.

capillary interstice An *interstice* small enough to hold water by surface tension at an appreciable height above a free water surface, yet large enough to prevent molecular attraction from extending across the entire opening. There are no definite size limitations (Meinzer, 1923, p. 18). Cf: *subcapillary interstice; supercapillary interstice.*

capillary migration The movement of water by capillarity. Syn: *capillary flow; capillary movement.*

capillary-moisture zone *capillary fringe.*

capillary movement (a) The rise of water in the subsoil above the water table by capillarity (Nelson, 1965, p. 66). (b) *capillary migration.*

capillary percolation *imbibition [water].*

capillary porosity The volume of interstices in a soil mass that hold water by capillarity (Jacks et al., 1960).

capillary potential A number representing the work required to move a unit mass of water from the soil to an arbitrary reference location and energy state (SSSA, 1965, p. 348). Symbol: M. Cf: *capillary head.*

capillary pressure The difference in pressure across the interface between two immiscible fluid phases jointly occupying the interstices of a rock. It is due to the tension of the interfacial surface, and its value depends on the curvature of that surface.

capillary pyrites *millerite.*

capillary ripple *capillary wave.*

capillary rise The height above the free water level to which water will rise as a result of capillarity.

capillary tension *moisture tension.*

capillary water (a) Water held in, or moving through, small interstices or tubes by capillarity. The term is considered obsolete by the Soil Science Society of America (1965, p. 332). Syn: *water of capillarity.* (b) Water of the capillary fringe.

capillary wave A wave whose wavelength is shorter than 1.7 cm, and whose propagation velocity is controlled mainly by the surface tension of the liquid in which the wave is traveling. Cf: *gravity wave.* Syn: *capillary ripple; ripple.*

capillary yield The amount of capillary water, in mm per day or liters per second per hectare, that rises through a plane parallel to the water table, at a given distance below the land surface (Schieferdecker, 1959, term 0301).

capitulum (a) A part of the carapace of a cirripede crustacean, enclosing trophic structures and commonly armored by calcareous plates. (b) An obsolete syn. of *gnathosoma.* (c) The head of a vertebrate rib, i.e. the articular surface in contact with the vertebral centrum.—Pl: *capitula.*

capped column A type of snow crystal in the shape of a hexagonal column with thin hexagonal plates or stars at each end.

capped deflection A sharp bend in the trend of a mountain range, in which the arcs meet approximately at right angles and the junction consists of a *cap range* of very high mountains (Wilson, 1950, p. 155).

capped quartz A variety of quartz containing thin layers of clay.

cappelenite A mineral: $(Ba,Ca,Na)(Y,La)_6B_6Si_{13}(O,OH)_{27}$.

capping (a) A syn. of *overburden [eco geol],* usually used for consolidated material. (b) A syn of *gossan.*

cap range A secondary mountain arc that curves around the junction of two primary arcs, as in a *capped deflection.*

capricorn An ammonoid shell resembling a goat's horn, encircled by widely spaced blunt ribs and subequal rounded areas between adjacent ribs.

cap rock [coast] Estuarine sandstones along the Yorkshire coast of England (Nelson & Nelson, 1967, p. 58).

cap rock [eco geol] (a) A syn. of *overburden [eco geol],* usually used for consolidated material. (b) A hard rock layer, usually sandstone, overlying the shale above a coal bed. Also spelled: *caprock.*

cap rock [tect] In a *salt dome,* an impervious body of anhydrite and gypsum, with minor calcite and sometimes with sulfur, that overlies the salt body, or plug. It probably results from accumulation of the less soluble minerals of the salt body during leaching in the course of its ascent.

caprock effect A sharp positive *gravity anomaly* superimposed on a broader negative, indicative of a salt dome. It is commonly produced by the dense caprock of the dome, but very shallow salt is denser than the surrounding sediments, so caprock is not essential in producing this effect. Syn: *cap effect.*

cap-rock fall A waterfall descending over a lip of strong and resistant rock.

captor stream *capturing stream.*

capture [chem] In a crystal structure, the substitution of a trace element for a major element of lower valence, e.g. Ba^{++} for K^+. Captured trace elements generally have a higher concentration relative to the major element in the mineral than in the fluid from which it crystallized. Cf: *admittance [chem]; camouflage.*

capture [streams] The natural diversion of the headwaters of one stream into the channel of another stream having greater erosional activity and flowing at a lower level; esp. diversion effected by a stream eroding headward at a rapid rate so as to tap and lead off the waters of another stream. See also: *abstraction; beheading; intercision.* Syn: *stream capture; river capture; piracy; stream piracy; river piracy; robbery; stream robbery.*

captured stream A stream whose former upper course has been diverted into the channel of another stream by capture. Syn: *pirated stream.*

capture theory The theory that holds that the Moon originated as an independent planet whose orbit around the Sun lay so close to the Earth that it strayed into the gravitational field of the Earth and was captured by it.

capturing stream A stream into which the headwaters of another stream have been diverted by capture. Syn: *pirate; pirate stream; captor stream.*

caracolite A monoclinic mineral: $Na_3Pb_2(SO_4)_3Cl$.

Caradocian European stage: Middle and Upper Ordovician (above Llandeilian, below Ashgillian). It is divided into a lower stage (Middle Ordovician) and an upper stage (Upper Ordovician).

carapace (a) A bony or chitinous case or shield covering the whole or part of the back of certain animals, such as the dorsal covering of the cephalothorax in arachnids. (b) The fossilized remains of an ostracode whose calcified cephalothorax covering has become divided laterally into two subsymmetrical parts (valves) joined at the dorsum.— The term has also been applied to trilobites, eurypterids, and crustaceans other than ostracodes, but for these organisms the term *dorsal exoskeleton* is preferred.

carapace carina A narrow ridge variously located on the carapace of a decapod crustacean.

carapace groove A furrow, generally dorsal, on the carapace of a decapod crustacean.

carapace horn The anteriorly dorsal termination of the valves of the carapace of phyllocarid crustaceans.

carapace spine A sharp projection of the carapace of a decapod crustacean.

carat A unit of weight for diamonds, pearls, and other gems. It formerly varied somewhat in different countries, but the *metric carat* or international carat equal to 0.2 gram, or 200 mg, was adopted in the U.S. in 1913 and is now standard in the principal countries of the world. Abbrev: c. Cf: *point [gem]; grain [gem].* Not to be confused with *karat.*

carbankerite Any coal microlithotype containing 20-60% by volume of carbonate minerals (calcite, siderite, dolomite, and ankerite) (ICCP, 1963).

carbapatite *carbonate-apatite.*

carbargillite Any coal microlithotype containing 20-60% by volume of clay minerals and mica, and lesser proportions of quartz, all with particle size averaging 1 to 3 microns (ICCP, 1963). Partial syn: *carbominerite.*

carbene An *asphaltene* that is insoluble in carbon tetrachloride, but soluble in carbon disulfide, benzene, or chloroform.

carbide A mineral compound that is a combination of carbon with a metal. An example is cohenite, $(Fe,Ni,Co)_3C$.

carbite A general term, now obsolete, applied to diamond and graphite.

carboborite A mineral: $Ca_2Mg(CO_3)(B_2O_5)\cdot 10H_2O$.

carbocer A pitchy, ocherous, and carbonaceous mineral substance containing rare-earth elements.

carbocernaite A mineral: $(Ca,Ce,Na,Sr)(CO_3)$.

carbohumin *ulmin.*

carbohydrate A polyhydroxy aldehyde or ketone, or a compound that can be hydrolyzed to such a form. Carbohydrates, of which sugars, starches, and cellulose are examples, are produced by all green plants and form an important animal food.

carboid A bitumen formed from asphaltenes at elevated temperatures. It may occur in cracked petroleum residues, and is insoluble in benzene.

carbominerite A microlithotype of coal containing different minerals in association with macerals (ICCP, 1971). Cf: *carbopyrite.* Partial syn: *carbargillite.*

carbon (a) A nonmetallic, chiefly tetravalent chemical element (atomic number 6; atomic weight 12.01115) occurring native in the crystalline form (as the diamond and as graphite) or amorphous, and forming a constituent of coal, petroleum, and asphalt, of limestone and other carbonates, and of all organic compounds. Symbol: C. (b) *carbonado.*

carbonaceous (a) Said of a rock or sediment that is rich in carbon; coaly. (b) Said of a sediment containing organic matter.

carbonaceous chondrite A group name for friable, dull-black, chondritic stony meteorites, characterized by the presence of hydrated clay-type silicate minerals (usually fine-grained serpentine or chlorite); by considerable amounts and a great variety of organic compounds (hydrocarbons, fatty and aromatic acids, porphyrins), believed to be of extraterrestrial origin; by a near or total absence of free nickel-iron; and by an abnormally high content of inert gases (esp. xenon). Much of the organic matter is a black insoluble complex of compounds of high molecular weight; the water content (usually water of hydration) is approximately 20% by weight. Carbonaceous chondrites are grouped in three types, each characterized by the amount of organic material and by other compositional features: type I contains the greatest amount of water and organic matter (3-5% combined carbon, 24-30% ignition loss); type II is intermediate in composition (12-24% ignition loss); and type III contains high-temperature minerals and some metallic constituents (2-12% ignition loss).

carbonaceous coal Coal that is intermediate in composition between metabituminous coal and anthracite (Tomkeieff, 1954, p.35).

carbonaceous rock A sedimentary rock that consists of, or contains an appreciable amount of, original or subsequently introduced organic material, including plant and animal residues and organic derivatives greatly altered (carbonized or bituminized) from the original remains; e.g. the coal series, black shale, asphaltic sediments, sapropel, certain clays, various solid substances derived from altered plant remains, and esp. *carbonaceous shale.* Syn: *carbonolite.*

carbonaceous shale A dark-gray or black shale with a significant content of carbon in the form of small disseminated particles or flakes; it is commonly associated with coal seams.

carbonado An impure opaque dark aggregate composed of minute diamond particles, forming a usually rounded mass with a granular to compact structure, and displaying a superior toughness as a result of its cryptocrystalline character and lack of cleavage planes. It is used as an *industrial diamond.* Cf: *bort; ballas.* Syn: *black diamond; carbon diamond; carbon.*

carbonate [mineral] A mineral compound characterized by a fundamental anionic structure of CO_3^{-2}. Calcite and aragonite, $CaCO_3$, are examples of carbonates. Cf: *borate; nitrate.*

carbonate [sed] A sediment formed by the organic or inorganic precipitation from aqueous solution of carbonates of calcium, magnesium, or iron; e.g. limestone and dolomite. See also: *carbonate rock.*

carbonate-apatite (a) A mineral of the apatite group: $Ca_5(PO_4,CO_3)_3(OH,F)$. It is the principal constituent of sedimentary phosphate rock. (b) An apatite mineral containing a considerable amount of carbonate.—Cf: *francolite.* Syn: *carbapatite, dahllite; podolite; collophane; tavistockite.*

carbonate-arenite *calclithite.*

carbonate buildup A carbonate rock mass that is thicker than, and different from, laterally equivalent strata, and probably stood above the surrounding sea floor during some or all of its depositional history. Cf: *reef; bank [sed]; submarine bank.* Syn: *limestone buildup.*

carbonate compensation depth In the ocean, that level below which the rate of solution of calcium carbonate exceeds the rate of its deposition. In the Pacific Ocean, this level is at 4000-5000 m; in the Atlantic it is somewhat shallower. Abbrev: CCD. Cf: *compensation depth [oceanog].* Syn: *calcium-carbonate compensation depth; depth of compensation [oceanog].*

carbonate-cyanotrichite A pale-blue mineral: $Cu_4Al_2(CO_3,SO_4)(OH)_{13}\cdot 2H_2O$.

carbonate cycle The biogeochemical pathways of carbonate, involving transformation from or to CO_2 and HCO_3^-, and its solution, deposition in minerals, and metabolism and regeneration in biological fixation.

carbonated spring A spring whose water contains carbon dioxide gas. This type of spring is especially common in volcanic areas (Comstock, 1878, p. 34).

carbonate-facies iron formation An *iron formation* characterized by alternating laminae of chert and iron-rich carbonate minerals (James, 1954, p.251-256).

carbonate-fluorapatite *francolite.*

carbonate hardness Hardness of water, expressed as $CaCO_3$, that is equivalent to the carbonate and bicarbonate alkalinity. When the total alkalinity expressed as $CaCO_3$ equals or exceeds the total hardness, all the hardness is carbonate. It can be removed by boiling and hence is sometimes called *temporary hardness,* although this synonym is becoming obsolete. Cf: *noncarbonate hardness; hardness [water].*

carbonate hydroxylapatite A mineral of the apatite group: $Ca_5(PO_4,CO_3)(OH)$.

carbonate rock A rock consisting chiefly of carbonate minerals, such as limestone, dolomite, or carbonatite; specif. a sedimentary rock composed of more than 50% by weight of carbonate minerals. See also: *calcareous rock; carbonate [sed].*

carbonate thermometer The temperature-dependent oxygen-18/oxygen-16 isotope ratio in the carbonate shells of fossil marine animals, as used to indicate the water temperature that existed at the time the shell was deposited (i.e. the paleotemperature). Accurate determination depends on the isotopic composition of the shell being in equilibrium with the surrounding water at the time of deposition, on knowledge of the isotopic composition of the water, and on the material being preserved without further isotopic fractionation or substitution. See also: *oxygen-isotope fractionation.*

carbonation (a) A process of chemical weathering involving the transformation of minerals containing calcium, magnesium, potassium, sodium, and iron into carbonates or bicarbonates of these metals by carbon dioxide contained in water (i.e. a weak carbonic-acid solution). Syn: *carbonatization.* (b) Introduction of carbon dioxide into a fluid.

carbonatite A carbonate rock of apparent magmatic origin, generally associated with *kimberlites* and alkalic rocks. The origin of carbonatites is controversial; they have been variously explained as derived from magmatic melt, solid flow, hydrothermal solution, and gaseous transfer. A carbonatite may be calcitic (*sövite*) or dolomitic (*rauhaugite*). See: Heinrich, 1966; Tuttle & Gittins, 1966.

carbonatization (a) Introduction of, or replacement by, carbonates. (b) *carbonation.*

carbon clock A popular syn. of *carbon-14,* used in radiometric dating of rocks.

carbon cycle The continued exchange and reactions of carbon in the biosphere, atmosphere, and hydrosphere (Pettijohn, 1949, p.363).

carbon dating *carbon-14 dating.*

carbon diamond *carbonado.*

carbon-hydrogen ratio The ratio of carbon to hydrogen in coal. It is the basis for a method of coal classification. Abbrev: C/H ratio.

Carbonic (a) *Pennsylvanian.* (b) *Carboniferous.*

Carboniferous The Mississippian and Pennsylvanian periods combined, ranging from about 345 to about 280 million years ago; also, the corresponding system of rocks. In European usage, the

Carboniferous is considered as a single period and is divided into upper and lower parts. The Permian is sometimes included. See also: *age of amphibians; age of coal.* Partial syn: *Carbonic.*

carbonification A syn. of *coalification,* suggested for standard use by the International Committee for Coal Petrology (ICCP, 1963).

carbonite *natural coke.*

carbonization (a) In the process of *coalification,* the accumulation of residual carbon by the changes in organic matter and decomposition products. (b) The accumulation of carbon by the slow, underwater decay of organic matter. (c) The conversion into carbon of a carbonaceous substance such as coal by driving off the other components, either by heat under laboratory conditions or by natural processes.

carbonolite *carbonaceous rock.*

carbonolith A term suggested by Grabau (1924, p. 298) for a carbonaceous sedimentary rock.

carbon ratio [coal] (a) The percentage of fixed carbon in a coal. (b) The ratio of fixed carbon in a coal to the fixed carbon plus the volatile hydrocarbons. Syn: *fixed-carbon ratio.*—Cf: *fuel ratio.*

carbon ratio [geochron] The ratio of the most common carbon isotope, carbon-12, which is nonradioactive, to either of the less common isotopes, carbon-13 (nonradioactive) or carbon-14 (radioactive), or the reciprocal of one of these ratios. If unspecified, the term generally refers to the ratio (carbon-12/carbon-13).

carbon-ratio theory The hypothesis that in any region, the specific gravity of oil varies inversely with the carbon ratio of the associated coals. As the percentage of fixed carbon in the coal increases as a result of metamorphism, the oil becomes lighter, i.e. higher in volatile hydrocarbons. The theory was most effectively stated by David White in 1915.

carbon spot (a) A black flecklike or flakelike carbon inclusion in the body of a diamond crystal. (b) A term in the jewelry trade referring to any apparently black inclusion or imperfection in a diamond. Under dark-field illumination, most "carbon spots" are found to be neither black nor carbon.

carbon trash Carbon remains of plant life found in sedimentary rocks and often associated with uranium and red-bed copper mineralization. See also: *tree ore.*

carbon-14 A heavy radioactive isotope of carbon having a mass number of 14 and a half-life of 5730 ± 40 years (Godwin, 1962). (The figure 5568 ± 30 is also used.) It is produced in nature by the reaction of atmospheric nitrogen with neutrons produced by cosmic-ray collisions and artificially by atmospheric nuclear explosions. Carbon-14 is useful in dating and tracer studies of materials directly or indirectly involved with the Earth's carbon cycle during the last 50,000 years. Symbol: ^{14}C. Partial syn: *radiocarbon.*

carbon-14 age A radiometric age expressed in years and calculated from the quantitative determination of the amount of carbon-14 remaining in an organic material. Syn: *radiocarbon age.* Popular syn: *carbon clock.*

carbon-14 dating A method of determining an age in years by measuring the concentration of carbon-14 remaining in an organic material, usually formerly living matter, but also water bicarbonate, etc. The method, worked out by Willard F. Libby, U.S. chemist, in 1946-1951, is based on the assumption that assimilation of carbon-14 ceased abruptly on removal of the material from the Earth's carbon cycle (i.e. on the death of an organism) and that it thereafter remained a closed system. Most carbon-14 ages are calculated using a half-life of 5730 ± 40 years or 5568 ± 30 years. Thus the method is useful in determining ages in the range of 500 to 30,000 or 40,000 years, although it may be extended to 70,000 years by using special techniques involving controlled enrichment of the sample in carbon-14. Syn: *radiocarbon dating; carbon dating.*

carbopyrite Any coal microlithotype containing 5-20% by volume of iron disulfide (pyrite and marcasite) (ICCP, 1963).

Carborundum Trade name for a synthetic substance (silicon carbide) used as an abrasive and as a refractory material. It is identical with the mineral moissanite.

carbuncle (a) A cabochon-cut red garnet. (b) An old name, now obsolete, for any of several precious stones of a fiery red color, such as ruby or spinel.

carbunculus A term applied to ruby, ruby spinel, almandine garnet, and pyrope.

carburan A pitchy hydrocarbon containing uranium.

cardella A *condyle* in the orifice of an autozooid in ascophoran cheilostome bryozoans. Pl: *cardellae.* Syn: *cardelle.*

cardhouse structure A structure found in certain marine sediments, consisting of platy aggregates of clay minerals arranged in edge-to-face fashion and resembling a "house" of playing cards (Lambe, 1953, p. 38). Cf: *bookhouse structure.*

cardiac lobe The median lobe of the prosoma and the opisthosoma in merostomes.

cardiac region The median part of the carapace of some decapods, behind the cervical groove or suture, between the urogastric and intestinal areas (TIP, 1969, pt. R, p. 92).

cardinal adj. Pertaining to the hinge of a bivalve shell; e.g. "cardinal extremity" (termination of posterior hinge margin of a brachiopod shell).—n. A cardinal part; e.g. a cardinal tooth.

cardinal angle The angle formed at each of the extremities of the hinge of a bivalve shell; e.g. the angle formed between the hinge line and the anterior or posterior free margin of an ostracode valve, or between the hinge line and the posterolateral margins of a brachiopod shell.

cardinal area (a) A flat or slightly concave, commonly triangular surface extending between the beak and the hinge margin in many bivalve mollusks, and partly or wholly occupied by ligament. It is set off from the remainder of the shell by a sharp angle. (b) The flattened, posterior sector of a valve of an articulate brachiopod, exclusive of the delthyrium or the notothyrium. It may be the interarea, a planarea, or the palintrope.—See also: *hinge area.*

cardinal axis The *hinge axis* in a bivalve mollusk.

cardinal fossula A *fossula* developed in the position of the cardinal septum in a rugose coral. It is most commonly due to abortion of the cardinal septum.

cardinalia A collective term for the varied internal outgrowths and structures of secondary shell located in the posterior and median region of the brachial valve near the beak of a brachiopod, and associated with articulation, support of the lophophore, and muscle attachment. It may include, for example, the cardinal processes, socket ridges, crural bases, and hinge plates.

cardinal margin The curved posterior margin of a brachiopod shell along which the valves are hinged, homologous with the *hinge line* of strophic shells but not parallel to the *hinge axis* (TIP, 1965, pt.H, p.141). See also: *hinge [paleont].*

cardinal muscle scar A posteriorly and laterally placed muscle scar in certain brachiopods (as in the superfamilies Acrotretacea and Obolellacea).

cardinal plate A plate extending across the posterior end of the brachial valve of a brachiopod, consisting laterally of outer hinge plates and medially of either conjunct inner hinge plates or a single plate, and commonly perforated posteriorly (TIP, 1965, pt.H, p.141).

cardinal platform The *hinge plate* in a bivalve mollusk.

cardinal point One of the four principal *points* of the compass (viz: the north, south, east, or west points) that lie in the direction of the Earth's two poles and of sunrise and sunset and that indicate the four principal astronomic directions on the Earth's surface, spaced at 90-degree intervals.

cardinal process A blade or variably shaped boss, ridge, or projection of secondary shell of a brachiopod, situated medially in the posterior end of the brachial valve, and serving for separation or attachment of diductor muscles.

cardinal septum The protoseptum lying in the plane of bilateral symmetry of a rugose corallite, distinguished from other protosepta by pinnate insertion of newly formed metasepta adjacent to it on both sides. Symbol: C. Cf: *alar septum; counter septum.*

cardinal tooth A *hinge tooth,* often relatively large, situated close to and directly beneath the beak of a bivalve mollusk. Its long axis is perpendicular or oblique to the hinge line. Cf: *lateral tooth.*

cardiophthalmic region The space between the ophthalmic ridges in merostomes.

Cardium clay A Pleistocene glacial clay of northern Europe, characterized by fossil shells of the genus *Cardium,* a marine bivalve mollusk of the family Cardiidae.

carex peat Peat that is derived mainly from *Carex,* a genus of sedges of the family Cyperaceae. Cf: *eriophorum peat.* Syn: *sedge peat.*

cargneule A French term for a porous or cavernous carbonate sedimentary rock (esp. a cellular dolomite), its cavities filled with soft, friable, evaporitic material that easily dissolves or falls out, leaving a rough, corroded surface. Syn: *cornieule.*

Cariboo orogeny A name proposed by W.H. White (1959) for an orogeny that is believed to have occurred during early Paleozoic time in the Cordillera of British Columbia, especially in the Selkirk and Omineca mountains, where Permian strata overlie deformed and metamorphosed Proterozoic and lower Paleozoic rocks.

caridoid adj. Said of a decapod crustacean of the infraorder Caridea containing most shrimps, prawns, and related forms; e.g. "caridoid facies", an aspect of primitive Eumalacostraca distinguished by enclosure of thorax by carapace, movably stalked eyes, biramous antennules, scaphocerite-bearing antennae, thoracopods with natatory exopods, ventrally flexed and powerfully muscled elongate abdomen, and caudal fan.—n. A caridoid crustacean.

caries texture In ore microscopy, a replacement pattern in which the younger mineral forms a series of scallop-shaped incursions into the host mineral, which resemble filled dental cavities. Syn: *cusp-and-caries texture.*

carina (a) An unpaired compartmental plate adjacent to the terga of a cirripede crustacean. (b) A flangelike elevation on the side of a septum of a rugose coral, formed by thickened trabecula. See also: *yardarm carina; zigzag carina.* (c) A keel-shaped structure or a flange going around the edge of some foraminiferal tests. (d) A prominent keel-like ridge on the exterior shell of a mollusk; e.g. an extended, somewhat angular linear elevation on the exterior of a whorl at the edge of a gastropod shell. (e) The central and denticulated, nodose, or smooth ridge extending down the middle of the platform or blade of a conodont. (f) A major angular elevation on the surface of a brachiopod valve, externally convex in transverse profile and radial from the umbo. Cf: *fold [paleont].* (g) A median ridge or keel-like structure on the frontal side of a bryozoan branch, chiefly in Cryptostomata. Syn: *keel.* (h) A frill-like or ridgelike structure more or less parallel to the free edge in an ostracode valve, usually situated above the velar structure; present in many genera of the order Palaeocopida.—Etymol: Latin, "keel". Pl: *carinae.*

carinal adj. Pertaining to a carina; e.g. a "carinal band" representing an imperforate marginal area between the carinae of a foraminiferal test.—n. One of a series of ossicles along the midline of the aboral surface of an asterozoan ray.

carinate (a) Said of a fold that is almost isoclinal. (b) Said of an antiform or synform that is confined to incompetent strata, so that adjacent strata are undisturbed.

carinolateral A compartmental plate in certain cirripede crustaceans, located on either side of the carina. Syn: *carinal latus.*

carletonite A mineral: $KNa_4Ca_4Si_8O_{18}(CO_3)_4(OH,F)\cdot 4H_2O$.

carlfriesite A monoclinic mineral: $CaTe_3O_5(OH)_4$.

carlinite A trigonal mineral: Tl_2S.

Carlin-type gold Gold occurring as microscopic particles (up to 30 microns) that must be identified by chemical analysis as it is not recoverable by panning. The term is taken from its occurrence at Carlin, Nevada. Syn: *invisible gold.*

Carlsbad B twin law A twin law that is now equated with the *x-pericline twin law.*

Carlsbad twin law A twin law in feldspar, especially orthoclase, that defines a penetration twin in which the twin axis is the *c* crystallographic axis and the composition surface is irregular. Also spelled: *Karlsbad twin law.*

carlsbergite A cubic mineral: CrN.

carmeloite A *basalt* or an *andesite,* depending on whether the plagioclase is andesine or labradorite, that contains iddingsite as an alteration product of olivine phenocrysts. Its name, given by Lawson in 1893, is derived from Carmel Bay, California. Not recommended usage.

carminite A carmine to tile-red mineral: $PbFe_2(AsO_4)_2(OH)_2$.

carnallite A milk-white to reddish orthorhombic mineral: $KMgCl_3\cdot 6H_2O$. It occurs as a saline residue and is a raw material of fertilizer manufacture in some European districts.

carnegieite A synthetic compound: $NaAlSiO_4$. It is the high-temperature equivalent of nepheline. It is triclinic at low temperatures, isometric at high temperatures.

carnelian A translucent red or orange-red variety of chalcedony, pale to deep in shade, containing iron impurities. It is used for seals and signet rings. Cf: *sard.* Also spelled: *cornelian.* Syn: *carneol.*

carneol *carnelian.*

Carnian European stage: Upper Triassic (above Ladinian, below Norian). Also spelled: *Karnian.*

carnivore A heterotrophic organism that nourishes itself mainly by feeding on other animals, living or dead. Adj: *carnivorous.* Cf: *herbivore.*

carnotite A strongly radioactive, canary-yellow to greenish-yellow secondary mineral: $K_2(UO_2)_2(VO_4)_2\cdot 3H_2O$. An ore of uranium and vanadium, and a source of radium, it occurs as a powdery incrustation or in loosely coherent masses, chiefly in sandstone (as in the western U.S.).

carobbiite A mineral: KF.

Carolina bay Any of various shallow, often oval or elliptical, generally marshy, closed depressions in the Atlantic coastal plain (from southern New Jersey to NE Florida, esp. developed in the Carolinas). They range from about 100 m to many kilometers in length, are rich in humus, and contain trees and shrubs different from those of the surrounding areas. Their origin is much debated and has been attributed to meteorites, upwelling springs, eddy currents, and solution. Syn: *bay.*

carpathite *karpatite.*

carpedolith *stone line.*

carpel An ovule-bearing locule of the ovary of seed plants, forming a simple *pistil* or a unit of a compound pistil. Said to represent a modified floral leaf (Swartz, 1971, p. 86).

carpholite A straw-yellow mineral: $MnAl_2Si_2O_6(OH)_4$.

carphosiderite A yellow mineral consisting of a basic hydrous iron sulfate. Much so-called carphosiderite is jarosite or natrojarosite. Cf: *hydronium jarosite.*

carpoid Any homalozoan echinoderm having an ambulacral groove and a stereome composed of crystalline calcite with reticular microstructure.

carpolite (a) A fossil fruit, nut, or seed. Syn: *carpolith.* (b) An ellipsoidal concretion or similar diagenetic structure, 1-2 cm in diameter, originally believed to be a fossil seed and assigned the generic name *Carpolites.*

carpopod The *carpus* of a malacostracan crustacean. Syn: *carpopodite.*

carpus (a) The fifth pereiopodal segment from the body of a malacostracan crustacean, located distal to the merus and proximal to the propodus. It comprises the third segment of a typical endopod. (b) The wrist of a tetrapod.—Pl: *carpi.* Syn: *carpopod; wrist.*

carr An isolated mass of rock found off the coast, esp. in the British Isles. Syn: *carrig; carrick.*

carrboydite A hexagonal mineral: $(Ni,Cu)_{14}Al_9(SO_4,CO_3)_6(OH)_{43}\cdot 7H_2O$.

carrig *carr.*

carrollite A light steel-gray mineral of the linnaeite group: $Cu(Co,Ni)_2S_4$. Syn: *sychnodymite.*

carrying capacity The natural *production* of a lake, as it influences the population of fish and other aquatic life that the lake will support.

carrying contour A single contour line representing two or more contours, used to show a vertical or near-vertical topographic feature such as a cliff.

carse A Scottish term for low, level, fertile land; esp. a tract of alluvial land or river bottom bordering an estuary or near a river mouth, representing a marine terrace or a raised beach. Example: Carse of Gowrie along the coast of Scotland. Syn: *carse land.*

carst *karst.*

carstone A British term for a hard, firmly cemented ferruginous sandstone, esp. one of Cretaceous age used as a building stone. Syn: *quernstone.*

cartilage-replacement bone Bone that replaces a cartilaginous precursor; includes the axial skeleton and most of the appendicular skeleton, but only limited parts of the skull. Syn: *endochondral bone.*

cartogram A small, abstracted, simplified map generally showing statistical data of various kinds in a diagrammatic way, usually by the use of shades, curves, or dots; e.g. a *dot map.* Syn: *diagrammatic map.*

cartographic unit A rock or group of rocks that is shown on a geologic map by a single color or pattern. The standard cartographic unit is a *formation.*

cartography (a) The art of map or chart construction, and the science on which it is based. It includes the whole series of map-making operations, from the actual surveying of the ground to the final printing of the map. Syn: *chartology.* (b) The study of maps as scientific documents and works of art.

cartology A graphic method of coal-seam correlation, involving the mapping and drawing of both vertical and horizontal sections. Cf: *composite map.*

cartouche A decorative frame or scroll-shaped embellishment on a map or chart, enclosing the title, scale, legend, and other descriptive matter. Syn: *title box.*

caryinite A monoclinic mineral: $(Na,Ca,Pb)_2(Mn,Mg,Fe^{+3})_3$

$(AsO_4)_3(?)$.

caryopilite A mineral of the kaolinite-serpentine group: $(Mn,Mg)_3 Si_2O_5(OH)_4(?)$.

caryopsis A small dry one-seeded indehiscent fruit that has a completely united seed coat and pericarp and forms a single grain, e.g. in grasses.

cascade [glaciol] *glacial stairway.*

cascade [streams] (a) A *waterfall,* esp. a small fall or one of a series of small falls descending over steeply slanting rocks; a shortened rapids. Also, a stepped series of small, closely spaced waterfalls or very steep rapids. Cf: *cataract.* (b) A short, rocky declivity in a stream bed over which water flows with greater rapidity and a higher fall than through a *rapids.*

cascade decay A little-used term for radioactive decay from a parent isotope through several daughter isotopes to a stable isotope.

cascade fold One of a series of folds that is formed by gravity collapse along the limb of larger folds.

cascade stairway *glacial stairway.*

Cascadia One of the *borderlands* proposed by Schuchert (1923), in this case along the western margin of North America, partly at sea, partly inland. Most of the evidence adduced for the existence of Cascadia can now be otherwise interpreted. Possibly there were minor offshore lands in places, and some former continental material may have disappeared by underthrusting at the edge of the continent, but the foundering of extensive lands into the Pacific Ocean basin is not considered a tenable concept.

Cascadian Revolution A name used by Schuchert and others for a supposed profound crustal disturbance in western North America which brought an end to the Tertiary. The concept is now known to be untenable. The type area (the Cascade Range) is unfortunate, as no notable crustal events seem to have occurred there at this time. The term should be abandoned.

cascading glacier A glacier passing over a steep, irregular bed, and therefore crossed by numerous crevasses and suggestive in appearance of a cascading stream. Cf: *icefall.*

cascadite A sodic *minette* containing biotite, olivine, and augite phenocrysts in a groundmass composed almost entirely of alkali feldspar. It was named by Pirsson in 1905 for Cascade Creek, Highwood Mountains, Montana. Not recommended usage.

cascajo Reef-derived material composed of coral detritus and other sediment, occurring in old deposits. Etymol: Spanish, "gravel". Pron: kahs-*kah*-ho.

cascalho A term used in Brazil for alluvial material, including gravel and ferruginous sand, in which diamonds are found. Etymol: Portuguese, "pebbles, small stones, coarse sand, gravel, grit".

case hardening The process by which the surface of a porous rock (esp. tuff and certain sandstones) is coated with a cement or desert varnish formed by evaporation of mineral-bearing solutions. Adj: *case-hardened.* Also spelled: *casehardening.*

casing Heavy metal pipe, lowered into a bore hole during or after drilling and cemented into place. It prevents the sides of the hole from caving, prevents loss of drilling mud or other fluids into porous formations, and prevents unwanted fluids from entering the hole. It consists of sections, usually about 30 ft. long, that are screwed together. A well may contain several strings of casing, the inner and smaller-diameter strings extending progressively deeper. See also: *surface pipe.* Syn: *well casing.*

casing-collar log A *well log* of relative magnetic intensity used in cased holes to identify the threaded junctions between consecutive lengths of well casing. It is used with a *gamma-ray log* run simultaneously to correlate between the geologic section and the sequence of collars, for depth control of perforating or other operations. Abbrev: CCL. Syn: *collar log.*

casing head A fitting attached to the top of the *casing* set in an oil or gas well, to which is attached the *christmas tree.* It may carry a valve to control pressure in the *annulus* between casing and *tubing.* Also spelled: casinghead.

casing-head gas Unprocessed natural gas produced from a reservoir containing oil. Such gas contains gasoline vapors and is so called because it is usually produced under low pressure through the casing head of an oil well.

casing point The depth in a drill hole to which a given string of *casing* extends.

Ca-spar *calcium feldspar.*

Cassadagan North American stage: Upper Devonian (above Chemungian, below Conewangoan).

Cassel brown *black earth [coal].* Etymol: source near Cassel, Germany.

Cassel earth *black earth [coal].* Etymol: source near Cassel, Germany.

Cassiar orogeny A name proposed by W.H. White (1959) for an orogeny that is believed to have occurred near the end of Paleozoic time in the Cordillera of British Columbia, especially in the Omineca and Cassiar districts. It was characterized by uplift, folding, and ultramafic intrusion.

cassidyite A mineral: $Ca_2(Mg,Ni)(PO_4)_2 \cdot 2H_2O$.

Cassinian curve An *isochromatic curve* in a biaxial crystal.

Cassini projection A map projection constructed by computing the lengths of arcs along a central meridian and along a great circle perpendicular to that meridian, and plotting these as rectangular coordinates on a plane. The scale is preserved along the two principal axes. It was formerly used as the base for much topographic and cadastral mapping. Named after C.F. Cassini de Thury (1714-1784), French astronomer, who introduced the projection in 1745.

cassinite (a) A bluish variety of orthoclase containing barium and occurring in Delaware County, Penna. (b) Perthitic intergrowths of hyalophane and plagioclase.

cassiterite A brown or black tetragonal mineral: SnO_2. It is the principal ore of tin. Cassiterite occurs in prismatic crystals of adamantine luster, and also in massive forms, either compact with concentric fibrous structure (*wood tin*) or in rolled or pebbly fragments (*stream tin*). Syn: *tinstone; tin ore; black tin.*

cast [paleont] Secondary rock or mineral material that fills a *natural mold;* specif. a replica or reproduction of the external details (size, shape, surface features) of a fossil shell, skeleton, or other organic structure, produced by the filling of a cavity formed by the decay or dissolution of some or all of the original hard parts of which the organism consisted. Cf: *mold [paleont].*

cast [sed] A sedimentary structure representing the infilling of an original mark or depression made on top of a soft bed and preserved as a solid form on the underside of the overlying and more durable stratum; e.g. a *flute cast* or a *load cast.* Cf: *mold.* Syn: *counterpart.*

castellated Said of a physiographic feature, such as a cliff, peak, or iceberg, that displays a towering or battlementlike structure.

casting [paleont] Something that is cast out or off, esp. a worm casting or a *fecal pellet.*

casting [sed] The process of forming a sedimentary cast or casts, or the configuration of a surface characterized by such casts; e.g. "load casting". Also, a cast so formed.

castle A natural rock formation bearing a fancied resemblance to a castle.

castle kopje *castle koppie.*

castle koppie A pointed *koppie,* or a small *bornhardt,* with a castellated profile, often occurring as a jumbled pile of joint-bounded granite blocks. Syn: *castle kopje.*

castor *castorite.*

castorite A transparent variety of *petalite.* Syn: *castor.*

casuzone A *biostratigraphic unit* defined as "a body of rock with upper and lower boundaries marked by reversible faunal changes that are essentially parallel to time planes" (Vella, 1964, p. 622).

caswellite A mineral consisting of a copper-red altered biotite, resembling clintonite.

cat *cat claw.*

cataclasis Rock deformation accomplished by fracture and rotation of mineral grains or aggregates without chemical reconstitution. See also: *cataclastic metamorphism; cataclastic rock.*

cataclasite *cataclastic rock.*

cataclasm A breaking down or rending asunder; a violent disruption.

cataclast A *megacryst* that is a remnant of incomplete cataclasis.

cataclastic (a) Pertaining to the structure produced in a rock by the action of severe mechanical stress during dynamic metamorphism; characteristic features include the bending, breaking, and granulation of the minerals. Also, said of the rocks exhibiting such structures. See also: *mortar structure.* (b) Pertaining to clastic rocks, the fragments of which have been produced by the fracture of pre-existing rocks by Earth stresses, e.g. crush breccia (Teall, 1887).—Also spelled: *kataclastic.*

cataclastic breccia *crush breccia.*

cataclastic conglomerate *crush conglomerate.*

cataclastic flow *Flow [exp struc geol]* involving intergranular movement, i.e. mechanical displacement of particles relative to each other.

cataclastic metamorphism A type of local metamorphism confined to the vicinity of faults and overthrusts, involving purely mechanical forces causing crushing and granulation of the rock fabric (*cataclasis*). Cf: *dislocation metamorphism; kinetic metamorphism.*

cataclastic rock A rock, such as a tectonic breccia, containing angular fragments that have been produced by the crushing and fracturing of preexisting rocks as a result of mechanical forces in the crust; a metamorphic rock produced by cataclasis. Its fabric is a structureless rock powder. Pettijohn (1957, p.281) would include glacial till as a "cataclastic deposit" as it is "an extensive gouge caused by the grinding along the base of an overthrust ice sheet". See also: *cataclasis; autoclastic rock; mylonite.* Syn: *cataclasite.*

cataclastic structure *mortar structure.*

cataclastic texture A texture in a dynamically metamorphosed rock produced by severe mechanical crushing and differential movement of the component grains and characterized by granular, fragmentary, deformed, or strained mineral crystals, commonly flattened in a direction at right angles to the mechanical stress. Syn: *pressure texture.*

cataclinal Said of a *dip stream* or of a valley that descends in the same direction as that of the general dip of the underlying strata it traverses. Term introduced by Powell (1875, p. 160). Ant: *anaclinal.* Syn: *conclinal; acclinal.*

catacline Said of the *inclination* of the cardinal area in either valve of a brachiopod lying at right angles to the orthocline position.

cataclysm (a) Any geologic event that produces sudden and extensive changes in the Earth's surface; e.g. an exceptionally violent earthquake. Adj: *cataclysmic; cataclysmal.* Cf: *catastrophe.* (b) Any violent, overwhelming flood that spreads over the land; a deluge.

catadupe An obsolete term for a cataract or waterfall.

catagenesis [evol] Evolution leading to decadence and decreased vigor. Also spelled: *katagenesis.*

catagenesis [sed] The changes occurring in an already formed sedimentary rock buried by a distinct (though sometimes thin) covering layer, characterized by pressure-temperature conditions that are much different from those of deposition (Fersman, 1922); specif. the breakdown of rocks. The term is more or less equivalent to *epigenesis* as applied by Russian geologists. Syn: *katagenesis.*

catagraphite A complex structure made up of traces of canals and cavities thought to be the result of the activity of algae and bacteria in the late Precambrian and Early Cambrian. Cf: *oncolite.*

catamorphism *katamorphism.*

catanorm Theoretical calculation of minerals in a metamorphic rock of the *katazone,* as indicated by chemical analyses. Cf: *mesonorm; epinorm.* See also: *Niggli molecular norm.*

cataphoresis *Electrophoresis* in which the movement of suspended positive particles in a fluid is toward the cathode. Cf: *anaphoresis.*

cataphorite *kataphorite.*

cataplasis In an evolutionary lineage, a late or final stage in which organisms become decadent and display decreasing vigor. Cf: *anaplasis; metaplasis.*

catapleiite A yellow or yellowish-brown hexagonal mineral: $(Na_2,Ca)ZrSi_3O_9 \cdot 2H_2O$. See also: *calcium-catapleiite.* Also spelled: *catapleite.*

cataract (a) A *waterfall,* esp. one of great volume in which the vertical descent has been concentrated in one sheer drop over a precipice. Cf: *cascade.* (b) A series of steep *rapids* in a large river, e.g. the Nile. (c) An overwhelming rush of water; a flood.

cataract lake A lake occupying the plunge basin of an extinct cataract whose stream has been diverted above the fall.

catarinite An obsolete term for an iron meteorite remarkable for high percentage of nickel.

cataspire A foldlike respiratory structure with thin calcite walls extending beneath the deltoid plate of a parablastoid, from its aboral margin to the edge of an adjacent ambulacrum. Seawater apparently flowed adorally through each cataspire (Sprinkle, 1973, p. 32).

catastable adj. Stable, with a tendency towards sinking. Cf: *anastable.*

catastrophe A sudden, violent disturbance of nature, ascribed to exceptional or supernatural causes, affecting the physical conditions and the inhabitants of the Earth's surface; e.g. the Noachian flood, or an extinction of an entire fauna. Cf: *cataclysm; paroxysm.* Syn: *convulsion.*

catastrophe theory A unique mathematical concept from topology, which allows quantification of the dynamics of abrupt and continuous changes within a system for modeling, e.g. some geological processes.

catastrophic advance *surge [glaciol].*

catastrophism (a) The doctrine that sudden violent, short-lived, more or less worldwide events outside our present experience or knowledge of nature have greatly modified the Earth's crust. (b) The doctrine that the present configuration of the Earth's crust, as well as the distribution of living beings, is mainly the result of "a great and sudden revolution" (Cuvier) of 5000 or 6000 years ago, and by extension that geologic processes of the past were of substantially greater intensity and number than those of the present. (c) The doctrine that changes in the Earth's fauna and flora are explained by recurring catastrophes, followed by creation of different organisms.—Cf: *uniformitarianism.* Syn: *convulsionism.*

catastrophist A believer in *catastrophism.*

catathermal n. A period of time during which temperatures are declining. The term was used by Emiliani (1955, p.547) for a part of a cycle as displayed in a deep-sea sediment core. Ant: *anathermal.*

catawberite A metamorphic rock of South Carolina that consists mainly of talc and magnetite.

catazone *katazone.*

catch basin (a) A reservoir or basin into which surface water may drain. (b) A basin to collect and retain material from a street gutter that would not readily pass through the sewer system.

catchment (a) A term used in Great Britain for an area that collects and drains rainwater; a *drainage basin.* (b) A depression that collects rainwater; a reservoir. (c) The act of catching water; also, the amount of water that is caught.

catchment area [grd wat] (a) The *recharge area* and all areas that contribute water to it. (b) An area paved or otherwise waterproofed to provide a water supply for a storage reservoir. Syn: *collecting area.*

catchment area [streams] *drainage basin.*

catchment basin *drainage basin.*

catchment glacier Nonpreferred syn. of *drift glacier.*

catchwater drain A ditch or surface drain designed to intercept flowing water on sloping land and to divert the flow so as to improve stability or to irrigate the soil. Syn: *catchwater; catchwork.*

cat claw A miner's term used in Peoria County, Ill., for an irregular protuberance (2.5-7.5 cm in height and width) in the lower surface of a bed of marcasite overlying a coal seam (Cady, 1921, p.164). Syn: *cat.*

cat coal Coal that contains pyrite.

catena [astrogeol] A term established by the International Astronomical Union for a chain of craters on Mars. Most are thought to be of volcanic origin, tectonically controlled (Mutch et al., 1976, p. 57). Etymol: Latin *catena,* a chain.

catena [soils] A sequence of soils of about the same age, derived from similar parent material and occurring under similar climatic conditions, but having different characteristics due to variation in relief and drainage. See also: *toposequence; soil association.*

catenary The curve assumed by a perfectly flexible, inextensible cord of uniform density and cross section when suspended freely from two fixed points both at the same level; e.g. such a curve as formed by a surveyor's tape hanging between adjacent supports. See also: *sag correction.*

catenary ripple mark "A ripple is described as catenary if the trace of its crestline has the pattern of a chain of catenary waves, such that the more pointed segments of the crestline face downcurrent" (Allen, 1968, p. 62-63). It differs from a lunate ripple mark in having lesser curvature of the crestline. Catenary ripple marks may be in or out of phase with the ripples immediately upstream or downstream.

catenicelliform Said of a jointed, delicately branching, erect colony in cheilostome bryozoans, the internodes of which consist of few zooids. It is attached by rootlets (Lagaaij & Gautier, 1965, p. 51).

cateniform Said of a tabulate corallum with the corallites united laterally as palisades that appear chainlike in cross section and commonly form a network.

catenulate colony Colonial growth of archaeocyathids in which the cups are contiguous and in a chainlike row and the outer wall does not develop between neighboring cups (TIP, 1972, pt. E, p. 6).

cat face A miner's term for glistening balls or nodules, or small discontinuous veinlets, of pyrite in the mining face of coal; also, lenticular deposits of pyrite associated with coal. Also spelled:

catface.

cathodoluminescence The emission of characteristic visible luminescence by a substance that is under bombardment by electrons.

cathole A local term used in southern Michigan for a shallow boggy depression less than an acre in extent, esp. one formed by a glacier in a till plain (Veatch & Humphrys, 1966, p. 59).

cat ice *shell ice.*

cation exchange The displacement of a cation bound to a site on the surface of a solid, as in silica-alumina clay-mineral packets, by a cation in solution. Syn: *base exchange.* See also: *ion exchange.*

catkin *ament.*

catlinite A red, siliceous, indurated clay from the upper Missouri River valley region (SW Minnesota), formerly used by the Dakota Indians for making tobacco pipes; a *pipestone.* Named after George Catlin (1796-1872), American painter of Indians.

catoctin A residual knob, hill, or ridge of resistant material rising above a peneplain and preserving on its summit a remnant of an older peneplain. Named after Catoctin Mountain, Maryland & Virginia. Cf: *monadnock.*

catogene Pertaining to sedimentary rocks, signifying that they were formed by deposition from above, as of suspended material. Cf: *katogene.*

catophorite *kataphorite.*

catoptrite *katoptrite.*

Ca-Tschermak molecule A synthetic pyroxene, $CaAl(AlSi)O_6$; a hypothetical component of natural pyroxenes. Not to be confused with *tschermakite.* Syn: *Tschermak molecule.*

cat's-eye [gem] Any gemstone that, when cut *en cabochon,* exhibits under a single strong point-source of light a narrow, well-defined chatoyant band or streak that moves across the summit of the gemstone, shifts from side to side as it is turned, and resembles a slit pupil of the eye of a cat. Internal reflection of light from parallel inclusions of tiny fibrous crystals or from long parallel cavities or tubes causes the "cat's-eye". Gemstones exhibiting this phenomenon include chrysoberyl, quartz, sillimanite, scapolite, cordierite, orthoclase, albite, beryl, and tourmaline. See also: *chatoyancy.*

cat's-eye [mineral] (a) A greenish gem variety of chrysoberyl that exhibits *chatoyancy.* Syn: *cymophane; oriental cat's-eye.* (b) A variety of minutely fibrous, grayish-green quartz (chalcedony) that exhibits an opalescent play of light. Syn: *occidental cat's-eye.* (c) A yellowish-brown silicified form of crocidolite. Cf: *tiger's-eye.* —The term, when used alone, is properly applied only to chrysoberyl.

catstep A *terracette,* esp. one produced by slumping of deep loess deposits as in western Iowa. Also spelled: *cat step.*

catter Shortened form of *bellicatter,* a syn. of *icefoot.*

cattierite A mineral with pyrite structure: CoS_2.

cauda The slender proximal portion of an autozooecium in some cheilostome bryozoans, comprising part of the gymnocyst and adjacent parts of lateral and basal walls (Thomas & Larwood, 1956, p. 370). Pl: *caudae.*

caudal fan A powerful swimming structure in malacostracan crustaceans, consisting of a combination of laterally expanded uropods and telson, and constituting a means of steering and balancing. Syn: *tail fan.*

caudal furca A crustacean *furca* consisting of a pair of caudal rami.

caudal process In certain ostracodes, a prolongation (usually pointed) of the posterior end of the carapace, commonly situated at or near mid-height.

caudal ramus One of a pair of appendages of the telson of a crustacean. It is usually rodlike or bladelike, but may be filamentous and multiarticular. Syn: *caudal appendage; cercus.*

Caudata A superorder of the lissamphibians, characterized by a tail and nonhopping gait. It includes the salamanders, newts, and caecilians.

caudex A basal part of the axis of an erect plant where it is neither clearly stem, rhizome, or root; the persistent base of an otherwise annual herbaceous stem (Fernald, 1950, p.1571).

cauk (a) A dialectal British term for a limestone or chalk. Syn: *calk.* (b) *cawk.*

cauldron [marine geol] *caldron [marine geol].*

cauldron [volc] As used by Smith & Bailey (1968), an inclusive term for all volcanic subsidence structures regardless of shape or size, depth of erosion, or connection with the surface. The term thus includes *cauldron subsidences,* in the classical sense, and *collapse calderas.*

cauldron subsidence A structure resulting from the lowering along a steep ring fracture of a more or less cylindrical block into a magma chamber; usually associated with ring dikes. Also, the process of forming such a structure. The ring fracture may or may not reach the surface of the Earth. See also: *ring-fracture stoping; surface cauldron subsidence; underground cauldron subsidence.*

caulescent Said of a plant that is more or less stemmed or stem-bearing; having an evident stem above ground (Lawrence, 1951, p.743).

caulk *cawk.*

caunter lode *cross vein.*

causse A French syn. of *karst,* in some usages implying a karstic plateau of relatively small size, or one restricted to southeastern France. Etymol: French dialect for *chaux,* "lime".

caustic n. (a) A curve representing a locus of points where a set of rays (as of light or sound) are brought exceptionally close together owing to the refractive properties of the medium through which they travel; specif. the curve to which adjacent orthogonals of waves, refracted by a bottom whose contour lines are curved, are tangents. Syn: *caustic curve.* (b) The envelope of the system or sequence of such rays. Syn: *caustic surface.*

caustic metamorphism The indurating, baking, burning, and fritting effects of lava flows and small dikes on the rocks with which they come in contact. The term was originated by Milch in 1922. Cf: *baking.* Syn: *optalic metamorphism.*

caustobiolith A combustible organic rock (Grabau, 1924, p.280). It is usually of plant origin. Cf: *acaustobiolith.*

caustolith A rock that has the property of combustibility (Grabau, 1924, p.280). It is usually of organic origin (e.g. coal and peat), but inorganic deposits (e.g. sulfur, asphalt, graphite) also occur.

caustophytolith A *caustobiolith* formed by the direct accumulation of vegetal matter; e.g. peat, lignite, and coal.

caustozoolith A rare *caustobiolith* formed by the direct accumulation of animal matter (Grabau, 1924, p.280); e.g. some oils.

cavaedium An irregular space within a sponge, communicating directly with the exterior, but morphologically outside the sponge in that it is lined by exopinacoderm. Pl: *cavaedia.*

cavalorite A granular plutonic rock containing more potassium feldspar than oligoclase. Named by Capellini in 1877 after Monte Cavaloro, Italy. Obsolete. Cf: *oligoclasite.*

cavansite An orthorhombic, greenish-blue mineral: $Ca(VO)(Si_4O_{10})\cdot 6H_2O$. It is dimorphous with pentagonite.

cavate (a) Descriptive of spores whose exine layers are separated by a cavity, including a rather slight separation as well as a more extensive separation producing a bladderlike protuberance (pseudosaccus). Syn: *camerate [palyn].* (b) Said of a dinoflagellate cyst with space or spaces of notable size between periphragm and endophragm (as in *Deflandrea phosphoritica*).

cave [coast] *sea cave.*

cave [speleo] (a) A natural underground open space, generally with a connection to the surface and large enough for a person to enter. The most common type of cave is formed in limestone by dissolution. Partial syn: *cavern.* (b) A similar feature that was formed artificially. (c) In informal use, any natural rock shelter, e.g. a *cliff overhang.*

cave balloon *cave blister.*

cave blister A partly or completely hollow hemispherical to nearly spherical speleothem, usually of gypsum or hydromagnesite, attached to a cave wall. Syn: *cave balloon.*

cave breakdown The collapse of the ceiling or walls of a cave; also, the accumulation of debris thus formed. See also: *cave breccia.* Syn: *breakdown.*

cave breathing The back-and-forth movement of air in the constricted passages of caves, usually with a cycle of a few minutes. See also: *blowing cave.*

cave breccia Angular fragments (*cave breakdown*) of limestone that have fallen to the floor from the roof and sides of a cave and that are cemented with calcium carbonate or occur in a matrix of cave earth. See also: *collapse breccia; solution breccia.*

cave bubble A nonattached hollow sphere, usually of calcite, that has formed around a gas bubble on the surface of a cave pool. Syn: *calcite bubble.*

cave coral A rough, knobby speleothem, usually of calcite, that resembles coral. Syn: *botryoid; cave popcorn; coralloid.*

cave cotton Thin flexible filaments of gypsum or epsomite projecting from a cave wall. Syn: *gypsum cotton.* See also: *cave flower.*

cave earth Fine-grained, generally unconsolidated detrital material partly filling a cave; also, similar material regardless of grain size. See also: *fill [speleo].* Syn: *cave soil.*

cave flower In a cave, gypsum or epsomite that occurs as a curved elongate deposit from a cave wall. Growth of the structure occurs at the attached end. Syn: *gypsum flower; oulopholite.* See also: *anthodite; cave cotton.*

cave formation *speleothem.*

cave ice (a) Naturally formed ice in a cave. (b) Formerly, calcium-carbonate deposits in a cave.

cave-in (a) The partial or complete collapse of earth material into a large underground opening, such as an excavation or a mine. (b) The sudden slumping of wall material into a pit. (c) A place where material has collapsed or fallen in or down.

cave-in lake A shallow body of water whose basin is produced by collapse of the ground following thawing of ground ice in regions underlain by permafrost; a lake occupying a *thaw depression.* Cf: *kettle lake.* Syn: *thaw lake [permafrost]; thermokarst lake; cryogenic lake.*

cave marble *cave onyx.*

cave of debouchure *outflow cave.*

cave onyx A compact banded deposit of calcite or aragonite found in caves, capable of taking a high polish, and resembling true onyx in appearance. See also: *dripstone; flowstone; onyx marble; speleothem; travertine.* Syn: *cave marble.*

cave pearl A nonattached rounded cave deposit, usually of calcite, formed by precipitation of concentric layers around a nucleus, and characterized by radial crystal structure. Syn: *cave pisolite.*

cave pisolite *cave pearl.*

cave popcorn *cave coral.*

caver One who engages in the hobby of cave exploration, or *caving [speleo].* Syn: *spelunker; potholer.* See also: *speleologist.*

cave raft A thin mineral film, usually of calcite, floating on a cave pool. Syn: *calcite raft; floe calcite.*

cavern A syn. of *cave [speleo],* with the implication of large size; a system or series of caves or cave chambers.

cavern flow Movement, often turbulent, of ground water through caves, coarse sorted gravel, or large open conduits, either by gravity or under pressure.

cavernous [speleo] Said of an area or geologic formation, e.g. limestone, that contains caverns, or caves.

cavernous [volc] Said of the texture of a volcanic rock that is coarsely *porous* or *cellular.*

cavernous rock Any rock that has many cavities, cells, or large interstices; e.g. a cliff face pitted with shallow holes resulting from cavernous weathering.

cavernous weathering Chemical and mechanical weathering on a cliff face, in which grains and flakes of rock are loosened so as to enlarge hollows and recesses "opened through a chemically hardened shell" on the surface of the cliff face (Cotton, 1958, p.15). It produces the *tafoni* in seaside cliffs. Cf: *honeycomb weathering.*

cavern porosity A pore system having large, cavernous openings. The lower size limit, for field analysis, is practically set at "about the smallest opening an adult person can enter" (Choquette & Pray, 1970, p. 244).

cavern system *cave system.*

cave soil *cave earth.*

cave system (a) A group of caves that are connected or hydrologically related. (b) A complex cave. Syn: *cavern system.*

caving [geomorph] (a) *bank caving.* (b) A falling in; the action of caving in.

caving [speleo] The exploration of caves, rather more as a hobby or sport than as a scientific study. See also: *caver; speleology.* Syn: *spelunking; potholing.*

cavings Rock fragments that fall from the walls of a borehole and contaminate the *well cuttings* or block the hole. They must be removed by drilling or *circulation* before the borehole can be deepened.

cavitation The collapse of bubbles in a fluid, caused by the static pressure being less than the fluid vapor pressure.

cavity (a) A solutional hollow in a limestone cave. (b) A small hollow in cavernous lava.

cavity dweller A *coelobitic* organism.

cawk A syn. of *barite;* esp. a white, massive, opaque variety of barite found in Derbyshire, England. Syn: *cauk; caulk; calk.*

c* axis That axis of a reciprocal crystal lattice which is perpendicular to (001). Cf: *a* axis; b* axis.*

c axis [cryst] One of the crystallographic axes used for reference in crystal description; it is oriented vertically. In tetragonal and hexagonal crystals, it is the unique symmetry axis. It is usually the principal axis. The letter *c* usually appears in italics. Cf: *a* axis; *b* axis.

***c* axis [struc petrol]** One of three orthogonal reference axes, *a, b,* and *c,* that are used in two distinct ways. (a) To help describe the geometry of a fabric that possesses monoclinic symmetry. The *c* axis lies in the unique symmetry plane at right angles to a prominent fabric plane; thus in many *tectonites* the *c* axis is normal to the schistosity. (b) In a kinematic sense, to describe a *deformation plan* that possesses monoclinic symmetry, such as a progressive simple shear. Here the *c* axis lies in the unique symmetry plane and normal to the movement plane. In a progressive simple shear the *c* axis lies normal to the shear plane. See also: *a* axis; *b* axis. Syn: *c* direction.

cay A small, low, coastal island or emergent reef of sand or coral; a flat mound of sand and admixed coral fragments, built up on a reef flat at or just above high-tide level. Term is used esp. in the West Indies where it is pronounced "key" and sometimes spelled *kay.* Etymol: Spanish *cayo,* "shoal or reef". Cf: *key.*

cayeuxite A nodular variety of pyrite containing silicon, arsenic, antimony, and germanium.

cay sandstone A friable to firmly cemented coral sand formed near the base of a coral-reef *cay* and reaching above high-tide level; it is cemented by calcium carbonate deposited from fresh water.

caysichite An orthorhombic mineral: $(Y,Ca)_4Si_4O_{10}(CO_3)_3 \cdot 4H_2O$.

Cayugan North American series: Upper Silurian (above Niagaran, below Helderbergian of Devonian).

Cazenovian North American stage: Middle Devonian (above Onesquethawan, below Tioughniogan).

CBL *cement-bond log.*

CBR *California bearing ratio.*

CCD *carbonate compensation depth.*

CCL *casing-collar log.*

CD interray Posterior interray in echinoderms situated between C ray and D ray and clockwise of C ray when viewed from the oral side. It differs frequently in shape from the other interrays and contains the anal opening; equal to ambulacrum IV of the Lovenian system.

***c* direction** *c axis [struc petrol].*

CDP *common depth point.*

cebollite A greenish to white fibrous mineral: $Ca_4Al_2Si_3O_{14}(OH)_2$ (?).

cecilite A basaltic rock with few phenocrysts, composed of leucite, augite, melilite, nepheline, olivine, anorthite, magnetite, and apatite. Leucite comprises about 50 percent of the total rock, followed by the mafic minerals, with melilite comprising about one-eighth, nepheline about eight percent, and anorthite four percent (Johannsen, 1939, p. 246). The rock is essentially a *leucitite.* It was named by Cordie in 1868 for the tomb of Cecilia Metella, Capo di Bove, Italy. Not recommended usage.

cedarite *chemawinite.*

cedar-tree structure A term applied to a laccolith or volcanic neck in which sill-like intrusive layers taper away from a central intrusive mass, the whole structure resembling the outline of a cedar tree in cross section. Cf: *compound laccolith.* Syn: *Christmas-tree laccolith.*

cedricite A *leucitite* that contains leucite and clinopyroxene phenocrysts in a very fine-grained groundmass containing phlogopite. Named by Wade and Prider in 1940 after Mount Cedric, Western Australia. Not recommended usage.

ceiling cavity A solutional hollow in the ceiling of a cave. Some are clearly joint-controlled; in others the control is problematical. See also: *joint cavity; pocket [speleo].*

ceiling channel A solutional groove on the ceiling of a cave that presumably was filled with water when it formed. Syn: *ceiling meander; upside-down channel.*

ceiling meander *ceiling channel.*

ceja A term used in the SW U.S. for the jutting edge along the top of a mesa or upland plain, and also for the cliff at this edge; an escarpment, esp. the steeper of the two slopes of a cuesta (if the slope is a cliff) or part of this slope that is a cliff. Etymol: Spanish, "eyebrow; mountain summit". Pron: *say*-ha.

celadonite A soft, green or gray-green, earthy, dioctahedral mineral of the mica group, consisting of a hydrous silicate of iron, magnesium, and potassium, and generally occurring in cavities in basaltic rocks. It has a structure very similar to that of glauconite, and has been regarded as a ferruginous glauconite, a mica near glauconite, and an aluminum-poor variety of glauconite, and also as "similar to illite" (Hey, 1962, 14.24.6). Not to be confused with *caledonite.* Syn: *kmaite; Verona earth.*

celerity The rate at which a small surface wave progresses outward in still water from the point of a slight disturbance.

celestial coordinate Any member of any system of coordinates used to locate a point on the celestial sphere; e.g. altitude, azimuth, declination, right ascension.

celestial equator The great circle on the celestial sphere whose plane is perpendicular to the Earth's axis of rotation. It is formed by the intersection of the celestial sphere and the extension of the Earth's equatorial plane. Often simply called the *equator.* Syn: *equinoctial circle.*

celestial horizon That circle of the celestial sphere formed by the intersection of the celestial sphere and a plane through the center of the Earth and perpendicular to the zenith-nadir line. Syn: *astronomic horizon; rational horizon.*

celestialite A variety of *ozocerite* found in some iron meteorites.

celestial latitude Angular distance north or south of the ecliptic; the arc of a circle of latitude between the ecliptic and a point on the celestial sphere, measured northward or southward from the ecliptic through 90°, and labeled N or S to indicate the direction of measurement. Cf: *latitude.* Syn: *ecliptic latitude.*

celestial longitude Angular distance east of the vernal equinox, along the ecliptic, between the circle of latitude of the vernal equinox and the circle of latitude of a point on the celestial sphere; measured eastward from the circle of latitude of the vernal equinox, through 360°. Cf: *longitude.* Syn: *ecliptic longitude.*

celestial mechanics The study of the theory of the motions of celestial bodies under the influence of gravitational fields (NASA, 1966, p.11).

celestial meridian One half of a great circle of the celestial sphere passing through the zenith of a given place and terminating at the celestial poles; the *hour circle* that contains the zenith, or the vertical circle that contains the celestial pole. Usually called simply the *meridian.*

celestial pole Either of the two points of intersection of the celestial sphere and the extended axis of rotation of the Earth, around which the diurnal rotation of the stars appears to take place; specif. *north pole* and *south pole.* The altitude of an observer's celestial pole is equal to his geographic latitude.

celestial sphere An imaginary sphere of infinite radius, described around an assumed center (usually the center of the Earth), and upon whose "inner surface" all the heavenly bodies (except the Earth) appear to be projected along radii passing through them. The apparent dome of the visible sky forms half of the celestial sphere. Since the radius of the Earth is negligible in comparison to the distances to most celestial bodies, the center of the celestial sphere is assumed for most surveying purposes to coincide with the point of observation on the Earth's surface.

celestine *celestite.*

celestite An orthorhombic mineral: $SrSO_4$. It is commonly white with an occasional pale-blue tint. It often occurs in residual clays and in deposits of salt, gypsum, and associated dolomite and shale. Celestite is the principal ore of strontium. Syn: *celestine; coelestine.*

celite *brownmillerite.*

cellariiform Said of a jointed, delicately or stoutly branching, erect colony in cheilostome bryozoans, the internodes of which consist of many zooids. It is attached by radicles on an encrusting base (Lagaaij & Gautier, 1965, p. 51).

celleporiform Said of a multilamellar, commonly massive colony in cheilostome bryozoans, formed by repeated self-overgrowth or budding in a frontal direction, from an encrusting basal layer of zooids (Lagaaij & Gautier, 1965, p. 51).

cell texture In mineral deposits, a network pattern formed by solution or replacement of organic structures, e.g. cell walls.

cellular Said of the texture of a rock (e.g. a cellular dolomite) characterized by openings or cavities, which may or may not be connected. Although there are no specific size limitations, the term is usually applied to cavities larger than pores and smaller than caverns. The syn. *vesicular* is preferred when describing igneous rocks. Cf: *porous; cavernous.*

cellular porosity A term applied originally by Howard & David (1936, p. 1406) to equidimensional openings formed by solution. The term has since been applied to intraparticle porosity formed organically within fossils. Choquette & Pray (1970, p. 245) suggest that the term be abandoned because of its infrequent use and diverse application.

cellular soil *polygonal ground.*

cellule A subdivision of a marginal chamberlet in the outer part of the marginal zone of a foraminifer (as in Orbitolinidae), formed by primary and secondary partitions.

cellulose A polymeric carbohydrate composed of glucose units, formula $(C_6H_{10}O_5)_x$, of which the permanent cell membranes of plants are formed, making it the most abundant carbohydrate.

cell wall A rigid wall outside the cytoplasmic membrane of the cells in most plants; commonly composed of cellulose but sometimes lignified.

cell-wall degradation matter *humic degradation matter.*

celsian A rare, colorless, monoclinic mineral of the feldspar group: $BaAl_2Si_2O_8$. It is the barium analogue of anorthite and is dimorphous with paracelsian.

celyphytic rim *kelyphytic rim.*

cement [mater] A manufactured gray powder which when mixed with water makes a plastic mass that will "set" or harden. It is combined with an *aggregate* to make *concrete.* Nearly all of today's production is *portland cement.* See also: *cement rock.*

cement [ore dep] Ore minerals, e.g. gold, that are part of or have replaced *cement* in the sedimentary use of the word.

cement [sed] Mineral material, usually chemically precipitated, that occurs in the spaces among the individual grains of a consolidated sedimentary rock, thereby binding the grains together as a rigid, coherent mass; it may be derived from the sediment or its entrapped waters, or it may be brought in by solution from outside sources. The most common cements are silica (quartz, opal, chalcedony), carbonates (calcite, dolomite, siderite), and various iron oxides. Others include clay minerals, barite, gypsum, anhydrite, and pyrite. Detrital clay minerals and other fine clastic particles may also serve as cements.

cementation [eng] *grouting.*

cementation [sed] The diagenetic process by which coarse clastic sediments become lithified or consolidated into hard, compact rocks, usually through deposition or precipitation of minerals in the spaces among the individual grains of the sediment. It may occur simultaneously with sedimentation, or at a later time. Cementation may occur by secondary enlargement. Syn: *agglutination.*

cementation [soil] The binding-together of the particles of a soil by such cementing agents as colloidal clay, hydrates of iron, or carbonates. Three degrees of cementation are recognized: weakly cemented, strongly cemented, and indurated.

cement-bond log An *acoustic log* of compressional wave amplitude run shortly after cementing well casing to evaluate the casing-to-cement and cement-to-formation bonds. Abbrev: CBL.

cement clay A clay with a variable amount of calcium carbonate, used in the manufacture of cement.

cement deposits In the Black Hills of the U.S., gold-bearing Cambrian conglomerates believed to be ancient beach or stream-channel deposits.

cement gravel Gravel that is consolidated by some binding material such as clay, silica, or calcite.

cementing The operation whereby cement slurry is pumped into a drill hole and forced up behind the casing, to seal the casing to the walls of the hole and to prevent unwanted leakage of fluids into the hole or migration of oil, gas, or water between formations.

cement rock Any rock that is capable of furnishing cement when properly treated, with little or no addition of other material; specif. a massive, sparsely fossiliferous, clayey limestone that contains the ingredients (alumina, silica, lime) for cement in approximately the required proportions. Example: the Blackjack Creek Limestone Member of the Fort Scott Limestone in Kansas. Syn: *cement stone.*

cement stone A syn. of *cement rock.* Also spelled: *cementstone.*

cemetery mound A small hillock caused by the melting of surrounding *ice wedges* in areas of patterned ground.

cenology An old term for the geology of surficial deposits.

Cenomanian European stage: lowermost Upper Cretaceous, or Middle Cretaceous of some authors (above Albian, below Turonian).

Cenophytic A paleobotanic division of geologic time, signifying the time since the development of the angiosperms in the middle or late Cretaceous. Cf: *Aphytic; Archeophytic; Eophytic; Paleophytic; Mesophytic.* Syn: *Neophytic.*

cenosis *coenosis.*

cenosite *kainosite.*

cenote In Yucatán, Mexico, a vertical shaft in limestone, open to the surface, that contains standing water. Partial syn: *vertical cave.* Etymol: Mayan, *tzonot.*

cenotypal Said of a fine-grained porphyritic igneous rock having the appearance of fresh or nearly fresh extrusive rocks, such as those of Tertiary and Holocene age. This term and the term *paleotypal* were introduced to distinguish *neovolcanic* and *paleovolcanic* fine-grained igneous rocks; both are little used.

Cenozoic An era of geologic time, from the beginning of the Tertiary period to the present. (Some authors do not include the Quaternary, considering it a separate era.) It is characterized paleontologically by the evolution and abundance of mammals, advanced mollusks, and birds; paleobotanically, by angiosperms. The Cenozoic is considered to have begun about 65 million years ago. Also spelled: *Cainozoic; Kainozoic.* See also: *age of mammals.*

cenozone A syn. of *assemblage-zone,* suggested by Moore (1957).

cenozoology Zoology of existing animals without regard to those that are extinct. Cf: *neontology; paleobiology.*

centare A metric unit of area equal to one square meter, 0.01 are, or 10.76 square feet. Abbrev: ca.

centered lattice A crystal lattice in which the axes have been chosen according to the rules for the crystal system, and in which there are lattice points at the centers of certain planes as well as at their corners; thus a centered lattice has two, three, or four lattice points per unit instead of one, as in a *primitive lattice*. See also: *base-centered lattice; one-face-centered lattice; face-centered lattice; body-centered lattice; side-centered lattice.*

center line A straight or curved line that continuously bisects a feature or figure (such as a stream, a strip of land, or the bubble tube in a spirit level); specif. the line connecting opposite corresponding corners of a quarter section or quarter-quarter section, or the line extending from the true center point of overlapping aerial photographs through each of the transposed center points.

center of gravity That point in a body or system of bodies through which the resultant attraction of gravity acts when the body or system is in any position; that point from which the body can be suspended or poised in equilibrium in any position.

center-of-gravity map A *vertical-variability map,* or *moment map,* that shows the relative, weighted mean position of a lithologic type in terms of its distance from the top of a given stratigraphic unit, expressed as a percentage of total thickness of the unit. Cf: *standard-deviation map.*

center of instrument The point on the vertical axis of rotation (of a surveying instrument) that is at the same elevation as that of the collimation axis when that axis is in a horizontal position. In a transit or theodolite, it is close to or at the intersection of the horizontal and vertical axes of the instrument.

center of symmetry A point in a crystal structure through which every aspect of an array is repeated by *inversion [cryst];* it is a symmetry element. Syn: *inversion center; symmetry center.*

center point (a) The *principal point* of a photograph. (b) The central point from which a map projection is geometrically based.

central body The main part of a pollen grain or spore; e.g. the main part of a vesiculate pollen grain, as distinct from the vesicles, or the compact central part of a dinoflagellate cyst from which the projecting structures extend.

central capsule The mucoid or chitinous perforated internal skeleton or sac of a radiolarian, enclosing the nucleus and intracapsular cytoplasm.

central cavity The opening or space enclosed by the inner wall along the axis of archaeocyathid cups (TIP, 1972, pt. E, p. 40). Cf: *internal cavity.*

central complex The core or central zone in which foraminiferal chamber passages bifurcate and anastomose in a reticulate pattern (as in Orbitolinidae).

central cylinder *stele.*

centrale (a) A prominent plate at the center of the aboral surface of the disc in many asterozoans. It is the central plate of the primary circlet. Syn: *centrodorsal.* (b) A noncirriferous thecal plate typically occurring inside the infrabasal circlet of some articulate crinoids (such as *Marsupites* and *Uintacrinus*). (c) One of the proximal bones of the wrist of primitive tetrapods.

central eruption Ejection of debris and lava flows from a central point, forming a more or less symmetrical *volcano.*

centrallasite *gyrolite.*

central lumen The opening in an edrioasteroid theca that extends from the proximal ends of the ambulacral tunnels down through the oral frame into the underlying thecal cavity (Bell, 1976).

central meridian The line of longitude at the center of a map projection; the meridian about which the geometric properties of a map projection are symmetric and which is a straight line on the map. It is used to determine the directions of axes of plane coordinates. See also: *principal meridian.*

central mound *central peak.*

central muscle One of an anteriorly or medially placed pair of muscles in lingulid brachiopods, originating on the pedicle valve, and passing anteriorly and dorsally to the brachial valve (TIP, 1965, pt.H, p.142).

central plate A small thin plate inside the marginal rim of some flattened early echinoderms, such as stylophorans, ctenocystoids, and some eocrinoids (TIP, 1968, pt. S, p. 537).

central projection *gnomonic projection.*

central-ring induction method An inductive electromagnetic method in which the transmitting and receiving coils are concentric.

central tendency Any measure or value representing or indicating the center of an entire statistical distribution; e.g. median, mode, and mean.

central uplift A central high area produced in an impact crater or in an explosion crater by inward and upward movement of material below the crater floor. It is formed at a relatively late stage during the crater-forming event and does not appear due to long-term slow adjustment. Central uplifts are characteristic of many cryptoexplosion structures believed to be produced by meteorite impact. Cf: *central peak.*

central valley *rift valley.*

central vent The opening at the Earth's surface of a volcanic conduit of cylindrical or pipelike form. Cf: *central eruption.*

centric Said of the texture of a rock in which the components are arranged about a center, either radially (e.g. *spherulitic*) or concentrically (e.g. *orbicular*).

centric diatom A diatom having basically radial symmetry; a member of the diatom order Centrales. Cf: *pennate diatom.*

centrifugal drainage pattern *radial drainage pattern.*

centrifugal replacement Mineral replacement in which the host mineral is replaced from its center outward. Cf: *centripetal replacement.*

centrifuge moisture equivalent *moisture equivalent.*

centripetal drainage pattern A drainage pattern in which the streams converge inward toward a central depression; it may be indicative of a volcanic crater or caldera, a structural basin, a breached dome, a sinkhole, or a bolson. Cf: *radial drainage pattern.* See also: *internal drainage.*

centripetal replacement Mineral replacement in which the host mineral is replaced from its periphery inward. Cf: *centrifugal replacement.*

centripetal selection Natural selection resulting in decreasing variation.

centroclinal adj. Said of strata and structures that dip towards a common center. Ant: *quaquaversal.* Cf: *periclinal [geol].*

centroclinal fold *centrocline.*

centrocline An equidimensional basin characteristic of cratonic areas, in which the strata dip towards a central low point. The term is little used in the U.S. Cf: *pericline.* Ant: *quaquaversal.* Syn: *centroclinal fold.*

centrodorsal n. (a) A commonly cirriferous crinoid columnal, or semifused to fused columnals attached to the theca of some articulate crinoids. (b) The *centrale* in many asterozoans.— adj. (a) Central and dorsal. (b) Pertaining to a centrodorsal; e.g. "centrodorsal cavity" consisting of a depression on the ventral surface of a crinoid centrodorsal and containing a chambered organ and accessory structures.

centrogenous skeleton The supporting rods generated at the cell center in acantharian radiolarians.

centronelliform Said of a simple spear-shaped loop of a brachiopod (as in the subfamily Centronellinae), suspended free of the valve floor, and commonly bearing a median vertical plate in addition to the echmidium (TIP, 1965, pt.H, p. 142).

centrosphere *barysphere.*

centrosymmetric Said of a crystal having a center of symmetry.

centrotylote A monaxonic sponge spicule (tylote) with a central swelling.

centrum [paleont] (a) A differentiated central part of a sponge spicule. (b) The substance of a stem plate (columnal or cirral) of a crinoid, including luminal flanges. (c) The cylindrical part of a vertebra that ossifies around the notochord.—Pl: *centra.*

centrum [seis] *focus [seis].*

cenuglomerate (a) A term proposed by Harrington (1946) for a rock resulting from the consolidation of mudflow material. (b) A coarse breccia formed by the accumulation of material resulting from rockfalls, landslides, or mudflows (Dunbar & Rodgers, 1957, p.171).—Etymol: Latin *coenum,* "mud", + *glomerare,* "to wind into a ball".

CEP *circular error probable.*

cephalic Pertaining to the head; esp. directed toward or situated on, in, or near the cephalon, such as the "cephalic shield" of a crustacean exoskeleton covering the head region and formed of fused tergites, or a "cephalic spine" carried by the cephalon of a trilobite.

cephalis The first or apical chamber of a nasselline radiolarian.

cephalon (a) The anterior or head region of an exoskeleton of a trilobite, consisting of several fused segments and bearing the eyes and mouth. (b) The most anterior tagma of a crustacean, bearing eyes, mouth, two pairs of antennae, mandibles, and two pairs of maxillae. See also: *head [paleont].*—Pl: *cephala.*

cephalopod Any marine mollusk belonging to the class Cephalopoda, characterized by a definite head, with the mouth surrounded by part of the foot that is modified into lobelike processes with tentacles or armlike processes with hooklets or suckers or both. The external shell, if present, as in nautiloids, is univalve and resembles a hollow cone, which may be straight, curved, or coiled and is divided into chambers connected by a siphuncle; the shell is internal in present-day cephalopods and their fossil ancestors, such as the belemnites. Nautiloids and ammonoids are extinct cephalopods, generally valuable as index fossils; octopuses, squids, and cuttlefishes are common living cephalopods. Range, Cambrian to present.

cephalothorax The fused *head* and *thorax* of certain arthropods; e.g. the anterior part of the body of a crustacean, composed of united cephalic and thoracic somites and covered by the carapace, or the fore part of the body of a merostome in front of the opisthosoma, or the anterior part of the body of an arachnid, bearing six pairs of appendages. Cf: *prosoma; gnathothorax.*

cerargyrite (a) *chlorargyrite.* (b) A group name for isomorphous isometric silver halides, mainly chlorargyrite, bromargyrite, and embolite.—Also spelled: *kerargyrite.*

ceratite Any ammonoid belonging to the order Ceratitida, characterized by a shell having sutures with serrate lobes and, in some groups, by an ornamented shell. Range, Permian to Triassic.

ceratitic suture A type of suture in ammonoids characterized by small, rounded, unbroken saddles and finely denticulate lobes developed on a major set; specif. a suture in ceratites. Cf: *ammonitic suture; goniatitic suture; pseudoceratitic suture.*

ceratoid Said of a very slenderly conical, horn-shaped corallite of a solitary coral.

ceratolith A horseshoe-shaped skeletal element of the coccolithophorid *Ceratolithus,* acting optically as a single unit of calcite.

cercopod *cercus.*

cercus Either of a pair of simple or segmented appendages situated at the posterior end of certain arthropods; e.g. a *caudal ramus* of a crustacean. Pl: *cerci.* Syn: *cercopod.*

ceresine A white wax that results from the bleaching of *ozocerite.*

cerianite A mineral: CeO_2. It usually contains some thorium.

cerine (a) *allanite.* (b) *cerite.*

cerioid Said of a massive corallum in which the walls of adjacent polygonal corallites are closely united.

cerite A trigonal mineral: $(Ce,Ca)_9(Mg,Fe)Si_7(O,OH,F)_{28}$. Syn: *cerine.*

cerolite A yellow or greenish waxlike mixture of serpentine and stevensite.

cerotungstite A monoclinic mineral: $CeW_2O_6(OH)_3$.

Cerozem *Sierozem.*

cerrito A small *cerro.* Syn: *cerrillo.*

cerro A term used in the SW U.S. for a hill, esp. a craggy or rocky eminence of moderate height. Etymol: Spanish.

ceruleite A turquoise-blue mineral: $Cu_2Al_7(AsO_4)_4(OH)_{13}\cdot 11.5H_2O$. Also spelled: *ceruléite.*

cerulene (a) A trade name for a form of calcite colored blue or green by azurite or malachite and used as a gemstone. (b) A term used less correctly for a blue variety of satin spar.

cerussite A colorless, white, yellowish, or grayish orthorhombic mineral of the aragonite group: $PbCO_3$. It is a common alteration product of galena and is a valuable ore of lead. Syn: *white lead ore; lead spar.*

cervantite A white or yellow orthorhombic mineral: $Sb^{+3}Sb^{+5}O_4$. It was formerly regarded as identical with stibiconite.

cervical groove In decapods, a transverse groove somewhat parallel to the postcervical groove and placed before it. It extends upward from the confluence of the hepatic and antennal grooves (Holthuis, 1974, p. 733).

cervical sinus An indentation at the front of the carapace of a cladoceran crustacean, exposing the rear part of the head.

cesarolite A steel-gray mineral: $H_2PbMn_3O_8$. It occurs in spongy masses.

cesbronite A green orthorhombic mineral: $Cu_5(TeO_3)_2(OH)_6\cdot 2H_2O$.

cesium kupletskite A mineral of the astrophyllite group: $(Cs,K,Na)_3(Mn,Fe)_7(Ti,Nb)_2Si_8O_{24}(O,OH,F)_7$. It forms a series with kupletskite.

cesium-vapor magnetometer A type of *optically pumped magnetometer* that measures the absolute total magnetic intensity with extreme sensitivity by determining the Larmor frequency of cesium atoms. Cf: *rubidium-vapor magnetometer.*

ceylonite A dark green, brown, or black variety of spinel containing iron. Syn: *pleonaste; candite; ceylanite; zeylanite.*

cf. (a) Used in paleontology to indicate that a specimen is very closely comparable to, but not certainly the same as, those of a named species; it implies more certain similarity than does *aff.* (b) Used in this glossary and other reference works to mean "compare".—Etymol: Latin *conferre,* "to compare".

chabazite A zeolite mineral: $CaAl_2Si_4O_{12}\cdot 6H_2O$. It sometimes contains sodium and potassium. Also spelled: *chabasite.*

chadacryst (a) The enclosed crystal in a poikilitic texture. (b) A syn. of *xenocryst.*—Also spelled: *cadacryst.*

Chadronian North American continental stage: Lower Oligocene (above Duchesnean, below Orellan).

chaemolith *humic coal.*

chaetetid Any organism characterized by massive coralla composed of very slender aseptate corallites with imperforate walls and complete tabulae. The chaetetids are currently placed in the tabulate-coral family Chaetetidae, but have been variously classified as hydrozoans, anthozoans, bryozoans, and sponges. Range, Ordovician to Permian.

chaff peat Peat that is derived from fragments of plants.

chagrenate Said of a smooth and translucent sculpture of pollen and spores.

chain [geomorph] A general term for any series or sequence of related natural features arranged more or less longitudinally, such as a chain of lakes, islands, seamounts, or volcanoes; esp. a *mountain chain* or other extended group of more or less parallel features of high relief.

chain [ore dep] adj. In mineral deposits, e.g. chromite, said of a crystal texture or structure in which a series of connected crystals resembles a linked or chainlike pattern.

chain [surv] (a) A measuring device used in land surveying, consisting of 100 links joined together by rings; specif. *Gunter's chain.* The term is commonly used interchangeably with *tape* although strictly a chain is a series of links and a tape is a continuous strip. (b) A unit of length prescribed by law for the survey of U.S. public lands and equal to 66 feet or 4 rods. It is a convenient length for land measurement because 10 square chains equals one acre.

chain coral Any coral (esp. one belonging to the family Halysitidae) characterized, in plan view, by cylindrical, oval, or subpolygonal corallites joined together on two or three sides to form a branching, chainlike network.

chain crater One of several small aligned depressions on the surface of the Moon, Mars, and Mercury, believed to be formed by either volcanic activity or by secondary impacts; more commonly applied to those of probable volcanic origin. See also: *crater chain.*

chain gage A type of *gage* used in determining water-surface elevation, consisting of a tagged or indexed chain, tape, or other form of line. It is used in situations in which the water surface is difficult to reach. Cf: *staff gage.*

chaining A term that was applied originally to measuring distances on the ground by means of a surveyor's chain, but later to the use of either a chain or a surveyor's tape. The term was formerly synonymous with *taping,* but "chaining" is now preferred (for historical and legal reasons) for surveys of the U.S. public-lands system and "taping" for all other surveys.

chainman A surveyor's assistant who measures distances, marks measuring points, and performs related duties; specif. one who marks the tape ends in chaining or who measures distances with a tape. See also: *rodman.* Syn: *tapeman.*

chain silicate *inosilicate.*

chalcanthite A blue triclinic mineral: $CuSO_4 \cdot 5H_2O$. It is a minor ore of copper. Syn: *blue vitriol; copper vitriol; bluestone; cyanosite.*

chalcedonic chert A transparent, translucent, vitreous, milky, smoky, waxy, or greasy variety of *smooth chert,* generally buff or blue-gray, sometimes mottled (Ireland et al., 1947, p. 1484); it breaks into splintery fragments with smooth conchoidal surfaces. In thin section, it is seen to consist of microcrystalline quartz crystals with a fibrous crystal habit. Cf: *novaculitic chert.*

chalcedonite *chalcedony.*

chalcedony (a) A cryptocrystalline variety of quartz. It is commonly microscopically fibrous, may be translucent or semitransparent, and has a nearly waxlike luster, a uniform tint, and a white, pale-blue, gray, brown, or black color; it has a lower density and lower indices of refraction than ordinary quartz. Chalcedony is the material of much chert, and often occurs as an aqueous deposit filling or lining cavities in rocks. In the gem trade, the name refers specif. to the light blue-gray or "common" variety of chalcedony. Varieties include carnelian, sard, chrysoprase, prase, plasma, bloodstone, onyx, and sardonyx. See also: *agate.* Var: *calcedony.* Syn: *chalcedonite.* (b) A general name for crystalline silica that forms concretionary masses with radial-fibrous and concentric structure and that is optically negative (unlike true quartz). (c) A trade name for a natural blue onyx.

chalcedony patch A milklike, semitransparent blemish in a ruby.

chalcedonyx An onyx with alternating stripes or bands of gray and white. It is valued as a semiprecious stone.

chalchihuitl A Mexican term for any green stone that has been carved into a decorative or useful object, and sometimes any stone, regardless of color, that has been carved. It refers esp. to jadeite or chalchuite (turquoise), but sometimes to porphyry, serpentine, or smithsonite. Syn: *chalchihuite; chalchiguite.*

chalchuite A blue or green variety of turquoise.

chalco- A prefix meaning "copper".

chalcoalumite A turquoise-green to pale-blue mineral: $CuAl_4(SO_4)(OH)_{12} \cdot 3H_2O$.

chalcocite A black or dark lead-gray mineral: Cu_2S. It has a metallic luster, occurs in orthorhombic crystals or massive, and is an important ore of copper. Syn: *copper glance; chalcosine; redruthite; beta chalcocite; vitreous copper.*

chalcocyanite A white mineral: $CuSO_4$. Syn: *hydrocyanite.*

chalcodite *stilpnomelane.*

chalcolite *torbernite.*

Chalcolithic *Copper Age.*

chalcomenite A blue mineral: $CuSeO_3 \cdot 2H_2O$.

chalconatronite A mineral: $Na_2Cu(CO_3)_2 \cdot 3H_2O$. It occurs as greenish-blue incrustations on ancient bronze objects from Egypt.

chalcopentlandite A hypothetical high-temperature sulfide of copper, nickel, and iron, now represented by mixtures of pentlandite and chalcopyrite.

chalcophanite A black triclinic mineral: $(Zn,Mn,Fe)Mn_3^{+4}O_7 \cdot 3H_2O$.

chalcophile (a) Said of an element concentrated in the sulfide rather than in the metallic and silicate phases of meteorites, and that is probably concentrated in the Earth's mantle relative to its crust and core (in Goldschmidt's tripartite scheme of element partition in the solid Earth). Cf: *lithophile; siderophile.* (b) Said of an element tending to concentrate in sulfide minerals and ores. Such elements have intermediate electrode potentials and are soluble in iron monosulfide.—(Goldschmidt, 1954, p. 24; Krauskopf, 1967, p. 580). Examples are: S, Se, As, Fe, Pb, Zn, Cd, Cu, Ag.

chalcophyllite A green mineral: $Cu_{18}Al_2(AsO_4)_3(SO_4)_3(OH)_{27} \cdot 33H_2O$. Syn: *copper mica.*

chalcopyrite A bright brass-yellow tetragonal mineral: $CuFeS_2$. It is generally found massive and constitutes the most important ore of copper. Syn: *copper pyrites; yellow copper ore; yellow pyrites; fool's gold.*

chalcosiderite A green mineral: $Cu(Fe,Al)_6(PO_4)_4(OH)_8 \cdot 4H_2O$. It is isomorphous with turquoise.

chalcosine *chalcocite.*

chalcosphere That zone or layer of the Earth containing heavy-metal oxides and sulfides; it is the equivalent of *stereosphere.*

chalcostibite A lead-gray mineral: $CuSbS_2$. Syn: *wolfsbergite.*

chalcothallite A mineral: Cu_3TlS_2.

chalcotrichite A capillary variety of cuprite occurring in fine slender interlacing fibrous crystals. Syn: *plush copper ore; hair copper.*

chalite A term, now obsolete, proposed by Pinkerton (1811, v.2, p.100) for a conglomerate containing pebbles intermediate between flint and chalcedony.

Chalk A stratigraphic term used in NW Europe for Upper Cretaceous. In Great Britain, it is divided into Lower Chalk (Cenomanian), Middle Chalk (Turonian), and Upper Chalk (Senonian).

chalk (a) A soft, pure, earthy, fine-textured, usually white to light gray or buff limestone of marine origin, consisting almost wholly (90-99%) of calcite, formed mainly by shallow-water accumulation of calcareous tests of floating microorganisms (chiefly foraminifers) and of comminuted remains of calcareous algae (such as coccoliths and rhabdoliths), set in a structureless matrix of very finely crystalline calcite. The rock is porous, somewhat friable, and only slightly coherent. It may include the remains of bottom-dwelling forms (e.g. ammonites, echinoderms, and pelecypods), and nodules of chert and pyrite. The best known and most widespread chalks are of Cretaceous age, such as those exposed in cliffs on both sides of the English Channel. Syn: *creta.* (b) A white, pure (or nearly pure), natural calcium carbonate, breaking into crumbly or powdery pieces. (c) *chalk rock.*—Etymol: Old English *cealc,* from Latin *calx,* "lime".

chalkland A region underlain by chalk deposits, characterized by rolling hills, undulating plateaus, open expanses of pastureland, and dry valleys; e.g. SE England.

chalk rock (a) A soft, milky-colored rock resembling white chalk, such as talc, calcareous tufa, diatomaceous shale, volcanic tuff, or a bed of white limestone. (b) A chalky rock; specif. the Chalk Rock, a bed of hard nodular chalk, in places containing green-colored calcareous or phosphatic nodules, occurring at or near the base of the Upper Chalk in England (Himus, 1954, p. 24).

chalk stream A stream flowing across or among the strata of a chalk deposit; a *bourne.*

chalky (a) Said of a soil or rock consisting of, rich in, or characterized by chalk. Syn: *cretaceous.* (b) Said of a limestone having the appearance of chalk. (c) Said of the porosity of such finely textured rocks as chalk and marl.

chalky chert A commonly dull or earthy, soft to hard, sometimes finely porous chert of essentially uniform composition, having an uneven or rough fracture surface, and resembling chalk (Ireland et al., 1947, p. 1487). It is common in insoluble residues. Cf: *smooth chert; granular chert.* Syn: *dead chert; cotton chert.*

chalky marl A grayish marly rock rich in chalk and containing up to 30% clayey material; specif. the Chalk Marl near the base of the English Chalk.

challantite A mineral: $6Fe_2(SO_4)_3 \cdot Fe_2O_3 \cdot 63H_2O$.

chalmersite *cubanite.*

chalybeate An adj. applied to water strongly flavored with iron salts or to a spring yielding such water. Etymol: Greek, an ancient tribe of ironworkers in Asia Minor.

chalybite British syn. of *siderite* (ferrous-carbonate mineral).

chamber [paleont] (a) The fundamental unit of a foraminiferal test, consisting of a cavity and its surrounding walls. It is a variously shaped enclosure that invariably is connected by pores, intercameral foramina, or other passages leading to similar enclosures or to the exterior. (b) One of the regular, juxtaposed, hollow structures formed by the skeleton of sphinctozoan sponges. Also, a term that is often used as an abbreviated form of *flagellated chamber;* this usage is not recommended. (c) A *camera* of a cephalopod. Also, the *body chamber* of a cephalopod.

chamber [speleo] A *room* of a cave.

chambered Said of a vein or lode of brecciated, irregular texture, e.g. a *stockwork.*

chambered level A spirit-level tube with a partition near one end which cuts off a small air reservoir, so arranged that the length of the bubble can be regulated.

chamberlet A small chamber in a foraminifer, created by subdivision of chambers by axial or transverse secondary septula.

chamber passage One of the radial corridors consisting of centrally directed extensions of marginal chamberlets of foraminifers (such as Orbitolinidae).

chambersite An orthorhombic mineral: $Mn_3B_7O_{13}Cl$.

chameolith *humic coal.*

chamosite A greenish-gray or black mineral of the chlorite group: $(Mg,Fe)_3Fe_3^{+3}(AlSi_3)O_{10}(OH)_8$. It is the monoclinic dimorph of orthochamosite, and is an important constituent of many oolitic and other bedded iron ores. See also: *berthierine.* Syn: *daphnite.*

Champlainian (a) North American provincial series: Middle Ordovician (above Canadian, below Cincinnatian). (b) Obsolete syn of *Ordovician.*

chance packing A random combination of systematically packed grains surrounded by or alternating with haphazardly packed grains (Graton & Fraser, 1935). The average porosity of a chance-packed aggregate of uniform solid spheres is slightly less than 40%.

Chandler motion *polar wandering.*

Chandler wobble An aspect of the Earth's rigid body motion departing from simple or pure spin, because its angular-momentum vector is not precisely colinear with a principal axis of inertia. The free *nutation,* a 428-day cycle, results in a variation of instantaneous astronomical latitude (as defined in accordance with an instantaneous angular-velocity vector or axis of rotation) of amplitude about 0.4 sec-arc.

change of color (a) An optical phenomenon consisting of a difference in color when a mineral or gemstone is moved about; specif. *labradorescence.* (b) An optical phenomenon consisting of a difference in color from daylight to artificial light, caused by selective absorption; e.g. that shown by alexandrite.—Cf: *play of color.*

channel [coast] (a) A relatively narrow sea or stretch of water, wider and larger than a *strait,* between two close landmasses and connecting two larger bodies of water (usually seas); e.g. the English Channel between England and France. (b) The deeper part of a moving body of water (as a bay, estuary, or strait) through which the main current flows or which affords the best passage through an area otherwise too shallow for navigation; it is often deepened by dredging. Also, a navigable waterway between islands or other obstructions, as on a lake.

channel [drill] A cavity or passage in a faulty *cementing* job behind the casing in a borehole.

channel [geophys] A system of interconnected devices through which data may flow from source to recorder; for example, geophone, cable, amplifier, recorder.

channel [ice] *lead [ice].*

channel [ore dep] *channelway [ore dep].*

channel [paleont] (a) A groove of an invertebrate, such as one that winds down the columella near its base in some gastropod shells and terminates in the siphonal notch or in the canal. (b) An area lying between septal pinnacles and the peripheral ends of septa in the zone between parent corallite and offset during the early stages of increase in corals (Fedorowski & Jull, 1976, p. 41).

channel [sed struc] (a) A linear current mark, larger than a *groove,* produced on a sedimentary surface, parallel to the current, and often preserved as a *channel cast*. It is 0.5-2 m wide, 20-50 cm deep, and up to 30 m long, and is best developed in a *turbidite* sequence. (b) An erosional feature "that may be meandering and branching and is part of an integrated transport system" (Pettijohn & Potter, 1964, p.288).

channel [streams] (a) The bed where a natural body of surface water flows or may flow; a natural passageway or depression of perceptible extent containing continuously or periodically flowing water, or forming a connecting link between two bodies of water; a watercourse. Syn: *channelway.* (b) The deepest or central part of the bed of a stream, containing the main current, and occupied more or less continuously by water; the *thalweg.* (c) A term used in quantitative geomorphology for a line or pattern of lines, without regard to width or depth, in the analysis of streams. Syn: *stream.* (d) An abandoned or buried watercourse represented by stream deposits of gravel and sand. (e) An artificial waterway, such as an open conduit, an irrigation ditch or canal, or a floodway. (f) An obsolete term for a stream or small river.

channel [volc] A narrow, sinuous flow channel, commonly formed in lava flows.

channel bar An elongate deposit of sand and gravel located in the course of a stream, esp. of a braided stream. Cf: *point bar.*

channel basin An obsolete term for a long, narrow proglacial valley, trench, or channel.

channel capacity The maximum flow that a given channel can transmit without overflowing its banks. See also: *bankfull stage.*

channel cast The cast of a *channel* that is generally cut in shale and filled with sand. Cf: *washout.* Syn: *channel fill; gouge channel.*

channeled scabland *Scabland* deeply eroded. On the Columbia Plateau of eastern Washington it represents intense scouring by glacial meltwater.

channeled upland *grooved upland.*

channel erosion Erosion in which material is removed by water flowing in well-defined courses; erosion caused by channel flow. Cf: *sheet erosion; rill erosion; gully erosion.*

channel fill (a) An alluvial deposit in a stream channel, esp. one in an abandoned cutoff channel or where the transporting capacity of the stream is insufficient to remove material supplied to it. (b) *channel cast.*—Syn: *channel filling.*

channel flow Movement of surface runoff in a long narrow troughlike depression bounded by banks or valley walls that slope toward the channel; specif. *streamflow.* Cf: *overland flow.* Syn: *concentrated flow.*

channel frequency *stream frequency.*

channel geometry The description of the shape (form) of a given cross section within a limited reach of a river channel. See also: *river morphology.*

channel-gradient ratio *stream-gradient ratio.*

channelization The straightening and deepening of a stream channel, to permit the water to move faster or to drain marshy acreage for farming.

channel length *stream length.*

channel line The line of the fastest current or the strongest flow of a stream; it generally coincides with (and is sometimes known as) the *thalweg.* Cf: *thread.*

channel maintenance constant Ratio of the area of a drainage basin to the total stream lengths of all the stream orders within the basin (Schumm, 1956, p. 607); approximately the reciprocal of the *drainage density* (Shreve, 1969, p. 412). It expresses the minimum limiting area required for the development of a drainage channel. Symbol: C.

channel morphology *river morphology.*

channel-mouth bar A bar built where a stream enters a body of standing water; it results from a decrease in the stream's velocity.

channel net The pattern of all stream channels within a drainage basin. Syn: *channel network.*

channel network *channel net.*

channel order *stream order.*

channel pattern The configuration in plan view of a limited reach of a river channel as seen from an airplane (Leopold & Wolman, 1957, p. 39-40). Recognized patterns include meandering, braided, sinuous, and relatively straight. See also: *river morphology.* Syn: *river pattern.*

channel porosity A system of pores in which the openings are markedly elongate and have developed independently of the textural or fabric elements of the rock (Choquette & Pray, 1970, p. 245).

channel precipitation Part of direct runoff; precipitation that falls directly onto lake and stream waters. It is usually considered with surface runoff (Chow, 1964, p. 14-2).

channel sample A composite rock sample, generally taken across the face of a formation or vein to give an average value.

channel sand A sand or sandstone deposited in a stream bed or other channel eroded into the underlying rocks. If exposed, such sands may contain gold or other valuable minerals; if buried, they may contain oil or gas. See also: *shoestring sand.*

channel segment *stream segment.*

channel splay *flood-plain splay.*

channel spring A type of *depression spring* issuing from the bank of a stream that has cut its channel below the water table.

channel storage In a stream channel, or over its flood plains, the volume of water at a given moment.

channel wave A type of *guided wave* that is propagated in a low-velocity layer within the Earth, or in the ocean or atmosphere. Cf: *Stoneley wave; sofar.*

channelway [ore dep] An opening or passage in a rock through which mineral-bearing solutions or gases may move. Syn: *channel [ore dep]; feeder [eco geol]; feeding channel.*

channelway [streams] *channel.*

channel width The distance across a channel or a stream, measured from bank to bank near bankfull stage. Symbol: w.

channery (a) Thin, flat *coarse fragments* of limestone, sandstone, or schist, having diameters as large as 150 mm (6 in.). (b) A term used in Scotland and Ireland for gravel.

chaoite A mineral: a hexagonal form of elemental carbon. Polymorphous with diamond, graphite, and lonsdaleite.

chaos [geol] A structural term proposed by Noble (1941, p. 963-977) for a gigantic breccia associated with thrusting, consisting of a mass of large and small blocks of irregular shape with very little fine-grained material, in a state of semidisorder. Type example:

the Amargosa chaos, a widespread deposit in the Death Valley area of California, consisting of an extremely complex mosaic of enormous, tightly packed, often unshattered but internally coherent, random blocks and masses of formations of different ages occupying a definite zone above a major thrust fault. The blocks range in size from pods a meter in diameter to blocks more than 800 m in length. Cf: *mélange [sed]; megabreccia.*

chaos [planet] The disorganized state of primordial matter and infinite space before the ordered universe was created. Ant: *cosmos.*

chaotic [geomorph] Said of a surface or land area consisting of short, jumbled ridges and valleys.

chaotic [petrology] Said of a massive, unstratified tuff consisting of a mixture of equally distributed fine and coarse material (such as the deposit of a *nuée ardente.*)

chaotic terrain Regions on Mars, first seen in 1969 on Mariner 6 images, that are topographically low and consist of irregular ridges, apparently formed at the expense of higher cratered terrain (Leighton et al., 1969). It has been interpreted as a feature of *thermokarst topography.*

chapeau de fer *gossan.*

chapeiro An isolated coral reef growing in small scattered patches, often rising like a tower to a height of 12-15 m, and sometimes spreading out in a mushroomlike top, as off the coast of Brazil. Etymol: Portuguese *chapeirão, "broad-brimmed hat".*

chapmanite A mineral: $Fe_2Sb(SiO_4)_2(OH)$.

char [coal] The solid carbonaceous residue that results from incomplete combustion of organic material. It can be burned for heat, or, if pure, processed for production of activated carbon for use as a filtering medium. See also: *charcoal; coke.*

char [streams] A term applied in India to a newly formed alluvial tract or flood-plain island formed of silt and sand deposited in the bed of a deltaic river, such as a sandbank left dry on the subsidence of a river after the flood season. Etymol: Hindu. Syn: *chur; diara.*

character [paleont] Any specifiable, definable, or recognizable attribute (often a morphologic feature) of an organism or taxon. A character may appear in several possible expressions or aspects (each termed a *character state*) in different organisms or taxa.

character [seis] A recognizable aspect of a seismic event or waveform that distinguishes it from others. It is usually a frequency or phasing effect, and is often not defined precisely and hence is dependent on subjective judgment.

characteristic fossil A fossil species or genus that is characteristic of a stratigraphic unit (formation, zone series, etc.) or time unit. It is either confined to the unit or is particularly abundant in it. Inappropriate syn: *index fossil.* Syn: *diagnostic fossil.*

characterizing accessory mineral *varietal mineral.*

character state A particular aspect or condition appearing in one organism or taxon, and recognizably different from its related expression in other organisms or taxa. See also: *character.*

charco (a) A term applied in SW U.S. to a small natural depression in which water collects, as in a desert alluvial plain; a *tank* or a water hole. Syn: *represo.* (b) A natural or artificial pool of water occupying a charco and supplied by desert floods. Also, a pool in a stream bed or a puddle in a playa.—Etymol: Spanish, "pond, small lake".

charcoal An impure carbon residue of the burning of wood or other organic material in the absence of air. It is black, often porous, and able to absorb gases. Like *coke,* it can be used as a fuel. See also: *char [coal].*

charge (a) The sediment that is carried into a channel, expressed as the ratio of the volume of sediment passing across a given cross section or portion of cross section of the channel in unit time to the portion of cross section of the channel in unit time. (b) In seismic work, the explosive combination employed for generating seismic energy, specified by the quantity and type of explosive used.

chargeability The primary unit of measurement in time-domain induced-polarization surveys. It is the area under the decay curve between two delay times after cessation of the transmitted current. Usually expressed in millivolt-seconds per volt.

Charmouthian Stage in Great Britain: Lower Jurassic (above Sinemurian, below Domerian).

Charnian orogeny An orogeny that supposedly occurred late in Precambrian time in the English Midlands; the dating is questionable, however, and the term has only local significance. The Charnian folds trend NW-SE and seem to have had a posthumous influence on structures in the surrounding Paleozoic. It is named for Charnwood Forest, Leicestershire, where small inliers of Precambrian sediments emerge.

charnockite An orthopyroxene-bearing *granite.* Most classifications require that quartz constitute at least 20% of the felsic constituents and that the ratio of alkali feldspar to total feldspar fall between 40% and 90%. Tobi (1971) places the quartz content at 10-60% and the ratio of alkali feldspar to total feldspar at 35-90% to correspond with Streckeisen's (1967) definition of granite. Although its origin (igneous or metamorphic) is controversial, charnockite is commonly found only in granulite-facies terranes, and high temperature and pressure are generally thought to be essential to its formation. The name is derived from that of Job Charnock (d.1693), the founder of Calcutta, India, from whose tombstone the rock was first described by Holland in 1893.

charnockite series A series of plutonic rocks compositionally similar to the granitic rock series but characterized by the presence of orthopyroxene.

charophyte One of a group of green algae corresponding to the order Charales and comprising the *stoneworts.*

chart (a) A special-purpose map; esp. one designed for purposes of navigation, such as a *hydrographic chart* or a *bathymetric chart.* (b) A base map conveying information about something other than the purely geographic. (c) *weather map.* (d) Obsolete syn. of *map.*

chart datum A standard water surface, usually low water, from which depths of soundings or tide heights are measured. When based on the tide, it may be called a *tidal datum.*

chartology *cartography.*

chartometer An instrument for measuring distances on charts or maps, such as the length of a stream in a drainage basin on a topographic map. See also: *opisometer.* Syn: *map measurer.*

chasm (a) A deep breach, cleft, or opening in the Earth's surface, such as a yawning fissure or narrow gorge; e.g. the Ausable Chasm near Keeseville, N.Y. (b) A deep recess extending below the floor of a cave.—Syn: *abyss.*

chasma A term established by the International Astronomical Union for a large canyon on Mars. Most are thought to be of structural origin. Generally used as part of a formal name for a Martian landform, such as Coprates Chasma (Mutch et al., 1976, p. 57). Etymol: Greek *chasma,* gulf, open mouth.

chasmophyte A plant growing in the crevices of a rock; a *saxifragous* plant.

chassignite An achondritic stony meteorite composed almost entirely (95%) of olivine, with accessory amounts of chromite, and lacking nickel-iron. It resembles terrestrial dunite.

chathamite A variety of nickel-skutterudite containing much iron.

chatoyancy An optical phenomenon, possessed by certain minerals in reflected light, in which a movable wavy or silky sheen is concentrated in a narrow band of light that changes its position as the mineral is turned. It results from the reflection of light from minute, parallel fibers, cavities or tubes, or needlelike inclusions within the mineral. The effect may be seen on a cabochon-cut gemstone, either distinct and well-defined (as the narrow, light-colored streak in a fine chrysoberyl cat's-eye) or less distinct (as in the usual tourmaline or beryl cat's-eye). Syn: *chatoyance.*

chatoyant adj. Said of a mineral or gemstone possessing *chatoyancy* or having a changeable luster or color marked by a narrow band of light.—n. A chatoyant gem.

chattermark [beach] A crescent-shaped mark on a wave-worn pebble, such as flint, caused by "hammering" of a beach by wave action.

chattermark [fault] Any mark, pit or scratch made on a rock surface by the surface of a mass that moves over it. Chattermarks can be caused by the material embedded in the bottom of a glacier, or they can occur on a fault surface. Cf: *slip-scratch; vibration mark.*

chattermark [glac geol] One of a series of small, closely spaced, short curved scars or cracks (smaller than a crescentic fracture) made by vibratory chipping of a firm but brittle bedrock surface by rock fragments carried in the base of a glacier. Each mark is roughly transverse to the direction of ice movement (although a succession of such marks is parallel to that direction), and usually convex toward the direction from which the ice moved (its "horns" point in the direction of ice movement). The term has been applied loosely to any glacial crescentic mark. Also spelled: *chatter mark.*

Chattian European stage: uppermost Oligocene (above Rupelian, below Aquitanian of Miocene).

Chautauquan North American provincial series: Upper Devonian (above Senecan, below Bradfordian).

Chayes point counter An instrument used for petrographic modal analysis. A pattern of regularly spaced traverses, along which are regularly spaced points, is placed over a thin section; at each point the mineral is identified and then mechanically tabulated.

Chazyan North American substage: Middle Ordovician (below Mohawkian, above Canadian).

check dam A dam designed to retard the flow of water in a channel, used esp. for controlling soil erosion.

checkerboard topography A landscape characterized by a repeating pattern in the relief, such as the "diaper pattern of rectangles" due to a fracture system as seen on the Elizabethtown topographic quadrangle, Adirondack Mountains, N.Y. (Hobbs, 1901, p. 150; 1911b, p. 131 & plate 9).

checker coal Rectangular grains of anthracite.

check shot A shot into a well seismometer to check the results of integrating a continuous velocity or sonic log. See also: *well shooting*.

cheek One of the two lateral or pleural parts of the cephalon of a trilobite, anterior to and typically much lower and flatter than the glabella. See also: *fixed cheek; free cheek*. Syn: *gena*.

cheesewring (a) A mushroom-shaped rock with a narrow stem and overhanging upper block; e.g. the Cheesewring on Bodmin Moor in Cornwall, England, a granite *tor* resembling an inverted muslin bag in which "curds from sour milk were once put by country people and the moisture wrung out so as to leave a white cream cheese" (Stamp, 1961, p. 102). (b) *gara*. See also: *mushroom rock*.

cheilostome Any ectoproct bryozoan belonging to the order Cheilostomata and characterized by the presence of a movable operculum over the orifice of the zooecia. Adj: *cheilostomatous*.

cheirographic coast "A coast of folded and faulted regions with complex submergences. It shows a series of deep gulfs and finger-like promontories" (Swayne, 1956, p. 33). Stamp (1961, p. 102) used "cheiragratic coast" and noted that the term is "apparently obsolete".

chela (a) The pincer-like claw or organ borne by certain of the limbs of arthropods; e.g. the distal part of a crustacean limb consisting of opposed movable and immovable fingers and usually involving dactylus and propodus, or the pincer of an arachnid appendage formed by a rigid process of the penultimate joint and a movable last joint. (b) A siliceous, monaxonic sponge spicule (microsclere) consisting of an arcuate shaft at each end of which is a recurved, cup-like expansion, either lobed or toothed. See also: *isochela; anisochela*.—Pl: *chelae*.

chelation Retention of a metallic ion by two atoms of a single organic molecule. An example is magnesium being retained by heme in hemoglobin.

chelicera One of the pre-oral appendages of all Chelicerata (subphylum of Arthropoda), corresponding to the second antennae of crustaceans, but modified for piercing or biting, and composed of two or three segments (as in arachnids) or of three or four(?) joints with the distal ones forming a chela (as in merostomes). Pl: *chelicerae*. Cf: *antenna*.

chelicerate Any terrestrial (*arachnid*) or aquatic (*merostome*) arthropod belonging to the subphylum Chelicerata, characterized chiefly by paired preoral appendages. Range, Cambrian to present. Cf: *pycnogonid*.

cheliped Any thoracopod bearing chelae; e.g. one of the pair of legs that bears the large chelae in decapod crustaceans.

chelkarite A mineral: $CaMgB_2O_4Cl_2 \cdot 7H_2O$ (?).

chelogenic Said of a cycle of continental evolution; shield-forming.

Chelonia An order of the reptilian subclass Anapsida, characterized by a carapace formed from modified ribs and dermal plates; it includes the turtles and tortoises. Range, Upper Triassic to Recent.

cheluviation *Eluviation* under the influence of chelating agents.

chemawinite A pale-yellow to dark-brown variety of retinite found in decayed wood at Cedar Lake in Manitoba. Syn: *cedarite*.

chemical activity *activity [chem]*.

chemical composition [mineral] *structural formula*.

chemical composition [petrology] The weight percent of the elements (generally expressed as certain oxide molecules) composing a rock. Syn: *composition [petrology]*.

chemical demagnetization A technique of partial *demagnetization* involving treatment by acid or other reagents to selectively remove one magnetically ordered mineral while leaving the remanent magnetization of another unaffected. Cf: *alternating-field demagnetization; thermal demagnetization*.

chemical equilibrium A state of balance between two opposing chemical reactions. The amount of any substance being built up is exactly counterbalanced by the amount being used up in the other reaction, so that concentrations of all participating substances remain constant.

chemical erosion *corrosion*.

chemical exfoliation A type of *exfoliation* caused by a volume increase induced by changes in the bulk chemical composition of the rock.

chemical fossil A chemical trace of an organism that has been mostly destroyed by diagenetic processes. Cf: *molecular paleontology*.

chemical gaging A type of *stream gaging* in which velocity of flow is measured by introducing a chemical of known saturation into the stream and then measuring the amount of dilution.

chemical limestone A limestone formed by direct chemical precipitation or by consolidation of calcareous ooze.

chemical magnetization *chemical remanent magnetization*.

chemical mining The extraction of valuable constituents of an orebody in place, by chemical methods such as leaching or dissolution. See also: *solution mining; in-situ mining*.

chemical oceanography The study of the chemistry and chemical changes of ocean water, its dissolved and suspended material, and the geographic and temporal variation of its chemical features.

chemical oxygen demand The amount of oxygen required for the oxidation of the organic matter in a water sample or a water body. Abbrev: COD. Cf: *biochemical oxygen demand*. Syn: *oxygen demand*.

chemical potential (a) An intensive quantity of a component that is defined as being equal to the change of the Gibbs free energy of the system, with the change in the number of moles *mi*, the temperature, the pressure, and the number of moles of the other components being kept constant. It is defined at each point of the system. (b) Partial molal free energy, usually symbolized by μ; the increase in energy of a system (due to an infinitesimal addition of an element or compound, without affecting thermal and mechanical energies) divided by the amount of the element or compound added.

chemical remanence *chemical remanent magnetization*.

chemical remanent magnetization A stable remanent magnetization caused by the slow growth of magnetically ordered mineral grains in the presence of a magnetic field, e.g. during such processes as oxidation, reduction, exsolution. Syn: *chemical remanence; chemical magnetization; crystallization remanent magnetization; crystallization magnetization*. Abbrev: CRM.

chemical residue A *residue* formed by chemical weathering in place; e.g. a deposit of sand resulting from the removal by solution of nitrates from a Chilean niter bed.

chemical rock (a) A sedimentary rock composed primarily of material formed directly by precipitation from solution or colloidal suspension (as by evaporation) or by the deposition of insoluble precipitates (as by mixing solutions of two soluble salts); e.g. gypsum, rock salt, chert, or tufa. It generally has a crystalline texture. (b) A sedimentary rock having less than 50% detrital material (Krynine, 1948, p. 134). Cf: *detrital rock*.

chemical unconformity An unconformity or stratigraphic boundary determined by chemical analysis, such as in the case of a limestone formation whose basal part has a higher concentration of impurities (silica, magnesia, sulfur) due to the presence of in-washed fine clastic and organic material (Landes, 1957).

chemical water *water of hydration*.

chemical weathering The process of weathering by which chemical reactions (hydrolysis, hydration, oxidation, carbonation, ion exchange, and solution) transform rocks and minerals into new chemical combinations that are stable under conditions prevailing at or near the Earth's surface; e.g. the alteration of orthoclase to kaolinite. Cf: *mechanical weathering*. Syn: *decomposition; decay [weath]*.

chemoautotrophic Said of an organism that obtains nourishment from chemical reactions of inorganic substances. Syn: *chemotrophic*.

chemocline The boundary between the circulating and the noncirculating water masses or layers of a lake; specif. the boundary separating the *mixolimnion* and the *monimolimnion* in a meromictic lake. Also, a zone of rapid change, with depth, in the chemical constituents of a lake.

chemofacies A term used by Keith & Degens (1959, p.40) to designate "all the chemical elements that are collected, precipitated, or adsorbed from the aqueous environment or fixed by chemical reactions within the bottom muds" and intended "only as a convenience in discussing chemical differences among environmental groups of sediments" (as in differentiating between marine and freshwater sediments).

chemogenic Said of a rock or mineral that was deposited directly from solution without biological mediation, e.g. travertine, in contrast to clastic, bioclastic, or organogenic limestones.

chemography The graphical representation of the compositions of minerals in terms of their components. Minerals in a binary system are depicted as points along a straight line, in a ternary system as points on or within a triangle, etc.

chemolithotrophic Said of an organism that obtains its nourishment by oxidation of inorganic compounds. Cf: *chemoorganotrophic.*

chemoorganotrophic Said of an organism that obtains its nourishment by the oxidation of organic compounds. Cf: *chemolithotrophic.*

chemotaxis *Taxis [ecol]* resulting from chemical stimuli. Cf: *chemotropism.*

chemotrophic *chemoautotrophic.*

chemotropism *Tropism* resulting from chemical stimuli. Cf: *chemotaxis.*

Chemungian North American stage: Upper Devonian (above Fingerlakesian, below Cassadagan).

chenevixite A dark-green to greenish-yellow mineral: $Cu_2Fe_2(AsO_4)_2(OH)_4 \cdot H_2O$ (?).

chengbolite *moncheite.*

chenier A long narrow wooded *beach ridge* or sandy hummock, 3 to 6 m high, forming roughly parallel to a prograding shoreline seaward of marsh and mud-flat deposits (as along the coast of SW Louisiana), enclosed on the seaward side by fine-grained sediments, and resting on foreshore or mudflat deposits. It is well drained and fertile, often supporting large evergreen oaks or pines on higher areas; its width ranges from 45 to 450 m and its length may be several tens of kilometers. Etymol: French *chêne,* "oak". Obsolete syn: *cheniere.*

chenier plain A strand plain, occupied by *cheniers* and intervening mud flats with marsh and swamp vegetation (Otvos and Price, 1979). Bight-coast chenier plains exist in SW Louisiana and in Guiana with a maximum length of 700 km. Chenier plains develop when (1) substantial quantities of river-supplied mud become available for nearshore marine transport and coastal mudflat deposition; (2) a balance exists between longshore sand transport, deposition, and erosional sand-winnowing; (3) these two conditions alternate.

cheralite A green monoclinic mineral: $(Ca,Ce,Th)(P,Si)O_4$. It is isostructural with monazite, and is essentially an intermediate member of a solid-solution series apparently extending between $CePO_4$ (monazite) and $CaTh(PO_4)_2$ (an artificial compound).

chernovite A mineral: $YAsO_4$.

Chernozem A great soil group of the 1938 classification system, a group of zonal soils whose surface horizon is dark and highly organic, below which is a lighter-colored horizon and an accumulation of lime. It is developed under conditions of temperate to cool subhumid climate (USDA, 1938). Most Chernozems are now classified as *Borolls* and *Ustolls.* Etymol: Russian *tschernosem,* "black earth". Also spelled: *Chernozyom; Tchornozem; Tschernosiom; Tschernosem.* Cf: *Chestnut soil.* Partial syn: *black earth.*

Chernozyom *Chernozem.*

chernykhite A mineral of the mica group: $(Ba,Na)(V^{+3},Al)_2(Si,Al)_4O_{10}(OH)_2$.

cherokite The dense brown residual sand constituting the cement of the chert breccias in the zinc-mining district of Joplin, Missouri.

cherry coal A soft, black, noncaking bituminous coal with a resinous luster that ignites and burns readily.

chert A hard, extremely dense or compact, dull to semivitreous, microcrystalline or cryptocrystalline sedimentary rock, consisting dominantly of interlocking crystals of quartz less than about 30 mμm in diameter; it may contain amorphous silica (opal). It sometimes contains impurities such as calcite, iron oxide, and the remains of siliceous and other organisms. It has a tough, splintery to conchoidal fracture, and may be white or variously colored gray, green, blue, pink, red, yellow, brown, and black. Chert occurs principally as nodular or concretionary segregations (*chert nodules*) in limestones and dolomites, and less commonly as areally extensive layered deposits (*bedded chert*); it may be an original organic or inorganic precipitate or a replacement product. The term *flint* is essentially synonymous, although it has been used for the dark variety of chert (Tarr, 1938). Cf: *jasper.* Syn: *hornstone; white chert; silexite.*

chert-arenite (a) A term used by McBride (1963, p.668) for a *quartzarenite* containing more than 25% chert. (b) A term used by Folk (1968, p.124) for a *litharenite* in which the main rock fragment is chert.

chertification A type of silicification in which fine-grained quartz or chalcedony is introduced into limestones, as in the Tri-State mining district of the Mississippi Valley (Fowler & Lyden, 1932).

chert nodule A dense, irregular, usually structureless, sometimes fossiliferous diagenetic segregation of *chert,* ranging from regular disks up to 5 cm in diameter to large, highly irregular, tuberous bodies up to 30 cm in length, frequently occurring distributed through calcareous strata. The larger nodules, of rounded contour, are marked by warty or knobby extensions. Examples include the cherts in the Mississippian limestones of the upper Mississippi Valley region, and the flint nodules of the Cretaceous chalk of England and France. See also: *nodular chert.*

cherty Containing chert; e.g. a "cherty limestone" so siliceous as to be worthless for the limekiln, or a "cherty iron carbonate" consisting of siderite intimately interbedded with chert.

chervetite A monoclinic mineral: $Pb_2V_2O_7$.

chessman spicule *discorhabd.*

chessylite A term commonly used in France and elsewhere for *azurite.* Syn: *chessy copper.*

Chesterian North American series: uppermost Mississippian (above Meramecian, below Morrowan of Pennsylvanian).

chesterlite Microcline feldspar from Chester County, Penna.

Chestnut soil A great soil group of the 1938 classification system, a group of zonal soils having a dark brown surface horizon, below which is a lighter-colored horizon and an accumulation of lime. It is developed under conditions of temperate to cool subhumid to semiarid climate, i.e. under slightly more arid conditions than that of *Chernozem.* Its characteristic vegetation is mixed tall and short grasses (USDA, 1938). Most Chestnut soils are now classified as *Xerolls, Ustolls,* and *Borolls.* Cf: *Reddish Chestnut soil.*

chevee A flat gem with a smooth, concave depression. Cf: *cuvette [gem].*

chevkinite A monoclinic mineral: $(Ca,Ce,Th)_4(Fe,Mg)_2(Ti,Fe)_3Si_4O_{22}$. It is dimorphous with perrierite.

chevron cast The cast of a chevron mark.

chevron cross-bedding Cross-bedding that dips in different or opposite directions in alternating or superimposed beds, forming a chevron or herringbone pattern. Syn: *herringbone cross-bedding; zigzag cross-bedding.*

chevron dune A V-shaped dune formed in a vegetated area where strong winds blow in a constant direction.

chevron fold A *kink fold,* the limbs of which are of equal length. Cf: *zigzag fold.*

chevron groove A V-shaped furrow on the cardinal area for the insertion of ligament in certain bivalve mollusks (as in some of the superfamily Arcacea and in early forms of the superfamilies Pteriacea and Pectinacea). Cf: *duplivincular.*

chevron halite Halite with enclosed impurities or bubbles arranged in a chevron (or rarely triradiate) pattern, believed to have formed as floating skeletal *hopper crystals* (Dellwig, 1955).

chevron mark A *tool mark* consisting of chevron-like depressions arranged in a row, the points of the chevrons generally but not always pointing downstream. Originally proposed by Dunbar & Rodgers (1957, p. 195). Cf: *reversed chevron mark; vibration mark.* Syn: *herringbone mark.*

Chézy equation An equation used to compute the velocity of uniform flow in an open channel: mean velocity of flow (V) equals the Chézy coefficient (C) times the square root of the product of hydraulic radius in feet (R) times the slope of the channel (S). Cf: *Manning equation.* See also: *Kutter's formula.*

chiastoclone A desma (of a sponge) in which several subequal, zygome-bearing arms radiate from a very short central shaft, giving the spicule an X-shaped profile.

chiastolite An opaque variety of andalusite containing black carbonaceous impurities arranged in a regular manner so that a section normal to the longer axis of the crystal shows a black Maltese cross formed as a result of the pushing aside of the impurities into definite areas as the crystal grew in metamorphosed shales. It has long been used for amulets, charms, and other inexpensive novelty jewelry. Syn: *cross-stone; crucite; macle [mineral].*

chiastolite slate A rock formed by contact metamorphism of carbonaceous shale, characterized by prominent cleavage or schistosity and the presence of conspicuous chiastolite crystals in a fine-grained groundmass.

chibinite A eudialyte-bearing *nepheline syenite* distinguished from *lujavrite* by its smaller amount of mafic components, which are in compact aggregates of thick rather than acicular crystals, and by having the eudialyte in patches in the interstices rather than as euhedral crystals. Also spelled: *khibinite.* The rock was named by Ramsay in 1898 for Khibina (Uruptek), Kola Peninsula, U.S.S.R. Not recommended usage.

Chickasawhay North American (Gulf Coast) stage: Oligocene (above Vicksburgian, below Anahuac).

chickenwire anhydrite An evaporite texture in which irregularly polygonal nodules of anhydrite (or pseudomorphous gypsum), 1 to 5 cm in diameter, are separated by thin darker stringers of other minerals, generally carbonates or clays. It is believed by some to be diagnostic of sabkha deposition; it may be the result of porphyroblastic recrystallization (Dean et al., 1975).

Chideruan European stage: Upper Permian (above Kazanian, below Triassic). Syn: Tatarian.

childrenite A pale-yellowish to dark-brown orthorhombic mineral: $(Fe,Mn)AlPO_4(OH)_2 \cdot H_2O$. It is isomorphous with eosphorite.

Chile-loeweite *humberstonite.*

Chile saltpeter Naturally occurring sodium nitrate; *soda niter* occurring in caliche in northern Chile. Cf: *saltpeter.* Syn: *Chile niter.*

chilidial plate One of a pair of posterior platelike extensions of the walls of the notothyrium of certain brachiopods, commonly forming lateral boundaries of the cardinal process.

chilidium The triangular plate covering the apex of the notothyrium of certain brachiopods, commonly convex externally and extending for a variable distance ventrally over the proximal end of the cardinal process.

chillagite A variety of wulfenite containing tungsten.

chilled border *chill zone.*

chilled contact That part of a mass of igneous rock, near its contact with older rocks, that is finer grained than the rest of the mass, because of its having cooled more rapidly.

chilled margin *chill zone.*

chill zone The border or marginal area of an igneous intrusion, characterized by finer grain than the interior of the rock mass, owing to more rapid cooling. Cf: *basic border.* Syn: *chilled border; chilled margin.*

Chimaeriformes The sole order of the chondrichthyan subclass *Holocephali,* including the living ratfishes or chimaerae and three extinct groups of uncertain relations. All are characterized by holostylic jaw suspension and teeth reduced to a few crushing plates.

chimney [coast] (a) An angular, columnar mass of rock, smaller than a *stack,* isolated on a wave-cut platform by differential wave erosion of a sea cliff. (b) A *blowhole;* a *spouting horn.*

chimney [ore dep] *pipe [ore dep].*

chimney [speleo] In a cave, a rounded vertical passage or opening. See also: *domepit.*

chimney [volc] A conduit through which magma reaches the Earth's surface. Cf: *vent; pipe [volc].* Syn: *feeder [volc].*

chimney rock A chimney-shaped column of rock rising above its surroundings or isolated on the face of a steep slope; a small, weathered outlier shaped like a sharp pinnacle; a *stack* formed by wave erosion. Syn: *pulpit rock.*

china clay A commercial term for *kaolin* obtained from china-clay rock after washing, and suitable for use in the manufacture of chinaware. Sometimes spelled: *China clay.*

china-clay rock Kaolinized *granite* composed chiefly of quartz and kaolin, with muscovite and tourmaline as possible accessories. The rock crumbles easily in the fingers. Also spelled: *China-clay rock.* Cf: *china stone [ign].*

Chinaman pebble A term used in New Zealand for a pebble derived from a conglomerate consisting of quartz pebbles cemented with chalcedony.

chinarump *shinarump.*

china stone [ign] Partially kaolinized *granite* containing quartz, kaolin, and sometimes mica and fluorite. It is harder than *china-clay rock* and is used as a glaze in the manufacture of china. Syn: *petunzyte; petuntse; porcelain stone.* Cf: *Cornish stone.*

china stone [sed] A fine-grained, compact Carboniferous mudstone or limestone found in England and Wales.

chine (a) A term used in England (esp. in Hampshire and in the Isle of Wight) for a narrow, deep ravine, gorge, or cleft, cut in a soft, earthy cliff by a stream descending steeply to the sea. (b) A ridge or crest of rocks.

Chinese-wall glacier A seldom used term for an ice sheet, such as along the coast of Greenland, whose front is a vertical or even overhanging cliff.

chink-faceting A term applied by Wentworth (1925, p. 260) to the localized grinding of smoothed, distinct, and often sharply limited facets on the surfaces of beach pebbles and cobbles that are lodged in crevices in such a way that they are subjected to recurrent rubbing and to-and-fro movement under the action of waves.

chinook A term used for a *foehn* occurring on the eastern slopes of the Rocky Mountains.

chiolite A snow-white tetragonal mineral: $Na_5Al_3F_{14}$.

chip A small fragment from a crystal; specif. a *diamond chip.*

chipping Abrasion of a rock fragment resulting in the flaking-off of its corners.

chip sample A series of chips of ore or rock taken at regular intervals across an exposure.

chip yard *forest bed.*

chiral twinning *optical twinning.*

chi-square test A statistical test that employs the sum of values given by the quotients of the squared difference between observed and expected (theoretical) frequencies divided by the expected frequency. It enables assessment of *goodness of fit,* association, or commonality in a population, and is used to determine equivalency of observed sample and expected population.

chitin A resistant organic compound with the same basic carbohydrate structure as cellulose, but nitrogenous because some hydroxyl groups are replaced by acetamide groups (i.e. it is a repeating unit of N-acetylglucosamine instead of glucose). It is a common constituent of various invertebrate skeletons such as insect exoskeletons and foraminiferal inner tests, and also occurs in hyphae and spores of fungi. Cf: *pseudochitin.*

chitinous Consisting of *chitin.*

chitinozoan A pseudochitinous marine microfossil of the extinct group Chitinozoa, having uncertain affinities but generally assumed to represent animal remains, shaped in general like a flask, occurring individually or in chains, and ranging primarily from uppermost Cambrian to Devonian. Chitinozoans have thin walls, which are usually black, structureless, and opaque but may be brown and translucent. Named by Eisenack (1931) who noted the resemblance of their walls to chitin.

chiton An invertebrate marine molluscan animal, class Amphineura, the shell of which consists of eight overlapping calcareous valves or plates. It is popularly called the "coat of mail" shell. Syn: *polyplacophoran.*

chkalovite A mineral: $Na_2BeSi_2O_6$.

chladnite [meteorite] A group name for achondritic stony meteorites (aubrites and diogenites) composed essentially of orthopyroxene. The term originally applied to achondrites composed essentially of enstatite.

chladnite [mineral] Pure meteoritic *enstatite.*

chlamydospore A thick-walled, nondeciduous spore, such as a unicellular *resting spore* in certain fungi, usually borne terminally on a hypha and rich in stored reserves; a *fungal spore* that may have chitinous walls and therefore occur as a microfossil in palynologic preparations.

chloanthite *nickel-skutterudite.*

chloraluminite A mineral: $AlCl_3 \cdot 6H_2O$.

chlorapatite (a) A rare mineral of the apatite group: $Ca_5(PO_4)_3Cl$. (b) An apatite mineral in which chlorine predominates over fluorine and hydroxyl.

chlorargyrite A white, pale yellow, or gray isometric waxlike mineral that darkens on exposure to light: AgCl. It occurs in the weathering zones of silver-sulfide deposits and it represents an important ore of silver. Syn: *cerargyrite; horn silver.*

chlorastrolite A mottled, green variety of *pumpellyite* used as a semiprecious stone, occurring as grains or small nodules of a radial, fibrous structure in geodes in basic igneous rocks. It resembles prehnite, and is found in the Lake Superior region (esp. on Isle Royale).

chlorides A miner's or prospector's term for ores containing silver chloride.

chlorine equivalent *chlorinity.*

chlorine log A *radioactivity log* designed to indicate chlorine content and hence salinity of pore water. Now largely replaced by the *pulsed-neutron-capture log.* Syn: *salinity log.*

chlorinity The chloride content of seawater, measured by mass, or grams per kilogram of seawater, and including the chloride equivalent of all the halides. Syn: *chlorine equivalent.*

chlorite (a) A group of platy, monoclinic, usually greenish minerals of the general formula: $(Mg,Fe^{+2},Fe^{+3})_6AlSi_3O_{10}(OH)_8$. It is characterized by prominent ferrous iron and by the absence of calcium and alkalies; chromium and manganese may be present. Chlorites are associated with and resemble the micas (the tabular crystals of chlorite cleave into small, thin flakes or scales that are flexible, but not elastic like those of mica); they may also be considered as clay minerals. The chlorites are widely distributed, esp. in low-grade metamorphic rocks, or as alteration products of ferromagnesian minerals. (b) Any mineral of the chlorite group, such as clinochlore, penninite, ripidolite, chamosite, thuringite, pennantite, and corundophilite.

chlorite schist A schist in which the main constituent, chlorite, imparts a schistosity by parallel arrangement of its flakes. Quartz, epidote, magnetite, and garnet may be accessories, the last two often as conspicuous porphyroblasts.

chloritic shale A poorly laminated shale containing a variety of angular to subrounded mineral particles of silt size, including unstable types, characterized by feldspar sometimes exceeding quartz in abundance, and by chlorite often abundant in the finer matrix. It is commonly associated with graywacke (high-rank or feldspathic graywacke) and represents accumulation of relatively finer detritus derived from rapidly eroded orogenic source areas and "poured" into rapidly subsiding depositional areas.

chloritization The replacement by, conversion into, or introduction of chlorite.

chloritoid A micaceous mineral: $Fe_2Al_4Si_2O_{10}(OH)_4$. It occurs in dull-green or dark-green to gray or grayish-black masses of brittle folia in metamorphosed argillaceous sedimentary rocks, and is related to the brittle micas. Magnesium may be present.

chlormanganokalite A yellow rhombohedral mineral: K_4MnCl_6. It is isomorphous with rinneite.

chlorocalcite *hydrophilite.*

chloromagnesite A mineral: $MgCl_2$.

chloromelanite (a) A dark-green to nearly black variety of jadeite. (b) A solid solution of roughly equal amounts of diopside, jadeite, and acmite.

chloropal (a) A name originally applied to a deep-green, opal-like mineral that was later shown to be a crystalline clay mineral and renamed *nontronite.* (b) A greenish variety of common opal from Silesia.

chlorophaeite A mineraloid closely related to chlorite in composition (hydrous silicate of magnesium, iron, and calcium) and found in the groundmass of tholeiitic basalts, where it occupies spaces between feldspar laths, forms pseudomorphs after olivine, or occurs in veinlets and amygdules. It is pale green when fresh, but may be dark green, brown, or red in weathered rocks.

chlorophane A variety of fluorite that emits a bright-green light when heated.

chlorophoenicite A gray-green monoclinic mineral: $(Mn,Zn)_5(AsO_4)(OH)_7$. It is isostructural with magnesium-chlorophoenicite.

chlorophyll Generally a mixture of two waxy pigments: chlorophyll a, $C_{55}H_{72}O_5N_4Mg$, blue-black, and chlorophyll b, $C_{55}H_{70}O_6N_4Mg$, yellow-green, which occurs in plasmic bodies (chloroplasts) of plants and serves as a catalyst in photosynthesis. Other forms of chlorophyll occur in diatoms, algae, etc.

chlorophyll *a* A pigment in phytoplankton that can be used to measure the abundance of phytoplankton.

chlorophyll coal A variety of *dysodile* which contains chlorophyll that can be extracted by alcohol.

chlorophyre A green porphyritic *quartz diorite.* Obsolete.

chlorospinel A grass-green variety of spinel containing some copper.

chlorothionite A bright-blue secondary mineral: $K_2Cu(SO_4)Cl_2$.

chlorotile A green orthorhombic mineral consisting of hydrated arsenate of copper. Cf: *mixite.*

chloroxiphite A dull-olive or pistachio-green monoclinic mineral: $Pb_3CuCl_2(OH)_2O_2$.

cho A rainy-season torrent carrying sand from the Himalayan foothills onto a plain below. Etymol: Panjabi, connoting "a bed of loose boulders, gravel and sand, indicating rapid erosion" (Stamp, 1961, p. 103). Also spelled: *choh.*

Choanichthyes *Sarcopterygii.*

choanocyte An endoderm cell of a sponge, bearing a distinct tubular collarlike contractile protoplasmic rim, surrounding the base of a flagellum. Choanocytes line the inner surfaces of canals and/or spongocoel. Adj: *choanocytal.* Syn: *collar cell.*

choanoderm A single layer of choanocytes in a sponge; a choanocytal membrane.

choanosome The inner layer of a sponge containing choanocyte-lined cavities (flagellated chambers).

choke [drill] An orifice or constriction in the *tubing* of an oil or gas well, for controlling the flow rate and producing pressure.

choke [speleo] An area in a cave that is blocked by debris.

chokedamp *blackdamp.*

choked stalagmite A stalagmite, the diameter of which is standard below and small above.

choma A ridgelike deposit of dense shell substance delimiting a tunnel in a fusulinid. Pl: *chomata.* Cf: *parachoma.*

chondri Chondrules; plural of *chondrus.*

Chondrichthyes A class of vertebrates including fish with skeletons of cartilage rather than bone; esp. the sharks.

chondrite [meteorite] A *stony meteorite* characterized by chondrules embedded in a finely crystalline matrix consisting of orthopyroxene, olivine, and nickel-iron, with or without glass. Chondrites constitute over 80% of meteorite falls and are usually classified according to the predominant pyroxene: e.g. "enstatite chondrite", "bronzite chondrite", and "hypersthene chondrite". Adj: *chondritic*. Cf: *achondrite.*

chondrite [paleont] A common *trace fossil* of the "genus" *Chondrites,* consisting of plantlike, regularly ramifying tunnel structures that neither cross each other nor anastomose but radiate around a central vertical tube. It is interpreted as a dwelling or feeding burrow, probably made by a marine worm. It is often called a *fucoid.*

chondrodite A dark-red, orange-red, or yellow monoclinic mineral of the humite group: $(Mg,Fe)_5(SiO_4)_2(OH,F)_2$. It commonly occurs in contact-metamorphosed dolomites. Also spelled: *condrodite.*

chondrophore A relatively prominent process with a hollowed-out surface for holding or attaching the internal ligament (resilium) of a bivalve mollusk. See also: *resilifer.*

Chondrostei An infraclass of ray-finned bony fish which includes the living sturgeons and a variety of Late Paleozoic forms near the stem of all ray-fins.

chondrule A spheroidal granule or aggregate, often radially crystallized and usually about one millimeter in diameter, consisting chiefly of olivine and/or orthopyroxene (enstatite or bronzite), and occurring embedded more or less abundantly in the fragmental bases of many stony meteorites (chondrites) and sometimes free in marine sediments. Most chondrules appear to have originated as molten silicate droplets. Syn: *chondrus; chondre.*

chondrus A syn. of *chondrule.* Pl: *chondri.*

chone An inhalant canal penetrating a cortex in a sponge, often leading from a vestibule to a subcortical crypt.

chonetid Any articulate brachiopod belonging to the suborder Chonetidina, characterized chiefly by a functional foramen located outside the delthyrium. Range, Lower Silurian (possibly Upper Ordovician) to Lower Jurassic.

chonolith An igneous intrusion whose form is so irregular that it cannot be classified as a laccolith, dike, sill, or other recognized body.

chop hill A term used in Nebraska for a *sand hill.*

chorate cyst A spiny encysted unicellular alga; esp. a condensed dinoflagellate cyst, bearing little morphologic resemblance to the motile theca. The ratio of the diameter of the main body to the total diameter of the cyst is 0.6 or less. Examples: *marginate chorate cyst; membranate chorate cyst; pterate chorate cyst; trabeculate chorate cyst.* See also: *proximochorate cyst; proximate cyst.*

chord The horizontal distance between ripple crests, measured normal to the crestlines. The term was introduced by Allen (1968, p. 61) to replace the term "wavelength," which is still commonly used. Syn: *ripple wavelength; ripple-mark wavelength; ripple spacing.*

Chordata A phylum of animals including those having a notochord, which in most taxa is represented by a bony spinal column. The phylum may or may not be considered to include the *Protochordata.*

chorismite Megascopically composite rock that consists of two or more petrogenetically dissimilar materials of any origin (Dietrich & Mehnert, 1961). The term was first introduced without genetic connotation, as a replacement for *migmatite.* Five types of chorismite were outlined. Not widely used.

choristid adj. Said of a sponge having a skeleton containing tetraxonic megascleres and lacking desmas.—n. A choristid sponge; specif. a sponge of the order Choristida, class Demospongiae.

choristoporate Pertaining to a highly specialized type of dasycladacean algae in which the sporangia are formed in specialized gametangia, which may be specialized rays of the second or third order.

C horizon A mineral horizon of a soil, lying beneath the A and/or B horizons, consisting of unconsolidated rock material that has been relatively little affected by pedogenic processes.

chorochromatic map A British term for a map in which broad distributions or variations are shown qualitatively over an area by means of different colors, tints, or shadings. Syn: *color-patch map.*

chorogram A generic term suggested by Wright (1944, p.653) for "any and all quantitative areal symbols" on a map.

chorographic Pertaining to *chorography;* specif. relating to an area of regional or continental extent, or said of a map representing a large region on a small scale (such as one between 1:500,000 and 1:5,000,000).

chorography (a) The art or practice of describing or mapping a particular region or district, esp. one larger than that considered by *topography* but smaller than that by geography. The term was widely used in the 17th and 18th centuries. (b) A broad account, description, map, or chart of a region considered by chorography. Also, the physical conformation or configuration, and the features, of such a region.—Syn: *Greek choros,* "place," + *graphein,* "to write".

chorology *biogeography.*

chott *shott.*

C/H ratio *carbon-hydrogen ratio.*

christensenite *tridymite.*

Christiansen effect In optical mineralogy, a dispersion phenomenon in which the boundary of a mineral grain that is immersed in a liquid of the same refractive index appears blue on one side and red to orange on the other.

christmas tree The assemblage of valves, pipes, gages, and fittings at the top of the casing of an oil or gas well, used to control the flow of fluids from the well and to prevent blowouts. See also: *casing head.*

christmas-tree laccolith *cedar-tree structure.*

christophite *marmatite.*

chromate A mineral containing the chromate ion CrO_4^{-2}. An example is potassium chromate, K_2CrO_4. Cf: *sulfate.*

chromatic aberration In crystal optics, the production of color fringes due to the failure of rays of different wavelengths to converge at the same point.

chromatite A citron-yellow mineral: $CaCrO_4$.

chromatography A general name for several processes of separating components of a sample by moving the sample in a mixture or solution over or through a medium using adsorption, partition, ion exchange, or other property in such a way that the different components have different mobilities and thus become separated. One of the earliest applications was in the separation of components of dye mixtures, giving rise to bands of different colors and hence to the name chromatography. See also: *column chromatography; electrochromatography; gas chromatography; liquid chromatography; paper chromatography; thin-layer chromatography.*

chrome A term commonly used to indicate ore of chromium, consisting esp. of the mineral chromite, or chromium-bearing minerals such as chrome mica or chrome diopside.

chrome-chert A cherty-looking rock formed by the replacement (by silica) of the silicate minerals of a chromite *peridotite,* the more resistant chromite grains remaining unaltered in the secondary siliceous matrix. The rock was named by Fermor in 1919.

chrome diopside A bright-green variety of diopside containing a small amount of Cr_2O_3.

chrome iron ore A syn. of *chromite.* Var: *chrome iron; chromic iron.*

chrome mica *fuchsite.*

chrome ocher A chromiferous clay; specif. a bright-green clay material containing 2-10.5% Cr_2O_3.

chrome spinel *picotite.*

chromic iron *chrome iron ore.*

chromite (a) A brownish-black to iron-black mineral of the spinel group: $(Fe,Mg)(Cr,Al)_2O_4$. It occurs in octahedral crystals as a primary accessory mineral in basic and ultrabasic igneous rocks; it also occurs massive, and it forms detrital deposits. Chromite is isomorphous with magnesiochromite, and is the most important ore of chromium. Syn: *chrome iron ore.* (b) A name applied to a series of isomorphous minerals in the spinel group, consisting of magnesiochromite and chromite.

chromitite (a) A rock composed chiefly of the mineral chromite. (b) A mixture of chromite with magnetite or hematite.

chromocratic *melanocratic.*

chromrutile *redledgeite.*

chron (a) The time span of a *chronozone.* (b) A term preferred by Dunbar & Rodgers (1957, p. 301) as a "reasonably unambiguous" and "mnemonic" syn. of *moment.* (c) A term used by Sutton (1940, p. 1404) for the time interval during which a "group" (now referred to as a "stage") of rocks was formed; i.e. used as a syn. of *age [geochron].*—The term was introduced by Williams (1901, p. 583-584) for an indefinite division of geologic time, and used by Wheeler et al. (1950, p. 2362) as a general geologic-time unit.

chronocline A gradational series of changes in the members of a natural group of organisms in successive stratigraphic units.

chronofauna A geographically restricted natural assemblage of interacting animal populations that maintained its basic structure over a geologically significant period of time.

chronogenesis The time sequence of appearance of organisms in stratified rocks.

chronohorizon A stratigraphic surface or interface that is everywhere of the same age. Although theoretically without thickness, it is commonly a very thin and distinctive interval that is essentially isochronous over its whole geographic extent and thus constitutes an excellent time-reference or time-correlation horizon (ISG, 1976, p. 67-68). Examples of horizons that may have strong chronostratigraphic significance include many *biohorizons,* bentonite beds, horizons of *magnetic reversal,* and coal beds. Cf: *buffer-zone; datum level [stratig].* See also: *moment; instant.* Syn: *chronostratigraphic horizon.*

chronolith *chronostratigraphic unit.*

chronolithologic unit *chronostratigraphic unit.*

chronology Arranging events in their proper sequence in time; also, considering or measuring time in discrete units. See also: *geochronology.*

chronomere Any interval of geological time, irrespective of duration (Challinor, 1978, p. 313). Cf: *stratomere.*

chronostratic unit *chronostratigraphic unit.*

chronostratigraphic classification The organization of rock strata into units on the basis of their age or time of origin (ISG, 1976, p. 66).

chronostratigraphic horizon *chronohorizon.*

chronostratigraphic unit A body of rock strata that is unified by having been formed during a specific interval of geologic time. It represents all the rocks formed during a certain time span of Earth history, and only the rocks formed during that time span. A chronostratigraphic unit is bounded by *isochronous* surfaces. Its rank and relative magnitude are a function of the length of the time interval that its rocks subtend, rather than of their thickness (ISG, 1976, p. 67). Chronostratigraphic units in order of decreasing rank: *eonothem, erathem, system, series, stage, substage.* Syn: *chronostratic unit; chronolithologic unit; time-stratigraphic unit; time-rock unit; chronolith.* See also: *chronozone.*

chronostratigraphic zone *chronozone.*

chronostratigraphy The branch of stratigraphy that deals with the age of strata and their time relations (IGS, 1976, p. 66). Syn: *time-stratigraphy.*

chronotaxial Pertaining to, characterized by, or exhibiting chronotaxy.

chronotaxis A term proposed by Henbest (1952, p.310) as a complementary term for "homotaxis". See: *chronotaxy.*

chronotaxy Similarity of time sequence; specif. correlation of fossil or stratigraphic sequences on identity in time, or the determination of age equivalence. The term was originally proposed as *chronotaxis* by Henbest (1952, p.310). Cf: *homotaxy.*

chronozone (a) A general term for "a zonal unit embracing all rocks formed anywhere during the time range of some geologic feature or some specified interval of rock strata" (ISG, 1976, p. 67). The basis of its time span may be any specified time interval, provided it has features allowing time-correlation with stratal sequences elsewhere. Thus we may speak of the chronozone of the ammonites, of *Exus albus,* or of the Sao Tome volcanic rocks. (b) A formal term for the lowest ranking division in the hierarchy of *chronostratigraphic units,* of lower rank than a stage but not necessarily an aliquot part of a stage (ISG, 1976, p. 68-69). Its time span is usually defined in terms of the time span of a previously designated formation, member, or biozone; e.g. a formal chronozone based on the time span of a biozone includes all strata equivalent in age to the total maximum time span of that biozone, regardless of the presence or absence of fossils diagnostic of the biozone (ISG, 1976, p. 69). Syn: *chronostratigraphic zone.*

chrysoberyl (a) A hard mineral: $BeAl_2O_4$. It is usually yellow, pale green, or brown, contains a small amount of iron, and is used as a gem. Principal varieties are cat's-eye and alexandrite. Syn: *chrysopal; gold beryl; cymophane.* (b) An obsolete syn. of *heliodor.*

chrysocolla (a) A blue, blue-green, or emerald-green mineral: $(Cu_2H_2(Si_2O_5)(OH)_4$. It is usually cryptocrystalline or amorphous, and it occurs as incrustations and thin seams in the zone of weathering of copper ores. Its chemical composition was formerly given as: $CuSiO_3 \cdot 2H_2O$. (b) An old name given to a mineral or minerals (such as chrysocolla, borax, and malachite) used for soldering gold (Hey, 1962, p. 384).

chrysocolla chalcedony Translucent to semitranslucent, vivid-blue to greenish-blue chalcedony. Minutely distributed chrysocolla causes the color.

chrysolite [gem] A misleading term applied in gemology to several yellow to yellowish-green gems. Correctly used it refers to the pale-yellow to yellowish-green gem variety of olivine.

chrysolite [mineral] (a) A yellowish-green, reddish, or brownish variety of olivine in which the ratio of magnesium to total magnesium plus iron is between 0.90 and 0.70 or in which the Fe_2SiO_4 content is 10-30 mole percent. The name has at times carried a wider meaning, as a syn. of *olivine.* (b) A name that has been applied at various times to topaz, prehnite, and apatite. This usage is obsolete.—Not to be confused with *chrysotile.*

chrysomonad Any of a group of microscopic organisms, either protozoans or algae, usually with flagella at some stage of their life history; many are autotrophic. *Holozoic* forms are assigned to the Protozoan order Chrysomonadina.

chrysopal (a) A translucent variety of common opal, colored apple-green by the presence of nickel. (b) *chrysoberyl.* (c) A gemstone-trade name for opalescent chrysolite (olivine).

chrysophyric Said of a *basalt* having olivine phenocrysts (Thrush, 1968, p. 208). Not recommended usage.

chrysoprase (a) An apple-green or pale yellowish-green variety of chalcedony containing nickel and valued as a gem. Syn: *green chalcedony.* (b) A misleading name used in the gem trade for green-dyed chalcedony having a much darker color than natural chrysoprase.

chrysoquartz Green *aventurine.*

chrysotile A white, gray, or greenish mineral of the serpentine group: $Mg_3Si_2O_5(OH)_4$. It is a highly fibrous, silky variety of serpentine, and constitutes the most important type of *asbestos.* Not to be confused with *chrysolite.* Cf: *antigorite.* Syn: *serpentine asbestos; clinochrysotile.*

chrystocrene A term introduced as *crystocrene* by Tyrrell (1904, p. 234) for a surface mass of ice formed each winter by the overflow of springs; also, the ice formed in the interstices of a mass of loose rock fragments (such as talus) by the freezing of a subjacent spring. The term is not synonymous with *rock glacier* (Tyrrell, 1910). Cf: *crystosphene.*

chthonic Said of deep-sea sediments and clastic debris derived from preexisting rocks. Ant: *halmeic.* Cf: *allogenic.*

chuco A term used in Chile for the upper part of a caliche deposit, composed mainly of sodium sulfate.

chudobaite A mineral: $(Na,K,Ca)(Mg,Zn,Mn)_2H(AsO_4)_2 \cdot 4H_2O$.

chukhrovite A mineral: $Ca_3(Y,Ce)Al_2(SO_4)F_{13} \cdot 10H_2O$.

chur *char [streams].*

churchite A monoclinic mineral: $YPO_4 \cdot 2H_2O$. Syn: *weinschenkite.*

churn drilling *cable-tool drilling.*

churn hole A *pothole* in a stream bed.

chute [geomorph] A term used in the Isle of Wight for a steep cutting affording a passage from the surface above a cliff to the lower undercliff ground (Stamp, 1961, p. 104).

chute [hydraul] An inclined water course, either natural or artificial.

chute [ore dep] Var. of *shoot,* as in *ore shoot.*

chute [speleo] An inclined channel or passage in a cave.

chute [streams] (a) A fall of water; a rapid or quick descent in a river; a steep channel by which water falls from a higher to a lower level; a rapids. See also: *shoot.* (b) A narrow channel through which water flows rapidly, esp. along an overflow river (such as the lower Mississippi River); specif. a *chute cutoff.*

chute cutoff A narrow "short cut" across a meander bend, formed at time of flood when the main flow of a stream is diverted to the inside of the bend, along or through a trough between adjacent parts of a point bar. Cf: *neck cutoff.* Syn: *chute.*

chymogenic Said of that portion of a composite rock formed by crystallization from an ionic- or molecular-dispersed phase or from a fluid, whether a gas, hydrothermal solution, or magma (Mehnert, 1968, p. 353). Also spelled: *chymogenetic.* Cf: *stereosome; metatect; mobilizate.* See also: *neosome.* Little used.

CI (a) *crystallization index.* (b) *contour interval.*

cicatricose Marked with scars; esp. said of sculpture of pollen and spores consisting of more or less parallel ridges.

cicatrisation Reconstruction of a broken or corroded crystal as a result of a secondary deposit of the same mineral in optical continuity.

cicatrix (a) A scar in an echinoderm; esp. the scar indicating the former position of the column in some echinoderms (such as cystoids) which apparently molt it. (b) A small groove or scar on the apex of some nautiloid conchs. (c) The impression on the inside of a bivalve shell caused by the insertion of the adductor muscle. See also: *adductor muscle scar.*—Pl: *cicatrices.* Syn: *cicatrice; scar.*

cienaga A marshy area where the ground is wet due to the presence of seepage or springs, often with standing water and abundant vegetation. The term is commonly applied in arid regions such as the southwestern U.S. Etymol: Spanish ciénaga, "marsh, bog, miry place". Also spelled: cienega.

cigar-shaped mountain An anticlinal ridge plunging at each end.

cilia Plural of *cilium.*

ciliate n. Any protozoan belonging to the class Ciliata and characterized by the presence of cilia throughout its life cycle. Known range, Upper Jurassic to the present.—adj. Possessing cilia.

cilifer Said of a variant of radulifer type of brachiopod crura, flattened in the plane of commissure, forming direct prolongations of horizontal hinge plates, then turning parallel to the plane of symmetry as slightly crescentic blades.

cilium One of numerous short hairlike processes found on the surface of cells, capable of rhythmic vibratory or lashing movement, and serving as organs of locomotion in free-swimming unicellular organisms and in some small multicellular forms and as producers of currents of water in higher animals. Pl: *cilia.* Cf: *flagellum.*

cima A mountain peak or dome. Etymol: Italian. Pron: *che*-ma.

ciminite A *trachydolerite* composed of olivine, clinopyroxene, and labradorite with alkali feldspar rims in a trachytic groundmass; named by Washington in 1896 for Monte Cimino, Italy. Not recommended usage.

Cimmerian orogeny One of the 30 or more short-lived orogenies during Phanerozoic time identified by Stille; in this case, two orogenies are included, the early Cimmerian late in the Triassic, between the Norian and Rhaetian stages, and the late Cimmerian at the end of the Jurassic.

cimolite A white, grayish, or reddish hydrous aluminum silicate mineral occurring in soft, claylike masses.

Cincinnatian North American provincial series: Upper Ordovician (above Champlainian, below Alexandrian of Silurian).

cinder A juvenile vitric vesicular pyroclastic fragment that falls to the ground in an essentially solid condition (Macdonald, 1972, p. 128). Cf: *block [volc]; scoria; lapilli; volcanic gravel.*

cinder coal *natural coke.*

cinder cone A conical hill formed by the accumulation of cinders and other pyroclasts, normally of basaltic or andesitic composition. Steepness of the slopes depends on coarseness of the ejecta, height of eruption, wind velocity, and other factors, but is normally greater than 10 degrees.

cinerite A deposit of volcanic cinders.

cingular archeopyle An *archeopyle* formed in a dinoflagellate cyst by breakage along and within the girdle.

cingular series The series of plates along the girdle in a dinoflagellate possessing a theca.

cingulate Having a girdle; esp. said of a spore possessing a cingulum.

cingulum (a) Either of the two *connecting bands* forming the sides of the two valves of a diatom; a *girdle [paleont].* (b) An annular, more or less equatorial extension of a spore in which the wall is thicker than that of the main body of the spore. Cf: *zone [palyn]; crassitude.* (c) Zooecial lining that results from secondary skeletal thickening in some stenolaemate bryozoans.—Pl: *cingula.*

cinnabar A rhombohedral mineral: HgS. It is dimorphous with metacinnabar and represents the most important ore of mercury.

Cinnabar occurs in brilliant red acicular crystals or in red, brownish, or gray masses in veins and alluvial deposits. Syn: *cinnabarite; vermilion.*

cinnamon stone Yellow-brown to reddish-brown *essonite.* Syn: *cinnamite.*

cipolin (a) *cipolino.* (b) A term used in France for any *crystalline limestone.*

cipolino A siliceous marble containing micaceous layers. Partial syn: *cipolin.*

CIPW classification A system for classifying and naming igneous rocks based on the *CIPW norm.* The initials represent the initial letters of the names of the men who devised the system, Cross, Iddings, Pirsson, and Washington (1902). Syn: *quantitative system; norm system.*

CIPW norm A norm in which the reported content of a mineral represents its weight percentage and in which the minerals are all anhydrous minerals of simplified composition (Cross et al., 1902). Cf: *norm.*

circadian Said of a time period approximately 24 hours in length, or of an event occurring at roughly 24-hour or daily intervals; e.g. "circadian rhythms". Cf: *circannian.*

circannian Said of a time period approximately one year in length, or of an event that occurs annually; e.g. "circannian rhythms". Cf: *circadian.*

circinate (a) Pertaining to the unrolling of a developing fern frond. (b) Pertaining to a protist that is curled downward from the apex.

circle [pat grd] A form of patterned ground whose horizontal mesh is dominantly circular. See: *sorted circle; nonsorted circle.*

circle [surv] The graduated disk of a surveying instrument, perpendicular to and centered about an axis of rotation, and calibrated to read the amount of rotation; e.g. a horizontal circle or a vertical circle of a theodolite or transit.

circle of influence *area of influence.*

circle of latitude A meridian of the terrestrial sphere, along which latitude is measured. Cf: *parallel of latitude.*

circlet A series of plates that forms a ring entirely or partially around the theca of an echinoderm.

circuit (a) A continuous series of connected survey lines that form a closed loop. (b) A line or series of lines connecting two fixed survey points.

circuit closure The *error of closure* of a level circuit, being the algebraic sum of all the junction closures in a circuit, usually reckoned counterclockwise around the circuit; hence, the accumulated error (before adjustment) of measured differences of elevation around the circuit, or the amount by which the last computed elevation fails to equal the initial elevation.

circular coal *eye coal.*

circular error probable The radius of a circle centered at the most probable point, such that half the measurements fall within the circle. Abbrev: CEP.

circularity ratio *basin-circularity ratio.*

circular level A *spirit level* having the inside surface of its upper part ground to a spherical shape, the outline of the bubble formed being circular, and the graduations being concentric circles. It is used where a high degree of precision is not required, as in setting an instrument in approximate position. Syn: *bull's-eye level; box level.*

circular normal distribution A *frequency distribution* of a polar variable analogous to a normal (Cartesian) distribution.

circular polarization In optics, circularly polarized light consisting of upward-spiraling vibration vectors that define a surface similar to the thread of a screw. It is caused by the interaction of mutually perpendicular wave motions whose path differences differ in phase by $(2n+1)/4\lambda$ on emergence from a crystal. Cf: *elliptical polarization.*

circular section [cryst] In a uniaxial crystal, an equatorial section perpendicular to the optic axis; in a biaxial crystal, one of two sections intersecting the beta axis of the biaxial indicatrix (Wahlstrom, 1948).

circular section [exp struc geol] One of the two circular cross sections through a strain ellipsoid.

circular slide A landslide whose slip surface follows the arc of a circle.

circulating fluid *drilling mud.*

circulation [drill] In *rotary drilling,* the continuous cycling of drilling mud (rarely air or foam) down the drill pipe, out through the drill bit, and up to the surface through the *annulus* between the drill pipe and the walls of the hole.

circulation [lake] The complete mixing of a lake or sea; generally it occurs when the waters are isothermal, often at the temperature of maximum density. See also: *overturn.*

circulation [oceanog] In oceanography, a general term for the flow of water in a large area, usually in a closed pattern or gyre, due to wind over the surface or to varying densities of water (resulting from differences in salinity and water temperature).

circulus A cameral deposit on the concave surface of a cyrtochoanitic septal neck of a nautiloid (TIP, 1964, pt.K, p.54).

circumdenudation The denudation or erosion of a landmass such that a part of the ground is left isolated and upstanding; e.g. denudation around a resistant rock mass. Syn: *circumerosion.*

circumdenudation mountain *mountain of circumdenudation.*

circumerosion *circumdenudation.*

circumferential wave An obsolete syn. of *surface wave.*

circumferentor A type of *surveyor's compass* having vertical slit sights on projecting arms.

circummural budding A type of *polystomodaeal budding* in which indirectly linked stomodaea are arranged around discontinuous collines or monticules of corallum. Cf: *intramural budding.*

circumoceanic basalt *Basalt* that issues from volcanoes on the margins of ocean basins (Chayes, 1964). A descriptive term, not part of a classification.

circumoral budding A type of *polystomodaeal budding* in which directly linked stomodaea are arranged concentrically around the central parent stomodaeum.

circum-Pacific belt The *great-circle belt* that borders the Pacific Ocean along the continental margins of Asia and the Americas, and meets the *Eurasian-Melanesian belt* in the Celebes.

circumvallation The process whereby mountains are formed by streams incising a featureless plain (Hobbs, 1912, p. 442).

cirque [geomorph] A term sometimes used for a semicircular, amphitheaterlike, or armchair-shaped hollow of nonglacial origin but resembling a glacial cirque; e.g. a *doline* in a limestone region, a *blowout* in an arid region, or a depression formed by landslide sapping. See also: *pseudocirque.*

cirque [glac geol] A deep steep-walled half-bowl-like recess or hollow, variously described as horseshoe- or crescent-shaped or semicircular in plan, situated high on the side of a mountain and commonly at the head of a glacial valley, and produced by the erosive activity of a mountain glacier. It often contains a small round lake, and it may or may not be occupied by ice or snow. Etymol: French, from Latin *circus,* "ring". Syn: *corrie; cwm; coire; kar; botn; amphitheater; combe; oule; van; zanoga.*

cirque [lunar] An obsolete syn. of *walled plain.*

cirque floor The nearly flat surface at the bottom of a cirque. See also: *cirque niveau.*

cirque glacier A small glacier occupying a *cirque,* or resting against the headwall of a cirque. It is the most common type of glacier in the mountains of the western U.S. Cf: *glacieret; niche glacier.* Syn: *corrie glacier.*

cirque lake A small, deep, commonly circular glacial lake occupying a cirque; it has no prominent inlet, being fed by runoff from the surrounding slopes and dammed by a lip of bedrock or by a small moraine. Syn: *tarn.*

cirque mountain *horn.*

cirque niveau The level of a *cirque floor* representing the surface of a terrace developed by preglacial erosion (Swayne, 1956, p. 34); it is the approximate altitude at which most cirques in a region have excavated their floors. Etymol: *cirque* + French *niveau,* "level".

cirque platform A relatively level surface formed by the coalescence of several cirques.

cirque stairway A succession of cirques situated in a row at different levels in the same glacial valley. Cf: *glacial stairway.* Syn: *cirque steps.*

cirral adj. Pertaining to a cirrus.—n. A single segment or plate of a crinoid cirrus.

cirri Plural of *cirrus.*

cirriped Any marine crustacean belonging to the class Cirripedia, characterized chiefly by the permanent attachment of the adult stage to some substrate; e.g. a barnacle. Range, Upper Silurian to present. Also spelled: *cirripede.*

cirrus (a) Any of the flexible rootlike jointed appendages attached to the side of the stem (and sometimes to the aboral surface) of a crinoid, exclusive of the radix. It is composed of small articulated plates, or cirrals. (b) A multiarticulate food-gathering thoracic appendage in a cirripede crustacean.—Pl: *cirri*.

cis link A *link* in a trunk stream channel bounded by tributaries that enter from the same side (James & Krumbein, 1969). The trunk channel is traced upstream by following the link of greater magnitude at each fork. Cf: *cis-trans link; trans link.*

cislunar Pertaining to phenomena, or to the space, between the Earth and the Moon or the Moon's orbit. Cf: *translunar.*

cistern (a) An artificial reservoir or tank for storing water. (b) A natural reservoir; a hollow containing water.

cis-trans link A link of magnitude μ that is formed at its upstream fork by the confluence of two links of unequal magnitude and that flows at its downstream fork into a link of a magnitude of less than 2μ (Mock, 1971, p. 1559). Symbol: CT. Cf: *cis link; trans link.*

citrine A transparent yellow to orange-brown variety of crystalline quartz closely resembling topaz in color. It can be produced by heating amethyst or dark smoky quartz. Syn: *topaz quartz; false topaz; Bohemian topaz; quartz topaz; yellow quartz.*

civil engineering A branch of engineering concerned primarily with the investigation, design, construction, operation, and maintenance of civil-works projects (public and private) such as highways, bridges, tunnels, waterways, harbors, dams, water supply, irrigation, railways, airports, buildings, sewage disposal, and drainage.

clade [evol] (a) A *monophyletic* higher taxon, one consisting of an ancestral species and all its known descendant species. (b) In *cladism,* all taxa distal to a given *node [evol]* in a *cladogram.*

clade [paleont] A branch at the extremity of an actine in an ebridian skeleton that may connect adjacent actines. See also: *proclade; opisthoclade; mesoclade.*

cladi Plural of *cladus.*

cladism That method in *systematics* wherein ancestor-descendant (phylogenetic or evolutionary) relationships among taxa are analyzed strictly on the basis of the distribution of *shared derived characters,* and consequently where genealogy is the sole criterion for the definition of taxa. Syn: *phylogenetic systematics.*

cladoceran Any crustacean belonging to the order Cladocera, characterized by a univalve carapace that is bent along the back giving a bivalve appearance. Cladocerans are commonly found in fresh- to brackish-water postglacial deposits. Range, Oligocene to present.

cladogenesis (a) Phylogenetic splitting or branching; speciation. (b) Progressive evolutionary specialization.

cladogram (a) A *dendrogram* expressing the classificatory relationships among a group of organisms, as based on their inferred phylogenetic or evolutionary relationships. Syn: *phylogenetic tree.* (b) In *cladism,* a dendrogram showing phylogenetic relationships among taxa in terms of recency of common ancestry. Identities of *nodes* (ancestors) are not specified, connecting lines usually represent *shared derived characters,* and all taxa are terminal in position. Cf: *phylogenetic tree.*

cladome The group of similar rays of a diaene, triaene, or tetraene sponge spicule.

Cladoselachii An order of marine elasmobranch fishes with fusiform body, broad-based fins, and terminal mouth; the most primitive true sharks. Stratigraphic range, Middle Devonian to Mississippian.

cladus One of the rays of a cladome; a branch of a ramose spicule. The term is usually used in the plural: *cladi.* Syn: *clad.*

Claibornian North American (Gulf Coast) stage: Eocene (above Wilcoxian, below Jacksonian).

claim In mining law, a portion of public land on which an individual may have mining rights; a *mining claim.* Size and other legal restrictions vary from country to country.

Clairaut's theorem An expression for the variation of normal gravity on the Earth that is the basis for standard gravity formulas such as the International Gravity Formula. It establishes the relationship between normal gravity and the flattening of the Earth, from which it becomes possible to compute the flattening from surface gravity observations.

clam A popular term for a bivalve mollusk, commonly applied to an edible one that lives partially or completely buried in sand or mud.

clamshell snapper A type of *grab sampler.*

clan [ecol] (a) A category in the hierarchy of classification used by some zoologists; it ranks below the *subfamily* and above the *genus.* (b) A small ecologic community, usually a *climax community,* that occupies only a few square meters of space and has only one dominant species.

clan [petrology] A group of igneous rocks that are closely related in chemical composition; a subdivision of a tribe. Clans are subdivided into families. See also: *tribe; family [petrology].* Syn: *igneous-rock clan.*

Clapeyron equation A statement in chemistry that the rate of change of pressure with temperature in a phase transition of a closed system is equal to the heat of the reaction divided by the product of the absolute temperature and the volume change of the reaction. It was developed by Clausius in 1850. Syn: *Clausius-Clapeyron equation.* See also: *Ehrenfest relation; Poynting's law.*

clarain A coal *lithotype* characterized macroscopically by semibright, silky luster and sheetlike, irregular fracture. It is distinguished from *vitrain* by containing fine intercalations of a duller lithotype, *durain.* Its characteristic microlithotype is *clarite.* Cf: *fusain.*

Clarendonian North American continental stage: Middle Miocene (above Barstovian, below Hemphillian).

clarinite The major maceral of clarain, according to the Stopes classification; the term is no longer in general use.

clarite A coal microlithotype that contains a combination of vitrinite and exinite totalling at least 95%, and containing more of each than of inertinite. Cf: *clarain.*

Clark degree A British unit for measuring hardness of water, equal to one grain per British gallon or 14.3 ppm as $CaCO_3$. Cf: *grain [water]; degree.*

clarke The average abundance of an element in the crust of the Earth. It is named in honor of F.W. Clarke. Cf: *clarke of concentration.* Syn: *crustal abundance.*

Clarke-Bumpus plankton sampler One of many plankton nets designed to include an opening–closing mechanism and a flow meter to record the volume of water filtered.

Clarke ellipsoid of 1866 The ellipsoid of reference for geodetic surveys in North and Central America, the Hawaiian Islands, and the Philippines. It was the basis of the *North American datum* of 1927. Cf: *ellipsoid.*

clarkeite A dark-brown or reddish-brown mineral: $(Na,Ca,Pb)_2U_2(O,OH)_7$.

clarke of concentration The concentration of an element in a mineral or rock relative to its crustal abundance. The term is applied to specific as well as average occurrences. Cf: *clarke.*

Clarkforkian North American continental stage: Upper Paleocene (above Tiffanian, below Wasatchian).

clarocollain A transitional lithotype of coal characterized by the presence of collinite with lesser amounts of other macerals. Cf: *colloclarain.* Syn: *clarocollite.*

clarocollite *clarocollain.*

clarodurain A transitional lithotype of coal characterized by vitrinite, but more of other macerals such as micrinite and exinite than of vitrinite; it corresponds to *semisplint coal.* Cf: *duroclarain.*

clarodurite A coal microlithotype containing at least 5% each of vitrinite, exinite, and inertinite, with more inertinite than vitrinite and exinite. It is a variety of *trimacerite,* intermediate in composition between clarite and durite, but closer to durite. Cf: *duroclarite.*

clarofusain A transitional lithotype of coal characterized by the presence of fusinite and vitrinite, with other macerals; fusinite is more abundant than it is in *fusoclarain.* Syn: *clarofusite.*

clarofusite *clarofusain.*

clarotelain A transitional lithotype of coal characterized by the presence of telinite, with lesser amounts of other macerals. Cf: *teloclarain.* Syn: *clarotelite.*

clarotelite *clarotelain.*

clarovitrain A transitional lithotype of coal characterized by the presence of vitrinite, with lesser amounts of other macerals. Cf: *vitroclarain.* Syn: *clarovitrite.*

clarovitrite *clarovitrain.*

clasmoschist A term suggested by W.D. Conybeare to replace "graywacke" (an arenaceous rock in the lower part of the Secondary strata) (Roberts, 1839, p.72).

clasolite A rock composed of fragments of other rocks; a clastic rock.

clasper (a) An appendage of a crustacean, modified for attachment in copulation or for fixation of parasites. (b) A part of the pelvic fin of many sharks, modified for copulation.

class [cryst] One of thirty-two possible combinations of the nontranslational elements of symmetry. Crystal classes are divided among the six crystal systems, and deal with outward symmetry. Syn: *point group.*

class [petrology] In the CIPW classification of igneous rocks, a subdivision based on the relative proportions of salic and femic standard minerals. The classes correspond approximately to the color-based divisions leucocratic, melanocratic, and mesocratic. The basic unit of the class is the *order [petrology].*

class [stat] A subdivision of the observed range of a variable, having stated limits.

class [taxon] A category in the hierarchy of classification of animals and plants, intermediate between *phylum* (or *division*) and *order.* Super- and sub-categories may be introduced as needed, provided they do not produce confusion (Blackwelder, 1967, p. 57). Cf: *subclass.*

Classic n. In New World archaeology, a cultural stage that follows the Formative and is characterized by the rise of civilizations such as the Mayan. It is followed by the Post-Classic.—adj. Pertaining to the Classic.

classical equilibrium constant An *equilibrium constant* that is defined by concentrations rather than by activities.

classification That part of *systematics* that deals chiefly with the grouping of like things within a system (Blackwelder, 1967, p. 4).

clast (a) An individual constituent, grain, or fragment of a sediment or rock, produced by the mechanical weathering (disintegration) of a larger rock mass; e.g. a *phenoclast.* (b) *pyroclast.* (c) *bioclast.*

clastation (a) The breaking-up of rock masses in situ by physical or chemical means (Grabau, 1924, p.17); *weathering.* (b) The disrupting of rocks to form clastic sediments (Galloway, 1922).

clastic adj. (a) Pertaining to a rock or sediment composed principally of broken fragments that are derived from preexisting rocks or minerals and that have been transported some distance from their places of origin; also said of the texture of such a rock. The term has been used to indicate a source both within and outside the depositional basin. (b) *pyroclastic.* (c) Said of a bioclastic rock. (d) Pertaining to the fragments (clasts) composing a clastic rock.—n. A clastic rock. Term is usually used in the plural; e.g. the commonest "clastics" are sandstone and shale.

clastic breccia A breccia formed by erosion (McKinstry, 1948, p.634).

clastic deformation One of the processes of dynamothermal metamorphism, which involves the fracture, rupture, and rolling-out of mineral and rock particles. In the extreme case, the rock may be thoroughly pulverized (Tyrrell, 1926). Cf: *blastic deformation; plastic deformation.*

clastic dike A *sedimentary dike* consisting of a variety of clastic materials derived from underlying or overlying beds; esp. a *sandstone dike* or a *pebble dike.*

clastichnic Said of a dolomite rock in which the original clastic texture of a limestone is preserved (Phemister, 1956, p. 72).

clasticity (a) The quality, state, or degree of being clastic. (b) The maximum apparent particle size in a sediment or sedimentary rock (Carozzi, 1957).

clastic pipe A cylindrical body of clastic material, having an irregular columnar or pillarlike shape, standing approximately vertical through enclosing formations (usually limestone), and measuring a few centimeters to 50 m in diameter and a meter to 60 m in height; esp. a *sandstone pipe.* Syn: *cylindrical structure.*

clastic ratio A term introduced by Sloss et al. (1949, p.100) for the ratio of the thickness or percentage of clastic material (conglomerate, sandstone, shale) to that of nonclastic material (limestone, dolomite, evaporites) in a stratigraphic section; e.g. a ratio of 5 indicates that the section contains an average of 5 m of clastics per meter of nonclastics. The ratio is a measure of the sediments carried into the environment as compared to those formed locally. Cf: *sand-shale ratio.* Syn: *detrital ratio.*

clastic rock (a) A consolidated sedimentary rock composed principally of broken fragments that are derived from preexisting rocks (of any origin) or from the solid products formed during chemical weathering of such rocks, and that have been transported mechanically to their places of deposition; e.g. a sandstone, conglomerate, or shale, or a limestone consisting of particles derived from a preexisting limestone. See also: *epiclastic rock.* Syn: *fragmental rock.* (b) *pyroclastic rock.* (c) *bioclastic rock* (d) *cataclastic rock.*

clastic sediment A sediment formed by the accumulation of fragments derived from preexisting rocks or minerals and transported as separate particles to their places of deposition by purely mechanical agents (such as water, wind, ice, and gravity); e.g. gravel, sand, mud, clay. Cf: *detrital sediment.* Syn: *mechanical sediment.*

clastic wedge The sediments of an *exogeosyncline,* derived from the tectonic land masses of the adjoining orthogeosynclinal belt (King, 1959, p. 59). Cf: *geosynclinal prism.*

clastizoic Said of a rock containing animal remains mainly in the form of angular, little-worn debris; esp. said of a fossiliferous-fragmental limestone that may often contain entire microfossils. Term introduced by Phemister (1956, p. 72).

clastizoichnic Said of a dolomite or recrystallized limestone that contains traces of original clastizoic features (Phemister, 1956, p. 72).

clathrate A term applied by Washington in 1906 to the texture commonly found in leucite-bearing rocks in which leucite crystals are surrounded by tangential augite crystals giving the appearance of a net or sponge, the augite representing the threads or walls and the leucite the holes (Johannsen, 1939, p.205).

clathrate wall The outer wall in archaeocyathids, in which vertical or longitudinal laths are applied to oblique annuli which are in turn attached to the outer edges of septa (TIP, 1972, pt. E, p. 11).

claudetite A monoclinic mineral: As_2O_3. It is dimorphous with arsenolite.

claugh *clough.*

Clausius-Clapeyron equation *Clapeyron equation.*

clausthalite A mineral: PbSe. It resembles galena in appearance.

clavalite A belonite with a globular enlargement at each end.

clavate adj. (a) Club-shaped, being slender at one end and gradually thickening near the other end, like a baseball bat; e.g. said of spores and pollen having sculpture consisting of processes that widen to a knob at the end. Cf: *pilate.* (b) Pertaining to a clavus of an ammonoid.—n. A club-shaped thecal form in edrioasteroids, in which the oral surface plates form an upper gibbous "head", a lower constricted pedunculate zone, and an outward flaring peripheral rim (Bell, 1976).

clavicle (a) A shelly buttress or heavy internal ridge supporting the chondrophore in some bivalve mollusks. (b) The collar-bone of Osteichthyes and tetrapods.

clavidisc A sponge spicule (microsclere) in the form of an ovate disk with a central perforation.

clavula A small ciliated spine in a fasciole of an echinoid. Pl: *clavulae.*

clavule A *sceptrule* in which the end bearing the axial cross is swollen or bears a ring of recurved teeth.

clavus An ammonoid tubercle elongated longitudinally in the direction of coiling. Pl: *clavi.*

clay [eng] Plastic material consisting mainly of particles having diameters less than 0.074 mm (passing U.S. standard sieve no.200). Cf: *silt [eng].*

clay [geol] (a) A rock or mineral fragment or a detrital particle of any composition (often a crystalline fragment of a clay mineral), smaller than a very fine silt grain, having a diameter less than 1/256 mm (4 microns, or 0.00016 in., or 8 phi units). This size is approximately the upper limit of size of particle that can show colloidal properties. See also: *coarse clay; medium clay; fine clay; very fine clay.* (b) A loose, earthy, extremely fine-grained, natural sediment or soft rock composed primarily of clay-size or colloidal particles and characterized by high plasticity and by a considerable content of *clay minerals* and subordinate amounts of finely divided quartz, decomposed feldspar, carbonates, ferruginous matter, and other impurities; it forms a plastic, moldable mass when finely ground and mixed with water, retains its shape on drying, and becomes firm, rocklike, and permanently hard on heating or firing. Some clays are nonplastic. Clay should have more than 50% clay-size particles (Twenhofel, 1937, p. 96), and clay minerals must form at least one-fourth of the total (Pettijohn, 1957, p. 341). Clays are classified by use, origin, composition, mineral constituents, and color; among their various uses are in the manufacture of tile, porcelain, and earthenware, and in filtration, oil refining, and paper manufacture. (c) A term that is commonly applied to any soft, adhesive, fine-grained deposit (such as loam or siliceous silt) and to earthy material, esp. when wet (such as mud or mire). (d) *clay mineral.*

clay [soil] (a) A term used in the U.S. and by the International Society of Soil Science for a rock or mineral particle in the soil, having a diameter less than 0.002 mm (2 microns). (b) *clay soil.* (c) *clay mineral.*

clay ball A chunk of clay released by erosion of a clayey bank and rounded by wave action; esp. an *armored mud ball.* Also spelled: *clayball.*

clay band A light-colored, argillaceous layer in clay ironstone. Also spelled: *clayband.*

clay-band ironstone A variety of *clay ironstone* characterized by abundant clay bands.

clay boil A *mud circle* that suggests a welling-up or heaving of the central core.

clay colloid (a) A clay particle having a diameter less than 1 micron (0.001 mm) (Jacks et al., 1960, p. 24). (b) A colloidal substance consisting of clay-size particles.

claycrete Weathered argillaceous material forming a layer immediately overlying bedrock.

clay dune A dune composed of clay fragments heaped up by the wind, as in the lower Rio Grande Valley, Tex. (Coffey, 1909).

C layer The seismic region of the Earth between 410km and 1000 km, equivalent to the *transition zone* of the upper mantle. It is a part of a classification of the Earth's interior made up of layers A to G.

clayey Abounding in, consisting of, characterized by, or resembling clay; *argillaceous.*

clayey breccia A term used by Woodford (1925, p.183) for a breccia containing at least 80% rubble and 10% clay.

clayey sand (a) An unconsolidated sediment containing 50-90% sand and having a ratio of silt to clay less than 1:2 (Folk, 1954, p.349). (b) An unconsolidated sand containing 40-75% sand, 12.5-50% clay, and 0-20% silt (Shepard, 1954).

clayey sandstone (a) A consolidated *clayey sand.* (b) A sandstone containing more than 20% clay (Krynine, 1948, p.141).—Cf: *argillaceous sandstone.*

clayey silt (a) An unconsolidated sediment containing 40-75% silt, 12.5-50% clay, and 0-20% sand (Shepard, 1954). (b) An unconsolidated sediment containing more particles of silt size than of clay size, more than 10% clay, and less than 10% of all other coarser sizes (Wentworth, 1922).

clay gall (a) A markedly flattened and somewhat rounded pellet or curled fragment, chip, or flake of clay, generally embedded in a sandy matrix and esp. abundant at the base of sandy beds. It may arise from the drying and cracking of a thin layer of coherent mud, the fragment commonly being rolled or blown into sand and buried, and forming a lenticular bleb upon wetting. (b) An ocherous, sometimes hollow inclusion of clay or mudstone, occurring esp. in oolitic limestones (Woodward, 1894, p.340). Syn: *crick.* —Syn: *gall.*

clay gouge (a) A clayey deposit in a fault zone; *fault gouge.* (b) A thin seam of clay separating masses of ore, or separating ore from country rock. See also: *gouge [ore dep].*

clay gravel Gravel containing fine-grained silica and clay, developing under puddling action (compaction) a dense, firm surface.

clay ironstone (a) A compact, hard, dark gray or brown fine-grained sedimentary rock, consisting of a mixture of argillaceous material (up to 30%) and iron carbonate (siderite), occurring in layers of nodules or concretions or as relatively continuous irregular thin beds, and usually associated with carbonaceous strata, esp. overlying a coal seam in the coal measures of the U.S. or Great Britain; a clayey iron carbonate, or an impure siderite ore occurring admixed with clays. The term has also been applied to an argillaceous rock containing iron oxide (such as hematite or limonite). See also: *blackband ironstone; clay-band ironstone.* (b) A sideritic concretion or nodule occurring in clay ironstone and other argillaceous rocks, often displaying septarian structure. (c) A sheet-like deposit of clay ironstone.—Syn: *ironstone.*

clayite A term proposed by Mellor (1908) for a hydrous aluminum silicate thought to be the true clay substance in kaolin and considered to be an amorphous (colloidal) material of the same chemical composition as kaolinite. Cf: *pelinite.*

clay loam A soil containing 27-40% clay, 20-45% sand, and the remainder silt. See also: *silty clay loam; sandy clay loam.*

clay marl A whitish, smooth, chalky clay; a marl in which clay predominates.

clay mineral (a) One of a complex and loosely defined group of finely crystalline, metacolloidal, or amorphous hydrous silicates, essentially of aluminum (and sometimes of magnesium and iron); they have a monoclinic crystal lattice of the two- or three-layer type, in which silicon and aluminum ions have tetrahedral coordination with respect to oxygen and in which aluminum, ferrous and ferric iron, magnesium, chromium, lithium, manganese, and other ions have octahedral coordination with respect to oxygen or hydroxyl. There may be exchangeable cations (usually calcium and sodium, sometimes potassium, magnesium, hydrogen, and aluminum) on the surfaces of the silicate layers, in amounts determined by the excess negative charge within the layer. Clay minerals are formed chiefly by alteration or weathering of primary silicate minerals such as feldspars, pyroxenes, and amphiboles, and are found in clay deposits, soils, shales, alteration zones of ore deposits, and other rocks, in flakelike particles or in dense, feathery aggregates of varying types. They are characterized by small particle size and ability to adsorb substantial amounts of water and ions on the surfaces of the particles. The most common clay minerals belong to the kaolin, montmorillonite, and illite groups. Syn: *clay; hydrosialite; sialite.* (b) Any crystalline substance occurring in the clay fraction of a soil or sediment.

claypan [geomorph] A term used in Australia for a shallow depression containing clayey and silty sediment, and having a hard, sun-baked surface; a *playa* formed by deflation of alluvial topsoils in a desert, in which water collects after a rain.

claypan [soil] A dense, heavy, relatively impervious subsurface soil layer that owes its hardness to a relatively higher clay content than that of the overlying material, from which it is separated by a sharply defined boundary. It is usually hard when dry and plastic when wet, and is presumably formed by the concentration of clay by percolating waters or by in-situ synthesis. Cf: *hardpan; iron pan.* Also spelled: *clay pan.*

clay parting (a) Clayey material between a vein and its wall. Syn: *parting.* (b) A seam of hardened carbonaceous clay between or in beds of coal, or a thin layer of clay between relatively thick beds of some other rock (such as sandstone).

clay plug A mass of silt, clay, and organic muck, deposited in and eventually filling an oxbow lake.

clay pocket A clay-filled cavity in rock; a mass of clay in rock or gravel.

clay rock An indurated clay, composed of argillaceous detrital material derived chiefly from the decomposition of feldspars, and sufficiently hardened to be incapable of being worked without grinding, but not chemically altered or metamorphosed; a *claystone.*

clay shale (a) A consolidated sediment consisting of no more than 10% sand and having a silt/clay ratio of less than 1:2 (Folk, 1954, p.350); a fissile claystone. (b) A shale that consists chiefly of clayey material and that becomes clay on weathering.

clay size A term used in sedimentology for a volume less than that of a sphere with a diameter of 1/256 mm (0.00016 in.). See also: *dust size.*

clay skin An *argillan* consisting of a coating of oriented clay minerals on the surface of a ped or on the wall of a void in a soil material.

clay slate (a) A low-grade, essentially unreconstituted slate, as distinguished from the more micaceous varieties that border on phyllite; specif. an *argillite,* less than 50% reconstituted, with a parting, slaty cleavage, or incipient foliation, or a weakly metamorphosed rock intermediate in character between a shale and a slate (Flawn, 1953, p.564). (b) A slate derived from an argillaceous rock, such as shale, rather than from volcanic ash; a metamorphosed clay, with cleavage developed by shearing or pressure, as distinguished from "mica slate". (c) An English term much used in the early 19th century for true slate.—Also spelled: *clayslate; clay-slate.*

clay soil A soil containing a high percentage of fine particles and colloidal substances, becoming sticky and plastic when wet and forming hard lumps or clods when dry; specif. a soil containing 40% or more of clay and not more than 45% of sand or 40% of silt. The term has also been used for a soil containing 30% or more of clay. Syn: *clay [soil].*

claystone [ign] (a) An obsolete term for a dull, altered, feldspathic igneous rock in which the groundmass or the whole rock has been reduced to a compact mass of earthy or clayey alteration products (Holmes, 1928, p. 61-62). (b) A term used in Australia for a soft earthy feldspathic rock occurring in veins and having the appearance of indurated clay.—Also spelled: *clay stone.*

claystone [sed] (a) An indurated clay having the texture and composition of shale but lacking its fine lamination or fissility; a massive *mudstone* in which clay predominates over silt; a nonfissile *clay shale.* Flawn (1953, p. 562-563) regards claystone as a weakly indurated sedimentary rock whose constituent particles have diameters less than 0.01 mm. Shrock (1948a) describes it as a somewhat unctuous, conchoidally fracturing sedimentary rock composed largely of clay material. Syn: *clay rock.* (b) A concretionary clay found in alluvial deposits in the form of flat rounded disks that are variously united to give rise to curious shapes (Fay, 1918, p.160). Also, a calcareous concretion frequently found in a bed of clay. (c) An old English term for an argillaceous limestone (Arkell

& Tomkeieff, 1953, p.24).—Also spelled: *clay stone.*

claystone porphyry An old and indefinite name for a *porphyry* whose fine groundmass is more or less kaolinized "so as to be soft and earthy, suggesting hardened clay" (Kemp, 1934, p. 203). Not recommended usage.

clay vein A body of clay, usually roughly tabular in form like a dike or vein, that fills a crevice in a coal seam. It is believed to originate where clay from the roof or floor has been forced into a small fissure, often altering or enlarging it. Cf: *horseback [coal]; spar [mining]; clastic dike.* Syn: *dirt slip.*

clay-with-flints (a) A term used in southern England (as on the North Downs and in Dorset) for a residual deposit of reddish-brown, tenacious clay containing mechanically unworn flint fragments, lying unevenly and directly on the surface of chalk or occurring in funnel-shaped pipes penetrating to considerable depths. It represents in part the insoluble residue of chalk subjected to prolonged subaerial weathering and in part admixed waste material derived from formerly overlying Tertiary rocks. There is much variation in the relative proportions of flints and clay. (b) A term applied loosely to any clay-flint drift deposit that rests on chalk.

clean (a) Said of a diamond or other gemstone that is free from noticeable internal flaws. (b) Said of a mineral that is virtually free of undesirable nonore or waste-rock material.

clean sandstone A relatively pure or well-washed sandstone containing little matrix; specif. an arenite with less than 10% argillaceous matrix (Williams, Turner & Gilbert, 1954) or an orthoquartzite with less than 15% detrital clay matrix (Pettijohn, 1954). The particles are held together by a mineral cement. It is usually deposited by fluids of low density. Cf: *dirty sandstone.*

clearing *polynya.*

cleat In a coal seam, a joint or system of joints along which the coal fractures. There are usually two cleat systems developed perpendicular to each other. See also: *face cleat; end cleat.* Also spelled: *cleet.*

cleating A syn. of *jointing,* used with reference to coal.

cleat spar Crystalline mineral matter occurring in the *cleat* planes of a coal seam; specif. *ankerite.*

cleavage [mineral] The breaking of a mineral along its crystallographic planes, thus reflecting crystal structure. The types of cleavage are named according to the structure, e.g. *prismatic cleavage.* Cf: *fracture [mineral]; parting [cryst].*

cleavage [struc geol] The property or tendency of a rock to split along secondary, aligned fractures or other closely spaced, planar structures or textures, produced by deformation or metamorphism. See also: *schistosity.* Obsolete syn: *secondary cleavage.*

cleavage banding A compositional banding that is parallel to the cleavage rather than to the bedding. It results from the mechanical movement of incompetent material, such as argillaceous rocks, into the cleavage planes in a more competent rock, such as sandstone. The argillaceous bands are commonly only a few millimeters thick (Billings, 1954). Cf: *segregation banding.*

cleavage face In a crystal, a smooth surface produced by cleavage. It may be almost planar, e.g. in mica.

cleavage fan *fan cleavage.*

cleavage fold A *shear fold* in which the shear occurs along cleavage planes of secondary foliation. Syn: *shear-cleavage fold.*

cleavage fragment A fragment of a crystal that is bounded by cleavage faces.

cleavage mullion A type of *mullion* formed by the intersection of cleavage planes with bedding (Wilson, 1953). Cf: *pencil cleavage; fold mullion.*

cleavage plane One of the surfaces along which a rock tends to split because of *cleavage.* Cleavage planes are parallel or subparallel.

cleavelandite A white, lamellar or leaflike variety of albite, having an almost pure Ab content and often forming fan-shaped aggregates of tabular crystals that show mosaic development and look as though bent. It is formed as a late-stage mineral in pegmatites, replacing other minerals. Also spelled: *clevelandite.*

cleavings The partings in a coal seam that separate it into beds.

cleet A less-preferred spelling of *cleat.*

cleft [geomorph] An abrupt chasm, cut, breach, or other sharp opening, such as a craggy fissure in a rock, a wave-cut gully in a cliff, a trench on the ocean bottom, a notch in the rim of a volcanic crater, or a narrow recess in a cave floor. Obsolete syn: *clift [geomorph].*

cleft [paleont] A tension fracture formed during growth on either side of the shell, anterior and/or posterior to the beak, of some rostroconch mollusks (Pojeta & Runnegar, 1976).

cleft deposit A *pocket,* specifically a fissure filling, in alpine regions.

cleft girdle On a fabric diagram, an annular maximum occupying a small circle of the net (Turner and Weiss, 1963, p. 58). Cf: *girdle; maximum.* Syn: *small-circle girdle.*

cleftstone *flagstone.*

cleme A long hexactinellid-sponge spicule (monactin) with alternating thornlike lateral spines arranged in two opposite rows. Cf: *uncinate.*

Clerici solution A solution of thallium malonate and thallium formate in water that is used as a *heavy liquid;* its specific gravity is 4.15. Cf: *Sonstadt solution; Klein solution; bromoform; methylene iodide.*

cleuch A Scottish var. of *clough.*

cleugh A Scottish var. of *clough.*

cleve (a) An English syn. of *cliff [geomorph].* (b) An English term for *brow* or steeply sloping ground.

cleveite A variety of uraninite containing rare earths (cerium).

clevelandite *cleavelandite.*

cliachite (a) A ferruginous bauxite. (b) A group name for colloidal aluminum hydroxides constituting most bauxite. Also spelled: *kliachite.* Syn: *alumogel.*

cliff [geomorph] (a) *sea cliff.* (b) Any high, very steep to perpendicular or overhanging face of rock; a precipice. A cliff is usually produced by erosion, less commonly by faulting. (c) A British term for a steep slope or declivity, or a hill. English syn: *cleve.* Dialectical var: *clift [geomorph].*

cliff [sed] (a) *clift.* (b) An old term used in SW England for rock lying directly above or between coal seams. Pl: *clives.*

cliffed headland A headland characterized by a cliff, such as one formed by erosion during the early development of an embayed coast.

cliff erosion *sapping [geomorph].*

cliff-foot cave A solution cave at the foot of a *karst tower* or cliff. Syn: *foot cave.*

cliff glacier A short glacier that occupies a niche or hollow on a steep slope and does not reach a valley, such as a glacier perched on a ledge or bench on the face of a cliff. Cf: *hanging glacier.* Nonpreferred syn: *cornice glacier.*

cliffline The *coastline* on a steep coast, represented by an imaginary line along the base of the cliffs.

cliff of displacement *fault scarp.*

cliffordite A mineral: UTe_3O_8.

cliff overhang A rock mass jutting out from a slope; esp. the upper part or edge of an eroded cliff projecting out over the lower, undercut part, as above a wave-cut notch.

cliffstone A hard chalk found in England and used in paint, as a filler for wood, and in the manufacture of rubber (Thrush, 1968, p. 218).

clift [geomorph] (a) Obsolete var. of *cleft.* (b) Dialectal var. of *cliff [geomorph].*

clift [sed] A term used in southern Wales for various kinds of shale, esp. a strong, usually silty, mudstone. Syn: *cliff.*

cliftonite A black polycrystalline aggregate of graphite, with cubic morphology, representing a minutely crystalline form of carbon occurring in meteorites. It is considered by some to be a pseudomorph after diamond.

climate The characteristic weather of a region, particularly as regards temperature and precipitation, averaged over some significant interval of time. See also: *climate classification; climatic province.*

climate classification An arrangement or description of the various climate types by particular descriptive factors, such as temperature, rainfall, vegetation, or position relative to land and sea; e.g. *Thornthwaite's classification; Köppen's classification.* See also: *climatic zone; temperature zone.*

climate-stratigraphic unit A term used by the American Commission on Stratigraphic Nomenclature (1959, p.669) for a time interval now known as a *geologic-climate unit.* It is not strictly a stratigraphic unit.

climatic Said of ecologic formations resulting from or influenced by differences in climate. Cf: *edaphic.*

climatic accident A departure from the normal cycle of erosion, caused by marked changes in the climate, such as those effected by glaciation or by a change to aridity, independent of the normal climatic change due to loss of relief from youth to old age. "The term has become virtually obsolete since the idea of humid temperate climate being 'normal' no longer holds credence" (Monkhouse, 1965, p. 3).

climatic amelioration A term designating a change to a warmer climate, applied specif. to the primary and secondary climatic trends of late glacial and Holocene time; an "improvement" of climate as seen by someone in high latitudes. Cf: *climatic deterioration.*

climatic deterioration A term designating a change to a colder climate, applied specif. to the primary and secondary climatic trends of late glacial and Holocene time such as occurred during the Little Ice Age; a "degeneration" or "worsening" of climate as seen by someone in high latitudes. Cf: *climatic amelioration.*

climatic optimum An informal term designating the postglacial interval of most equable climate, with warm temperatures and abundant rainfall. The concept is derived from the mid-Holocene warm interval and refers to *Atlantic* climatic intervals. As it began earlier in low latitudes, it is diachronous and should not be used as a precise time interval (Fairbridge, 1972). See also: *thermal maximum; Hypsithermal.*

climatic peat Peat that characteristically occurs in a certain climatic zone.

climatic province A region characterized by a particular *climate.*

climatic snowline (a) The average line or altitude above which horizontal surfaces have more than 50 percent snow cover, averaged over a long time period of climatic significance (e.g. 10 years). Cf: *snowline.* (b) The same line as observed in late summer so that it approximately coincides with the firn line or *equilibrium line* on glaciers.—See also: *regional snowline.*

climatic terrace A stream terrace whose formation is controlled by climatic changes that may induce aggradation or degradation of a valley.

climatic zone A general term for a latitudinal region characterized by a relatively homogeneous climate, e.g. any of the zones delimited by the tropics of Cancer and Capricorn and the Arctic and Antarctic circles, or a zone or province in some *climatic classification.* Cf: *temperature zone.*

climax In ecology, the final stable or equilibrium stage of development that a sere, community, species, flora, or fauna attains in a given environment. The major world climaxes correspond to *formations [ecol]* and *biomes.* Cf: *pioneer.*

climax avalanche A snow avalanche of maximum size, containing a large portion of old snow, and arising from conditions that developed over a period of time longer than one year. The term is part of an obsolete classification of avalanches.

climax community An assemblage of species that represents the permanent or usual long-term inhabitants of a region during ecologic or seral succession; it is normally preceded by one or more shorter-lived assemblages whose life activities prepare the habitat. Cf: *climax; pioneer.*

climbing bog An elevated boggy area on the margin of a swamp, usually in a region characterized by a short summer and considerable rainfall, caused by the upward growth of sphagnum from the original level of the swamp to higher ground.

climbing dune A dune formed by the piling-up of sand by wind against a cliff or mountain slope. Syn: *rising dune.*

climbing ripple One of a series of cross-laminae produced by superimposed migrating ripples, in which the crests of vertically succeeding laminae appear to be advancing upslope. See also: *ripple drift; cross-stratification.*

climograph A graphic representation of climatic data, esp. mean monthly values of precipitation and temperature.

cline A gradational series of variant forms (e.g. morphologic or physiologic variations) within a group of closely related (usually conspecific) organisms, generally developing as a result of environmental, geographic, chronological, or stratigraphic transition.

clinker [coal] (a) Coal that has been altered by igneous intrusion. Cf: *natural coke.* Syn: *scoria [coal].* (b) Masses of coal ash that are a byproduct of combustion. Cf: *coke.*

clinker [volc] A rough, jagged pyroclastic or autobrecciated fragment, such as *aa* that resembles the clinker or slag of a furnace. Adj: *clinkery.*

clinkertill Glacial till baked by the burning of lignite beds.

clinkery Adj. of *clinker [volc];* it is used to describe the surface of a lava flow.

clinkstone An obsolete syn. of *phonolite* (in its broadest sense).

clino adj. A term applied by Rich (1951, p. 2) to the environment of sedimentation that lies on the sloping part of the floor of a water body, extending from wave base down to the more or less level deeper parts. It may be used alone or as a combining form. See also: *clinoform; clinothem.* Cf: *unda; fondo.*

clinoamphibole (a) A group name for amphiboles crystallizing in the monoclinic system. (b) Any monoclinic mineral of the amphibole group, such as hornblende, cummingtonite, grunerite, tremolite, actinolite, riebeckite, glaucophane, and arfvedsonite.—Cf: *orthoamphibole.*

clinoaugite *clinopyroxene.*

clinoaxis In a monoclinic crystal, the lateral axis that is oblique to the vertical; it is the *a* axis [cryst]. Cf: *orthoaxis.*

clinobarrandite A variety of the mineral phosphosiderite containing aluminum.

clinobisvanite A monoclinic mineral: $BiVO_4$. It is dimorphous with pucherite.

clinobronzite A variety of clinoenstatite containing iron; a clinopyroxene intermediate in composition between clinoenstatite and clinohypersthene, having less than 20 mole percent of $FeSiO_3$.

clinochlore A greenish mineral of the chlorite group: $(Mg,Fe^{+2},Al)_3(Si,Al)_2O_5(OH)_4$. It is sometimes substantially free from iron. Cf: *ripidolite.*

clinochrysotile A syn. of *chrysotile.* The term is used to denote its monoclinic form. Cf: *orthochrysotile.*

clinoclase A dark-green mineral: $Cu_3(AsO_4)(OH)_3$. Syn: *clinoclasite.*

clinodome A *first-order prism* in the monoclinic system. Its indices are $\{0kl\}$ and its symmetry is $2/m$. Cf: *brachydome.*

clinoenstatite A mineral of the clinopyroxene group: $(Mg,Fe)SiO_3$; specif. the monoclinic magnesium silicate $MgSiO_3$.

clinoferrosilite A mineral of the clinopyroxene group: $(Fe,Mg)SiO_3$; specif. a mineral consisting of the monoclinic iron silicate $FeSiO_3$. See also: *ferrosilite.* Cf: *orthoferrosilite.*

clinoform The subaqueous land form analogous to the continental slope of the oceans or to the foreset beds of a delta (Rich, 1951, p. 2). It is the site of the *clino* environment of deposition. Cf: *undaform; fondoform.*

clinoform surface A sloping depositional surface, commonly associated with strata prograding into deep water (Mitchum, 1977, p. 205).

clinograde Pertaining to the decreasing concentration of oxygen or other chemicals in the *hypolimnion* of a lake. Cf: *orthograde.*

clinographic Pertaining to a representation of a crystal in which no crystal face is projected as a line.

clinographic curve A curve representing the slope or slopes of an area of the Earth's surface as it varies with altitude; in practice it is designed to show the actual variation of the average slope within each contour interval.

clinographic projection An oblique projection used for representing crystals in such a manner that no crystal face will be projected as a line.

clinohedral class *domatic class.*

clinohedrite A colorless, white, or purplish monoclinic mineral: $CaZnSiO_3(OH)_2$.

clinoholmquistite A monoclinic mineral of the amphibole group: $(Na,Ca)(Al,Li,Mg,Fe)_7Si_8O_{22}(OH,F)_2$.

clinohumite A monoclinic mineral of the humite group: $Mg_9Si_4O_{16}(F,OH)_2$.

clinohypersthene A mineral of the clinopyroxene group: $(Mg,Fe)SiO_3$. It has a higher iron content (20-50 mole percent of $FeSiO_3$) than that of clinoenstatite.

clinolimnion The upper part of a *hypolimnion,* where the rate of heat absorption falls off almost exponentially with depth. Cf: *bathylimnion.*

clinometer Any of various instruments used for measuring angles of slope, elevation, or inclination (esp. the dip of a geologic stratum or the slope of an embankment); e.g. a simple hand-held device consisting of a tube with cross hair, with a graduated vertical arc and an attached spirit level so mounted that the inclination of the line of sight can be read on the circular scale by centering the level bubble at the instant of observation. A clinometer is usually combined with a compass (e.g. the *Brunton compass*). See also: *inclinometer [drill].*

clinopinacoid In a monoclinic crystal, a pinacoid that is parallel to the mirror plane of symmetry and perpendicular to the axis of symmetry.

clinoplain An inclined plain projecting from the mountains and forming a low bluff on the side of a flood plain, as in the Rio Grande valley (Herrick, 1904, p. 379).

clinoptilolite A zeolite mineral: $(Na,K,Ca)_{2-3}Al_3(Al,Si)_2Si_{13}O_{36} \cdot 12H_2O$. It is a potassium-rich variety of *heulandite.*

clinopyroxene (a) A group name for pyroxenes crystallizing in the monoclinic system and sometimes containing considerable calcium with or without aluminum and the alkalies. (b) Any monoclinic mineral of the pyroxene group, such as diopside, hedenbergite, clinoenstatite, clinohypersthene, clinoferrosilite, augite, acmite, pigeonite, spodumene, jadeite, and omphacite.—Cf: *orthopyroxene.* Syn: *monopyroxene; clinoaugite.*

clinopyroxenite In the *IUGS classification,* a plutonic rock with M equal to or greater than 90, and cpx/(ol+opx+cpx) greater than 90.

clinosafflerite A monoclinic mineral: $(Co,Fe,Ni)As_2$. It is dimorphous with safflorite.

clinostrengite *phosphosiderite.*

clinothem Rock units formed in the *clino* environment of deposition (Rich, 1951, p. 2). Cf: *undathem; fondothem.*

clinounconformity An obsolete syn. of *angular unconformity.* Term proposed by Crosby (1912, p.296). Also spelled: *clinunconformity.*

clinoungemachite A monoclinic mineral consisting of a sulfate of ferric iron, sodium, and potassium. It is probably dimorphous with ungemachite.

clinozoisite A grayish-white, pink, or green mineral of the epidote group: $Ca_2Al_3Si_3O_{12}(OH)$. It is the monoclinic dimorph of *zoisite* and grades into, but is lighter in color than, epidote.

clint [geomorph] A Scottish term used in a general sense for any hard or flinty rock, such as a ledge projecting from a hillside or in a stream bed; also, a rocky cliff.

clint [karst] A slab of *limestone pavement* that is separated from adjacent clints by *solution fissures* along joints.

Clintonian Stage in New York state: Middle Silurian (above Medinan, below Lockportian).

clintonite (a) A mineral of the brittle-mica group: $Ca(Mg,Al)_3(Al_3Si)O_{10}(OH)_2$. It has a reddish-brown, copper-red, or yellowish color, and occurs in monoclinic crystals and foliated masses. Syn: *seybertite; xanthophyllite.* (b) A group name for the *brittle micas.*

Clinton ore A red, fossiliferous sedimentary iron ore, e.g. the Clinton formation (Middle Silurian) or correlative rocks of the east-central U.S., containing lenticular or oolitic grains of hematite. Cf: *fossil ore; flaxseed ore.*

clinunconformity *clinounconformity.*

clisere A *sere* that develops as a result of great physiographic changes.

clisiophylloid Said of an *axial structure* in a rugose coral (such as in *Clisiophyllum*) characterized by a short medial plate joining the cardinal and counter septa and resembling a spider web in transverse section.

clives Plural of *cliff [sed],* so called because of the "easy cleavage" of the rocks overlying the coal.

cloaca (a) An exhalant chamber in an invertebrate; esp. the large central cavity of a sponge into which pores and/or canals empty and which communicates through the osculum directly and externally with the surface of the sponge. Syn: *atrium; spongocoel; paragaster.* (b) The common excretory and reproductive passage of all vertebrates except marsupial and placental mammals.— Pl: *cloacae.*

clockwise inclination The inclination to the right of a heterococcolith suture as it proceeds to the periphery. Ant: *counterclockwise inclination.*

clod [mining] A miner's term applied to a soft, weak, or loosely consolidated shale (or to a hard, earthy clay), esp. one found in close association with coal or immediately overlying a coal seam. It is so called because it falls away in lumps when worked.

clod [soil] An artificially formed aggregate of soil particles. Cf: *ped.*

clone A group of individuals descended asexually from a single parent.

Cloosian dome An elliptical uplift or upwarping, described by Hans Cloos and exemplified in the East African and Rhine rift valleys (1939).

closed basin An enclosed area having no drainage outlet, from which water escapes only by evaporation, as in an arid region. Cf: *interior basin.*

closed bay A bay indirectly connected with the sea through a narrow pass.

closed-cavity ice Ice, commonly as large crystals of hoarfrost, formed in underground cavities within permafrost.

closed contour A contour line that forms a closed loop and does not intersect the edge of the map area on which it is drawn; e.g. a *depression contour* indicating a closed depression, or a normal contour indicating a hilltop.

closed depression An area of lower ground indicated on a topographic map by a hachured *depression contour* line forming a closed loop; e.g. a fault sag, or a hollow below the general land surface, with no surface outlet. Syn: *topographic depression.*

closed drainage *internal drainage.*

closed fold An old, rarely used syn. for *isoclinal fold.*

closed form A crystal form whose faces enclose space, e.g. a dipyramid. Cf: *open form.*

closed-in pressure *shut-in pressure.*

closed lake A lake that does not have a surface effluent and that loses water by evaporation (as in an arid or semiarid region, where the lake is usually saline or brackish) or by seepage (e.g. a *seepage lake*). Cf: *enclosed lake.* Ant: *open lake.*

closed ridge A circular, elliptical, or irregularly shaped ridge of glacial material surrounding a central depression (or sometimes a mound of glacial material or a moraine plateau), and resulting from the melting of a block of stagnant ice (Gravenor & Kupsch, 1959, p. 52-53).

closed structure A structure which, when represented on a map by contour lines, is enclosed by one or more closed contours; e.g. a closed anticline or a closed syncline. Ant: *open structure.*

closed system [chem] A chemical system in which, during the process under consideration, no transfer of matter (either into or out of the system) takes place. Cf: *open system [chem].*

closed system [permafrost] A condition of freezing of thawed ground within permafrost in which no additional ground water is available (Muller, 1947, p.214), exemplified by the pingos of the Mackenzie Delta, Canada. Ant: *open system [permafrost].*

closed traverse A surveying traverse that starts and terminates upon the same station or upon a station of known position. Cf: *open traverse.*

close fold A fold with an inter-limb angle between 30° and 70° (Fleuty, 1964, p. 470).

close-grained Said of a rock, and of its texture, characterized by fine, tightly packed particles.

close ice An obsolete term for sea-ice concentration of 8/10 to 10/10; now replaced generally by *close pack ice* and *very close pack ice.* Syn: *close pack; packed ice.*

close-joints cleavage An old term for both *slip cleavage* and *fracture cleavage.*

close-packed structure A type of crystal structure that provides the tightest possible packing: a first layer of atoms in which each atom has six similar atoms touching it, a second layer of atoms fitting into the indentations of the first, and a third layer either as a repetition of the first or in a third position. Cf: *open-packed structure.*

close pack ice Pack ice in which the ice cover or concentration is 7/10 to 9/10 and composed of floes mostly in contact. See also: *close ice; broken ice.*

close packing The manner of arrangement of uniform solid spheres packed as closely as possible so that the porosity is at a minimum; e.g. the packing of a face-centered cubic lattice or of a close-packed hexagonal lattice. See also: *rhombohedral packing.* Ant: *open packing.*

close suture A suture between immovably united but not fused crinoid ossicles. Cf: *loose suture.*

closing error *error of closure.*

closing the horizon Measuring the last angle of a series of adjacent horizontal angles at a station, required to make the series complete around the horizon. See also: *horizon closure.*

closterite A dense, laminated, brownish-red canneloid material from the Irkutsk River basin of Siberia. The organic matter is largely *Pila bibractensis* colonies (Twenhofel, 1950, p. 475).

closure [drill] The difference in the relative positions of the bottom and the surface location of a borehole, expressed in horizontal distance in a specific compass direction (Long, 1960).

closure [paleont] Calcification of the frontal membrane of an anascan cheilostome (bryozoan) autozooid, with loss of its lophophore and feeding function. Also, the calcified frontal structure so formed (Cook, 1965, p. 159).

closure [struc geol] In a subsurface fold, dome, or other structural trap, the vertical distance between the structure's highest point and its lowest closed *structure contour.* It is used in the estimation of petroleum reserves. Syn: *structural closure.*

closure [surv] A cumulative measure of the various individual errors in survey measurements; the amount by which a series of survey measurements fails to yield a theoretical or previously determined value for a survey quantity. See also: *error of closure; discrepancy.*

clot A group of ferromagnesian minerals in igneous rock, from a few inches to a foot or more in size, commonly drawn out longitudinally, that may be a segregation or an altered xenolith (Balk, 1937).

clotted *grumous.*

cloud A visible aggregate of minute water droplets or ice crystals above the Earth's surface. Water droplets form by condensation around nuclei such as particles of dust or pollen; ice crystallizes on dust particles or on other ice crystals. Clouds are classified according to height and shape.

cloudburst flood An ephemeral flood commonly occurring during an abrupt summer rain of high intensity, usually in an arid or semiarid region.

clouded agate A transparent or semitransparent light-gray *agate* with irregular, indistinct, or more or less rounded patches of darker gray resembling dark clouds. Syn: *cloud agate; cloudy agate.*

clouding The effect produced in crystals (as of plagioclase) by the presence of numerous minute dark particles (microlites, dustlike specks, short rods, thin hairlike growths, needles) distributed throughout. The particles consist of one or more minerals recognized with difficulty and seldom with certainty.

clough (a) A British term for a cleft in a hill; esp. a ravine, gorge, or glen with precipitous and rocky sides. (b) The cliff or precipitous face of a clough.—Pron: *kluf; klow* (as in "now"). Syn: *cleugh; cleuch; claugh.*

clove A term used in the Catskill Mountains, N.Y., for a narrow, deep valley; esp. a ravine or gorge. Etymol: Dutch *kloof,* "cleft".

Clovelly North American (Gulf Coast) stage: Miocene (above Duck Lakean, below Foleyan).

club moss One of a group of vascular cryptogams (chiefly the genera *Lycopodium* and *Selaginella*) that have small, simple leaves with a single midvein and have the sporangia borne in the axils of the leaves. Sometimes the sporophylls are modified and grouped into a terminal cone. See also: *lycopod.*

clunch (a) A term used in England for various stiff, tough, or indurated clays, esp. one forming the floor of a coal seam; a fireclay or underclay, or a bluish hard clay. Also, a soft, fine-grained, often clayey rock (shale) that breaks readily into irregular layers and that does not make a good roof during coal mining. See also: *stone clunch.* (b) A term used in England for a soft limestone; specif. a marly chalk.

cluse A narrow deep gorge, trench, or water gap, cutting transversely through an otherwise continuous ridge; esp. an antecedent valley crossing an anticlinal limestone ridge in the Jura Mountains of the European Alps. Etymol: French. Pron: *klooz.* Cf: *combe; val.*

cluster A group of criss-crossing dikes demonstrably related to an exposed pluton. Cf: *dike swarm; dike set.*

cluster analysis A procedure for arranging a number of objects in homogeneous subgroups based on their mutual similarities and hierarchical relationships.

clymenid Any ammonoid belonging to the order Clymeniida, characterized by the dorsal, rather than ventral, position of the marginal siphuncle. Clymenids are found only in the Upper Devonian.

clypeus (a) The labrum-carrying part of the cephalon of a crustacean. (b) The part of the carapace of an arachnid between its anterior edge and the eyes.—Pl: *clypei.*

CM diagram *CM pattern.*

CM pattern A sample-point pattern designed to distinguish different depositional environments of sediments and to define, compare, and correlate clastic sediments (Passega, 1957, p. 1952). It represents the variations in a sedimentary deposit of the maximum or one-percentile particle size (C) and the median particle size (M), as plotted on a diagram (C on the ordinate, M on the abscissa). Syn: *CM diagram.*

cnidarian Any coelenterate belonging to the subphylum Cnidaria, characterized mainly by the presence of nematocysts and simple muscles. All known fossil coelenterates belong to this subphylum.

cnidoblast A cell that produces a coral nematocyst or that develops into a nematocyst.

Coahuilan North American provincial series: Lower Cretaceous (above Upper Jurassic, below Comanchean).

coak An obsolete var. of *coke.*

coal A readily combustible rock containing more than 50% by weight and more than 70% by volume of carbonaceous material including inherent moisture, formed from compaction and induration of variously altered plant remains similar to those in peat. Differences in the kinds of plant materials (type), in degree of metamorphism (rank), and in the range of impurity (grade) are characteristic of coal and are used in classification (ASTM, 1970, p.70). Syn: *black diamond.*

coal age *age of coal.*

coal ball A concretion of mineralized plant debris, occurring in a coal seam or in adjacent rocks. Not to be confused with *ball coal.* Cf: *sulfur ball [coal].*

coal basin A coalfield with a basinal structure, e.g. the Carboniferous coal measures of England.

coal bed A *coal seam.* Also spelled: *coalbed.*

coal blende *coal brass.*

coal blossom *coal smut.*

coal brass Iron pyrites (pyrite) found in coal or coal seams. Syn: *coal blende.*

coal breccia Naturally fragmented coal in a seam. The fragments often show polished or slickensided surfaces (Stutzer & Noe, 1940, p.248).

coal classification (a) The analysis or grouping of coals according to a particular property, such as degree of metamorphism (*rank*), constituent plant materials (*type*), or degree of impurity (*grade*). (b) The analysis or grouping of coals according to the percentage of volatile matter, caking properties, and coking properties.

coal clay *underclay.*

coal equivalent The heat energy of fuels other than coal, expressed in terms of comparable heat energy of coal.

coalescing fan One of a series of confluent alluvial fans that form a *bajada.*

coalescing pediment One of a series of expanding pediments that join to produce a continuous pediment surrounding a mountain range or that merge over a broad region to ultimately reduce a desert mountain mass to an approximately continuous level; one of a number of pediments that make up a *pediplain.*

coalescive neomorphism A term introduced by Folk (1965, p. 22) for *aggrading neomorphism* in which small crystals are converted to large ones by gradual enlargement maintaining a uniform crystal size at all times (all crystals are consuming or being consumed); e.g. the process that forms most microspar calcite. Cf: *porphyroid neomorphism.*

coalfield A region in which coal deposits of known or possible economic value occur.

coal gas The fuel gas produced from *gas coal;* its average composition, by volume, is 50% hydrogen, 30% methane, 8% carbon monoxide, 4% other hydrocarbons, and 8% carbon dioxide, nitrogen, and oxygen (Nelson, 1965, p.89).

coal gravel A secondary coal deposit consisting of transported and redeposited fragments. Cf: *float coal.*

coalification The alteration or metamorphism of plant material into coal; the biochemical processes of diagenesis and the geochemical process of metamorphism in the formation of coal. See also: *carbonization; incorporation; vitrinization; fusinization; peat-to-anthracite theory; coalification break.* Syn: *carbonification; incarbonization; incoalation; bitumenization.*

coalification break That point in the process of *coalification* at which the exinites change from dark to pale gray in reflected light, owing to loss of methane. This change coincides with a significant change in the slope of curves of various coalification parameters, and corresponds to a volatile-matter content in vitrain of about 29% (on a dry, ash-free basis). Syn: *coalification jump.*

coalification jump *coalification break.*

coalingite A mineral: $Mg_{10}Fe_2(CO_3)(OH)_{24} \cdot 2H_2O$.

coal land An area containing coal beds which falls within the public domain.

Coal Measures A stratigraphic term used in Europe (esp. in Great Britain) for Upper Carboniferous, or for the sequence of rocks (typically, but not necessarily, coal-bearing) occurring in the upper part of the Carboniferous System. It is broadly synchronous with the Pennsylvanian of North America.

coal measures (a) A succession of sedimentary rocks (or *measures*) ranging in thickness from a meter or so to a few thousand meters, and consisting of claystones, shales, siltstones, sandstones, conglomerates, and limestones, with interstratified beds of coal. (b) A group of coal seams.

coal-measures unit A sequence (from oldest to youngest) of coal, shale, and sandstone, occurring in coal measures.

coal pipe A cylindrical extension from a coal seam into the overlying rock, representing a tree stump that was rapidly buried. Cf: *baum pot; bell; caldron bottom.*

coal plant A fossil plant found in association with, or contributing by its substance to the formation of, beds of coal, esp. in the coal measures.

coal seam A stratum or bed of coal. Syn: *coal bed.*

coal seat *seat earth.*

coal smut An earthy coal stratum near the surface; the weathered outcrop of a *coal seam.* Syn: *coal blossom.*

coal split *split.*

coaly Covered with coal, or containing or resembling coal; e.g. "coaly rashings", small pieces of soft, dark shale containing much carbonaceous material.

coarse Composed of or constituting relatively large particles; e.g. "coarse sandy loam". Ant: *fine [sed].*

coarse admixture A term applied by Udden (1914) to an *admixture* (in a sediment of several size grades) whose particles are coarser than those of the dominant or maximum grade; material coarser than that found in the maximum histogram class.

coarse aggregate The portion of an *aggregate* consisting of particles with diameters greater than approximately 1/4 inch or 4.76 mm. Cf: *fine aggregate.*

coarse clay A geologic term for a *clay* particle having a diameter in the range of 1/512 to 1/256 mm (2-4 microns, or 9 to 8 phi units). Also, a loose aggregate of clay consisting of coarse clay particles.

coarse fragment A rock or mineral particle in the soil with an equivalent diameter greater than 2 mm; it may be gravelly, cobbly, stony, flaggy, cherty, slaty, or shaly. See also: *channery.* Cf: *soil separate; fine earth.*

coarse-grained (a) Said of a crystalline rock, and of its texture, in which the individual minerals are relatively large; specif. said of an igneous rock whose particles have an average diameter greater than 5 mm (0.2 in.). Johannsen (1931, p. 31) earlier used a minimum diameter of 1 cm, and referred to igneous rocks having walnut-size to coconut-size grains as "very coarse-grained". Syn: *phaneritic.* (b) Said of a sediment or sedimentary rock, and of its texture, in which the individual constituents are easily seen with the unaided eye; specif. said of a sediment or rock whose particles have an average diameter greater than 2 mm (0.08 in., or granule size and larger). Cf: *coarsely crystalline.* The term is used in a relative sense, and various size limits have been suggested and used. Cf: *fine-grained; medium-grained.* (c) Said of a soil in which gravel and/or sand predominates. In the U.S., the minimum average diameter of the constituent particles is 0.05 mm (0.002 in.), or, as used by engineers, 0.074 mm (retained on U.S. standard sieve no.200); the International Society of Soil Science recognizes a diameter limit of 0.02 mm. Cf: *fine-grained.*

coarse gravel An engineering term for *gravel* whose particles have a diameter in the range of 19-76 mm (3/4 to 3 in.).

coarsely crystalline Descriptive of an interlocking texture of a carbonate sedimentary rock having crystals whose diameters are in the range of 0.25-1.0 mm (Folk, 1959), or exceed 0.2 mm (Carozzi & Textoris, 1967, p. 5) or 4 mm (Krynine, 1948, p. 143). Cf: *coarse-grained.*

coarse pebble A geologic term for a *pebble* having a diameter in the range of 16-32 mm (0.6-1.3 in., or -4 to -5 phi units) (AGI, 1958).

coarse sand (a) A geologic term for a *sand* particle having a diameter in the range of 0.5-1 mm (500-1000 microns, or 1 to zero phi units). Also, a loose aggregate of sand consisting of coarse sand particles. (b) An engineering term for a *sand* particle having a diameter in the range of 2 mm (retained on U.S. standard sieve no.10) to 4.76 mm (passing U.S. standard sieve no. 4) (c) A soil term used in the U.S. for a *sand* particle having a diameter in the range of 0.5-1 mm. The diameter range recognized by the International Society of Soil Science is 0.2-2 mm. (d) Soil material containing 85% or more of sand-size particles (percentage of silt plus 1.5 times the percentage of clay not exceeding 15), 25% or more of very coarse sand and coarse sand, and less than 50% of any other one grade of sand (SSSA, 1965, p. 347).

coarse silt A geologic term for a *silt* particle having a diameter in the range of 1/32 to 1/16 mm (31-62 microns, or 5 to 4 phi units). In Great Britain, the range 1/20 to 1/10 mm has been used. Also, a loose aggregate of silt consisting of coarse silt particles.

coarse-tail grading In a sedimentary bed, a progressive upward shift toward the finer grain sizes in the coarse tail (1-5%) of the distribution only (Middleton, 1967, p. 487). Cf: *distribution grading.*

coarse topography A topography with coarse *topographic texture,* characterized by low drainage density and widely spaced streams. It is common in regions of resistant rocks where the surface is incompletely dissected or the erosional features are on a large scale.

coast (a) A strip of land of indefinite width (may be many kilometers) that extends from the low-tide line inland to the first major change in landform features. (b) The part of a country regarded as near the coast, often including the whole of the *coastal plain;* a littoral district having some specific feature, such as the Gold *Coast.* Adj: *coastal.*

coastal Pertaining to a coast; bordering a coast, or located on or near a coast, as *coastal* waters, *coastal* zone management, or *coastal* shipping routes.

coastal area The areas of land and sea bordering the shoreline and extending seaward through the breaker zone (CERC, 1966, p. A6).

coastal desert Generally, any desert area bordering an ocean. See also: *west-coast desert.*

coastal dune A sand dune on low-lying land recently abandoned or built up by the sea; the dune may ascend a cliff and travel inland.

coastal energy The total energy, including that of wind, waves, tides, and currents, available for work along the coast. Most well-known coasts are characterized as dominated by *wave energy.*

coastal lake A lake produced by shoreline processes, as by the formation of a bar across a bay or by the joining of an offshore island to the mainland by a *double tombolo.*

coastal marsh A marsh bordering a seacoast, generally formed under the protection of a barrier beach, or enclosed in the sheltered part of an estuary. Cf: *open-coast marsh.*

coastal onlap *onlap.*

coastal plain (a) A low, generally broad plain that has its margin on an oceanic shore and its strata either horizontal or very gently sloping toward the water, and that generally represents a strip of recently prograded or emerged sea floor; e.g. the coastal plain of SE U.S. extending for 3000 km from New Jersey to Texas. (b) In less restricted usage any lowland area bordering a sea or ocean, extending inland to the nearest elevated land, and sloping very gently seaward; it may result from the accumulation of material, as along the Adriatic coast of northern Italy.—Not to be confused with *coast plain.*

coastland Land along a coast; esp. a section of seacoast.

coastline (a) The line that forms the boundary between land and water, esp. the water of a sea or ocean. (b) A general term to describe the appearance or configuration of the land along a coast, esp. as viewed from the sea; it includes bays, but crosses narrow inlets and river mouths. (c) A broad zone of land and water extending indefinitely both landward and seaward from a shoreline.—Cf: *shoreline.*

coast plain (a) A wave-cut plain of denudation. (b) A base level marking the sea level to which the land has been reduced by subaerial forces (Reusch, 1894, p. 349).—Obsolete. Not to be confused with *coastal plain.*

Coast Range orogeny (a) A name proposed by W.H. White (1959) for major deformation, metamorphism, and plutonism during Jurassic and Early Cretaceous time in the Coast Mountains of the Cordillera of British Columbia. It is broadly equivalent to the *Nevadan orogeny* of the U.S. (b) A term sometimes used for the late Cenozoic orogenic events in southern California; *Pasadenean orogeny.*

coast shelf *submerged coastal plain.*

coated grain A term used by Wolf (1960, p. 1414) for a sedimentary particle possessing concentric or enclosing layers of calcium carbonate; e.g. an oolith, a pisolith, or a skeletal grain encrusted by algae or foraminifers.

coated stone A gemstone partly or entirely covered by some transparent substance to heighten color, improve phenomenal effects, or conceal defects.

coaxial Said of a progressive deformation in which the principal directions of total and incremental strain are parallel at all times (Hsu, 1966, p. 217). Ant: *nonaxial.*

cob v. To carry on the process of *cobbing.*

coba Uncemented rock or gravel underlying the nitrate (caliche) deposits of Chile. Cf: *congela.*

cobalt bloom *erythrite.*

cobalt glance *cobaltite.*

cobaltite A grayish to silver-white isometric mineral with a reddish tinge: CoAsS. It usually occurs massive and in association with smaltite, and represents an important ore of cobalt. Syn: *cobalt glance; white cobalt; gray cobalt.*

cobaltocalcite (a) A red variety of calcite containing cobalt. (b) *spherocobaltite.*

cobalt ocher (a) *erythrite.* (b) *asbolite.*

cobaltomenite A mineral: $CoSeO_3 \cdot 2H_2O$.

cobalt pentlandite A mineral: $(Co,Ni)_9S_8$. It is the cobalt analogue of pentlandite.

cobalt pyrite A variety of pyrite containing cobalt.

cobalt pyrites *linnaeite.*

cobalt vitriol *bieberite.*

cobalt-zippeite An orthorhombic mineral: $Co(UO_2)_6(SO_4)_3(OH)_{10} \cdot 16H_2O$.

cobb *promontory.*

cobbing The separation, generally with a hand-held hammer, of worthless minerals from desired minerals in a mining operation, e.g. quartz from feldspar.

cobble [geomorph] A term used in the NE U.S. for a rounded hill of moderate elevation.

cobble [part size] (a) A rock fragment larger than a pebble and smaller than a boulder, having a diameter in the range of 64-256 mm (2.5-10 in., or -6 to -8 phi units, or a size between that of a tennis ball and that of a volleyball), being somewhat rounded or otherwise modified by abrasion in the course of transport; in Great Britain, the range of 60-200 mm has been used. Also, a similar rock fragment rounded in place by weathering at or somewhat below the surface of the ground; e.g. a "cobble of exfoliation" or a "cobble of spheroidal weathering". See also: *large cobble; small cobble.* (b) A rock or mineral fragment in the soil, having a diameter in the range of 20-200 mm (Atterberg, 1905). In the U.S., the term is used for a soil particle having a diameter in the range of 75-250 mm (3-10 in.) (SSSA, 1965, p. 333). Syn: *cobblestone.* (c) An engineering term for a particle having a diameter greater than 76 mm. (d) *cobblestone.*

cobble beach *shingle beach.*

cobble conglomerate A consolidated rock consisting mainly of cobbles.

cobble size A term used in sedimentology for a volume greater than that of a sphere with a diameter of 64 mm (2.5 in.) and less than that of a sphere with a diameter of 256 mm (10 in.).

cobblestone (a) A naturally rounded, usually waterworn stone suitable for use in paving a street or in other construction. Syn: *cobble; roundstone.* (b) A *cobble* in the soil. (c) A consolidated sedimentary rock consisting of cobble-size particles (Alling, 1943, p. 265).

cobbly Characterized by cobbles; e.g. a "cobbly soil" or a "cobbly land" containing an appreciable quantity of cobbles (SSSA, 1965, p. 333).

Coblenzian European stage: upper Lower Devonian (above Gedinnian, below Eifelian). It includes Siegenian and Emsian. Obsolete. Also spelled: *Coblencian.*

cocarde ore *cockade ore.*

coccinite A mercury mineral, supposedly HgI_2.

coccolite A granular variety of diopside of various colors.

coccolith (a) A general term applied to various microscopic calcareous structural elements or buttonlike plates having many different shapes and averaging about 3 microns in diameter (though some have diameters as large as 35 microns), constructed of minute calcite or aragonite crystals, and constituting the outer skeletal remains of a *coccolithophore.* Coccoliths are found in chalk and in deep-sea oozes of the temperate and tropical oceans, and were probably not common before the Jurassic. See also: *rhabdolith.* (b) Two shields connected by a central tube in a coccolithophore. (c) A term loosely applied to a coccolithophore.

coccolithophore Any of numerous minute mostly marine planktonic biflagellate algae, class Chrysophyceae, having brown pigment-bearing cells; at some phase of their life cycle they are encased in a sheath enclosing calcareous platelets, called coccoliths, which form a complex calcareous shell. Coccolithophores are autotrophic but they have been variously classified as algae and protozoans. Var: *coccolithophorid.*

coccosphere (a) The entire spherical or spheroidal test or skeleton of a coccolithophore, composed of an aggregate of interlocking coccoliths that are external to or embedded within an outer gelatinous layer of the cell. (b) A coccolithophore.

cocinerite A mixture of chalcocite and silver.

cocite A *lamprophyre* containing olivine, biotite, and clinopyroxene phenocrysts in a groundmass of leucite, alkali feldspar, biotite, and magnetite. Named by Lacroix in 1933 for Coc-Pia, Vietnam. Not recommended usage.

cockade ore An open-space vein filling in which the ore and gangue minerals are deposited in successive comblike crusts around rock fragments, e.g. around vein breccia fragments. Syn: *cocarde ore; ring ore; sphere ore.*

cockpit A steep-walled, star-shaped closed depression surrounded by conical hills in tropical karst areas, usually about ten times the size of a temperate-karst *sinkhole.* Type area: Jamaica.

cockpit karst A typical karst of the tropics, in which *cockpits* are separated by steep-walled rounded hills, forming a pattern that resembles a molded egg box. Syn: *cone karst; kegel karst; polygonal karst.* See also: *sinkhole karst; tower karst.*

cockscomb barite A comblike variety of barite displaying tabular crystals disposed roughly parallel to one another.

cockscomb pyrites A crestlike form of marcasite in twin crystals. Cf: *spear pyrites.*

cockscomb ridge A term used in South Africa for a wind-chiseled ridge, similar to a *yardang.*

cockscomb structure *hacksaw structure.*

coconinoite A light creamy-yellow secondary mineral: $Fe_2Al_2(UO_2)_2(PO_4)_4(SO_4)(OH)_2 \cdot 20H_2O$.

coconut-meat calcite Calcite forming crusts, a few millimeters to a few centimeters thick, made of white, very thin parallel fibers, perpendicular to the encrusted surface. It often has growth bands running athwart the fibers. Fibers commonly are optically length-slow. This calcite is found in modern and ancient cave deposits (Folk & Assereto, 1976, p. 486-496).

COD *chemical oxygen demand.*

cod A bag-shaped area of water or land; e.g. the inmost recess of a bay or meadow. Rare or colloquial.

coda The concluding part of a seismogram following the early, identifiable waves. Long trains of such waves may last for hours, especially if long oceanic paths are involved.

coefficient of acidity *oxygen ratio.*

coefficient of anisotropy The square root of the ratio of the true transverse resistivity to true longitudinal resistivity in an anisotropic material.

coefficient of earth pressure The principal-stress ratio at a given point in a soil.

coefficient of fineness The ratio of suspended solids to turbidity; a measure of the size of particles causing turbidity, the particle size increasing with coefficient of fineness (ASCE, 1962).

coefficient of kinematic viscosity *kinematic viscosity.*

coefficient of permeability *permeability coefficient.*

coefficient of storage *storage coefficient.*

coefficient of thermal expansion The relative increase of the volume of a system with increasing temperature in an isobaric process.

coefficient of transmissibility *transmissivity [hydraul].*

coefficient of variation The standard deviation of a set of data divided by its arithmetic mean. Syn: *coefficient of variability.*

coefficient of viscosity The ratio of the shear stress in a substance to the rate of shear strain; *viscosity.*

coefficient of volume compressibility The compression of a lithic unit per unit of original thickness per unit increase of pressure.

coelacanth A member of a suborder of crossopterygian fish that entered marine waters during the Mesozoic; it includes the sole living representative of the order *Crossopterygii,* the genus *Latimeria.* Range, Upper Devonian to Recent.

coelenterate Any multicelled invertebrate animal, solitary or colonial, belonging to the phylum Coelenterata, characterized by a body wall composed of two layers of cells connected by a structureless mesogloea, by a single body cavity with a single opening for ingestion and egestion, and by radial or biradial symmetry. Range, Precambrian to present.

coelenteron The spacious internal cavity enclosed by the body wall of a coelenterate and opening externally through the mouth. Pl: *coelentera.*

coelestine *celestite.*

coelobitic Pertaining to a *cavity dweller,* an organism (usually either a sessile or mobile invertebrate) that lives in a recessed cavity or void within reef-rock. Cf: *cryptic; endolithic; sciaphilic.* Syn: *cryptolithic.*

coeloconoid Said of a gastropod shell that approaches a conical shape but has concave sides. Cf: *cyrtoconoid.*

Coelolepida Order of diplorhinate jawless fishes characterized by flattened body form and armor consisting of a loosely articulated coat of small scales. Range, Upper Silurian to Lower Devonian.

coelom The general *body cavity* occurring in multicelled animals other than the sponges and coelenterates, as distinguished from all special cavities such as the *enteron.* Where well developed, the coelom forms a space between the alimentary viscera and the body walls. Adj: *coelomic.* Also spelled: *coelome.*

coelome *coelom.*

coenenchymal increase In some Paleozoic corals, esp. heliolitids, "offsets arise from the coenosteum that unites individual corallites in coenenchymal massif coralla. The offset cannot be related to any one parent" (Oliver, 1968, p. 20).

coenenchyme The complex mesogloea uniting the polyps of a compound coral; a collective term for both the coenosteum and the coenosarc. Adj: *coenenchymal.* Syn: *coenenchym; coenenchyma.*

coenobium A colony of independent organisms united by a common investment, usually having a definite arrangement (Jackson, 1928, p. 83).

coenocyte An organism such as certain filamentous green algae that consists of continuous, multinucleate protoplasm, lacking walls to separate protoplasts.

coenosarc Common soft tissue connecting coral polyps in a colony (TIP, 1956, pt.F, p.246).

coenosis A population that is held together by ecologic factors in a state of unstable equilibrium (Stamp, 1966, p. 115). Also spelled: *cenosis.* Plural: *coenoses.*

coenosteum (a) Calcareous skeletal deposits formed between the individual corallites of a colonial coral. (b) Vesicular or dense calcareous skeletal material between zooecia of some stenolaemate bryozoans, esp. in the exozone. Partial syn: *extrazooidal skeleton.* (c) The calcareous skeleton secreted by a millepore colony or a stromatoporoid.—Pl: *coenostea.*

coercive force The opposing applied magnetic field H required to reduce the remanent magnetization of a substance to zero. See also: *hysteresis.* Syn: *coercivity.*

coercivity *coercive force.*

coeruleolactite A mineral: $(Ca,Cu)Al_6(PO_4)_4(OH)_8 \cdot 4\text{-}5H_2O$. It has a milk-white to sky-blue color, and is related to turquoise.

coesite A monoclinic mineral: SiO_2. It is a very dense (2.93 g/cm^3) polymorph of *quartz* and is stable at room temperature only at pressures above 20,000 bars; the silicon has a coordination number of 4. Coesite is found naturally only in structures that are currently best explained as impact craters, or in rocks (such as suevite) associated with such structures. Cf: *stishovite.*

coetaneous A suggested replacement for the term *isochronous,* in the sense of being equal in duration or uniform in time. Rare.

coevolution A pattern of evolution in which two unrelated *lineages* profoundly influence each other's evolutionary directions and rates, so that the two evolve together as an integrated complex; e.g. the relation between flowering plants and pollinating insects through time.

coffinite A black mineral: $U(SiO_4)_{1-x}(OH)_{4x}$. An important ore of uranium, it occurs in many sandstone deposits and in hydrothermal veins. Syn: *nenadkevite.*

cogeoid *compensated geoid.*

cognate [pyroclast] Said of pyroclastics that are *essential* or *accessory.* Cf: *accidental.*

cognate [struc geol] Said of fractures or joints in a system that have the same time and deformational type of origin. Cf: *conjugate; complementary.*

cognate inclusion *autolith.*

cognate xenolith *autolith.*

cogwheel ore The mineral *bournonite* esp. when occurring in wheel-shaped twin crystals.

cohenite A tin-white isometric mineral: $(Fe,Ni,Co)_3C$. It occurs as an accessory mineral in iron meteorites.

coherence The property of wave trains being in phase. The coherence of adjacent seismic traces, along with an increase in amplitude, is the principal evidence for an *event.* Coherence is used qualitatively in picking events, but it can also be quantified. See: *semblance.*

coherent [bot] In plant morphology, pertaining to like parts that are partially united (Swartz, 1971, p. 113). Cf: *connate [bot]; adnate; adherent.*

coherent [geochem] Said of a group of elements which, owing to similarity in radius and valence, occur intimately associated in nature, such as entering into the same minerals at about the same stage of fractional crystallization; e.g. zirconium and hafnium form a "coherent pair" (Goldschmidt, 1937, p.662).

coherent [geol] Said of a rock or deposit that is consolidated, or that is not easily shattered.

cohesion *Shear strength* of a rock not related to interparticle friction. Cf: *adhesion.*

cohesionless Said of a soil that has relatively low shear strength when air-dried and low cohesion when wet, e.g. a sandy soil. Cf: *cohesive.* Syn: *noncohesive; frictional.*

cohesive Said of a soil that has relatively high shear strength when air-dried, and high cohesion when wet, e.g. a clay-bearing soil. Cf: *cohesionless.*

cohesiveness A mass property of unconsolidated, fine-grained sediments by which like or unlike particles (having diameters less than 0.01 mm) cohere or stick together by surface forces.

cohesive strength Inherent strength of a material when normal stress across the prospective surface of failure is zero.

coiling direction The direction (dextral or sinistral) in which a gastropod shell or a foraminiferal test is coiled. Changes in the coiling directions of planktonic foraminiferal tests are applied in stratigraphy to interpret paleoclimates or to determine correlations.

coire A var. of *corrie.* Etymol: Gaelic, "large kettle".

coke (a) A combustible material derived from *agglomerating* coal, consisting of mineral matter and fixed carbon fused together. It is produced by driving off by heat the coal's volatile matter, i.e., by carbonization. Coke is gray, hard, and porous, and as a fuel it is practically smokeless. It occurs in nature, but most is manufactured.—Also spelled: *coak* (obsolete). Cf: *clinker [coal]; natural coke; charcoal.*

coke coal *natural coke.*

cokeite *natural coke.*

coking coal A *caking coal* suitable for the production of coke for metallurgical use.

col (a) A high, narrow, sharp-edged *pass* or depression in a mountain range, generally across a ridge or through a divide, or between two adjacent peaks; esp. a deep pass formed by the headward erosion and intersection of two cirques, as in the French Alps. Also, the highest point on a divide between two valleys. (b) A marked, saddle-like depression in the crest of a mountain ridge; the lowest point on a ridge. Syn: *saddle.* (c) A short ridge or elevated neck of land connecting two larger and higher masses.—Etymol: French, from Latin *collum,* "neck". Cf: *gap [geomorph]; notch [geomorph].* Syn: *joch.*

cold avalanche A snow avalanche involving the movement of dry snow and occurring during the time of greatest cold, usually coinciding with a drop in temperature; e.g. a *dry-snow avalanche.* The term is part of an obsolete classification of avalanches.

cold desert A desert in a high latitude or at a high altitude, whose low temperature restricts or prohibits plant and animal life. The term is often used for *tundra* areas.

cold front The sloping boundary surface between an advancing mass of cold air and a warmer air mass beneath which it pushes like a wedge. Its passage is usually accompanied by a rise of pressure, a fall in temperature, a veer of wind, and a heavy shower. Cf: *warm front.*

cold fumarole A fumarole whose steam is less than 100°C in temperature.

cold glacier *polar glacier.*

cold ice Ice below the *pressure-melting temperature.*

cold lahar A flow of cooled volcanic materials down the slope of a volcano, produced by heavy rains, or by collapse of an unstable section of the volcano. Cf: *hot lahar.* Syn: *cold mudflow.*

cold loess Periglacial loess derived from glacial outwash and formed in garlands about the Pleistocene ice sheets, as in northern Europe and in north-central U.S. Cf: *warm loess.*

cold mudflow *cold lahar.*

cold region An area where the temperature is sufficiently low to affect engineering design, construction, and operation.

cold spring A spring whose water has a temperature appreciably below the mean annual atmospheric temperature in the area; also, a nonpreferred usage for any *nonthermal spring* in an area having thermal springs (Meinzer, 1923, p. 55).

colemanite A colorless or white monoclinic mineral: $Ca_2B_6O_{11} \cdot 5H_2O$. It is an important source of boron, occurring in massive crystals or as nodules in clay.

coleoid Any member of the subclass Coleoidea (=Dibranchiata) of the cephalopods, having a muscular mantle, internal shell, fins, ink bag, chromatophores, suction cups, closed funnel, and camera-like eyes (Jeletsky, 1966). See also: *decapod.* Syn: *dibranchiate; endocochlian.* Range, Lower Carboniferous to the present.

colina A term used in the SW U.S. for a hillock or other small *eminence.* Etymol: Spanish. Syn: *collado.*

colk A *pothole* in a stream bed. Etymol: Dutch, "hollow", such as a hole eroded by outrushing water at the base of a broken dike. Cf: *kolk.*

collabral Conforming to the shape of the outer lip of a gastropod shell, as indicated by growth lines.

collado (a) *colina.* (b) A term used in the SW U.S. for a saddle, gap, or pass.—Etymol: Spanish.

collain A kind of *euvitrain* that consists of ulmin compounds that are redeposited by precipitation from solution. Cf: *ulmain.*

collapse breccia A breccia formed by the collapse of rock overlying an opening, as by foundering of the roof of a cave or of the roof of country rock above an intrusion; e.g. a *solution breccia.* Syn: *founder breccia.*

collapse caldera A type of *caldera* produced by collapse of the roof of a magma chamber due to removal of magma by voluminous pyroclastic or lava eruptions or by subterranean withdrawal of magma. Most calderas are of this type. Cf: *erosion caldera; explosion caldera.*

collapse crater A large lunar crater believed to have formed by roof subsidence of lava-filled cavities. The type is not well established.

collapse depression An elliptical to elongate depression in the surface of a lava flow, resulting from partial or complete collapse of the roof of a long lava tunnel or of several short ones. Blocks from the roof may rest on a collapse-depression floor, and water and/or alluvium cover the blocks (Nichols, 1946, p. 1064). Not to be confused with a *kipuka.*

collapse fault A normal, gravity fault on the margin of a salt-withdrawal basin in the Gulf Coast (Seglund, 1974, p. 2389). Sediments overlying the salt collapse periodically as salt is withdrawn into clusters of domes or large intrusions.

collapse sinkhole A type of *sinkhole* that is formed by collapse of an underlying cave. See also: *solution sinkhole.*

collapse structure Any rock structure resulting from removal of support and consequent collapse by the force of gravity, e.g. gravitational sliding on fold limbs, salt solution causing collapse of overlying rocks in salt basins, sink-hole collapse, or collapse into mine workings.

collar [drill] (a) *drill collar.* (b) The mouth or upper end of a mine shaft.

collar [paleont] The smooth tapering part of an echinoid spine located above the milled ring.

collar cell *choanocyte.*

collar log *casing-collar log.*

collar pore A tiny aperture that occurs in a horizontal plate at the base of the cephalis in some nasselline radiolarians.

collecting area *catchment area [grd wat].*

collective group A zoological term for a collection of animals that can be divided into identifiable *species* but whose generic position is uncertain. No *type species* is required but the collective name is treated as having generic rank (ICZN, 1961, p. 43, 148; Cowan, 1968, p. 23).

collective species A botanical term for a collection of *species* that are individually recognizable but closely similar. The term is informal and has no nomenclatural status (Cowan, 1968, p. 23).

collector well A large-diameter well consisting of a concrete cylinder, sealed at the bottom, with perforated pipes extending radially into an aquifer. "The radial pipes are jacked hydraulically into the formation. Fine material around the pipes is removed by washing during construction. Collector wells are most often constructed in alluvial formations adjoining rivers. The radial pipes extend toward and under the river, thereby inducing movement of water downward through the stream bed to the pipes" (Chow, 1964, p. 13-31). This well was developed by the engineer Leo Ranney and is also known as the *Ranney collector.*

collenchyma A strengthening tissue in a plant, composed of cells with walls usually thickened at the angles of the walls (Fuller & Tippo, 1954, p. 954).

collencyte An amoebocyte of a sponge, often stellate or fusiform, that forms the cellular web of the mesohyle.

collenia A markedly convex, slightly arched or turbinate stromatolite, about 10 cm in diameter and less than 3 cm in height, produced by late Precambrian blue-green algae of the genus *Collenia* by the addition of external calcareous layers of varying thickness. It is associated with beds of flat-pebble conglomerate.

collimate (a) To make refracted or reflected rays of light parallel to a certain line or direction, such as by means of a lens or concave mirror. (b) To adjust the line of sight of a surveying instrument or the lens axis of an optical instrument so that it is in its proper position relative to other parts of the instrument, such as by means of a collimator. (c) To adjust the fiducial marks of a surveying camera so that they define the principal point.

collimating mark *fiducial mark.*

collimation axis The straight line passing through the rear nodal point of the objective lens, perpendicular to the axis of rotation of the telescope of a surveying instrument. It is perpendicular to the horizontal axis of the telescope in a transit or theodolite, and perpendicular to the vertical axis in a leveling instrument.

collimation error The angle by which the line of sight of an optical instrument differs from its *collimation axis.* Syn: *error of collimation.*

collimation line *line of collimation.*

collimation plane The plane described by the collimation axis of the telescope of a transit when the telescope is rotated about its horizontal axis.

collimator An optical device for producing a beam of parallel rays of light or for artificially creating an infinitely distant target that can be viewed without parallax, usually consisting of a tube having an objective converging lens with an arrangement of cross hairs placed in the plane of its principal focus. It is used in testing and adjusting certain optical surveying instruments. See also: *vertical collimator.*

colline A protuberant ridge of corallum surface between corallites of a scleractinian coral. Cf: *monticule [paleont].*

collinite (a) A maceral of coal within the *vitrinite* group, consisting of homogeneous jellified and precipitated plant material, lacking cell structure and of middle-range reflectance under normal reflected-light microscopy. Cf: *ulminite.* (b) A preferred syn. of *euvitrinite.*

collinsite A light-brown triclinic mineral: $Ca_2(Mg, Fe)(PO_4)_2 \cdot 2H_2O$. It is isomorphous with fairfieldite.

collite A microlithotype of coal; a variety of *vitrite.*

collobrierite A metamorphic rock composed of fayalite, garnet (almandine-spessartine), grunerite, magnetite, and some feldspar. The term was originated by LaCroix in 1917.

colloclarain A transitional lithotype of coal characterized by the presence of collinite, but in lesser amounts than other macerals. Cf: *clarocollain.* Syn: *colloclarite.*

colloclarite *colloclarain.*

colloclast A term used by Sander (1967, p. 327-328) for a weakly cemented, accretionary aggregate of mud or fine silt, typically possessing surficial irregularities with a lobate outline, and having a texture similar to the stratum in which it occurs; e.g. such an aggregate attaining silt or sand size, formed in place on the sea floor, or an aggregate attaining sand or gravel size, formed by the transport and redeposition of fragments made up of silt-, sand-, or gravel-size carbonates weakly cemented prior to movement.

collocryst A crystal formed by recrystallization of aggregated colloidal parent material, as in a mobilized sediment.

colloform Said of the rounded, finely banded kidneylike mineral texture formed by ultra-fine-grained rhythmic precipitation once thought to denote deposition of colloids. Cf: *botryoidal; reniform.*

colloid (a) A particle-size range of less than 0.00024 mm, i.e. smaller than clay size (U.S. Naval Oceanographic Office, 1966). (b) Originally, any finely divided substance that does not occur in crystalline form; in a more modern sense, any fine-grained material in suspension, or any such material that can be easily suspended (Krauskopf, 1967).

colloidal complex In a soil, a mixture of humus and clay.

colloidal dispersion (a) A suspension of particles of colloidal size in a medium, usually liquid; a *sol.* (b) An *aerosol.*

colloid plucking A mechanical-weathering process in which small fragments are pulled off or loosened from rock surfaces by soil colloids in contact with them (Reiche, 1945, p.14).

colloidstone A consolidated sedimentary rock consisting of colloid-size particles (Alling, 1943, p.265).

collophane Any of the massive cryptocrystalline varieties of apatite, often opaline, horny, dull, colorless, or snow-white in appearance, that constitute the bulk of phosphate rock and fossil bone and that are used as a source of phosphate for fertilizers; esp. *carbonate-apatite* or a hydroxylapatite containing carbonate, and sometimes francolite. The chemical formula $Ca_3P_2O_8 \cdot 2H_2O$ is sometimes given for collophane, but it is probably not a true mineral. Syn: *collophanite.*

collophanite *collophane.*

colluvial Pertaining to colluvium; e.g. "colluvial deposits".

colluviation The formation of colluvium.

colluvium (a) A general term applied to any loose, heterogeneous, and incoherent mass of soil material and/or rock fragments deposited by rainwash, sheetwash, or slow continuous downslope creep, usually collecting at the base of gentle slopes or hillsides. (b) Alluvium deposited by unconcentrated surface runoff or sheet erosion, usually at the base of a slope.—Cf: *slope wash.* Etymol: Latin *colluvies,* "collection of washings, dregs".

colmatage A term used in New Zealand for the artificial impounding of silt-laden water in order to build up the banks in the lower part of a river by deposition of alluvium; originally the term referred to the natural process of bank growth. Cf: *warping [sed].*

Cologne earth *black earth [coal].* Etymol: source near Cologne, Germany.

Cologne umber *black earth [coal].* Etymol: source near Cologne, Germany.

cololite A *trace fossil* now assigned to the "genus" *Lumbricaria,* consisting of a cylindrical, stringlike, and tortuous or convoluted body approximately 3 mm wide, probably representing a fossil cast of a worm, but formerly regarded as the petrified intestines of a fish or the contents of such intestines. It occurs esp. in *lithographic limestones* such as the Solenhofen stone.

colombianite A ball-like glass object found near Cali in Colombia, once identified as a tektite but now believed to be of volcanic origin; a variety of *amerikanite.*

colonial Said of an animal that lives in close association with others of the same species and that usually cannot exist as a separate individual; esp. a "colonial coral" in which the individuals are attached as a unit. Cf: *solitary coral.*

colonization A natural phenomenon wherein a species invades an area previously unoccupied by it and becomes established there.

colonnade In columnar jointing, the lower zone that has thicker and better-formed columns than the upper zone, or *entablature.*

colony (a) A physiologically and morphologically connected aggregation of organisms; e.g. certain bryozoans, graptolites, and anthozoan corals. (b) A group of living or fossil organisms found in an area or rock unit other than that of which they are characteristic, or that migrate into and become established in a barren area.

colophonite (a) A coarse, cloudy, yellow-brown variety of andradite garnet. (b) A nongem variety of vesuvianite.

color [mineral] A phenomenon of light or visual perception by which otherwise identical objects may be differentiated (Webster, 1967). In mineral analysis, color is an important diagnostic feature; it is affected or formed by the response of light to composition and structure. Color of a metallic mineral is usually constant, whereas that of a nonmetallic mineral is more apt to vary. The color of a mineral's powder (its *streak*) may differ from that of a massive sample.

color [sed] A mass property of a sediment, represented by the overall hue caused by combinations of the colors of the particles, surface coating, matrix, and cement, and controlled in part by the degree of fineness of the particles.

coloradoite A grayish-black isometric mineral: HgTe.

color center In crystal optics, a defect in the atomic structure that selectively absorbs a component of visible light. See also: *F center.*

color-composite image A color image prepared by projecting individual black-and-white multispectral images in color.

colored stone A gemstone of any species other than diamond. This usage illogically classifies all varieties of such species as colored stones, including colorless varieties, but does not include colored diamonds. However, it has proved a practicable and satisfactory classification.

color grade The grade or classification in which a gem is placed by examination of its color in comparison to the color of other gems of the same variety (Shipley, 1974).

colorimeter An instrument for measuring and comparing the intensity of color of a compound for quantitative chemical analysis, usually based on the relationship between concentration of a chemical solution and the amount of absorption of certain characteristic colors of light. See also: *Dubosq colorimeter; spectrocolorimeter.*

colorimetric Pertaining to *colorimetry.*

colorimetric analysis Quantitative chemical analysis performed by adding a certain amount of a substance to both an unknown and a standard solution and then comparing color intensities.

colorimetry The art or process of measuring and/or comparing colors, usually with a *colorimeter,* for quantitative chemical analysis.

color index In petrology, esp. in the classification of igneous rocks, a number that represents the percent, by volume, of dark-colored (i.e. mafic) minerals in a rock. According to this index, rocks may be divided into "leucocratic" (color index, 0-30), "mesocratic" (color index, 30-60), and "melanocratic" (color index, 60-100). Syn: *color ratio.*

color-patch map *chorochromatic map.*

color ratio *color index.*

color temperature (a) An estimate of the temperature of an incandescent body, determined by observing the wavelength at which it is emitting with peak intensity (its color) and applying *Wien's displacement law.* For an ideal blackbody, the temperature so estimated would be its true temperature and would agree with its *brightness temperature;* but for actual bodies, the color temperature is generally only an approximate value. (b) The temperature to which a blackbody radiator must be raised in order that the light it emits may match a given light source in color. It is usually expressed in degrees Kelvin.

colpa A nonrecommended syn. of *colpus.* Pl: *colpae.*

colpate Said of pollen grains having more or less elongated, longitudinal furrows (colpi) in the exine.

colpi Plural of *colpus.*

colporate Said of pollen grains having colpi in which there is a pore or some other modification of the exine, such as a transverse furrow, usually at the equator.

colpus A longitudinal furrowlike or groovelike modification in the exine of pollen grains, associated with germination. It either encloses a germ pore or serves directly as the place of emergence of the pollen tube, often with harmomegathic swelling. It may be distal (as in monocolpate pollen), meridional (as in tricolpate pollen), or otherwise disposed. The membrane of the colpus consists of exine in which ektexine and/or endexine are thinned or absent. Pl: *colpi.* Cf: *pore [palyn]; pseudocolpus.* Syn: *germinal furrow; sulcus [palyn].*

colpus transversalis *transverse furrow.*

columbite A black mineral: $(Fe,Mn)(Nb,Ta)_2O_6$. It is isomorphous with tantalite, occurs in granites and pegmatites, and is an ore of niobium as well as a source of tantalum. Syn: *niobite; dianite; greenlandite.*

columbotantalite A noncommittal term for minerals of the columbite-tantalite series.

columbretite A leucite *phonolite* composed of sanidine and altered hornblende laths in a dense groundmass of corroded oligoclase microlites with interstitial sanidine enclosing rounded leucite grains. The embayments of the microlites are filled with analcime, augite, and magnetite (Johannsen, 1939, p. 247). The name is from the Columbrete Islands, Spain. Not recommended usage.

columella [paleont] (a) A pillarlike calcareous axial structure of a corallite, formed by various modifications of the inner edges of septa. It commonly projects into the central part of the calice of many corals in the form of a sharp-pointed protuberance. See also: *trabecular columella; lamellar columella; fascicular columella; styliform columella.* (b) The medial pillar surrounding the axis of a spiral gastropod shell, formed by the coalescence of the inner (adaxial) walls of the whorls. (c) A vertical rod between two horizontal rings, or within the shell cavity, in certain radiolarians. (d) The stapes (columella auris) or middle-ear bone of lower tetrapods and birds, esp. when slender and rodlike as in lizards; also, the rodlike epipterygoid (columella cranii) of lower tetrapods.—Pl: *columellae.*

columella [palyn] One of the rodlets of ektexine that may branch and fuse distally to produce a tectum on pollen grains with complex exine structure. Pl: *columellae.*

columellar fold A *fold [paleont]* or spirally wound ridge on the columella of a gastropod, projecting into the shell interior.

columellar lip The adaxial part of the *inner lip* of a gastropod shell.

columellate Possessing or forming a columella or columellae; e.g. said of pollen grains with a complex ektexine structure consisting of columellae.

column [paleont] (a) A cylindrical structure consisting of a series of disklike plates mounted one on top of the other and attached to the aboral end of the theca of crinoids, blastoids, and most cystoids, and presumably used for anchoring or as a means of support. The distal end is known to be variously modified in some species. Syn: *stem.* (b) The smooth cylindrical body wall of a scleractinian coral polyp between the basal and oral disks.

column [speleo] A columnar deposit formed by the union of a *stalactite [speleo]* with its complementary *stalagmite [speleo].* Syn: *pillar [speleo]; stalactostalagmite.*

column [stratig] *geologic column.*

columnal One of the numerous individual vertical segments (ossicles or plates) that make up the column or stem of an echinoderm. Columnals are circular or polygonal, discoid, or button-shaped.

columnar Said of a crystal habit that is a subparallel arrangement of columnar individuals.

columnar coal Coal that has developed a columnar fracture, usually due to metamorphism by an igneous intrusion.

columnar facet A normally circular indentation in the basal plates of a cystoid theca to accommodate the proximal end of the column.

columnar ice Ice that has been built by *columnar ice crystals,* mostly broader in the lower part than in the upper part of the ice cover.

columnar ice crystal A vertical ice column. Massed together, columnar ice crystals form *columnar ice.*

columnar jointing Parallel, prismatic columns, polygonal in cross section, in basaltic flows and sometimes in other extrusive and intrusive rocks. It is formed as the result of contraction during cooling. Syn: *columnar structure; prismatic jointing; prismatic structure.*

columnar section A *vertical section,* or graphic representation on a vertical strip, of the sequence of rock units that occurs in an area or at a specific locality. Thicknesses are drawn to scale, and lithology is indicated by standard or conventional symbols, usually supplemented by brief descriptive notes indicating age, rock classification, fossil contents, etc. See also: *geologic column.*

columnar structure [mineral] A columnar, subparallel arrangement shown by aggregates of slender, elongate mineral crystals.

columnar structure [sed] A primary sedimentary structure found in some calcareous shales or argillaceous limestones, consisting of columns (9-14 cm in diameter, and 1-1.4 m in length) perpendicular to bedding and oval to polygonal in section (Hardy & Williams, 1959).

columnar structure [struc geol] *columnar jointing.*

column chromatography A chromatographic technique for separating components of a sample by moving it in a mixture or solution through tubular structures, packed with appropriate substrates, in such a way that the different components have different mobilities and thus become separated. See also: *chromatography.*

column crystal A snow crystal in the shape of a hexagonal prism, either solid or hollow.

colusite A bronze-colored tetrahedral mineral: $Cu_3(As,Sn,V,Fe,Te)S_4$.

comagmatic Said of igneous rocks that have a common set of chemical and mineralogic features, and thus are regarded as having been derived from a common parent magma. Also, said of the region in which such rocks occur. See also: *consanguinity.* Less preferred syn: *consanguineous.*

comagmatic region *petrographic province.*

Comanchean (a) North American provincial series: Lower and Upper Cretaceous (above Coahuilan, below Gulfian). (b) Obsolete name applied to a period (or system of rocks) between the Jurassic below and the Cretaceous above. Also spelled: *Comanchian.*

comb [geomorph] (a) The crest of a mountain or hill; a mountain ridge. Syn: *combe.* (b) A var. of *combe,* a valley. (c) A var. of *cwm.*

comb [ore dep] n. A vein filling in which subparallel crystals, generally of quartz, have grown perpendicular to the vein walls and thus resemble the teeth of a comb. — adj. Said of such a crystal texture or structure in a vein.

comb [paleont] (a) An arachnid structure resembling a comb; e.g. a row of serrated bristles on the fourth tarsi of an araneid, found only in the family Theridiidae, or a pair of abdominal appendages situated on the sternite following upon the genital opercula, present in all scorpions but not in any other arachnids. (b) A radial series of knobs or projections in acantharian radiolarians.

combe (a) A British term, also used in France, for a small, deep valley running down to the sea. Also, a bowl-shaped, generally unwatered valley or hollow on the flank of a hill, esp. a dry, closed-in valley on the the chalk downs of southern England. Etymol: Celtic. Syn: *comb [geomorph]; coom; coomb; coombe.* (b) A large longitudinal depression or valley along the crest or side of an anticline in the folded Jura Mountains of the European Alps, formed by downfaulting or more generally by differential erosion, often occurring along the line of junction of a hard crystalline rock with one that is soft. Cf: *cluse.* See also: *val; crêt.* (c) The amphitheaterlike steep bank of an incised meandering stream. (d) Var. of *comb [geomorph],* a mountain crest. (e) A term used in England and southern Scotland for a glacial valley and for a *cirque.* Also, a var. of *cwm.*

combeite A rhombohedral mineral: $Na_4Ca_3Si_6O_{16}(OH,F)_2$.

comber (a) A long, curling, deep-water ocean wave whose high, breaking crest (much larger than a whitecap) is pushed forward by a strong wind. (b) A long-period *spilling breaker.*—Cf: *roller.* Syn: *beachcomber.*

combination Any set that can be made by using all or part of a given collection of objects without regard to sequence. See also: *permutation.*

combination archeopyle An *archeopyle* formed by the release of a part of the dinoflagellate-cyst wall that corresponds to plates of more than one thecal plate series (such as combining the plates of the apical series and the precingular series).

combination trap A trap for oil or gas that has both structural and stratigraphic elements.

combination well An *open well* connected to one or more other wells.

combined twinning A rare type of twinning in quartz in which there appears to be a 180° rotation around *c* with reflection over $11\bar{2}0$ or over $\{0001\}$. The crystal axes are parallel but the electrical polarity of the *a* axes is not reversed in the twinned parts.

combined water "Water of solid solution and water of hydration which does not freeze even at the temperature of -78°C" (Muller, 1947, p. 214).

comb ridge A jagged, sharp-edged, steep-sided mountain ridge whose crest resembles a cockscomb because it bears pinnacles alternating with notches; an *arête* marked by a series of *aiguilles.* It commonly separates adjacent cirques in glaciated mountain regions.

combustible shale *tasmanite [coal].*

comendite A peralkaline *rhyolite* or *quartz trachyte,* less mafic than *pantellerite.* Macdonald and Bailey (1973) provide criteria for distinguishing comendite from pantellerite. Cf: *taurite.* The rock was named by Bertolio in 1895 from Le Comende, San Pietro Island, Sardinia. The spelling "commendite" (Irvine and Baragar, 1971) is in error.

comitalia Small megascleres (spicules) adherent to the rays of larger megascleres in lyssacine hexactinellid sponges.

commensal Said of organisms living in a state of *commensalism.*

commensalism The relationship that exists between two organisms in which the first benefits from the second, the second being neither benefitted nor harmed. Adj: *commensal.* Cf: *inquilinism; mutualism; symbiosis.*

commercial dust Impure *gold dust.*

commercial granite A general term for a decorative building stone that is hard and crystalline. It may be a granite, gneiss, syenite, monzonite, granodiorite, anorthosite, or larvikite. See also: *black granite.*

comminution (a) The gradual diminution of a substance to a fine powder or dust by crushing, grinding, or rubbing; specif. the reduction of a rock to progressively smaller particles by weathering, erosion, or tectonic movements. (b) The breaking, crushing, or grinding by mechanical means of stone, coal, or ore, for direct use or further processing.—Syn: *pulverization; trituration.*

comminution till Compact subglacial till manufactured more or less in place, composed of rock debris created by crushing and shearing, and made dense by water from melting of basal ice.

commissural plane The plane containing the cardinal margin of a brachiopod and either the commissure of a rectimarginate shell or points on the anterior commissure midway between crests of folds in both valves (TIP, 1965, pt.H, p.142).

commissure (a) The line of junction between the edges or margins of valves in a brachiopod or bivalve mollusk. (b) In a plant, the surface of joining of two mericarps, or of appressed stigmas or style branches. (c) The groove of the laesura along which an embryophytic spore germinates. It is essentially equivalent to *suture.*

common In the description of coal constituents, 5-10% of a particular constituent occurring in the coal. (ICCP, 1963). Cf: *rare; very common; abundant; dominant.*

common bud One of the confluent coelomic spaces and its protective outer membranous or membranous and skeletal wall, into which a new zooid is budded in stenolaemate bryozoans.

common depth point A portion of the subsurface that is involved in producing seismic reflections at different offset distances on several profiles. Abbrev: CDP. Syn: *common reflection point.*

common-depth-point shooting A type of seismic field layout designed for multiple subsurface coverage to yield reflections from *common depth points.*

common-depth-point stack A sum of seismic traces that have the same *common depth point.* The summing is done after appropriate statics and normal-moveout corrections have been applied to each trace. The objective is to attenuate noise and multiple reflections while accentuating reflection events.

common feldspar *orthoclase.*

common lead Any lead from a phase with a low value of U/Pb and/or Th/Pb such that no significant *radiogenic lead* has been generated in situ since the phase formed. Such phases include galena and other sulfides such as pyrite; feldspars, in particular K-feldspar; micas; and most abundant rock types of Cenozoic age. Data on common lead are used in determining ages, and, more important, in the solution of genetic problems (Doe, 1970, p. 35). Syn: *ordinary lead.*

common-lead age method The determination of an age in years for a lead by isotopically determining the lead ratios $^{206}Pb/^{204}Pb$, $^{207}Pb/^{204}Pb$, and $^{208}Pb/^{204}Pb$, which can be plotted and compared to proposed growth curves.

common mica *muscovite.*

common opal A variety of opal that never exhibits play of color, is found in a wide variety of colors and patterns, sometimes occurs as an earthy form, and generally is not suitable for gem use. Cf: *precious opal.*

common pyrites *pyrite.*

common reflection point *common depth point.*

common salt A colorless or white crystalline compound consisting of sodium chloride (NaCl), occurring abundantly in nature as a solid mineral (halite), or in solution (constituting about 2.6% of seawater), or as a sedimentary deposit (such as in *salt domes* or as a crust around the margin of a *salt lake*).

common strontium Strontium-87 in a rock or mineral which was present at the time the rock or mineral formed, and is not the result of in-situ decay of rubidium-87 after the formation of the rock or mineral. Cf: *radiogenic strontium.*

communication pore An opening in the wall or pore plate of a bryozoan zooid or extrazooidal structure.

community A group of organisms (living or fossil) occurring together because they possess an integrated system (*food chain* or *food web*) of energy transfer operating through several different feeding or *trophic* levels. Cf: *assemblage; association; biocoenosis.* Syn: *biotic community.*

community evolution Change in the composition and/or structure of a *community* of organisms through geologic time; it may appear as the development of new roles for the organisms to fill within the community, or as the development of new organisms to fill existing community roles.

compact (a) Said of any rock or soil that has a firm, solid, or dense texture, with particles closely packed. (b) Said of a *close-grained* rock in which no component particles or crystals can be recognized by the unaided eye. (c) Said of a finely textured rock with low matrix porosity.

compactability A property of a sedimentary material that permits it to decrease in volume or thickness under pressure; it is a function of the size, shape, hardness, and brittleness of the constituent particles.

compaction [sed] (a) Reduction in bulk volume or thickness of, or the pore space within, a body of fine-grained sediments in response to the increasing weight of overlying material that is continually being deposited or to the pressures resulting from earth movements within the crust. It is expressed as a decrease in porosity brought about by a tighter *packing* of the sediment particles. See also: *differential compaction.* (b) The process whereby fine-grained sediment is converted to consolidated rock, such as a clay lithified to a shale.—See: Rieke & Chilingarian, 1974; Chilingarian & Wolf, 1975, 1976.

compaction [soil] (a) Any process, such as burial or desiccation, by which a soil mass loses pore space and becomes more dense, thereby increasing its bearing capacity and general stability in construction. (b) The densification of a soil by mechanical means, accomplished by rolling, tamping, or vibrating. See also: *consolidation.*

compaction curve The curve showing the relationship between the density (dry unit weight) and the water content of a soil for a given compactive effort. Syn: *moisture-density curve.*

compaction fold A type of *supratenuous fold* developed by differential compaction of sedimentary material over more resistant rock, over a subsurface structure such as a buried hill, or over an active fault or fold.

compaction test A laboratory compacting procedure to determine the optimum water content at which a soil can be compacted so as to yield the maximum density (dry unit weight). The method involves placing (in a specified manner) a soil sample at a known water content in a mold of given dimensions, subjecting it to a compactive effort of controlled magnitude, and determining the resulting unit weight (ASCE, 1958, term 74). The procedure is repeated for various water contents sufficient to establish a relation between water content and unit weight. The maximum dry density for a given compactive effort will usually produce a sample whose saturated strength is near maximum. Syn: *moisture-density test.*

compartmental plate A rigid articulated skeletal element forming part of the shell wall in certain cirripede crustaceans; e.g. a lateral and a carinolateral. Syn: *mural plate.*

compass [paleont] A slender, arched radial rod in ambulacral position at the top of Aristotle's lantern in an echinoid.

compass [surv] (a) An instrument or device for indicating horizontal reference directions relative to the Earth by means of a magnetic needle or group of needles; specif. *magnetic compass.* Also, a nonmagnetic device that serves the same purpose; e.g. a *gyrocompass.* (b) A simple instrument for describing circles, transferring measurements, or subdividing distances, usually consisting of two pointed, hinged legs (one of which generally having a pen or pencil point) joined at the top by a pivot.

compass bearing A *bearing* expressed as a horizontal angle measured clockwise from north as indicated by a magnetic compass.

compass error The amount by which a compass direction differs from the true direction, usually expressed as the number of degrees east or west of true azimuth north and marked plus or minus according to whether the compass direction is less or greater than true azimuth. It combines the effects, or is the algebraic sum, of the deviation and variation of the compass.

compass rose A graduated circle, usually marked in 360 degrees, printed or inscribed on a nautical chart for reference, and indicating directions. It may be oriented with respect to true north, to magnetic north, or to both. Compass roses are used to give directions of prevailing winds. On geologic maps or diagrams, they may be adapted to show alignment of structures such as faults and joints. Syn: *rose.*

compass traverse A surveying traverse in which a number of straight lines are measured by tape or pace and their bearings are taken by a magnetic or prismatic compass. It is executed where lines are not shown or not clearly defined on a map, or when a base map of an area is not available.

compensated geoid One of various surfaces approximating the *geoid* and obtained from *Stokes' formula* or similar equations using gravity anomalies in the calculations. Syn: *cogeoid.* A different cogeoid is obtained for each system of reduction: free-air cogeoid, isostatic cogeoid, etc.

compensation depth [oceanog] Depth at which the rate of photosynthesis equals the rate of respiration. Cf: *carbonate compensation depth.* Syn: *depth of compensation [oceanog].*

compensation depth [tect] *depth of compensation [tect].*

compensation level [oceanog] The depth of the ocean at which the consumption and production of oxygen are equal. It is the deepest level at which phytoplankton, which produce oxygen, can exist; i.e., the bottom of the *euphotic zone.*

compensation level [tect] *depth of compensation [tect].*

compensation method A procedure for determining the voltage difference between two points in the ground by balancing against a voltage that is adjusted in phase and amplitude to effect the compensation. See also: *compensator [elect].*

compensation point In crystal optics, the point at which an interference color is compensated by introduction of a quartz wedge or by rotation of a Berek compensator.

compensation sac The *ascus* of an ascophoran cheilostome. Syn: *compensatrix.*

compensator [elect] An instrument to determine the voltage difference between two points in the ground by the *compensation method.*

compensator [optics] An apparatus in a polarizing microscope that measures the phase difference between two components of polarized light, e.g. a *Berek compensator* or a *biquartz plate.*

compensatrix *compensation sac.*

competence The ability of a current of water or wind to transport detritus, in terms of particle size rather than amount, measured as the diameter of the largest particle transported. It depends on velocity: a small but swift stream, for example, may have greater competence than a larger but slower-moving stream. Cf: *capacity [hydraul].* Adj: *competent [hydraul].*

competent [hydraul] Pertaining to the *competence* of a stream or of a current of air.

competent [struc geol] Said of a layer which, in contrast to adjacent layers, has formed more nearly *parallel folds* (the adjacent layers being more nearly in *similar folds*) and/or *boudins* or other brittle structures. Ant: *incompetent.*

competent rock A volume of rock which under a specific set of conditions is able to support a tectonic force. Such a volume may be competent or incompetent a number of times in its deformational history depending upon the environmental conditions, degree and time of fracturing, etc. Cf: *incompetent rock.*

compilation The selection and assembly of map detail from various source materials (such as existing maps, aerial photographs, surveys, and new data), and the preparation and production of a new or improved map (or a part of a map) based on this detail. See also: *delineation.*

compiled map A map (esp. a small-scale map of a large area) incorporating information collected from various source materials and not developed from original survey data for the map in question; a map prepared by *compilation.*

complementary [petrology] Said of different rocks or groups of rocks differentiated from the same magma, whose total composition is that of the parent magma.

complementary [struc geol] Said of sets of fractures that are considered to be *conjugate* although their origin is unknown. Cf: *cognate.*

complete caneolith A *caneolith* having upper and lower rim elements and a wall. Cf: *incomplete caneolith.*

complete flower A flower having all four types of floral appendages: sepals, petals, stamens, and carpels. Cf: *incomplete flower.*

complete overstep A term proposed by Swain (1949, p.634) for an *overstep* in which an unconformity (partly angular, partly parallel) is universal throughout a basin of deposition or in which the older rocks of the basin are entirely blanketed unconformably by younger rocks.

complete tabula A coral *tabula* consisting of a single platform. Cf: *incomplete tabula.*

complex n. (a) A large-scale field association or assemblage of different rocks of any age or origin, having structural relations so intricately involved or otherwise complicated that the rocks cannot be readily differentiated in mapping; e.g. a "volcanic complex". See also: *igneous complex; injection complex; metamorphic complex; basement complex.* (b) A rock-stratigraphic unit that includes a mass of rock "composed of diverse types of any class or classes or...characterized by highly complicated structure" (ACSN, 1961, art.6j). The term may be used as part of the formal name instead of a lithologic or rank term; e.g. Crooks Complex (Precambrian) of central Arizona. (c) A time-stratigraphic unit corresponding to the rocks formed during an eon (Kobayashi, 1944a, p. 478). This usage is not accepted by the ACSN or the IUGS Subcommission.

complex crater A meteorite impact crater of large diameter and relatively shallow depth, characterized by a central uplift and peripheral slumping or terracing, which apparently develop during the late stages of the crater-forming event as a result of yielding of the underlying rocks (Dence, 1968, p. 182). Cf: *simple crater.*

complex cuspate foreland A *cuspate foreland* in which erosion on one side of the cusp has truncated beach ridges and swales but which is later prograded so that beach ridges, swales, and other symmetrical lines of growth are parallel to the new shoreline (Johnson, 1919, p. 325); e.g. Cape Canaveral, Fla. Cf: *simple cuspate foreland.*

complex drainage pattern A drainage pattern that shows variations among component parts, such as one in an area of complicated geologic structure and geomorphic history (Thornbury, 1954, p. 123).

complex dune A dune formed by multidirectional winds, resulting in the intersection of two or more dunes.

complex fold A fold that contains a *cross fold.*

complex mountain A mountain that includes a combination of structures and a variety of landforms; a mountain whose structures defy simple classification.

complex ore (a) An ore that yields several metals. Cf: *simple ore.* (b) An ore that is difficult to utilize because it contains more than one metal or because of the presence of unusual metals.

complex prism A prism composed of fibrils radiating from an axis, found in many mollusks (MacClintock, 1967, p. 15).

complex resistivity Representation of apparent resistivity as having real and imaginary parts, to accommodate observed variations in resistivity with frequency in induced-polarization surveys.

complex ripple mark A syn. of *cross ripple mark.* Term used by Kelling (1958, p.121) for an "interference ripple-pattern" of any kind.

complex spit A large *recurved spit* with minor or secondary spits developed at the end of the main spit (as by minor currents). The lines of growth of the minor and the main spits do not merge or curve into each other, but intersect at distinct angles. Example: Sandy Hook, N.J. Cf: *compound spit.*

complex stream A stream that has entered a second or later cycle of erosion (Davis, 1889b, p. 218); e.g. the headwaters of a compound stream.

complex tombolo The system of islands and beaches that results when several islands are united with each other and with the mainland by a complicated series of *tombolos.* Syn: *tombolo series; tombolo cluster.*

complex twin A twin in feldspar that is the result of both normal twinning and parallel twinning.

complex valley A valley of which part may be parallel and part transverse to the general structure of the underlying strata. Term introduced by Powell (1874, p. 50). Cf: *simple valley; compound valley.*

component One of a set of chemical compositions of a thermodynamic system, the relative masses of which may be varied to describe all compositions within it. A component need not be physically realizable, and may be so defined as to have negative concentration for some components of the system. Components are the minimum number of chemical units required to describe the phase-rule behavior of a system. Cf: *exchange operator.*

componental movement In the deformation of a rock, the relative movements of component particles.

component-stratotype One of several specified intervals of strata making a *composite-stratotype* (ISG, 1976, p. 24).

composite n. A term occasionally used in ecology to denote an association of algae and fungi in an aqueous environment.

composite coast A term used by Cotton (1958, p. 441) for an initial coast resulting from deformation (either upwarping or subsidence) occurring along lines transverse to the coast, characterized by salients and embayments developed on a very large scale; a "coast of transverse deformation" consisting of alternate zones of submergence and emergence, as along the coast of New Zealand near Wellington.

composite cone *stratovolcano.*

composite fault scarp A fault scarp whose height results from the combined effects of differential erosion and fault movement.

composite fold *compound fold.*

composite gneiss Gneiss that is constituted of materials of at least two different phases (Dietrich & Mehnert, 1961). Syn: *mixed gneiss.* Cf: *arterite; phlebite; venite; veined gneiss.*

composite grain A sedimentary particle formed by aggregation of two or more discrete particles; esp. a carbonate particle resulting from clustering of lumps, pellets, coated grains, or detrital, skeletal, or algal particles (Bissell & Chilingar, 1967, p. 153).

composite intrusion Any igneous intrusion that is composed of two or more injections of different chemical and mineralogical composition. Cf: *multiple intrusion.* See also: *partial pluton.*

composite map In mining, a map that shows several levels of a mine on one sheet; a map that vertically projects data from different elevations in the mine to one level. Cf: *cartology.*

composite photograph A photograph made by assembling the separate photographs made by the several lenses of a multiple-lens camera in simultaneous exposure into the equivalent of a photograph taken with a single wide-angle lens.

composite profile A plot consisting of the highest points of a series of *profiles [geomorph]* drawn along several regularly spaced and parallel lines on a map (Monkhouse & Wilkinson, 1952); it represents the surface of any relief area as viewed in the horizontal plane of the summit levels from an infinite distance. Cf: *superimposed profile; projected profile.* Syn: *zonal profile.*

composite ripple mark A term used by Tanner (1960, p.484) for an *oscillation cross ripple mark* formed by two intersecting sets of waves neither of which is parallel with the crests of the ripple marks. It is best developed on shoals where wave refraction produces two sets of waves not quite at right angles to each other.

composite seam A coal seam composed of two or more distinct coal beds that come into contact where dirt bands or other intervening strata wedge out.

composite section A single inclined or vertical *section [geol],* prepared by projecting data from various more or less parallel sections.

composite sequence A term used by Duff & Walton (1962) for a sequence of beds that comprises all lithologic types (in a succession displaying cyclic sedimentation) in the order in which they tend to occur. It is constructed from statistical data based on actual rock successions.

composite set A term proposed by McKee & Weir (1953, p.384) for a large sedimentary unit of similar or gradational lithology, "compounded from both stratified and cross-stratified units", and consisting of horizontal strata together with *cosets* of cross-strata.

composite stone *assembled stone.*

composite-stratotype A *stratotype* formed by the combination of several specified type intervals of strata known as *component-stratotypes* (ISG, 1976, p. 24).

composite stream A stream whose drainage basin receives waters from areas of different geomorphologic structures (Davis, 1889b).

composite topography A landscape whose topographic features have developed in two or more cycles of erosion.

composite unconformity An unconformity representing more than one episode of nondeposition and possible erosion.

composite volcano *stratovolcano.*

compositing *mixing.*

composition [mineral] *structural formula.*

composition [petrology] (a) *chemical composition [petrology].* (b) The make-up of a rock in terms of the species and number of minerals present; mineralogic composition. Cf: *mode.*

compositional layering A synonym of *phase layering,* commonly used with reference to metamorphic rocks. Syn: *primary layering.*

compositional maturity A type of sedimentary *maturity* in which a clastic sediment approaches the compositional end product to which it is driven by the formative processes that operate upon it. It may be expressed as a ratio between chemical compounds (e.g. alumina/soda) or between mineral components (e.g. quartz/feldspar) (Pettijohn, 1957, p. 286 & 508). Cf: *mineralogic maturity; textural maturity.*

composition face *composition surface.*

composition plane In a crystal twin, a *composition surface* that is planar.

composition point In a plot of phase equilibria, that point whose coordinates represent the chemical composition of a phase or mixture.

composition surface The surface along which the individuals of a crystal twin are joined. It may or may not be a plane, and it is usually identical with the *twin plane.* See also: *composition plane.* Syn: *composition face.*

compositor-viewer A device in which black-and-white multispectral images are registered and projected with colored lights to produce a color image.

compound alluvial fan *bajada.*

compound conodont element A bladelike or barlike *conodont element,* commonly bearing denticles.

compound cooling unit An ash-flow unit or series of such units that record breaks in the history of cooling, compaction, and welding (Smith, 1960, p. 157).

compound coral The skeleton of a colonial coral.

compound cross-bedding A term used by Gilbert (1899, p.139) for a variety of ripple-drift cross-lamination. Not in common usage.

compound cross-stratification A complex type of cross-stratification that combines forms differing substantially in scale and orientation (Harms et al., 1975, p. 51). Cf: *compound foreset bedding.*

compound cuspate bar A bar formed by a *compound spit* that unites with the shore at its distal end (Johnson, 1919, p. 319).

compound eye An eye of an arthropod, consisting essentially of a great number of minute eyes crowded together; e.g. an array of contiguous visual units (ommatidia) of a crustacean having a common optic-nerve trunk, or the lateral eye of a merostome composed of many facets, or a *holochroal eye* of a trilobite. See also: *ocellus [paleont].*

compound fault *fault zone.*

compound fold A fold upon which minor folds with similar axes have developed. Cf: *simple fold.* Syn: *composite fold.*

compound foreset bedding Cross-bedding characterized by a concave base and by several foreset beds dipping in more than one direction, as where tangential foresets recently deposited may have their upper ends truncated and may then be overlain by new beds of similar nature dipping in the same or in a different direction (Lahee, 1923, p.80). It may develop by changes in the stream or by fluctuations of water level of a lake, such as interference among adjacent lobes on a delta front. Cf: *compound cross-stratification.*

compound laccolith A laccolith having several parts that are separated from each other by thick layers of country rock but that were formed by a single intrusion. Cf: *cedar-tree structure.*

compound leaf A leaf blade composed of several distinct parts or *leaflets.*

compound operculum A dinoflagellate operculum that is divided into two or more pieces that are separable from one another. Cf: *simple operculum.*

compound pellet A pellet of silt, sand, or granule size or larger, originating from pelleted limestone with micritic or sparry cement, and sometimes containing matrix or interstitial material (Bissell & Chilingar, 1967, p. 153).

compound plate An ambulacral-plate unit of an echinoid, composed of two or more individual ambulacral plates, each with a pore for a tube foot, bound together by a single large tubercle articulating with the *primary spine* (TIP, 1966, pt.U, p.253).

compound ripple mark (a) A *cross ripple mark* resulting from the simultaneous interference of wave oscillation with current action, and characterized by a systematic breaking or offsetting of the crests of the current ripples (Bucher, 1919, p.195). (b) A ripple mark produced by modification of a pre-existing set of ripples by a later set of either wave or current origin (Kelling, 1958, p.122); e.g. *wave-current ripple mark.*

compound shoreline A shoreline showing a very marked development of the features characteristic of at least two of the following: a *shoreline of emergence,* a *shoreline of submergence,* a *neutral shoreline* (Johnson, 1919, p. 172); e.g. where a formerly submerged shoreline is elevated slightly but not enough to destroy the effect of submergence, or where a dissected shoreline of emergence undergoes a slight submergence so that the coastal-plain valleys are drowned.

compound skeletal wall As seen in sections of stenolaemate bryozoans, a wall calcified by epidermis located on growing edges and both sides, generally producing a bilaterally symmetrical microstructural pattern (Boardman & Cheetham, 1969, p. 211). Cf: *simple skeletal wall.*

compound spit A *recurved spit* whose inner side shows a series of intermittent landward-deflected points representing successive recurved termini. Cf: *complex spit.*

compound stream A stream that is of different ages in its different parts (Davis, 1889b); e.g. a stream with old headwaters rising in the mountains and a young lower course traversing a coastal plain.

compound structure A term used by Cotton (1948) for an arrangement of rocks characterized by a simple *cover mass* resting unconformably on more complex *undermass.*

compound trabecula A *trabecula* of a scleractinian coral, composed of bundles of sclerodermites. Cf: *simple trabecula.*

compound valley A valley whose main course may be a *simple valley* or a *complex valley,* but whose tributary valleys are of a different kind. Term introduced by Powell (1874, p. 50).

compound valley glacier A glacier composed of two or more individual valley glaciers emanating from tributary valleys.

compound volcano A volcano that consists of a complex of two or more vents, or a volcano that has an associated volcanic dome, either in its crater or on its flanks. Examples are Vesuvius and Mont Pelée.

compreignacite A yellow secondary mineral: $K_2U_6O_{19} \cdot 11H_2O$.

compressed Said of the shape of the whorl section in a chambered cephalopod conch whose dorsoventral diameter is greater than its lateral diameter. Ant: *depressed.*

compressibility The reciprocal of *bulk modulus*; it equals 1/k where k is the bulk modulus. Its symbol is β. Syn: *modulus of compression.*
compressing flow A flow pattern on glaciers in which the velocity decreases with distance downstream; thus the longitudinal strain rate (velocity gradient) is compressive. This condition requires a transverse or vertical expansion, or a negative net balance on the surface, because ice is almost incompressible (Nye, 1952). Ant: *extending flow.*
compression [exp struc geol] A system of forces or stresses that tends to decrease the volume of, or shorten, a substance; also, the change of volume produced by such a system of forces.
compression [paleont] The remains of a fossil plant that has been flattened by the vertical pressure of overlying rocks.
compressional wave *P wave.*
compression fault A general term for a fault that has been produced by lateral crustal compression, e.g., a reverse fault. Little used. Cf: *tension fault.*
compression test *triaxial compression test.*
compressive strength The maximum *compressive stress* that can be applied to a material, under given conditions, before failure occurs.
compressive stress A *normal stress* that tends to push together material on opposite sides of a real or imaginary plane. See also: *compressive strength.* Cf: *tensile stress.*
compromise boundary A surface of contact, not corresponding to a crystal face, between two mutually growing but differently oriented crystals.
comptonite An opaque variety of thomsonite from the Lake Superior region.
computer An automatic electronic device capable of accepting information, applying prescribed processes to it, and supplying the results of these processes. The term is generally used for any type of computer. See also: *analog computer; digital computer; hybrid computer.*
computer-aided instruction Instruction where a computer controls the lesson material and many students are simultaneously accommodated on terminals interacting with the system (Merriam, 1976, p. 4). Abbrev: CAI. Syn: *computer-assisted instruction.*
computer-assisted instruction A syn. of *computer-aided instruction.* Abbrev: CAI.
computer graphics Maps, graphs, and diagrams plotted directly from data stored in computer banks and suitable for publication with minimal cartographic effort.
conate Said of sculpture of pollen and spores consisting of coni. Cf: *conus.*
concave bank The outer bank of a curved stream, with the center of the curve toward the channel; e.g. an *undercut* slope. Ant: *convex bank.*
concave cross-bedding (a) Cross-bedding with concave (downward-arching), generally tangential, foreset beds. This type of cross-bedding is very common and is used as a criterion for distinguishing top from bottom in sedimentary rocks. Cf: *convex cross-bedding.* (b) Cross-bedding deposited on a lower concave surface, as in *festoon cross-bedding.*
concave slope *waning slope.*
concavo-convex Concave on one side and convex on the other; e.g. said of a brachiopod shell having a concave brachial valve and a convex pedicle valve. Cf: *convexo-concave.*
concealed coalfield Deposits of coal that do not crop out. Cf: *exposed coalfield.*
concealed pediment A pediment that is buried by a thin layer of alluvium from an encroaching bajada; it is generally caused by a local rise of base level. Cf: *suballuvial bench; fan-topped pediment.*
concentrated flow *channel flow.*
concentrated wash *channel erosion.*
concentrates The valuable fraction of an ore that is left after worthless material is removed in processing. Cf: *tailings.*
concentration [ice] The ratio in eighths or tenths of the areal extent actually covered by sea ice to the total area of the sea surface, both covered by ice and ice free, at a specific location or over a defined area. See also: *ice cover.*
concentration [sed] *sediment concentration.*
concentration boundary "A line approximating the transition between two areas of pack ice with distinctly different concentrations" (U.S. Naval Oceanographic Office, 1968, p. B32).
concentration time (a) The time required for water to flow from the farthest point in a watershed to a gaging station. (b) That time at which the rate of runoff equals the rate of precipitation of a storm. Syn: *time of concentration.*
concentric fold *parallel fold.*
concentric shearing surface *surface of concentric shearing*
concentric weathering *spheroidal weathering.*
conceptual model *model.*
concertina fold *kink fold.*
conch (a) The portion of a cephalopod shell developed after the embryonic shell; e.g. the complete shell of an ammonoid exclusive of the protoconch, or all the hard calcareous parts of a nautiloid (including the external shell, septa, and siphuncle) exclusive of cameral deposits and any structures within the siphuncle. (b) Any of various large spiral-shelled marine gastropods, often of the genera *Strombus* or *Cassis.* Also, the shell of such a conch, often used for making cameos and formerly made into horns. (c) Any of various marine shells of invertebrates, including bivalve mollusks and brachiopods.—Pl: *conchs.*
conchal furrow A shallow, midventral groove on the inside wall of a nautiloid conch.
conchiform Shell-shaped; esp. shaped like one half of a bivalve shell.
conchilite A bowl-shaped body of limonite or goethite growing in an inverted position on mineralized bedrock and resembling the shell of an oyster or clam coated with a rusty deposit (Tanton, 1944). It is roughly oval or circular in plan, with a smooth or irregular and scalloped outline; it ranges from 2.5 cm to 1 m in diameter and from 2 cm to 7.5 cm in height.
conchiolin A fibrous protein, $C_3H_{48}N_9O_{11}$, that constitutes the organic basis of most mollusk shells; e.g. the material of which the periostracum and organic matrix of the calcareous parts of a bivalve-mollusk shell are composed. Syn: *conchyolin.*
conchitic Abounding in fossil shells; esp. said of a rock composed of or containing many shells.
conchoidal Said of a type of mineral or rock fracture that gives a smoothly curved surface. It is a characteristic habit of quartz and of obsidian. Etymol: like the curve of a conch (seashell).
conchology The study of shells of both fossil and existing animals. Cf: *malacology.*
conchorhynch The calcified hard tip of the horny lower jaw, probably only in nautiloids.
conchostracan Any branchiopod belonging to the order Conchostraca, characterized by a translucent bivalve shell with a clawlike furca developed at the posterior end. Range, Lower Devonian to present.
conclave A large or roomy atrium of a pollen grain, which in cross section appears to be filled with rodlike elements at an acute angle to the wall (Tschudy & Scott, 1969, p. 27).
conclinal A syn. of *cataclinal.* Term introduced by Powell (1874, p. 50).
concordance Parallelism of strata with sequence boundaries, with no terminations of strata against the boundary surfaces (Mitchum, 1977, p. 206). Cf: *discordance.*
concordance of summit levels *summit concordance.*
concordant [geochron] (a) Said of radiometric ages, determined by more than one method, that are in agreement within the analytical precision for the determining methods. (b) Said of radiometric ages given by coexisting minerals, determined by the same method, that are in agreement. (c) In a more restricted sense, the term has been used to indicate agreement of $^{238}U/^{206}Pb$, $^{235}U/^{207}Pb$, $^{207}Pb/^{206}Pb$, and $^{232}Th/^{208}Pb$ ages determined, within experimental error, for the same mineral sample. Ant: *discordant [geochron]*
concordant [hydraul] Said of flows, e.g. floods, at different points along a channel which have the same frequency of occurrence (Langbein & Iseri, 1960, p. 6).
concordant [intrus rocks] Said of a contact between an igneous intrusion and the country rock that parallels the foliation or bedding planes of the latter. Cf: *discordant [intrus rocks]; conformable [intrus rocks].*
concordant [stratig] Structurally *conformable;* said of strata displaying parallelism of bedding or structure. The term may be used where a hiatus cannot be recognized but cannot be dismissed. Ant: *discordant [stratig].*
concordant bedding A sedimentary structure marked by beds that are parallel and without angular junctions. Ant: *discordant bedding*. Syn: *parallel bedding.*
concordant coastline A coastline that is broadly parallel to the main trend of the land structure (such as mountain ranges or fold

belts) forming the margin of the ocean basin; it is generally linear and regular. Cf: *Dalmatian coastline.* Ant: *discordant coastline.* Syn: *Pacific-type coastline; longitudinal coastline.*

concordant drainage *accordant drainage.*

concordant junction *accordant junction.*

concordant summit level *accordant summit level.*

concordia The graphed curve formed when the $^{206}Pb/^{238}U$ ratio is plotted against the $^{207}Pb/^{235}U$ ratio as both increase in value due to nuclear decay of uranium to lead with passage of time, assuming a closed U-Pb system. The curve (concordia) is the locus of all concordant U-Pb ages, thus it is a time curve. Syn: *concordia plot; concordia diagram; concordia curve.*

concordia curve *concordia.*

concordia diagram *concordia.*

concordia intercept The intersections of the *concordia* curve (time curve) and a straight line (chord) which depicts the plot (locus) of discordant U-Pb ages. It may indicate the age of two important events in the life of the U-Pb system under study, the older age representing the start of the U-Pb system and the younger age the time when this system was disturbed.

concordia plot *concordia.*

concrescence A growing-together or coalescence of originally separate parts; e.g. union of radial spines in a radiolarian skeleton.

concrete A mixture of *cement [mater],* an *aggregate,* and water, which will "set" or harden to a rocklike consistency.

concretion (a) A hard, compact mass or aggregate of mineral matter, normally subspherical but commonly oblate, disk-shaped, or irregular with odd or fantastic outlines; formed by precipitation from aqueous solution about a nucleus or center, such as a leaf, shell, bone, or fossil, in the pores of a sedimentary or fragmental volcanic rock, and usually of a composition widely different from that of the rock in which it is found and from which it is rather sharply separated. It represents a concentration of some minor constituent of the enclosing rock or of cementing material, such as silica (chert), calcite, dolomite, iron oxide, pyrite, or gypsum, and it ranges in size from a small pellet-like object to a great spheroidal body as much as 3 m in diameter. Most concretions were formed during diagenesis, and many (especially in limestone and shale) shortly after sediment deposition. Cf: *nodule; secretion.* See also: *accretion; incretion; intercretion; excretion.* (b) A collective term applied loosely to various primary and secondary mineral segregations of diverse origin, including irregular nodules, spherulites, crystalline aggregates, geodes, septaria, and related bodies.

concretionary Characterized by, consisting of, or producing concretions; e.g. a "concretionary ironstone" composed of iron carbonate with clay and calcite, or a zonal "concretionary texture" (of an ore) characterized by concentric shells of slightly varying properties due to variation during growth.

concurrent-range zone A formal *range-zone* or body of strata defined by the overlapping ranges of two or more specified taxa from one or more of which it takes its name (ACSN, 1961, art.23); the part of the section where two range zones overlap. It is recognized by the lowest occurrence of certain fossils and the highest occurrence of others, and it is the biostratigraphic zone generally used in attempting time-correlation of strata. The term is equivalent to Oppel's (1856-1858) usage of *zone [stratig].* See also: *Oppel-zone; multifossil range-zone.* Syn: *overlap zone; range-overlap zone.*

concussion fracture One of a system of fractures in individual grains of a shock-metamorphosed rock which are generally radial to the grain surface and related to the contacts with adjacent grains. They are apparently formed by violent grain-to-grain contacts in the initial stages of passage of a shock wave and are formed by the resulting tensile stresses parallel to the surfaces of the impacting grains (Kieffer, 1971, p. 5456, 5468).

concyclothem A cyclic sequence of strata resulting from the local coalescence of two or more cyclothems. Term introduced by Gray (1955).

condensate Liquid hydrocarbons, generally clear or pale straw-colored and of high *API gravity* (above 60°), that are produced with *wet gas.* Syn: *distillate; natural gasoline.*

condensation [phys] The process by which a vapor becomes a liquid or solid; the opposite of *evaporation.*

condensation [stratig] A stratigraphic process in which the thinning of a sedimentary deposit or succession takes place contemporaneously with deposition, as by strong hydrostatic pressures resulting in solution along grain boundaries.

condensation room In a cave, an area in which water vapor condenses on the ceiling and walls because of a lower temperature of the wall rock than of the air that enters the room.

condensation zone A concentration of several *biozones* in a single bed (Heim, 1934).

condensed deposit A sedimentary deposit in a *condensed succession.*

condensed succession A relatively thin but uninterrupted stratigraphic succession of considerable time duration, in which the deposits accumulated very slowly; it is generally represented by a time-equivalent thick succession elsewhere in the same sedimentary basin or region. Ant: *extended succession.*

condensed system (a) A chemical system in which the vapor pressure is negligible, and can thus be ignored. (b) A chemical system in which the pressure maintained on the system is greater than the vapor pressure of any portion.

conditional resources *identified subeconomic resources.*

condrodite *chondrodite.*

conductance The inverse of electrical *resistance,* measured in mhos.

conductivity (a) *electrical conductivity.* (b) *thermal conductivity.*

conductivity log *induction log.*

conductivity-thickness product The product of average conductivity and thickness of a rock layer; it is used to characterize the response of an electromagnetic or direct-current method to the presence of a highly conductive layer. Measured in siemens units.

conduit A passage that is filled with water under hydrostatic pressure. See also: *siphon [speleo]; streamtube.*

condyle (a) In some cheilostome bryozoans, one of a pair of oppositely placed teeth or protuberances on which the operculum of an autozooid, the mandible of an avicularium, or the seta of a vibraculum pivots or is slung. Cf: *cardella.* (b) A swollen knob on the shell surface of an acantharian radiolarian. (c) A rounded articular prominence on a bone. See also: *occipital condyle.*

cone [bot] A fertile branch that bears an imbricated group of seed- or spore-bearing scales or sporophylls, as in pine or club moss. Cf: *strobilus.*

cone [geomorph] A mountain, hill, or other landform shaped like a cone, having relatively steep slopes and a pointed top; specif. an *alluvial cone.*

cone [glac geol] A steep-sided pile of sand, gravel, and sometimes boulders, with a fan-like outwash base, deposited against the front of a glacier by meltwater streams.

cone [marine geol] A type of *submarine fan.* The term is frequently used to describe a deep-sea fan associated with a major active *delta,* e.g. Mississippi, Nile, Ganges (Kelling & Stanley, 1976, p. 387).

cone [volc] *volcanic cone.*

cone cup The inside of an enclosing cone of a sedimentary cone-in-cone structure.

cone delta *alluvial cone.*

cone dike *cone sheet.*

cone-in-cone coal Coal exhibiting *cone-in-cone structure.* Syn: *crystallized coal.*

cone-in-cone structure [sed] (a) A minor sedimentary structure in thin, generally calcareous layers of some shales and in the outer parts of some large concretions, esp. septaria; it resembles a set of concentric, right circular cones fitting one into another in inverted positions (base upward, apex downward), commonly separated by clay films, and consisting usually of fibrous calcite and rarely of siderite or gypsum. The apical angles are between 30 and 60 degrees and the cone axes are normal to the bedding; the height of the cones usually ranges from 10 mm to 10 cm, and their sides are often ribbed, fluted, or grooved, and marked by annular depressions and ridges that are more pronounced near the bases and finer and more obscure near the apices. The structure appears to be due to pressure aided by crystallization and weathering (solution) along intersecting conical shear zones. (b) A similar structure developed in coal and consisting of a set of interpenetrating cones packed closely together.—Syn: *cone-in-cone.*

cone-in-cone structure [volc] A volcanic structure in which a younger cone or cones have developed within the original one; also, a similar pattern of crater development. Adj: *nested.*

cone karst *cockpit karst.*

conella A tiny prism or cone of spherulitic aragonite or secondary calcite, found occasionally on the surface of *steinkerns* of ammonoid conchs, less often on those of nautiloid conchs, produced by partial solution of the inner prismatic shell layer. Pl: *conellae.*

Conemaughian North American provincial stage: lower Upper Pennsylvanian (above Allegheny, below Monongahelan).

cone of dejection An *alluvial cone* consisting of coarse material, formed where a mountain torrent emerges from a narrow valley upon a plain or plunges over a valley-side bench. Syn: *dejection cone.*

cone of depression A depression in the potentiometric surface of a body of ground water that has the shape of an inverted cone and develops around a well from which water is being withdrawn. It defines the *area of influence* of a well. The shape of the depression is due to the fact that the water must flow through progressively smaller cross sections as it nears the well, and hence the hydraulic gradient must be steeper. Cf: *drawdown [grd wat]; cone of exhaustion.*

cone of detr[illegible] *alluvial cone.*

cone of exhaustion A cone-shaped depression in the potentiometric surface of a body of ground water that develops around a well when water is being withdrawn more rapidly than it can percolate laterally. Cf: *cone of depression.*

cone of pressure relief A cone of depression in the potentiometric surface of a body of confined ground water. As defined by Tolman (1937, p. 562), "an imaginary surface indicating pressure-relief conditions in a confined aquifer due to pumping".

cone penetration test A soil *penetration test* in which a steel cone of standard shape and size is pushed into the soil and the force required to advance the cone at a predetermined, usually slow and constant rate, or for a specified distance, or in some designs the penetration resulting from various loads, is recorded.

cone sheet A dike that is arcuate in plan and dips at 30°-45° toward the center of the arc. Cone sheets occur in concentric sets, which presumably converge at a magmatic center. They are associated with *ring dikes* to form a *ring complex.* Syn: *cone dike.*

Conewangoan *Bradfordian.*

conferva peat Peat that is derived mainly from filamentous algae of the genus *Tribonema.*

confidence interval The interval between *confidence limits.*

confidence limit Either the upper or the lower value of the range within which an actual measurement or parameter will fall with a stated probability.

configuration (a) The form or shape of a part of the Earth's surface with regard to its horizontal outline, its elevation, and its relative disposition or arrangement with other parts of the surface. (b) The topography of a region as shown by the typical contour map, where the contour interval is well suited to express the slope of the terrain of the area (ASCE, 1954, p. 39-40). Syn: *configuration of terrain.*

confined aquifer An aquifer bounded above and below by impermeable beds, or by beds of distinctly lower permeability than that of the aquifer itself; an aquifer containing confined ground water. Syn: *artesian aquifer.*

confined ground water Ground water under pressure significantly greater than that of the atmosphere. Its upper surface is the bottom of an impermeable bed or a bed of distinctly lower permeability than the material in which the water occurs. Ant: *unconfined ground water.* Syn: *artesian water; confined water; piestic water.*

confined water *confined ground water.*

confining bed A body of impermeable or distinctly less permeable material stratigraphically adjacent to one or more aquifers. Cf: *aquitard; aquifuge; aquiclude.*

confining pressure An equal, all-sided pressure, e.g. *geostatic pressure* or *hydrostatic pressure.*

confluence [glaciol] (a) A flowing-together of two or more glaciers. (b) A junction or place where confluence occurs.—Ant: *diffluence.*

confluence [streams] (a) A place of meeting of two or more streams; the point where a tributary joins the main stream; a fork. (b) A flowing together of two or more streams. (c) The stream or other body of water produced by confluence; a combined flood.—See also: *junction.*

confluence plain A plain formed by the merging of the valley floors of two or more streams.

confluence step A rock step that rises upstream toward the heads of two glacial valleys at their place of confluence. It is probably caused by the strengthening of glacial action downvalley from that point. Ant: *diffluence step.*

confluent adj. Said of a stream, glacier, vein, or other geologic feature that combines or meets with another like feature to form one stream, glacier, vein, etc. Ant: *diffluent.*—n. A confluent stream, usually a stream uniting with another of nearly equal size; a fork or branch of a river. The term is sometimes loosely applied to an *affluent.*

conformability The quality, state, or condition of being conformable, such as the relationship of conformable strata; *conformity.*

conformable [intrus rocks] Said of the contact of an intrusive body when it is aligned with the intrusion's internal structures. Cf: *disconformable [intrus rocks]; concordant [intrus rocks].*

conformable [stratig] Said of strata or stratification characterized by an unbroken sequence in which the layers are formed one above the other in parallel order by regular, uninterrupted deposition under the same general conditions; also said of the contacts (abrupt, gradational, or intercalated) between such strata. The term is often applied to a later formation having bedding planes that are parallel with those of an earlier formation and showing an arrangement in which disturbance or erosion did not take place at the locality during deposition. Dennis (1967, p.26) regards the term as descriptive, primarily referring to succession without disturbance and not necessarily implying parallelism of succeeding beds or continuity of deposition; e.g. cross-bedding without intervening erosion is conformable, but not parallel. Cf: *unconformable; concordant.*

conformably superimposed stream A term suggested by Kümmel (1893, p. 380) for a stream superimposed from a conformable cover.

conformality The unique property of a conformal map projection in which all small or elementary figures on the surface of the Earth retain their original shapes on the map. Cf: *equivalence [cart].* Syn: *orthomorphism.*

conformal projection A map projection on which the shape of any very small area of the surface mapped is preserved unchanged on the map and the scale at any point is the same in every direction (as along the meridian and the parallel at that point), although the scale may vary from point to point. It always shows a right-angle intersection of any parallel with any meridian. Examples include: stereographic projection, Mercator projection, and Lambert conformal conic projection. Cf: *equal-area projection.* Syn: *orthomorphic projection.*

conformity (a) The mutual and undisturbed relationship between adjacent sedimentary strata that have been deposited in orderly sequence with little or no evidence of time lapses; true stratigraphic continuity in the sequence of beds without evidence that the lower beds were folded, tilted, or eroded before the higher beds were deposited. Syn: *conformability.* (b) An uninterrupted sequence of strata displaying conformity. (c) A surface that separates younger strata from older ones, along which there is no physical evidence of erosion or nondeposition, and no significant hiatus (Mitchum, 1977, p. 206).—Cf: *unconformity.*

confused sea A highly disturbed sea surface with indeterminate direction and period of wave travel. It is caused by the superimposition of various wave trains.

congela A term used in Chile for *coba* with a high salt content.

congelation The change from a fluid to a solid state; freezing. Also, the product of such a change. Syn: *gelation [ice].*

congelifluction The progressive lateral flow of earth material under periglacial conditions; solifluction in a region underlain by frozen ground. Syn: *gelifluction; gelisolifluction.*

congelifract An angular rock fragment split off by frost action (*congelifraction*), ranging from very large blocks to finely comminuted material. Syn: *gelifract.*

congelifractate A mass of rock fragments (*congelifracts*) of different sizes produced by frost action.

congelifraction The mechanical disintegration, splitting, or breakup of a rock or soil due to the great pressure exerted by the freezing of water contained in cracks or pores, or along bedding planes; term introduced by Bryan (1946, p.640). Syn: *frost shattering; frost splitting; frost riving; frost bursting; frost weathering; frost wedging; gelivation; gelifraction.*

congeliturbate A mass of soil or other unconsolidated earth material moved or disturbed by frost action, and usually coarser than the underlying material; esp. a rubbly deposit formed by solifluction. See also: *head [mass move]; pocket [cryoped]; rubble drift; coombe rock.* Syn: *frost soil; cryoturbate; warp [mass move]; trail [mass move].*

congeliturbation A collective term suggested by Bryan (1946, p.640) to describe the stirring, churning, modification, and all other disturbances of soil, resulting from frost action; it involves frost heaving, solifluction, and differential and mass movements, and it produces patterned ground. Syn: *cryoturbation; frost stirring;*

frost churning; geliturbation.

congeneric Belonging to the same genus.

conglomerate A coarse-grained clastic sedimentary rock, composed of rounded to subangular fragments larger than 2 mm in diameter (granules, pebbles, cobbles, boulders) set in a fine-grained matrix of sand or silt, and commonly cemented by calcium carbonate, iron oxide, silica, or hardened clay; the consolidated equivalent of *gravel* both in size range and in the essential roundness and sorting of its constituent particles. The rock or mineral fragments may be of varied composition and range widely in size, and are usually rounded and smoothed from transportation by water or from wave action. Conglomerates may be classified according to nature or composition of fragments, proportion of matrix, degree of size sorting, type of cement, and agent or environment of formation. Etymol: Latin *conglomeratus,* "heaped, rolled, or pressed together". Cf: *breccia.* Syn: *puddingstone.*

conglomerated ice All forms of floating ice compacted into one mass. The term refers to the contents of an ice mass, not to the concentration.

conglomeratic Pertaining to a conglomerte; composed or having the properties of conglomerate.

conglomeratic mudstone A mudstone with a sparse to liberal sprinkling of pebbles or cobbles; e.g. a consolidated gravelly mud containing 5-30% gravel and having a ratio of sand to mud less than 1:1 (Folk, 1954, p.346), such as a *pebbly mudstone* or a *tilloid.* Pettijohn (1957, p.261) considered the term synonymous with *paraconglomerate.* Cf: *mudstone conglomerate.*

conglomeratic sandstone (a) A sandstone containing 5-30% gravel and having a ratio of sand to mud (silt + clay) greater than 9:1 (Folk, 1954, p.347); a consolidated gravelly sand. (b) A sandstone containing more than 20% pebbles (Krynine, 1948, p.141). Cf: *pebbly sandstone.*

conglomerite A term suggested by Willard (1930) for a conglomerate that has reached the same stage of induration as that of a quartzite, characterized by the welding together of matrix and clasts as evidenced by fractures passing through both. Examples include the partially metamorphosed or "stretched" Carboniferous conglomerates of Rhode Island, esp. those east of Newport and at Natick.

Congo copal A hard, yellowish to colorless *copal* derived from certain trees of the genus *Copaifera,* found as a fossil resin in the Congo, and used in making varnish. Syn: *Congo gum.*

congolite A trigonal mineral: $(Fe,Mg,Mn)_3B_7O_{13}Cl$. It is dimorphous with ericaite.

congregation Those fossil species occurring together in and characterizing the rocks of a particular *zone [stratig].* A congregation may include all or part of an *assemblage [paleoecol],* or more than one assemblage.

congressite A coarsely granular *urtite* consisting chiefly of nepheline, with minor amounts of sodalite, plagioclase, mica, and calcite. Cf: *craigmontite.* Named by Adams and Barlow in 1913 for Congress Bluff, Craigmont Hill, Ontario. Not recommended usage.

congruent Said of a parasitic fold, the axis and axial surface of which are parallel to the axis and axial surface of the main fold to which it is related. Var: *congruous.* Ant: *incongruous.*

congruent melting Melting of a substance directly to a liquid that is of the same composition as the solid. Cf: *incongruent melting.*

congruous *congruent.*

coni Plural of *conus.*

Coniacian European stage: Upper Cretaceous (above Turonian, below Santonian). See also: *Emscherian.*

coniatolite A hard, sheetlike crust of aragonite found in supratidal saline environments in the Persian Gulf area (Purser & Loreau, 1973, p. 375).

conical fold A fold model that can be described geometrically by the rotation of a line about one of its ends, which is fixed. Cf: *cylindrical fold.*

conical wave *head wave.*

conichalcite A pistachio-green to emerald-green mineral: $CaCu(AsO_4)(OH)$. It often contains phosphorus. Syn: *higginsite.*

conic projection One of a group of map projections produced by projecting the geographic meridians and parallels onto the surface of a cone that is tangent to, or intersects, the surface of the sphere, and then developing (unrolling and laying flat) the cone as a plane. True distances are measured along the line of tangency; everywhere else on the map the scale is too large or too small. Examples include: Lambert conformal conic projection, Albers projection, and Bonne projection. See also: *polyconic projection.* Syn: *conical projection.*

conidiophore A structure that bears conidia; specif. a specialized, typically erect *hypha* that produces successive conidia in certain fungi.

conidiospore A *conidium;* a *fungal spore* that may have chitinous walls and therefore occur as a microfossil in palynologic preparations.

conidium An asexual spore produced from the tip of a conidiophore; broadly, any asexual spore not borne within an enclosing structure, such as one not produced in a sporangium. Pl: *conidia.* See also: *conidiospore.*

conifer A gymnosperm, a member of the class Coniferae, having needlelike or scalelike leaves and naked seeds borne in cones. Conifers include pines, firs, and spruces.

coniferous Bearing cones, as in conifers.

coning (a) The cone-shaped rise of saltwater beneath fresh water in an aquifer as fresh water is produced from a well. Syn: *upconing.* (b) The cone-shaped rise of water underlying oil or gas in a reservoir as the oil or gas is withdrawn from a well.

conispiral adj. Said of a gastropod shell with a cone-shaped spire; said of a cephalopod shell characterized by a spiral coiled on the surface of a cone, the whorls not in a single plane.—n. A conispiral shell.

conjugate [fault] Said of faults that are of the same age and deformational episode.

conjugate [joint] Said of a joint system, the sets of which are related in deformational origin, usually compression. Also, said of the mineral deposits which may form in such joints. Cf: *cognate; complementary.*

conjugate fold system Two sets of minor folds, the axial surfaces of which are inclined towards each other. Cf: *kink band.*

conjugate solutions Two solutions coexisting in equilibrium whose compositions are separated by a miscibility gap in a potentially continuous compositional field. The possibility of a critical point at which the two phases would become identical is implicit.

conjugation line A *tie line [chem]* in a two-liquid field. The term is also sometimes used incorrectly as a synonym of *Alkemade line* and of *join.*

conjunct Said of a pore rhomb in which externally visible slits are continuous across the suture between plates bearing the rhomb. Cf: *disjunct.*

conker *kankar.*

connate [bot] In plant morphology, pertaining to the fusion of like parts. Cf: *adnate; adherent; coherent [bot].*

connate [petrology] Said of fluids derived from the same magma.

connate [sed] Originating at the same time as adjacent material; esp. pertaining to waters and volatile materials (such as carbon dioxide) entrapped in sediments at the time the deposits were laid down.

connate water Water entrapped in the interstices of a sedimentary rock at the time of its deposition; White (1957, p. 1661) has recommended that it be defined as "water that has been out of contact with the atmosphere for at least an appreciable part of a geologic period". Commonly misused by reservoir engineers and well-log analysts to mean *interstitial water* or *formation water.* Syn: *fossil water; fossilized brine; native water.*

connecting band (a) One of the two hooplike bands at the edge of the flange or valve mantle in a diatom frustule. See also: *girdle [paleont]; cingulum.* (b) A part of the loop of terebratellacean brachiopods that joins descending branches to the median septum or that joins ascending branches and descending branches posterior to their anterior curvature (TIP, 1965, pt. H, p. 142).

connecting bar A bar that is connected at both ends to a landmass; esp. a *tombolo.*

connecting lobe A rounded linear elevation of the valve surface of an ostracode, confluent with two or more subvertically trending *lobes* (TIP, 1961, pt.Q, p.49).

connecting ring The partly calcareous and partly conchiolinous delicate tubular membrane forming between septa the wall of a cephalopod siphuncle, such as one that connects the septal neck of an ellipochoanitic nautiloid conch with the septum immediately behind it. Vestiges of the connecting ring are also found in holochoanitic forms.

connecting tubule A subhorizontal tubular connection between neighboring corallites in a fasciculate corallum.

connective suture One of a pair of trilobite cephalic sutures, generally entirely on the ventral side, bounding the sides of the rostral plate. Cf: *median suture.*

connector bar A bar that joins the sagittal ring of a radiolarian skeleton to the lattice shell (as in the subfamily Trissocyclinae).

connellite A deep-blue mineral: $Cu_{19}Cl_4(SO_4)(OH)_{32}\cdot 3H_2O$. It is isomorphous with buttgenbachite. Syn: *footeite.*

conode *tie line [chem].*

conodont One of a large number of small, disjunct fossil elements assigned to the order Conodontophorida, phosphatic in composition, and commonly toothlike in form but not in function; produced in bilaterally paired, serial arrangement by small vagile marine animals of uncertain affinity. See: Hass (1962). Range, Cambrian (possibly Late Precambrian) to Upper Triassic; commonly abundant, widespread and useful biostratigraphically.

conodont element A unit or complete specimen of a conodont, e.g. *compound conodont element; platelike conodont element; lamellar conodont element; fibrous conodont element.*

conoplain A rarely used syn. of *pediment.* The term was used by Ogilvie (1905, p. 28) for an erosion surface peripheral to and sloping radially away in all directions from a laccolithic mountain mass.

conoscope A polarizing microscope using convergent light with the Bertrand lens inserted, used to test the interference figures of crystals. Cf: *orthoscope.*

conotheca Conoidal theca with a small circular aperture located at the terminus of a short neck developed irregularly in colonies of tuboid graptolithines (Whittington & Rickards, 1968).

Conrad discontinuity The seismic-velocity discontinuity within the Earth's crust that is equivalent to the boundary between the *sial* and the *sima,* at which velocities increase from ∾6.1 km/sec to 6.4-6.7 km/sec. It is not always discernible; i.e. there may be a general increase in velocity with depth in the crust, with no evidence of layer boundaries. See also: *Conrad layer.* A possible equivalent is the *Riel discontinuity.*

Conrad layer The seismic region of the Earth between the *Conrad discontinuity* and the Mohorovičić discontinuity. Also called the *intermediate layer* or the *lower crustal layer.*

consanguineous [petrology] The adj. of consanguinity and a less preferred syn. of *comagmatic.*

consanguineous [sed] Said of a natural group of sediments or sedimentary rocks related to one another by origin; e.g. a "consanguineous association" (such as flysch, molasse, or paralic sediments) interrelated by common ancestry, environment, and evolution, representing a "facies" in a broad sense but "not equivalent to a single lithologic type" (Fairbridge, 1958, p. 319).

consanguinity The genetic relationship that exists between igneous rocks that are presumably derived from the same parent magma. Such rocks are closely associated in space and time and commonly have similar geologic occurrence and chemical and mineralogic characteristics. Adj: *consanguineous.* See also: *comagmatic.*

consecutive-range zone "A body of strata within the range of a fossil group such that it forms the first part of the range of that fossil group before the first appearance of its immediate evolutionary descendant" (Geological Society of London, 1967, p. 85).

consequent [geomorph] Said of a geologic or topographic feature that originated as a result of and in harmony with preexisting conditions or features; e.g. a *consequent* ridge (such as an anticlinal arch that retained its axial eminence), or a *consequent* island in a lake basin (such as an elevation that remained above the water level at the time of the formation of the lake), or a *consequent* waterfall (such as one resulting from the irregularities of the surface over which the stream originally flowed). Cf: *subsequent; resequent; obsequent.*

consequent [streams] adj. Said of a stream, valley, or drainage system whose course or direction is dependent on or controlled by the general form and slope of an existing land surface. The term was first applied by Powell (1875, p. 163-166) to the valley thus formed.—n. *consequent stream.*

consequent divide A divide between two consequent streams.

consequent fault scarp A fault scarp whose face has been rapidly changed by mass-wasting shortly after the initial formation of the scarp.

consequent lake A lake occupying a depression that represents an original inequality in any new land surface; e.g. a lake existing in a depression in glacial deposits or among sand dunes, or in an irregularity of a recently uplifted sea floor. Syn: *newland lake.*

consequent stream A stream that originates on a newly exposed or recently formed surface and that flows along a course determined entirely by the initial slope and configuration of that surface; a stream whose direction of flow is directly related to and a necessary consequence of the original dip-slope surface of the land and the geologic structure of the area. Syn: *consequent; original stream.*

consertal A syn. of *sutured,* preferred in European usage but obsolescent in American usage.

conservative elements In seawater, elements whose total amount is invariant with time; they form most of the salts in seawater. Their abundance relative to each other remains constant; this principle is the basis for determining salinity. Cf: *nonconservative elements.* Syn: *conservative solutes; conservative ions.*

conservative ions *conservative elements.*

conservative solutes *conservative elements.*

consistency The relative ease with which a soil can be deformed. The term expresses the degree of firmness or cohesion of soil particles and their resistance to rupture or deformation, usually by hand pressure.

consistency index The ratio of the difference between the *liquid limit* and the natural water content of a soil to the *plasticity index.* Syn: *relative consistency.*

consistency limits *Atterberg limits.*

consociation An ecologic community, within an association, that has one dominant species, usually one of several dominants.

consolidated pack ice Pack ice in which the concentration is 8/8 (or 10/10) and the floes are frozen together. Obsolete syn: *consolidated pack; consolidated ice; field ice.*

consolidation (a) Any process whereby loosely aggregated, soft, or liquid earth materials become firm and coherent rock; specif. the *solidification* of a magma to form an igneous rock, or the *lithification* of loose sediments to form a sedimentary rock. (b) The gradual reduction in volume and increase in density of a soil mass in response to increased load or effective compressive stress; e.g. the squeezing of fluids from pore spaces. See also: *compaction [soil].* (c) A term sometimes used for diastrophic processes by which mobile belts are converted into rigid parts of the craton. The more specific terms *orogeny* and *orogenesis* are preferable.

consolidation grouting *grouting.*

consolute Said of liquids that are miscible in all proportions.

consolute point A point that represents the composition and maximum or minimum temperature or pressure of a miscibility gap. Syn: *critical solution point.*

consortium An intimately associated group of individuals of different kinds of organisms, generally belonging to different phyla, that live together.

conspecific Belonging to the same species.

constancy of composition *constancy of relative proportions.*

constancy of interfacial angles *law of constancy of interfacial angles.*

constancy of relative proportions The law that the ratios between the more abundant dissolved solids in seawater are virtually constant, regardless of the absolute concentration of the total dissolved solids. Syn: *law of constancy of relative proportions; constancy of composition.*

constant error A *systematic error* that is the same in both magnitude and sign throughout a given series of observations (the observational conditions remaining unchanged) and that tends to have the same effect upon all the observations of the series or part thereof under consideration; e.g. the *index error* of a precision instrument.

constant slope The straight part of a hillside surface, lying below the *free face* and determined by the angle of repose of the material eroded from it (Wood, 1942); it merges downslope into the *waning slope.* Cf: *gravity slope.* Syn: *debris slope.*

constitutional ash *inherent ash.*

constructional (a) Said of a landform that owes its origin or general character to the processes of upbuilding, such as accumulation by deposition (e.g. a plain or an alluvial terrace), volcanic eruption (e.g. a volcanic cone), and diastrophism or orogenic activity (e.g. a mountain, a fault block, or a tectonic valley); also said of a surface whose form was not acted upon by erosion. Ant: *destructional.* Cf: *initial landform.* (b) Said of a stream or drainage pattern that is formed by runoff from a constructional landform or surface (Davis, 1894, p. 74).

constructional void porosity Primary porosity in a carbonate framework (Murray, 1960, p. 61). Cf: *growth-framework porosity.*

constructive metamorphism An archaic syn. of *temperature-gradient metamorphism.*

constructive waterfall In a cave, a waterfall over a *rimstone* *dam* or a *canopy.*

constructive wave A wave that builds up a beach by moving material (esp. shingle) landward, as a gentle wave with a more powerful uprush than backwash. On sandy beaches, a wave with a steepness of less than 0.25 is considered a constructive wave. Ant: *destructive wave.*

consumer An organism that is unable to manufacture its food from nonliving matter but is dependent on the energy stored in other living things. Cf: *producer [ecol].*

consumptive use The difference between the total quantity of water withdrawn from a source for use and the quantity of water in liquid (and, rarely, solid) form returned to the source. It includes mainly water transpired by plants and evaporated from the soil, but may also include water diverted from one watershed to another, to the ocean, etc.

consumptive waste Water returned to the atmosphere without having benefited man (Langbein & Iseri, 1960).

contact n. (a) A plane or irregular surface between two types or ages of rock. (b) The surface between two fluids in a reservoir, i.e. oil and gas, oil and water, or gas and water. Syn: *interface.*—adj. Said of a mineral deposit that occurs at the contact of two unlike rock types.

contact area One of the areas of the proximal side of a spore or pollen grain that are in contact with the other members of the tetrad. Contact areas are seldom visible in mature pollen grains but are frequently apparent in spores. Trilete spores have three contact areas; monolete spores have two. Syn: *area contagionis.*

contact aureole *aureole.*

contact breccia A breccia around an igneous intrusion, caused by wall-rock fragmentation and consisting of both intrusive material and wall rock; *intrusion breccia.* Cf: *agmatite.*

contact erosion valley A valley eroded along a line of weakness at the contact between two different types of rock.

contact goniometer A *goniometer* that measures the solid angle between two crystal planes by contact with the surface. It is a 180° protractor, divided in degrees and accurate to half a degree, with a straightedge pivoted at the center. Cf: *two-circle goniometer; reflection goniometer.*

contactite A general term applied to rocks formed by contact metamorphism.

contact load The part of the *bed load* that is in substantially continuous contact with the stream bed.

contact log *wall-contact log.*

contact-metamorphic Said of a rock or mineral that has originated through the process of *contact metamorphism.* In older literature, used as a syn. of *contact-metasomatic* and *pyrometasomatic.*

contact metamorphism One of the principal local processes of thermal metamorphism, genetically related to the intrusion and extrusion of magmas and taking place in rocks or at near their contact with a body of igneous rock. Metamorphic changes are effected by the heat and materials emanating from the magma and by some deformation connected with the emplacement of the igneous mass (Holmes, 1920). Cf: *thermal metamorphism.* Adj: *contact-metamorphic.* See also: *exomorphism; endomorphism.*

contact-metasomatic Said of a rock or mineral that has originated through the process of *contact metasomatism.*

contact metasomatism A mass change in the composition of rocks in contact with an invading magma, from which "fluid" constituents are carried out to combine with some of the country-rock constituents to form a new suite of minerals. The term was originated by Barrell (1907). Cf: *regional metasomatism.*

contact mineral A mineral formed by contact metamorphism.

contact resistance The resistance observed between a grounded electrode and the ground, or between an electrode and a rock specimen.

contact spring A type of gravity spring whose water flows to the land surface from permeable strata over less permeable or impermeable strata that prevent or retard the downward percolation of the water (Meinzer, 1923, p. 51). Syn: *hillside spring; outcrop spring.* Cf: *barrier spring.*

contact twin A twinned crystal, the two individuals of which are symmetrically arranged about a twin plane. Syn: *juxtaposition twin.*

contact zone *aureole.*

contaminated *hybrid [ign petrol].*

contamination [ign] The process whereby the chemical composition of a magma is altered as a result of the *assimilation* of inclusions or country rock. Cf: *hybridization.*

contamination [water] The addition to water of any substance or property preventing the use or reducing the usability of the water for ordinary purposes such as drinking, preparing food, bathing, washing, recreation, and cooling. Sometimes arbitrarily defined differently from *pollution,* but generally considered synonymous.

contemporaneous Formed or existing at the same time. Said of lava flows interbedded in a single time-stratigraphic unit, and generally of any feature or facies that develops during the formation of the enclosing rocks. Cf: *penecontemporaneous.*

contemporaneous deformation Deformation that takes place in sediments during or immediately following their deposition. Includes many varieties of soft-sediment deformation, such as small-scale slumps, crumpling and brecciation, but in some areas (as the northern Apennines, Italy) features of large dimensions. Syn: *penecontemporaneous deformation.*

contemporaneous erosion Local erosion that goes on while elsewhere deposition is occurring generally or continuously.

contemporaneous fault *growth fault.*

contemporary carbon Carbon of living organisms. It has a carbon-14 activity of about 16 disintegrations per minute per gram carbon. Usage should be limited, as the isotopic composition of exchange reservoirs can change with time due to changes in proportions in the atmosphere. Also, carbon-14 activity may differ among organisms because they obtain carbon from different reservoirs, and among different parts of the same organism because of isotopic fractionation. Syn: *modern carbon.*

content graded bedding A type of graded bedding due to variation in the relative amounts ("contents") of different grains rather than the size: "although the mean size decreases upwards as in normal graded bedding, there is generally no gradation of the maximum size" (McBride, 1962, p. 50). The opposite of *coarse-tail grading,* in which the gradation is restricted to the maximum size.

continens An orange, yellow, or reddish area on the surface of Mars. Cf: *mare.* Pl: *continentes.*

continent (a) One of the Earth's major land masses, including both dry land and continental shelves. (b) An obsolete syn. of *terra.*

continental Formed on land rather than in the sea. Continental deposits may be of lake, swamp, wind, stream, or volcanic origin.

continental accretion (a) A theory, originally proposed by J.D. Dana in the 19th century, that continents have grown at the expense of the ocean basins by the gradual addition of new continental material around an original nucleus. Most of the new material was believed to have accumulated in concentric geosynclinal belts, each in turn consolidated by orogeny and succeeded by a new belt farther out. There is good evidence along some continental borders, as in western North America, that much new continental crust (now land area) was added to the continent during Phanerozoic time, but most geologists now reject the larger implications of the concept. It is in conflict with several other concepts, such as *oceanization* and *continental displacement.* (b) The outbuilding of continental crust at a subduction zone by adding offscraped ocean-floor material or newly deposited trench-slope sediment, usually deformed, to the upper plate. Syn: *accretion [struc geol]; tectonic accretion.*

continental alluvium Alluvium produced by the erosion of a highland area and deposited by a network of rivers to form an extensive plain.

continental apron *continental rise.*

continental basin A region, in the interior of a continent, that comprises one or more closed basins.

continental borderland That area of the *continental margin* between the shoreline and the continental slope which is topographically more complex than the *continental shelf.* It is characterized by ridges and basins, some of which are below the depth of the continental shelf. An example is the southern California continental borderland.

continental climate The climate of the interior of a continent, characterized by seasonal temperature extremes and by the occurrence of maximum and minimum temperature soon after summer and winter solstice, respectively. Cf: *marine climate.*

continental crust That type of the Earth's *crust* which underlies the continents and the continental shelves; it is equivalent to the *sial,* and ranges in thickness from about 35 km to as much as 60 km under mountain ranges. The density of the upper layer of the continental crust is $\sim$2.7 g/cm^3, and the velocities of compressional seismic waves through it are less than $\sim$7.0 km/sec. Cf: *oceanic crust.*

continental deposit A sedimentary deposit laid down on land (whether a true continent or only an island) or in bodies of water (whether fresh or saline) not directly connected with the ocean, as opposed to a marine deposit; a glacial, fluvial, lacustrine, or eolian deposit formed in a nonmarine environment. See also: *terrestrial deposit.*

continental displacement A general term, which can be used for many aspects of the theory originally propounded at length by Wegener (1912); also, less appropriately, called *continental drift.* Wegener postulated the displacement of large plates of continental (sialic) crust, moving freely across a substratum of oceanic (simatic) crust, but the mechanisms involved were so implausible to most geologists that the concept was generally discredited for many decades. New evidence has now been found and more acceptable mechanisms have been proposed, so that the original theory has gained a wider measure of acceptance: (1) the continents have remained relatively fixed but the Earth has expanded (see *expanding Earth*), leaving progressively wider gaps of oceanic areas between; (2) the continents have moved away from each other by *sea-floor spreading* along a median ridge or rift (see *world rift system*), producing new oceanic areas between the continents; or (3), the masses propelled away from the ridges consist of thick plates (see *plate tectonics*), composed of both continental and oceanic crust, which have moved in various directions independently of each other. Probably a true explanation of world tectonics will combine some or all of the newer explanations. Syn: *displacement theory; Wegener hypothesis; epeirophoresis theory; continental migration.*

continental divide A drainage *divide* that separates streams flowing toward opposite sides of a continent, often into different oceans; e.g. in North America, the divide separating the watersheds of the Pacific Ocean from those of the Atlantic Ocean, and extending from the Yukon Territory, along the British Columbia-Alberta boundary, through western Montana, Wyoming, Colorado, western New Mexico, and into Mexico.

continental drift *continental displacement.*

continental flexure A hinge-line structure along the contact of continent and sea floor, in which warping steepens the angle of slope of the continental shelf, causing relative uplift of the continent and eventual formation of coastal ranges.

continental glacier (a) A glacier of considerable thickness completely covering a large part of a continent or an area of at least 50,000 sq km, obscuring the relief of the underlying surface, such as the ice sheets covering Antarctica and Greenland. See also: *inland ice.* Syn: *continental ice sheet; continental ice; ice sheet.* (b) Any glacier in a continental, as opposed to a maritime, climatic environment. This usage is not recommended.

continental ice sheet *continental glacier.*

continental margin The ocean floor that is between the shoreline and the abyssal ocean floor, including various provinces: the *continental shelf, continental borderland, continental slope,* and the *continental rise.*

continental migration *continental displacement.*

continental nucleus *shield [tect].*

continental period The interval of time when a specific area was above sea level, forming part of a continent.

continental platform *continental shelf.*

continental rise That part of the *continental margin* that is between the continental slope and the abyssal plain, except in areas of an oceanic trench. It is a gentle incline with slopes of 1:40 to 1:2000, and generally smooth topography, although it may bear submarine canyons. Syn: *continental apron.*

continental river A river with no outlet to the sea, its water disappearing by percolation or evaporation.

continental sea *epicontinental sea.*

continental shelf That part of the *continental margin* that is between the shoreline and the continental slope (or, when there is no noticeable continental slope, a depth of 200 m). It is characterized by its very gentle slope of 0.1°. Cf: *insular shelf, marginal plateau; continental borderland.* Syn: *continental platform; shelf [marine geol].*

continental shield *shield [tect].*

continental slope That part of the *continental margin* that is between the continental shelf and the continental rise (or oceanic trench). It is characterized by its relatively steep slope of 3-6°. Cf: *insular slope.*

continental terrace The sediment and rock mass underlying the coastal plain, the continental shelf, and the continental slope (Curray, 1966, p. 207).

continental time A term used by Kobayashi (1944a, p. 477) for *fossil time* as indicated by nonmarine organisms. Cf: *marine time.*

continental transgression Any enlargement of the area of continental deposition (such as of piedmont deposition on a bajada) in which areas previously subjected to erosion, or areas of equilibrium between erosion and deposition, are covered with sediments. Cf: *transgression [stratig].*

continentes Plural of *continens.*

continuation A mathematical process of determining from a set of measurements of a potential field at one elevation the value of the field at another elevation. See also: *upward continuation; downward continuation.*

continuity equation An axiom stating that the rate of flow past one section of a conduit is equal to the rate of flow past another section of the same conduit plus or minus any additions or subtractions between the two sections (ASCE, 1962).

continuous-creation hypothesis *steady-state theory.*

continuous deformation Deformation by flow rather than by fracture. Cf: *discontinuous deformation.*

continuous permafrost A zone of permafrost that, for the most part, is uninterrupted by pockets or patches of unfrozen ground. Cf: *discontinuous permafrost; sporadic permafrost.*

continuous porosity A term proposed by Murray (1930, p. 452) for systems of interconnected pores, as opposed to *discontinuous porosity.* The term is little used and is not recommended (Choquette & Pray, 1970, p. 245). Cf: *effective porosity.*

continuous profiling A seismic method in which geophone groups are placed uniformly along the length of the line and so spread that a uniformly spaced set of points in the subsurface is sampled once. Cf: *correlation shooting.*

continuous reaction A metamorphic reaction that is continuously at equilibrium over a range of temperatures at constant pressure (or vice versa), owing to compositional variability among the minerals, for example in Fe/Mg ratio. A paragenetic diagram, e.g. the *AFM projection,* will show continuous variation in *tie line* orientations, but no topological change. Cf: *discontinuous reaction.*

continuous reaction series A *reaction series* in which reaction of early-formed crystals with later liquids takes place without abrupt phase changes; e.g. the plagioclase feldspars form a continuous reaction series. Cf: *discontinuous reaction series.*

continuous stream A stream that does not have interruptions in space; it may be perennial, intermittent, or ephemeral, but it does not have wet and dry reaches. Ant: *interrupted stream.*

continuous-velocity log *sonic log.*

contorted bedding *convolute bedding.*

contorted drift A glacial deposit that exhibits folding, thrusting, and other irregularities resulting from pressure during ice movement. The contortions are usually aligned in the direction of ice movement.

contortion (a) The intricate folding, bending, or twisting-together of laminated sediments on a considerable scale, the laminae being drawn out or compressed in such a manner as to suggest kneading more than simple folding; esp. *intraformational contortion.* Also, the state of being contorted. It occurs on a larger scale than *corrugation.* (b) A structure produced by contortion.

contour n. (a) An imaginary line, or a line on a map or chart, that connects points of equal value, e.g. elevation of the land surface above or below some reference value or datum plane, generally sea level. Contours are commonly used to depict topographic or structural surfaces; they can also readily show the laterally variable properties of sediments or any other phenomenon that can be quantified. Cf: *form line.* Partial syn: *isohypse; structure contour; topographic contour.* Syn: *isopleth; contour line.* (b) The outline or configuration of a surface feature seen two-dimensionally, e.g. the contour of a mountain pass or a coastline.—v. To provide a map with contour lines; to draw a contour line.

contour current An ocean current flowing along isopycnic lines approximately parallel to the bathymetric contours; e.g. a density current flowing parallel to the slope of a continental rise, such as the Western Boundary Undercurrent in the northwest Atlantic Ocean.

contour diagram An equal-area projection of structural data in which the poles have been contoured according to their density per unit area on the projection.

contour horizon *datum horizon.*

contour interval The difference in value between two adjacent contours; specif. the vertical distance between the elevations represented by two successive contour lines on a topographic map. It is generally a regular unit chosen according to the amount of vertical distance involved and the scale of the map, but it need not be constant over the entire map (a variable contour interval may be used for optimum portrayal of relief features). Syn: *vertical interval.*

contourite Any contour-current deposit, esp. a layer of coarse sediment in a marine-mud sequence, usually consisting of fine sand or coarse silt, deposited on the continental rise by contour-following bottom currents.

contour line (a) A line drawn on a map or chart representing a contour. Present usage makes contour and contour line synonymous: a line connecting points of equal value (generally elevation) above or below some reference value such as a datum plane. Contour lines are commonly used to depict topographic or structural shapes. The quantified properties of sediments or other phenomena can also be recorded by contour lines. Cf: *form line.* Syn: *isohypse; isopleth.* (b) A term used loosely in the general sense of an *isopleth;* e.g. a line (on a map) connecting points of equal magnitude of a mass property of a sediment (as of porosity, permeability, color, or thickness, or of size, shape, or roundness of sedimentary particles).

contour map A map that portrays surface configuration by means of contour lines; esp. a topographic map that shows surface relief by means of contour lines drawn at regular intervals above mean sea level, or a *structure-contour map* that shows the configuration of a specified rock surface underground and the inferred configuration of that surface where it has been removed by erosion.

contour sketching Freehand delineation, on a map, of the surface relief as seen in perspective view and controlled by map locations corresponding to salient points on the ground. See also: *field sketching.*

contour value A numerical value placed upon a contour line, such as a figure denoting elevation relative to mean sea level.

contraclinal A syn. of *anaclinal.* Term used by Powell (1873, p. 463).

contracting Earth A theory widely believed in the 19th and early 20th centuries that the orogenic and other structures of the Earth were produced by compression of the crust during its gradual contraction on the surface of a cooling but originally molten globe (a familiar textbook illustration of the time was a dried apple). The theory is now discredited, as the Earth is not cooling and contracting in the manner believed. Cf: *expanding Earth; tetrahedral hypothesis; wedge theory.*

contraction crack *frost crack.*

contraction fault A fault in sedimentary rocks along which there has been bed-parallel shortening (Norris, 1958, 1964), giving rise to tectonic thickening.

contraction fissure A fissure that is formed as a result of cooling or drying and consequent contraction of the rock. Cf: *cooling crack.*

contraction stripe One of the parallel lines on a playa or along its margin, consisting of vegetation growing along large-scale cracks caused by contraction of muds upon drying (Stone, 1967, p. 228).

contragradation Stream aggradation caused by an obstruction (Shaw, 1911). Syn: *dam gradation.*

contraposed shoreline A term introduced by Clapp in 1913 to denote a shoreline that has been cut through soft mantle until it reaches underlying hard rock. Thus the shoreline changes character as it evolves from a nearly straight, cliffed form to an irregular form following the shape of the hard-rock surface. As an example, Clapp cited the vicinity of Victoria, British Columbia, where the shoreline has been developed in "drift-covered crystalline rock".

contrast In photography, the difference in *density [optics]* between the shadows and the highlights (darkest parts) of a photographic negative, or the ratio of reflecting power between the shadows and the highlights (lightest parts) of a photographic print.

contrasted differentiation Differentiation of magma into basic and acidic magmas. Reactions between these contrasted magmas may produce intermediate rock types which resemble intermediate types usually considered as the product of progressive fractionation (Nockolds, 1934). Obsolete.

contratingent Said of a minor septum (of a rugose coral) that leans against the adjoining major septum on the side toward the counter septum.

contributory An obsolete syn. of *tributary.* Also spelled: *contributary.*

control [hydraul] (a) A section or reach of an open channel in which natural or artificial conditions make the water level above it a stable index of discharge. It may be either complete (i.e., water-surface elevation above the control is completely independent of downstream water-level fluctuations) or partial; it may also shift. (b) That waterway cross-section which is the bottleneck for a given flow and determines the energy head required to produce the flow. In an open channel, it is the point at which flow is at critical depth; in a closed conduit, it is the point at which hydrostatic pressure and cross-sectional area of flow are definitely fixed, except where the flow is limited at some other point by a hydrostatic pressure equal to the greatest vacuum that can be maintained unbroken at that point.—(ASCE, 1962).

control [surv] (a) The coordinated and correlated dimensional data that are used to establish the position, elevations, scale, and orientation of the detail of a map and that are responsible for the interpretations placed on the map. (b) The assemblage of accurately located points that determines the accuracy of a map and with which local secondary surveys may be tied in to insure their essential accuracy; a system of relatively precise field measurements of points, marks, or objects on the ground, whose horizontal and/or vertical positions have been (or will be) more or less accurately determined by surveying instruments. A map that includes many such points is said to have "good control".

control base A surface on which a map projection and ground control are plotted and on which templets have been assembled or aerotriangulation has been accomplished and the control points thus determined have been marked.

controlled mosaic A *mosaic [photo]* in which the photographs or images have been adjusted, oriented, and scaled to horizontal ground control in order to provide an accurate representation with respect to distances and distortions. It is usually assembled from photographs that have been corrected for tilt and for variations in flight altitude.

control point Any station in a horizontal and/or vertical control system that is identified on a photograph and used for correlating the data shown on that photograph (ASP, 1975, p. 2074). See also: *pass point.*

control station An accurately located point, mark, or object on the ground, whose horizontal and/or vertical position is used as a base for a dependent survey; any surveyed point used for horizontal and/or vertical control.

control survey A survey that provides horizontal and/or vertical position data for the support or control of subordinate surveys or for mapping; e.g. a survey that provides the geographic positions and/or plane coordinates of triangulation and traverse stations and the elevations of bench marks. Control surveys are classified according to their precision and accuracy: the highest prescribed order is designated first order, the next lower is second order, and so on.

conulariid Any marine fossil belonging to the order Conulariida (phylum Coelenterata?), characterized by a tetramerous cone-shaped to elongate, pyramidal or subcylindrical, chitinophosphatic periderm, which may be smooth or have longitudinal markings. Range, Middle Cambrian to Lower Triassic.

conule A cone-shaped projection from the body surface of certain sponges, generally over the end of a fiber.

conulite A hollow cylindrical or conical speleothem with the apex downward, formed as the lining of a drip-drilled hole in cave silt.

conus One of the pointed projections making up the sculpture of certain pollen and spores, being more or less rounded at the base and less than twice as high as the basal diameter. Pl: *coni.*

convection [eco geol] In hydrothermal systems, the flow of waters around and through heated zones adjacent to plutons in response to thermal gradients and controlled by porosity-permeability, salinity, fluid viscosity, and allied factors. The flow is generally down along the periphery, toward the system at depth, and upward along and through its central portions, possibly completing more than one loop.

convection [meteorol] (a) Transfer of heat by vertical movements within the Earth's atmosphere owing to density differences caused by heating from below. (b) The mixing and transport of properties other than heat.—Cf: *advection [meteorol].*

convection [oceanog] A general term for the density-driven movement and mixing of water masses in the ocean.

convection [tect] A supposed mass movement of subcrustal or mantle material, either laterally or in upward- or downward-directed *convection cells,* mainly as a result of heat variations.

Theories have been proposed utilizing convection currents to explain deep-sea trenches, island arcs, geosynclines, orogeny, and the like. Cf: *advection.*

convection cell In tectonics, a pattern of movement of mantle material in which the central area is uprising and the outer area is downflowing, due to heat variations. See also: *convection [tect].*

conventional age A potassium-argon age for which argon analysis is performed on one aliquot of sample and potassium analysis on a different aliquot of sample.

convergence [currents] The meeting of ocean currents or water masses having differing densities, temperatures, or salinities, resulting in the sinking of the denser, colder, or more saline water; also, the line or area in which convergence occurs. See also: *polar convergence.* Cf: *divergence [currents].*

convergence [evol] (a) The acquisition or possession of similar characteristics by animals or plants of different groups as a result of similarity in habits or environment. Cf: *adaptive radiation; parallelism; radiation [evol].* See also: *convergent evolution.* (b) In *cladism,* a syn. of *parallelism.*

convergence [meteorol] Contraction of an *air mass* toward a central region or zone, requiring vertical motion away from the region and horizontal motion toward it. Ant: *divergence [meteorol].*

convergence [petrology] The production, generally during metamorphism, of petrographically similar rocks from different original rocks.

convergence [stratig] The gradual decrease in the vertical distance or interval between two specified rock units or geologic horizons as a result of the thinning of intervening strata; e.g. the reduction in thickness of sedimentary beds (as measured in a given direction and at right angles to the bedding planes), caused by variable rates of deposition or by unconformable relationship.

convergence [surv] *convergence of meridians.*

convergence map *isochore map.*

convergence of meridians (a) The angular drawing-together of the geographic meridians in passing from the equator to the poles. (b) The difference between the two angles formed by the intersection of a great circle with two meridians. Also, the relative difference of direction of meridians at specific points on the meridians.—Syn: *convergence.*

convergent configuration A pattern in which a seismic unit thins laterally by nonsystematic reflection terminations, due to thinning of individual strata beyond the resolvable limit.

convergent evolution The development of similar-appearing forms in genetically unrelated lineages: *convergence.* Cf: *parallel evolution; homeomorphy.*

convergent plate boundary A boundary between two *plates* that are moving toward each other. It is essentially synonymous with *subduction zone,* but is used in different contexts (Dennis & Atwater, 1974, p. 1034).

converted wave A seismic wave that has been converted from a P-wave to an S-wave or vice versa by reflection or refraction at an interface. Such waves are sometimes designated PS or SP. Cf: *reflected wave.* Syn: *transformed wave.*

converter plant A plant that incorporates into its structure an insoluble element from the soil, and later, when the plant decays, returns that element to the soil in a soluble form.

convex bank The inner bank of a curved stream, with the center of the curve away from the channel; e.g. a *slip-off slope.* Ant: *concave bank.*

convex cross-bedding Cross-bedding with convex (upward-arching) foreset beds. Cf: *concave cross-bedding.*

convexo-concave Convex on one side and concave on the other; e.g. said of a resupinate brachiopod shell having a convex brachial valve and a concave pedicle valve. Cf: *concavo-convex.*

convexo-plane Convex on one side and flat on the other; e.g. said of a brachiopod shell having a convex brachial valve and a flat pedicle valve. Cf: *plano-convex.*

convex slope *waxing slope.*

convolute Coiled or wound together one part upon another; e.g. said of a coiled foraminiferal test in which the inner part of the last whorl extends to the center of the spiral and covers the inner whorls, or said of a coiled gastropod shell whose inner or earlier whorls are entirely concealed or embraced by the outer or later whorls. Cf: *involute; evolute; advolute.*

convolute bedding *convolute lamination.*

convoluted organ A loose, calcareous, spicular structure around the axial sinus in many camerate crinoids.

convolute lamination A descriptive term used by Kuenen (1953) for the wavy, extremely disorganized, and markedly and intricately crumpled, twisted, or folded laminae that are confined within a single, relatively thin, well-defined, undeformed layer, that die out both upward and downward, and that are overlain and underlain by parallel undisturbed layers. It is characteristic of some coarse-silt or fine-sand beds and involves only the internal laminae of the bed (and not the bed itself, which remains undeformed). The structure appears to result from deformation during deposition of sediments that become partially liquefied but still retain some cohesion. In some examples the anticlinal convolutions are deformed ripples. Convolutions may be truncated by erosion surfaces that are themselves convoluted, demonstrating the penecontemporaneous nature of the deformation. Axes of the convolutions generally have a preferred orientation normal to the paleocurrent. See also: *slip bedding; slump bedding; intraformational contortion.* Syn: *convolute bedding; convolution; contorted bedding; crinkled bedding; curly bedding; hassock structure; intrastratal flow structure.*

convolution [sed struc] (a) The process of producing convolute bedding; the state of being convoluted. (b) A structure produced by convolution, such as a small-scale but intricate fold. (c) *convolute bedding.*

convolution [seis] A change in wave shape as a result of passing through a linear *filter [seis];* the mathematical operation called "linear superposition". The Earth acts as a filter during the passage of seismic waves, and the movement that a seismic detector senses may be thought of as the result of the convolution of the Earth filter with the input seismic wave shape. Cf: *filtering.*

convolutional ball A "nearly closed, more or less elliptical, concentrically laminated structure" (Ten Haaf, 1956, p. 191) within a convolutely laminated bed. If the *convolute lamination* is not noticed, these features are easily mistaken for concretions or slump structures. Syn: *roll-up structure.*

convulsion *catastrophe.*

convulsionism *catastrophism.*

cookeite A mineral of the chlorite group: $LiAl_4(AlSi_3)O_{10}(OH)_8$.

coolgardite A mixture of coloradoite, sylvanite, and calaverite, found at Kalgoorlie in Western Australia.

cooling crack A joint that is formed as a result of cooling of an igneous rock. Cf: *contraction fissure.*

cooling unit An ash-flow unit or series of such units that were deposited rapidly enough to share a common history of cooling, compaction, and welding (Smith, 1960, p. 157).

coom (a) Var. of *cwm.* (b) An English term for a hollow in the side of a hill or mountain; a cove; a *combe.*

coomb (a) A var. of *combe,* a valley. (b) A var. of *cwm.*

coombe A var. of *combe,* a valley.

coombe rock An irregular mass of unstratified rock debris of any type accumulating as a result of solifluction; esp. the angular mass of unrolled and unweathered flints mixed with chalk rubble and other earthy matrix material, partly filling a dry valley (coombe), and spreading out onto the coastal plain, as in SW England. See also: *head [mass move]; rubble drift.* Syn: *elephant rock.*

coontail ore A term used in the Cave-in-Rock district of southern Illinois for a light- and dark-banded ore of sphalerite and fluorite.

cooperite A steel-gray tetragonal mineral: (Pt,Pd)S. It occurs in minute irregular grains in igneous rocks.

coordinate n. Any one of a set of numbers designating linear and/or angular quantities that specify the position of a point on a line, in space, or on a given plane or other surface in relation to a given reference system; e.g. latitude and longitude are coordinates of a point on the Earth's surface. The term is usually used in the plural, esp. to designate the particular kind of reference system (such as "spherical coordinates", "plane coordinates", and "polar coordinates").

coordinate system A reference system for defining points in space or on a particular surface by means of distances and/or angles with relation to designated axes, planes, or surfaces.

coordination number In crystallography, the number of nearest neighbor ions that surround a given ion in the crystal structure, e.g. four, six, or eight.

coorongite A soft, brown, rubbery variety of *elaterite,* originating from deposits of *Elaeophyton* algae in salt water bodies near the Coorong in Australia; it may represent a stage in the formation of boghead coal. See also: *n'hangellite.*

cop A term used in England for the steep-sided top or summit of a hill; also, a small hill with a rounded top.

copal An inclusive term for a wide variety of hard, brittle, semitransparent, yellowish to red fossil resins from various tropical trees (such as *Copaifera* and *Agathis*), being nearly insoluble in the usual solvents, and resembling amber in appearance; e.g. *Congo copal* and *kauri*. Copal also occurs as modern resinous exudations. Syn: *gum copal.*

copalite A clear, pale-yellow, dirty-gray, or dirty-brown fossil resin resembling copal in hardness, color, transparency, and difficult solubility in alcohol, containing succinic acid, and being much poorer in oxygen than most amber; e.g. the "Highgate resin" found in the blue Tertiary clay of Highgate Hill in London. Syn: *copaline; fossil copal.*

Copenhagen water *normal water.*

copepod Any crustacean belonging to the class Copepoda, characterized by the absence of both a carapace and compound eyes. The only known pre-Pleistocene fossil copepods have been found in Miocene lake deposits.

Copernican (a) Pertaining to the youngest lunar topographic features and lithologic map units constituting a system of rocks formed during the period of formation of ray craters (such as Copernicus). (b) Said of the stratigraphic period during which the Copernican System was developed.

Copernican system A concept of planetary motion according to which the Earth rotates on an axis once a day and revolves about the Sun once a year, and according to which all other planets have Sun-centered orbits. It is named after the Polish astronomer Copernicus (d.1543).

copiapite (a) A yellow mineral: $(Fe,Mg)Fe_4^{+3}(SO_4)_6(OH)_2 \cdot 20H_2O$. Syn: *ferrocopiapite; yellow copperas; ihleite; knoxvillite.* (b) A group of minerals containing hydrous iron sulfates, including copiapite, aluminocopiapite, calciocopiapite, cuprocopiapite, ferricopiapite, ferrocopiapite, and magnesiocopiapite.

coppaelite An olivine-free extrusive *melilitite* containing small clinopyroxene phenocrysts in a holocrystalline groundmass of melilite, clinopyroxene and phlogopite. The name, given by Sabatini in 1903, is for Coppaeli di Sotto, Umbria, Italy. Not recommended usage.

copper A reddish or salmon-pink isometric mineral, the native metallic element Cu. It is ductile and malleable, a good conductor of heat and electricity, usually dull and tarnished, and formerly an important ore. Copper is the only metal that occurs native abundantly in large masses; it frequently occurs in dendritic clusters or mossy aggregates, in sheets, or in plates filling narrow cracks or fissures. It has many uses, notably as an electric conductor and as the base metal in brass, bronze, and other alloys.

Copper Age In archaeology, a cultural level that is sometimes discernible between the *Bronze Age* and the *Iron Age* of the *three-age system.* It is characterized by experimentation with copper and its use for such technological purposes as weapons and tools. Correlation of relative cultural levels with actual age (and, therefore, with the time-stratigraphic units of geology) varies from region to region. Syn: *Chalcolithic; Eneolithic.*

copperas (a) *melanterite.* (b) A name sometimes applied to other sulfate minerals, such as copiapite ("yellow copperas") and goslarite ("white copperas").

copper glance *chalcocite.*

copper mica *chalcophyllite.*

copper nickel *nickeline.*

copper pyrites *chalcopyrite.*

copper uranite *torbernite.*

copper vitriol *chalcanthite.*

coppice *coral coppice.*

coppice mound A small mound of fine-grained desert material stabilized around shrubs or coppices.

coprocoenosis An accumulation of microvertebrate fossil remains that first passed through the digestive tracts of carnivores (mainly mammalian) and were deposited as fecal droppings in or near a stream or lake, where they were subsequently covered by sediment (Mellett, 1974). Etymol: Greek *copros,* "dung", + *koinos,* "common".

coprogenic Said of a deposit formed of animal excrement. Etymol: Greek *kopros,* "dung".

coprolite (a) The fossilized excrement of vertebrates such as fishes, reptiles, and mammals, larger than a *fecal pellet,* measuring up to 20 cm in length, characterized by an ovoid to elongate form, a surface marked by annular convolutions, and a brown or black color, and often composed largely of calcium phosphate; petrified excrement. The term is incorrectly used to refer to desiccated or fresh fecal remains. Term introduced by Buckland (1829a). (b) An English term applied commercially and popularly to any *phosphatic nodule* mined for fertilizer.

copropel A term introduced by Swain & Prokopovich (1954, p. 1184) to replace the vaguely applied term *gyttja,* and signifying a "dark-brown or gray coprogenic ooze, containing chitinous exoskeletons of benthonic arthropods in addition to reworked organic matter".

copula A short, hollow, funnel-like tube in the position of the mucron at the base of the chamber of chain-forming chitinozoans. Copulas connect to prosomes of preceding chambers to form chains.

coquimbite A white or slightly colored hexagonal mineral: $Fe_2(SO_4)_3 \cdot 9H_2O$. It sometimes contains appreciable aluminum, and it is dimorphous with paracoquimbite. Syn: *white copperas.*

coquina A detrital limestone composed wholly or chiefly of mechanically sorted fossil debris that experienced abrasion and transport before reaching the depositional site and that is weakly to moderately cemented but not completely indurated; esp. a porous light-colored limestone made up of loosely aggregated shells and shell fragments, such as the relatively recent deposits occurring in Florida and used for roadbeds and construction. Wentworth (1935, p. 244) recommended a particle size greater than 2 mm. Etymol: Spanish, "cockle, shellfish". Cf: *coquinoid limestone; microcoquina; coquinite.*

coquinite Compact, well-indurated, and firmly cemented equivalent of coquina.

coquinoid limestone A limestone consisting of coarse, unsorted, and often unbroken shelly materials that have accumulated in place without subsequent transportation or agitation, and generally having a fine-grained matrix. It is autochthonous, unlike the allochthonous *coquina;* under certain conditions it can develop into a biostrome. See also: *microcoquinoid limestone.* Syn: *coquinoid.*

coracite *uraninite.*

coral (a) A general name for any of a large group of bottom-dwelling, sessile, marine invertebrate organisms (polyps) that belong to the class Anthozoa (phylum Coelenterata), are common in warm intertropical modern seas and abundant in the fossil record in all periods later than the Cambrian, produce external skeletons of calcium carbonate, and exist as solitary individuals or grow in colonies. (b) A hard calcareous substance consisting of the continuous skeleton secreted by coral polyps for their support and habitation, and found in single specimens growing plantlike on the sea bottom or in extensive solidified accumulations (coral reefs). Also, any marine deposit like coral resulting from vital activities of various organisms (such as certain algae, or bryozoans and worms). (c) A piece of coral; e.g. "precious coral", a semitranslucent to opaque mass usually red to orange red, but sometimes white, cream, brown, blue, or black.

coral cap A thick deposit of coral-reef material overlying material of noncoral origin. Cf: *reef cap; coral crust.*

coral coppice A *coral thicket* within which broken coral branches have accumulated in significant quantities and thereby thickly carpet the sea floor around the bases of the colonies. Syn: *coppice.*

coral crust A thin layer of coral-reef material overlying material of noncoral origin. Cf: *coral cap.*

coralgal Said of a firm carbonate rock formed by an intergrowth of frame-building corals and algae (esp. coralline algae). The material so formed is an excellent sediment binder in a coral reef. Cf: *bryalgal.*

coralgal ridge A low ridge or elevated margin at the seaward (outer) edge of a reef flat, composed of the calcium-carbonate secretions of actively growing calcareous algae thickly encrusting and thus binding coral and shell rubble into a limestone mass. Cf: *algal ridge.*

coral head (a) A single massive, rounded coral colony. Syn: *niggerhead.* (b) A rounded, massive, often knobby or mushroom-shaped protuberance or growth of coral, usually forming on the submerged part of a coral reef, and frequently large enough to be dangerous to navigation; a small *reef patch* of coralline material. Syn: *coral knob; coral knoll.*

coral horse An elongate remnant of a former reef tract, characterized by a flat top (or by a flat side if it has been undercut and tilted); a *coral head* formed by solution and dismemberment of a former reef platform.

coral island (a) A coral reef that appears above sea level, situated far from any other kind of land. (b) An oceanic island formed from coral accumulations lying atop volcanic peaks. (c) A subaerial mound of sand, generally carbonate, resting on a flat coral reef.

coral knob *coral head.*

coral knoll *coral head.*

coralla Plural of *corallum.*

Corallian European stage: Upper Jurassic (above Callovian, below Kimmeridgian). Equivalent to Oxfordian.

coral limestone A limestone consisting of the calcareous skeletons of corals, often containing fragments of other organisms and often cemented by calcium carbonate. See also: *coral-reef limestone.*

coralline n. Any organism that resembles a coral in forming a massive calcareous skeleton or base, such as certain algae or stromatoporoids.—adj. Pertaining to, composed of, or having the structure of corals, as coralline limestone.

coralline algae *Calcareous algae* that form encrustations resembling coral.

corallite The calcareous exoskeleton formed by an individual coral polyp, consisting of walls, septa, and accessory structures such as tabulae and dissepiments. It is embedded in the general structure of the *corallum.*

corallith A subspheroidal, unattached coral colony that can be readily moved or rolled about (Glynn, 1974, p. 184). Cf: *rhodolith; oncolite.*

coralloid *cave coral.*

corallum The calcareous exoskeleton of a coral colony, or the *corallite* of a solitary coral; the entire skeleton of a coral. Pl: *coralla.*

coral pavement Within a *reef complex,* an extensive shallow bottom densely covered with coral colonies.

coral pinnacle *pinnacle [reef].*

coral rag A well-cemented, rubbly limestone composed largely of broken and rolled fragments of coral-reef deposits; e.g. the Coral Rag of the Jurassic, used locally in Great Britain as a building stone. Syn: *reef-rock breccia.*

coral reef (a) A coral-algal or coral-dominated *organic reef;* a mound or ridge of in-place coral colonies and accumulated skeletal fragments, carbonate sand, and limestone resulting from organic secretion of calcium carbonate that lithifies colonies and sands. A coral reef is built up around a potentially wave- and surf-resistant framework, especially of coral colonies but often including many algae; the framework may constitute less than half of the reef volume. Coral reefs occur today throughout the tropics, wherever the temperature is suitable (generally above about 18°C, a winter minimum). (b) A popular term for an organic reef of any type.

coral-reef coast A coast formed by deposits of coral and algae, partly exposed at low tide, and characterized by reefs built upward from a submarine floor or outward from the margin of a land area.

coral-reef limestone A *reef limestone* made up in large part of the skeletons of corals, but which may contain the remains of other organisms; a fossil coral reef. See also: *coral limestone.* Syn: *coral rock.*

coral rock *coral-reef limestone.*

coral thicket An aggregation of openly branching coral colonies, closely spaced and covering a portion of the sea bottom; esp. an aggregation in which the sea floor between the corals' bases is still largely exposed and not yet covered by coral debris. Such thickets range from shallow to deep, and from tropical to polar. Cf: *thicket reef; coral coppice.*

corange line A line on a chart joining points of equal tide range.

corbiculoid Said of heterodont dentition of a bivalve mollusk with three cardinal teeth in each valve, the middle tooth of the right valve occupying a median position below the beaks. Cf: *lucinoid.* Obsolete syn: *cyrenoid.*

cordate Said of structures, e.g. certain leaves, that are heart-shaped. Cf: *obcordate.*

corded pahoehoe The typical kind of *pahoehoe,* having a surface resembling coils of rope.

corderoite A cubic mineral: $Hg_3S_2Cl_2$.

cordierite A light-blue to dark-blue or violet-blue orthorhombic mineral: $(Mg,Fe)_2Al_4Si_5O_{18}$. It exhibits strong pleochroism, is easily altered by exposure, and is an accessory mineral in granites and a common constituent in metamorphic rocks formed under low pressure. Syn: *iolite; dichroite.*

cordierite-amphibolite facies A subdivision of the *amphibolite facies* in which pelitic rocks characteristically contain andalusite, cordierite, or sillimanite, but not kyanite or almandine (Winkler, 1967). It represents the low-pressure part of the amphibolite facies of Eskola. Syn: *hornblende-hornfels facies.*

cordillera (a) A comprehensive term for an extensive series or broad assemblage of more or less parallel ranges, systems, and chains of mountains (together with their associated valleys, basins, plains, plateaus, rivers, and lakes), the component parts having various trends but the mass itself having one general direction; esp. the main mountain axis of a continent, as the great mountain region of western North America from the eastern face of the Rocky Mountains to the Pacific Ocean, or the parallel chains of the Andes in South America; a mountain province. (b) An individual *mountain chain* with closely connected, distinct summits resembling the strands of a rope or the links of a chain; e.g. one of the parallel chains of the Rocky Mountains. (c) A term also used in South America for an individual *mountain range.*—Etymol: Spanish, "chain or range of mountains", from Latin *chorda,* "cord".

Cordilleran vein-type deposit One of a group of hydrothermal deposits of base and precious metals, chiefly Cu-W-Pb-Zn-Ag-Au, which are strongly vein-controlled and lack association with porphyry base-metal deposits. They may involve open-space filling, replacement or both; are typically mesothermal; and occur in Cordilleran continental margins. Examples: Mayflower, Utah; Magma, Arizona; Casapalca, Peru.

cordylite A colorless to wax-yellow mineral: $(Ce,La)_2Ba(CO_3)_3F_2$.

core [drill] n. A cylindrical section of rock, usually 5-10 cm in diameter and up to several meters in length, taken as a sample of the interval penetrated by a *core bit,* and brought to the surface for geologic examination and/or laboratory analysis. Cf: *core [oceanog].*—v. To obtain a core in drilling.

core [eng] A wall or structure of impervious material forming the central part of an embankment, dike, or dam, the outer parts of which are pervious. Syn: *core wall.*

core [fold] The inner or central part of a fold, especially of a folded structure that includes some sort of structural break. Cf: *envelope.*

core [interior Earth] The central zone or nucleus of the Earth's interior, below the *Gutenberg discontinuity* at a depth of 2900 km. It is divided into an *inner core* and an *outer core,* with a transition zone between, and is equivalent to the E, F, and G layers. Since only compressional waves propagate in the outer core, it is a fluid. The inner core, having a radius of approximately 1300 km, is solid, as shear waves have been observed to propagate through it. The magnetic field originates within the core.

core [oceanog] A relatively undisturbed, vertical section of ocean-bottom sediment collected by an oceanographic *corer.* See also: *box corer; piston corer; gravity corer.*

core-and-shell structure A term used by McKee (1954, p.65) for a sedimentary structure resembling a concretion, developed in massive silty mudstone, and characterized by a rounded lump (core), ovoid to elongate and 3 to 30 cm in diameter, surrounded by a series of concentric layers (shells) each with a thickness ranging from 3 to 12 mm and appearing to be due to shrinkage in the uniform structureless mud. It is characteristic of subaerial conditions following flooding in a region of considerable aridity.

core barrel (a) Two nested tubes above the bit of a core drill, the outer rotating with the bit, the inner receiving and preserving a continuous section or core of the material penetrated. (b) The tubular section of a *corer,* in which ocean-bottom sediments are collected either directly in the tube or in a plastic liner placed inside it.

core bit A hollow, cylindrical *drill bit* for cutting a core of rock in a drill hole; the cutting end of a *core drill.*

core box The wooden, metal, or cardboard box divided into narrow parallel sections, used to store the cores at the surface as they are extracted from a *core barrel* or *corer.*

cored bomb A type of *bomb [pyroclast]* that has a core of nonvolcanic rock or already solidified lava, around which the lava has molded itself. Syn: *perilith.*

core drill (a) A rotary *drilling rig* that cuts and brings to the surface a *core* from the drill hole. It is equipped with a *core bit* and a *core barrel.* (b) A lightweight, usually mobile drill that uses tubing instead of drill pipe and that can core down from the grass roots.

core hole Any hole drilled for the purpose of obtaining *cores;* loosely, a well, generally shallow, drilled for geological information only. Syn: *core test.*

core method A method for tracing a water mass or type from its origin across an area through which it is spreading, by variations in a characteristic parameter such as temperature, salinity or oxygen content.

corer An ocean-bottom sampler that is a metal or plastic cylinder or box, lowered by cable and driven into the ocean floor by impact of attached weights. See also: *piston corer; gravity corer; box corer.* Cf: *dredge; grab sampler.*

core record A record showing the depth, character, lithology, porosity, permeability, and fluid content of cores.

core recovery The amount of the drilled rock withdrawn as core in core drilling, generally expressed as a percentage of the total length of the interval cored.

core sample One or several pieces of whole or split parts of a *core,* selected for analysis; a sample obtained by coring.

core-stone An ellipsoidal or broadly rectangular joint block of granite formed by subsurface weathering in the same manner as a *tor,* but entirely separated from bedrock (Linton, 1955).

core test *core hole.*

core texture *atoll texture.*

core wall *core [eng].*

corindon *corundum.*

coring adj. Said of spicules, or foreign bodies such as sand grains, that occupy the axial region of a skeletal fiber of a sponge.

Coriolis acceleration The acceleration of a body in motion with respect to the Earth resulting from the rotation of the Earth, as seen by an observer on the Earth. A moving gravimeter has a Coriolis acceleration, which is involved in the *Eötvös effect.*

Coriolis effect The effect produced by the *Coriolis force;* the tendency of particles in motion on the Earth's surface to be deflected to the right in the Northern Hemisphere and to the left in the Southern Hemisphere. The magnitude of the effect is proportional to the velocity and latitude of the moving particles.

Coriolis force The apparent deflective component of the centrifugal force produced by the rotation of the Earth. It is named for Gustave Gaspard Coriolis (1792-1843), French mathematician who studied its effects. Cf: *pole-fleeing force.* See also: *Coriolis effect; Ferrel's law.* Also spelled: *coriolis force.* Syn: *geostrophic force.*

cork A protective layer of dead suberized tissue on the outside of older stems and roots, which is formed by a *cork cambium.* See also: *periderm [bot].* Cf: *phelloderm.*

cork cambium A *cambium* that functions in secondary growth by forming *cork* cells toward the outside of the stem and cork parenchyma or *phelloderm* toward the inside (Bold, 1967, p.521). See also: *periderm [bot].* Syn: *phellogen.*

corkite A rhombohedral mineral: $PbFe_3(PO_4)(SO_4)(OH)_6$. It is isomorphous with svanbergite, woodhouseite, and hinsdalite.

corkscrew flute cast A corkscrew-shaped flute cast, with a "twisted" beak at the upcurrent end.

cornelian *carnelian.*

corner (a) A point of intersection of two boundary lines of a tract of land; esp. a point on the Earth's surface, determined by surveying, that marks an extremity of a boundary of a subdivision of the public lands, usually at the intersection of two or more surveyed lines. Corners are described in terms of the points they represent; e.g. "township corner" located at the extremity of a township boundary. See also: *witness corner.* (b) A term that is often incorrectly used to denote the physical station, or *monument,* erected to mark the corner.

corner accessory A physical object adjacent to a corner, to which such a corner is referred for its future identification or restoration; e.g. a *bearing tree.* Corner accessories include mounds, pits, ledges, rocks and other natural features to which distances or directions, or both, from the corner or monument are recorded.

corner frequency For a seismic wave generated from an earthquake, the frequency at which the spectral field begins to decrease. It is related theoretically to the dimensions of the source.

corner reflector A special optical reflector used with electro-optical distance-measuring instruments to return a light beam to its source. A corner is cut from a highly polished cube of glass in which the three intersecting planes are precisely perpendicular. Light enters the cube through the cut surface and is reflected to the source by the polished cube faces. Syn: *corner prism; retrodirective prism.*

cornetite A peacock-blue mineral: $Cu_3(PO_4)(OH)_3$.

cornice An overhanging ledge or cantilevered mass of snow on the edge of a steep ridge or cliff face. Syn: *snow cornice.*

cornice glacier Nonpreferred syn. of *cliff glacier.*

cornieule *cargneule.*

Cornish diamond A rock crystal (clear quartz) from Cornwall, England.

Cornish stone A variety of *china stone* composed of feldspar, mica, and quartz and used as a bond in the manufacture of pottery. Syn: *Cornwall stone.*

corn snow *spring snow.*

cornstone (a) A calcareous concretion embedded in *marl* and grading into concretionary limestone. Its presence indicates fertile soil suitable for corn growing. (b) A calcareous conglomerate, consisting of fragments of marl and limestone embedded in a sandy or calcareous matrix.—The term is used in England and refers to two distinct rock types, both typically associated with the Old Red Sandstone and the New Red Sandstone. Allen (1960) suggests that the use of "concretionary cornstone" and "conglomeratic cornstone" would avoid confusion.

cornubianite A hornfels formed by contact metamorphism, and consisting of micas, quartz, and feldspar (Holmes, 1928, p.69). Cf: *leptynolite; proteolite.*

cornubite A mineral: $Cu_5(AsO_4)_2(OH)_4$. Cf: *cornwallite.*

cornulitid An invertebrate animal known only by a tapering, flexuous tube composed of calcium carbonate, with circular cross section, and with transverse rings developing in later growth stages. Cornulitids belong to the family Cornulitidae and are taxonomically unassigned in the Treatise on Invertebrate Paleontology (1962, pt. W, p.137), but were originally described as being closely related to certain annelids. Range, Middle Ordovician to Mississippian.

cornuspirine Having a tubelike planispirally coiled test; specif. pertaining to the foraminifer *Cyclogyra* ("*Cornuspira*").

cornutate Said of a diatom valve that has hornlike extensions.

cornwallite An emerald-green mineral: $Cu_5(AsO_4)_2(OH)_4 \cdot H_2O$. Cf: *cornubite.*

Cornwall stone *Cornish stone.*

corona [bot] In charophyte algae, the outer layer of the nucule, formed by cells cut off from the sheath cells of the female sex organ.

corona [paleont] The principal skeletal structure or main part of the calcareous test of an echinoid, including all ambulacra and interambulacra, but excluding the apical system of plates, the periproctal and peristomial systems, Aristotle's lantern, and appendages.

corona [palyn] A more or less equatorial extension of a spore, similar in disposition to a *zone [palyn]* but divided in fringelike fashion.

corona [petrology] A zone of minerals, usually with radial arrangement, around another mineral. It is a general term that has been applied to reaction rims, corrosion rims, and originally crystallized minerals. Cf: *kelyphytic rim.*

coronadite A black mineral: $Pb(Mn^{+2},Mn^{+4})_8O_{16}$. It is isostructural with hollandite and cryptomelane.

coronal (a) Pertaining to a corona. (b) Pertaining to certain openings of organisms, such as fringing an osculum of a sponge; e.g. "coronal pores" or tiny openings at the periphery of shields in acantharian radiolarians.

coronate Having or resembling a crown; esp. said of a cephalopod whorl section resembling a crown as viewed from the side, or said of a gastropod shell when the spire is surrounded by a row of spines or tubercles.

coronite A rock containing mineral grains surrounded by coronas.

coronula In charophyte oogonia, one or two tiers of small cells resting on the apical ends of the enveloping cells to form a more or less erect, elevated ring around the summit.

corrasion (a) A process of erosion whereby rocks and soil are mechanically removed or worn away by the abrasive action of solid materials moved along by wind, waves, running water, glaciers, or gravity; e.g. the wearing-away of the bed and banks of a stream by the cutting, scraping, scratching, and scouring effects of a sediment load carried by the stream, or the sawing and grinding action of sand, gravel, and boulders hurled by waves and currents against a shore. The term has also been used for the loosening of rock material by the impact of rushing water itself, and was used by Penck (1953, p. 112) for the "freeing of loosened rock fragments from their place of origin". The term *abrasion* is essentially synonymous. Syn: *mechanical erosion.* (b) A term sometimes used as a syn. of *attrition.* (c) A term formerly used as a syn. of *corrosion,* or as including the work of corrosion.—The term was first used by Powell (1875, p. 205) for channel cutting or the deepening of any valley floor, and extended by Gilbert (1877, p. 101) to the work of all running water (including lateral corrasion). Verb: *corrade.*

corrasion valley An elongated hollow or furrow excavated by the corrading action of a moving mass of material (Penck, 1953, p. 112).

correction In the analysis of physical measurements, a quantity that is applied to a measured quantity in order to negate the effects of a known interference, or to reduce the measurement to some arbitrary standard.

correction line A *standard parallel* in the U.S. Public Land Survey system.

correlate v. To show correspondence in character and stratigraphic position between such geologic phenomena as formations or fossil faunas of two or more separated areas.—adj. Belonging to the same stratigraphic position or level.

correlation [geol] (a) Demonstration of the equivalence of two or more geologic phenomena in different areas. "There are different kinds of correlation depending on the feature to be emphasized. Lithologic correlation demonstrates correspondence in lithologic character and lithostratigraphic position; a correlation of two fossil-bearing beds demonstrates correspondence in their fossil content and in their biostratigraphic position; and chronocorrelation demonstrates correspondence in age and in chronostratigraphic position" (ISG, 1976, p. 14). See also: *stratigraphic correlation; metamorphic correlation.* (b) The condition or fact of being correlated, such as the correspondence between two or more geologic phenomena, features, or events.

correlation [geomorph] The concept applied by Penck (1953, p. 419) which states that strata formed from the products of denudation are related to the period during which the denudation occurred.

correlation [seis] (a) The identification of a phase of a seismic record as representing the same phase on another record, thus relating reflections from the same stratigraphic sequence or refractions from the same marker. (b) The measurement of the degree of linear relationship between a pair of traces, or of the extent to which one can be considered as a linear function of the other (Sheriff, 1973, p. 39). Cf: *autocorrelation; crosscorrelation.*

correlation [stat] The intensity of association or interdependence between two or more mathematical variables; e.g. *linear correlation.*

correlation coefficient A number that expresses the degree of correlation between two mathematical variables. Many different correlation coefficients are in use. Symbol: *r.*

correlation log The generic term applied to any *well log* curve used in identifying equivalent subsurface geologic sections or individual lithologic units in wells. The *spontaneous-potential curve* and the *gamma-ray log* are often used because they can usually be compared directly to distinguish sandstone or limestone beds from shale intervals.

correlation shooting A seismic shooting method in which isolated profiles are shot and correlated to obtain relative structural positions of the horizons mapped. Cf: *continuous profiling.*

correlative estimate In hydraulics, a likely discharge value estimated by correlation for a particular span of time.

corrensite A clay-mineral mixture with a regular interstratification of chlorite and vermiculite. It shows swelling behavior in glycerol. See also: *swelling chlorite.*

correspondence analysis A method of factor analysis primarily designed for frequency data. Data are standardized by rows and columns in the same way before factoring.

corridor [karst] *solution corridor.*

corridor [speleo] In a cave, a traversable *passage [speleo]* that is generally narrow and straight.

corrie A term used in Scotland as a syn. of *cirque [glac geol],* esp. a *hanging cirque.* Etymol: Gaelic *coire,* "kettle". Also spelled: *coire; corry.*

corrie glacier *cirque glacier.*

corrom A Scottish term for a drainage divide developed on a delta by a stream whose drainage basin was altered by glacial action (Kendall & Bailey, 1908, p. 25).

corrosion [geomorph] (a) A process of erosion whereby rocks and soil are removed or worn away by natural chemical processes, esp. by the solvent action of running water, but also by other reactions such as hydrolysis, hydration, carbonation, and oxidation. Syn: *chemical erosion.* (b) A term formerly used interchangeably with *corrasion* for the erosion ("gnawing away") of land or rock, including both mechanical and chemical processes. The mechanical part is now properly restricted to "corrasion" and the chemical to "corrosion".—Verb: *corrode.*

corrosion [petrology] The partial resorption, dissolution, fusion, modification, or eating-away of the outer parts of early-formed crystals (such as quartz phenocrysts), or of xenoliths, by the solvent action of the residual magma in which they are contained. It sometimes results in the formation of corrosion borders. See also: *embayment [petrology].* Syn: *magmatic corrosion.*

corrosion border One of a series of borders of one or more secondary minerals around an original crystal, representing the modification of a phenocryst due to the corrosive action of its magma. Cf: *reaction rim.* Syn: *corrosion rim; corrosion zone; resorption border.*

corrosion rim A *corrosion border* as seen in section.

corrosion surface A pitted, irregular bedding surface found only in certain carbonate sediments, characterized by a black manganiferous stain, and presumed to result from cessation of lime deposition and from submarine solution or resorption of some of the previously deposited materials. Syn: *corrosion zone.*

corrosion zone (a) *corrosion surface.* (b) *corrosion border.*

corrugate Said of projections or ridges on pollen grains, with radial humps or bulges presenting a wrinkled appearance, as in *Riccia natan* (Tschudy & Scott, 1969, p. 25).

corrugated lamination A form of convolute bedding that differs from *convolute lamination* in that the lamination is intricately contorted and does not show a regular subparallel orientation of anticlines and synclines (cf. Davies, 1965).

corrugated ripple mark A *longitudinal ripple mark* with a sigmoidal profile, an equally rounded and usually symmetric crest and trough, and a ridge that rarely branches (Kelling, 1958, p.124). Cf: *mud-ridge ripple mark.*

corrugation (a) The process of deforming or crumpling sedimentary beds into small-scale folds, wrinkles, or furrows; esp. *intraformational corrugation.* Also, the state of being corrugated. It occurs on a smaller scale than *contortion.* (b) A structure produced by corrugation.

corry *corrie.*

corsite An orbicular *diorite* or *gabbro.* Cf: *esboite.* Syn: *napoleonite; miagite.* The name is from Corsica. Not recommended usage.

cortex (a) A layer of the ectosome of a sponge, consolidated by a distinctive skeleton, either organic or mineral, or both. Also, a layer of specialized sponge spicules or modified structure at the outer surface of the skeleton. (b) An outer coenenchymal layer of certain octacorals (esp. the Gorgonacea); the outer horny layer of the axis in the Holaxonia, as opposed to its medulla. (c) The dense, differentiated outer layer of an echinoid spine, usually bearing ornamentation. It is nonliving material on a mature spine. (d) The tissue, composed mostly of parenchyma cells, between the central vascular cylinder and the epidermis of a stem or root (Cronquist, 1961, p.873).

cortical Pertaining to, located in or on, or consisting of a cortex or outer part of an invertebrate; e.g. "cortical shell" (outermost of the concentric shells of a spumellinid radiolarian).

cortical fabric One of three major types of materials recognized in electron-microscope study of graptolithine periderm as a fundamental structural element in the periderm. Cortical fabric is layered, each layer being formed of relatively long fibers arranged in parallel; the fibers in one layer are commonly oriented at an angle to fibers in the subjacent and superjacent layers (Urbanek & Towe, 1974, p. 4-5). Cf: *fusellar fabric; sheet fabric.*

cortical tissue The outer part of the periderm in graptolithines; it is formed of layers of cortical fabric, sheet fabric, and, rarely, fusellar fabric (Urbanek & Towe, 1974, p. 5).

cortlandtite A *peridotite* that contains hornblende poikilitically enclosing olivine. Cf: *schriesheimite; scyelite.* Syn: *hudsonite.* The name, given by Williams in 1886, is from the Cortlandt complex near Peekskill, New York. Not recommended usage.

corundolite *emery rock.*

corundophilite A ferroan variety of clinochlore.

corundum A mineral: Al_2O_3. It occurs as shapeless grains and masses, or as variously colored rhombohedral crystals (such as prisms or tapering hexagonal pyramids), including the gem varieties such as ruby and sapphire. Corundum is extremely tough, has a hardness of 9 on the Mohs scale, and is used industrially as an abrasive. See also: *emery.* Syn: *adamantine spar; diamond spar; corindon.*

corvusite A blue-black, brown, or purplish mineral: $V_2O_4 \cdot 6V_2O_5 \cdot nH_2O$ (?). It is an ore of vanadium. Syn: *blue-black ore.*

corynite A variety of gersdorffite containing antimony.

cosalite A lead-gray or steel-gray mineral: $Pb_2Bi_2S_5$. It often contains copper.

cosedimentation Contemporaneous sedimentation, such as the precipitation of iron-bearing minerals at the same time as deposition of fine argillaceous sediment to produce a shale or mudstone with a high iron content (Pettijohn, 1957, p. 368).

coseismal line A line connecting points on the Earth's surface at which an earthquake wave has arrived simultaneously. Cf: *isoseismal line.* Syn: *homoseismal line.*

coset A term proposed by McKee & Weir (1953, p.384) for a sedimentary unit composed of two or more *sets [stratig],* either of strata or of cross-strata, "separated from other strata or cross-strata by original flat surfaces of erosion, nondeposition, or abrupt change in character". See also: *composite set.*

cosmecology A science that treats the Earth in its relation to cosmic phenomena (Webster, 1967, p. 514).

cosmic Of, from, or relating to the cosmos; esp. pertaining to phenomena or features that occur, exist, or originate beyond the Earth's atmosphere or in the universe in contrast to the Earth alone, such as "cosmic sediment" found in the oceans and containing particles of extraterrestrial origin.

cosmic dust (a) Very finely divided particles of solid matter moving about in interplanetary space or in any other part of the universe. (b) The smallest particles that invade the Earth's atmosphere from interplanetary space, reaching the Earth's surface (as on the sea floor or in polar ice) in an essentially unaltered state at an estimated accumulation rate of one thousand to one million tons a year. Their composition and structure are similar to those of meteorites, and they are believed to represent primordial condensation or sublimation products, cometary debris, or debris resulting from collisions between meteorites and asteroids or from collisions of meteorites, asteroids, and comets with the Moon and Earth. Cf: *meteoric dust; meteoritic dust.* Syn: *zodiacal dust.*

cosmic erosion The gradual degradation or catastrophic destruction of rocks in space on a planetary surface as a result of shock-wave interactions produced by hypervelocity impacts of particles against the exposed rock surfaces. The term includes the gradual wearing-away of rock surfaces due to spallation caused by interactions of the shock waves with the free surfaces of the rock, as well as the catastrophic rupture and breakup of entire rocks (Hö[illegible] et al., 1971, p. 5789).

cosmic radiation Very high-energy, subatomic particles from outer space, which bombard the Earth's atmosphere. Primary cosmic rays (atomic nuclei) are almost all absorbed in the upper atmosphere; secondary cosmic rays, which have less energy, may reach the Earth's surface, providing a part of natural background radiation.

cosmic spherule A small particle formed when a molten droplet ablates from a meteorite entering the Earth's atmosphere; first recognized in deep-sea sediments. See also: *magnetic spherule.*

cosmic water Juvenile water that is brought to the Earth from space in meteorites.

cosmochemistry The study of the origin, distribution, and abundance of elements in the universe.

cosmochlore *ureyite.*

cosmoclastic rock One of the original rocks of the Earth (Fairchild, 1904).

cosmogenic nuclide A *nuclide,* radiogenic or stable, that has been produced by the action of cosmic radiation.

cosmogony A scientific theory or a cultural mythology regarding the origin of the Universe, either as a whole or in more limited scope, such as the solar system of the Earth. Cf: *cosmology.*

cosmolite *meteorite.*

cosmological principle The principle that the universe presents essentially the same picture throughout all space and has done so throughout all time.

cosmology The study, both theoretical and observational, of the space-time structure of the Universe as a whole. Cf: *cosmogony.*

cosmopolitan Said of a kind of organism or a species that is widely distributed throughout the world in various geographic and ecologic provinces. Noun: *cosmopolite.*

cosmopolite A *cosmopolitan* organism.

cosmos The universe considered as an orderly and harmonious system. Ant: chaos.

cossyrite Sodium-rich variety of aenigmatite, occurring in minute black crystals in lava.

costa [bot] (a) In a simple leaf, the midrib. (b) In a pinnately compound leaf, the rachis. (c) In a diatom, a wall mark formed by two well defined ridges and containing fine pores.

costa [paleont] (a) One of the usually paired spines united medially and commonly laterally with neighboring spines to form a frontal shield overarching the frontal membrane in cribrimorph cheilostomes (bryozoans). Syn: *costula; costule.* (b) A round-topped elevation of moderate width and prominence disposed collabrally on the surface of a gastropod shell. (c) One of several moderately broad and prominent elevations of the surface of a bivalve-mollusk shell, directed radially or otherwise from the beak. Syn: *rib.* (d) A radial ridge on the external surface of a brachiopod shell, originating at the margin of the protegulal node. Also, any coarse rib of a brachiopod, without reference to origin (usually fewer than 15 costae in a width of 10 mm). Cf: *costella; capilla.* (e) A ridge on the external surface of a foraminiferal test. It may run along a suture or be transverse to it. (f) The prolongation of a septum on the outer side of a corallite wall. (g) A long narrow raised area or ridge of a conodont. (h) A transverse ridge, straight or slightly curved, on the surface of whorls of coiled cephalopod conchs.—Pl: *costae.*

costa [palyn] One of the riblike thickenings in the endexine of pollen, associated with colpi or pores. Costae are most often meridional and border colpi, but they may be transverse in association with transverse furrows.

costella A radial ridge on the external surface of a brachiopod shell, not extending to the margin of the protegulal node but arising by bifurcation of the existing costae or costellae or by intercalation between them. Also, a fine rib of a brachiopod, without reference to origin (usually 15-25 costellae in a width of 10 mm). Pl: *costellae.* Cf: *costa; capilla.*

costibite A mineral: CoSbS.

costula (a) A small ridge of an invertebrate; e.g. a marking that makes up the sculpture of a mollusk shell. (b) A *costa* of a cribrimorph cheilostome (bryozoan). Pl: *costulae.* Syn: *costule.*

coteau A French word used in parts of the U.S. for a variety of features: a range or sharp ridge of hills; a high plateau; a hilly upland, including the divide between two valleys; a morainal hill; an elevated pitted plain of rough surface (as in Missouri); a low, dry ridge within a swampy area (as in Louisiana); a side of a valley, esp. a prominent and dissected escarpment forming the edge of a plateau (as in Missouri). Etymol: Canadian French, "slope of a hill; hillock". Pron: co-*toe.*

cotectic Said of conditions of temperature, pressure, and composition under which two or more solid phases crystallize simultaneously and without resorption from a single liquid over a finite range of falling temperature; also, said of the geometric form (e.g. line or surface) representing the corresponding phase boundary on the liquidus of a phase diagram.

cotidal line A line on a chart connecting points at which high water occurs simultaneously. The lines show the time lapse, in lunar-hour intervals, between the Moon's passage over a reference meridian, usually Greenwich, and the succeeding high water at a specified location.

cotterite A variety of quartz having a peculiar metallic pearly luster.

cotton ball *ulexite.*

cotton chert *chalky chert.*

cotton rock (a) A term used in Missouri for a soft, fine-grained, siliceous, white to slightly gray or buff magnesian limestone having a chalky or porous appearance suggestive of cotton. (b) The white or light-colored decomposed exterior surrounding the dense black interior of a chert nodule.

cotton stone *mesolite.*

cotunnite A soft, white to yellowish, orthorhombic mineral: $PbCl_2$.

cotyledon The first leaf of the embryo of a seed plant, developed within the seed. It commonly serves as a storage organ. Syn: *seed leaf.*

Cotylosauria An order of anapsid reptiles of generalized structure and lizardlike or turtlelike habitus, mostly of Late Paleozoic age. Range, Lower Pennsylvanian to Upper Triassic.

cotype A term originally used for either a *syntype* or a *paratype* but recommended for rejection by the International Commission on Zoological Nomenclature because of its dual meaning (ICZN, 1961, p. 79, 148).

coulee [geomorph] (a) A term applied in western U.S. to a small stream, often intermittent. Also, the bed of such a stream when dry. (b) A term applied in NW U.S. to a dry or intermittent stream valley, gulch, or wash of considerable extent; esp. a long, steep-walled, trench-like gorge or valley representing an abandoned overflow channel that temporarily carried meltwater from an ice sheet, e.g. the Grand Coulee (formerly occupied by the Columbia River) in Washington State. (c) A small valley or a low-lying area.—Etymol: French coulée, "flow or rush of a torrent". Pron: *koo*-lee. Syn: *coulie.*

coulee [mass move] A tongue-like mass of debris moved by solifluction (Monkhouse, 1965, p. 81).
coulee [volc] A flow of viscous lava that has a blocky, steep-fronted form. Also spelled: *coulée.* Etymol: French, "outflowing".
coulee lake A lake produced by the damming of a water course by lava.
coulisse A term introduced by Scrivenor (1921, p. 354) for one of the prominent features, formed by the erosion of folded stratified rocks and of igneous intrusions, that are arranged en echelon on the Earth's surface "like the wings of the stage in a theatre". Etymol: French, "a side scene or wing on a theater stage". Pron: koo-*lease.*
couloir (a) A deep, narrow valley; esp. a gorge or gully on a mountain side in the Alps. (b) A French term for a passage in a cave, or a vertical cleft in a cliff.—Etymol: French, "passage". Pron: kool-*wahr.*
Coulomb's criterion A criterion of brittle shear failure based on the concept that shear failure will occur along a surface when the shear stress acting in that plane is large enough to overcome the cohesive strength of the material plus the frictional resistance to movement. Cohesive strength is equal to inherent shear strength when the stress normal to the shear surface is zero; frictional resistance to movement is equal to stress normal to the shear surface multiplied by the coefficient of internal friction of the material. See also: *Coulomb's equation.*
Coulomb's equation An equation describing the *Coulomb criterion,* or the failure of a material in shear fracture: critical shear stress for failure equals cohesion plus the coefficient of internal friction times normal stress across potential failure surface.
Coulomb's modulus *modulus of rigidity.*
coulometry A method of quantitative chemical analysis utilizing the number of coulombs necessary to release a substance during electrolysis to determine the amount of substance released. One faraday (96,500 coulombs) will release or deposit 1 gram-equivalent weight of an ion.
coulsonite A mineral of the spinel group: FeV_2O_4. Syn: *vanadomagnetite.*
counter In structural petrology, an instrument used to contour the density per unit area of a distribution of poles on an equal-area projection. It consists of a piece of plastic or similar material with a circular hole whose area usually equals 1.0% of that of the diagram Cf: *peripheral counter.*
counterclockwise inclination The inclination to the left of a heterococcolith suture as it proceeds to the periphery. Ant: *clockwise inclination.*
countercurrent A secondary current flowing in a direction opposite to that of the main or an adjacent current.
counter fossula A *fossula* developed in the position of the counter septum in a rugose coral.
counter-lateral septum One of two protosepta of a rugose corallite that adjoin the *counter septum* on either side. Symbol: KL.
counterlode *cross vein.*
counterpart *cast [sed].*
counterradiation The downward flux of infrared radiation reemitted by the atmosphere after prior absorption from the Earth's surface. Syn: *back radiation.*
counterscarp In landslides underlain by noncircular shear surfaces, a scarp that is parallel to the *crown scarp* and that occurs on the downslope side. Together, the scarps bound the structural trough or graben that is formed by such a landslide.
counter septum The protoseptum lying directly opposite the *cardinal septum* in the plane of bilateral symmetry of a rugose corallite. Symbol: K. Cf: *alar septum; counter-lateral septum.*
countervein *cross vein.*
country rock (a) The rock enclosing or traversed by a mineral deposit. Syn: *mother rock [eco geol].* Cf: *wall rock [eco geol].* (b) The rock intruded by and surrounding an igneous intrusion.
coupled wave A type of *surface wave* that is continuously generated by another wave which has the same phase velocity. Syn: *C wave.*
couplet Genetically related paired sedimentary laminae, generally occurring in repeating series, as *varves,* but applied to laminated nonglacial shales, evaporites, and other sediments as well.
course [stratig] (a) An old term used in Great Britain and applied to a stratum or outcrop, and to stratification. (b) A British term for a coal seam.
course [streams] The path followed by water, or the channel through which it flows; a *watercourse.*
course [surv] (a) A term used in surveying with several meanings: the bearing of a line; the length of a line; and the bearing (or azimuth) and length of a line, considered together. Also, the line connecting two successive stations in a traverse. (b) A term used in navigation for the azimuth or bearing of a line along which a ship or aircraft is to travel, without change of direction; the line drawn on a chart or map as the intended track.
course line A *slope line* linking *pits* via *passes* or *pales* (Warntz, 1975, p. 211).
cousinite A mineral: $MgU_2Mo_2O_{13}\cdot 6H_2O$(?).
Couvinian *Eifelian.*
covariance The arithmetic mean or the expected mean value of the product of the deviations of two variables from their respective mean values.
cove [coast] (a) A small narrow sheltered bay, inlet, creek, or recess in a coast, often inside a larger embayment; it usually affords anchorage to small craft. (b) A small, often circular, wave-cut indentation in a cliff; it usually has a restricted or narrow entrance. (c) A fairly broad, looped embayment in a lake shoreline (Veatch & Humphrys, 1966, p. 73). (d) A shallow tidal river, or the backwater near the mouth of a tidal river.
cove [geomorph] (a) Any precipitously walled and rounded or cirque-like opening, as at the head of a small steep valley; specif. a deep recess, hollow, or nook in a cliff or steep mountainside, or a small, straight valley extending into a mountain or down a mountainside. Examples: among the foothills of the Blue Ridge in Virginia, and in the English Lake District. Cf: *rincon.* (b) A term used in the southern Appalachian Mountains for a relatively open area sheltered by hills or mountains; e.g. Cades Cove in eastern Tennessee. (c) A gap or pass between hills. (d) A basin or hollow where the land surface has undergone differential weathering or subsidence, as from solution of underlying rock. (e) A Scottish term for a hollow or recess in rock, as a cave or cavern.
cove beach A *bayhead beach* formed in a cove.
covelline *covellite.*
covellite An indigo-blue hexagonal mineral: CuS. It is a common secondary mineral and represents an ore of copper. Syn: *covelline; indigo copper.*
cover (a) The sedimentary accumulation over the crystalline *basement.* See also: *cover mass.* (b) The vertical distance between any position in strata and the surface or any other position used as a reference.
covered flagellar field The part of a flagellate coccolithophore having a complete cover of coccoliths in the flagellar region. Cf: *naked flagellar field.*
covered karst Karst that forms below a soil cover that subdues its topographic features. See also: *naked karst; interstratal karst; paleokarst.*
covered plain A term introduced by Melton (1936, p. 594) for an alluvial flood plain through which a low-water channel does not actively meander; it usually has a natural levee and a thick alluvial cover derived from the suspended load. Cf: *meander plain; bar plain.*
cover head A thick accumulation of debris, consisting of talus cones and alluvial fans, resting on an elevated marine terrace; the material is deposited during and after the emergence of the terrace.
cover mass A *cover* of softer rock material with relatively simple structure, overlying the *undermass;* the material above the surface of an angular unconformity. See also: *compound structure.*
coverplate Any of small polygonal biserially arranged plates forming the walls and roof of the ambulacral groove of an echinoderm, such as a blastoid. See also: *anal coverplate.* Syn: *covering plate; ambulacral coverplate.* Also spelled: *cover plate.*
coverplate passageway A tubular canal in edrioasteroids that extends along lateral sutural faces of contiguous ambulacral and oral coverplates. These passageways connect the exterior to the interior of the theca (not in direct communication with the ambulacral tunnel) (Bell, 1976).
cover rock The thickness of rock between mine workings and the surface.
cover sand An eolian deposit of fine to very fine sand, usually containing more than 90% quartz, and believed to have been deposited by heavy snowstorms during the glacial epoch.
covite A *nepheline syenite* in which the mafic minerals are sodic clinopyroxene and amphibole. The term is more frequently used by Russian and European petrographers. The name, given by Washington in 1901, is for Magnet Cove, Arkansas. Not recom-

mended usage.

cow-dung bomb A type of volcanic bomb whose flattened shape is due to its impact while still viscous. Its surface is somewhat scoriaceous.

cowlesite A mineral of the zeolite group: $CaAl_2Si_3O_{10} \cdot 5\text{-}6H_2O$.

coxa The first segment of the leg of an arthropod, by which the leg articulates with the body; e.g. the proximal (basal) segment of the limb of a crustacean (except rarely where a precoxa is distinguishable), or the basal segment of all cephalothoracic appendages of an arachnid (the name is rarely used in the case of the chelicerae), or the proximal (basal) joint of the thoracic appendage of a merostome, directly attached to the body. Pl: *coxae.* Adj: *coxal.* See also: *coxopod; maxilla.*

coxopod The *coxa* of a crustacean limb. Syn: *coxopodite.*

crab The condition indicated on a vertical photograph by the lateral edges not being parallel to the air-base lines, caused by failure to orient the camera with respect to the track of the aircraft.

crab hole A low spot or depression in a *gilgai* Cf: *puff.*

crack [ice] Any *fracture* in sea ice, not sufficiently wide to be described as a *lead,* and usually narrow enough to jump across. See also: *tide crack.*

crack [struc geol] A partial or incomplete fracture.

cracking The breaking-up of more complex chemical compounds into simpler ones, usually by heating; i.e. the subjecting of compounds to pyrolysis.

crackle breccia (a) An incipient breccia having fragments parted by planes of rupture but showing little or no displacement (Norton, 1917, p.161). It is commonly a chemical deposit. (b) *shatter breccia.*

crag [geomorph] (a) A steep precipitous point or eminence of rock, esp. one projecting from the side of a mountain. Syn: *craig.* (b) An obsolete term for a sharp, rough, detached or projecting fragment of rock.

crag [sed] A shelly sandstone or a compacted fossiliferous sandy marl of marine origin and of Pliocene and Pleistocene age, found in eastern England (Norfolk, Suffolk, and Essex) and used as a fertilizer; e.g. the Coralline Crag.

crag and tail An elongate hill or ridge resulting from glaciation, having at the *stoss* end a steep, often precipitous, face or knob of ice-smoothed, resistant bedrock (the "crag") obstructing the movement of the glacier, and at the *lee* end a tapering, streamlined, gentle slope (the "tail") of intact weaker rock and/or drift protected by the crag. Cf: *knob and trail.*

craig A Scottish term for *crag;* e.g. Ailsa Craig. Etymol: Celtic, "rock".

craigmontite A light-colored nepheline *diorite* composed of nepheline, oligoclase, and muscovite with minor amounts of calcite, magnetite, corundum, and biotite. Craigmontite has more nepheline and less plagioclase and corundum than *congressite* or *raglanite.* It was named by Adams and Barlow in 1913 for Craigmont Hill, Ontario. Not recommended usage.

cranch In mining, a part of a vein that is left unworked (Nelson, 1965, p.110).

crandallite A white to light-gray mineral: $CaAl_3(PO_4)_2(OH)_5 \cdot H_2O$. It sometimes contains appreciable strontium, barium, iron, or rare earths. Syn: *pseudowavellite.*

craniacean Any inarticulate brachiopod belonging to the superfamily Craniacea, characterized generally by a strongly punctate, calcareous shell.

cranidium The central part of the cephalon of a trilobite, consisting of the glabella and its two fixed cheeks, and bounded by the facial sutures. Pl: *cranidia.* Adj: *cranidial.*

crassitude A more or less local exine thickening of a spore. Cf: *cingulum; zone [palyn].*

crater [geophys] A typically bowl-shaped or saucer-shaped pit or depression, generally of considerable size and with steep inner slopes, formed on a surface or in the ground by the explosive release of chemical or kinetic energy; e.g. an *impact crater* or an *explosion crater.*

crater [grd wat] *geyser crater.*

crater [lunar] An approximately circular or polygonal depression in the surface of the Moon, having a diameter that may range from a few centimeters to hundreds of kilometers and a depth that is small relative to its diameter. Large lunar craters often have lofty, rugged rim crests, terraced walls, and prominent central peaks. Lunar craters are believed to have formed in several ways, including meteoritic or cometary impact (e.g. Kepler and Tycho), secondary debris impact, volcanic activity, and subsidence. Syn: *lunar crater.*

crater [volc] A basinlike, rimmed structure that is usually at the summit of a volcanic cone. It may be formed by collapse, by an explosive eruption, or by the gradual accumulation of pyroclastic material into a surrounding rim. Cf: *caldera.*

crater chain A distinctive, chainlike, linear group of small craters first observed in most regions of the Moon's surface (such as east of the crater Copernicus). Lunar chains may be as much as 325 km long. Crater chains have also been found on Mars and Mercury. See also: *chain crater.*

crater depth (a) In an artificial crater, the maximum depth measured from the deepest point to the original ground surface (Flanders and Sauer, 1960, p. 5). (b) In a natural crater, in which the original ground level may be uncertain, the depth measured from the highest point on the rim crest to the deepest part of the observable depression.

crater fill Solidified lava at the bottom of a volcanic crater, with associated cinders and weathering debris.

cratering (a) The dynamic process or mechanism of formation of an individual crater. (b) The process of modification of a planetary or lunar surface by repeated crater formation.

crater lake A lake, usually of fresh water, that has formed in a volcanic crater or caldera by the accumulation of rain and ground water. See also: *caldera lake.*

crater radius (a) In an artificial crater, the average radius, measured at the level corresponding to the original ground surface (Flanders and Sauer, 1960, p. 6). (b) In a natural crater, in which the original ground level may be uncertain, the crater radius measured to the rim crest.

crater ring A low-relief rim of fragmental material surrounding a *maar.*

cratogene *craton.*

craton A part of the Earth's crust that has attained stability, and has been little deformed for a prolonged period. As originally defined, cratons included parts of both continents and ocean basins, but modern knowledge of the ocean basins indicates that existence of cratons there is unlikely, so the term is now restricted to continental areas. The extensive central cratons of the continents, including both *shields* and *platforms,* have been called *hedreocratons.* Parts of the more mature Phanerozoic fold belts have now achieved, or are approaching, a cratonic condition. Also spelled: *kraton.* See also: *thalassocraton.* Syn: *kratogen.*

cratonic basin *intracratonic basin.*

crawlway A cave passage that is traversable only by crawling.

C ray Right posterior ray in echinoderms situated clockwise of B ray when the echinoderm is viewed from the adoral side; equal to ambulacrum V in the Lovenian system.

craze plane A planar void in a soil material, having a highly complex conformation of the walls due to the interconnection of numerous short, flat and/or curved planes (Brewer, 1964, p.198).

cream ice *sludge [ice].*

crease [glac geol] A term used by Woodworth (1901, p. 665) for an *overflow channel* that formerly contained meltwater.

crease [mining] A limestone quarry in a mountainside.

creaseyite An orthorhombic mineral: $Pb_2Cu_2Fe_2Si_5O_{17} \cdot 6H_2O$.

crednerite A steel-gray to iron-black mineral: $CuMnO_2$.

creedite A white or colorless monoclinic mineral: $Ca_3Al_2(SO_4)(F,OH)_{10} \cdot 2H_2O$.

creek [coast] (a) A British term for a small inlet, narrow bay, or arm of the sea, longer than it is wide, and narrower and extending farther into the land than a cove. The term is used in the U.S. (as in Maryland and Virginia) in names given during the earliest period of English colonization for a narrow recess in the shore of the sea, a river, or a lake, and often offering port or anchorage facilities for vessels. (b) A small, narrow tidal inlet or estuary, esp. on a low-lying coast or on the lower reaches of a wide river; e.g. Napa Creek, Calif. Syn: *tidal creek.*

creek [streams] (a) A term generally applied over most of the U.S. (except New England) and in Canada and Australia to any natural stream of water, normally larger than a *brook* but smaller than a *river;* a branch or tributary of a main river; a lowland watercourse of medium size; a flowing rivulet. Also, a wide or short arm of a river, such as one filling a short ravine that joins the river. (b) A term used in the SW U.S. and in Australia for a long, shallow stream of intermittent flow; an *arroyo.*

creekology An ironical term for unscientific methods of choosing drilling sites or prospective oil or gas acreage and particularly applied to selection based on the general appearance of outcrops, topography, drainage, etc.

creep [mass move] The slow, more or less continuous downslope movement of mineral, rock, and soil particles under gravitational stresses. Many types of creep have been described, on the basis of material properties, stress level, stage and rate of deformation, fundamental mechanics of failure, geometric patterns, and cause of deformation. However, the term should not be limited by a presumption of mechanism, depth, velocity profile, thickness of creep zone, or lateral extent. Syn: *rock drift [mass move].*

creep [mining] Gradual strain failure of rock, as in mine pillars and roofs, owing to the pressure of superincumbent load. It generally involves lateral mass transfer in response to vertical stress. Cf: *flash [mining]; crown-in; inbreak.* Syn: *heave [mining].*

creep [struc geol] Continuously increasing, usually slow, deformation (strain) of solid rock resulting from a small, constant stress acting over a long period of time.

creeping The slow *shifting* of a divide from one position to another, as where a stream, because of its greater steepness and volume or the weaker nature of the rocks over which it flows, cuts down more rapidly than another stream on the opposite side of the divide (Cotton, 1958, p. 69-70). Cf: *leaping.*

creep limit The maximum stress that a material can withstand without observable creep.

creep recovery The gradual recovery of elastic strain when stress is released. Syn: *elastic aftereffect; elastic afterworking; afterworking; transient strain.*

creep strength The load per unit area leading to a specified steady creep strain rate at a given temperature.

creep wrinkle One of a series of small-scale corrugations of a bedding-plane surface, oriented at right angles to the direction of movement (slumping or creep) (McIver, 1961, p.227). See also: *crinkle mark; pseudo ripple mark.*

crenate Having the edge, margin, or crest cut into rounded scallops or shallow rounded notches; e.g. "crenate costae" of bivalve mollusks.

crenella (a) A narrow furrow between culmina of a crinoid columnal articulum. (b) A small radially disposed groove on the stem impression at the base of a blastoid theca and on the distal and proximal surfaces of columnals (TIP, 1967, pt.S, p.346).—Pl: *crenellae.*

crenula A ridge (culmen) combined with the adjacent furrow (crenella) of a crinoid columnal articulum. Pl: *crenulae.*

crenularium The entire area of a crinoid columnal articulum bearing crenulae. Pl: *crenularia.*

crenulate shoreline A minutely irregular shoreline, characterized by crenulate lines and sharp headlands, developed during a youthful stage of submergence by differential wave erosion acting upon less resistant rocks; e.g. the shoreline of SW Ireland.

crenulation Small-scale folding (wavelength up to a few millimeters) that is superimposed on larger-scale folding. Crenulations may occur along the cleavage planes of a deformed rock. Cf: *plication.*

crenulation cleavage *slip cleavage.*

creoline A purple epidotized *basalt.* Obsolete.

creolite (a) Red-and-white banded jasper from Shasta and San Bernardino counties, Calif. (b) A silicified rhyolite from Baja California.

crepis The initial fiber about which a desma of a sponge is secreted. Pl: *crepides.*

crescent beach A curving beach, concave toward the sea, formed along a hilly or mountainous coast at a bayhead or at the mouth of a stream entering a bay.

crescent cast *current crescent.*

crescentic crack *crescentic fracture.*

crescentic dune *barchan.*

crescentic fracture A *crescentic mark* in the form of a hyperbolic crack, of larger size (up to 10-12 cm long) than a *chattermark;* it is convex toward the direction from which the ice moved (its "horns" point in the direction of ice movement) and consists of a single fracture without removal of any rock. Cf: *lunate fracture; crescentic gouge.* Syn: *crescentic crack.*

crescentic gouge A *crescentic mark* in the form of a groove or channel with a somewhat rounded bottom, formed by glacial plucking on a bedrock surface; it is concave toward the direction from which the ice moved (its "horns" point away from the direction of ice movement) and it consists of two fractures from between which rock has been removed. Cf: *crescentic fracture.* Syn: *gouge [glac geol]; gouge mark; lunoid furrow.*

crescentic lake A lake occupying a crescent-shaped depression; e.g. an *oxbow lake.*

crescentic levee lake A lake confined between old natural levees inside a meander and resulting from the enlargement of the meander.

crescentic mark [glac geol] Any curved or lunate mark produced by a glacier moving over a bedrock surface; e.g. a *crescentic fracture* and a *crescentic gouge.* Syn: *lunate mark.*

crescentic mark [sed] A crescent-shaped scour mark; specif. *current crescent.*

crescent scour *current crescent.*

crescent-type cross-bedding *trough cross-bedding.*

crescumulate n. A general term for any plutonic rock formed by crystal accumulation and exhibiting *crescumulate texture.*

crescumulate texture The texture of certain igneous rocks in which large, elongated crystals are oriented roughly at right angles to cumulate layering in the rock. The term *harrisitic* was originally applied to this texture, which was thought to be limited to olivine crystals, as observed in the igneous rock harrisite; "crescumulate" has been suggested as a more general term without mineralogic restrictions (Wager, 1968, p.579).

crest [beach] (a) *berm crest.* (b) *beach crest.*

crest [geomorph] The highest point or line of a landform, from which the surface slopes downward in opposite directions; esp. the highest point of a mountain or hill, or the highest line or culminating ridge of a range of mountains or hills. See also: *summit.*

crest [struc geol] The highest point of a given stratum in any vertical section through a fold. Cf: *crest surface; crest line.*

crest line The line joining the crest points in a given stratum. Cf: *crest.*

crest plane Planar *crest surface.*

crest surface A surface that connects the *crest lines* of the beds of an anticline. Cf: *crest plane.*

crêt In the French Jura, an in-facing cliff or escarpment; the wall of a *combe.*

creta (a) *chalk.* (b) *fuller's earth.*—Etymol: Latin, "chalk".

Cretaceous The final period of the Mesozoic era (after the Jurassic and before the Tertiary period of the Cenozoic era), thought to have covered the span of time between 135 and 65 million years ago; also, the corresponding system of rocks. It is named after the Latin word for chalk ("creta") because of the English chalk beds of this age.

cretaceous A seldom-used syn. of *chalky.*

cretification The process or an instance of converting a rock into chalk, as by infiltration with calcium salts.

crevasse [geomorph] (a) A wide breach or crack in the bank of a river or canal; esp. one in a natural levee or an artificial bank of the lower Mississippi River. Etymol: American French. (b) A wide, deep break or fissure in the Earth after an earthquake. (c) A fissure in the surface of a glacier or icefall.

crevasse [glaciol] A deep, nearly vertical fissure, crack, or rift in a glacier or other mass of land ice, or in a snowfield, caused by stresses resulting from differential movement over an uneven surface. Crevasses may be concealed beneath *snowbridges,* and some are as much as 100 m in depth. Etymol: French. See also: *transverse crevasse; longitudinal crevasse; marginal crevasse; splaying crevasse.* Syn: *crevass; fissure [glaciol].*

crevasse filling A short, straight ridge of stratified sand and gravel believed to have been deposited in a crevasse of a wasting glacier and left standing after the ice melted; a variety of kame (Flint, 1928, p. 415). May also occur as long sinuous ridges and linear complexes of till or drift. Cf: *ice-channel filling; till crevasse filling.*

crevasse hoar A type of frost consisting of large leaf-, plate-, or cup-shaped ice crystals that form and grow below the surface of a snowfield or glacier in a crevasse or other large open space where water vapor can condense under calm, still conditions. Cf: *depth hoar.*

crevasse ridge A mass of fluvial material originally deposited in a crevasse and now forming a more or less straight ridge that stands above the general land surface and extends parallel to the direction of ice movement (Leighton, 1959, p. 340). May also occur in long sinuous ridges and linear complexes of till or drift in various directions as related to direction of ice movement.

crevice (a) A narrow opening or recess, as in a wave-eroded cliff. (b) A colloquial syn. of *crevasse.*

crevice karst A karst pattern of deep solution along closely spaced joints. See also: *limestone pavement.*

crib [eng] (a) A bin-type retaining wall consisting of interlocking members of steel, concrete, or wood, used to stabilize slopes and protect road cuts. (b) An engineering structure enclosing a water intake and filter offshore in a lake. (c) In mining, a structure, usually made of interlocking timbers, that forms a brace between the roof and floor. Also, short lengths of wood or other material placed in tunnels to transfer loads to steel load sets or other braces.

crib [glac geol] A Welsh term for *arête*.

cribellum A single or paired perforated plate in a small group of spiders, corresponding to the anterior and medial spinnerets of Liphistiina and serving as an outlet for special silk glands (TIP, 1955, pt.P, p.61). Pl: *cribella*.

cribrate (a) Like a sieve; e.g. said of a foraminifer perforated with round holes. (b) Said of an erect frondescent bryozoan colony with broad flattened anastomosing branches separated by large fenestrules.

cribrilith A discolith with numerous central perforations and a lamellar rim.

cribrimorph adj. Pertaining to the cheilostome bryozoan Cribrimorpha, characterized by a frontal shield of costae.—n. A cribrimorph cheilostome bryozoan.

crichtonite A mineral: $(Sr,La,Ce,Y)(Ti,Fe,Mn)_{21}O_{38}$. It was long supposed to be identical with ilmenite.

cricocalthrops A *calthrops* (sponge spicule) bearing a series of annular ridges on each ray.

cricolith A heterococcolith having units arranged in a simple ring; any elliptical ring coccolith.

crimp A marginal band on the aboral side of a plate in a platelike conodont element, representing the area covered by the last lamella accreted to the element.

crinanite An olivine-analcime *teschenite* in which the ophitic texture is well developed. It was named by Flett in 1909 for Loch Crinan, Scotland. Not recommended usage.

crinkled bedding (a) *convolute bedding.* (b) Bedding that displays minute wrinkles; in carbonate rocks, it is believed to be related to algal mats.

crinkle mark One of a series of subparallel corrugations of a bedding-plane surface, related to very small-scale and crumpled internal laminae, and produced by subaqueous solifluction (Williams & Prentice, 1957, p.289). See also: *creep wrinkle; pseudo ripple mark.*

crinoid Any pelmatozoan echinoderm belonging to the class Crinoidea, characterized by quinqueradiate symmetry, by a disk-shaped or globular body enclosed by calcareous plates from which appendages, commonly branched, extend radially, and by the presence of a stem, or column, more common in fossil than in living forms. Syn: *crinite; encrinite.* Range, Ordovician to present.

crinoidal limestone A limestone consisting almost entirely of the fossil skeletal parts of crinoids in which the plates, ossicles, or joints (representing single crystals of calcite) are often cemented with clear calcite in crystallographic continuity with the crinoid fragments. The hard parts are allochthonous and commonly show evidence of sorting. Examples are common in the Osagian (Lower Mississippian) of the Illinois basin. See also: *criquinite.* Syn: *encrinite; encrinal limestone.*

crinozoan Any attached echinoderm belonging to the subphylum Crinozoa, characterized by a partial meridional growth pattern tending to produce an aboral cup-shaped or globoid plated theca and a partial radially divergent growth pattern forming appendages. Among the major groups included in the subphylum are blastoids, crinoids, and cystoids.

cripple A swampy area in the Pine Barrens, N.J., supporting a growth of Atlantic white cedar. It is sometimes defined as always having flowing water. Cf: *spong.*

criquina A coquina composed of crinoid fragments.

criquinite Compact, well-indurated, and firmly cemented equivalent of *criquina.* See also: *crinoidal limestone.*

crisscross-bedding A kind of *cross-bedding* characteristic of eolian deposits, in which the layers dip in opposite directions.

crista One of the elevations making up the sculpture of certain pollen and spores, characterized by long, curved bases, sometimes irregularly fused, and by variously bumpy apices. Pl: *cristae*.

cristate Crested, or having a crest; esp. said of sculpture of pollen and spores consisting of cristae.

cristobalite A mineral: SiO_2. It is a high-temperature polymorph of quartz and tridymite, and occurs as white octahedrons in the cavities and fine-grained groundmasses of acidic volcanic rocks. Cristobalite is stable only above 1470°C; it has a tetragonal structure (alpha-cristobalite) at low temperatures and an isometric structure (beta-cristobalite) at higher temperatures. Cf: *tridymite.*

critical angle The least *angle of incidence* at which there is total reflection when an optic, acoustic, or electromagnetic wave passes from one medium to another medium that is less refractive.

critical damping *Damping* to the point at which the displaced mass just returns to its original position without oscillation.

critical density [chem] The density of a substance at its critical temperature and under its critical pressure.

critical density [exp struc geol] That density of a saturated, granular material below which, under rapid deformation, it will lose strength and above which it will gain strength.

critical depth In a channel of water, the depth at which flow is at a *critical velocity,* e.g. the depth at which a flow is at its minimum energy with respect to the channel bottom. Cf: *critical flow.*

critical distance In refraction seismic work, that distance at which the direct wave in an upper medium is matched in arrival time by that of the refracted wave from the medium below having greater velocity. Cf: *crossover distance.*

critical end point A point at which two of three or more phases participating in a univariant equilibrium become identical, thus terminating the line representing the equilibrium.

critical flow Fluid flow at a *critical velocity,* e.g. flow at the point at which it changes from laminar to turbulent. Cf: *critical depth.*

critical gradient *critical slope.*

critical height The maximum height at which a vertical or sloped bank of soil will stand unsupported under a given set of conditions (ASCE, 1958, term 98).

critical hydraulic gradient In a cohesionless soil, that *hydraulic gradient* at which intergranular pressure is reduced to zero by the upward flow of water (ASCE, 1962).

critical length Maximum horizontal distance over which sheet erosion does not occur, measured in the direction of overland flow from the drainage divide to a point at which the eroding stress becomes equal to the resistance of the soil to erosion (Horton, 1945, p.320); it determines the width of the *belt of no erosion.* Symbol: x_c. Cf: *length of overland flow.*

critically undersaturated Said of a rock having feldspathoids and olivine, but no hypersthene, in its norm. Cf: *undersaturated.*

critical mineral A mineral that is stable only under the conditions of one metamorphic facies or zone. Cf: *typomorphic mineral; index mineral.*

critical moisture In a soil, that degree of moisture below which, under constant load increase, deformation will increase, and above which it will decrease.

critical point A point representing a set of conditions (pressure, temperature, composition) at which two phases become physically indistinguishable; in a system of one component, the temperature and pressure at which a liquid and its vapor become identical in all properties.

critical pressure The pressure required to condense a gas at the critical temperature, above which, regardless of pressure, the gas cannot be liquefied.

critical resolved shear stress Shear stress on a crystallographic slip plane, acting in the slip direction, which is required for the initiation of slip.

critical slope The slope or grade of a channel that is exactly equal to the loss of head per foot resulting from flow at a depth that will give uniform flow at critical depth; the slope of a conduit which will produce critical flow (ASCE, 1962). Syn: *critical gradient.*

critical slope angle The local maximum slope inclination which the soil and rock materials underlying the slope can support without failure under existing climate, vegetation, and land use. Syn: *angle of ultimate stability.*

critical solution point *consolute point.*

critical temperature The temperature of a system at its critical point; for a one-component system, that temperature above which a substance can exist only in the gaseous state, no matter what pressure is exerted.

critical tractive force The minimum *tractive force* required to set sediment particles of a stream bed moving.

critical velocity (a) That velocity of fluid flow at which the flow changes from laminar to turbulent. (b) That velocity of fluid flow at which the flow changes from laminar to turbulent, and at which friction becomes proportional to a power of the velocity higher than the first power. Syn: *Reynolds critical velocity.* (c) That velocity of fluid flow at which the fluid's minimum energy value is attained. Syn: *Belanger's critical velocity.* (d) In an open channel, that velocity of fluid flow at which silt is neither picked up nor deposited. Syn: *Kennedy's critical velocity.* (e) In an open channel, that velocity of fluid flow at which the velocity head equals one half the mean depth, and at which the energy head is at a minimum. Syn: *Unwin's critical velocity.*—See also: *critical flow; critical depth.*

CRM *chemical remanent magnetization.*

crocidolite A lavender-blue, indigo-blue, or leek-green asbestiform variety of riebeckite, occurring in silky fibers and in massive and earthy forms. Also spelled: *krokidolite.* Syn: *blue asbestos; Cape blue.*

Crocodilia An order of the reptilian subclass Archosauria, characterized by persistently aquatic habit and conservative body form. Range, Upper Triassic to Recent.

crocoite A bright-red, yellowish-red, or orange monoclinic mineral: $PbCrO_4$. Syn: *red lead ore; crocoisite.*

crocydite Migmatite with a flakelike or flufflike light-colored part (Dietrich & Mehnert, 1961). Var: *krokydite.* Little used.

Croixian North American series: Upper Cambrian (above Albertan, below Canadian of Ordovician). Also spelled: *Croixan.*

cromaltite A *pyroxenite* in which sodic clinopyroxene is the predominant mineral, with melanite as a characteristic phase, and smaller amounts of biotite, perovskite, and oxide minerals. The name, given by Shand in 1906, is for the Cromalt Hills, Scotland. Not recommended usage.

Cromerian North European climatostratigraphic and floral stage: Middle Pleistocene (above Menapian, below Elsterian). Equivalent in time to Günz/Mindel interglacial. Also: British climatostratigraphic stage: Middle Pleistocene (above Beestonian, below Anglian). Correlation to northern Europe uncertain.

cromfordite *phosgenite.*

Cromwell current *equatorial undercurrent.*

cronstedtite A black to brownish-black mineral: $Fe_4{}^{+2}Fe_2{}^{+3}(Fe_2{}^{+3}Si_2)O_{10}(OH)_8$. It is structurally related to kaolinite; it is not a chlorite.

crooked hole A borehole that has deviated beyond the allowable limit from the vertical or from its intended course.

crookesite A lead-gray mineral: $(Cu,Tl,Ag)_2Se$.

crop n. Deprecated syn. of *outcrop.*—v. To appear at the surface of the ground; to *outcrop*.

crop coal (a) That part of a coal seam that is near the surface. (b) A coal deposit that crops out, as in an *exposed coalfield.* (c) Inferior, weathered coal occurring near the surface.—Cf: *deep coal.*

crop out v. *outcrop.*

cropping n. Deprecated syn. of *outcrop.*

cross-assimilation Simultaneous *assimilation* of country rock into magma and of magma into country rock, so that the same phases develop in both.

crossassociation A method for pairwise comparison of two strings or sequences of nonnumeric data (Sackin & Merriam, 1969, p. 7). Cf: *crosscorrelation.*

cross bar (a) A short, bifurcating "ridge" of a transverse ripple mark. (b) A low ridge trending across a *blind valley.*

crossbar The complete skeletal rim on which the fixed edge of the mandible of the avicularium pivots in some cheilostome bryozoans.

cross-bed A single bed, inclined at an angle to the main planes of stratification. The term is restricted by McKee & Weir (1953, p.382) to a bed that is more than 1 cm in thickness. See also: *cross-stratum; cross-lamina; cross-stratification.* Also spelled: *crossbed.*

cross-bedding (a) *Cross-stratification* in which the cross-beds are more than 1 cm in thickness (McKee & Weir, 1953, p. 382). (b) A cross-bedded structure; a cross-bed.—See also: *current bedding; inclined bedding; discordant bedding; crisscross-bedding.* Syn: *false bedding; diagonal bedding; oblique bedding; foreset bedding.* Also spelled: *crossbedding.*

cross channel A transverse drainageway cutting across an interstream area or connecting two successive low areas.

crosscorrelation (a) A measure of the similarity of two wave forms, the degree of linear relationship between them, or the extent to which one is a linear function of the other (Sheriff, 1973, p. 42). Cf: *autocorrelation.* (b) A method for comparison of two strings or sequences of numerical data (Sackin & Merriam, 1969, p. 7). Cf: *correlation [seis]; crossassociation.*

cross-coupling effect A term applied to the use of gravity meters aboard ship; the effect produced by simultaneous interactive accelerations in two directions.

cross course *cross vein.*

crosscut (a) A small passageway that may be driven at an angle to the main entry of a mine, to connect it with a parallel entry or an air course. (b) A level driven across the course of a vein or across the general direction of the workings; thus a mine opening that intersects a vein or ore-bearing structure at an angle.

crossed-lamellar Said of the type of mollusk shell structure composed of primary and secondary lamellae, the latter inclined in alternate directions in successive primary lamellae.

crossed nicols In a polarizing microscope, two Nicol prisms or Polaroid plates that are oriented so that the transmission planes of polarized light are at right angles; light that is transmitted from one will be intersected by the other, unless there is an intervening substance. Syn: *crossed polars.*

crossed polars *crossed nicols.*

crossed twinning *cross-hatched twinning.*

cross fault (a) A fault whose strike crosses at a high angle the strike of the constituent strata or the general trend of the regional structure. (b) A minor fault that intersects a major fault.

crossfeed Interference resulting from the unintentional pickup by one channel of information or noise on another channel. Syn: *crosstalk.*

cross fiber Veins of *fibrous* minerals, esp. asbestos, in which the fibers are at right angles to the walls of the vein. Cf: *slip fiber.*

cross fold A fold that intersects a pre-existing fold of different orientation; the resulting structure is a *complex fold.* Syn: *superimposed fold; transverse fold; subsequent fold.*

cross fracture A small-scale joint structure developed between *fringe joints.*

cross-grading The process of slope dissection by rills and gullies in which an original slope (alongside, and parallel with, a stream) is replaced by a new slope deflected toward the stream (Horton, 1945, p. 335). See also: *micropiracy.*

cross-hatched twinning Repeated twinning after two laws, e.g. microcline twin law, with intersecting composition planes. Syn: *gridiron twinning; crossed twinning.*

cross-hatching (a) The process of drawing or shading (on a map) with a pattern consisting of two sets of parallel lines crossing each other at a predetermined angle (obliquely or at right angles) so that the space between the lines in one set is identical to the space between the lines in the other set. (b) The effect produced by cross-hatching, such as a pattern indicating abrupt gradients.—See also: *hatching.*

crossing A term applied to the shallow part of a river channel, separating deeper pools of water (or *bendways*) at the bends of meanders.

crossing canal A proximal (prothecal) part of a graptoloid theca, growing across the sicula to develop on that side of the sicula opposite from the side from which it originated.

crossite A blue monoclinic mineral of the amphibole group, intermediate in composition between glaucophane and riebeckite; a variety of glaucophane rich in iron.

cross joint A joint that is perpendicular to the major lineation of the rock. Syn: *transverse joint; Q-joint; ac-joint.*

cross-joint fan In igneous rock, a fanlike pattern of cross joints that follow the arching of the flow lineation.

cross-lamina A *cross-bed*. The term is restricted by McKee & Weir (1953, p.382) to a *cross-stratum* that is less than 1 cm in thickness. Syn: *cross-lamination.*

cross-lamination (a) *Cross-stratification* characterized by cross-beds that are less than 1 cm in thickness (McKee & Weir, 1953, p. 382). (b) A cross-laminated structure; a *cross-lamina*.—See also: *flow-and-plunge structuree.* Syn: *oblique lamination; diagonal lamination.*

cross lode *cross vein.*

crossopodium A *trace fossil* of the "genus" *Crossopodia,* consisting of a sinuous or meandering marking about 1 cm wide with a median furrow, and believed to be a trail left by a creeping marine animal. Pl: *crossopodia.*

Crossopterygii An order of lobefinned bony fish characterized by a hinged braincase, amphistylic jaw suspension, and sharp, conical teeth with labyrinthodontine infolding of enamel. It includes the *coelacanths.*

crossover distance That distance at which a refracted wave becomes the *first arrival.* Cf: *critical distance.*

cross profile A plot of elevation drawn against distance along a line at right angles to the long direction of a valley, stream, or ridge. Cf: *longitudinal profile.* Syn: *transverse profile.*

cross ripple mark A ripple mark resulting from the interference of at least two sets of ripples, one set forming after the completion of, or simultaneously with, the other; e.g. *current cross ripple mark* and *oscillation cross ripple mark.* Term introduced by Bucher (1919, p.190) as "cross-ripple". See also: *compound ripple mark; tadpole nest.* Syn: *interference ripple mark; dimpled current mark; complex ripple mark.*

cross sea A confused, choppy state of the ocean, occurring where waves from two or more different directions meet. Wave direction may appear to be the same as one of the original directions, or it may be a new direction.

cross section (a) A diagram or drawing that shows features transected by a given plane; specif. a vertical section drawn at right angles to the longer axis of a geologic feature, such as the trend of an orebody, the mean direction of flow of a stream, or the axis of a fossil. Cf: *longitudinal section.* Syn: *transverse section.* (b) An actual exposure or cut that shows transected geologic features.—Adj: *cross-sectional.* Also spelled: *cross-section.*

cross spread (a) A seismic *spread* that makes a large angle to the line of traverse; it is used to determine the component of dip perpendicular to that line. (b) A seismic spread that is laid out in the pattern of a cross.

cross-stone (a) *chiastolite.* (b) *staurolite.* (c) *harmotome.*

cross-stratification Arrangement of strata inclined at an angle to the main stratification. In modern usage, following McKee and Weir (1953, p. 382), this is considered to be the general term, and to have two subdivisions: *cross-bedding,* in which the cross-strata are thicker than 1 cm, and *cross-lamination,* in which they are thinner than 1 cm. A single group of related cross-strata is a *set* and a group of similar, related sets is a *coset.* There are many types of cross-stratification; Allen (1963) recognized fifteen, to which he attached Greek-letter designations. Probably most cross-stratification is produced by the migration of bed forms, particularly ripples (which form small-scale cross-lamination) and dunes or megaripples (which form medium- to large-scale cross-lamination or cross-bedding). Syn: *false stratification; diagonal stratification.*

cross-stratum A *cross-bed.* McKee & Weir (1953, p.382) consider it a general term that includes cross-bed and *cross-lamina.*

crosstalk *crossfeed.*

cross valley *transverse valley.*

cross vein (a) A vein or lode that intersects a larger or more important one. Syn: *cross course; cross lode; countervein; counterlode; caunter lode.* (b) A vein that crosses the bedding planes in a sedimentary sequence.

crotovina *krotovina.*

croute calcaire Hardened *caliche [soil],* often found in thick masses or beds overlain by several centimeters of earth (SSSA, 1975, p. 5). Etymol: French *croûte calcaire,* "calcareous crust".

crowfoot Obsolete syn. of *stylolite.*

crown [gem] The portion of any faceted gemstone above the *girdle.* Cf: *pavilion.* Syn: *bezel; top [gem].*

crown [geomorph] The top or highest part of a mountain or an igneous intrusion; the summit.

crown [mass move] The practically undisturbed material still in place and adjacent to the highest parts of the scarp along which a landslide moved.

crown [paleont] The whole of a crinoid exclusive of the pelma; the part of a crinoid skeleton above the column, including the dorsal cup, tegmen, and arms.

crown-in In mining, a falling of the mine roof or a *heave* of the mine floor due to the pressure of overlying strata. Cf: *flash; inbreak; creep [mining].*

crown scarp The outward-facing scarp, bordering the upper portion of a landslide. It is almost always concave in a downslope direction. The scarp surface may be slickensided, indicating the downward displacement of the sublandslide shear surface; or it may have a rough, sugary surface, indicating a tension or cleavage fracture formed by downslope displacement of the slide mass (Varnes, 1958).

crowstone An English term for a very hard siliceous sandstone representing the floor of a coal seam; a *ganister.*

crucite (a) *chiastolite.* (b) Pseudomorph of hematite or limonite after arsenopyrite.

crude Said of a mineral material in its natural, unrefined state, e.g. crude oil, crude ore.

crude oil *Petroleum* in its natural state as it emerges from a well, or after passing through a gas-oil separator but before refining or distillation.

crumble coal An incoherent brown coal that lacks cementing material. Syn: *formkohle.*

crumble peat Peat that is friable and earthy.

crumb structure A type of soil structure in which the peds are spheroids or polyhedrons that have little or no accommodation to surrounding peds, are porous, and range in size from less than 1.0 mm to 5 mm. Cf: *granular structure.*

crumina A saclike semienclosed space developed in the ventral part of the carapace of female ostracodes belonging to some Paleozoic species. Pl: *cruminae.*

crump Ground movement, perhaps violent, caused by failure under stress of the ground surrounding an underground working (usually in coal), and so named because of the sound produced.

crumpled *plicated.*

crumpled ball A highly irregular, crumpled-up mass of laminated sandstone, measuring 5-25 cm across, and flattened parallel to the bedding (Kuenen, 1948, p.371). Cf: *slump ball.*

crumpled mud-crack cast A *mud-crack cast* that displays tortuous and contorted crumpling produced by the adjustment of the sand filling to the compaction of the enclosing mud matrix (Bradley, 1930).

crura Plural of *crus.*

crural base A part of the crus of a brachiopod, united to a hinge plate, and separating the inner and outer hinge plates when present.

cruralium A spoon-shaped structure of the brachial valve of a pentameracean brachiopod, formed by the dorsal union of outer plates (or homologues) and bearing adductor muscles.

crural plate A plate extending from the inner edge of an outer hinge plate or crural base to the floor of the brachial valve of a brachiopod (TIP, 1965, pt.H, p.143). See also: *septalial plate.*

crural process The pointed part of the crus of a brachiopod directed obliquely inward and ventrally.

crus Either of a pair of short, curved, calcareous basal processes that extend from the cardinalia or septum of a brachiopod to give support to the posterior end of the lophophore. The distal end may also be prolonged into primary lamella of spiralium or descending branch of loop (TIP, 1965, pt.H, p.142). Pl: *crura.*

crush belt A belt of crushed rock characterized by intense cataclasis and mylonitization.

crush border A microscopic, granular metamorphic structure sometimes characterizing adjacent feldspar particles in granite due to their having been crushed together during or subsequent to crystallization (Dale, 1933).

crush breccia (a) A breccia formed in place, or nearly in place, by mechanical fragmentation of rocks during crustal movements; a *tectonic breccia* associated with planes of movement and formed as a result of folding or faulting. (b) A term used by Norton (1917, p.186-188) for a tectonic breccia in which the brecciation was accomplished without either faulting or folding "except so far as the rupture planes of the breccia may be considered as minute faults"; e.g. a breccia produced by lateral pressure without any further mass deformation than that exhibited by gentle warpings.—Syn: *cataclastic breccia.*

crush conglomerate A rock formed essentially in place by deformation (folding or faulting) of brittle, closely jointed rocks, containing lozenge-shaped fragments produced by granulation of rotated joint blocks and rounded by attrition, and closely simulating a normal (sedimentary) conglomerate; a rock similar to a *crush breccia* but having fragments that are more rounded. It is characterized by similarity in composition (generally one rock type) of fragments and matrix. Term was introduced by Lamplugh (1895, p.563). Syn: *tectonic conglomerate; cataclastic conglomerate.*

crushing strength Unconfined compressive strength of a material.

crushing test *unconfined compression test.*

crush zone An area of fault breccia or fault gouge.

crust [ice] *ice rind.*

crust [interior Earth] The outermost layer or shell of the Earth, defined according to various criteria, including seismic velocity, density and composition; that part of the Earth above the Mohorovičić discontinuity, made up of the *sial,* or the sial and the *sima.* It represents less than 0.1% of the Earth's total volume. See also: *continental crust; oceanic crust.* Cf: *tectonosphere.*

crust [sed] A laminated, commonly crinkled deposit of algal dust, filamentous or bladed algae, or clots (from slightly arched forms to bulbous cabbage-like heads) of algae, formed on rocks, fossils, or other particulate matter by accretion, aggregation, or flocculation.

crust [snow] *snow crust.*
crustacean Any arthropod belonging to the superclass Crustacea, characterized chiefly by the presence of two pairs of antennae on the head. Most forms occur in marine environments. Crustaceans are second only to insects in numbers of individuals. Range, Cambrian to present.
crustaceous (a) Having, suggesting, or of the nature of a crust or shell. (b) Belonging to the Crustacea; crustacean.
crustal abundance *clarke.*
crusted strand *shelfstone.*
crustified Said of a vein in which the mineral filling is deposited in layers on the wall rock. Syn: *healed.*
crust reef A coral reef formed on a submerged bank.
cryergic A term recommended by Baulig (1956, paragraph 77) as a syn. of the broad meaning of *periglacial* as when the latter word is commonly but "rather incorrectly" applied to processes and deposits "in some regions not actually peripheral to glaciated regions". The term may also be used "to denote phenomena due to cold conditions" (Stamp, 1966, p.141).
cryic temperature regime A soil temperature regime in which the mean annual temperature (measured at 50cm depth) is more than 0°C but less than 8°C, with cold summer temperatures and a summer-winter variation of more than 5°C (USDA, 1975). Cf: *frigid temperature regime.*
cryochore A climatic term for the part of the Earth's surface covered with perpetual snow and ice.
cryoconite [glaciol] A dark, powdery dust, once thought to be of cosmic origin, transported by the wind and deposited on a snow or ice surface (e.g. on the Greenland Ice Cap). It is found mainly in *cryoconite holes,* but may form long stripes or an almost continuous cover. Syn: *kryokonite.*
cryoconite [mineral] A mixture of garnet, sillimanite, zircon, pyroxene, quartz, and other minerals.
cryoconite hole A cylindrical *dust well* containing particles of *cryoconite* that absorb solar radiation and cause increased ablation of glacier ice around and below them.
cryogenic lake *cave-in lake.*
Cryogenic period An informal designation for a period in geologic history "when large bodies of ice formed at or near the poles and the climate was generally suitable for the growth of continental glaciers" (ADTIC, 1955, p. 22).
cryogenics The branch of physics pertaining to the production and effects of very low temperatures; the science of extreme cold. Adj: *cryogenic.* Obsolete syn: *cryogeny.*
cryokarst *thermokarst.*
cryolaccolith *ice laccolith.*
cryolite A white or colorless monoclinic mineral: Na_3AlF_6. It may contain iron and has been found chiefly in a pegmatite at Ivigtut, Greenland, in cleavable masses of waxy luster. Natural and synthetic cryolite is used in the manufacture of aluminum. Syn: *Greenland spar; ice stone.*
cryolithionite A colorless isometric mineral: $Na_3Li_3Al_2F_{12}$.
cryolithology The study of the development, nature, and structure of underground ice, esp. ice in permafrost regions; a branch of *geocryology.*
cryology (a) In the U.S., cryology is the science of refrigeration, and although the term frequently has been proposed for the study of glaciers, it has not attained wide usage outside the field of low-temperature physics. Cf: *cryergy.* (b) *glaciology.*
cryomorphology The part of geomorphology "pertaining to the various processes and products of cold climates" (Black, 1966, p.332). See also: *periglacial geomorphology.*
cryonival (a) Pertaining to the combined action of frost and snow. (b) *periglacial* (Hamelin, 1961, p.200).
cryopedology The study of the processes of intensive frost action and the occurrence of frozen ground, esp. permafrost, including the civil-engineering methods used to overcome or minimize the difficulties involved; term introduced by Bryan (1946).
cryophilic Said of an organism that prefers low temperatures, esp. below 10°C. Syn: *psychrophilic.*
cryophyllite A variety of zinnwaldite with some deficiency in the (Li,Fe,Al) group and containing some ferric iron.
cryoplanation The reduction and modification of a land surface by processes associated with intensive frost action, such as solifluction, supplemented by the erosive and transport actions of running water, moving ice, and other agents (Bryan, 1946, p.640). Cf: *altiplanation.*
cryosphere The part of the Earth's surface that is perennially frozen; the zone of the Earth where ice and frozen ground are formed.
cryostatic Descriptive of frost-induced hydrostatic phenomena (Washburn, 1956, p. 842), e.g. the movement of water-saturated material confined between downward-advancing seasonal frost and an impermeable surface such as the permafrost table or bedrock. Such progressive downward freezing may generate large hydrostatic pressures.
cryotectonic Said of complicated and deranged features and deposits found at glacier borders, and consisting of material that has been overturned, inverted, folded, and transported by the shoving action of glaciers. Syn: *glaciotectonic.*
cryoturbate *congeliturbate.*
cryoturbation A syn. of *congeliturbation.* Also spelled: *kryoturbation.*
cryptacanthiiform Said of a brachiopod loop composed of descending branches fused distally to form an echmidium, which bears a hood on the ventral anterior end (TIP, 1965, pt.H, p.143). With continued growth, the echmidium becomes deeply cleft anteriorly but still connected with descending branches.
cryptalgal Said of rocks or rock structures formed "through the sediment-binding and/or carbonate-precipitating activities of nonskeletal algae" (Aitken, 1967, p. 1163). The influence of these organisms is more commonly inferred than observed, hence the etymol: Greek *kryptos,* "hidden, secret", + *algal.*
cryptalgalaminate Said of carbonate rocks "displaying a distinctive form of discontinuous, more or less planar lamination believed to have resulted from the activities upon and within the sediments of successive mats or films of blue-green and green algae" (Aitken, 1967, p. 1164). See also: *cryptalgal; stromatolite.*
crypthydrous Said of vegetal matter deposited on a wet substratum. Cf: *phenhydrous.*
cryptic Said of reef organisms, esp. invertebrates, that live inconspicuously under corals, shells, and rocks. Cf: *coelobitic; cryptofaunal; sciaphilic.*
cryptic layering That type of *layering* in an igneous intrusion in which there is a regular vertical change in chemical composition of the minerals; so named because it is less obvious than *rhythmic layering.* Typical element ratios defining cryptic layering are Fe/Mg in mafic minerals and Ca/Na in feldspars.
cryptic zoning Zoning in minerals that is not visible to the unaided eye. It may involve variation in major elements or in trace-element populations. The term is used with respect to igneous and hydrothermal systems.
cryptobatholithic Said of a mineral deposit occurring in the roof rocks of an unexposed batholith (Emmons, 1933). The term is little used. Cf: *acrobatholithic; embatholithic; endobatholithic; epibatholithic; hypobatholithic.*
cryptoclastic rock (a) A clastic rock whose extremely fine constituents can be seen only under the microscope. Ant: *macroclastic rock.* Cf: *microclastic rock.* (b) A carbonate sedimentary rock having an aphanic clastic texture and discrete particles whose diameters are less than 0.001 mm, and displaying little or no crystallinity under high-power magnification (Bissell & Chilingar, 1967, p.154); e.g. extremely finely comminuted carbonate dust. Cf: *cryptograined.*
cryptocrystalline (a) Said of the texture of a rock consisting of crystals that are too small to be recognized and separately distinguished even under the ordinary microscope (although crystallinity may be shown by use of the electron microscope); indistinctly crystalline, as evidenced by a confused aggregate effect under polarized light. Also, said of a rock with such a texture. Cf: *microcrystalline; dubiocrystalline.* Syn: *microaphanitic ; microcryptocrystalline; microfelsitic; felsophyric.* (b) Said of the texture of a crystalline rock in which the crystals are too small to be recognized megascopically. This usage is not recommended "since it cannot be known that an aphanitic rock is cryptocrystalline until the microscope has shown that it is actually microscopically crystalline" (Johannsen, 1939, p.206). (c)Descriptive of a crystalline texture of a carbonate sedimentary rock having discrete crystals whose diameters are less than 0.001 mm (Bissell & Chilingar, 1967, p.103) or less than 0.01 mm (Pettijohn, 1957, p.93). Some petrographers use an upper limit of 0.004 mm.
cryptocyst A frontal shield in cheilostome bryozoans, formed by calcification of an inner wall grown into the zooidal cavity subparallel to and beneath the frontal wall.
cryptodeltoid One of two plates on either side of the anal opening of a blastoid, generally overlapped aborally by the hypodeltoid and adjacent radial limbs, abutting against the superdeltoid plate

adorally, bordering the lancet plate laterally, and infolded into hydrospire folds on the inner side. By adoral extension and fusion above the anal opening, cryptodeltoids may form a horseshoe-shaped subdeltoid developed in some genera (TIP, 1967, pt. S, p. 346).

cryptodepression A lake basin whose bottom lies below sea level (although the water surface may be above sea level).

cryptodont Said of the dentition of certain bivalve mollusks of early origin lacking hinge teeth.

cryptoexplosion structure A nongenetic, descriptive term suggested by Dietz (1959, p. 496-497) to designate a roughly circular structure formed by the sudden, explosive release of energy and exhibiting intense, often localized rock deformation with no obvious relation to volcanic or tectonic activity. Such structures typically show some or all of the following: wide variation in diameter (less than 1.5 km to more than 50 km); a central dome-shaped uplift with intense structural deformation, often surrounded by a concentric ring depression; complex faulting and subordinate folding; widespread brecciation and shearing; and occurrence of shatter cones. Many cryptoexplosion structures are believed to be the result of hypervelocity impact of crater-forming meteorites of asteroidal dimensions; others may have been produced by obscure volcanic activity. The term, as presently used, largely replaces the earlier term *cryptovolcanic structure.* See also: *astrobleme.* Syn: *cryptoexplosive structure.*

cryptofaunal Pertaining to small mobile animals, such as crustaceans and polychaetes, that live on, in, and immediately around large sessile animals like sponges and corals. Cf: *coelobitic; cryptic.*

cryptogam A plant that lacks stamens and pistils and reproduces by spores rather than seeds. Examples include thallophytes, bryophytes, and pteridophytes. Cf: *phanerogam.*

cryptogene adj. Said of a rock whose origin cannot be determined.

cryptograined Said of the texture of a carbonate sedimentary rock having discrete clastic (or precipitated or flocculated) particles whose diameters are less than 0.001 mm (Bissell & Chilingar, 1967, p. 103; and DeFord, 1946) or less than 0.01 mm (Thomas, 1962). Some petrographers use an upper limit of 0.004 mm. Cf: *cryptoclastic rock.*

cryptographic *cryptocrystalline.*

cryptohalite A gray, cubic, high-temperature mineral: $(NH_4)_2SiF_6$. Cf: *bararite.*

cryptolite *monazite.*

cryptolithic *coelobitic.*

cryptomagmatic Said of a hydrothermal mineral deposit without demonstrable relationship to igneous processes. The term is little used. Cf: *apomagmatic; telemagmatic; perimagmatic.*

cryptomelane A mineral: $K(Mn^{+2},Mn^{+4})_8O_{16}$. It is isostructural with hollandite and coronadite.

cryptomere An *aphanitic* rock. Also spelled: *kryptomere.* The nonrecommended adjective forms "cryptomerous" and "kryptomerous" are therefore syns. of *aphanitic.*

cryptomphalus An umbilicus filled with callus in certain gastropods.

cryptonelliform Said of a long brachiopod loop (as in the superfamily Cryptonellacea) unsupported in adults by a median septum and having a narrow transverse band.

cryptoolitic Pertaining to an oolitic texture of such fine grain that it can be recognized only under the microscope. Also, said of a rock with such a texture.

cryptoperthite An extremely fine-grained variety of perthite in which the lamellae are of submicroscopic dimensions (1-5 microns wide) and are detectable only by X-rays or with the aid of the electron microscope. The potassium-rich host can be sanidine, orthoclase, or microcline; the sodium-rich lamellae can be analbite or albite. Cryptoperthite frequently displays a bluish to whitish milky luster. Cf: *microperthite.*

cryptorheic Said of drainage by subterranean streams. Also spelled: *cryptoreic.*

cryptorhomb In the cystoids, a specialized type of pore rhomb that has external openings of simple or compound pores. Cryptorhombs are found in the rhombiferan superfamily *Hemicosmitida* (Paul, 1968, p. 705).

cryptosiderite A stony meteorite poor in nickel-iron.

cryptostome Any ectoproct bryozoan belonging to the order Cryptostomata and resembling a trepostome but having a short endozone and an aperture at the bottom of the vestibule. Adj: *cryptostomatous.*

cryptovolcanic structure (a) A term introduced by Branco & Fraas (1905) and originally applied to a highly deformed, strongly brecciated, generally circular structure believed to have been produced by volcanic explosions, but lacking any direct evidence of volcanic activity, e.g. volcanic rocks, hydrothermal alteration, contact metamorphism, or mineralization; type example: Steinheim Basin, Germany. Many of these structures are now believed to have been formed by meteorite impact, and the nongenetic term *cryptoexplosion structure* is preferred. Also, the term cryptovolcanic structure "has tended to become a 'wastebasket' term and now includes many structures which are unquestionably of volcanic origin" (Dietz, 1959, p. 496). (b) A circular structure lacking evidence of shock metamorphism or of meteorite impact and therefore presumed to be of igneous origin, but lacking exposed igneous rocks or obvious volcanic features; a rock structure produced by concealed volcanic activity.

Cryptozoic That part of geologic time represented by rocks in which evidence of life is only slight and of primitive forms. Cf: *Phanerozoic.*

cryptozoon (a) A structure in Precambrian rocks, believed to be the remains of primitive life. (b) A hemispherical or cabbage-like algal structure of variable size, spreading somewhat above its base, composed of irregular and concentric laminae of calcite of very unequal thicknesses traversed by minute canals that branch irregularly, produced by the problematical Cambrian and Ordovician reef-forming calcareous alga of the genus *Cryptozoon.*—Pl: *cryptozoa.*

crystal A homogeneous, solid body of a chemical element, compound, or isomorphous mixture, having a regularly repeating atomic arrangement that may be outwardly expressed by plane faces.

crystal accumulation In a magma, the development of *layering* by the process of *crystal settling.*

crystal axial indices *indices of lattice row.*

crystal axis (a) *crystallographic axis.* (b) One of the three edges of the chosen unit cell in a crystal lattice. (c) Any *lattice row;* it can be considered a zone axis.—Syn: *axis [cryst].*

crystal-body playa A playa with one or more thick salt bodies at or near the surface, formed by the evaporation of a lake that once occupied the area (Stone, 1967, p. 220); e.g. Searles Lake in California.

crystal cast The filling of a crystal mold; e.g. *ice-crystal cast, salt-crystal cast.*

crystal chemistry The study of the relations among chemical composition, internal structure, and the physical properties of crystalline matter.

crystal class *class [cryst].*

crystal defect An imperfection in the ideal crystal structure. See also: *line defect; point defect; plane defect.* Syn: *lattice defect.*

crystal face (a) A planar face bounding a crystal. Syn: *face [cryst].* (b) A *rational face.* The terms are equivalent for igneous crystals, but the faces of crystals developed by recrystallization may not be rational.

crystal flotation In petrology, the floating of lighter-weight crystals in a body of magma. Cf: *crystal settling.* Syn: *flotation.*

crystal form (a) The geometric shape of a crystal. (b) An assemblage of symmetrically equivalent crystal planes making up a form which displays the symmetry of a crystal class. A crystal may be bounded by one or more forms, each consistent with the internal symmetry of the crystal. Crystal form may be symbolized by Miller indices enclosed in braces, e.g. $\{hkl\}$.

crystal fractionation Magmatic differentiation resulting from the floating or settling, under gravity, of mineral crystals as they form. Cf: *fractional crystallization.* Syn: *gravitational differentiation.*

crystal gliding Deformation of crystalline material by orderly displacement of atoms such that good crystal structure remains after the process is finished. It often produces crystal twins. See also: *twin gliding.* Syn: *gliding [cryst]; translation gliding; slip [cryst].*

crystal habit The general shape of crystals, e.g. cubic, prismatic, fibrous. For a given type of crystal, the habit may vary from locality to locality depending on environment of growth.

crystal indices *Miller indices.*

crystallaria A general term proposed by Brewer (1964, p.284) for a group of soil features consisting of single crystals or arrangements of crystals of relatively pure fractions of the soil plasma that do not enclose the matrix of the soil material but form cohesive masses whose shape and internal fabric are "consistent with their formation and present occurrence in original voids in the enclosing soil material"; e.g. spherulites, rosettes, crystal tubes, and intercalary crystals embedded in a dense soil matrix. See also: *crystal tube.*

crystal lattice The three-dimensional regularly repeating set of points that represent the translational periodicity of a crystal structure. Each lattice point has identical surroundings. There are fourteen possible lattice patterns. Syn: *Bravais lattice; space lattice; direct lattice; lattice [cryst]; translation lattice.*

crystalline [cryst] (a) Pertaining to or having the nature of a crystal, or formed by crystallization; specif. having a crystal structure or a regular arrangement of atoms in a space lattice. Ant: *amorphous.* (b) Said of a mineral particle of any size, having the internal structure of a crystal but lacking well-developed crystal faces or an external form that reflects the internal structure. (c) Resembling a crystal; clear, transparent, pure.

crystalline [petrology] adj. (a) Said of a rock consisting wholly of crystals or fragments of crystals; esp. said of an igneous rock developed through cooling from a molten state and containing no glass, or of a metamorphic rock that has undergone recrystallization as a result of temperature and pressure changes. The term may also be applied to certain sedimentary rocks (such as quartzite, some limestones, evaporites) composed entirely of contiguous crystals. (b) Said of the texture of a crystalline rock characterized by closely fitting or interlocking particles (many having crystal faces and boundaries) that have developed in the rock by simultaneous growth.—n. A crystalline rock. Term is usually used in the plural; e.g. the Precambrian "crystallines". This usage is not recommended.

crystalline carbonate A term used by Dunham (1962) for a carbonate sedimentary rock in which the depositional texture is not recognizable, owing to recrystallization and replacement; e.g. dolomite rock and dolomitic limestone.

crystalline chert Obsolescent syn. of *granular chert* (Ireland et al., 1947, p. 1486).

crystalline chondrite A hard, crystalline stony meteorite containing firm round radial chondrules that break with the matrix.

crystalline flake *flake graphite.*

crystalline-granular *granular.*

crystalline-granular texture (a) A primary granular texture of a sedimentary rock, produced by crystallization from an aqueous medium; it may be exhibited by rock salt, gypsum, or anhydrite. (b) A texture of a carbonate sedimentary rock produced by dolomitization of a limestone containing packed granules (Thomas, 1962, p.197).

crystalline limestone [meta] (a) A metamorphosed limestone; a *marble* formed by recrystallization of limestone as a result of metamorphism. Syn: *cipolin.* (b) A calcarenite with crystalline calcite cement formed in optical continuity with crystalline fossil fragments by diagenesis (Pettijohn, 1957, p.407-408). Syn: *sedimentary marble.*

crystalline limestone [sed] A limestone formed of abundant calcite crystals as a result of diagenesis; specif. a limestone in which calcite crystals larger than 20 microns in diameter are the predominant components (Schmidt, 1965, p. 128). Examples include the crinoidal limestones whose fragments (ossicles, plates, etc.) have been enlarged by growth of calcite. See also: *marble.*

crystalline rock (a) An inexact but convenient term designating an igneous or metamorphic rock, as opposed to a sedimentary rock. (b) A rock consisting wholly of relatively large mineral grains, e.g. a plutonic rock, an igneous rock lacking glassy material, or a metamorphic rock. (c) The term has also been applied to sedimentary rocks, e.g. some limestones, that are composed of coarsely crystalline grains or exhibit a texture formed by partial or complete recrystallization.

crystalline structure *crystal structure.*

crystallinity (a) The degree to which a rock (esp. an igneous rock) is crystalline (holocrystalline, hypocrystalline, etc.). (b) The degree to which the crystalline character of an igneous rock is developed (e.g. macrocrystalline, microcrystalline, or cryptocrystalline) or is apparent (e.g. phaneritic or aphanitic) (Challinor, 1978, p. 75).

crystallinoclastic rock A clastic rock containing abundant crystalline material, such as one having a crystalline cement.

crystallinohyaline *hyalinocrystalline.*

crystallite (a) A broad term applied to a minute body of unknown mineralogic composition or crystal form that does not polarize light. Crystallites represent the initial stage of crystallization of a magma or of a glass. Adj: *crystallitic.* Cf: *microlite; crystalloid.* (b) An obsolete syn. of *stylolite.*

crystal-lithic tuff A tuff that is intermediate between *crystal tuff* and *lithic tuff* or that is predominantly the latter. Cf: *lithic-crystal tuff.*

crystallitic Of, pertaining to, or composed of *crystallites.*

crystallization The process by which matter becomes crystalline, from a gaseous, fluid, or dispersed state.

crystallization banding *phase layering.*

crystallization differentiation The progressive change in composition of the liquid fraction of a magma as a result of the crystallization of mineral phases that differ in composition from the magma. The process may be *equilibrium crystallization* or *fractional crystallization* or some combination of the two.

crystallization fabric A term used by Friedman (1965, p.643) for the size and mutual relations of mineral crystals in sedimentary rocks such as evaporites, chemically deposited cements, and recrystallized limestones and dolomites. Cf: *crystallization texture.*

crystallization index In igneous petrology, specif. igneous differentiation, the number that is calculated from the system anorthite-diopside-forsterite and that represents the sum (in weight percent) of normative anorthite, magnesian diopside, normative forsterite, normative enstatite converted to forsterite, and magnesian spinel calculated from normative corundum in ultramafic rocks (Poldervaart & Parker, 1964, p. 281). Abbrev: CI. Cf: *petrogeny's primitive system.*

crystallization interval (a) The interval of temperature (or, less frequently, of pressure) between the formation of the first crystal and the disappearance of the last drop of a magma upon cooling, usually excluding late-stage aqueous fluids. (b) More specifically, with reference to a given mineral, the temperature range or ranges over which that particular phase is in equilibrium with liquid. In the case of equilibria along reaction lines or reaction surfaces, crystallization intervals defined in this way include temperature ranges in which certain solid phases are actually decreasing in amount with temperature decrease.— Syn: *freezing interval.*

crystallization magnetization *chemical remanent magnetization.*

crystallization remanent magnetization *chemical remanent magnetization.*

crystallization texture A term used by Friedman (1965, p.643) for the shape of mineral crystals in sedimentary rocks such as evaporites, chemically deposited cements, and recrystallized limestones and dolomites. Cf: *crystallization fabric.*

crystallized coal *cone-in-cone coal.*

crystallizing force The expansive force of a crystal that is forming within a solid medium. The force varies according to crystallographic direction. Syn: *force of crystallization.*

crystalloblast A crystal of a mineral produced entirely by metamorphic processes. See also: *idioblast; holoblast; hypidioblast; xenoblast.* Adj: *crystalloblastic.*

crystalloblastesis Deformation accomplished by metamorphic recrystallization (Knopf and Ingerson, 1938).

crystalloblastic (a) Pertaining to a *crystalloblast.* (b) Said of a crystalline texture produced by metamorphic recrystallization under conditions of high viscosity and directed pressure, in contrast to igneous rock textures that are the result of successive crystallization of minerals under conditions of relatively low viscosity and nearly uniform pressure (Becke, 1903). See also: *homeoblastic; heteroblastic.*

crystalloblastic order *crystalloblastic series.*

crystalloblastic series An arrangement of metamorphic minerals in order of decreasing *form energy,* so that crystals of any of the listed minerals tend to assume idioblastic outlines at surfaces of contact with simultaneously developed crystals of all minerals occupying lower positions in the series (Becke, 1913). Syn: *idioblastic series; crystalloblastic order.*

crystalloblastic strength *form energy.*

crystallogeny That branch of crystallography which deals with crystal growth.

crystallographic Pertaining to crystallography or to the properties of a crystal.

crystallographic axis One of three imaginary lines in a crystal (four in a hexagonal crystal) that pass through its center; it is used as a reference in describing crystal structure and symmetry. One or all of the crystallographic axes may coincide with axes of symmetry. Syn: *crystal axis.*

crystallographic orientation The relation of the axes or planes of a given crystal to some other established directions in space, e.g. geographic or geologic lines or planes.

crystallographic plane Any plane, crystal face, cleavage or lattice plane that can be described mathematically in terms of the lengths and directions of the crystallographic axes.

crystallographic texture A texture of mineral deposits formed by replacement or exsolution, in which the distribution and form of the inclusions are controlled by the crystallography of the host mineral.

crystallography The study of crystals, including their growth, structure, physical properties, and classification by form.

crystalloid n. A microscopic crystal which, when examined under a microscope, polarizes light but has no crystal outline or readily determinable optical properties. Cf: *crystallite; microlite.*

crystallolith A crystalline coccolith; e.g. a disciform holococcolith.

crystallothrausmatic A descriptive term applied to igneous rocks with an orbicular texture in which early phenocrysts form the nuclei of the orbicules (Eskola, 1938, p.476). Cf: *allothrausmatic; isothrausmatic; heterothrausmatic; homeothrausmatic.*

crystal mold A cavity left by solution or sublimation of a crystal (as of salt, ice, or pyrite) embedded in soft, fine-grained sediment.

crystal mush Partially crystallized magma; "an aggregate of solid crystals lubricated by compressed water vapor" (Krauskopf, 1967, p. 419).

crystal optics The study of the transmission of light in crystals; the concern of *optical crystallography.*

crystal pool In a cave, standing water lined with crystals of calcite.

crystal sandstone (a) A sandstone in which the quartz grains have been enlarged by deposition of silica so that the grains show regenerated crystal facets and sometimes nearly perfect quartz euhedra. Crystal sandstones of this nature sparkle in bright sunlight. (b) A sandstone in which calcite has been deposited in the pores in large patches or units having a single crystallographic orientation, resulting in a "poikiloblastic" or "luster-mottli g" effect. In some rare sandstones with incomplete cementation he carbonate occurs as sand-filled scalenohedra of calcite—*sand crystals* (Pettijohn, Potter & Siever, 1973, p. 164).

crystal sedimentation *crystal settling.*

crystal seeding The use of a *seed crystal* or foreign particle in a solution to initiate crystallization of the solute.

crystal settling In a magma, the sinking of crystals due to their greater density, sometimes aided by magmatic convection. It results in *crystal accumulation,* which develops *layering.* Cf: *crystal flotation.* Syn: *crystal sedimentation.*

crystal sorting The separation, by any process, of crystals from a magma, or of one crystal phase from another during crystallization of the magma.

crystal structure The regular, orderly, and repeated arrangement of atoms in a crystal, the translational properties of which are described by the crystal lattice or space lattice. Syn: *crystalline structure.*

crystal system One of six groups or classifications of crystals according to the symmetry of their crystal faces, and having characteristic dimensional equivalences in the lattices or axes of reference. The systems are: *isometric system, hexagonal system, tetragonal system, orthorhombic system, monoclinic system,* and *triclinic system.* Within the six systems there is a total of 32 crystal classes. Syn: *system [cryst].*

crystal tube A type of *crystallaria* consisting of masses of crystals filling or partly filling relatively large tube-shaped or acicular voids in soil material. It is usually formed by crystallization from the walls inward.

crystal tuff A tuff that consists predominantly of crystals and fragments of crystals. Cf: *crystal-vitric tuff; crystal-lithic tuff.*

crystal-vitric tuff A tuff that consists of fragments of crystals and volcanic glass. Cf: *crystal tuff; vitric tuff.*

crystal zone (a) Three or more nonparallel crystal faces, the edges of intersection of which are parallel to a common line or lattice row called the *zone axis.* (b) A result of *zoning [cryst].*

crystocrene *chrystocrene.*

crystosphene A buried mass or sheet of clear ice developed by a wedging growth between beds of other material (Tyrrell, 1904, p.234), such as the freezing of springwater rising and spreading laterally beneath alluvial deposits or under swamps in a tundra region. Cf: *chrystocrene.*

csiklovaite A trigonal mineral: $Bi_2Te(S,Se)_2$(?).

ctenodont Said of the dentition of certain bivalve mollusks of early origin having numerous short hinge teeth transverse to the margin.

ctenoid cast A very rare, toothlike *sole mark* having the form of an obliquely cut, longitudinally ribbed cylinder and probably representing a *bounce cast* made in mud by plant stems of *Equisetites* that intermittently touched bottom as they were carried along by a current of water. The term was introduced as "ctenoid marking" by Beasley (1914), who likened the structure to the large tortoise-shell comb worn in women's hair during the early Victorian period. Etymol: Greek *ktenos,* "comb".

ctenolium A comblike row of small teeth on the lower side of the byssal notch in some pectinacean mollusks.

ctenostome Any ectoproct bryozoan belonging to the order Ctenostomata and characterized by the presence of comblike protrusions at the mouth.

cubanite A bronze-yellow orthorhombic mineral: $CuFe_2S_3$. Syn: *chalmersite.*

cube A crystal form of six equivalent (not necessarily square) and mutually perpendicular faces, with indices of $\{100\}$.

cube ore *pharmacosiderite.*

cube spar *anhydrite.*

cubic cleavage Mineral cleavage parallel to the faces of a cube; e.g. in galena.

cubic close packing In a crystal, close packing of spheres by stacking close-packed layers in the sequence ABCABC etc. Cf: *hexagonal close packing.*

cubic coordination An atomic structure or arrangement in which an ion is surrounded by eight ions of opposite sign, whose centers form the points of a cube. An example is the structure of cesium chloride. Partial syn: *hexahedral coordination.*

cubic packing The "loosest" manner of systematic arrangement of uniform solid spheres in a clastic sediment or crystal lattice, characterized by a unit cell that is a cube whose eight corners are the centers of the spheres involved (Graton & Fraser, 1935). An aggregate with cubic packing has the maximum porosity (47.64%). Cf: *rhombohedral packing.* See also: *open packing.*

cubic plane In a crystal of the cubic system, any plane at right angles to a crystallographic axis.

cubic system *isometric system.*

cuboctahedron A cubic crystal form bounded by both the six equal squares of the cube and the eight equal triangles of the octagon, the latter cutting off the corners of the former.

cubo-dodecahedron A crystal in the cubic system that is bounded by cube and dodecahedron forms.

cucalite An obsolete term for chlorite-rich *diabase.* In the Rhaetian Alps it grades locally into chlorite schist.

cuchilla A term used in the SW U.S. for a sharply edged crest of a *sierra.* Etymol: Spanish, "large knife".

cuesta (a) A hill or ridge with a gentle slope on one side and a steep slope on the other; specif. an asymmetric ridge (as in the SW U.S.) with one face (dip slope) long and gentle and conforming with the dip of the resistant bed or beds that form it, and the opposite face (scarp slope) steep or even cliff-like and formed by the outcrop of the resistant rocks, the formation of the ridge being controlled by the differential erosion of the gently inclined strata. Originally, the term applied to the steep slope or scarp that terminates a gently sloping plain at its upper end; the term has also been used to denote the sloping plain itself, such as the top of a mesa. (b) A ridge or belt of low hills formed between lowlands in a region of gently dipping sedimentary rocks (as on a coastal plain), having a gentle slope conforming with the dip of the rocks and a relatively steep slope descending abruptly from its crest.—Etymol: Spanish, "flank or slope of a hill; hill, mount, sloping ground". Cf: *hogback [geomorph].* Syn: *wold; scarped ridge; escarpment.*

Cuisian European stage: Lower Eocene (above Ypresian, below Lutetian). See also: *Bruxellian.*

cul-de-sac [ice] *blind lead.*

cul-de-sac [karst] A closed abandoned *sinking stream* that has been partially filled.

cul-de-sac [speleo] A passage in a cave that has only one entrance.

culet The small facet that is polished parallel to the girdle plane across what would otherwise be the sharp point or ridge that terminates the pavilion of a diamond or other gemstone. Its principal function is to reduce the possibility of damage to the gem.

culm (a) *kolm.* (b) The anthracite contained in the series of shales and sandstones of North Devon, England, known as the Culm measures. (c) Fine particles of anthracite. (d) Coal dust or fine-grained waste from anthracite mines.

culmen A narrow ridge between adjoining crenellae on the articular surface of a columnal of a crinoid or blastoid. Pl: *culmina.*

culmination The highest point of a structural feature, e.g. of a dome, anticlinal crest, synclinal trough, or nappe. The axis of an anticline may have several culminations that are separated by saddles. See also: *crest.* Syn: *axis culmination; apex.*

cultural feature An artificial or man-made feature as shown on a map.

culture The details of a map, representing the works of man (such as roads, railroads, buildings, canals, trails, towns, and bridges), as distinguished from natural features; they are usually printed in black on a topographic map. The term also includes political boundary lines, meridians, parallels, place names, and the legends.

culvert Any covered structure, not classified as a bridge, that constitutes a transverse drain, waterway, or other opening under a road, railroad, canal, or similar structure.

cumacean Any malacostracan belonging to the order Cumacea, characterized chiefly by the long, slim, subcylindrical pleon which is usually strongly differentiated from the broad, commonly inflated pereion and cephalon (TIP, 1961, pt.R, p.368).

cumberlandite A coarse-grained ultramafic rock with olivine crystals (approximately 50 percent of the total rock) in a groundmass of ilmenite and magnetite (together, 40 percent), labradorite, and accessory spinel. The state rock of Rhode Island; the name, given by Wadsworth in 1884, is for the town of Cumberland in that state. Not recommended usage.

cumbraite A porphyritic extrusive rock composed of phenocrysts of basic plagioclase (bytownite, anorthite) in a groundmass of labradorite, orthopyroxene, clinopyroxene, and abundant glass; its chemical composition is andesitic rather than basaltic. See also: *inninmorite.* Named by Tyrrell in 1917 for Great Cumbrae, Scotland. Not recommended usage.

cumengite A deep-blue or light indigo-blue tetragonal mineral: $Pb_4Cu_4Cl_8(OH)_8 \cdot H_2O$. Also spelled: *cumengeite; cumengéite.*

cummingtonite A brownish monoclinic mineral of the amphibole group: $(Fe,Mg)_7Si_8O_{22}(OH)_2$. It contains more iron than anthophyllite, may contain zinc and manganese, and usually occurs in metamorphic rocks as lamellae or fibers. Cf: *grunerite.*

cumulate n. An igneous rock formed by the accumulation of crystals that settle out from a magma by the action of gravity. Syn: *accumulative rock.*

cumulative curve *cumulative frequency distribution.*

cumulative frequency distribution A *frequency curve* in which each group is added to the preceding one until the total number of observations is included; it commonly adds to 100%. Syn: *cumulative curve.*

cumulite A cloudy aggregate of globulites commonly found in glassy igneous rocks.

cumulo-dome *volcanic dome.*

cumulophyre An igneous rock characterized by *cumulophyric* texture. Not recommended usage.

cumulophyric Said of the texture of a porphyritic igneous rock in which the phenocrysts, not necessarily of the same mineral, are clustered in irregular groups; also, said of a rock exhibiting such texture, e.g. a *cumulophyre* (Cross et al., 1906, p. 703). Cf: *glomerophyric; synneusis; gregaritic.*

cumulose Pertaining to a soil material consisting chiefly of partly decomposed vegetable matter, accumulated in situ. An example is peat. Cf: *residual material.*

cumulo-volcano *volcanic dome.*

cumulus The accumulation of crystals that precipitated from a magma without having been modified by later crystallization. See also: *cumulus crystal.*

cumulus crystal A unit of the *cumulus.* Syn: *primary precipitate crystal.* See also: *primocryst.*

cuneate Said of structures, e.g. certain leaves, that are wedge-shaped or triangular, with the narrower end at the base.

cunette A small channel dug in the bottom of a larger channel or conduit for the purpose of concentrating the flow at low-water stages.

cuniculus A continuous tunnel-like cavity formed in foraminifers (such as Verbeekinidae) by strong septal fluting, the opposed folds of adjacent septa meeting to form continuous spiral sutures with vaulted arches between, and serving to connect adjoining chambers from one foramen to the next. Pl: *cuniculi.*

Cunnersdorf twin law A rare type of normal twin law in feldspar, in which the twin plane is $(\bar{2}01)$.

cup (a) A greatly expanded *basal cavity* beneath the anterior or posterior half of some conodont elements. (b) The calyx of a crinoid. (c) A skeleton in archaeocyathids, generally a double-walled inverted cone with an open central cavity and the space between the walls divided by septa, rods, tabulae or other structures (TIP, 1972, pt. E, p. 3).

cup and ball jointing *ball and socket jointing.*

cup coral *solitary coral.*

cup crystal An ice crystal with stepped surfaces in the form of a hollow hexagonal cup, one side of which may be undeveloped and appear to be rolled up. It is a common form of *depth hoar.*

cupid's dart *flèche d'amour.*

cupola [intrus rocks] An upward projection of an igneous intrusion into its roof. Cf: *roof pendant.*

cupola [paleont] A large vaulted dome in nassellarian and spumellarian radiolarians.

cupolate Said of a button-shaped scleractinian corallite with a flat base and a highly convex oral surface.

cupped pebble A pebble whose upper side has been subject to solution, often being so corroded that it becomes a mere shell (Scott, 1947). Cf: *pitted pebble.*

cup reef A small vase-shaped reef, often algal; in Bermuda, these reefs range up to 10 m high by 30 m across. Cf: *boiler; microatoll.*

cuprite A red (crimson, scarlet, vermilion, brownish-red) isometric mineral: Cu_2O. It is an important ore of copper, and occurs as a secondary mineral in the zone of weathering of copper lodes. Syn: *red copper ore; red oxide of copper; ruby copper; octahedral copper ore.*

cuproauride *auricupride.*

cuprobismutite A monoclinic mineral: $CuBiS_2$. It is dimorphous with emplectite, and was formerly regarded as a mixture of bismuthinite and emplectite.

cuprocopiapite A mineral of the copiapite group: $CuFe_4(SO_4)_6(OH)_2 \cdot 20H_2O$.

cuprodescloizite *mottramite.*

cuprorivaite A mineral: $CaCuSi_4O_{10}$.

cuproscheelite A mixture of scheelite and cuprotungstite.

cuprosklodowskite A strongly radioactive, greenish-yellow or grass-green, orthorhombic secondary mineral: $Cu(UO_2)_2Si_2O_7 \cdot 6H_2O$. It is isostructural with sklodowskite and uranophane.

cuprospinel A mineral of the spinel group: $(Cu,Mg)Fe_2O_4$.

cuprostibite A mineral: $Cu_2Sb(Te)$.

cuprotungstite A mineral: $Cu_2(WO_4)(OH)_2$.

cuprouranite *torbernite.*

cupule (a) A cuplike *involucre,* esp. of the acorn, whose bracts are adherent at the base and may or may not be free upwards. (b) A free sheathing structure from the peduncle, investing one or more seeds.

curie A unit of measurement of radioactivity, defined as the amount of a *radionuclide* in which the decay rate is 37 billion disintegrations per second, which is approximately equal to the decay rate of one gram of pure radium.

Curie balance A *magnetic balance* used to determine *saturation magnetization* as a function of temperature, hence to determine *Curie point.*

curienite An orthorhombic mineral: $Pb(UO_2)_2(VO_4)_2 \cdot 5H_2O$. It forms a series with francevillite.

Curie point The temperature above which thermal agitation prevents spontaneous magnetic ordering. Specifically, the temperature at which the phenomenon of ferromagnetism disappears and the substance becomes simply paramagnetic. Cf: *Néel point.* Syn: *Curie temperature.*

Curie's law The statement that magnetic susceptibility is inversely proportional to absolute temperature. It is applicable to substances which do not show spontaneous magnetic order at low temperatures. Cf: *Curie-Weiss Law.*

Curie temperature *Curie point.*

Curie-Weiss law The statement that the magnetic susceptibility of a ferromagnetic material above its Curie point is inversely proportional to the difference between actual temperature and the Curie point. Cf: *Curie's law.*

curio stone A stone that combines uniqueness or souvenir value with some degree of beauty and durability; e.g. fairy stone and Niagara spar. Cf: *ornamental stone.*

curite An orange-red, radioactive mineral: $Pb_2U_5O_{17} \cdot 4H_2O$. It is an alteration product of uraninite.

curl (a) A term used in southern U.S. for a bend in a stream. (b) An eddy in a stream.

curly bedding *convolute bedding.*
curly cannel A syn. of *cannel coal,* so named for its conchoidal fracture.
current (a) The part of a fluid body, esp. as air or water, that is moving continuously in a definite direction, often with a velocity much swifter than the average, or in which the progress of the fluid is principally concentrated. (b) A horizontal movement or continuous flow of water in a given direction with a more or less uniform velocity, producing a perceptible mass transport, set in motion by winds, waves, gravity, or differences in temperature and density, and of a permanent or seasonal nature; esp. an *ocean current.* (c) The velocity of flow of a fluid in a stream.
current bedding Any bedding or bedding structure produced by current action; specif. cross-bedding resulting from water or air currents of variable direction. The term is used (esp. in Great Britain) as a syn. of *cross-bedding,* but such usage is not recommended (Middleton, 1965, p.248). See also: *ripple bedding.* Syn: *false bedding.*
current crescent (a) A small, semicircular or U-shaped rounded ridge, convex upcurrent, commonly with a pit in the center, and developed on a muddy surface by current action (Peabody, 1947, p.73). (b) A flute cast of a horseshoe-shaped moat eroded on the upcurrent side of a pebble, shell, or other obstacle. Syn: *horseshoe flute cast; crescent cast; crescent scour; crescentic mark.*
current cross ripple mark A *cross ripple mark* resulting from the intersection at any angle of a preexisting current ripple mark by a later current moving in a different direction and being "sufficiently weak and only of very short duration" so as not to destroy the first set of ripples (Bucher, 1919, p.194-195).
current density The current per unit area perpendicular to the direction of current flow.
current direction *set [current].*
current electrode A metal contact with the ground used to facilitate current flow through the ground.
current-focused log *focused-current log.*
current lineation A term used by Stokes (1947) for *parting lineation.*
current mark (a) Any structure formed by the action of a current of water, either directly or indirectly, on a sedimentary surface; e.g. a *scour mark* made by the current itself, or a *tool mark* formed by solid objects swept along by the current (Dzulynski & Sanders, 1962). See also: *flow mark.* Syn: *current marking.* (b) An irregular structure made by a tidal current in the beach zone, consisting of a small depression extending toward the shore from the lee side of an obstruction. (c) A *current ripple mark*. Kindle (1917, p.36) used the term to designate the linguoid variety of current ripple mark.
current meter Any one of numerous instruments for measuring the speed alone, or both speed and direction, of flowing water, as in a stream or the ocean; it is usually activated by a wheel equipped with a set of revolving vanes or cups whose rate of turning is proportional to the velocity of the current.
current rip A *rip* consisting of small waves formed on the sea surface by the meeting of opposite currents.
current ripple *current ripple mark.*
current-ripple cast A term used by Kuenen (1957, fig.6) for a sedimentary structure now known as a *transverse scour mark.*
current ripple mark An *asymmetric ripple mark* with a sharp or rounded crest between rounded troughs, formed by currents of air or water moving more or less constantly in a uniform direction over a sandy surface (such as a stream bar, tidal flat, beach, or sand dune), the ripple slowly migrating downcurrent much like a miniature sand dune. See also: *ripple mark; linguoid ripple mark; rhomboid ripple mark; normal ripple mark.* Cf: *oscillation ripple mark.* Syn: *current ripple; current mark; parallel ripple mark.*
current rose A graphic representation of current direction for a given ocean area over a period of time; it indicates by means of proportional radiating arrows the direction toward which the prevailing current flows and the percentage frequency of any given direction of flow.
current scour A term introduced by Dzulynski and Walton (1965, p. 40) to refer to scour marks produced by a current, without the presence of an obstacle being essential to their formation. Cf: *scour mark; current mark; obstacle scour.*
curtain [geomorph] (a) A rock formation that connects two neighboring bastions. (b) One of a series of steps cut in a valley side and exaggerated by cultivation (Swayne, 1956, p. 45).
curtain [speleo] *drapery.*
curtain of fire A row of coalescing *lava fountains* along a fissure; a typical feature of a Hawaiian-type eruption (Macdonald, 1972, p. 214).
curtisite *idrialite.*
curvatura A visible line of some mid-Paleozoic trilete spores that connects the extremities of the ends of the laesura and outlines the contact areas; e.g. a "curvatura perfecta" having three lines complete all around the spore's proximal face, or a "curvatura imperfecta" having forklike projections from the radial ends of the laesura but not joining with their neighbors. Pl: *curvaturae.*
curvature (a) *terminal curvature.* (b) *outcrop curvature.* (c) *earth curvature.*
curvature correction An adjustment applied to an observation or computation (e.g. of difference in elevation) to allow for *earth curvature.* In geodetic spirit leveling, the effects of curvature and of atmospheric refraction are considered together, and tables have been prepared from which combined corrections can be taken.
curvature of gravity A vector of quantity calculated from torsion-balance data indicating the shape of the equipotential surface. It points in the direction of the longer radius of curvature.
curved fracture cleavage A pattern of cleavage surfaces in graded beds which cut more directly across the lower, coarser parts and curve to a more diagonal direction in the upper, finer parts, thus being convex outward from an anticlinal axis (Muller, 1965).
curved path The curvature of a seismic *raypath* because of variation in velocity. The curvature is usually concave upward because velocity increases with depth.
curve of erosion A theoretical profile of a surface shaped by erosion, portraying a a stream, valley, hill, coast, or skyline; esp. a single, continuously descending curve representing a stream course at grade, generally concave upward and flattening as it descends from the source to the sea.
curvette A misspelling of *cuvette.*
curviplanar Pertaining to a surface or form derived from the curving of a plane about one or more axes.
cuselite A biotite- and augite-bearing *lamprophyre* intermediate in composition between a minette and a vogesite and between a kersantite and a spessartite. The name, given by Rosenbusch in 1887, is for Cusel in the Saar Basin. Not recommended usage.
cushion An artificial pool designed to absorb the kinetic energy of falling water and so prevent erosion and reduce vibration.
cusp [coast] One of a series of sharp, seaward-projecting points of beach material, built by wave action and separated from its neighbors by smoothly curved shallow re-entrants. Cusps are spaced at more or less regular intervals along the shoreline, and generally at right angles to it; distance between them ranges from less than a meter to many kilometers. Cf: *beach cusp; storm cusp; giant cusp; cuspate spit; cuspate foreland.* The term "beach cusp" is frequently and loosely used as a synonym.
cusp [geomorph] A landform characterized by a projection with indentations of crescent shape on either side; e.g. a *meander cusp.*
cusp [paleont] A spinelike, fanglike, or cone-shaped structure (i.e., a large denticle) located above the basal cavity of conodont elements. It comprises the entire element in simple conodont elements. See also: *anticusp.*
cusp-and-caries texture *caries texture.*
cuspate bar A seaward-pointing, doubly crescentic bar uniting with the shore at each end. It may be formed by a single spit growing from the shore and then turning back to meet it, or by two spits growing obliquely from the shore and converging to form a bar of sharply cuspate form. Cf: *V-bar; looped bar.*
cuspate delta A tooth-shaped delta in which a single dominant river builds the delta forward into a lake or sea while vigorous wave action spreads the deltaic deposits uniformly on either side of the river mouth to form two curving beaches, each concave toward the water; e.g. the delta of the Tiber River on the Mediterranean Sea.
cuspate foreland The largest *cusp,* occurring as a cape or as a broadly triangular point of sand or shingle, with the apex pointing seaward, along an open coast. On some coasts, cuspate forelands measure many kilometers from apex to apex and extend seaward for several kilometers. They are formed by long-continued shore drifting of sediment, as by the convergence of separate spits or beach ridges from opposed directions, or by the progradation of cuspate bars. Term originated by Gulliver (1896, p. 401). Examples: Cape Canaveral, Fla., and Cape Hatteras, N.C.
cuspate-foreland bar (a) A transition form between a *compound cuspate bar* and a *cuspate foreland.* (b) A bar produced where a

cuspate bar enclosing a triangular lagoon or marsh is prograded by the addition of successive beach ridges on its seaward side (Johnson, 1919, p. 324-325).

cuspate reef A *wall reef* whose ends curve leeward and border the passages between adjacent reefs (Maxwell, 1968, p.99 & 101).

cuspate ripple mark One of a series of *linguoid ripple marks* arranged in phase, in rows parallel to the flow (Allen, 1968, p. 66).

cuspate sandkey A *cuspate bar* that has been built up above the water surface to form an island, as along the west coast of Florida.

cuspate spit A prominent point commonly extending from a barrier island into a bay or lagoon (Shepard, 1952, p. 1911); the distance between the crescentic tips is 2 km or more.

cusp cast A term introduced by Spotts & Weser (1964, p.199) for a crescentic, asymmetric sole mark that lacks a deeper upstream end and that is not elongated in the downstream direction. The original depression responsible for the cast may represent current scour.

cuspidine A mineral: $Ca_4Si_2O_7(F,OH)_2$. Syn: *custerite.*

cusplet A minor *beach cusp,* measuring 1.5 m or less between the tips of the crescent, occurring in the swash zone; it has a short life-span, appearing and disappearing with the turn of the tide. Syn: *beach cusplet.*

cusp-ripple A term used by McKee (1954, p.60) for a crescent-shaped current ripple resulting from an "irregular and fluctuating" stream. See also: *linguoid ripple mark.*

custerite *cuspidine.*

cut [gem] The style or form in which a gem has been fashioned; e.g. *brilliant cut* or *step cut.*

cut [geol] v. To excavate or hollow out a depression, channel, or furrow by erosion.—n. (a) A notch, depression, channel, inlet, or other incision produced by erosion or natural excavation, as by water or waves. (b) A passage or space from which material has been excavated, such as a road *cut.* Also, the material excavated.

cutan A modification of the texture, structure, or fabric of a soil material (such as a soil aggregate, ped, or skeleton grain) along a natural surface within it, caused by a concentration of a particular soil constituent. It can be composed of any of the component substances of the soil material. Examples: *argillan; mangan.* Etymol: Latin *cutis,* "a coating, surface, or skin".

cut and fill [eng] The excavating of material in one place and the depositing of it as compacted fill in an adjacent place, as in the building of a road, canal, or embankment, or in stope mining.

cut and fill [geomorph] A process of leveling whereby material eroded from one place by waves, currents, streams, or winds is deposited nearby until the surfaces of erosion and deposition are continuous and uniformly graded; esp. lateral erosion on the concave banks of a meandering stream accompanied by deposition within its loops. Cf: *scour and fill.*

cut and fill [sed struc] A sedimentary structure consisting of a small erosional channel that is subsequently filled; a small-scale *washout.*

cutbank A local term in the western U.S. for a steep bare slope formed by lateral erosion of a stream. Also spelled: *cut bank.*

cuticle (a) The layer, composed chiefly of *cutin,* that covers the outer walls of a plant's epidermal cells. (b) A secretion of epidermal cells comprising the noncalcified acellular parts of the body wall of a bryozoan.

cutin The waxy material of the *cuticle* covering external cell surfaces of vascular plants and some mosses (Scagel et al., 1965, p.614).

cutinite A maceral of coal within the *exinite* group, consisting of plant cuticles. Cf: *sporinite; resinite; alginite.*

cutoff [eng] An impermeable wall, collar, or other structure placed beneath the base or within the abutments of a dam to prevent or reduce losses by seepage along a construction interface or through porous or fractured strata. It may be made of concrete, compacted clay, interlocking sheet piling, or *grout* injected along a line of holes.

cutoff [stratig] A boundary, oriented normal to bedding planes, that marks the areal limit of a specific stratigraphic unit where the unit is not defined by erosion, pinch-out, faulting, or other obvious means. Cutoffs are applicable to map, cross-sectional, and three-dimensional views, and are in effect specialized facies boundaries. See also: *arbitrary cutoff.* Syn: *stratigraphic cutoff.*

cutoff [streams] (a) The new and relatively short channel formed when a stream cuts through a narrow strip of land and thereby shortens the length of its channel. See also: *neck cutoff; chute cutoff.* Syn: *meander cutoff; cutoff channel.* (b) A channel constructed to straighten a stream or to bypass large bends, thereby relieving an area normally subjected to flooding or channel erosion. See also: *pilot channel.* (c) The crescent-shaped body of water separated from the main stream by a cutoff. (d) The formation of a cutoff.

cutoff channel A meander *cutoff.*

cutoff grade In economic geology, the lowest grade of mineralized material that qualifies as ore in a given deposit; ore of the lowest assay value that is included in an ore estimate. Cf: *assay limit.*

cutoff lake *oxbow lake.*

cutoff limit *assay limit.*

cutoff meander A meander that has been abandoned by its stream after the formation of a neck cutoff. See also: *oxbow.* Syn: *abandoned meander.*

cutoff spur The remnant of a meander spur, formed when a vigorously downcutting stream breaks through a narrow strip of land between adjacent curves in the stream course; it usually stands as an isolated hill. Syn: *meander core.*

cutout A mass of shale, siltstone, or sandstone filling an erosional channel cut into a coal seam. Cf: *roll [coal]; washout [mining].* Syn: *horseback [coal]; want.*

cut plain A *stratum plain* on any hard rock that has been much dissected by erosion; the original surface is approximately represented by the summits of the least-eroded parts (Hill, 1900, p. 7).

cut platform *wave-cut platform.*

cutter A *solution fissure,* esp. as used in Tennessee for a solution crevice in limestone underlying a residual phosphate deposit.

cut terrace (a) *wave-cut terrace.* (b) *rock terrace.*

cuttings *well cuttings.*

cutty clay Plastic clay formerly used in England for making tobacco pipes; *pipe clay.*

cuvette [gem] An *intaglio* with a raised cameolike figure in a polished depression. Cf: *chevee.*

cuvette [sed] A large-scale *basin* in which sedimentation has occurred or is taking place, as distinguished from a tectonic basin due to folding of preexisting rocks; e.g. the Anglo-Parisian cuvette of SE England and NE France, in which Cenozoic rocks accumulated and were later folded into several distinct but smaller basins, such as the Paris Basin and the London Basin. Etymol: French, "small tub or vat". Sometimes misspelled *curvette.*

Cuvier's principle The theory that certain very different characteristics are commonly associated, e.g. kinds of feet and teeth among the vertebrates. Named after the French naturalist Georges Cuvier (1769-1832).

cuyamite A *teschenite* composed of labradorite, analcime, hauyne, hornblende, augite, and magnetite. Its name (Johannsen, 1939) is derived from Cuyamas Valley, California. Not recommended usage.

C wave *coupled wave.*

cwm Welsh term for *cirque [glac geol].* The term is occasionally used in Wales for a narrow, deep valley of nonglacial origin in a mountain region (Stamp, 1961, p. 142). Pron: *koom.* Pl: *cwms.* Syn: *coom; coomb; combe; comb.*

cyanite *kyanite.*

cyanochroite A blue mineral: $K_2Cu(SO_4)_2 \cdot 6H_2O$.

cyanosite A syn. of *chalcanthite.* Also spelled: *cyanose.*

cyanotrichite A bright-blue or sky-blue mineral: $Cu_4Al_2(SO_4)(OH)_{12} \cdot 2H_2O$. Syn: *lettsomite; velvet copper ore.*

cyatholith *placolith.*

cyathosponge *archaeocyathid.*

cycad *cycadophyte.*

cycadophyte A gymnosperm having compound leaves, and naked seeds borne separately on sporophylls or in simple cones. Cycadophytes include both cycadeoids and true cycads. They range from the Permian. Syn: *cycad.*

cycle [geol] (a) A series of events that are normally recurrent and return to a starting point, that are repeated in the same order at more or less regular intervals, and that end under conditions that are the same as they were at the beginning; e.g. the cycle of the seasons, a geochemical cycle, or a cycle of sedimentation. (b) An interval of time during which one sequence of a regularly recurring succession of events or phenomena runs to completion, with the last stage or event being quite different from the first; e.g. a cycle of erosion, or an *orogenic cycle.* (c) A group of rock units that occur in a certain order, with one unit being repeated frequently throughout the succession (Duff & Walton, 1962, p. 239); esp. a *cyclothem.*—Adj: *cyclic.*

cycle [paleont] (a) A ring of segments in heterococcoliths. (b) A set of septa or tentacles of like age in a coral.

cycle of denudation *cycle of erosion.*

cycle of erosion (a) The complete, progressive, and systematic sequence of natural changes or stages in a landscape from the start of its erosion on a newly uplifted or exposed surface through its dissection into mountains and valleys until it has been reduced in the final stage to a low, featureless plain or to a base level (such as sea level) that limits the activity of the agents involved; according to some authors, a complete cycle is from base level back to base level. The cycle, usually subdivided into youthful, mature, and old-age stages, is largely hypothetical because it is normally interrupted before it runs to completion; the landforms produced and destroyed during the sequence are a function of climate, geography, and geologic structure. The concept was first developed and formalized by Davis (1889b). See also: *normal cycle.* (b) The interval of time during which the cycle of erosion is completed; the time involved in the reduction of a newly uplifted land area to a base level.—Syn: *geomorphic cycle; geographic cycle; erosion cycle; cycle of denudation; physiographic cycle.*

cycle of fluctuation *phreatic cycle.*

cycle of relative change of sea level The interval of time during which a relative rise and fall of sea level taks place. Regional and global cycles are recognized (Mitchum, 1977, p. 206). See also: *paracycle of relative change of sea level; supercycle.*

cycle of sedimentation (a) A sequence of related processes and conditions, repeated in the same order, that is recorded in a sedimentary deposit; e.g. the processes and conditions that determine the ordered sequence of orthoquartzite, graywacke, and arkose. (b) The deposition of sediments in a basin between the beginnings of two successive marine transgressions, comprising the deposits formed initially on dry land, followed by shallow-water and then deep-water deposits that in turn gradually change to shallow-water and then dry-land type during a marine regression. (c) A cyclothem.—Syn: *sedimentary cycle.*

cycleology A term used by Elias (1965, p.339) to designate the detection and study of cycles in paleontologic and geologic phenomena. The basic unit in cycleology is the phase.

cycle skip An occurrence in acoustic or sonic logging, when the first arrival is strong enough to trigger the receiver closest to the transmitter but not the farthest receiver, which may then be triggered by a later cycle, resulting in an erroneously high transit time. The onset of cycle skip is characterized by an abrupt deflection corresponding to an added cycle of travel between receivers.

cyclic Adj. of *cycle;* recurrent rather than *secular.*

cyclic coverplate series A serially repeated set of coverplates in edrioasteroids, along each side of the ambulacral perradial line. Each set is a cycle; cycles on one side of the ambulacrum are a mirror image of those on the other side; opposing cycles alternate, offset by half a cycle (Bell, 1976).

cyclic crystallization A process of recurring crystallization of some mineral phases during magmatic settling that produces *rhythmic layering.* See also: *rhythmic crystallization.*

cyclic evolution Evolution, supposed by some to have occurred in many lineages, involving successively initial rapid and vigorous expansion, a long stable or slowly changing phase, and a final short episode in which overspecialized, degenerate, or inadaptive forms led to extinction.

cyclic salt Salt that is lifted from the sea as spray, is blown inland, and returns to the sea via drainage.

cyclic sedimentation A syn. of *rhythmic sedimentation.* The term is used esp. for sedimentation involving a circuitous sequence of conditions, such as found in a megacyclothem exhibiting asymmetric bedding.

cyclic terrace One of several stream terraces representing former valley floors formed during periods when downcutting had essentially stopped for a time and lateral erosion had become dominant; e.g. a *valley-plain terrace.* Terraces on opposite sides of the valley are paired or correspond in altitude along any given section of the valley. Cf: *noncyclic terrace.*

cyclic twinning *Repeated twinning* of three or more individual crystals according to the same twin law but with the twin axes or twin planes not parallel. Cyclic twinning often results in threefold, fourfold, fivefold, sixfold, or eightfold twins, which, if equally developed, display geometrical symmetry not formed in single crystals. Cf: *polysynthetic twinning.*

cyclocystoid Any small, discoid echinozoan belonging to the class Cyclocystoidea, characterized by a theca composed of calcareous plates separable into central oral and aboral discs, submarginal ring, and marginal ring, by a flat aboral surface, and by a multiple branching ambulacral system. Range, Middle Ordovician to Middle Devonian.

cyclodont Said of the dentition of bivalve mollusks characterized by arched hinge teeth curving out from below the hinge margin and by a small or absent hinge plate.

cyclographic projection A term used in structural geology for the representation of planes on a stereogram by means of great circles (Dennis, 1967, p.140).

cyclohexane A colorless liquid, a saturated homocyclic hydrocarbon of the cycloparaffin series, formula C_6H_{12}. It has a pungent odor and is found in petroleum. Syn: *hexamethylene.*

cyclolith Any elliptical or circular ring coccolith. The term should be restricted to circular forms.

cyclone An atmospheric low-pressure system with a closed, roughly circular wind motion that is counterclockwise in the Northern Hemisphere and clockwise in the Southern. See also: *tropical cyclone; anticyclone.*

cycloparaffin A saturated homocyclic hydrocarbon having the empirical formula C_nH_{2n}. Examples are cyclopentane and cyclohexane, both of which are found in petroleum. Syn: *naphthene.*

cyclopean texture *mosaic texture.*

cyclosilicate A class or structural type of *silicate* characterized by the linkage of the SiO_4 tetrahedra in rings, with a ratio of Si:O=1:3. An example of a cyclosilicate is beryl, $Be_3Al_2Si_6O_{18}$. Cf: *nesosilicate; sorosilicate; inosilicate; phyllosilicate; tectosilicate.* Syn: *ring silicate.*

Cyclostomata An order of monorhinate jawless fishes characterized by rasping and/or sucking mouth parts, predaceous or carrion-feeding habit, and lack of ossification. Lampreys and hagfishes are the sole surviving members. Range, Pennsylvanian to Recent.

cyclostome Any ectoproct bryozoan belonging to the order Cyclostomata and characterized by calcareous tubular zooecia with circular apertures and no operculum. Adj: *cyclostomatous.*

cyclosystem A gastrozooid of a hydrozoan, surrounded by a circular row of five to seven individual dactylozooids.

cyclothem (a) A term proposed by Weller (in Wanless & Weller, 1932, p.1003) for a series of beds deposited during a single sedimentary cycle of the type that prevailed during the Pennsylvanian Period. It is an informal lithostratigraphic unit equivalent to "formation". Because of extremely variable development, a cyclothem cannot be defined rigidly in terms of the members actually present at any locality (Weller, 1956, p.27-28). Cyclothems are typically associated with unstable shelf or interior basin conditions in which alternate marine transgressions and regressions occur; nonmarine sediments usually occur in the lower half of a cyclothem, marine sediments in the upper half. The term has also been applied to rocks of different ages and of different lithologies from the Pennsylvanian cyclothems. The cyclothem concept was developed by Weller (1930). See also: *ideal cyclothem; megacyclothem; rhythmite.* (b) A *cycle* applied to sedimentary rocks (Duff & Walton, 1962).

cyclowollastonite A triclinic mineral: $CaSiO_3$. It is trimorphous with wollastonite and parawollastonite.

cylindrical divergence The decrease in amplitude of seismic surface waves with distance from the source. The amplitude varies inversely as the distance. Cf: *spherical divergence.*

cylindrical fault A fault of which the plane is curved and on which the displacement is rotational about an axis parallel with the fault plane (Dennis, 1967).

cylindrical fold A fold model that can be described geometrically by the rotation of a line through space parallel to itself. Cf: *conical fold.* Syn: *cylindroidal fold.*

cylindrical projection A projection on the surface of a cylinder; esp. any of numerous map projections of the Earth, produced by projecting the geographic meridians and parallels onto the surface of a cylinder that is tangent to, or intersects, the surface of the sphere, and then developing (unrolling and laying flat) the cylinder as a plane. The principal scale is preserved along the line of tangency. Examples: *Mercator projection; Gall projection.*

cylindrical structure A vertical sedimentary structure with an irregular columnar or pillar-like shape; e.g. a *clastic pipe.*

cylindrite A blackish lead-gray mineral: $Pb_3Sn_4Sb_2S_{14}$. It occurs in cylindrical forms that separate under pressure into distinct shells or folia.

cylindroidal fold *cylindrical fold.*

cymatogeny Undulating movement or warping of the Earth's crust to produce regional, linear arching or doming, but with minimal deformation. The concept was introduced by Lester King (1959, p.117) and according to it, most mountain ranges are cymatogenic rather than orogenic. Cf: *orogeny; epeirogeny.*

cymoid adj. In economic geology, said of a vein that in cross section forms a reverse curve; a vein that swerves from its course, then returns. A pair of veins in such a pattern forms a cymoid loop.

cymophane A syn. of *chrysoberyl.* The name is applied esp. to chrysoberyl exhibiting a girasol or chatoyant effect, and more specif. to *cat's-eye* only.

cymrite An orthorhombic mineral: $Ba_2Al_5Si_5O_{19}(OH){\cdot}3H_2O$.

cyphonautes Pelagic bivalved larva of some bryozoans (such as *Membranipora*).

cyprine A light-blue variety of vesuvianite containing a trace of copper.

Cyprus-type deposit A pyritic copper deposit associated with underlying serpentinite and pillow basalt and with overlying cherts and ferruginous sediments. It is thought to form on oceanic crust of the sea floor.

cyrenoid Obsolete syn. of *corbiculoid.*

cyrilovite A brown mineral: $NaFe_3(PO_4)_2(OH)_4{\cdot}2H_2O$. Syn: *avelinoite.*

cyrtochoanitic Said of a comparatively short, retrochoanitic septal neck of a nautiloid, curved so as to be concave outward (TIP, 1964, pt.K, p.55). See also: *ellipochoanitic; orthochoanitic.*

cyrtocone A curved, slender cephalopod conch (like that of *Cyrthoceratites*) that completes less than one whorl. Syn: *cyrtoceracone.*

cyrtoconoid (a) Said of a gastropod shell in which the spire approaches a conical shape but has convex sides. Cf: *coeloconoid.* (b) Resembling a cyrtocone.

cyrtolite An altered variety of zircon containing uranium, beryllium, and rare earths.

cyrtolith A basket- or calotte-shaped heterococcolith with laths and a projecting central structure.

cyrtosome Any mollusk of the subphylum Cyrtosoma, characterized by a more or less bent gut and ordinarily a univalved shell. It includes the classes Monoplacophora, Gastropoda, and Cephalopoda.

cyst [paleont] A sac or capsule that is secreted by many protozoans and other minute animals as a prelude to a resting or a specialized reproductive phase and that envelops protoplasm and protects it from adverse environmental conditions.

cyst [palyn] A microscopic *resting spore* with a resistant wall, formed in dinoflagellates and in blue-green algae and desmids, by the breaking-up of parts of the filaments or by the enclosing of a cell group and their investment by a sheath or envelope. Dinoflagellate cysts exist abundantly as fossils. See also: *statospore.*

cystid (a) The combined cellular and skeletal layers of the body wall of a bryozoan zooid. (b) A *cystoid.*

cystiphragm A skeletal partition in stenolaemate bryozoans, extending from zooecial wall into zooecial cavity and recurved inwardly to form a compartment that is generally closed by abutting zooecial wall or diaphragm, or by another cystiphragm. Cystiphragms encircle the living chamber with a hollow collar or ring.

cystocarp In the red algae, the fertile structure and surrounding pericarp.

cystoid Any crinozoan belonging to the class Cystoidea, characterized by diplopores, pore rhombs, and brachioles. Range, Lower Ordovician to Upper Devonian. Var: *cystid.*

cystoporate An ectroproct bryozoan belonging to the order Cystoporata and characterized by calcareous tubular zooecia separated by vesicular tissue made up of small chambers (cystopores).

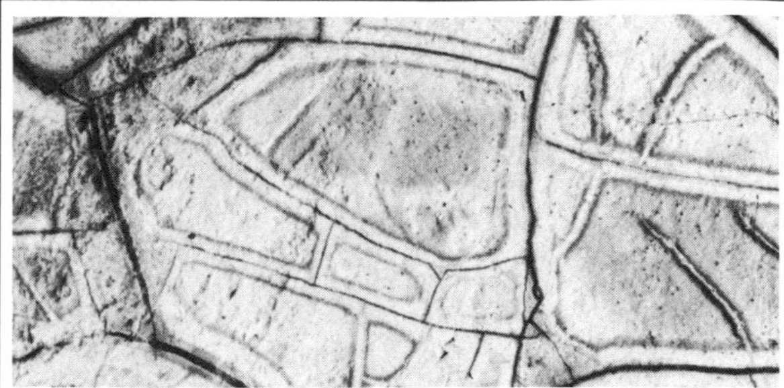

dachiardite A white to colorless zeolite mineral: $(Ca,Na_2,K_2)_5Al_{10}Si_{38}O_{96}\cdot25H_2O$.

Dacian Eastern European stage: Pliocene (above Pontian, below Rumanian or Akchagylian). See also: *Kimmerian*.

dacite A fine-grained extrusive rock with the same general composition as *andesite* but having a less calcic plagioclase and more quartz; according to many, it is the extrusive equivalent of *granodiorite*. Syn: *quartz andesite*. The name, given by Stache in 1863, is from the ancient Roman province of Dacia (now part of Romania).

dacitoid An extrusive rock with the same chemical composition as *dacite* but lacking modal quartz. Named by Lacroix in 1919. Not recommended usage.

dactylethra A polymorph in the stenolaemate bryozoan genus *Terebellaria* that is apparently the zooecium of a feeding zooid covered by a terminal diaphragm (Boardman & Cheetham, 1973, p. 155).

dactylite An igneous rock characterized by *dactylitic* texture. Not recommended usage.

dactylitic A term applied to a rock texture produced by a *symplectic* intergrowth, in which one mineral is penetrated by fingerlike projections from another mineral; also, said of a rock exhibiting such texture. Cf: *dactylotype intergrowth*.

dactylopod The *dactylus* of a malacostracan crustacean. Syn: *dactylopodite*.

dactylopore A relatively small or narrow tubular cavity of certain hydrozoans, occupied by a dactylozooid. Cf: *gastropore*.

dactylotype intergrowth A symplectic intergrowth in which fingerlike projections of one mineral penetrate another. See also: *dactylitic*.

dactylous (a) Pertaining to a dactylus of a crustacean. (b) Said of a pedicellaria of an echinoid having spoon-shaped jaws mounted on individual stalks. (c) In tetrapods, a suffix signifying "toes", e.g. pentadactylous, five-toed.

dactylozooid An elongate, slender, mouthless polyp housed in the dactylopore of a hydrozoan, equipped with numerous stinging cells, and performing protective, food-capturing, and tactile functions for the colony. Cf: *gastrozooid*.

dactylus The seventh and most distal segment of the pereiopod of a malacostracan crustacean. Pl: *dactyli*. Syn: *dactyl; dactylopod*.

dadsonite A monoclinic mineral: $Pb_{11}Sb_{12}S_{29}$.

dagala A syn. of *steptoe*, used in the region of Mt. Etna, Italy.

dahamite An albite-rich *paisanite* or *microgranite*. The name, given by Pelikan in 1902, is for Dahamis, Socotra Island, Yemen. Not recommended usage.

dahllite A resinous, yellowish-white carbonate-apatite mineral or association sometimes occurring as concretionary spherulites.

daily variation Daily fluctuations of the Earth's magnetic field. Typically, it has an amplitude of about 20 gamma and is most rapid near local noon, but it is subject to unpredictable fluctuations. Syn: *diurnal variation*.

daily wave *diurnal wave*.

dakeite *schroeckingerite*.

dal A Scandinavian term for a *valley*.

dale (a) A term used in northern England and southern Scotland for a broad, open river valley. (b) A poetic var. of *valley;* a *vale* or a small valley. Cf: *dell*. (c) A district whose *slope lines* run to the same *pit* (Warntz, 1975, p. 213).

dalles (a) The rapids in a deep, narrow stream confined between the rock walls of a canyon or gorge; e.g. The Dalles of the Columbia River where it flows over columnar basalt. (b) A steep-sided part of a stream channel, near the dalles proper, marked by clefts, ravines, or gorges; e.g. along the Wisconsin River, Wisc.—Etymol: French plural of *dalle,* "gutter". Pron: *dalz*. Syn: *dells*.

dallinid Any articulate brachiopod belonging to the family Dallinidae, characterized by loop development passing through precampagiform, campagiform, frenuliform, terebratelliiform, and dalliniform, or variations of these.

dalliniform Said of the loop, or of the growth stage in the development of the loop, of a dallinid brachiopod (as in *Dallina*), consisting of long descending branches recurved into ascending branches that meet in a transverse band, all free of the valve floor (TIP, 1965, pt.H, p.143). The dalliniform loop is morphologically similar to the *magellaniform* and *cryptonelliform* loops.

dallol A term applied in Nigeria to a flat-bottomed steep-sided dry valley, usually many kilometers wide; specif. a part of an old drainage system on the left bank of the Niger River.

Dalmatian coastline A *concordant coastline* developed where drowning or a rise of sea level has produced lines of narrow islands (representing the outer mountain ranges) separated by long inlets or straits (representing parallel valleys) lying peripheral, and trending roughly parallel, to the coast. Type region: the eastern coastline of the Adriatic Sea in the region of Dalmatia, Yugoslavia.

Dalradian A series of highly metamorphosed rocks of the Scottish Highlands. It is mostly late Precambrian in age but contains a Cambrian fauna in its upper part.

dalyite A mineral: $K_2ZrSi_6O_{15}$.

dam [eng] n. (a) An artificial barrier or wall constructed across a watercourse or valley for one or more of the following purposes: creating a pond or lake for storage of water; diverting water from a watercourse into a conduit or channel; creating a hydraulic head that can be used to generate power; improving river navigability; controlling floods; or retention of debris. It may be constructed of wood, earth materials, rocks, or solid masonry. Archaic syn: *barrage*. (b) A body of standing water confined or held by a dam.—v. To provide a dam; to obstruct or confine a flow of water.

dam [geomorph] An obstruction formed by a natural agent (such as a glacier, a lava flow, or the work of beavers) across a stream so as to produce a lake or pond; e.g. an *ice dam*.

damascened Said of the interwoven texture, observed in some volcanic glasses, that resembles the markings on Damascus sword blades.

dambo A term used in central Africa (esp. Zambia) for a small, ill-defined flood plain or drainage channel that is extremely flat with broad, grassy clearings, swampy during the wet season but dry for the greater part of the year. Etymol: Bantu.

dam gradation *contragradation*.

damkjernite A hypabyssal rock composed of phenocrysts of biotite and titanaugite in a fine-grained groundmass of pyroxene, biotite, perovskite, and magnetite, with interstitial nepheline, microcline, and calcite. The name, given by Brögger in 1921, is for the locality Damkjern (or Damtjern), Fen complex, Norway. Also spelled: damtjernite.

damourite A variety of muscovite, which gives off water more readily and has less elastic folia and a more pearly or silky luster. It is unctuous or talclike to the touch. Syn: *talcite*.

damp A general term for coal-mine gases.

damped-wave structure A series of ring-shaped uplifts and depressions of rapidly diminishing amplitude, surrounding the central uplift of a *cryptoexplosion structure* (Dietz, 1959, p. 496-497).

damping The loss of amplitude of an oscillation, owing to absorption. See also: *critical damping; damping factor*.

damping capacity The ability of a material to dissipate strain within itself.

damping factor The ratio of the observed damping to that required for *critical damping*.

danaite A variety of arsenopyrite containing cobalt.

danalite A mineral: $(Fe,Mn,Zn)_4Be_3(SiO_4)_3S$. It is the iron end-member isomorphous with helvite and genthelvite.

danburite An orange-yellow, yellowish-brown, grayish, or colorless transparent to translucent orthorhombic mineral: $CaB_2(SiO_4)_2$. It resembles topaz in crystal habit, physical properties, and appearance, and is used as a gemstone.

dancalite An extrusive rock containing phenocrysts of oligoclase, clinopyroxene and rare amphibole in a trachytic groundmass composed of plagioclase laths with interstitial analcime; an analcime *trachyandesite* or feldspar-bearing analcime *tephrite.* The name, given by De Angelis in 1925, is for Dancala, Ethiopia. Not recommended usage.

Danian European stage: lowermost Paleocene (above Maestrichtian of Cretaceous, below Montian). It is regarded by some authors as uppermost Cretaceous.

dannemorite A yellowish-brown to greenish-gray monoclinic mineral of the amphibole group: $(Fe,Mn,Mg)_7Si_8O_{22}(OH)_2$. It is a manganiferous cummingtonite.

dans A broad, shallow valley in South Africa. Etymol: Afrikaans.

d'ansite A mineral: $Na_{21}Mg(SO_4)_{10}Cl_3$.

daomanite An orthorhombic mineral: $(Cu,Pt)_2AsS_2$.

daphnite A mineral of the chlorite group: $(Mg,Fe)_3(Fe,Al)_3(Si,Al)_4O_{10}(OH)_8$. Syn: *chamosite.*

darapiosite A mineral of the osumilite group: $KNa_2LiMnZnZrSi_{12}O_{30}$.

darapskite A mineral: $Na_3(NO_3)(SO_4)\cdot H_2O$.

darcy A standard unit of permeability, equivalent to the passage of one cubic centimeter of fluid of one centipoise viscosity flowing in one second under a pressure differential of one atmosphere through a porous medium having an area of cross-section of one square centimeter and a length of one centimeter. Cf: *millidarcy.*

Darcy's law A derived formula for the flow of fluids on the assumption that the flow is laminar and that inertia can be neglected. The numerical formulation of this law is used generally in studies of gas, oil, and water production from underground formations. For example, in gas flow, the velocity of the flow is proportional to the pressure gradient multiplied by the ratio of permeability times density, divided by the viscosity of the gas.

Darcy-Weisbach formula A formula for determining the loss of head in flowing water: loss of head is equal to a coefficient that varies with the surface roughness of the conduit and the Reynolds number, times the length of the conduit, divided by its diameter, times the velocity head of the flowing fluid. In the case of a noncircular conduit or a circular conduit not flowing full, four times the hydraulic radius is substituted for the diameter (ASCE, 1962).

dark-colored Said of a rock-forming mineral having a dark color as viewed megascopically, but being transparent in thin section; also, said of the rock that such minerals form. By convention, dark-colored aphanites include those that are dark gray, dark green, black, and brownish black. Cf: *light-colored; mafic.* Syn: *melanocratic.*

dark field An image obtained in the transmission electron microscope by deliberately excluding all beams except the particular diffracted beam of interest. This is accomplished by placing an aperture in the back focal plane of the objective lens, permitting only a diffracted beam to form the image. Cf: *bright field.*

dark halo crater A small lunar crater surrounded by material with a lower albedo than that of the adjacent terrain.

dark mineral Any one of a group of rock-forming minerals that are dark-colored in thin section, e.g. biotite, hornblende, augite.

dark red silver ore *pyrargyrite.*

dark ruby silver *pyrargyrite.*

Darwin glass A highly siliceous vesicular frothy glass, found in the Mt. Darwin range in western Tasmania (near Queenstown) in the form of blobs, drops, and twisted shreds. It is probably an impactite, although believed by some to be a tektite or a glassy meteorite. Syn: *queenstownite.*

Darwinism The theory that evolution resulted from variation and the survival of favored individuals through *natural selection.* Named after the English naturalist Charles Darwin (1809-1882).

dashkesanite A monoclinic mineral of the amphibole group: $(Na,K)Ca_2(Fe,Mg)_5(Si,Al)_8O_{22}Cl_2$. It contains a high content (7.2%) of chlorine.

dascycladacean Pertaining to a type of green algae of the family Dascycladaceae, whose filaments are whorled about a central axis and often encrusted with calcium carbonate.

data-collection system On Landsat, the system that acquires information from seismometers, flood gauges, and other measuring devices. These data are relayed to an Earth receiving station.

date v. To assign a specific or approximate position on the geologic time scale to a past geologic event.

date line A longitudinal line on the Earth's surface that closely follows the 180° meridian, and is taken as the line along which the calendar day begins. Syn: *international date line.*

dating Age determination of naturally occurring substances or relicts by any of a variety of methods based on the amount of change, happening at a constant measurable rate, in a component. The changes may be chemical, or induced or spontaneous nuclear, and may take place over a period of time.

datolite A greenish monoclinic mineral: $CaBSiO_4(OH)$. It commonly occurs in cracks and cavities in diabase or basalt; it is sometimes used as a minor gem. Syn: *humboldtite; dystome spar.*

datum [geodesy] The astronomic latitude and longitude of an initial point, the astronomic azimuth of a line from this point, the *deflection of the vertical,* the *geoidal separation* (frequently these quantities are assumed to be zero for lack of more complete information), and the two constants necessary to define the reference ellipsoid on which horizontal control surveys are to be computed. See also: *geodetic coordinates.* Syn: *geodetic datum.*

datum [geol] (a) The top or bottom of a bed of rock, or any other surface, on which structure contours are drawn, e.g. a *datum horizon.* (b) *sea-level datum.*—Pl: *datums.*

datum [photo] A direction, level, or position from which angles, heights, depths, speeds, or distances are conventionally measured; e.g. in photographic mapping, the assumed horizontal plane on which the map is constructed.

datum [surv] Any numerical or geometric quantity or value that serves as a base or reference for other quantities or values; any fixed or assumed position or element (such as a point, line, or surface) in relation to which others are determined, such as a level surface to which depths or heights are referred in leveling. Pl: *datums*; the plural "data" is used for a group of statistical or inclusive references, such as "geographic data" for a list of latitudes and longitudes. See also: *datum plane; mean sea level.*

datum horizon A geologic horizon used as a reference plane for the position of rock strata or for the comparative measurement of the thickness of strata; the *key horizon* or bed on which elevations are taken or to which all elevations are finally referred in making a structure-contour map. Syn: *datum; structural datum; contour horizon.*

datum level [stratig] The base or top of a range of fossils that can be correlated in sections over a wide area. "The first evolutionary appearance of a fossil species immediately above a known ancestor may be accorded first order of reliability in defining a datum level" (Bolli, 1969, p. 200). Cf: *chronohorizon.* Syn: *datum plane [stratig].*

datum level [surv] Any level surface, such as mean sea level, used as a reference from which elevations are reckoned; a *datum plane.*

datum line *reference line.*

datum plane [seis] An arbitrary reference surface, used to minimize or eliminate local topographic and near-surface effects. Seismic times and velocity determinations are adjusted to the datum plane as if source and seismometers had been located on this plane and as if no low-velocity layer existed.

datum plane [stratig] *datum level [stratig].*

datum plane [surv] A permanently established horizontal plane, surface, or level to which soundings, ground elevations, water-surface elevations, and tidal data are referred; e.g. *mean sea level* is a common datum plane used in topographic mapping. See also: *tidal datum; chart datum; sounding datum.* Syn: *datum level [surv]; reference level; reference plane.*

datum point An assumed or fixed point used as a reference from which calculations or measurements may be taken.

daubrée A unit of intensity of wear of a sedimentary particle, equivalent to the removal of 0.1 gram from a 100-gram sphere of quartz (Wentworth, 1931, p. 25). Named in honor of Gabriel Auguste Daubrée (1814-1896), French mining engineer and mineralogist, who showed experimentally (Daubrée, 1879) that quartz grains lost only one part in 10,000 per kilometer of travel.

daubreeite A yellowish mineral: $BiO(OH,Cl)$. It is isomorphous with bismoclite. Also spelled: *daubréeite; daubreite.*

daubreelite A black meteorite mineral: $FeCr_2S_4$. Also spelled: *daubréelite.*

daughter A *nuclide* formed by disintegration of a radioactive precursor (*parent*). See also: *radioactive series; end product.*

daunialite A sedimentary rock consisting of a siliceous montmorillonitic clay, as distinct from bentonite of volcanic origin. It contains 25% organic silica (opal, chalcedony, quartz) and small amounts of sericite, chlorite, and kaolinite.

Dauphiné twin law A twin law in quartz in which two right-handed or two left-handed individuals form an interpenetration twin by a 180° rotation about the *c* crystal axis. The result is termed *electrical twinning.* Cf: *Brazil twin law.*

davainite An ultramafic rock (*hornblendite*) containing brown hornblende secondary after pyroxene, some orthopyroxene, and accessory plagioclase. According to Johannsen (1939, p. 248), it may be metamorphic. Not recommended usage.

davidite A dark-brown to brownish-black, uraniferous, iron-titanate mineral: $A_6B_{15}(O,OH)_{36}$, where A = Fe^{+2}, rare earths, U, Ca, Zr, & Th, and B = Ti, Fe^{+3}, V, & Cr. Ideal end-member is $FeTi_3O_7$. Davidite is a primary mineral in high-temperature hydrothermal lodes; it occurs in pegmatites and basic igneous rocks, and in all stages of intergrowth and exsolution with ilmenite and hematite. Syn: *ferutite; uferite.*

davidsonite A greenish or greenish-yellow variety of beryl.

Davisian Pertaining to the "American" school of geomorphology based on the teachings and writings of William Morris Davis (1850-1934), Harvard geologist and geographer; esp. said of the concepts of peneplanation and the cycle of erosion, and of the genetic method of landform description whereby differences are largely explainable in terms of geologic structure, geomorphic process, and stage of development.

davisonite A white mineral: $Ca_3Al(PO_4)_2(OH)_3 \cdot H_2O$. Syn: *dennisonite.*

davyne A chlorine-bearing feldspathoid mineral of the cancrinite group: $(Na,Ca,K)_8(Al_6Si_6O_{24})(Cl,SO_4,CO_3)_{2\text{-}3}$.

dawn stone *eolith.*

dawsonite A white bladed mineral: $NaAl(CO_3)(OH)_2$.

dayingite A cubic mineral: $CuCoPtS_4$.

daylighting In engineering geology, the exposure of a planar feature, such as bedding or a fault, by an open cut whose angle is steeper than that of the exposed feature. Such exposure may increase the likelihood of landsliding by removal of buttressing strata, but it may also reduce sliding tendencies by promoting drainage.

D-coal Microscopic coal particles that are predominantly durain, as found in miners' lungs. Cf: *F-coal; V-coal.*

dead (a) In economic geology, said of an economically valueless area, in contrast to a *quick* area or ore; barren ground. (b) In mining, said of an area of subsidence that is thought to be completely settled and will not move again; dead ground.

dead-burned dolomite A refractory product, CaO.MgO, produced by *calcination* of dolomite or dolomitic limestone.

dead cave A cave in which there is no longer any moisture or any growth of speleothems associated with the presence of moisture. See also: *live cave.* Syn: *dry cave.*

dead chert *chalky chert.*

dead cliff *abandoned cliff.*

dead fault A fault along which movmement has ceased. Cf: *active fault.*

dead glacier A glacier that is without an accumulation area or is no longer receiving material from one. It may continue to spread or creep downhill due to its bulk and topographic setting. Ant: *active glacier.* Cf: *dead ice.* Syn: *stagnant glacier.*

dead ground Rock in a mine that must be removed in order to get at productive ground.

dead ice (a) Ice that is not flowing forward and is not receiving material from an accumulation area; the ice of a *dead glacier.* (b) Detached blocks of ice left behind by a retreating glacier, usually buried in moraine and melting very slowly without the production of large quantities of water.—See also: *fossil ice.* Syn: *stagnant ice.*

dead lake An *extinct lake;* esp. one that has become filled with vegetation. See also: *senescent lake.*

dead line The level above which a batholith is metalliferous and below which it is economically barren. It is exposed during the epibatholithic stage of erosion (Emmons, 1933).

dead sea A body of water devoid of normal aquatic organisms, from which evaporites have been or are being precipitated. Type locality: Dead Sea in the Near East.

dead trace A trace on a seismic record that shows no appreciable deflection because of an instrument failure.

dead valley *dry valley.*

dead water In a stream, water that is or appears to be standing or still.

death assemblage *thanatocoenosis.*

debacle (a) A *breakup* on a river, esp. on the great rivers of the U.S.S.R. and of North America. (b) The rush of water, broken ice, and debris in a stream immediately following a breakup. Syn: *ice run.* (c) Any sudden, violent, destructive flood, deluge, or rush of water that breaks down opposing barriers and sweeps before it debris of all kinds.—Etymol: French *débâcle.*

debitumenization *devolatilization.*

debouchment (a) The issuing forth of a stream, as the *debouchment* of a tributary into the main stream. (b) The mouth of a river or channel. Syn: *debouchure.*—Etymol: French *débouchement.*

debouchure (a) *debouchment.* (b) The place where an underground stream reaches the surface; the opening from which a spring issues. See also: *resurgence.* (c) The point in a cave where a tubular passage connects with a larger passage or chamber.

debris [astron] Interplanetary material, ranging in size from particles less than one micron in diameter to chunks many kilometers across, including asteroids, comets, meteors, meteorites, and cosmic dust.

debris [geol] Any surficial accumulation of loose material detached from rock masses by chemical and mechanical means, as by decay and disintegration. It consists of rock fragments, soil material, and sometimes organic matter. The term is often used synonymously with *detritus,* although "debris" has a broader connotation. Etymol: French *débris*. Pl: *debris*. Syn: *rock waste.*

debris [glaciol] The rocks, earth, and other material lying on the surface, or incorporated in the body, of a glacier, or pushed ahead of the glacier front. Syn: *glacial debris.*

debris apron *bajada.*

debris avalanche The very rapid and usually sudden sliding and flowage of incoherent, unsorted mixtures of soil and weathered bedrock.

debris cone [geomorph] (a) *alluvial cone.* (b) A conical mound of tightly packed, fine-grained debris piled at the angle of repose atop certain boulders moved by a landslide.

debris cone [glaciol] A cone or mound of ice or snow on a glacier, covered with a veneer of *debris* thick enough to protect the underlying material from the ablation that has lowered the surrounding surface. Cf: *dirt cone; sand cone [glaciol].* Syn: *glacier cone.*

debris dam A mass of coarse alluvium deposited at the mouth of a tributary stream, commonly during a flash flood, and forming an obstruction in the main valley.

debris fall The relatively free collapse of predominantly unconsolidated or weathered mineral and rock material from a steep slope or cliff; it is esp. common along the undercut banks of streams. Cf: *soilfall.*

debris flood A disastrous flood, intermediate between the turbid flood of a mountain stream and a true mudflow, of the type that has occurred in southern California (Strahler, 1963, p. 465-466).

debris flow A moving mass of rock fragments, soil, and mud, more than half of the particles being larger than sand size. Slow debris flows may move less that 1 m per year; rapid ones reach 160 km per hour, as in the 1977 Huascaran flow in the Peruvian Andes. Cf: *mudflow [mass move]; sturzstrom; bentonite debris flow; rockfall avalanche.*

debris ice (a) Sea ice containing soil, mud, stones, shells, and other material. (b) *brash ice.*

debris island A *sorted circle* having a diameter of about 1 m and consisting of an isolated patch of fine-textured, compact material surrounded by frost-shattered boulders; term introduced by Washburn (1956, p.827). Syn: *earth island; rubble island.*

debris line A *swash mark* composed of debris washed up on a beach by storm waves. Cf: *trash line.*

debris plain A plain covered with rock waste.

debris slide A slide involving a slow to rapid downslope movement of comparatively dry and predominantly unconsolidated and incoherent earth, soil, and rock debris in which the mass does not show backward rotation (as in a slump) but slides or rolls forward, forming an irregular hummocky deposit resembling a moraine (Sharpe, 1938, p. 74). It is often called an "earth slide", but this is incorrect because the moving mass of a debris slide is greatly deformed or consists of many small units.

debris slope A *constant slope* with debris on it from the free face above. Cf: *talus slope.*

debris stream *debris flow.*

Debye-Scherrer method A method for recording the X-ray diffraction pattern of a crystalline material on film by using a specimen in the form of a powder. Syn: *powder method.*

decalcified Said of a soil that has been *leached* of its calcium carbonate.

decapod (a) Any *eumalacostracan* belonging to the order Decapoda, characterized by the presence of five pairs of uniramous limbs behind the maxillipeds. Range, Permian to present. (b) An early name for a *coleoid,* now discontinued.

decay [radioactivity] *radioactive decay.*

decay [waves] The attenuation or loss of energy from wind-generated ocean waves after they leave the generating area and pass into a region of lighter winds; it is accompanied by a gradual increase in wavelength and a gradual decrease in wave height.

decay [weath] The general weathering or wasting away of rock; specif. *chemical weathering.*

decay constant [elect] The time τ for an exponentially decaying quantity to decrease to 1/e of its initial value (e is the base of the Napierian logarithm).

decay constant [radioactivity] A constant, characteristic of a nuclear species, which expresses the probability that an atom of the species will decay in a given time interval. For a large number of atoms of a species, the decay constant is the ratio between the number of decaying atoms per unit of time and the existing number of atoms. Symbol: λ. See also: *branching ratio; mean life.* Syn: *disintegration constant; radioactive constant.*

Decca A brand name for an electronic positioning system for ships at sea, using a technique of comparison of low-frequency, continuous wave phases from two synchronized stations.

Deccan basalt A fine-grained nonporphyritic tholeiitic lava covering an area of about 200,000 square miles in the Deccan region of southeast India and consisting essentially of labradorite, clinopyroxene, and iron oxides. Olivine is generally absent, or is present in minor amount, usually near the bottom of flows. The rock corresponds to, among others, the plateau basalts of the Pacific Northwest of the U.S.A. and the Thulean province of western Scotland, northeast Ireland, and Iceland.

decementation The dissolving-out or leaching of the cement of a sedimentary rock, as in a sandstone whose void-filling fluids and solid grains do not form a closed system, thereby permitting fluids to move (or ions to diffuse) in and out (Pettijohn, 1957, p. 659); e.g. the removal of carbonates from a calcareous sandstone.

deciduous (a) Said of plants that shed their leaves annually. (b) Said of plant parts that are shed regularly within a year of their production.

decimate To resample digitized data using a longer sampling interval than formerly used.

decke The German equivalent of *nappe,* sometimes used in the English-language literature.

deckenkarren (a) Rounded *solution grooves* formed under a cover of vegetation or soil. See also: *karren.* (b) A group of cave *pendants.*—Etymol: German, "covered tracks", "ceiling tracks".

declination The horizontal angle in any given location between true north and magnetic north; it is one of the *magnetic elements*. Syn: *magnetic variation.*

declinator *declinometer.*

decline curve A graph of the decline in production of an oil or gas well or group of wells. Production rate (ordinate) is plotted against time (abscissa). It is used to predict *ultimate recovery.*

declined Said of a graptoloid rhabdosome with stipes hanging below the sicula and enclosing an angle less than 180 degrees between their ventral sides. Cf: *deflexed; reclined; reflexed.*

declining development The production of a landscape where the rate of downward erosion is more rapid than the rate of uplift or where valley widening exceeds valley deepening, characterized by a decrease of the relative relief and the formation of concave slopes. Cf: *accelerated development; uniform development.* Syn: *waning development; descending development.*

declinometer An instrument that measures magnetic declination. Syn: *declinator.*

declivity (a) A slope that descends from a point of reference; esp. a steep or overhanging slope, as of a cliff. Ant: *acclivity.* (b) A gradient of a surface; a deviation downward from the horizontal; an inclination.

décollement Detachment structure of strata owing to deformation, resulting in independent styles of deformation in the rocks above and below. It is associated with folding and with overthrusting. Etymol: French, "unsticking, detachment". Cf: *tectonic denudation; disharmonic folding.* See also: *bedding-plane slip.* Syn: *detachment.* Obsolete syn: *strip thrust.*

décollement fault *sole fault.*

décollement fold A fold in which the strata are independent of the basement, owing to décollement. Syn: *superficial fold; Jura-type fold.*

decomposers Organisms, usually microscopic, that break down organic matter and thus aid in recycling nutrients.

decomposition *chemical weathering.*

deconvolution A process designed to restore a wave shape to the form it is assumed to have had before it underwent a filtering action or *convolution.* It is a data-processing technique applied to seismic-reflection and other data for the purpose of improving the visibility and resolution of reflected events.

decorative stone A stone used for architectural decoration, as in mantels, columns, and store fronts.

decrement *ground-water discharge.*

decrepitation The breaking up of a mineral, usually violently and noisily, when it is heated.

decurrent Said of parts that extend downward below the point of insertion, as in decurrent leaves adnate to a stem.

decussate Said of plants with appendages opposite one another on an axis, alternating appendages in sequence being inserted at right angles (Swartz, 1971, p. 142). Cf: *distichous.*

decussate texture A microtexture in metamorphosed rocks, in which axes of contiguous crystals lie in diverse, criss-cross directions that are not random but rather are part of a definite mechanical expedient for minimizing internal stress. It is most noticeable in rocks composed largely of minerals with a flaky or columnar habit (Harker, 1939).

dedolomitization A process resulting from metamorphism, wherein part or all of the magnesium in a dolomite or dolomitic limestone is used for the formation of magnesium oxides, hydroxides, and silicates (e.g. brucite, forsterite) and resulting in an enrichment in calcite (Teall, 1903). The term was originally used by Morlot (1847) for the replacement of dolomite by calcite during diagenesis or chemical weathering.

deduction Reasoning from the general to the particular; inferring consequences from evidence; deriving applications from general principles. "It is a mental process not always given its proper priority in considering geological questions. An explanation or hypothesis in accord with deduction must be preferred to any other" (Challinor, 1978, p. 79). Ant: *induction.*

deep n. A clearly discernible depression of the ocean floor. Syn: *abyss.*

deep coal Coal that is far enough below the surface to require underground mining. Cf: *crop coal.*

deep earthquake *deep-focus earthquake.*

deeper-pool test A well located within the known limits of an oil or gas pool and drilled with the object of searching for new producing zones below the producing zone of the pool (Lahee, 1962, p. 134). Cf: *shallower-pool test.*

deep-focus earthquake An earthquake whose focus is at a depth of 300-700 km. Cf: *shallow-focus earthquake; intermediate-focus earthquake.* Syn: *deep earthquake.*

deep fore-reef *reef wall.*

deep lead A *lead [eco geol]* or alluvial placer that is buried under soil or rock. Pron: deep leed.

deep percolation Precipitation moving downward, below the plant-root zone, toward storage in subsurface strata. Cf: *shallow percolation.*

deep phreatic Said of cave formation at considerable depth below the top of the water-saturated zone. See also: *shallow phreatic.*

deep scattering layer A stratified area of marine organisms in the open ocean that scatter sound waves from an echo sounder. Cf: *shallow scattering layer; surface scattering layer.* Syn: *false bottom; phantom bottom.*

deep-sea channel A trough-shaped, low-relief valley on the deep-sea floor beyond the continental rise. It has few tributaries, and may be either parallel or at an angle to the continental margins. Syn: *mid-ocean canyon.*

deep-sea deposit Sediment found on the deep-sea floor. See: *contourite; turbidite; eupelagic deposit; hemipelagic deposit; pelagic deposit.*

deep-sea fan *submarine fan.*

deep-seated Said of geologic features and processes that originate or are situated at depths of one kilometer or more below the Earth's surface; *plutonic.*

deep soil (a) Generally, a soil deeper than 40 inches. (b) A soil with a deep black surface layer.

deep tow A submersible designed to be towed in deep water a short distance above the sea floor, which takes continuous oceanographic, geophysical and bathymetric measurements.

deep water A dense *water mass* that is formed by cooling, then sinking and spreading at great depth. Cf: *intermediate water; bottom water; surface water [oceanog].*

deep-water wave A wave on the surface of a body of water, the wavelength of which is less than twice the depth of the water, and for which the water depth is not an influence on the velocity or on the shape of the orbital. Cf: *shallow-water wave; transitional-water wave.* Syn: *short wave; surface wave.*

deep well A water well, generally drilled, extending to a depth greater than that typical of shallow wells in the vicinity. The term may be applied to a well 20 m deep in an area where shallow wells average 7 or 8 m deep, or to a much deeper well in an area where the shallowest aquifer supplies wells 100 m deep or more. Cf: *shallow well.*

deep-well disposal Disposal of liquid waste by injection into wells, usually constructed especially for the purpose, that penetrate deep, porous and permeable formations containing mineralized ground water and confined vertically by relatively impermeable beds. The method is used for disposal of saline water brought to the surface in oil wells, and for disposal of a variety of liquid wastes from industrial processes. Syn: *deep-well injection.*

deerite A monoclinic mineral: $(Fe^{+2},Fe^{+3},Mn)_{12}Si_8(O,OH)_{32}$.

Deerparkian North American stage: Lower Devonian (above Helderbergian, below Onesquethawan).

defeated stream A stream that, owing to uplift or other cause, is unable to degrade as fast as the land rises and thereby fails to maintain its original course, becomes ponded and diverted into a new course, and resumes as a consequent stream.

defect lattice A crystal lattice in which the expected systemic repetition is interrupted by an omission, an inclusion of an extra item, or the substitution of an unexpected item. Cf: *Schottky defect.*

defect-lattice solid solution *omission solid solution.*

deferred junction The place on a flood plain where the main stream is joined by a tributary whose course is prolonged downstream for a considerable distance by a barrier along the main stream; esp. the junction of a *yazoo stream* with the main stream, as along the convex side of a major meander. Syn: *yazoo; delayed junction; deferred tributary junction.*

deferred tributary *yazoo stream.*

defile A long, narrow pass or passage through hills or mountains, often forming the approach to a larger pass; esp. a passage enclosed between high, precipitous walls, as a gorge.

definition The degree of clarity and sharpness of an image in a photograph; also, the ability of a lens to record fine detail.

deflation The sorting out, lifting, and removal of loose dry fine-grained particles (clay and silt sizes) by the turbulent eddy action of the wind, as along a sand-dune coast or in a desert; a form of *wind erosion.*

deflation armor A *desert armor* whose surface layer of coarse particles is concentrated chiefly by deflation.

deflation basin A topographic basin excavated and maintained by wind erosion which removes unconsolidated material and commonly leaves a rim of resistant rock surrounding the depression. See also: *blowout [geomorph].* Syn: *wind-scoured basin.*

deflation hollow *blowout [geomorph].*

deflation lake A lake occupying a basin formed mainly by wind erosion, esp. in an arid or semiarid region; it is usually very shallow and may contain water only during certain seasons. See also: *dune lake.*

deflation residue *desert pavement.*

deflation ripple *granule ripple.*

deflection [drill] *deviation [drill].*

deflection [geodesy] *deflection of the vertical.*

deflection [mtns] A sharp change in the trend of a mountain chain. The term was introduced by Bucher (1933) as a translation of Staub's term *Beugung.* It differs from an *orocline* by not necessarily being a strain imposed on the completed orogen. See also: *capped deflection; fractured deflection.* Cf: *linkage [mtns].*

deflection [streams] A relatively spontaneous diversion of a stream, as by warping, alluviation, glaciation, lateral corrasion, volcanic action, or shoreline changes.

deflection angle [photo] A vertical angle, measured in the vertical plane containing the flight line, by which the datum of any model in a stereotriangulated strip departs from the datum of the preceding model.

deflection angle [surv] A horizontal angle measured from the forward prolongation of the preceding line to the following line; the angle between one survey line and the extension of another survey line that meets it. A deflection angle to the right is positive; one to the left is negative.

deflection basin A basin hollowed out by the erosive action of ice in front of a barrier obstructing the path of a glacier (Geikie, 1898, p. 297).

deflection of the vertical The angle at a given point on the Earth between the vertical, defined by gravity, and the direction of the normal to the reference ellipsoid through that point. It is sometimes referred to as *deviation of the vertical* or deflection of the plumb line. Cf: *datum; astronomic measurement.* Syn: *station error; deflection [geodesy].*

deflection pool A pool occupying a depression scooped out by a stream in its obstructing bed at a curve in its course (Miller, 1883, p. 275).

deflexed Said of a graptoloid rhabdosome with initial part of stipes hanging below the sicula and enclosing an angle less than 180 degrees between their ventral sides and distal parts of stipes tending to the horizontal. Cf: *declined; reflexed; reclined.*

defluent A stream that flows down, as from a lake or glacier.

deformation (a) A general term for the process of folding, faulting, shearing, compression, or extension of the rocks as a result of various Earth forces. (b) *strain.*

deformation band A planar region within a deformed mineral grain that has undergone a different kind of deformation from that in adjacent parts of the grain. The deformation may differ only in the amount of slip that has occurred on a particular slip plane, in which case the lattice is simply bent across the deformation-band boundary.

deformation ellipsoid *strain ellipsoid.*

deformation fabric A rock fabric that has resulted from deformation; the fabric of a *secondary tectonite.* Typical fabrics are composed of lineations, schistosities, cleavages, fold axes, and crystallographic preferred orientations. Cf: *depositional fabric.* Syn: *tectonic fabric.*

deformation lamella One of a series of narrow planar features developed by deformation within single grains; they have a slightly different refractive index than the host grain. Pl: *lamellae.*

deformation path The path taken by a deforming object in going from one strain state to another. The path is generally plotted on a Flinn or deformation diagram, in which one coordinate axis is the ratio of the major axis of the strain ellipsoid to the intermediate axis and the other coordinate axis is the ratio of the intermediate axis of the strain ellipsoid to the minor axis.

deformation plan The array of displacement vectors that connect points in the deformed state to those same points in the undeformed state. Syn: *movement picture.*

deformation plane In structural petrology, a term used in a kinematic sense to describe a deformation that has monoclinic symmetry. It is the unique symmetry plane and is parallel to the *a* axis and normal to the *b* axis. In a progressive simple shear the deformation plane is parallel to the direction of shear and normal to the planes of shear. It is also called the *a-c* plane.

deformation twin A crystal twin that is produced by gliding, i.e. deformation within a preexisting crystal. Syn: *glide twin; mechanical twin.*

deformation twinning In a crystal, twinning that is produced by gliding. Syn: *secondary twinning.*

deformed cross-bedding Cross-bedding in which the foreset beds are "overturned or buckled in the downcurrent direction usually prior to deposition of the overlying bed" (Pettijohn & Potter, 1964, p.299). The foreset dip angle may also be altered by subsequent tectonic folding.

deformed ice A term by the U.S. Naval Oceanographic Office (1968, p. B32) for *pressure ice.*

DeGeer moraine One of a series of recessional moraines produced individually during a single year as determined by varve chronology in Sweden.

degenerative recrystallization *degradation recrystallization.*

Deglacial n. A term used by Antevs (1953) for a time unit that covers in North America the time from the greatest extension of Wisconsin glaciation to the beginning of recession from the Cochrane moraines.

deglaciation The uncovering of a land area from beneath a glacier or ice sheet by the withdrawal of ice due to shrinkage by melting. As used in Great Britain, the term is restricted to a process that occurred in the past, in contrast to *deglacierization.* Also, the result of deglaciation.

deglacierization A term used in Great Britain for the gradual withdrawal, going on at the present time, of a glacier or an ice sheet from a land area. Cf: *deglaciation.*

degradation [geomorph] (a) The wearing down or away, and the general lowering or reduction, of the Earth's surface by the natural processes of weathering and erosion; e.g. the deepening by a stream of its channel. The term sometimes includes the process of transportation; and sometimes it is used synonymously with *denudation,* or used to signify the results of denudation. See also: *gradation.* (b) Less broadly, the vertical erosion or *downcutting* performed by a stream in order to establish or maintain uniformity of grade or slope. Cf: *aggradation.*

degradation [permafrost] The shrinkage or disappearance of permafrost due to natural or artificial causes. Ant: *aggradation [permafrost].* Cf: *depergelation.*

degradation [soil] A decrease in exchangeable bases in a soil, and destruction of layer-silicate clay, as a result of leaching (Jacks et al., 1960, p. 162).

degradation [stream] The lowering of a stream bed, due to such factors as increased scouring. Cf: *accretion [stream].*

degradation recrystallization Recrystallization resulting in a relative decrease in the size of crystals. Ant: *aggradation recrystallization.* Syn: *grain diminution; degenerative recrystallization.*

degradation vacuity The space-time value of the degradationally removed part of a transgressive-regressive depositional succession; e.g. the part of a *lacuna [stratig]* resulting from degradation of formerly existing rocks at an unconformity. The term was used by Wheeler (1964, p.602) to replace *erosional vacuity.* Cf: *hiatus.* Syn: *vacuity.*

Degraded Chernozem A great soil group in the 1938 classification system, a group of zonal soils having a dark brown to black surface horizon and an underlying gray leached horizon that in turn rests on a brown horizon. It is developed in an area in which forest vegetation has encroached upon grassland, and is intermediate between chernozemic and podzolic types (USDA, 1938). Most of these soils are now classified as *Boralfs* and *Borolls.*

degraded illite Illite that has had much of its potassium removed from the interlayer position as a result of prolonged leaching. Syn: *stripped illite.*

degrading neomorphism A kind of *neomorphism* in which the crystal size decreases (Folk, 1965, p. 23). Ant: *aggrading neomorphism.*

degrading stream (a) A stream that is actively cutting down its channel or valley and that is capable of transporting more load than it is supplied with. (b) A stream that is downcutting approximately at grade.

degree In hydrologic terminology, a unit for expressing the hardness of water. Cf: *Clark degree.*

degree-day A measure of the difference between the mean daily temperature and an arbitrary temperature such as 18.3°C (65°F), as used by heating engineers. It is normally applied to mean temperatures that are below the standard.

degree of curve A measure of the sharpness of curvature; e.g. the angle at the center of a circle subtended by a chord 100 ft long (as for U.S. railroads) or by an arc 100 ft long (as in highway surveying).

degree of freedom The capability of variation in a chemical system. The number of degrees of freedom in a system may be defined as the number of independent, intensive variables (e.g. temperature, pressure, and concentration in the various phases) necessary to define the system completely, or as the number of variables that may be changed independently without causing a change in phase. Cf: *phase rule.*

degree of meraspid period A developmental stage of a trilobite during the meraspid period, when the number of thoracic segments are increased. The degree is defined by the number of thoracic segments present and ranges from zero to the holaspid number of thoracic segments minus one.

degree of slope The angular measurement (expressed in degrees) of slope from a horizontal plane (Van Riper, 1962, p. 603).

degree of sorting A measure of the spread or range of variation of the particle-size distribution in a sediment. It is defined statistically as the extent to which the particles are dispersed on either side of the average: the wider the spread, the poorer is the sorting. It may be expressed by *sigma phi.*

dehiscent Said of certain fruits and sporangia that split open along definite seams at maturity.

dehrnite A grayish to greenish-white mineral of the apatite group: $(Ca,Na,K)_5(PO_4)_3(OH)$. Cf: *lewistonite.*

dehydration reaction A metamorphic reaction that results in the transfer of H_2O from a mineral into the fluid phase. Cf: *hydration reaction.*

DE interray Left posterior interray in echinoderms situated between D ray and E ray and clockwise of D ray when the echinoderm is viewed from the adoral side; equal to interambulacrum 1 in the Lovenian system.

dejection cone *cone of dejection.*

delafossite A mineral: $CuFeO_2$.

delatorreite *todorokite.*

delatynite A variety of amber rich in carbon, low in succinic acid, and lacking sulfur, found at Delatyn in the Carpathian Mountains of Galicia.

delawarite Pearly orthoclase from Delaware County, Penna.

delayed runoff Water from precipitation that sinks into the ground and discharges later into streams through seeps and springs (Tarr & Von Engeln, 1926, p. 70). As defined above, delayed runoff is a syn. of *ground-water runoff,* but it could also be defined as runoff delayed by any means, such as temporary storage of precipitation in the form of snow and ice.

delay time In seismic refraction work, the additional time required to traverse any raypath over the time that would be required to traverse the horizontal component at highest velocity encountered on the raypath, as it refers to either the source or receiver end of the trajectory. See also: *intercept time.*

deldoradite A light-colored cancrinite *syenite,* named by Johannsen (1939) for Deldorado Creek, Colorado. Not recommended usage.

delessite A mineral of the chlorite group; ferroan variety of clinochlore: $(Mg,Fe^{+2},Fe^{+3},Al)_6(Si,Al)_4O_{10}(OH)_8$. It occurs in cavities and seams in basic igneous rocks.

deleveling An alteration in the level of a part of the Earth's surface: it is positive when the land is depressed and negative when it is elevated, in relation to sea level.

delhayelite A mineral: $(Na,K)_{10}Ca_5Al_6Si_{32}O_{80}(Cl_2,F_2,SO_4)_3 \cdot 18H_2O$.

delineation A step in map *compilation* in which mapworthy features are distinguished and outlined on various possible source materials or are visually selected (as when operating a stereoscopic plotting instrument).

dell (a) A small, secluded wooded valley or natural hollow. The term is often used in a literary sense with no definite meaning. Cf: *dale.* (b) A depression upvalley from the source of a stream (Penck, 1953, p. 421). Etymol: German *Delle,* "dent".

dellaite A mineral: $Ca_6Si_3O_{11}(OH)_2$.

dellenite A term, equivalent to *rhyodacite,* proposed by Broegger in 1896, named for Dellen Lake, Sweden. Syn: *quartz latite.* Obsolescent; rhyodacite is the preferred term.

dells A corruption of *dalles,* as applied along the Wisconsin River, Wisc.

Delmontian North American stage: uppermost Miocene (above Mohnian, below Repettian).

delorenzite A syn. of *tanteuxenite.* It was originally described as: $(Y,U,Fe)(Ti,Sn)_3O_8$.

delphinite A yellowish-green epidote from France.

delrioite A mineral: $CaSrV_2O_6(OH)_2 \cdot 3H_2O$.

delta The low, nearly flat, alluvial tract of land at or near the mouth of a river, commonly forming a triangular or fan-shaped plain of considerable area, crossed by many distributaries of the main river, perhaps extending beyond the general trend of the coast, and resulting from the accumulation of sediment supplied by the river in such quantities that it is not removed by tides, waves, and currents. Most deltas are partly subaerial and partly below water. The term was introduced by Herodotus in the 5th century B.C. for the tract of land, at the mouth of the Nile River, whose outline broadly resembled the Greek capital letter "delta", Δ, with the apex pointing upstream.

delta bar A "bar" formed by a tributary stream that is building a delta into the channel of the main stream.

delta bedding The bedding characteristic of a delta, consisting of comparatively flat *topset beds* and *bottomset beds,* between which are steeper *foreset beds* leading from close to the delta surface to the bottom of surrounding water; the inclined bedding "presumed to originate as foresets of small deltas" (Pettijohn &

Potter, 1964, p.299).

delta cap An alluvial cone or fan built on a *delta plain,* and having an apex that migrates upstream (Russell, 1898b, p. 127).

delta cycle A term used by Barrell (1912, p. 397) for a two-phase tectonic cycle dependent on stream erosion and changing sea level, and involving deposition with a stationary crust followed by vertical movement (normally subsidence) of the bottom. For a geosynclinal fill that is essentially a deltaic accumulation, the cycle is one of increasing coarseness from the base upward, of an increasing volume of clastic material, and of eventual overtake of subsidence by sedimentation.

delta fan A deposit formed by the merging of an alluvial fan with a delta. Syn: *fan delta.*

delta front A narrow zone where deposition in deltas is most active, consisting of a continuous sheet of sand, and occurring at the effective depth of wave erosion (10 m or less). It is the break in slope separating the *prodelta* from the *intradelta,* and it may or may not be steep.

delta-front platform A zone of shallows, up to about 5 km wide, in front of the advancing distributaries of a delta.

delta-front trough A trough-shaped submarine valley formed off a large river delta on the continental shelf and slope. It has straight walls of soft rock, few if any tributaries, and a flat, seaward-sloping floor.

deltageosyncline *exogeosyncline.*

deltaic Pertaining to or characterized by a delta; e.g. "deltaic sedimentation". Also, constituting a delta; e.g. a "deltaic coast".

deltaic coastal plain A coastal plain composed of a series of coalescing deltas; it consists initially of natural levee ridges separated by basins.

deltaic deposit A sedimentary deposit laid down in a delta, characterized by well-developed local cross-bedding and by a mixture of sand, clay, and the remains of brackish-water organisms and of organic matter. Cf: *estuarine deposit.*

deltaic tract An extension of the *plain tract* of a stream, characterized by the formation of a delta and the deposition of fine sediments.

deltaite A mixture of crandallite and hydroxylapatite.

delta kame A flat-topped, steep-sided hill of well-sorted sand and gravel deposited by a meltwater stream flowing into a proglacial or other ice-marginal lake; the proximal margin of the delta was built in contact with glacier ice. Cf: *esker delta.* Syn: *kame delta; ice-contact delta; sand plateau.*

delta lake A lake formed along the margin of or within a delta, as by the building of bars across a shallow embayment or by the enclosure of part of the sea by the growth of deltaic deposits.

delta levee lake A lake on an advancing delta, formed between sandbars or natural levees deposited at the mouths of distributaries. Example: Lake Pontchartrain on the Mississippi River delta.

delta-mooreite *torreyite.*

delta plain The level or nearly level surface composing the landward part of a large delta; strictly, an alluvial plain characterized by repeated channel bifurcation and divergence, multiple distributary channels, and interdistributary flood basins.

delta plateau A raised or abandoned delta plain.

delta ray A secondary electron that was ejected by ionizing radiation.

delta shoreline A prograding shoreline produced by the advance of a delta into a lake or the sea.

delta structure The sedimentary structure produced by the three sets of beds in a delta: bottomset, foreset, and topset beds.

delta T The time difference between the arrivals of the same phase of a seismic event at two locations.

delta terrace A fan-shaped terrace composed of a delta that remained after the disappearance of the stream that produced it (Chamberlin, 1883, p. 304).

delta top *intradelta.*

delta unit *delta value.*

delta value The difference between the *isotope ratio* in a sample and that in a standard, divided by the ratio in the standard, and expressed as parts per thousand. Syn: *delta unit.*

delthyrial chamber The cavity beneath the umbo of the pedicle valve of a brachiopod, bounded by dental plates or by posterior and lateral shell walls if dental plates are absent. It corresponds to the *notothyrial chamber* of the brachial valve.

delthyrial plate A plate within the delthyrial chamber of some spiriferoid brachiopods, extending a variable distance from the apex between dental plates. It is probably homologous with the pedicle collar (TIP, 1965, pt.H, p.143).

delthyrium The median triangular or subtriangular opening beneath the beak of the pedicle valve of a brachiopod, bisecting the ventral cardinal area or pseudointerarea, and commonly serving as a pedicle opening. Pl: *delthyria.* Cf: *notothyrium.*

deltidial plate One of two plates growing medially (inward) from the margins of the delthyrium of a brachiopod and partly or completely closing it.

deltidium The cover of the delthyrium of a brachiopod, formed by conjunct deltidial plates (in contact anteriorly and dorsally of the pedicle) whose line of junction is visible. Pl: *deltidia.* Cf: *pseudodeltidium.*

deltohedron A *tristetrahedron* whose faces are quadrilateral, rather than triangular, as in the *trigonal tristetrahedron.* Syn: *deltoidal dodecahedron; tetragonal tristetrahedron.*

deltoid [lunar] A delta-shaped raised feature on the surface of the Moon, found in association with certain ring structures (Fielder, 1965, p. 170).

deltoid [paleont] *deltoid plate.*

deltoidal cast A term used by Birkenmajer (1958, p.143) for *frondescent cast.*

deltoidal dodecahedron *deltohedron.*

deltoid branch A branch of a stream, enclosing a whole delta (Jackson, 1834, p. 79).

deltoid island A *branch island* formed on a delta (Jackson, 1834, p. 79).

deltoid plate One of a circlet of interradial, subtriangular plates of a blastoid, situated near the summit (oral end) of the theca but aboral to the oral plates, between adjacent ambulacra, and above radial plates. Syn: *deltoid.*

deluge A great inundation or overflowing of the land by water; specif. The Deluge (the Noachian flood).

deluvium *diluvium.*

delvauxite A mineral, perhaps: $Fe_4{}^{+3}(PO_4)_2(OH)_6 \cdot n\,H_2O$.

delve A surface depression or hollow.

demagnetization The reduction of remanent magnetization, often selectively, i.e. only of its softer or less stable components. Techniques employed include *alternating-field demagnetization, thermal demagnetization,* and *chemical demagnetization.*

demantoid A bright-green to yellowish-green transparent gem variety of andradite garnet, characterized by a brilliant luster, a dispersion stronger than that of diamond, and a hardness less than that of other garnets. Syn: *Uralian emerald.*

demarcation line An imaginary line on the surface of a valve of a bivalve mollusk, originating at the beak and marking the locus of points on successive positions of the margin where the transverse-growth component has had maximum effect. It forms a dorsal/ventral profile when the valve is viewed from one end.

demersal *benthic.*

demesmaekerite A bottle-green to clear olive-green mineral: $Pb_2Cu_5(UO_2)_2(SeO_3)_6(OH)_6 \cdot 2H_2O$.

demic Said of a randomly interbreeding population smaller than the species population. The term is usually preceded by a prefix indicating the nature or cause of the separation of such a group; e.g., "topodemic," applied to a population restricted to a particular geographic area.

demiplate An ambulacral plate of an echinoid, touching an adradial suture but not touching a perradial suture.

demipyramid One of the ten elements that support the teeth in Aristotle's lantern of an echinoid. See also: *pyramid [paleont].*

demkha A term used in Algeria for an almond-shaped *dune massif* (Capot-Rey, 1945, p. 395).

demoiselle A term used in the French Alps for an *earth pillar* capped by a large boulder, esp. one formed by weathering of volcanic breccia or of glacial till. Etymol: French, "young lady". Pron: duh-mwa-*zel.*

demoiselle hill A symmetrical beehive-shaped elevation with a grassy surface, bordered by a cauldronlike depression, occurring on the Magdalen Islands in the Gulf of St. Lawrence (Clarke, 1911, p. 145); it may be a knob, knoll, or hill 175 m high.

demorphism An obsolete syn. of *weathering.*

demosponge Any sponge belonging to the class Demospongea and characterized chiefly by the presence of spongin as all or part of its skeleton. "Most demosponges contain siliceous spicules, with or without spongin. A few contain only spongin, and still fewer produce neither" (TIP, 1955, pt.E, p.34).

dendrite A mineral, e.g. a surficial deposit of an oxide of manganese, or an inclusion, that has crystallized in a branching pattern. Adj: *dendritic.* Syn: *dendrolite.*

dendritic Said of a mineral that has crystallized in a branching pattern; pertaining to a *dendrite.* Syn: *arborescent.*

dendritic drainage pattern A drainage pattern in which the streams branch randomly in all directions and at almost any angle, resembling in plan the branching habit of certain trees; it is produced where a consequent stream receives several tributaries which in turn are fed by smaller tributaries. It is indicative of insequent streams flowing across horizontal and homogeneous strata or complex crystalline rocks offering uniform resistance to erosion. Cf: *pinnate drainage pattern.*

dendritic glacier A *trunk glacier* joined by many tributary glaciers to form a pattern that resembles a branching tree.

dendritic ridge A type of *longitudinal ridge* that has a dendritic pattern. Bifurcation of most ridges is in the upcurrent direction (Dzulynski and Walton, 1963; 1965, p. 69).

dendritic tufa *dendroid tufa.*

dendrochore The part of the Earth's surface having a tree-sustaining climate. It constitutes the bulk of the *biochore.*

dendrochronology The study of annual *growth rings* of trees for dating of the recent past. Cf: *dendroclimatology.* Syn: *tree-ring chronologyy.*

dendroclimatology The study of the patterns and relative sizes of annual *growth rings* of trees for paleoclimatologic data of the recent past. Cf: *dendrochronology.*

dendroclone A desma (of a sponge) having a smooth, straight shaft with a tree-like, branching process at each end.

dendrodate A date calculated by *dendrochronology.*

dendrogram A treelike diagram depicting the mutual relationships of a group of items sharing a common set of variables, the variables representing either samples on which multiple measurements have been made or measured attributes for a group of samples; for example, a diagram illustrating the interrelations (based on degrees of similarity) throughout geologic time of fossil organisms connected by a common ancestral form. Cf: *dendrograph.*

dendrograph A treelike, two-dimensional correlation diagram depicting the mutual relationships between and within groups of items sharing a common set of variables. The added dimension of spacing, to separate the variables reflecting dissimilarities between groups to which the variables belong, permits a more accurate interpretation of data. Cf: *dendrogram.*

dendroid adj. (a) Said of certain invertebrates that form many-branched colonies or that have a treelike habit of growth, such as the irregularly branched bushy colonies formed by graptolites. (b) Said of an irregularly branching type of fasciculate corallum. (c) A syn. of *ramose* in stenolaemate bryozoans.—n. A sessile graptolite of the order Dendroidea ranging from Middle Cambrian to Carboniferous, characterized by a typically erect, dendroid rhabdosome having many stipes which are composed of autothecae, bithecae, and stolothecae arranged in regularly alternating triads along each branch.

dendroid colony An archaeocyathid colony in which each individual cup is isolated from others, except at its origin. It contrasts with massive or catenulate colonies (TIP, 1972, pt. E, p. 6).

dendroid tufa Gray tufa occurring as spheroidal, mushroom, or dome-shaped masses with concentric macrostructure and less pronounced internal dendritic structure; e.g. the tufa along the shore of the extinct Lake Lahontan in Nevada, where it constitutes the major part of the dome-like mass. Cf: *thinolitic tufa; lithoid tufa.* Syn: *dendritic tufa.*

dendrolite *dendrite.*

denivellation A variation in the level of a body of water, esp. of a lake; e.g. "wind denivellation" is a rise of water level due to wind drift, as along the windward shore of a lake. Etymol: French *dénivellation*, "difference of level".

denningite A colorless to pale-green tetragonal mineral: $(Mn,Zn)Te_2O_5$.

dennisonite *davisonite.*

dense [geol] (a) Said of a fine-grained, aphanitic igneous rock whose particles generally average less than 0.05 to 0.1 mm in diameter, or whose texture is so fine that the individual particles cannot be recognized by the unaided eye. (b) Said of a rock whose constituent grains are crowded close together. The rock may be fine- or coarse-grained. (c) Said of a rock or mineral possessing a relatively high *specific gravity.*

dense [optics] Said of a substance that is highly refractive.

dense [photo] Said of a photographic negative or a positive transparency in which the silver deposit per unit area is greater than normal owing to overexposure or overdevelopment.

density [oceanog] The mass of water per unit volume, usually stated in grams per cubic centimeter. Density of water ranges from about 1.0 for fresh water to 1.07 for water in the deep ocean. Density may be written as *sigma-t* or σ_T. See also: *in-situ density; potential density.*

density [optics] A measure of the degree of opacity of any translucent medium, defined strictly as the common logarithm of the opacity; e.g. the degree of blackening of an exposed photographic film, plate, or paper after development, or of the direct image. Cf: *contrast.*

density current A gravity-induced flow of one current through, over, or under another, owing to density differences. Factors affecting density differences include temperature, salinity, and concentration of suspended particles. See also: *salinity current; turbidity current; nuée ardente.*

density log The *well log* curve of induced radioactivity showing the bulk density of rocks and their contained fluids. It is a *porosity log* of the wall-contact type, indicating formation density by recording the *back scatter* of gamma rays. Syn: *gamma-gamma log.*

density profile A line of gravity readings taken over a topographic feature with appreciable relief that is not associated with variations in density or structure, the object of which is to determine the best density factor for elevation corrections. The most appropriate density is the one that minimizes the correlation of gravity values with elevation (Sheriff, 1973, p. 48).

density stratification The *stratification* of a lake produced as a result of density differences, the lightest layer occurring near the top and the heaviest layer at the bottom. It is usually brought about by temperature changes, but may also be caused by differences at different depths in the amount of suspended or dissolved material (e.g. where a surface layer of freshwater overlies salt water). See also: *thermal stratification.*

densofacies A term used by Vassoevich (1948) for *metamorphic facies.*

densospore A trilete spore, chiefly Paleozoic, with a pronounced cingulum that has a tendency to be "doubled", i.e. a thicker part toward the center of the spore and a thinner, more external part; e.g. the genus *Densosporites* and similar genera.

dental plate (a) One of variably disposed plates of secondary shell underlying and supporting the hinge teeth of a brachiopod and extending to the floor of the pedicle valve. (b) In gnathostomes, a large crushing element formed by the expansion of a single tooth or the aggregation of several teeth.

dental socket A shallow excavation in the posterior margin of the brachial valve of a brachiopod for the reception of a hinge tooth of the pedicle valve. Syn: *socket.*

dentate Toothed, or having small conical or toothlike projections; e.g. "dentate chela" of a sponge with toothed terminal expansions.

denticle (a) A small spinelike, needlelike, or sawtoothlike structure of compound and platelike conodont elements, similar to a cusp but commonly smaller. (b) One of the small sharp protruding ridges that alternate with complementary dental sockets located along the cardinal margin or the hinge line of both valves of a brachiopod. (c) A primary or secondary toothlet on the sutural edge of the radius of a compartmental plate of a cirripede crustacean, serving to strengthen the articulation of the plates; a small, delicate, spinelike or toothlike projection on the carapace of an ostracode. (d) In vertebrates, a small toothlike scale; also, a small toothlike structure attached to the gill bars.

denticulate (a) Having small teeth, or bearing a series of small spinelike or toothlike projections; e.g. said of a shell covered with such projections. (b) Minutely or finely dentate, as of a leaf.

denticulation The state of being denticulate; also, a series of small toothlike structures, such as the denticles on the anterior and posterior margins of the shell of a cytherid ostracode.

dentition The number, kind, and arrangement of teeth or toothlike structures in invertebrates, and of teeth in vertebrates.

denudation (a) The sum of the processes that result in the wearing away or the progressive lowering of the Earth's surface by various natural agencies, which include weathering, erosion, mass wasting, and transportation; also the combined destructive effects of such processes. The term is wider in its scope than *erosion,* although it is commonly used as a syn. of that term. It is also used as a syn. of *degradation,* although some authorities regard "denudation" as the actual processes and "degradation" as the results produced; Davis (1909, p. 408) distinguished between "denudation" as the active processes operating early in the cycle of landform development, and "degradation" as the more leisurely processes operating later. (b) Strictly, the laying bare, uncovering, or exposure of bedrock or a designated rock formation through the removal of overlying material by erosion. This is the original, and etymologically correct, meaning of the term, which was often applied in a catastrophic sense, e.g. the "Great Denudation" resulting from the universal flood.

denudation chronology The study of the timing and sequence of events leading to the formation and evolution of an existing landscape.

deoxidation sphere *bleach spot.*

departure The projection of a line onto an east-west axis of reference. The departure of a line is the difference of the meridian distances or longitudes of the ends of the line. It is east or positive (sometimes termed the *easting*) for a line whose azimuth or bearing is in the northeast or southeast quadrant; it is west or negative (sometimes termed the *westing*) for a line whose azimuth or bearing is in the southwest or northwest quadrant. Syn: *longitude difference.*

departure curve In well logging, a graph of correction factors to be applied to recorded (apparent) log measurements of particular sonde design to estimate "true" measurements under specific geometric and physical conditions. The term in the plural is usually applied to the collections of charts used in *resistivity log* analysis.

depauperate fauna (a) An *assemblage [paleoecol]* exhibiting substantially reduced *diversity.* Syn: *impoverished fauna.* (b) A *dwarf fauna.*

dependable yield The minimum water supply to a given area that is available on demand and which may decrease on the average of once every *n* number of years (Langbein & Iseri, 1960).

dependent variable A *variable* whose magnitude is plotted as a function of fixed consecutive values of a second or *independent variable.*

depergelation The act or process of thawing permafrost (Bryan, 1946, p.640). Cf: *degradation [permafrost].*

depletion [eco geol] The exhaustion of a natural resource, i.e. ore or oil, by commercial exploitation, measured quantitatively in monetary terms.

depletion [water] Loss of water from surface- or ground-water reservoirs at a rate greater than that of recharge.

depletion allowance A proportion of income derived from mining or oil production that is considered to be a return of capital not subject to income tax. It is a way of recognizing that mining or petroleum production ultimately exhausts the reserve.

depletion curve A *hydrograph* showing the loss of water from ground-water storage, by seepage or flowage into streams, or from any storage or channel. See also: *recession curve.*

depoaxis The axis of maximum sediment deposition during a geologic epoch.

depocenter An area or site of maximum deposition; the thickest part of any specified stratigraphic unit in a depositional *basin.*

deposit n. Earth material of any type, either consolidated or unconsolidated, that has accumulated by some natural process or agent. The term originally applied to material left by water, but it has been broadened to include matter accumulated by wind, ice, volcanoes, and other agents. Cf: *sediment.*—v. To lay down or let drop by a natural process; to become precipitated.

deposit feeder An animal that obtains its food from detrital material and associated microorganisms on the sea floor; e.g., sea cucumber, scaphopod. Syn: *detritovore; sediment feeder.*

deposition (a) The laying, placing, or throwing down of any material; specif. the constructive process of accumulation into beds, veins, or irregular masses of any kind of loose rock material by any natural agent, such as the mechanical settling of sediment from suspension in water, the chemical precipitation of mineral matter by evaporation from solution, or the accumulation of organic material on the death of plants and animals. See also: *sedimentation.* (b) Material that is deposited; a deposit or sediment.

depositional (a) Pertaining to the process of deposition; e.g. a "depositional basin" or a "depositional surface". (b) Formed by the process of deposition; e.g. a "depositional topography".

depositional dip *primary dip.*

depositional fabric Rock fabric or that fabric element or elements that result from deposition during the rock's formation, e.g. sedimentary grains in an unmetamorphosed, current-laid sediment or crystals deposited by crystal settling in a magma chamber; the fabric of a *primary tectonite.* Cf: *deformation fabric.*

depositional fault *growth fault.*

depositional interface The interface between the water and the bottom where sediments are deposited in relation to the energy level at the interface (Plumley et al., 1962, p. 86).

depositional magnetization *depositional remanent magnetization.*

depositional mark An irregularity formed on the bedding plane of a sediment during deposition; e.g. a scour mark or a tool mark. Syn: *depositional marking.*

depositional remanent magnetization Remanent magnetization resulting from mechanical orientation of ferrimagnetic mineral grains along the ambient field during sedimentation. Its inclination is generally less than that of the ambient field. Abbrev: DRM. Syn: *depositional magnetization; detrital remanent magnetization.*

depositional topography Topography formed as a result of sediments being dropped from a moving medium, e.g. coastal bars and barriers, kame terraces, or sand dunes.

depressed Said of the shape of a whorl section in a cephalopod conch whose lateral diameter is greater than its dorsoventral diameter.—Ant: *compressed.*

depressed flute cast A flat or weakly developed flute cast.

depressed moraine An irregular moraine "developed along the ice front in line with the normal morainal ridge, but failing to rise above the adjacent outwash", due more to "the nonaccumulation of marginal material than to an excess of outwash" (Fuller, 1914, p. 33-34); e.g. those on Long Island, New York.

depression [geomorph] Any relatively sunken part of the Earth's surface; esp. a low-lying area surrounded by higher ground and having no natural outlet for surface drainage, as an interior basin or a karstic sinkhole.

depression [meteorol] An atmospheric region of relatively low pressure. It may refer to the closed low-pressure area of cyclone type or to the open V-shaped trough of low pressure. Syn: *low [meteorol].*

depression [surv] (a) The angular distance of an object beneath the horizontal plane that passes through the observer. (b) The angular distance of a celestial object below the horizon.

depression [tect] A structurally low area in the crust, produced by negative movements that sink or downthrust the rocks. The term includes *basins* and *furrows.* Cf: *uplift.*

depression angle In *SLAR,* the angle between the horizontal plane passing through the antenna and the line connecting antenna and target.

depression contour A *closed contour,* inside of which the ground or geologic structure is at a lower elevation than that outside, and distinguished on a map from other contour lines by hachures marked on the downslope or downdip side. See also: *closed depression.*

depression spring A type of gravity spring, with its water flowing onto the land surface from permeable material as a result of the land surface sloping down to the water table (Meinzer, 1923, p. 51).

depression storage Accumulation of water from precipitation in depressions in the land surface (Langbein & Iseri, 1960); accumulation of rainwater or snowmelt in depressions when the soil has reached its infiltration capacity (Chow, 1964, p. 20-5). Cf: *detention.*

depressor muscle crest The elevated denticles, on the interior of a tergum of a cirripede crustacean, for attachment of muscles that depress or draw down.

depth The vertical distance from a specified datum to the bottom of a body of water. Syn: *water column.*

depth contour *isobath [oceanog].*

depth hoar Ice crystals formed below the surface and usually near the bottom of a snowpack by *temperature-gradient metamorphism.* The crystals are usually larger than the grains of the overlying snow mass, and when well developed may assume various complex shapes such as *cup crystals,* scrolls, or columns. Cf: *surface hoar; crevasse hoar.*

depth ice (a) *anchor ice.* (b) Small particles of ice formed below the surface of a sea churned by wave action (ADTIC, 1955, p. 24).

depth of compensation [oceanog] (a) *compensation depth [oceanog].* (b) *carbonate compensation depth.*

depth of compensation [tect] According to the concept of *isostasy,* the depth above which the rock material is brittle and below which there is a slow movement of plastic rock material to adjust to changes in load on the Earth's surface. Estimates of the depth of compensation range from 35 miles to 75 miles (Strahler, 1963, p.400). See also: *isostatic compensation.* Syn: *compensation level;*

compensation depth; level of compensation.

depth of exploration The maximum depth of the cause of a geophysical anomaly that is just recognizable above *noise.*

depth of focus The distance from the focus of an earthquake to the epicenter. Syn: *focal depth.*

depth of frictional influence *friction depth.*

depth of frictional resistance *friction depth.*

depth of penetration [elect] The depth in a dissipative medium at which the electric field intensity associated with a plane electromagnetic wave is smaller than its value at the surface of the medium by 1/e (e is the base of the Napierian logarithm). See also: *skin effect.*

depth of penetration [remote sensing] (a) The depth below the surface of a material by which the incident radiation has been attenuated to l/e, or to 37 percent. This is sometimes called *skin depth,* or *attenuation distance.* (b) The depth at which the integrated temperature differentials of adjacent spatial resolution cells are equivalent to the temperature resolution of the sensor. (c) The depth from which radiation may still reach the receiver. Such radiation is integrated over the distance downwards from the surface.

depth point In seismic work, a position at which a depth determination of a mapped horizon has been calculated. See also: *common depth point.*

depth rule An algorithm for calculating the depth to an anomalous mass, conductor, or magnetic body. Depth rules are often based on extreme values of an anomaly or its derivatives or on the distances between certain easily determined points on anomaly curves. See also: *index factor.*

depth section A seismic section plotted with its vertical scale in depth units rather than time units.

depth sounder *echo sounder.*

depth-velocity curve *vertical-velocity curve.*

depth zone [meta] One of the characteristic physicochemical environments at various depths in the Earth that give rise to different metamorphic phenomena (Grubenmann, 1904): *epizone; mesozone; katazone.*

depth zone [oceanog] One of four oceanic environments, or ranges of oceanic depths: the *littoral* zone, between high and low tides; the *neritic* zone, between low-tide level and 100 fathoms; the *bathyal* zone, between 100 and 500 fathoms; and the *abyssal* zone, 500 fathoms and deeper.

deranged drainage pattern A distinctively disordered drainage pattern in a recently glaciated area whose former surface and preglacial drainage have been remodeled and effaced, and in which the new drainage system shows a complete lack of underlying structural and bedrock control. It is characterized by irregular streams that flow into and out of lakes, by only a few short tributaries, and by swampy interstream areas.

derangement The process by which changes in a stream course are effected by agents other than streams, such as by glaciation, wind deposition, or diastrophism. Cf: *diversion.*

derbylite A black or brown monoclinic mineral: $Fe_4^{+3}Ti_3Sb^{+3}O_{13}(OH)$.

Derbyshire spar A popular name for fluorite from Derbyshire, England. Syn: *Derby spar.*

derbystone *blue john.*

derelict (a) A tract of dry land formed by *dereliction.* (b) Any property abandoned at sea, often of sufficient size as to be a menace to navigation.

dereliction A recession of water from the sea or other body of water so that land is left dry. Cf: *reliction.*

derivate A general term, now obsolete, for a rock derived from the products of destruction of older rocks; a sedimentary rock. Cf: *ingenite.*

derivative map [cart] A map derived from geologic data on several maps and based on interpretation of geologic structure, lithology, topography, rainfall, vegetation, ground-water levels, or other features.

derivative map [geophys] A map of one of the derivatives of a potential field, such as the Earth's gravity or magnetic field. It is usually of the second vertical derivative, or a *second-derivative map.*

derivative rock A rock composed of materials derived from the weathering of older rocks; a *sedimentary rock,* or a rock formed of material that has not been in a state of fusion immediately before its accumulation.

derivative structure A crystal structure having a multiple unit cell and/or a suppression of some symmetry elements, formed by the substitution of a simple fraction of one atom by another; e.g. the structure of chalcopyrite as derivative after sphalerite.

derived Said of geologic materials that are not native or that have been displaced or brought from elsewhere; esp. said of a fossil washed out from its original site and redeposited in a later formation at a different locality. Also said of geologic materials that are not primary or original. Cf: *reworked; remanié.*

derived character In *cladism,* a *character [paleont]* possessed by a descendant taxon and modified from a homologous ancestral *primitive character.* Syn: *apomorphy.*

derived till A till-like deposit "formed from the erosion of substantially older tillites, not necessarily with any glacial transport involved in the second formation"; e.g. the deposit of the so-called "Cretaceous glaciation" of South Australia (Harland et al., 1966, p. 232).

dermal [paleont] (a) Pertaining to the exopinacoderm, a cortex, or the ectosome of a sponge; e.g. "dermal skeleton" (differentiated skeleton at the outer surface), "dermal membrane" (roofing over a vestibule), or "dermal pores" (minute openings, or prosopores, in the surface). (b) Pertaining to the integument, ectoderm, or skin of vertebrates; also, originating there, as "dermal bone".

dermal [tect] Said of deformation or gliding in the upper part of the sialic crust. Cf: *epidermal; bathydermal.*

dermal bone Bone deposited in a more or less superficial layer of connective tissue, without an intervening phase of cartilage formation. It includes primarily the superficial flat bones of the skull, and the clavicle and interclavicle of the appendicular skeleton.

dermalium A usually specialized sponge spicule supporting all or part of the ectosome. Pl: *dermalia.*

dermolith An obsolete syn. of *pahoehoe.* Cf: *aphrolith.*

dermoskeleton *exoskeleton.*

derrick A framework tower of steel (formerly wood), erected over a deep drill hole (such as an oil well), used to support the various tools and tackle employed in hoisting and lowering the equipment used in drilling and completing the well. See also: *drilling rig.*

derriksite An orthorhombic mineral: $Cu_4(UO_2)(SeO_3)_2(OH)_6 \cdot H_2O$.

desalination The removal of dissolved salts from seawater in order to make it potable. The most common method is distillation.

desaulesite *pimelite.*

descendant A topographic feature carved from the mass beneath an older topographic form that has been removed (Willis, 1903, p. 74).

descending branch Either of two dorsal elements of a brachiopod loop, extending distally from crura and recurved ventrally at anterior ends. Cf: *ascending branch.*

descending development *declining development.*

descensional deposit A sedimentary deposit produced by the disintegration of rock and the aggregation of the resulting particles in beds. Little used.

descension theory A theory of formation of supergene mineral deposits involving the descent from above of mineral-bearing solutions. The theory originated with the Neptunian school of thought of the 18th Century, which postulated an aqueous origin for all rocks. Cf: *ascension theory.*

descloizite A brown to black mineral: $Pb(Zn,Cu)(VO_4)(OH)$. It is isomorphous with mottramite. Syn: *eusynchite.*

desert A region with a mean annual precipitation of 10 inches or less, and so devoid of vegetation as to be incapable of supporting any considerable population. Four kinds of deserts may be distinguished: (1) polar or high-latitude deserts, marked by perpetual snow cover and intense cold; (2) middle-latitude deserts, in the basinlike interiors of the continents, such as the Gobi, characterized by scant rainfall and high summer temperatures; (3) tradewind deserts, notably the Sahara, with negligible precipitation and a large daily temperature range; and (4) coastal deserts, as in Peru, where there is a cold current on the western coast of a large land mass. Adj: *desertic; eremic.*

desert arch An elongate *desert dome.*

desert armor A *desert pavement* whose surface of stony fragments protects the underlying finer-grained material from further wind erosion; a common feature of *stony deserts.* See also: *pebble armor.* Syn: *deflation armor.*

desert crust (a) A hard layer, containing calcium carbonate, gypsum, or other binding matter, exposed at the surface in a desert region. (b) *desert varnish.* (c) *desert pavement.*

desert dome A convex rock surface with uniform smooth slopes, representing the result of prolonged exposure of a mountain mass to desert erosion; e.g. Cima Dome in the Mojave Desert, Calif. See also: *desert arch; granite dome.* Syn: *pediment dome.*

desert lacquer *desert varnish.*

desert mosaic A *desert pavement* characterized by tightly interlocking and evenly set fragments, covering the surface in the manner of a mosaic; e.g. a *pebble mosaic.*

desert patina *desert varnish.*

desert pavement A natural residual concentration of wind-polished, closely packed pebbles, boulders, and other rock fragments, mantling a desert surface (such as an area of *reg*) where wind action and sheetwash have removed all smaller particles, and usually protecting the underlying finer-grained material from further deflation. The fragments commonly are cemented by mineral matter. See also: *desert armor; desert mosaic; lag gravel; boulder pavement; reg.* Syn: *desert crust; deflation residue.*

desert peneplain A syn. of *pediplain,* esp. in reference to a wind-erosion surface in southern Africa. The term is inappropriate because such a surface is produced under different conditions and by different processes than those of a humid-land peneplain.

desert plain (a) A general term used by Blackwelder (1931) for any plain commonly found in a desert; e.g. a flood plain, structural plain, playa, bajada, or pediment. (b) *pediplain.*

desert polish (a) A smooth, shiny or glossy surface imparted to rocks of desert regions by windblown sand and dust. Syn: *wind polish.* (b) A term sometimes used as a syn. of *desert varnish.*

desert rind *desert varnish.*

desert ripple One of a system of slightly arcuate ridges produced by the wind, arranged roughly en echelon about 15 m apart with crests supporting vegetation and troughs plated with caliche; it may be as high as 1 m and as long as 150 m.

desert rose A radially symmetric group of crystals with a fancied resemblance to a rose, formed in sand, soft sandstone, or clay. The crystals are usually of calcite, less commonly of barite, gypsum, or celestite.

Desert soil A great soil group in the 1938 classification system, a group of zonal soils having a light-colored surface horizon overlying calcareous material and, commonly, a hardpan. It is developed under conditions of aridity, warm to cool climate, and scant scrub vegetation (USDA, 1938). These soils are now classified as *Argids* and *Orthids.*

desert varnish A thin dark shiny film or coating, composed of iron oxide accompanied by traces of manganese oxide and silica, formed on the surfaces of pebbles, boulders, and other rock fragments in desert regions after long exposure, as well as on ledges and other rock outcrops. It is believed to be caused by exudation of mineralized solutions from within and deposition by evaporation on the surface. A similar appearance produced by wind abrasion is properly known as *desert polish.* Syn: *desert patina; desert lacquer; desert crust; desert rind; varnish.*

desiccation A complete or nearly complete drying-out or drying-up, or a deprivation of moisture or of water not chemically combined; e.g. the loss of water from pore spaces of soils or sediments as a result of compaction, or the formation of evaporites as a result of direct evaporation from bodies of water in an arid region, or the progressive increase in aridity of an area as a result of a climatic change (such as decreasing rainfall) or of accelerated erosion (such as deforestation). Cf: *dehydration; exsiccation.*

desiccation breccia A breccia formed where irregular dried-out and mud-cracked polygons have broken into angular fragments that have then been deposited with other sediments. Syn: *mud breccia.*

desiccation conglomerate A term used by Shrock (1948, p.208) for a conglomerate consisting of fragments eroded from a mud-cracked layer of sediment and rounded by transportation. Syn: *mudstone conglomerate.*

desiccation crack A crack in sediment, produced by drying; esp. a *mud crack.* Syn: *drying crack; desiccation fissure; desiccation mark; klizoglyph.*

desiccation mark A *desiccation crack;* esp. a *mud crack.*

desiccation polygon A small *nonsorted polygon* formed in a non-frigid environment (as on a mud flat) by drying of moist, clayey soil or sediment, thus producing contortion resulting in cracking. The polygon normally has three to five sides which may measure 2-30 cm in length. See also: *giant desiccation polygon.* Syn: *mud-crack polygon; mud-flat polygon; drought polygon; shrinkage polygon.*

design flood A flood against which protective measures are taken.

desilication [petrology] The removal of silica from a rock or magma by the breakdown of silicates and the resultant freeing of silica, or by reaction between a body of magma and the surrounding wall rock.

desilication [soil] The removal of silica from soils in a warm, humid climate by the percolation of large amounts of rainwater, resulting in a soil relatively rich in hydroxides of iron, aluminum, and manganese, i.e. an *Oxisol.*

desilting basin A *settling basin* consisting of an enlargement in a stream where silt carried in suspension may be deposited.

desma An irregularly branched, siliceous sponge spicule that bears knotty outgrowths (zygomes) which interlock with adjacent spicules. Pl: *desmas* or *desmata.*

desmid One of a family of unicellular, microscopic green algae, commonly composed of semicells that are mirror images of each other.

desmine *stilbite.*

desmite *Residuum [coal]* that is transparent; it is characteristic of higher coal grades.

desmodont adj. Said of the dentition of a bivalve mollusk characterized by the prominence of a large chondrophore inside the hinge line.—n. Any bivalve mollusk belonging to the order Desmodonta, characterized by the presence of two equal muscle scars and a pallial sinus and by the absence of hinge teeth or with irregular hinge teeth intimately connected with the chondrophore.

desmoid A siliceous sponge spicule bearing outgrowths like those of a desma but not interlocking with adjacent spicules.

Desmoinesian North American series: upper Middle Pennsylvanian (above Atokan, below Missourian). Syn: *Des Moines.*

desmosite A banded *adinole.*

despujolsite A hexagonal mineral: $Ca_3Mn^{+4}(SO_4)_2(OH)_6 \cdot 3H_2O$.

desquamation An obsolescent syn. of *exfoliation* characterized by the peeling-off or detachment of scaly rock fragments.

dess A term used in Morocco for silt deposited by a new stream flowing in an arid region (Termier & Termier, 1963, p.403).

destinezite *diadochite.*

destructional Said of a landform that owes its origin or general character to the removal of material by erosion and weathering; e.g. a mesa, canyon, cliff, or plain resulting from the wearing down or away of the land surface. Ant: *constructional.* Cf: *sequential landform.*

destructive metamorphism An archaic syn. of *equitemperature metamorphism.*

destructive wave A wave that erodes a beach by moving material seaward, as a storm wave with a more powerful backrush than uprush; on a sandy beach, a wave with a steepness greater than 0.25. Ant: *constructive wave.*

detached core The inner bed or beds of a fold that become separated or pinched off from their source due to extreme folding and compression.

detachment [fault] *décollement.*

detachment [soil] Separation of transportable particles from a soil layer, usually by running water, raindrop impact, or wind. Cf: *dispersion.*

detachment fault *sole fault.*

detachment thrust *sole fault.*

detail log Any well log plotted on a scale larger than the conventional scale (2 inches per 100 ft of depth) in order to portray minor variations in the formations penetrated by the borehole; specif. an electric log plotted on a scale of 5, 10, or 25 inches per 100 ft of depth.

detectability A measure of the smallest object that can be discerned on an image.

detectivity In infrared-detector terminology, the reciprocal of *noise-equivalent power.*

detector (a) The component of a remote-sensing system that converts electromagnetic radiation into a signal that can be recorded. Syn: *radiation detector.* (b) *seismic detector.*

detention The amount of water from precipitation existing as overland flow. *Depression storage* is not considered part of the detention (Rechard & McQuisten, 1968). Syn: *detention storage; surface detention.*

detention storage *detention.*

determinism "The theory that all occurrences in nature are determined by antecedent causes or take place in accordance with natural laws" (Webster, 1967, p. 616). It excludes chance and esp. probabilistic explanations.

deterministic process A process in which there is an exact mathematical relationship between the independent and dependent variables in the system. Ant: *stochastic process.*

detrital adj. Pertaining to or formed from *detritus;* said esp. of rocks, minerals, and sediments. The term may indicate a source outside the depositional basin (Krynine, 1948, p. 133) or a source within it.—n. A detrital rock. Term is usually used in the plural.

detrital fan *alluvial fan.*

detrital mineral Any mineral grain resulting from mechanical disintegration of parent rock; esp. a *heavy mineral* found in a sediment or weathered and transported from a vein or lode and found in a placer or alluvial deposit.

detrital ratio *clastic ratio.*

detrital remanent magnetization *depositional remanent magnetization.*

detrital rock A rock composed primarily of particles or fragments detached from pre-existing rocks either by erosion or by weathering; specif. a sedimentary rock having more than 50% detrital material (Krynine, 1948, p.134). Cf: *chemical rock.*

detrital sediment A sediment formed by the accumulation of detritus, esp. that derived from pre-existing rocks and transported to the place of deposition. Cf: *clastic sediment.* See also: *detrital deposit.*

detrition A general term for the processes involved in producing detritus; a wearing-away by breaking or rubbing of rock masses.

detritovore *deposit feeder.*

detritus A collective term for loose rock and mineral material that is worn off or removed by mechanical means, as by disintegration or abrasion; esp. fragmental material, such as sand, silt, and clay, derived from older rocks and moved from its place of origin. Cf: *debris [geol].* Pl: *detritus.* See also: *reef detritus.*

deuteric Referring to reactions between primary magmatic minerals and the water-rich solutions that separate from the same body of magma at a late stage in its cooling history. Syn: *paulopost; epimagmatic.* Cf: *multopost.* See also: *synantectic; synantexis; autometamorphism.*

deuteroconch The chamber in larger foraminifers immediately adjoining the proloculus and formed next after it.

deuteroforamen A secondary aperture independent of the tooth plate in some enrolled foraminifers. Cf: *protoforamen.*

deuterogene An old term for a secondary rock; a rock formed from a pre-existing rock. Adj: *deuterogenic.* Cf: *protogene.*

deuterogenic Adj. of *deuterogene.*

deuteroglacial Pertaining to the last great glaciation following the *proteroglacial* period (Hansen, 1894, p. 128).

deuterolophe A spirally coiled part of a brachiopod lophophore, bearing a double brachial fold and a double row of paired filamentary appendages. It is homologous with the side arms of *plectolophe.* Cf: *spirolophe.*

deuteromorphic A general term applied to crystals whose shapes have been acquired or modified by mechanical or chemical processes acting on the original forms (Loewinson-Lessing, 1899). Depending on the nature of the secondary agent, they may be described as *clastomorphic, lytomorphic, schizomorphic, tectomorphic,* or *neomorphic.* These terms are now obsolete.

deuteropore One of a group of *protopores* fusing into a single larger pore cavity in the outer wall of foraminifers.

deutonymph The second developmental stage in the arachnid order Acarida.

developed ore *developed reserves.*

developed reserves Ore that has been exposed on three sides and for which tonnage and quality estimates have been made; ore essentially ready for mining. Cf: *positive ore; proved reserves.* Syn: *developed ore; measured ore; ore in sight; blocked-out ore; assured mineral.*

development [eco geol] The preparation of a mining property or area so that an orebody can be analyzed and its tonnage and quality estimated. Development is an intermediate stage between *exploration* and *mining.*

development [grd wat] (a) In the construction of a water well, the removal of fine-grained material adjacent to a drill hole, enabling water to enter the hole more freely. (b) Exploitation of ground water.

development well A well drilled within the known or proved productive area of an oil field, with the expectation of obtaining oil or gas from the producing formation or formations in that field. Cf: *exploratory well; offset well.*

deviation [drill] (a) The departure of a drilled hole from being straight. The hole may be either vertical or inclined and the departure may be in any direction. Deviation may be intentional, as in *directional drilling,* or undesirable. Syn: *deflection [drill].* (b) In more general use, the angle of departure of a well bore from the vertical, without reference to direction. See also: *inclination [drill].*

deviation [geodesy] *deflection of the vertical.*

deviation [stat] (a) *mean deviation.* (b) *standard deviation.*

deviation of the vertical *deflection of the vertical.*

devilline A dark-green mineral: $Cu_4Ca(SO_4)_2(OH)_6 \cdot 3H_2O$. Syn: *devillite.*

devil's slide (a) An *avalanche track* down a steep slope. (b) A long narrow mass of talus material descending a steep mountain.

devitrification Conversion of glass to crystalline material.

devolatilization In coal, the loss of volatile constituents and resulting proportional increase in carbon content during coalification. It is a process of metamorphism; the higher the rank of coal, the higher the level of devolatilization. Syn: *debitumenization.*

Devonian A period of the Paleozoic era (after the Silurian and before the Mississippian), thought to have covered the span of time between 400 and 345 million years ago; also, the corresponding system of rocks. It is named after Devonshire, England, where rocks of this age were first studied. See also: *age of fishes.*

devonite A porphyritic diabase containing large potassium-rich labradorite phenocrysts in a groundmass of altered plagioclase and clinopyroxene. It was named by Johannsen in 1910 for Mt. Devon, Massachusetts. Not recommended usage.

dew Condensation of atmospheric water vapor on a surface whose temperature is below the dew point but above the freezing point. Cf: *frost.*

deweylite A mixture of clinochrysotile (or sometimes lizardite) and stevensite. It was formerly regarded as a mineral of the serpentine group. Syn: *gymnite.*

dewindtite A canary-yellow secondary mineral: $Pb(UO_2)_2(PO_4)_2 \cdot 3H_2O$.

dew point The temperature to which air must be cooled, at constant pressure and constant water-vapor content, in order for *saturation* to occur; the temperature at which the saturation pressure is the same as the existing vapor pressure.

dextral Pertaining, inclined, or spiraled to the right; specif. pertaining to the normal or clockwise direction of coiling of gastropod shells. A dextral gastropod shell in apical view (apex toward the observer) has the whorls apparently turning from the left toward the right; when the shell is held so that the axis of coiling is vertical and the apex or spire is up (as in orthostrophic shells) or down (as in hyperstrophic shells), the aperture is open toward the observer to the right of the axis. Actually, the definition depends on features of soft anatomy: with genitalia on the right side of the head-foot mass, the soft parts and shell are arranged such that the aperture is on the right when viewed with the apex (of orthostrophic shells) uppermost (TIP, 1960, pt.I, p.130). Ant: *sinistral.* Syn: *right-handed [paleont].*

dextral fault *right-lateral fault.*

dextral fold An asymmetric fold with the asymmetry of an S as opposed to that of a Z when seen in profile. The long limb is apparently offset to the right. Cf: *sinistral fold.*

dextral imbrication The condition in a heterococcolith in which each segment overlaps the one to the right when viewed from the center of the cycle. Ant: *sinistral imbrication.*

dextrorotatory *right-handed.*

DF *dark field.*

dhand A term used in Sind (region of West Pakistan) for a salt lake, esp. an alkali lake. Etymol: Sindhi.

D horizon An obsolete term for a soil horizon that may be present below a B or a C horizon, consisting of unweathered rock. It is currently designated as an R horizon.

DI (a) *durability index.* (b) *differentiation index.*

diabantite A mineral of the chlorite group: $(Mg,Fe^{+2},Al)_6(Si,Al)_4O_{10}(OH)_8$. It occurs in cavities in basic igneous rocks.

diabase In the U.S., an intrusive rock whose main components are labradorite and pyroxene and which is characterized by ophitic texture. As originally applied by Brongniart in 1807, the term corresponded to what is now recognized as *diorite.* "The word has come to mean a pre-Tertiary basalt in Germany, a decomposed basalt in England, and a dike-rock with ophitic texture in the United States and Canada" (Johannsen, 1939, p. 248). Cf: *traprock.* Syn: *dolerite.*

diabasic (a) A seldom used syn. of *nesophitic.* Kemp (1900, p.158-159) considered that "diabasic" applied to textures in which there was a predominance of plagioclase, with augite filling the interstices, while "ophitic" indicated a predominance of augite over plagioclase. (b) Composed of or resembling *diabase.*

diablastic Pertaining to a texture in metamorphic rock that consists of intricately intergrown and interpenetrating constituents, usually with rodlike shapes (Becke, 1903).

diaboleite A sky-blue mineral: $Pb_2CuCl_2(OH)_4$.

diabrochite Metamorphic rock that owes its mineralogical composition to intensive penetration by ascending solutions or vapors (*ichor*) or to partial fusion, but without injection of visible granitic material.

diachronic Pertaining to, or during, the time of the Earth's existence; considering events or changes as they happen or develop over time. Ant: *prochronic.* Cf: *synchronous.*

diachronism A term introduced by Wright (1926) for the transgression, across time planes or biozones, by a rock unit whose age differs from place to place; the state or condition of being diachronous.

diachronous Said of a rock unit that is of varying age in different areas or that cuts across time planes or biozones; e.g. said of a sedimentary formation related to a narrow depositional environment, such as a marine sand that was formed during an advance or recession of a shoreline and becomes younger in the direction in which the sea was moving. Syn: *time-transgressive.*

diachyte Rock product of marked mechanical and/or chemical contamination of anatectic magma by cognate basic material (Dietrich & Mehnert, 1961). Little used.

diaclinal Said of a stream or valley that passes through or across a fold, with a direction at right angles to the strike of the underlying strata. Also said of a region having diaclinal streams. Term introduced by Powell (1874, p. 50). Ant: *paraclinal.*

diacrystallic Pertaining to the texture of a diagenetically recrystallized and essentially monomineralic rock in which contiguous crystals interpenetrate in a complicated manner (Phemister, 1956, p.72).

diactine A sponge spicule having two rays, usually monaxonic. See also: *rhabdodiactine.* Syn: *diact; diactin.*

diad Said of a symmetry axis that requires a rotation of 180° to repeat the crystal's appearance. Also spelled: *dyad.* Cf: *triad.* Syn: *digonal.*

diadactic *diatactic.*

diadochite A brown or yellowish mineral: $Fe_2(PO_4)(SO_4)(OH)\cdot 5H_2O$. It is isomorphous with sarmientite. Syn: *destinezite.*

diadochy *ionic substitution.*

diadysite A migmatite consisting of granitic veins and metamorphic parent rock (Mehnert, 1968, p. 354). Cf: *arterite; phlebite; venite.* Little used.

diaene A sponge spicule with two rays of equal length and one of a different length, usually longer; a triaene with one ray reduced or absent.

diagenesis [mineral] Recombination or rearrangement of a mineral resulting in a new mineral; specif. the geochemical, mineralogic, or crystallochemical processes or transformations affecting clay minerals before burial in the marine environment of sedimentation, such as illitization, glauconitization, or any transformation affecting the lattice of a clay mineral before burial. For a discussion of diagenesis in clay minerals, see Keller (1963). Cf: *halmyrolysis.*

diagenesis [sed] All the chemical, physical, and biologic changes undergone by a sediment after its initial deposition, and during and after its lithification, exclusive of surficial alteration (weathering) and metamorphism. This is the definition as applied by most geologists in the U.S. (Twenhofel, 1939, p. 254-255) and Germany (Correns, 1950). It embraces those processes (such as compaction, cementation, reworking, authigenesis, replacement, crystallization, leaching, hydration, bacterial action, and formation of concretions) that occur under conditions of pressure (up to 1 kb) and temperature (maximum range of 100°C to 300°C) that are normal to the surficial or outer part of the Earth's crust; and it may include changes occurring after lithification under the same conditions of temperature and pressure. The father of this concept was Walther (1893-1894, p. 693-711), although the term "Diagenese" was first used by Gümbel (1868, p. 838) for a postsedimentary transformation of sediments into individual crystalline minerals, leading to the creation of metamorphic rocks such as gneiss and schist. Russian (and some U.S.) geologists restrict the term to the initial phase of postsedimentary changes, occurring in the zone where the sediment is still unconsolidated, the process being complete when the sediment has been converted to a more or less compact sedimentary rock (Fersman, 1922); in this usage, the term is equivalent to *early diagenesis* as used in the U.S. There is no universally accepted definition of the term, and no delimitation (such as the boundary with metamorphism). For a historical discussion and review, see Larsen & Chilingar (1967) and Dunoyer de Segonzac (1968). Cf: *epigenesis.* Syn: *diagenism.*

diagenetic Pertaining to or caused by diagenesis; e.g. a "diagenetic change" resulting from compaction, a "diagenetic structure" (such as a stylolite) formed after deposition, a "diagenetic deposit" (such as dolomitized limestone or one consisting of manganese nodules), or a "diagenetic environment" of rock consolidation. Syn: *postdepositional.*

diagenetic differentiation The redistribution of material within a sediment by solution and diffusion toward centers or nuclei where reprecipitation occurs, leading to segregation of minor constituents into diverse forms and structures, such as chert nodules in limestone or concretions in shale (Pettijohn, 1957, p. 672).

diagenetic facies A facies that includes all rocks or sedimentary materials that have, by a process of diagenesis, developed "mineral assemblages that are the result of adjustment to a particular diagenetic environment" (Packham & Crook, 1960, p. 400). A "low-rank" facies corresponds to an early stage of alteration, a "high-rank" facies to a late stage. Cf: *parfacies.*

diagenic metamorphism *diagenism.*

diagenism A term used by Grabau (1904, p. 235) as a syn. of *diagenesis* as defined by Walther. Cf: *static metamorphism.*

diagenite A diagenetic rock.

diagenodont Said of the dentition of a bivalve mollusk (e.g. *Astarte*) having differentiated cardinal teeth and lateral teeth located on the hinge plate, with the lateral teeth not exceeding two and the cardinal teeth not exceeding three in either valve. Cf: *teleodont.*

diaglomerate A conglomerate in which individual fragments are recognized as being related.

diaglyph A *hieroglyph* formed during diagenesis (Vassoevich, 1953, p.33).

diagnosis A statement of the characteristics of a *taxon* which in the opinion of its author distinguish it from others (McVaugh et al., 1968, p. 12).

diagnostic fossil *characteristic fossil.*

diagnostic mineral A mineral, such as olivine or quartz, whose presence in an igneous rock indicates whether the rock is undersaturated or oversaturated. Syn: *symptomatic mineral.*

diagnostic subsurface horizon A soil horizon that forms below the surface and is used to classify soils into orders, suborders, and great groups. It is usually a B horizon (e.g. argillic, spodic, cambic) but may be an A horizon (e.g. albic) or a C horizon (e.g. calcic, duripan). It occurs at the surface only if the soil has been truncated.

diagnostic surface horizon *epipedon.*

diagonal bedding An archaic syn. of *inclined bedding,* or bedding diagonal to the principal surface of deposition; specif. *cross-bedding.*

diagonal fault *oblique fault.*

diagonal joint A joint whose strike is oblique to the strike of the sedimentary strata, or to the cleavage plane of the metamorphic rocks, in which it occurs. Syn: *oblique joint; (hkO) joint.*

diagonal lamination *cross-lamination.*

diagonal scour mark One of a series of scour marks arranged diagonally to the main direction of flow and formed by concentration of smaller scour marks (usually longitudinal flutes) into distinct rows that alternate with areas where scour marks are absent or less abundant.

diagonal-slip fault *oblique-slip fault.*

diagonal stratification *cross-stratification.*

diagrammatic map *cartogram.*

dial n. A compass used for surface and underground surveying, fitted with sights, spirit levels, and *vernier,* and mounted on a tripod.—v. To survey or measure with a dial and chain.

diallage (a) A dark-green or grass-green, brown, gray, or bronze-colored clinopyroxene (usually a variety of augite or of aluminum-bearing diopside) occurring in lamellae or in foliated masses and often having a metallic or brassy luster. It is characterized by a conspicuous parting parallel to the front pinacoid, and is typically found in basic igneous rocks such as *gabbro.* (b) A term applied to various poorly defined alteration products of pyroxene.

diallagite A *pyroxenite* composed almost entirely of diallage. Other pyroxenes, hornblende, spinel, and garnet may be present as accessories. Not recommended usage.

dialogite A syn. of *rhodochrosite.* Also spelled: *diallogite.*

dialysis A method of separating compounds in solution or suspension by their differing rates of diffusion through a semipermeable membrane, some colloidal particles not moving through at all, some moving slowly, and others diffusing quite readily. Cf: *osmosis.* See also: *electrodialysis.*

diamagnetic Having a small, negative magnetic susceptibility. All materials that do not show paramagnetism or magnetic order are diamagnetic. Typical diamagnetic minerals are quartz and feldspar. Cf: *paramagnetic.*

diamantiferous *diamondiferous.*

diameter The maximum *link distance* in a drainage network.

diametral spine A basally fused spine opposite radial spines and passing through the diameter of the central capsule of an acantharian radiolarian.

diamict A general term proposed by Harland et al. (1966, p. 229) to include *diamictite* and *diamicton.*

diamictite A comprehensive, nongenetic term proposed by Flint et al. (1960b) for a nonsorted or poorly sorted, noncalcareous, terrigenous sedimentary rock that contains a wide range of particle sizes, such as a rock with sand and/or larger particles in a muddy matrix; e.g. a tillite or a pebbly mudstone. Cf: *diamicton.* Syn: *mixtite.* Originally termed *symmictite* by Flint et al. (1960a).

diamicton A general term proposed by Flint et al. (1960b) for the nonlithified equivalent of a *diamictite;* e.g. a till. Originally termed *symmicton* by Flint et al. (1960a).

diamond (a) An isometric mineral, representing a naturally occurring crystalline form of carbon dimorphous with graphite and being the hardest natural substance known (hardness of 10 on the Mohs scale). It often occurs in octahedrons with rounded edges or curved faces. Diamonds form under extreme temperatures and pressures and are found in ultrabasic breccias, pipes in igneous rocks, and alluvial deposits. Pure diamond is colorless or nearly so, color is imparted by impurities. When transparent and more or less free from flaws, it is the most cherished and among the most highly valued gemstones; its high refractive index and dispersive powers result in remarkable brilliance and play of prismatic color when faceted. Off-color or flawed diamonds are used for industrial purposes (such as in rock drills, abrasive powder, and cutting tools). (b) Artificially produced crystallized carbon similar to the native form. (c) A crystalline mineral that resembles diamond in brilliance, such as "Alençon diamond" (a smoky quartz sometimes valued as a jewel); esp. any of various kinds of rock crystal such as "Bristol diamond", "Herkimer diamond", "Lake George diamond", and "Arkansas diamond".

diamond bit A rotary-drilling bit studded with diamonds (usually bort). It is used for drilling and coring in extremely hard rock.

diamond chip A thin, tabular *chip* of an uncut diamond crystal, weighing less than 0.75 carat.

diamond drilling A variety of rotary drilling in which *diamond bits* are used as the rock-cutting tool. It is a common method of prospecting for mineral deposits, esp. in development work where core samples are desired.

diamond dust [ice] Minute ice crystals, usually columnar or bullet-shaped, precipitated out of clear air at very low temperatures as on the polar plateau of Antarctica.

diamond dust [mater] Powdered, crushed, or finely fragmented diamond material used as a cutting, grinding, and polishing abrasive medium.

diamondiferous Said of any substance (esp. rock or alluvial material) containing or yielding diamonds. Syn: *diamantiferous.*

diamond spar *corundum.*

diamond structure A type of crystal structure in which each atom or ion is four-coordinated; minerals having this structure are characteristically rigid and have low electrical conductivity.

diancistra A C-shaped siliceous sponge spicule (microsclere) having sharply recurved ends and bearing bladelike lamellae on the inner side so that it resembles a partly opened penknife. Pl: *diancistrae.*

dianite *columbite.*

diaphaneity The light-transmitting quality of a mineral.

diaphanotheca The relatively thick, light-colored to transparent, intermediate layer of the spirotheca next below the *tectum* in fusulinids.

diaphorite A gray-black orthorhombic mineral: $Pb_2Ag_3Sb_3S_8$. Syn: *ultrabasite.*

diaphragm Any of various more or less rigid partitions in the bodies or shells of invertebrates; e.g. a skeletal membranous partition extending across the zooidal chamber in stenolaemate bryozoans; a thin crescentic plate of secondary brachiopod shell developed around the visceral disc of the brachial valve; an imperforate partition crossing the siphuncle of a nautiloid; or a partial septum just below the aperture in a thecamoebian, perforated for protrusion of pseudopodia.

diaphthoresis *retrograde metamorphism.*

diaphthorite A crystalline rock in which minerals characteristic of a lower metamorphic grade have developed by retrograde metamorphism at the expense of minerals peculiar to a higher metamorphic grade. The term was originated by Becke in 1909.

diapir A dome or anticlinal fold in which the overlying rocks have been ruptured by the squeezing-out of plastic core material. Diapirs in sedimentary strata usually contain cores of salt or shale; igneous intrusions may also show diapiric structure. See also: *diapirism.* Syn: *piercement dome; diapiric fold; piercing fold.*

diapiric fold *diapir.*

diapirism The process of piercing or rupturing of domed or uplifted rocks by mobile core material, by tectonic stresses as in anticlinal folds, by the effect of geostatic load in sedimentary strata as in salt domes and shale diapirs, or by igneous intrusion, forming diapiric structures such as plugs. The concept was first applied to salt structures, which are the most common type of *diapir.* Obs. syn: *tiphon.*

diaplectic Said of glasslike mineralogic features produced by shock waves in such a way that the characteristics of the liquid state are lacking; e.g. a "diaplectic mineral" whose disordered and deformed crystals have been modified by shock waves without melting and have characteristics such as planar features, lowered refractive indices, and lowered birefringence, or a "diaplectic glass" (of quartz, feldspar, or other minerals) representing an amorphous phase produced by shock waves without melting. Diaplectic materials represent intermediate stages of structural order between the crystalline and the normal glassy phases. Term proposed by Wolf von Engelhardt in 1966 (Engelhardt & Stöffler, 1968, p. 163). Etymol: Greek *diaplesso,* "to destroy by beating or striking". Cf: *thetomorphic.*

diapositive A positive image on a transparent medium such as glass or film; a *transparency [photo].* The term originally was used primarily for a transparent positive on a glass plate used in a plotting instrument, a projector, or a comparator, but now is frequently used for any positive transparency (ASP, 1975, p. 2077).

diapsid adj. Pertaining to a reptilian skull characterized by two temporal fenestrae, upper and lower.—n. Loosely, any reptile with such a skull.

diara *char.*

diaresis A transverse groove on the posterior part of the exopod (rarely also the endopod) of a uropod of a malacostracan crustacean.

diarhysis A radial *skeletal canal* in dictyonine hexactinellid sponges, penetrating the body wall completely, open at each end, and containing a flagellated chamber. Pl: *diarhyses.*

diaschistic Said of the rock of a minor intrusion that consists of a differentiate, i.e. its composition is not the same as that of the parent magma. Cf: *aschistic.*

diasome Any benthic mollusk of the subphylum Diasoma, characterized by a more or less straight gut and one or two parts to the shell. It includes the classes Rostroconchia, Scaphopoda, and Pelecypoda.

diaspore A white, gray, yellowish, or greenish orthorhombic mineral: AlO(OH). It represents the alpha base dimorphous with boehmite. Diaspore is found in bauxite and is associated with corundum and dolomite; it occurs in lamellar masses with pearly luster or in prismatic crystals. Syn: *diasporite.*

diaspore clay A high-alumina refractory clay consisting essentially of the mineral diaspore. It has been interpreted as a desilication product of associated *flint clay* and other kaolinitic materials (Keller et al., 1954). Commercial diaspore of first-grade quality contains more than 68% alumina. See also: *burley clay.*

diastem A relatively short interruption in sedimentation, involving only a brief interval of time, with little or no erosion before deposition is resumed; a depositional break of lesser magnitude than a *paraconformity,* or a paraconformity of very small time value. Diastems are not ordinarily susceptible of individual measurement, even qualitatively, because the lost intervals are too short; they are often deduced solely on paleontologic evidence. The term was introduced by Barrell (1917, p.794). The synonymous term *non-sequence* is preferred in Great Britain. Adj: *diastemic.*

Etymol: Greek *diastema,* "interval".

diastrophic Adj. of *diastrophism.* Cf: *orographic.*

diastrophic eustatism The worldwide change in sea level produced by a change in the capacity of the ocean basins because of diastrophic changes (Thornbury, 1954, p.142). Cf: *glacio-eustatism; sedimento-eustatism.* See also: *eustasy.* Syn: *tectono-eustatism.*

diastrophic plateau A plateau formed by the upheaval of a plain, and cut, broken, or divided into parts by rivers, faults, or flexures (Powell, 1895, p. 39-40).

diastrophism A general term for all movement of the crust produced by tectonic processes, including the formation of ocean basins, continents, plateaus, and mountain ranges. *Orogeny* and *epeirogeny* are major subdivisions. The use of this general term for small-scale features (e.g. diastrophic event, diastrophic ridge, diastrophic structure) is vague and undesirable; more specific terms should be substituted. Adj: *diastrophic.* Syn: *tectonism.*

diatactic Said of a sedimentary structure, like that shown by varves, characterized by the repetition of a pair of unlike laminae showing a gradation in grain size from coarse below to fine above. Also spelled: *diadactic.*

diataphral Descriptive of a type of tectonics in which *syntaphral* folds and faults are refolded by upward diapiric regurgitation of the axial zone of a geosyncline (Carey, 1963, p. A6). Cf: *apotaphral.*

diatexis High-grade (i.e. near, but not complete) *anatexis,* involving rock components with high melting points (Dietrich & Mehnert, 1961). Cf: *metatexis.* Little used.

diatexite Rock formed by diatexis. Also spelled: *diatectite.* Little used.

diathermic Said of a substance or "wall" between two thermodynamic systems that is able to transmit heat. Two systems separated by a diathermic barrier will eventually reach thermal equilibrium.

diatom A microscopic, single-celled plant of the class Bacillariophyceae, which grows in both marine and fresh water. Diatoms secrete walls of silica, called *frustules,* in a great variety of forms. Frustules may accumulate in sediments in enormous numbers.

diatomaceous Composed of or containing numerous diatoms or their siliceous remains.

diatomaceous chert A diatomite that has a well-developed siliceous cement or groundmass.

diatomaceous earth *diatomite.*

diatomaceous shale An impure diatomite with much clayey matter and with shaly partings.

diatomite A light-colored soft friable siliceous sedimentary rock, consisting chiefly of opaline frustules of the diatom, a unicellular aquatic plant related to the algae. Some deposits are of lake origin but the largest are marine. Owing to its high surface area, high absorptive capacity, and relative chemical stability, diatomite has a number of uses, esp. as a filter aid and as an extender in paint, rubber, and plastics. The term is generally reserved for deposits of actual or potential commercial value. Syn: *diatomaceous earth; kieselguhr.* Obsolete syn: *infusorial earth; tripoli-powder.* See also: *tripoli; tripolite.*

diatom ooze A deep-sea pelagic sediment containing at least 30% diatom frustules; it is a siliceous ooze.

diatom-saprocol *dysodile.*

diatreme A breccia-filled volcanic *pipe* that was formed by a gaseous explosion.

diazo print A print on light-sensitized material, made directly by exposure to strong light from a positive transparency. Generally scale-stable when made in a vacuum frame, but elongated in direction of movement through a circular-drum printer. Syn: *ozalid.*

dibranchiate *coleoid.*

dicalycal Said of a graptoloid theca from which two others originate.

dice mineral A term used in Wisconsin for galena occurring in small cubes.

dicentric Said of a corallite formed by a polyp retaining a distomodaeal condition permanently.

dichotomous Said of a crinoid arm that is divided into two branches, which may be equal (isotomous) or unequal (heterotomous).

dichotomy A repeated, twofold equal branching of the main axis of a plant, e.g. in liverworts, seaweeds, and many pteridophytes.

dichotriaene A sponge *triaene* with dichotomously branched cladi.

dichroic Said of a mineral that displays *dichroism.*

dichroism *Pleochroism* of a crystal that is indicated by two different colors. A mineral showing dichroism is said to be *dichroic.* Cf: *trichroism.*

dichroite *cordierite.*

dichroscope An optical instrument that is used to analyze the colors of a pleochroic crystal; it consists of a calcite rhomb and a lens.

dickinsonite A green mineral: $H_2Na_6(Mn,Fe,Ca,Mg)_{14}(PO_4)_{12} \cdot H_2O$. It is isostructural with arrojadite.

dickite A well-crystallized clay mineral of the kaolin group: $Al_2Si_2O_5(OH)_4$. It is polymorphous with kaolinite and nacrite. Dickite is structurally distinct from other members of the kaolin group, having a more complex order of stacking in the *c*-axis direction than kaolinite. It usually occurs in hydrothermal veins.

dicolpate Said of pollen grains having two colpi.

dicolporate Said of pollen grains having two colpi, with at least one colpus provided with a pore or transverse furrow. Dicolporate pollen are rare.

dicot *dicotyledon.*

dicotyledon An angiosperm whose seeds contain an embryo with two embryonic leaves. Such a plant has flower parts in fours or fives, reticulate leaf venation, and tricolpate pollen. Examples include roses, thistles, and oaks. Dicotyledons range from the Jurassic. Cf: *monocotyledon.* Syn: *dicot.*

dicranoclone (a) A tuberculate monaxonic desma (of a sponge) of dipodal to polypodal form with rootlike terminal zygomes. (b) A desma (of a sponge) with arms that diverge from one side of a central point.

dictyonal framework The rigid interior skeleton of dictyonine sponges between the dermalia and gastralia, built of fused dictyonal strands, commonly but not always forming a cubic lattice.

dictyonalia The sponge spicules of a dictyonal framework.

dictyonal strand A linear series of hexactines (sponge spicules) in parallel orientation and fused to form a continuous strand.

dictyonine adj. Said of a hexactinellid sponge whose parenchymalia (spicules) form a rigid framework composed of dictyonal strands; more loosely, said of a sponge whose skeleton is composed of subparallel hexactines rigidly fused so that the limits of individual spicules are not apparent. Ant: *lyssacine.*—n. A dictyonine sponge.

dictyonite Migmatite with reticulated character, i.e. with a veinlet network (Dietrich & Mehnert, 1961). Var: *diktyonite.* Little used.

dictyospore A multicellular *fungal spore* that has both cross septa and longitudinal chitinous walls. Such spores may occur as microfossils in palynologic preparations.

dictyostele A *stele* consisting of separate vascular bundles, or of a network of bundles (Fuller & Tippo, 1954, p. 955). See also: *meristele.*

dicyclic (a) Said of a crinoid having two circlets of plates proximal to the radial plates or (in some inadunate crinoids that lack radial plates) proximal to the oral plates. (b) Said of the apical system of an echinoid in which ocular and genital plates are arranged in two concentric circles, the genital plates alone in contact with the periproctal margin.—Cf: *monocyclic.*

didodecahedron *diploid.*

diductor muscle A muscle that opens the valves in articulate brachiopods; commonly, one of a pair of two pairs of muscles attached to the brachial valve immediately anterior to the beak, usually to the cardinal process. The principal pair is commonly inserted in the pedicle valve on either side of the *adductor muscles* and the accessory pair is inserted posterior to them. Syn: *diductor; divaricator.*

didymoclone A desma (of a sponge) consisting of a short, straight shaft, from the enlarged ends of which several zygome-bearing arms project, predominantly on one side of the spicule.

didymolite *anorthite.*

die-back A large area of freshwater reed-swamp vegetation that is killed when the salinity of a coastal lagoon is increased by either natural or artificial means. These unprotected deposits may then be eroded by wave action.

diel Of or pertaining to a 24-hour day.

dielectric adj. Said of a material in which displacement currents predominate over conduction currents, i.e. an insulator.—n. Such a material.

dielectric loss The time rate of energy loss in a dielectric material due to conduction and hysteresis in polarization.

dielectric strength The maximum electric field that a dielectric can sustain without breakdown.

Dienerian European stage: Lower Triassic (above Griesbachian, below Smithian).

dienerite A gray-white isometric mineral: Ni_3As.

Diestian Northern European stage: Upper Miocene, approx. equivalent to Tortonian and Messinian.

dietella A small partly enclosed chamber near the base of the vertical walls of zooids in some cheilostome bryozoans. Its walls include one or more plates of interzooidal septulae. Syn: *pore chamber.*

dietrichite A mineral: $(Zn,Fe,Mn)Al_2(SO_4)_4 \cdot 22H_2O$.

dietzeite A dark golden-yellow mineral: $Ca_2(IO_3)_2(CrO_4)$.

differential compaction A kind of *compaction* produced by uneven settling of homogeneous earth material under the influence of gravity (as where thick sediments in depressions settle more rapidly than thinner sediments on hilltops) or by differing degrees of compactability of sediments (as where clay loses more interstitial water and comes to occupy less volume than sand).

differential curvature A quantity represented by the acceleration due to gravity times the difference in the curvatures in the two principal planes, that is, g(1/p1 - 1/p2), where p1 and p2 are the radii of curvature of the two principal planes.

differential entrapment The control of oil and gas migration and accumulation by selective trapping in interconnected reservoirs. A trap filled with oil is an effective gas trap but a trap filled with gas is not an effective oil trap. As a result, gas may be trapped downdip and oil updip (Gussow, 1954).

differential erosion Erosion that occurs at irregular or varying rates, caused by the differences in the resistance and hardness of surface materials: softer and weaker rocks are rapidly worn away, whereas harder and more resistant rocks remain to form ridges, hills, or mountains. Syn: *etching [geomorph].*

differential fault *scissor fault.*

differential leveling The process of measuring the difference of elevation between any two points by spirit leveling.

differential melting Partial melting of a rock, resulting from differences in melting temperatures of its constituent minerals.

differential pressure The difference in pressure between the two sides of an orifice; the difference between reservoir and sand-face pressure; between pressure at the bottom of a well and at the well head; between flowing pressure at the well head and that in the gathering line; any difference in pressure between upstream and downstream where a restriction to flow exists.

differential settlement Nonuniform *settlement;* the uneven lowering of different parts of an engineering structure, often resulting in damage to the structure.

differential solution *intrastratal solution.*

differential stress In experimental rock deformation, the maximum principal stress minus the least principal stress.

differential thermal analysis *Thermal analysis* carried out by uniformly heating or cooling a sample that undergoes chemical and physical changes, while simultaneously heating or cooling in identical fashion a reference material that undergoes no changes. The temperature difference between the sample and the reference material is measured as a function of the temperature of the reference material. Abbrev: DTA.

differential weathering Weathering that occurs at different rates, as a result of variations in composition and resistance of a rock or differences in intensity of weathering, and usually resulting in an uneven surface where more resistant material stands higher or protrudes above softer or less resistant parts. Syn: *selective weathering.*

differentiate n. A rock formed as a result of magmatic differentiation.

differentiated Said of an igneous intrusion in which there is more than one rock type, owing to *differentiation.*

differentiation [astrogeol] The processes by which planets and satellites develop concentric layers or zones of different chemical and mineralogical composition. An undifferentiated body would have a more homogeneous composition than a differentiated one (Lowman, 1976).

differentiation [intrus rocks] The process of developing more than one rock type, in situ, from a common magma. Cf: *assimilation.* Syn: *magmatic differentiation.*

differentiation [sed] (a) *sedimentary differentiation.* (b) *diagenetic differentiation.*

differentiation index In igneous petrology, the number that represents the sum of the weight percentages of normative quartz, orthoclase, albite, nepheline, leucite, and kalsilite (Thornton & Tuttle, 1960); a numerical expression of the extent of differentiation of a magma. Abbrev: DI. See also: *petrogeny's residua system.*

diffission Hobbs' term (1912, p. 204) for the natural process whereby rocks have been broken into fragments and blocks (as much as 8 m in diameter in western Texas) by a cloudburst or downpour of rain.

diffluence (a) A lateral branching or flowing-apart of a glacier in its ablation area. This separation may result from the glacier's spilling over a preglacial divide or through a gap made by basal sapping of a cirque wall, or from downvalley blocking at the junction of a tributary glacier. (b) A place at which diffluence occurs.—Ant: *confluence.*

diffluence pass The lower part of a trough end, where a distributary glacier has left the main valley.

diffluence step A rock step that rises downstream away from the main glacial valley at the place of diffluence. It is probably caused by the weakening of glacial action at that point. Ant: *confluence step.*

diffluent Said of a stream or glacier that flows away or splits into two or more branches. Ant: *confluent.*

diffraction [phys] The process by which the direction of wave motion in any medium is modified by bending around the edges of an obstacle, and the resultant formation of an interference pattern within the geometric shadow of the obstacle. See also: *electron diffraction; X-ray diffraction.*

diffraction [seis] (a) The generation and transmission in all directions of seismic wave energy in accordance with Huygens' principle. (b) An event observed on seismic data produced by diffracted energy. Such events result at the termination of reflectors and are characterized on seismic records and sections by a distinctive curved alignment.

diffraction [waves] The bending of a wave in a body of water around an obstacle, e.g. the interruption of a wave train by a breakwater or other barrier.

diffraction chart A chart showing the time-distance relationships that seismic diffractions should obey. It is used to identify diffractions, and for migration with certain techniques.

diffraction grating *grating.*

diffraction pattern The interference pattern of lines obtained when waves of rays, such as X-rays, light rays, or particle rays, are passed through a small opening or around the edge of a particle. Each substance has a characteristic diffraction pattern, which, when found, is taken to be evidence that that substance is present.

diffraction spacing In a crystal lattice, interplanar spacings given by a diffraction pattern.

diffractogram A record of diffraction of a crystalline sample, obtained by electronic detectors and recorded on a paper chart.

diffractometer In mineral analysis, an instrument that records either powder or single-crystal X-ray diffraction patterns.

diffuse layer The outer, mobile layer of ions in an electrolyte, required to satisfy a charge unbalance within a solid with which the electrolyte is in contact. It constitutes part of the *double layer* of charge adjacent to the electrolyte-solid interface. Cf: *fixed layer.*

diffuse-porous wood A type of wood in which the vessels are more or less uniform in size and distribution throughout each annual ring (Fuller & Tippo, 1954, p. 956).

diffusion metasomatism A process of mass transfer in which chemical components move by diffusion through a stationary aqueous solution occupying pores in rocks. Cf: *infiltration metasomatism.*

diffusivity *thermal diffusivity.*

digenite A blue to black mineral: Cu_9S_5. It is isometric and occurs with chalcocite. Syn: *blue chalcocite; alpha chalcocite.*

digestion Partial or complete assimilation of wall rock into a magma.

digital Said of the representation of measured quantities in discrete or quantized units. A digital system is one in which the information is stored and manipulated as a series of discrete numbers, as opposed to an *analog* system.

digital computer A *computer* that operates with numbers expressed directly as digits in a decimal, binary, or other system; a counting device that operates on discrete or discontinuous variables represented by digits of numbers and performs arithmetic by manipulating the digits and executing the basic arithmetic operations in a manner similar to a human mathematician. Cf: *analog computer; hybrid computer.*

digital log A *well log* whose curves have been discretely sampled and recorded on a magnetic tape for use in computer-processed interpretation and plotting.

digitate delta A *bird-foot delta* whose seaward-extending margin

has a fingerlike outline in plan.
digitation The emanation of subsidiary recumbent anticlines from a larger recumbent anticline.
digitization The process of converting analog data (such as an image on photographic film) into numerical format.
digitize To sample a continuous function at discrete time intervals and to record the values as a sequence of numbers.
digitized map (a) A map expressed or stored in digital form. (b) A map prepared from cartographic information that has been converted from analog to digital form for use in automatic plotters.
digonal *diad.*
digue *dike [eng].*
dihexagonal Said of a symmetrical twelve-sided figure, the alternate angles of which are equal. Such a figure is common in crystals of the hexagonal system.
dihexagonal dipyramid A crystal form that is a dipyramid of 24 faces, in which any section perpendicular to the sixfold axis is dihexagonal. Its indices are $\{hkl\}$ and its symmetry is $6/m\ 2/m\ 2/m$.
dihexagonal dipyramidal class That crystal class of the hexagonal system having symmetry $6/m\ 2/m\ 2/m$.
dihexagonal prism A crystal form of twelve faces parallel to the symmetry axis, in which any cross section perpendicular to the prism axis is dihexagonal. Its indices are $\{hk0\}$ with symmetry $6/m\ 2/m\ 2/m$.
dihexagonal pyramid A crystal form consisting of a pyramid of 12 faces, in which any cross section perpendicular to the sixfold axis is dihexagonal. Its indices are $\{hkl\}$ or $\{hk\bar{l}\}$ in symmetry $6mm$.
dihexagonal-pyramidal class That crystal class in the hexagonal system having symmetry $6mm$.
dihydrite *pseudomalachite.*
dike [eng] An artificial wall, embankment, ridge, or mound, usually of earth or rock fill, built around a relatively flat, low-lying area to protect it from flooding; a levee. A dike may also be constructed on the shore or border of a lake to prevent inflow of undesirable water. Syn: *digue; dyke.*
dike [intrus rocks] A tabular igneous intrusion that cuts across the bedding or foliation of the country rock. Also spelled: *dyke.* Cf: *sill [intrus rocks]; sheet [intrus rocks].* See also: *dikelet.*
dike [sed] *sedimentary dike.*
dike [streams] (a) An artificial watercourse; esp. a deep drainage ditch. The term has also been applied to any channel, including those formed naturally. (b) A pool or small pond.
dikelet A small *dike.* There is no agreement on specific size distinctions.
dike ridge (a) *dike wall.* (b) A small wall-like ridge (as one along a shore) produced by differential erosion.
dike rock The intrusive rock comprising a dike.
dike set A group of linear or parallel dikes. Cf: *dike swarm; cluster.*
dike spring A spring issuing from the contact between a dike composed of an impermeable rock, such as basalt or dolerite, and a permeable rock into which the dike was intruded.
dike swarm A group of dikes, which may be in radial, parallel, or en echelon arrangement. Their relationship with the parent plutonic body may not be directly observable. Cf: *cluster; dike set.*
dike wall A ridge, such as a *hogback,* consisting of a dike that formed in a more or less vertical crevice and was left standing after the rocks on either side were removed by erosion. Syn: *dike ridge.*
diktytaxitic Said of a rock texture, such as that of some olivine basalts in the NW U.S., that is characterized by numerous jagged, irregular vesicles bounded by crystals, some of which protrude into the cavities (Dickinson, 1965, p. 101). Not recommended usage.
dilatancy An increase in the bulk volume during deformation, caused by a change from close-packed structure to open-packed structure, accompanied by an increase in the pore volume. The latter is accompanied by rotation of grains, microfracturing, and grain boundary slippage.
dilatated septum A partly or wholly thickened septum of a rugose coral.
dilatation [exp struc geol] *dilation.*
dilatation [seis] *kataseism.*
dilatational transformation In a crystal, usually rapid thermal dilation and a rearrangement of the anion from cubic coordination to octahedral coordination, due to heating, e.g. the *transformation [cryst]* of CsCl to the NaCl structure at 460°C.
dilatational wave *P wave.*
dilatation theory The theory that attributed glacier movement to infiltration and freezing of water in cracks and other openings.
dilation Deformation by a change in volume but not shape. Also spelled: *dilatation.*
dilation vein A mineral deposit in a vein space formed by bulging of the walls, contrasted with veins formed by wall-rock replacement.
dillnite A fluorine-rich variety of zunyite.
Diluvial That period of geologic time since the appearance of man. Cf: *anthropozoic.*
diluvial (a) Pertaining to, produced by, or resembling a flood, esp. the Noachian flood; e.g. *diluvial* deposits. (b) Pertaining to *diluvium.*
diluvialist A believer in *diluvianism.* Cf: *fluvialist.*
diluvianism The doctrine that the widespread surficial deposits now known to be glacial drift, and other geologic phenomena, can be explained by a former worldwide flood or deluge, esp. the Noachian Flood.
diluvion (a) *diluvium.* (b) A term used in India as an ant. of *alluvion [law];* "it appears to mean loss of land by river erosion after flooding" (G.T. Warwick in Stamp, 1961, p. 157).
Diluvium A term used in continental Europe equivalent to Pleistocene.
diluvium (a) An archaic term applied during the early 1800s to certain widespread surficial deposits that could not be explained by the normal action of rivers and seas but were believed to be produced by extraordinary floods of vast extent, esp. the Noachian Flood; these deposits are now known to be mostly glacial drift. (b) A general term used in continental Europe for the older Quaternary, or Pleistocene, glacial deposits, as distinguished from the younger *alluvium.* Syn: *drift [glac geol].*—Syn: *deluvium; diluvion.*
dimensional orientation In structural petrology, a tendency for planar or linear fabric elements to be so arranged that a preferred orientation develops defined by an alignment of these elements.
dimension stone Building stone that is quarried and prepared in regularly shaped blocks according to specifications.
dimictic Said of a lake with two yearly overturns or periods of circulation, such as a deep freshwater lake in a temperate climate, with overturns in the spring and fall. Cf: *monomictic [lake].*
diminutive fauna *dwarf fauna.*
dimorph Either of two crystal forms displaying *dimorphism.* Syn: *allomorph.*
dimorphic [bot] Said of a plant or plant part that is normally produced in two forms, e.g. juvenile and adult types of foliage.
dimorphic [cryst] *dimorphous.*
dimorphism [biol] The characteristic of having two distinct forms in the same species, such as male and female, or megaspheric and microspheric stages.
dimorphism [cryst] That type of *polymorphism [cryst]* in which two crystalline species, known as *dimorphs,* occur. Adj: *dimorphous.* Cf: *trimorphism; tetramorphism.*
dimorphite An orange-yellow mineral: As_4S_3. It was originally described as one of two dimorphous substances (the other, however, being orpiment).
dimorphous Adj. of *dimorphism [cryst].* Syn: *dimorphic.*
dimple crater A small, almost circular craterlike feature restricted to mare regions of the Moon's surface and attributed to volcanic activity (possibly to withdrawal of molten subsurface lava). It lacks the raised rim of most lunar impact craters.
dimpled current mark An obsolete syn. of *cross ripple mark.*
dimyarian adj. Said of a bivalve mollusk or its shell with two adductor muscles, whether equal (isomyarian) or unequal (anisomyarian) in size. Cf: *monomyarian.*—n. A dimyarian mollusk.
Dinantian European stage: Lower Carboniferous. It includes Tournaisian and Viséan.
Dinarides The mountain belt that stretches from northeastern Italy to Greece, paralleling the eastern Adriatic coast. It constitutes the southwest-vergent branch of the Alpine mountain system.
Dinas rock A disintegrated sandstone of high silica content, formerly used for making refractory brick. Type locality: Craig-y-Dinas, a crag in south Wales. Syn: *Dinas clay.*
dinite A yellowish crystalline hydrocarbon mineral with a low melting point (30°C), found in lignite.
dinoflagellate A one-celled microscopic flagellated organism,

chiefly marine and usually solitary, with resemblances to both animal and plant kingdoms. It is characterized by one transverse flagellum encircling the body and usually lodged in the girdle and one posterior flagellum extending out from a similar median groove. Certain dinoflagellates have a theca or *test [paleont]* that is resistant to decay; it may be simple and smooth or variously sculptured and divided into characteristic plates and grooves. Others produce a resting stage or *cyst [palyn]* with a resistant organic wall that is often spiny and may differ markedly from the theca of the same species. Both thecae and cysts exist abundantly as fossils, and have a range primarily Triassic to present. Dinoflagellates are known from the Paleozoic, but are mainly important for correlating and dating Jurassic, Cretaceous, and Tertiary deposits. They inhabit all water types and are capable of extensive diurnal vertical migrations in response to light; they constitute a significant element in marine plankton, including certain brilliantly luminescent forms and those that cause *red tide.* See also: *hystrichosphaerid.*

dinosaur Any reptile of the subclass Archosauria distinguished from other reptiles especially by features of the pelvic bones. Dinosaurs were carnivorous or herbivorous, bipedal or quadripedal, land-dwelling, and of moderate to very large size. Range, Triassic to Cretaceous. See also: *Ornithischia; Saurischia.*

dinosaur leather A local term applied by Chadwick (1948) to complex sole marks, probably including both flute casts and load casts. Cf: *squamiform cast.*

Dinoseis A trade name for a seismic energy source in which a plate is driven against the ground by a confined explosion of gas.

dioctahedral Said of a layered-mineral structure in which only two of the three available octahedrally coordinated positions are occupied. Cf: *trioctahedral.*

dioecious Said of a taxonomic unit of plants with male and female reproductive organs entirely separated on different individuals.

diogenite An achondritic stony meteorite composed essentially of bronzite or hypersthene. Syn: *rodite.*

diopside A mineral of the clinopyroxene group: $CaMg(SiO_3)_2$. It contains little or no aluminum and may contain some iron. It ranges in color from white to green; transparent varieties are used in jewelry. Diopside occurs in some metamorphic rocks, and is found esp. as a contact-metamorphic mineral in crystalline limestones. Symbol: Di. Syn: *malacolite.*

diopside-jadeite *tuxtlite.*

diopsidite A *pyroxenite* composed almost entirely of diopside, with iron-titanium oxides, spinel, and garnet as common accessories. Not recommended usage.

dioptase A rare emerald-green hexagonal mineral: $CuSiO_3 \cdot H_2O$. It occurs in the zone of weathering of copper lodes in Chile and Siberia. Syn: *emerald copper.*

diorite (a) In the *IUGS classification,* a plutonic rock with Q between 0 and 5, P/(A+P) greater than 90, and plagioclase more sodic than An_{50}. (b) A group of plutonic rocks intermediate in composition between acidic and basic, characteristically composed of dark-colored amphibole (esp. hornblende), acid plagioclase (oligoclase, andesine), pyroxene, and sometimes a small amount of quartz; also, any rock in that group; the approximate intrusive equivalent of *andesite.* Diorite grades into *monzonite* with an increase in the alkali feldspar content. In typical diorite, plagioclase contains less than 50% anorthite, hornblende predominates over pyroxene, and mafic minerals total less than 50% of the rock. Etymol: Greek *diorizein,* "to distinguish", in reference to the fact that the characteristic mineral, hornblende, is usually identifiable megascopically. Cf: *dolerite; gabbro.* See also: *diabase.*

dioritoid In the *IUGS classification,* a preliminary term (for field use) for a plutonic rock with Q less than 20 or F less than 10, and P/(A+P) greater than 65.

dip [geomorph] (a) A low place or marked depression in the land surface; e.g. a steep-sided hollow among hills or a gap in a ridge. (b) A pronounced depression in a highway at the point of intersection with a dry stream bed, once common in the western U.S.

dip [magnet] *inclination [magnet].*

dip [seis] (a) The angle between a reflecting or refracting seismic wave front and the horizontal. (b) The angle between an interface associated with a seismic event and the horizontal.—Cf: *apparent dip [seis].*

dip [struc geol] n. The angle that a structural surface, e.g. a bedding or fault plane, makes with the horizontal, measured perpendicular to the *strike* of the structure and in the vertical plane. See also: *attitude; hade; inclination.* Syn: *true dip; angle of dip.*—v. To be tilted or inclined at an angle.

dip [surv] (a) The vertical angle, at the eye of an observer, between the plane of the horizon and the line of sight tangent to the apparent (visible or sensible) horizon; the angular distance of the apparent horizon below the horizontal plane through the observer's eye. See also: *dip angle.* Also called: "dip of horizon". (b) The apparent depression of the visible horizon due to the observer's elevation and to the convexity of the Earth's surface. (c) The first detectable decrease in the altitude of a celestial body after reaching its maximum altitude on or near the meridian transit.

dip angle The vertical angle, measured at an observation point in surveying or at an exposure station in photogrammetry, between the plane of the true horizon and a line of sight to the apparent horizon. See also: *dip.*

dip calculation Calculation of the dip of a reflecting interface from observations of the variation of the arrival time of seismic events as the observing point is moved. It is often associated with *migration.* See also: *moveout.*

dip circle An obsolete type of *inclinometer [magnet].*

dip-corrected map A map that shows strata in their original position before movement.

dip equator *magnetic equator.*

dip fault A fault that strikes parallel with the dip of the strata involved. Cf: *strike fault; oblique fault.*

diphyletic Said of a higher taxon that artificially includes descendants of two separate evolutionary lineages.

dip isogon A line joining points of equal dip. A classification of folds by Ramsay (1967, p. 363) is based on dip isogons.

dip joint A joint whose strike is approximately perpendicular to the bedding or cleavage of the containing rock. Cf: *strike joint.*

dipleural Said of the arrangement of the two rows of thecae (the stipes) in the biserial rhabdosome of a scandent graptoloid in which the rows are in contact back-to-back. Cf: *monopleural.*

diploblastic Said of the structure of lower invertebrates (sponges, coelenterates) having ectodermal and endodermal layers but lacking a true mesoderm.

diploclone A sublithistid desma (of a sponge) consisting of a shaft bearing expansions at each end but not articulating with neighboring spicules.

diploconical Said of a radiolarian shell formed by fusion of the bases of two cones opposite in one axis (TIP, 1954, pt.D, p.14).

diplodal Said of a flagellated chamber of a sponge that has both an aphodus and a prosodus.

dip log *dipmeter.*

diplogenetic Said of a mineral deposit that is in part *syngenetic* and in part *epigenetic* in origin (Lovering, 1963, p.315-316).

diplohedron *diploid.*

diploid n. A crystal form of the isometric system having 24 similar quadrilateral faces in a paired arrangement. Each face intersects the crystallographic axes at unequal lengths. Its indices are $\{hkl\}$ in symmetry $2/m\bar{3}$. Syn: *didodecahedron; diplohedron; dyakisdodecahedron.*

diploidal class That crystal class in the isometric system having symmetry $2/m\bar{3}$.

diplopore Any of double pores piercing a thecal plate in certain cystoids and mostly confined to that plate. It may be unbranched but it usually consists of a Y-shaped branching canal or tube that is oblique or perpendicular to the surface of the plate and that has two external openings. Cf: *pore rhomb; haplopore.*

Diplorhina A subclass of Agnatha characterized presumably by paired nostrils and assuredly by uniquely acellular bone.

diplorhysis The condition of a dictyonine hexactinellid sponge in which both epirhyses and aporhyses are present.

diploxylonoid Said of bisaccate pollen, in which the outline of the sacci in distal-proximal view is discontinuous with the body outline so that the grain appears to consist of three distinct, more or less oval figures. Cf: *haploxylonoid.*

dipmeter A 3- or 4-pad *wall-contact log* whose finely detailed *microresistivity log* curves are correlated to measure depth offsets relative to each other. In conjunction with simultaneous measurements of the caliper, inclination, and direction of the borehole, such measurements can be solved for dip and strike of the strata. Both the borehole curves as measured and the subsequent graphic plot of computed dip-strike symbols are called dipmeters, the former a "continuous dipmeter" or *dipmeter log,* the latter a "computed dipmeter" or "tadpole plot".

dipmeter log *dipmeter.*

dip needle An obsolete type of magnetometer used for mapping high-amplitude magnetic anomalies. It consists of a magnetized needle pivoted to rotate freely in a vertical plane, with an adjustable weight on the south side of the magnet.

Dipnoi An order of lobefinned bony fish characterized by non-hinged braincase, holostylic jaw suspension, and teeth distinctively modified into durophagous crushing plates, one up and one down on each side. Syn: *lungfish.* Range, Lower Devonian to Recent.

dipole Two poles of opposite charge an infinitesimal distance apart.

dipole-dipole array An electrode array in which one dipole provides current to the ground and an adjacent dipole is used to measure potential in the ground. The separation between dipoles is usually comparable to or greater than the spacing within each electrode pair constituting a dipole. The potential dipole lies entirely outside the current dipole.

dipole field A mathematically simple magnetic field, having an axis of symmetry, with magnetic-field lines pointing outward along one half of the axis (*positive pole*) and inward along the other half (*negative pole*). Most magnetic fields that are sufficiently remote from their source resemble a dipole field. See also: *axial dipole field.*

diporate Said of pollen grains having two pores.

dipping rod *divining rod.*

dip plain A *stratum plain* coincident in slope with the dip of the underlying resistant rock (Hill, 1891, p. 522).

dip pole One of the locations on the Earth where the horizontal magnetic field is zero and the magnetic inclination is $\pm 90°$. Approx. syn: *magnetic pole.*

dip reversal *rollover.*

dip separation The distance of *separation* of formerly adjacent beds on either side of a fault surface, measured along the dip of the fault. Cf: *dip slip; strike separation.* See also: *dip-separation fault.*

dip-separation fault A fault on which the displacement has been *dip separation.* Cf: *lateral fault.*

dip shift In a fault, the *shift* or relative displacement of the rock units parallel to the dip of the fault, but outside the fault zone itself. Cf: *dip slip; strike shift.*

dip shooting A system of seismic surveying in which the primary concern is determining the dip and position of reflecting interfaces rather than in tracing such interfaces continuously.

dip slip In a fault, the component of the movement or slip that is parallel to the dip of the fault. Cf: *dip separation; strike slip; oblique slip; dip shift.* Syn: *normal displacement.*

dip-slip fault A fault on which the movement is parallel to the dip of the fault. Cf: *strike-slip fault.*

dip slope A slope of the land surface, roughly determined by and approximately conforming with the direction and the angle of dip of the underlying rocks; specif. the long, gently inclined face of a cuesta. Cf: *scarp slope.* Syn: *back slope; outface.*

dip stream A consequent stream flowing in the general direction of dip of the strata it traverses; a *cataclinal* stream.

dip throw The component of the slip of a fault measured parallel with the dip of the strata.

dip valley A valley trending in the general direction of dip of the strata of a region; a valley at right angles to a subsequent stream.

dipyramid A closed crystal form consisting of two *pyramids* that are arranged base to base so that they appear as mirror images across the plane of symmetry. Adj. *dipyramidal.* Syn: *bipyramid.*

dipyramidal Having the symmetry of a *dipyramid.*

dipyre A syn. of *mizzonite;* specif. a term applied to a variety of scapolite with the components marialite and meionite in a ratio of about 3:1 to 3:2. Syn: *dipyrite.*

dipyrite (a) *dipyre.* (b) *pyrrhotite.*

direct-action avalanche A snow avalanche that occurs during or immediately after a storm and is the direct result of that storm. The term is part of an obsolete classification of avalanches.

direct angle An angle measured directly between two lines; e.g. a horizontal angle measured clockwise from a preceding surveying line to a following one.

direct intake Recharge to the aquifer directly through the zone of saturation.

direction (a) The position of one point relative to another without reference to the distance between them. It may be three-dimensional or two-dimensional. (b) The angle between a line or plane and an arbitrarily chosen reference line or plane; specif. the angle between a great circle passing through both the position of the observer and a given point on the Earth's surface and a true north-south line passing through the observer. When the reference line is north and the angle is designated east or west, the direction is called the *bearing;* when the reference line is south and the angle is reckoned clockwise, the direction is called the *azimuth.* (c) A syn. of *trend.*

directional drilling The intentional drilling of a well at controlled departures from the vertical and at controlled azimuths, often utilizing a *whipstock.* It is done to establish multiple wells from a single location such as an offshore platform, and for other purposes. Cf: *deviation; sidetracking.* Syn: *slant drilling.*

directional load cast A term originally applied to a structure interpreted as a flowage cast, but now regarded as a flute cast (Pettijohn & Potter, 1964, p.301).

directional log A *well log* that shows the inclination of a borehole, and the direction of the inclination. It is usually obtained with the *dipmeter* log.

directional structure Any sedimentary structure that indicates the direction of the current that produced it; e.g. cross-bedding, current marks, and ripple marks. Syn: *paleocurrent structure; aligned current structure; vector structure.*

directional survey (a) Determination of the direction and *deviation* from the vertical of a borehole by precise measurements at various points along its central axis. Also, the record of the information thus obtained. Syn: *borehole survey.*

directional well A well produced by *directional drilling.* Syn: *slant well.*

direction instrument theodolite A *theodolite* in which the graduated horizontal circle remains fixed during a series of observations, the telescope being pointed on a number of signals or objects in succession, and the direction of each read on the circle, usually by means of micrometer microscopes. Instrument theodolites are used almost exclusively in first- and second-order triangulation. Syn: *direction theodolite.*

direction of dip *line of dip.*

direction of the wind That point of the compass from which the wind blows, e.g. a "westerly" wind is blowing from the west. It may also be stated in degrees, measured clockwise from the north, e.g. an east wind has a direction of 90°. Syn: *wind direction.*

direction of tilt (a) The azimuth of the principal plane of a photograph. (b) The direction of the principal line on a photograph.

direction theodolite *direction instrument theodolite.*

directive couple A pair of mesenteries in the so-called dorsoventral plane (extending from the dorsal toward the ventral side) of a coral polyp, characterized by pleats on the opposite rather than the facing sides of the mesenteries.

directivity graph (a) A plot of relative intensity versus direction of an outgoing seismic wave, such as that resulting from a directional charge or from a source pattern. The directivity results from the interference of waves from the various components of the pattern. (b) A plot of the relative response of a geophone pattern or of directivity resulting from mixing.

direct lattice A syn. of *crystal lattice,* used when comparison is made with the *reciprocal lattice.*

direct leveling A type of *leveling* in which differences of elevation are determined by means of a continuous series of short horizontal lines, the vertical distances from these lines to adjacent ground marks being determined by direct observations on graduated rods with a leveling instrument equipped with a spirit level. Cf: *indirect leveling.*

direct linkage A type of *linkage* in scleractinian corals with mesenterial strands connecting the adjacent stomodaea. See also: *lamellar linkage.* Cf: *indirect linkage.*

direct runoff The *runoff* reaching stream channels immediately after rainfall or snow melting (Langbein & Iseri, 1960). Cf: *base runoff.* Syn: *direct surface runoff; immediate runoff; stormflow; storm runoff; storm water.*

direct stratification *primary stratification.*

direct surface runoff *direct runoff.*

direct tide An oceanic tide that is in phase with the apparent motions of the tide-producing body, so that high tide is directly under the tide-producing body and is accompanied by a high tide on the opposite side of the Earth. Cf: *reversed tide; opposite tide.*

Dirichlet's problem A well-known problem in geodesy: to determine a function that is harmonic outside of a given surface and assumes prescribed boundary values on the surface. Cf: *boundary-value problem; Neumann's problem.*

dirt band [coal] A thin stratum of shale or other inorganic rock material in a coal seam. Syn: *band [coal]; dirt bed [coal]; dirt* *parting; stone band.*

dirt band [glaciol] (a) Any dark layer in a glacier, usually the trace of silt or debris along a *summer surface.* (b) A dark band below an icefall that may be related to dirt collected in the broken ice of the icefall or between the ridge of one *wave ogive* and another.—Syn: *dust band.* (c) A term that was originally applied to a *Forbes band.* Cf: *dirt-band ogive.*

dirt-band ogive A curved band or *ogive* composed of debris-laden or dirt-laden ice that may be related to dirt collected in the broken ice of an icefall or between the ridge of one *wave ogive* and another. Cf: *dirt band [glaciol]; Forbes band.* See also: *Alaskan band.*

dirt bed [coal] *dirt band [coal].*

dirt bed [soil] A *paleosol* whose organic material is only partially decayed. It sometimes occurs in glacial drift.

dirt cone A cone or mound of ice or snow on a glacier, covered with a veneer of silt thick enough to protect the underlying material from the ablation that has lowered the surrounding surface. Cf: *debris cone [glaciol].*

dirt parting *dirt band [coal].*

dirt slip *clay vein.*

dirty arkose *impure arkose.*

dirty sand A term used in electrical prospecting for a sandstone that contains abundant clay and hence exhibits appreciable membrane polarization and abnormally high electrical conductivity because of surface conduction along the clay minerals.

dirty sandstone A sandstone containing much matrix; specif. a *wacke* with more than 10% argillaceous matrix (Williams, Turner & Gilbert, 1954) or a *graywacke* with more than 15% detrital clay matrix (Pettijohn, 1954). The particles are held together by primary, interstitial detritus of clay-like nature, or by authigenic derivatives of such material. It is usually deposited by fluids of high density or high viscosity. Cf: *clean sandstone.*

disaccate *bivesiculate.*

disaggregation Separation or reduction of an aggregate into its component parts; specif. *mechanical weathering.*

disappearing stream *sinking stream.*

disc (a) The central part of the body of an echinoderm, more or less distinctly separable from its arms. Sometimes spelled *disk.* (b) A discoidal, typically imperforate sclerite of a holothurian.

discharge [hydraul] The rate of flow at a given moment, expressed as volume per unit of time. See also: *specific discharge.*

discharge [sed] *sediment discharge.*

discharge area An area in which subsurface water, including both ground water and vadose water, is discharged to the land surface, to bodies of surface water, or to the atmosphere. Cf: *recharge area.*

discharge coefficient That coefficient by which a theoretical discharge must be multiplied to obtain the actual discharge. It is the product of the contraction coefficient and the velocity coefficient (ASCE, 1962).

discharge efficiency *drainage ratio.*

discharge-rating curve *stage-discharge curve.*

discharge velocity The rate of discharge of water through a porous medium, measured per unit of total area perpendicular to the direction of flow (ASCE, 1962).

disciform Of round or oval shape; e.g. "disciform holococcolith" having a discolithlike shape and a raised margin two or more cycles of microcrystals high.

discinacean Any inarticulate brachiopod belonging to the superfamily Discinacea, characterized by holoperipheral growth of the brachial valve, with the beak marginal to central.

discoaster One of the tiny star- or rosette-shaped calcareous plates, 10-35 microns in diameter, that are generally believed to be the remains of a planktonic organism and that may be isolated coccolithlike bodies of either a motile cell or a cyst. Discoasters are common in Tertiary deposits but are apparently absent in the Pleistocene; the level where they disappear has been suggested as a Pliocene-Pleistocene boundary. See also: *asterolith.*

discohexaster A hexactinal sponge spicule (microsclere) in which the ray tips bear branches terminated by umbels.

discoid adj. Having the shape of a disk; e.g. a solitary corallite. Syn: *discoidal.*—n. An object having such a shape.

discoidal *discoid.*

discolith A discoidal coccolith with a single, apparently imperforate elliptical or circular shield and a thickened margin. Cf: *tremalith.*

disconformable [intrus rocks] Said of the contact of an intrusive body that is not essentially parallel to the intrusion's internal structures. Cf: *conformable [intrus rocks]; discordant [intrus rocks].*

disconformable [stratig] Pertaining to a *disconformity.* Term proposed by Grabau (1905, p.534) to refer to formations that exhibit parallel bedding but "comprise between them a time break of greater or less magnitude".

disconformity An *unconformity* in which the bedding planes above and below the break are essentially parallel, indicating a significant interruption in the orderly sequence of sedimentary rocks, generally by a considerable interval of erosion (or sometimes of nondeposition), and usually marked by a visible and irregular or uneven erosion surface of appreciable relief; e.g. an unconformity in which the older rocks remained essentially horizontal during erosion or during simple vertical rising and sinking of the crust (without tilting or faulting). The tendency is to apply the term to breaks represented elsewhere by rock units of at least formation rank (Stokes & Varnes, 1955, p.157). The term formerly included what is now known as *paraconformity.* Syn: *parallel unconformity; erosional unconformity; nonangular unconformity; stratigraphic unconformity; paraunconformity.*

discontinuity [seis] A surface at which seismic-wave velocities abruptly change; a boundary between seismic layers of the Earth. Syn: *interface [seis]; seismic discontinuity; velocity discontinuity.*

discontinuity [stratig] Any interruption in sedimentation, whatever its cause or length, usually a manifestation of nondeposition and accompanying erosion; an unconformity. Syn: *break [stratig].*

discontinuity [struc geol] A surface separating two unrelated groups of rocks; e.g. a fault or an unconformity. See also: *discrete.*

discontinuity layer A *thermocline* in a lake or ocean.

discontinuous deformation (a) Deformation by fracture rather than flow. Cf: *continuous deformation.* (b) Deformation with development of kinematic discontinuities, e.g. fractures or cleavage planes.

discontinuous gully A gully with a vertical headcut at the upstream end and a fan at the point where its floor intersects the more steeply sloping plane of the original valley floor (Leopold et al., 1964, p. 448-449). The depth of its channel decreases downstream.

discontinuous permafrost A zone of permafrost containing patches of unfrozen ground, as beneath large rivers or lakes; it occurs in an intermediate zone between the northerly *continuous permafrost* and the southerly *sporadic permafrost.*

discontinuous porosity A term proposed by Murray (1930, p. 452) for poorly connected or isolated pores, as opposed to *continuous porosity.* The term is little used and is not recommended (Choquette & Pray, 1970, p. 245).

discontinuous reaction A metamorphic reaction that, despite compositional variability among the minerals, is at equilibrium, given a fixed pressure, at only one temperature. A paragenetic diagram, for example the *AFM projection,* will show a change in topology, such as "a *tie line* flip". Cf: *continuous reaction.*

discontinuous reaction series A *reaction series* in which reaction of early-formed crystals with later liquid represents an abrupt phase change; e.g., the minerals olivine, pyroxene, amphibole, and biotite form a discontinuous reaction series. Cf: *continuous reaction series.*

discordance (a) Lack of parallelism between adjacent strata. The term was used by Willis (1893, p.222) in cases where the cause is in doubt. Although the term has not been widely adopted, "it appears eminently suitable for descriptive use where there is insufficient evidence to decide between stratigraphic discordance (unconformity) and tectonic discordance (e.g. overthrust, slide, detachment fault)" (Dennis, 1967, p.36). Cf: *concordance.* (b) *angular unconformity.*

discordance index A numeric statistic used by Pearn (1964, p.401) to represent the amount of deviation of any actual rock sequence from the *ideal cyclothem.* It is defined as the minimum value of the number of missing lithologic units. Symbol: G.

discordant [geochron] (a) Said of radiometric ages, determined by more than one method for the same sample or coexisting minerals, that are in disagreement beyond experimental error. (b) Said of ages given by coexisting minerals determined by the same method that are in disagreement. (c) In a more restricted sense, the term has been used to indicate disagreement of $^{238}U/^{206}Pb$, $^{235}U/^{207}Pb$, $^{207}Pb/^{206}Pb$, and $^{232}Th/^{208}Pb$ ages determined for the same mineral sample. Discordant ages usually imply that one or more of the isotopic systems used for dating purposes has been disturbed by some geologic event (metamorphism, weathering) following the initial formation of the geologic material or by inadvertent laboratory procedures. Ant: *concordant [geochron].*

discordant [geomorph] Said of topographic features that do not have the same or nearly the same elevation; e.g. a *discordant* valley whose stream enters the main stream via a waterfall, or a *discordant* lip over which the floor of a hanging valley passes into the floor of the main valley. Ant: *accordant.*

discordant [intrus rocks] Said of a contact between an igneous intrusion and the country rock that is not parallel to the foliation or bedding planes of the latter. Cf: *concordant [intrus rocks]; disconformable [intrus rocks].*

discordant [stratig] Structurally *unconformable;* said of strata lacking conformity or parallelism of bedding or structure. Ant: *concordant [stratig].*

discordant bedding A sedimentary structure in which parallelism of beds is lacking or in which sedimentary layers are inclined to the major lines of deposition, such as bedding developed by rapid deposition of material from heavily laden currents of air or water; specif. *cross-bedding.* Ant: *concordant bedding.* See also: *inclined bedding.*

discordant coastline A coastline that develops where the general structural grain of the land (such as mountain chains or folded belts) is transverse to the margin of the ocean basin, and that represents rifting, faulting, subsidence, or other interruption of a formerly continuous and harmonious structure; it is generally irregular, with many inlets. Ant: *concordant coastline.* Syn: *Atlantic-type coastline; transverse coastline.*

discordant drainage Drainage that has not developed in a systematic relationship with, and is not consequent upon, the present geologic structure. Ant: *accordant drainage.*

discordant fold A fold whose axis is inclined to that of the *longitudinal fold* axes of the area.

discordant junction The joining of two streams or two valleys whose surfaces are at markedly different levels at the place of junction, as the abrupt entry of a tributary flowing at a high level into a main stream at a lower level. Ant: *accordant junction.*

discordant margin A margin of closed valves (of a bivalve mollusk) not in exact juxtaposition, but with one overlapping the other.

discordia A on line on a plot of $^{206}Pb/^{238}U$ versus $^{207}Pb/^{235}U$, formed by data on phases that have lost lead or gained uranium during a period of time that is short compared with the age of the phase.

discordogenic fault A fault in a tectonic belt that separates zones of uplift and subsidence, and that remains active during several geologic periods (Nikolaev, 1959).

discorhabd A sponge spicule (streptaster) consisting of a straight shaft bearing whorls of spines or transverse discoidal flanges. Syn: *chessman spicule.*

discotriaene A sponge *triaene* in which the cladome is represented by a transverse disk containing the axial canals of the three cladi.

discovery The actual finding of a valuable mineral, indicative of a deposit (lode, placer, or coal seam). Legally, a discovery is a prerequisite to making a mining claim on an area.

discovery claim A claim containing the original discovery of exploitable mineral deposits in a given locale, which may lead to claims being made on adjoining areas.

discovery vein The original mineral deposit on which a mining claim is based. Cf: *secondary vein; discovery claim.*

discovery well The first well to encounter gas or oil in a hitherto unproven area or at a hitherto unproductive depth; a successful *wildcat, outpost well, deeper-pool test,* or *shallower-pool test.*

discrepancy In surveying, the difference in computed values of a quantity obtained by different processes using data from the same survey; also, the difference between results of duplicate or comparable measures of a quantity. See also: *accumulated discrepancy.*

discrete [paleont] Said of conodont denticles that are not closely set, each denticle being separated from adjacent ones by open space. Cf: *appressed.*

discrete [struc geol] Said of any body of rock which has a definite boundary with adjacent rocks in space. See also: *discontinuity.*

discrete [weath] A term proposed by Gilbert (1898) to describe the surficial, weathered, and unconsolidated material composing the regolith.

discrete-film zone *belt of soil water.*

discriminant analysis A statistical procedure for classifying samples in categories previously defined and differentiated on the basis of samples from known populations. Syn: *discriminant function analysis.*

disembogue To discharge water through an outlet or into another body of water, such as a stream *disemboguing* into the ocean. Cf: *bogue.*

disequilibrium assemblage An association of minerals not in thermodynamic equilibrium.

disharmonic fold A fold that varies noticeably in profile form in the various layers through which it passes. Ant: *harmonic fold.*

disharmonic folding Folding in which there is an abrupt change in fold profile when passing from one folded surface or layer to another. It is characteristic of rock layers that have significant contrasts in viscosity (Whitten, 1966, p. 606). An associated structure is *décollement.* Ant: *harmonic folding.*

dish structure A primary sedimentary structure, generally found in sandstone, consisting of small meniscus-shaped lenses (4-50 cm long and one to a few centimeters thick) that are oval in plan, oriented parallel to the bedding, and defined by slightly finer-grained, concave-up bottoms each of which truncates the underlying lenses. It is thought to form as a result of elutriation of clay by pore water escaping soon after deposition of the sand. Term introduced by Wentworth (1967); see also Lowe and LoPiccolo (1974).

disintegration [coal] The decomposition of vegetable matter by slow combustion, in which there is no formation of carbon compounds and in which only volatile substances (carbon dioxide and water) are produced. Cf: *moldering; peat formation; putrefaction.*

disintegration [glaciol] *ice disintegration.*

disintegration [radioactivity] *radioactive decay.*

disintegration [weath] A syn. of *mechanical weathering.* Sometimes the term includes chemical action, in which case it is practically synonymous with *weathering.*

disintegration constant *decay constant [radioactivity].*

disjunct (a) Said of a cystoid pore rhomb in which externally visible slits forming parts of the rhomb are separated by solid areas of plates. Cf: *conjunct.* (b) Said of the apical system of an echinoid whose anterior part is separated from its posterior part.

disjunctive fold A fold in which the more brittle strata have fractured and separated and the more plastic beds have flowed under the forces of deformation.

disk [paleont] (a) The flattened circumoral part of a coelenterate (such as a sea anemone). See also: *oral disk; basal disk.* (b) *disc.*

disk [sed] A notably discoidal (flat and circular), or oblate or tabular, shape of a sedimentary particle, defined in *Zingg's classification* as having a width/length ratio greater than 2/3 and a thickness/width ratio less than 2/3.

disk hardness gage A device for measuring the penetration resistance of snow. Metal disks of various sizes are pressed against a snow surface by a calibrated spring until collapse occurs.

dislocation [cryst] *line defect.*

dislocation [struc geol] *displacement [struc geol].*

dislocation breccia *fault breccia.*

dislocation metamorphism A form of dynamic regional metamorphism concentrated along narrow belts of shearing or crushing without an appreciable rise in temperature. The term was originated by Lossen in 1883 and is considered to be equivalent to *dynamometamorphism.* Cf: *dynamic metamorphism; cataclastic metamorphism.*

dismal *pocosin.*

dismembered drainage A complex drainage system that has been altered by *dismembering,* thus creating a series of independent streams that enter the sea by separate mouths.

dismembered stream A tributary that is left as an independent stream after the lower part of the drainage system to which it formerly belonged was submerged by an invasion of the sea. Cf: *betrunked river.*

dismembering The making of a tributary into an independent stream by a change of geologic conditions, esp. by the submergence of the lower part of a valley by the sea. Cf: *betrunking.*

dismicrite A fine-textured limestone with less than 1% allochems, consisting mainly of lithified carbonate mud (micrite), and containing irregular patches or *bird's-eyes* of sparry calcite filling cavities caused by local disturbances (Folk, 1959, p. 28). Syn: *bird's-eye limestone.*

disomatic A term applied to a crystal now called a *xenocryst.*

disorder in minerals In a substitutional solid solution, the random occupation of one atom site in a crystal by two or more different atoms of similar size and charge, or of similar size and different charge if there is a concomitant substitution to balance charges, as in plagioclase, in which (Na and Si) in albite is substituted by (Ca and Al) as the composition approaches anorthite. Cf: *order in minerals.*

dispellet limestone A term used by Wolf (1960, p. 1416) for a pelleted limestone with tubules or irregular patches of sparite.

dispersal [ecol] The spreading of a species by migration into new areas having conditions favorable for its existence. Cf: *dispersion [ecol].*

dispersal [glac geol] *glacial dispersal.*

dispersal center The place on a delta at which the first stream distributary branches off from the main channel (Moore, 1966, p. 92).

dispersal map A stratigraphic map that shows the inferred source area and the direction or distance of transportation of clastic materials (Krumbein & Sloss, 1963, p.484).

dispersal shadow An accumulation of sediments formed down-current from a generating source (Pettijohn, 1957, p. 574); e.g. a boulder train on the lee side of a resistant knob overridden by ice. Cf: *sedimentary petrologic province.*

dispersed element An element that is generally too rare and unconcentrated to become an essential constituent of a mineral, and that therefore occurs principally as a substituent of the more abundant elements.

dispersed phase Colloidal material suspended in another phase, which in turn is known as the *dispersion medium.*

dispersion [ecol] The pattern of geographic or spatial distribution of individuals within a species; may be uniform, random or irregular, or aggregated or clumped. Cf: *dispersal.*

dispersion [gem] The property of a transparent gemstone to separate white light into the spectral colors.

dispersion [optics] The differences in the optical constants, e.g. wavelengths and indices of refraction, of a given mineral for different wavelengths of the spectrum. See also: *dispersion curve [optics].*

dispersion [seis] Distortion of the shape of a seismic-wave train because of variation of velocity with frequency. The peaks and troughs may advance toward or recede from the beginning of the wave as it travels (Sheriff, 1973, p. 57). See also: *normal dispersion; inverse dispersion.*

dispersion [soil] Breaking down or separation of soil aggregates into single grains. Cf: *detachment.*

dispersion [stat] The range or scatter of values about a central tendency; a statistical spread or variability. Common measures of dispersion are standard deviation and sorting.

dispersion curve [optics] The plotting on a logarithmic scale of a crystal's *dispersion.*

dispersion curve [seis] A plot of seismic-wave velocity versus frequency or period.

dispersion ellipse The ground area, usually elliptical in shape, covered by a meteorite shower. Syn: *strewn field.*

dispersion flow Flow of granular sediment in which collisions between particles maintain the fluidity of the material.

dispersion medium That material (solid, liquid, or gas) in which colloidal material, known as the *dispersed phase,* is suspended.

dispersion pattern The pattern of distribution of chemical elements, especially trace elements, in the wall rocks of an orebody or in the surface materials surrounding it. Cf: *halo.*

dispersion ratio In a soil, the ratio of the percentage of silt and clay that remains suspended (after a standard agitation procedure) to the percentage of the soil's clay and silt as analyzed mechanically.

dispersive power The refractive ability of a transparent substance, usually isotropic; it is symbolized by δ and equals n_F-n_C /n_{D-1}, in which n is the refractive index for Fraunhofer lines C, F, and D. (C = hydrogen discharge at wavelength 656.3, F = hydrogen discharge at wavelength 486.1, and D = sodium flame at wavelength 589.3, measured in nannometers.) Syn: *relative dispersion.*

dispersive stress *Bagnold dispersive stress.*

disphenoid A closed crystal form consisting of two *sphenoids,* in which the two faces of the upper sphenoid alternate with those of the lower. Adj: *disphenoidal.* Syn: *bisphenoid.*

disphenoidal Having the symmetry of a disphenoid, e.g. rhombic 222.

disphotic zone That part of the ocean, or of a cave, in which there is only dim light and little photosynthesis. It lies between the *aphotic zone* and the *euphotic zone;* or it may be considered as the lower part of the euphotic zone. Syn: *twilight zone.*

displacement [photo] Any shift in the position of an image on a photograph that does not alter the perspective characteristics of the photograph. It may be caused by the relief of the objects photographed, the tilt of the photograph, changes of scale, or atmospheric refraction. Cf: *distortion.*

displacement [struc geol] A general term for the relative movement of the two sides of a fault, measured in any chosen direction; also, the specific amount of such movement. Syn: *dislocation.*

displacement meter A seismometer designed to respond to the displacement of Earth particles.

displacement pressure The minimum capillary pressure required to force a nonwetting fluid into capillary openings in a porous medium saturated with a wetting fluid; specif., to force oil or gas from one water-filled pore to the next. Syn: *entry pressure.*

displacement shear A fracture surface that often occurs in the marginal zones of earthflows. It develops on shearing, parallel or subparallel to the direction of relative movement.

displacement theory *continental displacement.*

displacive transformation A high-low type of crystal *transformation* that involves no breaking of bonds, e.g. in high-low quartz at 573°C, involving rotation of SiO_4 tetrahedra. It is usually a rapid transformation. Cf: *dilatational transformation; reconstructive transformation; rotational transformation; substitutional transformation.*

disrupted Said of a metamorphic texture in which some phase or phases occur "in discontinuous stringers or groups of relatively small grains (partially separated by other minerals) that appear to have formed by disaggregation (with or without recrystallization) of initially larger grains. Although the disrupted grains may approach a lenticular shape, it is not intended to include this texture within the meaning of *laminated*" (Harte, 1977). In the past, the term has been used mainly with reference to the distribution and size of spinel or garnet in peridotites.

dissected pediment An eroded pediment. It is generally regarded as a product of second-cycle erosion of an originally nearly flat pediment, although it may also be "born dissected".

dissected peneplain An ancient and uplifted peneplain that has become the initial surface upon which erosion begins to cut the forms of a new cycle; a partially destroyed peneplain represented in a maturely dissected region by only a few remnants, such as plateaus or occasional flat-topped mountains and ridges.

dissected plateau A plateau in which a large part of the original level surface has been deeply cut into by streams.

dissection The process of erosion by which a relatively even topographic surface is gradually sculptured or destroyed by the formation of gullies, ravines, canyons, or other kinds of valleys; esp. the work of streams in cutting or dividing the land into hills and ridges, or into flat upland areas separated by fairly close networks of valleys. The process is applicable esp. to surfaces, such as plains and peneplains, that have been uplifted. Adj: *dissected.*

disseminated Said of a mineral deposit (esp. of metals) in which the desired minerals occur as scattered particles in the rock, but in sufficient quantity to make the deposit an ore. There is no genetic connotation. Cf: *impregnated.*

dissepiment (a) A small domed calcareous plate forming a *vesicle* or cystlike enclosure typically occurring between radiating septa in the peripheral region of a corallite. Its convex surface faces inward and upward. (b) A skeletal crossbar connecting branches of a fenestrate bryozoan colony. (c) An imperforate sagging or bubblelike plate in the intervallum or central cavity of an archaeocyathid. (d) A crossbar or strand of periderm uniting adjacent branches (stipes) in a dendroid graptolite colony or rhabdosome, as in *Dictyonema.* (e) A thin, upwardly or obliquely convex blisterlike internal structure partially filling the gallery space in most stromatoporoid coenostea or composing the entire coenosteum of a few stromatoporoids. Syn: *cyst; interlaminar partition; cyst plate.* Cf: *stromatoporoid; coenosteum; gallery.* Pl: *dissepiments.*

dissepimentarium The peripheral zone of the interior of a corallite, occupied by dissepiments. See also: *regular dissepimentarium.*

dissociation constant The equilibrium constant for a dissociation reaction, defined as the product of activities of the products of dissociation divided by the activity of the original substance. When used for ionization reactions, it is called an *ionization constant;* when it refers to a very slightly soluble compound, it is called a *solubility product.*

dissociation point That temperature at which a compound

breaks up reversibly to form two or more other substances, e.g. $CaCO_3$ becoming CaO plus CO_2. All variables should be stated in order to define the point precisely. The term dissociation refers to the breakup itself, and covers a wide variety of types, such as the breakup of molecular groupings in gases or liquids.

dissociation temperature A temperature point at which a given dissociation presumably occurs; in fact, it is usually a range of temperature owing to variations in composition or pressure, and may refer merely to the temperature at which the rate of a given dissociation becomes appreciable, under stated conditions.

dissoconch The postlarval shell of a bivalve mollusk.

dissolution *solution.*

dissolved-gas drive Energy within an oil pool, resulting from the expansion of gas liberated from solution in the oil. Cf: *gas-cap drive; water drive.*

dissolved load The part of the total *stream load* that is carried in solution, such as chemical ions yielded by erosion of the landmass during the return of rainwater to the ocean. Syn: *dissolved solids; solution load.*

dissolved oxygen The amount of oxygen, in parts per million by weight, dissolved in water, now generally expressed in mg/l. It is a critical factor for fish and other aquatic life, and for self-purification of a surface-water body after inflow of oxygen-consuming pollutants. Abbrev: DO.

dissolved solids (a) *dissolved load.* (b) A term that expresses the quantity of dissolved material in a sample of water, "either the residue on evaporation, dried at 180°C, or, for many waters that contain more than about 1000 parts per million, the sum of determined constituents" (USGS, 1958, p. 50).

distal [eco geol] Said of an ore deposit formed at a considerable distance, typically tens of kilometers, from a volcanic source to which it is related and from which its constituents have been derived. Cf: *proximal [eco geol].*

distal [paleont] Remote or away from the point of attachment, place of reference, or point of view. Examples in invertebrate morphology: "distal direction" away from a crinoid theca toward the holdfast or free lower extremity of the column; "distal portion" of the rhabdosome of a graptolite colony, farthest away from the point of origin; and "distal side" away from the ancestrula or origin of growth of a bryozoan colony. Ant: *proximal.*

distal [palyn] Said of the parts of pollen grains or spores away from the center of the original tetrad; e.g. said of the side of a monocolpate pollen grain upon which the colpus is borne, or said of the side of a spore opposite the laesura. Ant: *proximal [palyn].*

distal [sed] Said of a sedimentary deposit consisting of fine clastics and formed farthest from the source area; e.g. a "distal turbidite" consisting of thin silty varves, or the most remote foreland deposit derived from the borderland in a geosynclinal region. Cf: *proximal.*

distal downlap *Downlap* in the direction away from the source of clastic supply (Mitchum, 1977, p. 206).

distal onlap *Onlap* in the direction away from the source of clastic supply (Mitchum, 1977, p. 208). Cf: *proximal onlap.*

distance-function map A term used by Krumbein (1955) for a map now known as *facies-departure map.*

distance meter A device for measuring line-of-sight distances, generally by transmitting infrared light to a prism at a station and receiving the reflected light.

distant admixture A term applied by Udden (1914) to an *admixture* (in a sediment of several size grades) whose particles are most different in size from those of the dominant or maximum grade; material in one of the two classes at the extreme ends of a histogram.

disthene *kyanite.*

disthenite A metamorphic rock composed almost entirely of kyanite (disthene) and some quartz, often associated with magnetiferous quartzite and amphibolite (Lacroix, 1922, p.497).

distichous Said of a plant with two-ranked appendages, e.g. with leaves, leaflets, or flowers on opposite sides of the same point on a stem. Cf: *decussate.*

distillate *condensate.*

distillation [paleont] A process of fossilization whereby the liquid and/or gaseous components of an organic substance are removed leaving a carbonaceous residue.

distillation [water] Conversion of liquid to vapor by the addition of heat, and returning the vapor to a liquid by cooling, as in the purification of water.

distinctive mineral *varietal mineral.*

dististele (a) The distal region of a crinoid column. (b) The distal part of the stele of certain homalozoans.—Cf: *proxistele.*

distomodaeal budding A type of budding in scleractinian corals in which two stomodaea are developed within a common tentacular ring and two interstomodaeal couples of mesenteries are located between the original and each new stomodaeum.

distortion [cart] The change in shape and size of a land area on a map due to the flattening of the curved earth surface to fit a plane. Distortion is inevitable and is controlled in the development of a projection to produce the characteristics of equal area, conformality, or equidistance. Cf: *angular distortion.*

distortion [photo] (a) Any shift in the position of an image on a photograph that alters the perspective characteristics of the photograph. It may be caused by lens aberration, differential shrinkage of film or paper, or motion of the film or camera. (b) Compression or expansion of the scale of the imagery in the azimuth direction, perhaps caused by incorrect film speed. (c) A change in scale from one part of the imagery to another. Cf: *displacement.*

distortional wave *S wave.*

distributary [marine geol] A small ephemeral channel in the depositional regime of a deep-sea fan, usually at the mouth of a leveed fan valley.

distributary [streams] (a) A divergent stream flowing away from the main stream and not returning to it, as in a delta or on an alluvial plain. It may be produced by stream deposition choking the original channel. Ant: *tributary.* (b) One of the channels of a braided stream; a channel carrying the water of a stream distributary. Syn: *distributary channel.*

distributary glacier Any ice stream or lobe that flows away or forks off from the lower part of a glacier; a subsidiary terminus or outlet of a trunk glacier. Cf: *outlet glacier.* See also: *glacial lobe.*

distributed fault *fault zone.*

distribution [ecol] *range [ecol].*

distribution [stat] *frequency distribution.*

distribution coefficient *distribution ratio.*

distribution grading In a sedimentary bed, a progressive upward shift toward the finer grain sizes for almost all percentiles of the distribution (Middleton, 1967, p. 487). Cf: *coarse-tail grading.*

distribution ratio The ratio of concentrations of a solute in two immiscible solvents. Syn: *distribution coefficient.*

distribution scatter *scatter diagram [stat].*

distributive fault *step fault.*

distributive province The environment embracing all rocks that contribute to the formation of a contemporaneous sedimentary deposit and the agents responsible for their distribution (Milner, 1922, p. 366). Cf: *provenance.*

disturbance A term used by some geologists for a minor orogeny, e.g. the *Palisades disturbance.* Schuchert (1924) used *revolution* for a major orogeny at the end of an era, and *disturbance* for an orogeny within an era; this usage is obsolete. Cf: *event; pulsation.*

disturbed-neighborhood assemblage An *assemblage [paleoecol]* in which the specimens have been transported only a short distance from where they originally lived, and then deposited in essentially the same kind of sediment on which they had lived, in roughly the same proportions as when alive. Cf: *fossil community; winnowed community; transported assemblage; mixed assemblage.*

disturbing potential The difference between the *geopotential* and the *spheropotential* at a given point. Syn: *potential disturbance; potential of random masses; potential of disturbing masses.*

ditch A long, narrow excavation artificially dug in the ground; esp. an open and usually unpaved waterway, channel, or trench for conveying water for drainage or irrigation, and usually smaller than a canal. Some ditches may be natural watercourses.

ditetragonal Said of a crystal form having eight similar faces, the alternate interfacial angles of which are equal.

ditetragonal dipyramid A crystal form that is a dipyramid of 16 faces in which any section perpendicular to the fourfold axis is ditetragonal. Its indices are {hkl} in symmetry $4/m\ 2/m\ 2/m$.

ditetragonal-dipyramidal class That crystal class in the tetragonal system having symmetry $4/m\ 2/m\ 2/m$.

ditetragonal prism A crystal form of eight faces parallel to the symmetry axis in which any section perpendicular to the prism axis is ditetragonal. Its indices are {*hk0*} with symmetry $4/m\ 2/m\ 2/m$.

ditetragonal pyramid A crystal form consisting of eight faces in a pyramid, in which any section perpendicular to the fourfold symmetry axis is ditetragonal. Its indices are $\{hkl\}$ or $\{hk\bar{l}\}$ in symmetry $4mm$.

ditetragonal-pyramidal class That crystal class in the tetragonal system having symmetry $4mm$.

ditrigonal Said of a symmetrical, eight-sided figure, the alternate angles of which are equal. Such a figure is characteristic of certain crystal forms in the hexagonal system.

ditrigonal dipyramid A crystal form that is a dipyramid of twelve faces in which any section perpendicular to the threefold or sixfold symmetry axis is ditrigonal. Its indices are $\{hkl\}$ or $\{khl\}$ in symmetry $\bar{6}m2$.

ditrigonal-dipyramidal class That class in the hexagonal system having symmetry $\bar{6}m2$.

ditrigonal prism A crystal form of six faces parallel to the symmetry axis, in which any section perpendicular to the axis is ditrigonal. Its indices are $\{hk0\}$ or $\{kh0\}$ in symmetry $\bar{6}m2$ or $3m$.

ditrigonal pyramid A crystal form consisting of six faces in a pyramid, in which any section perpendicular to the symmetry 3 axis is ditrigonal. Its indices are $\{hkl\}$, $\{hk\bar{l}\}$, $\{khl\}$, or $\{kh\bar{l}\}$ in symmetry 3.

ditrigonal-pyramidal class That class in the rhombohedral division of the hexagonal system having symmetry $3m$.

ditrigonal-scalenohedral class *hexagonal-scalenohedral class.*

ditroite A *nepheline syenite* containing sodalite, biotite, and cancrinite. The term was introduced by Zirkel in 1866, and later Brögger proposed applying it to nepheline syenites having granular texture (Johannsen, 1939, p. 249). Its name is derived from Ditrau or Ditró, Romania. Cf: *foyaite.* Not recommended usage.

dittmarite An orthorhombic mineral: $(NH_4)Mg(PO_4)\cdot H_2O$.

Dittonian Series in the Old Red Sandstone of England: Lower Devonian (upper Gedinnian; above Downtonian).

diurnal current A tidal current that has only one flood period and one ebb period during a tidal day.

diurnal inequality The difference between the heights and durations of the two successive high waters or of the two successive low waters of a tidal day.

diurnal tide A tide with only one high water and one low water occurring during a tidal day, as in the Gulf of Mexico.

diurnal variation *daily variation.*

diurnal vertical migration A pattern of daily movement of certain marine organisms that is upward at sunset and downward at sunrise, in response to changes in light and other factors.

diurnal wave The daily cyclical heating and cooling of the upper 30 to 50 cm of the Earth in response to the daily solar cycle. Below this depth, daily temperature is constant. Cf: *annual wave.* Syn: *daily wave.*

divagation The lateral shifting of a stream course as a result of extensive deposition of alluvium in its bed, esp. accompanied by the development of meanders.

divariant Said of a chemical system having two *degrees of freedom.* Syn: *bivariant.*

divaricator A muscle that causes divergence or separation of parts; specif. a *diductor muscle* of a brachiopod. The term is rarely used.

divergence [currents] The separation of ocean currents by horizontal flow of water in different directions from a common source, usually upwelling; also, the area in which divergence occurs. Cf: *convergence [currents].*

divergence [evol] *adaptive radiation.*

divergence [glac geol] Interruption of a drainage pattern by the advance of a glacier or ice sheet.

divergence [meteorol] The spreading of air from a central region or zone. Vertically moving air replaces the air that moves outward horizontally. Ant: *convergence [meteorol].*

divergence [seis] The decrease in amplitude of a wave front because of geometrical spreading. See also: *spherical divergence; cylindrical divergence.*

divergent plate boundary *accreting plate boundary.*

diversion (a) The process by which a stream actively effects changes in the drainage or course of another stream, as by aggradation or capture. Cf: *derangement.* (b) The artificial draining, pumping, siphoning, or other removal of water from a stream, lake, or other body of water, into a canal, pipe, or other conduit. (c) A channel designed to divert water from a body of water for purposes such as prevention of flooding, reduction of erosion, or promotion of infiltration.

diversity The number of different kinds of organisms in an *assemblage.* It can be measured in various ways, some of which separate diversity in the strict sense (the number of species within the assemblage) from species evenness or *equitability* (the manner in which the individuals are distributed among the species present).

diversity stack A *stack [seis]* in which the components are weighted inversely as their mean power over certain intervals. It is used to prevent occasional large bursts of noise (such as traffic noise) from dominating the stacked record. See also: *mute.*

diverted stream A stream whose course or drainage has been affected by another stream; e.g. a captured stream.

diverter *diverting stream.*

diverting stream A stream that effects diversion; e.g. a capturing stream. Syn: *diverter.*

divide [grd wat] A ridge in the water table or other potentiometric surface from which the ground water represented by that surface moves away in both directions. Water in other aquifers above or below, and even in the lower part of the same aquifer, may have a potentiometric surface lacking the ridge, and so may flow past the divide. Syn: *water-table divide; ground-water divide; groundwater ridge.*

divide [streams] (a) The line of separation, or the ridge, summit, or narrow tract of high ground, marking the boundary between two adjacent drainage basins or dividing the surface waters that flow naturally in one direction from those that flow in the opposite direction; the line forming the rim of or enclosing a drainage basin; a line across which no water flows. An "anomalous" divide is one that does not follow the crest of the highest mountain range of a mountain chain. See also: *continental divide.* British syn: *watershed.* Syn: *drainage divide; water parting; height of land; topographic divide; watershed line.* (b) A tract of relatively high ground between two streams; a line that follows the summit of an interfluve. (c) The highest summit of a pass or gap. (d) *groundwater divide.*

dividing wall The wall dividing offset from parent corallite, jointly formed by both during increase in corals. It consists of two layers separated by a central epitheca (Fedorowski & Jull, 1976, p. 41). Cf: *partition.*

diviner *dowser.*

diving wave A seismic wave that is refracted in a strong velocity-gradient zone and returned to the surface, where it may be observed as a refraction arrival even though it has had no appreciable path through a distinctive refractor.

divining *dowsing.*

divining rod Traditionally, a forked wooden stick, cut from a willow or other water-loving plant, used in *dowsing.* It supposedly dips downward sharply when held over a body of ground water or a mineral deposit, thus revealing the presence of these substances. Syn: *witching stick; wiggle stick; dowsing rod; mineral rod; dipping rod; twig; dowser.* Cf: *water witch; waterfinder.*

division [bot] A category in the hierarchy of botanical classification intermediate in rank between *kingdom* and *subdivision* (ICBN, 1972, p. 17). It is generally regarded as equivalent to the rank of *phylum* in zoology.

division [stratig] A term proposed by Størmer (1966, p. 25) for a chronostratigraphic unit equivalent to stage (and possibly series), but having a regional or more limited geographic range.

divisional plane A general term that includes joints, cleavage, faults, bedding planes, and other surfaces of separation.

dixenite A black hexagonal mineral: $Mn_5(SiO_3)(AsO_3)_2(OH)_2$.

Dix equation A relationship that gives the velocity of an interval between two parallel seismic reflectors from measurements of their associated stacking velocities.

djalmaite *microlite [mineral].*

djebel *jebel.*

djerfisherite A meteorite mineral: $K_3(Cu,Na)(Fe,Ni)_{12}S_{14}$. It occurs in enstatite *chondrites.*

djurleite A mineral: $Cu_{1.96}S$. Its X-ray pattern is near to, but distinct from, that of chalcocite.

D layer The seismic region of the Earth between 1000 km and 2900 km, equivalent to the *lower mantle.* At a depth of 2700 km, there is a change from chemical homogeneity to inhomogeneity; the upper division is the D' layer, and the lower is the D" layer. It is a part of a classification of the Earth's interior made up of layers A to G.

DNA Deoxyribonucleic acid; concentrated mainly in the nuclear structures of organisms. See also: *gene.*

dneprovskite *wood tin.*

DO *dissolved oxygen.*

doab [sed] An Irish term for a dark sandy clay or shale found in the vicinity of bogs (Power, 1895).

doab [streams] (a) A term applied in the Indo-Gangetic Plain of northern India to the tongue of low-lying, alluvial land between two confluent rivers; specif. the Doab, the tract between the Ganges and Jumna rivers. The term is commonly restricted to the alluvial-plains portion characterized by very little relief. Cf: *interfluve.* (b) The confluence of two rivers.—Etymol: Persian, "two waters".

dock (a) The berthing space or waterway between two wharves or two piers, or cut into the land. Syn: *slip.* (b) A basin or enclosure (usually artificial) in connection with a harbor or river, designed to receive vessels and provided with means for controlling the water level. (c) An erroneous syn. of *pier.*

Docodonta One of two orders of mammals (the other being Triconodonta) of primitive structure and uncertain subclass assignment, mostly of Triassic and Jurassic age.

docrystalline A term, now obsolete, suggested by Cross et al. (1906, p.694) for porphyritic rocks that are dominantly crystalline, the ratio of crystals to glass being less than 7 to 1 but greater than 5 to 3.

dodd A term used in the English Lake District for a rounded summit, esp. a lower summit or blunt shoulder or *boss* attached to another hill. Syn: *dod.*

dodecahedral cleavage Mineral cleavage parallel to the faces of the dodecahedron (110); e.g. in sphalerite.

dodecahedron A crystal form with 12 faces that are either pentagonal or rhombic; if rhombic, the faces are equal, but if pentagonal, they are not regular. Each face is parallel to one crystallographic axis and intersects the other two. See also: *pyritohedron; rhombic dodecahedron.*

dodecant In the hexagonal crystal system, one of the 12 spatial divisions made by the four reference axes.

dodging The process of holding back light from certain areas of sensitized paper in making a print, in order to avoid overexposure.

dofemic One of five classes in the *CIPW classification* of igneous rocks, in which the ratio of salic minerals to femic is less than three to five but greater than one to seven. Cf: *salfemic; perfemic.*

dogger (a) A large, irregular nodule, usually of clay ironstone, sometimes containing fossils, found in a sedimentary rock, as in the Jurassic rocks of Yorkshire, England. (b) An English term for any large, lumpy mass of sandstone longer than it is broad, with steep rounded sides.

dogger stone A miner's term for a brown, compact, relatively pure, nonoolitic clay ironstone interbedded with oolitic ironstones in the British Middle Jurassic.

dogleg An abrupt angular change in course or direction, as of a borehole or in a survey traverse. Also, a deflected borehole, survey course, or anything with an abrupt change in direction resembling the hind leg of a dog.

dogtooth spar A variety of calcite in sharply pointed crystals of acute scalenohedral form resembling the tooth of a dog. Syn: *hogtooth spar.*

dohyaline In the *CIPW classification* of igneous rocks, those rocks in which the ratio of crystals to glassy material is greater than three to five but less than one to seven. Rarely used. Cf: *perhyaline; hyalocrystalline.*

dolarenaceous Said of the texture of a dolarenite.

dolarenite A dolomite rock consisting predominantly of detrital dolomite particles of sand size (Folk, 1959, p. 16); a consolidated dolomitic sand.

dolerite (a) In the U.S., a syn. of *diabase.* (b) In British usage, the preferred term for what is called *diabase* in the U.S.—Etymol: Greek *doleros* "deceitful", in reference to the fine-grained character of the rock which makes it difficult to identify megascopically. Cf: *diorite; traprock.*

doleritic (a) Of or pertaining to dolerite. (b) A preferred syn. of *ophitic* in European usage.

dolerophanite A brown monoclinic mineral: $Cu_2(SO_4)O$. Syn: *dolerophane.*

Dolgellian European stage: Upper Cambrian (above Festiniogian, below Tremadocian of Ordovician).

dolimorphic Said of an igneous rock in which *released minerals* are prominent; e.g. a lamprophyre composed chiefly of biotite and quartz, with a little hornblende.

dolina *doline.*

doline A syn. of *sinkhole.* Also spelled: *dolina.* Etymol: German transliteration from Slovene *dolina,* "valley".

Dollo's law A syn. of *irreversibility.* It is named after the Belgian paleontologist Louis Dollo (d.1931).

dolocast A cast or impression of a dolomite crystal, preserved in an insoluble residue. Adj: *dolocastic.* Cf: *dolomold.*

doloclast A lithoclast derived by erosion from an older dolomite rock; also, an intraclast disrupted from partly consolidated dolomitic mud on the bottom of a sea or lake.

dololithite A dolomite rock containing 50% or more of fragments of older dolomitic rocks that have been eroded and redeposited (Hatch & Rastall, 1965, p. 223).

dololutite A dolomitic rock consisting predominantly of detrital dolomite particles of silt and/or clay size (Folk, 1959, p. 16); a consolidated dolomitic mud. It is commonly interlayered with dense primary dolomites in evaporitic sequences. Cf: *dolosiltite.*

dolomicrite A sedimentary rock consisting of clay-sized dolomite crystals, interpreted as a lithified dolomite mud (analogous to calcite mud or micrite), and containing less than 1% allochems (Folk, 1959, p. 14). See also: *primary dolomite.* Syn: *dolomite mudstone.*

dolomilith *dolomith.*

dolomite [mineral] A common rock-forming rhombohedral mineral: $CaMg(CO_3)_2$. Part of the magnesium may be replaced by ferrous iron and less frequently by manganese. Dolomite is white, colorless, or tinged yellow, brown, pink, or gray; it has perfect rhombohedral cleavage and a pearly to vitreous luster, effervesces feebly in cold dilute hydrochloric acid, and forms curved, saddlelike crystals. Dolomite is found in extensive beds as dolomite rock; it is a common vein mineral, and is found in *serpentinite* and other magnesian rocks. Cf: *calcite.* Syn: *bitter spar; pearl spar; magnesian spar; rhomb spar.*

dolomite [sed] A carbonate sedimentary rock of which more than 50% by weight or by areal percentages under the microscope consists of the mineral dolomite, or a variety of limestone or marble rich in magnesium carbonate; specif. a carbonate sedimentary rock containing more than 90% dolomite and less than 10% calcite, or one having a Ca/Mg ratio in the range of 1.5-1.7 (Chilingar, 1957), or one having an approximate MgO equivalent of 19.5-21.6% or magnesium-carbonate equivalent of 41.0-45.4% (Pettijohn, 1957, p. 418). Dolomite occurs in crystalline and noncrystalline forms, is clearly associated and often interbedded with limestone, and usually represents a postdepositional replacement of limestone. Pure dolomite (unless finely pulverized) will effervesce very slowly in cold hydrochloric acid. Named after Déodat Guy de Dolomieu (1750-1801), French geologist, and first applied to certain carbonate rocks of the Tyrolean Alps. See also: *primary dolomite; magnesian limestone.* Syn: *dolostone; dolomite rock.*

dolomite limestone (a) *dolomitic limestone.* (b) A term suggested by Grout (1932, p.288) for a carbonate rock composed predominantly of the mineral dolomite. The term in this usage is not recommended (Pettijohn, 1957, p. 416).

dolomite mudstone *dolomicrite.*

dolomite rock *dolomite [sed].*

dolomith A term suggested by Grabau (1924, p. 298) for a dolomite rock. Syn: *dolomilith.*

dolomitic (a) Dolomite-bearing, or containing dolomite; esp. said of a rock that contains 5-50% of the mineral dolomite in the form of cement and/or grains or crystals. (b) Containing magnesium; e.g. "dolomitic lime" containing 30-50% magnesium.

dolomitic conglomerate (a) A conglomerate consisting of limestone pebbles and dolomite cement (Nelson & Nelson, 1967, p. 112). (b) A breccia-conglomerate of Keuper age in Somerset, England (Arkell & Tomkeieff, 1953, p. 38).

dolomitic limestone (a) A limestone in which the mineral dolomite is conspicuous, but calcite is more abundant; specif. a limestone containing 10-50% dolomite and 50-90% calcite and having an approximate magnesium-carbonate equivalent of 4.4-22.7% (Pettijohn, 1957, p. 418), or a limestone whose Ca/Mg ratio ranges from 4.74 to 60 (Chilingar, 1957). Cf: *calcitic dolomite; magnesian limestone.* Syn: *dolomite limestone.* (b) A limestone that has been incompletely dolomitized (Chilingar et al., 1967, p. 314).

dolomitic marble A variety of marble composed largely of dolomite and formed by the metamorphism of dolomitic or magnesian limestone. Cf: *magnesian marble.*

dolomitic mottling A textural feature resulting from incipient or arrested dolomitization of limestones, characterized by preferential alteration that leaves patches, blotches, bird's-eyes, laminae, allochems and/or other structures unaffected. Also, a similar phenomenon resulting from arrested or incomplete *dedolomitization.*

dolomitite A term used by Kay (1951) for a rock composed of the mineral dolomite; a *dolostone.*

dolomitization The process by which limestone is wholly or partly converted to dolomite rock or dolomitic limestone by the replacement of the original calcium carbonate (calcite) by magnesium carbonate (mineral dolomite), usually through the action of magnesium-bearing water (seawater or percolating meteoric water). It can occur penecontemporaneously or shortly after deposition of the limestone, or during lithification at a later period. Syn: *dolomization.*

dolomization *dolomitization.*

dolomold A rhombohedral opening in an insoluble residue, formed by the solution of a dolomite or calcite crystal. Adj: *dolomoldic.* Cf: *dolocast.*

dolomorphic A term used by Ireland et al. (1947, p. 1483) to describe an insoluble residue characterized by replacement or alteration of dolomite or calcite by an insoluble mineral that fills a dolomoldic cavity and assumes the crystal form of the dissolved mineral; it replaces "dolocastic" as used by Cloud et al. (1943, p. 135).

dolon In certain Paleozoic ostracodes, the strong curvature of the vela or frills such that the two meet outside the contact margin to form a false pouch; also the cavity formed by this curvature of the frills (TIP, 1961, pt. Q, p. 49). Syn: *false pouch.*

doloresite A dark-brown monoclinic mineral: $H_8V_6O_{16}$.

dolorudite A dolomite rock consisting predominantly of detrital dolomite particles larger than sand size (Folk, 1959, p. 15); a consolidated dolomitic gravel.

dolosiltite A dolomite rock consisting predominantly of detrital dolomite particles of silt size; a consolidated dolomitic silt. Cf: *dololutite.*

dolosparite A sparry dolomite crystal. Cf: *calcsparite.*

dolostone A term proposed by Shrock (1948a, p. 126) for the sedimentary rock *dolomite,* in order to avoid confusion with the mineral of the same name. Syn: *dolomitite.*

domain [magnet] A region within a grain of magnetically ordered mineral, within which the spontaneous magnetization has a constant value characteristic of the mineral composition and temperature. Syn: *magnetic domain.*

domain [meta] A macroscopically recognizable part of an altered rock, frequently of a mafic volcanic or plutonic igneous rock, that can be regarded as having a distinctive lithologic or bulk chemical composition (Smith, 1968).

domain [sed] The areal extent of a given lithology or environment; specif. the area in which a given set of physical controls combined to produce a distinctive sedimentary facies.

domain [struc petrol] *fabric domain.*

domal Said of a thecal shape in edrioasteroids, in which the oral surface is convex upward and is confined to the upper side of the theca; the distal edge of the oral surface forms the thecal ambitus. The nonplated aboral surface forms the entire lower side of the theca (Bell, 1976).

domatic class That crystal class in the monoclinic system having symmetry m. Syn: *clinohedral class.*

dome [beach] A miniature elevation (2.5 cm or more high, and 5 to 30 cm in diameter), composed of sand, with a hollow center, formed on beaches by the rush of waves entrapping and confining air (Shepard, 1967, p. 58). Syn: *air dome; sand dome.*

dome [cryst] An open crystal form composed of two nonparallel faces that intersect along and astride a symmetry plane, regardless of the orientation of the line of their intersection. Cf: *sphenoid.*

dome [fold] n. An uplift or anticlinal structure, either circular or elliptical in outline, in which the rocks dip gently away in all directions. A dome may be small, e.g. a Gulf Coast salt dome, or many kilometers in diameter. Domes include *diapirs,* volcanic domes, and cratonic uplifts. Type structure: Nashville Dome, Tennessee. See also: *pericline; arch.* Syn: *dome structure; structural dome; quaquaversal fold.* Less-preferred syn: *swell.* Ant: *basin.*—v. To bend, push, or thrust up into a dome, e.g. underlying magma doming the surface by upward pressure.

dome [geomorph] (a) A general term for any smoothly rounded landform or rock mass, such as a rock-capped mountain summit, that roughly resembles the dome of a building; e.g. the rounded granite peaks of Yosemite, Calif. The term is also applied to broadly up-arched regions, e.g. the English Lake District or the Black Hills of South Dakota. (b) A rounded snow peak, esp. in the French Alps.

dome [lunar] A small, almost circular surface bulge, generally several kilometers wide and a few hundred meters high, found in mare regions of the Moon's surface. Domes often have apparently smooth summits capped by craters; they are generally believed to be formed by local extrusive or intrusive igneous activity.

dome [marine geol] A general, nonrecommended term for such ocean-floor features as a *seamount* or a *knoll [marine geol].*

dome [petrology] A large magmatic or migmatitic intrusion whose surface is convex upward and whose sides slope away at low but gradually increasing angles. Intrusive igneous domes include laccoliths and batholiths; the term is used when the evidence as to the character of the lower parts of the intrusion is insufficient to allow more specific identification.

dome [volc] (a) *volcanic dome.* (b) *lava dome.*

dome mountain A mountain produced where a region of flat-lying sedimentary rocks is warped or bowed upward to form a structural dome; a mountain resulting from dissection of a structural dome. Examples: the Black Hills in South Dakota, and the Weald uplift in SE England. Syn: *domal mountain; domed mountain.*

domepit In a cave, a rounded vertical passage or high chamber, characterized by vertical solution grooves on its walls and usually by showering water. See also: *chimney [speleo].* Syn: *blind shaft; vertical shaft; aven; foiba.*

dome structure *dome.*

dome volcano *volcanic dome.*

domeykite A tin-white or steel-gray mineral: Cu_3As.

domiciliar dimorphism In certain ostracodes, a kind of dimorphism in which the adult form presumed to be the female has a much larger and more spacious posterior half of the carapace that the presumed adult male; also called "kloedenellid dimorphism", although not restricted to the family Kloedenellidae.

domicilium The main part of the carapace of an ostracode exclusive of alae or other accessory projecting structures.

dominant [coal] In the description of coal constituents, more than 60% of a particular constituent occurring in the coal (ICCP, 1963). Cf: *rare; common; very common; abundant.*

dominant [ecol] A species or group of species that is numerically very abundant, or largely controls the energy flow, or strongly affects the environment, within a community or association.

dominant discharge That discharge of a natural channel which determines the characteristics and principal dimensions of the channel. It depends on the sediment characteristics, the relationship between maximum and mean discharge, duration of flow, and flood frequency (ASCE, 1962).

domite An altered porphyritic oligoclase-biotite *trachyte* that contains tridymite in the groundmass. It is named for Puy de Dome in the Auvergne district of France. Not recommended usage.

domoikic In the *CIPW classification* of igneous rocks, those rocks in which the ratio of oikocrysts to chadacrysts is less than seven to one but greater than five to three. Rarely used. Cf: *peroikic; xenoikic.*

donathite A tetragonal mineral: $(Fe,Mg)(Cr,Fe)_2O_4$. It is a dimorph of chromite.

donga [glaciol] A small, steep-walled rift in a piedmont glacier or in an ice shelf. Cf: *barranca.*

donga [streams] (a) A term used in South Africa for a small narrow steep-sided ravine or gorge formed by turbulent water flow; it is usually dry except in the rainy season. A donga is similar to a wadi or a mullah. (b) A term used in South Africa for a gully formed by soil erosion.—Etymol: Afrikaans, from Bantu (Zulu).

doodlebug A popular term for any of various kinds of geophysical prospecting equipment.

dopatic In the *CIPW classification* of igneous rocks, those rocks in which the ratio of groundmass to phenocrysts is less than seven to one but greater than five to three. Rarely used. Cf: *perpatic; sempatic.*

doppler effect A change in the observed frequency of electromagnetic or sound waves, caused by relative motion between the source and the observer. See also: *doppler signal.*

dopplerite An amorphous brownish-black gelatinous calcium salt of a humic acid that is found at depth in marsh and bog deposits. It may represent an accumulation of *phytocollite* concentrated by ground water (Swain, 1963, p. 105).

doppler navigation *doppler positioning.*

doppler positioning A system which determines positions based on the *doppler effect* of satellite signals, radar, or sonar. Syn: *doppler navigation.*

doppler signal The difference in frequency of waves produced by the *doppler effect.*

dorbank A term used in southern Africa for a calcareous and siliceous concretion occurring beneath the surface layer of sandy loam. Etymol: Afrikaans, "dry layer".

dore A term used in the English Lake District for a narrow, doorlike opening or fissure between walls of rock, as a pass through a narrow gorge; it is often an open joint.

doreite An andesitic lava containing approximately equal amounts of potassium and sodium; the extrusive equivalent of *mangerite* (Streckeisen, 1967, p. 209). The name (for Mont Dore, Auvergne, France) was proposed by Lacroix in 1923 for olivine-bearing *trachyandesite*. Not recommended usage.

dorgalite A basalt in which the phenocrysts are exclusively olivine; an olivine basalt. It is named for Dorgali, Sardinia. Not recommended usage.

dormant volcano A volcano that is not now erupting but that has done so within historic time and is considered likely to do so in the future. There is no precise distinction between a dormant and an *active volcano*. Cf: *extinct volcano; inactive volcano*. Syn: *subactive volcano*.

dorr A glacial trough, open at both ends, across a ridge or mountain range, lying in a pass; it is formed through overdeepening by the crowding of ice through the pass. The term was introduced by Chadwick (1939, p. 362) to describe a fjord-like trough that may or may not have been submerged, and is named in honor of George B. Dorr (b.1853), executive of the Hancock County (Me.) Trustees of Public Reservations. Example: Somes Sound on Mount Desert Island, Me.

dorsal (a) Pertaining to, or situated near or on, the back or upper surface of an animal or of one of its parts; e.g. toward the brachial valve from the pedicle valve of a brachiopod, or pertaining to the hinge region of the shell of a bivalve mollusk, the side of the uniserial stipe opposite the thecal apertures of a graptoloid, or the spiral side of a trochoid foraminifer. (b) Referring to the direction or side of an echinoderm away from the mouth, normally downward and outward, e.g. toward the point of attachment of the column with blastoid theca or to the part of the crinoid or cystoid calyx located toward the column; aboral.—Ant: *ventral*.

dorsal area The part of the whorl of a coiled cephalopod conch in contact with the preceding whorl (TIP, 1964, pt.K, p.55).

dorsal cleft A triangular area on the dorsal side of a septum in which cameral deposits are absent (if such deposits are otherwise present); occurs in orthocerid cephalopods (Fischer & Teichert, 1969).

dorsal cup The cup-shaped part of a crinoid theca forming the aboral and lateral walls about the viscera. It does not include the free arms, tegmen, or column. Cf: *cup*.

dorsal exoskeleton (a) The resistant mineralized dorsal integument of a trilobite. The term is commonly used to include the reflexed border or doublure on the ventral side. (b) The commonly calcified part of the covering of a crustacean, including the cephalic shield and fold of integument arising from the posterior border of the maxillary somite and extending over the trunk, usually covering it laterally as well as dorsally and in many forms having a dorsal longitudinal hinge, and often fused to one or more thoracic somites.— Less-preferred syn: *carapace*.

dorsal furrow (a) An *axial furrow* of a trilobite. (b) A *septal furrow* of a nautiloid.

dorsal lobe The median primary lobe of a suture on the dorsum of a cephalopod shell. See also: *internal lobe*. Cf: *ventral lobe*.

dorsal shield (a) One of a series of ossicles along the midline of the aboral surface of an arm in an ophiuroid. Cf: *ventral shield*. (b) The entire dorsal test of a trilobite, including the cephalon, thorax, and pygidium.

dorsal valve The *brachial valve* of a brachiopod.

dorsomyarian Said of a nautiloid in which the retractor muscles of the head-foot mass are attached to the shell along the interior areas of the body chamber adjacent to, or coincident with, its dorsal midline (TIP, 1964, pt.K, p.55). Cf: *pleuromyarian; ventromyarian*.

dorsum [astrogeol] A term established by the International Astronomical Union for an elongated ridge on Mars. Generally used as part of a formal name for a Martian landform, such as Gordii Dorsum (Mutch et al., 1976, p. 57).

dorsum [paleont] The back or dorsal surface of an animal; e.g. the dorsal side of a cephalopod conch, opposite the ventral side and equivalent to the impressed area in slightly involute shells but referring only to that part of the conch adjacent to the venter of the preceding whorl in deeply involute shells; or the more or less flattened area of the carapace surface of an ostracode, adjacent to the hinge line and set off from the lateral surface of the valves. The term is somewhat loosely applied in fossil nautiloids, to the concave side of a whorl in coiled forms and to the side farthest removed from the siphuncle in straight or curved forms with eccentric siphuncle (TIP, 1964, pt. K, p. 55). Pl: *dorsa*. Cf: *venter*.

dosalic One of five classes in the *CIPW classification* of igneous rocks, in which the ratio of salic to femic minerals is less than seven to one but greater than five to three. Cf: *persalic; salfemic*.

dosemic In the *CIPW classification* of igneous rocks, those rocks in which the ratio of groundmass to phenocrysts is less than three to five but greater than one to seven. Rarely used. Cf: *sempatic; persemic*.

dot chart (a) A graphic aid used in the correction of station gravity for terrain effect, or for computing gravity effects of irregular masses. It can also be used in magnetic interpretation. (b) A transparent graph-type chart used in the calculation of the gravity effects of various structures. The dots on the chart represent unit areas.

dot map A *cartogram* utilizing dots (usually of uniform size), each dot representing a specific number of the objects whose distribution is being mapped.

double biseries Two distinct sets of coverplates in edrioasteroids, the pairs of one set alternating with those of the other to form an integrated system (Bell, 1976).

double Bouguer correction A *Bouguer correction* to sea level for measurements made on the ocean floor. It involves a correction to remove the upward attraction of the sea water above the meter, and another correction to replace the sea water with the replacement density. Similar "double" corrections are required for measurements made in mines or boreholes.

double layer A layer of ions in an electrolyte, required to satisfy a charge unbalance within a solid with which the electrolyte is in contact. See also: *diffuse layer; fixed layer*.

double-line stream A watercourse drawn to scale (on a map) by two lines representing the banks. Cf: *single-line stream*.

double-refracting spar *Iceland spar*.

double refraction *birefringence*.

double-serrate Said of a leaf having coarse marginal serrations with additional smaller teeth on the coarser serration lobes.

doublet A gem substitute composed of two pieces of gem material, or one of gem material and a second of glass or synthetic, fused or cemented together; e.g. a glass imitation with a thin layer of genuine garnet fused on the top. Cf: *triplet*.

double tide (a) A high tide consisting of two high-water maxima of nearly the same height separated by a slight lowering of water. (b) A low tide consisting of two low-water minima separated by a slight rise of water.

double tombolo Two separate bars or barriers connecting an island (usually of large extent and close to the shore) with the mainland. Cf: *tombolo*.

double valley A valley with a low divide on its floor, from which one stream flows in one direction and a second stream flows in another.

doublure An infolded margin of the exoskeleton of an arthropod, such as the reflexed ventral margin of the carapace integument of a crustacean or the inwardly deflected marginal part of the dorsal exoskeleton of a merostome; esp. a generally narrow band extending around the border of the dorsal exoskeleton of a trilobite, turned or bent under to the ventral side.

doubly plunging fold A fold, either an anticline or a syncline, that reverses its direction of plunge within the observed area (Billings, 1954, p. 46).

doughnut (a) A small circular *closed ridge* of glacial origin (Gravenor & Kupsch, 1959, p. 52). (b) *rock doughnut*.

douglasite A mineral: $K_2FeCl_4 \cdot 2H_2O$.

Douglas scale A series of numbers formerly used to indicate swell and the state of the sea, ranging from zero (calm) to 9 (confused). Named for H.P. Douglas, British naval officer, who devised the scale in 1921. Replaced in 1947 by World Meteorological Code 75.

doup A term used in northern England for a rounded depression or cavity in a rock or hillside.

dousing *dowsing*.

doverite *synchysite-(Y)*.

down An upland in southeastern England, generally treeless, underlain by chalk. Commonly used in the plural, e.g. South Downs.

downbuckle A compressional downfolding of sialic crust, associated with oceanic trenches. Syn: *tectogene* in its restricted sense.

downbuilding A theory of salt-dome formation, based on the fact that the top of the salt body in a dome is near the level at which the salt was originally deposited, and that the thick sequence of strata around it was formed by subsidence of the surroundings of the dome. While these facts are evident, the theory itself is mechanically implausible.
downcutting Stream erosion in which the cutting is directed in a downward direction (as opposed to lateral erosion). Cf: *degradation.* Syn: *vertical erosion; incision.*
downdip A direction that is downwards and parallel to the dip of a structure or surface. Cf: *updip.*
downdip block The rocks on the *downthrown* side of a fault. Cf: *updip block.*
downfaulted Said of the rocks on the *downthrown* side of a fault, or the *downdip block.* Cf: *upfaulted.*
downhole adj. In a borehole; e.g. "downhole equipment".—adv. Deeper; e.g. "to perforate downhole". Cf: *uphole.*
downlap A base-discordant relation in which initially inclined strata terminate downdip against an initially horizontal or inclined surface (Mitchum, 1977, p. 206). See also: *distal downlap.*
downslope n. A slope that lies downward; downhill.—adj. In a downward direction, or descending; e.g. a *downslope* ripple that migrated down a sloping surface.
downstream In a direction toward which a stream or glacier is flowing. Similarly, *downriver.* Ant: *upstream.*
down-structure method The method of examining structures on a geologic map by orienting the map so as to look down "into" it along the direction of pitch. The structures then appear in much the same attitude as they would in vertical cross section, since any plane, e.g. one parallel to the ground surface, that intersects plunging structures produces such a pattern, though with different proportions. The method is very useful for quick interpretation of complex folding and faulting (Mackin, 1950).
downthrow n. (a) The downthrown side of a fault. (b) The amount of downward vertical displacement of a fault.—Cf: *upthrow; heave.*
downthrown Said of that side of a fault that appears to have moved downward, compared with the other side. Cf: *upthrown.*
downthrown block *downthrow.*
down-to-basin fault A term used in petroleum geology for a fault whose downthrown side is toward the adjacent basin.
Downtonian Series in the Old Red Sandstone of England: Lower Devonian (lowermost Gedinnian; below Dittonian). It was originally assigned to uppermost Silurian (upper Ludlovian).
downward bulge *root [tect].*
downward continuation The process of determining, from values measured at one level, the value of a potential (e.g. gravitational) field at a lower level. See also: *continuation; upward continuation.*
downwarping Subsidence of a regional area of the Earth's crust, as in an *orogenic belt* or a *centrocline.* Cf: *upwarping.*
downwash Fine-grained surface material (such as soil) moved down a mountain slope or hillside by rain, esp. where there is little vegetation. Cf: *rainwash; sheetwash.*
downwasting (a) *mass wasting.* (b) The thinning of a glacier during ablation.—Cf: *backwasting.*
downwearing Erosion that causes the flattening-out of a hill or mountain and the decline of its slope; a process contributing to the development of a peneplain. Cf: *backwearing.*
downwelling *sinking.*
dowser (a) One who practices *dowsing.* Syn: *diviner.* (b) *divining rod.*—Syn: *water witch; waterfinder.*
dowsing The practice of locating ground water, mineral deposits, or other objects by means of a *divining rod* or a pendulum. A *dowser* may claim also to be able to diagnose diseases, determine the sex of unborn babies, etc. Syn: *dousing; divining; water witching.* Cf: *rhabdomancy; pallomancy.*
dowsing rod *divining rod.*
doxenic In the *CIPW classification* of igneous rocks, those rocks in which the ratio of oikocrysts to phenocrysts is less than three to five but greater than one to seven. Rarely used. Cf: *xenoikic; perxenic.*
Drachenfels trachyte A syn. of *drakonite.* Its name is derived from Drachenfels, Siebengebirge, Germany. Not recommended usage.
draft A term used in eastern U.S. for a gully or gorge, and for a small stream or creek.
drag [eco geol] *drag ore.*
drag [hydraul] (a) The friction of moving air against a water surface which tends to pull the water-surface layer in the direction of the wind. (b) The force exerted by a flowing fluid on an object in or adjacent to the flow. Cf: *push [hydraul].*
drag [struc geol] The bending of strata on either side of a fault, caused by the friction of the moving blocks along the fault surface; also, the bends or distortions so formed.
drag-and-slippage zone The zone bordering the crust and the Earth's interior, along which the entire crust may have shifted relative to the interior (Weeks, 1959, p.378). See also: *phorogenesis.*
drag cast A more appropriate term for *drag mark* as used by Kuenen (1957, p.243-245).
drag coefficient The ratio of the force per unit area exerted on a body by a flowing liquid to the pressure at the stagnation point (ASCE, 1962).
drag fold A minor fold, usually one of a series, formed in an incompetent bed lying between more competent beds, produced by movement of the competent beds in opposite directions relative to one another. Drag folds may also develop beneath a *thrust sheet.* They are usually a centimeter to a few meters in size. Cf: *subsidiary fold; intrafolial fold.*
drag groove A *drag mark* consisting of a long, narrow, even groove.
drag line A short, feeble glacial striation formed on the lee side of an older glacial groove.
dragma A siliceous monaxonic sponge spicule (microsclere) occurring in bundles and produced within a single sclerocyte. Pl: *dragmata.*
drag mark (a) A long, even *groove* or *striation* made by a solid body dragged over a soft sedimentary surface, as by a stone or shell pulled along the mud bottom by attached algae; it tends to be narrower and deeper than a typical *slide mark.* See also: *drag groove; drag striation.* (b) A term used by Kuenen (1957, p.243-245) for the structure called *groove cast* by Shrock (1948, p.162-163), being a broad and rounded or flat-topped or sharp-crested ridge, commonly with longitudinal striations, formed on the underside of an overlying bed by the filling of a *drag groove* probably under turbidity-current conditions. A more appropriate term for this feature would be *drag cast.*
Dragonian North American continental stage: Lower Paleocene (above Puercan, below Torrejonian).
dragonite A rounded quartz pebble, representing a quartz crystal that has lost its brilliancy and form; it was once once believed to be a fabulous stone obtained from the head of a flying dragon.
drag ore Crushed and broken fragments of rock or ore torn from an orebody and contained in and along a fault zone. See also: *trail of a fault.* Syn: *drag.*
drag striation A *drag mark* consisting of a short, narrow striation, curved or straight. Dzulynski & Slaczka (1958, p.234) used "drag stria" for a feature that is essentially a *striation cast.*
drain n. (a) A small, narrow natural watercourse. (b) A channel, conduit, or waterway, either natural or artificial, for draining or carrying off excess water from an area, such as a ditch designed to lower the water table so that land may be farmed; a sewer or trench.—v. To carry away the surface water or discharge of streams in a given direction or to an outlet.
drainage (a) The manner in which the waters of an area pass or flow off by surface streams or subsurface conduits. (b) The processes of surface discharge of water from an area by streamflow and sheet flow, and the removal of excess water from soil by downward flow. Also, the natural and artificial means for effecting this discharge or removal, such as a system of surface and subsurface conduits. (c) A collective term for the streams, lakes, and other bodies of surface water by which a region is drained; a *drainage system.* (d) The water features of a map, such as seas, lakes, ponds, streams, and canals. (e) An area or district drained of water, as by a stream; a *drainage area.* (f) The act or an instance of removing water from a previously marshy land area.
drainage area [petroleum] That area from which one well can produce the hydrocarbons contained in the reservoir rock.
drainage area [streams] The horizontal projection of the area whose surface directs water toward a stream above a specified point on that stream; a *drainage basin.*
drainage basin A region or area bounded by a drainage divide and occupied by a drainage system; specif. the tract of country that gathers water originating as precipitation and contributes it to a particular stream channel or system of channels, or to a lake,

reservoir, or other body of water. Cf: *river basin.* Syn: *basin; watershed; drainage area; catchment; catchment area; catchment basin; gathering ground; feeding ground; hydrographic basin.*

drainage coefficient The amount of runoff (expressed in water depth or other units) removed or drained from an area in 24 hours.

drainage composition Quantitative description of a drainage basin in terms of stream order, drainage density, bifurcation ratio, and stream-length ratio. Term introduced by Horton (1945, p.286) to imply "the numbers and lengths of streams and tributaries of different sizes or orders, regardless of their pattern".

drainage density Ratio of the total stream lengths of all the stream orders within a drainage basin to the area of that basin projected to the horizontal; approximately the reciprocal of the *channel maintenance constant* (Shreve, 1969, p. 412). It is an expression of *topographic texture:* high density values are favored in regions of weak or impermeable surface materials, sparse vegetation, mountainous relief, and high rainfall intensity. Term introduced by Horton (1932, p.357) to represent the average stream length within the basin per unit area. Symbol: D.

drainage district A governmental corporation or other public body created by a state to control drainage in a specified area; it functions under legal regulations for financing, constructing, and operating a drainage system.

drainage divide The boundary between adjacent drainage basins; a *divide.*

drainage lake An *open lake* that loses water through a surface outlet or whose level is largely controlled by the discharge of its effluent. Cf: *seepage lake.*

drainage line The course or channel of a major stream in a drainage system.

drainage network *drainage pattern.*

drainage pattern The configuration or arrangement in plan view of the natural stream courses in an area. It is related to local geologic and geomorphologic features and history. Syn: *drainage network.*

drainage ratio The ratio between runoff and precipitation in a given area for a given period of time. Syn: *discharge efficiency.*

drainage system A surface stream, or a body of impounded surface water, together with all other such streams and water bodies that are tributary to it and by which a region is drained. An artificial drainage system includes also surface and subsurface conduits.

drainage varve An abnormally thick and sandy varve formed by the drainage of lakes ponded between the ice edge and higher land, or behind dams of glacial deposits.

drainageway A channel or course along which water moves in draining an area.

drainage well A type of *inverted well* used to drain excess soil or surface water, where the aquifer penetrated is permeable enough, and has a head far enough below the land surface, to remove the water at a satisfactory rate. Drainage wells have been used to dispose of some untreated domestic and other wastes, but such uses are now largely prohibited. Syn: *absorbing well.* Cf: *relief well.*

drakonite An extrusive rock composed of phenocrysts of alkali feldspar, plagioclase, and biotite and/or hornblende, in a trachytic groundmass of alkali feldspar microlites and interstitial alkali amphibole or pyroxene. The plagioclase ranges from oligoclase to labradorite. Apatite, sphene, magnetite, and zircon may be present as accessories. Syn: *Drachenfels trachyte.* Not recommended usage.

drape fold (a) A *supratenuous fold* or *compaction fold.* (b) A fold produced in layered rocks by movement of an underlying brittle block at high angles to the layering (Friedman et al., 1976, p. 1049); a type of *forced fold.* In this usage, the term excludes supratenuous or compaction folds.

draper point The temperature, 977°F (525°C), at which red light first becomes visible from a heated object in darkened surroundings; hence, the minimum temperature of incandescent lava (Draper, 1847; Siegel & Howell, 1968).

drapery A thin translucent sheet of travertine formed when drops of water flow down an inclined cave ceiling and leave behind a sinuous trail of calcite. Syn: *curtain [speleo]; drip curtain.* Partial syn: *bacon-rind drapery.*

draping The general structural concordance of warped strata, lying above a limestone reef or other hard core, to the upper surface of that reef or core, due to initial dip, to differential compaction, or to both.

dravite A brown, magnesium-rich mineral of the tourmaline group: $NaMg_3Al_6(BO_3)_3(Si_6O_{18})(OH)_4$.

draw [geomorph] (a) A small natural watercourse or gully, generally shallower or more open than a ravine or gorge; a shallow gulch; a valley or basin. (b) A usually dry stream bed; a coulee whose water results from periodic rainfall. (c) A sag or troughlike depression leading up from a valley to a gap between two hills.

draw [mining] The horizontal distance, measured on the surface ahead of an underground coal face, over which the rocks are influenced by subsidence.

drawdown [grd wat] (a) The lowering of the water level in a well as a result of withdrawal. (b) The difference between the height of the water table and that of the water in a well. (c) Reduction of the pressure head as a result of the withdrawal of water from a well. Cf: *cone of depression.*

drawdown [hydraul] In a stream or conduit, the difference between the water-surface elevation at a constriction and what the elevation would be if there were no constriction (ASCE, 1962).

drawdown [water] The distance by which the level of a reservoir is lowered by the withdrawal of water.

draw slate In coal mining, shale that occurs above a coal seam and collapses during or shortly after removal of the coal.

draw works The powered winch used in *rotary drilling* for lifting and lowering the *drill string.* Also spelled: drawworks.

D ray Left posterior ray in echinoderms situated clockwise of C ray when the echinoderm is viewed from the adoral side; equal to ambulacrum I in the Lovenian system.

dredge [eng] (a) A large floating machine for scooping up or excavating earth material at the bottom of a body of water, raising it to the surface, and discharging it to the bank through a floating pipeline or conveyor, into a scow for removal, or, in the case of certain mining dredges, into the same body of water after removal of the ore mineral. A "hydraulic dredge" uses a centrifugal pump; other dredges use dippers, clamshells, bucket chains, and scrapers. They may or may not be self-propelled. Dredges are used to excavate or deepen harbor channels, to raise the level of lowland areas, to dig ditches and improve drainage, and to obtain sand, gravel, placer gold, and other materials. (b) A ship designed to remove sediment from a channel.

dredge [oceanog] An ocean-bottom sampler that scoops sediment and benthonic organisms as it is dragged behind a moving ship. It is usually a heavy, metal container; one variety is made of chain mail affixed to a metal collar. Cf: *corer; grab sampler.*

dredge peat *sedimentary peat.*

dreikanter (a) A doubly pointed ventifact or wind-worn stone, having three curved faces intersecting in three sharp edges, resembling a Brazil nut. Syn: *pyramid pebble.* (b) A term loosely applied as a syn. of *ventifact.*—Etymol: German *Dreikanter,* "one having three edges". Pl: *dreikanters; dreikanter.*

Dresbachian North American stage: Upper Cambrian (below Franconian, above Albertan).

dresserite A mineral: $Ba_2Al_4(CO_3)_4(OH)_8 \cdot 3H_2O$.

drewite A white neritic impalpable calcareous mud or ooze, consisting chiefly of minute aragonite needles a few microns in length, believed to have been precipitated directly from seawater through the action of nitrate- and sulfate-reducing bacteria. Named after George Harold Drew (1881-1913), British scientist, who studied the marine bacteria associated with this sediment in the shallow lagoons in the Bahamas (Drew, 1911).

driblet *spatter.*

driblet cone *hornito.*

dried ice Sea ice whose whitened surface contains cracks and thaw holes following the disappearance of meltwater.

dries n. "An area of a reef or other projection from the bottom of a body of water which periodically is covered and uncovered by the water" (Baker et al., 1966, p.51). Syn: *uncovers.*

drift [coast] Detrital material moved and deposited by waves and currents; e.g. *littoral drift.* Also, floating material (e.g. driftwood or seaweed) that has been washed ashore by waves and left stranded on a beach.

drift [drill] *inclination [drill].*

drift [geophys] A gradual change in a reference reading that is supposed to remain constant. An instrument such as a gravimeter may show drift as a result of elastic aging, long-term creep, hysteresis, or other factors. See also: *drift correction.*

drift [glac geol] A general term applied to all rock material (clay, silt, sand, gravel, boulders) transported by a glacier and deposited directly by or from the ice, or by running water emanating from a glacier. Drift includes unstratified material (till) that forms moraines, and stratified deposits that form outwash plains, eskers, kames, varves, glaciofluvial sediments, etc. The term is generally applied to Pleistocene glacial deposits in areas (as large parts of North America and Europe) that no longer contain glaciers. The term "drift" was introduced by Murchison (1839, v.1, p. 509) for material, then called *diluvium,* that he regarded as having drifted in marine currents and accumulated under the sea in comparatively recent times; this material is now known to be a product of glacial activity. Cf: *glacial drift; glaciofluvial drift; fluvioglacial drift.*

drift [hydraul] "The effect of the velocity of fluid flow upon the velocity (relative to a fixed external point) of an object moving within the fluid" (Huschke, 1959, p. 178).

drift [mining] A horizontal or nearly horizontal underground opening driven along a vein.

drift [oceanog] (a) One of the wide, slower movements of surface oceanic circulation under the influence of, and subject to diversion or reversal by, prevailing winds; e.g. the easterly drift of the North Pacific. Syn: *drift current; wind drift; wind-driven current.* (b) The slight motion of ice or vessels resulting from ocean currents and wind stress. (c) The speed of an ocean current or ice floe, usually given in nautical miles per day or in knots. (d) Sometimes used as a short form of *littoral drift.*

drift [photo] Apparent offset of aerial photographs with respect to the true flight line, caused by the displacement of the aircraft owing to cross winds, and by failure to orient the camera to compensate for the angle between the flight line and the direction of the aircraft's heading. The photograph edges remain parallel to the intended flight line, but the aircraft itself drifts farther and farther from that line.

drift [sed] n. (a) A general term, used esp. in Great Britain, for all surficial, unconsolidated rock debris transported from one place and deposited in another, and distinguished from solid bedrock; e.g. *river drift.* It includes loess, till, river deposits, etc., although the term is often used specif. for glacial deposits. (b) Any surface movement of loose incoherent material by the wind; also, an accumulation of such material, such as a *snowdrift* or a *sand drift.* —v. To accumulate in a mass or be piled up in heaps by the action of wind or water.

drift [speleo] *fill [speleo].*

drift [streams] In South Africa, a ford in a river. The term is used in many parts of Africa to indicate a ford or a sudden dip in a road over which water may flow at times (Stamp, 1961, p. 162). Syn: *drif (Afrikaans).*

drift [tect] *continental displacement.*

drift avalanche *dry-snow avalanche.*

drift-barrier lake A glacial lake formed upstream from a moraine that has blockaded a valley or a drainage course (Fairchild, 1913, p. 153). Cf: *valley-moraine lake.*

drift bed A layer of drift "of sufficient uniformity to be distinguished from associated ones of similar origin" (Fay, 1918, p. 231).

drift bedding An old term used by Sorby (1857, p.279) to replace *false bedding.* See also: *ripple drift.*

drift bottle A bottle containing a record of the date and place at which it was released into the sea and a card requesting return by the finder with the date and place of recovery. It is used in studying surface currents. Syn: *bottle post.*

drift clay *boulder clay.*

drift coal Coal formed according to the *drift theory;* allochthonous coal.

drift copper Native copper transported from its source by a glacier.

drift correction Adjustment to remove the effects of geophysical *drift,* usually by repeated observations at a base station.

drift current (a) *drift [oceanog].* (b) "A current defined by assuming that the wind stress is balanced by the sum of the Coriolis and frictional forces" (Baker et al., 1966, p. 52).—Cf: *stream current.*

drift curve A graph of a series of gravity values read at the same station at different times and plotted in terms of instrument reading versus time.

drift dam A dam formed by the accumulation of glacial drift in a pre-existing stream valley.

drift epoch A syn. of *glacial epoch;* specif. the "Drift epoch", also known as the Pleistocene Epoch. See also: *Drift period.*

drift glacier A small mass of flowing ice in a mountain area nourished primarily with windblown snow from adjacent snowfields, slopes, or ridges. Syn: *Ural-type glacier; snowdrift glacier; glacieret.* Nonpreferred syn: *catchment glacier.*

drift ice (a) Any ice that has been broken apart and drifted from its place of origin by winds and currents, such as a fragment of a floe or a detached iceberg; loose, unattached pieces of *floating ice* with open water exceeding ice; navigable with ease. (b) A syn. of *pack ice* as that term is used in a broad sense.

drifting ice station An oceanographic research base established on the ice in the Arctic Ocean.

drift lake A glacial lake occupying a depression left in the surface of glacial drift after the disappearance of the ice (White, 1870, p. 70).

driftless area A region that was surrounded, but presumably not covered, by continental ice sheets of the Pleistocene Epoch, and is supposedly devoid of glacial deposits; specif. the "Driftless Area" occupying SW Wisconsin and parts of Illinois, Iowa, and Minnesota.

drift line A line of drifted material washed ashore and left stranded. It marks the highest stage of water, such as of a flood.

drift map A British term for a geological map representing a true picture of the visible ground, including all surficial deposits and only those rock outcrops exposed at the surface (Nelson & Nelson, 1967, p.114). Cf: *solid map.*

driftmeter An instrument for determining the *inclination* of a drill pipe from the vertical and the depth of measurement.

drift mining (a) The extraction of *placer* ore by underground horizontal or inclined tunneling rather than by the use of water. Cf: *placer mining.* (b) The extraction of near-surface coal seams by underground inclined tunneling rather than by opencut mining or vertical-shaft methods.

drift peat Peat that occurs in association with glacial drift.

Drift period A term formerly used to designate the Pleistocene Epoch. See also: *drift epoch.*

drift plain A plain, e.g. a *till plain,* underlain by glacial drift.

drift scratch *glacial striation.*

drift sheet A widespread sheetlike body of glacial drift, deposited during a single glaciation (e.g. the Cary drift sheet) or during a series of closely related glaciations (e.g. the Wisconsin drift sheet).

drift terrace A term used in New England for an *alluvial terrace.*

drift theory [coal] The theory that coal originates from the accumulation of plant material that has been transported from its place of growth and deposited in another locality, where coalification occurs. Ant: *in-situ theory.* See also: *allochthony.*

drift theory [glac geol] A theory of the early 19th century which attributed the origin of widespread surficial deposits, including the erratic boulders, to the action of marine currents and floating ice. Cf: *glacier theory.*

driftwood Woody material, such as parts of trees, drifted or floated by water and cast ashore or lodged on beaches by storm waves.

drill n. A device with an edged or pointed end or compound contacts, used for making circular holes in rock or earth material by a succession of blows or by rotation of the cutting surface; specif. a *drill bit.* —v. To make a circular hole with a drill or other cutting tool.

drill bit Any device at the lower end of a *drill stem,* used as a cutting or boring tool in drilling a hole; the cutting edge of a *drill.* Cf: *core bit.* Syn: *bit; rock bit.*

drill collar A length of extra-heavy, thick-walled *drill pipe* in a rotary *drill string* directly above either the bit or the core barrel, to concentrate weight and give rigidity so that the bit will cut properly. Syn: *collar [drill].*

drill cuttings *well cuttings.*

drilled well A well constructed by cable-tool or rotary drilling methods in the search for water, oil, or gas.

driller's log The brief, often vernacular notations, included as part of a driller's *tour report,* that describe the gross characteristics of the well cuttings noted by the drilling crew as a well is drilled. It is useful only if a detailed sample log is not available.

drill hole A circular hole made by drilling; esp. one made by cable tools, or one made to explore for valuable minerals or to obtain geologic information. Cf: *borehole.* Syn: *hole [drill].*

drilling The act or process of making a circular hole with a drill or other cutting tool, for purposes such as blasting, exploration, prospecting, valuation, or obtaining oil, gas, or water.

drilling cable *cable [drill].*

drilling fluid *drilling mud.*

drilling in The process of completing a well by setting casing just above the producing oil-, water-, or gas-bearing formation and then drilling into it, leaving it uncased.

drilling mud A carefully formulated heavy suspension, usually in water but sometimes in oil, used in rotary drilling. It commonly consists of bentonitic clays, chemical additives, and weighting materials such as barite. It is pumped continuously down the drill pipe, out through openings in the drill bit, and back up in the annulus between the pipe and the walls of the hole to a surface pit where it is screened and reintroduced through the *mud pump.* The mud is used to lubricate and cool the bit; to carry the cuttings up from the bottom; and to prevent blowouts and cave-ins by plastering friable or porous formations with *mud cake,* and maintaining a hydrostatic pressure in the borehole offsetting pressures of fluids that may exist in the formation. See also: *oil-base mud.* Syn: *mud [drill]; drilling fluid; circulating fluid.*

drilling rig A general term for the *derrick,* power supply, *draw works,* and other surface equipment necessary in rotary or cable-tool drilling. Syn: *rig [drill].*

drilling time (a) The time required for a rotary drill bit to penetrate a specified thickness (usually one foot) of rock. (b) The elapsed time required to drill a well, excluding periods when not actually drilling.

drilling-time log A *strip log* showing the times required to drill an increment of depth, e.g. minutes per foot, in a borehole. Rapid drilling (a fast *break*) often indicates a porous reservoir rock.

drill pipe The heavy steel pipe that turns the drill bit in *rotary drilling* by transmitting the motion from the rotary table of the drilling rig to the bit at the bottom of the hole, and that conducts the *drilling mud* from the surface to the bottom. It is normally formed of 30-ft sections connected end to end. Cf: *drill collar.*

drill rod A thin, lightweight drill tubing, such as used in shallow-depth core or seismic shot-hole drilling.

drill stem (a) A term used in rotary drilling for the *drill string.* (b) A term used in cable-tool drilling for a solid shaft or cylindrical bar of steel or iron attached to the drill bit to give it weight.—Also spelled: *drillstem.* Syn: *stem [drill].*

drill-stem test A procedure for determining the potential productivity of an oil or gas reservoir by measuring *reservoir pressures* and flow capacities while the drill pipe is still in the hole, the well is still full of drilling mud, and usually the well is uncased. The tool consists of a packer to isolate the section to be tested and a chamber to collect a sample of fluid. If the formation pressure is sufficient, fluid flows into the tester. Abbrev: DST. Cf: *wire-line test.*

drill string (a) A term used in rotary drilling for the assemblage in a borehole of *drill pipe, drill collars, drill bit,* and *core barrel* (if in use), connected to and rotated by the *drilling rig* at the surface. Syn: *drill stem.* (b) A term used in cable-tool drilling for the assemblage in a borehole of *drill bit, drill stem, cable,* and other tools, connected to the *walking beam* at the surface.—Syn: *string [drill].*

drip curtain *drapery.*

driphole (a) A small hole or niche in clay or rock beneath a point where water drips. (b) The center hole in a feature built up beneath dripping water.

dripstone A general term for calcite or other mineral deposit formed in caves by dripping water, including stalactites and stalagmites, and also usually including similar deposits formed by flowing water. See also: *flowstone.* Syn: *dropstone.* Partial syn: *calcareous sinter; cave onyx; speleothem; travertine.*

driven well A shallow well, usually of small diameter (3-10 cm), constructed by driving a series of connected lengths of pipe into unconsolidated material to a water-bearing stratum, without the aid of any drilling, boring, or jetting device. Syn: *drivewell; tube well.*

drivewell *driven well.*

DRM *depositional remanent magnetization.*

drop [coal] In the roof of a coal seam, a funnel-shaped downward intrusion of sedimentary rock, usually sandstone. Cf: *stone intrusion.*

drop [hydraul] The difference in water-surface elevations measured upstream and downstream from a constriction in the stream.

drop-down curve The form of the water surface along a longitudinal profile, assumed by a stream or open conduit upstream from a sudden fall. In a uniform channel, the curve is convex upward (ASCE, 1962). Cf: *backwater curve.*

dropout A loss of information in reading from or recording on magnetic tape.

dropped coverage The portion of a seismic line that is not shot, perhaps because of permit problems, access difficulties, or danger of damage.

dropstone *dripstone.*

drought polygon *desiccation polygon.*

drowned Said of a land surface or land feature that has undergone *drowning;* e.g. a *drowned* coast.

drowned atoll An atoll occurring at great depth, so that further coral-algal reef growth is prevented; such atolls indicate relatively rapid subsidence. See also: *drowned reef.*

drowned reef A reef situated at such great depth that reef growth is prevented or greatly hampered. See also: *drowned atoll.*

drowned river mouth The lower end of a river that is widened or submerged by seawater invading the coast; an *estuary.* Example: Chesapeake Bay.

drowned valley A valley that is partly submerged by the intrusion of a sea or lake. Syn: *submerged valley.*

drowning The submergence of a land surface or topography beneath water, either by a rise in the water level or by a sinking or subsidence of the land.

druid stone A *sarsen* used in the building of ancient stone circles at Stonehenge and elsewhere in Great Britain. Syn: *druidical stone.*

drum (a) A Scottish term for a long narrow ridge. (b) *drumlin.*

drumlin (a) A low, smoothly rounded, elongate oval hill, mound, or ridge of compact glacial till or, less commonly, other kinds of drift (sandy till, varved clay), built under the margin of the ice and shaped by its flow, or carved out of an older moraine by readvancing ice; its longer axis is parallel to the direction of movement of the ice. It usually has a blunt nose pointing in the direction from which the ice approached, and a gentler slope tapering in the other direction. Height is 8-60 m, average 30 m; length is 400-2000 m, average 1500 m. Syn: *drum.* (b) *rock drumlin.*—Etymol: Irish & Gaelic, diminutive of *druim,* "back, ridge".

drumlin field *basket-of-eggs topography.*

drumlinoid A *rock drumlin* or a drift deposit whose form approaches that of a true drumlin but does not fully attain it even though it seemingly results from the work of moving ice.

drumloid An oval hill or ridge of glacial till whose shape resembles that of a *drumlin* but is less regular and symmetrical.

drupe A fruit with a fleshy pericarp and a stony pit, e.g. a peach. The pits are readily fossilized.

druse (a) An irregular cavity or opening in a vein or rock, having its interior surface or walls lined (encrusted) with small projecting crystals usually of the same minerals as those of the enclosing rock, and sometimes filled with water; e.g. a small solution cavity, a steam hole in lava, or a lithophysa in volcanic glass. Cf: *geode; vug; miarolitic cavity.* (b) A mineral surface covered with small projecting crystals; specif. the crust or coating of crystals lining a druse in a rock, such as sparry calcite filling pore spaces in a limestone. —Etymol: German. Adj: *drusy.*

drusy (a) Pertaining to a *druse,* or containing many druses. Cf: *miarolitic.* (b) Pertaining to an insoluble residue or encrustation, esp. of quartz crystals; e.g. a "drusy oolith" covered with subhedral quartz.

dry In climatology, *arid.*

dry assay Any type of assay procedure that does not involve liquid as a means of separation. Cf: *wet assay.*

dry avalanche *dry-snow avalanche.*

dry basin An *interior basin* (as in an arid region) containing no perennial lake because the drainage is "occasional only and not continuous" (Gilbert, 1890, p. 2).

dry-bone ore An earthy, friable, honeycombed variety of smithsonite, usually found in veins or beds in stratified calcareous rocks, accompanying sulfides of zinc, iron, and lead. The term is sometimes applied to hemimorphite. Syn: *dry bone.*

dry bulk density The specific gravity of a substance, e.g. a sediment, without interstitial water.

dry calving The breaking-away of a mass of ice from a glacier on dry land.

dry cave *dead cave.*

dry delta (a) *alluvial fan.* (b) *alluvial cone.*

dry digging *dry placer.*

drydock iceberg *valley iceberg.*

dry frozen ground Relatively loose and crumbly ground (or soil) that has a temperature below freezing but contains no ice.

dry gap A gap that is not occupied by a stream; specif. a *wind gap.*

dry gas Natural gas with a very low content of liquid hydrocarbons. Cf: *wet gas.*

dry hole The universal term in the petroleum industry for an unsuccessful well, i.e. one that does not produce oil or gas in commercial quantities.

dry ice (a) Ice at a temperature below the freezing point; specif. bare glacier ice on which there is no slush or standing water (ADTIC, 1955, p.26). (b) Solidified carbon dioxide.

drying crack *desiccation crack.*

dry lake (a) A lake basin that formerly contained a lake. (b) A *playa;* a tract of salt-encrusted land in an arid or semiarid region, occasionally covered by an intermittent lake.

dry peat Peat derived from humic matter and formed under drier conditions than those of a moor.

dry permafrost Loose and crumbly permafrost containing little or no ice or moisture.

dry placer A placer that cannot be mined owing to lack of the necessary water supply. Syn: *dry digging.*

dry playa A playa that is normally hard, buff in color, and smooth as a floor (Thompson, 1929); the water table is at a considerable distance beneath the dense, sun-baked surface. Cf: *wet playa.*

dry quicksand A sand accumulation that offers no support to heavy loads because of alternating layers of firmly compacted sand and loose, soft sand.

drysdallite A hexagonal mineral: $Mo(Se,S)_2$.

dry snow Deposited snow that has not been subject to melting or to infiltration of liquid water. Cf: *wet snow.*

dry-snow avalanche An avalanche composed of dry, loose or powdery snow that is set in motion by the wind and is sometimes drifted but not wind-packed; the driving-ahead of a column of compressed air creates a vacuum in its wake. It is the fastest-moving of the snow avalanches, capable of reaching a speed of 450 km/hr. Syn: *dry avalanche; drift avalanche; powder avalanche.*

dry-snow line The boundary on a glacier or ice sheet between the *dry-snow zone* and an area where surface melting occurs.

dry-snow zone The area on a glacier or ice sheet where no surface melting occurs even in summer, delimited by the *dry-snow line.*

dry steam Steam that has an enthalpy greater than that for equilibrium with water for the existing pressure.

dry unit weight The *unit weight* of soil solids per unit of total volume of soil mass. See also: *maximum unit weight.* Syn: *unit dry weight.*

dry valley A valley that is devoid or almost devoid of running water; a streamless valley. It may be the result of stream capture, a climatic change, or a fall in the water table. Dry valleys are common in areas underlain by chalk and limestone; other examples include wind gaps and glacial overflow channels. Syn: *dead valley.*

dry wash A *wash* that carries water only at infrequent intervals and for short periods, as after a heavy rainfall. Cf: *arroyo.*

dry weathering Mechanical weathering of a rock without the action of water, as in an arid region.

d-spacing In diffraction of X-rays by a crystal, the distance or separation between the successive and identical parallel planes in the crystal lattice. It is expressed as d in the *Bragg equation.*

DST *drill-stem test.*

DTA *differential thermal analysis.*

dubiocrystalline Said of the texture of a rock whose crystallinity can be determined only with difficulty or uncertainty; e.g. said of the texture of a porphyry whose groundmass is too fine to be resolved into its constituents under a microscope, but shows faint anisotropism or polarizes light like an aggregate. Also, said of a rock with such a texture. Term introduced by Zirkel (1893, p.455) as *dubiokrystallinisch.* Cf: *cryptocrystalline.*

dubiofossil A structure of undetermined or uncertain origin, possibly biogenic (Hofmann, 1972, p. 27); a *problematic fossil.* Etymol: Latin *dubius,* "doubtful", + fossil.

Dubosq colorimeter An instrument that compares visually the color intensity of a solution of unknown strength with that of a variable depth of standard solution. From the depth of standard solution required to obtain a visual match, the strength of the unknown can be determined. See also: *colorimeter.*

Duchesnian North American continental stage: Upper Eocene (above Uintan, below Chadronian).

Duck Lake North American (Gulf Coast) stage: Miocene (above Napoleonville, below Clovelly).

ducktownite A term used in Tennessee for an intimate mixture of pyrite and chalcocite, or for the matrix of a blackish copper ore containing grains of pyrite.

ductile Said of a rock that is able to sustain, under a given set of conditions, 5-10% deformation before fracturing or faulting. Cf: *brittle.*

ductility A measure of the degree to which a rock exhibits *ductile* behavior under given conditions, commonly expressed by the strain at which fracture or faulting commences.

ductolith A more or less horizontal igneous intrusion that resembles a tear drop in cross section.

duff A type of organic surface horizon of forested soils, consisting of matted, peaty organic matter that is only slightly decomposed. It is a constituent of the *forest floor.* Cf: *litter; leaf mold.*

dufrenite A blackish-green mineral: $Fe^{+2}Fe_4^{+3}(PO_4)_3(OH)_5 \cdot 2H_2O$. Syn: *kraurite; green iron ore.*

dufrenoysite A lead-gray orthorhombic mineral: $Pb_2As_2S_5$.

duftite An orthorhombic mineral: $PbCu(AsO_4)(OH)$.

dug well A shallow, large-diameter well constructed by excavating with hand tools or power machinery instead of by drilling or driving, such as a well for individual domestic water supplies.

Duhem's theorem The statement in chemistry that the state of any closed system is completely defined by the values of any two independent variables, extensive or intensive, provided the initial masses of each component are given. The choice of variables, however, must not conflict with the phase rule.

dull-banded coal *Banded coal* consisting mainly of vitrain and durain, some clarain, and minor fusain. Cf: *bright-banded coal.*

dull coal A type of banded coal defined microscopically as consisting mainly of clarodurain and durain and of 20% or less of bright materials such as vitrain, clarain, and fusain. Cf: *bright coal; semidull coal; semibright coal; intermediate coal.* Syn: *dulls.*

dull luster The luster of a mineral or rock surface that diffuses rather than reflects light, even though the surface may appear smooth. *Earthy* materials have a dull luster.

dulls *dull coal.*

dumalite A *trachyandesite* characterized by intersertal texture and a glassy mesostasis which possibly has the composition of nepheline. It was named by Loewinson-Lessing in 1905 for Dumala in the Caucasus. Not recommended usage.

dumbbell Two land areas connected by a relatively narrow isthmus of sand which is never below the high-water mark in any part of its length, and whose highest points are higher above sea level than any part of the isthmus (Schofield, 1920). Syn: *dumbbell island.*

dumontite A yellow orthorhombic mineral: $Pb_2(UO_2)_3(PO_4)_2(OH)_4 \cdot 3H_2O$.

dumortierite A bright-blue or greenish-blue mineral of the sillimanite group: $Al_7(BO_3)(SiO_4)_3O_3$. It may contain iron, and it occurs principally in schists and gneisses.

dumortierite-quartz A massive opaque blue variety of crystalline quartz, colored by intergrown crystals of dumortierite.

dumped deposit An unsorted sediment deposited directly below wave base or current base, or brought in at a rate too rapid for waves and currents to distribute it (Weeks, 1952, p. 2107); e.g. a shaly sand or a sandy shale.

dump moraine An end moraine consisting of englacial and superglacial material dropped by a glacier at its front.

dumpy level A leveling instrument in which the telescope is permanently attached (either rigidly or by a hinge) to the vertical spindle or leveling base and that is capable only of rotatory movement in a horizontal plane. The dumpy level takes its name from the dumpy appearance of the early type of this instrument, the telescope of which was short and had a large object glass. Cf: *wye level.*

dun An inconspicuous hill in the English Lake District. Etymol: Gaelic.

dundasite A white mineral: $PbAl_2(CO_3)_2(OH)_4 \cdot 2H_2O$.

dune [geomorph] A low mound, ridge, bank, or hill of loose, windblown granular material (generally sand, sometimes volcanic ash), either bare or covered with vegetation, capable of movement from place to place but always retaining its characteristic shape. Etymol: French. See also: *sand dune.*

dune [stream] A term used by Gilbert (1914, p.31) for a *sand wave* formed on a stream bed and usually transverse to the direction of flow, traveling downstream by the erosion of sand from the gentle upstream slope and its deposition on the steep downstream slope, and having an approximately triangular cross section in a vertical plane in the direction of flow; a large-scale mound or ridge of sand, similar to an eolian sand dune, but formed in a subaqueous environment. Most modern authors follow Simons and Richardson (1961) in considering that dunes are features greater than 5 cm in height (as compared with *ripples* which are less than 5 cm). Some authors insist that dunes must have a sinuous or irregular crest line, and be associated with deep scours; these characteristics distinguish dunes from *bars* and *sand waves.* Cf: *antidune.* Syn: *subaqueous sand dune.*

dune complex An aggregate of moving and fixed sand dunes in a given area, together with sand plains and the ponds, lakes, and swamps produced by the blocking of streams by the sand.

dune lake (a) A lake occupying a basin formed as a result of the blocking of the mouth of a stream by sand dunes migrating along the shore; e.g. Moses Lake, Wash. (b) A *deflation lake* occupying a blowout on a dune.

dune massif A large irregular cone- or pyramid-shaped dune with curved slopes and steep sides consisting of small hollows and terraces (Stone, 1967, p. 225).

dune movement In hydraulics, the movement of sediment along the bed of a stream in the form of a wave or dune which travels downstream. The upstream face of the wave is eroded and the eroded material is deposited on the downstream face of the wave. The water surface has only a slight undulation (ASCE, 1962).

dune phase The part of stream traction whereby a mass of sediment travels in the form of a small, dunelike body having a gentle upcurrent slope and a much steeper downcurrent slope (Gilbert, 1914, p. 30-34); it develops when the bed load is small or the current is weak. The dune form moves downstream. Cf: *smooth phase; antidune phase.*

dune ridge A series of parallel dunes, whose movements are arrested by the growth of vegetation, along the shore of a retreating sea. See also: *foredune.*

dune rock An *eolianite* consisting of dune sand.

dune sand A type of *blown sand* that has been piled up by the wind into a sand dune, usually consisting of rounded quartz grains having diameters ranging from 0.1 to 1 mm.

dune slack A damp *slack* or depression between dunes or dune ridges on a shore; a *dune valley.*

dune valley A hollow, furrow, or depression between dunes or dune ridges. Syn: *dune slack.*

dungannonite An alkalic corundum-bearing *diorite.* Its name, given by Adams and Barlow in 1910, is derived from Dungannon, Ontario. Not recommended usage.

dunite (a) In the *IUGS classification,* a plutonic rock with M equal to or greater than 90 and ol/(ol + opx + cpx + hbd) greater than 90. (b) *Peridotite* in which the mafic mineral is almost entirely olivine, with accessory chromite almost always present. Syn: *olivine rock.* Named by Hochstetter in 1864 from Dun Mountain, New Zealand.

Dunkardian North American provincial series: uppermost Pennsylvanian-Lower Permian (above Monongahelan).

duns A term used in SW England for a shale or massive clay associated with coal.

dunstone [ign] An amygdaloidal *spilite;* a local name in the Plymouth area of England. Not recommended usage.

dunstone [sed] (a) A term used near Matlock, England, for a hard granular yellowish or cream-colored magnesian limestone. (b) A term used in Wales for a hard fireclay or underclay, and in England for a shale.

duplex fault zone A structural complex consisting of a *roof thrust* at the top and a *floor thrust* at the base, within which a suite of more steeply dipping imbricate thrust faults thicken and shorten the intervening panel of rock (Dahlstrom, 1970, p. 418-421; Boyer, 1976).

duplexite *bavenite.*

duplicature A doubling or a fold; e.g. the distal portion of an ostracode carapace that is folded toward the interior of the valve around the free margin to form a doubling of the lamella (if fused, the radial pore canals can be seen extending from the inner to the outer margin, otherwise a cavity or vestibule is present, esp. in the anterior and posterior regions). See also: *skeletal duplicature.*

duplivincular Said of a type of ligament of a bivalve mollusk (e.g. *Arca*) in which the lamellar component is repeated as a series of bands, each with its two edges inserted in narrow grooves in the cardinal areas of the respective valves. Cf: *chevron groove.*

durability index The relative resistance to abrasion exhibited by a sedimentary particle in the course of transportation, represented by the ratio of the *reduction index* of a standard (such as quartz) to that of a given rock or mineral under the same conditions (Wentworth, 1931, p. 26). Abbrev: DI

durain A coal lithotype characterized macroscopically by dull, matte luster, grey to brownish black color, and granular fracture. It occurs in bands up to many centimeters in thickness; its characteristic microlithotype is *durite.* Cf: *vitrain; clarain; fusain.* Syn: *attritus.*

durangite An orange-red monoclinic mineral: $NaAl(AsO_4)F$.

Durangoan North American (Gulf Coast) stage: Lower Cretaceous (above LaCasitan of Jurassic, below Nuevoleonian).

duranusite An orthorhombic mineral: As_4S.

duration (a) The interval of time in which a tidal current is either ebbing or flooding, reckoned from the middle of slack water. (b) The interval of time from high water to low water (falling tide), or from low water to high water (rising tide).

duration-area curve A curve which shows the area beneath a *duration curve,* and any value of the flow, and is therefore the integral of duration with respect to stream flow. When the duration curve is plotted as a percentage of time, the resulting duration area shows the average flow available below a given discharge (ASCE, 1962).

duration curve A graphic illustration of how often a given quantity is equaled or exceeded during a given span of time, e.g. a *flow-duration curve.* It is used in hydraulics.

durbachite A plutonic rock composed chiefly of alkali feldspar, biotite, and hornblende, with a smaller amount of plagioclase and accessory quartz, apatite, sphene, zircon, and opaque oxides; a dark-colored biotite-hornblende *syenite.* The orthoclase phenocrysts form Carlsbad twins in a groundmass that is essentially an aggregate of biotite flakes and orthoclase (Johannsen, 1939, p. 249). Its name is derived from Durbach, in the Black Forest, Germany. Not recommended usage.

durdenite *emmonsite.*

duricrust A general term for a hard crust on the surface of, or layer in the upper horizons of, a soil in a semiarid climate. It is formed by the accumulation of soluble minerals deposited by mineral-bearing waters that move upward by capillary action and evaporate during the dry season. See also: *ferricrete; silcrete; calcrete; caliche [soil].* Etymol: Latin *durus,* "hard", + *crust.* Cf: *hardpan.*

durinite The major maceral of durain, according to the Stopes classification; the term is no longer in general use.

durinode In a soil, a nodule that has been cemented or indurated with silica.

duripan A diagnostic subsurface soil horizon that is characterized by cementation by silica (esp. opal or microcrystalline forms) and, possibly, by accessory cements. Duripans occur mainly in areas of volcanism that have arid or Mediterranean climates (USDA, 1975).

durite A coal microlithotype that contains a combination of inertinite and exinite totalling at least 95%, and containing more of each than of vitrinite. Cf: *durain.*

duroclarain A transitional lithotype of coal, characterized by the presence of vitrinite with lesser amounts of other macerals such as micrinite and exinite; it corresponds to *semisplint coal.* Cf: *clarodurain.*

duroclarite A coal microlithotype containing at least 5% each of vitrinite, exinite, and inertinite, with more vitrinite than inertinite and exinite. It is a variety of *trimacerite,* intermediate in composition between clarite and durite, but closer to clarite. Cf: *clarodurite.*

durofusain A coal lithotype transitional between fusain and durain, but predominantly fusain. Cf: *fusodurain.*

durotelain A coal lithotype transitional between telain and durain, but predominantly telain. Cf: *telodurain.*

durovitrain A coal lithotype transitional between vitrain and durain, but predominantly vitrain. Cf: *vitrodurain.*

dussertite A mineral: $BaFe_3(AsO_4)_2(OH)_5$.

dust [sed] (a) Dry, solid matter consisting of clay- and silt-size earthy particles (diameters less than 1/16 mm, or 62 microns), so finely divided or comminuted that they can be readily lifted and carried considerable distances in suspension by turbulent eddies in the wind, freely mixing with atmospheric gases, and staying aloft a long time but eventually falling back to the Earth's surface. Terrestrial sources of atmospheric dust include volcanic eruptions; salt spray from the seas; mineral particles; pollen and bacteria; and smoke and fly ash. See also: *volcanic dust.* (b) Small extraterrestrial particles that invade the Earth's atmosphere, such as *cosmic dust* and *meteoric dust.* (c) *gold dust.* (d) *diamond dust.* (e) A syn. of volcanic *ash,* esp. the finer fractions of ash.

dust [volc] A syn. of volcanic *ash,* esp. the finer fractions of ash.

dust band *dirt band [glaciol].*

dust basin A large, shallow *dust well.*

dust cloud *eruption cloud.*

dust-cloud hypothesis A theory of the formation of the planets by the accretion of a cloud of small, cold bodies which are sometimes called "planetesimals". Syn: *planetesimal hypothesis.*

dust fall (a) *dusting.* (b) *blood rain.*

dust hole A small *dust well.*

dusting The process by which dust and dustlike particles are deposited from the atmosphere; e.g. the deposition of opal phytoliths in the ocean. Syn: *dust fall.*

dust ring A ring of tiny inclusions seen in thin section, marking the original surface of a detrital sand grain that has grown by secondary enlargement.

dust sand A term used by Searle (1923, p. 1) for a material whose particles have diameters in the range of 0.025-0.04 mm and are washed out by a stream having a velocity of 1.5 mm/sec. The particle sizes correspond to medium silt and coarse silt.

dust size A term used in sedimentology for a volume less than that of a sphere with a diameter of 1/16 mm (0.0025 in.); it includes *silt size* and *clay size.*

dust tuff A tuff of very fine fraction; an indurated deposit of volcanic dust. Syn: *mud tuff.*

dust veil The stratospheric pall which results from ejection of volcanic dust into the stratosphere as an implicit consequence of paroxysmal eruptions (Lamb, 1970, p. 425-533).

dust-veil index A standard for classifying the degree by which dust in the atmosphere, esp. volcanic dust, forms a barrier to incoming solar radiation.

dust well A small hollow or pit on the surface of glacier ice or sea ice, produced by the gradual sinking into the ice of a patch of dark windblown particles that absorb solar radiation and cause the surrounding ice to melt more rapidly. Cf: *cryoconite hole; dust basin; meridian hole; dust hole.*

duttonite A pale-brown monoclinic mineral: $VO(OH)_2$.

duty of water The quantity (or depth) of irrigation water required for a given area for the purpose of producing a particular crop; it is commonly expressed in acre-inches or acre-feet per acre, or simply as depth in inches or feet. Syn: *duty.*

duxite An opaque, dark-brown variety of retinite containing about 0.5% sulfur, found in lignite at Dux in Bohemia, Czechoslovakia.

dwarf fauna A fossil *assemblage* consisting of specimens of small size. Many dwarf faunas result from sedimentary sorting, others from pathologies or environmentally influenced growth patterns. Also spelled: *dwarfed fauna.* Syn: *diminutive fauna; stunted fauna.* Partial syn: *depauperate fauna; impoverished fauna.*

dwip The basic unit of deposition in a tidal channel, consisting of a circular bank with a horseshoe outline and a hollow center, the toe pointing upstream (Strickland, 1940). It is caused primarily by channel bifurcation and reversing tidal currents.

dy A dark jellylike freshwater mud, consisting largely of unhumified or peaty organic matter, such as that derived from an acidic peat bog, which was brought to a nutrient-deficient lake in colloidal form and precipitated there. Etymol: Swedish, "silt". Cf: *gyttja; sapropel.*

dyad An uncommon grouping in which mature pollen grains are shed in fused pairs. Cf: *tetrad; polyad.*

dyakisdodecahedron *diploid.*

Dyassic An old equivalent of *Permian.*

dying lake A lake nearing extinction from any cause.

dyke *dike.*

dynamic correction In seismic work, a correction for *normal moveout.*

dynamic ellipticity A ratio expressed as the difference between the moments of inertia about the polar and equatorial axes, and the moment of inertia about the polar axis.

dynamic equilibrium A condition of a system in which there is a balanced inflow and outflow of materials. Cf: *stable equilibrium.*

dynamic geology A general term for the branch of geology that deals with the causes and processes of geologic phenomena; physical geology.

dynamic geomorphology The quantitative analysis of geomorphic processes treated as "gravitational or molecular shear stresses acting upon elastic, plastic, or fluid earth materials to produce the characteristic varieties of strain, or failure, that constitute weathering, erosion, transportation, and deposition" (Strahler, 1952a, p.923); the processes are considered in terms of steady-state operations that are self-regulatory to a large degree. Syn: *analytical geomorphology.*

dynamic head That head of fluid which would produce statically the pressure of a moving fluid (ASCE, 1962).

dynamic height The distance above the geoid of points on the same equipotential surface, expressed in linear units measured along a plumb line at a specified latitude, usually 45°. Cf: *height [geodesy]; orthometric height; geopotential number.* Syn: *geopotential height.*

dynamic magnetization A syn. in Russian literature of *shock remanent magnetization.*

dynamic metamorphism The total of the processes and effects of orogenic movements and differential stresses in producing new rocks from old, with marked structural and mineralogical changes due to crushing and shearing at low temperatures and extensive recrystallization at higher temperatures. It may involve large areas of the Earth's crust, i.e. be regional in character. Cf: *dynamothermal metamorphism; regional metamorphism; dislocation metamorphism.* Syn: *dynamometamorphism.*

dynamic pressure The pressure of a flowing fluid against a surface. Reaction to dynamic pressure affects direction and velocity of flow (ASCE, 1962).

dynamic range In seismic recording, the ratio of the maximum to minimum signal amplitude that can be handled by a recording system. It is usually specified over a certain frequency-band width and is measured in decibels.

dynamic recrystallization The formation of new grains in a rock during solid-state deformation. It is essentially a recovery process, in which simultaneous *nucleation* and *grain growth* compete, ultimately to yield a uniform grain size inversely proportional to the applied differential stress. See also: *annealing recrystallization.*

dynamic rejuvenation Renewal of the effectiveness of erosion processes, caused by uplift of a landmass with accompanying tilting and warping.

dynamic theory A theory of tides that considers the horizontal tide-producing forces to be the most important factor in causing movement of water and that regards the vertical tide-producing forces as small periodical variations in the acceleration of gravity (Baker et al., 1966, p. 53).

dynamic viscosity *viscosity coefficient.*

dynamofluidal Pertaining to a texture in dynamometamorphosed rocks showing parallel arrangement in one direction only.

dynamogranite A seldom-used term for an *augen gneiss* containing much microcline and orthoclase (Krivenko & Lapchik, 1934).

dynamometamorphism The equivalent of *dislocation metamorphism* and a syn. of *dynamic metamorphism.*

dynamometer In oceanography, an instrument that is used in conjunction with a bottom sampler to indicate that the bottom has been reached. Its operation is based on wire tension.

dynamo theory The statement that the Earth's main magnetic field is sustained by self-exciting dynamo action in the fluid core. The conducting liquid is supposed to flow in such a pattern that the electric current induced by its motion through the magnetic field sustains that field.

dynamothermal metamorphism A common type of metamorphism involving the effects of directed pressures and shearing stress as well as a wide range of confining pressures and temperatures. It is related both geographically and genetically to large orogenic belts, and hence is regional in character. Cf: *burial metamorphism; regional metamorphism; dynamic metamorphism.*

dypingite A mineral: $Mg_5(CO_3)_4(OH)_2 \cdot 5H_2O$.

dysanalyte A variety of perovskite containing niobium and tantalum.

dyscrasite A silver-white mineral: Ag_3Sb.

dyscrystalline *microcrystalline.*

dysgeogenous Not easily weathered; said of a rock that produces by weathering only a small amount of detritus. Ant: *eugeogenous.*

dysluite A brown variety of gahnite containing manganese and iron.

dysodile (a) A flexible, slightly elastic yellow or greenish-gray hydrocarbon, which burns with a highly fetid odor, from Melili, Sicily, and from certain German lignite deposits (Rice, 1945). See also: *chlorophyll coal.* (b) A sapropelic coal of lignitic rank, derived from diatomaceous sediments formed under anaerobic conditions. It burns readily and with a bad odor. Dysodile occurs in Tertiary limestones and lignites. Syn: *diatom-saprocol.*

dysodont Said of the dentition of a bivalve mollusk (e.g. some Mytilacea), characterized by small, weak hinge teeth close to the beaks.

dystome spar *datolite.*

dystrophic lake A lake that is characterized by a deficiency in nutrient matter and by a notably high oxygen consumption in the hypolimnion; its water is brownish or yellowish with much unhumified or dissolved humic matter and it has a small bottom fauna. It is often associated with acidic peat bogs. Juday and Birge rejected this term, and Hutchinson (1957, p. 380) called it "unfortunate". Cf: *oligotrophic lake.*

dysyntribite A hydrated aluminosilicate of sodium and potassium, probably a variety of pinite or an impure muscovite.

dzhalindite A yellow-brown mineral: $In(OH)_3$. It is an alteration product of indite.

dzhetymite A term proposed by Dzholdoshev (see Schermerhorn, 1966, p. 833) as a quantitative designation for a nonsorted rock composed of approximately equal proportions (25-35% each) of angular "gravel" (1-10 mm diameter in Russian literature), sand (0.1-1 mm), and mud (under 0.1 mm).

dzhezkazganite A mineral: a lead rhenium sulfide (?).

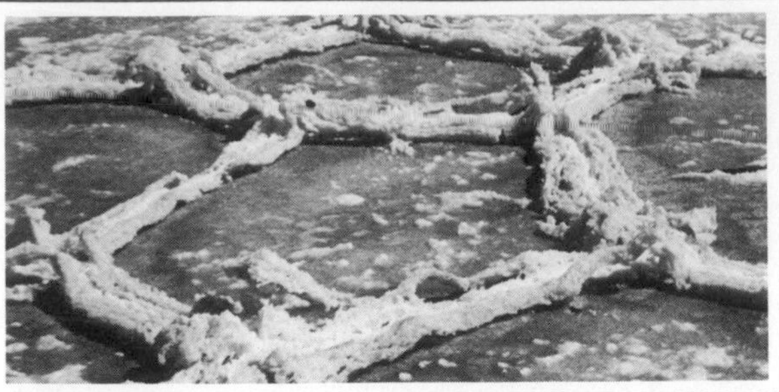

ea An English term for a stream.

Eaglefordian North American (Gulf Coast) stage: Upper Cretaceous (above Woodbinian, below Austinian).

eaglestone A walnut-sized concretionary nodule (usually of clay ironstone or flint), often containing a loose stone in its hollow interior, and believed by the ancients to be taken by an eagle to her nest to facilitate egg-laying.

EA interray Left anterior interray in echinoderms, situated between E ray and A ray and clockwise of E ray when the echinoderm is viewed from the adoral side; equal to interambulacrum 2 in the Lovenian system.

eakerite A mineral: $Ca_2SnAl_2Si_6O_{16}(OH)_6$.

ear The flattened or pointed extremity of a brachiopod shell, subtended between the hinge line and the lateral part of the commissure.

eardleyite A trigonal mineral: $(Ni,Zn)_6Al_2(CO_3)(OH)_{16}\cdot 4H_2O$(?).

earlandite A pale-yellow to white mineral consisting of a hydrous calcium citrate: $Ca_3(C_6H_5O_7)_2\cdot 4H_2O$. It is found in ocean-bottom sediments from the Weddell Sea in the south Atlantic Ocean.

early Pertaining to or occurring near the beginning of a segment of time. The adjective is applied to the name of a geologic-time unit (era, period, epoch) to indicate relative time designation and corresponds to *lower* as applied to the name of the equivalent time-stratigraphic unit; e.g. rocks of a Lower Jurassic batholith were intruded in Early Jurassic time. The initial letter of the term is capitalized to indicate a formal subdivision (e.g. "Early Devonian") and is lowercased to indicate an informal subdivision (e.g. "early Miocene"). The informal term may be used for eras and epochs, and for periods where there is no formal subdivision. Cf: *middle [geochron]; late.*

early diagenesis Diagenesis occurring immediately after deposition or immediately after burial. It is equivalent to *diagenesis* as interpreted by Russian geologists. Syn: *syngenesis; syndiagenesis.*

early wood Xylem formed in initial stages of a growth zone, which is not so dense as that produced later. Cf: *late wood.* Syn: *springwood.*

Earth That planet of the solar system which is fifth in size of the 9 major planets, and the third (between Venus and Mars) in order of distance from the Sun (about 150 x 10^6 km, or 93 million mi). Major data for the Earth: equatorial radius: 6,378 km (3,963.5 mi); polar radius: 6,357 km (3,941 mi); mean radius: 6,371 km (3,950 mi); equatorial circumference: 40,075 km (24,902 mi); surface area: 5.101 x 10^8 km^2 (197 x 10^6 sq mi).

earth [chem] (a) A difficultly reducible metallic oxide (such as alumina), formerly regarded as an element. (b) One of the four elements of the alchemists (the others: air, water, fire).—See also: *rare earths.*

earth [eng geol] Material that can be removed and handled economically with pick and shovel, or loosened and removed with a power shovel, scraper, or end loader.

earth [geog] A general term for the solid materials that make up the physical globe, as distinct from water and air. Also, the firm land or ground of the Earth's surface, uncovered by water.

earth [sed] (a) An organic deposit that has remained unconsolidated although it is no longer in the process of accumulation; e.g. *radiolarian earth* and *diatomaceous earth.* (b) An amorphous fine-grained material, such as a clay or a substance resembling clay; e.g. *fuller's earth.*

earth circle *soil circle.*

earth current Static or alternating electric current flowing through the ground and arising either in natural or artificial electric or magnetic fields. Syn: *ground current; telluric current.*

earth curvature The divergence of the surface of the Earth (spheroid) from a horizontal plane tangent at the point of observation. See also: *curvature correction.*

earth dam A dam constructed of earth material (such as gravel, broken weathered rock, sand, silt, or soil). It has a core of clay or other impervious material and a rock facing of riprap to protect against wave erosion.

earth dike An artificial levee or embankment built of earth fill.

earth finger A miniature "earth pillar" in a nearly horizontal attitude, produced by wind-driven rain falling upon clayey material (Cotton, 1958, p. 31 & 33).

earth flax A fine silky *asbestos.*

earthflow A mass-movement landform and process characterized by downslope translation of soil and weathered rock over a discrete basal shear surface (landslide) within well defined lateral boundaries. The basal shear surface is more or less parallel with the ground surface in the downslope portion of the flow, which terminates in lobelike forms. Overall, little or no rotation of the slide mass occurs during displacement, although, in the vicinity of the crown scarp, minor initial rotation is usually observed in a series of slump blocks. Earthflows grade into mudflows through a continuous range in morphology associated with increasing fluidity. Also spelled: *earth flow.*

earth foam Soft or earthy *aphrite.*

Earth history *geologic history.*

earth hummock A low, dome-shaped *frost mound,* consisting of a fine-textured earthen core covered by a tight mass of vegetation, esp. mosses, but also humus, grasses, sedges, and scrubby plants, and produced by hydrostatic pressure of ground water or by heaving from growth of ice lenses in arctic and alpine regions; the general height is 10-20 cm and the diameter ranges from 1/2 to 1 m. Earth hummocks form in groups to produce a nonsorted patterned ground. Cf: *turf hummock.* Syn: *earth mound; thufa.*

earth inductor A type of *inclinometer [magnet],* based on the principle that a voltage is induced in a coil rotating in the Earth's field whenever the rotation axis does not coincide with the field direction.

earth island *debris island.*

earthlight The faint illumination of the dark part of the Moon, produced by sunlight reflected onto the Moon from the Earth's surface and atmosphere. It is best seen during the Moon's crescent phases. Syn: *earthshine.*

earth mound *earth hummock.*

earth pillar A tall, conical column of unconsolidated to semiconsolidated earth materials (e.g. clay, till, or landslide debris), produced by differential erosion in a region of sporadic heavy rainfall (as in a badland or a high alpine valley), and usually capped by a flat, hard boulder that shielded the underlying softer material from erosion; it often measures 6-9 m in height, and its diameter is a function of the width of the protective boulder. Cf: *hoodoo.* Syn: *earth pyramid; demoiselle; fairy chimney; hoodoo column; penitent [geomorph].*

earth pitch *maltha.*

earth pressure The pressure exerted on a structure such as a retaining wall by earth materials, e.g. soil or sediments; the pressure exerted by soil on any boundary. See also: *active earth pressure; passive earth pressure; at rest.*

earth pyramid A less steep-sided and more conical *earth pillar,* occurring esp. where rainwash is the chief agent of erosion.

earthquake A sudden motion or trembling in the Earth caused by the abrupt release of slowly accumulated strain. Partial syn: *seismic event.* Syn: *shock; quake; seism; macroseism; temblor.*

earthquake engineering The study of the behavior of foundations and structures relative to seismic ground motion, and the attempt to mitigate the effect of earthquakes on structures.

earthquake intensity A measure of the effects of an earthquake at a particular place. Intensity depends not only on the *earthquake magnitude,* but also on the distance from earthquake to epicenter and on the local geology. See also: *intensity scale.*

earthquake magnitude A measure of the strength of an earthquake, or the strain energy released by it, as determined by seismographic observations. C. F. Richter first defined *local magnitude* as the logarithm, to the base 10, of the amplitude in microns of the largest trace deflection that would be observed on a standard torsion seismograph at a distance of 100 km from the epicenter. Magnitudes determined at teleseismic distances are called *body-wave magnitude* and *surface-wave magnitude.* The local, body-wave, and surface-wave magnitudes of an earthquake do not necessarily have the same numerical value. See also: *Richter scale.* Cf: *earthquake intensity.* Syn: *magnitude [seis].*

earthquake period That time during which a region affected by an earthquake continues to receive shocks without any long respite between them.

earthquake prediction That aspect of seismology which deals with the physical conditions or indications that precede an earthquake, in order to predict the size, time, and location of an impending shock.

earthquake record *seismogram.*

earthquake rent Popular syn. of *reverse scarplet.*

earthquake scarplet A low, nearly straight fault scarp or step, often continuous for many kilometers, formed simultaneously with an earthquake. Cf: *reverse scarplet.*

earthquake sea wave *tsunami.*

earthquake sound An *air wave* associated with an earthquake.

earthquake swarm A series of minor earthquakes, none of which may be identified as the main shock, occurring in a limited area and time. Syn: *swarm [seis].*

earthquake tremor *tremor.*

earthquake volume The volume of an earthquake's major potential energy content and chosen equal to the total volume of aftershocks following each shock (Båth, 1966, p.79). It is expressed in cm^3.

earthquake wave *seismic wave.*

earthquake zone An area of the Earth's crust in which fault movements and sometimes associated volcanism occur; a *seismic belt* or *seismic area.*

Earth radiation *terrestrial radiation.*

Earth rotation The turning of the Earth about its axis, described either as counterclockwise about the north pole or as eastward according to the movement of the equator. A proposed rate of rotation is 0.000072921 radians per second.

earth run A lobe of earth material that has flowed downhill beneath the vegetation cover, forming a sloping step whose front is often 1/3-2 m high (Sharpe, 1938, p. 42). Not preferred usage. Cf: *earthflow.*

earth science An all-embracing term for sciences related to the Earth (analogous, in educational parlance, to "life science"). It is occasionally used as a syn. for *geology* or *geological sciences,* but this usage is misleading because in its wider scope earth science may be considered to include such subjects as meteorology, physical oceanography, soil chemistry, and agronomy. The term is generally used in the singular.

Earth shell Any one of the concentric *shells [geol]* that constitute the structure of the Earth.

earthshine *earthlight.*

Earth's orbit The path through space of the Earth in its annual journey around the Sun. This path is an ellipse, with a semimajor axis of about 92,700,000 miles, an eccentricity of 0.03, and the Sun at one focus.

Earth stretching A method for calculating dispersion of surface waves in a spheroidal Earth using a computer program developed for flat anisotropic layers.

earth stripe *soil stripe.*

Earth tide The response of the solid Earth to the forces that produce the tides of the sea; semidaily earth tides have a fluctuation of between seven and fifteen centimeters (Strahler, 1963, p.110-111). Syn: *bodily tide.*

earth tremor *tremor.*

earth wave An obsolete syn. of *seismic wave.*

earth wax *ozocerite.*

earthwork (a) The operations involved in constructing earth embankments, preparing foundations for structures, and placing and compacting earth materials. (b) An embankment or other construction made of earth.

earthy [geol] Composed of or resembling earth, or having the properties or nature of earth or soil; e.g. an "earthy limestone" containing argillaceous material and characterized by high porosity, loosely aggregated particles, and close association with chalk.

earthy [mineral] (a) Said of minerals with a *dull luster;* the surface may feel rough to the touch. (b) Said of a type of fracture similar to that of a hard clay.

earthy breccia A term used by Woodford (1925, p.183) for a breccia in which rubble, sand, and silt + clay each constitute more than 10% of the rock.

earthy calamine *hydrozincite.*

earthy cobalt *asbolite.*

earthy lignite Lignite that is structurally amorphous and is soft and friable. Cf: *woody lignite.*

earthy manganese *wad [mineral].*

easting A *departure* (difference in longitude) measured to the east from the last preceding point of reckoning; e.g. a linear distance eastward from the north-south (vertical) grid line that passes through the origin of a grid system.

eastonite A variety of biotite: $K_2Mg_5Al_4Si_5O_{20}(OH)_4$.

ebb current The tidal current associated with the decrease in the height of a tide, generally moving seaward or down a tidal river or estuary. Cf: *flood current.* Erroneous syn: *ebb tide.*

ebbing-and-flowing spring *periodic spring.*

ebbing well A well in permeable rocks near the coast, in which the water level fluctuates with the tide.

ebb tide (a) *falling tide.* (b) An erroneous syn. of *ebb current.*

ebridian A marine protist characterized by the presence of flagella, a skeleton of solid silica rods, and the absence of chromatophores.

ecardinal Said of an inarticulate brachiopod (or of its shell) without a hinge.

eccentric *helictite.*

eccentricity (a) The condition, degree, amount, or instance of deviation from a center or of not having the same center; e.g. the horizontal displacement of a surveying instrument or signal from a triangulation station mark at the time an observation is made, or an effect seen in a surveyor's compass when the line of sight fails to pass through the vertical axis of the compass or when a straight line through the ends of the magnetic needle fails to pass through the center rotation of the needle. (b) The distance of the center of figure of a body from an axis about which it turns; e.g. the "eccentricity of alidade" represented by the distance between the center of figure of the index points on the alidade and the center of figure of the graduated circle. (c) The ratio of the distances from any point of a conic section to a focus and the corresponding directrix; e.g. the "eccentricity of ellipse" represented by the ratio of the distance between the center and a focus of an ellipse to the length of its semimajor axis, or $e^2 = (a^2 - b^2)/a^2$, where e = eccentricity, a = semimajor axis, and b = semiminor axis. Eccentricity is less than one in the ellipse, greater than one in the hyperbola, equal to one in the parabola, and equal to zero in the circle.

eccentricity reduction The correction that must be applied to an observed direction made with an eccentric instrument or signal, or both, in order to reduce the observed value to what it would have been if there had been no eccentricity. Also known as "eccentric correction" or "reduction-to-center".

eccentric signal A survey signal (target) which is not in the same vertical line with the station which it represents, such as a signal placed in triangulation at some point other than directly over the triangulation station and not in line with the station and the instrument.

eccentric station A survey point over which an instrument is centered and observations are made, and which is not in the same vertical line with the station which it represents and to which the observations will be reduced before being combined with observations at other stations (Mitchell, 1948, p. 26). It is established and occupied when it is not practicable to set up directly over the actual station center or when it becomes necessary in order to see points that are not visible from the station center.

ecdemite A bright-yellow to green mineral, perhaps: $Pb_6As_2O_7Cl_4$. Also spelled: *ekdemite.*

ecdysis The process of moulting of the exoskeleton in arthropods.

ecesis The establishment of a plant in a new location as a result of successful germination, growth, and reproduction. Syn: *establishment.*

echinate Spiny-surfaced, or densely covered with stiff, stout, or bluntish bristles, prickles, or spines; e.g. "echinate pollen" having a sculpture consisting of spines.

echinating Said of sponge spicules that protrude at an angle from the surface of a skeletal fiber.

echinoderm Any solitary marine benthic (rarely pelagic) invertebrate, belonging to the phylum Echinodermata, characterized by radial symmetry, an endoskeleton formed of plates or ossicles composed of crystalline calcite, and a water-vascular system. Echinozoans, asterozoans, crinozoans, and homalozoans are echinoderm subphyla.

echinoid Any echinozoan belonging to the class Echinoidea, characterized by a subspherical to modified spherical shape, interlocking calcareous plates, and movable appendages; e.g. a *sea urchin.*

echinozoan Any free-living globoid, discoid, or cylindroid echinoderm belonging to the subphylum Echinozoa, characterized by the absence of arms, brachioles, and outspread rays and by the dominance of a meridional growth pattern over bilateral symmetry. Range, Lower Cambrian to present.

echmidium A spear-shaped plate formed during ontogeny of a brachiopod loop by fusion of anterior ends of descending branches.

echodolite *phonolite.*

echogram The graphic record made by an *echo sounder,* in the form of a continuous profile. See also: *fathogram.*

echo sounder In oceanography, a *sounding* instrument that measures water depth by measurement of the time that it takes a sonic or supersonic sound signal to travel to and return from the sea floor. See also: *echogram; fathometer; pinger; precision depth recorder.* Syn: *sonic depth-finder; depth sounder.*

eckermannite A monoclinic mineral of the amphibole group: $Na_3(Mg,Li)_4(Al,Fe)Si_8O_{22}(OH,F)_2$.

Eckert projection One of a series of six map projections of the entire Earth in each of which the geographic poles are represented by parallel straight lines that are one half the length of the equator. The parallels are rectilinear and the meridians may be rectilinear or curved. They are broadly similar in appearance to the Mollweide projection. Named after Max Eckert (1868-1938), German cartographer, who developed the projections in 1906.

eckrite A monoclinic mineral of the amphibole group: $NaCa(Mg,Fe^{+2})_4Fe^{+3}Si_8O_{22}(OH)_2$.

eclipse One celestial body blocking out the light of another, e.g. the Moon obscuring the Sun.

ecliptic latitude *celestial latitude.*

ecliptic longitude *celestial longitude.*

eclogite A granular rock composed essentially of garnet (almandine-pyrope) and sodic pyroxene (omphacite). Rutile, kyanite, and quartz are typically present. Also spelled: *eklogite.*

eclogite facies The set of metamorphic mineral assemblages (facies) in which basic rocks are represented by omphacitic pyroxene and almandine-pyrope garnet. Also common, although not essential, is the association pyrope + olivine + diopside + enstatite. Phase-equilibrium work has shown that these high-density mineral associations indicate high pressure of crystallization, although the range of geologic environments in which the facies has been encountered, and the variation in mineral composition (Coleman et al., 1965), point to a broad range of possible pressure-temperature conditions. Many workers have suggested that low H_2O pressures are required, and that pressures and temperatures overlap those of several other metamorphic facies.

eclogitic (a) Pertaining to an *eclogite.* (b) Said of a rock having an association of clinopyroxene and garnet with a proportion of jadeite molecule in the pyroxene. In this case, no genetic connotation is implied or bulk composition considered (Church, 1968, p.757).

ecochronology A term introduced as *Oekochronologie* by Schindewolf (1950, p. 35) for *geochronology* based on the ecology of life forms.

ecocline A *cline* related to the gradation between two different *niches [ecol].*

ecography The strictly descriptive part of *ecology.* The term is seldom used.

ecologic facies *environmental facies.*

ecologic niche *niche [ecol].*

ecologic potential A term used by Lowenstam (1950) for the capability of an organism to directly control or modify its environment, such as that possessed by a reef-building organism that is able to erect a rigid and resistant framework in the zone of wave action.

ecologic reef A fossil *reef* recognizable as having been built and bound by organisms into a rigid, wave-resistant topographic high on the sea floor (Dunham, 1970, p. 1931). Cf: *stratigraphic reef.*

ecologic succession *succession [ecol].*

ecology The study of the relationships between organisms and their environment, including the study of communities, patterns of life, natural cycles, relationships of organisms to each other, biogeography, and population changes. See also: *paleoecology; ecography.* Adj: *ecologic; ecological.* Syn: *bionomics.*

economic geology The study and analysis of geologic bodies and materials that can be utilized profitably by man, including fuels, metals, nonmetallic minerals, and water; the application of geologic knowledge and theory to the search for and the understanding of mineral deposits.

economic limit *assay limit.*

economic yield The maximum estimated rate at which water may be withdrawn from an aquifer without creating a deficiency or affecting the quality of the supply. Cf: *safe yield.*

economy The input and consumption of energy within a system, and the changes that result; e.g. the *balance* of a glacier.

ecophenotype A variant of a species, produced by nongenetic modification of the *phenotype* by particular ecologic conditions.

ecospace As used by Valentine (1969, p.687), that volume within the *environmental hyperspace lattice* corresponding to the environmental conditions under which a particular organism may live. See also: *realized ecospace; prospective ecospace.*

ecosphere Portions of the universe favorable for the existence of living organisms; esp. the *biosphere.*

ecostratigraphic unit A stratigraphic unit based on the mode of origin or the environment of deposition of rocks (Hedberg, 1958, p.1893); e.g. a marine zone, a brackish-water zone, or a glacially deposited zone. The terminology of most ecostratigraphic units has not reached formal status. Syn: *ecozone.*

ecostratigraphy (a) A term introduced as *Oekostratigraphie* by Schindewolf (1950, p.35) for stratigraphy based on the ecology of life forms; the stratigraphic occurrence of local or regional faunal or floral assemblages that are valuable for ecologic considerations. (b) A term used by Hedberg (1958, p.1893) for the study and classification of stratified rocks according to their mode of origin or their environment of deposition.—Also spelled: *oecostratigraphy.*

ecosystem A unit in ecology consisting of the environment with its living elements, plus the nonliving factors that exist in and affect it.

ecotone A transition zone that exists between two ecologic communities. Members of both communities may compete within this zone, thus yielding an apparent enrichment known as the *edge effect.* Syn: *tension zone.*

ecotope The habitat of a particular organism. See also: *ecotopic.*

ecotopic Having the tendency to adjust to the specific conditions of the *ecotope.*

ecotype An ecologic variant of a species that has adapted to local environmental conditions.

écoulement A syn. of *gravitational sliding.* Etymol: French, "flowing".

ecozone (a) *ecostratigraphic unit.* (b) "A body of rock with upper and lower boundaries marked by reversible faunal changes that are essentially oblique to time planes. These faunal changes were probably caused by secular lateral shift of facies belts" (Vella, 1964, p. 622-623).

ectexine *ektexine.*

ectexis Migmatization with *in situ* formation of the mobile part (Dietrich & Mehnert, 1961). Var: *ektexis.* Cf: *entexis.*

ectexite Rock formed by ectexis. Also spelled: *ectectite.* Little used.

ectinite Rock formed as a result of essentially isochemical regional metamorphism, i.e. with no notable associated metasomatism (Dietrich & Mehnert, 1961). Little used.

ectocochleate Said of cephalopods whose body is lodged within a shell; sometimes grouped as Ectocochleata (e.g., *Nautilus*).

ectocyst The outer, cuticular layer of the body wall of a bryozoan, with or without incorporated calcification. Cf: *endocyst.*

ectoderm The outer body layer of an organism; e.g. the outer layer of the oral and basal disks, tentacles, and column wall of a coral polyp. In the vertebrates, the term refers exclusively to embryonic tissue. Cf: *endoderm; mesoderm.*

ectoderre The principal layer in the external covering of a chitinozoan. Cf: *endoderre; periderre.*

ectodynamic *ectodynamomorphic.*

ectodynamomorphic An old term applied to a soil whose characteristics were produced in part by external forces, e.g. vegetation, climate. Cf: *endodynamomorphic.* Syn: *ectodynamic.*

ectoexine *ektexine.*
ectogene Said of external factors that influence the texture of a rock (Sander, 1951, p.11). Cf: *entogene.*
ectooecium The outer, generally calcified part of the body-wall fold forming an ovicell in cheilostome bryozoans. Cf: *entooecium.*
ectophloic Pertaining to the siphonostele of certain vascular plants having phloem external to the xylem. Cf: *amphiphloic.*
ectophragm A thin membrane lying between distal ends of processes on a dinoflagellate cyst. Cf: *endophragm; periphragm.*
ectoproct Any bryozoan belonging to the subphylum Ectoprocta and characterized by a circular or horseshoe-shaped lophophore around the mouth but not the anus. Range, Ordovician (or possibly upper Cambrian) to the present. Cf: *entoproct.*
ectosiphuncle The wall of the siphuncle of certain cephalopods, consisting generally of septal necks and connecting rings. Cf: *endosiphuncle.* Syn: *ectosiphon.*
ectosolenian Said of a foraminifer (e.g. *Lagena*) having an external tubelike neck. Cf: *entosolenian.*
ectosome The peripheral region of a sponge beneath the inhalant surface and devoid of flagellated chambers; the cortical part of a sponge. Cf: *endosome.*
ectyonine Said of a sponge skeleton built of spiculofibers made up of both coring and echinating monaxons.
ecumeme That part of the Earth that is permanently inhabited.
edaphic Said of ecologic formations or effects resulting from or influenced by local conditions of the soil or substrate; also, an old term applied to any soil characteristic that affects plant growth, e.g. acidity, alkalinity. Cf: *climatic [ecol].*
edaphon All the animals and plants living in the soil.
Edaphosauria A suborder of pelycosaurian synapsid reptiles characterized by small heads, a tendency toward large size, and dentitions that suggest herbivorous habit. Stratigraphic range, Upper Pennsylvanian to Lower Permian.
eddy A circular movement of water that is generally in a different direction from that of the main current. It is a temporary current, usually formed at a point at which a current passes some obstruction, or between two adjacent currents flowing in opposite directions, or at the edge of a permanent current. Cf: *whirlpool; maelstrom.*
eddy-built bar A bar presumably built by currents rotating as an eddy in a tidal lagoon; e.g. one of the ridges surrounding some of the Carolina bays. Syn: *Neptune's racetrack.*
eddy coefficient *austausch.*
eddy conductivity *austausch.*
eddy diffusion Mixing by turbulent flow. Syn: *turbulent diffusion.*
eddy diffusivity The coefficient of the proportionality of the rate of transfer of mass to the gradient of the average concentration. It depends on the nature of the turbulent motion (Fairbridge, 1966, p. 230).
eddy flux The rate of transport or flux of fluid properties, e.g. momentum or suspended matter, by turbulent flow. Syn: *turbulent flux.*
eddy mark One of numerous superimposed or overlapping loops (0.3-1 m in diameter) forming a spiral impression on a sedimentary (sandstone) surface, believed to result from "dragging of a small limb of a larger floating log caught in a vortex or eddy current" of a stream, or from movement of a pebble or stick caught in circular winds of a "dust devil" after the sands had been exposed along the stream bank (Rigby, 1959).
eddy mill A *pothole* in a stream bed.
eddy spectrum Turbulent flow described in terms of the frequency distribution of eddy size, or of the partition of kinetic energy among eddies of various sizes. Syn: *turbulence spectrum.*
eddy viscosity The transfer coefficient for momentum, corresponding to *kinematic viscosity.*
Edenian North American stage: Upper Ordovician (above Mohawkian, below Maysvillian). Lowest substage of Cincinnatian.
edenite (a) A light-colored, iron-free variety of hornblende. (b) An end member in the amphibole mineral group: $NaCa_2Mg_5AlSi_7O_{22}(OH)_2$. Cf: *pargasite.*
edentulous Said of a bivalve mollusk lacking hinge teeth.
edge (a) A sharply pointed ridge; also the crest of such a ridge. (b) The escarpment that terminates a plateau; the extreme margin of a cliff. (c) The highest part of an elevated tract of land of great extent; esp. a ridge or divide between two streams.
edge coal An English and Scottish term for a steeply inclined to vertical coal seam.
edge dislocation In a crystal, a row of atoms marking the edge of a crystallographic plane and extending only part way; it is a type of *line defect.*
edge effect In ecology, the apparent increase in number of species inhabiting an *ecotone* as compared with the smaller number occupying either adjacent community.
edge line A heavy line on a relief map, depicting a sudden sharp change or break of slope.
edge water The water around the margins of an *oil pool* or a *gas pool.* Cf: *bottom water.* Also spelled: *edgewater.*
edgewise conglomerate A conglomerate exhibiting *edgewise structure;* e.g. an intraformational conglomerate containing elongated calcareous pebbles that are transverse to the bedding.
edgewise structure A primary sedimentary structure characterized by an arrangement of flat, tabular, or disc-shaped fragments whose long axes are set at varying steep angles to the bedding. It may be due to running water or to sliding or slumping soon after deposition. See also: *edgewise conglomerate.*
edge zone A fold of the body wall of a coral polyp, extending laterally and/or downward over the edge of the wall.
edingtonite A white or grayish-white zeolite mineral: $BaAl_2Si_3O_{10} \cdot 4H_2O$. It sometimes contains appreciable calcium.
EDM *electronic distance-measuring instrument.*
edolite A type of hornfels consisting mainly of feldspar and mica. There are varieties that also contain cordierite (aviolite) or andalusite (astite) (Holmes, 1928, p.87). Type locality: Edolo, Italian Alps.
edrioasteroid Any many-plated, attached echinozoan belonging to the class Edrioasteroidea, having a well-developed quinqueradial endothecal ambulacral system. Range, Lower Cambrian to Lower Carboniferous.
eel A series of hydrophones in a tube up to about 50 m in length, used in seismic reflection profiling in shallow water.
Eemian North European climatostratigraphic and floral stage: upper Pleistocene (above Saalian, below Weichselian). Equivalent in time to Riss/Würm interglacial.
effective diameter (a) The diameter of the particles in an assumed rock or soil that would transmit water at the same rate as the rock or soil under consideration, and that is composed of spherical particles of equal size and arranged in a specified manner. (b) The approximate diameter of a rock or soil particle equal to the sieve size that allows 10% (by weight) of the material to pass through; the particle diameter of the 90 percent line of a cumulative curve, or the maximum diameter of the smallest 10% of the particles of sediment.—Syn: *effective size.*
effective drainage porosity *effective porosity.*
effective force The force transmitted through the granular structure of a soil mass by effective stresses.
effective permeability The ability of a rock to conduct one fluid, e.g. gas, in the presence of other fluids, e.g. oil or water. See also: *absolute permeability; relative permeability.*
effective pore volume The pore space in rocks that is available for the free circulation of water. This excludes pore space taken up by air and pellicular water.
effective porosity The percent of the total volume of a given mass of soil or rock that consists of interconnecting interstices. The use of this term as a syn. of *specific yield* is to be discouraged. Syn: *effective drainage porosity.* Cf: *porosity [grd wat]; continuous porosity.*
effective precipitation (a) That part of precipitation producing runoff. (b) That part of precipitation falling on an irrigated area that meets the demands of consumptive use. Cf: *precipitation excess.*
effective pressure *effective stress.*
effective radiation *effective terrestrial radiation.*
effective size *effective diameter.*
effective stress The average normal force per unit area transmitted directly from particle to particle of a soil or rock mass. It is the stress that is effective in mobilizing internal friction. In a saturated soil in equilibrium, the effective stress is the difference between the total stress and the neutral stress of the water in the voids; it attains a maximum value at complete consolidation and before shear failure. Syn: *effective pressure; intergranular pressure.*
effective terrestrial radiation The difference between the outgoing infrared *terrestrial radiation* and the downward infrared *counterradiation* from the Earth's atmosphere. Syn: *effective radiation.*
effective unit weight The *unit weight* of a soil or rock mass that,

when multiplied by the height of the overlying column of soil or rock, yields the effective stress caused by the weight of the overburden.

effective velocity The actual velocity of ground water percolating through water-bearing material. "It is measured by the volume of ground-water passing through a unit cross-sectional area divided by effective porosity" (Tolman, 1937, p. 593). The velocity is the average for water moving through the interstices.

efficiency The *capacity [hydraul]* of a stream per unit discharge and unit gradient, or the quotient of capacity by the product of discharge and gradient (Gilbert, 1914, p. 36). It is a measure of the stream's potential work of transportation in relation to its potential energy. Symbol: E.

efflorescence (a) A whitish fluffy or crystalline powder, produced as a surface encrustation on a rock or soil in an arid region by evaporation of water brought to the surface by capillary action or by loss of water of crystallization on exposure to the air. It may consist of one or several minerals, commonly soluble salts such as gypsum, calcite, natron, and halite. Syn: *bloom [mineral].* (b) The process by which an efflorescent salt or crust is formed.

effluent adj. Flowing forth or out; emanating.—n. (a) A surface stream that flows out of a lake (e.g. an outlet), or a stream or branch that flows out of a larger stream (e.g. a distributary). Ant: *influent.* Cf: *effluent stream.* (b) A liquid discharged as waste, such as contaminated water from a factory or the outflow from a sewage works; water discharged from a storm sewer or from land after irrigation.

effluent cave *outflow cave.*

effluent flow Flow of water from the ground into a body of surface water; e.g. the flow of water to an effluent stream.

effluent lava flow A lava flow that is discharged from a volcano by way of a lateral fissure (Dana, 1890); an obsolete term. Cf: *interfluent lava flow; superfluent lava flow.*

effluent seepage Diffuse discharge of ground water to the land surface; *seepage* of water from out of the ground.

effluent stream (a) A stream or reach of a stream that receives water from the zone of saturation and provides *base flow;* its channel lies below the water table. Syn: *gaining stream.* (b) *effluent.*

efflux *outflow.*

effusion The emission of relatively fluid lava onto the Earth's surface; also, the rock so formed. Cf: *extrusion.*

effusive *extrusive.*

egeran A brown or yellowish-green variety of vesuvianite.

egg The female gamete of an embryophytic plant; a nonmotile gamete which can fuse with a sperm to form a zygote (Cronquist, 1961, p.874).

eggstone *oolite.*

eglestonite A brownish-yellow isometric mineral: $Hg_6Cl_3O_2H$.

egueiite A yellowish-brown material consisting of a hydrous phosphate of ferric iron with a little calcium and aluminum and occurring in small nodules in clay.

Egyptian jasper A brown or banded jasper occurring as pebbles or small boulders scattered over the desert surface between Cairo and the Red Sea. Syn: *Egyptian pebble.*

Eh The potential of a half-cell, measured against the standard hydrogen half-cell. Syn: *oxidation potential.*

Ehrenfest relation A modified *Clapeyron equation* that is used for second-order transitions.

ehrwaldite An *augitite* that contains both orthopyroxene and clinopyroxene. Obsolete.

EI *energy index.*

Eifelian European stage: Middle Devonian (above Emsian, below Givetian). Syn: *Couvinian.*

eightling A crystal twin, either cyclic or interpenetrating, that consists of eight individuals. Cf: *twoling; trilling; fourling; fiveling.*

einkanter A *ventifact* or wind-worn stone having only one face or a single sharp edge; it implies a steady, unchanging wind direction. Etymol: German *Einkanter,* "one having one edge". Pl: *einkanters; einkanter.*

eiscir An Irish term for ridge; esp. *esker.*

eitelite A hexagonal mineral: $Na_2Mg(CO_3)_2$.

ejecta [crater] Glass, shock-metamorphosed rock fragments, and other material thrown out of an explosion or impact crater during formation. Such material may be distributed around a crater in distinctive patterns, forming "ejecta rays" or "ejecta loops", as well as partially building the rim.

ejecta [pyroclast] Material thrown out by a volcano; *pyroclastics.* Syn: *ejectamenta.*

ejecta blanket A deposit surrounding an impact crater or explosion crater, consisting of material (such as *base-surge* deposits, *throwout,* and *fallout breccias*) ejected from the crater during formation; e.g. lunar-crater material, probably chiefly crushed rock with large blocks, occurring on a *mare* region and often forming hummocky to smooth layers ranging from about a meter to several hundred meters in thickness.

ejectamenta *ejecta.*

ekanite A green mineral: $(Th,U)(Ca,Fe,Pb)_2Si_8O_{20}$.

ekdemite *ecdemite.*

ekerite A *syenite* or *quartz syenite* that contains arfvedsonite as an essential component along with acmite, microperthite, and soda microcline, and with little quartz. The name was taken by Brögger in 1906 from Eker in the Oslo district, Norway. Not recommended usage.

eklogite *eclogite.*

Ekman layer A layer in the ocean, situated above a certain depth in the *Ekman spiral,* at which both the current and the frictional forces associated with it become negligibly small. The average flow of water is at right angles to the wind driving it (in the Northern Hemisphere). It may be produced near the surface by wind stresses (upper Ekman layer) or near the bottom by a pressure gradient (lower Ekman layer).

Ekman spiral A theoretical, graphic description of the way in which a wind, blowing uniformly and steadily over a homogeneous ocean of unlimited depth and extent and of constant viscosity, would cause currents in the surface layers to vary with depth, the water at the very surface drifting at an angle of 45° to the right of the wind direction in the Northern Hemisphere (and to the left in the Southern Hemisphere) and water at successive depths drifting in directions farther to the right (as a spiral) with a rapidly decreasing speed until at the *friction depth* it would move in the direction opposite to the wind; the net water transport (Ekman transport) is 90° to the right of the wind direction in the Northern Hemisphere. It is named for Vagn Walfrid Ekman, Swedish oceanographer, who in 1902 developed the theory of the spiral, which has also been applied to atmospheric motion.

eksedofacies Facies of the weathering environment (Vassoevich, 1948).

ektexine The outer of the two layers of the *exine* of spores and pollen, normally more densely or deeply staining than the *endexine,* and characterized by richly detailed external sculpture and often by complex internal structure of granules, columellae, and other elements. Also spelled: *ectexine.* Syn: *ectoexine; sexine.*

ektexis *ectexis.*

elaeolite *eleolite.*

élan vital In early evolutionary theories, the internal vital or driving force that supposedly stimulates the process of evolution. It was regarded as an inherent property of living matter. Etymol: French.

Elasmobranchii A subclass of cartilaginous fishes characterized by hyostylic or amphistylic jaw suspension and numerous teeth; if modified for crushing, the teeth are not reduced to a few plates as in the Bradydonti and Holocephali. The subclass includes sharks, skates, and rays.

elastic Said of a body in which strains are instantly and totally recoverable and in which deformation is independent of time. Cf: *plastic [struc geol].*

elastic aftereffect *creep recovery.*

elastic afterworking *creep recovery.*

elastic bitumen *elaterite.*

elastic compliance The reciprocal of *Young's modulus;* or more generally, a coefficient that relates a component of strain to a component of stress in an elastic material.

elastic constant One of various coefficients that define the elastic properties of a body, including the *Lamé constants, Poisson's ratio,* or one of the *moduli of elasticity.*

elastic deformation Deformation of a substance, which disappears when the deforming forces are removed. Commonly, that type of deformation in which stress and strain are linearly related, according to *Hooke's law.* Cf: *plastic deformation.*

elastic discontinuity A boundary between strata of different elastic moduli and/or density at which seismic waves are reflected and refracted.

elastic energy The energy stored within a solid by elastic deformation.

elastic limit The greatest stress that can be developed in a material without permanent deformation remaining when the stress is released.

elastic modulus *modulus of elasticity.*

elasticoplastic Said of deformation that has a perfectly elastic phase and a perfectly plastic phase. It is demonstrated by the model of a *Saint Venant substance.*

elasticoviscous Said of a material in which instantaneous elastic strain at a constant stress is followed by continuously developed permanent strain so long as the stress is maintained. See also: *Maxwell liquid.*

elastic rebound Elastic recovery from strain.

elastic-rebound theory The statement that movement along a fault is the result of an abrupt release of a progressively increasing elastic strain between the rock masses on either side of the fault. Such a movement returns the rocks to a condition of little or no strain. The theory was proposed by Harry Fielding Reid in 1911. Syn: *Reid mechanism.*

elastic strain The strain developed during the elastic behavior of a material.

elastic wave *seismic wave.*

elater The ribbonlike, filamentous appendage of certain spores (as of *Equisetum*), consisting of more or less coiled strips of exine. It aids in spore dispersal.

elaterite A brown asphaltic pyrobitumen, soft and elastic when fresh but hard and brittle on exposure to air. It is derived from the metamorphism of petroleum. See also: *coorongite.* Syn: *elastic bitumen; liverite; mineral caoutchouc.*

elatolite A supposedly high-temperature modification of calcite, but probably crystal casts of calcium carbonate after villiaumite (Hey, 1962, p.412).

E layer The seismic region of the Earth from 2900 km to 4710 km, equivalent to the *outer core.* It is divided into an upper (E') and a lower (E") part at 4560 km, at which level the velocity gradient of the P wave is reduced to zero. It is a part of a classification of the Earth's interior made up of layers A to G.

elb A *transverse dune* in the desert of Algeria.

elbaite A mineral of the tourmaline group: $Na(Li,Al)_3Al_6(BO_3)_3(SiO_3)_6(OH)_4$.

elbasin A term used by Taylor (1951, p. 613) for an "elevated basin, often wrongly called a plateau"; e.g. the British Columbia plateau.

Elbe Term formerly applied in northern Europe to the first glacial stage of the Pleistocene Epoch, followed by the Elster; probably equivalent to the *Günz* and *Nebraskan.* Superseded by *Menap.*

elbow of capture The point at which capture was effected along a stream course, characterized by an abrupt or sharp bend where the course turns from the captured part of its valley into the valley of the capturing stream.

elbow twin *geniculate twin.*

electrical conductivity A measure of the ease with which a conduction current can be caused to flow through a material under the influence of an applied electric field. It is the reciprocal of resistivity and is measured in mhos per meter.

electrical log *electric log.*

electrical method A geophysical prospecting method that depends on the electrical or electrochemical properties of rocks. The resistivity, spontaneous-polarization, induced-polarization, and inductive-electromagnetic methods are the principal electrical methods.

electrical resistivity The electrical resistance per unit length of a unit cross-sectional area of a material.

electrical-resistivity sounding A procedure for determining depths to geological interfaces, wherein separations of electrodes in an array are increased by increments. A plot of observed apparent resistivity versus electrode separation, when compared with similar plots for theoretically computed cases, yields estimates of the depths to the interfaces and the resistivities of the strata. See also: *geometric sounding; parametric sounding.* Syn: *resistivity sounding.*

electrical twinning Twinning in quartz according to the *Dauphiné twin law.*

electric calamine *hemimorphite.*

electric-field intensity The strength of an electric field at any point. It is measured by the force exerted on a unit positive charge placed at that point. Syn: *voltage gradient.*

electric log The generic term for a *well log* that displays electrical measurements of induced current flow (*resistivity log, induction log*) or natural potentials (*spontaneous-potential curve*) in the rocks of an uncased borehole. An electric log typically consists of the spontaneous-potential (SP) curve and one or more resistivity or induction curves. The *Archie equations* form the basis for interpretation of electric logs. Abbrev: E-log. Informal syn: *resistivity log.*

electrochemical induration A method of strengthening and consolidating saturated and unconsolidated soils or other granular earth materials by passing a direct current of electricity through probes placed in them (Titkov et al., 1965).

electrochromatography *Chromatography* wherein an applied electric potential is used to produce differential electrical migration.

electrode array A configuration of electrodes on or in the ground for the purpose of making an electrical survey.

electrodeless discharge Emission of light from matter energized by induced electrical currents.

electrodiagenesis Diagenesis affected or stimulated by electric currents and potentials.

electrodialysis *Dialysis* assisted by the application of an electric potential across the semipermeable membrane. Two important uses of electrodialysis are in water desalination and in removing electrolytes from naturally occurring colloids such as proteins. Cf: *electro-osmosis.*

electrofiltration *electrostatic precipitation.*

electrofiltration potential An electrical potential that is caused by movement of fluids through porous formations. Syn: *streaming potential; electrokinetic potential.*

electrographic Pertaining to a method for analyzing minerals and metals by transferring a small amount of the sample by electrical means to a prepared surface where the ions are identified.

electrokinetic potential *electrofiltration potential.*

electrolysis A method of breaking down a compound in its natural form or in solution by passing an electric current through it, the ions present moving to one electrode or the other where they may be released as new substances.

electromagnetic induction The generation of electric field or current in an electric conductor when that body is in a changing magnetic field or is moving through a magnetic field. It is a phenomenon of *electromagnetism.* Nonrecommended syn: *magnetic induction; induction [magnet].*

electromagnetic method An electrical exploration method based on the measurement of alternating magnetic fields associated with currents artificially or naturally maintained in the subsurface. If these currents are induced by a primary alternating magnetic field, the name inductive electromagnetic method applies, whereas if they are conducted into the ground via electrodes, the name conductive electromagnetic method applies. See also: *inductive method.*

electromagnetic radiation Energy propagated in the form of an advancing interaction between electric and magnetic fields. Abbrev: EMR.

electromagnetism The totality of electric and magnetic phenomena, or their study; particularly those phenomena with both electric and magnetic aspects, such as *electromagnetic induction.*

electromigration A method of separating isotopes or ions by their differing rates of movement during electrolysis.

electron capture A mode of radioactive decay in which an orbital electron is captured by the nucleus. The resulting nuclear transformation is identical with that in β^+ emission.

electron diffraction The *diffraction* of a beam of electrons, usually by the three-dimensional periodic array of atoms in a crystal that has periodic repeat distances (lattice dimensions) of the same order of magnitude as the wavelength of the electron beam. See also: *X-ray diffraction.*

electron diffraction analysis Analysis by *electron diffraction.*

electron diffraction pattern The interference pattern seen when a beam of electrons is sent through a substance, each pattern being characteristic for a particular substance. Electron diffraction patterns contain basic crystallographic information as well as information about orientation, defects, crystal size, and additional phases. See also: *X-ray diffraction pattern.*

electronic distance-measuring instrument A device that measures the phase differences between transmitted and returned (i.e. reflected or retransmitted) electromagnetic waves, of known frequency and speed, or the round-trip transit time of a pulsed signal, from which distance is computed. Commercial trade names are Electrotape, Geodimeter, and Tellurometer. Abbrev: EDM.

electron magnetic resonance *electron spin resonance.*

electron microprobe An analytical instrument that uses a finely focused beam of electrons to excite X-ray emission from selected portions of a sample. From the emitted X-ray spectrum the composition of the sample at the point of excitation can be determined. Spots as small as 1 micrometer in diameter can be analyzed, with sensitivities around 50ppm or less for most metals. Syn: *electron probe; microanalyzer.*

electron microscope An electron-optical instrument in which a beam of electrons, focused by systems of electrical or magnetic lenses, is used to produce enlarged images of minute objects on a fluorescent screen or photographic plate in a manner similar to that in which a beam of light is used in a compound microscope. The electron microscope, because of the very short wavelength of the electrons, is capable of resolving much finer structures than the optical instrument, and is capable of magnifications on the order of 100,000X. See also: *scanning electron microscope.*

electron microscopy Determining and identifying the structure of substances by using the *electron microscope.*

electron paramagnetic resonance A syn. of *electron spin resonance.* Abbrev: EPR.

electron probe *electron microprobe.*

electron spin resonance *Resonance* occurring when electrons that are undergoing transitions between energy levels in a substance are irradiated with electromagnetic energy of a proper frequency to produce maximum absorption. Abbrev: ESR. Syn: *electron magnetic resonance; electron paramagnetic resonance; paramagnetic resonance.*

electro-osmosis The motion of liquid through a membrane under the influence of an applied electric field. See also: *osmosis.*

electrophoresis The movement toward electrodes of suspended charged particles in a fluid by applying an electromotive force to the electrodes that are in contact with the suspension. See also: *cataphoresis; anaphoresis.*

electroprecipitation *electrostatic precipitation.*

electrostatic precipitation A method for removing suspended solid or liquid particles from a gas by applying a strong electric field to the mixture that charges the particles and precipitates them. Syn: *electrofiltration; electroprecipitation.*

electrostriction Deformation induced in materials on polarization by an applied electric field.

Electrotape A trade name for a precise electronic surveying device that transmits a radio-frequency signal to a responder unit which in turn transmits the signal back to the interrogator unit. The time lapse between original transmission and receipt of return signal is measured and displayed in a direct digital readout for eventual reduction to a precise linear distance. It operates on the same principle as the *Tellurometer.*

electroviscosity The viscosity of a fluid as influenced by electric properties, e.g. greater viscosity of a low-conductivity fluid than of a high-conductivity fluid flowing through narrow capillaries.

electrum (a) A naturally occurring, deep-yellow to pale-yellow alloy of gold with silver; argentiferous gold, containing more than 20% silver. Also spelled: *elektrum.* (b) An ancient Greek name, now obsolete, for *amber.* Also spelled: *elektron.*

electrum-tarnish method A technique for direct determination of sulfur fugacity in experimental study of sulfide systems. The method consists of determining the temperature at which electrum of a given gold-silver ratio tarnishes when equilibrated with a vapor in equilibrium with the sulfide assemblage being measured (Barton & Toulmin, 1964).

elements of symmetry *symmetry elements.*

eleolite A syn. of *nepheline,* esp. of a translucent, massive or coarsely crystalline, and dark variety (grayish, bright-green, or brown to brownish-red), having a greasy luster and sometimes used as an ornamental stone. Also spelled: *elaeolite; elaolite.*

eleolite syenite An obsolescent syn. of *nepheline syenite,* esp. of a coarse-grained variety containing eleolite.

elephant-head dune A *sand shadow* or small dune resembling the head of an elephant, having a rounded windward face covered with vegetation and a long, tapering snout of bare sand on the leeward side; examples occur in the Coachella Valley of the Colorado Desert in California.

elephant-hide pahoehoe A type of *pahoehoe* having a wrinkled and draped surface.

elephant rock A term used in SE Missouri for a *rocking stone,* not necessarily delicately balanced, formed in place by the weathering and removal of surrounding material.

eleutheromorph A new mineral in a metamorphic rock that has been freely developed and thus has gained its form independently. Cf: *pseudomorph.*

eleutherozoan n. Any echinoderm that does not live attached to a substrate.—adj. Said of an echinoderm having a free mode of life. Var: *eleutherozoic.*—Cf: *pelmatozoan.*

elevated shoreline A shoreline whose development has been interrupted by a relatively sudden rise of the coast or by a rapid lowering of the water level; it is not a true shoreline because it is no longer being shaped by waves and currents. Examples: a broad marine terrace (common along the continental and insular coasts of the Pacific Ocean), or a narrow strandline. Not to be confused with *shoreline of elevation.*

elevation [geomorph] A general term for a topographic feature of any size that rises above the adjacent land or the surrounding ocean bottom; a place or station that is elevated.

elevation [surv] The vertical distance from a datum (usually mean sea level) to a point or object on the Earth's surface; esp. the *height* of a ground point above the level of the sea. The term is used synonymously with *altitude* in referring to distance above sea level, but in modern surveying practice the term "elevation" is preferred to indicate heights on the Earth's surface whereas "altitude" is used to indicate the heights of points in space above the Earth's surface.

elevation correction The correction applied to time values observed in reflection or refraction seismic surveys due to difference of station elevation, in order to reduce the observations to an arbitrary reference datum.

elevation head The elevation of the point at which the hydrostatic pressure is measured, above or below an arbitrary horizontal datum. Syn: *potential head.*

elevation meter A mechanical or electromechanical device on wheels that measures slope and distance and that automatically and continuously integrates their product into difference of elevation.

elevation-relief ratio (Mean elevation - minimum elevation) ÷ (maximum elevation - minimum elevation) (Wood & Snell, 1960). Pike and Wilson (1971) have shown it to be identical to the *hypsometric integral.*

elevator tectonics A term used by Dietz and Holden (1966, p.353) for the rise and fall of blocks of sialic crust from atmospheric to abyssal levels.

elision An act or instance in which the continuity of the sedimentary record has been disturbed by the omission of sediments, such as produced by their removal and redeposition in adjacent depressions.

elkhornite A hypabyssal labradorite-bearing augite *syenite.* It was named by Johannsen in 1937 for the Elkhorn district of Montana. Not recommended usage.

ellestadite A mineral of the apatite group: $Ca_5(SiO_4,PO_4,SO_4)_3(OH,Cl,F)$.

ellipochoanitic Said of a relatively short, retrochoanitic septal neck of a nautiloid that does not reach as far as the preceding septum. See also: *cyrtochoanitic; orthochoanitic.*

ellipsoid A mathematical figure closely approaching the *geoid* in form and size. It is generally defined by its equatorial radius and by the reciprocal of the flattening, a/(a-b), where a and b are the equatorial and polar radii. A task of geodesists is the determination of more exact parameters of the ellipsoid. Cf: *spheroid [geodesy]; Clarke ellipsoid of 1866; reference ellipsoid.*

ellipsoidal lava (a) *pillow lava.* (b) An inclusive term for any lava flow with an ellipsoidal pattern, i.e. pillow lava and the *lava toes* of pahoehoe (Macdonald, 1953).

ellipsoid of revolution The simple mathematical figure that would be produced by an ellipse revolving around its minor axis. It is often used as reference surface for the Earth. See also: *spheroid [geodesy].*

elliptical polarization In optics, elliptically polarized light consisting of upward-spiraling vibration vectors, the surface of which is elliptical rather than circular, as in *circular polarization.* It is caused by the inconstant lengths of vibration vectors of mutually perpendicular plane-polarized waves whose path differences differ in phase by amounts other than $(n+1)/4\lambda$ on emergence from a crystal.

elliptical projection One of several map projections showing the Earth's surface upon the interior of an ellipse; e.g. *Mollweide projection; Aitoff projection.*

ellipticity (a) The degree of flattening of the reference ellipsoid as expressed by the equation $e=(a-b)/a$, where a and b are the equatorial and polar radii. Cf: *equatorial bulge.* (b) The ratio of minor to major axes of an ellipse.

ellipticone A coiled cephalopod shell having elliptic coiling of the last whorl or half whorl which breaks away from the spiral or slightly breaks the regularity of the spiral form.

ellsworthite *betafite.*

E-log *electric log.*

elongation *extension [exp struc geol].*

elongation ratio *basin-elongation ratio.*

elongation sign *sign of elongation.*

elpasolite A colorless isometric mineral: K_2NaAlF_6.

elphidiid Any foraminifer belonging to the family Elphidiidae, characterized by having a sutural canal system opening into rows of sutural pores. Range, Paleocene to present.

elpidite A white to brick-red mineral: $Na_2ZrSi_6O_{15}\cdot 3H_2O$.

Elsonian orogeny An orogenic event in the Canadian Shield about 1400 m.y. ago.

Elster The term applied in northern Europe to the second glacial stage of the Pleistocene Epoch, following the Elbe and preceding the Saale glacial stage; equivalent to the *Mindel* and *Kansan.*

Elsterian North European climatostratigraphic and floral stage: Middle Pleistocene (above Cromerian, below Holsteinian). Equivalent in time to Mindel glaciation.

Eltonian European stage: Upper Silurian (above Wenlockian, below Bringewoodian).

elutriation (a) A method of mechanical analysis of a sediment, in which the finer, lightweight particles are separated from the coarser, heavy particles by means of a slowly rising current of air or water of known and controlled velocity, carrying the lighter particles upward and allowing the heavier ones to sink. (b) Purification, or removal of material from a mixture or in suspension in water, by washing and decanting, leaving the heavier particles behind. (c) The washing away of the lighter-weight or finer particles in a soil by the splashing of raindrops.

eluvial [eco geol] Said of an incoherent ore deposit, such as a placer, resulting from the decomposition or disintegration of rock in place. The material may have slumped or washed downslope for a short distance but has not been transported by a stream.

eluvial [sed] Pertaining to or composed of wind-deposited eluvium; e.g. the "eluvial (or passive) phase" of a dune cycle, marked by sufficient vegetation to check deflation. Cf: *eolian.*

eluvial [weath] Pertaining to eluvium; *residual [weath].*

eluvial horizon A soil horizon from which material has been removed by the process of *eluviation.* Cf: *illuvial horizon.*

eluviated Said of a soil horizon or of materials that have been subjected to the process of *eluviation.*

eluviation The downward movement of soluble or suspended material in a soil, from the A horizon to the B horizon, by groundwater percolation. The term refers especially but not exclusively to the movement of colloids, whereas the term *leaching* refers to the complete removal of soluble materials. Adj: *eluvial; eluviated.* Cf: *illuviation.* See also: *cheluviation.*

eluvium [sed] Fine soil or sand moved and deposited by the wind, as in a sand dune. Cf: *alluvium.*

eluvium [weath] An accumulation of rock debris produced in place by the decomposition or disintegration of rock; a weathering product; a *residue.*

elvan A Cornish term (from Celtic, "white rock") for hypabyssal rocks having the composition of *granite,* esp. a *quartz porphyry.* Tourmaline, fluorite, and topaz may be accessories. Syn: *elvanite.* Not recommended usage.

elyite A monoclinic mineral: $Pb_4Cu(SO_4)(OH)_8$.

emanation *exhalation.*

emarginate Having a notched margin; e.g. said of a gastropod having a variously excavated margin of the outer lip, a bivalve mollusk whose margin is interrupted by a notch or sinus, the posteriorly deflected median segment of the anterior commissure of a brachiopod, or a leaf with a shallow notch at the apex.

embankment [coast] A narrow depositional feature, such as a spit, barrier, or bar, built out from the shore of a sea or lake by the action of waves and currents that deposit excess material at its deep end; it may be emerged or submerged. Syn: *bank.*

embankment [eng] A linear structure, usually of earth or gravel, constructed so as to extend above the natural ground surface and designed to hold back water from overflowing a level tract of land, to retain water in a reservoir, tailings in a pond, or a stream in its channel, or to carry a roadway or railroad; e.g. a dike, seawall, or fill.

embatholithic Said of a mineral deposit occurring in a batholith in which exposure of the batholith and of the country rock is about equal; also, said of that stage of batholith erosion (Emmons, 1933). The term is little used. Cf: *acrobatholithic; cryptobatholithic; endobatholithic; epibatholithic; hypobatholithic.*

embayed coast A coast with many projecting headlands, bays, and outlying islands, usually resulting from submergence.

embayed mountain A mountain that has been partly submerged, so that seawater enters the valleys; e.g. on the coast of SW Ireland.

embayment [coast] (a) The formation of a bay, as by the sea overflowing a depression of the land near the mouth of a river. (b) A *bay,* either the deep indentation or recess of a shoreline, or the large body of water (as an open bay) thus formed.

embayment [petrology] (a) Penetration of microcrystalline groundmass material into phenocrysts, making their "normal" euhedral boundaries incomplete. (b) An irregular *corrosion* or modification of the outline of a crystal by the magma from which it previously crystallized or in which it occurs as a foreign inclusion; esp. the deep corrosion into the sides of a phenocryst. (c) The penetration of a crystal by another, generally euhedral, crystal. Such a crystal is called an "embayed crystal".

embayment [struc geol] (a) A downwarped area containing stratified rocks, either sedimentary or volcanic or both, that extends into a terrain of other rocks, e.g. the Mississippi Embayment of the U.S. Gulf Coast. (b) *recess [fold].*

embedded Covered or enclosed by sediment in a matrix, such as gravel *embedded* in silt.

embolite A yellow-green isometric mineral: Ag(Cl,Br). It is intermediate in composition between chlorargyrite and bromargyrite.

embossed rock A term introduced by Hitchcock (1843, p. 180) as a syn. of *roche moutonnée.*

embouchure (a) The mouth of a river, or that part where it enters the sea. (b) An expansion of a river valley into a plain.—Etymol: French. Syn: *embouchement.*

embrechite Migmatite in which some textural components of the pre-existing rocks are preserved (Dietrich & Mehnert, 1961); a migmatite with preserved parallel layering, often including feldspar phenoblasts or granitic layers and lenses (Mehnert, 1968, p. 354). Little used.

embreyite A monoclinic mineral: $Pb_5(CrO_4)_2(PO_4)_2\cdot H_2O$.

embryo A young sporophytic plant; the *germ* of a *seed.*

embryonic Said of the earliest growth stage in the life history of an animal; the stage preceding the *nepionic* stage.

embryonic apparatus A group of chambers at the center of some megalospheric tests of foraminifers, larger in size and different in shape and arrangement from other chambers. See also: *juvenarium.* Syn: *nucleoconch.*

embryonic volcano A breccia-filled volcanic pipe without surface expression and considered to be produced by *phreatic explosions.* Examples of Permian age occur in Scotland.

Embryophyta The subkingdom of plants with embryos; includes all seed plants, pteridophytes, and bryophytes (Tippo, 1942, p. 204).

embryophytic Said of plants of the subkingdom *Embryophyta.*

emerald (a) A brilliant green variety of beryl, highly prized as a gemstone. The color, which is caused by the presence of chromium or possibly vanadium, ranges from medium-light or medium-dark tones of slightly bluish-green to those of slightly yellowish-green. Syn: *smaragd.* (b) Any of various gemstones having a green color, such as "oriental emerald" (sapphire), "copper emerald" (dioptase), "Brazilian emerald" (tourmaline), and "Uralian emerald" (demantoid). (c) Said of a gemmy and richly green-colored mineral, such as "emerald jade" (jadeite), "emerald spodumene" (hiddenite), and "emerald malachite" (dioptase).

emerald copper *dioptase.*

emerald cut A *step cut* in which the finished gem is square or rectangular and the rows (steps) of elongated facets on the crown and pavilion are parallel to the girdle with sets on each of the four sides and sometimes at the corners. It is commonly used on diamonds to emphasize the absence of color and on emeralds and other colored stones to enhance the color. See also: *square emerald cut.*

emerald nickel *zaratite.*

emerged bog A bog which tends to grow vertically, i.e. increase in thickness, by drawing water up through the mass of plants to above the water table where the growth takes place. Cf: *immersed bog.*

emerged shoreline *shoreline of emergence.*

emergence [bot] Any outgrowth of cortical and epidermal plant tissues that lack a vascular supply. See also: *enation.*

emergence [coast] A change in the levels of water and land such that the land is relatively higher and areas formerly under water are exposed; it results either from an uplift of the land or from a fall of the water level. Ant: *submergence*.

emergence [streams] The point where an underground stream appears at the surface to become a surface stream. Syn: *resurgence; rise; rising*.

emergence angle (a) *angle of emergence*. (b) *apparent dip [seis]*.

emergence velocity (a) A vertical component of glacier motion measured at the surface, representing the difference in vertical displacement of a stake or marker fixed in the ice and the product of horizontal displacement times the tangent of ice slope. (b) The rate the surface would rise if there were no ablation.

emergent Said of a plant that rises above its substrate; e.g., an emergent aquatic plant. Syn: *emersed*.

emergent aquatic plant A rooted plant growing in shallow water, with part of its stem and leaves above the water surface; e.g., bulrush, cattail.

emergent evolution Evolution characterized by the appearance of completely new and unpredictable characteristics or qualities at different levels due to a rearrangement of pre-existing entities.

emersed *emergent*.

emersio A gradual amplitude buildup of a seismic phase on a seismogram, without a clear onset. Cf: *impetus*.

emery (a) A gray to black granular impure variety of corundum, which contains varying amounts of iron oxides (usually magnetite or hematite). It is used in granular form for polishing and grinding. It occurs as masses in limestone and as segregations in igneous rocks. (b) A natural abrasive composed essentially of pulverized impure corundum. Also, the commercial product obtained by crushing emery rock. (c) *emery rock*.

emery rock A granular rock that is composed essentially of an impure mixture of corundum, magnetite, and spinel, and that may be formed by magmatic segregation or by metamorphism of highly aluminous sediments. Syn: *emery; corundolite*.

emigrant In ecology, a migrant plant or animal.

emildine A variety of spessartine garnet containing yttrium. Syn: *emilite*.

Emilian European stage: upper Lower Pleistocene (above Calabrian, below Sicilian).

eminence (a) An elevated area of any size, shape, or height; a mass of high land; a mountain or a hill. (b) The high point of an elevated feature.

eminent cleavage Perfectly displayed mineral cleavage, with smooth surfaces, as often seen in mica or calcite.

emission spectroscopy The observation of an *emission spectrum* and all processes of recording and measuring that go with it.

emission spectrum A general term for any spectrum issuing from a source. Cf: *arc spectrum; flame spectrum; spark spectrum; X-ray spectrum*.

emittance (a) The ratio of the emitted *radiant flux* per unit area of a substance to that of a blackbody radiator at the same temperature. See also: *spectral emittance*. (b) An obsolete term for the radiant flux per unit area emitted by a body.

emmonsite A yellow-green mineral: $Fe_2Te_3O_9 \cdot 2H_2O$. Syn: *durdenite*.

emplacement [intrus rocks] A term used to refer to the process of *intrusion*.

emplacement [ore dep] The localization of ore minerals, by whatever process; ore deposition.

emplectite A grayish or white orthorhombic mineral: $CuBiS_2$. It is dimorphous with cuprobismutite.

empolder v. To reclaim land by the creation of polders; to make low-lying or periodically flooded land cultivable by adequate drainage and the erection of dikes to prevent or control inundation. See also: *polderization*. Syn: *impolder*.—n. A tract of empoldered land; a polder.

empoldering *polderization*.

empressite A pale-bronze mineral: AgTe.

EMR *electromagnetic radiation*.

Emscherian An obsolete syn. of *Coniacian-Santonian*.

Emsian European stage: Lower Devonian (above Siegenian, below Eifelian).

emulsion stage That stage in the crystallization of some magmas in which the concentration of water exceeds the solubility and a new, water-rich phase is formed, either as a gas or as liquid droplets (Shand, 1947).

emulsion texture An ore texture showing minute blebs or rounded inclusions of one mineral irregularly distributed in another.

enalite A variety of thorite containing uranium.

enantiomorph Either of two crystals that display *enantiomorphism*.

enantiomorphism The characteristic of two crystals to be mirror images of each other, e.g. right-handed and left-handed quartz. Such crystals are called *enantiomorphs*. Enantiomorphism is produced by the improper crystallographic operations of reflection across a plane and inversion through a point. Adj: *enantiomorphous*.

enantiomorphous Adj. of *enantiomorphism*.

enantiotropy The relationship between polymorphs that possess a stable transition point, and that therefore can be stably interconverted by changes of temperature and/or pressure. Although the term was originally applied only in systems with a vapor present, modern usage seems to give the term the more general meaning above. Cf: *monotropy*.

enargite A grayish-black or iron-black orthorhombic mineral: Cu_3AsS_4. It is isomorphous with famatinite and dimorphous with luzonite. Enargite is an important ore of copper, occurring in veins in small crystals or granular masses, and often containing antimony (up to 6%) and sometimes iron and zinc.

enation An outgrowth from some organ of a plant; an *emergence*.

en cabochon adv. Cut in a style characterized by a smooth-domed, but unfaceted, surface; e.g. a ruby cut "en cabochon" in order to bring out the star. Etymol: French. See also: *cabochon*.

enclave An *inclusion*. This usage is not common in the U.S. (Holmes, 1928, p. 28).

enclosed lake A lake that has neither surface influent nor effluent and that never overflows the rim of its basin; e.g. a *kettle lake* or a *crater lake*. Cf: *closed lake*.

enclosed meander *inclosed meander*.

enclosure An *inclusion* in an igneous rock.

encrinal *encrinital*.

encrinal limestone A *crinoidal limestone*; specif. a limestone in which the crinoidal fragments constitute more than 10%, but less than 50%, of the bulk (Bissell & Chilingar, 1967, p. 156). Cf: *encrinite*.

encrinital Pertaining to, or made up of, encrinites; specif. said of a carbonate rock or sediment containing stem and/or plate fragments of crinoids. Syn: *encrinic; encrinal; encrinoid; encrinitic*.

encrinite [paleont] A syn. of *crinoid*, esp. a fossil crinoid belonging to the genus *Encrinus*.

encrinite [sed] A *crinoidal limestone;* specif. a limestone in which crinoidal fragments constitute more than 50% of the bulk (Bissell & Chilingar, 1967, p. 156). Cf: *encrinal limestone*.

encroachment [petroleum] The movement of *bottom water* or *edge water* into a petroleum reservoir as the oil and gas is removed.

encroachment [sed] Deposition of eolian sand from surface creep behind an obstruction of any sort, e.g. on the slip face of a dune (Bagnold, 1941, p. 127). Cf: *accretion [sed struc]*.

encroachment [stratig] The horizontal component of coastal *onlap* (Mitchum, 1977, p. 208). See also: *aggradation*.

encrustation (a) A crust or coating of minerals formed on a rock surface, e.g. calcite on cave objects or soluble salts on a playa. (b) A thin sheetlike organic growth, esp. a colonial invertebrate such as a bryozoan or coral, or a calcareous alga, closely adhering to the underlying solid substrate and mirroring its irregularities. (c) An *external mold* of a plant, usually in some incompressible rock such as sandstone or tufa (Walton, 1940); this usage is not recommended. (d) The process by which a crust or coating is formed.—Also spelled: *incrustation*.

endannulus An *annulus [palyn]* in the endexine of a pollen grain.

endarch Position of protoxylem indicating centrifugal sequence of maturation in primary wood (Foster and Gifford, 1974, p. 56).

end cleat The minor *cleat* system or jointing in a coal seam. See also: *end of coal*. Cf: *face cleat*. Syn: *butt cleat*.

endellionite *bournonite*.

endellite A name used in the the U.S. for a clay mineral: $Al_2Si_2O_5(OH)_4 \cdot 4H_2O$. It is the more hydrous form of halloysite, and is synonymous with *halloysite* of European authors. Syn: *hydrated halloysite; hydrohalloysite; hydrokaolin*.

endemic Said of an organism or group of organisms that is restricted to a particular region or environment. Syn: *indigenous; native*.

enderbite A plagioclase-rich member of the *charnockite* series containing quartz, plagioclase (commonly antiperthitic), hypersthene, and a small amount of magnetite. Most classification systems require that quartz constitute 10-65% of the felsic constituents and that the ratio of alkali feldspar to total feldspar be greater than 87.5%. Tobi (1971) has abandoned the term in favor of *alkali charnockite.* The name, proposed by Tilley in 1936, is for Enderby Land, Antarctica. Not recommended usage.

endexine The inner, usually homogeneous and smooth layer of the two layers of the *exine* of spores and pollen, normally less deeply staining than the *ektexine.* Syn: *intexine; nexine.*

endite One of the appendages of the inner side of the limb of an arthropod, such as the medially directed lobe of precoxa, coxa, basis, or ischium of a crustacean, or the median or inner lobe or segment of the biramous appendage of a trilobite. Cf: *exite.*

end lap *overlap [photo].*

endlichite A variety of vanadinite in which the vanadium is partly replaced by arsenic; a mineral intermediate in composition between mimetite and vanadinite.

end member (a) One of the two or more simple compounds of which an isomorphous (solid-solution) series is composed. For example, the end members of the plagioclase feldspar series are albite, $NaAlSi_3O_8$, and anorthite, $CaAl_2Si_2O_8$. Syn: *minal.* (b) One of the two extremes of a series, e.g. types of sedimentary rock or of fossils.

end moraine A ridgelike accumulation that is being produced at the margin of an actively flowing glacier at any given time; a moraine that has been deposited at the lower or outer end of a glacier. Cf: *terminal moraine.* Syn: *frontal moraine.*

endo- A prefix meaning "within".

endoadaptation Adjustment of one part of an organism to its other parts. Cf: *exoadaptation.*

endobatholithic Said of a mineral deposit occurring in or near an island or roof pendant of batholithic country rock; also, said of the stage of batholith erosion in which that area is exposed. (Emmons, 1933). The term is little used. Cf: *acrobatholithic; cryptobatholithic; embatholithic; epibatholithic; hypobatholithic.*

endobiontic Said of an organism living in bottom sediments. Cf: *epibiontic.*

endoblastesis Late or epimagmatic crystallization in an igneous rock, from *residual liquid.* Not in common use. Adj: *endoblastic.*

endoblastic The adj. of *endoblastesis.* Cf: *metablastic.*

endocast *steinkern.*

endochondral bone *cartilage-replacement bone.*

endocochleate Said of cephalopods with a shell or other hard parts lodged inside the soft body; sometimes grouped as Endocochleata (e.g., *Spirula*).

endocochlian *coleoid.*

endocoel A cavity in the capsule formed by the endophragm in a dinoflagellate cyst. Cf: *pericoel.*

endocone One of a series of concentric conical calcareous deposits or structures formed within the posterior or adapical part of the siphuncle of certain cephalopod conchs (e.g. *Endoceras*). The apices of the cones point toward the apex of the conch, and are usually perforated.

endocyclic Said of a *regular* echinoid whose periproct is located within the oculogenital ring. Ant: *exocyclic.*

endocyst The soft layer of the body wall of a bryozoan, lining the interior of a zooecium and enclosing the polypide, and giving rise to the *ectocyst* (TIP, 1953, pt.G, p.10).

endoderm The inner body layer of an organism; e.g. the inner layer of the outer body walls of a coral polyp, occurring as a double lamina in mesenteries. In the vertebrates the term refers exclusively to embryonic tissue. Cf: *ectoderm; mesoderm.*

endodermis A layer of specialized cells in many roots and some stems, delimiting the inner margin of the cortex (Cronquist, 1961, p.874). Cf: *epidermis [bot].*

endoderre The wall of the prosome of a chitinozoan. It is sometimes regarded as a third layer of the wall. Cf: *ectoderre; periderre.*

endodynamomorphic An old term applied to a soil whose characteristics reflect those of the parent material more than those of external agents. Cf: *ectodynamomorphic.*

end of coal The plane or surface of a coal seam in a direction at right angles to the *face of coal.*

endogastric (a) Said of a cephalopod shell that is curved or coiled, so that the venter is on or near the inner or concave side or area of whorls. (b) Said of a gastropod shell that is coiled so as to extend backward from the aperture over the extruded head-foot mass, as in most adult forms (TIP, 1960, pt.I, p.130).—Cf: *exogastric.*

endogene effect The contact-metamorphic effect of igneous intrusion on the margin of the intrusive body itself (Bateman, 1950). Cf: *exogene effect.*

endogenetic Derived from within; said of a geologic process, or of its resultant feature or rock, that originates within the Earth, e.g. volcanism, volcanoes, extrusive rocks. The term is also applied to chemical precipitates, e.g. *evaporites,* and to ore deposits that originate within the rocks that contain them. Cf: *exogenetic.* Syn: *endogenic; endogenous.*

endogenic *endogenetic.*

endogenous *endogenetic.*

endogenous dome A volcanic dome that has grown primarily by expansion from within and is characterized by a concentric arrangement of flow layers (Williams, 1932). Cf: *exogenous dome.*

endogenous inclusion *autolith.*

endoglyph A hieroglyph occurring within a single sedimentary bed (Vassoevich, 1953, p.37). Cf: *exoglyph.*

endokinematic Said of sedimentary operations in which "the largest displacement vectors occur between some matter within that part of the deposit destined to form the structure and the unmodified deposit" (Elliott, 1965, p.196); e.g. translational slumping, and horizontal or vertical transposition. Also, said of the sedimentary structures produced by endokinematic operations. Cf: *exokinematic.*

endokinetic Said of a fissure in a rock that is the result of strain within the rock unit itself. Cf: *exokinetic.*

endolistostrome A *broken formation* of sedimentary origin; an *olistostrome* without exotic blocks (Raymond, 1978).

endolithic Pertaining to organisms, generally microscopic algae or fungi, that live in minute burrows in corals, shells, or reef rock. Syn: *petricolous.* Cf: *epilithic.*

endolithic breccia A breccia formed by forces acting within the Earth's crust, as by tectonic movements, by swelling or hydration, or by foundering.

endometamorphism *endomorphism.*

endomorph A crystal that is surrounded by a crystal of a different mineral species. Adj: *endomorphic.*

endomorphic metamorphism *endomorphism.*

endomorphism Changes within an igneous rock produced by the complete or partial assimilation of country-rock fragments or by reaction upon it by the country rock along the contact surfaces. It is a form of *contact metamorphism* with emphasis on changes produced within the igneous body rather than in the country rock. The term was originated by Fournet in 1867. Cf: *exomorphism.* Partial syn: *endogene effects.* Syn: *endometamorphism; endomorphic metamorphism.*

endopelos Animals that lie on or burrow in soft mud.

endophragm (a) The complex internal skeletal structure of a crustacean, formed by the fusion of apodemes, and providing a framework for muscle attachment. Syn: *endophragmal skeleton.* (b) The inner-wall layer of a dinoflagellate cyst. Cf: *ectophragm; periphragm.*

endopinacoderm The *pinacoderm* lining the inhalant and exhalant systems of a sponge. Cf: *exopinacoderm.*

endoplicae Folds in the endexine of spores and pollen.

endopod The medial or internal ramus of a limb of a crustacean, arising from the basis. Cf: *exopod.* Syn: *endopodite.*

endopore The internal opening in the endexine of a pollen grain with a complex porate structure. See also: *vestibulum.* Cf: *exopore.* Syn: *os [palyn].*

endopsammon Animals that lie on or burrow in sand.

endopuncta A *puncta* of a brachiopod shell not extending to its external surface, occupied by a caecum. These pores are common over the whole inner surface of the shell but are not visible on the outer surface if the primary layer is intact. Cf: *exopuncta.* Pl: endopunctae. Syn: *endopunctum.*

endorheic Said of a basin or region characterized by internal drainage; relating to endorheism. Also spelled: *endoreic.*

endorheism (a) *internal drainage.* (b) The condition of a region in which little or no surface drainage reaches the ocean.—Ant: *exorheism.* Also spelled: *endoreism.*

endosiphotube A fine canal near the center of the siphuncle of certain eurysiphonate cephalopods, esp. endoceroids, in which the rest of the siphuncle is filled with organic deposits, e.g. *endocones* or *bullettes.*

endosiphuncle The space within the *ectosiphuncle* of certain cephalopods, including all organic tissues and calcareous structures. Syn: *endosiphon.*

endoskarn Skarn formed by replacement of intrusive or other aluminous silicate rock. Cf: *endomorphism.*

endoskeleton An internal *skeleton* in an animal, serving as a supporting framework; e.g. any internal hard parts serving for the attachment of muscles in a crustacean, or the internal system of articulated bones in a vertebrate. Cf: *exoskeleton.*

endosome The inner part of the body of various sponges; e.g. the choanosome with few if any supporting spicules, or the part of a sponge internal to a cortex, or the part surrounding the spongocoel and devoid of flagellated chambers. Because of conflicting usage, the term is not recommended. Cf: *ectosome.*

endosperm The food-storage tissue in a seed. In the gymnosperms, it is a part of the female gametophyte and is haploid; in the angiosperms, it results from the fusion of a sperm with two polar nuclei and is triploid (Fuller & Tippo, 1954, p. 957). Syn: *albumen.*

endosphere All that part of the Earth below the lithosphere.

endospore (a) A syn. of *intine.* The term is mostly applied to the sporoderm of spores, rather than to pollen. Syn: *endosporium.* (b) An asexual spore developed within the cell, esp. in bacteria. (c) A thin-walled spore of blue-green algae.—Cf: *exospore.*

endosternite A part of the endoskeleton of an arthropod; e.g. a tendinous endoskeletal plate in the cephalon of a crustacean.

endostratic Bedded within; e.g. said of the formation of bedding in clays as a result of "alternating desiccation and saturation by ground water" (Becker, 1932, p.85), or said of a breccia bedded within a distinct stratum (Norton, 1917).

endotheca A collective term for the dissepiments inside the wall of a scleractinian corallite. Cf: *exotheca.*

endothecal Said of edrioasteroid ambulacra between thecal plates, in contact with both the interior and the exterior of the theca (Bell, 1976).

endothermic Pertaining to a chemical reaction that occurs with an absorption of heat. Cf: *exothermic.*

endotomous Characterized by bifurcation in two main crinoid arms that give off branches only on their inner sides. Ant: *exotomous.*

endozone The inner part of a bryozoan colony, usually characterized by thin vertical walls, relative scarcity of intrazooidal skeletal structures, and a combination of zooidal growth directions at low angles to the colony growth directions or surfaces. Cf: *exozone.* Syn: *immature region.*

end peneplain *endrumpf.*

end product As applied to radioactivity, the stable nuclide at the end of a *radioactive series.* Cf: *parent; daughter.*

endrumpf A term proposed by W. Penck (1924) for the final landscape or plain that results from the erosion of a landmass that had high relief; it represents the end product of a period of degradation marked by waning uplift. Although Penck considered the Davisian *peneplain* as an equivalent term, *endrumpf* differs in that it does not imply a particular sequence of development leading up to leveling of the original relief. *Endrumpf* may be more likened to an extended use of *pediplain.* Etymol: German *Endrumpf,* "terminal torso". Cf: *primärrumpf.* Syn: *end peneplain.*

endurance limit That stress below which a material can withstand hundreds of millions of repetitions of stress without fracturing. It is considerably lower than rupture strength. Syn: *fatigue limit.*

endurance ratio The ratio of the *endurance limit* of a material to its static, tensile strength. Syn: *fatigue ratio.*

en echelon adj. Said of geologic features that are in an overlapping or staggered arrangement, e.g. faults. Each is relatively short but collectively they form a linear zone, in which the strike of the individual features is oblique to that of the zone as a whole. Etymol: French *en échelon,* "in steplike arrangement".

Eneolithic A syn. of *Copper Age.* Also spelled: *Aeneolithic.*

energy grade line *energy line.*

energy gradient The slope of the *energy line* of a body of flowing water, with reference to any plane. Syn: *energy slope.*

energy index [mineral] A term used by Gruner (1950) for a number that expresses the stability of a silicate. It is equal to the *bridging factor* multiplied by the electronegativity. Quartz has the highest energy index (1.80).

energy index [sed] The inferred degree of water agitation in the sedimentary environment of deposition. Abbrev: EI.

energy level The kinetic energy (due to wave or current action) that existed or exists in the water of a sedimentary environment, either at the interface of deposition or a meter or two above it. See also: *high-energy environment; low-energy environment.*

energy line In hydraulics, a line joining the elevations of the energy heads of a stream when referred to the stream bed. It lies above the water surface at any cross section; the vertical distance is equal to the velocity head at that cross section (ASCE, 1962). See also: *energy gradient.* Syn: *energy grade line.*

energy loss The difference between energy input and output as a result of transfer of energy between two points. In the flow of water, the rate of energy loss is represented by the slope of the *hydraulic grade line.*

energy of glacierization *activity index.*

energy slope *energy gradient.*

engineering geology Geology as applied to engineering practice, esp. mining and civil engineering. As defined by the Association of Engineering Geologists (1969), it is the application of geologic data, techniques, and principles to the study of naturally occurring rock and soil materials or ground water for the purpose of assuring that geologic factors affecting the location, planning, design, construction, operation, and maintenance of engineering structures, and the development of ground-water resources, are properly recognized and adequately interpreted, utilized, and presented for use in engineering practice. Syn: *geologic engineering.*

englacial Contained, embedded, or carried within the body of a glacier or ice sheet; said of meltwater streams, till, drift, moraine, etc. Syn: *intraglacial.*

englacial drift Rock material contained within a glacier or ice sheet.

englishite A white mineral: $K_2Ca_4Al_8(PO_4)_8(OH)_{10} \cdot 9H_2O$. Also spelled: *Englishite.*

engrafted stream A stream composed of the waters of several previously independent streams that unite before reaching the sea; esp. a main stream consisting of several separate extended streams flowing from an oldland and merging with each other on an uplifted coastal plain. Also spelled: *ingrafted stream.*

enhancement The process of altering the appearance of an image so that the interpreter can extract more information from it. Enhancement may be done by digital or photographic methods.

enhydrite (a) A mineral or rock having cavities containing water. (b) *enhydros.*

enhydros A hollow nodule or geode of chalcedony containing water, sometimes in large amount. Syn: *enhydrite; water agate.*

enhydrous Said of certain crystalline minerals containing water or having drops of included fluid; e.g. "enhydrous chalcedony". Not to be confused with *anhydrous.*

enigmatite *aenigmatite.*

enneri A term used in northern Africa (esp. Libya) for a wadi or dry river valley.

ennomoclone A desma (of a sponge) consisting of one short distal arm (brachyome) and three or six longer proximal arms directed symmetrically away from it; e.g. a *tricranoclone* or a *sphaeroclone.*

enrichment *supergene enrichment.*

enrockment A mass of large stones placed in water to form a base, as for a pier.

ensialic geosyncline A geosyncline, the geosynclinal prism of which contains clastics accumulating on sialic crust (Wells, 1949). Cf: *ensimatic geosyncline.* See also: *miogeosyncline.*

ensimatic geosyncline A geosyncline, the geosynclinal prism of which contains effusive rocks accumulating on simatic crust (Wells, 1949, p. 1927). Cf: *ensialic geosyncline.* See also: *eugeosyncline.*

enstatite A common rock-forming mineral of the orthopyroxene group: $MgSiO_3$. It is isomorphous with hypersthene, and may contain a little iron replacing the magnesium. Enstatite ranges from grayish white to yellowish, olive green, and brown. It is an important primary constituent of intermediate and basic igneous rocks. Symbol: En. Cf: *bronzite.* Syn: *chladnite.*

enstatolite A *pyroxenite* that is composed almost entirely of enstatite. Not recommended usage.

enstenite A group name for the orthopyroxenes of the $MgSiO_3$-$FeSiO_3$ isomorphous series. It includes enstatite, hypersthene, and orthoferrosilite.

entablature In columnar jointing, the upper zone that has thinner and less regular columns than the lower zone, or *colonnade.*

enterolithic (a) Said of a sedimentary structure consisting of ribbons of intestinelike folds that resemble those produced by tectonic deformation but that originate through chemical changes involving an increase or decrease in the volume of the rock; e.g. said of a small fold or local crumpling formed in an evaporite by flowage or by the swelling of anhydrite during hydration. See also: *tepee structure.* (b) Said of the deformation or folding that produces enterolithic structures.

enteron The digestive cavity or alimentary system of an animal, generally consisting of esophagus, stomach, and intestine. Cf: *coelom.*

entexis Migmatization with introduction from without of the more mobile part (Dietrich & Mehnert, 1961). Cf: *ectexis.* Little used.

entexite Rock formed by entexis. Also spelled: *entectite.* Little used.

enthalpy A thermodynamic quantity that is defined as the sum of a body's internal energy plus the product of its volume, multiplied by the pressure. Syn: *heat content.*

entire Said of a leaf with a continuous smooth margin, not lobed or dentate.

Entisol In U.S. Dept. of Agriculture soil taxonomy, a soil order characterized by dominance of mineral soil materials and absence of distinct pedogenic horizons. Entisols may have an ochric or anthropic epipedon. Other horizons are poorly formed because of lack of time, inert parent material, formation over slowly soluble rock, occurrence on steep, actively eroding slopes, or recent mechanical mixing (USDA, 1975). Suborders and great soil groups of this soil order have the suffix -ent. See also: *Aquent; Arent; Fluvent; Orthent; Psamment.* Cf: *azonal soil.*

entocoele The space within a pair of mesenteries of a coral. Cf: *exocoele.*

entogene Said of conditions within a depositional basin that influence the texture of a sedimentary rock formed in that basin (Sander, 1951, p.11). Cf: *ectogene.*

entomodont Said of a class of ostracode hinges intermediate in form between the merodont hinge and the amphidont hinge, having denticulate terminal and median elements with a partial subdivision of the median element.

entomophily Pollination by insects. Adj: *entomophilous.* Cf: *anemophily.*

entomostracan An obsolete term originally applied to insect-shelled crustaceans. Cf: *malacostracan.*

entooecium Calcified or uncalcified inner part of the body-wall fold forming an ovicell in cheilostome bryozoans. Cf: *ectooecium.*

entoolitic Pertaining to oolitic structures or grains that have formed or grown inward by the filling of small cavities, as by the deposition of successive coats on the cavity walls. Ant: *extoolitic.*

entoproct Any bryozoan belonging to the subphylum Entoprocta and lacking hard parts and a body cavity. These bryozoans are not found as fossils. Cf: *ectoproct.*

entoseptum A scleractinian-coral septum developed within an entocoele. Cf: *exoseptum.*

entosolenian Said of a foraminifer (e.g. *Oolina*) having an internal tubelike apertural extension (siphon). Cf: *ectosolenian.*

entotoichal Said of a cheilostome bryozoan ovicell that appears to be immersed in the distal zooid while opening independently to the exterior.

entozooecial *entozooidal.*

entozooidal Said of a cheilostome bryozoan ovicell that appears to be immersed in the distal zooid while opening below the operculum of the maternal zooid. Syn: *entozooecial.*

entrail pahoehoe A type of *pahoehoe* that has a surface resembling an intertwined mass of entrails, formed on steep slopes as dribbles around and through cracks in the flow crust.

entrainment The process of picking up and carrying along, as the collecting and movement of sediment by currents, or the incorporation of air bubbles into a cement *slurry.*

entrance angle *axil angle.*

entrapment burrow A term used by Kuenen (1957, p.253) for a burrow occupied by an animal buried below the sandy deposit of a passing turbidity current.

entrenched meander (a) An *incised meander* carved downward into the surface of the valley in which the meander originally formed; it exhibits a symmetric cross profile. Such a deepened meander, which preserves its original pattern with little modification, suggests rejuvenation of a meandering stream, as when there has been a rapid vertical uplift or a lowering of base level. Cf: *ingrown meander.* Syn: *inherited meander.* (b) A generic term used as a syn. of *incised meander.*—Syn: *intrenched meander.*

entrenched stream A stream, often meandering, that flows in a narrow trench or valley cut into a plain or relatively level upland; e.g. a stream that has inherited its course from a previous cycle of erosion and that cuts into bedrock with little modification of the original course. Also spelled: *intrenched stream.*

entrenchment The process whereby a stream erodes downward so as to form a trench or to develop an entrenched meander. Also, the results of such a process. Cf: *incision.* Also spelled: *intrenchment.*

entropy [phys] A macroscopic, thermodynamic quantity, ultimately reflecting the degree of microscopic randomness or disorder of a system. It was defined by Gibbs as the value of the integral of the quotient of the differential of the heat absorbed by the system divided by the absolute temperature for any reversible process by which the constituents of the system were brought from the states in which their entropies are zero into combination and to the state of interest. Symbol: S.

entropy [stratig] A measure of the degree of "mixing" of the different kinds of rock components in a stratigraphic unit (Pelto, 1954). The entropy value of a given component is the product of its proportion in the unit and the natural logarithm of that proportion. A stratigraphic unit with equal parts of each component has an entropy value of 100; as the composition approaches that of a single component, the entropy value approaches zero.

entropy [streams] A quantity that is expressed by the probability of a given distribution of energy utilization within or along a stream from headwaters to a downstream point, the most probable condition existing when the stream is graded or the energy is as uniformly distributed as may be permitted by physical constraints (Leopold & Langbein, 1962).

entropy map A *facies map* that is based on the degree of "mixing" of three end members (rock components) of a given stratigraphic unit, but that does not distinguish the natures of these end members. Cf: *entropy-ratio map.* Syn: *isentropic map.*

entropy-ratio map A *facies map* that is based on the degree of "mixing" of three end members (rock components) of a given stratigraphic unit and that indicates by map pattern the nature of the lithologic "mixture" through which a given end member is approached (Forgotson, 1960, p.93). Cf: *entropy map.*

entropy unit A unit of measurement defined as one calorie per mole-degree. It is essentially equivalent to the *gibbs.*

entry The mouth of a river.

entry pressure *displacement pressure.*

envelope The outer or covering part of a fold, especially of a folded structure that includes some sort of structural break. Cf: *core.*

enveloping surface An imaginary surface tangent to antiformal and synformal hinges in a single folded surface.

environment [biol] All those external factors and conditions which may influence an organism or a community. Syn: *habitat.*

environment [sed] A geographically restricted complex where a sediment accumulates, described in geomorphic terms and characterized by physical, chemical, and biological conditions, influences, or forces; e.g. a lake, swamp, or flood plain.

environmental assessment A detailed statement prepared by an organization for its own use to appraise the effect of a proposed project on the aggregate of social and physical conditions that influence a community or ecosystem. The assessment is often prepared to determine the need for a formal *environmental impact statement.*

environmental facies Facies that are concerned solely with environment or determined by the nature of environment; e.g. lithotopes, biotopes, and tectotopes (Weller, 1958, p.628). They are not material units or bodies of rock, but areas inferred from the results of a combination of mutually interacting influences and conditions as these are exhibited in the form of sedimentary types and organic communities. See also: *facies.* Syn: *ecologic facies.*

environmental geochemistry The effect on man of the distribution and interrelations of the chemical elements and radioactivity among surficial rocks, water, air, and biota.

environmental geology The application of geologic principles and knowledge to problems created by man's occupancy and exploitation of the physical environment. It involves studies of hydrogeology, topography, engineering geology, and economic geology, and is concerned with Earth processes, Earth resources, and engineering properties of Earth materials. It involves problems concerned with construction of buildings and transportation facilities, safe disposal of solid and liquid wastes, management of water resources, evaluation and mapping of rock and mineral resources, and long-range physical planning and development of the most efficient and beneficial use of the land. See also: *urban geology.* Syn: *geoecology.*

environmental hyperspace lattice In a geometric model of the environment and its ecologic units, a multidimensional space containing as many dimensions as there are possible environmental factors (Valentine, 1969, p.685). See also: *biospace; ecospace.*

environmental impact statement A document prepared by industry or a political entity on the environmental impact of its proposals for legislation and other major actions significantly affecting the quality of the human environment. Environmental impact statements are used as tools for decision-making and are required by the National Environmental Policy Act.

environmental resistance Factors in the environment that tend to restrict the development of an organism or group of organisms and to limit its numerical increase.

environmental science (a) Earth science applied to the human habitat: geomorphology, meteorology, climatology, soil science, and physical and applied oceanography. (b) A science that is involved with "all of nature we perceive or can observe, that is our physical environment—a composite of Earth, Sun, sea, and atmosphere, their interactions, and the hazards they present" (ESSA, 1968).

Eocambrian An approximate equivalent of *Riphean.* Syn: *Infracambrian.*

Eocene An epoch of the lower Tertiary period, after the Paleocene and before the Oligocene; also, the corresponding worldwide series of rocks. It is sometimes considered to be a period, when the Tertiary is designated as an era.

Eogene *Paleogene.*

eogenetic A term proposed by Choquette & Pray (1970, p. 219-220) for the period of time between final deposition of a sediment and burial of that sediment below the depth to which surface or near-surface processes are effective. The upper limit of the eogenetic zone is the land surface; the lower boundary, less clearly defined due to the gradual diminishing of surface-related processes, is the *mesogenetic* zone. Also applied to the porosity that develops during the eogenetic stage. Cf: *telogenetic.*

eohypse A contour line on a former land surface, reconstructed or "restored" on a map by plotting the contours of the surviving portions of the land and by extrapolation of those contours. Syn: *eohyps; eoisohypse.*

eoisohypse *eohypse.*

eolation The gradational work performed by the wind in modifying the land surface, such as the transportation of sand and dust, the formation of dunes, the effects of sandblasting, and, indirectly, the action of water waves generated by wind currents.

eolian (a) Pertaining to the wind; esp. said of such deposits as loess and dune sand, of sedimentary structures such as wind-formed ripple marks, or of erosion and deposition accomplished by the wind. (b) Said of the active phase of a dune cycle, marked by diminished vegetal control and increased dune growth. Cf: *eluvial [sed].*—Etymol: Aeolus, god of the winds. Syn: *aeolian; eolic.*

eolianite A consolidated sedimentary rock consisting of clastic material deposited by the wind; e.g. dune sand cemented below ground-water level by calcite. Also spelled: *aeolianite.* Syn: *dune rock.*

eolic *eolian.*

eolith The most primitive type of man-made stone implement. Syn: *dawn stone.*

Eolithic n. In archaeology, a cultural level that is pre-Paleolithic or at the very beginning of the *Paleolithic.* It is characterized by eoliths.—adj. Pertaining to the Eolithic.

eolomotion A relatively slow downwind or downhill movement of sand due to direct or indirect wind action on surface rock particles (Kerr & Nigra, 1952).

eometamorphism Early metamorphism, or the very beginnings of metamorphism, esp. as affecting hydrocarbons, which are highly vulnerable (Landes, 1967, p. 832).

eon (a) Any grand division or large part of geologic time; specif. the longest geologic-time unit, next in order of magnitude above *era,* e.g. the Phanerozoic Eon, which includes the Paleozoic Era, the Mesozoic Era, and the Cenozoic Era. (b) One billion (10^9) years.—Also spelled: *aeon.*

eonothem The chronostratigraphic equivalent of an *eon.*

Eophytic A paleobotanic division of geologic time, signifying that time during which algae were abundant. Cf: *Aphytic; Archeophytic; Cenophytic; Mesophytic; Paleophytic.*

eospar Primary sparry calcite deposited by direct precipitation in pore spaces, as opposed to *neospar* (Nichols, 1967, p. 1247-1248). Syn: *calcite cement.*

eosphorite A pink to rose-red mineral: $(Mn,Fe)AlPO_4(OH)_2 \cdot H_2O$. It is isomorphous with childrenite.

Eosuchia An order of lepidosaurian reptiles, of generally lizard-like habitus and Late Paleozoic and Early Mesozoic age, believed to include the ancestors of living lizards and snakes. Range, Upper Permian to Eocene.

Eötvös correction In gravity measurement, a correction for centripetal acceleration caused by east-west velocity over the surface of the rotating Earth. Syn: *Eötvös effect.*

Eötvös effect *Eötvös correction.*

Eötvös torsion balance *torsion balance.*

Eötvös unit A unit of gravitational gradient or curvature; 10^{-6} mgal/cm.

eozoan An obsolete term for a *protozoan*, or a one-celled animal of the subkingdom Eozoa. Syn: *eozoon.*

Eozoic An archaic term for part or all of the Precambrian.

eozoon (a) An inorganic banded structure of coarsely crystalline calcite and serpentine, occurring in the Grenville Series of Canada, originally interpreted as the remains of a gigantic foraminifer, *Eozoon canadense.* Pl: *eozoons; eozoa.* Adj: *eozoonal.* (b) *eozoan.*

epanticlinal fault A longitudinal or transverse fault that is associated with a doubly plunging minor anticline and formed concurrently with the folding (Irwin, 1926). Also spelled: *epi-anticlinal fault.*

epaulet A five-sided step cut of a gem, resembling a shoulder ornament (epaulet) in outline.

epeiric sea *epicontinental sea.*

epeirocratic (a) Adj. of *epeirocraton.* (b) Said of a period of low sea level in the geologic past. Cf: *thalassocratic.*

epeirocraton A craton of the continental block. Cf: *hedreocraton; thalassocraton.* Adj: *epeirocratic.*

epeirogenesis *epeirogeny.*

epeirogenetic *epeirogenic.*

epeirogenic Adj. of *epeirogeny.* Cf: *orographic.* Also spelled: *epeirogenetic.*

epeirogeny As defined by Gilbert (1890), a form of *diastrophism* that has produced the larger features of the continents and oceans, for example plateaus and basins, in contrast to the more localized process of *orogeny,* which has produced mountain chains. Epeirogenic movements are primarily vertical, either upward or downward, and have affected large parts of the continents, not only in the cratons but also in stabilized former orogenic belts, where they have produced most of the present mountainous topography. Some epeirogenic and orogenic structures grade into each other in detail, but most of them contrast strongly. Adj: *epeirogenic.* Syn: *epeirogenesis.* Cf: *bathygenesis; cymatogeny.*

epeirophoresis theory *continental displacement.*

epharmone An organism that has undergone morphologic change as a result of changes in the environment and therefore differs from the normal or usual form.

ephebic Said of the adult stage in the life history of an animal; i.e., the stage when the animal is normal in size and able to reproduce. The stage follows the *neanic* stage and precedes the *gerontic* stage.

ephemeral lake A short-lived lake. Cf: *intermittent lake; evanescent lake.*

ephemeral stream A stream or reach of a stream that flows briefly only in direct response to precipitation in the immediate locality and whose channel is at all times above the water table. The term "may be arbitrarily restricted" to a stream that does "not flow continuously during periods of as much as one month" (Meinzer, 1923, p. 58). Cf: *intermittent stream.*

ephemeris A publication giving coordinates of celestial bodies at uniform time intervals, generally for one calender year.

ephemeris second The fundamental invariable unit of time. It is defined as 1/31,556,925.9747 of the tropical year for 1900 January 0^d12^h ephemeris time. The ephemeris day is 86,400 seconds. The former unit of time was the mean solar second, defined as 1/86,400 of the mean solar day.

ephemeris time A uniform measure of time determined by relative changes in the positions of Earth, Moon, and stars. Cf: *ephemeris second.*

ephesite A mineral of the brittle-mica group: $NaLiAl_2(Al_2Si_2)O_{10}(OH)_2$. It is related to margarite.

ephippium The dorsal brood pouch of various cladoceran crustaceans that is shed with the eggs and serves for protection until hatching. Pl: *ephippia.*

epi- A prefix signifying "on" or "upon".

epi-anticlinal fault *epanticlinal fault.*

epibatholithic Said of a mineral deposit occurring in the peripheral area of a batholith; also, said of the stage of batholith erosion in which that area is exposed (Emmons, 1933). The term is little used. Cf: *acrobatholithic; cryptobatholithic; embatholithic; endobatholithic; hypobatholithic.* See also: *dead line.*

epibiontic Said of an organism living on the surface of bottom sediments, rocks, or shells. Cf: *endobiontic.*

epibole (a) A syn. of *acme-zone.* (b) The deposits accumulated during a *hemera* (Trueman, 1923, p. 200).

epibolite A term introduced by Jung & Roques for a migmatite with granitic layers concordant with the gneissosity of its nongranitic parent rock (Roques, 1961).

epicenter The point on the Earth's surface that is directly above the *focus* of an earthquake. Cf: *anticenter.* Syn: *epicentrum.*

epicentral distance The distance from the epicenter of an earthquake to the receiver. It may be measured in angular units (*angular distance*) or in linear distance, along a great-circle path.

epicentrum *epicenter.*

epiclastic rock A rock formed at the Earth's surface by consolidation of fragments of pre-existing rocks; a sedimentary rock whose fragments are derived by weathering or erosion. Cf: *autoclastic rock.*

epicontinental Situated on the continental shelf or on the continental interior, as an *epicontinental sea.*

epicontinental sea A sea on the continental shelf or within a continent. Cf: *mediterranean sea.* Syn: *inland sea; continental sea; epeiric sea.*

epicotyl The part of the embryo of the seed that lies above the cotyledon and becomes the growing point of the shoot. Syn: *plumule.*

epicycle A minor or secondary cycle within a major or primary cycle; specif. a subdivision of a cycle of erosion, initiated by a small change of base level, such as an episode of stillstand of sufficient duration to rank as a part of a cycle and recorded as a terrace.

epideltoid The anal deltoid between mouth and anus in blastoids that have only one or two anal deltoids.

epidermal Said of shallow or surficial deformation of the sialic crust. Cf: *dermal; bathydermal.*

epidermis [bot] The characteristic outermost tissue of a pla· , usually one cell thick, covering leaves, stems, and other parts (Cronquist, 1961, p. 874). Cf: *endodermis.*

epidermis [geol] The sedimentary part of the Earth's crust (Van Bemmelen, 1949, p. 285).

epidermis [paleont] Any of various animal integuments; e.g. the periostracum of a mollusk, the imperforate outer layer in foraminifera, or the external cellular layer in the body wall of a coelenterate. Cf: *hypodermis.*

epidesmine *stilbite.*

epidiabase A name proposed as a replacement for *epidiorite.*

epidiagenesis A term used by Fairbridge (1967) for the final emergent phase of diagenesis, in which sediments are lithified during and after uplift or emergence but before erosion. It is characterized by modification of connate solutions (by deeply penetrating and downward-migrating ground waters) and by reintroduction of oxidizing conditions; near the Earth's surface, it passes into a zone where weathering processes become dominant. It is equivalent to *late diagenesis.* See also: *syndiagenesis; anadiagenesis.* Cf: *epigenesis.* Adj: *epidiagenetic.*

epididymite A colorless, orthorhombic mineral: $NaBeSi_3O_7(OH)$. It is dimorphous with eudidymite.

epidiorite A metamorphosed gabbro or diabase in which generally fibrous amphibole (uralite) has replaced the original clinopyroxene (commonly augite). It is usually massive but may have some schistosity. See also: *epidiabase.*

epidosite A metamorphic rock consisting of epidote and quartz, and generally containing other secondary minerals such as uralite and chlorite.

epidote (a) A yellowish-green, pistachio-green, or blackish-green mineral: $Ca_2(Al,Fe)_3Si_3O_{12}(OH)$. It commonly occurs associated with albite and chlorite as formless grains or masses or as monoclinic crystals in low-grade metamorphic rocks (derived from limestones), or as a rare accessory constituent in igneous rocks, where it represents alteration products of ferromagnesian minerals. Syn: *pistacite; arendalite; delphinite; thallite.* (b) A mineral group, including minerals such as epidote, zoisite, clinozoisite, piemontite, and hancockite.

epidote-amphibolite facies *albite-epidote-amphibolite facies.*

epidotization The hydrothermal introduction of epidote into rocks or the alteration of rocks in which plagioclase feldspar is albitized, freeing the anorthite molecule for the formation of epidote and zoisite, often accompanied by chloritization. These processes are characteristically associated with metamorphism.

epieugeosyncline A deeply subsiding trough with limited volcanism, associated with rather narrow uplifts and overlying a deformed and intruded eugeosyncline (Kay, 1947, p. 1289-1293). Syn: *backdeep.* Cf: *secondary geosyncline.* See also: *nuclear basin.*

epifauna (a) Fauna living upon rather than below the surface of the sea floor. Cf: *infauna.* (b) Fauna living attached to rocks, seaweed, pilings, or to other organisms in shallow water and along the shore.

epigene (a) Said of a geologic process, or of its resultant features, occurring at or near the Earth's surface. Cf: *hypogene.* Syn: *epigenic.* (b) Pertaining to a crystal that is not natural to its enclosing material, e.g. a *pseudomorph.*

epigenesis [meta] The change in the mineral character of a rock as a result of external influences operating near the Earth's surface, e.g. mineral replacement during metamorphism.

epigenesis [sed] The changes, transformations, or processes, occurring at low temperatures and pressures, that affect sedimentary rocks subsequent to their compaction, exclusive of surficial alteration (weathering) and metamorphism; e.g. postdepositional dolomitization. The term is equivalent to *late diagenesis* (as used in the U.S.) and to *metharmosis,* but is considered by Russian geologists to include those changes occurring subsequent to *diagenesis* (as defined in a restricted sense). Cf: *epidiagenesis.* Syn: *metagenesis; metadiagenesis; catagenesis.*

epigenesis [streams] *superimposition.*

epigenetic [eco geol] Said of a mineral deposit of origin later than that of the enclosing rocks. Cf: *syngenetic [ore dep]; diplogenetic.* Syn: *xenogenous.*

epigenetic [sed] (a) Said of a sedimentary mineral, texture, or structure formed after the deposition of the sediment. (b) Pertaining to sedimentary epigenesis.—Cf: *syngenetic [sed].* Syn: *epigenic.*

epigenetic [streams] *superimposed.*

epigenetic ice Ice commonly in lenses in material that predates the growth of the ice. The common form of ice produced during the aggradation of permafrost; a form of *Taber ice.*

epigenic *epigene.*

epigenite A steel-gray mineral: $(Cu,Fe)_5AsS_6$ (?).

epiglacial bench A terrace cut by the lateral erosion of a supraglacial meltwater stream originating on a large glacier; also, the valley-side channel developed by such a stream (Stamp, 1961, p. 179).

epiglacial epoch A period of time that closes a "great glacial series", representing a "constant phase" of glacial activity (Hansen, 1894, p. 131). Not in current usage.

epiglyph A hieroglyph on the top of a sedimentary bed (Vassoevich, 1953, p.37). Cf: *hypoglyph.*

epigynous Said of a flower in which the sepals, petals, and stamens appear to arise from the top of the ovary. Such an ovary is called an *inferior ovary.* Cf: *hypogynous; perigynous.*

epiianthinite A yellow pseudomorphous alteration product of ianthinite, identical with schoepite.

epilimnetic Pertaining to an *epilimnion.* Syn: *epilimnial.*

epilimnial *epilimnetic.*

epilimnion The uppermost layer of water in a lake, characterized by an essentially uniform temperature that is generally warmer than elsewhere in the lake and by a relatively uniform mixing caused by wind and wave action; specif. the light (less dense), oxygen-rich layer of water that overlies the *metalimnion* in a thermally stratified lake. The oceanographic equivalent is *mixed layer.* Cf: *hypolimnion.*

epilithic Said of an organism that lives on or attached to rock or other stony matter. Syn: *petrophilous.* Cf: *endolithic.*

epimagma A vesicular magmatic residue that is relatively gas-free and of semisolid, pasty consistency, commonly formed by the cooling of lava in a lava lake. Cf: *hypomagma; pyromagma.*

epimagmatic *deuteric.*

epimatrix A term introduced by Dickinson (1970, p. 702) for "inhomogeneous interstitial materials grown in originally open interstices during diagenesis" of graywackes and arkoses.

epimere A lateral downfold of a crustacean tergite. Syn: *epimeron; pleurite; pleuron; tergal fold.*

epimerization A process in which there is an alteration of the configuration at only one asymmetric center in an organic compound containing more than one asymmetric center (Cram & Hammond, 1959, p. 131). Cf: *racemization age method.*

epinorm Theoretical calculation of minerals in a metamorphic rock of the *epizone,* as indicated by chemical analyses (Barth, 1959). Cf: *catanorm; mesonorm.* See also: *Niggli molecular norm.*

epipedon In U.S. soil classification, a diagnostic horizon that forms at the surface and is used to classify soils into orders, suborders, and great groups. It is not a syn. of *A horizon,* as its characteristics may extend well into the B horizon. See also: *mollic epipedon; umbric epipedon; ochric epipedon; anthropic epipedon; plaggen epipedon; histic epipedon.* Syn: *diagnostic surface horizon.*

epipelagic Pertaining to the *pelagic* environment of the ocean to a depth of 100 fathoms. Cf: *mesopelagic.*

epipelic Said of an organism growing on sediment, esp. soft mud.

epiphysis [intrus rocks] An *apophysis* or *tongue* of an intrusion which is detached from its source.

epiphysis [paleont] An interambulacral element at the top of Aristotle's lantern in echinoids.

epiphyte A plant not growing from the soil but living attached to another plant or some inanimate object such as a pole or wire. Adj: *epiphytic.*

epiplankton Organisms that are attached to floating vegetation or to mobile swimmers, esp. to vertebrates like turtles, sea snakes, and cetaceans. Syn: *pseudoplankton.*

epipsammon Animals that live on a sandy surface.

epirhysis A *skeletal canal* in dictyonine hexactinellid sponges corresponding to an inhalant canal. Pl: *epirhyses.*

episeptal deposit A proximal cameral deposit on the concave (adoral) side of the septum of a nautiloid. Ant: *hyposeptal deposit.*

episkeletal Above or outside the endoskeleton of an animal.

episode (a) A term used informally and without time implication for a distinctive and significant event or series of events in the geologic history of a region or feature; e.g. "glacial episode", "volcanic episode". (b) A term preferred by Størmer (1966, p. 25) for *subage,* or the geologic-time unit during which the rocks of a *substage* were formed. Syn: *phase [geochron] (a); time (d).*

episome The anterior part of the cell body above the girdle of an unarmored dinoflagellate. Cf: *hyposome.*

epispire A small round or elliptical sutural pore between the calyx plates of many primitive echinoderms such as eocrinoids, some edrioasteroids, and some stylophorans. Each epispire is thought to have housed an outward projecting soft-bodied papula used for respiration (Sprinkle, 1973, p. 28).

epistilbite A white or colorless zeolite mineral: $CaAl_2Si_6O_{16} \cdot 5H_2O$.

epistolite A triclinic mineral: $Na_2(Nb,Ti)_2Si_2O_9 \cdot nH_2O$.

epistoma *epistome.*

epistome (a) The region between the antennae and the mouth of a crustacean. Also, a plate covering this region, such as that between the labrum and the bases of the antennae in brachyuran decapods. (b) The *rostral plate* of a trilobite. (c) A small labiate organ overlapping the mouth in bryozoans of the class Phylactolaemata. (d) The median plate of the prosomal doublure of eurypterids.—Pl: *epistomes.* Syn: *epistoma.*

epitactic Recommended adj. of *epitaxy.*

epitaxial Adj. of *epitaxy.*

epitaxic Adj. of *epitaxy.*

epitaxy Orientation of one crystal with that of the crystalline substrate on which it grew. It is a type of overgrowth in which the two nets in contact share a common mesh. Adj: *epitactic; epitaxic; epitaxial.* Cf: *topotaxy; syntaxy.*

epitheca (a) An external calcareous layer or sheath of skeletal tissue surrounding a corallite and comprising an extension of the basal plate. (b) A dark secondary deposit in the inner wall of a fusulinid foraminifer; a tectorium. (c) The thin outermost calcareous layer of a thecal plate of a cystoid. It is thinner than the *stereotheca.* (d) The anterior part of a dinoflagellate theca, above the girdle. Cf: *hypotheca.* (e) *epivalve.*

epithelium (a) In an animal, a cellular tissue that forms a covering or lining and that may serve various functions such as protection or secretion. (b) In a plant, "any cellular tissue covering a free surface or lining a tube or cavity" (Swartz, 1971, p. 178); e.g. lining a resin duct and excreting resin.

epithermal Said of a hydrothermal mineral deposit formed within about 1 kilometer of the Earth's surface and in the temperature range of 50°-200°C, occurring mainly as veins (Park & MacDiarmid, 1970, p. 344). Also, said of that environment. Cf: *hypothermal; mesothermal; leptothermal; telethermal; xenothermal.*

epithermal-neutron log A *well log* of the *wall-contact log* type that measures radioactivity induced by neutron-neutron reactions. Fast neutrons of several Mev energy, emitted by a source in the *sonde,* may be slowed and backscattered as epithermal neutrons (0.4 to 100 ev) by collisions with hydrogen nuclei. Epithermal-neutron abundance is interpreted by assuming a constant lithology, usually limestone, that produces a continuous curve in porosity units. The epithermal-neutron log is largely insensitive to the differences in chlorinity of formation water that may hamper the interpretation of the *thermal-neutron log* and *neutron-gamma log.* Syn: *n-en log.*

epithet A term used in botanical nomenclature to designate the single word that names a species, which is written following the generic name as part of a *binomen.*

epithyridid Said of a brachiopod pedicle opening lying wholly within the ventral umbo and ventral from the beak ridges (TIP, 1965, pt.H, p.144).

epitract The part of a dinoflagellate cyst anterior to the girdle region. Ant: *hypotract.*

epityche An excystment aperture in the acritarch genus *Veryhachium.* It originates as an arched slit between two processes, in which rupture allows the folding-back of a relatively large flap.

epivalve The outer half of a diatom frustule. Cf: *hypovalve.* Syn: *epitheca.*

epixenolith A xenolith that is derived from the adjacent wall rock (Goodspeed, 1947, p. 1251). Cf: *hypoxenolith.*

epizoic Said of an organism growing on the body surface of an animal.

epizone According to Grubenmann's classification of metamorphic rocks (1904), the uppermost *depth zone* of metamorphism, characterized by low to moderate temperatures (less than 300°C) and hydrostatic pressures with low to high shearing stress. Mechanical and chemical metamorphism produce hydrous silicates (e.g. sericite, chlorite, talc) and carbonates (e.g. calcite, dolomite). Typical rocks are slate, phyllite, and sericite and chlorite schist. The concept was modified by Grubenmann and Niggli (1924) to include effects of low-temperature contact metamorphism and metasomatism. Modern usage stresses pressure-temperature conditions (low metamorphic grade) rather than the likely depth of zone. Cf: *mesozone; katazone.*

epizygal The distal brachial plate of a pair joined by syzygy in a crinoid. Cf: *hypozygal.*

epoch [geochron] (a) A geologic-time unit longer than an *age [geochron]* and shorter than a *period [geochron],* during which the rocks of the corresponding *series* were formed. (b) A term used informally to designate a length (usually short) of geologic time; e.g. *glacial epoch.*

epoch [paleomag] (a) A date to which measurements of a time-varying quantity are referred, e.g. "a chart of magnetic declination for epoch 1965.0". (b) *polarity epoch.*

epoch [tides] *tidal epoch.*

epontic Said of an organism that grows attached to some substratum.

EPR *electron paramagnetic resonance.*

epsomite [mineral] A mineral: $MgSO_4 \cdot 7H_2O$. It consists of native Epsom salts, and occurs in colorless prismatic crystals, botryoidal masses, incrustations in gypsum mines or limestone caves, or in solution mineral waters. Syn: *Epsom salt; bitter salts; hair salt.*

epsomite [sed] Obsolete syn. of *stylolite.*

Epsom salt (a) A chemical: $MgSO_4 \cdot 7H_2O$. It is a bitter, colorless or white, crystalline salt with cathartic qualities. (b) *epsomite.*—Syn: *Epsom salts.*

equal-area projection (a) A map projection on which a constant ratio of areas is preserved, so that any given part of the map has the same relation to the area on the sphere it represents as the whole map has to the entire area represented. Examples include: Bonne projection, Albers projection, and Mollweide projection. Cf: *conformal projection.* Syn: *authalic projection; equivalent projection; homolographic projection.* (b) *equiareal projection.*

equant (a) Said of a crystal having the same or nearly the same diameter in all directions. Cf: *tabular; prismatic.* Syn: *equidimensional; isometric.* (b) Said of a sedimentary particle whose length is less than 1.5 times its width (Krynine, 1948, p. 142). (c) Said of a rock in which the majority of grains are equant (Harte, 1977).

equant element A fabric element all of whose dimensions are approximately equal. Cf: *linear element; planar element.*

equation of state An equation interrelating the thermodynamic variables that define the state of a system. It is classically applied to simple gases and liquids in terms of pressure, volume, and temperature, but in modern geochemistry and petrology it is commonly extended to solids and to solutions, in which case the equations must contain terms describing the composition of the phase.

equator [astron] The great circle of a celestial sphere, having a plane that is perpendicular to the axis of the Earth.

equator [palyn] An imaginary line connecting points midway between the poles of a spore or pollen grain.

equator [surv] (a) The great circle formed on the surface of a sphere or spheroid by a plane drawn through its center and perpendicular to its polar axes, such as the great circle midway between the poles of rotation of a celestial body; specif. the Equator, or the great circle on the Earth's surface that is everywhere equally distant from the two poles and whose plane is perpendicular to the Earth's axis of rotation, that divides the Earth's surface into the northern and southern hemispheres, and that is the line from which latitudes are reckoned, its own latitude being everywhere zero degrees. It is the largest of the parallels of latitude, having a length on the Earth's surface of 40,075.76 km (24,901.92 miles). Syn: *terrestrial equator.* (b) *celestial equator.*

equatorial [clim] Said of a climate characterized by uniformly high temperature and humidity and heavy rainfall, and occurring in lowland areas within five to ten degrees of the equator. Cf: *tropical.*

equatorial [paleont] Pertaining to or located in the median plane normal to the axis of coiling or symmetry in a foraminifer; e.g. "equatorial section" representing a slice through a foraminiferal test passing through the proloculus. See also: *sagittal.*

equatorial aperture A symmetrical aperture in a planispiral foraminiferal test. It is commonly an *interiomarginal aperture,* but it may be areal or peripheral.

equatorial bulge An expression used to describe the *ellipticity* or flattening of the earth. Not recommended usage.

equatorial countercurrent A narrow and often variable ocean-surface current near the equator, flowing eastward between the westward-flowing equatorial currents to the north and south. Cf: *equatorial undercurrent.*

equatorial current (a) Any of the broad ocean-surface currents in the tropical areas just north or just south of the equator, driven southwest or west in the Northern Hemisphere by northeast trade winds (North Equatorial Current), or northwest or west in the Southern Hemisphere by southeast trade winds (South Equatorial Current). (b) A tidal current occurring twice monthly when the Moon is near or over the Earth's equator.

equatorial limb A term sometimes applied to the *amb* of a pollen grain or spore. It is undesirable because of possible confusion. Syn: *limb [palyn].*

equatorial projection One of a group of projections that have their center points on the equator and their polar axes vertical; e.g. an "equatorial cylindrical conformal map projection" (also known as the Mercator projection).

equatorial space A four-sided region resulting from formation of the *basal leaf cross* in an acantharian radiolarian.

equatorial spine A radial spine arising on the shell equator in an acantharian radiolarian.

equatorial tide A tide occurring twice monthly when the Moon is near or over the Earth's equator, and displaying the least diurnal inequality. Cf: *tropic tide.*

equatorial undercurrent A narrow *undercurrent* in the ocean, flowing from west to east beneath or sometimes embedded in the westward-flowing equatorial currents. Cf: *equatorial countercurrent.* Syn: *Cromwell current.*

equatorial view The view of a spore or pollen grain from an aspect more or less midway between the poles.

equator system of coordinates A system of curvilinear celestial coordinates (usually declination and right ascension) based on the celestial equator as the primary great circle. Cf: *horizon system of coordinates.*

equiareal projection A term used in structural petrology for an *equal-area projection* developed from the center of a sphere through points on its surface to a plane that is tangent at the south pole of the sphere and so constructed that areas between meridians and parallels on the plane are equal to corresponding areas on the surface of the sphere.

equidimensional *equant.*

equidistant projection A map projection in which distances are represented true to scale and without length distortion in all directions from a given point or along a given meridian or parallel.

equiform Said of crystals that have the same (or nearly the same) shape.

equiglacial line A line drawn on a map or chart to show coincidence of ice conditions, as in a lake or river, at a given time. See also: *isopag; isopectic; isotac.*

equigranular (a) Nonrecommended syn. of *homogranular.* The term was originally applied by Cross et al. (1906, p. 698) to igneous rocks. (b) Nonrecommended syn. of *granuloblastic.*

equilateral Bilaterally symmetrical; specif. said of a bivalve-mollusk shell whose parts anterior and posterior to the beaks are subequal or equal in length and nearly symmetrical. Cf: *equivalve.* Ant: *inequilateral.*

equilibrium (a) In geology, a balance between form and process, e.g. between the resistance of rocks along a coast and the erosional force of the waves. (b) That state of a chemical system in which the phases do not undergo any change of properties with the passage of time, provided they have the same properties when the same conditions are again reached by a different procedure.

equilibrium constant A number representing equilibrium of a chemical reaction, defined as the result of multiplying the activities of the equation, each raised to a power indicated by its coefficient in the equation, and dividing by a similar product of the activities of the reactants. It may be referred to as the *thermodynamic equilibrium constant;* when concentrations instead of activities are used, it may be called the *classical equilibrium constant.* See also: *van't Hoff equation.*

equilibrium crystallization Crystallization in which crystals formed on cooling continually react and re-equilibrate with the liquid. Cf: *fractional crystallization; crystallization differentiation.*

equilibrium diagram *phase diagram.*

equilibrium fusion Fusion in which the liquid produced on heating continually reacts and re-equilibrates with the crystalline residue (Presnall, 1969). Cf: *fractional fusion.*

equilibrium limit *equilibrium line.*

equilibrium line The level on a glacier where the net balance equals zero, and accumulation equals ablation; the line separating the superimposed ice zone of the accumulation area (above) from the ablation area (below). For some temperate mountain glaciers, it is very nearly coincident with the *firn line,* in which case it is common practice to use the latter term; but on subpolar glaciers, the equilibrium line is lower than the firn line because freezing of meltwater occurs below the firn line forming superimposed ice. Cf: *climatic snowline; snowline.* Syn: *equilibrium limit.*

equilibrium moisture content The moisture content of a soil mass at a time when there is no moisture movement (Nelson & Nelson, 1967, p. 127).

equilibrium path On a phase diagram, the crystallization sequence in which all crystals react continuously and completely with the liquid, so that adjustment of crystal composition throughout the crystallization interval is perfect.

equilibrium profile *profile of equilibrium.*

equilibrium shoreline A shoreline that has a local vertical profile of equilibrium and also an equilibrium shape in plan view; a *graded shoreline.*

equilibrium species A species that tends to be specialized in its adaptations, disperses and reproduces at moderate rates, but achieves large population size (close to the carrying capacity "K" in the logistic population-growth equation) because of the stable, predictable, or permanent nature of the environment occupied. Syn: *specialist species, K strategist.* Ant: *opportunistic species.*

equilibrium stage In hypsometric analysis of drainage basins, the stage in which a steady state is developed and maintained as relief slowly diminishes, and corresponding to maturity and old age in the geomorphic cycle (Strahler, 1952b, p.1130); the hypsometric integral is stable between 35% and 60%. Cf: *inequilibrium stage; monadnock phase.*

equilibrium theory A tidal hypothesis that assumes an ideal nonrotating Earth with no continental barriers and with a uniform deep ocean cover; the *equilibrium tide* would respond instantly to the gravitational forces of the Sun and the Moon, to form a surface in equilibrium moving around the Earth without friction.

equilibrium tide The hypothetical semidaily tide as described by the *equilibrium theory* for a frictionless, nonrotating water-covered Earth. Syn: *gravitational tide; astronomic tide.*

equinoctial circle *celestial equator.*

equinoctial tide A tide occurring when the Sun is near equinox, characterized by greater-than-average ranges of the spring tide.

equiplanation Those processes that operate at high latitudes and tend toward reduction of the land without reference to a base-level control and without involving any loss or gain of material

(Cairnes, 1912, p. 76). Cf: *altiplanation.*

equiplanation terrace A syn. of *altiplanation terrace* (G.T. Warwick in Stamp, 1961, p. 21).

equipotential line A contour line on the potentiometric surface; a line along which the pressure head of ground water in an aquifer is the same. Fluid flow is normal to these lines in the direction of decreasing fluid potential. Syn: *isopiestic line; isopotential line; piezometric contour.*

equipotential-line method An early electrical-survey method wherein lines of equal potential, near the current electrodes, were searched for and mapped using a pair of potential electrodes, one of which was held stationary for each line mapping.

equipotential surface A surface on which the potential is everywhere constant for the attractive forces concerned. The gravity vector is everywhere normal to a gravity equipotential surface; the geoid is an "equipotential". Syn: *gravity equipotential surface; niveau surface; level surface.*

equirectangular projection *plate carrée projection.*

equitability A partial syn. of *diversity.*

equitemperature metamorphism A process of modification of ice crystals in deposited snow, characterized by molecular transfer, mainly by vapor diffusion, from regions of high surface energy to regions of low surface energy at subfreezing temperatures. It tends to produce rounded grains of relatively uniform size, with ice bonds between grains. The resulting snow is cohesive. Cf: *temperature-gradient metamorphism.* Archaic syn: *destructive metamorphism.*

equivalence [cart] The unique property of an equal-area map projection in which the ratio between areas on the map is the same as that between the corresponding areas on the surface of the Earth. Cf: *conformality.* Syn: *orthembadism.*

equivalence [stratig] Geologic contemporaneity, esp. as indicated by identical fossil content. Syn: *equivalency.*

equivalent adj. Corresponding in geologic age or stratigraphic position; esp. said of strata or formations (in regions far from each other) that are contemporaneous in time of formation or deposition or that contain the same fossil forms.—n. A stratum that is contemporaneous or equivalent in time or character.

equivalent diameter Twice the *equivalent radius.*

equivalent grade A term used by Baker (1920, p. 367), and synonymous with *arithmetic mean diameter.*

equivalent projection *equal-area projection.*

equivalent radius A measure of particle size, equal to the computed radius of a hypothetical sphere of specific gravity 2.65 (quartz) having the same settling velocity and same density as those calculated for a given sedimentary particle in the same fluid; one half of the *equivalent diameter.* Cf: *nominal diameter; sedimentation diameter.*

equivalve Having valves equal in size and form; specif. said of a bivalve mollusk or its shell in which the right valve and the left valve are subequal or equal and symmetrical about the plane of commissure. Cf: *equilateral.* Ant: *inequivalve.*

equivoluminal wave *S wave.*

era A geologic-time unit next in order of magnitude below an *eon,* during which the rocks of the corresponding *erathem* were formed; e.g. the Paleozoic Era, the Mesozoic Era, and the Cenozoic Era. Each of these includes two or more *periods,* during each of which a *system* of rocks was formed. Long-recognized Precambrian eras are the Archeozoic (older) and Proterozoic (younger).

erathem The largest formal chronostratigraphic unit generally recognized, next in rank above *system;* the rocks formed during an *era* of geologic time, such as the Mesozoic Erathem composed of the Triassic System, the Jurassic System, and the Cretaceous System. See also: *supersystem.* Obsolete syn: *group; sequence.*

Eratosthenian (a) Pertaining to lunar topographic features and lithologic map units constituting a system of rocks formed during the period of formation of large craters (such as Eratosthenes) whose rays are no longer visible. Eratosthenian rocks are older than those of the Copernican System but younger than those of the Imbrian System. (b) Said of the stratigraphic period during which the Eratosthenian System was developed.

E ray [cryst] In a uniaxial crystal, the ray of light that vibrates in a plane containing the optic axis and at an angle with the basal pinacoid, and whose velocity or refraction approaches that of the O ray as the angle approaches zero; the *extraordinary ray.* Cf: *O ray.*

E ray [paleont] Left anterior ray in echinoderms situated clockwise of the D ray when the echinoderm is viewed from the adoral side; equal to ambulacrum II in the Lovenian system.

eremacausis The gradual transformation by oxidation of plant material into humus from exposure to air and moisture.

eremeyevite A syn. of *jeremejevite.* Also spelled: *eremeevite.*

eremic Pertaining to a *desert* or deserts, or to sandy regions.

eremology The scientific study of deserts and their phenomena.

eremophyte *xerophyte.*

erg A region in the Sahara, deeply covered with shifting sand and occupied by complex sand dunes; an extensive tract of *sandy desert;* a *sand sea.* Etymol: Hamitic. Pl: *areg; ergs.* See also: *koum; nefud.* Also spelled: *ergh.*

ergeron A French term for a very fine argillaceous sand, or variety of loess, containing a substantial amount of calcium carbonate, commonly occurring in northern France and the Belgian province of Hainaut.

ergh *erg.*

Erian North American provincial series: Middle Devonian (above Ulsterian, below Senecan).

Erian orogeny One of the 30 or more short-lived orogenies during Phanerozoic time identified by Stille, in this case at the end of the Silurian; the last part of the Caledonian orogenic era. Syn: *Hibernian orogeny.*

ericaite A mineral: $(Fe,Mg,Mn)_3B_7O_{13}Cl$.

ericophyte A plant growing on a heath or moor.

ericssonite A mineral: $BaMn_3Fe(Si_2O_7)(OH)$. It is dimorphous with orthoericssonite.

erikite A yellow-green, greenish-yellow, or brown mineral: $(La,Ce)_x(P,Si)O_4 \cdot H_2O$. It is perhaps a silicate-rich rhabdophane. The type erikite from Greenland has been shown to be monazite.

eriochalcite A bluish-green to greenish-blue mineral: $CuCl_2 \cdot 2H_2O$. Syn: *antofagastite.*

erionite A zeolite mineral: $(Ca,Na_2,K_2)_4(Al_8Si_{28})O_{72} \cdot 27H_2O$.

eriophorum peat Peat formed mainly from *Eriophorum,* also known as cotton grass, a genus of sedges of the family Cyperaceae. Cf: *carex peat.*

erlichmanite A mineral of the pyrite group: OsS_2.

ernstite A mineral: $(Mn^{+2}_{1-x}Fe^{+3}_x)Al(PO_4)(OH)_{2-x}O_x$. It is an oxidation product of eosphorite.

erode (a) To wear away the land, as by the action of streams, waves, wind, or glaciers. (b) To produce or modify a landform by the wearing-away of the land.

erodibility (a) The quality, degree, or capability of being eroded or yielding more or less readily to erosion. (b) The tendency of soil to be detached and carried away; the rate of soil erosion.—Cf: *erosiveness.* Adj: *erodible.* Syn: *erodability; erosibility.*

eroding channel A channel in which the energy of a stream is greater than that required to move the sediment available for transport.

eroding stress Shear stress of overland flow available to dislodge or tear loose soil material per unit area. Originally defined by Horton (1945, p.319) as the "eroding force" exerted parallel with the soil surface per unit of slope length and width. Symbol: F_1.

erosibility *erodibility.*

erosion (a) The general process or the group of processes whereby the materials of the Earth's crust are loosened, dissolved, or worn away, and simultaneously moved from one place to another, by natural agencies, which include weathering, solution, corrasion, and transportation, but usually exclude mass wasting; specif. the mechanical destruction of the land and the removal of material (such as soil) by running water (including rainfall), waves and currents, moving ice, or wind. The term is sometimes restricted by excluding transportation (in which case "denudation" is the more general term) or weathering (thus making erosion a dynamic or active process only). Cf: *denudation.* (b) An instance or product, or the combined effects, of erosion.

erosional Pertaining to or produced by the wearing-away of the land.

erosional flood plain A *flood plain* produced by the lateral erosion and gradual retreat of the valley walls.

erosional unconformity An unconformity made manifest by erosion, or a surface that separates older rocks that have been subjected to erosion from younger sediments that cover them; specif. *disconformity.*

erosional vacuity A term formerly used by Wheeler (1958, p.1057) and now replaced by *degradation vacuity.*

erosion caldera A type of *caldera* that is developed by the erosion and resultant widening of a caldera or by erosion of a volcanic cone, resulting in a large, central cirquelike depression. It is considered by some not to be a true caldera type, since it is not formed by volcanic processes. Cf: *explosion caldera; collapse caldera.*

erosion crater *makhtésh.*

erosion cycle *cycle of erosion.*

erosion fault scarp *fault-line scarp.*

erosion groove A sedimentary structure formed by "closely spaced lines of straight-sided scour marks" (Dzulynski & Sanders, 1962, p.66). The scouring may be initially concentrated by a pre-existing groove.

erosion integral An expression of the relative volume of a landmass removed by erosion at a given contour; the inverse of the *hypsometric integral.*

erosion intensity Quantity (or depth) of solid material actually removed from the soil surface by sheet erosion per unit of time and area; originally termed "erosion rate" by Horton (1945, p.324). Symbol: E_a.

erosionist A believer in the obsolete theory that the irregularities of the Earth's surface are mainly the result of erosion.

erosion lake A lake occupying a basin excavated by erosion.

erosion pavement A surficial concentration of pebbles, gravel, and other rock fragments that develops after sheet erosion or rill erosion has removed the finer soil particles and that tends to protect the underlying soil from further erosion.

erosion plain A general term for any plain produced by erosion, such as a peneplain, a pediplain, a panplain, or a plain of marine erosion.

erosion platform (a) A relatively level surface of limited extent formed by erosion. (b) A *wave-cut platform* along a coast. Cf: *abrasion platform.*

erosion proportionality factor Ratio of *erosion intensity* to *eroding stress.* It expresses the resistance of the ground surface to erosion by surface runoff by representing the quantity of solid material removed per unit of time and surface area. Symbol: k_3.

erosion ramp A sloping belt of reef rock immediately above the reef flat on an atoll islet, where marine erosion is active.

erosion remnant A topographic feature that remains or is left standing above the general land surface after erosion has reduced the surrounding area; e.g. a monadnock, a butte, or a stack. Syn: *residual [geomorph]; relic [geomorph]; remnant.*

erosion ridge One of a series of small ridges on a snow surface, formed by the corrasive action of windblown snow, either parallel or perpendicular to the direction of the wind (ADTIC, 1955, p.27). Cf: *sastrugi.*

erosion ripple A minor wavelike feature produced by the cutting action of the wind on a lower and somewhat more coherent layer of a sand dune.

erosion scarp A *scarp* produced by erosion; e.g. a *fault-line scarp* or a *beach scarp.*

erosion surface A land surface shaped and subdued by the action of erosion, esp. by running water. The term is generally applied to a level or nearly level surface; e.g. a *stripped structural surface.* Syn: *planation surface.*

erosion terrace A terrace produced by erosion; specif. a *rock terrace.*

erosion thrust A thrust fault on which the hanging wall moved across an erosion surface.

erosiveness (a) The quality, degree, or capability of accomplishing erosion; the power or tendency to effect erosion. (b) A term sometimes used as a syn. of *erodibility.* (c) The exposure of soil to erosion.—Adj: *erosive.*

erosive velocity That velocity of water in a channel above which erosion of the bed or banks will occur.

erpoglyph *worm casting.*

erratic n. A rock fragment carried by glacial ice, or by floating ice, deposited at some distance from the outcrop from which it was derived, and generally though not necessarily resting on bedrock of different lithology. Size ranges from a pebble to a house-size block. See also: *perched block; exotic.* Syn: *erratic block; erratic boulder; glacial erratic.*—adj. Transported by a glacier from its place of origin. Syn: *traveled.*

erratic block *erratic.*

erratic boulder *erratic.*

error (a) The difference between an observed, calculated, or measured value of a quantity and the ideal or true value of that quantity or of some conventional or standard value determined by established procedure or authority and used in lieu of the true value. (b) An inaccuracy or variation in the measurement, calculation, or observation of a quantity due to mistakes, imperfections in equipment or techniques, human limitations, changes of surrounding conditions, or other uncontrollable factors. (c) The amount of deviation of a measurement from some standard, arbitrary, estimated, or other reference value.—See also: *random error; systematic error; personal error.*

errorchron An isochron about which data are scattered, not only because of analytical error, but also because of departures of the geologic system investigated from an ideal model.

error of closure The amount by which a quantity obtained by a series of related measurements differs from the true or theoretical value or a fixed value obtained from previous determinations; esp. the amount by which the final value of a series of survey observations made around a closed loop differs from the initial value. The surveyed quantities may be angles, azimuths, elevations, or traverse-station positions. See also: *closure; circuit closure; triangle closure; horizon closure; mis-tie.* Syn: *misclosure; closing error.*

error of collimation *collimation error.*

ERTS Earth Resource Technology Satellite. See: *Landsat.*

erubescite *bornite.*

eruption The ejection of volcanic materials (lava, pyroclasts, and volcanic gases) onto the Earth's surface, either from a central vent or from a fissure or group of fissures. Cf: *central eruption; fissure eruption.*

eruption breccia *explosion breccia.*

eruption cloud A cloud of volcanic gases, with ash and other pyroclastic fragments, that forms by volcanic explosion; the ash may fall from it by *air-fall deposition.* Syn: *explosion cloud; ash cloud; dust cloud; volcanic cloud.* See also: *eruption column.*

eruption column The initial form that an *eruption cloud* takes at the time of explosion; the lower portion of such a cloud. It results from initial ejection velocity rather than buoyant rise of the hot gases.

eruption cycle The sequence of events that occurs during a volcanic eruption; the regular change in the behavior of the eruptions in a period of activity.

eruption rain A rain following a volcanic eruption that results from condensation of the eruption's associated steam. Syn: *volcanic rain.*

eruption rate *age-specific eruption rate.*

eruption-time In stochastic treatment of volcanologic data based on renewal theory, the age at which a repose period is ended by an outbreak (Wickman, 1966, p.293).

eruptive Said of a rock formed by the solidification of magma; i.e. either an *extrusive* or an *igneous* rock. Most writers restrict the term to its extrusive or volcanic sense.

eruptive evolution Evolution characterized by the relatively sudden appearance of varied new stocks from a common ancestral line.

erythraean Pertaining to the ancient sea that occupied the Arabian Sea, the Red Sea, and the Persian Gulf areas. Also spelled: *erythrean.*

erythrean *erythraean.*

erythrine *erythrite.*

erythrite A red or pink mineral: $Co_3(AsO_4)_2 \cdot 8H_2O$. It is isomorphous with annabergite, and may contain some nickel. Erythrite occurs in monoclinic crystals, in globular and reniform masses, or in earthy forms, as a weathering product of cobalt ores in the upper (oxidized) parts of veins. Syn: *erythrine; cobalt bloom; red cobalt; cobalt ocher; peachblossom ore.*

erythrosiderite A mineral: $K_2FeCl_5 \cdot H_2O$. It may contain some aluminum.

erythrozincite A variety of wurtzite containing manganese.

erzbergite Calcite and aragonite in alternate layers.

Erzgebirgian orogeny One of the 30 or more short-lived orogenies during Phanerozoic time identified by Stille, in this case in the early Upper Carboniferous (Namurian or Westphalian).

esboite An orbicular *diorite* in which andesine or oligoclase is the dominant plagioclase and forms the orbicules. Its name is derived from Esbo, Finland. Cf: *corsite.* Not recommended usage.

esboitic crystallization The process by which the orbicules of an esboite attain oligoclasitic composition (Eskola, 1938, p. 449).

escar *esker.*

escarpment (a) A long, more or less continuous cliff or relatively steep slope facing in one general direction, breaking the continuity of the land by separating two level or gently sloping surfaces, and produced by erosion or by faulting. The term is often used synonymously with *scarp,* although *escarpment* is more often applied to a cliff formed by differential erosion. (b) A steep, abrupt face of rock, often presented by the highest strata in a line of cliffs, and generally marking the outcrop of a resistant layer occurring in a series of gently dipping softer strata; specif. the steep face of a cuesta. See also: *scarp slope.* (c) A term used loosely in Great Britain as a syn. of *cuesta.*—Etymol: French *escarpment,* "steep face or slope".

eschar *esker.*

escharan *eschariform.*

eschariform Said of a rigid, bilamellar, foliaceous, erect colony; in cheilostome bryozoans, attached firmly by a calcareous base or loosely by radicles. Syn: *escharan.*

eschwegeite A syn. of *tanteuxenite.*

eschynite *aeschynite.*

escutcheon (a) The flat or simply curved, typically lozenge-shaped, dorsal differentiated area extending posteriorly from the beaks of certain bivalve mollusks and sometimes bordered by a ridge in each valve. It corresponds to the posterior part of the cardinal area and is separated from the remainder of the shell surface by a sharp change in angle. (b) An improper term for the *basal cavity* of a conodont.

eskar *esker.*

eskebornite A mineral: $CuFeSe_2$. It is isomorphous with chalcopyrite.

esker A long, narrow, sinuous, steep-sided ridge composed of irregularly stratified sand and gravel that was deposited by a subglacial or englacial stream flowing between ice walls or in an ice tunnel of a stagnant or retreating glacier, and was left behind when the ice melted. It may be branching and is often discontinuous, and its course is usually at a high angle to the edge of the glacier. Eskers range in length from less than 100 m to more than 500 km (if gaps are included), and in height from 3 to more than 200 m. Etymol: Irish *eiscir,* "ridge". Cf: *ice-channel filling; kame.* Syn: *ås; os [glac geol]; serpent kame; Indian ridge; morriner.* Also spelled: *eskar; eschar; escar.*

esker delta A flat-topped deposit of sand and gravel formed by, and at the mouth of, a glacial stream as it issued from an ice tunnel and flowed into a lake or sea; associated with an esker or ice-channel filling made at the same time. Cf: *kame delta; delta kame.* Syn: *sand plain [glac geol]; sand plateau.*

esker fan A small plain of gravel and sand built at the mouth of a subglacial stream, and associated with an esker formed at the same time.

eskerine Characteristic of an esker; e.g. *eskerine* topography.

esker lake A lake enclosed or dammed by an esker (Dryer, 1901, p. 129).

esker trough A term applied in Michigan to a shallow valley, cut in till, that contains an esker (Leverett, 1903, p. 118).

eskolaite A rhombohedral mineral: Cr_2O_3. It is isomorphous with hematite.

esmeraldite A coarse- to medium-grained rock having hypidiomorphic-granular texture, in which quartz and muscovite are the essential phases. Its name is derived from Esmeralda County, Nevada. Syn: *northfieldite; nordfieldite.* Not recommended usage.

espalier drainage pattern *trellis drainage pattern.*

esperite A mineral: $(Ca,Pb)ZnSiO_4$. Syn: *calcium-larsenite.*

espichellite A lamprophyric rock, similar to *camptonite,* in which hornblende, augite, olivine, magnetite, and pyrite phenocrysts are contained in a compact groundmass of magnetite, hornblende, augite, mica, and labradorite with orthoclase rims. Analcime may also occur in the groundmass. Espichellite resembles *teschenite* but has less analcime and is porphyritic rather than granular. It was named by Souza-Brandão in 1907 from Cape Espichel, Portugal. Not recommended usage.

esplanade (a) A term used in the SW U.S. for a rather broad bench or terrace bordering a canyon, esp. in a plateau region. (b) A level stretch of open and grassy or paved ground, often designed for providing a vista.

ESR *electron spin resonance.*

essential Said of pyroclastics that are formed from magma; it is part of a classification of ejecta based on mode of origin, and is equivalent to the terms *juvenile [volc]* and *authigenous.* Cf: *accessory; accidental; cognate.*

essential element An element whose presence is necessary in order for an organism to carry out its life processes.

essential mineral A mineral component of a rock that is necessary to its classification and nomenclature, but that is not necessarily present in large amounts. Cf: *accessory mineral.* Syn: *specific mineral.*

essexibasalt As defined by Lehmann in 1924, a nepheline *basanite* containing very calcic plagioclase (bytownite). Used by French petrologists as a synonym of *alkaline basalt.* Not recommended usage.

essexite (a) In the *IUGS classification,* a plutonic rock with F between 10 and 60, and P/(A+P) between 50 and 90. It is synonymous with both *foid monzodiorite* and *foid monzogabbro.* (b) An alkali gabbro primarily composed of plagioclase, hornblende, biotite, and titanaugite, with lesser amounts of alkali feldspar and nepheline. Essexite grades into theralite with a decrease in potassium feldspar and an increase in the feldspathoid minerals. Its name is derived from Essex County, Massachusetts, from where it was originally defined by Sears in 1891 (Johannsen, 1939, p. 250). Cf: *glenmuirite.*

essonite A yellow-brown or reddish-brown transparent gem variety of grossular garnet containing iron. Syn: *hessonite; cinnamon stone; hyacinth; jacinth.*

established dune *anchored dune.*

establishment *ecesis.*

estavelle A cave that is a spring during some periods and a sinking stream during others.

esterellite A porphyritic *quartz diorite* also containing zoned andesine and hornblende. It was named by Michel-Lévy in 1897 from Esterel, France. Not recommended usage.

Estérel twin law A twin law for parallel twins in feldspar with twin axis *a* and composition plane $(0kl)$ parallel to *a*.

estero (a) An estuary or inlet, esp. when marshy. (b) Land adjoining an estuary inundated by the tide.—Etymol: Spanish.

estuarine Pertaining to or formed or living in an estuary; esp. said of deposits and of the sedimentary or biological environment of an estuary. Syn: *estuarial.*

estuarine delta A long, narrow delta that has filled, or is in the process of filling, an estuary.

estuarine deposit A sedimentary deposit laid down in the brackish water of an estuary, characterized by fine-grained sediments (chiefly clay and silt) of marine and fluvial origin mixed with a high proportion of decomposed terrestrial organic matter; it is finer-grained and of more uniform composition than a *deltaic deposit.*

estuarine lagoon A lagoon produced by the temporary sealing of a river estuary by a storm barrier. Such lagoons are usually seasonal and exist until the river breaches the barrier; they occur in regions of low or spasmodic rainfall. Syn: *blind estuary.*

estuarine salinity Salinity that varies according to tidal or seasonal conditions, as in an estuary.

estuary (a) The seaward end or the widened funnel-shaped tidal mouth of a river valley where fresh water comes into contact with seawater and where tidal effects are evident; e.g. a *tidal river,* or a partially enclosed coastal body of water where the tide meets the current of a stream. Cf: *freshwater estuary; inverse estuary; positive estuary.* (b) A portion of an ocean, as a *firth* or an arm of the sea, affected by fresh water; e.g. the Baltic Sea. (c) A *drowned river mouth* formed by the subsidence of land near the coast or by the drowning of the lower portion of a nonglaciated valley due to the rise of sea level.—See also: *ria; branching bay; liman; fjord.* Etymol: Latin *aestus,* "tide".

étang A French term for a shallow pool, pond, or lake, esp. one lying among sand dunes, formed by the ponding of inland drainage by beach material thrown up by the sea, and gradually becoming filled with silt, like those along the Mediterranean coast of France (e.g. in Languedoc).

etched pothole *solution pan.*

etch figure A marking, usually in the form of minute pits, produced by a solvent on a crystal surface; the form varies with the mineral species and the solvent, but conforms to the symmetry of the crystal.

etching (a) The reduction of the Earth's surface by the slow processes of differential weathering, mass wasting (esp. creep), sheetwash, and deflation, so that areas underlain by more resistant rocks are brought into relief as the less resistant rocks are lowered (Rich, 1951). (b) A general term for the formation of a landform by erosion or chiseling, as the *etching* of a canyon by a stream.

etchplain A relatively inextensive erosion surface, believed to develop by the comparatively rapid but local differential lowering, during uplift, of a peneplain surface kept at or near base level by the removal of a deep overlying cover of weathered rock. The feature was originally described as an "etched plain" by Wayland (1934).

ethane A colorless, odorless, water-insoluble, gaseous paraffin hydrocarbon, formula C_2H_6, which occurs in natural gas or can be produced as a by-product in the cracking of petroleum.

ethmolith A discordant pluton that is funnel-like in cross section.

ethology The science concerned with animal behavior, some effects of which are preserved in the fossil record, esp. as *trace fossils.*

etindite A dark-colored extrusive rock intermediate in composition between *leucitite* and *nephelinite,* with phenocrysts of clinopyroxene in a dense groundmass of leucite, nepheline, and clinopyroxene. Named by Lacroix in 1923 for Etinde, Cameroon. Not recommended usage.

etnaite An alkali *olivine basalt* (Streckeisen, 1967, p. 185). Not recommended usage.

Etroeungtian European provincial stage: uppermost Devonian. See also: *Strunian.*

ettringite A mineral: $Ca_6Al_2(SO_4)_3(OH)_{12} \cdot 26H_2O$. Syn: *woodfordite.*

eu- A prefix meaning "well" or "well developed".

euaster A sponge spicule (microsclere) having the form of a modified aster in which the rays arise from a common center. Cf: *streptaster.*

euautochthony Accumulation of plant remains (such as roots, stumps, tree trunks) that are now found in the exact place, and more or less in the correct relative positions, in which they grew. Cf: *hypautochthony.*

eu-bitumen A collective name for those fluid, viscid, or solid bitumens that are easily soluble in organic solvents. Examples are petroleum, ozokerite, elaterite, and asphalt (Tomkeieff, 1954).

eucairite A silver-white to lead-gray isometric mineral: CuAgSe. Also spelled: *eukairite.*

euchlorin An emerald-green mineral: $(K,Na)_8Cu_9(SO_4)_{10}(OH)_6$. Also spelled: *euchlorine; euchlorite.*

euchroite An emerald-green or leek-green mineral: $Cu_2(AsO_4)(OH) \cdot 3H_2O$.

euclase A brittle monoclinic mineral: $BeAlSiO_4(OH)$. It occurs in pale tones of blue, green, yellow, or violet, and is sometimes colorless; the blue variety is greatly esteemed by gem collectors.

eucolite A variety of eudialyte that is optically negative. Also spelled: *eukolite.*

eucrite [ign] A very basic *gabbro* composed chiefly of calcic plagioclase (bytownite, anorthite) and clinopyroxene, with accessory olivine. Obsolescent. The name, given by Rose in 1864, is from Greek, "easily discerned".

eucrite [meteorite] An achondritic stony meteorite composed essentially of calcic plagioclase and pigeonite. It has a higher content of iron and calcium than that of howardite. Eucrites were originally regarded as anorthite-augite meteorites. Syn: *eukrite.*

eucryptite A colorless or white hexagonal mineral: $LiAlSiO_4$.

eucrystalline *macrocrystalline.*

eudiagnostic Said of the texture of a rock (esp. an igneous rock) in which all mineral components are of such size and shape as to be identifiable. The term includes both macrocrystalline and microcrystalline textures. It was originally used by Zirkel in German as *eudiagnostisch.* Ant: *adiagnostic.*

eudialyte A pale-pink to brownish-red mineral: $Na_4(Ca,Fe^{+2})_2ZrSi_6O_{17}(OH,Cl)_2$. It is optically positive. Cf: *eucolite.* Syn: *barsanovite.*

eudidymite A white, glassy, monoclinic mineral: $NaBeSi_3O_7(OH)$. It is dimorphous with epididymite.

eudiometer An instrument such as a graduated glass tube for measuring the amounts of different gases in a gas mixture by exploding the gases one at a time by passing an electric spark through the mixture.

eugenesis The period of development and death of the organic material found in coal-ball concretions (McCullough, 1977, p. 133). It is followed in order by syngenesis, diagenesis, and epigenesis.

eugeocline A term "tentatively used in the Great Basin to describe lower Paleozoic siliceous assemblage rocks that may be continental rise deposits" (Stewart & Poole, 1974, p. 29). Cf: *miogeocline.*

eugeogenous Easily weathered; said of a rock that produces by weathering a large amount of detritus. Ant: *dysgeogenous.*

eugeosyncline A geosyncline in which volcanism is associated with clastic sedimentation; the volcanic part of an *orthogeosyncline,* located away from the craton (Stille, 1940). Cf: *miogeosyncline.* Syn: *pliomagmatic zone.* See also: *ensimatic geosyncline.*

euglenoid One of a group of unicellular flagellates with a gullet and with the cell usually bounded by a fairly firm but flexible membrane instead of a cell wall.

eugranitic *granular.*

euhedral (a) Said of a mineral grain that is completely bounded by its own *rational faces,* and whose growth during crystallization or recrystallization was not restrained or interfered with by adjacent grains. (b) Said of the shape of such a crystal.—The term was proposed, originally in reference to igneous-rock components, by Cross et al. (1906, p. 698) in preference to the synonymous terms *idiomorphic* and *automorphic* (as they were originally defined). Cf: *anhedral; subhedral.*

euhedron Geometrical term for a solid figure completely bounded by plane surfaces. In petrology, it applies to those grains completely bounded by natural crystal faces. Pl: *euhedrons; euhedra.* Ant: *anhedron.* Syn: *idiomorph.*

eukairite *eucairite.*

eukaryote One of a major group of organisms, characterized by a complex protoplasmic organization, a vesicular nucleus, and various sorts of membrane-bounded cytoplasmic organelles. Cf: *prokaryote.*

eukolite A syn. of *eucolite.* Also spelled: *eukolyte.*

eukrite *eucrite [meteorite].*

euktolite *venanzite.*

eulerhabd A sinuous stout U-shaped oxea (sponge spicule); a more sharply curved variety of *ophirhabd.*

Eulerian (a) Pertaining to a system of coordinates or equations of motion in which the properties of a fluid are assigned to various points in space at each given time, without attempt to identify individual fluid parcels from one time to the next; e.g. a sequence of synoptic charts is an Eulerian representation. (b) Said of a direct method of measuring the speed and/or direction of an ocean current that flows past a geographically fixed point (as an anchored ship) where a current meter is stationed.—Cf: *Lagrangian.* Named in honor of Leonhard Euler (1707-1783), Swiss mathematician.

Euler number Inertial force divided by pressure-gradient force.

Euler's theorem A statement that any displacement of a spherical surface over itself leaves one point fixed. It is much used in *plate tectonics* because any displacement of a rigid body such as a lithospheric plate on the Earth's surface may be considered as a rotation about a properly chosen axis through a point on the sphere.

eulysite A *peridotite* containing manganese-rich fayalite, clinopyroxene, garnet, and magnetite and having a granular texture. Not recommended usage.

eulytite A mineral: $Bi_4Si_3O_{12}$. Syn: *eulytine; bismuth blende; agricolite.*

eumalacostracan A shrimplike crustacean belonging to the subclass Eumalacostraca, differing mainly from other malacostracans by the nonbivalve nature of the carapace and the presence of biramous thoracic appendages with a single joint in the protopod. Range, Middle Devonian to present.

eumorphism A property of an equal-area map projection in which undue distortion of shapes is not shown, as in the arithmetic mean of the sinusoidal and Mollweide projections (BNCG, 1966, p.17 & 47).

eumycete A plant of the subdivision Eumycetes, which comprises the true fungi. Cf: *myxomycete; schizomycete.*

euosmite A brownish-yellow resin, with a low oxygen content and a characteristic pleasant odor, found in brown coal.

eupelagic deposit Deep-sea sediment in which less than 25% of the fraction coarser than 5 microns is of terrigenous, volcanogenic, and/or neritic origin. Such deposits usually form far from the continents, beyond the continental margin and associated abyssal plain. They accumulate by vertical settling of particulate matter, are highly oxidized, and include pelagic clays and oozes.

eupholite A *euphotide* that contains talc. Not recommended usage.

euphotic zone That part of the ocean in which there is sufficient penetration of light to support photosynthesis. The depth varies, but averages about 80 m. Its lower boundary is the *compensation depth.* Cf: *disphotic zone; aphotic zone.* Syn: *photic zone.*

euphotide A *gabbro* in which the feldspar has been saussuritized. The term was originally applied by Haüy as a synonym of gabbro (Johannsen, 1939, p. 251). It is obsolete in the U.S.A. but is still used by French petrologists.

eupotamic Said of an aquatic organism adapted to living in both flowing and quiet fresh water. Cf: *autopotamic; tychopotamic.*

Eurasian-Melanesian belt The *great-circle belt* that extends from the Mediterranean across southern Asia to the Celebes, where it meets the *circum-Pacific belt.*

euretoid Said of the skeleton of a dictyonine hexactinellid sponge in which the dictyonal strands occur in more than one layer and are not parallel to the sponge surface. Cf: *farreoid.*

euripus A strait or narrow channel of water where the tide or a current flows and reflows with turbulent force. Pl: *euripi.* Etymol: Greek *euripos,* "strait", "channel".

eurite (a) A compact fine-grained porphyritic igneous rock that contains quartz phenocrysts. (b) Any fine-grained granitic rock. — Adj: *euritic.* Obsolescent.

euritic *microgranular.*

Euryapsida A subclass of reptiles characterized by a single upper temporal fenestra; it includes large marine forms such as plesiosaurs, nothosaurs, and placodonts, as well as smaller and less specialized terrestrial forms of uncertain relations. Range, Lower Permian to Cretaceous.

eurybathic Said of a marine organism that tolerates a wide range of depth. Cf: *stenobathic.*

eurybiontic *eurytropic.*

euryhaline Said of a marine organism that tolerates a wide range of salinities. Cf: *stenohaline.*

euryplastic Having great capacity for modification and adaptation to a wide range of environmental conditions; capable of major evolutionary differentiation. Cf: *stenoplastic.*

euryproct Said of a sponge in which the cloaca is conical, with the widest part forming the osculum.

eurypterid Any *merostome,* typically brackish or freshwater, belonging to the subclass Eurypterida, characterized by an elongate segmented lanceolate body that is only rarely trilobed, and a thin chitinous integument with tubercles or scalelike ornamentation. Range, Ordovician to Permian. Cf: *xiphosuran.*

eurypylous Said of a flagellated chamber of a sponge that has a very large apopyle.

eurysiphonate Said of nautiloids with relatively large siphuncles. Cf: *stenosiphonate.*

eurythermal Said of a marine organism that tolerates a wide range of temperatures. Cf: *stenothermal.*

eurytopic Said of an organism occurring in many different habitats. Cf: *eurytropic.*

eurytropic Said of an organism that can tolerate a wide range of a particular environmental factor. Cf: *eurytopic.* Syn: *eurybiontic.* Ant: *stenotropic.*

eusporangiate In a strict sense, said of isosporous ferns of subclass Eusporangiatae in which the sporangium wall is two or more cell layers thick. Also applied as a comparative term to similar sporangial walls in higher plants (Melchior and Werdermann, 1954, p. 287).

eustacy Nonrecommended var. of *eustasy.*

eustasy The worldwide sea-level regime and its fluctuations, caused by absolute changes in the quantity of seawater, e.g. by continental icecap fluctuations. See also: *glacio-eustatism; sedimento-eustatism; diastrophic eustatism.* Adj: *eustatic.* Nonrecommended spelling: *eustacy.* Syn: *eustatism.*

eustatic adj. Pertaining to worldwide changes of sea level that affect all the oceans. Eustatic changes may have various causes, but the changes dominant in the last few million years were caused by additions of water to, or removal of water from, the continental icecaps.—n. eustasy.

eustatic rejuvenation A renewal of the effectiveness of erosion processes resulting from causes that produce worldwide lowering of sea level.

eustatism *eustasy.*

eustratite A dense lamprophyric rock having rare phenocrysts of olivine, corroded hornblende, augite, and possibly oligoclase, in a groundmass composed of idiomorphic augite and magnetite with interstitial feldspar, mica, and colorless glass. The name, given by Ktenas in 1928, is for Haghios Eustratios Island in the Aegean Sea. Not recommended usage.

eusynchite *descloizite.*

eutaxic Said of a stratified mineral deposit. Cf: *ataxic.*

eutaxite A *taxite* whose components have aggregated into separate bands. Cf: *ataxite.*

eutaxitic Said of the *banded structure* of certain extrusive rocks, which results in a streaked or blotched appearance. Also, said of a rock exhibiting such structure, e.g. a *eutaxite.* The bands or lenses were originally ejected as individual portions of magma, were drawn out in a viscous state, and formed a heterogeneous mass in response to welding.

eutectic Said of a system consisting of two or more solid phases and a liquid whose composition can be expressed in terms of positive quantities of the solid phases, all coexisting at an (isobarically) invariant point, which is the minimum melting temperature for the assemblage of solids. Addition or removal of heat causes an increase or decrease, respectively, of the proportion of liquid to solid phases, but does not change the temperature of the system or the composition of any phases. See also: *eutectoid.*

eutectic point The lowest temperature at which a eutectic mixture will melt. Syn: *eutectic temperature.*

eutectic ratio The ratio of solid phases forming from the eutectic liquid at the eutectic point; it is such as to yield a gross composition for the crystal mixture that is identical with that of the liquid. It is most frequently stated in terms of weight percent.

eutectic temperature *eutectic point.*

eutectic texture A pattern of intergrowth of two or more minerals, formed as they coprecipitate during crystallization, e.g. the quartz and feldspar of *graphic granite.* See also: *exsolution texture.* Syn: *eutectoid texture.*

eutectofelsite *eutectophyre.*

eutectoid The equivalent of *eutectic,* when applied to a system all of whose participating phases are crystalline.

eutectoid texture *eutectic texture.*

eutectoperthite *mesoperthite.*

eutectophyre A light-colored tuffaceous igneous rock composed of interlocking quartz and orthoclase crystals. Syn: *eutectofelsite.* Obsolete.

Eutheria A subclass of mammals characterized by a placenta. It includes most living forms, which comprise about 24 orders. Range, Middle Cretaceous to Recent. See also: *placental; Marsupialia.*

eutrophic Said of a body of water characterized by a high level of plant nutrients, with correspondingly high primary productivity.

eutrophication The process by which waters become more eutrophic; esp. the artificial or natural enrichment of a lake by an influx of nutrients required for the growth of aquatic plants such as algae that are vital for fish and animal life.

eutrophic lake A lake that is characterized by an abundance of dissolved plant nutrients and by a seasonal deficiency of oxygen in the *hypolimnion;* its deposits usually have considerable amounts of rapidly decaying organic mud and its water is frequently shallow. Cf: *oligotrophic lake; mesotrophic lake.*

eutrophic peat Peat containing abundant plant nutrients, such as nitrogen, potassium, phosphorus, and calcium. Cf: *mesotrophic peat; oligotrophic peat.* Syn: *calcareous peat.*

eutrophy The quality or state of a *eutrophic lake.*

euvitrain Structureless, amorphous *vitrain.* It is the more common type. Cf: *provitrain.* See also: *collain; ulmain.* Syn: *xylovitrain.*

euvitrinite A variety of the maceral *vitrinite* characteristic of euvitrain and including the varieties ulminite and collinite. Plant material has been completely jellified and shows no cell structure. The term *collinite* has been proposed as a preferable synonym. Cf: *provitrinite.*

euxenite A brownish-black mineral: $(Y,Ca,Ce,U,Th)(Nb,Ta,Ti)_2O_6$. It is isomorphous with polycrase and occurs in granite pegmatites.

euxinic (a) Pertaining to an environment of restricted circulation and stagnant or anaerobic conditions, such as a fjord or a nearly isolated or silled basin with toxic bottom waters. Also, pertaining to the material (such as black organic sediments and hydrogen-sulfide muds) deposited in such an environment or basin, and to the process of deposition of such material (as in the Black Sea). (b) Pertaining to a rock facies that includes black shales and graphitic sediments of various kinds.—Etymol: Greek *euxenos,* "hospitable". Cf: *pontic.*

evaluation map A stratigraphic map that summarizes the results of stratigraphic analyses made for economic purposes (Krumbein & Sloss, 1963, p.484).

evanescent lake A short-lived lake formed after a heavy rain. Cf: *ephemeral lake.*

evansite A colorless or milky-white to brown or reddish-brown mineral: $Al_3(PO_4)(OH)_6 \cdot 6H_2O$ (?). It has a bluish, greenish, or yellowish tinge. Evansite may contain small amounts of uranium and thorium.

evapocryst An individual crystal of a primary mineral in an evaporite (Greensmith, 1957). Cf: *neocryst.*

evapocrystic texture A primary texture of an evaporite in which no lamination or linearity of evapocrysts is evident.
evapograph *atmometer.*
evapolensic texture A primary, nonporphyritic, roughly laminated texture of an evaporite.
evapoporphyrocrystic texture A texture of an evaporite in which large evapocrysts are embedded in a finer-grained matrix.
evaporates Goldschmidt's name for the evaporite group of sediments; sedimentary salts precipitated from aqueous solutions and concentrated by evaporation. The synonymous term *evaporite* is more commonly used. Cf: *reduzates; oxidates; resistates; hydrolyzates.*
evaporation The process, also called *vaporization,* by which a substance passes from the liquid or solid state to the vapor state. Limited by some to vaporization of a liquid, in contrast to *sublimation,* the direct vaporization of a solid. Also limited by some (e.g. hydrologists) to vaporization that takes place below the boiling point of the liquid. The opposite of *condensation* (Langbein & Iseri, 1960).
evaporation discharge The release of water from the zone of saturation by evaporation from the soil ("soil discharge") or by the transpiration of plants ("vegetal discharge").
evaporimeter *atmometer.*
evaporite A nonclastic sedimentary rock composed primarily of minerals produced from a saline solution as a result of extensive or total evaporation of the solvent. Examples include: gypsum, anhydrite, rock salt, primary dolomite, and various nitrates and borates. The term sometimes includes rocks developed by metamorphism or transport of other evaporites. Syn: *evaporate; saline deposit; saline residue.*
evaporite mineral A mineral precipitated as a result of evaporation; e.g. halite.
evaporite ratio A term used by Krumbein & Sloss (1963, p. 463) for the ratio of the thickness or percentage of evaporites (anhydrite, gypsum, salt) to that of carbonates (limestone, dolomite) in a stratigraphic section.
evaporite-solution breccia A term used by Sloss & Laird (1947, p.1422-1423) for a *solution breccia* formed where soluble evaporites (rock salt, anhydrite, gypsum, etc.) have been removed.
evapotranspiration Loss of water from a land area through transpiration of plants and evaporation from the soil. Also, the volume of water lost through evapotranspiration.
eveite A mineral: $Mn_2(AsO_4)(OH)$.
even-crested ridge One of the high fold ridges, as in the Appalachian Mountains of Pennsylvania, whose tops all rise to an approximately uniform elevation, indicating that a plain reconstructed by filling the valleys to the level of the ridgetops is an old peneplain. See also: *summit concordance; accordant summit level.*
even-grained *homogranular; granuloblastic.*
evening emerald Olivine (peridot or chrysolite) that loses some of its yellow tint in artificial light, appearing more greenish (like an emerald) and used as a gem. Syn: *night emerald.*
evenkite A hydrocarbon mineral: $C_{21}H_{44}$ (= n-tetracosane). It was formerly regarded as a paraffin wax: $C_{21}H_{42}$.
even-pinnate Said of a pinnately compound leaf having an even number of leaflets. Cf: *odd-pinnate.*
event [paleomag] *polarity event.*
event [seis] (a) An *earthquake.* (b) In seismic surveying, an *arrival,* denoted by a definite phase change or amplitude buildup on a seismic record.—Syn: *seismic event.*
event [tect] A noncommittal term used for any incident of probable tectonic significance that is suggested by geologic, isotopic or other evidence, but whose full implications are unknown. It is used especially for minor clusters of radiometric dates whose relations to geologic structures or processes have not been precisely evaluated. Cf: *pulsation; disturbance.*
everglade A term used esp. in the southern U.S. for a large expanse of marshy land, covered mostly with tall grass, e.g. the Florida Everglades.
evergreenite A quartz-bearing *syenite* that contains sulfide ore minerals, such as chalcopyrite and bornite. The name is from the Evergreen Mine, Colorado. Not recommended usage.
Evian water Noneffervescent alkaline mineral water. The term is derived from Évian-les-Bains, a town in southeastern France where it is found.
evisite A term proposed by Niggli in 1923, and never widely used, for a suite of alkalic *granites* and *syenites* at Evisa, Corsica. Obsolescent.
evolute Loosely coiled or tending to uncoil; e.g. said of a foraminiferal test with nonembracing chambers, a gastropod shell whose whorls are not in contact, or a cephalopod conch with little or no overlapping of the whorls. Cf: *involute; advolute; convolute.*
evolution (a) The development of a group of related organisms toward perfect or complete adaptation to the environmental conditions to which they have been exposed with the passage of time. (b) The theory that life on Earth has developed gradually, from one or a few simple organisms to more complex organisms. Syn: *organic evolution.* (c) The gradual permanent change in the form and function of organisms of successive ancestor-descendant generations or populations, over geologic time, so that the latest members of the succession differ significantly from the earliest.
evolutionary momentum The tendency of evolution to appear to continue along the same trend after the external stimulus provoking it has apparently diminished or disappeared. The term originated in the idea that organic function is activity, and activity is motion, which led to the notion that evolution may continue to inadaptive lengths and result in extinction.
evolutionary plexus A complex *lineage* of organisms that consists of minor lines that repeatedly divide and reunite. Syn: *plexus [evol].*
evolutionary series In paleontology, a morphologic series that corresponds with time to a significant degree. Cf: *lineage.*
evolutionary zone *lineage-zone.*
evorsion The formation of potholes in a stream bed by the erosional action of vortices and eddies. Etymol: Latin, *e,* "from", + *vortex,* "a whirlpool".
evorsion hollow A *pothole* in a stream bed.
ewaldite A mineral: $Ba(Ca,Y,Na,K)(CO_3)_2$.
Ewing corer The most commonly used variety of *piston corer.*
exaration The general process of glacial erosion. Grabau (1924, p. 263-264) suggested that the term be restricted to "glacial denudation, i.e. the removal and transport of weathered material" by glaciers.
excavation (a) The act or process of removing soil and/or rock materials from one location and transporting them to another. It includes digging, blasting, breaking, loading, and hauling, either at the surface or underground. (b) A pit, cavity, hole, or other uncovered cutting produced by excavation. (c) The material dug out in making a channel or cavity.
excentric Not centrally located; e.g. said of an ammonoid umbilicus characterized by an abrupt opening-up of the spiral described by the umbilical seam, or by a tendency toward closing of this spiral while the peripheral spiral is relatively unchanged (TIP, 1959, pt.L, p.4).
excess argon Argon-40 that is incorporated into rocks and minerals by processes other than in-situ radioactive decay of potassium-40. Cf: *atmospheric argon; radiogenic argon; inherited argon; extraneous argon.*
excess pore pressure Transient pore pressure at any point in an aquitard or aquiclude in excess of the pressure that would exist at that point if steady-flow conditions had been attained throughout the bed (Poland et al., 1972).
excess water *rainfall excess.*
exchange In glaciology, the arithmetic sum of accumulation plus ablation, averaged over a glacier for a balance year or a hydrologic year; a measure of the intensity of mass exchange with the atmosphere or hydrosphere. Usually given in millimeters or meters of water equivalent.
exchange capacity A quantitative measure of surface charge of a substance, reported in equivalents of exchangeable ions per unit weight of the solid. See also: *ion exchange.*
exchange coefficient *austausch.*
exchange component *exchange operator.*
exchange force A quantum-mechanical apparent interaction between electrons that is the cause of *magnetic order.*
exchange operator A *component* that changes, by addition, one compound or mineral into another or others, or that expresses ionic substitutions that can occur in a given mineral. Generally written as a formula containing negative coefficients, as $MgCa_{-1}$, and F_2O_{-1}, read "Mg, Ca minus one" and "F two, 0 minus one." Such operators can either be physically unrealizable (as in the above two examples) or may correspond to real substances, as $(OH)_2O_{-1}$ (equals water, H_2O). These components can be used to describe exchange processes, including diffusion, and reciprocal ternary chemical systems. Term originated by Burt (1974); concept originated earlier by J.B. Thompson, Jr. Syn: *exchange compo-*

nent (Brady, 1975).

excitation potential The characteristic minimum energy required to remove an electron from an atom.

excretion A term proposed by Todd (1903) for a *concretion* that grows progressively inward from the exterior; e.g. a shell of sand cemented by iron oxide and generally filled by unconsolidated sand or containing other shells of cemented sand.

excurrent (a) Said of a plant growth or structure that extends beyond the margin or tip, e.g. a midrib developing into a mucro or awn. (b) Said of the growth habit of plants having a continuous unbranched axis, e.g. the excurrent habit of spruces and firs (Lawrence, 1951, p.752).

excurrent canal *exhalant canal.*

exfoliation The process by which concentric scales, plates, or shells of rock, from less than a centimeter to several meters in thickness, are successively spalled or stripped from the bare surface of a large rock mass. It is caused by physical or chemical forces producing differential stresses within the rock, as by expansion of minerals as a result of near-surface chemical weathering, or by the release of confining pressure of a once deeply buried rock as it is brought nearer to the surface by erosion (*pressure-release jointing*). It often results in a rounded rock mass or dome-shaped hill. Cf: *spheroidal weathering; spheroidal parting.* Syn: *spalling; scaling; desquamation; sheeting; sheet jointing.*

exfoliation cave A cave formed by the partial destruction of a plate, sheet, or slab of rock produced by exfoliation, having a planar back wall and a continuation of the joint (of exfoliation) up into the roof of the cave (Bradley, 1963, p.525). Examples are found on the Colorado Plateau in SW U.S.

exfoliation dome A large dome-shaped form, developed in massive homogeneous coarse-grained rocks, esp. granite, by exfoliation; well-known examples occur in Yosemite Valley, Calif.

exhalant canal (a) Any canal forming part of the exhalant system of a sponge. Syn: *excurrent canal; apochete.* (b) A channel formed by the outer and the parietal lips in certain gastropods.

exhalant system The part of the *aquiferous system* of a sponge between the apopyles and the oscula, characterized by water flowing outward toward the oscula. Cf: *inhalant system.*

exhalation The streaming-forth of volcanic gases; also, the escape of gases from a magmatic fluid (Schieferdecker, 1959, terms 4462 & 4826). Syn: *emanation.*

exhalite A chemical sediment, usually containing oxide, carbonate, or sulfide as anions, and iron, manganese, base metals, and gold as cations, formed by the issuance of volcanically-derived fluids onto the sea floor or into the sea; thus, the product of *exhalation.*

exhumation The uncovering or exposure by erosion of a pre-existing surface, landscape, or feature that had been buried. See also: *resurrected.*

exhumed *resurrected.*

exilazooecium A polymorph in stenolaemate bryozoans, generally smaller than feeding zooecia in cross section, occurring in the exozone only, and containing few or no diaphragms so that an appreciable living chamber occurs (Boardman & Cheetham, 1973, p. 154). Cf: *mesopore.*

exine The outer, very resistant layer of the two major layers forming the wall (sporoderm) of spores and pollen, consisting of sporopollenin, and situated immediately outside the *intine.* It is divided into two layers (*ektexine* and *endexine*) on the basis of staining characteristics. See also: *perisporium.* Syn: *extine; exospore.*

exinite A coal maceral group including *sporinite, cutinite, alginite, resinite,* and *liptodetrinite,* derived from spores, cuticular matter, resins, and waxes. Exinite is relatively rich in hydrogen. It is a common component of attrital coal. Cf: *inertinite; vitrinite.* Syn: *liptinite.*

exinoid A maceral group that includes the macerals in the exinite series.

exinonigritite A type of *nigritite* that is derived from spore exines. Cf: *humonigritite; polynigritite; keronigritite.*

exinous Consisting of exine.

exite [coal] *liptite.*

exite [paleont] A movable lobe on the exterior side of the limb of an arthropod, such as a lateral ramus (e.g. an exopod) of the protopodal limb segments of a crustacean, or the lateral or outer lobe or joint of the biramous appendage of a trilobite. Cf: *endite.*

exo- A prefix meaning "outside" or "out of".

exoadaptation Adaptation of organisms to their external environments. Cf: *endoadaptation.*

exocast *external cast.*

exocoele The space between adjacent pairs of mesenteries of a coral. Cf: *entocoele.*

exocyathoid expansion Growth of an additional intervallum in an archaeocyathid cup, which appears to have had an adherent function (TIP, 1972, pt. E, p. 7).

exocyclic Said of an *irregular* echinoid whose periproct is located outside of the oculogenital ring. Ant: *endocyclic.*

exodiagenesis A term used by Shvetsov (1960) for diagenesis in subaerial environments and shallow stable seas. It is characterized by dehydration, coagulation of colloids, rapid growth of crystals (recrystallization), formation of concretions, and preservation of textural properties of sediments.

exogastric (a) Said of a cephalopod shell that is coiled so that the venter is on or near the outer or convex side of the whorls. (b) Said of a gastropod shell that is coiled so as to extend forward from the aperture over the front of the extruded head-foot mass, as in the earliest forms (TIP, 1960, pt.I, p.130).—Cf: *endogastric.*

exogene effect The effect of an igneous mass on the rock that it invades (Bateman, 1950). Cf: *endogene effect.*

exogenetic Said of processes originating at or near the surface of the earth, such as weathering and denudation, and to rocks, ore deposits, and landforms that owe their origin to such processes. Cf: *endogenetic.* Syn: *exogenic; exogenous.*

exogenic *exogenetic.*

exogenite A little-used term for an epigenetic mineral deposit differing in composition from the enclosing rock.

exogenous *exogenetic.*

exogenous dome A volcanic dome that is built by surface effusion of viscous lava, usually from a central vent or crater (Williams, 1932). Cf: *endogenous dome.*

exogenous inclusion *xenolith.*

exogeology A rarely used syn. of *astrogeology.*

exogeosyncline A *parageosyncline* accumulating clastic sediments from the uplifted orthogeosynclinal belt adjacent to it but outside the craton (Kay, 1947, p. 1289-1293). Syn: *deltageosyncline; foredeep; transverse basin.* Cf: *secondary geosyncline.*

exoglyph A hieroglyph occurring at the bounding surface of a sedimentary bed (Vassoevich, 1953, p.37). Cf: *endoglyph.*

exogyrate Shaped like the shell of *Exogyra* (a genus of bivalve mollusks having a thick shell and a spirally twisted beak); i.e. with the left valve strongly convex and its dorsal part coiled in the posterior direction, and with the right valve flat and spirally coiled.

exokinematic Said of sedimentary operations in which "the largest displacement vectors occur between matter outside the deposit and the unmodified deposit surrounding the structure produced" (Elliott, 1965, p.196); e.g. types of streamflow. Also, said of the sedimentary structures produced by exokinematic operations. Cf: *endokinematic.*

exokinetic Said of a fissure in a rock that is the result of strain in an adjacent rock unit. Cf: *endokinetic.*

exometamorphism *exomorphism.*

exomorphic metamorphism *exomorphism.*

exomorphic zone *aureole.*

exomorphism Changes in country rock produced by the intense heat and other properties of magma or lava in contact with them; *contact metamorphism* in the usual sense. The term was originated by Fournet in 1867. Cf: *endomorphism.* Syn: *exometamorphism; exomorphic metamorphism.*

exoolitic *extoolitic.*

exopinacoderm The *pinacoderm* covering the free surface of a sponge. Cf: *endopinacoderm.*

exopod The lateral or external ramus of a limb of a crustacean, arising from the basis. Cf: *endopod.* Syn: *exopodite.*

exopore The external opening in the ektexine of a pollen grain with a complex porate structure. See also: *vestibulum.* Cf: *endopore.*

exopuncta A *puncta* of the external shell surface of a brachiopod, commonly restricted to the primary layer, and never penetrating to the internal surface. Cf: *endopuncta.* Pl: exopunctae. Syn: *exopunctum.*

exorheic Said of a basin or region characterized by external drainage; relating to exorheism. Also spelled: *exoreic.*

exorheism (a) *external drainage.* (b) The condition of a region in which its water reaches the ocean directly or indirectly.—Ant: *endorheism.* Also spelled: *exoreism.*

exoseptum A scleractinian-coral septum developed within an exocoele. Cf: *entoseptum.*

exoskarn Skarn formed by replacement of limestone or dolomite. Cf: *exomorphism.*

exoskeleton An external *skeleton* of an animal, serving as a protective and supportive covering for its softer parts; e.g. the outer shell of a brachiopod or pelecypod, the system of sclerites covering the body of an arthropod, or the bony plates covering an armadillo. Cf: *endoskeleton.* Syn: *dermoskeleton.*

exosphere The outermost portion of a planet's atmosphere, in which the density is so low that an appreciable fraction of the molecules can escape into outer space. Its lower boundary for the Earth is estimated at 500 to 1000 km above the surface. It is also called "region of escape".

exospore (a) A syn. of *exine.* The term is mostly applied to the sporoderm of spores, rather than to pollen. Syn: *exosporium.* (b) One of the asexual spores formed by abstriction from a parent cell (as in certain fungi). (c) One of the spores formed from above downward and one at a time in certain blue-green algae.—Cf: *endospore.*

exostome The outer peristome of a protist.

exotheca A collective term for the dissepiments outside the wall of a scleractinian corallite. Cf: *endotheca.*

exothermic Pertaining to a chemical reaction that occurs with a liberation of heat. Cf: *endothermic.*

exotic [ecol] Said of an organism that has been introduced into a new area from an area where it grew naturally. Ant: *indigenous.*

exotic [struc geol] Applied to a boulder, block, or larger rock body unrelated to the rocks with which it is now associated, which has been moved from its place of origin by one of several processes. Exotic masses of tectonic origin are also *allochthonous;* those of glacial or ice-rafted origin are generally called *erratics.*

exotic block A mass of rock occurring in a lithologic association foreign to that in which the mass formed (Berkland et al., 1972, p. 2296).

exotic limonite Limonite precipitated in rock that did not formerly contain any iron-bearing sulfide. Cf: *indigenous limonite.*

exotic stream A stream that derives much of its waters from a drainage system in another region; e.g. a stream that has its source in a humid or well-weathered area but that flows across a desert before reaching the sea. Example: the Nile. Cf: *indigenous stream.*

exotomous Characterized by bifurcation in two main crinoid arms that give off branches only on their outer sides. Ant: *endotomous.*

exozone The outer parts of a stenolaemate bryozoan colony, characterized by thick vertical walls, concentrations of intrazooidal skeletal structures, and zooidal growth directions at high angles to the colony growth directions or surfaces. Cf: *endozone.* Syn: *mature region.*

expanded foot A broad, bulb-shaped lobelike or fanlike mass of ice formed where the lower part of a valley glacier leaves its confining walls and extends onto an adjacent lowland at the foot of a mountain slope. Syn: *piedmont bulb.* Nonpreferred syn: *ice fan; glacier bulb.*

expanded-foot glacier A small *piedmont glacier* consisting of an expanded foot. Syn: *foot glacier; bulb glacier.*

expanding Earth A theory, favored by some geologists, that the diameter of the Earth has grown larger, perhaps by a third or more during recorded geologic time, as a result of changes in atomic and molecular structure in the core and lower mantle, without change in actual mass. The theory has been linked with *continental displacement* and *sea-floor spreading,* although these have also been otherwise explained. Cf: *contracting Earth.*

expanding-lattice clay A clay mineral whose crystal lattice is expandable according to the amount of water it takes on; e.g. a three-layer clay (such as montmorillonite-smectite) in which diffuse negative charges originating in the central octahedral sheets result in less tendency for successive layers to be tightly bound by cations, thereby causing the layers to be readily pushed apart by adsorbed water.

expanding-Universe hypothesis Interpretation of the universally extragalactic red shift as a Doppler shift, suggesting that the Universe is expanding at a uniform rate.

expansion breccia A breccia formed by increase of volume due to chemical change, as by recrystallization or by hydration (Norton, 1917, p.191).

expansion fissure In petrology, one of a system of fissures that radiate irregularly through feldspar and other minerals adjacent to olivine crystals that have been replaced by serpentine. The alteration of olivine to serpentine involves considerable increase in volume, and the stresses so produced are relieved by the fissuring of the surrounding minerals. This phenomenon is common in norite and gabbro (Tyrrell, 1950).

expansion joint *sheeting.*

expectation The expected number of statistical occurrences of a given observation for a specified number of trials. It is expressed by the number of trials times the probability of occurrence of the given observation.

experimental petrology A branch of petrology dealing with the laboratory study of reactions designed to elucidate rock-forming processes. The term includes experiments relating to the physical properties or physical chemistry of minerals, rocks, rock melts, vapors, gases, or solutions coexisting with solid or molten rock materials.

experimental structural geology The study of high-pressure deformation of samples of rock; also, the construction of dynamic models that illustrate structural processes.

explanate Said of the flattened growth habit of erect bryozoan colonies.

explanation A term used by the U.S. Geological Survey in preference to *legend* or *key* (except on international maps).

exploding-bomb texture In mineral deposits, a pattern of pyrite replacement by copper sulfides, usually chalcopyrite-bornite, in which scattered, residual pyrite fragments are surrounded by the copper minerals in a manner suggesting a time-lapse photograph of an exploding bomb. Also spelled: *exploded-bomb texture.*

exploration (a) The search for deposits of useful minerals or fossil fuels; *prospecting.* It may include geologic reconnaissance, e.g. remote sensing, photogeology, geophysical and geochemical methods, and both surface and underground investigations. (b) Establishing the nature of a known mineral deposit, preparatory to *development.* In the sense that exploration goes beyond discovery, it is a broader term than prospecting.

exploration architecture A framework for the organization of an exploration program for mineral deposits. It includes regional, followup, and detailed levels (Fortescue, 1965, p. 6-7).

exploratory well A well drilled to an unexplored depth or in unproven territory, either in search of a new pool of oil or gas or with the expectation of greatly extending the known limits of a field already partly developed. Cf: *development well.* Syn: *test well; wildcat well.*

explorer's alidade A lightweight, compact alidade with a low pillar and a reflecting prism through which the ocular may be viewed from above. Syn: *Gale alidade.*

explosion breccia A type of volcanic breccia that is formed by a volcanic explosion. Syn: *eruption breccia; pyroclastic breccia.*

explosion caldera A type of *caldera* that is formed by explosive removal of the upper part of a volcanic cone. It is extremely rare, and is small in size. Cf: *collapse caldera; erosion caldera.*

explosion cloud *eruption cloud.*

explosion crater (a) A saucer-shaped to conical *crater* produced experimentally by detonation of a nuclear device or a chemical explosive. (b) A meteorite crater formed by hypervelocity impact. (c) A volcanic crater, e.g. a *maar.*

explosion seismology Obsolete term for *applied seismology.*

explosion tuff A tuff whose pyroclastic fragments are in the place in which they fell, rather than having been washed into place after they landed.

explosive bomb A *bread-crust bomb* that throws off fragments of its crust due to continued expansion of its interior after solidification of its crust.

explosive eruption An eruption or eruption phase that is characterized by the energetic ejection of pyroclastic material. Cf: *lava eruption.*

explosive evolution (a) Within a group or lineage of organisms, morphologic or ecologic change at an extremely rapid rate compared to the usual or normal rate. (b) Sometimes used to denote *adaptive radiation.*

explosive index The percentage of pyroclastics among the total products of a volcanic eruption. Syn: *explosivity index.*

explosive radiation An *adaptive radiation* that appears to have occurred very rapidly.

explosivity index *explosive index.*

exponential distribution A *frequency distribution* whose ordinate is proportional to the value of the dependent variable and plots as the variable exponent of a constant.

exponentiate To introduce an empirically derived exponential time-dependent gain in seismic playback.

exposed coalfield Deposits of coal that crop out at the surface, as along the rim of a coal basin. Cf: *concealed coalfield; crop coal.*
exposure [geol] (a) An area of a rock formation or geologic structure that is visible ("hammerable"), either naturally or artificially, i.e. is unobscured by soil, vegetation, water, or the works of man; also, the condition of being exposed to view at the Earth's surface. Cf: *outcrop.* (b) The nature and degree of openness of a slope or place to wind, sunlight, weather, oceanic influences, etc. The term is sometimes regarded as a syn. of *aspect.*
exposure [photo] (a) The total quantity of light received per unit area on a sensitized plate or film, usually expressed as the product of the light intensity and the time during which the light-sensitive material is subjected to the action of light. (b) A loosely used term "generally understood to mean the length of time during which light is allowed to act on a sensitive surface" (Smith, 1968, p.496). (c) The act of exposing a light-sensitive material to a light source. (d) An individual picture of a strip of photographs.
exposure interval The time interval between the taking of successive photographs.
exposure station *camera station.*
exsert adj. Projecting beyond an enclosing part or organ, such as having the ocular plates of an echinoid not in contact with the periproctal margin. Ant: *insert.* Syn: *exserted.*
exsiccation The drying-up of an area due to a change that drives out, or decreases the amount of, available moisture without reducing appreciably the average rainfall; e.g. the draining of a swamp or marsh, or the migration of sand dunes across cultivated ground. Cf: *desiccation.*
exsolution The process whereby an initially homogeneous solid solution separates into two (or possibly more) distinct crystalline phases without addition or removal of material, i.e., without change in the bulk composition. It generally, though not necessarily, occurs on cooling. Syn: *unmixing.*
exsolutional Pertaining to sediments or sedimentary rocks that solidified from solution either by precipitation or by secretion.
exsolution lamella A *lamella* produced by exsolution, such as one consisting of diopside associated with enstatite augen, or one contained within crystals of tremolite, hornblende, or cummingtonite.
exsolution texture In mineral deposits, a general term for the texture of any mineral aggregate or intergrowth formed by exsolution. It is generally fairly homogeneous, ranging from perthitic to geometrically regular. See also: *eutectic texture.*
exsudation A kind of *salt weathering* by which rock surfaces are scaled off owing to growth of salines by capillary action (Thornbury, 1954, p.39).
exsurgence The rising of a stream from a cave, the water having entered the cave by downward percolation through the overlying limestone. Syn: *karst spring.* See also: *resurgence.*
extended consequent stream A consequent stream that flows seaward across a newly emerged coastal plain and that forms an extension of an earlier, larger stream with headwaters in the older land behind the coastal plain.
extended stream A stream lengthened by the extension of its course downstream across newly emerged land (such as a coastal plain, a delta, or a plain of glacial deposition).
extended succession A relatively thick and uninterrupted stratigraphic succession in which the deposits accumulated rapidly. Ant: *condensed succession.*
extending flow A flow pattern on glaciers in which the velocity increases with distance downstream; thus the longitudinal strain rate (velocity gradient) is extending. This condition requires a transverse or vertical compression or a positive net balance on the surface to maintain the continuity of the ice (Nye, 1952). Ant: *compressing flow.*
extensiform Said of a graptoloid (such as Didymograptus) with two stipes that are horizontal.
extension [exp struc geol] (a) A strain term signifying increase in length. Cf: *tension,* which is a stress term. (b) A measure of the change in length of a line, specif. the ratio of the change in length to the original length. Cf: *stretch [exp struc geol].* Syn: *elongation.*
extension [streams] The lengthening of a stream by headward erosion and the multiplication of tributaries, or by regression of the sea or uplift of the coastal area.
extension fault A fault in sedimentary rocks along which there has been bed-parallel elongation (Norris, 1958, 1964), giving rise to tectonic thinning.
extension fracture A fracture that develops perpendicular to the direction of greatest stress and parallel to the direction of compression; a *tension fracture.* See also: *extension joint; tension crack.*
extension joint A joint that forms parallel to the direction of compression; a joint that is an *extension fracture.*
extension ore The *possible ore* ahead of or beyond an exposure in a mine. Cf: *probable ore.*
extension test *triaxial extension test.*
extension well Any well located as an outpost well or as a wildcat well that extends the productive area of a pool. The term cannot logically be applied until after the fact is demonstrated (Lahee, 1962, p. 134). Cf: *step-out.*
extensiveness In quantification of hydrothermal alteration, the degree to which susceptible minerals are converted to alteration phases; the amount of alteration in a host rock, regardless of mineralogy or distribution. It is analogous to extensive parameters in physical chemistry. Cf: *intensiveness; pervasiveness.*
extensive quantity A thermodynamic quantity such as volume or mass that depends on the total quantity of matter in the system.
extensometer An instrument used for measuring small deformations, as in tests of stress.
exterior link A *link* in a channel network emanating from a *source;* equivalent to a first-order stream segment of Strahler (Shreve, 1967). Cf: *link; stream order; interior link.*
exterior wall A body wall in bryozoans that extends the body of a zooid and of the colony. It includes an outermost cuticular layer.
extermination The local or even regional disappearance of a species that still exists elsewhere, as a result of changing environmental conditions, disease, competitors, or other adverse conditions. Cf: *extinction [evol].*
external cast An improper term sometimes used as a syn. of *external mold.* Syn: *exocast.*
external contact The planar or irregular surface between a pluton and the country rock (Compton, 1962, p. 277). Cf: *internal contact.*
external drainage Drainage whereby the water reaches the ocean directly or indirectly. Ant: *internal drainage.* Syn: *exorheism.*
external furrow One of the shallow, linear axial depressions or grooves on the outer surface of a fusulinid test, formed at the point of union between successive chambers and corresponding in position to a septum, and dividing the outer surface into melonlike lobes. Syn: *septal furrow.*
external lobe The *ventral lobe* in normally coiled cephalopod conchs.
external magnetic field The relatively small and varying portion of the natural *magnetic field* near the Earth's surface which is due to electric currents in the upper atmosphere.
external mold A *mold* or impression in the surrounding earth or rock, showing the surface form and markings of the outer hard parts of a fossil shell or other organic structure; also, the surrounding rock material whose surface receives the external mold. Cf: *internal mold; external cast.*
external rotation A change in orientation of structural features during deformation referred to coordinate axes external to the deformed body. Cf: *internal rotation.*
external suture The part of a suture of a coiled cephalopod conch exposed on the outside of whorls between the umbilical seams. Cf: *internal suture.*
externides Kober's term for the outer part of an orogenic belt, nearest to the craton or foreland, commonly the site of a miogeosyncline during early stages and of an exogeosyncline during late stages. It is usually subjected to marginal deformation (folding and lateral thrusting) during the orogenic phase. Cf: *internides.* See also: *secondary orogeny.* Syn: *secondary arc.*
extinction [evol] The total disappearance of a species or higher taxon, so that it no longer exists anywhere. Cf: *extermination.*
extinction [lake] The disappearance of a lake, by drying up (*temporary extinction*) or by destruction of the lake basin (*permanent extinction*).
extinction [optics] The more or less complete darkness obtained in a *birefringent* mineral at two positions during a complete rotation of a section between crossed nicols. Also, the darkness that persists for a complete rotation if the line of sight is parallel to the optic axis. See also: *extinction angle; extinction direction; inclined extinction; parallel extinction; undulatory extinction.*
extinction angle The angle through which a section of a *birefringent* mineral must be rotated from a known crystallographic plane or direction to the position at which it gives maximum *extinction* or darkness under a polarizing microscope. The extinction angle can be diagnostic in the identification of a mineral.

extinction coefficient In oceanography, a measure of the attenuation of downward radiation in the sea (U.S. Naval Oceanographic Office, 1968).

extinction direction One of the two positions at which a section of a birefringent crystal shows *extinction* between crossed nicols.

extinct lake (a) A lake that has lost all its water, either temporarily or permanently. (b) A lake whose open water has been replaced by vegetation and whose status has reached that of a bog, marsh, or swamp.—Syn: *dead lake.* See also: *senescent lake.*

extinct volcano A volcano that is not now erupting and that is not considered likely to erupt in the future. Cf: *active volcano; dormant volcano; inactive volcano.*

extine Var. of *exine.* The term is not in good usage in palynology.

extoolitic Pertaining to oolitic structures or grains that have formed or grown outward by deposition of material around a core or center, as in the formation of a small concretion. Ant: *entoolitic.* Syn: *exoolitic.*

extraclast A fragment of calcareous sedimentary material, produced by erosion of an older rock outside the area in which it accumulated; a component of calclithite. Cf: *intraclast.*

extragalactic nebula *galaxy.*

extraglacial Said of glacial deposits formed by meltwater beyond the farthest limit of the ice, or of glacial phenomena displayed in an area never covered by ice. Ant: *intraglacial.*

extramorainal Said of deposits and phenomena occurring outside the area occupied by a glacier and its lateral and end moraines. Ant: *intramorainal.* Syn: *extramorainic.*

extraneous argon *Inherited argon* and *excess argon* taken collectively.

extraneous ash *Ash* in coal that is derived from inorganic material introduced during formation of the seam, such as sedimentary particles, or filling cracks in the coal. Cf: *inherent ash.* Syn: *secondary ash; sedimentary ash.*

extraordinary ray *E ray [cryst].*

extrapolation Estimation of the value of a variate based on at least two known values on only one side of the unknown value. It is used to extend a line or curve by determining points on it beyond those for which data are available.

extra river A diamond (*river [gem]*) of the very highest grade.

extratentacular budding Formation of new scleractinian coral polyps by invagination of the edge zone or coenosarc outside of the ring of tentacles surrounding the mouth of the parent. Cf: *intratentacular budding.*

extraterrestrial Existing, occurring, or originating beyond the Earth or its atmosphere; e.g. "extraterrestrial radiation" or solar radiation received "on top of" the Earth's atmosphere.

extraterrestrial geology *astrogeology.*

extraumbilical aperture An aperture in the final chamber of a foraminiferal test not connecting with the umbilicus. It is commonly sutural midway between the umbilicus and the periphery. See also: *interiomarginal aperture.*

extraumbilical-umbilical aperture An aperture in the final chamber of a foraminiferal test that extends along its forward margin from the umbilicus toward the periphery, thus reaching a point outside the umbilicus (as in *Globorotalia*).

extravasation The eruption of molten or liquid material, e.g. lava, or water from a geyser, onto the surface of the Earth.

extrazooidal skeleton A protective or supportive colony structure in bryozoans which, once developed, remains outside zooidal boundaries throughout the life of the colony.

extremely coarsely crystalline Descriptive of an interlocking texture of a carbonate sedimentary rock having crystals whose diameters exceed 4 mm (Folk, 1959).

extremely finely crystalline *aphanocrystalline.*

extrusion The emission of relatively viscous lava onto the Earth's surface; also, the rock so formed. Cf: *effusion.*

extrusion flow A discredited hypothesis for a type of glacier flow in which the pressure of overlying ice is supposed to force the basal part of the glacier to flow faster than the upper part. *Gravity flow,* originally contrasted with extrusion flow, is now considered to account for all glacier flow.

extrusive adj. Said of igneous rock that has been erupted onto the surface of the Earth. Extrusive rocks include lava flows and pyroclastic material such as volcanic ash.—n. An extrusive rock.—Cf: *intrusive.* Syn: *effusive; volcanic; eruptive.*

extrusive ice Ice formed subaerially from water that emerges from the ground or below other ice, as in an icing.

exudation basin A spoon-shaped depression on the ice surface at the head of an *outlet glacier.* Examples are found on the Greenland and Antarctic ice sheets.

exudation vein *segregated vein.*

exuvia All or part of the exoskeleton of an arthropod that has been shed, molted, or cast off. Pl: *exuviae.*

exuviation The removal of the theca of a dinoflagellate, either plate by plate or as small groups of plates.

eye [grd wat] The opening from which the water of a spring flows out onto the land surface.

eye [meteorol] The approximately circular area of relatively light wind and good weather in the center of a tropical cyclone. Such an area may be 5 to 60 km in diameter.

eye [paleont] The ringlike part of a *hook* of a holothurian, sometimes partly closed by a bar. Also, a ringlike end of a *rod* of a holothurian.

eye agate Agate displaying concentric bands, usually of various colors, about a dark center, suggesting an eye. Syn: *Aleppo stone.*

eye-and-eyebrow structure A feature of certain rhyolites that contain crescent-shaped bodies of quartz with pealike bodies of quartz on the concave side; the convex side is toward the top of the flow.

eyebrow scarp A fault scarp that crosses a piedmont alluvial fan near its apex and that seldom maintains the dip of the fault surface in the unconsolidated gravels of the fan (Davis, 1927, p. 62).

eye coal Coal that contains structural disks in circular or elliptical shapes, either parallel or normal to the bedding, with concentric, bending rims and radiating striae. They reflect light in a mirror-like way. Syn: *circular coal.* Etymol: German *Augenkohle,* "eye coal".

eyed structure *augen structure.*

eyepiece The lens (or lenses) in a microscope or telescope through which the image formed by the *objective* is viewed. Syn: *ocular [optics].*

eye ridge A raised line or narrow band extending from the forward and inner part of a trilobite eye to the anterior part of the glabella.

eyestalk One of the movable peduncles in a decapod crustacean, carrying the eye at its distal extremity. Syn: *ophthalmite.*

eye tubercle A polished, transparent, rounded protuberance in the anterior and dorsal region of an ostracode valve, and on the cheek regions of some trilobites, forming the lens of the eye. See also: *tubercle.*

eylettersite A member of the crandallite group: $(Th,Pb)_{1-x}Al_3(PO_4,SiO_4)_2(OH)_6$(?).

eyot *ait.*

ezcurrite A triclinic mineral: $Na_4B_{10}O_{17}\cdot 7H_2O$. It is dimorphous with nasinite.

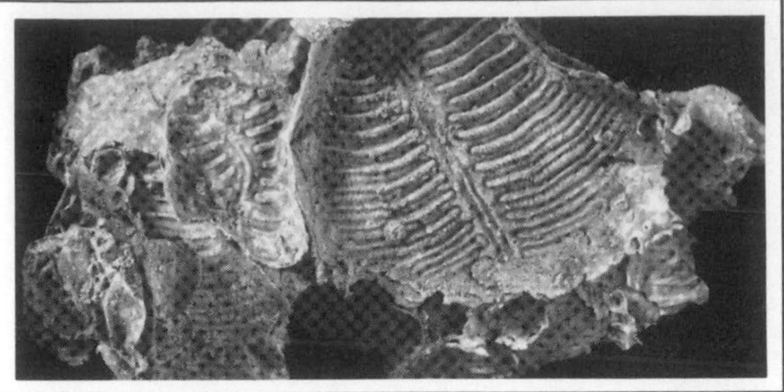

fabianite A monoclinic mineral: $CaB_3O_5(OH)$.
fabric [sed] The orientation (or lack of it) in space of the elements (discrete particles, crystals, cement) of which a sedimentary rock is composed. Cf: *packing.* See also: *crystallization fabric.*
fabric [soil] The physical nature of a soil according to the spatial arrangement of its particles and voids.
fabric [struc geol] The complete spatial and geometrical configuration of all those components that make up a deformed rock. It covers such terms as texture, structure, and preferred orientation, and so is an all-encompassing term that describes the shapes and characters of individual parts of a rock mass and the manner in which these parts are distributed and oriented in space (Hobbs, Means and Williams, 1976, p. 73). The term was first defined by Sander (1930), who used the German word Gefüge.
fabric analysis A term usually used as a synonym for structural petrology. In some instances, however, it may refer to the geometrical part of the much broader study of *structural analysis.*
fabric axis One of three orthogonal axes used in structural petrology as references in the orientation of fabric elements, and in the description of folding and of the movement symmetry of deformed rocks. Cf: *a* axis [struc petrol]; *b* axis [struc petrol]; *c* axis [struc petrol]. Syn: *reference axis; tectonic axis.*
fabric diagram In structural petrology, a stereographic or equal-area projection of fabric elements; an *orientation diagram.* See also: *point diagram; contour diagram.* Syn: *petrofabric diagram.*
fabric domain In a deformed rock, a volume defined by boundaries such as structural or compositional discontinuities, within which the fabric is more or less homogeneous. Syn: *domain.*
fabric element A component of a rock fabric. See also: *fabric; subfabric.*
face [cryst] *crystal face.*
face [geomorph] The principal side or surface of a landform; esp. *rock face.*
face [mining] Any surface on which mining operations are in progress. Syn: *breast; highwall.*
face [struc geol] n. A term used by Shrock (1948, p. 17-18) for the original top or upper surface of a layer of rock, esp. if it has been raised to a vertical or steeply inclined position.—v. To be directed toward or to present an aspect of. Sedimentary beds are said "to face" in the direction of the stratigraphic top of the succession (or to be directed toward the younger rocks or to the side that was originally upward), so that an overturned bed facing to the east may have a dip of 45 degrees to the west. Folds are said "to face" in the direction of the stratigraphically younger rocks along their axial surfaces and normal to their axes (Shackleton, 1958, p. 363); this coincides with the direction toward which the beds face at the hinge (a normal upright fold faces upward, an overturned anticline faces downward, and an asymmetric fold faces its steeper flank). Faults are said "to face" in the direction of the structurally lower unit. Syn: *young.*
face-centered lattice A type of *centered lattice* in which each unit cell has a lattice point at the center of each face as well as those at each corner, i.e. it has four lattice points per unit cell. Syn: *F-centered lattice.*
face cleat The major *cleat* system or jointing in a coal seam. Cf: *end cleat.* See also: *face of coal.*
facellite *kaliophilite.*
face of coal The plane or surface of the coal in situ at the advancing end of the working face. Cf: *end of coal.*
face pole In crystallography, a point on a projection surface that represents the intersection of a crystal *pole* with the crystal face.
facet [gem] One of the plane polished surfaces on a gemstone.
facet [geog] Any part of a landscape defined as a unit for geographic study on the basis of homogeneous topography.
facet [geomorph] (a) A nearly plane surface produced on a rock fragment by abrasion, as by wind sandblasting, by the grinding action of a glacier, or by a stream that differentially removes material from the upstream side of a boulder or pebble; inclined at an angle of 50 degrees or less to the direction of the impinging current (Maxson, 1940, p. 721). (b) Any plane surface produced by erosion or faulting, and intersecting a general slope of the land; e.g. a *triangular facet.*
facet [paleont] (a) A nearly flat surface on an echinoderm plate, serving for articulation with contiguous skeletal elements; e.g. *brachiolar facet.* (b) A small, circular or oval protuberance set within a channel of a cyclocystoid and connected with the ambulacral system. (c) A small, flat surface developed in trilobites on the anterolateral region of the pygidium and thoracic pleurae.
faceted boulder A boulder that has been ground flat on one or more sides by the action of natural agents, such as by glacier ice, streams, or wind. Cf: *faceted pebble.*
faceted pebble A pebble on which facets have been developed by natural agents, such as by wave erosion on a beach or by the grinding action of a glacier; specif. a *windkanter.* Cf: *faceted boulder.* Also spelled: *facetted.*
faceted spur A spur or ridge with an inverted-V face that was produced by faulting or by the trimming, beveling, or truncating action of streams, waves, or glaciers. See also: *truncated spur.*
facial (a) Pertaining to facies. The term is used by some non-English European geologists. (b) Pertaining to an outer surface.
facial suture One of the two symmetrical sutures that open at the time of molting of a trilobite, extending from the anterior margin of the cephalon around the eye and outward or backward to the lateral or posterior margin. It separates the free cheek from the fixed cheek.
faciation Part of an ecologic association, usually a large community, which is characterized by the dominance of two or more but not all of the more abundant organisms.
facieology The study of facies. Not recommended usage.
facies (a) The aspect, appearance, and characteristics of a rock unit, usually reflecting the conditions of its origin; esp. as differentiating the unit from adjacent or associated units. Cf: *stratigraphic facies; lithofacies; igneous facies.* (b) A mappable, areally restricted part of a lithostratigraphic body, differing in lithology or fossil content from other beds deposited at the same time and in lithologic continuity. Cf: *sedimentary facies.* (c) A distinctive rock type, broadly corresponding to a certain environment or mode of origin, e.g. "red-bed facies", "black-shale facies". Cf: *petrographic facies.* (d) A body of rock distinguished on the basis of its fossil content. Cf: *biofacies (a), (b).* (e) A local assemblage or association of living or fossil organisms, esp. one characteristic of some type of marine conditions. Cf: *biofacies (c).* (f) The environment or area in which a rock was formed, e.g. "sandy-bottom facies", "eolian facies", "volcanic facies". Cf: *environmental facies.* (g) Rocks broadly defined on a paleogeographic or paleotectonic basis. Cf: *geosynclinal facies; shelf facies.* (h) Rocks of any origin formed within certain pressure-temperature conditions. Cf: *mineral facies; metamorphic facies.*—The concept of stratigraphic facies was first defined by Gressly (1838, p. 10-12, 20-25). Usages and definitions have been reviewed by Moore (1949), Weller (1958), and Teichert (1958). "The general term 'facies' has been greatly overworked. . . . If the term is used, it is desirable to make clear the specific kind of facies to which reference is made" (ISG, 1976, p. 15). Etymol: Latin (and French), "face, form, aspect, condition". Pron *fay*sheez or *fay*seez. Pl: *facies.*
facies change A lateral or vertical variation in the lithologic or paleontologic characteristics of contemporaneous sedimentary deposits. It is caused by, or reflects, a change in the depositional

environment. Cf: *facies evolution.*

facies contour The trace (on a map) of a vertical surface that cuts a three-dimensional rock body into facies segments; a line indicating equivalence in lithofacies development.

facies-departure map A *facies map* based on the degree of similarity to some particular sedimentary rock composition (optimum facies or single-component end member). Term suggested by Forgotson (1960, p.94) to replace *distance-function map.*

facies evolution A gradual change of facies over a period of time, indicating gradually changing depositional conditions (Teichert, 1958, p.2723). Cf: *facies change.*

facies family A term used by Teichert (1958, p.2737) for several genetically interconnected *facies tracts;* e.g. coral-atoll deposits and desert deposits. See also: *facies suite.*

facies fauna A group of animals characteristic of a given stratigraphic facies or adapted to life in a restricted environment; e.g. the black-shale fauna of the Middle and Upper Devonian of the Appalachian region of U.S.

facies fossil A fossil, usually a single species or a genus, that is restricted to a defined stratigraphic facies or is adapted to life in a restricted environment. It prefers certain ecologic surroundings and may exist in them from place to place with little change for long periods of time.

facies map A broad term for a map showing the gross areal variation or distribution (in total or relative content) of observable attributes or aspects of different rock types occurring within a designated stratigraphic unit, without regard to the position or thickness of individual beds in the vertical succession; specif. a *lithofacies map.* Conventional facies maps are prepared by drawing lines of equal magnitude through a field of numbers representing the observed values of the measured rock attributes. Cf: *vertical-variability map.* See also: *biofacies map; isofacies map; isolith map; percentage map; ratio map; entropy map; entropy-ratio map; facies-departure map.*

facies plane A term used by Caster (1934, p.19 & 24) for the boundary between *magnafacies* or *parvafacies,* although it is usually not sharp enough to be recognizable as a plane in any section. Cf: *plane of contemporaneity.*

facies sequence A term used by Teichert (1958, p.2723) for a vertical succession of different (heteropic) facies formed at different times.

facies strike The compass direction of a facies contour at a given point on a map.

facies suite (a) A term suggested by Oriel (in Teichert, 1958, p.2737) for several genetically interconnected *facies families;* e.g. all marine deposits or all continental deposits. (b) A collection or group of rocks that shows variations within a single rock mass.

facies tract A system of different but genetically interconnected sedimentary facies of the same age (Teichert, 1958, p.2723); e.g. the outer-slope deposits of a coral atoll, or dry-channel deposits. It includes the areas of erosion from which the sediments of these facies are derived, so that an erosional interval represents part of a facies tract. The concept was developed by Walther (1893-1894). See also: *facies family.* Syn: *macrofacies.*

facing (a) The direction toward which a rock unit or layer *youngs.* (b) The direction along the axial plane of a fold in which it passes through younger layers. Syn: *fold facing.*

faciostratotype A supplemental stratotype designated for local reference or reference to different facies, and which distinguishes different ecologic conditions that existed during the time of the chronostratigraphic unit represented by the stratotype (Sigal, 1964).

facsimile crystallization *mimetic crystallization.*

factor analysis A method for identifying the minimum number of influences necessary to account for the maximum observed variation in a set of data and for indicating the extent to which each influence accounts for the variance observed in the data. See also: *Q-mode factor analysis; R-mode factor analysis.*

facultative Said of an organism capable of growth under a number of specific conditions; adaptable to alternate environments. Cf: *obligate.*

fadama A term used in western Africa for a flood plain in a wide river valley, subject to annual inundation and characterized by savanna vegetation (J.C. Pugh in Stamp, 1961, p. 187). Etymol: Hausa.

faecal pellet *fecal pellet.*

faheyite A white hexagonal mineral: $(Mn,Mg)Fe_2{}^{+3}Be_2(PO_4)_4 \cdot 6H_2O$.

fahlband A band of sulfide impregnation in metamorphic rocks. The sulfides are too abundant to be classed as accessory minerals, but too sparse to form an ore lens (Gammon, 1966, p. 177). Fahlbands have a characteristic rusty-brown appearance on weathering. Etymol: German.

fahlerz A syn. of *fahlore.* Etymol: German *Fahlerz,* "pale ore".

fahlore Any gray-colored ore mineral consisting essentially of sulfantimonides or sulfarsenides of copper; specif. tetrahedrite and tennantite. Syn: *fahlerz.*

fahlunite An altered form of cordierite.

faikes *fakes.*

failure Fracture or rupture of a rock or other material that has been stressed beyond its *ultimate strength.* Syn: *rock failure.*

fairchildite A mineral: $K_2Ca(CO_3)_2$. It is found in fused wood ash in partly burned trees. Cf: *buetschliite.*

fairfieldite A white, pale yellow or greenish-white triclinic mineral: $Ca_2(Mn,Fe)(PO_4)_2 \cdot 2H_2O$. It is isomorphous with collinsite.

fairway (a) The main navigable channel (usually buoyed) of a river or bay, through which vessels enter or leave a port or harbor; the part of a waterway that is kept open and unobstructed for navigation. (b) *midway.*

fair-weather runoff *base runoff.*

fairy-castle structure Intricate microtopography of the lunar surface, believed to be responsible for the Moon's optical properties.

fairy chimney A translation of the French term *cheminée de fée,* used in the French Alps for *earth pillar.*

fairy stone (a) A cruciform-twinned crystal of staurolite, used as a *curio stone* without fashioning for adornment. The term is also applied as a syn. of *staurolite,* and esp. to the variety occurring in the form of a twinned crystal. (b) Any of various odd or fantastically shaped calcareous or ferruginous concretions formed in alluvial clays. (c) A fossil sea urchin. (d) A stone arrowhead.

fakes A British vernacular term for a platy rock, such as a fissile sandy shale, a well-laminated siltstone, or a flaggy sandstone or micaceous flagstone. Adj: *fakey.* See also: *flaikes.* Also spelled: *faikes.*

falaise An old, low sea cliff, on an emergent coast, that re-establishes contact with the open sea; the type example is the coast of Normandy. Etymol: French, "cliff".

falcate Hooked or curved like a sickle; e.g. said of a sickle-shaped cephalopod rib. Syn: *falciform.*

falcifer A brachiopod crus that arises on the dorsal side of the hinge plate and projects into the brachial valve as a broad blade-like process.

falcondoite An orthorhombic mineral: $(Ni,Mg)_8Si_6O_{15}(OH)_2 \cdot 6H_2O$.

falcon's-eye *hawk's-eye.*

fall [mass move] (a) A very rapid downward movement of a mass of rock or earth that travels mostly through the air by free fall, leaping, bounding, or rolling, with little or no interaction between one moving unit and another; e.g.: *rockfall; debris fall.* (b) The mass of material moved by a fall.

fall [meteorite] One or more meteorites that are picked up immediately after they have reached the Earth's surface and for which information (place and time of fall) are known. Cf: *find.*

fall [slopes] (a) The descent of land or of a hill; a slope or declivity. (b) The distance to which a stream or physiographic feature slopes.

fall [streams] A *waterfall* or other precipitous descent of water. The plural "falls" is commonly used in place names, esp. where applied to a series of waterfalls.

fallback Fragmental material ejected from an impact or explosion crater during formation and redeposited within, and partly filling, the true crater almost immediately after formation. It includes slide-block deposits, talus material, and aerially transported dust. Cf: *fallout [crater]; throwout.*

fallback breccia An allochthonous breccia composed of *fallback.*

fall diameter The diameter of a sphere that has a specific gravity of 2.65 and the same uniform settling velocity as that of a given particle having any specific gravity "when each is allowed to settle alone in quiescent distilled water of infinite extent and at a temperature of 24°C" (Simons et al., 1961, p. vii).

falling dune An accumulation of sand that is formed as sand is blown off a mesa top or over a cliff face or steep slope, forming a solid wall, sloping at the angle of repose of dry sand, or a fan extending downward from a re-entrant in the mesa wall.

falling star *shooting star.*

falling tide That part of a tide cycle between high water and the following low water, characterized by seaward or receding movement of water. Also, an outgoing tidal river. Ant: *rising tide.* Syn: *ebb tide.*

fall line An imaginary line or narrow zone connecting the waterfalls on several adjacent near-parallel rivers, marking the points where these rivers make a sudden descent from an upland to a lowland, as at the edge of a plateau; specif. the Fall Line marking the boundary between the ancient, resistant crystalline rocks of the Piedmont Plateau and the younger, softer sediments of the Atlantic Coastal Plain in the eastern U.S. It also marks the limit of navigability of the rivers. Syn: *fall zone.*

fallout [crater] Fragmental material ejected from an impact or explosion crater during formation and eventually redeposited in and around the crater. It may have undergone considerable atmospheric sorting before deposition. Cf: *throwout; fallback.*

fallout [radioactivity] The descent of usually radioactive particles through the Earth's atmosphere, following a nuclear explosion; also, the particles themselves.

fallout breccia An allochthonous breccia composed of fallout from a crater. It is generally one of the last ejecta units to be deposited, and it characteristically contains small amounts of glass fragments and a limited range of fragment sizes.

fall overturn A seasonal *overturning* in lakes during fall and winter, when cool weather creates higher density at the surface of the lake.

fall velocity *settling velocity.*

fall zone *fall line.*

false beach A bar above water level, located a short distance offshore (Veatch & Humphrys, 1966, p. 111).

false Becke line A bright line, visible under the microscope, which moves in an opposite direction to that of the *Becke line* as the microscope tube is raised or lowered. It is especially conspicuous when the mineral fragment is thick and irregular or displays conspicuous cleavage; when the difference between the refractive indices of fragment and immersion medium is considerable; and when strongly converging light from the substage condensing lens passes through and near the fragment and then enters an objective lens of large angular aperture (Wahlstrom, 1969, p. 108). See also: *Becke line; Becke test.*

false bedding An old term for *cross-bedding* and *current bedding*, or bedding affected by currents that were often erratic and frequently changed direction. The term is not recommended in this sense because it might refer to *pseudobedding* (Hills, 1963, p.10). See also: *drift bedding.*

false body *thixotropic clay.*

false bottom [eco geol] An apparent bedrock underlying an alluvial deposit that conceals a lower alluvial deposit; e.g. a bed of clay or sand cemented by hydrous iron oxides, on which a gold placer deposit accumulates, and under which there is another alluvial deposit resting on bedrock.

false bottom [lake] The poorly defined bottom of a lake occurring where a firm bottom grades upward to a suspended or soupy mass of muck, colloidal sludge, soft marl, or organic matter, through which a weight easily sinks; e.g. in a bog lake.

false bottom [oceanog] *deep scattering layer.*

false cleavage A quarrymen's term for minor cleavage in a rock, e.g. *slip cleavage,* to distinguish it from the dominant or *true cleavage.* Geologically, the term is misleading and should be avoided.

false color Colors used to represent different frequency bands, or other measurable characteristics, where the colors are not those naturally characteristic of the frequencies. False color makes infrared differences evident and is useful in other applications.

false diamond A colorless mineral (such as zircon, white sapphire, white topaz, and quartz) that superficially resembles diamond when cut and polished.

false dip *apparent dip.*

false drumlin *rock drumlin.*

false esker A feature resembling an esker but "composed of till instead of water-laid drift" (Woodworth, 1894b).

false floor In a cave, a more or less horizontal layer of flowstone that has open space beneath it.

false folding Folding that is not generically related to lateral compression, e.g. shear folding, supratenuous folding. Ant: *true folding.* Syn: *bend folding.*

false form *pseudomorph.*

false galena *sphalerite.*

false gossan A laterally or vertically displaced, rather than indigenous, iron-oxide zone. It may be confused with the iron oxide of a *gossan,* which is weathered from underlying sulfide deposits.

false horizon (a) *artificial horizon.* (b) A line resembling the apparent horizon, but situated above or below it.

false lapis (a) *lazulite.* (b) Agate or jasper artificially dyed blue.

false mud crack A sedimentary structure resembling a mud crack, such as a polygonal pattern formed in soil or a fucoidal network. See also: *pseudo mud crack.*

false oolith *pseudo-oolith.*

false origin An arbitrary point to the south and west of the true origin of a grid system, from which grid distances are measured eastward and northward to insure that all points have positive coordinate values.

false pouch *dolon.*

false shoreline The line of contact between the open water of a lake and the front or edge of a floating mat of vegetation built out from the true shore (Veatch & Humphrys, 1966, p.111).

false stratification An old term for *cross-stratification*. It was used by Lyell (1838, p.38) for the diagonal arrangement of "minor layers placed obliquely to the general planes of stratification". The term is rarely used today. Cf: *pseudostratification.*

false stream An accumulation of water in a hollow along the side of a flood plain that slopes away from the main stream toward the side of the valley (Swayne, 1956, p. 58).

false synapomorphy *parallelism (b).*

false topaz (a) A yellow transparent variety of quartz resembling the color of topaz: specif. citrine. (b) A yellow variety of fluorite.

falun A French term for *shell marl* composed of an unconsolidated accumulation of sand-sized shell fragments.

famatinite A gray to copper-red orthorhombic mineral: Cu_3SbS_4. It is isomorphous with enargite.

Famennian European stage: uppermost Devonian (above Frasnian, below Tournaisian of Carboniferous).

family [ecol] An ecologic community composed of only one kind of organism, usually occupying a small area and representing an early stage in a succession.

family [petrology] (a) The basic unit of the *clan* of igneous rocks. (b) *clan.*

family [soil] In the U.S. Dept. of Agriculture soil taxonomy, a category that divides subgroups of soils into classes that have similar responses to use and management. Families have restricted ranges in particle size, mineralogy, temperature, and reaction. They are further subdivided into *soil series.* Syn: *soil family.*

family [taxon] A category in the hierarchy of zoological and botanical classification intermediate between *order* and *genus.* In zoology, the name of a family characteristically ends in -idae, e.g. Cytheridae; in botany the ending is -aceae, e.g. Rosaceae. Cf: *subfamily; superfamily.*

fan [geomorph] (a) A gently sloping, fan-shaped mass of detritus forming a section of a very low cone commonly at a place where there is a notable decrease in gradient; specif. an *alluvial fan.* (b) A fan-shaped mass of congealed lava that formed on a steep slope by the continually changing direction of flow.

fan [marine geol] *submarine fan.*

fan apron *bajada.*

fan bay The head of an alluvial fan that extends a considerable distance into a mountain valley. Cf: *fanhead.*

fan cleavage A type of *axial-plane cleavage* in which the cleavage planes fan out at small to large angles on each flank of the axial planes of folds. Syn: *cleavage fan.*

fan coral Any coral that forms flat, fanlike colonies.

fancy cut Any style of diamond cutting other than the round brilliant cut or single cut. It includes, among others, the marquise, emerald cut, pear-shape cut, baguette, and half moon. Syn: *modern cut.*

fancy diamond Any diamond with a natural body color strong enough to be attractive rather than off-color. Red, pink, blue, and green are very rare; orange and violet, rare; strong yellow, yellowish-green, brown, and black stones are more common.

fancy sapphire A *sapphire* of any hue other than blue, although colorless and red sapphires are sometimes included.

fan delta (a) A gently sloping alluvial deposit produced where a mountain stream flows out onto a lowland. (b) *delta fan.*—Cf: *arcuate delta.*

fan fold A fold with a broad *hinge* region and limbs that converge away from the hinge.

fanglomerate A sedimentary rock consisting of slightly waterworn, heterogeneous fragments of all sizes, deposited in an *alluvial fan* and later cemented into a firm rock; it is characterized by persistence parallel to the depositional strike and by rapid thinning downdip. The term was proposed by Lawson (1913, p.329) for the coarser, consolidated rock material occurring in the upper part of an alluvial fan. Cf: *bajada breccia.*

fanhead The area on an alluvial fan close to its apex. Cf: *fan bay.*

fanhead trench A linear depression formed by a drainage line that is incised considerably below the surface of an alluvial fan.

fan mesa An alluvial-fan remnant left standing after dissection of the fan.

fan scarp A *piedmont scarp* formed by faulting, occurring entirely within alluvium and not observed to cross bedrock in any part of its course (Billings, 1954, p. 156).

fan-shaped delta *arcuate delta.*

fan shooting A type of seismic shooting in which detectors are laid out along an arc so that each detector is in a different direction at roughly the same distance from a single shot point. It was used in the 1920's and 1930's to detect the presence of shallow salt domes intruding low-velocity sediments. Syn: *arc shooting.*

fan structure The fold structure of a *normal* anticlinorium.

fan system A fan-shaped pattern formed by diverging trabeculae in the plane of the septum of a scleractinian coral. See also: *axis of divergence.*

fan-topped pediment A pediment with a thin (15-60 m) covering of alluvial fans built upon it in response to some minor change of climate or of other controlling conditions (Blackwelder, 1931, p. 139). Cf: *concealed pediment.*

fan valley A valley in a submarine fan; a continuation of a submarine canyon. It is either V-shaped or trough-shaped, and has natural levees and distributaries.

faradaic path One of the two available paths for transfer of energy across an electrolyte-metal interface. Energy is carried by conversion of atom to ion or vice versa, due to electrochemical reaction and ion diffusion. Cf: *nonfaradaic path.*

faratsihite A pale-yellow clay mineral: $(Al,Fe)_2Si_2O_5(OH)_4$. It has been regarded as an iron-bearing variety of kaolinite, a mixture of kaolinite and nontronite, and identical with nontronite.

farinaceous Pertaining to a texture or structure of a rock or sediment that is mealy, soft, and friable, e.g. a limestone or a pelagic ooze; also, said of a lava flow in which the particles seem to be in a state of mutual repulsion. Syn: *mealy.*

far infrared Pertaining to the longer wavelengths of the *infrared* region, from 15 μm to 1 mm. The atmosphere transmits very little radiation at these wavelengths, so terrestrial use of this spectral band is severely limited. Cf: *reflected infrared.*

faro A small, atoll-shaped or oblong *reef* with a lagoon up to 30 m deep, forming part of the rim of a barrier reef or of an atoll. Etymol: political subdivision of an *atollon* in the Maldive Islands of the Indian Ocean.

far range The portion of a SLAR image farthest from the aircraft flight path.

farreoid Said of the skeleton of a dictyonine hexactinellid sponge in which the dictyonal strands occur in a single layer parallel to the sponge surface. Cf: *euretoid.*

farringtonite A colorless, wax-white, or yellow meteorite mineral: $Mg_3(PO_4)_2$.

farrisite A fine-grained hypabyssal rock composed primarily of a melilite-like mineral, making up approximately one third of the rock, and barkevikite, with smaller amounts of biotite, olivine, and magnetite. Feldspars and nepheline are almost entirely replaced by zeolites. The name, given by Brögger in 1898, is from Lake Farris in the Oslo district, Norway. Not recommended usage.

farsundite A hypersthene- and hornblende-bearing *granite* of the charnockite series. Cf: *opdalite.* The name, given by Kalderup in 1904, is for Farsund, Norway. Not recommended usage.

fascicular Said of an aggregate of *acicular* crystals.

fascicular columella A *columella* in scleractinian corals, formed by twisted vertical ribbons or rods resembling pali or paliform lobes.

fasciculate Arranged in fascicles; e.g. said of a corallum with cylindrical corallites that are somewhat separated from one another but may be joined by connecting tubules, or descriptive of ribbing in coiled ammonoid conchs having bunched or bundled ribs near the umbilical margin.

fasciole (a) A heavily ciliated tract in an echinoderm; esp. a narrow band of small tubercles bearing densely ciliated spines (clavulae) on the denuded test of certain echinoids. The term is also applied to a narrow band of such spines in which cilia beat to create currents. (b) A band generated on a gastropod shell by a narrow sinus or notch in, or a lamellose projection of, successive growth lines (TIP, 1960, pt.I, p.131).—See also: *anal fasciole.*

fashioning The sawing, cleaving, rounding up, facet-grinding and polishing, and other operations employed in preparing rough gem material for use in jewelry.

fasibitikite A medium-colored riebeckite-acmite *granite* that also contains eucolite and zircon. The name, given by Lacroix in 1915, is derived from the locality Ampasibitika, Malagasy. Not recommended usage.

fasinite A coarse-grained *melteigite* that contains titanaugite and nepheline as its main components along with alkali feldspar, olivine, and biotite. It has the same chemical composition as *berondrite* and differs from *bekinkinite* by the absence of hornblende and analcime. The name, given by Lacroix in 1916, is derived from Ampasindava, Malagasy. Not recommended usage.

fassaite A pale-green to dark-green variety of pyroxene containing considerable aluminum substituting for silicon: $(Ca,Mg,Fe^{+3},Al,Ti)(Si,Al)_2O_6$.

fast ice Any sea ice that forms along and remains attached to the coast (e.g. *icefoot*), or that forms between grounded icebergs, or is attached to the bottom in shallow water (e.g. *anchor ice*). Fast ice may form in situ from seawater or by freezing of *pack ice* to the shore. It may extend a few meters to several hundred kilometers from the coast.

fastigate Said of an ammonoid with a roof-shaped venter, the periphery of the shell being sharpened but not keeled.

fastland A *mainland;* esp. one that is high and dry near water, such as an *upland.*

fast ray In crystal optics, that component of light in any birefringent crystal section that travels with the greater velocity and has the lower index of refraction. Cf: *slow ray.*

fat clay A cohesive and compressible clay of high plasticity, containing a high proportion of minerals that make it greasy to the feel. It is difficult to work when damp, but strong when dry. Ant: *lean clay.* Syn: *long clay.*

fathogram The graphic record produced by a *fathometer;* a type of *echogram.*

fathometer A copyrighted name for a type of *echo sounder.* See also: *fathogram.*

fatigue Failure of a material after many repetitions of a stress that of itself is not strong enough to cause failure.

fatigue limit *endurance limit.*

fatigue ratio *endurance ratio.*

fatty acid Any one of a group of organic acids that occur in animal and vegetable oils and fats. Common examples are butyric acid ($C_4H_8O_2$); palmitic acid ($C_{16}H_{32}O_2$); stearic acid ($C_{18}H_{36}O_2$); and oleic acid ($C_{18}H_{34}O_2$).

fauces terrae A term used in international law for headlands and promontories that enclose territorial water that is not part of the high seas. Etymol: Latin, "gulf".

faujasite A cubic zeolite mineral: $(Na_2,Ca)Al_2Si_4O_{12}\cdot 6H_2O$. Cf: *gmelinite.*

fault [cryst] (a) A general term for a dislocation in a crystal. (b) A *stacking fault.*

fault [struc geol] A fracture or a zone of fractures along which there has been displacement of the sides relative to one another parallel to the fracture. Obsolete syn: *paraclase.*

fault apron A mass of rock waste deposited along the base of a fault scarp, formed by numerous coalescing alluvial cones.

fault basin A depression separated from the surrounding area by faults.

fault bench A small *fault terrace.*

fault block A crustal unit formed by *block faulting;* it is bounded by faults, either completely or in part. It behaves as a unit during block faulting and tectonic activity. An example is the Sierra Nevada of California. See also: *tilt block.* Syn: *block [tect].*

fault-block mountain *block mountain.*

fault breccia (a) A *tectonic breccia* composed of angular fragments resulting from the crushing, shattering, or shearing of rocks during movement on a fault, from friction between the walls of the fault, or from distributive ruptures associated with a major fault; a *friction breccia.* It is distinguished by its cross-cutting relations, by the presence of *fault gouge,* and by slickensided blocks. Syn: *dislocation breccia.* (b) A term sometimes used as a syn. of *fault rubble.*

fault cliff A cliff formed by faulting; esp. a *fault scarp.*

fault coast A coast formed directly by faulting (see Cotton, 1916), as one along a fault line, a fault scarp, or a narrow arm of the sea that floods a fault trough (e.g. the coast of the Red Sea). Cf: *fault-line coast.*

fault complex A group of faults that interconnect and intersect, having the same or different ages.

fault-dam spring *fault spring.*

fault embayment A fault trough, or other depressed region in a fault zone or between two faults, invaded by the sea; e.g. the Red Sea.

fault escarpment *fault scarp.*

fault fissure A fissure that is the result of faulting. It may or may not be filled with vein material.

fault-fold A structure that is associated with a combination of folding and nearly vertical faulting, in which crustal material that has been fractured into elongate strips tends to drape over the uplifted areas to resemble anticlines, and to crumple into the downthrown areas to resemble synclines. The structure has been described in parts of Germany (Hills, 1963).

fault gap A depression between the offset ends of a ridge, formed by a transverse fault that laterally displaces the ridge so that the two parts are no longer continuous (Lahee, 1961, p. 356). Cf: *fault-line gap.*

fault gouge Soft, uncemented pulverized clayey or claylike material, commonly a mixture of minerals in finely divided form, found along some faults or between the walls of a fault, and filling or partly filling a fault zone; a slippery mud that coats the fault surface or cements the *fault breccia.* It is formed by the crushing and grinding of rock material as the fault developed, as well as by subsequent decomposition and alteration caused by underground circulating solutions. Syn: *gouge; clay gouge; selvage.*

fault-graded beds Gradationally compacted marine muds that have been deformed by earthquake shock (Seilacher, 1969). A liquefied zone is underlain by a rubble zone and this by a step-faulted zone. See also: *seismite.*

fault growth Intermittent, small-scale movement along a fault surface that, accumulated, results in considerable displacement.

faulting The process of fracturing and displacement that produces a *fault.*

fault ledge *fault scarp.*

fault line The trace of a fault plane on the ground surface or on a reference plane. Syn: *fault trace.*

fault-line adj. Said of a secondary or subsequent landform or feature created solely by processes (such as erosion) acting upon faulted materials; e.g. a "fault-line outlier", an isolated hillock or ridge capped by resistant rock and created by differential erosion of a low-angle thrust mass (Sharp, 1954, p. 27).

fault-line coast A coast formed by the partial submergence of a *fault-line scarp* so that the waters of a sea or lake rest against the scarp.

fault-line gap A gap produced solely by erosion of a resistant ridge laterally offset by earlier faulting; e.g. such a gap located along the line of outcrop of a dip fault or of a diagonal fault that intersects the rock layer of the ridge (Lahee, 1961, p. 367). Cf: *fault gap.*

fault-line saddle A saddle created by rapid erosion of a ridge crest where it is crossed by a fault (Sharp, 1954, p. 27). Examples occur along the San Gabriel Fault within the San Gabriel Range in southern California. Cf: *fault saddle.*

fault-line scarp (a) A steep slope or cliff formed by differential erosion along a fault line, as by the more rapid erosion of soft rock on one side of a fault as compared to that of more resistant rock on the other side; e.g. the east face of the Sierra Nevada in California. See also: *obsequent fault-line scarp; resequent fault-line scarp.* Syn: *erosion fault scarp.* (b) A *fault scarp* that has been modified by erosion. This usage is not recommended because the scarp is usually not located on the fault line (Washburne, 1943, p. 496).

fault-line valley A valley that is formed along or follows a fault line; e.g. a subsequent valley developed by headward erosion in the soft, crushed, relatively weak material along a fault zone. Cf: *fault valley.*

fault-line-valley shoreline A shoreline formed by the partial submergence of a valley that has been eroded along the crushed zone of a fault or along a narrow strip of faulted weak rock, such as along the coast of northern Nova Scotia.

fault plane A *fault surface* that is more or less planar.

fault-plane solution Determination of the orientation of a fault plane and the direction of slip motion on it from an analysis of the sense of first motion of the *P waves* and/or the amplitudes of the P waves, *S waves,* and *surface waves.* The solution also gives the orientation of the principal axes of compression and tension. See also: *slip-vector analysis.* Syn: *focal mechanism.*

fault rubble An assemblage of detached, jumbled, and crushed or shattered angular fragments torn from the walls of a fault; an unconsolidated *fault breccia.*

fault saddle A particular type of *kerncol,* being a notch, col, or saddle in a ridge, created by actual displacement of the ridge crest by faulting (Sharp, 1954, p. 25). Cf: *fault-line saddle.*

fault sag A small, enclosed depression along an active or recent fault. It is caused by differential movement between slices and blocks within the fault zone or by warping and tilting associated with differential displacement along the fault, and it forms the site of a *sag pond.* Term introduced by Lawson et al. (1908, p. 33). Syn: *sag [struc geol].*

fault scarp (a) A steep slope or cliff formed directly by movement along a fault and representing the exposed surface of the fault before modification by erosion and weathering. It is an initial landform. Cf: *fault-line scarp.* Syn: *fault escarpment; fault cliff; fault ledge; cliff of displacement.* (b) A term used in England for any scarp that is due to the presence of a fault, even though the relief may be erosional.

fault-scarp shoreline A shoreline produced by recent faulting.

fault set A group of faults that are parallel or nearly so, and that are related to a particular deformational episode. Cf: *fault system.*

fault-slice ridge *slice ridge.*

fault splinter A narrow, ramplike connection between the opposite ends of two parallel normal faults. The feature occurs in major fault zones (Strahler, 1963, p. 596). Cf: *fault step.*

fault spring A spring flowing onto the land surface from a fault that brings a permeable bed into contact with an impermeable bed. Cf: *fracture spring; fissure spring.* Syn: *fault-dam spring.*

fault step Along a normal fault expressed at the surface, one of a series of thin rock slices along which the fault's total displacement is dispersed (Strahler, 1963, p. 596).

fault surface In a fault, the surface along which displacement has occurred. Cf: *fault plane.*

fault system (a) Two or more interconnecting *fault sets.* (b) A syn. of *fault set.*

fault terrace An irregular, terrace-like tract between two fault scarps, produced on a hillside by step faulting in which the downthrow is systematically on the same side of two approximately parallel faults. Cf: *fault bench.*

fault trace *fault line.*

fault trap A trap for oil or gas in which the *closure* results from the presence of one or more faults.

fault-trellis drainage pattern A *trellis drainage pattern* developed where a series of parallel faults have brought together alternating bands of hard and soft rocks (Thornbury, 1954, p. 123).

fault trench A cleft or crack formed on the Earth's surface as a result of faulting. It is a smaller-scale feature than a fault trough (rift valley).

fault trough *rift valley.*

fault-trough lake *sag pond.*

fault valley A linear depression produced by faulting; e.g. a small, narrow valley created within a major fault zone by relative depression of narrow slices, or a large graben situated between tilted block mountains, or a valley created by relative uplift on opposite sides of two parallel thrust faults. Cf: *fault trough; fault-line valley.*

fault wall *wall [fault].*

fault wedge A wedge-shaped rock mass bounded by two faults.

fault zone A fault that is expressed as a zone of numerous small fractures or of breccia or fault gouge. A fault zone may be as wide as hundreds of meters. Cf: *step fault.* Syn: *distributed fault; distributive fault.* Less-preferred syn: *shatter belt.*

fauna (a) The entire animal population, living or fossil, of a given area, environment, formation, or time span. Cf: *flora.* (b) Sometimes incorrectly used to include both the animal and plant fossils of a particular rock unit, i.e. the *biota.* Adj: *faunal.*

faunal break An abrupt change or *break [stratig]* from one fossil assemblage to another at a definite horizon in a stratigraphic sequence, usually produced by an unconformity or hiatus or sometimes by a change in bottom ecology without interruption of deposition; e.g. a gap in the orderly evolution of a single organism through a vertical series of beds.

faunal diversity (a) *Diversity* of a fauna. (b) Occasionally used more precisely to denote the number of species, either estimated or counted, whose combined totals comprise 95% of the total population.

faunal dominance That percentage of a population constituted by the most common species.
faunal evolution Change in the composition of a fauna with time.
faunal province A geographic region characterized by a specific assemblage of animals more or less widely distributed within it.
faunal stage A chronostratigraphic unit (stage) based on a faunizone.
faunal succession The observed chronologic sequence of life forms (esp. animals) through geologic time. See also: *law of faunal succession.*
faunal zone *faunizone.*
faunichron A term used by Dunbar & Rodgers (1957, p.300) for the geologic-time unit corresponding to faunizone of Buckman (1902).
faunizone (a) A biostratigraphic unit or body of strata characterized by a particular assemblage of fossils (specif. fossil faunas), regardless of whether it is inferred to have chronological or only environmental significance. (b) A term sometimes used for the strata equivalent in age to a certain overlap of "biozones" and having dominantly chronostratigraphic significance. (c) A term that has been used in the sense of "zone" regarded as a time unit of biochronologic significance.—The term was introduced by Buckman (1902) for "belts of strata, each of which is characterized by an assemblage of organic remains", and has been generally regarded as the animal-based variety of (biostratigraphic) *zone* of Oppel (1856-1858). The American Commission on Stratigraphic Nomenclature (1961, art.21d) states that the term is "not generally accepted" and that its correct definition is "in dispute". See also: *assemblage-zone.* Cf: *florizone.* Syn: *faunal zone.*
faunula (a) A set of animal species found in a relatively small and isolated region, and not peculiar to it. (b) *faunule.*
faunule (a) A collection of fossil animals obtained from a stratum over a very limited geographic area, esp. from only one outcrop. Syn: *local fauna.* (b) A term used by Fenton & Fenton (1928) for an assemblage of fossil animals associated in a single stratum or a few contiguous strata of limited thickness and dominated by the representatives of one community; the faunal assemblage of a *zonule.* Cf: *florule.* Syn: *faunula.*
faustite An apple-green mineral: $(Zn,Cu)Al_6(PO_4)_4(OH)_8 \cdot 5H_2O$. It is the zinc analogue of turquoise.
favositid Any tabulate coral belonging to the family Favositidae, characterized by massive colonies (usually without coenenchyme) of slender corallites with mural pores, short equal spinose septa, and complete tabulae. Range, Upper Ordovician to Permian, possibly Triassic.
***f* axis** A term used in crystal plasticity to denote a line in the crystal slip plane at right angles to the slip direction (*t* direction). It is commonly an axis of rotation of the crystal lattice during deformation.
fayalite A brown to black mineral of the olivine group: Fe_2SiO_4. It is isomorphous with forsterite, and occurs chiefly in igneous rocks. Symbol: Fa. Syn: *iron olivine.*
F center A type of *color center* in a crystal that is formed by a negative ion vacancy with two bound electrons. The F stands for Farbe, the German word for "color".
F-centered lattice *face-centered lattice.*
F-coal Microscopic coal particles that are predominantly fusain, as found in miners' lungs. Cf: *V-coal; D-coal.*
F-distribution test *F test.*
feather [gem] (a) A series of elongated and irregular liquid inclusions in a gemstone, grouped together in orderly proximity to each other so as to resemble the overall pattern of a bird's feather. (b) Any flaw inside a gemstone, such as a jagged fracture that is white in appearance. (c) In diamonds, a cleavage or fracture that has a feathery appearance when viewed at right angles to the separation plane.
feather [photo] v. To thin the edge of a photographic print, before assembly into a mosaic, by abrading the back surface with sandpaper or emery paper. Feathering is done to obtain a smooth mosaic surface and to reduce or eliminate shadows or sharp changes in contrast. Syn: *featheredging.*
feather alum (a) *halotrichite.* (b) *alunogen.*
feather amphibolite A metamorphic rock in which porphyroblastic crystals of amphibole (usually hornblende) tend to form stellate or sheaflike groups on the planes of foliation or schistosity. Cf: *amphibolite.* Syn: *garbenschiefer.*
feather edge The thin edge of a bed of sedimentary rock where it disappears by thinning, pinching, or wedging out. Also spelled: *featheredge.* Syn: *knife edge.*
featheredging *feathering.*
feather fracture A less-preferred syn. of *plume structure.* Although it was the original term, its use would lead to confusion with the term *feather jointing.*
feather ice *pipkrake.*
feathering [cart] The technique of progressively dropping contour lines to avoid congestion on steep slopes. A former practice, of tapering the line weight near the end of the contour line to be dropped, was abandoned with the advent of the scribing technique. Syn: *featheredging.*
feathering [seis] En-echelon arrangement of successive spreads, such as produced in marine shooting when a cross current causes the cable to drift at an angle to the seismic line.
feather jointing A joint pattern formed in a fault zone by shear and tension. The joints appear to the fault as the barbs of a feather to its shaft. Syn: *pinnate jointing.*
feather ore A capillary, fibrous, or feathery form of an antimony-sulfide mineral, such as stibnite or boulangerite; specif. jamesonite.
feather out To end irregularly. The term is applied to lenticular bodies of rock.
feather quartz Quartz in imperfect crystals, the bases of which meet at an angle along a crystal plane so that a cross section looks somewhat like a feather.
feather zeolite *hair zeolite.*
fecal pellet An organic excrement, mainly of invertebrates, occurring esp. in modern marine deposits but also fossilized in some sedimentary rocks, usually of a simple ovoid form less than a millimeter in length, or more rarely rod-shaped with longitudinal or transverse sculpturing, devoid of internal structure, and smaller than a *coprolite.* Also spelled: *faecal pellet.* Cf: *casting [paleont].*
federovskite An orthorhombic mineral: $Ca_2Mg_2B_4O_7(OH)_6$. It forms a series with roweite.
fedorite A mineral: $(Na,K)CaSi_4(O,OH)_{10} \cdot 1.5H_2O$.
Fedorov stage *universal stage.*
feedback Partial reversal of a certain process to its beginning or to a preceding stage as a means of reinforcement or modification, esp. in biologic, psychologic, and social systems.
feeder [eco geol] *channelway [ore dep].*
feeder [intrus rocks] The conduit through which magma passes from the magma chamber to some localized intrusion, e.g. a feeder dike.
feeder [streams] *tributary.*
feeder [volc] *chimney [volc].*
feeder beach An artificially widened beach serving to nourish downdrift beaches by natural littoral currents.
feeder current The part of a rip current that flows parallel to the shore (inside the breakers) before converging with other feeder currents to form the *neck* of the rip current.
feeding channel *channelway [ore dep].*
feeding esker A small esker joining a larger one.
feeding ground *drainage basin.*
feidj A term used in the Saharan region for a sand-covered interdune passage. Cf: *gassi.* Other spellings: *feidsh; feij; fejj.*
feitknechtite A mineral: β-MnO(OH). Cf: *manganite.*
fei ts'ui An emerald- or bluish-green variety of jadeite from Burma, esp. one resembling the color of the brilliant blue-green back of the kingfisher. Etymol: Chinese, "kingfisher jade".
felder Crustal blocks of a polygonal, mosaic pattern that are produced by *taphrogeny.* Syn: *tesserae.*
feldmark *fell-field.*
feldspar (a) A group of abundant rock-forming minerals of general formula: $MAl(Al,Si)_3O_8$, where M = K, Na, Ca, Ba, Rb, Sr, and Fe. Feldspars are the most widespread of any mineral group and constitute 60% of the Earth's crust; they occur as components of all kinds of rocks (crystalline schists, migmatites, gneisses, granites, most magmatic rocks) and as fissure minerals in clefts and druse minerals in cavities. Feldspars are usually white or nearly white and clear and translucent (they have no color of their own but are frequently colored by impurities), have a hardness of 6 on the Mohs scale, frequently display twinning, exhibit monoclinic or triclinic symmetry, and possess good cleavage in two directions (intersecting at 90° as in orthoclase and at about 86° as in plagioclase). On decomposition, feldspars yield a large part of the clay of soil and also the mineral kaolinite. (b) A mineral of the feldspar group, such as alkali feldspar (orthoclase, microcline), plagioclase (albite, anorthite), and celsian.—Syn: *felspar; feldspath.*

feldspath *feldspar.*

feldspathic Said of a rock or other mineral aggregate containing feldspar.

feldspathic arenite A sandstone containing abundant quartz, chert, and quartzite, less than 10% argillaceous matrix, and 10-25% feldspar (generally fresh and limpid), and characterized by an abundance of unstable materials in which the feldspar grains exceed the fine-grained rock fragments (Williams, Turner & Gilbert, 1954, p.294 & 316). It is less feldspathic and more mature than *arkosic arenite.* The rock is roughly equivalent to *subarkose.* See also: *feldspathic sandstone.*

feldspathic graywacke (a) A graywacke characterized by abundant unstable materials; specif. a sandstone containing generally less than 75% of quartz and chert and 15-75% detrital clay matrix, and having feldspar grains (chiefly sodic plagioclase, indicating a plutonic provenance) in greater abundance than rock fragments (indicating a supracrustal provenance) (Pettijohn, 1954; 1957, p.303). Williams, Turner and Gilbert (1954, p.294) give a feldspar content of 10-25% and an argillaceous matrix greater than 10%; the rock is less feldspathic than *arkosic graywacke.* It is equivalent to the *high-rank graywacke* of Krynine (1945). (b) A term used by Folk (1954, p.354) for a sandstone containing 25-90% micas and metamorphic rock fragments, 10-50% feldspars and igneous-rock fragments, and 0-65% quartz and chert. Cf: *impure arkose.* (c) A term used by Hubert (1960, p.176-177) for a sandstone containing 25-90% micas and micaceous metamorphic-rock fragments, 10-50% feldspars and feldspathic crystalline-rock fragments, and 0-65% quartz, chert, and metamorphic quartzite. Cf: *micaceous arkose.*—Cf: *lithic graywacke.* Syn: *lithic arkosic wacke.*

feldspathic litharenite (a) A term used by McBride (1963, p.667) for a litharenite containing appreciable feldspar; specif. a sandstone containing 10-50% feldspar, 25-90% fine-grained rock fragments, and 0-65% quartz, quartzite, and chert. (b) A term used by Folk (1968, p.124) for a sandstone containing less than 75% quartz and metamorphic quartzite and having a "F/R ratio" between 1:1 and 1:3, where "F" signifies feldspars and fragments of gneiss and granite, and "R" signifies all other fine-grained rock fragments.—Cf: *lithic arkose.*

feldspathic lithwacke Essentially a *lithic graywacke* (over 15 percent matrix) in which rock fragments exceed feldspar but the latter forms 10 percent or more of the sand fraction (Pettijohn, Potter & Siever, 1973, p. 164).

feldspathic polylitharenite A polylitharenite containing more than 10% feldspar (Folk, 1968, p. 135).

feldspathic quartzite (a) A term used by Hubert (1960, p.176-177) for a sandstone containing 70-95% quartz, chert, and metamorphic quartzite, 5-15% feldspars and feldspathic crystalline-rock fragments, and 0-15% micas and micaceous metamorphic-rock fragments. Cf: *micaceous quartzite.* (b) A term used by Pettijohn (1949, p.227) for a well-indurated *feldspathic sandstone,* and later (1954, p.364) as a syn. of *subarkose.*

feldspathic sandstone A feldspar-rich sandstone; specif. a sandstone intermediate in composition between an *arkosic sandstone* and a quartz sandstone, containing 10-25% feldspar and less than 20% matrix material of clay, sericite, and chlorite (Pettijohn, 1949, p.227). Petttijohn (1957, p.322) redefined the term and used it as a less-preferred syn. of *subarkose.* Krumbein & Sloss (1963, p.170) used the term for a quartzose sandstone with 10-25% feldspar (mainly potassic feldspar), and Williams, Turner and Gilbert (1954, p.316) used it as a general term to include *feldspathic arenite* and feldspathic wacke. See also: *feldspathic quartzite; arkose.*

feldspathic shale A shale characterized by a feldspar content greater than 10% in the silt size and by a finer matrix of kaolinitic clay minerals, commonly associated with arkose, and representing the removal of finer material from coarser arkosic debris (Krumbein & Sloss, 1963, p.175). Syn: *kaolinitic shale.*

feldspathic subgraywacke A term used by Folk (1954, p.354) for a sandstone composed of subequal amounts of rock fragments of igneous and metamorphic derivation; specif. a sandstone containing 10-25% feldspars and igneous-rock fragments, 10-25% micas and metamorphic-rock fragments, and 50-80% quartz and chert.

feldspathic sublitharenite *lithic subarkose.*

feldspathic wacke A sandstone containing abundant quartz, chert, and quartzite, more than 10% argillaceous matrix, and 10-25% feldspar (esp. sodic plagioclase), and characterized by an abundance of unstable materials in which the feldspar grains exceed the fine-grained rock fragments (Williams, Turner & Gilbert, 1954, p.292 & 316). It is less feldspathic and more mature than *arkosic wacke.* Syn: *subarkosic wacke.*

feldspathide *feldspathoid.*

feldspathization The formation of feldspar in a rock, usually as a result of metamorphism. Material for the feldspar may come from the country rock or be introduced by magmatic or other solutions.

feldspathoid (a) A group of comparatively rare rock-forming minerals consisting of aluminosilicates of sodium, potassium, or calcium and having too little silica to form feldspar. Feldspathoids are chemically related to the feldspars, but differ from them in crystal form and physical properties; they take the places of feldspars in igneous rocks that are undersaturated with respect to silica or that contain more alkalies and aluminum than can be accommodated in the feldspars. Feldspathoids may be found in the same rock with feldspars but never with quartz or in the presence of free magmatic silica. See also: *foid; lenad.* (b) A mineral of the feldspathoid group, including leucite, nepheline, sodalite, nosean, hauyne, lazurite, cancrinite, and melilite.—Syn: *felspathoid; feldspathide.*

feldspathoidite A rarely used name applied to a group of igneous rocks that contains the most feldspathoid-rich of the *foidites.* Proposed by Johannsen (1939) for monomineralic feldspathoid rocks.

felite *belite.*

fell (a) A term used in Scotland and northern England for a bare, uncultivated, open hillside or mountain. (b) A term used in Great Britain for an elevated tract of wasteland or a mountain moorland; a *fell-field.*—Etymol: Scandinavian. See also: *fjeld; fjäll.*

fell-field An open, treeless, rock-strewn area that is above the timberline or in a high latitude and that has a sparse ground cover of low plants or grasses and sedges. Syn: *fell; feldmark; fjeldmark.*

fellside A hillside or mountainside. Rarely used.

feloid A group name for the feldspar and feldspathoid minerals.

fels An unfortunate term applied to massive metamorphic rock lacking schistosity or foliation, e.g. calcsilicate fels (Winkler, 1967). Cf: *granofels.*

felsenmeer *block field.*

felsic A mnemonic adjective derived from *fe*ldspar + *l*enad (feldspathoid) + *si*lica + *c,* and applied to an igneous rock having abundant light-colored minerals in its mode; also, applied to those minerals (quartz, feldspars, feldspathoids, muscovite) as a group. It is the complement of *mafic.*

felsic index A chemical parameter of igneous rocks, equal to 100 $\times$ $(Na_2O+K_2O)/(Na_2O+K_2O+CaO)$. It ranges from about 25 (basalt) to 100 (rhyolite). It is most commonly plotted as the abscissa on variation diagrams, on which the ordinate represents the mafic index. It reflects changes produced by fractional crystallization of the felsic minerals (Simpson, 1954). Cf: *mafic index.* Abbrev: FI.

felside An informal term proposed by Johannsen (1938) for field use, to be applied to any fine-grained light-colored nonporphyritic igneous rock, e.g. nonporphyritic rhyolite, trachyte, phonolite, latite, and light-colored andesite.

felsiphyric A syn. of *cryptocrystalline* originally used by Cross et al. (1906, p. 703).

felsite A general term for any light-colored, fine-grained or aphanitic extrusive or hypabyssal rock, with or without phenocrysts and composed chiefly of quartz and feldspar; a rock characterized by *felsitic* texture. Syn: *felstone.* Cf: *aphanite; felsitoid; felsophyre; mafite.*

felsitic A syn. of *aphanitic* applied to the light-colored dense rocks, with "aphanitic" being then reserved for the dark-colored; of or pertaining to a *felsite.* The term is not recommended because in its original use it was applied to a mineral substance now known to be a mixture of quartz and feldspar.

felsitoid An informal term applied to any light-colored igneous rock in which the mineral grains are too small to be distinguished by the unaided eye. Cf: *felsite.* Syn: *aphanite.*

felsoandesite An *andesite* having a felsitic groundmass; a felsitic andesite. Not recommended usage.

felsöbanyite A snow-white mineral: $Al_4(SO_4)(OH)_{10}\cdot 5H_2O$. It has the same formula as, but a different X-ray pattern from, basaluminite. Incorrectly spelled: *felsobanyite.*

felsophyre A general term for any porphyritic *felsite.* Syn: *aphanophyre.* Cf: *vitrophyre; granophyre.*

felsophyric A syn. of *microcrystalline* originally used by Cross et al. (1906, p. 703).

felsosphaerite A spherulite composed of a felsitic substance. Obsolete.

felspar A chiefly British spelling of *feldspar.*

felspathoid *feldspathoid.*

felstone An obsolete syn of *felsite.*

felty *pilotaxitic.*

femag *mafic.*

femic Said of an igneous rock having one or more normative, dark-colored iron-, magnesium-, or calcium-rich minerals as the major components of the norm; also, said of such minerals. Etymol: a mnemonic term derived from *fe*rric + *m*agnesium + *ic.* Cf: *basic; salic; mafic; felsic.*

femur (a) The thigh-bone of a tetrapod. (b) The third segment of a leg of an arachnid, forming the "hip" articulation with the preceding segment (trochanter) and the "knee" articulation with the following segment (patella) (TIP, 1955, pt.P, p.61). (c) A joint belonging to the proximal part of a prosomal appendage of a merostome.

fen Waterlogged, spongy ground containing alkaline decaying vegetation, characterized by reeds, that may develop into peat. It sometimes occurs in the sinkholes of karst regions. Cf: *bog.*

fenaksite A pale-rose monoclinic mineral: $(K,Na)_4(Fe,Mn)_2(Si_4O_{10})_2(OH,F)$. Not to be confused with *phenakite.*

fence diagram [geochem] A diagram of chemical factors, such as Eh and pH, that influence mineral stability, having discrete fields defined by boundaries between phases in an assemblage of minerals, rocks, or compounds. Cf: *geochemical fence.*

fence diagram [geol] A drawing in perspective of three or more geologic sections, showing their relationships to one another.

fenestra [paleont] (a) A small opening in an invertebrate; e.g. an open space in a reticulate or anastomosing bryozoan colony, or an open or closed window in the wall or lorica of a tintinnid. Pl: *fenestrae.* Syn: *fenestrule.* (b) An opening in a bone that does not serve as a passage for blood vessel or nerve.

fenestra [sed] A term used by Tebbutt et al. (1965, p.4) for *shrinkage pore*, or "primary or penecontemporaneous gap in rock framework, larger than grain-supported interstices". It may be an open space in the rock, or be completely or partly filled with secondarily introduced sediment or cement. Also used to describe the porosity and fabric of rocks with fenestral features (Choquette & Pray, 1970, p.246). Pl: *fenestrae*.

fenestrate Having openings or transparent areas; perforated or reticulated. The term has been applied esp. to bryozoans possessing small windowlike openings between branches, arranged in a reticulate or anastomosing pattern; to corals having regularly perforated septa; and to pollen exhibiting large geometrically arranged holes in the exine. Syn: *fenestrated.*

fenestrated *fenestrate.*

fenestrule *fenestra [paleont].*

fenêtre A syn. of *window.* Etymol: French, "window".

fengluangite An antimonian variety of guanglinite.

fenite A quartzo-feldspathic rock that has been altered by alkali metasomatism at the contact of a carbonatite intrusive complex. The process is called *fenitization.* Fenite is mostly alkalic feldspar, with some aegirine, subordinate alkali-hornblende, and accessory sphene and apatite.

fenitization As generally used today, widespread alkali metasomatism of quartzo-feldspathic country rocks in the environs of carbonatite complexes. The name fenite for the altered rock was originated by Brögger (Turner & Verhoogen, 1960).

fen peat *lowmoor peat.*

fenster A syn. of *window.* Etymol: German, "window".

feral Said of an "unsubdued" landform or landscape in early maturity, when the crests of ridges and spurs are shaped by the intersection of the slopes of valley sides that are for the most part still steep, so that the ridges are sharp and serrate (Cotton, 1958, p. 110). Cf: *subdued.*

ferberite A grayish to black mineral of the wolframite series: $FeWO_4$. It is isomorphous with huebnerite, and may contain up to 20% manganese tungstate.

ferdisilicite A mineral: $FeSi_2$.

ferghanite A sulfur-yellow secondary mineral, possibly: $(UO_2)_3(VO_4)_2 \cdot 6H_2O$. It is perhaps a leached or weathered product of tyuyamunite. Also spelled: *ferganite.*

fergusite (a) In the *IUGS classification,* a plutonic rock in which F is between 60 and 100, M is between 30 and 50, and potassium exceeds sodium. Cf: *ijolite.* (b) A plutonic foidite containing leucite and 30 to 60 percent mafic minerals, such as olivine, apatite, and biotite, with accessory opaque oxides. Its name, given by Pirsson in 1905, is derived from Fergus County, Montana. Cf: *arkite; missourite; italite.*

fergusonite A brownish-black mineral: $Y(Nb,Ta)O_4$. It is isomorphous with formanite and dimorphous with beta-fergusonite. It may contain erbium, cerium, iron, titanium, and uranium.

Fermat's principle The statement that a seismic wave will follow the path between two points that takes less time than variations of this path. Such a path is called a *minimum-time path.*

fermorite A white mineral of the apatite group: $(Ca,Sr)_5[(As,P)O_4]_3(OH)$.

fern A vascular, nonflowering plant of the class Filicineae. Many modern ferns have complex fronds growing from an underground rhizome, and sporangia grouped on the surface of the leaf.

fernandinite A dull-green mineral: $CaV_2^{+4}V_{10}^{+5}O_{30} \cdot 14H_2O$ (?).

feroxyhyte A hexagonal mineral: $FeO(OH)$. It is a polymorph of goethite, lepidocrocite, and akaganeite.

ferralite (a) A term used in the formerly French parts of North Africa for a soil that originated from basic crystalline rocks that have undergone chemical change and that consist of a mixture of hydrates of iron, aluminum, and sometimes manganese and titanium. (b) A humid, tropical soil formed by the leaching of silica and bases by mildly acidic or neutral solutions and characterized by a large content of iron oxide. A "ferralitic soil" has a silica/sesquioxide ratio of less than 2 (Van Riper, 1962, p.83). Also spelled: *ferrallite.*

ferrazite A mineral: $(Pb,Ba)_3(PO_4)_2 \cdot 8H_2O$ (?).

Ferrel's law The statement that the centrifugal force produced by the rotation of the Earth (*Coriolis force*) causes a *rotational deflection* of currents of water and air to the right in the Northern Hemisphere and to the left in the Southern Hemisphere.

ferretto zone A term used by early geologists for a reddish or reddish-brown B horizon produced under conditions of free subsurface drainage in permeable near-surface material such as loess or sand and gravel extending to the surface or overlain by thin till (Flint, 1957, p.213). It is the well-drained equivalent of *gumbotil.*

ferrian Containing ferric iron. Cf: *ferroan.*

ferricopiapite A mineral of the copiapite group: $Fe_5^{+3}(SO_4)_6O(OH) \cdot 20H_2O$.

ferricrete (a) A term suggested by Lamplugh (1902) for a *conglomerate* consisting of surficial sand and gravel cemented into a hard mass by iron oxide derived from the oxidation of percolating solutions of iron salts. (b) A ferruginous *duricrust.*—Etymol: *ferr*uginous + con*crete.* Cf: *calcrete; silcrete.*

ferricrust (a) A general term for an indurated soil horizon cemented with iron oxide, mainly hematite. (b) The hard crust of an iron concretion.

ferrierite A zeolite mineral: $(Na,K)_2MgAl_3Si_{15}O_{36}(OH) \cdot 9H_2O$.

ferriferous Iron-bearing; said esp. of a mineral containing iron, or of a sedimentary rock that is richer in iron than is usually the case, such as a shale whose iron-oxide content is greater than 15%. Cf: *ferruginous.*

ferrihydrite A hexagonal mineral: $Fe_{10}O_{15} \cdot 9H_2O$.

ferrilith A term suggested by Grabau (1924, p. 298) for an iron-rich sedimentary rock (ironstone). Syn: *ferrilyte.*

ferrimagnetism A type of *magnetic order*, macroscopically resembling *ferromagnetism*. Magnetic ions at different crystal sites are opposed, i.e., antiferromagnetically coupled. There is nevertheless a net magnetization because of inequality in the number or magnitude of atomic magnetic moments at the two sites. This type of magnetic order occurs in magnetite. Cf: *antiferromagnetism.*

ferrimolybdite A yellowish mineral: $Fe_2(MoO_4)_3 \cdot 8H_2O$ (?). It occurs as an earthy powder or incrustation, or as silky, fibrous, and radiating crystals, and is formed by the oxidation of molybdenite. Cf: *molybdite.* Syn: *molybdic ocher.*

ferrinatrite A grayish-white or whitish-green mineral: $Na_3Fe^{+3}(SO_4)_3 \cdot 3H_2O$.

ferrisicklerite A dark-brown mineral: $Li(Fe^{+3},Mn^{+2})PO_4$. It is isomorphous with sicklerite.

ferrisymplesite An amber-brown mineral: $Fe_3(AsO_4)_2(OH)_3 \cdot 5H_2O$. Cf: *symplesite.*

ferrite [ign] A general term applied to grains, scales, and threads of unidentifiable, more or less transparent or amorphous, red, brown, or yellow iron oxide in the groundmass of a porphyritic rock (Johannsen, 1939, p. 177). Cf: *opacite; viridite.*

ferrite [sed] A term used by Tieje (1921, p. 655) for a cemented iron-rich sediment whose particles do not interlock.

ferritungstite A mineral: $Ca_2Fe_2^{+2}Fe_2^{+3}(WO_4)_7 \cdot 9H_2O$. It occurs as a pale-yellowish to brownish-yellow earthy powder. Syn: *tung-*

stic ocher.

ferriturquoise A variety of turquoise containing 5% Fe_2O_3.

ferroactinolite A monoclinic mineral component representing a theoretical end-member of the amphibole group: $Ca_2Fe_5{}^{+2}Si_8O_{22}(OH)_2$. It is a variety of actinolite containing no magnesium, and is isomorphous with tremolite. Syn: *ferrotremolite.*

ferroan Containing ferrous iron. Cf: *ferrian.*

ferroan dolomite A mineral that is intermediate in composition between dolomite and ferrodolomite; specif. *ankerite.*

ferroaxinite A mineral: $Ca_2(Fe,Mn)Al_2BSi_4O_{15}(OH)$.

ferrobasalt A lava marked by strong relative and absolute enrichment in iron. "Total iron normally exceeds 12 or 13%, and MgO is less than 6%. Silica, which shows little enrichment with respect to primitive tholeiites, ranges from about 48 to 50%" (McBirney & Williams, 1969, p. 144).

ferrobustamite A mineral of the pyroxenoid group: $Ca(Fe,Ca,Mn)Si_2O_6$. It is isostructural with bustamite.

ferrocarpholite A mineral: $(Fe,Mg)Al_2Si_2O_6(OH)_4$.

ferrocolumbite An orthorhombic mineral: $FeNb_2O_6$. It forms a series with ferrotantalite and manganocolumbite.

ferrocopiapite *copiapite.*

Ferrod In U.S. Dept. of Agriculture soil taxonomy, a suborder of the soil order *Spodosol,* characterized by having at least six times as much elemental iron as organic carbon in its spodic horizon. A Ferrod has none of the characteristics associated with wetness (USDA, 1975). Cf: *Aquod; Humod; Orthod.*

ferrodiorite A dioritic rock "in which the actual (not normative) plagioclase is less calcic than about An_{50}, and the ferromagnesian minerals iron-rich" (Wager & Brown, 1967, p. 78).

ferrodolomite A mineral component: $CaFe(CO_3)_2$. It is isomorphous with dolomite, but probably does not occur naturally except in ankerite.

ferrogabbro A name given by Wager & Deer in 1939 to igneous rocks in the upper zone of the Skaergaard intrusion, East Greenland, containing iron-rich pyroxenes and olivine. The term was withdrawn by Wager & Brown (1967), in favor of *ferrodiorite.*

ferrohexahydrite A monoclinic mineral: $FeSO_4 \cdot 6H_2O$.

ferrohortonolite A mineral of the fosterite-fayalite solid-solution series, containing 70 to 90% of the fayalite component.

ferromagnesian Containing iron and magnesium; applied to *mafic* minerals.

ferromagnetism A type of *magnetic order* in which all magnetic atoms in a domain have their moments aligned in the same direction; loosely, any type of magnetic order. Cf: *ferrimagnetism; antiferromagnetism.*

ferroselite An orthorhombic mineral: $FeSe_2$. It resembles marcasite.

ferrosilite (a) A mineral component in the orthopyroxene group: $FeSiO_3$. It is the iron analogue of enstatite and occurs in hypersthene, but it does not exist separately in nature. Symbol: Fs. Syn: *iron hypersthene.* (b) A mineral group consisting of *clinoferrosilite* and *orthoferrosilite.*

ferrospinel (a) *hercynite.* (b) A synthetic magnetic substance of spinel structure, containing iron, and being a poor conductor of electricity.

ferrotantalite An orthorhombic mineral: $FeTa_2O_6$. It forms a series with ferrocolumbite and manganotantalite.

ferrotremolite *ferroactinolite.*

ferruccite An orthorhombic mineral: $NaBF_4$.

ferruginate adj. A term restricted by Allen (1936, p. 22) to designate the iron-bearing cement of a sedimentary rock.—v. To stain a rock with an iron compound.

ferruginous (a) Pertaining to or containing iron, e.g. a sandstone that is cemented with iron oxide. Cf: *ferriferous; siderose.* (b) Said of a rock having a red or rusty color due to the presence of ferric oxide (the quantity of which may be very small).

fersilicite A mineral: FeSi.

fersmanite A brown triclinic mineral: $(Ca,Na)_4(Ti,Nb)_2Si_2O_{11}(F,OH)_2$. Sometimes misspelled: *fersmannite.*

fersmite A black mineral: $(Ca,Ce,Na)(Nb,Ti,Fe,Al)_2(O,OH,F)_6$.

ferutite *davidite.*

fervanite A golden-brown mineral: $Fe_4(VO_4)_4 \cdot 5H_2O$. It occurs with radioactive minerals but is not itself radioactive.

Festiniogian European stage: Upper Cambrian (above Maentwrogian, below Dolgellian).

festoon The upfolded or pointed part of a layer in a congeliturbate. Ant: *pocket [cryoped].*

festoon cross-bedding A variety of *trough cross-bedding* described by Knight (1929); it consists of elongate, semi-ellipsoidal, eroded, plunging troughs or scooplike structures that are filled by sets of thin laminae conforming in general to the shapes of the troughs, and that crosscut each other so that only parts of each unit are preserved, resulting in a festoonlike (looped or curved) appearance in section. The cross-beds are deposited on concave surfaces so that both the lower bounding surfaces and the cross-beds are trough-shaped.

festooned pahoehoe A type of *pahoehoe,* the ropy surface of which has been dragged by flow of underlying molten lava into festoon patterns.

fetch (a) A term used in wave-forecasting for the area of the open ocean over the surface of which the wind blows with constant speed and direction, thereby creating a wave system. Syn: *generating area.* (b) The extent of the fetch, measured horizontally in the direction of the wind. Syn: *fetch length.*

fetch length *fetch.*

feuerstein A syn. of *firestone.* Etymol: German *Feuerstein,* "flint".

FFI log *free-fluid index log.*

FI *felsic index.*

fiamme Dark, vitric lenses in welded tuffs, averaging a few centimeters in length, perhaps formed by the collapse of fragments of pumice. The presence of fiamme may be called *flame structure [pyroclast].* Etymol: Italian *fiamma,* "flame". Cf: *piperno.*

fiard Anglicized variant of *fjard.*

fiasconite An anorthite-bearing *leucitite-basanite* that also contains augite, olivine, nepheline, and iron oxides. Its name (Johannsen, 1939) is derived from Montefiascone, Italy. Not recommended usage.

fiber A strengthening cell, usually elongated, tapering, and thick-walled, occurring in various parts of vascular plants.

fiber tracheid A type of *tracheid* that occurs in angiospermous secondary wood, with a thick secondary wall, pointed ends, and small bordered pits having lenticular to slitlike, usually extended, inner apertures (Record, 1934, p.30).

Fibrist In U.S. Dept. of Agriculture soil taxonomy, a suborder of the soil order *Histosol,* characterized by containing much undecomposed identifiable plant fiber that accumulated in bogs that were saturated most of the time. Some Fibrists may be artificially drained to permit cultivation. Fibrists usually have a moist bulk density of less than 0.1 (USDA, 1975). Cf: *Folist; Hemist; Saprist.*

fibroblastic Pertaining to a *homeoblastic* type of texture of a metamorphic rock due to the development during recrystallization of minerals with a fibrous habit. Cf: *nematoblastic.*

fibrocrystalline Characterized by the presence of fibrous crystals.

fibroferrite A yellowish mineral: $FeSO_4(OH) \cdot 5H_2O$.

fibrolite *sillimanite.*

fibrous conodont element A *lamellar conodont element* in which the lamellae are thick and "white matter" is absent or reduced to a thin column along cusp and denticle axes, and whose broken edges are typically frayed.

fibrous habit The tendency of certain minerals, e.g. asbestos, to crystallize in needlelike grains or fibers.

fibrous ice *acicular ice.*

fibrous layer The *secondary layer* in many articulate brachiopods, secreted intracellularly as fibers bounded by cytoplasmic sheaths.

fibrous ligament The part of a ligament of a bivalve mollusk characterized by fibrous structure and in which conchiolin is commonly impregnated with calcium carbonate. It is secreted by the epithelium of the mantle and is elastic chiefly to compressional stresses. Cf: *lamellar ligament.*

fibrous peat Peat in which original plant structures are only slightly altered by degradation of cellulose matter. It is tough and nonplastic. Cf: *pseudofibrous peat; amorphous peat.* Syn: *woody peat.*

fibrous texture In mineral deposits, a pattern of finely acicular, rodlike crystals, e.g. in chrysotile and amphibole asbestos. See also: *cross-fiber.*

fichtelite A white, translucent, crystalline, non-aromatic hydrocarbon, with an approximate hydrogen:carbon ratio of 1:6. It is a resin acid found in fossil wood.

fiducial mark An index or point used as a basis of reference; one of usually four index marks connected with a camera lens (as on the metal frame that encloses the negative) that form an image on the negative or print such that lines drawn between opposing points intersect at and thereby define the *principal point* of the

photograph. Syn: *collimating mark.*
fiducial time A time marked on a record to correspond to some arbitrary time. Such marks may aid in synchronizing different records or may indicate a reference, such as a *datum plane [seis].*
fiedlerite A colorless monoclinic mineral: $Pb_3(OH)_2Cl_4$.
field [eco geol] A region or area that possesses or is characterized by a particular mineral resource, e.g. gold field, coal field.
field [geol] A broad term for the area, away from the laboratory and esp. outdoors, in which a geologist makes firsthand observations and collects data, rock and mineral samples, and fossils.
field [geophys] That space in which an effect, e.g. gravity or magnetism, occurs and is measurable. It is characterized by continuity, i.e. there is a value associated with every location within the space.
field [ice] (a) *ice field [ice].* (b) A very large floe or other unbroken area of sea ice.
field capacity The quantity of water held by soil or rock against the pull of gravity. It is sometimes limited to a certain drainage period, thereby distinguishing it from *specific retention* which is not limited by time. Syn: *field-moisture capacity; normal moisture capacity.*
field classification A preliminary analysis of fossils or hand specimens of rocks or minerals in the field, usually with the aid of a hand lens.
field coefficient of permeability *field permeability coefficient.*
field completion Obtaining additional information in the field in order to edit a topographic map from a compiled manuscript or to fill in and confirm that part of a map manuscript prepared by stereocompilation. It includes: a comprehensive examination of the compilation for completeness, quality, and topographic expression; the addition, deletion, or correction of map features; the classification of buildings, roads, and drainage; the mapping of public-land subdivision lines and civil boundaries; obtaining place names; and checking the map for compliance with vertical-accuracy standards.
field contouring Contouring of a topographic map by field methods accomplished by planetable surveys. It is usually done for terrain unsuitable for contouring by photogrammetric methods. Cf: *field sketching.*
field focus The total area or volume that is the source of an earthquake, inferred from the area of shaking as observed in the field. The concept is inexact and the term is not commonly used.
field geology Geology as practiced by direct observation in the field; original, primary reconnaissance; *field work.*
field ice (a) An obsolete term for *consolidated pack ice* consisting of very large, relatively flat floes many kilometers across. (b) A general term used for all types of *sea ice* except that newly formed.
field intensity The force of attraction exerted on a unit mass particle at a point by the matter causing the force field.
field map A preliminary or original map of the geology of an area, made in the field, on which a final map may be based.
field moisture Water present in the ground above the water table.
field-moisture capacity *field capacity.*
field-moisture deficiency The amount of water required to raise the moisture content of the soil to field capacity.
field-moisture equivalent The minimum water content of a soil mass, expressed as a percentage of its dry weight, at which a drop of water placed on a smoothed surface of the soil will not be absorbed but will spread out, giving a shiny appearance to the soil.
field of force A region of space at each point in which there exists a value of some force.
field of view (a) *angular field of view.* (b) *instantaneous field of view.*
field permeability coefficient The *permeability coefficient* defined for prevailing conditions rather than for a temperature of 60°F. Syn: *coefficient of field permeability.*
field reversal *geomagnetic reversal.*
field sketching The art of drawing contours based on the elevations of selected features located on a planetable sheet. Cf: *field contouring.* See also: *contour sketching.*
field well A well drilled for oil or gas within the area of a pool that has already been essentially proved for production (Lahee, 1962, p. 132-133).
field work *field geology.*
figure of the Earth The *geoid,* or surface of the Earth, as approximated by mean sea level over the oceans and the sea-level surface extended continuously through the continents. Irregularities or undulations of the geoid have been called humps and hollows, but the surface is nowhere concave.
figure stone *agalmatolite.*
filamented pahoehoe A type of *pahoehoe,* the surface of which displays threadlike strands that are formed by escaping gas bubbles and are recumbent and aligned with the direction of flow. It is a common type and is often found superimposed on other forms.
filiform *capillary.*
filiform lapilli *Pele's hair.*
fill [eng geol] (a) Man-made deposits of natural earth materials (e.g. rock, soil, gravel) and waste materials (e.g. tailings or spoil from dredging), used to fill an enclosed space such as an old stope or chamber in a mine, to extend shore land into a lake or bay, or in building dams. See also: *backfill; made land.* (b) Soil or loose rock used to raise the surface of low-lying land, such as an embankment to fill a hollow or ravine in railroad construction. Also, the place filled by such an embankment. (c) The depth to which material is to be placed to bring the surface to a predetermined grade.
fill [sed] Any sediment deposited by any agent so as to fill or partly fill a valley, sink, or other depression.
fill [speleo] Detrital material partly or completely filling a cave. Syn: *drift [speleo]; wash [speleo].* See also: *cave earth.*
filled-lake plain A swampy plain formed by the filling of a lake by sediments, aided by the growth of plants (Tarr, 1902, p. 82).
filled valley A wide-basin valley, in an arid or semiarid region, that contains abundant alluvium in the form of fans, flood plains, and lake deposits.
filler [mater] *mineral filler.*
filler [streams] A stream that empties into and fills a lake.
fill-in fill terrace A terrace left by a stream that, having incised its valley fill, partly fills up the new valley and incises anew (Schieferdecker, 1959, term 1514).
fillowite A brown, yellow, or colorless mineral: $H_2Na_6(Mn,Fe,Ca)_{14}(PO_4)_{12} \cdot H_2O$ (?).
fillstrath terrace A *fill terrace* whose surface has been eroded to a level below the original depositional surface (Howard, 1959, p. 242); it consists of alluvial material as contrasted with a *strath terrace* formed in bedrock.
fill terrace (a) A term used by Bucher (1932, p. 131) for a remnant, resulting from stream rejuvenation, of a flat valley bottom (or of an alluvial plain) that had been produced by stream aggradation; e.g. an *alluvial terrace* or a glacial terrace. (b) The part of a former alluvial valley floor built upward by deposition of valley-filling sediments (Howard, 1959, p. 242); it includes *filltop terrace* and *fillstrath terrace.*—Leopold et al. (1964, p. 460) find that the term is "confusing" and "should probably be abandoned".
filltop terrace A *fill terrace* whose flat surface is the original depositional surface (Howard, 1959, p. 242).
film water *pellicular water.*
filter [photo] Any transparent material that, by absorption or reflection, selectively modifies the radiation transmitted through an optical system; specif. a glass or gelatin plate placed in front of, in, or behind a camera lens to reduce or eliminate the effect of light of a certain color or colors on the film or plate.
filter [seis] A device or system that changes the wave form or amplitude of a signal. The discrimination is most commonly on the basis of frequency, but dip, wavelength, amplitude and other bases are sometimes used. The device may be electrical or mechanical, or it may be a computer. The Earth acts as a filter to seismic waves. Cf: *convolution [seis].*
filter bridge A narrow *land bridge* that permits the selective migration of some organisms.
filter cake *mud cake [drill].*
filter feeder An animal that obtains its food by straining out organic matter from water as it passes through some part of its body; e.g. a bryozoan. Cf: *suspension feeder.*
filtering The attenuation of certain frequency components of a signal and the enhancement of others. It may be done electrically, or numerically in a digital computer after the signal has been recorded. Cf: *convolution [seis].*
filter pressing A process of magmatic differentiation wherein a magma, having crystallized to a "mush" of interlocking crystals in liquid, is compressed by earth movements and the liquid moves toward regions of lower pressure, thus becoming separated from the crystals. Syn: *filtration differentiation.*
filtration Removal of suspended and/or colloidal material from a liquid by passing it through a relatively fine porous medium.
filtrational Said of an ore-forming fluid or mineralizer that is a

nonmagmatic underground water (Smirnov, 1968). Cf: *juvenile [ore dep]; assimilated.*

filtration differentiation *filter pressing.*

filtration spring A spring whose water percolates from numerous small openings in permeable material. It may have either a small or a large discharge (Meinzer, 1923, p. 50). Cf: *fracture spring; seepage spring.*

filum aquae The thread of a stream. Etymol: Latin, "thread of water". Pl: *fila aquarum.*

fimmenite A peat that is derived mainly from spores.

finandranite A coarse-grained potassium-rich *syenite* composed of alkali feldspar, amphibole, and some biotite, ilmenite, and apatite; described from Malagasy in 1922 by Lacroix. Not recommended usage.

find A meteorite not seen to fall, but recognized as such by its composition and structure. Cf: *fall [meteorite].*

fine Composed of or constituting relatively small particles; e.g. "fine sandy loam". Ant: *coarse [sed].*

fine admixture A term applied by Udden (1914) to an *admixture* (in a sediment of several size grades) whose particles are finer than those of the dominant or maximum grade; material finer than that found in the maximum histogram class.

fine aggregate The portion of an *aggregate* consisting of particles with diameters smaller than approximately 1/4 inch or 4.76 mm. Cf: *coarse aggregate.*

fine clay A geologic term for a *clay* particle having a diameter in the range of 1/2048 to 1/1024 mm (0.5-1 micron, or 11 to 10 phi units). Also, a loose aggregate of clay consisting of fine clay particles.

fine earth (a) That part of a soil that can be passed through a No. 10 (2.0mm) sieve. Cf: *coarse fragment.* (b) A general term for loose earth.

fine-grained (a) Said of a crystalline or glassy rock, and of its texture, in which the individual minerals are relatively small; specif. said of an igneous rock whose particles have an average diameter less than 1 mm (0.04 in.). Syn: *aphanitic.* (b) Said of a sediment or sedimentary rock, and of its texture, in which the individual constituents are too small to distinguish with the unaided eye; specif. said of a sediment or rock whose particles have an average diameter less than 1/16 mm (62 microns, or silt size and smaller). Cf: *finely crystalline.* The term is used in a relative sense, and various size limits have been suggested and used. Cf: *coarse-grained; medium-grained.* (c) Said of a soil in which silt and/or clay predominate. In the U.S., the maximum average diameter of the constituent particles is 0.05 mm (0.002 in.), or as used by engineers, 0.074 mm (passing U.S. standard sieve No. 200); the International Society of Soil Science recognizes a diameter limit of 0.02 mm. Cf: *coarse-grained; fine earth.*

fine-granular *microgranular [ign].*

fine gravel (a) A soil term used in the U.S. for *gravel* whose particles have a diameter in the range of 2-12.5 mm (1/12 to 1/2 in.); it was formerly applied to soil particles (now called *very coarse sand*) having diameters of 1-2 mm. (b) An engineering term for *gravel* whose particles have a diameter in the range of 4.76 mm (retained on U.S. standard sieve no.4) to 19 mm (3/4 in.).

finely crystalline Descriptive of an interlocking texture of a carbonate sedimentary rock having crystals whose diameters are in the range of 0.016-0.062 mm (Folk, 1959), 0.01-0.1 mm (Carozzi & Textoris, 1967, p. 5), 0.01-0.05 mm (Bissell & Chilingar, 1967, p. 103), or less than 1.0 mm (Krynine, 1948, p. 143). Cf: *fine-grained.*

fineness The state of subdivision of a substance; the size of the constituent particles of a substance. The term is applied in describing sedimentary texture.

fineness factor A measure of the average particle size of clay and ceramic material, computed by summing the products of the reciprocal of the size-grade midpoints and the weight percentage of material in each class (expressed as a decimal part of the total frequency) (Purdy, 1908). The measure is based on the assumption that the surface areas of two powders are inversely proportional to their average particle sizes. Syn: *surface factor.*

fineness modulus A means of evaluating sand and gravel deposits, consisting of passing samples through standardized sets of sieves, accumulating percentages retained, dividing by 100, and comparing the resultant fineness-modulus number to various specification requirements.

fine pebble A geologic term for a *pebble* having a diameter in the range of 4-8 mm (1/6 to 0.3 in., or -2 to -3 phi units) (AGI, 1958).

fines [mining] Finely crushed or powdered material, e.g. of coal, crushed rock, or ore, as contrasted with the coarser fragments; esp. material smaller than the minimum specified size or grade, such as coal with a maximum particle size less than 8 mm, or ores too pulverulent to be smelted in the ordinary way; or material passing through a given screen or sieve.

fines [sed] (a) Very small particles, esp. those smaller than the average in a mixture of particles of various sizes; e.g. the silt and clay fraction in glacial drift, or the fine-grained sediment that settles slowly to the bottom of a body of water. (b) An engineering term for the clay- and silt-sized soil particles (diameters less than 0.074 mm) passing U.S. standard sieve no. 200.

fine sand (a) A geologic term for a *sand* particle having a diameter in the range of 0.125-0.25 mm (125-250 microns, or 3 to 2 phi units). Also, a loose aggregate of sand consisting of fine sand particles. (b) An engineering term for a *sand* particle having a diameter in the range of 0.074 mm (retained on U.S. standard sieve no.200) to 0.42 mm (passing U.S. standard sieve no. 40). (c) A soil term used in the U.S. for a *sand* particle having a diameter in the range of 0.10-0.25 mm. The diameter range recognized by the International Society of Soil Science is 0.02-0.2 mm. (d) Soil material containing 85% or more of sand-size particles (percentage of silt plus 1.5 times the percentage of clay not exceeding 15) and 50% or more of fine sand or less than 25% of very coarse sand, coarse sand, and medium sand together with less than 50% of very fine sand (SSSA, 1965, p.347).

fine silt A geologic term for a *silt* particle having a diameter in the range of 1/128 to 1/64 mm (8-16 microns, or 7 to 6 phi units). In Great Britain, the range 1/100 to 1/20 mm has been used. Also, a loose aggregate of silt consisting of fine silt particles.

fine topography A topography with fine *topographic texture,* characterized by high drainage density and closely spaced streams, and common in regions of weak rocks. An "ultra-fine" topography is characterized by the extremely fine dissection of badlands topography.

finger One of two pincerlike blades of the distal end of a cheliped of a crustacean or of a chela of an arachnid. One finger is movable and the other is fixed.

finger bar *bar finger.*

finger coal *Natural coke* occurring as small, hexagonal columns associated with igneous intrusion.

finger gully One of a group of very small gullies that forms a fan-shaped extension at the head of a system of gullies.

finger lake A long, relatively narrow lake, usually of glacial origin, which may occupy a rock basin in the floor of a glacial trough or be held in by a morainal dam across the lower end of the valley; esp. one of a group of such lakes disposed somewhat like the fingers of a hand, such as the Finger Lakes in central New York State.

Fingerlakesian North American stage: lower Upper Devonian (above Erian, below Chemungian). Syn: *Finger Lakes.*

fingertip channel One of the smaller, unbranched stream channels at the head of a drainage network.

finite-strain theory A theory of material deformation which considers displacements and strains too large to be evaluated through *infinitesimal-strain theory.*

finnemanite A gray, olive-green, or black hexagonal mineral: $Pb_5(AsO_3)_3Cl$.

fiord Anglicized variant of *fjord.*

fiorite *siliceous sinter.*

fire Flashes of the different spectral colors seen in diamonds and other gemstones as the result of *dispersion.* Cf: *play of color.*

fire assay Any type of assay procedure that involves the heat of a furnace.

fireball A bright or brilliant meteor with luminosity that equals or exceeds that of the brightest planets. Cf: *bolide.*

fireball hypothesis *"big bang" hypothesis.*

fireblende A syn. of *pyrostilpnite.* Also spelled: *fire blende.*

fireclay (a) A siliceous clay rich in hydrous aluminum silicates, capable of withstanding high temperatures without deforming (either disintegrating or becoming soft and pasty), and useful for the manufacture of refractory ceramic products such as crucibles and firebrick. It is deficient in iron, calcium, and alkalies, and approaches kaolin in composition, the better grades containing at least 35% alumina when fired. (b) A term formerly, but inaccurately, used for *underclay.* Although many fireclays commonly occur as underclays, not all fireclays carry a roof of coal and not all underclays are refractory.—Also spelled: *fire clay.* Syn: *firestone; refractory clay; sagger.*

fireclay mineral A disordered variety of kaolinite. See also: *mellorite.*

firedamp A coal-mine gas that is explosive and consists mainly of *methane.* Cf: *blackdamp; whitedamp; afterdamp.*

fire fountain Rise-and-fall eruption of incandescent lava, either from a central volcanic vent or along a fissure, as a "jet" of molten material that breaks into a spray of melt droplets and *bombs.*

fire marble *lumachelle.*

fire opal A transparent to translucent and yellow, orange, red, or brown variety of opal that gives out fiery reflections in bright light and that may or may not have play of color. See also: *gold opal.* Syn: *sun opal; pyrophane.*

firestone (a) Any fine-grained siliceous stone formerly used for striking fire; specif. *flint.* Syn: *feuerstein.* (b) A nodule of pyrite formerly used for striking fire. (c) A fine-grained siliceous rock that can resist or endure high heat and that is used for lining furnaces and kilns, such as certain Cretaceous and Jurassic sandstones in southern England. (d) *fireclay.*

firmatopore A polymorph that is slender and proximally directed on the reverse sides of the colonies of some stenolaemate bryozoans. Cf: *hematopore.*

firm ground A tunnelman's term for materials in which a tunnel can be advanced without the aid of initial roof support, and the permanent lining can be constructed before the crown or walls begin to deform. Firm ground may become unstable if construction methods are changed or tunnel size is increased. See also: *flowing ground; raveling ground; running ground; squeezing ground; swelling ground.*

firmoviscosity The *elasticoviscous* state, as modeled by a Kelvin body. In response to a given stress, elastic strain is produced only over a finite period of time; unloading is time-dependent, although all of the strain is recoverable. At a constant strain, stress may be supported indefinitely (Turner and Weiss, 1963, p. 279).

firn (a) A material that is transitional between snow and glacier ice, being older and denser than snow, but not yet transformed into glacier ice. Snow becomes firn after existing through one summer melt season; firn becomes glacier ice when its permeability to liquid water drops to zero. The term has also been defined, although rarely, on the basis of certain physical properties, such as density (snow becomes firn at a density greater than 0.4 g/cc, as in the older literature, or 0.55 g/cc, the greatest density of ice if its grains were shifted around so that they fit most snugly), but this criterion is difficult to measure and variable in specific areas. Syn: *névé; firn snow.* (b) A geographic term applied to the accumulation area or upper region of a glacier. This usage is being supplanted by *firn field,* or by *névé* (in Great Britain).—Etymol: German, adjective meaning "old, of last year".

firn basin *firn field.*

firn edge The boundary on a glacier between glacier ice and firn during the ablation season.

firn field The *accumulation area* of a glacier; a broad expanse of glacier surface over which snow accumulates and firn is created; an area of *firn.* Syn: *firn basin; névé.*

firn ice *iced firn.*

firnification The process whereby snow is transformed into firn and then into ice in a glacier.

firn limit *firn line.*

firn line (a) The highest level to which the winter snow cover retreats on a glacier; the *snowline.* (b) The edge of the snow cover at the end of the summer season, thus the boundary between the *superimposed ice zone* below and the *soaked zone* above. See also: *equilibrium line.* Syn: *firn limit.*

firn snow (a) *firn.* (b) *old snow.*

firnspiegel A thin sheet or film of clear ice on a snow surface, bridging hollows in the snow, formed when surface meltwater is immediately refrozen as a thin ice film, and the snow below continues to melt by radiation passing through the transparent ice sheet. Etymol: German *firn,* "old, last year's", + *Spiegel,* "mirror".

first antenna *antennule.*

first arrival The first energy to arrive from a seismic source. First arrivals on reflection records are used for information about a surficial low-velocity or weathering layer; refraction studies are often based on first arrivals. Syn: *first break.*

first bottom The normal flood plain of a river. Cf: *second bottom.*

first break *first arrival.*

first-class ore An ore of sufficiently high grade to be acceptable for shipment to market without preliminary treatment. Cf: *second-class ore.* Syn: *shipping ore.*

first law of thermodynamics The statement describing the *internal energy* of a system, which says that the change of energy of a system equals the amount of energy received from the external world, which in turn equals the heat taken in by the system and the work done on the system.

first maxilla *maxillule.*

first meridian *prime meridian.*

first-order leveling Leveling of high precision and accuracy in which, for a section of 1-2 km in length, the maximum allowable difference obtained by running the line first forward to the objective point and then backward to the starting point is 4.0 mm times the square root of the distance in kilometers separating the ends of the line (or 0.017 ft times the square root of the distance in miles). Cf: *second-order leveling; third-order leveling.*

first-order pinacoid In a triclinic crystal, the $\{0kl\}$ pinacoid and the $\{0\bar{k}l\}$ pinacoid. Cf: *second-order pinacoid; third-order pinacoid; fourth-order pinacoid.*

first-order prism A crystal form: in a tetragonal crystal, the $\{110\}$ prism; in a hexagonal crystal, the $\{10\bar{1}0\}$ prism; in an orthorhombic crystal, any $\{0kl\}$ prism; and in a monoclinic crystal, any $\{0kl\}$ prism. Cf: *second-order prism; third-order prism; fourth-order prism.* See also: *clinodome; brachydome.*

first water A term occasionally used, esp. in England, for the highest quality of a gemstone, such as that of a diamond that is flawless, perfectly clear and transparent, and colorless or almost blue-white. Cf: *second water; third water.*

first-year ice Sea ice, not more than one winter's growth, developing from young ice; it is subdivided on the basis of thickness: "thin" (30-70 cm; also known as *white ice*); "medium" (70-120 cm); and "thick" (120 cm to 2 m). See also: *winter ice; one-year ice.*

firth A long, narrow arm of the sea; also, the opening of a river into the sea. Along the Scottish coast, it is usually the lower part of an *estuary* (e.g. Firth of Forth), but sometimes it is a fjord (e.g. Firth of Lorne) or a strait (e.g. Pentland Firth). Etymol: Scottish. Syn: *frith.*

fischerite A mineral, consisting of a green hydrous aluminum phosphate, that is probably identical with *wavellite.*

fischesserite A cubic mineral: Ag_3AuSe_2.

fish [drill] (a) Broken or lost equipment in a well bore, recoverable only by *fishing.* (b) Any foreign material, in a well, that cannot be removed at will.

fish [oceanog] Any oceanographic sensing device that is towed behind a ship.

fish [seis] A sensor that is towed in the water, such as that used with side-scan sonar.

fish-eye stone *apophyllite.*

fishhook dune A dune consisting of a long, sinuous, sigmoidal ridge forming the "shaft" and a well-defined crescent forming the "hook" (Stone, 1967, p. 228). Syn: *hooked dune.*

fishing Searching for and attempting to recover, by the use of specially prepared tools, a piece or pieces of drilling equipment (such as sections of pipe, cables, or casing) that have become detached, broken, or lost from the drill string or have been accidentally dropped into the hole.

fish kill Destruction of fish in lakes or ponds, due to a decrease in oxygen resulting from snow or from excessive amounts of suspended organic matter; to toxic pollutants; or to the total freezing of shallow lakes or ponds.

fishtail structure The ragged lateral termination of a coal seam, produced where wedges of clastic sediment entered the parent peat deposit parallel to the bedding.

fissiculate Said of a blastoid having exposed or partly exposed hydrospire slits or spiracular slits.

fissile (a) Capable of being easily split along closely spaced planes; exhibiting *fissility.* (b) Said of bedding that consists of laminae less than 2 mm in thickness (Payne, 1942).

fissility A general term for the property possessed by some rocks of splitting easily into thin layers along closely spaced, roughly planar, and approximately parallel surfaces, such as bedding planes in shale or cleavage planes in schist; its presence distinguishes shale from mudstone. The term includes such phenomena as *bedding fissility* and *fracture cleavage.* Etymol: Latin *fissilis,* "that which can be cleft or split". Adj: *fissile.*

fission [evol] Asexual reproduction occurring when a single cell or polyp divides into two theoretically equal parts.

fission [isotope] The spontaneous or induced splitting, by particle collision, of a heavy nucleus into a pair (only rarely more) of nearly equal fission fragments plus some neutrons. Fission is accompanied by the release of a large amount of energy. Cf: *fusion [isotope].* See also: *spallation.* Syn: *nuclear fission.*

fissionable Said of nuclei, such as uranium and plutonium, that are capable of *fission.*

fission-track dating A method of calculating an age in years by determining the ratio of the spontaneous fission-track density to induced fission tracks. The method, which has been used for ages from 20 years to 1.4 x 10^9 years, works best for micas, tektites, and meteorites, and is also useful for determining the amount and distribution of the uranium in the sample. Syn: *fission-track method; spontaneous fission-track dating.*

fission-track method *fission-track dating.*

fission tracks The paths of *radiation damage* made by nuclear particles in a mineral or glass by the spontaneous fission of uranium-238 impurities. They are similar in occurrence and formation to *alpha-particle recoil tracks,* but are larger and less numerous. Fission-track density is established by etching and subsequent microscopic examination.

fissure [geol] A surface of fracture or a crack in rock along which there is a distinct separation. It is often filled with mineral-bearing material. Syn: *joint fissure; open joint.*

fissure [glaciol] *crevasse [glaciol].*

fissure eruption An eruption that takes place from an elongate fissure, rather than from a central vent. Cf: *central eruption; plateau eruption.* See also: *fissure flow; fissure vent.*

fissure flow A flow formed as a result of a *fissure eruption.*

fissure-flow volcano *fissure volcano.*

fissure polygon A *nonsorted polygon* marked by intersecting grooves or fissures producing a gently convex polygonal surface pattern and by the absence of a well-defined stone border. The term is inadequate because some polygons with fissures have stone borders coincident with the fissures (Washburn, 1956, p.825). The feature is typical of broad areas of NW Canadian lowlands. See also: *mud polygon; ice-wedge polygon.*

fissure spring A spring issuing from a crack or joint. Several springs of this type may flow out along the same fissure line. Cf: *fault spring; fracture spring.*

fissure system A group of fissures of the same age and of more or less parallel strike and dip.

fissure theory An early theory, now partially discredited, that oil and gas migrate extensively through fissures resulting from the arching of beds into anticlines.

fissure vein A type of mineral deposit of veinlike shape with the implication of clearly defined walls rather than extensive host-rock replacement.

fissure vent The opening at the Earth's surface of a volcanic conduit having the form of a crack or fissure. Cf: *fissure eruption.*

fissure volcano One of a series of volcanic vents in a pattern of eruption along a fissure. Syn: *fissure-flow volcano.*

fistulose Said of a foraminifer having tubular irregular growth in the apertural region.

fitzroyite A *lamproite* composed of phenocrysts of leucite and phlogopite in a very fine-grained groundmass. Named by Wade and Prider in 1940 for the Fitzroy Basin, Western Australia.

fiveling A crystal twin consisting of five individuals. Cf: *twoling; trilling; fourling; eightling.*

fix (a) A position determined from terrestrial, electronic, or astronomical data. Also, the point thus established. (b) The act of determining a fix.

fix-bitumens All authigenic, nonfluid bitumens. They are divided into stabile protobitumens and stabile metabitumens (Tomkeieff, 1954).

fixed ash *inherent ash.*

fixed carbon In coal, coke, and bituminous materials, the solid combustible matter remaining after removal of moisture, ash, and volatile matter. It is expressed as a percentage.

fixed-carbon ratio *carbon ratio.*

fixed cheek The part of the *cheek* of a trilobite inside the facial suture, remaining attached to the glabella at the time of molting. Cf: *free cheek.* Syn: *fixigena.*

fixed dune *anchored dune.*

fixed elevation An elevation that has been adopted either as a result of tide observations or previous adjustment of spirit leveling and that is held at its accepted value in any subsequent adjustment (Mitchell, 1948, p.27).

fixed form A crystal form whose indices are fixed relative to length, e.g. cube {100}, or octahedron {111} of the cubic system. Syn: *singular crystal form.*

fixed ground water Ground water in material having interstices so small that the water is held permanently to the walls of the interstices, or moves so slowly that it is not available for withdrawal at useful rates. Outside the zone of saturation material with infinitely small openings can hold water indefinitely against the pull of gravity, whereas within the zone of saturation there is apparently always movement, even though at very low rates.

fixed layer The inner, relatively immobile layer of ions in an electrolyte, required to satisfy a charge unbalance within a solid with which the electrolyte is in contact. It constitutes part of the *double layer* of charge adjacent to the electrolyte-solid interface. Cf: *diffuse layer.*

fixed moisture Moisture retained in the soil in a quantity that is less than the *hygroscopic coefficient.*

fixigena *fixed cheek.*

fixity of species The 18th-century theory that a species, once created, remained constant in its characteristics throughout its span of existence.

fizelyite A lead-gray mineral: $Pb_5Ag_2Sb_8S_{18}$. It is closely related to andorite and ramdohrite. Also spelled: *fizélyite.*

fjäll A Swedish word for a mountain rising above the timberline and having flat undissected areas (Stamp, 1961, p. 193). See also: *fjeld; fell.*

fjard A small, narrow, irregular inlet or bay, typically formed by submergence of a glacial valley excavated in a lowland along the margin of a flat rocky coast, such as that of SE Sweden; often accompanied by numerous fringing islands. A fjard is shorter, shallower, and broader in profile than a *fjord,* but deeper than a ria. Pron: *fyard.* Etymol: Swedish *fjärd,* a large continuous area of water surrounded by *skerry-guard* islands (Stamp, 1961, p. 193); usage in English therefore has taken a more specialized meaning not apparent in Sweden. Syn: *fiard.*

fjeld A Norwegian word for "field" having a wide meaning, but when used in English refers to an elevated, rocky, almost barren plateau above the timberline, covered with snow during the winter, as in the Scandinavian upland. See also: *fell; fjäll.* Syn: *fjell.*

fjeldbotn A Norwegian term for a cirque carved by an ice field (Termier & Termier, 1963, p. 405).

fjeldmark *fell-field.*

fjord (a) A long narrow winding inlet or arm of the sea, U-shaped and steep-walled, generally several hundred meters deep, between high rocky cliffs or slopes along a mountainous coast; typically with a shallow sill or threshold of solid rock or earth material submerged near its mouth, and becoming deeper inland. A fiord usually represents the seaward end of a deeply excavated glacial-trough valley that is partially submerged by drowning after the melting of the ice. Examples: along the glaciated coasts of Alaska, Greenland, and Norway. (b) Any embayment of the seacoast in a Scandinavian country regardless of the adjacent topography, as a *fjard* in the low flat Swedish coast or a *förde* in eastern Denmark.—Pron: *fyord.* Etymol: Norwegian. Cf: *estuary; ria.* Syn: *fiord; sea loch.*

fjord coast A deeply indented, glaciated coast characterized by a partial submergence of glacial troughs and by the presence of steep parallel walls, truncated spurs, and hanging valleys. Its development is favored by marine west-coast climates combined with strong relief, as in Alaska and southern Chile.

fjord ice Ice formed during the winter in a fjord and melting in place during the summer. Cf: *sikussak.*

fjord lake A lake in a glacially excavated rock basin of a U-shaped valley near sea level.

fjord shoreline A shoreline of submergence characterized by the development of numerous *fjords;* e.g. along the west coast of Norway. Cf: *fiord coast.*

fjord strait A strait between two fjords opening in opposite directions; e.g. the Straits of Magellan.

f,k space A domain in which the independent variables are frequency (f) and wavenumber (k). Seismic data analysis sometimes involves f-k plots, in which energy density within a given time interval is plotted and contoured on a frequency-versus-wavenumber basis.

flabellate Fanlike; e.g. said of a fan-shaped corallite, or of a meandroid corallum with a single continuous, laterally free, linear series of corallites.

flabellum A body part resembling a fan; esp. the thin, most distal exite of the limb of a branchiopod crustacean. Pl: *flabella.*

Fladen (a) A distinctive, pancake-shaped body resembling a volcanic bomb and composed of glass and fragments of rocks and minerals, found in the *suevites* at the Ries basin in Germany. It exhibits flow structure and surface sculpturing apparently produced by aerodynamic forces. See also: *impact bomb.* (b) Any similar glass-rich, aerodynamically shaped body, formed by meteorite impact, and found associated with other meteorite impact structures.—Etymol: German, "flat cake". Pl: *Fladen.* Syn: *Flädle.*

flag A syn. of *flagstone.* The term is often used in the plural, such as the "Lingula flags" of the European Upper Cambrian.

flagellar field The area around the flagella of a coccolithophore; e.g. *covered flagellar field* and *naked flagellar field.*

flagellar pore One of the pores in a dinoflagellate for extrusion of flagella, usually located at the anterior or the posterior junction of girdle and sulcus.

flagellate n. An organism, esp. a protozoan or alga, that bears flagella.—adj. Possessing flagella.

flagellated chamber Any cavity, in a sponge, lined by choanocytes. See also: *chamber [paleont].*

flagellum (a) Any of various threadlike appendages of animals, such as the multiarticulate distal part of the limb of a crustacean, or a long whiplike extension in a sponge. (b) A long, whiplike protoplasmic process that projects singly or in groups from a cell or microorganism, is possibly equivalent to a much enlarged *cilium,* and is the primary organ controlling the movement (through water) of a flagellated protozoan and of many algae, bacteria, and zoospores.—Pl: *flagella.*

flaggy (a) Splitting or tending to split into layers of suitable thickness for use as flagstones; specif. descriptive of a sedimentary rock that splits into layers from 1 cm to 5 cm in thickness (McKee & Weir, 1953, p.383). (b) Said of bedding that consists of layers from 1 cm to 10 cm in thickness (Payne, 1942). (c) Pertaining to a flag or flagstone. (d) Said of a soil full of flagstone fragments.

flagstaffite A mineral: $C_{10}H_{22}O_3$. It occurs in colorless, transparent, orthorhombic crystals identical with *cis*-terpin hydrate, and is found with resin in the radial cracks of fossil pine trees.

flagstone (a) A hard sandstone, usually micaceous and fine-grained, that occurs in extensive thin beds with shale partings; it splits uniformly along bedding planes into thin slabs suitable for use in terrace floors, retaining walls, and the like. Cf: *bluestone [rock]; freestone [rock].* (b) A flat slab of flagstone used for paving; esp. a thin piece split from flagstone. Also, a surface of such stone. (c) A relatively thin flat fragment (of limestone, sandstone, shale, slate, or schist) occurring in the soil, having a length in the range of 15-38 cm (6-15 in.) (SSSA, 1965, p.336).—Syn: *flag; slabstone; grayband; cleftstone.*

flaikes A Scottish term for a shaly or fissile sandstone that splits along the grain. See also: *fakes.* Syn: *flakes.*

flajolotite *tripuhyite.*

flake graphite In economic geology, graphite disseminated in metamorphic rock as thin, visible flakes that are separable from the rock by mechanical means. Syn: *crystalline flake.*

flake mica Finely divided mica recovered from mica schist or sericite schist or obtained as a byproduct of beneficiation of feldspar or kaolin.

flame emission spectrometry *flame photometry.*

flame photometer A *spectrophotometer* using flame excitation of samples, usually compounds in solution, to provide spectra for analysis. See also: *photometer.*

flame photometry Measurement of the intensity of the lines in a flame spectrum by a flame photometer. Syn: *flame emission spectrometry.*

flame spectroscopy The observation of a *flame spectrum* and all processes of recording and measuring that go with it.

flame spectrum The spectrum of light emitted by a substance by heating it in a flame.

flame structure [pyroclast] The presence of *fiamme* in a welded tuff, e.g. piperno.

flame structure [sed] A term introduced by Walton (1956, p.267) for a sedimentary structure consisting of wave- or flame-shaped plumes of mud that have been squeezed irregularly upward into an overlying layer. It is probably formed by load casting accompanied by horizontal slip or drag. The term *antidune* as used by Lamont (1957) is synonymous. See also: *load wave; streaked-out ripples.*

flame test A qualitative analysis of a mineral made by intensely heating a sample in a flame and observing the flame's color, which will be indicative of the element involved, e.g. green from copper.

Flandrian British climatostratigraphic stage: Holocene (post-10,000 ^{14}C years; above Devensian).

flange (a) A projecting rim or edge of an organism; e.g. a platelike marginal extension along foraminiferal chambers (as in *Sphaeroidinella*), the part of a coccolith that spreads out like a rim, a shelflike structure along the inner or outer side of a blade or bar of a conodont, a lateral projection from a brachiopod crus, or the outwardly bent rim of the aperture in some cephalopod conchs. (b) An equatorial extension of the exine of a spore. It is a less precisely defined term than *cingulum* or *zona.*

flank [paleont] (a) The lateral side of a coiled cephalopod conch between the venter and the umbilical seam; the *whorl side.* (b) The sloping surface between the venter and the lateral margin of a brachiopod valve. (c) In a bivalve mollusk, the median part of the surface of the valve.

flank [struc geol] *limb.*

flank eruption An eruption on the side of a volcano. Cf: *summit eruption.*

flanking moraine A moraine left by a glacial lobe or by a tongue-like projection of an ice sheet (Fairchild, 1932, p. 629). Cf: *lateral moraine [glac geol].*

flaser The streaky layers of parallel, scaly aggregates surrounding the lenticular bodies of granular material in *flaser structure.* Etymol: German, "streak".

flaser gabbro Coarse-grained *blastomylonite* formed by dislocation metamorphism of a gabbro. Flakes of mica or chlorite sweep around augen of feldspar and/or quartz with much recrystallization and neomineralization (Joplin, 1968, p.21). Cf: *zobtenite.*

flaser structure [meta] A structure in dynamically metamorphosed rock in which lenses and layers of original or relatively unaltered granular minerals are surrounded by a matrix of highly sheared and crushed material, giving the appearance of a crude flow structure, e.g. flaser gabbro. Cf: *mylonitic structure; augen structure.* Syn: *phacoidal structure.*

flaser structure [sed] *Ripple cross-lamination* in which mud streaks are preserved in the troughs but incompletely or not at all on the crests. A detailed classification is given by Reineck and Wunderlich (1968). Cf: *wavy bedding.*

flash [mining] (a) A subsidence of the surface resulting from underground mining, esp. from the working of rock salt and pumping of brine. Cf: *inbreak; heave [mining]; crown-in.* (b) A small lake or shallow reach of water occupying a hollow produced by such a subsidence.

flash [water] (a) A sudden rise of water in a stream, as where water is held back by a dam. (b) A pool of water; a marsh or marshy place.—Cf: *flush.*

flash figure An optic analysis of uniaxial crystals under the conoscope, a vague black cross that appears when the optic axis is parallel to the plane of polarization of either prism. Only slight rotation from this position causes the cross to separate into two hyperbolic segments and leave the field (Wahlstrom, 1948).

flash flood A local and sudden flood or torrent of relatively great volume and short duration, overflowing a stream channel in a usually dry valley (as in a semiarid area), carrying an immense load of mud and rock fragments, and generally resulting from a rare and brief but heavy rainfall over a relatively small area having steep slopes. It may also be caused by ice jams and by dam failure. See also: *freshet.* Syn: *flashy stream.*

flash opal Opal in which the play of color is pronounced only in one direction (Shipley, 1951, p. 82).

flashy stream *flash flood.*

flat [eco geol] n. A horizontal orebody, regardless of genetic type.

flat [geog] (a) A tract of low-lying, level wetland; e.g. a marsh or swamp in a river valley. (b) A term used in northern and central U.S. as a syn. of *bottom,* or low-lying land along a watercourse.

flat [geomorph] adj. Having or marked by a continuous surface or stretch of land that is smooth, even, or horizontal, or nearly so, and that lacks any significant curvature, slope, elevations, or depressions.—n. A general term for a level or nearly level surface or small area of land marked by little or no relief, as a plain; specif: *mud flat; valley flat.* Also, a nearly level region that visibly displays lower relief than its surroundings.

flat [lake] (a) The low-lying, exposed, flat land of a lake delta or of a lake bottom. (b) The flat bottom of a desiccated lake in the arid parts of western U.S.—Commonly used in the plural.

flatiron One of a series of short, triangular hogbacks forming a spur or ridge on the flank of a mountain, having a narrow apex and a broad base, resembling (when viewed from the side) a huge flatiron; it usually consists of a plate of steeply inclined resistant rock on the *dip slope.*

flat joint In igneous rock, joint dipping at an angle of 45° or less. Rarely applied to joints dipping more than 20°.

flatland A region, or tract of land, characterized by predominant levelness or by no significant variation in elevation, as along a river or a coast.

flatness A term used by Wentworth (1922a) to express the shape of a pebble, defined as the ratio of the radius of curvature of the most convex portion of the flattest developed face to the mean radius of the pebble. Wentworth (1922b) also expressed flatness as the arithmetic mean of the long and intermediate diameters (length and width) of the pebble divided by the short diameter (thickness). Cf: *roundness.*

flats and pitches (a) A phrase descriptive of the structure of the lead and zinc deposits in dolomite of the Upper Mississippi Valley region of the U.S., esp. in Wisconsin. The "flats" are nearly horizontal solution openings; the "pitches" are the inclined, interconnecting joints. (b) A slump structure of both horizontal and steeply inclined cracks in sedimentary strata. Syn: *pitches and flats.*

flattening *ellipticity.*

flat-topped ripple mark A ripple mark with a flat, wide crest between narrow troughs; e.g. a shallow-water ripple mark whose crest was planed off during ebb tide or fall in water level.

flaw [gem] A general term for any internal or external *imperfection* of a fashioned diamond or other gemstone. It includes cracks, inclusions, visibly imperfect crystallization, internal twinning, and cleavage.

flaw [ice] (a) A narrow separation zone (*fracture [ice]*) between pack ice and fast ice, formed when pack ice at this boundary undergoes shearing due to a strong wind or current, and characterized by pieces of ice in a chaotic state; it is not wide enough to permit passage of a navigable vessel. (b) Obsolete syn. of *flaw lead.*

flaw [struc geol] An old term for a strike-slip fault.

flaw lead A navigable passage between pack ice and fast ice. See also: *shore lead.* Obsolete syn. *flaw [ice].* Pron: flaw leed.

flawless Said of a diamond or other gemstone that is free from *flaws* of any description when observed by a trained eye under efficient illumination with a fully corrected magnifier of not less than 10 power.

flaxseed ore An iron-bearing sedimentary deposit, e.g. the *Clinton ore,* composed of disk-shaped hematitic oolites that have been somewhat flattened parallel to the bedding plane. Cf: *fossil ore.*

F layer [seis] The seismic region of the Earth from 4710 km to 5160 km, equivalent to the *transition zone* between the *outer core* and the *inner core.* It is a part of a classification of the Earth's interior made up of layers A to G. Together with the G layer, it is the equivalent of the *lower core.*

F layer [soil] A surficial layer of humus or partially decomposed organic matter over a forest soil. It lies beneath the *L layer* and above the *H layer.* See also: *O horizon.*

flèche d'amour An acicular, hairlike crystal of rutile embedded in sagenitic quartz. The term is used loosely as a syn. of "sagenitic quartz", and was used formerly for amethyst containing brown needles of goethite. Etymol: French, "arrow of love". Syn: *cupid's dart; love arrow.*

fleckschiefer A type of *spotted slate* characterized by minute flecks or spots of indeterminate material. Etymol: German. Cf: *knotenschiefer; garbenschiefer; fruchtschiefer.*

fleet (a) A term used in England for a small, shallow inlet, estuary, creek, or arm of the sea. Also, a place where water flows; a small rapid stream. (b) A small, usually salty or brackish lagoon behind the coastline, separated from the open sea by a long bank of sand or shingle parallel to the coast (Monkhouse, 1965, p. 125).

fleischerite A mineral: $Pb_3Ge(SO_4)_2(OH)_6 \cdot 3H_2O$.

fleshy sponge Any sponge that lacks a skeleton, i.e. one of the demosponges.

fleur-de-lis A sedimentary structure consisting of frondescent spatulate elevations and resembling the appearance of an iris. Etymol: French *fleur de lis,* "lily flower". Syn: *fleur-de-lys.*

flexible Said of a mineral which can be bent without breaking but will not return to its original form; e.g. talc.

flexible crinoid Any crinoid belonging to the subclass Flexibilia, characterized by the incorporation, but not firm attachment, of the lower brachials in the dorsal cup and by a flexible tegmen.

flexible sandstone A fine-grained, thin-layered variety of *itacolumite.*

flexible silver ore *sternbergite.*

flexostyle The tubular enrolled chamber of a foraminiferal test immediately following the proloculus (as in *Amphisorus*).

flexural fold A general term for *flexure-flow folds* and *flexure-slip folds.* Cf: *passive fold.*

flexural slip *bedding-plane slip.*

flexural-slip thrust fault A syn. of *uplimb thrust fault.* See: Price (1965).

flexure *hinge.*

flexure correction In pendulum observations of gravity, a necessary correction for the influence of the rather complex coupled vibration phenomena caused by the oscillation of the receiver case, the pillar, and the surface soil. Such vibrations cause the period of the pendulum itself to change.

flexure-flow fold A flexural fold in which the mechanism of folding includes displacement parallel to layer boundaries and some flow within layers, resulting in thickening of hinge areas and thinning of limbs. Cf: *flexure-slip fold.*

flexure line A line, extending from the beak to the anterior border of both ventral propareas in some lingulacean brachiopods, marked by deflection of growth lines (TIP, 1965, pt.H, p.145).

flexure-slip fold A *flexure fold* in which the mechanism of folding is slip along bedding planes or along surfaces of foliation. There is no change in thickness of individual strata, and the resultant folds are parallel. Cf: *flexure-flow fold.*

flight *terrace flight.*

flight altitude The vertical distance above a given datum, usually mean sea level, of an aircraft in flight or during a specified portion of a flight (ASP, 1975, p. 2082). See also: *flight height.*

flight height A term used in aerial photography for the *flight altitude* when the datum is mean ground level of the area being photographed.

flight line A line drawn on a map or chart to represent the planned or actual track of an aircraft during the taking of aerial photographs.

flight map A map that shows the desired flight lines and/or air stations before the taking of aerial photographs, or a map on which are plotted, after photography, selected air stations and the tracks between them.

flight path The line on the ground directly beneath a remote sensing aircraft or satellite.

flight strip A succession of overlapping aerial photographs taken along a single course. Syn: *strip [photo].*

flimmer *mastigoneme.*

flinkite A greenish-brown mineral: $Mn_3(AsO_4)(OH)_4$.

flint [mineral] (a) A term that has been considered as a mineral name for a massive, very hard, somewhat impure variety of chalcedony, usually black or of various shades of gray, breaking with a conchoidal fracture, and striking fire with steel. See also: *chert.* Syn: *firestone.* (b) Pulverized quartz of any kind; e.g. "potter's flint" in the form of powdered quartz, made by pulverizing flint pebbles.

flint [sed] A term that is widely used as a syn. of *chert* or for the homogeneous, dark-gray or black variety of chert. According to Tarr (1938), the term "flint" should either be discarded or be reserved for siliceous artifacts (such as the "flint arrowheads" used by primitive man) because rocks described as flint are identical with chert in texture and composition, despite the fact that the term "flint" has been in use since about A.D. 700 for "anything hard" and since A.D. 1000 for "a variety of stone" and that it antedates "chert" by almost 1000 years. Flint has been described as having a denser texture, a more perfect (smooth) conchoidal or less splintery fracture, a smaller quartz content, and greater infusibility than chert, and as having thin translucent splinters or sharp cutting edges. The term is commonly used in southern England for one of the siliceous nodules occurring in the Cretaceous chalk beds, and elsewhere in England for any hard rock (such as in Shropshire for a fine-grained sandstone suitable for building). Syn: *black chert; silex; hornstone; petrosilex.*

flint clay A smooth, flintlike microcrystalline clay rock composed dominantly of kaolin, which breaks with a pronounced conchoidal fracture and resists slaking in water. It becomes plastic upon prolonged grinding in water, as in an industrial wet-pan unit.

flint curtain A concentration of silica derived from flints, occurring along a vertical joint plane in the chalk beds of eastern Denmark; it results from dehydration of silica gel and flowage of silica, immediately subsequent to jointing.

flint meal Fine, flourlike material consisting primarily of minute fossils (such as sponge spicules) and occurring in an enclosed cavity of a flint nodule from the chalk beds of southern England.

flinty (a) Composed of flint, or containing more than the normal

percentage of silica; e.g. a "flinty slope". (b) Resembling flint in hardness or fracture; flintlike.
flinty crush rock *ultramylonite.*
flinty slate A *touchstone* consisting of siliceous slate.
float A general term for isolated, displaced fragments of a rock, esp. on a hillside below an outcropping ledge or vein. Cf: *floating reef.* Syn: *floater.*
float coal Small isolated bodies of coal in sandstone or shale, probably deposited as pieces of peat that were eroded and transported from the original deposit. Cf: *coal gravel.* Syn: *raft [coal].*
float coccolith A modified coccolith serving as a suspension organ in nonmotile coccolithophores exhibiting dimorphism (such as *Thorosphaera*).
floater *float.*
floating ice Any form of ice floating in water, including ice that is stranded or grounded and ice formed on land but drifting in the sea. The term formerly excluded icebergs and other forms of land ice. See also: *drift ice.*
floating island A mass or mat of vegetation with little or no soil, floating freely in a lake or tropical sea, usually due to detachment from a marshy or boggy shore during a storm or a rise in the water level.
floating marsh *flotant.*
floating peat Peat that is derived from floating plants.
floating reef An isolated, displaced rock mass in alluvium. Cf: *float.*
floating sand grain An isolated grain of quartz sand that is not, or does not appear to be, in contact with neighboring sand grains scattered throughout the finer-grained matrix of a sedimentary rock, esp. of a limestone; e.g. a grain surrounded on all sides by coarse mosaic of calcite cement.
float ore Scattered fragments of vein material broken from outcrops and dispersed in soil; a type of *float.*
floatstone [mineral] A lightweight, porous, friable variety of opal that floats on water and occurs in white or grayish, spongy, and concretionary or tuberous masses. Also spelled: *float stone.* Syn: *swimming stone.*
floatstone [mining] A miner's term for cellular or honeycomb quartz detached from a lode.
floc (a) A loose, open-structured mass formed in a suspension by *flocculation;* e.g. a small aggregate of tiny sedimentary grains or colloidal clay particles. (b) A term used by Brewer (1964, p.367) for soil *plasma* that has a relatively low luster and a rough surface, giving it the appearance of clusters of silt-size grains in reflected light up to magnifications of approximately 20 times. Cf: *lac [soil].*
flocculation The process by which a number of individual minute suspended particles are tightly held together in clotlike masses, or are loosely aggregated or precipitated into small lumps, clusters, or granules; e.g. the joining of soil colloids into a small group of soil particles, or the deposition or settling out of suspension of clay particles in salt water. See also: *floc.*
floe A piece of floating ice other than fast ice or glacier ice, larger than an *ice cake* and smaller than an *ice field.* Floes are subdivided according to horizontal extent and many varying size limits have been used; the U.S. Naval Oceanographic Office (1968, p. A27) gives the following dimensions: "giant" (over 10 km); "vast" (2-10 km); "big" (500 m to 2 km); "medium" (100-500 m); "small" (20-200 m). Syn: *ice floe.*
floeberg (a) A massive piece of sea ice composed of a *hummock [ice]* or a group of hummocks, frozen together and separated from any ice surroundings, and floating with its highest point up to 5 m above sea level (U.S. Naval Oceanographic Office, 1968, p. B33). It resembles a small iceberg. (b) In the older literature, a thick mass of well-hummocked sea ice originating from an ice floe, and sometimes projecting more than 15 m above sea level.
floe calcite *cave raft.*
floe till *berg till.*
Floetz A name applied by A.G. Werner in the 1790's to the group or series of rocks that included most of the obviously stratified, comparatively flat, fossiliferous rocks (and certain associated trap rocks) and that were believed to represent the emergence of mountains from beneath the receding ocean, with products of the resulting erosion deposited on the mountain flanks. The rocks succeeded the *Transition* series and included the whole range of strata from the Devonian through the Tertiary. Etymol: German *Flötz* (now *Flöz*), "flat layer, stratum, seam, bed". Syn: *Secondary.*
floitite A rock consisting of biotite and those minerals that are typical of the greenschist facies. The term was originated by Becke in 1922.
flokite *mordenite.*
flood [sed] A term used by Milner (1940, p. 457) to describe the occurrence, in a sedimentary rock, of a mineral species "so far in excess of all others as to constitute almost a pure concentrate".
flood [water] (a) A rising body of water (as in a stream, lake, or sea, or behind a dam) that overtops its natural or artificial confines and that covers land not normally under water; esp. any relatively high streamflow that overflows its banks in any reach of the stream, or that is measured by gage height or discharge quantity. (b) A flood of special severity or local interest; specif. the Flood, or the *Noachian flood.* (c) An archaic term for a large body of moving water, such as a river.
flood absorption A reduction in discharge resulting from the storage of flowing water in a reservoir, channel, or lake (ASCE, 1962).
flood basalt *plateau basalt.*
flood basin (a) The tract of land covered by water during the highest known flood. (b) The broad, flat area between a sloping, low plain and the natural levee of a river, "occupied by heavy soils and commonly having either no vegetation or strictly swamp vegetation" (Bryan, 1923b, p. 39). Syn: *tule land.*
flood control The prevention or reduction of damage caused by flooding, as by containing water in reservoirs removed from areas where it would do damage, improving channel capacity to convey water past or through critical areas with the least amount of damage, and diverting excess water into bypasses or floodways.
flood crest The highest stage of a flood. The term is nearly synonymous with *flood peak,* but does not refer to discharge since it connotes the top of the flood wave (Langbein & Iseri, 1960, p. 10).
flood current The tidal current associated with the increase in the height of a tide, generally set toward the shore or up a tidal river or estuary. Cf: *ebb current.* Erroneous syn: *flood tide.*
flood dam A dam to store floodwaters temporarily, or to supply a surge of water as for clearing a channel. It is often specifically designed to withstand rapid changes in reservoir water level. Syn: *flooding dam.*
flooded stream *drowned stream.*
flood frequency The average occurrence of flooding of a given magnitude, over a period of years.
flood-frequency curve A graphic illustration of the number of times per year that a flood of a given magnitude is equaled or exceeded.
flood fringe *pondage land.*
floodgate (a) A gate for shutting out, admitting, releasing, or otherwise regulating a body of water, such as excess water in times of flood; specif. the lower gate of a lock. See also: *sluice.* (b) A stream stopped by or allowed to pass by a floodgate.
flood icing *icing.*
flooding (a) The covering or causing to be covered with a fluid, such as the covering of flat lands with a thin sheet of water; the filling or becoming full with water, esp. to excess. (b) A method of injecting water via select wells into the *pay zone* of an oil pool in order to force the oil toward producing wells; *water flooding.*
flooding ice *icing.*
floodland The land along a river that may be submerged by floodwaters; a *flood plain.*
flood map A map that depicts the extent of former floods or the anticipated extent of any particular magnitude of flood.
flood peak The highest discharge or stage value of a flood. Cf: *flood crest.* Syn: *peak discharge.*
flood plain (a) The surface or strip of relatively smooth land adjacent to a river channel, constructed by the present river in its existing regimen and covered with water when the river overflows its banks. It is built of alluvium carried by the river during floods and deposited in the sluggish water beyond the influence of the swiftest current. A river has one flood plain and may have one or more *terraces* representing abandoned flood plains. Cf: *valley flat; erosional flood plain.* (b) Any flat or nearly flat lowland that borders a stream and that may be covered by its waters at flood stages; the land described by the perimeter of the maximum probable flood. Syn: *floodland.* (c) The part of a lake-basin plain between the shoreline and the shore cliff, subject to submergence during a high stage of the lake.—Also spelled: *floodplain; flood-plain.*
flood-plain deposit Sandy and clayey sediment deposited by river water that was spread out over a flood plain; a deposit beneath and forming a flood plain, being thickest near the river and thinning out toward the valley slopes. See also: *overbank deposit.* Syn: *vertical-accretion deposit.*

flood-plain icing *icing.*
flood-plain lobe The part of a flood plain enclosed by a stream meander.
flood-plain meander scar A crescentic mark indicating the former position of a river meander on a flood plain.
flood-plain scroll One of a series of short, crescentic, slightly sinuous strips or patches of coarse alluvium formed along the inner bank of a stream meander and representing the beginnings of a flood plain. Syn: *flood scroll.*
flood-plain splay A small alluvial fan or other outspread deposit formed where an overloaded stream breaks through a levee (artificial or natural) and deposits its material (often coarse-grained) on the flood plain. Syn: *sand splay; channel splay; splay [geomorph].*
flood plane (a) The position occupied by the water surface of a stream during a specific flood. (b) Loosely, the elevation of the water surface at any of various points along the stream during a specific flood.
flood probability The probability, determined statistically, that a flood of a given size will be equaled or exceeded in a given period, e.g. a 10% probability would be called a ten-year flood.
flood-prone area Land on a flood plain that is subject to inundation during a flood of a designated frequency or magnitude; for example, a 100-year flood or a *standard project flood.*
flood routing Progressive determination of the timing and shape of a flood wave at successive points along a river (Langbein & Iseri, 1960, p. 10).
flood scroll *flood-plain scroll.*
flood series A listing of flood events for a given period of time, arranged in order of magnitude.
flood stage (a) The height of the gage at the lowest bank of the reach (other than an unusually low place or break). (b) The stage at which stream overflow begins to cause damage.
flood tide (a) *rising tide.* (b) A tide at its greatest height. (c) An erroneous syn. of *flood current.*
flood tuff *ignimbrite.*
floodwall A wall, often of reinforced concrete, built to prevent flooding. Cf: *levee [streams].*
floodwater (a) Water that has overflowed its confines; the water of a flood. (b) The flooded area behind a dam; an impoundment.
flood wave A rise in the stage of a stream that culminates in a crest before receding.
floodway (a) A large-capacity channel constructed to divert floodwaters or excess streamflow from populous or flood-prone areas, such as a bypass route bounded by levees. (b) The part of a flood plain kept clear of encumbrances and reserved for emergency diversion of floodwaters. (c) *flowage land.*
flood zone [stratig] *acme-zone.*
flood zone [streams] (a) The land bordering a stream, subject to floods of about equal frequency; e.g. a strip of the flood plain, subject to flooding more often than once but not as frequently as twice in a century (Langbein & Iseri, 1960, p. 11). (b) The land bordering a reservoir or stream impoundment, subject to inundation above the normal operating level.
floor [eco geol] The *footwall* of a horizontal orebody.
floor [geomorph] (a) The *bed* of any body of water; esp. the continuous and gently curved or essentially horizontal surface of the ground beneath the water of a stream, lake, or ocean. Syn: *bottom.* (b) *valley floor.*
floor [intrus rocks] The country rock bordering the lower surface of an igneous intrusion. Cf: *roof [intrus rocks].*
floor [stratig] A rock surface, usually an eroded surface, on which sedimentary strata have been deposited.
floorplate Any plate, in a double or single row, that forms an ambulacral groove in an echinoderm. Syn: *ambulacral floorplate.*
floorplate passageway A tubular canal in edrioasteroids that extends along lateral sutural faces of contiguous ambulacral floorplates. Floorplate passageways connect the ambulacral tunnel to the interior of the theca (Bell, 1976).
floor thrust The lower boundary of a *duplex fault zone* (Dahlstrom, 1970, p. 418).
flora The entire plant population of a given area, environment, formation, or time span. Cf: *fauna.*
floral stage A chronostratigraphic unit (stage) based on a florizone or commonly on a floral assemblage; e.g. the Ravenian (upper Eocene) of Washington State.
floral zone *florizone.*
florencite A pale-yellow mineral: $CeAl_3(PO_4)_2(OH)_6$.
floricome A sponge spicule (hexaster) with petal-like terminations on the rays.
Florida earth A variety of fuller's earth from Florida (esp. from Quincy and Jamieson) or resembling that from Florida.
florizone A biostratigraphic unit or body of strata characterized by a particular assemblage of fossil floras, regardless of whether it is inferred to have chronological or only environmental significance. Although the term, like *faunizone,* has been given different meanings, it is close in concept to *assemblage-zone* and has been generally regarded as the plant-based variety of (biostratigraphic) zone. The American Commission on Stratigraphic Nomenclature (1961, art.21d) states that the term is "not generally accepted" and that its correct definition is "in dispute". Syn: *floral zone.*
florule (a) An assemblage of fossil plants obtained from a stratum over a very limited geographic area, esp. from only one exposure. (b) A term used by Fenton & Fenton (1928,p.15) for an assemblage of fossil plants associated in a single stratum or a few contiguous strata of limited thickness and dominated by the representatives of one community; the floral assemblage of a *zonule.* Cf: *faunule.* Syn: *florula.*
floscelle A star-shaped area around the peristome of an echinoid, formed by phyllodes and bourrelets.
flos ferri An arborescent variety of aragonite occurring in delicate white coralloid masses that commonly encrust hematite, forming picturesque snow-white pendants and branches.
floss A British term for a stream.
flotant A coastal marsh formed along an abandoned channel or in a low basin between natural levees of active and inactive stream channels, as in Louisiana south of New Orleans; it is not as firm as marshland. Syn: *floating marsh.*
flotation *crystal flotation.*
flour A finely powdered rock or mineral mass, resulting from pulverization and grinding; e.g. *rock flour,* of glacial origin, or chalky-looking finely comminuted carbonate mud formed under intense wave or current action in shoal areas.
flour sand *very fine sand.*
flow [coast] A Scottish term for an arm of the sea; used chiefly in place names, e.g. Scapa Flow.
flow [exp struc geol] Any rock deformation that is not instantly recoverable without permanent loss of cohesion. Various types of flow in which the mechanism is known include *cataclastic flow, gliding flow,* and *recrystallization flow.* Syn: *rock flowage.*
flow [glaciol] *glacier flow.*
flow [hydraul] The movement of water, and the moving water itself; also, the rate of movement.
flow [mass move] (a) A mass movement of unconsolidated material that exhibits a continuity of motion and a plastic or semifluid behavior resembling that of a viscous fluid; e.g. *creep; solifluction; earthflow; mudflow; debris flow; sturzstrom.* Water is usually required for most types of flow movement. (b) The mass of material moved by a flow.
flow [volc] *lava flow.*
flowage (a) An act of flowing or flooding, such as the overflowing of a stream onto adjacent land; the state of being flooded. (b) A body of water resulting from flowage; the floodwater of a stream. The term is used locally in Wisconsin for the backwater of an artificial lake. (c) The area affected by a previous flooding.
flowage cast A term used by Birkenmajer (1958, p.141) for a sole mark believed to result from the flowage of mobile, hydroplastic sand over the uneven bottom in the direction of slope; it may be transverse, longitudinal, or multidirectional. See also: *flow cast.*
flowage differentiation The tendency of suspended crystals to concentrate in the high-velocity zone of a magma that is moving by laminar flow.
flowage fold *flow fold.*
flowage land The part of a flood plain that will be covered by the water impounded by a proposed dam, exclusive of the river channel; the principal flow-carrying part of the natural cross section of a stream. Syn: *floodway.*
flowage line *flow line.*
flow-and-plunge structure A variety of *cross-lamination,* consisting of short, obliquely laminated beds deposited irregularly at various angles of slope, resulting from tidal action accompanied by plunging waves.
flow bog A peat bog whose surface level fluctuates with rain and the tides.
flow breccia A breccia that is formed contemporaneously with the movement of a lava flow; the cooling crust becomes fragmented while the flow is still in motion. It is a type of *autobreccia.*

flow cast (a) A term introduced by Shrock (1948, p.156) for a *sole mark* consisting of a lobate ridge or other raised feature and representing the filling of a depression produced by the flowage or warping of the soft and hydroplastic underlying sediment. Kuenen (1953, p.1058) applied the term *load cast* to such a structure produced by vertical adjustments. Prentice (1956) revived "flow cast" for a sole mark resulting from a combination of load casting and current-oriented flow, such as a load cast modified by horizontal flowage of sediment during or after settling. See also: *flowage cast.* (b) *flute cast.*

flowchart A graphic representation or schematic diagram of steps in a sequence of operations that are represented by symbols, as for a computer program. Also spelled: *flow chart.*

flow cleavage A syn. of *slaty cleavage,* so called because of the assumption that recrystallization of the platy minerals is accompanied by rock flowage.

flow direction The axis parallel to the direction of relative displacement in both igneous and metamorphic rocks. In the former it is a *flow line [petrology].* In the latter, it is usually subparallel to lineations visible on hand specimens, and corresponds to the average orientation of the slip direction of individual crystals, even in conditions of *dynamic recrystallization.*

flow-duration curve A type of *duration curve* showing how often a particular stream discharge is equaled or exceeded.

flow earth *solifluction mantle.*

flower The reproductive structure of an angiosperm. In a morphologic sense, it is considered to be a specialized branch system.

flowering plant An informal designation of an *angiosperm.*

flow failure A form of slope movement involving the transport of earth materials in a fluid manner over distances of at least several tens of feet.

flow fold A fold composed of relatively plastic rocks that have flowed towards the synclinal trough. In this type of deformation, there are no apparent surfaces of slip. Syn: *flowage fold.* Cf: *reverse-flowage fold; rheid fold.*

flow gneiss Gneiss whose structure was produced by flowage in an igneous mass before complete solidification.

flowing artesian well An *artesian well* whose head is sufficient to raise the water above the land surface. Cf: *nonflowing artesian well; flowing well.* Syn: *blow well.*

flowing ground A tunnelman's term for soil that flows into a tunnel from floor, walls, or roof, driven by water seepage. Cohesionless soil below the water table is typical flowing ground. See also: *firm ground; raveling ground; running ground; squeezing ground; swelling ground.*

flowing well (a) A well that yields water at the land surface without pumping. It is distinguished from a *flowing artesian well* by the possibility that the flow may be due to gas rather than artesian pressure. (b) *flowing oil well.*

flow joint A joint parallel to the flow layers of a plutonic rock (Tomkeieff, 1943).

flow layer A layer in an igneous rock that is characterized by *flow layering.* Cf: *flow line [petrology].*

flow layering The structure of an igneous rock, characterized by alternating layers of color, mineralogic composition, and/or texture, formed as a result of the flow of magma or lava. Syn: *flow banding; fluxion banding.* See also: *banding [ign].*

flow line [hydraul] (a) The position of the surface of a flowing fluid. (b) A water-level contour around a body of water, e.g. maximum or mean flow line of a lake; *flowage line.* (c) In an open channel, the *hydraulic grade line.*

flow line [petrology] A lineation of crystals, mineral streaks, or inclusions in an igneous rock, indicating the direction of flow before consolidation. Cf: *flow layer.*

flow mark (a) A small channel or gouge cut in a sedimentary surface by a current of water; a flute. (b) A cast of a flow mark, preserved in overlying sediment (Rich, 1950); specif. a *flute cast.* (c) A small ridge formed on the upper surface of a muddy sediment by a current of water (McKee, 1954, p.63).—See also: *current mark.* Syn: *flow marking.*

flow net In the study of seepage phenomena, a graph of flow lines and equipotential lines.

flow plane The plane along which displacement occurs in both igneous and metamorphic rocks. In the former it is a *flow layer.* In the latter, it is generally subparallel to the *foliation* visible in hand specimens and corresponds to the average orientation of slip plane of individual crystals, even in conditions of *dynamic recrystallization.*

flow profile The form of the water surface of a *gradually varied flow;* it is commonly known as the *backwater curve.*

flow regime A range of streamflows with similar bed forms, resistance to flow, and mode of sediment transport (Middleton, 1965, p. 249).

flow roll A rounded, pillow-like body or mass of sandstone occurring within or just above finer-grained sediment or commonly within the basal part of a sandstone overlying shale or mudstone, having a shape approaching that of an elongate, flattened ellipsoid (short axis more or less vertical), and presumed to form by deformation, as by large-scale load casting or mud flowage accompanied by subaqueous slump or foundering of sand channels. The term was used by Pepper et al. (1954, p.88) in reference to the characteristic rolled appearance of the structure and because deformation of strata occurred prior to complete lithification of the rocks. See also: *ball-and-pillow structure; pseudonodule; storm roller.*

flow slide *liquefaction slide.*

flowstone A general term for any deposit of calcium carbonate or other mineral formed by flowing water on the walls or floor of a cave. See also: *dripstone; travertine; cave onyx.*

flow stress A general term for the stress required to sustain *flow [exp struc geol].*

flow structure A primary sedimentary structure resulting from subaqueous slump or flow (Cooper, 1943, p. 190).

flow system In hydrodynamics, a set of flow lines in which any two are always adjacent, and can be intersected in one direction only by an uninterrupted surface across which flow takes place.

flow texture A texture characterized by a wavy or swirling pattern in which platy or prismatic minerals are oriented along planes of lamellar flowage or in flow lines in fine-grained and glassy igneous rocks. Syn: *fluidal texture; fluxion texture; rhyotaxitic texture.*

flowtill A superglacial till that is modified and transported by plastic mass flow (Hartshorn, 1958, p. 481).

flow unit A group of sheets or beds of lava or pyroclasts that were formed by a single eruption or outpouring.

flow velocity A vector point function used to indicate rate and direction of movement of water through soil, per unit of time and perpendicular to the direction of flow.

fluctuation The alternate rising and lowering of the water table either regularly or periodically.

flue [intrus rocks] A pipelike igneous intrusion.

flue [sed] A hard, sandy shale in the Lancashire (England) coalfield, probably so named in reference to its splitting or fissile quality.

fluellite A colorless or white mineral: $Al_2(PO_4)F_2(OH)\cdot 7H_2O$. It was previously formulated: $AlF_3\cdot H_2O$.

fluent An obsolete term for a stream or other current of water.

fluidal texture (a) A syn. of *flow texture.* (b) A metamorphic texture in which narrow stripes or lenticles of a mineral, present as grains approx. 0.01 mm in diameter, are connected with *porphyroclasts* of the same mineral and extend across regions in which another mineral shows a dominantly *mosaic* texture. This texture has been given a specific genetic connotation relating to *superplasticity* (Harte, 1977).

fluid dynamics That aspect of *fluid mechanics* which deals with motion of fluids.

fluid escape structure A general category of sedimentary feature produced by the escape of fluids from a bed of sediment after deposition. It includes *dish structure, pillar structure* and *vertical sheet structure.* If the escaping fluid is water, the feature may be called a *water escape structure* (Lowe, 1975).

fluid inclusion In a mineral, a tiny cavity, 1.0-100.0 microns in diameter, containing liquid and/or gas, formed by the entrapment in crystal irregularities of fluid, commonly that from which the rock crystallized. Partial syn: *liquid inclusion.*

fluidity index The ratio of sand detritus to the interstitial detrital matrix of a sandstone (Pettijohn, 1954, p. 362-363). It is a measure of the fluidity (density and viscosity) of the depositing medium and it partly determines the sorting of transported sediment; e.g. a high ratio indicates a poorly sorted sediment deposited from a medium of high density and high viscosity. Syn: *fluidity factor.*

fluidization The mixing process of gas and loose fine-grained material so that the whole flows like a liquid, e.g. the formation of an ash flow or nuée ardente during a volcanic eruption.

fluid mechanics The study of the mechanics or behavior of liquids and gases. It is broad in scope, and includes such disciplines as hydraulics and aerodynamics. See also: *fluid dynamics.*

fluid potential The mechanical energy per unit mass of a fluid, e.g. water or oil, at any given point in space and time, with respect to an arbitrary state and datum. The fluid potential is proportional to the *total head;* it is the *head* multiplied by the acceleration due to gravity. Syn: *potential.*

fluke One of two or more recurved components of an *anchor* of a holothurian.

flume [eng] An artificial inclined channel used for conveying water for industrial purposes, such as irrigation, transportation, mining, logging, or power production; or for diverting the water of a stream from its channel for the purpose of washing or dredging the sand and gravel in the dry bed or to aid in engineering construction.

flume [geomorph] A ravine, gorge, or other deep narrow valley, with a stream flowing through it in a series of cascades; e.g. in the White Mountains, N.H.

fluoborite A colorless hexagonal mineral: $Mg_3(BO_3)(F,OH)_3$. Syn: *nocerite.*

fluocerite A pale-yellow or reddish-yellow hexagonal mineral: $(Ce,La,Nd)F_3$. Syn: *tysonite.*

fluolite *pitchstone.*

fluor (a) The original form of *fluorite,* still used chiefly in Great Britain. (b) An obsolete name for a mineral belonging to a group including fluorite and characterized by the alchemists as resembling gems and usable as metallurgic fluxes (Webster, 1967, p. 877). Etymol: Latin *fluere,* "to flow".

fluorapatite (a) A very common mineral of the apatite group: $Ca_5(PO_4)_3F$. It is a common accessory mineral in igneous rocks. Syn: *apatite.* (b) An apatite mineral in which fluorine predominates over chlorine and hydroxyl.

fluorescence A type of *luminescence* in which the emission of light ceases when the external stimulus ceases; also, the light so produced. Cf: *phosphorescence.*

fluorescence spectrum The *emission spectrum* produced by irradiation of a material with radiation of higher energy, as in *X-ray fluorescence spectroscopy.*

fluorine dating Determination of relative age of Pleistocene or Recent bones on the basis of fluorine content. The method depends on the gradual combination with time of fluorine in ground water with the calcium phosphate of bone. In areas where this method has been used, the fluorine content averages 2% in lower Pleistocene bone, 1% in middle Pleistocene bone, 0.5% in upper Pleistocene bone, and 0.3% in Recent bone.

fluorite A transparent to translucent mineral: CaF_2. It is found in many different colors (often blue or purple) and has a hardness of 4 on the Mohs scale. Fluorite occurs in veins, usually as a gangue mineral associated with lead, tin, and zinc ores, and is commonly found in crystalline cubes with perfect octahedral cleavage. It is the principal ore of fluorine, and is used as a flux, in the preparation of glass and enamel, in the manufacture of hydrofluoric acid, and for carved ornamental objects. Syn: *fluorspar; fluor; Derbyshire spar.*

flurospar *fluorite.*

flurosion A term proposed by Glock (1928, p. 477-478) for the work of transportation and erosion carried on by streams.

flush (a) A sudden increase in the volume of a stream; a sudden flow or rush of water, as down a stream and filling the channel or overflowing the banks. Syn: *fresh.* (b) A British term for a pool or for a low swampy place.—Cf: *flash.*

flushing period The interval of time necessary for an amount of water equal to the volume of a lake to pass through its outlet, computed by dividing lake volume by flow rate (usually mean flow) of the outlet.

flustriform Said of a flexible, erect colony in cheilostome bryozoans, with lightly calcified zooids and attached by radicles or direct adherence.

flute [geomorph] A small, shallow channel formed by differential weathering and erosion, running nearly vertically down the face of a rock surface.

flute [sed] (a) A primary sedimentary structure, commonly seen as a *flute cast,* consisting of a discontinuous scoop-shaped or lobate depression or groove generally 2-10 cm in length, usually formed by the scouring action of a turbulent, sediment-laden current of water flowing over a muddy bottom, and having a steep or abrupt upcurrent end where the depth of the mark tends to be the greatest. Its long axis is generally parallel to the current. The term was first used by Maxson and Campbell (1935) for flutes cut on boulders in the Colorado River. (b) A term that is sometimes used loosely as a syn. of *flute cast.* (c) A scalloped or rippled rock surface. The term is usually used in the plural.—See also: *fluting.*

flute [speleo] (a) An incised vertical channel or groove developed in a cave shaft by solution. See also: *rillenkarren.* (b) A syn. of *scallop [speleo].*

flute cast A term suggested by Crowell (1955, p.1359) for a spatulate or lingulate *sole mark* consisting of a raised, oblong, and subconical bulge on the underside of a siltstone or sandstone bed, characterized by a steep or blunt bulbous or beaked upcurrent end from which the structure flattens or flares out in the downcurrent direction and merges with the bedding plane. It is formed by the filling of a *flute.* See also: *lobate rill mark.* Syn: *fluting; flute; flow cast; flow mark; scour cast; vortex cast; linguoid sole mark; turboglyph; lobate plunge structure.*

fluted moraine *fluted moraine surface.*

fluted moraine surface Moraine surface in front of a glacier containing parallel ridges that have a more or less constant height over distances of the order of tens or hundreds of meters. The ridge axes are parallel to the flow direction of the glacier (Schytt, 1963). Syn: *fluted moraine.*

fluted till *Lodgment till* that has been eroded into grooves, or in which positive constructional linear ridges have been built.

fluting [geomorph] A process of differential weathering and erosion by which an exposed well-jointed coarse-grained rock, such as granite or gneiss, develops a corrugated surface of *flutes;* esp. the formation of small-scale ridges and depressions by wave action.

fluting [glac geol] (a) The formation by glacial action of smooth deep gutterlike channels or furrows on the stoss side of a rocky hill obstructing the advance of a glacier; the furrows are larger than glacial grooves, and they do not extend around the hill to the lee side (Chamberlin, 1888, p. 246). Also, a furrow so formed. (b) Lineations or streamline grooves and ridges parallel to the direction of ice movement, formed in newly deposited till or older drift. They range in height from a few centimeters to 25 m, and in length from a few meters to more than 20 km.

fluting [sed] (a) The process of forming a *flute* by the cutting or scouring action of a current of water flowing over a muddy surface. (b) Scalloped or rippled rock surfaces. (c) *flute cast.*

Fluvent In U.S. Dept. of Agriculture soil taxonomy, a suborder of the soil order *Entisol,* characterized by formation in recent alluvial deposits that are flooded frequently and usually stratified. Most sediments are derived from eroding soils or streambanks. Organic carbon is mainly in the clay fraction and decreases irregularly with depth (USDA, 1975). Cf: *Aquent; Arent; Orthent; Psamment.*

fluvial (a) Of or pertaining to a river or rivers. (b) Existing, growing, or living in or about a stream or river. (c) Produced by the action of a stream or river. See also: *fluviatile.*—Etymol: Latin *fluvius,* "river".

fluvial cycle of erosion *normal cycle.*

fluvial denudation Erosional reduction of a land surface by streams, assisted by weathering, mass wasting, and overland flow.

fluvial deposit A sedimentary deposit consisting of material transported by, suspended in, or laid down by a stream. Syn: *fluviatile deposit.*

fluvialist A believer in the doctrine that the widespread surficial deposits now known to be glacial drift, and other geologic phenomena, can be explained by ordinary stream action. Cf: *diluvialist.*

fluvial lake (a) A lake with a perceptible flow of water, e.g. a body of water connecting two larger bodies whose differences in elevation may be sufficient to create a flow from one to another. (b) A slowly moving part of a river, as where its width has been expanded. See also: *fluviatile lake.*

fluviatile A syn. of *fluvial* (Fowler, 1937, p. 184). Geologists tend to use the term for the results of river action (e.g. *fluviatile* dam, or *fluviatile* sands) and for river life (e.g. *fluviatile* fauna). Syn: *fluviatic.*

fluviatile lake A lake formed as a result of the action of a river or stream, or a lake occupying a basin produced by running water capable of erosion or deposition; e.g. an *oxbow lake* on the flood plain of a meandering river, or a lake formed by the damming action of excess sediment at the confluence of a tributary and the main river. See also: *fluvial lake.*

fluviation The activities engaged in, and the processes employed, by streams (Glock, 1928, p. 477). Syn: *stream action.*

fluvicoline Said of an animal that lives in or frequents streams.

fluvioclastic rock A hydroclastic rock containing current- or river-worn fragments (Grabau, 1924, p.295). Syn: *potamoclastic rock.*

fluvioeolian Pertaining to the combined action of streams and wind; e.g. a *fluvioeolian* deposit.

fluvioglacial A syn. of *glaciofluvial.* The term "glaciofluvial" is preferred in U.S. "since logically the *glac-* precedes the *fluv-*" (Monkhouse, 1965, p. 127).

fluvioglacial drift *glaciofluvial drift.*

fluviograph A device for measuring and recording automatically the rise and fall of a river. Syn: *fluviometer.*

fluviokarst Karst formed near the margin of a soluble-rock terrain.

fluviolacustrine Pertaining to sedimentation partly in lake water and partly in streams, or to sediments deposited under alternating or overlapping lacustrine and fluvial conditions.

fluviology The science of rivers.

fluviomarine Said of marine sediments that contain resorted and redistributed fluvial material along with the remains of marine organisms.

fluviometer *fluviograph.*

fluviomorphology *river morphology.*

fluvioterrestrial Consisting of or pertaining to the land and its streams.

fluviraption A term introduced by Malott (1928a) for *hydraulic action* performed by rivers.

flux (a) A stream of flowing water; a flood or an outflow. (b) The number of radioactive particles in a given volume times their mean velocity.

fluxgate magnetometer An electrical instrument that measures the change in magnetic field along the axis of its sensor with a sensitivity of one gamma or more. Used as an *airborne magnetometer* it measures relative total magnetic intensity; used as a ground magnetometer it measures relative vertical magnetic intensity.

fluxie A field term generally used by sedimentologists for *fluxoturbidite.*

fluxion Obsolescent British usage for "flow". Cf: *flow layering; flow texture.*

fluxion banding British usage for *flow layering.*

fluxion texture British usage for *flow texture.*

flux ore A material containing an appreciable amount of valuable metal but smelted mainly because it contains fluxing agents useful in the reduction of richer ores. Cf: *fluxstone.*

fluxoturbidite A term proposed by Dzulynski et al. (1959, p. 1114) for a sediment produced by a mechanism related both to deposition from turbidity currents and to submarine slumping or sliding. It is characterized by coarse grain, thick bedding, and poor development of grading and of sole marks. Cf: *undaturbidite.* Syn: *fluxie.*

fluxstone Limestone, dolomite, or other rock or mineral used in metallurgical processes to lower the fusion temperature of the ore, combine with impurities, and make a fluid *slag.*

fly ash All particulate matter that is carried in a gas stream, esp. in stack gases at a coal-fired plant for generation of electric power.

flyer A number of geophones permanently connected at intervals along a short cable. A typical flyer might contain, for example, six geophones connected in series at 20-ft intervals, with clips at one end for joining it to the main cable that carries the signal to the recording equipment.

flying bar A looped bar or spit formed on the landward side of an island that is subsequently reduced below sea level by wave erosion before the bar or spit is destroyed. Term originated by Gulliver (1899, p. 190).

flying magnetometer *airborne magnetometer.*

flying veins A pattern of mineral-deposit veins overlapping and intersecting in a branchlike pattern.

fly leveling Spirit leveling in which some of the restrictions of precise leveling (such as limiting the length of sight, or balancing backsights and foresights) are relaxed in order to obtain elevations of moderate accuracy more rapidly; e.g. running a level line (a line over which leveling operations are accomplished) at the close of a working day in order to check the results of an extended line run in one direction only. Also called "flying levels".

flysch (a) A marine sedimentary facies characterized by a thick sequence of poorly fossiliferous, thinly bedded, graded deposits composed chiefly of marls and sandy and calcareous shales and muds, rhythmically interbedded with conglomerates, coarse sandstones, and graywackes. See also: *wildflysch; macigno.* (b) An extensive, preorogenic sedimentary formation representing the totality of the flysch facies deposited in different troughs, during the later stages of filling of a geosynclinal system, by rapid erosion of an adjacent and rising mountain belt at a time directly previous to the main paroxysmal (diastrophic) phase of the orogeny or when initial diastrophism had already developed interior ridges exposed to erosion; specif. the Flysch strata of late Cretaceous to Oligocene age along the borders of the Alps, deposited in the foredeeps in front of northward-advancing nappes rising from beneath the sea, before the main phase (Miocene) of the Alpine orogeny. (c) A term that has been loosely applied to any sediment with nearly all of the lithologic and stratigraphic characteristics of a flysch, such as almost any turbidite.—Etymol: dialectal term of German origin used in Switzerland for a crumbly or fissile material that slides or flows. Pron: *flish.* Pl: *flysches.* Adj: *flyschoid.* Cf: *molasse.*

fm *formation.*

f-number The ratio of the equivalent focal length of a photographic lens to the *relative aperture,* or a number expressing the relative aperture of the lens; e.g. the f-number of a lens with a relative aperture of $f/4.5$ is 4.5. The smaller the number, the brighter the image and therefore the shorter the exposure required. Etymol: *f,* symbol for focal length. See also: *speed.*

foam *pumice.*

foam crust A snow-surface feature produced during ablation and resembling small overlapping waves, like sea foam on a beach. Cf: *plowshare.*

foaming earth Soft or earthy aphrite.

foam mark A surface sedimentary structure consisting of a pattern of almost imperceptible ridges and hollows, formed where foam produced by wind action on seawater is driven over a surface of moist or wet sand.

focal depth *depth of focus.*

focal distance *focal length.*

focal length A general term for the distance from the principal point or center of a lens to the principal focus. In photogrammetry, the term "equivalent focal length" is the distance measured along the lens axis from the rear nodal point of the lens to the position of the focal plane that provides the best average definition in the aerial negative, and the term "calibrated focal length" is an adjusted value of the equivalent focal length computed to distribute the effect of lens distortion over the entire field of the negative. Symbol: f. Syn: *focal distance.*

focal mechanism *fault-plane solution.*

focal plane The plane, perpendicular to the axis of a lens, in which images of points in the object field are brought to a focus; a plane that passes through a principal focus.

focal point *principal focus.*

focal sphere An arbitrary reference sphere drawn about the hypocenter or focus of an earthquake, to which body waves recorded at the Earth's surface are projected for studies of earthquake mechanisms.

focus [photo] (a) The point at which rays of light converge to form an image after passing through a lens or optical system or after reflection by a mirror. See also: *principal focus.* (b) The condition of sharpest imagery.—Pl: *foci.*

focus [seis] The initial rupture point of an earthquake, where strain energy is first converted to elastic wave energy; the point within the Earth which is the center of an earthquake. Cf: *epicenter.* Syn: *hypocenter; seismic focus; centrum [seis].*

focused-current log The *resistivity log* curves from a multi-electrode *sonde* designed to focus the surveying current radially through the rocks in a horizontal, disk-shaped pattern. This permits sharp definition of bed boundaries and improved measurement of resistivity. Focused-current logs are marketed under several trade names, e.g. Laterolog, Guard Log. Syn: *current-focused log; guard-electrode log.*

foehn A warm, dry *katabatic wind* that descends the leeward side of a mountain ridge and warms from compression. The type locality is the Alps but a variety of local names are used elsewhere, e.g. *chinook.* Also spelled: *föhn.*

fog A cloud at the Earth's surface, formed by condensation of atmospheric water vapor into tiny droplets of water, about 40 microns in diameter, or rarely into crystals of ice.

fog desert A west-coast desert having a cold-water marine current just offshore of a warm continental land mass; the coolness and moist air combine to produce fog but little precipitation.

foggara A term used in the Saharan desert region (esp. in Morocco and Mauritania) for a gently inclined, underground conduit or tunnel designed to intercept ground water near the foot of mountains and to conduct it by gravity to a neighboring lowland for irrigation; a horizontal well. Etymol: Arabic. Cf: *qanat; karez.*

foggite An orthorhombic mineral: $CaAl(PO_4)(OH)_2 \cdot H_2O$.
föhrde *förde.*
foiba *domepit.*
foid A collective term coined by Johannsen (1917, p. 69-70) to denote the *feldspathoid* group of minerals. Etymol: *f*eldspath*oid.*
foid-bearing In the *IUGS classification,* a modifier preceding a rock name (syenite, monzonite, etc.) indicating a plutonic rock with F between 0 and 10. The rock name is determined by P/(A+P) and, for some categories, the plagioclase composition. For a specific rock, the term "foid" is replaced by the feldspathoid that is present. For example, nepheline-bearing gabbro contains nepheline and has F between 0 and 10, P/(A+P) greater than 90, and plagioclase more calcic than An_{50}. Cf: *foid.*
foid diorite In the *IUGS classification,* a plutonic rock with F between 10 and 60, P/(A+P) greater than 90, and plagioclase more sodic than An_{50}. Cf: *theralite.*
foid gabbro In the *IUGS classification,* a plutonic rock with F between 10 and 60, P/(A+P) greater than 90, and plagioclase more calcic than An_{50}. Syn: *theralite.*
foidite A plutonic or volcanic rock in which feldspathoids constitute 60-100 percent of the light-colored components; e.g. *urtite, ijolite, melteigite, italite.* Sometimes restricted to those igneous rocks in which feldspathoids represent 90-100 percent of the light-colored constituents. Cf: *feldspathoidite.* According to Sørensen (1974, p. 564), foidite should be used for volcanic rocks, *foidolite* for their coarser-grained equivalents.
foid monzodiorite In the *IUGS classification,* a plutonic rock with F between 10 and 60, P/(A+P) between 50 and 90, and plagioclase more sodic than An_{50}. Syn: *essexite.*
foid monzogabbro In the *IUGS classification,* a plutonic rock with F between 10 and 60, P/(A+P) between 50 and 90, and plagioclase more calcic than An_{50}. Syn: *essexite.*
foid monzosyenite In the *IUGS classification,* a plutonic rock with F between 10 and 60, and P/(A+P) between 10 and 50. Less preferred syn: *foid plagisyenite.*
foidolite In the *IUGS classification,* a plutonic rock with F greater than 60; a general term that includes *melteigite, missourite, ijolite, fergusite, urtite,* and *italite.*
foid plagisyenite Less preferred synonym of *foid monzosyenite.*
foid syenite In the *IUGS classification,* a plutonic rock with F between 10 and 60, and P/(A+P) less than 10.
fold [geomorph] A British term for an undulation in the land surface, either a low rounded hill or a shallow depression.
fold [paleont] (a) A major rounded elevation on the surface of a brachiopod valve (generally the brachial valve), externally convex in transverse profile and radial from the umbo, and usually median in position. It is typically associated with the *sulcus.* Cf: *carina.* (b) A spirally wound ridge on the interior of the wall of a gastropod shell; e.g. *columellar fold* and *parietal fold.*
fold [struc geol] n. A curve or bend of a planar structure such as rock strata, bedding planes, foliation, or cleavage. A fold is usually a product of deformation, although its definition is descriptive and not genetic and may include primary structures.
fold belt A widely used syn. of *orogenic belt.* Also spelled: *foldbelt; fold-belt.*
fold breccia A local *tectonic breccia* composed of angular fragments resulting from the sharp folding of thin-bedded, brittle rock layers between which are incompetent ductile beds; e.g. a breccia formed where interbedded chert and shale are sharply folded. Syn: *reibungsbreccia.*
fold coast A coast whose configuration is controlled by folded rocks.
folded fault (a) Any fault that has been deformed by folding. (b) A thrust fault, the hanging wall of which has become slightly folded due to the development of step thrusting or *step faults* beneath it (Jones, P.B., 1971); a *warped fault.*
fold facing *facing [struc geol].*
fold fault An overfold, the middle limb of which is replaced by a fault surface. Cf: *lag fault; slide [fault].*
folding The formation of *folds [struc geol]* in rocks.
fold mountains Mountains that have been formed by the large-scale folding and later uplift of stratified rocks.
fold mullion A type of *mullion* formed by the cylindrical undulations of bedding; internal structure displays conformable bedding laminations (Wilson, 1953). Cf: *cleavage mullion.*
fold system A group of congruent folds that are produced by the same tectonic episode.
Foleyan North American (Gulf Coast) stage: Pliocene (above Clovelly, below Pleistocene).
folgerite *pentlandite.*
folia [meta] Thin, leaflike layers or laminae, specif. cleavable folia of gneissic or schistose rocks. Singular: *folium.*
folia [speleo] *shelfstone.*
foliate Adj. of *foliation.*
foliated Adj. of *foliation.* Obsolete syn: *parafoliate.*
foliated ground ice A large mass of ice commonly occupying thermal contraction cracks in permafrost, and characterized by parallel or subparallel structures marked by air bubbles, films of organic or inorganic matter, or boundary surfaces between ice layers of different composition. It is usually, but not always, wedge-shaped. Syn: *wedge ice.*
foliation [glaciol] The planar or layered structure produced in the ice of a glacier by plastic deformation, manifest as alternating layers of coarse-grained and fine-grained ice, or bubbly and clear ice. Nonpreferred syn: *banding [glaciol].*
foliation [struc geol] A general term for a planar arrangement of textural or structural features in any type of rock; esp. the planar structure that results from flattening of the constituent grains of a metamorphic rock. Adj: *foliate; foliated.*
foliose Leafy, or resembling a leaf; esp. said of a corallum with laminar branches.
Folist In U.S. Dept. of Agriculture soil taxonomy, a suborder of the soil order *Histosol,* characterized by O horizons derived from leaf litter, twigs, and branches resting on rocks or from other fragmental material whose interstices contain organic material (USDA, 1975). Cf: *Fibrist; Hemist; Saprist.*
follicle A simple dry dehiscent fruit that produces seeds and that has one carpel and splits along one seam only.
following wind (a) A wind whose direction is the same as that of ocean waves. (b) A tailwind.
fondo adj. A term applied by Rich (1951, p. 2) to the environment of sedimentation that lies on the deep floor of a water body. It may be used alone or as a combining form. See also: *fondoform; fondothem.* Cf: *clino; unda.* Etymol: Spanish *fondo,* "bottom".
fondoform The subaqueous land form constituting the main floor of a water body (Rich, 1951, p. 2). It is the site of the *fondo* environment of deposition. Cf: *clinoform; undaform.*
fondothem Rock units formed in the *fondo* environment of deposition (Rich, 1951, p. 2). Cf: *clinothem; undathem.*
Foner magnetometer *vibration magnetometer.*
font An archaic term for a stream or a spring, fountain, or source of a stream; it forms part of place names, such as Chalfont.
Fontainebleau sandstone (a) A desilicified quartz sandstone (or an uncemented quartz sand) whose calcareous cement forms a crystalline aggregate of calcite in which sand grains are embedded; the cement is continuous and the easily fractured surfaces of the rock indicate that the calcite is in crystallographic continuity. Type locality: Fontainebleau, in the Paris Basin of France. Cf: *meulerization.* (b) A name given to a variety of calcite; specif. *sand-calcite.* Syn: *Fontainebleau limestone.*
food chain The passage of energy and materials from *producers* through a progressive, essentially linear sequence of plant-eating and meat-eating *consumers.* Cf: *food cycle; food web.*
food cycle All the *food chains* in an association of organisms; the food relations between the members of a population that make it possible for the population to survive.
food groove (a) An *ambulacral groove* in an echinoderm; e.g. a furrow running along the adoral surface of crinoid ray ossicles and traversing the tegmen to the mouth. (b) In a brachiopod, a trough running the length of the frontal surface of the brachia, bounded by a muscular lip and the base of the filaments.
food web The passage of energy and materials from *producers* through a progressive, many-stranded, anastamosing network of plant-eating and meat-eating *consumers.* Cf: *food chain; food cycle.*
fool's gold A popular term for pyrites resembling gold in color; specif. pyrite and chalcopyrite.
foot [geol] (a) The bottom of a slope, grade, or declivity. Cf: *head [geomorph].* (b) In a landslide or slump, the line of intersection (usually buried, but inferred) between the surface of rupture and the original ground surface (Varnes, 1958). (c) The lower bend of a fold or structural terrace. Syn: *lower break.*
foot [paleont] (a) The ventral part of the body of a mollusk, consisting chiefly of a muscular surface or process, and used for locomotion; e.g. a broad, flattened muscular sole used for creeping in most gastropods, or a protrusible, tapering, bladelike muscular struc-

ture extending from the midline of the body of a bivalve mollusk (anteriorly and ventrally in more typical bivalves), and used for burrowing or locomotion. (b) A limb of an arthropod. (c) One of the radial appendages extending from the ultimate joint of the shell of certain radiolarians (as in the suborders Nassellina and Phaeodarina).—Cf: *head [paleont].*

foot cave *cliff-foot cave.*

footeite *connellite.*

foot glacier *expanded-foot glacier.*

foothills A region of relatively low, rounded hills at the base of or fringing a mountain range; e.g. the low, undulating region along the western base of the Sierra Nevada in California.

footing A relatively shallow foundation by which concentrated loads of a structure are distributed directly to the supporting soil or rock through an enlargement of the base of a column or wall. Its ratio of base width to depth of foundation commonly exceeds unity. Cf: *pier [eng].*

foot layer A downward extension of ektexine of a pollen grain, partly surrounded by endexine.

footprint *track [paleont].*

foot slope A general term for a hillside surface whose top part is the *wash slope* and that includes "all the slopes of diminishing gradient" (Penck, 1953, p. 419). Syn: *fusshang.*

footwall The underlying side of a fault, orebody, or mine working; esp. the wall rock beneath an inclined vein or fault. Syn: *heading wall; heading side; lower plate.* Cf: *hanging wall.*

footwall cutoff A syn. of *trailing edge.* See: Royse et al. (1975).

foralite An inorganic structure resembling a worm tube, found in stratified rock.

foram *foraminifer.*

foramen (a) A small opening, perforation, pore, or orifice, such as the *pedicle foramen* of a brachiopod or the *septal foramen* of a nautiloid; esp. an opening that connects adjacent chambers in the test of a foraminifer, located at the base of septa or areal in position, and often representing a previous aperture or a secondarily formed aperture (but not equivalent to the pore of a perforate test). (b) A small opening in a bone, esp. one that gives passage to a blood vessel or nerve.—Pl: *foramina.*

foramina Plural of *foramen.*

foraminifer Any protozoan belonging to the subclass Sarcodina, order Foraminifera, characterized by the presence of a test of one to many chambers composed of secreted calcite (rarely silica or aragonite) or of agglutinated particles. Most foraminifers are marine but freshwater forms are known. Range, Cambrian to the present. Syn: *foram; foraminiferan; foraminifera.*

foraminiferal Pertaining to or derived from the Foraminifera or their shells; e.g. "foraminiferal test".

foraminiferal limestone A limestone composed chiefly of the remains of bottom-dwelling and floating foraminifers, and commonly lacking a fine-grained matrix; e.g. *fusulinid limestone* and *nummulitic limestone.*

foraminiferal ooze An *ooze* whose skeletal remains are the tests of foraminifera; it is a calcareous ooze. See also: *globigerina ooze.*

foraminite A sedimentary rock composed predominantly of the remains of foraminifers.

foram number In quantitative foraminiferal studies, the total number of all foraminiferal species in a one-gram sample (dry weight) of material greater than 0.1 mm in diameter.

forb A noncultivated dicotyledonous herbaceous plant; a herb other than grass; a weed. The term appears in some palynologic literature dealing with Quaternary sediments.

Forbes band An obsolescent term for one of a group of bands forming a type of *ogive* pattern that occurs on valley glaciers below icefalls and is characterized by alternating dark and light curved bands that cross the glacier. These bands normally occur in a regular succession at roughly equal intervals. This type of band was described by James D. Forbes (1809-1868), English physicist, who originally called it a *dirt band.* Syn: *Forbes ogive.* Cf: *dirt-band ogive; wave ogive.*

forbesite A mixture of annabergite and arsenolite.

force [phys] That which tends to put a stationary body in motion or to change the direction or speed of a moving body.

force [streams] A name given in northern England to a waterfall or cascade. Etymol: Old Norse *fors.* Cf: *fors.* Syn: *foss; fosse [streams].*

force couple Two equally intense forces acting in opposite directions but not in the same line on a body, creating a tendency for the body to rotate. Syn: *couple.*

forced-cut meander A meander in which deposition on the inner bank proceeds at the same rate as lateral erosion on the outer bank, thereby maintaining a channel of constant width (Melton, 1936, p. 596-597). Cf: *advance-cut meander.* Syn: *scroll meander; forced meander.*

forced fold A fold whose final overall shape and trend are dominated by the shape of some forcing member below (Stearns, 1978). See also: *forced folding; drape fold.*

forced folding Deformation of the sedimentary strata above the crystalline basement by dominantly vertical movement along faults, many of which are curved in cross section. This structural style is well developed in the Rocky Mountains foreland (Stearns, 1978). See also: *drape fold; forced fold.*

forced oscillation An oscillation that is imposed on a body, e.g. the Earth, by an external force. Cf: *free oscillation.*

forced wave A wave that is generated and maintained by a continuous force, e.g. wind. Cf: *free wave.*

force of crystallization *crystallizing force.*

forceps A C-shaped siliceous sponge spicule (sigma) having the form of tongs, with subparallel extremities.

forcherite An orange-yellow variety of opal colored with orpiment.

forcible intrusion Emplacement of magma that forcibly created the space into which it moved; also, the magma or rock body so emplaced. Cf: *permissive intrusion.* Syn: *aggressive intrusion.*

ford (a) A shallow and usually narrow part of a stream, estuary, or other body of water that may be crossed, as by wading or by a wheeled land vehicle. It usually has a firm, level, and relatively boulder-free bottom. Syn: *fording; wath.* (b) An archaic term for a stream or other body of water.

förde A Danish term for a long, narrow straight-sided inlet of the sea in a coastline consisting of till or surrounded by terminal moraines, and typically produced by drowning of a subglacial valley along a low-lying coast, like that of the SE Jutland Peninsula. Pl: *förden.* Cf: *fjord; bodden.* Syn: *föhrde.*

foredeep (a) An elongate depression bordering an island arc or other orogenic belt. Cf: *trench.* (b) A syn. of *exogeosyncline,* so named because of its relative position, near the craton.

foredune A coastal dune or *dune ridge* oriented parallel to the shoreline, occurring at the landward margin of the beach, along the shoreward face of a beach ridge, or at the landward limit of the highest tide, and more or less completely stabilized by vegetation.

foreign inclusion A fragment of country rock enclosed in an igneous rock. Syn: *xenolith.*

foreland A stable area marginal to an orogenic belt, toward which the rocks of the belt were thrust or overfolded. Generally the foreland is a continental part of the crust, and is the edge of the craton or platform area.

foreland [coast] (a) An extensive area of land, either high ground or low land, jutting out from the coast into a large body of water (usually the sea); a *headland;* a *promontory.* See also: *cuspate foreland.* (b) A prograded strip of low, flat land built by waves and currents at the base of a cliff; an initial stage in the development of a *strand plain.* (c) A part of the natural shore, located seaward of an embankment, that receives the shock of sea waves and deadens their force.

foreland [geog] The land lying in front of or adjoining other land and physiographically related to it.

foreland [glac geol] A lowland area, now or formerly covered by *piedmont glaciers.*

foreland facies *shelf facies.*

forelimb The steeper of the two limbs of an asymmetrical, anticlinal fold. Cf: *backlimb.*

forelimb thrust *front-limb thrust fault.*

forellenstein *troctolite.*

Forel scale A color scale of yellows, greens, and blues that is used against a white Secchi disk to measure the color of seawater.

forensic geology Application of the earth sciences to the law (Murray & Tedrow, 1975). Syn: *legal geology.*

forepoling A method of advancing an excavation in loose, caving, or watery ground, such as quicksand, by driving sharp-pointed poles, timbers, sections of steel, or slabs into the ground ahead of, or simultaneously with, the excavating; a method of supporting a very weak roof. It is useful in tunneling and in extracting coal from under shale or clay.

fore reef The seaward side of a reef; in places a steep slope covered with deposits of reef talus, elsewhere an organism-constructed vertical wall. Cf: *back reef; off-reef.* Also spelled: *forereef.*

forerunner A low, long-period ocean swell representing the gradual change of water level that commonly begins several hours before the arrival of the main swell from a distant storm, esp. a tropical hurricane.

foreset n. A *foreset bed*.—adj. Pertaining to or forming a steep and advancing frontal slope, or the sediments deposited on such a slope.

foreset bed One of the inclined, internal, and systematically arranged layers of a cross-bedded unit; specif. one of the gently inclined layers of sandy material deposited upon or along an advancing and relatively steep frontal slope, such as the outer margin of a delta or the lee side of a dune, and progressively covering the *bottomset bed* and in turn being covered or truncated by the *topset bed*. Foreset beds represent the greater part of the bulk of a delta. Also spelled: *fore-set bed*. Syn: *foreset*.

foreset bedding A syn. of *cross-bedding;* an internal bedding inclined to the principal surface of accumulation. See also: *compound foreset bedding*.

foreshock A small tremor that commonly precedes a larger earthquake or *main shock* by an interval ranging from seconds to weeks and that originates at or near the focus of the larger earthquake. Cf: *aftershock*.

foreshore (a) The lower or outer, gradually seaward-sloping, zone of the shore or beach, lying between the crest of the most seaward berm on the *backshore* (or the upper limit of wave wash at high tide) and the ordinary low-water mark; the zone regularly covered and uncovered by the rise and fall of the tide, or the zone lying between the ordinary tide levels. Sometimes referred to as the *shore*. Syn: *beach face*. (b) A term loosely applied to a strip of ground lying between a body of water, as a lake or stream, and land that is cultivated or otherwise occupied.

foreside A stretch of country fronting the sea, as Falmouth Foreside, Maine.

foresight (a) A sight on a new survey point, taken in a forward direction and made in order to determine its bearing and elevation. Also, a sight on a previously established survey point, taken to close a circuit. (b) A reading taken on a level rod to determine the elevation of the point on which the rod rests. Syn: *minus sight*.—Abbrev: FS. Ant: *backsight*.

foreslope The steeply sloping part of an *organic reef*, extending from its outer margin to an arbitrary depth of 10 fathoms.

forest bed An interglacial deposit consisting of soil and woody remains of trees and other vegetation. Syn: *black drift; chip yard; woodyard*.

forest floor The highly organic surface layer of a forest soil, including *litter, duff*, and *leaf mold*.

forest marble *landscape marble*.

forest-moss peat Peat formed in forested swamps.

forest peat A highmoor peat formed of the remains of trees.

fork (a) A place where two or more streams join to form a larger waterway; a confluence. (b) The smaller of two streams that unite at a fork; any uniting stream at a fork where the resulting stream is much larger; a branch. (c) The land bounded by, adjoining, or lying in the angle made by, a fork. The term is often used in the plural in place names.

form [geomorph] *landform*.

form [sed struc] Those aspects of a particle's shape that are not expressed by sphericity or roundness (Sneed and Folk, 1958, p. 123). It can be described by the use of ratios of the long, intermediate, and short axes, which can be combined into various "form indices", and by such terms as platy, bladed, elongate, and compact.

forma The smallest category used in ordinary taxonomic work. It is generally applied to trivial variations occurring among individuals of any population (Lawrence, 1951, p.56).

formal unit A stratigraphic unit that is "defined and named in accordance with the rules of an established or conventional system of classification and nomenclature" (ISST, 1961, p.18); e.g. a unit that is established in conformance with Article 13 of the Code of Stratigraphic Nomenclature prepared by the American Commission on Stratigraphic Nomenclature (1961). The initial letter of each word used in forming the name of a formal unit should be capitalized (although a name is not necessarily formal because it is capitalized). Cf: *informal unit*.

formanite A black mineral: $Y(Ta,Nb)O_4$. It is isomorphous with fergusonite, and may contain uranium, thorium, calcium, titanium, and zirconium.

format A term introduced by Forgotson (1957, p.2110) for an informal, laterally continuous lithostratigraphic unit that includes two or more lithologically dissimilar units but is suitable for regional mapping, "defined in general terms as a rock unit which is related at one point (or in one area) to a unit of formal stratigraphy, but which crosses facies boundaries and cutoffs to reach other areas where other formal units are employed". The term is applied to an operational unit representing strata sandwiched between observable *markers* that are believed to be isochronous surfaces. Formats are useful for correlation (particularly in the subsurface) between areas where the stratigraphic section is divided into different formations that do not correspond in time value. See also: *assise*.

formation [cart] A persistent body of igneous, sedimentary, or metamorphic rock, having easily recognizable boundaries that can be traced in the field without recourse to detailed paleontologic or petrologic analysis, and large enough to be represented on a geologic map as a practical or convenient unit for mapping and description; the basic *cartographic unit* in geologic mapping.

formation [drill] A general term applied by drillers without stratigraphic connotation to a sedimentary rock that can be described by certain drilling or reservoir characteristics; e.g. "hard formation", "cherty formation", "porous formation".

formation [ecol] A group of *associations* that exist together as a result of their closely similar life pattern, habits, and climatic requirements.

formation [geomorph] A naturally formed topographic feature, commonly differing conspicuously from adjacent objects or material, or being noteworthy for some other rason; esp. a striking erosional form on the land surface.

formation [speleo] *speleothem*.

formation [stratig] (a) A body of rock strata, of intermediate rank in the hierarchy of *lithostratigraphic units*, which is unified with respect to adjacent strata by consisting dominantly of a certain lithologic type or combination of types or by possessing other unifying lithologic features. Thickness may range from less than a meter to several thousand meters, depending on the size of units locally required to best express the lithologic development of a region. The formation is the fundamental unit of *lithostratigraphic classification;* it is the only *formal unit* for subdividing the whole stratigraphic column all over the world into named units on the basis of lithology (ISSC Rept. 3, 1970, p. 9). Most formations have a prevailingly tabular shape, and are mappable at the Earth's surface at scales on the order of 1:25,000 or are traceable in the subsurface. A formation is a genetic unit, or a product of uniform or uniformly alternating conditions, and may contain rock of one lithologic type, repetitions of two or more types, or extreme heterogeneity that in itself may constitute a form of unity compared to the adjacent strata. Also, it may represent a long or short time interval, be composed of materials from one or several sources, and include breaks in the chronostratigraphic sequence; its age or time value may not necessarily be the same wherever it is recognized. Formations may be combined into *groups* or subdivided into *members*. A formation name should preferably consist of a geographic name followed by a descriptive lithologic term (usually the dominant rock type) or by the word "formation" if the lithology is so variable that no single lithologic distinction is appropriate. Abbrev: fm. (b) A lithologically distinctive, mappable body of igneous or metamorphic rock. (c) A term that has been used for a chronostratigraphic unit representing a rock of constant time span; specif. a stage, or the rocks corresponding to the time interval of an age. This usage is not recommended. In Germany, the term "Formation" is equivalent to the time-stratigraphic term "system".—Syn: *geologic formation*.

formation evaluation The process of evaluating gas- or oil-bearing formations penetrated by a well or wells, and of appraising their commercial significance.

formation factor The ratio of the conductivity of an electrolyte to the conductivity of a rock saturated with that electrolyte. Symbol: F. Syn: *resistivity factor; formation resistivity factor*.

formation pressure *bottom-hole pressure*.

formation resistivity factor *formation factor*.

formation-volume factor The factor applied to convert a barrel of gas-free oil in a stock tank at the surface into an equivalent amount of oil in the reservoir. It generally ranges between 1.14 and 1.60. Cf: *shrinkage factor*.

formation water Water present in a water-bearing formation under natural conditions, as opposed to introduced fluids, such as *drilling mud*. Syn: *native water*. Cf: *connate water*.

Formative n. In New World archaeology, the cultural stage that follows the Archaic and is characterized by the development of agriculture and a settled population. It is followed by the Classic. Correlation of relative cultural levels with actual age (and, therefore, with the time-stratigraphic units of geology) varies from region to region. Syn: *Pre-Classic.*—adj. Pertaining to the Formative.

form contour A topographic contour determined by stereoscopic examination of aerial photographs without ground control or by some other means not involving conventional surveying.

form energy The potentiality of a mineral to develop its own crystal form against the resistance of the surrounding solid medium (Eskola, 1939). Syn: *power of crystallization; crystalloblastic strength.*

form genus (a) A *taxon* primarily for convenience in classifying fossils of problematic relationship that show similarity in morphology. Among plant fossils a form genus is one that is unassignable to a family but may be referable to a taxon of higher rank (ICBN, 1972, p. 17). Cf: *parataxon.* (b) Less commonly applied informally to a genus in a series of related genera which have resulted from the splitting up of an old familiar genus. (c) Also, applied informally to a genus containing several species with the same general morphology but suspected of having unrelated ancestors.—Also spelled: *form-genus.*

formkohle *crumble coal.*

formkreis (a) One of a series of related landforms that owe their existence to the same natural agent (such as running water or moving ice). (b) *morphogenetic region.*—Etymol: German *Formkreis,* "form cycle". Pl: *formkreise.*

form line A line (usually broken) on a map, sketched or drawn by visual observation, depicting the general surface configuration or shape of the terrain without regard to a true vertical datum and regular spacing and usually without indicating elevations; an uncontrolled or interpolated *contour line,* or one that is not instrumentally or accurately surveyed. Also spelled: *formline.*

form ratio The ratio of mean stream depth to width of stream measured from bank to bank (Gilbert, 1914, p. 35-36); it expresses the deepness or shallowness of a stream channel. The *hydraulic radius* is sometimes substituted for the mean depth if the stream is broad and shallow. Symbol: R.

form set A *set* of cross-stratification with an upper surface that preserves the shape of the original bed form (McKee and Weir, 1953). See also: *fossil ripple.*

fornacite An olive-green mineral: $Pb_2Cu(CrO_4)[(As,P)O_4](OH)$. It is isomorphous with vauquelinite.

fors A Swedish term for a rapids or cataract, or a waterfall of low inclination. Cf: *force.*

forsterite A whitish or yellowish mineral of the olivine group: Mg_2SiO_4. It is isomorphous with fayalite, and occurs chiefly in metamorphosed dolomites and crystalline limestones. Symbol: Fo. Syn: *white olivine.*

fortification agate An agate having angular markings or parallel zigzag lines resembling the plan of a fortification.

fortunite A *trachyte* that contains phenocrysts of olivine and phlogopite in a very fine-grained groundmass that is megascopically undeterminable but under the microscope can be identified as a combination of ortho- and clinopyroxene, mica, feldspar, and some glass. Its name, given by Osann in 1889, is derived from Fortuna, Spain. Cf: *verite.* Not recommended usage.

forward lap *overlap [photo].*

forward scatter The scattering of radiant energy into the hemisphere of space bounded by a plane normal to the direction of the incident radiation and lying on the side toward which the incident radiation was advancing; the opposite of *back scatter.*

foshagite A white mineral: $Ca_4Si_3O_9(OH)_2$.

foshallassite A snow-white mineral: $Ca_3Si_2O_7 \cdot 3H_2O$ (?). Also spelled: *foschallasite.*

foso A term used in SW U.S. for a stream channel without conspicuous banks or bluffs. Etymol: Spanish, "moat, ditch".

foss (a) *fosse [streams].* (b) *force [streams].*

fossa [astrogeol] A term established by the International Astronomical Union for a linear topographic depression on Mars, similar to a terrestrial graben. Generally used as part of a formal name for a Martian landform, such as Claritas Fossae (Mutch et al., 1976, p. 57). Etymol: Latin *fossa,* ditch.

fossa [paleont] (a) A depression on an articular face of a crinoid ossicle for the attachment of muscles or ligaments. (b) A depression in the surface of a bone that serves for the attachment of a muscle or houses a gland, cartilage, or other soft part.—Pl: *fossae.*

fosse [glac geol] A long, narrow depression or troughlike hollow between the edge of a retreating glacier and the wall of its valley, or between the front of a moraine and its outwash plain. It may result from local acceleration of melting due to absorbed or reflected heat from the valley sides.

fosse [streams] (a) A long, narrow waterway; a canal, ditch, or trench. Etymol: Latin *fossa,* "ditch". Syn: *foss.* (b) *force [streams].*

fosse lake A glacial lake occupying a fosse.

fossette (a) One of the slitlike pits, depressions, or grooves paralleling the periphery on the surface of the tests of some foraminifers (such as *Elphidium*). (b) A depression for the resilium in the shell of a bivalve mollusk.

fossil n. Any remains, trace, or imprint of a plant or animal that has been preserved in the Earth's crust since some past geologic or prehistoric time; loosely, any evidence of past life.—adj. Said of any object that existed in the geologic past and of which there is still evidence. Cf: *subfossil.*

fossil assemblage *assemblage [paleoecol].*

fossil association *association [ecol].*

fossil community An *assemblage [paleoecol]* in which the individuals lived in the same place where their fossils are found, are present in approximately the same numbers and sizes as when alive, and thus have experienced no post-mortem transport. Cf: *winnowed community; disturbed-neighborhood assemblage; transported assemblage; mixed assemblage; paleobiocoenosis.* Syn: *in-place assemblage.*

fossil copal *copalite.*

Fossildiagenese A German term applied to the branch of paleoecology concerned with the history of organic remains after burial. Cf: *biostratonomy; taphonomy.*

fossil erosion surface An *erosion surface* that was buried by younger sediments and was later exposed by their removal. Sometimes used as a syn. of *buried erosion surface.*

fossil flood plain A flood plain that is beyond the reach of the highest flood (Bryan, 1923a, p. 88). Cf: *living flood plain.*

fossil flour *diatomaceous earth.*

fossil fuel A general term for any hydrocarbon that may be used for fuel: chiefly petroleum, natural gas, and coal.

fossil geochronometry Measurement of *growth lines* on fossil shells as a means of estimating the length of days and lunar months in *geologic time.* Cf: *lichenometry.* See also: *geochronometry.*

fossil ice (a) Ice formed in, and remaining from, the geologically recent past. It is preserved in cold regions, such as the coastal plains of northern Siberia, where remains of Pleistocene ice have been found. See also: *dead ice.* (b) Relatively old *ground ice [permafrost]* in a permafrost region. Also, underground ice in a region where present-day temperatures are not low enough to create it (Huschke, 1959, p. 230).

fossil ice wedge A sedimentary structure formed by the filling of the space formerly occupied by an *ice wedge* that had melted; the sediment fill may be wedge-shaped or very irregular. Syn: *ice-wedge pseudomorph; ice-wedge fill; ice-wedge cast.*

fossiliferous Containing fossils.

fossilization All processes involving the burial of a plant or animal in sediment and the eventual preservation of all, part, or a trace of it.

fossilized brine *connate water.*

fossil karst (a) *relict karst.* (b) *paleokarst.*

fossil meteorite crater *astrobleme.*

fossil ore An iron-bearing sedimentary deposit, e.g. *Clinton ore,* in which shell fragments have been replaced and cemented together by hematite and carbonate. Cf: *flaxseed ore.*

fossil patterned ground Patterned ground that is inactive or no longer developing; e.g. ice-wedge casts and involution structures in a region of mild climate that formed during colder periods of the Pleistocene epoch when the region was under periglacial conditions. Ant: *active patterned ground.*

fossil peneplain A peneplain that was buried by younger sediments and was later exposed by their removal. Syn: *stripped peneplain.*

fossil permafrost (a) Permafrost left from colder times and commonly below the average depth reached by seasonal frost. (b) Permafrost that, having formed during an earlier colder period, will not, under present climatic conditions, reform after it is once disturbed or destroyed.—Ant: *active permafrost.* Syn: *passive permafrost.*

fossil pingo The remains of a pingo. See also: *pingo-remnant.*

fossil plain A plain that was buried by younger sediments and was later exposed by their removal.
fossil record The record of life in the geologic past as indicated by fossils. Cf: *stratigraphic record.* Syn: *paleontologic record.*
fossil resin Any of various natural *resins* found in geologic deposits as exudates of long-buried plant life; e.g. amber, retinite, and copal.
fossil ripple A *ripple mark* preserved on a sedimentary rock surface. See also: *form set.*
fossil soil *paleosol.*
fossil time *Geologic time* estimated on the basis of organic evolution (Kobayashi, 1944a, p. 476). See also: *marine time; continental time.* Cf: *relative time.*
fossil turquoise *odontolite.*
fossil water *connate water.*
fossil wax *ozocerite.*
fossula (a) An unusually wide or relatively prominent space between septa of a rugose coral, distinguished by its shape and size, and caused by failure of one or more septa to develop as rapidly as others. See also: *cardinal fossula; counter fossula; alar fossula.* (b) A small pitlike depression that may occur in the cephalic axial or posterior border furrow in trilobites.
fossulate Said of sculpture of pollen and spores consisting of grooves that anastomose.
fouling The attachment and growth of aquatic plants and/or animals on submerged surfaces.
foundation (a) The lower, supporting part of an engineering structure, in contact with the underlying soil or rock and transmitting the weight of the structure and its included loads to the underlying earth material (Ireland, 1969). It is usually below ground level. (b) A term that is sometimes applied to the upper part of the soil or rock mass in contact with, and supporting the loads of, an engineering structure; the subsoil.
foundation bed The rock or soil layer immediately beneath the foundation, which receives the load of an engineering structure.
foundation coefficient A coefficient expressing how many times stronger is the effect of an earthquake in a given rock than would have been the case in an undisturbed crystalline rock under the same conditions (Schieferdecker, 1957, p.197).
founder breccia *collapse breccia.*
fount A *fountain* or spring of water.
fountain (a) A spring of water issuing from the Earth. Syn: *fount.* (b) The *source* or head of a stream.
fountain head According to Tolman (1937, p. 559), "The elevation of water surface in a conduit if the overlying confining stratum extends above the water table, or elevation of water table above the upper termination of the confining stratum where the latter is below the water table".
fountainhead The fountain or spring that is the source of a stream. Syn: *springhead; wellspring.*
fourchite An olivine-free *monchiquite.* Its name, given by Williams in 1890, is derived from the Fourche Mountains, Arkansas.
Fourier analysis A method for representing a periodic mathematical function as an infinite series of summed sine and cosine terms. It involves comparison of observed periodic data with this theoretical form, and also all harmonic (period, amplitude, phase) relationships of the series. Named after Jean Baptiste Joseph Fourier (1768-1830), French physicist. Syn: *harmonic analysis.*
fourling A crystal twin consisting of four individuals. Cf: *twoling; trilling; fiveling; eightling.*
fourmarierite An orange-red to brown secondary mineral: $PbU_4O_{13}\cdot 4H_2O$.
fourth-order pinacoid In a triclinic crystal, the $\{\bar{h}kl\}$, $\{h\bar{k}l\}$, or $\{hk\bar{l}\}$ pinacoid. Cf: *first-order pinacoid; second-order pinacoid; third-order pinacoid.*
fourth-order prism A crystal form in monoclinic crystals with two pairs of parallel faces, rhombic in cross section and parallel to an inclined direction. Its indices are $\{hkl\}$ or $\{\bar{h}kl\}$. Cf: *first-order prism; second-order prism; fourth-order prism.* Obsolescent syn: *hemipyramid.*
four-way dip In seismic prospecting, dip determined by in-line and cross spreads placed at approximately right angles to each other.
foveolate Pitted; e.g. said of sculpture of pollen and spores consisting of pits in the ektexine.
fowlerite A variety of rhodonite containing zinc.
foyaite A *nepheline syenite* containing a predominance of potassium feldspar. Originally described by Blum as synonymous with nepheline syenite, and later applied by Brögger to nepheline syenite with trachytic texture (Johannsen, 1939, p. 252). Its name is derived from Foya, Portugal. Cf: *ditroite.* Not recommended usage.
fractional crystallization [petrology] Crystallization in which the early-formed crystals are prevented from equilibrating with the liquid from which they grew, resulting in a series of residual liquids of more extreme compositions than would have resulted from *equilibrium crystallization.* Cf: *crystallization differentiation.* Syn: *fractionation.*
fractional crystallization [salt] Controlled precipitation from a saline solution of salts of different solubilities, as affected by varying temperatures or by the presence of other salts in solution (Bateman, 1950, p. 183).
fractional fusion Fusion in which the liquid produced on heating is isolated from the system as soon as it is formed and is thereby prevented from further reaction with the crystalline residue (Presnall, 1969). Cf: *equilibrium fusion.*
fractional section A *section* containing appreciably less than 640 acres, usually due to invasion by a body of water or by land which cannot properly be surveyed as part of that section, or due to closing of the public-land survey on the north and west boundaries of the township.
fractional township A *township* containing appreciably less than 36 normal sections, usually due to invasion by a body of water or by land which cannot properly be surveyed as part of that township, or due to closing of the public-land survey on a State boundary or other limiting line.
fractionation [geochem] Separation of chemical elements in nature, by processes such as preferential concentration of an element in a mineral during magmatic crystallization, or differential solubility during rock weathering.
fractionation [petrology] *fractional crystallization.*
fractoconformity The relation between conformable strata where faulting of the older beds proceeds contemporaneously with deposition of the newer.
fractography The study of the surfaces of fractures, esp. microscopic study.
fracture [exp struc geol] Deformation due to a momentary loss of cohesion or loss of resistance to differential stress and a release of stored elastic energy. Cf: *flow [exp struc geol].* Syn: *rupture.*
fracture [ice] Any break or rupture through fast ice, a single floe, or highly concentrated pack ice, resulting from deformation caused by tides, temperature changes, currents, or wind. Its length may vary from a few meters to many kilometers, and its width from zero to more than 500 m. Includes: *crack [ice]; flaw [ice]; lead [ice].*
fracture [mineral] The breaking of a mineral other than along planes of *cleavage.* A mineral can be described in part by its characteristic fracture, e.g. uneven, fibrous, conchoidal, or hackly.
fracture [struc geol] A general term for any break in a rock, whether or not it causes displacement, due to mechanical failure by stress. Fracture includes cracks, joints, and faults.
fracture cleavage A type of cleavage that occurs in deformed but only slightly metamorphosed rocks and that is based on closely spaced, parallel joints and fractures. Syn: *close-joints cleavage.*
fractured deflection A marked change in the trend of a mountain range, where arcs meet at large obtuse angles, and from the junction two lineaments appear to cross the ranges and mark major structural changes (Wilson, 1950, p. 151).
fracture porosity Porosity resulting from the presence of openings produced by the breaking or shattering of an otherwise less pervious rock.
fracture spring A spring whose water flows from joints or other fractures, in contrast to the numerous small openings from which a *filtration spring* flows (Meinzer, 1923, p. 50). Cf: *fissure spring; fault spring.*
fracture strength In experimental structural geology, the differential stress at the moment of fracture. Syn: *fracture stress; breaking strength.*
fracture stress *fracture strength.*
fracture system A set or group of contemporaneous fractures related by stress.
fracture zone On the deep-sea floor, an elongate zone of unusually irregular topography that often separates regions of different depths. Such zones commonly cross and apparently displace the mid-oceanic ridge by faulting.
fracturing *hydraulic fracturing.*

fragipan A dense subsurface layer of soil whose hardness and relatively slow permeability to water are chiefly due to extreme compactness rather than to high clay content (as in *claypan*) or cementation (as in *hardpan*). It appears indurated when dry, but shows a moderate to weak brittleness when moist; it contains much silt and sand, but little clay and organic matter.

fragment (a) A rock or mineral *particle* larger than a *grain.* (b) A piece of rock that has been detached or broken from a pre-existing mass; e.g. a clast produced by volcanic, dynamic, or weathering processes.

fragmental rock (a) *clastic rock.* (b) *pyroclastic rock.* (c) *bioclastic rock.*

fragmental texture (a) A texture of sedimentary rocks, characterized by broken, abraded, or irregular particles in surface contact, and resulting from the physical transport and deposition of such particles; the texture of a clastic rock. The term is used in distinction to a "crystalline" texture. (b) The texture of a pyroclastic rock, such as that of a tuff or a volcanic breccia.

fragmentary Consisting of or characterized by clastic or detrital material; fragmental. The term was formerly applied to rocks composed of fragments of older rocks, or to rocks having an inhomogeneous texture; as used in geology, the term is obsolete.

fragmentation The act or process of breaking into pieces or fractionating, or the state of being fragmentated or fractionated; esp. the breaking-up of a sponge into several others without concomitant formation of specialized reproductive bodies.

fraipontite A mineral of the kaolinite-serpentine group: $(Zn,Al)_6(Si,Al)_4O_{10}(OH)_8$.

Fra Mauro basalt A class of basaltic rocks found in the lunar highlands, differing from mare basalts primarily by their higher plagioclase content. A number of varieties, e.g. *KREEP,* are distinguished on the basis of chemical composition (Taylor, 1975, p. 234). Named from the Apollo 14 landing site near the crater Fra Mauro, where this basalt was first collected.

framboid A microscopic aggregate of pyrite grains in shale, often in spheroidal clusters resembling raspberry seeds. It was considered to be the result of colloidal processes but is now linked with the presence of organic materials; sulfide crystals fill chambers or cells in bacteria (Park & MacDiarmid, 1970, p.133). Adj: framboidal. Etymol: French *framboise,* "raspberry".

frame-builders Organisms, generally calcareous and sessile-benthonic, that collectively construct a reef frame or *growth lattice.*

framesite A South African variety of bort showing minute brilliant points.

framework [paleont] The coarsely porous underlying wall of a double wall in archaeocyathids (TIP, 1972, pt. E, p. 40).

framework [sed] (a) The rigid arrangement created in a sediment or sedimentary rock by particles that support one another at their points of contact; e.g. the clasts of a fragmental rock (esp. a sandstone), constituting a mechanically firm structure capable of supporting open pore spaces, although interstices may be occupied by cement or matrix. (b) The rigid, wave-resistant, calcareous structure built by sedentary organisms (such as sponges, corals, and bryozoans) in a high-energy environment.

framework [tect] *tectonic framework.*

framework silicate *tectosilicate.*

francevillite A yellow secondary mineral: $(Ba,Pb)(UO_2)_2(VO_4)_2 \cdot 5H_2O$.

franckeite A dark-gray or black mineral: $Pb_5Sn_3Sb_2S_{14}$.

francoanellite A hexagonal mineral: $H_6K_3Al_5(PO_4)_{18} \cdot 13H_2O$.

francolite A colorless mineral of the apatite group: $Ca_5(PO_4,CO_3)_3(F,OH)$. Cf: *carbonate-apatite.* Syn: *carbonate-fluorapatite; staffelite; kurskite.*

Franconian North American stage: Upper Cambrian (above Dresbachian, below Trempealeauan).

frangite A comprehensive term proposed by Bastin (1909, p. 450) for all sedimentary rocks (unconsolidated or cemented), and their dynamically metamorphosed representatives, formed from the disintegration of igneous rocks without extensive decomposition or mechanical sorting; e.g. arkose, graywacke, gneiss. Etymol: Latin *frangere,* "to break". Adj: *frangitic.*

frankdicksonite A mineral: BaF_2. It is found in Nevada.

franklinite An iron-black mineral of the magnetite series in the spinel group: $(Zn,Mn^{+2},Fe^{+2})(Fe^{+3},Mn^{+3})_2O_4$. It resembles magnetite but is less strongly magnetic. It is an ore of zinc.

Frasch process A process for mining native sulfur, in which superheated water is forced into the deposits for the purpose of melting the sulfur. The molten sulfur is then pumped to the surface.

Frasnian European stage: Upper Devonian (above Givetian, below Famennian).

Fraunhofer line Any of the *absorption lines* in the spectrum of the sun corresponding to the *absorption spectra* of the gases around it.

frazil (a) A group of individual ice crystals, having the form of small discoids or spicules, which are formed in supercooled turbulent water. Syn: *frazil ice.* (b) *frazil crystal.*—Etymol: Canadian French *frasil,* from French *fraisil,* "forge cinders".

frazil crystal A small discoid or needlelike spicule of ice formed by freezing of supercooled turbulent water. Syn: *frazil; ice crystal.*

frazil ice A spongy, slushy, cinderlike mass or aggregate of frazil crystals collected by adhesion or regelation and suspended in supercooled turbulent water, esp. common in a rapidly flowing stream, but also found in turbulent seawater, where it is called *lolly ice.* Syn: *frazil; needle ice.*

frazil slush An agglomerate of loosely packed frazil floating on the water surface which can accumulate under the ice cover. Cf: *slush [snow].*

freboldite A hexagonal mineral: CoSe.

Fredericksburgian North American (Gulf Coast) stage: Lower Cretaceous (above Trinitian, below Washitan).

free Said of a *native element,* e.g. free gold.

free-air anomaly A gravity anomaly calculated from a theoretical model and elevation above sea level, but without allowance for the attractive effect of topography and isostatic compensation.

free-air correction A correction for the elevation of a gravity measurement, required because the measurement was made at a different distance from the center of the Earth than the datum. The first term of the free-air correction if 0.09406 mgal/ft or 0.3086 mgal/m.

free arm The freely mobile part of a crinoid ray not incorporated in the theca.

free blade The portion of a blade of a platelike conodont element not flanked by platforms.

freeboard (a) The additional height above the recorded or design high-water mark of an engineering structure, such as a dam, seawall, flume, or culvert, that represents an allowance against overtopping by transient disturbances, including waves induced by winds or landslides. (b) The vertical distance between the water level at a given time and the top of an engineering structure, such as the vertical distance between the normal operating level of a reservoir and the crest of the associated dam.

free-burning coal *noncaking coal.*

free cheek A lateral part of the cephalon of a trilobite outside the facial suture, separated from the cranidium at the time of molting and including the visual surface of the eye. See also: *cheek [paleont].* Cf: *fixed cheek.* Syn: *librigena.*

free corer A type of *gravity corer* that operates from a float rather than from a ship.

free degradation Degradation of a slope from the foot of which no debris is removed, e.g. an abandoned cliff (Hutchinson, 1967).

free energy A thermodynamic function of the state of a system, providing a measure of the maximum work obtainable from the system under specified conditions. The functions most commonly used are the *Helmholtz free energy* and the *Gibbs free energy.* Because these functions (among others) also measure the driving forces for processes occurring under specified conditions, they are often referred to as *thermodynamic potentials.*

free enthalpy *Gibbs free energy.*

free face The part of a hillside surface consisting of an outcrop of bare rock (such as a scarp or cliff) that stands more steeply than the angle of repose of the *constant slope* immediately below (Wood, 1942); a rock wall from which weathered material falls to the slope below.

free flow In hydraulics, flow that is not disturbed by submergence or backwater (Brown & Runner, 1939, p. 155).

free-fluid index log A syn. of *nuclear-magnetism log.* Abbrev: FFI log.

free ground water *unconfined ground water.*

free margin The peripheral distal border of an ostracode valve, exclusive of the hinge.

free meander A stream meander that displaces itself very easily by lateral corrasion, esp. where vertical corrasion is of no importance. Syn: *free-swinging meander.*

free moisture *free water.*

free operculum The part of a dinoflagellate cyst that is completely surrounded by archeopyle sutures, with no unsutured connection to the rest of the cyst. Cf: *attached operculum.* Syn: *free opercular piece.*

free oscillation An oscillation of a body, e.g. the Earth, that occurs without external influence other than the initiating force, and that has its own natural frequency. Such oscillations follow major earthquakes. Cf: *forced oscillation.*

free period The time for one complete swing of the seismograph mass when all damping is removed and no driving force is present.

freestone [rock] Any stone (esp. a thick-bedded, even-textured, fine-grained sandstone) that breaks freely and can be cut and dressed with equal ease in any direction without splitting or tending to split. The ease with which it can be shaped into blocks makes it a good building stone. The term was originally applied to limestone, and is still used for such rock. See also: *konite.* Cf: *flagstone.*

freestone [water] Water containing little or no dissolved material. Syn: *freestone water.*

free surface The upper surface of a layer of fluid where the pressure on it is equal to the external atmospheric pressure.

free water (a) Water in the soil in excess of field capacity that is free to move in response to the pull of gravity. Syn: *gravity water; gravitational water; infiltration water.* Cf: *gravity ground water.* (b) Water that can be removed from another substance, as in ore analysis, without changing the structure or composition of the substance. Ant: *bound water.* Syn: *free moisture.*

free-water content The fraction of the total mass (or volume) of wet snow that is liquid. Not to be confused with *water equivalent.* Syn: *snow moisture; liquid water content; water saturation.* See also: *water content [snow].*

free-water elevation *water table.*

free-water level *free-water surface.*

free-water surface The surface of a body of water at which the pressure is atmospheric and below which the pressure is greater than atmospheric; the surface of any pond, reservoir, etc., that is open to the atmosphere, or a *water table.* Syn: *free-water level.*

free wave A wave that is created by a sudden rather than a continuous impulse and that continues to exist after the generating force is gone, influenced by friction and basin characteristics. Cf: *forced wave.*

freeze-out lake A shallow lake subject to being frozen for long periods.

freeze-thaw action *frost action.*

freezeup (a) The formation of a continuous ice cover, generally restricted to the hardening of locally formed young ice, but sometimes including the freezing together of pieces of drift ice. (b) The period during which a body of water in an area is frozen over, esp. when marking the beginning of winter.

freezing The process by which a liquid becomes a solid, involving the removal of heat. Commonly applied to water but also used for solidification on cooling of molten metals and magma.

freezing interval *crystallization interval.*

freibergite A mineral: $(Ag,Cu)_{12}(Sb,As)_4S_{13}$. It is isomorphous with tetrahedrite.

freieslebenite A steel-gray to dark-gray mineral: $PbAgSb_3$.

freirinite *lavendulan.*

fremontite *natromontebrasite.*

French chalk A soft, white variety of talc, steatite, or soapstone, finely ground into powder and used for crayons, as a grease remover in dry cleaning, or for other special purposes.

Frenkel defect In a crystal lattice, the displacement of an atom from its original position to an interstice; it is a type of *point defect.* Cf: *interstitial defect; Schottky defect.*

frenuliniform Said of the loop, or of the growth stage in the development of the loop, of a dallinid brachiopod (as in the subfamily Frenulininae), marked by lateral resorption gaps (lacunae) occurring in the hood but before resorption of the shell occurs posterior to the gaps (TIP, 1965, pt.H, p.145). It is subsequent to the *campagiform* stage.

frenulum A small cylinder connected to the internal part of the nasal tube of a style near the base of a galea in a phaeodarian radiolarian. Pl: *frenula.*

frequency curve A curve that graphically represents a *frequency distribution;* e.g. a smooth line drawn on a histogram if the class interval is made smaller and the steps between several bars grow smaller.

frequency distribution A systematic arrangement of statistical data (such as a graphic or tabular display of the number of observations on a variable) that exhibits the division of the values of the variable into mutually exclusive (but closely related), usually ranked, and exhaustive discrete categories or classes, and that indicates the frequencies or relative frequencies that correspond to each of the categories or classes. It is generally selected on the basis of some progressively variable physical character, such as the diameter of sedimentary particles. Syn: *distribution [stat].*

frequency domain (a) Measurements as a function of frequency, or operations in which frequency is the variable, in contrast to the *time domain.* (b) Transmission of a continuous wave (usually sinusoidal) and simultaneous reception of electromagnetic energy as a function of frequency. It is used with induced electrical polarization and electromagnetic methods.

frequency response Attenuation as a function of frequency produced by passage of information through an element such as a geophone or filter.

fresh [water] adj. Said of water that does not contain or is not composed of salt water.—n. (a) An increased, sudden flow or rush of water; a *freshet* or a *flush.* (b) A stream, spring, or pool of fresh water. (c) A freshwater stream flowing into the sea, or the part of a stream or its shores above the flow of tidal seawater. (d) The mingling of fresh water and salt water.

fresh [weath] Said of a rock or rock surface that has not been subjected to or altered by surface weathering, such as a rock newly exposed by fracturing. Syn: *unweathered.*

freshening Making water less salty; separating water from saline constituents.

freshet (a) A great rise in, or a sudden overflowing of, a small stream, usually caused by heavy rains or rapidly melting snow in the highlands at the head of the stream; a rapidly rising flood, usually of minor severity and short duration. See also: *flash flood.* Syn: *fresh; spate; high water.* (b) A small clear freshwater stream or current flowing swiftly into the sea; an area of comparatively fresh water at or near the mouth of a stream flowing into the sea. (c) A small stream flowing swiftly into a lake (as in the spring) and often carrying a heavy silt load during its peak flow.

fresh ice (a) *young ice.* (b) *freshwater ice.* (c) Ice that was formed on salt water but is now salt-free.

fresh water Water containing only small quantities of dissolved minerals, such as the water of streams and lakes unaffected by salt water or salt-bearing rocks; water that lacks a saline or mineral taste, though it may contain suspended sediment, pathogenic organisms, and/or small quantities of dissolved constituents not detectable by taste but nevertheless toxic, and may have harmless but objectionable taste, odor, or color. Cf: *potable water.* Syn: *sweet water.* Also spelled: *freshwater; fresh-water.*

freshwater estuary (a) An estuary into which river water pours with sufficient volume to exclude salt water. See also: *positive estuary.* (b) In the Great Lakes and other large lakes, the lower *reach* of a tributary to the lake that has a *drowned river mouth,* shows a zone of transition from stream water to lake water, and is influenced by changes in lake level as a result of *seiches* or *wind tides.*

freshwater ice Ice formed by the freezing of fresh water in lakes or streams, or in the ground. Syn: *fresh ice.*

freshwater limestone A limestone formed by accumulation or precipitation in a freshwater lake, a stream, or a cave. It is often algal and sometimes nodular. See also: *underclay limestone.*

freshwater sediment A sediment that accumulates, or has accumulated, in a freshwater environment; e.g. a sediment resulting from lacustrine, fluvial, or glaciofluvial activity.

freshwater swamp A swamp that depends on nontidal fresh water rather than a saltwater source.

Fresnian North American stage: Upper Eocene (above Narizian, below Refugian).

fresnoite A mineral: $Ba_2TiSi_2O_8$.

fret A spot worn or eroded by *fretting,* as on a limestone surface.

fretted upland A preglacial upland surface completely consumed by the intersection of cirques from opposite sides; the "ultimate product of cirque sculpture by glaciers" (Hobbs, 1912, p. 373). Cf: *grooved upland; scalloped upland.*

fretting (a) *honeycomb weathering.* (b) The wearing-away of a rock surface, as by a stream cutting its channel. (c) Agitation or disturbance of running water, such as the rippling of a brook over rocks.

fretum An arm of the sea; a strait. Pl: *freta.*

fretwork weathering *honeycomb weathering.*

freudenbergite A black hexagonal mineral: $Na_2(Ti,Fe)_8O_{16}$.

freyalite A variety of thorite high in rare earths (esp. cerium).

friability The condition of being friable.

friable (a) Said of a rock or mineral that crumbles naturally or is easily broken, pulverized, or reduced to powder, such as a soft or poorly cemented sandstone. (b) Said of a soil consistency in which moist soil material crushes easily under gentle to moderate pressure (between thumb and forefinger) and coheres when pressed together.

friction Mechanical resistance to the relative motion of contiguous bodies or of a body and a medium. Cf: *internal friction.*

frictional As applied to a soil, a syn. of *cohesionless.*

frictional layer The layer of the ocean that is affected by the action of the wind on the water surface; also, the zone of friction between bottom water and submerged rock or sediment.

friction breccia A breccia composed of broken or crushed rock fragments resulting from friction, such as a "volcanic friction breccia" formed where a rising column of nearly congealed lava was shattered against the walls of the volcanic vent and later cemented by newly rising magma; specif. a *fault breccia* produced by friction of the two walls of the fault rubbing against each other.

friction crack A short *crescentic mark* that is transverse to the direction of ice movement and that includes a distinct fracture that dips forward into the bedrock, indicating the direction of ice movement (Harris, 1943). It presumably results from local increase in frictional pressure between ice and bedrock.

friction depth The depth in the ocean where the velocity vector in an *Ekman spiral* is exactly opposite to the wind direction; commonly about 100 m, rarely deeper than 200 m. Syn: *depth of frictional influence; depth of frictional resistance.*

friction head That head of fluid which is lost because of friction. See also: *friction slope.* Syn: *head loss; friction loss.*

friction loss *friction head.*

friction slope The *friction head* or loss per unit length of conduit. For most conditions of flow the friction slope coincides with the energy gradient, but where a distinction is made between energy losses due to bends, expansions, impacts, etc., a distinction must also be made between the friction slope and the energy gradient. In uniform channels, the friction slope is equal to the bed or surface slope only for uniform flow (ASCE, 1962).

friedelite A rose-red mineral: $Mn_8Si_6O_{18}(OH,Cl)_4 \cdot 3H_2O$.

friendly ice A submariner's term for an ice canopy containing more than 10 large skylights (or other features that permit a submarine to surface) per 30 nautical miles (56 km) along the submarine's track (U.S. Naval Oceanographic Office, 1968, p. B33). Ant: *hostile ice.*

frigid climate A type of climate characteristic of a region in which there is a more or less permanent cover of snow and ice over the permanently frozen surface. Cf: *polar climate.*

frigid temperature regime A soil temperature regime in which the mean annual temperature (measured at 50cm depth) is more than 0°C but less than 8°C, with a summer-winter variation of more than 5°C and with warm summer temperatures (USDA, 1975). Cf: *isofrigid temperature regime; cryic temperature regime.*

frill (a) A relatively large lamella projecting well beyond the general contour of a brachiopod valve, deposited by the margin of highly retractile mantle (TIP, 1965, pt.H, p.145). Cf: *growth lamella.* (b) The velum in an ostracode.

fringe [coast] The line beyond which detritus from the delta-forming river is no longer the major fraction in the sea-floor sediment (Bayly, 1968, p. 151-152).

fringe [glac geol] A thin sprinkling of erratics in front of the end moraine of a glacier.

fringe [paleont] The peripheral pitted bilaminar portion of the cephalon in trinucleid and harpetid trilobites.

fringe joint A small-scale joint peripheral to a master joint, usually at a 5-25° angle from the face of the main joint. It is formed by tension or shear. See also: *cross fracture.*

fringe ore Ore at the outer limits of the mineralization pattern or halo. Syn: *halo ore.*

fringe water Water of the capillary fringe. Syn: *anastatic water.*

fringing reef An *organic reef* that is directly attached to or borders the shore of an island or continent, having a rough, tablelike surface that is exposed at low tide; it may be more than 1 km wide, and its seaward edge slopes sharply down to the sea floor. There may be a shallow channel or lagoon between the reef and the adjacent mainland. Cf: *barrier reef.* Syn: *shore reef.*

frith *firth.*

fritting The partial melting of grains of quartz and other minerals, so that each grain becomes surrounded by a zone of glass. Fritting results from the contact action of basalt and related lavas on other rocks (Johannsen, 1931).

fritzscheite A mineral of the autunite group: $Mn(UO_2)_2(VO_4)_2 \cdot 10H_2O$ (?).

frohbergite A mineral: $FeTe_2$. It is isomorphous with marcasite.

frolovite A white mineral: $CaB_2O_4 \cdot 4H_2O$.

frond The expanded compound foliage of ferns or a similar leaflike structure.

frondelite A mineral: $MnFe_4{}^{+3}(PO_4)_3(OH)_5$.

frondescent cast A term used by Ten Haaf (1959, p.30) for a feather-like *sole mark* resembling certain shrubs or large cabbage leaves, the spreading "foliage" always directed downcurrent; it is usually several decimeters in length. Syn: *cabbage-leaf mark; deltoidal cast.*

frondose Said of the flattened, frond-shaped growth habit of erect stenolaemate bryozoan colonies that lack median laminae.

front [geomorph] (a) The more or less linear outer slope of a mountain range that rises above a plain or plateau. (b) Land that faces or abuts, esp. on a body of water. Syn: *frontage.*

front [glaciol] (a) *ice front.* (b) *snout [glaciol].*

front [meteorol] The contact at the Earth's surface between two different air masses, commonly cold and warm, that generally moves in an easterly direction. See also: *cold front; warm front.*

front [paleont] The part of the carapace of a brachyuran decapod crustacean (crab) between the orbits.

front [petrology] A metamorphic zone of changing mineralization developed outward from an igneous mass.

frontal adj. Pertaining or belonging to the front part; esp. pertaining to the orifice-bearing side of a bryozoan zooid or colony, or to the interorbital portion of the vertebrate skull. Syn: *obverse.*

frontal apron *apron [geomorph].*

frontal kame A kame that consists of a steep alluvial fan against the edge of an ice sheet.

frontal membrane Uncalcified part of the frontal wall in anascan cheilostome bryozoans, to which parietal muscles are attached. It may be exposed or overarched by a frontal shield.

frontal moraine (a) *end moraine.* (b) "A moraine rampart at the front of a former glacier" (Schieferdecker, 1959, term 0918).

frontal plain *outwash plain.*

frontal plate The modified rostrum of a brachyuran decapod crustacean, bearing a process that projects ventrally between antennules to unite with the epistome.

frontal pore A pore on the front of the lattice shell of a radiolarian and adjacent to the basal ring. It is similar in appearance to the sternal pore.

frontal region The anteromedian part of the carapace of some decapods, including the rostrum and the area behind it (TIP, 1969, pt. R, p. 92).

frontal scar The scar on the interior of the carapace of an ostracode, just anterior and dorsal to the adductor muscle scars.

frontal shield The calcareous part of the frontal surface of a cheilostome bryozoan. It is developed in different patterns with varying relationships to the frontal wall of the zooid. Cf: *cryptocyst; gymnocyst; pericyst; umbonuloid.*

frontal terrace *outwash terrace.*

frontal wall The exterior zooidal wall in cheilostome and some stenolaemate bryozoans, attached to the wall containing the orifice and providing a front side to the zooid.

front bay A large, irregular, shallow bay connected with the sea through a pass between barrier islands, as along the coast of Texas. Cf: *back bay.*

front-limb thrust fault A thrust fault developed on the steeper "front limb" of an asymmetric anticline, in which the direction of tectonic transport is toward the adjacent syncline; the fault dips in the same direction as the anticlinal axis, elongating the limb of the anticline (Douglas, 1950). See also: *back-limb thrust fault; extension fault.* Syn: *forelimb thrust.*

front pinacoid In an orthorhombic, monoclinic, or triclinic crystal, the {100} pinacoid. Cf: *basal pinacoid; side pinacoid.* Syn: *macropinacoid; orthopinacoid.*

front range The outermost range of a mountain system; e.g. the Front Range of the Rocky Mountains, extending southward from Casper, Wyo., through Colorado and into New Mexico, including the Sangre de Cristo Mountains.

front slope The *scarp slope* of a cuesta.

froodite A monoclinic mineral: $PdBi_2$. Cf: *michenerite.*

frost A granular or flaky deposit of ice crystals caused by the sublimation of water vapor on a surface whose temperature is below the freezing point. Cf: *dew.* Syn: *hoarfrost.*

frost action (a) The mechanical weathering process caused by alternate or repeated cycles of freezing and thawing of water in pores, cracks, and other openings, usually at the surface. It includes *congelifraction* and *congeliturbation.* (b) The resulting effects of frost action on materials and structures.—Syn: *freeze-thaw action.*

frost-active soil A fine-grained soil that undergoes changes in volume and bearing capacity due to frost action (Nelson & Nelson, 1967, p.151).

frost belt A ditch designed to assist the early and rapid freezing of the soil in order to obstruct seepage of shallow ground water. It is commonly placed upslope from foundations in permafrost areas. Syn: *frost dam.*

frost blister A low *frost mound,* usually no more than a few meters high, caused by growth of segregated layers of ice or by hydrostatic pressure of ground water. Cf: *gravel mound.* Syn: *ice laccolith.*

frost boil (a) An accumulation of excess water and mud liberated from ground ice by accelerated spring thawing, commonly softening the soil and causing a quagmire. (b) A low mound developed by local differential frost heaving at a place most favorable for the formation of segregated ice and accompanied by an absence of an insulating cover of vegetation (Taber, 1943, p.1458-1459). (c) A break in a surface pavement due to swelling frost action; as the ice melts, soupy subgrade materials issue from the break.

frost bursting *congelifraction.*

frost churning *congeliturbation.*

frost circle A term used by Williams (1936) in referring to a *sorted circle* developed in horizontal thin-bedded limestones in Ontario.

frost crack A nearly vertical fracture developed by thermal contraction in rock or in frozen ground with appreciable ice content. Frost cracks commonly intersect to form polygonal patterns in plan view. Syn: *ice crack; contraction crack; thermal contraction crack.*

frost cracking The contraction cracking of frozen ground, and ice on lakes and rivers, at very low temperatures; the formation of *frost cracks.*

frost-crack polygon A *nonsorted polygon* formed by intersecting *frost cracks.* It is similar to an *ice-wedge polygon,* but lacks a border underlain by ice wedges and therefore is not necessarily associated with permafrost.

frost creep Soil creep resulting from frost action (Kerr, 1881).

frost dam *frost belt.*

frost drift The movement, by frost action, of debris on a slope (Kerr, 1881). Obsolete.

frost flower A type of *surface hoar,* usually growing on an ice surface, characterized by leafy or dendritic blades oriented at various angles to the surface. Syn: *ice flower.*

Frost gravimeter An astatic gravity meter of the balance type, consisting of a mass at the end of a nearly vertical arm, supported by a main spring inclined to the vertical at about a 45° angle. The beam rises and falls with gravity variation, but is restored to its normal position by a sensitive weighing spring tensioned by a micrometer screw.

frost heave *frost heaving.*

frost heaving The uneven lifting or upward movement, and general distortion, of surface soils, rocks, vegetation, and structures such as pavements, due to subsurface freezing of water and growth of ice masses (esp. ice lenses); any upheaval of ground caused by freezing. Syn: *frost heave.*

frost hillock The marked upward bulging sometimes present in the center of a *mud polygon* (Sharpe, 1938, p.36).

frosting (a) A lusterless ground-glass or mat surface on rounded mineral grains, esp. of quartz. It may result from innumerable impacts of other grains during wind action, or from deposition of many microscopic crystals, e.g. fine silica secondarily deposited on quartz grains. (b) The process that produces such a surface.

frost line (a) The maximum depth of frozen ground in areas where there is no permafrost; it may be expressed for a given winter, as the average of several winters, or as the greatest depth on record. Cf: *frost table.* (b) The bottom limit of permafrost. Cf: *permafrost table.* (c) The altitudinal limit below which frost never occurs; applied esp. in tropical regions.

frost mound A general term for a knoll, hummock, or conical mound in a permafrost region, containing a core of ice, and representing a generally seasonal and localized upwarp of the land surface, caused by frost heaving and/or hydrostatic pressure of ground water. See also: *pingo; palsa; hydrolaccolith; ice laccolith; earth hummock; frost blister; ground-ice mound; ice mound; gravel mound.* Syn: *soil blister; suffosion knob.*

frost-pattern soil A term used by Troll (1944) for what is now known as *patterned ground;* it is a misleading term because patterned ground need not consist of soil, nor need it involve a periglacial origin.

frost point That temperature to which a sample of moist air must be cooled (at constant pressure and water-vapor content) in order to cause the sublimation of ice.

frost polygon One of the network polygons forming *polygonal ground.*

frost riving *congelifraction.*

frost scar A *nonsorted circle* or irregular form representing a small patch of bare soil produced by local frost heaving intense enough to disturb the vegetation cover. See also: *mud circle.* Syn: *mud spot; spot medallion.*

frost shattering *congelifraction.*

frost soil *congeliturbate.*

frost splitting *congelifraction.*

frost stirring A syn. of *congeliturbation* involving no mass movement.

frost table An irregular surface that represents, at any given time, the penetration of thawing in seasonally or perennially frozen ground; the upper limit of frozen ground. Cf: *frost line; permafrost table.*

frost weathering *congelifraction.*

frost wedge A term used loosely for any *ice wedge,* whether in perennially or seasonally frozen ground or in fossil form; any wedge-shaped mass whose origin involves cold or freezing conditions.

frost wedging A type of *congelifraction* by which jointed rock is pried and dislodged by ice acting as a wedge.

frost zone *seasonally frozen ground.*

Froude number A dimensionless numerical quantity used as an index to characterize the type of flow in a hydraulic structure that has the force of gravity (as the only force producing motion) in conjunction with the resisting force of inertia. It is the ratio of inertia forces to gravity forces, and is equal to the square of a characteristic velocity (mean, surface, or maximum velocity) of the system divided by the product of a characteristic linear dimension (e.g. diameter or depth) and the gravity constant, acceleration due to gravity, all of which are expressed in consistent units in order that the combinations will be dimensionless. The number is used in open-channel flow studies or where the free surface plays an essential role in influencing motion (ASCE, 1962).

frozen [coal] *burned.*

frozen [ore dep] Said of the contact between the wall of a vein and the mineral deposit filling it, in which the vein material adheres closely to the wall; also, said of the vein material and of the wall.

frozen ground Ground that has a temperature below freezing and generally contains a variable amount of water in the form of ice. Terms inadvisedly used as syns: *frost; ground frost; permafrost.* Syn: *tjaele; gelisol; merzlota; tele; taele.*

fruchtschiefer A type of *spotted slate* characterized by concretionary spots having shapes suggestive of grains of wheat. Etymol: German. Cf: *fleckschiefer; garbenschiefer; knotenschiefer.*

fructification A reproductive organ or fruiting structure of a plant.

fruit (a) In a strict sense, the pericarp and its seeds, the fertilized and developed ovary. (b) More broadly, the matured pericarp and its contents, with any integral external part (Jackson, 1928, p.153).

frustule The siliceous cell wall of a *diatom,* consisting of two halves, the epivalve and the hypovalve. It is ornate, microscopic, and boxlike.

FS *foresight.*

F test A statistical test for equality or comparison of sample variances, expressed as the ratio between sample variances. Syn: *F-distribution test.*

fuchsite A bright-green, chromium-rich variety of muscovite. Syn: *chrome mica.*

fucoid n. (a) An informal name now applied loosely to any indefinite trail-like or tunnel-like sedimentary structure identified as a *trace fossil* but not referred to a described genus. It was once considered to be the remains of the marine alga *Fucus,* and later was regarded as a cylindrical, U-shaped, regularly branching feeding burrow of a marine animal and assigned to the plantlike "genus" *Fucoides.* The term has been broadly applied to crustacean tracks, worm burrows, molluscan trails, marks made by the tide or waves, and rill marks. Fucoids have been defined as being within sedimentary layers and formed of material more or less unlike the matrix (Vassoevich, 1953, p. 21); but in most usage fucoids are surface features formed of the same material as the matrix. See also: *chondrite [paleont].* (b) A fossil of an alga, or a fossil resembling an alga or the remains or impression of a seaweed. (c) A seaweed of the order Fucales (brown algae).—adj. Pertaining to or resembling a fucoid. Syn: *fucoidal.*

fuel ratio In coal, the ratio of fixed carbon to volatile matter. It is sometimes a factor in the analysis or classification of coals. Cf: *carbon ratio [coal].*

fugacity A thermodynamic function defined by the equation $dG = R T d \ln f$, where G is the *Gibbs free energy,* R is the gas constant, and T is absolute temperature. Fugacity is expressed in units of pressure. Fugacities are used in calculations of chemical equilibrium because they can be calculated in practical applications whereas G cannot. For further information, see: Klotz, I. M., 1972, *Chemical Thermodynamics.*

fugacity coefficient The ratio of the fugacity of a gas to its potential pressure; hence, a measure of the nonideality of the gas.

fugitive In ecology, said of an organism, esp. a plant species, that is not permanently established and is likely to disappear.

fugitive constituent A substance that was originally present in a magma but was lost during crystallization, so that it does not commonly appear as a rock constituent (Shand, 1947, p. 34). Syn: *volatile component.*

fugitive species *opportunistic species.*

fukuchilite A mineral: Cu_3FeS_8.

fulcral plate A small plate raised above the floor of the brachial valve of a brachiopod, extending between the posterior margin and the brachiophore base, and bounding the dental socket anteriorly and laterally (TIP, 1965, pt.H, p.145).

fulcral ridge A linear elevation on an articular face of a crinoid ossicle, serving as an axis of differential movement.

fulcrum The intersection of the end of a recurved spit with the next succeeding stage in development of a compound spit.

fulgurite An irregular, glassy, often tubular or rod-like structure or crust produced by the fusion of loose sand (or rarely, compact rock) by lightning, and found esp. on exposed mountain tops or in dune areas of deserts or lake shores. It may measure 40 cm in length and 5-6 cm in diameter. Etymol: Latin *fulgur,* "lightning". Syn: *lightning stone; lightning tube; sand tube.*

fulji A term used in northern Arabia for a depression between *barchans,* occurring esp. where the dunes are pressing closely on one another; it has a steep slope on the windward side and a gentle slope on the lee side. Etymol: Arabic. Pl: *fuljis.* Syn: *fulje.*

full A British term for *beach ridge.*

full-cut brilliant A brilliant-cut diamond or colored stone with the usual total of 58 facets, consisting of 32 facets and a table above the girdle, and 24 facets and a culet below. The girdle is usually polished on colored stones, but seldom on diamonds.

full-depth avalanche A *snow avalanche* that glides over a rock or ground surface; formerly termed *ground avalanche.*

full dip *true dip.*

fuller's earth A very fine-grained, naturally occurring clay or claylike material possessing a high adsorptive capacity, consisting largely of hydrated aluminum silicates (chiefly the clay minerals montmorillonite and palygorskite). Used originally in England for whitening, degreasing, or fulling (shrinking and thickening by application of moisture) woolen fabrics, fuller's earth is now extensively used as an adsorbent in refining and decolorizing oils and fats; it is a natural bleaching agent. Its color ranges from light brown through yellow and white to light and dark green, and it differs from ordinary clay by having a higher percentage of water and little or no plasticity, tending to break down into a muddy sediment in water. Fuller's earth probably forms as a residual deposit by decomposition of rock in place, as by devitrification of volcanic glass. The term is applied without reference to any particular chemical or mineral composition, texture, or origin. Syn: *creta; walker's earth.*

full meander A stream meander consisting of two loops, one in a clockwise direction and the other in a counterclockwise direction.

fully arisen sea *fully developed sea.*

fully developed sea A sea in which all possible wave frequencies in the wave spectrum for prevailing wind speed have developed the maximum energies. The ocean waves are at the maximum height for a given wind force blowing over sufficient fetch regardless of duration. Syn: *fully arisen sea.*

fülöppite A lead-gray mineral with a bluish or bronze tarnish: $Pb_3 Sb_8S_{15}$. Also spelled: *fuloppite.*

fulvic acid That organic matter of indefinite composition that remains in solution when an aqueous alkaline extract of soil is acidified.

fulvurite An old synonym for *brown coal.*

fumarole A vent, usually volcanic, from which gases and vapors are emitted; it is characteristic of a late stage of volcanic activity. It is sometimes described by the composition of its gases, e.g. chlorine fumarole. Fumaroles may occur along a fissure or in apparently chaotic clusters or fields. See also: *solfatara; fumarolic stage; fumarole field.* Also spelled: fumerole.

fumarole field A group of cool fumaroles (Rittmann, 1962, p. 10). Cf: *solfatara field.*

fumarole mound A small mound from which fumarole or solfatara gases escape.

fumarolic stage A late or decadent type of volcanic activity characterized by the emission of gases and vapors from a vent or *fumarole.* Cf: *solfataric stage.*

fume cloud A vaporous cloud of volcanic gases from a body of molten lava.

fumerole *fumarole.*

functional morphology The study of the form and structure of an organism in relation to its adaptation to a specific environment and/or survival under specific conditions; the *morphology* of an animal or plant as it responds or responded to environmental changes and conditions.

fundamental complex *basement.*

fundamental jelly *ulmin.*

fundamental strength The maximum stress that a body can withstand, under given conditions but regardless of time, without creep.

fundamental substance *ulmin.*

fungal spore A spore of the usually multicellular, nonvascular, heterotrophic plants (fungi). Such spores include a wide variety of types, from simple unicellular to multicellular sclerotia; they have a range of Precambrian to Holocene, and those that are preserved in sediments and studied in coal petrology and palynology are chitinous. Examples: *basidiospore; chlamydospore; conidiospore; dictyospore; phragmospore; teleutospore; urediospore.*

Fungi Class of heterotrophic thallophytes; multicelled, commonly microscopic plants that may be saprophytic, symbiotic, or parasitic on green plants. They are probably polyphyletic. Range is from at least Late Precambrian. Informal sing.: *fungus.*

fungus Informal term for a member of the class *Fungi.*

funicle A spirally wound narrow ridge extending upward from the inner lip of a gastropod shell into the umbilicus.

funicular water Capillary water contained in a cluster of rock or soil particles in the zone of aeration, the interstices of the cluster being completely filled with water bounded by a single closed capillary meniscus (Smith, W.O., 1961, p. 2). Cf: *pendular water; pellicular water; sejunction water; capillary condensation.*

funiculus [bot] The stalk by which an ovule is attached to the ovary wall or placenta in angiospermous plants.

funiculus [paleont] A strand of tissue connecting the polypide with the body wall or septulae in a bryozoan. Pl: *funiculi.*

funnel filling The red-brown to dark-brown, opaque to translucent, coarsely laminated portion of the basal plate of some conodont elements, occupying the cavity in the basal funnel.

funnel intrusion An igneous intrusion with an inverted conical shape; typically layered, and mafic or ultramafic in composition.

funnel joint A joint in a *joint set* that is *concentric,* with the joints dipping towards a common center.

funnel sea A gulf or bay that is narrow at its head and wide at its mouth, and that deepens rapidly from head to mouth, thus resembling one half of a funnel split lengthwise; e.g. the Gulf of California.

furca A two-forked last abdominal segment of certain crustaceans; specif. *caudal furca.* Pl: *furcae.*

fur-cap rock *mushroom rock.*

furcula (a) The wishbone, or fused clavicles, of a bird. (b) A wishbone-shaped sponge spicule.—Pl: *furculae.*

furious cross-bedding Bedding that is doubly cross-bedded, characterized by foreset beds that are themselves cross-bedded (Reiche, 1938, p.926).

furrow (a) A linear depression produced by the removal of rock material, as by glacial action; e.g. a *groove.* (b) A nongeneric term used by Bucher (1933) for a depressed part of the crust of any size with a distinct linear development. Cf: *welt.* (c) A term applied in Africa to a natural or artificial watercourse used for drainage or irrigation. (d) *swale.* (e) *colpus.*

furrow cast A term introduced by McBride (1962, p.58) for a *sole mark* consisting of a cast of a group of closely spaced, parallel, and linear indentations separated by long, narrow, round or flat-topped, slightly sinuous septa which appear as depressions in the cast; it lacks the steep or blunt upcurrent end of a flute cast. The term was suggested hesitantly by Kuenen (1957, p.244) to replace the ambiguous "groove cast". Cf: *furrow flute cast.*

furrow flute cast A *furrow cast* with an upcurrent termination similar to that of a flute cast. Syn: *sludge cast; rill cast.*

fusain A coal lithotype characterized macroscopically by its silky luster, fibrous structure, friability, and black color. It occurs in strands or patches and is soft and dirty when not mineralized. Its characteristic microlithotype is *fusite.* Cf: *vitrain; clarain; durain.* Syn: *mineral charcoal; mother of coal.* Obsolete syn: *motherham.*

fusainisation *fusinization.*

fusellar fabric One of the three major types of materials recognized in electron-microscope study of graptolithine periderm as a fundamental structural element of the periderm. Fusellar fabric is formed from slightly wavy, and commonly branching, fibrils that are so interwoven that they form a three-dimensional mesh (Urbanek & Towe, 1974, p. 4). Cf: *cortical fabric; sheet fabric.*

fusellar tissue Fusellar tissue in graptolithines is composed of spongy-appearing fusellar fabric bounded or enclosed by thin, membranous sheet fabric; it forms the inner part of the graptolithine periderm, the outer part being formed from cortical tissue (Urbanek & Towe, 1974, p. 5).

fusibility A characteristic of minerals by which they can be ranked on a temperature scale. See also: *fusibility scale.* Cf: *infusible.*

fusibility scale A temperature scale based on the *fusibility* of a standard group of minerals, with which other minerals may be compared. An analysis that can be made with a burner and a blowpipe is based on the following series: stibnite, 550°C; chalcopyrite, 800°C; almandine garnet, 1050°C; actinolite, 1200°C; orthoclase, 1300°C; enstatite (bronzite), 1400°C; and quartz, infusible.

fusiform Shaped like a spindle, i.e. tapering toward each end from a swollen middle.

fusiform bomb A volcanic bomb that tapers at both ends from a swollen middle; it includes both *rotational bombs* and *spindle-shaped bombs.*

fusil A spindle-shaped siliceous concretion.

fusinite A maceral of coal within the inertinite group, with cellular structure of high reflectance and relatively high carbon content; it consists of carbonized woody material. Fusinite is characteristic of fusain.

fusinization A process of *coalification* in which fusain is formed. Cf: *incorporation; vitrinization.* Also spelled: *fusainisation.*

fusinoid Fusinite that has a reflectance distinctly higher than that of associated xylinoids, vitrinoids, or anthrinoids, and that has well-developed cellular structure (ASTM, 1970, p.175).

fusion [isotope] The combination, or fusion, of two light nuclei to form a heavier nucleus. The reaction is accompanied by the release of a large amount of energy as in the hydrogen bomb. Cf: *fission [isotope].* Syn: *nuclear fusion.*

fusion [petrology] (a) The process whereby a solid becomes liquid by the application of heat; melting. (b) The unification or mixing of two or more substances, as by melting together.

fusion [photo] *stereoscopic fusion.*

fusion crust A thin glassy coating, usually black and rarely exceeding one millimeter in thickness, formed on the surface of a meteorite by frictional heating during atmospheric flight. Owing to differing effects of the atmosphere upon different meteorite surfaces, fusion crusts may be knobby, striated, ribbed, net, porous, warty, or scoriaceous.

fusion tectonite An igneous rock whose alignment of early-formed crystals was caused by continuous movement in an enclosing melt; a type of *primary tectonite* (Turner and Weiss, 1963, p.39).

fusite A coal microlithotype that contains at least 95% fusinite. It is a variety of *inertite.* Cf: *fusain.*

fusoclarain A transitional lithotype of coal, characterized by the presence of fusinite and vitrinite with other macerals; fusinite is less abundant than it is in *clarofusain.* Syn: *fusoclarite.*

fusoclarite *fusoclarain.*

fusodurain A coal lithotype transitional between durain and fusain, but predominantly durain. Cf: *durofusain.*

fusotelain A coal lithotype transitional between telain and fusain, but predominantly telain. Cf: *telofusain.*

fusovitrain A coal lithotype transitional between vitrain and fusain, but predominantly vitrain. Cf: *semifusain.* Syn: *fusovitrite.*

fusovitrite *fusovitrain.*

fusshang A syn. of *foot slope.* Etymol: German *Fusshang,* "foot slope".

fusulinacean Any fusulinid belonging to the superfamily Fusulinacea, characterized by a spindle-shaped, spheroidal, or discoid test with a complex internal structure.

fusuline *fusulinid.*

fusulinid Any foraminifer belonging to the suborder Fusulinina, family Fusulinidae, characterized by a multichambered elongate calcareous micrygranular test, commonly resembling the shape of a grain of wheat. Range, Ordovician to Triassic. Syn: *fusuline.* See also: *alveolinid.*

fusulinid limestone A *foraminiferal limestone* composed chiefly of fusulinid tests; e.g. the numerous Missourian and Virgilian (Upper Pennsylvanian) limestones of midwestern U.S.

future ore *possible ore.*

gabbride A term used in the field for any igneous rock having pyroxene as the only dark mineral, which forms over 50 percent of the rock, with a smaller amount of feldspar; e.g. augite *diorite, gabbro, norite.* Not recommended usage.

gabbro (a) In the *IUGS classification,* a plutonic rock with Q between 0 and 5, P/(A+P) greater than 90, and plagioclase more calcic than An_{50}. (b) A group of dark-colored, basic intrusive igneous rocks composed principally of basic plagioclase (commonly labradorite or bytownite) and clinopyroxene (augite), with or without olivine and orthopyroxene; also, any member of that group. It is the approximate intrusive equivalent of *basalt.* Apatite and magnetite or ilmenite are common accessory minerals. Gabbro grades into *monzonite* with increasing alkali-feldspar content. According to Streckeisen (1967, p. 171, 198), plagioclase with more than 50% anorthite distinguishes gabbro from *diorite;* quartz is 0-20% of the light-colored constituents, and the plagioclase/total feldspar ratio is 90/100.—The name, introduced by Buch in 1810, is apparently after the town of Gabbro in Tuscany, Italy.

gabbroic layer *basaltic layer.*

gabbroid (a) In the *IUGS classification,* a preliminary term (for field use) for a plutonic rock with Q less than 20 or F less than 10, P/A+P) greater than 65, and pl/(pl+px+ol) between 10 and 90. Cf: *leucogabbroid; melagabbroid.* (b) Said of a rock resembling gabbro. (c) A nonpreferred syn. of *ophitic.*

gabbronorite In the *IUGS classification,* a plutonic rock satisfying the definition of *gabbro,* in which pl/(pl+px+ol) and pl/(pl+px+hbl) are between 10 and 90, and ol/(pl+px+ol) and hbl/(pl+px+hbl) are less than 5.

gabbrophyre A porphyritic hypabyssal rock composed of phenocrysts of labradorite and augite in a groundmass of calcic plagioclase and hornblende. Not recommended usage.

gabion A specially designed container, cylinder, or box of corrosion-resistant wire used to hold coarse rock aggregate, as in forming a groin or seawall, or to assist in developing a bar or dike in a harbor. Gabions are also often placed at the toe of slide-prone slopes to improve stability.

gabrielsonite A mineral: $PbFe(AsO_4)(OH)$.

gadolinite A black, greenish-black, or brown mineral: $Be_2FeY_2Si_2O_{10}$. It is a source of rare earths.

gagarinite A creamy, yellowish, or rosy hexagonal mineral: $NaCaY(F,Cl)_6$.

gagatite Coalified woody material, resembling jet. See also: *gagatization.*

gagatization In coal formation, the impregnation of wood fragments with dissolved organic substances. See also: *gagatite.*

gage n. In hydraulics, a device for measuring such factors as water-surface elevation, velocity of flow, water pressure, and precipitation. See also: *staff gage; chain gage.*

gage height *stage [hydraul].*

gageite A mineral: $(Mn,Mg,Zn)_7Si_2O_7(OH)_8$.

gaging *stream gaging.*

gaging station A particular site on a stream, canal, lake, or reservoir where systematic observations of gage height, discharge, or water quality (or any combination of these) are obtained.

gahnite A dark-green to yellowish, gray, or black mineral of the spinel series: $ZnAl_2O_4$. It often contains some magnesium. Syn: *zinc spinel.*

gahnospinel A blue or greenish variety of spinel containing zinc.

gaidonnayite An orthorhombic mineral: $Na_2ZrSi_3O_9 \cdot 2H_2O$. It is dimorphous with catapleiite.

gain control In a seismic amplifier, a device to change the amplification with time. It may be automatic by individual channel, ganged for all channels together, programmed prior to the shot, or otherwise pre-arranged. Syn: *volume control.*

gaining stream *effluent stream.*

gaize A porous fine-grained micaceous glauconitic sandstone, containing much soluble silica, occurring among the Cretaceous rocks of France and Belgium; a calcareous clastic sediment cemented by chert or flint. See also: *opoka.*

gal A unit of acceleration, used in gravity measurements. One gal = 1 cm/sec^2 = 10^{-2}m/sec^2. The Earth's normal gravity is 980 gal. The term is not an abbreviation: it was invented to honor the memory of Galileo. See also: *milligal; microgal.*

galactic cluster *star cluster.*

galactite (a) A variety of white natrolite, occurring in colorless acicular crystals. (b) An obsolete syn. of *novaculite.* (c) An unidentified stone (possibly of calcium nitrate) whose milky solution gave rise to several medieval legends and superstitions.

galaxite A black mineral of the spinel series: $MnAl_2O_4$. The manganese is often replaced in part by ferrous iron or magnesium, and the aluminum by ferric iron.

galaxy One of billions of large systems of stars, nebulae, star clusters, globular clusters, and interstellar matter that make up the Universe. When the term is capitalized, it refers to the Milky Way stellar system. Syn: *extragalactic nebula.*

galea (a) A conical process in the skeleton of a phaeodarian radiolarian. (b) The spinning tube on the movable finger of the chelicera of certain arachnids (pseudoscorpions). (c) The outer distal hoodlike lobe of the second segment of the maxillule of a crustacean, adjacent to the *lacinia* (TIP, 1969, pt.R, p. 96).

Gale alidade A syn. of *explorer's alidade.* Named for Hoyt S. Gale (1876-1952), American geologist.

galeite A trigonal mineral: $Na_{15}(SO_4)_5F_4Cl$. Cf: *schairerite.*

galena A bluish-gray to lead-gray mineral: PbS. It frequently contains included silver minerals. Galena occurs in cubic or octahedral crystals, in masses, or in coarse or fine grains; it is often associated with sphalerite as disseminations in veins in limestone, dolomite, and sandstone. It has a shiny metallic luster, exhibits highly perfect cubic cleavage, and is relatively soft and very heavy. Galena is the most important ore of lead and one of the most important sources of silver. Syn: *galenite; lead glance; blue lead.*

galenite *galena.*

galenobismutite A lead-gray or tin-white mineral: $PbBi_2S_4$.

galkhaite A cubic mineral: $(Hg,Cu,Zn)(As,Sb)S_2$.

gall [sed] (a) *clay gall.* (b) A sand pipe.

gall [soil] A small barren or infertile surface spot or area from which the original surface soil has been removed by erosion or excavation.

gallery [grd wat] *infiltration gallery.*

gallery [paleont] A laterally continuous internal open space between adjacent laminae in most stromatoporoid coenostea, partially filled by pillars and dissepiments.

gallery [speleo] A large, more or less horizontal passage in a cave.

galliard A hard, smooth, close-grained, siliceous sandstone; a *ganister.* Also spelled: *calliard.*

gallite A tetragonal mineral: $CuGaS_2$.

galloping glacier A popular term for *surging glacier.*

Gall projection A stereographic, modified-cylindrical map projection in which the cylinder intersects the globe along the parallels 45°N and 45°S. The scale is preserved along these parallels, but is too small between them and too large poleward of them; there is less distortion of areas and shapes in high latitudes than in the Mercator projection. It was introduced in 1885. Named after Johann G. Galle (1812-1910), German astronomer.

galmei *hemimorphite.*

galt *gault.*

gamagarite A dark-brown monoclinic mineral: $Ba_4(Fe,Mn)_2V_4O_{15}(OH)_2$.

gametophyte The individual or sexual generation of a plant that produces gametes; e.g. the haploid generation of an embryophytic plant, produced by germination of the spores. In lower vascular plants and bryophytes, the gametophyte is a separate plant, but in seed plants, it is confined to the several cells of the *microgametophyte* in the pollen grain and the multicellular *megagametophyte* in the ovule, with the seed developing from the fertilized ovule. Cf: *sporophyte.* See also: *prothallus.*

gamius A characteristic anomaly on a gamma-ray log in uranium exploration that occurs in unaltered sandstone and is an indicator of a nearby *roll-front orebody* (Bailey & Childers, 1977, p. 413). Etymol: *gam*ma *i*n *u*naltered *s*andstone.

gamma [cryst] (a) In a biaxial crystal, the largest *index of refraction.* (b) The interaxial angle between the *a* and *b* crystallographic axes.—Cf: *alpha [cryst]; beta [cryst].*

gamma [magnet] The cgs unit of magnetic field intensity commonly used in magnetic exploration. It is equal to 10^{-5} oersted. Syn: *nanotesla.*

gamma [mineral] adj. Of or relating to one of three or more closely related minerals and specifying a particular physical structure (esp. a polymorphous modification); specif. said of a mineral that is stable at a temperature higher than those of its *alpha* and *beta* polymorphs (e.g. "gamma quartz" or "γ-quartz").

gamma* angle The angle of the reciprocal lattice between the *a* axis* and the *b* axis,* which is equal to the interfacial angle between (100) and (010). Cf: *alpha* angle; beta* angle.*

gamma decay De-excitation of an atomic nucleus without a change in atomic number or mass number, usually by emission of *gamma radiation.*

gamma-gamma log *density log.*

gamma-MnO_2 *nsutite.*

gamma radiation Electromagnetic radiation from an atomic nucleus, often accompanying emission of *alpha particles* and *beta particles.* Cf: *gamma ray.*

gamma ray A photon from an atomic nucleus. See: *gamma radiation.*

gamma-ray log The *radioactivity log* curve of the intensity of broad-spectrum, undifferentiated natural gamma radiation emitted from the rocks in a cased or uncased borehole. It is used for correlation, and for distinguishing shales (which are usually richer in naturally radioactive elements) from sandstones, carbonates, and evaporites. Cf: *spectral gamma-ray log.*

gamma-ray spectrometer An instrument for measuring the energy distribution, or spectrum, of gamma rays, whether from natural or artificial sources. It is used in airborne remote sensing for potassium, thorium, and uranium. Cf: *scintillation spectrometer.*

gamma-ray spectrometry Determination of gamma-ray energies and states of polarization, and also studies of the correlations between gamma rays emitted in sequence from a nucleus.

gamma-ray spectroscopy The observation of a gamma-ray spectrum and all processes of recording and measuring that go with it.

gamma-sulfur *rosickyite.*

gangmylonite An ultramylonite or mylonite that shows intrusive relations with the adjacent rock with no evidence of fusion (Hammer, 1914).

gangue The valueless rock or mineral aggregates in an ore; that part of an ore that is not economically desirable but cannot be avoided in mining. It is separated from the ore minerals during concentration. Syn: *matrix [ore dep].* Cf: *ore mineral.*

ganister (a) A hard, fine-grained quartzose sandstone or quartzite, used in the manufacture of silica brick. It is composed of subangular quartz particles (0.15-0.5 mm in diameter, although some authors use a lower limit of 0.05 mm, thereby extending into the silt-size range), cemented with secondary silica, and possessing a characteristic splintery fracture that gives rise to smooth, subconchoidal surfaces and sharp edges. Ganister is distinguished from chert by its more granular texture and by the relatively small quantity of chalcedonic or amorphous silica. (b) In England, a highly siliceous *seat earth* of coal seams, e.g. the Sheffield ganister of the Lower Coal Measures of Yorkshire. See also: *pencil ganister; bastard ganister; silica rock; crowstone; galliard.* (c) A mixture of ground quartz and fireclay used as a furnace lining.—Also spelled: gannister.

ganoid adj. Pertaining to fish scales of rhomboid shape, with a heavy outer coat of enamel.—n. (obsolete) Any bony fish with this type of scale, mostly chondrosteans and holosteans but including acanthodians and some crossopterygians.

ganomalite A colorless to gray tetragonal mineral: $Ca_2Pb_3Si_3O_{11}$.

ganophyllite A brown mineral: $(Na,K)(Mn,Fe,Al)_5(Si,Al)_6O_{15}(OH)_5 2H_2O$.

gap [coast] (a) A narrow passage or channel between an island and the shore. (b) A break in a levee through which a stream distributary may flow; a tidal inlet.

gap [fault] In a fault, the horizontal component of separation measured parallel to the strike of the strata, with the faulted bed absent from the measured interval. Cf: *overlap [fault].* Obsolete syn: *stratigraphic heave.*

gap [geomorph] (a) A term used in Pennsylvania and farther south for a sharp break or opening in a mountain ridge, or for a short *pass* through a mountain range; e.g. a *wind gap.* Cf: *notch [geomorph]; col.* (b) A ravine or gorge cut deeply through a mountain ridge, or between hills or mountains; e.g. a *water gap.*—Cf: *gate [geomorph].*

gap [marine geol] *abyssal gap.*

gap [stratig] *break [stratig].*

gape (a) A localized opening remaining between the margins of a bivalve shell of a mollusk or ostracode when the valves are shut or drawn together by adductor muscles. (b) In brachiopods, the anterior and lateral space between the valves when they are open.

gara A mushroom-shaped rock occurring in arid or desert regions, resulting from the undercutting of soft rock by wind-driven sand, esp. if the soft rock is overlain by more resistant strata. Pl: *gour.* Syn: *cheesewring.* See also: *mushroom rock.*

G/A ratio The principle of the geometric increase in ore tonnage relative to the arithmetic decrease of ore grade, as applied mainly to sedimentary, some residual, and disseminated orebodies (Lasky, 1950).

garbenschiefer (a) A type of *spotted slate* characterized by concretionary spots whose shape resembles that of a caraway seed. Etymol: German. Cf: *knotenschiefer; fleckschiefer; fruchtschiefer.* (b) *feather amphibolite.*

gardening A phenomenon in which the lunar regolith is constantly and very slowly churning due to successive impacts whereby bottom material works its way to the top and surface material gets buried.

garéwaïte A nearly feldspar-free *lamprophyre,* in the same series with vogesite, spessartite, and odinite, composed of corroded clinopyroxene phenocrysts in a fine-grained holocrystalline groundmass of olivine, pyroxene, chromite, and magnetite, with green spinel and labradorite as accessories. Described by Duparc and Pearce in 1904 from the northern Urals. Not recommended usage.

garganite A *vogesite* that contains both augite and hornblende. Not recommended usage.

Gargasian Substage in Switzerland: Lower Cretaceous (upper Aptian; above Bedoulian Substage).

gargulho A Brazilian term used in the plateau region of Bahia for a comparatively coarse, clay-cemented ferruginous conglomerate in which diamonds are found.

garland *stone garland.*

garnet (a) A group of minerals of formula: $A_3B_2(SiO_4)_3$, where A = Ca, Mg, Fe^{+2}, and Mn^{+2}, and B = Al, Fe^{+3}, Mn^{+3}, V^{+3}, and Cr. (b) Any of the minerals of the garnet group, such as the end members almandine (Fe-Al), andradite (Ca-Fe), grossular (Ca-Al), pyrope (Mg-Al), spessartine (Mn-Al), uvarovite (Ca-Cr), and goldmanite (Ca-V).—Garnet is a brittle and transparent to subtransparent mineral, having a vitreous luster, no cleavage, and a variety of colors, dark red being the most common. It occurs as an accessory mineral in a wide range of igneous rocks, but is most commonly found as distinctive euhedral isometric crystals in metamorphic rocks (gneiss, mica schist, eclogite); it may also be massive or granular. Garnet is used as a semiprecious stone and as an abrasive.

garnetiferous Containing garnets; e.g. "garnetiferous peridotite".

garnetite A metamorphic rock consisting chiefly of an aggregate of interlocking garnet grains. Cf: *tactite.*

garnetization Introduction of, or replacement by, garnet. This process is commonly associated with contact metamorphism.

garnet jade A light-green variety of grossular garnet, closely approaching fine jadeite in appearance, esp. that found in Transvaal, South Africa.

garnetoid A group name for minerals with structures similar to that of garnet; e.g. griphite and berzeliite.

garnierite (a) A group name for poorly defined hydrous nickel-magnesium silicates. (b) An apple-green or pale-green mineral, probably: $(Ni,Mg)_3Si_2O_5(OH)_4$. The ratio Ni:Mg is highly variable. Garnierite has been regarded as having no crystal structure, but it may be monoclinic; it is sometimes considered to be a nickel-rich antigorite. It is an important ore of nickel and is used as a gemstone.—See also: *genthite.* Syn: *noumeite; nepouite.*

garrelsite A monoclinic mineral: $Ba_3NaSi_2B_7O_{16}(OH)_4$.

garronite A mineral: $Na_2Ca_5Al_{12}Si_{20}O_{64} \cdot 27H_2O$.

gas *natural gas.*

gas barren An area, as large as several acres, that is characterized by a lack of vegetation due to fumarolic activity and acid leaching of the surface rocks.

gas cap Free gas overlying liquid hydrocarbons in a reservoir under trap conditions.

gas-cap drive Energy within an oil pool, supplied by expansion of an overlying volume of compressed free gas as well as by expansion of gas dissolved in the oil. Cf: *dissolved-gas drive; water drive.*

gas chromatography A process for separating gases or vapors from one another by passing them over a solid (*gas-solid chromatography*) or liquid (*gas-liquid chromatography*) phase. The gases are repeatedly adsorbed and released at differential rates resulting in separation of their components. Abbrev: GC. See also: *liquid chromatography; chromatography.*

gas coal Bituminous coal that is suitable for the manufacture of flammable gas because it contains 33-38% volatile matter. Cf: *high-volatile bituminous coal.* See also: *coal gas.*

gas-cut mud Drilling mud returned from the bottom of a drill hole, characterized by a fluffy texture, gas bubbles, and reduced density due to the retention of entrained natural gas rising from the strata traversed by the drill.

gaseous transfer Separation from a magma of a gaseous phase that moves relative to the magma and releases dissolved substances, usually in the upper levels of the magma, when it enters an area of reduced pressure. See also: *pneumatolytic differentiation.* Syn: *volatile transfer.*

gas field (a) A *gas pool.* (b) Two or more gas pools on a single geologic feature or otherwise closely related.

gas fluxing A rapid upward streaming of free juvenile gas through a column of molten magma in the conduit of a volcano. The gas acts as a flux to promote melting of the wall rocks. Syn: *volcanic blowpiping.*

gash breccia A term used in Pembrokeshire, England, for a rock believed to have originated from the collapse of walls and roofs of caves that had been eroded by the solvent action of underground water.

gas heave The distortion of prodelta sediments, produced by the weight of a distributary-mouth bar compacting the underlying sediment and causing carbon dioxide to escape (Moore, 1966, p.101). See also: *air heave.*

gas-heave structure A sedimentary structure, restricted to the smaller distributaries of the Mississippi River delta, produced by gas heave. See also: *air-heave structure.* Cf: *mudlump.*

gash fracture A small-scale tension fracture that occurs at an angle to a fault and tends to remain open. Syn: *open gash fracture.*

gash vein A nonpersistent vein that is wide above and narrow below, and that terminates within the formation it traverses. The term was originally applied to vein fillings of solution joints in limestone.

gas inclusion A gas bubble within a gemstone, often visible to the unaided eye as in a three-phase inclusion characteristic of some South American emeralds; esp. one within a synthetic stone, often enabling it to be distinguished from a natural gem.

gas-liquid chromatography Process in which a gas, such as helium, argon, hydrogen, or nitrogen, carrying a gaseous mixture to be resolved, is passed over a nonvolatile liquid coated on a porous inert solid support (May & Cuttitta, 1967, p.115) where the components are separated by differential mobility rates. Abbrev: GLC. See also: *gas-solid chromatography; gas chromatography.*

gasoclastic sediment A sediment resulting from sedimentary volcanism, such as mud ejected by enormous volumes of gas (Bucher, 1952, p.87).

gas-oil contact The boundary surface between an accumulation of oil and an overlying accumulation of natural gas. Syn: *gas-oil interface.*

gas-oil interface *gas-oil contact.*

gas-oil ratio (a) The quantity of gas produced with oil from an oil well, usually expressed as the number of cubic feet of gas per barrel of oil. Abbrev: GOR. (b) *reservoir gas-oil ratio.*

gaspeite A mineral: $(Ni,Mg,Fe)CO_3$. A member of the calcite group.

gas phase That stage in a volcanic eruption that is characterized by the release of large amounts of *volcanic gases.*

gas pit A circular pit, 2.5-30 cm in diameter and from less than 3 cm to more than 30 cm deep, surrounded by a mound of mud and produced by the escape of gas bubbles (as of methane generated during the decomposition of organic matter) rising from the surface of a mud bar (Maxson, 1940a).

gas pool A subsurface accumulation of natural gas that will yield gas in economic quantities. Cf: *gas field.*

gas sand A sand or sandstone containing a large quantity of natural gas.

gassi A term used in the Saharan region for a sand-free interdune passage, in some cases traversing an entire *erg* from end to end. Etymol: Arabic, "neck, closed ground". Pl: *gassis.* Cf: *feidj.*

gas skin The envelope of hot volcanic gas surrounding each of the particles in an ash flow or nuée ardente.

gas-solid chromatography Process in which a gas, such as helium, argon, hydrogen, or nitrogen, carrying a gaseous mixture to be resolved, is passed over a porous adsorbing solid (May & Cuttitta, 1967, p. 114), where the components are separated by differential mobility rates. Abbrev: GSC. See also: *gas-liquid chromatography; gas chromatography.*

gas streaming A process of magmatic differentiation involving the formation of a gaseous phase, usually during a late stage in consolidation of the magma, that results in partial expulsion, by escaping gas bubbles, of residual liquid from the crystal network.

gasteropod *gastropod.*

gastral Pertaining to the surface of the spongocoel of a sponge; e.g. "gastral membrane" (endopinacoderm lining the spongocoel) and "gastral cavity" (the spongocoel itself).

gastralium (a) A specialized sponge spicule lining the cloaca. (b) The abdominal rib of a lower tetrapod; usually used in the plural as a collective term.— Pl: *gastralia.*

gas trap A *trap [petroleum]* that contains gas.

gastric region The median part of the carapace in some decapods, in front of the cervical groove and behind the frontal region. It is divided by some authors into epigastric, mesogastric, metagastric, protogastric, and urogastric subregions (TIP, 1969, pt. R, p. 92).

gastrolith A highly polished, rounded stone or pebble from the stomach of some vertebrates (esp. reptiles), thought to have been used in grinding up their food. Syn: *stomach stone; gizzard stone.*

gastro-orbital groove A groove in decapods that extends forward from the upper part of the cervical groove in the direction of the orbit. It forms the upper border of the elevated area behind the antennal spine (Holthuis, 1974, p. 733).

gastropod Any mollusk belonging to the class Gastropoda, characterized by a distinct head with eyes and tentacles and, in most, by a single calcareous shell that is closed at the apex, sometimes spiralled, not chambered, and generally asymmetrical; e.g. a snail. Range, Upper Cambrian to present. Also spelled: *gasteropod.*

gastropore A relatively large tubular cavity of certain hydrozoans, providing lodgment for a gastrozooid. Cf: *dactylopore.*

gastrovascular cavity The interior space (coelenteron) of a coral polyp, radially partitioned by septa, and functioning in both digestion and circulation.

gastrozooid A short, cylindrical feeding and digestive polyp housed in a gastropore of a hydrozoan. Cf: *dactylozooid.*

gas-water contact The boundary surface between an accumulation of gas and the underlying *bottom water.* Syn: *gas-water interface.*

gas-water interface *gas-water contact.*

gas well A *well* that is capable of producing natural gas or that produces chiefly natural gas. Some statutes define the term on the basis of the *gas-oil ratio.*

gat (a) A natural or artificial opening, as a strait, channel, or other passage, extending inland through shoals, or between sandbanks, or in the cliffs along a coast. (b) A strait or channel from one body of water to another, as between offshore islands or shoals, or connecting a lagoon with the sea.—Syn: *gate.*

gate [coast] (a) An entrance to a bay or harbor, located between promontories; e.g. the Golden Gate in San Francisco, Calif. (b) *gat.*

gate [geomorph] (a) A mountain pass affording an entrance into a country. (b) A broad, low valley or opening between highlands, generally wider than a *gap.* Syn: *geocol.* (c) A restricted passage along a river valley; e.g. the Iron Gate on the River Danube.

gate [paleont] A large opening or fissure in the skeleton of a spumellarian radiolarian.

gather A display of input data, arranged so that all the seismic traces corresponding to some criterion, such as shot-detector distance, are displayed side by side. It is used for checking corrections and evaluating the components of a *stack*. A *CDP* gather displays data for the same reflecting point after correction for normal moveout and statics; a "common-range" or "common-offset" gather displays data for the same offset but for a number of nearby reflection points.

gathering ground (a) An area over which water is collected from precipitation, springs, and surface drainage; esp. an area that supplies water to a reservoir. (b) *drainage basin.*

gathering zone A term suggested for the area between the land surface and the water table.

gating In marine seismic profiling and echo sounding, a method of eliminating near-surface background noise from bottom and sub-bottom echoes.

gaudefroyite A mineral: $Ca_4Mn_{3-x}{}^{+3}(BO_3)_3(CO_3)(O,OH)_3$.

Gault European (Great Britain) stage: Lower Cretaceous (above Wealden). Equivalent to Albian and Aptian.

gault A stiff, firm, compact clay, or a heavy, thick clay soil; specif. the Gault, a Lower Cretaceous clay formation in Great Britain. Also spelled: *galt.*

Gause's principle In ecology, the statement that two identical species cannot coexist in the same area of space at the same time. Named after G.F. Gause (d. 1855), German geneticist. Var: *Gause's rule; Gause's axiom.*

gauss The cgs unit for magnetic induction (flux density), the magnetic field conventionally symbolized by B. The field one cm from a straight wire carrying 5 amps is one gauss.

gaussbergite An extrusive *leucitite* similar to *orendite* but with a glassy groundmass and containing phenocrysts of leucite, and of clinopyroxene and olivine in place of phlogopite. Its name is derived from Gaussberg volcano, Kaiser Wilhelm II Land, Antarctica. Not recommended usage.

Gaussian distribution A syn. of *normal distribution.* Named after Karl Friedrich Gauss (1777-1855), German mathematician.

Gauss projection Any of several conformal map projections used or developed by Karl Friedrich Gauss (1777-1855), German mathematician; esp. the "Gauss-Krüger projection", a special case of the *transverse Mercator projection* derived by the direct conformal representation of the spheroid on a plane.

gauteite A porphyritic hypabyssal rock, probably *trachyte* or *trachyandesite,* characterized by a bostonitic groundmass predominantly of plagioclase, more than in a typical *bostonite,* along with magnetite, clinopyroxene, hornblende, and biotite and with phenocrysts of hornblende, clinopyroxene, some biotite, and abundant plagioclase. Analcime is also a common constituent. The name, given by Hibsch in 1898, is for Gaute (now Kout), Czechoslovakia. Not recommended usage.

gaylussite A yellowish-white to gray mineral: $Na_2Ca(CO_3)_2 \cdot 5H_2O$.

GC *gas chromatography.*

geanticlinal n. The original, now obsolete, form of *geanticline.* adj. Pertaining to a geanticline.

geanticline (a) A mobile upwarping of the crust of the Earth, of regional extent. Ant: *geosyncline.* (b) More specifically, an anticlinal structure that develops in geosynclinal sediments, due to lateral compression.—Var: *geoanticline.*

gearksutite An earthy mineral: $CaAl(OH)F_4 \cdot H_2O$. It occurs with cryolite. Syn: *gearksite.*

gedanite A brittle, wine-yellow variety of amber with very little succinic acid.

Gedinnian European stage: Lower Devonian (above Ludlovian of Silurian, below Siegenian).

gedrite An aluminum-bearing variety of anthophyllite. A member of the amphibole group.

gedroitzite A clay mineral of the vermiculite group: $6(K,Na)_2O \cdot 5Al_2O_3 \cdot 14SiO_2 \cdot 12H_2O$. It is characteristic of many alkali soils from the Ukraine. Also spelled: *gedroizite.*

geest (a) Alluvial material that is not of recent origin lying on the surface. An example is the sandy region of the North Sea coast in Germany. (b) *saprolite.*

gehlenite A mineral of the melilite group: $Ca_2Al_2SiO_7$. It is isomorphous with akermanite. Syn: *velardenite.*

Geiger counter *Geiger-Müller counter.*

Geiger-Müller counter An instrument consisting of a *Geiger-Müller tube* plus a voltage source and the electronic equipment necessary to record the tube's electric pulses. Syn: *G-M counter; Geiger counter.*

Geiger-Müller tube A radiation detector consisting of a gas-filled tube with a cathode envelope and an axial wire anode. It functions by producing momentary current pulses caused by ionizing radiation. It is a part of the *Geiger-Müller counter.* Syn: *G-M tube.*

geikielite A bluish-black or brownish-black mineral: $MgTiO_3$. It is isomorphous with ilmenite, and often contains much iron. It is usually found in rolled pebbles.

GEK *geomagnetic electrokinetograph.*

gel (a) A translucent to transparent, semisolid, apparently homogeneous substance in a colloidal state, generally elastic and jellylike, offering little resistance to liquid diffusion, and containing a dispersion or network of fine particles that have coalesced to some degree. (b) A nonhomogeneous gelatinous precipitate; e.g. a coagel. (c) A liquified mud, which became firm and then reabsorbed most of the water released earlier.—A gel is in a more solid form than a *sol,* and can sustain limited shear stress.

gelation [chem] The formation of a *gel* from a *sol,* as by coagulation or by precipitation with an electrolyte.

gelation [ice] *congelation.*

geli- Of or pertaining to cold, frost action, or permafrost; e.g. *gelivation.*

Ge:Li detector *lithium-drifted germanium detector.*

gelifluction A syn. of *congelifluction.* It is commonly used as a modifying or combining term with *bench, lobe, sheet, slope,* and *stream* to indicate periglacial origin by soil flow. Also spelled: *gelifluxion.*

gelifraction *congelifraction.*

gelisol *frozen ground.*

gelisolifluction *congelifluction.*

gelite A name for opal (or chalcedony?) as a secondary deposit in rocks (Hey, 1962, p. 435).

geliturbation *congeliturbation.*

gelivation *congelifraction.*

gelivity The sensitivity or susceptibility of rock to congelifraction; the property of being readily split by frost (Hamelin & Clibbon, 1962, p.219).

gel mineral *mineraloid.*

gélose *ulmin.*

gélosic coal Coal that is rich in gélose or ulmin; a syn. of algal coal or *boghead coal.*

gelosite A microscopic constituent of torbanite, consisting of squashed, translucent, pale-yellow spheres of birefringent material (Dulhunty, 1939).

gem n. (a) A cut-and-polished stone that has intrinsic value and possesses the necessary beauty, durability, rarity, and size for use in jewelry as an ornament or for personal adornment; a jewel whose value is not derived from its setting. (b) An especially fine or superlative specimen (as compared to others cut from the same species), generally of superb color, unusual internal quality, and fine cut; e.g. a gem turquoise having a pure uniform blue of the highest possible intensity. In this usage, the meaning depends on the ethics and the range of qualities handled by the seller.—adj. Said of a rough diamond that has the necessary shape, purity, and color to allow it to be used for cutting and polishing into a diamond usable in jewelry. Only about 20% of all diamonds mined meet this requirement.

gem color The finest or most desirable color for a particular gem variety. Cf: *perfection color.*

gem crystal A crystal from which a gem can be cut.

gem gravel A gravel *placer* containing an appreciable concentration of gem minerals.

gemma An asexual propagule sometimes appearing as, but not homologous with, a vegetative bud (Lawrence, 1951, p.753).

gemmary (a) The science of gems. (b) A collection of gems; gems considered collectively. (c) A house or receptacle for gems and jewels.

gemmate adj. (a) Said of sculpture of pollen and spores consisting of more or less spherical projections. (b) Having gemmae.—v. To produce or propagate by a bud.

gem material Any rough material, either natural or artificial, that can be fashioned into a jewel.

gemmation Asexual reproduction involving the origination of a new organism as a localized area of growth on or within the body of the parent and subsequently differentiating into a new individual, as in some corals.

gemmiferous Producing or containing gems.

gem mineral Any mineral species that yields varieties with sufficient beauty and durability to be classed as gemstones.

gemmology British var. of *gemology.*

gemmule An internal resistant asexual reproductive body of a sponge, consisting of a mass of archaeocytes charged with reserves and enclosed in a noncellular protective envelope.

gemmy Having the characteristics (such as hardness, brilliance, and color) desired in a gemstone.

gemologist One who has successfully completed recognized courses in *gemology* and has demonstrated his competence in identification and evaluation of gem materials.

gemology The science and study of gemstones, including their source, description, origin, identification, grading, and appraisal. British spelling: *gemmology.*

gemstone Any mineral, rock, or other natural material (including organic materials such as pearl, amber, jet, shell, ivory, and coral) that, when cut and polished, has sufficient beauty and durability for use as a personal adornment or other ornament.

gem variety The variety of a mineral species that yields gemstones.

gena A *cheek* of a trilobite. Pl: *genae.*

genal angle The posterior lateral corner of the cephalon of a trilobite (typically terminating in a genal spine, but may be rounded), or of the prosoma of a merostome.

genal spine A spine extending backward from the genal angle of the cephalon of a trilobite, produced on the outer posterior margin of the free cheek.

gendarme A sharp rock pinnacle on an arête, such as will retard or prevent progress along the crest of the arête; it is less pointed and more towerlike than an *aiguille,* and is commonly found in the Alps. Pron: zhan-*darm.* Etymol: French, "policeman".

gene The fundamental unit governing the transmission and development or determination of hereditary characteristics. Genes occur in a linear sequence on the chromosomes of a cell nucleus, and are now thought to originate in the deoxyribonucleic acid (DNA) component in the chromosomes.

gene complex The system comprising all the interacting genetic factors of an organism.

genera The plural of *genus.*

general age equation The relationship between radioactive decay and geological time which takes into account the presence of daughter atoms initially. Mathematically expressed it is $t = 1/\lambda \ln(D_t - D_o/P + 1)$, where t is time, λ is the decay constant for the isotope in question, D_t is the number of daughter atoms at time (t), D_o is the number of daughter atoms initially present, and P is the number of parent atoms at time (t). Cf: *age equation.*

general base level *ultimate base level.*

general form The crystal form in each crystal class that has different intercepts on each crystal axis, and has the largest number of equivalent crystal faces for the symmetry present. The form displays and is characteristic of the point group symmetry; other forms may display more symmetry.

generalist species *opportunistic species.*

general-purpose geologic map A map designed to provide a large amount of geological information for many purposes (land-use planning, petroleum geology, highway location, etc.). In the U.S. its scale is usually 1/24,000 and not smaller than 1/62,500.

generating area *fetch.*

generation All the crystals of the same mineral species that appear to have crystallized at essentially the same time; e.g., if there are olivine phenocrysts in a groundmass containing olivine, there are said to be two "generations" of olivine.

generic Pertaining to a genus.

genetic Pertaining to relationships due to a common origin, or to features involving *genes* or *gene complexes.*

genetic drift Gradual change with time in the genetic composition of a continuing population resulting from the elimination of some genetic features and the appearance of others, and appearing to be unrelated to the environmental benefits or detriments of the genes involved.

genetic pan A natural *pan [soil]* of low permeability, with a high concentration of small particles, and differing in certain physical and chemical properties from the soil immediately above or below it (SSSA, 1975, p. 12); e.g. *claypan; fragipan.* Cf: *pressure pan.*

genetic physiography *geomorphogeny.*

genetics The science that deals with the materials and processes of inheritable characteristics or features from generation to generation.

genetic type A sedimentary body representing a complex of genetically related facies formed in the same environment (on land or in the sea) and mostly as the effect of a single leading process (Krasheninnikov, 1964, p.1245); e.g. alluvial deposits, deltaic deposits, lagoonal deposits, or marine deposits. The concept of genetic types, long established in Russian geology, was introduced by Pavlov (1889) to demonstrate the diversity of continental deposits according to their origin.

genicular spine A spine originating from the geniculum of a graptolite.

geniculate Bent abruptly at an angle; specif. said of a brachiopod shell characterized by an abrupt and more or less persistent change in direction of valve growth, producing an angular bend in lateral profile.

geniculate twin A type of crystal twin that bends abruptly, e.g. crystals of rutile that are twinned on the second-order pyramid (101). Syn: *elbow twin.*

geniculum A small knee-shaped structure or abrupt bend in an organism; e.g. a distinct change in the direction of growth of the free ventral wall of a graptolithine theca, or between two successive segments of erect jointed coralline algae. Pl: *genicula.*

genital plate (a) One of the inner circlet of primordial plates of the *apical system* of an echinoid, arranged in an interradial position and usually perforated by one or more pores for discharge of reproductive products. Cf: *ocular plate.* (b) One of the circlet of five plates on the aboral surface immediately around the centrale of an asterozoan.

genobenthos All terrestrial organisms.

genocline A *cline* caused by hybridization between adjacent populations that are genetically distinct.

genotype (a) *type species.* (b) In genetics, the genetic constitution of an organism or a species in contrast to its observable physical characteristics. See also: *biotype.*

genthelvite A mineral: $(Zn,Fe,Mn)_4Be_3(SiO_4)_3S$. It is the zinc end member, isomorphous with helvite and danalite.

genthite A soft, amorphous, pale-green or yellowish mixture of hydrous nickel-magnesium silicates, representing a mineral near: $(Ni,Mg)_4Si_3O_{10}\cdot 6H_2O$. See also: *garnierite.*

gentle fold A fold with an interlimb angle between 120° and 180° (Fleuty, 1964, p. 470).

gentnerite A mineral: $Cu_8Fe_3Cr_{11}S_{18}$.

genus A category in the hierarchy of plant and animal classification intermediate in rank between *family* and *species.* Adj: *generic.* Plural: *genera.* Cf: *subgenus.*

genus zone The *taxon-range-zone* of a genus (ISG, 1976, p. 54).

geo (a) A term used in northern Scotland for a long deep narrow coastal inlet or cove, walled in by steep, rocky cliffs, and formed by marine erosion of the cliffs along a line of weakness, esp. in a well-jointed rock, as in the Old Red Sandstone of Caithness and Orkney. Syn: *gio; goe; gja.* (b) Broadly, an opening to the sea.—Etymol: Scandinavian, akin to Old Norse *gja,* "chasm".

geo- *A prefix meaning "Earth".*

geoacoustics The study of the sounds of the Earth, e.g. those of a volcanic eruption.

geoanticline Var. of *geanticline.*

geoastronomy (a) *astrogeology.* (b) *geocosmology.*

geobarometer A mineral pair the composition of which is pressure-dependent.

geobarometry Any method, such as the interpretation of pressure-sensitive chemical or mineral reactions, for the direct or indirect determination of the pressure conditions under which a rock or mineral formed. The term is analogous to *geothermometry.*

geobasin A structure of geosynclinal type but without folding of its sedimentary filling (Rich, 1938).

geobiology The study of the biosphere, esp. through geologic time. Cf: *biogeology; paleobiology; paleontology.*

geobios That area of the Earth occupied by terrestrial plants and animals. Cf: *hydrobios.*

geobleme A term used by Bucher (1963) for a cryptoexplosion structure caused from within the Earth. Cf: *astrobleme.*

geobotanical prospecting The visual study of plants, their morphology, and their distribution as indicators of such things as soil composition and depth, bedrock lithology, the possibility of ore bodies, and climatic and groundwater conditions. Cf: *biogeochemical prospecting.* See also: *botanical anomaly.*

geocentric Referring to the center of mass of the Earth, in defining coordinate systems. When combined with latitude, as in "geocentric latitude", it is the angle at the center of the Earth between the plane of the *celestial equator* and a line to a point on the surface of the Earth.

geocentric horizon The plane passing through the center of the

Earth, parallel to the apparent horizon.

geocentric latitude (a) The *latitude* or angle at the center of the Earth between the plane of the celestial equator and a line (radius) to a given point on the Earth's surface. It would be identical with *geodetic latitude* only for a truly spherical Earth. Symbol: ψ. (b) The celestial latitude of a body based on or as seen from the Earth's center.

geocentric longitude (a) *geodetic longitude.* (b) The celestial longitude of a body based on or as seen from the Earth's center.

geocerain *geocerite.*

geocerite A white, flaky, waxlike resin of approximate composition: $C_{27}H_{53}O_2$. It is found in brown coal. Syn: *geocerain; geocerin.*

geochemical analysis The application of chemical analysis to geological problems.

geochemical anomaly A concentration of one or more elements in rock, soil, sediment, vegetation, or water that is markedly higher than background. The term may also be applied to hydrocarbon concentrations in soils.

geochemical balance The study of the proportional distribution, and the rates of migration, in the global fractionation of a particular element, mineral, or compound; e.g. the distribution of quartz in igneous rocks, its liberation by weathering, and its redistribution to sediments and, in solution, to terrestrial waters and the oceans.

geochemical cycle The sequence of stages in the exchange of elements among major geochemical reservoirs. Rankama and Sahama (1950) distinguish a major cycle, proceeding from magma to igneous rocks to sediment to sedimentary rocks to metamorphic rocks, and possibly through migmatites and back to magma; and a minor or exogenic cycle proceeding from sediment to sedimentary rocks to weathered material and back to sediments again.

geochemical exploration The search for economic mineral deposits or petroleum by detection of abnormal concentrations of elements or hydrocarbons in surficial materials or organisms, usually accomplished by instrumental, spot-test, or "quickie" techniques that may be applied in the field. Syn: *geochemical prospecting.*

geochemical facies Any areal geological entity that is distinguishable on the basis of trace-element composition, radioactivity, or other geochemical property. A lithofacies could be defined chemically, but a geochemical facies is delimited by features that remain imperceptible lithologically.

geochemical fence In a *fence diagram [geochem],* a boundary between discrete fields that represent different phases.

geochemical prospecting *geochemical exploration.*

geochemical survey The mapping of geochemical facies.

geochemistry As defined by Goldschmidt (1954, p. 1), the study of the distribution and amounts of the chemical elements in minerals, ores, rocks, soils, water, and the atmosphere, and the study of the circulation of the elements in nature, on the basis of the properties of their atoms and ions; also, the study of the distribution and abundance of isotopes, including problems of nuclear frequency and stability in the universe. A major concern of geochemistry is the synoptic evaluation of the abundances of the elements in the Earth's crust and in major classes of rocks and minerals.

geochron An interval of geologic time corresponding to a *rock-stratigraphic unit.* It may, like the age of that unit, vary from place to place.

geochrone A standard unit of geologic time, now obsolete. It was proposed by Williams in 1893 and was set equal to the duration of the Eocene.

geochronic An obsolete syn. of *geochronologic.*

geochronologic Pertaining or relating to *geochronology.* Obsolete syn: *geochronic.*

geochronologic interval The time span between two geologic events (ISG, 1976, p. 14). Syn: *interval [stratig].*

geochronologic unit *geologic-time unit.*

geochronology Study of time in relationship to the history of the Earth, esp. by the *absolute age determination* and *relative dating* systems developed for this purpose. Cf: *geochronometry.* See also: *chronology.* Syn: *geologic chronology; geochrony.*

geochronometer A physical feature, material, or element whose formation, alteration, or destruction can be calibrated or related to a known interval of time. See also: *radioactive clock.*

geochronometry Measurement of *geologic time* by geochronologic methods, esp. radiometric dating. Cf: *geochronology.* See also: *fossil geochronometry.*

geochrony An obsolete syn. of *geochronology.*

geocline A *cline* related to a geographic transition.

geocol A term introduced by Taylor (1951, p. 614) for a broad, low gap between highlands; a *gate.*

geocosmogony The science of the origin of the Earth. See also: *geogony.*

geocosmology The science that deals with the origin and geologic history of the Earth, including its planetary attributes (shape, mass, density, physical fields, rotation, location of poles); the influence of the solar system, the galaxy, and the universe on the geologic development of the Earth; and the material interaction (exchange) between the Earth and the universe (Galkiewicz, 1968). Syn: *geoastronomy.*

geocronite A lead-gray mineral: Pb_5SbAsS_8.

geocryology The study of ice and snow on the Earth, esp. the study of permafrost. See also: *cryolithology.*

geocyclic (a) Pertaining to or illustrating the rotation of the Earth. (b) Circling the Earth periodically.

geocyclicity The quality or state of being geocyclic, as in defining all events of cyclic nature recorded in sedimentary rocks for which a specific causal mechanism could not be inferred.

geode (a) A hollow or partly hollow and globular or subspherical body, from 2.5 cm to 30 cm or more in diameter, found in certain limestone beds and rarely in shales; it is characterized by a thin and sometimes incomplete outermost primary layer of dense chalcedony, by a cavity that is partly filled by an inner drusy lining of inward-projecting crystals (often perfectly formed and usually of quartz or calcite and sometimes of barite, celestite, and various sulfides) deposited from solution on the cavity walls, and by evidences of growth by expansion in the cavities of fossils or along fracture surfaces of shells. Unlike a *druse,* a geode is separable (by weathering) as a discrete nodule or concretion from the rock in which it occurs and its inner crystals are not of the same minerals as those of the enclosing rock. A geode tends to be slightly flattened with its equatorial plane parallel to the bedding. (b) The crystal-lined cavity in a geode. (c) A term applied to a rock cavity and its lining of crystals that is not separable as a discrete nodule from the enclosing rock.—Cf: *vug.*

geodepression Haarmann's term for a long, narrow depression on the scale of a geosyncline, but not necessarily filled with sediments (Glaessner & Teichert, 1947, p. 476).

geodesic n. A line of shortest distance between any two points on a mathematically defined surface.

geodesy (a) The science concerned with the determination of the size and shape of the Earth and the precise location of points on its surface. (b) The determination of the gravitational field of the Earth and the study of temporal variations such as Earth tides, polar motion, and rotation of the Earth.

geodetic azimuth The angle between the tangent to the meridian at a given point and the tangent to the geodesic line from the given point to the azimuth point. See also: *Laplace azimuth.*

geodetic coordinates Quantities defining the horizontal position of a point on an ellipsoid of reference with respect to a specific geodetic *datum,* usually expressed as latitude and longitude. These may be referred to as *geodetic positions* or *geographic coordinates.* The elevation of a point is also a geodetic coordinate and may be referred to as a height above sea level.

geodetic datum *datum [geodesy].*

geodetic engineering *geodetic surveying.*

geodetic equator The circle on the spheroid midway between its poles of revolution and connecting points of zero degrees geodetic latitude. Its plane, which contains the center of the spheroid and is perpendicular to its axis, cuts the celestial sphere in a line coinciding with the celestial equator because the axis of the spheroid is by definition parallel to the axis of rotation of the Earth. Cf: *astronomic equator.*

geodetic latitude The *latitude* or angle which the normal at a point on the spheroid makes with the plane of the geodetic equator (the equatorial plane of the spheroid). It is equivalent to the *astronomic latitude* corrected for station error and to true *geographic latitude.* Latitude as shown on topographic maps is geodetic latitude. Symbol: ϕ. Cf: *geocentric latitude.*

geodetic leveling Spirit leveling of a high order of accuracy, usually extended over large areas (and with proper applications of orthometric corrections), to furnish accurate vertical control for surveying and mapping operations.

geodetic longitude The *longitude* or angle between the plane of the geodetic meridian and the plane of an arbitrarily chosen initial meridian (generally the Greenwich meridian). It is equivalent to the *astronomic longitude* corrected for station error and to true *geographic longitude.* Symbol: λ. Syn: *geocentric longitude.*

geodetic meridian A line (ellipse) on the spheroid having the same geodetic longitude at every point. Syn: *geographic meridian.*

geodetic parallel A line or circle on the spheroid having the same geodetic latitude at every point; a small circle whose plane is parallel with the plane of the geodetic equator.

geodetic position The latitude and longitude of a point on the Earth's surface, determined by *triangulation, trilateration,* or traverse, and referred to an adopted ellipsoid of reference. Cf: *geodetic coordinates.*

geodetic sea level *mean sea level.*

geodetic survey (a) A survey of a large land area accomplished by the processes of geodetic surveying and used for the precise location of basic points suitable for controlling other surveys. (b) An organization engaged in making geodetic surveys.

geodetic surveying Surveying in which account is taken of the figure and size of the Earth and corrections are made for earth curvature; the applied science of geodesy. It is used where the areas or distances involved are so great that results of desired accuracy and precision cannot be obtained by *plane surveying.* Syn: *geodetic engineering.*

geodic Pertaining to a geode; e.g. a "geodic cavity or vein" in which iron sulfide may precipitate.

geodiferous Containing geodes.

Geodimeter Trade name of an electronic-optical device that measures ground distances precisely by electronic timing and phase comparison of modulated light waves that travel from a master unit to a reflector and return to a light-sensitive tube where an electric current is set up. It is normally used at night and is effective with first-order accuracy up to distances of 5-40 km (3-25 miles). Etymol: acronym for *geo*detic-*di*stance *meter*.

geodynamic Pertaining to physical processes within the Earth as they affect the features of the crust.

geodynamics That branch of science which deals with the forces and processes of the interior of the Earth.

Geodynamics Project An international program of research (1971-1977) on the dynamics and dynamic history of the Earth with emphasis on deep-seated geological phenomena, esp. movements and deformations of the lithosphere. The program is coordinated by the Inter-Union Commission on Geodynamics which was established by the International Council for Scientific Unions (ICSU) at the request of the International Union of Geodesy and Geophysics (IUGG) and the International Union of Geological Sciences (IUGS).

geoecology *environmental geology.*

geoelectricity The Earth's natural electric fields and phenomena. It is closely related to geomagnetism. See also: *telluric current.*

geoevolutionism A term introduced by Goode (1969) for the "idea of evolution of geologic processes" in which some geologic changes foster new processes that in turn bring about new changes.

geofact A supposed *artifact* that may actually be of geological origin (Haynes, 1973).

geoflex An early syn. of *orocline.*

geogeny *geogony.*

geognosy An 18th-century term for a science accounting for the origin, distribution, and sequence of minerals and rocks in the Earth's crust. It developed out of mineralogy and was distinct from *oryctology.* The term was superseded by *geology* as early ideas were abandoned. It has become restricted to absolute knowledge of the Earth, as distinct from the theoretical and speculative reasoning of geology.

geogony The science or theory of the formation of the Earth, esp. a speculative study of its origin. Syn: *geogeny.* See also: *geocosmogony.*

geogram A term used by Marr (1905, p.lxii) for a hypothetical geologic column, connoting principally lateral variations in lithology and in organic assemblages that could be traced in a deposit over an area of any width up to that of the circumference of the Earth.

geographic Pertaining or relating to geography.

geographical position An inclusive term used to designate both geodetic and astronomic coordinates.

geographic center The point on which an area on the Earth's surface would balance if it were a plate of uniform thickness (i.e. the center of gravity of such a plate). The geographic center of the conterminous U.S. is in the eastern part of Smith County, Kansas (lat. 39°50′N, long. 98°35′W); the geographic center of North America is in Pierce County, N.D., a few miles west of Devils Lake.

geographic coordinates *geodetic coordinates.*

geographic cycle *cycle of erosion.*

geographic grid A system of parallels and meridians used to locate points on the Earth's surface.

geographic horizon The boundary line of that part of the Earth's surface visible from a given point of view; the *apparent horizon.*

geographic latitude A general term, applying alike to *astronomic latitude* and *geodetic latitude.*

geographic longitude A general term, applying alike to *astronomic longitude* and *geodetic longitude.*

geographic meridian *geodetic meridian.*

geographic name The proper name or expression by which a particular geographic entity is known; esp. the name of a river, town, or other natural or artificial feature at or near which a rock-stratigraphic unit is typically developed.

geographic north *true north.*

geographic parallel A general term for a line or circle on the Earth's surface having the same latitude at every point. It is applied to astronomic parallel and geodetic parallel.

geographic pole Either of the two *poles* or points of intersection of the Earth's surface and its axis of rotation; specif. *North Pole* and *South Pole.* Syn: *terrestrial pole.*

geographic province An extensive region all parts of which are characterized by similar geographic features. Cf: *physiographic province.*

geographic race Part of a biologic species coinciding with a particular geographic area and probably resulting from specific characteristics of the environment; a geographic subspecies or variety.

geography The study of all aspects of the Earth's surface including its natural and political divisions, the distribution and differentiation of areas and, often, man in relationship to his environment. See also: *physical geography.*

geohistory *geologic history.*

geohydrologic unit An aquifer, a confining unit (aquiclude or aquitard), or a combination of aquifers and confining units, comprising "a framework for a reasonably distinct hydraulic system" (Maxey, 1964, p. 126). Cf: *hydrostratigraphic unit.*

geohydrology A term, often used interchangeably with *hydrogeology,* referring to the hydrologic or flow characteristics of subsurface waters. Also used in reference to all hydrology on the Earth without restriction to geologic aspects (Stringfield, 1966, p. 3). The term was first suggested in 1942 by Meinzer for the branch of hydrology dealing with subterranean waters.

geoid The *figure of the Earth* considered as a sea-level surface extended continuously through the continents. It is a theoretically continuous surface that is perpendicular at every point to the direction of gravity (the plumb line). It is the surface of reference for astronomical observations and for geodetic leveling. See also: *compensated geoid; datum [geodesy].*

geoidal horizon A circle on the celestial sphere formed by the intersection of the celestial sphere and a plane tangent to the sea-level surface of the Earth at the zenith-nadir line.

geoidal map The contoured development of the separations of the geoid and a specified reference ellipsoid.

geoidal section A section profile of the geoid with relation to the reference ellipsoid.

geoidal separation The distance between the reference ellipsoid and the geoid. Cf: *datum [geodesy].*

geoisotherm *isogeotherm.*

geolith *lithostratigraphic unit.*

geologese (a) Literary style or jargon peculiar to geologists. (b) Geological language that is "progressing rapidly" toward the construction of "sentences in such a way that their meaning is not apparent on first reading" (Vanserg, 1952, p.221).

geologic *geological.*

geologic age (a) The age of a fossil organism or of a particular geologic event or feature referred to the geologic time scale and expressed in terms either of years or centuries (*absolute age*) or of comparison with the immediate surroundings (*relative age*); an age datable by geologic methods. (b) The term is also used to emphasize the long-past periods of time in geologic history, as distinct from present-day or historic times. See also: *age [geochron].*

geologic-age determination Determination of the *relative age* or *absolute age* of a geologic event or feature.

geological Pertaining to or related to geology. The choice between this term and *geologic* is optional, and may be made according to the sound of a spoken phrase or sentence. *Geological* is generally preferred in the names of surveys and societies, and in English and

Canadian usage.

geological oceanography That aspect of the study of the ocean that deals specifically with the ocean floor and the ocean—continent border, including submarine relief features, the geochemistry and petrology of the sediments and rocks of the ocean floor, and the influence of seawater and waves on the ocean bottom and its materials. Syn: *marine geology; submarine geology.*

geological ore *possible ore.*

geological science Any of the subdisciplinary specialties that are part of the science of *geology*; e.g. geophysics, geochemistry, paleontology, petrology, etc. The term is commonly used in the plural. See also: *geoscience.* Cf: *Earth science.*

geologic chronology *geochronology.*

geologic climate *paleoclimate.*

geologic-climate unit A term used by the American Commission on Stratigraphic Nomenclature (1961, art.39) for "an inferred widespread climatic episode defined from a subdivision of Quaternary rocks"; e.g. glaciation, interglaciation, stade, and interstade. It is strictly not a stratigraphic unit. The different stratigraphic boundaries that define the limits of the geologic-climate unit in different latitudes are not likely to be isochronous. Syn: *climate-stratigraphic unit.*

geologic column (a) A composite diagram that shows in a single column the subdivisions of part or all of geologic time or the sequence of stratigraphic units of a given locality or region (the oldest at the bottom and the youngest at the top, with dips adjusted to the horizontal) so arranged as to indicate their relations to the subdivisions of geologic time and their relative positions to each other. See also: *columnar section.* (b) The vertical or chronologic arrangement or sequence of rock units portrayed in a geologic column. See also: *geologic section.*—Syn: *stratigraphic column; column [stratig].*

geologic cycle *orogenic cycle.*

geologic engineering *engineering geology.*

geologic erosion A syn. of *normal erosion,* or erosion caused naturally by geologic processes.

geologic formation *formation [stratig].*

geologic hazard A naturally occurring or man-made geologic condition or phenomenon that presents a risk or is a potential danger to life and property. Examples include landsliding, flooding, earthquakes, ground subsidence, coastal and beach erosion, faulting, dam leakage and failure, mining disasters, pollution and waste disposal, and seawater intrusion.

geologic high An oil-field term for a structure on which rocks occur at a higher position than in the surrounding area. Cf: *high [struc geol].*

geologic history The history of the Earth and its inhabitants throughout geologic time, often considered for a certain area or duration of time, or in certain aspects. It comprises all chemical, physical, and biologic conditions that have existed on and in the Earth, all processes that have operated to make and modify these conditions, and all events that have affected any part of the Earth, including its inhabitants, from the beginning of the planet to the present, and is circumscribed "in no way by what we think we know about it" (Moore, 1949, p.3). Syn: *geohistory; Earth history.*

geologic horizon *horizon [geol].*

geologic low An oil-field term for a structure on which rocks occur at a lower position than in the surrounding area. Cf: *low [struc geol].*

geologic map A map on which is recorded geologic information, such as the distribution, nature, and age relationships of rock units (surficial deposits may or may not be mapped separately), and the occurrence of structural features (folds, faults, joints), mineral deposits, and fossil localities. It may indicate geologic structure by means of formational outcrop patterns, by conventional symbols giving the direction and amount of dip at certain points, or by structure-contour lines.

geologic name The name of a *formation [stratig]* or other lithostratigraphic unit.

geologic norm The condition resulting from *normal erosion* of the land, undisturbed by the activity of man and his agents.

geologic province An extensive region characterized throughout by similar geologic history or by similar structural, petrographic, or physiographic features. Cf: *physiographic province.*

geologic range *stratigraphic range.*

geologic record The "documents" or "archives" of the history of the Earth, represented by bedrock, regolith, and the Earth's morphology; the rocks and the accessible solid part of the Earth. Also, the geologic history based on inferences from a study of the geologic record. See also: *stratigraphic record.*

geologic section (a) Any sequence of rock units found in a given region either at the surface or below it (as in a drilled well or mine shaft); a local *geologic column.* Syn: *stratigraphic section.* (b) *section [geol].*

geologic thermometer *geothermometer.*

geologic time The period of time dealt with by historical geology, or the time extending from the end of the formative period of the Earth as a separate planetary body to the beginning of written or human history; the part of the Earth's history that is represented by and recorded in the succession of rocks. The term implies extremely long duration or remoteness in the past, although no precise limits can be set. See also: *time (a).*

geologic time scale An arbitrary chronologic arrangement or sequence of geologic events, used as a measure of the relative or absolute duration or age of any part of geologic time, and usually presented in the form of a chart showing the names of the various rock-stratigraphic, time-stratigraphic, or geologic-time units, as currently understood; e.g. the geologic time scales published by Holmes (1959), Kulp (1961), and Harland et al. (1964). See also: *atomic time scale; relative time scale; biologic time scale.* Syn: *time scale.*

geologic-time unit A span of continuous time in geologic history, during which a corresponding *time-stratigraphic unit* was formed; a division of time "distinguished on the basis of the rock record, particularly as expressed by time-stratigraphic units" (ACSN, 1961, art. 36). It is not a material unit or body of strata, and therefore it is strictly not a stratigraphic unit. Geologic-time units in order of decreasing magnitude are eon, era, period, epoch, and age. The names of periods, epochs, and ages are identical with those of the corresponding time-stratigraphic units; the names of eras and eons are independently formed. Cf: *biochronologic unit.* Syn: *geochronologic unit; time unit.*

geologist One who is trained in and works in any of the geological sciences.

geologize v. To participate in or talk about geology; to practice geology.

geology The study of the planet Earth—the materials of which it is made, the processes that act on these materials, the products formed, and the history of the planet and its life forms since its origin. Geology considers the physical forces that act on the Earth, the chemistry of its constituent materials, and the biology of its past inhabitants as revealed by fossils. Clues on the origin of the planet are sought in a study of the Moon and other extraterrestrial bodies. The knowledge thus obtained is placed in the service of man—to aid in discovery of minerals and fuels of value in the Earth's crust, to identify geologically stable sites for major structures, and to provide foreknowledge of some of the dangers associated with the mobile forces of a dynamic Earth. See also: *geological science; Earth science; geoscience; historical geology; physical geology.*

geomagnetic axis The axis of the dipole magnetic field most closely approximating the actual magnetic field of the Earth.

geomagnetic electrokinetograph A shipboard instrument containing electrodes and designed to measure ocean-surface currents in depths greater than 100 fathoms by recording the electric current generated by the movement of a high-conductivity electrolyte (ocean water) through the Earth's magnetic field. Abbrev: GEK.

geomagnetic equator The great circle of the Earth whose plane is perpendicular to the geomagnetic axis; the line connecting points of zero geomagnetic latitude.

geomagnetic field The Earth's magnetic field.

geomagnetic latitude The magnetic latitude that a location would have if the Earth's field were replaced by the dipole field most closely approximating it. It is latitude reckoned relative to the geomagnetic axis instead of to the Earth's rotational axis. Cf: *geomagnetic longitude.*

geomagnetic longitude Longitude reckoned around the geomagnetic axis instead of around the Earth's rotational axis. The prime geomagnetic meridian is the one extending to the geographic south from the north geomagnetic pole. Cf: *geomagnetic latitude.*

geomagnetic meridian A great circle of the Earth through the geomagnetic poles.

geomagnetic polarity epoch *polarity epoch.*

geomagnetic polarity event *polarity event.*

geomagnetic polarity reversal *geomagnetic reversal.*

geomagnetic-polarity time scale A chronology based on counting reversals of the Earth's magnetic field. From youngest to oldest, named epochs are Brunhes normal (now to 0.7 m.y.), Matuyama reversed (0.7 my to 2.5 m.y.), Gauss normal (2.5 m.y. to 3.3 m.y.) and Gilbert reversed (3.3 m.y. to 5 m.y.). See also: *polarity epoch.*

geomagnetic poles The points of emergence at the Earth's surface of the axis of the geocentric magnetic dipole that most closely approximates the Earth's magnetic field. Partial syn: *magnetic poles.*

geomagnetic reversal A change of the Earth's magnetic field between *normal polarity* and *reversed polarity*. Syn: *field reversal; geomagnetic polarity reversal; magnetic polarity reversal; magnetic reversal; reversal; polarity reversal.*

geomagnetic secular variation *secular variation.*

geomagnetism The magnetic phenomena exhibited by the Earth and its atmosphere; also, the study of such phenomena. Syn: *terrestrial magnetism.*

geomagnetochronology Establishment of a time scale of Earth history on the basis of *paleomagnetism.*

geomalism Equal lateral growth of an organism in response to gravitational force. Cf: *geotaxis; geotropism.*

geomathematics All applications of mathematics to studies of the Earth's crust. See also: *mathematical geology.*

geomechanics A branch of geology that embraces the fundamentals of structural geology and a knowledge of the response, or strain, of natural materials to deformation or changes caused by the application of stress and/or strain energy.

geometric factor The factor by which the ratio of voltage to current is multiplied in order to obtain the apparent resistivity. It depends on electrode array and spacing.

geometric grade scale A *grade scale* having a constant ratio between size classes; e.g. the Wentworth grade scale, each size class of which differs from its predecessor by the constant ratio 1/2.

geometric horizon A term originally applied to the *celestial horizon,* but now more commonly to the intersection of the celestial sphere and an infinite number of straight lines tangent to the Earth's surface and radiating from the eye of the observer. It would coincide with the *apparent horizon* if there were no terrestrial refraction.

geometric mean The *n*th root of the product of the values of *n* positive numbers; the antilogarithm of the mean of the logarithms of individual values. See also: *mean; harmonic mean.*

geometric mean diameter An expression of the average particle size of a sediment or rock, obtained by taking the antilogarithm of the *phi mean diameter;* the diameter equivalent of the arithmetic mean of the logarithmic frequency distribution.

geometric projection *perspective projection.*

geometrics Measurements of the Earth through space and time.

geometric sounding That form of electromagnetic *sounding* in which separation between transmitting and receiving coils is the variable. Cf: *parametric sounding.*

geometry number A dimensionless constant representing the ratio of the product of maximum basin relief and drainage density within a given drainage basin to the tangent of the stream gradient; it summarizes the essentials of landform geometry, particularly the relations between planimetric and relief aspects of a drainage basin (Strahler, 1958, p.287 & 295). Symbol: N_G.

geomonocline A broad flank of a geosyncline, the beds of which have a uniform dip; a unilateral, marginal geosyncline.

geomorphic (a) Pertaining to the form of the Earth or of its surface features; e.g. a *geomorphic* province. (b) Pertaining to geomorphology; geomorphologic.

geomorphic cycle *cycle of erosion.*

geomorphic geology *geomorphology.*

geomorphogeny The part of geomorphology that deals with the origin, development, and changes of the Earth's surface features or landforms. Cf: *geomorphography.* Syn: *genetic physiography.*

geomorphography The part of geomorphology that deals with the description of the Earth's surface features or landforms. Cf: *geomorphogeny.*

geomorphology (a) The science that treats the general configuration of the Earth's surface; specif. the study of the classification, description, nature, origin, and development of present landforms and their relationships to underlying structures, and of the history of geologic changes as recorded by these surface features. The term is esp. applied to the genetic interpretation of landforms, but has also been restricted to features produced only by erosion or deposition. The term was applied widely in Europe before it was used in the U.S., where it has come to replace the term *physiography* and is usually considered a branch of geology; in Great Britain, it is usually regarded as a branch of geography. See also: *physical geography.* Syn: *physiographic geology; geomorphic geology; geomorphy.* (b) Strictly, any study that deals with the form of the Earth (including geodesy, and structural and dynamic geology). This usage is more common in Europe, where the term has even been applied broadly to the science of the Earth. (c) The features dealt with in, or a treatise on, geomorphology; e.g. the *geomorphology* of Texas.

geomorphy (a) *geomorphology.* (b) A syn. of *topography* "in the broader application of that term" (Lawson, 1894, p. 241).

geomyricite A white, waxy resin of approximate composition: $C_{32}H_{62}O_2$. It is found in brown coal. Syn: *geomyricin.*

geomythology "The geological application of euhemerism" (Vitaliano, 1968; 1973)—i.e. of the theory that myths may be interpreted as accounts of historical personages and events. Geomythology was first used for myths or legends embodying the memory of some real geologic event, but has been expanded to include any folklore or myths having a geological basis.

geonomy A term variously recommended as a synonym for *geology,* as the science of the dynamic Earth, as the science concerned exclusively with the physical forces relating to the Earth, and to denote the study of the Earth's upper mantle. The word has not been given broad recognition or acceptance, and the variety of proposed definitions suggests that it should be used with care, if at all.

geop A surface within the gravity field of the Earth in which all points have equal and constant *geopotential.* Syn: *geopotential surface.*

geopetal Pertaining to any rock feature that indicates the relation of top to bottom at the time of formation of the rock; e.g. a "geopetal fabric", the internal structure or organization that indicates original orientation of a stratified rock, such as cross-bedding or grains on a boundary surface. Term introduced by Sander (1936, p.31; 1951, p.2).

geophagous Said of an organism that feeds on soil.

geophone A *seismic detector* that produces a voltage proportional to the displacement, velocity, or acceleration of ground motion, within a limited frequency range. Syn: *jug; seismometer; pickup.* Cf: *seismograph; transducer.*

geophysical exploration The use of geophysical techniques, e.g. electric, gravity, magnetic, seismic, or thermal, in the search for economically valuable hydrocarbons, mineral deposits, or water supplies, or to gather information for engineering projects. Syn: *applied geophysics; geophysical prospecting.*

geophysical log *well log.*

geophysical prospecting *geophysical exploration.*

geophysical survey The use of one or more geophysical techniques in *geophysical exploration,* such as earth currents, electrical, gravity, heat flow, magnetic, radioactivity, and seismic. Geochemical techniques may sometimes be included.

geophysicist One who studies the physical properties of the Earth, or applies physical measurements to geological problems; a specialist in *geophysics.* A geophysicist may also study the Moon and the other planets.

geophysics Study of the Earth by quantitative physical methods. Basic divisions include solid-earth geophysics, physics of the atmosphere and hydrosphere, and solar-terrestrial physics. There are numerous specialties within the field, e.g. seismology, tectonophysics, engineering geophysics. The term is sometimes used to include instrumental study of the Moon and planets.

geophysiography The synthesis of all knowledge available about the Earth; "the combination...of relevant parts of geophysics, geochemistry, geomorphology, and geoecology" (Strøm, 1966, p. 8). Apparently a unique usage.

geopiezometry Measurement of pressure and compressibility of rocks in metamorphism (Chinner, 1966).

geoplanetology A term proposed by Nayak (1970, p. 1279) for the study of "geological and related aspects of the Moon and other planets which are likely to be investigated in the future".

geopolar Pertaining or relating to a pole of the Earth.

geopolitics A pejorative term suggesting organizational manipulation within government, society, or university circles dealing with earth science.

geopotential The potential energy of a unit mass relative to sea level, numerically equal to the work that would be done in lifting the unit mass from sea level to the height at which the mass is located. It is commonly expressed in terms of *dynamic height.* Cf:

disturbing potential; spheropotential.

geopotential height *dynamic height.*

geopotential number The numerical value, C, assigned to a given *geop* or geopotential surface when expressed in geopotential units (g.p.u.) where 1 g.p.u. = 1 m x 1 kgal. Since g = 1.98 k gal, C = gH = 0.98 H. Thus the geopotential number in g.p.u. is almost equal to the height above sea level in meters. Geopotential numbers are generally preferred to *dynamic heights* because the physical meaning is more correctly expressed.

geopotential surface *geop.*

geopressured aquifer A term used for an aquifer, esp. in the Gulf Coast, in which fluid pressure exceeds normal hydrostatic pressure of 0.465 pound per square inch per foot of depth (Jones, 1969, p. 34).

georgiadesite A white or brownish-yellow orthorhombic mineral: $Pb_3(AsO_4)Cl_3$.

georgiaite A greenish tektite from Georgia, U.S.

Georgian *Waucoban.*

geoscience (a) A short form, sometimes used in the plural, denoting the collective disciplines of the *geological sciences.* The term, as such, is synonymous with *geology.* (b) A syn. of *earth science.*

geoselenic Relating to the Earth and the Moon.

geosere A series of *climax communities* that succeed each other as a result of changes in the physical and climatic characteristics of the environment.

geosophy A term coined by Wright (1947, p. 12) for "the study of geographical knowledge from any or all points of view", dealing with "the nature and expression of geographical knowledge both past and present", and covering "the geographical ideas, both true and false, of all manner of people".

geosphere (a) The *lithosphere.* (b) The lithosphere, hydrosphere, and atmosphere combined. (c) Any of the so-called spheres or layers of the Earth.

geostatic pressure The vertical pressure at a point in the Earth's crust, equal to the pressure caused by the weight of a column of the overlying rock or soil. Syn: *lithostatic pressure; overburden pressure; rock pressure.*

geostatistics (a) Statistics as applied to geology; the application of statistical methods or the collection of statistical data for use in geology. (b) Statistical techniques developed for mine evaluation by the French school of G. Matheron.

geostratigraphic Pertaining to worldwide stratigraphy; e.g. "geostratigraphic standards" or "geostratigraphic stage".

geostrome A term, now obsolete, proposed by Patrin "to denote the strata of the Earth" (Pinkerton, 1811, v.l, p.542).

geostrophic current A wind or ocean current in which the horizontal pressure force is exactly balanced by the equal but opposite *Coriolis force.* The geostrophic current is neither accelerated nor affected by friction. It flows to the right of the pressure gradient force along the isobars in the northern hemisphere. Cf: *gradient current.*

geostrophic cycle A term suggested by Tomkeieff (1946, p.326) for the Huttonian concept of one great cycle of dynamic changes occurring in the Earth; the cycle embraces both the organic and the inorganic spheres, and consists alternately of the complementary processes of destruction and construction. Cf: *orogenic cycle.*

geostrophic force *Coriolis force.*

geosuture In Dietz's use, a boundary zone between contrasting tectonic units of the crust; in many places a fault which probably extends through the entire thickness of the crust. In Wilson's use, a place where two continents have come together.

geosynclinal n. The original, now obsolete, term for *geosyncline,* first used by Dana in 1873.—adj. Pertaining to a geosyncline.

geosynclinal couple Aubouin's concept of the true geosyncline as miogeosynclinal and eugeosynclinal furrows linked by geanticlinal ridges (1965, p. 34); an *orthogeosyncline.*

geosynclinal cycle *tectonic cycle.*

geosynclinal facies A sedimentary facies characterized by great thickness, predominantly argillaceous character, and paucity of carbonate rocks; it consists of uniform, rhythmic, and graded beds of shale or silty shale regularly interbedded with graywackes, deposited rapidly in a strongly subsiding geosyncline of a deep-water marine environment. Cf: *shelf facies.* See also: *orogenic facies; graptolitic facies.*

geosynclinal prism The load of sediments that accumulates, often to great thicknesses, in the downwarped part of a geosyncline, having a shape similar to that of a long, plano-convex prism whose convexity is at the floor. Cf: *nepton; clastic wedge.*

geosynclinal trough A linear depression or basin that subsides as it receives clastic material, located not far from the source supplying the sediment.

geosyncline A mobile downwarping of the crust of the Earth, either elongate or basinlike, measured in scores of kilometers, in which sedimentary and volcanic rocks accumulate to thicknesses of thousands of meters. A geosyncline may form in part of a tectonic cycle in which orogeny follows. The concept was presented by Hall in 1859, and the term *geosynclinal* was proposed by Dana in 1873. Differing opinions of the origin, mechanics, and essential features of geosynclines are reflected in various schemes that have been used to define their aspects. Some are based on the tectonic relationship of crustal units, some emphasize mountain-building processes, and others are concerned with the relationship of geosynclinal sedimentation to subsidence. Recognition of the plate structure of the lithosphere has led to appreciation that nearly all geosynclinal phenomena are related to ocean opening and closing (Wilson, 1968). Cf: *mobile belt.* See also: *synclinorium.* Ant: *geanticline.*

geotaxis *Taxis [ecol]* resulting from gravitational attraction. Cf: *geotropism; geomalism.*

geotechnical Pertaining to the broad field of *geotechnics.*

geotechnics The application of scientific methods and engineering principles to the acquisition, interpretation, and use of knowledge of materials of the Earth's crust for the solution of engineering problems; the applied science of making the Earth more habitable. It embraces the fields of soil mechanics and rock mechanics, and many of the engineering aspects of geology, geophysics, hydrology, and related sciences. Syn: *geotechnique.*

geotechnique A syn. of *geotechnics.* Etymol: French.

geotechnology The application of scientific methods and engineering techniques to the exploitation and use of natural resources.

geotectocline The geosynclinal accumulation of sediments formed above a *downbuckle;* the basin between the limbs of a downbuckle (Hess, 1938, p. 79). Syn: *tectocline.*

geotectogene *tectogene.*

geotectonic *tectonic.*

geotectonic cycle A sequence of geosynclinal, orogenic, and cratonic stages. Such a cycle may be repeated (Stille, 1940).

geotectonics *tectonics.* Cf: *megatectonics.*

geotexture The texture of the Earth's surface as manifested by the largest features of relief (such as continental massifs and ocean basins), the formation of which is connected with worldwide processes (I.P. Gerasimov & J.A. Mescherikov in Fairbridge, 1968, p. 731).

geotherm *isogeotherm.*

geothermal Pertaining to the heat of the interior of the Earth. Syn: *geothermic.*

geothermal brine A brine that is overheated with respect to its depth, as a result of association with an anomalous heat source.

geothermal energy Energy that can be extracted from the Earth's internal heat.

geothermal flux *Geothermal heat flow* per unit time. Syn: *heat flux.*

geothermal gradient The rate of increase of temperature in the Earth with depth. The gradient differs from place to place depending on the heat flow in the region and the thermal conductivity of the rocks. The average geothermal gradient in the Earth's crust approximates 25°C/km of depth.

geothermal heat flow The amount of heat energy leaving the Earth per cm^2/sec, measured in calories/cm^2/sec. The mean heat flow for the Earth is about 1.5 ± 0.15 microcalories/cm^2/sec, or about 1.5 heat-flow units. Heat-flow measurements in igneous rocks have shown a linear correlation between heat production in rocks and surface heat flow. The heat production is due to the presence of uranium, potassium, and thorium. See also: *geothermal flux.* Syn: *heat flow.*

geothermal metamorphism A type of deep-seated static metamorphism in which a regular downward increase in temperature attributed to deep burial by overlying rocks is the controlling factor. Cf: *thermal metamorphism; load metamorphism; static metamorphism.*

geothermal prospecting Exploration for sources of *geothermal energy.* Syn: *thermal prospecting.*

geothermic *geothermal.*

geothermometer A mineral or mineral assemblage whose composition, structure, or inclusions are fixed within known thermal limits under particular conditions of pressure and composition and whose presence thus denotes a limit or a range for the temperature of formation of the enclosing rock. Examples are the composition of co-existing minerals that undergo solid solution, the filling temperatures of fluid inclusions, and the thermal discoloration of spores and pollen.

geothermometry (a) The study of the Earth's heat, including the temperature of the Earth, the effect of temperature on physical and chemical processes, the source of the Earth's heat, and volcanology. (b) Determination of the temperature of chemical equilibration of a rock, mineral, or fluid.

geotomical axis A minor axis with small spines in the shell of an acantharian radiolarian. Cf: *hydrotomical axis.*

geotropism *Tropism* resulting from gravitational attraction. Cf: *geotaxis; geomalism.*

geotumor An uplift of regional extent (Haarmann, 1930).

geoundation An upward or downward warping of the Earth's crust, on a continental or oceanic scale (Van Bemmelen, 1932).

gerasimovskite A mineral: $(Mn,Ca)_2(Nb,Ti)_5O_{12} \cdot 9H_2O$.

gerhardtite An emerald-green mineral: $Cu_2(NO_3)(OH)_3$.

germ (a) In botany, an obsolete syn. of *embryo.* (b) A common term used in reference to bacteria.

germanite A reddish-gray mineral: $Cu_3(Ge,Ga,Fe)(S,As)_4$.

germanotype tectonics The tectonics of the cratons and stabilized foldbelts, typified by structures in Germany north of the Alps. The milder phases of germanotype tectonics are epeirogenic, but they also include broad folds dominated by vertical uplift and high-angle faults, block-faulted terranes, and sedimentary basins deformed within a frame of surrounding massifs. Cf: *alpinotype tectonics.* Syn: *paratectonics.*

germinal aperture An *aperture [palyn],* such as a colpus or germ pore, through which the pollen nuclei emerge on germination of the grain. The term is sometimes used to include the laesura of spores.

germinal furrow *colpus.*

germ pore A *pore [palyn]* or thin area in the exine of a pollen grain through which the pollen tube emerges on germination.

geröllton A term used by Pettijohn (1957, p.265) for a nonglacial conglomeratic mudstone (or as a syn. of *tilloid*). It was introduced as *Geröllton* (German for "gravel clay") by Ackerman (1951) who applied it to a pebble-bearing clay in which the pebbles and clay were deposited simultaneously.

gerontic The senile or old-age growth stage in ontogeny; the stage following the *ephebic* stage.

gerontomorphosis Evolutionary changes involving modifications of the adult characteristics of organisms.

gersdorffite A silver-white to steel-gray isometric mineral: NiAsS. It closely resembles cobaltite, and may contain some iron and cobalt. Syn: *nickel glance.*

gerstleyite A red mineral: $(Na,Li)_4As_2Sb_8S_{17} \cdot 6H_2O$.

gerstmannite An orthorhombic mineral: $(Mg,Mn)_2Zn(SiO_4)(OH)_2$.

getchellite A mineral: $AsSbS_3$.

geversite A mineral of the pyrite group: $PtSb_2$.

geyser A type of hot spring that intermittently erupts jets of hot water and steam, the result of ground water coming into contact with rock or steam hot enough to create steam under conditions preventing free circulation; a type of intermittent spring. Syn: *pulsating spring; gusher [grd wat].*

geyser basin A valley that contains numerous springs, geysers, and steaming fissures fed by the same ground-water flow (Schieferdecker, 1959, term 4524).

geyser cone A low hill or mound built up of siliceous sinter around the orifice of a geyser. The term is sometimes mistakenly applied to an algal growth on objects (such as wooden snags) occurring along the shores of some Tertiary lakes. Syn: *geyser mound.*

geyser crater The bowl- or funnel-shaped opening of the geyser pipe, which often contains a *geyser pool.* Syn: *crater [grd wat].*

geyserite A syn. of *siliceous sinter,* used esp. for the compact, loose, concretionary, scaly, or filamentous incrustation of opaline silica deposited by precipitation from the waters of a geyser.

geyser jet The plume of water and steam emitted during the eruption of a geyser.

geyser mound *geyser cone.*

geyser pipe The narrow tube or well of a geyser extending downward from the *geyser pool.* Syn: *pipe [grd wat]; geyser shaft.*

geyser pool The comparatively shallow pool of heated water ordinarily contained in a *geyser crater* at the top of a *geyser pipe.*

geyser shaft *geyser pipe.*

GF *grading factor.*

ghat A term used in India originally for a mountain pass, or a path leading down from a mountain, but now commonly and erroneously applied (by Europeans) to a mountain range, specif. the mountain ranges parallel to the east and west coasts of India. Etymol: Hindi *ghāt.* Syn: *ghaut.*

ghizite An analcime- and olivine-bearing *andesite* characterized by the presence of biotite. Cf: *scanoite.* The name, given by Washington in 1914, is for Ghizo, Sardinia, Italy. Not recommended usage.

G horizon An obsolete term for soil horizon of intense reduction characterized by ferrous iron and gray or olive colors. Cf: *gley soil.*

ghost [petrology] A visible outline of a former crystal shape, fossil, or other rock structure that has been partly obliterated (as by diagenesis or replacement) and that is bounded by inclusions and outlined by bubbles or foreign material. Syn: *phantom.*

ghost [seis] Seismic energy that travels upward from a seismic profiling shot and then is reflected downward at the base of the weathering or at the surface. Such energy usually unites with a downward-traveling wave train, although it may sometimes be distinguished as a separate wave. Ghosts may also occur where the detectors (e.g. marine *streamers*) are below the water surface and respond to events reflected from it.

ghost coal A coal that burns with a bright white flame.

ghost crystal *phantom crystal.*

ghost member Any part of the ideal or typical cyclothem that is absent in a particular cyclothem. Cf: *phantom.*

ghost stratigraphy Traces, in highly metamorphosed strata, of the original lithology and stratification.

Ghyben-Herzberg ratio A ratio describing the static relation of fresh ground water and saline ground water in coastal areas. For each foot of freshwater head above sea level, the salt-water surface is displaced to 40 feet below sea level, i.e. in a ratio of 1:40. The static relationship is modified by dynamic factors that cause mixing of fresh and salt water, esp. the seaward flow of fresh water and tide-induced fluctuations of the interface. The ratio was formally defined by Badon-Ghyben in 1889 and independently by Herzberg in 1901. Syn: *Ghyben-Herzberg relation; Ghyben-Herzberg principle; Ghyben-Herzberg formula.* See also: *salt-water encroachment.*

ghyll Obsolescent spelling of *gill.*

giant cusp A slightly protruding *cusp,* commonly 300-500 m from adjacent ones, with a submarine ridge continuing seaward as a transverse bar along one or both sides of which is a deep channel. Giant cusps appear to be characteristic of areas or times of relatively strong littoral currents.

giant desiccation polygon A *desiccation polygon* formed on a playa by contraction of muds upon drying, bounded by fissures or cracks measuring several meters in depth and up to 1 m in width, and extending over an area of several hundred square meters (Stone, 1967, p. 228).

giant granite *pegmatite.*

giantism The tendency for an evolving *lineage* to develop extremely large body size compared to that in other closely related lineages. Cf: *gigantism.*

giant ripple A *ripple* that is more than 30 m in length; it usually shows superimposed *megaripples* (Reineck & Singh, 1973, p. 37-39).

giant's cauldron *giant's kettle.*

giant's kettle A cylndrical hole bored in bedrock beneath a glacier by water falling through a deep *moulin* or by boulders rotating in the bed of a meltwater stream; it may contain a pond or marsh. Syn: *moulin pothole; glacial pothole; pothole [glac geol]; giant's cauldron.*

giant stairway *glacial stairway.*

gibber An Australian term for a pebble or boulder; esp. one of the wind-polished or wind-sculptured stones that compose a desert pavement or the lag gravels of an arid region. It is pronounced with a hard "g".

gibber plain A desert plain strewn with wind-abraded pebbles, or *gibbers;* a gravelly desert in Australia.

gibbs A unit of measurement of entropy, heat capacity, and various commonly used thermodynamic functions that is essentially equivalent to *entropy unit.*

Gibbs free energy The Gibbs free energy G is defined by the equation $G = H - TS$, where H is *enthalpy,* T is absolute temperature, and S is *entropy.* With respect to any process at constant temperature and pressure, the state of equilibrium is defined by $dG = O$. Cf:

Helmholtz free energy. Syn: *Gibbs function; free enthalpy.*

Gibbs function *Gibbs free energy.*

gibbsite A white or tinted monoclinic mineral: $Al(OH)_3$. It is polymorphous with bayerite and nordstrandine. Gibbsite is formed by weathering of igneous rocks and is the principal constituent of bauxite; it occurs in micalike crystals or in stalactitic and spheroidal forms. Syn: *hydrargillite.*

Gibbs phase rule *phase rule.*

gibelite A *trachyte* characterized by the presence of abundant large phenocrysts of alkali feldspar, sometimes microperthite, and a small amount of sodic clinopyroxene in a groundmass composed of alkali feldspar and a little quartz, and exhibiting flow texture. Amphibole is present as an accessory. The type locality, described by Washington in 1913, is on the island of Pantelleria in the western Mediterranean. Not recommended usage.

Gibraltar stone A light-colored *onyx marble* found at Gibraltar and elsewhere.

giessenite An orthorhombic mineral: $Pb_9CuBi_6Sb_{1.5}S_{30}$.

gigantism (a) In animals, development to abnormally large size as a result of the excessive growth of certain hard parts, often accompanied by structural weakness and sexual impotence. Cf: *nanism.* (b) In plants, excessive vegetative growth.

gigayear A term proposed by Rankama (1967) for one billion (10^9) years. Syn: *eon (b).*

gilgai n. The microrelief of heavy clay soils with high coefficients of expansion and contraction according to changes in moisture. Gilgai is typical of *Vertisols.*—adj. Said of a soil that displays gilgai.—See also: *crab hole; puff.*

gill [paleont] An organ for obtaining oxygen from water; e.g. a *branchia* of a crustacean or mollusk.

gill [streams] (a) A term used in the English Lake District for a deep narrow rocky valley, esp. a wooded ravine with a rapid stream running through it. (b) A narrow mountain stream or brook flowing swiftly through a gill. Also, a term used in Yorkshire, England, for a stream flowing in a shallow valley, sometimes ending in a pothole.—Also spelled: *ghyll; ghyl; gil.*

gill bar One of several serially arranged bars of bone or cartilage, jointed and movable in gnathostomes but fixed in Agnatha, which support the gills behind or below the head; collectively the visceral skeleton.

gill chamber *branchial chamber.*

gillespite A red mineral: $BaFeSi_4O_{10}$.

gill slit (a) An opening in an echinoderm, such as a fissure in the disc of an ophiuroid along the side of the base of an arm and leading into the bursa, or an indentation of the peristomial margin of echinoid interambulacra for the passage of the stem of an external branchia. (b) One of several openings behind or below the head of an aquatic vertebrate, for the escape of water being passed over the gills.— Syn: *branchial slit.*

gilpinite *johannite.*

gilsonite *uintahite.*

ginkgo A member of the gymnosperm subclass Ginkgoales, characterized by dichotomous venation, fan-shaped or regularly bifurcating leaves, and terminally borne seeds. Ginkgoes range from the Permian.

ginorite A white monoclinic mineral: $Ca_2B_{14}O_{23}\cdot 8H_2O$.

ginzburgite A group name for iron-rich clay minerals of the kaolin group.

gio *geo.*

giobertite *magnesite.*

gipfelflur (a) Uniformity or accordance of high mountain summits, presumed to develop independent of geologic structure and of rock type, at a level of uplift limited by isostatic control or because the degrading agencies above that level keep pace with continuing uplift (Von Engeln, 1942, p. 100). (b) An imaginary, relatively smooth surface touching the tops of the accordant summits of a region. See also: *summit plane; peak plain.*—Etymol: German *Gipfelflur,* "summit plain". The term was originated by A. Penck (1919).

girasol adj. Said of any gem variety, such as sapphire or chrysoberyl, that exhibits a billowy, gleaming, round or elongated area of light that "floats" or moves about as the stone is turned or as the light source is moved.—n. A name that has been applied to many gemstones with a girasol effect, such as moonstone; specif. a translucent variety of *fire opal* with reddish reflections in a bright light and a faint bluish-white floating light emanating from the center of the stone.

girdle [gem] The outer edge or periphery of a fashioned gemstone; the portion that is usually grasped by the setting or mounting; the dividing line between the *crown* and the *pavilion.*

girdle [paleont] (a) The region of overlap of the two valves of a diatom frustule, consisting of the *connecting band* from each valve. Also, either of the two connecting bands forming the girdle. Syn: *cingulum.* (b) A transverse furrow around the theca or body of a dinoflagellate; the part of the shell lying between epivalve and hypovalve in certain dinoflagellates. (c) The muscular, flexible marginal band of uniform width of the mantle of a chiton, encircling the shell plates and differentiated from the central portion of the back. It belongs with the soft parts of the chiton, although its covering may be studded with needlelike or scalelike calcareous spicules. (d) A spiral or annular shelf in the skeleton of a spumellarian radiolarian. (e) In the vertebrates, that part of the skeleton that connects front or hind limbs to the axial skeleton.—See also: *perignathic girdle.*

girdle [struc petrol] On an equal-area projection, a belt of concentration of points representing orientations of fabric elements. If this belt coincides approximately with a great circle of the projection, then it is referred to as a great-circle girdle. If the belt of concentration coincides approximately with a small circle of the projection, then it is called a small-circle or *cleft girdle* (Turner and Weiss, 1963, p. 58). Cf: *maximum.*

girdle axis On a *fabric diagram* in which a girdle is produced, the girdle axis coincides with that pole which is normal to the plane best approximating a great-circle girdle and is the axis of the cone corresponding to a small-circle girdle.

girdle facet One of the 32 triangular facets that adjoin the girdle of a round brilliant-cut gem; there are 16 facets above the girdle and 16 below.

girdle list A high membranous ridge perpendicular to the wall and bordering the girdle of a dinoflagellate.

girdle view The side view of a diatom frustule, showing only the edges of the valves.

girdle zone The circular central region with shelves in a radiolarian shell.

Girondian European stage: Lower Miocene (above Chattian, below Langhian). It includes Burigalian and Aquitanian.

girt *bayou.*

girvanella An *algal biscuit* characterized by a complex of microscopic filaments.

gisement The angle between the *grid meridian* and the *geographic meridian*. It is a term used primarily in connection with military grids. Cf: *grid azimuth.* Etymol: French. Syn: *mapping angle.*

Gish-Rooney method In electrical prospecting, the use of a double commutator to reverse periodically the direction of flow of current in both power and potential leads, to eliminate Earth-current potentials.

gismondine A zeolite mineral: $CaAl_2Si_2O_8\cdot 4H_2O$. It sometimes contains potassium. Syn: *gismondite.*

gitology A term increasingly in use, esp. in Europe, to describe the study of ore-deposit genesis in the broadest sense, including chemical, thermodynamic, petrological, and economic disciplines. Etymol: French.

giumarrite An amphibole-bearing *monchiquite,* named for Giumarra, Sicily. Not recommended usage.

Givetian European stage: Middle Devonian (above Eifelian, below Frasnian).

gizzard (a) The last part of the foregut (anterior part of the alimentary canal) of an arachnid, developed as a pumping organ. Its dorsal dilative muscle is attached to an apodeme visible on the external surface of the carapace (TIP, 1955, pt.P, p. 62). (b) The posterior end of the stomach of birds, esp. as characterized by extreme development of muscular walls.

gizzard stone *gastrolith.*

gja [coast] *geo.*

gja [volc] A gaping, dilation fissure of the Icelandic rift system, from which volcanic eruptions take place. Etymol: Icelandic, "chasm". Pl: *gjár.*

glabella The raised axial part of the cephalon of a trilobite. It represents the anterior part of the axis or axial lobe. In some forms the term includes the occipital ring. Pl: *glabellae.*

glabellar furrow A narrow groove extending transversely across the glabella of a trilobite. It is commonly incomplete or interrupted. Also called "lateral glabellar furrow".

glabellar lobe A transverse lobe on the glabella of a trilobite, more or less bounded by complete or partial glabellar furrows. It represents a remnant of the original segments fused in the cephalon. Also called "lateral glabellar lobe".

glacial adj. (a) Of or relating to the presence and activities of ice or glaciers, as *glacial* erosion. (b) Pertaining to distinctive features and materials produced by or derived from glaciers and ice sheets, as *glacial* lakes. (c) Pertaining to an ice age or region of glaciation. (d) Suggestive of the extremely slow movement of glaciers. (e) Used loosely as descriptive or suggestive of ice, or of below-freezing temperature.—n. A *glacial age,* or *glacial stage,* of a glacial epoch, esp. of the Pleistocene Epoch; e.g. the Wisconsin *glacial.*

glacial action All processes due to the agency of glacier ice, such as erosion, transportation, and deposition. The term sometimes includes the action of meltwater streams derived from the ice. See also: *glacial erosion.*

glacial advance *advance [glaciol].*

glacial age A subdivision of a glacial epoch, esp. of the Pleistocene Epoch. Syn: *glacial.*

glacial basin A *rock basin* caused by erosion of the floor of a glacial valley.

glacial block A large, markedly angular rock fragment that has not been greatly modified during glacial transport.

glacial boulder A *boulder* or large rock fragment that has been moved for a considerable distance by a glacier, being somewhat modified by abrasion but not always "rounded". Cf: *erratic.* Syn: *ice boulder [glac geol].*

glacial boundary The position occupied, in a given region or during a given glacial stage, by the outer or lower margin of an ice sheet; this may extend beyond the terminal moraine.

glacial canyon A canyon eroded by a glacier, usually occupying the site of an older stream valley and having a U-shaped cross profile.

glacial chute A term suggested by Harvey (1931, p. 231) for one of a group of narrow, closely spaced, steeply plunging glacial troughs, with vertical or nearly vertical walls and gently curving U-shaped bottoms, that acted as valves in passing or stopping snow, ice, and rock material. Examples occur on the flanks of Mount Puy Puy in Peru.

glacial-control theory A theory of coral-atoll and barrier-reef formation according to which marine erosion and lowering of the sea level during the Ice Age destroyed existing coral reefs and left an extensive, level rock surface from which coral reefs were built up continuously during the gradual postglacial rise of sea level when the oceans became rapidly warmed. Theory was proposed by Reginald A. Daly in 1910. Cf: *antecedent-platform theory; subsidence theory.*

glacial cycle (a) A term used by Davis (1911, p. 56) for the ideal case of glaciation operating for a long period of time under fixed climatic conditions such that glacial erosion would be complete and replaced by normal erosion. (b) "A major global climatic oscillation of the order of 10^5 yr, developed within an 'ice age' *sensu lato,* which may last 10^6-10^7 yr and which recurs at widely spaced intervals in geologic time (i.e. 2×10^8yr) (Fairbridge, 1972).

glacial debris (a) *glacial drift.* (b) *debris [glaciol].*

glacial deposit *glacial drift.*

glacial-deposition coast A coast with partly submerged moraines, drumlins, and other glacial deposits.

glacial dispersal The fan-shaped deployment of *erratics* by glacial ice.

glacial drainage (a) The flow system of glacier ice. (b) The system of meltwater streams flowing from a glacier or ice sheet.

glacial drainage channel A "safer expression" for *overflow channel,* because "the exact mode of origin of many channels supposedly due to overflow is uncertain" (Challinor, 1978, p. 212); a channel formed by an ice-marginal, englacial, or subglacial stream.

glacial drift A general term for *drift* transported by glaciers or icebergs, and deposited directly on land or in the sea. Cf: *glaciofluvial drift.* Syn: *glacial deposit; glacial debris.*

glacial epoch Any part of geologic time, from Precambrian onward, in which the climate was notably cold in both the northern and southern hemispheres, and widespread glaciers moved toward the equator and covered a much larger total area than those of the present day; specif. the latest of the glacial epochs, known as the Pleistocene Epoch. Syn: *glacial period; ice age; drift epoch.*

glacial erosion The grinding, scouring, plucking, gouging, grooving, scratching, and polishing effected by the movement of glacier ice armed with rock fragments frozen into it, together with the erosive action of meltwater streams. See also: *glacial action.*

glacial erratic *erratic.*

glacial eustasy *glacio-eustatism.*

glacial flour *rock flour.*

glacial geology (a) The study of the geologic features and effects resulting from the erosion and deposition caused by glaciers and ice sheets. Cf: *glaciology.* (b) The features of a region that has undergone glaciation.—Syn: *glaciogeology.*

glacial groove A deep, wide, usually straight furrow cut in bedrock by the abrasive action of a rock fragment embedded in the bottom of a moving glacier; it is larger and deeper than a *glacial striation,* ranging in size from a deep scratch to a glacial valley.

glacialism (a) The *glacier theory.* (b) *glaciation.*

glacialist (a) A believer in the *glacier theory.* (b) A student of glaciology; a glaciologist.—Syn: *glacierist.*

glacial lake (a) A lake that derives much or all of its water from the melting of glacier ice, e.g. fed by meltwater, or lying on glacier ice and due to differential melting. (b) A lake occupying a basin produced by glacial deposition, as one held in by a morainal dam. (c) A lake occupying a basin produced in bedrock by glacial erosion (scouring, quarrying), as a *cirque lake* or *fjord lake*. (d) A lake occupying a basin produced by collapse of outwash material surrounding masses of stagnant ice. (e) *glacier lake.*

glacial lobe A large, rounded, tonguelike projection from the margin of the main mass of an ice cap or ice sheet; a short, broad *distributary glacier.* Cf: *tongue [glaciol]; outlet glacier.* Syn: *lobe [glaciol].*

glacial-lobe lake A lake occupying a depression that was excavated by a glacial lobe as it advanced over the drainage basin of a former river.

glacial marine Said of marine sediments that contain glacial material. Syn: *glaciomarine.*

glacial maximum The time or position of the greatest advance of a glacier, or of glaciers (such as the greatest extent of Pleistocene glaciation). Ant: *glacial minimum.* Syn: *glaciation limit.*

glacial meal *rock flour.*

glacial milk *glacier milk.*

glacial mill *moulin.*

glacial minimum The time or position of the greatest retreat of a glacier. Ant: *glacial maximum.*

glacial pavement A polished, striated, relatively smooth, planed-down rock surface produced by glacial abrasion. Cf: *boulder pavement.* Syn: *ice pavement.*

glacial period (a) A syn. of *glacial epoch;* specif. the Pleistocene Epoch. Usage of this term is not strictly correct "as glacial intervals during earth history were not of the period rank, but of shorter duration" (ADTIC, 1955, p. 34). (b) A geologic *period*, such as the Quaternary Period, that embraced an interval of time marked by one or more major advances of ice.

glacial plain A plain formed by the direct action of the glacier ice itself. Cf: *outwash plain.*

glacial polish A smoothed surface produced on bedrock by glacial abrasion.

glacial pothole A syn. of *giant's kettle.* The term is misleading because the feature is not produced by the direct scouring action of glacier ice, but by a stream of water falling through a *moulin,* by lateral or subglacial meltwater, or perhaps by a viscous mixture of water, ice, and rock fragments.

glacial pressure ridge *ice-pushed ridge.*

glacial recession *recession [glaciol].*

glacial refuge *refugium.*

glacial retreat *recession [glaciol].*

glacial scour The eroding action of a glacier, including the removal of surficial material and the abrasion, scratching, and polishing of the bedrock surface by rock fragments dragged along by the glacier. Cf: *grinding [glac geol].* Syn: *scouring.*

glacial scratch *glacial striation.*

glacial spillway *overflow channel.*

glacial stage A major subdivision of a glacial epoch, esp. one of the cycles of growth and disappearance of the Pleistocene ice sheets; e.g. the "Wisconsin Glacial Stage". Syn: *glacial.*

glacial stairway A glacial valley whose floor is shaped like a broad staircase composed of a series of irregular steplike benches (*treads*) separated by steep *risers.* Cf: *cirque stairway.* Syn: *giant stairway; cascade stairway; cascade [glac geol].*

glacial stream A flow of water that is supplied by melting glacier ice; a *meltwater* stream.

glacial stria A syn. of *glacial striation.* Pl: *striae.*

glacial striation One of a series of long, delicate, finely cut, commonly straight and parallel furrows or lines inscribed on a bedrock surface by the rasping and rubbing of rock fragments embedded at the base of a moving glacier, and usually oriented in the direction of ice movement; also formed on the rock fragments transported by the ice. Cf: *glacial groove.* Syn: *glacial scratch; glacial stria; drift scratch.*

glacial terrace A terrace formed by glacial action, either by rearranging glacial materials in terrace form (such as a remnant of a valley train), or by cutting into bedrock. See also: *kame terrace.*

glacial theory *glacier theory.*

glacial till *till.*

glacial trough A deep, steep-sided *U-shaped valley* leading down from a cirque, and excavated by an alpine glacier that has widened, deepened, and straightened a preglacial river valley; e.g. Yosemite Valley, Calif.

glacial valley A U-shaped, steep-sided valley showing signs of glacial erosion; a glaciated valley, or one that has been modified by a glacier.

glaciated Said of a formerly glacier-covered land surface, esp. one that has been modified by the action of a glacier or an ice sheet, as a *glaciated* rock knob. Cf: *glacier-covered.*

glaciated coast A coast whose features were modelled by continental glaciers of the Pleistocene epoch, or a coast covered by glaciers at the present time.

glaciation (a) The formation, movement, and recession of glaciers or ice sheets. (b) The covering of large land areas by glaciers or ice sheets. Syn: *glacierization.* (c) The geographic distribution of glaciers and ice sheets. (d) A collective term for the geologic processes of glacial activity, including erosion and deposition, and the resulting effects of such action on the Earth's surface. (e) Any of several minor parts of geologic time during which glaciers were more extensive than at present; a *glacial epoch,* or a *glacial stage.* "A climatic episode during which extensive glaciers developed, attained a maximum extent, and receded" (ACSN, 1961, art.40).—The verb *glaciate* is apparently derived from *glaciation.* Syn: *glacialism.*

glaciation limit (a) The lowest altitude in a given locality at which glaciers can develop, usually determined as below the minimum summit altitude of mountains on which glaciers occur but above the maximum summit altitude of mountains having topography favorable for glaciers but on which none occur. (b) *glacial maximum.*

glacier (a) A large mass of ice formed, at least in part, on land by the compaction and recrystallization of snow, moving slowly by creep downslope or outward in all directions due to the stress of its own weight, and surviving from year to year. Included are small mountain glaciers as well as ice sheets continental in size, and ice shelves which float on the ocean but are fed in part by ice formed on land. See also: *ice stream [glaciol]; ice sheet; ice cap.* (b) Nonpreferred syn. of *alpine glacier.* (c) A term used in Alaska for flood-plain *icing* or a mass of *ground ice.* (d) A streamlike landform having the appearance of, or moving like, a glacier; e.g. a *rock glacier.*—Etymol: French *glace,* "ice", from Latin *glacies.*

glacier advance *advance [glaciol].*

glacier band The appearance of one of a series of more or less extensive zones, layers, or lenses, on or within a glacier, that differ visibly (such as in color or texture) from the adjacent material. It may consist of ice, firn, snow, rock, debris, dirt, organic matter, or any mixture of these materials, and it may originate by infilling, plastic flow, or shear. Syn: *band [glaciol].*

glacier bed The surface under a glacier.

glacier berg *glacier iceberg.*

glacier breeze *glacier wind.*

glacier bulb *expanded foot.*

glacier burst *glacier outburst flood.*

glacier cap *ice cap*

glacier cave (a) A cave that is formed within a glacier. (b) A nonrecommended syn. of *ice cave.*

glacier cone A rarely used syn. of *debris cone.*

glacier corn Glacier ice broken into irregular crystals of various sizes (Brigham, 1901, p.92).

glacier cornice A mass of glacier ice projecting into an open crevasse. It was formerly underlain by ice containing numerous rock fragments which, when warmed by solar radiation, melted the ice around them and caused the ice above them to project like a *cornice* (Hobbs, 1912, p.397).

glacier-covered Said of a land surface overlain by glacier ice at the present time. Cf: *glaciated.* Syn: *glacierized; ice-covered.*

glacière *ice cave.*

glacieret (a) A very small glacier on a mountain slope or in a cirque, as in the Sierra Nevada, Calif.; a miniature alpine glacier. Cf: *cirque glacier.* (b) A tiny mass of ice or firn in high mountains, resembling a glacier but defying a precise definition; a *drift glacier.*—Also spelled: *glacierette.*

glacier flood *glacier outburst flood.*

glacier flow The slow downward or outward movement of the ice in a glacier, due to the force of gravity (*gravity flow*). Deformation within the ice, by intragranular gliding, grain-boundary migration, and recrystallization, is involved, together with sliding of the glacier on its bed in some situations. Usually expressed in meters per day or year. Syn: *ice flow; flow [glaciol].*

glacier grain (a) An individual ice crystal in a glacier. (b) A mechanically separate particle of ice in a glacier.—Cf: *grain [glaciol].*

glacier ice Any ice that forms in or was once a part of a glacier, including land ice that is flowing or that shows evidence of having flowed, and glacier-derived ice floating in the sea.

glacier iceberg A greenish, irregularly shaped iceberg consisting of ice detached from a coastal glacier, typically found in the Arctic. Syn: *glacier berg.*

glacierist *glacialist.*

glacierization A term used in Great Britain for *glaciation* in the sense of the gradual covering or "inundation" of a land surface by glaciers or ice sheets.

glacierized A term used in Great Britain for *glacier-covered.*

glacier lake A lake held in place by the damming of natural drainage by the edge or front of a glacier or ice sheet, as a lake ponded by glacier ice advancing across a valley, or a lake occurring along the margin of a continental ice sheet. Cf: *proglacial lake.* Syn: *glacial lake; marginal lake; ice-dammed lake.*

glacier meal *rock flour.*

glacier milk A stream of turbid, whitish meltwater containing *rock flour* in suspension. Syn: *glacial milk.*

glacier mill *moulin.*

glacier mouse A small, rounded, moss-covered stone, 7-10 cm in diameter, found on certain glaciers, that either lies on morainal material or has rolled off it onto the adjacent ice. Glacier mice were described and named by Eythórsson (1951, p.503). Cf: *polster.*

glacier outburst flood A sudden, often annual, release of meltwater from a glacier or glacier-dammed lake, sometimes resulting in a catastrophic flood, formed by melting of a drainage channel or buoyant lifting of ice by water or by subglacial volcanic activity. Syn: *jökulhlaup; outburst [glaciol]; glacier burst; glacier flood.*

glacier pothole *moulin.*

glacier recession *recession [glaciol].*

glacier remanié A glacier formed by regelation of accumulated ice blocks brought down by avalanches and icefalls from the ends of glaciers at higher levels. Etymol: French, "reworked glacier". Syn: *recemented glacier; reconstructed glacier; regenerated glacier; reconstituted glacier.* See also: *remanié [glaciol].*

glacier snout *terminus.*

glacier surge *surge [glaciol].*

glacier table A boulder or large block of rock supported by an *ice pedestal* that rises from the surface of a glacier. It occurs where the melting of the glacier is retarded by the insulation effect of the rock.

glacier theory The theory, first propounded about 1840 and now universally accepted, that the drift was deposited through the agency of glaciers and ice sheets moving slowly from higher to lower latitudes during the Pleistocene Epoch. Cf: *drift theory [glac geol].* See also: *glacialist.* Syn: *glacial theory; glacialism.*

glacier tongue *tongue [glaciol].*

glacier tongue afloat The *tongue* of a glacier that extends so far into the sea that its lower end floats on the ocean. It is esp. common in the Antarctic. Nonpreferred syn: *ice tongue afloat.*

glacier wave A syn. of *wave ogive;* also used loosely and nonspecifically for *kinematic wave* and the active rapid flow phase of a *surging glacier.*

glacier well *moulin.*

glacier wind (a) A cold *katabatic wind* blowing off a glacier. (b) A cold wind blowing out of ice caves in a glacier front, due to the difference in density between the colder air inside and the warmer air outside.—Syn: *glacier breeze.*

glacioaqueous Pertaining to or resulting from the combined action of ice and water; the term is often used as a syn. of *glaciofluvial.* Syn: *aqueoglacial.*

glacio-eustatic Pertaining to *glacio-eustatism.* Also spelled: *glacioeustatic.*

glacio-eustatism The worldwide changes in sea level produced by the successive withdrawal and return of water in the oceans accompanying the formation and melting of ice sheets. Adj: *glacioeustatic.* Also spelled: *glacioeustatism.* Syn: *glacial eustasy; glacio-eustasy.*

glaciofluvial Pertaining to the meltwater streams flowing from wasting glacier ice and esp. to the deposits and landforms produced by such streams, as kame terraces and outwash plains; relating to the combined action of glaciers and streams. Syn: *fluvioglacial; glacioaqueous.*

glaciofluvial drift A general term for *drift* transported and deposited by running water emanating from a glacier. Cf: *glacial drift.* Syn: *fluvioglacial drift.*

glaciogeology *glacial geology.*

glacio-isostasy The state of hydrostatic equilibrium in the Earth's crust as influenced by the weight of glacier ice.

glaciokarst *alpine karst.*

glaciolacustrine Pertaining to, derived from, or deposited in glacial lakes; esp. said of the deposits and landforms composed of suspended material brought by meltwater streams flowing into lakes bordering the glacier, such as deltas, kame deltas, and varved sediments.

glaciology (a) The study of all aspects of snow and ice; the science that treats quantitatively the whole range of processes associated with all forms of solid existing water. Syn: *cryology.* (b) The study of existing glaciers and ice sheets, and of their physical properties. This definition is not internationally accepted.

glaciomarine *glacial marine.*

glacionatant Relating to or derived from floating ice of glacial origin, as *glacionatant* till.

glaciosolifluction Gravitative sliding, on the surface of a melting glacier, of heterogeneous material mixed with water, generally into crevasses or potholes.

glaciospeleology The study of *glacier caves.*

glaciotectonic *cryotectonic.*

glacis A gently inclined slope or bank, less steep than talus; e.g. a piedmont slope, or an easy slope on a mountain side. Etymol: from its resemblance to a glacis used in fortifications as a defense against attack; originally from Middle French *glacer,* "to freeze". A British usage.

glacon A fragment of sea ice ranging in size from brash ice (2 m across) to a floe of medium to big dimensions (about 1 km across). Etymol: French glaçon, "piece of ice".

glade [geog] A term that usually indicates a clearing between slopes; it can be a high meadow, sometimes marshy and forming the headwaters of a stream, or it can be a low, grassy marsh, which is periodically inundated.

glade [ice] A *polynya,* esp. a stretch of open water in the ice of a lake or of a river. Rarely used.

glade [karst] A Jamaican term for a closed depression formed by the coalescence of several *cockpits.* See also: *karst valley.*

gladite A lead-gray mineral: $PbCuBi_5S_9$.

gladkaite A fine-grained gray hypabyssal rock composed of a granular aggregate of sodic plagioclase, abundant quartz, hornblende, and some biotite; a quartz *lamprophyre.* Named by Duparc and Pearce in 1905 for Gladkaia Sopka, northern Urals. Not recommended usage.

glady [clay] A term used in Devon, England, for a variegated black and white clay having a slippery or smooth texture and often associated with stoneware clays. Also spelled: *gladii.*

glady [soil] Said of a limestone-outcrop area having shallow soil.

glaebule A term proposed by Brewer (1964, p.259-260) for a three-dimensional unit, usually prolate to equant in shape, within the matrix of a soil material, recognizable by its greater concentration of some constituent, by its difference in fabric as compared with the enclosing soil material, or by its distinct boundary with the enclosing soil material; e.g. a nodule, concretion, septarium, pedode, or papule. Etymol: Latin *glaebula,* "a small clod or lump of earth". Pl: *glaebules.* See also: *papule.*

glamaigite An intrusive breccia composed of dark patches of *marscoite* in a lighter groundmass. The name, given by Harker in 1904, is for Glamaig, Isle of Skye, Scotland, and was intended for use only at that locality.

glance A mineral that has a resplendent luster; e.g. chalcocite, or copper glance.

glance coal *pitch coal.*

glance pitch A variety of *asphaltite* with a brilliant conchoidal fracture, sometimes called *manjak.* It is similar to gilsonite, but has a higher specific gravity and fixed-carbon content. It fuses between 230°F and 250°F.

gland duct *setal duct.*

glare ice A smooth, glassy or bright, highly reflective sheet of ice on a surface of water, land, or glacier.

glareous Said of an organism that lives in gravelly soil. Syn: *glareal.*

glaserite *aphthitalite.*

glass [chem] A state of matter intermediate between the close-packed, highly ordered array of a crystal and the poorly packed, highly disordered array of a gas. Most glasses are supercooled liquids, i.e. are metastable, but there is no true break in the change in properties between the metastable and stable states. The distinction between glass and liquid is made solely on the basis of viscosity, and is not necessarily related, except indirectly, to the difference between metastable and stable states.

glass [ign] An amorphous product of the rapid cooling of a magma. It may constitute the whole rock (e.g. obsidian) or only part of a groundmass.

glass porphyry *vitrophyre.*

glass rock A term used in northern Illinois and southern Wisconsin for a pure cryptocrystalline limestone of Trentonian age.

glass sand A sand that is suitable for glassmaking because of its high silica content (93-99 + %) and its low content of iron oxide, chromium, cobalt, and other colorants.

glass schorl *axinite.*

glass sponge *hyalosponge.*

glassy Said of the texture of certain extrusive igneous rocks, which is similar to that of broken glass or quartz and developed as a result of rapid cooling of the lava, without distinct crystallization. Also, said of any of the other properties of a volcanic rock that resemble those of glass, such as hardness, luster, or composition. Syn: *hyaline [ign]; vitreous.*

glassy feldspar *sanidine.*

glassy luster *vitreous luster.*

glauberite A brittle, light-colored, monoclinic mineral: $Na_2Ca(SO_4)_2$. It has a vitreous luster and saline taste, and occurs in saline residues.

Glauber's salt A syn. of *mirabilite.* Named after Johann R. Glauber (1604-1668), German chemist. Also spelled: *Glauber salt.*

glaucocerinite *glaucokerinite.*

glaucochroite A bluish-green, violet, or pale-pink orthorhombic mineral: $CaMnSiO_4$.

glaucodot A mineral: (Co,Fe)AsS. Syn: *glaucodote.*

glaucokerinite A sky-blue mineral: $(Cu,Zn)_{10}Al_4(SO_4)(OH)_{30}\cdot 2H_2O$ (?). Also spelled: *glaucocerinite.*

glauconarenite *glauconitic sandstone.*

glauconite (a) A dull-green earthy or granular mineral of the mica group: $(K,Na)(Al,Fe^{+3},Mg)_2(Al,Si)_4O_{10}(OH)_2$. It has often been regarded as the iron-rich analogue of illite. Glauconite occurs abundantly in greensand, and seems to be forming in the marine environment at the present time; it is the most common sedimentary (diagenetic) iron silicate and is found in marine sedimentary rocks from the Cambrian to the present. Glauconite is an indicator of very slow sedimentation. (b) A name applied to a group of green minerals consisting of hydrous silicates of potassium and iron.

glauconitic Said of a mineral aggregate that contains glauconite, resulting in the characteristic green color, e.g. glauconitic shale or clay.

glauconitic sand *greensand.*

glauconitic sandstone A sandstone containing sufficient grains of glauconite to impart a marked greenish color to the rock; *greensand; glauconarenite.*

glauconitization A submarine replacement process whereby a mineral, e.g. biotite, is converted to glauconite under very slow rates of sedimentation and at depths of 100 to 300 m.

glaucophane A blue, bluish-black, or grayish-blue monoclinic mineral of the amphibole group: $Na_2(Mg,Fe^{+2})_3Al_2Si_8O_{22}(OH)_2$. It is a fibrous or prismatic mineral that occurs only in certain crystalline schists resulting from regional metamorphism of sodium-rich igneous rocks (such as *spilites*).

glaucophane schist A type of amphibole schist in which glaucophane rather than hornblende is an abundant mineral. Epidote frequently occurs, and there are quartz and mica varieties (Holmes, 1928, p.106). Cf: *blueschist.*

glaucophane-schist facies The set of metamorphic mineral assemblages (facies) in which basic rocks are represented by combinations of sodic amphibole (e.g. glaucophane, crossite), lawsonite, sodic pyroxene, aragonite, epidote, and garnet. The mineral pair jadeite + quartz is also diagnostic. Exact definitions of the facies and its subdivisions vary (Turner, 1968). It is believed to represent lower temperatures and higher pressures than the *greenschist facies.* It is characteristic of metamorphism in subduction zones, with their unusually low geothermal gradients. Syn: *blueschist facies.*

glaucopyrite A variety of loellingite containing cobalt.

glaukosphaerite A green monoclinic mineral: $(Cu,Ni)_2(CO_3)(OH)_2$. Cf: *rosasite.*

G layer The seismic region of the Earth below 5160 km, equivalent to the *inner core.* It is a part of a classification of the Earth's interior made up of layers A to G. Together with the F layer, it is the equivalent of the *lower core.*

GLC *gas-liquid chromatography.*

glei *gley.*

gleization The formation of a *gley soil.* Syn: *gleying.*

glen A narrow, steep-sided secluded valley, usually wooded, often containing a stream or lake at its bottom; esp. a narrow-floored, glaciated mountain valley in Scotland and Ireland. It is narrower and more steep-sided than a *strath.* Syn: *glyn.*

Glenarm A provincial series of the late Precambrian in New Jersey, Pennsylvania, Delaware, Maryland, and Virginia.

Glencoe-type caldera *Cauldron subsidence* resulting from the collapse of the roof of a magma chamber along ring fractures, i.e. stoping of a cylindrical block of the crust (Williams, 1941, p. 246).

glendonite A pseudomorph of a carbonate (calcite or esp. siderite) after glauberite.

Glen flow law An empirical relation of the shear strain rate (ϵ) of ice to the shear stress (σ), as $\epsilon = k\sigma^n$, where the parameter k varies with temperature, type of ice, and geometry of stress, and n is a number between 1.5 and 4.5 (Glen, 1955, p.528). This relation is basic to most analyses of glacier flow.

glenmuirite Similar to *essexite,* differing only in having analcime, rather than nepheline, as the principal feldspathoid. Named by Johannsen (1938) for Glenmuir Water, Ayrshire, Scotland. Not recommended usage.

glessite A brown variety of retinite found on the shores of the Baltic Sea.

gletscherschlucht A gorge sculptured by meltwater streams, often initiated where a moulin empties onto jointed rock. Etymol: German *Gletscherschlucht,* "glacier gorge".

gley A syn. of *gley soil.* Also spelled: *glei.*

gleying *gleization.*

gley soil Soil developed under conditions of poor drainage, resulting in reduction of iron and other elements and in gray colors and mottles. The term is obsolete in the U.S. Cf: *G horizon.*

glide A gently flowing, calm reach of shallow water in a stream.

glide bedding A variety of *convolute bedding* produced by subaqueous gliding. The term is little used. Cf: *slip bedding.*

glide breccia A breccia formed by subaqueous gravitational movements that deform, shatter, or crush newly formed or partly consolidated bottom sediments deposited under somewhat unstable conditions at higher levels. It may be produced by overloading, earthquakes, or deformation.

glide direction The direction of gliding along *glide planes* in a mineral.

glide fold *shear fold.*

glide plane A symmetry element in a crystal that relates parts on opposite sides by reflection plus translation parallel to the plane. The possible translation components associated with a glide plane must correspond to one half of a lattice translation. Syn: *glide reflection; translation plane; gliding plane; slip plane [cryst].*

glide reflection *glide plane.*

glide twin *deformation twin.*

gliding [cryst] *crystal gliding.*

gliding [tect] *gravitational sliding.*

gliding flow *Flow [exp struc geol]* involving gliding parallel to the preferred crystallographic orientation, e.g. intragranular deformation in a crystal by twin gliding or translation gliding.

gliding plane *glide plane.*

gliding surface *slip surface [mass move].*

glimmer A syn. of *mica.* Etymol: German *Glimmer.*

glimmergabbro A biotite-bearing *gabbro.* Not recommended usage.

glimmerite *biotitite.*

glimmerton An early name for illite.

glinite A group name for clay minerals from clay deposits.

glint An escarpment or steep cliff, esp. one produced by erosion of a dipping resistant formation. See also: *klint [coast].* Etymol: Norwegian, "boundary".

glint lake A lake formed along a *glint line,* esp. a long, narrow glacial lake occupying a basin excavated in bedrock where a glacier is dammed by an escarpment ("glint"), e.g. certain lakes in Norway and Scotland. Syn: *glint-line lake.*

glint line An extensive erosional escarpment produced by the denudation of a very gently dipping resistant formation, as the Silurian limestone of the Great Lakes region. The term is used specif. for the boundary between an ancient shield and younger rocks, e.g. in Russia where Paleozoic rocks rise above the Baltic Shield.

global scale A map scale (smaller than 1/5,000,000) involving all or a major part of the Earth's surface.

global tectonics Tectonics on a global scale, such as tectonic processes related to very large-scale movement of material within the Earth; specif. *new global tectonics.* Cf: *megatectonics.*

globe (a) A body having the form of a sphere; specif. a spherical, typically hollow ball that has a map of the Earth drawn on it and that is usually rotatable at an angle corresponding to the inclination of the Earth's axis. Also, a chart of the celestial sphere, depicted on a sphere. (b) A planet; esp. the Earth.

globigerina ooze A deep-sea pelagic sediment containing at least 30% foraminiferal tests, predominantly of the genus *Globigerina.* It is calcareous, and a particular type of *foraminiferal ooze.*

globigerinid Any planktonic foraminifer belonging to the superfamily Globigerinacea, characterized by a perforate test with bilamellid septa and walls of radial calcite crystals. Range, Middle Jurassic to present. Var: *globigerine.*

globosphaerite A more or less spherical cumulite in which the globulites have a somewhat radial arrangement.

globular *spherulitic.*

globular projection A map projection (neither conformal nor equal-area) representing a hemisphere upon a plane parallel to its base, the point of projection being removed to a point outside of the opposite surface of the sphere. The equator and central meridian are straight lines intersecting at right angles; all other meridians and parallels are circular arcs. The projection is an arbitrary distribution of curves conveniently constructed; distance and directions can neither be measured nor plotted. It is commonly used in pairs in atlases.

globulite A spherical crystallite commonly found in volcanic glass.

globulith An intrusive body, or group of associated bodies, having a globular or botryoidal shape and almost concordant contacts, resulting from the effects of the intrusion(s) on the immediate surroundings (Berthelsen, 1970, p. 73).

glomeroclastic Pertaining to particles grouped together in clusters in a carbonate sedimentary rock. Also said of the texture characterized by lumps.

glomerocryst An aggregate of crystals of the same mineral. Cf: *polycrystal.*

glomerophyric An obsolete term applied to the texture of porphyritic igneous rocks containing closed clusters of equant crystals of the same mineral. Cf: *cumulophyric; gregaritic.* Syn: *glomeroporphyritic.*

glomeroplasmatic An obsolete term applied to the *synneusis* texture of granites and gneisses that contain open clusters of individual crystals or grains of the same mineral. Cf: *glomerophyric.*

glomeroporphyritic *glomerophyric.*

glomospirine Having an irregularly wound coiled tubular chamber; specif. pertaining to the foraminifer *Glomospira.*

glory hole A large open pit from which ore is being or has been extracted.

glory-hole mining A type of opencut mining in which the orebody is worked from the top down in a conical excavation and the ore is removed by an underground system beneath the orebody. Syn: *mill-hole mining.*

gloss *polish.*

gloss coal The highest ranking lignite. It is deep black, compact, with definite conchoidal fracture and glossy luster. Cf: *subbituminous coal.*

glossopterid n. The informal name for the fossil gymnosperm genus *Glossopteris* and its allies, whose foliage is common in the Permian.—adj. Pertaining to such a plant or plant assemblage.

glossothyropsiform Said of a brachiopod loop developed from the cryptacanthiiform stage by final resorption of the posterior part of the echmidium and consisting of two descending branches unconnected posteriorly, bearing two broad ascending elements joined by a wide transverse band (TIP, 1965, pt.H, p.145).

glowing avalanche *ash flow.*

glowing cloud *nuée ardente.*

glucine A mineral: $CaBe_4(PO_4)_2(OH)_4 \cdot 1/2H_2O$.

glyn A Welsh syn. of *glen.*

glyptogenesis The *sculpture* of the Earth's surface by erosion.

glyptolith A term proposed by Woodworth (1894a, p. 70) for a wind-cut stone or *ventifact.* Etymol: Greek *glyptos,* "carved", + *lithos,* "stone".

G-M counter *Geiger-Müller counter.*

gmelinite A hexagonal zeolite mineral: $(Na_2,Ca)Al_2Si_4O_{12} \cdot 6H_2O$. Cf: *faujasite.*

GMR *grain-micrite ratio.*

G-M tube *Geiger-Müller tube.*

gnamma hole A term used in the deserts of Western Australia for a rounded hollow eroded or indented in solid rock, usually at the intersections of joints in granite, and frequently containing water; it has a narrow orifice, but widens out below.

gnathal lobe The masticatory endite of the mandible of a crustacean.

gnathobase The serrate oral margin on the coxa of a eurypterid.

gnathosoma The anterior part of the body in the arachnid order Acarida, bearing the mouth parts. Obsolete syn: *capitulum.* Syn: *gnathosome.*

gnathostome Any vertebrate with anterior visceral skeleton modified into jaws; a member of any vertebrate class except the *Agnatha.*

gnathothorax The thorax and the part of the head bearing the feeding organs of an arthropod; e.g. the tagma of a crustacean resulting from fusion of mandibular and two maxillary somites with one or more thoracic somites, limbs of which are modified to act as mouth parts (TIP, 1969, pt.R, p.96). Cf: *cephalothorax.*

gneiss A foliated rock formed by regional metamorphism, in which bands or lenticles of granular minerals alternate with bands or lenticles in which minerals having flaky or elongate prismatic habits predominate. Generally less than 50% of the minerals show preferred parallel orientation. Although a gneiss is commonly feldspar- and quartz-rich, the mineral composition is not an essential factor in its definition. Varieties are distinguished by texture (e.g. augen gneiss), characteristic minerals (e.g. hornblende gneiss), or general composition and/or origins (e.g. granite gneiss). See also: *gneissic; gneissoid; gneissose.*

gneissic Pertaining to the texture or structure typical of gneisses, with foliation that is more widely spaced, less marked, and often more discontinuous than that of a *schistose* texture or structure (Johannsen, 1931). Cf: *gneissoid; gneissose.*

gneissic structure In a metamorphic rock, commonly *gneiss,* the coarse, textural lineation or banding of the constituent minerals into alternating silicic and mafic layers. Syn: *gneissosity; gneissose structure.*

gneissoid Pertaining to a gneisslike structure or texture that is the result of nonmetamorphic processes, e.g. viscous magmatic flow forming a gneissoid granite. Cf: *gneissic; gneissose.*

gneissose (a) Said of a rock, or of its structure, that resembles gneiss but that is not the result of metamorphic processes. Cf: *gneissoid.* (b) Said of a rock whose structure is composite, having alternating schistose and granulose bands and lenses which differ in mineral composition and texture. Cf: *gneissic.*—Use of this ambiguous term is discouraged.

gneissose structure *gneissic structure.*

gneissosity *gneissic structure.*

gnocchi Ovoid or nut-shaped cavities, usually 1-3 cm in diameter, filled partially to completely with white calcite, dolomite, or other minerals (Folk and Assereto, 1974, p. 34-35). Pron: *nyok*-kee. Plural form, commonly used; sing: gnoccho. Origin: In allusion to the Italian potato dumpling of the same size, shape and color.

gnomonic projection (a) A perspective azimuthal map projection (of a part of a hemisphere) on a plane tangent to the surface of the sphere, having the point of projection at the center of the sphere. All straight lines on the tangent plane represent arcs of great circles on the Earth's surface; all great circles appear as straight lines. The point of tangency may be at a pole, on the equator, or at any point in between (oblique gnomonic projection). It is used, in conjunction with the Mercator projection, to plot great-circle courses in navigation. Syn: *central projection; great-circle projection; great-circle chart.* (b) A similar projection used in optical mineralogy to plot data obtained by measurements of crystals with a two-circle goniometer, characterized by a plane of projection that is tangent to the north pole of the sphere, with the poles of the faces parallel to the vertical axis of the sphere lying at infinity.

gobi (a) A Mongolian term introduced by Berkey & Morris (1924, p. 105) for a small, open, level-surfaced basin within a *tala.* (b) A lenticular mass of sedimentary deposits occupying a gobi.

goblet valley *wineglass valley.*

gob pile A heap of mine refuse on the surface.

godlevskite A mineral: $(Ni,Fe)_7S_6$.

goe *geo.*

goedkenite A monoclinic mineral: $(Sr,Ca)_2Al(PO_4)_2(OH)$.

goethite A yellowish, reddish, or brownish-black mineral: α-FeO (OH). It is trimorphous with lepidocrocite and akaganeite. Goethite is the commonest constituent of many forms of natural rust or of limonite, and it occurs esp. as a weathering product in the gossans of sulfide-bearing ore deposits. Also spelled: *göthite.* Syn: *allcharite; xanthosiderite.*

gold A soft, heavy, yellow, isometric mineral, the native metallic element Au. It is often naturally alloyed with silver or copper and occasionally with bismuth, mercury, or other metals, and is widely found in alluvial deposits (as nuggets and grains) or in veins associated with quartz and various sulfides. Gold is malleable and ductile, and is used chiefly for jewelry and as the international standard for world finance.

gold amalgam A variety of native gold containing mercury; a naturally occurring *amalgam* composed of gold, silver, and mercury, the gold averaging about 40%. It is usually associated with platinum, and occurs in yellowish-white grains that crumble readily.

gold beryl A syn. of *chrysoberyl.* Not to be confused with *golden beryl.*

gold dust Fine particles, flakes, or pellets of gold, such as those obtained in placer mining. Cf: *commercial dust.*

golden beryl A clear, golden-yellow or yellowish-green gem variety of beryl. See also: *heliodor.* Not to be confused with *gold beryl.*

goldfieldite A dark lead-gray mineral: $Cu_{12}(Sb,As)_4(Te,S)_{13}$. It is the tellurium analogue of tetrahedrite.

goldichite A pale-green monoclinic mineral: $KFe(SO_4)_2 \cdot 4H_2O$.

goldmanite A mineral of the garnet group: $Ca_3(V,Al,Fe)_2(SiO_4)_3$.

gold opal A *fire opal* that exhibits only an overall color of golden yellow.

gold quartz Milky quartz containing small inclusions of gold.

goldschmidtine *stephanite.*

goldschmidtite *sylvanite.*

Goldschmidt's phase rule *mineralogical phase rule.*

goldstone A translucent, reddish-brown glass containing a multitude of tiny thin tetrahedra or hexagonal platelets of metallic copper, which exhibit bright reflections and produce a popular but poor imitation of *aventurine.* See also: *sunstone.* Syn: *aventurine glass.*

goletz terrace A terrace "in which bedrock is exposed in the scarp and is close to the surface of the bench" (Bird, 1967, p. 248).

gompholite *nagelfluh.*

gonal spine A spine situated at plate corners on a dinoflagellate cyst.

gonatoparian adj. Of or concerning a trilobite having facial sutures, the posterior sections of which reach the cephalic margin at the genal angles; also, said of the sutures themselves.

gondite A metamorphic rock consisting of spessartine and quartz, probably derived from manganese-bearing sediment. It is named after Gonds and the Gondite Series, central India. Cf: *collobrierite; eulysite.*

Gondwana The Late Paleozoic continent of the Southern Hemisphere. It was named by Suess for the Gondwana system of India, which has an age range from Carboniferous to Jurassic and contains glacial tillite in its lower part and coal measures higher up. Similar sequences of the same age are found in all the continents of the hemisphere; this similarity, along with much compelling evidence of other sorts, indicates that all these continents were once joined into a single larger mass. The preponderance of modern evidence indicates that the present continents are fragments that have been separated from each other by *continental displacement.* The counterpart of Gondwana in the Northern Hemisphere was *Laurasia;* the supercontinent from which both were derived was *Pangea.* Var: *Gondwanaland.*

Gondwanaland Var. of *Gondwana* introduced by Suess. Gondwana is preferred as it means "land of the Gonds" and Gondwanaland is tautological.

gongylodont Said of a class of ostracode hinges consisting of three elements wherein the terminal elements are opposites in the same valve, as in the genus *Loxoconcha.*

goniatite Any ammonoid cephalopod belonging to the order Goniatitida, characterized generally by a shell having sutures of angular appearance with eight undivided lobes. Range, Middle Devonian to Upper Permian.

goniatitic suture A type of suture in ammonoids characterized by simple fluting in which most or all of the lobes and saddles are entire or plain (not denticulate or frilled), the only common exception being the ventral lobe, which is subdivided and may be denticulate; specif. a suture in goniatites. Cf: *ammonitic suture; ceratitic suture.*

goniometer (a) An instrument used in optical crystallography for measuring the angles between crystal faces. Types are the *contact goniometer,* the *reflection goniometer,* and the *two-circle goniometer.* (b) An instrument that measures X-ray diffractions; a *diffractometer.*

gonnardite A zeolite mineral: $Na_2CaAl_4Si_6O_{20} \cdot 7H_2O$. It occurs in finely fibrous, radiating spherules.

gonopore (a) A simple opening that serves as an exit from the genital system of an echinoderm, e.g. a cystoid or an edrioasteroid; a genital pore. In certain cystoids it is combined with a *hydropore.* (b) The outlet of the genital ducts in crustaceans; a sexual pore.

gonozooid A zooid modified as a brood chamber in the stenolaemate bryozoans.

gonyerite A mineral of the chlorite group: $(Mn,Mg)_6Si_4O_{10}(OH)_8$.

gooderite A plutonic *nepheline syenite* with a predominance of albite rather than potassium feldspar. Named by Johannsen (1938) for Gooderham Township, Ontario. Not recommended usage.

goodness of fit A statistical test used to ascertain agreement of observed data with theoretical distributions, with other observed data, or with some mathematical functions. See also: *chi-square test.*

Goodsprings twin law A rare type of normal twin in feldspar, in which the twin plane is ($\bar{1}12$).

goongarrite A lead-bismuth sulfide consisting of a mixture of cosalite and galena, found at Lake Goongarrie in Western Australia.

gooseberry stone A syn. of *grossular,* esp. used for the yellowish-green varieties.

gooseneck The part of a winding valley resembling in plan the curved neck of a goose; esp. a part formed by an entrenched meander.

GOR *gas-oil ratio.*

gorceixite A brown mineral: $BaAl_3(PO_4)_2(OH)_5 \cdot H_2O$.

gordonite A colorless mineral: $MgAl_2(PO_4)_2(OH)_2 \cdot 8H_2O$.

gordunite A *peridotite* composed chiefly of olivine and less pyroxene, with small amounts of pyrope, picotite, and opaque oxides; a garnet-bearing *wehrlite.* Not recommended usage.

gore [cart] One of the series of related and triangular or lune-shaped sections of a map or chart, usually bounded by meridians and tapering to the poles, which can be applied to the surface of a sphere (with a negligible amount of distortion) to form a globe.

gore [surv] A small irregularly shaped tract of land, generally triangular, left between two adjoining surveyed tracts, often because of inaccuracies in the boundary surveys or as a remnant of a systematic survey. It is an officially recognized tract in some States (e.g. Maine and Vermont).

gorge (a) A narrow, deep valley with nearly vertical rocky walls, enclosed by mountains, smaller than a *canyon [geomorph],* and more steep-sided than a *ravine;* esp. a restricted, steep-walled part of a canyon. (b) A narrow defile or passage between hills or mountains.—Etymol: French, "throat".

görgeyite A mineral: $K_2Ca_5(SO_4)_6 \cdot H_2O$.

goshenite A colorless, white, or bluish beryl from Goshen, Mass.

goslarite A white mineral: $ZnSO_4 \cdot 7H_2O$. It forms by oxidation of sphalerite and usually occurs massive. Syn: *white vitriol; zinc vitriol; white copperas.*

gossan An iron-bearing weathered product overlying a sulfide deposit. It is formed by the oxidation of sulfides and the leaching-out of the sulfur and most metals, leaving hydrated iron oxides and rarely sulfates. Syn: *capping; leached capping; iron hat; chapeau de fer.* Also spelled: *gozzan.* Cf: *oxidized zone; false gossan.*

gossany Pertaining to or comprising gossan.

gote A British term for a watercourse.

göthite *goethite.*

Gothlandian *Gotlandian.*

Gotlandian An alternative name of the Silurian, specif. the late Silurian, used in Europe. Also spelled: *Gothlandian.*

götzenite A mineral: $(Ca,Na)_3TiSi_2O_7(F,OH)_2$.

gouffre A French term for a gulf, chasm, or pit, sometimes applied in English-language publications to a natural gorge, a karstic depression, or a pit cave.

gouge [glac geol] *crescentic gouge.*

gouge [ore dep] (a) A thin layer of soft, earthy fault-comminuted rock material along the wall of a vein or between the country rock and the vein, so named because a miner is able to "gouge" it out and thereby facilitate the mining of the vein itself. Syn: *selvage [ore dep]; pug.* See also: *clay gouge.* (b) A term used in Nova Scotia for a narrow band of gold-bearing slate next to a vein, extractable by a thin pointed stick.

gouge channel A term used by Kuenen (1957, p.242) for a large sole mark (larger than a flute cast) now known as a *channel cast.* Syn: *megaflow mark.*

gouge mark *crescentic gouge.*

gouging [glac geol] (a) The formation of *crescentic gouges.* (b) The local basining of a bedrock surface by the action of glacier ice (Thornbury, 1954, p. 48).

gouging [mining] The working of a mine without plan or system, by which only the high-grade ore is mined. Syn: *high-grading.*

gour [geomorph] Plural of *gara.*

gour [speleo] A syn. of *rimstone dam* and *rimstone pool;* it is taken from the French. See also: *microgour.*

gowerite A monoclinic mineral: $CaB_6O_{10} \cdot 5H_2O$.

goyazite A yellowish-white mineral: $SrAl_3(PO_4)_2(OH)_5 \cdot H_2O$. Syn: *hamlinite.*

goyle An English term for a ravine or other steep, narrow valley.

goz A term used in Sudan for a long, gentle, dunelike accumulation of sand, ranging in thickness from a few decimeters to tens of meters; also, a large-scale, undulating tract containing such accumulations. Etymol: Arabic. Pl: *gozes.*

gozzan *gossan.*

gr *group.*

graben An elongate, relatively depressed crustal unit or block that is bounded by faults on its long sides. It is a structural form that may or may not be geomorphologically expressed as a *rift valley.* Etymol: German, "ditch". Cf: *horst.* Syn: *trough [fault].*

grab sampler An ocean-bottom sampler that commonly operates by enclosing material from the seafloor between two jaws upon contact with the bottom. See also: *Shipek bottom sampler; Peterson grab; Van Veen grab; clamshell snapper.* Cf: *dredge; corer.* Syn: *snapper.*

gradation [geomorph] (a) The leveling of the land, or the bringing of a land surface or area to a uniform or nearly uniform grade or slope through erosion, transportation, and deposition; specif. the bringing of a stream bed to a slope at which the water is just able to transport the material delivered to it. See also: *gradation; degradation; aggradation.* (b) Often used as a syn. of *degradation.*

gradation [part size] The proportion of material of each particle size, or the frequency distribution of various sizes, constituting a particulate material such as a soil, sediment, or sedimentary rock. The limits of each size are chosen arbitrarily. Cf: *sorting; grading [part size].*

gradation period "The entire time during which the base level remains in one position; that is, the interval between two elevations of the Earth's surface of sufficient magnitude to produce a marked change in the position of sea level" (Hayes, 1899, p. 22).

gradation zone "A body of rock with upper and lower boundaries defined by selected evolutionary stages of a gradational bioseries of planktonic Foraminifera or of any similarly well distributed bioseries" (Vella, 1964, p. 621).

grade [coal] A *coal classification* based on degree of purity, i.e., quantity of inorganic material or ash left after burning. Cf: *type [coal]; rank [coal].*

grade [eng] (a) A degree of inclination, or a rate of ascent or descent, with respect to the horizontal, of a road, railroad, embankment, conduit, or other engineering structure; it is expressed as a ratio (vertical to horizontal), a fraction (such as m/km or ft/mi), or a percentage (of horizontal distance). (b) A graded part of a road, embankment, or other engineering structure that is ascending, descending, or level.—The synonymous term *gradient* is used in geomorphology.

grade [evol] (a) A group of organisms, all at the same or a similar level of organization or advancement; some such groups may be formalized as higher taxa, which then are usually *polyphyletic,* not generally considered desirable in modern systematics. (b) The common level of evolutionary development that has been independently attained by two or more separate but related evolving *lineages.* (c) A stage of evolution of an animal in which one or more features have undergone a *chronocline.*

grade [meta] *metamorphic grade.*

grade [ore dep] The relative quantity or the percentage of ore-mineral content in an orebody. Syn: *tenor.*

grade [part size] A particular size (diameter), size range, or size class of particles of a soil, sediment, or rock; a unit of a grade scale, such as "clay grade", "silt grade", "sand grade", or "pebble grade".

grade [streams] (a) The condition of balance, achieved by a stream, between erosion and deposition, brought about by the adjustments between the capacity of the stream to do work and the quantity of work that the stream has to do (Davis, 1902, p. 86). It is represented by the continuously descending curve (the longitudinal profile of the stream) which everywhere is just steep enough to allow the stream to transport the load of sediment made available to it. Grade involves an equilibrium among slope, load, volume, velocity and channel characteristics. The term was used by Gilbert (1876) but the concept was first formally introduced by Davis (1894) who admits that it "cannot be understood without rather careful thinking"; although a precise definition is difficult, the concept is useful as it implies both an adjustability of the channel to changes in independent variables and a stability in form and profile. (b) A term sometimes used as a syn. of *gradient* of a given length of a stream. This usage is confusing and not recommended.

grade [surv] (a) A datum level; a level of reference. (b) Height above sea level; actual elevation. Also, the elevation of the finished surface of an engineering project (such as of a canal bed, embankment top, or excavation bottom). (c) Rate of slope; degree of inclination.

grade correction A correction applied to a distance measured on a slope to reduce it to a horizontal distance between the vertical lines through its end points. Syn: *slope correction.*

graded [geomorph] Said of a surface or feature when neither degradation nor aggradation is occurring, or when both erosion and deposition are so well balanced that the general slope of equilibrium is maintained. Cf: *in regime.* Syn: *at grade.*

graded [part size] (a) A geologic term pertaining to an unconsolidated sediment or to a cemented detrital rock consisting of particles of essentially uniform size or of particles lying within the limits of a single grade. Syn: *sorted [part size].* (b) An engineering term pertaining to a soil or an unconsolidated sediment consisting of particles of several or many sizes or having a uniform or equable distribution of particles from coarse to fine; e.g. a "graded sand" containing coarse, medium, and fine particle sizes. See also: *well-graded.*—The term is "rarely used in geology to refer to the sorting of the sediment, although this is common among engineers" (Middleton, 1965, p. 249). Ant: *nongraded.*

graded bed A sedimentary bed, usually thin, exhibiting *graded bedding,* generally having an abrupt contact with the fine material of the underlying bed but a gradational or indefinite contact near the top; e.g. sand or coarse silt grading upward into shaly material.

graded bedding A type of bedding in which each layer displays a gradual and progressive change in particle size, usually from coarse at the base of the bed to fine at the top. It may form under conditions in which the velocity of the prevailing current declined in a gradual manner, as by deposition from a single short-lived turbidity current. See also: *sorted bedding.*

graded profile *profile of equilibrium.*

graded reach (a) A part of a stream characterized by a condition of balance between erosion and deposition, as where a stream crossing an outcrop of weak rocks is in equilibrium while its profile across resistant rocks remains for a long time irregular and steep. (b) A reach of a graded stream.

graded shoreline A shoreline that has been straightened or simplified by the formation of barriers across embayments and by the cutting-back of headlands, and that possesses a broadened surface profile so adjusted in slope that the energy of incoming waves is completely absorbed and the shifting of the shoreline is reduced to a very slow rate; a shoreline with a vertical *profile of equilibrium,* typical of an advanced stage of development. Syn: *equilibrium shoreline.*

graded slope The downstream gradient of a graded stream; it permits the most effective transport of load and is represented by the *profile of equilibrium.*

graded stream (a) A stream in equilibrium, showing a balance between its transporting capacity and the amount of material supplied to it, and thus between degradation and aggradation in the stream channel. "A graded stream is one in which, over a period of years, slope and channel characteristics are delicately adjusted to provide, with available discharge, . . . just the velocity required for the transportation of the load supplied from the drainage basin. . . . Its diagnostic characteristic is that any change in any of the controlling factors will cause a displacement of the equilibrium in a direction that will tend to absorb the effect of the change" (Mackin, 1948, p. 471, revised by Leopold & Maddock, 1953, p. 51). A graded stream is not a stream that is loaded to capacity (streams probably never attain this condition), and neither is it a stream that is neither eroding nor depositing (erosion may occur in one part of the channel and deposition in another part). The term is not to be confused with *gradient,* which is possessed by all streams. Syn: *steady-state.* (b) A stream characterized by the absence of waterfalls and rapids (Kesseli, 1941).—Cf: *poised stream; regime stream.*

graded unconformity *blended unconformity.*

grade level The level attained by a stream when its "whole course" has been reduced to a uniform gradient, or when its longitudinal profile is a straight line (Park, 1914, p. 42).

grade scale A systematic, arbitrary division of an essentially continuous range of particle sizes (of a soil, sediment, or rock) into a series of classes or scale units (or grades) for the purposes of standardization of terms and of statistical analysis; it is usually logarithmic. Examples include: *Udden grade scale; Wentworth grade scale; phi grade scale; Atterberg grade scale; Tyler Standard grade scale; Alling grade scale.* See also: *geometric grade scale.*

gradient [geomorph] (a) A degree of inclination, or a rate of ascent or descent, of an inclined part of the Earth's surface with respect to the horizontal; the steepness of a slope. It is expressed as a ratio (vertical to horizontal), a fraction (such as m/km or ft/mi), a percentage (of horizontal distance), or an angle (in degrees). Syn: *slope [geomorph].* (b) A part of a surface feature that slopes upward or downward; a slope, as of a stream channel or of a land surface.—The synonymous term *grade* is used in engineering.

gradient [geophys] The first derivative or change in value of one variable with respect to another such as a change in gravity, temperature, magnetic susceptibility, or electrical potential with respect to horizontal or vertical distance. See also: *gradiometer.*

gradient [hydraul] *hydraulic gradient.*

gradient [streams] A *stream gradient.* Cf: *grade [streams].*

gradient array An electrode array used in resistivity and induced-polarization surveys. Both current electrodes are fixed, while the two potential electrodes are close enough together to measure the gradient of the potential. The potential probes are moved along traverse lines normal to geologic lineation and parallel to a line joining the current electrodes. A square area, of dimensions one-third of the separation of the current electrodes and situated midway between them, is surveyed.

gradient current A wind or ocean current in which the horizontal pressure force is exactly balanced by the sum of the *Coriolis force* and the surface or bottom frictional forces. It flows to the right of the pressure gradient force in the Northern Hemisphere, but not along the isobars. Cf: *geostrophic current.*

gradienter An attachment to a surveyor's transit with which an angle of inclination is measured in terms of the tangent of the angle instead of in degrees and minutes. It may be used as a telemeter in observing horizontal distances.

gradient of the head *Hydraulic gradient* for which the specified direction is that of maximum rate of increase in head.

grading [geomorph] The reduction of the land to a level surface or equilibrium slope, such as erosion to base level by streams.

grading [part size] The gradual reduction, in a progressively upward direction within an individual stratification unit, of the upper particle-size limit. It implies pulsatory turbulent-fluid deposition. Cf: *gradation [part size].*

grading factor A *sorting index* developed by Baker (1920, p. 368) and defined as the difference between unity and the quotient of mean deviation divided by arithmetic mean diameter (equivalent grade); a measure of how nearly the degree of sorting approaches perfection. Abbrev: GF.

gradiometer Any instrument that is used to measure the *gradi-*

ent of a physical quantity, e.g. a device consisting of two magnetometers, one above the other, that measures the difference in the magnetic field at two locations.

gradualistic speciation *phyletic gradualism.*

gradually varied flow The flow in an open channel where the velocity changes slowly along the channel and the flow is assumed uniform for an increment of length. See also: *flow profile.*

graduation (a) The method or system of placing degrees or other equally spaced intermediate marks on an instrument or device (such as a thermometer or tape) to represent standard or conventional values. (b) A mark or the marks so placed; one of the equal divisions or dividing lines on a graduated scale.

graemite An orthorhombic mineral: $CuTeO_3 \cdot H_2O$.

Graf sea gravimeter A balance-type gravity meter that is heavily damped in order to attenuate shipboard vertical accelerations. It consists of a mass at the end of a horizontal arm, supported by a torsion-spring rotational axis. The mass rises and falls with gravity variation, but is restored to near its null position by a horizontal reading spring, tensioned with a micrometer screw. The difference between actual beam position and null position gives indication of gravity value after the micrometer screw position has been taken into account (U.S. Naval Oceanographic Office, 1966, p.72).

graftonite A salmon-pink monoclinic mineral: $(Fe,Mn,Ca)_3(PO_4)_2$. It occurs in laminated intergrowths with triphylite.

grahamite [meteorite] *mesosiderite.*

grahamite [mineral] A black asphaltite with a variable luster, black streak, high specific gravity, and high fixed-carbon content.

grail Coarse- or medium-grained sediment particles; specif. gravel or sand.

grain [eco geol] A quarrymen's term for a plane of parting in a metamorphic rock, e.g. slate, that is perpendicular to the flow cleavage; or for a direction of parting in massive rock, e.g. granite. Cf: *rift [eco geol].*

grain [gem] (a) A unit of weight commonly used for pearls and sometimes for other gems, equal to 1/4 *carat*, or 0.0500 gram. (b) A diamond-cutting term referring to cleavage direction.

grain [geomorph] (a) The broad, linear arrangement of the topographic features (such as mountain ranges and valleys) or underlying geologic structures (such as folds and bedding) of a country or region; e.g. the arrangement of roughly parallel ridges and valleys often displayed in regions of tilted strata. (b) The general direction or trend of such physical or structural features; e.g. the *grain* of northern Scotland runs NE and SW.—Syn: *grain of the country.*

grain [glaciol] An individual particle in snow, ice, or glacier material, consisting of a single ice crystal or a mechanically separate particle of ice. Cf: *snow grain; glacier grain.*

grain [palyn] *pollen grain.*

grain [petrology] (a) A mineral or rock *particle,* smaller than a *fragment,* having a diameter of less than a few millimeters and generally lacking well-developed crystal faces; esp. a small, hard, more or less rounded mineral particle, such as a sand grain. Also, a general term for sedimentary particles of all sizes (from clay to boulders), as used in the expressions "grain size", "fine-grained", and "coarse-grained". (b) The factor of rock texture that depends on the absolute sizes (fineness or coarseness) of the distinct particles composing the rock. Also, the factor of rock texture that is due to the arrangement or trend of constituent particles, such as a lineation or stratification; e.g. the "magnetic grain" in the crustal structure of a region.

grain [water] A unit of hardness of water, expressed in terms of equivalent $CaCO_3$. A hardness of one grain per U.S. gal equals 17.1 ppm by weight as $CaCO_3$. Cf: *Clark degree.*

grain boundary In a polycrystalline solid, the boundary between two crystals. See also: *plane defect.*

grain density (a) Specific gravity of the grains composing a sediment or sedimentary rock. (b) Syn. of *packing density.*

grain diminution (a) *degradation recrystallization.* (b) *micritization.*

grain growth (a) The growth of a crystal, as from solution on the walls of a container, in open pore space or in a magma chamber; *crystal growth.* (b) Applied by Bathurst (1958, p.24) to carbonate sediments, e.g. calcite mud or fibers changing to calcite mosaic with a coarser texture; in this sense it is equivalent to recrystallization; Folk (1965, p.16-20) objects to the usage of this term in carbonate petrology. (c) A metallurgical term for the solid-state growth, coalescence or enlargement of a crystal at the expense of another, occurring between unstrained or undeformed grains.

grain-micrite ratio A ratio that expresses the relative proportion of larger to smaller particles in a carbonate sedimentary rock. It is defined as the sum of the percentages of grains (detrital grains, skeletal grains, pellets, lumps, coated grains, and mineral grains) divided by the percentage of micrite (calcareous mud or its consolidated equivalent). It excludes diagenetic or postdepositional features such as cement, vugs, fractures, vein fillings, and recrystallized areas. Abbrev: GMR.

grain shape *particle shape.*

grain size (a) *particle size.* (b) *granularity.*

grainstone A term used by Dunham (1962) for a mud-free (less than 1% of material with diameters less than 20 microns), grain-supported, carbonate sedimentary rock. It may be current-laid or formed by mud being washed out from previously deposited sediment, or it may result from mud being bypassed while locally produced particles accumulated. Cf: *packstone; mudstone.*

grain-supported A term used by Dunham (1962) to describe a sedimentary carbonate rock with little or no muddy matrix, whose sand-size particles are so abundant that they are in three-dimensional contact and able to support one another. Cf: *mud-supported.*

gralmandite A garnet intermediate in chemical composition between grossular and almandine (almandite).

gramenite *nontronite.*

graminivore A *herbivore* that eats primarily grains or grasses.

granat (a) A term used in Ireland for quartzose grit. It is presumably an obsolete variant of "granite" (Arkell & Tomkeieff, 1953, p.50). (b) An obsolete form of "garnet". Etymol: German *Granat,* "garnet".

Grand Canyon A provincial series of the Proterozoic in Arizona.

grandidierite A mineral: $(Mg,Fe)Al_3(BO_4)(SiO_4)O$.

grandite A garnet intermediate in chemical composition between grossular and andradite.

granide A syn. of *granitic rock,* proposed by Johannsen (1939, p. 253) for light-colored medium- to coarse-grained quartz- and feldspar-bearing plutonic rocks also containing biotite or hornblende.

granilite An obsolete term formerly applied to a crystalline igneous rock having more than three phases.

graniphyric Said of the texture of a porphyritic igneous rock having a microlitic groundmass (Cross et al., 1906, p.704). Cf: *granophyric.*

granite [mater] *commercial granite.*

granite [petrology] (a) In the *IUGS classification,* a plutonic rock with Q between 20 and 60 and P/(A+P) between 10 and 65. (b) A plutonic rock in which quartz constitutes 10 to 50 percent of the felsic components and in which the alkali feldspar/total feldspar ratio is generally restricted to the range of 65 to 90 percent. Rocks in this range of composition are scarce, and sentiment has been growing to expand the definition to include rocks designated as *adamellite* or *quartz monzonite,* which are abundant in the U.S. (c) Broadly applied, any holocrystalline, quartz-bearing plutonic rock. Syn: *granitic rock.*—Etymol: Latin *granum,* "grain".

granite [seis] In early seismologic work, any rock in which velocity of the compressional wave is about 5.5-6.2 km/sec.

granite dome A term introduced by Davis (1933) for what he later termed a *desert dome* (Davis, 1938), because the feature was not always developed across granites.

granite gneiss (a) A gneiss derived from a sedimentary or igneous rock and having the mineral composition of a granite. (b) A metamorphosed granite.

granitelle An obsolete term formerly applied to a compound of quartz and feldspar; originally used as a syn. of *two-mica granite.*

granitello An obsolete term formerly applied to fine-grained *granite.*

granite-pebble conglomerate A term used by Krumbein & Sloss (1963, p.164) for *arkosic conglomerate.*

granite porphyry A hypabyssal rock differing from a *quartz porphyry* by the presence of sparse phenocrysts of mica, amphibole, or pyroxene in a medium- to fine-grained groundmass.

granite series A sequence of products that evolved continuously during crustal fusion, earlier products tending to be deep-seated, syntectonic, and granodioritic, and later products tending to be shallower, late syntectonic or post-tectonic, and more potassic (Turner & Verhoogen, 1960, p. 388).

granite tectonics The study of the structural features, such as foliation, lineation, and faults, in plutonic rock masses, and the reconstruction of the movements that created them.

granite wash A driller's term for material eroded from outcrops of granitic rocks and redeposited to form a rock having approximately the same major mineral constituents as the original rock (Taylor & Reno, 1948, p. 164); e.g. an arkose consisting of granitic detritus. Cf: *basic wash.*

granitic (a) Pertaining to or composed of granite. (b) A nonrecommended syn. of *granular,* or, more restrictively, of *subautomorphic.* —Syn: *granitoid; eugranitic.*

granitic layer A syn. of *sial,* so named for its supposed petrologic composition. A layer is sometimes called "granitic layer" if it possesses the appropriate seismic velocity ($\sim$ 6.0 km/s), although nothing may be known about its composition. Cf: *basaltic layer.*

granitic rock A term loosely applied to any light-colored coarse-grained plutonic rock containing quartz as an essential component, along with feldspar and mafic minerals. Syn: *granite [petrology]; granitoid; granide.*

granitification *granitization.*

granitine A crystalline rock containing any three minerals other than those of a *granite.* Obsolete.

granitite A *granite* that contains biotite but no other ferromagnesian mineral or muscovite. The term, now obsolete, was mainly used by French petrologists; it has also been used for a granite rich in oligoclase or containing hornblende or accessory minerals (Johannsen, 1939, p. 254).

granitization An essentially metamorphic process or group of processes by which a solid rock is converted or transformed into a granitic rock by the entry and exit of material, without passing through a magmatic stage. Some authors include in this term all granitic rocks formed from sediments by any process, regardless of the amount of melting or evidence of movement. The precise mechanism, frequency, and magnitude of the processes are still in dispute. See also: *transformism.* Cf: *magmatism; transfusion.* Syn: *transformation; granitisation; granitification.*

granitizer *transformist.*

granitogene Said of a sediment composed of granitic fragments.

granitoid n. (a) In the *IUGS classification,* a preliminary term (for field use) for a plutonic rock with Q between 20 and 60. (A *granitic rock.* —adj. A syn. of *granitic.*

granitotrachytic *ophitic.*

granoblastic (a) Pertaining to a *homeoblastic* type of texture in a nonschistose metamorphic rock upon which recrystallization formed essentially equidimensional crystals with normally well sutured boundaries (Harker, 1939). See also: *sutured.* Cf: *granuloblastic.* (b) Pertaining to a secondary texture due to diagenetic change, either by crystallization or recrystallization in the solid state, in which the grains are of equal size (Pettijohn, 1949). This usage is discouraged on the basis that diagenesis is not a metamorphic process.

granodiorite (a) In the *IUGS classification,* a plutonic rock with Q between 20 and 60, and P/(A+P) between 65 and 90. (b) A group of coarse-grained plutonic rocks intermediate in composition between *quartz diorite* and *quartz monzonite* (U.S. usage), containing quartz, plagioclase (oligoclase or andesine), and potassium feldspar, with biotite, hornblende, or, more rarely, pyroxene, as the mafic components; also, any member of that group; the approximate intrusive equivalent of *rhyodacite.* The ratio of plagioclase to total feldspar is at least two to one but less than nine to ten. With less alkali feldspar it grades into quartz diorite, and with more alkali feldspar, into granite or quartz monzonite. The term first appeared in print in 1893 in a paper by Lindgren and was applied to all rocks intermediate in composition between *granite* and *diorite.* The term has the connotation that the rock is a diorite with granitic characteristics, i.e. with quartz and a certain amount of alkali feldspar (Johannsen, 1939, p. 254).

granodolerite A *dolerite* that contains interstitial quartz and alkali feldspar, usually as granophyric intergrowths. Not recommended usage.

granofels A field name for a medium- to coarse-grained granoblastic metamorphic rock with little or no foliation or lineation (Goldsmith, 1959). Cf: *fels.*

granogabbro A *gabbro* containing quartz and alkali feldspar, analogous to *granodiorite* but having plagioclase more calcic than An_{50}. Proposed by Johannsen in 1917. Not recommended usage.

granolite A general term for any plutonic rock characterized by granitic rather than porphyritic texture. The term, proposed by Pirsson in 1899, has never been widely used.

granomerite A holocrystalline rock that contains no kryptomerous groundmass. Obsolete.

granophyre (a) An irregular microscopic intergrowth of quartz and alkali feldspar. (b) As defined by Rosenbusch in 1872, a porphyritic extrusive rock characterized by a micrographic holocrystalline groundmass, or a fine-grained granitic rock having a micrographic texture. Syn: *pegmatophyre.* (c) As defined by Vogelsang in 1867, a porphyritic rock of granitic composition characterized by a crystalline-granular groundmass.—See: Barker, 1970.—Adj: *granophyric.* Cf: *felsophyre; vitrophyre.*

granophyric (a) As defined by Rosenbusch, a term applied to the texture of a porphyritic igneous rock in which the phenocrysts and groundmass penetrate each other, having crystallized simultaneously; of or pertaining to a *granophyre* (Cross et al., 1906, p.703). (b) As defined by Vogelsang, a term applied to a porphyritic igneous rock having a microgranular groundmass (Johannsen, 1939, p.214). Cf: *graniphyric.*

granoschistose Pertaining to a structure of a monomineralic metamorphic rock produced by the parallel elongation of grains of a mineral that is normally equidimensional or nearly so.

granosphaerite A spherulite composed of radially or concentrically arranged grains.

granosyenite A term used esp. in the Russian literature for a plutonic igneous rock with a composition intermediate between that of granite and that of syenite.

grantsite A dark olive-green to greenish-black mineral: $Na_4Ca_xV_{2x}{}^{+4}V_{12-2x}{}^{+5}O_{32}\cdot 8H_2O$.

granular [geol] (a) Said of the texture of a rock that consists of mineral grains of approximately equal size. The term may be applied to sedimentary rocks, e.g. sandstones, but is esp. used to describe holocrystalline igneous rocks whose major-phase grain size ranges from 2 to 10 mm. Nonrecommended syn: *granitic.* The syn. *granoblastic* is used for metamorphic rocks. (b) An element in compound adjectives, esp. used to differentiate a texture (e.g. *xenomorphic-granular*) from the corresponding crystal shape (e.g. *xenomorphic*). This usage is obsolescent.

granular [paleont] Covered with very small grains or having numerous small protuberances; e.g. "granular pattern" of ornamentation on the walls of spores and pollen grains, or "granular hyaline wall" representing a perforate and lamellar part of a foraminiferal test composed of minute, equidistant, and variously oriented grains of calcite and seen between crossed nicols as a multitude of tiny flecks of color (TIP, 1964, pt.C, p.60).

granular cementation Chemical deposition of material from solution onto a free surface between detrital grains of a sediment, resulting in outward growth of crystalline material adhering to that surface (Bathurst, 1958, p. 14); e.g. growth of calcite in the pores of an unconsolidated sand. Cf: *rim cementation.*

granular chert A compact, homogeneous, hard to soft chert, common in insoluble residues, composed of distinguishable and relatively uniform-sized grains, characterized by an uneven or rough fracture surface and by a dull to glimmering luster (Ireland et al., 1947, p. 1486); it may appear saccharoidal. See also: *granulated chert.* Cf: *smooth chert; chalky chert.* Syn: *crystalline chert.*

granular disintegration A type of weathering consisting of grain-by-grain breakdown of rock masses composed of discrete mineral crystals that separate from one another along their natural contacts. It produces a coarse mineral debris, each grain having much the same shape and size as in the original rock. It develops esp. in coarse-grained rocks (such as granite, gneiss, sandstone, and conglomerate) occurring in regions of great temperature extremes. Syn: *mineral disintegration; granular exfoliation.*

granular exfoliation *granular disintegration.*

granular ice Ice made of small crystals of irregular form but having somewhat rounded forms like sand particles.

granularity The quality, state, or property of being granular; specif. one of the component factors of the texture of a crystalline rock, including both grain size and grain-size distribution. Granularity and grain relationships define textures, which, together with orientation, are included in the definition of *fabric* (in Sander's sense). Friedman (1965, p.647) used the term to refer to the "size and mutual relations of crystals" in a sedimentary rock such as an evaporite, a chemically deposited cement, or a recrystallized limestone or dolomite.

granular snow Cohesionless coarse-grained snow.

granular structure A type of soil structure in which the peds are spheroids or polyhedrons that have little or no accommodation to surrounding peds, are relatively nonporous, and range in size from less than 1.0mm to more than 10.0mm. Cf: *crumb structure.*

granular texture A rock texture resulting from the aggregation of mineral grains of approximately equal size. The term may be applied to a sedimentary or metamorphic rock, but is esp. used to describe an equigranular, holocrystalline igneous rock whose particles range in diameter from 0.05 to 10 mm. See also: *granitic.*

granulated chert A type of *granular chert* composed of rough, irregular grains or granules of chert tightly or loosely held together in small masses or fragments (Hendricks, 1952, p. 12 & 18).

granulation The act or process of being formed into grains, granules, or other small particles; specif. the crushing of a rock under such conditions that no visible openings result. Also, the state or condition of being granulated.

granule [paleont] A minute, more or less spherical skeletal element situated on the surface of asterozoan ossicles, generally in a pit or distributed in covering skin.

granule [sed] (a) A term proposed by Wentworth (1922, p. 380-381) for a rock fragment larger than a very coarse sand grain and smaller than a pebble, having a diameter in the range of 2-4 mm (1/12 to 1/6 in., or -1 to -2 phi units, or a size between that of the head of a small wooden match and that of a small pea) being somewhat rounded or otherwise modified by abrasion in the course of transport. The term *very fine pebble* has been used as a synonym. (b) A little grain or small particle, such as one of a number of the generally round or oval, nonclastic (precipitated), internally structureless grains of glauconite or other iron silicate in iron formation; a pseudo-oolith.

granule gravel An unconsolidated deposit consisting mainly of granules.

granule ripple A large *wind ripple* consisting in part of granule-sized particles. Syn: *deflation ripple.*

granule texture A texture of *iron formation* in which precipitated or nonclastic granules are separated by a fine-grained matrix.

granulite [ign] A muscovite-bearing *granite,* esp. in the French literature. Obsolescent; not recommended usage.

granulite [meta] (a) A metamorphic rock consisting of even-sized, interlocking mineral grains less than 10% of which have any obvious preferred orientation. (b) A relatively coarse, granular rock formed at the high pressures and temperatures of the *granulite facies,* which may exhibit a crude gneissic structure due to the parallelism of flat lenses of quartz and/or feldspar. The texture is typically granuloblastic.

granulite [sed] A sedimentary rock composed of sand-sized aggregates of constructional (nonclastic) origin, simulating in texture an arenite of clastic origin; e.g. a rock formed of lapilli or of oolitic grains. The term was introduced by Grabau (1911, p. 1007). Syn: *granulyte.*

granulite facies The set of metamorphic mineral assemblages (facies) in which basic rocks are represented by diopside + hypersthene + plagioclase, with amphibole generally minor in amount. Almandine is characteristic of basic and pelitic rocks. Pelitic assemblages show the association of sillimanite or kyanite with perthitic feldspar and almandine, often also with cordierite; muscovite is absent and biotite small in amount. The facies is typical of deep-seated regional dynamothermal metamorphism, at temperatures in excess of 650°C. Cf: *pyroxene-hornfels facies.*

granulitic [ign] (a) Of, pertaining to, or composed of *granulite.*—All other usages are now discouraged, including the following: (b) A syn. of *granoblastic;* (c) a syn. of *granuloblastic;* (d) a descriptive term for a fine-grained texture produced by crushing and *cataclasis;* (e) a syn. of *xenomorphic* as used by Michel-Lévy in 1874 and 1889; (f) a term applied by Judd in 1886 to the texture of basaltic or doleritic rocks (Johannsen, 1939, p. 215) (preferred syn: *intergranular*); (g) a syn. of *microgranular.*

granulitic [meta] (a) Pertaining to a *granoblastic* texture having xenoblastic crystal development. (b) Pertaining to a structure resulting from the production of granular fragments in a rock by crushing.—Use of this term requires precise definition of its meaning.

granulitic [sed] Said of the rock structure resulting from the production of flattened or granular fragments in a rock by crushing (Stokes & Varnes, 1955, p. 66).

granulitization In regional metamorphism, reduction of the components of a solid rock such as a gneiss to grains. The extreme result of the process is the development of *mylonite.*

granuloblastic Said of a metamorphic homogranular texture "in which mineral grains largely lack rational faces but have straight or smoothly curving grain boundaries and approximately polygonal shapes" (Harte, 1977). This texture is typically fine-grained (2 mm or less). Granuloblastic and coarser-grained *granoblastic* textures are common in rocks of the granulite facies and in monomineralic rocks (Joplin, 1968). Syn: *homogranular; even-grained.* Nonrecommended syn: *equigranular; granulitic [meta].*

granulometric facies An interpretative term introduced by Rivière (1952) for semilogarithmic cumulative curves representing grain-size analyses of sediments. The facies is subdivided as "linear", "parabolic", "logarithmic", and "hyperbolic" according to the shape of the curve. "The use of the term facies to describe such statistical representations of one single property of a sedimentary rock seems inadmissible" (Teichert, 1958, p. 2726).

granulometry The measurement of grains, esp. of grain sizes.

granulose [meta] Pertaining to the structure that is typical of granulite and that is due to the presence of granular minerals, e.g. quartz, feldspars, garnet, pyroxene, in alternating streaks and bands developed on a megascopic or microscopic scale. No typical foliation is developed due to the absence of lamellar or prismatic minerals.

granulose [paleont] Having a surface roughened with granules; e.g. having very small grains on the tests of certain foraminifers or on the epitheca or tabulae in some corals.

grapestone A term used by Illing (1954) for a cluster of small calcareous pellets or other grains, commonly of sand size, stuck together by incipient cementation shortly after deposition. The cluster has a lumpy outer surface that resembles a bunch of grapes. Grapestones occur in modern carbonate environments, such as on the Bahama Banks. See also: *bahamite.*

grapevine drainage pattern *trellis drainage pattern.*

graphic Said of the texture of an igneous rock that results from the regular intergrowth of quartz and feldspar crystals. The quartz commonly occupies triangular areas, producing the effect of cuneiform writing on a background of feldspar. Similar intergrowths of other minerals, e.g. ilmenite-pyroxene, are less common. See also: *graphic intergrowth.* Syn: *runic.*

graphic granite A *pegmatite* characterized by graphic intergrowths of quartz and alkali feldspar. See: Barker, 1970. Syn: *Hebraic granite; runite.* See also: *pegmatite.*

graphic intergrowth An intergrowth of crystals, commonly feldspar and quartz, that produces a type of poikilitic texture in which the larger crystals have a fairly regular geometric outline and orientation, resembling cuneiform writing. See also: *graphic.*

graphic log *sample log.*

graphic tellurium An old name for sylvanite occurring in monoclinic crystals that are arranged in more or less regular lines, having a fanciful resemblance to writing (such as to runic characters).

graphiocome A sponge spicule (hexaster) with fine brushlike terminations on the rays. Syn: *graphiohexaster.*

graphiphyre A rock having a granophyric groundmass in which the constituents are of microscopic size (Cross et al., 1906, p. 704). Cf: *graphophyre.* Adj: *graphiphyric.* Not recommended usage.

graphite A hexagonal mineral, a naturally occurring crystalline form of carbon dimorphous with diamond. It is opaque, lustrous, greasy to the touch, and iron black to steel gray in color; it occurs as crystals or as flakes, scales, laminae, or grains in veins or bedded masses or as disseminations in metamorphic rocks. Graphite conducts electricity well, and is soft and unctuous, immune to most acids, and extremely refractory. It is used in lead pencils, paints, and crucibles, as a lubricant and an electrode, and as a moderator in nuclear reactors. Syn: *plumbago; black lead.*

graphitic Pertaining to, containing, derived from, or resembling graphite; e.g. "graphitic rock".

graphitite A variety of shungite or type of graphitic rock that does not give the so-called nitric-acid reaction (Tomkeieff, 1954, p. 52).

graphitization The formation of graphitic material from organic compounds.

graphitoid n. (a) A variety of shungite that will burn in the Bunsen flame. It may represent merely impure graphite. (b) A term applied to meteoritic graphite.—adj. Resembling graphite.

graphocite The end product of coal metamorphism, comparable to *meta-anthracite* and composed mainly of graphitic carbon (ASTM, 1970, p.185).

graphoglypt A *trace fossil,* consisting of a presumed worm trail, appearing as a relief on the undersurface of *flysch* beds (mostly sandstones), and having a meandering, spiral, or netlike pattern related to a highly organized foraging behavior; e.g. *Paleodictyon.* It was interpreted by Fuchs (1895) as a string of spawn of gastropods. Cf: *rhabdoglyph; vermiglyph.*

graphophyre A rock having a granophyric groundmass in which the constituents are of megascopic size (Cross et al., 1906, p. 704). Cf: *graphiphyre.* Adj: *graphophyric.* Not recommended usage.

graptolite Any colonial marine organism belonging to the class Graptolithina, variously assigned to the phylum Coelenterata or to the Hemichordata, characterized by a cup- or tube-shaped, highly resistant exoskeleton of organic composition, arranged with other individuals along one or more branches (*stipes*) to form a colony (*rhabdosome*). Graptolites commonly occur in black shales. Range, Middle Cambrian to Carboniferous. Adj: *graptolithine; graptolitic.*

graptolitic facies A term applied to a *geosynclinal facies* containing an abundance of graptolites.

graptoloid Any graptolite belonging to the order Graptoloidea, characterized by a planktonic or epiplanktonic mode of life and by a colony consisting of a few branches with only one kind of theca, the autotheca. Range, Lower Ordovician to Lower Devonian.

grass opal An *opal phytolith* derived from a grass.

grat A term used in the Alps for a small, lateral *arête.* Etymol: German *Grat,* "ridge".

graticule [cart] The network of lines representing meridians of longitude and parallels of latitude on a map or chart, upon which the map or chart was drawn. Not to be confused with *grid.*

graticule [geophys] A template divided into blocks or cells that is used to integrate graphically a geophysical quantity such as gravity. It is used in computing terrain corrections and gravitational or magnetic effects of bodies of irregular shape.

graticule [optics] An accessory to an optical instrument such as a microscope to aid in measurement of the object under study; it is a thin glass disk bearing a scale which is superimposed upon the object.

grating (a) In optical spectroscopy, equidistant and parallel lines that are used in producing spectra by diffraction. Syn: *diffraction grating.* (b) The gratelike pattern of lines observed in some serpentinized hornblende crystals, resulting from the occurrence of the initial alteration along cleavage cracks.

gratonite A rhombohedral mineral: $Pb_9As_4S_{15}$.

graupel A soft, usually spherical snow crystal that has been enveloped by frozen water droplets. Etymol: German *Graupel,* "sleet, soft hail". Syn: *pellet snow; soft hail.*

grauwacke Var. of *graywacke.* Etymol: German *Grauwacke.*

gravel (a) An unconsolidated, natural accumulation of rounded rock fragments resulting from erosion, consisting predominantly of particles larger than sand (diameter greater than 2 mm, or 1/12 in.), such as boulders, cobbles, pebbles, granules, or any combination of these fragments; the unconsolidated equivalent of conglomerate. In Great Britain, the range of 2-10 mm has been used. Cf: *rubble; pebble.* (b) A popularly used term for a loose accumulation of rock fragments, such as a detrital sediment associated esp. with streams or beaches, composed predominantly of more or less rounded pebbles and small stones, and mixed with sand that may compose 50-70% of the total mass. (c) A soil term for rock or mineral particles having a diameter in the range of 2-20 mm (Jacks et al., 1960, p. 14); in this usage, the term is equivalent to *pebbles.* The term has also been used in Great Britain for such particles having a diameter in the range of 2-50 mm. In the U.S., the term is used for rounded rock or mineral soil particles having a diameter in the range of 2-75 mm (1/6 to 3 in.); formerly the term applied to fragments having diameters ranging from 1 to 2 mm. See also: *fine gravel.* (d) An engineering term for rounded fragments having a diameter in the range of 4.76 mm (retained on U.S. standard sieve no. 4) to 76 mm (3 in.). See also: *fine gravel; coarse gravel.* (e) A stratum of gravel. (f) An obsolete term for sand. (g) *volcanic gravel.*

gravel deposit In economic geology, an alluvial deposit consisting mainly of gravel but commonly including sand and clay. The gravel and sand may be used as a construction material, either directly as fill or as aggregate in concrete. See also: *fineness modulus.*

gravel desert *reg.*

gravelly mud An unconsolidated sediment containing 5-30% gravel and having a ratio of sand to mud (silt + clay) less than 1:1 (Folk, 1954, p.346).

gravelly sand (a) An unconsolidated sediment containing 5-30% gravel and having a ratio of sand to mud (silt + clay) greater than 9:1 (Folk, 1954, p.346). (b) An unconsolidated sediment containing more particles of sand size than of gravel size, more than 10% gravel, and less than 10% of all other finer sizes (Wentworth, 1922, p.390).

gravelly soil A soil that contains an abundance of gravel, usually between 35 and 60% by volume. The remainder of the soil volume is *fine earth.*

gravel mound Any mound of gravel; restricted by Muller (1947, p.217) to a low *frost mound* of sand and gravel, formed by hydrostatic pressure of ground water. Cf: *frost blister.*

gravel pack Gravel or coarse sand placed opposite an oil-producing sand in a well, to prevent or retard the movement of loose sand grains (along with the oil) into the well bore. It is usually forced through perforations under pressure.

gravel piedmont A term used by Hobbs (1912, p. 214) for a feature now known as a *bajada.*

gravel pipe A *pipe [sed]* filled predominantly with gravel.

gravel rampart A *rampart* of loosely compacted reef rubble built along the seaward edge of a reef; the rubble pieces average smaller than those in a *boulder rampart.* Syn: *gravel ridge.*

gravel ridge *gravel rampart.*

gravelstone (a) A rounded rock fragment or constitutent of a gravel. (b) Consolidated gravel; a conglomerate.

gravel train A *valley train* composed chiefly of gravel.

gravimeter An instrument for measuring variations in the gravitational field, generally by registering differences in the weight of a constant mass as the gravimeter is moved from place to place. Syn: *gravity meter.*

gravimetric (a) Pertaining or relating to measurement by weight, e.g. *gravimetric analysis.* (b) Pertaining to measurements of variations of the gravitational field.

gravimetric analysis Quantitative chemical analysis in which the different substances of a compound are measured by weight.

gravimetry The measurement of gravity or gravitational acceleration, especially as used in geophysics and geodesy.

graviplanation A term used by Coleman (1952, p. 455) for the process whereby lunar material is transported and deposited under the "morphological influence" of the Moon's gravitational force.

gravitation The mutual attraction between two masses. See also: *law of universal gravitation.*

gravitational acceleration The mutual acceleration between any two masses resulting from gravitational attraction.

gravitational constant The constant γ in the law of universal gravitation: its value is $6.670 \pm 0.005 \times 10^{-11}$ newton m^2/kg^2.

gravitational differentiation *crystal fractionation.*

gravitational field A region associated with any mass distribution that gives rise to forces of gravitational attraction. See also: *gravity field.*

gravitational gliding *gravitational sliding.*

gravitational intensity The measure of gravitational force, expressed as a vector quantity, exerted on a unit mass at a particular point.

gravitational method *gravity prospecting.*

gravitational separation (a) The stratification of gas, oil, and water in a subsurface reservoir according to their specific gravities. (b) The separation of these fluids in a gravity separator after production.

gravitational sliding Downward movement of rock masses on slopes by the force of gravity, e.g. along a thrust-fault plane. See also: *gravity tectonics.* Syn: *gravity sliding; gravity gliding; gravitational gliding; gliding; écoulement; sliding.*

gravitational tide *equilibrium tide.*

gravitational water *free water.*

gravitational wave A hypothetical wave that travels at the speed of light, by which gravitational attraction is propagated.

gravitite A bed of unsorted clastics deposited by a sedimentary flow impelled only by gravitational forces. There is no internal bedding, and particle arrangement is random (Natland, 1976).

gravity [geophys] (a) The effect on any body in the universe of the inverse-square-law attraction between it and all other bodies and of any centrifugal force that may act on the body because of its motion in an orbit. (b) The resultant force on any body of matter at or near the Earth's surface due to the attraction by the Earth and to its rotation about its axis. (c) The force exerted by the Earth and by its rotation on unit mass, or the acceleration imparted to a freely falling body in the absence of frictional forces.

gravity [petroleum] A general term for *API gravity* or *Baumé gravity* of crude oil.

gravity anomaly The difference between the observed value of gravity at a point and the theoretically calculated value. It is based on a simple gravity model, usually modified in accordance with some generalized hypothesis of variation in subsurface density as related to surface topography.

gravity compaction Compaction of sediment resulting from pressure of overburden.

gravity corer An oceanographic *corer* that penetrates the ocean floor solely by its own weight. It is less efficient than a *piston corer.* There are several varieties, including the *Phleger corer* and the *free corer.*

gravity correction *gravity reduction.*

gravity dam A dam so proportioned that it will resist reservoir-water-induced overturning moments and sliding forces by its own weight; e.g. Grand Coulee Dam on the Columbia River and Aswan Dam on the Nile. All earth dams are gravity dams; the term is applied principally to concrete dams.

gravity equipotential surface *equipotential surface.*

gravity erosion *mass erosion.*

gravity fault *normal fault.*

gravity field A term used instead of *gravitational field* when other influences are also involved, such as centrifugal force.

gravity flow Movement of glacier ice as a result of the inclination of the slope on which the glacier rests; *glacier flow.* See also: *extrusion flow.*

gravity fold A fold that is genetically related to isostatic movements.

gravity formula A formula expressing normal gravity on the surface of a specified reference ellipsoid as a function of latitude.

gravity gliding *gravitational sliding.*

gravity gradient The partial derivative of the acceleration of gravity with respect to distance in a particular direction, for which purpose the acceleration of gravity is considered as a scalar.

gravity ground water The water that would be withdrawn from a body of rock or soil by the influence of gravity should the zone of saturation and capillary fringe be moved downward entirely below that body, remaining there for a specific length of time, no water being lost or received by the body except through the force of gravity (Meinzer, 1923, p. 27). Cf: *free water.*

gravity meter *gravimeter.*

gravity orogenesis A concept proposed by Bucher (1956) for mountain building that results entirely from gravitational stresses. Others believe that such forces may account for folding and buckling but not for entire mountains. Cf: *sedimentary tectonics.*

gravity prospecting The determination of specific-gravity differences of rock masses by mapping the force of gravity of an area, using a gravimeter. Syn: *gravitational method.*

gravity reduction Applying the free-air, Bouguer, isostatic, or other corrections to gravity measurements. Syn: *gravity correction; reduction [geophys].*

gravity separation Separation of mineral particles, with the aid of water or air, according to the differences in their specific gravities.

gravity sliding *gravitational sliding.*

gravity slope The upper, relatively steep slope of a hillside, commonly lying at the angle of repose of the material eroded from it; it is steeper than the *wash slope* below. Term introduced by Meyerhoff (1940). Cf: *constant slope.* Syn: *steilwand; böschung.*

gravity solution A solution used to separate the different mineral particles of rock by exploiting their differences in specific gravity; e.g. a solution of mercuric iodide in potassium iodide, having a maximum specific gravity of 3.19.

gravity spring A spring issuing from the point where the water table and the land surface intersect; an outcrop of the water table.

gravity standard The value of gravity at the Pendelsaal of the Geodetic Institute in Potsdam, East Germany. It is defined as 981.27400 gal. (This value may be about 13 mgal too large.) Syn: *Potsdam gravity.*

gravity survey Measurements of the gravitational field at a series of different locations. The object is to associate variations with differences in the distribution of densities and hence of rock types. Occasionally the whole gravitational field is measured, as with a pendulum, or derivatives of the field, as with a torsion balance, but usually the difference between the gravity field at two points is measured, with a gravimeter. Gravity data usually are displayed as Bouguer or free-air anomaly maps.

gravity tectonics Tectonics in which the dominant propelling mechanism is down-slope gliding under the influence of gravity. In general, the extent of structures produced mainly by gravity remains controversial; probably all gravity movements were triggered by deeper-seated crustal forces, and probably many structures produced by dominantly deep-seated forces were modified to some extent by gravity. See also: *gravitational sliding.*

gravity unit One tenth of a *milligal.* Abbrev: G unit.

gravity water (a) *free water.* (b) Water delivered in canals or pipelines by gravity instead of by pumping, as for irrigation or a public water supply.

gravity wave A wave whose propagation velocity is controlled mainly by gravity, and whose wavelength is 1.7 cm or more. Cf: *capillary wave.*

gravity wind *katabatic wind.*

gray antimony (a) *stibnite.* (b) *jamesonite.*

grayband Sandstone used for sidewalks; flagstone.

Gray-Brown Podzolic soil A great soil group of the 1938 classification system, one of a group of zonal soils characterized by a thin A1 horizon, a grayish-brown A2 horizon, and a brown illuvial B horizon in which silicate clays have accumulated. They form under deciduous forest in a humid, temperate climate (USDA, 1938). Most of these soils are now classified as *Udalfs.*

gray cobalt (a) *smaltite.* (b) *cobaltite.*

gray copper ore (a) *tetrahedrite.* (b) *tennantite.* Syn: *gray copper.*

Gray Desert soil *Sierozem.*

gray earth *Sierozem.*

gray hematite *specularite.*

gray ice A type of *young ice* (10-15 cm thick) that is less elastic than *nilas;* it breaks on swell and usually rafts under pressure. Cf: *gray-white ice.*

grayite A yellow powdery mineral: $(Th,Pb,Ca)PO_4 \cdot H_2O$.

gray manganese ore (a) *manganite.* (b) *pyrolusite.*

gray mud A type of *mud [marine geol]* that is intermediate in composition between globigerina ooze and red clay.

Gray Podzolic soil *Gray Wooded soil.*

gray scale A monochrome strip of continuous tones ranging from white to black with intermediate tones of gray, used to determine the density of a color photograph. Cf: *step wedge.*

graystone A dense gray-green rock, resembling *basalt* and composed of feldspar and augite (Thrush, 1968, p. 508). Not recommended usage.

graywacke An old rock name that has been variously defined but is now generally applied to a dark gray firmly indurated coarse-grained sandstone that consists of poorly sorted angular to subangular grains of quartz and feldspar, with a variety of dark rock and mineral fragments embedded in a compact clayey matrix having the general composition of slate and containing an abundance of very fine-grained illite, sericite, and chloritic minerals; e.g. the Jackfork Sandstone (Mississippian) in Oklahoma, parts of the Franciscan Formation (Mesozoic) in western California, and certain Ordovician rocks in the Taconic region of New York and Vermont. This description is similar to Naumann's (1858, p.663) definition of the type graywacke, the Tanner Graywacke (Upper Devonian and Lower Carboniferous) of the Harz Mountains, Germany. Graywacke is abundant within the sedimentary section, esp. in the older strata, usually occurring as thick, extensive bodies with sole marks of various kinds and exhibiting massive or obscure stratification in the thicker units but marked graded bedding in the thinner layers. It generally reflects an environment in which erosion, transportation, deposition, and burial were so rapid that complete chemical weathering did not occur, as in an orogenic belt where sediments derived from recently elevated source areas were "poured" into a geosyncline. Graywackes are typically interbedded with marine shales or slates, and associated with submarine lava flows and bedded cherts; they are generally of marine origin and are believed to have been deposited by submarine turbidity currents (Pettijohn, 1957, p.313). Selected modern definitions have been given by Allen (1936, p.22); Twenhofel (1939, p.289); Krynine (1948); Folk (1954); Williams, Turner & Gilbert (1954, p.293-297); Pettijohn (1957); McBride (1962a); and Krumbein & Sloss (1963, p.171-172). The first recorded use of the term was by Lasius (1789, p.132-152) who referred to "Grauewacke" as a German miner's term for barren country rock of certain ore veins in the Harz Mountains, and who described the rock as a gray or dark quartz "breccia" with mica flakes and fragments of chert or sandstone in a clay cement (see Dott, 1964). The term "greywacke" was probably first used in English by Jameson (1808). Early usage was wide and vague: "geologists differ much respecting what is, and what is not, Grey Wacce" (Mawe, 1818, p.92), and "it has already been amply shown that this word should cease to be used in geological nomenclature, and . . . is mineralogically worthless" (Murchison, 1839). In view of the diversity of usage, the term "graywacke" should not be used formally without either a specific definition or a reference to a readily available published definition. Folk (1968, p.125) advocates discarding the term for any precise petrographic usage, and relegating it to nonquantitative field usage for a hard, dark, clayey, impure sandstone "that you can't tell much about in the field". Etymol: German *Grauwacke*, "gray stone", probably so named because the original graywackes resembled partly weathered basaltic residues (wackes). See also: *wacke.* Cf: *arkose; subgraywacke.* Also spelled: *greywacke; grauwacke.* Syn: *apogrit.*

graywether Var. of *greywether.* Also spelled: *gray weather.*

gray-white ice A type of *young ice* (15-30 cm thick) that is more likely to be ridged than rafted under pressure. Cf: *gray ice.*

Gray Wooded soil A great soil group, introduced in 1949, characterized by a thin 01 horizon and a thin A1 horizon, a relatively thick, bleached A2 horizon, and a brown clayey B horizon. They occur in cool to temperate, subhumid to semiarid parts of Western Canada and U.S. (Thorp and Smith, 1949). These soils are now classified under the *Alfisols.* Syn: *Gray Podzolic soil.*

grazing The feeding of zooplankton upon phytoplankton.

grease ice A soupy layer of *new ice* formed on a water surface (esp. in the sea) by the coagulation of frazil crystals; it reflects little light, giving the sea a matte, greasy appearance. Syn: *ice slush.*

greasy Said of minerals that appear oily to the touch or to the sight.

greasy quartz A type of *milky quartz* with a greasy luster.

great circle A circle on the Earth's surface, the plane of which passes through the center of the Earth and an arc of which constitutes the shortest distance between any two terrestrial points. Cf: *small circle.* Syn: *orthodrome.*

great-circle belt The distribution pattern of *primary arcs* on the Earth's surface in belts of major tectonic activity, i.e. the *circum-Pacific belt* and the *Eurasian-Melanesian belt* (Strahler, 1963, p.403).

great-circle chart (a) A chart on a gnomonic projection, on which a great circle appears as a straight line. (b) *gnomonic projection.*

great-circle projection *gnomonic projection.*

great divide A drainage divide between major drainage systems; specif. the Great Divide (Continental Divide) of the North American continent.

great elliptic The line of intersection of the ellipsoid and a plane containing two points on the ellipsoid and its center.

Great Ice Age The Pleistocene Epoch.

great soil group (a) In the 1938 soil classification system, the most widely used of the higher categories. Soils in each class of this category have common internal characteristics, i.e. kind and sequence of horizons, chemical and physical properties; for example, Podzols and Chernozems. (b) In the current U.S. Dept. of Agriculture soil taxonomy, a category, properly called merely a great group, that divides soil suborders. Great groups are differentiated by properties of the whole soil, not just diagnostic horizons, and by moisture and temperature regimes. Names are formed by prefixing a syllable to suborder names, e.g. Hapludalf, Argiaquoll.

green algae A group of algae corresponding to the phylum Chlorophyta, that owes its grassy green color to the dominance of chlorophyll pigmentation. Such algae occur in a great variety of forms, from unicellular and microscopic types to large complex seaweeds. Cf: *brown algae; blue-green algae; red algae; yellow-green algae.*

greenalite An earthy- or pale-green mineral: $(Fe^{+2},Fe^{+3})_{5\text{-}6}Si_4O_{10}(OH)_8$. It is a member of the kaolinite-serpentine group. It occurs in small ellipsoidal granules in cherty rock associated with the iron ores of the Mesabi district, Minn. Greenalite resembles glauconite in appearance, but contains no potassium.

greenalite rock A dull, dark-green rock, uniformly fine-grained with conchoidal fracture, containing grains of greenalite in a matrix of chert, carbonate minerals, and ferruginous amphiboles (Van Hise & Leith, 1911, p. 165 & 474).

green beryl A term applied to the light-green or pale-green gem variety of beryl, as distinguished from the full-green or richly green-colored emerald and the light blue-green aquamarine.

green chalcedony (a) Chalcedony that has been artificially colored green. (b) *chrysoprase.*

green earth Any of various naturally occurring silicates (esp. of iron) used chiefly as bases for green basic dyes and for green pigments; specif. glauconite and celadonite. Syn: *terre verte; terra verde.*

greenhouse effect The heating of the Earth's surface because outgoing long-wavelength terrestrial radiation is absorbed and re-emitted by the carbon dioxide and water vapor in the lower atmosphere and eventually returns to the surface.

green iron ore *dufrenite.*

green john A green variety of fluorite.

greenlandite *columbite.*

Greenland spar *cryolite.*

green lead ore *pyromorphite.*

Green Mountains disturbance A name used by Schuchert (1924) for a supposed time of deformation at the end of the Cambrian. Evidence for this disturbance is unconvincing, even in the type area of the Green Mountains of Vermont, and the term should be discarded.

green mud A type of *mud [marine geol]* whose greenish color is due to the presence of chlorite or glauconite minerals.

greenockite A yellow or orange hexagonal mineral: CdS. It is dimorphous with hawleyite and usually occurs as an earthy incrustation or coating on sphalerite and other zinc ores. Syn: *cadmium blende; cadmium ocher; xanthochroite.*

greenovite A red, pinkish, or rose-colored variety of sphene containing manganese.

greensand (a) A sand having a greenish color, specif. an unconsolidated marine sediment consisting largely of dark greenish grains of glauconite, often mingled with clay or sand (quartz may form the dominant constituent), found between the low-water mark and the inner mud line. The term is loosely applied to any glauconitic sediment. Syn: *glauconitic sand.* (b) A sandstone consisting of greensand that is often little or not at all cemented, having a greenish color when unweathered but an orange or yellow color when weathered, and forming prominent deposits in Cretaceous and Eocene beds (as in the Coastal Plain areas of New Jersey and Delaware); specif. either or both of the Greensands (Lower and Upper) of the Cretaceous System in England, whether containing glauconite or not. Syn: *glauconitic sandstone.*—Also spelled: *green sand.*

greensand marl A marl containing sand-size grains of glauconite.

greenschist A schistose metamorphic rock whose green color is due to the presence of chlorite, epidote, or actinolite. Cf: *greenstone.*

greenschist facies The set of metamorphic mineral assemblages (facies) in which basic rocks are represented by albite + epidote + chlorite + actinolite (Eskola, 1939). Chlorite, white mica, biotite, and chloritoid are typical minerals in pelitic rocks. The facies includes the common products of low-grade regional metamorphism in all parts of the world. It is believed to correspond to temperatures in the range 300°-500°C. It is distinguished from the *prehnite-pumpellyite facies* by the absence of prehnite and pumpellyite, from the *zeolite facies* by the absence of zeolites, and from the *blueschist facies* by the absence of sodic amphiboles, omphacite, jadeite, lawsonite, and aragonite.

green snow A general name for snow colored by a growth of green microscopic algae, such as *Stichococcus* and *Chlamydomonas.* Cf: *red snow; yellow snow.*

greenstone [ign] In Scotland, any intrusion of igneous rock in the Coal Measures.

greenstone [meta] A field term applied to any compact dark-green altered or metamorphosed basic igneous rock (e.g. spilite, basalt, gabbro, diabase) that owes its color to the presence of chlorite, actinolite, or epidote.

greenstone [mineral] (a) *nephrite.* (b) An informal name for a greenish gemstone, such as fuchsite or chiastolite.

greenstone [sed] A compact, nonoolitic, relatively pure chamosite mudstone interbedded with oolitic ironstone in the Lower Jurassic of Great Britain.

greenstone belt Term applied to elongate or beltlike areas within Precambrian shields that are characterized by abundant *greenstone [meta].* An individual belt may contain the deformed and metamorphosed rocks of one or more volcano-sedimentary piles, in each of which there is typically a trend from mafic to felsic volcanics. The resultant volcano-sedimentary complexes are of economic interest as host rocks for presumably volcanogenic metal deposits.

green vitriol *melanterite.*

Greenwich meridian The astronomic meridian that passes through the original site of the Royal Astronomical Observatory at Greenwich, near London, England. Its adoption as the world-wide reference standard, or *prime meridian,* was approved almost unanimously at an International Meridian Conference in Washington, D.C., in 1884. Cf: *national meridian.*

greet stone A term used in Yorkshire, England, for a coarse-grained or gritty sandstone.

gregaritic Said of the texture of a porphyritic igneous rock in which independently oriented grains of the same mineral (esp. augite) in the groundmass occur in clusters. Cf: *synneusis; cumulophyric; glomerophyric.*

greigite A dark mineral with spinel-like structure: Fe_3S_4. Syn: *melnikovite.*

greisen A pneumatolytically altered granitic rock composed largely of quartz, mica, and topaz. The mica is usually muscovite or lepidolite. Tourmaline, fluorite, rutile, cassiterite, and wolframite are common accessory minerals. See also: *greisenization.*

greisenization A process of hydrothermal alteration in which feldspar and muscovite are converted to an aggregate of quartz, topaz, tourmaline, and lepidolite (i.e., *greisen*) by the action of water vapor containing fluorine. Syn: *greisenisation; greisening.*

grenatite (a) *staurolite.* (b) *leucite.*

grennaite A fine-grained porphyritic *nepheline syenite* containing catapleiite and eudialyte. The name, given by Adamson in 1944, is for Grenna farm, Norra Kärr complex, Sweden. Not recommended usage.

Grenville A provincial series of the Precambrian of Canada and New York.

Grenville orogeny A name that is widely used for a major plutonic, metamorphic, and deformational event during the Precambrian, dated isotopically as between 880 and 1000 m.y. ago, which affected a broad province along the southeastern border of the Canadian Shield. Originally, the name Grenville was used for a metasedimentary series in the southern part of the province, and the name *Laurentian* was used for the associated plutonic rocks. Pertinent objections have been raised (Osborne, 1956; Gilluly, 1966) to the use of Grenville for the orogeny, the province, and its northwestern structural "front", but these uses will be continued until generally acceptable alternatives are proposed.

grenz A horizon in coal beds resulting from a temporary halt in the accumulation of vegetal material. It is frequently marked by a bed of clay or sand. Etymol: German "Grenzhorizont", recurrence horizon.

greywacke *graywacke.*

greywether A popular term for a *sarsen* on the English chalk downs, so named from its fancied resemblance to a sheep lying down on a distant hillside. Also spelled: *graywether; gray weather.*

grid (a) A network composed of two sets of uniformly spaced parallel lines, usually intersecting at right angles and forming squares, superimposed on a map, chart, or aerial photograph, to permit identification of ground locations by means of a system of coordinates and to facilitate computation of direction and distance. The term is frequently used to designate a plane-rectangular coordinate system superimposed on a map projection, and usually carries the name of the projection; e.g. "Lambert grid". Not to be confused with *graticule.* (b) A systematic array of points or lines; e.g. a rectangular pattern of pits or boreholes used in alluvial sampling.

grid azimuth The angle at a given point in the plane of a rectangular coordinate system between the central meridian or a line parallel to it, and a straight line to the azimuth point. Cf: *gisement.*

gridiron twinning *cross-hatched twinning.*

grid line One of the lines used to establish a grid.

grid meridian A line through a point parallel to the central meridian or Y axis of a system of plane-rectangular coordinates. Cf: *gisement.*

grid method A method of plotting detail from oblique photographs by superimposing a perspective of a map grid on a photograph and transferring the detail by eye (using the corresponding lines of the map grid and its perspective as placement guides).

grid north The northerly or zero direction indicated by a meridional line of a rectangular map grid. It is coincident with true north only along the meridian of origin.

grid residual A method of emphasizing anomalies of a certain size in a potential-field map, e.g. gravity or magnetic. A grid (usually square or trigonal) is drawn on a contour map and values are determined at the grid intersections by interpolation. The residual at a grid intersection is the value at that point less the average at other intersections a fixed distance away. Averages at several distances may be used and weighted to approximate second-derivative values or other operators. The process of making grid residuals is called *map convolution* because it represents map data convolved with a residualizing operator (or template). See also: *residualizing.*

grid smoothing A method of smoothing sharp irregularities in potential-field measurements that arise from very shallow disturbances. A grid is drawn on a contour map and the smoothed value at a grid intersection is the average of values a fixed small distance away.

Griesbachian European stage: Lowermost Triassic (above Permian, below Dienerian).

griffithite An iron-rich clay mineral of the montmorillonite group; specif. a variety of saponite containing ferrous iron. It was formerly regarded as identical with nontronite.

grike A syn. of *solution fissure.* Also spelled: *gryke.*

grimaldite A trigonal mineral: $CrO(OH)$. It is polymorphic with bracewellite and guyanaite.

grimselite A yellow hexagonal mineral: $K_3Na(UO_2)(CO_3)_3 \cdot H_2O$.

grinder A spherical or discus-shaped stone rotated by the force of helical water currents in a stream pothole, the rotation producing a deepening of the pothole.

grinding [geomorph] The process of erosion by which rock fragments are worn down, crushed, sharpened, or polished through the frictional effect of continued contact and pressure by larger fragments.

grinding [glac geol] Abrasion by rock fragments embedded in a glacier and dragged along the bedrock floor; it produces gouges and grooves, and chips out fragments of bedrock. Cf: *glacial scour.*

griotte A French quarrymen's term for a marble or fine-grained limestone of red color, often variegated with small dashes of purple and spots or streaks of white or brown. It includes goniatite shells and is often used as an ornamental stone. Etymol: French, "morello cherry".

griphite A mineral: $(Na,Al,Ca,Fe)_6Mn_4(PO_4)_5(OH)_4$. Its crystal structure is related to that of garnet.

griquaite A coarse-grained garnet and clinopyroxene rock that may or may not contain olivine or phlogopite. It occurs as nodular xenoliths in *kimberlite* pipes and dikes; a garnet *pyroxenite.* Cf: *ariegite.* Named by Beck in 1907 for Griqualand, S. W. Africa. Not recommended usage.

grit (a) A coarse-grained sandstone, esp. one composed of angular particles; e.g. a breccia composed of particles ranging in diameter from 2 mm to 4 mm (Woodford, 1925, p.183). (b) A sand or sandstone made up of angular grains that may be coarse or fine. The term has been applied to any sedimentary rock that looks or feels gritty on account of the angularity of the grains. (c) *gritstone.* (d) A sandstone composed of particles of conspicuously unequal sizes (including small pebbles or gravel). (e) A sandstone with a calcareous cement. The term has been applied incorrectly to any nonquartzose rock resembling a grit; e.g. *pea grit* or a calcareous grit. (f) A small particle of a stone or rock; esp. a hard, angular granule of sand. Also, an abrasive composed of such granules. (g) The structure or "grain" of a stone that adapts it for grinding or sharpening; the hold of a grinding substance. Also, the size of abrasive particles, usually expressed as their *mesh number.* (h) An obsolete term for sand or gravel, and for earth or soil.—The term is vague and has been applied widely with many different connotations. Allen (1936, p.22) proposed to restrict the term to a coarse-grained sandstone composed of angular particles varying in diameter from 0.5 mm to 1 mm. Etymol: Old English *greot,* "gravel, sand".

gritrock *gritstone.*

gritstone A hard, coarse-grained, siliceous sandstone; esp. one used for millstones and grindstones. Syn: *grit; gritrock.*

gritty (a) Said of the feel of a soil or of a loose or cemented sediment containing enough angular particles of sand to impart a roughness to the touch. The actual quantity of sand in such a soil is usually small. (b) Containing or resembling sand or grit. Syn: *arenose.*

grivation The angular difference in direction between grid north and magnetic north at any point, measured east or west from grid north.

groin A low, narrow jetty constructed of timber, stone, concrete, or steel, usually extending roughly perpendicular to the shoreline, designed to protect the shore from erosion by currents, tides, or waves, or to trap sand and littoral drift for the purpose of building up or making a beach. It may be permeable or impermeable. Syn: *groyne.*

grønlandite A hypersthene *hornblendite* containing more hornblende than hypersthene. Not recommended usage.

groove [fault] One of a series of parallel scratches developed along a fault surface. A groove is a larger structure than a *striation.* Cf: *slickensides; mullion structure; slip-scratch.*

groove [glac geol] *glacial groove.*

groove [sed] A long, straight narrow depression, with an almost uniform depth and cross section, on a sedimentary surface (as of mud or shale). It is thought to have been produced by a simultaneous rectilinear advance of objects propelled by a continuous current. Often preserved as a *groove cast,* it is larger and wider than a *striation* but smaller than a *channel.* See also: *drag mark; slide mark.*
groove-and-spur structure *spur-and-groove structure.*
groove cast A term used by Shrock (1948, p.162-163) for a rounded or sharp-crested rectilinear ridge, a few millimeters high and many centimeters in length and width, produced on the underside of a sandstone bed by the filling of a *groove* on the surface of an underlying mudstone. This structure was called a *drag mark* by Kuenen (1957, p.244) who considered "groove cast" as a general term including drag marks and slide marks. Cf: *striation cast; mud furrow.* See also: *ruffled groove cast.* Syn: *proglyph.*
grooved lava Lava with grooves or striations made when it was plastic or viscous, by the sliding of one block of lava over another, by the squeezing of viscous lava through cracks in the lava crust, or by impact of volcanic fragments on viscous lava (Nichols, 1938, p. 601). See also: *squeeze-up.*
grooved upland An upland surface largely unaffected by feeble cirque-cutting that has left extensive undissected remnants of the preglacial surface (Hobbs, 1911a, p. 30). Cf: *fretted upland.* Syn: *channeled upland.*
groove lake A lake occupying a *glacial groove.*
groove spine One of a cluster or row of short, blunt, generally recumbent spines bordering ambulacral grooves in many asteroids.
grooving (a) The formation of *furrows* on a rock surface. (b) A furrow, or a set of furrows.
grorudite A hypabyssal rock composed of phenocrysts of microcline or microcline-perthite, acmite, and less kataphorite in a tinguaitic groundmass of microcline or microperthite, acmite, and abundant quartz; an acmite-rich sodic *granite.* Named by Brögger in 1894 for Grorud, Oslo district, Norway. Not recommended usage.
grospydite An ultramafic rock, occurring in nodules in kimberlite pipes of Yakutia, U.S.S.R., containing garnets (esp. grossular), plagioclase, pyroxene, and, at high pressures, kyanite, rather than spinel or olivine. Its name, given by Sobolev and others in 1966, is derived from the initial letters of *gros*sular, *py*roxene, and *di*sthene (kyanite).
gross calorific value (a) A *calorific value* calculated on the assumption that the water in the products is completely condensed. Cf: *net calorific value.* (b) For solid fuels and liquid fuels of low volatility, the heat produced by combustion of unit quantity, at constant volume, in an oxygen-bomb calorimeter under specified conditions.—Syn: *gross heat of combustion.*
gross heat of combustion *gross calorific value.*
gross primary production Amount of organic matter produced by living organisms within a given volume or area in a given time, including that consumed by the respiratory processes of the organisms (Odum, 1959). Cf: *net primary production.*
grossular The calcium-aluminum end member of the garnet group, usually characterized by a green color: $Ca_3Al_2(SiO_4)_3$. It may be colorless, yellow, orange, brown, rose, or red, and it often occurs in contact-metamorphosed impure limestones. The principal variety is essonite. Syn: *grossularite; grossularia; gooseberry stone.*
grossularite *grossular.*
grothite *sphene.*
grotto A small cave, or one of the rooms of a cave.
ground [elect] (a) The voltage level of the ground. (b) The reference voltage level of an electrical system or instrument.
ground [geog] (a) The surface or upper part of the Earth. (b) Land, particularly a region or area.
ground air Air in the ground, principally in the zone of aeration, but including any bubbles trapped in the zone of saturation. Cf: *subsurface air; soil atmosphere; included gas.*
ground avalanche An obsolete syn. of *full-depth avalanche.*
ground control Accurate data on the horizontal and/or vertical positions of identifiable ground points so that they may be recognized in aerial photographs.
ground current *earth current.*
ground data Information collected on or near the surface of the Earth, in conjunction with an aerial or remote-sensing survey. Nonrecommended syn: *ground truth.*
grounded hummock A hummocked formation of *grounded ice,* appearing singly or in a line or chain. Cf: *stamukha.*
grounded ice Floating ice that is aground in shallow water. Cf: *stranded ice; grounded hummock.*
ground failure A permanent differential ground movement capable of damaging or seriously endangering a structure.
ground fog *radiation fog.*
ground frost [meteorol] An occurrence of below-freezing temperature on the surface of the ground, while the air temperature remains above the freezing point.
ground frost [permafrost] (a) Any frozen soil including permafrost; a deprecated syn. of *frozen ground.* (b) *ground ice.*
ground ice [ice] (a) A deprecated syn. of *anchor ice.* (b) Glacier ice, sea ice, or lake ice that has been covered with soil (ADTIC, 1955, p. 37). (c) Ice formed on the ground by freezing of rain or snow, or by compaction of a snow layer.
ground ice [permafrost] All ice, of whatever origin or age, found below the surface of the ground, esp. a lens, sheet, wedge, seam, or irregular mass of clear nonglacial ice enclosed in perennially or seasonally frozen ground, often at considerable depth. Syn: *fossil ice; subsurface ice; subsoil ice; subterranean ice; underground ice; ground frost.*
ground-ice layer *ice layer [permafrost].*
ground-ice mound Any *frost mound, ice laccolith,* or *pingo* containing bodies of ice.
ground-ice wedge *ice wedge.*
grounding The temporary dropping and lodgement of sedimentary particles carried in saltation, typically in sandbars, natural levees, or gravel beds (McGee, 1908, p. 199).
ground magnetometer A magnetometer primarily suitable for making static observations of magnetic-field intensity on the surface of the Earth.
groundmass [ign] The material between the phenocrysts of a porphyritic igneous rock. It is relatively finer grained than the phenocrysts and may be crystalline, glassy or both. Cf: *mesostasis.* Syn: *matrix [ign].*
groundmass [sed] A term sometimes used for the *matrix* of a sedimentary rock.
ground mix The use of a pattern of shots or geophones distributed over a sizable surface area. The objective is to have vertically reflected energy add up in-phase while horizontally traveling energy partially cancels. The term is sometimes reserved for situations where adjacent geophone or source patterns overlap.
ground moraine (a) An accumulation of till after it has been deposited or released from the ice during ablation, to form an extensive area of low relief devoid of transverse linear elements. Syn: *bottom moraine.* (b) The rock debris dragged along in and beneath a glacier or ice sheet (a less desirable usage in Europe, where *moraine* is sometimes used as a synonym of *till*).
ground-moraine shoreline An irregular shoreline formed where masses of glacial drift abut against the sea as a result of submergence.
ground motion A general term for all seismic motion, including ground acceleration, velocity, displacement, and strain. See also: *strong motion.*
ground noise In exploration seismology, ground motion that is not caused by the shot.
ground range On *SLAR* images, the distance from the flight path to an object.
ground receiving station A facility that records image data transmitted by *Landsat.*
ground resolution cell The area on the terrain that is covered by the *instantaneous field of view* of a detector. Its size is determined by the altitude of the remote-sensing system and the field of view of the detector.
ground roll A seismic surface wave, generally of low frequency and velocity.
ground slope *valley-side slope.*
ground-surge deposit A well-bedded, heterogeneous pyroclastic flow deposit; includes *base surge* deposits and some deposits of *nuées ardentes* (Sparks & Walker, 1973, p. 62).
ground survey A survey made by ground methods, as distinguished from an aerial survey.
ground swell A long, high ocean *swell.*
ground truth A nonrecommended syn. of *ground data.* There is no reason to infer that data acquired on the ground are more "true" than those acquired by remote sensing.
ground water (a) That part of the subsurface water that is in the

zone of saturation, including underground streams. See also: *phreatic water.* Syn: *plerotic water.* (b) Loosely, all *subsurface water* (excluding internal water) as distinct from surface water.—Also spelled: groundwater; ground-water. Syn: *subterranean water; underground water.*

ground-water artery A roughly tubular body of permeable material surrounded by impermeable or less permeable material and saturated with water confined under artesian pressure. "The term is especially applicable to deposits of gravel along ancient stream channels that have become buried in less permeable alluvial material under alluvial fans" (Meinzer, 1923, p. 42).

ground-water barrier A natural or artifical obstacle, such as a dike or fault gouge, to the lateral movement of ground water, not in the sense of a confining bed. It is characterized by a marked difference in the level of the ground water on opposite sides. Syn: *barrier [grd wat]; hydrologic barrier; ground-water dam.* Cf: *ground-water cascade; interrupted water table.*

ground-water basin (a) A subsurface structure having the character of a basin with respect to the collection, retention, and outflow of water. (b) An aquifer or system of aquifers, whether basin-shaped or not, that has reasonably well defined boundaries and more or less definite areas of recharge and discharge. Cf: *basin; artesian basin.*

ground-water budget A numerical account, the *ground-water equation,* of the recharge, discharge, and changes in storage of an aquifer, part of an aquifer, or system of aquifers. Syn: *ground-water inventory.*

ground-water cascade The near-vertical or vertical flow of ground water over a *ground-water barrier*. Cf: *interrupted water table.*

ground-water cement A secondary concentration of calcium carbonate or calcium sulfate, usually in the desert, resulting from evaporation of ground water at the surface or in shallow soil; a type of *water-table rock.* Syn: *water-table cement.*

ground-water dam *ground-water barrier.*

ground-water decrement *ground-water discharge.*

ground-water discharge (a) Release of water from the zone of saturation. (b) The water or the quantity of water released. Syn: *ground-water decrement; decrement; phreatic-water discharge.*

ground-water divide *divide [grd wat].*

ground-water equation (a) The equation that balances the *ground-water budget;* $R = E + S + I$, where R is rainfall, E is evaporation and transpiration loss, S is water discharged from the area as streamflow, and I is the recharge. (b) A mathematical statement of ground-water losses and gains in a specified area.—(Tolman, 1937, p. 560).

ground-water flow (a) *ground-water movement.* (b) *ground-water runoff.*

ground-water geology The science of subsurface water, with emphasis on the geologic aspects; *hydrogeology.*

ground-water hydrology *geohydrology.*

ground-water increment *recharge.*

ground-water inventory *ground-water budget.*

ground-water lake A body of surface water that represents an exposure of the upper surface of the zone of saturation, or of the water table.

Ground-Water Laterite soil A great soil group in the 1938 classification system, an intrazonal, hydromorphic group of soils having an A2 horizon containing concretions that is underlain by a hardpan composed of iron and aluminum compounds. These soils are formed in warm-temperate to tropical climates in response to a fluctuating water table (USDA, 1938). They are now classed as *Aquults, Udults,* and *Ustults,* and all contain *plinthite.*

ground-water level (a) A syn. of *water table.* (b) The elevation of the water table or another potentiometric surface at a particular place or in a particular area, as represented by the level of water in wells or other natural or artificial openings or depressions communicating with the zone of saturation.

ground-water mining The process, deliberate or inadvertent, of extracting ground water from a source at a rate so in excess of the replenishment that the ground-water level declines persistently, threatening exhaustion of the supply or at least a decline of pumping levels to uneconomic depths.

ground-water mound A mound-shaped elevation in a water table or other potentiometric surface that builds up as a result of the downward percolation of water, through the zone of aeration or an overlying confining bed, into the aquifer represented by the potentiometric surface. Syn: *water-table mound.*

ground-water movement The movement, or flow, of water in the zone of saturation, whether naturally or artificially induced. Syn: *ground-water flow.*

ground-water outflow The discharge from a drainage basin, or from any area, occurring as ground water.

Ground-Water Podzol soil A great soil group in the 1938 classification system, an intrazonal, hydromorphic group of soils having a prominent light-colored leached A2 horizon overlain by thin organic material and underlain by a dark brown B horizon, irregularly cemented with iron or organic compounds. It develops under various types of forest vegetation, in humid climates of varying temperature (USDA, 1938). These soils are now classified as *Spodosols.*

ground-water province An area or region in which geology and climate combine to produce ground-water conditions consistent enough to permit useful generalizations.

ground-water recession curve The part of a stream hydrograph supposedly representing the inflow of ground water at a decreasing rate after surface runoff to the channel has ceased. Because the base runoff to the stream may include some water that had been stored in lakes and swamps rather than in the ground, the lower recession curve cannot be assumed to represent ground water only.

ground-water recharge *recharge.*

ground-water replenishment *recharge.*

ground-water reservoir (a) *aquifer.* (b) A term used to refer to all the rocks in the zone of saturation, including those containing permanent or temporary bodies of perched ground water.—Syn: *ground-water zone; reservoir [grd wat].*

ground-water ridge (a) A linear elevation in the water table that develops beneath an influent stream. Cf: *interstream ground-water ridge.* (b) *divide [grd wat].*

ground-water runoff The *runoff* that has entered the ground, become ground water, and been discharged into a stream channel (Langbein & Iseri, 1960). Cf: *surface runoff; storm seepage; delayed runoff.* Syn: *ground-water flow.*

ground-water storage (a) The quantity of water in the zone of saturation. (b) Water available only from storage as opposed to capture.

ground-water surface *water table.*

ground-water table *water table.*

ground-water trench A troughlike depression in the water table or potentiometric surface, caused by flow of ground water into a stream or drainage ditch.

ground-water wave A "high" in the water table or other potentiometric surface that moves laterally, with a wavelike motion, away from a place where a substantial quantity of water has been added to the zone of saturation within a brief period. Syn: *phreatic wave.*

ground-water withdrawal The process of withdrawing ground water from a source; also, the quantity of water withdrawn. Syn: *offtake; recovery [grd wat].*

ground-water zone *ground-water reservoir.*

ground wave A seismic wave whose path of propagation is through both the material beneath the ocean floor and the ocean water.

group (a) The formal *lithostratigraphic unit* next in rank above *formation.* A group includes two or more contiguous or associated formations with significant lithologic features in common. The type or reference sections of a group are those of its component formations. The proposal for recognition of a group should outline clearly the unifying characteristics on which it is based and the formations of which it is composed (ISG, 1976, p. 34). A group name customarily combines a geographic name with the word "group", and no lithologic designation is included. Abbrev: gr. See also: *subgroup; supergroup; synthetic group.* (b) A stratigraphic sequence that will probably be divided in whole or in part into formations in the future (ISG, 1976, p. 34). See also: *analytic group.* (c) A general term for an assemblage or consecutive sequence of related layers of rock, such as of igneous rocks or of sedimentary beds. (d) A term proposed at the 2nd International Geological Congress in Bologna in 1881 as the chronostratigraphic equivalent of an era, and subsequently used quite widely for the rocks formed during an era. This usage is not recommended, the synonymous term *erathem* having been formally adopted (ISST, 1961, p.28). (e) An obsolete term for a chronostratigraphic unit representing a local or provincial subdivision of a system (usually less than a standard series, or the equivalent of "stage" as that term is presently used) and containing two or more formations.

group velocity The velocity with which seismic energy moves through a medium. Where velocity varies with frequency, individual phases will appear to travel at different *phase velocities.* See also: *dispersion [seis]; particle velocity.*

grout (a) A cement *slurry* of high water content, fluid enough to be poured or injected into spaces and thereby fill or seal them (such as the fissures in the foundation rock of a dam, or the space between the lining of a tunnel and the surrounding earth). (b) The stony waste material, of all sizes, obtained in quarrying.

grouting The injection of *grout* into fissured, jointed, or permeable rocks in order to reduce their permeability or increase their strength. Syn: *cementation [eng]; consolidation grouting.*

groutite A jet-black mineral: $HMnO_2$. It is polymorphous with manganite and feitknechtite.

grovesite A mineral of the grovesite group: $(Mn,Fe,Al)_6(Al,Si)_4O_{10}(OH)_8$.

growan (a) An old English term for a coarse-grained granite, grit, or sandstone. (b) A *grus* developed by the disintegration of a granite.—Syn: *grouan.*

growler A small fragment of massive floating ice of glacier or sea-ice origin, extending less than 1 m above sea level and smaller than a *bergy bit.*

growth axis The line formed by tips of lamellae in cusps and denticles of conodont elements and commonly emphasized by concentration of "white matter".

growth band A *growth line* on the surface of a bivalve-mollusk shell.

growth curve The curve along which lead ratios change with time by the addition of radiogenic lead from a source in which the U/Pb and Th/Pb ratio is changed only by radioactive decay.

growth fabric Orientation of fabric elements independent of the influences of stress and deformation, i.e. characteristic of the manner in which the rock was formed.

growth fault A fault in sedimentary rock that forms contemporaneously and continuously with deposition, so that the throw increases with depth and the strata of the downthrown side are thicker than the correlative strata of the upthrown side. Such a structure occurs in the Gulf Coast region. See also: *hinge-line fault.* Syn: *contemporaneous fault.* Less-preferred syn: *depositional fault; flexure fault; Gulf Coast-type fault; progressive fault; sedimentary fault; slump fault; synsedimentary fault.*

growth-framework porosity Primary porosity developed from organic and/or inorganic processes during the in-place growth of a carbonate-rock framework (Choquette & Pray, 1970, p. 246-247). Intraparticle porosity of individual organisms or of particles that were clastic components of the rock is excluded, thus giving the term a more restrictive meaning than that of *constructional void porosity,* which includes these openings.

growth habit The general form, shape and internal structure developed by a bryozoan colony, and its relationship to the substrate.

growth island An irregular layer or patch on a crystal face due to *spiral growth* along an internal screw dislocation.

growth lamella A concentric outgrowth of a brachiopod shell, smaller than a *frill*, deposited by the margin of retractile mantle (TIP, 1965, pt.H, p.145).

growth lattice The rigid, reef-building, in-situ framework of an *organic reef,* consisting of the skeletons of sessile calcareous organisms and excluding reef-flank and associated fragmental deposits (MacNeil, 1954, p.390). Syn: *organic lattice; reef frame; lattice [reef].*

growth layer *growth ring [geochron].*

growth line (a) One of a series of fine to coarse ridges on the outer surface of a brachiopod shell, concentric about the beak and parallel or subparallel to the margins of the valves, and indicating the former positions of the margins when the anterior and lateral growth of the shell temporarily was in abeyance. (b) One of a usually irregularly arranged and more or less obscure series of comarginal lines on the surface of a bivalve-mollusk shell, approximately parallel to the borders of the valve, and representing successive advances of the shell margin at earlier growth stages. Cf: *growth ruga.* Syn: *growth band.* (c) One of a series of lines on the surface of a cephalopod conch, denoting periodic increases in size and hence former positions of the aperture. (d) One of a series of collabrally disposed surface markings (low ridges) on the outer surface of a gastropod shell, parallel to and indicating the former positions of the outer lip. (e) An irregular marking on the epitheca of rugose corallites, such as a slight ridge or depression parallel to the upper edge of the corallite, defining a former position of this margin during growth. Syn: *growth ring.*

growth ring [geochron] Layer of wood produced in a tree or woody plant during its annual growth period. It is seen in cross section as a ring. Growth rings can be analyzed for chronologic and climatic data based on number and relative sizes. See also: *dendrochronology; dendroclimatology.* Syn: *annual growth ring; growth layer; tree ring.*

growth ring [paleont] A *growth line* on a rugose coral.

growth ruga An irregular *ruga* or wrinkle on the surface of a bivalve-mollusk shell, having an origin similar to that of a *growth line* but corresponding to a more pronounced halt in growth.

growth twin A twinned crystal that developed as a result of change in lattice orientation during growth.

groyne *groin.*

grumous Formed of clustered, aggregated, or flocculated grains; esp. said of a secondary texture in a microcrystalline, carbonate sedimentary rock that has experienced pervasive recrystallization (such as a diagenetic dolomite). It is characterized by patches of coarse crystals or limy particles irregularly invading shell fragments, ooliths, and matrix, and by dark, dense, fine-grained unrecrystallized areas that are ultimately surrounded by sparry calcite. Syn: *clotted.*

Grumusol Formerly, a general U.S. term for a *Vertisol* (Oakes and Thorp, 1951).

grunerite A monoclinic mineral of the amphibole group: $Fe_7Si_8O_{22}(OH)_2$. Cf: *cummingtonite.* Also spelled: *grünerite.*

grünlingite A trigonal mineral: Bi_4TeS_3(?).

grus The fragmental products of in-situ granular disintegration of granite and granitic rocks. Syn: *residual arkose; slack [weath]; growan.* Etymol: German *Grus,* "grit, fine gravel, debris". Also spelled: *gruss; grush.*

grush *grus.*

gruss *grus.*

gryke *grike.*

gryphaeate Shaped like the shell of *Gryphaea* (a genus of fossil bivalve mollusks); i.e. with the left valve strongly convex and its dorsal part incurved, and with the right valve flat.

GSC *gas-solid chromatography.*

guadalcazarite A variety of metacinnabar containing zinc.

Guadalupian North American series: Lower and Upper Permian (above Leonardian, below Ochoan).

guanajuatite A bluish-gray mineral: Bi_2Se_3.

guanglinite An orthorhombic mineral: Pd_3As.

guanine A monoclinic mineral: $C_5H_3(NH_2)N_4O$ (2-amino-6-hydroxypurine).

guano (a) A phosphate or nitrate deposit formed by the leaching of bird excrement accumulated in arid regions, e.g. islands of the eastern Pacific Ocean and the West Indies. It is processed for use as a fertilizer. Syn: *ornithocopros.* (b) Similar deposits of bat excrement, found in caves and worked for phosphate or nitrate, as in Malaya.

guard The thick hard cigar-shaped calcareous structure that ensheathes the phragmocone of a belemnite and is located in the rear portion of the body. Syn: *rostrum.*

guard cells Specialized epidermal cells, two of which by varying turgor regulate each epidermal pore or stoma (Esau, 1965, p. 160-161).

guard-electrode log *focused-current log.*

guayaquilite A soft, pale-yellow, amorphous fossil resin with a high (15%) oxygen content, soluble in alcohol and alkalies, and found near Guayaquil, Ecuador. Its approximate formula: $C_{40}H_{26}O_6$. Also spelled: *guayaquillite; guyaquillite.*

gudmundite A silver-white to steel-gray orthorhombic mineral: $FeSbS$.

guerinite A mineral: $Ca_5H_2(AsO_4)_4 \cdot 9H_2O$.

guern *khurd.*

guest A mineral introduced into and usually replacing a pre-existent mineral or rock; a *metasome [geol].* Ant: *host.*

guest element *trace element.*

guettardite A mineral: $Pb_9(Sb,As)_{16}S_{33}$.

gugiaite *meliphanite.*

guhr (a) A white (sometimes red or yellow), loose, earthy, water-laid deposit of a mixture of clay or ocher, occurring in the cavities of rocks. (b) *kieselguhr.*

guidebook (a) A *road log* of a field trip, summarizing the geology. (b) A guide to the minerals or fossils available in an area.

guided wave Any seismic wave that is propagated in a single layer or along some surface or discontinuity, e.g. a *surface wave, Stoneley wave,* or *channel wave.*

guide fossil (a) Any fossil that has actual, potential, or supposed value in identifying the age of the strata in which it is found or in indicating the conditions under which it lived; a fossil used esp. as an index or guide in the local correlation of strata. (b) A fossil that is most characteristic of an *assemblage-zone,* but that is not necessarily restricted to the zone or found throughout every part of it (ACSN, 1961, art. 21e).—See also: *zonal guide fossil.* Cf: *index fossil.*

guide meridian A north-south line used for reference in surveying; specif. one of a set of auxiliary governing lines of the U.S. Public Land Survey system, projected north or south from points established on the base line or a standard parallel, usually at intervals of 24 miles east or west of the principal meridian, and on which township, section, and quarter-section corners are established.

guild A group of closely related but distinct species that have very similar ecological requirements and also occur together in particular habitats.

guildite A dark chestnut-brown mineral: $CuFe(SO_4)_2(OH)\cdot 4H_2O$.

guilielmite A subaqueous sedimentary structure formed in mud by collapse around a fossil and characterized by small, polished slip surfaces arranged with radial or orthorhombic symmetry around the fossil (Wood, 1935). It is esp. common in the shales of coal-bearing sections.

guilleminite A canary-yellow secondary mineral: $Ba(UO_2)_3(SeO_3)_2(OH)_4\cdot 3H_2O$.

gula A projecting, rather ornate extension of the trilete laesura of fossil megaspores. Pl: *gulae.* Cf: *apical prominence; acrolamella.*

gulch A term used esp. in the western U.S. for a narrow, deep ravine with steep sides, larger than a *gully;* esp. a short, precipitous cleft in a hillside, formed and occupied by a torrent, and containing gold (as in California).

gulf [coast] A relatively large part of an ocean or sea extending far into the land, partly enclosed by an extensive sweep of the coast, and opened to the sea through a strait; the largest of various forms of inlets of the sea. It is usually larger, more enclosed, and more deeply indented than a bay.

gulf [geomorph] A deep, narrow hollow, gorge, or chasm; e.g. one of the long precipitous stream-worn excavations west of the Adirondack Mountains in northern New York State.

gulf [karst] A *sinkhole,* commonly containing flowing water on its floor.

Gulf Coast-type fault *growth fault.*

gulf-cut island An island formed by the cutting backward of two parallel inlets into opposite sides of a piece of subsiding land (Powell, 1895, p. 64).

Gulfian North American provincial series: Upper Cretaceous (above Comanchean, below Paleocene of Tertiary).

Gulf-type gravimeter A gravity meter consisting of a mass suspended at the end of a spring, the latter so designed that its extension will cause the mass to rotate. By this means the linear displacement of the spring is converted into an angular deflection which is more easily measured (Wyckoff, 1941, p.13). The design also minimizes the sensitivity to seismic disturbances, and the basic instrument is therefore well suited for underwater observations (Pepper, 1941, p. 34) Syn: *Hoyt gravimeter.*

gull A structure formed by mass-movement processes, consisting of widened, steeply inclined tension fissures or joints, resulting from lateral displacement of a slide mass and filled with debris derived from above. Gulls generally trend parallel to surface contours, and are usually associated with *camber* (Hollingworth et al., 1950). Primarily a British usage.

gullet [paleont] (a) A variably tubular invagination of the cytoplasm of various protists (such as tintinnids) that sometimes functions in the intake of food. (b) A longitudinal groove present in certain algae (such as some Cryptophyceae and Euglenophyceae).

gullet [streams] A narrow opening or depression, such as a defile or ravine; a *gully* or other channel for water.

gull hummock A conical or dome-shaped peaty mound, formed by accretion of well-manured grasses, sedges, and other vascular plants near the nest of the great black-backed gull, on islands in the Arctic (ADTIC, 1955, p. 61). Owls and other perching birds form similar hummocks in the tundra regions. Cf: *peat mound.*

gully [coast] A wave-cut chasm in a cliff, or a minor channel incised in a mud flat below the high-water level (Schieferdecker, 1959, terms 1149 & 1233).

gully [geomorph] (a) A very small valley, such as a small *ravine* in a cliff face, or a long, narrow hollow or channel worn in earth or unconsolidated material (as on a hillside) by running water and through which water runs only after a rain or the melting of ice or snow; it is smaller than a *gulch.* Syn: *gulley; gullet.* (b) Any erosion channel so deep that it cannot be crossed by a wheeled vehicle or eliminated by plowing, esp. one excavated in soil on a bare slope. (c) A small, steep-sided wooded hollow.

gully erosion Erosion of soil or soft rock material by running water that forms distinct, narrow channels that are larger and deeper than rills and that usually carry water only during and immediately after heavy rains or following the melting of ice or snow. Cf: *sheet erosion; rill erosion; channel erosion.* Syn: *gullying; ravinement.*

gully gravure A term used by Bryan (1940) for the process or processes whereby the steep slopes of hills and mountains retreat by "repeated scoring or graving", each groove (gully) "so disposed as to reduce rather than emphasize inequalities" (p. 92); the development of rills into gullies.

gullying *gully erosion.*

gum An organic, viscid juice extracted from, or exuded by, certain trees and plants. It hardens in the air and is soluble in water.

gumbo A term used locally in the U.S. for a clay soil that becomes sticky, impervious, and plastic when wet.

gumbotil (a) A gray to dark-colored, leached, deoxidized clay representing the B horizon of fully mature soils, developed from profoundly weathered clay-rich till under conditions of low relief and poor subsurface drainage (as beneath broad, flat uplands). It consists chiefly of beidellite and/or illite, and may contain altered rock fragments originally mixed with the clay; it is very sticky and plastic when wet, extremely firm when dry. Term introduced by Kay (1916). Cf: *silttil; mesotil.* See also: *ferretto zone.* (b) A term used for a fossilized soil beneath a deposit of later till.

gum copal An inferior resin or amber; copal.

gummite A general term for yellow, orange, red, or brown secondary minerals consisting of a mixture of hydrous oxides of uranium, thorium, and lead, and occurring as alteration products of uraninite and not otherwise identified. It includes silicates, phosphates, and oxides; much of the material is probably mixtures or amorphous gels, but some consists perhaps largely of curite. Syn: *uranium ocher.*

G unit *gravity unit.*

gunite n. A mixture of portland cement, sand, and water applied by pneumatic pressure through a specially adapted hose and used as a fireproofing agent and as a sealing agent to prevent weathering of mine timbers and roadways. Etymol: *Gunite,* a trademark. Cf: *shotcrete.*—v. To apply gunite; to cement by spraying gunite.

gunningite A monoclinic mineral: $(Zn,Mn)(SO_4)\cdot H_2O$.

Gunnison River A provincial series of the Precambrian in Colorado.

Gunter's chain A surveyor's *chain* that is 66 feet long, consisting of a series of 100 metal links each 7.92 inches long and fastened together with rings. It served as the legal unit of length for surveys of U.S. public lands, but has been superseded by steel or metal tapes graduated in chains and links. Named after Edmund Gunter (1581-1626), English mathematician and astronomer, who invented the device about 1620. Syn: *pole chain.*

Günz (a) A European stage of the Pleistocene, above the Donau gravels, below the Mindel. (b) Long considered the first glacial stage of the Pleistocene Epoch in the Alps; it is now known that there were earlier ones. See also: *Nebraskan.*—Etymol: Günz River, Bavaria. Adj: *Günzian.*

Günz-Mindel The term applied in the Alps to the first classical interglacial stage of the Pleistocene Epoch, following the Günz and preceding the Mindel glacial stages. See also: *Aftonian.*

gurhofite A snow-white variety of dolomite mineral, containing a large proportion of calcium. Syn: *gurhofian.*

gusher *geyser.*

gushing spring *vauclusian spring.*

gustavite A mineral: $PbAgBi_3S_6$.

gut (a) A very narrow passage or channel connecting two bodies of water; e.g. a contracted strait, or a small creek in a marsh or tidal flat, or an inlet. Also, "a channel in otherwise shallow water, generally formed by water in motion" (CERC, 1966, p. A14). (b) A tidal stream connecting two larger waterways. (c) A term used in the Virgin Islands and elsewhere for a gully, ravine, small valley, or narrow passage on land.

Gutenberg discontinuity The seismic-velocity discontinuity at

2900 km, marking the mantle-core boundary, at which the velocities of *P* waves are reduced and *S* waves disappear. It probably reflects the change from a solid to a liquid phase and a change in composition. It is named after Beno Gutenberg, seismologist. Syn: *Oldham-Gutenberg discontinuity; Wiechert-Gutenberg discontinuity.*

Gutenberg low-velocity zone The *low-velocity zone* of the upper mantle.

gutsevichite A mineral: $(Al,Fe)_3(PO_4,VO_4)_2(OH)_3 \cdot 8H_2O$ (?).

guttation The process by which water in liquid form is exuded from an uninjured surface of a plant. Cf: *transpiration.*

gutter [ore dep] The lowest and usually richest portion of an alluvial placer. The term is used in Australia for the dry bed of a buried Tertiary river. Syn: *bottom [ore dep].*

gutter [streams] (a) A shallow, natural channel, furrow, or gully worn by running water. (b) A shallow, steep-sided valley that drains a marshy upland; it usually marks an area where the drainage is about to be rejuvenated. (c) An artificially paved watercourse, such as a roadside ditch for carrying off excess surface water to a sewer. (d) An archaic term for a brook.

gutter cast A down-bulge on the bottom of a sedimentary bed, of great length (usually one meter or more) compared with its width and depth (a few centimeters to several decimeters). In cross section it has the form of a small channel (Whitaker, 1973, p. 405). Cf: *channel; washout.*

guyanaite An orthorhombic mineral: CrO(OH). It is polymorphous with grimaldite and bracewellite.

guyaquillite *guayaquilite.*

guyot A type of *seamount* that has a platform top. Etymol: Arnold Guyot, nineteenth-century Swiss-American geologist. Syn: *tablemount; tableknoll.*

G wave A long-period (1-4 min) *Love wave* in the upper mantle, usually restricted to an oceanic path. The G stands for Gutenberg.

gymnite *deweylite.*

gymnocyst A frontal shield in cheilostome bryozoans, formed by calcification of part of the frontal wall. In anascan cheilostomes it is principally developed on the proximal and lateral margins of the frontal membrane.

Gymnophiona *Caecilia.*

gymnosolen A fingerlike or digitate form of stromatolite, splitting off in two or more upward directions from algal structures, resembling a series of stacked inverted thimbles or (if large) of soup bowls similarly arranged (Pettijohn, 1957, p.222). It is produced by blue-green algae of the genus *Gymnosolen.*

gymnosperm A plant whose seeds are commonly in cones and never enclosed in an ovary. Examples include cycad, ginkgo, pine, fir, and spruce. Such plants range from the Late Devonian. Cf: *angiosperm.*

gymnospore A naked spore, or one not developing in a sporangium. The term is not in good usage in palynology.

gyp A syn. of *gypsum.* Also spelled: *gyps.*

gyparenite A sandstone composed of discrete, wind-drifted particles of gypsum.

gypcrete A gypsum-cemented crust or rock, found in some playa-lake beachrock environments in an arid climate (Fairbridge, 1968, p. 555).

gyprock *rock gypsum.*

gypsey A syn. of *bourne.* Also spelled: gipsy; gypsy.

gypsic horizon A diagnostic subsurface soil horizon at least 15 cm thick that is characterized by enrichment in calcium sulfate.

gypsiferous Gypsum-bearing, as *gypsiferous* shales.

gypsification Development of, or conversion into, gypsum; e.g. the hydration of anhydrite.

gypsinate Cemented with gypsum.

gypsite (a) An earthy variety of gypsum containing dirt and sand, found only in arid regions as an efflorescent deposit occurring over the ledge outcrop of gypsum or of a gypsum-bearing stratum. Syn: *gypsum earth.* (b) *gypsum.*

gypsolith A term suggested by Grabau (1924, p. 298) for a gypsum rock.

gypsum A widely distributed mineral consisting of hydrous calcium sulfate: $CaSO_4 \cdot 2H_2O$ It is the commonest sulfate mineral, and is frequently associated with halite and anhydrite in evaporites, forming thick, extensive beds interstratified with limestone, shale, and clay (esp. in rocks of Permian and Triassic age). Gypsum is soft (hardness of 2 on the Mohs scale); it is white or colorless when pure, but commonly has tints of gray, red, yellow, blue, or brown. It occurs massive (alabaster), fibrous (satin spar), or in monoclinic crystals (selenite) Gypsum is used chiefly as a soil amendment, as a retarder in portland cement, and in making plaster of Paris. Etymol: Greek *gypsos,* "chalk". Syn: *gypsite; gyp; plaster stone; plaster of Paris.*

gypsum cave (a) A cave that is formed in gypsum rock by solution. (b) A cave containing abundant gypsum incrustations.

gypsum cotton *cave cotton.*

gypsum earth *gypsite.*

gypsum flower *cave flower.*

gypsum plate In a polarizing microscope, a plate of clear gypsum (selenite) that gives a first-order red interference color; it is used to determine optical sign with crystals or interference figures and to determine the position of vibration-plane traces in crystal plates.

gyral *gyre.*

gyrate Winding or coiled round; convolute, like the surface of the brain.

gyre A circular motion of water in each of the major ocean basins, centered on a subtropical high-pressure region; its movement is generated by convective flow of warm surface water poleward, by the deflective effect of the Earth's rotation, and by the effects of prevailing winds. The water within each gyre turns clockwise in the Northern Hemisphere and counterclockwise in the Southern Hemisphere. Syn: *gyral.*

gyrocompass A nonmagnetic *compass* that functions by virtue of the couples generated in a rotor when the latter's axis of rotation is displaced from parallelism with that of the Earth and that consists of a continuously driven gyroscope whose supporting ring confines the spring axis to a horizontal plane. It automatically aligns itself in the celestial meridian (thus pointing to the true north) by the Earth's rotation which causes it to assume a position parallel to the Earth's axis. The gyrocompass is used in underground and borehole surveying. Syn: *gyroscopic compass; gyrostatic compass.*

gyrocone A loosely coiled cephalopod shell in which the successive whorls are not in contact with each other, or in which only a single whorl is approximately completed. Syn: *gyroceracone.*

gyrogastric Said of a gastropod shell that is coiled toward the posterior of the body; caused by rotation (torsion) of the larval exogastric shell (Pojeta & Runnegar, 1976).

gyrogonite A dispersed fossil oogonium of charophytes.

gyroid An isometric crystal form consisting of 24 crystal faces with indices $\{hkl\}$ and symmetry 432. A gyroidal crystal may be right- or left-handed.

gyroidal class That crystal class in the isometric system having symmetry 432.

gyrolite A white mineral with a micaceous cleavage: $Ca_2Si_3O_7(OH)_2 \cdot H_2O$. Syn: *centrallasite.*

gyroscopic compass (a) *gyrocompass.* (b) A magnetic compass whose equilibrium is maintained by the use of gyroscopes.

gyrostatic compass *gyrocompass.*

gyttja A dark, pulpy, freshwater mud characterized by abundant organic matter that is more or less determinable, and deposited or precipitated in a marsh or in a lake whose waters are rich in nutrients and oxygen. It is an anaerobic sediment laid down under conditions ranging from aerobic to anaerobic and is capable of supporting aerobic life. Cf: *dy; sapropel.* Etymol: Swedish.

Gzhelian Stage in Russia: upper Upper Carboniferous (above Moscovian, below Orenburgian).

Haalck gravimeter A gravimeter in which the change in weight of a mercury column is balanced by a gas spring (Heiland, 1940, p.124).

haapalaite A hexagonal mineral: $2(Fe,Ni)_2S_2 \cdot 3(Mg,Fe)(OH)_2$.

habit (a) The characteristic crystal form or combination of forms of a mineral, including characteristic irregularities. (b) A general term for the outward appearance of a mineral or rock. (c) The characteristic appearance of an organism, esp. those aspects that most affect its mode of life. Syn: *habitus.*

habitat The particular *environment* or place where an organism or species tends to live; a more locally circumscribed portion of the total environment.

habitus *habit.*

hachure n. One of a series of short, straight, evenly spaced, parallel lines used on a topographic map for shading and for indicating surfaces in relief (such as steepness of slopes), drawn perpendicular to the contour lines; e.g. an inward-pointing "tick" trending downslope from a depression contour. Hachures are short, broad (heavy), and close together for a steep slope, and long, narrow (light), and widely spaced for a gentle slope, and they enable minor details to be shown but do not indicate elevations above sea level. Etymol: French. Syn: *hatching; hatchure.*—v. To shade with or show by hachures.

hackly fracture The property shown by certain minerals or rocks of fracturing or breaking on jagged surfaces.

hackmanite A variety of sodalite containing a little sulfur and usually fluorescing orange or red under ultraviolet light.

hacksaw structure The irregular, *saw-toothed* or saw-shaped termination of a crystal (such as of augite) or mineral particle due to intrastratal solution. Syn: *hacksaw termination; cockscomb structure.*

hadal Pertaining to the deepest oceanic environment, specifically that of oceanic trenches, i.e., over 6.5 km in depth.

hade n. In structural geology, the complement of the *dip;* the angle that a structural surface makes with the vertical, measured perpendicular to the strike. It is little used. Syn: *rise; underlay.*—v. To incline from the vertical.

Hadley cell A thermally driven unit of atmospheric circulation that extends in both directions from the equator to about 30°. Air rises at the equator, flows poleward, descends, and then flows toward the equator. It is named after G. Hadley who described it in 1735.

Hadrynian In a three-part division of the Proterozoic of Canada, the latest division, above the *Helikian.* Cf: *Aphebian.*

Haeckel's law *recapitulation theory.*

haematite Original spelling of *hematite.*

haff A shallow freshwater coastal lagoon separated from the open sea by a sandspit (*nehrung*) across a river mouth; esp. such a lagoon on the East German coast of the Baltic Sea. Pl: *haffs; haffe.* Etymol: German *Haff,* "lagoon".

hafnon A tetragonal mineral: $HfSiO_4$.

hagendorfite A greenish-black mineral: $(Na,Ca)(Fe,Mn)_2(PO_4)_2$.

häggite A black monoclinic mineral: $V_2O_2(OH)_3$.

haidingerite A white or colorless mineral: $CaHAsO_4 \cdot H_2O$.

hail Precipitation in the form of spheroidal layered ice pellets that usually fall from cumulonimbus clouds of a thunderstorm. The layered structure is produced by successive accretions of clear and frothy ice.

hail imprint A small, shallow depression or crater-like pit formed by a hailstone falling on a soft sedimentary surface. It is generally larger, deeper, and more irregular than a *rain print.* Syn: *hail pit; hailstone imprint.*

hail pit *hail imprint.*

hailstone imprint *hail imprint.*

hair ball *lake ball.*

hair copper *chalcotrichite.*

hairpin dune A greatly elongated *parabolic dune* that has migrated downwind, its horns drawn out parallel to each other, formed where a constant wind is in conflict with vegetation.

hair pyrites (a) *millerite.* (b) Capillary pyrite.

hair salt (a) *alunogen.* (b) Silky or fibrous *epsomite.*

hairstone A variety of clear crystalline quartz thickly penetrated with fibrous, threadlike, or acicular inclusions of other minerals, usually crystals of rutile or actinolite; esp. sagenitic quartz. See also: *Venus hairstone; Thetis hairstone.* Also spelled: *hair stone.* Syn: *needle stone.*

hair zeolite A group of fibrous zeolite minerals, including natrolite, mesolite, scolecite, thomsonite, and mordenite. See also: *needle zeolite.* Syn: *feather zeolite.*

haiweeite A pale-yellow to greenish-yellow secondary mineral: $Ca(UO_2)_2Si_6O_{15} \cdot 5H_2O$.

hakite A mineral of the tetrahedrite group: $(Cu,Hg)_{12}Sb_4(S,Se)_{13}$.

haldenhang A syn. of *wash slope.* Etymol: German *Haldenhang,* "under-talus rock slope of degradation" (Penck, 1924).

half-blind valley A *blind valley* whose stream may overflow during floods, when the sinking stream cannot manage all the water.

Half-Bog soil A great soil group in the 1938 classification system, an intrazonal, hydromorphic group of soils with mucky or peaty surface soil underlain by a gray mineral soil (USDA, 1938). Most of these soils are now classified as *Aquepts.* Cf: *Bog soil.*

half-life The time necessary for a radioactive substance to lose half of its radioactivity (provided there are a large number of atoms involved). Each *radionuclide* has a characteristic half-life.

half-moon cut A style of gem cutting that produces a stone in the shape of a half circle.

half-round Part of a rounded pebble or cobble that was deposited, sheared by faulting, and then redeposited in two or more pieces (Tanner, 1976, p. 80).

half section A half of a normal *section* of the U.S. Public Land Survey system, representing a piece of land containing 320 acres as nearly as possible; any two quarter sections within a section which have a common boundary. It is usually identified as the north half, south half, east half, or west half of a particular section.

half-tide level *mean tide level.*

half-tube A remnant or trace of a *tube [speleo]* visible on the roof or walls of a cave.

half-value thickness The thickness of an absorbing medium which will reduce any incident radiation to half its initial intensity.

half width Half the width of a simple *anomaly* (esp. a gravity or magnetic anomaly) at the point of half its maximum value. For simple models the maximum depth at which the body causing the anomaly can lie can be calculated from the half width.

halide A mineral compound characterized by a halogen such as fluorine, chlorine, iodine, or bromine as the anion. Halite, NaCl, is an example. Syn: *halogenide.*

halilith A term suggested by Grabau (1924, p. 298) for rock salt (a sedimentary rock).

halite A mineral: NaCl. It is native salt, occurring in massive, granular, compact, or cubic-crystalline forms, and having a distinctive salty taste. Symbol: Hl. Syn: *common salt; rock salt.*

halitic Pertaining to halite; esp. said of a sedimentary rock containing halite as cementing material, such as "halitic sandstone".

hälleflinta Fine-grained, compact granular hornfels formed by contact metamorphism of an acid igneous rock such as rhyolite, quartz porphyry, or acid tuff resulting in a banded and/or blastoporphyritic rock. Cf: *leptite.*

hälleflintgneiss An obsolete syn. of *leptite.*

hallimondite A yellow secondary mineral: $Pb_2(UO_2)(AsO_4)_2$.

halloysite (a) A name used in the U.S. for a porcelainlike clay mineral: $Al_2Si_2O_5(OH)_4 \cdot 2H_2O$. It is synonymous with *metahalloysite* of European authors. Halloysite is made up of minute slender tubes, as shown by the electron microscope. The term has also been used to designate a nonhydrated mineral having the chemical composition of, but structurally distinct from, kaolinite. (b) A name used in Europe for a clay mineral: $Al_2Si_2O_5(OH)_4 \cdot 4H_2O$. It is synonymous with *endellite* of U.S. authors, and designates a more highly hydrated mineral than metahalloysite. (c) A general term proposed by MacEwan (1947) and used by Grim (1968, p. 39) for all the naturally occurring halloysite minerals (hydrated, nonhydrated, and intermediate) and for artificially prepared complexes. It is a group name used with qualifications to denote the type that is being described, (e.g. "glycerol-halloysite", "fully hydrated halloysite").

halmeic Said of a deep-sea sediment formed directly from solution or around an organic nucleus, e.g. barite, phosphorite, manganese nodules. Cf: *authigenic.* Ant: *chthonic.*

halmyrogenic *halmeic.*

halmyrolysis The geochemical reaction of sea water and sediments in an area of little or no sedimentation. Examples include modification of clay minerals, formation of glauconite from feldspars and micas, and the formation of phillipsite and palagonite from volcanic ash. Alternate spelling: halmyrosis. Cf: *diagenesis.* Syn: *submarine weathering.*

halo (a) A circular or crescentic distribution pattern about the source or origin of a mineral, ore, mineral association, or petrographic feature. It is encountered principally in magnetic and geochemical surveys. Cf: *dispersion pattern.* (b) Discoloration of a mineral, viewed in thin section, in the form of a ring. Most haloes of this sort are caused by radiation damage by alpha particles emitted from uranium- and thorium-bearing mineral inclusions.

halocline A steep downward increase in salinity of seawater. Cf: *thermocline.*

halogenic *halmeic.*

halogenide *halide.*

halokinesis *salt tectonics.*

halomorphic soil A type of intrazonal soil (1938 classification system) whose characteristics have been strongly influenced by the presence of neutral or alkali salts or both.

halo ore *fringe ore.*

halophilic Said of an organism that prefers a saline environment. Noun: *halophile.* Cf: *haloxene.*

halophreatophyte A plant receiving its water supply from saline ground water.

halophyte A plant growing in soil or water with a high content of salts.

halosere A *sere* that develops in a saline environment.

halotrichite (a) A mineral: $FeAl_2(SO_4)_4 \cdot 22H_2O$. It occurs in yellowish fibrous crystals. Syn: *feather alum; iron alum; mountain butter; butter rock.* (b) Any of several sulfates similar to halotrichite in construction and habit; e.g. alunogen.

haloxene Said of an organism that can tolerate saline conditions but does not prefer them. Cf: *halophilic.*

hals A British term for a pass or col. Cf: *hause.* Syn: *halse.*

halurgite A mineral: $Mg_2B_8O_{14} \cdot 5H_2O$.

hamada *hammada.*

hambergite A grayish-white or colorless mineral: Be_2BO_3OH.

hamlinite *goyazite.*

hammada An extensive, nearly level upland desert surface that is either bare bedrock or bedrock thinly veneered by pebbles, smoothly scoured and polished and generally swept clear of sand and dust by wind action; a *rock desert* of the plateaus, esp. in the Sahara. The term is also used in other regions, as in Western Australia and the Gobi Desert. Etymol: Arabic, *hammadah.* See also: *reg; serir.* Also spelled: *hamada; hammadah; hammadat; hamadet.* Syn: *nejd.*

hammarite A reddish steel-gray mineral: $Pb_2Cu_2Bi_4S_9$ (?).

Hammer-Aitoff projection An equal-area map projection derived from the equatorial aspect of the Lambert azimuthal equal-area projection by doubling the horizontal distances along each parallel from the central meridian until the entire spherical surface can be represented within an ellipse whose major axis (equator) is twice the length of its minor axis (central meridian). It resembles the Mollweide projection, but all parallels (except the equator) are represented by curved lines and there is less angular distortion near the margins. The projection was introduced in 1892 by H.H. Ernst von Hammer (1858-1925), German geodesist, but is often attributed to David Aitoff (d.1933), Russian geographer, who previously introduced a similar-appearing projection based on the azimuthal equidistant projection. Incorrect syn: *Aitoff projection.*

Hammer chart A template for making gravity terrain corrections. Named for Sigmund Hammer.

hammock (a) *hummock [geog].* (b) A term applied in the SE U.S. to a fertile area of deep, humus-rich soil, generally covered by hardwood vegetation and often rising slightly above a plain or swamp; esp. an island of dense, tropical undergrowth in the Florida Everglades. Syn: *hummock.*

hammock structure The intersection of two vein or fracture systems at an acute angle.

hampshirite Steatite pseudomorphous after olivine.

hamrongite A dark violet-gray fine-grained *lamprophyre* containing phenocrysts of black mica in a groundmass characterized by intersertal texture and composed of mica, andesine, and some quartz; a quartz *kersantite.* The name, given by Eckermann in 1928, is from Hamrånge, Sweden. Not recommended usage.

hamulus A hook-shaped secondary deposit on the chamber floor in foraminifers of the family Endothyridae. The point of the hook is directed toward the aperture of the test. Pl: *hamuli.*

hancockite A mineral of the epidote group: $(Pb,Ca,Sr)_2(Al,Fe)_3(SiO_4)_3(OH)$.

hand lens A small magnifying glass (usually X6 to X10) for use in the field or in other preliminary investigations of a mineral, fossil, or rock. Syn: *pocket lens.*

hand level A small, hand-held leveling instrument in which the spirit level is so mounted that the observer can view the bubble at the same time that he sights an object through the telescope. The viewing of the bubble is accomplished by means of a prism or mirror in the telescope tube: when the cross hair bisects the bubble and the object in view, that object is on a level with the eye. The hand level is used where a high degree of precision and accuracy is not required, such as in reconnaissance surveys. See also: *Abney level; Locke level.*

hand specimen A piece of rock of a size that is convenient for megascopic study and for preserving in a study collection.

hanger *hanging wall.*

hanging Situated on steeply sloping ground (such as a *hanging* meadow) or on top of other ground (such as a *hanging wall*), or jutting out and downward (such as a *hanging* rock), or situated at or having a discordant junction (such as a *hanging valley*).

hanging cirque A cirque on a mountainside, excavated by a former hanging glacier and not continued in a valley. Cf: *valley-head cirque.* See also: *corrie.*

hanging drumlin A drumlin on a valley slope, consisting of subglacial debris pushed laterally or molded into its present position by a thin overriding glacier.

hanging glacier A glacier, generally small, protruding from a basin or niche on a mountainside above a cliff or very steep slope, from which ice may break off occasionally and abruptly to form an ice avalanche. Cf: *cliff glacier.*

hanging side *hanging wall.*

hanging tributary A tributary stream or tributary glacier occupying a *hanging valley.*

hanging trough A glacial *hanging valley.*

hanging valley [glac geol] A glacial valley whose mouth is at a relatively high level on the steep side of a larger glacial valley. The larger valley was eroded by a trunk glacier and the smaller one by a tributary glacier, and the discordance of level of their floors, as well as their difference in size, is due to the greater erosive power of the trunk glacier. Syn: *hanging trough; perched glacial valley.*

hanging valley [streams] (a) A tributary valley whose floor at the lower end is notably higher than the floor of the main valley in the area of junction, produced where the more rapid deepening of the main valley results in the creation of a cliff or steep slope over which a waterfall may develop. (b) A coastal valley whose lower end is notably higher than the shore to which it leads, produced where betrunking or rapid cliff recession causes the mouths of streams to "hang" along the cliff front. Syn: *valleuse.*

hanging wall The overlying side of an orebody, fault, or mine working; esp. the wall rock above an inclined vein or fault. Syn: *hanging side; hanger.* Cf: *footwall; upper plate.*

hanging-wall cutoff A syn. of *leading edge.* See: Royse et al. (1975).

hanksite A white or yellow hexagonal mineral: $Na_{22}K(SO_4)_9$

$(CO_3)_2Cl$.

hannayite A mineral: $Mg_3(NH_4)_2H_4(PO_4)_4 \cdot 8H_2O$. It occurs as slender yellowish crystals in guano.

hanusite A mixture of stevensite and pectolite. It was formerly regarded as a mineral: $Mg_2Si_3O_7(OH)_2 \cdot H_2O$.

haplite *aplite.*

haplome A "more correct", but "not generally accepted", spelling of *aplome* (Hey, 1962, p. 446).

haplophyre A *granite* found in the Alps and characterized by large quartz and feldspar grains in a mortar structure (Thrush, 1968, p. 526). Not recommended usage.

haplopore An unbranched pore lying normally within one thecal plate of a cystoid. If connected in pairs, haplopores are designated *diplopores.*

haplotabular archeopyle An *apical archeopyle* in a dinoflagellate cyst consisting of a single plate.

haploxylonoid Said of bisaccate pollen, in which the outline of the sacci in distal-proximal view is more or less continuous with the outline of the body, the sacci appearing more or less crescent-shaped, and the outline of the whole grain presenting generally smooth ellipsoidal forms. Cf: *diploxylonoid.*

haplozoan Any one of a small group of supposedly free-living echinoderms belonging to the subphylum Haplozoa, comprising only two genera, and characterized by a thick calcareous skeleton composed of a few plates arranged around a median, craterlike depression. They are known only from the Middle Cambrian.

hapteron An attaching structure in some of the larger seaweeds, esp. in the brown algae; it is usually multicellular, branched, and rootlike.

haptonema A threadlike to clublike part of a cell in a coccolithophorid, located between, but more rigid than, the flagella. It can be contracted into a spiral or extended and used as an organ of attachment to substrate material.

haptotypic character A feature of spores and pollen grains that is a product of its contact with other members of the tetrad in which it was formed; e.g. the laesura and contact areas of spores.

haradaite A mineral: $SrVSi_2O_7$.

harbor (a) A small bay or a sheltered part of a sea, lake, or other large body of water, usually well protected either naturally or artificially against high waves and strong currents, and deep enough to provide safe anchorage for ships; esp. such a place in which port facilities are furnished. (b) An inlet; e.g. Pearl Harbor, Hawaii.—British spelling: *harbour.*

harbor bar A bar built across the exit to a harbor.

hardcap A term used in bauxite mining for the uppermost foot or two of a bauxite deposit. Since it is harder and tougher than the material below it, it is usually used as a roof during mining.

hard coal (a) A syn. of *anthracite.* Cf: *soft coal.* (b) Outside the U.S., the term is sometimes used for any coal with a calorific value higher than 5700 kcal/kg (10,260 BTU/lb) on a moist, mineral-matter-free basis; this includes bituminous and certain subbituminous coals.

hardebank Unaltered *kimberlite* below the zone of *blue ground.*

hardground A zone at the sea bottom, usually a few cm thick, the sediment of which is lithified to form a hardened surface, often encrusted, discolored, case-hardened, bored, and solution-ridden. It implies a gap in sedimentation and may be preserved stratigraphically as an unconformity. Also spelled: *hard ground; hard-ground.*

hardhead A syn. of *niggerhead.* Also spelled: *hard head.*

hard magnetization Magnetization that is not easily destroyed; specifically, remanent magnetization with a large *coercive force.* Cf: *soft magnetization; stable magnetization.*

hard mineral A mineral that is as hard as or harder than quartz, i.e. ranking seven or higher on the Mohs scale. Cf: *soft mineral.*

hardness [mineral] The resistance of a mineral to scratching; it is a property by which minerals may be described, relative to a standard scale of ten minerals known as the *Mohs scale,* to the *technical scale* of fifteen minerals, or to any other standard.

hardness [water] A property of water causing formation of an insoluble residue when the water is used with soap, and forming a scale in vessels in which water has been allowed to evaporate. It is primarily due to the presence of ions of calcium and magnesium, but also to ions of other alkali metals, other metals (e.g. iron), and even hydrogen. Hardness of water is generally expressed as parts per million as $CaCO_3$ (40 ppm Ca produces a hardness of 100 ppm as $CaCO_3$); also as milligrams per liter; and as the combination of *carbonate hardness* and *noncarbonate hardness.* Syn: *total hardness.* Cf: *soft water.*

hardness points Small, pointed pieces of minerals of different hardness, affixed to metal handles and used for testing the hardness of another mineral by ascertaining which point will scratch it. Minerals of hardness 6 to 10 on the Mohs scale are usually used as the points for testing gemstones.

hard ore A term used in the Lake Superior region for a compact, massive iron ore mainly composed of specular hematite and/or magnetite and containing more than 58% iron. Cf: *soft ore.*

hardpan (a) A general term for a relatively hard, impervious, and often clayey layer of soil lying at or just below the surface, produced as a result of cementation of soil particles by precipitation of relatively insoluble materials such as silica, iron oxide, calcium carbonate, and organic matter. Its hardness does not change appreciably with changes in moisture content, and it does not slake or become plastic when mixed with water. The term is not properly applied to hard clay layers that are not cemented, or to layers that may seem indurated but which soften when soaked in water. See also: *lime pan; iron pan; ortstein.* Cf: *duricrust; claypan [soil]; fragipan.* (b) A layer of gravel encountered in the digging of a gold placer, occurring one or two meters below the ground surface and partly cemented with limonite. (c) A term commonly applied in NW U.S. to a compact, subglacial till that must be drilled or blasted before removal. Also, a cemented layer of sand or gravel enclosed within till, or a capping of partly cemented material at the top of a water-bearing layer of sand or gravel in till (Wentworth, 1935, p.243). (d) A popular term used loosely to designate any relatively hard layer that is difficult to excavate or drill; e.g. a thin resistant layer of limestone interbedded with easily drilled soft shales. (e) *caliche [soil].* (f) *plow sole.*—Legget (1962, p.798) suggests that the term be avoided "in view of its wide and essentially popular local use for a wide range of materials".

hard rock (a) A term used loosely for igneous or metamorphic rock, as distinguished from sedimentary rock. (b) A rock that is relatively resistant to erosion. (c) Rock that requires drilling and blasting for its economical removal. (d) A term used loosely by drillers for a pre-Cretaceous sedimentary rock that is drilled relatively slowly and that produces samples usually readily identified as to depth interval.—Cf: *soft rock.*

hard-rock geology A colloquial term for geology of igneous and metamorphic rocks, as opposed to *soft-rock geology.*

hard-rock phosphate A term used in Florida for pebbles and boulders of a hard massive homogeneous light-gray phosphorite, showing irregular cavities that are usually lined with secondary mammillary incrustations of calcium phosphate. It is essentially equivalent to the term *white-bedded phosphate* that is used in Tennessee.

hard shore A shore composed of sand, gravel, cobbles, boulders, or bedrock. Ant: *soft shore.*

hard spar A name applied to corundum and andalusite.

hardware The physical equipment or components involved in computing. Cf: *software.*

hard water Water that does not lather readily when used with soap, and that forms a scale in containers in which it has been allowed to evaporate; water with more than 60 mg/l of hardness-forming constituents, expressed as $CaCO_3$ equivalent. See also: *hardness [water].* Cf: *soft water.*

hardwood The wood of an angiospermous tree. Actually, such wood may be either hard or soft. Cf: *softwood.*

hardystonite A white mineral: $Ca_2ZnSi_2O_7$.

Harker diagram A *variation diagram* on which silica content is shown on the abscissa and other oxides on the ordinate.

harkerite A colorless mineral: $Ca_{48}Mg_{16}Al_3(BO_3)_{15}(CO_3)_{18}(SiO_4)_{12}(OH,Cl)_8 \cdot 3H_2O$.

Harlechian European stage: Lower Cambrian.

harlequin opal Opal with small, close-set, angular (mosaiclike) patches of play of color of similar size. Cf: *pinfire opal.*

harmomegathus The *membrane* of the pore or colpus of a pollen grain when it serves to accommodate, by expansion, an increase in volume of the grain; this usually results from the taking-up of water. Pl: *harmomegathi.* Adj: *harmomegathic.*

harmonic analysis *Fourier analysis.*

harmonic fold A fold whose form is constant throughout its constituent strata. Ant: *disharmonic fold.*

harmonic folding Folding in which the strata remain parallel or concentric, without structural discordances between them, and in which there are no sudden changes in the form of the folds at depth. Ant: *disharmonic folding.*

harmonic mean *n* numbers divided by the sum of the reciprocals of the variables (Freund, 1960, p. 67). See also: *mean; geometric mean.*

harmotome A zeolite mineral: $(Ba,K)_{1-2}(Al,Si)_{8}O_{16}\cdot 6H_2O$. It forms cruciform twin crystals. Syn: *cross-stone.*

harpolith (a) A large, sickle-shaped igneous intrusion that was injected into previously deformed strata and was subsequently deformed with the host rock by horizontal stretching or orogenic forces. (b) Essentially a *phacolith* with a vertical axis.

harrisite [mineral] Chalcocite pseudomorphous after galena.

harrisite [petrology] A granular igneous rock composed chiefly of olivine and a smaller amount of anorthite, and characterized by *harrisitic* texture. Named by Harker in 1908 for Harris, Isle of Rhum. Cf: *troctolite.* Not recommended usage.

harrisitic Said of the texture observed in certain olivine-rich rocks (esp. *harrisite*) in which the olivine crystals are oriented at approximately right angles to the cumulate layering of the rock. This phenomenon is now known to occur with other minerals and is called *crescumulate* texture (Wager, 1968, p.579).

harrow mark One of a group of parallel fine-grained ridges of sand, silt, and clay, from about 1 to 10 cm high and 5 to 50 cm apart, with intervening trough-like strips of coarser sediments, occurring in stream channels and extending for distances as great as 100 m. It has been ascribed to the action of regular longitudinal helical flow patterns with alternating senses of rotation (Karcz, 1967).

harstigite A mineral: $Ca_6(Mn,Mg)Be_4Si_6(O,OH)_{24}$.

hartite A white, crystalline, fossil resin found in lignites. Syn: *bombiccite; branchite; hofmannite; josen.*

Hartley gravimeter An early form of *stable gravimeter* consisting of a weight suspended from a spiral spring, a hinged lever, and a compensating spring for restoring the system to a null position.

Hartmann's law The statement that the acute angle between two sets of intersecting shear planes is bisected by the axis of greatest principal stress, and the obtuse angle by the axis of least principal stress.

hartschiefer A metamorphic rock of compact, dense, cherty, or felsitic texture, having a banded structure in which the bands are of approximately even thickness, have rigid parallelism, and differ considerably in mineral and chemical composition. It is formed by intense dynamic metamorphism from ultramylonites and is associated with other rocks of mylonitic habit (Holmes, 1920, p.116). The term was originated by Quensel in 1916. Etymol: German.

harzburgite (a) In the *IUGS classification,* a plutonic rock with M equal to or greater than 90, ol/(ol+opx+cpx) between 40 and 90, and cpx/ol+opx+cpx) less than 5. (b) A *peridotite* composed chiefly of olivine and orthopyroxene. Named by Rosenbusch in 1887 for Harzburg, Germany. See also: *saxonite.*

hassock A term used in England for a soft, somewhat calcareous sandstone containing glauconite. Syn: *calkstone.*

hassock structure A variety of *convolute bedding* in which the laminae resemble tufts of grass or sedge. Syn: *hassock bedding.*

hastate Said of a leaf that is arrow-shaped, as in leaves with a sharp tip and basal lobes that point away from the petiole.

hastingsite A monoclinic mineral of the amphibole group: $NaCa_2(Fe,Mg)_5Al_2Si_6O_{22}(OH)_2$. It generally contains a little potassium.

hastite An orthorhombic mineral: $CoSe_2$. It is dimorphous with trogtalite.

hatchettine A soft yellow-white paraffin wax, perhaps $C_{38}H_{78}$, having a melting point of 55-65°C in the natural state and 79°C after purification. It occurs as veinlike masses in ironstone nodules associated with coal-bearing strata (as in south Wales) or in cavities in limestone (as in France). Syn: *hatchettite; adipocire; adipocerite; mineral tallow; mountain tallow; naphthine.*

hatchettolite *betafite.*

hatching (a) The drawing of hachures on a map to give an effect of shading. See also: *cross-hatching.* (b) *hachure.*

hatchite A lead-gray triclinic mineral: $(Pb,Tl)_2AgAs_2S_5$.

hatchure Var. of *hachure.*

hatherlite An anorthoclase-bearing biotite-hornblende *syenite.* Syn: *leeuwfonteinite.* Obsolete.

hauchecornite A tetragonal mineral: $Ni_9(Bi,Sb,As,Te)_2S_8$.

hauerite A reddish-brown or brownish-black mineral: MnS_2. It occurs in octahedral or pyritohedral crystals.

haughtonite A black, iron-rich variety of biotite.

hause An English term for a pass, or a ridge connecting two higher elevations, or a narrow gorge (Whitney, 1888, p. 137). Cf: *hals.* Syn: *haws.*

hausmannite A brownish-black, opaque mineral: Mn_3O_4.

haustorium A food-absorbing organ of a fungus or other parasitic plant.

Hauterivian European stage: Lower Cretaceous (above Valanginian, below Barremian).

hauyne A blue feldspathoid mineral of the sodalite group: $(Na,Ca)_{4-8}(Al_6Si_6O_{24})(SO_4,S)_{1-2}$. It is related to nosean and occurs in rounded and subangular grains embedded in various volcanic rocks. Pron: ah-*ween.* Also spelled: *haüyne.* Syn: *hauynite.*

hauynite A syn. of *hauyne.* Also spelled: *haüynite.*

hauynitite A plutonic or hypabyssal rock composed chiefly of hauyne and pyroxene, usually titanaugite. Small amounts of feldspathoids and sometimes plagioclase and/or olivine are present. Apatite, sphene, and opaque oxides occur as accessories. See also: *hauynophyre.*

hauynolith A monomineralic extrusive rock composed entirely of hauyne.

hauynophyre An extrusive rock similar in composition to a *leucitophyre* but containing hauyne in place of some of the leucite. Other possible phases include nepheline, augite, magnetite, apatite, melilite, and mica. A partial syn. of *hauynitite;* some rocks are called "hauynophyre" when hauyne is a conspicuous mineral but not necessarily a major constituent.

Haüy's law *law of rational indices.*

haven A small bay, recess, or inlet of the sea affording anchorage and protection for ships; a harbor.

havsband A Swedish term for the outermost or seaward part of a *skerry-guard,* constituting bare skerries and the smallest rock islets (Stamp, 1961, p. 230 & 419).

Hawaiian-type bomb A type of volcanic bomb formed when a still-plastic clot of lava strikes the ground, so that its shape is controlled by impact, not by its flight through the air. Cf: *pancake bomb; Strombolian-type bomb.*

Hawaiian-type eruption A type of volcanic eruption in which great quantities of extremely fluid basaltic lava are poured out, mainly issuing in lava fountains from fissures on the flanks of a volcano. Explosive phenomena are rare, but much spatter and scoria are piled into cones and mounds along the vents. Characteristic of shield volcanoes. Cf: *Peléean-type eruption; Strombolian-type eruption; Vulcanian-type eruption.*

hawaiite [mineral] A pale-green, iron-poor gem variety of olivine from the lavas of Hawaii.

hawaiite [petrology] As defined by Iddings in 1913, an olivine "basalt" with andesine as the normative plagioclase (thus differing from true *basalt,* in which the normative plagioclase is more calcic). Macdonald (1960, p. 175) revived the term and extended the definition: "a rock with moderate to high colour index, and frequently basaltic habit, in which the normative and modal feldspar is andesine, and with soda:potash ratio greater than 2:1. It generally, but not always, lacks normative quartz, and commonly contains normative and modal olivine. Hawaiite is intermediate in composition between *alkali olivine basalt* and *mugearite,* and grades into both."

hawk's-eye A transparent to translucent colorless variety of quartz, which contains minute parallel closely packed fibrous crystals of partly replaced crocidolite (from which a pale-blue to greenish-blue sheen is produced by reflection of light) and that resembles the eye of a hawk when cut cabochon; a blue variety of tiger's-eye. Cf: *sapphire quartz.* Syn: *hawk-eye; falcon's-eye.*

hawleyite A yellow isometric mineral: CdS. It is dimorphous with greenockite.

haws *hause.*

haxonite A meteorite mineral: $(Fe,Ni)_{23}C_6$.

haycockite An orthorhombic mineral: $Cu_4Fe_5S_8$.

Hayford zone A subdivision of the globe, used in the calculation of topographic and isostatic reductions around a gravity station. It is named after the U.S. geodesist J.F. Hayford. See also: *Bullard's method.*

haystack hill *karst tower.*

hazardous waste Waste that poses a present or potential danger to human beings or other organisms because it is toxic, flammable, radioactive, explosive, or has some other property that produces substantial risk to life.

haze (a) Fine particles of dust, salt, smoke, or water dispersed through a part of the atmosphere, diminishing transparency of the air, causing colors to assume a characteristic subdued opalescent appearance, and reducing the horizontal visibility to more than one, but less than two, kilometers. (b) The obscuration, or a lack of transparency, of the atmosphere near the Earth's surface, caused by haze or by heat refraction (shimmering).

HDM *humic degradation matter.*

HDR *hot dry rock.*

HE *horizontal equivalent.*

head [coast] (a) A *headland,* usually coupled with a specific place name, e.g. Diamond Head, Hawaii. (b) The inner part of a bay, creek, or other coastal feature extending farthest inland. (c) The apex of a triangular-shaped delta or fan.

head [geomorph] (a) The influent end of a lake; the end opposite the outlet. (b) The source, beginning, or upper part of a stream. (c) The farthest upstream point reached by vessels; the limit of river navigation. (d) The upper part or end of a slope or valley; *valley head.*

head [hydraul] (a) The elevation to which water rises at a given point as a result of reservoir pressure. Cf: *fluid potential.* (b) Water-level elevation in a well, or elevation to which the water of a flowing artesian well will rise in a pipe extended high enough to stop the flow. (c) When not otherwise specified, it usually refers to *static head.*

head [mass move] A term used in southern England for a thick, poorly stratified, compact mass of locally derived angular rubble mixed with sand and clay, formed by solifluction under periglacial conditions, and mantling the high ground or occurring on slopes and in valley bottoms; a *congeliturbate.* See also: *coombe rock.* Syn: *rubble drift.*

head [paleont] (a) The anterior tagma of a crustacean, consisting of the *cephalon* alone or comprising the cephalon and one or more anterior thoracomeres (having limbs that are modified as mouth parts) fused to it. See also: *cephalothorax.* (b) The anterior dorsal part of the body of a mollusk, bearing the mouth, sensory organs, and major nerve ganglia. (c) In the vertebrates, the term is often used to designate the convex proximal end of ribs, humerus, and femur. (d) *algal head.*—Cf: *foot [paleont].*

head [sed] *algal head.*

head [struc geol] The upper bend of a fold or structural terrace. Cf: *foot [geol].* Syn: *upper break.*

headcut A vertical face or drop on the bed of a stream channel, occurring at a knickpoint.

head dune A dune that accumulates on the windward side of an obstacle. Cf: *tail dune.*

headed dike A dike which has a terminal expansion of teardrop shape.

head erosion *headward erosion.*

heading [grd wat] A horizontal tunnel into an aquifer that taps ground water penetrating fissures, for the purpose of supplying wells and reservoirs.

heading [surv] The compass direction (azimuth) of the longitudinal axis of a ship or aircraft.

heading side *footwall.*

heading wall *footwall.*

headland [coast] (a) An irregularity of land, esp. of considerable height with a steep cliff face, jutting out from the coast into a large body of water (usually the sea or a lake); a bold *promontory* or a high *cape.* Syn: *head; mull.* (b) The high ground flanking a body of water, such as a cove. (c) The steep crag or cliff face of a promontory.

headland [soil] A term used in soil conservation for the source of a stream.

headland beach A narrow beach formed at the base of a *cliffed headland.*

headland mesa A part of a general plateau that projects into a meander loop of a large river (Lee, 1903, p. 73). See also: *island mesa.*

head loss *friction head.*

headpool A pool near the head of a stream.

heads Low-grade material overlying an alluvial *placer.*

headstream A stream that is the source or one of the sources of a larger stream or river.

headwall A steep slope at the head of a valley; esp. the rock cliff at the back of a cirque. Syn: *backwall.*

headwall recession The steepening and backward movement of the headwall of a cirque, caused in part by alternate thawing and refreezing.

headward erosion The lengthening and cutting upstream of a young valley or gully above the original source of its stream, effected by erosion of the upland at the valley head; it is accomplished by rainwash, gullying, spring sapping, and the slumping of material into the head of the growing valley. Syn: *head erosion; headwater erosion; retrogressive erosion.*

headwater (a) The *source* (or sources) and upper part of a stream, esp. of a large stream or river, including the upper drainage basin; a stream from this source. The term is usually used in the plural. Syn: *waterhead.* (b) The water upstream from a structure, as behind a dam.

headwater erosion *headward erosion.*

headwater opposition The position of, or relationship shown by, two valleys facing in opposite directions, each growing upstream by headward erosion and separated from the other by a ridgelike divide (Fenneman, 1909, p. 35-36).

head wave A seismic wave travelling downward at the critical angle to a high-velocity layer, moving along the top of that layer, and later emerging at the critical angle. The term is sometimes restricted to the part of such a wave that is also a *first arrival.* Syn: *refracted wave; conical wave; von Schmidt wave.*

healed *crustified.*

heaped dune *star dune.*

heat Energy in transit from a higher-temperature system to a lower-temperature system. The process ends in thermal equilibrium.

heat balance (a) Equilibrium that exists on the average between the radiation received by the Earth and its atmosphere from the Sun and that emitted by the Earth and its atmosphere. That the equilibrium does exist in the mean is demonstrated by the observed long-term constancy of the Earth's surface temperature. On the average, regions of the Earth nearer the equator than about 35° latitude receive more energy from the Sun than they are able to radiate, whereas latitudes higher than 35° receive less. The excess of heat is carried from low latitudes to higher latitudes by atmospheric and oceanic circulations and is reradiated. (b) The equilibrium known to exist when all sources of heat gain and heat loss for a given region or body are accounted for. This balance includes advective and evaporative aspects as well as radiation.—Cf: *heat budget.*

heat budget (a) The amount of heat required to raise the water of a lake from its minimum winter temperature to its maximum summer temperature; it is usually expressed as gram calories of heat per square centimeter of lake surface. (b) The accounting for the total amount of heat received and lost by a particular system, such as a lake, a glacier, or the entire Earth during a specific period. See also: *heat balance.*

heat capacity That quantity of heat required to increase the temperature of a system by one degree at constant pressure and volume. It is usually expressed in calories per degree Celsius. Syn: *thermal capacity.*

heat conduction The process of heat transfer through solids, from a higher-temperature to a lower-temperature region, by molecular impact without transfer of the matter itself, i.e. without convection. Syn: *thermal conduction.*

heat conductivity *thermal conductivity.*

heat content *enthalpy.*

heated stone A gemstone that has been heated to change its color, such as blue zircon, or to improve its color, such as many aquamarines. Cf: *stained stone; burnt stone.*

heat flow *geothermal heat flow.*

heat-flow measurement Measurement of the amount of heat leaving the Earth. It involves measuring the *geothermal gradient* of rocks by accurate resistance thermometers in drill holes (preferably more than 300 meters deep), and the *thermal conductivity* of rocks, usually in the laboratory, on samples from the drill holes. Heat-flow measurements on the ocean floors use slightly different techniques.

heat-flow unit A measurement of terrestrial heat flow equivalent to 10^{-6} cal/cm^2/sec.

heat flux *geothermal flux.*

heat-generation unit A measure of the rate of heat production per unit volume of material due to radioactive decay within the earth. One heat-generation unit (abbrev: HGU) is 10^{-13} cal/cm^3-sec.

heath peat *calluna peat.*

heave [mining] *creep [mining].*

heave [soil] A predominantly upward movement of a surface, caused by expansion or displacement, as from swelling clay, removal of overburden, seepage pressure, or frost action; esp. *frost heaving.* See also: *air heave; gas heave.* Syn: *heaving.*

heave [struc geol] In a fault, the horizontal component of separation or displacement. It is an old, imprecise term. Cf: *throw.* Syn:

horizontal throw.

heaving shale An incompetent or hydrating shale that runs, falls, swells, or squeezes into a borehole.

heavy-bedded Said of a shale whose splitting property is intermediate between that of a thin-bedded shale (easy to split) and that of a platy or flaggy shale (hard to split) (Alling, 1945, p.753).

heavy crop A collective term used in Great Britain for the *heavy minerals* of a sedimentary rock.

heavy gold Gold occurring as large particles. Cf: *nugget.*

heavy isotope An isotope of an element having a greater than normal mass; e.g. deuterium.

heavy liquid In analysis of minerals, a liquid of high density, such as *bromoform,* in which specific-gravity tests can be made, or in which mechanically mixed minerals can be separated. When a mineral grain is placed in the liquid, the liquid's specific gravity is adjusted by the addition of a lighter or heavier liquid until the mineral neither rises nor sinks; the specific gravity of the liquid and of the mineral are then equal. See also: *Klein solution; Sonstadt solution; Clerici solution; Westphal balance; methylene iodide.* Syn: *specific-gravity liquid.*

heavy metal Any of the metals that react readily with dithizone (C_6H_5N), e.g. zinc, copper, lead, and many others.

heavy mineral [petrology] A rock-forming mineral generally having a specific gravity greater than 2.9; e.g. a mafic mineral.

heavy mineral [sed] A *detrital mineral* from a sedimentary rock, having a specific gravity higher than a standard (usually 2.85), and commonly forming as a minor constituent or *accessory mineral* of the rock (less than 1% in most sands); e.g. magnetite, ilmenite, zircon, rutile, kyanite, garnet, tourmaline, sphene, apatite, biotite. Cf: *light mineral.* See also: *heavy crop.*

heavy oil Crude oil that has a low *API gravity* or *Baumé gravity.* Cf: *light oil.*

heavy spar *barite.*

heazlewoodite A mineral: Ni_3S_2.

Hebraic granite *graphic granite.*

Hebridean *Lewisian.*

hebronite *amblygonite.*

hecatolite Orthoclase *moonstone.*

hectare A metric unit of land area equal to 10,000 square meters, 100 ares, or 2.471 acres. Abbrev: ha.

hectorite A trioctahedral, lithium-rich clay mineral of the montmorillonite group: $Na_{0.33}(Mg,Li)_3Si_4O_{10}(F,OH)_2$. It represents an end-member, in which the replacement of aluminum by magnesium and lithium in the octahedral sheets is essentially complete.

hedenbergite A black mineral of the clinopyroxene group: $CaFeSi_2O_6$. It occurs as a *skarn* mineral at the contact of limestones with granitic masses.

hedgehog stone Quartz with needle-shaped inclusions of goethite.

hedleyite A mineral: Bi_7Te_3. It is an alloy consisting of a solid solution of Bi_5 in Bi_2Te_3.

hedreocraton A stable, continental craton, including both continental shield and platform. Cf: *thalassocraton; epeirocraton.*

hedrumite A coarse-grained, light-colored porphyritic hypabyssal rock characterized by trachytoid texture and containing accessory nepheline; a *pulaskite* porphyry. Its name, given by Brögger in 1890, is derived from Hedrum, Norway. Not recommended usage.

hedyphane A yellowish-white mineral of the apatite group: $(Ca,Pb)_5(AsO_4)_3Cl$. It may contain barium.

Heersian European stage: Lower Paleocene (above Danian, below Landenian).

heideite A monoclinic mineral: $(Fe,Cr)_{1+x}(Ti,Fe)_2S_4$.

heidornite A monoclinic mineral: $Na_2Ca_3B_5O_8(SO_4)_2Cl(OH)_2$.

height [geodesy] The distance between an equipotential surface through a point and a reference surface, measured along a line of force or along its tangent. Cf: *dynamic height; orthometric height.*

height [geomorph] (a) A landform or area that rises to a considerable degree above the surrounding country, such as a hill or plateau. The term is often used in the plural. (b) The highest part of a ridge, plateau, or other high land.

height [paleont] (a) The maximum distance, measured normal to the length in the plane of symmetry, between a concavo-convex or convexo-concave shell of a brachiopod and the line joining the beak and the anterior margin. Also, the *thickness* of the shell of a brachiopod. (b) The distance between two planes parallel to the hinge axis of a bivalve-mollusk shell and perpendicular to the plane of symmetry, which just touch the most dorsal and ventral parts of the shell. Cf: *length.* (c) In coiled cephalopod conchs, the linear distance between the venter and the umbilical seam (TIP, 1964, pt. K, p. 22).

height [surv] The vertical distance above a datum (usually the surface of the Earth); *altitude* or *elevation* above a given level or surface.

height of instrument A surveying term used in spirit leveling for the height of the line of sight of a leveling instrument above the adopted datum, in trigonometric leveling for the height of the center of the theodolite above the ground or station mark, in stadia surveying for the height of the center of the telescope of the transit or telescopic alidade above the ground or station mark, and in differential leveling for the height of the line of sight of the telescope at the leveling instrument when the instrument is level. Abbrev: HI.

height of land The highest part of a plain or plateau; specif. a drainage *divide,* or a part thereof.

heinrichite A yellow to green secondary mineral: $Ba(UO_2)_2(AsO_4)_2 \cdot 10\text{-}12H_2O$. Cf: *metaheinrichite.*

heintzite *kaliborite.*

hekistotherm A plant that can grow at low temperatures, esp. in areas where the warmest month has a mean temperature below 10°C.

helatoform Shaped like a nail; e.g. "helatoform cyrtolith" having a nail-shaped central structure.

Helderbergian North American stage: lowermost Devonian (above Upper Silurian, below Ulsterian).

held water Water retained within the soil as a liquid or a vapor.

helen The long narrow curved part of the *hyolithid* shell that projects laterally from the junction of the operculum and cone; there are right and left helens (Runnegar et al., 1975, p. 181). Etymol: the genus *Helenia,* named by Walcott.

helenite A variety of *ozocerite.*

helical flow *helicoidal flow.*

helicitic Pertaining to a metamorphic-rock texture consisting of bands of inclusions that indicate original bedding or schistosity of the parent rock and cut through later-formed crystals of the metamorphic rock. The relict inclusions commonly occur in porphyroblasts as curved and contorted strings. The term was originally, but is no longer, confined to microscopic texture. Also spelled: *helizitic.* Cf: *poikiloblastic.*

helicoid Forming or arranged in a spiral; specif. said of a gastropod shell having the form of a flat coil or flattened spiral, or sometimes of an ammonoid coiled in regular three-dimensional spiral form with a constant angle. See also: *torticone.*

helicoidal flow At the bend of a river, a coiling type of flow motion that results in erosion of the concave, outer bank and deposition along the convex, inner bank. Syn: *helical flow.*

helicoplacoid Any echinozoan belonging to the class Helicoplacoidea, characterized by a fusiform to pyriform placoid body with a spirally pleated expansible and flexible test (TIP, 1968, pt.U, p.131). They are known only from the Lower Cambrian.

helictite A curved twiglike cave deposit, usually of calcite, that grows at the free end by deposition from water emerging there from a nearly microscopic central canal. See also: *anemolite.* Syn: *eccentric.*

heligmite A *helictite* that grows from the floor.

Helikian In a three-part division of the Proterozoic of Canada, the middle division, between the *Aphebian* below and the *Hadrynian* above.

heliodor A golden, greenish, or brownish-yellow transparent gem variety of beryl found in southern Africa. See also: *golden beryl.* Obsolete syn: *chrysoberyl.*

heliolite *sunstone.*

heliolith (a) A *coccolith* constructed of many tiny calcite crystals, commonly in radial arrangement. (b) An individual of the Heliolithae, a subdivision of the family Coccolithophoridae.—Cf: *ortholith.*

heliolitid Any coral belonging to the family Heliolitidae, characterized by massive coralla with slender tabularia separated by coenenchyme and commonly having 12 equal spinose septa and complete tabulae. Heliolitids are considered by some workers to be tabulates. Range, Middle Ordovician to Middle Devonian.

heliophyllite A mineral representing an orthorhombic polymorph of ecdemite: $Pb_6As_2O_7Cl_4$ (?).

heliotrope [mineral] *bloodstone.*

heliotrope [surv] An instrument used in geodetic surveying to aid in making long-distance (up to 320 km) observations and composed of one or more plane mirrors so mounted and arranged that a beam of sunlight may be reflected toward a distant survey station where it can be observed with a theodolite.

heliozoan Any actinopod protozoan belonging to the subclass Heliozoa, characterized by pseudopodia that are not stiff or rigid, being strengthened only by an axial rod of fibrils.

helium age method Determination of the age of a mineral in years, based on the known radioactive decay rates of uranium and thorium isotopes to helium. This oldest method of radioactive age measurement is not in common use, as both uranium and helium are known for their ease of movement through geologic materials and this affects the determination of reliable ages. Syn: *helium dating.*

helium dating *helium age method.*

helium index An obsolete term for the experimental age obtained by substituting helium and radioactivity values in the age equation.

helizitic *helicitic.*

hellandite A mineral: $(Ca,Y)_2(Si,B,Al)_3O_8 \cdot H_2O$.

helluhraun An Icelandic term for *pahoehoe.* Cf: *apalhraun.*

hellyerite A mineral: $NiCO_3 \cdot 6H_2O$.

Helmholtz coil In a magnetometer, a pair of coaxial coils with their distance apart equal to their radius. Electric current in the coils produces an unusually uniform magnetic field between the coils.

Helmholtz free energy A *thermodynamic potential* that is a function of temperature and volume. It is one of the commonly used functions describing *free energy,* and is useful in determining the course of constant-volume isothermal processes. Cf: *Gibbs free energy.*

helminthite An obsolete term for a doubtfully distinguished *trace fossil* consisting of a long, sinuous surface trail or filled-up burrow of a supposed marine worm, without impressions of lateral appendages. Syn: *helmintholite.*

heloclone A sinuous, monaxonic sponge spicule of irregular outline, often bearing articulatory notches along its length or at its end.

helophyte A perennial marsh plant that has its overwintering buds beneath the water. See also: *hydrophyte.*

helsinkite A hypidiomorphic-granular hypabyssal rock composed primarily of albite and epidote. Its name is derived from Helsinki, Finland. Not recommended usage.

Helvetian European stage: Miocene (above Serravallian, below Tortonian).

helvite A mineral: $(Mn,Fe,Zn)_4Be_3(SiO_4)_3S$. It is the manganese end member isomorphous with danalite and genthelvite. Syn: *helvine.*

hemachate A light-colored agate spotted with red jasper. Syn: *blood agate.*

hemafibrite A brownish to garnet-red mineral: $Mn_3(AsO_4)(OH)_3 \cdot H_2O$.

hematite A common iron mineral α-Fe_2O_3. It is dimorphous with maghemite. Hematite occurs in splendent, metallic-looking, steel-gray or iron-black rhombohedral crystals, in reniform masses or fibrous aggregates, or in deep-red or red-brown earthy forms: it has a distinctive cherry-red to reddish-brown streak and a characteristic brick-red color when powdered. It is found in igneous, sedimentary, and metamorphic rocks, both as a primary constituent and as an alteration product. Hematite is the principal ore of iron. Symbol: Hm. See also: *specularite.* Also spelled: *haematite.* Syn: *red hematite; red iron ore; red ocher; rhombohedral iron ore; oligist iron; bloodstone.*

hematite schist *itabirite.*

hematolite A trigonal mineral: $(Mn,Mg)_{13}Al_2As^{+3}(AsO_4)_2(OH)_{21}O_4$.

hematophanite A mineral: $Pb_4Fe_3O_8(Cl,OH)$.

hematopore A polymorph that is slender and distally directed on the reverse sides of the colonies of some stenolaemate bryozoans. Cf: *firmatopore.*

hemera (a) The geologic-time unit corresponding to *acme-zone;* the time span of the acme or greatest abundance, in a local section, of a taxonomic entity. Also, the period of time during which a race of organisms is at the apex of its evolution. The term was proposed by Buckman (1893, p.481-482) for the time of acme of development of one or more species, but later used by him (Buckman, 1902) in the sense of *moment* or the time during which a biostratigraphic zone was deposited, and by Jukes-Browne (1903, p.37) for the duration of a subzone. (b) A term sometimes incorrectly applied to a biostratigraphic zone (body of strata) comprising the time range of a particular fossil spacies.—Etymol: Greek, "day". Pl: *hemerae.* Adj: *hemeral.*

hemichoanitic Said of a retrochoanitic septal neck of a nautiloid that extends one-half to three-fourths of the distance to the preceding septum.

Hemichordata A subdivision of the *Protochordata* or of the *Chordata,* including animals with a pre-oral notochord and three primary coelom segments in the adult.

hemicone *alluvial cone.*

hemicrystalline *hyalocrystalline.*

hemicyclothem Half of a cyclothem. The term is generally applied either to the lower nonmarine part, or to the upper marine part, of a Pennsylvanian cyclothem.

hemidisc A sponge spicule consisting of an unequal-ended amphidisc.

hemihedral Said of the *merohedral* crystal class (or classes) in a system, the general form of which has half the number of equivalent faces of the corresponding holohedral form. Syn: *hemisymmetrical.*

hemihedrite A mineral: $Pb_{10}Zn(CrO_4)_6(SiO_4)_2F_2$.

hemimorph A crystal having polar symmetry, i.e. displaying *hemimorphism.*

hemimorphism The characteristic of a crystal that has *polar symmetry,* so that its two ends have different forms. Such a crystal is a *hemimorph.* Adj: *hemimorphic.*

hemimorphite (a) A white or colorless to pale-green, blue, or yellow orthorhombic mineral: $Zn_4Si_2O_7(OH)_2 \cdot H_2O$. It is similar to smithsonite, but is distinguished from it by strong pyroelectric properties. Hemimorphite is a common secondary mineral, and is an ore of zinc. Syn: *calamine; electric calamine; galmei.* (b) A term sometimes used (esp. in the gem trade) as a syn. of *smithsonite.*

Hemingfordian North American continental stage: Lower Miocene (above Arikareean, below Barstovian).

hemiopal *semiopal.*

hemipelagic deposit Deep-sea sediment in which more than 25% of the fraction coarser than 5 microns is of terrigenous, volcanogenic, and/or neritic origin. Such deposits usually accumulate near the continental margin and its adjacent abyssal plain, so that continentally derived sediment is more abundant than in eupelagic sediments, and the sediment has undergone lateral transport. Cf: *terrigenous deposit; pelagic deposit.*

hemipelagite Sediments formed by the slow accumulation on the sea floor of biogenic and fine terrigenous particles deposited on top of the *pelitic* interval of a *turbidite.* Fossil species are *indigenous* (Natland, 1976, p. 702).

hemiperipheral growth Growth of brachiopod shells in which new material is added anteriorly and laterally but not posteriorly.

hemiphragm A transverse calcareous shelflike platform extending from the zooecial wall part way across the zooecial chamber in some stenolaemate bryozoans. Hemiphragms commonly alternate in ontogenetic series from opposite sides of chamber walls.

hemipyramid An old term for *fourth-order prism.*

hemiseptum (a) A transverse calcareous shelflike platform extending from the zooecial wall part way across the zooecial chamber in some stenolaemate bryozoans. They generally occur singly on the proximal sides of zooecia or in one or two pairs in alternate positions on proximal and distal sides. A "superior hemiseptum" may be present on the proximal wall and an "inferior hemiseptum" on the distal wall. (b) A partial septum, between normal ones, subdividing a foraminiferal chamber, as in some Lituolacea.—Pl: hemisepta.

hemisphere Half a sphere; usually refers to half of the Earth, as divided by the Equator into northern and southern hemispheres, or by the 20°W and 160°E meridians into eastern (Old World) and western (New World) hemispheres.

Hemist In U.S. Dept. of Agriculture soil taxonomy, a suborder of the soil order *Histosol,* characterized by organic materials most of which have been decomposed beyond recognition, by a bulk density of 0.1 to 0.2, and by ground water at or very near the surface almost all the time unless artificial drainage has been provided (USDA, 1975). Cf: *Fibrist; Folist; Saprist.*

hemisymmetrical *hemihedral.*

Hemphillian North American continental stage: Upper Miocene—Lower Pliocene (above Clarendonian, below Blancan).

hemusite A cubic mineral: Cu_6SnMoS_8.

hendersonite A black mineral: $Ca_2V^{+4}V_8^{+5}O_{24} \cdot 8H_2O$.

hendricksite A mineral of the mica group: $K(Zn,Mn)_3(Si_3Al)O_{10}(OH)_2$.

Hennigian Referring to *cladism* (or *phylogenetic systematics* in its restricted sense), in honor of its principal proponent, W. Hennig.

henritermierite A mineral of the hydrogarnet group: $Ca_3(Mn,Al)_2(SiO_4)_2(OH)_4$.

hepatic cinnabar A liver-brown or black variety of cinnabar. Syn: *liver ore.*

hepatic groove A groove in decapods that continues from the posterior end of the antennal groove posteriorly, and unites, with a looplike curve, the lower ends of the cervical and postcervical grooves (Holthuis, 1974, p. 733).

hepatic region Part of the carapace of some decapods that may touch the antennal, cardiac, and pterygostomial regions (TIP, 1969, pt. R, p. 92).

hepatite A variety of barite that emits a fetid odor when rubbed or heated.

heptane Any of nine colorless liquid isomeric paraffin hydrocarbons of formula C_7H_{16}. One of these, n-heptane, $CH_3(CH_2)_5CH_3$, occurs in crude oils and in some pine oils.

heptaphyllite (a) A group of mica minerals that contain seven cations per ten oxygen and two hydroxyl ions. (b) Any mineral of the heptaphyllite group, such as muscovite and other light-colored micas; a dioctahedral clay mineral.—Cf: *octaphyllite.*

heptorite A dark-colored *lamprophyre* composed of barkevikite, titanaugite, and hauyne phenocrysts in a glassy groundmass containing labradorite microlites. Essentially a hauyne *monchiquite.* Not recommended usage.

herb Any vascular plant of low stature whose stem does not become woody; either annual or growing from a perennial root or rhizome.

herbaceous Said of green, vascular plants of low stature, either annual or perennial.

herbivore A *heterotrophic* organism that feeds on plants. Cf: *carnivore.*

Hercules stone A syn. of *lodestone.* Also called: *Heraclean stone.*

Hercynian orogeny By present usage, the late Paleozoic orogenic era of Europe, extending through the Carboniferous and Permian, hence synonymous with the *Variscan orogeny;* European usage today is about equally divided between the two terms. Many German geologists regard "Hercynian" as a NW orographic direction without time significance, as proposed by von Buch; hence they prefer the name Variscan. Many French and Swiss geologists, following M. Bertrand (1892), prefer "Hercynian" in the time sense; thus, the crystalline massifs of the northern Alps are said to be Hercynian, rather than Variscan. Cf: *Armorican orogeny.*

Hercynides A name used for the fold belt created by the *Hercynian orogeny,* extending from southern Ireland and Wales to northern France, Belgium, and northern Germany. Approx. syn: *Variscides.*

hercynite A black mineral of the spinel series: $FeAl_2O_4$. It often contains some magnesium. Syn: *iron spinel; ferrospinel.*

herderite A colorless to pale-yellow or greenish-white monoclinic mineral: $CaBe(PO_4)(F,OH)$. It is isomorphous with hydroxyl-herderite.

heredity All the qualities and potentialities that an individual has acquired genetically from its ancestors.

Herkimer diamond A quartz crystal from Herkimer County, N.Y. See also: *Lake George diamond.*

Hermann-Mauguin symbols An internationally accepted shorthand notation system of the elements of symmetry of crystal classes, which expresses both outward and inward symmetry. An example is $4/m\ \bar{3}2/m$ for the hexoctahedral class of the isometric system, in which the numbers with a bar are axes of rotoreflection, m is a symmetry plane, and a number over m indicates an axis of symmetry with a plane of symmetry perpendicular to it. Cf: *Schoenflies notation.*

hermatobiolith An organic *reef rock.*

hermatolith *reef rock.*

hermatopelago A submerged *reef cluster.* Etymol: Greek *hermato,* "sunken reef", + *pelagos,* "sea".

hermatypic coral A reef-building coral; a coral characterized by the presence within its endodermal tissue of many symbiotic algae; a coral incapable of adjusting to aphotic conditions. Ant: *ahermatypic coral.* Syn: *hermatype.*

heronite A hypabyssal rock composed of spheroidal phenocrysts of alkali feldspar in a groundmass composed of radiating bundles of labradorite and acmite with interstitial analcime. The rock appears to be an altered *tinguaite* (Johannsen, 1939, p. 256). Its name is derived from Heron Bay, Ontario. Not recommended usage.

herrerite A blue and green variety of smithsonite containing copper.

herringbone cross-bedding *chevron cross-bedding.*

herringbone mark *chevron mark.*

herringbone texture In mineral deposits, a pattern of alternating rows of parallel crystals, each row in a reverse direction from the adjacent one. It resembles the "herringbone" textile fabric.

herschelite A zeolite mineral: $(Na,Ca,K)AlSi_2O_6 \cdot 3H_2O$.

hervidero A syn. of *mud volcano.* Etymol: Spanish *hervir,* "to boil".

herzenbergite A mineral: SnS. Syn: *kolbeckine.*

hessite A lead-gray cubic mineral: Ag_2Te. It is sectile, usually massive, and often auriferous.

hessonite *essonite.*

hetaerolite A black mineral: $ZnMn_2O_4$. It is found with chalcophanite.

heteractine A spicule in a heteractinid sponge. It is generally octactine-based in more advanced forms, but may be polyactine as well.

heterad crystallization Adcumulus growth in which cumulus crystals and poikilitic crystals of the same composition continue to develop until little or no interstitial liquid remains.

heteradcumulate A cumulate in which cumulus crystals and unzoned poikilitic crystals have the same composition.

hetero- A prefix meaning "different".

heteroblastic Pertaining to a type of *crystalloblastic* texture in a metamorphic rock in which the essential mineral constituents are of two or more distinct sizes. The term was originated by Becke (1903). Cf: *homeoblastic.*

heterochronism The phenomenon by which two analogous geologic deposits may not be of the same age although their processes of formation were similar.

heterochronous [evol] Said of a fauna or flora appearing in a new region at a time that is quite different from the time it appeared in the region which it previously inhabited.

heterochronous [stratig] Said of a sequence of sediments representing lateral development of a similar lithofacies in successively younger stages. Term introduced by Nabholz (1951).

heterochronous homeomorph A homeomorph from a later geologic time that resembles one from an earlier geologic time. Cf: *isochronous homeomorph.*

heterochthonous (a) Said of a transported rock or sediment, or one that was not formed in the place where it now occurs. Also, said of fossils removed by erosion from their original deposition site and re-embedded. Cf: *allochthonous.* (b) Said of a fauna or flora that is not indigenous.

heterococcolith A coccolith constructed of differing elements. Cf: *holococcolith.*

heterocoelous Said of sponges whose cloacae are not lined with choanocytes; specif. pertaining to syconoid or leuconoid sponges having calcium-carbonate spicules. Cf: *homocoelous.*

heterocolpate Said of pollen grains having large elongate holes (pseudocolpi) geometrically arranged in the exine.

heterocyst A differentiated cell, usually large, produced by some blue-green algae; its function is uncertain (Fogg et al., 1973).

heterodesmic Said of a crystal or other material that is bonded in more than one way. Cf: *homodesmic.*

heterodont adj. (a) Said of the dentition of a bivalve mollusk having a small number of distinctly differentiated cardinal and lateral teeth that fit into depressions on the opposed valve. (b) Said of the hingement of ostracode valves effected by a combination of tooth-and-socket and ridge-and-groove types, characterized by pointed or slightly crenulate teeth in one or both valves associated with a ridge in one valve and a groove in the other (TIP, 1961, pt.Q, p. 50). (c) Said of vertebrates in which the teeth are markedly nonuniform from front to back, esp. mammals.—n. A heterodont mollusk; specif. a bivalve mollusk of the order Heterodonta, having few hinge teeth but usually with both lateral and cardinal teeth and with unequal adductor muscles. Cf: *taxodont.*

heterogeneous equilibrium Equilibrium in a system consisting of more than one phase. Cf: *homogeneous equilibrium.*

heterogenite A black mineral occurring in mammillary masses: $CoO(OH)$. It occurs in two forms: heterogenite-3R and heterogenite-2H. It may contain some copper and iron. Syn: *stainierite.*

heterogony *alternation of generations.*

heterogranular (a) Said of the texture of a rock having crystals of significantly different sizes. In igneous rocks, the grain-size distribution curve may be relatively flat, illustrating a large range of sizes being nearly equally represented. In metamorphic rocks, the grain size is characterized by several overlapping log-normal distribution curves. (b) Said of a rock with such a texture. Cf: *seriate; hiatal [ign].* Ant: *homogranular.* Syn: *inequigranular.*

heterolithic unconformity A term proposed by Tomkeieff (1962, p.412) to replace *nonconformity* in the sense of an unconformity developed between "unlike rocks".

heteromorph (a) An organism or part that differs from the normal form, specif. an ammonoid or ammonoid shell that deviates from the normal (planispiral) mode of coiling and/or in which the walls of some or all of the coils are not in contact. (b) In dimorphic ostracodes, the adult form that differs in general shape from the juvenile instars. It is generally presumed to be the female of the species. Cf: *tecnomorph.*

heteromorphic [evol] Deviating from the usual form, or having diversity of form. Syn: *heteromorphous.* Cf: *isomorphic.*

heteromorphic [petrology] Said of igneous rocks having similar chemical composition but different mineralogic composition.

heteromorphism The crystallization of two magmas of nearly identical chemical composition into two different mineral aggregates as a result of different cooling histories.

heteromorphite A mineral: $Pb_7Sb_8S_{19}$.

heteromorphosis The production by an organism of an abnormal or misplaced part, esp. as the result of regeneration.

heteromyarian adj. Said of a bivalve mollusk, or of its shell, having the anterior adductor muscle conspicuously smaller than the posterior adductor muscle.—n. A heteromyarian mollusk. Cf: *anisomyarian; monomyarian.*

heterophragm A small skeletal cystoidal structure that projects from the body wall into the zooecial chamber in some stenolaemate bryozoans.

heteropic Said of sedimentary rocks of different facies, or said of facies characterized by different rock types. The rocks may be formed contemporaneously or in juxtaposition in the same sedimentation area or both, but the lithologies are different; e.g. facies that replace one another laterally in deposits of the same age. Also, said of a map depicting heteropic facies or rocks. Cf: *isopic.*

heteropod Any prosobranch gastropod belonging to the suborder Heteropoda, a group of pelagic forms with shells of aragonite.

heteropygous Said of a trilobite having cephalon and pygidium of unequal size. Ant: *isopygous.*

heterosis The high capacity for growth and activity frequently displayed by crossbred organisms as compared with those that are inbred.

heterosite A mineral: $(Fe^{+3},Mn^{+3})PO_4$. It is isomorphous with purpurite.

heterosporous Characterized by heterospory; specif. said of plants that produce both microspores and megaspores.

heterospory The condition in embryophytic plants in which spores are of two types: microspores and megaspores. Cf: *homospory.*

Heterostraci An order of diplorhinate jawless fishes, characterized by flattened dorsal and ventral head-trunk plates articulated with each other by narrow plates covering the gill apertures. It includes the oldest known vertebrates. Range, Lower Ordovician to Upper Devonian.

heterostrophy The quality or state of being coiled in a direction opposite to the usual one; specif. the condition of a gastropod protoconch in which the whorls appear to be coiled in a direction opposite to those of the teleoconch. Adj: *heterostrophic.*

heterotactic [stratig] *heterotaxial.*

heterotactic [struc petrol] Said of the symmetry of a fabric in which not all the subfabrics agree in symmetry. Cf: *homotactic [struc petrol].*

heterotaxial Pertaining to, characterized by, or exhibiting heterotaxy. Syn: *heterotactic; heterotactous; heterotaxic.*

heterotaxis An erroneous transliteration of *heterotaxy.*

heterotaxy Abnormal or irregular arrangement; specif. the condition of strata that are widely separated and not equivalent as to their relative positions in the geologic sequence, or that are lacking uniformity in stratification or arrangement. Ant: *homotaxy.* Syn: *heterotaxis; heterotaxia.*

heterotherm *poikilotherm.*

heterothermic *poikilothermic.*

heterothrausmatic A descriptive term applied to igneous rocks with an orbicular texture in which the nuclei of the orbicules are composed of various kinds of rock or mineral fragments. Cf: *allothrausmatic; crystallothrausmatic; isothrausmatic; homeothrausmatic.*

heterotomous Said of a crinoid arm characterized by division into unequal branches. Ant: *isotomous.*

heterotrichy The most advanced type of growth habit in filamentous algae, in which both a prostrate or creeping system and a projecting or erect system of filaments are formed (Fritsch, 1961, p. 20).

heterotrophic Said of an organism that nourishes itself by utilizing organic material to synthesize its own living matter. Most animals are heterotrophic. Noun: *heterotroph.* Cf: *autotrophic.* Syn: *metatrophic; allotrophic; zootrophic.*

heterozooid A bryozoan polymorph differing from an autozooid in lacking some or all organs concerned with feeding.

Hettangian European stage: lowermost Jurassic (above Rhaetian of Triassic, below Sinemurian).

heubachite A nickel-containing variety of heterogenite.

heulandite A zeolite mineral: $(Na,Ca)_{4-6}Al_6(Al,Si)_4Si_{26}O_{72}\cdot 24H_2O$. It often occurs as foliated masses or as coffin-shaped monoclinic crystals in cavities in decomposed basic igneous rocks. See also: *clinoptilolite; stilbite.*

heumite A dark-colored, fine-grained hypabyssal rock characterized by granular texture and composed of alkali feldspar, barkevikite, biotite, and smaller amounts of nepheline, sodalite, clinopyroxene, and minor accessories. Its name, given by Brögger, 1898, is derived from Heum, Norway. Not recommended usage.

hewettite A deep-red mineral: $CaV_6O_{16}\cdot 9H_2O$. It occurs in aggregates of silky, slender orthorhombic crystals.

hexacoral *scleractinian.*

hexactine A siliceous sponge spicule having six rays arising from a common center at right angles to one another.

hexactinellid A nonpreferred syn. of *hyalosponge.*

hexadisc A hexactinellid-sponge spicule (microsclere) composed of three interpenetrating amphidiscs at right angles to one another about a common center. Cf: *staurodisc.*

hexagonal close packing In a crystal, close packing of spheres by stacking close-packed layers in the sequence ABAB etc. Cf: *cubic close packing.*

hexagonal cross ripple mark An *oscillation cross ripple mark* formed by parallel ripples arranged in zigzag fashion and characterized by obtuse angles in adjoining ripples facing in opposite directions, by crossbars connecting apexes on opposite sides of the ripple, and by an enclosed pit that tends to be bounded by six sides. It appears to be formed by waves that oscillate at some angle between 45° and 90° to the direction of the original ripple mark.

hexagonal dipyramid A crystal form of 12 faces consisting of two hexagonal pyramids repeated across a mirror plane of symmetry. A cross section perpendicular to the sixfold axis is hexagonal. Its indices are $\{h0l\}$ or $\{hhl\}$ with symmetry $6/m\ 2/m\ 2/m$, and 622, $\{hhl\}$ only with symmetry $\bar{6}m2$, also $\{hkl\}$ with symmetry $6/m$.

hexagonal-dipyramidal class That class of the hexagonal system having symmetry $6/m$.

hexagonal indices *Miller-Bravais indices.*

hexagonal prism A crystal form of six faces parallel to the symmetry axis, whose interfacial angles are 60°. Its indices are $\{110\}$ or $\{100\}$ in several hexagonal classes, or $\{hk0\}$ in symmetry $6/m$, 6, and $\bar{3}$.

hexagonal pyramid A crystal form consisting of six faces in a pyramid, in which any cross section perpendicular to the sixfold axis is hexagonal. Its indices are $\{h0l\}$ and $\{hhl\}$ in classes 6mm and 6, only $\{hhl\}$ in 3m, and $\{hkl\}$ in 6.

hexagonal-pyramidal class That crystal class in the hexagonal system having symmetry 6.

hexagonal-scalenohedral class That crystal class in the rhombohedral division of the hexagonal system having symmetry $\bar{3}2/m$. Syn: *trigonal-scalenohedral class; ditrigonal-scalenohedral class.*

hexagonal scalenohedron A *scalenohedron* of twelve faces and having symmetry $\bar{3}2/m$. It resembles a ditrigonal pyramid. Cf: *tetragonal scalenohedron.*

hexagonal system One of the six *crystal systems,* characterized by one unique axis of threefold or sixfold symmetry that is perpendicular and unequal in length to three identical axes that intersect at angles of 120°. This definition includes the *trigonal system* of threefold symmetry; however, the two systems of threefold and sixfold symmetries may be defined separately. Cf: *isometric system; tetragonal system; orthorhombic system; monoclinic system; triclinic system.*

hexagonal-trapezohedral class That crystal class in the hexagonal system having symmetry 622.

hexagonal trapezohedron A crystal form with 12 faces, a sixfold axis, and three twofold axes, but neither mirror planes nor a center of symmetry. It is composed of top and bottom hexagonal pyramids, one of which is twinned less than 30° about c with respect to the other. It may be right-handed or left-handed, and its indices are $\{hkl\}$ or $\{hk\bar{l}\}$ in symmetry 622.

hexahedral Adj. of *hexahedron.*

hexahedral coordination An atomic structure or arrangement in which an ion is surrounded by eight nearest-neighbor ions of opposite sign, whose centers form the points of a hexahedron (which may or may not be a cube). It may be synonymous with *cubic coordination.*

hexahedrite An *iron meteorite* made up of large single crystals or coarse aggregates of kamacite, usually containing 4-6% nickel in the metal phase, and characterized upon etching by the presence of Neumann bands caused by twinning parallel to the octahedral planes. Symbol: *H.* Cf: *octahedrite; ataxite.*

hexahedron A polyhedron of six equivalent faces, e.g. a cube or a rhombohedron. Adj: *hexahedral.*

hexahydrite A white or greenish-white monoclinic mineral: $MgSO_4 \cdot 6H_2O$.

hexamethylene *cyclohexane.*

hexane Any of five colorless liquid volatile paraffin hydrocarbons of formula C_6H_{14}. The hexanes, especially n-hexane, $CH_3(CH_2)_4CH_3$, occur in crude oil.

hexarch Said of a stele having six strands or origins (Jackson, 1928, p. 180).

hexastannite *stannoidite.*

hexaster A sponge spicule (microsclere) having the form of a hexactin with anaxial branches or extensions at the ray tips.

hexatestibiopanickelite A hexagonal mineral: $(Ni,Pd)_2SbTe$.

hexatetrahedron *hextetrahedron.*

hexoctahedral class That crystal class in the isometric system having symmetry $4/m\ 3\ 2/m$.

hexoctahedron An isometric crystal form of 48 equal triangular faces, each of which cuts the three crystallographic axes at different distances. Its indices are $\{hkl\}$ and its symmetry is $4/m\ \bar{3}\ 2/m$.

hextetrahedral class That crystal class in the isometric system having symmetry $\bar{4}\ 3m$.

hextetrahedron An isometric crystal form of 24 faces, with indices $\{hkl\}$ and symmetry $\bar{4}\ 3m$. Also spelled: *hexatetrahedron.*

heyite A monoclinic mineral: $Pb_5Fe_2(VO_4)_2O_4$.

heyrovskite A mineral: $(Pb,Ag,Bi)_6Bi_2S_9$.

HGU *heat-generation unit.*

HI *height of instrument.*

hiatal [ign] Said of the texture of an igneous rock in which the sizes of the crystals are not in a continuous series but are broken by hiatuses, or in which there are grains of two or more markedly different sizes, as in *porphyritic* rocks (Johannsen, 1939, p. 216). Cf: *seriate; heterogranular.*

hiatal [stratig] Pertaining to or involving a stratigraphic hiatus.

hiatus (a) A break or interruption in the continuity of the geologic record, such as the absence in a stratigraphic sequence of rocks that would normally be present but either were never deposited or were eroded before deposition of the overlying beds. (b) A lapse in time, such as the time interval not represented by rocks at an unconformity; the *time value* of an episode of nondeposition or of nondeposition and erosion together. Cf: *lacuna [stratig].*

hibernaculum A dormant zooid with protective walls, produced mainly at the onset of winter by freshwater and brackish-water bryozoans (such as those in the class Gymnolaemata) and developed as part of a colony in the spring. Pl: *hibernacula.* See also: *statoblast.*

Hibernian orogeny *Erian orogeny.*

hibonite A dark-brown mineral: $(Ca,Ce)(Al,Ti,Mg)_{12}O_{18}$.

hibschite *hydrogrossular.*

hidalgoite A white mineral: $PbAl_3(SO_4)(AsO_4)(OH)_6$.

hidden encruster A *cryptic* species that is sessile and encrusting in habit, such as a cheilostome bryozoan or serpulid worm.

hiddenite An intensely green transparent gem variety of spodumene containing chromium.

hielmite *hjelmite.*

hieratite A grayish, cubic, high-temperature mineral of fumaroles: K_2SiF_6.

hieroglyph Any sedimentary mark or structure found on a bedding plane; esp. a *sole mark.* A classification of hieroglyphs has been proposed by Vassoevich (1953). The term was first used by Fuchs (1895) for a problematic fossil whose appearance is suggestive of a drawing or ornament.

higginsite *conichalcite.*

high n. A general term for such features as a crest, culmination, anticline, or dome. Cf: *low [struc geol].* Syn: *structural high.*

high albite High-temperature albite, stable above 450°C. Natural high albite almost always contains appreciable amounts of potassium and calcium in solid solution. Cf: *low albite.*

high-alumina basalt Nonporphyritic *basalt* distinguished by "higher content of Al_2O_3 (generally higher than 17 per cent and rarely as low as 16 per cent) than that of the *tholeiite* with the corresponding SiO_2 and total alkalies, and by lower alkali content than that of the *alkali basalt,* provided only aphyric rocks are compared" (Kuno, 1960, p. 122). In Japan, high-alumina basalt is transitional between alkaline and tholeiitic basalts geographically, mineralogically, and chemically (except for the higher aluminum content). Kuno's suggestion that it represents a primary magma, independent of alkaline and tholeiitic liquids, has not been widely accepted, and Yoder and Tilley (1962) stated that high-alumina basalt occurs in both the tholeiitic and alkaline basalt suites.

high-angle cross-bedding Cross-bedding in which the cross-beds have an average maximum inclination of 20° or more (McKee & Weir, 1953, p.388). Cf: *low-angle cross-bedding.*

high-angle fault A fault with a dip greater than 45°. Cf: *low-angle fault.*

high-calcium limestone A limestone that contains very little magnesium; specif. a limestone in which the approximate MgO equivalent is less than 1.1%, or in which the approximate magnesium-carbonate equivalent is less than 2.3% (Pettijohn, 1957, p. 418). Dolomite is not present (the magnesium carbonate is in solid solution in calcite). According to Cooper (1945, p. 9), the calcium-carbonate content is greater than 95%. Cf: *magnesian limestone.*

high chalcocite Hexagonal chalcocite, stable above 105°C.

high-energy coast A coast exposed to ocean swell and stormy seas and characterized by average breaker heights of greater than 50 cm. Cf: *moderate-energy coast; low-energy coast.*

high-energy environment An aqueous sedimentary environment characterized by a high *energy level* and by turbulent action (such as that created by waves, currents, or surf) that prevents the settling and accumulation of fine-grained sediment; e.g. a beach or a river channel. Cf: *low-energy environment.*

high-grade adj. Said of an ore with a relatively high ore-mineral content. Cf: *low-grade.* See also: *grade [ore dep].*

high-grading (a) Larceny of valuable ore or mineral specimens by employees in a mine. (b) *gouging [mining].*

high island In the Pacific Ocean, a volcanic rather than a coralline island. Cf: *low island.*

highland (a) A general term for a relatively large area of elevated or mountainous land standing prominently above adjacent low areas; a mountainous region. The term is often used in the plural in a proper name; e.g. the Highlands of Scotland. (b) A relative term denoting the higher land of a region; it may include mountains, valleys, and plains. Cf: *upland.* (c) A lofty headland, cliff, or other high landform.

highland glacier A semicontinuous ice cap or glacier system covering the highest or central position of a mountainous area, partly reflecting irregularities of the land surface beneath; e.g. a *plateau glacier.* Syn: *highland ice.* Cf: *ice field [glaciol].*

highlands The heavily cratered, telescopically bright regions of the Moon. They contrast with the *maria,* which are telescopically darker, topographically lower, and geologically younger.

high-level ground water Ground water occurring above the basal water table and separated from it by impermeable or less permeable material such as ash beds, intrusive igneous rocks, or ice. It makes its way through, over, and around the low-permeability materials to join the basal water (Stearns & Macdonald, 1942, p. 132).

high marsh A syn. of *salting.* The term is a "less correct and rather obsolete" syn. of *salt marsh* (Schieferdecker, 1959, term 1243). Cf: *low marsh.*

highmoor bog A bog, often on the uplands, whose surface is largely covered by sphagnum mosses which, because of their high degree of water retention, make the bog more dependent on rainfall than on the water table. The bog often occurs as a *raised peat bog* or *blanket bog.* Cf: *lowmoor bog.*

highmoor peat Peat occurring on high moors and formed

predominantly of moss, such as sphagnum. Its moisture content is derived from rain water rather than from ground water, and is acidic. Ash and nitrogen content is low, and cellulose content is high. Cf: *lowmoor peat.* Syn: *moorland peat; moor peat; sphagnum peat; bog peat; moss peat.*

high oblique n. An *oblique* that includes the apparent horizon. Cf: *low oblique.*

high plain An extensive area of comparatively level land not situated near sea level; e.g. the High Plains, a relatively undissected section of the Great Plains of the U.S., extending along the eastern side of the Rocky Mountains at elevations above 600 m.

high-polar glacier A *polar glacier* in whose accumulation area the firn is at least 100 m thick, and which does not melt appreciably even during the summer (Ahlmann, 1933); e.g. most of the glaciers in Antarctica. Cf: *subpolar glacier.*

high quartz High-temperature quartz; specif. beta quartz.

high-rank graywacke A term introduced by Krynine (1945) for a graywacke containing abundant (20%) feldspar, usually a sodic plagioclase. It is formed in eugeosynclines. The rock is equivalent to *feldspathic graywacke* of Pettijohn (1954) and is regarded as "graywacke proper" by Pettijohn (1957, p.320). Cf: *low-rank graywacke.*

high-rank metamorphism Metamorphism that is accomplished under conditions of high temperature and pressure. Cf: *low-rank metamorphism.*

high-speed layer A subsurface layer in which the speed of seismic-wave propagation is appreciably greater than that in the layer just above it. See also: *stringer [seis].*

highstand The interval of time during one or more *cycles of relative change of sea level* when sea level is above the shelf edge in a given local area (Mitchum, 1977, p. 207). Cf: *lowstand.*

high tide The tide at its highest; the accepted popular syn. of *high water* in the sea.

high-volatile A bituminous coal A bituminous coal, commonly agglomerating, that has more than 31% volatile matter (on a dry, mineral-matter-free basis) and 14,000 or more BTU/lb (moist, mineral-matter-free). Cf: *high-volatile B bituminous coal; high-volatile C bituminous coal.*

high-volatile B bituminous coal A bituminous coal, characteristically agglomerating, that has 13,000 to 14,000 BTU/lb (moist, mineral-matter-free). Cf: *high-volatile A bituminous coal; high-volatile C bituminous coal.*

high-volatile bituminous coal Bituminous coal that contains more than 31% volatile matter, analyzed on a dry, mineral-matter-free basis. See also: *high-volatile A bituminous coal; high-volatile B bituminous coal; high-volatile C bituminous coal.* Cf: *low-volatile bituminous coal; medium-volatile bituminous coal; gas coal.*

high-volatile C bituminous coal An agglomerating or nonagglomerating bituminous coal that has 11,500 to 13,000 BTU/lb (moist, mineral-matter-free); or an agglomerating, high-volatile bituminous coal that has 10,500 to 11,500 BTU/lb (moist, mineral-matter-free). Cf: *high-volatile A bituminous coal; high-volatile B bituminous coal.*

highwall The working face of a surface mine or quarry, esp. of an open-pit coal mine.

high water Water at the maximum level reached during a tidal cycle. Abbrev: HW. Cf: *low water.* Syn: *high tide.*

high-water platform A *wave-cut bench* or a *solution platform* developed a little below high-water level, commonly on a rock surface.

highway geology Engineering geology as applied to the planning, design, construction, and maintenance of public roads.

highwoodite A dark-colored intrusive rock composed of alkali feldspar, labradorite, pyroxene, biotite, iron oxides, apatite, and possibly a small amount of nepheline. It is essentially a *monzonite.* Its name, given by Johannsen in 1938, is derived from the Highwood Mountains, Montana. Not recommended usage.

hilairite [ign] A porphyritic *nepheline syenite* composed of large albite, nepheline, sodalite, acmite, and eudialyte phenocrysts in a trachytic groundmass of acmite, nepheline, albite, and orthoclase. The name (Johannsen, 1938) is for Mont St. Hilaire, Quebec. Not recommended usage.

hilairite [mineral] A trigonal mineral: $Na_2ZrSi_3O_9 \cdot 3H_2O$.

hilate Said of a spore or pollen grain possessing a *hilum [palyn].*

hilgardite A mineral: $Ca_2B_5ClO_8(OH)_2$.

hill (a) A natural elevation of the land surface, rising rather prominently above the surrounding land, usually of limited extent and having a well-defined outline (rounded rather than peaked or rugged), and generally considered to be less than 300 m (1000 ft) from base to summit; the distinction between a hill and a *mountain* is arbitrary and dependent on local usage. See also: *mount.* (b) Any slightly elevated ground or other conspicuous elevation in a relatively flat area. (c) An eminence of inferior elevation in an area of rugged relief. (d) A range or group of hills, or a region characterized by hills or by a highland. Term usually used in the plural; e.g. the Black Hills of South Dakota. (e) A district whose *slope lines* run to the same *peak* (Warntz, 1975, p. 213).

hill creep Slow downhill movement, on a steep hillside and under the influence of gravity, of soil and rock waste flowing toward the valleys; it is an important factor in the wasting of hillsides during dissection, as in the Alps. See also: *creep; terminal creep.* Syn: *hillside creep.*

hillebrandite A white mineral: $Ca_2SiO_3(OH)_2$.

hill-island A glacial moraine, mainly of sand and of variable size, rising as a mature hill from an outwash plain of a later glacial epoch (P.R. Barham in Stamp, 1966, p. 235).

hillock A small, low hill; a mound. Adj: *hillocky.*

hillock moraine A moraine consisting of a series of hillocks.

hill of planation A term applied by Gilbert (1877, p. 130-131) to a bedrock erosion surface now described as a *pediment* (although such a surface is not a hill in any sense).

hill peat Peat occurring in cold, temperate areas and derived from mosses, heather, pine trees, and related plant forms. Syn: *subalpine peat.*

hill shading (a) A method of showing relief on a map by simulating the appearance of sunlight and shadows, assuming an oblique light from the NW so that slopes facing south and east are shaded (the steeper slopes being darker), thereby giving a three-dimensional impression similar to that of a relief model. The method is widely used on topographic maps in association with contour lines. (b) The pictorial effect (of contoured topographic features) emphasized by hill shading, in which the features are shown by the shadows they cast.—Syn: *hillwork; relief shading; plastic shading; shading.*

hillside A part of a hill between its crest and the drainage line at the foot of the hill. Syn: *hillslope.*

hillside creep *hill creep.*

hillside spring *contact spring.*

hillslope *hillside.*

hillwash The process of *rainwash* operating on a hillslope. Also spelled: *hill wash.*

hillwork *hill shading.*

Hilt's law The generalization that, in a vertical succession at any point in a coal field, coal rank increases with depth.

hilum [bot] A scar on a seed coat that marks the place of attachment of the seed stalk to the seed.

hilum [palyn] A germinal aperture of a spore or pollen grain, formed by the breakdown of the exine in the vicinity of one of the poles. The hilum in the spore *Vestispora* is associated with an operculum that may become separated from the spore.

hinge [fold] The locus of maximum curvature or bending in a folded surface, usually a line. Cf: *hinge line.* Syn: *flexure.*

hinge [paleont] A collective term for the structures of the dorsal region that function during the opening and closing of the valves of a bivalve shell; esp. a flexible ligamentous joint. The term is often used loosely for *hinge line,* and for the *cardinal margin* of a brachiopod (TIP, 1965, pt.H, p.145).

hinge area (a) The flattened area marginal to the hinge of a brachiopod or pelecypod shell. See also: *cardinal area.* (b) The surface involved in the hingement of ostracode valves, commonly differentiated into anterior and posterior areas containing more complex elements and between these an interterminal area with simpler structures (TIP, 1961, pt.Q, p.50-51).

hinge axis (a) An imaginary straight line about which the two valves of a bivalve shell are hinged. Syn: *cardinal axis.* (b) The line joining the points of articulation about which the valves of a brachiopod rotate when opening and closing.—Cf: *hinge line [paleont].*

hinge fault A fault on which the movement of one side hinges about an axis perpendicular to the fault plane; displacement increases with distance from the hinge. It is a questionable term. Cf: *scissor fault; rotational fault; pivotal fault.*

hinge line [fault] That line along a fault surface at which the direction of apparent displacement changes. See also: *scissor fault.*

hinge line [fold] A line connecting the points of flexure or maximum curvature of the bedding planes in a fold. See also: *hinge.*

hinge line [paleont] (a) A line along which articulation takes place; e.g. the middorsal line of junction of two valves of a crustacean carapace, permitting movement between them, or the line along which the two valves of an ostracode articulate, seen when the carapace is complete. (b) The straight posterior margin of a brachiopod shell, parallel to the *hinge axis.* The term is also used as a syn. of *cardinal margin.* (c) A term applied loosely to the part of a bivalve-mollusk shell bordering the dorsal margin and occupied by or situated close to the hinge teeth and ligament. The term is sometimes used as a syn. of *hinge axis.*—Syn: *hinge.*

hinge line [struc geol] A line or boundary between a stable region and a region undergoing upward or downward movement. In Pleistocene geology, it is the boundary between regions undergoing postglacial uplift and those of no uplift (for example, in the Great Lakes area).

hinge-line fault A fault that is caused by sedimentary overload, as in an area of embayment, with downthrow and sedimentary thickening in a seaward direction. It is an alternative interpretation to *growth fault* for the Gulf Coast region of the U.S.

hingement The area of the juncture and articulation of the two halves or valves of the carapace of an ostracode.

hinge node A localized thickened part of the hinge of the right valve of phyllocarid crustaceans, serving to strengthen the hinge.

hinge plate (a) The shelly internal plate bearing the hinge teeth in a bivalve mollusk, situated below the beak and the adjacent parts of the dorsal margin of each valve, and lying in a plane parallel to that of commissure. Syn: *cardinal platform.* (b) A plate, simple or divided, typically nearly parallel to the plane between the valves of a brachiopod, lying along the hinge line in the interior of the brachial valve and bearing its dental sockets, and joined to crural bases. See also: *inner hinge plate; outer hinge plate.*

hinge tooth An articulating projection of one valve of a bivalve shell, located near the hinge line or adjacent to the dorsal margin, and fitting into an accompanying socket in the opposite valve for the purpose of holding the valves in position when closed; e.g. a *cardinal tooth* or a *lateral tooth* of a bivalve mollusk, or one of a pair of small or stout, wedge-shaped processes situated at the base of the delthyrium of the pedicle valve of a brachiopod and articulating with the dental sockets in the brachial valve. Syn: *tooth.*

hinge trough A V- or U-shaped depression formed by fusion of the bifurcated median septum with combined socket ridges and crural bases of some terebratellacean brachiopods. Syn: *trough [paleont].*

hinsdalite A dark-gray or greenish rhombohedral mineral: $(Pb,Sr)Al_3(PO_4)(SO_4)(OH)_6$. It is isomorphous with svanbergite, corkite, and woodhouseite.

hinterland An area bordering, or within, an orogenic belt on the internal side, away from the direction of overfolding and thrusting; it is related to the *internides* and to the discredited *borderlands* of Schuchert. Syn: *backland.*

hinterland sequence A depositional sequence that consists entirely of nonmarine deposits laid down at a site interior to the coastal area, where depositional mechanisms are controlled only indirectly or not at all by the position of sea level (Mitchum, 1977, p. 207).

hintzeite *kaliborite.*

hiortdahlite A pale-yellow triclinic mineral: $(Ca,Na)_3ZrSi_2O_7(O,OH,F)_2$.

hipotype *hypotype.*

hirnantite A *keratophyre* composed of lath-shaped sodic plagioclase (albitized andesine) with interstitial chlorite, clinopyroxene, and small amounts of quartz, leucoxene, hematite, and calcite. Named for Hirnant, North Wales. Not recommended usage.

hirst *hurst.*

hisingerite A black or brownish-black monoclinic mineral: $Fe_2^{+3}Si_2O_5(OH)_4 \cdot 2H_2O$.

hislopite A bright grass-green variety of calcite in which the color is due to admixed glauconite.

hispid Rough, or covered with minute hairlike spines.

histic epipedon A diagnostic surface soil horizon, consisting of *organic soil material,* which unless drained is saturated for 30 or more consecutive days (USDA, 1975).

histium The near-ventral ridge confluent with the connecting lobe of the carapace of some heteromorphic ostracodes.

histogram A vertical-bar graph representing a *frequency distribution,* in which the height of bars is proportional to frequency of occurrence within each class interval and, due to the subdivision of the x-axis into adjacent class intervals, there are no empty spaces between bars when all classes are represented in a sample so graphed. Histograms are used to depict particle-size distribution in sediments.

histometabasis Preservation of structure by the minerals that have replaced organic tissues, as in silicified wood and many other fossils.

historical geology A major branch of *geology* that is concerned with the evolution of the Earth and its life forms from its origins to the present day. The study of historical geology therefore involves investigations into stratigraphy, paleontology, and geochronology, as well as the consideration of paleoenvironments, glacial periods, and plate-tectonic motions. It is complementary to *physical geology.* Not to be confused with *history of geology.*

historical geomorphology A branch of geomorphology concerned with the series of events within a particular geologic period and in a particular geographic region.

history of geology That branch of the history of science that treats the development of geologic knowledge, including the history of observations of geologic features, the history of theories to explain their origin, and the history of the organization and development of geologic institutions and societies. The biographical study of geologists is included. Not to be confused with *historical geology.*

Histosol In U.S. Dept. of Agriculture soil taxonomy, a soil order characterized by being more than half organic in its upper 80cm. Most Histosols are saturated or nearly saturated most of the year unless they have been artificially drained (USDA, 1975). Suborders and great soil groups of this order have the suffix -ist. Cf: *Bog soil.* See also: *Fibrist; Folist; Hemist; Saprist.*

hjelmite A black, often metamict mineral of formula: AB_2O_6 or $A_2B_3O_{10}$, where A = Y, Fe^{+2}, U^{+4}, Mn, or Ca, and B = Nb, Ta, Sn, or W. It may be equivalent to pyrochlore + tapiolite. Also spelled: *hielmite.*

***hkl* indices** The *Miller indices,* in general terms that represent integral numbers.

(*hkO*) joint A partial syn. of *diagonal joint.*

H layer In a forest soil, a layer of amorphous organic material below the litter and the partially decomposed *F layer [soil].* Syn: *humus layer.* See also: *O horizon.*

hoarfrost *frost.*

hocartite A tetragonal mineral: Ag_2FeSnS_4.

Hochmoor *raised bog.*

hodgkinsonite A bright-pink to reddish-brown mineral: $MnZb_2SiO_5 \cdot H_2O$.

hodochrone A rarely used syn. of *traveltime curve.*

hodograph [oceanog] The locus of one end of a variable vector as the other end remains fixed, and representing the linear velocity of a moving point; used specif. in oceanography to describe the Ekman spiral and the motion of the mean tidal-current cycle.

hodograph [seis] A rarely used syn. of *traveltime curve.*

hodrushite A mineral: $Cu_4Bi_6S_{11}$.

hoe A promontory or a point of land stretching into the sea; a spur of a hill, or a projecting ridge of land; a cliff. Term is obsolete except in English place names; e.g. Plymouth Hoe. Syn: *howe.*

hoegbomite *högbomite.*

hoelite A yellow mineral: $C_{14}H_8O_2$ (anthraquinone).

hoernesite A white monoclinic mineral: $Mg_3(AsO_4)_2 \cdot 8H_2O$. Its crystals resemble those of gypsum. Also spelled: *hörnesite.*

hofmannite *hartite.*

hogback [geomorph] Any ridge with a sharp summit and steep slopes of nearly equal inclination on both flanks, and resembling in outline the back of a hog; specif. a sharp-crested ridge formed by the outcropping edges of steeply inclined resistant rocks, and produced by differential erosion. The term is usually restricted to ridges carved from beds dipping at angles greater than 20° (Stokes & Varnes, 1955, p. 71). Cf: *cuesta.* See also: *dike wall; razorback.* Also spelled: *hog-back.* Syn: *hog's back; stone wall.*

hogback [glac geol] A term applied in New England to a *drumlin* (western Massachusetts) and to a *horseback* or esker (Maine).

högbomite A black mineral: $(Mg,Fe)_2(Al,Ti)_5O_{10}$. Also spelled: *hoegbomite.*

högbomitite A högbomite-rich *magnetitite.* Not recommended usage.

hog's back Var. of *hogback.* Also spelled: *hog's-back; hogsback.*

hogtooth spar *dogtooth spar.*

hog wallow (a) A faintly rolling land surface characterized by many low, coalescent or rounded mounds (such as *Mima mounds*) that are slightly higher than the basin-shaped depressions between them. (b) A *wallow* made by swine. Also, a similar depression believed to be formed by heavy rains. Also spelled: *hogwallow.*

hohmannite A mineral: $Fe_2(SO_4)_2(OH)_2 \cdot 7H_2O$.

holacanth A trabecula of a rugose coral, seemingly consisting of a clear rod of calcite, as in septa of *Tryplasma* (TIP, 1956, pt.F, p.248).

holarctic Pertaining to the arctic regions as a whole.

holaspis A trilobite at the developmental stage having the number of thoracic segments typical of the species. Pl: *holaspides.*

holdenite A red orthorhombic mineral: $(Mn,Zn)_6(AsO_4)(OH)_5O_2$.

holdfast Something that supports or holds in place; e.g. a basal discoid or rootlike structure by which the thallus of many algae is attached to a solid object in water, any structure at the distal extremity of a crinoid column that serves for fixation, or a multiplated cylindrical or globular structure that attaches a primitive crinoid or crinoid calyx to an object on the substrate.

hole [coast] A term used in New England for a small bay, cove, or narrow waterway; e.g. Woods Hole, Mass.

hole [cryst] *vacancy.*

hole [drill] (a) *borehole.* (b) *drill hole.* (c) A mine, well, or shaft dug in earth material.

hole [geomorph] (a) A term used in the western U.S. for a comparatively level, grassy valley shut in by mountains; e.g. Jackson Hole, Wyo. Cf: *park.* (b) An abrupt hollow in the ground, such as a pothole, a kettle, or a cave.

hole [streams] (a) A deep place in a stream. (b) A *water hole* in the bed of an intermittent stream.

hole fatigue A delay in the effective initiation of a seismic impulse from a shot because of changes in the shot environment (usually cavity formation) produced by an earlier shot in the same hole.

holism The theory that in nature organisms develop from individual structures acting as "whole" units. Cf: *vitalism.*

holisopic Said of an *isopic* condition that includes both lithologic and paleontologic similarities.

hollaite A hybrid rock produced by interaction between *sövite* and rocks of the *ijolite-melteigite* series. The name, given by Brögger in 1921, is for Holla Church, Fen Complex, Norway. Not recommended usage.

hollandite A silvery-gray to black mineral: $Ba(Mn^{+2},Mn^{+4})_8O_{16}$. It is isostructural with coronadite and cryptomelane.

hollingworthite A mineral: (Rh,Pt,Pd)AsS.

hollow (a) A low tract of land surrounded by hills or mountains; a small, sheltered valley or basin, esp. in a rugged area. (b) A term used in the Catskill Mountains of New York for a notch or pass. (c) A landform represented by a depression, such as a cirque, a cave, a large sink, or a blowout.

Holmes' classification A classification of igneous rocks based chiefly on the degree of saturation of a rock, with other aspects of the mineralogy as secondary considerations. The system was proposed in 1928 by Arthur Holmes (1890-1965).

Holmes effect The effect leading to overestimation of the relative area of an opaque grain in a thin section of a rock viewed in transmitted light. The apparent area of an opaque grain will always be that of its maximum cross section in the slide, whereas the desired reference is the surface of the thin section. The term was coined by F. Chayes (1956, p. 95) for the effect described by the petrographer Arthur Holmes (1890-1965) in 1927.

holmite An igneous rock similar to *monchiquite* but with a groundmass of melilite rather than analcime; a pyroxene-rich *alnoite.* The name, given by Johannsen in 1938, is for Holm in the Orkney Islands. Not recommended usage.

holmquistite A bluish-black orthorhombic mineral of the amphibole group: $(Na,K,Ca)Li(Mg,Fe)_3Al_2Si_8O_{22}(OH)_2$. It is related to anthophyllite.

holo- A prefix meaning "completely, wholly".

holoaxial *holohedral.*

holoblast A *crystalloblast* that is newly and completely formed during metamorphism. The term was first used by Sander (1951).

Holocene An epoch of the Quaternary period, from the end of the Pleistocene, approximately 8 thousand years ago, to the present time; also, the corresponding series of rocks and deposits. When the Quaternary is designated as an era, the Holocene is considered to be a period. Syn: *Recent.*

Holocephali A subclass of cartilaginous fishes consisting entirely of the order *Chimaeriformes.* Syn: *ratfish.*

holochoanitic Said of a retrochoanitic septal neck of a nautiloid that extends backward through the length of one camera.

holochroal eye A trilobite compound eye consisting of numerous adjoining lenses covered by a continuous cornea. Cf: *schizochroal eye.* Syn: *compound eye.*

holoclastic rock A sedimentary clastic rock, as distinguished from a *pyroclastic rock.*

holococcolith A coccolith consisting entirely of microcrystals of usual crystallographic shape, whether or not they are identical. Cf: *heterococcolith.*

holocrystalline Said of the texture of an igneous rock composed entirely of crystals, i.e. having no glassy part. Also, said of a rock with such a texture. Syn: *pleocrystalline.*

holocyst *olocyst.*

holohedral Said of that crystal class having the maximum symmetry possible in each crystal system. Cf: *merohedral; tetartohedral.* Syn: *holosystematic; holosymmetric; holoaxial.*

holohedron Any crystal form in the holohedral class of a crystal system.

holohyaline Said of an igneous rock that is composed entirely of glass.

holokarst Karst that is completely developed, characterized by thick limestone bedrock, little or no surface drainage, and a bare surface with well-formed depressions and caves. See also: *merokarst.*

hololeims Coalified remains of entire plants (Kristofovich, 1945, p.138). Cf: *meroleims.* See also: *phytoleims.*

holomictic lake A lake that undergoes a complete mixing of its waters during periods of circulation or *overturn.* Cf: *meromictic lake.*

holomixis The process leading to, or the condition of, a *holomictic lake.*

holoperipheral growth Increase in size of a brachiopod valve all around the margins (in posterior, anterior, and lateral directions). Cf: *mixoperipheral growth.*

holophyte A plant that derives its nourishment entirely from its own organs. Adj: *holophytic.*

holophytic The obsolescent adj. of *holophyte;* denoting nutrition of the plant type. Cf: *holozoic.* Syn: *phototrophic; photoautotrophic.*

holoplankton Plankton that are planktonic during their complete life cycles, as opposed to the temporarily planktonic *meroplankton.*

holosiderite A meteorite consisting of metallic iron without stony matter. Cf: *oligosiderite.*

holosome A term introduced by Wheeler (1958, p.1061) for an intertongued chronostratigraphic unit that may be either depositional (comprising one or more contiguous holostromes) or hiatal (consisting of combined contiguous hiatuses). Cf: *lithosome; biosome.*

Holostei An infraclass of ray-finned bony fishes that includes the living bowfin (*Amia*) and gar (*Lepisosteus*), as well as a variety of Mesozoic forms structurally more or less intermediate between the Paleozoic chondrosteans and the Tertiary teleosts. It is probably a grade in the evolution of the Osteichthyes.

holostomatous Said of a gastropod shell with a more or less circular apertural margin uninterrupted by a siphonal canal or notch. Cf: *siphonostomatous.*

holostratotype The original stratotype designated by the author at the time of establishment of a stratigraphic unit or boundary (ISG, 1976, p. 26).

holostrome A term introduced by Wheeler (1958, p.1055-1056) for a chronostratigraphic unit "embodying the space-time value of a complete (restored) transgressive-regressive depositional sequence", including strata that may later have been removed by erosion. Cf: *lithostrome.*

holostylic Pertaining to a jaw suspension in which the cartilaginous or cartilage-replacement skeleton of the upper jaw is immovably fused to the neurocranium.

holosymmetric *holohedral.*

holosystematic *holohedral.*

holothuroid Any cylindroid echinozoan, usually free-living, belonging to the class Holothuroidea, characterized by the absence of an articulated test and by the reduction of skeletal elements to microscopic sclerites; e.g. a *sea cucumber.* Var: *holothurian.*

holotomous Said of a crinoid arm characterized by a division on each successive brachial plate; typically, of a crinoid arm with pinnules.

holotype The one specimen or other element designated by the author as the nomenclatural type in describing a new *species.* As long as the holotype is extant, it automatically fixes the application of the name concerned (ICBN, 1972, p. 18). Cf: *lectotype; neotype.*

holozoic Said of an organism that is nourished by the ingestion of organic matter; denoting nutrition of the animal type. Cf: *holophytic.*

Holsteinian North European climatostratigraphic and floral stage: Middle Pleistocene (above Elsterian, below Saalian). Equivalent in time to Mindel/Riss interglacial.

holtite A mineral that is related to dumortierite: $(Al,Sb,Ta)_7(B,Si)_4O_{18}$ (?).

Holweck-Lejay inverted pendulum An instrument for measuring differences in gravity, in which a mass is suspended from below by a weak flat-leaf spring. The instrument is used near its instability configuration and, as used, its period of oscillation varies with changes in gravity by a much greater percentage than is the case in the gravity pendulum (Nettleton, 1940, p. 30).

holyokeite An *albitite* characterized by ophitic texture, named by Emerson in 1902 for Holyoke, Massachusetts. Not recommended usage.

homalographic projection *homolographic projection.*

homalozoan Any echinoderm belonging to the subphylum Homalozoa, characterized by the absence of radial symmetry and having a basically asymmetrical body. The subphylum includes the carpoids and possibly the Machaeridia.

homeoblastic Pertaining to a type of *crystalloblastic* texture in a metamorphic rock in which the essential mineral constituents are approximately of equal size. Depending on the habit of the minerals involved, this texture may also be called more specifically *granoblastic, lepidoblastic, nematoblastic,* or *fibroblastic.* The term was originated by Becke (1903). Cf: *heteroblastic.*

homeochilidium An externally convex triangular plate closing almost all, or only the apical part, of the notothyrium in the brachiopod order Paterinida. Cf: *homeodeltidium.* Also spelled: *homoeochilidium.*

homeocrystalline *homogranular.*

homeodeltidium An externally convex triangular plate closing almost all, or only the apical part, of the delthyrium in the brachiopod order Paterinida. Cf: *homeochilidium.* Also spelled: *homoeodeltidium.*

homeomorph [cryst] A crystal that displays *homeomorphism* with another.

homeomorph [evol] An individual that bears a close resemblance to another organism although the two have different ancestors.

homeomorphic [cryst] Adj. of *homeomorphism.*

homeomorphic [paleont] Said of a crinoid column consisting of columnals that are all alike in size and shape.

homeomorphism The characteristic of crystalline substances of dissimilar chemical composition to have similar crystal form and habit; such crystals are known as *homeomorphs.* Adj: *homeomorphic; homeomorphous.*

homeomorphous Adj. of *homeomorphism.*

homeomorphy The phenomenon in which species having superficial resemblance are unlike in structural details; general similarity but dissimilarity in detail. The term is sometimes used as a syn. of *convergent evolution.* Not to be confused with *homomorphy.*

homeostasis The trend toward a relatively stable internal condition in the bodies of the higher animals as a result of a sequence of interacting physiologic processes; e.g., the ability to maintain relatively constant body heat during widely varying external temperatures.

homeothrausmatic A genetic term applied to igneous rocks with an orbicular texture in which the nuclei of the orbicules are formed of inclusions of the same generation as the groundmass (Eskola, 1938, p. 476). Cf: *isothrausmatic; allothrausmatic; heterothrausmatic; crystallothrausmatic.*

Homerian Floral stage in Alaska: Miocene and Pliocene(?).

homilite A black or blackish-brown mineral: $Ca_2(Fe,Mg)B_2Si_2O_{10}$.

Hommel's classification A classification of igneous rocks in which a rock is represented by a two-part formula, one giving the molecular proportions of the oxides and the other the percentages of the normative minerals. The system was proposed in 1919 by W. Hommel.

homo- A prefix meaning "same".

homoclinal Adj. of *homocline.*

homoclinal shifting A term used by Cotton (1922, p. 392) as a syn. of *monoclinal shifting.*

homocline A general term for a rock unit in which the strata have the same dip, e.g. one limb of a fold, a tilted fault block, a monocline, or an isocline. Cf: *monocline.* Adj: *homoclinal.*

homocoelous Said of sponges whose flagellated chambers are also spongocoels; specif. pertaining to asconoid sponges having calcium-carbonate spicules. Cf: *heterocoelous.*

homodesmic Said of a crystal or other material that is bonded in only one way. Cf: *heterodesmic.*

homogeneous equilibrium Equilibrium in a system consisting of only one phase, typically liquid or gaseous. Cf: *heterogeneous equilibrium.*

homogeneous strain A state of *strain* of an extended body in which all initially straight lines remain straight, because the components of strain are identical at every point. More properly termed a *homogeneous strain field.* Syn: *uniform strain.*

homogeny *homology.*

homogranular (a) Said of the texture of a rock having crystals of the same or nearly the same size. In metamorphic rocks, this unimodal grain-size distribution typically is log-normal as a result of competing growth and nucleation processes. (b) Said of a rock with such a texture. Ant: *heterogranular.* Syn: *equigranular; even-grained; homeocrystalline .*

homoiolithic Said of a sedimentary rock containing fragments of similar rock material or composed of two similar rock materials, and having a structure that indicates "contemporary erosion and redistribution" (Phemister, 1956, p.73).

homoiothermic Said of an organism whose body temperature remains relatively uniform and independent of the temperature of the environment; warm-blooded. Syn: *homothermous; homothermal; homeothermic.* Cf: *poikilothermic.*

homologous [geol] (a) Said of strata, in separated areas, that are correlatable (contemporaneous) and that are of the same general lithologic character or facies and/or occupy analogous structural positions along the strike. (b) Said of faults, in separated areas, that have the same relative position or structure.

homologous [paleont] The adj. of *homology.*

homolographic projection An *equal-area projection.* The term is sometimes given to a particular map projection, such as the "Mollweide homolographic projection". Syn: *homalographic projection.*

homologue An organism or part of an organism exhibiting homology. Also spelled: *homolog.*

homology (a) In biology, similarity but not identity between parts of different organisms as a result of evolutionary differentiation from the same or corresponding parts of a common ancestor. Syn: *true homology; homogeny.* Cf: *homoplasy.* (b) Similarity of position, proportion, structure, etc. without restriction to common ancestry. Cf: *analogy.* (c) In *cladism,* a *character* possessed in common by two or more organisms or taxa and also inherited from a common ancestor, remote or immediate. See also: *shared character; shared primitive character; shared derived character.*—Adj: *homologous.*

homolosine projection An equal-area map projection consisting of a sinusoidal projection (between parallels 40°N and 40°S) combined with a Mollweide homolographic projection (between these parallels and the poles); specif. an *interrupted projection* that allows continental masses to be recentered on several meridians in order that they be shown with a minimum of shape distortion, leaving gaps in the interrupted ocean areas between each section to accommodate errors. Also, a similar projection showing the oceans to best advantage.

homomorphosis The regeneration by an organism of a part having a form similar to that of a part that has been lost.

homomorphy Superficial similarity in the morphology of members of different phyla, e.g. the rudistid lamellibranchs and the corals (Whitten, 1972, p. 227). Not to be confused with *homeomorphy.*

homomyarian *isomyarian.*

homonym Any one of two or more identical names used to identify different organisms or objects. In zoologic and botanic nomenclature, it refers to duplication of name for a *taxon* of the same rank based on a different *type* (McVaugh et al., 1968, p. 15). See also: *homonymy.*

homonymy In zoological and botanical nomenclature, identity in spelling of the names applied to different taxa of the same rank. See also: *law of homonymy.*

homoplasy Similarity or correspondence of parts or organs that developed as a result of convergence or parallelism, rather than from a common ancestry. Cf: *homology.* Adj: *homoplastic.*

homopolar Of uniform polarity; not separated or changed into ions; not polar in activity.

homopycnal inflow Flowing water of the same density as the body of water it enters, resulting in easy mixing (Moore, 1966, p. 89). Cf: *hyperpycnal inflow; hypopycnal inflow.*

homoseismal line *coseismal line.*

homospore One of the spores of an embryophytic plant that reproduced by homospory; a plant spore functioning as either male or female in reproduction. Its range is Silurian to Holocene. Syn: *isospore.*

homosporous Characterized by homospory.

homospory The condition in embryophytic plants in which all spores produced are of the same kind; the production by various plants of homospores. Cf: *heterospory.* Syn: *isospory.*

homotactic [stratig] *homotaxial.*

homotactic [struc petrol] Said of the symmetry of a fabric in which all the subfabrics agree in symmetry. Cf: *heterotactic [struc petrol].*

homotaxial Pertaining to, characterized by, or exhibiting homotaxy; e.g. said of rock-stratigraphic units or biostratigraphic units that have a similar order of arrangement in different locations but are not necessarily contemporaneous (ACSN, 1961, art.2a). Syn: *homotaxeous; homotactic.*

homotaxis An erroneous transliteration of *homotaxy.*

homotaxy Similarity of serial arrangement; specif. taxonomic similarity between stratigraphic or fossil sequences in separate regions, or the condition of strata characterized by similar fossils occupying corresponding positions in different vertical sequences, without connotation of similarity of age. The term was originally proposed as *homotaxis* by Huxley (1862, p.xlvi) to avoid the common fallacy of confusing taxonomic similarity with synchroneity. Etymol: Greek. Cf: *chronotaxy.* Ant: *heterotaxy.* Syn: *homotaxia.*

homothetic Said of geomorphologic features that show geometric similarity (similar in shape though perhaps differing in size) and that have corresponding points that are colinear (Strahler, 1958, p.291).

hondo A term used in the SW U.S. for a broad, low-lying arroyo. Etymol: Spanish, "bottom".

hondurasite *selen-tellurium.*

honessite A mineral consisting of a basic sulfate of iron and nickel.

honeycomb structure [ice] A sea-ice consisting of soft, spongy ice filled with pockets of meltwater or seawater, and characteristic of *rotten ice.*

honeycomb structure [weath] A rock or soil structure having cell-like forms suggesting a honeycomb; e.g. a *stone lattice.* See also: *tafone.*

honeycomb weathering A type of chemical weathering in which innumerable pits are produced on a rock exposure. The pitted surface resembles an enlarged honeycomb and is characteristic of finely granular rocks, such as tuffs and sandstones, in an arid region. Cf: *cavernous weathering.* Syn: *fretwork weathering; alveolar weathering.*

honey stone A syn. of *mellite.* Also spelled: *honeystone.*

hongquiite A cubic mineral: TiO.

hongshiite A hexagonal mineral: PtCuAs.

Honkasalo correction A term added to the conventional Earth-tide correction formula to reduce observed gravity values at a point to a common average value, instead of eliminating the lunar-solar effect altogether.

hood (a) An arched plate of secondary shell of a brachiopod, arising from the echmidium of *Cryptacanthia* or from the median septum of dallinids and terebratellids. (b) The tough fleshy structure located above the head of *Nautilus* and covering the aperture when the head is withdrawn into the living chamber. (c) Ventral flattened part of a *rhyncholite,* commonly of rhomboidal shape (TIP, 1964, pt. K, p. 475). (d) Curved lamellose plates of rostroconch mollusks connected to carinae in the Conocardiacea; growing edges of these plates form a tubular extension of the ventral orifice.—Syn: *collar; schleppe; eventail; fringe; kragen.*

hoodoo (a) A fantastic column, pinnacle, or pillar of rock produced in a region of sporadic heavy rainfall by differential weathering or erosion of horizontal strata, facilitated by joints and by layers of varying hardness, and occurring in varied and often eccentric or grotesque forms. Cf: *earth pillar.* Syn: *rock pillar.* (b) *hoodoo column.*—Etymol: African; from its fancied resemblance to animals and embodied evil spirits.

hoodoo column A term sometimes applied to an *earth pillar.* Syn: *hoodoo.*

hoodoo rock One of several topographic forms of bizarre shape, developed or modified by differential weathering; e.g. *pedestal rock; earth pillar; hoodoo.*

hook [geomorph] (a) A sandy or gravelly barrier spit or narrow cape turned sharply landward at the outer end, so as to resemble a hook in plan view; e.g. a low peninsula or barrier ending in a recurved spit and formed at the end of a bay. Also, a *recurved spit.* (b) A sharp bend, curve, or angle in a stream.

hook [paleont] A holothurian sclerite in the form of a fishhook, consisting of an *eye,* a *shank,* and a *spear.*

Hookean substance A material that deforms in accordance with *Hooke's law.*

hooked bay A bay similar to a *bight* but having a headland at only one end.

hooked dune *fishhook dune.*

hooked spit *recurved spit.*

Hooke's law A statement of *elastic deformation,* that the strain is linearly proportional to the applied stress. See also: *Hookean substance.*

hook valley The valley containing a barbed tributary.

hope [geog] A piece of dry, arable land surrounded by swamp or marsh.

hope [geomorph] A British term (used esp. in southern Scotland) for a small enclosed valley; esp. the broad upper end of a narrow mountain valley, or a blind valley branching from a larger or wider valley. It is usually rounded and often has a stream flowing through it.

hopeite A gray orthorhombic mineral: $Zn_3(PO_4)_2 \cdot 4H_2O$. It is dimorphous with parahopeite.

hopper crystal A cubic crystal of salt in which the faces of the cube have grown more at the edges than in the center, giving each face a centrally depressed or hopper-shaped form.

horizon [geol] An interface indicative of a particular position in a stratigraphic sequence. In practice it is commonly a distinctive very thin bed (ISG, 1976, p. 14). See also: *biohorizon; lithostratigraphic horizon; chronostratigraphic horizon.* Syn: *geologic horizon.*

horizon [soil] *soil horizon.*

horizon [surv] One of several lines or planes used as reference for observation and measurement relative to a given location on the Earth's surface and referred generally to a horizontal direction (Huschke, 1959, p. 283); esp. *apparent horizon.* The term is also frequently applied to *celestial horizon, actual horizon,* and *artificial horizon.*

horizon A The uppermost *reflecting horizon* of the ocean floor. Cf: *horizon beta; horizon B.*

horizon B The lowermost *reflecting horizon* of the ocean floor. Cf: *horizon A; horizon beta.*

horizon beta A *reflecting horizon* of the ocean floor, occurring between *horizon A* and *horizon B.*

horizon circle A circle, in an azimuthal projection, defined by points equidistant from the center of the projection. The maximum horizon circle on a polar projection is the equator.

horizon closure The amount by which the sum of a series of adjacent measured horizontal angles around a point fails to equal exactly the theoretical sum of 360 degrees; the *error of closure* of horizon. Also known as "closure of horizon". See also: *closing the horizon.*

horizon system of coordinates A set of celestial coordinates, usually altitude and azimuth or azimuth angle, based on the celestial horizon as the primary great circle. Cf: *equator system of coordinates.*

horizontal adj. In geodesy, said of a direction that is tangent to the geop at a given point. Cf: *vertical [geophys].*

horizontal angle An angle in a horizontal plane. Cf: *vertical angle.*

horizontal axis The axis about which the telescope of a theodolite or transit rotates when moved vertically. It is the axis of rotation that is perpendicular to the *vertical axis* of the instrument.

horizontal circle A graduated disk affixed to the lower plate of a transit or theodolite by means of which horizontal angles can be measured.

horizontal control A system of points whose horizontal positions and interrelationships have been accurately determined for use as fixed references in positioning and correlating map features.

horizontal dip slip *horizontal slip.*

horizontal direction A direction in a horizontal plane; an observed horizontal angle at a triangulation station, reduced to a common initial direction.

horizontal displacement *strike slip.*

horizontal equivalent The distance between two points on a land surface, projected onto a horizontal plane; e.g. the shortest distance between two contour lines on a map. Abbrev: HE. Cf: *vertical interval.*

horizontal fault A fault with no dip. Cf: *vertical fault.*

horizontal-field balance An instrument that measures the horizontal component of the magnetic field by means of the torque that the field component exerts on a vertical permanent magnet. The two most common types are the *Schmidt field balance* and the *torsion magnetometer.* Cf: *vertical-field balance.*

horizontal fold *nonplunging fold.*

horizontal form index A term used by Bucher (1919, p.154) to express the degree of asymmetry of a current ripple mark, defined as the ratio of the horizontal length of the steep (downcurrent) side to that of the gentle (upcurrent) side. Twenhofel (1950, p.568) used the ratio of the length of the upcurrent side to that of the downcurrent side. Allen (1963) used the ratio of the *span* (crest length) to the *chord* (ripple wavelength). Cf: *vertical form index.* See also: *ripple symmetry index.*

horizontal inclined fold A fold with a horizontal axis and an inclined axial plane (Turner & Weiss, 1963, p. 119).

horizontal intensity The horizontal component of the vector magnetic-field intensity; it is one of the *magnetic elements*, and is symbolized by H. Cf: *vertical intensity.*

horizontal-loop method An electromagnetic method in which the planes of the transmitting and receiving coils are horizontal.

horizontal normal fold A fold with a horizontal axis and vertical axial plane (Turner & Weiss, 1963, p. 119).

horizontal separation In faulting, the distance between the two parts of a disrupted unit (e.g. bed, vein, or dike), measured in any specified horizontal direction.

horizontal slip In a fault, the horizontal component of the net slip. Cf: *vertical slip.* Syn: *horizontal dip slip.*

horizontal throw The *heave* of a fault.

hormite A group name suggested (but not approved) for the sepiolite and palygorskite clay minerals.

hormogonium A multicellular trichome of a filamentous blue-green alga, which may become fragmented at random or at a *heterocyst.*

horn [geog] (a) A body of land (such as a spit), or of water, shaped like a horn. (b) The pointed end of a dune or beach cusp; e.g. the forward, outer end of a barchan crescent. Syn: *wing.*

horn [glac geol] A high rocky sharp-pointed mountain peak with prominent faces and ridges, bounded by the intersecting walls of three or more cirques that have been cut back into the mountain by headward erosion of glaciers; e.g. the Matterhorn of the Pennine Alps. See also: *tind; arête.* Syn: *matterhorn; cirque mountain; pyramidal peak; monumental peak; horn peak.*

hornblende (a) The commonest mineral of the amphibole group: $Ca_2Na(Mg,Fe^{+2})_4(Al,Fe^{+3},Ti)$ (Al,Si_8OË IF22) $(O,OH)_2$. It has a variable composition, and may contain potassium and appreciable fluorine. Hornblende is commonly black, dark green, or brown, and occurs in distinct monoclinic crystals or in columnar, fibrous, or granular forms. It is a primary constituent of many acid and intermediate igneous rocks (granite, syenite, diorite, andesite) and less commonly of basic igneous rocks, and it is a common metamorphic mineral in gneiss and schist. Symbol: Ho. (b) A term sometimes used (esp. by the Germans) to designate the *amphibole* group of minerals. The term "Hornblende" is an old German name for any dark, prismatic crystal found with metallic ores but containing no valuable metal (the word "Blende" indicates "a deceiver").—Obsolete syn: *hornstone.*

hornblende andesite An *andesite* containing abundant hornblende. Obsolete syn: *hungarite.*

hornblende gabbro In the *IUGS classification,* a plutonic rock satisfying the definition of *gabbro,* in which pl/(pl+hbl+px) is between 10 and 90, and px/(pl+hbl+px) is less than 5.

hornblende-hornfels facies The set of metamorphic mineral assemblages (facies) in which basic rocks are represented by hornblende + plagioclase, with epidote and almandine excluded (Turner, 1968). Pelitic assemblages consist of micas accompanied by andalusite, cordierite, or sillimanite; almandine and staurolite are uncommon, kyanite absent. The facies is typical of the middle grades of thermal (contact) metamorphism, and of low-pressure regional dynamothermal metamorphism of the Abukuma or Buchan type. Pressures are estimated to be less than 4000 bars, temperatures in the range 400°-650°C. Cf: *amphibolite facies; pyroxene-hornfels facies; albite-epidote-hornfels facies.* Syn: *cordierite-amphibolite facies.*

hornblende peridotite In the *IUGS classification,* a plutonic rock with M equal to or greater than 90, ol/(ol+hbl+px) between 40 and 90, and px/(ol+hbl+px) less than 5.

hornblende pyroxenite In the *IUGS classification,* a plutonic rock with M equal to or greater than 90, ol/(ol+hbl+px) less than 5, and px/(px+hbl) between 50 and 90.

hornblende schist A schistose metamorphic rock consisting principally of hornblende, with little or no quartz. Unlike *amphibolite,* it does not need to contain plagioclase.

hornblendite (a) In the *IUGS classification,* a plutonic rock with M equal to or greater than 90 and hbl/(hbl+px+ol) greater than 90. (b) An igneous rock composed almost entirely of hornblende. The term has been equated incorrectly by some authors with the metamorphic rock *amphibolite.*

horn coral *solitary coral.*

hörnesite *hoernesite.*

hornfels A fine-grained rock composed of a mosaic of equidimensional grains without preferred orientation and typically formed by contact metamorphism. Porphyroblasts or relict phenocrysts may be present in the characteristically granoblastic (or decussate) matrix (Winkler, 1967). Cf: *skleropelite.*

hornfels facies A loosely defined term used to denote the physical conditions involved, or the set of mineral assemblages produced, by thermal (contact) metamorphism at relatively shallow depths in the Earth's crust. It encompasses the *albite-epidote-hornfels facies,* the *hornblende-hornfels facies,* the *pyroxene-hornfels facies,* and the *sanidinite facies.*

hornito A small mound of *spatter* built on the back of a lava flow (generally pahoehoe), formed by the gradual accumulation of clots of lava ejected through an opening in the roof of an underlying lava tube. Syn: *driblet cone.*

horn lead *phosgenite.*

horn mercury *calomel.*

horn peak *horn [glac geol].*

horn quicksilver *calomel.*

horn silver *chlorargyrite.*

hornstein A syn. of *hornstone [rock].* Etymol: German *Hornstein,* "chert".

hornstone [mineral] (a) A compact, flinty, brittle variety of chalcedony; "a siliceous mineral substance, sometimes approaching nearly to flint, or common quartz" (Lyell, 1854, p. 807). (b) An obsolete name formerly applied to *hornblende.*

hornstone [rock] A general term for a compact tough siliceous rock having a splintery or subconchoidal fracture (Holmes, 1920). It has been used to describe flint or chert as well as hornfels, and has also been confused with hornblende. The term should be abandoned. Syn: *hornstein; petrosilex [sed].*

horny sponge Any demosponge that possesses spicules but has a skeleton composed of spongin.

horotely A phylogenetic phenomenon characterized by a normal or average rate of evolution. Cf: *bradytely; tachytely.*

horse [coal] *horseback [coal].*

horse [fault] A displaced rock mass that has been caught between the walls of a fault.

horse [ore dep] A miners' term for a barren mass of country rock occurring within a vein.

horseback [coal] (a) A syn. of *cutout.* (b) A bank or ridge of foreign matter in a coal seam. (c) A large *roll* in a coal seam. (d) A *clay vein* in a coal seam.—Syn: *kettleback; horse; symon fault; washout.*

horseback [glac geol] A low, sharp ridge of sand, gravel, or rock; specif. an esker or eskerlike deposit, or a kame, in northern New England, esp. Maine. Syn: *hogback; boar's back.*

horseflesh ore Cornish syn. of *bornite.*

horse latitudes Zones of oceanic calms at about 30-35°N and S, characterized by warmth and dryness. These belts move north and south about 5°, following the Sun, and are zones of *divergence* that mark the average region of origin of the *trade winds* and *prevailing westerlies.*

horseshoe A topographic feature, such as a valley or a mountain range, shaped like a horseshoe.

horseshoe bend An *oxbow* in the course of a stream.

horseshoe dissepiment One of a single vertical series of dissepiments of a rugose coral, characterized by a horizontal base and a strongly arched top part. Cf: *lateral dissepiment.*

horseshoe dune *barchan.*

horseshoe flute cast *current crescent.*

horseshoe lake A lake occupying a horseshoe-shaped basin; specif. an *oxbow lake.*

horseshoe moraine A terminal moraine, markedly convex on the down-valley side, usually formed at the end of a valley glacier that never advanced beyond the mountain front.

horseshoe reef A horseshoe-shaped reef that develops from a reef *pinnacle* or *table reef* parallel to the dominant wave action on a platform reef. Its cusps extend downwind and its interior parts often become densely vegetated to produce a small wooded island of low relief.

horsetail [bot] *sphenopsid.*

horsetail [ore dep] adj. Said of a major vein dividing or fraying into smaller fissures; also, said of an ore comprising a series of such veins.

horsetailing A feathery or frondlike fluting, grooving, or other structure developed on the surfaces of shatter-coned rocks by the distinctive striations that radiate from the apex of each *shatter cone* and extend along its length. The presence of multiple nested and parasitic shatter cones produces a distinctive horsetail-like effect.

horsfordite A silver-white mineral: Cu_5Sb.

horst An elongate, relatively uplifted crustal unit or block that is bounded by faults on its long sides. It is a structural form and may or may not be expressed geomorphologically. Etymol: German, no direct English equivalent. Cf: *graben.*

hortite An obsolete name applied to a dark-colored *syenite* that may have formed from gabbro as a result of the assimilation of limestone.

Horton number A dimensionless number formed by the product of *runoff intensity* and the *erosion proportionality factor.* It expresses the relative intensity of the erosion process on the slopes of a drainage basin. Named in honor of Robert E. Horton (1875-1945), U.S. hydraulic engineer. Symbol: N_H.

hortonolite A mineral of the olivine group: $(Fe,MgMn)_2SiO_4$. It is a variety of fayalite containing magnesium and manganese.

hoshiite A nickel-containing variety of magnesite.

host A rock or mineral that is older than rocks or minerals introduced into it or formed within or adjacent to it, such as a *host rock,* or a large crystal with inclusions of smaller crystals of a different mineral species; a *palasome.* Ant: *guest.*

host element An essential element replaced by a *guest element* in a mineral.

hostile ice A submariner's term for an ice canopy containing no large skylights or other features that permit a submarine to surface (U.S. Naval Oceanographic Office, 1968, p. B33). Ant: *friendly ice.*

host rock A body of rock serving as a *host* for other rocks or for mineral deposits; e.g. a pluton containing xenoliths, or any rock in which ore deposits occur.

hot As related to radioactivity, said of a highly radioactive substance.

hot brine Warm and very saline water such as is found on the bottom of the Red Sea. Temperature may be as high as 56°C, and salinity as high as 256%.

hot desert An arid area where the mean annual temperature is higher than 18°C (64.4°F) (Stone, 1967, p. 230).

hot dry rock A potential source of heat energy within the Earth's crust: rocks at depths less than 10 km and at temperatures above 150°C. They are related to two types of heat source: igneous magmas, and conduction from the Earth's deeper interior. Abbrev: HDR.

hot lahar A flow of hot volcanic materials down the slope of a volcano, produced by heavy rains after an eruption or by intrusion of lava into ice, snow, or water-saturated soil. Cf: *cold lahar.* Syn: *hot mudflow.*

hot mudflow *hot lahar.*

hot spot A volcanic center, 100 to 200 km across and persistent for at least a few tens of millions of years, that is thought to be the surface expression of a persistent rising *plume* of hot mantle material. Hot spots are not linked with arcs, and may or may not be associated with oceanic ridges. Some 200 late Cenozoic hot spots have been identified (Cloud, 1974, p. 879). See also: *melting spot.*

hot spring A *thermal spring* whose temperature is above that of the human body (Meinzer, 1923, p. 54). Cf: *warm spring.*

hourglass structure A type of zoning, especially common in clinopyroxenes and chloritoids, in which a "core", distinguished from the outer part by a difference of color or optical properties, has a cross section resembling that of an hourglass.

hourglass valley (a) A valley whose pattern in plan view resembles an hourglass; e.g. a valley extending without interruption across a former divide, toward which it narrows from both directions (Von Engeln, 1942, p. 377). (b) *wineglass valley.*

hover A floating island of vegetation.

how An English term for a low, small hill in a valley or dale; a mound or hillock.

howardite An achondritic stony meteorite consisting largely of calcic plagioclase and orthopyroxene (commonly hypersthene). It has a lower content of iron and calcium than that of eucrite.

howe (a) A Scottish term for a hollow or depression, esp. one on the Earth's surface, as a basin or a valley. (b) *hoe.*

howieite A mineral: $Na(Fe,Mn)_{10}(Fe,Al)_2Si_{12}O_{31}(OH)_{13}$.

howlite A white nodular or earthy mineral: $Ca_2B_5SiO_9(OH)_5$.

Hoxnian British climatostratigraphic stage: Middle Pleistocene (above Anglian, below Wolstonian). Interglacial; probably correlative with Holsteinian.

hoya A stream bed, valley, or basin in a rugged mountainous region, as the Peruvian Andes. Etymol: Spanish, "large hole, cavity, pit".

Hoyt gravimeter *Gulf-type gravimeter.*

hsianghualite An isometric mineral: $Ca_3Li_2Be_3(SiO_4)_3F_2$.

huangho deposit A general term applied by Grabau (1936, p. 253) to a coastal-plain deposit that consists of alluvium that is spread out over a level surface (as a flood plain or a delta) above the normal reach of the sea but that passes laterally into marine beds of equivalent age. Type locality: the loess-derived alluvial deposits at the mouth of the Huang Ho (Yellow River) in northern China. See also: *shantung.*

huanghoite A hexagonal mineral: $BaCe(CO_3)_2F$.

hub The cylindrical or hemispherical projection on the central part (and usually on the lower surface) of a *wheel* of a holothurian.

Hubble constant The amount by which the distance to a galaxy must be multiplied in order to get its velocity or red shift: 75 km sec^{-1} 10^{-6} PSC^{-1}. It represents the present expansion rate of the Universe. It has the unit of reciprocal time and would represent the age of the Universe were there no gravitational deceleration.

hübnerite *huebnerite.*

Hudsonian orogeny A name proposed by Stockwell (1964) for a time of plutonism, metamorphism, and deformation during the Precambrian in the Canadian Shield (especially in the Churchill, Bear, and Southern provinces), dated isotopically as between 1640 and 1820 m.y. ago.

hudsonite A syn. of *cortlandtite,* nonpreferred since "hudsonite" had been used earlier for a variety of pyroxene.

huebnerite A brownish-red to black mineral of the wolframite series: $MnWO_4$. It is isomorphous with ferberite, and may contain up to 20% iron tungstate. Also spelled: *hübnerite.*

huemulite A triclinic mineral: $Na_4MgV_{10}O_{28}\cdot 24H_2O$.

huerfano A term used in the SW U.S. for a hill or mountain of older rock entirely surrounded, but not covered, by any kind of later sedimentary material; esp. a solitary eminence separated by erosion from the mass of which it once formed a part. Etymol: Spanish *huérfano,* "orphan". Pron: *ware*-fah-no. Cf: *lost mountain; tejon.*

hügelite A brown to orange-yellow secondary mineral: $Pb_2(UO_2)_3(AsO_4)_2(OH)_4\cdot 3H_2O$.

Hugoniot The locus of points describing the pressure-volume-energy relations or states that may be achieved within a material by shocking it from a given initial state. Named after Pierre Henri Hugoniot (1851-1887), French physicist. Pron: ewe-*go*-neeoh. Syn: *Hugoniot curve.*

hühnerkobelite A mineral: $(Na_2,Ca)(Fe^{+2},Mn^{+2})_2(PO_4)_2$. It is isomorphous with varulite.

hullite A soft, black, waxy-appearing aluminosilicate of ferric iron, magnesium, calcium, and alkalies, occurring as interstitial matter and amygdaloidal infillings in certain basalts. It is perhaps identical with chlorophaeite.

hulsite A black mineral: $(Fe^2,Mg)_2(Fe^3,Sn)(BO_3)O_2$.

hum A syn. of *karst tower.* Etymol: from a town of that name in Yugoslavia.

human paleontology That branch of paleontology concerned with the fossil record and evolution of early man and his immediate predecessors. Syn: *paleoanthropology.*

humanthracite Humic coal of anthracitic rank; it is the highest stage in the *humolith series.* Cf: *sapanthracite.*

humanthracon Humic coal of bituminous rank; it is the fifth stage in the *humolith series.* Cf: *sapanthracon.*

humate A salt or ester of *humic acid.*

humatipore In cystoids, an exothecal dipore in a single calyx plate having compound calcified canals. These canals carry coelomic fluids for respiration and are found in the rhombiferan family Holocystitidae (Paul, 1972, p. 7).

humatirhomb In cystoids, an exothecal pore structure having calcified thecal canals extending across the suture between adjacent calyx plates. These canals carry coelomic fluids for respiration and are found in the rhombiferan superfamily Caryocystitidae (Paul, 1972, p. 2).

humberstonite A mineral: $Na_7K_3Mg_2(SO_4)_6(NO_3)_2 \cdot 6H_2O$. Syn: *Chile-loeweite.*

Humble gravimeter A gravimeter consisting of a mass, hinged lever, and several springs. The gravity force is balanced by an elastic force. The instrument depends for its sensitivity on proximity to an instability configuration (Bryan, 1937).

humboldtine A mineral: $FeC_2O_4 \cdot 2H_2O$. It occurs in capillary or botryoidal forms in brown coal and black shale. Syn: *humboldtite; oxalite.*

humboldtite (a) *datolite.* (b) *humboldtine.*

humic Pertaining to or derived from *humus.*

humic acid Black acidic organic matter extracted from soils, low-rank coals, and other decayed plant substances by alkalis. It is insoluble in acids and organic solvents.

humic-cannel coal *pseudocannel coal.*

humic coal Coal that is derived from peat by the process of *humification.* Most coal is of this type. Cf: *sapropelic coal.* See also: *humolith series.* Syn: *cahemolith; chameolith; chaemolith; humulith; humus coal; humulite; humolite; humite [coal]; humolith.*

humic decomposition Chemical breakdown of rocks and minerals by the action of vegetable acids.

humic degradation matter Organic degradation matter that is cellulosic and in which the individual particles are still recognizable; it is similar to *anthraxylon.* It is classified according to the type of constituent plant material. See also: *translucent humic degradation matter.* Syn: *cell-wall degradation matter; brown matter.* Abbrev: HDM.

Humic Gley soil A great soil group in the 1949 revised classification system, an intrazonal, hydromorphic group of mineral soils having a dark organic-rich surface horizon underlain by a gleyed horizon. It occurs in wet meadows and in swamps with forest vegetation (Thorp and Smith, 1949). Most of these soils are now classified as *Aquolls, Aquepts,* or *Aquults.* Syn: *Weisenboden.*

humidity The water-vapor content of the atmosphere. The unmodified term often signifies *relative humidity.* See also: *mixing ratio; absolute humidity; specific humidity.*

humification The process of development of humus or humic acids, essentially by slow oxidation. Adj: *humified.* See also: *mor.*

humin *ulmin.*

huminite (a) A group of macerals in brown coal, consisting of humic matter derived mainly from lignin and cellulose (ICCP, 1971). The group is distinguished microscopically by having a middle level of reflectance and a brownish yellow to reddish brown color in transmitted light. It is the precursor of the *vitrinite* group in bituminous coals. (b) A variety of oxidized bitumen, resembling brown coal, found in a granite-pegmatite vein in Sweden.

humite [coal] *humic coal.*

humite [mineral] (a) A white, yellow, brown, or red orthorhombic mineral: $Mg_7Si_3O_{12}(F,OH)_2$. It sometimes contains appreciable iron, and it is found in the masses ejected from volcanoes. (b) A group of isomorphous magnesium-silicate minerals frequently containing fluorine and closely resembling one another in chemical composition, physical properties, and crystallization. It consists of olivine, humite, clinohumite, chondrodite, and norbergite.

hummerite A mineral: $KMgV_5O_{14} \cdot 8H_2O$.

hummock [geog] A rounded or conical knoll, mound, hillock, or other small elevation. Also, a slight rise of ground above a level surface. Syn: *hammock.*

hummock [ice] A mound, hillock, or pile of broken floating ice, either fresh or weathered, that has been forced upward by pressure, as in an ice field or ice floe. Cf: *bummock.* Syn: *ice hummock.*

hummock [permafrost] A small irregular knob of earth or turf. Neither *earth hummock* nor *turf hummock* is diagnostic of permafrost, but both are most common in subpolar and alpine regions. Both require vegetation (Washburn, 1973, p. 126).

hummocked ice Sea ice having a rugged, uneven surface due to the formation of hummocks; it has the appearance of smooth hillocks when weathered. A form of *pressure ice.*

hummocking Pressure process by which floating ice becomes broken up into hummocks.

hummocky Abounding in hummocks, or uneven; said of topographic landforms, as a hummocky dune, and of hummocked ice.

hummocky cross-stratification A type of cross-stratification in which lower bounding surfaces of sets are erosional and commonly slope at angles less than 10°, though dips can reach 15°; laminae above these erosional set boundaries are parallel to that surface, or nearly so; laminae can systematically thicken laterally in a set so that their traces on a vertical surface are fan-like and dip diminishes regularly; and the dip directions of erosional set boundaries and of the overlying laminae are scattered (Harms et al., 1975, p. 87). It is thought to be formed by storm-wave surges on the shoreface. Cf: *truncated wave-ripple laminae.*

hummocky moraine An area of *knob-and-kettle topography* that may have been formed either along a live ice front or around masses of stagnant ice (Gravenor & Kupsch, 1959, p. 52).

humocoll Humic material of the rank of peat; it is the second stage in the *humolith series.* Cf: *saprocol.*

Humod In U.S. Dept. of Agriculture soil taxonomy, a suborder of the soil order *Spodosol,* characterized by large accumulations of organic matter relative to iron in its spodic horizon. The albic horizon may be either intermittent or continuous, and the upper part of the spodic horizon is nearly black (USDA, 1975). Cf: *Aquod; Ferrod; Orthod.*

humodil Humic coal of lignitic rank; it is the third stage in the *humolith series.* Cf: *saprodil.*

humodite Humic coal of subbituminous rank; it is the fourth stage in the *humolith series.* Cf: *saprodite.*

humodurite *translucent attritus.*

humogelite *ulmin.*

humolite *humic coal.*

humolith *humic coal.*

humolith series Humic material and coals in order of metamorphic rank: *humopel, humocoll, humodil, humodite, humanthracon* and *humanthracite* (Heim & Potonié, 1932, p.146). Cf: *sapropelite series; humosapropelic series; saprohumolith series.* See also: *humic coal.*

humonigritite A type of *nigritite* that occurs in sediments. Cf: *polynigritite; exinonigritite; keronigritite.*

humopel Organic matter, or *ulmin,* of humic coals; it is the first stage in the *humolith series.* Cf: *sapropel.*

humosapropelic series Organic materials and coals intermediate between the *humolith series* and the *sapropelite series,* with humolithic materials predominating. Cf: *saprohumolith series.*

humosite A dark brownish-red microscopic constituent of torbanite, translucent and isotropic (Dulhunty, 1939).

humovitrinite Vitrinite in vitrain of humic coal. Cf: *saprovitrinite.*

Humox In U.S. Dept. of Agriculture soil taxonomy, a suborder of the soil order *Oxisol,* characterized by formation in relatively cool humid climates at high altitudes. Most have a high content of organic matter within the first meter of depth, a low supply of bases, a udic moisture regime, and a thermic or hyperthermic temperature regime (USDA, 1975). Cf: *Aquox; Orthox; Torrox; Ustox.*

humpy A small morainal mound with a central depression (Gravenor & Kupsch, 1959, p. 53).

humulite *humic coal.*

humulith *humic coal.*

Humult In U.S. Dept. of Argiculture soil taxonomy, a suborder of the soil order *Ultisol,* characterized by having a high organic-carbon content. They occur mainly in mountainous areas that have high but seasonally deficient rainfall. Because of the slope, cultivated soils often have lost their surface horizons, and the argillic horizon is at the surface (USDA, 1975). Cf: *Aquult; Udult; Ustult; Xerult.*

humus The generally dark, more or less stable part of the organic matter of the soil, so well decomposed that the original sources cannot be identified. The term is sometimes used incorrectly for the total organic matter of the soil, including relatively undecomposed material. Syn: *soil ulmin.*

humus coal *humic coal.*

humus layer *H layer.*

hungarite An obsolete syn. of *hornblende andesite.* The name,

given by Lang in 1877, is from Hungary.
hungchaoite A mineral: $MgB_4O_7 \cdot 9H_2O$.
hungry (a) Said of a rock, lode, or belt of country that is barren of ore minerals or of geologic indications of ore, or that contains very low-grade ore. Ant: *likely.* (b) Said of a soil that is poor or not fertile.
huntite A white mineral: $CaMg_3(CO_3)_4$.
Hunt-Wentworth recording micrometer An instrument used for petrographic modal analysis. A series of linear traverses across a thin section are made. The traverse lengths of the various mineral grains are recorded. With a sufficient number of traverses, the total linear distance traversed for a given mineral when divided by the total distance traversed is proportional to its volume in the rock (Heinrich, 1956, p. 8).
hureaulite A monoclinic mineral: $Mn_5(PO_4)_2[PO_3(OH)]_2 \cdot 4H_2O)$. It occurs in yellow, reddish, or gray prismatic crystals, or in massive form. Syn: *bastinite; palaite.*
hurlbutite A mineral: $CaBe_2(PO_4)_2$.
Huronian A division of the Proterozoic of the Canadian Shield.
hurricane A *tropical cyclone,* esp. in the West Indies, in which the wind velocity equals or exceeds 64 knots (73 mph).
hurricane delta A delta formed by storm waves carrying sand across a reef or barrier island and depositing it in a lagoon.
hurricane surge *storm surge.*
hurricane tide (a) *storm surge.* (b) The height of a storm surge above the astronomically predicted level of the sea.
hurricane wave *storm surge.*
hurst (a) A wooded knoll, hill, or other small eminence; a grove or a thick wood; a copse. (b) A bank or piece of rising ground; esp. a sandbank in or along a river.—The term is very common in place names. Also spelled: *hirst; hyrst.*
husebyite A plagioclase-bearing *nepheline syenite,* named by Brögger in 1933 for Huseby, Oslo district, Norway. Not recommended usage.
hushing *hydraulic prospecting.*
hussle A term used in England for soft clay associated with coal and for contorted carbonaceous shale immediately below a coal seam. Etymol: Yorkshire dialect for "rubbish".
hutchinsonite A scarlet to deep cherry-red orthorhombic mineral: $(Pb,Tl)_2(Cu,Ag)As_5S_{10}$.
Huttonian Of or relating to James Hutton (1726-1797), Scottish geologist, who advocated the theory of *plutonism,* introduced the concepts of uniformitarianism and the geologic cycle, and emphasized the indefinite length of geologic time.
huttonite A colorless to pale-cream monoclinic mineral: $ThSiO_4$. It is dimorphous with thorite and isostructural with monazite.
Huygens' principle The statement that any particle excited by wave energy becomes a new point source of wave energy. It is named after the Dutch astronomer and mathematician Christian Huygens (d.1695).
HW *high water.*
H wave *hydrodynamic wave.*
hyacinth (a) A transparent orange, red, or reddish-brown zircon, sometimes used as a gem. The term has been used interchangeably with *jacinth,* and loosely to signify any zircon. (b) Yellow, orange, or brownish essonite used as a gem. Syn: *hyacinth garnet; hyacinthoid.* (c) A term applied as a syn. of various orange-red to orange minerals, and also of minerals such as harmotome, vesuvianite, and meionite. (d) A precious stone believed by the ancients to be the sapphire.
hyaline [ign] A syn. of *glassy;* sometimes used as a prefix ("hyalo") to names of volcanic rocks with a glassy texture, e.g. "hyalobasalt".
hyaline [mineral] Said of a mineral that is amorphous.
hyaline [paleont] Said of the glassy clear or transparent fine-textured outer wall of a foraminifer.
hyalinocrystalline Obsolescent term applied to the texture of a porphyritic igneous rock in which the phenocrysts are in a glassy groundmass irrespective of the relative proportions. Also, said of a rock having such texture. Syn: *crystallohyaline.*
hyalite A colorless variety of common opal that is sometimes clear as glass and sometimes translucent or whitish and that occurs as globular concretions (resembling drops of melted glass) or botryoidal crusts lining cavities or cracks in rocks. Syn: *water opal; Müller's glass.*
hyalithe An opaque glass resembling porcelain and frequently black, green, brown, or red.
hyalo- A prefix meaning "glassy". Cf: *vitr-.*
hyalobasalt *tachylyte.*
hyaloclastite A deposit formed by the flowing or intrusion of lava or magma into water, ice, or water-saturated sediment, and its consequent granulation or shattering into small angular fragments. Syn: *aquagene tuff.* Cf: *palagonite tuff.*
hyalocrystalline Said of the texture of a porphyritic igneous rock in which crystals and glassy groundmass are equal or nearly equal in volumetric proportions, the ratio of phenocrysts to groundmass being between 5:3 and 3:5 (Cross et al., 1906, p. 694). Cf: *intersertal.* Syn: *semicrystalline.*
hyalomelane A term given by Hausmann in 1847 to a volcanic glass, commonly porphyritic, differing from *tachylyte* in its purported insolubility in acids; a basaltic *vitrophyre.* Obsolete.
hyalomylonite A glassy rock formed by fusion of granite, arkose, etc. by frictional heat in zones of intense differential movement. Cf: *buchite.*
hyalophane A colorless monoclinic mineral of the feldspar group: $(K,Ba)Al(Al,Si)_3O_8$. It is intermediate in composition between celsian and orthoclase.
hyalophitic Said of the texture of an igneous rock in which the mesostasis is glassy and makes up a proportion of the rock intermediate in texture between *hyalopilitic* and *hyalocrystalline.* Cf: *intersertal.*
hyalopilitic Said of the *intersertal* texture of an igneous rock in which needlelike microlites of the groundmass are set in a glassy mesostasis, the phenocrysts forming less than one-eighth of the rock. Cf: *hyalophitic; sporophitic.*
hyalopsite *obsidian.*
hyalosiderite A rich olive-green variety of olivine containing considerable iron (30-50 mole percent of Fe_2SiO_4).
hyalosponge Any sponge belonging to the class Hyalospongea, characterized chiefly by a skeleton composed of six-rayed siliceous spicules, without calcium carbonate or spongin. Syn: *hexactinellid; glass sponge.*
hyalotekite A white or gray mineral: $(Pb,Ca,Ba)_4BSi_6O_{17}(OH,F)$.
hybrid [evol] An individual having parents belonging to different species.
hybrid [ign petrol] adj. Pertaining to a rock whose chemical composition is the result of *assimilation.* Syn: *contaminated.*—n. A rock whose composition is the result of assimilation.—See also: *hybridization.*
hybrid age The radiometric age given by an isotopic system that has lost radiogenic isotopes owing to thermal, igneous, or tectonic activity some time after the start of the isotopic system. See also: *overprint [geochron]; updating; mixed ages.*
hybrid computer A computer with combined capabilities of the *analog computer* and *digital computer.*
hybridism (a) *hybridization.* (b) The condition of being *hybrid.*
hybridization The process whereby rocks of different composition from that of the parent magma are formed, by *assimilation.* Cf: *contamination [ign].* Syn: *hybridism.*
hydatogenesis The crystallization or precipitation of salts from normal aqueous solutions; the formation of an evaporite.
hydatogenic Said of a rock or mineral deposit formed by an aqueous agent, e.g. a mineral deposit in a vein from a magmatic solution or an evaporite from a body of salt water. Cf: *pneumatogenic; hydatopneumatogenic.* Syn: *hydatomorphic.*
hydatomorphic *hydatogenic.*
hydatopneumatogenic Said of a rock or mineral deposit formed by both aqueous and gaseous agents. Cf: *hydatogenic; pneumatogenic.*
hydatopyrogenic *aqueo-igneous.*
hydnophorid Said of a scleractinian corallum with corallite centers arranged around protuberant collines or monticules (TIP, 1956, pt.F, p.248).
hydrarch adj. Said of an ecologic succession (i.e. a *sere*) that develops under *hydric* conditions. Cf: *mesarch; xerarch.* See also: *hydrosere.*
hydrargillite (a) *gibbsite.* (b) A name that has been applied to various aluminum-bearing minerals, including aluminite, wavellite, and turquoise.
hydrate n. A mineral compound that is produced by hydration, or one in which water is part of the chemical composition.—v. To cause the incorporation of water into the chemical composition of a mineral.
hydrated halloysite *endellite.*
hydration reaction A metamorphic reaction that results in the transfer of H_2O from the fluid phase into the structure of a miner-

al. Cf: *dehydration reaction.*

hydration rind dating *obsidian hydration dating.*

hydration shattering The process of grain loosening and rock disintegration by the wedging pressure of water in films of varying thickness on silicate mineral surfaces. This water is drawn between the grains by electro-osmosis and exerts differential pressures up to 2000 kg/cm^2, strong enough to loosen and separate the grains. Such a process may be significant in all climates, without the aid of freezing and thawing. It produces loosened and separated grains, the accumulation being *grus.*

hydration water *water of hydration.*

hydraulic [eng] Conveyed, operated, effected, or moved by means of water or other fluids, such as a "hydraulic dredge" using a centrifugal pump to draw sediments from a river channel.

hydraulic [hydraul] Pertaining to a fluid in motion, or to movement or action caused by water.

hydraulic [mater] Hardening or setting under water; e.g. "hydraulic lime" or "hydraulic cement".

hydraulic action The mechanical loosening and removal of weakly resistant material solely by the pressure and *hydraulic force* of flowing water, as by a stream surging into rock cracks or impinging against the bank on the outside of a bend, or by ocean waves and currents pounding the base of a cliff. See also: *fluviraption.*

hydraulic conductivity *permeability coefficient.*

hydraulic current A local current produced by differences in water level at the two ends of a channel, set up by the rising and falling tide at constrictions in a baymouth or in the narrow strait connecting two bodies of water having tides that differ in time or range; e.g. in Hell Gate, where Long Island Sound joins the East River, N.Y.

hydraulic diffusivity In ground water, the *transmissivity* divided by the *storage coefficient,* or T/S. It is the conductivity of the saturated medium when the unit volume of water moving is that involved in changing the head a unit amount in a unit volume of medium (Lohman et al., 1970, p. 23).

hydraulic element A quantity pertaining to a particular stage of flowing water in a particular cross section of a conduit or stream channel, e.g. depth of water, cross-sectional area, hydraulic radius, wetted perimeter, mean depth of water, velocity, energy head, friction factor (ASCE, 1962).

hydraulic equivalent The number of Udden size grades between the size of a given mineral grain of a sediment and the size of the quartz grain with which it was deposited or to which it is hydraulically equivalent (the size of a larger or smaller grain that settles with the given mineral grain under the same conditions) (Rittenhouse, 1943). See also: *hydraulic ratio.*

hydraulic fill Earth or waste material that has been excavated, transported, and flushed into place by moving water.

hydraulic-fill dam A dam composed of *hydraulic fill,* in which the sorting of particle sizes into an impervious central core supported by outer zones of coarser material is accomplished by arrangement of peripheral discharge outlets and flow in the central pool.

hydraulic force The eroding and shearing force of flowing water, involving no sediment load and resulting in *hydraulic action.*

hydraulic fracturing A general term, for which there are numerous trade or service names, for the fracturing of rock in an oil or gas reservoir by pumping in water (or other fluid) and sand under high pressure. The purpose is to produce artificial openings in the rock in order to increase permeability. The pressure opens cracks and bedding planes, and sand introduced into these serves to keep them open when pressure is reduced. Syn: *fracturing; hydrofracturing.*

hydraulic friction The resistance to flow exerted on the perimeter or contact surface between a stream and its containing conduit, due to the roughness characteristic of the confining surface, which induces a loss of energy. Energy losses arising from excessive turbulence, impact at obstructions, curves, eddies, and pronounced channel changes are not ordinarily ascribed to hydraulic friction (ASCE, 1962).

hydraulic geometry The description, at a given cross section of a river channel, of the graphical relationships among plots of hydraulic characteristics (such as width, depth, velocity, channel slope, roughness, and bed particle size, all of which help to determine the shape of a natural channel) as simple power functions of river discharge (Leopold & Maddock, 1953).

hydraulic grade line In a closed channel, a line joining the elevations that water would attain in atmospheric pressure; in an open channel, the free water surface or *flow line [hydraul].* Its slope represents *energy loss.* See also: *hydraulic gradient; hydraulic head.*

hydraulic gradient (a) In an aquifer, the rate of change of *total head* per unit of distance of flow at a given point and in a given direction. Cf: *pressure gradient.* See also: *gradient of the head.* Syn: *potential gradient.* (b) In a stream, the slope of the *hydraulic grade line.* See also: *critical hydraulic gradient.* Syn: *gradient [hydraul].*

hydraulic head (a) The height of the free surface of a body of water above a given subsurface point. (b) The water level at a point upstream from a given point downstream. (c) The elevation of the *hydraulic grade line* at a given point above a given point of a pressure pipe.

hydraulic jump In fluid flow, a change in flow conditions accompanied by a stationary, abrupt turbulent rise in water level in the direction of flow. It is a type of stationary wave.

hydraulic limestone An impure limestone that contains silica and alumina (usually as clay) in varying proportions and that yields, upon calcining, a cement that will harden under water. See also: *cement rock.* Syn: *waterlime.*

hydraulic mean depth *hydraulic radius.*

hydraulic mining The extraction of ore by means of strong jets of water. The ore may be surficial, such as a placer, or subterranean, such as a soft coal. Cf: *placer mining.*

hydraulic permeability The ability of a rock or soil to transmit water under pressure. It may vary according to direction.

hydraulic plucking A process of stream erosion by which rock fragments are forcibly removed by the impact of water entering cracks in a rock. Syn: *quarrying.*

hydraulic profile A vertical section of the potentiometric surface of an aquifer.

hydraulic prospecting The use of water to clear away surficial deposits and debris to expose outcrops, for the purpose of exploring for mineral deposits. Syn: *hushing.*

hydraulic radius In a stream, the ratio of the area of its cross section to its *wetted perimeter.* Symbol: R. Syn: *hydraulic mean depth.* See also: *form ratio.*

hydraulic ratio A value expressing the quantity of any given heavy mineral in a sediment, equal to the weight of a heavy mineral in a given size class divided by the weight of light minerals in the *hydraulic-equivalent* class (Rittenhouse, 1943). The value is commonly multiplied by 100 to reduce the number of decimal places.

hydraulics The aspect of engineering that deals with the flow of water or other liquids; the practical application of *hydromechanics.*

hydraulic wedging Pressure produced in a cavity within a reef or other body of rock by pounding surf (Cloud, 1957, p.1016).

hydric Said of a habitat that has or requires abundant moisture; also, said of an organism or group of organisms occupying such a habitat. Cf: *xeric; mesic.* See also: *hydrarch.*

hydroamphibole A mixture of hornblende and chlorite.

hydroastrophyllite A triclinic mineral: $(H_3O,K,Ca)_3(Fe,Mn)_{5\text{-}6}Ti_2Si_8(O,OH)_{31}$. It is a member of the astrophyllite group.

hydrobasaluminite A mineral: $Al_4(SO_4)(OH)_{10}\cdot 36H_2O$.

hydrobiology The biology of bodies of water, esp. of lakes and other bodies of fresh water. Cf: *biohydrology.*

hydrobios That area of the Earth occupied by aquatic plants and animals. Cf: *geobios.*

hydrobiotite (a) A light-green, trioctahedral, mixed-layer clay mineral composed of interstratification of biotite and vermiculite. (b) A term applied originally to a biotitelike material high in water.

hydroboracite A white mineral: $CaMgB_6O_{11}\cdot 6H_2O$.

hydrocalcite (a) A mineral name applied to material that is perhaps $CaCO_3\cdot 2H_2O$ or $CaCO_3\cdot 3H_2O$. (b) A mineral name used by Marschner (1969) for a compound now known as *monohydrocalcite.*

hydrocalumite A colorless to light-green mineral: $Ca_2Al(OH)_7\cdot 3H_2O$ or $Ca_4Al_2O_7\cdot 12H_2O$.

hydrocarbon Any organic compound, gaseous, liquid, or solid, consisting solely of carbon and hydrogen. They are divided into groups of which those of especial interest to geologists are the *paraffin, cycloparaffin, olefin,* and *aromatic* groups. Crude oil is essentially a complex mixture of hydrocarbons.

hydro cast *hydrographic cast.*

hydrocerussite A colorless hexagonal mineral: $Pb_3(CO_3)_2(OH)_2$. It occurs as a secondary product as an encrustation on native lead or on galena.

hydrochemical facies (a) The diagnostic chemical character of ground-water solutions occurring in hydrologic systems (Back, 1966, p.11). It is determined by the flow pattern of the water and by the effects of chemical processes operating between the ground water and the minerals within the lithologic framework. (b) A term used by Chebotarev (1955, p.199) to indicate concentration of dissolved solids (facies may be low-, transitional-, or high-saline).

hydrochloroborite A monoclinic mineral: $Ca_4B_8O_{15}Cl_2 \cdot 22H_2O$.

hydrochore A plant whose seeds or spores are distributed by water.

hydroclast A rock fragment that is transported and deposited in an aqueous environment.

hydroclastic rock (a) A clastic rock deposited by the agency of water. (b) A rock broken by wave or current action. (c) A volcanic rock broken or fragmented during chilling under water or ice.

hydroclimate The physical and often the chemical factors that characterize a particular aquatic environment.

hydrocyanite *chalcocyanite.*

hydrodialeima A term proposed by Sanders (1957, p.295) for an unconformity caused by subaqueous processes.

hydrodolomite A mixture of hydromagnesite and calcite.

hydrodynamic jetting Directional ejection of molten or vaporized material at very high velocities as a result of shock-wave interactions at the interface between projectile and target in the early stages of *hypervelocity impact* (Gault et al., 1968, p. 90). Such a process is believed to form tektites by meteorite impact.

hydrodynamics The aspect of *hydromechanics* that deals with forces that produce motion. Cf: *hydrostatics; hydrokinetics.*

hydrodynamic wave An obsolete term for a type of *surface wave* that is similar to a Rayleigh wave but has an opposite particle motion. Syn: *H wave.*

hydroelectric power Electrical energy generated by means of a power generator coupled to a turbine through which water passes. Cf: *waterpower; hydropower; white coal.*

hydroexplosion General term for a volcanic explosion caused by the generation of steam from any body of water. It includes phreatic, phreatomagmatic, submarine, and littoral explosions. Cf: *littoral explosion.*

hydrofracturing *hydraulic fracturing.*

hydrogarnet (a) A group of garnet minerals of the general formula: $A_3B_2(SiO_4)_{3-x}(OH)_{4x}$. (b) A mineral of the hydrogarnet group, such as hydrogrossular.

hydrogenesis The natural condensation of moisture in the air spaces of surficial soil or rock material.

hydrogenic Said of a soil whose dominant formative influence is water, as in a cold, humid area.

hydrogenic rock A sedimentary rock formed by the agency of water. The term was restricted by Grabau (1924, p. 280) to a *hydrolith* "wholly of chemical origin", such as a precipitate from solution in water.

hydrogen-index log *neutron log.*

hydrogen-ion concentration *pH.*

hydrogenous (a) Said of coals high in moisture, such as brown coals. (b) Said of coals high in volatiles, such as sapropelic coals.

hydrogen-sulfide mud *black mud.*

hydrogeochemistry The chemistry of ground and surface waters, particularly the relationships between the chemical characteristics and quality of waters and the areal and regional geology.

hydrogeology The science that deals with subsurface waters and with related geologic aspects of surface waters. Also used in the more restricted sense of *ground-water geology* only. The term was defined by Mead (1919, p. 2) as the study of the laws of the occurrence and movement of subterranean waters. More recently it has been used interchangeably with *geohydrology.*

hydroglauberite A mineral: $Na_4Ca(SO_4)_3 \cdot 2H_2O$.

hydrograph A graph showing stage, flow, velocity, or other characteristics of water with respect to time (Langbein & Iseri, 1960). A stream hydrograph commonly shows rate of flow; a ground-water hydrograph, water level or head. See also: *depletion curve.*

hydrographic basin (a) The *drainage basin* of a stream. (b) An area occupied by a lake and its drainage basin.

hydrographic cast A hydrographic survey station at which temperature and salinity measurements are made at standard ocean depths, in order to compute densities. Also, the process itself. Cf: *Nansen cast.* Syn: *hydro cast; oceanographic cast.*

hydrographic chart A map used in navigation, showing water depth, bottom relief, tides and currents, adjacent land, and distinguishing surface features. Syn: *nautical chart.*

hydrography (a) The science that deals with the physical aspects of all waters on the Earth's surface, esp. the compilation of navigational charts of bodies of water. (b) The body of facts encompassed by hydrography.

hydrogrossular A mineral of the hydrogarnet group: $Ca_3Al_2(SiO_4)_{3-x}(OH)_{4x}$, with x near 1/2. Syn: *hibschite; plazolite; hydrogrossularite.*

hydrohalite A mineral: $NaCl \cdot 2H_2O$. It is formed only from salty water at or below the freezing temperature of pure water.

hydrohalloysite *endellite.*

hydrohematite *turgite.*

hydroherderite *hydroxyl-herderite.*

hydrohetaerolite A black mineral of uncertain composition: $Zn_2Mn_4O_8 \cdot H_2O$ (?).

hydroid Any one of a group of hydrozoans belonging to the order Hydroida, among which the polypoid (usually colonial) generation is dominant, and the skeleton is commonly composed of a hornlike material. Cf: *millepore; stylaster.* Range, Cambrian to present.

hydrokaolin (a) *endellite.* (b) A fibrous variety of kaolinite from Saglik in Transcaucasia, U.S.S.R.

hydrokinetics The aspect of *hydromechanics* that deals with forces that cause change in motion. Cf: *hydrodynamics; hydrostatics.*

hydrolaccolith An approx. syn. of *ice laccolith.* The term is used often as a syn. of *pingo,* but a hydrolaccolith can be a seasonal mound, whereas a pingo is perennial.

hydrolite A term variously applied to enhydros, to siliceous sinter, and to the zeolite mineral gmelinite.

hydrolith (a) A term proposed by Grabau (1904) for a rock that is chemically precipitated from solution in water, such as rock salt or gypsum; a *hydrogenic rock.* (b) A rock that is "relatively free from organic material" (Nelson & Nelson, 1967, p. 185). (c) A *hydroclastic rock* consisting of carbonate fragments (Bissell & Chilingar, 1967, p. 158).

hydrologic balance *hydrologic budget.*

hydrologic barrier *ground-water barrier.*

hydrologic budget An accounting of the inflow to, outflow from, and storage in a hydrologic unit such as a drainage basin, aquifer, soil zone, lake, or reservoir (Langbein & Iseri, 1960); the relationship between evaporation, precipitation, runoff, and the change in water storage, expressed by the *hydrologic equation.* Syn: *water balance; water budget; hydrologic balance.*

hydrologic cycle The constant circulation of water from the sea, through the atmosphere, to the land, and its eventual return to the atmosphere by way of transpiration and evaporation from the sea and the land surfaces. Syn: *water cycle.*

hydrologic properties Those properties of a rock that govern the entrance of water and the capacity to hold, transmit, and deliver water, e.g. porosity, effective porosity, specific retention, permeability, and direction of maximum and minimum permeability.

hydrologic regimen (a) *regimen [water].* (b) *regimen [lake].*

hydrologic system A complex of related parts — physical, conceptual, or both — forming an orderly working body of hydrologic units and their man-related aspects such as the use, treatment and reuse, and disposal of water and the costs and benefits thereof, and the interaction of hydrologic factors with those of sociology, economics, and ecology.

hydrology (a) The science that deals with global water (both liquid and solid), its properties, circulation, and distribution, on and under the Earth's surface and in the atmosphere, from the moment of its precipitation until it is returned to the atmosphere through evapotranspiration or is discharged into the ocean. In recent years the scope of hydrology has been expanded to include environmental and economic aspects. At one time there was a tendency in the U.S. (as well as in Germany) to restrict the term "hydrology" to the study of subsurface waters (DeWiest, 1965, p. 1). Syn: *hydroscience.* (b) The sum of the factors studied in hydrology; the hydrology of an area or district.

hydrolysates *hydrolyzates.*

hydrolysis A decomposition reaction involving water. In geology, it commonly indicates reaction between silicate minerals and either pure water or aqueous solution. In such reactions, H^+ ions or OH^- ions are consumed, thus changing the H^+/OH^- ratio. In the hydrolysis of a silicate mineral, an amount of cation chemically equivalent to the quantity of H^+ consumed must be released to the solution. An example is the hydrolysis of potassium feldspar to muscovite plus quartz: $1.5KAlSi_3O_3 + H^+ \rightleftarrows 0.5KAl_3Si_3O_{10} + K^+$

+ $3SiO_2$. (See Hemley & Jones, 1964).

hydrolyzates Sediments characterized by elements that are readily hydrolyzed, concentrate in the fine-grained alteration products of primary rocks, and are thus abundant in clays, shales, and bauxites. Hydrolyzate elements are aluminum and associated silicon, potassium, and sodium. It is one of Goldschmidts' groupings of sediments as analogues of differentiation stages in rock analysis. Also spelled: *hydrolysates.* Cf: *resistates; oxidates; reduzates; evaporates.*

hydromagnesite A white, earthy mineral: $Mg_5(CO_3)_4(OH)_2 \cdot 4H_2O$. It occurs in small monoclinic crystals (as in altered ultrabasic rocks) or in amorphous masses or chalky crusts (as in the temperate caves of eastern U.S.).

hydromagniolite A general term for hydrous magnesium silicates.

hydromarchite A tetragonal mineral: $Sn_3O_2(OH)_2$.

hydromechanics The theoretical, experimental, or practical study of the action of forces on water. See also: *hydrodynamics; hydrokinetics; hydrostatics; hydraulics.*

hydrometamorphism Alteration of rock by material that is added, removed, or exchanged by water solutions, without the influence of high temperature and pressure. Syn: *hydrometasomatism.*

hydrometasomatism *hydrometamorphism.*

hydrometeor A minute droplet of water or crystal of ice falling through or suspended in the atmosphere.

hydrometer An instrument that is used to measure the specific gravity of a liquid such as seawater.

hydrometry (a) The use of the hydrometer to measure specific gravity of a fluid. (b) The study of the flow of water, esp. measurement.

hydromica Any of several varieties of muscovite that are less elastic and more unctuous than mica, that have a pearly luster, and that sometimes contain less potash and more water than ordinary muscovite; e.g. a common micaceous clay mineral resembling sericite but having weaker double refraction. The term is practically synonymous with *illite.* Syn: *hydrous mica.*

hydromolysite A mineral: $FeCl_3 \cdot 6H_2O$.

hydromorphic soil A type of intrazonal soil (1938 classification system) having characteristics that were developed in the presence of excess water all or part of the time; e.g. a *Bog soil.*

hydromuscovite A term applied loosely to any fine-grained, muscovite-like clay mineral commonly but not always high in water content and deficient in potassium. It is probably an illite.

hydronium jarosite A mineral: $(H_3O)Fe_3(SO_4)_2(OH)_6$. Cf: *carphosiderite.*

hydrophane A white, yellow, brown, or green variety of common opal that becomes more translucent when immersed in water.

hydrophilite A white mineral: $CaCl_2$. Syn: *chlorocalcite.*

hydrophone A pressure-sensitive detector that responds to sound transmitted through water. It is used in marine seismic surveying, or as a seismometer in a well. Syn: *pressure detector.*

hydrophyte (a) A plant growing in water, either submerged, emergent, or floating; esp. a *helophyte.* (b) A plant that requires large quantities of water for its growth. Syn: *hygrophyte.* —Cf: *mesophyte; xerophyte.*

hydroplasticity Plasticity that results from the presence of pore water and absorbed water films in a sediment, so that it yields easily to changes of pressure.

hydroplutonic *aqueo-igneous.*

hydropore A pore, slit, or small external opening that serves as an adit to the water-vascular system of an echinoderm. It may be covered by the madreporite. Cf: *gonopore.*

hydropore oral An unpaired oral plate in edrioasteroids, in the right posterior part of the oral region; it commonly forms the posterior edge of the hydropore (Bell, 1976).

hydropore structure A group of plates that externally surround the hydropore in edrioasteroids (Bell, 1976).

hydropower Literally, *waterpower,* but now generally considered a syn. of *hydroelectric power.*

hydropsis *synoptic oceanography.*

hydroscarbroite A mineral: $Al_{14}(CO_3)_3(OH)_{36} \cdot n\,H_2O$.

hydroscience *hydrology.*

hydroscopic water *hygroscopic water.*

hydrosere A *sere* that develops in an aquatic environment; a *hydrarch* sere. Cf: *mesosere; xerosere.*

hydrosialite A syn. of *clay mineral.* Also spelled: *hydrosyalite.*

hydrosilicate inclusion A fluid inclusion in a crystal representing the late-silicate fraction of magmatic crystallization.

hydrospace *inner space.*

hydrosphere The waters of the Earth, as distinguished from the rocks (lithosphere), living things (biosphere), and the air (atmosphere). Includes the waters of the ocean; rivers, lakes, and other bodies of surface water in liquid form on the continents; snow, ice, and glaciers; and liquid water, ice, and water vapor in both the unsaturated and saturated zones below the land surface. Included by some, but excluded by others, is water in the atmosphere, which includes water vapor, clouds, and all forms of precipitation while still in the atmosphere.

hydrospire An infolded, thin-walled elongate calcareous structure in the interior of a blastoid theca, beneath and parallel to the ambulacral border. Its function is apparently respiratory.

hydrospire pore One of numerous minute rounded *pores* between the side plates or outer side plates near the margin of an ambulacrum, leading into a hydrospire of a blastoid or connecting the space enclosed by a hydrospire with the exterior.

hydrospire slit A longitudinal opening of a hydrospire in some fissiculate blastoids, excavated in the surface of deltoid and radial plates.

hydrostatic equilibrium In a fluid, the horizontal coincidence of the surfaces of constant pressure and constant mass; gravity and pressure are in balance.

hydrostatic head The height of a vertical column of water whose weight, if of unit cross section, is equal to the hydrostatic pressure at a given point; *static head* as applied to water. See also: *artesian head.*

hydrostatic level The level to which the water will rise in a well under its full pressure head. It defines the potentiometric surface. Syn: *static level.*

hydrostatic pressure [exp struc geol] Stress that is uniform in all directions, e.g. beneath a homogeneous fluid, and causes dilation rather than distortion in isotropic materials.

hydrostatic pressure [hydraul] The pressure exerted by the water at any given point in a body of water at rest. The hydrostatic pressure of ground water is generally due to the weight of water at higher levels in the zone of saturation (Meinzer, 1923, p. 37) See also: *artesian pressure.*

hydrostatics The aspect of *hydromechanics* that deals with forces that produce equilibrium. Cf: *hydrodynamics; hydrokinetics.*

hydrostatic stress A state of stress in which the normal stresses acting on any plane are equal and where shearing stresses do not exist in the material.

hydrostratigraphic unit A term proposed by Maxey (1964, p. 126) for a body of rock having considerable lateral extent and composing "a geologic framework for a reasonably distinct hydrologic system". Cf: *geohydrologic unit.*

hydrotachylyte A volcanic glass similar to *tachylyte* but containing as much as 13% water (Johannsen, 1937, p. 290).

hydrotalcite A pearly-white rhombohedral mineral: $Mg_6Al_2(CO_3)(OH)_{16} \cdot 4H_2O$. It is dimorphous with manasseite.

hydrothermal Of or pertaining to hot water, to the action of hot water, or to the products of this action, such as a mineral deposit precipitated from a hot aqueous solution, with or without demonstrable association with igneous processes; also, said of the solution itself. "Hydrothermal" is generally used for any hot water but has been restricted by some to water of magmatic origin.

hydrothermal alteration Alteration of rocks or minerals by the reaction of hydrothermal water with pre-existing solid phases.

hydrothermal deposit A mineral deposit formed by precipitation of ore and gangue minerals in fractures, faults, breccia openings, or other spaces, by replacement or open-space filling, from watery fluids ranging in temperature from 50° to 700°C but generally below 400°C, and ranging in pressure from 1 to 3 kilobars. The fluids are of diverse origin. Alteration of host rocks is common.

hydrothermal metamorphism A local type of metamorphism caused by the percolation of hot solutions or gases through fractures, causing mineralogic changes in the neighboring rock. Syn: *hydrothermal metasomatism.*

hydrothermal metasomatism *hydrothermal metamorphism.*

hydrothermal processes Those processes associated with igneous activity that involve heated or superheated water, esp. alteration, space filling, and replacement.

hydrothermal stage That stage in the cooling of a magma during which the residual fluid is strongly enriched in water and other volatiles. The exact limits of the stage are variously defined by different authors, in terms of phase assemblage, temperature, composition, and/or vapor pressure; most definitions consider it as the last stage of igneous activity, coming at a later time, and hence at a lower temperature, than the *pegmatitic stage.*

hydrothermal synthesis Mineral synthesis in the presence of water at elevated temperatures.

hydrothermal water Subsurface water whose temperature is high enough to make it geologically or hydrologically significant, whether or not it is hotter than the rock containing it. It may include magmatic and metamorphic water, water heated by radioactive decay or by energy release associated with faulting, meteoric water that descends slowly enough to acquire the temperature of the rocks in accordance with the normal geothermal gradient but then rises more quickly so as to retain a distinctly above-normal temperature as it approaches the surface, meteoric water that descends to and is heated by cooling intrusive rocks, water of geopressured aquifers, and brine that accumulates in an area of restricted circulation at the bottom of a sea.

hydrotomical axis A major axis with large spines in the skeleton of an acantharian radiolarian. Cf: *geotomical axis.*

hydrotroilite A black, finely divided colloidal material: $FeS \cdot nH_2O$. It is perhaps formed by bacteria on bottoms of marine basins characterized by reducing conditions and restricted circulation; it quickly changes to more stable pyrite.

hydrotungstite A mineral: $H_2WO_4 \cdot H_2O$.

hydrougrandite A mineral of the garnet group: $(Ca,Mg,Fe)_3(Fe,Al)_2(SiO_4)_{3-x}(OH)_{4x}$.

hydrous Said of a mineral compound containing water.

hydrous mica *hydromica.*

hydroxide A type of *oxide* characterized by the linkage of a metallic element or radical with the ion OH, such as brucite, $Mg(OH)_2$.

hydroxyapatite *hydroxylapatite.*

hydroxylapatite (a) A rare mineral of the apatite group: $Ca_5(PO_4)_3(OH)$. (b) An apatite mineral in which hydroxyl predominates over fluorine and chlorine.—Syn: *hydroxyapatite.*

hydroxyl-bastnaesite A wax-yellow to dark-brown mineral: $(Ce,La)CO_3(OH,F)$.

hydroxyl-ellestadite A mineral of the apatite group: $Ca_{10}(SiO_4)_3(SO_4)_3(OH,Cl,F)_2$.

hydroxyl-herderite A monoclinic mineral: $CaBe(PO_4)(OH)$. It is isomorphous with herderite. Syn: *hydroherderite.*

hydrozincite A white, grayish, or yellowish mineral: $Zn_5(CO_3)_2(OH)_6$. It is a minor ore of zinc and is found in the upper (oxidized) zones of zinc deposits as an alteration product of sphalerite. Syn: *zinc bloom; calamine; earthy calamine.*

hydrozoan Any coelenterate belonging to the class Hydrozoa, characterized by forms, usually colonial and more specialized than sponges, that are both polypoid and medusoid or exclusively medusoid; and by the absence of nematocysts and a stomodaeum. Range, Precambrian or Lower Cambrian to present.

hyetal Pertaining to rain, rainfall, or rainy regions; e.g. a *hyetal* interval, or the difference in rainfall between two *isohyets.* Cf: *pluvial [meteorol].*

hyetometer *rain gage.*

hygrograph A self-recording *hygrometer.*

hygrometer An instrument that is used to measure the humidity of the air. See also: *hygrograph.*

hygrophilous Said of an organism that lives in moist areas. Syn: *hygrophile; hygrophilic.*

hygrophyte *hydrophyte.*

hygroscopic capacity *hygroscopic coefficient.*

hygroscopic coefficient The ratio of the weight of water that a completely dry mass of soil will absorb if in contact with a saturated atmosphere until equilibrium is reached, to the weight of the dry soil mass, expressed as a percentage. See also: *hygroscopic water.* Syn: *hygroscopic capacity.*

hygroscopicity "The quantity of water absorbed by dry soil in a secluded space above 10 percent sulphuric acid at room temperature (about 18°C), expressed as a percentage of the weight of dry soil" (Schieferdecker, 1959, term 0356); the ability of a soil to absorb and retain water.

hygroscopic moisture *hygroscopic water.*

hygroscopic water Moisture held in the soil that is in equilibrium with that in the atmosphere to which the soil is exposed. Syn: *hygroscopic moisture; hydroscopic water.* See also: *hygroscopic coefficient.*

hyoid apparatus In the vertebrates, that part of the visceral skeleton that supports the tongue and related structures; elements of the hyoid apparatus serve to brace the jaws and gill cover in fish, and become the stapes of tetrapods.

hyolithid An invertebrate animal assigned to the order Hyolithida, characterized by bilateral symmetry, with a closed tapering shell that is triangular in cross section, has a conical embryonic chamber not separated from the rest of the shell, and has an operculum possessing two pairs of bilaterally symmetrical muscle scars. Hyolithids are questionably assigned to the mollusks, but have been variously identified as worms, pteropods, cephalopods, and *incertae sedis.* Range, Lower Cambrian to Middle Permian.

hyostylic Adj. Pertaining to a jaw suspension in which the cartilaginous or cartilage-replacement skeleton of the upper jaw is movably articulated to the neurocranium, primarily via elements of the visceral skeleton.

hypabyssal Pertaining to an igneous intrusion, or to the rock of that intrusion, whose depth is intermediate between that of *abyssal* or *plutonic* and the surface. This distinction is not considered relevant by some petrologists (Stokes and Varnes, 1955). Syn: *subvolcanic.*

hypautochthony (a) Accumulation of plant remains that no longer occur in the exact place of their growth, but still within the same general area, as in a peat bog. Cf: *euautochthony.* (b) A term sometimes used as a syn. of *allochthony.*

hypautomorphic A syn. of *subautomorphic* and *hypidiomorphic.* The term *hypautomorphisch* was proposed by Rohrbach (1885, p. 17-18) and has priority, but is rarely used. See also: *subhedral.*

hypautomorphic-granular *hypidiomorphic; subautomorphic.*

hyper- A prefix meaning "very, greatly".

hyperborean Pertaining or relating to the far north; of a frigid northern region.

hypercline Said of the dorsal and anterior *inclination* of the cardinal area in the brachial valve of a brachiopod, lying in the top right or second quadrant moving clockwise from the orthocline position (TIP, 1965, pt.H, p.60, fig.61).

hypercyclothem A term proposed by Weller (1958a, p.203-204) for a great cyclic sequence consisting of four *megacyclothems* and an alternating detrital sequence "of more than ordinary thickness and complexity".

hyperfusible n. Any substance capable of lowering the melting ranges in end-stage magmatic fluids. Syn: *hyperfusible component.*

hypergene *supergene.*

hypergenesis A term introduced by Fersman (1922), and persisting to the present day in Russian geology, for surficial alteration (weathering) of sedimentary rocks. Syn: *retrograde diagenesis; regressive diagenesis; retrodiagenesis.*

hyperglyph A hieroglyph formed during weathering (Vassoevich, 1953, p.33).

hyperite A plutonic rock composed of orthopyroxene, plagioclase, olivine, and clinopyroxene, being intermediate in composition between *gabbro* and *norite.* See also: *hyperite texture.* Not recommended usage.

hyperite texture The texture characteristic of *hyperite,* in which a fibrous amphibole reaction rim is formed at the contacts between olivine and plagioclase grains (Johannsen, 1939, p.257).

hypermelanic Said of igneous rocks that consist of 90-100% mafic minerals. Cf: *melanocratic; ultramafic.*

hypermorphosis *anaboly.*

hyperpiestic water A class of *piestic water* including waters that rise above the land surface. Cf: *hypopiestic water; mesopiestic water.*

hyperpycnal inflow Flowing water that is denser than the body of water it enters, resulting in formation of a turbidity current. Its flow pattern is that of a *plane jet* (Moore, p. 89). Cf: *hypopycnal inflow; homopycnal inflow.*

hypersaline Excessively saline; with a salinity substantially greater than that of normal sea water. Specif., having a salinity above the lowest at which halite can be precipitated.

hypersolvus Said of those granites, syenites, and nepheline syenites that are characterized by the absence of plagioclase except as a component of perthite (Tuttle and Bowen, 1958). Cf: *solvus; subsolvus.*

hypersthene A common rock-forming mineral of the orthopyroxene group: $(Mg,Fe)SiO_3$. It is isomorphous with enstatite. Hypersthene is grayish, greenish, black, or dark brown, and often has a bronze or greenish-brown play of color (*schiller*) on the cleavage surface. It is an essential constituent of many igneous rocks (gabbros, andesites). Symbol Hy.

hypersthene basalt *Basalt* that is silica-saturated, containing normative hypersthene and diopside, with neither quartz nor olivine. The term was defined by Yoder and Tilley in 1962.

hypersthenfels *norite.*

hypersthenite Originally defined as a syn. of *norite,* but now commonly used to mean a rock composed entirely of hypersthene; an *orthopyroxenite.* Not recommended usage.

hyperstomial Said of a cheilostome bryozoan ovicell that rests on or is partly embedded in the distal zooid and that opens above the operculum of the maternal zooid.

hyperstrophic Said of a rare gastropod shell in which the whorls are coiled on an inverted cone so that the apex points forward rather than backward, and the spire is depressed instead of elevated. A hyperstrophic shell is not easily distinguished from an *orthostrophic* shell unless the aperture shows the siphon pointed in the same direction as the apex (Beerbower, 1968, p.341).

hypersubsidence A phase of subsidence that is attained "when the algebraic sum of terms in the epeirogenic cycle is equivalent to a final basement depression of 15,000 feet, an amount which may be definitive in the quantitative classification of major epeirogenic sedimentary basins" (Kamen-Kaye, 1967, p. 1838).

hypertely Evolution to the extreme of being a disadvantage.

hyperthermic temperature regime A soil temperature regime in which the mean annual temperature (measured at 50cm) is at least 22°C, with a summer-winter variation of at least 5°C (USDA, 1975). Cf: *isohyperthermic temperature regime.*

hypertrophy Excessive growth of a species.

hypervelocity impact The impact of a projectile onto a surface at a velocity such that the stress waves produced on contact are orders of magnitude greater than the static bulk compressive strength of the target material. The minimum required velocities vary for different materials, but are generally 1-10 km/sec, and about 4-5 km/sec for most crystalline rocks. In such an impact, the kinetic energy of the projectile is transferred to the target material in the form of intense shock waves, whose interactions with the surface produce a crater much larger in diameter than the projectile. Meteorites striking the Earth at speeds in excess of about 5 km/sec give examples of large hypervelocity impacts and produce correspondingly large craters (Dietz, 1959, p. 499).

hypha One of the individual tubular filaments or threads that make up the *mycelium* of a fungus; e.g. a *conidiophore.* Pl: *hyphae.*

hypidioblast A mineral grain that is newly formed by metamorphism and is bounded only in part by its characteristic crystal faces. It is a type of *crystalloblast.* Cf: *idioblastic; xenoblastic.* Syn: *subidioblast.*

hypidioblastic Pertaining to a *hypidioblast* of a metamorphic rock; also, said of such a texture. It is analogous to the term *hypidiomorphic* in igneous rocks. Cf: *idioblastic.*

hypidiomorphic A syn. of *subautomorphic,* originally proposed by Rosenbusch (1887, p. 11) to describe in an igneous rock the individual mineral crystals (now termed *subhedral*) that are bounded only in part by their own *rational faces.* Syn: *subidiomorphic.*

hypidiomorphic-granular *hypidiomorphic; subautomorphic; subhedral.*

hypidiotopic Intermediate between *idiotopic* and *xenotopic;* esp. said of the fabric of a crystalline sedimentary rock in which the majority of the constituent crystals are subhedral. Also, said of the rock (such as an evaporite, a chemically deposited cement, or a recrystallized limestone or dolomite) with such a fabric. The term was proposed by Friedman (1965, p.648).

hypo- A prefix meaning "under, nearly".

hypobatholithic Said of a mineral deposit occurring in the deeply eroded region of a batholith; also, said of that stage of batholith erosion (Emmons, 1933). The term is little used. Cf: *acrobatholithic; cryptobatholithic; embatholithic; endobatholithic; epibatholithic.*

hypocenter *focus [seis].*

hypocotyl The portion of the axis of a plant embryo or seedling below the attachment of the cotyledons (seed leaves) and above the root.

hypocrystalline Said of the texture of an igneous rock that has crystalline components in a glassy groundmass, the ratio of crystals to glass being between 7:1 and 5:3.

hypocrystalline-porphyritic Said of the porphyritic texture of an igneous rock having a hypocrystalline groundmass.

hypodeltoid The anal deltoid on the aboral side of the anus in many blastoids. In some Pennsyvanian and Permian fissiculate blastoids, it is either small and loosely attached or completely atrophied.

hypodermalium A specialized sponge spicule of the cortex, lying largely beneath the exopinacoderm. Cf: *autodermalium.*

hypodermis (a) A reticulate layer beneath the *epidermis* in the walls of certain foraminifera. (b) The cellular layer that underlies and secretes the external chitinous membrane of arthropods.

hypodigm A group consisting of all the specimens used by an author as the basis for his description of a species, and all specimens later referred to it (Simpson, 1940, p. 18). Cf: *type material.*

hypogastralium A specialized sponge spicule lying largely beneath the endopinacoderm of the spongocoel. Cf: *autogastralium.*

hypogeal *hypogene.*

hypogeic *hypogene.*

hypogene (a) Said of a geologic process, and of its resultant features, occurring within and below the crust of the Earth. Cf: *epigene.* Syn: *hypogenic; hypogeal; hypogeic.* (b) Said of a mineral deposit formed by ascending solutions; also, said of those solutions and of that environment. Cf: *supergene; mesogene.* (c) A rarely used syn. of *plutonic.*

hypogenesis The direct development of an organism without alternation of generations.

hypogenic *hypogene.*

hypoglyph A hieroglyph on the bottom of a sedimentary bed (Vassoevich, 1953, p.37). Cf: *epiglyph.*

hypogynous Said of flowers in which the ovary surmounts a receptacle, so that the sepals, petals, and stamens radiate from below the locules. Cf: *epigynous; perigynous.*

hypohyaline Said of the texture of an igneous rock that has crystalline components in a glassy groundmass, with a ratio of crystals to glass between 3:5 and 1:7.

hypolimnetic Pertaining to a *hypolimnion.* Syn: *hypolimnial.*

hypolimnion The lowermost layer of water in a lake, characterized by an essentially uniform temperature (except during a *turnover*) that is generally colder than elsewhere in the lake, and often by relatively stagnant or oxygen-poor water; specif. the dense layer of water below the *metalimnion* in a thermally stratified lake. See also: *clinolimnion; bathylimnion.* Cf: *epilimnion.*

hypolithic Said of a plant that grows beneath rocks.

hypomagma Relatively immobile, viscous lava that forms at depth beneath a shield volcano, is undersaturated with gases, and gives rise to volcanic activity. Cf: *epimagma; pyromagma.*

hyponome The muscular tube, nozzle, or swimming funnel just below the head of a cephalopod, extending externally from the mantle cavity, through which water is expelled from the mantle cavity.

hyponomic sinus The large, concave ventral notch or re-entrant in the middle of the aperture of a cephalopod, marking the location through which the hyponome protrudes.

hyponym The name of a plant *taxon* that cannot be recognized because of the vagueness or inadequacy of its description (Cowan, 1968, p. 48).

hypoparian adj. Of or concerning a trilobite that lacks facial sutures and that is generally blind.—n. A hypoparian trilobite; specif. a trilobite of the order Hypoparia (now obsolete).

hypopiestic water A class of *piestic water* including waters that rise above the bottom of the upper confining bed but not as high as the water table. Cf: *hyperpiestic water; mesopiestic water.*

hypopycnal inflow Flowing water that is less dense than the body of water it enters, e.g. a river entering the ocean. Its flow pattern is that of an *axial jet.* (Moore, 1966, p. 89). Cf: *hyperpycnal inflow; homopycnal inflow.*

hyposeptal deposit A distal cameral deposit on the convex (adapical) side of septum of a nautiloid. Ant: *episeptal deposit.*

hyposome The posterior part of the cell body below the girdle of an unarmored dinoflagellate. Cf: *episome.*

hypostega A part of the body cavity of a cheilostome bryozoan zooid between a cryptocyst or an umbonuloid frontal shield and the overlying membranous wall.

hypostegal coelom The confluent coelomic spaces beyond the ends of zooidal walls in stenolaemate bryozoans. Also, a syn. of *hypostega.*

hypostoma A syn. of *hypostome.* Pl: *hypostomata.*

hypostomal suture The line of junction in a trilobite between the anterior margin of the hypostome and the posterior margin of the frontal doublure or rostral plate.

hypostome (a) A ventral plate of the head region behind and above which the mouth of a trilobite is located. Cf: *metastoma.* Syn:

labrum. (b) A vase-shaped or conical process bearing the mouth of a hydrozoan.—Syn: *hypostoma.*

hypostracum A term used originally for the inner layer of the shell wall of a bivalve mollusk, secreted by the entire epithelium of the mantle, but also applied in a later sense to the *myostracum.* Cf: *ostracum.*

hypostratotype A stratotype designated to extend knowledge of an established unit or boundary to other geographical areas or to other facies. It is always subordinate to the *holostratotype* (ISG, 1976, p. 26). See also: *reference section.*

hypotheca (a) The posterior part of a dinoflagellate theca, below the girdle. Cf: *epitheca.* (b) *hypovalve.*

Hypothermal n. A term proposed by Cooper (1958, p.944) for a postglacial interval (the last 2600 years) characterized by a moderate decrease in temperature, some limited glacial expansions, and inferred eustatic lowerings of sea level.

hypothermal Said of a hydrothermal mineral deposit formed at great depth and in the temperature range of 300°-500°C (Park & MacDiarmid, 1970, p. 293). Also, said of that environment. Cf: *mesothermal; epithermal; leptothermal; telethermal; xenothermal.* Syn: *katathermal.*

hypothesis A conception or proposition that is tentatively assumed, and then tested for validity by comparison with observed facts and by experimentation; e.g. the "planetesimal hypothesis" and the "nebular hypothesis" to explain the evolution of the planets. It is less firmly founded than a *theory.*

hypothetical resources Undiscovered mineral resources that we may still reasonably expect to find in known mining districts (Brobst & Pratt, 1973, p. 4). Cf: *identified resources; speculative resources.*

hypothyridid Said of a brachiopod pedicle opening that is located below or on the dorsal side of the beak ridges with the umbo intact (TIP, 1965, pt.H, p.146).

hypotract The part of a dinoflagellate cyst posterior to the girdle region. Ant: *epitract.*

hypotype A described or figured specimen used in extending or correcting the knowledge of a *species,* or in other publications regarding it. See: Frizzell, 1933, p. 653.

hypovalve The inner valve of a diatom frustule. Cf: *epivalve.* Syn: *hypotheca.*

hypoxenolith A xenolith derived from a source more remote than the adjacent wall rock (Goodspeed, 1947, p. 1251). Cf: *epixenolith.*

hypozygal The proximal brachial plate of a pair joined by syzygy in a crinoid. Cf: *epizygal.*

Hypsithermal n. A term proposed by Deevey & Flint (1957) as a substitute for *climatic optimum* and *thermal maximum.* It represents the Holocene interval when "most of the world entered a period when mean annual temperatures exceeded those of the present". The Hypsithermal was taken to include the Boreal, Atlantic, and Subboreal climatic intervals, or from about 9000 to 2500 years ago. See also: *Megathermal; Xerothermic.*—adj. Pertaining to the postglacial Hypsithermal interval and to its climate, deposits, biota, and events.

hypsographic curve A cumulative-frequency profile representing the statistical distribution of the absolute or relative areas of the Earth's solid surface (land and sea floor) at various elevations above, or depths below, a given datum, usually sea level. Syn: *hypsometric curve.*

hypsography (a) A branch of geography dealing with the observation and description of the varying elevations of the Earth's surface with reference to a given datum, usually sea level. Cf: *hypsometry.* (b) Topographic relief. Also, the parts of a map, collectively, that represent topographic relief. (c) The portrayal of topographic relief on maps.

hypsometer An instrument used in estimating the elevation of a point on the Earth's surface in relation to sea level by determining atmospheric pressure through observation of the boiling point of water at that point. It is useful in mountainous or high-altitude regions. Syn: *thermobarometer.*

hypsometric Pertaining to hypsometry or to elevation above a datum; esp. relating to elevations above sea level determined with a hypsometer.

hypsometric analysis The measurement of the distribution of ground surface area (or horizontal cross-sectional area) of a landmass with respect to elevation (Strahler, 1952b, p.1118). Syn: *area-altitude analysis.* Cf: *hypsometry.*

hypsometric curve *hypsographic curve.*

hypsometric integral Proportionate area below the *percentage hypsometric curve;* it expresses the relative volume of a landmass at a given contour. In the study of drainage basins, Strahler (1952b, p.1121) used this term to express the ratio of the volume of earth material to the volume of the solid reference figure having a base equal to basin area and a height equal to maximum basin relief. Symbol: I. Cf: *erosion integral.* Pike and Wilson (1971) have shown that the hypsometric integral is identical to the *elevation-relief ratio* (E) (Wood & Snell, 1960) and is very easily calculated: E = (mean elevation - minimum elevation) ÷ (maximum elevation - minimum elevation).

hypsometric map Any map showing relief by means of contours, hachures, shading, tinting, or any other convention.

hypsometric tint A color applied to the area between two selected contour lines on a map of an area whose relief is depicted by layer tinting. Syn: *layer tint.*

hypsometry The science of determining, by any method, height measurements on the Earth's surface with reference to sea level; e.g. "barometric hypsometry" in which elevations are determined by means of mercurial or aneroid barometers. Cf: *hypsography; hypsometric.*

hyrst *hurst.*

hysteresis (a) A lag in the return of an elastically deformed body to its original shape after the load has been removed. (b) The property that a rock is said to exhibit when its magnetization is nonreversible. Syn: *magnetic hysteresis.* (c) A phase lag of dielectric displacement behind electric-field intensity, due to energy dissipation in polarization processes.

hysterobase A *diabase* composed of plagioclase, quartz, biotite, and brown hornblende that is paramorphic after augite. Not recommended usage.

hystero-brephic stage The earliest stage of development of the offset during increase in colonial corals. It commences with the initial modification of the parent corallite and normally includes insertion of at least some of the primary septa in the offset (Fedorowski & Jull, 1976, p. 42). Cf: *hystero-neanic stage; late neanic stage.*

hysterocrystalline An obsolete term applied to a mineral produced in an igneous rock as a result of secondary crystallization.

hysterogenetic Said of the last crystallization products of a magma; e.g. dikes.

hysterogenous A little-used term pertaining to a mineral deposit on the Earth's surface formed from the debris of other rocks. Syn: *hysteromorphous.* Cf: *idiogenous; xenogenous.*

hysteromorphous *hysterogenous.*

hystero-neanic stage The stage of development of the offset following the hystero-brephic stage during increase in colonial corals. It usually is taken to commence with the appearance of the first metaseptum in the offset and commonly includes completion of the partition or dividing wall separating offset from parent corallite (Fedorowski & Jull, 1976, p. 42). Cf: *hystero-brephic stage; late neanic stage.*

hystero-ontogeny Changes that occur in the offset, or asexually developed corallite, during its development (Fedorowski & Jull, 1976, p. 40). Cf: *blastogeny; astogeny.*

hysterosoma That section of the body of an acarid that is behind the second pair of legs.

hystrichosphaerid A general term formerly used for a great variety of resistant-walled organic microfossils, ranging from Precambrian to Holocene and characterized by spherical to ellipsoidal, usually spinose, remains found among fossil microplankton. These are now divided among the *acritarchs* and *dinoflagellate* cysts. The term has no formal taxonomic status. Syn: *hystrichosphere.*

hystrichosphere *hystrichosphaerid.*

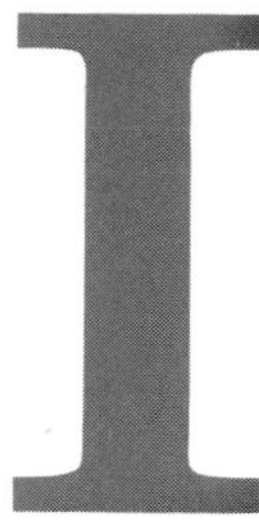

ianthinite (a) A violet-black orthorhombic secondary mineral: $UO_2 \cdot 5UO_3 \cdot 10H_2O$). (b) A mineral name that was erroneously given to *wyartite.*

IASY *International Active Sun Years.*

ice (a) Water in the solid state; specif. the dense substance formed in nature by the freezing of liquid water, by the condensation of water vapor directly into ice crystals, or by the recrystallization or compaction of fallen snow. It is colorless to pale blue or greenish blue, usually white from included gas bubbles. At standard atmospheric pressure, it is formed at and has a melting point of 0°C; in freezing it expands about one eleventh in volume. Ice commonly occurs as hexagonal crystals, and in large masses is classed as a rock. (b) A term often substituted for *glacier,* as in "continental ice".

ice age A loosely used syn. of *glacial epoch,* or time of extensive glacial activity; specif. the latest of the glacial epochs, also known as the Pleistocene Epoch.

ice apron The thin mass of snow and ice attached to the headwall of a cirque above the *bergschrund* (but not present at a *randkluft*). Syn: *apron [glaciol].*

ice avalanche A sudden fall, down a steep slope, of ice broken from an ice sheet or glacier (most commonly a hanging glacier). Syn: *icefall.*

ice bar An ice edge consisting of floes compacted by wind, sea, and swell, and difficult to penetrate.

ice barchan A small crescentic dune composed of ice crystals. Rarely used.

ice barrier [glaciol] A syn. of *ice front* and *ice shelf.* The term was introduced by Sir James C. Ross in 1841 to designate the high, steep, seaward cliff face or edge of a great ice mass in Antarctica because it obstructed navigation; the syn. "ice front" is now preferred. The term ice barrier was later applied to the entire mass of ice, and was widely adopted for similar polar morphologic features, esp. for an immense ice formation of very great extent; the term "ice shelf" is now used for this type of ice mass. Syn: *barrier [glaciol].*

ice barrier [ice] A rarely used syn. of *ice dam.*

ice-barrier lake A lake formed in a mountain valley whose lower end is dammed by a glacier descending another valley.

ice-basin lake A lake, pond, or pool on sea ice or on glacier ice (ADTIC, 1955, p. 41). Rarely used.

ice bay *bight [ice].*

iceberg A large, massive piece of floating or stranded glacier ice of any shape, detached (calved) from the front of a glacier into a body of water. An iceberg extends more than 5 m above sea level and has the greater part of its mass (4/5 to 8/9) below sea level. It may reach a length of more than 80 km. Syn: *berg [glaciol].* Cf: *floeberg.*

iceberg tongue "A major accumulation of icebergs projecting from the coast, held in place by grounding, and joined together by fast ice" (U.S. Naval Oceanographic Office, 1968, p.B34).

ice blade A "crest" or "spire" of ice, 0.5-1.5 m high, rising from a surface of firn, and formed by unequal melting (Russell, 1885, p.318).

iceblink [glaciol] A cliff extending along the seaward margin of a mass of inland ice. Examples are found on the coast of Greenland. Syn: *isblink.*

iceblink [meteorol] A relatively bright, usually yellowish or whitish glare in the sky near the horizon or on the underside of a cloud layer, produced in a polar region by light reflected from a large ice-covered surface (as an ice sheet) that may be too far away to be visible; not as bright as *snowblink.* Also spelled: *ice blink.*

ice blister *ice mound.*

ice-block ridge A ridge, either closed or linear, surrounding or separating depressions in a moraine (Deane, 1950, p. 14).

ice boulder [glac geol] An obsolete syn. of *glacial boulder.*

ice boulder [ice] A large fragment of sea ice shaped by wave action into a nearly spherical form and then stranded on the shore.

ice boundary "The demarcation at any given time between fast ice and pack ice or between areas of pack ice of different concentrations" (U.S. Naval Oceanographic Office, 1968, p. B34). Cf: *ice edge.*

ice breccia Pieces of ice of different ages, or angular fragments of glacier ice of comparable age, frozen together.

ice cake A *floe* or piece of floating sea ice less than 10 m across. Syn: *block [ice]; cake [ice].*

ice canopy A submariner's term for *pack ice.*

ice cap A dome-shaped or platelike cover of perennial ice and snow, covering the summit area of a mountain mass so that no peaks emerge through it, or covering a flat landmass such as an Arctic island; spreading due to its own weight outwards in all directions; and having an area of less than 50,000 sq km. An ice cap is considerably smaller than an *ice sheet.* Cf: *ice field [glaciol]; ice sheet; glacier.* Nonpreferred syn: *ice carapace; cap [glaciol]; glacier cap.* Also spelled: *icecap.*

ice carapace Nonpreferred syn. of *ice cap.*

ice cascade *icefall.*

ice cast A shell of ice formed around a beach pebble as a result of the wetting action of spray, tides, and waves, and subsequent freezing; the ice is sometimes separated from the pebble.

ice cauldron A wide area in a valley, upon which glacier ice once piled up so high as to flow radially outward through pre-existing passes that were deepened by glacial scour.

ice cave (a) An artificial or natural cave, in a temperate climate, in which ice forms and persists throughout all or most of the year. Syn: *ice grotto; glacière.* (b) A nonrecommended syn. of *glacier cave.*

ice-channel filling An inclusive term for *esker* and *crevasse filling,* to eliminate arbitrary and commonly unfounded distinctions between them. Emphasizes that linear ridges of stratified debris may form in tunnels under the ice, in valleys open to the sky either superglacial or on the substrate, as beaded or continuous deltaic or fluvial forms in standing water at mouths of tunnels ("DeGeer eskers"), or as kettle rims between closely spaced kettle holes in outwash (Jahns, 1953).

ice clearing (a) The end phase of breakup. (b) *polynya.*—Rarely used.

ice cliff Any vertical wall of ice; e.g. a very steep surface bounding a glacier or a mass of shelf ice. Nonrecommended syn: *ice front; ice wall.* Syn: *ice face.*

ice cluster A concentration of sea ice covering hundreds of square kilometers and found in the same region every summer. See also: *pack ice.* Syn: *ice massif.*

ice column *ice pillar.*

ice concrete A dense frozen mixture of sand, rock fragments, and ice. Syn: *icecrete.*

ice cone *ice pyramid.*

ice-contact delta A delta built by a stream flowing into a lake between a valley slope and the margin of glacier ice. Syn: *delta kame; kame delta.* Obsolete syn: *morainal delta.*

ice-contact deposit Stratified drift deposited in contact with melting glacier ice, such as an esker, a kame, a kame terrace, or a feature marked by numerous kettles.

ice-contact plain *kame plain.*

ice-contact slope The steep slope of sediment that was deposited against a wall of glacier ice, marking the position of the ice margin; an irregular scarp against which glacier ice once rested.

ice-contact terrace *kame terrace.*

ice-contorted ridge *ice-pushed ridge.*

ice cover (a) The extent of glacier ice on a land surface at the present time, with special reference to its thickness. (b) The ratio of an area of sea ice of any *concentration* to the total area of sea surface within some large geographic locale that may be global, hemispheric, or specific to a given study (U.S. Naval Oceanographic Office, 1968, p. B34). (c) The extent of ice on a lake surface.

ice-covered *glacier-covered.*

ice crack *frost crack.*

ice-crack moraine A linear ridge consisting of very sandy, unstratified drift believed to have been deposited between blocks of dead ice in a disintegrating glacier (Sproule, 1939, p. 104).

ice creek A narrow inlet or rift in the seaward part of an *ice shelf,* sometimes extending many kilometers in from the *ice front.* Ice creeks are ephemeral, but may recur in the same place (Armstrong et al., 1977).

icecrete *ice concrete.*

ice crust [glaciol] A type of *snow crust,* formed when meltwater or rainwater freezes to form a continuous layer of ice on the surface of deposited snow. See also: *ice layer [glaciol].*

ice crust [ice] (a) *ice rind.* (b) A thin layer of ice on a rock surface, formed by freezing of water condensed from the air. (c) A layer of ice 2 to 4 mm thick on a snow pack.

ice crystal (a) A macroscopic particle of ice with a regular structure, usually hexagonal; it is anisotropic. (b) *frazil crystal.*

ice-crystal cast A *crystal cast* formed by the filling of an ice-crystal mark with mud or sand; it commonly appears as a straight, slightly raised ridge on the underside of a sandstone bed.

ice-crystal mark A crack formed on a sedimentary surface by the sublimation of a crystal of ice.

ice dam A river obstruction formed of floating blocks of ice that may cause ponding and widespread flooding during spring and early summer. Syn: *ice barrier [ice].*

ice-dammed lake *glacier lake.*

ice dendrite A thin branching ice crystal. Dendritic ice sometimes forms the first skim of ice over still water, and also may grow under water from an existing ice surface.

iced firn A mixture of ice and firn; firn permeated with meltwater and then refrozen. Syn: *firn ice.*

ice dike A secondary formation of ice, usually made up of columnar crystals, filling a crevasse or other gash in glacier ice.

ice disintegration The process of breaking up a stagnant and wasting glacier into numerous small blocks: it is said to be "controlled" where the blocks are separated along fractures or other lines of weakness to form linear or lobate features, and "uncontrolled" where equal forces break up the glacier along cracks extending in all directions to produce round, oval, or rudely polygonal features (Gravenor & Kupsch, 1959, p.48-49). Syn: *disintegration [glaciol].*

ice dome (a) A rounded, gently sloping elevation in the surface of an inland *ice sheet.* Ice domes do not have precisely defined margins and may cover large areas — 100,000 sq km or more (Armstrong et al., 1977). (b) An accumulation of glacier ice in a caldera.

ice edge The boundary at any given time between open water and sea, lake, or river ice of any kind, whether fast or drifting. It may be "compacted" by wind or current, or it may be "diffuse" or "open" when dispersed or poorly defined. Cf: *ice boundary; ice limit.*

ice face *ice cliff.*

icefall (a) The part of a glacier that is highly crevassed because of a very steep slope of the glacier bed. Syn: *ice cascade.* Cf: *cascading glacier.* (b) *ice avalanche.*—Also spelled: *ice fall.*

ice fan Nonpreferred syn. of *expanded foot.*

ice field [glaciol] (a) An extensive mass of land ice covering a mountain region, consisting of many interconnected alpine and other types of glaciers, covering all but the highest peaks and ridges. Cf: *ice cap; highland glacier.* (b) A general, but not recommended, designation for a large and irregular body of glacier ice.—Also spelled: *icefield.*

ice field [ice] An extensive area of pack ice, consisting of floes, and greater than 10 km (6 mi) across; the largest areal subdivision of sea ice. Ice fields are subdivided according to horizontal extent as follows: "large" (over 20 km); "medium" (15-20 km); "small" (10-15 km). Cf: *ice patch.* Syn: *field [ice].*

ice field [permafrost] *icing.*

ice floe (a) A large fragment or extensive sheet of ice, detached and floating freely in open water. (b) *floe.*

ice flow *glacier flow.*

ice flower *frost flower.*

ice foot (a) The ice at the lower end or front of a glacier. (b) A mass or wall of ice formed by the freezing of snow that accumulated along the foot of a mountain slope. It is not formed from converging glaciers. (c) *icefoot.*

icefoot A narrow strip, belt, or fringe of ice formed along and firmly attached to a coast, unmoved by tides, and remaining after the fast ice has broken away; it is usually formed by the freezing of wind-driven spray, or of seawater during ebb tide. A true icefoot has its base at or below the low-water mark. Also used for ice at the foot of lake bluffs. Also spelled *ice foot.* Syn: *ballycadder; bellicatter; cadder; catter; kaimoo; shore-ice belt.*

ice-foot niche A hollow created at the base of a soft cliff (as of limestone) by floating sea ice; during spring it permits the collapse of the overhanging wall (Hamelin & Cook, 1967, p. 101).

ice free "No sea ice present. There may be some ice of land origin" (U.S. Naval Oceanographic Office, 1968, p. B34). Cf: *open water [ice].*

ice fringe (a) A very narrow *ice piedmont,* extending less than about one kilometer inland from the sea. (b) A belt of sea ice that extends a short distance offshore.

ice front (a) The floating vertical cliff forming the seaward edge of an ice shelf or other glacier that enters water, ranging in height from 2 to more than 50 m above sea level. Cf: *ice wall.* Syn: *ice barrier [glaciol]; front [glaciol].* (b) Nonrecommended syn. of *ice cliff* in the sense of any vertical wall of ice. (c) The *snout* of a glacier.

ice gang A rush of water following a breakup; a *debacle.*

ice gland A rudely cylindrical vertical column of ice or of iced firn in a firn field.

ice gneiss Frozen ground with ice segregated in laminae so as to resemble a gneissic rock; term used by Taber (1943).

ice gorge The vertical-walled opening left after an *ice jam* has broken through.

ice gouge A general term for the processes of ice interacting with the sea floor. See: Reimnitz et al., 1973, 1977.

ice grass *ice stalk.*

ice grotto *ice cave.*

ice gruel A type of *sludge* floating on the sea surface and formed by the irregular freezing together of frazil crystals.

ice hummock *hummock [ice].*

ice island A form of large *tabular iceberg* broken away from an ice shelf and found in the Arctic Ocean, having a thickness of 15 to 50 m and an area between a few thousand square meters and 500 sq km or even more. The surface of an ice island is usually marked by broad, shallow, regular undulations that give it a ribbed appearance from the air.

ice jam (a) An accumulation of broken river ice lodged in a narrow or obstructed part of the channel; it frequently produces local floods during a spring breakup. Cf: *ice gorge.* (b) An accumulation of large fragments of lake ice or sea ice thawed loose from the shore during early spring and subsequently piled up on or blown against the shore by the wind, often exerting great pressures.

ice keel The submerged mass of broken ice under a *pressure ridge,* forced downward by pressure, and extending as much as 50 m below sea level.

ice laccolith A small mound with a solid ice core resembling a laccolith in section. It is formed in the active layer by the freezing of water forced into a plano-convex lens where it is trapped between the downward-freezing active layer and the underlying permafrost. Ice laccoliths are produced in a few days or weeks during fall freeze-up in areas of continuous permafrost. Most are less than a meter high and a few meters to 10 m across. They may not melt for several years. They differ from many *hydrolaccoliths* that are not active-layer phenomena (Black, 1954, p. 848). Syn: *cryolaccolith; frost blister.*

ice-laid drift *till.*

Iceland agate A syn. of *obsidian,* applied to gem-quality varieties.

icelandite A term proposed by Carmichael in 1964 for intermediate lavas of Thingmuli volcano, Iceland, that, compared to calc-alkaline *andesites,* are low in aluminum, high in iron, and have fewer mafic phenocrysts.

Iceland spar A very pure and transparent variety of calcite, the best of which is obtained in Iceland, that cleaves easily and perfectly into rhombohedrons and that exhibits strong double refraction; an *optical calcite.* It occurs in vugs and cavities in volcanic rocks and as nodules in residual clays in limestone regions. Syn: *Iceland crystal; double-refracting spar.*

ice layer [glaciol] A layer of solid ice or iced firn in a mass of snow or firn, either a remnant *ice crust* that has been covered by snow or a layer formed by the freezing of water trapped on a relatively impermeable horizon such as a snow crust.

ice layer [permafrost] An approximately horizontal layer of ground ice, sometimes lenticular (see *ice lens [permafrost]*). Syn: *ground-ice layer.*

ice lens [glaciol] A horizontal ice layer that tapers out in all directions.

ice lens [permafrost] A discontinuous layer of *ground ice [permafrost]* tapering at the periphery; ice lenses in soil commonly occur parallel to each other in repeated layers.

ice limit (a) The extreme minimum or the extreme maximum extent of the *ice edge* in any given time period, based on observations over several years. The term should be prefaced by "minimum" or "maximum". (b) Obsolete syn. of *mean ice edge.*

ice mantle *ice sheet.*

ice-marginal drainage Stream drainage along the side or front of a glacier.

ice-marginal lake A lake of meltwater dammed between an ice barrier and the valley wall, or between ice lobes. Syn: *proglacial lake.*

ice-marginal terrace *kame terrace.*

ice-marginal valley A valley parallel to the margin of a glacier. It may serve for the draining-away of meltwater.

ice massif *ice cluster.*

ice mound (a) A *frost mound* containing bodies of ice. (b) *icing mound.* Syn: *ice blister.*

ice mountain A popular term for a large iceberg.

I-centered lattice *body-centered lattice.*

ice pack *pack ice.*

ice pan A large flat piece of first-year ice protruding several centimeters to a meter above the sea surface. Syn: *pan [ice].*

ice patch An area of pack ice smaller than an *ice field.* Syn: *patch [ice].*

ice pavement *glacial pavement.*

ice pedestal A pinnacle, column, or cone of ice projecting from the surface of a glacier and supporting, or formerly supporting, a large rock (*glacier table*) or mass of debris which protects the ice underneath it from solar radiation, so that it ablates less rapidly than the ice around it. See also: *mushroom ice.* Syn: *ice pillar; pedestal [glaciol].*

ice penitente A *nieve penitente* consisting of glacier ice.

ice period The period of time from freezeup to breakup of ice.

ice piedmont A mass of ice, sloping gently seaward, that covers a coastal strip of low-lying land backed by mountains (Armstrong & Roberts, 1956, p.7). It may be anywhere from one kilometer to 50 km wide, and its outer edge may be marked by a line of ice cliffs. Ice piedmonts frequently merge into *ice shelves.* See also: *ice fringe.*

ice pillar Any tall, narrow mass of ice, such as an *ice pedestal* or *mushroom ice*. Syn: *ice column.*

ice pipe An ice mass of cylindrical shape.

ice plateau (a) An ice-covered highland area whose upper surface is nearly level and whose sides descend steeply to lowlands or to the ocean. See also: *plateau glacier.* (b) Any ice sheet with a level or gently rounded surface, as the Polar Plateau in Antarctica.

ice pole The approximate center of the most consolidated part of the arctic pack ice, and therefore a difficult point to reach by surface travel; it is near lat. 83°-84°N and long. 160°W. Syn: *pole of inaccessibility.*

ice potential "The potential amount of ice that would be formed in a given water mass if surface heat loss provided the thermohaline circulation" (Baker et al., 1966, p. 84).

ice push (a) The lateral pressure exerted by the expansion of shoreward-moving ice, esp. of lake ice. Syn: *ice shove; ice thrust.* (b) The ridge of material formed by an ice push. Syn: *lake rampart; ice-push ridge.*

ice-pushed ridge An asymmetric ridge of local, essentially nonglacial material (such as deformed bedrock, with some drift incorporated in it) that has been pressed up by the shearing action of an advancing glacier. It is typically 10-60 m high, about 150-300 m wide, and as much as 5 km long. Examples are common on the Great Plains of North America, where such ridges occur on the sides of escarpments formed of relatively incompetent rocks that face the direction from which the ice moved. The term is sometimes used "not quite correctly" as a synonym of *push moraine,* which is an accumulation of glacial drift (Schieferdecker, 1959, term 0924). Syn: *ice-contorted ridge; ice-thrust ridge; glacial pressure ridge; pressure ridge [glaciol].*

ice-push ridge *lake rampart.*

ice-push terrace A terracelike accumulation of coarse material pushed up along a shore; esp. a terrace consisting of successive *lake ramparts.* Syn: *ice-pushed terrace.*

ice pyramid A roughly conical mound of ice on the surface of a glacier, formed by differential ablation; e.g. an ice pedestal whose sides have been melted back, making a cone shape. Syn: *ice cone.*

ice-rafted Said of material such as boulders or till, which is deposited by the melting of floating ice containing it; esp. said of clasts and till distributed widely in marine sediments.

ice-rafting The transporting of rock fragments of all sizes on or within icebergs, ice floes, or other forms of floating ice. See also: *rafting.*

ice ramp An accumulation of snow forming a slope connecting two levels, including those which may build up on *fast ice* against *ice fronts, ice walls,* or rock cliffs. Ice ramps can also be artificially constructed or excavated (Armstrong et al., 1977).

ice rampart A syn. of *lake rampart.* The term is misleading because the rampart is a wall of boulders and other coarse material, not of ice.

ice-receiving area The portion of a surging glacier, generally near the terminus, that is periodically refilled by glacier surges.

ice-reservoir area That portion of a glacier that is periodically drained by glacier surges. This reservoir, which is refilled by direct snow accumulation or by normal ice flow between surges, may be located in nearly any part of the glacier system.

ice ribbon A thin, white, curly deposit of ice growing to a length of 10 cm or more, formed by the freezing of moisture exuded from the dead stem of a plant just above the ground in the early frosty period of winter before the ground is thoroughly frozen.

ice rind A brittle, thin sea ice formed on a quiet surface by direct freezing or from grease ice, usually in water of low salinity; its thickness is generally less than 5 cm, and it is easily broken into rectangular pieces by wind or swell (U.S. Naval Oceanographic Office, 1968, p. B34). Syn: *ice crust [ice]; crust [ice].*

ice rise An ice mass, usually dome-shaped, resting on unexposed rock, and surrounded by an ice shelf, or in part by an ice shelf and in part by sea or ice-free land or both. Cf: *ice rumples.*

ice rumples A locally grounded area of *ice shelf* which is overridden by an *ice sheet.* Ice rumples are distinguished by crevassing, and a rise of as much as 50 m above the ice-shelf surface level. Ice may be deflected or even halted by ice rumples (Armstrong et al., 1977).

ice run (a) Movement of ice floes with the current in a river at breakup. The ice run may be characterized as thin, close, or compact. (b) A rush of water following a breakup; a *debacle.*

ice-scour lake A glacial lake occupying a rock basin eroded by a glacier; e.g. a *finger lake.*

ice sheet A *glacier* of considerable thickness and more than 50,000 sq km in area, forming a continuous cover of ice and snow over a land surface, spreading outward in all directions and not confined by the underlying topography; a *continental glacier.* Ice sheets are now confined to polar regions (as on Greenland and Antarctica), but during the Pleistocene Epoch they covered large parts of North America and northern Europe. Armstrong & Roberts (1956, p.7) also apply the term to any extensive body of floating sea ice, but it should be restricted to land ice. Not to be confused with *sheet ice.* See also: *inland ice.* Cf: *ice cap.* Syn: *ice mantle.*

ice shelf A sheet of very thick ice, with a level or gently undulating surface, which is attached to the land along one side but most of which is afloat and bounded on the seaward side by a steep cliff (*ice front*) rising 2 to 50 m or more above sea level. Ice shelves have been formed along polar coasts (e.g. those of Antarctica, the Canadian Arctic islands, and Greenland), and they are generally of great breadth, some of them extending several hundreds of kilometers seaward from the coastline. They are nourished by annual snow accumulation and by seaward extension of land glaciers; limited areas may be aground. Term used by Sir Douglas Mawson in 1912. Nonrecommended syn: *shelf ice; barrier [glaciol]; ice barrier [glaciol]; barrier ice.*

ice shove *ice push.*

ice slush *grease ice.*

ice spar A white or colorless, glassy, transparent variety of orthoclase; specif. *sanidine.*

ice spicule A small needlelike ice crystal that grows in water.

ice stalk A fibrous or spiky efflorescence of *Taber ice,* developed on the surface of freezing sediments. Syn: *ice grass.*

ice stone *cryolite.*

ice stream (a) A current of ice in an ice sheet or ice cap that flows more rapidly than the surrounding ice, usually flowing to the ocean or to an ice shelf and not constrained by exposed rock. Cf: *outlet glacier.* See also: *glacier.* (b) One component of an *alpine glacier;* e.g. an *inset ice stream.* (c) An obsolete syn. of outlet glacier and alpine glacier. (d) A term sometimes popularly applied to a glacier of any kind, esp. an alpine glacier. Syn: *stream [glaciol]; valley glacier.*

ice table A mass of *level ice.* Cf: *sheet ice.*

ice thrust *ice push.*

ice-thrust ridge (a) *ice-pushed ridge.* (b) *lake rampart.*

ice tongue Nonpreferred syn. of *tongue [glaciol].*

ice tongue afloat Nonpreferred syn. of *glacier tongue afloat.*

ice vein *ice wedge.*

ice wall (a) A cliff of ice forming the seaward margin of a glacier that is not afloat, such as of an ice sheet, an ice piedmont, or an ice rise. It is aground, the rock floor being at or below sea level. Cf: *ice front.* (b) Nonrecommended syn. of *ice cliff* in the sense of any vertical wall of ice.

ice-walled channel A term used by Gravenor & Kupsch (1959, p. 56) for a meltwater channel, either an open trench or a *tunnel valley,* containing a stream that may have flowed beneath a glacier.

ice wedge Wedge-shaped, foliated ground ice produced in permafrost, occurring as a vertical or inclined sheet, dike, or vein tapering downward, and measuring from a few millimeters to as much as 6 m wide and from 1 m to as much as 30 m high. It originates by the growth of hoar frost or by the freezing of water in a narrow crack or fissure produced by thermal contraction of the permafrost. See also: *fossil ice wedge.* Syn: *ground-ice wedge; ice vein.*

ice-wedge cast *fossil ice wedge.*

ice-wedge fill *fossil ice wedge.*

ice-wedge polygon A large *nonsorted polygon* characterized by borders of intersecting ice wedges, found only in permafrost regions and formed by contraction of frozen ground. The fissured borders delineating the polygon may be ridges (low-centered polygon in which sediments are being upturned) or shallow troughs (high-centered polygon in which erosion and thawing are prevalent), and are underlain by ice wedges. Diameter: up to 150 m, averaging 10-40 m. In plan, the pattern tends to be three- to six-sided. See also: *fissure polygon.* Cf: *frost-crack polygon.* Syn: *tundra polygon; Taimyr polygon.*

ice-wedge pseudomorph *fossil ice wedge.*

ice-worn Abraded by ice; rubbed, striated, grooved, polished, or scoured by glacial action.

ice yowling A long, high-pitched sound accompanying the formation of contraction cracks in ice. Syn: *yowling.*

ichn A combining form signifying preservation of an original feature after alteration, e.g. a "clastichnic rock".

ichnite A fossil footprint or track. Syn: *ichnolite.*

ichnocoenosis An association of trace fossils.

ichnofacies A sedimentary facies characterized by particular *trace fossils* or by evidences of the life activities of fossil animals.

ichnofossil *trace fossil.*

ichnolite (a) *ichnite.* (b) The rock containing an ichnite.—Obsolete.

ichnology The study of *trace fossils;* esp. the study of fossil tracks. It includes both *palichnology* and *neoichnology.*

ichor A fluid thought to be responsible for such processes as granitization. Originally the term carried the connotation of derivation from a magma (Dietrich & Mehnert, 1961). Syn: *residual liquid.*

ichthammol A brownish-black viscous liquid that is a distillation product of bituminous schists and is used as an emollient and antiseptic.

ichthyodorulite A spine, commonly a cutwater or support for a fin, of any nontetrapod vertebrate, esp. such a spine used as a basis for a form taxon.

Ichthyosauria The sole order of the reptilian subclass Ichthyopterygia, of uncertain ancestry but of porpoiselike or sharklike body form as adaptation for life in the sea. Range, Middle Triassic to Upper Cretaceous.

Ichthyostegalia An order of labyrinthodont amphibians characterized by a fishlike skull pattern and in at least one form by retention of bony supports for tail-fin rays. It includes the earliest known tetrapod. Range, Upper Devonian to Lower Mississippian.

icicle A narrow cone-shaped spike or shaft of clear ice, hanging with its point downward, formed by the freezing of dripping water; its length ranges from finger length to 8 m.

icing (a) A surface ice mass formed during the winter in a permafrost area by successive freezing of sheets of water that may seep from the ground, or from a spring or river. Syn: *aufeis; flood icing; river icing; flooding ice; floodplain icing; glacier (colloquial in Alaska); ice field [permafrost].* (b) The accumulation of an ice deposit on exposed objects.

icing mound A thick, localized surface mound on an *icing;* it may form by the upwarp and breaching of a layer of ice (as in a river) by the hydrostatic pressure of water, which then issues from the mound as a freezing spring. Syn: *ice blister; ice mound.*

icosacanthic law *Müllerian law.*

icosahedron A solid figure having 20 faces.

icositetrahedron A term for the isometric *trapezohedron.*

Idahoan-type facies series Rocks produced in a type of dynamothermal regional metamorphism characteristic of Boehls Butte, Idaho, in which the polymorphs andalusite, kyanite, and sillimanite occur together with some staurolite and cordierite. Thus the Al_2SiO_5 triple point is characteristically involved in this type of regional dynamothermal metamorphic series. At pressures between 3000 and 6000 bars, it lies approximately between the *Pyrenean-type facies series* and the *Barrovian-type facies series* (Hietanen, 1967, p.195).

idaite A mineral: Cu_3FeS_4, but perhaps Cu_5FeS_6.

idd A term applied in northern Sudan to a place in the bed of an intermittent stream where water may be obtained in shallow wells for most if not all of the dry season (J.H.G. Lebon in Stamp, 1961, p. 254).

Iddings' classification A classification of igneous rocks in which the mineralogic classifications of Rosenbusch and Zirkel are correlated with the CIPW or *normative* classification. The system was proposed in 1913 by J. P. Iddings.

iddingsite A reddish-brown mixture of silicates (of ferric iron, calcium, and magnesium) formed by the alteration of olivine. It forms rust-colored patches in basic igneous rocks.

ideal cut *Tolkowsky theoretical brilliant cut.*

ideal cyclothem A theoretical *cyclothem* that represents, in a given region and within a given stratigraphic interval, the optimum succession of deposits during a complete sedimentary cycle. It is constructed from theoretical considerations and from accumulated data from modern environments and experimental evidence. An ideal cyclothem of ten members for western Illinois consists of the following sequence (Weller & others, 1942, p.10): (10) marine shale with ironstone concretions; (9) clean marine limestone; (8) black laminated shale with limestone concretions or layers; (7) impure, lenticular, fine-grained marine limestone; (6) gray marine shale with pyritic nodules; (5) coal; (4) underclay; (3) freshwater, usually nonfossiliferous limestone; (2) sandy shale; and (1) fine-grained micaceous sandstone, locally unconformable on underlying beds. See also: *discordance index.*

ideal section A geologic cross section that combines observed evidence on stratigraphy and/or structure with interpretation of what is not present. It may be the summation or average of several successive cross sections.

ideal solution A solution in which the molecular interaction between components is the same as that within each component; a solution that conforms to *Raoult's law.* Cf: *nonideal solution.*

identified resources Specific bodies of mineral-bearing rock whose existence and location are known (Brobst & Pratt, 1973, p. 3). They may or may not be evaluated as to extent and grade. Identified resources include *reserves* and *identified subeconomic resources.* Cf: *hypothetical resources; speculative resources.*

identified subeconomic resources Mineral resources that are not *reserves,* but that may become reserves as a result of changes in economic and legal conditions (Pratt & Brobst, 1974, p. 2). Syn: *conditional resources.* See also: *identified resources.*

idioblast A mineral constituent of a metamorphic rock formed by recrystallization and bounded by its own crystal faces. It is a type of *crystalloblast.* The term was originated by Becke (1903). Cf: *hypidioblast; xenoblast.*

idioblastic Pertaining to an *idioblast* of a metamorphic rock. It is analogous to the term *idiomorphic* in igneous rocks. Cf: *hypidioblastic.*

idioblastic series *crystalloblastic series.*

idiochromatic Said of a mineral whose color is a result of its chemical composition, i.e. is inherent. Cf: *allochromatic.*

idiogenous A little-used syn. of *syngenetic [ore dep].* Cf: *hysterogenous; xenogenous.*

idiogeosyncline A type of late-cycle geosyncline between stable and mobile areas of the crust, the sediments of which are only weakly folded, such as the marginal basins of the East Indian island arc (Umbgrove, 1933, p. 33-43). Cf: *parageosyncline (b).*

idiomorph *euhedron.*

idiomorphic A synonym of *automorphic,* originally proposed by Rosenbusch (1887, p. 11) to describe individual *euhedral* crystals. Though the term lacks priority, it is now commonly applied to the igneous-rock texture characterized by such euhedral crystals, especially in American usage.

idiomorphic-granular *idiomorphic.*

idiophanous Said of a crystal that exhibits an interference figure to the naked eye, without the help of optical instruments.

idiotopic Said of the fabric of a crystalline sedimentary rock in which the majority of the constituent crystals are euhedral. Also, said of the rock (such as an evaporite, a chemically deposited cement, or a recrystallized limestone or dolomite) with such a fabric. The term was proposed by Friedman (1965, p.648). Cf: *xenotopic; hypidiotopic.*

idocrase *vesuvianite.*

idrialite A hydrocarbon mineral: $C_{22}H_{14}$. It was previously formulated $C_{24}H_{18}$. Idrialite is often found mixed with cinnabar and clay. Syn: *curtisite.*

igdloite *lueshite.*

igneous Said of a rock or mineral that solidified from molten or partly molten material, i.e. from a magma; also, applied to processes leading to, related to, or resulting from the formation of such rocks. Igneous rocks constitute one of the three main classes into which rocks are divided, the others being metamorphic and sedimentary. Etymol: Latin *ignis,* "fire". See also: *magmatic; plutonic; hypabyssal; extrusive.* Obsolete syn: *pyrogenous.* Deprecated syn: *eruptive.*

igneous breccia (a) A breccia that is composed of fragments of igneous rock. (b) Any breccia produced by igneous processes, e.g. volcanic breccia, intrusion breccia. Cf: *agmatite.*

igneous complex An assemblage of intimately associated and roughly contemporaneous igneous rocks differing in form or in petrographic type; it may consist of plutonic rocks, volcanic rocks, or both.

igneous-contact shoreline A shoreline formed by the partial submergence of the relatively steep slope left by the removal of weak beds from one side of a straight igneous contact (Johnson, 1925, p. 24-27).

igneous cycle The sequence of events in which volcanic activity is followed by major plutonic intrusions, and then minor intrusions (e.g. dikes).

igneous facies A part or variety of a single igneous rock body, differing in some attribute (structure, texture, or mineralogic or chemical composition) from the "normal" or typical rock of the main mass; e.g. a granite mass may grade into a porphyritic "igneous facies" near its borders. It is due to differentiation in place.

igneous lamination In plutonic rocks, the arrangement of tabular crystals parallel to each other and to any layering the rocks may have.

igneous metamorphism A high-temperature metamorphic process that includes the effects of magma on adjacent rocks as well as those due to injection pegmatitization (Lindgren, 1933). The term is no longer in common use. Cf: *pyrometamorphism.*

igneous province *petrographic province.*

igneous quartz (a) *quartzolite.* (b) *silexite [ign].*

igneous-rock clan *clan.*

igneous-rock series An assemblage of temporally and spatially related igneous rocks of the same general form of occurrence (plutonic, hypabyssal, or volcanic), characterized by possessing in common certain chemical, mineralogic, and textural features or properties so that the rocks together exhibit a continuous variation from one extremity of the series to the other. Syn: *series [ign]; rock series.*

ignimbrite The rock formed by the widespread deposition and consolidation of ash flows and nuées ardentes. The term originally implied dense welding but there is no longer such a restriction, so that the term includes rock types such as *welded tuff* and nonwelded *sillar.* See also: *tufflava; ash-flow tuff.* Syn: *flood tuff.*

ignimbrite shield A *shield volcano* built of rhyolitic ash flows, with a *collapse caldera* at its summit. Esp. well developed in the Tibesti region of Africa (Macdonald, 1972, p. 267).

ignispumite A type of rhyolite characterized by lenticles and banding, which is believed to have been deposited as an acid, foamy lava and to be transitional with true ignimbrite. Cf: *tufflava.*

IGY *International Geophysical Year.*

ihleite *copiapite.*

iimoriite A triclinic mineral: $Y_5(SiO_4)_3(OH)_3$.

ijolite (a) In the *IUGS classification,* a plutonic rock in which F is between 60 and 100, M is between 30 and 70, and sodium exceeds potassium. Cf: *fergusite.* (b) A series of plutonic rocks containing nepheline and 30-60% mafic minerals, generally clinopyroxene, and including sphene, apatite, and melanite; also, any rock of that series. *Melteigite* and *jacupirangite* are more mafic members of the series; *urtite* is a type rich in nepheline. Named by Ramsay in 1891 for Ijola (Iivaara), Finland. Cf: *fergusite; tawite.*

ijussite A mafic *teschenite* composed of abundant titanaugite and barkevikite with smaller amounts of bytownite, anorthoclase, and analcime. The name is for the Ijuss River, Siberia, U.S.S.R. Not recommended usage.

ikaite A chalky mineral: $CaCO_3 \cdot 6H_2O$.

ikunolite A mineral: $Bi_4(S,Se)_3$. Cf: *laitakarite.*

ilesite A green mineral: $(Mn,Zn,Fe)SO_4 \cdot 4H_2O$.

ilimaussite A mineral: $Ba_2Na_4CeFeNb_2Si_8O_{28} \cdot 5H_2O$.

illidromica A hydromica, low in potassium and high in water; *illite.* Also, a clay mineral intermediate in composition between illite and montmorillonite.

Illinoian Pertaining to the classical third glacial stage of the Pleistocene Epoch in North America, between the Yarmouthian and Sangamonian interglacial stages. See also: *Riss.* Also spelled: *Illinoisan.* Obsolete.

illite (a) A general name for a group of three-layer, micalike clay minerals that are widely distributed in argillaceous sediments (esp. in marine shales and in soils derived from them). They are intermediate in composition and structure between muscovite and montmorillonite, have 10-angstrom *c*-axis spacings with substantially no expanding-lattice characteristics, and have the general formula: $(H_3O,K)_y(Al_4 \cdot Fe_4 \cdot Mg_4 \cdot Mg_6)(Si_{8-y} \cdot Al_y)\ O_{20}(OH)_4$, with *y* less than 2 and frequently 1 to 1.5. The term is generally used for "clay-mineral micas of both dioctahedral and trioctahedral types and of muscovite and biotite crystallizations" (Grim, 1968, p. 42). Illite contains less potassium and more water than true micas, and more potassium than kaolinite and montmorillonite; it appears intermediate between kaolin and montmorillonite clays in cation-exchange capacity, in ability to absorb and retain water, and in physical characteristics (such as plasticity index). Much so-called illite may be a mechanical mixture of fine-grained montmorillonite and muscovite, or a clay containing alternate layers having a montmorillonite and a muscovite structure. (b) A mineral of the illite group; esp. the mineral having the complex chemical composition of muscovite or of a hydrated muscovite but giving a line-poor X-ray powder pattern (Hey, 1962, p. 16.3.20).—The term was proposed by Grim et al. (1937) as a general term (not a specific mineral name) in recognition of the state (Illinois) in which clay study has received much encouragement. See also: *bravaisite.* Syn: *hydromica; illidromica; glimmerton.*

illuvial horizon A soil horizon to which material has been added by the process of *illuviation.* Cf: *eluvial horizon.*

illuviation The accumulation, in a lower soil horizon, of soluble or suspended material that was transported from an upper horizon by the process of *eluviation.* Adj: *illuvial.*

ilmajokite A mineral: $(Na,Ca,Ba)_{10}Ti_5Si_{14}O_{22}(OH)_{44} \cdot nH_2O$.

ilmenite An iron-black, opaque, rhombohedral mineral: $FeTiO_3$. It is the principal ore of titanium. Ilmenite occurs as a common accessory mineral in basic igneous rocks (esp. gabbros and norites), and is also concentrated in mineral sands. See also: *menaccanite [mineral].* Syn: *titanic iron ore; mohsite.*

ilmenitite A hypabyssal rock composed almost entirely of ilmenite, with accessory pyrite, chalcopyrite, pyrrhotite, hypersthene, and labradorite. Cf: *nelsonite.*

ilmenomagnetite (a) Magnetite with microintergrowths of ilmenite. (b) Titanian maghemite with exsolution ilmenite.—Cf: *magnetoilmenite.*

ilmenorutile A black mineral: $(Ti,Nb,Fe)_3O_6$.

ilsemannite A black, blue-black, or blue mineral: $Mo_3O_8 \cdot nH_2O$ (?).

ilvaite A brownish to black orthorhombic mineral: $CaFe_2{}^{+2}Fe^{+3}(SiO_4)_2(OH)$. It is related to epidote, and usually contains manganese in small amounts. Syn: *lievrite; yenite.*

ILW *international low water.*

image [photo] (a) The recorded representation of a scene or object produced by an optical-mechanical or electronic scanner that converts electromagnetic radiation (EMR) outside the visible part of the spectrum into visible EMR. Cf: *photograph.* (b) The optical counterpart of an object, produced by the reflection or refraction of light when focused by a lens or mirror.

image [sed] (a) A term introduced by Wadell (1932, p. 449) for a binomial expression of the shape of a sedimentary particle, expressed as a fraction giving the roundness of the particle in the numerator and the sphericity in the denominator. (b) A two-dimensional (plane) projection or cross section of a sedimentary particle, obtained by photography or by tracing; it is useful in determining *roundness.*

image motion The smearing or blurring of imagery on an aerial photograph due to the relative movement of the camera with respect to the ground. An "image-motion compensator" installed with the camera intentionally imparts movement to the film at such a rate as to compensate for the forward motion of the aircraft during exposure time.

imager A nonrecommended term for a remote-sensing system.

imagery (a) The process of producing *images [photo];* the term is analogous to *photography.* (b) A collective term for the images so produced.

imandrite A rock composed chiefly of quartz and albite, formed by the combination of a nepheline syenite magma with a graywacke. A genetic term; not recommended usage.

imatra stone *marlekor.*

imbibition [rock] Formation of feldspathic minerals by the penetration of alkaline solutions of magmatic origin into aluminum-rich metamorphic rocks.

imbibition [water] (a) The absorption of a fluid, usually water, by a granular rock or any other porous material, under the force of capillary attraction, and in the absence of any pressure. Syn: *capillary percolation.* (b) Absorption of water by plants.

imbibometry A method of analysis involving measurement of the uptake of water or other fluids in a solid substance.

Imbrian (a) Pertaining to lunar topographic features and lithologic map units constituting a system of rocks formed during the period of formation of the Mare Imbrium basin and of deposition of mare material of the Procellarum Group, or during any time between these two events. Imbrian rocks are older than the post-mare craters and associated ejecta of the Eratosthenian and Copernican systems. (b) Said of the stratigraphic period during which the Imbrium System was developed.

imbricated Overlapping, as tiles on a roof or scales on a bud.

imbricated fault zone A zone of closely spaced faults exhibiting *imbricate structure;* it is often developed above a décollement, and is underlain by a sole fault.

imbricated texture A texture resembling overlapping plates, seen in certain minerals (such as tridymite) under the microscope.

imbricate structure [sed] A sedimentary structure characterized by *imbrication* of pebbles all tilted in the same direction, with their flat sides commonly displaying an upstream dip. Syn: *shingle structure.*

imbricate structure [tect] A tectonic structure displayed by a series of nearly parallel and overlapping minor thrust faults, high-angle reverse faults, or slides, and characterized by rock slices, sheets, plates, blocks, or wedges that are approximately equidistant and have the same displacement and that are all steeply inclined in the same direction (toward the source of stress). See also: *imbricated fault zone.* Syn: *schuppen structure; shingle-block structure.*

imbrication [sed] (a) A sedimentary fabric characterized by disk-shaped or elongate fragments dipping in a preferred direction at an angle to the bedding. It is commonly displayed by pebbles on a stream bed, where flowing water tilts the pebbles so that their flat surfaces dip upstream. (b) Formation of *imbricate structure* .—Syn: *shingling.*

imbrication [tect] (a) The steeply inclined, overlapping arrangement of thrust sheets in imbricate structure. (b) Formation of such structure.

imerinite A colorless to blue monoclinic mineral of the amphibole group: $Na_2(Mg,Fe)_6Si_8O_{22}(O,OH)_2$. It is related to richterite.

imgreite A hexagonal mineral: NiTe (?).

imhofite A mineral: $Tl_6CuAs_{16}S_{40}$.

imitation Any material that simulates a genuine, natural gem; specif. glass, plastic, other amorphous material, or certain crystalline materials. Cf: *synthetic stone; assembled stone.* Syn: *simulated stone.*

immature [geomorph] Said of a topography or region, and of its features, that have not attained maturity; esp. said of a valley or drainage system that is well above base level and whose side slopes descend steeply to the riverbanks, or of a drainage system that is expanding into an incompletely drained upland.

immature [sed] (a) Pertaining to the first stage of textural maturity (Folk, 1951); said of a clastic sediment that has been differentiated or evolved from its parent rock by processes acting over a short time and/or with a low intensity and that is characterized by relatively unstable minerals (such as feldspar), abundance of mobile oxides (such as alumina), presence of weatherable material (such as clay), and poorly sorted and angular grains. Example: an "immature sandstone" containing over 5% clay and commonly occurring in deeper marine, flood-plain, swamp, and mudflow deposits. Cf: *submature; mature; supermature.* (b) Said of an argillaceous sedimentary material intermediate in character between a clay and a shale; e.g. an "immature shale".

immature region *endozone.*

immature soil *azonal soil.*

immaturity A stage in the cycle of erosion or the sequential development of land forms that is characterized by *immature* features.

immediate runoff *direct runoff.*

immersed bog A bog which tends to expand horizontally by growth of plants under water. Cf: *emerged bog.*

immersion cell A cuplike accessory to a microscope or polariscope, in which a gemstone may be immersed in a liquid, generally of high refractive index, in order to eliminate surface reflection and refraction and thus allow efficient observation of the interior of the stone.

immersion liquid A liquid of known refractive index, used in the *immersion method* of determining a mineral's refractive index. An example is acetone. Syn: *index liquid.*

immersion method A method of determining the relative refractive index of a mineral in order to identify it, by immersing the sample in a liquid of known refractive index (an *immersion liquid*).

immigrant In ecology, an organism that becomes established in a region where it was previously unknown.

immiscibility gap A term that is incorrectly used for *miscibility gap.*

immiscible Said of two or more phases that, at mutual equilibrium, cannot dissolve completely in one another, e.g. oil and water. Cf: *miscible.*

imogolite A thread-shaped paracrystalline clay mineral having the general formula $1.1SiO_2 \cdot Al_2O_3 \cdot 2.3\text{-}2.8H_2O(+)$.

impact A forceful contact or collision between bodies, such as that involved in the production of a meteorite crater or cryptoexplosion structure. Also, the degree or concentration of force in a collision.

impact bomb A porous mass of *impactite* formed by splattering, and exhibiting aerodynamic sculpturing. See also: *Fladen.*

impact cast *prod cast.*

impact crater A *crater* formed on a surface by the impact of an unspecified projectile; esp. a terrestrial, lunar, or Martian crater in which the nature of the impacting body (meteorite, asteroid, comet, etc.) is not known. See also: *meteorite crater; penetration funnel; primary crater; secondary crater.*

impact erosion The wearing-away of rocks or fragments through the effect of definite blows of relatively large fragments.

impact glass Glassy *impactite.*

impactite (a) A vesicular, glassy to finely crystalline material produced by fusion or partial fusion of target rock by the heat generated from the impact of a large meteorite, and occurring in and around the resulting crater, typically as individual bodies composed of mixtures of melt and rock fragments, often with traces of meteoritic material; a rock (such as *suevite*) from a presumed impact site. Syn: *impact slag; impact glass.* (b) A term used incorrectly for any shock-metamorphosed rock.

impact lava *impact melt.*

impact law A physical law governing the settling of coarse particles, in which, for a given particle density, fluid density, and fluid viscosity, the settling velocity is directly proportional to the square root of the particle diameter. Cf: *Stokes' law.*

impact mark *prod mark.*

impact melt Molten material produced by fusion of target rock during a meteorite impact, and emplaced in and around the resulting crater as discrete, partly to completely crystalline dikelike or sill-like bodies, as the matrix of fragmental breccias, or as discrete fragments ejected from the crater. Syn: *impact lava.*

impact metamorphism A type of *shock metamorphism* in which the shock waves and the observed changes in rocks and minerals result from the hypervelocity impact of a body such as a meteorite (Chao, 1967, p. 192). The term was used by Dietz (1961) to describe a field of geology involving the study of minerals (such as coesite and minute diamonds) created by meteorite impact.

impact slag *impactite.*

impact structure A generally circular or craterlike structure produced by impact (usually extraterrestrial) on a planetary surface. The stage of erosion of the structure and the nature of the impacting body need not be specified.

imparipinnate *odd-pinnate.*

impedance [elect] The complex ratio of voltage to current in an electrical circuit, or the complex ratio of electric-field intensity to magnetic-field intensity in an electromagnetic field. It is the reciprocal of *admittance.*

impedance [seis] *acoustic impedance.*

imperfect flower A flower having either stamens or carpels but not both. Cf: *perfect flower.*

imperfection *flaw [gem].*

imperforate (a) Not perforated, or lacking a normal opening; esp. descriptive of foraminiferal-test walls without pores. (b) Said of a spiral mollusk shell in which the first whorl is tightly coiled upon itself, leaving no central opening.—Cf: *perforate.*

Imperial jade A translucent to semitransparent variety of jadeite characterized by the finest, highly intense emerald-green color; "true jade". Also spelled: *imperial jade.*

impermeability The condition of a rock, sediment, or soil that renders it incapable of transmitting fluids under pressure. Syn: *imperviousness.* Ant: *permeability [geol].* Adj: *impermeable.*

imperviousness *impermeability.*

impetus A sharp onset or arrival of a seismic phase on a seismogram. Cf: *emersio.*

impingement The mechanism or process in dolomitization whereby dolomite crystals replace calcareous particles (commonly skeletal particles such as crinoid ossicles and plates) but not in optical continuity with the calcite of the original particle (Lucia, 1962).

impolder *empolder.*

impoverished fauna *depauperate fauna.*

impregnated Said of a mineral deposit (esp. of metals) in which the minerals are epigenetic and diffused in the host rock. Cf: *disseminated; interstitial.*

impressed area The concave dorsum of a coiled cephalopod conch in contact with the venter of the preceding (next-older) whorl and tending to overlap it. Syn: *impressed zone.*

impression (a) The form, shape, or indentation made on a soft sedimentary surface (as of mud or sand) by an organic or inorganic body (usually a harder structure, such as a fossil shell or the strengthened surface of a leaf) that has come in contact with it; a shallow *mold.* It usually occurs as a negative or concavity found on the top of a bed, and a cast of it may then be found on the base of the overlying bed. (b) A small circular pit formed by rain, hail, drip, or spray. (c) A fossil footprint, trail, track, or burrow.—Syn: *imprint.*

imprint An *impression,* esp. one made by a thin object such as a leaf or by a falling hailstone or raindrop.

imprisoned Said of a boulder or block of rock resting intimately against others with common or closely fitting interfaces, such as those found along a rocky coast like that of Victoria, Australia. Term used by Baker (1959, p. 206).

imprisoned lake A term used by Dana (1895, p.199) for a lake occupying a basin that had been cut off from a river system, or a crater of an extinct volcano, or a basin or depression formed by earth movements or glacial action.

impsonite A dull black, nearly infusible asphaltic pyrobitumen, with a hackly fracture and high fixed-carbon content. It closely resembles albertite, but is almost insoluble in turpentine; it is derived from the metamorphism of petroleum.

impulse An energy pulse of such short time-duration that its wave shape is of no consequence. Theoretically it is of infinite magnitude, infinitesimal time duration, and finite energy.

impulsive Said of an *impetus,* or sharp arrival of a seismic phase. Cf: *emergent [seis].*

impunctate Lacking pores, perforations, or punctae; specif. said of a brachiopod shell without endopunctae or pseudopunctae, in which the shell substance is dense. Impunctate brachiopods are by far the most numerous of all brachiopods. Cf: *punctate.*

impure arkose A term commonly applied to a sandstone (esp. a graywacke) that is highly feldspathic but is not an arkose; specif. a sandstone containing 25-90% feldspar and igneous-rock fragments, 10-50% mica and metamorphic-rock fragments, and 0-65% quartz and chert (Folk, 1954, p.354). The term is roughly equivalent to *micaceous arkose* of Hubert (1960), and was used by Krynine (1948, p.137) for a transitional rock between arkose and high-rank graywacke. Cf: *lithic arkose; feldspathic graywacke.* Syn: *dirty arkose.*

IMW *International Map of the World.*

inactive volcano A volcano that has not been known to erupt. Cf: *active volcano; dormant volcano; extinct volcano.*

inadunate Any crinoid belonging to the subclass Inadunata, characterized by firmly jointed calyx plates, a mouth concealed by the tegmen, and arms that are free above the radials. Range, Ordovician to Triassic.

in-and-out channel A crescentic valley excavated on a hillside by meltwater flowing around a projecting glacial lobe (Kendall, 1902, p. 483).

inaperturate Said of pollen and spores having no germinal, harmomegathic, or other openings. Cf: *acolpate; alete.*

inarticulate n. Any brachiopod belonging to the class Inarticulata, characterized by valves that are calcareous or composed of chitinophosphate and commonly held together by muscles rather than hinge teeth and dental sockets.—adj. Said of a brachiopod having such valves, or of the valves themselves. Cf: *articulate.*

inbreak A subsidence of the surface over a mine due to subterranean shattering of rock material. Cf: *crown-in; flash [mining]; creep [mining].*

incaite A monoclinic mineral: $Pb_4FeSn_4Sb_2S_{15}$.

incandescent Said of an ash flow, nuée ardente, or any pyroclastic matter that is glowing.

incandescent tuff flow A term essentially synonymous with *ash flow;* it was originally used to describe the outbursts of fine-grained fragmental rhyolitic material in the Arequipa region of Peru (Fenner, 1948, p. 879).

incarbonization *coalification.*

Inceptisol In U.S. Dept. of Agriculture soil taxonomy, a soil order characterized by having one or more horizons in which mineral materials other than carbonates or amorphous silica have been altered or removed but not accumulated to a significant degree. These soils contain a preponderance of weatherable minerals and do not have argillic, oxic, or spodic horizons. Inceptisols commonly have an ochric or umbric epipedon over a cambic horizon, and sometimes a fragipan (USDA, 1975). Suborders and great soil groups of this order have the suffix -ept. See also: *Andept; Aquept; Ochrept; Plaggept; Tropept; Umbrept.*

incertae sedis A term applied to a fossil or modern specimen whose classification is regarded as uncertain. Etymol: Latin, "of uncertain place".

inch-scale layering In igneous rocks, a type of rhythmic layering consisting of layers slightly less than one inch thick, which may be repeated hundreds of times (Hess, 1960, p. 133).

incidental vein A vein discovered after the original vein on which a claim is based.

incipient peneplain A syn. of *strath* in the restricted sense proposed by Bucher (1932, p. 131), of a fluvial-degradation surface consisting of a broad valley floor and extensive valley-floor side strips. Syn: *partial peneplain; local peneplain.*

incised In geomorphology, said of a stream meander, channel, or notch that has been downcut or entrenched into the surface during, and because of, rejuvenation of the stream, whether due to relative uplift of the surface or to other cause. See also: *incision.*

incised meander (a) A generic term for an old stream meander that has become deepened by rejuvenation and that is more or less closely bordered or enclosed by valley walls. Two types are usually recognized: *entrenched meander* and *ingrown meander.* Syn: *inclosed meander.* (b) Used in a more restricted sense as a syn. of *entrenched meander.*

incision (a) The process whereby a downward-eroding stream deepens its channel or produces a narrow, steep-walled valley; esp. the downcutting of a stream during, and as a result of, rejuvenation, whether due to relative movement (uplift) of the crust or to other cause. Also, the product of such a process, e.g. an incised notch or meander. Cf: *entrenchment.* (b) The process whereby a steep-sided trench or notch intersects a plane surface or slope; e.g.

current erosion of the continental slope to produce a submarine canyon.

incisor (a) A process with a biting surface on the gnathal lobe of the mandible of a crustacean. (b) One of the premaxillary (anterior) teeth of mammals, which are relatively unspecialized and in most forms are replaced in an individual's lifetime.—Cf: *molar.*

inclination [drill] The angle of the axis of a *well bore* measured from the vertical at a stated depth. See also: *deviation [drill].* Syn: *drift [drill].*

inclination [magnet] The angle at which magnetic-field lines dip; it is one of the *magnetic elements.* Syn: *dip [magnet]; magnetic dip; magnetic inclination.*

inclination [paleont] The attitude of the cardinal area (or pseudointerarea) in either valve of a brachiopod, based on the convention of viewing the specimen in lateral profile with beaks to the left and brachial valve uppermost, referring the cardinal area to its position within one of four quadrants defined by the commissural plane and the plane normal to it and the plane of symmetry, touching the base of the cardinal areas (TIP, 1965, pt.H, p.146). See also: *orthocline; catacline; anacline; hypercline; apsacline; procline.*

inclination [slopes] (a) A deviation from the true vertical or horizontal. Also, the amount of such deviation; the rate of slope, or grade. (b) An inclined surface; a slope.

inclination [struc geol] A general term for the slope of any geological body or surface, measured in the upward or downward direction and from the horizontal or the vertical. It is often used synonymously with *dip.*

inclinator *inclinometer [magnet].*

inclined bedding (a) An inclusive term for bedding inclined to the principal surface of deposition. The term is not recommended for use as a syn. of *cross-bedding* because it may "equally refer ... to any initial dip" (Hills, 1963, p.10). See also: *discordant bedding.* Archaic syn.: *diagonal bedding; oblique bedding.* (b) Bedding laid down with primary or initial dip (Dennis, 1967, p.12).

inclined extinction A type of *extinction* seen in birefringent crystal sections in which the vibration directions are inclined to a crystal axis or direction of cleavage. Cf: *parallel extinction; undulatory extinction.* Syn: *oblique extinction.*

inclined fold A fold whose axial surface is inclined from the vertical, and in which one limb may be steeper than the other. The term sometimes includes the restriction that the steeper of the two limbs not be overturned.

inclinometer [drill] Any of various instruments for measuring *inclination [drill];* a *driftmeter.*

inclinometer [magnet] An instrument that measures magnetic inclination. See also: *earth inductor; dip circle.* Syn: *inclinator.*

inclosed meander A syn. of *incised meander;* it was proposed as a generic term by Moore (1926). Also spelled: *enclosed meander.*

included gas Gas in isolated interstices in either the zone of aeration or the zone of saturation (Meinzer, 1923, p. 21). The term may also be applied to bubbles of air or other gas, not in isolated interstices, that are surrounded by water in either zone and that act as obstacles to water flow until the gas disappears by dissolving in the water. Cf: *ground air; subsurface air.*

inclusion A fragment of older rock within an igneous rock to which it may or may not be genetically related. See also: *xenolith; autolith.* Syn: *enclave; enclosure.*

incoalation *coalification.*

incoherent Said of a rock or deposit that is loose or unconsolidated.

incompetent Said of rocks that have deformed in a ductile manner compared to adjacent more brittle rocks, e.g. the matrix around *boudins;* or of layers that have formed more nearly *similar folds* in contrast to competent layers which have formed more nearly *parallel folds.* It is a relative term. Ant: *competent.*

incompetent rock A volume of rock which at a specific time and under specific conditions is not able to support a tectonic force. Cf: *competent rock.*

incomplete caneolith A *caneolith* having upper and lower rim elements but lacking a wall. Cf: *complete caneolith.*

incomplete flower A flower which lacks one or more of the four floral appendages (stamens, carpels, corolla or petals, and calyx or sepals). Cf: *complete flower.*

incomplete ripple mark A ripple-marked surface characterized by isolated crests of ripple marks, such as one receiving an insufficient supply of sand. Syn: *starved ripple mark.*

incomplete tabula A coral *tabula* consisting of several *tabellae* joined together. Cf: *complete tabula.*

incompressibility modulus *bulk modulus.*

incongruent melting Melting accompanied by decomposition or by reaction with the liquid, so that one solid phase is converted into another; melting to give a liquid different in composition from the original solid. An example is orthoclase melting incongruently to give leucite and a liquid richer in silica than the original orthoclase. Cf: *congruent melting.*

incongruent solution Dissolution accompanied by decomposition or by reaction with the liquid so that one solid phase is converted into another; dissolution to give dissolved material in different proportions from those in the original solid.

incongruous Said of a parasitic fold, the axis and axial surface of which are not parallel to the axis and axial surface of the main fold to which it is related. Ant: *congruent.*

inconsequent A syn. of *insequent.* The term was used by Gilbert (1877, p. 143-144), but is not now in general use.

incorporation A process of *coalification* in which there is no modification of material. Cf: *vitrinization; fusinization.*

increase The addition of corallites to colonies by *offset* formation resulting from parent polyps. Examples: *axial increase; lateral increase; intermural increase; peripheral increase.*

increment *recharge.*

incremental strain A finite or infinitesimal strain relating two sequential configurations of a body. The total finite strain of a rock body is the cumulative result of a number of incremental strains (Means, 1976, p. 224).

incretion (a) A term proposed by Todd (1903) for a cylindrical *concretion* with a hollow core; e.g. a rhizocretion. (b) A concretion whose growth has been directed inward from without.

incrop A former outcrop concealed by or buried beneath younger unconformable strata.

incrustation *encrustation.*

incumbent Lying above; said of an overlying or superimposed stratum.

incurrent canal *inhalant canal.*

indehiscent Said of a fruit or other plant structure that does not open spontaneously to distribute its contents (seeds or spores).

indelta A term used in Australia for an inland area where a river subdivides (Taylor, 1951, p. 615).

independent ovicell A bryozoan ovicell that develops independently of the distal zooid. Syn: *recumbent ovicell.*

independent variable A *variable* whose magnitude changes systematically. Cf: *dependent variable.*

inderborite A monoclinic mineral: $CaMg[B_3O_3(OH)_5]_2 \cdot 6H_2O$.

inderite A mineral: $Mg_3O_3(OH)_5 \cdot 5H_2O$. Syn: *lesserite.*

index bed *key bed.*

index contour A contour line shown on a map in a distinctive manner for ease of identification, being printed more heavily than other contour lines and generally labeled with a value (such as figure of elevation) along its course. It appears at regular intervals, such as every fifth or sometimes every fourth contour line (depending on the contour interval). Syn: *accented contour.*

index ellipsoid The *indicatrix* of an anisotropic crystal.

index error (a) An instrument error, constant in behavior, caused by the displacement of the zero or index mark of a vernier; e.g. an error resulting from inclination of the upper plate in a transit having a fixed vertical vernier. (b) An instrument error in the magnetic bearing given by readings of the needle of a compass, such as an error arising from oblique magnetization of the needle or from the disturbance of the line of sight.

index factor A constant which, when multiplied by certain measurements made on potential-field anomalies, gives an estimate as to the depth to an anomalous mass (sometimes the maximum depth at which the mass could be located). It is used in magnetic and gravity interpretation. See also: *depth rule.*

index fossil A fossil that identifies and dates the strata or succession of strata in which it is found; esp. any fossil taxon (generally a genus, rarely a species) that combines morphologic distinctiveness with relatively common occurrence or great abundance and that is characterized by a broad, even worldwide, geographic range, and by a narrow or restricted stratigraphic range that may be demonstrated to approach isochroneity. The best index fossils include swimming or floating organisms that evolved rapidly and were distributed widely, such as graptolites and ammonites. The fossil need not necessarily be either confined to, or found throughout every part of, the strata for which it serves as an index. Cf: *characteristic fossil; guide fossil.* Syn: *key fossil; type fossil.*

index horizon A structural surface used as a reference in analyzing the geologic structure of an area. Syn: *index plane.*
index liquid *immersion liquid.*
index map (a) A map, usually of small scale, that depicts the location of, or specific data regarding, one or more small areas in relation to (or within) a larger area and that typically points up special features in the larger area; e.g. a map showing a mine property in relation to the main surface features (towns, roads, streams, etc.) of the surrounding area. It is often shown in a small rectangle on a large map. (b) A map showing the location and numbers of flight strips and aerial photographs; a map showing the outline of the area covered by each aerial photograph. Cf: *photoindex.*
index mineral A mineral developed under a particular set of temperature and pressure conditions, thus characterizing a particular degree of metamorphism. When dealing with progressive metamorphism, it is a mineral whose first appearance (in passing from low to higher grades of metamorphism) marks the outer limit of the zone in question (Turner and Verhoogen, 1960, p.491). Cf: *critical mineral; typomorphic mineral.*
index of refraction In crystal optics, a number that expresses the ratio of the velocity of light in vacuo to the velocity of light within the crystal. Its conventional symbol is n. Modifying factors include wavelength, temperature, and pressure. *Birefringent* crystals have more than one index of refraction. See also: *relative index of refraction; alpha [cryst]; beta [cryst]; gamma [cryst].* Syn: *refractive index.*
index plane *index horizon.*
index species (a) A species of plant or animal that is characteristic of a particular set of environmental conditions and therefore whose presence in a particular area indicates the existence of those conditions in that area. (b) An *index fossil* of species rank.
index surface A two-shelled geometric surface that represents the indices of refraction in a biaxial crystal in the direction of propagation. Cf: *indicatrix.*
index zone A stratum or body of strata, recognizable by paleontologic or lithologic characters, that can be traced laterally and identifies a reference position in a stratigraphic section.
indialite A hexagonal mineral: $Mg_2Al_4Si_5O_{18}$.
indianaite A white, porcelainlike clay mineral representing an impure variety of halloysite from Lawrence County, Indiana.
Indiana limestone *Bedford limestone.*
indianite A variety of anorthite occurring as gangue of the corundum of the Carnatic of India.
Indian ridge A term used in New England for a sinuous *esker*. Syn: *serpent kame.*
indicated ore Ore for which there are quantitative estimates of tonnage and grade, made partly from inference and partly from specific sampling. Cf: *inferred ore; possible ore; potential ore.* Syn: *probable ore.*
indicator [eco geol] A geologic or other feature that suggests the presence of a mineral deposit, e.g. a geochemical anomaly, a carbonaceous shale indicative of coal, or a pyrite-bearing bed that may lead to gold ore at its intersection with a quartz vein.
indicator [ecol] A plant or animal peculiar to a specific environment, which can therefore be used to identify that environment.
indicator [glac geol] *indicator stone.*
indicator fan A pattern formed by the distribution of *indicator stones* derived from a restricted source. Cf: *boulder train.*
indicator horizon *marker band.*
indicator plant (a) A plant whose occurrence is broadly indicative of the soil of an area, e.g. its salinity or alkalinity, level of zone of saturation, and other soil conditions. (b) A plant that grows exclusively or preferentially on soil rich in a given metal or other element.—Syn: *plant indicator.*
indicator stone A glacial erratic whose source and direction of transportation are known because of its identity with bedrock in a certain small or restricted area. Syn: *indicator [glac geol].*
indicatrix In optics, a geometric figure that represents the refractive indices of a crystal: it is formed by drawing, from a central point representing the center of the crystal, lines in all directions, whose lengths represent the refractive indices for those vibration directions. The figure for an isotropic crystal is a sphere; for a uniaxial crystal, an ellipsoid of revolution; and for a biaxial crystal, a triaxial ellipsoid (Berry and Mason, 1959, p.192). Cf: *index surface.* Partial syn: *index ellipsoid.* Syn: *optic indicatrix.*
indices of lattice row Integral numbers, enclosed in square brackets, symbolized by [*uvw*], and determined, with any lattice point on the *lattice row* as origin, by the coordinates ua, vb, and wc, of the next lattice point on the row in terms of the unit cell edges a, b, and c. Indices enclosed in the following manner indicate a symmetrical set of axes: < *uvw* >. See also: *Miller indices; zone symbol.* Syn: *crystal axial indices.*
indicolite An indigo-blue (light-blue to bluish-black) variety of tourmaline, used as a gemstone. Syn: *indigolite.*
indifferent point In a system having two or more components, that point at which two phases become identical in composition, with the result that the system loses one degree of freedom, e.g. the maximum and minimum in a solid-solution series, or the melting point of a congruently melting compound (Levin et al., 1964, p. 6).
indigene An indigenous organism. Var: *indigen.*
indigenous Said of an organism originating in a specific place; native. Syn: *endemic.* Ant: *exotic [ecol].*
indigenous coal Coal formed according to the *in-situ theory;* autochthonous coal.
indigenous limonite Sulfide-derived limonite that remains fixed at the site of the parent sulfide, often as boxworks or other incrustation. Cf: *exotic limonite; relief limonite.*
indigenous stream A stream that lies wholly within its drainage basin. Cf: *exotic stream.*
indigirite A mineral: $Mg_2Al_2(CO_3)_4(OH)_2 \cdot 15H_2O$.
indigo copper *covellite.*
indigolite *indicolite.*
indirect effect *Bowie effect.*
indirect intake Recharge to the aquifer by way of another body of rock.
indirect leveling A type of *leveling* in which differences of elevation are determined indirectly, as from vertical angles and horizontal distances (trigonometric leveling), from atmospheric pressures (barometric leveling), or from the boiling point of water (thermometric leveling). Cf: *direct leveling.*
indirect linkage A type of *linkage* in scleractinian corals with one or more couples of mesenteries between each pair of neighboring stomodaea. See also: *trabecular linkage.* Cf: *direct linkage.*
indirect stratification *secondary stratification.*
indite An iron-black mineral: $FeIn_2S_4$.
indium A silvery-white tetragonal mineral, the native metallic element In. It is soft and malleable, and occurs in very small quantities in ores of zinc and other metals.
indochinite A tektite from southeast Asia (Cambodia, Laos, Vietnam, China, Thailand).
Induan European provincial stage: Lower Triassic (above Olenekian, below Virglorian).
induced infiltration Recharge to ground water by infiltration, either natural or man-made, from a body of surface water as a result of the lowering of the ground-water head below the surface-water level. Syn: *induced recharge.*
induced magnetization The magnetic field spontaneously induced in a volume of rock by the uniform action of an applied field. Its direction and magnitude are parallel and proportional, respectively, to the applied field. In the absence of *remanent magnetization,* induced magnetization is the magnetic moment per unit volume.
induced meander *advance-cut meander.*
induced polarization The production of a double layer of charge at a mineral interface, or production of changes in double-layer density of charge, brought about by application of an electric or magnetic field (induced electrical or magnetic polarization). Induced electrical polarization is manifested either by a decay of voltage in the Earth following the cessation of an excitation current pulse, or by a frequency dependence of the apparent resistivity of the Earth. Abbrev: *IP.*
induced radioactivity Radioactivity that is produced by *activation.* Cf: *artificial radioactivity.*
induced recharge *induced infiltration.*
inductance That property of an electric circuit by which an electromotive force is induced in it by a current variation, either in it or in a neighboring circuit.
induction [magnet] (a) *magnetic induction*. (b) *electromagnetic induction*.
induction [philos] Reasoning from the particular to the general, or from the individual to the universal; deriving general principles from the examination of separate facts. Ant: *deduction.*
induction log An *electric-log* curve obtained, without the use of electrodes, by raising through the uncased borehole transmitting coils (fed with a constant alternating current) that induce, in the

rocks surrounding the borehole, concentric eddy currents. These in turn induce fields that are detected by receiver coils. The magnitude of the fields is proportional to the conductivity of the surrounding rocks, and the log gives a continuous record of conductivity with depth. It is appropriate for measurements in empty holes, or in holes drilled with oil or with oil-base or freshwater muds. Syn: *conductivity log.*

induction number A distance in a conductive material measured as a number of radian wave lengths (real wave length/2π). It is a dimensionless quantity used to characterize the dimensions of a conductive body in an electromagnetic field.

inductive coupling The mutual impedance between a transmitting wire and a potential wire, arising in induction. This effect can lead to fictitious anomalies in induced-electrical-polarization surveys.

inductive method An electrical exploration method in which electric current is introduced into the ground by means of electromagnetic induction and in which the magnetic field associated with the current is determined.

inductura The smooth shelly layer of a gastropod shell secreted by the general surface of the mantle, commonly extending from the inner side of the aperture over the parietal region, columellar lip, and part or all of the shell exterior (TIP, 1960, pt.I, p.131).

indurated Said of a rock or soil hardened or consolidated by pressure, cementation, or heat.

indurated soil A very strongly cemented soil horizon. Cf: *cementation [soil].*

induration (a) The hardening of a rock or rock material by heat, pressure, or the introduction of cementing material; esp. the process by which relatively consolidated rock is made harder or more compact. See also: *lithification.* (b) The hardening of a soil horizon by chemical action to form a *hardpan.*

indusium The covering of a *sorus* on the leaf of a fern.

industrial diamond A general term for diamonds used in drilling, in wire drawing, and as a general abrasive. See also: *ballas; bort; carbonado.*

industrial lithostratigraphic unit A lithostratigraphic body recognized primarily for utilitarian purposes, such as an aquifer, oil sand, quarry layer, or ore-bearing bed. It is considered to be an *informal unit* even if named (ISG, 1976, p. 35).

industrial mineral Any rock, mineral, or other naturally occurring substance of economic value, exclusive of metallic ores, mineral fuels, and gemstones; one of the *nonmetallics.*

inequigranular A nonrecommended syn. of *heterogranular.* The term was originally applied by Cross et al. (1906, p. 698) to igneous rocks, but it has also been used for sedimentary rocks (such as recrystallized carbonate rocks) whose constituent crystals vary in size.

inequilateral Having the two ends unequal; specif. said of a bivalve-mollusk shell whose parts anterior and posterior to the beaks differ appreciably in length. Cf: *inequivalve.* Ant: *equilateral.*

inequilibrium stage In hypsometric analysis of drainage basins, the stage of early development corresponding to youth in the geomorphic cycle (Strahler, 1952b, p.1130); the hypsometric integral is greater than 60%. Cf: *equilibrium stage.*

inequivalve Having valves unequal in size and form; specif. said of a bivalve mollusk or its shell in which one valve is flatter (and often smaller) than the other. Cf: *inequilateral.* Ant: *equivalve.*

inert component A component whose amount in a rock after a metasomatic process depends on its initial concentration rather than on its chemical potential as externally fixed by the environment. Cf: *mobile component; perfectly mobile component.* Syn: *initial value component.*

inert gas Any of the six elements that have no tendency to react with any of the other elements: helium, neon, argon, krypton, xenon, and radon. They are all gases under usual conditions. Syn: *noble gas.*

inertial surveying instrument A device, mounted in a helicopter or truck, which combines gyroscopes, accelerometers, and electronic computers, and is used to extend positions from a known position or a basic net.

inertial surveying system A complex system of precise gyroscopes and accelerometers combined with electronics to sense and measure movement. Originally designed for navigation and guidance of military aircraft and ships, through continued development and refinement the systems are competitive in remote areas with conventional ground surveying methods for extending horizontal control.

inertinite A coal maceral group including *micrinite, macrinite, sclerotinite, fusinite, semifusinite,* and *inertodetrinite.* They are characterized by a relatively high carbon content and a reflectance higher than that of vitrinite. They are relatively inert during the carbonization process. Cf: *exinite; vitrinite.* Syn: *inerts.*

inertite A microlithotype of coal that contains a combination of inertinite macerals totalling at least 95% (ICCP, 1971). Cf: *fusite; liptite; vitrite.*

inertodetrinite A maceral of coal within the *inertinite* group, having an angular outline but no internal structure (ICCP, 1971). Particles are usually less than 30 microns across.

inertodetrite A microlithotype of coal consisting of at least 95% of *inertodetrinite* (ICCP, 1971).

inerts An informal term, synonymous with *inertinite.*

inesite A rose-red to flesh-red mineral: $Ca_2Mn_7Si_{10}O_{28}(OH)_2 \cdot 5H_2O$.

inface The steeper of the two slopes of a cuesta; the *scarp slope.* The term is an abbreviation of "inward-facing escarpment", referring to the cliff portion (of a cuesta) facing the oldland, as on a coastal plain.

infancy The initial or very early stage in any developmental sequence, esp. in the cycle of erosion, commencing when a region is freshly exposed to the action of surface waters. Regional infancy is characterized by smooth, nearly level erosion surfaces, imperfectly dissected by narrow stream gorges; numerous original and slight depressions, occupied by marshy lakes and ponds; shallow streams; and imperfect drainage systems. Cf: *youth.* Syn: *topographic infancy.*

infantile Pertaining to the initial or very early stage of the cycle of erosion, either of a stream that has just begun its work of erosion, or of a landscape with a smooth surface and numerous shallow lakes. Cf: *youthful.*

infauna Those aquatic animals that live within rather than on the bottom sediment. Cf: *epifauna.*

inferior groove A groove in decapods that extends downward from the posterior part of the hepatic groove; it might be considered a continuation of the postcervical groove (Holthuis, 1974, p. 733).

inferior ovary An *epigynous* plant ovary.

inferred ore Ore for which there are quantitative estimates of tonnage and grade made in only a general way, based on geologic relationships and on past mining experience, rather than on specific sampling. Cf: *indicated ore; possible ore; potential ore.*

infilling A process of deposition by which sediment falls or is washed into depressions, cracks, or holes, as the filling-in of crevasses upon the melting of glacier ice.

infiltration The flow of a fluid into a solid substance through pores or small openings; specif. the movement of water into soil or porous rock. Cf: *percolation.*

infiltration capacity The maximum or limiting infiltration rate. The term is considered an obsolete syn. of *infiltration rate* by the Soil Science Society of America (1965, p. 338). Symbol: f.

infiltration coefficient The ratio of infiltration to precipitation for a specific soil under specified conditions (Nelson & Nelson, 1967, p. 192).

infiltration front *pellicular front.*

infiltration gallery A horizontal conduit constructed for the purpose of intercepting ground water. The galleries often parallel rivers, which provide a perennial water supply to the conduit. Syn: *gallery [grd wat].*

infiltration index The average rate of infiltration, expressed in inches per hour. It equals the average rate of rainfall "such that the volume of rainfall at greater rates equals the total direct runoff" (Langbein & Iseri, 1960, p. 12).

infiltration metasomatism A process of mass transfer in which chemical components are transported by a stream of aqueous solutions percolating through pores in rocks. Cf: *diffusion metasomatism.*

infiltration rate The rate at which a soil under specified conditions can absorb falling rain or melting snow; expressed in depth of water per unit time (cm/sec; in/hr). Syn: *infiltration velocity.* Cf: *infiltration capacity.*

infiltration vein An interstitial mineral deposit formed by the action of percolating waters. Cf: *segregated vein.*

infiltration velocity *infiltration rate.*

infiltration water Nonpreferred syn. of *free water.*

infiltrometer An instrument used to measure the infiltration of water into soil.

infinitesimal-strain theory A theory of material deformation in which small displacements and small strains —in geology, less than about 1%—are analyzed. Cf: *finite-strain theory.*

inflammable cinnabar A mixture of cinnabar, idrialite, and clay.

inflation [paleont] The distance, measured normal to the plane of symmetry, between the right and left sides of a bivalve mollusk; the length of the middorsal-midventral line, or the "width" of the shell. Syn: *thickness [paleont].*

inflation [volc] *tumescence.*

inflection angle The angle at which a contour line diverges downstream from a stream channel. Symbol: ψ.

inflorescence A cluster of flowers. See also: *panicle.*

inflow (a) The act or process of flowing in; e.g. the flow of water into a lake. Syn: *influx.* (b) Water that flows in; e.g. ground water and rainfall flowing into the streams of a drainage basin. Also, the amount of water that has flowed in.

inflow cave A cave into which a stream flows, or is known to have flowed. See also: *outflow cave; sinking stream.* Syn: *influent cave.*

influent adj. Flowing in.—n. (a) A surface stream that flows into a lake (e.g. an inlet), or a stream or branch that flows into a larger stream (e.g. a tributary). Ant: *effluent.* Cf: *influent stream.* Syn: *affluent.* (b) A stream that flows into a cave.

influent cave *inflow cave.*

influent flow Flow of water into the ground from a body of surface water; e.g. the flow of water from an influent stream.

influent seepage Movement of gravity water in the zone of aeration, from the ground surface toward the water table; *seepage* of water into the ground.

influent stream (a) A stream or reach of a stream that contributes water to the zone of saturation and develops bank storage; its channel lies above the water table. Syn: *losing stream.* (b) *influent.*

influx *inflow.*

informal unit A lithologic body to which casual reference is made but for which there is insufficient need, insufficient information, or inappropriate basis to justify designation as a *formal unit.* It may be referred to informally as a lithozone, e.g. shaly zone or coal-bearing zone; or as a bed or member, e.g. sandy beds, shaly member (ISG, 1976, p. 35). Other entities considered to be informal units include *industrial lithostratigraphic units;* units such as "formation A" or "map unit 2"; zones as applied to rock-stratigraphic units (e.g. "producing zone" or "mineralized zone"); marker horizons established on electric and other mechanically recorded logs; soils in a rock-stratigraphic classification; several minor units in one vertical sequence having identical geographic names; the several units constituting a cycle of sedimentation; general rock units (e.g. "Lower Cambrian strata"); and a unit within another (e.g. "Chinle sandstone" referring to a sandstone of the Chinle Formation). Names of informal units follow the same rules of capitalization as ordinary common nouns (although failure to capitalize a unit name does not necessarily render the name informal).

infrabasal plate Any plate of the proximal circlet in a crinoid dorsal cup that has two circlets of plates (dicyclic) below the radial plates. Syn: *infrabasal.*

Infracambrian *Eocambrian.*

infracrustal Below the surface (i.e., within the crust) and, in general, below the zone of weathering and diagenesis. Cf: *supracrustal.*

infraglacial *subglacial.*

infralaminal accessory aperture An *accessory aperture* in the test of a planktonic foraminifer that leads to a cavity beneath accessory structures and that is at the margin of these structures (as in *Catapsydrax*). Cf: *intralaminal accessory aperture.*

inframarginal sulcus *scrobis septalis.*

infrared Pertaining to or designating that part of the electromagnetic spectrum ranging in wavelength from 0.7 μm to about 1 mm.

infrared absorption spectroscopy The observation of an *absorption spectrum* in the infrared frequency region and all processes of recording and measuring that go with it. Absorption of radiant energy for transitions within molecules at the lowest energy levels produces an infrared absorption spectrum.

infrared atmospheric transmission window In the atmosphere, a spectral band in which there is minimal absorption, hence maximum transmission of radiation. Water vapor is the single most important attenuator of infrared radiation. The largest transmission window in the middle infrared region occurs between 8 and 14 m. Syn: *transmission window.*

infrared color photograph A color photograph in which the red-imaging layer is sensitive to photographic *infrared* wavelengths, the green-imaging layer to red light, and the blue-imaging layer to green light.

infrared image An image produced by the *thermal infrared* spectral band.

infrasculpture A kind of structure of spores and pollen consisting of organized internal modifications of exine.

infrastructure Structure produced at a deep crustal level, in a plutonic environment, under conditions of elevated temperature and pressure, which is characterized by plastic folding, and the emplacement of granite and other migmatitic and magmatic rocks. This environment occurs in the internal parts of most orogenic belts, but the term is used especially where the infrastructure contrasts with an overlying, less disturbed layer, or *superstructure.*

infundibulum (a) A deep indentation of the *scrobis septalis* or a basal indentation of the apertural face of a foraminiferal test (as in *Alabamina*) (TIP, 1964, pt.C, p.61). (b) The apertural region of a tintinnid. (c) In the vertebrates, a neural component of the pituitary gland.—Pl: *infundibula.*

infusible Said of a mineral that ranks with quartz on the fusibility scale; i.e., that will not fuse in temperatures up to about 1500°C. Cf: *fusibility.*

infusorial earth An obsolete syn. of *diatomite.* Syn: *infusorial silica.*

infusorial silica *infusorial earth.*

ingenite A general term, now obsolete, for a rock originating below the Earth's surface; an igneous or metamorphic rock. Cf: *derivate; plutonic rock.*

ingrafted stream *engrafted stream.*

ingression The entering of the sea at a given place, as the drowning of a river valley (Schieferdecker, 1959, terms 1260 & 1840).

ingrown meander A term proposed by Rich (1914, p. 470) for a continually growing or expanding *incised meander* formed during a single cycle of erosion by the enlargement or accentuation of an initial minor sinuosity while the stream was actively downcutting; a meander that "grows in place". It exhibits a pronounced asymmetric cross profile (a well-developed, steep undercut slope on the outside of the meander, a gentle slip-off slope on the inside) and is produced when the rate of downcutting is slow enough to afford time for lateral erosion. Cf: *entrenched meander.*

ingrown stream A stream that has enlarged its original course by undercutting the outer (concave) banks of its curves.

inhalant canal Any canal forming part of the inhalant system of a sponge. Syn: *incurrent canal; prosochete.*

inhalant system The part of the *aquiferous system* of a sponge between the ostia and the prosopyles, characterized by water flowing inward from the ostia. Cf: *exhalant system.*

inherent ash *Ash* in coal derived from inorganic material that was structurally part of the original plant material. It cannot be separated mechanically from the coal, of which it commonly constitutes not more than 1%. Cf: *extraneous ash.* Syn: *constitutional ash; fixed ash; intrinsic ash; plant ash.*

inherent mineral matter *Mineral matter* in coal that was structurally part of the original organic material.

inherent moisture (a) In coal, that fraction of the *moisture content* that is structurally contained in the material. Syn: *bed moisture.* (b) The maximum moisture that a sample of coal will hold at 100% humidity and atmospheric pressure.

inherited (a) Said of a geologic structure, feature, or landscape that owes its character to conditions or events of a former period; esp. said of a *superimposed* stream, valley, or drainage system. (b) Also, said of a soil or sediment characteristic that is directly related to the nature of the parent material rather than to formative processes. See also: *lithomorphic.*

inherited argon Argon-40 that is produced within mineral grains by the decay of potassium-40 before the event being dated. It may be argon-40 generated during the premetamorphic history of a rock, which has survived the metamorphic event, or argon-40 due to the incorporation of older contaminating mineral grains in a dated sample. Cf: *excess argon; extraneous argon; radiogenic argon.*

inherited flow control The control of glacial drainage by the disposition of blocks of dead ice that had broken along thrust planes or open crevasses when the ice was in motion (Gravenor & Kupsch, 1959, p. 49).

inherited meander *entrenched meander.*
inherited septum A septum originally in the parent corallite, which is inherited by the offset during increase in corals (Fedorowski & Jull, 1976, p. 41).
inhomogeneity breccia A term used by Sander (1951, p.2) for a breccia that forms paradiagenetically by the rupture of relatively friable layers occurring within a plastic sediment. It contains sharp fragments with broken borders that sometimes can be matched with each other.
initial bud (a) An outgrowth through a hole in the wall of a graptolithine sicula, producing the first theca of the rhabdosome. (b) The first youthful polyp formed asexually from the protopolyp in a colony of corals.
initial dip (a) A syn. of *primary dip.* (b) The dip that a bedded deposit attains due to compaction after sedimentation, but before tectonic deformation.
initial landform A landform that is produced directly by epeirogenic, orogenic, or volcanic activity, whose original features are only slightly modified by erosion; it is dominant in the initial and youthful stages of the erosion cycle. Cf: *sequential landform; ultimate landform.*
initial meridian *prime meridian.*
initial point The point from which any survey is initiated; esp. the point from which a survey within a given area of the U.S. public-land system begins and from which a base line is run east and west and a principal meridian is run north and south. Syn: *point of origin.*
initial shoreline A shoreline brought about by regional tectonic activity (subsidence, uplift, faulting, folding), by volcanic accumulation, or by glacial action; it may have any slope, from almost vertical to nearly horizontal, and may be either smooth or irregular.
initial value component *inert component.*
injection [ign] *intrusion [ign].*
injection [sed] (a) The forcing, under abnormal pressure, of sedimentary material (downward, upward, or laterally) into a preexisting deposit or rock, either along some plane of weakness or into a crack or fissure; e.g. the transformation of wet sands and silts to a fluid state and their emplacement in adjacent sediments, producing structures such as sandstone dikes or sand volcanoes. See also: *intrusion [sed].* (b) A sedimentary structure or rock formed by injection.—Syn: *sedimentary injection.*
injection breccia A fragmental rock formed by the introduction of largely foreign rock fragments into veins and fractures in the host rock (Speers, 1957). Some examples (notably the Sudbury breccias at Sudbury, Canada) are associated with structures of probable meteorite-impact origin and may have formed as a result of the impact process itself.
injection complex An assemblage or association of rocks consisting of igneous intrusions in intricate relationship to sedimentary and metamorphic rocks, such as the ancient rocks underlying the oldest sedimentary formations in eastern U.S.
injection dike A sedimentary dike formed by abnormal pressure of injection from below or above or from the side. Cf: *neptunian dike.*
injection fold An *incompetent* fold.
injection gneiss A composite rock whose banding is wholly or partly caused by *lit-par-lit* injection of granitic magma into layered rock (Holmes, 1928, p. 124). Cf: *arterite; phlebite; venite.*
injection ice Shallow ground ice, either seasonal or perennial, formed when water intrudes beneath layers of rock or soil; it can occur in beds several meters thick and hundreds of meters long.
injection metamorphism Metamorphism accompanied by intimate injection of sheets and streaks of liquid magma (usually granitic) in zones near deep-seated intrusive contacts. Cf: *plutonic metamorphism; lit-par-lit.*
injection well (a) In water supply, a *recharge well* or *input well.* (b) A well in an oil or gas field through which water, gas, steam, or chemicals are pumped into the reservoir formation for *pressure maintenance* or *secondary recovery,* or for storage or disposal of the injected fluid.
inland Pertaining to or lying in the interior part of a country or continent, or not bordering on the sea; e.g. an *inland* lake.
inland basin *interior basin.*
inland drainage *internal drainage.*
inland ice (a) The ice forming the inner part of a *continental glacier* or large *ice sheet.* The term is applied esp. to the ice on Greenland. (b) A continental glacier or ice sheet in its entirety.
inland sea *epicontinental sea.*
inland waters The *territorial waters* (such as lakes, canals, rivers, inlets, bays) within the territory of a state, but excluding high seas and marginal waters that are subject to sovereign rights of bordering states; waters that are above the rise and fall of the tides.
inland waterway One of a system of navigable inland bodies of water (such as a river, canal, or sound).
inlet (a) A small, narrow opening, recess, indentation, or other entrance into a coastline or a shore of a lake or river, through which water penetrates into the land. Cf: *pass.* (b) A waterway entering a sea, lake, or river; a creek; an inflowing stream. (c) A short, narrow waterway between islands, or connecting a bay, lagoon, or similar body of water with a larger body of water, such as a sea or lake; e.g. a waterway through a coastal obstruction (such as a reef or a barrier island) leading to a bay or lagoon. Syn: *tongue.* (d) *tidal inlet.*
inlier An area or group of rocks surrounded by rocks of younger age, e.g. an eroded anticlinal crest. Cf: *outlier.*
in-line offset The component of the distance from a geophone to the shotpoint in the direction of the line of the spread. Cf: *perpendicular offset.*
innate Said of certain igneous rocks that have undergone transformation without intrusion or other change of position, such as rocks formed by simple fusion in place (Medlicott & Blanford, 1879, p. 752).
innelite A mineral: $Na_2(Ba,K)_4(Ca,Mg,Fe)Ti_3Si_4O_{18}(OH,F)_{1.5}(SO_4)$.
inner bar A bar formed at the upper bend of a flood channel, or where the waters of a river are checked by a flood tide. Ant: *outer bar.*
inner beach The part of a sandy beach that is covered by the wash of gentle waves and is ordinarily saturated. Cf: *foreshore.*
inner core The central part of the Earth's *core,* extending from a depth of about 5100 km to the center (6371 km) of the Earth; its radius is about one third of the whole core. The inner core is probably solid, as evidenced by the observation of *S* waves that are propagated in it, and because compressional waves travel noticeably faster through it than through the outer core. Density ranges from 10.5 to 15.5 g/cm^3. It is equivalent to the *G layer.* Cf: *outer core.* Partial syn: *lower core.* Syn: *siderosphere.*
inner hinge plate Either of a pair of subhorizontal *hinge plates* in the cardinalia of some brachiopods (such as rhynchonelloids, spiriferoids, and terebratuloids), located median of the crural bases and fused laterally with them. Cf: *outer hinge plate.*
inner lamella The thin layer covering an ostracode body in the anterior, ventral, and posterior parts of the carapace, chitinous except for calcified marginal parts forming the duplicature (TIP, 1961, pt.Q, p.51). Cf: *outer lamella.*
inner lamina The inner shell layer of a compartmental plate of certain cirripede crustaceans, separated from an outer lamina by parietal tubes.
inner lead An area of calm water between a line of parallel offshore islands (such as a string of *skerries*) and the mainland.
inner lip The adaxial (inner) margin of the aperture of a gastropod shell, extending from the foot of the columella to the suture. It consists of the *columellar lip* and the *parietal lip.* Cf: *outer lip.*
inner lowland The innermost of the lowland belts of a *belted coastal plain,* formed in less-resistant rocks that separate the oldland from the cuesta landscape (the first cuesta scarp descending to the bottom of the lowland). Syn: *inner vale.*
inner mantle *lower mantle.*
inner plate One of a pair of subvertical plates in the cardinalia of some pentameracean brachiopods, lying on the ventral side of the base of the brachial process and fused dorsally with it. Cf: *outer plate.*
inner reef One of the reefs comprising the landward or shelfward part of a *reef complex* or *reef tract.* These are often smaller and less developed than *outer reefs* in the same region. Cf: *leeward reef; windward reef.*
inner-shelf shoal An arcuate and linear shelf sand body such as occurs on the U. S. inner continental shelf, which is formed by nearshore processes (Swift et al., 1972, Ch. 23).
inner side The portion of a conodont element on the concave side of the anterior-posterior midline. Ant: *outer side.*
inner space The region involved in marine research; the ocean or the ocean environment. Cf: *outer space.* Syn: *hydrospace.*
inner vale *inner lowland.*
inner vesicle A membranous expansion from the distal part of a

maternal zooid in some cheilostome bryozoans, which partly fills the opening of a brood chamber (Ryland, 1970, p. 96).

inner wall The innermost wall in double-walled archaeocyathids, surrounding the central cavity (TIP, 1972, pt. E, p. 19-20).

inninmorite An igneous rock composed of augite and plagioclase (anorthite to labradorite) in a groundmass of sodic plagioclase, augite, and abundant glass. It is similar in composition to *cumbraite.* The name is for Inninmore, Scotland. Not recommended usage.

inoperculate adj. Having no operculum; e.g. said of an irregular tear that serves as the opening of a sporangium through which spores are discharged.—n. An inoperculate animal or shell; e.g. an inoperculate gastropod shell.

inorganic Pertaining or relating to a compound that contains no carbon. Cf: *organic.*

inosculation The union of tributaries to form a main stream.

inosilicate A class or structural type of *silicate* characterized by the linkage of the SiO_4 tetrahedra into linear chains by the sharing of oxygens. In a simple chain, e.g. pyroxenes, two oxygens are shared; in a double chain or band, e.g. amphiboles, half the SiO_4 tetrahedra share three oxygens and the other half share two. The Si:O ratio of the former type is 1:3 and for the latter it is 4:11. Cf: *nesosilicate; sorosilicate; cyclosilicate; phyllosilicate; tectosilicate.* Syn: *chain silicate.*

in-place assemblage *fossil community.*

input well *injection well.*

inquilinism A form of *commensalism* in which one organism lives inside another, usually in the digestive tract or respiratory chamber. Adj: *inquiline.* Cf: *parasitism.*

in regime Said of a stream or channel that has attained an average equilibrium or that is capable of adjusting its cross-sectional form or longitudinal profile by means of alterations imposed by the flow, and in which the average values of the quantities that constitute regime show no definite trend over a period of years (such as 10-20 years). Cf: *graded [geomorph].*

inselberg A prominent isolated residual knob, hill, or small *mountain of circumdenudation,* usually smoothed and rounded, rising abruptly from and surrounded by an extensive lowland erosion surface in a hot, dry region (as in the deserts of southern Africa or Arabia), generally bare and rocky although partly buried by the debris derived from and overlapping its slopes; it is characteristic of an arid or semiarid landscape in a late stage of the erosion cycle. The term was originated by W. Bornhardt. Etymol: German *Inselberg,* "island mountain". Pl: *inselbergs; inselberge.* Cf: *monadnock; bornhardt.* Syn: *island mountain.*

insequent adj. Said of a stream, valley, drainage system, or type of dissection that is seemingly uncontrolled by the associated rock structure or surface features, being determined by minor inequalities not falling into any larger-scale pattern. Etymol: *in* + con*sequent.* Syn: *inconsequent.*—n. *insequent stream.*

insequent stream A stream developed on the present surface but not consequent upon it and apparently not controlled or adjusted by the rock structure and surface features; a self-guided stream that develops under accidental or chance controls and whose resulting drainage pattern is dendritic, as a young stream wandering irregularly on a nearly level plain underlain by homogeneous or horizontally stratified rocks. The term was proposed by Davis (1897, p. 24). Syn: *insequent.*

insert adj. Having the ocular plates of an echinoid in contact with the periproctal margin. Ant: *exsert.* Syn: *inserted.*

inset [cart] *inset map.*

inset [petrology] A term proposed by Shand (1947) to replace the term *phenocryst.*

inset [streams] A channel where water flows in.

inset ice stream An *ice stream* from a tributary glacier that is set into the surface of a larger glacier and does not extend to the bed; e.g. a *superimposed ice stream* set into the surface of a trunk glacier a short distance from their confluence. Cf: *juxtaposed ice stream.*

inset map A small, separate map that is positioned within the *neat line,* and in an unimportant part, of a larger map for economy of space or for legibility; e.g. a map of an area geographically outside a map sheet but included therein for convenience of publication, or a part of a larger map drawn at an enlarged or reduced scale. It may or may not be at the same scale as the larger map. Syn: *inset.*

inset terrace A *stream terrace* formed during successive periods of vertical and lateral erosion such that remnants of the former valley floor are left on both sides of the valley (Schieferdecker, 1959, term 1512).

inshore (a) Situated close to the shore or indicating a shoreward position; specif. said of a zone of variable width extending from the low-water shoreline through the breaker zone. See also: *offshore; nearshore.* (b) In a narrow sense, said of a zone that is equivalent to the *shoreface.*

inshore water (a) Water that is adjacent to land, with physical properties influenced considerably by continental conditions. Ant: *offshore water.* (b) A strip of open water located seaward of an icefoot or land, produced by the melting of *fast ice* along the shore.

inside pond A body of fresh water enclosed or partly enclosed by sediment deposited from bifurcating distributaries, and lying inland from the delta front of the Mississippi River delta. Cf: *outside pond.*

in-situ combustion A technique used for recovering oil of low gravity and high viscosity from a reservoir when primary methods have failed. The method involves heating the oil in the formation by igniting it (burning it in place), keeping combustion alive by pumping air into the formation. As the front of burning oil advances, the heat breaks down the oil into coke and light oil, and the latter is pushed ahead to producing wells.

in-situ density The *density* of a unit of water, measured at its actual depth. Cf: *potential density.*

in-situ mining Removal of the valuable components of a mineral deposit without physical extraction of the rock, e.g. by *solution mining.*

in-situ modeling The interpretation of gravity or magnetic data in which the calculated field points are the same as those of actual field stations, as opposed to points located on a datum or reference plane.

in-situ temperature The temperature of a unit of water, measured at its actual depth. Cf: *potential temperature [oceanog].*

in-situ theory The theory that coal originates at the place where its constituent plants grew and decayed. Ant: *drift theory [coal].* See also: *autochthony.* Syn: *swamp theory.*

insizwaite A mineral of the pyrite group: $Pt(Bi,Sb)_2$.

insoak The absorption of free surface water by unsaturated soil.

insolation [meteorol] The combined direct solar and sky radiation reaching a given body, e.g. the Earth; also, the rate at which it is received, per unit of horizontal surface. Cf: *solar constant.*

insolation [weath] (a) Exposure to the Sun's rays. (b) The geologic effect of the Sun's rays on the Earth's surficial materials; specif. the effect of changes of temperature on the mechanical weathering of rocks. See also: *shadow weathering.*

insolilith A relatively rounded pebble with a rough or cracked surface produced by exfoliation or granular disintegration resulting from insolation.

insoluble residue The material remaining after the more soluble part of a rock sample has been dissolved in hydrochloric acid or acetic acid. It is chiefly composed of siliceous material (e.g. chert or quartz) and various detrital minerals (e.g. anhydrite, glauconite, pyrite, and sphalerite). Abbrev: IR. See also: *siliceous residue.*

inspissation The alteration of an oil deposit, as by evaporation or oxidation during long exposure, in which gases and lighter fractions escape, and *asphalt* and *heavy oils* remain.

instant The geochronologic equivalent of a *chronohorizon* that has no appreciable time duration (ISG, 1976, p. 68). It is marked by a *time plane.* Cf: *moment.*

instantaneous field of view The solid angle through which a detector is sensitive to radiation. In a scanning system this refers to the solid angle subtended by the detector when the scanning motion is stopped. Cf: *angular field of view.*

instant rock A colloquial term for a fragile rock produced from originally fragmental materials by the shock waves associated with explosions or meteorite impacts, i.e. by *shock lithification.*

instar (a) The ontogenetic stage in the life of an arthropod occurring between two successive molts (periods of ecdysis); an immature molted or shed carapace of an ostracode. Also, an individual in a specified instar. (b) A single episode of shell formation in foraminifers, commonly of a single chamber.

instrument error A *systematic error* resulting from imperfections in, or faulty adjustment of, the instrument or device used. Such an error "may be accidental or random in nature and result from the failure of the instrument to give the same indication when subjected to the same input signal" (ASP, 1966, p. 1138).

instrument station A *station* at which a surveying instrument is set up for the purpose of making measurements; e.g. the point over which a leveling instrument is placed for the purpose of taking a backsight or foresight. Syn: *setup.*

insular [clim] Said of a climate in which there is little seasonal temperature variation, e.g. a marine climate.

insular [ecol] Said of an organism that has a limited or isolated range or habitat, esp. if on an island or group of islands.

insular shelf An area of the ocean floor analagous to the *continental shelf,* but surrounding an island. Syn: *island shelf.*

insular slope An area of the ocean floor analogous to a *continental slope,* but surrounding an island. Syn: *island slope.*

insulated stream A stream or reach of a stream that neither contributes water to the zone of saturation nor receives water from it (Meinzer, 1923, p. 56); it is separated from the zone of saturation by an impermeable bed.

insulosity The percentage of the area of a lake that is occupied by islands.

insurgence *sinking stream.*

intaglio A carved gem that may be used as a seal, in which the design has been engraved into the stone. Pron: in-*tal*-yoe. Cf: *cameo.* See also: *cuvette [gem].*

intake (a) *recharge.* (b) The openings in water-bearing materials through which water passes into a well.

intake area *recharge area.*

intectate Said of a pollen grain lacking a tectum.

integraph Any device used in carrying out a mathematical integration by graphical means, e.g., a *dot chart* for computing terrain effects used in the reduction of gravity data.

integrate Said of a type of wall structure in trepostome bryozoans in which the zooecial boundaries appear in tangential section as narrow, well-defined lines. Cf: *amalgamate.*

integrated drainage Drainage developed during maturity in an arid region, characterized by coalescence (across intervening ridges and mountains) of drainage basins as a result of headward erosion in the lower basins or of spilling over from the upper basins due to aggradation (Lobeck, 1939, p. 12-13); drainage developed where various higher local base levels are replaced by a single lower base level.

integrated geophysics The combination of seismic, gravity, magnetic, electrical, and/or well-log data to effect a more accurate and complete interpretation than any one data set could provide if used alone.

integration In petrology, the formation of larger crystals from smaller crystals by recrystallization. See also: *regenerated crystal.*

integripalliate Said of a bivalve mollusk devoid of a pallial sinus in the pallial line. Cf: *sinupalliate.*

integument In a flowering plant, the covering layer of an ovule. It contributes to the coat or *testa* of a mature seed.

intensity *earthquake intensity.*

intensity scale A standard of relative measurement of *earthquake intensity.* Three such systems are the *Mercalli scale,* the *modified Mercalli scale,* and the *Rossi-Forel scale.*

intensiveness In quantification of hydrothermal alteration, the result of fixing intensive variables, which in turn set mineral stabilities. It is described in terms of stable alteration-mineral assemblages, such as "potassic", "quartz-sericite", or "greisen". Cf: *pervasiveness; extensiveness.*

intensive variable A thermodynamic variable that is independent of the total amount of matter in the system. Examples include temperature, pressure, and mole fraction.

interambulacral adj. Situated between ambulacra of echinoderms, such as an "interambulacral ray" of a crinoid; esp. referring to the thecal plates making up the area between ambulacra. Cf: *ambulacral.*—n. Any plate situated between the ambulacral plates of an echinoderm.

interambulacrum One of the areas between two ambulacra in an echinoderm; any of the five interradial sections of most echinoderms.

interamnian Situated between or enclosed by rivers. See also: *interfluvial.*

interantennular septum A plate that separates two antennular cavities in some malacostracan crustaceans. Syn: *proepistome.*

interarea The posterior sector of a brachiopod shell with the growing edge at the hinge line. The term is more commonly used for any plane or curved surface lying between the beak and the posterior margin of a brachiopod valve and bisected by the delthyrium or the notothyrium (TIP, 1965, pt.H, p.146). The interarea is generally distinguished by a sharp break in angle from the remainder of the valve and by the absence of costae, plicae, or coarse growth lines. Cf: *pseudointerarea; planarea.*

interbasin area A roughly triangular area, located between adjacent tributary drainage basins, that has not developed a drainage channel but contributes drainage directly into a higher-order channel (Schumm, 1956, p.608). Symbol: A_0.

interbasin length The maximum horizontal length of the *interbasin area,* measured from the apex of the triangular ground surface to the adjacent channel (Strahler, 1964). Symbol: L_0.

interbed A bed, typically thin, of one kind of rock material occurring between or alternating with beds of another kind.

interbedded Said of beds lying between or alternating with others of different character; esp. said of rock material laid down in sequence between other beds, such as a contemporaneous lava flow "interbedded" with sediments. Cf: *intercalated.* Syn: *interstratified.*

interbiohorizon-zone *interval-zone.*

interbrachial adj. Situated between arms; e.g. an "interbrachial margin" of the disc of an asterozoan.—n. A crinoid plate occurring in the dorsal cup between brachial plates of adjacent ambulacral rays or between brachs of any single ray.

intercalary adj. Interposed or inserted between, or introduced or existing interstitially, e.g. "intercalary apical system" of an echinoid, in which ocular plates II and IV meet at midline so as to separate anterior and posterior portions; hooplike "intercalary bands" located between the valves and connecting bands in a diatom frustule; or "intercalary cuticle" of a cheilostome bryozoan formed by the outermost layers of lateral walls of two contiguous lineal series of zooids.—n. One of many thecal plates occurring between radial and basal circlets of some crinoids (e.g. *Acrocrinus*).

intercalated Said of layered material that exists or is introduced between layers of a different character; esp. said of relatively thin strata of one kind of material that alternate with thicker strata of some other kind, such as beds of shale that are intercalated in a body of sandstone. Cf: *interbedded.*

intercalation (a) The existence of one or more layers between other layers; e.g. the presence of sheets of lava between sedimentary strata, the occurrence of a particular fossil horizon between fossil zones of a different character, or the inclusion of lamellar particles of one mineral in another in such a way that the inclusions are oriented in planes related to the crystal structure of the host mineral. (b) The introduction of a layer between other layers already formed, such as the insertion of rock by intrusion or thrusting in a preexisting series of stratified rock. (c) An intercalated body of material, such as a bed in an intertongued zone or a lens of volcanic ash in a sedimentary deposit.

intercamarophorial plate A short, low median septum on the posterior midline of the camarophorium in brachiopods of the superfamily Stenoscismatacea, extending to the underside of the hinge plate but independent of the median septum duplex (TIP, 1965, pt.H, p.146).

intercameral Located between the chambers of a foraminiferal test; e.g. "intercameral foramen" representing a primary or secondary opening between successive chambers.

intercardiophthalmic region A small rectangular area of the prosoma of a merostome, embracing the cardiac lobe and a minor part of the interophthalmic region.

intercellular space The space resulting from a separation of adjacent plant cell walls from each other along more or less extended areas of contact. In some cases it may result from the splitting of the *middle lamella* in plant cell walls (Esau, 1965, p.62).

intercept [sed] One of the three linear dimensions or diameters of a sedimentary particle: the longest dimension is the "maximum" intercept, and the shorter dimensions are the "intermediate" and "short" intercepts.

intercept [surv] The part of the rod seen between the upper and lower stadia hairs of a transit or telescopic alidade; e.g. a stadia interval.

interception The process by which water from precipitation is caught and stored on plant surfaces and eventually returned to the atmosphere without having reached the ground. Also, the amount of water intercepted. Cf: *throughfall.*

intercept time The time obtained by extrapolating a seismic refraction alignment on a time-distance curve back to zero shot-to-geophone distance; the sum of the *delay times* at the shot and receiver ends of the path.

intercision (a) A type of *capture* characterized by sidewise swinging of mature streams (Lobeck, 1939, p. 201). (b) A type of diversion accomplished by the cutting back of bluffs along a lake shore such that the lake advances inland and cuts into a bend of a river valley some distance above its mouth (Goldthwait, 1908).

intercretion A term proposed by Todd (1903) for a *concretion* that grows by accretion (on the exterior) and by irregular and interstitial addition, causing a circumferential expansion and resultant cracking and wedging apart of the interior of the concretion; e.g. a *septarium.*

intercrystal porosity Porosity between equant, equal-sized crystals (Choquette & Pray, 1970, p. 247).

intercumulus The space between crystals of a cumulus.

intercumulus liquid Magmatic liquid that surrounds the crystals of a cumulus, i.e. that occupies the intercumulus. See also: *intercumulus material.* Syn: *interprecipitate liquid; interprecipitate material.*

intercumulus material Material that crystallized from *intercumulus liquid.* Syn: *interprecipitate material.*

interdigitation *intertonguing.*

interdistributary bay A pronounced indentation of the delta front between advancing stream distributaries, occupied by shallow water, and either open to the sea or partly enclosed by minor distributaries.

interdune Pertaining to the relatively flat surface, whether sand-free or sand-covered, between dunes; e.g. said of the long, trough-like, wind-swept passage between parallel longitudinal dunes, such as a *gassi* or a *feidj.*

interestuarine Situated between two estuaries.

interface [petroleum] A syn. of *contact (b).*

interface [sed] A depositional boundary separating two different physicochemical regions; specif. the surface separating the top of the uppermost layer of sediment and the medium (usually water) in which the sedimentation is occurring.

interface [seis] *discontinuity [seis].*

interfacial angle In crystallography, the angle between two faces of a crystal.

interference The condition occurring when the area of influence of a water well comes into contact with or overlaps that of a neighboring well, as when two wells are pumping from the same aquifer or are located near each other.

interference colors In crystal optics, the colors displayed by a *birefringent* crystal in crossed polarized light. Thickness and orientation of the sample and the nature of the light are factors that affect the colors and their intensity.

interference figure The pattern or figure that a crystal displays in polarized light under the conoscope. It is a combination of the *isogyre* and the *isochromatic curve,* and is used to distinguish axial from biaxial crystals and to determine optical sign. See also: *axial figure; biaxial figure.*

interference ripple mark *cross ripple mark.*

interfingering *intertonguing.*

interflow [ore dep] adj. Occurring in or between volcanic flows, as of an exhalite zone or tuff bed; specif., said of metalliferous sediments or pyroclastics intercalated with flows of Keewatin age at Cobalt, Ontario.

interflow [water] *storm seepage.*

interfluent lava flow A lava flow that is discharged into and through subterranean fissures and cavities in a volcano and may never reach the surface (Dana, 1890); an obsolete term. Cf: *effluent lava flow; superfluent lava flow.*

interfluminal *interfluvial.*

interfluve The area between rivers; esp. the relatively undissected upland or ridge between two adjacent valleys containing streams flowing in the same general direction. Cf: *doab.* Syn: *interstream area.*

interfluve hill A relatively flat-topped remnant of an antecedent slope on which gradation was arrested when the adjacent streams and valleys were developed (Horton, 1945, p. 360); it occurs along the divides in drainage basins approaching maturity. Syn: *interfluve plateau.*

interfluvial Lying between streams; pertaining to an interfluve. See also: *interstream; interamnian.* Syn: *interfluminal.*

interfolding The simultaneous development of discrete fold systems with different orientations.

interformational Formed or existing between one *formation [stratig]* and another; e.g. "interformational unconformity".

interformational conglomerate A conglomerate that is present within a formation, the constituents of which have a source external to the formation. Cf: *intraformational conglomerate.*

intergelisol *pereletok.*

interglacial adj. Pertaining to or formed during the time interval between two successive *glacial epochs* or between two *glacial stages.* The term implies both the melting of ice sheets to about their present level, and the maintenance of a warm climate for a sufficient length of time to permit certain vegetational changes to occur (Suggate, 1965, 619).—n. *interglacial stage.*

interglacial stage A subdivision of a glacial epoch separating two glaciations, characterized by a relatively long period of warm or mild climate during which the temperature rose to at least that of the present day; esp. a subdivision of the Pleistocene Epoch, as the "Sangamon Interglacial Stage". Syn: *interglacial; thermal [glac geol].*

interglaciation A climatic episode "during which the climate was incompatible with the wide extent of glaciers that characterized a glaciation" (ACSN, 1961, art. 40).

intergrade n. A soil that is transitional between two other soils. The subgroup category of the U.S. Dept. of Agriculture soil taxonomy is set up specifically to accommodate soils that are intergrades between two orders, suborders, or great groups.

intergranular Said of the ophitic texture of an igneous rock in which the augite occurs as an aggregation of grains, not in optical continuity, in the interstices of a network of feldspar laths that may be diverse, subradial, or subparallel (Johannsen, 1939, p. 219). The interstitial augite forms a relatively small proportion of the rock. This texture is distinguished from *intersertal* texture by the absence of interstitial glass. Nonrecommended syn: *granulitic [ign].*

intergranular movement A process that goes on within a glacier when grains of ice rotate and slide over each other like grains of corn in a chute. It is a significant factor in glacier flow only near the surface of a glacier. Cf: *intragranular movement.*

intergranular porosity The porosity between the grains or particles of a rock, such as that between the lithoclasts or the bioclasts of a carbonate sedimentary rock. Cf: *interparticle porosity.*

intergranular pressure *effective stress.*

intergrowth The state of interlocking of grains of two different minerals as a result of their simultaneous crystallization. Cf: *graphic intergrowth.*

interio-areal aperture An aperture in the face of the final chamber of a foraminiferal test, not at its base.

interiomarginal aperture A basal aperture in a foraminiferal test at the margin of the final chamber and along the final suture. In coiled forms it may be an *equatorial aperture* or an *extraumbilical aperture.* See also: *spiroumbilical aperture.*

interior basin (a) A depression entirely surrounded by higher land and from which no stream flows outward to the ocean. Cf: *closed basin.* Syn: *inland basin.* (b) *intracratonic basin.*

interior drainage *internal drainage.*

interior link A *link* in a channel network emanating from a *fork* (Shreve, 1967). Cf: *link; exterior link.*

interior plain A plain that is situated far from the borders of a continent, as contrasted with a *coastal plain.*

interior valley A large flat-floored closed depression in a karst area. Its drainage is subsurface, its size is measured in kilometers or tens of kilometers, and its floor is commonly covered by alluvium. Interior valleys may become intermittent lakes during periods of heavy rainfall, when the sinking streams that drain them cannot manage the runoff. Syn: *polje.* See also: *karst valley.*

interior wall A body wall in bryozoans that partitions pre-existing body cavity into zooids, parts of zooids, or extrazooidal parts.

interlacing drainage pattern *braided drainage pattern.*

interlacustrine Situated between lakes; e.g. an "interlacustrine overflow stream" spilling over from one lake to another.

interlaminated Said of laminae occuring between or alternating with others of different character; *intercalated* in very thin layers. Syn: *interleaved.*

interlayer A layer placed between others of a different nature; e.g. an interbed.

interlayering The regular or random arrangement of structural units of clay minerals, each unit differing from the adjacent unit either in composition or in crystallographic orientation.

interleaved *interlaminated.*

interlobate deposit Drift lying between two adjacent glacial lobes.

interlobate moraine A lateral or end moraine formed along the line of junction and roughly parallel to the axes of two adjacent glacial lobes that have pushed their margins together. Syn: *intermediate moraine.*

interlocking Pertaining to two seismic records made with the end geophone of one record occupying the shotpoint location of the other, so that a common raypath and a common arrival time are involved.

interlocking spur One of several projecting ridges extending alternately from the opposite sides of the wall of a young, V-shaped valley down which a river with a winding course is flowing, each lateral spur extending into a concave bend of the river so that viewed upstream the spurs seem to "interlock" or "overlap". Syn: *overlapping spur.*

interlocking texture A rock texture in which particles with irregular boundaries interlock by mutual penetration, as in a crystalline limestone.

interloculum A distinct interspace between the ektexine and endexine of certain triporate pollen grains (Kremp, 1965, p. 74).

intermediate Said of an igneous rock that is transitional between *basic* and *silicic* (or between *mafic* and *felsic*), generally having a silica content of 54 to 65 percent; e.g. syenite and diorite. "Intermediate" is one subdivision of a widely used system for classifying igneous rocks on the basis of their silica content; the other subdivisions are *acidic, basic,* and *ultrabasic.* Syn: *mediosilicic.*

intermediate belt That part of the *zone of aeration* that lies between the capillary fringe and the belt of soil water. Syn: *intermediate zone.*

intermediate coal A type of banded coal defined microscopically as consisting of between 60% and 40% of bright ingredients such as vitrain, clarain, and fusain, with clarodurain and durain composing the remainder. Cf: *semibright coal; semidull coal; bright coal; dull coal.*

intermediate contour A contour line drawn between index contours.

intermediate-focus earthquake An earthquake whose focus occurs between a depth of about 60 km and 300 km. Cf: *shallow-focus earthquake; deep-focus earthquake.*

intermediate layer *sima.*

intermediate moraine *interlobate moraine.*

intermediate plain A plain intermediate in altitude between the highest summits of an erosion surface and the bottoms of the deepest valleys (Trowbridge, 1921, p. 31-33). The term is misleading because it implies a plain intermediate in position between two other plains.

intermediate-scale map A map at a scale (in the U.S. Geological Survey, 1/50,000, 1/62,500, 1/100,000, and 1/125,000) that can show on quadrangle or county format most details of roads and culture, together with topography by appropriate contour interval. The Defense Mapping Agency defines intermediate scale as from 1/200,000 to 1/500,000. Cf: *large-scale map; small-scale map.*

intermediate vadose water Water of the intermediate belt. Syn: *argic water.*

intermediate water A cold *water mass* of relatively low *salinity,* originating at arctic and antarctic convergences. It lies above *deep water* and *bottom water.* Cf: *surface water [oceanog].*

intermediate wave *transitional-water wave.*

intermediate zone *intermediate belt.*

intermineral Pertaining to a time interval between periods of mineralization (Kirkham, 1971, p. 1245-1246); also, pertaining to those features, e.g. dikes, that were emplaced during such an interval. The term has been applied to a time interval within a period of mineralization and to its resulting features, but *intramineral* is preferred for this meaning. Cf: *premineral; postmineral.*

intermittent island A patch of the shallow bottom of a lake, exposed during a period of low lake level and covered during periods of higher level. See also: *shoal.*

intermittent lake A lake that normally contains water for only part of the year or that is only seasonally dry; e.g. a *deflation lake.* Cf: *ephemeral lake; playa lake.* Syn: *temporary lake.*

intermittent spring A spring that discharges only periodically. A geyser is a special type of intermittent spring (Meinzer, 1923, p. 54). Syn: *intermitting spring.* Cf: *perennial spring; periodic spring.*

intermittent stream (a) A stream or reach of a stream that flows only at certain times of the year, as when it receives water from springs or from some surface source. The term "may be arbitrarily restricted" to a stream that flows "continuously during periods of at least one month" (Meinzer, 1923, p. 68). (b) A stream that does not flow continuously, as when water losses from evaporation or seepage exceed the available streamflow.—Cf: *ephemeral stream.* Syn: *temporary stream; seasonal stream.*

intermitting spring *intermittent spring.*

intermont adj. *intermontane.*—n. A hollow between mountains.

intermontane Situated between or surrounded by mountains, mountain ranges, or mountainous regions; e.g. the Great Basin of western U.S., between the Sierra Nevada and the Wasatch Mountains. Syn: *intermont; intermountain.*

intermontane glacier A glacier formed by the confluence of several alpine glaciers and occupying a depression between mountain ranges or ridges.

intermontane plateau A plateau that is partly or completely enclosed by mountains, and that is formed in association with them; e.g. the Tibetan plateau.

intermontane trough (a) A subsiding area in an island-arc region of the ocean, lying between stable or uprising regions. (b) A basin-like area between mountain ranges, sometimes occupied by an intermontane glacier.

intermorainal Situated between moraines, as an *intermorainal* lake occupying a narrow depression between parallel moraines of a retreating glacier. Syn: *intermorainic.*

intermountain *intermontane.*

intermural increase A type of *increase* (offset formation of corallites) in cerioid coralla occurring at or within the intercorallite wall, so that no single corallite is recognizable as the parent.

internal cast A syn. of *steinkern.* The term should not be used for an *internal mold.*

internal cavity The central opening enclosed in single-walled archaeocyathids, comparable to the central cavity in double-walled forms (TIP, 1972, pt. E, p. 40). Cf: *central cavity.*

internal contact The planar or irregular surface between separately emplaced portions of a composite pluton (Compton, 1962, p. 277). Cf: *external contact.*

internal drainage Surface drainage whereby the water does not reach the ocean, such as drainage toward the lowermost or central part of an interior basin. It is common in arid and semiarid regions, as in western Utah. Ant: *external drainage.* See also: *centripetal drainage pattern.* Syn: *interior drainage; inland drainage; closed drainage; endorheism.*

internal energy That energy of a system described by the *first law of thermodynamics.*

internal erosion Erosion effected within a compacting sediment by movement of water through the larger pores (Bathhurst, 1958, p. 33). Cf: *internal sedimentation.*

internal friction That part of the shear strength of a rock or other intact solid that depends on the magnitude of the normal stress on a potential shear fracture. Cf: *cohesion.*

internal lobe The *dorsal lobe* in normally coiled cephalopod conchs. See also: *annular lobe.*

internal mold A *mold* or impression showing the form and markings of the inner surfaces of a fossil shell or other organic structure; it is made on the surface of the rock material filling the hollow interior of the shell or organism. It is sometimes called incorrectly a "cast of the interior", but can be so called only if the shell or structure itself be regarded as a mold. Cf: *external mold; internal cast.* See also: *steinkern.*

internal oblique muscle One of a pair of muscles in some inarticulate brachiopods, originating on the pedicle valve between the anterior adductor muscles, and passing posteriorly and laterally to insertions on the brachial valve located anteriorly and laterally from the posterior adductor muscles (TIP, 1965, pt.H, p.146). Cf: *lateral oblique muscle.*

internal rotation In structural petrology, a change in the orientation of structural features during deformation, referred to coordinate axes internal to the deformed body.

internal sedimentation Accumulation of clastic or chemical sediments derived from the surface of, or within, a more or less consolidated carbonate sediment (mud or silt), and deposited in secondary cavities formed in the host rock (after its deposition) by bending of laminae or by *internal erosion* or solution (Bathurst, 1958, p. 31).

internal seiche A free oscillation of a submerged layer in a stratified body of water occupying an enclosed or semi-enclosed basin; esp. an oscillation of the thermocline in a lake. It is believed to be initiated by the same factors that produce a surface *seiche.*

internal suture The part of a suture of a coiled cephalopod conch situated on the dorsum (or within the impressed area), extending between umbilical seams, and hidden from view unless the conch is broken. Cf: *external suture.*

internal tide Submerged vertical oscillations on density surfaces in the sea with tidal periods.

internal water Water in the interior of the Earth, below the zone of saturation, where interstices cannot exist due to the pressure of overlying rocks (Meinzer, 1923, p. 22).

internal wave A submerged wave occurring on a density surface, e.g. the thermocline, in density-stratified water. Because of the small density gradients involved in internal waves compared with external or surface waves, the internal wave heights, periods, and wavelengths are usually large. See also: *lee wave.*

International Active Sun Years An international cooperative program of studying solar-terrestrial phenomena during an active-sun, i.e. sunspot-maximum, period. It is related to the *International Geophyscial Year* and to the *International Years of the Quiet Sun.* Abbrev: IASY.

international date line *date line.*

International Geophysical Year An international cooperative program of observation of geophysical phenomena from July 1, 1957 to December 31, 1958, a period that was near a maximum in sunspot activity. See also: *International Active Sun Years; International Years of the Quiet Sun.* Abbrev: IGY.

international gravity formula The theoretical gravity at the latitude ϕ is 978,049 $(1 + 0.0052884 \sin^2\phi - 0.0000059 \sin^2 2\phi)$ mgal. (A better value may be 978,031.8 $(1 + 0.0053024 \sin^2\phi - 0.0000058 \sin^2 3\phi)$mgal.) See also: *latitude correction.*

International Hydrological Decade A ten-year program, 1965-74, patterned after the International Geophysical Year, aimed at training hydrologists and technicians and at establishment of networks for measuring hydrologic data. The idea originated in the United States, but the program was sponsored by UNESCO, and a large proportion of the membership of the United Nations participated.

international low water A datum plane so low that the tide will seldom fall below it: a reference plane below mean sea level by an amount equal to half the range between mean lower low water and mean higher high water multiplied by 1.5 (Baker et al., 1966, p. 87). Abbrev: ILW.

International Map of the World A map series at a scale of 1/1,000,000 (one inch to 15.78 miles), having a uniform set of symbols and conventional signs, using the metric system for measuring distances and elevations, and printed in modified polyconic projection on 840 sheets, each covering an area of 4°lat. and 6° long. except above the 60th parallel where the longitude covered is 12° on each sheet. The International Map of the World on the Millionth Scale was first suggested at the 5th International Geographical Congress in 1891 and was accepted in principle in 1909. It consists of an incomplete series of map sheets (many needing revision) generally published by national mapping agencies of concerned countries under the auspices of the United Nations. Abbrev: IMW. Syn: *millionth-scale map of the world.*

International Years of the Quiet Sun An international cooperative program during 1964-1965 of studying solar-terrestrial phenomena during a quiet-sun, i.e. sunspot-minimum, period. It is related to the *International Geophysical Year* and to the *International Active Sun Years.* Abbrev: IQSY.

internides Kober's term for the internal part of an orogenic belt, farthest away from the craton, commonly the site of a eugeosyncline during its early phases, and subjected later to plastic folding and plutonism. Cf: *externides.* See also: *primary orogeny; hinterland.* Syn: *primary arc.*

internodal A crinoid columnal disposed between nodals, generally smaller than nodals and lacking cirri.

internode An interval or part between two successive nodes; e.g. a segment of a colony of cheilostome bryozoans between articulations, or a segment of a jointed algal thallus.

interophthalmic region The space between the cardiac lobe and ophthalmic ridge of a merostome.

interparticle porosity The porosity between particles in a rock; e.g. *breccia porosity.* Choquette & Pray (1970, p. 247) recommend use of this term rather than the term *intergranular porosity* which suggests limitation to grain-size particles.

interpenetration twin A twinned crystal, in which the individuals appear to have grown through one another. Syn: *penetration twin.*

interpleural furrow A transversely directed furrow on the pleural regions of the pygidium of a trilobite, indicating the join between adjacent fused pleurae. Syn: *interpleural groove; rib furrow.*

interpluvial adj. Said of an episode of time that was drier than the *pluvial* periods between which it occurred.—n. Such an episode or period of time.

interpolation Estimation of the value of a variate based on two or more known surrounding values; a method used to determine intermediate values between known points on a line or curve.

interpositum In certain receptaculitids, a pentagonal plate which, in conjunction with a triangulum, is associated with an increase in the number of meromes per whorl.

interprecipitate liquid *intercumulus liquid.*

interprecipitate material *intercumulus material.*

interpretative log *interpretive log.*

interpretive log A *sample log* based on rotary well cuttings, in which the geologist attempts to show only the rock encountered by the bit at each sampled depth, ignoring the admixed material from higher levels. Cf: *percentage log.*

interpretive map As used in environmental geology, a map prepared for the general public that classifies the suitability of land for a particular use on the basis of geologic characteristics. Examples: general construction, sand and gravel development, land burial of waste, ground-water development.

interradial [paleont] adj. Situated midway between the axes of adjacent rays of an echinoderm; e.g. an "interradial suture" representing a common line or division between adjacent radial plates of a blastoid.—n. (a) A structure in the interradial area; e.g. a crinoid plate above a basal. (b) *interray.*

interradial [palyn] Pertaining to areas of the proximal face or equator of trilete spores, lying between the arms of the laesura. Cf: *radial [palyn].*

interradial loculus A subdivision of intervallar space bounded by septae, rods, pillars, or other radial skeletal elements in double-walled archaeocyathids (TIP, 1972, pt. E, p. 13). Pl: *loculi.*

interray The area between two adjacent rays of an echinoderm; e.g. the part of a theca between any two adjacent crinoid rays. Syn: *interradial [paleont].*

interreef Situated between reefs; e.g. the "interreef region" characterized by relatively unfossiliferous rock, or "interreef sediments" deposited between reefs. Also spelled: *inter-reef.* Cf: *off-reef.*

interrupted profile A normal profile that has been altered by an *interruption;* e.g. a longitudinal profile of a stream, where, after rejuvenation, the head of the second-cycle valley touches the first-cycle valley.

interrupted projection A map projection lacking continuous outlines, having several central meridians instead of one, or whose origin is repeated, in order to reduce the peripheral shape distortion and the linear scale discrepancy; e.g. a *homolosine projection* split along several meridians. Syn: *recentered projection.*

interrupted stream A stream that contains perennial reaches with intervening intermittent or ephemeral reaches, or a stream that contains intermittent reaches with intervening ephemeral reaches (Meinzer, 1923, p. 58). Ant: *continuous stream.*

interrupted water table A water table that slopes steeply over a *ground-water barrier,* with pronounced difference in elevation above and below the barrier, but not as steep as a *ground-water cascade.*

interruption A break in, or the cutting short of, the cycle of erosion, characterized by a change in the position of base level relative to a landmass or terrain, and resulting in the initiation of a new erosion cycle. It may be dynamic, caused by earth movements (involving deformation, dislocation, or tilting) or fluctuations of sea level; or static, caused by climatic change or piracy. Cf: *accident.*

intersecting peneplain One of two peneplains forming a *morvan* landscape.

intersection (a) A method in surveying by which the horizontal position of an unoccupied point is determined by drawing lines to that point from two or more points of known position. Cf: *resection.* (b) Determination of positions by triangulation.

intersection shoot An ore shoot located at the intersection of one vein or vein system with another. It is a common type of ore deposit.

intersept That portion of an archaeocyathid wall between the edges of two adjacent septae (TIP, 1972, pt. E, p. 40).

interseptal ridge A longitudinal elevation on the outer surface of the wall of a corallite, corresponding in position to the space between a pair of adjacent septa on the inner surface of the wall. Cf: *septal groove.*

intersequent stream A stream following a consequent course in a depression between the margins of opposing alluvial fans, as on a bajada.

intersertal Said of the texture of a porphyritic igneous rock in which the groundmass, composed of a glassy or partly crystalline material other than augite, occupies the interstices between unoriented feldspar laths, the groundmass forming a relatively small proportion of the rock. Cf: *hyalopilitic; hyalophitic; hyalocrystalline.*

interstade A warmer substage of a glacial stage, marked by a temporary retreat of the ice; "a climatic episode within a glaciation during which a secondary recession or a stillstand of glaciers took place" (ACSN, 1961, art. 40). Example: the Alleröd interstade of Denmark. Syn: *interstadial; oscillation [glac geol].*

interstadial adj. Pertaining to or formed during an *interstade.*—n. *interstade.*

interstice An opening or space, as in a rock or soil. On the basis of origin, it may be classified as an *original interstice* or a *secondary interstice;* on the basis of size, as a *capillary interstice,* a *subcapillary interstice,* or a *supercapillary interstice.* Syn: *void; pore [geol].* Adj: *interstitial.*

interstitial Said of a mineral deposit in which the minerals fill the pores of the host rock. Cf: *impregnated.*

interstitial defect In a crystal structure, the filling of a normally void interstice with an extra atom; it is a type of *point defect.* See also: *addition solid solution.* Cf: *Frenkel defect; Schottky defect.*

interstitial solid solution *addition solid solution.*

interstitial water Subsurface water in the voids of a rock. Syn: *pore water.* Cf: *connate water.*

interstratal karst Karst that forms at depth below a resistant nonsoluble rock. Syn: *subjacent karst.* See also: *covered karst; paleokarst.*

interstratification (a) The state or condition of being interstratified or occurring between strata of a different character. (b) *interlayering.*

interstratified *interbedded.*

interstream Said of an area, divide, or topographic feature situated or lying between streams, such as an *interfluve.* See also: *interfluvial.*

interstream ground-water ridge A residual ridge in the water table that develops between two effluent streams as a result of the percolation of ground water toward the streams. Cf: *ground-water ridge.*

intertentacular Situated between tentacles; e.g. "intertentacular organ", a flask-shaped, tubular bryozoan structure providing passageway for extrusion of ova between two tentacles on the distal side of the lophophore near the midline.

intertextic Said of an arrangement in a soil fabric whereby the skeleton grains are linked by intergranular braces or are embedded in a porous matrix (Brewer, 1964, p.170). Cf: *porphyroskelic; agglomeroplasmic.*

interthecal Between thecae; e.g. "interthecal septum" separating adjacent thecal cavities in graptoloids.

intertidal *littoral.*

intertidalite A *tidalite* that is known to be deposited by tidal processes in the intertidal zone.

intertongued lithofacies A *lithofacies* whose irregular boundaries separate intertonguing stratigraphic bodies of contrasting characteristics (such as shale and sandstone) (Weller, 1958, p.633). It is not a unit that can be mapped in the normal manner. The term *lithosome,* as originally defined, is synonymous. Cf: *statistical lithofacies.*

intertonguing The disappearance of sedimentary bodies in laterally adjacent masses owing to splitting into many thin *tongues [stratig],* each of which reaches an independent pinch-out termination; the intergradation of markedly different rocks through a vertical succession of thin interlocking or overlapping wedge-shaped layers. Syn: *interfingering; interdigitation.*

intertrappean Pertaining to a deposit that occurs between two lava flows.

intertrough The median, narrowly triangular furrow dividing the pseudointerarea of the pedicle valve of some acrotretacean brachiopods.

interval [cart] *contour interval.*

interval [geog] A term used in New England as a syn. of *bottom,* or a tract of low-lying, alluvial land along a watercourse, between the river and the hills or higher ground by which the valley floor is bounded. Also spelled: *intervale.*

interval [glac geol] An informal term for a subdivision of an interstade.

interval [stratig] (a) *stratigraphic interval.* (b) *geochronologic interval.* (c) *polarity interval.*

interval change A lateral increase or decrease of the time interval between two seismic reflection events.

interval correlation Stratigraphic correlation based on well-defined stratigraphic intervals identified by their positions between marker horizons (Krumbein & Sloss, 1963, p.343).

interval density In a well bore, the density of an interval integrated from gamma-gamma log data or determined by a borehole gravity meter. See also: *apparent density.*

intervale *interval.*

interval-entropy map A multicomponent *vertical-variability map* that expresses the degree of vertical alternations (homogeneity or heterogeneity) of rock types in a given succession of beds or within a given stratigraphic unit. The term was introduced by Forgotson (1960).

intervallar coefficient The ratio of the width of the intervallum to the width of the central cavity in double-walled archaeocyathids (TIP, 1972, pt. E, p. 11).

intervallum The space between the outer and inner walls of an archaeocyathid (TIP, 1955, pt.E, p.7). It may contain various structures, esp. the septa. Adj: *intervallar.*

intervalometer A timing device on an aerial camera that automatically operates the shutter at predetermined intervals.

interval velocity The distance across a stratigraphic interval divided by the time for a seismic wave to traverse it; the *average velocity [seis]* measured over a depth interval, e.g. in a sonic log or borehole survey. It usually refers to compressional velocity and implies measurement perpendicular to bedding.

interval-zone An interval of strata between two distinctive biostratigraphic horizons. The name given to an interval-zone may be derived from the names of the boundary horizons, the name for the basal boundary preceding that for the upper, e.g. *Globigerinoides sicanus/Orbulina suturalis* interval-zone. Syn: *biointerval-zone; interbiohorizon-zone.*

interzonal time *Geologic time* represented by a diastem or stratigraphic hiatus (Kobayashi, 1944, p. 745).

interzone A term used by Henningsmoen (1961, p.83) for the barren or nonfossiliferous rocks between two local range zones.

interzooecial Used loosely to indicate skeletal zooecial walls between adjacent zooidal chambers in stenolaemate bryozoans.

interzooidal Existing between or among zooids; e.g. "interzooidal avicularia" of cheilostome bryozoans, which are intercalated in lineal series in spaces smaller than those occupied by autozooids.

intexine A syn. of *endexine.* Also spelled: *intextine.*

intine The thin, inner layer of the two major layers forming the wall (sporoderm) of spores and pollen, composed of cellulose and pectates, and situated inside the *exine,* surrounding the living cytoplasm. Syn: *endospore.*

intra-ambulacral extension Part of an ambulacral coverplate or oral plate of an edrioasteroid that is produced inward into the ambulacral tunnel (Bell, 1976).

intracapsular Said of cell materials within the central capsule of a radiolarian; e.g. "intracapsular layer" consisting of protoplasm exclusive of nucleus.

intraclast A broad, general term introduced by Folk (1959, p. 4) for a component of a limestone, representing a torn-up and reworked fragment of a penecontemporaneous sediment (usually weakly consolidated) that has been eroded within the basin of deposition (such as the nearby sea floor or an exposed carbonate mud flat) and redeposited there to form a new sediment; an *allochem* derived from the same formation. The fragment may range in size from fine sand to gravel (smaller grains are *pellets*), and is generally rounded but may be equant to discoidal. Cf: *protointraclast; extraclast.*

intracoastal Being within or near the coast; esp. said of inland waters near the coast.

intracontinental geosyncline *intrageosyncline.*

intracratonic basin A basin on top of a *craton.*

intracrystal porosity The porosity within individual crystals, pores in large crystals of echinoderms, and fluid inclusions (Choquette & Pray, 1970, p. 247).

intracyclothem A cyclic sequence of strata resulting from the splitting of a cyclothem. Term introduced by Gray (1955).

intradeep A geosynclinal trough appearing within a geosynclinal belt at the end of or following uplift of the belt; a type of *secondary geosyncline.* See also: *foredeep; backdeep.*

intradelta The landward part of a delta, largely subaerial but extending for a short distance below the water level, marked by a great diversity of environments and commonly covered by marshes and swamps; it contains the distributary channels, flanked by levees. Cf: *delta front; prodelta.* Syn: *delta top.*

intrafacies A term used by Cloud & Barnes (1957, p.169) to denote a minor or subordinate facies occurring within a differing major facies.

intrafolial fold A minor fold involving only a few layers in an otherwise unfolded rock.

intraformational (a) Formed within a geologic formation, more or less contemporaneously with the enclosing sediments. The term is esp. used in regard to syndepositional folding or slumping, e.g. "intraformational deformation" or "intraformational breccia". (b) Existing within a formation, with no necessary connotation of time of origin. See also: *intrastratal.*

intraformational breccia A rock formed by brecciation of partly consolidated material, followed by practically contemporaneous sedimentation. It is similar in nature and origin to an *intraformational conglomerate* but contains fragments showing greater angularity.

intraformational conglomerate (a) A conglomerate in which the clasts are essentially contemporaneous with the matrix in origin, developed by the breaking-up and rounding of fragments of a newly formed or partly consolidated sediment (usually shale or limestone) and their nearly immediate incorporation in new sedimentary deposits; e.g. an *edgewise conglomerate.* Fragmentation is commonly caused by shoaling and temporary withdrawal of water, followed by desiccation and mud cracking. Examples abound in the lowest Paleozoic limestones and dolomites of the Appalachian region of eastern U.S. (b) A conglomerate occurring in the midst of a geologic formation, such as one formed during a brief interruption in the orderly deposition of strata. It may contain clasts external to the formation. The term is used in this sense esp. in England.—Cf: *interformational conglomerate.*

intraformational contortion Intricate and complicated folding, as exhibited in *convolute bedding;* esp. such deformation resulting from the subaqueous slumping or sliding of unconsolidated sediments under the influence of gravity. See also: *intraformational corrugation.* Syn: *intrastratal contortion.*

intraformational corrugation A term applied to *intraformational contortion* on a small scale.

intraformational fold A minor fold confined to a sedimentary layer lying between undeformed beds; it results from processes, such as sliding or slumping, that took place prior to complete lithification.

intrageosyncline DuToit's term for a *parageosyncline* (1937).

intraglacial (a) Said of glacial deposits formed on ground actually covered by the ice, or of glacial phenomena pertaining to a region covered by the ice at any given time. Ant: *extraglacial.* (b) *englacial.*

intragranular movement A gliding movement by which favorably oriented ice crystals are deformed by slip along layers, without breaking the continuity of the crystal lattice. It is an important mechanism in glacier flow. Cf: *intergranular movement.*

intragranular porosity The porosity existing within individual grains or particles of a rock, esp. within skeletal material of a carbonate sedimentary rock. Cf: *intraparticle porosity.*

intralaminal accessory aperture An *accessory aperture* in the test of a planktonic foraminifer that leads through accessory structures into a cavity beneath them and not directly into the chamber cavity (as in *Rugoglobigerina*). Cf: *infralaminal accessory aperture.*

intramicrite A limestone containing at least 25% intraclasts and in which the carbonate-mud matrix (micrite) is more abundant than the sparry-calcite cement (Folk, 1959, p. 14).

intramicrudite An intramicrite containing gravel-sized intraclasts.

intramineral Pertaining to the time interval of a period of mineralization (Kirkham, 1971, p. 1245); also, pertaining to those features, e.g. a breccia mineralized during its formation, that were emplaced during such an interval. The term has been applied to a time interval between distinct periods of mineralization and to its resulting features, but *intermineral* is preferred for this meaning.

intramontane space *Zwischengebirge.*

intramorainal Said of deposits and phenomena occurring within a lobate curve of a moraine. Ant: *extramorainal.* Syn: *intramorainic.*

intramural budding A type of *polystomodaeal budding* in which the stomodaea are directly or indirectly linked in a single linear series. Cf: *circummural budding.*

intraparticle porosity The porosity within individual particles of a rock. Choquette & Pray (1970, p. 247) recommend use of this term rather than the term *intragranular porosity* which suggests limitation to grain-size particles.

intrapermafrost water Unfrozen ground water in layers or lenses within permafrost.

intrapositional deposit Sediments deposited by the process of *stratigraphic leak,* e.g. as crevice or erosional-channel fillings (Foster, 1966).

intrasparite A limestone containing at least 25% intraclasts and in which the sparry-calcite cement is more abundant than the carbonate-mud matrix (micrite) (Folk, 1959, p. 14). It is common in environments of high physical energy, where the spar usually represents pore-filling cement.

intrasparrudite An intrasparite containing gravel-sized intraclasts.

intrastratal Formed or occurring within a stratum or strata; e.g. formation of iron-rich authigenic clay by "intrastratal alteration" of hornblende. See also: *intraformational.*

intrastratal contortion *intraformational contortion.*

intrastratal flow structure A variety of *convolute bedding* formed by flowage.

intrastratal solution Removal by chemical solution of certain mineral species from within a sedimentary bed following deposition. Syn: *differential solution.*

intratabular Said of features of a dinoflagellate cyst that correspond to the central parts of thecal plates rather than to the lines of separation between them. Cf: *nontabular; peritabular.*

intratelluric (a) Said of a phenocryst, of an earlier generation than its groundmass, that formed at depth, prior to extrusion of a magma as lava. (b) Said of that period of crystallization occurring deep within the Earth just prior to the extrusion of a magma as lava. (c) Located, formed, or originating deep within the Earth.

intratentacular budding Formation of new scleractinian coral polyps by invagination of the oral disk of the parent inside the ring of tentacles surrounding its mouth. Cf: *extratentacular budding.*

intrathecal extension Part of an external thecal plate in an edrioasteroid that is produced inward into the interior of the theca (Bell, 1976).

intraumbilical aperture An aperture in a foraminiferal test located in the umbilicus but not extending outside of it.

intrazonal soil In the 1938 classification system, one of the *soil orders* including soils with more or less well-developed soil characteristics that reflect the dominating influence of some local factor of relief, parent material, or age over the normal effects of the climate and vegetation; also, any soil belonging to the intrazonal soil order. Cf: *zonal soil; azonal soil.*

intrazonal time *Geologic time* represented by a biostratigraphic zone (Kobayashi, 1944, p. 745); the *hemera* of Buckman (1902).

intrazooidal Said of a structure occurring within a zooid, e.g. "intrazooidal septulae" that connect the hypostega with the main body cavity in some ascophoran cheilostomes (bryozoans) (Banta, 1970, p. 39). Also spelled: *intrazoidal.*

intrenched meander *entrenched meander.*

intrenched stream *entrenched stream.*

intrinsic ash *inherent ash.*

intrinsic geodesy The study of the gravity field of the Earth by considering only natural quantities capable of being observed and measured.

intrinsic ionic conduction Electrical conduction arising in transport of charge through a solid as a result of a movement of ions through a crystal lattice. It usually takes place at elevated temperatures and does not depend on the presence of impurities or vacancies.

intrinsic permeability *specific permeability.*

introgression The evolutionary process in which a *hybrid [evol]* crosses with a member of one of its parental species, thus introducing some genes from the other parental species into the next generation of the first species. Syn: *introgressive hybridization.*

introgressive hybridization *introgression*

intrusion [grd wat] *salt-water encroachment.*

intrusion [ign] The process of emplacement of magma in pre-existing rock; magmatic activity; also, the igneous rock mass so formed within the surrounding rock. See also: *pluton.* Syn: *injection [ign]; emplacement [intrus rocks]; invasion [ign]; irruption [intrus rocks].*

intrusion [sed] (a) A sedimentary *injection* on a relatively large scale; e.g. the forcing upward of clay, chalk, salt, gypsum, or other plastic sediment, and its emplacement under abnormal pressure in the form of a diapiric plug. See also: *autointrusion.* (b) A sedimentary structure or rock formed by intrusion. (c) *stone intrusion.* —Syn: *sedimentary intrusion.*

intrusion breccia (a) *contact breccia.* (b) *intrusive breccia.*

intrusion displacement Faulting due to an igneous intrusion. See also: *marginal thrust; trap-door fault.*

intrusive adj. Of or pertaining to *intrusion,* both the processes and the rock so formed. n. An *intrusive* rock. — Cf: *extrusive.* Syn: *irruptive.*

intrusive breccia A heterogeneous mixture of angular to rounded fragments in a matrix of clastic material, which has been mobilized and intruded into its present position along pre-existing structures (Bryant, 1968, p. 4). It is commonly hydrothermally altered. Syn: *intrusion breccia.*

intrusive ice Ice that results from the freezing of injected water, as in a domal or tabular body, forming a raised surface, such as an *ice laccolith* or a *pingo.*

intrusive tuff *tuffisite.*

intrusive vein An igneous intrusion resembling a *sheet,* apparently formed from a magma rich in volatiles.

intumescence The property that some minerals have of swelling or frothing when heated, owing to the release of gases.

inundation A rising of water and its spreading over land not normally submerged.

invar An alloy of nickel and iron, containing about 36% nickel, and having an extremely low coefficient of thermal expansion. It is used in the construction of surveying instruments such as pendulums, level rods, first-order leveling instruments, and tapes. Etymol: *Invar,* a trademark.

invariant equilibrium Equilibrium of a phase assemblage that has zero degrees of freedom. See also: *invariant point.*

invariant point A point representing the conditions of *invariant equilibrium.*

invasion [ign] *intrusion [ign].*

invasion [stratig] *transgression.*

inver (a) A place where a river flows into the sea or into an arm of the sea; e.g. Inverness, Scotland. (b) The confluence of two streams.—Etymol: Gaelic.

invernite A granitelike holocrystalline intrusive rock characterized by phenocrysts of orthoclase and plagioclase in a groundmass of euhedral feldspar and rare hornblende or mica with interstitial quartz. Not recommended usage.

inverse dispersion The *dispersion* of seismic surface waves in which the recorded wave period decreases with time. Cf: *normal dispersion.*

inverse estuary An "estuary" in which evaporation exceeds the influx of fresh water (land drainage and precipitation) so that the salinity rises above that of seawater. Ant: *positive estuary.* Syn: *negative estuary.*

inverse filter A filter with characteristics complementary to another filter in such a way that when used in series with the other filter no frequency-selective filtering occurs (except for overall time delay).

inverse problem The problem of gaining knowledge of the physical features of a disturbing body by analysis of its effects, e.g., fields and potentials; finding the model from observed data. It is in contrast to the direct, forward, or normal problem, which is calculating what would have been observed from a given model (Sheriff, 1973, p. 116).

inverse projection *transverse projection.*

inverse thermoremanent magnetization An artificial *remanent magnetization* acquired during a temperature increase from subzero. Cf: *partial thermoremanent magnetization.* Abbrev: ITRM.

inverse zoning *reversed zoning.*

inversion [cryst] *Transformation* [cryst] through a center of symmetry.

inversion [geomorph] (a) The development of *inverted relief* whereby anticlines are transformed into valleys and synclines into mountains; e.g. the formation of a deep basin in an area formerly occupied by land that produced quantities of sediment. (b) The occupancy by a lava flow of a former ravine or valley in the side of a volcano, thereby producing a divide over the former valley and forcing the stream to develop a new valley on its former divide (Cotton, 1958, p. 366-367).—Syn: *inversion of relief.*

inversion [geophys] Construction of a general geophysical model from an array of local data points; for example, using numerous gravity measurements to infer subsurface density distributions, or using slip vectors and spreading rates to define global plate motions.

inversion [meteorol] In meteorology, a reversal of the gradient of a meteorologic element, e.g. an increase rather than a decrease of temperature with height.

inversion center *center of symmetry.*

inversion layer In a body of water, a water layer whose temperature increases rather than decreases with depth.

inversion point (a) A point representing the temperature at which one polymorphic form of a substance, in equilibrium with its vapor, reversibly changes into another under invariant conditions. (b) The temperature at which one polymorphic form of a substance inverts reversibly to another under univariant conditions and a specific pressure. (c) More loosely, the lowest temperature at which a monotropic phase inverts at an appreciable rate into a stable phase, or at which a given phase dissociates at an appreciable rate, under given conditions. (d) A single point at which different phases are capable of existing together at equilibrium.—Syn: *transition point; transition temperature.*

invert The floor or bottom of the internal cross section of a closed conduit, such as an aqueduct, tunnel, or drain. The term originally referred to the inverted arch used to form the bottom of a masonry-lined sewer or tunnel.

invertebrate n. An animal belonging to the Invertebrata, i.e. without a backbone, such as the mollusks, arthropods, and coelenterates.—adj. Of or pertaining to an animal that lacks a backbone.

invertebrate paleontology The branch of paleontology dealing with fossil invertebrates. Syn: *invertebrate paleozoology.*

invertebrate paleozoology *invertebrate paleontology.*

inverted *overturned.*

inverted pendulum *pendulum (b).*

inverted plunge The plunge of folds, or sets of folds, whose inclination has been carried past the vertical, so that the plunge is now less than 90° in a direction opposite from the original attitude. It is a rather common feature in excessively folded or refolded terranes.

inverted relief A topographic configuration that is the inverse of the geologic structure, as where mountains occupy the sites of synclines and valleys occupy the sites of anticlines. See also: *inversion [geomorph].* Ant: *uninverted relief.*

inverted siphon A portion of a water conduit that is depressed in a U shape. Cf: *siphon [hydraul].*

inverted stream (a) A beheaded stream whose drainage turns back into the capturing stream (Lobeck, 1939, p. 199). (b) A term proposed by Davis (1889b, p. 210) but later abandoned (Davis, 1895, p.134) in favor of *obsequent stream.*

inverted tide *reversed tide.*

inverted unconformity (a) An unconformity in which the younger strata end abruptly against the older rocks, such as one produced by intense folding of a complex region (Grabau, 1924, p.826). (b) Truncation of the upper parts of laminae in sediment, shown in some load casts where underlying plastic material has been squeezed upward and intruded laterally into overlying sediment (Kuenen, 1957, p.250).

inverted well A well that takes in water near its top and discharges it at lower levels, into permeable material; e.g. a *drainage well.*

invisible gold A syn. used in the popular press for *Carlin-type gold.*

involucre One or more whorls of bracts subtending a flower or inflorescence. See also: *cupule.*

involute Coiled or rolled inward, e.g. said of a foraminiferal test having closely coiled and strongly overlapping whorls, in which the inner part of the last whorl extends in toward the center of the coil to cover part of the adjacent inner whorl; or a coiled gastropod shell having the last whorl enveloping the earlier whorls, which are more or less visible in the umbilici; or a coiled cephalopod conch with considerably overlapping whorls and a narrow umbilicus. Cf: *convolute; evolute; advolute.*

involution [sed] (a) A highly irregular, aimlessly contorted sedimentary structure consistig of local folds and interpenetrations o fine-grained material in clayey strata, and developed by the formation, growth, and melting of ground ice (*congeliturbation*) in the active layer overlying permafrost. Syn: *Brodelboden.* (b) An irregularly contorted and penetrating structure, as a wave or fold, in a soil deposit.

involution [struc geol] The refolding of nappes, resulting in complex patterns of association.

inwash Alluvium deposited against the margin of a glacier by a stream of nonglacial origin.

inyoite A colorless monoclinic mineral: $Ca_2B_6O_{11} \cdot 13H_2O$.

iodargyrite A yellowish or greenish hexagonal mineral: AgI. Syn: *iodyrite.*

iodate A mineral compound that is characterized by the radical IO_3^-. An example is salesite, $Cu(IO_3)(OH)$.

iodobromite An isometric mineral: Ag(Br,Cl,I).

iodyrite *iodargyrite.*

iolanthite A banded reddish jasperlike mineral from Oregon.

iolite A syn. of *cordierite,* esp. the gem variety.

ion exchange Reversible exchange of ions contained in a crystal for different ions in solution without destroying crystal structure or disturbing electrical neutrality. It is accomplished by diffusion and occurs most easily in crystals having one- or two-dimensional channelways where ions are relatively weakly bonded; it also takes place at higher temperatures in network silicates, involving the most weakly bonded cations such as those of potassium and sodium. Ion exchange is also common in resins consisting of three-dimensional hydrocarbon networks to which many ionizable groups are attached. See also: *exchange capacity; anion exchange; cation exchange.*

ionic substitution The replacement of one or more kinds of ion in a crystal structure by other kinds of generally similar size and charge. Syn: *diadochy; proxying; substitution.*

ionite (a) *anauxite.* (b) An earthy, resinous, brownish-yellow fossil hydrocarbon in the lignite of Ione Valley, Amador County, Calif. It is not a recognized mineral species.

ionium An old but still-used name for thorium-230, a member of the uranium series and daughter of uranium-234.

ionium-deficiency method Calculation of an age in years for fossil coral or shell from 10,000 to 250,000 years old, based on the growth of ionium (thorium-230) toward equilibrium with uranium-238 and uranium-234 which entered the carbonate shortly after its formation or burial. The age depends on the departure from the equilibrium ratio, which is directly related to the passage of time. See also: *uranium-series age method; thorium-230/protactinium-231 deficiency method.*

ionium-excess method The calculation of an age in years for deep-sea sediments formed during the last 300,000 years, based on the assumptions that the initial ionium (thorium-230) content of accumulating sediments has remained constant for the total section of sediments under study and that this initial ionium content is "ionium-excess", i.e. uranium-unsupported thorium-230. The age depends on this ionium excess content, which decreases with the passage of time. See also: *ionium-thorium age method; uranium-series age method.*

ionium-thorium age method The calculation of an age in years for deep-sea sediments formed during the last 300,000 years, based on the assumption that the initial thorium-230 ratio for accumulating sediments has remained constant for the total section of sediments under study. The age depends on the thorium-230 to thorium-232 ratio, which gradually decreases with the passage of time. See also: *ionium-excess method; uranium-series age method.* Syn: *thorium-230/thorium-232 age method.*

ionization constant As applied to ionization reactions, a syn. of *dissociation constant.*

ionization potential The voltage required to drive an electron from an atom or molecule, leaving a positive ion.

ionizing radiation Any electromagnetic or particulate radiation that displaces electrons within a medium.

ion microprobe An instrument, similar in principle to an *electron microprobe,* with which a focused beam of ions is made to strike the surface of a sample, with resultant emission of ions from the impact area. The emitted ions are characteristic of the isotopes of the elements present in the area of impact.

ionography Ion-exhange *electrochromatography* wherein the ions migrate by electrostatic attraction, usually over or through an ion-exchange resin.

iowaite A mineral: $Mg_4Fe^{+3}(OH)_8OCl \cdot 2\text{-}4H_2O$.

Iowan Originally defined as a separate stage between the Illinoian and Wisconsinan, and later as the earliest substage of the Wisconsinan. The area of Iowan drift in northeastern Iowa is now recognized as an erosional surface cut into the Kansan till plain (Ruhe, 1969).

iozite *wüstite.*

IP *induced polarization.*

IQSY *International Years of the Quiet Sun.*

IR (a) *insoluble residue.* (b) *infrared.*

iranite A saffron-yellow triclinic mineral: $Pb_{10}Cu(CrO_4)_6(SiO_4)_2(F,OH)_2$. Syn: *khuniite.*

iraqite A tetragonal mineral: $K(La,Ce,Th)_2(Ca,La,Na)_5Si_{16}O_{40}$.

irarsite A mineral: (Ir,Ru,Rh,Pt)AsS. It forms a series with hollingworthite.

irhtemite A monoclinic mineral: $Ca_4MgH_2(AsO_4)_4$.

irhzer A term used in northern Africa for a straight groove carved in a mountainside by a stream (Termier & Termier, 1963, p. 408). Etymol: Berber.

iridescence The exhibition of prismatic colors (producing rainbow effects) in the interior or on the surface of a mineral, caused by interference of light from thin films or layers of different refractive index.

iridosmine A tin-white or steel-gray rhombohedral mineral: (Os, Ir). It is a native alloy containing 20-68% iridium and 32-80% osmium; it usually contains some rhodium, platinum, ruthenium, iron, and copper. Cf: *osmiridium.* Syn: *iridosmium.*

iridosmium *iridosmine.*

iriginite A canary-yellow secondary mineral: $(UO_2)Mo_2O_7 \cdot 3H_2O$.

iris (a) A transparent quartz crystal containing minute air-filled or liquid-filled internal cracks that produce iridescence by interference of light. The cracks may occur naturally or be caused artificially by heating and sudden cooling of the specimen. Syn: *iris quartz; rainbow quartz.* (b) An iridescent mineral; e.g. "California iris" (kunzite).

IRM *isothermal remanent magnetization.*

iron [meteorite] *iron meteorite.*

iron [mineral] A heavy, magnetic, malleable and ductile, and chemically active mineral, the native metallic element Fe. It has a silvery or silver-white color when pure, but readily oxidizes in moist air. Native iron occurs rarely in terrestrial rocks (such as disseminated grains in basalts), but is common in meteorites; it occurs combined in a wide range of ores and in most igneous rocks. Iron is the most widely used of the metals.

Iron Age In archaeology, a cultural level that is the final age in the *three-age system,* and is characterized by the technology of iron. Correlation of relative cultural levels with actual age (and, therefore, with the time-stratigraphic units of geology) varies from region to region; e.g. the Iron Age began in Europe about 1100 B.C. at the earliest, but in the Americas, there was no iron technology until contact with European culture was made. In general, the Iron Age falls within historic time (Bray and Trump, 1970, p.115).

iron alum *halotrichite.*

iron bacteria Anaerobic bacteria that precipitate iron oxide from solution, either by oxidizing ferrous salts or by releasing oxidized metals from organic compounds. Accumulations of iron developed in this way are *bacteriogenic* ore deposits. Cf: *sulfur bacteria.*

iron ball A term used in Lancashire, England, for an ironstone nodule. See also: *ballstone.*

iron-bearing formation *iron formation.*

iron cordierite *sekaninaite.*

iron-cross twin law A twin law according to which crystals of the diploidal class of the isometric system are formed by interpenetration twinning of two pyritohedrons. The twin axis is perpendicular to a face of the rhombic dodecahedron.

iron formation A chemical sedimentary rock, typically thin-bedded and/or finely laminated, containing at least 15% iron of sedimentary origin, and commonly but not necessarily containing layers of chert (James, 1954, p.239). Various primary facies (usually not weathered) of iron formation are distinguished on the basis of whether the iron occurs predominantly as oxide, silicate, carbonate, or sulfide. Most iron formation is of Precambrian age. In mining usage, the term refers to a low-grade sedimentary iron ore with the iron mineral(s) segregated in bands or sheets irregularly mingled with chert or fine-grained quartz (Thrush, 1968, p. 590). Cf: *ironstone; jaspilite.* See also: *oxide-facies iron formation; carbonate-facies iron formation; silicate-facies iron formation; sulfide-facies iron formation.* Essentially synonymous terms: *itabirite; banded hematite quartzite; taconite; quartz-banded ore; banded ironstone; calico rock; jasper bar; iron-bearing formation.* Also spelled: *iron-formation.*

iron froth A fine, spongy or micaceous variety of hematite.

iron glance A variety of hematite; specif. *specularite.*

iron hat *gossan.*

iron hypersthene (a) An iron-rich hypersthene. (b) *ferrosilite.*

iron meteorite A general name for meteorites consisting essentially of nickeliferous iron (solid solution of iron with 4% to 30% or more of nickel); e.g. a *hexahedrite,* an *octahedrite,* and an *ataxite.* Syn: *iron [meteorite]; siderite [meteorite]; meteoric iron.*

iron mica (a) *lepidomelane.* (b) *biotite.* (c) Micaceous hematite.

iron-monticellite *kirschsteinite.*

iron olivine *fayalite.*

iron ore Ferruginous rock containing one or more distinct natural chemical compounds from which metallic iron may be profitably extracted. The chief ores of iron consist mainly of the oxides: hematite (Fe_2O_3); goethite (α-FeO(OH)); magnetite (Fe_3O_4); and the carbonate, siderite or chalybite ($FeCO_3$).

iron pan A general term for a *hardpan* in a soil in which iron oxides are the principal cementing agents; several types of iron pans are found in dry and wet areas and in soils of widely varying textures. Also spelled: *ironpan.* See also: *moorpan.* Cf: *claypan [soil].*

iron pitch A variety of Lake Trinidad *land asphalt* that has overflowed the lake onto the land and hardened on weathering to such an extent that it resembles refined *lake asphalt* (Abraham, 1960, p.177).

iron pyrites (a) *pyrite.* (b) *marcasite [mineral].*—Sometimes incorrectly spelled: *iron pyrite.*

iron range A term used in the Great Lakes region of the U.S. and Canada for a productive belt of iron formations. The term implies a linear region rather than a topographic elevation.

iron sand A sand containing particles of iron ore (usually magnetite), as along a coastal area.

iron shale A material, usually with a laminated structure, consisting of iron oxides and produced by the weathering of an iron meteorite.

ironshot adj. (a) Said of a mineral that is streaked, speckled, or marked with iron or an iron ore. (b) Containing small nodules or oolitic bodies of limonite or hematite; e.g. an "ironshot rock" in which the ooliths are essentially composed of limonite.—n. A limonitic oolith in an ironshot rock.

iron spar *siderite [mineral].*

iron spinel *hercynite.*

ironstone (a) Any rock containing a substantial proportion of an iron compound, or any iron ore from which the metal may be smelted commercially; specif. an iron-rich sedimentary rock, either deposited directly as a ferruginous sediment or resulting from chemical replacement. The term is customarily applied to a hard, coarsely banded or nonbanded, and noncherty sedimentary rock of post-Precambrian age, in contrast with *iron formation.* The iron minerals may be oxides (limonite, hematite, magnetite), carbonate (siderite), or silicate (chamosite); most ironstones containing iron oxides or chamosite are oolitic. (b) *clay ironstone.* (c) *banded ironstone.*

ironstone cap A surficial or near-surface sheet or cap of concretionary *clay ironstone.*

iron-stony meteorite *stony-iron meteorite.*

iron talc *minnesotaite.*

iron vitriol *melanterite.*

irradiance The radiant energy per unit time per unit area incident upon a surface.

irregular Pertaining to an echinoid of the order Exocycloida displaying an *exocyclic* test in which the periproct is located outside of the oculogenital ring or in a posterior or oral position. Cf: *regular.*

irreversibility In evolution, the theory that an evolving group of organisms, or part of an organism, does not return to the ancestral condition. Syn: *Dollo's law.*

irreversible process Any process which proceeds in one direction spontaneously, without external interference.

irrotational strain Strain at a point, in which the orientation of the principal axes of strain remains unchanged. Cf: *rotational strain.* Syn: *nonrotational strain.* Not to be confused with noncoaxial *progressive deformation.*

irrotational wave *P wave.*

irruption [ecol] An abrupt, sharp, but temporary increase in a natural population, usually connected with exceptionally favorable environmental conditions. Adj: *irruptive.*

irruption [intrus rocks] A nearly obsolete syn. of *intrusion.*

irruptive *intrusive.*

Irvingtonian North American (California) continental stage: Pleistocene (above Blancan, below Rancholabrean).

isallobar A line on a weather map connecting points of equal barometric tendencies.

isallotherm A line connecting points of equal temperature variation in a given time interval. Cf: *isotherm.*

isanomaly *isoanomaly.*

isarithm An *isopleth,* esp. one drawn through points on a graph at which a given quantity has the same numerical value.

isblink A term used in Greenland for *iceblink* or seaward cliff of ice. Etymol: Danish.

ischium (a) The posteroventral of the three bones of the vertebrate pelvic girdle. (b) The third pereiopodal segment from the body of a malacostracan crustacean, distal to the basis and proximal to the merus. It comprises the first segment of the endopod. Syn: *ischiopod; ischiopodite.*—Pl: *ischia.*

isenite A reputedly feldspathoid-bearing hornblende *trachyandesite* containing phenocrysts of andesine, soda microcline, hornblende, and biotite in a groundmass of oligoclase, orthoclase, and nosean, with smaller amounts of augite, apatite, and iron ore. Tröger in 1935 stated that apatite was misidentified as nosean. The term is obsolete.

isentrope A line or surface that represents the locus of points of a given entropy.

isentropic Said of a process that is at constant entropy.

isentropic map *entropy map.*

iserine A variety of ilmenite found as loose rounded crystals or grains in the sands at Iserwiese in Bohemia. It was formerly regarded as probably a ferruginous rutile. Syn: *iserite.*

iserite (a) A doubtful variety of rutile with considerable amounts of FeO. (b) *iserine.*

I-shaped valley An extremely young valley (such as a canyon) in which downcutting greatly exceeds lateral erosion (Lane, 1923).

ishikawaite A black mineral: $(U,Fe,Y,Ce)(Nb,Ta)O_4$.

ishkulite A chromium-bearing variety of magnetite.

ishkyldite A mineral: $Mg_{15}Si_{11}O_{27}(OH)_{20}$. It may be a variety of chrysotile high in silica. Also spelled: *ishkildite.*

Ising gravimeter An astatic, balance-type gravity meter consisting of a quartz frame and thread, with an attached mass at the end of a short vertical beam attached to the center of the thread, the size of the mass and the stiffness of the thread being chosen so that the system is in nearly neutral equilibrium. When the frame is given a small measured tilt, the change in the equilibrium position of the beam is dependent on the tilt and on the value of gravity.

isinglass A syn. of *mica,* esp. muscovite in thin transparent sheets.

island (a) A tract of land smaller than a continent, surrounded by the water of an ocean, sea, lake, or stream. The term has been loosely applied to land-tied and submerged areas, and to land cut off on two or more sides by water, such as a peninsula. (b) An elevated piece of land surrounded by a swamp, marsh, or alluvial land, or isolated at high water or during floods. (c) Any isolated and distinctive tract of land surrounded by terrain with other characteristics; e.g. a woodland surrounded by prairie or flat open country.

island arc A chain of islands, e.g. the Aleutians, rising from the deep-sea floor and near to the continents; a *primary arc* expressed as a curved belt of islands. Its curve is generally convex toward the open ocean. According to Bucher (1965), an island-arc pattern results from shrinkage. This is a typical tension pattern which would result in the brittle crust by rotating a polar circumference into an equitorial circumference such as by sliding the Earth's crust on the interior. Syn: *volcanic arc.*

island hill An isolated, partly buried bedrock hill standing islandlike in the midst of alluvium (as of a *sandy*) or in the silt of a lake plain (Shaw, 1911, p. 489). Type example: the hill bearing the town of Island in Kentucky.

island mesa A *headland mesa* that has been cut off from the main plateau by a river, so that it stands as an isolated mass (Lee, 1903, p. 73).

island mountain (a) A mountain more or less completely encircled by valleys that separate it from other mountains or drainage-divide ridges. (b) *inselberg.*

island shelf *insular shelf.*

island slope *insular slope.*

island-tying The process of *tombolo* formation.

island volcano *volcanic island.*

isle An *island,* generally but not necessarily of small size; e.g. the British Isles. The term is a diminutive of "island".

islet A small or minor island.

iso- A prefix meaning "equal", e.g. in *isopach,* equal thickness, or *isotherm,* equal temperature.

isoanomalous line A syn. of *isoanomaly*. Also spelled: *isanomalous line.*

isoanomaly A line connecting points of equal geophysical *anomalies*. Syn: *isanomaly; isoanomalous line.*

isoanomaly curve A curve representing equal gravity anomaly values.

isoanthracite line (a) On a map or diagram, a line connecting points of equal carbon-hydrogen ratio in anthracite. (b) A non-preferred synonym of *isovol* as applied to anthracite.

isobar A line on a map or chart connecting points of equal pressure.

isobaric surface A surface, all points of which have equal pressure; it is not necessarily horizontal.

isobase A term used for a line that connects all areas of equal uplift or depression; it is used especially in Quaternary geology as a means for expressing crustal movements related to postglacial uplift.

isobath [grd wat] An imaginary line on a land surface along which all points are the same vertical distance above the upper or lower surface of an aquifer or above the water table.

isobath [oceanog] A line on a map or chart that connects points of equal water depth. Syn: *bathymetric contour; depth contour.*

isobed map *isostratification map.*

isobiolith A *para-time-rock unit* defined by fossils (Wheeler et al., 1950, p. 2362).

isocal On a map or diagram, a line connecting points of equal calorific value in coal. Cf: *isocarb; isodeme; isohume; isovol.*

isocarb On a map or diagram, a line connecting points of equal fixed-carbon content in coal. Cf: *isocal; isodeme; isohume; isovol.* See also: *isocarbon map.*

isocarbon map A coal-deposit map showing points of equal fixed-carbon content by contour lines, or *isocarbs.*

isocenter (a) The unique point common to the principal plane of a tilted photograph and the plane of an assumed truly vertical photograph taken from the same camera station and having an equal principal distance. It is the center of radial displacement of images due to tilt. (b) The point on an aerial photograph intersected by the bisector of the angle between the plumb line and the perpendicular to the photograph. (c) The point of intersection (on a photograph) of the principal line and the isometric parallel.

isochela A sponge *chela* having equal or similar ends. Cf: *anisochela.*

isochemical metamorphism Metamorphism that involves no change in bulk chemical composition (Eskola, 1939). It is a theoretical concept that is probably only approached in nature. Syn: *treptomorphism.*

isochemical series Rocks displaying the same bulk chemical composition throughout a sequence of mineralogic or textural changes, as in a sequence of metamorphic rocks of varying grade.

isochore [chem] In a phase diagram, a line connecting points of constant volume.

isochore [stratig] A line drawn on a map through points of equal drilled thickness for a specified subsurface unit. Thickness figures are uncorrected for dip in vertical wells, and corrected for hole angle, but not for dip, in deviated wells. Cf: *isopach.*

isochore map (a) A map showing drilled thickness of a given stratigraphic unit by means of *isochores.* Syn: *convergence map.* (b) A map showing by contours the thickness of the pay section of an oil pool between the oil-water contact and the roof rock. It is used for making calculations of reservoir volume (Levorsen, 1967, p. 616).—Cf: *isopach map.*

isochromatic curve In optics of biaxial and uniaxial crystals, a band of color indicating the emergence of those components of light having equal path difference. It is a part of the *interference figure.* Cf: *isogyre.* See also: *Cassinian curve.*

isochrome map A contour map that depicts the continuity and extent of color stains on geologic formations.

isochron [geochron] If isotope ratios are determined separately on several systems with the same history (for instance, different minerals in a rock), and if the ratio of a daughter isotope (D_r) to a related non-radiogenic isotope (D_n) was initially the same for all samples, the n/a plot of D_r/D_n versus P/D_r (where P is the parent of D_r) will yield a straight line, or isochron. The slope of an isochron increases with the age of the systems investigated. See also: *isochrone.*

isochron [seis] A line on a map connecting points at which a characteristic time or interval has the same value; e.g. in seismology, a line passing through points at which the difference between arrival times of seismic waves from two reflecting surfaces is equal. Syn: *isotime line.*

isochronal (a) *isochronous.* (b) *isochronic.*

isochrone A line, on a map or chart, connecting all points at which an event or phenomenon occurs simultaneously or which represent the same time value or time difference; e.g. a line along which duration of travel is constant, or a line indicating the places at which rain begins at a specified time. See also: *isochron.*

isochroneity The state or quality of being *isochronous;* equivalence in duration. Syn: *isochronism.*

isochronic Having isochrones; e.g. an isochronic map. Syn: *isochronal.*

isochronism *isochroneity.*

isochronous (a) Equal in duration or uniform in time; e.g. an "isochronous interval" between two synchronous surfaces, or an "isochronous unit" of rock representing the complete rock record of an isochronous interval. Mann (1970, p. 750) has proposed that the word *coetaneous* would be more appropriate for use in this sense. (b) A term frequently applied in the sense of *synchronous,* such as an "isochronous surface" having everywhere the same age or time value within a body of strata (Hedberg, 1958, p. 1890; and ACSN, 1961, art. 28c).—Syn: *isochronal.*

isochronous homeomorph A homeomorph that develops at the same geologic time as another and therefore poses special problems in identification. Cf: *heterochronous homeomorph.*

isoclasite A white mineral: $Ca_2(PO_4)(OH) \cdot 2H_2O$.

isoclinal Adj. of *isocline.*

isocline A fold whose limbs are parallel. Adj: *isoclinal.*

isoclinic line An *isomagnetic line* connecting points of equal magnetic inclination.

isocommunity A natural community that closely resembles another community in morphology and ecology.

isocon A line connecting points of equal geochemical concentration, e.g. salinity.

isodeme On a map or diagram, a line connecting points of equal swelling characteristics. Cf: *isocal; isocarb; isohume; isovol.*

isodesmic Said of a crystal or other material having ionic bonding of equal strength, e.g. NaCl. Cf: *anisodesmic.*

isodiff A line on a map or chart connecting points of equal correction or difference in datum; e.g. *isolat* and *isolong.*

isodimorphism The characteristic of two crystalline substances to be both dimorphous and isomorphous, e.g. calcite and aragonite. Adj: *isodimorphous.*

isodimorphous Said of two crystalline substances displaying *isodimorphism.*

isodont Said of the dentition of a bivalve mollusk (e.g. *Spondylus* and *Plicatula*) characterized by a small number of symmetrically arranged hinge teeth.

isodynamic line *isogam.*

isofacial [petrology] Pertaining to rocks belonging to the same metamorphic facies and having reached equilibrium under the same set of physical conditions. Cf: *allofacial.* Syn: *isogradal.*

isofacial [stratig] Pertaining to rocks belonging to the same facies; e.g. an "isofacial line" on a map, along which the thickness of stratum of the same lithologic composition is constant.

isofacies map A map showing the distribution of one or more facies within a designated stratigraphic unit. See also: *facies map.*

isoferroplatinum A cubic mineral: Pt_3Fe (approximately).

isofract A graphic representation of the locus of all compositions in a system having a given value for the index of refraction.

isofrigid temperature regime A soil temperature regime having the same temperature range as the *frigid temperature regime,* but with a summer-winter variation of less than 5°C (USDA, 1975).

isogal A contour line of equal gravity values. Cf: *gal.*

isogam An *isomagnetic line* connecting points of equal magnetic-field intensity. It is used for maps of total, horizontal, or vertical magnetic intensity. Syn: *isodynamic line.*

isogeolith A *para-time-rock unit* defined by lithology (Wheeler et al., 1950, p.2362).

isogeotherm A line or surface within the Earth connecting points of equal temperature. Syn: *geotherm; geoisotherm.*

isogon *isogonic line.*

isogonic line An *isomagnetic line* connecting points of equal magnetic declination. See also: *agonic line.* Syn: *isogon.*

isograd A line on a map joining points at which metamorphism proceeded at similar values of pressure and temperature as indicated by rocks belonging to the same metamorphic facies. Such a line represents the intersection of an inclined surface with the Earth's surface corresponding to the boundary between two contiguous facies or zones of metamorphic grade, as defined by the appearance of specific index minerals, e.g. garnet isograd, staurolite isograd.

isogradal Pertaining to rocks which have reached the same grade of metamorphism irrespective of their initial compositions. Cf: *isophysical series.* Syn: *isograde; isofacial [petrology].*

isograde *isogradal.*

isogram A general term proposed by Galton (1889, p.651) for any line on a map or chart connecting points having an equal numerical value of some physical quantity (such as temperature, pressure, or rainfall); an *isopleth.*

isogranular An obsolescent syn. of *subautomorphic,* restricted to igneous rocks having pyroxene grains in the interstices between, and of the same size as, plagioclase crystals.

isogriv A line on a map or chart connecting points of equal grivation.

isogyre In crystal optics, a black or shadowy part of an *interference figure* that is produced by extinction and indicates the emergence of those components of light having equal vibration direction. It may look like one arm of a black cross. Cf: *isochromatic curve.* Syn: *polarization brush.*

isohaline adj. Of equal or constant salinity.—n. A line on a chart that connects points of equal salinity in the ocean.—Cf: *isopycnic.*

isoheight *isohypse.*

isohume On a map or diagram, a line connecting points of equal moisture content in coal. Cf: *isocal; isocarb; isodeme; isovol.*

isohyet A line connecting points of equal precipitation.

isohyperthermic temperature regime A soil temperature regime having the same temperature range as the *hyperthermic temperature regime,* but with a summer-winter variation of less than 5°C (USDA, 1975).

isohypse An isopleth for height or elevation; a *contour* on a topographic map. Syn: *isoheight.*

isokite A white monoclinic mineral: $CaMg(PO_4)F$. It is isomorphous with tilasite.

isolat An *isodiff* connecting points of equal latitude correction.

isolated porosity The property of rock or soil of containing noncommunicating interstices, e.g. vesicles in lava, expressed as the percent of bulk volume occupied by such interstices; the numerical difference between total porosity and effective porosity.

isolation In biology, any process or condition by which a group of individuals is separated for a considerable length of time from other groups, as a result of geographic, behavioral, or ecologic factors.

isolation of outcrops A method of geologic mapping that outlines all areas of exposed rock to distinguish them from areas where the rock is buried or otherwise concealed. Syn: *multiple-exposure method.*

isoline *isopleth.*

isolith (a) An imaginary line connecting points of similar lithology and separating rocks of differing nature, such as of color, texture, or composition (Kay, 1945a, p.427). The term "isolithic boundary" was used by Grossman (1944, p.48) for a zone of lithofacies change separating rocks of different grain sizes. (b) An imaginary line of equal aggregate thickness of a given lithologic facies or particular class of material within a formation, measured perpendicular to the bedding at selected points (which may be on outcrops or in the subsurface).

isolith map A map that depicts isoliths; esp. a *facies map* showing the net thickness of a single rock type or selected rock component in a given stratigraphic unit.

isolong An *isodiff* connecting points of equal longitude correction.

isomagnetic line A line connecting points of equal value of some *magnetic element,* e.g. *isoclinic line; isogonic line; isodynamic line; isopor.*

isomegathy A term introduced by Shepard & Cohee (1936) for a line, on a map, connecting points of equal median size of sedimentary particles.

isomertieite A cubic mineral: $(Pd,Cu)_5(Sb,As)_2$. It is polymorphous with mertieite.

isomesic temperature regime A soil temperature regime having the same temperature range as the *mesic temperature regime,* but with a summer-winter variation of less than 5°C (USDA, 1975).

isometric *equant.*

isometric line A term introduced by Wright (1944) for a line, drawn on a map, representing a constant value obtained from measurement at a series of points along its course; an *isopleth.*

isometric projection A projection in which the plane of projection is equally inclined to the three spatial axes of a three-dimensional object, so that equal distances along the axes are drawn equal. It gives a bird's-eye view, combining the advantages of a ground plan and elevation; e.g. as in a block diagram showing three faces.

isometric system One of the six crystal systems, characterized by four threefold axes of symmetry as body diagonals in a cubic unit cell of the lattice. It comprises five crystal classes or point groups. Cf: *hexagonal system; tetragonal system; orthorhombic system; monoclinic system; triclinic system.* Syn: *cubic system.*

isometry The constancy of shape or proportions of a system (or organism) as the magnitude of the system changes. Cf: *allometry.*

isomicrocline An optically positive variety of microcline.

isomodal layering Layering in a cumulate in which the layers are characterized by a uniform proportion of one or more cumulus minerals.

isomorph An organism, or a part of an organism, that is similar to another but unrelated to it.

isomorphic Having identical or similar form. Cf: *heteromorphic [evol].* Syn: *isomorphous.*

isomorphism [cryst] The characteristic of two or more crystalline substances to have similar chemical composition, axial ratios, and crystal forms, and to crystallize in the same crystal class. Such substances form an *isomorphous series.* Adj: *isomorphous.* Cf: *isostructural.* Syn: *allomerism.*

isomorphism [evol] The similarity that develops in organisms of different ancestry as a result of *convergence [evol].*

isomorphous Adj. of *isomorphism.* Syn: *isomorphic; allomeric.*

isomorphous mixture *isomorphous series.*

isomorphous series Two or more crystalline substances that display *isomorphism;* their physical properties vary along a smooth curve. An example is olivine, usually found in nature as a solid solution of Mg_2SiO_4 and Fe_2SiO_4, i.e. an isomorphous series between forsterite and fayalite. The exact lattice dimensions and other physical properties vary with change of the Mg/Fe ratio. Syn: *solid-solution series.*

isomyarian adj. Said of a bivalve mollusk or its shell having two adductor muscles of equal or nearly equal size. Syn: *homomyarian.* —n. An isomyarian mollusk.

isontic line Obsolete syn. of *isopleth.* Term proposed by Lane (1928, p.37).

iso-orthoclase An optically positive variety of orthoclase. It has been found in granitic gneiss. Syn: *isorthoclase; isorthose.*

isopach A line drawn on a map through points of equal true thickness of a designated stratigraphic unit or group of stratigraphic units. Cf: *isochore.* Syn: *isopachyte; thickness line; thickness contour.*

isopach map A map that shows the thickness of a bed, formation, sill, or other tabular body throughout a geographic area by means of isopachs at regular intervals. Cf: *isochore map.* Syn: *thickness map.* Nonrecommended syn: *isopachous map.*

isopachous Of, relating to, or having an isopach; e.g. an "isopachous contour". Not recommended usage.

isopachous cement A crust of carbonate cement, often fibrous, of equal thickness around all clastic grains; common as a beachrock or submarine cement.

isopachous map A nonrecommended syn. of *isopach map.*

isopach strike The compass direction of an isopach line at a given point on a map.

isopachyte British term for *isopach.*

isopag An *equiglacial line* connecting points where ice is present for approximately the same number of days per year.

isopectic An *equiglacial line* connecting points where ice begins to form at the same time in winter. Cf: *isotac.*

isoperimetric curve A line on a map or map projection (as on an equal-area projection) along which there is no variation from exact scale.

isoperthite A variety of alkali feldspar consisting of perthitic intergrowths of the same kind of feldspar or of two kinds of feldspar belonging to the same isomorphous series.

isophysical series A series of rocks of different chemical composition that were metamorphosed under identical physical conditions. Cf: *isogradal.*

isopic Said of sedimentary rocks of the same facies, or said of facies characterized by identical or closely similar rock types. The rocks may be formed in different sedimentation areas or at different times or both, but the lithologies are the same; e.g. a facies repeated in vertical succession. Also, said of a map depicting isopic facies or rocks. Cf: *heteropic; holisopic.*

isopiestic line *equipotential line.*

isopleth [geochem] In a strict sense, a line or surface on which some mathematical function has a constant value. It is sometimes distinguished from a *contour* by the fact that an isopleth need not refer to a directly measurable quantity characteristic of each point in the map area, e.g. maximum temperature of a particular point. More generally, the term is used as a synonym of "isocomposition-al section".

isopleth [phys sci] (a) A general term for a line, on a map or chart, along which all points have a numerically specified constant or equal value of any given variable, element, or quantity (such as abundance or magnitude), with respect to space or time; esp. a *contour.* Etymol: Greek *isos,* "equal", + *plethos,* "fullness, quantity, multitude". Syn: *isogram; isoline; isontic line; isometric line.* (b) A line drawn through points on a graph at which a given quantity has the same numerical value (or occurs with the same frequency) as a function of two coordinate variables. It is often used of a meteorologic element that varies with the time of the year (month) and the time of day (hour).—Syn: *isarithm.*

isopleth map A general term for any map showing the areal distribution of some variable quantity in terms of lines of equal or constant value; e.g. an *isopach map.*

isopod Any malacostracan crustacean belonging to the order Isopoda, characterized generally by the absence of a carapace and the presence of sessile eyes and a compressed body. Range, Triassic to present. Cf: *amphipod.*

isopollen A line on a map connecting locations with samples having the same percentage or amount of pollen of a given kind. Syn: *isopoll.*

isopor An *isomagnetic line* of equal secular change, e.g. equal annual change of isogonic or isoclinic lines.

isopotal Having equal infiltration capacities; e.g. an isopotal area in a watershed.

isopotential line *equipotential line.*

isopotential surface A surface on which points of equal *fluid potential* lie.

isopycnic adj. Of constant or equal density, measured in space or in time.—n. A line on a chart that connects points of equal density.—Cf: *isohaline.* See also: *isostere.*

isopygous Said of a trilobite having cephalon and pygidium of subequal size. Ant: *heteropygous.*

isorad A line connecting points of equal radioactivity.

isorat Line connecting points of equal isotope ratios.

isoseism *isoseismal line.*

isoseismal n. A syn. of *isoseismal line.*

isoseismal line A line connecting points on the Earth's surface at which earthquake intensity is the same. It is usually a closed curve around the *epicenter.* Cf: *coseismal line.* Syn: *isoseism; isoseismal.*

isosinal map A *slope map* whose contour lines are lines of equal slope represented by sines of slope angles read from a topographic map.

isospore *homospore.*

isospory *homospory.*

isostannite *kesterite.*

isostasy The condition of equilibrium, comparable to floating, of the units of the lithosphere above the asthenosphere. Two differing concepts of the mechanism of isostasy are called the *Airy hypothesis* and the *Pratt hypothesis.* See also: *isostatic compensation; depth of compensation.*

isostatic adjustment *isostatic compensation.*

isostatic anomaly A gravity anomaly calculated on a hypothesis that the gravitational effect of masses extending above sea level is approximately compensated by a deficiency of density of the material beneath those masses; the effect of deficiency of density in ocean waters is compensated by an excess of density in the material under the oceans.

isostatic compensation The adjustment of the lithosphere of the Earth to maintain equilibrium among units of varying mass and density; excess mass above is balanced by a deficit of density below, and vice versa. See also: *depth of compensation; isostasy.* Syn: *isostatic adjustment.*

isostatic correction The adjustment made to values of gravity, or to deflections of the vertical, observed at a point to take account of the assumed mass deficiency under topographic features for which a topographic correction is also made.

isostatic isocorrection-line map A contour map showing lines connecting places for which the isostatic correction has the same value.

isostere A line connecting points of equal density of the Earth's atmosphere; an *isopycnic* of the atmosphere.

isostratification map A map that shows the number or thickness of beds in a stratigraphic unit by means of contour lines representing equal *stratification indices* (Kelley, 1956, p.299). Syn: *isobed map.*

isostructural Said of two or more chemical compounds with similar crystal structures but with little tendency to show *isomorphism.*

isotac An *equiglacial line* connecting points where ice melts at the same time in spring. Cf: *isopectic.*

isotach A line connecting points of equal wind velocity.

isotangent map A *slope map* whose contour lines are lines of equal slope represented by tangents of slope angles read from a topographic map.

isotaque In crystal optics, one of several curves representing equal wave-normal velocities; in a uniaxial crystal it is a circle concentric with the optic axis, and in a biaxial crystal, it is a spherical ellipse.

isotherm A line connecting points of equal temperature. Isotherm maps are often used to portray surface temperature patterns of water bodies. Cf: *isallotherm.*

isothermal Pertaining to the process of changing the thermodynamic state of a substance, e.g. its pressure and volume, while maintaining the temperature constant.

isothermal remanent magnetization Remanent magnetization due solely to application of a magnetic field, without change of temperature. Abbrev: IRM.

isothermic temperature regime A soil temperature regime that has the characteristics of a *thermic temperature regime* except for a summer-winter variation of less than 5°C (USDA, 1975).

isothrausmatic A descriptive term applied to igneous rocks with an orbicular texture in which the nuclei of the orbicules are composed of the same rock as the groundmass (Eskola, 1938, p.476). Cf: *allothrausmatic; crystallothrausmatic; homeothrausmatic; heterothrausmatic.*

isotime line *isochron [seis].*

isotomous Said of a crinoid arm characterized by division into equal branches. Ant: *heterotomous.*

isotope One of two or more species of the same chemical element, i.e. having the same number of protons in the nucleus, but differing from one another by having a different number of neutrons. The isotopes of an element have slightly different physical and chemical properties, owing to their mass differences, by which they can be separated. See also: *radioisotope.*

isotope dilution An analytical method in which a known quantity of an element with an isotopic composition different from that of the natural element (a *spike*) is mixed with the sample being analyzed. Measurement of the isotopic composition of the mixture allows calculation of the amount of the natural element in the sample.

isotope effect *Isotopic fractionation* in a variety of chemical and physical processes, resulting from slight differences in the properties (heat capacity, vapor pressure, density, free energy, etc.) of the various isotopic forms of the elements involved.

isotope geochemistry *isotope geology.*

isotope geology The application of the study of radioactive and stable isotopes, especially their abundances, to geology. It includes the calculation of geologic time, and the determination of the origin, mechanisms, and conditions of geologic processes by isotopic means. Syn: *isotope geochemistry; nuclear geology; nuclear geochemistry; radiogeology.*

isotope ratio The ratio of abundance of any two isotopes of a given element, e.g. ^{18}O to ^{16}O. It is conventionally written as the ratio of the heavy isotope to the light isotope. Cf: *delta value.*

isotopic (a) Pertaining to an *isotope.* (b) Said of rocks formed in the same environment, such as in the same sedimentary basin or geologic province. Ant: *heterotopic.*

isotopic age *radiometric age.*

isotopic age determination *radiometric dating.*

isotopic fractionation The relative enrichment of one isotope of an element over another, owing to slight variations in their physical and chemical properties. It is proportional to differences in their masses. See also: *isotope effect.*

isotopic number The number of excess neutrons, i.e. the number of neutrons minus the number of protons, in an atomic nucleus. It is usually an indication of the radioactivity of the nucleus.

isotropic Said of a medium whose properties are the same in all directions; in crystal optics, said of a crystal whose physical properties do not vary according to crystallographic direction, e.g. one in which in which light travels with the same speed in any direction. Cubic crystals and amorphous substances are usually *isotropic.* Ant: *anisotropic.*

isotropization The solid-state conversion of an originally birefringent mineral such as quartz or feldspar into a more or less isotropic phase at temperatures below the melting point, as a result of destruction of the crystallinity by such processes as shock-wave action or neutron bombardment.

isotropy The condition of having properties that are uniform in all directions. Adj: *isotropic.*

isotypic Said of crystalline substances that have analogous crystal structures and chemical compositions, e.g. zircon and xenotime.

isovol On a map or diagram, a line connecting points of equal volatile content in coal. Cf: *isocarb; isocal; isodeme; isohume.* See also: *isoanthracite line.*

issite A dark-colored hypabyssal rock characterized by xenomorphic-granular texture and composed chiefly of hornblende with smaller amounts of green pyroxene and even less labradorite and accessory magnetite and apatite. Its name is derived from the Issa River, Penza oblast, U.S.S.R. Not recommended usage.

issue The place where a stream flows out into a larger body of water.

isthmus A narrow strip or neck of land, bordered on both sides by water, connecting two larger land areas, such as a peninsula and the mainland (e.g. Isthmus of Suez) or two continents (e.g. Isthmus of Panama). Etymol: Greek *isthmos.* See also: *submarine isthmus.*

itabirite A laminated, metamorphosed *oxide-facies iron formation* in which the original chert or jasper bands have been recrystallized into megascopically distinguishable grains of quartz and in which the iron is present as thin layers of hematite, magnetite, or martite (Dorr & Barbosa, 1963, p. 18). The term was originally applied in Itabira, Brazil, to a high-grade massive specular-hematite ore (66% iron) associated with a schistose rock composed of granular quartz and scaly hematite. The term is now widely used outside Brazil. Cf: *jacutinga; canga.* Syn: *banded-quartz hematite; hematite schist.*

itacolumite A micaceous sandstone or a schistose quartzite that contains interstitial, loosely interlocking grains of mica, chlorite, and talc, and that exhibits flexibility when split into thin slabs. Type locality: Itacolumi Mountain in the state of Minas Gerais, Brazil. Syn: *flexible sandstone; articulite.*

italite (a) In the *IUGS classification,* a plutonic rock in which F is between 60 and 100, M is 10 or less, and potassium exceeds sodium. Cf: *urtite.* (b) A volcanic *foidite* rich in leucite and containing up to 30 percent mafic minerals, such as melilite, biotite, and apatite. Its name, given by Washington in 1920, is derived from Italy. Cf: *fergusite; missourite.*

iterative evolution Repeated development of new forms from the same ancestral stock; repeated, independent evolution.

itoite An orthorhombic mineral: $Pb_3GeO_2(SO_4)_2(OH)_2$.

ITRM *inverse thermoremanent magnetization.*

itsindrite A potassium-rich hypabyssal *nepheline syenite* containing microcline, nepheline, biotite, acmite, and zoned melanite. Its name, given by Lacroix in 1922, is derived from the Itsindra Valley, Malagasy. Not recommended usage.

IUGS classification An internationally adopted classification of plutonic rocks, presented in 1973 by the Subcommission on the Systematics of Igneous Rocks of the International Union of Geological Sciences, A. L. Streckeisen, chairman. The classification is based on modal proportions of minerals in five groups: Q = quartz and other polymorphs of SiO_2; A = alkali feldspars; P = plagioclase more calcic than An_5, and scapolite; F = feldspathoids (foids); M = all other phases (mafites). If the percentage of mafites M is less than 90, a rock is classified according to its position in a double triangle QAPF. Ultramafic rocks (M=90 to 100) are classified according to the proportions of mafites. A more detailed classification of gabbroic rocks is also provided, and a "preliminary system (for field use)" was suggested. No agreement was reached on nomenclature of charnockitic rocks. See: IUGS Subcommission on the Systematics of Igneous Rocks, 1973, p. 26-30.

ivorite A black tektite from the Ivory Coast, western Africa.

ivory The fine-grained creamy-white dentine forming the tusks of elephants, and the teeth or tusks of certain other large animals such as the walrus; it has long been esteemed for a wide variety of ornamental articles. "Vegetable ivory" is produced from the corozo nut and the Doum palm nut.

I wave A longitudinal or *P wave* in the Earth's inner core. Cf: *K wave.*

ixiolite A mineral: $(Ta,Nb,Sn,Fe,Mn)_4O_8$. It was previously considered to be a mixture of cassiterite with columbite or tapiolite, and to be a manganese-tantalate isomorph of tapiolite and of mossite.

jacinth (a) A zircon; specif. a yellow or brown zircon. The term was originally an alternate spelling of *hyacinth,* and has been used to designate a red or orange zircon and sometimes a gem zircon more nearly pure orange in color than a hyacinth. (b) An orange-red to orange essonite.—The term, "having become meaningless", is obsolete in the American gem trade (Shipley, 1951, p. 116).
jack [coal] (a) *Cannel coal* interstratified with shale. (b) Coaly, often canneloid, shale. (c) A large ironstone nodule in the coal measures of Wales.
jack [mineral] A zinc ore; specif. *sphalerite.*
jack iron A term used in the zinc-mining area of Missouri for a solid flint rock containing disseminated sphalerite, or blackjack.
Jacksonian North American (Gulf Coast) stage: Eocene (above Claibornian, below Vicksburgian).
jackstraw texture A texture of metamorphic rocks characterized by a criss-cross arrangement of an elongate mineral. It is often exhibited by olivine in a matrix of talc in metamorphosed ultramafic rocks (Snoke and Calk, 1978).
jacobsite A black magnetic mineral of the magnetite series in the spinel group: $(Mn^{+2},Fe^{+2},Mg)(Fe^{+3},Mn^{+3})_2O_4$.
Jacob's staff A single, straight rod, staff, or pole, pointed and shod with iron at its lower end for insertion in the ground, and fitted with a ball-and-socket joint at its upper end for adjustment to a level position, used instead of a tripod for mounting and supporting a surveyor's compass or other instrument. Named after Jacob (St. James), symbolized in religious art by a pilgrim's staff.
jacupirangite An ultramafic plutonic rock that is part of the *ijolite* series, composed chiefly of titanaugite and magnetite, with a smaller amount of nepheline; a nepheline-bearing *clinopyroxenite.* Its name, given by Derby in 1891, is derived from Jacupiranga, Brazil.
jacutinga A term used in Brazil for disaggregated, powdery *itabirite,* and for variegated thin-bedded high-grade hematite iron ores associated with and often forming the matrix of gold ore. Etymol: from its resemblance to the colors of the plumage of *Pipile jacutinga,* a Brazilian bird.
jade (a) A hard, extremely tough, compact gemstone consisting of either the pyroxene mineral jadeite or the amphibole mineral nephrite, and having an unevenly distributed color ranging from dark or deep green to dull or greenish white. It takes a high polish, and has long been used for jewelry, carved articles, and various ornamental objects. Syn: *jadestone.* (b) A term that is often applied to various hard green minerals; e.g. "California jade" (or californite, a green compact variety of vesuvianite), "Mexican jade" (or tuxtlite, and also green-dyed calcite), saussurite, and green varieties of sillimanite, pectolite, garnet, and serpentine.
jade-albite *Maw-sit-sit.*
jadeite A high-pressure mineral of the clinopyroxene group, essentially: $Na(Al,Fe)Si_2O_6$. It occurs in various colors (esp. green) and is found chiefly in Burma; when cut, it furnishes the most valuable and desirable variety of jade and is used for ornamental purposes.
jadeitite A metamorphic rock consisting principally of jadeite, commonly associated with small amounts of feldspar or feldspathoids. It is probably derived from an alkali-rich igneous rock by high-pressure metamorphism.
jadeolite A deep-green chromiferous *syenite* cut as a gemstone and resembling jade in appearance. Obsolete.
jadestone *jade.*
jager A high-quality bluish-white diamond.
jagoite A yellow-green mineral: $Pb_3FeSi_3O_{10}(OH,Cl)$.
jagowerite A triclinic mineral: $BaAl_2(PO_4)_2(OH)_2$.
jahnsite A monoclinic mineral: $CaMn(Mg,Fe)_2Fe_2{}^{+3}(PO_4)_4(OH)_2 \cdot 8H_2O$.
jahresringe An annual lamination, e.g. a varve; also used in the more general sense of any sedimentary couplet that may have specific time significance, though this is generally implied rather than documented.
jalpaite A lead-gray mineral: Ag_3CuS_2.
jamborite A hexagonal mineral: $(Ni^{+2},Ni^{+3},Fe)(OH)_2(OH,S,H_2O)$ (?).
jamesonite A lead-gray to gray-black orthorhombic mineral: $Pb_4FeSb_6S_{14}$. It is a minor ore of lead, and sometimes contains copper and zinc. Jamesonite has a metallic luster and commonly occurs in acicular crystals with fibrous or featherlike forms. Syn: *feather ore; gray antimony.*
Jamin effect The restrictive force exerted upon the flow of fluids through narrow tubes or passages by successive bubbles of air or other gas. If a narrow tube is expanded at several places, and each bulb or enlargement contains a gas, the liquid can support a pressure of several atmospheres before it begins to flow.
Jänecke diagram A square phase diagram whose corners represent two reciprocal salt pairs (e.g., NaCl - KCl - NaBr - KBr) and on which is plotted the configuration of the surface representing the aqueous solution saturated with the salts. It is particularly useful in the study of phase equilibria relevant to evaporites. Syn: *reciprocal salt-pair diagram.*
janggunite An orthorhombic mineral: $Mn_{4.85}{}^{+4}(Mn_{0.90}{}^{+2}Fe_{0.30}{}^{+3})O_{8.09}(OH)_{5.91}$.
Japanese twin law A twin law in quartz that governs twinning of two individuals with a composition plane of (1122); four varieties are possible.
jardang *yardang.*
jargoon A colorless, pale-yellow, or smoky gem variety of zircon from Ceylon. Syn: *jargon.*
jarlite A colorless to brownish mineral: $NaSr_3Al_3F_{16}$.
jarosite (a) An ocher-yellow or brown mineral of the alunite group: $KFe_3(SO_4)_2(OH)_6$. Syn: *utahite.* (b) A group of minerals consisting of hydrous iron sulfates, including jarosite, natrojarosite, ammoniojarosite, argentojarosite, plumbojarosite, and hydronium jarosite.
jars A mechanical device included in the *drill string* of a cable-tool well to allow a sharp vertical stress to be applied if the string should become stuck while drilling, or in *fishing* for "stuck tools".
jaspachate *jaspagate.*
jaspagate A syn. of *agate jasper,* esp. that in which jasper predominates. Also spelled: *jaspachate.*
jasper A variety of *chert* associated with iron ores and containing iron-oxide impurities that give it various colors, characteristically red, although yellow, green, grayish-blue, brown, and black cherts have also been called jasper. The term has also been applied to any red chert or chalcedony irrespective of associated iron ore. Syn: *jasperite; jaspis; jasperoid.*
jasper bar A term used in Australia for *iron formation.* Syn: *bar; jaspilite.*
jasperine Banded jasper of varying colors.
jasperite *jasper.*
jasperization The conversion or alteration of igneous or sedimentary rocks into banded rocks like jaspilite by metasomatic introduction of iron oxides and cryptocrystalline silica.
jasperoid n. (a) A dense, usually gray, chertlike siliceous rock, in which chalcedony or cryptocrystalline quartz has replaced the carbonate minerals of limestone or dolomite; a silicified limestone. It typically develops as the gangue of metasomatic sulfide deposits of the lead-zinc type, such as those of Missouri, Oklahoma, and Kansas. (b) *jasper.* — adj. Resembling jasper.
jasper opal A yellow or yellow-brown, almost opaque common opal containing iron oxide and other impurities, having the color

of yellow jasper but the luster of common opal. Some varieties are almost reddish brown to red. Syn: *jaspopal; opal jasper.*

jaspery Resembling or containing jasper; e.g. "jaspery iron ore" (impure hematite interbedded with jasper), or "jaspery chert" (a silicified radiolarian ooze associated with volcanic rocks in Ordovician strata of southern England). Syn: *jaspidean.*

jaspidean Resembling or containing jasper; jaspery.

jaspilite (a) A banded compact siliceous rock containing at least 25% iron, occurring with iron ores, and resembling jasper; e.g. the rock of the Precambrian iron-bearing district of the Lake Superior region. (b) A general term (used esp. in Australia) for banded *iron formation.* Syn: *jasper bar.* Also spelled: *jaspilyte.* See also: *jasperization.*

jaspis A syn. of *jasper.* Etymol: German *Jaspis.*

jaspoid (a) Resembling jasper (Thrush, 1968, p. 598). (b) A syn. of *tachylyte* (Hey, 1962, p. 470).

jasponyx An opaque onyx, part or all of whose bands consist of jasper.

jaspopal A syn. of *jasper opal.* Also spelled: *jasp-opal.*

javaite An Indonesian tektite from Java. Syn: *javanite.*

jebel A hill, mountain, or mountain range in northern Africa. Etymol: Arabic. Syn: *jabal; djebel.*

jefferisite A variety of vermiculite.

jeffersonite A dark-green or greenish-black mineral of the clinopyroxene group: $Ca(Mn,Zn,Fe)Si_2O_6$.

jelly *ulmin.*

jenkinsite A variety of antigorite containing iron.

jennite A mineral: $Na_2Ca_8(SiO_3)_3(Si_2O_7)$.

jeremejevite A colorless to pale yellowish-brown hexagonal mineral: $Al_6B_5O_{15}(OH)_3$. Syn: *eremeyevite.*

jeromite A mineral: $As(S,Se)_2$ (?).

jet [coal] A hard, lustrous pure black variety of lignite; it has a conchoidal fracture and will take a high polish. It occurs as isolated masses in bituminous shale and is probably derived from waterlogged pieces of driftwood. Jet is used for jewelry and other ornamentation. Syn: *black amber.* See also: *jet shale; pitch coal.*

jet [hydraul] A sudden and forceful rush or gush of fluid through a narrow or restricted opening, either in spurts or in a continuous flow; e.g. a stream of water or air used to flush cuttings from a borehole.

jet flow A type of streamflow characterized by water moving in plunging, jetlike surges, produced where a stream reaches high velocity along a sharply inclined stretch or moves swiftly over a waterfall, or where a turbulent stream enters a body of standing water. Syn: *shooting flow.*

jetonized wood Lamellae of vitrain in coal.

jet rock *jet shale.*

jet shale Bituminous shale containing *jet* [coal]. Syn: *jet rock.*

jetted well A shallow water well, constructed by a high-velocity stream of water directed downward into the ground.

jetting The process of sinking a borehole, or of flushing cuttings or loosely consolidated materials from a borehole, by using a directed, forceful stream (jet) of drilling mud, air, or water.

jetty (a) An engineering structure (such as a breakwater, groin, seawall, or small pier) extending out from the shore into a body of water, designed to direct and confine the current or tide, to protect a harbor, to induce scouring, or to prevent shoaling of a navigable passage by littoral materials. Jetties are often built in pairs on either side of a harbor entrance, or at the mouth of a river. (b) A British term for a landing wharf or pier used as a berthing place for vessels.

jew's-stone [mineral] A piece of marcasite used in making ornaments (esp. costume jewelry).

jew's-stone [paleont] A large fossil clavate spine of a sea urchin.

jezekite *morinite.*

jheel A term applied in the Ganges flood plain of India to a backwater, such as a pool, marsh, or lake, remaining from inundation, existing during the cold weather at about the same level as that of the river and rising with the river during the rainy season. Etymol: Hindi. Cf: *bhil.* Pron: *jeel.* Also spelled: *jhil.*

jhil *jheel.*

jimboite An orthorhombic mineral: $Mn_3(BO_3)_2$. It is isostructural with kotoite.

joaquinite A honey-yellow mineral: $NaBa_2Ce_2Fe(Ti,Nb)_2Si_8O_{26}(OH,F)$.

Job's tears Rounded grains of olivine (peridot) found associated with garnet in Arizona and New Mexico.

joch A mountain pass with a long, approximately level summit between two parallel slopes (Stamp, 1961, p. 491); a *col.* Etymol: German *Joch,* "yoke". Pron: *yuhkh.* Syn: *yoke-pass.*

joesmithite A mineral: $PbCa_2(Mg,Fe)_4Fe^3Si_6O_{12}(OH)_4(O,OH)_8$.

johachidolite A colorless and transparent mineral: $Na_2Ca_3Al_4B_6O_{14}(F,OH)_{10}$.

johannite A green secondary mineral: $Cu(UO_2)_2(SO_4)_2(OH)_2 \cdot 6H_2O$. Syn: *gilpinite.*

johannsenite A clove-brown, grayish, or greenish mineral of the clinopyroxene group: $CaMnSi_2O_6$.

Johannsen number A number, composed of three or four digits, that defines the position of an igneous rock in *Johannsen's classification.* The first digit represents the class, the second the order, and the third and fourth the family.

Johannsen's classification A quantitative mineralogic classification of igneous rocks developed by the petrographer Albert Johannsen (1939). Cf: *Johannsen number.*

johnstrupite A brownish-green mineral, approximately: $(Ca,Na)_3(Ce,Ti,Zr)(SiO_4)_2F$. Cf: *mosandrite.*

join The line or plane drawn between any two or three composition points in a phase diagram. There is no special phase significance to a join; it need not be a limiting binary or ternary subsystem. Incorrect syn: *conjugation line.*

joint [paleont] (a) An articulation in a crustacean; commonly, the movable connection of an individual *segment* of an appendage with its neighbors or with the body, or the movable connection of body parts. (b) A connection between any pair of contiguous crinoid ossicles. (c) A segment of the shell of a nasselline radiolarian. (d) For vertebrates, see: *articular; articulation.*

joint [struc geol] A surface of fracture or parting in a rock, without displacement; the surface is usually plane and often occurs with parallel joints to form part of a *joint set.* See also: *jointing.*

joint block A body of rock that is bounded by joints; the rock that occurs between adjacent joints.

joint-block separation A type of mechanical weathering in which the rock breaks down or comes apart along well-defined joint planes. Syn: *block disintegration.*

joint cavity A solutional hollow whose position is controlled by a joint on the inner surface of a cave. See also: *ceiling cavity.*

joint frequency *joint spacing.*

jointing n. The condition or presence of *joints* in a body of rock Partial syn: *cleating.*

joint plane The surface of a joint.

joint-plane fall A waterfall whose crest is irregularly broken by the falling away of joint blocks (Tarr & Von Engeln, 1926, p. 83

joint set A group of more or less parallel joints. See also: *joint sys-tem.*

joint spacing The interval between joints of a particular joint set measured on a line perpendicular to the joint planes. Syn: *joint frequency.*

joint system Two or more *joint sets* that intersect. They may be of the same age or of different ages.

joint valley A valley whose drainage pattern is controlled by master joint systems, e.g. in a *rectangular drainage pattern.*

jokul An Icelandic mountain permanently covered with ice an snow; an Icelandic snow-capped peak. The term has been applie as well to an Icelandic glacier or small ice cap. Etymol: Icelandi *jökull,* "icicle, glacier". Pron: *yo*-kool. Pl: *jokuls.* Syn: *jöku jökul.*

jökulhlaup An Icelandic term for *glacier outburst flood.* Pron: *yo*-kool-loup (the last syllable as in "out").

joliotite An orthorhombic mineral: $(UO_2)CO_3 \cdot nH_2O$, with n a proximately equal to 2.

Jolly balance In mineral analysis, a delicate spring balance use to measure specific gravity.

Joplin-type lead *J-type lead.*

jordanite A lead-gray mineral: $Pb_{14}As_6S_{23}$.

Jordan's law A theory in evolutionary biology stating that close related organisms tend to occupy adjacent rather than identic or distant ranges. Named after the American biologist David Jo dan (1851-1931). Not to be confused with *Jordan's rule.*

Jordan's rule The empirical relationship, noted by the biologi David Jordan, that the number of vertebrae in a freshwater fi depends on the temperature of the water in which the fish develo ed. Not to be confused with *Jordan's law.*

jordisite An amorphous mineral: MoS_2. Cf: *molybdenite.*

josefite An altered hypabyssal rock having microgranular textu and composed of augite, olivine, serpentine, and calcite. N recommended usage.

joseite A mineral: $Bi_3Te(Se,S)$.
josen *hartite.*
josephinite A mineral consisting of a natural alloy of iron and nickel occurring in stream gravel from Josephine County, Oregon; nickel-iron.
jotunite A plutonic rock of the *charnockite series,* intermediate between *monzonite* and *norite,* containing orthopyroxene, plagioclase, and microperthite, and attributed in part to monzonite, in part to monzonite-diorite, and in part to monzonite-gabbro (Streckeisen, 1967, p. 169). The name, given by Goldschmidt in 1916, is for Jotunheim, Norway. Syn: *jotun-norite.* Not recommended usage.
jotun-norite *jotunite.*
jouravskite A mineral: $Ca_6Mn_2(SO_4,CO_3)_4(OH)_{12}\cdot 24H_2O$.
J-type lead *Anomalous lead* that gives model ages younger than the age of the enclosing rock, in some cases even negative model ages. Cf: *B-type lead.* Syn: *Joplin-type lead.*
juanite An orthorhombic mineral: $Ca_{10}Mg_4Al_2Si_{11}O_{39}\cdot 4H_2O$ (?).
jug A colloquial syn. of *geophone.*
jugum (a) A medially placed connection of secondary shell between two primary lamellae of brachiopod spiralia; a more or less complex skeletal crossbar linking the right and left halves of the brachidium of certain brachiopods. (b) A transverse structure crossing the center of a heterococcolith and connecting one side of the cycle with the other.—Pl: *juga* or *jugums.* Adj. *jugal.*
julgoldite A mineral: $Ca_2Fe^{+2}(Fe,Al)_2(SiO_4)(Si_2O_7)(OH)_2\cdot H_2O$. It is related to pumpellyite.
julienite A blue mineral occurring in needlelike crystals: $Na_2Co(SCN)_4\cdot 8H_2O$ (?).
jumillite An extrusive rock, commonly fine-grained, composed of phenocrysts of barium-bearing sanidine, olivine, and phlogopite in a fine-grained groundmass of altered leucite, sanidine, and iron-rich diopside mantled with acmite-augite, with interstitial kataphorite. A variety of *leucitite,* named by Osann in 1906 for Jumilla, Spain. Cf: *orendite.* Not recommended usage.
jump correlation Identification of events on noncontiguous seismic records as involving the same interfaces in the Earth.
junction [streams] The meeting of two or more streams; also, the place of such a meeting; a *confluence.* Examples: *accordant junction; deferred junction.*
junction [surv] A point common to two or more survey lines.
junction closure The amount by which a new survey line into a junction fails to give the previously determined position or elevation for the junction point.
jungle A equatorial region of wild, tangled, dense vegetation.
junitoite An orthorhombic mineral: $CaZn_2Si_2O_7$.
junoite A monoclinic mineral: $Pb_3Cu_2Bi_8(S,Se)_{16}$.
Jura *Jurassic.*
Jurassian relief A type of relief found in young mountains that consist of many parallel anticlines and synclines, characterized by primary structural forms or by features upon which erosion has had relatively little influence (Schieferdecker, 1959, term 1944). Type example: the relief of the Jura Mountains in Switzerland. Cf: *Appalachian relief.*
Jurassic The second period of the Mesozoic era (after the Triassic and before the Cretaceous), thought to have covered the span of time between 190 and 135 million years ago; also, the corresponding system of rocks. It is named after the Jura Mountains between France and Switzerland, in which rocks of this age were first studied. See also: *age of cycads.* Syn: *Jura.*
Jura-Trias The Jurassic and Triassic periods, combined.
Jura-type fold *décollement fold.*
jurbanite A monoclinic mineral: $Al(SO_4)(OH)\cdot 5H_2O$.
jurupaite A variety of xonotlite containing magnesium.
juvenarium The proloculus and first few chambers of a foraminifer. See also: *embryonic apparatus.*
juvenile [geomorph] *youthful.*
juvenile [ore dep] Said of an ore-forming fluid or mineralizer that is derived from a magma, via fractional crystallization or other plutonic mechanism, as opposed to fluids of surface, connate, or meteoric origin. Cf: *assimilated; filtrational.*
juvenile [volc] In the classification of pyroclastics, the equivalent of *essential;* derived directly from magma reaching the surface.
juvenile [water] A term applied to water and gases that are known to have been derived directly from magma and are thought to have come to the Earth's surface for the first time. Cf: *resurgent.*
juvite A light-colored *nepheline syenite* in which the feldspar is exclusively or predominantly potassium feldspar and the potassium-oxide content is higher than the sodium oxide. Not recommended usage.
juxta-epigenesis Post-diagenetic changes that affect sediments while they are near the original environment of deposition, either under a relatively thin overburden or exposed above sea level (Chilingar et al., 1967, p. 316). Cf: *apo-epigenesis.*
juxtaposed ice stream Ice from a tributary glacier that is set into the surface of a glacier and extends to the bed. Cf: *inset ice stream.*
juxtaposition twin *contact twin.*
J wave A transverse or *S wave* traveling through the Earth's inner core.

kaersutite A black variety of hornblende containing titanium.

kafehydrocyanite A mineral: $K_4Fe(CN)_6 \cdot 3H_2O$.

kahlerite A yellow to yellow-green secondary mineral of the autunite group: $Fe(UO_2)_2(AsO_4)_2 \cdot nH_2O$.

kaimoo A stratified ice and sediment rampart built during the autumn on an Arctic beach by wave action. Etymol: Eskimo.

kainite A usually whitish monoclinic mineral: $MgSO_4 \cdot KCl \cdot 3H_2O$. It is a natural salt occurring in irregular granular masses, and is used as a source of potassium and magnesium compounds.

kainosite A yellowish-brown mineral: $CaIF2(Ce,Y)_2(SiO_4)_3(CO_3) \cdot H_2O$. Syn: *cenosite.*

Kainozoic *Cenozoic.*

kaiwekite A trachytic extrusive rock composed of phenocrysts of barkevikite, small acmite-augite crystals, and anorthoclase with inclusions of acmite-augite and other minerals, and with a few pseudomorphs of serpentine after olivine, in a groundmass composed chiefly of oligoclase with some augite and magnetite; the approximate extrusive equivalent of *larvikite.* Essentially a *trachyte.* Named by Marshall in 1906 for Kaiweke, New Zealand. Not recommended usage.

kajanite A feldspar-free extrusive rock composed of phenocrysts of bronze-colored mica and olivine in a groundmass of leucite, diopside, and titaniferous iron oxide; a *leucitite.* The name, given by Lacroix in 1926, is for Kajan River, Borneo. Not recommended usage.

kakirite A megascopically sheared and brecciated rock in which fragments of original material are surrounded by gliding surfaces along which intense granulation and some recrystallization has occurred. It was named by Svenonius after Lake Kakir, Swedish Lapland.

kakortokite A layered *nepheline syenite* of varied composition, having light-colored layers rich in feldspar and nepheline (white) or in eudialyte and nepheline (red) and dark-colored layers rich in acmite and arfvedsonite (black). The rock was originally described by Ussing in 1912 from Kakortok in the Olimaussaq complex, SW Greenland. Not recommended usage.

kalahari A term used in SW Africa for a salt pan; specif. in the Kalahari Desert of Botswana.

kali- A prefix which, when in an igneous rock name, signifies an absence of plagioclase or a plagioclase content of less than 5.0%.

kalialaskite An *alaskite* with no modal albite; the intrusive equivalent of *kalitordrillite.* Cf: *birkremite.* Not recommended usage.

kaliborite A colorless to white mineral: $HKMg_2B_{12}O_{21} \cdot 9H_2O$. Syn: *heintzite; hintzeite; paternoite.*

kalicinite A colorless to white or yellowish mineral: $KHCO_3$. Syn: *kalicine; kalicite.*

kaligranite A plagioclase-free *granite* that frequently contains soda pyriboles; the intrusive equivalent of *kalirhyolite.* Cf: *alaskite.* Syn: *orthogranite.* Not recommended usage.

kalikeratophyre A *keratophyre* containing potassium feldspar instead of albite. Not recommended usage.

kaliliparite A potassic *rhyolite* having a chemical composition of approximately 68 percent SiO_2, 16 percent Al_2O_3, 1 percent CaO, 1 percent MgO, 1 percent Fe_2O_3, 11 percent K_2O, and 2 percent Na_2O. It is used for low-alkali glass (Thrush, 1968, p. 605). Not recommended usage.

kalinite A mineral of the alum group: $KAl(SO_4)_2 \cdot 11H_2O$. Cf: *alum.* Syn: *potash alum.*

kaliophilite A hexagonal mineral of volcanic origin: $KAlSiO_4$. It is dimorphous with kalsilite. Syn: *facellite; phacellite.*

kalirhyolite An extrusive rock composed of quartz, alkali feldspar, and a ferromagnesian mineral; the extrusive equivalent of *kaligranite.* Not recommended usage.

kalistrontite A hexagonal mineral: $K_2Sr(SO_4)_2$. It is isostructural with palmierite.

kalisyenite A *syenite* containing less than 5 percent of the total feldspar as plagioclase. Not recommended usage.

kalitordrillite A plagioclase-free hypabyssal or extrusive rock composed chiefly of quartz and alkali feldspar; the extrusive equivalent of *kalialaskite.* Not recommended usage.

kalitrachyte An alkalic *trachyte* in which potassium feldspar is the only feldspar and the dark minerals are usually alkalic. Not recommended usage.

kalkowskite A very rare, brownish or black mineral: $Fe_2Ti_3O_9$ (?). It may be ilmenite. It usually contains small amounts of rare-earth elements, niobium, and tantalum. Cf: *arizonite [mineral].* Syn: *kalkowskyn.*

kalmafite A general term for igneous rocks composed of *kal*silite (or, presumably, any other polymorph of $KAlSiO_4$) and *maf*ic minerals, proposed by Hatch, Wells and Wells (1961). Syn: *mafurite; katungite.*

kalsilite A mineral: $KAlSiO_4$. It is dimorphous with kaliophilite, and sometimes contains sodium.

kamacite A meteorite mineral consisting of the body-centered cubic alpha-phase of a *nickel-iron* alloy, with a fairly constant composition of 5-7% nickel. It occurs in iron meteorites as bars or "girders" flanked by lamellae of taenite.

kamafugite A general term for potassium-rich, silica-undersaturated igneous rocks, proposed by Sahama in 1974. The term, embracing *katungite, mafurite,* and *ugandite,* is derived from the first letters of these. Not recommended usage.

kamarezite *brochantite.*

kame A low mound, knob, hummock, or short irregular ridge, composed of stratified sand and gravel deposited by a subglacial stream as a fan or delta at the margin of a melting glacier; by a superglacial stream in a low place or hole on the surface of the glacier; or as a ponded deposit on the surface or at the margin of stagnant ice. The term has undergone several changes in meaning, but can still be usefully applied to a deposit of glaciofluvial and glaciolacustrine sand and gravel whose precise mode of formation is uncertain (Thornbury, 1954, p. 378-379). Etymol: a Scottish variant of "comb", a long, steep-sided ridge. Cf: *esker.*

kame-and-kettle topography *knob-and-kettle topography.*

kame complex An assemblage of kames, constituting a hilly landscape.

kame delta *delta kame.*

kame field A group of closely spaced kames, interspersed in places with kettles and eskers, and having a characteristic hummocky topography.

kame moraine (a) An *end moraine* that contains numerous kames. (b) A group of kames along the front of a stagnant glacier, commonly comprising the slumped remnants of a formerly continuous outwash plain built up over the foot of rapidly wasting or stagnant ice.—See also: *moraine kame.*

kamenitza *solution pan.*

kame plain A flat-topped outwash plain originally entirely bounded by ice-contact slopes. Syn: *ice-contact plain.*

kame terrace A terracelike ridge consisting of stratified sand and gravel formed as a *glaciofluvial* or *glaciolacustrine* deposit between a melting glacier or a stagnant ice lobe and a higher valley wall or lateral moraine, and left standing after the disappearance of the ice; a filling of a *fosse.* A kame terrace terminates a short distance downstream from the terminal moraine; it is commonly pitted with kettles and has an irregular ice-contact slope. Cf: *glacial terrace.* Syn: *ice-contact terrace; ice-marginal terrace.*

kämmererite (a) A reddish variety of penninite containing chromium. Its formula is near: $Mg_5(Al,Cr)_2Si_3O_{10}(OH)_8$. (b) A hypothetical end member of the chlorite group: $Mg_2Cr_2SiO_5(OH)_4$.

kamperite A fine-grained black hypabyssal rock composed of small euhedral alkali feldspar crystals and a small amount of oligoclase in a groundmass of dark-colored mica. The name, given by Brögger in 1921, is for Kamperhaug in the Fen complex, Norway. Not recommended usage.

kanat *qanat.*

kandite A name suggested (but not approved) for the kaolin group of clay minerals, including kaolinite, nacrite, dickite, and halloysite.

kanemite An orthorhombic mineral: $NaH(Si_2O_4)(OH)_2 \cdot 2H_2O$.

kankan-ishi A black resinous flinty *andesite,* composed of hypersthene, oligoclase, and hornblende microphenocrysts in a groundmass of colorless glass and a network of acicular crystals of colorless bronzite (Johannsen, 1939, p. 259). Not recommended usage.

kankar (a) A term used in India for masses or layers of concretionary calcium carbonate, usually occurring in nodules, found embedded in the older alluvium or stiff clay of the Indo-Gangetic plain, or for precipitated calcium carbonate in the form of cement in porous sediments or as a coating on pebbles. (b) A limestone containing kankar and used for making lime and building roads.—Etymol: Hindi. The term was used originally for gravel, stone, or any small rock fragments, whether rounded or not; it is occasionally applied in the U.S. to a residual calcareous deposit, such as *caliche.* Inappropriate syn: *travertine.* Also spelled: *kunkur; kunkar; conker.*

kankite A monoclinic mineral: $FeAsO_4 \cdot 3.5H_2O$.

Kansan Pertaining to the classical second glacial stage of the Pleistocene Epoch in North America, after the Aftonian interglacial stage and before the Yarmouthian. See also: *Mindel.*

kansite *mackinawite.*

kantography The depiction of *edge lines* on relief maps. Etymol: German *Kantographie.*

kaoleen A colloquial term used in south-central Missouri for a chalky, porous, weathered chert with a white to tan or buff color. Etymol: corruption of *kaolin,* to which the material bears a slight resemblance.

kaolin [mineral] (a) A group of clay minerals characterized by a two-layer crystal structure in which each silicon-oxygen sheet is alternately linked with one aluminum-hydroxyl sheet and having approximate composition: $Al_2Si_2O_5(OH)_4$. The kaolin minerals include kaolinite, nacrite, dickite, and anauxite; although the minerals halloysite, endellite, and allophane are structurally and chemically different, they are sometimes included. The kaolin minerals are generally derived from alteration of alkali feldspars and micas. They have lower base-exchange capacities than montmorillonite and illite, and they absorb less water and thus have lower plasticity indexes, lower liquid limits, and less shrinkage when drying from a wet state. See also: *kandite.* Syn: *kaolinite.* (b) A mineral of the kaolin group; specif. *kaolinite.* The term was once applied to a single clay mineral which later was known to include at least four minerals of the kaolin group.

kaolin [petrog] A soft, fine, earthy, nonplastic, usually white or nearly white clay (rock) composed essentially of clay minerals of the kaolin group, principally kaolinite, derived from in-situ decomposition (extreme weathering or pneumatolysis) of aluminous minerals (such as feldspars in a granitic rock), containing a variable proportion of other constituents (quartz, mica flakes) derived from the parent rock, and remaining white or nearly white on firing; a *porcelain clay,* or natural (unwashed) *china clay.* It is used in the manufacture of ceramics, refractories, and paper. Type locality: Kao-ling (meaning "high hill"), hill in Kiangsi province, SE China. Syn: *kaoline; white clay; bolus alba.*

kaolinic Pertaining to or resembling kaolin.

kaolinite (a) A common clay mineral of the kaolin group: $Al_2Si_2O_5(OH)_4$. It is the characteristic mineral of most kaolins, and is polymorphous with dickite and nacrite. Kaolinite consists of sheets of tetrahedrally coordinated silicon joined by an oxygen shared with octahedrally coordinated aluminum; it is a high-alumina clay mineral that does not appreciably expand under varying water content and does not exchange iron or magnesium. The mineral was formerly known as kaolin. (b) A name sometimes applied to the kaolin group of clay minerals, and formerly applied to individual minerals of that group (such as to dickite and nacrite).

kaolinitic shale *feldspathic shale.*

kaolinization Replacement or alteration of minerals, esp. feldspars and micas, to form kaolin as a result of weathering or hydrothermal alteration. Also spelled: *kaolinisation.* Cf: *argillation; argillization.*

kaolinton An obsolescent term used by ceramists (esp. in Europe) for the portion of a clay that is soluble in sulfuric acid but not soluble in hydrochloric acid. Cf: *allophaneton.*

kar A syn. of *cirque [glac geol].* Etymol: Swiss-German *Kar,* "cirque".

K-Ar age method *potassium-argon age method.*

karang A term used in Indonesia for an emerged terrace composed of ancient fringing-reef material, and also for the coral limestone itself (Termier & Termier, 1963, p.408). Etymol: Malay, "reef, coral reef".

karat The proportion of pure gold in an alloy. Pure or fine gold is 24 karat; 10-karat gold is 10/24 pure, or 10 parts of pure gold by weight mixed with 14 parts of other metals. Not to be confused with *carat.* Abbrev: k.

karelianite A black mineral: V_2O_3. It may contain some iron, chromium, and manganese.

karewa A term applied in Kashmir to the level surface between the incised streams dissecting a terrace. Etymol: Kashmiri.

karez A term used in Pakistan for a gently inclined, underground channel dug so as to conduct ground water by gravity from alluvial gravels and the foot of hills to an arid lowland or basin; a horizontal well. Etymol: Baluchi. Pl: *karezes.* Cf: *qanat; foggara.*

karibibite An orthorhombic mineral: $Fe_2^{+3}As_4^{+3}(O,OH)_9$.

karite A *grorudite* containing approximately 50 percent quartz. Named for the Kara River, U.S.S.R. Not recommended usage.

karling (a) A high, dissected region, characterized by cirques, as Mount Anne in Tasmania. (b) A cluster or group of cirques.

Karlsbad twin law *Carlsbad twin law.*

kärnäite A rock found on the island of Kärnä in Lake Lappajärvi, central Finland, originally described as a volcanic rock with glassy groundmass and numerous inclusions, consisting of agglomerate-like tuff, and having a composition similar to dacite, the principal feldspar phenocrysts being monoclinic (probably sanidine). The rock has also been interpreted as an *impactite* with bedrock fragments (Svensson, 1968). Not recommended usage.

karnasurtite A honey to pale-yellow metamict mineral: $(Ce,La,Th)(Ti,Nb)(Al,Fe)(Si,P)_2O_7(OH)_4 \cdot 3H_2O$. It gives a monazitelike X-ray pattern when heated, but it may be equivalent to rhabdophane.

Karnian *Carnian.*

karoo *karroo.*

karpatite A hydrocarbon mineral: $C_{24}H_{12}$. Syn: *carpathite; pendletonite.*

karpinskite A greenish-blue mineral: $(Mg,Ni)_2Si_2O_5(OH)_2$ (?). Not to be confused with *karpinskyite.*

karpinskyite A mixture of leifite and Zn-bearing clay, formerly thought to be a distinct mineral species. Not to be confused with *karpinskite.*

karren In karst topography, a general term for solution grooves ranging in width from a few millimeters to more than a meter, and commonly separated by knifelike ridges. Karren that originate under a soil cover are rounded and average about 50 cm wide, whereas those that originate at the surface are sharp and typically 1 cm wide. Some have been through a sequence of covering and uncovering, and pure and hybrid varieties are distinguished; see *rillenkarren; rinnenkarren; deckenkarren; spitzkarren; trittkarren.* Etymol: German, "wheel tracks". See also: *karrenfeld.* Syn: *solution grooves; lapiaz.*

karrenfeld A karstic surface on limestone, characterized by solution grooves. See also: *limestone pavement; karren.*

karroo A tableland, found esp. in South Africa, that commonly rises to a considerable height in terraces. It does not support vegetation in the dry season, but becomes a grassy plain or pastureland during the wet season. Also spelled: *karoo.*

karst A type of topography that is formed on limestone, gypsum, and other rocks by dissolution, and that is characterized by sinkholes, caves, and underground drainage. Etymol: German, from the Yugoslavian territory Krš; type locality, a limestone plateau in the Dinaric Alps of northwestern Yugoslavia and northeastern Italy. First published on a topographic map, *Ducatus Carnioliae,* in 1774. Also spelled: *carst.* Adj: *karstic.* Syn: *karst topography; causse.* See also: *pseudokarst; thermokarst.*

karst base level A level below which karstification is presumed to cease.

karst bridge A *natural bridge* in a soluble rock.

karst cone One of a group of rounded hills separated by cockpits

in a *cockpit karst.* See also: *karst tower.*
karst corridor *solution corridor.*
karst fenster *karst window.*
karst hydrology The drainage pattern and features that are characteristic of karst.
karstic Adj. of *karst.*
karstification The formation of karst features by the solutional, and sometimes mechanical, action of water in a region of limestone, gypsum, or other bedrock.
karst lake *karst pond.*
karstland An area characterized by karst.
karst-margin plane *marginal karst plane.*
karst plain A plain, usually of limestone, on which karst features are developed. See also: *marginal karst plain; labyrinth karst.* Syn: *karst plateau.*
karst plateau *karst plain.*
karst pond A body of standing water in a closed depression of a karst region. Syn: *karst lake; sink lake; solution lake.*
karst spring *exsurgence.*
karst street *solution corridor.*
karst topography *karst.*
karst tower An isolated hill in a karst region surrounded by a plain that is commonly alluviated. Syn: *haystack hill; hum; mogote; pepino.* See also: *karst cone.*
karst valley A closed depression formed by the coalescence of several *sinkholes.* Its drainage is subsurface, its size is measured in hundreds of meters to a few kilometers, and it usually has an irregular floor and a scalloped margin inherited from the sinkholes. Syn: *nested sinkholes; solution valley; valley sink; uvala.* See also: *sinkhole; interior valley; glade; solution corridor.*
karst window An unroofed cave, at the bottom of which can be seen a subterranean stream. Syn: *karst fenster.*
kåsenite A pyroxene-rich, nepheline-bearing *carbonatite,* named by Brögger in 1921 at the Fen complex, Norway. Not recommended usage.
kasoite A variety of celsian containing potassium.
kasolite A yellow to brown monoclinic mineral: $Pb(UO_2)SiO_4 \cdot H_2O$.
kassaite A fine-grained hypabyssal rock composed of phenocrysts of hauyne, labradorite with oligoclase rims, barkevikite, and augite in a holocrystalline tinguaitic groundmass of acicular hastingsite crystals and andesine with oligoclase and orthoclase rims. Described by Lacroix in 1918 from the Los Archipelago, Guinea. Not recommended usage.
kassite A mineral: $CaTi_2O_4(OH)_2$.
kata- A prefix meaning "down", "beneath".
katabatic wind A local wind that moves down a slope, e.g. as a result of surface cooling during the night. Ant: *anabatic wind.* See also: *mountain wind; foehn; glacier wind.* Syn: *gravity wind.*
kataclastic Var. of "cataclastic". Also spelled: *kataklastic.*
katagenesis *catagenesis.*
kataglacial The part of a paleoclimatic cycle marking the transition from a full glacial (pleniglacial) to an interglacial. It is generally much shorter than an *anaglacial.*
kataglyph A hieroglyph formed during catagenesis, or under a covering set of beds (Vassoevich, 1953, p.33).
katamorphic zone The shallow zone in the Earth's crust in which *katamorphism* takes place. The term was originated by Van Hise. Cf: *anamorphic zone.*
katamorphism Destructive metamorphism in the *katamorphic zone,* at or near the Earth's surface, in which complex minerals are broken down and altered through oxidation, hydration, solution, and allied processes to produce simpler and less dense minerals. The term was introduced by Van Hise in 1904. Also spelled: *catamorphism.* Cf: *anamorphism.*
kataphorite A brownish monoclinic mineral of the amphibole group: $Na_2Ca(Fe^{+3},Al)_5AlSi_7O_{22}(OH)_2$. Also spelled: *catophorite; cataphorite.*
kataseism Earth movement toward the focus of an earthquake. Cf: *anaseism.* Syn: *dilatation [seis]; rarefaction.*
katatectic layer A sedimentary layer that is "built downward"; specif. a distinct, generally horizontal or slightly dipping layer of solution residue (gypsum and/or anhydrite) formed by intermittent compaction of sulfate accumulating on top (in the cap rock) of a salt stock by solution of salt. The term was introduced by Goldman (1933, p.84). Etymol: Greek *kata,* "down", + *tekton,* "builder".
katatectic surface A surface separating two katatectic layers (Goldman, 1952, p.v).
Katathermal n. The interval of gradually cooling climates since the *climatic optimum.* See also: *Little Ice Age.*
katathermal *hypothermal.*
katazone According to Grubenmann's classification of metamorphic rocks (1904), the lowermost *depth zone* of metamorphism, which is characterized by high temperatures (500°-700°C), mostly strong hydrostatic pressure, and low or no shearing stress. Long-continued reconstitution and recrystallization, often without deformation, and deep-seated metamorphism associated with igneous action, produce such rocks as high-grade schists and gneisses, granulites, eclogites, and amphibolites. The concept was modified by Grubenmann and Niggli (1924) to include effects of high-temperature contact metamorphism and metasomatism. Modern usage stresses temperature-pressure conditions (highest metamorphic grade) rather than the likely depth of zone. Cf: *mesozone; epizone.* Also spelled: *catazone.* See also: *katamorphism.*
Katmaian-type eruption The violently explosive ejection of huge amounts of pumice and ash, followed by an *ash flow* and extensive fumarole activity (Macdonald, 1972, p. 241). Type area: vicinity of Mount Katmai, Alaska, including Valley of Ten Thousand Smokes. Cf: *Peléean-type eruption; Vulcanian-type eruption.*
katogene Pertaining to the breaking-down of a rock by atmospheric or other agents, or to shallow-depth replacement. Obsolete.
katoptrite A black monoclinic mineral: $(Mn,Mg,Fe)_{14}(Al,Fe)_4Sb_2Si_2O_{29}$. Also spelled: *catoptrite.*
katothermal Said of a lake (such as a polar lake) whose temperature increases with depth. The term is "apparently defunct" (Stamp, 1961, p.278).
katungite An extrusive rock composed chiefly of melilite, with a smaller amount of olivine and magnetite and minor leucite and perovskite; a pyroxene-free *melilitite.* Named by Holmes in 1937 for Katunga, Uganda.
katzenbuckelite A hypabyssal rock with tinguaitic texture and composed of phenocrysts of nepheline, biotite, olivine, nosean, leucite, and apatite in a fine-grained groundmass of nepheline, leucite, and acmite. Named by Osann in 1903 for Katzenbuckel, Germany. Not recommended usage.
kauaiite An orthoclase-bearing olivine-augite *diorite* in which the feldspar is zoned, with calcic labradorite in the inner zones grading outward into alkali feldspar. Its name, given by Iddings in 1913, is derived from the Hawaiian island of Kauai. Not recommended usage.
kauri A light-colored, whitish-yellow, or brown *copal,* usually found as a fossil resin from the kauri pine (a tree of the genus *Agathis*), esp. from *Agathis australis,* a tall timber tree of New Zealand. Etymol: Maori *kawri.* Syn: *kauri resin; kauri gum; kauri copal; agathocopalite.*
kavir (a) A term used in Iran for a *salt desert;* specif. the Great Kavir of inner Iran, a series of closed basins noted for marsh conditions and high salinities. (b) A *playa* on a kavir.—Also spelled: *kewire; kevir.*
kawazulite A trigonal mineral: $BiTe_2Se$.
kaxtorpite A *nepheline syenite* containing pectolite, eckermannite, and sodic clinopyroxene. The name, given by Adamson in 1944, is for Kaxtorp, Norra Kärr complex, Sweden. Not recommended usage.
kay A variant of *key* and *cay.*
kazakovite A trigonal mineral: $Na_6H_2TiSi_6O_{18}$.
Kazanian European stage: Upper Permian (above Kungurian, below Tatarian).
kazanskite A hypabyssal plagioclase-bearing *dunite,* with accessory hornblende, plagioclase, magnetite (approximately one fourth of the rock), and green spinel. Not recommended usage.
KB *kelly bushing.*
K-bentonite *potassium bentonite.*
keatite A tetragonal polymorph of SiO_2, synthesized hydrothermally at high pressures.
keazoglyph A hieroglyph consisting of small transverse displacements along cracks (Vassoevich, 1953, p.64).
kedabekite An igneous rock similar to *eucrite* and composed of bytownite, calcium-iron garnet, and hedenbergite. The name, given by Fedorov in 1901, is for Kedabek, Caucasus, U.S.S.R. Not recommended usage.
keel (a) A costa or riblike structure on the aboral (under) side of platelike conodont elements. (b) A continuous sharp ridge along the venter of a coiled nautiloid or ammonoid conch. (c) A *carina* of a bryozoan; also that part of a distal zooecial wall in the end

zone of a stenolaemate that is between sinuses, and proximally is the recumbent basal zooecial wall and distally is the flattened median part of the vertical zooecial wall. (d) A vertical sail-like plate in a radiolarian; a keel-like ridge along the outer margin of the test of a foraminifer. (e) A canal or cleft in the valve of some pennate diatoms. (f) A spiral ridge on a gastropod shell.

keeleyite *zinkenite.*

Keewatin A division of the Archeozoic rocks of the Canadian Shield. Also spelled: *Keewatinian.*

Keewatinian *Keewatin.*

kegelite A pseudohexagonal mineral: $Pb_{12}(Zn,Fe)_2Al_4(Si_{11}S_4)O_{54}$.

kegel karst A syn. of *cockpit karst.* Etymol: German.

kehoeite An amorphous mineral: $(Zn, Ca)_4 Al_8 (PO_4)_8 (OH)_8 \cdot 20H_2O$ (?).

keilhauite A radioactive variety of sphene containing aluminum, iron, and yttrium and other rare earths.

keldyshite A mineral: $(Na,H)_2ZrSi_2O_7$.

kellerite A cuprian variety of pentahydrite.

kelly A steel pipe of square or hexagonal cross section, 40 ft (12 m) long, forming the top section of the rotary *drill string.* It is fitted into and passes through the *rotary table* and is turned by it during drilling, thereby transmitting the rotary motion of the table to the drill pipe. Syn: *kelly joint.*

kelly bushing The journal-box insert in the *rotary table* of a rotary drilling rig through which the *kelly* passes. Its upper surface is commonly used as the zero-depth reference for well-log and other downhole measurements in a well bore. Abbrev: KB.

kellyite A mineral of the kaolinite-serpentine group: $(Mn^{+2},Mg)_2 Al(Si,Al)O_5(OH)_4$.

Kelvin wave A tide progression developing in a relatively confined body of water (such as the English Channel or the North Sea) in which, because of the Coriolis force, the tide range in the Northern Hemisphere increases to the right and decreases to the left of the direction of travel.

kelyphitic border *kelyphitic rim.*

kelyphytic rim (a) In some igneous rocks, a peripheral zone of pyroxene or amphibole developed around olivine where it would otherwise be in contact with plagioclase, or around garnet where it would otherwise be in contact with olivine or other magnesium-rich minerals. Cf: *reaction rim; corona [petrology].* (b) A secondary *reaction rim.* (c) *reaction rim.*—Also spelled: *celyphytic rim.* Syn: *kelyphytic border.*

kemmlitzite A mineral: $SrAl_3(AsO_4)(SO_4)(OH)_6$.

kempite An emerald-green orthorhombic mineral: $Mn_2(OH)_3Cl$.

kennedyite A mineral: $MgFe_2{}^{+3}Ti_3O_{10}$.

Kennedy's critical velocity *critical velocity (d).*

kennel (a) A Scottish term for a hard sandstone, often with calcareous cement; *kingle.* (b) *kennel coal.*

kennel coal *cannel coal.*

Kenoran orogeny A name proposed by Stockwell (1964) for a time of plutonism, metamorphism, and deformation during the Precambrian of the Canadian Shield (especially in the Superior and Slave provinces), dated radiometrically at 2390-2600 m.y. ago, or at the end of the Archean of the present Canadian classification. It is synonymous with *Algoman orogeny* of Minnesota.

kenozooid A bryozoan *heterozooid* without a polypide and usually without either orifice or muscles.

kentallenite A dark-colored *monzonite* composed of approximately equal amounts of augite, olivine, potassium feldspar, and plagioclase, with smaller amounts of biotite, apatite, and opaque oxides; an olivine-bearing monzonite. Its name is derived from Kentallen, Argyllshire, Scotland. Not recommended usage.

kentrolite A dark reddish-brown mineral: $Pb_2Mn_2Si_2O_9$.

kentsmithite A local name used in the Paradox Valley, Colorado, for a black vanadium-bearing sandstone.

kenyaite A mineral: $Na_2Si_{22}O_{41}(OH)_8 \cdot 6H_2O$.

kenyte An olivine-bearing *phonolite* composed of anorthoclase, nepheline, acmite-augite, sodic amphibole, olivine, apatite, and opaque oxides. The groundmass may have a trachytic or hyalopilitic texture. Its name, given by Gregory in 1900, is derived from Mount Kenya, Kenya. Not recommended usage.

Kepler's laws of planetary motion The statements that each planet moves in an elliptical orbit with the Sun at one focus of the ellipse; that the line from the Sun to any planet sweeps out equal areas of space in equal intervals of time; and that the squares of the sidereal periods of the several planets are proportional to the cubes of their mean distances from the Sun. Although Kepler's laws are a mathematical consequence of Newton's laws, which are more fundamental, they preceded Newton's laws, were empirically based on the observations of Tycho Brahe, and were a significant extension of Copernican philosophy.

kerabitumen *kerogen.*

keralite A quartz-biotite hornfels. The term was originated by Cordier in 1868.

kerargyrite *cerargyrite.*

keratophyre A name originally applied by Gümbel (1874, p. 43) to trachytic rocks containing highly sodic feldspars, but now more generally applied to all salic extrusive and hypabyssal rocks characterized by the presence of albite or albite-oligoclase and chlorite, epidote, and calcite, generally of secondary origin. Originally the term was restricted to lavas of pre-Tertiary age but this distinction is not recognized in current usage. Some varieties of keratophyre contain sodic orthoclase and sodic amphiboles and pyroxenes. Keratophyres commonly are associated with spilitic rocks and interbedded with marine sediments.

keratose Said of a horny sponge whose skeleton consists entirely of organic fibers without spicules (although it sometimes may contain foreign particles including spicules of other sponges).

keriotheca The relatively thick shell layer with honeycomblike structure in the wall of some fusulinids (such as schwagerinids), occurring next below the tectum, and forming part of the spirotheca. It may be divisible into *lower keriotheca* and *upper keriotheca.*

kermesite A cherry-red mineral: Sb_2S_2O. It usually occurs as tufts of capillary crystals resulting from the alteration of stibnite. Syn: *antimony blende; red antimony; purple blende; pyrostibite.*

kernbut A projecting ridge or buttress created by displacement on a fault traversing a hillslope and separated from the hill by a *kerncol;* the outer ridgelike edge of a fault terrace or fault bench. The term was introduced by Lawson (1904, p. 332) for a primary feature occurring in Kern Canyon, Calif., but the type locality has since been shown to be one of fault-line forms (Webb, 1936). Etymol: *Kern* Canyon + *but*tress.

kerncol A low sag or trough separating a *kernbut* from the hillside, occurring where a faulted block joins the hill. The term was introduced by Lawson (1904, p. 332) for a primary feature occurring in Kern Canyon, Calif., but the type locality has since been shown to be one of fault-line forms (Webb, 1936). Etymol: *Kern* Canyon + *col.*

kernite A colorless to white monoclinic mineral: $Na_2B_4O_7 \cdot 4H_2O$. Syn: *rasorite.*

kerogen Fossilized insoluble organic material found in sedimentary rocks, usually shales, which can be converted by distillation to petroleum products. Syn: *kerabitumen.*

kerogenite *oil shale.*

kerogen shale *oil shale.*

kerolite A mixture of serpentine and stevensite.

keronigritite A type of *nigritite* that is derived from kerogen. Cf: *polynigritite; humonigritite; exinonigritite.*

kerosene shale *kerosine shale.*

kerosine shale (a) A syn. of *torbanite.* (b) Any bituminous oil shale.—Also spelled: *kerosene shale.*

kersantite A *lamprophyre* containing biotite and plagioclase (usually oligoclase or andesine), with or without clinopyroxene and olivine. Defined by Delesse in 1851, named for the village of Kersanton, France. Obsolete syn: *kersanton.*

kersanton An obsolete syn. of *kersantite.*

kertschenite An oxidation product of vivianite.

kerzinite In the Urals, lignite that is impregnated with hydrated nickel silicate and thus is mined for nickel.

kess-kess An Arabic term used in Morocco for a fossil reef knoll isolated by erosion. Cf: *klint [reef].*

kesterite A mineral: $Cu_2(Zn,Fe)SnS_4$. It is the zinc analogue of stannite. Also spelled: *kësterite.* Syn: *isostannite.*

kettle [glac geol] A steep-sided, usually basin- or bowl-shaped hole or depression, commonly without surface drainage, in glacial-drift deposits (esp. outwash and kame fields), often containing a lake or swamp; formed by the melting of a large, detached block of stagnant ice (left behind by a retreating glacier) that had been wholly or partly buried in the glacial drift. Kettles range in depth from about a meter to tens of meters, and in diameter to as much as 13 km. Thoreau's Walden Pond is an example. Cf: *pothole [glac geol].* Syn: *kettle hole; kettle basin; potash kettle.*

kettle [streams] A *pothole* in a stream bed.

kettleback *horseback [coal].*

kettle basin *kettle.*

kettle bottom *caldron bottom.*
kettle drift Drift within and adjacent to a kettle.
kettle hole *kettle.*
kettle lake A body of water occupying a kettle, as in a pitted outwash plain or in a kettle moraine. Cf: *cave-in lake.* Syn: *kettle-hole lake; pit lake.*
kettle moraine A terminal moraine whose surface is marked by many kettles.
kettle plain A *pitted outwash plain* marked by many kettles.
kettnerite A brown to yellow tetragonal mineral: $CaBi(CO_3)OF$.
Keuper European stage (esp. in Germany): Upper Triassic (above Muschelkalk, below Jurassic).
kevir *kavir.*
Keweenawan A provincial series of the Precambrian in Michigan and Wisconsin.
kewire *kavir.*
key [cart] A *legend* on a map.
key [coast] A *cay,* esp. one of the coral islets or barrier islands off the southern coast of Florida. See also: *sandkey.* Syn: *kay.*
key [taxon] An artificial analytic device or arrangement for use in identification of plants or animal forms whereby a choice is provided between two contradictory propositions resulting in the acceptance of one and rejection of the other (Lawrence, 1951, p. 225).
key bed (a) A well-defined, easily identifiable stratum or body of strata that has sufficiently distinctive characteristics (such as lithology or fossil content) to facilitate correlation in field mapping or subsurface work. (b) A bed the top or bottom of which is used as a datum in making structure-contour maps.—Syn: *key horizon; index bed; marker bed [stratig].*
keyed *sutured.*
key fossil *index fossil.*
key horizon (a) The top or bottom of an easily recognized, extensive bed or formation that is so distinctive as to be of great help in stratigraphy and structural geology; e.g. a *datum horizon.* (b) A term that is used interchangeably with *key bed.*
keyite A monoclinic mineral: $(Cu,Zn,Cd)_3(AsO_4)_2$.
keystone fault A grabenlike structure developed on the crest of an anticline.
K-feldspar *potassium feldspar.*
K-feldspar-cordierite-hornfels facies A name given to the *pyroxene-hornfels facies* (Winkler, 1967) in order to indicate a more diagnostic mineral assemblage that would distinguish it from the hornblende-hornfels facies. Characteristic orthopyroxenes form in the higher-temperature parts of this facies (in excess of about 700°C).
khadar A term used in India for a low-lying area, e.g. an alluvial plain, that is liable to be flooded by the waters of a river. Etymol: Urdu-Hindi *khādar.* Cf: *bhangar.* Syn: *khaddar; khuddar.*
khademite An orthorhombic mineral: $Al(SO_4)(OH)\cdot 5H_2O$.
khagiarite A black *pantellerite* characterized by a glassy microlitic groundmass exhibiting flow texture. Defined by Washington in 1913. Not recommended usage.
khal (a) A term used in East Pakistan for a narrow stream channel. (b) A sluggish creek on the lower delta of the Ganges.—Etymol: Bengali.
kharafish A limestone plateau in the Libyan desert, formed by wind erosion.
khari A term used in East Pakistan for a small, deep stream of local origin. Etymol: Bengali.
kheneg A term used in the Atlas Mountains of northern Africa for a canyon. Etymol: Arabic.
khibinite [mineral] *mosandrite.*
khibinite [petrology] *chibinite.*
khibinskite A monoclinic, pseudotrigonal mineral: $K_2ZrSi_2O_7$.
khlopinite A tantalian variety of samarskite.
khoharite A hypothetical end member of the garnet group: $Mg_3Fe_2(SiO_4)_3$.
khondalite A group of metamorphosed aluminous sediments consisting of garnet-quartz-sillimanite rocks with garnetiferous quartzites, graphite schists, and marbles (Walker, 1902, p.11). It is named after Khonds and the Khondalite series, India.
khor (a) A term used in Sudan for an intermittent stream. (b) A term used in northern Africa for a watercourse or ravine, esp. one that is dry.—Etymol: Arabic *khawr,* "wadi, dry wash".
khud A term used in India for a ravine or precipice. Etymol: Hindi *khad.*
khuniite *iranite.*
khurd A term used in Algeria for a pyramid-shaped sand dune, 80 to 150 m high, with curved slopes, formed by the intersection of seif dunes (Capot-Rey, 1945, p. 393). Cf: *rhourd.* Syn: *guern.*
kick [drill] (a) A surge against the normal fluid circulation in an oil well, caused by the formation pressure in the well exceeding the pressure exerted by the drilling mud. (b) A quick snap of the drill stem caused by the core breaking in a blocked core barrel. (c) A small sidewise displacement in a borehole, caused by the deflection of the drill bit when entering a hard, dipping stratum underlying softer rock.
kick [seis] *arrival.*
kickout The lateral distance from a drilling site reached by a *directional well.*
kidney ore A variety of hematite occurring in compact kidney-shaped masses, concretions, or nodules, together with clay, sand, calcite, or other impurities; concretionary ironstone. Syn: *kidney iron ore.*
kidneys A miner's term for a mineral zone that contracts, expands, and again contracts downwards.
kidney stone [mineral] *nephrite.*
kidney stone [sed] A pebble or nodule roughly resembling the shape of a kidney; e.g. a small hard red-coated ironstone nodule common in the Oxford Clay of England.
kies A general term for the sulfide ores. Etymol: German *Kies,* "finer gravel".
kieselguhr A syn. of *diatomite.* Etymol: German *Kieselguhr.* Syn: *guhr; kieselgur.*
kieserite A white monoclinic mineral: $MgSO_4\cdot H_2O$. It occurs in saline residues.
kiirunavaarite *magnetitite.*
Kikuchi lines Pairs of parallel lines that appear in transmission-electron-microscope diffraction patterns and are formed as a result of coherent scattering of inelastically scattered electrons in the upper parts of the sample foil.
kilchoanite A mineral: $Ca_3Si_2O_7$. It is dimorphous with rankinite.
kilkenny coal *anthracite.*
kill A creek, channel, stream, or river. The term is used chiefly in place names in Delaware and New York State; e.g. Peekskill, N.Y. Etymol: Dutch *kil.*
killalaite A monoclinic mineral: $Ca_6Si_4O_{14}\cdot H_2O$.
Killarney Revolution A name proposed by Schuchert (1924) for a supposed major orogeny at the end of Precambrian time in North America; based on the Killarney Granite north of Lake Huron in Ontario, supposed to be of post-Keweenawan age. Radiometric data now indicate that the Keweenawan is 1000 m.y. old, and that the Killarney Granite is older, and probably equivalent to the Penokean Granite (see *Penokean orogeny*) of Michigan and Minnesota. Actually, no notable tectonic events are now known to have occurred in this part of North America at the end of the Precambrian. The term Killarney Revolution is obsolete, and should be abandoned.
killas A name used in Devon and Cornwall for any rock that has been metamorphosed by contact with granite.
kimberlite A porphyritic alkalic *peridotite* containing abundant phenocrysts of olivine (commonly serpentinized or carbonatized) and phlogopite (commonly chloritized), and possibly geikielite and chromian pyrope, in a fine-grained groundmass of calcite and second-generation olivine and phlogopite and with accessory ilmenite, serpentine, chlorite, magnetite, and perovskite. The name, derived from Kimberley, South Africa, was proposed by Lewis in 1887.
Kimeridgian *Kimmeridgian.*
Kimmerian Eastern European stage (Black Sea area): Upper Miocene to Lower Pliocene (above Pontian, below Akchagylian). See also: *Dacian.*
Kimmerian orogeny One of 30 or more short-lived orogenies during Phanerozoic time postulated by Stille, in this case consisting of three subdivisions occurring at the start, in the early stages, and at the end of Jurassic time.
Kimmeridgian European stage: Upper Jurassic (above Oxfordian, below Portlandian). The spelling *Kimeridgian* was used by Arkell (1956, p.20) on the basis that the type locality, the village of Kimmeridge in the Isle of Purbeck, southern England, was spelled with one "m" until 1892. See also: *LaCasitan.*
kimzeyite A mineral of the garnet group: $Ca_3(Zr,Ti)_2(Al,Si)_3O_{12}$.
kin A headland. Also, a term used in Ireland for the highest point of anything. Etymol: Gaelic.
kindchen A nodule or concretion that resembles the head of a

child; specif. *loess kindchen.* Etymol: German *Kindchen,* "little child, infant, baby".

Kinderhookian North American series: lowermost Mississippian (above Conewangoan of Devonian, below Osagian).

K index A measure of intensity of magnetic disturbance. It is a figure ranging from zero to nine, indicative of range of magnetic intensity in a three-hour interval, after subtraction of normal daily variation.

kindly *likely.*

kindred *rock association.*

kinematic viscosity The ratio of the viscosity coefficient (in poises) to density at room temperature (in q/cu cm). See also: *eddy viscosity.* Syn: *coefficient of kinematic viscosity.*

kinematic wave A bulge on a glacier surface that travels down the glacier with a velocity three to four times faster than the surface velocity of the glacier. Cf: *glacier wave.* Syn: *traveling wave.*

kinetic metamorphism A type of metamorphism that produces deformation of rocks without chemical reconstitution or recrystallization to form new minerals (Turner and Verhoogen, 1951, p.370). Cf: *cataclastic metamorphism.* Syn: *mechanical metamorphism.*

kingdom (a) The highest category in the hierarchy of classification of animals and plants that is subject to formal regulation in nomenclature. See: ICZN, 1964; ICBN, 1972. Cf: *subkingdom.* (b) Any one of the three major divisions into which all natural objects are traditionally classified, viz. animal kingdom, plant kingdom, mineral kingdom.

kingeniform Said of the loop of an adult dallinid brachiopod (as in the subfamily Kingeninae) in which the tendency to retain the campagiform hood during loop development leads to a broad sheet-like transverse band with connecting bands leading to the septum "in addition to normal ones joining descending branches with septum" (TIP, 1965, pt. H, p. 147).

kingite A white mineral: $Al_3(PO_4)_2(OH,F)_3 \cdot 9H_2O$.

kingle A Scottish term for a very hard rock, esp. a siliceous or calcareous sandstone occurring in oil shales but without bituminous matter; e.g. *kennel.*

kink band A type of *deformation band* occurring microscopically in crystals and megascopically in foliated rocks, in which the orientation of the lattice or of the foliation is changed or deflected by gliding or slippage. Kink bands are associated with shock-wave action as well as with normal deformation. They commonly occur as conjugate systems. Cf: *conjugate fold system.* Syn: *knick band; knick zone.*

kink fold A fold with planar limbs and sharp angular hinge.

kinoite A mineral: $Ca_2Cu_2Si_3O_{10} \cdot 2H_2O$.

kinoshitalite A mineral of the mica group: $(Ba,K)(Mg,Mn,Al)_3(Al_2Si_2)O_{10}(OH)_2$.

kinradite A name used in California and Oregon for jasper containing spherical inclusions of colorless or nearly colorless quartz.

kinzigite A coarse-grained metamorphic rock of pelitic composition occurring in the granulite facies. Essential minerals are garnet and biotite, with which occur varying amounts of quartz, K-feldspar, oligoclase, muscovite, cordierite, and sillimanite. The term was originated in 1860 by Fischer, who named it after Kinzig, Schwarzwald, Germany.

kipuka An area surrounded by a lava flow. Etymol: Hawaiian, "opening".

kirovite A mineral: $(Fe,Mg)SO_4 \cdot 7H_2O$. It is a variety of melanterite containing magnesium.

kirschsteinite A mineral: $Ca(Fe,Mg)SiO_4$. It is isomorphous with monticellite. Syn: *iron-monticellite.*

kitkaite A mineral: NiTeSe.

kivite A dark-colored olivine-bearing leucite *basanite.* Its name, given by Lacroix in 1923, is derived from Lake Kivu in east-central Africa. Not recommended usage.

kjelsasite A *monzodiorite* similar to *larvikite* but with more calcium oxide and fewer alkalic oxides. Brögger in 1933 derived the name from Kjelsås, Oslo district, Norway. Not recommended usage.

kladnoite A mineral: $C_6H_4(CO)_2NH$. It occurs as monoclinic crystals formed in burning waste heaps in the Kladno coal basin of Bohemia.

klapperstein A *rattle stone* that results from the weathering of a box-stone. Etymol: German *Klapperstein.*

klebelsbergite A mineral consisting of a basic antimony sulfate (?) and occurring in the interstices between crystals in columnar aggregates of stibnite.

kleinite A yellow to orange mineral: $Hg_2N(Cl,SO_4) \cdot nH_2O$. Cf: *mosesite.*

Klein solution A solution of cadmium borotungstate that is used as a *heavy liquid;* its specific gravity is 3.6. Cf: *bromoform; Clerici solution; Sonstadt solution; methylene iodide.*

kliachite *cliachite.*

klimakotopedion A term introduced by Schwarz (1912, p. 95) as a syn. of *stepped plain.* Etymol: translation into Greek of "stepped plain".

klingstein An obsolete syn. of *phonolite.*

klinker bed *clinker bed.*

klint [coast] A term used in Denmark and Sweden for a vertical mountain wall or abrasion precipice, several meters high and 100 m or more long; esp. a steep cliff along the shore of the Baltic Sea. Pl: *klintar.* See also: *glint.*

klint [reef] An exhumed fossil *bioherm* or coral reef, its surrounding rocks having been eroded, leaving the reef core standing in relief as a prominent knob, ridge, or hill. Pl: *klintar.* Cf: *tepee butte.* Not to be confused with *clint.*

klintite The rock composing a klint (Pettijohn, 1957, p. 397); e.g. a loosely knit and cavernous or reticulating network of dense, hard, tough dolomite, which because of its rigid framework gives to the massive reef core of an exhumed bioherm its strength and resistance to denudation.

klippe An isolated rock unit that is an erosional remnant or outlier of a *nappe*. The original sense of the term was merely descriptive, i.e. included any isolated rock mass such as an erosional remnant. Plural: *klippen.* Etymol: German, "rock protruding from a sea or lake floor".

klizoglyph *desiccation crack.*

klockmannite A reddish-violet to slate-gray mineral: CuSe. It tarnishes blue-black and is found in granular aggregates.

kloof A term used in South Africa for a deep rugged gorge, ravine, glen, or other short steep-sided valley, and also for a mountain pass. In some place names, the term may refer to a wide, open valley. Etymol: Afrikaans.

kluftkarren A group or series of *solution fissures.*

kmaite *celadonite.*

knap (a) A crest or summit of a hill. (b) A small hill or slight rise of ground.

kneaded (a) Said of a vague sedimentary structure resembling kneaded dough, such as a variety of flow roll or ball-and-pillow structure, or a structure formed by intrastratal slippage. (b) Said of a sediment or sedimentary particles transported by mudflows; e.g. "kneaded gravel".

knebelite A mineral: $(Fe,Mn)_2SiO_4$. It is a manganoan fayalite.

knee fold A *zigzag fold* occurring in gravity-collapse structures.

knick (a) A *knickpoint;* esp. the place of junction where a gently inclined pediment and the adjacent mountain slope meet at a sharp angle. Syn: *knickpunkt.* (b) *nick.*

knick band *kink band.*

knickline A line formed by the angle of a *knick* in a slope, esp. in a desert region where there is an abrupt transition from a pediment surface to the mountain slope.

knickpoint Any interruption or break of slope; esp. a point of abrupt change or inflection in the longitudinal profile of a stream or of its valley, resulting from rejuvenation, glacial erosion, or the outcropping of a resistant bed. Etymol: German *Knickpunkt,* "bend point". Also spelled: *knick point.* Syn: *knick; nick; nick-point; knickpunkt; break; rejuvenation head; rock step.*

knickpunkt (a) A *knickpoint* in a stream profile, esp. one resulting from rejuvenation or from an uplift. (b) The sharp angle made by the *haldenhang* and the *steilwand;* a *knick.*—Pl. *knickpunkte.* Etymol: German *Knickpunkt,* "bend point".

knick zone *kink band.*

knife edge (a) A narrow ridge of rock or sand. (b) *feather edge.*

knipovichite A mineral consisting of a chromium-containing variety of alumohydrocalcite.

knitted texture A texture that is typical of the mineral serpentine in a rock when it replaces a clinopyroxene. Cf: *lattice texture.*

knob (a) A rounded eminence, as a knoll, hillock, or small hill or mountain; esp. a prominent or isolated hill with steep sides, commonly found in the southern U.S. See also: *knobs.* (b) A peak or other projection from the top of a hill or mountain. Also, a boulder or group of boulders or an area of resistant rocks protruding from the side of a hill or mountain.

knob-and-basin topography *knob-and-kettle topography.*

knob-and-kettle topography An undulating landscape in which a disordered assemblage of knolls, mounds, or ridges of glacial drift is interspersed with irregular depressions, pits, or kettles that are commonly undrained and may contain swamps or ponds. See also: *hummocky moraine.* Syn: *knob-and-basin topography; kame-and-kettle topography.*

knob and trail A structure, found in glaciated areas, that is made up of a protruding mass of resistant rock (the "knob") and a ridge of softer rock (the "trail") extending from the lee side of the knob (Chamberlin, 1888, p. 244-245). Cf: *crag and tail.*

knobs An area marked by a group of rounded, isolated hills ("knobs").

knock A hill in the English Lake District or in Scotland. Etymol: Gaelic.

knocker A colloquial field term denoting a resistant, rounded monolith, a few feet to several hundred feet across, that stands out prominently above the level of the surrounding terrain. It is a convenient term for both *tectonic blocks* and *exotic blocks* (Berkland et al., 1972, p. 2296).

knoll [geomorph] (a) A small, low, rounded hill; a hillock or mound. (b) The rounded top of a hill or mountain.—Syn: *knowe; knowle.*

knoll [marine geol] A mound-like relief form of the sea floor, less than 1000 m in height. Syn: *seaknoll.*

knoll reef *reef knoll.*

knopite A variety of perovskite containing cerium.

knorringite A mineral of the garnet group: $Mg_3Cr_2(SiO_4)_3$. It forms a series with pyrope.

knot [geomorph] (a) A term used in the English Lake District for a hill of moderate height; esp. one having a bare-rock surface. (b) An elevated land area formed by the meeting of two or more mountainous regions; e.g. the structural-junction area of ridges of folded mountains.

knot [mining] A miner's term for small concretions, e.g. galena in sandstone, or for segregations of darker minerals in granite and gneiss.

knotenschiefer A type of *spotted slate* characterized by conspicuous subspherical or polyhedral clots that are often individual minerals, e.g. cordierite, biotite, chlorite, andalusite (Holmes, 1928). Etymol: German. Cf: *garbenschiefer; fruchtschiefer; fleckschiefer.*

knotted With reference to metamorphic rocks, a syn. of *maculose.*

knotted-hornfels facies Metamorphic rocks formed in the lowest grades of thermal (contact) metamorphism at temperatures between 200° and 350°C and at pressures not exceeding 2500 bars (Hietanen, 1967). Syn: *albite-epidote-hornfels facies.*

knotted schist *spotted slate.*

knotted slate *spotted slate.*

knowe A Scottish syn. of *knoll.* Syn: *know.*

knoxvillite *copiapite.*

Knudsen formula A formula that expresses the relationship between salinity and chlorinity of sea water: salinity=0.030+1.8050(chlorinity). See also: *Knudsen's tables.*

Knudsen's tables Hydrographic tables to aid in the computation of salinity, density, and sigma-t from chlorinity titrations and hydrometer readings, based on the *Knudsen formula.*

koashvite An orthorhombic mineral: $Na_6(Ca,Mn)(Ti,Fe)Si_6O_{18}\cdot H_2O$.

kobeite A black mineral: $(Y,U)(Ti,Nb)_2(O,OH)_6$. Also spelled: *kobeïte.*

kobellite A blackish-gray mineral: $Pb_5Bi_8S_{17}$.

kodurite A coarsely crystalline rock of doubtful origin consisting of potassium feldspar, garnet (spessartine, andradite), and apatite. Named by Fermor in 1907 for Kodur Mine, Vizagapatam, India. Obsolete.

koechlinite A greenish-yellow orthorhombic mineral: Bi_2MoO_6.

koenenite A very soft mineral: $Na_4Mg_9Al_4Cl_{12}(OH)_{22}$.

Koenigsberger ratio The ratio of the *remanent magnetization* to magnetization induced by the Earth's field. Its symbol is Q.

koenlinite *könlite.*

koettigite A carmine mineral: $Zn_3(AsO_4)_2\cdot 8H_2O$. Also spelled: *köttigite.*

köfelsite A frothy, pumiceous high-silica glass occurring as small veins in fractured gneisses in the Köfels structure, Austria, and apparently formed by vesiculation of an impact melt. The material is extremely heterogeneous, ranging in color from white to dark brown, and contains shock-metamorphosed mineral fragments and shock-melted glasses which provide definite evidence of origin by meteorite impact.

köflachite A dark-brown variety of retinite found in brown coal at Köflach in Styria, Austria.

kogarkoite A monoclinic mineral: $Na_3(SO_4)F$.

kohalaite An *andesite* that contains normative oligoclase and may or may not contain modal olivine. Its name, given by Iddings in 1913, is derived from the Kohala Mountains, Hawaii. Cf: *mugearite.* Not recommended usage.

koktaite A mineral: $(NH_4)_2Ca(SO_4)_2\cdot H_2O$.

kolbeckine *herzenbergite.*

kolbeckite A blue to gray mineral: $ScPO_4\cdot 2H_2O$. It was formerly described as a hydrous phosphate and silicate of aluminum, beryllium, and calcium. Syn: *sterrettite.*

kolk A deep, isolated hole or depression, scoured out by eddying water in soft rock. Etymol: German *Kolk,* "deep pool, eddy, scour". Cf: *colk.*

kollanite A term, now obsolete, proposed by Pinkerton (1811, v.2, p.98) to distinguish the English *puddingstones* from other conglomerates.

kolm Nodular or concretionary bodies of coal found in the Paleozoic alum shales of Sweden and that contain rare metals, especially uranium. Syn: *culm.*

kolovratite A mineral which is a hydrous vanadate of nickel and zinc.

kolskite A mixture of lizardite and sepiolite.

komarovite An orthorhombic mineral: $(Ca,Mn)Nb_2Si_2O_9(O,F)\cdot 3.5H_2O$.

komatiite (a) An igneous suite, analogous to the ophiolitic, tholeiitic, calc-alkaline, and alkaline suites, distinguished by the presence of ultramafic lavas. (b) Ultramafic lava. The term was originally applied to basaltic and ultramafic lavas by Viljoen and Viljoen (1969), but was extended to include the associated rocks by Arndt and others (1977). Their definition of the komatiite suite includes "noncumulate rocks ranging in composition from peridotite ($\approx$30 per cent MgO, 44 per cent SiO_2) to basalt (8 per cent MgO, 52 per cent SiO_2) or andesite (12 per cent MgO, 56 per cent SiO_2), and cumulate rocks ranging from peridotite (up to 40 per cent MgO) to mafic gabbro ($\approx$12 per cent MgO)." The lavas commonly exhibit *spinifex texture.* All rocks of the series have low Ti and Fe/(Fe+Mg), and high Mg, Ni, and Cr. The name is for the Komati River, Barberton Mountain Land, Transvaal, South Africa.

kona A term used in Hawaii for the leeward side, i.e. one away from the trade winds (Stamp, 1966, p. 282).

kongsbergite A silver-rich variety of native amalgam, containing about 95% silver and 5% mercury.

koninckite A yellow mineral: $FePO_4\cdot 3H_2O$ (?).

konite A term, now obsolete, introduced by Pinkerton (1811, v.1, p. 429) for a *freestone* consisting of limestone. Etymol: Greek *konia,* a term used by Theophrastus for "lime".

könlite A brown to yellow hydrocarbon found in brown coal and having an approximate composition of 91.75% carbon, 7.50% hydrogen, and 0.75% oxygen. Syn: *könleinite; koenlinite.*

kop A mountain or large hill that stands out prominently. Etymol: Dutch, "head".

kopje A var. of *koppie.* Etymol: Dutch, "small head".

Köppen's classification A *climate classification,* formulated by W. Köppen in 1918, that is based on the climatic requirements of certain types of vegetation. Cf: *Thornthwaite's classification.*

koppie A small but prominent hill occurring on the veld of South Africa, sometimes reaching 30 m above the surrounding land; esp. an isolated, elongate, scrub-covered hillock or knob composed of igneous rock and representing an erosion remnant, such as a small inselberg. See also: *castle koppie.* Etymol: Afrikaans, from Dutch *kopje,* "small head". Syn: *kopje.*

koppite A variety of pyrochlore containing iron, potassium, and cerium and lacking titanium.

koris A term used in northern Africa for a dry valley.

korite A variety of *palagonite* (Hey, 1962, p. 485). Not recommended usage.

kornelite A colorless to brown mineral: $Fe_2(SO_4)_3\cdot 7H_2O$.

kornerupine A colorless, yellow, brown, or sea-green mineral: $Mg_3Al_6(Si,B,Al)_5O_{21}(OH)$. It resembles sillimanite in appearance.

korzhinskite A mineral: $CaB_2O_4\cdot H_2O$.

kosmochlor *ureyite.*

kostovite A mineral: $CuAuTe_4$.

koswite A *peridotite* composed of olivine, clinopyroxene, and hornblende in a groundmass of magnetite; a magnetite peridotite. Named by Duparc in 1902 for Koswinsky in the northern Urals. Not recommended usage.

kotoite An orthorhombic mineral: $Mg_3(BO_3)_2$. It is isostructural with jimboite.

kotschubeite A rose-red variety of clinochlore containing chromium.

köttigite *koettigite.*

kotulskite A mineral: $Pd(Te,Bi)_{1-2}$.

koum A *sandy desert* or continuous tract of sand dunes in central Asia, equivalent to an *erg.* Etymol: French. See also: *kum.*

koutekite A hexagonal mineral: Cu_5As_2.

kozulite A manganese-rich mineral of the amphibole group: $(Na,K)_3(Mn,Mg,Fe)_5Si_8O_{22}(OH,F)_2$.

krablite A rhyolitic *crystal tuff* containing plagioclase grains enclosed in orthoclase phenocrysts, along with smaller amounts of augite and quartz. The rock was originally identified as the mineral feldspar. It occurs as ejecta from Mount Krafla, Iceland. Syn: *baulite; kraflite.* Obsolete.

kraflite *krablite.*

kragerite *krageröite.*

krageröite A rutile-bearing albite *aplite,* with minor amounts of quartz, potassium feldspar, and ilmenite. Its name, given by Brögger in 1904, is derived from Kragerö, Norway. Syn: *kragerite.* Not recommended usage.

Krakatoan caldera A type of caldera formed in the summit region of a volcano following evacuation of the underlying magma chamber by a voluminous outpouring of pyroclasts, normally of silicic composition.

kramerite *probertite.*

krans *krantz.*

krantz A term used in southern Africa for a precipitous rock face or sheer cliff. Etymol: Afrikaans *krans,* "wreath". Pl: *krantzes.* Syn: *krans.*

krantzite A variety of retinite found in small yellowish grains disseminated in brown coal.

Krasnozem A Russian term for a zonal red soil developed in a Mediterranean climate.

kratochvilite A hydrocarbon mineral: $C_{13}H_{10}$ (fluorene).

kratogen *craton.*

kraton *craton.*

kraurite *dufrenite.*

krausite A yellowish-green mineral: $KFe(SO_4)_2 \cdot H_2O$.

krauskopfite A mineral: $BaSi_2O_5 \cdot 3H_2O$.

krautite A monoclinic mineral: $MnHAsO_4 \cdot H_2O$.

KREEP An acronym for a basaltic lunar rock type, first found in Apollo 12 fines and breccias, characterized by an unusually high content of potassium (K), rare-earth elements (REE), phosphorus (P), and other trace elements, in comparison to other lunar rock types. The material, which is found in a variety of crystalline and glassy (shock-melted?) rock types, is distinctly different from the iron-rich mare basalts. Syn: *nonmare basalt.*

kreittonite A black variety of gahnite containing ferrous iron or ferric iron, or both.

kremastic water A syn. of *vadose water* proposed by Meinzer (1939, p. 676), including *rhizic water, argic water,* and *anastatic water.*

kremersite A ruby-red mineral: $[(NH_4),K]_2FeCl_5 \cdot H_2O$. It is a variety of erythrosiderite containing ammonium.

krennerite A silver-white to pale-yellow mineral: $AuTe_2$. It often contains silver. Syn: *white tellurium.*

kribergite A white, chalklike mineral: $Al_5(PO_4)_3(SO_4)(OH)_4 \cdot 2H_2O$ (?).

krinovite A meteorite mineral: $NaMg_2CrSi_3O_{10}$.

kröhnkite An azure-blue monoclinic mineral: $Na_2Cu(SO_4)_2 \cdot 2H_2O$. Also spelled: *kroehnkite.*

krokidolite *crocidolite.*

krokydite *crocydite.*

krotovina An irregular tubular or tunnel-like structure in soil, made by a burrowing animal and subsequently filled with material from another horizon. Also spelled: *crotovina.* Etymol: Russian.

krupkaite An orthorhombic mineral: $PbCuBi_3S_6$.

krutaite A mineral of the pyrite group: $CuSe_2$.

krutovite A cubic mineral: $NiAs_2$.

kryokonite *cryoconite [glaciol].*

kryomer A relatively cold period within the Pleistocene Epoch, such as a *glacial stage* (Lüttig, 1965, p. 582). Ant: *thermomer.*

kryoturbation *cryoturbation.*

kryptomere *cryptomere.*

krystic Pertaining to ice in all its forms, as a surface feature of the Earth. Rarely used.

kryzhanovskite A mineral: $MnFe_2(PO_4)_2(OH)_2 \cdot H_2O$. Also spelled: *kruzhanovskite.*

K selection The evolutionary process favoring development of *K strategists* or *equilibrium species* (Wilson & Bossert, 1971, p. 110-111).

ksimoglyph A hieroglyph produced by a solid object that is dragged or propelled by a continuous current over a sedimentary surface (Vassoevich, 1953, p.61 & 72); e.g. a drag mark or a groove cast.

K-spar (a) *potassium feldspar.* (b) *potash spar.*

K strategist *equilibrium species.*

ktenasite A blue-green mineral: $(Cu,Zn)_3(SO_4)(OH)_4 \cdot 2H_2O$.

ktypéite A mineral substance intermediate between calcite and aragonite.

kuemmerform n. An organism that is small and stunted as a result of environmental stress. The term is commonly applied to planktonic foraminifers but may also refer to other forms. Erroneous spelling: *kummerform.*

kukersite An organic sediment rich in the alga *Gloexapsamorpha prisca,* found in the Ordovician of Estonia.

kulaite An extrusive nepheline *trachybasalt* or nepheline *basanite* containing potassium feldspar and calcic plagioclase, small amounts of nepheline and olivine, and hornblende as the dominant mafic mineral. Washington in 1894 derived the name from Kula Devit, Turkey. Not recommended usage.

kulanite A triclinic mineral: $Ba(Fe^{+2},Mn,Mg)_2Al_2(PO_4)_3(OH)_3$.

kullaite A porphyritic *diorite* containing phenocrysts of altered plagioclase and alkali feldspar in an ophitic groundmass of plagioclase, alkali feldspar, and pseudomorphs of chlorite after augite, with minor amounts of quartz, apatite, and opaque oxides. The name, given by Hennig in 1899, is derived from Kullen, Sweden. Not recommended usage.

Kullenberg corer A type of *piston corer* that has many modifications or varieties.

kullerudite A mineral: $NiSe_2$.

kum A Turkish term for "sand", applied to the sandy deserts of central Asia; e.g. Kizil Kum. See also: *koum.*

Kungurian European stage: Lower or Middle Permian of some authors (above Artinskian, below Kazanian).

kunkar (a) Var. of *kankar.* (b) A term used in Australia for a calcareous duricrust, or caliche.

kunkur *kankar.*

kunzite A pinkish, light-violet, or lilac-colored transparent gem variety of spodumene.

kupfernickel *nickeline.*

kupletskite A mineral: $(K,Na)_3(Mn,Fe)_7(Ti,Nb)_2Si_8O_{24}(O,OH)_7$.

kuranakhite An orthorhombic mineral: $PbMn^{+4}Te^{+6}O_6$.

kurchatovite A mineral: $Ca(Mg,Mn)B_2O_4$.

kurgantaite A mineral: $(Sr,Ca)_2B_4O_8 \cdot H_2O$ (?).

kurnakovite A white mineral: $Mg_2B_6O_{11} \cdot 15H_2O$.

kuroko deposit A type of massive base-metal sulfide deposit in Japan. Kuroko deposits are typically zoned and strata-bound. They are volcanogenic deposits of Miocene age, precipitated on the sea floor adjacent to fumaroles or hot springs on the flanks of submarine dacite domes during the late stages of explosive felsic volcanic cycles. See: Matsukuma & Horikoshi, 1970. Cf: *Cyprus-type deposit.*

kurskite An alkali-bearing francolite.

kurtosis (a) The quality, state, or condition of peakedness or flatness of the graphic representation of a statistical distribution. (b) A measure of the peakedness of a *frequency distribution;* e.g. a measure of concentration of sediment particles about the median diameter: $(Q_3\text{-}Q_1)/2(P_{90}\text{-}P_{10})$, where Q_3 and Q_1 are the particle diameters, respectively, at the 75% and 25% intersections on the cumulative frequency distribution, P_{90} is the particle diameter such that 90% of the particles are larger and 10% smaller, and P_{10} is the particle diameter such that 10% of the particles are larger and 90% smaller. Various approximations or coefficients of kurtosis have been devised in an attempt to assign genetic significance to sediment distributions. Abbrev: K.—Cf: *skewness.*

kurumsakite A mineral: $(Zn,Ni,Cu)_8Al_8V_2Si_5O_{35} \cdot 27H_2O$ (?).

kuskite A light-colored hypabyssal rock originally thought to contain scapolite, quartz, and some decomposed plagioclase phenocrysts in a fine-grained groundmass of quartz, orthoclase, and muscovite; later the "scapolite" was identified as quartz and the name was withdrawn (Johannsen, 1939, p. 261). Its name is derived from the Kuskokwim River, Alaska. Obsolete.

kusuite A tetragonal mineral: $(Ce^{+3},Pb^{+2},Pb^{+4})VO_4$.

kutinaite A mineral: Cu_2AgAs.
kutnahorite A mineral: $Ca(Mn,Mg,Fe)(CO_3)_2$. It is isomorphous with dolomite. Also spelled: *kutnohorite.*
Kutter's formula A formula that expresses the value of the Chézy coefficient in the *Chézy equation* in terms of the friction slope, hydraulic radius, and a roughness coefficient (Brown & Runner, 1939, p. 199).
kvellite A very dark-colored hypabyssal rock containing phenocrysts of biotite, olivine, barkevikite, apatite, ilmenite, and magnetite in a groundmass of lath-shaped anorthoclase and nepheline. The name, given by Brögger in 1898, is derived from Kvelle, Oslo district, Norway. Not recommended usage.
K wave A longitudinal or *P wave* in the Earth's outer core. Cf: *I wave.*
kX unit A unit in which X-ray wave lengths were generally given prior to World War II. 1 kX = 1.002056 $\pm$ 0.000005 A.
kyanite A blue or light-green triclinic mineral: Al_2SiO_5. It is trimorphous with andalusite and sillimanite. Kyanite occurs in long, thin, bladed crystals and crystalline aggregates in schists, gneisses, and granite pegmatites, and has a hardness of 4-5 along the length of the crystal and 6-7 across it. It forms at medium temperatures and high pressures in regionally metamorphosed sequences. Also spelled: *cyanite.* Syn: *sappare; disthene.*
kyanophilite A mineral: $(K,Na)Al_2Si_2O_7(OH)$ (?).
kyle A Scottish term for a narrow channel, sound, or strait between two islands or an island and the mainland, or for a narrow inlet into the coast. Etymol: Gaelic.
kylite An olivine-rich *theralite,* named by Tyrrell in 1912 for the Kyle district, Ayrshire, Scotland. Not recommended usage.
kymoclastic rock A hydroclastic rock containing marine or other wave-formed fragments (Grabau, 1924, p.295).
kyphorhabd A curved monaxonic sponge spicule with transverse swellings on the convex side.
kyr A term used in central Asia for flat land, a plateau, or the top of a small hill or mountain, and indiscriminately for a low hill or small mountain, but applied specif. in Turkmenia to stony, hard ground as contrasted to *adyr* (Murzaevs & Murzaevs, 1959, p. 131).
kyriosome A term used by Niggli (1954, p. 191) for the fundamental mass or framework fraction of a complex rock; the major part of a migmatite. Cf: *akyrosome.* Little used.
kyrtome A triradiate, more or less thickened area associated with and bordering the laesura of a trilete spore. Cf: *torus.*
kyschtymite A medium- to fine-grained hypabyssal rock composed of euhedral corundum crystals and some biotite and green spinel in a groundmass of calcic plagioclase. The name, given by Moroziewicz in 1897, is from Kyschtym in the Urals. Not recommended usage.

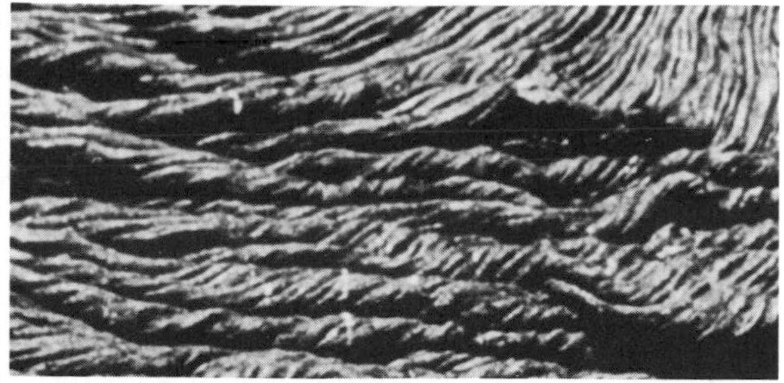

laagte A broad, almost level drainage course in the veld of southwestern Africa, less well-defined than a valley, and dry for most of the year. Etymol: Afrikaans. Pron: *loch*-ta. Syn: *leegte.*

laanilite A coarse-grained *pegmatoid* composed chiefly of garnet, biotite, quartz, and iron oxides. Named by Hackman in 1905 for Laanila, Lapland. Not recommended usage.

laavenite *lavenite.*

labial aperture An *accessory aperture* formed in a foraminiferal test by the free parts of the apertural lip, not leading directly to a chamber.

labial pore A median or submedian pore in the proximal wall of the peristome in ascophoran cheilostomes (bryozoans), resulting from closure of a notch in the secondary orifice.

labiate Having lips; e.g. said of an exaggerated marginate foramen of a brachiopod in which the dorsal edge is prolonged liplike (TIP, 1965, pt. H, p. 147).

labiatiform cyrtolith A cyrtolith coccolith with a central structure shaped like a double lip (as in *Anthosphaera robusta*).

labile (a) Said of rocks and minerals that are mechanically or chemically unstable; e.g. a "labile sandstone" or "labile graywacke" containing abundant unstable fragments of rocks and minerals and less than 75% matrix of fine silt and clay (Packham, 1954), or "labile constituents" (such as feldspar and rock fragments in a sandstone) that are easily decomposed. Cf: *unstable [sed].* (b) Said of protobitumen that represents easily decomposable plant and animal products (such as fat, oil, or protein) in peat and sapropel. Ant: *stabile.*

labilizing force In an unstable gravimeter, a force acting in the same direction as the force being measured, and therefore opposite to the direction of the ordinary restoring force.

labite A mineral: $MgSi_3O_6(OH)_2 \cdot H_2O$. It may be chrysotile.

labium (a) The lower lip of an arthropod (such as of an insect); the *metastoma* of a crustacean. Cf: *labrum.* (b) The columellar part of the aperture of a gastropod shell.—Pl: *labia.*

labor A Spanish term used in early land surveys in Texas for unit of area equal to about 177.14 acres (representing a tract 100 varas square). Pron: la-*bore.*

labradite *labradoritite.*

labradophyre An *anorthosite* composed of labradorite phenocrysts in a groundmass of the same mineral. Not recommended usage.

labradorescence An optical phenomenon consisting of flashes of a laminated iridescence of a single bright hue that changes gradually as a mineral or gemstone is moved about in reflected light, caused by internal structures that selectively reflect only certain colors; specif. the light-interference effect exhibited by labradorite and set up in thin plates of feldspar (produced by repeated twinning or by exsolution), resulting in a series of vivid colors (usually brilliant blue or green) spread over large areas. Syn: *change of color.*

labradorfels *labradoritite.*

labradorite [mineral] A dark mineral of the plagioclase feldspar group with composition ranging from $Ab_{50}An_{50}$ to $Ab_{30}An_{70}$. It commonly shows a rich, beautiful play of colors (commonly blue or green), and is therefore much used for ornamental purposes. Labradorite is common in igneous rocks of intermediate to low silica content. Pron: *lab*-ra-daw-rite. Syn: *Labrador spar.*

labradorite [rock] A name applied by French petrologists to light-colored labradorite-rich *basalt* and by Soviet petrologists to a light-colored *gabbro* or *norite* (i.e. *anorthosite*). Not recommended usage.

labradoritite An *anorthosite* composed almost entirely of the mineral labradorite. Syn: *labradorfels; labradite.* Not recommended usage.

Labrador spar A syn. of *labradorite [mineral].* Also called: *Labrador stone; Labrador rock.*

labrum (a) An unpaired outgrowth of an arthropod, consisting of a single median piece or flap immediately in front of or above the mandibles and more or less covering the mouth; e.g. the upper lip in front of the mouth of a crustacean or of a merostome, or the *hypostome* of a trilobite. Cf: *labium.* (b) The external margin of a gastropod shell. (c) A more or less enlarged and modified liplike primordial plate of an echinoid, bordering the peristome in the interambulacrum 5.—Adj: *labral.*

labuntsovite A mineral: $(K,Ba,Na)(Ti,Nb)(Si,Al)_2(O,OH)_7 \cdot H_2O$. It was originally described as titaniferous elpidite but is now shown to contain only a trace of ZrO_2.

labyrinthic Said of some agglutinated foraminifers having a complex spongy wall with interlaced dendritic channels perpendicular to the surface.

labyrinth karst A *karst plain* characterized by intersecting *solution corridors* (Brook & Ford, 1978, p. 493).

labyrinthodont adj. Pertaining to conical teeth in which the outer layer of enamel is complexly folded to form a labyrinthine pattern in cross section. Such teeth occur in crossopterygian fish and labyrinthodont amphibians.—n. A member of the amphibian subclass *Labyrinthodonta.*

Labyrinthodonta A subclass of amphibians, characterized by teeth with labyrinthine infolding of the enamel and by composite vertebral centra.

labyrinthus A term established by the International Astronomical Union for a region on Mars of intersecting linear depressions; one such region has been named, the Labyrinthus Noctis at the west end of the Valles Marineris. These depressions are generally considered to be graben (Mutch et al., 1976, p. 57, 77).

lac [geog] The French word for *lake.* It appears in proper names in parts of the U.S. where the influence of early French settlement remains, and in French-speaking regions of Canada.

lac [soil] A term used by Brewer (1964, p.366) for soil *plasma* that has a high luster and smooth surface in reflected light under a hand lens up to magnifications of approximately 20 times. Cf: *floc.*

LaCasitan North American (Gulf Coast) stage: Upper Jurassic (above Zuloagan, below Durangoan of Cretaceous; it is equivalent to European Portlandian and Kimmeridgian) (Murray, 1961).

laccolite The original term, now superseded, for a *laccolith.*

laccolith A concordant igneous intrusion with a known or assumed flat floor and a postulated dikelike feeder commonly thought to be beneath its thickest point. It is generally planoconvex in form and roughly circular in plan, less than five miles in diameter, and from a few feet to several hundred feet in thickness. See also: *bysmalith.* Syn: *laccolite.*

lacine One of a series of detached, tongue-shaped, ridgelike *meander scrolls,* frequently found spread apart like the rays of a fan. Lacines are not as long, as smoothly curved, or as closely spaced as the more symmetric meander scrolls. Term introduced by Melton (1936, p. 599). Etymol: Latin *lacinia,* "flap, tongue".

lacine meander A detached *scroll meander* in which lateral erosion of the outer bank is somewhat retarded, thereby producing a low-water channel of unequal width (Melton, 1936, p. 599-600). Examples occur in the Mississippi River between Cairo, Ill., and Baton Rouge, La.

lacinia The inner distal spiny lobe of the second segment of the maxillule of a crustacean, adjacent to the *galea* (TIP, 1969, pt.R, p.97).

lacinia mobilis A small, generally toothed process articulated with the incisor process of the mandible of a malacostracan crustacean.

laciniate Deeply incised, as in a leaf having narrow, pointed lobes, as if slashed.

LaCoste-Romberg gravimeter A long-period vertical-seismograph suspended system adapted to the measurement of gravity differences. Sensitivity is achieved by adjusting the system to proximity to an instability configuration. Syn: *zero-length-spring gravimeter.*

lacroixite A pale yellowish-green mineral, approximately: $NaAl(PO_4)(F,OH)$. It often contains considerable fluorine.

lacullan *anthraconite.*

lacuna [paleont] (a) A pore or slitlike opening between costae making up the frontal shield in cribrimorph cheilostomes (bryozoans). (b) A space, in the lorica of a tintinnid, lacking reticulation or other surface marking. (c) A lateral hole or gap in the hood of some terebratellacean brachiopods, produced by resorption during loop development. (d) In receptaculitids, a circular area devoid of meromes, in the apical portion of the thallus.—Pl: *lacunae* or *lacunas.*

lacuna [palyn] A rarely used term for a depressed space, pit, or hole on the outer surface of a pollen grain.

lacuna [stratig] A chronostratigraphic unit representing a gap in the stratigraphic record; specif. the missing interval at an unconformity, representing the interpreted space-time value of both *hiatus* (period of nondeposition) and *degradation vacuity* (period of erosion) (Wheeler, 1964, p.599).

lacuster The central part of a lake (Veatch & Humphrys, 1966, p.162).

lacustrine (a) Pertaining to, produced by, or formed in a lake or lakes; e.g. "lacustrine sands" deposited on the bottom of a lake, or a "lacustrine terrace" formed along its margin. (b) Growing in or inhabiting lakes; e.g. a "lacustrine fauna". (c) Said of a region characterized by lakes; e.g. a "lacustrine desert" containing the remnants of numerous lakes that are now dry.—Cf: *limnic.* Syn: *lacustral; lacustrian.*

lacy residue An insoluble residue containing irregular openings and having constituent material comprising less than 25% of the volume (Ireland et al., 1947, p. 1482). Cf: *skeletal residue.*

ladder lode *ladder vein.*

ladder reef *ladder vein.*

ladder vein One of a series of mineral deposits in transverse, roughly parallel fractures that have formed along foliation planes perpendicular to the walls of a dike during its cooling, or along shrinkage joints in basaltic rocks or dikes. Syn: *ladder lode; ladder reef.*

lade (a) The mouth of a river. (b) A watercourse.

Ladinian European stage: upper Middle Triassic (above Anisian, below Carnian).

ladu The basal avalanchelike part of a pyroclastic flow.

laesura The line or scar on the proximal face of an embryophytic spore, marking the contact with other members of the tetrad. It may be trilete or monolete. Pl: *laesurae.* See also: *Y-mark; suture [palyn].* Syn: *tetrad scar.*

laevigate A syn. of *psilate.* The term is more often applied to spores than to pollen.

laffittite A monoclinic mineral: $AgHgAsS_3$.

lag (a) *lag gravel.* (b) *sedimentary lag.*

lag deposit *lag gravel.*

lag fault An overthrust, the thrusted rocks of which move differentially so that the upper part of the geologic section is left behind; the replacement of the upper limb of an overturned anticline by a *fold fault.* Syn: *tectonic gap.*

lagg The depressed marginal areas, often covered with water, that surround some gently domed bogs. Etymol: Swedish. Syn: *bog moat.*

lag gravel (a) A residual accumulation of coarse, usually hard rock fragments remaining on a surface after the finer material has been blown away by winds. See also: *desert pavement.* (b) Coarse-grained material that is rolled or dragged along the bottom of a stream at a slower rate than the finer material, or that is left behind after currents have winnowed or washed away the finer material.—Syn: *lag; lag deposit.*

lag mound A remnant of thin unconsolidated surface material left on a limestone pavement when the cover is partly removed at underlying solution fissures by *piping.*

lagoon [coast] (a) A shallow stretch of seawater, such as a sound, channel, bay, or saltwater lake, near or communicating with the sea and partly or completely separated from it by a low, narrow, elongate strip of land, such as a reef, barrier island, sandbank, or spit; esp. the sheet of water between an offshore coral reef and the mainland. It often extends roughly parallel to the coast, and it may be stagnant. (b) A shallow freshwater pond or lake near or communicating with a larger lake or a river; a stretch of freshwater cut off from a lake by a barrier, as in a depression behind a shore dune; a barrier lake. (c) A shallow body of water enclosed or nearly enclosed within an atoll. (d) The term has been widely applied to other coastal features, such as an estuary, a slough, a bayou, a marsh, and a shallow pond or lake into which the sea flows.—Etymol: Latin *lacuna,* "pit, pool, pond". Syn: *lagune; laguna.*

lagoon [geog] A closed depression in a high, grass-covered tableland of the cordilleras of the western U.S.

lagoon [water] (a) The basin of a hot spring; also, the pool formed by a hot spring in such a basin. Etymol: Italian *lagone,* "large lake". (b) A perennial brine pool near the margin of an alkaline lake; e.g. near Lake Magadi in southern Kenya.

lagoon [water res] Any shallow artificial pond or other water-filled excavation for the natural oxidation of sewage or disposal of farm manure, or for some decorative or aesthetic purpose.

lagoonal Pertaining to a lagoon, esp. *lagoonal* deposition.

lagoon atoll *pseudoatoll.*

lagoon beach The sandy fringe on the inner, protected side of a reef island, facing toward the lagoon.

lagoon channel (a) The stretch of deep water separating a reef from the neighboring land (mainland or island). (b) A *pass* through a reef, and into and through a lagoon.

lagoon cliff A very steep to vertical *lagoon slope.* Syn: *lagoon scarp.*

lagoon cycle The sequence of events, and the interval of time, involved in the filling of a lagoon with sediments followed by eventual erosion by wave action and refilling by deposition.

lagoon flat The nearly horizontal reef flat located lagoonward of the *lagoon beach.*

lagoon floor The undulating to nearly level bottom of a lagoon, often encircled by the *lagoon slope.*

lagoon island (a) One of many scattered islets rising from within the lagoon of a composite *atoll* or a large *barrier reef,* generally marking former fringing reefs that grew up with the postglacial eustatic rise. (b) *atoll.*

lagoonlet A small *lagoon;* esp. a shallow pool of water on a platform reef. Syn: *miniature lagoon.*

lagoon margin The lagoonward margin of the *lagoon shelf* or of the reef flat along those parts of the reef that lack islands.

lagoon phase The strata or stratigraphic facies formed by accumulation of sediment in a shallow coastal water body that is separated from the open sea by a barrier.

lagoon plain A flat landform produced by the filling of lagoons with sediments.

lagoon scarp *lagoon cliff.*

lagoon shelf The part of a *reef* that borders the lagoon side of a reef island; the sand-covered, lagoonward-sloping shelf commonly found where sedimentation conspicuously exceeds organic growth.

lagoonside The land bordering on a lagoon.

lagoon slope The seaward border zone or side of a lagoon, sloping downward from the *lagoon margin* to the *lagoon floor;* it may be gentle or steep. Cf: *lagoon cliff.*

Lagorio's rule An approximate rule according to which quartz usually begins to crystallize early from highly siliceous rhyolites and porphyries, and late from the less siliceous ones. The rule was proposed in 1887 by A. Lagorio.

Lagrangian (a) Pertaining to a system of coordinates or equations of motion in which the properties of a fluid are identified for all time by assigning them coordinates that do not vary with time. (b) Said of a direct method of measuring the speed and/or direction of an ocean current by tracking the movement of the same water mass through the ocean by means of tracers, drift bottles, buoys, deep drogues, current poles, and other devices.—Cf: *Eulerian.* Named in honor of Joseph Louis Lagrange (1736-1813), French mathematician.

laguna [coast] A var. of *lagoon.* Etymol: Spanish and Italian.

laguna [lake] A term used in areas of Spanish influence, including the southwestern U.S., for a lake or lagoon; esp. a shallow ephemeral lake in the lower part of a *bolson,* fed by streams rising in the neighboring mountains and flowing only as a result of rainstorms. Etymol: Spanish, "pond, small lake".

lagunar Var. of *lagoonal,* sometimes used in Britain.

lagune (a) Var. of *lagoon.* Etymol: French. (b) A term used in the SW U.S. for a small lake.

lahar A *mudflow [mass move]* composed chiefly of volcaniclastic materials on the flank of a volcano. The debris carried in the flow includes pyroclasts, blocks from primary lava flows, and epiclastic material. Etymol: Indonesian. Syn: *mudflow [volc].*

laihunite An orthorhombic mineral: $Fe_3{}^{+3}Fe_2{}^{+2}(SiO_4)_3$ (?).

laitakarite A mineral: $Bi_4(Se,S)_3$. Cf: *ikunolite.* Syn: *selenjoseite.*

lakarpite A subsolvus *nepheline syenite* composed of orthoclase or microcline, calcic plagioclase, and an amphibole resembling arfvedsonite, accessory acmite, altered nepheline, rosenbuschite, and secondary natrolite. The name, proposed by Törnebohm in 1906, is from Lakarp, Norra Kärr complex, Sweden. Not recommended usage.

lake [coast] A term loosely applied to a sheet of water lying along a coast and connected with the sea; e.g. one of the shallow (1-2 m deep), interconnected bodies of water in the Florida Bay area, Fla. See also: *seashore lake.*

lake [ice] A submariner's term for a *polynya* during the summer. Cf: *skylight.*

lake [lake] (a) Any inland body of standing water occupying a depression in the Earth's surface, generally of appreciable size (larger than a *pond*) and too deep to permit vegetation (excluding subaqueous vegetation) to take root completely across the expanse of water; the water may be fresh or saline. The term includes an expanded part of a river, a reservoir behind a dam, or a lake basin intermittently or formerly covered by water. (b) An inland area of open, relatively deep water whose surface dimensions are sufficiently large to sustain waves capable of producing somewhere on its periphery a barren wave-swept shore (Welch, 1952).—Syn: *lac; lago; loch; lough.*

lake [streams] An English term for a brook or small stream; also, a channel.

lake asphalt Soft asphalt, rich in bitumen, from the pitch lake of Trinidad. See also: *land asphalt.* Syn: *lake pitch.*

lake ball A spherical mass of tangled, waterlogged fibers and other filamentous material of living or dead vegetation (such as blue-green algae, moss, spruce needles, and fragments of peat, grass, or twigs), produced mechanically along a lake bottom by wave action, and usually impregnated with sand and fine-grained mineral fragments. Lake balls may range in size up to that of a man's head. See also: *aegagropile; peat ball.* Cf: *sea ball.* Syn: *hair ball; burr ball.*

lake basin (a) The depression in the Earth's surface occupied or formerly occupied by a lake and containing its shore features. (b) The area from which a lake receives drainage.

lakebed (a) The flat to gently undulating ground underlain by fine-grained sediments deposited in a former lake. (b) The ground on which a lake rests at present; the bottom of a lake; a lake basin.

lake biscuit *algal biscuit.*

lake delta A delta, usually arcuate with a steep front, built out by a river into a freshwater lake; e.g. the delta of the Rhône River in Lake Geneva.

lake deposit A sedimentary deposit laid down conformably on the floor of a lake, usually consisting of coarse material near the shore and sometimes passing rapidly into clay and limestone in deeper water; most of it is of fluvial or glacial origin mixed with freshwater or terrestrial organic matter. It may show a clearly marked seasonal layering, as in *varved clays.*

lake district A region marked by the grouping together of lakes, e.g. in northern England.

lake gage A gage for measuring the elevation of the water surface of a lake.

Lake George diamond Colorless, doubly terminated quartz crystal from Herkimer County, N.Y. See also: *Herkimer diamond.*

lake gun A term used on Seneca Lake, New York, for a phenomenon that produces sounds like distant thunder or gunfire. It is known to occur on several European lakes and has a variety of names (Hutchinson, 1957, p.361-362).

lake-head delta A delta built by a river at the head of a lake. Cf: *lake delta.*

lake ice Ice formed on a lake, regardless of observed location; it is usually *freshwater ice.*

lakelet A small lake.

lake loam A term applied to loess that may have been formed by deposition in lakes (Veatch & Humphrys, 1966, p. 171).

lake marl *bog lime.*

lake marsh (a) A part of the bottom of a lake, supporting a dense growth of emergent aquatic plants. (b) A marsh that occupies the site of a former lake.—See also: *marsh lake.*

lake ocher Ocherous deposits formed on the bottom of a lake by bacteria capable of precipitating ferric hydroxide, or found in a marsh or swamp that was formerly the site of a lake.

lake ore (a) A disklike or irregular concretionary mass of ferric hydroxide less than a meter thick, or a layer of porous, yellow bedded limonite, formed along the borders of certain lakes. Cf: *bog ore.* (b) *bog iron ore.*

lake peat *sedimentary peat.*

lake pitch *lake asphalt.*

lake plain (a) The nearly level surface marking the floor of an extinct lake, filled in by well-sorted deposits from inflowing streams. (b) A flat lowland or a former lake bed bordering an existing lake. See also: *lake terrace.*

lake rampart A conspicuous wall-like ridge composed of unconsolidated coarse material along a lake shore. It is produced by shoreward movement of lake ice, as by winds, waves, or currents, and esp. by expansion of ice against yielding lake-shore deposits, or by pushing and then stranding bottom-lake deposits as it overrides the shore. Lake ramparts may reach 2 m in height. Examples occur along the shores of the Great Lakes. See also: *walled lake.* Syn: *ice rampart; rampart [lake]; ice-push ridge; ice-thrust ridge.*

lakescape All or part of a lake, including water surface, islands, and shoreline features, that can be viewed from an observation point.

lakeshore The narrow strip of land in contact with or bordering a lake; esp. the beach of a lake. Also spelled: *lake shore.* Syn: *lakeside.*

lakeside *lakeshore.*

lake terrace A narrow shelf, partly cut and partly built, produced along a lake shore in front of a nip or line of low cliffs, and later exposed when the water level falls (Cotton, 1958, p. 489). See also: *lake plain.*

lallan Scottish var. of *lowland.* Syn: *lalland.*

Lamarckism A 19th-century theory of evolution stating that changes in the environment cause structural changes in an organism especially by inducing new or increased use of organs or parts as a result of adaptive modification or greater development, and also cause disuse and eventual atrophy of other parts, and that these changes are passed on to offspring. This theory is named after the French naturalist J.B. de Monet Lamarck (1744-1829).

Lambert azimuthal equal-area projection An azimuthal map projection having the pole of the projection at the center of the area mapped, the azimuths of great circles radiating from this pole (center) and being truly represented on the map but the scale along such great-circle lines so varying with distance from the center that an equal-area projection is produced. The pole (center) of the projection may be at the pole of the sphere, on its equator, or at any point in between. The projection is useful for representing a single hemisphere or continental masses, but extreme distortion of areas is encountered near the map periphery. See also: *Schmidt projection.*

Lambert conformal conic projection A conformal conic map projection on which all meridians are represented by equally spaced straight lines that radiate from a common point outside the map limits and the parallels (of which one or two are standard parallels along which the scale is exact) are represented by circular arcs having this common point for a center and intersecting the meridians at right angles. The scale is the same in every direction at any point on the map, but increases north and south from the standard parallel(s); where there are two standard parallels, the scale is too small between them and too large beyond them. The projection is used for maps of middle latitudes (for maps of the conterminous U.S., smallest distortion occurs when the standard parallels represent latitudes 33°N and 45°N) and as a base for sets of large-scale aeronautical charts produced by the National Geodetic Survey. Named after Johann H. Lambert (1728-1777), German physicist, who introduced the projection in 1772.

Lambert's law The statement that the intensity of blackbody radiation emerging from an aperture is greatest in the direction perpendicular to the plane of the aperture, and decreases with the cosine of the angle between the perpendicular and the direction of observation. Such a reflective body is called a *perfectly diffuse reflector;* real bodies seldom approach this condition.

Lamb's problem An investigation in seismology that is concerned with disturbances initiated at a point or along a line on the surface of a semi-infinite, perfectly elastic medium.

Lamé constants Two *elastic constants* or parameters, λ and μ, which express the relationships between the components of stress and strain for linear elastic behavior of an isotropic solid; λ is identical with rigidity, and μ is equivalent to the bulk modulus minus $2\mu/3$.

lamella (a) A thin scale, leaf, lamina, or layer, e.g. one of the units of a polysynthetically twinned mineral, such as plagioclase. See also: *deformation lamella; exsolution lamella.* (b) An organ, process, or part of an organism resembling a leaf or thin plate, e.g. a *primary lamella* of a brachiopod.—Pl: *lamellae.*

lamellar Composed of or arranged in lamellae; disposed in layers like the leaves of a book. Syn: *lamellate.*

lamellar columella A platelike coral *columella.* In rugose corals, it is generally in the plane of the cardinal septum and the counter septum; in scleractinian corals, it is oriented parallel with the longer axis of the calice.

lamellar conodont element A *conodont element* consisting of numerous thin layers or sheaths, being most obvious in specimens that also contain opaque "white matter". See also: *fibrous conodont element.*

lamellar flow Flow of a liquid in which layers glide over one another. Cf: *laminar flow [hydraul].*

lamellar layer The *primary layer* of a brachiopod.

lamellar ligament The part of a ligament of a bivalve mollusk characterized by lamellar structure and containing no calcium carbonate. It is secreted at the edge of the mantle and is elastic to both compressional and tensional stresses. Cf: *fibrous ligament.*

lamellar linkage The joining, by lamellar septal plates, of corallite centers in scleractinian corals, corresponding to *direct linkage* of stomodaea.

lamellar pyrites *marcasite [mineral].*

lamellar wall A foraminiferal test constructed of thin platelike layers of aragonite or calcite, one layer being formed with addition of each new chamber, and covering the whole previously formed test.

lamellibranch *pelecypod.*

lamina [bot] The blade or expanded portion of a leaf.

lamina [paleont] A thin platelike, scalelike, or sheetlike structure in an organism; e.g. a uniform thin sheet of wall substance in the *lorica* of a tintinnid, or a sheetlike structure in some corals, formed by the juxtaposition of two layers of skeletal material in septa and the column. Cf: *lamella.* Pl: *laminae.*

lamina [sed] The thinnest recognizable unit layer of original deposition in a sediment or sedimentary rock, differing from other layers in color, composition, or particle size; specif. such a sedimentary layer less than 1 cm in thickness (commonly 0.05-1.00 mm thick). It may be parallel or oblique to the general stratification. Several laminae may constitute a *bed [stratig]* (Payne, 1942, p.1724) or a *stratum* (McKee & Weir, 1953, p.382). Pl: *laminae.* Syn: *lamination; straticule.*

laminar Consisting of, arranged in, or resembling laminae; e.g. "laminar structure" produced by alternation of thin sedimentary layers of differing composition.

laminar flow [glaciol] A type of glacier flow in which the surface, bed, and flow vectors are all parallel; there is neither extending flow nor compressing flow.

laminar flow [hydraul] Water flow in which the stream lines remain distinct and in which the flow direction at every point remains unchanged with time. It is characteristic of the movement of ground water. Cf: *turbulent flow; mixed flow; lamellar flow.* Syn: *streamline flow; sheet flow [hydraul].*

laminarian Pertaining to a large family of kelps, Laminariaceae, of the order Laminariales.

laminar layer The secondary shell layer in the majority of strophomenid brachiopods, consisting of flat-lying blades or laths of calcite amalgamated laterally and disposed subparallel with the surface of deposition (Williams, 1970).

laminar velocity That velocity of water in a stream below which the flow is laminar and above which it may be either laminar or turbulent. Cf: *turbulent velocity.*

laminar wall A single layer of silica in the wall of a diatom frustule. It may be either of uniform thickness or have local thickenings that form ribs or costae.

laminate adj. Consisting of or containing laminae. Syn: *laminated.*

laminated [petrology] (a) Said of the texture imparted to a lava by *lamination [ign];* also, said of a rock with such a texture. (b) Said of the texture of a metamorphic rock showing layers or lenticles, generally less than 2.5 mm thick, that are associated with modal variations, e.g. olivine and orthopyroxene in peridotite. The laminations have the characteristics of mylonitic compositional layering (Vernon, 1974). Also, said of a rock having such a texture. Syn: *blastolaminar.*

laminated [sed struc] (a) Said of a rock (such as shale) that consists of laminae or that can be split into thin layers. Syn: *laminate.* (b) Said of a substance that exhibits lamination; e.g. "laminated clay" formed in a lake. (c) Said of the sedimentary structure possessed by a laminated rock.

laminated quartz Vein quartz containing slabs, blades or laminar films of other material.

lamination [ign] The spreading-out of the constituents of a lava parallel to the underlying rocks.

lamination [sed] (a) *lamina [sed].* (b) The formation of a lamina or laminae. (c) The state of being laminated; specif. the finest stratification or bedding, typically exhibited by shales and fine-grained sandstones. (d) A laminated structure.

laminite (a) A term used by Lombard (1963) for a finely laminated detrital rock of the flysch lithofacies, frequently occurring in geosynclinal successions in natural sequences complementary to typical turbidites. It is finer-grained and thinner-bedded than a turbidite, ranging in thickness from a few millimeters to 30 cm, and is believed to form seaward from turbidites as a bottomset bed of a large delta. (b) A term suggested by Adolph Knopf (in Sander, 1951, p. 135) to replace *rhythmite* in order to avoid the positive implication of perfect periodicity in the recurrence of laminae.

laminoid Laterally elongate parallel to stratification; e.g. "laminoid-fenestral fabric" of a limestone, characterized by particulate carbonate interrupted by horizontally elongate gaps (fenestrae) that tend to outline lamination (Tebbutt et al., 1965, p.4).

laminoid-fenestral structure A sedimentary fabric in limestone or dolomite, consisting of irregular patches or laminae, once voids, now filled by clear calcite. Patches are crudely oriented parallel with bedding. The structure was formed by rot-out of algal or other tissues, burrowing, gas escape, etc. Syn: *bird's-eye; dismicrite.*

lampadite A variety of wad containing as much as 18% copper oxide and often containing cobalt. The term is often used for all hydrous manganese oxides containing copper.

lamprobolite *basaltic hornblende.*

lamproite A group name for dark-colored hypabyssal or extrusive rocks rich in potassium and magnesium; also, any rock in that group, such as *madupite, orendite, fitzroyite, verite, cedricite,* or *wyomingite.* The term was proposed by Niggli in 1923.

lamprophyllite A mineral: $Na_2(Sr,Ba)_2Ti_3(SiO_4)_4(OH,F)_2$.

lamprophyre A group of dark-colored, porphyritic, hypabyssal igneous rocks characterized by panidiomorphic texture, a high percentage of mafic minerals (esp. biotite, hornblende, and pyroxene), which form the phenocrysts, and a fine-grained groundmass with the same mafic minerals in addition to feldspars and/or feldspathoids; also, any rock in that group, e.g. *minette, vogesite, kersantite, spessartite, camptonite, monchiquite, fourchite, alnoite.* Most lamprophyres are highly altered. They are commonly associated with *carbonatites.* Cf: *leucophyre.* Adj: *lamprophyric.*

lamprophyric Said of the *holocrystalline-porphyritic* texture exhibited by *lamprophyres,* in which phenocrysts of mafic minerals are contained in a fine-grained crystalline groundmass.

lamproschist Metamorphosed lamprophyre with a schistose structure containing brown biotite and green hornblende.

lamp shell A syn. of *brachiopod,* esp. a terebratuloid.

lanarkite A white, greenish, or gray monoclinic mineral: Pb_2SO_5 or $PbO \cdot PbSO_4$.

lanceolate Spear-shaped, or shaped like a lance head, such as a leaf or prism that is much longer than broad, widening above the base and tapering to a point at the apex; e.g. said of the form of a lobe of an ammonoid suture, or of a nautiloid whorl section with an acute periphery, or of a bryozoan colony.

lancet plate An elongate and spear-shaped or triangular plate located along the midline of the ambulacrum of blastoids.

land In a general sense, that part of the Earth's surface that stands above mean sea level. The inclusion of Antarctica's permanent ice in calculating the land surface of the Earth is controversial. Not to be confused with the term "soil".

land accretion Reclamation of land from the sea or other low-lying or flooded areas by draining and pumping, dumping of fill, or planting of marine vegetation.

land asphalt Hard asphalt, containing less bitumen and more impurities than *lake asphalt,* from areas outside the pitch lake of Trinidad. It is divided into several varieties depending on the depth at which it is found. Syn: *land pitch.*

landauite A mineral: $(Zn,Mn,Fe)Ti_3O_7$.

landblink A yellowish reflection on the underside of a cloud layer over snow-covered land in a polar region; yellower than *iceblink [meteorol].* Also spelled: *land blink.*
land bridge A land connection between continents or landmasses, often subject to temporary or permanent submergence, that permits the migration of organisms; e.g. the Bering Land Bridge. See also: *neck; filter bridge.*
land compass *surveyor's compass.*
Landenian European stage: Upper Paleocene (above Montian-Heersian, below Ypresian of Eocene). It includes Thanetian and Sparnacian.
landerite A pink to rose-pink variety of grossular garnet. Syn: *rosolite; xalostocite.*
landesite A brown mineral: $Mn_3{}^{+2}Fe^{+3}(PO_4)_2(OH)_3 \cdot 3H_2O$ (?). It occurs as an alteration product of reddingite. Cf: *salmonsite.*
landfill *sanitary landfill.*
landflood An overflowing of inland water onto the land.
landform Any physical, recognizable form or feature of the Earth's surface, having a characteristic shape, and produced by natural causes; it includes major forms such as plain, plateau, and mountain, and minor forms such as hill, valley, slope, esker, and dune. Taken together, the landforms make up the surface configuration of the Earth. Also spelled: *land form.* See also: *physiographic form; topographic form.* Syn: *relief feature.*
landform map *physiographic diagram.*
land hemisphere That half of the Earth containing the bulk (about six-sevenths) of the dry land surface; it is mostly north of the Equator, with Paris as its approx. center. Cf: *water hemisphere.*
land ice Any ice mass formed from snow, rain, or other freshwater on land, as an ice shelf or a glacier, even though it may be floating in the sea, as an iceberg. Ant: *sea ice.*
landlocked Said of a body of water that is nearly or entirely enclosed by land; e.g. a landlocked bay separated from the main body of water by a bar, or a landlocked lake having no surface outlet.
landmark (a) Any conspicuous object, natural or artificial, located near or on land and of sufficient interest or prominence in relation to its surroundings to make it outstanding or useful in determining a location or a direction. (b) Any monument, material mark, or fixed object (such as a river, tree, or ditch) used to designate the location of a land boundary on the ground.
landmass A land area studied as a unit, without regard necessarily to size or relief, on the basis of the sediments derived from it or the paleogeographic evidence indicated by the change in shorelines (Eardley, 1962, p. 6).
landmass volume The volume, beneath a land surface, of a body with vertical sides and a base equal to *basin area* at the elevation of the stream's mouth (Strahler, 1952b, p.1120). Symbol: V.
land pebble *land-pebble phosphate.*
land-pebble phosphate A term used in Florida for a *pebble phosphate* occurring as pellets, pebbles, and nodules in gravelly beds a few feet below the ground surface. It is extensively mined. Cf: *river-pebble phosphate.* Syn: *land pebble; land rock; matrix (b).*
land pitch *land asphalt.*
land rock A syn. used in South Carolina for *land-pebble phosphate.*
Landsat An unmanned earth-orbiting NASA satellite that transmits multispectral images in the 0.4 to 1.1 μm region to Earth receiving stations. It was formerly called Earth Resource Technology Satellite, or ERTS.
landsbergite *moschellandsbergite.*
landscape The distinct association of landforms, esp. as modified by geologic forces, that can be seen in a single view, e.g. glacial landscape.
landscape marble A close-grained limestone characterized by dark conspicuous dendritic markings that suggest natural scenery (woodlands, forests); e.g. the argillaceous limestone in the Cotham Marble near Bristol, England. Syn: *forest marble.*
land sculpture *sculpture.*
landside (a) That part of a near-water feature that is facing toward the land. (b) An obsolete term for *shore.*
land sky Dark or gray streaks or patches in the sky near the horizon or on the underside of low clouds, caused by the absence of reflected light from ground that is not snow-covered; not as dark as *water sky.* See also: *blink.*
landslide A general term covering a wide variety of mass-movement landforms and processes involving the downslope transport, under gravitational influence, of soil and rock material en masse. Usually the displaced material moves over a relatively confined zone or surface of shear. The wide range of sites and structures, and of material properties affecting resistance to shear, result in a great range of landslide morphology, rates, patterns of movement, and scale. Landsliding is usually preceded, accompanied, and followed by perceptible *creep* along the surface of sliding and/or within the slide mass. Terminology designating landslide types generally refers to the landform as well as the process responsible for it, e.g. rockfall, translational slide, block glide, avalanche, mudflow, liquefaction slide, and slump. Syn: *landsliding; slide [mass move].*
landslide breccia A breccia that is largely fragmented and wholly assembled by the force of gravity, as by a rockfall or a rockslide.
landslide lake (a) A lake resulting from the damming of a stream valley by the material in a landslide. (b) A long, narrow lake between the back slope of a landslide terrace and a valley wall.
landslide sapping The process of causing landslides by a stream undermining a canyon wall (Freeman, 1925, p. 78).
landslide scar A bare or relatively bare surface or niche on the side of a mountain or other steep slope, left by the removal of earth material from the place where a landslide started.
landslide shear surface *slip surface [mass move].*
landslide terrace A short, rough-surfaced terrace resulting from a landslide.
landslide track The exposed path in rock or earth formed by a landslide. Syn: *slide.* Not to be confused with *avalanche track.*
landsliding The downward movement of a landslide.
land survey A survey made to determine boundaries and areas of tracts of land, esp. of privately owned parcels of land. Cf: *cadastral survey; boundary survey.*
land-tied island A *tied island* connected with the mainland by a tombolo.
lane (a) A narrow, not necessarily navigable, fracture or channel of water through sea ice; it may widen into a *lead.* (b) A syn. of *lead [ice].* Rarely used.
langbanite An iron-black hexagonal mineral: $(Mn^{+2},Ca)_4(Mn^{+3},Fe^{+3})_9SbSi_2O_{24}$. Not to be confused with *langbeinite.*
langbeinite A colorless to reddish isometric mineral: $K_2Mg_2(SO_4)_3$. It is much used in the fertilizer industry as a source of potassium compounds. Not to be confused with *langbanite.*
Langhian European stage: Middle Miocene (above Burdigalian, below Serravallian).
langisite A hexagonal mineral: $(Co,Ni)As$.
langite A blue to green mineral: $Cu_4(SO_4)(OH)_6 \cdot 2H_2O$.
Langmuir cell A unit of water circulation, at or adjacent to the surface, driven by the wind, having its long axis essentially parallel with the wind direction, having opposite senses of spiral flow in adjacent cells, and having pronounced ellipticity in shallow water, where it may influence sediment transport (Campbell et al., 1977).
lansfordite A colorless mineral: $MgCO_3 \cdot 5H_2O$. It alters to nesquehonite on exposure to air.
lantern *Aristotle's lantern.*
lantern-node The central octahedron of a *lychnisc* in a sponge.
lanthanite A colorless, white, pink, or yellow mineral: $(La,Ce)_2(CO_3)_3 \cdot 8H_2O$.
lapiaz A syn. of *karren.* Etymol: Pyrennées dialect. Also spelled: *lapiés.*
lapidary (a) A cutter, grinder, and polisher of colored stones, or of precious stones other than diamonds. Syn: *lapidist.* (b) The art of cutting gems. (c) An obsolete term for a short treatise on metals, stones, and gems, describing their supposed medicinal, magical, or mythical characteristics.
lapidification An obsolete term signifying conversion into stone or stony material, such as the process of petrifaction or lithification.
lapidofacies Facies related to diagenesis (Vassoevich, 1948).
lapiés *lapiaz.*
lapilli Pyroclastics that may be either essential, accessory, or accidental in origin, of a size range that has been variously defined within the limits of 2 and 64 mm. The fragments may be either solidified or still viscous when they land (though some classifications restrict the term to the former); thus there is no characteristic shape. An individual fragment is called a *lapillus.* Cf: *volcanic gravel; block [volc]; cinder.*
lapillistone A pyroclastic rock composed largely of *lapilli.*
lapillite *lapilli tuff.*
lapilli tuff An indurated deposit that is predominantly lapilli,

with a matrix of tuff. Syn: *lapillite*.

lapillus The singular form of *lapilli*.

lapis lazuli (a) A blue, semitranslucent to opaque, granular crystalline rock used as a semiprecious stone for ornamental purposes and composed essentially of lazurite and calcite but also containing hauyne, sodalite, pyrite inclusions, and other minerals. It usually has a rich azure-blue color, but may be other shades of blue, depending on the amount of inclusions. It is probably the original sapphire of the ancients. Syn: *lazuli.* (b) An old name for lazurite, still used esp. for the gem variety. (c) An ultramarine-colored serpentine from India.

Laplace azimuth A *geodetic azimuth* derived from an astronomic azimuth by means of the *Laplace equation,* expressing the relationship between astronomic and geodetic azimuths in terms of astronomic and geodetic longitudes and geodetic latitude. Cf: *azimuth.*

Laplace equation An equation, used to derive the *Laplace azimuth,* that expresses the relationship between astronomic and geodetic azimuths in terms of astronomic and geodetic longitudes and geodetic latitude. See also: *Laplace station.*

Laplace station A triangulation or traverse station at which the *Laplace equation* can be formulated through observation of astronomic longitude and azimuth.

laplandite An orthorhombic mineral: $Na_4CeTiPSi_7O_{22}\cdot 5H_2O$.

lap-out map A map showing the areal distribution of formations immediately overlying an unconformity. Syn: *worm's-eye map.*

lapparentite *tamarugite.*

lappered ice *anchor ice.*

lappet Projection of the peristome on whorl sides or venter of nautiloid and ammonoid conchs (TIP, 1957, pt. L, p. 4).

lapse rate The rate at which some atmospheric property, usually temperature, decreases with height; the vertical gradient. For temperature, the average is about 0.6°C/100 meters; however, it varies according to the moisture content of the air.

laqueiform Said of the loop pattern in a dallinid brachiopod (as in the family Laqueidae) in which posterior connecting bands from the ascending branches to the descending branches "are retained during enlargement and proportional thinning during change from frenuliniform to terebrataliiform loop" (TIP, 1965, pt.H, p.147).

Laramian orogeny *Laramide orogeny.*

Laramic orogeny *Laramide orogeny.*

Laramide orogeny A time of deformation, typically recorded in the eastern Rocky Mountains of the United States, whose several phases extended from late Cretaceous until the end of the Paleocene. Intrusives and accompanying ore deposits emplaced about this time in the mountain states are commonly called Laramide (e.g. the Boulder batholith, Montana). Geologists differ as to whether to restrict the Laramide closely in time and space, as to a single event near the end of the Cretaceous, and to deformations near the type area, or to apply the term broadly to all orogenies from early in the Cretaceous through the Eocene or later, and to deformations in the whole Cordilleran belt of western North America. Best usage is probably somewhere between the extremes, and the Laramide can properly be considered as an orogenic era in the sense of Stille. It is named for the Laramie Formation of Wyoming and Colorado, probably a synorogenic deposit. Also spelled: *Laramic orogeny; Laramian orogeny.* Syn: *Laramide Revolution.*

Laramide Revolution *Laramide orogeny.*

lardalite *laurdalite.*

larderellite A white mineral: $(NH_4)B_5O_8\cdot 2H_2O$.

lardite (a) White hydrated silica, probably a variety of opal, occurring in clay in central Russia. (b) Massive talc; steatite. (c) *agalmatolite.*

lard stone Massive talc; steatite.

large boulder A *boulder* having a diameter in the range of 1024-2048 mm (40-80 in., or -10 to -11 phi units).

large cobble A geologic term for a *cobble* having a diameter in the range of 128-256 mm (5-10 in., or -7 to -8 phi units).

larger foraminifera An informal term generally used to designate those foraminifers that are studied without the aid of thin sectioning. Cf: *smaller foraminifera.*

large-scale map A map at a scale (in the U.S., 1/25,000 and larger) such that a small area can be shown in fine detail and with great accuracy; a map whose representative fraction has a small denominator (such as 1/10,000). The Defense Mapping Agency defines large-scale maps as those at 1/75,000 or larger. Cf: *intermediate-scale map; small-scale map.*

large wave An obsolete syn. of *surface wave.*

larnite A gray mineral: β-Ca_2SiO_4. It is a metastable monoclinic phase of calcium orthosilicate, stable from 520° to 670°C, and tending to break down to the stable calcio-olivine. Cf: *bredigite.* Syn: *belite.*

larosite An orthorhombic mineral: $(Cu,Ag)_{21}(Pb,Bi)_2S_{13}$.

larsenite A colorless or white orthorhombic mineral: $PbZnSiO_4$.

Larsen method *Lead-alpha age method.* This method was suggested and developed under the guidance of E. S. Larsen, Jr. (1879-1961), U.S. mineralogist and petrologist.

Larsen variation diagram A diagram in which the weight percent of each oxide constituent of a rock is plotted as the ordinate against the abscissa, which is one-third of the $SiO_2 + K_2O$ - FeO - MgO - CaO. The diagram was devised by E. S. Larsen (1938).

larvikite An alkalic *syenite,* grading to *monzonite,* composed of phenocrysts of two feldspars (esp. oligoclase and alkali feldspar), often intimately intergrown, which comprise up to 90% of the rock, with diopsidic augite and titanaugite as the chief mafic minerals, and accessory apatite (generally abundant), ilmenite, and titaniferous magnetite, and less commonly olivine, bronzite, lepidomelane, and quartz or feldspathoids (less than 10 percent by volume). Its name, given by Brögger in 1890, is derived from Larvik, Norway. Also spelled: *laurvikite.* Syn: *blue granite.*

lassenite A name, now obsolete, that was formerly applied to a volcanic glass thought to have the composition of *trachyte* but now known to be dacitic. Cf: *metabolite.*

late Pertaining to or occurring near the end of a segment of time. The adjective is applied to the name of a geologic-time unit (era, period, epoch) to indicate relative time designation and corresponds to *upper* as applied to the name of the equivalent time-stratigraphic unit; e.g. rocks of an Upper Jurassic batholith were intruded in Late Jurassic time. The initial letter of the term is capitalized to indicate a formal subdivision (e.g. "Late Devonian") and is lowercased to indicate an informal subdivision (e.g. "late Miocene"). The informal term may be used for eras and epochs, and for periods where there is no formal subdivision. Cf: *middle [geochron]; early.*

late diagenesis Deep-seated diagenesis, occurring a long time after deposition, when the sediment is more or less compacted into a rock, but still in the realm of pressure-temperature conditions similar to those of deposition; it represents a transition from diagenesis to metamorphism. Syn: *epigenesis; epidiagenesis; metharmosis.*

late-glacial Pertaining to the time of the waning of the last glaciation; specif. the "Late Glacial period" of the Pleistocene Epoch, immediately preceding the Preboreal phase. Also spelled: *Lateglacial.*

late neanic stage During increase in colonial corals, the stage following the hystero-neanic stage. It is taken to start when the offset possesses all or nearly all the characters of the adult stage except that it has smaller dimensions and fewer septa (Fedorowski & Jull, 1976, p. 42). Cf: *hystero-brephic stage; hystero-neanic stage.*

latent magma A highly viscous magma that exists under high pressure beneath the Earth's crust and reacts as a solid body, e.g. with respect to the propagation of earthquake waves. With a decrease in pressure, the magma becomes sufficiently fluid to flow (Schieferdecker, 1959, term 3827).

latera Plural of *latus.*

lateral [paleont] n. (a) A compartmental plate in certain cirripede crustaceans, bounded by a carinolateral and a rostrolateral. In other cirripedes, the term is synonymous with *latus.* (b) One of a series of ossicles along the side of an arm in an ophiuroid. Also, one of a circlet of five plates in certain cystoids. (c) A lateral part; e.g. a lateral tooth. (d) In coiled cephalopod conchs, those parts of the shell wall lying between the ventral and dorsal sides.

lateral [streams] n. A *lateral stream.*

lateral [volc] *parasitic [volc].*

lateral accretion Outward or horizontal sedimentation; e.g. digging away of the outer bank of a stream meander and building up of the inner bank to water level by deposition of material brought there by rolling or pushing along the bottom. Cf: *vertical accretion.*

lateral bud In seed plants, an axillary bud.

lateral channel A channel formed by a meltwater stream flowing laterally away from a glacier through a notch in bordering hills (Rich, 1908, p. 528).

lateral consequent stream A *secondary consequent stream* flowing down the flank of an anticline or syncline.

lateral corrasion Corrasion of the banks of a stream.

lateral crevasse *marginal crevasse.*

lateral depression series A group of small, basinlike depressions on the exterior lateral edges of the coverplates of lebetodiscid edrioasteroids. They flank the suture lines between contiguous ambulacral coverplates, and are associated with the external foramina of the coverplate passageways (Bell, 1976).

lateral depressor pit A small hollow near one or both basal angles of the scutum of a cirripede crustacean, serving for attachment of the lateral muscle that depresses or draws down.

lateral dissepiment A dissepiment of a rugose coral, characterized by a blisterlike form and developed in isolated manner on the sides of septa. Cf: *horseshoe dissepiment.*

lateral dune A sand dune flanking a larger dune, formed around an obstacle.

lateral erosion The wearing-away of its banks by a meandering stream as it swings from side to side, impinging against and undercutting the banks as it flows downstream; it results in *lateral planation.*

lateral fault A fault along which there has been strike separation. Cf: *strike-slip fault; dip-separation fault.* See also: *wrench fault; right-lateral separation; left-lateral separation.* Syn: *strike-separation fault.*

lateral increase A type of *increase* (offset formation of corallites) in fasciculate and massive coralla characterized by sideward outgrowth.

lateral lake A *fluviatile lake* formed in the valley of a tributary stream by the silting-up of the channel of the main stream, thereby producing embankments or levees that impound the water of the tributary.

lateral levee lake A lake occupying a depression behind a *natural levee.*

lateral lobe Any adapical inflection of a suture of a nautiloid or ammonoid shell between the ventral and dorsal lobes; in coiled conchs, the lateral lobes may be external or internal according to whether they are on the flanks or dorsal areas (TIP, 1964, pt. K, p. 57).

lateral log An obsolete 3-electrode *resistivity log,* designed to record resistivity in a porous formation well beyond the radius of invasion by drilling mud. Cf: *focused-current log.*

lateral migration Movement of oil or gas through permeable zones parallel to the stratification.

lateral moraine (a) A low ridgelike moraine carried on, or deposited at or near, the side margin of a mountain glacier. It is composed chiefly of rock fragments loosened from the valley walls by glacial abrasion and plucking, or fallen onto the ice from the bordering slopes. (b) An end moraine built along the side margin of a glacial lobe occupying a valley. Cf: *flanking moraine.*—Syn: *side moraine; valley-side moraine.*

lateral oblique muscle One of a pair of muscles in some inarticulate brachiopods, originating on the pedicle valve anteriorly and laterally from the posterior adductor muscles, and passing anteriorly and dorsally to insertions either on the brachial valve and anterior body wall against the anterior adductor muscles (as in the family Discinidae) or entirely on the anterior body wall (as in the family Craniidae) (TIP, 1965, pt.H, p.147). Cf: *internal oblique muscle.*

lateral planation The reduction of the land in an interstream area to a plain or a nearly flat surface by the *lateral erosion* of a meandering stream; the creation and development by a stream of its flood plain.

lateral saddle An adoral inflection of a suture of a nautiloid or ammonoid shell between the ventral and dorsal lobes; in coiled conchs, lateral saddles may be external or internal according to whether they are on the flanks or dorsal areas (TIP, 1964, pt. K, p. 57).

lateral search *profiling.*

lateral secretion A theory of ore genesis formulated in the 18th century and passing in and out of use since. It postulates the formation of ore deposits by the leaching of adjacent wall rock. In current usage, convectively driven fluids associated with cooling plutons are thought to have abstracted metals from adjacent host rocks and transported them to new sites of deposition, as in the formation of certain porphyry base-metal deposits. See also: *lithogene; segregated vein.*

lateral shared coverplate One of a pair of oral plates in edrioasteroids that flank the transverse oral midline. They are proximal to the lateral ambulacral bifurcation plate and distal to the central primary orals (Bell, 1976).

lateral sinus A notch or re-entrant in the lateral part of the apertural margin (peristome) of a cephalopod.

lateral spread Lateral movements in a fractured mass of rock or soil, which result from liquefaction or plastic flow of subjacent materials.

lateral storage *bank storage.*

lateral stream A stream situated on, directed toward, or coming from the side; e.g. a stream flowing along the edge of a lava flow that recently filled part of a valley. Syn: *lateral.*

lateral tooth A *hinge tooth* located some distance from the beak of a bivalve mollusk, situated anterior or posterior to the middle of the hinge, and lying ahead of or behind the *cardinal teeth.* Its long axis is parallel to the hinge line.

lateral valley A *longitudinal valley* developed parallel to the regional structure.

lateral variation A change in the sedimentary characteristics of a formation in a horizontal direction. It is particularly significant in petroleum geology when porosity and permeability are affected, as in a *facies change* from sandstone to siltstone or shale (a *shaleout*).

lateral wall One of a pair of oppositely placed *vertical walls* bounding the sides of a cheilostome bryozoan zooid and commonly developed as an *exterior wall.*

laterite An older term for a highly weathered red subsoil or material rich in secondary oxides of iron, aluminum, or both, nearly devoid of bases and primary silicates, and commonly with quartz and kaolinite. It develops in a tropical or forested warm to temperate climate, and is a residual product of weathering. Laterite is capable of hardening after a treatment of wetting and drying, and can be cut and used for bricks; hence its etymology: Latin, *latericius,* "brick". See also: *laterite soil; plinthite.*

laterite soil A great soil group in the 1938 classification system, characterized by a thin A horizon and a reddish leached B horizon overlying laterite (USDA, 1938). This group was dropped in the 1949 revision and these soils were included with *Latosols.* Laterite soils are now classified as *Orthox.*

lateritization *laterization.*

laterization A general term for the process that converts a rock or soil to *laterite.* Also spelled: *lateritization.*

Laterolog A trade name for a *focused-current log.* See also: *Microlaterolog.*

late wood Xylem formed in late stages of a growth zone, which is more dense than the wood produced earlier. Cf: *early wood.* Syn: *summerwood.*

lath The part of a heterococcolith with one large dimension, one intermediate, and one very small.

lath-shaped Said of the habit of a crystal that is long and thin, and of moderate to narrow width. In thin section, lath-shaped crystals often cross sections of platy or tabular crystals.

latilamina A layer, 1-10 mm thick, composed of several laminae or dissepiments in some stromatoporoids, the boundaries of which are marked as planes of weakness in the coenosteum and by crowding of the laminae. Pl: *latilaminae.*

Latin square An array of size *n* x *n,* containing *n* letters displayed *n* times in such a manner that each letter occurs once only in each row and in each column of the array.

latite A porphyritic extrusive rock having phenocrysts of plagioclase and potassium feldspar (probably mostly sanidine) in nearly equal amounts, little or no quartz, and a finely crystalline to glassy groundmass, which may contain obscure potassium feldspar; the extrusive equivalent of *monzonite.* Latite grades into *trachyte* with an increase in the alkali feldspar content, and into *andesite* or *basalt,* depending on the presence of sodic or calcic plagioclase, as the alkali feldspar content decreases. It is usually considered synonymous with *trachyandesite* and *trachybasalt,* depending on the color. The name, given by Ransome in 1898, is derived from Latium, Italy.

latitude (a) Angular distance from some specified circle or plane of reference, such as the angle measured at the center of a sphere or spheroid between the plane of the equator and the radius to any point on the surface; specif. angular distance of a point on the Earth's surface north or south of the equator, measured along a meridian through 90 degrees (the equator being latitude zero degrees, the North Pole lat. 90°N, and the South Pole lat. 90°S). A degree of latitude on the Earth's surface ranges in length from 68.704 statute miles at the equator to 69.407 statute miles at the poles. Abbrev: lat. Symbol: ϕ. See also: *astronomic latitude; geodetic latitude; geocentric latitude; geographic latitude; celestial latitude.* Cf: *parallel.* (b) The projection on the meridian of a given *course* in a plane survey equal to the length of the course multiplied by the cosine of its bearing. (c) A linear coordinate distance measured north or south from a specified east-west line of reference; e.g. *northing* and *southing.*—Cf: *longitude.*

latitude correction (a) The north-south correction made to observed magnetic-field intensities in order to remove the Earth's normal field (leaving, as the remainder, the anomalous field). Cf: *longitude correction.* (b) A correction of gravity data with latitude, because of variations in centrifugal force owing to the Earth's rotation and because of differences in the radius owing to polar flattening. The correction for latitude ϕ amounts to 1.308 sin 2 ϕ mgal/mi = 0.813 sin 2 ϕ mgal/km. See also: *international gravity formula.*

latitude difference The length of the projection of a line onto a meridian of reference in a plane survey, being equal to the length of the line multiplied by the cosine of its bearing; e.g. *northing* (positive difference) and *southing* (negative difference). Cf: *meridional difference; departure.*

latiumite A mineral: $(Ca,K)_8(Al,Mg,Fe)(Si,Al)_{10}O_{25}(SO_4)$.

Latosol A great group of zonal soils introduced in 1950 and characterized by deep weathering and abundant hydrous-oxide material. They are developed under forested humid tropical conditions (Kellogg, 1950). Most of these soils are now classified as *Humults* and *Tropepts.*

latrappite A mineral: $(Ca,Na)(Nb,Ti,Fe)O_3$. Cf: *perovskite.*

lattice [cryst] *crystal lattice.*

lattice [reef] *growth lattice.*

lattice bar A bar composing the lattice shell of a radiolarian.

lattice constant In a crystal lattice, the length of one edge of its unit cell; also, the angle between two edges of the cell. Syn: *lattice parameter; parameter.*

lattice defect *crystal defect.*

lattice drainage pattern *rectangular drainage pattern.*

lattice parameter *lattice constant.*

lattice point A point representing the translational periodicity of a crystal structure. See also: *crystal lattice.*

lattice pore One of the open spaces surrounded by lattice bars in a radiolarian skeleton of the subfamily Trissocyclinae.

lattice-preferred orientation The preferred orientation of crystallographic axes or planes. In metamorphic rocks, it results from *plastic deformation* (crystal gliding and/or dynamic recrystallization) and is dependent on the mineral structure and on the physical conditions (pressure, temperature, stress) during deformation. In igneous rocks, it is mainly related to the original shape of the crystals during settling or flow. Cf: *shape-preferred orientation.*

lattice row A row of *lattice points.* See also: *crystal axis.*

lattice shell A porous sheath that surrounds all or a part of the sagittal ring of some radiolarians, or is divided into symmetric halves by the sagittal ring in other radiolarians; a meshwork radiolarian skeleton.

lattice spine One of the spines that project from the lattice bar in a radiolarian skeleton, either distributed randomly or confined to the junctions of two or more lattice bars.

lattice texture [eco geol] In mineral deposits, a texture produced by exsolution in which elongate crystals are arranged along structural planes.

lattice texture [meta] A texture that is typical of the mineral serpentine in a rock when it replaces an amphibole. Cf: *knitted texture.*

Lattorfian European stage: Lower Oligocene (above Priabonian, below Rupelian). Also spelled: *Latdorfian.*

latus (a) Any of paired plates forming part of the shell in certain cirripede crustaceans, but not including tergum and scutum; e.g. "carinal latus" (carinolateral) or "rostral latus" (rostrolateral). Syn: *lateral [paleont].* (b) The surface of a crinoid columnal or cirral, exclusive of articular facets.—Pl: *latera.*

laubmannite A mineral: $Fe_3^{+2}Fe_6^{+3}(PO_4)_4(OH)_{12}$. It may contain a little manganese or calcium.

Laue camera The instrument used in the *Laue method* of X-ray diffraction analysis. A pinhole defines an X-ray beam perpendicular to a flat film. When the back-reflection method is used, a pinhole is made through the film which is then mounted between the X-ray tube and the crystal.

Laue equations Three simultaneous equations that represent the necessary conditions for radiation to be diffracted by a three-dimensional crystal structure.

Lauegram The diagram of X-ray diffraction made according to the *Laue method.* See also: *Laue spot.* Syn: *Laue pattern.*

laueite A honey-brown triclinic mineral: $MnFe_2(PO_4)_2(OH)_2 \cdot 8H_2O$. It is polymorphous with strunzite. Cf: *pseudolaueite.*

Laue method A technique of X-ray diffraction analysis using a single fixed crystal irradiated by a beam of a continuous spectrum of X-rays. The patterns, or *Lauegram,* are observed after X-ray transmission or by reflection back to their source, called back reflection. See also: *Laue camera.*

Laue pattern *Lauegram.*

Laue spot A single spot on a *Lauegram.*

laugenite An oligoclase *diorite.* The name, proposed by Iddings in 1913, is for Laugendal, Norway. Not recommended usage.

laumontite A white zeolite mineral: $CaAl_2Si_4O_{12} \cdot 4H_2O$. It sometimes contains appreciable sodium, and on exposure to air it loses water, becomes opaque, and crumbles. It occurs as prismatic crystals in veins in schist and slate, and in cavities in igneous rocks. Also spelled: *laumonite; lomonite; lomontite.*

laumontite-prehnite-quartz facies A term introduced by Winkler (1967) to replace the term *zeolite facies.*

launayite A mineral: $Pb_{22}Sb_{26}S_{61}$.

launder A trough, channel, gutter, flume, or chute by which water or powdered ore is conveyed in a mining operation.

lauoho o pele *Pele's hair.*

Laurasia The protocontinent of the Nortern Hemisphere, corresponding to *Gondwana* in the Southern Hemisphere, from which the present continents of the Northern Hemisphere have been derived by separation and continental displacement. The hypothetical supercontinent from which both Laurasia and Gondwana were derived is *Pangea.* The protocontinent Laurasia included most of North America, Greenland, and most of Eurasia, excluding India. The main zone of separation was in the North Atlantic, with a branch in Hudson Bay, and geologic features on opposite sides of these zones are very similar. Etymol: a combination of Laurentia, a paleographic term for the Canadian Shield and its surroundings, and Eurasia.

laurdalite An alkalic *syenite* containing more than 10 percent modal feldspathoids and characterized by porphyritic texture. Also spelled: *lardalite.* The name, given by Brögger in 1890, is for Laurdal, Norway.

Laurentian A name that is widely and confusingly used for granites and orogenies of Precambrian age in the Canadian Shield. It is named for the Laurentian Highlands northwest of the St. Lawrence River in eastern Canada (a part of the Grenville province of current usage), where Logan (1863) recognized the Laurentian granites, now dated radiometrically at about 1000 m.y. The term was misapplied by Lawson (1885) to the oldest granites near the U.S.-Canadian border northwest of Lake Superior, from which Schuchert subsequently derived his Laurentian Revolution, or orogeny, that was supposed to have terminated the Archeozoic. Modern work shows that Lawson's Laurentian is older than the 2400-m.y.-old Algoman orogeny and granites, at the end of the Archean of the present Canadian classification, but no radiometric dates for it survive, and its significance and extent are uncertain. It has been suggested that the term Laurentian be restored to Logan's original meaning. See also: *Grenville orogeny.*

laurionite A colorless mineral: $Pb(OH)Cl$. It is dimorphous with paralaurionite.

laurite An iron-black mineral: RuS_2. It often contains osmium. Laurite is found in association with platinum in placer deposits and usually occurs in minute octahedrons resembling those of magnetite.

laurvikite *larvikite.*

lausenite A white silky or fibrous mineral: $Fe_2(SO_4)_3 \cdot 6H_2O$.

lautarite A monoclinic mineral: $Ca(IO_3)_2$.

lautite A mineral: CuAsS.

lava A general term for a molten *extrusive;* also, for the rock that is solidified from it.

lava ball A globular mass of lava that is scoriaceous inside and compact on the outside; it is formed by the coating of a fragment of scoria by fluid lava. Syn: *pseudobomb; volcanic ball.*

lava blister *blister [volc].*

lava breccia (a) Autoclastic breccia produced by fragmentation during movement of the upper and lower crusts of lava flows. (b) Volcanic breccia.

lava cascade A cascade of fluid, incandescent lava, formed when a lava river passes over a cliff or steep part of its course.

lava cave *lava tube.*

lava column The column of fluid or solidified lava in a volcanic conduit. Syn: *magma column.*

lava-dam lake A lake formed where a lava flow obstructs or has obstructed a watercourse. Syn: *lava-dammed lake.*

lava dome (a) A dome-shaped mountain of solidified lava in the form of many individual flows, formed by the extrusion of highly fluid lava, e.g. Mauna Loa, Hawaii. Cf: *volcanic dome.* Syn: *dome [volc].* (b) *shield volcano.* (c) A *tumulus* developed on a lava flow.

lava eruption A volcanic phase that is characterized by the emission of lava, with few if any explosive phenomena. Cf: *explosive eruption; mixed eruption.*

lava field A more or less well-defined area that is covered by lava flows. Cf: *ash field; volcanic field.*

lava flow A lateral, surficial outpouring of molten lava from a vent or a fissure; also, the solidified body of rock that is so formed. Syn: *flow [volc]; nappe [volc].*

lava flow-unit A separate, distinct lobe of lava that issues from the main body of a lava flow (Nichols, 1936).

lava fountain A jet of incandescent lava, shot into the air as the magma reaches the surface by the hydrostatic pressure on the liquid and the expansion of gas bubbles forming in it. Fountains usually range from about 10 to 100 m in height, but occasionally reach 300 m. They are characteristic of *Hawaiian-type eruptions.* See also: *curtain of fire.*

lava lake A lake of molten lava, usually basaltic, in a volcanic crater or depression. The term refers to solidified and partly solidified stages as well as to the molten, active lava lake.

lava levee The scoriaceous sheets of lava that overflowed their natural channels and solidified to form a levee, similar to levees formed by an overflowing stream of water.

lava plain A broad stretch of level or nearly level land, usually many hundreds of square kilometers in extent, underlain by a relatively thin succession of lava flows, most of which are basaltic and the product of *fissure eruptions.*

lava plateau A broad, elevated tableland or flat-topped highland, usually many hundreds or thousands of square kilometers in extent, underlain by a thick succession of lava flows, most of which are tholeiitic basalts and the product of fissure eruption. Syn: *basaltic plateau.*

lava rag A bit of scoriaceous material ejected from a volcano.

lava shield A *shield volcano* of the basaltic type.

lava toe One of a series of small, bulbous projections that develop at the front of a moving pahoehoe flow, formed by the breaking-open of the crust and the emergence of fluid lava. Syn: *toe [volc].*

lava tree A half-cylindrical projection above the surface of a lava flow, formed when fluid lava flows against the upstream side of a tree; the tree then burns, forming a *lava-tree mold.* An unusual form, seen only on recent flows.

lava-tree mold A cylindrical hollow in a lava flow formed by the envelopment of a tree by the flow, solidification of the lava in contact with the tree, and disappearance of the tree by burning and subsequent removal of the charcoal and ash. The inside of the mold preserves the surficial features of the tree. See also: *lava tree.*

lava trench A collapsed *lava tube.*

lava tube A hollow space beneath the surface of a solidified lava flow, formed by the withdrawal of molten lava after the formation of the surficial crust. See also: *lava trench; volcanic flow drain.* Syn: *lava cave; lava tunnel.*

lava tunnel *lava tube.*

lavendulan A lavender mineral: $NaCaCu_5(AsO_4)_4Cl \cdot 5H_2O$. Syn: *freirinite.*

lavenite A mineral: $(Na,Ca)_3Zr(Si_2O_7)(O,OH,F)_2$. Cf: *wöhlerite.* Also spelled: *laavenite.*

lavialite A metamorphosed basaltic rock with relict phenocrysts of labradorite in an amphibolitic groundmass. The term was originated by Sederholm in 1899, who named it after Lavia, Finland.

lavrovite A green variety of diopside containing small amounts of vanadium and chromium. Syn: *lavroffite.*

law [geomorph] A Scottish term for a more or less rounded or conical hill or mound. Syn: *low [geomorph].*

law [philos] A formal statement of the invariable and regular manner in which natural phenomena occur under given conditions; e.g. the "law of superposition" or a "law of thermodynamics".

law of acceleration A theory in biology stating that the order of development of a structure or organ is directly related to its importance to the organism.

law of accordant junctions *Playfair's law.*

law of basin areas A general law expressing the direct geometric relation between stream order and the mean basin area of each order in a given drainage basin, originally stated by Schumm (1956, p.606). The law is expressed as a linear regression of logarithm of mean basin area on stream order, the positive regression coefficient being the logarithm of the basin-area ratio.

law of constancy of interfacial angles The statement in crystallography that the angles between corresponding faces on different crystals of one substance are constant. It was first noted by the Danish scientist Nicolaus Steno in 1669. Syn: *constancy of interfacial angles.*

law of constancy of relative proportions *constancy of relative proportions.*

law of correlation of facies A ruling principle in stratigraphy, enunciated by Walther (1893-1894): within a given sedimentary cycle, the same succession of facies that occurs laterally is also present in vertical succession.

law of crosscutting relationships A stratigraphic principle whereby relative ages of rocks can be established: a rock (esp. an igneous rock) is younger than any other rock across which it cuts.

law of equal declivities Where homogeneous rocks are maturely dissected by consequent streams, all hillside slopes of the valleys cut by the streams tend to develop at the same slope angle, thereby producing symmetric profiles of ridges, spurs, and valleys. The principle was formulated by Gilbert (1877, p. 141).

law of equal volumes *Lindgren's volume law.*

law of extralateral rights *apex law.*

law of faunal assemblages A general law of geology: Similar assemblages of fossil organisms (faunas and floras) indicate similar geologic ages for the rocks that contain them.

law of faunal succession A general law of geology: Fossil organisms (faunas and floras) succeed one another in a definite and recognizable order, each geologic formation having a different total aspect of life from that in the formations above it and below it; or, the relative age of rocks can be determined from their fossil content.

law of homonymy A principle in taxonomy stating that any name that is a junior homonym of another name must be rejected and replaced. See also: *homonymy; law of priority.*

law of minimum lateral thrust The statement that the relative displacement of overlying strata to underlying strata surrounding an inclined, concordant intrusion can be expressed as B-A$\cot\Theta$, in which B = the horizontally measured width of the inclined part of the intrusive body, A = the vertically measured width of the horizontal part of the intrusive body, and Θ = the angle of inclination (DuToit, 1920, p. 28).

law of nature A generalization of science, representing an intrinsic orderliness of natural phenomena or their necessary conformity to reason. Syn: *natural law.*

law of original continuity A general law of geology: A water-laid stratum, at the time it was formed, must continue laterally in all directions until it thins out as a result of nondeposition or until it abuts against the edge of the original basin of deposition. The law was first clearly stated by Steno (1669).

law of original horizontality A general law of geology: Water-laid sediments are deposited in strata that are horizontal or nearly horizontal, and parallel or nearly parallel to the Earth's surface. The law was first clearly stated by Steno (1669).

law of priority A principle in taxonomy stating that nomenclature of a taxonomic group is based on priority of publication (ICBN, 1972, p. 16). See also: *priority.*

law of rational indices The statement in crystallography that crystal faces make simple rational intercepts on suitable crystal axes, i.e. the axes of reference or the three axes forming the edges of the unit cell of each crystal lattice. Syn: *Haüy's law; law of rational intercepts.*

law of rational intercepts *law of rational indices.*

law of reflection The statement in physics that the angle between the reflected ray and the normal to the reflecting surface is the same as the angle between this normal and the incident ray, provided the wave travels with the same velocity as the incident wave. See also: *reflection.*

law of refraction The statement in physics that when a wave crosses a boundary between two isotropic substances, the wave normal changes direction in such a manner that the sine of the angle of incidence between wave normal and boundary normal divided by the velocity in the first medium equals the angle of refraction divided by the velocity in the second medium. Syn: *Snell's law.*

law of stream gradients A general law expressing the inverse geometric relation between stream order and the mean stream gradient of a given order in a given drainage basin, originally stated by Horton (1945, p.295).

law of stream lengths A general law expressing the direct geometric relation between stream order and the main stream lengths of each order in a given drainage basin, originally stated by Horton (1945, p.291). The law is expressed as a linear regression of logarithm of mean stream length on stream order, the positive regression coefficient being the logarithm of the stream-length ratio.

law of stream numbers A general law expressing the inverse geometric relation between stream order and the number of streams of each order in a given drainage basin, originally stated by Horton (1945, p.291). The law is expressed as a linear regression of logarithm of number of streams on stream order, the negative regression coefficient being the logarithm of the bifurcation ratio. See also: *number of streams.*

law of superposition A general law upon which all geologic chronology is based: In any sequence of sedimentary strata (or of extrusive igneous rocks) that has not been overturned, the youngest stratum is at the top and the oldest at the base; i.e., each bed is younger that the bed beneath, but older than the bed above it. The law was first clearly stated by Steno (1669).

law of surface relationships A principle developed by Wheeler (1964, p.602-603): "time as a stratigraphic dimension has meaning only to the extent that any given moment in the Earth's history may be conceived as precisely coinciding with a corresponding worldwide lithosphere surface and all simultaneous events either occurring thereon or directly related thereto". The "lithosphere surfaces" (surfaces of deposition or surfaces of erosion) are envisioned as "the only universal physical geologic 'datum' surfaces with direct stratigraphic implication".

law of unequal slopes A stream flowing down the steeper slope of an asymmetric ridge or divide erodes its valley more rapidly than one flowing down the gentler slope, thereby causing the crest of the divide to migrate away from the more actively eroding stream toward the less actively eroding one. The principle was first recognized by Gilbert (1877, p. 140).

law of universal gravitation The statement that every mass particle in the universe attracts every other mass particle with a force directly proportional to the product of the two masses and inversely proportional to the square of the distance between them, the direction of the force being in the line joining the two particles. The law applies only to particles, and not to bodies of finite size. See also: *gravitation.*

lawrencite A green or brown meteorite mineral: $(Fe,Ni)Cl_2$. It occurs as an abundant accessory mineral in iron meteorites.

lawsonite A colorless to grayish-blue orthorhombic mineral: $CaAl_2(Si_2O_7)(OH)_2 \cdot H_2O$.

lawsonite-albite facies A term introduced by Winkler (1967) for rocks formed by *burial metamorphism* at the same temperatures as those of the *lawsonite-glaucophane-jadeite facies* but at lower pressures, e.g. 6000 to 7500 bars.

lawsonite-glaucophane-jadeite facies A term introduced by Winkler (1967) for that part of the *glaucophane-schist facies* formed at 250° to 400°C. The rocks are formed by *burial metamorphism* at very high pressures (above 7500 bars), and the coexistence of lawsonite and glaucophane as index minerals is required. Cf: *lawsonite-albite facies.*

laxite An old name for unconsolidated fragmental rocks.

layer [seis] One of a series of concentric zones or belts of the Earth, delineated by seismic discontinuities. A classification of the interior of the Earth designates layers A to G from the surface inward.

layer [stratig] A general term for any tabular body of rock (igneous, metamorphic, or sedimentary), of ice, or of unconsolidated material, lying in a position essentially parallel to the surface or surfaces on or against which it was formed, and more or less distinctly limited above and below; specif. a *bed* or *stratum* of rock, with no limitation as to thickness.

layer-cake Said of the geologic concept of successive layers of strata, each separated by an unconformity and completely independent of or structurally different from other layers above and below (Levorsen, 1943, p.907-912).

layer depth In the ocean, the depth to the top of the *thermocline;* i.e., to the bottom of the *mixed layer.*

layered intrusion An intrusive body in which there are layers of varying mineralogical composition, e.g. the Bushveld Igneous Complex of South Africa. See also: *banded differentiate.* Syn: *stratiform intrusion.*

layered permafrost Ground consisting of perennially frozen layers alternating with unfrozen layers or *taliks* (Muller, 1947, p.218).

layered series A body of igneous rocks showing banding that simulates the stratification of a sedimentary sequence and having a parallelism of platy minerals that resembles the bedding of sediments (Wager & Deer, 1939, p. 36).

layering [cart] *layer tinting.*

layering [petrology] A tabular succession of the different components (mineralogic, textural, structural) of igneous, sedimentary, or metamorphic rocks, or the formation of layers of material, one upon the other, in a particular rock; e.g. the phenomenon in plutonic rocks resulting from *crystal settling* in magma. Wager & Brown (1967, p. v) suggest that the term "be kept for the high-temperature sedimentation features of igneous rocks", leaving *bedding* and *stratification* for use with sedimentary rocks and *banding* for use with metamorphic rocks. However, "layering" is now preferred for metamorphic rocks as it implies three dimensions rather than two. Cf: *rhythmic layering; phase layering.*

layering [stratig] *bedding [stratig].*

layer silicate *phyllosilicate.*

layer stripping A procedure that removes the effects of an upper crustal layer. It modifies seismic and gravity data to make them appear as though the seismic source and receivers, or the gravity stations, were at the base of the stripped layer and unaffected by it. Syn: *stripping.*

layer structure A type of crystal structure built up by distinct layer units, as in the micas, clay, and graphite. See also: *two-layer structure; three-layer structure.*

layer tint *hypsometric tint.*

layer tinting A method of depicting relief on a map by the distinctive shading or coloring of the areas between contour lines in a manner suggestive of progressive change, so that the pattern of distribution of high and low areas is revealed or emphasized at a glance. Syn: *layering.*

lay of the land *topography.*

layover In *SLAR* images, the geometric displacement of the top of objects toward the near infrared range, relative to their base.

lazarevicite *arsenosulvanite.*

lazuli A syn. of *lapis lazuli.* Also spelled: *lazule.*

lazulite An azure-blue to violet-blue mineral: $(Mg,Fe^{+3})Al_2(PO_4)_2(OH)_2$. It is isomorphous with scorzalite, and occurs in small masses or in monoclinic crystals. Syn: *blue spar; false lapis; berkeyite.* Not to be confused with *lazurite.*

lazurite An intense blue or violet-blue feldspathoid mineral of the sodalite group: $(Na,Ca)_{7-8}(Al,Si)_{12}(O,S)_{24}[SO_4,Cl_2,(OH)_2]_2$. It is the principal constituent of *lapis lazuli.* See also: *ultramarine.* Not to be confused with *lazulite.*

leachate A solution obtained by leaching; e.g. water that has percolated through soil containing soluble substances and that contains certain amounts of these substances in solution. Syn: *lixivium.*

leached Said of a soil in which *leaching* has taken place. Partial syn: *decalcified.*

leached capping *gossan.*

leach hole *sinkhole.*

leaching (a) The separation, selective removal, or dissolving-out of soluble constituents from a rock or orebody by the natural action of percolating water. (b) The removal in solution of nutritive or harmful constituents (such as mineral salts and organic matter) from an upper to a lower soil horizon by the action of percolating water, either naturally (by rainwater) or artificially (by irrigation). Cf: *eluviation.* (c) The extraction of soluble metals or salts from an ore by means of slowly percolating solutions; e.g. the separation of gold by treatment with a cyanide solution.—Syn: *lixiviation.*

lead [eco geol] (a) A syn of *lode [eco geol].* (b) A placer deposit. See also: *back lead; blue lead; deep lead.*—Pron: leed.

lead [ice] Any *fracture [ice], water opening,* or long narrow strip of ocean water through sea ice (esp. pack ice), navigable by surface vessels, and sometimes covered by young ice; wider than a *lane.* Cf: *polynya.* Syn: *channel [ice]; lane.* Pron: leed.

lead [mineral] (a) A soft, heavy, malleable and ductile isometric mineral, the native metallic element Pb. It is silvery when freshly cut, but tarnishes readily in moist air to dull gray. Lead rarely occurs in native form, being found mostly in combination (as in galena, cerussite, and anglesite). (b) A term sometimes applied to graphite.—Pron: *led.*

lead [seis] A syn. of *time lead.* Pron: leed.

lead [streams] An open watercourse, usually artificial, leading to or from a mill, mine, reservoir, etc. Syn: *leat.*

lead-alpha age method A method of calculating an age in years by spectrographically determining the total lead content and the alpha-particle activity of a zircon, monazite, or xenotime concentrate, the alpha-particle activity representing the uranium-thorium content. This age method is less precise than the potassium-argon or rubidium-strontium age methods, and is best used for rocks younger than Precambrian. Syn: *Larsen method; lead-alpha dating.*

lead-alpha dating *lead-alpha age method.*

lead glance *galena.*

leadhillite A yellowish or greenish-white monoclinic mineral. $Pb_4(SO_4)(CO_3)_2(CH)_2$. It is dimorphous with susannite.

leading edge The frontal edge of a thrust sheet (Dahlstrom, 1970) with respect to a given stratigraphic unit. Syn: *hanging-wall cutoff.* Cf: *trailing edge.*

leading stone *lodestone.*

lead-isotope age *lead-lead age.*

lead-lead age An age in years calculated from the ratio of lead-207 to lead-206, a by-product of the *uranium-thorium-lead age method.* Syn: *lead-isotope age.*

lead line A weighted line of wire or cord that is used in *sounding.* The line is lowered from a ship until it reaches bottom; then its length is measured. Syn: *sounding line.*

lead ocher A yellow or red, scaly or earthy lead monoxide; specif. massicot and litharge. Syn: *plumbic ocher.*

lead ratio The ratio of one lead isotope to another. The ratios normally encountered are $^{206}Pb/^{204}Pb$, $^{207}Pb/^{204}Pb$, $^{208}Pb/^{204}Pb$, and $^{207}Pb/^{206}Pb$.

lead spar (a) *cerussite.* (b) *anglesite.*

lead-uranium age method *uranium-lead age method.*

lead-uranium ratio The ratio of lead-206 to uranium-238 and/or lead-207 to uranium-235, formed by the radioactive decay of uranium within a mineral. The ratios are frequently used as part of the *uranium-thorium-lead age method.*

lead vitriol *anglesite.*

lead-210 age method A method of age determination based on the activity of ^{210}Pb (half-life = 22.2 years), and comparing the measured activity with the activity of present-day samples. It has been applied to studies of past precipitation rates in Antarctica.

leaf The principal photosynthetic appendage of higher plants. Typically leaves have a *petiole* or stalk and a veined widened portion or *lamina.*

leaf clay *book clay.*

leaf gap A parenchymatous opening into a stele, left by the departure of a leaf trace (Cronquist, 1961, p.877).

leaf gold Gold occurring naturally as thin flakes or sheets; not to be confused with man-made gold leaf.

leaflet One of the parts of a *compound leaf.*

leaf mold A general term for an accumulation on the soil surface that is composed chiefly of partially decomposed vegetable matter (usually fallen leaves and the remains of herbaceous plants). It is a constituent of the *forest floor.* Cf: *litter; duff.*

leaf peat *paper peat.*

leaf primordium The *meristem* within a bud which develops into a leaf.

leaf scar A scar on a twig following the abscission of a leaf. See also: *bundle scar.*

leaf trace Vascular tissue extending from a stem into a leaf. Cf: *branch trace.*

leafy Pertaining to a sedimentary structure resembling a leaf, or said of a rock containing such a structure; e.g. "leafy post", a thinly laminated sandstone containing micaceous layers (as in Durham, England).

league (a) Any of various linear units of distance, ranging from about 2.42 to 4.6 statute miles; esp. "land league" (an English land unit equal to 3 statute miles) and "marine league" (a marine unit equal to 3 nautical miles). (b) Any of various units of land area equal to a square league; esp. an old Spanish unit for the area of a tract 5000 varas square, equal to 4428.4 acres (1792.1 hectares) in early Texas land descriptions or equal to 4439 acres (1796 hectares) in old California surveys.

leaked fossil A fossil deposited as a result of a *stratigraphic leak.*

leaking mode A seismic wave that is imperfectly trapped, so that its energy leaks or escapes across a layer boundary, causing some attenuation.

leaky aquifer A confined aquifer whose confining beds will conduct significant quantities of water into or out of the aquifer. Cf: *leaky confining bed.*

leaky confining bed A confining bed through which water can move into or out of the adjacent aquifer. Cf: *aquitard; leaky aquifer.*

lean *low-grade.*

lean cannel coal Cannel coal that is low in hydrogen and transitional to bituminous coal in rank. Cf: *subcannel coal.* Syn: *semicannel coal.*

lean clay A clay of low to medium plasticity owing to a relatively high content of silt or sand. Ant: *fat clay.*

leaping The sudden and radical *shifting* of a divide from one position to another, as where the valley system drained by a captured stream is "transferred and added in a moment to that of the master stream" (Cotton, 1958, p. 69). Cf: *creeping.*

least squares *method of least squares.*

least-time path *minimum-time path.*

leat An English dialectal syn. of *lead [streams].*

leatherstone *mountain leather.*

lebensspur A sedimentary structure left by a living organism; a fossil form is a *trace fossil.* The term is also applied to a Holocene track or burrow. Etymol: German *Lebensspur,* "life mark". Pl: *lebensspuren.*

lechatelierite Naturally fused amorphous silica, occurring in fulgurites and impact craters as a vitreous or glassy product formed by the melting of quartz sand as a result of lightning or of the heat generated by the impact of a meteorite; a natural *silica glass* formed at high temperatures. Also spelled: *lechateliérite.*

Le Chatelier's rule The statement in chemistry that, if conditions of a system that is initially at equilibrium are changed, the equilibrium will shift in such a direction as to tend to restore the original conditions.

lechosos opal A variety of precious opal exhibiting a deep-green play of color; esp. a Mexican opal exhibiting emerald-green play of color and flashes of carmine, dark violet, dark blue, and purple.

lecontite A colorless mineral found in bat guano: $(NH_4,K)Na(SO_4)\cdot 2H_2O$.

lectoparatype *paralectotype.*

lectostratotype A stratotype selected later in the absence of an adequately designated original stratotype (ISG, 1976, p. 26). Etymol: Latin *lectus,* "bed", + stratotype.

lectotype A *syntype,* chosen if needed after the original description, to take the place of the *holotype* (Frizzell, 1933, p. 655). Cf: *holotype; neotype.*

ledge (a) A narrow shelf or projection of rock, much longer than wide, formed on a rock wall or cliff face, as along a coast by differential wave erosion of softer rocks. (b) A rocky outcrop; solid rock. (c) An underwater ridge of rocks, esp. near the shore; also, a nearshore reef. (d) A quarry exposure or natural outcrop of a mineral deposit.

Ledian European stage: Middle Eocene (equivalent to Middle Lutetian).

ledikite A clay mineral: $K(Fe,Mg)_3(Si,Al)_8O_{20}(OH)_4$. It is the trioctahedral analogue of illite.

ledmorite A melanite-bearing *nepheline syenite* or *malignite* with more pyroxene and less melanite than *borolanite* with which it occurs (Johannsen, 1939, p. 262). Its name, given by Shand in 1910, is derived from the Ledmore River, Scotland. Not recommended usage.

lee n. The part or side of a hill or prominent object that is sheltered or turned away from the wind.—adj. Said of a side or slope of a hill or knob that faces away from an advancing glacier or ice sheet; facing the downstream side of a glacier and relatively protected from its abrasive action. Ant: *stoss.*

Lee configuration A configuration employing electrodes, the outer two of which are the current and the inner three of which are the potential electrodes. Syn: *partitioning method.*

lee dune A general term for a dune formed to the leeward of a source of loose sand or of an obstacle of any kind, and generally under a wind of constant direction. See also: *umbracer dune; umbrafon dune.*

leegte *laagte.*

lee shore A shore crossed by wind from the land and thereby protected from strong wave action. Ant: *weather shore.*

lee-source dune *umbrafon dune.*

leeuwfonteinite A syn. of *hatherlite.* This name was suggested by Brouwer in 1903 because the rock does not occur at Hatherley factory but at Leeuwfontein, Transvaal, South Africa. The rock is a *syenite* rich in anorthoclase and containing sodic amphibole. Not recommended usage.

leeward adj. Said of the side or slope (as of a hill or prominent rock) sheltered or located away from the wind; downwind.—n. The *lee* side, or the lee direction.—Ant: *windward.*

leeward reef In a *reef complex* or *reef tract,* a reef on the side opposite that from which the prevailing winds blow. Leeward reefs tend to be less well-developed than *windward reefs,* as on Eniwetok Atoll. See also: *outer reef; inner reef.*

lee wave An *internal wave* occurring on the downstream side of a submarine ridge.

left bank The bank of a stream situated to the left of an observer who is facing downstream.

left-handed [cryst] Said of an optically active crystal that rotates the plane of polarization of light to the left. Cf: *right-handed.* Syn: *levorotatory.*

left-handed [paleont] *sinistral.*

left-handed separation *left-lateral separation.*

left-lateral fault A fault on which the displacement is *left-lateral separation.* Syn: *sinistral fault; left-lateral slip fault; left-slip fault.*

left-lateral separation Displacement along a fault such that, in plan view, the side opposite the observer appears displaced to the left. See also: *left-lateral fault.* Cf: *right-lateral separation.* Syn: *left-handed separation.*

left-lateral slip fault *left-lateral fault.*

left-slip fault *left-lateral fault.*

left valve The valve lying on the left-handed side of a bivalve mollusk when the shell is placed with the anterior end pointing away from the observer, the commissure being vertical and the hinge being uppermost. Abbrev: LV. Ant: *right valve.*

leg On a seismogram, a single cycle in a wave train. Cf: *leggy.*

legal geology *forensic geology.*

legend A brief explanatory list of the symbols, cartographic units, patterns (shading and color hues), and other cartographic conventions appearing on a map, chart, or diagram. On a geologic map, it shows the sequence of rock units, the oldest at the bottom and the youngest at the top. The legend formerly included a textual inscription of, and the title on, the map or chart. Syn: *explanation; key.*

leggy Said of a wave train that contains a number of cycles or *legs.* Syn: *tailing.*

legrandite A yellow to nearly colorless mineral: $Zn_2(AsO_4)(OH)\cdot H_2O$.

legume A dry, dehiscent fruit that is produced from a single carpel that splits along two seams, as in pods of peas and beans.

lehiite A white mineral: $(Na,K)_2Ca_5Al_8(PO_4)_8(OH)_{12}\cdot 6H_2O$ (?).

lehm A term used in Alsace, France, for *loess.* Etymol: German *Lehm,* "loam".

lehmanite An obsolete term, proposed by Pinkerton but never widely used, for an igneous rock containing feldspar and quartz (Johannsen, 1939, p. 262).

leidleite A glassy variety of *dacite* or *rhyodacite* containing microlites, not phenocrysts, of calcic plagioclase and pyroxene, with accessory apatite and opaque oxides. Named by Thomas and Bailey in 1915 for Glen Leidle, Mull, Scotland. Not recommended usage.

leifite A colorless mineral: $Na_2(Si,Al,Be)_7(O,OH,F)_{14}$.

leightonite A pale-blue mineral: $K_2Ca_2Cu(SO_4)_4\cdot 2H_2O$.

Leintwardinian European stage: Upper Silurian (above Bringewoodian, below Whitecliffian).

leiosphaerid A thin-walled, more or less spherical body of probable algal relationship, characterized by the genus *Leiosphaeridia,* and usually referred to the acritarchs. It is mostly Ordovician to Silurian in age.

leiteite A monoclinic mineral: $(Zn,Fe)As_2O_4$.

lekolith A term proposed by Coats (1968, p.71) for "a mass of extrusive igneous rock more or less equant in plan, with a nearly level upper surface, commonly a lower surface determined by the shape of the basin that it filled, and a diameter greater than its depth"; e.g. a mass formed by a congealed lava lake. Etymol: Greek *lekos,* "dish", + *lithos,* "stone".

Lemberg's stain A test used to distinguish calcite from dolomite. A solution of logwood in an aqueous solution of aluminum chloride is used to stain the minerals by boiling; calcite and aragonite become violet, whereas dolomite does not change color. Cf: *Meigen's solution.*

Lemnian bole A gray to yellow or red clay obtained from the island of Lemnos in the Aegean Sea, once used for medicinal purposes. Syn: *Lemnian earth; terra Lemnia.*

lemoynite A mineral: $(Na,K)_2CaZr_2Si_{10}O_{26}\cdot 5\text{-}6H_2O$.

Lemuria An imaginary continent, also known as Amosnuria, beloved by science-fiction writers, that is alleged to have occupied most of the central Pacific Ocean until historic time, when it sank, leaving only the Pacific islands as tiny remnants. The dispersal of Polynesian peoples and cultures is supposed to have been facilitated by the existence of Lemuria, but this dispersal is easily explained otherwise. Geologically, the existence of such a continent, either modern or ancient, is impossible.

lenad (a) A group name for the *feldspathoid* standard minerals. (b) A mnemonic term for leucite and nepheline.—Etymol: *le*ucite + *n*epheline + *ad.*

Lenan European stage: Middle Cambrian (above Aldanian, below Amgan).

lengenbachite A steel-gray mineral: $Pb_6(Ag,Cu)_2As_4S_{13}$.

length [lake] The shortest distance, through the water or on the water surface, between the most distant points on a lake shore.

length [paleont] (a) The distance from the most posterior part (normally the umbo) of a brachiopod valve to the farthest point on the anterior margin, measured on or parallel with the commissural plane in the plane of symmetry. (b) The distance between two planes perpendicular to the hinge axis of a bivalve-mollusk shell and just touching the anterior and posterior extremities of the shell. Cf: *height.*

length distortion The ratio of scaled map distance to true ground distance.

length of overland flow Distance along the ground surface, projected to the horizontal, of nonchannel flow from a point on the drainage divide to a point of contact with a definite stream channel; the length is always measured at right angles to the contour lines in the drainage basin. Symbol: L_g. Cf: *critical length.* Syn: *slope length.*

lennilite (a) A green variety of feldspar (orthoclase) from Lenni Mills, Delaware County, Penna. (b) A vermiculite mineral.

lenoblite A mineral: $V_2O_4\cdot 2H_2O$.

lens n. A geologic deposit bounded by converging surfaces (at least one of which is curved), thick in the middle and thinning out toward the edges, resembling a convex lens. A lens may be double-convex or plano-convex. See also: *lentil.* —v. To disappear laterally in all directions; e.g. a unit is said to "lens out" within a mapped area.

lensing The thinning-out of a stratum in one or more directions; e.g. the disappearing laterally of a stratum.

lentelliptical Lenticular and elliptical; e.g. said of a lens-shaped radiolarian shell with elliptical outline.

lenticel In stems and other plant parts, a pore through which the exchange of gases occurs. In a woody stem, lenticels occur in the bark (Fuller & Tippo, 1954, p. 962).

lenticle (a) A large or small lens-shaped stratum or body of rock; a *lentil.* (b) A lens-shaped rock fragment of any size.

lenticular (a) Resembling in shape the cross section of a lens, esp. of a double-convex lens. The term may be applied, for example, to a body of rock, a sedimentary structure, or a mineral habit. (b) Pertaining to a stratigraphic lens or lentil.—Syn: *lentiform.*

lenticular bedding A form of interbedded mud and ripple cross-laminated sand, in which "the ripples or lenses are discontinuous not only in the vertical but also more or less in the horizontal direction" (Reineck & Wunderlich, 1968, p. 102). Cf: *flaser structure; wavy bedding.*

lenticule A small *lentil.*

lentiform *lenticular.*

lentil (a) A minor rock-stratigraphic unit of limited geographic extent, being a subdivision of a formation and similar in rank to a *member,* and thinning out in all directions; "a geographically restricted member that terminates on all sides within a formation" (ACSN, 1961, art.7). Term originated by Keith (1895). Cf: *tongue [stratig].* (b) A lens-shaped body of rock, enclosed by strata of different material; a geologic *lens.* See also: *lenticule; lenticle.*

lentil ore *liroconite.*

lentocapillary point An obsolete term for the moisture content at which water movement through the soil becomes slow (Jacks et al., 1960, p. 54).

Leonardian North American series: Lower Permian (above Wolfcampian, below Guadalupian).

leonhardite A zeolite mineral: $Ca_2Al_4Si_8O_{24} \cdot 7H_2O$. It is a variety of laumontite altered by partial loss of water. Not to be confused with *leonhardtite.*

leonhardtite A synonym of *starkeyite.* Not to be confused with *leonhardite.*

leonite A colorless, white, or yellowish monoclinic mineral: $K_2Mg(SO_4)_2 \cdot 4H_2O$.

leopardite An igneous rock composed of small quartz phenocrysts in a microgranitic groundmass of quartz, orthoclase, albite, and mica. Iron and manganese hydroxide stains give the rock a characteristic streaked or spotted appearance. Not recommended usage.

leopoldite *sylvite.*

leperditiid Any ostracode belonging to the order Leperditicopida, characterized by a large, strongly calcified, thick-walled shell that is usually smooth, but sometimes finely ornamented to nodose, and that has a long straight hinge, large muscle-scar pattern, and secondary shell layers. Leperditiids are commonly four or five times larger than other ostracodes. Range, Lower Ordovician (or possibly Upper Cambrian) to Upper Devonian.

lepidoblastic Pertaining to a *homeoblastic* type of texture of a foliated or schistose rock that is due to the parallel orientation during recrystallization of minerals with a flaky or scaly habit, e.g. mica, chlorite.

lepidocrocite A ruby-red or blood-red to reddish-brown mineral: γ-FeO(OH). It is trimorphous with akaganeite and goethite, and is associated with limonite in iron ores.

lepidodendrid n. An arborescent lycopsid of the family Lepidodendraceae, well known from Carboniferous deposits.—adj. Pertaining to the genus *Lepidodendron* or to related genera.—Cf: *sigillarian.*

lepidolite A mineral of the mica group: $K(Li,Al)_3(Si,Al)_4O_{10}(F,OH)_2$. It commonly occurs in rose or lilac-colored masses made up of small scales, as in pegmatites. Syn: *lithium mica; lithia mica; lithionite.*

lepidolith A thin, apparently homogeneous, elliptical coccolith; e.g. a surface plate of the coccolithophorid *Thorosphaera flagellata.*

lepidomelane A black variety of biotite with a high content of ferric iron. Syn: *iron mica.*

lepidote Said of a plant part that is covered with fine scales. Cf: *squamose.*

lepisphere A micron-sized spheroidal diagenetic body, usually composed of a silica mineral, with radial crystal orientation and scaly crystal terminations on the outer surface.

Lepospondyli A subclass of amphibians, characterized by unitary vertebral centra and distinctive skull pattern, and by restriction to the Paleozoic.

leptite A Fennoscandian term for a granular quartzofeldspathic metamorphic rock formed by regional metamorphism of the highest grade (*granulite facies*). The terms *granulite (b), leptynite,* and *hälleflinta* are approximately synonymous.

leptochlorite (a) A group name for chlorites of indistinct crystallization. (b) A group name for chlorites with a composition corresponding to: $(Mg,Fe^{+2},Al)_n(Si,Al)_4O_{10}(OH)_8$, where n is less than 6 (Hey, 1962, p. 495).—Cf: *orthochlorite.*

leptogeosyncline An oceanic trough containing only a minor sedimentary accumulation and associated with volcanism (Trümpy, 1955).

leptokurtic (a) Said of a frequency distribution that has a concentration of values about its mean greater than for the corresponding normal distribution. (b) Said of a narrow frequency distribution curve that is more peaked than the corresponding normal distribution curve.—Cf: *platykurtic; mesokurtic.*

leptoma A thin region of exine situated at the distal pole of a pollen grain and usually functioning as the point of emergence of the pollen tube. See also: *pseudopore [palyn].*

leptomorphic An obsolete syn. of *xenomorphic.*

leptopel Finely particulate, mainly colloidal organic and inorganic matter (such as silicates, hydrous oxides, or insoluble carbonates) occurring suspended in natural waters (Fox, 1957, p. 383). Cf: *pelogloea.*

leptosporangiate A fern of the subclass Leptosporangiatae, in which the sporangium wall is only one cell-layer thick and the sporangial structure arises from only the outer segment of an initial cell (Melchior and Werdermann, 1954, p. 287).

leptothermal Said of a hydrothermal mineral deposit formed at temperature and depth conditions intermediate between *mesothermal [eco geol]* and *epithermal;* also, said of that environment. Cf: *hypothermal; xenothermal; telethermal.*

leptynite A French term for *leptite.*

leptynolite A fissile or schistose variety of hornfels containing mica, quartz, and feldspar, with or without accessories such as andalusite and cordierite. The term was originated by Cordier in 1868 (Holmes, 1928, p.139). Cf: *cornubianite.*

lermontovite A mineral: $(U,Ca,Ce)_3(PO_4)_4 \cdot 6H_2O$(?).

lesserite *inderite.*

lestiwarite A *syenite-aplite* composed chiefly of microperthite, acmite, arfvedsonite, and accessory sphene. Its name, given by Brögger in 1898, is derived from Lestiware, Kola Peninsula, U.S.S.R. Not recommended usage.

letdown The natural lowering (in the stratigraphic section) of slabs and fragments of a resistant formation by weathering and erosion of more vulnerable underlying rock; e.g. the stage of brecciation in which bedding is but little disturbed (Landes, 1958, p. 125).

letovicite A mineral: $(NH_4)_3H(SO_4)_2$.

lettsomite *cyanotrichite.*

leucaugite A white or grayish variety of augite resembling diopside: $CaMgSi_2O_6$.

leuchtenbergite A mineral (clinochlore) of the chlorite group, often resembling talc and containing little or no iron.

leucite A white or gray mineral of the feldspathoid group: $KAlSi_2O_6$. It is an important rock-forming mineral in alkalic rocks (esp. lavas), and usually occurs in trapezohedral crystals with a glassy fracture. Syn: *amphigene; grenatite; white garnet; Vesuvian garnet; vesuvian.*

leucitite A fine-grained or porphyritic extrusive or hypabyssal igneous rock chiefly composed of pyroxene (esp. titanaugite) and leucite, with little or no feldspar and without olivine.

leucitohedron *trapezohedron.*

leucitolith An extrusive rock composed almost entirely of leucite.

leucitophyre A porphyritic extrusive rock composed chiefly of leucite, nepheline, and clinopyroxene. Cf: *hauynophyre.*

leuco- A prefix meaning "light-colored"; in the *IUGS classification,* it is used to designate a rock that is more felsic than the specified range, e.g. leucogabbro. Cf: *mela-.*

leucochalcite *olivenite.*

leucocratic Light-colored; applied to light-colored igneous rocks that are relatively poor in mafic minerals. The percentage of mafic minerals necessary for a rock to be classified as leucocratic varies among petrologists, but is usually given as less than 30 to 37.5 percent. Cf: *melanocratic; mesocratic.* Noun: *leucocrate.* Syn: *light-colored.*

leucogabbroid In the *IUGS classification,* a plutonic rock satisfying the definition of *gabbroid,* and in which pl/(pl+px+ol) is between 65 and 90.

leucon A sponge or sponge larva in which the flagellated chambers are connected to both exhalant and inhalant canals and do not open directly either to the spongocoel or to the exterior except through a canal. Cf: *ascon; sycon.* See also: *rhagon.* Adj: *leuconoid.*

leucophanite A glassy greenish to pale-yellow mineral: $(Na,Ca)_2BeSi_2(O,F,OH)_7$. Syn: *leucophane.*

leucophoenicite A light purplish-red mineral: $Mn_7Si_3O_{12}(OH)_2$.

leucophosphite A white mineral: $K_2Fe_4(PO_4)_4(OH)_2 \cdot 9H_2O$.

leucophyre A term originally applied to altered *diabase* in which the feldspar has been altered to saussurite, kaolin, and chlorite. This usage is obsolete, but the term is occasionally used for a light-colored hypabyssal rock, being the antithesis of *lamprophyre.* Not recommended usage.

leucophyride A term used in the field for any light-colored porphyritic igneous rock with a fine-grained groundmass. Cf: *felsite; felsitoid.* Not recommended usage.

leucopyrite *loellingite.*

leucosome The light-colored part of a migmatite, usually rich in quartz and feldspar (Mehnert, 1968, p. 355). Cf: *melasome.*

leucosphenite A white to grayish-blue monoclinic mineral: $BaNa_4Ti_2B_2Si_{10}O_{30}$.

leucotephrite (a) As defined by Fouqué and Michel-Lévy in 1879, a *tephrite* containing leucite as the only feldspathoid. (b) In current usage (Sørensen, 1974, p. 568), the term means "light-colored tephrite". Not recommended usage.

leucoxene A general term for fine-grained, opaque, whitish alteration products of ilmenite, commonly consisting mostly of rutile and partly of anatase or sphene, and occurring in some igneous rocks. The term has also been applied to designate a variety of sphene.

leumafite A general term for igneous rocks composed of *leu*cite and *maf*ic minerals, proposed by Hatch, Wells, and Wells in 1961. Syn: *leucitite; ugandite.*

leurodiscontinuity A term proposed by Sanders (1957, p.295) for an unconformity characterized by a regular surface. Cf: *trachydiscontinuity.* Etymol: Greek *leuros,* "smooth, even", + discontinuity.

levee [marine geol] An embankment of sediment, bordering one or both sides of a submarine canyon, fan valley, or deep-sea channel. It is similar to a river-channel levee in the subaerial environment.

levee [streams] (a) *natural levee.* (b) An artificial embankment built along the bank of a watercourse or an arm of the sea, to protect land from inundation or to confine streamflow to its channel. Cf: *floodwall.* Syn: *earth dike.* (c) A landing place along a river; a pier or quay. (d) *mudflow levee.*—Etymol: French *levée.*

levee delta A delta having the form of a long narrow ridge, resembling a natural levee (Dryer, 1910).

levee-flank depression *backswamp depression.*

levee lake A lake formed by a *natural levee* that acts as a barrier or enclosure for holding water. See also: *lateral levee lake; delta levee lake.*

levee ridge The elevated strip of land on which a river flows, produced by the building-up of the stream bed and the natural levees on each side.

level [geog] n. Any large expanse of relatively flat, usually low-lying country, unbroken by noticeable elevations or depressions; specif. any alluvial tract of recent formation, such as the Bedford Level in Lincolnshire, England.

level [surv] n. (a) A *leveling instrument.* (b) A device or attachment for finding a horizontal line or plane or for adjusting an instrument to the horizontal; specif. a *spirit level.* (c) A measurement of the difference of altitude of two points on the Earth's surface by means of a level.—v. To find the heights of different points by means of a level.

level [water] (a) An open reach of water in a stream or canal, such as between two canal locks. (b) The elevation of the surface of a body of water; a water table.

level fold *nonplunging fold.*

level ice Sea ice that has not been deformed, displays a flat surface, and typically occurs in undisturbed waters. Ant: *pressure ice.*

leveling The operation of determining the comparative altitude of different points on the Earth's surface, usually by sighting through a leveling instrument at one point to a level rod at another point. Also, the finding of a horizontal line or the establishing of grades (such as for a railway roadbed) by means of a level. See also: *spirit leveling; direct leveling; indirect leveling.* Also spelled: *levelling.*

leveling instrument An instrument for establishing a horizontal line of sight, usually by means of a spirit level or a pendulum device; e.g. a *surveyor's level* and a *pendulum level.* It is used, with a level rod, to determine differences in elevation between two widely separated points on the Earth's surface. Syn: *level [surv].*

leveling rod A syn. of *level rod.* Also known as a "leveling pole" or "leveling staff".

leveling screw One of three or more adjusting screws for bringing an instrument (such as a surveyor's level) to the horizontal.

levelman A surveyor who operates a leveling instrument.

level of compensation *depth of compensation [tect].*

level of saturation *water table.*

level of zero amplitude The maximum depth below the Earth's surface reached by seasonal temperature changes.

level rod A straight rod or bar, with a flat face graduated in plainly visible linear units with zero at the bottom, used in measuring the vertical distance between a point on the Earth's surface and the line of sight of a leveling instrument that has been adjusted to a horizontal position. It is usually made of metal or well-seasoned wood. See also: *target rod; self-reading leveling rod.* Syn: *rod; leveling rod; surveyor's rod.*

level surface *equipotential surface.*

level trier An apparatus for use in measuring the angular value of the divisions of a spirit level.

leverrierite A discredited name for a clay mineral known to be kaolinite, or a mixture of alternating plates of kaolinite and muscovite, or a mixture of kaolinite and illite.

levorotatory *left-handed.*

levynite A white or light-colored zeolite mineral: $(Ca,Na_2,K_2)_3(Al_6Si_{12})O_{36}\cdot 18H_2O$. Syn: *levyne; levyine; levyite.*

Lewisian Precambrian gneisses of the Northwest Highlands of Scotland; the oldest rocks in Britain. Syn: *Hebridean.*

lewisite A mineral: $(Ca,Fe,Na)_2(Sb,Ti)_2O_7$. It is a titanian romeite.

lewistonite A white mineral of the apatite group: $(Ca,K,Na)_5(PO_4)_3(OH)$. It is a potassium-rich variety of hydroxylapatite. Cf: *dehrnite.*

lexicon An alphabetic compilation of geologic names, accompanied by formal definitions that state the lithology, thickness, age, underlying and overlying formations, type locality, and original reference; e.g. the "Wilmarth lexicon" containing 13,090 names (Wilmarth, 1938) and the "Keroher lexicon" containing 14, 634 names (Keroher et al., 1966).

Lg wave A short-period, higher-mode *surface wave,* with a group velocity of about 3.5 km/sec, that travels over long paths in the continental crust only. The "g" refers to the granitic layer. Cf: *Rg wave.*

lherzite A *hornblendite* composed chiefly of brown hornblende, with minor amounts of biotite, ilmenite, and occasionally garnet. The name, given by Lacroix in 1917, is from Lherz in the Pyrenees. Not recommended usage.

lherzolite (a) In the *IUGS classification,* a plutonic rock with M equal to or greater than 90, ol/(ol+opx+cpx) between 40 and 90, and both opx/(ol+cpx+opx) and cpx/(ol+cpx+opx) greater than 5. (b) *Peridotite* composed chiefly of olivine, orthopyroxene, and clinopyroxene, in which olivine is generally most abundant; a two-pyroxene peridotite. Cf: *bielenite.*—The name, dating from 1797, is for Lherz in the Pyrenees.

Lias Middle European series: Lower Jurassic (above Triassic, below Dogger). Syn: *Liassic.*

lias A bluish or whitish, compact, argillaceous limestone or cement rock, typically interbedded with shale or clay; esp. such a limestone quarried in Somerset and other parts of SW England. Syn: *lyas.*

Liassic *Lias.*

liberite A mineral: Li_2BeSiO_4.

libethenite An olive-green to dark-green orthorhombic mineral: $Cu_2(PO_4)(OH)$.

libolite A variety of *albertite* from Angola.

libration The small, angular change in the face that a body, e.g. the Moon, presents toward the Earth. Only a tiny part is due to dynamic rotational motion (physical libration). In the case of the Moon, it is due primarily to the fact that although the Moon is in synchronous rotation, its rotation is uniform while its rate of revolution varies due to orbital eccentricity, producing longitudinal geometric libration; and also the fact that its rotational axis is not exactly perpendicular to the plane of its orbit (producing latitudinal geometric libration).

libriform Said of wood fibers that are thick-walled and elongate, and have simple pits.

librigena The *free cheek* of a trilobite. Syn: *librigene.*

Libyan Desert glass Silica glass from the Libyan Desert, possibly a tektite or an impactite.

lichen A thallophytic plant of the subdivision Lichenes that is composed of a fungus and an alga living in symbiotic relationship. The alga is protected by the fungus, which in turn relies upon the alga for the production of food (Melchior & Werdermann, 1954, p. 204).

lichenometry Measurement of the diameter of lichens growing on exposed rock surfaces as a method of dating geomorphic features. Cf: *fossil geochronometry.*

lichen polygon A highly specialized *vegetation polygon,* in which the pattern appears to be confined to thick reindeer moss, as in northern Quebec (Rousseau, 1949, p.50). The polygon sides are 30-50 cm in length.

lick *salt lick.*

lido (a) An Italian term for a barrier beach; e.g. the one protecting the lagoon of Venice. (b) A bathing beach at a seaside resort, but now extended to include those at freshwater and artificial-lake resorts. Type example: the Lido near Venice. Syn: *plage.*

lie A British term for the disposition of topographic features or for the slope of the land surface.

liebenbergite A mineral of the olivine group: $(Ni,Mg)_2SiO_4$.

liebenerite A variety of pinite containing alkalies, iron, and calcium.

liebigite An apple-green or yellow-green mineral: $Ca_2U(CO_3)_4 \cdot 10H_2O$. It occurs as secondary concretions or coatings. Syn: *uranothallite.*

Liesegang banding *Liesegang rings.*

Liesegang rings Secondary, nested rings or bands caused by rhythmic precipitation within a fluid-saturated rock. Syn: *Liesegang banding.*

lievrite *ilvaite.*

life assemblage *biocoenosis.*

life cycle The phases, changes, or stages that an individual organism passes through during its lifetime. Syn: *ontogeny.*

lift A slight rise or elevation of the ground.

lift joint A horizontal tension joint in massive rock such as granite, probably formed as a result of reduction of load pressure during quarrying; a type of *strain break.*

ligament (a) A tough structure of connecting tissue in an animal; esp. a horny elastic band in bivalve mollusks connecting the valves of the shell dorsally along a line adjacent to the umbones and acting as a spring to open the valves when the adductor muscles relax. (b) In the vertebrates, a fibrous band or cord of tissue connecting skeletal elements, usually without direct involvement of muscle.

ligamentary articulation A type of articulation of crinoid ossicles effected solely by ligaments but sometimes supplemented by calcareous deposition. Cf: *muscular articulation.*

ligament field A concave or flat part of a crinoid articular face for the attachment of ligaments.

ligament groove A narrow depression in the cardinal area of a bivalve mollusk for the attachment of the fibers of a ligament. Cf: *ligament pit.*

ligament pit (a) A relatively broad depression in the cardinal area of a bivalve mollusk for the attachment of the ligament. Cf: *ligament groove.* (b) A generally steep-sided small depression in a crinoid dorsal-ligament fossa adjoining the center of the transverse ridge.

light-colored Said of a rock-forming mineral that is light in color and generally also light in weight; also, said of the rock such minerals form; *leucocratic.* By convention, light-colored aphanites include those that are white, light and medium gray, yellow, light and medium green, red, purple, and brown. Cf: *dark-colored.*

lighthouse A term used in Kentucky for a *natural bridge.*

light mineral (a) A rock-forming mineral of a detrital sedimentary rock, having a specific gravity lower than a standard (usually 2.85); e.g. quartz, feldspar, calcite, dolomite, muscovite, feldspathoids. Cf: *heavy mineral.* (b) A *light-colored* mineral.

lightning stone *fulgurite.*

lightning tube A tubular *fulgurite.*

light oil Crude oil that has a high *API gravity* or *Baumé gravity.* Cf: *heavy oil.*

light red silver ore *proustite.*

light ruby silver *proustite.*

lightweight aggregate An *aggregate* with a relatively low specific gravity, e.g. pumice, volcanic cinders, expanded shale, foamed slag, or expanded perlite or vermiculite.

lightweight concrete A concrete made with *lightweight aggregate.*

lignilite Obsolete syn. of *stylolite.*

lignin A complex amorphous polysaccharide containing methoxyl and phenyl propane units, always occurring with cellulose and forming an important chemical constituent of wood (Treiber, 1957, p. 446).

lignite (a) A brownish-black coal that is intermediate in coalification between peat and subbituminous coal; consolidated coal with a calorific value less than 8300 BTU/lb, on a moist, mineral-matter-free basis. Cf: *brown coal; lignite A; lignite B.*

lignite A *Lignite* that contains 6,300 or more BTU/lb but less than 8,300 BTU/lb (moist, mineral-matter-free). Cf: *lignite B.* Syn: *black lignite.*

lignite B *Lignite* that contains less than 6,300 BTU/lb (moist, mineral-matter-free); essentially synonymous with *brown lignite* or *brown coal.*

ligule A term used for various straplike plant structures, e.g. a membranous structure internal to the leaf base in the heterosporous lycopsids *Isoetes, Selaginella,* and *Lepidodendron;* or the limb of the ray flowerets in a member of the family Compositae.

ligurite An apple-green variety of sphene.

likasite A sky-blue orthorhombic mineral: $Cu_{12}(NO_3)_4(PO_4)_2(OH)_{14}$.

likely Said of a rock, lode, or belt of ground that gives indications of containing valuable minerals. Syn: *kindly.* Ant: *hungry.*

LIL An acronym for large ion lithophile elements, such as potassium, rubidium, and uranium, which have large ionic radii. First coined to describe the composition of lunar samples (Taylor, 1975).

lillianite A steel-gray mineral: $Pb_3Bi_2S_6$.

lily pad *shelfstone.*

lily-pad ice A term used for *pancake ice* consisting of circular pieces of ice that are not more than about 50 cm in diameter. Rarely used.

liman (a) A shallow muddy lagoon, bay, or marshy lake, formed at the mouth of a river behind the seaward deposits of a delta and protected by a barrier or a spit; an *estuary* or broad freshwater bay of the sea. Etymol: Russian, from Greek *limen,* "harbor". (b) An area of mud or slime deposited near the mouth of a river.

liman coast A coast with many lagoons (*limans*) and drowned valleys, protected from the open sea by a barrier or a spit; e.g. the northern coast of the Black Sea.

limb [astron] The outer edge of a lunar or planetary disk.

limb [fold] That area of a fold between adjacent fold hinges. It generally has a greater radius of curvature than the hinge region and may be planar. Syn: *flank.* Obsolete syn: *shank.*

limb [palyn] *equatorial limb.*

limb [surv] (a) The graduated margin of an arc or circle in an instrument for measuring angles, such as the part of a marine sextant carrying the altitude scale. (b) The graduated staff of a leveling rod.

limbate Having a thickened border or edge of a foraminiferal chamber, commonly at the suture but sometimes elevated.

limb darkening A photometric function of the Moon in which, at full moon, the center of the disk is much brighter than the limbs, evidently because of the Moon's roughness at optical dimensions.

limburgite A dark-colored, porphyritic extrusive igneous rock having olivine and clinopyroxene as phenocryst minerals in an alkali-rich glassy groundmass that may have microlites of clinopyroxene, olivine, and opaque oxides; some nepheline and/or analcime may be present, and feldspars are typically absent. Its name, given by Rosenbusch in 1872, is derived from Limburg, Kaiserstuhl, Germany. Partial syn: *magma basalt.* Not recommended usage.

limbus A crease at the edge of the vesicle of a vesiculate pollen grain, or at the edge of the pseudosaccus of a pseudosaccate grain, in which the proximal and distal exine layers are more or less fused.

lime (a) Calcium oxide, CaO; specif. quicklime and hydraulic lime. The term is used loosely for calcium hydroxide (as in *hydrated lime*) and incorrectly for calcium carbonate (as in agricultural lime). (b) A cubic mineral: CaO. (c) A term commonly misused for calcium in such deplorable expressions as "carbonate of lime" or "lime feldspar". (d) A limestone. The term is sometimes used by drillers for any rock consisting predominantly of calcium carbonate.

limeclast A lithoclast derived by erosion from an older limestone; also, an intraclast disrupted from partly consolidated calcareous mud on the bottom of a sea or lake.

lime concretion A concretion in soil, having a variable shape and size and consisting of an aggregate of precipitated calcium carbonate or of other material cemented by it.

lime feldspar A misnomer for *calcium feldspar.*

lime mica *margarite [mineral].*

lime mud The unconsolidated micritic component of a limestone.

lime mudstone A term proposed by Dunham (1962) for a fairly pure (93-99% calcium carbonate), mainly nonporous and impermeable, texturally uniform limestone whose main constituent (75-85%) is calcite mud (micrite). See also: *micritic limestone.*

lime olivine *calcio-olivine.*

lime pan [geomorph] A playa with a smooth, hard surface of calcium carbonate, commonly tufa.

lime pan [soil] A type of *hardpan* cemented chiefly with calcium carbonate. Also spelled: *limepan.*

lime rock A term used in SE U.S. (esp. Florida and Georgia) for an unconsolidated or partly consolidated form of limestone, usually containing shells or shell fragments, with a varying percentage of silica. It hardens on exposure and is sometimes used as *road metal.* Also spelled: *limerock.*

lime-silicate rock *calc-silicate rock.*

lime-soda feldspar A misnomer for *sodium-calcium feldspar.*

limestone (a) A sedimentary rock consisting chiefly (more than 50% by weight or by areal percentages under the microscope) of calcium carbonate, primarily in the form of the mineral calcite, and with or without magnesium carbonate; specif. a carbonate sedimentary rock containing more than 95% calcite and less than 5% dolomite. Common minor constituents include silica (chalcedony), feldspar, clays, pyrite, and siderite. Limestones are formed by either organic or inorganic processes, and may be detrital, chemical, oolitic, earthy, crystalline, or recrystallized; many are highly fossiliferous and clearly represent ancient shell banks or coral reefs. Limestones include chalk, calcarenite, coquina, and travertine, and they effervesce freely with any common acid. Abbrev: ls. (b) A general term used commercially (in the manufacture of lime) for a class of rocks containing at least 80% of the carbonates of calcium or magnesium and which, when calcined, gives a product that slakes upon the addition of water.

limestone buildup *carbonate buildup.*

limestone log An obsolete *resistivity log* device of short *lateral log* type that used 5 electrodes.

limestone pavement (a) A limestone bedding-plane or glaciated surface in a karst area that is divided into *clints* by *solution fissures.* See also: *crevice karst.* (b) A solution-grooved surface on limestone. See also: *karrenfeld.*

lime uranite *autunite.*

limewater Natural water with large amounts of dissolved calcium bicarbonate or calcium sulfate.

limiting beds The oldest strata immediately above and the youngest strata immediately below an angular unconformity; they are used to date the folding and erosion (Spieker, 1956).

limnal Pertaining to a body or bodies of fresh water, esp. to a lake or lakes.

limnetic (a) Relating to the pelagic or open part of a body of fresh water. (b) Said of lake-dwelling organisms and communities that are free from direct dependence on the bottom or shore.—Syn: *limnic [lake].*

limnic [coal] (a) Said of coal deposits formed inland in freshwater basins, peat bogs, or swamps, as opposed to *paralic* coal deposits. (b) Said of peat formed beneath a body of standing water. Its organic material is mainly planktonic.

limnic [lake] (a) Pertaining to a body of fresh water. Cf: *lacustrine.* (b) *limnetic.*

limnite *bog iron ore.*

limnogenic rock A sedimentary rock formed by precipitation from fresh water, esp. that of a lake (Grabau, 1924, p. 329).

limnogeology The geology of lakes.

limnogram A record of lake-level variations as recorded by a water-level gage, such as a record made on a limnimeter. Syn: *limnograph.*

limnograph *limnogram.*

limnokrene *spring lake.*

limnology The scientific study of the physical, chemical, meteorological, and esp. biological and ecological conditions and characteristics of pools, ponds, and lakes; and, by extension, of all inland waters. Etymol: Greek *limne*, "marsh, lake, pool".

limon (a) Viscous mud deposited during floods by rivers of the Mediterranean basin, the Atlantic coast of Morocco, and western Africa, and characterized by a binder of fine iron-hydroxide grains. (b) A widespread, fine-grained, surficial deposit of periglacial loam in northern France, from which brown loamy soils have developed. It is probably of windblown and wind-deposited origin, but different from loess in that it is formed under a more humid climate. (c) A term sometimes used as a French syn. of *loess*.—Etymol: French, "loam, silt, ooze, mud". Pron: lee-*mone.*

limonite A general field term for a group of brown, amorphous, naturally occurring hydrous ferric oxides whose real identities are unknown. Limonite was formerly thought to be a distinct mineral ($2Fe_2O_3 \cdot 3H_2O$), but is now considered to have a variable composition (and variable chemical and physical properties) and to consist of any of several iron hydroxides (commonly goethite) or of a mixture of several minerals (such as hematite, goethite, and lepidocrocite) with or without presumably adsorbed additional water. It is a common secondary material formed by oxidation (weathering) of iron or iron-bearing minerals, and it may also be formed as an inorganic or biogenic precipitate in bogs, lakes, springs, or marine deposits; it occurs as coatings (such as ordinary rust), as loose or dense earthy masses, as pseudomorphs after other iron minerals, and in a variety of stalactitic, fibrous, reniform, botryoidal, or mammillary forms, and it represents the coloring material of yellow clays and soils. Limonite is commonly dark brown or yellowish brown, but may be yellow, red, or nearly black; it is a minor ore of iron. See also: *bog iron ore.* Syn: *brown iron ore; brown hematite; brown ocher.*

limpid dolomite Mineral dolomite that occurs as exceptionally clear, perfectly formed euhedra with mirrorlike faces and gemlike luster; it is most easily recognized under the binocular microscope on acid-etched surfaces. Thought to form by slow crystallization in fresh or brackish water (Folk and Siedlecka, 1972).

limurite A metasomatic rock found at the contact of calcareous rocks with intruded granite and consisting of over 50% axinite. Other minerals include diopside, actinolite, zoisite, albite, and quartz. The term was originated by Zirkel in 1879.

limy (a) Containing a significant amount of lime or limestone; e.g. "limy soil". (b) Containing calcite; e.g. "limy dolomite" (a calcitic dolomite rock).

lin *linn.*

linarite A deep-blue monoclinic mineral: $PbCu(SO_4)(OH)_2$.

lindackerite A light-green or apple-green mineral: $H_2Cu(AsO_4)_4 \cdot 8\text{-}9H_2O$. It may contain a little nickel or cobalt.

Lindblad-Malmquist gravimeter *Boliden gravimeter.*

lindgrenite A green mineral: $Cu_3(MoO_4)_2(OH)_2$.

Lindgren's volume law The principle that during formation of ore by replacement, there is no change in rock volume or form (Lindgren, 1933). Syn: *law of equal volumes.*

lindinosite A dark *granite* composed of more than 50 percent riebeckite, with quartz and microcline. The name was derived by Lacroix in 1922 from Lindinosa, Corsica. Not recommended usage.

lindoite A light-colored hypabyssal rock characterized by bostonitic texture and similar in composition to *solvsbergite,* but being rich in quartz and poor in dark-colored minerals; the extrusive equivalent of an alkalic *granite.* The name, given by Brögger in 1894, is for the island of Lindö, Oslo district, Norway. Not recommended usage.

lindströmite A lead-gray to tin-white mineral: $Pb_3Cu_3Bi_7S_{15}$. Also spelled: *lindstromite.*

line [cart] A mark on a map, indicating a boundary, division, or contour.

line [seis] A linear array of seismologic observation points.

linea A linear marking on the carapace of some crustaceans, typically weakly calcified (TIP, 1969, pt. R, p. 97).

lineage In evolution, a line of descent, usually expressed as a chronological succession of ancestor-descendant species (or genera). Although it is sometimes used synonymously with *evolutionary series,* it usually refers to a particular line of descent within the evolutionary plexus.

lineage boundary The surface along which *plane defects* in a crystal occur.

lineage-zone A type of *range-zone* consisting of the body of strata containing specimens representing a segment of an evolutionary or developmental line or trend (ISG, 1976, p. 58-59). Syn: *evolutionary zone; morphogenetic zone; phylogenetic zone; phylozone.*

lineal series A single line of connected zooids in a cheilostome bryozoan colony, sequentially related by direct asexual descent, and commonly in contact with zooids of adjacent series along exterior walls (Boardman & Cheetham, 1969, p. 233).

lineament A linear topographic feature of regional extent that is believed to reflect crustal structure (Hobbs et al., 1976, p. 267). Examples are fault lines, aligned volcanoes, and straight stream courses. Nonrecommended syn: *linear.*

linear adj. Arranged in a line or lines; pertaining to the linelike character of some object or objects.—n. A nonrecommended syn. of *lineament.*

linear correlation The *correlation [stat]* of two or more variables measured by their straight-line relationship. If the correlation is perfect (unity or negative unity), the value of one variable is proportional to that of the others; if the correlation is absent (zero), there is no predictability of one value given that for any other.

linear element A fabric element having one dimension that is much greater than the other two. Lineations are the common linear elements. Cf: *planar element; equant element.*

linear flow structure *platy flow structure.*

linear scale ratio In model analysis, a ratio of the length in the prototype to the length in the model (Strahler, 1958, p.291). Symbol: λ.

linear selection Natural selection favoring variation in a particular direction.

lineation [sed] Any linear structure, of megascopic or microscopic nature, on or within a sedimentary rock, and esp. characterizing a bedding plane; e.g. a ripple mark, a sole mark, or a linear parallelism in fabric caused by preferred alignment of long axes of clasts or fossils at the time of deposition. It is largely the product of current action. See also: *parting lineation.*

lineation [struc geol] A general, nongeneric term for any linear structure in a rock; e.g. flow lines, slickensides, linear arrangements of components in sediments, or axes of folds. Lineation in metamorphic rocks includes mineral streaking and stretching in the direction of transport, crinkles and minute folds parallel to fold axes, and lines of intersection between bedding and cleavage, or of variously oriented cleavages. See: O'Leary et al. (1976); El-Etr (1976).

line defect A type of *crystal defect* occurring along certain lines in the crystal structure. Cf: *plane defect; point defect.* See also: *screw dislocation; edge dislocation.* Syn: *dislocation [cryst].*

line map *planimetric map.*

line of collimation The *line of sight* of the telescope of a surveying instrument, defined as the line through the rear nodal point of the objective lens of the telescope and the center of the *reticle* when they are in perfect alignment. Syn: *collimation line.*

line of concrescence In an ostracode, the proximal line of junction or fusion of the duplicature with the outer lamella.

line of dip The direction of the angle of dip, measured in degrees by compass direction. It generally refers to true dip, but can be said of apparent dip as well. Syn: *direction of dip.*

line of force [magnet] *magnetic-field line.*

line of force [phys] In a *field of force,* a line that is perpendicular to every equipotential surface it intersects; a line designating the direction of the force at every point along it.

line of induction *magnetic-field line.*

line of section A line on a map, indicating the position of a *profile section.* It is the *profile line* of the section as seen in plan.

line of seepage *seepage line.*

line of sight (a) A line extending from an observer's eye or an observing instrument to a distant point (such as on the celestial sphere) toward which the observer is looking or directing the instrument; e.g. *line of collimation.* (b) The straight line between two points. It is in the direction of a great circle but does not follow the curvature of the Earth. (c) A line joining the Earth or the Sun and a distant astronomic body.

line of strike *strike.*

line rod *range rod.*

line scanner A nonrecommended syn. of *scanner.*

line source A straight current element of infinite extent but infinitesimal cross section.

line spectrum The array of intensity values in the spectrum that occurs in very short, distinct ranges (i.e. only certain wavelengths) of the ordering variable, so that the spectrum appears to be a number of discrete lines with spaces between. An optical line spectrum results from electron transitions within atoms. Cf: *band spectrum.*

line-up n. On a seismogram trace, alignment in phase; an *event [seis].*

lingulacean n. Any inarticulate brachiopod belonging to the superfamily Lingulacea, characterized by subequal, generally phosphatic valves, with the pedicle valve being slightly larger. Range, Lower Cambrian to present.—adj. Said of a brachiopod having subequal phosphatic valves, or of the valves themselves. Var: *linguloid.*

lingulid Any lingulacean brachiopod belonging to the family Lingulidae, characterized mainly by an elongate oval to spatulate outline and a biconvex shell. Their stratigraphic range is Silurian (possibly Ordovician) to present. The genus *Lingula* belongs to this family and has frequently been used loosely for any Ordovician species in the family.

linguloid *lingulacean.*

linguloid ripple mark *linguoid ripple mark.*

linguoid current ripple *linguoid ripple mark.*

linguoid ripple mark An aqueous *current ripple mark* characterized by a tongue-shaped outline or having a barchanlike shape whose horns point into the current; it is best developed on the bottoms of shallow streams where it shows a highly irregular pattern with a wide variety of forms. The term "linguoid" applied to a ripple or ripple mark was introduced by Bucher (1919, p.164). See also: *cusp-ripple.* Syn: *linguoid current ripple; linguloid ripple mark; cuspate ripple mark.*

linguoid sole mark *flute cast.*

lining (a) A casing of brick, concrete, cast iron, or steel, placed in a tunnel or shaft to provide support. (b) A cover of clay, concrete, synthetic film, or other material, placed over all or part of the perimeter of a conduit or reservoir, to resist erosion, minimize seepage losses, withstand pressure, and improve flow.

lining pole *range rod.*

link [geomorph] An unbroken section of stream channel between two nodes in a *drainage network* (Shreve, 1966). The upstream node may be either a *source* or *fork.* The downstream node may be a fork or the *mouth* of the network. Cf: *bifurcating link; cis link; trans link; exterior link; interior link.*

link [paleont] A radial lath of skeletal material connecting the walls or vertical pillars in cups of dictyocyathid archaeocyathids (TIP, 1972, pt. E, p. 40).

link [surv] (a) One of the 100 standardized divisions of a surveyor's chain, each consisting of iron rods or heavy steel wire looped at both ends and joined together by three oval rings, and measuring 7.92 inches in length. (b) A unit of linear measure equal to 7.92 inches or one one-hundredth of a chain.

linkage [mtns] The joining at a sharp angle of two branches of a mountain range, as shown in plan view. Cf: *deflection.*

linkage [paleont] A type of intratentacular budding in scleractinian corals, characterized by development of two or more mouths with stomodaea inside the same tentacular ring. See also: *direct linkage; indirect linkage.*

link distance A dimensionless integer assigned to each *link* in a channel network, denoting the number of links in the flow path from the link to the mouth of the network (Jarvis, 1972). It is a measure of the relative position of the link in the drainage network. Cf: *diameter.* Syn: *topologic path length.*

linked veins An ore-deposit pattern in which adjacent, more or less parallel veins are connected by diagonal veins or veinlets.

links [coast] A Scottish term for a narrow area of flat or undulating land built up along a coast by drifting sand, and covered with turf or coarse grass; in Scotland, such land is often used as a golf course.

links [streams] A winding course of a river. Also, the ground along such a winding course.

linn (a) A pool of water, esp. a deep one below a fall. (b) A torrent running over rocks; a waterfall, cataract, or cascade. (c) A precipice or a steep ravine.—The term is used chiefly in Scotland and northern England. Etymol: Gaelic *linne,* "pool". Also spelled: *lin; lyn; lynn.*

Linnaean Conforming to the principles of *binomial nomenclature* as advocated by the Swedish botanist Carl von Linné, who Latinized his name to Carolus Linnaeus.

Linnaean species A species defined, usually in a broad sense, on the basis of its morphology. Also spelled: *Linnean species.*

linnaeite (a) A pale steel-gray isometric mineral: $(Co,Ni)_3S_4$. It has a coppery-red tarnish and constitutes an ore of cobalt. Syn: *linneite; cobalt pyrites.* (b) A group of isomorphous nickel-bearing sulfides, including linnaeite, carrollite, siegenite, violarite, and polydymite.

linophyre An igneous rock characterized by *linophyric* texture. Not recommended usage.

linophyric A term, now obsolete, applied to porphyritic igneous rocks with the phenocrysts arranged in lines or streaks (Cross et al., 1906, p.703); of or pertaining to a *linophyre.*

linosaite A basaltic rock with alkalic affinities, indicated by the presence of sodic pyribole or minor feldspathoid or by being associated with feldspathoid-bearing rocks. Its name (Johannsen, 1938) is derived from Linosa, one of the Pelagie Islands in the Mediterranean Sea. Not recommended usage.

linsey A term used in Lancashire, NW England, for a strong, striped shale and a streaky, banded sandstone or siltstone, interbedded in such a manner as to resemble a mixed linen and woollen fabric ("linsey-woolsey").

lintonite A greenish, agatelike variety of thomsonite from the Lake Superior region.

lip [eng] A low parapet erected on the downstream edge of a millrace or dam apron, to direct discharged water so that scouring of the river bottom is minimized.

lip [geomorph] (a) A projecting or overhanging edge, rim, or margin, such as of a rock on a mountainside. (b) A steep slope or abyss. (c) *crater lip.*

lip [paleont] (a) A margin of the aperture of a gastropod shell; e.g. *inner lip* and *outer lip.* (b) An elevated border of the aperture of a foraminiferal test. It may be small and at one side of the aperture, or completely surround it. (c) The labrum (upper lip) or labium (lower lip) of an arthropod.

Lipalian A name formerly used for the interval of time represented by a widespread unconformity separating Precambrian and Cambrian strata.

liparite A syn. of *rhyolite* used by German and Soviet authors. Its name, given by Roth in 1861, is derived from the Lipari Islands, in the Tyrrhenian Sea. Not recommended usage; the much more widely used synonym *rhyolite* has priority by one year.

lipid Any of several saponifiable oxygenated fats or fatty-acid-containing substances such as waxes, exclusive of hydrocarbon and certain other nonsaponifiable ether-soluble compounds, which in general are soluble in organic solvents, but barely soluble in water. They, along with proteins and carbohydrates, are the principal structural components of living cells. Also spelled: *lipide*.

lipide *lipid*.

lipogenesis In evolution, accelerated development as a result of the omission of certain ancestral stages. Cf: *bradytely; tachytely*.

lipotexite Nonliqufied basic material within anatectic magma (Dietrich & Mehnert, 1961). Also spelled: *lipotectite*. Little used.

lipscombite A mineral: $(Fe,Mn)Fe_2(PO_4)_2(OH)_2$.

liptinite *exinite*.

liptite A microlithotype of coal that contains a combination of exinite macerals totalling at least 95%. Cf: *vitrite; inertite*. Syn: *exite*.

liptobiolite (a) A resistant plant material that is left behind after the less resistant parts of the plant have wholly decomposed and that is characterized by relative stability of composition; e.g. resin, gum, wax, amber, copal, and pollen. (b) *liptobiolith*.

liptobiolith A combustible organic rock formed by an accumulation of liptobiolites; e.g. spore coal and pollen peat. Syn: *liptobiolite*.

liptocoenosis In paleontology, an assemblage of dead organisms together with the traces and products of their life prior to burial. Its syn., *necrocoenosis,* is used more commonly in biology.

liptodetrinite A maceral of coal within the *exinite* group, having no recognizable structure and low reflectance and fluorescence; because of its finely detrital condition it cannot be assigned with certainty to any of the other macerals of the group (ICCP, 1971).

liquation (a) In a magma, the separation of the residual liquid from earlier formed crystals. Rarely used. (b) In Russian usage, a syn. of *liquid immiscibility*.

liquefaction [sed] The transformation of loosely packed sediment into a fluid mass preliminary to movement of a turbidity current by subaqueous slumping or sliding.

liquefaction [soil] In cohesionless soil, the transformation from a solid to a liquid state as a result of increased pore pressure and reduced effective stress.

liquefaction slide The rapid and often catastrophic failure of a loose mass of predominantly cohesionless material which is generally at or near full saturation. The essential mechanism of such a slide is the sudden transfer of load from the particle contacts to the pore fluid, with resultant high transient pore-fluid pressures and consequent loss of strength. Liquefaction slides usually follow upon a disturbance (e.g. by earthquake or conventional slide) and can occur both subaqueously and subaerially (Koppejan et al., 1948). Syn: *flow slide*.

liquefied petroleum gas A compressed hydrocarbon gas obtained through distillation and usable as a motor fuel, for heating, or in certain industrial processes. Abbrev: LPG.

liquid chromatography A process for separating components in a liquid phase from one another by passing them over a solid or liquid stationary phase where the components are separated by their differential mobility rates. The technique used, based on the nature of the stationary phase, is often *column chromatography, paper chromatography*, or *thin-layer chromatography*. Cf: *gas chromatography*. See also: *chromatography*.

liquid flow Movement of a liquid that is usually of low viscosity, involving laminar and/or turbulent flow. Cf: *viscous flow; solid flow*.

liquid immiscibility A process of magmatic differentiation involving division of the magma into two or more liquid phases that may be physically separated from each other by gravity or other processes.

liquid inclusion A partial syn. of *fluid inclusion*.

liquidity index An expression of the consistency of a soil at its natural moisture content: its water content minus the water content at the plastic limit, all divided by the plasticity index at the liquid limit (Nelson and Nelson, 1967). Syn: *water-plasticity ratio; relative water content*.

liquid limit The water-content boundary between the semiliquid and the plastic states of a sediment, e.g. a soil. It is one of the *Atterberg limits*. Cf: *plastic limit*.

liquidus The locus of points in a temperature-composition diagram representing the maximum solubility (saturation) of a solid component or phase in the liquid phase. In a binary system it is a line; in a ternary system, a curved surface; in a quaternary system, a volume. In an isopleth study, at temperatures above the liquidus, the system is completely liquid, and at the intersection of the liquidus and the isopleth, the liquid is in equilibrium with one crystalline phase.

liquid-water content *free-water content*.

lira A fine raised line or linear elevation on the surface of some shells, resembling a thread or a hair; e.g. one of the parallel fine ridges on the surface of a nautiloid conch separated by *striae* and not easily discernible with the naked eye, or a fine linear elevation within the outer lip or on the shell surface of a gastropod. Pl: *lirae* or *liras*.

liroconite A blue to green monoclinic mineral: $Cu_2Al(AsO_4)(OH)_4 \cdot 4H_2O$. It usually contains some phosphorus. Syn: *lentil ore*.

liskeardite A soft, white mineral: $(Al,Fe)_3(AsO_4)(OH)_6 \cdot 5H_2O$.

lissamphibian Member of a subclass of amphibians characterized by loss or extreme reduction of dermal armor. The subclass was established primarily to accommodate living forms. Cf: *Anura*.

listric surface A curvilinear, usually concave-upward surface of fracture that curves, at first gently and then more steeply, from a horizontal position. Listric surfaces bound wedge-shaped masses, appearing to be thrust against or along each other. Etymol: Greek, *listron,* "shovel".

listrium A plate closing the anterior end of the pedicle opening that has migrated posteriorly in some discinacean brachiopods.

litchfieldite A *nepheline syenite* composed of albite, with smaller amounts of potassium feldspar, nepheline, biotite, cancrinite, and sodalite. Its name, given by Bayley in 1892, is derived from Litchfield, Maine. Not recommended usage.

-lite A rock-name suffix derived from the Greek word *lithos,* "stone".

lith A combining form, prefix or suffix, which means stone or rock.

litharenite (a) A term introduced by McBride (1963, p.667) as a shortened form of *lithic arenite* and used by him for a sandstone containing more than 25% fine-grained rock fragments, less than 10% feldspar, and less than 75% quartz, quartzite, and chert. See also: *sublitharenite*. (b) A general term used by Folk (1968, p.124) for a sandstone containing less than 75% quartz and metamorphic quartzite and more than 25% fine-grained volcanic, metamorphic, and sedimentary rock fragments, including chert; or whose content of such rock fragments is at least three times that of feldspar and plutonic-rock fragments.

litharge A red or yellow tetragonal mineral: PbO. Cf: *massicot*. Syn: *lead ocher*.

lithia mica *lepidolite*.

lithia water Mineral water containing lithium salts (e.g. lithium bicarbonate, lithium chloride).

Lithic n. In New World archaeology, the basal prehistoric cultural stage, characterized by the migration of man into the New World and the hunting of big game. It is followed by the Archaic. Correlation of relative cultural levels with actual age (and, therefore, with the time-stratigraphic units of geology) varies from region to region.—adj. Pertaining to the Lithic.

lithic (a) A syn. of *lithologic,* as in "lithic unit". (b) Said of a medium-grained sedimentary rock, and of a pyroclastic deposit, containing abundant fragments of previously formed rocks; also, said of such fragments. (c) Pertaining to or made of stone; e.g. "lithic artifacts" or "lithic architecture".

lithic arenite (a) A term used by Williams, Turner and Gilbert (1954, p.294 & 304) for a sandstone containing abundant quartz, chert, and quartzite, less than 10% argillaceous matrix, and more than 10% feldspar, and characterized by an abundance of unstable materials in which the fine-grained rock fragments exceed feldspar grains. It is better sorted and more porous and permeable, and contains better-rounded grains, than lithic wacke. The rock is roughly equivalent to "subgraywacke" as redefined by Pettijohn (1957). See also: *litharenite*. (b) A term used by Pettijohn (1954, p.364) as a syn. of *lithic sandstone*.

lithic arkose (a) A term used by McBride (1963, p.667) for an arkose containing appreciable rock fragments; specif. a sandstone containing 10-50% fine-grained rock fragments, 25-90% feldspar, and 0-65% quartz, quartzite, and chert. (b) A term used by Folk (1968, p.124) for a sandstone containing less than 75% quartz and metamorphic quartzite and having an "F/R ratio" between 1:1 and 3:1, where "F" signifies feldspars and fragments of gneiss and granite, and "R" signifies all other fine-grained rock fragments.—Cf: *feldspathic litharenite; impure arkose.*

lithic arkosic wacke A graywacke in which feldspar exceeds rock particles (Pettijohn, Potter & Siever, 1973, p. 167). Syn: *feldspathic graywacke.*

lithic-crystal tuff A tuff that is intermediate between *crystal tuff* and *lithic tuff* but is predominantly the latter. Cf: *crystal-lithic tuff.*

lithic graywacke A graywacke characterized by abundant unstable materials; specif. a sandstone containing a variable content (generally less than 75%) of quartz and chert and 15-75% detrital-clay matrix, and having rock fragments (primarily of sedimentary or low-rank metamorphic origin) in greater abundance than feldspar grains (chiefly sodic plagioclase, indicating a plutonic provenance) (Pettijohn, 1957, p.304). Example: some of the gray sandstones of the Siwalik Series (India), with little or no feldspar and 40-45% metamorphic-rock fragments (mainly phyllite or schist). The rock is equivalent to *low-rank graywacke* of Krynine (1945) and to *subgraywacke* as originally defined by Pettijohn (1949). The term was introduced by Pettijohn (1954, p.364) and by Williams, Turner and Gilbert (1954, p.294). Cf: *feldspathic graywacke.*

lithiclast *lithoclast.*

lithic sandstone A sandstone containing rock fragments in greater abundance than feldspar grains. The term was used by Pettijohn (1954, p.364) for such a sandstone with less than 15% detrital clay matrix (e.g. subgraywacke and protoquartzite), by Williams, Turner and Gilbert (1954, p.310) to include lithic arenite and lithic wacke, and by Hatch & Rastall (1965, p.111-112) to include the sublitharenite of McBride (1963). See also: *lithic arenite.*

lithic subarkose A term used by McBride (1963, p.667) for a sandstone composed of subequal amounts of feldspar and rock fragments; specif. a sandstone containing 10-25% feldspar, 10-25% rock fragments, and 50-80% quartz, quartzite, and chert. Syn: *feldspathic sublitharenite.*

lithic subarkosic wacke A wacke with subequal proportions of feldspar and rock fragments but no more than 25 percent of either (Pettijohn, Potter & Siever, 1973, p. 167).

lithic tuff An indurated deposit of volcanic ash in which the fragments are composed of previously formed rocks, e.g. accidental particles of sedimentary rock, accessory pieces of earlier lavas in the same cone, or small bits of new lava (essential ejecta) that first solidify in the vent and are then blown out. Cf: *crystal-lithic tuff; crystal tuff; lithic-crystal tuff.*

lithic wacke (a) A sandstone containing abundant quartz, chert, and quartzite, more than 10% argillaceous matrix, and more than 10% feldspar (esp. sodic plagioclase), and characterized by an abundance of unstable materials in which the fine-grained rock fragments exceed feldspar grains (Williams, Turner & Gilbert, 1954, p.291-292 & 301). (b) A quartz wacke containing abundant (up to 40-50%) fine-grained rock fragments (bits of shale, coal, etc.) (Krumbein & Sloss, 1963, p.172-173).

lithidionite A mineral: $KNaCuSi_4O_{10}$.

lithifaction *lithification.*

lithification [coal] A compositional change in a coal seam from coal to bituminous shale or other rock; the lateral termination of a coal seam due to a gradual increase in impurities.

lithification [sed] (a) The conversion of a newly deposited, unconsolidated sediment into a coherent, solid rock, involving processes such as cementation, compaction, desiccation, crystallization. It may occur concurrent with, soon after, or long after deposition. (b) A term sometimes applied to the *solidification* of a molten lava to form an igneous rock.—See also: *consolidation; induration.* Syn: *lithifaction.*

lithify To change to stone, or to petrify; esp. to consolidate from a loose sediment to a solid rock.

lithionite *lepidolite.*

lithiophilite A salmon-pink or clove-brown orthorhombic mineral: $Li(Mn^{+2},Fe^{+2})PO_4$. It is isomorphous with triphylite.

lithiophorite A mineral: $(Al,Li)MnO_2(OH)_2$.

lithiophosphate A white or colorless mineral: Li_3PO_4. It is a hydrothermal alteration product of montebrasite. Syn: *lithiophosphatite.*

lithistid adj. Said of a stonelike or stony sponge whose rigid skeletal framework consists of interlocking or fused siliceous spicules (desmas).—n. Any demosponge belonging to the order Lithistida and characterized by the presence of desmas, interlocked and cemented to form a rigid framework.

lithium-drifted germanium detector A *semiconductor radiation detector* containing germanium rather than silicon, in which lithium is diffused into the semiconductor to compensate for impurities. Syn: *Ge:Li detector.*

lithium mica *lepidolite.*

lithizone A *para-time-rock unit* representing a zone or succession of strata possessing common lithologic characteristics (Wheeler et al., 1950, p.2364). Cf: *monothem.* Syn: *lithozone.*

litho- A prefix meaning "rock" or "stone".

lithocalcarenite A calcarenite containing abundant limeclasts.

lithocalcilutite A calcilutite containing abundant limeclasts.

lithocalcirudite A calcirudite containing abundant limeclasts.

lithocalcisiltite A calcisiltite containing abundant limeclasts.

lithoclast A mechanically formed and deposited fragment of a carbonate rock, normally larger than 2 mm in diameter, derived from an older, lithified limestone or dolomite within, adjacent to, or outside the depositional site. Syn: *lithiclast.*

lithodesma A small calcareous plate reinforcing the internal ligament (resilium) in many shells of bivalve mollusks. Pl: *lithodesmata.* Syn: *ossiculum.*

lithodolarenite A dolarenite containing abundant doloclasts.

lithodololutite A dololutite containing abundant doloclasts.

lithodolorudite A dolorudite containing abundant doloclasts.

lithodolosiltite A dolosilitite containing abundant doloclasts.

lithodomous *lithotomous.*

lithofacies (a) A lateral, mappable subdivision of a designated stratigraphic unit, distinguished from adjacent subdivisions on the basis of lithology, including all mineralogic and petrographic characters and those paleontologic characters that influence the appearance, composition, or texture of the rock; a *facies* characterized by particular lithologic features. Laterally equivalent lithofacies may be separated by vertical arbitrary-cutoff planes, by intertonguing surfaces, or by gradational changes. See also: *statistical lithofacies; intertongued lithofacies; operational facies.* (b) A term used by Moore (1949, p.17 & 32) to signify any particular kind of sedimentary rock or distinguishable rock record formed under common environmental conditions of deposition, without regard to age or geologic setting or without reference to designated stratigraphic units, and represented by the sum total of the lithologic characteristics of the rock. This usage closely parallels Wells' (1947) definition of *lithotope.* (c) A term that has been applied to "lithology", "lithologic type", and the "manifestation" of lithologic characters.

lithofacies map A *facies map* based on lithologic attributes, showing areal variation in the overall lithologic character of a given stratigraphic unit. The map may emphasize the dominant, average, or specific lithologic aspect of the unit, and it gives information on the changing composition of the unit throughout its geographic extent.

lithofraction The fragmentation of rocks during transportation in streams or by wave action on beaches.

lithogene adj. Said of a mineral deposit formed by the process of mobilization of elements from a solid rock and their transportation and redeposition elsewhere. On a local scale the process may be called a product of *lateral secretion;* on a larger scale, the deposit may be called a product of regional metamorphism (Lovering, 1963, p.315-316).

lithogenesis (a) The origin and formation of rocks, esp. of sedimentary rocks. Also, the science of the formation of rocks. Cf: *petrogenesis.* Syn: *lithogeny.* (b) The first stage of mountain building, during which sediment is accumulated in the sea (esp. in a sinking geosyncline) and later compacted to form sedimentary rock.—Adj: *lithogenetic.*

lithogenetic unit A term used by Schenck & Muller (1941) for a local mappable assemblage of rock strata (such as a formation, member, or bed), considered without regard to time; a cartographic unit. See also: *lithostratigraphic unit.*

lithogenous Said of stone-secreting organisms, such as a coral polyp.

lithogeny *lithogenesis.*

lithogeochemical Said of "geochemical exploration techniques dealing with sampling and studying the mineral fraction of the lithosphere (rock, soil, stream and lake sediments)" (Beus, 1978, p. 110).

lithogeochemistry The chemistry of the mineral fraction of the lithosphere, i.e. rocks, soils, and stream and lake sediments (Beus, 1978, p. 110; Govett, 1978, p. 109). Cf: *biogeochemistry; hydrogeochemistry.*

lithographic limestone A compact, dense, homogeneous, exceedingly fine-grained limestone having a pale creamy yellow or grayish color and a conchoidal or subconchoidal fracture; a *micritic limestone.* It was formerly much used in lithography for engraving and the reproduction of colored plates. See also: *Solenhofen stone.* Syn: *lithographic stone.*

lithographic stone *lithographic limestone.*

lithographic texture A sedimentary texture of certain calcareous rocks, characterized by uniform particles of clay size and by an extremely smooth appearance resembling that of the stone used in lithography.

lithoherm A deep-water mound (up to several hundred meters long by 50 m high) of limestone, apparently formed by submarine lithification of carbonate mud, sand, and skeletal debris; e.g., in the Florida Straits (James, 1977). Cf: *bioherm; bank [sed].*

lithohorizon A surface of lithostratigraphic change or of distinctive lithostratigraphic character, pre-eminently valuable for correlation (not necessarily time-correlation); commonly the boundary of a lithostratigraphic unit, though also often a lithologically distinctive horizon or very thin marker bed within a lithostratigraphic unit (ISG, 1976, p. 32). Cf: *biohorizon; chronohorizon.* Syn: *lithostratigraphic horizon.*

lithoid Pertaining to or resembling a rock or stone, e.g. lithoid tufa.

lithoidal Said of the texture of some dense, microcrystalline igneous rocks, or of devitrified glass, in which individual constituents are too small to be distinguished with the unaided eye.

lithoidite A nonporphyritic, cryptocrystalline *rhyolite* composed of felsitic minerals. Obsolete.

lithoid tufa Gray, compact, bedded tufa, occasionally containing gastropod shells, occurring in the core of domelike masses in the desert basins of NW Nevada, as along the shore of the extinct Lake Lahontan. It is older and more stonelike than the overlying *thinolitic tufa* and *dendroid tufa.*

lithologic Adj. of *lithology.* Syn: *lithic.*

lithologic correlation A kind of *stratigraphic correlation* based on the correspondence in lithologic characters such as particle size, color, mineral content, primary structures, thickness, weathering characteristics, and other physical properties.

lithologic guide In mineral exploration, a kind of rock known to be associated with an ore. Cf: *stratigraphic guide.* See also: *ore guide.*

lithologic log *sample log.*

lithologic map A type of geologic map showing the rock types of a particular area.

lithologic unit *lithostratigraphic unit.*

lithology (a) The description of rocks, esp. in hand specimen and in outcrop, on the basis of such characteristics as color, mineralogic composition, and grain size. As originally used, "lithology" was essentially synonymous with *petrography* as currently defined. (b) The physical character of a rock. —Adj: *lithologic.* Cf: *petrology.*

lithomarge A smooth, indurated variety of common kaolin, consisting at least in part of a mixture of kaolinite and halloysite.

lithomorphic Said of a soil whose characteristics are mainly *inherited.*

lithophagous Said of an organism that feeds on rock material.

lithophile (a) Said of an element that is concentrated in the silicate rather than in the metal or sulfide phases of meteorites. Such elements concentrate in the Earth's silicate crust in Goldschmidt's tripartite division of elements in the solid Earth. Cf: *chalcophile; siderophile.* (b) Said of an element with a greater free energy of oxidation per gram of oxygen than iron. It occurs as an oxide and more often as an oxysalt, esp. in silicate minerals.—(Goldschmidt, 1954, p.24). Examples are: Se, Al, B, La, Ce, Na, K, Rb, Ca, Mn, U. Syn: *oxyphile.*

lithophilous *rupestral.*

lithophysa A hollow, bubblelike structure composed of concentric shells of finely crystalline alkali feldspar, quartz, and other minerals; found in silicic volcanic rocks, e.g. rhyolite and obsidian. Pl: *lithophysae.* Syn: *stone bubble.*

lithophyte A plant living on the surface of a rock. Adj: *lithophytic.*

lithorelic Said of a soil feature that is derived from the parent material. Cf: *pedorelic.*

lithosere A *sere* that develops on a rock surface.

lithosiderite *stony-iron meteorite.*

Lithosol A great soil group in the 1938 classification system, an azonal group of soils characterized by shallow depth to bedrock and by recent and imperfect weathering. It usually develops on steep slopes (USDA, 1938). Most of these soils are now classified as *Orthents.* See also: *mountain soil.* Syn: *skeletal soil.*

lithosome (a) A rock mass of essentially uniform or uniformly heterogeneous lithologic character, having intertonguing relationships in all directions with adjacent masses of different lithologic character. The term was introduced by Wheeler & Mallory (in Fischer et al., 1954, p.929) and defined by them (1956, p.2722) as a lithostratigraphic body or vertico-laterally segregated unit that is "mutually intertongued with one or more bodies of differing lithic constitution". It is essentially identical with Caster's (1934) *magnafacies.* Cf: *biosome; holosome; intertongued lithofacies.* (b) The sedimentary record of a physicochemical environment or of a more or less uniform lithotope; a body of sediment deposited under uniform physicochemical conditions (Sloss, in Weller, 1958, p.624). (c) A term defined by Moore (1957a, p.1787-1788) as "an independent body of genetically related sedimentary deposits of any sort", or, alternatively, "a spatially segregated part of any genetically related body of sedimentary deposits".—Cf: *lithostrome.*

lithospar A naturally occurring mixture of spodumene and feldspar.

lithosphere (a) The solid portion of the Earth, as compared with the *atmosphere* and the *hydrosphere.* (b) In plate tectonics, a layer of strength relative to the underlying *asthenosphere* for deformation at geologic rates. It includes the *crust* and part of the *upper mantle* and is of the order of 100 km in thickness (Dennis & Atwater, 1974, p. 1031).

lithostatic pressure *geostatic pressure.*

lithostratic unit *lithostratigraphic unit.*

lithostratigraphic classification The organization of rock strata into units on the basis of their lithologic character (ISG, 1976, p. 30). The fundamental unit is the *formation [stratig].*

lithostratigraphic horizon *lithohorizon.*

lithostratigraphic unit A body of rock that is unified by consisting dominantly of a certain lithologic type or combination of types, or by possessing other unifying lithologic features. It may consist of sedimentary, igneous, or metamorphic rocks, or of two or more of these. It may or may not be consolidated. The critical requirement is a substantial degree of overall lithologic homogeneity. Lithostratigraphic units are recognized and defined by observable physical features and not by inferred geologic history or mode of genesis. Fossils may be important as physical constituents or because of their rock-forming character, as in coquina, diatomite, coal beds, etc. Only major lithologic features readily recognizable in the field (or subsurface) should serve as the basis for recognition (ISG, 1976, p. 31). A lithostratigraphic unit has a binomial designation, preferably consisting of a geographic name, derived from the type area, combined with a descriptive lithologic term (e.g. Ohio Shale), or with the appropriate rank term (e.g. Rome Formation). See also: *lithogenetic unit.* Syn: *rock-stratigraphic unit; lithostratic unit; lithologic unit; rock unit; geolith.*

lithostratigraphic zone *lithozone.*

lithostratigraphy The element of stratigraphy that deals with the lithology of strata and with their organization into units based on lithologic character (ISG, 1976, p. 30). Syn: *petrostratigraphy; rock stratigraphy.*

lithostrome A term introduced by Wheeler & Mallory (1956, p.2721-2722) for a sedimentary unit "consisting of one or more beds of essentially uniform or uniformly heterogeneous lithologic character" and representing the "three-dimensional counterpart of a lithotope"; esp. an individual tongue projecting from a lithosome. The term is regarded as essentially synonymous with *lithosome* as defined by some, and with *lithotope* as defined by Wells (1947). Weller (1958, p.636) would reject the term because it represents "nothing more than a rock stratigraphic unit" such as "bed", "member", "tongue", "stratum", or "layer". Cf: *holostrome.*

lithothamnion A plant of the genus *Lithothamnion,* an encrusting or nodular red calcareous alga of the family Corallinaceae, abundant in post-Jurassic rocks, and reported as a living form from considerable depths and very cold waters. It is most abundant on the seaward edge of reef flats, where it acts as a cementing medium of some coral reefs.

lithothamnion ridge An *algal ridge* built by *Lithothamnion* and other red calcareous algae, rising about 1 m above the surrounding reef and extending to depths of 6-7 m below sea level.

lithotomous Said of an organism that bores into rock. Also spelled: *lithodomous.*

lithotope (a) An area or surface of uniform sediment, sedimentation, or sedimentary environment, including associated organisms. (b) A paleoecologic term originally proposed by Wells (1944, p.284) for "the sedimentary rock record of a biotope" whose life community or biocoenosis was preserved; it was later defined by Wells (1947, p.119) as "the rock record of the environment" (including both its physical and biologic expressions). The term has subsequently been used for a stratigraphic unit, a part of a stratigraphic section, a particular kind of sediment or rock, and a body of uniform sediments formed by persistence of the depositional environment, and also in an intangible sense for a sedimentary rock environment and a physical environment. Cf: *lithofacies; lithostrome; biotope [stratig].*

lithotype A macroscopically visible band in humic coals, analyzed by physical characteristics rather than by botanical origin. The four lithotypes of banded bituminous coal are *vitrain, clarain, durain,* and *fusain.* These were originally described by Stopes in 1919. Beds or bands of cannel and boghead coal are sometimes included as lithotypes of coal. See also: *banded ingredients.*

lithoxyl A term applied to *wood opal* in which the original woody structure is observable. The term is also used to designate petrified (opalized) wood. Syn: *lithoxyle; lithoxylite; lithoxylon.*

lithozone (a) An informal term to indicate a body of strata that is unified in a general way by lithologic features but for which there is insufficient need or information to justify its designation as a formal unit, e.g. the shaly zone in the lower part of the Parker Formation, the coal-bearing zone exposed south of Ravar, the Burgan oil-producing zone (ISG, 1976, p. 30). Syn: *lithostratigraphic zone.* (b) A "more euphonious" syn. of *lithizone* (P. F. Moore, 1958, p. 449).

lit-par-lit adj. Having the characteristic of a layered rock, the laminae of which have been penetrated by numerous thin, roughly parallel sheets of igneous material, usually granitic. Etymol: French, bed-by-bed. Cf: *injection gneiss.*

litter In forestry, a general term for the layer of loose organic debris, composed of freshly fallen or only slightly decayed material, that accumulates in wooded areas. It is a constituent of the *forest floor.* Cf: *duff; leaf mold.*

Little Ice Age A cool, brief interval in an otherwise warm interglacial stage. Originally employed for a mid-Holocene event in the Yosemite area, California, about 3000 years B.P. (Matthes, 1930), it is also widely used for the 16th- and 18th-century cool phases (Pittock et al., 1978).

littoral (a) Pertaining to the benthic ocean environment or *depth zone* between high water and low water; also, pertaining to the organisms of that environment. Syn: *intertidal.* (b) Pertaining to the depth zone between the shore and a depth of about 200 m. In this meaning, which is obsolete and nonrecommended, the term includes the *neritic* zone. See also: *sublittoral.*—Cf: *supralittoral.*

littoral cone An adventive or accidental ash or tuff cone formed on a lava flow when it runs into a body of water, usually the sea. Such cones are the result of steam explosions that hurl into the air large amounts of ash, lapilli, and small bombs derived from the new lava.

littoral current An ocean current caused by the approach of waves to a coast at an angle. It flows parallel to and near to the shore. See also: *littoral drift.* Syn: *longshore current.*

littoral drift Material (such as shingle, gravel, sand, and shell fragments) that is moved along the shore by a *littoral current.* Syn: *longshore drift; shore drift.*

littoral explosion An explosion that is the result of the contact of a flow of molten lava with the edge of a body of water; a *hydroexplosion.*

littoral shelf A shallow, nearshore, terracelike part of a submerged lake bed, produced by the combined effects of wave erosion and current deposition, and often extending a considerable distance lakeward from the beach.

lituicone A nautiloid conch (like that of *Lituites*) that is coiled in the early stages of development and then becomes straight in the mature stages. Syn: *lituiticone.*

live cave A cave in which there is moisture and growth of speleothems associated with the presence of moisture. See also: *dead cave.* Syn: *active cave.*

liveingite A mineral: $Pb_9As_{13}S_{28}$. Syn: *rathite-II.*

liverite Name used locally in Utah for *elaterite.*

liver opal *menilite.*

liver ore *hepatic cinnabar.*

liver rock A sandstone that breaks or cuts as readily in one direction as in another and that can be worked without being affected by stratification; a dense *freestone* that lacks natural division planes.

liverwort A member of the bryophyte class Hepaticae, characterized by a creeping or dichotomously branched thallus, no roots or leaves, and unicellular rhizoids. Liverworts range from the Devonian. Cf: *moss [bot].*

livesite A clay mineral intermediate between kaolinite and halloysite; a disordered kaolinite.

live speleothem A moist-cave mineral deposit that is actively forming. Syn: *active speleothem.*

live stream *perennial stream.*

living chamber (a) The *body chamber* housing the soft parts of a cephalopod. (b) The outermost part of the zooidal body cavity in stenolaemate bryozoans, which contains the functional organs.

living flood plain A flood plain that is overflowed in times of high water (Bryan, 1923a, p. 88). Cf: *fossil flood plain.*

living fossil An animal or plant that lives at the present time, is also known as a fossil from an earlier geologic time, and has undergone relatively little modification since that earlier time.

livingstonite A lead-gray mineral: $HgSb_4S_8$.

lixiviation *leaching.*

lixivium *leachate.*

lizardite A platy mineral of the serpentine group: $Mg_3Si_2O_5(OH)_4$. It is a polymorph of chrysotile, distinct from clinochrysotile, orthochrysotile, and parachrysotile.

L-joint A syn. of *primary flat joint.* Rarely used.

Llandeilian European stage: Middle Ordovician (above Llanvirnian, below lower Caradocian). Also spelled: *Llandeillian.*

Llandoverian European stage: Lower Silurian (above Ashgillian of Ordovician, below Wenlockian). Syn: *Valentian.*

llanite A hypabyssal porphyritic *rhyolite* composed of phenocrysts of red alkali feldspar and blue quartz in a fine-grained groundmass of quartz, microcline, albite, and biotite. The name was derived by Iddings in 1904 from Llano County, Texas. Not recommended usage.

Llano A provincial series of the Precambrian in Texas.

llano A term for an extensive tropical plain, with or without vegetation, applied esp. to the generally treeless plains of northern South America and the southwestern U.S. Etymol: Spanish.

Llanoria One of the *borderlands* proposed by Schuchert (1923), in this case south of North America, between the Ouachita geosyncline and the Gulf of Mexico. Evidence proposed for Llanoria was more tenuous than for Appalachia and Cascadia, and modern knowledge of the substructure of the Gulf Coastal Plain and Gulf of Mexico virtually precludes its former existence.

Llanvirnian European stage: Middle Ordovician (above Arenigian, below Llandeilian).

L layer A surficial layer of leaf litter over a soil. See also: *F layer [soil]; O horizon.*

llyn A Welsh term for a pool or lake.

load (a) The material that is moved or carried by a natural transporting agent, such as a stream, a glacier, the wind, or waves, tides, and currents; specif. *stream load.* (b) The quantity or amount of such material at any given time.—Syn: *sediment load.*

load cast A *sole mark,* usually measuring less than a meter in any direction, consisting of a swelling in the shape of a slight bulge, a deep or shallow rounded sack, a knobby excrescence, a highly irregular protuberance, or a bulbous, mammillary, or papilliform protrusion of sand or other coarse clastics, extending downward into finer-grained, softer, and originally hydroplastic underlying material, such as wet clay, mud, or peat, that contained an initial depression. It is produced by the exaggeration of the depression as a result of unequal settling and compaction of the overlying material and by the partial sinking of such material into the depression, as during the onset of deposition of a turbidite on unconsolidated mud. A load cast is more irregular than a flute cast (it is usually not systematically elongated in the current direction), and is characterized by an absence of a distinction between the upcurrent and downcurrent ends. The term was proposed by Kuenen (1953, p.1058) to replace *flow cast* used by Shrock (1948, p.156), although Kuenen excluded the phenomenon of warping of underlying laminae and applied the term to a feature resulting from vertical adjustment only. See also: *load-flow structure.* Syn: *load casting; teggoglyph.*

load-casted Said of a current mark (such as a groove or flute) that is exaggerated, misshapen, or obscured by the development of a load cast. Also said of a sole mark (such as a flute cast or groove cast) that is similarly modified by load casting.

load-casted ripple A term apparently first used by Dzulynski and Kotlarczyk (1962) for a structure formed by sinking of ripples into the underlying mud during deposition of the ripple. See also Dzulynski and Walton (1965, p. 146). Syn: *ripple load cast.*

load casting (a) The formation or development of a load cast or load casts; also, the configuration of the underside of a stratum characterized by load casts. (b) *load cast.*

load-cast lineation A small-scale, poorly defined, irregular linear structure that appears as a cast on the underside of a sandstone bed, and that is attributed to a dense, sluggish turbidity current moving over soft mud (Crowell, 1955, p.1358).

load-cast striation A rill-like sedimentary structure of uncertain origin (Pettijohn & Potter, 1964, p.319).

loaded stream A stream that has all the sediment it can carry. A partly loaded stream is one carrying less than full capacity. See also: *overloaded stream; underloaded stream.*

load-flow structure A term sometimes used for *load cast* because the structure forms by downsinking of overlying material and not by infilling of a depression (as implied by the term "cast").

load fold A plication of an underlying stratum, believed to result from unequal pressure and settling of overlying material (Sullwold, 1959).

load metamorphism A type of *static metamorphism* in which pressure due to deep burial has been a controlling influence, along with high temperature. Cf: *geothermal metamorphism; thermal metamorphism.*

load mold The mold of a load cast; the depression in an underlying stratum, occupied by a load pocket, such as the sea-floor surface beneath a depositing turbidite (Sullwold, 1960).

load pocket The material within a load cast, consisting of a "bulge of sand" pressing into an underlying stratum (Sullwold, 1959, p.1247).

loadstone *lodestone.*

load wave The "salient curved unevenness" of underlying material that appears to have been "squirted up" into a superjacent turbidity-current deposit as a result of unequal settling of the overlying material (Sullwold, 1959); it resembles a ripple mark or other wavelike structure. The term "refers to smooth upward bulges as well as to tenuous breaking wave or flame shapes" (Sullwold, 1960, p.635). See also: *flame structure.*

loam (a) A rich, permeble soil composed of a friable mixture of relatively equal and moderate proportions of clay, silt, and sand particles, and usually containing organic matter (humus); specif. a soil consisting of 7-27% clay, 28-50% silt, and 23-52% sand. It has a somewhat gritty feel yet is fairly smooth and slightly plastic. Loam may be of residual, fluvial, or eolian origin, and includes many loesses and many of the alluvial deposits of flood plains, alluvial fans, and deltas. It usually implies fertility, and is sometimes called *topsoil* in contrast to the subsoils that contain little or no organic matter. (b) A term used in the old English literature for a mellow soil rich in organic matter, regardless of texture. (c) An obsolete term formerly used in a broad sense for clay, impure clay, clayey earth, and mud.

loaming A method of geochemical prospecting in which samples of soil or other surficial material are tested for traces of the metal desired, its presence presumably indicating a near-surface orebody.

loamy sand A soil containing 70-90% sand, 0-30% silt, and 0-15% clay, or a soil containing 85-90% sand at the upper limit and having the percentage of silt plus 1.5 times the percentage of clay not less than 15, or a soil containing 70-85% sand at the lower limit and having the percentage of silt plus twice the percentage of clay not exceeding 30 (SSSA, 1965, p.347); specif. such a soil containing at least 25% very coarse sand, coarse sand, and medium sand, and less than 50% fine sand or very fine sand. It is subdivided into loamy coarse sand, loamy fine sand, and loamy very fine sand. Cf: *sandy loam.*

loamy soil A soil (such as a clay loam, silt loam, or sandy loam) whose texture and properties are intermediate between those of a coarse-textured or sandy soil and a fine-textured or clayey soil.

lobate delta *arcuate delta.*

lobate plunge structure *flute cast.*

lobate rill mark A term used by Clarke (1918) and Shrock (1948, p.131) for a spatulate or lingulate sedimentary structure (cast) resembling the bowl of an inverted spoon and believed to develop on a beach by ebbing tidal currents or retreating storm waves of the intertidal zone. The structure is now considered to be a *flute cast* formed by current action. See also: *rill mark.*

lobate soil *step [pat grd].*

lobe [bot] Any protuberant part or segment of a plant organ; specifically a segment of a petal or calyx or leaf that represents division to about the middle (Lawrence, 1951, p.759).

lobe [glaciol] (a) A rounded, tonguelike projection of glacial drift beyond the main mass of drift. (b) *glacial lobe.*

lobe [lake] A long, rounded indentation of a lake.

lobe [paleont] (a) An element or undulation of a suture line in a cephalopod shell that forms an angle or curve whose convexity is directed backward or away from the aperture (or toward the apex). Ant: *saddle.* (b) One of the longitudinal divisions of the body, or one of the lateral divisions of the glabella, in a trilobite. (c) A rounded major protuberance of the valve surface of an ostracode, generally best developed in the dorsal part of the carapace (TIP, 1961, pt.Q, p.52); e.g. *connecting lobe.*

lobe [streams] (a) *meander lobe.* (b) *flood-plain lobe.*

lobefin *Sarcopterygii.*

local base level *temporary base level.*

local current A natural *earth current* of any origin, e.g. from the oxidation of a sulfide deposit.

local fauna *faunule.*

local-gravity map A gravity map from which regional changes of gravity have been eliminated.

local horizon (a) *apparent horizon; visible horizon.* (b) The actual lower boundary of the observed sky or the upper outline of terrestrial objects including nearby obstructions or irregularities.

local magnitude A measure of the strain energy released by an earthquake within 100 km of the epicenter. See: *earthquake magnitude.*

local metamorphism Metamorphism caused by a local process, e.g. contact metamorphism or metasomatism near an igneous body, hydrothermal metamorphism, or dislocation metamorphism in a fault zone. Cf: *regional metamorphism.*

local peat Peat developed by ground water. Syn: *basin peat; azonal peat.*

local peneplain (a) *incipient peneplain.* (b) *partial peneplain.*

local range-zone The *range-zone* of a specified taxon in some particular section or local area. The sum of all the local range-zones is the true range-zone of the taxon. Syn: *teilzone; topozone; partial range-zone.*

local relief The vertical difference in elevation between the highest and the lowest points of a land surface within a specified horizontal distance or in a limited area. Syn: *relative relief.* Cf: *available relief.*

local sorting The action responsible for the size-sorting (homogeneity) of sedimentary particles at a given place (Pettijohn, 1957, p. 540-541). Cf: *progressive sorting.*

local unconformity An unconformity that is strictly limited in geographic extent and that usually represents a relatively short period, such as one developed around the margins of a sedimentary basin or along the axis of a structural trend that rose intermittently while continuous deposition occurred in an adjacent area. It may be similar in appearance to, but lacks the regional importance of, a disconformity. Cf: *regional unconformity.*

location (a) The spot or place where a borehole is to be drilled; e.g. a *well site.* (b) The spacing unit between two boreholes or wells, e.g. a 40-acre location.

lochan A Scottish term for a small lake (loch) or pond, usually lying in a *cirque.*

lock A stretch of water in a canal, stream, or dock, enclosed by gates at each end, and used in raising or lowering boats as they pass from one water level to another.

Locke level A *hand level* with a fixed bubble tube that can be used only for approximate horizontal sighting.

Lockportian Stage in New York State: upper Middle Silurian (above Clintonian, below Salinan).

locomorphic stage A term introduced by Dapples (1962) for the middle geochemical stage of diagenesis characterized by prominent mineral replacement (without reactions). It is typical of lithification of a clastic sediment, and is more advanced than the *redoxomorphic stage* and precedes the *phyllomorphic stage.*

locular dimorphism Development of a *loculus* in certain ostracodes.

locular wall (a) A wall of a diatom frustule having separate inner and outer laminae connected by vertical partitions that form areolae. (b) The wall of a seed or spore chamber (locule) in an ovary or sporangium.

locule A compartment, cavity or chamber; in flowering plants, a cavity in the ovary containing ovules, or in the anther containing pollen grains (Scagel et al., 1965, p. 622). Syn: *loculus [bot].*

loculus [bot] *locule.*

loculus [paleont] One of the chambers in a foraminiferal test. (b) The space between two adjacent septa in the intervallum of archaeocyathids. (c) In certain dimorphic ostracodes, a pocketlike structure developed in the presumed female (heteromorph) between the frill and the marginal ridge, usually several in each valve (as in *Ctenoloculina, Abditoloculina, Tetrasacculus,* and others).—Pl: *loculi.* Adj: *locular.*

lode [eco geol] A mineral deposit consisting of a zone of veins, veinlets, disseminations, or planar breccias; a mineral deposit in consolidated rock as opposed to *placer* deposits. Syn: *lead [eco geol].* Cf: *vein [ore dep]; vein system.*

lode [streams] A local English term for a channel or watercourse, usually partly artificial and embanked above the surrounding country.

lode claim A mining claim on an area containing a known vein or lode. Cf: *placer claim.*

lode country *ore channel.*

lodestone (a) A magnetic variety of natural iron oxide (Fe_3O_4) or of the mineral magnetite; specif. a piece of magnetite possessing polarity like a magnet or magnetic needle and hence one that, when freely suspended, will attract iron objects. Also spelled: *loadstone.* Syn: *leading stone; Hercules stone.* (b) An intensely magnetized rock or ore deposit.

lodestuff Both the gangue and the economically valuable minerals of a lode; the contents of an *ore channel.*

lode tin Cassiterite occurring in veins, as distinguished from *stream tin.*

lodge moraine A terminal moraine of billowy relief, consisting of subglacial debris lodged under a thin margin of a glacier; widespread in North America. Syn: *submarginal moraine.*

lodgment The plastering beneath a glacier of successive layers of basal till upon bedrock or other glacial deposits. Cf: *plastering-on.* Also spelled: *lodgement.*

lodgment till A *basal till* commonly characterized by compact fissile structure and containing stones oriented with their long axes generally parallel to the direction of ice movement. Also spelled: *lodgement till.* See also: *fluted till.*

lodranite A stony-iron meteorite composed of a mixture of bronzite and olivine, enclosed within a fine network of nickel-iron.

loellingite An orthorhombic mineral: $FeAs_2$. Also spelled: *löllingite; lollingite.* Syn: *leucopyrite.*

loess A widespread, homogeneous, commonly nonstratified, porous, friable, slightly coherent, usually highly calcareous, fine-grained blanket deposit (generally less than 30 m thick), consisting predominantly of silt with subordinate grain sizes ranging from clay to fine sand. It covers areas extending from north-central Europe to eastern China as well as the Mississippi Valley and Pacific Northwest of the U.S. Loess is generally buff to light yellow or yellowish brown, often contains shells, bones, and teeth of mammals, and is traversed by networks of small narrow vertical tubes (frequently lined with calcium-carbonate concretions) left by successive generations of grass roots, which allow the loess to stand in steep or nearly vertical faces. Loess is now generally believed to be windblown dust of Pleistocene age, carried from desert surfaces, alluvial valleys, and outwash plains, or from unconsolidated glacial or glaciofluvial deposits uncovered by successive glacial recessions but prior to invasion by a vegetation mat. The mineral grains, composed mostly of silica and associated heavy minerals, are fresh and angular, and are generally held together by calcareous cement. Etymol: German *Löss,* from dialectal (Switzerland) *lösch,* "loose", so named by peasants and brickworkers along the Rhine valley where the deposit was first recognized. Pron: *luehss.* Cf: *limon; adobe.* Syn: *löss; lehm; bluff formation.*

loessal Pertaining to or consisting of loess. Syn: *loessial.*

loess doll A compound nodule or concretion of calcium carbonate found in loess and resembling a doll, a potato, or a child's head. It is often hollow but may contain a loose stone. Syn: *loess nodule; loess kindchen; puppet.*

loess flow A fluid suspension of dry porous silt in air, moving downslope, such as occurred following the 1920 earthquake in Kansu Province, China (Close & McCormick, 1922).

loessification Formation and development of loess.

loess kindchen A *loess doll* resembling the head of a child. Etymol: German *Lösskindchen.*

loessland Land whose surface is underlain by loess.

loess nodule *loess doll.*

loessoïde A Dutch term for deposits in southern Limburg (a province in southern Netherlands), believed to be of loessal origin, but reworked and redeposited by streams, possibly with an admixture of residual material from in-situ decomposition.

loeweite A white to pale-yellow mineral: $Na_{12}Mg_7(SO_4)_{13} \cdot 15H_2O$. Also spelled: *löweite.*

Loewinson-Lessing classification A chemical classification of igneous rocks (into the four main types—acid, intermediate, basic, and ultrabasic) based on silica content.

loferite A term suggested by Fischer (1964, p. 124) for a limestone or dolomite riddled by shrinkage pores, such as the carbonate sediments in the Triassic Dachstein Formation (Lofer facies) in Salzburg, Austria. The term is partly synonymous with *bird's-eye limestone.*

log A continuous record as a function of depth, usually graphic and plotted to scale on a narrow paper strip, of observations made on the rocks and fluids of the geologic section exposed in a well bore; e.g., *well log, sample log, strip log, drilling-time log.*

logan [geog] Shortened form of *pokelogan.*

logan [geomorph] *logan stone.*

Logan's Line A structural discontinuity along the northwestern edge of the Northern Appalachians, between complexly deformed rocks on the southeast and undisturbed rocks on the northwest. The name commemorates its discovery by Logan (1863) near Quebec City. For part of its distance the line is a major low-angle thrust fault, but northeast of Quebec City it is beneath the St. Lawrence Estuary and its nature is undetermined; southward in Vermont the frontal fault changes into a succession of discontinuous breaks. It is interpreted by many geologists as having been formed during the Taconic orogeny of early Paleozoic time.

logan stone An English name for a *rocking stone* consisting of a large mass of granite or gneiss chemically weathered along horizontal joints and so balanced on its base as to "log" or rock from side to side; e.g. the stone weighing about 80 tons near Land's End in Cornwall. Syn: *logan [geomorph]; loggan stone; logging stone.*

logarithmic mean diameter An expression of the average particle size of a sediment or rock, obtained by taking the arithmetic mean of the particle-size distribution in terms of logarithms of the class midpoints. Cf: *phi mean diameter.*

loggan stone *logan stone.*

logging (a) The act or process of making or recording a *log.* (b) The method or technique by which subsurface formations are characterized relative to depth by measurements or observations on the rocks of a borehole.

logging stone *logan stone.*

logging tool *sonde.*

lognormal distribution A *frequency distribution* whose logarithm follows a normal distribution.

log strip A long, narrow piece of paper on which a *strip log* is plotted.

loipon A term proposed by Shrock (1947) for a residual surficial layer produced by intense and prolonged chemical weathering and composed largely of certain original constituents of the source rock. Typical accumulations of loipon are the gossans over orebodies, bauxite deposits in Arkansas, terra rossa deposits of Europe, and duricrust of Australia. Etymol: Greek, "residue". Pron: *loy-pon.* Adj: *loiponic.*

lokkaite A mineral: $(Y,Ca)_2(CO_3)_3 \cdot 2H_2O$.

löllingite *loellingite.*

lolly ice Soft *frazil ice* formed in turbulent seawater. Syn: *lolly.* Rarely used.

loma A term used in the SW U.S. for an elongated, gentle swell or rise of the ground (as on a plain), or a rounded, broad-topped, inconspicuous hill. Etymol: Spanish, "hillock, rising ground, slope".

lomita A small, low *loma.*

lomonite Original spelling of *laumontite.*

lomonosovite A dark cinnamon-brown to black or rose-violet mineral: $Na_2Ti_2Si_2O_9 \cdot Na_3PO_4$. Cf: *murmanite.*

lomontite *laumontite.*

lonchiole A *sceptrule* with a single spine opposite the single ray.

Longaxones A group of primitive, usually lightly sculptured, tricolpate, Cretaceous and younger angiosperm pollen in which the polar axis is as long as, or longer than, the equatorial diameter. Cf: *Brevaxones.*

long clay A highly plastic clay; a *fat clay.*

longicone A long, slender, conical, gradually tapering shell characteristic of certain orthoconic cephalopods. Also, a fossil animal having such a shell. Cf: *brevicone.*

longitude (a) An angular distance between the plane of a given meridian through any point on a sphere or spheroid and the plane of an arbitrary meridian selected as a line of reference, measured in the plane of a great circle of reference or in a plane parallel to that of the equator; specif. the length of the arc or portion of the Earth's equator or of a parallel of latitude intersected between the meridian of a given place and the prime meridian (or sometimes a national meridian), expressed either in time or in degrees east or west of the prime or national meridian (which has longitude zero degrees) to a maximum value of 180 degrees. A degree of longitude on the Earth's surface varies in length as the cosine of the latitude, being 69.95 statute miles at the equator, 53.43 miles at lat. 40°, and zero at the poles; it represents 4 minutes of time, so that 15 degrees of longitude is equivalent to a difference of one hour of local time. Longitude may also be measured as the angle at the poles lying between the two planes that intersect along the Earth's axis to produce the two meridians. Abbrev: long. Symbol λ. See also: *astronomic longitude; geodetic longitude; geographic longitude; celestial longitude.* Cf: *meridian.* (b) A linear coordinate distance measured east or west from a specified north-south line of reference; e.g. *easting* and *westing.*—Cf: *latitude.*

longitude correction The east-west corrections made to observed magnetic-field intensities by subtracting the Earth's normal field. Cf: *latitude correction [magnet].*

longitude difference *departure.*

longitudinal Said of an entity that is extended lengthwise; esp. said of a topographic feature that is oriented parallel to the general strike or topographic trend of a region. Ant: *transverse [geomorph].*

longitudinal band *Foliation* in a glacier that is parallel to the direction of ice movement.

longitudinal coastline *concordant coastline.*

longitudinal conductance The product of the average conductivity and thickness of a rock layer, measured in siemens units.

longitudinal consequent stream A consequent stream whose direction is determined by the plunge of a fold; esp. a stream flowing in a synclinal trough.

longitudinal crevasse A crevasse roughly parallel to the direction of ice movement. This type of *crevasse* in a valley glacier is longitudinal only in the center of the glacier; away from the center it becomes a *splaying crevasse* (the preferred term).

longitudinal drift A long, tapered, sharp-crested *sand drift* formed on the lee side of a narrow gap in a ridge or scarp oriented transversely to the prevailing wind, esp. in a desert or steppe region where ridges interrupt flat plains or plateaus; it may be 500 m in length.

longitudinal dune A long, narrow sand dune, usually symmetrical in cross profile, oriented parallel with the direction of the prevailing wind; it is wider and steeper on the windward side but tapers to a point on the leeward side, and commonly forms behind an obstacle in an area where sand is abundant and the wind is strong and constant. Such dunes may be a few meters high and up to 100 km long. See also: *seif.*

longitudinal fault A fault whose strike is parallel with that of the general structural trend of the region.

longitudinal flagellum A thread-shaped flagellum in a dinoflagellate, trailing after the body and arising from the posterior pore in the sulcus if two are present, its proximal part lying in the ventral sulcus near the major axis.

longitudinal fold A fold whose axis trends in accordance with the general strike of the area's structures. Cf: *discordant fold.* Syn: *strike fold.*

longitudinal furrow One of many closely spaced continuous furrows elongated parallel to the current and developed rhythmically on all or part of a stream bed. Furrows are separated by regularly spaced *longitudinal ridges,* the separation ranging from 3 mm to as much as 5 cm (Dzulynski and Walton, 1965, p. 61). The furrows are rounded in section; they begin upstream in a convex "beak" and are broken along their length by occasional cuspate bars. They are commonly preserved as *sole marks.* The term was first defined by Dzulynski and Walton (1963, p. 285). Cf: *furrow cast; furrow flute cast; rib and furrow.*

longitudinal joint A steeply dipping joint plane in a pluton that is oriented parallel to the lines of flow. Syn: *S-joint; (hO1) joint; bc-joint.*

longitudinal moraine A moraine rampart consisting of a medial moraine and an englacial moraine of a former glacier (Schieferdecker, 1959, term 0920).

longitudinal profile (a) The profile of a stream or valley, drawn along its length from the source to the mouth of the stream; it is the straightened-out, upper edge of a vertical section that follows the winding of the stream or valley. See also: *thalweg.* Cf: *cross profile.* Syn: *long profile; valley profile; stream profile; river profile.* (b) A similar profile of a landform, such as a pediment.

longitudinal resistivity Resistivity of rock measured along the direction of bedding. Cf: *transverse resistivity.*

longitudinal ridge One of many closely spaced continuous ridges elongated parallel to the current and developed rhythmically on all or part of a stream bed. Ridges are separated by regularly spaced *longitudinal furrows,* the separation ranging from 3 mm to as much as 5 cm (Dzulynski and Walton, 1965, p. 61). Cf: *rib and furrow.*

longitudinal ripple mark A ripple mark with a relatively straight crest, formed parallel to the direction of the current, such as one related to oscillatory wave action (Straaten, 1951); its profile may be asymmetric or symmetric. See also: *corrugated ripple mark; mud-ridge ripple mark.*

longitudinal section A diagram drawn on a vertical or inclined plane and parallel to the longer axis of a given feature; e.g. a section drawn parallel to the strike of a vein, the length of a valley, or the axis of a fossil. Cf: *cross section.*

longitudinal septum A septum in certain cirripede crustaceans disposed normal to the inner and outer laminae of a compartmental plate and separating the parietal tubes. Syn: *parietal septum.*

longitudinal stream A *subsequent stream* that follows the strike of the underlying strata.

longitudinal valley (a) A *subsequent valley* developed along or in the same direction as the general strike of the underlying strata; a valley at right angles to a consequent stream. This is the current usage of the term, as used by Powell (1873, p. 463). (b) A term originally applied by Conybeare & Phillips (1822, p. xxiv) to a long valley developed parallel to the general trend of a ridge, range, or chain of mountains or hills. According to current usage, the term is correctly used only "where the mountain or hill ranges are parallel to the strike" (Stamp, 1961, p. 300). Syn: *lateral valley.*—Cf: *transverse valley.*

longitudinal wave *P wave.*

Longmyndian A division of the Precambrian in Shropshire, England.

long period A period of seismic activity that is more than six seconds in duration.

long profile *longitudinal profile.*

long-range fossil A fossil taxon that possesses an extensive vertical range and that may be expected to occur through a great thickness of strata.

long-range order (a) A strong tendency for the random atoms in a random solid solution to become ordered as the solution cools from the elevated temperature at which it was formed. (b) That state of a crystal structure in which a given atomic species occupies the same specific site in each unit cell, so that the probability of finding atoms on wrong sites is zero. Cf: *short-range order.*

longshore Pertaining or belonging to the shore or coast, or a seaport; littoral. Syn: *alongshore.*

longshore bar A low, elongate sand ridge, built chiefly by wave action, occurring at some distance from, and extending generally parallel with, the shoreline, being submerged at least by high tides, and typically separated from the beach by an intervening *trough.* Syn: *ball; offshore bar; submarine bar; barrier bar.*

longshore current *littoral current.*

longshore drift *littoral drift.*

longulite A cylindrical or conical belonite thought to have formed by the coalescence of globulites. See also: *bacillite.*

longwall Said of a method of underground mining in flat-lying strata, esp. of coal. Parallel entries are driven into the seam, to the limit of the block to be mined; from the end of these entries, workings are driven at right angles in both directions. A longwall face is produced as these workings are widened back toward the point of entry. Working space is provided by timbers or other supports; the roof caves as mining progresses. Almost all of the coal or other desired mineral is recovered, in contrast to the *room-and-pillar* method.

long wave [seis] An obsolete syn. of *surface wave [seis].*

long wave [water] *shallow-water wave.*

lonsdaleite A meteorite mineral consisting of a form of carbon. It is hexagonal and is polymorphous with diamond, graphite, and chaoite.

lonsdaleoid dissepiment In rugose corals, a dissepiment lying between the peripheral ends of septa and the outer wall, as in *Lonsdaleia.* They are usually larger than other types of dissepiments (Jull, 1967, p. 622). See also: *lonsdaleoid septum.*

lonsdaleoid septum A rugose corallite septum characterized by discontinuity toward the peripheral edge of the septum, as in *Lonsdaleia.* See also: *lonsdaleoid dissepiment.*

look direction The direction in which pulses of microwave energy are transmitted by a *SLAR* system. It is normal to the flight direction. Syn: *range direction.*

loop A pattern of field observations that begin and end at the same point, with a number of intervening observations. Such a pattern is useful in correcting for drift in gravity-meter observations, for diurnal variation in magnetometer surveys, and for faults or other cause of misclosure in seismic shooting.

loop [coast] *looped bar.*

loop [geophys] A pattern of field observations that begins and ends at the same point with a number of intervening measurements. Such a pattern is useful in correcting for drift in gravity-meter observations and for diurnal variation in magnetometer surveys, and in detecting faults or other causes of misclosure in seismic shooting.

loop [glac geol] *loop moraine.*

loop [paleont] A support (brachidium) for a brachiopod lophophore, composed of secondary shell and extending anteriorly from crura as a closed apparatus, variably disposed and generally ribbonlike, with or without a supporting septum from the floor of the brachial valve (TIP, 1965, pt. H, p. 147).

loop [waves] *antinode.*

loop bar *looped bar.*

loop bedding Bedding characterized by small groups of laminae that are sharply constricted or that end abruptly at intervals, giving the effect of long, thin loops or links of a chain; it is found in fine calcareous sediments and in oil shale.

looped bar A curved bar on the leeward or landward side of an offshore island undergoing wave erosion, formed by the union of two separate spits that have trailed off behind and joined together to form a loop that encloses or nearly encloses a body of water. Cf: *cuspate bar.* Syn: *loop; loop bar.*

loop lake *oxbow lake.*

loop moraine An end moraine of a valley glacier, shaped like an arc or half-loop, concave toward the direction from which the ice approached; it is usually steep on both sides and extends across the valley. Syn: *valley-loop moraine; moraine loop; loop [glac geol].*

loose ice A rarely used syn. for *broken ice.*

loose-snow avalanche A snow avalanche that starts at a point and widens downhill, in snow lacking cohesion. Cf: *wind-slab avalanche; slab avalanche.*

loose suture An externally visible suture between movably united crinoid ossicles. Cf: *close suture.*

lopadolith A basket-shaped coccolith opening distally.

loparite A brown to black mineral: $(Ce,Na,Ca)_2(Ti,Nb)_2O_6$. It was formerly regarded as a variety of perovskite containing alkalies and cerium.

lopezite An orange-red mineral: $K_2Cr_2O_7$.

lophophore A feeding organ in bryozoans and brachiopods, usually consisting of a circular or horseshoe-shaped fleshy ridge surrounding the mouth and bearing the tentacles that serve to engulf food particles and provide a respiratory current. In brachiopods, it is a feeding organ with filamentous appendages, symmetrically disposed about the mouth, suspended from the anterior body wall or attached to the dorsal mantle, and occupying the mantle cavity (TIP, 1965, pt.H, p.147). See also: *brachia.*

lophophytous Said of a sponge that is fastened to the substrate by a tuft of spicules.

lophotrichous Said of a bacterial cell having a tuft of flagella at one or both ends. Cf: *monotrichous; peritrichous.*

lopolith A large, concordant, typically layered igneous intrusion, of plano-convex or lenticular shape, that is sunken in its central part owing to sagging of the underlying country rock.

Lorac A hyperbolic radio location system similar to loran, in which two or more fixed transmitters emit continuous waves and the position of a mobile receiver in the resulting standing-wave pattern is determined by measuring the phase difference of the waves emanating from two of the transmitters. The useful range is about 200 nautical miles. A trade name. Etymol: *lo*ng-*r*ange *ac*curacy.

loran Any of various long-range radio position-fixing systems by which hyperbolic lines of position are determined by measuring the difference in arrival times of synchronized pulse signals from two or more fixed transmitting radio stations of known geographic position. Loran fixes may be obtained at a range of 1400 nautical miles at night. Cf: *shoran.* Etymol: *lo*ng-*ra*nge *n*avigation.

lorandite A vivid red or dark lead-gray monoclinic mineral: $TlAsS_2$.

loranskite A black mineral: $(Y,Ce,Ca,Zr)TaO_4$ (?).

lorenzenite A dark-brown to black mineral: $Na_2Ti_2Si_2O_9$. Syn: *ramsayite.*

lorettoite A honey-yellow mineral: $Pb_7O_6Cl_2$.

lorica (a) A hard, protective organic covering, commonly tubular or vaselike, secreted or built with agglutinated foreign matter by tintinnids, thecamoebians, certain algae, and other protists, and having a calcareous or siliceous composition. (b) The cell wall or two valves of a diatom.—Pl: *loricae.*

loseyite A bluish-white mineral: $(Mn,Zn)_7(CO_3)_2(OH)_{10}$.

losing stream *influent stream.*

löss Var. of *loess.* Etymol: German *Löss.*

lost circulation The condition during *rotary drilling* when the drilling mud escapes into porous, fractured, or cavernous rocks penetrated by the borehole and does not return to the surface.

lost mountain An isolated mountain standing in a desert and so far removed from the main mass of mountains as to have no apparent connection with them; e.g. an outlier or a monadnock that has resisted erosion more effectively than the surrounding land. A smaller feature is called a "lost hill". Cf: *huerfano.*

lost stream (a) A dried-up stream in an arid region. (b) A *sinking stream.*

Lotharingian European stage: uppermost lower Lower Jurassic (above Sinemurian, below Pliensbachian).

lotrite *pumpellyite.*

lottal A field term used by King (1962, p. 179) for the aqueous clayey mixtures formed by mass movement down hillslopes. Etymol: Richard Armour's verse, "Shake and shake the catsup bottle. None will come, and then a lot'll."

louderback A remnant of a lava flow appearing in a tilted fault block and bounded by a dip slope. It is named after George D. Louderback, a North American geologist, who used it as evidence of block faulting in basin-and-range topography.

louderbackite *roemerite.*

loughlinite A pearly-white, asbestiform mineral: $Na_2Mg_3Si_6O_{16} \cdot 8H_2O$.

loupe Any small magnifying glass mounted for use in the hand or so that it can be held in the eye socket or attached to spectacles, and used to study gemstones.

louver A transverse wall plate in archaeocyathids, commonly developed between the edges of adjacent septa or longitudinal ribs and usually tilted with reference to the wall surface (TIP, 1972, pt. E, p. 41).

lovchorrite *mosandrite.*

lovdarite An orthorhombic mineral: $(Na,K,Ca)_4(Be,Al)_2Si_6O_{16} \cdot 4H_2O$.

love arrow *flèche d'amour.*

Lovenian system A numbering system in which the individual ambulacral areas of the tests of echinoids and edrioasteroids are designated by Roman numerals (I-V), and the interambulacral areas by Arabic numerals (1-5). It is based on bilateral symmetry with respect to a plane passing through the apical system, peristome, and periproct in irregular echinoids, and chiefly by the position of the madreporite in regular echinoids. Named after Sven L. Lovén (1809-1895), Swedish zoologist.

Love wave A type of *surface wave* having a horizontal motion that is shear or transverse to the direction of propagation. Its velocity depends only on density and rigidity modulus, and not on bulk modulus. It is named after A.E.H. Love, the English mathematician who discovered it. See also: *G wave.* Syn: *Q wave;*

Querwellen wave.

lovozerite A mineral: $(Na,Ca)_3(Zr,Ti)Si_6(O,OH)_{18}$.

low [beach] (a) *swale.* (b) *trough.*

low [geomorph] *law [geomorph].*

low [meteorol] *depression [meteorol].*

low [struc geol] n. A general term for such features as a structural basin, a syncline, a saddle, or a sag. Cf: *high.* Syn: *structural low.*

low albite Low-temperature albite common in nature, stable below 450°C. It takes almost no calcium or potassium into solid solution, and has a completely ordered structure. Cf: *high albite.*

low and ball A descriptive name for a *longshore bar* ("ball") separated by a distinct longitudinal *trough* ("low") lying parallel to the shoreline in the shoreface or offshore region along a seashore or a lake shore.

low-angle cross-bedding Cross-bedding in which the cross-beds have an average maximum inclination of less than 20° (McKee & Weir, 1953, p.388). Cf: *high-angle cross-bedding.*

low-angle fault A fault with a dip of 45° or less. Cf: *high-angle fault.*

low-angle thrust *overthrust.*

low chalcocite Orthorhombic chalcocite, stable below 105°C.

löweite *loeweite.*

low-energy coast A coast protected from strong wave action by headlands, wide gently sloping bottom, dominance by winds from the land, or other factors, and characterized by average breaker heights of less than 10 cm. Cf: *high-energy coast; moderate-energy coast; zero-energy coast.*

low-energy environment An aqueous sedimentary environment characterized by a low *energy level* and by standing water or a general lack of wave or current action, thereby permitting very fine-grained sediment to settle and accumulate; e.g. a coastal lagoon or an alluvial swamp. Cf: *high-energy environment.*

lower Pertaining to rocks or strata that are normally below those of later formations of the same subdivision of rocks. The adjective is applied to the name of a chronostratigraphic unit (system, series, stage) to indicate position in the geologic column and corresponds to *early* as applied to the name of the equivalent geologic-time unit; e.g. rocks of the Lower Jurassic System were formed during the Early Jurassic Period. The initial letter of the term is capitalized to indicate a formal subdivision (e.g. "Lower Devonian") and is lowercased to indicate an informal subdivision (e.g. "lower Miocene"). The informal term may be used where there is no formal subdivision of a system or of a series. Cf: *upper; middle [stratig].*

lower break *foot [geol].*

Lower Carboniferous In European usage, the approximate equivalent of the *Mississippian.* Cf: *Upper Carboniferous.*

lower core A term that includes the *inner core* and the transitional zone of the *outer core,* i.e. the equivalent of the *F layer* and the *G layer.*

lower crustal layer *Conrad layer.*

lower keriotheca The adaxial (lower) part of *keriotheca* in the wall of a fusulinid, characterized by coarse alveolar structure (as in *Schwagerina*). Cf: *upper keriotheca.*

lower low-water datum An approximation to the plane of *mean lower low water,* adopted as a standard reference plane for a specific area (as the Pacific coast of the U.S.) and retained for an indefinite period although it may differ slightly from a later (and better) determination. Cf: *low-water datum.*

lower mantle That part of the *mantle* that lies below a depth of about 1000 km and has a density of 4.7 g/cm^3, in which the seismic velocity increases slowly with depth. It is equivalent to the *D layer.* Syn: *inner mantle; mesosphere.*

lower Paleolithic n. The first and oldest division of the *Paleolithic,* characterized by *Australopithecus* and *Homo erectus.* Correlation of cultural levels with actual age (and, therefore, with the time-stratigraphic units of geology) varies from region to region. Cf: *middle Paleolithic; upper Paleolithic.*—adj. Pertaining to the lower Paleolithic.

lower plate The footwall of a fault. Cf: *upper plate.*

Lower Silurian An old syn. of *Ordovician.*

lower tectorium The adaxial secondary dark layer of spirotheca in the wall of a fusulinid, next below the diaphanotheca or tectum, as in *Profusulinella.* Cf: *tectorium; upper tectorium.*

Lower Volgian European stage: Upper Jurassic (above Kimmeridgian, below Upper Volgian). See also: Portlandian.

low-flow frequency curve A graphic illustration of both the magnitude and frequency of minimum flows in a given time span.

low-grade Said of an ore with a relatively low ore-mineral content. Syn: *lean.* Cf: *high-grade.* See also: *grade [ore dep].*

low island In the Pacific Ocean, a coralline rather than a volcanic island. Cf: *high island.*

lowland (a) A general term for low-lying land or an extensive region of low land, esp. near the coast and including the extended plains or country lying not far above tide level. (b) The low and relatively level ground of a region, in contrast with the adjacent, higher country; e.g. a vale between two cuestas. (c) A low or level tract of land along a watercourse; a *bottom.*—The term is usually used in the plural. Ant: *upland.*

low-latitude desert *tropical desert.*

low marsh The flat, usually bare ground situated seaward of a *salt marsh* and regularly covered and uncovered by the tides (Carey & Oliver, 1918, p. 166); e.g. a *mud flat.* Syn: *slob land.*

lowmoor bog A bog that is at or only slightly above the water table, on which it depends for accumulation and preservation of peat (chiefly the remains of sedges, reeds, shrubs, and various mosses). Cf: *highmoor bog.*

lowmoor peat Peat occurring on low-lying moors or swamps and containing little or no sphagnum. Its moisture is standing surface water and is low in acidity. Ash and nitrogen content is high; cellulose content is low. Cf: *highmoor peat.* Syn: *fen peat.*

low oblique n. An *oblique* that does not include the apparent horizon. Cf: *high oblique.*

low quartz Low-temperature quartz; specif. *alpha quartz.*

low-rank graywacke A term introduced by Krynine (1945) for a graywacke in which feldspar is almost absent. It is related to miogeosynclines. The rock is equivalent to *subgraywacke* as originally defined by Pettijohn (1949) and to *lithic graywacke* of Pettijohn (1954). Cf: *high-rank graywacke.*

low-rank metamorphism Metamorphism that is accomplished under conditions of low to moderate temperature and pressure. Cf: *high-rank metamorphism.*

lowstand The interval of time during one or more *cycles of relative change of sea level* when sea level is below the shelf edge (Mitchum, 1977, p. 208). Cf: *highstand.*

low tide The tide at its lowest; the accepted popular syn. of *low water* in the sea.

low-tide delta A delta formed at the *step* by drainage of water from the beach onto the tidal flat. It is associated with enlarged and accentuated rills.

low-tide platform *solution platform.*

low-tide terrace A relatively horizontal zone of the foreshore near the low-water line.

low-velocity-layer correction *weathering correction.*

low-velocity zone (a) *low-velocity layer.* (b) The zone in the upper mantle, variously defined as from 60 to 250 km in depth, in which velocities are about 6% lower than in the outermost mantle. It is probably caused by the near-melting-point temperature of the material. Syn: *Gutenberg low-velocity zone; B layer.* (c) A region inside the core boundary below a depth of 2900 km which produces a *shadow zone [seis]* at the Earth's surface.

low-volatile bituminous coal Bituminous coal, characteristically agglomerating, that contains 15-22% volatile matter, analyzed on a dry, mineral-matter-free basis. Cf: *high-volatile bituminous coal; medium-volatile bituminous coal.*

low water Water at the minimum level reached during a tide cycle. Abbrev: LW. Cf: *high water.* Syn: *low tide.*

low-water datum An approximation to the plane of *mean low water,* adopted as a standard reference plane for a specific area (as the Atlantic coast of the U.S.) and retained for an indefinite period although it may differ slightly from a later (and better) determination. Cf: *lower low-water datum.*

loxochoanitic Said of a short, straight retrochoanitic septal neck of a nautiloid that points obliquely toward the interior of the siphuncle.

loxoclase A variety of orthoclase containing considerable sodium: $(K,Na)AlSi_3O_8$. It has a green tinge due to small inclusions of diopside. The loxoclase series ranges from Or_1Ab_1 to Or_1Ab_4, with K_2O in the range of 4-7%. Syn: *soda orthoclase.*

loxodrome *rhumb line.*

loxodromic curve *rhumb line.*

LPG *liquefied petroleum gas.*

L-tectonite A tectonite whose fabric is dominated by the presence of lineations, such as a deformed conglomerate in which the pebbles are strongly elongate. Cf: *S-tectonite.*

lublinite A very soft, cheesy or spongy mixture of calcite and water. See also: *moonmilk.* Syn: *rock milk; mountain milk.*

lubricating layer In a décollement, that stratum which acted as a lubricant for the gliding of the overthrust. See also: *sole [fault].*

lucinite *variscite.*

lucinoid Said of heterodont dentition of a bivalve mollusk with two cardinal teeth in each valve, the anterior tooth in the left valve occupying a median position below the beaks. Cf: *corbiculoid.*

luclite A fine-grained *diorite* composed chiefly of plagioclase, hornblende, and occasionally a small amount of quartz. It is somewhat coarser grained than *malchite,* which it otherwise resembles. Obsolete.

Lüder's lines Planar deformation features, wider than ordinary shear fractures, inclined along planes of high shear stress, on which plastic or cataclastic deformation is concentrated.

Ludian European stage: uppermost Eocene (above Bartonian, below Tongrian of Oligocene).

ludlamite A green monoclinic mineral: $(Fe,Mg,Mn)_3(PO_4)_2 \cdot 4H_2O$.

ludlockite A triclinic mineral: $(Fe,Pb)As_2O_6$.

Ludlovian European stage: Upper Silurian (above Wenlockian, below Gedinnian of Devonian).

ludwigite A blackish-green orthorhombic mineral: $(Mg,Fe^{+2})_2Fe^{+3}BO_5$. It is isomorphous with vonsenite. Syn: *magnesioludwigite.*

lueneburgite A colorless mineral: $MgIF3B_2(PO_4)_2(OH)_6 \cdot 5H_2O$.

lueshite An orthorhombic mineral: $NaNbO_3$. It has structure of the perovskite type and is dimorphous with natroniobite. Syn: *igdloite.*

lugarite A coarse-grained porphyritic *ijolite* that contains analcime in place of nepheline. Phenocrysts of barkevikite, titanaugite, and zoned labradorite occur in an analcime groundmass. The name, given by Tyrrell in 1912, is for Lugar, Ayrshire, Scotland. Not recommended usage.

luhite A calcite-rich dike rock with olivine and titanaugite in a groundmass of clinopyroxene, melilite, hauyne, nepheline, and calcite; a hauyne-melilite *damkjernite* or hauyne-nepheline *alnoite.* The rock was named by Scheumann in 1913 for Luh, Czechoslovakia. Not recommended usage.

Luisian North American stage: Miocene (above Relizian, below Mohnian).

lujavrite A coarse-grained trachytic eudialyte-bearing *nepheline syenite,* containing thin parallel feldspar crystals with interstitial nepheline grains and acicular acmite crystals. The rock was originally described by Brögger in 1890 from Luijaur (Lujavr), now Lovozero, Kola Peninsula, U.S.S.R. Also spelled: *lujaurite; luijaurite; lujauvrite.* Cf: *chibinite.* Not recommended usage.

lumachelle (a) A compact, dark-gray or dark-brown limestone or marble, composed chiefly of fossil mollusk shells, and characterized by a brilliant iridescence or chatoyant reflection from within. Syn: *fire marble.* (b) Any accumulation of shells (esp. oysters) in stratified rocks.—Etymol: French, "coquina, oyster bed", from Italian *lumachella,* "little snail".

lumen (a) A small, round central open space through a columnal of a crinoid, blastoid, or cystoid; e.g. the wide cavity in a short, ringlike, proximal columnal of many cystoids. (b) One of the spaces between muri of pollen and spores exhibiting reticulate sculpture.—Pl: *lumina.* Adj: *luminal.*

lumen pore A pseudopore on the frontal surface of a costa in some cribrimorph cheilostomes (bryozoans), opening into the costal lumen. Cf: *pelma; pelmatidium.*

lumina Plural of *lumen.*

luminescence The emission of light by a substance that has received energy or electromagnetic radiation of a different wavelength from an external stimulus; also, the light so produced. It occurs at temperatures lower than those required for incandescence. See also: *phosphorescence; fluorescence.*

lump (a) A descriptive term applied to a composite, lobate grain in recent carbonate sediments, believed to have formed by aggregation, flocculation, or clotting of two or more pellets, ooliths, skeletons, etc., or fragments thereof, or by disruption of newly deposited or partly indurated carbonate mud. It typically possesses surficial re-entrants. See also: *megalump.* (b) *mudlump.*

lump graphite Cryptocrystalline or very finely crystalline natural graphite from vein deposits, occurring in particle sizes ranging from that of walnuts to finer than 60-mesh.

lumping In taxonomy, the practice of ignoring minor differences in the recognition or definition of species and genera. A taxonomist known for his frequent lumping of taxa is called a "lumper" Cf: *splitting.*

lump limestone A limestone containing numerous lumps (such as aggregates of pellets or ooliths) in a matrix of micrite, e.g. some of the Cenozoic limestones of the western interior of the U.S.

lumpy Said of a gemstone characterized by a thick cut (too great a depth in proportion to its width).

lunabase Dark lunar surface rocks of basic (basaltic?) composition; e.g. *marebase.* An obsolete term, introduced by Spurr in 1944. Cf: *lunarite.*

lunar (a) Pertaining to or occurring on the Moon, such as a "lunar probe" designed to pass close to the Moon, or "lunar dust" consisting of fine-grained material produced by meteoritic bombardment. (b) Resembling the surface of the Moon, such as the "lunar landscape" of certain glaciers.

lunar crater *crater [lunar].*

lunar day The time required for the Earth to rotate once with respect to the Moon, or the interval between two successive upper transits of the Moon across a local meridian; it is about 24.84 hours (24 hours and 50 minutes) of solar time. Cf: *tidal day.*

lunar geology A science that applies geologic principles and techniques to the study of the Moon, esp. its composition and the origin of its surface features. See also: *selenology.*

lunarite A general term for light-toned, brightly reflecting surface rocks of the lunar highlands (terrae). Term introduced by Spurr (1944). Cf: *lunabase.*

lunarium A skeletal structure in some stenolaemate bryozoans that is part of the zooecial wall, occurs on the proximal or lateral side of a feeding zooid, and generally produces a troughlike depression along the length of the zooidal chamber. It commonly projects beyond the remainder of the zooecial wall. Pl: *lunaria.*

lunar microcrater *micrometeorite crater.*

lunar playa A relatively small, level area of the Moon's surface, as much as a few kilometers long, occupying a low place in the ejecta blankets surrounding lunar craters such as Tycho and Copernicus. It is believed to be either a *fallback* deposit or a small lava flow.

lunar regolith A thin, gray layer on the surface of the Moon, perhaps several meters deep, consisting of partly cemented or loosely compacted fragmental material ranging in size from microscopic particles to blocks more than a meter in diameter. It is believed to be formed by repeated meteoritic and secondary fragment impact over a long period of time. Syn: *lunar soil; soil [lunar].*

lunar soil *lunar regolith.*

lunar tide The part of the tide caused solely by the tide-producing force of the Moon. Cf: *solar tide.* Syn: *moon tide.*

lunar transient phenomenon A temporarily abnormal appearance of a small area of the Moon, generally involving brightening, darkening, obscuration of surface features, or significant color changes, especially in the red and blue. Durations range from a few seconds to about a day, with a typical duration of about 30 minutes. The areas involved are generally part of a single crater or an isolated mountain but may occasionally include a whole crater or larger areas. These phenomena are strongly distributed around mare margins, as in the occurrence of dark-floored and dark-haloed craters, domes, and sinous rilles. They are generally thought to arise from some type of internal volcanic action involving outgassing or extrusion of lava (Cameron, 1972). Abbrev: *LTP.*

lunar varnish A hypothetical substance forming a dark coating on lunar particles and suggested as responsible for the low albedo of shallow subsurface lunar regolith.

lunate bar A crescent-shaped bar commonly found off a pass between barrier islands, at the entrance to a harbor, or at a stream mouth.

lunate fracture A *crescentic mark [glac geol]* that is similar to a *crescentic fracture* but consists of two fractures from between which rock has been removed.

lunate mark *crescentic mark [glac geol].*

lunate ripple mark A type of ripple mark in which the crest lines are strongly curved and open out downcurrent: the form is similar to that of a *barchan* and has the opposite orientation to that of a *linguoid ripple mark.* Cf: *cuspate ripple mark.*

lunate sandkey A *lunate bar* that has been built up above the water surface to form a crescent-shaped island, as along the west coast of Florida.

lundyite An intrusive *granite* or *quartz syenite* characterized by orthophyric texture, a high content of alkali minerals, and a kataphoritelike amphibole. The name, given by Hall in 1914, is for Lundy Island in the Bristol Channel, England. Not recommended usage.

lunette A term proposed by Hills (1940) for one of the broad low even-crested crescentic mounds or ridges, rarely more than 6-9 m high, of clay loam or silty clay bordering the leeward (eastern) shore of almost every lake and swamp in the plains of northern Victoria, Australia; it is produced by dust-laden winds.

lungfish *Dipnoi.*

lunitidal interval The interval between the Moon's transit over the local or Greenwich meridian and the time of the following high water or low water. Syn: *retardation [tides].*

lunker A Scottish term for a lenticular mass of sandstone or clay ironstone; a big nodule.

lunoid furrow *crescentic gouge.*

lunula A crescentic marking on the selenizone of certain gastropods, formed by growth increments.

lunule (a) A flat or curved, commonly cordate, area in front of the beak on the outside of many bivalve (pelecypod) shells, corresponding to the anterior part of the cardinal area, and distinguished from the remainder of the shell surface by a sharp change in angle. (b) One of the openings in an echinoid test from the aboral surface through the oral surface at a perradial or interradial suture.

lunulitiform Said of conical or discoid colonies of cheilostome bryozoans in late stages of growth.

lurain Lunar terrain.

lusakite A variety of staurolite containing cobalt.

luscladite An olivine *theralite* or *essexite* having hyperite texture and characterized by the absence of hornblende and by the presence of biotite as well as olivine. Alkali feldspar forms the reaction rim around the plagioclase. Nepheline is not abundant and fills the interstices. Cf: *berondrite; kylite.* Named by Lacroix in 1920 for Ravin de Lusclade, Mont Dore, France. Not recommended usage.

Lusitanian European stage: Upper Jurassic (above Oxfordian, below Kimmeridgian). It includes the Argovian, Rauracian, and Sequanian substages.

lusitanite (a) In the *IUGS classification,* a plutonic rock in which Q is less than 5 or F is less than 10, P/(A+P) is less than 10, and M is between 45 and 75. (b) A dark-colored albite *syenite* composed of riebeckite, acmite, alkali feldspar and minor amounts of quartz. Its name, given by Lacroix in 1916, is derived from Lusitania (i.e. Portugal).

luster The reflection of light from the surface of a mineral, described by its quality and intensity; the appearance of a mineral in reflected light. Terms such as metallic or resinous refer to general appearance; terms such as bright or dull refer to intensity.

luster mottling [ign] The macroscopic appearance of *poikilitic* rocks. The term was originated by Raphael Pumpelly (Johannsen, 1939, p.183).

luster mottling [sed] The shimmering appearance of a broken surface of a sandstone cemented with calcite, produced by the brilliant reflection of light from the cleavage faces of conspicuously large and independently oriented calcite crystals, a centimeter or more in diameter, incorporating colonies of detrital sand grains; e.g. the luster displayed by *Fontainebleau sandstone.* It may also develop locally in barite, gypsum, or dolomite cements.

lusungite A rhombohedral mineral: $(Sr,Pb)Fe_3(PO_4)_2(OH)_5 \cdot H_2O$.

lutaceous Said of a sedimentary rock formed from mud (clay- and/or silt-size particles) or having the fine texture of impalpable powder or rock flour; pertaining to a lutite. Also said of the texture of such a rock. Term introduced by Grabau (1904, p.242). Cf: *argillaceous; pelitic.*

lutalite A glassy leucite *nephelinite* or olivine *leucitite,* with more than 50 percent mafic minerals and a higher sodium-to-potassium ratio than most olivine leucitites. Named by Holmes in 1937 at Bufurnbira, Zaire. Not recommended usage.

lutecite Fibrous chalcedony characterized by inclined extinction and by fibers that are seemingly elongated about 30° to the *c*-axis. Syn: *lutecin.*

luteous Having an essential proportion of muddy sediment in a limestone, recognized by the presence of many particles of clastic quartz of silt size. Term introduced by Phemister (1956, p. 73).

Lutetian European stage: Eocene (above Ypresian, below Priabonian). See also: *Bruxellian.*

lutetium-hafnium age method The determination of an age in years based on the known radioactive decay rate of lutetium-176 (half-life approximately 2.2 x 10^{10} years) to hafnium-176. The method can be used, under favorable conditions, for dating minerals containing rare earths.

lutite A general name used for consolidated rocks composed of silt and/or clay and of the associated materials which, when mixed with water, form mud; e.g. shale, mudstone, and calcilutite. The term is equivalent to the Greek-derived term, *pelite,* and was introduced as *lutyte* by Grabau (1904, p.242) who used it with appropriate prefixes in classifying fine-grained rocks (e.g. "anemolutyte", "anemosilicilutyte", "hydrolutyte", and "hydrargillutyte"). Etymol: Latin *lutum,* "mud". See also: *rudite; arenite.*

lutyte Var. of *lutite.*

luxullianite A *granite* characterized by phenocrysts of potassium feldspar and quartz which enclose clusters of radially arranged acicular tourmaline crystals in a groundmass of quartz, tourmaline, alkali feldspar, brown mica, and cassiterite. Its name is derived from Luxulyan, Cornwall. Also spelled: *luxulianite; luxulyanite.* Var: *luxuliane.* Not recommended usage.

luzonite An isometric mineral: Cu_3AsS_4. It is dimorphous with enargite, and was formerly regarded as a variety of famatinite containing arsenic. Cf: *sinnerite.*

LV *left valve.*

LVL *low-velocity layer.*

LW *low water.*

L wave *surface wave [seis].*

lyas *lias.*

lychnisc A dictyonal hexactin (sponge spicule) in which the central crossing of the rays is replaced by an open octahedron with the rays arising from the octahedral angles.

lycopod A pteridophyte characterized by dense, simple, spirally arranged leaves, and by spore-bearing organs situated in the leaf axils, which in some forms produce stromboli or cones; a member of the Lycopodineae, coextensive with the class Lycopsida (Melchior and Werdermann, 1954, p. 273). Lycopods range from the Devonian, and include the club mosses. Syn: *lycopsid.*

lycopsid A member of the subdivision Lycopsida; a *lycopod.*

Lydian stone A *touchstone* consisting of a compact, extremely fine-grained, velvet- or gray-black variety of *jasper.* Etymol: Greek *Lydia,* ancient country in Asia Minor. Syn: *lydite; basanite.*

lydite A syn. of *Lydian stone.* Also spelled: *lyddite.*

lyell (a) A block of rock transported and released by an iceberg; an ice-rafted block (Hamelin, 1961, p. 202). (b) Imperfectly stratified deposit resembling till, but transitional between till and clay, produced by sedimentation of till into water through the agency of floating ice; sufficient depth of water is required for flotation of bergs (Lougee & Lougee, 1976).

lyn *linn.*

lynchet A bank of earth that accumulates on the downhill side of an ancient ploughed field as the disturbed soil moves downslope under gravity (Bray & Trump, 1970, p. 137).

lyndochite A variety of aeschynite, relatively high in calcium and thorium.

lyrula In some acrophoran cheilostomes (bryozoans), a median skeletal protuberance projecting over the operculum from the proximal lip of the orifice. Cf: *mucro.* Syn: *lyrule.*

lysimeter A structure used to measure quantities of water used by plants, evaporated from soil, and lost by deep percolation. It consists of a basin, having closed sides and a bottom fitted with a drain, in which soil is placed and plants are grown. Quantities of natural and/or artificial precipitation are measured, the deep percolate is measured and analyzed, water taken up by plants is weighed, etc.

lysocline The level or ocean depth at which the rate of solution of calcium carbonate just exceeds its combined rate of deposition and precipitation.

lyssacine adj. Said of a hexactinellid sponge whose megascleres (spicules) are unfused or separate, incompletely fused, or so fused that their individual boundaries are apparent. Ant: *dictyonine.*—n. A lyssacine sponge.

maar A low-relief, broad volcanic crater formed by multiple shallow explosive eruptions. It is surrounded by a *crater ring,* and may be filled by water. Type occurrence is in the Eifel area of Germany.

Maastrichtian *Maestrichtian.*

macallisterite A mineral: $Mg_2B_{12}O_{20}\cdot15H_2O$. Also spelled: *mcallisterite.*

macaluba A syn. of *mud volcano.* The name is taken from that of a low mud volcano, Macaluba, in Sicily.

macconnellite *mcconnellite.*

macdonaldite A mineral: $BaCa_4Si_{15}O_{35}\cdot11H_2O$.

macedonite [mineral] A mineral: $PbTiO_3$.

macedonite [rock] A fine-grained rock composed of orthoclase, sodic plagioclase, biotite, olivine, and rare pyriboles; an olivine-biotite *trachyte* or *trachyandesite.* Named for Mt. Macedon, Victoria, Australia by Skeats and Summers in 1909. Cf: *woodendite.* Not recommended usage.

maceral One of the organic constituents that comprise the coal mass; all petrologic units seen in polished or thin sections of coal. Macerals are to coal as minerals are to inorganic rock. Maceral names bear the suffix "-inite" (*vitrinite, exinite,* etc.). Cf: *phyteral.* Syn: *micropetrological unit.*

maceration The act or process of disintegrating sedimentary rocks (such as coal and shale) by various chemical and physical techniques in order to extract and concentrate acid-insoluble microfossils (including palynomorphs). It includes mainly chemical treatment by oxidants and alkalies and use of other separating techniques that will remove extraneous mineral and organic constituents. Maceration is widely used in palynology.

macfarlanite A silver ore consisting of a mixture of sulfides, arsenides, etc., and containing cobalt, nickel, and lead. Cf: *animikite.*

macgovernite A mineral: $(Mn,Mg,Zn)_{15}As_2Si_2O_{17}(OH)_{14}$. Also spelled: *mcgovernite.*

Machaeridia A class questionably assigned to the homalozoans and characterized by an elongate, bilaterally symmetrical test composed of an even number of longitudinal columns of plates. They have been variously identified as mollusks, annelids, and arthropods.

macigno The classical *flysch* facies in the northern Apennines, consisting of alternating strata of sandstone and mudstone, and showing graded bedding. Etymol: Italian, "millstone". Pron: mah-*cheen*-yo.

mackayite A green mineral: $FeTe_2O_5(OH)$ (?).

mackelveyite A dark-green or black mineral, approximately: $Na_2Ba_4Ca(Y,U)_2(CO_3)_9\cdot5H_2O$. Also spelled: *mckelveyite.*

Mackereth sampler A variety of *piston corer* that operates by air hoses. It is used in shallow water.

mackinawite A tetragonal mineral: $(Fe,Ni)_{1.1}S$. It occurs as a corrosion product on iron pipes. Syn: *kansite.*

mackinstryite A mineral: $(Ag,Cu)_2S$. Also spelled: *mckinstryite.*

mackintoshite *thorogummite.*

macle [cryst] A twinned crystal; esp. a flat, often triangular, diamond composed of two flat crystals. Etymol: French, "wide-meshed net".

macle [mineral] (a) A dark or discolored spot in a mineral. (b) *chiastolite.*

macled [cryst] Said of a crystal having a twin structure.

macled [mineral] (a) Said of a mineral that is marked like chiastolite. (b) Said of a mineral that is spotted.

maconite A vermiculite from North Carolina.

macrinite A maceral of coal within the *inertinite* group, characterized by a lack of cell structure (ICCP, 1971). It occurs in particles mostly larger than 10 microns across, as a groundmass or in isolated forms.

macro- A prefix meaning "large" or "great". Cf: *micro-.* Syn: *mega-.*

macro-axis The longer lateral axis of an orthorhombic or triclinic crystal; it is usually the *b* axis. Cf: *brachy-axis.*

macrochoanitic Said of a retrochoanitic septal neck of a nautiloid that reaches backward beyond the preceding septum and is invaginated into the preceding septal neck.

macroclastic Said of coal that contains many recognizable fragments. Cf: *microclastic.*

macroclastic rock A clastic rock whose constituents are visible to the unaided eye. Ant: *cryptoclastic rock.*

macroclimate The general climate of an extensive region or a country. See also: *mesoclimate; microclimate.*

macrococcolith One of the larger coccoliths in coccolithophores exhibiting dimorphism but with the dimorphic coccoliths irregularly placed. Cf: *micrococcolith.*

macroconch A mature conch of a chambered cephalopod which, in all respects except size and occasional modification of the aperture, resembles smaller conchs (*microconchs*) found in the same fossil association. Macroconchs are now generally regarded as representing females. Macroconchs and microconchs have in the past been described as different species or even placed in different genera (Callomon, 1963).

macrocrystalline Said of the texture of a rock consisting of or having crystals that are large enough to be distinctly visible to the unaided eye or with the use of a simple lens; also, said of a rock with such a texture. Howell (1922) applied the term to the texture of a recrystallized sedimentary rock having crystals whose diameters exceed 0.75 mm, and Bissell & Chilingar (1967, p.103) to the texture of a carbonate sedimentary rock having crystals whose diameters exceed 1.0 mm. Syn: *megacrystalline; eucrystalline.* See also: *macromeritic; phaneritic.*

macrodome A crystal form of either two or four faces which are parallel to the macroaxis in an orthorhombic crystal. A macrodome with four faces is a rhombic prism, or a *second-order prism* of the orthorhombic system.

macroevolution (a) The evolution or origin of higher taxa (i.e., above species rank; esp. orders or classes), as contrasted to *microevolution.* Syn: *megaevolution.* (b) Evolution occurring in large, complex stages, such as the development of one species from another. Cf: *microevolution.*

macrofabric *megafabric.*

macrofacies *facies tract.*

macrofauna (a) Living or fossil animals large enough to be seen with the naked eye. (b) An obsolete term for the animals occupying a broad area of uniform characteristics; a large or widespread group of animals. Cf: *microfauna; megaflora.* Syn: *megafauna.*

macroflora *megaflora.*

macrofossil A fossil large enough to be studied without the aid of a microscope. Cf: *microfossil.* Syn: *megafossil.*

macrofragmental Said of a coal composed of recognizable fragments or lenses of vegetal matter. Cf: *microfragmental.*

macrograined Said of the texture of a carbonate sedimentary rock having clastic particles whose diameters are greater than one millimeter (Bissell & Chilingar, 1967, p. 103). See also: *megagrained.*

macrohabitat A *habitat* large enough for a human observer to move around in, i.e. tens or hundreds of feet in dimensions. Cf: *microhabitat.*

macroite A microlithotype of coal, a variety of *inertite,* consisting of 95% or more of macrinite (ICCP, 1971).

macrolinear "Any lineation . . . which is two to less than ten kilometers long" (El-Etr, 1976, p. 485).

macrolithology The study of rocks considered as a part of the stratigraphic column in a given area; also, the collective characteristics of such rocks. Cf: *microlithology.*

macromeritic *macrocrystalline; phaneritic.*

macromutation Mutation that is large and easily observed.

macronucleus A relatively large vegetative nucleus that is believed to exert a controlling influence over the trophic activities in the body of a tintinnid. Cf: *micronucleus.*

macrophagous Said of an organism that feeds on relatively large particles. Cf: *microphagous.*

macrophyric *megaphyric.*

macrophyte A megascopic plant, esp. in an aquatic environment.

macropinacoid *front pinacoid.*

macroplankton Plankton of the size range 1 mm to 1 cm. They are larger than *ultraplankton, nannoplankton,* and *microplankton,* but smaller than *megaloplankton.*

macropolyschematic Said of mineral deposits having megascopically distinguishable textural elements. The term is little used.

macropore A pore too large to hold water by capillarity. Syn: *megapore.*

macroporphyritic *megaphyric.*

macropygous Said of a trilobite with a pygidium larger than the cephalon. Cf: *isopygous; micropygous.*

macrorelief A term applied to surface irregularities only when it is necessary to distinguish them from *microrelief.*

macrosclere *megasclere.*

macroscopic (a) *megascopic.* (b) According to Dennis (1967, p.152), a term introduced to describe tectonic features that are too large to be observed directly in their entirety. Cf: *mesoscopic.*

macroseism A syn. of *earthquake,* as opposed to *microseism.*

macrospore A seldom-used and unsatisfactory syn. of *megaspore.*

macrotectonics A term used by Tomkeieff (1943, p.348) as a syn. of *megatectonics.*

macrotherm *megatherm.*

macula [intrus rocks] A local pocket of magma, formed by the fusion of shale, that acts as a type of *magma chamber.* Pl: *maculae.*

macula [paleont] One of the clusters of small, modified zooecia that may be regularly spaced throughout a Paleozoic bryozoan colony, appearing at the surface as a flattened or slightly depressed area typically surrounded by unusually large zooecia. Pl: *maculae.*

maculose Said of a group of contact-metamorphic rocks, e.g. spotted slates, that have a spotted or knotted character; also, said of the structure itself. Syn: *spotted; knotted.*

madeirite A porphyritic *gabbro* composed of large phenocrysts of augite and olivine in a fine-grained groundmass that is less abundant than the phenocrysts and is composed of plagioclase microlites in secondary calcite and magnetite. Its name, given by Gagel in 1913, is derived from Madeira, Spain. Not recommended usage.

made land Man-made land; an area artificially filled with earth materials more or less mixed with refuse. Made land is common along marshy shorelines and at many former *sanitary landfill* sites. See also: *fill [eng geol].*

madocite A mineral: $Pb_{17}(Sb,As)_{16}S_{41}$.

madreporite A porous or sievelike structure that is situated at the distal end of the stone canal in an echinoderm and that provides access to the water-vascular system from the exterior; e.g. a conspicuous plate on the aboral surface of an asteroid or in the right anterior genital plate of an echinoid.

madupite An extrusive rock composed of phlogopite, clinopyroxene, and perovskite phenocrysts in a brown glassy groundmass with the composition of leucite and nepheline; a *lamproite.* The name was coined by Cross in 1897 from an Indian word for "sweetwater", after Sweetwater County, Wyoming, where the Leucite Hills are located. Not recommended usage.

maelstrom A rapid, confused, and often destructive current, formed by the combination of strong wind-generated waves and a strong opposing tidal current; it may display *eddy*-type or *whirlpool*-type characteristics. It typically occurs along the south shore of the Lofoten Islands of Norway.

maenaite A hypabyssal plagioclase-bearing *bostonite* differing from normal bostonite in being richer in calcium and poorer in potassium; an altered *trachyte.* It was named by Brögger in 1894 for Lake Maena, Oslo district, Norway. Not recommended usage.

Maestrichtian European stage: Upper Cretaceous (above Campanian, below Danian of Tertiary). Also spelled: *Maastrichtian.*

mafelsic Said of an igneous rock in which the felsic and mafic minerals are present in approximately equal amounts.

mafic Said of an igneous rock composed chiefly of one or more ferromagnesian, *dark-colored* minerals in its mode; also, said of those minerals. The term was proposed by Cross, et al. (1912, p. 561) to replace the term *femag,* which they did not consider to be euphonious. Etymol: a mnemonic term derived from *ma*gnesium + *f*erric + *ic.* It is the complement of *felsic.* Cf: *femic; salic; basic.* Partial syn: *ferromagnesian.*

mafic front A term preferred by some petrologists to its synonym *basic front.*

mafic index A chemical parameter of igneous rocks, equal to $100 \times (FeO + Fe_2O_3) / MgO + FeO + Fe_2O_3$. It is most commonly plotted as the ordinate on variation diagrams, on which the abscissa represents the *felsic index.* It reflects changes produced by fractional crystallization of the mafic minerals (Simpson, 1954). Abbrev: MI.

mafic margin *basic border.*

mafite (a) A mafic mineral. (b) A dark-colored *aphanite.*

mafraite A hypabyssal *theralite* containing labradorite with alkali feldspar rims, pyroxene, magnetite, and euhedral hornblende, and without modal nepheline although nepheline does occur in the norm. Cf: *berondrite.* Named by Lacroix in 1920 for Mafra, Portugal. Not recommended usage.

mafurite A variety of olivine *leucitite* in which kalsilite is present instead of leucite. The name, proposed by Holmes in 1945, is for Mafura, Uganda. Not recommended usage.

magadiite A mineral: $NaSi_7O_{13}(OH)_3 \cdot 4H_2O$. It is found in lake beds at Lake Magadi, Kenya.

magadiniform Said of the loop, or of the growth stage in the development of the loop, of a terebratellid brachiopod (as in the subfamily Magadinae), marked by completed descending branches from the cardinalia to the median septum, with a ringlike structure on the septum representing an early ascending portion of the loop (TIP, 1965, pt.H, p.147). Cf: *premagadiniform.*

magamius In uranium exploration, "a massive gamma in unaltered sandstone", i.e. a *ma*ssive *gamius* (Bailey & Childers, 1977, p. 413).

magbasite A mineral: $KBa(Al,Sc)(Mg,Fe^{+2})_6Si_6O_{20}F_2$.

magellaniform Said of the free loop, or of the growth stage in the development of the loop, of a terebratellid brachiopod (as in *Magellania*), consisting of long descending branches recurved into ascending branches that meet in transverse band (TIP, 1965, pt.H, p.147). The magellaniform loop is morphologically similar to the *dalliniform* loop.

magelliform Said of the loop, or of the growth stage in the development of the loop, of a terebratellid brachiopod (as in *Magella*), in which the bases of the septal ring on the median septum meet and fuse with the attachments of completed descending branches (TIP, 1965, pt.H, p.147).

maghemite A strongly magnetic mineral of the magnetite series in the spinel group: γ-Fe_2O_3. It is dimorphous with hematite. Syn: *oxymagnite.*

magma Naturally occurring mobile rock material, generated within the Earth and capable of intrusion and extrusion, from which igneous rocks are thought to have been derived through solidification and related processes. It may or may not contain suspended solids (such as crystals and rock fragments) and/or gas phases. Adj: *magmatic.*

magma basalt A partial syn. of *limburgite,* also applied to a porphyritic, glassy basaltic rock with more resemblance to ordinary basalt. Obsolescent.

magma blister A pocket of magma whose formation has raised the overlying land surface.

magma chamber A reservoir of magma in the shallow part of the lithosphere (to a few km. or tens of km.), from which volcanic materials are derived; the magma has ascended into the crust from an unknown source. See also: *macula; dike chamber.* Syn: *magma reservoir.*

magma column *lava column.*

magmagranite A *granite* produced by crystallization of a magma.

magma province *petrographic province.*

magma reservoir *magma chamber.*

magmatic Of, pertaining to, or derived from *magma.* Syn: *orthotectic.*

magmatic assimilation *assimilation.*

magmatic corrosion *corrosion [petrology].*

magmatic deposit *magmatic ore deposit.*

magmatic differentiation *differentiation.*

magmatic dissolution The solution of country rock by magma; *assimilation.* Syn: *magmatic solution.*

magmatic emanation A combination of gases and liquids given off by a magma, e.g. aqueous, pegmatitic, or hydrothermal fluids. Cf: *mineralizer.*

magmatic evolution The continuing change in composition of a magma as a result of magmatic differentiation, assimilation, or mixing of magmas.

magmatic ore deposit An ore deposit formed by *magmatic segregation,* generally in mafic rocks and layered intrusions, as crystals of metallic oxides or from an immiscible sulfide liquid. Syn: *magmatic segregation deposit; magmatic deposit.*

magmatic segregation Concentration of crystals of a particular mineral (or minerals) in certain parts of a magma during its cooling and crystallization. Some economically valuable ore deposits (i.e. *magmatic ore deposits*) are formed in this way. See also: *differentiation.* Syn: *segregation [petrology].*

magmatic segregation deposit *magmatic ore deposit.*

magmatic solution *magmatic dissolution.*

magmatic stoping A term suggested by Daly for a process of magmatic emplacement or intrusion that involves detaching and engulfing pieces of the country rock. The engulfed material presumably sinks downward and/or is assimilated. See also: *piecemeal stoping; ring-fracture stoping.*

magmatic water Water contained in or expelled from magma. Cf: *juvenile water; plutonic water.*

magmatism (a) The development and movement of magma, and its solidification to igneous rock. (b) The theory that much granite has formed through crystallization from magma rather than through *granitization;* opposed to *transformism.* A proponent of this theory is called a *magmatist.*

magmatist A proponent of the theory of *magmatism.*

magmatite A rock formed from magma.

magma type Categorization of magma having a distinctive chemical composition.

magmosphere *pyrosphere.*

magnacycle A term proposed by Merriam (1963, p.106) for a large, complex rock unit that follows a repetitious pattern and that can be considered cyclic in nature.

magnacyclothem A magnacycle that is larger than a *megacyclothem* (Merriam, 1963, p.106).

magnafacies A term proposed by Caster (1934, p.19) for a major, continuous, and homogeneous belt of deposits that is distinguished by similar lithologic and paleontologic characters and that extends obliquely across time planes or through several defined chronostratigraphic units; a complete "lithic member" or perfect lithostratigraphic unit of the same facies but formed at different times; an *isopic* facies of European usage. It represents a distinct depositional environment that persisted with more or less shifting of geographic placement during time, and it may be divisible into, or assignable to, several noncontemporaneous *parvafacies.* The term is very nearly synonymous with *lithosome* as defined by Wheeler & Mallory (1956). Etymol: Latin *magna,* "great", + *facies.* See also: *megafacies.*

magnesia alum *pickeringite.*

magnesia mica (a) *phlogopite.* (b) *biotite.*

magnesian calcite A variety of calcite: $(Ca,Mg)CO_3$. It consists of randomly substituted magnesium ions in solid solution for calcium in the calcite structure. Low-magnesian calcite has less than 4% $MgCO_3$ in solid substitution, and is essentially the common form of calcite. High-magnesian calcite has 4-19% $MgCO_3$ in solid substitution; it is metastable and during limestone formation converts to low-magnesian calcite or to dolomite. Syn: *magnesium calcite.*

magnesian dolomite A dolomite rock with an excess of magnesium; specif. a dolomite rock whose Ca/Mg ratio ranges from 1.0 to 1.5 (Chilingar, 1957), or a dolomite rock containing 50-75% dolomite and 25-50% magnesite (Bissell & Chilingar, 1967, p. 108).

magnesian limestone (a) A limestone that contains appreciable magnesium; specif. a limestone having at least 90% calcite, no more than 10% dolomite, an approximate MgO equivalent of 1.1-2.1%, and an approximate magnesium-carbonate equivalent of 2.3-4.4% (Pettijohn, 1957, p. 418); or a limestone whose Ca/Mg ratio ranges from 60 to 105 (Chilingar, 1957); or a limestone containing 5-15% magnesium carbonate but in which dolomite cannot be detected (Holmes, 1928, p. 149). Some petrographers use the term for a limestone with some MgO but no dolomite; others for a rock with all possible mixtures of dolomite and calcite. Cf: *high-calcium limestone; dolomitic limestone.* (b) A dolomitic limestone; specif. the Magnesian Limestone, a facies of the Permian of NE England. (c) A term commonly but loosely used to indicate *dolomite* rock.

magnesian marble A type of metamorphosed magnesian limestone containing some dolomite (generally less than 15%). Cf: *dolomitic marble.*

magnesian spar *dolomite [mineral].*

magnesioarfvedsonite A mineral of the amphibole group: $(Na,Ca)_3(Mg,Fe,Al)_5(Si,Al)_8O_{22}(OH,F)_2$.

magnesioaxinite A triclinic mineral: $Ca_2MgAl_2BSi_4O_{15}(OH)$.

magnesiochromite (a) A mineral of the spinel group: $(Mg,Fe)(Cr,Al)_2O_4$. It is isomorphous with chromite. Syn: *magnochromite.* (b) *picrochromite.*

magnesiocopiapite A mineral of the copiapite group: $MgFe_4(SO_4)_6(OH)_2 \cdot 20H_2O$. It is a magnesium-rich variety of copiapite.

magnesioferrite A mineral of the spinel group: $(Mg,Fe)Fe_2O_4$. It is strongly magnetic and usually black. Syn: *magnoferrite.*

magnesioludwigite *ludwigite.*

magnesioriebeckite A mineral of the amphibole group: $Na_2(Mg,Fe^{+2},Fe^{+3})_5Si_8O_{22}(OH)_2$.

magnesite A white to grayish, yellow, or brown mineral: $MgCO_3$. It is isomorphous with siderite. Magnesite is generally found as earthy masses or irregular veins resulting from the alteration of dolomite rocks, or of rocks rich in magnesium silicates, by magmatic solutions. It is used chiefly in making refractories and magnesia. Syn: *giobertite.*

magnesium astrophyllite A mineral of the astrophyllite group: $(K,Na)_4(Fe,Mg,Mn)_7Ti_2Si_8O_{24}(O,OH,F)_7$.

magnesium calcite *magnesian calcite.*

magnesium-chlorophoenicite A monoclinic mineral: $(Mg,Mn)_5(AsO_4)(OH)_7$. It is isostructural with chlorophoenicite.

magnesium front *basic front.*

magnesium-zippeite An orthorhombic mineral: $Mg_2(UO_2)_6(SO_4)_3(OH)_{10} \cdot 16H_2O$.

magnet A magnetized body, especially a *permanent magnet.*

magnetic aftereffect *magnetic viscosity.*

magnetic anisotropy (a) *susceptibility anisotropy.* (b) *magnetocrystalline anisotropy.*

magnetic azimuth The *azimuth [surv]* measured clockwise from magnetic north through 360 degrees; the angle at the point of observation between the vertical plane through the observed object and the vertical plane in which a freely suspended magnetized needle, influenced by no transient artificial magnetic disturbance, will come to rest.

magnetic balance (a) An instrument in which the translational force on a magnetic moment in a nonuniform magnetic field is balanced against a spring, torsional, or gravitational force. See also: *Curie balance.* (b) *Schmidt field balance.*

magnetic basement The upper surface of extensive rocks having relatively large magnetic susceptibilities compared with those of sediments; often but not necessarily coincident with the geologic *basement.* It generally excludes magnetic sediments and thin volcanic and other high-susceptibility rocks intruded into the sedimentary section, but thick volcanic rocks in the sedimentary section would be classed as magnetic basement where the magnetic effects of deeper bodies would not be resolvable.

magnetic bearing The *bearing* expressed as a horizontal angle between the local magnetic meridian and a line on the Earth; a bearing measured clockwise from magnetic north. It differs from a *true bearing* by the amount of magnetic declination at the point of observation.

magnetic cleaning Partial demagnetization of natural remanent magnetization, by the removal of less stable, secondary components of magnetization or viscous remanent magnetization. Syn: *magnetic washing.*

magnetic compass A *compass* whose operation depends upon an element that senses the Earth's magnetic field; e.g. an instrument having a magnetic needle that turns freely on a pivot in a horizontal plane and that always swings to such a position that one end points to magnetic north. See also: *prismatic compass.*

magnetic dip *inclination [magnet].*

magnetic domain *domain [magnet].*

magnetic elements The characteristics of a magnetic field that can be expressed numerically. The seven magnetic elements are *declination* D, *inclination* I, *total intensity* F, *horizontal intensity* H, *vertical intensity* Z, *north component* X, *east component* Y. Only three elements are needed to give a complete vector specification of the magnetic field.

magnetic epoch *polarity epoch.*

magnetic equator The line on the Earth's surface at which magnetic inclination is zero; the locus of points with zero magnetic latitude. Syn: *aclinic line; dip equator.*

magnetic field (a) A region in which *magnetic forces* would be exerted on any magnetized bodies or electric currents present; the region of influence of a magnetized body or an electric current. See also: *external magnetic field.* (b) *magnetic-field intensity.* (c) *magnetic induction.*

magnetic-field intensity The force exerted by the magnetic field on a magnetic material at a point in space. Commonly used as a syn. for *total intensity.* It is expressed in SI as *teslas* and in cgs units as *gauss* or *gammas.* Syn: *magnetic field; magnetic field strength.* Nonrecommended syn: *magnetic force.*

magnetic-field line A curve whose tangent at any point is in the magnetic-field direction at that point. Syn: *line of force; line of induction; magnetic-flux line.*

magnetic-field strength *magnetic-field intensity.*

magnetic flux The surface area times the normal component of magnetic induction B; the number of magnetic-field lines crossing the surface of a given area. Expressed in *webers* in SI or in *maxwells* in the cgs system.

magnetic-flux line *magnetic-field line.*

magnetic force (a) The physical force experienced by a magnetic substance when placed in a *magnetic field* or between magnetized bodies and electric currents. (b) A nonrecommended syn. of *magnetic-field intensity.*

magnetic hysteresis *hysteresis (b).*

magnetic inclination *inclination [magnet].*

magnetic induction (a) Magnetic-flux density, symbolized by B. In a magnetic medium, it is the vector sum of the inducing field H and the magnetization M. B is expressed in *teslas* in SI and in *gauss* or *gammas* in the cgs system. Syn: *magnetic field.* (b) A nonrecommended syn. of *electromagnetic induction.* (c) The process of magnetizing a body by applying a magnetic field. This usage is not recommended.

magnetic interval The time interval of constant polarity of the Earth's magnetic field.

magnetic iron ore A syn. of *magnetite.* Var: *magnetic iron.*

magnetic latitude The angle whose tangent is one-half the tangent of the magnetic inclination. It would equal geographic latitude if the Earth's actual magnetic field were an axial dipole field.

magnetic meridian *magnetic north.*

magnetic moment A vector quantity characteristic of a magnetized body or an electric-current system; it is proportional to the magnetic-field intensity produced by this body and also to the force experienced in the magnetic field of another magnetized body or electric current. The magnetic moment per unit volume is the *magnetization.*

magnetic needle A short, slender, wire-like length of magnetic material (such as a bar magnet) that is used as a compass and that is so suspended at its midpoint as to indicate the direction of the magnetic field in which it is placed by orienting itself toward the Earth's magnetic north. Usually referred to as *needle.*

magnetic north The uncorrected direction indicated by the north-seeking end of the needle of a magnetic compass; the direction from any point on the Earth's surface of the horizontal component of the Earth's magnetic lines of force connecting the observer with the north magnetic pole; the northerly direction of the magnetic meridian at any given point. It is the common zero-degree (or 360-degree) reference in much of navigational practice. Cf: *true north.* Syn: *magnetic meridian.*

magnetic order A repetitive arrangement of the magnetic moments of ions in mineral crystals, analogous to the repetitive arrangement of the positions of the ions. It is applicable only for ions with an intrinsic magnetic moment, such as Fe^{+3}, Fe^{+2}, or Mn^{+2}. See also: *exchange force; ferromagnetism; ferrimagnetism; antiferromagnetism.*

magnetic permeability The ratio of the magnetic induction B to the inducing field strength H. It is dimensionless.

magnetic polarity reversal *geomagnetic reversal.*

magnetic poles (a) Two areas near opposite ends of a magnet where the magnetic intensity is greatest. The magnetic lines of force leave the magnet at the positive or north-seeking pole and enter at the negative or south-seeking pole. See also: *negative pole; positive pole.* (b) *dip poles.* (c) An approx. syn. for *geomagnetic poles.*

magnetic potential A scalar quantity whose negative gradient is the vector magnetic-field intensity H. It is often symbolized by W.

magnetic prospecting A technique of applied geophysics: a survey is made with a magnetometer, on the ground or in the air, which yields local variations, or anomalies, in magnetic-field intensity. These anomalies are interpreted as to the depth, size, shape, and magnetization of geologic features causing them.

magnetic pyrites *pyrrhotite.*

magnetic resonance Interaction between the magnetic motion, electron spin, and nuclear spin of certain atoms with an external magnetic field.

magnetic reversal *geomagnetic reversal.*

magnetic signature A shape of a magnetic anomaly, useful for comparison with known or model anomalies.

magnetic spherule A black *cosmic spherule* consisting of magnetite and sometimes including a metal core.

magnetic storm A world-wide disturbance of the Earth's magnetic field, commonly with amplitude of 50 to 200 gammas. It generally lasts several days, and is thought to be caused by charged particles ejected by solar flares. Magnetic prospecting usually has to be suspended during such periods.

magnetic stratigraphy *paleomagnetic stratigraphy.*

magnetic survey Measurement of a component or element of the geomagnetic field at different locations. It is usually made to map either the broad patterns of the Earth's main field or local anomalies due to variation in rock magnetization. See also: *aeromagnetic survey; magnetic prospecting.*

magnetic susceptibility *susceptibility [magnet].*

magnetic variation (a) Changes of the magnetic field in time or in space. (b) Magnetic *declination.*

magnetic viscosity A slow change of magnetization towards the direction of the ambient magnetic field. See also: *viscous magnetization.* Syn: *magnetic aftereffect.*

magnetic washing *magnetic cleaning.*

magnetism A class of physical phenomena associated with moving electricity, including the mutual mechanical forces among magnets and electric currents.

magnetite (a) A black, isometric, strongly magnetic, opaque mineral of the spinel group: $(Fe,Mg)Fe_2O_4$. It often contains variable amounts of titanium oxide, and it constitutes an important ore of iron. Magnetite commonly occurs in octahedrons and also granular or massive; it is a very common and widely distributed accessory mineral in rocks of all kinds (in orebodies as a magmatic segregation, in lenses enclosed in schists and gneisses, in igneous rocks as a primary mineral or as an alteration product, in placer deposits, and as a *heavy mineral* in sands). Syn: *magnetic iron ore; octahedral iron ore.* (b) A name applied to a series of isomorphous minerals in the spinel group, consisting of magnetite, magnesioferrite, franklinite, jacobsite, trevorite, and maghemite.—Symbol: Mt.

magnetitite An igneous rock composed chiefly of the mineral magnetite. It has an iron content of at least 65%. Apatite may be present. Syn: *kiirunavaarite.*

magnetization The magnetic moment per unit volume; a vector quantity symbolized by M, I or J. The magnetization of a rock is the sum of its two types: *induced magnetization* and *remanent magnetization*. Syn: *volume magnetization.* Nonrecommended syn: *polarization [magnet].*

magnetocrystalline anisotropy Dependence of the electronic energy of a magnetically ordered crystal upon the direction in which the atomic magnetic moments are aligned. Those crystallographic directions for which the energy is lowest are called "easy" directions. In magnetite these are the (111) directions. Syn: *magnetic anisotropy.*

magnetogram A continuous record produced by a *magnetograph* of temporal variations in magnetic elements.

magnetograph An instrument to record, automatically and continuously, temporal variations in the magnetic elements; the record it produces is a *magnetogram.*

magnetohydrodynamics The study of the relationship between a magnetic field and an electrically conducting fluid. It is relevant to studies of the Earth's core.

magnetoilmenite A high-temperature solid solution of magnetite in ilmenite. Cf: *ilmenomagnetite.*

magnetometer An instrument that measures the Earth's magnetic field and its changes, or the magnetic field of a particular rock (from which its magnetization is deduced).

magnetometric resistivity method A method of electrical surveying in which the ground is energized with direct current through a pair of electrode contacts, and the behavior of the current is surveyed by measuring the resulting magnetic field. Abbrev: MMR.

magnetoplumbite A black hexagonal mineral: $Pb(Fe^{+3},Mn^{+3})_{12}O_{19}$. Cf: *plumboferrite.*

magnetopolarity unit A body of rock unified by its remanent magnetic polarity (Oriel et al., 1976, p. 275). Syn: *polarity rock-stratigraphic unit; magnetostratigraphic polarity unit.*

magnetosphere The confines of the Earth's magnetic field, due to interaction between the solar wind and the geomagnetic field. On the sunlit side, the magnetosphere is approximately hemispherical, with a radius of about ten Earth radii under quiet conditions; it may be compressed to about six Earth radii by magnetic storms. Opposite the sunlit side, the magnetosphere extends in a "tail" of several hundred Earth radii.

magnetostratigraphic polarity-change horizon *polarity-change horizon.*

magnetostratigraphic polarity unit *magnetopolarity unit.*

magnetostratigraphic polarity zone *polarity zone.*

magnetostratigraphic unit A body of rock characterized by remanent magnetic properties (Oriel et al., 1976, p. 275). Cf: *magnetopolarity unit.*

magnetostratigraphy All parts of stratigraphy based on paleomagnetic signatures (*remanent magnetization*) (Oriel et al., 1976, p. 275).

magnetostriction Elastic strain or deformation accompanying magnetization. Cf: *piezomagnetism.*

magnetotelluric method An electromagnetic method of surveying in which natural electric and magnetic fields are measured. Usually the two horizontal electric-field components plus the three magnetic-field components are recorded; orthogonal pairs yield elements of the tensor impedance of the Earth. This impedance is measured at frequencies within the range 10^{-5}hz to 10hz.

magnioborite *suanite.*

magniophilite *beusite.*

magniotriplite A mineral: $(Mg,Fe,Mn)_2(PO_4)F$. It is a magnesium-rich variety of triplite.

magniphyric An obsolete term for the texture of a microphyric igneous rock in which the greatest dimension of the phenocrysts is between 0.2 mm and 0.4 mm (Cross et al., 1906, p.702); also, said of a rock having such texture. Cf: *mediophyric; magnophyric.*

magnitude [geomorph] A dimensionless integer assigned to each *link* in a channel network denoting the number of *sources* ultimately tributary to the link. The magnitude of a network is given by the magnitude of its outlet link, and thus denotes the number of *exterior links* in the network (Shreve, 1967). Symbol: N_1. Cf: *bifurcating link.*

magnitude [seis] *earthquake magnitude.*

magnocalcite Dolomitic calcite; a mixture of dolomite and calcite.

magnochromite *magnesiochromite.*

magnocolumbite A black orthorhombic mineral: $(Mg,Fe,Mn)(Nb,Ta)_2O_6$. It is the magnesium analogue of columbite.

magnoferrite *magnesioferrite.*

magnophorite A monoclinic mineral of the amphibole group: $NaKCaMg_5Si_8O_{23}OH$.

magnophyric An obsolete term for the texture of a porphyritic igneous rock in which the greatest dimension of the phenocrysts is more than 5 mm (Cross et al., 1906, p. 702); also, said of a rock having such texture. Cf: *magniphyric; mediophyric.*

magnussonite A green isometric mineral: $Mn_5(AsO_3)_3(OH,Cl)$. It may contain some magnesium and copper.

main joint *master joint.*

mainland A continuous body of land constituting the chief part of a country; e.g. a continent, or a main island relative to an adjacent smaller island. Syn: *fastland.*

main partition A radial wall of a foraminiferal test, extending from the marginal zone toward the center of the chamber (as in Orbitolinidae). It may be a simple transverse septum.

main shock The largest earthquake in a sequence. See also: *aftershock; foreshock.* Syn *principal earthquake.*

main stem The principal course of a stream.

main stream The principal, largest, or dominating stream of any given area or drainage system. Syn: *master stream; trunk stream.*

maitlandite *thorogummite.*

majakite A hexagonal mineral: PdNiAs.

major earthquake An earthquake having a *surface-wave magnitude* of seven or greater on the Richter scale. Such a limit is arbitrary, and may vary according to the user. Cf: *microearthquake; ultramicroearthquake.*

major fold A large-scale or dominant fold in an area, with which *minor folds* are usually associated.

majorite A meteorite mineral of the garnet group: $Mg_3(Fe,Al,Si)_2Si_3O_{12}$.

major joint *master joint.*

major septum One of the initial or secondary septa of a corallite; specif. a protoseptum or a metaseptum. Major septa are of subequal length and extend most of the distance from the wall to the axis. Cf: *minor septum.*

makatea A Polynesian term used in the south Pacific Ocean for a raised rim of a *coral reef,* for a broad uplifted coral reef surrounding an island, or for an *atoll* uplifted so that its lagoon waters drained away and left the atoll exposed as a large carbonate island. Etymol: Tuamotu.

makatite A mineral: $Na_2Si_4O_9 \cdot 5H_2O$.

make n. (a) A formation or accumulation of ore in a vein; esp. the wide or thick part of a lode or orebody. Cf: *pinch.* (b) The output, actual yield, or amount produced by an oil or gas well or a mine over a specified period. The term is colloquial.

makhtésh A term used in Israel for a huge, cirque-like hollow somewhat resembling an elongated meteorite crater, produced by erosion of a structural dome (Amiran, 1950-1951). Etymol: Hebrew, "mortar". Pl: *makhtéshim.* Syn: *erosion crater.*

mäkinenite A mineral: gamma-NiSe.

making hole The act of, or the portion of work time spent in, actual drilling and deepening of a drill hole or well.

malachite A bright-green monoclinic mineral: $Cu_2CO_3(OH)_2$. It is an ore of copper and is a common secondary mineral associated with azurite in the upper (oxidized) zones of copper veins. Malachite occurs in masses having smooth mammillated or botryoidal surfaces, and it is often concentrically banded in different shades of colors. It is used to make ornamental objects.

malacolite A syn. of *diopside.* The term originally designated a light-colored (pale-green or yellow) translucent variety of diopside from Sweden.

malacology The study of mollusks. Cf: *conchology.*

malacoma Collective name for the soft parts of radiolarians.

malacon A brown altered or hydrated variety of zircon. Also spelled: *malakon; malacone.*

malacostracan (a) Any crustacean belonging to the class Malacostraca, characterized by compound eyes, a thorax composed of eight somites, typically with a carapace, and by an abdomen composed of six or seven somites. Range, Lower Cambrian to present. (b) In very early usage, a soft-shelled crustacean. Cf: *entomostracan.*

malanite A cubic mineral: $(Cu,Pt,Ir)S_2$.

malayaite A mineral: $CaSnSiO_5$.

malaysianite A tektite from the Malay Peninsula.

malchite A fine-grained *lamprophyre,* generally porphyritic, with small phenocrysts of hornblende, labradorite, and sometimes biotite, in a groundmass of hornblende, andesine, and a small amount of quartz. Its name, given by Osann in 1892, is derived from Malchen, Germany. Cf: *luclite.* Not recommended usage.

maldonite A mineral: approximately Au_2Bi. It is a pinkish to silvery-white alloy of gold and bismuth. Syn: *black gold; bismuth gold.*

malenclave A body of contaminated or unusable ground water surrounded by uncontaminated water. Classification of malenclaves depends on whether their volume expands, diminishes, or is constant with time (Legrand, 1965, p. 88).

malezal swamp A swamp due to drainage of water over an extensive plain which has only a slight, almost imperceptible slope.

malignite (a) In the *IUGS classification,* a plutonic rock in which F is between 10 and 60, $P/A+P$ is 10 or less, and M is between 30 and 60. (b) A mafic *nepheline syenite* which has more than 5% nepheline and roughly equal amounts of pyroxene and potassium feldspar. The name, given by Lawson in 1896, is derived from the Maligne River, Ontario, Canada.

malinowskite A variety of tetrahedrite containing lead.

malladrite A hexagonal, low-temperature mineral of fumaroles: Na_2SiF_6. Not to be confused with *mallardite.*

mallardite A pale-rose monoclinic mineral: $MnSO_4 \cdot 7H_2O$. Not to be confused with *malladrite.*

Mallard's constant In *Mallard's law,* the constant for any combination of lenses on a given microscope; it is written as K.

Mallard's law (a) A statement in optics that relates to the determination of 2V, the optic axial angle. The formula is $D=K \sin E$, in which D equals half the distance between the points of emergence of the optic axes, E equals one half the optic axial angle in air, and K equals *Mallard's constant.* (b) An empirical relationship in twinning, which states that if a unit cell has a symmetry or pseudosymmetry greater than that of the crystal structure, then twinning is likely to occur, with the additional symmetry or pseudosymmetry acting as the twin element.—Named after the French crystallographer and mineralogist of the nineteenth century, Ernest Mallard.

malleable Said of a mineral, e.g. gold, silver, copper, platinum, which can be plastically deformed under compressive stress, e.g. hammering.

malloseismic Said of an area that is likely to be visited several times in a century by destructive earthquakes. Obsolete.

Malm Middle European series: Upper Jurassic (above Dogger, below Cretaceous).

malmstone (a) A hard, cherty, grayish-white sandstone whose matrix contains minute opaline globules derived from sponge spicules that once filled now-empty molds; specif. the Malmstone from the upper part of the Upper Greensand (Cretaceous) of Surrey and Sussex in England, used as a building and paving material. (b) A marly or chalky rock.—Syn: *malm rock.*

malpais A term used in the southwestern U.S. and Mexico for a region of rough and barren lava flows. The connotation of the term varies according to the locality. Etymol: Spanish, *mal país,* "bad land".

maltha A term used to designate the softer, more viscid varieties of native asphalt. In Trinidad, it is called *brea.* Syn: *earth pitch; mineral tar; malthite.*

malthacite A scaly, sometimes massive, white or yellowish clay related to fuller's earth, having a Si/Al ratio of about 4.

malthite *maltha.*

Malthusian principle The concept that all animals, including man, potentially outbreed the food supply; conversely, the food supply is the primary limiting factor on population. Thus most populations, if allowed a free breeding range, maintain themselves at the point of starvation.

mamelon [paleont] (a) A raised, rounded top of an echinoid tubercle, on which the spine articulates. (b) A member of a group of similarly shaped and regularly arranged domelike elevations on the surface of some stromatoporoids, the summits of which commonly mark the points of divergence of astrorhizae. Cf: *stromatoporoid; astrorhiza.*

mamelon [volc] A small, rounded volcano formed over a vent by slow extrusion of viscous, siliceous lava.

mamlahah A term used on the Arabian peninsula for an interior salt-encrusted playa. Cf: *sabkha.*

mammal Any vertebrate of the class Mammalia: warm-blooded, clothed in hair, bringing forth their young alive and nursing them. Range, Jurassic to present.

mammillary Forming smoothly rounded masses resembling breasts or portions of spheres; said of the shape of some mineral aggregates, as malachite or limonite.

mammillary hill A smooth, rounded, more or less elongate drumlin having an elliptical base.

mammillary structure *pillow structure [sed].*

mammillated surface A hummocky rock surface characterized by smoothed and rounded mounds alternating with hollows, esp. a streamlined surface formed by glacial erosion in mountainous areas, as in the Adirondack Mountains, N.Y.

manaccanite *menaccanite [mineral].*

manandonite A white mineral: $LiAl_4BSi_3O_{10}(OH)_8$. A member of the chlorite group.

manasseite A hexagonal mineral: $Mg_6Al_2(CO_3)(OH)_{16}\cdot 4H_2O$. It is dimorphous with hydrotalcite.

mancusanite *amerikanite.*

mandchurite A glassy nepheline *basanite,* named by Lacroix in 1923 for Manchuria. Not recommended usage.

mandible (a) In cheilostome bryozoans, the relatively enlarged, generally intricately reinforced but uncalcified, movable part of an *avicularium.* It is generally hinged on condyles or a crossbar. (b) Any of various invertebrate mouth parts serving to hold or bite into food materials and/or to move food into the mouth; e.g. one of the third pair of cephalic appendages of a crustacean. (c) An obsolete term used by some arachnologists for chelicera and by others for pedipalpal coxa. (d) The lower jaws of a gnathostome, esp. if they are fused as in *Homo.* Obsolescent except in human anatomy.

mandibular joint covering A point near the end of the cervical groove in decapods (Holthuis, 1974, p. 735).

mandibular muscle scar The place of attachment of the muscle leading to the mandible of an ostracode from the inner surface of the carapace just anterior and ventral to the adductor muscle scars.

mandibular palp The distal articulated part of the mandible of a crustacean, which aids in feeding and cleaning.

Manebach-Ala twin law A complex twin law in triclinic feldspar according to which the twin axis is perpendicular to [001] and the composition plane is (001). Cf: *Ala-A twin law.* Syn: *acline-A twin law.*

Manebach pericline twin law A complex twin law in feldspars, in which the twin axis is at right angles to [010], and the composition plane is (001).

Manebach twin law A twin law in feldspars, both monoclinic and triclinic, usually simple, with the twin plane and composition plane of (001).

mangan A *cutan* consisting of manganese oxides or hydroxides (Brewer, 1964, p.215).

manganandalusite *viridine.*

manganapatite A variety of apatite containing managnese in solid solution for calcium.

manganaxinite A mineral: $Ca_2(Mn,Fe)Al_2BSi_4O_{15}(OH)$.

manganbabingtonite A mineral: $Ca_2(Mn,Fe^{+2})Fe^{+3}Si_5O_{14}(OH)$.

manganbelyankinite A mineral: $(Mn,Ca)(Ti,Nb)_5O_{12}\cdot 9H_2O$.

manganberzeliite A mineral: $(Mn,Mg)_2(Ca,Na)_3(AsO_4)_3$. It is isomorphous with berzeliite.

manganblende *alabandite.*

manganese alum *apjohnite.*

manganese epidote *piemontite.*

manganese-hoernesite A mineral: $(Mn,Mg)_3(AsO_4)_2\cdot 8H_2O$. Also spelled: *manganese-hörnesite.*

manganese nodule A small, irregular, black to brown, friable, laminated concretionary mass consisting primarily of manganese salts and manganese-oxide minerals (Mn content is 15-30%), alternating with iron oxides. These nodules are abundant on the floors of the world's oceans (and also of the Great Lakes) as a result of pelagic sedimentation or precipitation, esp. in an area of slow deposition, and occur on or in sediments (esp. red clay and sometimes organic ooze). Manganese nodules range from a few mm to 25 cm in diameter (generally 3-5 cm) and have an average weight of 115 grams, although larger ones exist (a nodule weighing 770 kg has been found).

manganese-shadlunite A mineral of the pentlandite group: $(Mn,Pb,Cd)(Fe,Cu)_8S_8$.

manganese spar (a) *rhodonite.* (b) *rhodochrosite.*

manganite A brilliant steel-gray or iron-black orthorhombic mineral: γ-MnO(OH). It is trimorphous with groutite and feitknechtite, and is an ore mineral of manganese. Syn: *gray manganese ore.*

mangan-neptunite A dark-red mineral: $Na_2KLi(Mn,Fe)_2Ti_2Si_8O_{24}$. Cf: *neptunite.*

manganocalcite (a) A variety of rhodochrosite containing calcium. (b) A variety of calcite containing manganese.

manganocolumbite A mineral: $(Mn,Fe^{+2})(Nb,Ta)_2O_6$.

manganolangbeinite A rose-red isometric mineral: $K_2Mn_2(SO_4)_3$.

manganolite [mineral] *rhodonite.*

manganolite [rock] A general term for rocks composed of manganese minerals, esp. manganese oxides such as wad and psilomelane.

manganomelane A field term used synonymously for *psilomelane* to designate hard, massive, botryoidal, colloform manganese oxides not specifically identified. The term was rejected by the International Mineralogical Association.

manganophyllite (a) A manganoan variety of biotite: $K(Mn,Mg,Al)_{2-3}(Al,Si)_4O_{10}(OH)_2$. (b) A hypothetical biotite end member: $K_2Mn_5Al_4Si_5O_{10}(OH)_4$.

manganosiderite A variety of siderite containing manganese. It is an intermediate member of the isomorphous series siderite-rhodochrosite.

manganosite An isometric mineral: MnO. It occurs in small emerald-green octahedrons that turn black on exposure.

manganostibite An orthorhombic mineral: $(Mn,Fe)_7Sb^{+5}As^{+5}O_{12}$.

manganotantalite A mineral: $(Mn,Fe)(Ta,Nb)_2O_6$, with Mn > Fe. Cf: *tantalite.*

manganpyrosmalite A mineral: $(Mn,Fe)_8Si_6O_{15}(OH,Cl)_{10}$. Cf: *pyrosmalite.*

mangerite A plutonic rock of the *charnockite series,* corresponding to *monzonite.* Typically it contains microperthite as the dominant feldspar, with varying amounts of mafic minerals, esp. hypersthene; a hypersthene-bearing alkalic *monzonite* containing a predominance of perthitic feldspars. It is the intrusive equivalent of *doreite* (Streckeisen, 1967, p. 209). See also: *pyroxene monzonite.* The name was given by Kolderup in 1903 for Manger, Norway. Not recommended usage.

mangrove coast A tropical or subtropical low-energy coast with a shore zone overgrown by mangrove vegetation. Such coasts are common in Indonesia, Papua New Guinea, and other tropical regions. The Everglades region of Florida is the only significant U.S. example.

mangrove swamp A tropical or subtropical *marine swamp* characterized by abundant mangrove trees.

maniculifer Said of brachiopod crura derived from the radulifer type, with handlike processes at the end of straight, ventrally directed crura.

manjak A variety of asphaltite found in Barbados, which contains 0.7% to 0.9% sulfur and 1% to 2% mineral matter. See also: *glance pitch.*

manjakite An igneous rock exhibiting equigranular texture and containing garnet, biotite, pyroxene, and variable amounts of feldspar, magnetite, hypersthene, and labradorite. It resembles *kentallenite* but contains less calcium (Thrush, 1968, p. 678). Obsolete.

manjiroite A tetragonal mineral: $(Na,K)Mn_8O_{16}\cdot nH_2O$.

man-made shoreline A shoreline consisting of the works of man, such as harbor areas, breakwaters, causeways, piers, seawalls, and docks.

Manning equation An equation used to compute the velocity of uniform flow in an open channel: $V=1.486/n\ R^{2/3}\ S^{1/2}$, where V is the mean velocity of flow (in cfs units), R is the hydraulic radius in feet, S is the slope of the channel or sine of the slope angle, and n is the Manning roughness coefficient. Cf: *Chézy equation.*

mansfieldite A white to pale-gray orthorhombic mineral: $AlAsO_4\cdot 2H_2O$. It is isomorphous with scorodite.

mantle [cryst] The outer zone in a zoned crystal; *an overgrowth.*

mantle [geol] A general term for an outer covering of material of one kind or another, such as a *regolith;* specif. *waste mantle.*

mantle [interior Earth] The zone of the Earth below the crust and above the core, which is divided into the *upper mantle* and the *lower mantle,* with a transition zone between.

mantle [paleont] (a) The fold, lobe, or pair of lobes of the body wall in a mollusk or brachiopod, lining the shell and bearing the shell-secreting glands, and usually forming a mantle cavity; e.g. a prolongation of the body wall of a brachiopod, such as the two folds of ectodermal epithelium lying above and below the viscera and lining the inner surface of each valve, or the integument surrounding the vital organs of a bivalve mollusk. Syn: *pallium.* (b) The fleshy structure of cirripede crustaceans, strengthened by five calcified plates (carina, terga, and scuta) (TIP, 1969, pt.R, p.98). (c) Variously formed covering or coat in a radiolarian.

mantle canal Any of the flattened tubelike branching extensions of the body cavity into the mantle of a brachiopod, through which fluids circulate in the mantle. Syn: *pallial sinus.*

mantle cavity The cavity, between the mantle and the body proper, holding the respiratory organs of a mollusk or brachiopod; e.g. the anterior space between brachiopod valves, bounded by the mantle and the anterior body wall, and containing the lophophore. Syn: *pallial chamber.*

mantle-crust mix Rock whose properties are between those of the crust and those of the mantle, e.g. having *P*-wave velocities between 7.4 and 7.7 km/sec.

mantled Covered, as by an *ash fall* that conforms to the underlying surface.

mantled gneiss dome A term used by Eskola (1948) for a dome in metamorphic terranes that has a core of gneiss that was remobilized from an original basement and has risen through a cover of younger rocks, also metamorphosed. The gneiss is surrounded by a concordant sheath of the basal part of the overlying metamorphic sequence.

mantle rock A syn. of *regolith.* Also spelled: *mantlerock.*

mantle source volume The region in the asthenosphere in which basalt is melted from its parent rock (Dalrymple et al., 1974, p. 30).

manto A flat-lying, bedded deposit; either a sedimentary bed or a replacement strata-bound orebody. Etymol: Spanish, "vein, stratum".

manus The broad proximal part of a cheliped propodus (i.e., this propodus minus the fixed *finger*) (TIP, 1969, pt. R, p. 98).

map n. A diagram, drawing, or other graphic representation, usually on a flat surface, of selected physical features (natural, artificial, or both) of a part or the whole of the surface of the Earth, some other planet, the Moon, or any desired surface or subsurface area, by means of signs and symbols and with the means of orientation indicated, so that the relative position and size of each feature on the map corresponds to its correct geographic situation according to a definite and established scale and projection. The type of information that a map is primarily designed to convey is frequently designated by a descriptive adjective, e.g. "geologic map", "topographic map", or "structure map". Etymol: Latin *mappa,* "napkin, cloth". Cf: *chart; plan.* —v. To produce or prepare a map; to represent or delineate on a map; to engage in a mapping operation.

map collar *marginalia.*

map convolution The process of making *grid residuals.*

map face The area on a map, enclosed by the *neat line.*

map measurer *chartometer.*

mapping The process of making a map of an area; esp. the field work necessary for the production of a map.

mapping angle *gisement.*

map projection (a) Any orderly system or arrangement of lines drawn on a plane surface and representing a corresponding system of imaginary lines on an adopted terrestrial or celestial datum surface; esp. a graticule formed by two intersecting systems of lines (representing parallels of latitude and meridians of longitude) that portray upon a flat surface the whole or any part of the curved surface of the Earth, or a grid based on such parallels and meridians. It is frequently referred to as a *projection.* (b) Any systematic method by which a map projection is made; the process of transferring the outline of surface features of the Earth onto a plane. (c) The mathematical concept of a map projection.

map reading The interpretation of the information shown on a map.

map scale *scale [cart].*

map series A group of maps generally conforming to the same cartographic specifications or having some common unifying characteristic, such as the same scale or the same size of area covered. It usually has a uniform format and is identified by a name, number, or a combination of both. Examples are the National Topographic Map Series and the Geologic Quadrangle Map Series published by the U.S. Geological Survey. Syn: *series.*

map sheet An individual map, including *marginalia [cart],* either complete in itself or part of a map series.

mar A Swedish term for a bay or creek whose entrance is filled with silt so that the water is almost fresh (Stamp, 1961, p. 308). Pl: *marer.*

marais A French term for swamp used in place names in certain localities of the U.S.

marble (a) A metamorphic rock consisting predominantly of fine- to coarse-grained recrystallized calcite and/or dolomite, usually with a granoblastic, saccharoidal texture. (b) In commerce, any crystallized carbonate rock, including true marble and certain types of limestone (*orthomarble*), that will take a polish and can be used as architectural or ornamental stone. (c) *verd antique.*

marcasite [gem] A popular term used in the gemstone trade to designate any of several minerals with a metallic luster (esp. crystallized pyrite, as used in jewelry) and also polished steel and white metal.

marcasite [mineral] A common light yellow or grayish orthorhombic mineral: FeS_2. It is dimorphous with pyrite and resembles it in appearance, but marcasite has a lower specific gravity, less chemical stability, and usually a paler color. Marcasite often occurs in sedimentary rocks (such as chalk) in the form of nodules or concretions with a radiating fibrous structure. Syn: *white iron pyrites; iron pyrites; white pyrite; white pyrites; cockscomb pyrites; spear pyrites; lamellar pyrites.*

marchite A *pyroxenite* composed of enstatite and diopside. The name, given by Kretschmer in 1918, is for the March River, Czechoslovakia. Not recommended usage.

mare (a) One of the several dark, low-lying, level, relatively smooth, plainslike areas of considerable extent on the surface of the Moon, having fewer large craters than the *highlands,* and composed of mafic or ultramafic volcanic rock; e.g. Mare Imbrium (a circular mare) and Mare Tranquillitatis (a mare with an irregular outline). It is completely waterless. Cf: *terra.* (b) A dark area on the surface of Mars, whose origin is not definitely known. Cf: *continens.*—Etymol: Latin, "sea", from Galileo's belief that lunar maria represented great seas of water. Pron: *mah*-rey. Pl: *maria.* Syn: *sea.*

marebase Lunar rock of basic composition specific to the maria.

See also: *lunabase.*
mare basin A large, approximately circular or elliptical topographic depression in the lunar surface, filled or partly filled with mare material; e.g. the Imbrium basin. See also: *thalassoid.*
marekanite *Obsidian* that occurs as rounded to subangular bodies, usually less than two inches in diameter and having indented surfaces. These bodies occur in masses of *perlite* and are of special interest because of their low water content as compared with the surrounding perlite. The name is from the Marekanka River, Okhotsk, Siberia, U.S.S.R. Cf: *obsidianite.*
mare material Dark, relatively smooth, heavily cratered igneous rock, chiefly of mafic or ultramafic composition, underlying the lunar maria.
maremma A low, marshy or swampy tract of coastland. Etymol: Italian.
Maremmian European stage: Middle Miocene (above Vindobonian, below Vallesian).
mareogram *marigram.*
mareograph *marigraph.*
mare ridge *wrinkle ridge.*
mareugite A bytownite- and hauyne-bearing plutonic rock; a hauyne *gabbro.* Its name, given by Lacroix in 1917, is derived from Mareuges, Auvergne, France, where the rock forms inclusions in *ordanchite.* Not recommended usage.
margarite [ign] A beadlike string of globulites, commonly found in glassy igneous rocks.
margarite [mineral] A mineral of the brittle-mica group: $CaAl_2(Al_2Si_2)O_{10}(OH)_2$. It has a pale pink or yellowish color, and is marked by a pearly luster. Syn: *lime mica; calcium mica; pearl mica.*
margarodite A pearly variety of muscovite, resembling talc and giving a small percentage of water on ignition.
margarosanite A colorless or snow-white triclinic mineral: $Pb(Ca,Mn)_2(SiO_3)_3$.
marginal carina The ridge forming the posterior and lateral margins of the carapace on some decapods (Holthuis, 1974, p. 735).
marginal chamberlet A simple subdivision of a primary chamber of a foraminiferal test, located in the marginal zone of the chamber, and formed by main partitions only (as in Orbitolinidae).
marginal channel A channel formed by a meltwater stream flowing along the margin of a glacier or an ice sheet (Rich, 1908, p. 528).
marginal conglomerate A conglomerate that forms along a shore, on the landward margins of sediments of other types into which it grades (Twenhofel, 1939, p.30). It lies at different stratigraphic levels in the section (as seen over a large area) and thereby diagonally transects time intervals. If sea level is rising, the conglomerate is a *basal conglomerate.*
marginal cord A thick spiral structure beneath the surface at the periphery of a foraminiferal test (as in Nummulitidae) (TIP, 1964, pt.C, p.61).
marginal crevasse A *crevasse* near the margin of a glacier. It normally extends obliquely upstream from either side toward its middle at an angle of about 45° (as seen in plan). Cf: *transverse crevasse; splaying crevasse.* Syn: *lateral crevasse.*
marginal fault *boundary fault.*
marginal fissure A fracture, bordering an igneous intrusion, that has become filled with magma.
marginal granule A dotlike body in a lamella of a tintinnid.
marginal groove The groove on decapods that extends along the inner side of the marginal carina of the carapace (Holthuis, 1974, p. 733).
marginalia [cart] All printed or other material outside the *neat line* of a map.
marginalia [paleont] Sponge spicules (prostalia) around or on an oscular margin.
marginal karst plain That part of a *karst plain* flanked on one side by higher karst terrain and on the other by nonsoluble rock. Syn: *karst margin plain.*
marginal lagoon A lagoon that is adjacent to a shore or coastline.
marginal lake *glacier lake.*
marginal moraine A term formerly used as a syn. of *terminal moraine* (Hobbs, 1912, p. 279).
marginal nunatak A *nunatak* that is partly bounded by the sea or by land; e.g. Jensen Nunatak of western Greenland.
marginal plain An obsolete term for an *outwash plain* flanking the margin of a terminal moraine. Also, a vague term loosely applied to various topographic features around the margins of glaciers.
marginal plate A large elongate plate forming part of the massive marginal rim of some flattened early echinoderms, such as stylophorans, ctenocystoids, and some eocrinoids (TIP, 1968, pt. S, p. 538).
marginal plateau A relatively flat shelf adjacent to a continent and similar topographically to, but deeper than, a *continental shelf.* The Blake Plateau is an example.
marginal ring The distal part of a cyclocystoid, bordering the *submarginal ring* and composed of small imbricating plates that distally decrease in size.
marginal salt pan A natural salt pan along a coast, such as the Great Rann of Kutch in the Gujarat region of western India; a salt marsh along a coast.
marginal sea A semi-enclosed sea adjacent to a continent, floored by submerged continental mass. See also: *shelf sea.*
marginal spine In some anascan cheilostomes (bryozoans), one of a series of spines, some uncalcified at the base, placed on the inner margin of the gymnocyst around the frontal membrane; e.g. a *scutum.*
marginal suture (a) The ecdysial (molting) junction between exoskeleton elements at the prosomal margin in a merostome. (b) A suture running along the edge of the cephalon of certain trilobites (TIP, 1959, pt.O, p.122).
marginal trench *trench [marine geol].*
marginal zone The peripheral portion of foraminiferal chambers where chamberlets are subdivided by primary and secondary partitions (as in Orbitolinidae).
marginarium The peripheral part of the interior of a corallite, characterized by generally abundant dissepiments or by a dense deposit of skeletal tissue producing a stereozone. Adj: *marginarial.* Cf: *tabularium.*
marginate chorate cyst A dinoflagellate *chorate cyst* whose outgrowths are characteristically localized on the lateral margins, leaving the dorsal and more often the ventral surfaces free of outgrowths.
margination texture An obsolescent genetic term used for the *sutured* texture of a granite in which the sinuous contacts between quartz and feldspar grains result from the corrosion of earlier formed crystals in some later magmatic process.
margo (a) A modified margin of the colpus of a pollen grain, consisting of a thickening or thinning in the ektexine. Cf: *annulus [palyn].* (b) A term sometimes used for similar marginal features associated with the laesura of spores.
maria Plural of *mare.*
marialite A mineral of the scapolite group: $3NaAlSi_3O_8 \cdot NaCl$ (or three albite plus sodium chloride). It is isomorphous with meionite. Symbol: Ma.
marienbergite A plagioclase-bearing *phonolite* containing natrolite instead of nepheline. Named by Johannsen in 1938 for Marienberg (now Mariánske Lázně), Czechoslovakia. Not recommended usage.
marignacite A variety of pyrochlore containing an appreciable amount of rare earths, esp. cerium.
marigram A *tide curve;* esp. the autographic record traced by a *marigraph.* Syn: *mareogram.*
marigraph A self-registering *tide gage,* usually actuated by a float in a tube or pipe communicating with the sea through a small hole that filters out short-period waves. See also: *marigram.* Syn: *mareograph.*
marine abrasion (a) Erosion of the ocean floor by sediment that is moved by wave energy. Syn: *wave erosion.* (b) Erosion of submarine canyons by downslope movement of sediments under the influence of gravity.
marine arch *sea arch.*
marine bank *submarine bank.*
marine bench *marine-cut bench.*
marine biology The study of marine fauna and flora.
marine bridge *sea arch.*
marine-built Constructed or built up by the action of waves and currents of the sea. See also: *wave-built.*
marine-built platform A syn. of *marine-built terrace.* The term is inconsistent because a platform is usually regarded as an erosional feature.
marine-built terrace A *wave-built terrace* produced by marine processes.
marine cave (a) *sea cave.* (b) A cave formed on the bottom of the sea.
marine cliff *sea cliff.*

marine climate The climate of islands and of land areas bordering the ocean, characterized by only moderate diurnal and annual temperature ranges and by the occurrence of maximum and minimum temperatures longer after the summer and winter solstices, respectively, than in a *continental climate.* Syn: *oceanic climate; maritime climate.*

marine-cut Carved or cut away by the action of waves and currents of the sea. See also: *wave-cut.* Syn: *sea-cut.*

marine-cut bench A *wave-cut bench* of marine origin. Syn: *marine bench.*

marine-cut platform A *wave-cut platform* produced by marine processes.

marine-cut terrace A syn. of *marine-cut platform.* The term is inconsistent because a terrace is usually regarded as a constructional feature.

marine delta plain A nearly flat plain built in a bay by stream deposits at the place where the current is checked upon entering quiet water (Tarr, 1902, p. 73-74); it is built a slight distance above sea level.

marine-deposition coast A coast whose configuration results chiefly from marine deposition, such as one straightened by the formation of spits or bars, or prograded by wave and current deposits.

marine ecology The study of the relationships between marine organisms and their environment, including associated organisms.

marine-erosion coast A coast whose configuration results chiefly from marine erosion, as in the straightening of sea cliffs by waves.

marine erratic A sedimentary particle of anomalous size or lithology, transported and deposited in marine sediments by ice rafting, plants, or animals.

marine geodesy The precise determination of positions at sea and the establishing of boundaries and boundary markers at sea. It also includes the measurement of gravity at sea and the study of all the physical characteristics of the sea environment that effect such measurements.

marine geology *geological oceanography.*

marine invasion The spreading of the sea over a land area.

marine limit The present or former limit of the sea; a shoreline. Commonly used in areas of postglacial isostatic uplift for the highest record of late-glacial submergence. See also: *marin gräns.*

marine marsh A flat vegetated savannalike land surface at the edge of the sea, usually covered by water during high tide. Cf: *salt marsh.*

marine onlap A term proposed by Melton (1947, p.1869) for *onlap* in connection with marine strata that are progressively pinched out landward above an unconformity. Example: the relations of the Cambrian rocks of the Grand Canyon.

marine peneplain An abrasion platform of large areal extent, uplifted above the reach of the waves before wave erosion had succeeded in perfecting a smooth plane; an almost-plane surface of uncompleted marine denudation. Cf: *plain of marine erosion.*

marine plain (a) *plain of marine erosion.* (b) A coastal plain of marine sediments.

marine plane A hypothetical wave-cut surface produced during the ultimate stage of marine erosion; a *plain of marine erosion.*

marine platform *marine-cut platform.*

marine salina A body of salt water along an arid coast, separated from the sea by a sand or gravel barrier through which seawater enters, and having little or no inflow of fresh water; e.g. at Larnaca on Cyprus. Some salt may be deposited in it.

marine snow *sea snow.*

marine stack *stack [coast].*

marine swamp A low area of salty or brackish water along the seashore, characterized by an abundant growth of grass, reeds, mangrove trees, and similar types of vegetation. See also: *mangrove swamp.* Syn: *paralic swamp.*

marine terrace (a) A narrow constructional coastal strip, sloping gently seaward, veneered by a marine deposit (typically silt, sand, fine gravel). See also: *wave-built terrace.* (b) A narrow coastal plain whose margin has been strongly cliffed by marine erosion. (c) Loosely, a wave-cut platform that has been exposed by uplift along a seacoast or by the lowering of the sea level, and from 3 m to more than 40 m above mean sea level; an elevated marine-cut bench. Cf: *raised beach.* (d) A terrace formed along a seacoast by the merging of a wave-built terrace and a wave-cut platform.—Syn: *sea terrace; shore terrace.*

marine time A term used by Kobayashi (1944a, p. 477) for *fossil time* as indicated by marine organisms. Cf: *continental time.*

marine transgression *transgression [stratig].*

marin gräns Any maximum stand of the sea against the coast; esp. the highest *marine limit* or coastline of the postglacial sea. Etymol: Swedish, "marine border (or limit)". Abbrev: MG.

marining A term proposed by Grabau (1936, p. 254) for a temporary or short-lived flooding of a level coastal plain or deltaic deposits by an epicontinental sea; e.g. a momentary submergence accompanying a tsunami.

mariposite A bright-green, chromium-rich variety of muscovite (or phengite), having a high silica content.

maritime Bordering on the sea, as a *maritime* province.

maritime climate *marine climate.*

maritime plant A plant growing in salty conditions of the foreshore.

mariupolite An albite-nepheline *syenite* containing acmite and biotite, with zircon and beckellite as the main accessories. Named by Morozewicz in 1902 for Mariupol (now Oktj' abv), Ukraine, U.S.S.R. Not recommended usage.

mark A sedimentary structure along a bedding plane. The term usually signifies a *mold* or primary sedimentary structure (depression), such as a slide mark or a tool mark, but is also frequently applied to a cast (filling), such as a sole mark or a drag mark. Syn: *marking.*

marker [seis] (a) A layer that accounts for a characteristic segment of a seismic-refraction time-distance curve and can be followed over reasonably extensive areas. (b) A layer that yields characteristic reflections over a more or less extensive area.

marker [stratig] (a) An easily recognized stratigraphic feature having characteristics distinctive enough for it to serve as a reference or datum or to be traceable over long distances, esp. in the subsurface, as in well drilling or in a mine working; e.g. a stratigraphic unit readily identified by characteristics recognized on an electric log, or any recognizable rock surface such as an unconformity or a corrosion surface. See also: *format.* Syn: *marker bed [stratig]; marker horizon [stratig].* (b) A term used in South Africa for an outcrop.

marker band An identifiable thin bed that has the same stratigraphic position throughout a considerable area (Wills, 1956, p.14). Syn: *indicator horizon.*

marker bed (a) A geologic formation serving as a *marker.* (b) *key bed.*

marker horizon A *marker* represented by a rock surface or stratigraphic level, such as a vertical or lateral boundary based on electric or other mechanically recorded logs, that may serve to delineate lithostratigraphic units.

markfieldite A hypabyssal *granite* containing plagioclase phenocrysts in a granophyric groundmass. Its name, given by Hatch in 1909, is derived from Markfield, England. Not recommended usage.

marking *mark.*

Markov process A *stochastic process* in which the state of a system at time $t(n)$ depends on the state of the system at time $t(n-1)$. It assumes that in a sequence of random events, the outcome or probability of each event is influenced by or depends upon the outcome of the immediately preceding event. Process introduced by Andrei A. Markov (1856-1922), Russian mathematician. Syn: *Markov chain; Markoff process.*

marl (a) An old term loosely applied to a variety of materials, most of which occur as loose, earthy deposits consisting chiefly of an intimate mixture of clay and calcium carbonate, formed under marine or esp. freshwater conditions; specif. an earthy substance containing 35-65% clay and 65-35% carbonate (Pettijohn, 1957, p. 410). Marl is usually gray; it is used esp. as a fertilizer for acid soils deficient in lime. In the Coastal Plain area of SE U.S., the term has been used for calcareous clays, silts, and sands, esp. those containing glauconite (greensand marls); and for newly formed deposits of shells mixed with clay. The term has also been used to designate a soft, friable clay with very little calcium carbonate, and a very fine, loose, almost pure calcium carbonate with little clay or silt. Syn: *calcareous clay.* (b) A soft, grayish to white, earthy or powdery, usually impure calcium carbonate precipitated on the bottoms of present-day freshwater lakes and ponds largely through the chemical action of aquatic plants, or forming deposits that underlie marshes, swamps, and bogs that occupy the sites of former (glacial) lakes. The calcium carbonate may range from 90% to less than 30%. Syn: *bog lime.* (c) A term occasionally used (as in Scotland) for a compact, impure, argillaceous limestone. (d) A term loosely applied to any soil that falls readily to pieces on exposure to air. (e) A literary term for clay or earthy material.—Etymol: French *marle.*

marlaceous Resembling or abounding with marl.

marl ball *marl biscuit.*

marl biscuit An *algal biscuit* found on the shore or shallow bottom of a lake (esp. in northern U.S. and southern Canada), consisting of a hard, flattish, rounded concretion of marl formed around a shell fragment or other nucleus. Syn: *marl ball; marl pebble.*

marlekor A calcareous concretion of certain glacial clays, as of the varved lake clays of Scandinavia and in the Connecticut River valley of New England. Syn: *imatra stone.*

marlite (a) A hardened marl resistant to the action of air; *marlstone.* (b) A semi-indurated sheet or crust formed on the bottoms and shores of lakes by the intergrowth or cementation of a considerable number of *marl biscuits.*—Syn: *marlyte.*

marl lake (a) A lake whose bottom deposits contain large quantities of marl. (b) A lake that has been mined or dredged as a commercial source of marl, esp. for the manufacture of portland cement.—Syn: *merl.*

marloesite A pale-gray, fine-grained extrusive rock, apparently an altered *andesite,* composed of phenocrysts of plagioclase and lath-shaped pseudomorphs of mica after olivine in a groundmass characterized by glomerophyric texture and composed of augite, sodic plagioclase, and iron oxides. Its name, given by Thomas in 1911, is derived from Marloes, Pembrokeshire, Wales. Cf: *skomerite.* Not recommended usage.

marl pebble *marl biscuit.*

marl slate An English term for fissile calcareous rock (shale); it is not a true slate.

marlstone (a) An indurated rock of about the same composition as *marl,* more correctly called an earthy or impure argillaceous limestone. It has a blocky subconchoidal fracture, and is less fissile than shale. Syn: *marlite.* (b) A hard ferruginous rock (ironstone) of the Middle Lias in England, worked as an iron ore; specif. the Marlstone, a calcareous and sideritic oolite made up of ooliths, shell chips, and crinoid ossicles, set in a carbonate cement. (c) A term originally applied by Bradley (1931) to slightly magnesian calcareous mudstones or muddy limestones in the Green River Formation of the Uinta Basin, Utah, but subsequently applied to associated rocks (including conventional shales, dolomites, and oil shales) whose lithologic characters are not readily determined. Picard (1953) recommends abandonment of the term as used in the Uinta Basin.

marly Pertaining to, containing, or resembling marl; e.g. "marl limestone" containing 5-15% clay and 85-95% carbonate, or "marly soil" containing at least 15% calcium carbonate and no more than 75% clay (in addition to other constituents).

marlyte (a) *marlite.* (b) An obsolete term for a shale having imperfect lamination and so slightly indurated as to be fragile (Dana, 1874).

marmarization *marmorization.*

marmarosis *marmorosis.*

marmatite A dark-brown to black, iron-rich variety of sphalerite. Syn: *christophite.*

marmolite A thinly laminated, usually pale-green serpentine mineral; a variety of chrysotile.

marmoraceous Pertaining to or resembling marble.

Marmorian North American stage: Middle Ordovician (subdivision of Chazyan, above Whiterock, below Ashby) (Cooper, 1956).

marmorization The conversion of limestone into marble by any process of metamorphism. Also spelled: *marmarization.* Syn: *marmorosis.*

marmorosis A syn. of *marmorization.* Also spelled: *marmarosis.*

marne A French term for a marl or calcareous clay containing more than 50% clay and not less than 15% calcium carbonate.

marokite A black orthorhombic mineral: $CaMn_2O_4$.

marosite A *shonkinite* composed of biotite, augite with hornblende rims, sanidine, calcic plagioclase, nepheline, sodalite, apatite, and iron oxides. The name was derived by Iddings in 1913 from Pic de Maros, Celebes, Indonesia. Not recommended usage.

marquise cut A doubly pointed, elongate variation of the *brilliant cut* in which the girdle outline is boat-shaped.

marrite A monoclinic mineral: $PbAgAsS_3$.

marscoite A hybrid intrusive rock containing quartz and feldspar phenocrysts in a gabbroid groundmass. The term, proposed by Harker in 1904, was intended for use only in the Marsco area of Skye, Scotland.

Marsden chart A chart that is used to show meteorologic data over the oceans. It is based on a Mercator map projection with systematically numbered squares.

Marsden square One of a system of numbered areas each 10 degrees latitude by 10 degrees longitude, based on the Mercator projection, and used chiefly for identifying geographic positions and showing distribution of worldwide oceanographic and meteorologic data on a chart. Each square is subdivided into 100 one-degree subsquares which are numbered from 00 to 99 starting with 00 nearest the intersection of the equator and the Greenwich meridian. The system was introduced in 1831 by William Marsden (1754-1836), Irish orientalist.

marsh A water-saturated, poorly drained area, intermittently or permanently water-covered, having aquatic and grasslike vegetation, essentially without the formation of peat. Cf: *swamp; bog.*

Marshall line A syn. of *andesite line,* named after the New Zealand geologist P. Marshall.

marsh bar A narrow ridge of sand piled up at the seaward edge of a marsh undergoing wave erosion, as along the Delaware Bay shores of New Jersey.

marsh basin The depression occurring between stream banks in a *salting.*

marsh creek A drainage channel developed on a salt marsh.

marsh gas *Methane* produced during the decay of vegetable substances in stagnant water.

marshite A reddish, oil-brown isometric mineral: CuI.

marsh lake (a) An area of open water in a marsh, surrounded by wide expanses of marshland. (b) A lake covered completely or nearly so by emergent aquatic plants, esp. sedge and grasses.—See also: *lake marsh.*

marsh ore *bog iron ore.*

marsh pan A *salt pan* in a marsh.

marsh peat Peat that is derived from both plant debris and sapropelic matter. Cf: *banded peat.*

marsh shore A lake shore consisting of marsh vegetation, which often merges with the emergent aquatic vegetation of the lake.

Marsupialia An order of mammals, characterized by lack of placenta and consequent birth of young at a very immature state, with later development taking place in a specialized pouch. See also: *Eutheria.*

marthozite A mineral: $Cu(UO_2)_3(SeO_3)_3(OH)_2 \cdot 7H_2O$.

martinite [mineral] A variety of whitlockite containing carbonate.

martinite [rock] A leucite-bearing orthoclase-labradorite extrusive rock, a *tephrite* characterized by fine-grained, vesicular texture and composed of phenocrysts of leucite, plagioclase, alkali feldspar, and augite in a felty groundmass of labradorite, alkali feldspar, augite, leucite, olivine, magnetite, and apatite. Johannsen (1938) derived the name from Croce di San Martino, Vico crater, Italy. Not recommended usage.

martite Hematite occurring in iron-black octahedral crystals pseudomorphous after magnetite.

masafuerite A hypabyssal *picrite* consisting of olivine phenocrysts, comprising over 50 percent of the rock, in a groundmass of pleochroic augite and calcic plagioclase, ilmenite, and magnetite. The name, given by Johannsen in 1937, is for Masafuera in the Juan Fernandez islands, Chile. Not recommended usage.

masanite A *quartz monzonite* containing phenocrysts of zoned plagioclase and corroded quartz in a granophyric groundmass. The name, given by Koto in 1909, is from Masan-po, Korea. Not recommended usage.

masanophyre A *masanite* containing phenocrysts of oligoclase with orthoclase rims in a groundmass of blue-green hornblende and sphene. Not recommended usage.

mascagnite A yellowish-gray mineral: $(NH_4)_2SO_4$. It occurs as powdery crusts in volcanic districts and with other ammonium sulfates in guano deposits.

mascon A large-scale, high-density, lunar mass concentration below a ringed mare (Muller & Sjogren, 1968, p. 680). Etymol: *mass* + *con*centration.

maskeeg *muskeg.*

maskelynite Thetomorphic plagioclase glass; a colorless meteorite mineral consisting of a shock-formed noncrystalline phase that results from vitrification of plagioclase in rocks transfigured by shock waves and that retains the external features of crystalline plagioclase.

masonite A variety of chloritoid occurring in broad dark-green plates.

mass balance *balance.*

mass budget *balance.*
mass defect The difference between the sum of the atomic weights of the constituent particles of an isotope and its atomic weight as a whole.
mass erosion A term which includes all processes by which soil and rock materials fail and are transported downslope predominantly en masse by the direct application of gravitational body stresses. Syn: *gravity erosion.*
mass heaving The all-sided, general expansion of the ground during freezing, involving significant horizontal forces over a considerable area (Washburn, 1956, p.840). Syn: *mass heave.*
massicot A yellow, orthorhombic mineral: PbO. Cf: *litharge.* Syn: *lead ocher.*
massif A massive topographic and structural feature, especially in an orogenic belt, commonly formed of rocks more rigid than those of its surroundings. These rocks may be protruding bodies of basement rocks, consolidated during earlier orogenies, or younger plutonic bodies. Examples are the crystalline massifs of the Helvetic Alps, whose rocks were deformed mainly during the Hercynian orogeny, long before the Alpine orogeny.
massive [eco geol] Said of a mineral deposit (esp. of sulfides) characterized by a great concentration of ore in one place, as opposed to a disseminated or veinlike deposit.
massive [ign] (a) Said of granite, diorite, and other igneous rocks that have a more or less homogeneous texture (fabric) over wide areas and that display an absence of layering, foliation, cleavage, or similar features. Also, said of the fabric of such rocks. (b) Said of a pluton that is not tabular in shape.
massive [meta] Said of a metamorphic rock whose constituents are neither oriented in parallel position nor arranged in layers; that is, a rock that does not have schistosity, foliation, or any similar structure.
massive [mineral] (a) Said of a mineral that is physically isotropic, e.g. lacking a platy, fibrous, or other structure. (b) Said of an amorphous mineral, or one without apparent crystalline structure. This usage is not recommended.
massive [paleont] Said of a corallum composed of corallites with or without intercorallite walls closely in contact with one another, or of a bryozoan colony consisting of a thick heavy zoarium, generally hemispherical or subglobular in shape.
massive [rock mech] Said of a durable rock that is considered to be essentially isotropic and homogeneous and is free of fissures, bedding, foliation, and other planar discontinuities. Massive rock possesses a strength that does not vary appreciably from point to point.
massive [sed] (a) Said of a stratified rock that occurs in very thick, homogeneous beds, or of a stratum that is imposing by its thickness; specif. said of a bed that is more than 10 cm (4 in.) in thickness (Payne, 1942) or more than 1.8 m (6 ft) in thickness (Kelley, 1956, p. 294). (b) Said of a stratum or stratified rock that is obscurely bedded, or that is or appears to be without internal structure (such as a rock free from minor joints, fissility, or lamination), regardless of thickness. The massive appearance may be deceptive, as many "massive" beds display laminae and other structures when X-rayed. See also: *unstratified.* (c) Descriptive of a sedimentary rock that is difficult to split, or that splits into layers greater than 120 cm (4 ft) in thickness (McKee & Weir, 1953, p.383).
massive sulfide deposit Any mass of unusually abundant metallic sulfide minerals, e.g. a *kuroko deposit.*
mass movement A unit movement of a portion of the land surface; specif. *mass wasting* or the gravitative transfer of material down a slope. Cf: *mass transport.*
mass property A property of a sediment considered as an aggregate, e.g. porosity, color, density, plasticity.
mass spectrograph Strictly, a recording *mass spectrometer,* but commonly used only for those instruments that record on a photographic plate as contrasted to those that record numerically or graphically. The latter are usually simply called mass spectrometers.
mass spectrometer An instrument for producing and measuring, usually by electrical means, a *mass spectrum.* It is especially useful for determining molecular weights and relative abundances of isotopes within a compound. See also: *mass spectrograph.*
mass spectrometry The art or process of using a *mass spectrometer* to study mass spectra.
mass spectroscopy The observation of a *mass spectrum* and all processes of recording and measuring that go with it.
mass spectrum The pattern of relative abundances of ions of different atomic or molecular mass (mass-to-charge ratio) within a sample. It frequently refers to the measured relative abundances of various isotopes of a given element.
mass susceptibility *specific susceptibility.*
mass transfer The redistribution of matter, typically involving chemical interaction between water and rock, and generally also including diffusion. It is now used chiefly to analyze major and minor element redistribution in hydrothermal alteration systems and in modelling of those systems.
mass transport [oceanog] The movement of water.
mass transport [sed] The carrying of material in a moving medium such as water, air, or ice. Cf: *mass movement.*
massula (a) A more or less irregular, coherent mass of many pollen grains shed from the anther and fused together. Cf: *pollinium.* (b) A term sometimes applied to a structure associated with the laesura and the attached nonfunctional spores of certain megaspores.—Pl: *massulae.*
mass wasting A general term for the dislodgement and downslope transport of soil and rock material under the direct application of gravitational body stresses. In contrast to other erosion processes, the debris removed by mass wasting is not carried within, on, or under another medium. The mass properties of the material being transported depend on the interaction of the soil and rock particles and on the moisture content. Mass wasting includes slow displacements, such as creep and solifluction, and rapid movements such as rockfalls, rockslides, and debris flows. Cf: *mass erosion.* Syn: *mass movement.*
master cave An area in or a portion of a cave that seems to be the largest, most level part, to which the auxiliary passages seem to lead.
master curve One of a set of theoretical curves calculated from models, against which an observed curve is matched in an effort to find a fit sufficiently close that the model parameters can be considered applicable to the actual situation. It is used in resistivity surveying and in gravity and magnetic interpretation. Syn: *type curve.*
master joint A persistent joint plane of greater-than-average extent. Syn: *main joint; major joint.*
master map An original map, usually of large scale, containing all the information from which other maps showing specialized information can be compiled; a primary source map. Syn: *base map.*
master stream *main stream.*
mastigoneme One of the delicate, hairlike lateral threads, filaments, or processes along the length of some flagella. Syn: *flimmer.*
mastodon One of a group of extinct, elephantlike mammals widely distributed in the Northern Hemisphere in the Oligocene and Pleistocene. It differs from mammoths and other true elephants in that teeth are low-crowned, with closed roots.
masutomilite A mineral of the mica group: $K(Li,Mn,Al,Fe)_3(Si,Al)_4O_{10}(F,OH)_2$.
masuyite An orthorhombic mineral: an oxide of Pb and U (?).
matched terrace *paired terrace.*
maternal zooid A zooid in cheilostome bryozoans that deposits eggs in a brood chamber enclosed by body walls of the same or adjacent zooids (Woollacott & Zimmer, 1972, p. 165).
mathematical geography That branch of geography that is concerned with the representation of the Earth on maps and charts using various projection methods.
mathematical geology Mathematics as applied to geology; in particular, "the discipline devoted to the investigation of probability distributions of values of random variables, with the object of obtaining information concerning geological processes" (Vistelius, 1967, p.9). See also: *geomathematics.*
matildite A gray mineral: $AgBiS_2$. Syn: *schapbachite; plenargyrite.*
matlockite A mineral: PbFCl.
matraite A mineral: ZnS.
matrix [gem] A gemstone cut from material consisting of a mineral and the surrounding rock material, e.g. opal matrix or turquoise matrix.
matrix [ign] *groundmass [ign].*
matrix [ore dep] (a) A syn. of *gangue.* (b) A local term for the phosphate-bearing gravel in the land-pebble deposits of Florida.
matrix [paleont] The natural rock or earthy material in which a fossil is embedded, as opposed to the fossil itself.
matrix [sed] The finer-grained material enclosing, or filling the interstices between, the larger grains or particles of a sediment or

sedimentary rock; the natural material in which a sedimentary particle is embedded. The term refers to the relative size and disposition of the particles, and no particular particle size is implied. In carbonate sedimentary rocks, the matrix usually consists of clay minerals or micritic components surrounding coarser material; although the term should be used in a descriptive, nongenetic, and noncompositional manner, it has been applied inappropriately as a syn. of *micrite*. Syn: *groundmass*.

matrix limestone *micritic limestone*.

matrix porosity The porosity of the matrix or finer part of a carbonate rock, as opposed to the porosity of the coarser constituents (Choquette & Pray, 1970, p. 247).

matrix rock *matrix [ore dep] (b)*.

matrosite Black, opaque microscopic material forming the matrix of torbanite (Dulhunty, 1939).

mattagamite A mineral of the marcasite group: $(Co,Fe)Te_2$.

matterhorn A glacial *horn* resembling the Matterhorn, a peak in the Pennine Alps. Syn: *Matterhorn peak*.

matteuccite A mineral: $NaHSO_4 \cdot H_2O$.

Matura diamond Colorless to faintly smoky gem-quality zircon from the Matara (Matura) district of southern Ceylon. When it does not occur colorless, the color may be removed by heating.

mature [geomorph] Pertaining to the stage of *maturity* of the cycle of erosion; esp. said of a topography or region, and of its landforms (such as a plain or plateau), having undergone maximum development and accentuation of form; or of a stream (and its valley) with a fully developed profile of equilibrium; or of a coast that is relatively stable.

mature [sed] Pertaining to the third stage of textural maturity (Folk, 1951); said of a clastic sediment that has been differentiated or evolved from its parent rock by processes acting over a long time and with a high intensity and that is characterized by stable minerals (such as quartz), deficiency of the more mobile oxides (such as soda), absence of weatherable material (such as clay), and well-sorted but subangular to angular grains. Example: a clay-free "mature sandstone" on a beach, with a standard deviation of less than 0.5 phi units (a range of less than 1 phi unit between the 16th and 84th percentiles of the particle-size distribution). Cf: *immature; submature; supermature*.

matureland The land surface of the mature stage of the cycle of erosion, ranging from surfaces having attained maximum relief to those of reduced "but not low" relief (Maxson & Anderson, 1935, p. 90). The term was introduced by Willis (1928, p. 493) in a broader sense to include eroded surfaces "qualified as vigorous, advanced, or subdued, according to the stage of development", a subdued matureland approaching the flatness of a peneplain. Davis (1932, p. 429), noting that a "subdued" surface is neither mature nor old, but senescent, proposed the term *senesland* for this kind of matureland.

mature region *exozone*.

mature soil *zonal soil*.

mature stream A stream developed during the stage of *maturity [streams]*; a graded stream.

maturity [coast] A stage in the development of a shore, shoreline, or coast that begins when a profile of equilibrium is attained, and that is characterized by decrease of wave energy; creation of beaches; disappearance of lagoons and marshes; straightening of the shoreline by bridging of bays and cutting back of headlands so as to produce a smooth, regular shoreline consisting of sweeping curves; and—eventually—retrogradation of the shore beyond the bayheads so that it lies against the mainland as a line of eroded cliffs throughout its course. This process does not necessarily occur at the same rate everywhere, owing to varying rock resistance. See also: *secondary [coast]*.

maturity [sed] The extent to which a clastic sediment texturally and compositionally approaches the ultimate end product to which it is driven by the formative processes that operate upon it (Pettijohn, 1957, p. 508 & 522). See also: *textural maturity; mineralogic maturity; compositional maturity*.

maturity [streams] The stage in the development of a stream at which it has reached its maximum efficiency, having attained a profile of equilibrium and a velocity that is just sufficient to carry the sediment delivered to it by tributaries. It is characterized by: a load that is just about equal to the ability of the stream to carry it; lateral erosion predominating over downcutting, with the formation of a broad, open, flat-floored valley having a regular and moderate or gentle gradient and gently sloping, soil-covered walls with few outcrops; absence of waterfalls, rapids, and lakes; a steady but deliberate current, and muddy water; numerous and extensive tributaries, some of whose headwaters may still be in the youthful stage; development of flood plains, alluvial fans, deltas, and meanders, as the stream begins to deposit material; and a graded bed.

maturity [topog] The second of the three principal stages of the cycle of erosion in the topographic development of a landscape or region, intermediate between youth and old age (or following adolescence), lasting through the period of greatest diversity of form or maximum topographic differentiation, during which nearly all the gradation resulting from operation of existing agents has been accomplished. It is characterized by numerous, closely spaced mature streams; disappearance of initial level surfaces, as the land is completely dissected and reduced to slopes; large, well-defined drainage systems with numerous and extensive tributaries and sharp, narrow divides, and an absence of swamps or lakes on the uplands; greatest degree of ruggedness possible, with a new plain of erosion just beginning to appear; and pedimentation (in an arid cycle). Syn: *topographic maturity*.

maturity index A measure of the progress of a clastic sediment in the direction of chemical or mineralogic stability; e.g. a high ratio of alumina/soda, of quartz/feldspar, or of quartz + chert/feldspar + rock fragments, indicates a highly mature sediment (Pettijohn, 1957, p. 509).

maucherite A reddish silver-white mineral: $Ni_{11}As_8$. It tarnishes to gray copper-red. Syn: *temiskamite*.

maufite A mineral: $(Mg,Ni)Al_4Si_3O_{13} \cdot 4H_2O$ (?).

Maui-type well A type of *basal tunnel* characterized by a vertical or inclined shaft dug from the land surface to the basal water table, and by one or more tunnels dug along the water table to skim off the uppermost basal ground water to avert possible salt-water encroachment (Stearns & Macdonald, 1942, p. 126). This procedure was first used on the island of Maui, Hawaii.

mauzelite A variety of romeite containing Pb.

Maw-sit-sit A jade simulant from Burma, consisting of an intimate mixture of chrome-rich jadeite and albite. Syn: *jade-albite*.

mawsonite A mineral: $Cu_6Fe_2{}^{+3}SnS_8$.

maxilla (a) The main tooth-bearing bone in the upper jaws of Osteichthyes and tetrapods. (b) One of the first or second pairs of mouth parts posterior to the mandibles in various arthropods; e.g. the last cephalic appendage of a crustacean, following the maxillule and serving for feeding and respiration, or the *coxa* of a pedipalpus of an arachnid. Pl: *maxillae*. Adj: *maxillary*.

maxilliped One of the three pairs of appendages of a crustacean, situated next behind the maxillae; an anterior thoracopod modified to act as a mouth part, its somite usually fused to the cephalon.

maxillule The fourth cephalic appendage of a crustacean, between the mandible and the maxilla, serving as a mouth part. Syn: *first maxilla; maxillula*.

maximum [geophys] n. An anomaly characterized by values greater than those in neighboring areas; e.g. a gravity maximum or a geothermal maximum. Cf: *minimum [geophys]*.

maximum [glac geol] *glacial maximum*.

maximum [struc petrol] On a fabric diagram, a single area of concentration of poles representing the orientations of fabric elements (Turner and Weiss, 1963, p.58). Cf: *girdle; cleft girdle*. Syn: *point maximum*.

maximum projection sphericity The maximum projection area of a sphere of the same volume as a sedimentary particle, divided by the maximum projection area of the particle (Sneed and Folk, 1958, p. 118). It is a measure of shape more closely related to the hydraulic resistance of the particle than is *sphericity* as defined by Waddell.

maximum slope A slope that is steeper than the slope units above or below it.

maximum unit weight The *dry unit weight* defined by the peak of a compaction curve.

maximum water-holding capacity The average moisture content of a disturbed soil sample, one centimeter high, after equilibration with a water table at its lower surface (Jacks et al., 1960, p. 45). The retained water represents the lower part of the capillary fringe.

maxwell The cgs (centimeter-gram-second) unit of magnetic flux. One maxwell = 10^{-8} *weber*, or the flux through one square centimeter normal to a field of magnetic induction of one gauss.

Maxwell liquid A model of *elasticoviscous* behavior. During the application of stress the body deforms both elastically and viscously. When the stress is released the elastic strain is recovered, releasing the stored energy. If the body is retained in a strained condition, the stress is relaxed as the elastic strain is slowly recovered.

mayaite A white to gray-green or yellow-green material grading from tuxtlite to a nearly pure albite, found in the ancient tombs of the Mayans and elsewhere in Central America.

Mayan European stage: Middle Cambrian (above Amgan, below Tuorian).

mayenite An isometric mineral: $Ca_{12}Al_{14}O_{33}$.

Maysvillian North American stage: Upper Ordovician (above Edenian, below Richmondian).

maze cave Any cave consisting of repeatedly rejoined passages. Syn: *network cave.* See also: *branchwork cave.*

mazzite A hexagonal mineral of the zeolite family: $K_2CaMg_2(Al,Si)_{36}O_{72}\cdot 28H_2O$.

mboziite A mineral of the amphibole group: $Na_2CaFe_3{}^{+2}Fe_2{}^{+3}Al_2Si_6O_{22}(OH)_2$.

mbr *member.*

mbuga A term used in SW Africa for a temporary swamp or "black claypan" (playa) marking the last stand of a now desiccated lake.

mcallisterite *macallisterite.*

mcconnellite A trigonal mineral: $CuCrO_2$. Also spelled: *macconnellite.*

mcgovernite *macgovernite.*

m-charnockite A name proposed by Tobi (1971, p. 202) in his classification of the *charnockite series* for that member which contains mesoperthite as the only feldspar.

mckelveyite *mackelveyite.*

mckinstryite *mackinstryite.*

md *millidarcy.*

M-discontinuity Syn. of *Mohorovičić discontinuity,* suggested by Vening Meinesz (1955, p. 321). Also spelled: *M discontinuity.*

meadow ore *bog iron ore.*

meadow peat Peat derived from grasses.

meadow soil *Weisenboden.*

mealy *farinaceous.*

mean An arithmetic average of a series of values; esp. *arithmetic mean.* See also: *geometric mean; harmonic mean.* Cf: *mode [stat]; median.*

mean depth The cross-sectional area of a stream divided by its width at the surface. Cf: *mean hydraulic depth.*

meander [streams] n. (a) One of a series of regular freely developing sinuous curves, bends, loops, turns, or windings in the course of a stream. It is produced by a mature stream swinging from side to side as it flows across its flood plain or shifts its course laterally toward the convex side of an original curve. Etymol: Greek *maiandros,* from Maiandros River in western Asia Minor (now known as Menderes River in SW Turkey), proverbial for its windings. (b) *valley meander.*—v. To wind or turn in a sinuous or intricate course; to form a meander.

meander [surv] v. To survey on or along a meander line.—n. *meander line.*

meander amplitude The distance between points of maximum curvature of successive meanders of opposite phase, measured in a direction normal to the general course of the meander belt (Langbein & Iseri, 1960, p. 14).

meander bar A deposit of sand and gravel located on the inside of, and extending into the curve of, a meander; specif. a *point bar.*

meander belt The zone along a valley floor across which a meandering stream shifts its channel from time to time; specif. the area of the flood plain included between two lines drawn tangentially to the extreme limits of all fully developed meanders. It may be from 15 to 18 times the width of the stream.

meander breadth The distance between the lines used to define the *meander belt* (Langbein & Iseri, 1960, p. 14).

meander core (a) A central hill encircled or nearly encircled by a stream meander. Syn: *rock island.* (b) *cutoff spur.*

meander cusp A projection on the eroded edge of a meander-scar terrace, formed by the intersection of two or more meander scars. See also: *two-swing cusp; three-swing cusp; two-sweep cusp.* Syn: *terrace cusp.*

meander cutoff A *cutoff* formed when a stream cuts through a meander neck.

meandering stream A stream having a pattern of successive meanders. Syn: *snaking stream.*

meandering valley A valley having a pattern of successive windings broadly resembling the trace of a meandering stream. The windings, or *valley meanders,* are of the same general order of size.

meander length (a) The distance between corresponding parts of successive meanders of the same phase, measured along the general course of the meanders (Langbein & Iseri, 1960, p. 14). (b) Twice the distance between successive points of inflection of the meander (Leopold & Wolman, 1957, p. 55).

meander line A surveyed line, usually of irregular course, that is not a boundary line; esp. a metes-and-bounds traverse of the margin or bank of a permanent natural body of water, run approximately along the mean-high-water line for the purpose of defining the sinuosities of the bank or shoreline and as a means of providing data for computing the area of land remaining after the water area has been segregated. Syn: *meander.*

meander lobe The more or less elevated, tongue-shaped area of land enclosed within a stream meander. Syn: *tongue.*

meander neck The narrow strip of land, between the two limbs of a meander, that connects a meander lobe with the mainland.

meander niche On the wall of a cave, a crescentic opening formed by stream erosion. Syn: *wall niche.*

meander plain A term introduced by Melton (1936, p. 594) for a plain built by the meandering process, or a plain of lateral accretion; it is seldom or never subject to overbank floods and thus lacks any alluvial cover. Cf: *covered plain; bar plain.*

meander scar (a) A crescentic, concave mark on the face of a bluff or valley wall, produced by the lateral planation of a meandering stream which undercut the bluff, and indicating the abandoned route of the stream. See also: *flood-plain meander scar.* Syn: *meander scarp.* (b) An abandoned meander, often filled in by deposition and vegetation, but still discernible (esp. from the air).

meander scarp *meander scar.*

meander-scar terrace A local terrace formed by the shifting of meanders during the slow and continuous excavation of a valley (Schieferdecker, 1959, term 1519). Syn: *alternate terrace.*

meander scroll (a) One of a series of long, parallel, closely fitting, arcuate ridges and troughs formed along the inner bank of a stream meander as the channel migrated laterally down-valley and toward the outer bank. Cf: *point bar; lacine.* (b) A small, elongate lake occurring on a flood plain in a well-defined part of an abandoned stream channel, commonly in an oxbow.

meander spur An undercut projection of high land extending into the concave part of, and enclosed by, a meander.

meander terrace A small, relatively short-lived *stream terrace* formed by a freely swinging meander cutting into a former and higher flood plain; an *unpaired terrace.*

mean deviation The arithmetic mean of the absolute deviations of observations from their mean. Syn: *average deviation.*

mean diameter (a) *arithmetic mean diameter.* (b) *geometric mean diameter.* (c) *logarithmic mean diameter.* (d) *phi mean diameter.*

meandroid Said of a corallum characterized by meandering rows of confluent corallites with walls only between the rows.

mean earth ellipsoid The hypothesized ellipsoid that coincides most closely with the actual figure of the Earth at sea level, i.e. with the geoid.

mean higher high water The average height of all the higher high waters recorded at a given place over a 19-year or computed equivalent period. Abbrev: MHHW.

mean high water The average height of all the high waters recorded at a given place over a 19-year period or a computed equivalent period. Abbrev: MHW.

mean high-water neap The average high-water height during *quadrature [astron],* recorded over a 19-year or computed equivalent period. Abbrev: MHWN. Cf: *mean low-water neap.*

mean high-water spring The average high-water height at *syzygy [astron],* recorded over a 19-year or computed equivalent period. Abbrev: MHWS. Cf: *mean low-water spring.*

mean hydraulic depth The cross-sectional area of a stream divided by the length of its wetted perimeter. Cf: *mean depth.*

mean ice edge The average position of the *ice edge* in any given time period (usually a month), based on observations over several years. Formerly known as *ice limit.*

mean life In a *radionuclide* with a large number of atoms, the average of the lives of the individual atoms (each life ending in *radioactive decay*). For a single atom, the mean life is the time interval for which the probability of decay within the interval is 1/2. Mean life is the reciprocal of the *decay constant.*

mean lower low water The average height of all the lower low waters recorded at a given place over a 19-year or computed equivalent period. Abbrev: MLLW. See also: *lower low-water datum.*

mean low water The average height of all the low waters recorded at a given place over a 19-year or computed equivalent period. Abbrev: MLW. See also: *low-water datum.*

mean low-water neap The average low-water height during *quadrature [astron],* recorded over a 19-year or computed equivalent period. Abbrev: MLWN. Cf: *mean high-water neap.*

mean low-water spring The average low-water height at *syzygy [astron],* recorded during a 19-year or computed equivalent period. Abbrev: MLWS. Cf: *mean high-water spring.*

mean range The difference in height between mean high water and mean low water. Abbrev: Mn. Cf: *tide range.*

mean refractive index (a) The median *index of refraction* for any crystalline substance, with variation due to zoning. (b) The median index of refraction in any microcrystalline substance for which specific index values related to crystal directions are not determinable. (c) In a biaxial crystal, the beta, N_y or N_m index of refraction (in which y or m = mean). This is not the average index.

mean sea level The average height of the surface of the sea for all stages of the tide over a 19-year period, usually determined from hourly height observations on an open coast or in adjacent waters having free access to the sea; the assumed or actual sea level at its mean position midway between mean high water and mean low water. It is adopted as a *datum plane,* i.e. *sea-level datum,* for the measurement of elevations and depths. Cf: *mean tide level.* Abbrev: MSL. Popular syn: *sea level.* Syn: *geodetic sea level.*

mean spring range The average semidiurnal range of the tide at *syzygy [astron].* Abbrev: Sg. Syn: *spring range.*

mean stress The algebraic average of the three *principal stresses.*

mean tide level The plane or surface that lies exactly midway between mean high water and mean low water; the average of observed heights of high water and low water. Abbrev: MTL. Cf: *mean sea level.* Syn: *ordinary tide level; half-tide level.*

mean velocity *average velocity [hydraul].*

mean velocity curve *vertical-velocity curve.*

mean water level The average height of the surface of water, determined at equal (usually hourly) intervals over a considerable period of time. Abbrev: MWL.

measured ore *developed reserves.*

measures A group or series of sedimentary rocks having some characteristic in common, specif. *coal measures.* The term apparently refers to the old practice of designating the different seams of a coalfield by their "measure" or thickness.

meat earth A term sometimes used in mining for the topsoil of an opencut mine, which may be saved and used for restoration of the area.

mechanical analysis Determination of the particle-size distribution of a soil, sediment, or rock by screening, sieving, or other means of mechanical separation; "the quantitative expression of the size-frequency distribution of particles in granular, fragmental, or powdered material" (Krumbein & Pettijohn, 1938, p. 91). It is usually expressed in percentage by weight (and sometimes by number or count) of particles within specific size limits. See also: *particle-size analysis.*

mechanical clay A clay formed from the products of abrasion of rocks.

mechanical erosion *corrasion.*

mechanical metamorphism *kinetic metamorphism.*

mechanical sediment *clastic sediment.*

mechanical stage A microscope *stage* that allows exact recording of the position of the object, e.g. a thin section, and that has a device for moving the object sideways, forward, and backward.

mechanical twin *deformation twin.*

mechanical weathering The process of weathering by which frost action, salt-crystal growth, absorption of water, and other physical processes break down a rock to fragments, involving no chemical change. Cf: *chemical weathering.* Syn: *physical weathering; disintegration [weath]; disaggregation.*

mechanoglyph A hieroglyph of mechanical origin (Vassoevich, 1953, p.38).

medano A Spanish term for a *sand dune,* esp. one occurring along a seashore, as in Chile or Peru.

medial *middle [geochron].*

medial moraine (a) An elongate moraine carried in or upon the middle of a glacier and parallel to its sides, usually formed by the merging of adjacent and inner lateral moraines below the junction of two coalescing valley glaciers. (b) A moraine formed by glacial abrasion of a rocky protuberance near the middle of a glacier and whose debris appears at the glacier surface in the ablation area. (c) The irregular ridge left behind in the middle of a glacial valley, when the glacier on which it was formed has disappeared.—Syn: *median moraine.*

median The value of the middle item in a set of data arranged in rank order. If the set of data has an even number of items, the median is the arithmetic mean of the middle two ranked items. Cf: *mean; mode [stat].*

median carina In some decapods, a longitudinal ridge extending over the full length of the middorsal area of the carapace from the tip of the rostrum to the middle of the posterior margin (Holthuis, 1974, p. 735).

median diameter An expression of the average particle size of a sediment or rock, obtained graphically by locating the diameter associated with the midpoint of the particle-size distribution; the middlemost diameter that is larger than 50% of the diameters in the distribution and smaller than the other 50%.

median dorsal plate An elongate plate that posteriorly and dorsally separates the carapace valves of a phyllocarid crustacean.

median lamina In many stenolaemate bryozoans, the erect median colony wall from which zooids grow in two back-to-back layers to form bifoliate colonies. Syn: *mesotheca.*

median mass *Zwischengebirge.*

median moraine *medial moraine [glac geol].*

median muscle An anterior or posterior pedal retractor muscle inserted across the dorsal midline of the shells of primitive rostroconchs (Pojeta & Runnegar, 1976, p. 15).

median section A slice in the central sagittal part and perpendicular to the axis of coiling of a foraminiferal test.

median septum (a) A calcareous ridge built along the midline of the interior of a brachiopod valve (Beerbower, 1968, p.284). (b) In biserial scandent graptoloid graptolithines, a partition that separates two rows of thecae.

median sulcus A prominent vertical depression in the anterior and median surface of an ostracode valve.

median suture A suture along the sagittal line on the ventral cephalic doublure of a trilobite, developed in many forms where no rostral plate is present. Cf: *connective suture.*

median valley *rift valley.*

medical geology The application of geology to medical and health problems, involving such studies as the occurrence of toxic elements in unusual quantities in parts of the Earth's crust, the distribution of trace elements as related to nutrition, or the geographic patterns of disease. The medical syn. is "regional pathology". See also: *environmental geochemistry.*

medicinal spring A spring of reputed therapeutic value due to the substances contained in its waters. Cf: *spa.*

mediglacial Relating to or formed between glaciers, or situated in the midst of glaciers.

mediiphyric An obsolete term applied by Cross et al. (1906, p.702) to porphyritic rocks in which the longest dimension of the phenocrysts is between 0.04 mm and 0.008 mm.

Medinan (a) Obsolete syn. of *Alexandrian.* (b) North American stage: Lower Silurian (above Richmondian of Upper Ordovician, below Clintonian).

mediophyric An obsolete term applied to the texture of a porphyritic rock in which the longest dimension of the phenocrysts is between 1 mm and 5 mm (Cross et al., 1906, p.702); also, said of a rock having such texture. Cf: *magnophyric; magniphyric.*

mediosilicic A term proposed by Clarke (1908, p. 357) to replace *intermediate.* Cf: *subsilicic; persilicic.*

mediterranean n. *mesogeosyncline.*

Mediterranean climate A climate characterized by hot, dry summers and mild, rainy winters.

mediterranean delta A delta built out into a landlocked sea that is tideless or has a low tidal range (Lyell, 1840, v.1. p. 422).

mediterranean sea A type of *epicontinental sea* that is deep and that connects with the ocean by a narrow opening.

Mediterranean suite A major group of igneous rocks, characterized by high potassium content. This suite was so named because of the predominance of potassium-rich lavas around the Mediterranean Sea; specif. those of Vesuvius and Stromboli. Cf: *Atlantic suite; Pacific suite.*

Medithermal A term used by Antevs (1948, p.176) for a period of time in the late Holocene marked by decreasing temperatures.

medium bands In banded coal, vitrain bands from 2.0 to 5.0 mm thick (Schopf, 1960, p.39). Cf: *thin bands; thick bands; very thick* *bands.*

medium-bedded A relative term applied to a sedimentary bed whose thickness is intermediate between *thin-bedded* and *thick-bedded.* See also: *stratification index.*

medium boulder A *boulder* having a diameter in the range of 512-1024 mm (20-40 in., or -9 to -10 phi units).

medium clay A geologic term for a *clay* particle having a diameter in the range of 1/1024 to 1/512 mm (1-2 microns, or 10 to 9 phi units). Also, a loose aggregate of clay consisting of medium clay particles.

medium-crystalline Descriptive of an interlocking texture of a carbonate sedimentary rock having crystals whose diameters are in the range of 0.062-0.25 mm (Folk, 1959) or 0.1-0.2 mm (Carozzi & Textoris, 1967, p. 5) or 1-4 mm (Krynine, 1948, p. 143). Cf: *medium-grained.*

medium-grained (a) Said of an igneous rock, and of its texture, in which the individual crystals have an average diameter in the range of 1-5 mm (0.04-0.2 in.). Johannsen (1931, p.31) earlier used the range of 1-10 mm. (b) Said of a sediment or sedimentary rock, and of its texture, in which the individual particles have an average diameter in the range of 1/16 to 2 mm (62-2000 microns, or sand size). Cf: *medium-crystalline.*—The term is used in a relative sense to describe rocks that are neither *coarse-grained* nor *fine-grained.*

medium pebble A geologic term for a *pebble* having a diameter in the range of 8-16 mm (0.3-0.6 in., or -3 to -4 phi units) (AGI, 1958).

medium sand (a) A geologic term for a *sand* particle having a diameter in the range of 0.25-0.5 mm (250-500 microns, or 2 to 1 phi units). Also, a loose aggregate of sand consisting of medium sand particles. (b) An engineering term for a *sand* particle having a diameter in the range of 0.42 mm (retained on U.S. standard sieve no. 40) to 2 mm (passing U.S. standard sieve no. 10). (c) A soil term used in the U.S. for a *sand* particle having a diameter in the range of 0.25-0.5 mm.

medium silt A geologic term for a *silt* particle having a diameter in the range of 1/64 to 1/32 mm (16-31 microns, or 6 to 5 phi units). Also, a loose aggregate of silt consisting of medium silt particles.

medium-volatile bituminous coal Bituminous coal, characteristically agglomerating, that contains 23-31% volatile matter, analyzed on a dry, mineral-matter-free basis. Cf: *high-volatile bituminous coal; low-volatile bituminous coal.*

medmontite A mixture of chrysocolla and mica.

medulla (a) The central zone of certain octocorals; e.g. the central chord of the axis of the Holaxonia (TIP, 1956, pt. F, p. 174). (b) The internal part of some protozoans.

medullary shell The internal concentric shell of spumellarian radiolarians.

meerschaum Massive *sepiolite.* Etymol: German *Meerschaum,* "sea froth".

mega- A prefix signifying "large" or "great". Syn: *macro-.*

megabarchan A giant *barchan,* up to 100 m or more in height (Stone, 1967, p. 232).

megabreccia (a) A term used by Landes (1945) for a rock produced by brecciation on a very large scale, containing blocks that are randomly oriented and invariably inclined at angles from 6° to 25° and that range from a meter to more than 100 m in horizontal dimension. (b) A term used by Longwell (1951) for a coarse breccia containing individual blocks as much as 400 m long, developed downslope from large thrusts by gravitational sliding. It is partly tectonic and partly sedimentary in origin, containing blocks that are shattered but little rotated.—Cf: *chaos.*

megacanthopore The larger of two sizes of *acanthopore* in stenolaemate bryozoans, which increases in size through the exozone until comparable in diameter to mesozooecia and feeding zooecia. Cf: *micracanthopore.*

megaclast (a) One of the larger fragments in a variable matrix of a sedimentary rock (Crowell, 1964). Cf: *phenoclast.* (b) A constituent of a mixtite (Schermerhorn, 1966).

megaclone A large, smooth monaxonic desma (of a sponge), having branches that bear cuplike articular facets, mostly terminal.

megacryst A nongenetic term introduced by Clarke (1958, p. 12) for "any crystal or grain" in an igneous or metamorphic rock that is "significantly larger" than the surrounding groundmass or matrix; e.g. a large microcline crystal in porphyritic granite. It may be a phenocryst, a xenocryst, a porphyroblast, or a porphyroclast.

megacrystalline *macrocrystalline.*

megacyclothem A term introduced by Moore (1936, p.29) to designate a combination of related *cyclothems,* or a cycle of cyclothems, such as in the Pennsylvanian of Kansas. Also, a cyclothem on a large scale, comprising minor cyclothems. Cf: *hypercyclothem; magnacyclothem.*

megaevolution *macroevolution.*

megafabric The fabric of a rock as seen in hand specimen or outcrop, without the aid of a microscope. Cf: *microfabric.* Syn: *macrofabric.*

megafacies (a) A term used by Cooper & Cooper (1946, p.68) apparently for a large intertonguing lithologic body. (b) A term used mistakenly for *magnafacies.*

megafauna *macrofauna.*

megaflora (a) Plants large enough to be seen with the naked eye. (b) An obsolete term for the plants of a large habitat; a large, widespread group of plants. Cf: *microflora; macrofauna.* Syn: *macroflora.*

megaflow mark A term used by Kuenen (1957, p.243) as a syn. of *gouge channel.*

megafossil *macrofossil.*

megagametophyte The multicellular female *gametophyte* or haploid generation that develops from the megaspore of a heterosporous embryophytic plant. In lower vascular plants, it may be a small free-living plant bearing archegonia, but in seed plants it is contained within the ovule, and the egg is produced in it. Cf: *microgametophyte.*

megagrained Said of the texture of a carbonate sedimentary rock having clastic particles whose diameters are greater than one millimeter (DeFord, 1946). See also: *macrograined.*

megagroup A term used by Swann & Willman (1961) for a rock-stratigraphic unit that is next higher in rank than group and that represents a major event in the course of geologic history. It is not recognized as a formal unit by the American Commission on Stratigraphic Nomenclature (1961). Cf: *supergroup.*

megalineament "Any lineation more than 100 kilometers long" (El-Etr, 1976, p. 485).

megaloplankton The largest plankton; they are more than 1 cm in size. Cf: *ultraplankton; nannoplankton; microplankton; macroplankton.*

megalospheric Said of a foraminiferal test or shell produced asexually and characterized by a large initial chamber (proloculus), relatively few chambers, small size of the adult test, and incomplete ontogeny. Cf: *microspheric.*

megalump A gravel-sized *lump* in a limestone. It usually originates by disruption (by high-energy waves or currents or possibly by turbidity currents) of newly deposited or partly indurated carbonate mud, which is then incorporated within the sedimentary unit from which it was derived.

megaphyric Said of the texture of a porphyritic igneous rock in which the greatest dimension of the phenocrysts is more than 2 mm (Cross et al., 1906, p.702); also, said of a rock having such texture. Cf: *microphyric.* Syn: *macroporphyritic; megaporphyritic; macrophyric.*

megapore (a) *macropore.* (b) In the pore-size classification of Choquette & Pray (1970, p.233), an equant to equant-elongate pore or a tubular or platy pore with an average diameter or thickness greater than 4 mm.—Cf: *mesopore [petrology]; micropore.*

megaporphyritic *megaphyric.*

megarhizoclone A large *rhizoclone* approaching the form of a megaclone.

megaripple A large *sand wave* or ripplelike feature having a wavelength greater than 1 m (Van Straaten, 1953) or a ripple height greater than 10 cm (Imbrie & Buchanan, 1965, p.155), composed of sand, and formed in a subaqueous environment. Not to be confused with *metaripple.* See also: *giant ripple; sand wave.*

megasclere A large *sclere;* specif. one of the primary spicules forming the principal skeletal support in a sponge. It usually differs in form from a *microsclere.* Syn: *macrosclere.*

megascopic Said of an object or phenomenon, or of its characteristics, that can be observed with the unaided eye or with a hand lens. Syn: *macroscopic.*

megashear A strike-slip fault with a horizontal displacement that exceeds significantly the thickness of the crust (Carey, 1958, p. 196).

megasporangium A *sporangium* that develops or bears megaspores; e.g. the nucellus in a gymnospermous seed plant. Cf: *microsporangium.*

megaspore (a) One of the spores of a heterosporous embryophytic plant that germinates to produce a megagametophyte and that is ordinarily larger than the *microspore.* Its range is mid-Devonian to Holocene. (b) A term arbitrarily defined in paleopalynology as a spore or pollen grain greater than 200 microns in diameter; it may not have been biologically a megaspore in function. Cf: *miospore.*—Syn: *macrospore.*

megatectonics The tectonics of the very large structural features of the Earth, or of the whole Earth. Similar terms are *geotectonics* and *global tectonics,* but all these large, vague words seem superfluous since the subject of *tectonics* itself differs from the subject of *structural geology* in dealing only with the very large structural features. Cf: *microtectonics.* Syn: *macrotectonics.*

megatherm A plant requiring high temperatures and large quantities of water for its existence. Syn: *macrotherm.* Cf: *microtherm; mesotherm.*

Megathermal n. A term proposed by Judson (1953, p.59) for *Altithermal.* See also: *Hypsithermal.*

megathermal Pertaining to a climate characterized by high temperature. Cf: *mesothermal; microthermal.*

megayear A term proposed by Rankama (1967) for one million (10^6) years.

megazone *superzone.*

megerliiform Said of the loop of a terebratellacean brachiopod with descending branches joining anterior projections from the large ring on a low median septum, differing from similar dallinid and terebratellid loops by the appearance of a well-developed ring before growth of descending branches (TIP, 1965, pt.H, p.148).

Meigen's reaction A test used to distinguish calcite from aragonite. A cobalt-nitrate solution is used to stain the minerals by boiling; aragonite becomes lilac and retains this color in thin section, whereas calcite and dolomite become pale blue but do not show the color in thin section. Cf: *Lemberg's stain.*

meimechite *meymechite.*

meinzer A syn. of *permeability coefficient,* named for O.E. Meinzer (1876-1948), a hydrogeologist with the U.S. Geological Survey. Syn: *Meinzer unit.*

meiofauna Organisms ranging in size from 1 mm to 0.1 mm that live within sediments; size class between *megafauna* and *microfauna* (Hulings & Gray, 1971).

meionite A mineral of the scapolite group: $3CaAl_2Si_2O_8 \cdot CaCO_3$ (or three anorthite plus calcium carbonate). It is isomorphous with marialite, and may contain other anions (sulfate, chloride). Symbol: Me.

meixnerite A trigonal mineral: $Mg_6Al_2(OH)_{18} \cdot 4H_2O$.

meizoseismal Pertaining to the maximum destructive force of an earthquake.

mela- A prefix meaning "dark-colored"; in the *IUGS classification,* it is used to designate a rock that is more mafic than the specified range, e.g. melagranite. Cf: leuco-.

melaconite *tenorite.*

melagabbroid In the *IUGS classification,* a plutonic rock satisfying the definition of *gabbroid,* and in which pl/(pl+px+ol) is between 10 and 35.

melanchym A complex humic substance separated into two fractions by alcohol: insoluble *melanellite* and soluble *rochlederite.* It is found in the brown coal of Bohemia.

melanellite The insoluble portion remaining when *melanchym* is treated with alcohol. See also: *rochlederite.*

mélange [gem] An assortment of mixed sizes of diamonds larger than those of a *mêlée,* i.e. weighing more than 1/4 carat.

mélange [sed] A mappable body of rock characterized by the inclusion of fragments and blocks of all sizes, both exotic and native, embedded in a fragmented and generally sheared matrix of more tractable material (Berkland et al., 1972, p. 2296). Mode of origin is not a part of this definition. "Where the origin of a particular mélange is known, the mélange may be designated either an *olistostrome* (sedimentary mélange) or a tectonic mélange" (Raymond, 1975, p. 8). The term was introduced by Greenly in 1919; etymol: French, "mixture". Cf: *chaos [geol].*

melanic *melanocratic.*

melanite A black variety of andradite garnet containing titanium. Cf: *schorlomite.* Syn: *pyreneite.*

melanized Said of a soil whose dark color is due to its content of humus.

melanocerite A brown or black rhombohedral mineral: $(Ce,Ca)_5(Si,B)_3O_{12}(OH,F) \cdot nH_2O$ (?).

melanocratic Dark-colored; applied to dark-colored igneous rocks rich in mafic minerals. The percentage of mafic minerals required for a rock to be classified as melanocratic varies among petrologists; the lower limit ranges from 60 to 67%. Cf: *leucocratic; mesocratic.* Syn: *chromocratic; melanic; dark-colored.* Noun: *melanocrate.*

melanophlogite A mineral consisting of silicon dioxide (SiO_2) and containing carbon and sulfur. It was formerly believed to be a partly oriented pseudomorph of alpha quartz after cristobalite containing H_2SO_4.

melanophyre A broad term used in the field for any dark-colored porphyritic igneous rock having a fine-grained groundmass.

melanosome *melasome.*

melanostibite A mineral: $Mn(Sb,Fe)O_3$.

melanotekite A black or dark-gray mineral: $Pb_2Fe_2^{+3}Si_2O_9$.

melanovanadite A black mineral: $Ca_2V_4^{+4}V_6^{+5}O_{25} \cdot nH_2O$.

melanterite A green or greenish-blue monoclinic mineral: $FeSO_4 \cdot 7H_2O$. It usually results from the decomposition of iron sulfides. Syn: *copperas; green vitriol; iron vitriol.*

melaphyre A term originally applied to any dark-colored porphyritic igneous rock but later restricted to altered *basalt,* esp. of Carboniferous and Permian age. Obsolete.

melasome The dark-colored part of a migmatite, rich in mafic minerals (Mehnert, 1968, p. 355). Cf: *leucosome.* Also spelled: *melanosome.*

melatope In an *interference figure,* a point indicating the crystal's optic axis.

mêlée (a) A collective term for small round faceted diamonds, such as those mounted in jewelry. The term is sometimes applied to colored stones of the same size and shape as the diamonds. (b) A small diamond cut from a fragment of a larger size. (c) In diamond classification, a term for small round cut diamonds weighing more than 1/4 carat. Cf: *mélange [gem].*—Etymol: French.

melikaria (a) Skeletal structures of quartz formed in place by deposition of silica from rising waters in the bottoms of deep shrinkage cracks in septaria or other concretions, the enclosing rock having been removed by solution (Burt, 1928). They resemble septarian veins in form, and may be as large as 45 x 20 x 10 cm (as in the Quaternary alluvial deposits of Brazos County, Tex.). Cf: *septarium.* (b) A term applied to the vein skeletons of septaria (Twenhofel, 1939, p.552).—Etymol: Greek, "honeycombs".

melilite (a) A group of minerals of general formula: $(Na,Ca)_2(Mg,Al)(Si,Al)_2O_7$. It consists of an isomorphous solid-solution series, and may contain some iron. (b) A tetragonal, often honey-yellow mineral of the melilite group, such as the end members gehlenite and akermanite. It occurs as a component of certain recent basic volcanic rocks.—The melilites of volcanic rocks are usually classed as feldspathoids, but have also been considered as "undersaturated pyroxenes". Also spelled: *mellilite.*

melilite basalt An obsolete synonym of *melilitite.*

melilitholith An extrusive rock composed entirely of melilite. Johannsen (1938, p. 337) proposed this mellifluous term.

melilitite A generally olivine-free extrusive rock composed of melilite and clinopyroxene (or other mafic mineral) usually comprising more than 90 percent of the rock, with minor amounts of feldspathoids and sometimes plagioclase. Syn: *melmafite.*

melilitolite A group of rare plutonic mafic rocks with a predominance of melilite (Streckeisen, 1967, p. 174); also, any rock in that group, e.g. *uncompahgrite.* The intrusive equivalent of *melilitite.*

meliphanite A yellow, red, or black mineral: $(Ca,Na)_2Be(Si,Al)_2(O,OH,F)_7$. Syn: *meliphane; gugiaite.*

melkovite A mineral: $CaFeH_6(MoO_4)_4(PO_4) \cdot 6H_2O$.

mellilite (a) *melilite.* (b) *mellite.*

mellite A honey-colored mineral: $Al_2[C_6(COO)_6] \cdot 18H_2O$. It has a resinous luster, usually occurs as nodules in brown coal, and is in part a product of vegetable decomposition. Syn: *honey stone; mellilite.*

mellorite A name suggested for a poorly crystallized material of the kaolin group of clay minerals in which randomness of stacking of the layer packets in the *c*-axis direction is present. Because there is considerable range of disorder in the less well-crystallized kaolinites, there is no need for a specific mineral name such as "mellorite"; the term in general use is *fireclay mineral.*

melmafite A general term for igneous rocks composed of *mel*ilite and other *maf*ic minerals, proposed by Hatch, Wells, and Wells in 1961. Syn: *melilitite; melilite basalt.*

melnikovite *greigite.*

melonite A reddish-white mineral: $NiTe_2$.

melonjosephite A mineral: $CaFe^{+2}Fe^{+3}(PO_4)_2(OH)$.

melt n. In petrology, a liquid, fused rock.

melteigite (a) In the *IUGS classification,* a plutonic rock in which F is between 60 and 100, M is between 70 and 90, and sodium exceeds potassium. Cf: *missourite.* (b) A dark-colored plutonic rock that is part of the *ijolite* series and contains nepheline and 60-90% mafic minerals, esp. green pyroxene. Cf: *turjaite; urtite; algarvite; micromelteigite.*—The name is from Melteig farm, Fen complex, Norway.

melting spot A region in the mantle within which *tholeiite* magma is generated and whose vertical projection on the Earth's surface is an area within which tholeiitic eruptions have occurred or may occur (Dalrymple et al., 1974, p. 31). See also: *hot spot.*

melt metamorphism The process of snow-grain growth and rounding in the presence of the melt, caused by an inverse relationship between grain radius and melting temperature. It usually results in an increase in density and in modification of other snow properties.

melt-out till Till derived from slow melting of thick masses of debris-rich stagnant ice buried beneath sufficient overburden to inhibit deformation under gravity, thus preserving structures derived from the parent ice (Boulton, 1970).

meltwater Water derived from the melting of snow or ice, esp. the stream flowing in, under, or from melting glacier ice. Also spelled: *melt water.*

member A lithostratigraphic unit of subordinate rank, comprising some specially developed part of a *formation.* It may be formally defined and named, informally named, or unnamed. It is not necessarily mappable, and a named member may extend from one formation into another. Laterally equivalent parts of a formation that differ recognizably may be considered members; e.g. the gravel member and the silt member of the Bonneville Formation. A member name combines a geographic name followed by the word "member"; where a lithologic designation is useful, it should be included (e.g. the Wedington Sandstone Member of the Fayetteville Shale). It is higher in rank than a *bed.* Abbrev: mbr. Cf: *lentil; tongue.*

membranate chorate cyst A dinoflagellate *chorate cyst* with a prominent membrane, e.g. *Membranilarnacia.*

membrane The thinned, generally delicate and elastic exinous floor of a pore or colpus of a pollen grain; e.g. *harmomegathus.*

membranelle A flattened, bladelike vibrating organ in a tintinnid, consisting of a row of fused cilia and fringed with lamellae, and used for locomotion. Syn: *membranella.*

membranimorph adj. Pertaining to generally simple anascan cheilostomes (bryozoans), characterized by extensive frontal membranes, slightly to moderately developed gymnocysts with or without marginal spines, and small or no cryptocysts. —n. An anascan cheilostome having such a structure.

membraniporiform Said of a generally unilaminate encrusting colony in cheilostome bryozoans, firmly attached by calcified or membranous walls of zooids (Lagaaij & Gautier, 1965, p. 51).

membranous sac In living stenolaemate bryozoans, the membrane surrounding the digestive and reproductive organs in retracted position.

menaccanite A variety of *ilmenite* found as a sand near Manaccan (Menachan) in Cornwall, England. Also spelled: *menachanite; manaccanite.*

Menap North European glacial stage: Pleistocene (below Elster). Equivalent to the Günz of the Alpine sequence. Replaces *Elbe* in glacial chronology.

mendeleyevite A titanium- and rare-earth-bearing betafite. Also spelled: *mendelyeevite; mendeleevite.*

m-enderbite A name proposed by Tobi (1971, p. 202) in his classification of the *charnockite series* for that member in which mesoperthite and free plagioclase are both present.

mendip (a) A buried hill that is exposed (by the cutting of a valley across a cuesta) as an inlier. (b) A coastal-plain hill that at one time was an offshore island.—Type locality: Mendip Hills in Somerset, England.

mendipite A white orthorhombic mineral: $Pb_3Cl_2O_2$.

mendozite A monoclinic mineral of the alum group: $NaAl(SO_4)_2 \cdot 11H_2O$ (?). Cf: *soda alum.*

meneghinite A blackish lead-gray mineral: $CuPb_{13}Sb_7S_{24}$.

Menevian European stage: Middle Cambrian (above Solvan, below Maentwrogian).

mengwacke A wacke with 33-90% unstable materials (Fischer, 1934). Etymol: German *Mengwacke,* "mixed wacke".

menilite An opaque, impure, dull-grayish or brown variety of opal found in rounded or flattened concretions at Menilmontant near Paris, France. Syn: *liver opal.*

meniscus A liquid surface curved by capillarity. Pl: *menisci.*

Meotian Eastern European stage (Black Sea area): Upper Miocene (above Sarmatian, below Pontian). It has also been regarded as lowermost Pliocene, and is equivalent in age to Pannonian.

Meramecian North American series: Upper Mississippian (above Osagian, below Chesterian).

meraspid Adj. of *meraspis.*

meraspis A juvenile trilobite that has a distinct cephalon and pygidium but does not yet have the number of thoracic segments typical of the species; a late trilobite larva in which the thorax progressively develops. Pl: *meraspides.*

Mercalli scale An arbitrary scale of *earthquake intensity,* ranging from I (detectable only instrumentally) to XII (causing almost total destruction). It is named after Giuseppi Mercalli (d.1914), the Italian geologist who devised it in 1902. Its adaptation to North American conditions is known as the *modified Mercalli scale.*

mercallite A sky-blue mineral: $KHSO_4$.

Mercator chart A chart or map drawn on the Mercator projection. It is commonly used for marine navigation.

Mercator equal-area projection *sinusoidal projection.*

Mercator projection An equatorial, cylindrical, conformal map projection derived by mathematical analysis (not geometrically) in which the equator is represented by a straight line true to scale, the meridians by parallel straight lines perpendicular to the equator and equally spaced according to their distance apart at the equator, and the parallels by straight lines perpendicular to the meridians and parallel with (and the same length as) the equator. The parallels are spaced so as to achieve conformality, their spacing increasing rapidly with their distance from the equator so that at all places the degrees of latitude and longitude have the same ratio to each other as to the sphere itself, resulting in great distortion of distances, areas, and shapes in the polar regions (above 80° lat.), the scale increasing poleward as the secant of the latitude. Because any line of constant direction (azimuth) on the sphere is truly represented on the projection by a straight line, the Mercator projection is of great value in navigation and is used for hydrographic charts, and also to show geographic variations of some physical property (such as magnetic declination) or to plot trajectories of Earth satellites in oblique orbits. Named after Gerhardus Mercator (1512-1594), Flemish mathematician and geographer, whose world map of 1569 used this projection. See also: *transverse Mercator projection.*

Mercator track A *rhumb line* constructed on a Mercator projection.

mercury A heavy, silver-white to tin-white hexagonal mineral, the native metallic element Hg. It is the only metal that is liquid at ordinary temperatures. Native mercury is found as minute fluid globules disseminated through cinnabar or deposited from the waters of certain hot springs, but it is unimportant as a source of the metal. It usually contains small amounts of silver. Mercury combines with most metals to form alloys or amalgams. Syn: *quicksilver.*

mercury barometer A type of *barometer* that measures barometric change by its effect on the mercury or other liquid in a U-shaped glass tube closed at one end. Cf: *aneroid barometer.*

mer de glace A general term applied to any of the large glaciers or ice sheets of the Pleistocene Epoch. Type example: Mer de Glace, the largest glacier on the Mont Blanc massif in the Alps. Etymol: French, "sea of ice".

mere [coast] An obsolete term for an estuary, creek, inlet, or other arm of the sea, and for the sea itself.

mere [lake] (a) A sheet of standing water; esp. a large pond or a small, shallow lake, occupying a hollow among drumlins, or often occupying a basin resulting from subsidence caused by the removal of subsurface salt or by the solvent action of ground water on the salt. (b) A levee lake behind a barrier consisting of sediment carried upstream by the tide.

merenskyite A mineral: $(Pd,Pt)(Te,Bi)_2$. Syn: *biteplapallidite.*

mergifer Said of a variant of the radulifer type of long brachiopod crura, very close together and parallel, arising directly from the swollen edge of a high dorsal median septum.

meridian (a) An imaginary great circle on the surface of the Earth passing through the poles and perpendicular to the equator, connecting all points of equal *longitude;* a north-south line of constant longitude, or a plane, normal to the geoid or spheroid and passing through the Earth's axis, defining such a line. Also, a half of such a great circle included between the Earth's poles. Syn: *terrestrial meridian.* (b) Any one of a series of lines, corresponding to meridians, drawn on a globe, map, or chart at intervals due north and south and numbered according to the degrees of longitude east or west from the prime meridian. (c) *celestial meridian*. —Cf: *parallel*.

meridian hole A term introduced by Agassiz (1866, p.293-294) for a shallow, crescent-shaped *dust well* that accurately registers on the surface of a glacier the position of the Sun during the day. In the northern hemisphere it has a steeper wall on its southern side than on its northern side.

meridian line A line running accurately north and south through any given point on or near the Earth's surface; specif. a line used in plane surveying and defined by the intersection of the plane of the celestial meridian and the plane of the horizon.

meridional (a) Pertaining to a movement or direction between the poles of an object, e.g. the Earth's north-south water or air circulation patterns, or the alignment of colpi on a pollen grain. (b) Southern.

meridional difference The difference (distance) between the meridional parts of any two given parallels of latitude. It is found by subtraction if the two parallels are on the same side of the equator, and by addition if they are on opposite sides. Cf: *latitude difference; departure*. Also called: *meridional difference of latitude*.

meridional part The linear length of the arc of a meridian between the equator and a given parallel of latitude on a Mercator chart, expressed in units of one minute of longitude at the equator.

meridional projection A projection of a sphere onto a plane that is parallel to the plane of a meridian passing through the point of projection; e.g. "meridional orthographic map projection" or "Lambert meridional equal-area projection".

merismite Chorismite in which there is irregular penetration of the diverse units (Dietrich & Mehnert, 1961). Little used.

meristele A strand of vascular tissue enclosed in a sheath of endodermis forming part of a *dictyostele* (Swartz, 1971, p. 289).

meristem A plant tissue consisting of undifferentiated formative or generative cells that give rise to daughter cells capable of differentiation, as found in the *cambium* and other plant tissues and organs (Swartz, 1971, p. 289).

mero- A prefix signifying "part" or "portion".

merocrystalline *hypocrystalline*.

merodont Said of a class of ostracode hinges having three elements and characterized by crenulate terminal elements with either a positive or a negative, crenulate or smooth, median element.

merohedral Said of crystal classes in a system, the general form of which has only one half, one fourth, or one eighth the number of equivalent faces of the corresponding form in the *holohedral* class of the same system. This condition is known as *merohedrism*. Cf: *tetartohedral*. Syn: *merosymmetric*. See also: *hemihedral*.

merohedrism The condition of being *merohedral*. Syn: *merohedry*.

merohedry *merohedrism*.

merokarst Karst that is imperfectly developed, characterized by thin or impure limestone bedrock and by the presence of surface drainage. See also: *holokarst*.

meroleims Coalified remains of plant debris (Kristofovich, 1945, p.138). Cf: *hololeims*. See also: *phytoleims*.

merome In receptaculitids, an elongate calcified branch (lateral) originating on the main axis.

meromictic lake A lake that undergoes incomplete mixing of its waters during periods of circulation; specif. a lake in which the bottom, noncirculating water mass (*monimolimnion*) is adiabatically isolated from the upper, circulating layer (*mixolimnion*). Cf: *holomictic lake*.

meromixis The process leading to, or the condition of, a *meromictic lake*.

meroplankton Plankton that are temporarily planktonic, e.g. eggs and larvae of benthic and nektonic organisms. Cf: *holoplankton*. Syn: *temporary plankton*.

meropod The *merus* of a malacostracan crustacean. Syn: *meropodite*.

merostome Any aquatic *arthropod* belonging to the class Merostomata, characterized by the presence of one pair of preoral appendages with three, possibly four, joints. Cf: *eurypterid; arachnid*.

merosymmetric *merohedral*.

merosyncline Bubnoff's term for that part of a geosynclinal belt having independent mobility (Glaessner & Teichert, 1947, p. 588).

meroxene A variety of biotite with its axial plane parallel to the crystallographic *b*-axis.

Merriam effect The relationship between mountain mass and the vertical distribution of animals and plants. The term was designated by Lowe (1961, p. 45-46) for the influence of factors such as the elevation of a mountain, its size or mass, and the elevation of the basin or plain from which it rises, on the vertical placement of species and communities of plants and animals. Named after Clinton Hart Merriam (1855-1942), U.S. biologist, who first recognized the relationship (Merriam, 1890).

merrihueite A mineral: $(K,Na)_2(Fe,Mg)_5Si_{12}O_{30}$.

merrillite A syn. of *whitlockite*. It is found in meteorites.

Mersey yellow coal *tasmanite [coal]*.

mertieite A pseudohexagonal mineral: $Pd_5(Sb,As)_2$. It is polymorphous with isomertieite.

merumite A mixture of eskolaite and several hydrous chromium oxides.

merus The fourth pereiopodal segment from the body of a malacostracan crustacean, bounded proximally by the ischium and distally by the carpus. Pl: *meruses*. Syn: *meropod*.

merwinite A colorless to pale-green monoclinic mineral: $Ca_3Mg(SiO_4)_2$.

merzlota A Russian term for *frozen ground*.

mesa (a) An isolated, nearly level landmass standing distinctly above the surrounding country, bounded by abrupt or steeply sloping erosion scarps on all sides, and capped by layers of resistant, nearly horizontal rock (often lava). Less strictly, a very broad, flat-topped, usually isolated hill or mountain of moderate height bounded on at least one side by a steep cliff or slope and representing an erosion remnant. A mesa is similar to, but has a more extensive summit area than, a *butte*, and is a common topographic feature in the arid and semiarid regions of the U.S. See also: *table mountain*. (b) A broad terrace or comparatively flat plateau along a river valley, marked by an abrupt slope or escarpment on one side. See also: *bench [geomorph]*.—Etymol: Spanish, "table". Pron: *may*-sa.

mesabite An ocherous variety of goethite from the Mesabi Range in Minnesota.

mesa-butte A *butte* formed by the erosion and reduction of a mesa. Cf: *volcanic butte*.

mesa plain The flat summit of a hilly mountain or plateau (Hill, 1900, p. 6). Cf: *plateau plain*.

mesarch adj. Said of an ecologic succession (i.e. a sere) that develops under *mesic* conditions. Cf: *hydrarch; xerarch*. See also: *mesosere*.

mesa-terrace An obsolete term used by Lee (1900, p. 504-505) for an alluviated, planate rock surface contained within a valley, lying between the flood plain of a nearby stream and the steeper slope leading up to a mesa.

mesenchyme (a) The *mesohyle* of a sponge. (b) A term used by zoologists for the fleshy connective tissue in coelenterates, but applied by paleontologists to the stony skeletal structures between corallites secreted by the common fleshy connective tissue (Shrock & Twenhofel, 1953, p.133). Cf: *sclerenchyme*.—Also spelled: *mesenchyma*.

mesentery One of several radially disposed fleshy laminae or sheets of soft tissue that are attached to the inner surface of the oral disk and column wall of a coral polyp, and that partition the internal body cavity by extending inward from the body wall. Adj: *mesenterial*.

meseta (a) A small *mesa*. (b) An extensive plateau or flat upland, often with an uneven or eroded surface, forming the central physical feature of a region; e.g. the high, dissected tableland of the interior of Spain.—Etymol: Spanish, "tableland". Syn: *mesita; mesilla*.

mesh [part size] One of the openings or spaces between the wires of a sieve or screen. See also: *mesh number*.

mesh [pat grd] The unit component of patterned ground (excepting steps and stripes), as a circle, a polygon, or an intermediate form (Washburn, 1956, p.825).

mesh number The size of a sieve or screen, or of the material passed by a sieve or screen, in terms of the number of *meshes* per linear inch; e.g. mesh number 20 indicates that the sieve or screen has 20 holes per linear inch (this takes no account of the diameter of the wire, so that the mesh number does not always have a definite relation to the size of the hole).

mesh texture A rock texture that is *reticulate*.

mesic Said of a habitat receiving a moderate amount of moisture; also, said of an organism or group of organisms occupying such a habitat. Cf: *hydric; xeric.* See also: *mesarch.*

mesic temperature regime A soil temperature regime in which the mean annual temperature (measured at 50cm) is at least 8°C but less than 16°C, with a summer-winter variation of more than 5°C (USDA, 1975). Cf: *isomesic temperature regime.*

mesilla A term used in SW U.S. for a small *mesa.* Etymol: Spanish, "small table". Syn: *mesita; meseta.*

mesistele The intermediate part of a crinoid column, between proxistele and dististele; it is doubtfully distinguishable in pluricolumnals. Also, the medial part of the stele in certain homalozoans.

mesita *mesilla.*

mesitis Transformation, tending to promote homogenization, between chemically different rocks under the same temperature and pressure (Dietrich & Mehnert, 1961, p. 61).

mesitite A white variety of magnesite containing 30-50% iron carbonate. Syn: *mesitine; mesitine spar.*

meso- A prefix meaning "middle".

mesoclade One of the median *clades* or skeletal branches that connect the bifurcating parts of the actines in an ebridian skeleton.

mesoclimate The climate of a small area, for example a valley or a densely forested area, that may differ from the general climate of the region. See also: *macroclimate; microclimate.*

mesoconch The part of a dissoconch of a bivalve mollusk formed at an intermediate stage of growth and separated from earlier- and later-formed parts by pronounced discontinuities.

mesocoquina A term used by Bissell & Chilingar (1967, p. 153) for a detrital limestone composed of weakly cemented shell detritus of sand size (2 mm in diameter) or less. Cf: *microcoquina.*

mesocratic Composed of almost equal amounts of light and dark constituents; applied to igneous rocks intermediate in color between *leucocratic* and *melanocratic.* The percentage of mafic minerals required for a rock to be classified as mesocratic varies among petrologists; the lower limit ranges from 30 to 37%, the upper limit from 60 to 67%.

mesocrystalline Said of the texture of a rock consisting of or having crystals whose diameters are intermediate between those of a microcrystalline and a macrocrystalline rock; also, said of a rock with such a texture. Howell (1922) applied the term to the texture of a recrystallized sedimentary rock having crystals whose diameters are in the range of 0.20-0.75 mm, and Bissell & Chilingar (1967, p.103) to the texture of a carbonate sedimentary rock having crystals whose diameters are in the range of 0.05-1.0 mm.

mesocumulate A cumulate containing a small amount of intercumulus material; a cumulate intermediate between an *orthocumulate* and an *adcumulate.*

mesoderm The middle layer of cell wall in an embryo, which gives rise to muscular, vascular, and connective tissues.

Mesogea A name, used mainly in France, for the sea usually called *Tethys.*

mesogene Said of a mineral deposit or enrichment of mingled *hypogene* and *supergene* solutions; also, said of such solutions and environment.

mesogenetic A term proposed by Choquette & Pray (1970, p. 220) for the period between the time when newly buried deposits are affected mainly by processes related to the depositional interface (*eogenetic* stage) and the time when long-buried deposits are affected by processes related to the erosional interface (*telogenetic* stage). Also applied to the porosity that develops during the mesogenetic stage.

mesogeosyncline A geosyncline between two continents and receiving clastics from both of them (Schuchert, 1923). Syn: *mediterranean.*

mesogloea A gelatinous substance between endoderm and ectoderm of certain invertebrates; e.g. an extracellular gel, containing proteins and carbohydrates, found in the mesohyle of many sponges, or a noncellular jellylike middle layer of the outer walls and mesenteries of coral polyps. Syn: *mesoglea.*

mesograined Said of the texture of a carbonate sedimentary rock having clastic particles whose diameters are in the range of 0.05-1.0 mm (Bissell & Chilingar, 1967, p. 103) or 0.1-1.0 mm (DeFord, 1946).

mesogyrate Said of the umbones (of a bivalve mollusk) curving toward the center. Cf: *orthogyrate.*

mesohyle Loosely organized material constituting a sponge between the pinacoderm and the choanoderm, commonly consisting of spongin, spicules, and various types of cells (mainly amoebocytes), embedded in mesogloea, although one or more of these elements may be missing. Syn: *mesenchyme; parenchyma.*

mesokurtic Closely resembling a normal frequency distribution; e.g. said of a distribution curve that is neither *leptokurtic* (very peaked) nor *platykurtic* (flat across the top).

mesolimnion A *metalimnion* in a lake. Adj: *mesolimnetic.*

mesolite A zeolite mineral: $Na_2Ca_2Al_6Si_9O_{30}\cdot 8H_2O$. It is intermediate in chemical composition between natrolite and scolecite, and is usually found in white or colorless tufts of very delicate acicular crystals in amygdaloidal basalts. Syn: *cotton stone.*

Mesolithic n. In archaeology, the middle division of the *Stone Age,* characterized by the change from glacial to postglacial climate and the absence of agriculture. Correlation of relative cultural levels with actual age (and, therefore, with the time-stratigraphic units of geology) varies from region to region. Cf: *Paleolithic; Neolithic.* Syn: *Transitional; Middle Stone Age.*—adj. Pertaining to the Mesolithic.

mesolithion Animals that live in cavities in rock.

mesomicrocline A pseudomonoclinic mineral of the alkali feldspar group: $KAlSi_3O_8$. It is intermediate in degree of ordering between microcline and orthoclase.

mesonorm Theoretical calculation of normative minerals in metamorphic rocks of the *mesozone* from chemical analyses (Barth, 1959). Cf: *catanorm; epinorm.* See also: *Niggli molecular norm.*

mesopelagic Pertaining to the *pelagic* environment of the ocean between 100 and 500 fathoms. Cf: *epipelagic.*

mesopeltidium A sclerite (commonly one of a pair) of the segmented carapace of an arachnid, situated immediately behind the *propeltidium* and in front of the *metapeltidium.*

mesoperthite A variety of perthitic feldspar consisting of an intimate mixture of about equal amounts of potassium feldspar and plagioclase (usually albite, sometimes oligoclase). It is intermediate in composition between perthite and antiperthite. Syn: *eutectoperthite.*

mesophilic Said of an organism that prefers a moderate environment; e.g. mesothermal conditions. Noun: *mesophile.*

mesophyll The chlorophyllous tissues in the interior of a leaf.

mesophyte A plant that cannot survive extreme conditions of temperature or water supply. Cf: *hydrophyte; xerophyte.* Adj: *mesophytic.*

Mesophytic A paleobotanic division of geologic time, signifying the time between the first occurrence of gymnosperms and that of angiosperms. Cf: *Aphytic; Archeophytic; Eophytic; Paleophytic; Cenophytic.*

Mesophyticum A paleobotanic division of geologic time, corresponding approximately to, and characterized by the plant life of, the Mesozoic. Cf: *Palaeophyticum; Cainophyticum.*

mesopiestic water A class of *piestic water* including water that rises above the water table but not to the land surface. Cf: *hypopiestic water; hyperpiestic water.*

mesoplankton (a) Plankton of the size range 0.5-1.0 mm; a type of *microplankton.* (b) Plankton that live at middle depths. The term is rarely used because it is confusing.

mesopore [paleont] A polymorph in stenolaemate bryozoans, generally polygonal and smaller in cross section than feeding zooecia, occurring between feeding zooecia in exozones only, and containing closely spaced diaphragms throughout its length so that no appreciable living chamber occurs. Cf: *exilazooecium.* Syn: *mesozooecium.*

mesopore [petrology] In the pore-size classification of Choquette & Pray (1970, p.233), an equant to equant-elongate pore or a tubular or platy pore with an average diameter or thickness between 4 and 1/16 mm. Cf: *megapore; micropore.*

mesopsammon Animals that live in cavities in sand.

Mesosauria An order of presumably anapsid reptiles, aquatic fishcatchers, known only from black shales of Early Permian age in South Africa and Brazil.

mesoscopic According to Dennis (1967, p.152), a term introduced to describe a tectonic feature large enough to be observed without the aid of a microscope yet small enough that it can still be observed directly in its entirety. Cf: *macroscopic.*

mesosere A *sere* that develops in an environment having a moderate amount of moisture, i.e. in a mesic environment; a *mesarch* sere. Cf: *hydrosere; xerosere.*

mesosiderite A *stony-iron meteorite* in which the silicates are mainly pyroxene (usually orthopyroxene) and calcic plagioclase. Mesosiderites often appear to be breccias made up of fragments of widely different chemical and mineralogical composition, cemented together by a nickel-iron matrix. Olivine is sometimes present, generally as separately enclosed crystals of fairly large size. Syn: *grahamite [meteorite].*

mesosilexite A *silexite* in which the dark-colored components constitute more than five percent of the rock (Johannsen, 1932, p. 24). Not recommended usage.

mesosoma The middle region of the body of some invertebrates, esp. when this cannot be readily analyzed into its primitive segmentation (as in arachnids and most mollusks); specif. the anterior part of a merostome opisthosoma carrying appendages. Cf: *metasoma.* Also spelled: *mesosome.*

mesosphere The *lower mantle;* it is probably not involved in the Earth's tectonic processes.

mesostasis The last-formed interstitial material, either glassy or aphanitic, of an igneous rock. Cf: *groundmass [ign].* Syn: *basis [ign]; base [ign].*

mesotheca *median lamina.*

mesotherm A plant that requires moderate temperatures for successful growth. Cf: *microtherm; megatherm.*

mesothermal [clim] Pertaining to a climate characterized by moderate temperature. Cf: *megathermal; microthermal.*

mesothermal [eco geol] Said of a hydrothermal mineral deposit formed at considerable depth and in the temperature range of 200°-300°C (Park & MacDiarmid, 1970, p.317). Also, said of that environment. Cf: *hypothermal; epithermal; leptothermal; telethermal; xenothermal.*

mesothermal [ecol] Said of an organism that prefers moderate temperatures, i.e. in the 25-37°C range.

mesothyridid Said of a brachiopod pedicle opening when the foramen is located partly in the ventral umbo and partly in the delthyrium, with the beak ridges appearing to bisect the foramen (TIP, 1965, pt.H, p.148).

mesotil A semiplastic or semifriable derivative of chemically weathered till, developed beneath a partially drained area, and intermediate in texture between *gumbotil* and *silttil* (Leighton & MacClintock, 1930, p. 42-43).

mesotourmalite A *tourmalite* in which tourmaline comprises 5-50 percent of the rock (Johannsen, 1932, p. 24). Not recommended usage.

mesotrophic lake A lake that is characterized by a moderate supply of nutrient matter, neither notably high nor low in its total production; it is intermediate between a *eutrophic lake* and an *oligotrophic lake.*

mesotrophic peat Peat containing a moderate amount of plant nutrients. Cf: *oligotrophic peat; eutrophic peat.*

mesotrophy The quality or state of a *mesotrophic lake.*

mesotype (a) A group of zeolite minerals, including natrolite, mesolite, and scolecite. Syn: *needle zeolite.* (b) A term used, mainly in France, in the restricted meaning of *natrolite,* because its form is intermediate between those of stilbite and analcime.

Mesozoic An era of geologic time, from the end of the Paleozoic to the beginning of the Cenozoic, or from about 225 to about 65 million years ago. See also: *age of gymnosperms; age of reptiles.* Obsolete syn: *Secondary.*

mesozone According to Grubenmann's classification of metamorphic rocks (1904), the intermediate *depth zone* of metamorphism, which is characterized by temperatures of 300°-500°C, moderate hydrostatic pressure and shearing stress. Chemical and regional metamorphism predominate; association of some epizone and katazone minerals is characteristic. The concept was modified by Grubenmann and Niggli (1924) to include effects of intermediate-temperature contact metamorphism. Modern usage stresses temperature-pressure conditions (medium to high metamorphic grade) rather than the likely depth of zone. Cf: *katazone; epizone.*

mesozooecium *mesopore.*

messelite A mineral: $Ca_4Fe_2(PO_4)_4 \cdot 5H_2O$. Syn: *neomesselite.*

messenger A sliding metal weight on the cable of an oceanographic vessel, which activates an oceanographic device.

Messinian European stage: uppermost Miocene (above Tortonian, below Zanclean). See also: *Diestian.*

meta- A prefix that, when used with the name of a sedimentary or igneous rock, indicates that the rock has been metamorphosed, e.g. metabasalt, metaquartzite.

meta-aluminite A mineral: $Al_2(SO_4)(OH)_4 \cdot 5H_2O$.

meta-alunogen A mineral: $Al_4(SO_4)_6 \cdot 27H_2O$.

meta-ankoleite A yellow secondary mineral: $K_2(UO_2)_2(PO_4)_2 \cdot 6H_2O$.

meta-anthracite Coal having a fixed-carbon content of 98% or more; the highest rank of anthracite. Cf: *graphocite.* Syn: *superanthracite; subgraphite.*

meta-argillite An *argillite* that has been metamorphosed.

meta-arkose Arkose that has been "welded" or recrystallized by metamorphism so that it resembles a granite or a granitized sediment (Pettijohn, 1957, p.325). Cf: *recomposed granite.*

meta-autunite A yellow secondary mineral: $Ca(UO_2)_2(PO_4)_2 \cdot 2\text{-}6H_2O$. It is apparently not formed directly in nature, but most field and museum specimens of autunite have been partly dehydrated to this phase. Cf: *para-autunite.*

metabasite A collective term, first used by Finnish geologists, for metamorphosed mafic rock that has lost all traces of its original texture and mineralogy owing to complete recrystallization.

metabentonite (a) Metamorphosed, altered, or somewhat indurated bentonite, characterized by clay minerals (esp. illite) that no longer have the property of absorbing or adsorbing large quantities of water; nonswelling bentonite, or bentonite that swells no more than do ordinary clays. The term has been applied to certain Ordovician clays of the Appalachian region and upper Mississippi River valley. See also: *potassium bentonite.* Syn: *subbentonite.* (b) A mineral of the montmorillonite group with SiO_2 layers in the montmorillonite structure.

metabituminous coal Coal that contains 89-91.2% carbon, analyzed on a dry, ash-free basis. Cf: *semibituminous coal.*

metablastesis (a) Recrystallization and growth of a preferred mineral or group of minerals. (b) Essentially isochemical recrystallization without evidence of a separate mobile phase.— (Dietrich & Mehnert, 1961, p. 61).

metaboghead coal High-ranking torbanite.

metabolism of rocks A term proposed by Barth (1962) for the redistribution of granitizing materials within sediments by mobilization, transfer, and reprecipitation, as opposed to metasomatism involving addition of new materials.

metabolite [ecol] An excretion or external secretion (e.g., an enzyme, hormone, or vitamin) of an organism that affects the associated organisms by inhibiting their activities or even killing them.

metabolite [meteorite] An iron meteorite showing metamorphic effects due to reheating.

metabolite [rock] A term proposed, but not used, for altered *lassenite.* Obsolete.

metaboly The capability of an organism to change its shape.

metaborite A white mineral: HBO_2. It is the cubic modification of metaboric acid.

metacalciouranoite A mineral: $(Ca,Na,Ba)U_2O_7 \cdot 2H_2O$. Syn: *metacaltsuranoite.*

metacaltsuranoite *metacalciouranoite.*

metacannel coal Cannel coal of high metamorphic rank. Cf: *subcannel coal.*

metacinnabar A black isometric mineral: HgS. It is dimorphous with cinnabar and represents an ore of mercury. Syn: *metacinnabarite; saukovite.*

metaclase Leith's term for a rock possessing *secondary cleavage,* or cleavage in its modern meaning (1905, p. 12). Cf: *protoclase.*

metacolloid An originally colloidal substance that has become crystalline, e.g. serpophite.

metacryst Any large crystal developed in a metamorphic rock by recrystallization, such as garnet or staurolite in mica schist; a syn. of *porphyroblast.* Syn: *metacrystal.*

metacrystal *metacryst.*

metadelrioite A triclinic mineral: $CaSrV_2O_6(OH)_2$.

metadiagenesis *epigenesis [sed].*

metagenesis [evol] *alternation of generations.*

metagenesis [sed] A term applied by Russian geologists to *epigenesis* (changes occurring in a more or less compact sedimentary rock) or to late epigenesis.

metagenic Said of a sediment or sedimentary rock formed through diagenetic alteration of other sediments (Grabau, 1920, p. 2).

metaglyph A hieroglyph formed during metamorphism (Vassoevich, 1953, p.33).

metahaiweeite A secondary mineral: $Ca(UO_2)_2Si_6O_{15} \cdot nH_2O$, where *n* is less than 5. It is apparently a dehydration product of haiweeite.

metahalloysite A name used in Europe for the less hydrous form of halloysite. It is synonymous with *halloysite* of U.S. authors. The term has also been used to designate the nonhydrated form of halloysite.

metaharmosis Var. of *metharmosis.*

metaheinrichite A yellow to green secondary mineral: $Ba(UO_2)_2$

$(AsO_4)_2 \cdot 8H_2O$. Cf: *heinrichite.*

metahewettite A red mineral: $CaV_6O_{16} \cdot 9H_2O$. It resembles hewettite but differs slightly from it in its behavior during hydration; it is found in highly oxidized ore as coatings and fracture fillings.

metahohmannite An orange mineral: $Fe_2(SO_4)_2(OH)_2 \cdot 3H_2O$. It constitutes a partly dehydrated hohmannite.

metajennite A mineral: $Na_2Ca_8Si_5O_{19} \cdot 7H_2O$.

metakahlerite A yellow to yellowish-green secondary mineral: $Fe(UO_2)_2(AsO_4)_2 \cdot 8H_2O$.

metakaolinite An intermediate product obtained when kaolinite is heated between about 500°C and 850°C; artificially dehydrated kaolinite. Syn: *metakaolin.*

metakirchheimerite A pale-rose mineral: $Co(UO_2)_2(AsO_4)_2 \cdot 8H_2O$.

metal In the older geologic literature, a now obsolete term for any hard rock.

metal factor A derived parameter used to represent induced polarization anomalies. Abbrev: MF.

metalignitous coal Coal that contains 80-84% carbon, analyzed on a dry, ash-free basis. Not listed by ASTM as a rank classification. Cf: *subbituminous coal.*

metalimnion The horizontal layer of a thermally stratified lake in which the temperature decreases rapidly with depth. The metalimnion lies between the *epilimnion* and the *hypolimnion,* and includes the thermocline. Less preferred syn: *thermocline [lake] (b); mesolimnion.*

metallic (a) Pertaining to a metal. (b) Said of a type of luster that is characteristic of metals. Cf: *nonmetallic; submetallic luster.*

metalliferous Metal-bearing; specif., pertaining to a mineral deposit from which a metal or metals can be extracted by metallurgical processes.

metallization The process or processes by which metals are introduced into a rock, resulting in an economically valuable deposit; the mineralization of metals.

metallized hood The upper shell or roof of a batholith, which is the first area to solidify after intrusion and may contain virtually all the hydrothermal metalliferous lodes of the intrusion (Emmons, 1933). Syn: *hood [intrus rocks].*

metallogenetic *metallogenic.*

metallogenic Adj. of *metallogeny.* Syn: *metallogenetic; minerogenic; minerogenetic.*

metallogenic element An element that occurs as a native element or that occurs in sulfides, selenides, tellurides, arsenides, antimonides, or sulfosalts. It is one of H.S. Washington's bipartate groupings of elements of the lithosphere, now obsolete. Cf: *petrogenic element.*

metallogenic epoch A unit of geologic time favorable for the deposition of ores, or characterized by a particular assemblage of mineral deposits. Several metallogenic epochs may be represented within a single area, or *metallogenic province.*

metallogenic map A map, usually on a regional scale, on which is shown the distribution of particular assemblages or provinces of mineral deposits and their relationship to such geologic features as tectonic trends and petrographic types.

metallogenic province An area characterized by a particular assemblage of mineral deposits, or by one or more characteristic types of mineralization. A metallogenic province may have had more than one episode of mineralization, or *metallogenic epoch.* Syn: *metallographic province.*

metallogeny The study of the genesis of mineral deposits, with emphasis on their relationship in space and time to regional petrographic and tectonic features of the Earth's crust. The term has been used for both metallic and nonmetallic mineral deposits. Adj: *metallogenic.*

metallographic province A little-used syn. of *metallogenic province.*

metallo-organic Said of a compound in which an atom of a metal is bound to an organic compound through an atom other than carbon, such as oxygen, nitrogen, or sulfur, to form a coordination compound. Cf: *organometallic.*

metallotect A term used in metallogenic studies for any geologic feature (tectonic, lithologic, geochemical, etc.) considered to have influenced the concentration of elements to form mineral deposits; an *ore control,* but without the implication of economic value.

metallurgy The science and art of separating metals and metallic minerals from their ores by mechanical and chemical processes; the preparation of more metalliferous materials from raw ore.

metalodevite A mineral of the meta-autunite group: $Zn(UO_2)_2(AsO_4)_2 \cdot 10H_2O$.

metaluminous Said of an igneous rock in which the molecular proportion of aluminum oxide is greater than that of sodium and potassium oxides combined but generally less than of sodium, potassium, and calcium oxides combined; one of Shand's (1947) groups of igneous rocks, classified on the basis of the degree of aluminum-oxide saturation. Cf: *peralkaline; peraluminous; subaluminous.*

metamarble A term proposed by Brooks (1954) for metamorphic carbonate rock that is commercially valuable because it will take a polish, e.g. the Vermont metamarble. Cf: *orthomarble.*

metamict Said of a mineral containing radioactive elements in which various degrees of lattice disruption and changes have taken place as a result of radiation damage while its original external morphology has been retained. Examples occur in zircon, thorite and several other minerals. Not all minerals containing radioactive elements are metamict; e.g. xenotime and apatite.

metamorphic adj. Pertaining to the process of metamorphism or to its results. n. A *metamorphic rock.*

metamorphic assemblage (a) A metamorphic *mineral assemblage.* (b) *metamorphic complex.*

metamorphic aureole *aureole.*

metamorphic complex The metamorphic rocks constituting a whole group closely related on a regional and/or stratigraphic basis, e.g. the Dalradian metamorphic complex of Scotland. Syn: *metamorphic assemblage.*

metamorphic convergence A term to indicate two metamorphic processes converging from opposite directions but resulting in the same metamorphic product, e.g. at the same temperature a diorite may be converted retrogressively and a dolomitic marl progressively into the identical epidote-chlorite-actinolite rock.

metamorphic correlation The determination of equivalence of metamorphic features, either between the metamorphic grades of rocks of different original composition, or between a metamorphic unit and its unmetamorphosed representative elsewhere.

metamorphic differentiation A collective term for the various processes by which minerals or mineral assemblages are locally segregated from an initially uniform parent rock during metamorphism, e.g. garnet porphyroblasts in fine-grained mica schist.

metamorphic diffusion Migration, by diffusion, of materials from one part of a rock mass to another during metamorphism. Diffusion may involve chemically active fluids from magmatic sources, hot pore fluids, or fluids released from hydrous minerals or carbonates. Ionic diffusion in the solid state may also occur. Cf: *solid diffusion.*

metamorphic facies A set of metamorphic mineral assemblages, repeatedly associated in space and time, such that there is a constant annd therefore predictable relation between mineral composition and chemical composition. The concept was introduced by Eskola (1915, 1939). It is generally assumed that the metamorphic facies represent the results of equilibrium crystallization of rocks under a restricted range of externally imposed physical conditions, e.g. temperature, lithostatic pressure, H_2O-pressure. Syn: *mineral facies; densofacies.* See also: *metamorphic facies series; metamorphic subfacies.*

metamorphic facies series A group of *metamorphic facies* characteristic of an individual area or terrane, and represented by a curve or a group of curves in a pressure-temperature diagram illustrating the range of the different types of metamorphism and metamorphic facies (Hietanen, 1967). The term was introduced by Miyashiro (1961).

metamorphic grade The intensity or rank of metamorphism, measured by the amount or degree of difference between the original parent rock and the metamorphic rock. It indicates in a general way the P-T environment or facies in which the metamorphism took place. For example, conversion of shale to slate or phyllite would be low-grade dynamothermal metamorphism (greenschist facies), whereas its continued alteration to a garnet-sillimanite schist would be high-grade metamorphism (almandine-amphibolite facies). Syn: *metamorphic rank.*

metamorphic overprint *overprint.*

metamorphic rank *metamorphic grade.*

metamorphic rock (a) In its original usage (Lyell, 1833), the group of gneisses and crystalline schists. (b) In current usage, any rock derived from pre-existing rocks by mineralological, chemical, and/or structural changes, essentially in the solid state, in response to marked changes in temperature, pressure, shearing stress, and chemical environment, generally at depth in the Earth's crust. Syn: *metamorphic (n.); metamorphite.*

metamorphic subfacies A subdivision of a *metamorphic facies* based on minor but significant differences in mineral assemblages. Such subdivisions should be used with care, since they tend to have only local signifcance. The recent trend is away from the use of subfacies.

metamorphic zoning *zoning [meta].*

metamorphism The mineralogical, chemical, and structural adjustment of solid rocks to physical and chemical conditions which have generally been imposed at depth below the surface zones of weathering and cementation, and which differ from the conditions under which the rocks in question originated. In an older and now obsolete sense, the scope of the term included *katamorphism,* i.e. the processes of cementation and weathering (Van Hise, 1904).

metamorphite *metamorphic rock.*

metamorphosis (a) In biology, a process involving marked or abrupt reorganization of an animal during post-embryonic development, such as the transformation of a larva into a succeeding stage of development and growth. (b) Any change in form, structure, substance, etc.

metanauplius A postnaupliar crustacean larva with the same general body and limb morphology as a *nauplius,* but having additional limbs (about seven pairs).

metanovacekite A yellow mineral: $Mg(UO_2)_2(AsO_4)_2 \cdot 4\text{-}8H_2O$. It is a partly dehydrated form of novacekite.

metaparian Of or concerning a trilobite that appears to have nonfunctional facial sutures both beginning and ending at the posterior margin of the cephalon.

metapeltidium The last sclerite (usually single, rarely one of a pair) of a segmented carapace of an arachnid, following upon the *mesopeltidium.*

metaphyte A multicellular plant. Cf: *protophyte.*

metaplasis That middle stage in an evolutionary *lineage* in which organisms attain maximum vigor and diversification. Cf: *anaplasis; cataplasis.*

metapodosoma A section of the body of an acarid arachnid, bearing the third and fourth pairs of legs. Cf: *propodosoma.*

metaprotaspis A large trilobite protaspis in which the protopygidium is well defined. Cf: *anaprotaspis.* Pl: *metaprotaspides.*

metaquartzite A quartzite formed by metamorphic recrystallization, as distinguished from an *orthoquartzite,* whose crystalline nature is of diagenetic origin.

metaripple A term introduced by Bucher (1919, p.190) for a large asymmetric ripplelike feature whose surface configuration and internal structure show that the final form was produced under conditions quite distinctive from those under which it was initiated; e.g. a ripple that is transformed into another pattern by waves acting in the same direction as the preceding current when the velocity of the current is changed. Not to be confused with *megaripple.* Cf: *para-ripple.*

metarossite A light-yellow or pale greenish-yellow mineral: $CaV_2O_6 \cdot 2H_2O$. It is a dehydration product of rossite.

metaschoderite A monoclinic mineral: $Al_2(PO_4)(VO_4) \cdot 6H_2O$. It is a dehydration product of schoderite.

metaschoepite A mineral: $UO_3 \cdot nH_2O$, where n is less than 2. It is a dehydration product of *schoepite.*

metasediment A sediment or sedimentary rock that shows evidence of having been subjected to metamorphism.

metaseptum One of the main septa of a corallite other than a *protoseptum,* generally distinguished by its extension axially much beyond that of minor septa (TIP, 1965, pt.F, p.249).

metasicula The distal part of the *sicula* of a graptolithine, formed of normal growth increments of fusellar tissue overlain by cortical tissue. Cf: *prosicula.*

metasideronatrite A yellow mineral: $Na_4Fe_2^{+3}(SO_4)_4(OH)_2 \cdot 3H_2O$. It is a partly dehydrated form of sideronatrite.

metasilicate According to the now obsolete classification of silicates as oxyacids of silicon, a salt of the hypothetical metasilicic acid, H_2SiO_3. Cf: *orthosilicate.*

metasom *metasome [geol].*

metasoma The hind region of the body of some invertebrates, esp. when this cannot be readily analyzed into its primitive segmentation (as in some mollusks and arachnids); specif. the posterior part of a merostome opisthosoma lacking appendages, or the *metasome* of a copepod crustacean. Cf: *mesosoma.*

metasomasis *metasomatism.*

metasomatic Pertaining to the process of metasomatism and to its results. The term is especially used in connection with the origin of ore deposits.

metasomatic rock A rock whose chemical composition has been substantially changed by the metasomatic alteration of its original constituents; a *metasomatite.*

metasomatism The process of practically simultaneous capillary solution and deposition by which a new mineral of partly or wholly different chemical composition may grow in the body of an old mineral or mineral aggregate (Lindgren, 1928). In current usage, the presence of interstitial, chemically active pore liquids or gases contained within the rock body or introduced from external sources are essential for the replacement process, which often, though not necessarily, occurs at constant volume with little disturbance of textural or structural features. Cf: *pyrometasomatism.*

metasomatite *metasomatic rock.*

metasome [geol] (a) A replacing mineral, which grows in size at the expense of another mineral (the host or *palasome*); a mineral grain formed by metamomatism. Syn: *guest.* (b) The newly formed part of a migmatite or composite rock, introduced during metasomatism. Cf: *neosome.* —Also spelled: *metasom.*

metasome [paleont] The posterior part of the prosome of a copepod crustacean, consisting of free thoracic somites in front of the major articulation. Also spelled: *metasoma.*

metaspondyl In dasycladacean algae, one of the primary branches or rays that occur in clusters of three or six and that are regularly arranged in whorls.

metastable (a) Said of a phase that is stable with respect to small disturbances but that is capable of reaction with evolution of energy if sufficiently disturbed. (b) Said of a phase that exists in the temperature range in which another phase of lower vapor pressure is stable. A vapor phase need not be present. The metastable phase is not to be confused with instability. In general, metastability is due to the reluctance of a system to initiate the formation of a new, stable phase.

metastable relict *unstable relict.*

metastasis [meta] Changes of a paramorphic character, such as the recrystallization of a limestone or the devitrification of a glassy rock (Bonney, 1886).

metastasis [tect] *metastasy.*

metastasy A term used by Gussow (1958) for lateral adjustments of the Earth's crust, as opposed to vertical movements (isostasy). Syn: *metastasis.*

metaster The portion of a migmatite that remained solid (immobile or less mobile) during migmatization. Cf: *restite; stereosome.* See also: *paleosome.* Little used.

metastibnite A noncrystalline mineral: Sb_2S_3.

metastoma A median platelike process behind the mouth in certain arthropods; e.g. a plate at the posterior edge of the mouth of a merostome. It is possibly represented in some trilobites by a plate posterior to the hypostome. Also, the lower lip behind the mandibles of a crustacean, usually cleft into paragnaths. Pl: *metastomata.* Cf: *hypostome.* Syn: *labium.* Also spelled: *metastome.*

metastome *metastoma.*

metastrengite *phosphosiderite.*

metatarsus (a) Collectively, the foot bones of a tetrapod. (b) The proximal (typically the sixth) segment of a leg of an arachnid, following the tibia and preceding the tarsus. Pl: *metatarsi.*

metatect The fluid or more mobile part of a migmatite. Cf: *chymogenic; mobilizate.* See also: *neosome.* Little used.

metatectite Lipotexite whose mineralogy and texture have been changed mainly through metasomatism accompanying anatexis. Synonymous with the *metatexite* of some workers (Dietrich & Mehnert, 1961). Little used.

metatexis Low-grade *anatexis,* i.e. partial or differential melting (Mehnert, 1968, p. 355). Cf: *diatexis; anamigmatization.*

metatexite The rock resulting from metatexis. Synonymous with the *metatectite* of some workers (Dietrich & Mehnert, 1961). Little used.

metatheca The distal part of a graptoloid theca. It is equivalent to the *autotheca* in those graptolites with more than one type of theca.

metathenardite A mineral representing a high-temperature polymorph (perhaps hexagonal) of thenardite and occurring in fumaroles on Martinique Island.

metatorbernite A green secondary mineral: $Cu(UO_2)_2(PO_4)_2 \cdot 8H_2O$. It contains less water than torbernite.

metatrophic *heterotrophic.*

metatyuyamunite A yellow secondary mineral: $Ca(UO_2)_2(VO_4)_2 \cdot 3\text{-}5H_2O$.

meta-uranocircite A mineral of the meta-autunite group: $Ba(UO_2)_2(PO_4)_2 \cdot 8H_2O$.

meta-uranopilite A yellow, grayish, brown, or green mineral: $(UO_2)_6(SO_4)(OH)_{10} \cdot 5H_2O$. It is partly dehydrated uranopilite.

meta-uranospinite A yellow secondary mineral: $Ca(UO_2)_2(AsO_4)_2 \cdot 8H_2O$. It is partly dehydrated uranospinite.

metavandendriesscheite A mineral: $PbU_7O_{22} \cdot n\,H_2O$, with n less than 12.

metavanuralite A mineral: $Al(UO_2)_2(VO_4)_2(OH) \cdot 8H_2O$.

metavariscite A green monoclinic mineral: $AlPO_4 \cdot 2H_2O$. It is dimorphous with variscite and isomorphous with phosphosiderite.

metavauxite A colorless mineral: $Fe^{+2}Al_2(PO_4)_2(OH)_2 \cdot 8H_2O$. It has more water than vauxite but less than paravauxite.

metavivianite A mineral: $Fe_3^{+2}(PO_4)_2 \cdot 8H_2O$. It is dimorphous with vivianite.

metavoltine A mineral: $K_2Na_6Fe^{+2}Fe_6^{+3}(SO_4)_{12}O_2 \cdot 18H_2O$.

metaxite [mineral] A fibrous serpentine mineral; a variety of chrysotile.

metaxite [sed] *micaceous sandstone.*

metaxylem Primary xylem which matures after the *protoxylem,* concomitantly with or after the surrounding tissues (Cronquist, 1961, p.877).

metazellerite A yellow secondary mineral: $Ca(UO_2)(CO_3)_2 \cdot 3H_2O$.

metazeunerite A green secondary mineral: $Cu\,(UO_2)_2\,(AsO_4)_2 \cdot 8H_2O$. It has less water than zeunerite.

Metazoa The large group of multicellular animals in which the cells are arranged in two layers in the embryonic gastrula stage.

meteor (a) The visible streak of light resulting from the entry into the atmosphere of a solid particle from space. (b) Any physical object or relatively small fragment of solid material associated with a meteor and made luminous as a result of friction during its passage through the Earth's atmosphere; a *meteoroid.* Syn: *shooting star.*

meteor crater *meteorite crater.*

meteoric [meteorite] Relating to or composed of meteors or meteoroids.

meteoric [water] (a) Pertaining to water of recent atmospheric origin. (b) Pertaining to, dependent on, derived from, or belonging to the Earth's atmosphere; e.g. "meteoric erosion" caused by rain, wind, or other atmospheric forces.

meteoric dust Small particles (diameters ranging up to 100 microns) representing the product of melting and oxidation of meteors in the Earth's atmosphere. Cf: *meteoritic dust; cosmic dust.*

meteoric iron (a) Iron of meteoric origin. (b) An *iron meteorite.*

meteoric stone (a) A stone of meteoric origin; a *stony meteorite.* (b) A meteorite having the appearance of a stone.

meteorite Any *meteoroid* that has fallen to the Earth's surface in one piece or in fragments without being completely vaporized by intense frictional heating during its passage through the atmosphere; a stony or metallic meteoroid large enough to survive passage through the Earth's atmosphere and reach the ground. Most meteorites are believed to be fragments of asteroids and to consist of primitive solid matter similar to that from which the Earth was originally formed. Adj: *meteoritic.* Syn: *cosmolite; skystone.*

meteorite crater An *impact crater* formed by the falling of a large meteorite onto a surface; e.g. Barringer Crater (Meteor Crater) in Coconino County, Ariz., and Chubb Crater in Quebec, Canada. Cf: *penetration funnel.* Syn: *meteor crater; meteorite impact crater.*

meteorite impact crater *meteorite crater.*

meteoritic dust Small angular or flat particles representing the product of fragmentation or crushing of meteorites. The particles maintain the composition and structure peculiar to meteorites. Cf: *meteoric dust; cosmic dust.*

meteoritics A science that deals with meteors and meteorites. Cf: *aerolithology.*

meteoroid One of the countless solid objects moving in interplanetary space, distinguished from asteroids and planets by its smaller size but considerably larger than an atom or molecule. Syn: *shooting star; falling star.* Cf: *meteor; meteorite.*

meteorolite An obsolete term for a meteorite, esp. a *stony meteorite.* Syn: *meteorlithe.*

meteorologic tide A change in water level due to such factors as strong winds or barometric pressure. See also: *wind set-up.*

meteorology The study of the Earth's atmosphere, including its movements and other phenomena, especially as they relate to weather forecasting.

meteor shower A large concentration of falling meteors; also, the phenomenon observed when members of a meteor swarm encounter the Earth's atmosphere and their luminous paths appear to diverge from a single point in the sky.

meteor swarm A group of meteoroids that have closely similar orbits around the sun.

meter rod A precise leveling rod graduated in whole and fractional meters.

metes and bounds The boundaries or limits of a tract of land; esp. the boundaries of irregular pieces of land (such as claims, grants, and reservations) in which the bearing and length of each successive line is given and in which the lines may be described by reference to local natural or artificial monuments along it (such as a stream, ditch, road, or fence). Such boundaries have been established for much of the land in non public-land surveys, and are distinguished from those established by beginning at a fixed starting point and running therefrom by stated compass courses and distances.

methane A colorless odorless inflammable gas, the simplest paraffin hydrocarbon, formula CH_4. It is the principal constituent of natural gas and is also found associated with crude oil. See also: *marsh gas; firedamp.*

methane series The homologous series of saturated aliphatic hydrocarbons, empirical formula C_nH_{2n+2}, of which methane is the lowest and representative member, followed by ethane, propane, the butanes, etc. Syn: *paraffin series.*

metharmosis The changes occurring in a sediment after its burial (after uplift or consolidation) but before weathering begins; in this usage, the term is equivalent to *late diagenesis* or *epigenesis.* The term was proposed by Kessler (1922) in a less restricted sense to designate all changes that a sediment may undergo, including diagenesis proper and metamorphism. Syn: *metaharmosis.*

method of least squares Any of several statistical methods for fitting a line, curve, or higher-degree surface to a set of data such that the sum of the squares of the distances of points to that fitted surface is minimized. Syn: *least squares.*

methylene iodide A liquid compound that is used as a *heavy liquid;* its specific gravity is 3.33. Cf: *Clerici solution; Sonstadt solution; Klein solution; bromoform.*

metric carat *carat.*

meulerization Local cementation, and replacement (in part) by opaline or chalcedonic silica carried by ground water, of a carbonate sandstone or a limestone, such as the reaction occurring in certain sedimentary rocks of the Paris Basin. Etymol: French *meule,* "millstone". Cf: *Fontainebleau sandstone.*

Mexican onyx Yellowish-brown or greenish-brown *onyx marble,* found chiefly in Tecali, Mexico.

meyerhofferite A colorless triclinic mineral: $Ca_2B_6O_{11} \cdot 7H_2O$. It is an alteration product of inyoite.

meymacite A resinous, light-brown mineral: $WO_3 \cdot 2H_2O$.

meymechite An ultramafic igneous rock composed of abundant olivine phenocrysts (usually altered) in a serpentine-rich or glassy groundmass; according to Russian petrologists, the extrusive equivalent of *kimberlite.* Also spelled: *meimechite.* Kotulsky in 1943 derived the name from Meimecha Kotuj, Siberia, U.S.S.R.

MF *metal factor.*

MG *marin gräns.*

mgal *milligal.*

MHHW *mean higher high water.*

MHW *mean high water.*

MHWN *mean high-water neap.*

MHWS *mean high-water spring.*

MI *mafic index.*

miagite *corsite.*

mianthite Dark-colored enclosures, patches, or streaks in an anatexite (Dietrich & Mehnert, 1961). Little used.

miargyrite An iron-black to steel-gray monoclinic mineral: $AgSbS_2$. It has a cherry-red powder.

miarolithite A chorismite having miarolitic cavities or remnants thereof; a variety of *ophthalmite.*

miarolitic A term applied to small irregular cavities in igneous rocks, esp. "granites", into which small crystals of the rock-forming minerals protrude; characteristic of, pertaining to, or occurring in such cavities. Also, said of a rock containing such cavities.

miaskite A biotite-bearing *nepheline syenite* containing oligoclase and microperthite. Its name, given by Rose in 1839, is derived from Miask, in the Urals, U.S.S.R. Also spelled: *miascite.* Not recommended usage.

mica (a) A group of minerals of general formula: (K,Na,Ca) (Mg,Fe, $Li,Al)_{2\text{-}3}(Al,Si)_4O_{10}(OH,F)_2$. It consists of complex *phyllosilicates* that crystallize in forms apparently orthorhombic or hexagonal (such as tabular six-sided prisms) but really monoclinic; that are characterized by low hardness and by perfect basal cleavage, readily splitting into thin, tough, somewhat elastic laminae or plates with a splendent pearly luster; and that range in color from colorless, silvery white, pale brown, or yellow to green or black. Micas are prominent rock-forming constituents of igneous and metamorphic rocks, and commonly occur as flakes, scales, or shreds. Sheet muscovite is used in electric insulators; ground mica in paint and as a dusting agent. Cf: *brittle mica.* Syn: *isinglass; glimmer.* (b) Any mineral of the mica group, including muscovite, biotite, lepidolite, phlogopite, zinnwaldite, roscoelite, paragonite, and sericite.

mica book A crystal of mica, usually large and irregular. It is so named because of the resemblance of its cleavage plates to the leaves of a book. Syn: *book.*

micaceous (a) Consisting of, containing, or pertaining to mica; e.g. a "micaceous sediment". (b) Resembling mica; e.g. a "micaceous mineral" capable of being easily split into thin sheets, or a "micaceous luster".

micaceous arkose A term used by Hubert (1960, p.176-177) for a sandstone containing 25-90% feldspars and feldspathic crystalline-rock fragments, 10-50% micas and micaceous metamorphic-rock fragments, and 0-65% quartz, chert, and metamorphic quartzite. The term is roughly equivalent to *impure arkose* of Folk (1954). Cf: *feldspathic graywacke.*

micaceous iron ore A soft, unctuous variety of hematite having a foliated structure resembling that of mica.

micaceous quartzite A term used by Hubert (1960, p.176-177) for a sandstone containing 70-95% quartz, chert, and metamorphic quartzite, 5-15% micas and micaceous metamorphic-rock fragments, and 0-15% feldspars and feldspathic crystalline-rock fragments. Cf: *feldspathic quartzite.*

micaceous sandstone A sandstone containing conspicuous layers or flakes of mica, usually muscovite. Syn: *metaxite [sed].*

micaceous shale A gray or brownish-gray shale, usually well-laminated, containing abundant muscovite flakes along its lamination planes and finer-grained sericite in its clay matrix; it is commonly associated with subgraywacke, and represents detrital deposition under moderately unstable conditions.

mica plate In a polarizing microscope, a phase plate consisting of a sheet of muscovite that is used to determine optical sign from interference figures. Its interference color in white light is a light, neutral gray. Syn: *quarter-wave plate.*

micarelle Mica pseudomorphous after scapolite.

mica schist A schist whose essential constituents are mica and quartz, and whose schistosity is mainly due to the parallel arrangement of mica flakes.

michenerite An isometric mineral: (Pd,Pt)BiTe. Cf: *froodite.*

micracanthopore An acanthopore in stenolaemate bryozoans that belongs to the smaller of two distinct sizes occurring in the same zoarium. Cf: *megacanthopore.*

micrinite A maceral of coal within the *inertinite* group that is granular but shows no plant-cell structure. It is opaque, of high reflectance, and generally less than 10 microns in grain diameter. Larger particles are commonly termed *macrinite.* See also: *residuum [coal].*

micrinoid A maceral group that includes the macerals in the micrinite series.

micrite (a) A descriptive term used by Folk (1959) for the semi-opaque crystalline matrix of limestones, consisting of chemically precipitated carbonate mud with crystals less than 4 microns in diameter, and interpreted as a lithified ooze. The term is now commonly used in a descriptive sense without genetic implication. Leighton & Pendexter (1962) used a diameter limit of 31 microns. Chilingar et al. (1967, p. 317) and Bissell & Chilingar (1967, p.161) extended usage of the term to include unconsolidated material that may be of either chemical or mechanical origin (and possibly biologic, biochemical, or physicochemical). Micrite is finer-textured than *sparite.* See also: *matrix [sed].* (b) A limestone with less than 1% allochems and consisting dominantly of micrite matrix (Folk, 1959, p. 14); e.g. lithographic limestone. See also: *micritic limestone.*

micrite envelope A thin coating of micrite around allochems, particularly skeletal grains. It is produced by coating or boring algae, or perhaps by mechanical adhesion of carbonate mud (Bathurst, 1966). Syn: *algal circumcrust* (K. H. Wolf); *dust ring* (H. C. Sorby).

micritic limestone A limestone consisting of more than 90% micrite (Leighton & Pendexter, 1962, p. 60) or less than 10% allochems (Wolf, 1960, p. 1415); a *micrite.* See also: *calcilutite; lithographic limestone; lime mudstone; calcimicrite.* Syn: *micrite limestone; matrix limestone.*

micritization Decrease in the size of sedimentary carbonate particles, possibly due to boring algae. Syn: *grain diminution.*

micro- A prefix meaning "small". When modifying a rock name, it signifies fine-grained hypabyssal, as in *microgranite.* Cf: *macro-* .

microaerophilic Said of an organism that can exist with very little free oxygen present. Noun: *microaerophile.*

microanalyzer *electron microprobe.*

microaphanitic *cryptocrystalline.*

microatoll (a) A ring-shaped growth of corals or serpulids, surrounding a central dead area or depression, with a width of 1 to 6 m. They are commonly found in the intertidal belt of relatively warm seas or scattered across a reef flat. Cf: *cup reef.* Syn: *miniature atoll.* (b) A small atoll-like reef or knoll, developed within the lagoon or shallows of a reef complex like an atoll or platform reef, characterized by a rim of coral growth surrounding a central sandy depression. Smaller than a *faro,* such a microatoll is on the order of 100 m across and 10 m high, with a depression of 1 to 3 m in the center (Kornicker & Boyd, 1962).

microbiofacies The biologic aspect of a *microfacies* (Fairbridge, 1954).

microbiostratigraphy Biostratigraphy based on microfossils.

microbreccia (a) A poorly sorted sandstone containing relatively large and sharply angular particles of sand set in a very fine silty or clayey matrix; e.g. a graywacke. It is somewhat less micaceous than a siltstone. (b) A breccia within fragments of a coarser breccia (Sander, 1951, p.28).

microchemical test A chemical test made on minute grains or polished surfaces under a microscope. It is often combined, in identifying a substance, with observations on form, color, and optical properties.

microclastic Said of coal that is composed mainly of fine particles, e.g. cannel coal. Cf: *macroclastic.*

microclastic rock A clastic rock whose constituents are very minute. Cf: *cryptoclastic rock.*

microclimate The climatic structure close to the Earth's surface, affected by the character of the surface materials; for example, over a snow surface, lake, or cornfield. See also: *macroclimate; mesoclimate.*

microcline A clear, white to gray, brick-red, or green mineral of the alkali feldspar group: $KAlSi_3O_8$. It is the fully ordered, triclinic modification of potassium feldspar and is dimorphous with orthoclase, being stable at lower temperatures; it usually contains some sodium in minor amounts. Microcline is a common rock-forming mineral of granitic rocks and pegmatites, and is often secondary after orthoclase. It is generally characterized by cross-hatch twinning.

microcline-perthite A perthite consisting of an intergrowth of microcline and plagioclase.

microclinite A *syenite* composed entirely of microcline. Not recommended usage.

micrococcolith One of the smaller coccoliths in coccolithophores exhibiting dimorphism but with the dimorphic coccoliths irregularly placed. Cf: *macrococcolith.*

microconch A mature conch of a chambered cephalopod which, in all respects except size and occasional modification of the aperture, resembles larger conchs (*macroconchs*) found in the same fossil association. Microconchs are now generally regarded as representing males (Callomon, 1963).

microconglomerate A poorly sorted sandstone containing relatively large rounded particles of sand set in a very fine silty or clayey matrix.

microcontinent A submarine plateau that is an isolated fragment of continental crust. Cf: *aseismic ridge.*

microcoquina (a) A detrital limestone composed wholly or chiefly of weakly cemented shell detritus of sand size (2 mm in diameter) or less. (b) A variety of chalk (Bissell & Chilingar, 1967, p. 153).—Cf: *coquina; mesocoquina.*

microcoquinoid limestone A *coquinoid limestone* composed of small shells. Syn: *microcoquinoid.*

microcosmic salt *stercorite.*

microcrater *micrometeorite crater.*

micro cross-lamination A small but distinctive cross-lamination, similar to a small-scale trough cross-bedding. See also: *rib and furrow.*

microcryptocrystalline *cryptocrystalline.*

microcrystal A crystal, the crystalline nature of which is discernible only under the microscope; such crystals form a *microcrystalline* substance.

microcrystalline Said of the texture of a rock consisting of or having crystals that are small enough to be visible only under the microscope; also, said of a rock with such a texture. In regard to carbonate sedimentary rocks, various diameter ranges are in use: 0.01-0.20 mm (Pettijohn, 1957, p. 93); less than 0.01 mm (Carozzi & Textoris, 1967, p. 6); and 0.001-0.01 mm (Bissell & Chilingar, 1967, p. 161, who note that some petrographers use 0.004-0.062 mm). Cf: *cryptocrystalline; felsophyric.* See also: *microcrystal.* Syn: *micromeritic.*

microdelta A small-scale delta or bar, generally not more than a few meters across, with a slip face on which foreset beds are deposited. It resembles a dune or megaripple, but does not show a repetitive, wavelike form, and lacks a well-developed inclined stoss side. Also spelled: *micro-delta.*

microdistributive fracture One of a pattern of numerous tiny fractures along which slight movement has taken place. Such movement in the aggregate can have a strong effect on the form and structure of a large rock body (Rubey & Hubbert, 1959, p.197).

microearthquake An earthquake having a *body-wave magnitude* of two or less on the Richter scale. Such a limit is arbitrary, and may vary according to the user. Cf: *major earthquake; ultramicroearthquake.*

microelement *trace element.*

microeutaxitic Said of the texture of certain extrusive igneous rocks that are microscopically *eutaxitic.*

microevolution (a) The evolution or origin of species, as contrasted to that of higher taxa. (b) Evolution that occurs within a continuous population but does not result in the development of genetic discontinuities; the changes, brought about by selective accumulation of minute variations, are thought to be chiefly responsible for evolutionary differentiation.—Cf: *macroevolution,* from which it probably differs only in degree.

microfabric The fabric of a rock as seen under the microscope. Cf: *megafabric.*

microfacies Those characteristic and distinctive aspects of a sedimentary rock that are visible and identifiable only under the microscope (low-power magnification). See also: *microbiofacies; microlithofacies.*

microfauna (a) Living or fossil animals too small to be seen with the naked eye. (b) An obsolete term for a very localized or small group of animals; animals occupying a small habitat. Cf: *microflora; macrofauna.*

microfelsitic *cryptocrystalline.*

microflora (a) A community of microorganisms assigned to the plant kingdom. The term is commonly misapplied to the *microfossil* remains of higher plants. (b) An obsolete term for a very localized or small group of plants; plants occupying a very small habitat. Cf: *microfauna; megaflora.*

microfluidal Said of the *flow texture* of an igneous rock that is visible only with the aid of a microscope.

microfluxion Obsolescent British usage for *microflow.*

microforaminifera (a) The chitinous inner tests of certain foraminifers, almost always spiral, frequently found in palynologic preparations of marine sediments; they are generally much smaller than "normal" whole foraminifers but display recognizable characteristics of "normal" species. (b) Foraminifers much smaller than those generally observed and studied.

microfossil A fossil too small to be studied without the aid of a microscope, e.g. an invertebrate such as a *foraminifer* or an *ostracode.* It may be the remains of a microscopic organism or a part of a larger organism. Cf: *macrofossil; nannofossil.*

microfragmental Said of a coal composed of macerated vegetal matter. Cf: *macrofragmental.*

microgal A unit of acceleration commonly used in borehole gravity work; 10^{-6} gal.

microgametophyte The male *gametophyte* or haploid generation that develops from the microspore of a heterosporous embryophytic plant. In lower vascular plants, a few-celled microgametophyte as well as the sperm cells are produced entirely within the microspore; in seed plants, the microgametophyte plus the surrounding microspore wall is the pollen grain, in which the microgametophyte is further reduced, consisting of only three cells in the angiosperms. Cf: *megagametophyte.*

micro-gas survey Soil analysis to determine the presence of hydrocarbon gases that have presumably seeped upwards into the overburden from buried sources.

microgeography The detailed analysis of the natural features of a very limited area.

microgeology (a) Study of the geologic and geochemical role of microorganisms (Ehrenberg, 1854). (b) Study of microscopic features of rocks.

microgour A *gour* or rimstone dam or pool on the scale of a few centimeters.

micrograined (a) Said of the texture of a carbonate sedimentary rock having clastic particles whose diameters are in the range of 0.001-0.01 mm (Bissell & Chilingar, 1967, p. 103) or 0.001-0.004 mm (DeFord, 1946). Some petrographers use the limits of 0.004-0.062 mm. (b) Said of the texture of a carbonate sedimentary rock wherein the particles are mostly 0.01-0.06 mm in diameter, are poorly sorted, and are admixed with clay-sized calcareous mud (Thomas, 1962). Also said of a sedimentary rock with such a texture. Cf: *microgranular.*

microgranitic *microgranular [ign].*

microgranitoid *microgranular [ign].*

microgranular [ign] Said of the texture of a *microcrystalline, xenomorphic* igneous rock. Also, said of a rock with such a texture. Syn: *fine-granular.* Nonrecommended syns: *euritic; microgranitic; microgranitoid.*

microgranular [paleont] Said of a foraminiferal wall (as in Endothyracea) composed of minute calcite crystals, probably originally granular but possibly recrystallized. The granules may be aligned in rows perpendicular to the outer wall, resulting in fibrous structure.

microgranular [sed] Minutely granular; specif. said of the texture of a carbonate sedimentary rock wherein the particles are mostly 10-60 microns in diameter and are well-sorted, and the finer clay-sized matrix is absent (Thomas, 1962). Also said of a sedimentary rock with such a texture. Cf: *micrograined.*

microgranulitic Obsolete and nonrecommended syn. of (a) *microgranular* and (b) *intergranular.*

micrograph A graphic recording of something seen through the microscope, e.g. a *photomicrograph* of a petrologic thin section.

micrographic Said of the *graphic* texture of an igneous rock that is distinguishable only with the aid of a microscope; also, said of a rock having such texture.

microgroove cast A term used by McBride (1962, p.56) for a *striation cast* of a striation less than 2.5 cm in length.

microhabitat A very small *habitat* (dimensions measurable in mm or cm). Cf: *macrohabitat.*

microhill A very rough, miniature sand column raised by the formation of *pipkrakes,* ranging from a few millimeters to several centimeters high and having a height-diameter ratio of 2:5 (Otterman & Bronner, 1966, p.56).

Microlaterolog Trade name for a *microresistivity log,* obtained with a miniaturized *focused-current log* electrode arrangement, designed to measure a shallow volume of rock at the borehole face. Cf: *Microlog.* Syn: *trumpet log.*

microlinear "Any lineation that is invisible to the unaided eye" (El-Etr, 1976, p. 485).

microlite [cryst] A microscopic crystal that polarizes light and has some determinable optical properties. Cf: *crystallite; crystalloid.* Syn: *microlith.*

microlite [mineral] A pale-yellow, reddish, brown, or black isometric mineral of the pyrochlore group: $(Na,Ca)_2(Ta,Nb)_2O_6(O,OH,F)$. It is isomorphous with pyrochlore, with Ta greater than Nb, and it often contains small amounts of other elements (including uranium and titanium). Microlite occurs in granitic pegmatites and in pegmatites related to alkalic igneous rocks, and it constitutes an ore of tantalum. Syn: *djalmaite.*

microlith *microlite.*

microlithofacies The lithologic aspect of a *microfacies* (Fairbridge, 1954).

microlithology The study, or characteristics, of rocks as they appear under the microscope. Cf: *macrolithology.*

microlithon The rock material between cleavage planes that is folded, kinked, or flattened (DeSitter, 1954).

microlithotype A typical association of macerals in coals, occurring in bands at least 50 microns wide. Microlithotype names bear the suffix "-ite". Cf: *lithotype.*

microlitic Said of the texture of a porphyritic igneous rock in which the groundmass is composed of an aggregate of differently oriented or parallel microlites in a glassy or cryptocrystalline mesostasis. Hyalopilitic, pilotaxitic, orthophyric, and trachytic are microlitic textures.

Microlog Trade name for a *well log* consisting of two microresistivity curves: the micronormal curve (see *normal log*) and a very short lateral curve (see *lateral log*). Response is dominated by the presence of *mud cake,* which causes separation between otherwise virtually coincident curves and thus indicates porous zones. Use of logs of this type for quantitative estimation of porosity is obsolete, although such logs with accompanying caliper curves are still valued for precise location of filterable rock layers of implied reservoir quality. Cf: *Microlaterolog.*

micromeritic An obsolete syn. of *microcrystalline.*

micromeritics The study of the characteristics and behavior of small particles. It is applicable to soil physics.

micrometeorite A meteorite or meteoritic particle with a diameter generally less than a millimeter; a meteorite so small that it undergoes atmospheric entry without vaporizing or becoming intensely heated and hence without disintegration.

micrometeorite crater A small crater produced by hypervelocity impact of primary micrometeorite particles on exposed surfaces of lunar rocks on the lunar surface. The craters are typically less than a few millimeters in diameter and are characterized by a central glass-lined pit, a concentric lightened area of shock-fractured minerals, and a roughly circular spall area approximately 4.5 times larger in diameter than the central pit. The informal term *zap crater* is equivalent (Hörz et al., 1971, p. 5785). Syn: *microcrater; lunar microcrater.*

micromineral Any crystalline matter of the clay fraction in sediments and soils, including the iron and aluminum oxides, allophane, fine-grained carbonates, etc., in addition to the phyllosilicates, to which the term *clay mineral* is usually restricted (Yaalon, 1965).

micromoon The amount of mass responsible for a lunar *mascon;* equal to 10^{-6} mass of the moon (Sjogren, 1974, p. 115).

micronucleus A small nucleus that is concerned with reproductive functions in the body of a tintinnid. Cf: *macronucleus.*

micro-oil A term used by Vernadskiy for hydrocarbons occurring in a diffused state in sedimentary rock; the nascent oil, still within and sorbed to its source rock (Vassoevich, 1965, p.510).

micro-ophitic Said of the ophitic texture of an igneous rock that is distinguishable only with the aid of a microscope. Also, said of a rock with such texture.

micropaleontology A branch of paleontology that deals with the study of fossils too small to be observed without the aid of a microscope; the study of microfossils.

micropegmatite A less-preferred syn. of *granophyre.*

micropegmatitic A nonrecommended syn. of *graphic* and *micrographic.*

micropellet A pellet or pelletlike sedimentary particle of a fine to very fine grade size, "possibly smaller than 0.01 mm in diameter" (Bissell & Chilingar, 1967, p. 161). Adj: *micropelletoid.*

microperthite A variety of perthite in which the lamellae (5-100 microns wide) are visible only with the aid of the microscope. Cf: *cryptoperthite.*

microperthitite A *syenite* composed entirely of microperthite. Not recommended usage.

micropetrological unit *maceral.*

microphagous Said of an organism that feeds on relatively minute particles. Cf: *macrophagous.*

microphotograph A less-preferred syn. of *photomicrograph.*

microphyllous Having small leaves with one vein and leaf trace, as in Psilophyta and Lycopodiophyta (Scagel et al., 1965, p. 623).

microphyric Said of the texture of a porphyritic igneous rock in which the phenocrysts are of microscopic size, i.e. their longest dimension does not exceed 0.2 mm (Cross et al., 1906, p.702); also, said of a rock having such texture. Cf: *macrophyric.* Syn: *microporphyritic.*

micropiracy The overtopping and breaking-down of the narrow ridge between adjacent rill channels, and diversion of flow from the higher, shorter, and shallower channel to the lower, longer, and deeper one closer to the initial rill (Horton, 1945, p. 335). See also: *cross-grading.*

microplankton Plankton of the size range 60 microns to 1 mm, e.g. most phytoplankton. They are larger than *ultraplankton* and *nannoplankton,* but smaller than *macroplankton* and *megaloplankton.* Syn: *net plankton.*

micropoikilitic Said of the poikilitic texture of an igneous rock that can be distinguished only with the aid of a microscope, because of the small size of both *oikocrysts* and crystalline inclusions (*chadacrysts*). Also, said of a rock having such texture.

micropore (a) A pore small enough to hold water against the pull of gravity and to inhibit the flow of water. (b) In the pore-size classification of Choquette & Pray (1970, p.233), an equant to equant-elongate pore or a tubular or platy pore with an average diameter or thickness of less than 1/16 mm.—Cf: *mesopore [petrology]; megapore.*

microporphyritic *microphyric.*

micropulsation An oscillatory geomagnetic variation in the frequency range from 0.01 to 3 Hz, commonly with amplitudes less than 1 gamma (Sumner, 1976, p. 243).

micropygous Said of a trilobite with a pygidium smaller than the cephalon. Cf: *isopygous; macropygous.*

micropyle The minute opening in the integument of an ovule or seed, through which a pollen tube grows to reach the female gametophyte (Fuller & Tippo, 1954, p. 963).

microquartz Nonclastic anhydrous crystalline silica occurring in sediments and having particle diameters usually less than 20 microns.

microradiograph A picture produced by X-rays or rays from a radioactive source showing the minute internal structure of a substance.

microrelief (a) Local, slight irregularities of a land surface, including such features as low mounds, swales, and shallow pits, generally about a meter in diameter, and causing variations amounting to no more than 3 m. (b) Relief features that are too small to show on a topographic map; e.g. gullies, mounds, boulders, pinnacles, or other features less than 60 m in diameter and less than 6 m in elevation, in an area for which the topographic map has a scale of 1:50,000 or smaller and a contour interval of 3 m (10 ft) or larger.—Cf: *macrorelief.*

microresistivity log Generic term for any *resistivity log* curve obtained from measurements between electrodes spaced a few inches apart and held in direct contact with the wall of the borehole. *Microlog* and *Microlaterolog* are examples of such logs used to determine porosity in potential reservoir rocks flushed by mud filtrate.

microrhabd A rod-shaped monaxonic sponge spicule (microsclere).

Microsauria An order of lepospondylous amphibians characterized by small size and salamanderlike or snakelike body form. The older literature includes (in error) early and primitive cotylosaurian reptiles in this group. Range, Lower Mississippian to Lower Permian.

microsclere A small *sclere;* specif. one of the minute secondary spicules scattered throughout a sponge or concentrated in the cortex or elsewhere. It usually differs in form from a *megasclere.*

microscope An optical instrument that is used to produce an enlarged image of a small object; it consists of the lens (or lenses) of the objective and of the eyepiece set into a tube, and held by an adjustable arm over a stage on which the object is placed. Types of microscopes vary according to intended use and to type of energy used, e.g. natural light, polarized light, transmitted light, reflected light, electrons, or X-rays.

microscopic (a) Said of an object or phenomenon or of its characteristics that cannot be observed without the aid of a microscope. (b) Of or pertaining to a microscope.

microsection (a) Any thin section used in microscopic analysis. (b) A *polished section.* (c) A *polished thin section.*

microseism A collective term for small motions in the Earth that are unrelated to an earthquake and that have a period of 1.0-9.0 sec. They are caused by a variety of natural and artificial agents, esp. atmospheric events. Cf: *macroseism.* Syn: *seismic noise.*

microsere A *sere* of a very small habitat, usually failing to attain climax and ending with the loss of identity of the habitat. Syn: *serule.*

microsolifluction The frost movements that produce patterned ground (Troll, 1944).

microsommite A mineral of the cancrinite group: $(Na,Ca,K)_9(Si,Al)_{12}O_{24}(Cl,SO_4,CO_3)_{2\text{-}3}$.

microspar Calcite matrix in limestones, occurring as uniformly sized and generally loaf-shaped crystals ranging from 5 to more than 20 microns in diameter. It develops by recrystallization or neomorphism of carbonate mud (micrite) (Folk, 1959). Not to be used as a synonym for fine pore-filing spar. Cf: *microsparite.*

microsparite (a) A term used by Folk (1959, p. 32) for a limestone whose carbonate-mud matrix has recrystallized to microspar. (b) A term used by Chilingar et al. (1967, p. 320) for a sparry crystal of calcite whose diameter ranges from 5 to 20 microns. Cf: *microspar.*

microspheric Said of a foraminiferal test or shell produced sexually and characterized by a very small initial chamber (proloculus), many chambers, often large size of the adult test, and more complete ontogeny. Cf: *megalospheric.*

microspherulitic Said of the spherulitic texture of an igneous rock that is distinguishable only with the aid of a microscope, owing to the small size of the spherules. Also, said of a rock having such texture.

microsporangium A *sporangium* that develops or bears microspores; e.g. the anther in an angiosperm or the pollen sac in all other seed plants. Cf: *megasporangium.*

microspore One of the spores of a heterosporous embryophytic plant that germinates to produce a microgametophyte and that is ordinarily smaller than the *megaspore* of the same species. In seed plants, pollen grains consist of a microspore wall or exine with a microgametophyte contained inside. See also: *small spore.*

microstriation A microscopic scratch developed on the polished surface of a rock or mineral as a result of abrasion.

microstructure (a) The internal structure and character of plant and animal tissues, esp. skeletal tissues, as revealed by the microscope. (b) Structural features of rocks that can be discerned only with the aid of the microscope.

microstylolite A *stylolite* in which the relief along the surface is less than a millimeter, such as one indicating differential solution between two mineral grains.

microtectonics A syn. of *structural petrology.*

microtektite A small glassy object, less than one millimeter in diameter and usually spherical, found in some deep-sea sediments, similar to and possibly related to tektites in outward form and composition.

microtherm A plant that requires low temperatures for successful growth. Cf: *mesotherm; megatherm.*

microthermal Pertaining to a climate characterized by low temperature. Cf: *mesothermal; megathermal.*

microtinite A light-colored, coarse-grained igneous rock characterized by monzonitic texture and by the presence of vitreous plagioclase. Named by Lacroix in 1901 for inclusions in lavas of the Auvergne district, France. Obsolete.

microtopography Topography on a small scale. The term has been applied to features having relief as small as 1-10 cm as well as to those involving amplitudes of 50-100 meters and wavelengths of a few kilometers.

microvermicular Said of the texture of a rock having wormlike intergrowths of crystals that are visible in thin section under the polarizing microscope. Also said of the texture of the rock with such intergrowths. Cf: *micrographic.*

microvitrain Vitrain bands occurring in clarain, .05-2.0 mm thick.

microwave The region of the electromagnetic spectrum in the approximate wavelength range from 1 mm to beyond 1 m. *Passive remote sensing* systems operating at these wavelengths are called microwave systems; *active remote sensing* systems are called radar.

micstone A very fine-grained sedimentary rock, containing more than 65% by volume material less than 5 microns in diameter, and consisting predominantly of carbonate minerals; a carbonate mudstone (Lewan, 1978). Etymol: *mic-*rite + mud-*stone.*

mictite Coarsely composite rock formed as the result of contamination, by the incorporation and partial or complete assimilation of country-rock fragments, of a magma under conditions of relatively low temperature and probably at relatively high levels in the crust (Dietrich & Mehnert, 1961). Little used.

midalkalite An obsolete syn. of *nepheline syenite.*

mid-bay bar A bar built across a bay at some point between its mouth and its head.

midden [arch] A heap or stratum of refuse (broken pots and tools, ashes, food remains, etc.) normally found on the site of an ancient settlement (Bray & Trump, 1970, p. 147).

midden [sed] A mound-like accumulation of calcareous sediment trapped or bound together by algal growth.

midden [soil] A mass of highly organic soil formed by an earthworm around its burrow; also, any organic debris on soil, deposited by an animal.

middle [geochron] Pertaining to a segment of time intermediate between *late* and *early.* The adjective is applied to the name of a geologic-time unit (era, period, epoch) to indicate relative time designation and corresponds to *middle* as applied to the name of the equivalent time-stratigraphic unit; e.g. rocks of a Middle Jurassic batholith were intruded in Middle Jurassic time. The initial letter of the term is capitalized to indicate a formal subdivision (e.g. "Middle Devonian") and is lowercased to indicate an informal subdivision (e.g. "middle Miocene"). The informal term may be used for eras and epochs, and for periods where there is no formal subdivision. Syn: *medial.*

middle [stratig] Pertaining to rocks or strata that are intermediate between *upper* and *lower.* The adjective is applied to the name of a time-stratigraphic unit (system, series, stage) to indicate position in the geologic column and corresponds to *middle* as applied to the name of the equivalent geologic-time unit; e.g. rocks of the Middle Jurassic System were formed during the Middle Jurassic Period. The initial letter of the term is capitalized to indicate a formal subdivision (e.g. "Middle Devonian") and is lowercased to indicate an informal subdivision (e.g. "middle Miocene"). The informal term may be used where there is no formal subdivision of a system or of a series.

middle diagenesis *anadiagenesis.*

middle ground A bar deposit or shoal formed in the middle of a channel or fairway at the entrance and exit of a constricted passage (as a strait) by the rise and fall of the tide, and characterized by a flow of water on either side of the deposit.

middle lamella In plants, the intercellular substance between the primary walls of two contiguous cells, composed chiefly of calcium pectate (Esau, 1965, p.34). See also: *intercellular space.*

middle lateral muscle One of a pair of muscles in some lingulid brachiopods, originating on the pedicle valve between the central muscles, and diverging slightly posteriorly before insertion on the brachial valve (TIP, 1965, pt.H, p.148). Cf: *outside lateral muscle.*

middle latitude n. The latitude of the point situated midway on a north-south line between two parallels; half the arithmetic sum of the latitudes of two places on the same side of the equator.

middle-latitude desert A vast desert area occurring within lat. 30°-50° north or south of the Equator in the interior of a large continental mass, usually situated in the lee of high mountains that stand across the path of prevailing winds (thus, a *rain-shadow desert*), and commonly characterized by a cold, dry climate.

middle Paleolithic n. The second division of the *Paleolithic,* characterized by Neanderthal man in Eurasia. Correlation of cultural levels with actual age (and, therefore, with the time-stratigraphic units of geology) varies from region to region. Cf: *lower Paleolithic; upper Paleolithic.*—adj. Pertaining to the middle Paleolithic.

middle Stone Age *Mesolithic.*

midfan The area between the fanhead and the outer, lower margins of an alluvial fan.

midfan mesa A much eroded, islandlike remnant of an old upfaulted alluvial fan, commonly a primary product of piedmont faulting (Eckis, 1928, p. 243-246).

mid-ocean canyon *deep-sea channel.*

mid-oceanic ridge A continuous, seismic, median mountain range extending through the North and South Atlantic Oceans, the Indian Ocean, and the South Pacific Ocean. It is a broad, fractured swell with a central rift valley and usually extremely rugged topography; it is 1-3 km in elevation, about 1500 km in width, and over 84,000 km in length. According to the hypothesis of sea-floor spreading, the mid-oceanic ridge is the source of new crustal material. See also: *rift valley; sea-floor spreading.* Syn: *mid-ocean rise; oceanic ridge.*

mid-ocean rift *rift valley.*

mid-ocean rise *mid-oceanic ridge.*

midrange The arithmetic mean of the smallest and largest values in a sample. Syn: *range midpoint.*

midrib The central rib of leaf venation. It is a continuation of the petiole.

midstream (a) The part of a stream well removed from both sides or from the source and the mouth. (b) A line along a stream course, midway between the sides of the stream.

midwater trawl A towed, netlike device that is used to gather marine organisms anywhere between the bottom and the water surface.

midway The middle, deepest, or best navigable channel used in defining water boundaries between states, as in a sound, bay, strait, estuary, or other arm of the sea, and in a lake or landlocked sea. Syn: *fairway; thalweg.*

Midwayan North American (Gulf Coast) stage: Paleocene (above Navarroan of Cretaceous, below Sabinian).

miemite A yellowish-brown, fibrous dolomite mineral occurring at Miemo in Tuscany, Italy.

miersite A canary-yellow isometric mineral: (Ag,Cu)I.

Mie scattering Multiple reflection of light waves by atmospheric particles with the general dimensions of the wavelength of light.

Mie theory A theory of the scattering of electromagnetic radiation by spherical particles, developed by G. Mie in 1908. In contrast to *Rayleigh scattering,* the Mie theory embraces all ratios of diameter to wavelength. Mie theory is important in meteorological optics, where diameter-to-wavelength ratios of the order of unity and larger are characteristic of many problems regarding haze and cloud scattering.

migma Mobile, or potentially mobile, mixture of solid rock material(s) and magma, the magma having been injected into or melted out of the rock material (Dietrich & Mehnert, 1961). Etymol: Greek, "mixture".

migmatite A composite rock composed of igneous or igneous-appearing and/or metamorphic materials, which are generally distinguishable megascopically (Dietrich, 1960, p. 50). The term was introduced by Sederholm in 1907 (p. 88-89). Cf: *chorismite.*

migmatitization *migmatization.*

migmatization Formation of a *migmatite.* The more mobile, typically light-colored, part of a migmatite may be formed as the result of anatexis, lateral secretion, metasomatism, or injection. Also spelled: *migmatitization.*

migrating dip A dipping event in a reflection seismogram that is mapped to its true position in space. See also: *migration [seis].* Syn: *swinging dip.*

migrating dune *wandering dune.*

migrating inlet A tidal inlet, such as that connecting a coastal lagoon with the open sea, that shifts its position laterally in the direction in which the dominant longshore current flows. It results from deposition on one side of the inlet, accompanied by erosion on the other.

migration [ecol] A broad term applied to the movements of plants and animals from one place to another over long periods of time.

migration [geomorph] The movement of a topographic feature from one locality to another by the operation of natural forces; specif. the movement of a dune by the continual transfer of sand from its windward to its leeward side.

migration [petroleum] The movement of liquid and gaseous hydrocarbons from their source or generating beds through permeable formations into reservoir rocks.

migration [seis] The process by which events on a reflection seismogram are mapped in an approximation of their true spatial positions. It requires knowledge of the velocity distribution along the raypath. Cf: *migrating dip.*

migration [streams] (a) *shifting [streams].* (b) The slow downstream movement of a system of meanders, accompanied by enlargement of the curves and widening of the meander belt.

miharaite A basalt in which the groundmass is olivine-free and contains free silica; a *subalkaline basalt* or *tholeiite.* The name, given by Tsuboi in 1918, is for Miharayama, Japan. Not recommended usage.

mijakite A manganese-rich *basalt* composed of phenocrysts of augite and bytownite, and sometimes biotite, hypersthene, and apatite, in a groundmass having intersertal texture and composed of lath-shaped feldspar, magnetite grains, and a nearly opaque red-brown mineral identified as pyroxene. Also spelled: *miyakite.* Named by Petersen in 1891 for Mijakeshima in the Bonin Islands, Japan. Not recommended usage.

Milankovitch curve That curve drawn for any latitude (but usually 65°N) and any geologic time that represents the amount of solar energy received by that latitude. It is computed from an analysis of the earth's orbital variations.

Milankovitch theory An astronomical theory of glaciation, formulated by Milutin Milankovitch (1879-1958), Yugoslav mathematician, in which climatic changes result from fluctuations in the seasonal and geographic distribution of insolation, determined by variations of the Earth's orbital elements, namely eccentricity, tilt of rotational axis, and longitude of perihelion (Milankovitch, 1941). It is supported by recent radiometrically dated reconstructions of ocean temperature and glacial sequences.

milarite A colorless to greenish, glassy, hexagonal mineral: $K_2Ca_4Be_4Al_2Si_{24}O_{60}\cdot H_2O$.

Milazzian European stage: Upper Pleistocene (above Sicilian, below Tyrrhenian).

mile Any of various units of distance that were derived from the ancient Roman marching unit of 1000 double paces (a double pace = 5 ft) and that underwent many changes as the term came into use among the western nations (e.g. a mile = 1620 English yards or 1482 meters); specif. *statute mile* and *nautical mile.* Etymol: Latin *mille,* "thousand".

milieu A French term used in paleontology, sedimentation, and stratigraphy for environment, surroundings, or setting; e.g. the environment characteristic of a stratigraphic facies.

miliolid A foraminifer belonging to the family Miliolidae, characterized by a test that usually has a porcelaneous and imperforate wall and has two chambers to a whorl variably arranged about a longitudinal axis.

milioline Pertaining or belonging to or resembling the foraminiferal genus *Miliola* or suborder Miliolina; e.g. formed as in the foraminiferal test of the superfamily Miliolacea, commonly with narrow elongate chambers (two to a whorl) added in differing planes of coiling (TIP, 1964, pt.C, p.61).

miliolite A fine-grained limestone of eolian origin, consisting chiefly of the tests of *Miliola* and other foraminifers.

military geology Those branches of the earth sciences, especially geomorphology, soil science, and climatology, that are applied to such military concerns as *terrain analysis,* water supply, cross-country movement, location of construction materials, and the building of roads and airfields.

milk opal A translucent and milk-white to green, yellow, or blue variety of common opal.

milky quartz A milk-white, nearly opaque variety of crystalline quartz often having a greasy luster. The milkiness is usually due to the presence of innumerable very small cavities containing fluids. Syn: *greasy quartz.*

milled ring A flange near the base of an echinoid spine for the attachment of muscles controlling the movement of the spine.

millepore Any one of a group of hydrozoans belonging to the order Milleporina, characterized by a calcareous skeleton and free-swimming sexual individuals. Cf: *hydroid; stylaster.*

Miller-Bravais indices A four-index type of *Miller indices,* useful but not necessary in order to define planes in crystal lattices in the hexagonal system; the symbols are $hk\bar{\imath}l$, in which $\bar{\imath}=-(h+k)$. Syn: *hexagonal indices.*

Miller indices A set of three or four symbols (letters or integers) used to define the orientation of a crystal face or internal crystal plane. The indices are determined by expressing, in terms of lattice constants, the reciprocals of the intercepts of the face or plane on the 3 crystallographic axes, and reducing (clearing fractions) if necessary to the lowest integers retaining the same ratio. When the exact intercepts are unknown, the general symbol (hkl) is used for the indices, where h, k, and l are respectively the reciprocals of rational but undefined intercepts along the a, b, and c crystallographic axes. In the hexagonal system, the Miller indices are $(hk\bar{\imath}l)$; these are known as the *Miller-Bravais indices.* Indices designating individual crystal faces are enclosed in parentheses; complete crystal forms, in braces; crystal zones, in square brackets; and crystallographic lines, in greater-than/less-than symbols. To denote the interception at the negative end of an axis, a line is placed over the appropriate index, as $(1\bar{1}1)$. The indices were proposed by William H. Miller (1801-1880), English mineralogist. See also: *indices of lattice row.* Syn: *crystal indices; hkl indices.*

millerite A brass-yellow to bronze-yellow rhombohedral mineral: NiS. It usually has traces of cobalt, copper, and iron, and is often tarnished. Millerite generally occurs in fine hairlike or capillary crystals of extreme delicacy, chiefly as nodules in clay ironstone. Syn: *capillary pyrites; nickel pyrites; hair pyrites.*

millet-seed sand Sand that consists essentially of smoothly and conspicuously rounded grains about the size of a millet seed; specif. a desert sand whose grains have a surface like that of ground glass and are very perfectly rounded as a result of wind action that caused them to be constantly impacting against each other.

mill-hole mining *glory-hole mining.*

millidarcy The customary unit of measurement of fluid permeability, equivalent to 0.001 *darcy.* Abbrev: md.

milligal A unit of acceleration used with gravity measurements; 10^{-3} gal = 10^{-5}m/sec^2. Abbrev: mgal.

milling ore *second-class ore.*

millionth-scale map of the world *International Map of the*

World. Abbrev: IMW.

millisite A white mineral: $(Na,K)CaAl_6(PO_4)_4(OH)_9 \cdot 3H_2O$.

mill ore Var. of *milling ore.*

millosevichite A mineral: $(Al,Fe)_2(SO_4)_3$.

mill-rock (a) A type of coarse acidic pyroclastic breccia found in or close to the volcanic units in which Canadian massive sulfide ore deposits occur (Sangster, 1972, p. 3). (b) More generally, any proximal, typically explosive rhyolite pyroclastic breccia.

millstone A *buhrstone;* e.g. a coarse-grained sandstone or a fine-grained quartz conglomerate. Also, one of two thick disks of such material formerly used for grinding grain and other materials, which were fed through a center hole in the upper stone.

millstone grit Any hard, siliceous rock suitable for use as a material for millstones; specif. the Millstone Grit of the British Carboniferous, a coarse conglomeratic sandstone.

Mima mound A term used in the NW U.S. for one of numerous low, circular or oval domes composed of loose, unstratified, gravelly silt and soil material, built upon glacial outwash on a *hog-wallow* landscape; the basal diameter varies from 3 m to more than 30 m, and the height from 30 cm to about 2 m. The mounds are probably built by pocket gophers (Arkley & Brown, 1954). Named after the Mima Prairie in western Washington State. Pron: *my*-ma. Cf: *pimple mound.* Also spelled: *mima mound.*

mimetene *mimetite.*

mimetesite *mimetite.*

mimetic [cryst] Pertaining to a twinned or malformed crystal that appears to have a higher grade of symmetry than it actually does.

mimetic [evol] Said of an organism that exhibits or is characterized by *mimicry.*

mimetic [struc petrol] Said of a tectonite whose deformation fabric, formed by recrystallization or neomineralization, reflects and is influenced by pre-existing anisotropic structure; also, said of the fabric itself.

mimetic crystallization Recrystallization and/or neomineralization in metamorphism that reproduces any pre-existent anisotropy, bedding, schistosity, or other structures (Knopf and Ingerson, 1938). Syn: *facsimile crystallization.*

mimetite A yellow to yellowish-brown mineral of the apatite group: $Pb_5(AsO_4)_3Cl$. It is isomorphous with pyromorphite, and commonly contains some calcium or phosphate. Mimetite usually occurs in the oxidized zone of lead veins, and is a minor ore of lead. Syn: *mimetene; mimetesite.*

mimicry The superficial similarity that exists between organisms, or between an organism and its surroundings, as a means of concealment, protection, or other advantage. See also: *mimetic.*

mimosite A dark-colored *dolerite* containing abundant augite and ilmenite. Cf: *soggendalite.* Obsolete.

minable Said of a mineral deposit for which extraction is technically feasible and economically worthwhile.

minal *end member.*

minasragrite A blue efflorescent mineral: $VO(SO_4) \cdot 5H_2O$.

Mindel (a) European stage: Pleistocene (above Günz, below Riss). (b) The second classical glacial stage of the Pleistocene Epoch in the Alps, after the Günz-Mindel interglacial stage. See also: *Kansan; Elster.*—Etymol: Mindel River, Bavaria. Adj: *Mindelian.*

Mindel-Riss The term applied in the Alps to the second classical interglacial stage of the Pleistocene Epoch, after the Mindel glacial stage and before the Riss. See also: *Yarmouth.*

mine n. (a) An underground excavation for the extraction of mineral deposits, in contrast to surficial excavations such as quarries. The term is also applied to various types of open-pit workings. (b) The area or property of a mineral deposit that is being excavated; a mining claim.—v. To excavate for and extract mineral deposits or building stone.

mineragraphy An obsolescent syn. of *ore microscopy.*

mineral (a) A naturally occurring inorganic element or compound having an orderly internal structure and characteristic chemical composition, crystal form, and physical properties. Those who include the requirement of crystalline form in the definition would consider an amorphous compound such as opal to be a *mineraloid.* (b) Any naturally formed inorganic material, i.e. a member of the mineral kingdom as opposed to the plant and animal kingdoms.

mineral aggregate An *aggregate* or assemblage of more than one crystal grain (which may be of one or several mineral species) and containing more than one crystal lattice. It can occur as sediment if loosely bound, or as rock if tightly bound.

mineral assemblage (a) The minerals that compose a rock, esp. an igneous or metamorphic rock. The term includes the different kinds and relative abundances of minerals, but excludes the texture and fabric of the rock. See also: *metamorphic assemblage.* (b) *mineral association.*

mineral association A group of minerals found together in a rock, esp. in a sedimentary rock. Syn: *mineral assemblage.*

mineral belt An elongated region of mineralization; an area containing several mineral deposits.

mineral blossom drusy quartz.

mineral caoutchouc *elaterite.*

mineral charcoal *fusain.*

mineral deposit A mass of naturally occurring mineral material, e.g. metal ores or nonmetallic minerals, usually of economic value, without regard to mode of origin. Accumulations of coal and petroleum may or may not be included; usage should be defined in context. Cf: *mineral occurrence.*

mineral disintegration *granular disintegration.*

mineral facies (a) An approx. syn. of *metamorphic facies.* (b) Rocks of any origin whose constituents have been formed within the limits of a certain pressure-temperature range characterized by the stability of certain index minerals.

mineral filler A finely pulverized inert mineral or rock that is included in a manufactured product, e.g. paper, rubber, and plastics, to impart certain useful properties, such as hardness, smoothness, or strength. Common mineral fillers include asbestos, kaolin, and talc.

mineralization [ore dep] The process or processes by which a mineral or minerals are introduced into a rock, resulting in a valuable or potentially valuable deposit. It is a general term, incorporating various types, e.g. fissure filling, impregnation, replacement.

mineralization [paleont] A process of fossilization whereby the organic components of an organism are replaced by inorganic ("mineral") material.

mineralize To convert to a mineral substance; to impregnate with mineral material. The term is applied to the processes of ore formation and also to the process of fossilization.

mineralizer (a) A gas or fluid that dissolves, receives by fractionation, transports, and precipitates ore minerals. A mineralizer is typically aqueous, with various hyperfusible gases (CO_2, CH_4, H_2 S, HF), simple ions (H^+, HS, Cl^-, K, Na, Ca), complex ions (esp. chloride complexes), and dissolved base and precious metals. Syn: *ore-forming fluid.* (b) A gas that is dissolved in a magma and that aids in the concentration, transport, and precipitation of certain minerals and in the development of certain textures as it is released from the magma by decreasing temperature and/or pressure. Cf: *fugitive constituent; volatile component.*

mineral lands Legally, areas considered more valuable for their ore deposits or mineral potential than for agriculture or other purpose. Cf: *stone land.*

mineral matter The inorganic material in coal. See also: *inherent mineral matter.*

mineral occurrence Any ore or economic mineral in any concentration found in bedrock or as float; esp. a valuable mineral in sufficient concentration to suggest further exploration. Cf: *mineral deposit.*

mineralogic Adj. of *mineralogy.*

mineralogical Adj. of *mineralogy.*

mineralogical phase rule Any of several modifications of the fundamental Gibbs *phase rule,* taking into account the number of degrees of freedom consumed by the fixing of physical-chemical variables in the natural environment. The most famous such rule, that of Goldschmidt, assumes that two variables (taken as pressure and temperature) are fixed externally and that consequently the number of phases (minerals) in a system (rock) will not generally exceed the number of components. The Korzhinskiy-Thompson version takes into account the external imposition of chemical potentials of perfectly mobile components, and thereby reduces the maximum expectable number of minerals in a given rock to the number of inert components. Syn: *Goldschmidt's phase rule.*

mineralogic maturity A type of sedimentary *maturity* in which a clastic sediment approaches the mineralogic end product to which it is driven by the formative processes that operate upon it (Pettijohn, 1957). The ultimate sand is a concentration of pure quartz, and the mineralogic maturity of sandstones is commonly expressed by the quartz/feldspar ratio; this ratio is not so appropriate for sand derived from feldspar-poor rocks and the ratio of quartz + chert/feldspar + rock fragments may be substituted as more generally applicable. Cf: *compositional maturity; textural* *maturity.*

mineralogist One who studies the formation, occurrence, properties, composition, and classification of minerals; a geologist whose field of study is *mineralogy.*

mineralography A syn. of *mineragraphy;* both are obsolescent terms for *ore microscopy.*

mineralogy (a) The study of minerals: formation, occurrence, properties, composition, and classification. See also: *mineralogist.* Adj: *mineralogic; mineralogical.* Obsolete syn: *oryctology; oryctognosy.* (b) An obsolete use of the term is for the general geology of a region.

mineraloid A naturally occurring, usually inorganic substance that is not considered to be a *mineral* because it is amorphous and thus lacks characteristic crystal form; e.g. opal. Syn: *gel mineral.*

mineral pathology Study of the changes undergone by unstable minerals in an environment whose conditions of temperature, pressure, and composition are different from those under which the minerals originally formed (Pettijohn, 1957, p. 502).

mineral pigment An inorganic pigment, either natural or synthetic, used to give color, opacity, or body to a paint, stucco, plaster, or similar material. See also: *ocher; sienna.*

mineral pitch An obsolete syn. of *asphalt.*

mineral reserves *reserves.*

mineral resin Any of a group of resinous, usually fossilized, mineral hydrocarbon deposits; e.g. bitumen and asphalt. See also: *resin.*

mineral resources *resources.*

mineral rod *divining rod.*

mineral sands *beach placer.*

mineral sequence *paragenetic sequence.*

mineral soap *bentonite.*

mineral soil A soil that is composed mainly of mineral matter but having some organic material also.

mineral spring A spring whose water contains enough mineral matter to give it a definite taste, in comparison to ordinary drinking water, esp. if the taste is unpleasant or if the water is regarded as having therapeutic value. This type of spring is often described in terms of its principal characteristic constituent; e.g. *salt spring.*

mineral streaking In metamorphic rocks, lineation of grains of a mineral. Cf: *stretching.* Syn: *streaking.*

mineral survey The marking of legal boundaries of ore deposits or mineralized formations on public land, when such boundaries are not the normal land subdivisions.

mineral tallow *hatchettine.*

mineral tar *maltha.*

mineral time *Geologic time* estimated on the basis of radioactive minerals (Kobayashi, 1944a, p. 476). Cf: *absolute time.*

mineral water Water that contains naturally or artificially supplied mineral salts or gases (e.g. carbon dioxide).

mineral wax *ozocerite.*

mineral wool A generic term for felted or matted fibers manufactured by blowing or spinning threads of molten rock, slag, or glass. The material is used for thermal insulation. Syn: *rock wool.*

mineral zone An informal term for a stratigraphic unit classified on the basis of mineral content (usually detrital minerals) and usually named from characteristic minerals (ISST, 1961, p.29).

mineral zoning *zoning of ore deposits.*

minerocoenology The study of mineral associations in the broadest sense, such as the correlation of igneous rocks or magmatic provinces with their ore deposits (Thrush, 1968, p. 712).

minerogenetic *metallogenic.*

minerogenic *metallogenic.*

miner's inch A measure of water flow equal to 1.5 cu ft/min.

minette A *lamprophyre* primarily composed of biotite phenocrysts in a groundmass of alkali feldspar and biotite.

minguzzite A green monoclinic mineral (oxalate): $K_3Fe(Cr_2O_4)_3 \cdot 3H_2O$.

miniature atoll *microatoll.*

miniature lagoon (a) *lagoonlet.* (b) *pseudolagoon.*

minimicrite Abnormally fine micrite, generally 0.5 to 1.5 microns in grain size. It often originated as magnesium calcite, e.g. in blue-green algal mats, micrite envelopes, porcelaneous forams, and fecal pellets. Minimicrite is very "dense" and opaque in thin section (Folk, 1974).

minimum [geophys] n. An anomaly characterized by values smaller than those in neighboring areas; e.g. a gravity minimum or a geothermal minimum. Cf: *maximum [geophys].*

minimum [glac geol] *glacial minimum.*

minimum detectable power In infrared detector technology, the incident power that will give a signal-to-noise ratio equal to unity at the output of the detector (Smith et al., 1968, p.250). Syn: *noise-equivalent power.*

minimum pendulum A pendulum used in gravity measurements, so designed that changes in period resulting from small changes in length are at a minimum. Among factors that may tend to change the length are temperature, creep, and knife-edge wear.

minimum phase A characteristic of a waveform that has its energy concentrated in its front portion.

minimum slope A slope that is flatter in gradient than the slope units above or below it.

minimum-time path The path between two points along which the time of travel is less than on neighboring paths. See also: *Fermat's principle.* Syn: *least-time path; brachistochrone.*

mining The process of extracting metallic or nonmetallic mineral deposits from the Earth. The term may also include preliminary treatment, e.g. cleaning or sizing. Cf: *mining geology; mining engineering.*

mining claim A *claim* on mineral lands.

mining engineering The planning and design of mines, taking into account economic, technical, and geologic factors; also supervision of the extraction, and sometimes the preliminary refinement, of the raw material. Cf: *mining; mining geology.*

mining geology The study of the geologic aspects of mineral deposits, with particular regard to problems associated with mining. Cf: *mining; mining engineering.*

miniphyric An obsolete term applied to the texture of a porphyritic igneous rock in which the greatest dimension of the phenocrysts does not exceed 0.008 mm (Cross et al., 1906, p.702); also, said of a rock having such texture.

minium A bright-red, scarlet, or orange-red mineral: Pb_3O_4. Syn: *red lead.*

minnesotaite A mineral: $(Fe,Mg)_3Si_4O_{10}(OH)_2$. It is probably isomorphous with talc. It occurs abundantly in the iron ores of Minnesota. Syn: *iron talc.*

minophyric An obsolete term applied to the texture of a porphyritic igneous rock in which the greatest dimension of the phenocrysts is between 0.2 mm and 1 mm (Cross et al., 1906, p.702); also, said of a rock with such texture.

minor element (a) A syn. of *trace element.* (b) A term that is occasionally used for an element that normally comprises between one and five percent of a rock; it is not quantitatively defined.

minor fold A small-scale fold that is associated with or related to the *major fold* of an area. Cf: *subsidiary fold.*

minor planet *asteroid.*

minor septum One of the relatively short, third-cycle septa of a corallite, commonly inserted between, and much shorter than, adjacent *major septa.*

minus-cement porosity The porosity that a sedimentary material would have if it contained no chemical cement.

minus sight *foresight.*

minute (a) A unit of time equal to 1/60 of an hour and containing 60 seconds. Abbrev: min; m (in physical tables). (b) A unit of angular measure equal to 1/60 of a degree and containing 60 seconds of arc. Symbol: ′.

minverite A *diabase* containing hornblende and albite. According to Johannsen (1939, p. 267), the albite is in part primary and in part secondary, and the rock may be metamorphic. The name is from St. Minver, Cornwall, England. Not recommended usage.

minyulite A white mineral: $KAl_2(PO_4)_2(OH,F) \cdot 4H_2O$.

Miocene An epoch of the upper Tertiary period, after the Oligocene and before the Pliocene; also, the corresponding worldwide series of rocks. It is considered to be a period when the Tertiary is designated as an era.

miocrystalline *hyalocrystalline.*

miogeocline A prograding wedge of shallow-water sediment at the continental margin (Dietz & Holden, 1966) or along a geosynclinal seaway. Cf: *eugeocline.*

miogeosyncline A geosyncline in which volcanism is not associated with sedimentation; the nonvolcanic aspect of an orthogeosyncline, located near the craton (Stille, 1940). Syn: *miomagmatic zone.* Cf: *eugeosyncline.* See also: *ensialic geosyncline.*

miomagmatic zone *miogeosyncline.*

miomirite A variety of davidite containing lead.

miospore A term arbitrarily defined in paleopalynology as a spore or pollen grain less than 200 microns in diameter. Cf: *megaspore.* See also: *small spore.*

miothermic Pertaining to or characterized by prevailing temperature conditions on the Earth as opposed to exceptionally warmer or colder periods. Cf: *pliothermic.* Rarely used.

mirabilite A white or yellow monoclinic mineral: $Na_2SO_4 \cdot 10H_2O$. It occurs as a residue from saline lakes, playas, and springs, and as an efflorescence. Syn: *Glauber's salt.*

mire (a) A small piece of marshy, swampy, or boggy ground; wet spongy earth. (b) Soft, heavy, often deep mud or slush.—Obsolete syn: *slough [geog].*

mirror glance *wehrlite [mineral].*

mirror plane of symmetry *plane of mirror symmetry.*

mirror stone *muscovite.*

miry ground Ground that is deeply wet, generally sticky, and not having sufficient bearing strength to support loads.

mischungskorrosion *mixture dissolution.*

miscibility gap A compositional range intermediate between phases of variable composition, in which the assemblage of those phases is stable relative to a single phase. It is sometimes incorrectly called an *immiscibility gap.*

miscible Said of two or more phases that, when brought together, have the ability to mix and form one phase. Cf: *immiscible.*

misclosure *error of closure.*

mise a la masse A drill-hole resistivity or induced-polarization survey technique in which a buried conductor is directly energized and serves as a large buried electrode. Potentials are measured on the surface, in bore holes, or underground. A great depth of exploration is expected with the technique. Etymol: French.

misenite A white mineral: $KHSO_4$.

miserite A pink mineral: $K(Ca,Ce)_4Si_5O_{13}(OH)_3$.

misfit stream (a) A stream whose meanders are obviously not proportionate in size to the meanders of the valley or to the meander scars preserved in the valley wall; a stream that is either too large (an *overfit stream*) or too small (an *underfit stream*) to have eroded the valley in which it flows. (b) A term that is often incorrectly used as a syn. of *underfit stream.*

mispickel *arsenopyrite.*

Mississippian A period of the Paleozoic era (after the Devonian and before the Pennsylvanian), thought to have covered the span of time between 345 and 320 million years ago; also, the corresponding system of rocks. It is named after the Mississippi River valley, in which there are good exposures of rocks of this age. It is the approximate equivalent of the *Lower Carboniferous* of European usage.

Mississippi Valley-type deposit A strata-bound deposit of lead and/or zinc minerals in carbonate rocks, together with associated fluorite and barite. These deposits characteristically have relatively simple mineralogy, occur as veins and replacement bodies, are at moderate to shallow depths, show little post-ore deformation, are marginal to sedimentary basins, and are without an obvious source of the mineralization. Examples: Wisconsin-Illinois lead deposits; Kentucky-Illinois fluorspar deposits; Appalachian zinc and barite deposits.

Missourian North American series: lower Upper Pennsylvanian (above Desmoinesian, below Virgilian).

missourite (a) In the *IUGS classification,* a plutonic rock in which F is between 60 and 100, M is between 70 and 90, and potassium exceeds sodium. Cf: *melteigite.* (b) A plutonic rock containing a potassium feldspathoid (leucite) and 60 to 90 percent mafic minerals, such as pyroxene and olivine.—Its name, proposed by Weed and Pirsson in 1896, is derived from the Missouri River in Montana. Cf: *fergusite; italite.*

mis-tie A term used in surveying for the failure of the first and final observations around a closed loop to be identical, or for the failure of the values at identical points on intersecting loops to be the same. See also: *error of closure.*

misy A term for various poorly defined iron sulfates.

mitridatite A mineral: $Ca_3Fe_4{}^{+3}(PO_4)_4OH_6 \cdot 3H_2O$.

mitscherlichite A greenish-blue tetragonal mineral: $K_2CuCl_4 \cdot 2H_2O$.

mix-crystal *solid solution.*

mixed ages Discordant ages given by various dating methods (e.g. potassium-argon or rubidium-strontium) for the same igneous or metamorphic body. They are the result of thermal and/or dynamic changes that affected the body at some time after its formation. See also: *hybrid age; overprint [geochron]; updating.*

mixed assemblage An *assemblage [paleoecol]* composed of some specimens representing a *fossil community* or a *winnowed community,* plus others representing one or more *transported assemblages* brought into the locality where found. See also: *disturbed-neighborhood assemblage; thanatocoenosis.* Syn: *remanié assemblage.*

mixed-base crude A crude oil in which both paraffinic and naphthenic hydrocarbons are present in approximately equal proportion. Cf: *paraffin-base crude; asphalt-base crude.*

mixed crystal *solid solution.*

mixed current A tidal current characterized by a conspicuous difference in velocity between the two flood periods or ebb periods usually occurring each tidal day.

mixed cut A combination of *brilliant cut* above the girdle, usually with 32 facets, and *step cut* below, with the same number of facets. It is often used for colored stones, esp. fancy sapphires, to improve color and retain brilliancy.

mixed eruption A volcanic phase that includes both the emission of lava and the explosive ejection of pyroclasts. Cf: *explosive eruption; lava eruption.*

mixed flow Water flow that is partly *turbulent flow* and partly *laminar flow.*

mixed gneiss *composite gneiss.*

mixed layer The layer of ocean water above the thermocline; it is mixed by wind action. It is equivalent to the *epilimnion* in a lake.

mixed-layer mineral A mineral whose structure consists of alternating layers of clay minerals and/or mica minerals; e.g. chlorite, made up of alternating biotite and brucite sheets.

mixed ore An ore of both oxidized and unoxidized minerals.

mixed peat Peat that is stratified according to plant associations. Cf: *banded peat.*

mixed tide A tide with two high waters and two low waters occurring during a tidal day, and having a marked diurnal inequality (as in parts of the Pacific and Indian oceans). The term is usually applied to a tide that is intermediate between a predominantly diurnal tide and a predominantly semidiurnal tide, or to a tide with alternating periods of diurnal and semidiurnal components.

mixed volatile Pertaining to more than one species or component in the fluid phase of natural, experimental, or hypothetical rock-fluid systems. In metamorphic petrology, mixed volatile reactions are considered in particular in the progressive metamorphism of impure limestones and dolostones.

mixed water A term used by White (1957, p. 1639) for any mixture of volcanic and meteoric waters in any proportion. White recommended discontinuing use of the term for chloride- and sulfide-rich acid waters.

mixing Summing the output of different channels to attenuate noise. Syn: *compositing.*

mixing coefficient *austausch.*

mixing length The length, normal to the flow direction, over which a small volume of fluid is assumed to retain its identity in the mass exchange process in turbulent flow. It is related to the coefficient of eddy viscosity and the rate of change of velocity normal to the line of flow (Middleton, 1965, p. 250).

mixing ratio The ratio of the mass of water vapor to the mass of dry air with which it is associated. It is an important atmospheric quantity. Cf: *humidity; absolute humidity; relative humidity; specific humidity.*

mixite An emerald-green or blue-green to whitish mineral: $Bi_2Cu_{12}(AsO_4)_6(OH)_{12} \cdot 6H_2O$.

mixolimnion The upper, low-density, freely circulating layer of a *meromictic lake.* Cf: *monimolimnion; chemocline.*

mixoperipheral growth Growth of a brachiopod valve in which the posterior part increases in size anteriorly and toward the other valve. Cf: *holoperipheral growth.*

mixotrophic Said of an organism that is nourished by both *autotrophic* and *heterotrophic* mechanisms.

mixtite A descriptive group term proposed by Schermerhorn (1966, p.834) for a coarsely mixed, nonsorted or poorly sorted, clastic sedimentary rock, without regard to composition or origin; e.g. a tillite. Syn: *diamictite.*

mixtum A term proposed by Schermerhorn (1966, p. 834) for an unconsolidated *mixtite.*

mixture dissolution The ability of two calcite-saturated waters of different carbon-dioxide content to dissolve additional calcite when mixed. The CO_2 limits the solubility at low CO_2 content, whereas the water limits it at high CO_2 content, and the two factors together provide optimal solubility at an intermediate content of CO_2. Consequently, despite the saturated condition of the two initial waters, their mixture can dissolve more calcite (Moore & Sullivan, 1978, p. 147). Syn: *mischungskorrosion.*

miyakite *mijakite.*

mizzonite A mineral of the scapolite group intermediate between meionite and marialite, and containing 54-57% silica; esp. such a variety of scapolite occurring in clear crystals in ejected masses on volcanoes. Syn: *dipyre.*

MLLW *mean lower low water.*

MLW *mean low water.*

MLWN *mean low-water neap.*

MLWS *mean low-water spring.*

MMR *magnetometric resistivity method.*

MM scale *modified Mercalli scale.*

Mn *mean range.*

mo A Swedish term for "glacial silts or rock flour having little plasticity" (Stokes & Varnes, 1955, p. 93). Pron: *moo.*

moat [glac geol] A glacial channel resembling a moat; e.g. a deep trench in glacier ice, surrounding a *nunatak,* and produced by ablation; or a channel at the margin of a dwindling glacier.

moat [marine geol] A ringlike depression around the base of many seamounts. It may be discontinuous. Syn: *sea moat.*

moat [reef] An elongate water-filled channel, on or adjacent to a reef flat, and only a few meters deep and wide. Cf: *boat channel.*

moat [streams] A syn. of *oxbow lake.* The term is used in New England and was also applied by Shaler (1890, p. 277) to the waters in abandoned channels in the Mississippi River flood plain.

moat [volc] A valleylike depression around the inner side of a crater or caldera, between its rim and a resurgent dome or cone constructed within it.

moat lake A *senescent lake* characterized by a peripheral or outer ring of water enclosing a filled interior (Veatch & Humphrys, 1966, p.202). See also: *atoll moor.*

mobile belt A long, relatively narrow crustal region of tectonic activity, measured in scores of miles. The term *geosyncline* is applied to its phase of sedimentation and subsidence. See also: *orogenic belt; orogenic cycle.*

mobile component (a) A component whose amount in a system changes during a given process. Cf: *perfectly mobile component; inert component.* (b) An element (or group of elements) that can migrate beyond the limits of a single mineral (Mehnert, 1968, p. 356).

mobility A term used by W. Penck (1924) for the concept that the relative rate of uplift of the Earth's crust primarily determines the nature of the landforms produced by erosional processes.

mobilizate English translation of the German word *Mobilisat,* introduced to refer to the mobile phase, of any consistency, that existed during migmatization. Cf: *chymogenic; metatect.* See also: *neosome.*

mobilization (a) Any process that renders a solid rock sufficiently plastic to permit it to flow or to permit geochemical migration of the mobile components. Cf: *rheomorphism.* (b) Any process that redistributes and concentrates the valuable constituents of a rock into an actual or potential ore deposit.

Mocha stone A white, gray, or yellowish form of moss agate containing brown to red iron-bearing or black manganese-bearing dendritic inclusions. The term is also used as a syn. of *moss agate.* Named for the city of Mocha (Al Mukhā) in Yemen. Also spelled: *mocha stone; mochastone.* Syn: *Mocha pebble.*

mock lead *sphalerite.*

mock ore *sphalerite.*

moctezumite A bright-orange mineral: $Pb(UO_2)(TeO_3)_2$.

modal The adj. of *mode.*

modal analysis A statement of the composition of a rock in terms of the relative amounts of minerals present; also, the procedure (usually *point counter analysis* or *Rosiwal analysis*) that yields such a statement (Chayes, 1956, p. 1).

modal cycle A term proposed by Duff & Walton (1962) for a particular group of beds that occurs most frequently through a succession displaying cyclic sedimentation.

modal diameter An expression of the average particle size of a sediment or rock, obtained graphically by locating the highest point of the frequency curve or by finding the point of inflection of the cumulative curve; the diameter that is most frequent in the particle-size distribution.

mode [petrology] The actual mineral composition of a rock, usually expressed in weight or volume percentages. Adj: *modal.* Cf: *norm.*

mode [stat] The value or group of values that occurs with the greatest frequency in a set of data; the most typical observation. Cf: *mean; median.*

model A working hypothesis or precise simulation, by means of description, statistical data, or analogy, of a phenomenon or process that cannot be observed directly or that is difficult to observe directly. Models may be derived by various methods, e.g. by computer, from stereoscopic photographs, or by scaled experiments. Syn: *conceptual model.* See: Wolf, 1976, vol. 1, chap. 1.

model scale The relationship existing between a distance measured in a model (such as in a stereoscopic image) and the corresponding distance on the Earth.

moder Plant material in a state intermediate between living and decayed.

moderate-energy coast A coast protected from strong wave action by headlands, wide gently sloping bottom, dominance of winds from the land, or other factors, and characterized by average breaker heights of 10-50 cm. Cf: *high-energy coast; low-energy coast.*

moderately sorted Said of a *sorted* sediment that is intermediate between a well-sorted sediment and a poorly sorted sediment and that has a *sorting coefficient* (Trask's So) in the range of 2.5 to 4.0. Based on the phi values associated with the 84 and 16 percent lines, Folk (1954, p. 349) suggests sigma phi limits of 0.50-1.00 for moderately sorted material.

modern carbon *contemporary carbon.*

modern cut A syn. of *fancy cut.* Also spelled: *moderne cut.*

modified Mercalli scale An earthquake *intensity scale,* having twelve divisions ranging from I (not felt by people) to XII (damage nearly total). It is a revision of the *Mercalli scale* made by Wood and Neumann in 1931. Cf: *Rossi-Forel scale.* Abbrev: MM scale.

modified polyconic projection A projection used for maps of large areas, derived from the regular *polyconic projection* by so altering the scale along the central meridian that the scale is exact along two standard meridians, one on either side of the central meridian and equidistant therefrom. Scale is preserved along these two meridians.

modlibovite A monticellite-free *polzenite* with biotite phenocrysts in a groundmass of olivine, melilite, lazurite, phlogopite, biotite, and interstitial nepheline. Cf: *vesecite.* The name, given by Scheumann in 1922, is for Modlibov, Czechoslovakia. Not recommended usage.

modulus of compression *compressibility.*

modulus of elasticity The ratio of stress to its corresponding strain under given conditions of load, for materials that deform elastically, according to Hooke's law. It is one of the *elastic constants.* See also: *Young's modulus; modulus of rigidity; modulus of deformation; static modulus; bulk modulus.* Syn: *elastic modulus; modulus of volume elasticity.*

modulus of incompressibility *bulk modulus.*

modulus of rigidity A *modulus of elasticity* in shear. Symbol: μ or G. Syn: *torsion modulus; shear modulus; rigidity modulus; Coulomb's modulus.*

modulus of volume elasticity *modulus of elasticity.*

modumite A light-colored *essexite* or *anorthosite* containing bytownite, pyroxene, barkevikite, and biotite. The name, given by Brögger in 1933, is from Modum in the Oslo district, Norway. Not recommended usage.

moel A term used in Wales for a rounded hill with a vegetation-clad summit (Marr, 1901). Etymol: Welsh, "bare field".

mofette The exhalation of carbon dioxide in an area of late-stage volcanic activity; also, the small opening from which the gas is emitted. Examples are in Yellowstone National Park in the U.S. Etymol: French, "noxious gas".

mofettite A natural carbon-dioxide gas.

mogote *karst tower.*

mohavite *tincalconite.*

Mohawkian North American stage: Middle Ordovician (above Chazyan, below Cincinnatian). See also: *Trentonian.*

Mohnian North American stage: Miocene (above Luisian, below Delmontian).

Moho Abbreviated form of *Mohorovičić discontinuity,* suggested by Birch (1952, p. 229).

Mohole project A now discontinued project to penetrate the Earth's crust and sample the mantle, i.e. to drill through the Mohorovičić discontinuity. The drill hole itself may be called the mohole.

Mohorovičić discontinuity The boundary surface or sharp seismic-velocity discontinuity that separates the Earth's crust from the subjacent mantle. It marks the level in the Earth at which *P*-wave velocities change abruptly from 6.7-7.2 km/sec (in the lower crust) to 7.6-8.6 km/sec or average 8.1 km/sec (at the top of the upper mantle); its depth ranges from about 5-10 km beneath the ocean floor to about 35 km below the continents, although it may reach 60 km or more under some mountain ranges. The discontinuity probably represents a chemical change from basaltic or simatic materials above to peridotitic or dunitic materials below, rather than a phase change (basalt to eclogite); however, the discontinuity should be defined by seismic velocities alone. It is variously estimated to be between 0.2 and 3 km thick. It is named in honor of its discoverer, Andrija Mohorovičić (1857-1936), Croatian seismologist. Syn: *Moho; M-discontinuity.*

Mohr circle A graphic representation of the state of stress at a particular point at a particular time. The coordinates of each point on the circle are the shear stress and the normal stress on a particular plane. See also: *Mohr envelope.*

Mohr envelope An envelope of a series of *Mohr circles*; the locus of points whose coordinates represent the stresses at failure. Syn: *rupture envelope.*

mohrite A mineral: $(NH_4)_2(Fe,Mg)(SO_4)_2 \cdot 6H_2O$.

Mohr-Knudsen method In oceanography, a chemical method for estimating the chlorinity of seawater.

mohsite *ilmenite.*

Mohs scale A standard of ten minerals by which the *hardness* of a mineral may be rated. The scale includes, from softest to hardest and numbered one to ten: talc; gypsum; calcite; fluorite; apatite; orthoclase; quartz; topaz; corundum; and diamond. Cf: *technical scale.*

Moinian A highly metamorphosed series of Precambrian rocks in the Northwest Highlands of Scotland.

moire Said of feldspars having the appearance of watered silk.

moissanite A meteoritic mineral: SiC. It is identical with the manufactured product Carborundum.

moist playa *wet playa.*

moisture Water diffused in the atmosphere or the ground, including soil water.

moisture content [coal] In coal, both the surface or free moisture that can be removed by natural drying, and the *inherent moisture* that is structurally contained in the substance.

moisture content [soil] The amount of moisture in a given soil mass, expressed as weight of water divided by weight of oven-dried soil, multiplied by 100 to give a percentage. See also: *water content [sed].*

moisture-density curve *compaction curve.*

moisture-density test *compaction test.*

moisture equivalent The ratio of weight of water that a saturated soil will retain against a centrifugal force 1000 times the force of gravity to the weight of dry soil (Meinzer, 1923, p. 25). Syn: *centrifuge moisture equivalent.*

moisture index A means for classifying climates devised by Thornthwaite; wet-season surplus minus 0.6 times dry-season deficiency, divided by total need—all expressed in the same unit, such as inches—multiplied by 100 to get a percentage.

moisture meter An instrument for determining the percentage of moisture in a substance such as timber or soil, usually by measuring its electrical resistivity.

moisture tension In a soil, negative pressure or suction of the water, equal to the equivalent pressure necessary to bring the soil water to hydraulic equilibrium through a porous wall, with a pool of water of equivalent composition (SSSA, 1970). Syn: *soil-moisture tension; capillary tension.*

molar n. (a) One of the posterior grinding or shearing teeth of mammals, which are not normally replaced in an individual's lifetime. (b) A process with a grinding surface on the gnathal lobe of the mandible of a crustacean.— Cf: *incisor.*

molasse (a) A paralic (partly marine, partly continental or deltaic) sedimentary facies consisting of a very thick sequence of soft, ungraded, cross-bedded, fossiliferous conglomerates, sandstones, shales, and marls, characterized by primary sedimentary structures and sometimes by coal and carbonate deposits. It is more clastic and less rhythmic than the preceding flysch facies. (b) An extensive, postorogenic sedimentary formation representing the totality of the molasse facies resulting from the wearing down of elevated mountain ranges during and immediately succeeding the main paroxysmal (diastrophic) phase of an orogeny, and deposited considerably in front of the preceding flysch; specif. the Molasse strata, mainly of Miocene and partly of Oligocene age, deposited on the Swiss Plain and Alpine foreland of southern Germany subsequent to the rising of the Alps.—Etymol: French *mollasse,* "soft". Pron: mo-*laas.* Adj: *molassic.* Cf: *flysch.*

molasse sandstone A sandstone of the *molasse* facies, characterized by Cayeux as a poorly rounded, poorly sorted, coarse sand rich in rock fragments and generally calcareous; generally a lithic arenite, in places arkosic. It formed by the demolition of a newly elevated orogenic belt (Pettijohn, Potter & Siever, 1973, p. 168).

mold [paleont] (a) An *impression* made in the surrounding earth or rock material by the exterior or interior of a fossil shell or other organic structure. A complete mold would be the hollow space with its boundary surface. Cf: *cast [paleont].* See also: *external mold; internal mold.* (b) *natural mold.* (c) A cast of the inner surface of a fossil shell or other organic structure.—Also spelled: *mould.*

mold [sed] An original mark or primary depression made on a sedimentary surface; e.g. a flute, striation, or groove. The filling of such a depression produces a *cast;* unfortunately "some authors reverse the usage and regard the structures on the bottoms of beds as molds" (Middleton, 1965, p. 247), and others regard "cast" and "mold" as synonymous. Syn: *mark.*

mold [soil] (a) An old term for a soft, friable soil rich in humus and suited to plant growth, e.g. leaf mold. (b) An old term for surface soil; the surface of the Earth; the ground.—Etymol: Old English *molde,* "earth; dust; soil".

moldavite [astron] A translucent, olive- to brownish-green or pale-green tektite from western Czechoslovakia (southern Bohemia and southern Moravia), characterized by marked sculpturing on its surface due to solution etching. Named after the Bohemian river Moldau (German name for Vltava), in whose valley moldavites are found. Syn: *moldauite; vltavite; pseudochrysolite.*

moldavite [mineral] A variety of *ozocerite* from Moldavia.

moldering The decomposition of organic matter under conditions of insufficient oxygen, so that a carbon-rich residue is formed. Cf: *disintegration [coal]; peat formation; putrefaction.*

moldic porosity Porosity resulting from the removal, usually by solution, of an individual constituent of a rock, such as a shell (Choquette & Pray, 1970, p. 248-249).

mole A massive solid-fill protective structure extending from the shore into deep water, formed of masonry and earth or large stones, and serving as a breakwater or a pier.

molecular paleontology The study of the molecular-scale aspects of fossils, such as skeletal mineralogy and geochemistry, or chemical traces of fossils that have been largely destroyed by diagenesis. Cf: *chemical fossil.*

molecular proportion The ratio of the weight percentage of a particular rock component, esp. an oxide, to its molecular weight.

mole fraction The number of moles of a given component in a phase, divided by the total number of moles of all components in the phase. Mole fractions are thus useful in defining the composition of a phase.

mole track A small, geologically short-lived ridge, 30-60 cm high, formed by the humping up and cracking of the ground where movement along a large strike-slip fault occurred in heavily alluviated terrain. It resembles the track of a gigantic mole, or a line of disturbed earth turned by a great plowshare.

mollic epipedon A diagnostic soil horizon that is dark and thick and has at least 0.6% organic carbon, a base saturation of at least 50% when measured at a pH of 7, and less than 250 ppm P_2O_5 soluble in citric acid (USDA, 1975). Mollic epipedons are normally formed under grass vegetation. Cf: *umbric epipedon; ochric epipedon.* See also: *anthropic epipedon.*

Mollisol In U.S. Dept. of Agriculture soil taxonomy, a soil order characterized by a *mollic epipedon* with an underlying horizon having a base saturation of 50% or more. It has no oxic or spodic horizon but may contain a histic epipedon or a natric, albic, argillic, cambic, gypsic, calcic, or petrocalcic horizon. Mollisols characteristically form under grass in climates that have a moderate to severe seasonal moisture deficit. They are dark-colored soils with a relatively high cation-exchange capacity dominated by calcium. They are most extensive in midlatitudes between the *Andisols* of arid climates and the *Spodosols* or *Alfisols* of humid climates. Many are very productive agricultural soils (USDA, 1975). Suborders and great soil groups of this soil order have the suffix -oll. See also: *Alboll; Aquoll; Boroll; Rendoll; Udoll; Ustoll.*

mollisol *active layer.*

molluscoid In some classifications, any invertebrate animal possessing a lophophore; i.e. a brachiopod or bryozoan.

mollusk A solitary invertebrate belonging to the phylum Mollusca, characterized by a nonsegmented body that is bilaterally symmetrical and by a radially or biradially symmetrical mantle and shell. Among the classes included in the mollusks are the gastropods, pelecypods, and cephalopods. Also spelled: *mollusc.* Adj: *molluscan.*

Mollweide projection An equal-area map projection on which the entire surface of the Earth is enclosed within an ellipse whose major axis (the equator, representing 360° of longitude) is twice the length of the minor axis (the central meridian, representing 180° of latitude). All parallels are represented by straight lines at right angles to the central meridian and more widely spaced at the equator than at the poles, and all meridians are represented by equally spaced elliptical arcs with the exception of the central meridian (a straight line) and the meridian 90° from the center (a full circle, representing the hemisphere centered at the origin of the projection). The meridional curvature increases away from the central meridian. There is excessive angular distortion (shearing) at the margins of the map. Named after Karl B. Mollweide (1774-1825), German mathematician and astronomer, who introduced the projection in 1805. Also known as "Mollweide homolographic projection".

molten In the fluid state as a result of heating; fused; melted.

moluranite A black amorphous mineral: $UO_2 \cdot 3UO_3 \cdot 7MoO_3 \cdot 20H_2O$.

molybdate A mineral compound characterized by the radical MoO_4, in which the hexavalent molybdenum ion and the four oxygens form a flattened square rather than a tetrahedron. Tungsten and molybdenum may substitute for each other. An example of a molybdate is wulfenite, $PbMoO_4$. Cf: *tungstate.*

molybdenite A lead-gray hexagonal mineral: MoS_2. It is the principal ore of molybdenum. Molybdenite generally occurs in foliated masses or scales, and is found in pegmatite dikes and quartz veins or disseminated in porphyry; it resembles graphite in appearance and to the touch, but has a bluer color. Cf: *jordisite.*

molybdenite-3R A trigonal mineral: MoS_2. It is polymorphous with molybdenite and jordisite.

molybdic ocher (a) *ferrimolybdite.* (b) *molybdite.*

molybdite A mineral: MoO_3. Much so-called molybdite is *ferrimolybdite.* Syn: *molybdine; molybdic ocher.*

molybdomenite A colorless to yellowish-white mineral: $PbSeO_3$.

molybdophyllite A colorless, white, or pale-green mineral: $Pb_2Mg_2Si_2O_7(OH)_2$.

molysite A brownish-red or yellow mineral: $FeCl_3$.

moment (a) The geochronologic equivalent of a *chronohorizon* (ISG, 1976, p. 68). (b) A term recommended by Teichert (1958a, p. 113-115, 117) for the time interval during which a biostratigraphic zone was deposited; the geologic-time unit corresponding to Oppel's (1856-1858) "zone". It is, for all practical purposes, the shortest perceptible interval into which geologic time can be subdivided.—Cf: *instant; hemera.* Syn: *phase [geochron]; secule; zone time.*

moment map A stratigraphic map that expresses the positional relations of beds as a continuous variable (Krumbein & Libby, 1957, p.200); e.g. a *center-of-gravity map* and a *standard-deviation map.*

moment measure The expected value of each of the powers of a random variable that has a given distribution; a weighted measure of central tendency. It is used to describe the character of a distribution curve. The first moment measure is the *mean;* the second is the *standard deviation;* the third is the *skewness;* and the fourth is the *kurtosis* (Cole & King, 1968, p. 660). In sedimentology, moment measures are related to the center of gravity of the particle-size distribution curve and are defined about the mean value of the variable.

monacanth A trabecula in a rugose coral in which the fibers are related to a single center of calcification and radiate upward and outward from the axis formed by upward shifting of the center (TIP, 1956, pt.F, p.235). Cf: *rhabdacanth; rhipidicanth.*

monactine A sponge spicule having a single ray. Syn: *monact; monactin.*

monadnock An upstanding rock, hill, or *mountain of circumdenudation* rising conspicuously above the general level of a peneplain in a temperate climate, representing an isolated remnant of a former erosion cycle in a mountain region that has been largely beveled to its base level. Type locality: Mount Monadnock in New Hampshire. Cf: *catoctin; unaka; inselberg.* Syn: *torso mountain.*

monadnock phase In hypsometric analysis of drainage basins, the transitory stage characterized by abnormally low hypsometric integrals (less than 35%): removal of the monadnock by fluvial erosion will restore the distorted hypsographic curve to equilibrium form (Strahler, 1952b, p.1130). Cf: *equilibrium stage.*

monalbite Monoclinic albite; a monoclinic, high-temperature modification of sodium feldspar. It forms a complete solid-solution series with sanidine. Formerly called: *barbierite.*

monaxial Having one axis.

monaxon A simple uniaxial sponge spicule with a single axial filament or canal, or one developed by growth along a single axis. It may be curved or straight and may bear expansions at one or both ends. Obsolete syn: *rhabd.*

monazite A yellow, brown, or reddish-brown monoclinic mineral: $(Ce,La,Nd,Th)(PO_4,SiO_4)$. It is a rare-earth phosphate with appreciable substitution of thorium for rare earths and silicon for phosphorus; thorium-free monazite is rare. It is widely disseminated as an accessory mineral in granites, gneisses, and pegmatites, and it is often naturally concentrated in detrital sand, gravel, and alluvial tin deposits. Monazite is a principal ore of the rare earths and the main source of thorium. Syn: *cryptolite.*

monazite-(La) A monoclinic mineral: $(La,Ce,Nd)PO_4$.

moncheite A steel-gray hexagonal mineral: $(Pt,Pd)(Te,Bi)_2$. Syn: *biteplatinite; chengbolite.*

monchiquite A *lamprophyre* containing phenocrysts of olivine, clinopyroxene, and typically biotite or amphibole (barkevikite), in a groundmass of glass or analcime, often highly altered. Nepheline or leucite may be present. Its name, given by Hunter and Rosenbusch in 1890, is derived from Serra de Monchique, Portugal. Cf: *fourchite.*

mondhaldeite A hypabyssal rock similar in composition to *camptonite* and characterized by the presence of long acicular hornblende phenocrysts along with augite, bytownite, and leucite, in a glassy groundmass with felty texture. The name, given by Graeff in 1900, is for Mondhalde, Kaiserstuhl, Germany. Not recommended usage.

mondmilch *moonmilk.*

monetite A yellowish-white mineral: $CaHPO_4$.

monheimite Smithsonite with iron in solid solution for zinc.

moniliform Beadlike, or jointed at regular intervals so as to resemble a string of beads; e.g. "moniliform antennae". Syn: *nummuloidal.*

moniliform wall A vertical zooecial wall in stenolaemate bryozoans that is thickened in transverse annular ridges so that in longitudinal section it looks beaded.

monimolimnion The deep, usually salty layer of a *meromictic lake,* of high density and perennially stagnant or noncirculating. Cf: *mixolimnion; chemocline.*

monimolite A yellowish, brownish, or greenish mineral: $(Pb,Ca)_3Sb_2O_8$ (?). It may contain ferrous iron.

monk rock *penitent rock.*

monmouthite An *urtite* containing hastingsite in place of pyroxene. It was named by Adams in 1904 after Monmouth Township, Ontario. Not recommended usage.

monocentric Said of a corallite formed by a monostomodaeal polyp.

monochromatic illuminator *monochromator.*

monochromatic light Electromagnetic radiation of a single wavelength or frequency. It is used in crystal optics to determine indices of refraction.

monochromator An instrument for selecting a narrow portion of a spectrum. In optics, a variable filter, grating, or prism which can isolate light of only one wavelength (color) or of a very narrow range of wavelengths. Syn: *monochromatic illuminator.*

monoclinal Adj. of *monocline.*

monoclinal coast A coast resulting from monoclinal flexure at the shoreline (Cotton, 1958, p. 475); e.g. the west coast of South Island, New Zealand.

monoclinal scarp A scarp resulting from a steep downward flexure between an upland block and a tectonic basin (Cotton, 1958, p. 174).

monoclinal shifting The downdip migration of a divide (and of a stream channel) resulting from the tendency of streams in a region of inclined strata to flow along the strike of less resistant strata, as where differential erosion proceeds more rapidly along the steeper slope of a cuesta or monoclinal ridge. The process was first noted by Gilbert (1877, p. 135-140). See also: *shifting.* Syn: *homoclinal shifting; uniclinal shifting.*

monocline A local steepening in an otherwise uniform gentle dip. Cf: *homocline; flexure.* Adj: *monoclinal.* Obsolete syn: *unicline.*

monoclinic system One of the six *crystal systems,* characterized by either a single twofold axis of symmetry, a single plane of symmetry, or a combination of the two. Of the three nonequivalent axes, one is perpendicular to the plane formed by the other two. Cf: *isometric system; hexagonal system; tetragonal system; orthorhombic system; triclinic system.*

monocolpate Said of pollen grains having a single, normally distal colpus. Syn: *monosulcate.*

monocot *monocotyledon.*

monocotyledon An angiosperm whose seeds contain a single, parallel-veined embryonic leaf. Such a plant usually has flowering parts in threes, parallel leaf venation, and monocolpate pollen. Examples include grasses, palms, and lilies. Monocotyledons range from the Cretaceous. Cf: *dicotyledon.* Syn: *monocot.*

monocrepid Said of a desma (of a sponge) with a monaxial crepis.

monocyclic (a) Said of a crinoid having only a single circlet of plates proximal to the radial plates. (b) Said of the apical system of an echinoid in which genital plates and ocular plates are arranged in a single ring around the periproct.—Cf: *dicyclic.*

monoecious Said of a plant that has both staminate and pistillate flowers, or both male and female gametangia.

monofacies A term used by Bailey & Childers (1977, p. 27) for *roll-front orebodies* of uranium that "occur in formations which are uniformly reducing". Cf: *bifacies.*

monogene adj. (a) *monogenetic.* (b) A term applied specif. by Naumann (1850, p. 433) to an igneous rock (such as dunite) composed essentially of a single mineral. Cf: *polygene; monomineralic; monomictic [sed].* Syn: *monogenic.*

monogenetic (a) Resulting from one process of formation or derived from one source, or originating or developing at one place and time; e.g. said of a volcano built up by a single eruption. (b) Consisting of one element or type of material, or having a homogeneous composition; e.g. said of a gravel composed of a single type of rock. —Cf: *polygenetic.* Syn: *monogene; monogenic.*

monogenic (a) *monogenetic.* (b) *monogene.*

monogeosyncline A single geosynclinal trough along the continental margin and receiving sediments from a borderland on its oceanic side (Schuchert, 1923). Cf: *polygeosyncline.*

monoglacial theory The belief that the Pleistocene ice sheet made only one general advance and one general recession, without any substantial "interglacial" recession and readvance (Wright, 1914, p. 124-125).

monograptid n. Any *graptoloid* belonging to the family Monograptidae, characterized by scandent uniserial rhabdosomes with thecae of variable form and "monograptid" growth. Range, Silurian and Lower Devonian.—adj. Said of the upward direction of growth in graptoloid thecae.

monohydrocalcite A rare mineral: $CaCO_3 \cdot H_2O$. It was first observed in lake-bottom sediments, and it may be formed by precipitation from cold water in contact with air. Cf: *hydrocalcite.*

monolete adj. Said of an embryophytic spore having a laesura consisting of a single line or mark. Cf: *trilete.*—n. A monolete spore. The usage of this term as a noun is improper.

monolith (a) A piece of unfractured bedrock, generally more than a few meters across; e.g. an unweathered joint block moved by a glacier. (b) A large upstanding mass of rock, such as a volcanic *spine.* (c) One of many large blocks of stone or concrete forming the component parts of an engineering structure, such as a dam. (d) A vertical soil section, taken to illustrate the *soil profile.*

monomaceral Said of a coal microlithotype consisting of a single maceral. Cf: *bimaceral; trimaceral.*

monomict breccia A brecciated meteorite in which all the fragments have essentially the same composition. Cf: *polymict breccia.*

monomictic [lake] Said of a lake with only one yearly *overturn. Tropical lakes* overturn in the winter and *polar lakes* in the summer. Cf: *dimictic.*

monomictic [sed] Said of a clastic sedimentary rock composed of a single mineral species. Cf: *oligomictic; polymictic.* Syn: *monomict.*

monomineralic Said of a rock composed wholly or almost wholly of a single mineral; esp. said of an igneous rock (such as anorthosite or dunite) consisting of one essential mineral. The amounts of other minerals tolerated under the definition vary with different authors. Cf: *polymineralic; anchimonomineralic; monogene.*

monomorphic Said of a bryozoan colony in which all zooids in *zones of astogenetic repetition* are feeding zooids of similar morphology.

monomyarian adj. Said of a bivalve mollusk or its shell with only the posterior adductor muscle. Cf: *dimyarian; anisomyarian.*—n. A monomyarian mollusk, such as an oyster or scallop.

Monongahelan North American provincial stage: Upper Pennsylvanian (above Conemaughian, below Dunkardian).

monophyletic Evolving from a single ancestral stock. Cf: *nonmonophyletic; polyphyletic.*

monoplacophoran Any mollusk belonging to the class Monoplacophora, characterized by nearly bilateral symmetry and internal serial repetition. Originally thought to be represented only by Paleozoic forms, it is now known to exist in present-day marine environments.

monopleural Said of the arrangement of the two rows of thecae in the biserial rhabdosome of a scandent graptoloid in which the rows are in side-by-side contact. Cf: *dipleural.*

monopodial Having one main axis of growth.

monoporate Said of pollen grains provided with a single pore, as in grasses.

monopyroxene *clinopyroxene.*

Monorhina A subclass of the Agnatha characterized by a single midline nostril associated with an externalized hypophysis; bone, if present, is cellular and histologically comparable to that of gnathostomes.

monosaccate Said of pollen with a single vesicle, usually extending all around the pollen grain more or less at the equator.

monoschematic Said of mineral deposits having a uniform texture. Cf: *polyschematic.*

monosomatic chondrule A chondrule consisting of a single crystal. Cf: *polysomatic chondrule.*

monostatic radar A *radar* system with the transmitter and receiver located at the same place.

monostomodaeal Said of the stomodaea of a scleractinian coral polyp, each having its own tentacular ring after originating by distomodaeal or tristomodaeal budding.

monostratum A simple layer, as in a first-order *laminite* (Lombard, 1963, p.14).

monostromatic Said of foliaceous tissue that is composed of only one layer of cells, such as leaves of mosses or algal thalli that are only one cell thick. Cf: *oligostromatic; polystromatic.*

monosulcate *monocolpate.*

monothalamous *unilocular.*

monothem A term proposed by Caster (1934, p.18) for a noncyclic, or not obviously cyclic, chronostratigraphic unit of genetically related strata, representing a "more ordinary, and perhaps more normal major subdivision" of a stage; but interpreted by Moore (1949, p.19) as a "local deposit having essentially uniform lithologic character" and corresponding to formation or member of lithostratigraphic classification. Weller (1958, p.636) regards the term as "superfluous because its meaning is the same as 'substage'". Cf: *lithizone.*

monothermite A clay-mineral material that shows a single high-temperature endothermal reaction at about 550°C. It appears to be a mixture in which illite and kaolinite are important components (Grim, 1968, p. 48).

monotopism The origin of a species or other systematic group in only one geographic area. Cf: *polytopism.*

Monotremata The sole order of the mammalian subclass Prototheria, characterized by egg-laying reproduction, lack of nipples, and near or complete loss of teeth.

monotrichous Said of a bacterial cell with a single flagellum occurring at one pole. Cf: *lophotrichous; peritrichous.*

monotrophic Said of an organism that feeds on one kind of food only.

monotropy The relationship between two different forms of the same substance, e.g. pyrite and marcasite, that have no definite transition point, since only one of the forms, e.g. pyrite, is stable; and in which the change from the unstable to the stable form is irreversible. Cf: *enantiotropy.*

monotypic Said of a taxon that includes only one taxon of the next lower rank, e.g. a genus or subgenus with only one originally included species.

monroe A small mound of mud occurring on tidal flats in cold regions. They generally occur in groups, have a rounded conical shape with occasionally a nipple at the top, have steep slopes, and are 5 to 25 cm in diameter. The name is "a descriptive term referring to a world-known American movie star that may easily be used in many languages" (Dionne, 1973, p. 848). They are thought to form under an *ice foot* as a result of load pressure that expels air and water trapped under mud and ice.

mons A term established by the International Astronomical Union for a large isolated mountain on Mars. Most are of volcanic origin. Generally used as part of a formal name for a Martian landform such as Olympus Mons (Mutch et al., 1976, p. 57). Etymol: Latin *mons,* mountain.

monsmedite A mineral: $H_8K_2Tl_2^{+3}(SO_4)_8 \cdot 11H_2O$.

monsoon A type of wind system whose direction changes with the seasons, for example over the Arabian Sea, where the winds are from the northeast for six months and then from the southeast for the next six months.

monster An organism with extreme departure in form or structure, generally pathologic, from the usual type of its species.

monstrosity A part of an organism exhibiting considerable deviation in structure or form, which may be injurious or unuseful to the species and is usually not propagated.

montane Of, pertaining to, or inhabiting cool upland slopes below the timber line, characterized by the dominance of evergreen trees. Cf: *alpine [ecol].* Syn: *subalpine; alpestrine.*

montanite A mineral: $Bi_2O_3 \cdot TeO_3 \cdot 2H_2O$.

montan wax A solid bitumen that may be extracted by solvents from certain lignites or brown coals. It is white to brown and melts at 77° to 93°C.

Mont Blanc ruby Reddish quartz or *rubasse* from Mont Blanc, SE France.

montbrayite A tin-white triclinic mineral: Au_2Te_3.

montebrasite A mineral: $LiAlPO_4(OH)$. It is isomorphous with amblygonite and natromontebrasite.

Monte Carlo method A random-sampling process for generating uniformly distributed pseudorandom numbers and using these to "draw" random samples from known frequency distributions (Harbaugh & Bonham-Carter, 1970, p. 74).

monteponite A black mineral: CdO.

montesite A mineral: $PbSn_4S_5$. It is plumboan herzenbergite.

montgomeryite A green to colorless mineral: $Ca_4MgAl_4(PO_4)_6(OH)_4 \cdot 12H_2O$.

Montian European stage: Paleocene (above Danian, below Thanetian).

monticellite A colorless or gray mineral related to olivine: $CaMgSiO_4$. It is isomorphous with kirschsteinite, and usually occurs in contact-metamorphosed limestones.

monticle *monticule [geomorph].*

monticule [geomorph] (a) A little mound; a hillock, mound, knob, or other small elevation. (b) A small, subordinate volcanic cone developed on the flank or about the base of a larger volcano. —Etymol: French. Syn: *monticle.*

monticule [paleont] (a) A protuberant part of the corallum surface of a scleractinian coral, produced in circummural budding. Cf: *colline.* (b) One of the small rounded nodes or swellings of a brachiopod shell, commonly bearing spines. (c) One of the clusters of small, modified zooecia that may be regularly spaced throughout a stenolaemate bryozoan colony, appearing at the surface as a small protuberance or elevation. Cf: *macula.*

montiform Having the shape of a mountain; mountainlike.

montmartrite A variety of gypsum from the Montmartre section of Paris, France.

montmorillonite (a) A group of expanding-lattice clay minerals of general formula: $R_{0.33}Al_2Si_4O_{10}(OH)_2 \cdot nH_2O$, where R includes one or more of the cations Na^+, K^+, Mg^{+2}, Ca^{+2}, and possibly others. The minerals are characterized by a three-layer crystal lattice (one sheet of aluminum and hydroxyl between two sheets of silicon and oxygen); by deficiencies in charge in the tetrahedral and octahedral positions balanced by the presence of cations (most commonly calcium and sodium) subject to base exchange; and by swelling on wetting (and shrinking on drying) due to introduction of considerable interlayer water in the *c*-axis direction. Magnesium or iron may proxy for aluminum, and aluminum for silica. The montmorillonite minerals are generally derived from alteration of ferromagnesian minerals, calcic feldspars, and volcanic glasses; they are the chief constituents of *bentonite* and *fuller's earth,* and are common in soils, sedimentary rocks, and some mineral deposits. Syn: *smectite.* (b) A dioctahedral clay mineral of the montmorillonite group: $Na_{0.33}Al_{1.67}Mg_{0.33}Si_4O_{10}(OH)_2 \cdot nH_2O$. It is usually white, grayish, pale red, or blue, and represents a high-alumina end-member that has some slight replacement of Al^{+3} by Mg^{+2} and substantially no replacement of Si^{+4} by Al^{+3}. Cf: *beidellite.* (c) Any mineral of the montmorillonite group, such as montmorillonite, nontronite, saponite, hectorite, sauconite, beidellite, volkonskoite, or griffithite.

montmorillonite-saponite An alternative name preferred by the International Mineralogical Association for the smectite group of clay minerals.

montrealite An ultramafic rock (*hornblende peridotite*) containing, in order of decreasing abundance, clinopyroxene, amphibole, and olivine, with little or no feldspar or nepheline. The name, given by Adams in 1913, is for Montreal (Mount Royal), Quebec. Not recommended usage.

montroseite [mineral] A black mineral: (V,Fe)O(OH).

montroseite [sed] A uranium-bearing sandstone.

montroydite A mineral: HgO.

monument [geomorph] (a) An isolated pinnacle, column, or pillar of rock resulting from erosion and resembling a man-made monument or obelisk, usually extremely regular in form and of grand dimensions. (b) *tind.*

monument [surv] A natural or artificial physical structure that marks the location on the ground of a *corner* or other survey point; e.g. a pile of stones indicating the boundary of a mining claim, or a road or fence marking the boundary of real property. See also: *boundary monument.*

monumental peak *horn [glac geol].*

monumented upland A term proposed by Hobbs (1921, p. 373) for "the extreme type of mountain sculpture ... believed to be due to continued glacial action upon a fretted upland like that of the Alps", characterized by enlargement of cirques and reduction of horns. Example: Glacier National Park, Mont.

monzodiorite In the *IUGS classification,* a plutonic rock with Q between 0 and 5, P/(A+P) between 65 and 90, and plagioclase more sodic than An_{50}.

monzogabbro In the *IUGS classification,* a plutonic rock with Q between 0 and 5, P/(A+P) between 65 and 90, and plagioclase more calcic than An_{50}.

monzonite (a) In the *IUGS classification,* a plutonic rock with Q between 0 and 5, and P/(A+P) between 35 and 65. (b) A group of plutonic rocks intermediate in composition between *syenite* and *diorite,* containing approximately equal amounts of alkali feldspar and plagioclase, little or no quartz, and commonly augite as the main mafic mineral; also, any rock in that group; the intrusive equivalent of *latite*. With a decrease in the alkali feldspar content, monzonite grades into diorite or *gabbro,* depending on the composition of the plagioclase; with an increase in alkali feldspar, it grades into syenite. Syn: *syenodiorite.*—The name, given by de Lapparent in 1864, is for Monzoni, Tyrolean Alps.

monzonitic (a) Said of the texture of an igneous rock containing euhedral plagioclase crystals and some interstitial potassium feldspar. (b) Of, pertaining to, or composed of *monzonite.*

monzonorite In Tobi's classification (1971, p. 202) of the *charnockite series,* a quartz-poor member containing more plagioclase than microperthite.

Moody diagram A diagram showing the variation of the Darcy-Weisbach coefficient against the Reynolds number.

mooihoekite A tetragonal mineral: $Cu_9Fe_9S_{16}$.

moon Any natural satellite of a planet; specif. the Moon, the Earth's only known natural satellite and next to the Sun the most conspicuous object in the sky, deriving its light from the Sun and reflecting it to the Earth. The Moon revolves about the Earth from west to east in about 29.53 days with reference to the Sun (interval from new moon to new moon) or about 27.32 days with reference to the stars; it has a mean diameter of 3475.9 km (2160 miles, or about 27% that of the Earth), a mean distance from the Earth of about 384,400 km (238,857 miles), a mass of 7.354×10^{25} g (about 1/81 that of the Earth), a volume about 1/49 that of the Earth, and a mean density of 3.34 g/cm^3. The Moon rotates once on its axis during each revolution in its orbit and therefore it always presents nearly the same face (41%) to the Earth; it has essentially no atmosphere, no water, and no life forms or organic matter.

moonmilk (a) A soft white plastic calcareous deposit, which occurs on the walls of limestone caves. It may consist of calcite, hydromagnesite, nesquehonite, huntite, aragonite, magnesite, or dolomite. Etymol: Swiss dialect *moonmilch,* "elf's milk". Syn: *mountain milk; mondmilch; rock milk; rock meal; bergmehl; agaric mineral.* Partial syn: *lublinite.*

moonquake An agitation or disturbance of the Moon's surface, analogous to a terrestrial earthquake.

moonscape The surface of the Moon, as observed in photographs or through a telescope or as delineated on the basis of photographic or telescopic evidence.

moonstone (a) A semitransparent to translucent alkali feldspar (adularia) or cryptoperthite that exhibits a bluish to milky-white pearly or opaline luster; an opalescent variety of orthoclase. Flawless moonstones are used as gemstones. Cf: *sunstone.* Syn: *hecatolite.* (b) A name incorrectly applied to peristerite or to opalescent varieties of plagioclase (esp. albite). (c) A name incorrectly applied (without proper prefix) to milky or *girasol* varieties of chalcedony, scapolite, corundum, and other minerals.

moon tide *lunar tide.*

moor coal A lignite or brown coal that is friable.

mooreite A glassy white mineral: $(Mn,Zn,Mg)_8(SO_4)(OH)_{14} \cdot 4H_2O$. Cf: *torreyite.*

moorhouseite A mineral: $(Co,Ni,Mn)(SO_4) \cdot 6H_2O$.

moorland pan *moorpan.*

moorland peat *highmoor peat.*

moorpan An *iron pan* occurring in a peaty soil or forming at the bottom of a bog, containing compact redeposited iron and humus compounds. Syn: *moorland pan.*

moor peat *highmoor peat.*

mor A raw type of humus, usually on the surface with a sharp lower boundary, and developed under cool, moist conditions. Cf: *mull [soil].* Syn: *raw humus.*

moraesite A white mineral: $Be_2(PO_4)(OH) \cdot 4H_2O$.

morainal Of, relating to, forming, or formed by a moraine. Syn: *morainic.*

morainal apron *outwash plain.*

morainal channel A meltwater-stream channel formed during the construction of a moraine (Rich, 1908, p. 528).

morainal-dam lake A glacial lake impounded by a *drift dam* left in a pre-existing valley by a retreating glacier.

morainal delta An obsolete syn. of *ice-contact delta.*

morainal lake A glacial lake occupying a depression resulting from irregular deposition of drift in an end moraine or ground moraine of a continental glacier.

morainal plain *outwash plain.*

morainal stuff An obsolete term for the material carried upon the surface of a glacier.

morainal topography An irregular landscape produced by deposition of drift and characterized by irregularly scattered hills and undrained depressions.

moraine [glac geol] A mound, ridge, or other distinct accumulaion of unsorted, unstratified glacial drift, predominantly till, deposited chiefly by direct action of glacier ice, in a variety of topographic landforms that are independent of control by the surface on which the drift lies. The term was probably used originally, and is still often used in European literature, as a petrologic name for *till* that is being carried and deposited by a glacier; but it is now more commonly used as a geomorphologic name for a landform composed mainly of till that has been deposited by either a living or an extinct glacier. Etymol: French, a term used by Alpine peasants in the 18th century for any heap of earth and stony debris; neither the exact origin of the term nor its first use in glacial geology can be traced.

moraine [volc] Solidified volcanic debris carried on the surface of a lava flow.

moraine bar A terminal moraine serving as a bar, rising out of deep water some distance from the shore (Tarr & Martin, 1914, p. 294).

moraine kame A term applied by Salisbury et al. (1902, p. 118) to a kame that forms one of a group having the characteristics of, and "the same general significance" as, a terminal moraine. See also: *kame moraine.*

moraine loop *loop moraine.*

moraine plateau A relatively flat area within a hummocky moraine, generally at the same elevation as, or a little higher than, the summits of surrounding knobs (Gravenor & Kupsch, 1959, p. 50).

moraine rampart An elongated ridge or row of lateral and terminal moraines, sometimes forming an amphitheatrical arrangement (Schieferdecker, 1959, terms 0916 & 0922).

morainic *morainal.*

moralla (a) Poorly crystallized or massive opaque-appearing greenish material from Colombian emerald mines. (b) Any of the poorer grades of emerald.—Syn: *morallion; morallon.*

morass ore *bog iron ore.*

mordenite A zeolite mineral: $(Ca,Na_2,K_2)_4Al_8Si_{40}O_{96} \cdot 28H_2O$. Syn: *ashtonite; flokite; arduinite; ptilolite.*

morel basin One of a series of cavities about 1 cm in diameter, which resemble the pits on a morel mushroom, produced on impure limestone by dissolution (Scott, 1947, p. 147). See also: *solution pan; solution morel.*

morencite *nontronite.*

morenosite An apple-green or light-green mineral: $NiSO_4 \cdot 7H_2O$. It may contain appreciable magnesium, and it occurs in secondary incrustations. Syn: *nickel vitriol.*

morganite *vorobyevite.*

morinite A mineral: $Na_2Ca_4Al_4(PO_4)_4O_2F_6 \cdot 5H_2O$. Syn: *jezekite.*

Morin transition A magnetic transition in hematite, occurring at about 250°K, below which "the feeble ferromagnetism with a molecular saturation moment of about one-hundredth of a Bohr magnetron disappears" (Runcorn et al., 1967, p.867-868).

morion A nearly black, opaque variety of smoky quartz or cairngorm.

morlop A mottled variety of jasper found in New South Wales, Australia. It often occurs as pebbles associated with diamonds.

morphocline In *cladism,* the ordered spectrum of variation displayed by a homologous character. Syn: *transformation series.*

morphogenesis The origin and subsequent early growth or development of landforms or of a landscape.

morphogenetic Pertaining to the origin of morphological features.

morphogenetic region A climatic zone in which the predominant geomorphic processes produce distinctive regional landscape characteristics that contrast with those of other areas developed under different climatic conditions. Peltier (1950) has postulated nine morphological regions based on temperature and moisture conditions. See also: *formkreis.*

morphogenetic zone *lineage-zone.*

morphogeny The interpretative morphology of a region; specif. *geomorphogeny.*

morphographic map *physiographic diagram.*

morphography The descriptive morphology of a region, or the phenomena so described; specif. *geomorphography.*

morphologic region A region delimited according to its distinctive landforms, rock structure, and evolutionary history. Cf: *physiographic province.*

morphologic series A graded series of fossils showing variation either in individuals as a whole or in some particular variable feature.

morphologic species A *species* based solely on morphologic characteristics.

morphologic unit [geomorph] A surface, either depositional or erosional, that is recognized by its topographic character.

morphologic unit [stratig] *morphostratigraphic unit.*

morphology [geomorph] (a) The shape of the Earth's surface; *geomorphology,* or "the morphology of the Earth" (King, 1962). (b) The external structure, form, and arrangement of rocks in relation to the development of landforms.

morphology [meteorite] The study of the dimensions, form, surface relief, fusion crust, and inner macrostructure of meteorites.

morphology [paleont] (a) A branch of biology or paleontology that deals with the form and structure of animals and plants or their fossil remains; esp. a study of the forms, relations, and phylogenetic development of organs apart from their functions. See also: *paleomorphology; functional morphology.* (b) The features included in the form and structure of an organism or any of its parts.

morphology [soil] The study of the distribution patterns of horizons in a soil profile, and of the soil's properties.

morphometrics Statistical analysis of morphological measurements in organisms.

morphometry [geomorph] "The measurement and mathematical analysis of the configuration of the Earth's surface and of the shape and dimensions of its landforms. The main aspects examined are the area, altitude, volume, slope, profile, and texture of the land as well as the varied characteristics of rivers and drainage basins" (Clarke, 1966, p.235).

morphometry [lake] (a) The measurement of the form characteristics (area, depth, length, width, volume, bottom gradients) of lakes and their basins. (b) The branch of limnology dealing with such measurements.

morphosculpture A topographic feature smaller than, and often developed within or on, a *morphostructure;* e.g. a ripple mark, ledge, or knoll on the ocean floor.

morphosequent Said of a surface feature that does not reflect the underlying geologic structure. Ant: *tectosequent.*

morphostratigraphic unit A distinct stratigraphic unit, defined by Frye & Willman (1960, p.7) as "a body of rock that is identified primarily from the surface form it displays"; it may or may not be distinctive lithologically from contiguous units, and may or may not transgress time throughout its extent. The term is used in stratigraphic classification of surficial deposits such as glacial moraines, alluvial fans, beach ridges, and other such deposits where landforms serve to give identity to a body of clastic sediments. Syn: *morphologic unit.*

morphostructure A major topographic feature that coincides with or is an expression of a geologic structure (e.g. a trench or ridge on the ocean floor) or that is formed directly by tectonic movements (e.g. a basin or dome). It is produced by the interaction of endogenic and exogenic forces, the former being predominant. Cf: *morphosculpture.*

morphotectonics The tectonic interpretation of the morphological or topographic features of the Earth's surface; it deals with their tectonic or structural relations and origins, rather than their origins by surficial processes of erosion and sedimentation. Cf: *orogeny.*

morriner *esker.*

morro A term used in Latin America for an isolated hill or ridge, which may or may not be on a coastal plain near the present shoreline; esp. a headland or bluff. Etymol: Spanish and Portuguese.

Morrowan North American series: lowermost Pennsylvanian (above Chesterian of Mississippian, below Atokan).

mortar bed A valley-flat deposit, occurring in Nebraska and Kansas, consisting of sand, or of a mixture of clay, silt, sand, and gravel, firmly cemented by calcium carbonate and resembling hardened mortar; a type of *caliche.*

mortar structure A structure in crystalline rocks characterized by a mica-free aggregate of small grains of quartz and feldspar occupying the interstices between, or forming borders on the edges of, much larger and rounded grains of the same minerals. Long considered a product of *cataclasis,* the structure may actually be the result of plastic deformation and dynamic recrystallization (Harte, 1977). See also: *pseudogritty structure.* Syn: *cataclastic structure; murbruk structure; porphyroclastic structure.*

mortar texture A texture found in crystalline sedimentary rocks in which relatively large crystalline grains are separated by a microcrystalline mosaic.

mortlake A British syn. of *oxbow lake.* The term is "practically obsolete" in Britain (Stamp, 1961, p. 327). Etymol: probably from Mortlake, a parish in a SW suburb of London, situated near a drained oxbow lake of the River Thames.

morvan (a) The intersection of two peneplains, as where an exhumed, tilted peneplain is cut across obliquely by a younger surface that has more nearly retained its original horizontal attitude; e.g. the intersection between the stripped and distinctly sloped Fall Zone peneplain (along the eastern Piedmont Plateau of the U.S.) with the late Tertiary Harrisburg peneplain. Also, the "problem" of the intersection of two peneplains. (b) A region that exhibits a *morvan* relationship, marked by a hard-rock upland bordered by a sloping land of older rock. The term was introduced by Davis (1912, p. 115); "a region of composite structure, consisting of an older undermass, usually made up of deformed crystalline rocks, that had been long ago worn down to small relief and that was then depressed, submerged, and buried beneath a heavy overmass of stratified deposits, the composite mass then being uplifted and tilted, the tilted mass being truncated across its double structure by renewed erosion, and in this worn-down condition rather evenly uplifted into a new cycle of destructive evolution". Type locality: Morvan region of central France. Syn: *skiou.*

mosaic [geomorph] *desert mosaic.*

mosaic [paleont] (a) A pattern formed on the interior of a brachiopod valve by outlines of the adjacent fibers of the secondary layer of shell. (b) Arrangement of plates in edrioasteroids and cyclocystoids, more or less in plane and not imbricating, and presumably rather rigid. Syn: *tessellate.*

mosaic [petrology] A textural subtype in which individual mineral grains are approximately equant (Harte, 1977).

mosaic [photo] An assembly of aerial or space photographs or images whose edges have been *feathered* and matched to form a continuous photographic representation of a part of the Earth's surface; e.g. a composite photograph formed by joining together parts of several overlapping vertical photographs of adjoining areas of the Earth's surface. See also: *controlled mosaic; uncontrolled mosaic.* Syn: *aerial mosaic; photomosaic.*

mosaic breccia A breccia having fragments that are largely but not wholly disjointed and displaced.

mosaic evolution The pattern seen within an evolving *lineage,* in which various morphologic characters of the organisms change at different rates; as a result, a particular species within the lineage appears to possess some advanced, some primitive, and some intermediate characters simultaneously.

mosaic structure Slight irregularity of orientation of small, angular, and granular fragments of varying sizes in a crystal, the fragments appearing in polarized light like pieces of a mosaic.

mosaic texture [meta] A granoblastic texture in a dynamically metamorphosed rock in which the individual grains meet with straight or only slightly curved, but not interlocking or sutured, boundaries. Syn: *cyclopean texture.*

mosaic texture [sed] A texture in a crystalline sedimentary rock characterized by more or less regular grain-boundary contacts; e.g. a texture in a dolomite in which the mineral dolomite forms rhombs of uniform size so that in section contiguous crystals appear to dovetail, or a texture in an orthoquartzite in which secondary quartz is deposited in optical continuity on detrital grains.

mosandrite A reddish-brown or yellowish-brown mineral: $(Na,Ca,Ce)_3Ti(SiO_4)_2F$. Cf: *johnstrupite.* Syn: *rinkite; rinkolite; lovchorrite; khibinite.*

moschellandsbergite A mineral: Ag_2Hg_3. It consists of a naturally occurring alloy of silver with mercury, and was formerly included with *amalgam.* Syn: *landsbergite.*

Moscovian Stage in Russia: middle Upper Carboniferous (above Namurian, below Gzhelian).

moscovite *muscovite.*

mosesite A yellow mineral: $Hg_2N(SO_4,MoO_4)\cdot H_2O$. Cf: *kleinite.*

mosor A monadnock that has survived because of remoteness from the main drainage lines; esp. a *hum* in a karstic region. Etymol: originally a German term named by Penck (1900) after the Mosor Mountains in Dalmatia, Yugoslavia. Pl: *mosore.*

moss [bot] A bryophyte of the class Musci, characterized by a leafy upright gametophyte, which bears the sporophyte in haustorial connection. Cf: *liverwort.*

moss [eco geol] adj. A syn. of *capillary [mineral],* e.g. moss gold.

moss [gem] A fracture, fissure, or other flaw in a gemstone, having the appearance of moss; specif. such a fracture in an emerald.

moss agate (a) A general term for any translucent chalcedony containing inclusions of any color arranged in dendritic patterns resembling trees, ferns, leaves, moss, and similar vegetation; specif. an agate containing brown, black, or green mosslike markings due to visible inclusions of oxides of manganese and iron. See also: *Mocha stone; tree agate.* (b) A moss agate containing green inclusions of actinolite or of other green minerals.

moss animal *bryozoan.*

Mössbauer effect The almost recoil-free emission and absorption of nuclear gamma rays by atoms tightly bound in a solid. It is a special case of *nuclear resonance* characterized by an extremely sharply defined resonant frequency as the atom is tightly bound in a crystal lattice, atomic recoil is minimized, and the emitted gamma ray is limited to a narrow frequency range. The Mössbauer effect was named in honor of Rudolf L. Mössbauer (1929-), German physicist. So sharp is the resonance that very slight Doppler shifts and even gravitational shifts of the emitted frequency can be detected.

Mössbauer spectrometry The art or process of using a spectrometer to analyze a *Mössbauer spectrum,* mainly for determining chemical structure. Measurement is made of the nuclear resonant absorption of gamma rays passing from a radioactive source to an absorber, usually the material being studied (DeVoe & Spijkerman, 1966, p. 382R).

Mössbauer spectrum The spectrum seen when Mössbauer gamma-ray intensity is plotted as a function of relative velocity between the radioactive source and the absorber, usually the material being studied (DeVoe & Spijkerman, 1966, p.382R).

moss coral *bryozoan.*

mossite A mineral: $Fe(Nb,Ta)_2O_6$. It is isomorphous with tapiolite.

moss land An area with abundant moss, yet not wet enough to be a bog.

moss peat *highmoor peat.*

moss polyp *bryozoan.*

mother cell A cell from which new cells are formed; e.g. *spore mother cell* and *pollen mother cell.*

mother crystal A mass of raw quartz (faced or rough) as found in nature.

mother geosyncline Stille's term for a geosyncline that matured by evolving into a folded mountain system (Glaessner & Teichert,

1947, p. 588). See also: *orogeosyncline.*
motherham An old syn. of *fusain.*
mother liquor In crystallization, the liquid that remains after the substances readily and regularly crystallizing have been removed. Syn: *mother liquid.*
mother lode (a) A main mineralized unit that may not be economically valuable in itself but to which workable veins are related, e.g. the Mother Lode of California. (b) An ore deposit from which a placer is derived; the *mother rock* of a placer.
mother of coal A syn. of *fusain.* Also spelled: *mother-of-coal.*
mother-of-emerald (a) *prase.* (b) Green fluorite.
mother-of-pearl The *nacre* of a pearl-bearing mollusk, extensively used for making small ornamental objects.
mother rock [eco geol] (a) A general term for the rock in which a secondary or transported ore deposit originated; *mother lode.* (b) A syn. of *country rock.*
mother rock [sed] *parent rock [sed].*
motile Exhibiting or capable of movement, as by cilia; e.g. the flagellate stage or "motile phase" in the life cycle of a coccolithophorid. Ant: *nonmotile.*
mottle (a) A spot, blotch, or patch of color occurring on the surface of a sediment or soil. (b) A small, irregular body of material in a sedimentary matrix of different texture (difference in color not being essential) (Moore & Scruton, 1957, p.2727).
mottled [sed] Said of a sediment or sedimentary rock marked with spots of different colors, usually as a result of oxidation of iron compounds. Cf: *variegated.*
mottled [soil] Said of a soil that is irregularly marked with spots or patches of different colors, usually indicating poor aeration or seasonal wetness.
mottled limestone Limestone with narrow, branching, fucoidlike cylindrical masses of dolomite, often with a central tube or hole (Van Tuyl, 1916, p. 345); it may be organic or inorganic in origin.
mottled structure Discontinuous lumps, tubes, pods, and pockets of a sediment, randomly enclosed in a matrix of contrasting textures, and usually formed by the filling of animal borings and burrows (Moore & Scruton, 1957, p.2727). Syn: *mottling.*
mottling (a) Variation of color in sediments and soils, as represented by localized spots, patches, or blotches of color or shades of color. Also, the formation of mottles or of a mottled appearance. (b) *mottled structure.* (c) *luster mottling.* (d) *dolomitic mottling.*
mottramite A mineral: $Pb(Cu,Zn)(VO_4)(OH)$. It is isomorphous with descloizite. Syn: *cuprodescloizite; psittacinite.*
Mott Smith gravimeter An instrument in which the moving system consists entirely of fused quartz, the restoring and labilizing forces being provided by quartz fibers.
motu A Polynesian term for a small coral island with vegetation. Pl: *motu; motus.*
mould *mold [paleont].*
moulin A roughly cylindrical, nearly vertical, well-like opening, hole, or shaft in the ice of a glacier, scoured out by swirling meltwater as it pours down from the surface. Etymol: French, "mill", so called because of the loud roaring noise made by the falling water. Pron: moo-lanh. Syn: *glacier mill; glacial mill; glacier pothole; pothole [glaciol]; glacier well.*
moulin kame A conical hill of glaciofluvial material formed in a large circular hole (*moulin*) in glacier ice.
moulin pothole *giant's kettle.*
mounanaite A mineral: $PbFe_2(VO_2)_2(OH)_2$.
mound (a) A low rounded natural hill, generally of earth; a knoll. (b) A small man-made hill, composed either of debris accumulated during successive occupations of the site (syn: *tell; teppe*) or of earth heaped up to mark a burial site. (c) An organic structure built by fossil colonial organisms, such as crinoids.
mound spring A spring characterized by a mound at the place where it flows onto the land surface. According to Meinzer (1923, p. 55) "mound springs may be produced, wholly or in part, by the precipitation of mineral matter from the spring water; or by vegetation and sediments blown in by the wind—a method of growth common in arid regions". Cf: *pool spring.* See also: *spring mound.*
mount (a) An abbreviated form of the term *mountain,* esp. used preceding a proper name and usually referring to a particular summit within a group of elevations; e.g. Mount Marcy in the Adirondack Mountains. Abbrev: mt. (b) A high *hill;* esp. an eminence rising abruptly above the surrounding land surface, such as Mount Vesuvius. (c) *seamount.*
mountain (a) Any part of the Earth's crust higher than a *hill,* sufficiently elevated above the surrounding land surface of which it forms a part to be considered worthy of a distinctive name, characterized by a restricted summit area (as distinguished from a plateau), and generally having comparatively steep sides and considerable bare-rock surface; it can occur as a single, isolated eminence, or in a group forming a long chain or range, and it may form by earth movements, erosion, or volcanic action. Generally, a mountain is considered to project at least 300 m (1000 ft) above the surrounding land, although older usage refers to an altitude of 600 m (2000 ft) or more above sea level. When the term is used following a proper name, it usually signifies a group of elevations, such as a range (e.g. the Adirondack Mountains) or a system (e.g. the Rocky Mountains). Abbrev: mt.; mtn. Syn: *mount.* (b) Any conspicuous or prominent elevation in an area of low relief, esp. one rising abruptly from the surrounding land and having a rounded base. (c) A region characterized by mountains; term usually used in the plural.
mountain and bolson desert A desert area made up of elongated mountain ranges and intervening alluvium-filled fault basins or *bolsons* (Stone, 1967, p. 233).
mountain apron *bajada.*
mountain blue A blue copper mineral; specif. azurite and chrysocolla.
mountain building *orogeny.*
mountain butter A term used for various salts; esp. *halotrichite.*
mountain chain A complex, connected series of several more or less parallel *mountain ranges* and *mountain systems* grouped together without regard to similarity of form, structure, and origin, but having a general longitudinal arrangement or well-defined trend; e.g. the Mediterranean mountain chain of southern Europe. See also: *cordillera.*
mountain climate A climate of high altitudes, characterized by extremes of surface temperature, low atmospheric temperature, strong winds, and rarefied air (Swayne, 1956, p.98).
mountain cork (a) A white or gray variety of *asbestos* consisting of thick interlaced fibers and resembling cork in texture and lightness (it floats on water). Syn: *rock cork.* (b) A fibrous clay mineral such as sepiolite or palygorskite.
mountain flax A fine silky *asbestos.*
mountain glacier An *alpine glacier;* a glacier formed on a mountain slope.
mountain green A green mineral; specif. malachite, green earth, and chrysocolla.
mountain group An assemblage of several mountain peaks or of short mountain ridges; e.g. the Catskill Mountains, N.Y.
mountainite A monoclinic mineral: $(Ca,Na_2,K_2)_2Si_4O_{10}\cdot 3H_2O$. Cf: *rhodesite.*
mountain leather (a) A tough variety of *asbestos* occurring in thin flexible sheets made of interlaced fibers. Syn: *rock leather; mountain paper.* (b) A fibrous clay mineral such as sepiolite or palygorskite.—Syn: *leatherstone.*
mountain limestone A term used in England for a Carboniferous limestone occurring in the hills and mountains; specif. the Early Carboniferous limestone forming the Pennine Chain of northern England.
mountain mahogany *obsidian.*
mountain milk *moonmilk.*
mountain of accumulation A symmetric mountain, frequently of great height, formed by the accretion of material on the Earth's surface, esp. by the ejection of material from a volcano; it tends to occur as an isolated peak. Syn: *accumulation mountain.*
mountain of circumdenudation A mountain consisting of resistant rock that remains after the surrounding, less resistant rock has been worn away, or a mountain representing the remains of a pre-existing plateau; e.g. a *monadnock or an inselberg.* Syn: *relict mountain; remainder mountain; circumdenudation mountain.*
mountain of denudation A remnant of "undisturbed and otherwise continuous strata, that have been in part removed by erosion" (Gilbert, 1875, p. 21).
mountain of dislocation A mountain "due to the rearrangement of strata, either by bending or fracture" (Gilbert, 1875, p. 21); a "fold mountain" or "fault mountain".
mountainous (a) Descriptive of a region characterized by conspicuous peaks, ridges, or mountain ranges. (b) Resembling a mountain, such as a *mountainous* dome that is strongly elevated and around whose flanks the strata are steeply dipping.
mountain paper A paperlike variety of *asbestos* occurring in thin sheets; specif. *mountain leather.*

mountain pediment (a) A term introduced by Bryan (1923a. p. 30, 52-58, 88) for a plain of combined erosion and transportation at the foot of a desert mountain range, similar in form to an alluvial plain, and surrounding a mountain in such a manner that at a distance the plain appears to be a broad triangular mass (resembling a pediment or gable of a low-pitched roof) above which the mountain projects. This usage is similar to *piedmont pediment.* (b) A pediment occurring within a mountain mass as a relatively high-altitude surface truncating a mountain structure (Tator, 1953, p. 51).

mountain range A single, large mass consisting of a succession of mountains or narrowly spaced mountain ridges, with or without peaks, closely related in position, direction, formation, and age; a component part of a *mountain system* or of a *mountain chain.*

mountainside A part of a mountain between the summit and the foot. Syn: *mountain slope.*

mountain slope *mountainside.*

mountain soap A dark clay mineral having a greasy feel and streak; specif. *saponite.* Syn: *rock soap.*

mountain soil An old term for a skeletal soil or *Lithosol,* formed by physical weathering processes in mountainous areas.

mountain system A group of *mountain ranges* exhibiting certain unifying features, such as similarity in form, structure, and alignment, and presumably originating from the same general causes; esp. a series of ranges belonging to an *orogenic belt.* Cf: *mountain chain.*

mountain tallow *hatchettine.*

mountain-top detritus *block field.*

mountain tract The narrow, upper part of a stream near its source in the mountains, characterized by a steep gradient and a narrow, V-shaped valley in which the water is flowing swiftly. Cf: *valley tract; plain tract.* Syn: *torrent tract.*

mountain wall A very steep mountainside.

mountain wind A nighttime *katabatic wind,* flowing down a mountain slope. It often alternates with a daytime *valley wind.*

mountain wood (a) A compact, fibrous, gray to brown variety of *asbestos* resembling dry wood in appearance. Syn: *rock wood.* (b) A fibrous clay mineral such as sepiolite or palygorskite.

mourite A violet mineral consisting of a hydrous uranium molybdate: $U^{+4}Mo^{+6}{}_5O_{12}(OH)_{10}$.

mouth [geol] (a) The place of discharge of a stream, as where it enters a larger stream, a lake, or the sea. (b) *baymouth.* (c) The entrance or opening of a geomorphic feature such as a cave, valley, or canyon. (d) The surface outlet of an underground conduit, as of a volcano.

mouth [paleont] The entrance to the digestive tract through which food passes into the body of an animal; e.g. the central opening at the summit of the theca leading to the alimentary system of a blastoid, the external opening of the body cavity of a coelenterate through which indigestible material is also discharged, or the open end or aperture of the body chamber of an ammonoid shell.

mouth frame The angulated girdle of ossicles surrounding the mouth of an asterozoan.

moutonnée A French adjective meaning "fleecy" or "curled", but often used (incorrectly) as a shortened form of *roche moutonnée.* The term was introduced into geologic literature by Saussure (1786, par. 1061, p. 512-513) in describing an assemblage of rounded Alpine hills whose contiguous and repeated curves, taken as a whole and as seen from a distance, resemble a thick fleece and also a curly or wavy wig ("perruque moutonnée") that was fashionable in the late 18th century. The term later implied a fancied resemblance between the general form of a roche moutonnée and that of a grazing sheep whose head is represented by the stoss side.

movable bed A stream bed consisting of readily transportable materials.

movement picture *deformation plan.*

movement plan *deformation plan.*

moveout The difference in arrival times of a reflection event on adjacent traces of a seismic record, esp. resulting from the dip of the reflecting interface. Cf: *normal moveout.* Syn: *stepout.*

moyite A granite in which quartz comprises more than 50 percent of the light-colored phases, and potassium feldspar is the only feldspar present. Its name, given by Johannsen in 1932, is derived from the Moyie Sill, British Columbia. Not recommended usage.

mozarkite The state rock of Missouri: a varicolored, easily polished Ordovician chert.

mpororoite A monoclinic mineral: $(Al,Fe)_2W_2O_9 \cdot 6H_2O$.

mroseite An orthorhombic mineral: $CaTe^{+4}O_2(CO_3)$.

MSL *mean sea level.*

MTL *mean tide level.*

muck [mining] n. A syn. of *waste rock.* —v. To remove waste rock.

muck [sed] Dark finely divided well decomposed organic material, intermixed with a high percentage of mineral matter, usually silt; it forms surface deposits in some poorly drained areas, e.g. areas of permafrost and lake bottoms.

muckite A yellow variety of retinite found in minute particles in coal in a region of central Europe about the upper valley of the Oder River. Named after its discoverer, H. Muck, 19th-century German mineralogist.

muck soil A soil that contains at least 50% organic matter that is well decomposed (SSSA, 1970, p. 11).

mucro An abrupt, sharp terminal point or process of an animal part or a plant part; e.g. a projection of the frontal shield over the proximal lip of the orifice in some ascophoran cheilostome bryozoans (cf: *lyrula*), or the terminal segment of the springing appendage of certain arthropods, or a short abrupt spur or spiny tip of some leaves. Pl: *mucrones* or *mucros.* Syn: *mucron.*

mucron (a) A perforate central scar or buttonlike projection on the aboral end of a chitinozoan test, serving for attachment. (b) *mucro.*

mucronate Ending in an abrupt, sharp terminal point or process, such as by a distinct and obvious mucro; e.g. said of the cardinal margin of a brachiopod in which the posterolateral extremities extend into sharp points.

mud [drill] *drilling mud.*

mud [marine geol] A sticky, fine-grained, marine detrital sediment, either pelagic or terrigenous. Muds are usually described by color: *blue mud; black mud; gray mud; green mud; red mud.*

mud [sed] (a) A slimy, sticky, or slippery mixture of water and silt- or clay-sized earth material, with a consistency ranging from semifluid to soft and plastic; a wet, soft soil or earthy mass; mire, sludge. (b) An unconsolidated sediment consisting of clay and/or silt, together with material of other dimensions (such as sand), mixed with water, without connotation as to composition; e.g. a recently exposed lake-bottom clay in a soft, ooze-like condition. (c) A mixture of silt and clay; the silt-plus-clay portion of a sedimentary rock, such as the finely divided calcareous matrix of a limestone.

mud aggregate An aggregate of mud grains, commonly having the size of a sand or silt particle, and usually mechanically deposited.

mud ball (a) A spherical mass of mud or mudstone in a sedimentary rock, developed by weathering and breakup of clay deposits. It may measure as much as 20 cm in diameter. (b) *armored mud ball.*

mudbank (a) A submerged or partly submerged ridge of mud along a shore or in a river, usually exposed during low tide. (b) A *submarine bank* consisting largely of carbonate mud (*micrite*); some mudbanks preserve traces of plants or sessile animals that helped to stabilize the loose sediment.—Also spelled: *mud bank.* See also: *bank [sed].*

mud breccia A term used by Ransome & Calkins (1908, p.31) for a *desiccation breccia* containing angular or slightly rounded fragments of "fine-grained argillite embedded in somewhat coarser-grained and more arenaceous material".

mud-buried ripple mark Ripple mark covered by mud settling out of water, characterized by the filling up of troughs and little or no accumulation on the crests (Shrock, 1948, p.109-110). Cf: *flaser structure.*

mud cake [drill] A clay lining or layer of concentrated solids adhering to the walls of a well or borehole, formed where the *drilling mud* lost water by filtration into a porous formation during rotary drilling. Syn: *filter cake; cake [drill].*

mud cake [sed] A clast formed by desiccation and occurring in an intraformational breccia.

mud circle A *nonsorted circle* characterized by a central core of upwardly injected clay, silt, or sometimes fine sand, surrounded by vegetation; the center is round and generally 10 cm to 2 m in diameter. See also: *frost scar; plug [pat grd].* Syn: *clay boil; tundra ostiole.*

mud column The height, measured from the bottom of a borehole, of *drilling mud* standing in the hole.

mud cone A small cone of sulfurous mud built around the opening of a *mud volcano* or mud geyser.

mud crack (a) An irregular fracture in a crudely polygonal pattern, formed by the shrinkage of clay, silt, or mud, generally in the course of drying under the influence of atmospheric surface conditions. Also referred to as a *sun crack,* a *shrinkage crack,* and a *desiccation crack.* (b) *mud-crack cast.*—Also spelled: *mudcrack.*

mud-crack cast A mud crack after it has been filled and the filling material (generally sand) has been hardened into rock; often occurs on the underside of a bed immediately overlying a mudstone. Syn: *mud crack.*

mud-crack polygon *desiccation polygon.*

mudding off Blockage of the flow of reservoir fluids, e.g. oil, into a well bore, resulting from the formation of *mud cake* and/or loss of permeability from invasion by *drilling mud.* It may require remedial treatment to establish production.

muddy [geomorph] n. A topographic feature consisting of a very fine-grained stream deposit produced when water has been ponded, and marked by a top that is "nearly horizontal though more or less concave" (Shaw, 1911, p. 489). Type example: the bottom of Muddy River valley in southern Illinois. Cf: *sandy [geomorph].*

muddy [sed] adj. Pertaining to or characterized by mud; esp. said of water made turbid by sediment, or of sediment consisting of mud.

muddy gravel An unconsolidated sediment containing 30-80% gravel and having a ratio of sand to mud (silt + clay) less than 1:1 (Folk, 1954, p.346).

muddy sand An unconsolidated sediment containing 50-90% sand and having a ratio of silt to clay between 1:2 and 2:1 (Folk, 1954, p.349).

mud engineer A specialist who studies and prescribes the materials, chemicals, and proprietary additives to make up and maintain the properties of the *drilling mud* used in rotary drilling.

mud field An area saturated with ground water owing to the presence of fumaroles (Schieferdecker, 1959, term 4511).

mud flat A relatively level area of fine silt along a shore (as in a sheltered estuary) or around an island, alternately covered and uncovered by the tide, or covered by shallow water; a muddy *tidal flat* barren of vegetation. Cf: *sand flat.* Also spelled: *mudflat.* Syn: *flat.*

mud-flat polygon *desiccation polygon.*

mudflow [mass move] A general term for a mass-movement landform and a process characterized by a flowing mass of predominantly fine-grained earth material possessing a high degree of fluidity during movement. The degree of fluidity is revealed by the observed rate of movement or by the distribution and morphology of the resulting deposit. If more than half of the solid fraction of such a mass consists of material larger than sand size, the term *debris flow* is preferable (Sharp & Nobles, 1953; Varnes, 1958). Mudflows are intermediate members of a gradational series of processes characterized by varying proportions of water, clay, and rock debris. The water content of mudflows may range up to 60%. The degree of water bonding, determined by the clay content and mineralogy, critically affects the viscosity of the matrix and the velocity and morphology of the flow. With increasing fluidity, mudflows grade into loaded and clear streams; with a decrease in fluidity, they grade into earthflows. Also spelled: *mud flow.*

mudflow [sed] A minor sedimentary structure found in fine-grained rocks and indicative of local flowage while the material was still soft. Not in common usage. Syn: *mud mark.*

mudflow [volc] *lahar.*

mudflow levee A sharp linear ridge marking the edge of a narrow debris flow or mudflow, and consisting of boulders shoved aside by the force of the flow.

mud geyser A geyser that erupts sulfurous mud; a type of mud volcano.

mud glacier A viscous mass of surficial material moving slowly downslope by solifluction.

mud lava [grd wat] The sulfurous and sometimes carbonaceous material contained in mud pots or erupted from mud volcanoes or mud geysers.

mud lava [volc] A term that has been used to describe some materials in Japan which are now known to be *ash flow* deposits (Ross & Smith, 1961, p. 5).

mud log A continuous analysis of the drilling mud and well cuttings during *rotary drilling,* for entrained oil or gas. Visual observation, ultraviolet fluoroscopy, partition gas chromatograph, and hydrogen-flame ionization analyzer may be used. A *drilling-time log* is kept concurrently.

mudlump A diapiric sedimentary structure that forms a small short-lived island, some 4000 square meters in area, near the mouth of a major distributary of the Mississippi River; it consists of a broad mound or swelling of silt or thick plastic clay that stands 2 to 4 m above sea level. It is created by the loading of rapidly deposited delta-front sands upon lighter-weight prodelta clays, causing the clays to be intruded or thrust upward into and through the overlying sandbar deposits. Cf: *gas-heave structure.* Also spelled: *mud lump.*

mud mark *mudflow [sed].*

mud pellet A small, flattened to rounded, irregularly shaped mass of mud or mudstone, 3 to 13 mm in diameter, in a sedimentary rock. Pellets are pieces of compacted mud produced by breakup, transported short distances, and redeposited.

mud pit *slush pit.*

mud polygon (a) A *nonsorted polygon* whose center is bare of vegetation but whose outlining reticulate fissures contain peat and plants. The term was suggested by Elton (1927, p.165) to replace *fissure polygon,* but it is ambiguous because forms without a stone border do not invariably consist of "mud" but may consist of sand, gravel, or a nonsorted mixture of sand, clay, and silt with stones (Washburn, 1956, p.825-826). See also: *frost hillock.* (b) *desiccation polygon.*

mud pot A type of hot spring containing boiling mud, usually sulfurous and often multicolored, as in a *paint pot.* Mud pots are commonly associated with geysers and other hot springs in volcanic areas, esp. Yellowstone National Park, Wyo. Syn: *sulfur-mud pool.*

mud pump The reciprocating pump used to impel *drilling mud* through the essentially closed circulating system used in rotary drilling. Syn: *slush pump.*

mud-ridge ripple mark A *longitudinal ripple mark* with a regular profile, a usually symmetric crest, and a narrow and angular ridge, that is situated between much wider and relatively flat troughs and that frequently branches (always converging downcurrent). Cf: *corrugated ripple mark.*

mud rock A syn. of *mudstone.* Also spelled: *mudrock.*

mud-rock flood A violent and destructive rush of water generated by a cloudburst and laden with rocks, mud, and debris engulfed along its path.

mudrush The sudden inflow of mud into shallow mine workings. See also: *running ground.*

mud shale A consolidated sediment consisting of no more than 10% sand and having a silt/clay ratio between 1:2 and 2:1 (Folk, 1954, p.350); a fissile mudstone.

mudslide A relatively slow-moving type of mudflow in which movement occurs predominantly by sliding upon a discrete boundary shear surface (Hutchinson & Bhandari, 1971). Cf: *earthflow.*

mud spot *frost scar.*

mud stalagmite A stalagmite that is composed of mud cemented by calcite.

mudstone (a) An indurated mud having the texture and composition of shale, but lacking its fine lamination or fissility; a blocky or massive, fine-grained sedimentary rock in which the proportions of clay and silt are approximately equal; a nonfissile mud shale. Shrock (1948a) regards mudstone as a partly indurated mud that slakes upon wetting. See also: *claystone; siltstone.* (b) A general term that includes clay, silt, claystone, siltstone, shale, and argillite, and that should be used only when the amounts of clay and silt are not known or specified or cannot be precisely identified; or "when a deposit consists of an indefinite mixture of clay, silt, and sand particles, the proportions varying from place to place, so that a more precise term is not possible" (Twenhofel, 1937, p.98); or when it is desirable to characterize the whole family of finer-grained sedimentary rocks (as distinguished from sandstones, conglomerates, and limestones). Syn: *mud rock.* (c) A term used by Dunham (1962) for a mud-supported carbonate sedimentary rock containing less than 10% grains (particles with diameters greater than 20 microns); e.g. a calcilutite. The term specifies neither mineralogic composition nor mud of clastic origin. Cf: *wackestone; packstone; grainstone.*—The term was apparently first used by Murchison (1839) for certain massive dark-gray fine-grained Silurian shales of Wales, which on exposure and wetting rapidly disintegrate into mud.

mudstone conglomerate (a) A conglomerate containing mudstone clasts, such as one produced by penecontemporaneous compaction and induration of muds. (b) *desiccation conglomerate.*—Cf: *conglomeratic mudstone.*

mudstone ratio A uranium prospector's term, esp. on the Colorado Plateau, for the ratio of the total thickness of red mudstone to that of green mudstone within an assumed stratigraphic interval. "Its value is based upon the premise that uranium-bearing solutions will bleach red mudstone containing ferric iron to green mudstone containing ferrous iron" (Ballard & Conklin, 1955, p. 194), in the course of depositing uranium minerals.

mud-supported A term used by Dunham (1962) to describe a sedimentary carbonate rock whose sand-size particles (at least 10% of the total bulk) are embedded or "floating" in, and supported by, the muddy matrix. Cf: *grain-supported.*

mud tuff *dust tuff.*

mud volcano An accumulation, usually conical, of mud and rock ejected by volcanic gases; also, a similar accumulation formed by escaping petroliferous gases. The term has also been used for a *mud cone* not of eruptive origin. Syn: *hervidero; macaluba.* Cf: *air volcano; salinelle.*

mugearite An extrusive or hypabyssal igneous rock, a member of the *alkali basalt* suite, consisting of oligoclase with subordinate alkali feldspar and mafic minerals. In many examples, olivine is more abundant than clinopyroxene. Although generally nepheline-normative, mugearite may contain normative hypersthene, or even quartz. It has been defined (Baker et al., 1974) as having a differentiation index between 45 and 65 and normative plagioclase more sodic than An_{30}. Cf: *hawaiite; benmoreite; basalt.* The term, proposed by Harker in 1904, is from the village of Mugeary, Isle of Skye, Scotland. Usage was revived by Macdonald (1960), and by Muir and Tilley (1961).

muirite A mineral: $Ba_{10}Ca_2MnTiSi_{10}O_{30}(OH,Cl,F)_{10}$.

mukhinite A mineral of the epidote group: $Ca_2(Al_2V)(SiO_4)_3(OH)$.

mull [coast] A Scottish term for a *headland;* e.g. Mull of Galloway.

mull [soil] A type of humus, usually developed in the forest, that is incorporated with underlying mineral matter. Cf: *mor.*

Müllerian law The law that expresses the regularity in distribution of 20 radial spines on the shells of radiolarians of the suborder Acantharina (four spines on each of the five circles that are comparable to the equatorial, two tropical, and two polar circles of the terrestrial globe): "between two poles of a spineless axis are regularly disposed five parallel zones, each with four radial spines; the four spines of each zone are equidistant one from another, and also equidistant from each pole; and the four spines of each zone are so alternating with those of each neighboring zone, that all twenty spines together lie in four meridian planes, which transect one another at an angle of 45°" (translated from Haeckel, 1862, p.40). Named by Haeckel in honor of Johannes Müller (1801-1858), German physiologist and zoologist, who first recognized the regularity in the disposition of the 20 radial spines (Müller, 1858, p.12 & 37). Syn: *icosacanthic law.*

Müller's glass *hyalite.*

mullicite A variety of vivianite occurring in cylindrical masses.

mullion A columnar structure in folded sedimentary and metamorphic rocks in which columns of rock form a coarse lineation. Mullions may be formed parallel to the direction of movement, as along fault planes, or perpendicular to it, as in *fold mullions* and *cleavage mullions* (Wilson, 1961, p. 510). Cf: *rodding.*

mullion structure A wavelike pattern of parallel grooves and ridges, measuring as much as several feet from crest to crest, and formed on a folded surface or along a fault surface. Etymol: Old French *moienel,* "medial". Cf: *striation; groove [fault]; slickenside; slip-scratch.*

mullite A rare orthorhombic mineral: $Al_6Si_2O_{13}$. Synthetic mullite is a valuable refractory material. Syn: *porcelainite.*

mullitization The formation of mullite from minerals of the sillimanite group by heating.

mullock *waste rock.*

multiband A nonrecommended syn. of *multispectral.*

multichannel *multispectral.*

multicycle adj. Said of a landscape or landform produced during or passing through more than one cycle of erosion, and bearing the traces of the former condition(s); e.g. a *multicycle* coast with a series of elevated sea cliffs separated from each other in stair-like fashion by narrow wave-cut benches, each sea cliff representing a separate shoreline cycle (Cotton, 1922, p. 426); or a *multicycle* valley showing on its sides a series of straths resulting from successive uplifts. Syn: *multicyclic; multiple-cycle, polycyclic.*

multifossil range-zone A lithostratigraphic unit marked by a concentration of the *range-zones* of a number of different fossils (ACSN Rept. 5, 1957, p. 1884). See also: *concurrent-range zone.*

multigelation Often-repeated freezing and thawing (Washburn, 1956, p.838). See also: *regelation.*

multilocular Divided into or composed of many small chambers or vesicles; specif. said of a many-chambered test of a unicellular organism such as a foraminifer. See also: *polythalamous.*

multipartite map A *vertical-variability map* that shows the degree of distribution of one lithologic type within certain parts (such as the top, middle, and bottom thirds) of a given stratigraphic unit. The map was introduced by Forgotson (1954).

multiple biseries Two or more sets of coverplates in edrioasteroids, each of which forms an alternating biseries of pairs of plates, the pairs of each set alternating with those of the other sets to form an integrated system (Bell, 1976).

multiple-cycle *multicycle.*

multiple detectors Two or more seismic detectors whose combined outputs are fed into a single amplifier-recorder channel in order to reduce undesirable noise.

multiple-exposure method *isolation of outcrops.*

multiple fault *step fault.*

multiple glaciation The alternating advance and recession of glacier ice during the Pleistocene Epoch.

multiple intrusion Any type of igneous intrusion that has been produced by several injections separated by periods of crystallization. Chemical composition of the various injections is approximately the same. Cf: *composite intrusion.*

multiple reflection A seismic wave that has been reflected more than once. Syn: *secondary reflection.*

multiple tunnel One of a series of openings in the chamber of a fusulinid test, produced by resorption of the lower (adaxial) parts of septa.

multiple twin A twinned crystal that is formed by *repeated twinning.*

multiple working hypotheses The name given by Chamberlin (1897) to a method of "mental procedure" applicable to geologic studies, in which several rational and tenable explanations of a phenomenon are developed, coordinated, and evaluated simultaneously in an impartial manner.

multiplex n. A stereoscopic (anaglyphic) plotting instrument used in preparing topographic maps from aerial photographs.—v. To transmit several channels of seismic information over a single channel without crossfeed. Usually different input channels are sampled in sequence at regular intervals and the samples are fed into a single output channel. Digital tapes are sometimes multiplexed in this way (Sheriff, 1973, p. 148).

multiringed basin A large circular depression, hundreds of kilometers in diameter, with two or more zones of concentrically arranged mountains, e.g. Orientale basin on the Moon. Multiringed basins are probably formed by meteoroid impact.

multisaccate Said of pollen with more than two vesicles.

multiserial Arranged in or consisting of several or many rows or series; e.g. "multiserial ambulacrum" of an echinoid with pore pairs arranged in more than two longitudinal series, or said of a protist composed of numerous rows of cells or other structural features.

multispectral Said of a remote-sensing system that simultaneously records information at several wavelength regions of the electromagnetic spectrum. Multispectral images may be acquired with cameras, scanners, or antennas. Syn: *multichannel.* Nonrecommended syn: *multi-band.*

multisystem A set of phases more numerous than can coexist stably under any set of conditions, thus formally possessing, in the phase-rule sense, a negative number of degrees of freedom; the equilibrium relationships among the phases of a multisystem can be represented by an array of invariant points connected by univariant, bivariant, etc., equilibria, some of which will in general be stable and some metastable.

Multituberculata *Allotheria.*

multivariate Pertaining to, having, or involving two or more independent mathematical variables; e.g. "multivariate analysis" that separates and defines the effects of a number of statistically independent variables.

multivincular Said of a type of ligament of a bivalve mollusk (e.g. *Isognomon*) consisting of serially repeated elements of alivincular type.

multi-year ice *Old ice* up to 3 m thick that has survived at least two summers' melt. The hummocks are smooth, the ice is almost salt-free, and the color when bare is usually blue. The melt pattern consists of interconnecting *puddles* and a well-developed drainage system.

multopost Said of a process involving an igneous rock that occurs some time after consolidation of the magma, i.e. later than a *deuteric* process. Rarely used.

mundic A syn. of *pyrite.* Drillers often use "mundick" to designate pyrite.

muniongite A hypabyssal rock resembling *tinguaite,* being composed of alkali feldspar, nepheline, acmite, and sometimes cancrinite; a nepheline-rich *phonolite.* The name is for Muniong, New South Wales, Australia. Not recommended usage.

munro A Scottish term for a hill more than 900 m in height, separated from another by a "dip" of more than 150 m (Darling & Boyd, 1964, p. 23). Named after H.T. Munro, Scottish mountaineer.

Munsell color system A system of color classification that is applied to rocks and soils. Color is defined by its hue, value (brilliance), and chroma (purity).

muntenite A variety of amber from Rumania.

mural deposit A *cameral deposit* along the wedgelike extension of each septum attached to the wall of a nautiloid conch.

mural escarpment A rocky cliff with a face nearly vertical, like a wall (Lee, 1840, p. 350).

muralite A *phyteral* of coal that represents the structure of plant cell walls and that occurs in some types of vitrain.

mural joint structure A pattern of cubical or rectangular blocks of rock formed by numerous right-angle joints. Syn: *rectangular joint structure.*

mural plate *compartmental plate.*

mural pore (a) A small hole in the wall between adjoining corallites, as in some tabulates. (b) An opening in the shell wall of a foraminifer, as distinguished from a *septal pore.* (c) A *communication pore* in a bryozoan.

mural rim The line of attachment of the frontal membrane to the vertical walls or the inner margin of the gymnocyst in anascan cheilostomes (bryozoans). It commonly forms a ridge, bears marginal spines, or borders the outer margin of a cryptocyst.

murambite A leucite *basanite* containing abundant mafic minerals, named by Holmes in 1936. Not recommended usage.

murataite A cubic mineral: $(Na,Y)_4(Zn,Fe)_3(Ti,Nb)_6O_{18}(F,OH)_4$.

murbruk structure *mortar structure.*

murchisonite (a) A flesh-red perthitic variety of orthoclase with good cleavage and often gold-yellow reflections in a direction perpendicular to (010). (b) A name applied to moonstone and the iridescent feldspar from Frederiksvaern, Norway.

Murderian Stage in New York State: Upper Silurian (upper Cayugan; above Canastotan).

murdochite A black isometric mineral: $PbCu_6O_8$.

muri The walls of the positive reticulate sculpture in pollen and spores.

murite A dark-colored feldspathoid-rich *phonolite* in which the mafic minerals comprise about 50 percent of the rock. Named by Lacroix in 1927 for Cape Muri, Rarotonga, Cook Islands. Not recommended usage.

murmanite A violet mineral: $Na_2(Ti,Nb)_2Si_2O_9 \cdot nH_2O$. Cf: *lomonosovite.*

muromontite A mineral: $Be_2FeY_2(SiO_4)_3$ (?). It is perhaps identical with gadolinite or is a member of the clinozoisite group.

murram Deposits of *bog iron ore* in the tropics of Africa.

murus reflectus A sutural indentation of the apertural face of a foraminiferal test, longitudinally and obliquely folded below the aperture (as in *Osangularia*). Etymol: Latin.

Muschelkalk European stage (esp. in Germany): Middle Triassic (above Bunter, below Keuper).

muscle field (a) An area of a brachiopod valve where muscle scars are concentrated. (b) A concave or flat area on the ventral (inner) side of an articular face of a muscularly articulated plate of a crinoid ray, serving for the attachment of muscle fibers.

muscle platform A relatively broad and solid or undercut elevation of the inner surface on either valve of some brachiopods, to which muscles are attached. Syn: *platform [paleont].*

muscle scar (a) One of the differentiated, more or less well-defined impressions or elevations on the inner surface of a bivalve shell (as in an ostracode, rostroconch, brachiopod, or pelecypod), or on a bone, marking the former place of attachment of a muscle; e.g. an *adductor muscle scar.* Syn: *muscle mark; scar.* (b) Smooth or slightly depressed paired areas in the external surface of the axial region of a trilobite exoskeleton, interpreted as areas of muscle attachment (TIP, 1959, pt.O, p.123).

muscle track The path of successive muscle impressions formed in a brachiopod, rostroconch, or pelecypod by migration of the muscle base during growth. Syn: *track [paleont].*

muscovadite A cordierite-biotite *norite,* formed as a result of the partial absorption of fragments of the country rock by the magma. Named by A. N. Winchell in 1900. Obsolete.

muscovado A term applied in Minnesota to rusty-colored outcropping rocks, such as gabbros and quartzites, that resemble brown sugar. Etymol: Spanish, "brown sugar".

muscovite (a) A mineral of the mica group: $KAl_2(AlSi_3)O_{10}(OH)_2$. It is colorless to yellowish or pale brown, and is a common mineral in gneisses and schists, in most acid igneous rocks (such as granites and pegmatites), and in many sedimentary rocks (esp. sandstones). See also: *sericite.* Also spelled: *moscovite.* Syn: *white mica; potash mica; common mica; Muscovy glass; mirror stone.* (b) A term applied in clay mineralogy to *illite.*

Muscovy glass *muscovite.*

muscular articulation A type of articulation of crinoid ossicles effected by muscle fibers in addition to ligaments. Cf: *ligamentary articulation.*

mush *brash ice.*

mush frost *pipkrake.*

mushroom ice An *ice pedestal* with a round and expanded top. See also: *ice pillar.*

mushroom rock A tablelike rock mass formed by wind abrasion or differential weathering in an arid region, consisting of an upper layer of resistant rock underlain by a softer, partially eroded layer, thereby forming a thin "stem" supporting a wide mass of rock, the whole feature resembling a mushroom in shape. See also: *pedestal rock; cheesewring; zeuge; gara.* Syn: *toadstool rock; fur-cap rock.*

musical sand A *sounding sand* that emits a definite musical note or tone when stirred, trodden on, or otherwise disturbed; esp. *whistling sand.*

muskeg (a) A bog, usually a *sphagnum bog,* frequently with grassy *tussocks,* growing in wet, poorly drained boreal regions, often areas of permafrost. Tamarack and black spruce are commonly associated with muskeg areas. Syn: *maskeeg.* (b) A term sometimes used in Michigan for a *bog lake.*

muskoxite A mineral: $Mg_7Fe_4{}^{+3}O_{13} \cdot 10H_2O$.

mussel (a) Any of the common freshwater pelecypods belonging to the superfamily Unionacea. (b) *mytilid.*

mustard gold A spongy type of free gold found in the gossan above gold-silver-telluride deposits.

mutant The offspring bearing a *mutation.*

mutation A spontaneously occurring, fundamental change in heredity, which results in the development of new individuals that are genetically unlike their parents and therefore can be acted upon by natural selection to effect desirable changes and eventually to establish new species. Mutations are now thought to be chemical changes in the DNA of a chromosome; some are visible, but most are not; many are deleterious. Mutations are the raw material of evolution. See also: *mutant.*

mute v. To change the relative contribution of the components of a record *stack* with record time. The long offset traces in the early part of the record may be muted or excluded because they are dominated by refraction arrivals or because their frequency content after correction for normal moveout is appreciably lower than that of other traces. Muting may also be done over certain time intervals to keep ground roll, air waves, or noise bursts out of the stack. Cf: *diversity stack.*

muthmannite A gray-white mineral: (Ag,Au)Te.

mutual-boundary stratotype A *boundary stratotype* that serves as the top of one stage and the bottom of the next younger stage (ISG, 1976, p. 84-86).

mutual inductance The complex ratio of voltage in one circuit to current in another to which it is inductively coupled.

mutualism A relationship between two organisms in which both are benefitted. Cf: *commensalism; symbiosis.*

MWL *mean water level.*

mycelium The entire filamentous vegetative growth of fungi, composed of *hyphae.*

mycobiont The fungal partner, or component, of a lichen. Cf: *phycobiont.*

mycorrhiza A symbiotic association of fungus with a root and/or rhizome of a higher plant.

mylonite As introduced by Lapworth in 1885, a compact, chertlike rock without cleavage, but with a streaky or banded structure, produced by the extreme granulation and shearing of rocks that have been pulverized and rolled during overthrusting or intense dynamic metamorphism. Mylonite may also be described as a microbreccia with flow texture (Holmes, 1920). See also: *protomylo-*

nite; ultramylonite; blastomylonite.

mylonite gneiss A metamorphic rock that is intermediate in character between mylonite and schist. Felsic minerals show cataclastic phenomena with little or no recrystallization, and commonly occur as *augen* surrounded by and alternating with schistose streaks and lenticles of recrystallized mafic minerals (Holmes, 1928, p.164).

mylonitic structure A structure characteristic of mylonites, produced by intense microbrecciation and shearing which gives the appearance of a flow structure. Cf: *flaser structure.*

mylonitization Deformation of a rock by extreme microbrecciation, due to mechanical forces applied in a definite direction, without noteworthy chemical reconstitution of granulated minerals. Characteristically the mylonites thus produced have a flinty, banded, or streaked appearance, and undestroyed augen and lenses of the parent rock in a granulated matrix (Schieferdecker, 1959). Also spelled: *mylonization.*

mylonization *mylonitization.*

myocyte A fusiform contractile cell in sponges.

myodocope Any ostracode belonging to the order Myodocopida, characterized by a shell with subequal valves that may be ornamented or smooth and by a well developed rostrum. Most planktonic marine ostracodes are myodocopes. Range, Ordovician to present.

myophore A part of a shell adapted for the attachment of a muscle; e.g. a process for attachment of an adductor muscle of a pelecypod, or the distal expanded part of the differentiated cardinal process of a brachiopod to which the diductor muscles were attached.

myophragm A median ridge of secondary shell of a brachiopod, secreted between muscles and not extending beyond the muscle field.

myostracum The part of the shell wall of a mollusk secreted at the attachments of the muscles or mantle. Syn: *hypostracum.*

myriapod Any terrestrial arthropod belonging to the superclass Myriapoda, which includes insects, centipedes, and millipedes. They are characterized by a body that is divided into a head and trunk with one pair of antennae on the head and uniramous appendages. They are rarely preserved as fossils but are known from the Upper Silurian to the present.

myrickite (a) A white or gray chalcedony, opal, or massive quartz unevenly colored by or intergrown with pink or reddish inclusions of cinnabar, the color of which tends to become brown. The opal variety is known as *opalite.* (b) Cinnabar intergrown with common white opal or translucent chalcedony.

myrmekite An intergrowth of plagioclase feldspar (generally oligoclase) and vermicular quartz, generally replacing potassium feldspar, formed during the later stages of consolidation in an igneous rock or during a subsequent period of plutonic activity (Barker, 1970). The quartz occurs as blobs, drops, or vermicular shapes within the feldspar.

myrmekite-antiperthite A myrmekitelike intergrowth of predominant plagioclase and vermicular orthoclase (Schieferdecker, 1959, term 5177).

myrmekite-perthite A myrmekitelike intergrowth of microcline and vermicular plagioclase (Schieferdecker, 1959, term 5176).

myrmekitic (a) Said of a *symplectic* texture characterized by intergrowths of feldspar and vermicular quartz. (b) Pertaining to or characteristic of *myrmekite.*

mytilid Any bivalve mollusk belonging to the family Mytilidae, characterized by an equivalve, inequilateral shell with prosogyrate umbones. Syn: *mussel.*

mytiliform Said of a slipper-shaped shell of a bivalve mollusk; specif. shaped like a marine mussel shell, such as the elongated and equivalve shell of *Mytilus* (a genus of marine bivalve mollusks).

myxomycete An organism of the class Myxomycetes, commonly called the slime molds, of uncertain systematic position but usually associated with the fungi. It exists as complex, mobile plasmodia and reproduces by spores. Cf: *eumycete; schizomycete.*

myxosponge Any *demosponge* whose only skeleton is mesogloea. It is without spicules or spongin.

nab A British term for a projecting part of an eminence; e.g. a headland or promontory (a *ness*), or a spur of an escarpment.

nacre The hard, iridescent internal layer of various mollusk shells, having unusual luster and consisting chiefly of calcium carbonate in the form of aragonite deposited as thin tablets normal to the surface of the shell and interleaved with thin sheets of organic matrix. Cf: *calcitostracum; treppen.* Syn: *mother-of-pearl.*

nacreous luster A type of mineral luster resembling that of mother-of-pearl. Syn: *pearly luster.*

nacrite A well-crystallized clay mineral of the kaolin group: $Al_2Si_2O_5(OH)_4$. It is polymorphous with kaolinite and dickite. Nacrite is structurally distinct from other members of the kaolin group, being the most closely stacked in the *c*-axis direction.

NAD *North American datum.*

nadir (a) The point on the celestial sphere that is directly beneath the observer and directly opposite the *zenith.* (b) The point on the ground vertically beneath the perspective center of an aerial-camera lens.

nadorite A brownish-yellow mineral: $PbSbO_2Cl$.

naëgite A variety of zircon containing thorium and uranium. Also spelled: *naegite.*

nafud *nefud.*

nagatelite A phosphatian variety of allanite.

nagelfluh A massive and variegated Miocene conglomerate accompanying the *molasse* of the Alpine region in Switzerland. It contains pebbles that appear like flights or swarms of nailheads. Etymol: German *Nagel,* "nail", + *Fluh,* "mass of rock, stratum, layer". Syn: *gompholite.*

nagyagite A dark lead-gray mineral: $Pb_5Au(Te,Sb)_4S_{5\text{-}8}$. Syn: *black tellurium; tellurium glance.*

nahcolite A white monoclinic mineral: $NaHCO_3$.

naif Said of a gemstone having a true or natural luster when uncut, e.g. of the natural, unpolished faces of a diamond crystal. Syn: *naife.*

nailhead spar A variety of calcite in crystals showing a combination of hexagonal prisms with flat rhombohedrons.

nailhead striation A glacial striation with a definite or blunt head or point of origin, generally narrowing or tapering in the direction of ice movement and coming to an indefinite end. Syn: *nailhead scratch.*

Nain province A Precambrian province in the Canadian Shield (Stockwell, 1964).

nakauriite An orthorhombic mineral: $Cu_8(SO_4)_4(CO_3)(OH)_6 \cdot 48H_2O$.

naked flagellar field The area around the flagella in which no coccoliths are present in coccolithophores lacking a complete cover of coccoliths. Cf: *covered flagellar field.*

naked karst Karst that is developed in a region without soil cover, so that its topographic features are well exposed. Syn: *bare karst.* See also: *covered karst.*

naked pole The end in a nonflagellate coccolithophore that is free of coccoliths.

nakhlite An achondritic stony meteorite consisting of a holocrystalline aggregate of diopside (75%) and olivine.

naled A Russian term for *icing; aufeis.* Pl: *naledi.*

nallah *nullah.*

nambulite A mineral of the pyroxenoid group: $NaLiMn_8Si_{10}O_{28}(OH)_2$.

Namurian European stage: Lower and Upper Carboniferous (above Viséan, below Westphalian). It is divided into a lower stage (Lower Carboniferous or Upper Mississippian) and an upper stage (Upper Carboniferous or Lower Pennsylvanian).

nanism The development of abnormally or exceptionally small size; dwarfishness. Cf: *gigantism.*

nanlingite A trigonal mineral: $CaMg_3(AsO_3)_2F$.

nannofossil (a) A collective term for fossil *discoasters* and *coccoliths,* both primarily calcareous microfossils, mostly rather near the limit of resolution of the light microscope and hence best studied with electron microscopy. (b) A term sometimes used in a more general sense for other extremely small marine (usually algal) fossils, smaller than *microfossils.*

nanno ooze Unconsolidated pelagic deposit, consisting of more than 69% $CaCO_3$, in which nannoplankton make up more than 30% of the recognizable skeletal remains.

nannoplankton Passively floating unicellular organisms less than about 35 microns in diameter. They are larger than *ultraplankton* but smaller than *microplankton, macroplankton,* and *megaloplankton.*

nanotesla 10^{-9} *tesla,* the SI unit for magnetic-field strength equivalent to the *gamma* in the cgs system.

nanozooid A small polymorph with a single tentacle and no digestive system in a few stenolaemate bryozoans.

Nansen bottle A device used in oceanography to obtain subsurface seawater samples and in-situ temperature measurements. It is a bottle, open at both ends, to which are attached a pair of *reversing thermometers.* When the device is in place, it is reversed, which encloses the water sample and records its temperature. Nansen bottles are usually used in a series on a line.

Nansen cast The use of a series of *Nansen bottles* to obtain seawater samples and measurements. Cf: *hydrographic cast.*

nant A little valley with a stream. Etymol: Celtic, "brook".

nantokite A colorless, white, or grayish isometric mineral: CuCl.

naotic septum A rugose corallite septum characterized by development peripherally in a series of closely spaced dissepimentlike plates, as in *Naos.*

NAP *nonarborescent pollen.*

naphtha An archaic term for liquid petroleum. It is now used to designate those hydrocarbons of the lowest boiling point (under 250°C) that are liquid at standard conditions, but easily vaporize and become inflammable. They are used as cleaners and solvents.

naphthalene A white crystalline bicyclic aromatic hydrocarbon, formula $C_{10}H_8$, that has a characteristic odor and occurs in coal tar and some crude oils.

naphthene *cycloparaffin.*

naphthene-base crude *asphalt-base crude.*

naphthine A syn. of *hatchettine.* Also spelled: *naphtine; naphtein.*

napoleonite *corsite.*

Napoleonville North American (Gulf Coast) stage: Miocene (above Anahauc, below Duck Lake).

nappe [hydraul] A sheet of water overflowing a dam.

nappe [struc geol] A sheetlike, allochthonous rock unit, which has moved on a predominantly horizontal surface. The mechanism may be thrust faulting, recumbent folding, or both. The term was first used as "nappe de recouvrement" (Schardt, 1893) for the large allochthonous sheets of the western Alps, and it has been adopted into English. The German equivalent, *Decke,* is also sometimes used in English. Etymol: French, "cover sheet, tablecloth". See also: *klippe.*

nappe [volc] *lava flow.*

nappe outlier *klippe.*

nari A variety of *caliche* that forms by surface or near-surface alteration of permeable calcareous rocks (dissolution and redeposition of calcium carbonate) and that occurs in the drier parts of the Mediterranean region. It is characterized by a fine network of veins surrounding unreplaced remnants of the original rock, and it often contains clastic particles (rocks and shells). Etymol: Arabic *nar,* "fire", in allusion to its use in limekilns.

Narizian North American stage: Middle Eocene (above Ulatisian, below Refugian).

narrow A constricted section of a mountain pass, valley, or cave; a gap or narrow passage between mountains. Commonly used in the plural, e.g. the *Narrows* of New River.

narrow-band filter A filter that transmits only frequencies or wavelengths within a narrow band.

narsarsukite A yellow mineral: $Na_2(Ti,Fe)Si_4(O,F)_{11}$.

NASA National Aeronautics and Space Administration.

nasal tube A curved cylinder or prismatic tube in a phaeodarian radiolarian, embracing the central capsule on one side and a galea on the other side. Syn: *rhinocanna.*

nase *naze.*

nasinite A monoclinic mineral: $Na_4B_{10}O_{17} \cdot 7H_2O$. It is dimorphous with ezcurrite.

nasledovite A mineral: $PbMn_3Al_4(CO_3)_4(SO_4)O_5 \cdot 5H_2O$.

nasonite A white mineral: $Ca_4Pb_6Si_6O_{21}Cl_2$.

Na-spar (a) *sodium feldspar.* (b) *soda spar.*

nasselline Any radiolarian belonging to the suborder Nassellina, characterized by a central capsule perforated only at one pole and enclosed by a single membrane.

nasturan *pitchblende.*

national meridian A meridian chosen in a particular nation as the reference datum for determining longitude for that nation. It is commonly defined in several European countries with respect to a key point in the capital city. Cf: *Greenwich meridian.*

natisite A tetragonal mineral: $Na_2(TiO)(SiO_4)$.

native *endemic.*

native asphalt Liquid or semiliquid asphalt in exudations or seepages including surface flows and lakes. Syn: *natural asphalt.*

native coke *natural coke.*

native element Any element found uncombined in a nongaseous state in nature. Nonmetallic examples are carbon, sulfur, and selenium; semimetal examples are antimony, arsenic, bismuth, and tellurium; *native metals* include silver, gold, copper, iron, mercury, iridium, lead, palladium, and platinum.

native metal A metallic *native element.*

native mud The aqueous suspension consisting of pulverized *well cuttings* without additives, produced by the bit when drilling a rotary borehole.

native paraffin *ozocerite.*

native water (a) *connate water.* (b) *formation water.*

natric horizon A diagnostic subsurface soil horizon that has the same properties as an argillic horizon but also displays a blocky, columnar, or prismatic structure and has a subhorizon with an exchangeable-sodium percentage of over 15 (USDA, 1975).

natroalunite A mineral of the alunite group: $NaAl_3(SO_4)_2(OH)_6$. It is isomorphous with alunite. Syn: *almeriite.*

natroborocalcite A syn. of *ulexite.* Also spelled: *natronborocalcite.*

natrochalcite An emerald-green mineral: $NaCu_2(SO_4)_2(OH) \cdot H_2O$.

natrofairchildite A mineral: $Na_2Ca(CO_3)_2$. It may be identical with nyerereite.

natrojarosite A yellowish-brown to golden-yellow mineral of the alunite group: $NaFe_3(SO_4)_2(OH)_6$. Syn: *utahite.*

natrolite A zeolite mineral: $Na_2Al_2Si_3O_{10} \cdot 2H_2O$. It sometimes contains appreciable calcium, and usually occurs in slender, acicular or prismatic crystals. Partial syn: *mesotype [mineral]; needle zeolite.*

natromontebrasite A mineral: $(Na,Li)AlPO_4(OH,F)$. It is isomorphous with amblygonite and montebrasite. Syn: *fremontite.*

natron A white, yellow, or gray monoclinic mineral: $Na_2CO_3 \cdot 10H_2O$. It is very soluble in water, and occurs mainly in solution (as in the soda lakes of Egypt and the western U.S.) or in saline residues.

natroniobite A monoclinic mineral: $NaNbO_3$. It is dimorphous with lueshite.

natron lake *soda lake.*

natrophilite A mineral: $NaMn(PO_4)$.

natrophosphate A cubic mineral: $Na_6H(PO_4)_2F \cdot 17H_2O$.

natrosilite A monoclinic mineral: $Na_2Si_2O_5$.

natural arch (a) A *natural bridge* resulting from erosion. (b) A landform similar to a natural bridge but not formed by erosive agencies (Cleland, 1910, p. 314). (c) *sea arch.*—Syn: *arch [geomorph].*

natural area (a) An area of land or water that has retained its wilderness character, although not necessarily completely natural and undisturbed, or that has rare or vanishing flora, fauna, archaeological, scenic, historical, or similar features of scientific or educational value (Ohio Legislative Service Commission, 1969, p. 3); e.g. a "research natural area" where "natural processes are allowed to predominate and which is preserved for the primary purposes of research and education" (U.S. Federal Committee on Research Natural Areas, 1968, p. 2). (b) Any outdoor site that contains an unusual biologic, geologic, or scenic feature or that illustrates "common principles of ecology uncommonly well" (Lindsey et al., 1969, p. 4).—See also: *wilderness area.*

natural asphalt *native asphalt.*

natural bridge (a) Any archlike rock formation created by erosive agencies and spanning a ravine or valley; an opening found where a stream abandoned a meander and broke through the narrow meander neck, as at Rainbow Bridge, Utah. (b) In a limestone terrane, the remnant of the roof of an underground cave or tunnel that has collapsed. Syn: *karst bridge.* (c) *sea arch; natural arch.* See also: *lighthouse.*

natural brine *brine [geol].*

natural coke Coal that has been naturally carbonized by contact with or proximity to an igneous intrusion, or by natural combustion. Syn: *carbonite; coke coal; cokeite; native coke; finger coal; blind coal; cinder coal.* Cf: *clinker [coal]; coke.*

natural gas (a) Hydrocarbons that exist as a gas or vapor at ordinary pressures and temperatures. Methane is the most important, but ethane, propane, and others may be present. Common impurities include nitrogen, carbon dioxide, and hydrogen sulfide. Natural gas may occur alone or associated with oil. Syn: *gas.* (b) Gaseous hydrocarbons trapped in the zone of ground-water saturation, under pressure from, and partially dissolved in, underlying water or petroleum (Meinzer, 1923, p. 21). Cf: *subsurface air; included gas.*

natural-gas liquids Hydrocarbons that occur naturally in gaseous form or in solution with oil in the reservoir, and that are recoverable as liquids by condensation or absorption; e.g. *condensate* and *liquefied petroleum gas.*

natural gasoline *condensate.*

natural glass A vitreous, amorphous, inorganic substance that has solidified from magma too quickly to crystallize. Granitic or acid natural glass includes pumice and obsidian; an example of a basaltic natural glass is tachylite.

natural history The study of the nature and history of all animal, vegetable, and rock and mineral forms. The term is sometimes considered old-fashioned except when applied to the animal world; but it is still used by some in its geologic sense, e.g. in environmental geology.

natural horizon *apparent horizon.*

natural landscape A landscape that is unaffected by the activities of man (in contrast to the "cultural landscape" resulting from man's settlement); it includes landforms and their natural plant cover, and the contrast between land and water. Syn: *physical landscape.*

natural law *law of nature.*

natural levee (a) A long broad low ridge or embankment of sand and coarse silt, built by a stream on its flood plain and along both banks of its channel, esp. in time of flood when water overflowing the normal banks is forced to deposit the coarsest part of its load. It has a gentle slope (about 60 cm/km) away from the river and toward the surrounding flood plain, and its highest elevation (about 4 m above the flood plain) is closest to the river bank, at or near normal flood level. Syn: *levee; raised bank; spill bank.* (b) Any naturally produced low ridge resembling a natural levee; e.g. a lava levee, or a sediment ridge bordering a fan-valley.

natural load The quantity of sediment that a stable stream carries.

natural mold The empty space or cavity left after solution of an original fossil shell or other organic structure, bounded by the external impression (external mold) and the surface of the internal filling (*steinkern*) (Shrock & Twenhofel, 1953, p.19). See also: *mold [paleont]; cast [paleont].*

natural region (a) A part of the Earth's surface characterized by relatively uniform and distinctive physical features (relief, structure, climate, vegetation) within its borders, and therefore possessing to a certain extent a uniformity in human activities. (b) A region that possesses a unity based on significant geographic characteristics (physical, biological, cultural), in contrast to an area marked out by boundaries imposed for political or administrative purposes.—The term provides a convenient regional basis for nomenclature and integration of the whole landscape.

natural remanence *natural remanent magnetization.*

natural remanent magnetism *natural remanent magnetization.*

natural remanent magnetization The entire *remanent magnetization* of a rock in situ. Abbrev: NRM. Syn: *natural remanence; natural remanent magnetism.*

natural resin An unmodified *resin* from a natural source (such as a tree) and distinguished from synthetic resin; e.g. a copal.

natural scale (a) The *scale* of a map, expressed in the form of a fraction or ratio, independent of the linear units of measure; specif. *representative fraction.* The term is not recommended. (b) True scale, as it exists in nature, without magnification or reduction.

natural selection The process by which less vigorous or less well-adapted individuals tend to be eliminated from a population, so that they tend to leave fewer descendants to perpetuate their stock. See also: *struggle for existence.* Syn: *selection.*

natural slope (a) The slope assumed by a mass of loose heaped-up material, such as earth. (b) *angle of repose.*

natural stone A gemstone that occurs in nature, as distinguished from a man-made substitute.

natural tunnel A cave that is nearly horizontal and that is open at both ends. It may contain a stream. Syn: *tunnel cave; tunnel [speleo].*

natural well A sinkhole or other natural opening resembling a well that extends below the water table and from which ground water can be withdrawn.

naujaite A coarse, hypidiomorphic-granular sodalite-rich *nepheline syenite* that contains microcline and small amounts of albite, analcime, acmite, and sodium amphiboles and is characterized by poikilitic texture; the sodalite is surrounded by the other phases. The rock was first described by Ussing in 1912 from Naujakasik, Ilimaussaq complex, on the southwest coast of Greenland. Not recommended usage.

naujakasite A silvery-white or grayish mineral: $Na_6(Fe^{+2},Mn)Al_4Si_8O_{26}$.

naumannite An iron-black isometric mineral: Ag_2Se.

naupliar eye An unpaired median eye appearing in the nauplius and retained in some mature stages.

nauplius A crustacean larva in the early stage after leaving the egg, having only three pairs of limbs (corresponding to antennules, antennae, and mandibles), a median (naupliar) eye, and little or no segmentation of the body. Pl: *nauplii.* Cf: *metanauplius.*

nautical chart *hydrographic chart.*

nautical distance The length in nautical miles of the *rhumb line* joining any two places on the Earth's surface.

nautilicone A strongly involute nautiloid conch (like that of *Nautilus*) coiled in a plane spiral with the outer whorls embracing the inner whorls.

nautiloid Any cephalopod belonging to one of the subclasses Nautiloidea, Endoceratoidea, or Actinoceratoidea, characterized by a centrally located siphuncle and by a straight, curved, or coiled chambered external shell with less elaborate sutural flexures than in ammonoids. Nautiloids, known today only from the genus *Nautilus,* reached their peak in the Ordovician and Silurian. Range, Upper Cambrian to present.—adj. Pertaining to Nautiloidea.

navajoite A dark-brown mineral: $V_2O_5 \cdot 3H_2O$.

Navarroan North American (Gulf Coast) stage: Upper Cretaceous (above Tayloran, below Midwayan).

Navier-Stokes equations Equations of motion for a viscous fluid.

navite A dark-colored porphyritic igneous rock containing phenocrysts of calcic plagioclase (labradorite) and olivine frequently altered to iddingsite, and some augite and rare enstatite in a holocrystalline groundmass composed chiefly of feldspar and minor second-generation augite. The name, given by Rosenbusch in 1887, is after Nave, Germany. Obsolete.

naze A promontory or headland; a *ness.* Etymol: perhaps from the Naze, a promontory in Essex, England. Syn: *nase.*

neanic Said of a youthful or immature growth stage of an organism; the stage following the *nepionic* stage and preceding the *ephebic stage.*

neap tide A tide occurring at the first and third quarters of the Moon when the gravitational pull of the Sun opposes (or is at right angles to) that of the Moon, and having an unusually small or reduced tide range (usually 10-30% less than the mean range). Cf: *spring tide.*

near earthquake An earthquake whose epicenter is within 1000-1200 km of the detector.

near infrared *reflected infrared.*

nearshore Extending seaward or lakeward an indefinite but generally short distance from the shoreline; specif. said of the indefinite zone extending from the low-water shoreline well beyond the breaker zone, defining the area of *nearshore currents,* and including the *inshore* zone and part of the *offshore* zone. Depths are generally less than 5 fathoms (10 m).

neat line The line that surrounds a map, separating the map from the margin. Also spelled: *neatline.*

Nebraskan Pertaining to the first classical glacial stage of the Pleistocene Epoch in North America, followed by the Aftonian interglacial stage. See also: *Günz.*

nebula Historically, any faintly luminous, diffuse object seen in the heavens. In modern usage, an interstellar cloud of gas or dust. The other diffuse objects are clusters of stars, or galaxies. Galaxies are still often referred to as extragalactic nebulae (Stokes & Judson, 1968, p. 512).

nebular hypothesis A model for the origin of the Universe (by Laplace in 1796) which supposes a rotating, primeval nebula of gas and dust which increased its rotation as it contracted. This led to a flattening of the mass and, as centrifugal forces exceeded gravity, ejections of matter from its equator. These castoffs formed planets around the original mass, the Sun. The model has been abandoned since it was discovered that the angular momentum of the Sun is too low.

nebulite A migmatite characterized by indistinct inclusions (schlieren or skialiths) (Dietrich & Mehnert, 1961). Not widely used.

neck [bot] The tapering apical portion of an *archegonium.*

neck [currents] The narrow band or "rip" of water forming the part of a rip current where *feeder currents* converge and flow swiftly through the incoming breakers or surf and out to the *head.*

neck [geog] (a) A narrow stretch or strip of land connecting two larger areas; e.g. the lowest part of a level mountain pass between two ridges, or a narrow isthmus joining a peninsula with the mainland. See also: *land bridge.* (b) Any narrow strip of land such as a cape, promontory, peninsula, bar, or hook. (c) *meander neck.*

neck [ore dep] *pipe [ore dep].*

neck [paleont] (a) The constricted anterior part of the body chamber in specialized brevicones between the flared aperture and the inflated portion (TIP, 1964, pt.K, p.57). (b) *septal neck.* (c) The often constricted, usually flexible part of the tetrapod body between head and shoulders; neck vertebrae are morphologically distinct from trunk vertebrae in all except snakelike forms.

neck [volc] A vertical, pipelike intrusion that represents a former volcanic *vent.* The term is usually applied to the form as an erosional remnant. Cf: *plug [volc].*

neck cutoff A meander *cutoff* formed where a stream breaks through or across a narrow meander neck, as where downstream migration of one meander has been slowed and the next meander upstream has overtaken it. Cf: *chute cutoff.*

neck ring *occipital ring.*

necrocoenosis *liptocoenosis.*

necronite A blue pearly variety of orthoclase that emits a fetid smell upon hammering. It occurs in limestone near Baltimore, Md.

necrophagous Said of an organism that feeds on dead matter.

Nectridia An order of lepospondylous amphibians characterized by fan-shaped neural and haemal spines on caudal vertebrae, elongate body form, and limb reduction. Range, Lower Pennsylvanian to Upper Permian.

needle [cryst] A needle-shaped or acicular mineral crystal.

needle [geol] A pointed, elevated, and detached mass of rock formed by erosion, such as an *aiguille* or a *stack.*

needle [surv] *magnetic needle.*

needle ice (a) *pipkrake.* (b) *frazil ice.* (c) *candle ice.*

needle ironstone A variety of goethite occurring in fibrous aggregates of acicular crystals. Syn: *needle iron ore.*

needle ore (a) Iron ore of very high metallic luster, found in small quantities, which may be separated into long slender filaments resembling needles. (b) *aikinite.*

needle stone (a) *needle zeolite.* (b) *hairstone.*—Also spelled: *needlestone.*

needle tin ore A variety of cassiterite with acute pyramidal forms.

needle zeolite A syn. of *mesotype [mineral];* specif. a syn. of *natrolite.* See also: *hair zeolite.* Syn: *needle stone.*

Néel point (a) The temperature at which the susceptibility of an antiferromagnetic mineral has a maximum. Above this point, thermal agitation prevents antiferromagnetic ordering. (b) The temperature at which thermal agitation overcomes magnetic order in a ferrimagnetic mineral. *Curie point* is also used with this meaning.—Syn: *Néel temperature*.

Néel temperature *Néel point*.

neftdegil *neft-gil*.

neft-gil A mixture of paraffins and a resin found in the Caspian area on Cheleken Island; it is related to *pietricikite*. Syn: *neftdegil*.

nefud (a) A deep or large sandy desert in Arabia, equivalent to an *erg*. Syn: *nafud*. (b) A high sand dune in the Syrian desert (Stone, 1967, p. 267).

negative [optics] (a) Said of anisotropic crystals: of a uniaxial crystal, in which the extraordinary index of refraction is greater than the ordinary index; and of a biaxial crystal in which the intermediate index of refraction β is closer to γ than to α. Cf: *positive [optics]*. (b) Said of a crystal containing a cavity, the form of which is one of the possible crystal forms of the mineral.

negative [photo] (a) A photographic image on film, plate, or paper that reproduces the bright parts of the subject as dark areas and the dark parts as light areas. Cf: *positive [photo]*. (b) A film, plate, or paper containing such an image.

negative area *negative element*.

negative confining bed The upper confining bed of an aquifer whose head is below the upper surface of the zone of saturation, i.e. below the water table. Little used.

negative delta A term used by Playfair (1802, p. 430) for an *estuary*.

negative element A large structural feature or portion of the Earth's crust, characterized through a long period of geologic time by frequent and conspicuous downward movement (subsidence, submergence) or by extensive erosion, or by an uplift that is considerably less rapid or less frequent than those of adjacent *positive elements*. Syn: *negative area*.

negative elongation In a section of an anisotropic crystal, a *sign of elongation* that is parallel to the faster of the two plane-polarized rays. Cf: *positive elongation*.

negative estuary *inverse estuary*.

negative landform A relatively depressed or low-lying topographic form, such as a valley, basin, or plain, or a volcanic feature formed by a lack of material (as a caldera). Ant: *positive landform*.

negative movement (a) A downward movement of the Earth's crust relative to an adjacent part of the crust, such as produced by subsidence; a negative movement of the land may result in a *positive movement* of sea level. (b) A relative lowering of the sea level with respect to the land, such as produced by a *positive movement* of the Earth's crust or by a retreat of the sea.

negative pole The south-seeking member of the *magnetic poles*. Cf: *positive pole*. See also: *dipole field*.

negative shoreline *shoreline of emergence*.

negative skin friction Frictional forces developed on the sides of a pile, or other structure embedded in soil, that tends to cause it to be dragged downward. It is often developed in a consolidating soil mass. See also: *skin friction*.

negative snowflake *Tyndall figure*.

negative strip *Vening Meinesz zone*.

negrohead *niggerhead*.

nehrung A long, narrow sandspit, sandbar, or barrier beach enclosing or partially enclosing a lagoon (*haff*), formed across a river mouth by longshore drifting of sand; esp. such a feature along the East German coast of the Baltic Sea. Pl: *nehrungs; nehrungen*. Etymol: German *Nehrung*, "sandbar, spit".

neighborite An orthorhombic mineral: $NaMgF_3$.

nejd A syn. of *hammada*. Var: *nijd*.

nek A term used in South Africa for a low place in a mountain range; a saddle or col. Etymol: Afrikaans, "neck".

nekoite A triclinic mineral: $Ca_3Si_6O_{15}\cdot 8H_2O$. Cf: *okenite*.

nektobenthos Those forms of marine life that live just above the ocean bottom and occasionally rest on it.

nekton Aquatic animals that are actively free-swimming, e.g. cephalopods, fish. Adj: *nektonic*.

nektonic Said of that type of *pelagic* organism which actively swims; adj. of *nekton*. Cf: *planktonic*.

nelsonite A group of hypabyssal rocks composed chiefly of ilmenite and apatite, with or without rutile. The ratio of ilmenite to apatite varies widely. Cf: *ilmenitite*. It was named by Watson in 1907 after Nelson County, Virginia.

nema A hollow threadlike prolongation of the apex of the prosicula of a graptolite. The term is used where the prolongation is "exposed", as in all except scandent rhabdosomes. Cf: *virgula*.

nemafite A general term for igneous rocks composed of *nep*heline and *maf*ic minerals, proposed by Hatch, Wells, and Wells in 1961. Syn: *nephelinite*.

nemalite A fibrous variety of brucite containing ferrous oxide.

nematath A term used by Carey (1958) for a submarine ridge across an Atlantic-type ocean basin, which is not an orogenic structure but is composed of continental crust that has been stretched across a *sphenochasm* or *rhombochasm*. Carey cites as an example the Lomonosov ridge that extends across the Arctic Ocean basin from North America to Asia.

nematoblastic Pertaining to a *homeoblastic* type of texture of a metamorphic rock due to the development during recrystallization of slender parallel prismatic crystals. Cf: *fibroblastic*.

nematocyst One of the minute stinging cells or organs of hydrozoans, scyphozoans, and anthozoans; e.g. a "thread cell" formed within a *cnidoblast* of a coral.

nenadkevichite An orthorhombic mineral: $(Na,Ca,K)(Nb,Ti)Si_2O_7\cdot 2H_2O$.

nenadkevite *coffinite*.

n-en log *epithermal neutron log*.

neoautochthon A stable basement or autochthon formed where a nappe has ceased movement and has become defunct. Cf: *paleoautochthon*.

neoblast A grain, esp. in a metamorphic rock, that is of more recent formation than other grains of the same or other mineral species in the rock. Some neoblasts consist wholly or in part of introduced material; others represent late-stage recrystallization of original rock components. Cf: *paleoblast*.

Neocene An obsolete syn. of *Neogene*.

Neocomian European stage: Lower Cretaceous (above Tithonian-Upper Volgian of Jurassic, below Aptian). It includes Berriasian (lowermost Cretaceous), Valanginian, Hauterivian, and Barremian (some authors omit Barremian).

neocryst An individual crystal of a secondary mineral in an evaporite (Greensmith, 1957). Cf: *evapocryst*.

neocrystallization Crystallization or recrystallization that involves the development of new minerals among the fabric elements.

neocrystic texture A secondary, nonlaminated texture of an evaporite.

neo-Darwinism *Darwinism* modified or rephrased in accordance with modern genetics.

neoformation *neogenesis*.

Neogene An interval of time incorporating the Miocene and Pliocene of the Tertiary period; the later Tertiary. When the Tertiary is designated as an era, then the Neogene, together with the *Paleogene*, may be considered to be its two periods. Obsolete syn: *Neocene*.

neogenesis The formation of new minerals, as by diagenesis or metamorphism. Cf: *authigenesis*. Syn: *neoformation*.

neogenic Said of newly formed minerals; pertaining to neogenesis.

neoglaciation The readvance of mountain glaciers during a *Little Ice Age* interval of the late Holocene (Moss, 1951, p. 62).

Neognathae A superorder of the avian subclass Neornithes which includes, in some 25 orders, all birds except ground-dwelling forms and toothed birds of the Mesozoic. See also: *Paleognathae*.

neoichnology The study of Holocene tracks, burrows, and other structures left by living organisms, as opposed to *palichnology*. See also: *ichnology*.

neokaolin Kaolinite artificially produced from nepheline.

neolensic texture A secondary, nonporphyritic, roughly laminated texture of an evaporite.

Neolithic n. In archaeology, the last division of the *Stone Age*, characterized by the development of agriculture and the domestication of farm animals. Correlation of relative cultural levels with actual age (and, therefore, with the time-stratigraphic units of geology) varies from region to region. Syn: *New Stone Age*.—adj. Pertaining to the Neolithic.

neomagma A magma formed by partial or complete fusion of preexisting rock under conditions of plutonic metamorphism. Cf: *anatexis; palingenesis*.

neomesselite *messelite*.

neomineralization Chemical interchange within a rock whereby its mineral constituents are converted into new mineral species; a type of *recrystallization*.

neomorphism An inclusive term suggested by Folk (1965, p. 20-21) for all transformations between one mineral and itself or a polymorph, whether the new crystals are larger or smaller or simply differ in shape from the previous ones, or represent a new mineral species. It includes the processes of inversion, recrystallization, and strain recrystallization, in which the gross composition remains essentially constant. The term is appropriate where it is not possible to distinguish between recrystallization and inversion, or where the mechanism of change is not known. See also: *aggrading neomorphism; degrading neomorphism.*

neontology The study of existing organisms, as opposed to *paleontology.* Approx. syn: *biology.* Cf: *cenozoology.*

Neophytic *Cenophytic.*

neoporphyrocrystic texture A texture of an evaporite in which large neocrysts are embedded in a finer-grained matrix.

Neornithes A subclass of birds that includes all except the Upper Jurassic genus *Archaeopteryx.*

neosome A geometric element of a composite rock or mineral deposit, appearing to be younger than the main rock mass (or *paleosome*); e.g. an injection in country rock, or the introduced or newly formed material of a migmatite. Sometimes used in place of the term *metatect, mobilizate,* or *chymogenic.* Cf: *metasome [geol].*

neospar Sparry calcite formed by *neomorphism* of earlier finer-textured carbonate, e.g. micrite (Nichols, 1967, p. 1247-1248). Syn: *pseudosparite.*

neostratotype A stratotype established after the holostratotype has been destroyed or is otherwise not usable (Sigal, 1964).

neotectonic map A map portraying post-Miocene geologic structures. Cf: *neotectonics.*

neotectonics The study of the post-Miocene structures and structural history of the Earth's crust.

neoteny (a) Arrested development such that youthful characteristics are retained by the adult organism. Syn: *paedomorphism; paedomorphosis; proterogenesis.* (b) Acceleration in the attainment of sexual maturity relative to general body development. Syn: *paedogenesis.*

Neothermal n. A term introduced by Antevs (1948, p.176) for the climatic interval from the end of the Wisconsinan glaciation to the present (approximately the past 10,000 years).—adj. Pertaining to the postglacial Neothermal interval and to its climate, deposits, biota, and events.

Neotian *Pannonian.*

neotocite A mineral consisting of a hydrous silicate of manganese and iron of uncertain formula. It may be an alteration product of rhodonite, possibly an opal with disseminated manganese and iron oxides.

neotype A single specimen designated as the *type specimen* of a species or subspecies when the *holotype* (or *lectotype*) and all paratypes or all syntypes have been lost or destroyed (ICZN, 1964, p. 150).

neovolcanic Said of extrusive rocks that are of Tertiary or younger age. Cf: *paleovolcanic.*

NEP *noise-equivalent power.*

nepheline A hexagonal mineral of the feldspathoid group: $(Na,K)AlSiO_4$. It occurs as glassy crystals or colorless grains, or as coarse crystals or green to brown masses of greasy luster without cleavage, in alkalic igneous rocks, and it is an essential constituent of some sodium-rich rocks. Syn: *nephelite; eleolite.*

nepheline basalt *olivine nephelinite.*

nepheline syenite A plutonic rock composed essentially of alkali feldspar and nepheline. It may contain an alkali ferromagnesian mineral, e.g. an amphibole (riebeckite, arfvedsonite, barkevikite) or a pyroxene (acmite or acmite-augite); the intrusive equivalent of *phonolite.* Sodalite, cancrinite, hauyne, and nosean, in addition to apatite, sphene, and opaque oxides, are common accessories. Rare minerals are also frequent accessories. Cf: *foyaite; ditroite; foid syenite.* Obsolete syn: *eleolite syenite; midalkalite.*

nephelinite A fine-grained or porphyritic extrusive or hypabyssal rock, of basaltic character, but primarily composed of nepheline and clinopyroxene, esp. titanaugite, and lacking olivine and feldspar. Cf: *olivine nephelinite.* Syn: *nemafite.*

nephelinitoid A nepheline-rich groundmass in an igneous rock; the glassy groundmass in nepheline rocks.

nephelite *nepheline.*

nepheloid layer A layer of water in the deep ocean basin that contains significant amounts of suspended sediment. It is from 200 m to 1000 m thick.

nephelometer An instrument used in nephelometry, designed to measure the amount of cloudiness of a medium.

nephelometry The measurement of the cloudiness of a medium; esp. the determination of the concentration or particle sizes of a suspension by measuring, at more than one angle, the scattering of light transmitted or reflected by the medium. Cf: *turbidimetry.*

nephlinolith An extrusive igneous rock composed entirely of nepheline.

nephrite An exceptionally tough, compact, fine-grained, greenish or bluish amphibole (specif. tremolite or actinolite) constituting the less rare or valuable kind of jade and formerly worn as a remedy for kidney diseases. Syn: *kidney stone; greenstone.*

nepioconch The earliest-formed part of a dissoconch of a bivalve mollusk, separated from the later part by a pronounced discontinuity. Cf: *mesoconch.*

nepionic Said of the stage or period in which the young shell of an invertebrate does not yet show distinctive specific characteristics, i.e. of the stage following the *embryonic* stage and preceding the *neanic* stage.

nepouite A monoclinic mineral: $Ni_3Si_2O_5(OH)_4$. It is a member of the kaolinite-serpentine group of two-layer phyllosilicates.

nepton A term used by Makiyama (1954) for a body of sedimentary rock filling a basin; e.g. a *geosynclinal prism.*

Neptune's racetrack *eddy-built bar.*

neptunian adj. (a) Pertaining to neptunism and the rocks whose origin was explained by it. (b) Formed by the agency of water.—n. *neptunist.*

neptunian dike A sedimentary dike formed by infilling of sediment, generally sand, in an undersea fissure or hollow. Cf: *injection dike.*

neptunic rock (a) A rock formed in the sea. (b) A general term proposed by Read (1944) for all sedimentary rocks. Cf: *plutonic rock; volcanic rock.*

neptunism The theory, advocated by A. G. Werner and long since obsolete, that the rocks of the Earth's crust all consist of material deposited sequentially from, or crystallized out of, water. Etymol: Neptune, Roman god of waters. See also: *Wernerian.* Ant: *plutonism.* Syn: *neptunianism; neptunian theory.*

neptunist A believer in the theory of neptunism. Ant: *plutonist.* Syn: *neptunian.*

neptunite A black mineral: $Na_2KLi(Fe,Mn)_2Ti_2Si_8O_{24}$. Cf: *mangan-neptunite.*

nereite A *trace fossil* of the "genus" *Nereites,* consisting of a meandering feeding trail 1 to 2 cm wide with a narrow central axis and regularly spaced lateral, leaf-shaped, or lobelike projections, and formed perhaps by a worm or gastropod.

neritic Pertaining to the ocean environment or *depth zone* between low-tide level and 100 fathoms, or between low-tide level and approximately the edge of the continental shelf; also, pertaining to the organisms living in that environment. It is called by some the *sublittoral zone*, i.e. is considered to be a part of the *littoral* zone.

Nernst distribution law The statement that the ratio of the molar concentration of a substance dissolved in two immiscible liquids is constant and depends only on temperature. The ratio is called the *partition coefficient.*

nesophitic Said of the ophitic texture of igneous rocks, especially diabases and some gabbros, in which pyroxene is interstitial to plagioclase and occurs in isolated areas (Walker, 1957, p.2). Cf: *sporophitic.* Syn: *diabasic.*

nesosilicate A class or structural type of *silicate* characterized by isolated SiO_4 tetrahedra, rather than by linkage of tetrahedra by the sharing of common oxygens. An example of a nesosilicate is olivine, $(Mg_2SiO_4{-}Fe_2{}^{+2}SiO_4)$. Cf: *sorosilicate; cyclosilicate; inosilicate; phyllosilicate; tectosilicate.*

nesquehonite A colorless or white mineral: $Mg(HCO_3)(OH)\cdot 2H_2O$. It occurs in radiating groups of prismatic crystals.

ness A British term used esp. in Scotland for a *promontory,* headland, or cape, or any point or projection of the land into the sea; commonly used as a suffix to a place name, e.g. Fife*ness.* Syn: *naze; nose; nore; nab.*

nest A concentration of some relatively conspicuous element of a geologic feature, such as a "nest" of pebbles in a sand layer or inclusions in an igneous rock; esp. a small, pocketlike mass of ore or mineral within another formation.

nested (a) Said of volcanic cones, craters, or calderas that occur one within another, i.e. show *cone-in-cone structure [volc].* (b) Said of two or more calderas that intersect, having been formed at different times or by different explosions.

nested sinkholes *karst valley.*

net [pat grd] A form of horizontal patterned ground whose mesh is intermediate between a *circle* and a *polygon.* See: *sorted net; nonsorted net.*

net [struc petrol] In structural petrology, a stereographic or an equal-area projection of a sphere in which the network of meridians and parallels forms a coordinate system. The meridians and parallels are normally projected at 2° intervals and the net is used to plot points that represent lineations, the normals to foliations, or crystallographic directions. Syn: *projection net; stereographic net.*

net [surv] A series of surveying or gravity stations that have been interconnected in such a manner that closed loops or circuits have been formed or that are so arranged as to provide a check on the consistency of the measured values; e.g. a *base net* and a *triangulation net.* Syn: *network.*

net ablation A nonrecommended term with various meanings, such as *summer balance* and net balance of the ablation area.

net accumulation A nonrecommended term with various meanings, such as *winter balance* and *net balance* of the accumulation area of a glacier.

net balance The change in mass of a glacier from the time of minimum mass in one year to the time of minimum mass in the succeeding year (*balance year*); the mass change between one *summer surface* and the next. It can be determined at a point, as an average for an area, or as a total mass change for the glacier. Units of millimeters, meters, or cubic meters of water equivalent are generally used. Syn: *net budget.* Cf: *annual balance; balance.*

net budget *net balance.*

net calorific value A *calorific value* calculated from *gross calorific value* under conditions such that all the water in the products remains in the form of vapor. Syn: *net heat of combustion.*

net heat of combustion *net calorific value.*

net plankton *microplankton.*

net primary production The amount of organic matter produced by living organisms within a given volume or area in a given time, minus that which is consumed by the respiratory processes of the organisms. Cf: *gross primary production.*

net slip On a fault, the distance between two formerly adjacent points on either side of the fault, measured on the fault surface or parallel to it. It defines both the direction and relative amount of displacement. Syn: *total slip.*

net texture A network system of nickel sulfides in peridotite, originally a heavy interstitial melt in which lighter solid elements floated, or at least levitated (Naldrett, 1973).

net venation In a leaf, a type of *venation* in which the veins branch repeatedly to form a network. Cf: *parallel venation.* See also: *pinnate venation; palmate venation.*

network *net [surv].*

network cave (a) Any cave consisting of repeatedly rejoined passages. Syn: *maze cave.* See also: *branchwork cave.* (b) A cave in which the passages intersect as a grid. See also: *anastomotic cave; spongework cave.*

network deposit *stockwork.*

neudorfite A waxy, pale-yellow variety of retinite containing a little nitrogen, found in coal at Neudorf in Moravia, Czechoslovakia.

Neumann bands Fine straight lines observed on etched surfaces of iron meteorites (hexahedrites), caused by mechanical twinning on (211) planes in kamacite. Named after Franz E. Neumann (1798-1895), German mineralogist. Syn: *Neumann lines; Neumann lamellae.*

Neumann's problem A well-known problem in geodesy: to determine a function that is harmonic outside of a given surface and whose normal derivatives assume prescribed boundary values on the surface. Cf: *boundary-value problem; Dirichlet's problem.*

neurocranium The cartilage or cartilage-replacement component of the vertebrate braincase.

neuromotorium A ganglionlike granular body forming the dynamic center of ciliates (as in tintinnids).

neuston Plankton that are suspended on the surface film of water owing to surface tension. See also: *pleuston.*

neutral axis In a two-dimensional structural model, the equivalent of a *neutral surface.*

neutral depth *normal depth.*

neutral dune A small, irregular sand dune (Wolfe et al., 1966, p. 614).

neutral estuary An estuary in which neither freshwater inflow nor evaporation dominates.

neutral pressure (a) *neutral stress.* (b) The lateral earth pressure when the soil is *at rest.*

neutral shoreline A shoreline whose essential features are independent of either the submergence of a former land surface or the emergence of a former underwater surface (Johnson, 1919, p. 172 & 187); a shoreline resulting without a change in the relative level of land and water. It includes shorelines of deltas, alluvial and outwash plains, volcanoes, and coral reefs, as well as those produced by faulting.

neutral soil A soil whose pH value is 7.0. In practice, the pH value of a neutral soil ranges from 6.6 to 7.3.

neutral stress The stress transmitted by the fluid that fills the voids between particles of a soil or rock mass; e.g. that part of the total normal stress in a saturated soil caused by the presence of interstitial water. Syn: *pore pressure; pore-water pressure; neutral pressure.*

neutral surface *surface of no strain.*

neutron activation *Activation analysis* using neutrons to irradiate the sample.

neutron-activation log A *radioactivity log* of neutron-spectral gamma type, usually run in cased wells, in which high-energy neutrons (about 14 Mev) bombard well-bore rocks and transmute natural elements to gamma-ray-emitting isotopes of characteristic identity. Behavior of calcium versus silicon permits lithology interpretation, and that of carbon versus oxygen may distinguish oil from water. See also: *spectral gamma-ray log.*

neutron-gamma log The *well log* curve of induced gamma radioactivity that results from bombardment of rocks near the well bore by fast neutrons. A low count rate implies near-source dissipation in high-porosity rocks, esp. capture by chlorine. See also: *neutron log.* Syn: *n-g log.*

neutron log A *radioactivity log* curve that indicates the intensity of radiation (neutrons or gamma rays) produced when the rocks in a borehole are bombarded by neutrons from a *sonde.* It indicates the presence of fluids (but does not distinguish between oil and water) in the rocks, and is used with the *gamma-ray log* to differentiate porous from nonporous formations. See also: *neutron-gamma log; neutron-neutron log; epithermal-neutron log.* Syn: *hydrogen-index log; nuclear log.*

neutron-neutron log Any of the several *neutron log* curves that measure the abundance of neutrons of a discrete energy range. Neutrons arrive at the detector after "random walk" scattering and slowing, most effectively by hydrogen nuclei. Depending on the neutron-energy selectivity level of the detector, these curves may be subdivided into *epithermal-neutron log* and *thermal-neutron log* types. See also: *neutron log.* Syn: *n-n log.*

neutron soil-moisture meter An instrument for measuring water content of soil and rocks as indicated by the scattering and absorption of neutrons emitted from a source, and resulting gamma radiation received by a detector, in a probe lowered into an access hole.

Nevadan orogeny A time of deformation, metamorphism, and plutonism during Jurassic and Early Cretaceous time in the western part of the North American Cordillera, typified by relations in the Sierra Nevada, California. In that area, deformation of the supracrustal rocks can be closely dated by limiting fossiliferous strata as late in the Jurassic (between the Kimmeridgian and Portlandian Stages), but earlier and later Nevadan deformation occurs elsewhere. In the Sierra Nevada itself, the emplacement of granite and other plutonic activity were more prolonged than the deformation, and have been dated radiometrically between 180 m.y. and 80 m.y., or from Early Jurassic to Early Cretaceous. Geologists differ as to whether to restrict the Nevadan closely in time and space, or to use it broadly; it can most properly be considered as an orogenic era, in the sense of Stille. Also spelled: *Nevadian orogeny; Nevadic orogeny.* Cf: *Coast Range orogeny.*

Nevada twin law A rare, parallel twin law in feldspar, with a twin axis of [112].

Nevadian orogeny *Nevadan orogeny.*

Nevadic orogeny *Nevadan orogeny.*

nevadite A term, now obsolete, that was applied to *rhyolite* containing abundant large phenocrysts of quartz, feldspar, biotite, and hornblende in a small amount of groundmass. The name, given by Richtofen in 1868, was derived from Nevada.

névé A French term meaning a mass of hardened snow at the source or head of a glacier; it refers to the overall snow cover that exists during the melting period and sometimes from one year to another. The term was originally used in English as an exact equivalent of *firn* (the material), and is still frequently so used, but it is perhaps best to restrict it, as proposed by Bristish glaciologists, to a geographic meaning, such as an area covered with perennial snow or an area of firn (a *firn field*), or more generally the *accumulation area* above or at the head of a glacier.

nevyanskite A tin-white variety of iridosmine containing 35-50% osmium or more than 40% iridium and occurring in flat scales.

newberyite A white orthorhombic mineral: $HMgPO_4 \cdot 3H_2O$.

new global tectonics A general term introduced by Isacks et al. (1968) for *global tectonics* based on the related concepts of continental drift, sea-floor spreading, transform faults, and underthrusting of the lithosphere (crust and uppermost mantle) at island arcs, as they are jointly applied to an integrated global analysis of the relative motions of crustal segments delineated by the major seismic belts.

new ice A general term for recently formed ice (esp. floating sea ice) less than 5 cm thick, composed of ice crystals that may be weakly frozen together and that have a definite form only while they are afloat; e.g. *frazil ice, grease ice, sludge, shuga, ice rind, nilas, and pancake ice.*

newlandite A *griquaite* containing garnet, enstatite, and chrome diopside. The name, given by Bonney in 1899, is for the Newlands diamond pipe, South Africa. Not recommended usage.

newland lake A term used by Hobbs (1912, p.401) for a *consequent lake,* esp. one occupying a depression on a newly emerged ocean bottom.

New Red Sandstone The red sandstone facies of the Permian and Triassic systems, well-developed in NW England. See William Buckland in Phillips (1818, p.71-79).

new snow (a) Fallen snow in which the original crystalline structure is still recognizable. Ant: *old snow.* (b) Snow that has fallen very recently, as in the past 24 hours.

new Stone Age *Neolithic.*

Newtonian flow In experimental structural geology, flow in which the rate of shear strain is directly proportional to the shear stress; flow of a *Newtonian liquid.* Cf: *non-Newtonian flow.* Syn: *viscous flow.*

Newtonian liquid A substance in which the rate of shear strain is proportional to the shear stress. This constant ratio is the *viscosity* of the liquid. See also: *Newtonian flow.*

Newton's law of gravitation The statement that every particle of matter attracts every other particle with a force whose magnitude is proportional to the product of their masses and inversely proportional to the square of the distance between them.

nexine The inner division of the exine of pollen, more or less equivalent to *endexine.* Cf: *sexine.*

neyite A mineral: $Pb_7(Cu,Ag)_2Bi_6S_{17}$.

ngavite A chondritic stony meteorite composed of bronzite and olivine in a friable, breccialike mass of chondrules.

n-g log *neutron-gamma log.*

ngurumanite A hypabyssal nepheline-clinopyroxene rock *(ankaratrite* or *melteigite)* with an iron-rich mesostasis of phyllosilicates, calcite, and analcime. The name, given by Saggerson and Williams in 1964, is for the Nguruman Escarpment, Kenya. Not recommended usage.

n'hangellite A green, elastic bitumen, similar to *coorongite,* that represents deposits of the alga *Coelosphaerium.*

Niagaran North American series: Middle Silurian (above Alexandrian, below Cayugan).

Niagara spar A name applied in the vicinity of Niagara Falls, N.Y., to fibrous gypsum imported through Canada from England, and to fibrous calcite found in veins in limestone near Niagara Falls, Ontario (Shipley, 1951, p. 152).

niccolite A syn. of *nickeline.* Also spelled: *nicolite.*

niche [ecol] The position of an organism or a population in the environment as determined by its mode of life, needs, contributions, potential, and interaction with other organisms or populations. Syn: *ecologic niche.*

niche [geomorph] A shallow cave or re-entrant produced by weathering and erosion near the base of a rock face or cliff, or beneath a waterfall.

niche glacier A common type of small mountain glacier, occupying a funnel-shaped hollow or irregular recess in a mountain slope. Cf: *cirque glacier.*

nick (a) A place of abrupt inflection in a stream profile; a *knickpoint.* (b) A sharp angle cut by waves, currents, or ice at the base of a cliff.—Syn: *knick.*

nickel A nearly silver-white hard mineral, the metallic element Ni. It occurs native esp. in meteorites and also alloyed with iron in meteorites. Nickel is used chiefly in alloys and as a catalyst.

nickel-antimony glance *ullmannite.*

nickel bloom A green hydrated and oxidized patina or incrustation on outcropping rocks, indicating the existence of primary nickel minerals; specif. annabergite (a nickel arsenate). The term is also applied to zaratite (a nickel carbonate) and to morenosite (a nickel sulfate).

nickel glance *gersdorffite.*

nickelhexahydrite A mineral: $(Ni,Mg,Fe)SO_4 \cdot 6H_2O$.

nickeline A pale copper-red hexagonal mineral: NiAs. It is one of the chief ores of nickel, and may contain antimony, cobalt, iron, and sulfur. Syn: *niccolite; arsenical nickel; copper nickel; kupfernickel.*

nickel-iron An alloy of nickel and iron (Ni,Fe) occurring in pebbles and grains (as in stream gravel), and also in meteorites. See also: *kamacite; taenite.* Syn: *awaruite; josephinite.*

nickel ocher *annabergite.*

nickel pyrites *millerite.*

nickel-skutterudite A tin-white to steel-gray isometric mineral: $(Ni,Co)As_3$. It may contain iron, and it represents a valuable ore of nickel, often associated with smaltite and skutterudite. Syn: *chloanthite; white nickel.*

nickel vitriol *morenosite.*

nickel-zippeite An orthorhombic mineral: $Ni_2(UO_2)_6(SO_4)_3(OH)_{10} \cdot 16H_2O$.

nickpoint A syn. of *knickpoint.* Also spelled: *nick point.*

nicol (a) *Nicol prism.* (b) Any apparatus that produces polarized light, e.g. Nicol prism or Polaroid; a *polarizer.*

nicolite *niccolite.*

nicolo A variety of onyx with a black or brown base and a bluish-white or faint-bluish top layer.

Nicol prism In a polarizing microscope, a pair of prisms that polarize and analyze the light used for illumination of the thin section under study. The lower nicol, or *polarizer,* is located below the stage; it consists of a rhombohedron of optically clear calcite so cut and recemented that the ordinary ray produced by double refraction in the calcite is totally reflected and the extraordinary ray transmitted. The upper nicol, or *analyzer,* is located above the objective and receives the polarized light after it has passed through the object under study. Its vibration direction is normally set at right angles to that of the polarizer. Partial syn: *nicol.* Syn: *polarizing prism.*

nicopyrite *pentlandite.*

nieve penitente (a) A jagged pinnacle or spike of snow or firn, up to several meters in height, resulting from differential ablation under conditions of strong insolation, especially in high altitude-low latitude environments; an advanced stage of *sun cup* development. (b) An assemblage of nieve penitentes.—Etymol: Spanish, "penitent snow", shortened from "nieve de los penitentes", from the illusion of human figures hanging their heads in penitence; first used in South America. Syn: *penitent [glaciol]; penitente; ice penitente; snow penitente; sun spike.* Cf: *serac.*

nifontovite A mineral: $Ca_3B_6O_6(OH)_{12} \cdot 2H_2O$.

nigerite A dark-brown mineral: $(Zn,Mg,Fe^{+2})(Sn,Zn)_2(Al,Fe^{+3})_{12}O_{22}(OH)_2$.

niggerhead (a) A large block or boulder of coral torn from the outer face of a reef and thrown by storm waves onto the reef flat, where it is overgrown by a crust of algae that become black on dying, as on the Great Barrier Reef, Australia. (b) An isolated erosion remnant of an emerged reef, rising above an eroded reef flat. Syn: *coral head; bommy.* (c) A growing coral head. (d) Any dark round shaggy tussock, hummock, peat mass, or clump of organic or soil material found in far northern regions, resulting from the differential frost heaving of wet soil, and often covered with mosses, lichens, or other low plant growth; specif. one of black, matted sedge tussocks with a peaty base, formed by the growth of cotton grass, and separated from other tussocks by wet hollows filled with snow, esp. noticeable in many permafrost regions. Also, any more or less uniform hummock of tundra.—Syn: *negrohead; hardhead.*

niggliite A silver-white mineral: PtSn.

Niggli molecular norm A *norm* in which the reported content of a mineral represents the percentage of the total cations present in the rock that are tied up in that particular mineral. Three variant mineral associations are possible: the *catanorm,* the *mesonorm,* and the *epinorm.* Syn: *molecular norm.*

Niggli number *Niggli value.*

Niggli's classification A chemical classification of igneous rocks that is essentially a modification and simplification of *Osann's classification.* This system was proposed in 1920 by the Swiss mineralogist Paul Niggli (1888-1953).

Niggli value A parameter produced from the chemical analysis of a rock reported as molecular proportions. The proportions of oxides other than SiO_2 are summed, and this sum is divided into SiO_2 to produce *si;* into Al_2O_3 to produce *al;* into the sum $FeO+2Fe_2O_3+MgO$ to produce *fm;* into CaO to produce *c;* and into the sum Na_2O+K_2O to produce *alk.* Syn: *Niggli number.*

night emerald *evening emerald.*

nigrine A black variety of rutile containing iron.

nigritite Coalified, carbon-rich bitumens. See also: *polynigritite; humonigritite; exinonigritite; keronigritite.*

nijd *nejd.*

niklesite A *pyroxenite* purportedly containing three pyroxenes: diopside, enstatite, and diallage. Named by Kretschmer in 1918 for Niklesgraben, Moravia, Czechoslovakia. Obsolete.

nilas A thin elastic crust of gray-colored ice formed on a calm sea, having a matte surface, and easily bent by waves and thrust into a pattern of interlocking "fingers"; it is subdivided by color into "dark nilas" (less than 5 cm thick) and "light nilas" (5-10 cm thick). Etymol: Russian.

niligongite A plutonic *foidite* intermediate in composition between *fergusite* and *ijolite,* containing approximately equal amounts of nepheline and leucite and 30 to 60 percent mafic minerals. Its name, given by Lacroix in 1933, is derived from Niligongo, Zaire. Not recommended usage.

nimesite A mineral of the kaolinite-serpentine group: $(Ni,Mg)_2Al(Si,Al)O_5(OH)_4$.

nimite A mineral of the chlorite group: $(Ni,Mg,Fe,Al)_6AlSi_3O_{10}(OH)_8$.

ningyoite A brownish-green to brown mineral: $(U,Ca,Ce)_2(PO_4)_2\cdot1\text{-}2H_2O$. It occurs as coatings or cavity fillings in uranium ore.

niningerite A meteorite mineral: (Mg,Fe,Mn)S.

niobite *columbite.*

niobo-aeschynite An orthorhombic mineral: $(Ce,Ca,Th)(Nb,Ti)_2(O,OH)_6$. It forms a series with aeschynite.

niobophyllite A mineral: $(K,Na)_3(Fe^{+2},Mn)_6(Nb,Ti)_2Si_8(O,OH,F)_{31}$.

niocalite A pale-yellow orthorhombic mineral: $Ca_4NbSi_2O_{10}(O,F)$.

nip [coal] A *pinch* or thinning of a coal seam, esp. as a result of tectonic movements. Cf: *want.*

nip [coast] A small, very low cliff or break in slope produced at the high-water mark by wavelets, and often cited as an initial feature in the development of a shoreline of emergence. The term has also been applied in a broader sense to the small *notch* resulting from the formation of such a cliff.

nip [speleo] *solution nip.*

nip [streams] The place on the bank of a meander lobe where erosion occurs as a result of the crowding of the stream current toward the lobe (Tower, 1904, p. 593).

nisbite A mineral: $NiSb_2$.

nissonite A mineral: $Cu_2Mg_2(PO_4)_2(OH)_2\cdot5H_2O$.

niter (a) A white orthorhombic mineral: KNO_3. It is a soluble crystalline salt that occurs as a product of nitrification in most arable soils in hot, dry regions, and in the loose earth forming the floors of some natural caves. Cf: *soda niter.* Syn: *saltpeter.* (b) A term that was formerly used for a variety of saline efflorescences, including natron and soda niter.—Also spelled: *nitre.*

nitrate A mineral compound characterized by a fundamental anionic structure of NO_3^-. Soda niter, $NaNO_3$, and niter, KNO_3, are nitrates. Cf: *carbonate; borate.*

nitratine *soda niter.*

nitre *niter.*

nitride A mineral compound that is a combination of nitrogen with a more positive element. An example is osbornite, TiN.

nitrification The formation of nitrates by the oxidation of ammonium salts to nitrites (usually by bacteria) followed by oxidation of nitrites to nitrates. It is one of the processes of soil formation.

nitrobarite A colorless mineral: $Ba(NO_3)_2$.

nitrocalcite A mineral: $Ca(NO_3)_2\cdot4H_2O$. It occurs as an efflorescence, as on walls and in limestone caves. Syn: *wall saltpeter.*

nitrogen fixation In a soil, the conversion of atmospheric nitrogen to a combined form by the metabolic processes of some algae, bacteria, and actinomycetes.

nitroglauberite A mixture of darapskite and soda niter.

nitromagnesite A mineral: $Mg(NO_3)_2\cdot6H_2O$. It occurs as an efflorescence in limestone caverns.

nitrophyte A plant that requires nitrogen-rich soil for growth.

nival Characterized by or living in or under snow, or pertaining to a snowy environment; e.g. *nival* fauna or climate.

nival gradient The angle between a *nival plane* and the horizon (Young, 1910, p.252).

nival karst *alpine karst.*

nival plane The imaginary planar surface containing all the different snowlines of the same time period (Young, 1910, p.252).

nivation (a) The process of excavation of a shallow depression or *nivation hollow* in a mountainside by removal of fine material around the edge of a shrinking snow patch or snowbank, chiefly through sheetwash, rivulet flow, and solution in meltwater (Thorn, 1976). Freeze-thaw action is apparently insignificant. Syn: *snow-patch erosion.* (b) More generally, the work of snow and ice beyond the limits of glacier action.

nivation cirque *nivation hollow.*

nivation glacier A small, "new-born" glacier, representing the initial stage of glaciation. Syn: *snowbank glacier.*

nivation hollow A shallow depression or hollow in a mountainside, permanently or intermittently occupied by a snowbank or snow patch and produced by *nivation.* If the snow completely melts each summer the hollow is deepened; otherwise not (Thorn, 1976). It has been suggested that deepening of a nivation hollow produces a *cirque,* but this is not proven. Syn: *nivation cirque; snow niche.*

nivation ridge A low convex accumulation of fine sediment downslope from a *nivation hollow,* consisting of fine material carried by sheetwash and rivulet flow from beneath the melting edge of a snow patch or snowbank. Cf: *nivation.*

niveal Said of features and effects "due to the action of snow and ice" (Scheidegger, 1961, p. 24). See also: *niveoglacial.*

niveau surface *equipotential surface.*

nivenite A velvet-black variety of uraninite containing rare earths (cerium and yttrium).

niveo-eolian *niveolian.*

niveoglacial Pertaining to the combined action of snow and ice. See also: *niveal.*

niveolian Pertaining to simultaneous accumulation and intermixing of snow and airborne sand at the side of a gentle slope; e.g. said of material deposited by snowstorms under periglacial conditions. Syn: *niveo-eolian.*

nivo-karst "A characteristic of periglacial areas" (Hamelin & Cook, 1967, p.73) whereby differential chemical weathering beneath snowbanks produces a karstlike topography, as the solution of limestone fragments by snowmelt containing carbonic acid.

NML *nuclear-magnetism log.*

NMO *normal moveout.*

NMR *nuclear magnetic resonance.*

n-n log *neutron-neutron log.*

Noachian flood The flood, described in Genesis 5:28-10:32, during which the patriarch Noah was said to have saved his family and representative creatures. Early writers believed that the waters of this flood deposited material now known as *drift.* Also known as "The Deluge".

no-basement interpretation *thin-skinned structure.*

noble gas *inert gas.*

nobleite A monoclinic mineral: $CaB_6O_{10}\cdot4H_2O$.

noble metal Any metal or alloy of comparatively high economic value, or one that is superior in certain desired properties, e.g. gold, silver, or platinum. Cf: *base metal.*

nocerite *fluoborite.*

nodal A crinoid columnal that is generally larger than adjacent columnals and that commonly bears cirri.

nodal line A line in an oscillating area of water along which there is little or no rise or fall of the tide. Cf: *node.*

nodal point *amphidromic point.*

node [bot] The place on a plant stem from which a leaf and bud normally emerge.

node [evol] A branching point on a *dendrogram,* especially on a *cladogram* or *phylogenetic tree.*

node [fault] That point along a fault at which the direction of apparent displacement changes. It can occur, for instance, at the intersection of a lateral fault with a fold. See also: *scissor fault.*

node [paleont] (a) The uncalcified proximal extremity of a branch forming an articulation in a jointed colony of cheilostome bryozoans. (b) A knob, protuberance, or thickened or swollen body part of an animal, such as a small boss at the end of a foraminiferal pillar.

node [waves] That point on a standing wave at which the vertical motion is least and the horizontal velocity is greatest. It is also associated with seiches. Cf: *nodal line; partial node.* Ant: *antinode.*

nodular (a) Composed of nodules; e.g. "nodular bedding" consisting of scattered to loosely packed nodules in matrix of like or unlike character. (b) Having the shape of a nodule, or occurring in the form of nodules; e.g. "nodular ore" such as a colloform mineral aggregate with a bulbed surface. Syn: *nodulated.* (c) *orbicular.*

nodular anhydrite *chickenwire anhydrite.*

nodular chert (a) Chert in the form of *chert nodules.* (b) A term used in Missouri for chalky chert containing small irregular grains (Grohskopf & McCracken, 1949, pl.3).

nodulated Occurring in the form of nodules; *nodular.*

nodule [ign] A fragment of a coarse-grained igneous rock, apparently crystallized at depth, occurring as an *inclusion* in an extrusive rock; e.g. a "peridotite nodule" in a flow of olivine basalt. Cf: *xenolith.* Syn: *plutonic nodule.*

nodule [sed] (a) A small, irregularly rounded knot, mass, or lump of a mineral or mineral aggregate, normally having a warty or knobby surface and no internal structure, and usually exhibiting a contrasting composition from the enclosing sediment or rock matrix in which it is embedded; e.g. a nodule of pyrite in a coal bed, a *chert nodule* in limestone, or a *phosphatic nodule* in marine strata. Most nodules appear to be secondary structures: in sedimentary rocks they are primarily the result of postdepositional replacement of the host rock and are commonly elongated parallel to the bedding. Nodules can be separated as discrete masses from the host material. (b) One of the widely scattered concretionary lumps of manganese, cobalt, iron, and nickel found on the floors of the world's oceans; esp. a *manganese nodule.*—Etymol: Latin *nodulus,* "small knot". Cf: *concretion.*

noise (a) Any undesired sound, and, by extension, any unwanted disturbance, within a useful frequency band. Cf: *signal.* (b) An erratic, intermittent, or statistically random oscillation. (c) That portion of the unwanted signal that is statistically random, as distinguished from "hum", which is an unwanted signal occurring at multiples of the power-supply frequency.

noise-equivalent power A syn. of *minimum detectable power.* Abbrev: NEP. Cf: *detectivity.*

nolanite A black hexagonal mineral: $Fe_3V_7O_{16}$.

nomen ambiguum A name used in different senses (e.g. the same name applied to different taxa) so that it has become "a long-persistent source of error" (McVaugh et al., 1968, p. 18). Etymol: Latin, "uncertain name". Pl: *nomina ambigua.*

nomenclature The practice of naming allied groups of plants and animals (*taxa*) according to the hierarchical system and formal procedure prescribed by accepted authoritative codes, i.e. the International Code of Botanical Nomenclature and the International Code of Zoological Nomenclature.

nomen confusium A name applied to a *taxon* of dubious integrity, e.g. one based on a *type* consisting of discordant elements (Cowan, 1968, p. 70; McVaugh et al., 1968, p. 19). Etymol: Latin, "confused name". Pl: *nomina confusia.*

nomen conservandum Any name that must be adopted as a correct name in accordance with a special regulation, although otherwise contrary to nomenclatural rules (McVaugh et al., 1968, p. 10; ICBN, 1972, appendix III). Etymol: Latin, "preserved name". Plural: *nomina conservanda.*

nomen nudum A scientific name published without description or *diagnosis* (McVaugh et al., 1968, p. 10). Etymol: Latin, "nude name, mere name". Plural: *nomina nuda.*

nominal In zoological nomenclature, a term applied to a particular *taxon* whose rank has been objectively defined by its *type;* e.g. "the nominal genus *Musca* is always that to which its type species, *Musca domestica,* belongs" (ICZN, 1964, p.153).

nominal diameter The computed diameter of a hypothetical sphere having the same volume as that calculated for a given sedimentary particle; it is a true measure of particle size independent of either the shape or the density of the particle. Cf: *equivalent radius; sedimentation diameter.*

nomogenesis A theory of evolution stating that evolutionary change is governed by predetermined natural processes and is independent of environmental influences.

nomogram A type of line chart that graphically represents an equation of three variables, each of which is represented by a graduated straight line. It is used to avoid lengthy calculations; a straight line connecting values on two of the lines automatically intersects the third line at the required value. Syn: *nomograph.*

nomograph *nomogram.*

nonangular unconformity *disconformity.*

nonarborescent pollen Pollen of herbs and shrubs. Abbrev: NAP. Syn: *nontree pollen.*

nonartesian ground water *unconfined ground water.*

nonasphaltic pyrobitumen Any of a group of pyrobitumens, including peat, coal, and nonasphaltic pyrobitumenous shales, that are dark-colored relatively hard nonvolatile solids, composed of hydrocarbons containing oxygenated bodies. They are sometimes associated with mineral matter, the nonmineral constituents being infusible and largely insoluble in carbon disulfide (Abraham, 1960, p. 57).

nonassociated gas Natural gas that occurs in a reservoir without oil. Cf: *associated gas.*

nonbanded coal Coal without bands of vitrain or lustrous material, consisting mainly of clarain, durain, or intermediate material. Cf: *sapropelic coal.*

noncaking coal Coal that does not cake or agglomerate when heated; it is usually a hard or dull coal. Syn: *free-burning coal.*

Noncalcic Brown soil A great soil group in the 1938 classification system, for a group of zonal soils having a slightly acidic, light pink or reddish brown A horizon and a light brown or dull red B horizon. It is developed under a mixture of grass and forest vegetation, in a subhumid climate (USDA, 1938). Most of these soils are now classified as *Xeralfs.* Syn: *Shantung soil.*

noncapillary porosity The volume of large interstices in a rock or soil that do not hold water by capillarity (Jacks et al., 1960). Cf: *aeration porosity.*

noncarbonate hardness Hardness of water, expressed as $CaCO_3$, that is in excess of the $CaCO_3$ equivalent of the carbonate and bicarbonate alkalinity. It cannot be removed by boiling and hence is sometimes called *permanent hardness,* although this synonym is becoming obsolete. Cf: *carbonate hardness; hardness [water].*

nonclastic (a) Said of a sedimentary texture showing no evidence that the sediment was derived from a pre-existing rock or was deposited mechanically. (b) Pertaining to a chemically or organically formed sediment or sedimentary rock.—Syn: *nonmechanical.*

noncoaxial Antonym of *coaxial* (Hsu, 1966, p. 217).

noncognate *accidental.*

noncohesive *cohesionless.*

nonconformable Pertaining to a nonconformity or to the stratigraphic relations shown by a nonconformity.

nonconformity (a) An *unconformity* developed between sedimentary rocks and older rocks (plutonic igneous or massive metamorphic rocks) that had been exposed to erosion before the overlying sediments covered them. The restriction of the term to this usage was proposed by Dunbar & Rodgers (1957, p.119). Although the term is "well known in the classroom", it is "not commonly used in practice" (Dennis, 1967, p.160). Syn: *heterolithic unconformity.* (b) A term that formerly was widely, but now less commonly, used as a syn. of *angular unconformity,* or as a generic term that includes angular unconformity.—Term proposed by Pirsson (1915, p.291-293).

nonconservative elements In seawater, elements whose total quantities may vary with time. Cf: *conservative elements.*

noncyclic terrace One of several stream terraces representing former valley floors formed during periods when continued valley deepening accompanied lateral erosion. Terraces on opposite sides of the valley are unpaired. Cf: *cyclic terrace.*

nondepositional unconformity A term used by Tomkeieff (1962, p.412) for a surface of nondeposition in marine sediments. It is equivalent to *paraconformity.*

nondetrital Pertaining to sedimentary material derived from solution by physicochemical or biochemical means, including authigenic minerals formed in the sediment after deposition. In the next erosion cycle, nondetrital material may become detrital.

noneroding velocity The velocity of water in a channel that will maintain silt in movement but will not scour the bed. Cf: *transporting erosive velocity.*

nonesite A porphyritic *basalt* composed of enstatite, labradorite, and augite phenocrysts in a groundmass of plagioclase and augite. The name, given by Lepsius in 1878, is for Nonsberg, Austria. Obsolete.

nonfaradaic path One of the two available paths for transfer of energy across an electrolyte-metal interface. Energy is carried by capacitive transfer, i.e. charging and discharging of the double-layer capacitance. Cf: *faradaic path.*

nonferrous Said of metals other than iron, usually the base metals.

nonflowing artesian well An *artesian well* whose head is not sufficient to lift the water above the land surface. Cf: *flowing artesian well; nonflowing well.*

nonflowing well A well that yields water at the land surface only by means of a pump or other lifting device. It may be either a *water-table well* or a *nonflowing artesian well.*

nonfoliate Pertaining to a metamorphic rock lacking foliation on the scale of hand specimens.

nongraded (a) A rarely used geologic term pertaining to an unconsolidated sediment or to a cemented detrital rock consisting notably of particles of more than one size or of particles lying within the limits of more than one grade; e.g. a loam or a till. Syn: *poorly sorted.* (b) An engineering term pertaining to a soil or an unconsolidated sediment consisting of particles of essentially the same size. See also: *poorly graded.*—Ant: *graded [part size].*

nonideal solution A solution in which the molecular interaction between components is not the same as that within each component. Cf: *ideal solution.*

nonmare basalt *KREEP.*

nonmechanical *nonclastic.*

nonmetal (a) A naturally occurring substance that does not have metallic properties, such as high luster, conductivity, opaqueness, and ductility. (b) In economic geology, any rock or mineral mined for its nonmetallic value, such as stone, sulfur, or salt. Syn: *nonmetallic; industrial mineral.*

nonmetallic adj. (a) Of or pertaining to a nonmetal. (b) Said in general of mineral lusters other than *metallic* luster. Cf: *submetallic.* —n. A *nonmetal* or *industrial mineral;* usually used in the plural.

nonmonophyletic In *cladism,* referring to a taxon which is either *paraphyletic, polyphyletic,* or both. Cf: *monophyletic.*

nonmotile Not *motile;* e.g. the nonflagellate stage or "nonmotile phase" in the life cycle of a coccolithophorid.

non-Newtonian flow Flow in which the relationship of the shear stress to the rate of shear strain is nonlinear, i.e. flow of a substance in which viscosity is not constant. Cf: *Newtonian flow.*

nonparametric statistics Statistics that do not assume specific distributions. Cf: *parametric statistics.*

nonpareil A large, specially cut gemstone; esp. a *solitaire.*

nonpenetrative Said of a texture of deformation that affects only part of a rock, e.g. kink bands. Cf: *spaced cleavage.*

nonplunging fold A fold whose hinge line is horizontal. Cf: *plunging fold.* Syn: *horizontal fold; level fold.*

non-point-source pollution Pollution from sources that cannot be defined as discrete points, such as areas of crop production, timber, surface mining, disposal of refuse, and construction. Cf: *point-source pollution.*

nonrotational strain *irrotational strain.*

nonsaline alkali soil *nonsaline sodic soil.*

nonsaline sodic soil A soil with a content of exchangeable sodium greater than 15%, which gives a conductance of less than 4 mmhos/cm. The pH values are usually between 8.5 and 10.0. Cf: *saline sodic soil.* Syn: *nonsaline alkali soil.*

non-sequence A term used in Great Britain for a *diastem,* or for a break or gap in the continuity of the geologic record, representing a time during which no permanent deposition took place. A non-sequence usually can be detected only by a study of successive fossil contents. Cf: *paraconformity.*

nonsilting velocity The velocity of water in a channel that maintains silt in movement. Syn: *transportation velocity.*

nonsorted Said of a nongenetic group of *patterned ground* features that do not have a border of stones surrounding or alternating with finer material, as in *sorted* patterned ground; often there is a border of vegetation between areas of relatively bare ground or finer material.

nonsorted circle A form of patterned ground "whose mesh is dominantly circular and has a nonsorted appearance due to the absence of a border of stones" (Washburn, 1956, p.829); developed singly or in groups. Vegetation characteristically outlines the pattern by forming a bordering ridge. When well-developed, it has a distinctly domed central area. Diameter: commonly 0.5 to 3 m. Examples: *mud circle; frost scar; peat ring; tussock ring.*

nonsorted crack A rare form of patterned ground representing the boulder-free variant of a *sorted crack.*

nonsorted net A form of patterned ground "whose mesh is intermediate between that of a nonsorted circle and a nonsorted polygon and has a nonsorted appearance due to the absence of a border of stones" (Washburn, 1956, p.830); e.g. an *earth hummock.*

nonsorted polygon A form of patterned ground "whose mesh is dominantly polygonal and has a nonsorted appearance due to the absence of a border of stones" (Washburn, 1956, p.831-832); never developed singly. Its borders commonly, but not invariably, are marked by wedge-shaped fissures narrowing downward; it typically results from infilling of these fissures. Diameter: a few centimeters to tens of meters. See also: *fissure polygon; mud polygon; ice-wedge polygon; vegetation polygon; sand-wedge polygon; frost-crack polygon; desiccation polygon.*

nonsorted step A form of patterned ground "with a steplike form and a nonsorted appearance due to a downslope border of vegetation embanking an area of relatively bare ground upslope" (Washburn, 1956, p.834); formed in groups. See also: *sorted step; turf-banked terrace.*

nonsorted stripe One of the alternating bands comprising a form of patterned ground characterized by "a striped pattern and a nonsorted appearance due to parallel lines of vegetation-covered ground and intervening strips of relatively bare ground oriented down the steepest available slope" (Washburn, 1956, p.837). Vegetation characteristically outlines the pattern, as the absence of lines of stones is an essential feature; the bare ground consists of finer-grained material or a nonsorted mixture of fines and stones. See also: *solifluction stripe; vegetation stripe; stripe hummock.*

nonsteady flow *unsteady flow.*

nonstrophic Said of a brachiopod shell whose posterior margin is not parallel with the hinge axis. Cf: *strophic.*

nonsystematic joints Joints that are not part of a set. They do not cross other joints, they often terminate at bedding surfaces, their surfaces may be strongly curved, and the structures on their faces are not oriented. Cf: *systematic joints.*

nontabular Said of projecting surface features of a dinoflagellate cyst that are neither sutural nor *intratabular* and have a random arrangement. Cf: *peritabular.*

nontectonite Any rock whose fabric shows no influence of movement of adjacent grains, e.g. a rock formed by mechanical settling. Some rocks are transitional between a tectonite and a nontectonite (Turner and Weiss, 1963, p.39).

nonthermal spring A spring in which the temperature of the water is not appreciably above the mean atmospheric temperature in the vicinity. A spring whose temperature approximates the mean annual temperature, or a *cold spring,* is considered a nonthermal spring (Meinzer, 1923, p. 55).

nontree pollen A syn. of *nonarborescent pollen.* Abbrev: NTP.

nontronite A pale yellow to green dioctahedral iron-rich clay mineral of the montmorillonite group: $Na_{0.33}Fe_2{}^{+3}(Al_{0.33}Si_{3.67})O_{10}(OH)_2 \cdot nH_2O$. It represents an end member in which the replacement of aluminum by ferric iron in the octahedral sheets is essentially complete. Nontronite commonly occurs in weathered basaltic rocks, where it may occupy vesicles or veins or occur between lava flows. Syn: *chloropal; gramenite; morencite; pinguite.*

nonuniform flow In hydraulics, a type of steady flow in an open channel in which velocity varies at different points along the channel.

nonuniformist One who believes that past changes in the Earth's structure have proceeded from cataclysms or processes more violent than are now operating; a believer in the doctrine of catastrophism. Syn: *nonuniformitarian.*

nonvascular plant A plant without a vascular system or well differentiated roots, stems, and leaves, e.g. a thallophyte or bryophyte.

nonwetting sand Sand that resists infiltration of water, consisting of angular particles of varying sizes, and occurring as a tightly packed mass, generally lenticular.

nook An obsolete syn. of *promontory.*

norbergite A yellow or pink orthorhombic mineral of the humite group: $Mg_3SiO_4(F,OH)_2$.

nordenskiöldine A mineral: $CaSn(BO_3)_2$.

nordfieldite *esmeraldite.*

nordite A pale brown mineral: $(La,Ce)(Sr,Ca)Na_2(Na,Mn)(Zn,Mg)Si_6O_{17}$.

nordmarkite [mineral] A variety of staurolite containing manganese.

nordmarkite [rock] A quartz-bearing alkali *syenite* that has microperthite as its main constituent with smaller amounts of oligoclase, quartz, and biotite, and is characterized by granitic or trachytoid texture. The name, given by Brögger in 1890, is for Nordmark, Oslo district, Norway. Not recommended usage.

nordsjoite A *nepheline syenite* that contains melanite and calcite. Named by Johannsen in 1938 for Nordsjö, the lake bordering the Fen complex, Norway. Not recommended usage.

nordstrandite A mineral: $Al(OH)_3$ or $Al_2O_3 \cdot 3H_2O$. It is trimorphous with gibbsite and bayerite.

nore *ness.*

Norian European stage: Upper Triassic (above Carnian, below Rhaetian).

norilskite A mineral consisting of platinum with a high content of iron and nickel.

norite (a) In the *IUGS classification,* a plutonic rock satisfying the definition of *gabbro,* in which pl/(pl + px + ol) is between 10 and 90 and opx/(opx + cpx) is greater than 95. (b) A coarse-grained plutonic rock containing basic plagioclase (labradorite) as the chief constituent and differing from gabbro by the presence of orthopyroxene (hypersthene) as the dominant mafic mineral. Cf: *hypersthenite.* Syn: *hypersthenfels.*—The name was first used by Esmark in 1823.

norm The theoretical mineral composition of a rock expressed in terms of *normative mineral* molecules that have been determined by specific chemical analyses for the purpose of classification and comparison; "the theoretical mineral composition that might be expected had all chemical components crystallized under equilibrium conditions according to certain rules" (Stokes & Varnes, 1955, p. 94). Adj: *normative.* Cf: *mode.* See also: *CIPW classification.*

normal [fold] Said of an anticlinorium in which the axial surfaces of the subsidiary folds converge downwards; said of a synclinorium in which the axial surfaces of the subsidiary folds converge upwards. Cf: *abnormal [fold].* See also: *fan structure.*

normal [meteorol] n. The average value of a meteorological element (such as pressure, temperature, rainfall, or duration of sunshine) over any fixed period of years that is recognized as standard for a given country or element. The period 1901-1930 was selected by the International Meteorological Organization at a Warsaw conference in 1935 as the international standard period for climatological normals; the U.S. Weather Bureau's temperature and precipitation normals, however, are computed from the period 1921-1950.—adj. Approximating the statistical norm or average, such as the "normal rainfall" of a region for a definite time.

normal consolidation Consolidation of sedimentary material in equilibrium with overburden pressure. Cf: *overconsolidation.*

normal curve A bell-shaped curve that graphically represents a *normal distribution.*

normal cycle A *cycle of erosion* in which the complete reduction or lowering of a region to base level is effected largely by running water, specif. the action of rivers as the dominant erosion agent. Cf: *arid cycle.* Syn: *fluvial cycle of erosion.*

normal depth (a) Water depth in an open channel that corresponds to uniform velocity for a given flow. It is the hypothetical depth in a steady, nonuniform flow; the depth for which the surface and bed are parallel. Syn: *neutral depth.* (b) Water depth measured perpendicular to the bed.—(ASCE, 1962).

normal dip *regional dip.*

normal dispersion The *dispersion* of seismic surface waves in which the recorded wave period increases with time. Cf: *inverse dispersion.*

normal displacement A syn. of *dip slip.* Cf: *total displacement.*

normal distribution A *frequency distribution* whose plot is a continuous, infinite, bell-shaped curve that is symmetrical about its arithmetic mean, mode, and median (which in this distribution are numerically equivalent). Syn: *Gaussian distribution; bell-shaped distribution.*

normal erosion (a) The wearing-away of topographic features that is effected by prevailing agencies and that is mainly responsible for the present modification of the habitable land surface; specif. subaerial erosion by running water, rain, and certain physical and organic weathering processes. The term, used originally for stream erosion in a temperate climate, is open to criticism because erosion as found in temperate areas may in fact be "abnormal" (esp. in regard to past geologic conditions) or because one mode of erosion is just as "normal" as another. Cf: *special erosion.* (b) Erosion of rocks and soil under natural environmental conditions, undisturbed by human activity. It includes erosion by running water, rain, wind, ice, waves, gravity, and other geologic agents. Cf: *accelerated erosion.* See also: *geologic norm.* Syn: *geologic erosion.*

normal fault A fault in which the hanging wall appears to have moved downward relative to the footwall. The angle of the fault is usually 45-90°. There is dip separation but there may or may not be dip slip. Cf: *thrust fault.* Syn: *gravity fault; normal slip fault; slump fault.*

normal fold *symmetrical fold.*

normal geopotential number *spheropotential number.*

normal gradient *normal gravity.*

normal gravity The gravity caused by the attraction of the equipotential *mean earth ellipsoid* combined with the centrifugal force due to the Earth's rotation. Syn: *normal gradient.*

normal horizontal separation *offset.*

normal hydrostatic pressure In porous strata or in a well, pressure at a given point that is approximately equal to the weight of a column of water extending from that point to the surface.

normal log The *resistivity log* curve derived from a simple 2-electrode array, with spacing that ranges from 2 inches in the micronormal (see *Microlog*) to 16 and 64 inches for the short and long normal curves (see *electric log*) and to as much as 1,000 feet in the *ultra-long-spaced electric log.* See also: *lateral log; focused-current log; induction log.*

normal magnetic field A smooth component of the Earth's magnetic field, without anomalies of exploration interest.

normal moisture capacity *field capacity.*

normal moveout The increase in arrival time of a seismic-reflection event resulting from an increase in the distance from source to detector, or from dip of the reflector. Seismic data must be corrected for normal moveout. Abbrev: NMO. See also: *moveout; dynamic correction.*

normal-moveout velocity The constant velocity for an overlying medium that would most nearly give the observed normal moveout for a horizontal reflector. It is determined from normal moveout values in *velocity analysis.* Syn: *stacking velocity.*

normal polarity (a) A natural remanent magnetization closely parallel to the present ambient geomagnetic-field direction. See also: *geomagnetic reversal.* (b) A configuration of the Earth's magnetic field with the magnetic negative pole, where field lines enter the Earth, located near the geographic north pole.—Cf: *reversed polarity.*

normal pore canal A tubule or *pore canal* piercing an ostracode carapace at right angles, and believed to serve as a receptor of sensory setae. Cf: *radial pore canal.*

normal-pressure surface A potentiometric surface that coincides with the upper surface of the zone of saturation (Meinzer, 1923, p. 39). It is usually the same as the water table. Cf: *subnormal-pressure surface; artesian-pressure surface.*

normal projection (a) A projection in which a three-dimensional object is projected onto two mutually perpendicular planes. (b) A projection whose surface axes coincide with those of the sphere.

normal ripple mark An aqueous *current ripple mark* consisting of a "simple asymmetrical ridge" that may have "various ground plans" (Shrock, 1948, p.101).

normal sandstone A term used by Shrock (1948a) for a sandstone composed almost exclusively of quartz, with subordinate amounts of other minerals.

normal section A line between two points on the surface of an ellipsoid, formed by the intersection of the ellipsoid and plane containing the normal at one point and the other point.

normal shift In a fault, the horizontal component of the shift, measured perpendicular to the strike of the fault. Cf: *offset.*

normal slip fault *normal fault.*

normal soil A soil whose profile is more or less in equilibrium with the environment, and which shows the effects of the environment on its development from the parent material.

normal strain Change of length per unit length in a given direction.

normal stress That component of *stress* which is perpendicular to a given plane. It may be either *tensile stress* or *compressive stress.* Symbol: σ. Cf: *shear stress.*

normal twin A twinned crystal, the twin axis of which is perpendicular to the composition surface. Cf: *parallel twin.*

normal water A standardized seawater; its chlorinity is between 19.30 and 19.50 parts per thousand and it has been analyzed to within 0.001 part per thousand. Syn: *Copenhagen water; standard seawater.*

normal zoning *Zoning* in a crystal of plagioclase, in which the zones become progressively more sodic outward. Cf: *reversed zon-*

ing.

Normapolles A group of Cretaceous and lower Paleogene porate pollen with a complex pore apparatus (e.g. an *oculus*) and sometimes other peculiarities such as double *Y-marks.* Cf: *Postnormapolles.*

normative The adj. of *norm.*

normative mineral A mineral whose presence in a rock is theoretically possible on the basis of certain chemical analyses. A normative mineral may or may not be actually present in the rock. See also: *norm.* Syn: *standard mineral.*

norm system *CIPW classification.*

norsethite A rhombohedral mineral: $BaMg(CO_3)_2$.

North American datum First known as the United States standard datum, in 1913 it was adopted by Canada and Mexico and renamed North American datum (NAD). It was defined by the data of the station at Meade's Ranch in Kansas: latitude 39°13′ 28.686″N., longitude 98°32′30.506″W., azimuth to Waldo 75°28′14.52″; and by the *Clarke ellipsoid of 1866.* When the entire triangulation network of the United States was readjusted between 1925 and 1930, the azimuth to Waldo was corrected to 75°28′09.64″. The triangulation networks of Canada and Mexico were adjusted to this and the result was the North American datum of 1927. The datum is being readjusted.

northfieldite *esmeraldite.*

north geographic pole *north pole.*

northing A *latitude difference* measured toward the north from the last preceding point of reckoning; e.g. a linear distance northward from the east-west line that passes through the origin of a grid.

north pole [astron] The north *celestial pole,* representing the zenith of the heavens as viewed from the north geographic pole.

north pole [geog] The *geographic pole* in the northern hemisphere of the Earth at lat. 90°N, representing the northernmost point of the Earth or the northern extremity of its axis of rotation. Also spelled: *North Pole.* Syn: *north geographic pole.*

northupite A colorless, white, yellow, or gray isometric mineral: $Na_3Mg(CO_3)_2Cl$.

nose [fold] A short, plunging anticline without closure. Syn: *structural nose; anticlinal nose.*

nose [geomorph] (a) A projecting and generally overhanging buttress of rock. (b) The projecting end of a hill, spur, ridge, or mountain. (c) The central forward part of a parabolic dune. (d) *ness.*

nose [sed] The forward part of a turbidity current, which is more dense than the *tail* and carries coarser material.

nosean A feldspathoid mineral of the sodalite group: $Na_8Al_6Si_6O_{24}(SO_4)$. It is gray, blue, or brown, and is related to hauyne. Syn: *noselite.*

noseanite A feldspar- and olivine-free basalt that contains abundant nosean. Obsolete.

noseanolith An extrusive rock composed almost entirely of nosean (Johannsen, 1938).

noselite *nosean.*

noselitite An extrusive rock composed chiefly of nosean, with less than about 10% pyroxene, amphibole, or both (Johannsen, 1938).

notch [coast] A deep, narrow cut or hollow along the base of a sea cliff near the high-water mark, formed by undercutting due to wave erosion and/or chemical solution, and above which the cliff overhangs. See also: *nip.*

notch [geomorph] (a) A term used in the NE U.S. for a narrow passageway or short defile between mountains or through a ridge, hill, or mountain; a deep, close *pass.* Also, the narrowest part of such a passage. Cf: *gap [geomorph]; col.* (b) A breached opening in the rim of a volcanic crater.

notite A variety of *palagonite* (Hey, 1962, p. 541). Not recommended usage.

notochord A flexible or elastic rod that provides a supporting and stiffening structure in an animal's body; in the true vertebrates it is replaced by a backbone composed of bone or cartilage.

notothyrial chamber The cavity in the umbo of the brachial valve of a brachiopod, bounded laterally by brachiophore bases (or homologues) or by posterior and lateral shell walls if brachiophore bases are absent. It corresponds to the *delthyrial chamber* of the pedicle valve.

notothyrial platform Umbonal thickening of the floor of the brachial valve of a brachiopod between brachiophore bases (or homologues).

notothyrium The median subtriangular opening in the brachial valve of a brachiopod, bisecting the dorsal cardinal area or pseudointerarea. Pl: *notothyria.* Cf: *delthyrium.*

noumeite A syn. of *garnierite,* esp. of a dark-green unctuous variety.

nourishment [beach] The replenishment of a beach, either naturally (as by littoral transport) or artificially (as by the deposition of dredged materials).

nourishment [glaciol] *accumulation.*

novacekite A yellow secondary mineral of the autunite group: $Mg(UO_2)_2(AsO_4)_2 \cdot 9H_2O$.

novaculite (a) A dense hard even-textured light-colored cryptocrystalline siliceous sedimentary rock, similar to chert but characterized by dominance of microcrystalline quartz over chalcedony. It was formerly believed to be the result of primary deposition of silica, but in the type occurrence (Lower Paleozoic of the Ouachita Mountains, Arkansas and Oklahoma) it appears to be a thermally metamorphosed bedded chert, distinguished by characteristic polygonal triple-point texture (Keller et al., 1977). The origin of novaculite has also been ascribed to crystallization of opaline skeletal material during diagenesis. The rock is used as a *whetstone.* See also: *Arkansas stone; Washita stone.* Syn: *razor stone; Turkey stone; galactite.* (b) A term used in southern Illinois for an extensive *bedded chert* (J.E. Lamar, in Tarr, 1938, p. 19). (c) A general name formerly used in England for certain argillaceous stones that served as whetstones.

novaculitic chert A generally gray chert that breaks into slightly rough, splintery fragments; it is less vitreous and somewhat coarser-grained than *chalcedonic chert.*

novakite A tetragonal mineral: $(Cu,Ag)_4As_3$.

nowackiite A mineral: $Cu_6Zn_3As_4S_{12\text{-}13}$.

NQR *nuclear quadrupole resonance.*

NRM *natural remanent magnetization.*

nsutite A mineral: $Mn_{1-x}{}^{+4}Mn_x{}^{+2}O_{2-2x}(OH)_{2x}$. Syn: *gamma-*$MnO_2$.

n-tn log *thermal neutron log.*

NTP *nontree pollen.*

nubbin (a) One of the isolated bedrock knobs or small hills forming the last remnants of the crest of a mountain or mountain range that has succumbed to desert erosion (backwearing). The term was introduced by Lawson (1915) and extended by Cotton (1942) to include small remnants of spurs and ridges. (b) A residual boulder, commonly granitic, occurring on a desert dome or broad pediment (Stone, 1967, p. 235).

nuclear age determination *radiometric dating.*

nuclear basin A postorogenic basin in a mobile belt; a contemporary *epieugeosyncline.*

nuclear clock *radioactive clock.*

nuclear fission *fission.*

nuclear fusion *fusion.*

nuclear geochemistry *isotope geology.*

nuclear geology *isotope geology.*

nuclear log (a) *neutron log.* (b) *radioactivity log.*

nuclear magnetic resonance The selective absorption of electromagnetic radiation at the appropriate resonant frequency by nuclei undergoing precession in a strong magnetic field. Abbrev: NMR.

nuclear-magnetic-resonance spectrometer An instrument for scanning and measuring the *nuclear magnetic resonance* spectrum of nuclei.

nuclear-magnetism log The open-hole *well log* that measures the free protons present in a few inches of rock near the well bore. It is in limited commercial use to measure permeability in sandstones and effective porosity in carbonates. Abbrev: NML. Syn: *free-fluid index log.*

nuclear quadrupole resonance Resonance of an atomic nucleus whose electric charge distribution deviates from a spherical distribution. Abbrev: NQR.

nuclear radiation Radiation from an atomic nucleus, commonly alpha or beta particles and gamma rays.

nuclear reaction (a) A change in identity of an atomic nucleus, brought about by interaction with an elementary particle or with another nucleus. (b) Any interaction of a nucleus with a particle or photon that involves the nuclear potential. Syn: *reaction [radioactivity].*

nuclear resonance *Resonance* occurring when a nucleus is irradiated with gamma rays of exactly the same frequency as those that the nucleus naturally tends to radiate.

nuclear-resonance magnetometer A type of magnetometer that measures total magnetic-field intensity by means of the precession of magnetic nuclei, precession frequency being proportional to field intensity. In practice, only the *proton-precession magnetometer* has been used.

nuclear snow gage Any type of gage using a radioactive source and a detector to measure, by the absorption or scattering of radiation, the mass or density of snow.

nucleation The beginning of crystal growth at one or more points.

nucleic acid A complex organic substance that is the genetic material in all known organisms. Cf: *DNA.*

nucleoconch *embryonic apparatus.*

nucleogenesis The origin of the chemical elements of the universe.

nucleosynthesis The generation of elements from hydrogen nuclei or protons by nuclear processes under the high-temperature, high-pressure conditions common in the life of a star.

nucleus The earliest-formed part of the shell or operculum of a gastropod. The term should not be used synonymously with *protoconch.*

nuclide A species of atom characterized by the number of neutrons and protons in its nucleus. See also: *radionuclide; cosmogenic nuclide.*

nucule (a) In fossil seeds, the central cavity formerly occupied by nucellus and gametophyte. (b) The female reproductive structure of a charophyte. It includes the oogonium and the outer protective cells.

nuculoid Any bivalve mollusk belonging to the order Nuculoida, characterized by a taxodont, equivalve, isomyarian shell with closed margins.

nudibranch Any *opisthobranch* belonging to the order (or suborder) Nudibranchia, characterized chiefly by the absence of a shell in the adult stage and by the absence of gills, or their replacement by secondary gills.

nuée ardente A swiftly flowing, turbulent gaseous cloud, sometimes incandescent, erupted from a volcano and containing ash and other pyroclastics in its lower part; a *density current* of pyroclastic flow. This lower part of the nuée ardente is comparable to an *ash flow,* and the terms are sometimes used synonymously in this sense. Etymol: French, "glowing cloud". Syn: *Peléan cloud; glowing cloud.*

nuevite *samarskite.*

Nuevoleonian North American (Gulf Coast) stage: Lower Cretaceous (above Durangoan, below Trinitian).

nuffieldite An orthorhombic mineral: $Pb_2Cu(Pb,Bi)Bi_2O_7$.

nugget A large lump of placer gold or other metal. Cf: *heavy gold.*

nullah (a) A term used in the desert regions of India and Pakistan for a sandy river bed or channel, or a small ravine or gully, that is normally dry except after a heavy rain. (b) The small, intermittent, generally torrential stream that flows through a nullah.—Etymol: Hindi *nala.* Pron: *nala.* See also: *wadi; arroyo.* Also spelled: *nulla; nallah; nalla.*

null hypothesis The assumption that no significant difference exists between two items or samples that are being compared statistically, and that any observed difference is purely by chance and not due to a systematic cause.

nullipore A coralline alga, formerly thought to be an animal.

number of streams Total number of stream segments of a specified order or orders in a given drainage basin. The symbol N_u refers to the total number of stream segments of a given order *u* within a specified drainage basin. See also: *law of stream numbers.*

numerical aperture A measurement or indicator of a microscope's resolving power.

numerical taxonomy The use of statistics in classifying and analyzing fossils and their paleoecologic implications.

nummulite Any foraminifer belonging to the family Nummulitidae, characterized by a test that is usually planispiral. Range, Upper Cretaceous to present. Adj: *nummulitic.* Var: *nummulitid.*

Nummulitic A syn. of *Paleogene,* used in Europe.

nummulitic limestone A *foraminiferal limestone* composed chiefly of nummulite shells; specif. the "Nummulite Limestone", a thick, distinctive, and widely distributed Eocene formation stretching from the Alps and northern Africa to China and eastern and southern Asia, composed esp. of the remains of the genus *Nummulites.*

nummuloidal *moniliform.*

nunakol A *nunatak* rounded by glacial erosion; a rounded "island" of rock in a glacier. Etymol: Eskimo. Syn: *rognon.*

nunatak An isolated hill, knob, ridge, or peak of bedrock that projects prominently above the surface of a glacier and is completely surrounded by glacier ice. Nunataks are common along the coast of Greenland. Etymol: Eskimo, "lonely peak". Swedish plural: *nunatakker.* Cf: *rognon; nunakol; marginal nunatak.* Also spelled: *nunatag.*

Nusselt's number The ratio of convective to conductive heat transfer.

nut An indehiscent, one-celled, and one-seeded hard and bony fruit, even if resulting from a compound ovary (Lawrence, 1951, p.762).

nutation The motion of the true axis of rotation of the Earth about its mean position with a principal term of about 18.6 years.

nutrient In oceanography, any inorganic or organic compound used to sustain plant life; e.g. silica for diatoms.

nyerereite A hexagonal mineral: $Na_2Ca(CO_3)_2$.

nymph (a) One of the narrow, thickened lunate processes or platforms of many bivalve mollusks extending posteriorly from the beak along the dorsal margin and serving for attachment of the ligament. Syn: *nympha.* (b) An immature stage in the life cycle of an acarid arachnid; e.g. protonymph, deutonymph, and tritonymph.

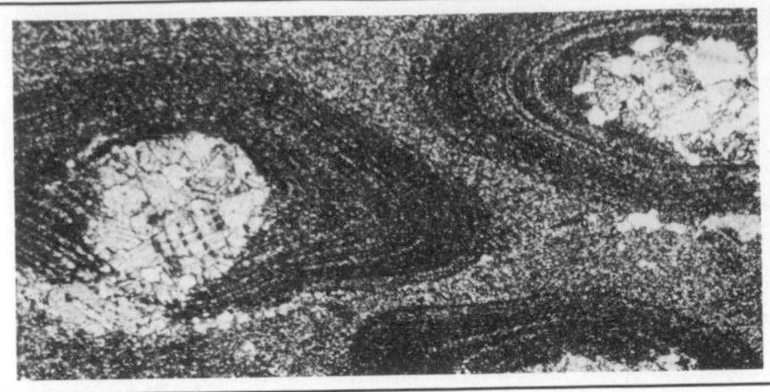

oasis [astrogeol] Any of numerous small dark spots occurring at the intersection of the supposed canals on the planet Mars. An astronomical term, now obsolete.

oasis [geog] A fertile, vegetated area in the midst of a desert, where the water table has come close enough to the surface for wells and springs or seepages to exist, thus making it suitable for human habitation.

obcordate Inversely heart-shaped, as in a leaf that is deeply lobed at the base (Swartz, 1971, p. 320). Cf: *cordate.*

obduction The overriding or overthrusting of oceanic crust onto the leading edges of continental lithospheric *plates;* plate accretion (Coleman, 1971).

object glass *objective.*

objective The lens (or lenses) that gives an image of an object in the focal plane of a microscope's or telescope's *eyepiece.* Syn: *objective lens; object glass.*

objective lens *objective.*

objective synonym In taxonomy, one of two or more names based on the same *type.* Cf: *subjective synonym.*

oblate Flattened or depressed at the poles; e.g. "oblate pollen" whose equatorial diameters are longer than the dimensions from pole to pole. Ant: *prolate.*

oblate ellipsoid An ellipsoid that is flattened at its poles.

oblate-prolate index A numerical index devised by Dobkins and Folk (1970, p. 1188) to describe the shape of a particle. It is based mainly on the value (L-I/L-S), which defines whether the intermediate axis I is closer in length to the short axis S or to the long axis L.

obligate adj. Said of an organism that can grow only under certain restricted conditions. Cf: *facultative.* Syn: *obligative.*

oblique n. An aerial photograph taken with the camera axis intentionally inclined. It combines the ground view with the pattern obtained from a height. See also: *high oblique; low oblique.*

oblique bedding An archaic syn. of *inclined bedding,* or bedding oblique to the principal surface of deposition; specif. *cross-bedding.*

oblique extinction *inclined extinction.*

oblique fault A fault that strikes oblique to, rather than parallel or perpendicular to, the strike of the constituent rocks or dominant structure. Cf: *oblique-slip fault; strike fault; dip fault.* Syn: *diagonal fault.*

oblique joint *diagonal joint.*

oblique lamination (a) *cross-lamination.* (b) *transverse lamination.*

oblique projection A projection that is not centered on a pole or on the equator and that does not use the equator or a meridian as a center line of orientation, or that has an axis inclined at an oblique angle to the equatorial plane; e.g. "oblique stereographic projection" or "oblique Mercator projection".

oblique section A slice through a foraminiferal test cut in a direction neither parallel nor normal to the axis of coiling.

oblique slip In a fault, movement or slip that is intermediate in orientation between the *dip slip* and the *strike slip.*

oblique-slip fault A fault on which the slip is oblique to, rather than parallel or perpendicular to, the dip of the constituent rocks or dominant structure. Cf: *oblique fault.* Syn: *diagonal-slip fault.*

obovate Inversely egg-shaped, as in a leaf whose terminal end is broader than its basal end. Cf: *ovate.*

obovoid Inversely ovoid, as in a fruit whose terminal portion is broader than its basal portion. Cf: *ovoid.*

obruchevite A brown mineral of the pyrochlore group: (Y,Na,Ca,U)(Nb,Ta,Ti,Fe)$_2$(O,OH)$_7$.

obsequent [geomorph] Said of a geologic or topographic feature that does not resemble or agree with a *consequent* feature from which it developed at a later date; esp. said of a tilt-block mountain (or of a rift-block mountain) that was formerly the floor of the original valley (or graben) but that was left standing as a result of differential erosion, or said of a tilt-block valley (or of a rift-block valley) that occupies the site of the former mountain (or horst) after the original topography was modified by differential erosion. Ant: *resequent.*

obsequent [streams] adj. Said of a stream, valley, or drainage system whose course or direction is opposite to that of the original consequent drainage. The term was proposed by Davis (1895, p. 134). Etymol: "opposite to consequent". See also: *anaclinal.*—n. *obsequent stream.*

obsequent fault-line scarp A *fault-line scarp* that faces in the opposite direction from the original fault scarp (i.e. faces the upthrown block) or in which the structurally downthrown block is topographically higher than the upthrown block. Cf: *resequent fault-line scarp.*

obsequent stream A stream that flows in a direction opposite to that of an original consequent stream and that is a tributary to a subsequent stream developed along the strike of weak beds; e.g. a short stream flowing down the scarp slope of a cuesta, or a stream flowing in a direction opposite to that of the dip of the local strata or the tilt of the land surface. See also: *scarp stream; antidip stream; reversed consequent stream.* Syn: *obsequent; anticonsequent stream; inverted stream.*

observation well A special well drilled in a selected location for the purpose of observing parameters such as fluid levels and pressure changes, e.g. within an oil reservoir as production proceeds.

observed gravity Gravity value obtained by either relative or absolute measurements.

obsidian A black or dark-colored volcanic glass, usually of rhyolite composition, characterized by conchoidal fracture. It is sometimes banded or has microlites. Usage of the term goes back as far as Pliny, who described the rock from Ethiopia. Obsidian has been used for making arrowheads, other sharp implements, jewelry, and art objects. Syn: *Iceland agate; hyalopsite; mountain mahogany.*

obsidian dating *obsidian hydration dating.*

obsidian hydration dating A method of calculating an age in years for an obsidian artifact or Holocene volcanic glass by determining the thickness of the hydration rim which has been produced by water vapor slowly diffusing into a freshly chipped surface and producing a hydrated layer or rind. It is applicable to glasses 200 to 200,000 years old. Syn: *hydration rind dating; obsidian dating.*

obsidianite An obsolete term, proposed by Walcott (1898) for a small rounded glassy object now known as a *tektite.* Most stones originally described as "obsidianite" were later shown to be true obsidian and not tektites.

obstacle mark *obstacle scour.*

obstacle scour A term introduced by Dzulynski and Walton (1965, p. 40) to refer to a scour mark produced by the interaction of an obstacle on the bed and a current flowing around it. Cf: *scour mark; current mark; current scour.* Syn: *obstacle mark.*

obstructed stream A stream whose valley has been blocked, as by a landslide, glacial moraine, sand dune, or lava flow; it frequently consists of a series of ponds or small lakes.

obstruction cave A *glacier cave* formed as the result of interrupted ice flow due to a bedrock or other subglacial protuberance. Entrance to such caves is generally impossible without tunneling.

obstruction moraine A moraine formed where the movement of ice is obstructed, as by a ridge of bedrock.

obtuse Blunt or rounded, as in a leaf.

obtuse bisectrix The *bisectrix* of the obtuse angle between the optic axes of a biaxial crystal. Cf: *acute bisectrix.*

obverse Pertaining to the *frontal* side of a bryozoan zooid or colony.—Cf: *reverse.*

occidental (a) Said of a gemstone of inferior quality, e.g. "occidental agate" (poorly marked and not very translucent). (b) Misrepresenting a substitute as being the genuine gem it represents; e.g. "occidental turquoise" (odontolite), or "occidental topaz" (citrine). (c) Said of a gemstone found in any part of the world other than the Orient.—Cf: *oriental.*

occidental cat's-eye A syn. of quartz *cat's-eye.*

occipital condyle An articular surface on the bone along the back part of the head of a tetrapod, by which the skull articulates with the first cervical vertebra (atlas). See also: *condyle.*

occipital furrow (a) The transverse groove on the cephalon of a trilobite running from axial furrow to axial furrow and forming the posterior boundary of the *glabella.* (b) A groove in front of the rim along the posterior border of the prosoma of a merostome (TIP, 1955, pt.P, p.8).

occipital ring The axial region of the most posterior element of the cephalon of a trilobite, generally delimited frontally by a prominent occipital furrow. Syn: *neck ring.*

occludent margin The margin of scutum and tergum forming the aperture in a cirripede crustacean and occluding it with comparable margins of opposed scutum and tergum.

occlusion A syn. of *absorption [chem].* Adj: *occluded.*

occlusor lamina In some cheilostome bryozoans, one of a pair of oppositely placed calcareous partitions arising from the lateral walls to provide the origin for the occlusor muscles.

occlusor muscle One of a pair of oppositely placed muscles that originate on the walls of a cheilostome bryozoan zooid and insert on the operculum (Ryland, 1970, p. 34).

occult mineral A mineral that might be expected to be present in a rock (as from the evidence of chemical analysis) but which is not identifiable, even with the aid of a microscope. Common examples are quartz and orthoclase in the glassy or cryptocrystalline groundmasses of certain lavas.

occupy To set a surveying instrument over a point for the purpose of making observations or measurements.

ocean The continuous salt-water body that surrounds the continents and fills the Earth's great depressions; also, one of its major geographic divisions. See also: *sea.*

ocean-basin floor The area of the sea floor between the base of the continental margin, usually the foot of the continental rise, and the mid-ocean ridge (Heezen and Menard, 1963, p. 236).

ocean current (a) A permanent predominantly horizontal movement of the surface water of the ocean, constituting part of its general circulation. (b) Broadly, any current in the ocean—tidal or nontidal, permanent or seasonal, horizontal or vertical—characterized by regularity, either as a continuous stream flowing along a definable path, or less commonly of a cyclic nature. It may be produced by wind stresses (drift), long-wave motions (tidal current), or density gradients due to variations in temperature and salinity (density or geostrophic currents).

ocean-floor spreading *sea-floor spreading.*

ocean hole *blue hole.*

oceanic (a) Pertaining to those areas of the ocean that are deeper than the littoral and neritic zones. (b) Pertaining to the ocean.

oceanic climate *marine climate.*

oceanic crust That type of the Earth's *crust* which underlies the ocean basins; it is equivalent to the *sima,* i.e. is characterized by the absence of the sialic layer. The oceanic crust is about 5-10 km thick; it has a density of 3.0 g/cm^3, and compressional seismic-wave velocities travelling through it exceed 6.2 km/sec. Cf: *continental crust.*

oceanic delta A delta built out into a tide-influenced sea and characterized by a concave delta plain (Moore, 1966, p. 87).

oceanic formation Name proposed for a formal or informal ocean-basin lithostratigraphic unit; the word "oceanic" should precede the formal or informal name (ACSN, Note 39, 1971, p. 1870).

oceanic ridge *mid-oceanic ridge.*

oceanic tholeiite The principal igneous rock encrusting the deeper parts of the ocean basins (Engel et al., 1965, p. 720). It is believed to be parental to *alkaline basalt.* Syn: *abyssal tholeiite.*

oceanic trench *trench [marine geol].*

oceanite A picritic *basalt* with more than 50 percent olivine in the mode. Named by Lacroix in 1923 for its occurrence at Piton de la Fournaise, Réunion Island, Indian Ocean. Not recommended usage.

oceanization The conversion of continental crust into oceanic crust.

oceanographic cast *hydrographic cast.*

oceanographic equator The zone of maximum temperature of the surface of the ocean; also defined as the zone in which the temperature of the surface of the ocean is higher than 28°C. Its position may vary seasonally, but it is always near to the geographic equator. Syn: *thermal equator.*

oceanography (a) The study of the ocean, including its physical, chemical, biologic, and geologic aspects. (b) In a narrower sense, the study of the marine environment.—Syn: *oceanology.*

oceanology *oceanography.*

ocellar Said of the texture of an igneous rock, esp. one with nepheline, in which the "phenocrysts" consist of aggregates of smaller crystals (e.g. of biotite or acmite) arranged radially or tangentially around larger, euhedral crystals (e.g. of leucite or nepheline) or form rounded eyelike branching forms. Also, said of a rock having such a texture.

ocellus [paleont] (a) A minute simple eye in an arthropod; e.g. the only type of eye found in an arachnid, the median visual organ located on the prosoma of a merostome, or an unpaired median eye common in some branchiopod and copepod crustaceans. Also, one of the elements of a *compound eye.* (b) A short hyaline process on the frustule in some diatoms (as in Auliscus).—Pl: *ocelli.*

ocellus [petrology] A phenocryst in an *ocellar* rock. Plural: *ocelli.*

ocher (a) An earthy, pulverulent, red, yellow, or brown iron oxide that is used as a pigment; e.g. "yellow or brown ocher" (limonite) and "red ocher" (hematite). Also, any of various clays strongly colored by iron oxides. (b) A similar earthy and pulverulent metallic oxide used as a pigment; e.g. "antimony ocher" (stibiconite and cervantite), "lead ocher" (massicot and litharge), and "tungstic ocher" (tungstite and ferritungstite).—Cf: *umber; sienna.* Also spelled: *ochre.*

ocherous Pertaining to, containing, or resembling ocher; e.g. "ocherous iron ore", a red, powdery or earthy hematite. Var: *ochreous; ochrous.*

Ochoan North American series: uppermost Permian (above Guadalupian, below Lower Triassic).

ochre *ocher.*

ochreous *ocherous.*

Ochrept In U.S. Dept. of Agriculture soil taxonomy, a suborder of the soil order *Inceptisol* that includes light-colored, brownish, more or less freely drained soils of mid to high latitudes. Most are formed in Holocene deposits or on Holocene surfaces. Ochrepts commonly have an ochric epipedon and a cambic horizon, and some also have a calcic horizon, fragipan, or duripan (USDA, 1975). Cf: *Andept; Aquept; Plaggept; Tropept; Umbrept.*

ochric epipedon A diagnostic surface horizon that is thinner, lighter in color, and lower in content of organic matter than a *mollic epipedon* or an *umbric epipedon* (USDA, 1975).

ochrous *ocherous.*

Ockham's razor The "principle of parsimony", which enjoins economy in explanation, i.e. use of the minimum number of assumptions. It was set forth by William of Ockham, English philosopher, d. ca. 1349.

Ocoee A provincial series of the Precambrian in Virginia, Tennessee, North Carolina, and Georgia.

ocrite A group name for powdery ochers.

octactine A sponge spicule having six equidistant rays in one plane and two rays at right angles to them.

octahedral Pertaining to an *octahedron.*

octahedral borax A variety of tincalconite occurring in crystals that simulate octahedrons, as from the lagoons of Tuscany, Italy.

octahedral cleavage Mineral cleavage parallel to the faces of the octahedron (111); e.g. in diamond.

octahedral coordination An atomic arrangement in which an ion is surrounded by six ions of opposite sign, whose centers form the corners of an octahedron. An example is the structure of NaCl. Cf: *six coordination.*

octahedral copper ore *cuprite.*

octahedral iron ore *magnetite.*

octahedral planes Those planes in a cubic crystal lattice having three equivalent Miller indices.

octahedrite [meteorite] The commonest *iron meteorite* containing 6-18% nickel in the metal phase and, on etching, showing Widmanstätten structure due to the presence of intimate intergrowths (of plates of kamacite with narrow selvages of taenite) oriented parallel to the octahedral planes. Symbol: *O.* Cf: *hexahedrite; ataxite.*

octahedrite [mineral] A syn. of *anatase.* The term is a misnomer because anatase crystallizes in tetragonal dipyramids and not in octahedrons.

octahedron An isometric crystal form of eight faces that are equilateral triangles. Its indices are {111} and its symmetry is $4/m\ \bar{3}\ 2/m$. Adj: *octahedral.*

octane Any of the several isomeric liquid paraffin hydrocarbons having the formula C_8H_{18}, including n-octane $CH_3(CH_2)_6CH_3$ which is found in petroleum.

octaphyllite (a) A group of mica minerals that contain eight cations per ten oxygen and two hydroxyl ions. (b) Any mineral of the octaphyllite group, such as biotite; a trioctahedral clay mineral.—Cf: *heptaphyllite.*

octocoral Any anthozoan belonging to the subclass Octocorallia, characterized by exclusively polypoid forms with pinnate tentacles and colonial growth. Range, Silurian (questionably) to present. See also: *alcyonarian.*

ocular [optics] *eyepiece.*

ocular [paleont] n. *ocular plate.*

ocular plate One of the outer circlet of primordial plates of the *apical system* of an echinoid, located at the aboral terminus of an ambulacrum and perforated by an ocular pore. Cf: *genital plate.* Syn: *ocular.*

ocular pore A perforation in an ocular plate of an echinoid for the passage of a *terminal tentacle.*

ocular sinus One of a pair of small, shallow sinuses at the sides of the aperture in the position of the eyes in *Nautilus.*

oculogenital ring A ring formed in echinoids by a circlet of ocular plates surrounding a circlet of genital plates in the center of the aboral surface at the apical end of ambulacral and interambulacral areas. It surrounds the periproct in regular echinoids, and it represents the initial plates of an echinoid skeleton. See also: *apical system.*

oculus A much enlarged part of the pore structure of triporate pollen, consisting of a thick protrusion of ektexine. Pl: *oculi.*

OD *ordnance datum.*

Oddo-Harkins rule A statement in geochemistry that, with four exceptions, the cosmic abundances of elements of even atomic number exceed those of adjacent elements of odd atomic number. This relationship was perceived by both Oddo and Harkins.

odd-pinnate Said of a pinnately compound leaf having an odd number of leaflets. Cf: *even-pinnate.* Syn: *imparipinnate.*

odenite A variety of biotite supposed to contain a new element (odenium).

odinite A greenish-gray dike rock composed of phenocrysts of labradorite and clinopyroxene, sometimes hornblende, in a groundmass of fine lath-shaped or equigranular feldspar and a felty mesh of acicular hornblende crystals. Named by Chelius in 1892 for the Odenwald, Germany. Obsolete.

odograph An instrument that automatically plots the course and distance traveled by a vehicle and that draws directly on paper by electronic or photoelectric methods a continuous map of the route taken.

odometer An instrument attached to a wheel of a vehicle to count the number of turns made by the wheel and used to measure the approximate distance traveled as a function of the number of revolutions and the circumference of the wheel.

odometry Mechanical measurement of distances.

Odontognathae A superorder of the avian subclass Neornithes that is restricted to the toothed forms of the late Mesozoic.

odontolite A fossil bone or tooth colored deep blue by iron phosphate (vivianite), and rarely green by copper compounds, and resembling turquoise, such as that from the tusks of mammoths found in Siberia. It is cut and polished for jewelry. Syn: *bone turquoise; fossil turquoise.*

odontology The study of teeth, including their structure, development, and diseases.

oeciopore The skeletal aperture of a brood chamber of stenolaemate bryozoans through which larvae escape.

oecostratigraphy *ecostratigraphy.*

oersted (a) Commonly used as the cgs (centimeter-gram-second) unit of *magnetic-field intensity.* Except in magnetized media, a magnetic field with an intensity H of one oersted has an induction B of one gauss. (b) The cgs unit of *magnetic force.*

offlap (a) The progressive offshore regression of the updip terminations of the sedimentary units within a conformable sequence of rocks (Swain, 1949, p.635), in which each successively younger unit leaves exposed a portion of the older unit on which it lies. Also, the successive contraction in the lateral extent of strata (as seen in an upward sequence) due to their being deposited in a shrinking sea or on the margin of a rising landmass. Ant: *onlap.* Syn: *regressive overlap.* (b) The progressive withdrawal of a sea from the land. Cf: *regression [stratig].* (c) A term commonly used by seismic interpreters for reflection patterns generated from strata prograding into deep water (Mitchum, 1977, p. 208).

off-lying *offshore.*

off-reef Pertaining to the sea floor away from the margin of a reef; e.g. the "off-reef facies" or the "off-reef sediments" deposited immediately surrounding a reef. Cf: *fore reef; interreef.* Also spelled: *offreef.*

offretite A zeolite mineral: $(K_2,Ca)(Al_{10}Si_{26})O_{72}\cdot 30\text{-}32H_2O$.

offset [cart] The small distance added (during construction of a map projection) to the length of meridians on each side of the central meridian in order to determine the top latitude of the constructed chart.

offset [coast] The migration of an upcurrent part of a shore to a position a little farther seaward than a downcurrent part; esp. the offset of a spit across a coastal inlet. Cf: *overlap.*

offset [fault] In a fault, the horizontal component of displacement, measured perpendicular to the disrupted horizon. Cf: *normal shift.* Syn: *normal horizontal separation.*

offset [geomorph] (a) A spur or minor branch from a range of hills or mountains. (b) A level terrace on a hillside.

offset [paleont] A new corallite formed in a corallum by budding; a corallite formed directly or indirectly from a protocorallite.

offset [seis] n. (a) The horizontal distance from energy source to detector, esp. to the nearest detector. (b) The horizontal distance from a shothole to the line of profile, measured perpendicularly to the line. (c) The horizontal displacement, measured from the detector, of a point for which a calculated depth applies.—v. To make such an adjustment of position or depth.

offset [surv] (a) A short line measured perpendicular to a traverse course or a surveyed line or principal line of measurement, for the purpose of locating a point with respect to a point on the course or line; e.g. a perpendicular distance measured from a great-circle line to a parallel of latitude in order to locate a section corner on that parallel in the U.S. Public Land Survey system. (b) A jog in a survey line which has approximately the same direction both before and after passing the jog.

offset deposit (a) A mineral deposit, esp of sulfides, formed partly by magmatic segregation and partly by hydrothermal solution, near the source rock. (b) At Sudbury, Ontario, the term refers to dikelike bodies radiating from the Sudbury Complex, thought to have been filled from above by xenolithic rock fragments and massive pyrrhotite-chalcopyrite-pentlandite.

offset line A supplementary line established close to and roughly parallel with the main survey line to which it is referenced by measured offsets. Offset lines are used where it is convenient to avoid obstructions, over which it would be difficult to make measurements, located along the main line.

offset ridge A ridge that is discontinuous on account of faulting.

offset shotpoint In seismic shooting, a shotpoint at some distance from the line of detectors. See also: *perpendicular offset; in-line offset.*

offset stream A stream displaced laterally or vertically by faulting.

offset well (a) An oil well drilled near the boundary of a property and opposite to a producing or completed well on an adjoining property, for the purpose of preventing the drainage of oil or gas by the earlier well. (b) Any *development well.*

offshore (a) Situated off or at a distance from the shore; specif. said of the comparatively flat, always submerged zone of variable width extending from the breaker zone to the seaward edge of the continental shelf. Ten meters is a suggested minimal depth. The offshore zone is seaward of the *inshore* or *nearshore* zone or the *shoreface* (CERC, 1966, p. A43; and Johnson, 1919, p. 161), although it is often regarded (e.g. Shepard, 1967, p.43) as the zone extending seaward from the low-water shoreline. (b) Pertaining to a direction seaward or lakeward from the shore; e.g. an *offshore* wind or one that blows away from the land, or an *offshore* current or one moving away from the shore.—Ant: *onshore.* See also: *nearshore.* Syn: *off-lying.*

offshore bar (a) *longshore bar.* (b) A catchall term used by Johnson (1919) for features now known as a *barrier beach* and a *barrier* *island.*—The term is undesirable as it has been applied both to a submerged feature (a *bar*) and an emergent feature (a *barrier*).

offshore barrier *barrier beach.*

offshore beach A syn. of *barrier beach.* The term was used by Gilbert & Brigham (1902, p. 306) for a long narrow sandy beach with a belt of quiet water separating it from the mainland.

offshore slope The frontal slope below the outer edge of the wave-built terrace.

offshore terrace A wave-built terrace in the offshore zone, composed of gravel and coarse sand. See also: *shoreface terrace.*

offshore water Water lying seaward of *inshore water,* whose physical properties are influenced only slightly by continental conditions. Ant: *inshore water.*

offtake *ground-water withdrawal.*

oghurd A term used in the Saharan region for a massive, mountainous dune, formed by some underlying rocky topographic feature, and rising considerably above the general dune level.

ogive A dark, curved, arcuate structure, one of a series repeated periodically down a glacier, generally formed at the base of an icefall, and resembling the pointed arch or rib across a Gothic vault (the "arch" is convex downslope due to faster flow in the middle of the glacier); esp. *dirt-band ogive.* Pron: *o*-jive. Cf: *Forbes band; wave ogive.*

O horizon An organic layer formed above a mineral soil incorporating one or both of two subdivisions: the O1 horizon, in which the original forms of plant remains are recognizable, and the O2, in which these forms cannot be recognized. Cf: *L layer; F layer [soil]; H layer.*

-oid A suffix (derived from the Greek) meaning "like, having the form of". A rock name or geologic feature thus qualified (e.g. granitoid, gneissoid) resembles but is not the same as the name or feature to which it is attached.

oikocryst In poikilitic fabric, the enclosing crystal.

oil *petroleum.*

oil accumulation *oil pool.*

oil-base mud A *drilling mud* with clay particles suspended in oil rather than in water.

oil-cut mud Drilling mud unintentionally admixed with crude oil; may result from oil entering the mud while drilling or from a *drill-stem test* of an oil reservoir.

oil field (a) An *oil pool.* (b) Two or more oil pools on a single geologic feature or otherwise closely related.

oil-field brine Water found by the drill in rocks penetrated at depth. It usually has a high concentration of dissolved salts.

oil pool A subsurface accumulation of petroleum that will yield crude oil in economic quantities. Cf: *oil field.* Syn: *oil accumulation.*

oil sand A term applied loosely to any porous stratum containing petroleum or impregnated with hydrocarbons; specif. a sandstone or unconsolidated sand from which oil is obtained by drilled wells. The term is also applied to productive limestone and dolomite. See also: *gas sand; tar sand; sand [drill].* Cf: *water sand.*

oil seep The emergence of liquid petroleum at the surface as a result of slow upward migration from its buried source. Syn: *seepage [petroleum].*

oil shale A *kerogen*-bearing, finely laminated brown or black sedimentary rock that will yield liquid or gaseous hydrocarbons on distillation. Syn: *kerogen shale; kerogenite.*

oil show *show (a).*

oil trap A *trap [petroleum]* that contains oil.

oil-water contact The boundary surface between an accumulation of oil and the underlying *bottom water.* Syn: *oil-water interface.*

oil-water interface *oil-water contact.*

oil well A *well* from which petroleum is obtained by pumping or by natural flow. Some statutes define the term on the basis of the gas-oil ratio.

ojo A term used in SW U.S. for a very small lake or a pond. Etymol: Spanish, "eye". Pron: *o*-ho.

okaite Hauyne *melilitolite* with accessory biotite, perovskite, apatite, calcite, and opaque oxides. It resembles *turjaite* except that the feldspathoid is hauyne rather than nepheline. Named by Stansfield in 1923 for Oka, Quebec. Not recommended usage.

okenite A whitish mineral: $CaSi_2O_4(OH)_2 \cdot H_2O$. Cf: *nekoite.*

old age [coast] A hypothetical stage in the development of a shore, shoreline, or coast, characterized by a wide wave-cut platform, a faintly sloping sea cliff pushed far inland, and a coastal region approaching peneplanation. The stage is probably a theoretical abstraction, since it is doubtful whether stability of sea level is maintained long enough for the land to be so reduced (Dietz, 1963).

old age [streams] The stage in the development of a stream at which erosion is decreasing in vigor and efficiency, and aggradation becomes dominant as the gradient is greatly reduced. It is characterized by: a load that exceeds the stream's ability to carry it, and is therefore readily deposited; a very broad, shallow, open valley with gently sloping sides and a nearly level floor (flood plain) that may be 15 times the width of the meander belt; numerous oxbows, meander scars, levees, yazoos, bayous, and swamps and lakes on valley floors; a sluggish current; graded or mature tributaries, few in number; and slow erosion, effected chiefly by mass-wasting at valley sides.

old age [topog] The final stage of the cycle of erosion of a landscape or region, in which the surface has been reduced almost to base level and the landforms are marked by simplicity of form and subdued relief. It is characterized by a few large meandering streams flowing sluggishly across broad flood plains, separated by faintly swelling hills, and having dendritic distributaries; and by peneplanation. Cf: *senescence; senility.* Syn: *topographic old age.*

Older Dryas n. A term used primarily in Europe for an interval of late-glacial time (centered about 11,500 years ago) following the Bølling and preceding the Allerød. It was characterized by tundra vegetation, and the climate, as inferred from stratigraphic and pollen data (Iversen, 1954), deteriorated so as to favor either expansion or retarded retreat of glaciers.—adj. Pertaining to the late-glacial Older Dryas interval and to its climate, deposits, biota, and events.

Oldest Dryas n. A term used primarily in Europe for a late-glacial interval (about 14,000-13,000 years ago) preceding the Bølling, characterized by tundra vegetation. The climate as inferred from stratigraphic and pollen data (Iversen, 1954) was colder than in the succeeding Bølling.—adj. Pertaining to the late-glacial Oldest Dryas interval and to its climate, deposits, biota, and events.

old-from-birth peneplain A term used by Davis (1922) for a peneplain presumably formed during an uplift of such extreme slowness over a long period of time that vertical corrasion was outpaced by valley-side grading and by general downwearing of the interstream uplands, thereby producing a landscape that will at once be "old" or that lacks any features characterizing youth or maturity; it is essentially a *primärrumpf.*

Oldham-Gutenberg discontinuity *Gutenberg discontinuity.*

oldhamite A pale-brown meteorite mineral: CaS.

old ice (a) Floating sea ice more than two years old (Armstrong et al., 1966, p. 30). It may be as much as 3 m or more thick, and it shows features that are smoother than those in second-year ice. (b) A term formerly applied to sea ice that has survived at least one summer's melt and that shows features that are smoother than those in first-year ice; e.g. second-year ice and multi-year ice. (c) A term loosely applied to a deposit of ice in permafrost.

old lake (a) A lake in an advanced stage of filling by sediments or vegetation. See also: *aging.* (b) A eutrophic or dystrophic lake. See also: *senescent lake.* (c) A lake whose shoreline exhibits an advanced stage of development.

oldland (a) Any ancient land; specif. an extensive area (such as the Canadian shield) of ancient crystalline rocks reduced to low relief by long-continued erosion and from which the materials of later sedimentary deposits were derived. (b) A region of older land, projected above sea level behind a coastal plain, that supplied the material of which the coastal-plain strata were formed; the land adjoining a new land surface that has just been brought above sea level. (c) A term proposed by Maxson & Anderson (1935, p. 90) for the land surface of the old-age stage of the cycle of erosion, characterized by subdued relief. Maxson later adopted (1950, p. 101) the earlier term *senesland* for this feature.

old mountain A mountain that was formed prior to the Tertiary period, esp. a *fold mountain* produced before the Alpine orogeny. Ant: *young mountain.*

Old Red Sandstone A thick sequence of nonmarine, predominantly red sedimentary rocks, chiefly sandstones, conglomerates, and shales, representing the Devonian System in parts of Great Britian and elsewhere in NW Europe. See Miller (1841). Abbrev: ORS.

old snow Fallen snow that has lost most traces of its original snow-crystal shapes, as *firn* or *settled snow,* through metamorphism. Ant: *new snow.* Syn: *firn snow.*

old Stone Age *Paleolithic.*

old stream A stream developed during the stage of *old age.*

olefin An unsaturated *aliphatic hydrocarbon,* empirical formula C_nH_{2n}, which contains at least one double bond. The olefins form a series, analogous to the methane series, of which ethylene, C_2H_4, a sweet-smelling gas present in common gas, is the lowest member. Also spelled: *olefine.*

olefine *olefin.*

Olenekian European provincial stage: lowermost Triassic (above Thuringian, below Induan).

olenellid Any trilobite belonging to the family Olenellidae, characterized generally by a subovate to elongate exoskeleton, the absence of dorsal sutures on the cephalon, numerous segments in the thorax, and well-developed pleural spines or acutely terminating, falcate distal portions (TIP, 1959, pt.O, p.191). Range, Lower Cambrian only.

oligist iron A syn. of *hematite.* Also spelled: *oligiste iron.*

oligo- A prefix meaning "small", "a little".

Oligocene An epoch of the early Tertiary period, after the Eocene and before the Miocene; also, the corresponding worldwide series of rocks. It is considered to be a period when the Tertiary is designated as an era.

oligoclase A mineral of the plagioclase feldspar group with composition ranging from $Ab_{90}An_{10}$ to $Ab_{70}An_{30}$. It is common in igneous rocks of intermediate to high silica content.

oligoclasite A granular *diorite* composed almost entirely of oligoclase. When the term was first defined, by Bombicci in 1868, it was applied to a rock containing more alkali feldspar than oligoclase, i.e. what was later called *cavalorite* (Johannsen, 1939, p. 246). Syn: *oligosite.* Not recommended usage.

oligomictic [lake] Said of a lake that circulates only at rare intervals when abnormally cold spells occur; e.g. a lake of small or moderate area or of very great depth, or in a region of high humidity, in which a small temperature difference between surface and bottom suffices to maintain stable stratification (Hutchinson, 1957, p.462). Cf: *polymictic [lake].*

oligomictic [sed] Said of a clastic sedimentary rock composed of a single rock type, such as an orthoquartzitic conglomerate; also, said of the clasts of such a rock. Oligomictic rocks are characteristic of stable conditions such as are found in epicontinental seas. Cf: *monomictic; polymictic.* Syn: *oligomict.*

oligonite A variety of siderite containing up to 40% manganese carbonate. Syn: *oligon spar.*

oligopelic Said of a lake-bottom deposit that contains very little clay (Veatch & Humphrys, 1966, p. 218).

oligophyre A light-colored *diorite* containing oligoclase phenocrysts in a groundmass of the same mineral. The name was proposed by Coquand in 1857. Not recommended usage.

oligosiderite A meteorite containing only a small amount of metallic iron. Cf: *holosiderite.*

oligosite *oligoclasite.*

oligostromatic Said of a plant part that is composed of only a few layers of cells. Cf: *monostromatic; polystromatic.*

oligotrophic lake A lake that is characterized by a deficiency in plant nutrients and usually by abundant dissolved oxygen in the *hypolimnion;* its bottom deposits have relatively small amounts of organic matter and its water is often deep. Cf: *dystrophic lake; mesotrophic lake; eutrophic lake.*

oligotrophic peat Peat containing a small amount of plant nutrients. Cf: *mesotrophic peat; eutrophic peat.*

oligotrophy The quality or state of an *oligotrophic lake.*

olistoglyph A hieroglyph produced by sliding or interlaminar gliding (Vassoevich, 1953, p.61); specif. a *slide mark.*

olistolith An *exotic block* or other rock mass transported by submarine gravity sliding or slumping and included within the binder of an olistostrome. Term introduced by G. Flores in Beneo (1955, p. 122).

olistostrome A sedimentary deposit consisting of a chaotic mass of intimately mixed heterogeneous materials (such as blocks and muds) that accumulated as a semifluid body by submarine gravity sliding or slumping of unconsolidated sediments. It is a mappable, lens-like stratigraphic unit lacking true bedding but intercalated among normally bedded sequences, as in the Tertiary basin of central Sicily. Raymond (1978) gives "olistostrome" as a general term for either a *broken formation* or a *mélange* of sedimentary origin. Cf: *allolistostrome; endolistostrome.* Term introduced by G. Flores in Beneo (1955, p. 122). Etymol: Greek *olistomai,* "to slide", + *stroma,* "bed".

olivenite An olive-green, dull-brown, gray, or yellowish orthorhombic mineral: $Cu_2(AsO_4)(OH)$. Syn: *leucochalcite; wood copper.*

olivine (a) An olive-green, grayish-green, or brown orthorhombic mineral: $(Mg,Fe)_2SiO_4$. It consists of the isomorphous solid-solution series forsterite-fayalite. Olivine is a common rock-forming mineral of basic, ultrabasic, and low-silica igneous rocks (gabbro, basalt, peridotite, dunite); it crystallizes early from a magma, weathers readily at the Earth's surface, and metamorphoses to serpentine. (b) A name applied to a group of minerals forming the isomorphous system $(Mg,Fe,Mn,Ca)_2SiO_4$, including forsterite, fayalite, tephroite, and a hypothetical calcium orthosilicate. Also, any member of this system.—See also: *peridot; chrysolite.* Syn: *olivinoid.*

olivine basalt (a) *Basalt* that lies on the plane of critical silica saturation, containing normative olivine and diopside with neither nepheline nor hypersthene (Yoder and Tilley, 1962). (b) The basalt of the island of Mull (Kennedy, 1933).

olivine clinopyroxenite In the *IUGS classification,* a plutonic rock with M equal to or greater than 90, ol/(ol+opx+cpx) between 5 and 40, opx/(ol+opx+cpx) less than 5, and cpx/(ol+opx+cpx) less than 90.

olivine gabbronorite In the *IUGS classification,* a plutonic rock satisfying the definition of *gabbro* and in which pl/(pl+px+ol) is between 10 and 90, px/(pl+px+ol) is greater than 5, and ol/(pl+px+ol) is greater than 5.

olivine-hornblende pyroxenite In the *IUGS classification,* a plutonic rock with M equal to or greater than 90, ol(ol+hbl+px) between 5 and 40, and pyroxene more abundant than amphibole.

olivine hornblendite In the *IUGS classification,* a plutonic rock with M equal to or greater than 90, ol/(ol+hbl+px) between 5 and 40, px/(ol+hbl+px) less than 5, and hbl/(ol+hbl+px) less than 90.

olivine leucitite *ugandite.*

olivine nephelinite An extrusive igneous rock differing in composition from nephelinite only by the presence of olivine. Syn: *nepheline basalt; ankaratrite.*

olivine orthopyroxenite In the *IUGS classification,* a plutonic rock with M equal to or greater than 90, ol/(ol+opx+cpx) between 5 and 40, cpx/(ol+opx+cpx) less than 5, and opx/(ol+opx+cpx) less than 90.

olivine-pyroxene hornblendite In the *IUGS classification,* a plutonic rock with M equal to or greater than 90, ol/(ol+hbl+px) between 5 and 40, and amphibole more abundant than pyroxene.

olivine pyroxenite In the *IUGS classification,* a plutonic rock with M equal to or greater than 90, ol/(ol+hbl+px) between 5 and 40, hbl/(ol+hbl+px) less than 5, and px/(ol+hbl+px) less than 90.

olivine rock *dunite.*

olivine tholeiite *Basalt* that is silica-undersaturated, containing normative olivine, hypersthene, and diopside, with neither quartz nor nepheline (Yoder and Tilley, 1962).

olivine websterite In the *IUGS classification,* a plutonic rock with M equal to or greater than 90, ol/(ol+opx+cpx) between 5 and 40, opx/(ol+opx+cpx) greater than 5, and cpx/(ol+opx+cpx) greater than 5.

olivinite (a) In the *IUGS classification,* a syn. of *dunite.* (b) An olivine-rich ore-bearing igneous rock that also contains other pyroxenes and/or amphiboles.

olivinoid (a) An olivinelike substance found in meteorites. (b) *olivine.*

ollenite A type of hornblende schist characterized by abundant epidote, sphene, and rutile. Garnet is one of the accessories (Holmes, 1928, p.170).

olmsteadite An orthorhombic mineral: $K_2Fe_4{}^{+2}(Nb,Ta)_2(PO_4)_4 \cdot 4H_2O$.

olocyst A term used in the older literature for a thin, smooth calcareous layer thought to form the initial deposit in frontal shields of certain ascophoran cheilostomes (bryozoans). Syn: *holocyst.*

olsacherite A mineral: $Pb_2(SeO_4)(SO_4)$.

olshanskyite A mineral: $Ca_3B_4(OH)_{18}$.

olynthus (a) The first stage in the development of a sponge, in which the initial functional aquiferous system has a single flagellated chamber. (b) Newly attached sponge larva resembling a vase in form and having a simple and asconoid body wall.

ombrogenous Said of peat deposits whose moisture content depends on rainfall. Cf: *soligenous; topogenous.*

ombrophilous Said of a plant adapted to extremely rainy conditions. Cf: *ombrophobous.* Noun: *ombrophile.*

ombrophobous Said of an organism that cannot tolerate extremely rainy conditions. Cf: *ombrophilous.* Noun: *ombrophobe.*

ombrotiphic Pertaining to a short-lived pond whose water is derived from rainfall. Cf: *tiphic.*

omission The elimination or nonexposure of certain stratigraphic beds at the surface or in any specified section owing to disruption and displacement of the beds by faulting. Ant: *repetition.*

omission solid solution A crystal in which there is incomplete filling of particular atomic sites. Cf: *substitution solid solution.* Syn: *defect-lattice solid solution.*

ommatidium One of the basic visual units of the compound eye of an arthropod. Pl: *ommatidia.*

omphacite A grass-green to pale-green granular aluminous clinopyroxene found as a common constituent in the rock *eclogite;* ideally $CaNaMgAlSi_4O_{12}$. In thin section it is colorless, superficially resembling olivine.

omuramba A term used in central and NE Namibia for the clearly defined dry bed of an intermittent stream, carrying water sometimes only as a series of shallow lakes and vleis. Etymol: Bantu (Herero). Pl: *omirimbi.* Cf: *oshana.*

oncoid An algal biscuit that resembles an ancient oncolite.

oncolite A small, variously shaped, concentrically laminated, calcareous sedimentary structure, resembling an oolith, and formed by the accretion of successive layered masses of gelatinous sheaths of blue-green algae. It is smaller than a *stromatolite* and generally does not exceed 10 cm in diameter. Also spelled: *onkolite.* Cf: *catagraphite.*

one-face-centered lattice A type of *centered lattice* in which the unit cell has one pair of faces centered, i.e. there are two lattice points per unit cell. If the (100) plane is centered, the symbol A is used; if the (010) plane is centered, the symbol B; and if the (001) plane is centered, the symbol C. In orthorhombic and monoclinic crystal lattices, all types are possible; in tetragonal crystal lattices, only the C type of centering is possible.

onegite A pale amethyst-colored sagenitic quartz penetrated by needles of goethite.

Onesquethawan North American stage: Lower and Middle Devonian (above Deerparkian, below Cazenovian).

one-year ice Sea ice of not more than one winter's growth, and a thickness of 70 cm to 2 m; it includes the "medium" and "thick" subdivisions of *first-year ice.*

ongonite Topaz-bearing quartz *keratophyre,* enriched in Li and F. Named by Kovalenko and co-workers in 1971 for a region in Mongolia.

onion-skin weathering *spheroidal weathering.*

onkilonite An olivine-leucite *nephelinite* that also contains augite and perovskite, but no feldspar. Backlund in 1915 named the rock for the Onkilones, the inhabitants of the type locality, Wilketski Island, U.S.S.R. Not recommended usage.

onkolite *oncolite.*

onlap (a) An *overlap* characterized by the regular and progressive pinching out, toward the margins or shores of a depositional basin, of the sedimentary units within a conformable sequence of rocks (Swain, 1949, p.635), in which the boundary of each unit is transgressed by the next overlying unit and each unit in turn terminates farther from the point of reference. Also, the successive extension in the lateral extent of strata (as seen in an upward sequence) due to their being deposited in an advancing sea or on the margin of a subsiding landmass. Ant: *offlap.* Cf: *overstep.* See also: *marine onlap; apparent onlap; proximal onlap; distal onlap.* Syn: *transgressive overlap; coastal onlap.* (b) The progressive submergence of land by an advancing sea. Cf: *transgression [stratig].*

onofrite A mineral: Hg(S,Se). It is a variety of metacinnabar containing selenium, and is a source of selenium.

onokoid A small or microcrystalline, dense, nodular, pealike body in ophthalmitic rocks (Niggli, 1954, p. 191). Little used.

Onondagan North American provincial stage: Lower Devonian (above Deerparkian, below Onesquethawan.

onoratoite A triclinic mineral: $Sb_8O_{11}Cl_2$.

onset *arrival.*

onset-and-lee topography *stoss-and-lee topography.*

onshore (a) Pertaining to a direction toward or onto the shore; e.g. an *onshore* wind or one that blows landward from a sea or lake, or an *onshore* current or one moving toward the shore. (b) Situated on or near the shore, as *onshore* oil reserves.—Ant: *offshore.*

Ontarian (a) Stage in New York State: Middle Silurian (middle and lower parts of Clinton Group). (b) An obsolete name for the Middle and Upper Ordovician in New York State.

ontogenetic stage Developmental stage in the growth of an individual organism.

ontogeny Development of an individual organism in its various stages from initiation through maturity. Adj: *ontogenetic.* Cf: *phylogeny.* Syn: *life cycle.*

ontozone A term used by Henningsmoen (1961) for a *biozone* that signifies the range of a taxon.

onychium The distal subsegment of the tarsus carrying the claws, found in some arachnids but wanting in others. Pl: *onychia.*

onyx (a) A variety of chalcedony that is like *banded agate* in consisting of alternating bands of different colors but unlike it in that the bands are always straight and parallel. Onyx is used esp. in making cameos. Cf: *agate; sardonyx; jasponyx.* (b) A name applied incorrectly to dyed, unbanded, solid-colored chalcedony; esp. *black onyx.* (c) *onyx marble.*—adj. (a) Parallel-banded; e.g. "onyx marble" and "onyx obsidian". (b) Jet black.

onyx agate A banded agate with straight parallel bands of white and different tones of gray.

onyx marble A compact, usually banded, generally translucent variety of calcite (or rarely of aragonite) resembling true onyx in appearance; esp. parallel-banded *travertine* capable of taking a good polish, and used as a decorative or architectural material or for small ornamental objects. It is usually deposited from cold-water solutions, often in the form of stalagmites and stalactites in caves. See also: *cave onyx.* Syn: *onyx; Mexican onyx; alabaster; oriental alabaster; Gibraltar stone; Algerian onyx.*

onyx opal Common opal with straight parallel markings.

oocast *oolicast.*

ooecium The *ovicell* in cheilostome bryozoans. Pl: *ooecia.*

oogonium The female sex organ in the thallophytes, containing one or more oospheres (Swartz, 1971, p. 324).

ooid (a) An individual spherite of an oolitic rock; an *oolith.* The term has been used in preference to "oolith" to avoid confusion with "oolite". Syn: *ooide.* (b) A general, nongeneric term for a particle that resembles an oolith in outer appearance and size (Henbest, 1968, p. 2). Cf: *pseudo-oolith.*—Adj: *ooidal.*

oolicast One of the small, subspherical openings found in an oolitic rock, produced by the selective solution of ooliths without destruction of the matrix. The term is inappropriate unless the opening is subsequently filled. See also: *oomold.* Syn: *oocast.*

oolicastic porosity The porosity produced in an oolitic rock by removal of the ooids and formation of oolicasts (Imbt & Ellison, 1947, p. 369-370).

oolite (a) A sedimentary rock, usually a limestone, made up chiefly of ooliths cemented together. The rock was originally termed *oolith.* Syn: *roestone; eggstone.* (b) A term often used for *oolith,* or one of the ovoid particles of an oolite.—Etymol: Greek *oon,* "egg". Pron: *oh-oh*-lite, not *oo*-lite. Cf: *pisolite.* Also spelled: *oölite.*

oolith One of the small round or ovate accretionary bodies in a sedimentary rock, resembling the roe of fish, and having diameters of 0.25 to 2 mm (commonly 0.5 to 1 mm). It is usually formed of calcium carbonate, but may be of dolomite, silica, or other minerals, in successive concentric layers, commonly around a nucleus such as a shell fragment, an algal pellet, or a quartz-sand grain, in shallow, wave-agitated water; it often shows an internal radiating fibrous structure indicating outward growth or enlargement at the site of deposition. Ooliths are frequently formed by inorganic precipitation, although many noncalcareous ooliths are produced by replacement, in which case they are less regular and spherical, and the concentric or radial internal structure is less well-developed, than in accretionary oolites. The term was originally used for a rock composed of ooliths (an *oolite*), and is sometimes so used today. Pron: *oh-oh*-lith. Cf: *pisolith.* Also spelled: *oölith.* Syn: *ooid; oolite; ovulite.*

oolitic Pertaining to an oolite, or to a rock or mineral made up of ooliths; e.g. an "oolitic ironstone", in which iron oxide or iron carbonate has replaced the calcium carbonate of an oolitic limestone. Also spelled: *oölitic.*

oolitic limestone An even-textured limestone composed almost wholly of relatively uniform calcareous ooliths, with virtually no interstitial material. It is locally an important oil reservoir (such as the Smackover Formation in Arkansas), and is also quarried for building stone.

oolitic texture The texture of a sedimentary rock consisting largely of ooliths showing tangential contacts with one another.

oolitization The act or process of forming ooids or an oolitic rock. Also, the result of such action or process.

oolitoid A sedimentary particle similar in size and shape to an oolith, but lacking its internal structure (Bissell & Chilingar, 1967, p. 162). Cf: *pseudo-oolith.*

ooloid A term used by Martin (1931, p.15) for a tiny, elliptically shaped, concretionlike siliceous form constructed of thin concentric layers around a central siliceous mass. It is found singly or cemented together in irregularly shaped clusters embedded in silicified shells of bryozoans, brachiopods, etc., and even composing the entire pseudomorphic shell. See also: *beekite.*

oomicrite A limestone containing at least 25% ooliths and no more than 25% intraclasts and in which the carbonate-mud matrix (micrite) is more abundant than the sparry-calcite cement (Folk, 1959, p. 14). It generally represents a mixing of two environments, such as where ooliths are swept from a bar into a muddy lagoon.

oomicrudite An oomicrite containing ooliths that are more than one millimeter in diameter.

oomold A spheroidal opening in a sedimentary rock or insoluble residue, produced by solution of an oolith. Adj: *oomoldic.* See also: *oolicast.*

oopellet A spherical or subspherical grain displaying characteristics of both an oolith and a pellet. "The internal part is pelletoidal, and thus may be ovoid in shape, but it has an accretionary coating, the thickness of all layers being equal to or slightly greater than the diameter of the pellet which they enclose" (Bissell & Chilingar, 1967, p. 162). Cf: *superficial oolith.*

oophasmic Said of a dolomite or recrystallized limestone that contains vague but unmistakable traces of oolitic texture (Phemister, 1956, p. 73).

oospararenite An oosparite containing sand-sized ooliths; an oolitic sandstone.

oosparite A limestone containing at least 25% ooliths and no more than 25% intraclasts and in which the sparry-calcite cement is more abundant than the carbonate-mud matrix (micrite) (Folk, 1959, p. 14). It is common in environments of high wave or current energy, where the spar usually represents pore-filling cement. Cf: *pisosparite.*

oosparrudite An oosparite containing ooliths that are more than one millimeter in diameter.

oosterboschite A mineral: $(Pd,Cu)_7Se_5$.

oovoid A void in the center of an incompletely replaced oolith.

ooze [geog] A piece of soft, muddy ground, such as a mudbank; a marsh, fen, or bog resulting from the flow of a spring or brook.

ooze [marine geol] A pelagic sediment consisting of at least 30% skeletal remains of pelagic organisms (either calcareous or siliceous), the rest being clay minerals. Grain size is often bimodal (partly in the clay range, partly in the sand or silt range). Oozes are further defined by their characteristic organisms: *diatom ooze; foraminiferal ooze; globigerina ooze; pteropod ooze; radiolarian ooze.* See also: *calcareous ooze; siliceous ooze.*

ooze [sed] (a) A soft, soupy mud or slime, typically found covering the bottom of a river, estuary, or lake. (b) Wet earthy material that flows gently, or that yields easily to pressure.

oozy Pertaining to or composed of ooze; e.g. "oozy fraction" of soils in which mineral grains are less than one micron in diameter.

opacite A general term applied to swarms of opaque, microscopic grains in rocks, esp. as rims that develop mainly on biotite and hornblende phenocrysts in volcanic rocks, apparently as a result of post-eruption oxidation and dehydration. Opacite is generally supposed to consist chiefly of magnetite dust. Cf: *viridite; ferrite [ign].*

opal A mineral or mineral gel: $SiO_2 \cdot nH_2O$. It has been shown by electron diffraction to consist of packed spheres of silica; some so-called opal gives weak X-ray patterns of cristobalite or tridymite. Opal has a varying proportion of water (as much as 20% but usually 3 to 9%); it occurs in nearly all colors, is transparent to nearly opaque and typically exhibits a marked iridescent *play of color.* It differs from quartz in being isotropic, and has a lower refractive index than quartz and is softer and less dense. Opal usually occurs massive and frequently pseudomorphous after other minerals, and is deposited at low temperatures from silica-bearing water. It is found in cracks and cavities of igneous rocks, in flintlike nodules in limestones, in mineral veins, in deposits of thermal springs, in siliceous skeletons of various marine organisms (such as diatoms and sponges), in serpentinized rocks, in weathering products, and in most chalcedony and flint. The transparent colored varieties exhibiting opalescence are valued as gemstones. Syn: *opaline.*

opal-agate A variety of banded opal having different shades of color and an agatelike structure, consisting of alternate layers of opal and chalcedony. Cf: *agate opal.*

opalescence A milky or somewhat pearly appearance or luster of a mineral, such as that shown by opal and moonstone. Cf: *play of color.*

opaline n. (a) Any of several minerals related to or resembling opal; e.g. a pale-blue to bluish-white opalescent or girasol corundum, or a brecciated impure opal pseudomorphous after serpentine. (b) *opal.* (c) An earthy form of gypsum. (d) A rock with a groundmass or matrix consisting of opal.—adj. Resembling opal, esp. in appearance; e.g. "opaline feldspar" (labradorite) or "opaline silica" (tabasheer).

opalite An impure, colored variety of common opal; e.g. *myrickite.*

opalized wood *silicified wood.*

opal jasper *jasper opal.*

opal phytolith A discrete, distinctively shaped *phytolith [paleont],* or solid body of isotropic silica, usually less than 80 microns in diameter, precipitated by terrestrial plants (sedges, reeds, some woods, and esp. grasses) as excess material or as reinforcement of cell structures. Such bodies may represent recent or fossil forms, can be transported by wind, and may be deposited in the ocean. Syn: *plant opal; grass opal.*

opaque Said of a material that is impervious to visible light, or of a material that is impervious to radiant energy other than visible light, e.g. radiation. Cf: *transopaque; translucent; transparent.*

opaque attritus Attrital material in coal, consisting of abundant particles of *inertinite,* opaque in transmitted light. Cf: *translucent attritus.*

opdalite A hypersthene-biotite *granodiorite.* It was named by Goldschmidt in 1916 after Opdal, Norway. Cf: *farsundite.* Not recommended usage.

open bay An indentation between two capes or headlands, so broad and open that waves coming directly into it are nearly as high near its center as on adjacent parts of the open sea; a *bight.*

opencast mining *opencut mining.*

open-cavity ice Ice that results by condensation from water vapor in an open cavity or crack in the ground. It is similar to hoarfrost, except that the ice crystals grow in cavities rather than on the surface. It is common in thermal contraction cracks and mine workings in permafrost.

open channel A conduit in which water flows with a free surface (ASCE, 1962).

open coast A coast exposed to the full action of waves and currents.

open-coast marsh A *salt marsh* found along an open coast. Cf: *coastal marsh.*

opencut mining Surficial mining, in which the valuable rock is exposed by removal of overburden. Coal, numerous nonmetals, and metalliferous ores (as of iron and copper) are worked in this way. Cf: *quarrying.* Syn: *strip mining; opencast mining; openpit mining.*

open fold A fold with an inter-limb angle between 70° and 120° (Fleuty, 1964, p. 470).

open form A crystal form whose faces do not enclose space, e.g. a trigonal prism. Cf: *closed form.*

open gash fracture *gash fracture.*

open hole (a) An uncased well or borehole, or that portion extending below the depth at which *casing* has been set. (b) A borehole free of any obstructing object or material.

open ice (a) Ice that is sufficiently broken up to permit passage of vessels. (b) *broken ice.*

open joint *fissure.*

open lake (a) A lake that has an effluent; e.g. a *drainage lake.* Ant: *closed lake.* (b) A lake having open water, free of ice or emergent vegetation.

open-packed structure In crystal structure, a pattern of stacking of equal spheres in an orthogonal arrangement such that each sphere is in contact with six others. Cf: *close-packed structure.*

open pack ice Pack ice in which the concentration is 4/10 through 6/10 with many leads and polynyas; the floes are generally not in contact with one another. See also: *broken ice; scattered ice.*

open packing The manner of arrangement of uniform solid spheres packed as loosely as possible so that the porosity is at a maximum; e.g. *cubic packing.* Ant: *close packing.*

openpit mining *opencut mining.*

open rock Any stratum sufficiently open or porous to contain a significant amount of water or to convey it along its bed.

open sound A sound with large openings between the protecting islands.

open-space structure A structure in a carbonate sedimentary rock, formed by the partial or complete occupation by internal sediments and/or cement (Wolf, 1965).

open structure A geologic structure which, when represented on a map by contour lines, is not surrounded by closed contours. Ant: *closed structure.*

open system [chem] A chemical system in which, during the process under consideration, material is either added or removed. Cf: *closed system.*

open system [permafrost] A condition of freezing of the ground in which additional ground water is available through either free percolation or capillary movement (Muller, 1947, p.219), exemplified by the pingos of East Greenland. Ant: *closed system [permafrost].*

open traverse A surveying traverse that starts from a station of known or adopted position but does not terminate upon such a station and therefore does not completely enclose a polygon. Cf: *closed traverse.*

open valley (a) A broad band of lowland, with relatively straight and parallel valley sides, through which a stream swings from side to side in broad, open curves (Rich, 1914, p. 469). (b) A strath produced by progressive widening of a valley by lateral stream cutting (Bucher, 1932, p. 131).

open water [ice] A relatively large area of freely navigable water in an ice-filled region; specif. water in which the concentration of floating ice is less than 1/8 (or 1/10). Cf: *ice free; polynya.*

open water [lake] (a) Lake water that remains unfrozen or uncovered by ice during the winter. (b) Lake water that is free of emergent vegetation or artificial obstructions and of dense masses of submerged vegetation at very shallow depths.

open well (a) A well large enough (one meter or more in diameter) for a man to descend to the water level. See also: *combination well.* (b) An artificial pond formed where a large excavation into the zone of saturation has been filled with water to the level of the water table (Veatch & Humphrys, 1966, p. 351).

openwork Said of a gravel with unfilled voids.

operational facies A term used by Krumbein & Sloss (1963, p.328) for stratigraphic *facies* designating lateral variations of any characteristic of a defined stratigraphic unit, occupying mutually exclusive areas bounded by arbitrarily (or preferably, quantitatively) determined limits, and usually comprising one or several lithosomes and biosomes that occur in vertical succession or are intertongued.

operational unit A term used by Sloss et al. (1949) for an arbitrary stratigraphic unit that is distinguished by objective criteria for some practical purpose (such as regional facies mapping or analysis); e.g. a unit delimited by easily recognizable and traceable markers, or a unit defined by the velocity of transmission of seismic or sonic energy. Its boundaries do not necessarily correspond with those of any conventional stratigraphic unit. Syn: *parastratigraphic unit.*

operator variation An effect of all experimental procedures, arising from a constant bias characteristic of each operator and inconsistent differences within one operator and among a group of operators.

operculate adj. Having an operculum; e.g. said of a pollen grain having pore membranes with an operculum, or said of an archeopyle covered by a lid.—n. An operculate gastropod.

operculum [bot] A lid or cover, as in a protistan, a moss capsule, or the fungi; it may be part of a cell wall.

operculum [paleont] (a) A corneous or calcareous plate that develops on the posterior dorsal surface of the foot of a gastropod and that serves to close the aperture. (b) A generally uncalcified lamina or flap, hinged or pivoting on condyles, that closes the zooidal orifice in cheilostome bryozoans. (c) The valves (terga or scuta) and associated membranes forming an apparatus that guards the aperture of cirripede crustaceans. (d) A lid, usually disklike and composed of one or two parts, that closes an opening (such as the anus or genital opening) of an arachnid; a plate adjoining the appendages of the genital segment of a merostome. (e) A lidlike covering of the calice in some solitary corals, formed of one or more independent plates. (f) A structure that may serve to close the pseudostome of chitinozoans. It may be external in position or sunken within the neck. (g) The flat pore-bearing base of the podoconus in nassellarian radiolarians; the central part of the astropyle of phaeodarian radiolarians. (h) In the vertebrates, a flat scalelike bone that serves as a controllable cover of an opening, e.g. in Osteichthyes the gill cover, in frogs and toads the small bone that partially closes the lateral opening of the inner-ear capsule.—Pl: *opercula.*

operculum [palyn] (a) A lid consisting of the plate or plates that originally filled the archeopyle of a dinoflagellate or the pylome of an acritarch. (b) A thicker central part of a pore membrane of a pollen grain, or a large section or cap of exine completely surrounded by a single circular colpus. For certain hilate spores and pollen, the operculum is a less well-defined lid of exine associated with the formation of the hilum.

opesiula One of the small notches or pores in a cryptocyst for the passage of parietal muscles attached to the frontal membrane of some anascan cheilostomes (bryozoans). Pl: *opesiulae.* Syn: *opesiule.*

opesium In some anascan cheilostomes (bryozoans), an opening defined by the inner margin of the cryptocyst, serving as a passage for the lophophore. It also serves as a passageway for parietal muscles in some anascans lacking opesiulae. Pl: *opesia.*

opferkessel A *solution pan* formed on silicate rocks and commonly stained by reddish-brown iron minerals. Etymol: German, "sacrificial basin".

Ophiacodontia A suborder of pelycosaurian synapsid reptiles, characterized by generally conservative structure; later forms appear to be more highly aquatic than most pelycosaurs. Range, Lower Pennsylvanian to Lower Permian.

ophicalcite A recrystallized metamorphic rock composed of calcite and serpentine, commonly formed by dedolomitization of a siliceous dolostone. Some ophicalcites are highly veined and brecciated and are associated with serpentinite.

ophicarbonate A metamorphic rock composed of serpentine and a carbonate mineral (calcite, dolomite, or magnesite). Cf: *ophicalcite.*

ophiocistioid Any quinqueradiate, free-living echinozoan belonging to the class Ophiocistioidea, having a depressed, dome-shaped body covered entirely or on one side only by plates. Range, Lower Ordovician to Upper Silurian (possibly Middle Devonian).

ophiolite A group of mafic and ultramafic igneous rocks ranging from spilite and basalt to gabbro and peridotite, including rocks rich in serpentine, chlorite, epidote, and albite derived from them by later metamorphism, whose origin is associated with an early phase of the development of a geosyncline. The term was originated by Steinman in 1905 (Miyashiro, 1968, p.826).

ophiolitic suite The association of ultramafic rocks, coarse-grained gabbro, coarse-grained diabase, volcanic rocks, and radiolarian chert that characteristically occurs in eugeosynclinal sequences.

ophirhabd A sinuous oxea (sponge spicule). Cf: *eulerhabd.*

ophite A general term for diabases that have retained their ophitic structure although the pyroxene is altered to uralite. The term was originated by Palasson in 1819.

ophitic Said of the holocrystalline, hypidiomorphic-granular texture of an igneous rock (esp. diabase) in which lath-shaped plagioclase crystals are partially or completely included in pyroxene crystals, typically augite. Also, said of a rock exhibiting ophitic texture, or, rarely, of a rock involving other pairs of minerals. The term *diabasic* was distinguished from "ophitic" by Kemp (1900, p. 158-159), who considered the latter as requiring an excess of augite over plagioclase, and the former as having a predominance of plagioclase, with augite filling the interstices. Cf: *poikilitic; poikilophitic.* Nonpreferred syns: *basiophitic; granitotrachytic; gabbroid.* Syn: *doleritic.*

ophiuroid Any asterozoan echinoderm belonging to the subclass Ophiuroidea, characterized by slender, elongate arms that are distinct from the disc in almost all cases; e.g. starfishlike animals such as brittle stars and basket stars. Var: *ophiurid; ophiuran.*

ophthalmic ridge A longitudinal ridge above the compound eye of a merostome and extending forward and backward from it.

ophthalmite [migma] Chorismite characterized by augen and/or other lenticular aggregates of minerals (Dietrich & Mehnert, 1961). Little used.

ophthalmite [paleont] *eyestalk.*

opisometer A *chartometer* consisting of a small toothed wheel, geared to a pointer moving over a graduated recording dial, used for measuring distances on a map. The wheel is run along a given line, which may be curved or irregular (such as one representing a stream, road, or railway).

opisthobranch Any marine gastropod belonging to the subclass Opisthobranchia, characterized by the reduction or absence of the shell. Range, Lower Carboniferous to present.

opisthoclade A *clade* or bar in the ebridian skeleton that arises from an upper actine and is directed toward the posterior. In the triaene ebridian skeleton, it may rejoin the distal extremity of the rhabde. Cf: *proclade.*

opisthocline (a) Said of the body of the shell (and in some genera, of the hinge teeth) of a bivalve mollusk, sloping (from the lower end) in the posterior or backward direction. (b) Said of the growth lines that incline backward relative to the growth direction of a gastropod shell.—Cf: *prosocline.*

opisthodetic Said of a ligament of a bivalve mollusk situated wholly posterior to (or behind) the beaks. Cf: *amphidetic.*

opisthogyrate Said of the umbones (of a bivalve mollusk) curved so that the beaks point in the posterior or backward direction. Ant: *prosogyrate.* Syn: *opisthogyral.*

opisthoparian adj. Of or concerning a trilobite whose facial sutures extend backward from the eyes to the posterior margin of the cephalon; e.g. an "opisthoparian facial suture" that crosses a cheek, passes along the medial edge of the eye, and intersects the posterior border of the cephalon medial to the genal angle. Cf: *proparian.* Syn: *opisthoparous.*—n. An opisthoparian trilobite; specif. a trilobite of the order Opisthoparia including those in which the genal angles or genal spines are borne by the free cheeks.

opisthosoma The posterior part of the body of an arthropod; esp. the *abdomen* behind the *prosoma* of a merostome or following the fourth pair of legs of an arachnid. Syn: *thoraceton.*

opisthosome A dark fusiform body at the base of the body chamber of a chitinozoan, usually convex upward or even spherical, and commonly having a ragged appearance (as if burst open) and a longitudinally striate surface. It is not always present, and may at times be mistaken for a fold of the body-chamber wall. Cf: *prosome.*

opoka A porous, flinty, and calcareous sedimentary rock, with conchoidal or irregular fracture, consisting of fine-grained opaline silica (up to 90%), and hardened by the presence of silica of organic origin (silicified residues of radiolaria, sponge spicules, and diatoms). It differs from *gaize* in its absence of quartz grains and the rarity of glauconite, although a distinction is not drawn in Russia between the two rocks (P.G.H. Boswell in Allen, 1936, p.11). Etymol: Polish.

Oppel-zone A *biozone* characterized by an association of selected taxons of restricted and largely concurrent range, chosen as indicative of approximate contemporaneity. Not all of the taxons considered diagnostic need be present at any one place for the zone to be legitimately identified. The lower part is commonly marked largely by first appearance, and the upper part by last appearance, of certain taxons; the body of the zone is marked largely by concurrence of the diagnostic taxons. The Oppel-zone is a more subjective, more loosely defined, and more easily applied biozone than the *concurrent-range zone.* The term has not been widely used, but the concept corresponds to a widespread practice in biostratigraphic zonation (ISG, 1976, p. 57-58). Named for Albert Oppel (1831-1865), German stratigrapher.

opportunistic species A species that tends to be generalized in its adaptations, disperses and reproduces rapidly (a high "r" in the logistic population-growth equation), but remains at small population size because of the unstable, unpredictable, or temporary nature of the environment occupied. Syn: *fugitive species; generalist species; r strategist.* Ant: *equilibrium species.*

opposite tide The high tide at a corresponding place on the opposite side of the Earth accompanying a *direct tide.*

optalic metamorphism *caustic metamorphism.*

optical activity The property or ability of a mineral, e.g. quartz, to rotate the plane of polarization of light. Such a mineral is said to be optically active. Syn: *rotary polarization.*

optical axis In an optical system, the line passing through the nodal points of a lens.

optical calcite Crystalline calcite so clear that it has value for optical use. It is usually *Iceland spar.*

optical center That point on the axis of an optical system at which light rays cross.

optical character In optical crystallography, the designation positive or negative, depending on the values of the different indices of refraction of a mineral. For uniaxial crystals with two indices of refraction, if the index of the extraordinary ray exceeds that of the ordinary ray, the mineral has a positive optical character. For biaxial crystals with three indices of refraction, the intermediate index is nearer in value to the smaller index than to the larger one for optically positive crystals.

optical constant Any characteristic optical property of a crystal, e.g. index of refraction, optic angle.

optical crystallography That branch of crystallography that deals with the optical properties of crystals; *crystal optics.* Cf: *optical mineralogy.*

optical emission spectrometry Chemical analysis performed by heating the sample to high temperatures whereupon the atoms emit light of definite wavelengths characteristic of the specific elements or molecules present (May & Cuttitta, 1967, p. 130). See also: *emission spectrum; arc spectrum; flame spectrum; spark spectrum.*

optical emission spectroscopy The observation of an optical emission spectrum and all processes of recording and measuring that go with it.

optical glass Glass that is suitable for use as prisms, lenses, and other optical items.

optically pumped magnetometer A type of magnetometer that measures total magnetic-field intensity by means of the precession of magnetic atoms, with precession frequency proportional to field intensity. The magnetic atoms are usually gaseous rubidium, cesium, or helium, which are magnetized by optical pumping, i.e. by irradiation by circularly polarized light of suitable wavelength. See also: *cesium-vapor magnetometer; rubidium-vapor magnetometer.*

optical microscope A microscope that utilizes visible light for illumination.

optical mineralogy That branch of science dealing with the optical properties of minerals. Cf: *optical crystallography.*

optical oceanography That aspect of physical oceanography which deals with the optical properties of seawater and natural light in seawater.

optical path The path along which light rays travel through the *optical system* of a microscope or other optical apparatus. Syn: *path [optics].*

optical pyrometer A type of *pyrometer* that measures high temperature by comparing the intensity of light of a particular wavelength from the hot material with that of a filament of known temperature. It is used to determine the temperature of incandescent lavas.

optical rotation The angle of rotation, measured in degrees, of plane-polarized light as it passes through an optically active crystal.

optical square A small hand instrument used in surveying for accurately setting off a right angle by means of two plane mirrors placed at an angle of 45 degrees to each other or by means of a single plane mirror so placed that it makes an angle of 45 degrees with a sighting line.

optical system The lenses, prisms, and mirrors of an optical apparatus such as a microscope, through which goes the *optical path.* Syn: *optical train.*

optical train *optical system.*

optical twinning A type of twinning in quartz, the individuals of which are alternately right-handed and left-handed, e.g. Brazil twinning. Syn: *chiral twinning.*

optical wedge (a) A refracting prism of very small angle, inserted in an optical train to introduce a small bend in the ray path. It is used in the eyepiece of certain stereoscopes. (b) A strip of film or a glass plate used to reduce the intensity of light or radiation (gradually or in steps, as in determining the density of a photographic negative), and having a layer of neutral or colored substance varying progressively in transmittance with distance along the wedge; e.g. a *step wedge.*

optic angle The acute angle between the two optic axes of a biaxial crystal; its symbol is 2V. See also: *apparent optic angle.* Syn: *axial angle; optic-axial angle.*

optic-axial angle *optic angle.*

optic axis A direction in an anisotropic crystal along which there is no double refraction. In tetragonal and hexagonal crystals it is parallel to the threefold, fourfold, or sixfold symmetry axis; in orthorhombic, monoclinic, and triclinic crystals there are two optic axes, which are determined by the indices of refraction. See also: *primary optic axis; secondary optic axis.*

optic ellipse Any noncircular section of an *index ellipsoid.*

optic indicatrix *indicatrix.*

optic normal The axis of a crystal that is perpendicular to the optic axis.

optimum moisture content The water content at which a specified force can compact a soil mass to its maximum dry unit weight.

oral adj. (a) Said of the surface on which the mouth of an invertebrate is situated, such as the upward-directed *actinal* surface of the theca of an edrioasteroid. Also, relating to or located on an oral surface, or situated at, near, or toward the mouth or peristome (such as of an echinoderm); e.g. an "oral pole" representing the end theca containing the mouth in a cystoid. Ant: *aboral.* (b) Pertaining to the *orifice* (not the mouth) of a bryozoan zooid. (c) Toward the upper side of a conodont element.—n. An *oral plate* of an echinoderm.

oral disk The fleshy, more or less flattened wall closing off the upper or free end of the cylindrical column that forms the sides of a scleractinian coral polyp, its center containing the mouth. Cf: *basal disk.* Also spelled: *oral disc.*

oral frame A structure in edrioasteroids underlying the external oral area, formed by proximal ambulacral floorplates and commonly other elements. It surrounds the central lumen or stomial chamber (Bell, 1976).

oral margin The trace of the oral side of a conodont element in lateral (side) view. The term has also been used for the oral side itself.

oral membrane A sheet of cilia in the gullet of a tintinnid.

oral pinnule Any proximal pinnule of a crinoid, differentiated from distal pinnules in function, structure, or both.

oral plate (a) Any of five interradially disposed plates forming a circlet surrounding or covering the mouth of an echinoderm. Syn: *oral.* (b) One of the elements in edrioasteroids that form the external covering of the oral area and roof the central lumen; distally continuous with the ambulacral coverplate series (Bell, 1976).

oral pole The end of a flask-shaped chitinozoan that includes the neck and the mouth. Cf: *aboral pole.*

oral side The upper side of a conodont element opposite that toward which the basal cavity opens. It commonly supports denticles, nodes, and ridges in compound and platelike conodont elements. Cf: *aboral side.*

oral tooth One of the sharp triangular projections around the basal shell opening in phaeodarian radiolarians.

orangite A bright orange-yellow variety of thorite.

oranite A lamellar intergrowth of a potassium feldspar and a plagioclase near anorthite.

orate Said of a porate pollen grain having an internal opening in the endexine.

O ray In uniaxial crystals, the ray that vibrates perpendicular to the optic axis; the *ordinary ray.* Cf: *E ray [cryst].*

orbicular (a) Said of the structure of a rock containing numerous orbicules; also, said of a rock having such structure. (b) Having the shape of an orbicule. Cf: *centric; nodular; spheroidal; spherulitic.*

orbiculate Said of a circular or disk-shaped leaf.

orbicule A more or less spherical body, from microscopic size to several centimeters in diameter, whose components are arranged in concentric layers. Their centers may or may not exhibit xenolithic nuclei. Cf: *spherulite [petrology].*

orbit [paleont] (a) A circular opening in the anterior part of the carapace of a decapod crustacean, enclosing the eyestalk. (b) In the vertebrates, a circular opening between skull roof and cheek that houses the eye.

orbit [waves] The path of a water particle affected by wave motion, being almost circular in deep-water waves and almost elliptical in shallow-water waves. Orbits are generally slightly open in the direction of wave motion, giving rise to *wave drift.*

orbital carina The ridge forming the orbital margin on some decapods (Holthuis, 1974, p. 735).

orbital region Part of the carapace of some decapods behind the eyes, bordered by the frontal and antennal regions (TIP, 1969, pt. R, p. 92).

orbite A porphyritic *gabbro* containing large phenocrysts of hornblende, or plagioclase and hornblende. Chelius in 1892 named the rock for Orbishöhe, Odenwald, Germany. Obsolete.

orbitoid Any foraminifer belonging to the superfamily Orbitoidacea, characterized by large discoidal saddle-shaped or stellate tests with walls composed of radially arranged calcite crystals and with bilamellid septa. Range, Cretaceous to present.

orbitolinid Any foraminifer belonging to the family Orbitolinidae, characterized by a relatively large conical test ranging from a high pointed cone to a broad shield or disc. Range, Lower Cretaceous to Eocene.

orbitolite Any foraminifer belonging to the genus *Orbitolites* of the suborder Miliolina, characterized by a discoidal test containing numerous small chambers in annular series. Range, Upper Paleocene to Eocene.

orcelite A mineral: $Ni_{5-x}As_2$.

ordanchite An extrusive rock containing phenocrysts of sodic plagioclase, hauyne, hornblende, augite, and some olivine; an olivine-bearing hauyne *trachyandesite.* Named by Lacroix in 1917 for Banne d'Ordanche, Auvergne, France. Not recommended usage.

order [geomorph] (a) *stream order.* (b) *basin order.*

order [petrology] In the CIPW classification of igneous rocks, the basic unit of the *class [petrology].*

order [taxon] A category in the hierarchy of classification of plants and animals intermediate between *class* and *family.* In botany, the name of an order characteristically ends in -ales; e.g. Filicales. Cf: *suborder.*

order-disorder inversion *substitutional transformation.*

order-disorder polymorphs Two crystal substances of the same composition but of different atomic arrangement. In the higher-temperature or disordered form, two or more elements are randomly distributed over a particular set of atom sites; in the lower-temperature or ordered form, the atoms become ordered with respect to the same sites. The ordered form usually has lower symmetry.

order-disorder transformation A *transformation [cryst]* between two polymorphic forms, one of which has a more ordered structure than the other. In general, if the ordered, low-symmetry, low-temperature form is heated, a point is reached at which some portion of the structure becomes disordered, or random, usually with an increase in crystal symmetry, to produce the high-temperature form. Cf: *substitutional transformation.*

order in minerals The ordered substitution of one ion for another in the crystal structure, e.g. in microcline, in which one fourth of the Si positions are occupied by Al. Cf: *disorder in minerals.* See also: *short-range order; long-range order.*

order of crystallization The apparent chronologic sequence in which crystallization of the various minerals of an assemblage takes place, as evidenced mainly by textural features. Syn: *sequence of crystallization.*

ordinary chert A generally homogeneous *smooth chert.* It has an even fracture surface, is nearly opaque, has slight granularity or crystallinity, and may be of any color (chiefly white, gray, or brown, or sometimes mottled) (Ireland et al., 1947, p. 1485).

ordinary coccolith One of the unmodified coccoliths in a coccolithophore exhibiting dimorphism.

ordinary lead *common lead.*

ordinary ray *O ray.*

ordinary tide level *mean tide level.*

ordnance datum A name given to several horizontal datums to which heights have been referred on official maps of the British Ordnance Survey; specif. in Great Britain (but not Ireland) the mean sea level at Newlyn in Cornwall. Abbrev: OD.

ordonezite A brown tetragonal mineral: $ZnSb_2O_6$. Also spelled: *ordoñezite.*

ordosite A dark-colored *syenite,* containing about 60 percent sodic clinopyroxene. The name, given by Lacroix in 1925, is for the Ordos Plateau, China. Obsolescent.

Ordovician The second earliest period of the Paleozoic era (after the Cambrian and before the Silurian), thought to have covered the span of time between 500 and 440 million years ago; also, the corresponding system of rocks. It is named after a Celtic tribe called the Ordovices. In the older literature the Ordovician is sometimes known as the *Lower Silurian.* See also: *age of marine invertebrates.* Obsolete syn: *Champlainian.*

ore (a) The naturally occurring material from which a mineral or minerals of economic value can be extracted at a reasonable profit. Also, the mineral(s) thus extracted. The term is generally but not always used to refer to metalliferous material, and is often modified by the name of the valuable constituent, e.g., "iron ore". See also: *mineral deposit; orebody; ore mineral.* (b) The term "ores" is sometimes applied collectively to opaque accessory minerals, such as ilmenite and magnetite, in igneous rocks.

ore beds Metal-rich layers in a sequence of sedimentary rocks.

ore block A section of an orebody, usually rectangular, that is used for estimates of overall tonnage and quality. See also: *blocking out.*

ore blocked out *developed reserves.*

orebody A continuous, well-defined mass of material of sufficient *ore* content to make extraction economically feasible. See also: *mineral deposit.*

ore channel A little-used term for the orebody or lode, including both gangue and economically valuable minerals. See also: *lodestuff.* Syn: *lode country.*

ore chimney *pipe [ore dep].*

ore cluster A genetically related group of orebodies that may have a common root or source rock but that may differ structurally or otherwise.

ore control Any tectonic, lithologic, or geochemical feature considered to have influenced the formation and localization of ore. Cf: *metallotect.*

ore-forming fluid *mineralizer.*

oregonite A hexagonal mineral: Ni_2FeAs_2.

ore guide Any natural feature, such as alteration products, geochemical variations, local structures, or plant growth, known to be indicative of an orebody or mineral occurrence. See also: *lithologic guide; stratigraphic guide.*

ore in sight *developed reserves.*

Orellan North American continental stage: Middle Oligocene (above Chadronian, below Whitneyan).

ore magma A term proposed by Spurr (1923) for a magma that may crystallize into an ore; the sulfide, oxide, or other metallic facies of a solidified magma.

ore microscopy The study of opaque ore minerals in polished section with a reflected-light microscope. Syn: *mineragraphy; mineralography.*

ore mineral The part of an *ore,* usually metallic, which is economically desirable, as contrasted with the *gangue.*

Orenburgian Stage in Russia: uppermost Upper Carboniferous (above Gzhelian, below Permian Asselian).

orendite A porphyritic leucite *lamproite* containing phlogopite phenocrysts in a nepheline-free reddish-gray groundmass of leucite, sanidine, phlogopite, amphibole, and diopside; a phlogopite-leucite *trachyte.* Its name, given by Cross in 1897, is derived from Orenda Butte, Leucite Hills, Wyoming. Not recommended usage.

ore of sedimentation *placer.*

oreography *orography.*

ore pipe *pipe [ore dep].*

ore roll *roll orebody.*

ore shoot An elongate pipelike, ribbonlike, or chimneylike mass of ore within a deposit (usually a vein), representing the more valuable part of the deposit. Syn: *shoot [ore dep].*

organ genus A genus name used for groups of fossil plants that are assignable to a family with minimal distinction from genera of plants as normally considered.

organic adj. Pertaining or relating to a compound containing carbon, especially as an essential component. Organic compounds usually have hydrogen bonded to the carbon atom. Cf: *inorganic.* —n. A substance containing carbon, as in such expressions as "organic-rich shale".

organic bank *bank [sed].*

organic evolution *evolution.*

organic geochemistry That branch of chemistry concerned with naturally occurring carbonaceous and biologically derived substances of geological interest.

organic hieroglyph *bioglyph.*

organic lattice *growth lattice.*

organic mound *bioherm.*

organic reef A *bioherm* of sufficient size to develop associated facies. It is erected by, and composed mostly of the remains of, sedentary or colonial and sediment-binding organisms, generally marine: chiefly corals and algae, less commonly crinoids, bryozoans, sponges, mollusks, and other forms that live their mature lives near but below the surface of the water (although they may have some exposure at low tide). Their exoskeletal hard parts remain in place after death, and the deposit is firm enough to resist wave erosion. An organic reef may also contain still-living organisms. See also: *coral reef; algal reef.* Cf: *bank [sed].*

organic rock A sedimentary rock consisting primarily of the remains of organisms (plant or animal), such as of material that originally formed part of the skeleton or tissues of an animal. Cf: *biogenic rock.*

organic soil A general term applied to a soil that consists primarily of organic matter such as peat soils and muck soils. Cf: *Histosol.*

organic soil material Water-saturated soil material that contains a certain minimum percentage of organic carbon, depending on the clay content of the mineral fraction: for 60% or more clay, 18% or more carbon; for no clay, 12% or more carbon; and for intermediate clay contents, a proportionate amount of carbon. Soil material that has never been saturated with water must have 20% or more organic carbon (USDA, 1975).

organic texture A sedimentary texture resulting from the activity of organisms.

organic weathering Biologic processes and changes that assist in the breakdown of rocks; e.g. the penetrating and expanding force of roots, the presence of moss and lichen causing humic acids to be retained in contact with rock, and the work of animals (worms, moles, rabbits) in modifying surface soil. Syn: *biologic weathering.*

organized elements A term used by Claus & Nagy (1961) for circular, microscopic organic particles observed in carbonaceous chondrites, having diameters of about 4 to 30 microns, and resembling fossil algae, dinoflagellates, and other terrestrial microorganisms. They are believed by some to be extraterrestrial microfossils and by others to be terrestrial microbiologic contaminations or inorganic microstructures (such as mineral grains or sulfur droplets).

organogenic Said of a rock or sediment made up of products of organic activity; e.g. a crinoidal limestone (termed an "organogenic conglomerate" by Hadding, 1933).

organolite Any rock consisting mainly of organic material, esp. one derived from plants; e.g. coal, resin, and bitumen.

organometallic Said of a compound in which an atom of a metal is bound to an organic compound directly through a carbon atom. Cf: *metallo-organic.*

organosedimentary Pertaining to sedimentation as affected by organisms; e.g. said of a stromatolite, a sedimentary structure produced by the life processes of blue-green algae.

organotrophic Relating to the development and nourishment of living organs.

orido An Italian term for a gorge cut through a rock barrier holding a lake in a glaciated region, like those around Lago di Como in Italy.

orient [gem] (a) The minute play of color on, or just below, the surface of a gem-quality pearl, caused by diffraction and interference of light from the irregular edges of the overlapping crystals or platelets of aragonite that comprise the nacre of the pearl. (b) A pearl of great luster.

orient [surv] (a) To place or set a map so that the map symbols are parallel with their corresponding ground features. (b) To turn a plane table in a horizontal plane until all lines connecting positions on the plane-table sheet have the same azimuths as the corresponding lines connecting ground objects. (c) To turn a transit so that the direction of the zero-degree line of its horizontal circle is parallel to the direction it had in the preceding (or in the initial) setup or parallel to a standard line of reference.—Etymol: Latin *oriens,* "rising", used originally in connection with the rising of the Sun in the east.

oriental (a) Said of a genuine gemstone; e.g. "oriental ruby" (the true ruby), or "oriental sapphire" (the true sapphire). (b) Indicating a finer variety of a gem, or one having superior grade, luster, or value; e.g. "oriental chalcedony" (fine, translucent, gray to white chalcedony), or "oriental carnelian" (deep, bright red, translucent carnelian). When applied to minerals, the term is frequently used in the same sense as *precious,* e.g. "oriental opal" (fire opal). (c) Being corundum or sapphire but simulating another gem in color; e.g. "oriental amethyst", "oriental aquamarine", "oriental beryl", "oriental emerald", and "oriental topaz", are all varieties of corundum. (d) Said of natural pearls as opposed to cultured pearls. (e) Said of a gem originating in the Orient.—Cf: *occidental.*

oriental alabaster *onyx marble.*

oriental amethyst (a) Violet to purple variety of sapphire. (b) Any amethyst of exceptional beauty.

oriental cat's-eye A syn. of chrysoberyl *cat's-eye.*

oriental chrysolite Greenish-yellow chrysoberyl.

oriental jasper *bloodstone.*

orientation [cryst] In describing crystal form and symmetry, the placing of the crystal so that its crystallographic axes are in the conventional position.

orientation [photo] The direction in which an aerial photograph is turned with respect to observer or map. A single photo is best oriented for study when turned so that the shadows are cast toward the observer.

orientation [surv] The assignment or imposition of a definite direction in space; the act of establishing the correct relationship in direction, usually with reference to the points of the compass. Also, the state of being in such relationship.

orientation diagram In structural petrology, a general term for a fabric diagram.

oriented Said of a specimen or thin section that is so marked as to show its original position in space.

oriented core A *core* that can be positioned on the surface as it was in the borehole prior to extraction.

orientite A brown to black orthorhombic mineral: $Ca_2Mn^{+3}{}_3(SiO_4)_3(OH)$.

orifice (a) An opening in the zooid wall through which the lophophore and tentacles of a bryozoan are protruded. Cf: *aperture.* (b) An opening in the upper part of a crustacean shell, containing the operculum. (c) Any major opening through the outer covering of an echinoderm. (d) An aperture or other opening in a foraminiferal test.

origin (a) A point in a coordinate system that serves as an *initial point* in computing its elements or in prescribing its use; esp. the point defined by the intersection of coordinate axes, from which the coordinates are reckoned. The term has also been applied to the point to which the coordinate values of zero and zero are assigned (regardless of its position with reference to the axes) and to the point from which the computation of the elements of the coordinate system, or projection, proceeds. Syn: *origin of coordinates.* (b) Any arbitrary zero or starting point from which a magnitude is reckoned on a scale or other measuring device.

original dip *primary dip.*

original horizontality The state of strata being horizontal or nearly horizontal at the time they were deposited. See also: *law of original horizontality.*

original interstice An *interstice* that formed contemporaneously with the enclosing rock. Cf: *secondary interstice.* Syn: *primary interstice.*

original stream *consequent stream.*

original valley A valley formed by *hypogene* action, or by *epigene* action other than that of running water (Geikie, 1898, p. 347).

origofacies Facies of the primary sedimentary environment (Vassoevich, 1948); the sedimentary "facies" of most western authors (Teichert, 1958, p.2736).

Oriskanian North American provincial stage: Lower Devonian (above Helderbergian, below Deerparkian).

orizite *epistilbite.*

ornament A pattern of diagonal lines, plus-signs, curlicues, or the like, printed in black, grey, a contrasting color, or negative (white), over a color hue on a geologic map, distinguishing one cartographic unit from another of basically the same hue. The pattern may suggest lithology and/or internal arrangement of rock-unit constituents.

ornamental stone An attractive natural stone, usually opaque, that is not practical for jewelry but is useful for fashioning into ornamental and decorative objects, such as figurines, ash trays, and lamp bases; e.g. onyx marble, agate, or malachite. Cf: *curio stone.*

ornamentation [paleont] The characteristic markings or patterns on the body of an animal; e.g. the external surface features of preserved hard parts (ridges, grooves, granules, spines, etc.) that may interrupt the smooth surface of a shell. See also: *sculpture [paleont]; prosopon.*

ornamentation [palyn] *sculpture [palyn].*

Ornithischia One of the two orders of archosaurian reptiles commonly treated as *dinosaurs,* characterized by a bifurcate or reduced pubis, a tendency toward elaboration of the dental battery and development of a beak, and herbivorous habit. Range, Upper Triassic to Upper Cretaceous. Cf: *Saurischia.*

ornithocopros *guano.*

ornoite A variety of hornblende *diorite.* "There seems to be no necessity for a new name for this rock" (Johannsen, 1937, p. 115-116).

orocline An orogenic belt with an imposed curvature or sharp bend, interpreted by Carey (1958) as a result of horizontal bending of the crust, or "deformation in plan". Syn: *geoflex.*

oroclinotath An orogenic belt, interpreted by Carey (1958) as having been subjected both to substantial horizontal bending and to stretching along the strike.

orocratic Pertaining to a period of time in which there is much diastrophism. Cf: *pediocratic.*

orogen *orogenic belt.*

orogene *orogenic belt.*

orogenesis *orogeny.*

orogenetic *orogenic.*

orogenic Adj. of *orogeny.* Cf: *orographic.*

orogenic belt A linear or arcuate region that has been subjected to folding and other deformation during an *orogenic cycle.* Orogenic belts are *mobile belts* during their formative stages, and most of them later became mountain belts by postorogenic processes. Syn: *fold belt; orogen; orogene.*

orogenic cycle The interval of time during which an originally *mobile belt* evolved into a stabilized *orogenic belt,* passing through a *preorogenic phase,* an *orogenic phase,* and a *postorogenic phase.* The concept has been rendered obsolete by the recognition of the plate structure of the Earth (Coney, 1970). Syn: *geologic cycle; tectonic cycle; geotectonic cycle.*

orogenic facies A term applied to a *geosynclinal facies* when emphasizing its tectonic environment.

orogenic phase The median part of an *orogenic cycle,* characterized by a climax of crustal mobility and orogenic activity. It is commonly shorter than the *preorogenic phase* and the *postorogenic phase,* and may be less than a geologic period in length, although it is commonly prolonged by a succession of *pulsations.*

orogenic sediment Any sediment that is produced as the result of an orogeny or that is directly attributable to the orogenic region in which it later becomes involved; e.g. a clastic sediment such as flysch or molasse.

orogenic unconformity An *angular unconformity* produced locally in a region affected by mountain-building movements.

orogeny Literally, the process of formation of mountains. The term came into use in the middle of the 19th Century, when the process was thought to include both the deformation of rocks within the mountains, and the creation of the mountainous topography. Only much later was it realized that the two processes were mostly not closely related, either in origin or in time. Today, most geologists regard the formation of mountainous topography as postorogenic. By present geological usage, orogeny is the process by which structures within fold-belt mountainous areas were formed, including thrusting, folding, and faulting in the outer and higher layers, and plastic folding, metamorphism, and plutonism in the inner and deeper layers. Only in the very youngest, late Cenozoic mountains is there any evident causal relation between rock structure and surface landscape. Little such evidence is available for the early Cenozoic, still less for the Mesozoic and Paleozoic, and virtually none for the Precambrian—yet all the deformational structures are much alike, whatever their age, and are appropriately considered as products of orogeny. See also: *diastrophism.* Cf: *epeirogeny; tectogenesis; cymatogeny; morphotectonics.* Syn: *orogenesis; mountain building; tectogenesis.* Adj: *orogenic; orogenetic.*

orogeosyncline Kober's term for a geosyncline that later became an area of orogeny (Glaessner & Teichert, 1947, p. 588). See also: *mother geosyncline.*

orographic [geog] (a) Pertaining to mountains, esp. in regard to their location and distribution. (b) Said of the precipitation that results when moisture-laden air encounters a high barrier and is forced to rise over it, such as the precipitation on the windward slopes of a mountain range facing a steady wind from a warm ocean. Also, said of the lifting of an air current caused by its passage up and over a mountain. (c) Pertaining to a *rain-shadow desert.*

orographic [tect] A term, now little used, for features relating to mountain structure and topography. More explicit adjectives such as *diastrophic, epeirogenic,* and *orogenic,* are now preferred.

orographic desert *rain-shadow desert.*

orographic snowline The lower elevation limit of patches of perennial snow between glaciers. Term was introduced by Ratzel in 1886 (Flint, 1971, p. 64). Cf: *regional snow line.*

orography (a) The branch of physical geography that deals with the disposition, character, formation, and structure of mountains and of chains, ranges, and systems of mountains. (b) Broadly, the description or depiction of the relief of the Earth's surface or of a part of it, or the representation of such relief on a map or model; the land features of a specified region.—Etymol: Greek *oros,* "mountain". Syn: *orology; oreography.*

orohydrography A branch of hydrography dealing with the relations of mountains to drainage.

orology A syn. of *orography;* esp. the study of mountain building and mountain formation.

orophilous Said of an organism that lives in *montane* conditions.

orophyte A plant growing in *montane* regions.

orotath An orogenic belt, interpreted by Carey (1958) as having been substantially stretched in the direction of its length.

orotvite A *diorite* composed of hornblende, biotite, plagioclase, nepheline, and cancrinite, with accessory sphene, ilmenite, and apatite. The name was given by Streckeisen in 1939 for the Orotva Valley, Detro, Romania. Not recommended usage.

orpheite A trigonal mineral: $H_6 Pb_{10} Al_{20} (PO_4)_{12} (SO_4)_5 (OH)_{40} \cdot 11H_2O$.

orpiment A lemon-yellow to orange monoclinic mineral: As_2S_3. It is generally foliated or massive, and is frequently associated with realgar. Orpiment occurs as a deposit from some hot springs and as a sublimate from some volcanoes. Syn: *yellow arsenic.*

ORS *Old Red Sandstone.*

orthembadism A term used in cartography as a synonym of *equivalence.* An "orthembadic projection" is an equal-area projection.

Orthent In U.S. Dept. of Agriculture soil taxonomy, a suborder of the soil order *Entisol,* characterized by soils that form on recent erosion surfaces. They are either shallow to bedrock or contain more than 35% coarse fragments, and they are either on steep slopes or the organic carbon content decreases regularly with depth (USDA, 1975). Cf: *Aquent; Arent; Fluvent; Psamment.*

Orthid In U.S. Dept. of Agriculture soil taxonomy, a suborder of the soil order *Aridisol,* characterized by the presence of a cambic, calcic, petrocalcic, gypsic, or salic horizon or a duripan and by the absence of an argillic or natric horizon. They form in sediments or on erosion surfaces of late Pleistocene or younger age and are products of the present environment (USDA, 1975). Cf: *Argid.*

orthid Any articulate brachiopod belonging to the order Orthida, characterized chiefly by an impunctate shell or one with endopuncta and by an open delthyrium and brachiophores. Range, Lower Cambrian to Upper Permian.

orthite A syn. of *allanite,* applied esp. when occurring in slender prismatic or acicular crystals.

ortho- In petrology, a prefix that, when used with the name of a metamorphic rock, indicates that it was derived from an igneous rock, e.g. orthogneiss, orthoamphibolite; it may also indicate the primary origin of a crystalline, sedimentary rock, e.g. "orthoquartzite" as distinguished from "metaquartzite".

orthoamphibole (a) A group name for amphiboles crystallizing in the orthorhombic system. (b) Any orthorhombic mineral of the amphibole group, such as anthophyllite, gedrite, and holmquistite.—Cf: *clinoamphibole.*

orthoandesite An *andesite* containing orthopyroxene (Johannsen, 1937, p. 173). Syn: *sanukitoid.* Cf: *sanukite.* Not recommended usage.

orthoantigorite A mineral of the serpentine group: $Mg_3Si_2O_5(OH)_4$. It is a six-layer orthorhombic form of antigorite.

orthoapsidal projection A map projection produced by means of the orthographic projection of a graticule from some solid body other than the sphere or spheroid.

ortho-arenite An arenite with detrital matrix under 15 percent (Pettijohn, Potter & Siever, 1973, p. 168).

orthoaxis In a monoclinic crystal, the lateral axis that has twofold symmetry and/or is perpendicular to the mirror plane of symmetry; it is the *b* axis [cryst]. Cf: *clinoaxis.*

orthobituminous Said of bituminous coal containing 87-89% carbon, analyzed on a dry, ash-free basis. Not listed by ASTM as a rank classification. Cf: *parabituminous; perbituminous.*

orthoceratite Any nautiloid belonging to the genus *Orthoceras,* characterized by the presence of three longitudinal furrows on the body chamber.

orthochamosite A mineral of the chlorite group: $(Fe^{+2},Mg,Fe^{+3})_6AlSi_3O_{10}(OH)_8$. It is the orthorhombic dimorph of chamosite.

orthochem An essentially normal precipitate formed by direct chemical action within a depositional basin or within the sediment itself, as distinguished from material transported in a solid state (Folk, 1959, p. 7); e.g. aphanocrystalline calcareous ooze (micrite), intergranular cement, recrystallized sedimentary material, and replacement minerals such as dolomite. Adj: *orthochemical.* Cf: *allochem.*

orthochlorite (a) A group name for distinctly crystalline forms of chlorite (such as clinochlore and penninite). (b) A group name for chlorites conforming to the general formula: $(R^{+2},R^{+3})_6(Si,Al)_4O_{10}(OH)_8$ (Hey, 1962, p. 546).—Cf: *leptochlorite.*

orthochoanitic Said of a straight, cylindrical, retrochoanitic septal neck of a nautiloid that extends only a short distance to the preceding septum.

orthochronology *Geochronology* based on a standard succession of biostratigraphically significant faunas or floras, or on irreversible evolutionary processes. Ideally, it is based on a stratigraphic succession of species "where each successive species is the descendent of the one which immediately precedes it stratigraphically" (Teichert, 1958a, p. 106). Cf: *parachronology.* See also: *biochronology.*

orthochrysotile A mineral of the serpentine group: $Mg_3Si_2O_5(OH)_4$. It is an orthorhombic form of chrysotile. Cf: *clinochrysotile.*

orthoclase (a) A colorless, white, cream-yellow, flesh-pink, or gray mineral of the alkali feldspar group: $KAlSi_3O_8$. It is the partly ordered, monoclinic modification of potassium feldspar and is dimorphous with microcline, being stable at higher temperatures; it usually contains some sodium in minor amounts. Ordinary or common orthoclase is a common rock-forming mineral; it occurs esp. in granites, acid igneous rocks, and crystalline schists, and is usually perthitic. Syn: *common feldspar; orthose; pegmatolite.* (b) A general term applied to any potassium feldspar that is or appears to be monoclinic; e.g. sanidine, submicroscopically twinned microcline, adularia, and submicroscopically twinned analbite.—Cf: *plagioclase; anorthoclase.*

orthoclasite An orthoclase-bearing porphyritic intrusive rock, such as granite or syenite. The term is sometimes restricted to rocks containing more than 90 percent orthoclase. Not recommended usage.

orthocline (a) Said of the *inclination* of the cardinal area in either valve of a brachiopod, lying on the continuation of the commissural plane. (b) Said of the body of the shell (and in some genera, of the hinge teeth) of a bivalve mollusk, oriented perpendicular or nearly perpendicular to the hinge axis. Syn: *acline [paleont].* (c) Said of the growth lines that traverse the whorl at right angles to the growth direction of a gastropod shell.

orthocone A straight, slender nautiloid conch, resembling that of *Orthoceras.* Syn: *orthoceracone.*

orthoconglomerate A term used by Pettijohn (1957, p.256) for a conglomerate with an intact gravel framework, characterized by a mineral cement, and deposited by "ordinary" but highly turbulent water currents (either high-velocity streams or the surf); e.g. *orthoquartzitic conglomerate* and *arkosic conglomerate.* It is strongly current-bedded and is associated with coarse cross-bedded sandstones. Cf: *paraconglomerate.*

orthocumulate A cumulate composed chiefly of one or more cumulus minerals plus the crystallization products of the intercumulus liquid. Cf: *mesocumulate.*

Orthod In U.S. Dept. of Agriculture soil taxonomy, a suborder of the order *Spodosol,* characterized by a *spodic horizon* in which aluminum, iron, and organic carbon are all present, but none dominates. Orthods are the most common Spodosols in northern parts of Europe and the United States. They formed mainly in coarse, acid parent materials under coniferous forests. Undisturbed soils usually have an O horizon and a thin albic horizon over the spodic horizon, but natural mixing or cultivation may obliterate most or all of the albic horizon (USDA, 1975). Cf: *Aquod; Ferrod; Humod.*

orthodolomite (a) A *primary dolomite,* or one formed by sedimentation. (b) A term used by Tieje (1921, p. 655) for a dolomite rock so well-cemented that the particles are interlocking.

orthodome An old term for a monoclinic crystal form whose faces parallel the orthoaxis. Its indices are $\{h0l\}$.

orthodont Said of the hinge of a bivalve mollusk in which the direction of the hinge teeth is parallel or nearly parallel to the cardinal margin.

orthodrome *great circle.*

orthodromic projection A map projection, derived from the gnomonic projection, in which angles are correct at two points and all great circles are straight lines.

orthoericssonite A mineral: $BaMn_2(FeO)Si_2O_7(OH)$.

orthofelsite *orthophyre.*

orthoferrosilite A mineral of the orthopyroxene group: $(Fe,Mg)SiO_3$; specif. a mineral consisting of the orthorhombic iron silicate $FeSiO_3$. See also: *ferrosilite.* Cf: *clinoferrosilite.*

orthogenesis Evolution that follows a single direction or specific trend continuously for many generations of an evolving *lineage,* and often appears to be independent of the effects of natural selection or other external factors. Cf: *rectilinear evolution.* Syn: *straight-line evolution.*

orthogeosyncline A geosyncline between continental and oceanic cratons, containing both volcanic (eugeosynclinal) and nonvolcanic (miogeosynclinal) belts (Stille, 1935, p. 77-97). Syn: *primary geosyncline; geosynclinal couple.* See also: *eugeosyncline; miogeosyncline.*

orthogonal n. A curve that is everywhere perpendicular to the wave crests on a refraction diagram. Syn: *wave ray.*

orthogonal projection A projection in which the projecting lines (straight and parallel) are perpendicular to the plane of the projection; e.g. an *orthographic projection.*

orthograde Pertaining to the uniform distribution of dissolved oxygen in the hypolimnion of a lake, dependent only "on conditions at circulation and on subsequent physical events" (Hutchinson, 1957, p.603). Cf: *clinograde.*

orthogranite The original name for *kaligranite;* withdrawn by the term's originator (Johannsen, 1932, p. 51).

orthographic projection (a) A perspective azimuthal map projection produced by straight parallel lines from a point at an infinite distance from the sphere to points on the sphere and perpendicular to the plane of projection. The largest area depicted is that of a hemisphere, and the projection is true to scale only at the center. The plane of projection may be perpendicular to the Earth's axis of rotation (polar orthographic projection, with the center at a pole) or parallel to the plane of some selected meridian (meridional orthographic projection, with the center on the equator). It is used for star charts and for pictorial world maps. (b) A similar projection used in optical mineralogical study of the origin of interference phenomena under the polarizing microscope, obtained by dropping perpendiculars from the poles (in the projection of the sphere) to the plane of projection which is normal to the north-south axis of the sphere. (c) *orthogonal projection.*

orthogyrate Said of the umbones (of a bivalve mollusk) curved so that each beak points neither anteriorly nor posteriorly but directly toward the other valve. Cf: *mesogyrate.*

orthohexagonal Pertaining to an orthorhombic lattice having a hexagonal array of points. Such a lattice is c-centered and has $b=\sqrt{3}a$.

orthohydrous (a) Said of coal containing 5-6% hydrogen, analyzed on a dry, ash-free basis. (b) Said of a maceral of normal hydrogen content, e.g. vitrinite.—Cf: *subhydrous; perhydrous.*

ortholignitous Said of coal containing 75-80% carbon, analyzed on a dry, ash-free basis. Not listed by ASTM as a rank classification.

ortholimestone A term proposed by Brooks (1954) for sedimentary limestone. Cf: *metalimestone; orthomarble.*

ortholith (a) A coccolith composed of one or very few crystals, as in the coccolithophorid *Braarudosphaera.* (b) An individual of the Ortholithae, a subdivision of the family Coccolithophoridae.—Cf: *heliolith.*

orthomagmatic stage The main stage in the crystallization of silicates from a typical magma; the stage during which as much as 90 percent of a magma may crystallize. Syn: *orthotectic stage.*

orthomarble A term proposed by Brooks (1954) for sedimentary carbonate rock that is commercially valuable because it will take a polish, e.g. the Holston orthomarble of Tennessee. Cf: *ortholimestone; metamarble.*

orthomatrix A term introduced by Dickinson (1970, p. 702) for recrystallized detrital clay or *protomatrix* in graywackes and arkoses.

orthometric correction A systematic correction that must be applied to a gravity-oriented measured difference of elevation because level surfaces at different elevations are not exactly parallel.

orthometric height The distance of a point above the geoid expressed in linear units measured along the plumb line at the point. Orthometric corrections are applied to measurements of precise leveling because level surfaces at different elevations are not parallel. Cf: *height.*

orthomicrite A genetic term applied to unaltered or primary calcareous micrite (Chilingar et al., 1967, p. 318). It includes *allomicrite* and *automicrite.* Cf: *pseudomicrite.*

orthomicrosparite A genetic term applied to microsparite that has developed by precipitation in open voids (Chilingar et al., 1967, p. 228). Cf: *pseudomicrosparite.*

orthomimic feldspar A group of feldspars that by repeated twinning simulate a higher degree of symmetry with rectangular cleavages. Also spelled: *orthomic feldspar.*

orthomorphic projection *conformal projection.*

orthomorphism *conformality.*

orthophotograph A photographic copy, prepared from a photograph formed by a perspective projection, in which the displacements due to tilt and relief have been removed; a photograph that has been transformed to an orthographic projection.

orthophotomap An orthophotograph, or a mosaic of orthophotographs, in standard quadrangle format, printed in colors to approximate ground conditions and enhanced with cartographic symbols including contours, elevations, boundaries, roads, and drainage. Syn: *photomap.*

orthophotomosaic A uniform-scale photographic mosaic consisting of an assembly of orthophotographs.

orthophotoquad A monocolor orthophotograph, or a mosaic of orthophotographs, in standard quadrangle format with little or no cartographic treatment.

orthophyre An obsolete term for a porphyritic rock containing phenocrysts of orthoclase. Syn: *orthofelsite.*

orthophyric Said of the texture of the groundmass in certain holocrystalline porphyritic igneous rocks in which the feldspar crystals have quadratic or short, stumpy rectangular cross sections, rather than the lath-shaped outline observed in *trachytic* texture. Also, said of a groundmass with this texture, or of a rock having an orthophyric groundmass.

orthopinacoid *front pinacoid.*

orthopinakiolite A black orthorhombic mineral $(Mg,Mn^{+2})_2Mn^{+3}BO_5$. It is a polymorph of pinakiolite.

orthopyroxene (a) A group name for pyroxenes crystallizing in the orthorhombic system and usually containing no calcium and little or no aluminum. (b) Any orthorhombic mineral of the pyroxene group, such as enstatite, bronzite, hypersthene, and orthoferrosilite.—Cf: *clinopyroxene.*

orthopyroxenite In the *IUGS classification,* a plutonic rock with M equal to or greater than 90, and opx/(ol+opx+cpx) greater than 90.

orthoquartzite A clastic sedimentary rock that is made up almost exclusively of quartz sand (with or without chert), that is relatively free of or lacks a fine-grained matrix, and that is derived by secondary silicification; a *quartzite* of sedimentary origin, or a "pure quartz sandstone". The term generally signifies a sandstone with more than 90-95% quartz and detrital chert grains that are well-sorted, well-rounded, and cemented primarily with secondary silica (sometimes with carbonate) in optical and crystallographic continuity with the grains. The rock is characterized by stable but scarce heavy minerals (zircon, tourmaline, magnetite), by lack of fossils, and by prominence of cross-beds and ripple marks. It commonly occurs as thin but extensive blanket deposits associated with widespread unconformities (e.g. an epicontinental deposit developed by an encroaching sea) and it represents intense chemical weathering of original minerals other than quartz, considerable transport and washing action before final accumulation (the sand may experience more than one cycle of sedimentation), and stable conditions of deposition (such as the peneplanation stage of diastrophism). Example: St. Peter Sandstone (Middle Ordovician) of midwestern U.S. The term was introduced by Tieje (1921, p.655) for a quartz sandstone whose interlocking particles were cemented by infiltration and pressure (in contrast to *paraquartzite*), and was used by Krynine (1948, p.149) in contrast to *metaquartzite,* but the term is objectionable because it is an exception to the use of "ortho-" for a metamorphic rock indicating an igneous origin and because "quartzite" is traditionally applied to quartzose rocks that break across instead of between grains. See also: *quartzose sandstone.* The term is essentially equivalent to *quartzarenite* and *quartzitic sandstone.* Syn: *sedimentary quartzite; orthoquartzitic sandstone.*

orthoquartzitic conglomerate A well-sorted, lithologically homogeneous, light-colored *orthoconglomerate* consisting of mature or supermature quartzose residues (chiefly vein quartz, chert, and quartzite, in fine to medium pebble size) that represent relatively stable material derived from eroded granitic or metamorphic terrain, with removal of finer material and less-stable lithologic types by weathering or by long transport. It is commonly interbedded with pure quartz sandstone. Syn: *quartz-pebble conglomerate.*

orthoquartzitic sandstone *orthoquartzite.*

orthorhombic system One of the six *crystal systems,* characterized by three axes of symmetry that are mutually perpendicular and of unconstrained relative lengths. Cf: *isometric system; tetragonal system; hexagonal system; monoclinic system; triclinic system.* Syn: *rhombic system.*

orthoscope A polarizing microscope in which light is transmitted by the crystal parallel to the microscope axis, in contrast to the *conoscope,* in which a converging lens and Bertrand lens are used.

orthose (a) A syn. of *orthoclase,* esp. yellow orthoclase. (b) An obsolete term introduced by Haüy (1801) for the feldspar group of minerals.—Etymol: French.

orthoselection The continuous action of *natural selection* in the same direction over a long period of time.

orthosilicate According to the now obsolete classification of silicates as oxyacids of silicon, a salt of the hypothetical orthosilicic acid, H_4SiO_4. Cf: *metasilicate.*

orthosite A light-colored coarse-grained *syenite* composed almost entirely of orthoclase, described by Turner in 1900. Not recommended usage.

orthosparite A sparite cement developed by physicochemical precipitation in open voids (Chilingar et al., 1967, p. 320). Cf: *pseudosparite.*

orthostratigraphy Standard or "main" stratigraphy based on fossils identifying recognized biostratigraphic zones (such as trilobites in the Cambrian and graptolites in the Silurian) (Schindewolf, 1957, p.397). Cf: *parastratigraphy.*

orthostrophic Having harmonious coiling throughout; specif. said of the common gastropod shell in which the whorls are coiled on an erect cone so that the apex points backward rather than forward, and the spire is slightly to strongly elevated. Cf: *hyperstrophic.*

orthotectic *magmatic.*

orthotectic stage *orthomagmatic stage.*

orthotectonics A syn. of *alpinotype tectonics.* It is used by Dewey (1969) for orogenic belts of the Andean type.

orthotill A till formed by immediate release from the transporting ice, as by ablation and melting (Harland et al., 1966, p. 231). Ant: *paratill.*

orthotriaene A sponge triaene in which the cladi are oriented close to 90 degrees to the rhabdome. Cf: *protriaene.*

Orthox In U.S. Dept. of Agriculture soil taxonomy, a suborder of the soil order *Oxisol,* characterized by formation under equatorial rainforests of the Amazon and Congo basins. They have essentially no dry season. Colors are generally yellowish or reddish, and the oxic horizon lacks any evidence of wetness (USDA, 1975). Cf: *Aquox; Humox; Torrox; Ustox.*

orthozone A term suggested by Kobayashi (1944, p.742) to replace *zone* as defined by Oppel (1856-1858).

ortlerite An obsolete term for a porphyritic hypabyssal diorite resembling greenstone and containing minor accessory orthoclase. Its name is derived from the Ortler Alps, Italy. Cf: *suldenite.*

ortstein A cemented *spodic horizon* in which the cementing material consists of illuviated sesquioxides, mostly iron and organic matter; the hardened B horizon of a Podzol. See also: *hardpan.*

orvietite An extrusive *trachybasalt* composed of approximately equal amounts of plagioclase and sanidine, along with leucite, augite, minor biotite and olivine, and accessory apatite and opaque oxides. The name, proposed by Niggli in 1923, is for Orvieto, Italy. Not recommended usage.

oryctocoenosis That part of a *thanatocoenosis* that has been preserved as fossils. Var: *oryctocoenose.*

oryctognosy An obsolete syn. of *mineralogy.*

oryctology (a) An obsolete syn. of *mineralogy.* (b) A term used in the middle of the 18th century for the study of fossils, at that time including almost anything dug out of the ground. The term was restricted at the same time "fossil" was restricted, and was later replaced by "paleontology" (Challinor, 1978, p. 211).

os [glac geol] Anglicized spelling of the Swedish term *ås,* meaning *esker.* Pl: *osar.* Syn: *ose.*

os [palyn] *endopore.*

Osagian North American series: Lower Mississippian (above Kinderhookian, below Meramecian). Also spelled: *Osagean.*

Osann's classification A purely chemical classification of igneous rocks. "The system is based on certain definite characteristics of the mineral combinations formed from the magmas, namely, on the combination of the alkalies with Al_2O_3 in definite proportions in the feldspars and feldspathoids, and on the union of lime with alumina in the anorthite molecule of the plagioclase and with iron and magnesia in the ferromagnesian minerals. The rock is classified from the amounts of these combinations, the percentage of silica, the silica coefficient, and the ratio of soda to the sum of the alkalies" (Johannsen, 1931, v.1, p. 68). See also: *Niggli's classification.*

osar Plural of *os.* The term is often mistakenly used as a singular noun.

osarizawaite A yellow mineral: $PbCuAl_2(SO_4)_2(OH)_6$.

osarsite A monoclinic mineral: (Os,Ru)AsS.

osbornite A meteorite mineral: TiN.

oscillation [glac geol] *interstade.*

oscillation [stratig] A term used by Ulrich (1911) for the repeated transgressions and regressions of the seas in constantly shifting patterns, bringing about changes in the character of the sediments being deposited. Cf: *pulsation [stratig].*

oscillation cross ripple mark A *cross ripple mark* resulting from the concurrent or successive action of two sets of waves or from the intersection of a set of waves with a pre-existing current ripple mark; e.g. *rectangular cross ripple mark* and *hexagonal cross ripple mark.* See also: *composite ripple mark.* Syn: *wave cross ripple mark; wave interference ripple mark.*

oscillation ripple *oscillation ripple mark.*

oscillation ripple mark A *symmetric ripple mark* with a sharp, narrow, relatively straight crest between broadly rounded troughs, formed by the orbital or to-and-fro motion of water agitated by oscillatory waves. Cf: *current ripple mark.* Syn: *oscillation ripple; oscillatory ripple mark; wave ripple mark.*

oscillation theory A theory, proposed by Haarman (1930), that cosmic energy produces the Earth's major tectonic features, and that secondary features are the result of gravitational sliding, compressional settling, or subsidence.

oscillatory extinction *undulatory extinction.*

oscillatory ripple mark *oscillation ripple mark.*

oscillatory twinning Repeated, parallel twinning.

oscillatory wave A water wave in which the individual particles move in closed vertical orbits about a point with little or no change in position, although the wave form itself advances; e.g. an ocean wave in deep water. Cf: *wave of translation.* Syn: *wave of oscillation.*

oscillatory-wave theory A modern theory of the tides (replacing the *progressive-wave theory*) involving the assumption that the basic tidal movement in the open ocean consists of a system of waves oscillating within subdivided units of the ocean surface, each unit having a fixed node from which the height of tidal rise increases outward; the oscillations vary with the relative positions of the Earth, Moon, and Sun, together with the shape, size, and depth of the body of water within the unit. Progressive-wave movement is of secondary importance. Syn: *stationary-wave theory.*

osculum A large opening from the internal cavity of a sponge to the exterior, through which water leaves the sponge. Pl: *oscula.* Cf: *ostium; pore [paleont].* Syn: *oscule.*

ose Var. of *os.* Pl: *oses.*

oshana A poorly defined stream channel in the flat-lying Ovamboland region of Namibia, containing water only during the highest floods and usually in the form of a chain of standing pools that quickly dry away. Cf: *omuramba.* Etymol: Afrikaans.

osmiridium A white to gray cubic mineral: (Ir,Os). It is a native alloy containing 25-40% osmium and 50-60% iridium, and is often found with platinum. The name has also been used as a syn. of *iridosmine.*

osmium A mineral: (Os,Ir).

osmosis The movement at unequal rates of a solvent through a semipermeable membrane, which usually separates the solvent and a solution, or a dilute solution and a more concentrated one, until the solutions on both sides of the membrane are equally strong. Cf: *dialysis.* See also: *electro-osmosis.*

osseous amber Opaque or cloudy amber containing numerous minute bubbles. Syn: *bone amber.*

osseous breccia *bone breccia.*

ossicle (a) Any of the numerous individual calcified elements or pieces of the skeleton of many echinoderms; e.g. a *plate.* The term is normally used for the larger of such elements. (b) A tiny bone, esp. of the middle ear.— Syn: *ossiculum.*

ossiculum (a) *ossicle.* (b) *lithodesma.*—Pl: *ossicula.*

ossipite A coarse-grained variety of *troctolite* containing labradorite, olivine, magnetite, and a small amount of clinopyroxene. Also spelled: *ossypite.* The rock was named by Dana in 1872 for the Ossipee Indians of New Hampshire. Not recommended usage.

Osteichthyes A class of vertebrates, the bony fishes. Range, Devonian to present.

osteolite A massive, earthy mineral (apatite) consisting of an impure, altered calcium phosphate.

osteolith (a) A femur-shaped heterococcolith built up of lamellae as in the coccolithophorid *Ophiaster hydroideus.* (b) A fossil bone.

Osteostraci An order of monorhinate jawless vertebrates characterized by expanded head and trunk shield and flattened body form. Range, Upper Silurian to Upper Devonian.

ostiole [geomorph] *tundra ostiole.*

ostiole [paleont] One of the small inhalant openings of a sponge; an *ostium.*

ostium Any opening through which water enters a sponge. The term is sometimes applied only to an opening larger than a *pore,* and it was used in the older literature as a synonym of *posticum.* Pl: *ostia.* Cf: *osculum.* Syn: *ostiole [paleont].*

ostracode Any aquatic crustacean belonging to the subclass Ostracoda, characterized by a bivalve, generally calcified carapace with a hinge along the dorsal margin. Most ostracodes are of microscopic size (0.4-1.5 mm long) although freshwater forms up to 5 mm long and marine forms up to 30 mm long are known. Range, Lower Cambrian to present. Also spelled: *ostracod.*

ostracum A term used originally for the outer part of the calcareous wall of the shell of a bivalve mollusk, secreted at the edge of the mantle, but also applied by some later authors to the entire calcareous wall. Cf: *hypostracum; periostracum.* Pl: *ostraca.*

ostraite A *jacupirangite* containing abundant green spinel. Named by Duparc in 1913 for Ostraia Sopka, Urals, U.S.S.R. Not recommended usage.

Ostwald's rule The statement in phase studies that an unstable phase does not necessarily transform directly to the truly stable phase, but rather it may first pass through successive intermediate phases, presumably due to lower activation energy barriers via that route.

osumilite A hexagonal mineral: $(K,Na)(Mg,Fe_{+2})_2(Al,Fe)_3(Si,Al)_{12}O_{30}\cdot H_2O$. It is commonly mistaken for cordierite.

osumilite-(K,Mg) A mineral of the osumilite group: $(K,Na)(Mg,Fe^{+2})_2(Al,Fe)_3(Si,Al)_{12}O_{30}\cdot H_2O$.

otavite A hexagonal mineral: $CdCO_3$. It is isostructural with calcite.

otolith Ear bone of a fish.

ottajanite A leucite *tephrite* or *trachybasalt* having the chemical, but not mineralogic, composition of a *sommaite,* being composed of augite and leucite phenocrysts in a groundmass of calcic plagioclase, leucite, and augite, with some sanidine, nepheline, olivine, opaque oxides, hornblende, biotite, and apatite. Its name, given by Lacroix in 1917, is derived from Ottajano, Vesuvius, Italy. Not recommended usage.

ottemannite A mineral: Sn_2S_3.

ottrelite A gray to black variety of chloritoid containing manganese.

Ouachita stone *Washita stone.*

ouachitite An olivine-free biotite *lamprophyre* having a glassy or analcime groundmass. The name, given by Kemp in 1891, is for the Ouachita River, Arkansas. Not recommended usage.

oued A var. of *wadi.* Etymol: French. Pron: *wed.* Pl: *oueds; ouadi.*

ouenite A fine-grained *diabase* resembling *eucrite* and containing chrome diopside, anorthite, and smaller amounts of hypersthene and olivine. The name, given by Lacroix in 1911, is for Ouen, New Caledonia. Not recommended usage.

ouklip A term used in southern Africa for a conglomerate. Etymol: Afrikaans, "old rock".

oule A term used in the Pyrenees for *cirque [glac geol].* Etymol: Spanish *olla,* "pot, kettle".

oulopholite *cave flower.*

outburst *glacier outburst flood.*

outcrop n. That part of a geologic formation or structure that appears at the surface of the Earth; also, bedrock that is covered only by surficial deposits such as alluvium. Cf: *exposure.* Syn: *crop* (deprecated); *cropping* (deprecated); *outcropping.* —v. To appear exposed and visible at the Earth's surface; *crop out.*

outcrop area The area occupied by a particular rock unit.

outcrop curvature *settling [mass move].*

outcrop map A type of geologic map that shows the distribution and shape of actual outcrops, leaving blank those areas without outcrops. It often includes measured data for specific places, such as specimen or fossil collections, or strike and dip of beds.

outcropping n. *outcrop.*

outcrop spring *contact spring.*

outcrop water Rain and surface water that seeps downward through outcropping porous and fissured rock, fault planes, old shafts, or surface drifts.

outer bar A bar formed at the mouth of an ebb channel of an estuary. Ant: *inner bar.*

outer bark For stems and roots of dicotyledons and gymnosperms, a nontechnical term incorporating the rough corky tissue developed from the cork cambium, outside of the inner bark which is developed from the stelar cambium (Swartz, 1971, p. 328). The technical term for outer bark is *rhytidome.*

outer core The outer or upper zone of the Earth's *core,* extending from a depth of 2900 km to 5100 km, and including the *transition zone;* it is equivalent to the *E layer* and the *F layer.* It is presumed to be liquid because it sharply reduces compressional-wave velocities and does not transmit shear waves. Its density ranges from 9 to 11 g/cm^3. Cf: *inner core.*

outer epithelium The ectodermal epithelium adjacent to the shell of a brachiopod and responsible for its secretion.

outer hinge plate Either of a pair of concave or subhorizontal *hinge plates* in the cardinalia of some brachiopods, separating inner socket ridges and crural bases. Cf: *inner hinge plate.*

outer lamella The relatively thick mineralized shell layer of an ostracode, enclosed between thin chitinous layers, and serving to conceal and protect the soft parts of the body and appendages (TIP, 1961, pt.Q, p.53). Cf: *inner lamella.*

outer lip The abaxial (lateral) margin of the aperture of a gastropod shell, extending from the suture to the foot of the columella. Cf: *inner lip.*

outer mantle *upper mantle.*

outer mantle lobe The outer peripheral part of the mantle of a brachiopod, separated by a mantle groove from an inner lobe, and responsible (in articulate brachiopods) for the secretion of the primary shell layer (TIP, 1965, pt.H, p.149).

outer plate One of a pair of subvertical plates in the cardinalia of pentameracean brachiopods, with the ventral surface fused to the base of the brachial process and the dorsal edge attached to the floor of the valve (TIP, 1965, pt.H, p.149). Cf: *inner plate.*

outer reef One of the reefs comprising the seaward or basinward part of a *reef complex* or *reef tract.* They tend to be larger and better developed than *inner reefs* in the same region. Cf: *leeward reef; windward reef.*

outer side The portion of a conodont element on the convex side of the anterior-posterior midline. Ant: *inner side.*

outer wall The exterior layer in the skeleton of double-walled archaeocyathids. It is homologous with the single wall in one-walled forms (TIP, 1972, pt. E, p. 7).

outface *dip slope.*

outfall (a) The mouth of a stream or the outlet of a lake; esp. the narrow end of a watercourse or the lower part of any body of water where it drops away into a larger body. (b) The vent or end of a drain, pipe, sewer, ditch, or other conduit that carries waste water, sewage, storm runoff, or other effluent into a stream, lake, or ocean.

outflow (a) The act or process of flowing out; e.g. the discharge of water from a river into the sea. Syn: *efflux.* (b) Water that flows out; e.g. ground-water seepage and stream water flowing out of a drainage basin. Also, the amount of water that has flowed out. (c) An *outlet* where water flows out of a lake.

outflow cave A cave from which a stream issues, or is known to have issued. See also: *inflow cave; through cave.* Syn: *effluent cave; cave of debouchure.*

outgassing The removal of occluded gases, usually by heating; e.g. the process involving the release of gases and water vapor from molten rocks, leading to the formation of the Earth's atmosphere and oceans.

outlet (a) The relatively narrow opening at the lower end of a lake through which water is discharged into an outflowing stream or other body of water. Syn: *outflow.* (b) A stream flowing out of a lake, pond, or other body of standing water; also, the channel through which such a stream flows. (c) The lower end of a watercourse where its water flows into a lake or sea; e.g. a channel, in or near a delta, diverging from the main river and delivering water into the sea. (d) A crevasse in a levee.

outlet glacier A glacier issuing from an ice sheet or ice cap through a mountain pass or valley, constrained to a channel or path by exposed rock. Cf: *ice stream [glaciol]; glacial lobe; distributary glacier.*

outlet head The place where water leaves a lake and enters an effluent.

outlier An area or group of rocks surrounded by rocks of older age, e.g. an isolated hill or butte. Cf: *inlier.*

outline map A map that presents minimal geographic information, usually only coastlines, principal streams, major civil boundaries, and large cities, leaving as much space as possible for the reception of additional particular data. See also: *base map.*

outpost well A hole drilled for oil or gas with the thought that it will probably extend, by a considerable distance, a pool already partly developed. It is far enough from the limits of the pool to make its outcome uncertain, but not far enough to be designated a *wildcat well* (Lahee, 1962, p. 133).

outside lateral muscle One of a pair of muscles in some lingulid brachiopods, originating on the pedicle valve laterally to the central muscles, and extending posteriorly to insertions behind the *middle lateral muscles* on the brachial valve (TIP, 1965, pt.H, p.149).

outside pond A body of water enclosed or partly enclosed by sediment deposited by bifurcating distributaries in the outermost extension of the Mississippi River delta; it is usually connected with the Gulf of Mexico. Cf: *inside pond.*

outwash [glac geol] (a) Stratified detritus (chiefly sand and gravel) removed or "washed out" from a glacier by meltwater streams and deposited in front of or beyond the end moraine or the margin of an active glacier. The coarser material is deposited nearer to the ice. Syn: *glacial outwash; outwash drift.* (b) The meltwater from a glacier.

outwash [sed] Soil material washed down a hillside by rainwater and deposited upon more gently sloping land.

outwash apron *outwash plain.*

outwash cone A steeply sloping, cone-shaped accumulation of outwash deposited by meltwater streams at the margin of a shrinking glacier. Syn: *wash cone.*

outwash drift A deposit of *outwash.*

outwash fan A fan-shaped accumulation of outwash deposited by meltwater streams in front of the end moraine of a glacier. Coalescing outwash fans form an *outwash plain.*

outwash plain A broad, gently sloping sheet of outwash deposited by meltwater streams flowing in front of or beyond a glacier, and formed by coalescing *outwash fans;* the surface of a broad body of outwash. Cf: *valley train; glacial plain.* See also: *sand plain.* Syn: *outwash apron; morainal apron; frontal plain; wash plain; marginal plain; sandur; morainal plain.*

outwash-plain shoreline A prograding shoreline formed where the outwash plain in front of a glacier is built out into a lake or the sea.

outwash terrace A dissected and incised valley train or benchlike deposit extending along a valley downstream from an outwash plain or terminal moraine; a flat-topped bank of outwash with an abrupt outer face. Syn: *frontal terrace; overwash terrace.*

outwash train *valley train.*

ouvarovite *uvarovite.*

ovary In a flower, the basal, enlarged part of the pistil, in which seeds develop.

ovate Shaped like an egg in sectional view, as a leaf whose basal end is broader than its terminal end. Cf: *obovate.*

oven (a) A rounded, sacklike pit or hollow in a rock, esp. in a granitic rock, having an arched roof and resembling an oven (Bell, 1894, p. 358). It is produced by chemical weathering. Cf: *weather pit.* (b) *spouting horn.*

oven-dry soil A soil sample that has been dried at 105°C.

overbank deposit Fine-grained sediment (silt and clay) deposited from suspension on a flood plain by floodwaters that cannot be contained within the stream channel. See also: *flood-plain deposit.*

overburden [eco geol] Barren rock material, either loose or consolidated, overlying a mineral deposit, which must be removed prior to mining. Syn: *top [ore dep]; baring. Caprock [eco geol]* and *capping* are synonyms that are usually used for consolidated material.

overburden [sed] (a) The upper part of a sedimentary deposit, compressing and consolidating the material below. (b) The loose soil, silt, sand, gravel, or other unconsolidated material overlying bedrock, either transported or formed in place; *regolith.*

overburden pressure *geostatic pressure.*

overconsolidation Consolidation (of sedimentary material) greater than that normal for the existing overburden; e.g. consolidation resulting from desiccation or from pressure of overburden that has since been removed by erosion. Ant: *underconsolidation.* Cf: *normal consolidation.*

overdeepened valley The degraded channel or valley of an alpine glacier, now occupied by an aggrading stream.

overdeepening The process by which an eroding glacier excessively deepens and broadens an inherited preglacial valley to a level below that of the original base level. Undrained depressions and rock-floored lakes are common. Cf: *oversteepening.*

overdraft Withdrawal of ground water in excess of replenishment.

overfall [currents] A turbulent, disturbed surface of water (such as a breaking wave) caused by the meeting of strong currents, by winds moving against a current, or by a current setting over a submerged ridge or shoal; a *rip.* Term is usually used in the plural.

overfall [eng] A place provided on a dam or weir for the overflow of surplus water.

overfall [streams] An obsolete term for a waterfall.

overfit stream A *misfit stream* that is too large to have eroded the valley in which it flows, or whose flood plain is too small for the size of the stream. There is some doubt as to whether such a stream exists.

overflow v. To flow over the margin of; to cover with water.—n. A flowing over the banks of a stream or river; an inundation.

overflow channel A channel or notch cut by the overflow waters of a lake, esp. the channel draining meltwater from a glacially dammed lake; an outlet of a proglacial lake. See also: *glacial drainage channel.* Syn: *spillway [glac geol]; glacial spillway; sluiceway; crease [glac geol].*

overflow ice Ice formed during high spring tides by water rising through cracks in the surface ice and then freezing (Swayne, 1956, p. 104).

overflow spring A type of contact spring that develops where a permeable deposit dips beneath an impermeable mantle. Ground water overflows onto the land surface at the edge of the impermeable stratum.

overflow stream (a) A stream containing water that has overflowed the banks of a river. Syn: *spill stream.* (b) An effluent from a lake, carrying water to a stream, sea, or another lake.

overfold An *overturned* fold.

overgrowth [cryst] (a) Secondary material deposited in optical and crystallographic continuity around a crystal grain of the same composition, as in the diagenetic process of secondary enlargement. (b) A deposit of one mineral growing in oriented crystallographic directions on the surface of another mineral; e.g. hematite on quartz, or chalcopyrite on galena. See also: *mantle [cryst].*

overgrowth [paleont] A subsequent encrusting growth of zooids and possible extrazooidal parts separated from the supporting stenolaemate bryozoan colony by a basal lamina. The overgrowth can be intracolonial, intercolonial and conspecific, or interspecific.

overhand stoping The working of a block of ore from a lower level to a level above.

overhang (a) *cliff overhang.* (b) A part of the mass of a *salt dome* that projects out from the top of the dome much like the cap of a mushroom.

overhanging ripple *rhomboid ripple mark.*

overite A pale-green to colorless mineral: $CaMgAl(PO_4)_2(OH) \cdot 4H_2O$.

overland flow That part of surface runoff flowing over land surfaces toward stream channels; specif. *sheet flow [geomorph].* After it enters a stream, it becomes a part of the total runoff (Langbein & Iseri, 1960). Syn: *unconcentrated flow.* Cf: *channel flow; streamflow.*

overlap [coast] The migration of an upcurrent part of a shore to a position that extends seaward beyond a downcurrent part; esp. the lapping-over of an inlet by a spit. Cf: *offset.*

overlap [paleont] In ostracodes, the closure of the two valves in such a manner that the contact margin of the larger valve extends over the margin of the smaller.

overlap [photo] The area common to two successive aerial or space photographs or images along the same flight strip, expressed as a percentage of the photo area. Cf: *sidelap.* Syn: *end lap; forward lap.*

overlap [stratig] A general term referring to the extension of marine, lacustrine, or terrestrial strata beyond underlying rocks whose edges are thereby concealed or "overlapped", and to the unconformity that commonly accompanies such a relation; esp. the relationship among conformable strata such that each successively younger stratum extends beyond the boundaries of the stratum lying immediately beneath. The term is often used in the sense of *onlap,* and sometimes in the sense of *overstep* (as by De la Beche, 1832); because of such conflicting usage, Melton (1947, p.1869) and Swain (1949, p.634) urged that the term be abandoned. See also: *replacing overlap.*

overlap fault (a) *thrust fault.* (b) A fault structure in which the displaced strata are doubled back upon themselves.

overlapping pair Two photographs taken at different camera stations in such a manner that part of one photograph shows the same terrain as shown on a part of the other photograph; e.g. *stereoscopic pair.*

overlapping spur *interlocking spur.*

overlap zone *concurrent-range zone.*

overlay Graphic data on a transparent or translucent sheet to be superimposed on another sheet (such as a map or photograph) to show details not appearing, or requiring special emphasis, on the original; a *template.* Also, the medium or sheet containing an overlay.

overlie To lie or be situated over or upon, or to occupy a higher position than. The term is usually applied to certain rocks (usually sedimentary or volcanic) resting or lying upon certain older rocks. Ant: *underlie.*

overload The amount of sediment that exceeds the ability of a stream to transport it and that is therefore deposited.

overloaded stream A stream that is so heavily loaded with sediment that it is forced to deposit a part of its load; e.g. the Platte River in Nebraska.

overpressure Pressure in excess of lithostatic pressure, e.g. from tectonic stress.

overprint [geochron] A complete or partial disturbance of an isolated radioactive system, by thermal, igneous, or tectonic activities, which results in loss or gain of radioactive or radiogenic isotopes, and hence a change in the radiometric age that will be given by the disturbed system. See also: *updating; mixed ages; hybrid age.*

overprint [struc petrol] The superposition of a new set of structural features on an older set. Syn: *superprint; metamorphic overprint.*

oversaturated Said of an igneous rock or magma that contains silica (or occasionally alumina) in excess of the amount required to form *saturated* minerals from the bases present. Syn: *silicic.* Cf: *undersaturated; unsaturated.*

oversteepened valley "An ice-free valley in which one side is higher and steeper than the other, a condition caused by the swing of a former glacier directed against that side" (Swayne, 1956, p. 105).

oversteepened wall A *trough end* having an almost vertical slope due to glacial action.

oversteepening The erosive process by which an alpine glacier excessively steepens the sides of an inherited preglacial valley. Cf: *overdeepening.*

overstep n. (a) An *overlap* characterized by the regular truncation of older units of a complete sedimentary sequence by one or more later units of the sequence (Swain, 1949, p.635). The term, which is more commonly used in Great Britain than in U.S., refers to the progressive burial of truncated edges of underlying strata below an unconformity (esp. when an unconformity is not very obvious but is made evident by detailed mapping). Cf: *onlap.* See also: *strike-overlap; complete overstep; regional overstep.* (b) A stratum laid down on the upturned edges of underlying strata.—v. To transgress; e.g. an unconformable stratum that truncates the upturned edges of the underlying older rocks is said to "overstep" each of them in turn.

overthrust A low-angle *thrust fault* of large scale, with displacement generally measured in kilometers. Cf: *underthrust fault.* Syn: *low-angle thrust; overthrust fault.*

overthrust block *overthrust nappe.*

overthrust fault *overthrust.*

overthrust nappe The body of rock that forms the hanging wall of a large-scale overthrust; a *thrust nappe*. Syn: *overthrust block; overthrust sheet; overthrust slice.*

overthrust sheet *overthrust nappe.*

overthrust slice *overthrust nappe.*

overturn The circulation, esp. in the fall and spring, of the layers of water in a lake or sea, whereby surface water sinks and mixes with bottom water; it is caused by changes in density differences due to changes in temperature, and is esp. common wherever lakes are icebound in winter. See also: *turnover [lake]; circulation [lake].*

overturned Said of a fold, or the limb of a fold, that has tilted beyond the perpendicular. Sequence of strata thus appears reversed. Such a fold may be called an *overfold.* Syn: *inverted; reversed.*

overturning The rising movement of bottom waters to the surface, either *upwelling* in the ocean or the slow seasonal movement in lakes known as *fall overturn.*

overwash (a) A mass of water representing the part of the uprush that runs over the berm crest (or other structure) and that does not flow directly back to the sea or lake. (b) The flow of water in restricted areas over low parts of barriers or spits, esp. during high tides or storms.

overwash mark A narrow, tongue-like ridge of sand formed by overwash on the landward side of a berm.

overwash pool A *tide pool* between a berm and a beach scarp, which water enters only at high tide.

ovicell (a) The structure for brooding embryos present in most cheilostome bryozoans. It generally consists of a fold of body wall, one or both walls of which are calcified, enclosing the water-filled brood chamber. Syn: *ooecium.* (b) A term used loosely for any skeletal structure that houses bryozoan larvae during their development.

ovoid Egg-shaped, as in a fruit whose basal portion is broader than its terminal portion. Cf: *obovoid.*

ovulate Containing or bearing an egg or ovule (Swartz, 1971, p. 329).

ovule The unfertilized young seed in the ovary of a seed plant (Swartz, 1971, p. 329).

ovulite *oolith.*

owyheeite A steel-gray to silver-white mineral: $Ag_2Pb_5Sb_6S_{15}$. It occurs in metallic fibrous masses and acicular crystals. Syn: *silver jamesonite.*

oxalite *humboldtine.*

oxammite A yellowish-white transparent orthorhombic mineral (ammonium oxalate): $(NH_4)_2C_2O_4 \cdot H_2O$.

oxbow (a) A closely looping stream meander resembling the U-shaped frame embracing an ox's neck, having an extreme curvature such that only a neck of land is left between two parts of the stream. Syn: *horseshoe bend.* (b) A term used in New England also for the land enclosed, or partly enclosed, within an oxbow (bend of a stream). (c) The abandoned, bow- or horseshoe-shaped channel of a former meander, left when the stream formed a cutoff across a narrow meander neck. See also: *cutoff meander.* Syn: *abandoned channel.* (d) *oxbow lake.*

oxbow lake The crescent-shaped, often ephemeral, body of standing water situated by the side of a stream in the abandoned channel (oxbow) of a meander after the stream formed a neck cutoff and the ends of the original bend were silted up. Examples are common along the banks of the Mississippi River, where they are often known as *bayous.* See also: *billabong.* Syn: *oxbow; loop lake; mort-lake; moat; horseshoe lake; cutoff lake; crescentic lake.*

oxea A needle-shaped monaxonic sponge spicule tapering to a sharp point at each end. Pl: *oxeas* or *oxeae.* Cf: *tornote.*

Oxfordian European stage: Upper Jurassic (above Callovian, below Kimmeridgian). Syn: *Corallian.*

oxic horizon A diagnostic subsurface soil that is composed primarily of the hydrated oxides of iron and aluminum. The original weatherable materials are almost completely gone. It is at least 30cm thick (USDA, 1975).

oxidates Sediments composed of the oxides and hydroxides of iron and manganese, crystallized from aqueous solution. It is one of Goldschmidt's groupings of sediments or analogues of differentiation stages in rock analysis. Cf: *resistates; evaporates; reduzates; hydrolyzates.*

oxidation potential *Eh.*

oxide A mineral compound characterized by the linkage of oxygen with one or more metallic elements, such as cuprite, Cu_2O, rutile, TiO_2, or spinel, $MgAl_2O_4$. See also: *hydroxide.*

oxide-facies iron formation An *iron formation* in which the principal iron-rich minerals are oxides, typically hematite or magnetite (James, 1954, p.256-263). It is thought to be the shoreward facies of iron formation. See also: *specular schist; itabirite.*

oxidite *shale-ball.*

oxidized zone An area of mineral deposits modified by surface waters, e.g. sulfides altered to oxides and carbonates. See also: *supergene enrichment.* Cf: *sulfide zone; gossan; protore.*

oxidizing flame In blowpiping, the outer, almost invisible, and less intense part of the flame, from which oxygen may be added to the compound being tested. Cf: *reducing flame.*

Oxisol In U.S. Dept. of Agriculture soil taxonomy, a soil order characterized by mixtures of quartz, kaolin, free oxides and organic matter, lacking clearly marked horizons. Oxisols have either an oxic horizon within 2 m, or a continuous phase of plinthite within 30 cm, of the surface. There is no underlying spodic or argillic horizon. Oxisols are deeply weathered soils on stable surfaces in tropical to subtropical regions (USDA, 1975). See also: *Aquox;*

Humox; Orthox; Torrox; Ustox.

oxoferrite A variety of native iron with some FeO in solid solution.

oxyaster A stellate sponge spicule (aster) having acute, sharp rays.

oxybasiophitic Said of an ophitic rock that may be either *basiophitic* or *oxyophitic.*

oxycone A laterally compressed, coiled cephalopod conch with an acute periphery and a usually narrow or occluded umbilicus, as in *Oxynoticeras.*

oxygen deficit The difference between the actual amount of dissolved oxygen in lake or sea water and the saturation concentration at the temperature of the water mass sampled.

oxygen demand *chemical oxygen demand.*

oxygen-isotope fractionation Fractionation of oxygen isotopes (oxygen-18/oxygen-16) in oxygen-bearing geologic materials, e.g. carbonate shells of marine organisms, which may be used as an indication of the temperature of formation of the materials. See also: *carbonate thermometer.*

oxygen ratio The ratio of the number of atoms of oxygen in the basic oxides of a mineral or rock to the number of atoms of oxygen in SiO_2 (Johannsen, 1939, v.1, p. 164). Syn: *acidity coefficient; acidity quotient; coefficient of acidity.*

oxyhexaster A hexaster whose simple terminal rays end in sharp points.

oxyhornblende *basaltic hornblende.*

oxylophyte A plant preferring or restricted to acid soil.

oxymagnite *maghemite.*

oxymesostasis A mesostasis composed of quartz, orthoclase, or both.

oxyophitic Said of the texture of an ophitic rock with an oxymesostasis; also, said of a rock with such texture. Cf: *basiophitic.*

oxyphile *lithophile.*

oxysphere A term that was proposed as a replacement for *lithosphere;* that zone or layer of the Earth whose constituent rocks are 60% oxygen.

oxytylote A sponge spicule shaped like a common pin.

oyster reef An *organic reef* or *bank [sed]* composed mostly of oyster shells attached upon one another in growth position; living examples tend to be small (a hundred meters or so across, by a few meters high) and to occur in estuarine waters.

ozalid *diazo print.*

Ozarkian A now obsolete term for the time represented by rocks formed between the Cambrian and the Canadian. It was so named after the Ozark uplift of Missouri.

ozarkite White massive thomsonite from Arkansas.

ozocerite A brown to jet black paraffin wax. It occurs in irregular veins, is soluble in chloroform, has a variable melting point, and yields *ceresine* on heating with a 20-30% solution of concentrated H_2SO_4 at 120° to 200° C. Varieties: *baikerite; celestialite; helenite; moldavite; pietricikite.* Also spelled: *ozokerite.* Syn: *ader wax; earth wax; fossil wax; mineral wax; native paraffin.*

ozokerite Original spelling of *ozocerite.*

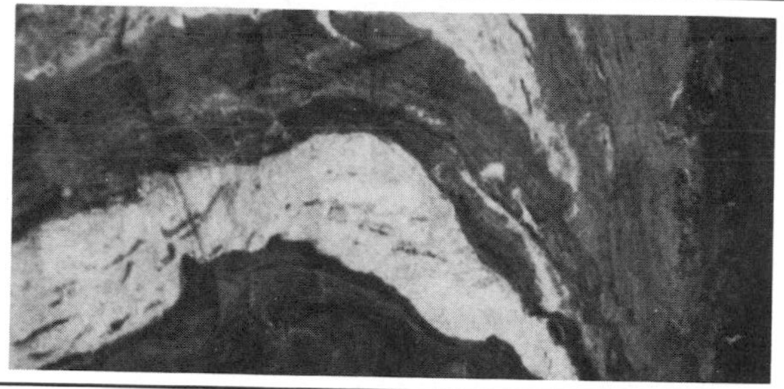

paar A depression produced by the moving-apart of crustal blocks rather than by subsidence within a crustal block. It is floored with upper-mantle igneous rocks and is essentially devoid of crustal material. Examples are the Gulf of California and the Dead Sea. Etymol: Hebrew. Cf: *rift [tect].*

pabstite A mineral: $Ba(Sn,Ti)Si_3O_9$.

pachnolite A colorless to white monoclinic mineral: $NaCaAlF_6 \cdot H_2O$.

pachyodont Said of the dentition of a bivalve mollusk characterized by hinge teeth that are large, heavy, blunt, thick, and amorphous.

Pacific suite One of two large groups of igneous rocks, characterized by calcic and calc-alkalic rocks. Harker (1909) divided all Tertiary and Holocene igneous rocks of the world into two main groups, the *Atlantic suite* and the Pacific suite, the latter being so named because of the predominance of calcic and calc-alkalic rocks in the area of the circum-Pacific orogenic belt. Because there is such a wide variation in tectonic environments and associated rock types in the areas of Harker's Atlantic and Pacific suites, the terms are now seldom used to indicate kindred rock types. Cf: *Mediterranean suite.* Syn: *anapeirean.* See also: *andesite line.*

Pacific-type coastline A *concordant coastline,* esp. one as developed around the Pacific Ocean (e.g. the coastline of British Columbia), reflecting the continuous linear trends of the circum-Pacific fold-mountain system. A "modified" Pacific-type coastline develops behind festoons of island arcs and adjacent foredeep trenches, such as the coastline of Asia. Ant: *Atlantic-type coastline.*

pack *pack ice.*

packed biomicrite A *biomicrite* in which the skeletal grains make up over 50% of the rock. Cf: *sparse biomicrite.*

packed ice *close ice.*

packer A short expansible-retractable device deliberately set in a cased or uncased *well bore* to prevent upward or downward fluid movement; generally for temporary use.

packet texture An obsolete term for the close grouping of quartz crystals in pegmatite (Knopf & Ingerson, 1938, p. 170).

pack ice (a) A term used in a broad sense to include any area of sea ice (other than fast ice) regardless of its form or disposition, composed of a heterogeneous mixture of ice of varying sizes and ages, and formed by the jamming or crushing together of pieces of floating ice; the mass may be either loosely or tightly packed but it covers the sea surface with little or no open water. See also: *drift ice.* (b) The ice material in an area of pack ice, or forming an ice pack.—Syn: *ice canopy; ice pack; pack.* "The terms 'pack ice' and 'ice pack' have been used indiscriminately for both the sea area containing floating ice, and the material itself" (Huschke, 1959, p. 410). See also: *ice cluster.*

packing The manner of arrangement or spacing of the solid particles in a sediment or sedimentary rock, or of the atoms or ions in a crystal lattice; specif. the arrangement of clastic grains, entirely apart from any authigenic cement that may have crystallized between them. Cf: *fabric; compaction.*

packing density A measure of the extent to which the grains of a sedimentary rock occupy the gross volume of the rock in contrast to spaces between the grains, equal to the cumulated grain-intercept length along a traverse in a thin section (Kahn, 1956). Syn: *grain density.*

packing index The ratio of the ion volume to the volume of the unit cell in a crystal (Fairbairn, 1943).

packing proximity An estimate of the number of grains (in a sedimentary rock) that are in contact with their neighbors, equal to the total percentage of grain-to-grain contacts along a traverse measured on a thin section (Kahn, 1956).

packing radius Half the distance of closest approach of like atoms or ions in a crystal.

packsand A very fine-grained sandstone that is so loosely consolidated by a little calcareous cement as to be readily cut by a spade.

packstone A term used by Dunham (1962) for a sedimentary carbonate rock whose granular material is arranged in a self-supporting framework, yet also contains some matrix of calcareous mud. Cf: *mudstone; grainstone; wackestone.*

paddle (a) The flat distal part of the last prosomal appendage (toward the rear) in a merostome. (b) A tetrapod limb modified into a rowing organ for life in the sea, as in plesiosaurs and sea turtles.

padmaragaya A light orange or reddish-yellow variety of sapphire; a synthetic corundum of various shades of yellow or orange. Etymol: Sinhalese, "lotus color". Syn: *padmaradschah; padparadscha.*

padparadscha A syn. of *padmaragaya.* Also spelled: *padparadschah.*

paedogenesis *neoteny.*

paedomorphism *neoteny.*

paedomorphosis (a) *neoteny.* (b) Evolution as a result of modification in the immature growth stages.

pagoda stone (a) A Chinese limestone showing in section fossil orthoceratites arranged in pagoda-like designs. (b) An agate whose markings resemble pagodas. (c) *pagodite.*

pagodite Massive pinite or *agalmatolite* carved by the Chinese into miniature pagodas. Syn: *pagoda stone.*

paha A low, elongated, rounded glacial ridge or hill consisting mainly of drift, rock, or windblown sand, silt, or clay but capped with a thick cover of loess; found esp. in NE Iowa. Height varies between 10 and 30 m. Etymol: Dakota *pahá,* "hill". Pl: *paha; pahas.*

pahoehoe A Hawaiian term for a type of basaltic lava flow typified by a smooth, billowy, or ropy surface. Varieties include corded, elephant-hide, entrail, festooned, filamented, sharkskin, shelly, and slab pahoehoe. Cf: *aa.* Pron: pa-*ho*-e-*ho*-e. Obsolete syn. *dermolith.* Syn: *ropy lava.*

Pahrump A provincial series of the Precambrian in California.

paigeite *vonsenite.*

painite A mineral: $CaZrBAl_9O_{18}$.

paint A term used in SW U.S. for an earthy, pulverulent variety of cinnabar (Bureau of Mines, 1968, p. 788).

paint pot A type of *mud pot* containing multicolored mud. Also spelled: *paintpot; paint-pot.*

paired terrace One of two *stream terraces* that face each other at the same elevation from opposite sides of the stream valley and that represent the remnants of the same flood plain or valley floor. Cf: *unpaired terrace.* Syn: *matched terrace.*

paisanite A light-colored peralkaline hypabyssal or extrusive *rhyolite* characterized by small sanidine and quartz phenocrysts, few in number, and aggregates of sodic amphibole in a groundmass of quartz and alkali feldspar, frequently intergrown. Its name, given by Osann in 1893, is derived from Paisano Pass, Texas. Cf: *comendite; pantellerite.* Not recommended usage.

pakihi A term used in New Zealand for a waterlogged gravel flat (Stamp, 1966, p. 352). Etymol: Maori.

palaeo- *paleo-.*

palaeocope Any ostracode belonging to the order Palaeocopida, characterized by a shell with a long straight dorsal margin, commonly with lobes, sulci, and ventral structures. Range, Lower Ordovician to Middle Permian, with some questionably identified in present-day waters.

palaeoethnobotany The study of plants found in archeological sites (Renfrew, 1973).

Palaeophyticum A paleobotanic division of geologic time, corresponding approximately to, and characterized by the plant life of, the Paleozoic. Cf: *Mesophyticum; Cainophyticum.*

palaetiology Explanation of past changes in the Earth's condition as being governed by the laws of cause and effect. Var: *paletiology.*

palagonite An altered *tachylyte,* brown to yellow or orange and found in pillow lavas as interstitial material or in amygdules. Named by Waltershausen in 1845, for Palagonia, Sicily.

palagonite tuff A pyroclastic rock consisting of angular fragments of hydrothermally altered or weathered basaltic glass. Cf: *hyaloclastite.*

palagonitization Formation of palagonite by hydration of tachylyte.

palaite *hureaulite.*

palasome A syn. of *host,* used in economic geology. Also spelled: *palosome.* Cf: *paleosome; metasome.*

palate (a) In the vertebrates, esp. Osteichthyes and tetrapods, the bony roof of the mouth and its soft covering. (b) In some cheilosome bryozoans, the membranous or partly calcified wall of an avicularium, on which the mandible occludes.

Palatinian orogeny *Pfalzian orogeny.*

palatinite An old term for basalts and diorites that contain orthopyroxenes. Obsolete.

pale The self-crossing point of some *contour line* that forms two loops around adjacent *pits* (Warntz, 1975, p. 210).

paleic surface A smooth preglacial erosion surface, such as widely developed in Scandinavia.

paleo- (a) A combining form denoting the attribute of great age or remoteness in regard to time (*Paleo*zoic), or involving ancient conditions (*paleo*climate), or of ancestral origin, or dealing with fossil forms (*paleo*anthropic). Sometimes given as *pale-* before vowels (*pale*vent). Also spelled: *palaeo-; palaio-.* (b) A prefix indicating pre-Tertiary origin, and generally altered character, of a rock to the name of which it is added, e.g. *paleo*picrite; by some the prefix has been applied to pre-Carboniferous rocks or features, e.g. the *Paleo*atlantic Ocean.

paleoagrostology The study of fossil grasses.

paleoaktology Study of ancient nearshore and shallow-water environments.

paleoalgology The study of fossil algae. Syn: *paleophycology.*

paleoanthropology *human paleontology.*

paleoaquifer A stratigraphic unit or subterranean channel system that functioned as an aquifer at some time in the geologic past.

paleoautochthon The original autochthon or basement of a tectonic region, esp. where overlain successively by *neoautochthon* and *allochthon.* It may be folded and faulted by later movements, but it has not been greatly displaced in a horizontal direction.

paleobiochemical Relating to ancient biochemical products, such as amino acids, fatty acids, and sugars, isolated from geological specimens, that have undergone little change since they were produced.

paleobiocoenosis An assemblage of organisms that lived together in the geologic past as an interrelated community. Syn: *paleocoenosis.* Cf: *fossil community.*

paleobiogeography That branch of paleontology that treats the geographic distribution of plants and animals in past geologic time, esp. with regard to the ecologic, climatologic, and evolutionary factors involved. See also: *paleophytogeography; paleozoogeography.*

paleobiology A branch of paleontology dealing with the study of fossils as organisms rather than as features of historical geology. Cf: *biogeology; geobiology; cenozoology.*

paleobiotope A term sometimes used in paleoecology to designate a region of unspecified size that is characterized by essentially uniform environmental conditions and by a correspondingly uniform population of animals or plants or both. See also: *biotope [ecol].*

paleoblast A crystal, or a remnant of a crystal, esp. in a metamorphic rock, that is older than other grains of the same or other mineral species in the rock. Such relicts represent former conditions of equilibration of the rock. Because of the kinetics of solid-state *recrystallization,* paleoblasts have high dislocation densities and narrow subgrain-boundary spacings. Cf: *neoblast; porphyroclast.*

paleobotanic province A large region characterized and defined by similar fossil floras.

paleobotany The study of the plant life of the geologic past (Arnold, 1947, p.1). Syn: *phytopaleontology; paleophytology.*

Paleocene An epoch of the early Tertiary period, after the Gulfian of the Cretaceous period and before the Eocene; also, the corresponding worldwide series of rocks.

paleochannel A remnant of a stream channel cut in older rock and filled by the sediments of younger overlying rock; a buried stream channel.

paleoclimate The climate of a given interval of time in the geologic past. Syn: *geologic climate.*

paleoclimatologic map A map that depicts paleoclimatic data of a given time interval.

paleoclimatology The study of *paleoclimates* throughout geologic time, and of the causes of their variation, on either a local or a worldwide basis. It involves the interpretation of glacial deposits, fossils, and sedimentologic and other types of data.

paleocoenosis *paleobiocoenosis.*

paleocrystic ice Old sea ice, esp. well-weathered polar ice, generally considered to be at least 10 years old; it is often found in floebergs and in the pack ice of the central Arctic Ocean.

paleocurrent An ancient current (generally of water) that existed in the geologic past, whose direction is inferred from the sedimentary structures and textures of the rocks formed at that time.

paleocurrent structure *directional structure.*

paleodepth The depth at which an ancient organism or group of organisms lived.

paleodrainage pattern A drainage pattern representing the distribution of a valley system as it existed at a given moment of geologic time (Andresen, 1962).

paleoecology The study of the relationships between ancient organisms and their environments, the death of organisms, and their burial and postburial history in the geologic past, based on fossil faunas and floras and their stratigraphic position. See also: *ecology.*

paleoenvironment An environment in the geologic past.

paleoequator The position of the Earth's equator in the geologic past as defined for a specific geologic period and based on geologic evidence such as paleomagnetic measurements, oxygen-isotope ratios, fauna and flora, distribution of evaporites, reefs, coal deposits, and tillites; e.g. the Ordovician paleoequator for North America, running from the southern tip of Baja California to the north end of Greenland. Paleoequators are great circles that were formerly normal to the axis of rotation but are now displaced and vary from continent to continent and plate to plate.

paleofloristics The composition and distribution of ancient floras.

paleofluminology The study of ancient stream systems. Cf: *paleohydrology.*

Paleogene An interval of geologic time incorporating the Paleocene, Eocene, and Oligocene of the Tertiary; the earlier Tertiary. When the Tertiary is designated as an era, then the Paleogene, together with the *Neogene,* may be considered to be its two periods. Syn: *Eogene; Nummulitic.*

paleogeographic event *palevent.*

paleogeographic map A map that shows the reconstructed physical geography at a particular time in the geologic past, including such information as the distribution of land and seas, geomorphology of the land, depth of the sea, directions of currents in water and air, distribution of bottom sediments, and climatic belts. Cf: *paleotectonic map.*

paleogeographic stage *palstage.*

paleogeography The study and description of the physical geography of the geologic past, such as the historical reconstruction of the pattern of the Earth's surface or of a given area at a particular time in the geologic past, or the study of the successive changes of surface relief during geologic time. Syn: *paleophysiography.*

paleogeologic map A map that shows the areal geology of an ancient surface at some time in the geologic past; esp. such a map of the surface immediately below an unconformity, showing the geology as it existed at the time the surface of unconformity was completed and before the overlapping strata were deposited. Paleogeologic maps were introduced by Levorsen (1933). Cf: *subcrop map.* Syn: *peel map.*

paleogeomorphology A branch of geomorphology concerned with the recognition of ancient erosion surfaces and with the study of ancient topographies and topographic features that are now concealed beneath the surface or have been removed by erosion. Syn: *paleophysiography.*

Paleognathae A superorder of the avian subclass Neornithes which encompasses the ground-dwelling orders of birds: ostriches, cassowaries, rheas, emus, tinamous, moas, elephant birds, and kiwis. See also: *Neognathae.*

paleohydrology (a) The study of the earliest uses and management of water. (b) The study of ancient hydrologic features preserved in rock, e.g. paleokarst. Cf: *paleofluminology.*

paleoichnology *palichnology.*

paleoisotherm The locus of points of equal temperature at some time in the geologic past.

paleokarst A rock or area that has been karstified and subsequently buried under sediments. See also: *covered karst; interstratal karst; relict karst.* Syn: *buried karst; fossil karst.*

paleolatitude The latitude of a specific area on the Earth's surface in the geologic past; specif. the distance measured in degrees from the *paleoequator* or *paleopole.*

paleolimnology (a) The study of the conditions and processes in ancient lakes; interpretation of the accumulated sediments and the geomorphology and geologic history of ancient lake basins. (b) The study of the sediments and history of existing lakes.

Paleolithic n. In archaeology, the first division of the *Stone Age,* characterized by the appearance of man and man-made implements. Correlation of relative cultural levels with actual age (and, therefore, with the time-stratigraphic units of geology) varies from region to region; however, the age generally given for the Paleolithic more or less coincides with the Pleistocene. Cf: *Mesolithic; Neolithic.* See also: *lower Paleolithic; middle Paleolithic; upper Paleolithic; Eolithic.* Syn: *Old Stone Age.*—adj. Pertaining to the Paleolithic.

paleolithologic map A *paleogeologic map* that shows lithologic variations at some buried horizon or within some restricted zone at a particular time in the geologic past.

paleomagnetic pole *virtual geomagnetic pole.*

paleomagnetic stratigraphy The use of natural remanent magnetization to identify stratigraphic units. It depends on the temporal variation of the ambient magnetic field, which is due to geomagnetic secular variation and reversals. Syn: *magnetic stratigraphy.*

paleomagnetism The study of natural remanent magnetization in order to determine the intensity and direction of the Earth's magnetic field in the geologic past.

paleometeoritics The study of variation of extraterrestrial debris as a function of time over extended parts of the geologic record, esp. in deep-sea sediments and possibly in sedimentary rocks, and for more recent periods in ice.

paleomorphology The *morphology* or study of form and structure of fossil remains (hard parts) in order to determine the original anatomy (soft parts) of an organism; e.g. the study of a brachiopod muscle scar whose depth may indicate the strength of the muscle.

paleomycology The study of fossil fungi.

paleontography The formal, systematic description of fossils. Adj: *paleontographic.*

paleontological resource A locality, esp. one of such limited size that it can be destroyed by human activity, that has yielded type fossils or described fossils of unusually high quality, or is documented for its preservation of paleoenvironmental features. Also, the fossils, described or not, from such a locality.

paleontologic facies A term recommended by Teichert (1958, p.2734) to replace *biofacies* as used in stratigraphy, signifying the paleontologic characteristics of a sedimentary rock.

paleontologic record A syn. of *fossil record.* Cf: *stratigraphic record.*

paleontologic species A *morphologic species* based on fossil specimens. It may include specimens that would be considered specifically distinct if living individuals could be observed.

paleontologist One who studies the fossilized remains of animals and/or plants.

paleontology The study of life in past geologic time, based on fossil plants and animals and including phylogeny, their relationships to existing plants, animals, and environments, and the chronology of the Earth's history. Cf: *neontology.* See also: *historical geology.*

paleopalynology A division of *palynology* concerned with the study of fossil spores and pollen. It is now interpreted broadly to include study of a wide range of fossil microscopic bodies in addition to spores and pollen: animal remains such as chitinozoans, as well as fungal spores, dinoflagellates, acritarchs, and other organisms resistant to acids and found in sedimentary rocks of all ages (nannofossils and diatoms are sometimes included). The usual criteria for inclusion are that the bodies be microscopic in size and composed of a resistant organic substance (usually sporopollenin, chitin, or pseudochitin) that results in their being preserved in sedimentary rocks and available for separation by maceration from such rocks.

paleopedology The study of soils of past geologic ages, including determination of their ages.

paleophycology *paleoalgology.*

paleophyre A reddish porphyritic *andesite* intruded into Silurian strata in the Fichtelgebirge, German-Czechoslovakian border. Also spelled: *palaeophyre.* Obsolete.

paleophysiography (a) *paleogeomorphology.* (b) *paleogeography.*

Paleophytic A paleobotanic division of geologic time, signifying that time during which pteridophytes were abundant, between the development of algae and the appearance of the first gymnosperms. Cf: *Aphytic; Archeophytic; Eophytic; Mesophytic; Cenophytic.* Syn: *Pteridophytic.*

paleophytogeography The branch of *paleobiogeography* dealing with the distribution of plants in past geologic time. Cf: *paleozoogeography.*

paleophytology An obsolescent syn. of *paleobotany.*

paleoplain A term introduced by Hill (1900, p. 5) for an ancient degradational plain that is now buried beneath later deposits.

paleopole A pole of the Earth, either magnetic or geographic, at some former geologic time.

paleosalinity The salinity of a body of water in the geological past, as evaluated on the basis of chemical analyses of sediment or formation water.

paleosere A sequence of ecologic communities in the geologic past that led to a *climax community;* a *sere* in the geologic past.

paleoslope The direction of initial dip of a former land surface; esp. the regional slope of a large, ancient physiographic unit, such as a flood plain or a continental slope.

paleosol A buried soil horizon of the geologic past. When uncovered, it is said to be exhumed. Syn: *buried soil; fossil soil.*

paleosome A geometric element of a composite rock or mineral deposit, appearing to be older than an associated younger rock element (or *neosome*); e.g. wall rock in a vein or replacement deposit, or the unaltered and relatively immobile pre-existing part of a migmatite. Sometimes used in place of the term *stereosome, metaster,* or *restite.* Cf: *host.*

paleostructure The geologic structure of a region or sequence of rocks at some time in the geologic past; the structure of a paleogeologic area.

paleostructure map A map that shows, by thickness contour lines, the geologic structure that existed at the time when a surface of unconformity was completed and before the overlapping strata were deposited (Levorsen, 1960, p. 4).

paleotectonic map A map intended to show geologic and tectonic features as they existed at some time in the geologic past, rather than the sum of all the tectonics of the region, as portrayed on a *tectonic map.* It is similar to a *paleogeographic map* but more emphasis is placed on the tectonic features than on the distribution of lands and seas. See also: *neotectonic map.*

paleotemperature The mean temperature at a given time or place in geologic history; esp. the paleoclimatic temperature of the sea.

paleothanatocoenosis A group of organisms buried together in the geologic past.

paleothermal Pertaining to or characteristic of warm climates of the geologic past; e.g. a "paleothermal fauna". Syn: *paleothermic.*

paleothermometry Measurement or estimation of paleotemperatures; esp. the determination of the temperature of a shallow sea of the geologic past, based on the mass-spectrometric measurement of the relative abundance of oxygen isotopes in the carbonates of marine fossil shells. See also: *geologic thermometry.*

paleotopographic map A map that shows the relief of a surface of unconformity, or the relief believed to have existed at some time in the geologic past.

paleotopography The topographic relief of an area at a particular time in the geologic past; the topography of a paleogeologic area, such as the configuration of the surface of an unconformity at the time it was overlapped.

paleotypal Said of a fine-grained porphyritic igneous rock having the characteristics of altered extrusive or hypabyssal rocks such as those of pre-Tertiary age. This term and the term *cenotypal* were introduced to distinguish Tertiary (*neovolcanic*) and pre-

Tertiary (*paleovolcanic*) fine-grained igneous rocks; both are little used.

paleovolcanic Said of extrusive rocks that are of pre-Tertiary age. Cf: *neovolcanic*.

paleovolcanology The study of the processes and products of volcanic activity in the geologic past.

paleowind A wind of the geologic past. Its direction is recorded by distribution of volcanic ash falls, growth patterns of coral reefs, orientation of sand dunes, etc.

Paleozoic An era of geologic time, from the end of the Precambrian to the beginning of the Mesozoic, or from about 570 to about 225 million years ago. Obsolete syn: *Primary*.

paleozoogeography The branch of *paleobiogeography* dealing with the distribution of animals in past geologic time. Cf: *paleophytogeography*.

paleozoology That branch of paleontology dealing with the study of *subfossil* and *fossil* animals, both invertebrate and vertebrate.

palermoite A mineral: $(Li,Na)_2(Sr,Ca)Al_4(PO_4)_4(OH)_4$.

palette *shield [speleo]*.

palevent A relatively sudden and short-lived paleogeographic happening, such as the brief existence of a particular depositional environment, or a rapid geographic change separating two *palstages* (Wills, 1956, p.14). Syn: *paleogeographic event*. Rarely used.

pali [geog] An Hawaiian term for a steep slope; e.g. the Nuuanu Pali, a steep-faced scarp on the NE side of Oahu.

pali [paleont] Plural of *palus*.

palichnology The study of *trace fossils*, as opposed to *neoichnology*. Also spelled: *paleoichnology*. See also: *ichnology*.

paliform Resembling a *palus*; specif. "paliform lobes" of the septa in corals, formed by detached trabecular offsets from the inner edges of the septa, appearing in vertical succession, and differing from pali in not being formed as a result of substitution.

palimpsest [marine geol] Said of relict sediments of the *continental shelf*, reworked by physical or biological processes.

palimpsest [meta] adj. Said of a structure or texture in a metamorphic rock in which remnants of some pre-existing structure or texture, perhaps even the original one, are preserved, and sometimes even megascopically visible. The term was first used by Sederholm (1891). Cf: *relict [meta]*.

palimpsest [streams] Said of a kind of drainage in which a modern, anomalous drainage pattern is superimposed upon an older one, clearly indicating different topographic and possibly structural conditions at the time of development.

palimpsest surface *Continental shelf* morphology that results from shore-face retreat and modification (but not obliteration) by subsequent processes (Swift et al., 1972, p. 537).

palingenesis [paleont] Recapitulation, without change, in the young stages of an organism of the characteristics of its ancestors. See also: *recapitulation theory*.

palingenesis [petrology] Formation of a new magma by the melting of pre-existing magmatic rock in situ. Considered incorrectly by some workers as a syn. of *anatexis*. Adj: *palingenic*.

palingenetic *resurrected*.

palingenic Formed by or involving *palingenesis*.

palinspastic map A name coined by Kay (1937) for a paleogeographic or paleotectonic map in which the features represented have been restored as nearly as possible to their original geographic positions, before the rocks of the crust were shortened by folding, or telescoped by thrusting.

palinspastic section A cross section in which the features represented have been restored as nearly as possible to their original geographic positions, before the rocks of the crust were shortened by folding or telescoped by thrusting (Kay, 1937).

palintrope A term used initially for the morphologically posterior sector of either valve of some brachiopod shells, which was reflexed to grow anteriorly (mixoperipheral growth), but more recently for the curved surface of the shell, bounded by beak ridges and cardinal margin of nonstrophic shells (TIP, 1965, pt.H, p.149). It differs from a *planarea* in being curved in all directions.

palisade A picturesque rock cliff or line of bold cliffs, rising precipitously from the margin of a stream or lake; esp. one consisting of basalt with columnar structure, such as the Palisades along the Hudson River of New York and New Jersey. Term is usually used in the plural.

palisade mesophyll *palisade tissue*.

Palisades disturbance A time of deformation, or orogeny, supposed by Schuchert (1924) to have closed the Triassic Period in eastern North America and elsewhere. It is based on the block-faulted structure of the Upper Triassic Newark series in the Appalachian area, which was truncated before younger Mesozoic (mainly Cretaceous) strata were laid over it. The concept of a distinct orogeny at this time is dubious and has only local application at most. Named for the Palisades of New York and New Jersey, the edge of a diabase sill intruded at this time.

palisade tissue Tissue composed of long cylindrical chlorophyllous cells oriented normal to the lamina beneath the upper epidermis of angiospermous leaves. Syn: *palisade mesophyll*.

palladium A soft silver-white or steel-white isometric mineral, the native metallic element Pd. It is one of the platinum metals, and it resembles and occurs with platinum, usually occurring in grains and frequently alloyed with platinum and iridium.

palladium amalgam *potarite*.

palladium gold *porpezite*.

palladoarsenide A monoclinic mineral: Pd_2As.

palladobismutharsenide An orthorhombic mineral: $Pd_2(As,Bi)$.

pallasite [ign] Any ultramafic rock, whether of meteoric or terrestrial origin, that contains approximately 60 percent iron if meteoric or more iron oxides than silica if terrestrial; e.g. *cumberlandite* (Thrush, 1968, p. 789).

pallasite [meteorite] A *stony-iron meteorite* composed essentially of large single glassy crystals of olivine embedded in a network of nickel-iron. Pallasites are believed to have been formed at the interface of the stony mantle and metal core of a layered planetoid. Syn: *pallas iron*.

pallial chamber *mantle cavity*.

pallial line A line or narrow band on the inner surface of a valve of a bivalve-mollusk or rostroconch shell, close to and more or less parallel with the margin, and marking the line of attachment of the marginal muscles of the mantle. It is typically distinguished by a groove or ridge and by a change in texture of shell material. Syn: *pallial impression*.

pallial sinus (a) An often conspicuous embayment or inward bend in the posterior and ventral part of the pallial line of a bivalve mollusk, marking the point of attachment of the siphonal retractor muscles. See also: *sinus [paleont]*. (b) A *mantle canal* of a brachiopod.

pallite A ferrian variety of millisite.

pallium The *mantle* of a mollusk or brachiopod. Pl: *pallia*.

pallomancy A form of *dowsing* using a pendulum. Cf: *rhabdomancy*.

palmate Having lobes or divisions radiating from a common center in a handlike fashion; digitate.

palmate venation In a leaf, a type of *net venation* in which the main veins branch out from the stalk apex like the fingers of a hand. Cf: *pinnate venation*.

palmierite A white hexagonal mineral: $(K,Na)_2Pb(SO_4)_2$. It is isostructural with kalistrontite.

palmitic acid A long-chain waxlike fatty acid, formula $C_{16}H_{32}O_2$, present in numerous plant and animal fats as glycerides.

palosome *palasome*.

palp A reduced distal portion of the limb of a crustacean, usually only one of its rami, but sometimes comprising both rami and basis (TIP, 1969, pt.R, p.99). See also: *palpus*.

palpebral lobe (a) An elevated portion of the fixed cheek of a trilobite, extending laterally from the axial furrow to the upper and inner margin of the visual surface of an eye. (b) The elevated portion of the lateral eye of a eurypterid.

palpi Plural of *palpus*.

palpus A term applied either to a *pedipalpus* (including pedipalpal coxae) or more properly to one of the five segments following the coxa in an arachnid (TIP, 1955, pt.P, p.62). Pl: *palpi*. Adj: *palpal*. See also: *palp*.

pals *palsa*.

palsa An elliptical domelike *frost mound* containing ice lenses in peat, commonly 3-6 m high and 2-25 m long, occurring in subarctic bogs of the tundra, esp. in Scandinavia, and often surrounded by shallow open water. Etymol: Swedish, "elliptical". Pl: *palsen*. See also: *frost mound; peat mound; peat hummock*. Also spelled: *pals*.

palsen Plural of *palsa*.

palstage A period of time when paleogeographic conditions were relatively static, or were changing gradually and progressively, with relation to such factors as sea level, surface relief, or distance from shore (Wills, 1956, p.14). Cf: *palevent*. Syn: *paleogeographic stage*. Rarely used.

paludal Pertaining to a marsh. Syn: *paludous; palustral*.

paludification *ulmification*.

paludous *paludal.*

palus Any of several slender vertical calcareous lamellae, plates, or pillars, developed along the inner edge of certain entosepta of a coral and comprising the remnant part of a pair of exosepta joined at their inner margins. Pl: *pali.* See also: *paliform.*

palustrine Pertaining to material growing or deposited in a marsh or marsh-like environment.

palygorskite A chain-lattice clay mineral: $(Mg,Al)_2Si_4O_{10}(OH) \cdot 4H_2O$. The term has also been used as a group name for light-weight fibrous clay minerals showing a considerable amount of substitution of aluminum for magnesium and characterized by distinctive rodlike shapes under the electron microscope. Syn: *attapulgite.*

palyniferous Bearing pollen. The term in palynology usually refers to rocks or sediment samples that yield pollen, spores, or other palynomorphs on maceration.

palynofacies A term used in paleopalynology for an assemblage of palynomorphs in a portion of a sediment, representing local environmental conditions and not typical of the regional palynoflora.

palynoflora The whole suite of palynomorphs from a given rock unit. The term *microflora* is sometimes used as a synonym but should be avoided as it better applies to assemblages of extant microscopic algae and fungi.

palynology A branch of science concerned with the study of pollen of seed plants and spores of other embryophytic plants, whether living or fossil, including their dispersal and applications in stratigraphy and paleoecology. Term suggested by Hyde & Williams (1944, p.6). Etymol: Greek παλυνω, *sprinkle,* suggestive of Latin *pollen,* "fine flour, mill dust". See also: *paleopalynology; pollen analysis.*

palynomorph A microscopic, resistant-walled organic body found in palynologic maceration residues; a palynologic study object. Palynomorphs include pollen, spores of many sorts, acritarchs, chitinozoans, dinoflagellate thecae and cysts, certain colonial algae, and other acid-insoluble microfossils. Cf: *sporomorph.*

palynostratigraphy The stratigraphic application of palynologic methods.

pamet A dry valley formed in glacial deposits on the outer part of Cape Cod, Mass.

pampa A vast treeless grassy plain of temperate regions, esp. as used in Argentina and adjacent parts of Uruguay. It is comparable to the *prairies* of North America, the *steppes* of the U.S.S.R., and the *veld* of South Africa.

pan [geomorph] (a) A shallow, natural depression or basin, esp. one containing a lake, pond, or other body of standing water; e.g. a shallow depression holding a temporary or permanent pool in a tidal marsh along the Atlantic coast of the U.S. (b) A term used in South Africa for a hollow in the ground where the neck of a volcano formerly existed.

pan [ice] (a) Shortened form of *pancake ice.* (b) An individual piece of pancake ice. (c) *ice pan.* (d) A large fragment of flat, relatively thin ice, having a diameter about 60 m, formed in a bay or fiord or along the shore and subsequently loosened to drift about the sea. Syn: *pan ice.*

pan [salt] (a) A *salt pan,* esp. in South African usage. Cf: *vloer.* (b) An artificial basin for producing salt by evaporation of salt water or brine. Also, a vessel for evaporating salt water or brine.

pan [soil] A hard, cementlike layer, crust, or horizon within or just beneath the surface soil, being strongly compacted, indurated, or high in clay content, and usually impeding the movement of water and air and the growth of plant roots; specif. *hardpan.* See also: *genetic pan; pressure pan.*

panabase *tetrahedrite.*

panautomorphic *automorphic.*

panautomorphic-granular *automorphic.*

pancake bomb A type of volcanic bomb whose flattened shape is due to impact. Cf: *Hawaiian-type bomb.*

pancake ice One or more small, predominantly circular pieces of newly formed sea ice (diameter ranging from about 30 cm to about 3 m) with slightly raised rims caused by the pieces rotating and striking against one another; it often forms during the early fall in polar regions. See also: *lily-pad ice.* Syn: *pan [ice]; pancake.*

pandaite A mineral of the pyrochlore group: $(Ba,Sr)_2(Nb,Ti)_2(O,OH)_7$.

pandemic Said of conditions that occur over a broad geographic area and affect a major part of the population; also said of a widely dispersed population.

panderian opening A small opening in the fixigenal or thoracic doublure of trilobites, situated immediately behind the panderian protuberance. Its function is not known, but it may have been a pore leading to some internal organ. Cf: *panderian protuberance.*

panderian protuberance A small protuberance on the fixigenal or thoracic doublure of trilobites, which functioned as a stopping device to prevent segments overlapping too far on enrollment. Cf: *panderian opening.*

pandermite *priceite.*

panethite A meteorite mineral: $(Na,Ca,K)_2(Mg,Fe,Mn)_2(PO_4)_2$.

panfan A graded bedrock surface consisting of a series of coalescing pediments and representing the penultimate stage of an arid cycle of erosion. The synonymous term *pediplain* is preferred because the feature does not involve alluvial fans, although the term "panfan" was proposed by Lawson (1915, p. 33) for a vast alluvial fan representing the end stage in the process of geomorphic development in a desert region.

Pangaea *Pangea.*

Pangea A supercontinent that existed from about 300 to about 200 m.y. ago and included most of the continental crust of the Earth, from which the present continents were derived by fragmentation and *continental displacement.* During an intermediate stage of the fragmentation, between the existence of Pangea and that of the present continents, Pangea is believed to have split into two large fragments, *Laurasia* on the north and *Gondwana* on the south. The proto-ocean around Pangea has been termed *Panthalassa.* Some geologists, while believing in the former existence of Laurasia and Gondwana, are reluctant to concede the existence of an original Pangea. Also spelled: *Pangaea.*

Pang-Yang depression A large erosional basin with a flat bottom and steep sides, developed on a rocky plain or plateau (Stone, 1967, p. 236). Type locality: Pang Yang, Burma.

panhole *solution pan.*

panicle A compound *inflorescence* with several main branches, each of which bears pedicelled flowers arranged along its axis; in grass panicles, the flowers are borne in spikelets on the pedicels (Fuller & Tippo, 1954, p. 965).

panidiomorphic *idiomorphic.*

panidiomorphic-granular A syn. of *idiomorphic.* Johannsen (1939, p. 226) states that Rosenbusch, who first used the term in 1887, incorrectly applied it to the *microgranular xenomorphic* texture of aplite, although he had defined it as involving euhedral components; this usage was corrected in the last edition of Rosenbusch's book, by its editor, Osann.

pan lake A lake occupying a shallow natural depression, or pan.

panmixis The free interchange of genes within an interbreeding population.

panning A technique of prospecting for heavy metals, e.g. gold, by washing placer or crushed vein material in a pan. The lighter fractions are washed away, leaving the heavy metals behind in the pan.

Pannonian Eastern European stage: Upper Miocene (above Sarmatian, below Pontian). Also used as equivalent to lower Pontian in earlier usage; considered as lower Pliocene by some authors. Equivalent to Meotian. See also: *Turolian.*

panplain (a) A term introduced by Crickmay (1933, p. 344-345) for a very broad plain formed by the coalescence of several adjacent flood plains, each resulting from long-continued lateral erosion by meandering streams; it represents the end stage of an erosion cycle. Cf: *peneplain; plain of lateral planation.* Syn: *panplane.* (b) A very level plain with a general seaward inclination (Von Engeln, 1942).

panplanation The action or process of formation and development of a *panplain.*

panplane *panplain.*

pantellerite A peralkaline *rhyolite* or *quartz trachyte* with normative quartz exceeding 10%. It is more mafic than *comendite* (Macdonald & Bailey, 1973). Its name, given by Förstner in 1881, is derived from Pantelleria, an island in the Mediterranean Sea south of Sicily.

Panthalassa The ocean surrounding *Pangea.*

pantograph An instrument for copying a map or drawing on any predetermined scale of reduction or enlargement. It consists of four bars hinged to form an adjustable parallelogram, so that as one tracing stylus is moved over the material to be copied the other makes the desired copy. Specif: a mechanical attachment to a stereo-plotting instrument to permit tracing map detail at model scale while reproducing on the compilation manuscript at a predetermined reduction.

pantonematic Said of a featherlike flagellum (as in Euglenophyta) provided throughout its length with a single row of tiny cilia.

paolovite An orthorhombic mineral: Pd_2Sn.

papa A soft, bluish clay, mudstone, siltstone, or sandstone found in North Island, N.Z., and used for whitening fireplaces. Etymol: Polynesian.

papagoite A blue monoclinic mineral: $CaCuAlSi_2O_6(OH)_3$.

paper chromatography A chromatographic technique for separating components of a sample by moving it in a mixture of solution by gravity or capillarity through a paper substrate in such a way that the different components have different mobilities and thus become separated. The technique usually involves partition procedures (May & Cuttitta, 1967, p.116). See also: *chromatography.*

paper clay A fine-grained, white, kaolinic clay with high retention and suspending properties and a very low content of free silica, used for coating or filling paper.

paper peat Thinly laminated peat. Syn: *leaf peat.*

paper shale A shale that easily separates on weathering into thin layers or laminae suggesting sheets of paper; it is often highly carbonaceous.

paper spar A crystallized variety of calcite occurring in thin lamellae or paperlike plates.

papery Descriptive of a fine-grained sedimentary rock that splits into laminae less than 2 mm in thickness (McKee & Weir, 1953, p.383).

papilla (a) A surficial mound associated with a pore in cystoids. (b) A minute scalelike ossicle or projection in ophiuroids. (c) *apical papilla.*—Pl: *papillae.*

papillus polaris A swelling or protuberance in the polar region ("polpapillus") of pollen grains (Tschudy and Scott, 1969, p. 27).

papula A short protuberance of integument between ossicles of the aboral or oral surface of an asteroid and functioning as an external gill. Pl: *papulae.*

papule A prolate to equant, somewhat rounded *glaebule* composed dominantly of clay minerals with a continuous lamellar fabric and having sharp external boundaries (Brewer, 1964, p.274-275); e.g. a clay gall in soil material.

para- (a) A prefix that, when used with a metamorphic rock name, indicates that the rock was derived from a sediment, e.g. paragneiss. (b) When used with the name of a clastic sedimentary rock, a prefix signifying matrix-rich (Pettijohn, Potter & Siever, 1973, p. 168), e.g. *pararenite.*

para-autunite An artificial compound: $Ca(UO_2)_2(PO_4)_2$. It represents the complete dehydration product of autunite. Cf: *meta-autunite.*

parabiont Any one of the organisms involved in *parabiosis.*

parabiosis The condition in which members of two or more species maintain colonies close to one another without conflict. See also: *parabiont.*

parabituminous Said of bituminous coal containing 84-87% carbon, analyzed on a dry, ash-free basis. Not listed by ASTM as a rank classification. Cf: *perbituminous; orthobituminous.*

parabolic dune (a) A sand dune with a long, scoop-shaped form, convex in the downwind direction so that its horns point upwind, whose ground plan, when perfectly developed, approximates the form of a parabola. It is characteristically covered with sparse vegetation, and is often found along the coast where strong onshore winds are supplied with abundant sand. (b) A term used loosely as a syn. of *upsiloidal dune.*—Cf: *barchan.*

parabutlerite An orange orthorhombic mineral: $FeSO_4(OH) \cdot 2H_2O$. Cf: *butlerite.*

paracelsian A pale-yellow orthorhombic mineral: $BaAl_2Si_2O_8$. It is dimorphous with celsian.

parachoma A ridge of dense calcite developed between adjacent foramina in some fusulinacean foraminiferal tests having multiple foramina (as in Verbeekinidae and Neoschwagerininae). Pl: *parachomata.* Cf: *choma.*

parachronology (a) Practical dating and correlation of stratigraphic units. (b) Geochronology based on fossils that supplement, or are used instead of, biostratigraphically significant fossils. Cf: *orthochronology.*

parachrysotile A mineral of the serpentine group: $Mg_3Si_2O_5(OH)_4$. It is a polymorph of chrysotile, distinct from clinochrysotile, orthochrysotile, and lizardite.

paraclase An obsolete term for a *fault.*

paraclavule An apparently monaxonic sponge spicule (microsclere) consisting of a short straight shaft pointed at one end and bearing an umbel at the other end. It resembles an amphidisc with one umbel missing.

paraclinal Said of a stream or valley that is oriented in a direction parallel to the fold axes of a region. Also said of a region having paraclinal streams. Term introduced by Powell (1874, p. 50). Ant: *diaclinal.*

paracme The period in the phylogeny of a group of organisms that follows the *acme* and is marked by decadence or decline.

paraconformable Not really or not quite conformable; esp. said of strata exhibiting paraconformity.

paraconformity A term introduced by Dunbar & Rodgers (1957, p.119) for an obscure or uncertain *unconformity* in which no erosion surface is discernible or in which the contact is a simple bedding plane, and in which the beds above and below the the break are parallel. This type of unconformity was formerly classed by Pirsson (1915, p.291-293) as a kind of *disconformity,* and is recognized in Great Britain as a *non-sequence* "of major time-significance" rather than as an unconformity (Challinor, 1978, p.321). Not to be confused with *paraunconformity.* Cf: *diastem.* Syn: *nondepositional unconformity.*

paraconglomerate A term proposed by Pettijohn (1957, p.261) for a conglomerate that is not a product of normal aqueous flow but is deposited by such modes of mass transport as subaqueous turbidity slides and glacier ice; it is characterized by a disrupted gravel framework (stones not generally in contact), is often unstratified, and is notable for containing more matrix than gravel-sized fragments (pebbles may form less than 10% of the rock). Examples include tillites, pseudotillites, pebbly mudstones, and relatively structureless clay or shale bodies in which pebbles or cobbles are randomly distributed. Cf: *orthoconglomerate.* Syn: *conglomeratic mudstone.*

paracontinuity A small-scale *disconformity* that is also a small-scale *faunal break.* It is "a kind of geographically widespread diastem" and "offers a reliable means by which significant time-stratigraphic boundaries may be determined precisely" (Conkin & Conkin, 1975, p. 1-2).

paracoquimbite A pale-violet rhombohedral mineral: $Fe_2(SO_4)_3 \cdot 9H_2O$. It is dimorphous with coquimbite.

paracostibite A mineral: CoSbS. It is dimorphous with costibite.

paracycle of relative change of sea level The interval of time occupied by one regional or global relative rise and stillstand of sea level, followed by another relative rise, with no intervening relative fall (Mitchum, 1977, p. 209). See also: *cycle of relative change of sea level; supercycle.*

paradamite A triclinic mineral: $Zn_2(AsO_4)(OH)$. It is isomorphous with tarbuttite and dimorphous with adamite.

paradelta A term proposed by Strickland (1940, p. 10) for the landward or upper part of a delta, or that part undergoing degradation.

paradiagenetic Signifying a close relation with sedimentary diagenesis; e.g. "paradiagenetic movement", or deformation that is precrystalline in relation to spathization (Sander, 1951, p. 52).

paradocrasite A mineral: $Sb_2(Sb,As)_2$.

paradoublural line Furrow, flexure, or weak ridge developed on the dorsal exoskeleton of a trilobite directly above the inner margin of the doublure.

paradox of anisotropy In a sequence of rocks that exhibits anisotropy because of layering, a measurement of apparent resistivity made with a four-terminal electrode array laid out in the direction of maximum resistivity (that is, perpendicular to the bedding) will yield the minimum value of resistivity, i.e. for current flowing along the bedding planes. This is known as the "paradox of anisotropy".

para-ecology *taphonomy.*

paraffin-base crude Crude oil which will yield large quantities of paraffin in the process of distillation. Cf: *asphaltic-base crude; mixed-base crude.*

paraffin coal A type of light-colored bituminous coal from which oil and paraffin are produced.

paraffin hydrocarbon Any of the hydrocarbons of the *methane series.*

paraffinic Pertaining or relating to a *paraffin hydrocarbon* or paraffin wax.

paraffin series *methane series.*

paraffin wax A colorless, odorless, tasteless, amorphous solute of complex hydrocarbons with a high methane-series composition.

paraflagellar boss A swelling near the base of a flagellum in some

Euglenophyta. It possibly serves as a photoreceptor.

parafoliate An obsolete syn. of *foliate.*

paragaster The *spongocoel* of a sponge.

paragenesis A characteristic association or occurrence of minerals or mineral assemblages in ore deposits, connoting contemporaneous formation. Cf: *paragenetic sequence.*

paragenetic (a) Pertaining to paragenesis. (b) Pertaining to the genetic relations of sediments in laterally continuous and equivalent facies.

paragenetic sequence The sequential order of mineral deposition, as individual phases or assemblages, in an ore deposit. Cf: *paragenesis.* Syn: *mineral sequence.*

parageosyncline (a) A geosyncline within a craton or stable area; an epeirogenic basin rather than an orogenic belt (Stille, 1935, p. 77-97). Syn: *intrageosyncline.* (b) A contemporary oceanic depression marginal to the craton (Schuchert, 1923, p. 151-260). Cf: *idiogeosyncline.*

paraglacial *periglacial.*

paragnath One of a pair of leafy lobes of the metastoma lying behind the mandibles in most crustaceans. Syn: *paragnathus.*

paragon A perfect diamond of 100 carats or more.

paragonite A yellowish or greenish mineral of the mica group: $NaAl_2(AlSi_3)O_{10}(OH)_2$. It corresponds to muscovite but with sodium instead of potassium, and it usually occurs in metamorphic rocks. Syn: *soda mica.*

paraguanajuatite A rhombohedral mineral: $Bi_2(Se,S)_3$.

parahilgardite A triclinic mineral: $Ca_2B_5O_8Cl(OH)_2$. It is dimorphous with hilgardite.

parahopeite A colorless triclinic mineral: $Zn_3(PO_4)_2 \cdot 4H_2O$. It is dimorphous with hopeite.

parajamesonite A mineral: $Pb_4FeSb_6S_{14}$. It is dimorphous with jamesonite.

parakeldyshite A triclinic mineral: $Na_2ZrSi_2O_7$.

paralaurionite A white mineral: Pb(OH)Cl. It is dimorphous with laurionite.

paralectotype Any of the *syntypes* other than the one designated as *lectotype.* Syn: *lectoparatype.*

paraliageosyncline A geosyncline developing along a present-day continental margin, e.g. the Gulf Coast geosyncline (Kay, 1945).

paralic [coal] Said of coal deposits formed along the margin of the sea, as opposed to *limnic* coal deposits.

paralic [sed] By the sea, but nonmarine; esp. pertaining to intertongued marine and continental deposits laid down on the landward side of a coast or in shallow water subject to marine invasion, and to the environments (such as lagoonal or littoral) of the marine borders. Also said of basins, platforms, marshes, swamps, and other features marked by thick terrigenous deposits intimately associated with estuarine and continental deposits, such as deltas formed on the heavily alluviated continental shelves. Etymol: Greek *paralia,* "seacoast".

paralic swamp *marine swamp.*

paralimnion The littoral part of a lake, extending from the margin to the deepest limit of rooted vegetation. Adj: *paralimnetic.*

parallax [surv] (a) The apparent displacement of the position of an object, with respect to a reference point or system, caused by an actual shift in the point of observation; e.g. "instrument parallax" in which an imperfect adjustment of a surveying instrument or a change in the position of the observer causes a change in the apparent position of an object with respect to the reference mark(s) of the instrument. (b) The difference in the apparent direction of an object as seen from two different points not on a straight line with the object (such as the apparent difference in position of a point on two consecutive photographs, or the apparent difference in direction between objects on the Earth's surface due to their difference in elevation); the angular distance between two straight lines drawn to an object from two different points of view.

parallax [tides] The ratio of the mean radius of the Earth to the distance of a tide-producing body (usually the Moon), represented by the angle at the center of the Moon between a line to the center of the Earth and a line tangent to the Earth's surface. The term is used in regard to the variation in tide range or in tidal-current speed resulting from the continually changing distance of the Moon from the Earth.

parallax bar *stereometer.*

parallel (a) One of the imaginary circles on the surface of the Earth, parallel to the equator and to one another and connecting all points of equal *latitude;* a circle parallel to the primary great circle of a sphere or spheroid, or a closed curve approximating such a circle; an east-west line of constant latitude. Each parallel is a small circle except for the equator. (b) A line, corresponding to a parallel, drawn on a globe, map, or chart.—Cf: *meridian.* Syn: *parallel of latitude.*

parallel bedding *concordant bedding.*

parallel cleavage An obsolete syn. of *bedding-plane cleavage.*

parallel-displacement fault A little-used term for a fault on which the linear features that were parallel before displacement are still parallel afterwards.

parallel drainage pattern A drainage pattern in which the streams and their tributaries are regularly spaced and flow parallel or subparallel to one another over a considerable area. It is indicative of a region having a pronounced, uniform slope and a homogeneous lithology and rock structure.

parallelepiped A closed crystal form bounded by three pairs of parallelograms.

parallel evolution The development of similar forms by related but distinct phylogenetic lineages. See also: *parallelism.* Cf: *convergent evolution.*

parallel extinction A type of *extinction* in anisotropic crystals parallel to crystal outlines or traces of cleavage planes. Cf: *inclined extinction; undulatory extinction.*

parallel fold A fold in which the orthogonal thickness of the layers is constant. Syn: *concentric fold.*

parallel growth *parallel intergrowth.*

parallel intergrowth Intergrowth of two or more crystals in which one or more axes in each crystal are almost parallel. Syn: *parallel growth.*

parallelism (a) The development or possession of similar characteristics by two or more related organisms in separate lineages, often as a result of similar environmental conditions acting upon similar heredities derived from a long-distant common ancestor. See also: *parallel evolution.* Cf: *convergence [evol].* (b) In *cladism,* a character shared by two taxa but mistakenly considered a *derived character.* Syn: *false synapomorphy; convergence [evol] (b).*

parallelkanter An elongated *ventifact* having parallel faces or edges. Etymol: German *Parallelkanter,* "one having parallel edges".

parallel of latitude A *parallel* or line of latitude. Cf: *circle of latitude.*

parallel retreat of slope (a) The recession of a scarp or of the side of a hill or mountain (once the angle of slope is established) without change in declivity, the slope at any given time retreating parallel to its former positions. (b) The concept or principle of backwearing of a slope as proposed by W. Penck (1924).

parallel ripple mark A ripple mark with a relatively straight crest and an asymmetric profile; specif. a *current ripple mark.*

parallel roads A series of horizontal beaches or wave-cut terraces occurring parallel to each other at different levels on each side of a glacial valley, as those at Glen Roy in Scotland. Each beach, or *road [glac geol],* represents a former shoreline that corresponds with a temporary level of overflow from a proglacial lake formed by ice-damming.

parallel section A slice through a foraminiferal test in a plane normal to the axis of coiling but not through the proloculus.

parallel shot In seismic prospecting, a test shot made with all the amplifiers connected in parallel and activated by a single geophone, in order to check for lead, lag, polarity, and phasing in the amplifier-to-oscillograph circuits. Syn: *bridle.*

parallel texture A general term to include *tabular* texture and *prismatic* texture.

parallel twin A twinned crystal, the twin axis of which is parallel to the composition surface. Cf: *normal twin.*

parallel unconformity *disconformity.*

parallel venation In a leaf, a type of *venation* in which the main veins are parallel with each other and with the longitudinal axis of the leaf. Cf: *net venation.*

parallochthon Rocks that were brought from intermediate distances and deposited on or near an allochthonous mass during transit. Also spelled: *para-allochthon.*

paramagnetic Having a small positive magnetic susceptibility. A paramagnetic mineral such as olivine, pyroxene, or biotite contains magnetic ions that tend to align along an applied magnetic field but do not have a spontaneous magnetic order. Cf: *diamagnetic.* See also: *superparamagnetism.*

paramagnetic resonance *electron spin resonance.*

paramarginal resources Low-grade *resources* that are recoverable at prices as much as 1.5 times those prevailing now (Brobst & Pratt, 1973, p. 3).

paramelaconite A black tetragonal mineral: $Cu_{1-2x}{}^{+2},Cu_{2x}{}^{+1})O_{1-x}$.

parameter [cryst] (a) Any of the axial lengths or interaxial angles that define a unit cell. Syn: *lattice constant.* (b) On a crystal face, the rational multiple of the axial length intercepted by a plane, which determines the position of the plane relative to the crystal lattice. (c) The proportions (x, y, z) of the unit-cell axial lengths that define the position of an atom relative to any lattice point.

parameter [stat] (a) A number describing a population (Davis, 1973, p. 62). Cf: *statistics.* (b) A constant or variable in a mathematical expression that distinguishes various specific situations (Krumbein & Graybill, 1965, p. 53). (c) Any of a set of physical properties whose values determine the characteristics or behavior of a system. (d) An approx. syn. of *variable* (Cole & King, 1968, p. 661).

parametric hydrology That branch of hydrology dealing with "the development and analysis of relationships among the physical parameters involved in hydrologic events and the use of these relationships to generate, or synthesize, hydrologic events" (Hofmann, 1965, p. 120). Cf: *stochastic hydrology; synthetic hydrology.*

parametric latitude *reduced latitude.*

parametric sounding That form of electromagnetic *sounding* in which frequency is the variable. Cf: *geometric sounding.*

parametric statistics Statistics in which assumptions are made regarding distributions. Cf: *nonparametric statistics.*

paramontroseite An orthorhombic mineral: VO_2.

paramorph A *pseudomorph* with the same composition as the original crystal, as calcite after aragonite.

paramorphism The property of a mineral to change its internal structure without changing its external form or chemical composition. Cf: *paramorph.* Syn: *allomorphism.*

paramoudra A flint nodule of exceptionally large size (up to a meter in length and one-third meter in diameter), shaped like a barrel, pear, or cylinder, standing erect in the chalk beds of NE Ireland and the eastern coast of England. It appears to be a gigantic fossil zoophyte allied to the sponges. Term introduced by Buckland (1817). Etymol: vernacular Irish. Pl: *paramoudras.* Syn: *potstone.*

paraphyletic In *cladism,* pertaining to a higher taxon that does not contain all the species known to be descended from a given ancestral species.

parapierrotite A monoclinic mineral: $Tl(Sb,As)_5S_8$.

parapyla An accessory tubular aperture of the central capsule (in addition to the astropyle) of a phaeodarian radiolarian. Pl: *parapylae.*

pararammelsbergite A mineral: $NiAs_2$. It is dimorphous with rammelsbergite.

pararenite An arenite with detrital matrix between 15% and 70% (Pettijohn, Potter & Siever, 1973, p. 168).

para-ripple A term introduced by Bucher (1919, p.262-263) for a large symmetric or nearly symmetric ripple having gentle surface slopes and "showing no assortment of grains". Cf: *metaripple.*

paraschachnerite An orthorhombic mineral: Ag_3Hg_2.

paraschoepite A mineral: $UO_3{\cdot}2H_2O$ (?). It is closely related to *schoepite.*

parasitic [ecol] Said of an organism that lives by *parasitism.*

parasitic [volc] Said of a volcanic cone, crater, or lava flow that occurs on the side of a larger cone; it is a subsidiary form. Syn: *lateral [volc]; adventive.*

parasitic ferromagnetism *weak ferromagnetism.*

parasitic fold A fold on the limb or hinge of a larger fold with which it is *congruent.* The small fold is said to be parasitic on the larger. Syn: *subsidiary fold.*

parasitism The relationship that exists when one organism derives its food, and usually other benefits, from another living organism without killing it, but usually causing it some harm. Cf: *commensalism; inquilinism; mutualism; symbiosis.* Adj: *parasitic.*

parastratigraphic unit *operational unit.*

parastratigraphy (a) Supplemental stratigraphy based on fossils other than those governing the prevalent *orthostratigraphy* (Schindewolf, 1957, p.397). (b) Stratigraphy based on *operational units.*

parastratotype A supplementary stratotype used in the original definition by the original author to aid in elucidating the *holostratotype* (ISG, 1976, p.26).

parasymplesite A monoclinic mineral: $Fe_3(AsO_4)_2{\cdot}8H_2O$. Cf: *symplesite.*

paratacamite A rhombohedral mineral: $Cu_2(OH)_3Cl$. It is trimorphous with atacamite and botallackite.

parataxitic Said of an extremely well layered *eutaxitic* texture (Whitten, 1972).

parataxon An informal designation for those fossilized remains, esp. of animals, that are only part of the whole individual and that usually occur widely separated from other parts; e.g. a particular conodont. Cf: *form genus.* See also: *taxon.*

paratectonic Pertaining to orogenic belts characterized by steep cleavages in low-grade metamorphic rocks (Dewey, 1969).

paratectonics *germanotype tectonics.*

paratellurite A tetragonal mineral: TeO_2. It is dimorphous with tellurite.

paratheca A wall of a scleractinian corallite, formed by closely spaced rows of dissepiments. Cf: *septotheca; synapticulotheca.*

paratill A till formed by ice-rafting in a marine or lacustrine environment; it includes deposits from ice floes and icebergs (Harland et al., 1966, p. 232). Ant: *orthotill.*

para-time-rock unit A term introduced by Wheeler et al. (1950, p.2364) for a working *time-stratigraphic unit* that is biostratigraphic and lithostratigraphic in character and that therefore is intrinsically transgressive with respect to time; e.g. zone (lithizone, radiozone), stage, isobiolith, and isogeolith. It "approaches synchrony", whereas a true chronostratigraphic unit (such as system and series) expresses "absolute synchrony".

paratype Any of the specimens, other than the *holotype,* on which the original description of a species or subspecies is based. See: Frizzell, 1933, p. 660.

paraunconformity *parunconformity.*

parautochthon A tectonic unit only slightly displaced with respect to the autochthon and usually lying between autochthon and allochthon.

parautochthonous Said of a rock unit that is intermediate in tectonic character between *autochthonous* and *allochthonous.*

paravane A device with vanes, used in marine seismic surveying, that is towed through the water. The force of the flowing water on the adjustable vanes causes the device to dive, maintain a particular orientation, or move to one side.

paravauxite A colorless mineral: $Fe^{+2}Al_2(PO_4)_2(OH)_2{\cdot}10H_2O$. It has more water than vauxite and metavauxite.

paraveatchite *p-veatchite.*

parawollastonite A monoclinic mineral: $CaSiO_3$. It is dimorphous with wollastonite.

parchettite A leucite *trachyandesite* similar in composition to leucite *tephrite* but containing more leucite than feldspar and more plagioclase than alkali feldspar. The name was derived by Johannsen (1938) from Fosso della Parchetta, San Martina, Italy. Not recommended usage.

parenchyma [bot] A plant tissue composed of thin-walled cells that are relatively undifferentiated or unspecialized, vacuolate, and isodiametric or polyhedral.

parenchyma [paleont] (a) The *mesohyle* of a sponge. (b) The endoplasm of a protozoan.

parenchymalium One of the spicules of the interior of a hexactinellid sponge, excluding specialized dermal and gastral spicules. Pl: *parenchymalia.*

parenchymella Sponge larva composed of an envelope of uniflagellate cells surrounding more or less completely an internal mass of cells. See also: *parenchymula.*

parenchymula (a) The *planula* of a coelenterate. (b) An alternate, but not recommended, spelling of *parenchymella.*

parent A *radionuclide* regarded in relation to the nuclide or nuclides into which it is transformed by decay. Cf: *daughter; end product.*

parental magma The magma from which a particular igneous rock solidified or from which another magma was derived. It is sometimes used as a syn. of *primary magma.*

parent material The unconsolidated material, mineral or organic, from which the *solum* develops. See also: *parent rock [soil]; residual material; transported soil material.*

parent rock [sed] A rock from which other sediments or rocks are derived. Syn: *mother rock [sed]; source rock [sed].*

parent rock [soil] The rock mass from which a soil's *parent material* is derived.

parfacies A subfacies of a *diagenetic facies,* based on pH-Eh limits (Packham & Crook, 1960, p. 400).

pargasite (a) A monoclinic mineral of the amphibole group: $NaCa_2Mg_4Al_3Si_6O_{22}(OH)_2$. Cf: *edenite.* (b) A green or blue-green variety of hornblende containing sodium and found in contact-metamorphosed rocks.

parichno One of a pair of scars located adjacent to the leaf-trace scar of a lepidophytic leaf bolster that reflects the presence of aerenchymatous tissue extending into the cortex.

paries The triangular middle part of a compartmental plate of a cirripede crustacean. Pl: *parietes.*

parietal (a) Pertaining to the walls of a part or cavity of an organism. (b) Said of a plant part that is peripheral in position or orientation.

parietal fold A *fold* or spirally wound ridge on the parietal region of a gastropod, projecting into the shell interior.

parietal gap An opening from the spongocoel to the exterior of a lyssacine hexactinellid sponge, extending completely through the body wall and interrupting the regular skeletal framework.

parietal lip The part of the *inner lip* of a gastropod shell situated on the parietal region.

parietal muscle One of a pair or of multiple pairs of oppositely placed muscles that originate on the lateral or basal walls of a cheilostome bryozoan zooid and insert on the frontal membrane or the floor of the ascus (Ryland, 1970, p. 34).

parietal pore (a) A *parietal tube* in a cirripede crustacean. (b) A gonopore in a cystoid.

parietal region (a) The basal surface of a gastropod shell just within and immediately outside the aperture. (b) In gnathostomes, the roof of the skull behind the orbits.

parietal septum (a) A *longitudinal septum* in a cirripede crustacean. (b) A seldom preserved longitudinal wall extending inward in the posterior region of certain echinoderms.

parietal tube One of the longitudinal tubes in certain cirripede crustaceans situated between the inner and outer laminae of a compartmental plate and separated by longitudinal septa. Syn: *parietal pore.*

parietes Plural of *paries.*

pariety A name formerly applied to the *septum* of an archaeocyathid.

parisite A brownish-yellow secondary mineral: $(Ce,La)_2Ca(CO_3)_3F_2$. It is related to *synchysite.*

parivincular Said of a longitudinally elongated type of ligament of a bivalve mollusk, located posterior to the beaks and comparable to a cylinder split on one side with severed edges attached respectively along the dorsal margin of two valves.

park (a) A term used in the Rocky Mountain region of Colorado and Wyoming for a wide, grassy open valley lying at a high altitude and walled in by wooded mountains; e.g. South Park in central Colorado. It is more extensive than a *hole.* Also, a level valley between mountain ranges. (b) A relatively large, open, grassy area surrounded by woodland, or interrupted by scattered clumps of trees and shrubby vegetation; e.g. a tropical grassland in Africa.

parkerite A bright-bronze mineral: $Ni_3(Bi,Pb)_2S_2$.

parmal pore One of the pores piercing the shield of an acantharian radiolarian and bordered only by united branches of apophyses. Cf: *sutural pore.*

parna A term used in SE Australia for an eolian clay occurring in sheets. Etymol: an aboriginal word for "sandy and dusty ground".

parogenetic Formed before the enclosing rock; esp. said of a concretion formed in a different (older) rock from its present (younger) host. Term introduced by Bates (1938, p. 91). Etymol: Greek *paros,* "before, formerly".

paroxysm Any sudden and violent action of physical forces occurring in nature, such as the explosive eruption of a volcano or the convulsive "throes" of an earthquake; specif. the most violent and explosive action during a volcanic eruption, sometimes leading to the destruction of the volcano and generally preceded and followed by smaller explosions. Cf: *catastrophe.*

paroxysmal eruption An eruption of the Katmaian, Peléean, Plinian, or Vulcanian type.

parricidal budding Formation of a new scleractinian coral polyp from the inner surface of a wedge-shaped fragment split off lengthwise from the parent.

parrot coal A syn. of *cannel coal,* so named because of the crackling noises it makes while burning.

parsettensite A copper-red mineral: $Mn_5Si_6O_{13}(OH)_8$ (?). It often contains appreciable aluminum and potassium.

parsonite A pale-yellow to pale-brown mineral: $Pb_2(UO_2)(PO_4)_2 \cdot 2H_2O$.

partial-duration flood A flood peak that exceeds a given base stage or discharge. Syn: *basic-stage flood; flood above a base.*

partial node That part (a point, line, or surface) of a *standing wave* where some characteristic of the wave field has a minimum amplitude differing from zero. Cf: *node.*

partial pediment (a) A term proposed by Mackin (1937, p. 877) for a broadly planate gravel-capped interstream bench or terrace. (b) A broad, planate erosion surface formed by the coalescence of contemporaneous, valley-restricted benches developed at the same elevation in proximate valleys, which would produce a pediment if uninterrupted planation were to continue at this level (Tator, 1953, p. 52-53).

partial peneplain (a) *incipient peneplain.* (b) A planation surface intermediate in development between a berm (or a strath terrace) and a peneplain; a base-leveled area that need not be limited to the confines of a valley. It can cross divides on rocks of medium resistance or on decayed resistant rocks (Bascom, 1931, p. 173). Syn: *local peneplain.*

partial pluton That part of a *composite intrusion* that represents a single intrusive episode.

partial range-zone *local range-zone.*

partial thermoremanent magnetization The thermoremanent magnetization acquired by cooling in an ambient field only over a restricted temperature interval, as opposed to the entire temperature range from *Curie point* to room temperature. Cf: *inverse thermoremanent magnetization.* Abbrev: PTRM.

particle A general term, used without restriction as to shape, composition, or internal structure, for a separable or distinct unit in a rock; e.g. a "sediment particle", such as a *fragment* or a *grain,* usually consisting of a mineral.

particle diameter The length of a straight line through the center of a sedimentary particle considered as a sphere; a common expression of *particle size.*

particle shape The spatial or geometric form of the particles in a sediment or rock; a fundamental property of a particle that determines the relation between its mass and surface area. It depends upon the *sphericity* and *roundness* of the particle, although the term is frequently applied to sphericity as distinguished from roundness. Syn: *grain shape.*

particle size The general dimensions (such as average diameter or volume) of the particles in a sediment or rock, or of the grains of a particular mineral that make up a sediment or rock, based on the premise that the particles are spheres or that the measurements made can be expressed as diameters of equivalent spheres. It is commonly measured by sieving, by calculating settling velocities, or by determining areas of microscopic images. See also: *particle diameter.* Syn: *grain size.*

particle-size analysis Determination of the statistical proportions or distribution of particles of defined size fractions of a soil, sediment, or rock; specif. *mechanical analysis.* Syn: *size analysis; size-frequency analysis.*

particle-size distribution The percentage, usually by weight and sometimes by number or count, of particles in each size fraction into which a powdered sample of a soil, sediment, or rock has been classified, such as the percentage of sand retained on each sieve in a given size range. It is the result of a *particle-size analysis.* Syn: *size distribution; size-frequency distribution.*

particle velocity The velocity with which an individual particle of a medium moves under the influence of wave motion. Cf: *group velocity; phase velocity.*

parting [cryst] The breaking of a mineral along planes of weakness caused by deformation, twinning, or exsolution; e.g. in garnet. Cf: *cleavage [mineral].*

parting [ore dep] (a) A band or bed of waste material between veins or beds of ore. (b) *clay parting.*

parting [stratig] A lamina or very thin sedimentary layer, following a surface of separation between thicker strata of different lithology; e.g. a *shale break* in sandstone, or a thin bed of shale or slate in a coal bed.

parting [struc geol] A joint or fissure; specif. a plane or surface along which a rock is readily separated or is naturally divided into layers, e.g. a *bedding-plane parting.* See also: *splitting.*

parting cast A sand-filled tension crack produced by creep along the sea floor (Birkenmajer, 1959, p.111). Syn: *pseudo mud crack.*

parting lineation A term introduced by Crowell (1955, p. 1357) for "faint irregularities or streaks on lamination planes", parallel to the direction in which the current flowed. It consists of parallel ridges and grooves a few millimeters wide and many centimeters long, esp. in thin-bedded sandstone. This feature has been termed *current lineation* by Stokes (1947), *parting-plane lineation* by

McBride and Yeakel (1963), and *streaming lineation* by Conybeare and Crook (1968).

parting-plane lineation A *parting lineation* on a laminated surface, consisting of subparallel linear shallow grooves and ridges of low relief, generally less than 1 mm (McBride & Yeakel, 1963).

parting-step lineation A *parting lineation* characterized by subparallel, step-like ridges where the parting surface cuts across several adjacent laminae (McBride & Yeakel, 1963).

partition [paleont] The wall separating an offset from its parent corallite during increase in corals. It is constructed of some combination of the thickened ends of septa, septal and/or pseudoseptal pinnacles, sclerenchyme, and horizontal skeletal elements. An epitheca is absent. It is a permanent structure in some species, and temporary in others (Fedorowski & Jull, 1976, p. 41). Cf: *dividing wall.*

partition [speleo] In a cave, a solutional remnant of rock that spans a passage from floor to ceiling. See also: *bridge; wall [speleo].*

partition coefficient The ratio of the molar concentration of a substance dissolved in two immiscible liquids, as described by the *Nernst distribution law.*

partitioning method *Lee configuration.*

partiversal Said of a series of local dips in different directions ranging through about 180° in compass direction, occurring at or near the end of a plunging anticlinal axis.

partridgeite *bixbyite.*

partzite A mineral: $Cu_2Sb_2(O,OH)_7$ (?).

parunconformity Var. of *paraunconformity.*

parvafacies A term proposed by Caster (1934, p.19) for a body of rock that comprises the part of any *magnafacies* lying between designated chronostratigraphic planes or key beds traced across the magnafacies; a laterally limited or grading chronostratigraphic unit of different facies but formed at the same time; a *heteropic* facies of European usage. Etymol: Latin *parva,* "small", + *facies.*

parwelite A mineral: $(Mn,Mg)_5SbSiAsO_{12}$.

Pasadenan orogeny The youngest of 30 or more short-lived orogenies during Phanerozoic time recognized by Stille (1936), in this case in the middle of the Pleistocene, based on relations in southern California between Pliocene and lower Pleistocene strata, and the unconformably overlying upper Pleistocene; named for Pasadena, California. Syn: *Coast Range orogeny.*

pascoite An orange mineral: $Ca_3V_{10}O_{28}\cdot 17H_2O$.

pass [coast] (a) A relatively permanent channel through which a distributary on a delta flows to the sea; specif. a navigable channel in the Mississippi River delta. (b) A navigable channel connecting a body of water with the sea; e.g. a narrow opening between two closely adjacent islands or through a coastal obstruction such as a barrier reef, a barrier island, a bar, or a shoal. Cf: *inlet.* Syn: *passage.* (c) An expanse of open water in a marsh.

pass [geomorph] (a) A natural passageway through high, difficult terrain; e.g. a break, depression, or other relatively low place in a mountain range, affording a passage across, or an opening in a ridge between two peaks, usually approached by a steep valley. Cf: *col; gap [geomorph]; notch [geomorph].* (b) "The self-crossing point of some contour line that forms two loops, one around each of two adjacent *peaks*" (Warntz, 1975, p. 210).

pass [remote sensing] The frequency transmitted by a filter. High-pass filters transmit high-frequency data, low-pass filters low-frequency data.

pass [streams] A river crossing; a ford. Syn: *passage [streams].*

passage [coast] *pass.*

passage [speleo] An opening between the rooms of a cave. Partial syn: *corridor; crawlway; squeezeway.*

passage [streams] *pass [streams].*

passage bed A stratum that is transitional in lithologic or paleontologic character between rocks below and above or between rocks of two geologic systems; e.g. a deposit formed during the period of transition from one set of geographic conditions to another, such as a stratum of the Rhaetian Stage intermediate in character and position between the continental deposits of the Upper Triassic and the marine clays of the Lower Jurassic. See also: *transitional series.*

pass band In seismic profiling, the range of frequencies transmitted without attenuation.

passing Transportation of sediment; e.g. *bypassing* and *total passing.*

passive earth pressure The maximum value of lateral *earth pressure* exerted by soil on a structure, occurring when the soil is compressed laterally causing its internal shearing resistance along a potential failure surface to be completely mobilized; the maximum resistance of a vertical earth face to deformation by a horizontal force. Cf: *active earth pressure.*

passive fold A fold model in which the folded rocks are believed to have behaved in a purely passive way during folding.

passive glacier A glacier with sluggish movement, generally occurring in a continental environment at a high latitude, where accumulation and ablation are both small. Ant: *active glacier.*

passive method (a) A seismic method that involves monitoring naturally produced ground motions. (b) A construction method in permafrost areas, in which the frozen ground near the structure is not disturbed or altered, and the foundations are provided with additional insulation to prevent thawing of the underlying ground.—Cf: *active method.*

passive permafrost *fossil permafrost.*

passive remote sensing Remote-sensing methods that utilize energy naturally reflected or radiated from the terrain, e.g. photography which records available light. Cf: *active remote sensing.* Syn: *passive system.*

passive system *passive remote sensing.*

pass point A point whose horizontal and/or vertical position is determined from photographs by photogrammetric methods and which is intended for use as a supplemental *control point* in the orientation of other photographs.

paste The clay-like matrix of a "dirty" sandstone; e.g. the microcrystalline matrix of a graywacke, consisting of quartz, feldspar, clay minerals, chlorite, sericite, and biotite.

pastplain A plain that has been uplifted and dissected; thus "it is no longer a true plain" (Davis, 1890, p. 88).

pat (a) A term used in Pakistan for an arid plain formed by deposits of fine light-colored muddy clay that accumulates on evaporation of shallow pools of water. (b) A term used in Chota Nagpur, India, for a small, steep-sided plateau.—Etymol: Sindhi.

patagium A spongy *veil* between the arms in the skeleton of a spumellarian radiolarian. Pl: *patagia.*

patch [geog] A small, isolated piece of ground distinguished from that about it by its appearance or by the vegetation it bears.

patch [ice] (a) *ice patch.* (b) An irregular small mass of floating sea-ice fragments of any concentration.

patch reef (a) A moundlike or flat-topped *organic reef,* generally less than a kilometer across, less extensive than a *platform reef,* and frequently forming a part of a larger reef complex. (b) A small, thick, generally unbedded lens of limestone or dolomite, more or less isolated and surrounded by rocks of unlike facies.—Cf: *reef patch; table reef.*

patella (a) The fourth segment in the pedipalpus or in the leg of an arachnid, following upon and forming the "knee" articulation with the femur (TIP, 1955, pt.P, p.62). (b) A joint forming the "knee" in the prosomal appendage of a merostome.—Pl: *patellae* or *patellas.*

patellate (a) Pertaining to the tetrapod kneecap. (b) Said of a low solitary corallite with sides expanding from the apex at an angle of about 120 degrees. Cf: *trochoid; turbinate.*

patera A term established by the International Astronomical Union for an irregular crater with scalloped edges on Mars. Most are believed to be calderas. Generally used as part of a formal name for a Martian landform, such as Tyrrhenum Patera (Mutch et al., 1976, p. 57, 66). Etymol: Latin *patera,* saucer.

paternoite *kaliborite.*

paternoster lake One of a chain or series of small circular lakes occupying rock basins, usually at different levels, in a glacial valley, separated by morainal dams or *riegels,* but connected by streams, rapids, or waterfalls, to resemble a rosary or a string of beads. Syn: *rock-basin lake; step lake; beaded lake.*

path [optics] *optical path.*

path [seis] *raypath.*

pathfinder In geochemical exploration, a relatively mobile element or gas that occurs in close association with an element or commodity being sought, but can be more easily found because it forms a broader halo or can be detected more readily by analytical methods. A pathfinder serves to lead investigators to a deposit of a desired substance.

pathline In computer modeling of hydrothermal systems, the path along which a fluid packet moves in response to physicochemical gradients near an intruded stock (Norton, 1978). See also: *sourceline.*

patina [geol] (a) A colored film or thin outer layer produced on the surface of a rock or other material by weathering after long exposure; e.g. a *desert varnish,* or a case-hardened layer on a chert nodule. (b) Strictly, the greenish film formed naturally on copper and bronze after long exposure to a moist atmosphere, and consisting of a basic carbonate. Etymol: Italian. Pron: *pat*-in-a.

patina [palyn] A thickening of the exine of spores that extends over approximately half of the surface, i.e. over the entire surface of one hemisphere. Pron: *pat*-i-na. Adj: *patinate.*

patination The quality or state of being coated with a patina, or the act or process of coating with a patina.

patronite A black mineral consisting of an impure vanadium sulfide whose exact composition is not known. It is mined as an ore of vanadium at Minasragra, Peru.

patterned ground A group term suggested by Washburn (1950, p.7-8) for certain well-defined, more or less symmetrical forms, such as circles, polygons, nets, steps, and stripes, that are characteristic of, but not necessarily confined to, surficial material subject to intensive frost action. It is classified according to type of pattern and presence or absence of sorting. Patterned ground occurs principally in polar, subpolar, and arctic regions, but also includes features in tropical and subtropical areas. Previous terms more or less synonymous: *structure ground; soil structure [pat grd]; Strukturboden; frost-pattern soil; soil patterns.*

patterned sedimentation Sedimentation characterized by a systematic sequence of beds; e.g. recurrent sedimentation (interbedding, interdigitation, etc.), repetition of beds, or rhythmic or cyclic sedimentation.

pattern shooting In seismic prospecting, the use of a number of energy sources arranged in a definite geometric pattern.

Patterson function *Patterson synthesis.*

Patterson map *Patterson projection.*

Patterson projection A projection of the *Patterson synthesis* on a section through a crystal. See also: *Patterson vectors.*

Patterson synthesis A type of Fourier synthesis whose coefficients are corrected for absorption and other factors; it is used in direct determination of crystal structure. See also: *Patterson vectors.* Syn: *Patterson function.*

Patterson vectors In analysis of crystal structure, the vectors of peaks relative to the origin in a *Patterson synthesis* or *Patterson projection.*

paulingite An isometric zeolite mineral consisting of an aluminosilicate of potassium, calcium, and sodium.

paulopost *deuteric.*

paurocrystalline Descriptive of an interlocking texture of a carbonate sedimentary rock having crystals whose diameters are in the range of 0.01-0.1 mm (Bissell & Chilingar, 1967, p. 163).

paurograined Said of the texture of a carbonate sedimentray rock having clastic particles whose diameters are in the range of 0.01-0.1 mm (Bissell & Chilingar, 1967, p. 163-164) or 0.004-0.1 mm (DeFord, 1946).

pavement A bare rock surface that suggests a paved road in smoothness, hardness, horizontality, surface extent, or close packing of its units. Examples: *boulder pavement; glacial pavement; desert pavement; limestone pavement; erosion pavement.*

pavilion The portion of a faceted gemstone below the *girdle.* Cf: *crown [gem].* Syn: *base [gem].*

pavilion facet A main facet on the pavilion of any fashioned gemstone; e.g. a large facet extending from the girdle to the culet of a brilliant-cut gem, or a facet in the center row of facets on the pavilion of a step-cut gem.

pavonite A mineral: $(Ag,Cu)(Bi,Pb)_3S_5$.

pawdite A dark-colored fine-grained granular hypabyssal rock composed of magnetite, titanite, biotite, hornblende, calcic plagioclase, and traces of quartz. The plagioclase is abundant and approximates labradorite in composition, having bytownite centers surrounded by oligoclase rims (Johannsen, 1939, p. 273). The rock, apparently a *diabase,* was named by Duparc and Grosset in 1916 for Pawdinskaya Datcha, Urals, U.S.S.R. Not recommended usage.

paxilla A pillarlike ossicle of an asterozoan, having a flattened summit bearing a tuft of spinelets or granules. Pl: *paxillae.*

paxillose (a) Resembling a little stake. (b) Bearing paxillae.

paxite A mineral: Cu_2As_3. It is probably orthorhombic.

pay adj. Said of a structure or stratum that contains a mineral deposit (pay gravel, pay streak) or oil and gas (pay sand); also, said of a mineral deposit or part of it that is especially profitable, e.g. pay ore.—n. A *reservoir rock* containing oil or gas.—The term is colloquial.

pay zone The vertical interval(s) of the stratigraphic section in an oil or gas field that will yield oil or gas in economic quantities.

PBM *permanent bench mark.*

PDB standard Carbon dioxide prepared from belemnites (*Belemnitella americana*) collected from the Peedee Formation (Upper Cretaceous) of South Carolina. It is used as a standard of comparison in determining the isotopic composition of carbon and of oxygen. It was first used by H. C. Urey in 1951 to determine paleotemperatures on the basis of the fractionation of oxygen isotopes. The initials stand for "Peedee belemnite".

PDR *precision depth recorder.*

peachblossom ore *erythrite.*

peacock copper *peacock ore.*

peacock ore Informal name for an iridescent copper mineral having a lustrous, tarnished surface exhibiting variegated colors, such as chalcopyrite and esp. bornite. Syn: *peacock copper.*

pea gravel Clean gravel, the particles of which are similar in size to that of peas.

pea grit A limestone containing calcareous pisoliths; a *pisolite.*

pea iron ore *pea ore.*

peak [coast] A headland or promontory; a jut of land.

peak [geomorph] (a) The more or less conical or pointed top of a hill or mountain; one of the crests of a mountain; a prominent summit or the highest point. (b) An individual mountain or hill taken as a whole, esp. when isolated or having a pointed, conspicuous summit.

peak [paleont] A skeletal extension of the wall, shaped like the brim of a baseball cap, from the upper half of the rim in skeletal pores in archaeocyathids (TIP, 1972, pt. E, p. 41).

peak diameter The dominant or modal particle diameter as determined on a particle-size distribution curve.

peak discharge *flood peak.*

peak plain A high-level plain formed by a series of accordant summits, often explained as an uplifted and fully dissected peneplain. See also: *gipfelflur.* Syn: *summit plain.*

peak runoff The maximum rate of runoff at a given point or from a given area, during a specified period.

peak-zone *acme-zone.*

pea ore A variety of pisolitic limonite or *bean ore* occurring in small, rounded grains or masses. Syn: *pea iron ore.*

pearceite A black mineral: $Ag_{16}As_2S_{11}$. It may contain copper. Cf: *polybasite.*

pearl A dense spherical calcareous concretion, usually white or light-colored, consisting of occasional layers of conchiolin and predominant nacreous layers of aragonite (or rarely calcite), deposited concentrically about a foreign particle within or beneath the mantle of various marine and freshwater mollusks, either free from or attached to the shell.

pearlite *perlite.*

pearl mica *margarite [mineral].*

pearl opal *cacholong.*

pearl spar A crystalline carbonate mineral, such as ankerite, having a pearly luster; specif. *dolomite [mineral].* Also spelled: *pearlspar.*

pearlstone *perlite.*

pearly luster *nacreous luster.*

pear-shaped cut A variation of the *brilliant cut,* usually with 58 facets, having a pear-shaped girdle outline. See also: *pendeloque; briolette.*

peastone A rock whose texture resembles an aggregate of peas; a *pisolite.*

peat An unconsolidated deposit of semicarbonized plant remains in a watersaturated environment, such as a bog or fen, and of persistently high moisture content (at least 75%). It is considered an early stage or rank in the development of coal; carbon content is about 60% and oxygen content is about 30% (moisture-free). Structures of the vegetal matter can be seen. When dried, peat burns freely.

peat ball A *lake ball* containing an abundance of peaty fragments.

peat bed *peat bog.*

peat bog A *bog* in which peat has developed, under conditions of acidity, from the characteristic vegetation, esp. sphagnum. Syn: *peat moor; peat bed.*

peat breccia Peat that has been broken up and then redeposited by water. Syn: *peat slime.*

peat coal (a) A coal transitional between peat and brown coal or lignite. (b) Artificially carbonized peat that is used as a fuel.

peat flow A flow of peat produced in a peat bog by a *bog burst.* Cf: *bog flow.*

peat formation The decomposition of vegetable matter under conditions intermediate between those of *moldering* and those of *putrefaction,* in stagnant water with small amounts of oxygen. Cf: *disintegration [coal].*

peat hummock A hummock of peat rather than soil. Cf: *earth hummock.*

peat moor *peat bog.*

peat moss Moss from which peat has formed, usually *sphagnum moss.*

peat mound Any mound composed largely of peat, ranging in size from a small hummock to a *palsa.* Cf: *gull hammock.*

peat ring A *nonsorted circle* in peat.

peat-sapropel Organic degradation matter that is transitional between peat and sapropel. Syn: *sapropel-peat.*

peat slime *peat breccia.*

peat soil An organic soil that consists mainly of peat.

peat-to-anthracite theory A theory of coal formation as a process in which the progressive ranks of coal are indicative of the degree of *coalification* and, by inference, of the relative geologic age of the deposit. Peat, as the initial stage of coalification, is of recent geologic age; lignite, as an intermediate stage, is usually Tertiary or Mesozoic, and bituminous coal and anthracite, as the more advanced stages of coalification, are usually Carboniferous (Nelson & Nelson, 1967, p.271).

pebble [gem] (a) A rough gem occurring in the form of a pebble, as in a stream. (b) Transparent, colorless quartz, or *quartz crystal;* e.g. "Brazilian pebble".

pebble [part size] (a) A general term for a small, roundish, esp. waterworn stone; specif. a rock fragment larger than a granule and smaller than a cobble, having a diameter in the range of 4-64 mm (1/6 to 2.5 in., or -2 to -6 phi units, or a size between that of a small pea and that of a tennis ball), being somewhat rounded or otherwise modified by abrasion in the course of transport. In Great Britain, the range of 10-50 mm has been used. The term has been used to include fragments of cobble size; it is frequently used in the plural as a syn. of *gravel.* See also: *very coarse pebble; coarse pebble; medium pebble; fine pebble.* Syn: *pebblestone.* (b) A rock or mineral fragment in the soil, having a diameter in the range of 2-20 mm (Atterberg, 1905). The U.S. Bureau of Soils has used a range of 2-64 mm. Cf: *gravel.*

pebble armor A *desert armor* consisting of rounded pebbles, as on a *serir.*

pebble bed Any pebble conglomerate, esp. one in which the pebbles weather conspicuously and fall loose; e.g. the Bunter pebble beds of Devon and Somerset in England. Syn: *popple rock.*

pebble coal *ball coal.*

pebble conglomerate A consolidated rock consisting mainly of pebbles.

pebble dent A depression formed by a pebble on an unconsolidated sedimentary surface, represented by a downward curvature of laminae beneath the pebble.

pebble dike (a) A *clastic dike* composed largely of pebbles. (b) A tabular body containing sedimentary fragments in an igneous matrix, as from the Tintic district in Utah (Farmin, 1934); e.g. one whose fragments were broken from underlying rocks by gaseous or aqueous fluids of magmatic origin and injected upward into country rock, becoming rounded due to the milling and/or corrosive action of the hydrothermal fluids.

pebble gravel An unconsolidated deposit consisting mainly of pebbles.

pebble mosaic A *desert mosaic* consisting of pebbles.

pebble peat Peat that is formed in a semiarid climate by the accumulation of moss and algae, no more than 1/4 inch in thickness, under the surface pebbles of well-drained soils.

pebble phosphate A secondary phosphorite of either residual or transported origin, consisting of pellets, pebbles and nodules of phosphatic material mixed with sand and clay, as in Florida; e.g. *land-pebble phosphate* and *river-pebble phosphate.*

pebble pup (a) A geologist's assistant. (b) A student of geology. (c) An inexperienced *rock hound.*

pebble size A term used in sedimentology for a volume greater than that of a sphere with a diameter of 4 mm (1/6 in.) and less than that of a sphere with a diameter of 64 mm (2.5 in.).

pebblestone *pebble [part size].*

pebbly mudstone A delicately laminated *conglomeratic mudstone* in which thinly scattered pebbles are embedded among somewhat distorted bedding planes. The term is advocated by Crowell (1957, p.1003) as a descriptive name, without regard to manner of origin, for a poorly sorted, till-like rock composed of dispersed pebbles in an abundant mudstone matrix. See also: *tilloid; pseudotillite.*

pebbly sand An unconsolidated sediment containing at least 75% sand and a conspicuous number of pebbles that do not exceed 25% of the total aggregate (Willman et al., 1942, p.343-344). Cf: *sandy gravel.*

pebbly sandstone (a) A consolidated *pebbly sand.* (b) A sandstone containing 10-20% pebbles (Krynine, 1948, p.141). Cf: *conglomeratic sandstone.* (c) A term used in Scotland for a conglomerate.

pecoraite A mineral of the serpentine group: $Ni_3Si_2O_5(OH)_4$.

pectinacean Any bivalve mollusk assigned to the superfamily Pectinacea, characterized by an orbicular, monomyarian, subequilateral shell with winglike extensions from the hinge margin; e.g. a *scallop [paleont].*

pectinate Comblike, with close narrow divisions or parts; also said of spine connections in cacti when small lateral spines radiate like comb teeth from the areole.

pectinate tabula A transverse, porous skeletal element in archaeocyathids, shaped like the teeth of two combs, each projecting toward the other from adjoining septa (TIP, 1972, pt. E, p. 41).

pectinirhomb A specialized type of *pore rhomb* found in cystoids, consisting of a compact rhomboidal structure of closely spaced comblike grooves. It is typically set in a distinct depressed area on thecal plates.

pectolite A whitish or grayish monoclinic mineral: $NaCa_2Si_3O_8(OH)$. It occurs in compact masses of divergent or parallel fibers, commonly in cavities in basalts and scoriaceous lavas. Cf: *serandite.*

ped A naturally formed unit of soil structure, e.g. granule, block, crumb, aggregate. Cf: *clod [soil].*

pedal elevator muscle A thin bundle of muscle fibers attached to the bivalve-mollusk shell in the umbonal cavity and serving to raise the foot.

pedalfer An old, general term for a soil in which there is a concentration of sesquioxides. It is the characteristic type of soil in a humid region. Cf: *pedocal.*

pedal gape An opening between margins of the shell of a bivalve mollusk for the protrusion of the foot.

pedality The physical nature of a soil as expressed by the features of its constituent peds.

pedal levator muscle A pedal muscle serving to retract the foot of a bivalve mollusk.

pedal muscle One of a pair or several pairs of muscles connecting the foot of a mollusk to the interior surface of the shell; e.g. "pedal protractor muscle" serving to extend the foot of a bivalve mollusk or gastropod, and "pedal retractor muscle" serving to retract it.

pedcal *pedocal.*

pedestal [geomorph] A relatively slender neck or column of rock capped by a wider mass of rock and produced by undercutting as a result of wind abrasion (as in the SW U.S.) or by differential weathering. See also: *pedestal rock.* Syn: *rock pedestal.*

pedestal [glaciol] *ice pedestal.*

pedestal rock (a) An isolated and residual or erosional mass of rock supported by or balanced on a *pedestal.* The term is also applied to the entire feature. See also: *balanced rock; mushroom rock.* Syn: *pedestal boulder.* (b) *perched block.*

pedia Plural of *pedion.*

pedial class That class in the triclinic system having symmetry 1 (no symmetry).

pedicel [bot] In an inflorescence, the stalk of an individual flower.

pedicel [paleont] (a) The greatly modified first segment of the abdomen in arachnids of the subclass Caulogastra, reaching its extreme development in spiders. (b) A small foot or footlike organ of an invertebrate, such as a tube foot of an echinoderm. (c) The area of attachment of the body of a tintinnid to the lorica. (d) A small or short stalk or stem in an animal; esp. a narrow basal part by which a larger part or body is attached, such as the *pedicle* of a brachiopod.

pedicellaria Any of various minute organs resembling forceps that are borne in large numbers on certain echinoderms, e.g. a minute stalked organ in the external integument of an echinoid, used for grasping or defending; or a minute pincerlike or valvate calcareous appendage on or in the skin, ossicles, or spines of an asteroid. Pl: *pedicellariae.*

pedicle A variably developed, cuticle-covered, fleshy or muscular appendage of a brachiopod, commonly protruding from the pedicle valve and serving to attach the animal to the substratum. See also: *pedicel [paleont].*

pedicle callist A localized thickening of secondary-shell layer in the apex of the pedicle valve of a brachiopod, representing the track of anterior migration of junction between pedicle epithelium and outer epithelium.

pedicle collar The complete or partial ringlike thickening of the inner surface of the ventral beak of a brachiopod, "continuous laterally with internal surface of deltidial plates, sessile, with septal support, or free anteriorly and secreted by anteriorly migrating outer epithelium at its junction with pedicle epithelium" (TIP, 1965, pt.H, p.149).

pedicle epithelium The ectodermal epithelium investing the pedicle of a brachiopod.

pedicle foramen The subcircular to circular perforation of the shell, adjacent to the beak of the pedicle valve, through which the pedicle of a brachiopod passes. Cf: *pedicle opening.* Syn: *foramen.*

pedicle groove A commonly subtriangular groove dividing the ventral pseudointerarea medially and affording passage for the pedicle in many lingulids.

pedicle opening *pedicle foramen.*

pedicle sheath An externally directed tube projecting posteriorly and ventrally from the ventral umbo of a brachiopod, probably enclosing the pedicle in young stages of development of some shells with supra-apical pedicle opening (TIP, 1965, pt.H, p.150).

pedicle tube An internally directed tube of the secondary shell of a brachiopod, continuous with the margin of the pedicle foramen and enclosing the proximal part of the pedicle (TIP, 1965, pt.H, p.150).

pedicle valve The valve of a brachiopod through which the pedicle commonly emerges. It is usually larger than the *brachial valve*, and it contains the teeth by which the valves are hinged. Syn: *ventral valve.*

pediment A broad gently sloping rock-floored erosion surface or plain of low relief, typically developed by subaerial agents (including running water) in an arid or semiarid region at the base of an abrupt and receding mountain front or plateau escarpment, and underlain by bedrock (occasionally by older alluvial deposits) that may be bare but are more often partly mantled with a thin discontinuous veneer of alluvium derived from the upland masses and in transit across the surface. The longitudinal profile of a pediment is normally slightly concave upward, and its outward form may resemble a *bajada* (which continues the forward inclination of a pediment). The term was first applied to a landform by McGee (1897, p. 92), although Gilbert (1877, p. 130-131) recognized and described the feature as a *hill of planation.* Etymol: from an architectural pediment, a triangular feature crowning a portico of columns in front of a Grecian-style building; in this sense the term is not appropriate for a gently sloping surface commonly forming a broad approach to a mountain range. Cf: *rock fan; plain of lateral planation; peripediment.* Syn: *piedmont interstream flat; conoplain; rock pediment.*

pedimentation The action or process of formation and development of a *pediment* or pediments. The two processes recognized as being most active in pediment formation are lateral planation by steep-gradient streams, and backwearing and removal of debris by rill wash and unconcentrated flow; the latter process appears to be the more widely accepted. Cf: *pediplanation.*

pediment dome *desert dome.*

pediment gap A term applied by Sauer (1930) to a broad opening formed by the enlargement of a *pediment pass.*

pediment pass A term applied by Sauer (1930) to a narrow, flat, rock-floored tongue extending back from a pediment and penetrating sufficiently along a mountain to join another pediment extending into the mountain front from the other side; the pediments are frequently at different levels. Cf: *pediment gap.*

pediocratic Pertaining to a period of time in which there is little diastrophism. Cf: *orocratic.*

pedion An open crystal form having only a single face, with no symmetrically equivalent face. Pl: *pedia.*

pedipalpus One of the second pair of cephalothoracic appendages that lie on each side of the mouth of an arachnid and that are subject to many variations in structure, such as being the largest and most conspicuous appendages in scorpions (ending in a powerful chela), stout and conspicuous in whip scorpions and the order Phrynichida (but ending in a pointed joint), and the least conspicuous appendages in the order Architarbida. Pl: *pedipalpi.* See also: *palpus.* Syn: *pedipalp.*

pediplain A term proposed by Maxson & Anderson (1935, p. 94) for an extensive thinly alluviated erosion surface formed in a desert region by the coalescence of two or more adjacent pediments and occasional desert domes, and representing the end result (the "peneplain") of the mature stage of the arid erosion cycle. Howard (1942) objected to the term because the surface to which it is applied is not wholly at the base of a slope and is not a "plain" in the true geomorphic sense. Cf: *pediplane; coalescing pediment.* Syn: *panfan; desert peneplain; desert plain.*

pediplanation (a) A general term for all the processes by which *pediplanes* are formed (Howard, 1942, p. 11). (b) The action or process of formation and development of a *pediplain* or pediplains; *pedimentation* of regional magnitude, assisted by slope retreat.

pediplane (a) A general term proposed by Howard (1942, p.11) for any planate erosion surface, such as a pediment or a peripediment, produced in the piedmont area of an arid or semiarid region, either exposed or covered with a veneer of alluvium no greater than the depth of effective scour (the thickness that can be moved during floods). (b) A term sometimes used as a syn. of *pediplain.*

pedocal An old, general term for a soil in which there is an accumulation or concentration of carbonates, usually calcium carbonate. It is the characteristic type of soil in an arid or semiarid region. Cf: *pedalfer.* Also spelled: *pedcal.*

pedode A term proposed by Brewer (1964, p.271) for a spheroidal, discrete glaebule with a hollow interior, often with a drusy lining of crystals like that of a geode. It may have an outside layer of chalcedony.

pedogenesis *soil genesis.*

pedogenic Pertaining to soil formation.

pedogeography The study of the geographic distribution of soils.

pedography The systematic description of soils; an aspect of *soil science.*

pedolith A surface formation that has undergone one or more pedogenic processes (Dewolf, 1970).

pedologic age The relative maturity of a soil profile.

pedologic unit A soil considered without regard to its stratigraphic relations (ACSN, 1961, art.18b). Cf: *soil-stratigraphic unit.*

pedology One of the disciplines of soil science, the study of soil morphology, genesis, and classification. It is sometimes used as a syn. of *soil science.*

pedometer A pocket-size, watch-shaped instrument that registers the linear distance a pedestrian covers by responding to his body motion at each step. It is carried in an upright position attached to the body or to a leg, and it can be adjusted to the length of the pace of the person carrying it.

pedon The smallest unit or volume of soil that represents or exemplifies all the horizons of a soil profile. It is usually a horizontal, more or less hexagonal area of about one square meter, but may be larger. The term is part of the soil classification system of the National Cooperative Soil Survey.

pedorelic Said of a soil feature that is derived from a pre-existing soil horizon. Cf: *lithorelic.*

pedosphere That shell or layer of the Earth in which soil-forming processes occur.

pedrosite A synonym for *hornblendite.* The name, given by Osann in 1923, is for Pedrosa, Portugal. Not recommended usage.

peduncle [bot] A plant stalk that bears an inflorescence or a strobilus.

peduncle [paleont] (a) A narrow part by which some larger part or the whole body of an animal is attached, such as the pedicle of a brachiopod, the column of an echinoderm, or the basal portion of certain crustacean appendages. (b) The fleshy stalklike portion of the body of certain cirripedes. (c) The mass of cyptoplasm projecting from a thecamoebian-test aperture, giving rise to pseudopodia; pseudopodial trunk.

peel A thin film of acetate or other plastic substance, obtained in the *peel technique* for rock study. In some fields, e.g. paleobotany, peels are superior to thin sections for microscopic study, esp. because they cover much larger areas.

peel map A *paleogeologic map* on which the formations overlying an unconformity are, in effect, "peeled off" and the pre-unconformity distribution of the underlying formations is uncovered and mapped (Levorsen, 1960, p.4).

peel-off time In seismic prospecting, the time correction to be applied to observed data to adjust them to a depressed reference datum.

peel technique A method of preparing rock sections for study. A smooth surface is differentially etched with acid, and then covered with cellulose acetate or a similar liquid that will dry to a transparent film. When this film is peeled off, it brings with it a thin layer of rock substance, preserving this material in its original space relations. It may then be studied microscopically. See also: *peel.*

peel thrust A sedimentary sheet peeled off a sedimentary sequence, essentially along a bedding plane. A series of peel thrusts may be imbricated above a décollement (Bucher, 1955).

peep-sight alidade An *alidade* used with a plane table, consisting of a rear (open) sight mounted on a straightedge.

peg adjustment The adjustment of a spirit-leveling instrument of the dumpy-level type in which the line of collimation is made parallel with the axis of the spirit level by means of two stable marks (pegs) the length of one instrument sight apart.

pegma In rostroconchs, a plate connecting right and left valves in the umbonal part of the shell; it supports the large muscle in primitive rostroconchs (Pojeta & Runnegar, 1976, p. 47). Etymol: Greek, "fastened".

pegmatite An exceptionally coarse-grained igneous rock, with interlocking crystals, usually found as irregular dikes, lenses, or veins, esp. at the margins of batholiths. Most grains are one cm or more in diameter. Although pegmatites having gross compositions similar to other rock types are known, their composition is generally that of *granite;* the composition may be simple or complex and may include rare minerals rich in such elements as lithium, boron, fluorine, niobium, tantalum, uranium, and rare earths. Pegmatites represent the last and most hydrous portion of a magma to crystallize and hence contain high concentrations of minerals present only in trace amounts in granitic rocks. The first use of the term "pegmatite" is attributed to Haüy, who used it in 1822 as a syn. of *graphic granite.* Haidinger in 1845 introduced the present meaning. Cf: *pegmatoid; symplectite.* Adj: *pegmatitic.* Syn: *giant granite.*

pegmatitic (a) Said of the texture of an exceptionally coarsely crystalline igneous rock. (b) Occurring in, pertaining to, or composed of *pegmatite.*

pegmatitic stage A stage in the normal sequence of crystallization of a magma containing volatiles, at which time the residual fluid is sufficiently enriched in volatile materials to permit the formation of coarse-grained rocks (i.e. pegmatites). The relative amounts of silicate and volatile materials in the fluid, the temperature range, and the relationship of these fluids to hydrothermal fluids are in dispute. Cf: *hydrothermal stage.*

pegmatitization Replacement by pegmatite.

pegmatoid n. An igneous rock that has the coarse-grained texture of a *pegmatite* but lacks graphic intergrowths and/or typically granitic composition.—adj. Said of the texture of a pegmatitic rock lacking graphic intergrowths and/or typical granitic composition. Introduced by Evans in 1912. Not recommended usage.

pegmatolite *orthoclase.*

pegmatophyre *granophyre.*

pegmatophyric *granophyric.*

peiroglyph A cross-cutting sedimentary structure, such as a sandstone dike (Vassoevich, 1953, p.37).

pekoite An orthorhombic mineral: $PbCuBi_{11}(S,Se)_{18}$.

pelagic [lake] Pertaining to the deeper part of a lake (10-20 m or more), characterized by deposits of mud or ooze and by the absence of aquatic vegetation.

pelagic [oceanog] (a) Pertaining to the water of the ocean as an environment. See also: *epipelagic; mesopelagic.* (b) Said of marine organisms whose environment is the open ocean, rather than the bottom or shore areas. Pelagic organisms may be either *nektonic* or *planktonic.*

pelagic deposit Marine sediment in which the fraction derived from the continents indicates deposition from a dilute mineral suspension distributed throughout deep-ocean water (Arrhenius, 1963, p. 655). Cf: *terrigenous deposit; hemipelagic deposit; deep-sea deposit.* Syn: *abyssal deposit.*

pelagic limestone A fine-textured limestone formed chiefly by the accumulation of the calcareous tests of floating organisms (esp. of post-Jurassic foraminifers). It is characteristic of geosynclinal belts, but can also occur as a shelf sediment.

pelagochthonous Said of coal derived from a submerged forest or from driftwood.

pelagosite A term used by Revelle & Fairbridge (1957, p. 258) for a superficial calcareous crust a few millimeters thick, generally white, gray, or brownish, with a pearly luster, formed in the intertidal zone by ocean spray and evaporation (alternate solution and evaporation), and composed of calcium carbonate accompanied by larger amounts of magnesium carbonate, strontium carbonate, calcium sulfate, and silica than are found in normal limy sediments. It was originally described as an impure calcite mineral. (See Purser and Loreau, 1973.)

peldon An English term for a very hard, smooth, compact sandstone with conchoidal fracture, occurring in coal measures.

Peléan cloud A syn. of *nuée ardente,* so named because it is a characteristic type of eruption of Mt. Pelée on the island of Martinique in the West Indies.

pelecypod Any benthic aquatic mollusk belonging to the class Pelecypoda, characterized by a bilaterally symmetrical bivalve shell, a hatchet-shaped foot, and sheetlike gills. Syn: *lamellibranch.* Partial syn: *bivalve.* Range, Ordovician to present.

Peléean-type eruption A type of volcanic eruption characterized by gaseous clouds (*nuées ardentes*) and/or the development of *volcanic domes.* Etymol: Mont Pelée, island of Martinique. Cf: *Hawaiian-type eruption; Strombolian-type eruption; Vulcanian-type eruption.*

Pele's hair A natural spun glass formed by blowing-out during quiet fountaining of fluid lava, cascading lava falls, or turbulent flows, sometimes in association with *Pele's tears [pyroclast].* A single strand, with a diameter of less than half a millimeter, may be as long as two meters. Etymol: Pele, Hawaiian goddess of fire. Syn: *lauoho o pele; filiform lapilli; capillary ejecta.*

Pele's tears [gem] Hawaiian name for a clear chalcedony or opal in cabochon cut.

Pele's tears [pyroclast] Small, solidified drops of volcanic glass behind which trail pendants of *Pele's hair.* They may be tear-shaped, spherical, or nearly cylindrical. Etymol: Pele, Hawaiian goddess of fire. Cf: *tear-shaped bomb.*

pelhamine A light gray-green serpentine mineral from Pelham, Mass. It may be altered chrysotile.

pelinite A term proposed by Searle (1912, p. 148) for a hydrous aluminum silicate thought to be the true clay substance in clays other than the kaolins and considered to be an amorphous (colloidal) and plastic material of varying composition but of generally higher silica content than that in *clayite* and also with appreciable alkalies and/or alkaline earths.

pelionite A name proposed by W.F. Petterd for a bituminous coal resembling English cannel coal, occurring near Monte Pelion in Tasmania (Thrush, 1968, p. 802).

pelite (a) A sediment or sedimentary rock composed of the finest detritus (clay- or mud-size particles); e.g. a *mudstone,* or a calcareous sediment composed of clay and minute particles of quartz. The term is equivalent to the Latin-derived term, *lutite.* (b) A fine-grained sedimentary rock composed of more or less hydrated aluminum silicates with which are mingled small particles of various other minerals (Twenhofel, 1937, p.90); an aluminous sediment. (c) A term regarded by Tyrrell (1921, p.501-502) as the metamorphic derivative of lutite, such as the metamorphosed product of a siltstone or mudstone. "As commonly used, a pelite means an aluminous sediment metamorphosed, but if used systematically, it means a fine-grained sediment metamorphosed" (Bayly, 1968, p.230).—Etymol: Greek *pelos,* "clay mud". See also: *psammite; psephite.* Also spelled: *pelyte.*

pelitic (a) Pertaining to or characteristic of pelite; esp. said of a sedimentary rock composed of clay, such as a "pelitic tuff" representing a consolidated volcanic ash consisting of clay-size particles. (b) Said of a metamorphic rock derived from a pelite; e.g. a "pelitic hornfels" or a "pelitic schist", derived by metamorphism of an argillaceous or a fine-grained aluminous sediment.—Cf: *argillaceous; lutaceous.*

pelitomorphic Pertaining to clay-size carbonate particles in a limestone or dolomite rock. Also, said of a limestone or dolomite consisting of an aggregate of pelitomorphic particles or having a matrix of such particles.

pellet (a) A small, usually rounded aggregate of accretionary material, such as a *lapillus* or a *fecal pellet;* specif. a spherical to elliptical homogeneous clast made up almost exclusively of clay-sized calcareous (micritic) material, devoid of internal structure, and contained in a well-sorted carbonate rock. Folk (1959; 1962) suggested that the term apply to allochems less than 0.15-0.20 mm in diameter. Pellets appear to be mainly the feces of mollusks and worms; other types include pseudo-ooliths and aggregates produced by gas bubbling, by algal "budding" phenomena, or by other intraformational reworking of lithified or semilithified carbonate mud. (b) A small rounded aggregate (0.1-0.3 mm in diameter) of clay minerals and fine quartz found in some shales and clays, separated from a matrix of the same materials by a shell of organic material, and ascribed to the action of water currents (Allen & Nichols, 1945).

pelleted limestone A limestone characterized by abundant pellets, such as some lower Paleozoic limestones whose major constituents are fecal pellets, or carbonate muds that display rounded or ellipsoidal aggregates of "grains of matrix" material. The adjectival term is sometimes given as "pellet", "pelletoid", "pelletoidal", "pelletal", and "pelleted".

pelletoid Pelleted, or containing abundant pellets; e.g. a "pelletoid limestone". Alternate form: *pelletoidal.*

pellet snow *graupel.*

pellicle An imperforate *pellis.*

pellicular envelope The delicate outer covering of soft parts in a tintinnid.

pellicular front The even front, developed only in pervious granular material, on which pellicular water depleted by evaporation, transpiration, or chemical action is regenerated by influent seepage (Tolman, 1937, p. 593). Syn: *infiltration front; wetting front.*

pellicular water Water in layers more than one or two molecules thick that adheres to the surfaces of soil and rock particles in the zone of aeration. Layers more than a few microns in thickness are short-lived, owing to the requirement that free energy and capillary surface be at a minimum when moisture is at equilibrium (Smith, W.O., 1961, p. 11). Syn: *adhesive water; film water; sorption water.* Cf: *pendular water; funicular water; attached ground water.*

pellis A thin, commonly imperforate sheet or sheath outside the cup in archaeocyathids (TIP, 1972, pt. E, p. 41).

pellodite A term used by Schuchert (1924, p. 441) for a water-laid sandy varved clay, and by Pettijohn (1957, p.273) for the lithified equivalent of a varved clay. It is apparently a syn. of *pelodite.*

pellyite An orthorhombic mineral: $Ba_2Ca(Fe^{+2},Mg)_2Si_6O_{17}$.

pelma (a) An entire crinoid column with attached cirri and holdfast structure, if present. (b) A relatively large *lumen pore* in a costa of a cribrimorph cheilostome (bryozoan).—Pl: *pelmata.*

pelmatidium A relatively small *lumen pore* in a costa of a cribrimorph cheilostome (bryozoan).

pelmatozoan n. Any echinoderm, with or without a stem, that lives attached to a substrate.—adj. Said of an echinoderm having an attached mode of life. Var: *pelmatozoic.*—Cf: *eleutherozoan.*

pelmicrite A limestone consisting of a variable proportion of pellets and carbonate mud (micrite); specif. a limestone containing less than 25% intraclasts and less than 25% ooliths, with a volume ratio of pellets to fossils and fossil fragments greater than 3 to 1, and the carbonate-mud matrix more abundant than the sparry-calcite cement (Folk, 1959, p. 14).

pelodite A term proposed by Woodworth (1912, p. 78) for a lithified glacial rock flour, composed of glacial pebbles in a silty or clayey matrix, formed by redeposition of the fine fraction of a till. Syn: *pellodite.*

pelogloea Organic matter, mainly colloidal, occurring adsorbed on sedimentary particles in natural waters (Fox, 1957, p. 384). Cf: *leptopel.*

peloid An allochem composed of micrite, irrespective of size or origin. Includes both pellets and intraclasts; useful where exact origin is unknown.

pelolithic *argillaceous.*

pelphyte A lake-bottom deposit consisting mainly of fine, nonfibrous plant remains (Veatch & Humphrys, 1966, p.227). Cf: *psephyte [lake].*

pelsparite A limestone consisting of a variable proportion of pellets and clear calcite (spar); specif. a limestone containing less than 25% intraclasts and less than 25% ooliths, with a volume ratio of pellets to fossils and fossil fragments greater than 3 to 1, and the sparry-calcite cement more abundant than the carbonate-mud matrix (micrite) (Folk, 1959, p. 14).

pelta A lidlike flap, dorsally closing or partly closing the internal cavity in one-walled archaeocyathids; it may or may not be porous (TIP, 1972, pt. E, p. 41).

peltate Shield-shaped, as in a leaf that is attached to its stalk inside the margin (Lawrence, 1951, p.764).

pelvis The proximal or dorsal part of a blastoid theca from the aboral tips of the ambulacra to the proximal pole.

Pelycosauria An order of synapsid reptiles of generally reptilian grade, as indicated by crocodiliform or lacertiform habitus. Range, Lower Pennsylvanian to ?Upper Permian.

pelyte *pelite.*

pen A British term variously used for a hill, mountain, highland, or headland. Etymol: Celtic.

peña A rock; a rocky point; a needlelike eminence; a cliff. Term is used in the SW U.S. Etymol: Spanish, "rock".

peñasco A term used in the SW U.S. for a projecting rock, esp. one isolated by the recession of a cliff or of a mountain slope. Etymol: Spanish, "large rock".

pencil cleavage Cleavage in which fracture produces long, slender pieces of rock. It is produced by the intersection of a direction of cleavage with the stratification of the rock, usually in weakly metamorphosed rocks.

pencil ganister A variety of *ganister* characterized by fine carbonaceous streaks or markings, and so called from the likeness of these to pencil lines. The carbonaceous traces are often recognizable as roots and rootlets of plants.

pencil gneiss A gneiss that breaks into roughly cylindrical, pencil-like quartz-feldspar crystal aggregates, often mantled by mica flakes. Syn: *stengel gneiss.*

pencil ore Hard, fibrous masses of hematite that can be split up into thin rods.

pencil stone A compact pyrophyllite once used for making slate pencils.

pendant [intrus rocks] *roof pendant.*

pendant [speleo] One of a closely spaced group of solutional remnants hanging from the ceiling of a cave. Syn: *rock pendant; solution pendant.* See also: *deckenkarren.*

pendeloque A modification of the round *brilliant cut,* having a drop-shaped outline similar to that of the *pear-shaped cut,* but with the narrow end longer and more pointed. Cf: *briolette.*

pendent [geomorph] Said of a landform that slopes steeply down (as a hillside) or overhangs (as a cliff).

pendent [paleont] Said of a graptoloid rhabdosome with approximately parallel stipes that hang below the sicula.

pendletonite *karpatite.*

pendular water Capillary water ringing the contact points of adjacent rock or soil particles in the zone of aeration (Smith, W.O., 1961, p. 2). Cf: *funicular water; pellicular water; capillary condensation.*

pendulum (a) A body so suspended from a fixed point as to swing freely to and fro under the combined action of gravity and momentum. Syn: *physical pendulum.* (b) A vertical bar so supported from below by a stiff spring as to vibrate to and fro under the combined action of gravity and the restoring force of the spring (Dobrin, 1952, p. 50-51). Syn: *inverted pendulum.*

pendulum level A *leveling instrument* in which the line of sight is automatically maintained horizontal by means of a built-in pendulum device (such as a horizontal arm and a plumb line at right angles to the arm).

penecontemporaneous Formed or existing at almost the same time; e.g. said of a structure or mineral that was formed immediately after deposition of a sediment but before its consolidation into rock. Cf: *contemporaneous.*

penecontemporaneous deformation *contemporaneous deformation.*

penecontemporaneous faulting A deformation occurring in soft rock, soon after the deposition of the strata involved, and caused by gravitational sliding or slump.

penecontemporaneous fold A fold that develops in sediment shortly after deposition.

peneloken *pokelogan.*

peneplain n. A term introduced by Davis (1889a, p. 430) for a low, nearly featureless, gently undulating land surface of considerable area, which presumably has been produced by the processes of long-continued subaerial erosion (primarily mass-wasting of and sheetwash on interstream areas of a mature landscape, assisted by stream erosion) almost to base level in the penultimate stage of a humid, fluvial geomorphic cycle; also, such a surface uplifted to form a plateau and subjected to dissection. The term has been extended in the literature of geomorphology to include surfaces produced by marine, eolian, and even glacial erosion; and many topographic surfaces interpreted as peneplains are actually pediplains, panplains, or plains of marine erosion. A peneplain may be characterized by gently graded and broadly convex interfluves sloping down to broad valley floors, by truncation of strata of varying resistance and structure, by accordant levels, and by isolated erosion remnants (monadnocks) rising above it. Etymol: Latin *pene-,* "almost", + *plain.* Cf: *endrumpf; base-level plain; ma-*

rine peneplain; panplain; rumpffläche. Syn: *peneplane; base-level peneplain.*—v. *peneplane.*

peneplanation The act or process of formation and development of a *peneplain;* esp. the decline and flattening out of hillsides during their retreat and the accompanying downwasting of divides and residual hills.

peneplane n. A term suggested by Johnson (1916) to replace *peneplain* since the latter, when first introduced by Davis (1889a), was not intended to signify "almost a plain" (i.e. a region of nearly horizontal structure) but a region with almost a flat surface; the suggestion has received little support.—v. To erode to a peneplain. Syn: *peneplain.*

penesaline (a) Said of an environment intermediate between normal marine and *hypersaline,* characterized by evaporitic carbonates often interbedded with gypsum or anhydrite. According to Sloss (1953, p.145), the upper limit is characterized by a salinity of 352 parts per thousand (the maximum salinity at which sodium chloride is not precipitated) and the lower limit by a salinity high enough to be toxic to normal marine organisms. Cys (1963, p.162) proposed a lower-limit salinity of 72 parts per thousand which is "the lowest salinity at which calcium carbonate will be precipitated from solution by purely chemical processes with no influence of or catalyzing action by organisms or organic matter". (b) Said of the water in the back-reef zone characterized by a salinity too great to sustain organisms (Lang, 1937, p. 887).

penetration funnel An *impact crater,* generally funnel-shaped, formed by a small meteorite striking the Earth at a relatively low velocity and containing nearly all the impacting mass within it (Cassidy, 1968, p. 117). Cf: *meteorite crater.*

penetration test A test to determine the relative densities of noncohesive soils, sands, or silts; e.g. the "standard penetration test" that determines the number of blows required by a standard weight, when dropped from a standard height (30 in. per blow), to drive a standard sampling spoon a standard penetration (12 in.), or the "dynamic penetration test" that determines the relative densities of successive layers by recording the penetration per blow or a specified number of blows. See also: *cone penetration test.*

penetration twin *interpenetration twin.*

penetrometer An instrument for measuring the consistency of materials (such as soil, snow, asphalt, or coal) by indicating the pressure necessary to inject a rigid weight-driven rod or needle of specified shape to a specific depth.

penfieldite A white hexagonal mineral: $Pb_2(OH)Cl_3$.

penikkavaarite An intrusive rock, probably an *essexite,* composed chiefly of augite, barkevikite, and green hornblende in a groundmass of andesine. Named by Johannsen (1938) for Penikkavaara, Finland. Not recommended usage.

peninsula (a) An elongated body or stretch of land nearly surrounded by water and connected with a larger land area, usually by a neck or an isthmus. (b) A relatively large tract of land jutting out into the water, with or without a well-defined isthmus; e.g. the Italian peninsula.—Etymol: Latin *paeninsula,* "almost island".

penitent [geomorph] A term used in the French Alps for an *earth pillar.*

penitent [glaciol] n. A syn. of *nieve penitente.*—adj. A term used to refer to a nieve penitente; e.g. "penitent ice" is a nieve penitente consisting mainly of ice, and "penitent snow" is one consisting mainly of snow.

penitent rock A variety of tor formed on rock with a dipping foliation, joint pattern, or bedding (Ackermann, 1962). Syn: *monk rock.*

penkvilksite A monoclinic or orthorhombic mineral: $Na_4Ti_2Si_8O_{22}\cdot 5H_2O$.

pennantite An orange, manganese-bearing mineral of the chlorite group: $(Mn,Al)_6(Si,Al)_4O_{10}(OH)_8$. Some ferric iron may be present. It is isomorphous with thuringite.

pennate diatom A diatom having elongate form and bilateral symmetry; a member of the diatom order Pennales. Cf: *centric diatom.*

pennine *penninite.*

penninite A variety of clinochlore: $(Mg,Fe^{+2},Al)_6(Si,Al)_4O_{10}(OH)_8$. It has an emerald, olive-green, pale-green, or bluish color. Syn: *pennine.*

Pennsylvanian A period of the Paleozoic era (after the Mississippian and before the Permian), thought to have covered the span of time between 320 and 280 million years ago; also, the corresponding system of rocks. It is named after the state of Pennsylvania in which rocks of this age are widespread and yield much coal. It is the approximate equivalent of the *Upper Carboniferous* of European usage. See also: *age of ferns.* Syn: *Carbonic.*

Penokean orogeny A time of deformation and granite emplacement during the Precambrian in Minnesota and Michigan, dated radiometrically at about 1700 m.y. ago, which occurred between the formation of the Animikie ("Huronian") Series and the Keweenawan Series. It is perhaps the same as the *Hudsonian orogeny* of the Canadian Shield. See also: *Killarney Revolution.*

peñon A high, rocky point. Etymol: Spanish, "large rock, rocky mountain". Obsolete syn: *peñol.*

penroseite A lead-gray mineral: $(Ni,Co,Cu)Se_2$. It may contain some lead and silver. Penroseite is isostructural with pyrite and occurs in radiating columnar masses. Syn: *blockite.*

pentactine A sponge spicule having five rays; specif. a hexactine with one ray suppressed.

pentagonal dodecahedron *pyritohedron.*

pentagonite An orthorhombic mineral: $Ca(VO)Si_4O_{10}\cdot 4H_2O$. It is polymorphous with cavansite.

pentahydrite A triclinic mineral: $MgSO_4\cdot 5H_2O$. It is isostructural with chalcanthite. Syn: *allenite.*

pentahydroborite A mineral: $CaB_2O(OH)_6\cdot 2H_2O$.

pentalith A coccolith formed of five crystal units diverging at 72 degrees.

pentameracean Any articulate brachiopod assigned to the superfamily Pentameracea; characterized in general by a large, strongly biconvex shell with a smooth, costellate, or costate exterior, and by a spondylium in the pedicle valve. Range, Upper (possibly Middle) Ordovician to Upper Devonian. Var: *pentamerid.*

pentamerous *quinqueradiate.*

pentane Any of three isomeric, low-boiling paraffin hydrocarbons, formula C_5H_{12}, found in petroleum and natural gas.

pentlandite A pale bronze to light-brown isometric mineral: $(Fe,Ni)_9S_8$. It is commonly intergrown with pyrrhotite, from which it is distinguished by its octahedral cleavage and lack of magnetism. Pentlandite is the principal ore of nickel. Syn: *folgerite; nicopyrite.*

penumbra (a) The partly shadowed region of an eclipse. (b) The outer, lighter region of a sunspot.—Cf: *umbra.*

Penutian North American stage: Lower Eocene (above Bulitian, below Ulatisian).

Peorian A term previously used for an interglacial stage between the now-discarded Iowan (earlier) and the Wisconsinan (later) glacial stages. Named for exposures near Peoria, Ill.

peperino (a) An unconsolidated, gray tuff of the Italian Albano hills, containing crystal fragments of leucite and other minerals. (b) An indurated pyroclastic deposit containing fragments of various sizes and types.

peperite A breccialike material in marine sedimentary rock, interpreted by some as a mixture of lava with sediment, and by others as shallow intrusions of magma into wet sediment.

pepino *karst tower.*

pepper-and-salt texture Said of disseminated ores, esp. with dark grains in a light matrix.

peracidite *silexite [ign].*

peraeopod *pereiopod.*

peralkaline Said of an igneous rock in which the molecular proportion of aluminum oxide is less than that of sodium and potassium oxides combined; one of Shand's (1947) groups of igneous rocks, classified on the basis of the degree of aluminum-oxide saturation. Cf: *peraluminous; metaluminous; subaluminous.*

peraluminous Said of an igneous rock in which the molecular proportion of aluminum oxide is greater than that of sodium and potassium oxides combined; one of Shand's (1947) groups of igneous rocks, classified on the basis of the degree of aluminum-oxide saturation. Cf: *peralkaline; metaluminous; subaluminous.*

perbituminous Said of bituminous coal containing more than 5.8% hydrogen, analyzed on a dry, ash-free basis. Not listed by ASTM as a rank classification. Cf: *orthobituminous; parabituminous.*

percentage hypsometric curve Hypsographic curve using dimensionless parameters independent of an absolute scale of topographic features by relating the area enclosed between a given contour and the highest contour to the height of the given contour above a basal plane.

percentage log A *sample log* in which the percentage of each type of rock (except obvious *cavings*) present in each sample of well cuttings is estimated and plotted. Cf: *interpretive log.*

percentage map A *facies map* that depicts the relative amount (thickness) of a single rock type in a given stratigraphic unit.

percent slope The direct ratio between the vertical distance and the horizontal distance for a given slope; e.g. a 3-meter rise in 10 meters horizontal distance would be a 30 *percent slope.*

perch A unit of length, varying locally in different countries, but by statute in Great Britain and U.S. equal to 16.5 ft. It was used extensively in the early public-land surveys and is equivalent in length to a *rod* or *pole.*

perched aquifer An aquifer containing *perched ground water.*

perched block (a) A large, detached rock fragment, generally of boulder size, believed to have been transported and deposited by a glacier, and lying in a conspicuous and relatively unstable or precariously poised position on a hillside. Cf: *erratic.* Syn: *perched boulder; perched rock; balanced rock; pedestal rock.* (b) A rock forming a glacier table in a glacier. (c) A rock capping an earth pillar.

perched boulder *perched block.*

perched glacial valley *hanging valley.*

perched ground water Unconfined ground water separated from an underlying main body of ground water by an unsaturated zone. See also: *perched aquifer; perching bed.* Syn: *perched water.*

perched lake A perennial lake whose surface level lies at a considerably higher elevation than those of other bodies of water, including aquifers, directly or closely associated with it; e.g. a lake on a bench that borders the shore of a larger lake.

perched rock *perched block.*

perched spring A spring whose source of water is a body of perched ground water.

perched stream (a) A stream or reach of a stream whose upper surface is higher than the water table and that is separated from the underlying ground water by an impermeable bed in the zone of aeration (Meinzer, 1923, p. 57). (b) A stream flowing on an antecedent hillside along a graded valley of a higher-order stream into which it flows at nearly right angles (Horton, 1945, p. 352).

perched water *perched ground water.*

perched water table The water table of a body of perched ground water. Syn: *apparent water table.*

perching bed A body of rock, generally stratiform, that supports a body of *perched ground water.* Its permeability is sufficiently low that water percolating downward through it is not able to bring water in the underlying unsaturated zone above atmospheric pressure. At a given place there may be two or more perching beds and bodies of perched ground water, separated from each other and from the main zone of saturation by unsaturated zones.

percolating water (a) A legal term for water that oozes, seeps, or filters through the soil without a definite channel in a course that is unknown or not discoverable. Cf: *underground stream.* (b) Water involved in percolation.

percolation Slow laminar movement of water through small openings within a porous material. Also used as a syn. of "infiltration". Flow in large openings such as caves is not included. Cf: *infiltration.*

percolation rate The rate, expressed as either velocity or volume, at which water percolates through a porous medium.

percolation test A term used in sanitary engineering for a test to determine the suitability of a soil for the installation of a domestic sewage-disposal system. A hole is dug and filled with water and the rate of water-level decline measured; the dimensions of the hole and acceptable rate of decline differ from one jurisdiction to another.

percolation zone The area on a glacier or ice sheet where a limited amount of surface melting occurs, but the meltwater refreezes and the snow layer is not completely soaked or brought up to the melting temperature. The percolation zone may be bordered at higher altitudes by the *dry-snow line* and at lower altitudes by the *saturation line.* Cf: *soaked zone.*

percrystalline A term, now obsolete, suggested by Cross et al. (1906, p.694) for porphyritic rocks that are extremely crystalline with only a little glass, the ratio of crystals to glass being greater than 7 to 1.

percussion figure A pattern of radiating lines produced on a section of a crystal by a blow from a sharp point. Cf: *pressure figure.*

percussion mark A crescentic scar produced on a hard, dense pebble (esp. one of chert or quartzite) by a sharp blow, as by the violent impact of one pebble on another; it may be indicative of high-velocity flow. Syn: *percussion scar.*

percussion scar *percussion mark.*

percylite A pale-blue mineral: $PbCuCl_2(OH)_2$ (?).

pereion The thorax of a malacostracan crustacean, exclusive of the somites bearing maxillipeds. It is usually provided with locomotory appendages (pereiopods). Pl: *pereia*. Also spelled: *pereon.*

pereionite A somite of the pereion of a malacostracan crustacean. Also spelled: *pereonite.*

pereiopod A locomotory thoracopod of a malacostracan crustacean; an appendage of the pereion. Also spelled: *peraeopod; pereopod.* Syn: *walking leg.*

pereletok A frozen layer of ground, between the *active layer* above and the *permafrost* below, that remains unthawed for one or several years. Etymol: Russian, "survives over the summer". Syn: *intergelisol.*

perennial lake A lake that retains water in its basin throughout the year, and that usually is not subject to extreme fluctuations in level.

perennially frozen ground *permafrost.*

perennial spring A spring that flows continuously, as opposed to an *intermittent spring* or a *periodic spring.*

perennial stream A stream or reach of a stream that flows continuously throughout the year and whose upper surface generally stands lower than the water table in the region adjoining the stream. Syn: *permanent stream; live stream.*

pereon *pereion.*

pereonite *pereionite.*

pereopod *pereiopod.*

perezone A depositional zone containing nonfossiliferous sediments and occurring mostly between low tide and low-lying land undergoing active erosion; it includes lagoons and brackish-water bays.

perfect crystal A crystal without lattice defects. It is an unattained ideal or standard.

perfect flower A flower having both stamens and carpels. Cf: *imperfect flower.*

perfect fractionation path On a phase diagram, a line or a path representing a crystallization sequence in which any crystal that has been formed remains inert, i.e. does not have its composition altered.

perfection color As applied to gem diamonds, the finest color grade; i.e. a total absence of color. Cf: *gem color.*

perfectly diffuse reflector A surface that reflects radiant energy in such a manner that the reflected energy may be treated as if it were being radiated in accordance with *Lambert's law.* The energy reflected in any direction from a unit area of such a reflector varies as the cosine of the angle between the normal to the surface and the direction of the reflected radiant energy.

perfectly mobile component A component whose amount in a system is determined by its externally imposed chemical potential rather than by its initial amount in the system. Cf: *inert component; mobile component.* Syn: *boundary value component.*

perfect solution A solution that is ideal throughout its compositional range.

perfect stone (a) A *flawless* gemstone. (b) A colored stone in which small inclusions or structural faults are less undesirable than in a flawless gemstone and may even be desirable.

perfemic One of five classes in the *CIPW classification* of igneous rocks, in which the ratio of salic to femic minerals is less than one to seven. Cf: *dofemic.*

perfoliate Said of a sessile leaf or bract that surrounds the stem so that the stem seems to pass through it.

perforate (a) Said of the wall between corallites of some colonies, characterized by the presence of many irregularly arranged small openings. Also, pertaining to Perforata, a division of corals whose skeleton has a porous texture. (b) Descriptive of foraminiferal-test walls punctured or pierced by numerous pores or small openings that are distinct from apertures, foramina, and canals. Perforate walls are esp. characteristic of calcareous hyaline tests. Also, pertaining to Perforata, a division of foraminifers whose shells have small perforations for the protrusion of pseudopodia. (c) Said of an echinoid tubercle with a small depression in the top for the ligament connecting a spine with the tubercle. (d) Said of a spiral mollusk shell having an opening in the center of the umbilicus. Cf: *imperforate.*

perforated crust A type of *snow crust* containing pits and hollows produced by ablation.

perforation Puncturing of well casing opposite an oil- or gas-bearing zone to permit oil or gas to flow into a cased borehole.

perforation deposit A term suggested by Cook (1946) for an isolated kame consisting of material that accumulated in a vertical shaft piercing a glacier and affording no outlet for water at the bottom.

pergelation The formation of permafrost in the present or in the past (Bryan, 1946, p.640).

pergelic temperature regime A soil temperature regime in which the mean annual temperature is less than 0°C, and there is permafrost (USDA, 1975).

pergelisol A term introduced by Bryan (1946) for *permafrost*.

perhyaline In the *CIPW classification* of igneous rocks, those rocks in which the ratio of crystals to glassy material is greater than one to seven. Rarely used. Cf: *dohyaline*.

perhydrous (a) Said of coal containing more than 6% hydrogen, analyzed on a dry, ash-free basis. (b) Said of a maceral of high hydrogen content, e.g. exinite, resinite.—Cf: *orthohydrous; subhydrous*.

peri- A prefix meaning "around", "near".

perianth A collective term for the corolla and the calyx of a flower when considered together or when structurally undifferentiated.

periblain A kind of *provitrain* in which the cellular structure is derived from cortical material. Cf: *suberain; xylain*.

periblinite A variety of provitrinite characteristic of periblain and consisting of cortical tissue. Cf: *xylinite; suberinite; telinite*.

pericarp (a) The wall of a fruitified ovary. (b) The wall of the capsule in mosses. (c) Improperly used for the protective husks surrounding certain fruits (Jackson, 1928, p.273).

periclase An isometric mineral: MgO. It alters easily to brucite. Syn: *periclasite*.

periclinal [bot] Parallel to the surface or circumference of a plant organ. Cf: *anticlinal [bot]*.

periclinal [geol] Said of strata and structures that dip radially outward from, or inward towards, a center, to form a dome or a basin. Cf: *quaquaversal; centroclinal*.

pericline [fold] A general term for a fold in which the dip of the beds has a central orientation; beds dipping away from a center form a *dome*, and beds dipping towards a center form a *basin*. The term is generally British in usage. See also: *centrocline; quaquaversal*.

pericline [mineral] A variety of albite elongated in the direction of the *b*-axis and often twinned with this as the twinning axis. It occurs in veins as large milky-white opaque crystals. Pericline is probably an albitized oligoclase.

pericline ripple mark A term used by Ten Haaf (1959, p.22) for a ripple mark arranged in an orthogonal pattern either parallel or transverse to the current direction and having a wavelength up to 80 cm and amplitude up to 30 cm.

pericline twin law A parallel twin law in triclinic feldspars, in which the twin axis is the crystallographic *b* axis and the composition surface is a rhombic section. It occurs alone or with the albite twin law.

pericoel The space between the periphragm and endophragm in a cavate dinoflagellate cyst. Cf: *endocoel*.

pericolpate Said of pollen grains having more than three colpi, not meridionally arranged.

pericolporate Said of pollen grains having more than three colpi, not meridionally arranged, with at least part of the colpi provided with pores or transverse furrows.

pericycle In roots and stems, a layer (or layers) of cells immediately outside the phloem and inside the endodermis, from which branch roots develop (Fuller & Tippo, 1954, p. 966).

pericyst The frontal shield of a cribrimorph cheilostome (bryozoan), composed of costae united medially and commonly laterally.

perideltaic Adjacent to or surrounding a delta.

perideltidium One of a pair of slightly raised triangular parts of the interarea of a brachiopod, flanking the pseudodeltidium or lateral to it, and characterized by vertical striae in addition to horizontal growth lines parallel to the posterior margin (TIP, 1965, pt.H, p.150).

periderm [bot] A collective name for *cork, cork cambium,* and *phelloderm*. It constitutes a major portion of what is commonly called bark in woody plants.

periderm [paleont] The protein substances that compose the rhabdosome of a graptolite, consisting of an inner part (fusellar tissue) with growth increments and an outer part (cortical tissue) of layers of material (cortical fabric).

periderre A thinner layer located outside the *ectoderre* in the wall of a chitinozoan. Cf: *endoderre*.

peridot (a) A transparent to translucent green gem variety of olivine. Syn: *peridote*. (b) A yellowish-green or greenish-yellow variety of tourmaline, approaching olivine in color. It is used as a semiprecious stone. Syn: *peridot of Ceylon*.

peridotite (a) In the *IUGS classification,* a plutonic rock with M equal to or greater than 90 and ol/(ol+opx+cpx) greater than 40. (b) A general term for a coarse-grained plutonic rock composed chiefly of olivine with or without other mafic minerals such as pyroxenes, amphiboles, or micas, and containing little or no feldspar. Peridotites encompass the more specific terms *saxonite, harzburgite, lherzolite, wehrlite, dunite*. Accessory minerals of the spinel group are commonly present. Peridotite is commonly altered to *serpentinite*.

peridotite shell A syn. of the outer or *upper mantle,* so named because its composition may be peridotitic.

periembryonic chamber An immature (nepionic) part of a foraminiferal test formed on the ventral side and partly surrounding the proloculus (as in Orbitolinidae).

perigean tide A tide of increased range (e.g. a spring tide) occurring monthly when the Moon is at or near the perigee of its orbit. Ant: *apogean tide*.

perigee That point on the orbit of an Earth satellite, natural or man-made, which is nearest to the Earth. Cf: *apogee*.

perigenic Said of a rock constituent or mineral formed at the same time as the rock of which it constitutes a part "but not at the specific location in which it is now found in that rock" (Lewis, 1964, p.875); e.g. said of a glauconite grain formed from an agglutinated clay pellet and subjected to "short" transportation prior to its final incorporation into the sediment.

periglacial (a) Said of the processes, conditions, areas, climates, and topographic features at the immediate margins of former and existing glaciers and ice sheets, and influenced by the cold temperature of the ice. (b) By extension, said of an environment in which frost action is an important factor, or of phenomena induced by a periglacial climate beyond the periphery of the ice. Syn: *cryergic; cryonival; paraglacial; subnival*.—Term introduced by Lozinski (1909).

periglacial geomorphology "The study of all processes and phenomena found in cold regions" (Hamelin & Cook, 1967, p.11). See also: *cryomorphology*.

perignathic girdle A continuous or discontinuous ring of internal processes around the peristomial opening of an echinoid, serving for the attachment of muscles supporting and controlling Aristotle's lantern. See also: *girdle [paleont]*.

perigynous Said of a plant in which the stamens and petals are borne on a ring of the receptacle surrounding a pistil and usually adnate to the calyx. Cf: *epigynous; hypogynous*.

perilith *cored bomb*.

perilumen A raised inner border of a columnal articulum of a crinoid, developed as a smooth-topped, granulose, tuberculate or vermiculate ridge or field surrounding a lumen.

perine *perisporium*.

perinium A sometimes present, more or less sculptured outer coat of a pollen grain; *perisporium*. Pl: *perinia*. Adj: *perinate*.

period [geochron] (a) A geologic-time unit longer than an *epoch [geochron]* and a subdivision of an *era,* during which the rocks of the corresponding *system* were formed. It is the fundamental unit of the standard geologic time scale. (b) A term used informally to designate a length of geologic time; e.g. *glacial period*.

period [phys] The interval of time required for the completion of a cyclic motion or recurring event, such as the time between two consecutive like phases of the tide or a current.

periodic current *tidal current*.

periodic spring A spring that ebbs and flows, owing to natural siphon action. Such springs issue mainly from carbonate rocks, in which solution channels form the natural siphons. It is distinguished from a geyser by its temperature—that of ordinary ground water—and general lack of gas emission. Syn: *ebbing-and-flowing spring*. Cf: *perennial spring; intermittent spring*.

periostracum The thin organic layer covering the exterior of the shell of brachiopods and many mollusks, such as the thin coat covering the calcareous part of the shell of a bivalve mollusk, or the outer horny shell layer of a gastropod, composed dominantly of conchiolin. Cf: *ostracum*.

peripediment A term proposed by Howard (1942, p. 11) for the segment of a *pediplane* extending across the younger rocks or alluvium of a basin which is always beyond but adjacent to the segment (termed *pediment* by Howard) developed on the older upland rocks.

peripheral counter In structural petrology, an instrument used to prepare density contours for the marginal areas of a fabric diagram. It consists of a strip of plastic or similar material with a circular hole in either end, the area of each of which is equivalent to 1.0% of the area of the total diagram. Cf: *counter.*

peripheral depression *ring depression.*

peripheral fault An *arcuate fault* bounding an elevated or depressed area such as a diapir. Partial syn: *border fault.*

peripheral increase A type of *increase* (offset formation of corallites) characterized by offsets that arise in marginarial or coenenchymal tissue. A number of offsets are usually formed, and increase characteristically involves termination of development of the parietal corallite.

peripheral moraine A term proposed by Chamberlin (1879, p. 14) and now considered an obsolete syn. of *recessional moraine.*

peripheral rim A quasi-regularized series of plates in edrioasteroids that form the margin of the oral surface; it includes several circlets of plates, which diminish in size distally (Bell, 1976).

peripheral-rim transition plate One of the plates in the peripheral rim of edrioasteroids that separate externally elongate plates concentric with the thecal margin from those that are radial (Bell, 1976).

peripheral sink *rim syncline.*

peripheral stream A stream that flows parallel with the edge of a glacier, usually just beyond the moraine (Todd, 1902, p. 39).

periphery The part of a gastropod shell or any particular whorl that is farthest from or most lateral to the axis of coiling; also, the outer margin of a coiled foraminiferal test.

periphract A continuous band, composed of muscles and of fibrous tissues (aponeuroses) providing means of linear attachment to the muscles, that encircles the body of a nautiloid (TIP, 1964, pt.K, p.57). Adj: *periphractic.* Syn: *annulus [paleont].*

periphragm The outer layer of a dinoflagellate cyst, usually carrying extensions in the form of spines, and projecting to the position of the former thecal wall. It may have served as a support during the period of cyst formation. Cf: *ectophragm; endophragm.*

periphyton Micro-organisms that coat rocks, plants, and other surfaces on the water bottom. Cf: *aufwuchs.*

peripolar space A three-sided pyramidal space resulting from formation of *basal leaf cross* in an acantharian radiolarian. Cf: *perizonal space.*

periporate Said of pollen grains having many pores scattered over the surface.

periproct The membranous or irregularly polyplated area surrounding the anal opening of an echinoderm, such as the space in the CD interray of an echinoid containing the anus and covered in life by skin in which small plates are embedded; or the irregular circlet of plates around the anus of lebetodiscid edrioasteroids. Adj: *periproctal.*

perisome The *body wall* of an invertebrate, esp. of an echinoderm. Syn: *perisoma.*

perispore *perisporium.*

perisporium An additional wall layer external to the *exine* in certain spores and pollen. It is composed of thin and loosely attached sporopollenin and is therefore not usually encountered in palynomorphs. Syn: *perine; perinium; perispore.*

peristerite A gem variety of albite with blue or bluish-white luster characterized by sharp internal reflections of blue, green, and yellow; an inhomogeneous, unmixed sodic plagioclase with a composition ranging between An_2 and An_{24}. It resembles moonstone, and is falsely called *moonstone* by jewelers.

peristome (a) The region around the mouth in various invertebrates; e.g. the space containing the oral plates at the summit of the theca of a blastoid, the edge of the aperture of the body chamber of a cephalopod, the frontal depression above the mouth of a tintinnid, or the raised rim around the aperture of a foraminiferal test. (b) An elevated rim surrounding the primary orifice in a cheilostome bryozoan, or the raised terminal portion of the zooid in a stenolaemate bryozoan. Syn: *peristomie.*

peristomial Pertaining to the peristome; e.g. "peristomial ovicell" formed as a dilatation of the peristome of a bryozoan.

peristomice *secondary orifice.*

peristomie *peristome.*

peritabular Said of the surface features of a dinoflagellate cyst that originate immediately interior to the margins of reflected plate areas (as in *Areoligera* and *Eisenackia*). Cf: *intratabular; nontabular.*

perite An orthorhombic mineral: $PbBiO_2Cl$.

peritectic point *reaction point.*

peritidal Referring to depositional environments in a zone from somewhat above highest storm or spring tides to somewhat below lowest tides; a broader term than "intertidal" (Folk, 1973).

peritreme The edge or margin of the aperture of a shell; e.g. the peristome of a gastropod.

peritrichous Said of a bacterial cell having flagella uniformly distributed over all the surface. Cf: *lophotrichous; monotrichous.*

perizonal space A four-sided region resulting from the formation of *basal leaf cross* in an acantharian radiolarian. Cf: *peripolar space.*

perizonium An outer, silicified membrane on a diatom frustule, formed during auxospore development, from which the new hypovalve and epivalve are produced.

perknide An informal field term for any holocrystalline igneous rock composed almost entirely of dark-colored minerals; e.g. *hornblendite, pyroxenite, peridotite.* Cf: *perknite.* Introduced by Johannsen (1931). Not recommended usage.

perknite Any of a group of igneous rocks containing as their main constituents clinopyroxene and amphibole, with accessory orthopyroxene, biotite, iron oxides, and little or no feldspar. Included in this group are pyroxenite and hornblendite. The term, proposed by Turner in 1901 and derived from a Greek work meaning "dark", is not commonly used, having been replaced by "ultramafic rocks" and "ultramafites". Cf: *perknide.*

perlite A volcanic glass having the composition of *rhyolite,* a perlitic texture, and a generally higher water content than *obsidian.* Syn: *pearlite; pearlstone.*

perlitic (a) Said of the texture of a glassy igneous rock that has cracked due to contraction during cooling, the cracks forming small spheruloids. It is generally confined to natural glass, but occasionally found in quartz and other noncleavable minerals and as a relict structure in devitrified rocks. (b) Pertaining to or characteristic of *perlite.*

perloffite A monoclinic mineral: $Ba(Mn,Fe)_2^{2+}Fe_2^{3+}(PO_4)_3(OH)_3$.

permafrost Any soil, subsoil, or other surficial deposit, or even bedrock, occurring in arctic, subarctic, and alpine regions at a variable depth beneath the Earth's surface in which a temperature below freezing has existed continuously for a long time (from two years to tens of thousands of years). This definition is based exclusively on temperature, and disregards the texture, degree of compaction, water content, and lithologic character of the material. The thickness of permafrost ranges from over 1000 m in the north to 30 cm in the south; it underlies about one-fifth of the world's land area. Etymol: *perma*nent + *frost.* See also: *tjaele.* Cf: *pereletok.* Term introduced by Muller (1947) who included as synonyms the terms "frozen ground" or "frozen soil" preceded by any of the following modifiers: "constantly", "eternally", "ever", "perennially", "permanently", "perpetually", and "stable". Syn: *pergelisol; perennially frozen ground.*

permafrost island A small, shallow, isolated patch of permafrost surrounded by unfrozen ground; occurs on protected north-facing slopes in regions of *sporadic permafrost.*

permafrost line A line on a map representing the geographic limits of permafrost.

permafrost table The upper limit of permafrost, represented by an irregular surface dependent on local factors. Cf: *frost table; frost line.* Syn: *pergelisol table.*

permanence of continents An hypothesis, propounded as a virtual dogma by many North American geologists during the 19th and early part of the 20th centuries, that the continents (and by implication the intervening ocean basins) have been fixed in their present positions throughout geologic time. Even these geologists found it necessary, in order to explain at least some of the intercontinental resemblances, to resort to borderlands, isthmian links, and other supposed land features, now foundered beneath the oceans. The hypothesis was severely challenged later in the 20th Century by opposing hypotheses, such as *continental displacement* and *oceanization,* and is now out of favor in anything like its original form.

permanent axis The axis of the greatest moment of inertia of a rigid body, about which it can rotate in equilibrium.

permanent bench mark A readily identifiable, relatively permanent, recoverable *bench mark* that is intended to maintain its elevation with reference to an adopted datum without change over a long period of time and located where disturbing influences are believed to be negligible. Abbrev: PBM. Cf: *temporary bench mark.*

permanent extinction The *extinction* of a lake by destruction of the lake basin, as by deposition of sediments, erosion of the basin rim, or filling with vegetation.

permanent hardness *noncarbonate hardness.*

permanent icefoot An icefoot that does not melt completely during the summer.

permanently frozen ground *permafrost.*

permanent magnet A *magnet* having a large, hard remanent magnetization.

permanent set The amount of permanent deformation of a material that has been stressed beyond its elastic limit. Syn: *set [exp struc geol].*

permanent stream A syn. of *perennial stream.* The term should be avoided because a stream is not permanent (unchanging) in course, volume, or velocity.

permanent water A source of water that remains constant throughout the year.

permanent wilting A degree of wilting from which a plant can recover only if water is added to the soil. Cf: *wilting point; temporary wilting.*

permeability [geol] The property or capacity of a porous rock, sediment, or soil for transmitting a fluid; it is a measure of the relative ease of fluid flow under unequal pressure. The customary unit of measurement is the *millidarcy.* Cf: *absolute permeability; effective permeability; relative permeability.* Syn: *perviousness.* Adj: *permeable.* Ant: *impermeability.*

permeability [magnet] The ratio of magnetic induction B to inducing field strength H. Syn: *magnetic permeability.*

permeability coefficient The rate of flow of water in gallons per day through a cross section of one square foot under a unit hydraulic gradient, at the prevailing temperature (*field permeability coefficient*) or adjusted for a temperature of 60°F (Stearns, 1927, p. 148). Cf: *capillary conductivity.* Syn: *hydraulic conductivity; coefficient of permeability; meinzer.*

permeability trap A *trap* for oil or gas formed by lateral variation of permeability within a reservoir bed.

permeameter An instrument for measuring *permeability.*

permeation [grd wat] Penetration by passing through the interstices, as of a rock or soil, without causing physical change.

permeation [meta] The intimate penetration of country rock by metamorphic agents, such as granitizing solutions, particularly of an already metamorphosed rock so that it becomes more or less completely recrystallized.

permeation gneiss A gneiss formed as a result of or modified by the passage of geochemically mobile materials through or into solid rock.

permesothyridid Said of a brachiopod pedicle foramen located mostly within the ventral umbo (TIP, 1965, pt.H, p.150). Cf: *submesothyridid.*

Permian The last period of the Paleozoic era (after the Pennsylvanian), thought to have covered the span of time between 280 and 225 million years ago; also, the corresponding system of rocks. The Permian is sometimes considered part of the Carboniferous, or is divided between the Carboniferous and Triassic. It is named after the province of Perm, USSR, where rocks of this age were first studied. See also: *age of amphibians.* Syn: *Dyassic.*

per mille Per thousand.

permineralization A process of fossilization whereby the original hard parts of an animal have additional mineral material deposited in their pore spaces.

permingeatite A mineral: Cu_3SbSe_4.

permissive intrusion Emplacement of magma in spaces created by forces other than its own, e.g. orogenic forces; sometimes termed "phacolithic injection." Also, the magma or rock body so emplaced. Cf: *forcible intrusion.*

Permo-Carboniferous (a) The entire Permian and Carboniferous, considered as a unit. (b) The Permian and Pennsylvanian periods combined. (c) An age, or corresponding rock unit, transitional between the uppermost Pennsylvanian and lowermost Permian.

permutation Any different ordered subset, or arrangement, of a given set of objects. See also: *combination.*

perofskite *perovskite.*

peroikic In the *CIPW classification* of igneous rocks, those rocks in which the ratio of oikocrysts to chadacrysts is greater than seven to one. Rarely used. Cf: *domoikic.*

perovskite A yellow, brown, or grayish-black mineral: $CaTiO_3$. It sometimes has cerium and other rare-earth elements. Cf: *latrappite.* Also spelled: *perofskite.*

perpatic In the *CIPW classification* of igneous rocks, those rocks in which the ratio of groundmass to phenocrysts is greater than seven to one. Rarely used. Cf: *dopatic.*

perpendicular n. A very steep slope or precipitous face, as on a mountain.

perpendicular offset In seismic surveying, the right-angle distance from a shotpoint to the spread line. Cf: *in-line offset.*

perpendicular separation The separation of a fault as measured at right angles to the fault plane.

perpendicular slip The component of the slip of a fault that is measured perpendicular to the trace of the fault on any intersecting surface (Dennis, 1967, p. 138).

perpendicular throw In a faulted bed, vein, or other planar feature, the distance between two formerly adjacent points, measured perpendicular to the surface.

perpetual frost climate A type of *polar climate* having a mean temperature in the warmest month of less than 0°C. Cf: *tundra climate.*

perpetually frozen ground *permafrost.*

perradial Having a meridional position; e.g. a "perradial suture" situated at the midline between two columns of an ambulacrum in an echinoid, a "perradial position" precisely along one of the radii of a crinoid, or a "perradial plane" occupying a meridional position in acantharian radiolaria.

perradial line The junction in edrioasteroids between closed opposing members of coverplate pairs or series; it extends approximately along the midline of each ambulacrum (Bell, 1976).

Perret phase That stage of a volcanic eruption characterized by the emission of much high-energy gas that may significantly enlarge the volcanic conduit.

perrierite A mineral: $(Ca,Ce,Th)_4(Mg,Fe)_2(Ti,Fe)_3Si_4O_{22}$. It is a polymorph of chevkinite.

perryite A meteorite mineral: $(Ni,Fe)_5,(Si,P)_2$.

persalic One of five classes in the *CIPW classification* of igneous rocks, in which the ratio of salic to femic minerals is greater than seven to one. Cf: *dosalic.*

persemic In the *CIPW classification* of igneous rocks, those rocks in which the ratio of groundmass to phenocrysts is less than one to seven. Rarely used. Cf: *dosemic.*

persilicic A term proposed by Clarke (1908, p. 357) to replace "acidic". Syn: *silicic.* Cf: *subsilicic; mediosilicic.*

personal error An *error,* either random or systematic, caused by an observer's personal habits in making observations, by his mental or physical reactions, or by his inability to perceive or measure dimensional values exactly.

perspective The appearance to the eye of objects in respect to their relative distance and positions. Also, a picture (or other representation) in perspective.

perspective center The point of origin or termination of bundles of rays directed to a point object such as a photographic image.

perspective plane Any plane containing the perspective center; its intersection with the ground always appears as a straight line on an aerial photograph.

perspective projection A projection of points by straight lines drawn through them from some given point to an intersection with the plane of projection; e.g. a photograph is formed by a perspective projection of light rays from a point within the lens. The point of projection (unless otherwise indicated) is understood to be within a finite distance from the plane of projection. Examples include: stereographic projection, orthographic projection, and gnomonic projection. Syn: *geometric projection.*

perthite A variety of alkali feldspar consisting of parallel or subparallel intergrowths in which the potassium-rich phase (usually microcline) appears to be the host from which the sodium-rich phase (usually albite) exsolved. The exsolved areas are visible to the naked eye, and typically form strings, lamellae, blebs, films, or irregular veinlets. Cf: *antiperthite; mesoperthite.*

perthitic (a) Said of a texture produced by parallel or subparallel intergrowths of sodium-rich feldspar (typically albite) occurring as small stringers or irregular veinlets in potassium-rich feldspar (typically microcline). Cf: *perthitoid.* (b) Pertaining to or characteristic of *perthite.*

perthitoid Said of a perthitelike texture produced by minerals other than the feldspars.

perthophyte A plant living on a dead plant or on the decaying portions of a live plant.

perthosite A light-colored *syenite* composed almost entirely of *perthite,* with less than three percent mafic minerals. The name was introduced by Phemister in 1926. Not recommended usage.

Peru saltpeter Naturally occurring sodium nitrate; *soda niter* occurring in Peru. Cf: *saltpeter.* Syn: *Peruvian saltpeter.*

pervalvar axis The axis connecting the midpoints of the two valves in a diatom frustule. Cf: *apical axis; transapical axis.*

pervasiveness In quantification of hydrothermal alteration, the degree to which alteration is strictly veinlet-controlled at one extreme and generally distributed without regard to veinlets at the other. Cf: *extensiveness; intensiveness.*

perviousness *permeability [geol].*

perxenic In the *CIPW classification* of igneous rocks, those rocks in which the ratio of oikocrysts to chadacrysts is less than one to seven. Rarely used. Cf: *doxenic.*

petal [bot] A member of the second set of floral leaves, i.e. the set just internal to the sepals (Cronquist, 1961, p.879).

petal [paleont] An expanded, differentiated, petal-shaped segment of the ambulacrum situated toward the apical system of an echinoid, characterized by tube feet more or less specialized for respiration and by typically unequal or enlarged pore pairs.

petalite A white, gray, or colorless monoclinic mineral: $LiAlSi_4O_{10}$.

Peters length A measurement made on profiles across potential-field (especially magnetic) anomalies, the objective of which is to determine the depth to an anomalous tabular body or dike model. Peters' rule gives the depth as 5/8 the horizontal distance between the points on the side of an anomaly (Peters, 1949).

Peterson grab *Van Veen grab sampler.*

petiole The stalk of a leaf.

Petoskey stone A waterworn fragment of Devonian colonial coral from the beach of Lake Michigan at Petoskey, Mich. It is the "state rock" of Michigan.

petra A term introduced by Swain (1958, p. 2876) for "the rock materials produced in specific sedimentary organic environments"; a petrologic type. Pl: *petrai.* Etymol: Greek, a small mass of naturally occurring rock.

petraliiform Said of a unilamellar encrusting colony in cheilostome bryozoans, loosely attached by protuberances from basal walls of zooids or by radicles.

petricole *petrocole.*

petricolous *endolithic.*

petrifaction A process of fossilization whereby organic matter is converted into a stony substance by the infiltration of water containing dissolved inorganic matter (e.g. calcium carbonate, silica) which replaces the original organic materials, sometimes retaining the structure. Syn: *petrification.*

petrification *petrifaction.*

petrified moss A moss-like coating of *tufa* deposited on growing plants.

petrified rose *barite rosette.*

petrified wood *silicified wood.*

petro- A prefix meaning "rock".

petroblastesis Formation of rocks chiefly as the result of crystallization of diffusing ions (Dietrich & Mehnert, 1961). Not widely used.

petrocalcic horizon A diagnostic subsurface soil horizon that is characterized by an induration with calcium carbonate, sometimes with magnesium carbonate (USDA, 1975).

petrochemistry The study of the chemical composition of rocks; it is an aspect of geochemistry, and is not equivalent to petroleum chemistry.

petroclastic rock *detrital rock.*

petrocole An organism that lives in rocky areas. Also spelled: *petricole.*

petrofabric *fabric [struc petrol].*

petrofabric analysis An equivalent term for *structural petrology*, used by Knopf and Ingerson (1938, p.13).

petrofabric diagram *fabric diagram.*

petrofabrics *structural petrology.*

petrofacies *petrographic facies.*

petrogenesis A branch of petrology that deals with the origin and formation of rocks. Cf: *lithogenesis.* Adj: *petrogenetic; petrogenic.* Syn: *petrogeny.*

petrogenetic grid A diagram whose coordinates are intensive parameters characterizing the rock-forming environment (e.g. pressure, temperature) on which may be plotted equilibrium curves delimiting the stability fields of specific minerals and mineral assemblages.

petrogenic element An element that occurs mainly as an oxide, silicate, fluoride, or chloride, and is therefore a characteristic occurrence in ordinary rocks. It is one of H.S. Washington's two major groupings of elements of the lithosphere, now obsolete. Cf: *metallogenic element.*

petrogeny *petrogenesis.*

petrogeny's primitive system The system $CaAl_2Si_2O_8$ - $CaMgSi_2O_6$ - Mg_2SiO_4, in whose constituents the residual liquids produced by crystallization differentiation are invariably depleted (Poldervaart & Parker, 1965). Cf: *crystallization index; petrogeny's residua system.*

petrogeny's residua system The system $NaAlSiO_4$ - $KAlSiO_4$ - SiO_2, whose constituents are invariably concentrated in the final residual melts produced by crystallization differentiation. It includes as phases quartz, albite, the potassium feldspars, leucite, nepheline, and kaliophilite (kalsilite). The term was first used by N. L. Bowen. Cf: *petrogeny's primitive system; differentiation index.*

petroglyph Literally, a rock carving; it usually excludes writing and therefore is of prehistoric or protohistoric age. Cf: *petrogram.*

petrogram A picture painted or otherwise drawn on rocks. Cf: *petroglyph; pictograph.*

petrographer One versed in the science of petrography.

petrographic Adj. of *petrography.*

petrographic facies Facies distinguished primarily on the basis of appearance or composition without respect to form, boundaries, or mutual relations (Weller, 1958, p.627). They consist of actual large bodies of rock occurring in certain areas and in more or less restricted parts of the stratigraphic section (e.g. "red-bed facies", "paralic facies", "geosynclinal facies", "evaporite facies"), or they may consist of all rocks of a single kind (e.g. "black-shale facies", "graywacke facies"). See also: *facies.* Cf: *stratigraphic facies.* Syn: *petrofacies.*

petrographic microscope *polarizing microscope.*

petrographic period The extension in time of a rock association. Cf: *petrographic province.*

petrographic province A broad area in which similar rocks are considered to have been formed during the same period of igneous activity. Cf: *petrographic period.* Syn: *magma province; comagmatic region; igneous province.*

petrography That branch of geology dealing with the description and systematic classification of rocks, esp. igneous and metamorphic rocks and esp. by means of microscopic examination of thin sections. Petrography is more restricted in scope than *petrology.* Adj: *petrographic.* See also: *lithology; sedimentary petrography.*

petroleum (a) A naturally occurring complex liquid hydrocarbon, which after distillation and removal of impurities yields a range of combustible fuels, petrochemicals, and lubricants. Syn: *crude oil; oil.* (b) A general term for all naturally occurring hydrocarbons, whether gaseous, liquid, or solid (Levorsen, 1967, p. 3).

petroleum coke (a) The residue obtained in the distillation of petroleum. (b) A cokelike substance found in cavities of igneous intrusions into carbonaceous sediments (Tomkeieff, 1954).

petroleum geologist A geologist engaged in exploration for, or production of, oil or gas. See also: *petroleum geology.*

petroleum geology The branch of economic geology that relates to the origin, migration, and accumulation of oil and gas, and to the discovery of commercial deposits. Its practice involves the application of geochemistry, geophysics, paleontology, structural geology, and stratigraphy to the problems of finding hydrocarbons. See also: *petroleum geologist.*

petrologic Adj. of *petrology.*

petrologic province *sedimentary petrologic province.*

petrologist One who is engaged in the study of *petrology.*

petrology That branch of geology dealing with the origin, occurrence, structure, and history of rocks, esp. igneous and metamorphic rocks. Petrology is broader in scope than *petrography.* Adj: *petrologic.* See also: *sedimentary petrology.*

petromictic Said of a sedimentary deposit characterized by an assortment of metastable rock fragments; e.g. a "petromictic conglomerate" containing a mixture of pebbles or cobbles of plutonic, eruptive, sedimentary, and/or metamorphic rocks. McElroy (1954, p.151) proposed the term "petromictic sandstone" to replace "greywacke" for certain Permian and Triassic sedimentary rocks of New South Wales, having the general granular composition of classical graywackes but being light-colored, well-sorted, and mildly indurated, with a matrix that may consist of an introduced mineral cement. Syn: *petromict.*

petromorph A feature, such as a calcite vein, that stands in relief on a cave wall because limestone was dissolved from around it. Etymol: Greek, "rock form".

petrophilous *epilithic.*

petrophysics Study of the physical properties of *reservoir rocks.*

petrosilex [petrology] (a) An old name for an extremely fine crystalline porphyry or quartz porphyry, and for the groundmass of such porphyries; also, a finely crystalline aggregate now known to be devitrified glass. "It was practically a confession by the older petrographers that they did not know of what the rock consisted" (Kemp, 1896, p. 156). (b) A term applied by Lyell (1839, p. 99) to igneous rocks. (c) An obsolete term formerly applied to clinkstone, to fusible hornstone, to felsite, and to compact feldspar.

petrosilex [sed] (a) An obsolete term applied to *flint* occurring in the form of a rock mass, or as part of the rock, as distinguished from detached nodular flint in chalk beds; also, a rock partly converted to flint. (b) A French term for *hornstone [rock],* or flinty slate without slaty cleavage. (c) *amausite.*

petrostratigraphy *lithostratigraphy.*

petrous Said of a material that resembles stone in its hardness; e.g. petrous phosphates. Little used.

petrovicite An orthorhombic mineral: $PbHgCu_3BiSe_5$.

Petschau twin law A rare type of parallel twin law in feldspar, having a twin axis of [$\bar{1}$10].

Pettersson theory An astronomical theory of climatic change, proposed by Sven Otto Pettersson (1848-1941), Swedish oceanographer, in which climatic changes are related to tidal cycles produced by systematic orbital variations of Earth and Moon about the Sun.

petuntse *china stone [ign].*

petunzyte *china stone [ign].*

petzite A steel-gray to iron-black mineral: Ag_3AuTe_2.

pezograph *regmaglypt.*

Pfalzian orogeny One of the 30 or more short-lived orogenies during Phanerozoic time identified by Stille, in this case at the end of the Permian. Syn: *Palatinian orogeny.*

pH The negative $\log_{10}$ of the hydrogen-ion activity in solution; a measure of the acidity or basicity of a solution.

phacellite *kaliophilite.*

phaceloid Said of a fasciculate corallum having subparallel corallites. Also spelled: *phacelloid.*

phacoidal structure An infrequently used term for a lenticular metamorphic structure, e.g. *flaser structure; augen structure.*

phacolite A variety of chabazite, characterized by colorless lenticular crystals.

phacolith A minor syntectonic concordant concavo-convex intrusion within folded strata. Cf: *harpolith.*

phaeodarian Any radiolarian belonging to the suborder Phaeodarina, characterized mainly by a central capsule enclosed by a double-walled membrane.

phagotrophic Said of an organism that is nourished by the ingestion of solid matter.

phaneric Said of the texture of a carbonate sedimentary rock (esp. limestone) characterized by individual crystals or clastic grains whose diameters are greater than 0.01 mm (Bissell & Chilingar, 1967, p. 163). The term was proposed by DeFord (1946) who used 0.004 mm as the limiting diameter. Cf: *aphanic.*

phanerite An igneous rock having the grains of its essential minerals large enough to be seen macroscopically. Obsolete syn: *phaneromere.*

phaneritic Said of the texture of an igneous rock in which the individual components are distinguishable with the unaided eye, i.e. megascopically crystalline. Also, said of a rock having such texture. Cf: *aphanitic.* Syn: *macromeritic; phanerocrystalline; phenocrystalline.*

phanerocryst *phenocryst.*

phanerocrystalline *phaneritic.*

phanerogam (a) A seed-bearing plant, as opposed to a spore-bearing plant or *cryptogam.* Cf: *spermatophyte.* (b) A flowering plant in which the stamens and pistils are distinctly developed (Jackson, 1928, p.279).

phanerogenic Of known origin; e.g. said of a species that is proved to descend from a known species found in an older geologic formation.

phaneromere *phanerite.*

phaneromphalous Said of a gastropod shell with a completely open umbilicus. Cf: *anomphalous.*

phanerophyte A perennial plant whose overwintering buds are above the ground surface.

Phanerozoic That part of geologic time represented by rocks in which the evidence of life is abundant, i.e. Cambrian and later time. Cf: *Cryptozoic.*

phanoclastic rock An "even-grained or uniformly sized" clastic rock (Pettijohn, 1949, p.30).

phantom (a) A bed or member that is missing from a given stratigraphic section although it elsewhere occupies a characteristic position in a sequence of similar age. Cf: *ghost member.* (b) *ghost [petrology].*

phantom bottom *deep scattering layer.*

phantom crystal A crystal within which an earlier stage of crystallization or growth is outlined by dust, tiny inclusions, or bubbles, e.g. serpentine containing a ghost or phantom of original olivine. Syn: *ghost crystal.*

phantom horizon In seismic reflection prospecting, a line drawn on seismic sections so that it is parallel to nearby dip segments thought to indicate structural attitude. It is used where actual events are not continuous enough to be used alone.

pharaonite A mineral that may be magnesium-rich davyne.

pharetrone Any calcisponge having spicules that have the shape of a tuning fork.

pharmacolite A monoclinic mineral: $CaH(AsO_4)\cdot 2H_2O$. It occurs in white or grayish silky fibers. Syn: *arsenic bloom.*

pharmacosiderite A green or yellowish-green mineral: $Fe_3(AsO_4)_2(OH)_3\cdot 5H_2O$. It commonly occurs in cubic crystals. Syn: *cube ore.*

pharynx (a) A differentiated part of the alimentary canal in many invertebrates; e.g. the tubular passageway between the mouth and gastrovascular cavity of an octocoral, or the internal oral tube in a phaeodarian radiolarian. (b) In the vertebrates, that part of the alimentary tract that connects the buccal cavity to the esophagus. Its walls support the gills in gilled vertebrates and in many lower chordates as well.

phase [chem] A homogeneous, physically distinct portion of matter.

phase [geochron] (a) A term approved by the 8th International Geologic Congress in Paris in 1900 for the geologic-time unit next in order of magnitude below *age,* during which the rocks of a *substage* (then referred to as a "zone") were formed; a *subage.* The term was seldom used and is now obsolete in this usage. Syn: *episode; time (d).* (b) *moment [stratig].*

phase [glac geol] An informal subdivision of a *stage.* The term has been used by Flint (1957) for the deposits of various glacial Great Lakes at different levels and dates.

phase [ign] An interval in the development of a given process; esp. a chapter in the history of the igneous activity of a region, such as the "volcanic phase" and major and minor "intrusive phases".

phase [phys] (a) A stage in periodic motion (rotation, oscillation, etc.), measured with respect to a given initial point and expressed in angular measure. (b) A peak, trough, zero-crossing, or other identifiable point of a travelling wave. See also: *Airy phase; T-phase.*

phase [sed] (a) A product of "deposition during a single fluctuation in the competency of the transporting agent" (Apfel, 1938). Such a subunit is probably a *lamina.* (b) A transitory or minor fluctuation in the velocity of a depositing current, resulting in the formation of a lamina.

phase [stratig] (a) A lithologic facies, esp. on a small scale, such as a minor variety within a dominant or normal facies, or a facies of short duration or local occurrence; e.g. the "marine phase" and the "fluviatile phase" of the Pocono Formation (Barrell, 1913, p.465). The term was used by McKee (1938, p.13-14) for a lateral subdivision (or facies) of a formation. (b) A term defined by Fenton & Fenton (1930, p.150) as "a local or regional aspect or condition of a stratum or group of strata, as determined both by original nature and secondary change; the latter being the determining factor"; e.g. a change arising from faulting, folding, secondary dolomitization, or erosion. (c) Characteristic strata repeated at various positions in a stratigraphic section that record the recurrence of a particular kind of environment (Allan, 1948, p.8). (d) A term that has been applied variously in stratigraphy to (see Weller, 1958, p.620): a part of a depositional cycle; a part of a cyclothem; sediments consisting of identical constituents deposited under relatively uniform conditions; a time-stratigraphic division; a type of lithologic development; a sedimentary province; and a major biotope.—Because the term has been used widely and vaguely in stratigraphy and because its technical application is not needed, "this word may well be left without restriction as to special meaning" (Moore, 1957a, p.1787).

phase [surv] The apparent displacement of a surveying object or signal caused by one side being more strongly illuminated than the other and resulting in an error in sighting.

phase boundary *boundary line.*

phase diagram A graph designed to show the boundaries of the fields of stability of the various phases of a system. The coordinates are usually two or more of the intensive variables temperature, pressure, and composition, but are not restricted to these. Syn: *equilibrium diagram.*

phase equilibria In physical chemistry, the study of those phases which, under specified conditions, may exist in equilibrium.

phase lag *tidal epoch.*

phase layering (a) The mineralogical *layering* of plutonic rocks caused by the generally sudden appearance and progressive disappearance (from bottom to top) of some mineral phase. Cf: *cryptic layering; rhythmic layering.* (b) A large-scale structure in metamorphic rocks produced by subparallel layers of various origins that are mineralogically distinct from the host rocks such as gneisses and peridotites. Syn: *primary banding; compositional layering; primary layering; crystallization banding.*

phase microscope A microscope that utilizes the phase differences of light rays transmitted by different portions of an object to create an image in which the details of the object are distinct, despite their near uniformity of refractive index.

phase plate In a polarizing microscope, a plate of doubly refracting material, e.g. mica of a quarter-wave plate, that changes the relative phase of the polarized light's components.

phase response *phase spectrum.*

phase rule The statement that for any system in equilibrium, the number of *degrees of freedom* is two greater than the difference between the number of components and the number of phases. It may be symbolically stated as F = (C-P) + 2. Syn: *Gibbs phase rule.* See also: *mineralogical phase rule.*

phase spectrum In seismology, the shift of phase with frequency produced by passage through a filter. Syn: *phase response.*

phase velocity The velocity with which an observable, individual wave or wave crest is propagated through a medium; the velocity of a point of constant phase. It is the product of wavelength and frequency. Symbol: c. Cf: *group velocity; particle velocity.*

phassachate A lead-colored agate.

phelloderm Secondary tissue of woody plant stems, produced by the cork cambium on its inner surface, as opposed to the cork, which is produced by the cork cambium on its outer surface (Cronquist, 1961, p.879). See also: *periderm [bot].*

phellogen *cork cambium.*

phenacite *phenakite.*

phenakite A colorless, white, or pale varicolored glassy rhombohedral mineral: Be_2SiO_4. It is sometimes confused with quartz. Phenakite is used as a minor gemstone. Not to be confused with *fenaksite.* Also spelled: *phenacite.*

phenetic system A classification system for living organisms that is based on morphologic, anatomic, physiologic, or any other available biochemical, criteria. By definition the term does not have evolutionary implications (Cowan, 1968, p. 78).

phengite (a) A variety of muscovite with a high silica content. (b) A transparent or translucent stone (probably crystallized gypsum) used by the ancients for windows.

phenhydrous Said of vegetal matter deposited under water. Cf: *cryptohydrous.*

phenicochroite *phoenicochroite.*

phenoclast One of the larger and more conspicuous fragments in a sediment or sedimentary rock composed of various sizes of material, such as a cobble or pebble (*spheroclast*) embedded in a fine-textured matrix of a conglomerate, or a fragment (*anguclast*) of a breccia.

phenoclastic rock A "nonuniformly sized" clastic rock containing phenoclasts (Pettijohn, 1949, p.30).

phenocryst A term suggested by J.P. Iddings, and widely used, for a relatively large, conspicuous crystal in a porphyritic rock. The term *inset* has been suggested as an alternative.

phenocrystalline *phaneritic.*

phenogenesis The development of the *phenotype.*

phenomenal gem A gemstone exhibiting an optical phenomenon, such as asterism, chatoyancy, or play of color.

phenomenology The science that treats of the description and classification of phenomena.

phenoplast A large rock fragment (in a *rudaceous* rock) that was plastic at the time of its incorporation in the matrix.

phenotype The visible characters of an organism that reflect the interaction of genotype and environment.

phi A logarithmic transformation of the ratio of a grain diameter in mm to a standard grain diameter of 1 mm. It is not a dimension but a ratio (McManus, 1962, p. 673). Symbol: ϕ.

phialine Said of an everted apertural rim (as on the neck of a vial or bottle) of some foraminiferal tests.

phi deviation measure A graphic measure of dispersion of particle size shown on a plot of phi units.

phi grade scale A logarithmic transformation of the *Wentworth grade scale* in which the negative logarithm to the base 2 of the particle diameter (in millimeters) is substituted for the diameter value (Krumbein, 1934); it has integers for the class limits, increasing from -5 for 32 mm to +10 for 1/1024 mm. The scale was developed specifically as a statistical device to permit the direct application of conventional statistical practices to sedimentary data. Syn: *phi scale.*

philippinite A tektite from the Philippine Islands. See also: *rizalite.*

philipstadite A monoclinic mineral of the amphibole group, approximately: $Ca_2(Fe,Mg)_5(Si,Al)_8O_{22}(OH)_2$.

phillipsite A white or reddish zeolite mineral: $(K_2,Na_2,Ca)Al_2Si_4O_{12}\cdot 4\text{-}5H_2O$. It sometimes contains no sodium, but always contains considerable potassium. It commonly occurs in complex (often cruciform) fibrous crystals, and makes up an appreciable part of the red-clay sediments in the Pacific Ocean.

phi mean diameter A *logarithmic mean diameter* obtained by using the negative logarithms of the class midpoints to the base 2. See also: *geometric mean diameter.*

phi scale *phi grade scale.*

phi standard deviation The *standard deviation* of a grain-size distribution expressed in *phi units.* It may be determined by calculation or graphically. Syn: *phi deviation measure.*

phi unit The unit interval on the curve of values in phi transformation. It should be used in the same context as intervals on the *Wentworth grade scale* (McManus, 1962, p. 673).

phlebite Metamorphite or migmatite with roughly banded or veined appearance (Dietrich & Mehnert, 1961). Originally proposed, without genetic connotation, to replace the term *veined gneiss* (Mehnert, 1968, p. 17). Cf: *venite; arterite; composite gneiss; injection gneiss; diadysite.* Little used.

Phleger corer A type of *gravity corer* that has a check valve and/or core catcher to retain the sample.

phloem The food-conducting tissue of vascular plants, consisting of various types of cells, such as sieve tubes, phloem parenchyma, fibers, and companion cells. Syn: *sieve tissue.*

phlogopite A magnesium-rich mineral of the mica group: $K(Mg,Fe)_3AlSi_3O_{10}(OH,F)_2$. It is yellowish brown to brownish red or copper-colored, and usually occurs in crystalline limestones as a result of dedolomitization. Phlogopite is near biotite in composition, but contains little iron. Syn: *magnesia mica; amber mica; brown mica.*

phobotaxis *Taxis [ecol]* in which an organism avoids a concentration or intensity of something. Cf: *strophotaxis; thigmotaxis.*

phoenicochroite A red mineral: Pb_2CrO_5. Also spelled: *phenicochroite.* Syn: *phoenicite; berezovite; scheibeite.*

pholad Any bivalve mollusk belonging to the family Pholadidae, characterized by an equivalve shell of variable size, commonly gaping open at the posterior end.

pholerite A discredited name for a clay mineral identical with kaolinite or one of the other kaolin minerals.

phonolite (a) In the strictest sense, a group of fine-grained extrusive rocks primarily composed of alkali feldspar (esp. anorthoclase or sanidine), and with nepheline as the main feldspathoid (Streckeisen, 1967, p. 185); also, any rock in that group; the extrusive equivalent of *nepheline syenite.* (b) In the broadest sense, any extrusive rock composed of alkali feldspar, mafic minerals and any feldspathoid, such as nepheline, leucite, or sodalite. Syn (all obsolete): *clinkstone; klingstein; echodolite.*— Streckeisen (1967, p. 186) suggests that "phonolite" be preceded by the name of the main feldspathoid mineral (e.g. "leucite phonolite", "analcime phonolite", etc.). Etymol: Greek *phone,* "sound", in reference to the allegedly characteristic ringing sound emitted by a phonolite when struck with a hammer.

phorogenesis The shifting or slipping of the Earth's crust relative to the mantle. See also: *drag-and-slippage zone.*

phosgenite A white, yellow, or grayish tetragonal mineral with adamantine luster: $Pb_2Cl_2(CO_3)$. Syn: *horn lead; cromfordite.*

phosinaite An orthorhombic mineral: $H_2Na_3(Ca,Ce)(SiO_4)(PO_4)$.

phosphammite A mineral: $(NH_4)_2HPO_4$.

phosphate A mineral compound containing tetrahedral PO_4^{-3} groups. An example is pyromorphite, $Pb_5(PO_4)_3Cl$. Phosphorus, arsenic, and vanadium may substitute for each other in the tetrahedron. Cf: *arsenate; vanadate.*

phosphate rock Any rock that contains one or more phosphatic minerals of sufficient purity and quantity to permit its commercial use as a source of phosphatic compounds or elemental phosphorus. About 90% of the world's production is sedimentary phosphate rock, or *phosphorite;* the remainder is igneous rock rich in apatite.

phosphatic Pertaining to or containing phosphates or phosphoric acid; said esp. of a sedimentary rock containing phosphate minerals, such as a "phosphatic limestone" produced by secondary enrichment of phosphatic material, or a "phosphatic shale" representing mixtures of primary or secondary phosphate and clay minerals.

phosphatic nodule A black, gray, or brown rounded mass or "pebble", ranging in diameter from a few millimeters to more than 30 cm, consisting of coprolites, corals, shells, bones, sand grains, mica flakes, or sponge spicules, more or less enveloped in *collophane* (calcium phosphate). They occur in marine strata (as in Permian beds of western U.S. and in the Cretaceous chalk of England), and are forming at present on the sea floor (as off the coast of California). See also: *coprolite.*

phosphatite *phosphorite.*

phosphatization Conversion to a phosphate or phosphates; e.g. the diagenetic replacement of limestone, mudstone, or shale by phosphate-bearing solutions, producing phosphates of calcium, aluminum, or iron. Cf: *phosphorization.*

phosphide A mineral compound that is a combination of phosphorus with a metal. An example is schreibersite, $(Fe,Ni)_3P$.

phosphochalcite *pseudomalachite.*

phosphoferrite A white, pale-green, or yellow orthorhombic mineral: $(Fe,Mn)_3(PO_4)_2 \cdot 3H_2O$. It is isomorphous with reddingite.

phosphophyllite A colorless or pale blue-green monoclinic mineral with perfect micaceous cleavage: $Zn_2(Fe,Mn)(PO_4)_2 \cdot 4H_2O$.

phosphorescence A type of *luminescence* in which the stimulated substance continues to emit light after the external stimulus has ceased; also, the light so produced. The duration of the emission is temperature-dependent, and has a characteristic rate of decay. Cf: *fluorescence.*

phosphorite A sedimentary rock with a high enough content of phosphate minerals to be of economic interest. Most commonly it is a bedded primary or reworked secondary marine rock composed of microcrystalline carbonate fluorapatite in the form of laminae, pellets, oolites, nodules, and skeletal, shell, and bone fragments. Aluminum and iron phosphate minerals (wavellite, millisite) are usually of secondary formation. See also: *brown rock; bone phosphate; pebble phosphate; guano.*

phosphorization Impregnation or combination with phosphorus or a compound of phosphorus; e.g. the diagenetic process of *phosphatization.*

phosphorochalcite *pseudomalachite.*

phosphorroesslerite A monoclinic mineral: $MgH(PO_4) \cdot 7H_2O$. It is isomorphous with roesslerite. Also spelled: *phosphorrösslerite.*

phosphosiderite A pinkish-red monoclinic mineral: $FePO_4 \cdot 2H_2O$. It is dimorphous with strengite and isomorphous with metavariscite. Syn: *clinostrengite; metastrengite.*

phosphuranylite A golden-yellow secondary mineral: $Ca(UO_2)_3(PO_4)_2 \cdot 6H_2O$. It exhibits phosphorescence upon exposure to radium emanations.

photic zone *euphotic zone.*

photoalidade A photogrammetric instrument having a telescopic alidade, a plateholder, and a hinged ruling arm, mounted on a tripod frame, and used for plotting lines of direction and measuring vertical angles to selected features appearing on oblique and terrestrial photographs.

photoautotrophic *holophytic.*

photobase The length of the *air base* as represented on a photograph.

photoclinometer An obsolete device containing a sensitized graduated paper disk and a light source, used to determine the *inclination* from the vertical of a well bore. It hangs freely on a light cable, and upon being lowered to a determined depth, the energy source is activated to register a "burn" on the sensitized disk.

photoclinometry A technique for ascertaining slope information from a distribution of image brightness; it is used esp. for studying the amount of slope of a lunar crater wall or ridge by measuring the density of its shadow.

photoconductive detector A *detector* in which the conductivity varies in response to the electromagnetic radiation impinging on it.

photoelasticity The property of a transparent, isotropic solid to become doubly refracting under nonhydrostatic states of stress. This makes it possible to study stress-distribution patterns under the polariscope.

photogeologic guide Any photographic element that assists in the interpretation of the geology of a given area.

photogeologic map A geologic map based on information derived from the interpretation of aerial photographs.

photogeology The identification, recording, and study of geologic features and structures by means of photography; specif. the geologic interpretation of aerial and space photographs and images and the presentation of the information so obtained. It now includes the interpretation of second-generation photographs obtained by photographing images recorded on television-type tubes (the images recording wavelengths outside the visible spectrum).

photogrammetry The art and science of obtaining reliable measurements from photographic images. Measurements relate not only to size, shape, and position, but also to color or tone, texture, and patterns of distribution of these elements.

photograph The recorded representation of a scene or object made by the action of light on a base material coated with a sensitized solution that is chemically treated to fix the image points at the desired density. The term is generally taken to mean the direct action of electromagnetic radiation on the sensitized material. Cf: *image [photo].*

photograph center The center of a photograph as indicated by the images of the fiducial marks of the camera. For a perfectly adjusted camera, it is identical to the *principal point* of the photograph.

photographic interpretation *photointerpretation.*

photography (a) The process of producing *photographs.* Cf: *imagery.* (b) A collective term for the photographs so produced.

photohydrology The science involving extraction of hydrologic data from aerial photographs and Landsat imagery.

photoindex A mosaic made by assembling individual photographs (with accompanying designations) into their proper relative positions and copying the assembly photographically at a reduced scale. It is not an *index map.* Also spelled: *photo index.*

photointerpretation The extraction of information from aerial photographs and images for a particular purpose, such as mapping the geologic features of an area. Also spelled: *photo interpretation.* Syn: *photographic interpretation.*

photomap An aerial photograph or a controlled mosaic of rectified photographs to which have been added a reference grid, scale, place names, marginal information, and other pertinent data or map symbols; e.g. an *orthophotomap.*

photometer An instrument for measuring the intensity of light. See also: *spectrophotometer; flame photometer.*

photometry (a) Study of ways and means to measure the intensity of light. (b) The art or process of using a *photometer.*

photomicrograph A photographic enlargement of a microscopic image such as a petrologic thin section; a type of *micrograph.* See also: *photomicrography.* Less-preferred syn: *microphotograph.*

photomicrography The preparation of *photomicrographs,* performed by the projection of the image through the eyepiece of the microscope onto the photographic recording medium.

photomosaic *mosaic [photo].*

photoperiod The relative number of alternating daylight and dark hours in a 24-hour period. The photoperiod has a significant effect on the development of certain organisms, esp. flowering plants.

photoreceptor A sense organ that undergoes specific stimulation when exposed to light, such as (perhaps) a paraflagellar body in some Euglenophyta.

photorelief map (a) A map consisting of a photograph of a relief model of the area under study and showing salient physical features. (b) A diagrammatic map that simulates or gives the impression of a photograph of a relief model of the area under study.—Also spelled: *photo-relief map.*

phototaxis *Taxis* resulting from stimulation by light.
phototheodolite A ground-surveying instrument used in terrestrial photogrammetry, combining the functions of a theodolite and a camera mounted on the same tripod.
phototriangulation The addition of horizontal and/or vertical control points by photogrammetric methods, whereby "the measurements of angles and/or distances on overlapping photographs are related into a spatial solution using the perspective principles of the photographs" (ASP, 1966, p.1148); esp. *aerotriangulation.*
phototrophic *holophytic.*
photovoltaic detector A *detector* in which the voltage varies in response to the electromagnetic radiation impinging on it.
phragmites peat Peat that is derived mainly from the reed genus *Phragmites.*
phragmocone (a) The chambered part of the conch of a nautiloid or an ammonoid. (b) The conical, chambered internal shell of a belemnite or aulacocerid, fitted into a deep anterior cavity (the *alveolus*) in the rostrum or telum.
phragmospore A plant spore having two or more septa; e.g. a septate *fungal spore* that may have a chitinous wall and therefore be preserved as a palynomorph.
phreatic cycle The period of time during which the water table rises and falls. It may be a daily, annual, or other cycle. Syn: *cycle of fluctuation.*
phreatic explosion A volcanic eruption or explosion of steam, mud, or other material that is not incandescent; it is caused by the heating and consequent expansion of ground water due to an underlying igneous heat source. Cf: *phreatomagmatic.*
phreatic gas A gas formed by the contact of atmospheric or surface water with ascending magma. Cf: *juvenile [water]; resurgent gas.*
phreatic ground water *phreatic water.*
phreatic line *seepage line.*
phreaticolous Said of an organism or of the fauna inhabiting the interstices of mixtures of sand and gravel.
phreatic solution The solution action by ground water below the water table. Cf: *vadose solution.*
phreatic surface *water table.*
phreatic water A term that originally was applied only to water that occurs in the upper part of the zone of saturation under water-table conditions (syn. of *unconfined ground water,* or *well water*), but has come to be applied to all water in the zone of saturation, thus making it an exact syn. of *ground water* (Meinzer, 1923, p. 5). In 1939 Meinzer used "phreatic water" (in the sense of "unconfined ground water") as a class of *plerotic water.* Syn: *phreatic ground water.*
phreatic-water discharge *ground-water discharge.*
phreatic wave *ground-water wave.*
phreatic zone *zone of saturation.*
phreatomagmatic explosion A volcanic explosion that extrudes both magmatic gases and steam; it is caused by the contact of magma with ground water or shallow surface water. Cf: *phreatic explosion.*
phreatophyte A plant that obtains its water supply from the *zone of saturation* or through the *capillary fringe* and is characterized by a deep root system.
phrenotheca One of the thin dense diaphragmlike partitions that extend across the chamber of a foraminiferal test at various angles and in various parts of the chamber (as in *Pseudofusulina*).
Phthinosuchia A suborder of predaceous synapsid reptiles, brigaded in the order Therapsida but showing many features transitional between sphenacodont pelycosaurs and synapsids proper. Range, early Upper Permian.
phycobiont The algal partner or component of a lichen. Cf: *mycobiont.*
phyla The plural of *phylum.*
phyletic *phylogenetic.*
phyletic evolution Evolution involving changes in *lineages* but little or no increase in the number of taxonomic groups.
phyletic gradualism That evolutionary process or pattern in which morphologic change occurs gradually and continuously throughout the time duration of several successive species within an evolutionary *lineage,* so that an ancestral species appears to grade imperceptibly into its immediately descendent species. Syn: *gradualistic speciation.* Ant: *punctuated equilibria.*
phyllarenite A term used by Folk (1968, p.124) for a *litharenite* composed chiefly of foliated, phyllosilicate-rich metamorphic-rock fragments (as of slate, phyllite, and schist). It may have any particle size from silt through gravel, and any clay content, sorting, or rounding. See also: *subphyllarenite.*
phyllite [mineral] (a) A general term used by some French authors for the scaly minerals, such as the micas, chlorites, clays, and vermiculites. (b) A general term for minerals with a layered crystal structure.
phyllite [petrology] A metamorphosed rock, intermediate in grade between slate and mica schist. Minute crystals of sericite and chlorite impart a silky sheen to the surfaces of cleavage (or schistosity). Phyllites commonly exhibit corrugated cleavage surfaces. Cf: *phyllonite.*
phyllite-mylonite *phyllonite.*
phyllocarid Any malacostracan belonging to the subclass Phyllocarida, characterized by a relatively large bivalve carapace that may or may not be hinged along the dorsal margin. Range, Lower Cambrian to present.
phylloclade A somewhat flattened branch or stem that functions as a leaf, e.g. in Christmas cactus. Cf: *phyllode [bot].*
phyllode [bot] A flattened and expanded petiole that functions as a leaf. Cf: *phylloclade.*
phyllode [paleont] The more or less depressed area of enlarged pores in the adoral part of an ambulacrum in an echinoid. It bears specialized podia. Cf: *bourrelet.*
phyllofacies A facies differentiated on the basis of stratification characteristics, esp. the *stratification index* (Kelley, 1956, p.299).
phylloid Leaf-shaped, or resembling a leaf; esp. said of the minor elements or endings of the saddles of an ammonoid suture (as in *Phylloceras*).
phyllomorphic stage A term introduced by Dapples (1962) for the latest (most advanced) geochemical stage of diagenesis characterized by authigenic development of micas, feldspars, and chlorites at the expense of clays (unidirectional reactions). It follows the *locomorphic stage.* See also: *redoxomorphic stage.*
phyllonite A rock that macroscopically resembles *phyllite [petrology]* but that is formed by mechanical degradation (mylonitization) of initially coarser rocks (e.g. graywacke, granite, or gneiss). Silky films of recrystallized mica or chlorite, smeared out along schistosity surfaces, and formation by dislocation metamorphism are characteristic. The term was originated by Sander (1911). Syn: *phyllite-mylonite.*
phyllonitization The processes of mylonitization and recrystallization to produce a phyllonite.
phyllopodium A broad flat leaflike thoracic appendage of a crustacean. Pl: *phyllopodia.* Cf: *stenopodium.*
phyllosilicate A class or structural type of *silicate* characterized by the sharing of three of the four oxygens in each tetrahedron with neighboring tetrahedra, to form flat sheets; the Si:O ratio is 2:5. An example is the micas. Cf: *nesosilicate; sorosilicate; cyclosilicate; inosilicate; tectosilicate.* Syn: *layer silicate; sheet mineral; sheet silicate.*
phyllosilicate cement A term applied by Dickinson (1970, p. 702) to a cement in graywackes and arkoses that is clear, monomineralic, and crystalline, and has certain other features that distinguish it from inhomogeneous matrix.
phyllotaxy The arrangement of leaves or floral parts on an axis. It is generally expressed numerically by a fraction, the numerator representing the number of revolutions of a spiral made in passing from one leaf past each successive leaf to reach the leaf directly above the initial leaf, and the denominator representing the number of leaves passed in the spiral thus made (Lawrence, 1951, p.765).
phyllotriaene A sponge *triaene* in which the cladi are expanded into flattened, sometimes digitate, leaflike structures.
phyllovitrinite *provitrinite.*
phylogenetic The adj. of *phylogeny.* Syn: *phyletic.*
phylogenetic evolution Evolution within a single *lineage.*
phylogenetic systematics (a) The study of *systematics* with the ultimate goal of inferring *phylogeny* of the organisms investigated. (b) In more restricted usage, the study of systematics by means of the methods of *cladism;* this usage results from the work of W. Hennig, and thus may be termed *Hennigian* systematics.
phylogenetic tree (a) A *dendrogram* or other diagrammatic portrayal expressing ancestor-descendant or genealogical relationships among several taxa, as inferred from various types of evidence. Syn: *phylogram; cladogram.* (b) More specifically, in *cladism,* a *dendrogram* showing phylogenetic relationships among taxa in terms of inferred genealogical history; identities of *nodes* (ancestors) are specified, connecting lines represent ancestral

taxa, and not all the taxa included are terminal in position. Cf: *cladogram*.

phylogenetic zone *lineage-zone*.

phylogeny (a) The line, or lines, of direct descent in a given group of organisms, as opposed to the development of an individual organism. Cf: *ontogeny*. (b) The study or history of such relationships.—Adj: *phylogenetic*.

phylogerontism The condition of apparent deterioration and probable approaching extinction of an evolving *lineage*. Syn: *racial senescence*. The term has been overapplied.

phylogram *phylogenetic tree (a)*.

phylozone *lineage-zone*.

phylum A category in the hierarchy of zoological classification intermediate between *kingdom* and *class*. It is generally regarded as equivalent to the rank of *division* in botanical classification. Cf: *subphylum*.

-phyre A suffix which, in a rock name, signifies "porphyry".

physical exfoliation A type of *exfoliation* caused by physical forces, such as by the freezing of water in fine cracks in the rock or by the removal of overburden concealing deeply buried rocks.

physical geography That branch of *geography* which is the descriptive study of the Earth's surface as man's physical environment, dealing with the classification, form, and extent of the natural phenomena directly related to the exterior physical features and changes of the Earth, including land, water, and air. It differs chiefly from *geology* in that it considers the present rather than the past conditions of the Earth, and it is more inclusive than *geomorphology*, dealing not only with landforms but also climate, oceans, atmosphere, soils, geologic processes, natural resources, and sometimes the biogeographical distribution of animal and plant life. In the 18th century, the term was applied in a broader sense, commonly including the races of men and their physical works on the Earth. Cf: *physiography*.

physical geology A broad division of *geology* that concerns itself with the processes and forces involved in the inorganic evolution of the Earth and its morphology, and with its constituent minerals, rocks, magmas, and core materials. Cf: *historical geology*.

physical landscape *natural landscape*.

physical oceanography The study of such physical aspects of the ocean as optical and acoustic properties; temperature; density; and currents, waves, and tides.

physical pendulum *pendulum* (a).

physical residue A *residue* formed by mechanical weathering in place; e.g. a deposit of gravel resulting from the removal of finer particles by water or wind, as on the floor of a desert valley.

physical stratigraphy Stratigraphy based on the physical aspects of rocks (esp. the sedimentologic aspects); e.g. *lithostratigraphy*.

physical time A term used by Jeletzky (1956, p. 682) to designate time as measured by any physical phenomenon or process (such as by radioactive decay of elements), and proposed by him to replace *absolute time* as used in the geologic sense.

physical weathering *mechanical weathering*.

physicogeographical Pertaining to *physical geography*.

physiofacies A term suggested by Moore (1949, p.17) for "the total inorganic characteristics of a sedimentary rock", or that part of lithofacies not represented by biofacies. The term is essentially identical with *lithofacies* as that term has been interpreted by some. Moore (1957a, p.1784-1785) later wrote that "the concept of physiofacies ... may well be forgotten". Cf: *physiotope*.

physiographic cycle *cycle of erosion*.

physiographic diagram A small-scale map showing landforms by the systematic application of a standardized set of simplified pictorial symbols that represent the appearances such forms would have if viewed obliquely from the air at an angle of about 45°. The first major map of this kind was published by Lobeck (1921). Syn: *morphographic map; landform map*.

physiographic feature A prominent or conspicuous *physiographic form* or noticeable part thereof (Mitchell, 1948, p. 64). Cf: *topographic feature*.

physiographic form A *landform* considered with regard to its origin, cause, or history (Mitchell, 1948, p. 64). Cf: *topographic form*.

physiographic geology A branch of geology that deals with topography; *geomorphology*. The term was previously used as a syn. of *physiography*.

physiographic pictorial map *trachographic map*.

physiographic province A region of which all parts are similar in geologic structure and climate and which has consequently had a unified geomorphic history; a region whose pattern of relief features or landforms differs significantly from that of adjacent regions (see Fenneman, 1914). Examples: the Valley and Ridge, Blue Ridge, and Piedmont provinces in eastern U.S., and the Basin and Range, Rocky Mountains, and Great Plains provinces in western U.S. Cf: *geologic province; geographic province; morphologic region; structural province*.

physiography Originally, a description of the physical nature (form, substance, arrangement, changes) of objects, esp. of natural features; the term was introduced into geography in 1869 by Huxley for the study or description of "natural phenomena in general". The term later came to mean, esp. in the U.S., "a description of the surface features of the Earth, as bodies of air, water and land" (Powell, 1895), with an emphasis on mode of origin; i.e. it became synonymous with *physical geography*, and embraced geology, meteorology, and oceanography. Still later, esp. in the U.S., the term was restricted to a part of physical geography, namely the description and origin of landforms; in this sense, it is obsolescent and is replaced by *geomorphology*, although there is a general tendency to regard "physiography" as the descriptive, and "geomorphology" as the interpretative, study of landforms. See also: *physiographic geology*. Etymol: Greek *physis*, "nature", *graphein*, "to write".

physiotope A term defined by Moore (1949, p.17) as "designation of all purely physiochemical elements of an environment", but intended by him to represent the sedimentary environment of a *physiofacies*. Weller (1958, p.616) notes that if the term is to be accepted as having a meaning similar to "biotope" or "lithotope", it should be defined as an "area".

phytal zone The part of a lake bottom covered by water shallow enough to permit the growth of rooted plants. Cf: *aphytal zone*.

phytem A major Precambrian biostratigraphic unit defined largely on the basis of *stromatolites* (Keller et al., 1968, p. 190).

phyteral Vegetal matter in coal that is recognizable as morphologic forms, e.g. cuticle, spore coats, or wax, as distinguished from the *macerals*, or organic material forming the coal mass. See also: *muralite*.

phytoclast An organic particle of roughly the same size as a mineral clast of the containing rock. Phytoclasts make up 0.1-0.5% of most shale and are less abundant in sandstone; they also occur widely in metasediments (Bostick, 1970, p.74).

phytocoenosis The plant population of a particular habitat.

phytocollite A black gelatinous nitrogenous humic body occurring beneath or within peat deposits. Cf: *dopplerite*.

phytoecology The branch of ecology concerned with the relationships between plants and their environment. Cf: *zooecology*.

phytogenic dam A natural dam consisting of plants and plant remains. Such dams may account for ponds and lakes in tundra regions.

phytogenic dune Any dune in which the growth of vegetation influences the form of the dune, as by arresting the drifting of sand; e.g. a *foredune*.

phytogenic rock A *biogenic rock* produced by plants or directly attributable to the presence or activities of plants; e.g. algal deposits, peat, coal, some limestones, and lithified ooze containing diatoms. Cf: *phytolith*. Syn: *phytogenous rock*.

phytogenous rock *phytogenic rock*.

phytogeography The branch of *biogeography* dealing with the geographic distribution of plants. Cf: *zoogeography*.

phytokarst A type of solution landscape in which the major morphology is produced by the attack of boring algae and/or fungi on limestone. It is characterized by jagged, grotesque sculpture; it differs from ordinary solution karst by a random orientation of the sculpture with respect to gravity, and a black algal coat (Folk et al., 1973).

phytoleims Coalified remains of plants (Kristofovich, 1945, p.138). See also: *meroleims; hololeims*.

phytolith [paleont] A stony or mineral structure, generally microscopic, secreted by a living plant; often composed of calcium oxalate or opaline silica. Cf: *opal phytolith*.

phytolith [sed] A *biolith* formed by plant activity or composed of plant remains; specif. *phytogenic rock*.

phytopaleontology *paleobotany*.

phytophagous Said of an organism that feeds on plants.

phytoplankton The plant forms of *plankton*, e.g. diatoms. Cf: *zooplankton*.

phytozoan *zoophyte*.

Piacenzian *Plaisancian*.

pi axis *pi pole.*
picacho A term used in the SW U.S. for a large, sharply pointed, isolated hill or mountain; a peak. Etymol: Spanish.
pi circle On a *pi diagram,* a girdle of points representing poles to folded surfaces.
pick In the interpretation of seismograph records, the selection of an event; also, any selected event on a seismic record.
pickeringite A mineral: $MgAl_2(SO_4)_4 \cdot 22H_2O$. It occurs in white to faintly colored fibrous masses. Syn: *magnesia alum.*
pickup *geophone.*
picotite A dark-brown, chromium-bearing variety of hercynite (spinel). Much so-called picotite is ceylonite or magnesiochromite. Syn: *chrome spinel.*
picotpaulite A mineral: $TlFe_2S_3$.
picrite A dark-colored, generally hypabyssal rock containing abundant olivine along with pyroxene, biotite, possibly amphibole, and less than 10 percent plagioclase. The term was first used by Tschermak who applied it to a rock composed chiefly of olivine, titanaugite, and barkevikite, with or without biotite; later the term was used by Rosenbusch for a rock composed chiefly of olivine and augite, with or without hornblende and biotite (Streckeisen, 1967, p. 176). Not recommended usage.
picritic Said of an olivine-rich igneous rock.
picrochromite An end-member of the spinel group: $MgCr_2O_4$. It is produced synthetically. Syn: *magnesiochromite.*
picrocollite A hypothetical end-member of the palygorskite group: $MgSi_3O_5(OH)_4 \cdot 2H_2O$.
picrolite A term that has been applied to a fibrous or columnar variety of serpentine mineral. It is now regarded as a syn. of *antigorite.*
picromerite A white mineral: $K_2Mg(SO_4)_2 \cdot 6H_2O$. Syn: *schoenite.*
picropharmacolite A mineral: $H_2Ca_4Mg(AsO_4)_4 \cdot 12H_2O$.
pictogram *pictograph.*
pictograph A picture painted on a rock by primitive peoples and used as a sign. Cf: *petrogram.* Syn: *pictogram.*
picture element In a digitized image, the area on the ground represented by each digital value. Because the analog signal from the detector of a scanner may be sampled at any desired interval, the size of the picture element may be different from the *ground resolution cell* of the detector. Commonly abbreviated to *pixel.*
picurite *pitch coal.*
pi diagram In structural petrology, a fabric diagram in which the poles representing normals to fabric planes have been plotted. In an area of simple cylindrical folding, the pi diagram for bedding would be a great-circle girdle at right angles to the fold axis. Also written: *π diagram.* Cf: *beta diagram.*
piecemeal stoping *Magmatic stoping* in which relatively small blocks of roof rock are detached. Cf: *ring-fracture stoping.*
piedmont adj. Lying or formed at the base of a mountain or mountain range; e.g. a *piedmont* terrace or a *piedmont* pediment.—n. An area, plain, slope, glacier, or other feature at the base of a mountain; e.g. a foothill or a bajada. In the U.S., the Piedmont is a plateau extending from New Jersey to Alabama and lying east of the Appalachian Mountains. Etymol: from Piemonte, a region of NW Italy at the foot of the Alps.
piedmont alluvial plain *bajada.*
piedmont angle The sharp break of slope between a hill and a plain, such as the angle at the junction of a mountain front and the pediment at its base. Cf: *knick.*
piedmont bench (a) An upfaulted alluvial fan or pediment surface at the base of a mountain, bounded on its outer side by a scarplet (piedmont scarp) (Sharp, 1954, p. 23). Such features are displayed along the south sides of the San Gabriel and San Bernardino ranges in southern California. (b) *piedmont step.*
piedmont benchland One of several successions or systems of *piedmont steps.* Syn: *piedmonttreppe; piedmont stairway.*
piedmont bulb *expanded foot.*
piedmont flat *piedmont step.*
piedmont glacier A thick continuous sheet of ice at the base of a mountain range, resting on land, formed by the spreading out and coalescing of valley glaciers from the higher elevations of the mountains. Cf: *expanded-foot glacier.*
piedmont gravel Coarse gravel derived from high ground by mountain torrents and spread out on relatively flat ground where the velocity of the water is decreased.
piedmont interstream flat A term used by Tator (1949) for a planate rock surface along the east flank of the Colorado Front Range, and regarded by him (1953, p. 47) as a syn. of *pediment.*
piedmontite *piemontite.*
piedmont lake An oblong lake occupying a partly overdeepened basin excavated in rock by, or dammed by a moraine of, a piedmont glacier.
piedmont pediment A term used by Davis (1930, p. 154) for a pediment peripheral to, and along the base of, a mountainous area. Cf: *mountain pediment.*
piedmont plain *bajada.*
piedmont plateau A plateau lying between the mountains and the plains or the ocean; e.g. the plateau of Patagonia in southern Argentina and southern Chile, between the Andes and the Atlantic Ocean.
piedmont scarp A small, low cliff occurring in alluvium on a piedmont slope at the foot of, and essentially parallel to, a steep mountain range (as in the western U.S.), resulting from dislocation of the surface, esp. by faulting; term proposed by Gilbert (1928, p. 34). See also: *fan scarp.* Syn: *scarplet.*
piedmont slope (a) *bajada.* (b) A gentle slope at the base of a mountain in a semiarid or desert region, composed of a pediment (upper surface of eroded bedrock) and a bajada (lower surface of aggradational origin).
piedmont stairway *piedmont benchland.*
piedmont step An extensive or regional terrace- or bench-like feature sloping outward or down-valley (as in the Black Forest or Schwarzwald region of SW Germany), assumed by W. Penck (1924) to develop in response to a continually accelerated uplift of a rising or expanding dome. See also: *treppen concept; piedmont benchland.* Syn: *piedmont bench; piedmont flat.*
piedmonttreppe A syn. of *piedmont benchland.* Etymol: German *Piedmonttreppe,* "piedmont staircase".
piemontite A dark-red or reddish-brown manganese-bearing mineral of the epidote group: $Ca_2(Al,Mn^{+3},Fe)_3Si_3O_{12}(OH)$. Cf: *withamite.* Also spelled: *piedmontite.* Syn: *manganese epidote.*
pienaarite A sphene-rich *malignite* in which the feldspar is anorthoclase. The name, given by Brouwer in 1910, is for Pienaar Creek, Transvaal, South Africa. Not recommended usage.
pier [coast] (a) A long, narrow *wharf* extending out from the shore into the water, serving as a berthing or landing place for vessels or as a recreational facility. Cf: *jetty; quay.* Erroneous syn: *dock.* (b) A breakwater, groin, mole, or other structure used to protect a harbor or shore, and serving also as a promenade or as a landing place for vessels.
pier [eng] (a) An underground structural member that transmits a concentrated load to a stratum capable of supporting it without danger of failure or excessive settlement. Its ratio of base width to depth of foundation is usually less than 1:4. Cf: *footing.* (b) A rectangular or circular column, usually of concrete or masonry, designed to support heavy concentrated loads from arches or the superstructure of a bridge.
piercement dome *diapir.*
piercing fold *diapir.*
piercing point In a quaternary chemical system, the point at which a univariant curve (representing the compositions of liquids that can exist in equilibrium with three particular solid phases) and a ternary *join* intersect at some point other than a ternary univariant point.
pier dam An engineering structure, such as a groin, built from shore to deepen a channel, to divert logs, or to direct the water flow. Syn: *wing dam.*
pierre-perdue Blocks of stone or concrete heaped loosely in the water to make a foundation. Etymol: French, "lost stone".
pierrepontite Iron-rich variety of tourmaline.
pierrotite A mineral: $Tl_2(Sb,As)_{10}S_{17}$.
piestic interval *potential drop.*
piestic water A term proposed by Meinzer (1939) as a syn. of *confined ground water* and one of two classes of *plerotic water.* It includes *hyperpiestic water, hypopiestic water,* and *mesopiestic water.* This classification is not commonly used in the U.S.
pietricikite A variety of *ozocerite.* Originally incorrectly spelled *zietrisikite.* See also: *neft-gil.*
piezocrystallization Crystallization of a magma under pressure, such as pressure associated with orogeny.
piezoelectric crystal A crystal, e.g. of quartz or tourmaline, that displays the *piezoelectric effect.* Any nonconducting crystal lacking a center of symmetry may be piezoelectric.
piezoelectric effect In certain crystals, the development of an electric potential in certain crystallographic directions when mechanical strain is applied, or, the development of a mechanical

strain, hence vibration, when an electric potential is applied. Quartz and tourmaline are examples of naturally *piezoelectric crystals.*

piezogene Pertaining to the formation of minerals primarily under the influence of pressure (Kostov, 1961). Cf: *thermogene.*

piezoglypt *regmaglypt.*

piezomagnetism Stress dependence of magnetic properties. It is the inverse of *magnetostriction.*

piezometric contour *equipotential line.*

piezometric surface *potentiometric surface.*

pigeonite A mineral of the clinopyroxene group: $(Mg,Fe^{+2},Ca)(Mg,Fe^{+2})Si_2O_6$. It is intermediate in composition between clinoenstatite and diopside, and has little calcium, little or no aluminum or ferric iron, and less ferrous iron than magnesium. Pigeonite is characterized optically by a small and variable axial angle ($2V = 0\text{-}30°$). It is found in basic igneous rocks at Pigeon Point in Minnesota. Cf: *augite.*

pigeon's-blood ruby A gem variety of ruby of the finest color: intense, clear, dark red to slightly purplish red. It is found almost exclusively in upper Burma. Also spelled: *pigeon-blood ruby.*

pigment mineral A mineral having economic value as a coloring agent. The most important are hematite and limonite.

pike A term used in England for any summit or top of a mountain or hill, esp. one that is peaked or pointed. Also, a mountain or hill having a peaked summit.

pikeite An obsolete name for an augite-bearing phlogopite *peridotite.* Its name (Johannsen, 1938) is derived from Pike County, Arkansas. Cf: *kimberlite.*

pila Plural of *pilum.*

pilandite A hypabyssal *syenite* containing abundant anorthoclase phenocrysts in a groundmass of the same mineral. The name, given by Henderson in 1898, is for the Pilanesberg complex, Transvaal, South Africa. Not recommended usage.

pilar A term used in the SW U.S. for a large pillarlike or projecting rock. Etymol: Spanish, "pillar". Pron: pe-*lahr.*

pilate Said of spores and pollen having sculpture that is similar to that of *clavate* forms but that consists of smaller hairlike processes (pila) with more or less spherical knobs. Syn: *piliferous.*

pile A long, relatively slender structural foundation element, usually made of timber, steel, or reinforced or prestressed concrete, that is driven or jetted into the ground or cast in place in a borehole. Piles are used to support vertical or lateral loads, to form a wall to exclude water or soft material or to resist their pressure, to compact the surrounding ground, or rarely to restrain the structure from uplift forces. See also: *sheet pile.*

piliferous A less-preferred syn. of *pilate.*

piling A structure or group of piles.

pilite (a) Actinolite pseudomorphous after olivine. (b) *tinder ore.*

pill An English term for a pool, and for a small stream or creek.

pillar [geomorph] A natural formation shaped like a pillar; specif. an *earth pillar* and a *rock pillar.*

pillar [paleont] (a) A tiny rodlike structure, larger and straighter than a *trabecula,* connecting discrete layers of sclerite in holothurians. (b) An elongate peglike structure produced near the center or axis of coiling in certain foraminifera by thickening of the wall. The ends of the pillars appear as small bosses or nodes on the ventral side of the test. (c) In archaeocyathids, longitudinal cylindrical to laterally flattened rodlike skeletal elements, which may bridge upward and outward from the inner wall to the outer wall. (d) A rodlike internal structure, 0.02 to 0.25 mm in diameter, between adjacent laminae, which partly fills the gallery space in most stromatoporoid coenostea.

pillar [speleo] (a) Bedrock support remaining after removal of surrounding rock in solution. (b) A syn. of *column [speleo].*

pillar [struc geol] A joint block produced by columnar jointing.

pillar reef A *coral reef* consisting of large massive corals grown on one another to form tall sturdy pillars, separated by spacious caverns that tend to become filled by secondary internal encrustations and fine carbonate sediments. Cf: *thicket reef.*

pillar structure A columnar structure, normal or oblique to the bedding, consisting of massive or "swirled" sand that cuts across laminae or other primary structures in a sand bed. It is thought to be formed by the escape of water after deposition of the bed. The term was first published by Lowe and LoPiccolo (1974). Cf: *vertical sheet structure.*

pillow breccia A deposit of pillows and fragments of lava in a matrix of tuff.

pillow lava A general term for those lavas displaying *pillow structure* and considered to have formed in a subaqueous environment; such lava is usually basaltic or andesitic. Syn: *ellipsoidal lava.*

pillow structure [ign] A structure, observed in certain extrusive igneous rocks, that is characterized by discontinuous pillow-shaped masses ranging in size from a few centimeters to a meter or more in greatest dimension (commonly between 30 and 60 cm). The pillows are close-fitting, the concavities of one matching the convexities of another. The spaces between the pillows are few and are filled either with material of the same composition as the pillows, with clastic sediments, or with scoriaceous material. Grain sizes within the pillows tend to decrease toward the exterior. Pillow structures are considered to be the result of subaqueous extrusion, as evidenced by their association with sedimentary deposits, usually of deep-sea origin. See also: *pillow lava.*

pillow structure [sed] A primary sedimentary structure resembling the size and shape of a pillow; it is most characteristic of the basal parts of a sandstone overlying shale. See also: *ball-and-pillow structure.* Syn: *mammillary structure.*

pilotaxitic Said of the texture of the groundmass of a holocrystalline igneous rock in which lath-shaped microlites (typically plagioclase) are arranged in a glass-free mesostasis and are generally interwoven in irregular unoriented fashion. Cf: *trachytic.* Syn: *felty.*

pilot channel One of a series of *cutoffs* for converting a meandering stream into a straight channel of greater slope. It is built only large enough to start flow along the new course, since erosion during floods is expected to create channels of adequate capacity.

pilum One of the small, spinelike rods on the exine of pollen and spores, characterized by rounded or swollen knoblike ends. Pl: *pila.*

pimelite An apple-green clay mineral of the montmorillonite group: $(Ni,Mg)_3Si_4O_{10}(OH)_2 \cdot 4H_2O$. Syn: *desaulesite.*

pimple *pimple mound.*

pimple mound A term used along the Gulf Coast of eastern Texas and SW Louisiana for one of hundreds of thousands of low, flattened, rudely circular or elliptical domes composed of sandy loam that is coarser than, and distinct from, the surrounding soil; the basal diameter ranges from 3 m to more than 30 m, and the height from 30 cm to more than 2 m. Cf: *Mima mound.* Syn: *pimple.*

pimple plain A plain characterized by numerous conspicuous *pimple mounds.*

pin [geomorph] An Irish term for a mountain peak. Etymol: Gaelic *beann* or *beinn,* "peak". See also: *ben.*

pin [sed] (a) A thin, irregular bed, band, or seam of ironstone or other hard rock in the coal measures of south Wales. (b) A cylindrical nodule, usually of clay ironstone, in the coal measures of south Wales.

pin [surv] A sharp pointed metal rod with a ring at the end, used for marking taped measurements on the ground. Syn: *arrow.*

pinacocyte One of the cells, generally flat, of the pinacoderm of a sponge.

pinacoderm An unstratified layer of cells (pinacocytes), other than the choanaderm, constituting the soft parts of a sponge and delimitimg it from the external milieu; e.g. *endopinacoderm* and *exopinacoderm.*

pinacoid An open crystal form consisting of two parallel faces. Adj: *pinacoidal.*

pinacoidal class That crystal class in the triclinic system having only a center of symmetry.

pinacoidal cleavage Mineral cleavage parallel to one of the crystal's pinacoidal surfaces; e.g. the (010) cleavage of gypsum.

pinakiolite A black mineral: $(Mg,Mn^{+2})_2Mn^{+3}BO_5$. It is polymorphous with orthopinakiolite.

pinch n. (a) A marked thinning or squeezing of a rock layer; e.g. a coming-together of the walls of a vein, or of the roof and floor of a coal seam, so that the ore or coal is more or less completely displaced. See also: *nip [coal].* (b) A thin place in, or a narrow part of, an orebody; the part of a mineral zone that almost disappears before it widens out in another place to form an extensive orebody, or swell. Cf: *make.* —v. *pinch out.*

pinch-and-swell structure A structural condition commonly found in quartz veins and pegmatites in metamorphosed rocks, in which the vein is pinched and thinned at frequent intervals, leaving expanded parts between (Ramberg, 1955).

pinchite An orthorhombic mineral: $Hg_5O_4Cl_2$.

pinch out To taper or narrow progressively to extinction; to *thin out.* See also: *pinch-out.*

pinch-out The termination or end of a stratum, vein, or other body of rock that narrows or thins progressively in a given horizontal direction until it disappears and the rocks it once separated are in contact; esp. a *stratigraphic trap* formed by the thinning-out of a porous and permeable sandstone between two layers of impermeable shale. The lithologic character of the stratum is typically maintained to the feather edge. Cf: *shale-out.* See also: *wedge-out.*

pinfire opal Opal in which the patches of *play of color* are very small and close together and usually less regularly spaced than in *harlequin opal.*

pinger A battery-powered, low-energy source for an *echo sounder.*

pingo (a) A large *frost mound,* esp. a relatively large conical mound of soil-covered ice (commonly 30-50 m high and up to 400 m in diameter), raised in part by hydrostatic pressure of water within or below the permafrost of Arctic regions (esp. Canada), and of more than one year's duration; an intrapermafrost ice-cored hill or mound. Its crest is sometimes ruptured or collapsed due to melting of the ice, thus forming a star-shaped crater; the term has also been applied to such a depression (Monkhouse, 1965, p.237). The mound itself often resembles a small volcano. The term was introduced for this feature by Porsild (1938, p.46). Pl: pingos. See also: *ground-ice mound; ice laccolith.* Syn: *boolgoonyakh.* (b) The term has been used in several related senses, as a conical hill or mound, or as a hill completely covered by an ice sheet but revealing its presence by surface indications (ADTIC, 1955, p.61).—Etymol: Eskimo, "conical hill". Syn: *pingok.*

pingok *pingo.*

pingo remnant A rimmed depression, as in the northern Netherlands, where it was previously regarded as a kettle (Gravenor & Kupsch, 1959, p. 62). It is formed by rupturing of a pingo summit resulting in exposure of the ice core to melting followed by partial or total collapse. See also: *fossil pingo.* Syn: *pseudokettle.*

pinguite *nontronite.*

pinhole chert Chert containing weathered pebbles pierced by minute holes or pores.

pinite A compact, fine-grained, usually amorphous mica (chiefly muscovite) of a dull-gray, green, or brown color, derived from the alteration of other minerals (such as cordierite, nepheline, scapolite, spodumene, and feldspar).

pink snow *red snow.*

pinna A primary subdivision of a pinnately compound leaf or frond; a leaflet. See also: *pinnule [bot].* Pl: *pinnae.*

pinnacle [geomorph] (a) A tall slender tapering tower or spire-shaped pillar of rock, either isolated or at the summit of a mountain or hill; esp. a lofty peak. (b) A hill or mountain with a pointed summit.

pinnacle [reef] A small, isolated spire or column of rock or coral, either slightly submerged or awash; specif. a small reef patch, consisting of coral growing sharply upward (with slopes ranging from 45° to nearly vertical), usually within an atoll lagoon, often rising close to the water surface. Syn: *pinnacle reef; reef pinnacle; coral pinnacle.*

pinnacled iceberg An irregular iceberg, shaped and weathered in such a way as to be topped with spires and pinnacles.

pinnacle reef (a) A syn. of *pinnacle [reef].* (b) A term used in the Michigan Basin to apply to an isolated stromatoporoid-algal reef mound, now dolomitized, in the Middle Silurian rocks of the subsurface; many are productive of oil. They range up to 500 acres in area and 500 feet in relief, with slopes rarely exceeding 15 degrees. They are mounds rather than true pinnacles.

pinnate (a) Having leaflets or veins on each side of a common axis in a featherlike arrangement (Swartz, 1971, p. 358). (b) Referring to an erect growth habit of colonies of stenolaemate bryozoans, in which lateral branches grow in the same plane from opposite sides of a main axial branch.

pinnate drainage pattern A *dendritic drainage pattern* in which the main stream receives many closely spaced, subparallel tributaries that join it at acute angles, resembling in plan a feather; it is believed to indicate unusually steep slopes on which the tributaries developed.

pinnate jointing *feather jointing.*

pinnate venation A type of *net venation* in which the secondary veins branch from the midrib in parallel pattern. Cf: *palmate venation.*

pinnoite A yellowish tetragonal mineral: $MgB_2O_4 \cdot 3H_2O$.

pinnular A plate forming part of a pinnule of a crinoid.

pinnule [bot] A subdivision of a *pinna,* or a secondary subdivision of a pinnately compound leaf or frond; a secondary leaflet.

pinnule [paleont] (a) One of several generally slender, unbifurcated, uniserial branchlets of the food-gathering system of a crinoid arm, typically borne on alternate sides of successive brachial plates. (b) A secondary branch of a plumelike organ, such as a digitate lateral branch of a tentacle of an octocoral polyp or one of the biserial branches of the cystoid *Caryocrinites.*—Also spelled: *pinule.*

pinolite A metamorphic rock containing magnesite (breunnerite) as crystals and granular aggregates in a matrix of phyllite or talc schist. It is so named because the magnesite inclusions commonly resemble pine cones in shape (Holmes, 1928, p.184).

pintadoite A green mineral: $Ca_2V_2O_7 \cdot 9H_2O$.

pinule (a) Var. of *pinnule.* (b) *pinulus.*

pinulus A sponge spicule (usually a pentactine or hexactine) in which one ray (such as the unimpaired one in a pentactine) is enlarged and projects either internally or externally from the sponge and bears numerous small oblique spines giving the spicule the appearance of a pine tree. Pl: *pinuli.* Syn: *pinule.*

pinwheel garnet A porphyroblastic garnet with inclusions so arranged as to suggest rotation during formation.

pioneer In ecology, a community, species, flora, fauna, or individual that establishes itself in a barren area, initiating a new ecologic cycle or *sere.* Cf: *climax.*

piotine *saponite.*

pipe [grd wat] *geyser pipe.*

pipe [intrus rocks] A discordant pluton of tubular shape.

pipe [ore dep] A cylindrically shaped, more or less vertical orebody. The ore may be a vein deposit, a breccia column, or a diamond-bearing volcanic breccia. Syn: *ore pipe; ore chimney; chimney [ore dep]; neck [ore dep]; stock [ore dep].*

pipe [sed] (a) A tubular cavity from several centimeters to a few meters in depth, formed esp. in calcareous rocks, and often filled with sand and gravel; e.g. a vertical joint or sinkhole in chalk, enlarged by solution of the carbonate material and filled with clastic material. See also: *sand pipe.* (b) *clastic pipe.*

pipe [volc] A vertical conduit through the Earth's crust below a volcano, through which magmatic materials have passed. It is usually filled with volcanic breccia and fragments of older rock. As a zone of high permeability, it is commonly mineralized. Cf: *plug [volc]; chimney [volc]; vent; diatreme.* Syn: *breccia pipe.*

pipe amygdule An elongate amygdule that occurs in a lava, towards the base of the flow, probably formed by the generation of gases or vapor from the underlying material.

pipe clay (a) A white to grayish-white, highly plastic clay, practically free from iron, suitable for use in making tobacco pipes. The term has been extended to include any white-burinng clay of considerable plasticity. Syn: *ball clay; cutty clay.* (b) A mass of fine clay, generally of lenticular form, forming the surface of bedrock, on which the gravel of deep leads (old river beds) frequently rests.—Also spelled: *pipeclay.*

piperno A welded tuff characterized by *fiamme,* or flame structure. Such a rock is said to be pipernoid. Etymol: Italian.

pipernoid Said of the eutaxitic texture of certain extrusive igneous rocks, e.g. *piperno,* in which dark patches and stringers occur in a light-colored groundmass. Also, said of a rock exhibiting such texture.

pipe-rock A marine sandstone containing abundant *scolithus* structures.

pipe-rock burrow *scolithus.*

pipestone A pink or mottled argillaceous stone, carved by the Indians into tobacco pipes; esp. *catlinite.*

pipette analysis A kind of particle-size analysis of fine-grained sediment, made by removing samples from suspension with a pipette.

pipe vesicle Slender vertical cavities a few centimeters or tens of centimeters in length extending upward from the base of a lava flow. Most are formed by water vapor, derived from the underlying wet ground, that streamed upward into the lava. Cf: *spiracle.*

piping Erosion by percolating water in a layer of subsoil, resulting in caving and in the formation of narrow conduits, tunnels, or "pipes" through which soluble or granular soil material is removed; esp. the movement of material, from the permeable foundation of a dam or levee, by the flow or seepage of water along underground passages. See also: *water creep.* Syn: *tunnel erosion.*

pipkrake (a) A small, thin spike or needlelike crystal of ground ice, from 2.5-6 cm in length, formed just below, and growing perpendicular to, the surface of the soil in a region where the daily temperatures fluctuate across the freezing point. It is common in periglacial areas, where it contributes to the sorting of material in patterned ground and to downslope movement of surface material. (b) A bundle, cluster, or tuft of pipkrakes.—Etymol: Swedish, "needle ice". Syn: *needle ice; feather ice; mush frost; spew frost.*

pi pole In structural petrology, that pole which represents the normal to a fabric plane. Commonly written: π pole. There is confusion in the literature, and some workers have used the term as equivalent to *beta axis;* but the π pole for a particular plane is always at right angles to the β axis defined by the intersection of that plane with some other plane.

piracy *capture [streams].*

pirate (a) *capturing stream.* (b) *pirate valley.*

pirated stream *captured stream.*

pirate stream *capturing stream.*

pirate valley A valley that appropriates the waters of another valley; a valley containing a capturing stream. Syn: *pirate.*

pirssonite A white to colorless orthorhombic mineral: $Na_2Ca(CO_3)_2 \cdot 2H_2O$.

pisanite A blue mineral: $(Fe,Cu)SO_4 \cdot 7H_2O$. It is a variety of melanterite containing copper.

pisiform Resembling the size and shape of a pea; e.g. a "pisiform concretion" (or pisolith).

pisolite [mineral] A variety of calcite or aragonite.

pisolite [sed] (a) A sedimentary rock, usually a limestone, made up chiefly of pisoliths cemented together; a coarse-grained oolite. Syn: *peastone; pea grit.* (b) A term often used for a *pisolith,* or one of the spherical particles of a pisolite.—Etymol: Greek *pisos,* "pea". Cf: *oolite.*

pisolite [volc] An individual unit in a mass of *accretionary lapilli.*

pisolith One of the small, round or ellipsoidal accretionary bodies in a sedimentary rock, resembling a pea in size and shape, and constituting one of the grains that make up a *pisolite.* It is often formed of calcium carbonate, and some are thought to have been produced by a biochemical algal-encrustation process. A pisolith is larger and less regular in form than an *oolith,* although it has the same concentric and radial internal structure. The term is sometimes used to refer to the rock made up of pisoliths. Syn: *pisolite [sed].*

pisolitic [ign] Said of a tuff composed of accretionary lapilli.

pisolitic [sed] Pertaining to *pisolite [sed],* or to the texture of a rock made up of pisoliths or pealike grains; e.g. "pisolitic bauxite" or "pisolitic limestone".

pisosparite A limestone containing at least 25% pisoliths and no more than 25% intraclasts and in which the sparry-calcite cement is more abundant than the carbonate-mud matrix (micrite) (Folk, 1959, p. 22). Cf: *oosparite.*

pistacite A syn. of *epidote,* esp. the pistachio-green variety rich in ferric iron. Also spelled: *pistazite.*

pistil In a flower, the female reproductive organ consisting of an ovary, style (when present), and stigma. It may consist of a single *carpel* (simple pistil) or of two or more carpels (compound pistil).

pistillate Said of a flower which has a pistil but no stamens. Cf: *staminate.*

piston corer An oceanographic *corer* containing a piston inside the cylinder which reduces friction by creating suction. There are several varieties, including the *Ewing corer,* the *Mackereth sampler,* and the *Kullenberg corer.* Cf: *gravity corer.*

piston organelle A moundlike structure rising from the floor of peristome in a tintinnid.

pit [bot] A thin place in a cell wall. See also: *simple pit; bordered pit.*

pit [geol] A small indentation or depression left on the surface of a rock or particle (esp. of a clastic particle) as a result of some eroding or corrosive process, such as etching, differential solution, or impact.

pit [geomorph] A local minimum of elevation, as at the lowest point of a closed depression (Warntz, 1975, p. 210).

pit [speleo] *vertical cave.*

pit-and-mound structure A sedimentary structure consisting of a small blister-like mound, 1 mm high and 3-12 mm in diameter, that surrounds or contains at its summit a tiny crater-like central pit up to 1 mm in diameter simulating a rain print. It is formed during rapid settling of low-viscosity mud, by gas bubbles or water currents emerging at the surface. The term was introduced by Kindle (1916).

pitch [coast] An obsolete term for the tip of a piece of land, such as a cape, extending into a body of water.

pitch [petroleum] *asphalt.*

pitch [slopes] A steep place; a declivity.

pitch [speleo] A vertical shaft in a cave.

pitch [struc geol] The angle between the horizontal and any linear feature, e.g. an ore shoot or lineation, measured in the plane containing the linear feature (Phillips, 1954, p. 10). Syn: *rake.*

pit chamber The space between the pit chamber and the overarching border of a *bordered pit* (Swartz, 1971, p. 359).

pitchblende A massive brown to black variety of uraninite, found in hydrothermal sulfide-bearing veins. It is fine-grained (colloform), amorphous, or microcrystalline, and has a distinctive pitchy to dull luster. Pitchblende contains a slight amount of radium; thorium and the rare earths are generally absent. Syn: *pitch ore; nasturan.*

pitch coal (a) A brittle, lustrous bituminous coal or lignite, with conchoidal fracture. Syn: *bituminous lignite; bituminous brown coal; glance coal; picurite; specular coal.* (b) A kind of *jet [coal].*

pitches and flats *flats and pitches.*

pitching fold A fold of which the hinge line is inclined to the horizontal. See *pitch.*

pitch opal A yellow to brown inferior quality of common opal displaying a pitchy luster.

pitch ore (a) *pitchblende.* (b) *pitchy copper ore.*

pitch peat Peat that resembles asphalt.

pitchstone A volcanic glass with a waxy dull resinous luster. Its color and composition vary widely; it contains a higher percentage of water than *obsidian.* Crystallites are detectable in thin section. Syn: *fluolite.*

pitchy copper ore A dark, pitchlike oxide of copper; a mixture of chrysocolla and limonite. Syn: *pitch ore.*

pitchy iron ore (a) *pitticite.* (b) *triplite.*

pit crater A *sink [volc]* or a small *caldera.*

pith The spongy central tissue of an exogenous stem, consisting chiefly of parenchyma (Jackson, 1928, p. 287).

pith rays Primary bands of parenchyma cells extending from the pith to the pericycle in herbaceous and young woody stems (Fuller & Tippo, 1954, p. 967).

pit lake A *kettle lake* in a pitted outwash plain.

piton A term commonly used for volcanic peaks, especially steep-sided domes, in the West Indies and other French-speaking regions.

pit run *bank gravel.*

pitted outwash Outwash with pits or kettles, produced by the partial or complete burial of glacial ice by outwash and the subsequent thaw of the ice and collapse of the surficial materials. Cf: *pitted outwash plain.*

pitted outwash plain An *outwash plain* marked by many irregular depressions such as kettles, shallow pits, and potholes; many are found in Wisconsin and Minnesota. See also: *pitted outwash; kettle plain.* Syn: *pitted plain.*

pitted pebble A pebble having marked concavities not related to the texture of the rock in which it appears or to differential weathering (Kuenen, 1943). The depressions range in size from minute pits caused by sand particles to cups a few centimeters across and a centimeter deep; they are common at the contacts between adjacent pebbles, and have been explained as the result of pressure-induced solution at points of contact. The term has also been applied to cobbles. Cf: *cupped pebble.* Syn: *scarred pebble.*

pitted plain *pitted outwash plain.*

pitticite A brown to yellowish or reddish mineral found in reniform masses, consisting of a hydrous arsenate and sulfate of iron, and having a highly variable composition. Syn: *pittizite; pitchy iron ore.*

pivotability A measure of roundness of sedimentary particles, expressed by the ease with which a particle can be dislodged from a surface or by the tendency of a particle to start rolling on a slope. The term was introduced by Shepard & Young (1961, p.198) who assigned the highest values of roundness to the particles that "could be most easily pivoted".

pivotal fault A partial syn. of *hinge fault.* Cf: *scissor fault.* See also: *trochoidal fault.*

pixel *picture element.*

placanticline A gentle, anticlinelike uplift of the continental platform, usually asymmetric and without a typical outline. There is no corresponding synclinelike structure. The term is used mainly in the Russian literature of the Volga-Urals region (Shatsky, 1945). A corresponding term in western literature is *plains-type fold.*

placenta The part of the ovary to which the ovules are attached, usually the margin of the carpels (Swartz, 1971, p. 360).

placental A member of the mammalian subclass Eutheria, characterized by bearing young in a relatively advanced state of development by means of a placenta. Range, Cretaceous to present.

placer A surficial mineral deposit formed by mechanical concentration of mineral particles from weathered debris. The common types are *beach placers* and *alluvial* placers. The mineral concentrated is usually a heavy mineral such as gold, cassiterite, or rutile. Cf: *lode [eco geol].* Syn: *lead [eco geol]; ore of sedimentation.* Pron: plasser.

placer claim A *claim* on a placer deposit in which a discovery has been made. Cf: *lode claim.*

placer mining The extraction and concentration of heavy metals or minerals from placer deposits by various methods, generally using running water. Cf: *hydraulic mining; drift mining.*

place value The value that a mineral deposit has by virtue of its location.

placic horizon A diagnostic subsurface soil horizon that is black to dark red, thin, usually cemented with iron, and not very permeable (USDA, 1975).

placochela A sponge chela in which both shaft and recurved ends are broadly expanded and flattened.

Placodermi A class of jawed vertebrates characterized by development of dermal armor, very heavy in primitive forms, in which it forms elaborately jointed head and trunk shields; ossification of the internal skeleton is poor. Range, Lower to Upper Devonian.

placolith A perforate coccolith having two shields connected by a central tube. See also: *tremalith.* Syn: *cyatholith.*

placon A layer of iron oxide, 1 to 30 mm thick, that forms 20 to 150 cm below the surface of a *placosol* (Tilsley, 1977, p. 21). It is essentially impervious to water and divides the soil into two chemically dissimilar environments. Conditions above the placon are acid, and base metals are leached; conditions below are reducing, and metals are relatively stable.

placosol An accumulation of humified organic material above a leached near-surface mineral-soil horizon underlain by an impermeable layer of iron oxide (Tilsley, 1977, p. 21). Placosols develop in regions of cool moist climate. The iron-oxide layer, or *placon,* is their most important feature.

pladdy A term used in Northern Ireland for a "residual island drumlin awash at high tide" (Stamp, 1961, p. 365).

pladorit A hornblende-bearing *granitite* or a magnesium- and mica-bearing hornblende *granite* (Johannsen, 1939, p. 275). Obsolete.

plage A French term for a sandy beach, esp. at a seaside resort. The term is being supplanted by *lido.*

plaggen epipedon A diagnostic surface soil horizon that is man-made by manuring and mixing, and that is more than 50cm in thickness (USDA, 1975).

Plaggept In U.S. Dept. of Agriculture soil taxonomy, a suborder of the soil order *Inceptisol,* characterized by the presence of a *plaggen epipedon.* Plaggepts are known to occur only in the British Isles and northwestern Europe (USDA, 1975). Cf: *Andept; Aquept; Ochrept; Tropept; Umbrept.*

plagiaplite An *aplite* composed chiefly of plagioclase (oligoclase to andesine), possibly green hornblende, and accessory quartz, biotite, and muscovite; a fine-grained *diorite.* Not recommended usage.

plagioclase (a) A group of triclinic feldspars of general formula: $(Na,Ca)Al(Si,Al)Si_2O_8$. At high temperatures it forms a complete solid-solution series from Ab ($NaAlSi_3O_8$) to An ($CaAl_2Si_2O_8$). The plagioclase series is arbitrarily subdivided and named according to increasing mole fraction of the An component: albite (An 0-10), oligoclase (An 10-30), andesine (An 30-50), labradorite (An 50-70), bytownite (An 70-90), and anorthite (An 90-100). The Al/Si ratio ranges with increasing An content from 1:3 to 1:1. Plagioclase minerals are among the commonest rock-forming minerals, have characteristic twinning, and commonly display zoning. (b) A mineral of the plagioclase group; e.g. albite, anorthite, peristerite, and aventurine feldspar.—The term was introduced by Breithaupt (1847, p. 490) who applied it to all feldspars having an oblique angle between the two main cleavages. Cf: *alkali feldspar; orthoclase.* Syn: *sodium-calcium feldspar.*

plagioclase arenite A term used by McBride (1963, p.668) for an arkose containing more than 25% plagioclase, and by Folk (1968, p.124) for an arkose in which plagioclase is the main feldspar.

plagioclase arkose An arkose of which the chief feldspar is plagioclase (Folk, 1968, p. 130).

plagioclase rock *anorthosite.*

plagioclasite *anorthosite.*

plagiogranite A term commonly used by Russian petrologists for igneous rock having a low potassium content. It includes rocks ranging in composition from *quartz diorite* to *trondhjemite.* The original plagiogranite of Kruschov, described in 1931, had an average modal composition of 56% plagioclase, 27% quartz, 12% biotite, 5% amphibole. Not recommended usage.

plagionite A blackish lead-gray mineral: $Pb_5Sb_8S_{17}$.

plagiophyre A porphyritic igneous rock similar to an *orthophyre* but containing plagioclase rather than orthoclase phenocrysts. The term was introduced by Tyrrell in 1912. Not recommended usage.

plagiostome An asymmetrically placed aperture or pseudostome in a thecamoebian test (as in *Centropyxis* or *Plagiopyxis*).

plain (a) Any flat area, large or small, at a low elevation; specif. an extensive region of comparatively smooth and level or gently undulating land, having few or no prominent surface irregularities but sometimes having a considerable slope, and usually at a low elevation with reference to surrounding areas. A plain may be either forested or bare of trees, and may be formed by deposition or by erosion. (b) An extensive tract of level or rolling, almost treeless country with a shrubby vegetation; a *prairie.* In Australia, "plain" implies treelessness. The term is usually used in the plural. (c) A region underlain by horizontal strata or characterized by horizontal structure, which may be dissected into hills and valleys by stream erosion; Davis (1885) introduced this concept of a "plain" but the term should be used without regard to the underlying geologic structure.—Cf: *plateau [geomorph].*

plain of denudation A surface that has been reduced to or nearly to sea level by the agents of erosion (usually considered to be of subaerial origin); it is relatively flat but may be marked by residual hills of resistant rock rising somewhat above the general level. See also: *plain of marine denudation.*

plain of lateral planation An extensive smooth apronlike surface developed at the base of a mountain or escarpment (as at Book Cliffs, Utah) by the widening of valleys and the coalescence of flood plains as a result of *lateral planation.* It resembles the landscape form of a *pediment* and is often so called. Cf: *panplain.*

plain of marine denudation A plane or nearly plane surface worn down by the gradual encroachment of ocean waves upon the land; or an imaginary plane representing such a surface after uplift and partial subaerial erosion. The concept was first recognized by Ramsay (1846); however, the term has often been applied to a plain produced by subaerial or nonmarine agents, which was subsequently submerged beneath the sea, only a minor role being assigned to wave erosion. Syn: *plain of submarine denudation.*

plain of marine erosion A largely theoretical platform, representing a plane surface of unlimited width, produced below sea level by the cutting-away of the land by marine processes acting over a very long period of stillstand; the ultimate abrasion platform. Cf: *plain of marine denudation; marine peneplain.* Syn: *marine plain; marine plane; sea plain; submarine plain.*

plain of submarine denudation *plain of marine denudation.*

plains-type fold An anticlinal or domelike structure of the continental platform that has no typical outline and for which there is no corresponding synclinal structure. It is associated with the vertical uplift of normal faulting. The corresponding term used in the Russian literature is *placanticline.* See also: *drape fold.*

plain tract The lower part of a stream, characterized by a low gradient and a wide flood plain. Cf: *mountain tract; valley tract.* See also: *deltaic tract.*

Plaisancian European stage: Lower Pliocene (above Zanclean, below Astian). Also spelled: *Plaisanzian; Piacenzian.*

plaiting A texture seen in some schists that results from the intersection of relict bedding planes with well developed cleavage planes. Syn: *gaufrage.*

Plait point A point representing conditions at which two conjugate phases become identical; a critical point.

plan A drawing, sketch, or diagram of any object or structure, made by horizontal projection upon a plane or flat surface; esp. a very large-scale and considerably detailed *map* of a small area, such as one showing underground mine workings.

planaas A Danish term for an outwash plain formed as a flat-topped delta in standing water between two walls of stagnant ice. Pron: pla-*nouse.*

planar Lying or arranged as a plane or in planes, usually implying more or less parallelism, as in bedding or cleavage. It is a two-dimensional arrangement, in contrast to the one-dimensional *lin-*

ear arrangement.

planar cross-bedding (a) Cross-bedding in which the lower bounding surfaces are planar surfaces of erosion (McKee & Weir, 1953, p.385); it results from beveling and subsequent deposition. (b) Cross-bedding characterized by planar foreset beds.

planarea One of two flattened areas developed on either side of the posterior part of a brachiopod shell in place of the more common single median *interarea.* Cf: *palintrope.*

planar element A fabric element having two dimensions that are much greater than the third. Examples are bedding, cleavage, and schistosity. Cf: *linear element; equant element.*

planar features Closely spaced parallel microscopic planes, distinct from cleavage planes, that occur in shock-metamorphosed minerals (particularly quartz and feldspar) and are regarded as unique and important indicators of shock metamorphism. The structures are characteristically multiple (often more than five distinct sets per grain) and are oriented parallel to specific planes in the host-crystal lattice. They have been produced experimentally by shock pressures of 80 to 250 kb. Syn: *shock lamellae.*

planar flow structure *platy flow structure.*

planate adj. Said of a surface that has been flattened or leveled by *planation;* e.g. a pediment is a nearly planate erosion surface.—v. To reduce to a plain or other flat surface.

planation (a) The process or processes of erosion whereby the surface of the Earth or any part of it is reduced to a fundamentally even, flat, or level surface; specif. *lateral planation* by a meandering stream. The term also includes erosion by waves and currents, and abrasion by glaciers or wind, in producing a flat surface. The term was originated by Gilbert (1877, p. 126-127) who considered alluviation of the flattenend surface as part of the planation process; however, this condition is not necessary. (b) A broad term for the general lowering of the land; e.g. *peneplanation; panplanation; pediplanation; cryoplanation; altiplanation.*

planation stream piracy Capture effected by the lateral planation of a stream invading and diverting the upper part of a smaller stream.

planation surface *erosion surface.*

planchéite A blue mineral: $Cu_8Si_8O_{22}(OH)_4 \cdot H_2O$. Cf: *shattuckite.*

planchette *planetable board.*

plane A two-dimensional form that is without curvature; ideally, a perfectly flat or smooth surface. In geology the term is applied to such features as a *bedding plane* or a *planation surface* . Adj: *planar.* Cf: *surface.*

plane bed A sedimentary bed "without elevations or depressions larger than the maximum size of the bed material" (Simons et al., 1961, p. vii). It is characteristic of the lower part of the upper *flow regime.*

plane coordinates (a) Two coordinates that represent the perpendicular distances of a point from a pair of axes that intersect at right angles, reckoned in the plane of those axes. (b) A coordinate system in a horizontal plane, used to describe the positions of points with respect to an arbitrary origin by means of plane coordinates. It is used in areas of such limited extent that the errors introduced by substituting a plane for the curved surface of the Earth are within the required limits of accuracy.—Syn: *rectangular coordinates; plane-rectangular coordinates.*

plane correction A correction applied to observed surveying data to reduce them to a common reference plane.

plane defect A type of *crystal defect* that occurs along the boundary plane (*lineage boundary*) between two regions of a crystal, or along the *grain boundary* between two crystals. Cf: *line defect; point defect.* See also: *stacking fault.*

plane fault A fault with a surface that is planar rather than curved. Cf: *arcuate fault.*

plane group One of 17 kinds of two-dimensional patterns that can be produced by one asymmetric motif that is repeated by symmetry operations to produce a unit of pattern, which then is repeated by translation to build up an ordered pattern that fills any two-dimensional area. Cf: *space group.* Syn: *plane symmetry group.*

plane jet A flow pattern characteristic of *hyperpycnal inflow,* in which the inflowing water spreads as a parabola whose width is about three times the square root of the distance downstream from the mouth (Moore, 1966, p. 87). Cf: *axial jet.*

plane of composition A term used by Cullison (1938, p.983) for the plane of contact between the part of an internal mold composed of material similar to the matrix in which the fossil shell was embedded and the material (such as secondary calcite) that partially filled the hollow interior of the shell. It defines the horizontal plane at the time the filling became hardened. Not to be confused with *composition plane.*

plane of contemporaneity A term used by Caster (1934, p.19 & 24) for the horizontal or nearly horizontal surface between stratigraphic units (primarily formations) as seen in section; e.g. the surface separating parvafacies belonging to the same magnafacies. Cf: *facies plane.*

plane of incidence A plane that contains an incident ray and the normal to the surface at the point of incidence.

plane of maximum shear stress Either of two planes that lie on opposite sides of the maximum principal stress axis at angles of 45° to it and that are parallel to the intermediate principal stress axis.

plane of mirror symmetry A symmetry element in a crystal that is a plane dividing the crystal into halves, one of which is the mirror image of the other. Syn: *mirror plane of symmetry; plane of symmetry [cryst]; symmetry plane; reflection plane.*

plane of polarization *vibration plane.*

plane of saturation *water table.*

plane of stratification *bedding plane.*

plane of symmetry [cryst] *plane of mirror symmetry.*

plane of symmetry [paleont] The plane that bisects a shell symmetrically.

plane of vibration *vibration plane.*

plane-polarized Said of a moving wave, e.g. of light, that has been polarized so that it vibrates in a single plane.

planerite A variety of coeruleolactite containing copper, or a variety of turquoise containing calcium.

plane strain A state of strain in which all displacements that arise from deformation are parallel to one plane, and the longitudinal strain is zero in one principal direction.

plane stress A state of stress in which one of the principal stresses is zero.

plane surveying Ordinary field and topographic surveying in which earth curvature is disregarded and all measurements are made or reduced parallel to a plane representing the surface of the Earth. The accuracy and precision of results obtained by plane surveying will decrease as the area surveyed increases in size. Cf: *geodetic surveying.*

plane symmetry group *plane group.*

planet (a) One of the nine celestial bodies of the solar system that revolve around the Sun in elliptical orbits and in the same direction. A planet shines only by reflected light. (b) A similar body in another solar system.

planetable A simple surveying instrument for graphically plotting the lines of a survey directly from field observations. It consists essentially of a small drawing board mounted on a tripod and fitted with a compass and a straightedge ruler (alidade) that is pointed at the object observed usually by means of a telescope or other sighting device. Also spelled: *plane table.*

planetable board The drafting board of a planetable instrument. Means are provided on the upper side for attachment of drawing paper, and a plate on the lower side is threaded for attachment to the tripod head. Syn: *planchette.*

planetable map A map made by planetable surveying methods. It includes maps made by complete field mapping on a base projection and by field contouring on a planimetric base map.

planetabling Plotting with, or making use of, a planetable.

planetary (a) Pertaining to the planets of the solar system. Cf: *terrestrial.* (b) Pertaining to the Earth as a whole.

planetary geology A science that applies geologic principles and techniques to the study of planets and their natural satellites. The term is frequently used as a syn. of *astrogeology.* Syn: *planetary geoscience.*

planetary wave A major, prominent atmospheric wave characterized by a long wavelength, a significant amplitude, and a velocity directed always to the west. Also, a similar free progressive wave in the model ocean, not yet convincingly demonstrated to exist there, but believed to be about 1600 km (1000 mi) long and to be caused by the gravitational attraction of the Sun and Moon on the Earth but largely governed by the depth of the water and by the Earth's rotational effects.

planetary wind Any major part of the worldwide system of large-scale wind circulation, e.g. the *trade winds.*

planetesimal hypothesis *dust-cloud hypothesis.*

planetography The descriptive science of the physical features of planets.

planetoid *asteroid.*

planetology A term originally applied to the study and interpretation of surface markings of planets and their natural satellites, and later to the study of the condensed matter of the solar system, including planets, satellites, asteroids, meteorites, and interplanetary material. The term is frequently used as a syn. of *astrogeology,* and was redefined by Rankama (1962, p. 519) as "the universal science that studies the configuration and movements of matter and the accompanying energy transformations in planets, their natural satellites, and other cosmic bodies of a similar nature" in our solar system and other possible planetary systems.

planèze An erosional relief form consisting of a lava flow protecting the underlying volcanic cone. It may be a wedge-shaped unit on the slope of an erosionally dissected volcano, or a lava-capped plateau. Etymol: French, "lava plateau". Also spelled: *planeze.*

planimeter A mechanical instrument for measuring the area of any plane figure by means of a pointer or moving arm that traces its boundary or perimeter. It is used esp. for measuring irregular areas on a chart or map.

planimetric map A map that presents only the relative horizontal positions of natural or cultural features, by lines and symbols. It is distinguished from a *topographic map* by the omission of relief in measurable form. Syn: *line map.*

planimetry (a) The measurement of plane surfaces; e.g. the determination of horizontal distances, angles, and areas on a map. (b) The plan details of a map; the natural and cultural features of a region (excluding relief) as shown on a map.

planispiral adj. Having the shell coiled in a single plane; esp. said of gastropod and cephalopod shells formed by a spiral coiled in a single plane and ideally symmetric in that plane, and said of a coiled foraminiferal test with whorls of the coil in a single plane. Also spelled: *planospiral.*—n. A planispiral shell or test.

planitia A term established by the International Astronomical Union for a plain lower than the surrounding terrain on Mars, generally of regional extent. Some are probably areas of lava flows. Generally used as part of a formal name for a Martian landform, such as Acidalium Planitia (Mutch et al., 1976, p. 57).

plankter An individual planktonic organism.

planktivorous Said of an organism that feeds on plankton.

plankton Aquatic organisms that drift, or swim weakly. See also: *phytoplankton; zooplankton.* Adj: *planktonic.*

plankton bloom *water bloom.*

plankton equivalent A quantitative chemical relationship between one aspect of plankton and another, e.g. in phytoplankton, 1 mg carbon = 2.3 mg dry organic matter.

planktonic Said of that type of *pelagic* organism which floats; adj. of *plankton.* Cf: *nektonic.*

plankton snow *sea snow.*

planoconformity A term used by Crosby (1912, p.297) for the relation between conformable strata that are approximately uniform in thickness and sensibly parallel throughout.

plano-convex Flat on one side and convex on the other; e.g. said of a brachiopod shell having a flat brachial valve and a convex pedicle valve. Cf: *convexo-plane.*

planophyre A porphyritic rock characterized by *planophyric* texture. Obsolete.

planophyric A term, now obsolete, applied by Cross et al. (1906, p.703) to the texture of a porphyritic igneous rock in which the phenocrysts are arranged in layers or laminae in the groundmass; also, characteristic of or pertaining to a *planophyre.*

planorasion The process by which wind, working in conjunction with other erosional agents in a desert, "acts as an abrading and eroding agent" that works uphill (Hobbs, 1917, p. 48); the process may grade a slope as steep as 4 degrees. See also: *antigravitational gradation.*

Planosol A great soil group in the 1938 classification system, an intrazonal, hydromorphic group of soils having a leached surface layer above a definite clay pan or hardpan. These soils develop on nearly flat upland surfaces under grass or trees in a humid to subhumid climate (USDA, 1938). Most are now classified as *Aqualfs* or *Albolls.*

planospiral *planispiral.*

plan-position indicator A radar-display device that consists of an oscilloscope (radarscope) equipped with a cathode-ray tube and a rotating antenna for scanning horizontally all or part of a complete circle, and that presents visually or graphically in plan (map) position the range and direction (azimuth or bearing) of an object (such as a ship, building, cliff, or mountain), from which echoes are reflected in the form of spots of light whose brightness corresponds to the strength of the target signal detected by radar. The position of the radar itself is displayed in the center of the indicator. Abbrev: PPI.

plant [bot] Any member of the vegetable group (plant kingdom) of living organisms.

plant [seis] n. (a) The manner in which a geophone is placed on or in the Earth. Syn: *seismometer plant.* (b) The coupling to the ground.—v. To place a geophone in its proper place on the ground.

plant ash *inherent ash.*

plant indicator *indicator plant.*

plant opal *opal phytolith.*

planula The very young free-swimming larva of a coelenterate (such as of a coral polyp), consisting of an outer layer of ciliated ectoderm cells and an internal mass of endoderm cells. Pl: *planulae.* Syn: *parenchymula.*

planulate Said of a moderately evolute and compressed cephalopod shell with an open umbilicus and a bluntly rounded venter.

planum A term established by the International Astronomical Union for a relatively smooth, cratered plateau on Mars. Generally used as part of a formal name for a Martian landform or region, such as Lunae Planum (Mutch et al., 1976, p. 57). Etymol: Latin *planum,* level.

plash A shallow short-lived pool or small pond resulting from a flood, heavy rain, or melting snow; a puddle.

plasma [mineral] A semitranslucent green variety of chalcedony, sometimes having white or yellowish spots. The green color is attributed to chlorite. Cf: *bloodstone.*

plasma [soil] The part of a soil material that is capable of being, or has been, moved, reorganized, and/or concentrated by soil-forming processes (Brewer & Sleeman, 1960); e.g. all mineral or organic material of colloidal size, and the relatively soluble material that is not bound up in *skeleton grains.* See also: *lac [soil]; floc.*

plaster conglomerate A conglomerate composed entirely of boulders derived from a partially exhumed monadnock and forming a wedgelike mass on its flank.

plastering-on The addition of material to a ground moraine by the melting of ice at the base of a glacier (Gravenor & Kupsch, 1959, p. 60). Cf: *lodgment.*

plaster stone *gypsum.*

plastic [biol] Having the capability of variation and phylogenetic change.

plastic [struc geol] Said of a body in which strain produces continuous, permanent deformation without rupture. Cf: *elastic.*

plastic deformation (a) Permanent deformation of the shape or volume of a substance, without rupture. It is mainly accommodated by *crystal gliding* and/or *dynamic recrystallization.* (b) Deformation by one or both of two grain-scale mechanisms: slip, and twinning. This is a metallurgical definition, increasingly used by geologists. Sometimes called "crystal plasticity." (c) Rheological term for deformation characterized by a *yield stress,* which must be exceeded before flow begins.

plastic equilibrium State of stress within a soil mass or a portion thereof that has been deformed to such an extent that its ultimate shearing resistance is mobilized (ASCE, 1958, term 263). Plastic equilibrium is "active" if it is obtained by lateral expansion of a soil mass, and "passive" if obtained by lateral compression of a soil mass.

plasticity index The water-content range of a soil at which it is plastic, defined numerically as the liquid limit minus the plastic limit.

plasticlast An intraclast consisting of calcareous mud that has been torn up while still soft (Folk, 1962). Cf: *protointraclast.*

plastic limit The water-content boundary of a sediment, e.g. a soil, between the plastic and semisolid states. It is one of the *Atterberg limits.* Cf: *liquid limit.*

plastic relief map A topographic map printed on plastic and then molded by heat and pressure into a three-dimensional form to emphasize the relief.

plastic shading Archaic syn. of *hill shading* (BNCG, 1966, p.31).

plastic strain *plastic deformation.*

plastic zone A region adjacent to the *rupture zone* of an explosion crater and at an increased distance from the shock site, differing from the rupture zone by having less fracturing and only small permanent deformations.

plastotype An artificial specimen molded or cast directly from a type specimen.

plastron (a) The more or less inflated and enlarged adoral segment of the posterior interambulacral area of certain echinoids. (b) A firmly sutured, rigid belly-shield, formed from flattened and expanded girdle bones (in plesiosaurs) or from dermal scales combined with gastralia and girdle elements (in turtles).—Adj: *plastral.*

plat [cart] (a) A diagram drawn to scale, showing boundaries and subdivisions of a tract of land as determined by survey, together with all essential data required for accurate identification and description of the various units shown and including one or more certificates indicating due approval. It differs from a map in that it does not necessarily show additional cultural, drainage, and relief features. (b) A precise and detailed plan or map representing a township, private land claim, mineral claim, or other surveyed area, and showing the actual or proposed divisions, special features, or uses of the land.—See also: *plot.*

plat [geog] An obsolete term for a plateau, tableland, or other expanse of open level land.

plate [geol] (a) A thin flat fragment of rock, such as a slab or flagstone. (b) A torsionally rigid thin segment of the Earth's lithosphere, which may be assumed to move horizontally and adjoins other plates along zones of seismic activity (Dennis & Atwater, 1974, p. 1032). See also: *plate tectonics.* (c) A snow crystal in the form of a flat hexagonal plate.

plate [paleont] (a) Any discrete, normally flat or tabular *ossicle* in the skeleton of an echinoderm, composed of a single crystal of calcium carbonate. The term is sometimes used only for external plates, "but all calcareous bodies formed serve as framework of support for soft parts and constitute plates" (TIP, 1967, pt.S, p.113). (b) A structure consisting of inner and outer platforms and adjoining a portion of the axis of a platelike conodont. The term is used incorrectly when referring to a *platform.* (c) A lamina that forms part of an animal body, such as a valve of a mollusk or crustacean, or a flat octocorallian sclerite too thick to be called a scale. (d) A flat, calcified, generally rhomboidal structure, forming the most distal portion of a receptaculitid merome.

plateau [geomorph] (a) Broadly, any comparatively flat area of great extent and elevation; specif. an extensive land region considerably elevated (more than 150-300 m in altitude) above the adjacent country or above sea level; it is commonly limited on at least one side by an abrupt descent, has a flat or nearly smooth surface but is often dissected by deep valleys and surmounted by high hills or mountains, and has a large part of its total surface at or near the summit level. A plateau is usually higher and has more noticeable relief than a *plain* (it often represents an elevated plain), and is usually higher and more extensive than a mesa; it may be tectonic, residual, or volcanic in origin. See also: *tableland.* (b) A flat, upland region underlain by horizontal strata or characterized by horizontal structure, which may be highly dissected; Davis (1885) introduced this concept of a "plateau", but the term should be used without regard to the underlying geologic structure.—Etymol: French. Pl: *plateaus; plateaux.*

plateau [marine geol] A broad, more or less flat-topped and ill-defined elevation of the sea floor, generally over 200 m in elevation. Syn: *submarine plateau.*

plateau age In the *argon-40/argon-39 age method* and the *xenon-xenon age method*, an age calculated by averaging results from gas fractions released continuously that have obviously similar apparent ages.

plateau basalt A term applied to those basaltic lavas that occur as vast composite accumulations of horizontal or subhorizontal flows, which, erupted in rapid succession over great areas, have at times flooded sectors of the Earth's surface on a regional scale. They are generally believed to be the product of fissure eruptions. Cf: *shield basalt.* Syn: *flood basalt.*

plateau eruption Successive lava flows that spread in sheets over a large area. Cf: *fissure eruption.*

plateau glacier A *highland glacier* overlying a relatively flat mountain tract, and usually overflowing its edges in hanging glaciers. See also: *ice plateau.*

plateau gravel A sheet, spread, or patch of surficial gravel, often compacted, occupying a flat area on a hilltop, plateau, or other high region at a height above that normally occupied by a stream-terrace gravel. It may represent a formerly extensive deposit that has been raised by earth movements and largely removed by erosion.

plateau mountain A *pseudomountain* produced by the dissection of a plateau; e.g. the Catskill Mountains, N.Y.

plateau plain An extensive plain surmounted by a sublevel summit area and bordered by escarpments (Hill, 1900, p. 8). Cf: *mesa plain.*

plate boundary Zone of seismic and tectonic activity along the edges of lithosphere plates, presumed to indicate relative motion between plates (Dennis & Atwater, 1974, p. 1033). Syn: *plate juncture; plate margin.*

plate carrée projection A simple cylindrical projection with an evenly spaced network of horizontal parallels (spaced at their correct meridional distance from the equator) and vertical meridians (spaced at their correct equatorial distances). Only cardinal directions are true, and scale is true on all meridians and the standard parallel but is greatly distorted away from the center. If the equator is the standard parallel, the network consists of squares; any other standard parallel will produce rectangles with the north-south dimension the longer. The projection is neither equal-area nor conformal, and is used in geologic mapping for small areas and in geographic referencing for large-scale city maps. A modified version has the scale preserved along two parallels of latitude other than the equator. Etymol: French, "regular-square projection". Syn: *equirectangular projection.*

plate-equivalent Said of the part of the dinoflagellate-cyst wall judged to occupy a position equivalent to that occupied by a plate of the theca.

plate juncture *plate boundary.*

platelet A small ice crystal which, when united with other platelets, forms a layer of floating ice, esp. sea ice, and serves as seed crystals for further thickening of the ice cover. Platelets in sea ice retain their identity for some time because they are bounded by rows or layers of brine cells.

platelike conodont element A *conodont element* having platforms or a greatly expanded basal cavity.

plate margin *plate boundary.*

plate-scale One of the oval to elliptic organic scales embedded in the surface layer of the periplast (cell membrane) of a coccolithophorid.

plate tectonics A theory of *global tectonics* in which the lithosphere is divided into a number of *plates* whose pattern of horizontal movement is that of torsionally rigid bodies that interact with one another at their boundaries, causing seismic and tectonic activity along these boundaries (Dennis & Atwater, 1974, p. 1031).

platform [coast] A flat or gently sloping underwater erosional surface extending seaward or lakeward from the shore; specif. a *wave-cut platform* or an abrasion platform. See also: *wave-built platform.*

platform [geomorph] (a) A general term for any level or nearly level surface; e.g. a terrace or bench, a ledge or small space on a cliff face, a flat and elevated piece of ground such as a tableland or plateau, a peneplain, or any beveled surface. (b) A small plateau.

platform [paleont] (a) A laterally broadened structure or shelf along the inner side or the outer side of the anterior-posterior axis of a conodont. The term is also used (incorrectly) for *plate.* (b) The *muscle platform* of a brachiopod. (c) The flat bottom or floor of the calyx of a coral.

platform [tect] That part of a continent that is covered by flat-lying or gently tilted strata, mainly sedimentary, which are underlain at varying depths by a *basement* of rocks that were consolidated during earlier deformations. A platform is a part of the *craton.*

platform facies *shelf facies.*

platform reef An *organic reef,* more extensive than a *patch reef* (i.e., several km across), with a flat upper surface, and sometimes forming an island. Platform reefs are common off the coast of Australia. Cf: *table reef.*

platidiiform Said of a brachiopod loop consisting of descending branches from the cardinalia to the median septum, with only rudimentary prongs on the septum representing the ascending part of the loop (TIP, 1965, pt.H, p.150).

platina Crude native platinum.

platiniridium A silver-white cubic mineral: (Ir,Pt). It is a native alloy of iridium with platinum and other related metals.

platinite *platynite.*

platinum A very heavy, steel-gray to silvery-white, isometric mineral, the native metallic element Pt, commonly containing palladium, iridium, iron, and nickel. It occurs as grains and nuggets in alluvial deposits (often associated with nickel sulfide and gold ores), and disseminated in basic and ultrabasic igneous rocks. Platinum is a highly corrosion-resistant, ductile, and malleable metal, and is the most abundant metal of the platinum group. Syn: *polyxene.*

platte A resistant knob of rock in a glacial valley or rising in the midst of an existing glacier, often causing a glacier to split near its snout. Etymol: German *Platte,* "slab". Pl: *platten.*

platting The action or process of mapping a surveyed area; the making of a plat.

plattnerite An iron-black mineral: PbO_2.

platy (a) Said of a sedimentary particle whose length is more than three times its thickness (Krynine, 1948, p. 142). Cf: *acicular [sed].* (b) Said of a sandstone or limestone that splits into laminae having thicknesses in the range of 2 to 10 mm (McKee & Weir, 1953, p. 383).

platycone A coiled cephalopod conch with a flattened form, without implication as to the width of the umbilicus or the shape of the venter.

platy flow structure An igneous rock structure of tabular sheets suggesting stratification. It is formed by contraction during cooling; the structure is parallel to the surface of cooling, and is commonly accentuated by weathering. Syn: *platy structure; linear flow structure; planar flow structure.*

platykurtic (a) Said of a *frequency distribution* that has a concentration of values about its mean less than for the corresponding normal distribution. (b) Said of a broad, flat-topped frequency-distribution curve that is less peaked than the corresponding normal curve.—Cf: *leptokurtic; mesokurtic.*

platynite An iron-black mineral: $PbBi_2(Se,S)_3$. It occurs in thin metallic plates resembling graphite. Syn: *platinite.*

platyproct Said of a sponge in which the exhalant surface is nearly or quite flat.

platy structure *platy flow structure.*

plauenite A plagioclase-rich *quartz syenite,* grading to *quartz monzonite.* The name was given by Brögger in 1895, from Plauenscher Grund, near Dresden, Germany. Obsolete.

playa [coast] (a) A small, generally sandy land area at the mouth of a stream or along the shore of a bay. (b) A flat, alluvial coastland, as distinguished from a beach.—Etymol: Spanish, "beach, shore, strand, coast". Pron: *ply*-ah.

playa [geomorph] (a) A term used in SW U.S. for a dry, vegetation-free, flat area at the lowest part of an undrained desert basin, underlain by stratified clay, silt, or sand, and commonly by soluble salts. The term is also applied to the basin containing an expanse of playa, which may be marked by *ephemeral lakes.* See also: *salina; alkali flat; salt flat; salt pan; salar; salada.* Syn: *dry lake; vloer; sabkha; kavir; takir.* (b) A term that is often used for *playa lake.*

playa basin *bolson.*

playa furrow A shallow but distinct indentation or trail left on a playa by a *playa scraper* moving across the surface before the playa is completely desiccated.

playa lake A shallow, intermittent lake in an arid or semiarid region, covering or occupying a playa in the wet season but drying up in summer; an *ephemeral lake* that upon evaporation leaves or forms a playa. Syn: *playa.*

playa scraper An object (usually a cobble or boulder) that produces a *playa furrow* by moving across the moist sediment surface.

playfairite A mineral: $Pb_{16}Sb_{18}S_{43}$.

Playfair's law A generalized statement about the relation of stream systems to their valleys in areas of uniform bedrock and structure that have been subject to stream erosion for a long period of time; viz. that streams cut their own valleys, which are proportional in size to the streams they contain, and that the stream junctions in these valleys are accordant in level. For a quantitative statement of a major part of Playfair's law, see Horton (1945, p. 293, eq. 17). The law was enunciated by John Playfair (1747-1819), professor of natural philosophy at the University of Edinburgh: "Every river appears to consist of a main trunk, fed from a variety of branches, each running in a valley proportioned to its size, and all of them together forming a system of valleys, communicating with one another, and having such a nice adjustment of their declivities, that none of them join the principal valley, either on too high or too low a level; a circumstance which would be infinitely improbable, if each of these valleys were not the work of the stream which flows in it" (Playfair, 1802, p. 102). Syn: *law of accordant junctions.*

play of color An optical phenomenon consisting of flashes of a variety of prismatic colors, seen in rapid succession as certain minerals (esp. opal) or cabochon-cut gems are moved about; e.g. *opalescence.* It is caused by diffraction of light from innumerable minute regularly arranged optically transparent uniform spherical particles of amorphous silica, and from the spaces between them, stacked in an orderly three-dimensional pattern that behaves like a diffraction grating. Cf: *fire; change of color.* Syn: *schiller.*

plaza A term used in the SW U.S. for the exceptionally wide floor of a flat, open valley; the flat bottom of a shallow canyon. Etymol: Spanish, "square, marketplace".

plazolite *hydrogrossular.*

pleat A longitudinal fold of retractile muscle fibers with associated mesogloea on the side of a coral mesentery.

plectolophe A brachiopod lophophore in which each brachium consists of a U-shaped side arm bearing a double row of paired filamentary appendages "but terminating distally in medially placed plano-spire normal to commissural plane and bearing single row of paired appendages" (TIP, 1965, pt.H, p.150). Cf: *deuterolophe; spirolophe.*

Pleistocene An epoch of the Quaternary period, after the Pliocene of the Tertiary and before the Holocene; also, the corresponding worldwide series of rocks. It began two to three million years ago and lasted until the start of the Holocene some 8,000 years ago. When the Quaternary is designated as an era, the Pleistocene is considered to be a period. Syn: *Ice Age; Great Ice Age; glacial epoch; Diluvium.*

plenargyrite *matildite.*

pleniglacial The full glacial phase of a paleoclimate cycle, preceded by the *anaglacial* phase and followed by the *kataglacial* phase.

pleochroic Said of a mineral that displays *pleochroism.*

pleochroic formula An expression of a crystal's pleochroism, or the color of transmitted light. Cf: *absorption formula.*

pleochroic halo A minute zone of color or darkening surrounding and produced by a radioactive mineral crystal or inclusion.

pleochroic halo dating Determination of geologic age by measuring the increase in color darkening of the *pleochroic halo* of alpha-particle radiation damage around a zircon, monazite, xenotine, or apatite crystal as a function of time and alpha-particle activity. So many variables (e.g. mica sensitivity to alpha radiation, thermal annealing, color reversal) have been discovered that this dating method has only limited application.

pleochroism The ability of an anisotropic crystal to differentially absorb various wavelengths of transmitted light in various crystallographic directions, and thus to show different colors in different directions. This property is more easily seen under polarized light than by the unaided eye. A mineral showing pleochroism is said to be *pleochroic.* Cf: *bireflectance.* Syn: *polychroism.* See also: *dichroism; trichroism.*

pleocrystalline An obsolete syn. of *holocrystalline.*

pleomere A somite of the abdomen of a malacostracan crustacean. Syn: *pleonite.*

pleomorph *polymorph [evol].*

pleomorphism *polymorphism [evol].*

pleomorphous *polymorphic.*

pleon The abdomen of a malacostracan crustacean.

pleonaste *ceylonite.*

pleonastite A rock composed of spinel and clinochlore surrounding corundum crystals (Thrush, 1968, p. 836). Obsolete.

pleonite *pleomere.*

pleopod An abdominal limb of a crustacean; specif. any appendage of the pleon of a malacostracan crustacean, excluding caudal ramus and uropod. Syn: *pleopodite.*

pleosponge *archaeocyathid.*

pleotelson A structure of a malacostracan crustacean resulting from fusion of one or more abdominal somites (pleomeres) with the telson.

plerotic water A syn. of *ground water* proposed by Meinzer (1939), including *piestic water* and *phreatic water.*

plesiomorphy *primitive character.*

plesiostratotype A complementary stratotype (Sigal, 1964).

plessite A fine-grained intergrowth of kamacite and taenite. It occurs as triangular or polygonal areas in iron meteorites exhibiting Widmanstätten structure.

pleura (a) A laterally located part of the body of an invertebrate; e.g. a lateral part of the opisthosoma of a merostome. (b) One of the two lateral parts of each exoskeletal segment of a trilobite that extend outward from its axis; the part of the thoracic segment or pygidium that is lateral to the axial lobe. In this use, the term has been used as both a syn. and a plural of *pleuron.* (c) An *epimere* of a crustacean.—Pl: *pleurae.*

Pleuracanthodii An order of elasmobranch fishes, almost exclusively freshwater, characterized by lobate fins, amphistylic jaw suspension, and distinctive double-cusped teeth. Range, Upper Devonian to Upper Triassic.

pleural angle The angle between two straight lines lying tangential to the last two whorls on opposite sides of a gastropod shell.

pleural furrow (a) A groove along the surface of a pleura of a trilobite. (b) A groove crossing the pleura of a merostome.

pleuralia Sponge spicules (prostalia) on the sides of the body.

pleural spine A pointed or sharply rounded extension of the distal end of a pleura of a trilobite. It is narrower than the medial part of the pleura.

pleural suture The line of splitting-apart in molting of the carapace of a decapod crustacean. It is present in all brachyurans.

pleurite *epimere.*

pleurocyst A term used in the older literature for a generally granular imperforate calcareous layer thought to form a secondary deposit over an *olocyst* in frontal shields of certain ascophoran cheilostomes (bryozoans).

pleurodont Pertaining to vertebrate teeth that have undergone *ankylosis* to the medial side of a groove in the upper and lower jaws.

pleuromyarian Said of a nautiloid in which the retractor muscles of the head-foot mass are attached to the shell along the lateral areas of the interior of the body chamber (TIP, 1964, pt.K, p.57). Cf: *dorsomyarian; ventromyarian.*

pleuron A term that has been used as both a synonym and a singular form of *pleura.*

pleuston *Plankton* that float because of low specific gravity.

plexus [evol] *evolutionary plexus.*

plexus [glac geol] An area in a subglacial deposit that encloses a *giant's kettle* (Stone, 1899).

plica (a) One of the strong parallel ridges and depressions involving the entire thickness of a bivalve shell (mollusk or brachiopod), extending radially from beak to shell margin, and appearing as corrugations on the inner as well as the outer surface of the shell; e.g. a major undulation of the commissure of a brachiopod, with crest directed dorsally, commonly but not invariably associated with the dorsal fold and the ventral sulcus (TIP, 1965, pt.H, p.150). Syn: *plication [paleont].* (b) A term used, irrespective of commissure, for a small carina or fold in the surface of a brachiopod valve. (c) A small projection on the surface of an opisthosomal segment of a eurypterid.—Pl: *plicae.*

plicated Adj. of *plication.* Syn: *crumpled.*

plication [paleont] A coarse radial corrugation in the surface of a bivalve-mollusk or brachiopod shell; specif. a *plica.*

plication [struc geol] Intense, small-scale folding. Adj: *plicated.* Cf: *crenulation.*

Pliensbachian European stage: Lower Jurassic (above Sinemurian, below Toarcian).

Plinian eruption An explosive eruption in which a steady, turbulent stream of fragmented magma and magmatic gas is released at a high velocity from a vent. Large volumes of *tephra* and tall eruption columns are characteristic (Wilson, 1976, p. 543). Etymol: Pliny the Younger, A.D. 79.

plinth A term suggested by Bagnold (1941, p. 229) for the lower and outer part of a seif dune, beyond the slip-face boundaries, that has never been subjected to sand avalanches.

plinthite In a soil, a material consisting of a mixture of clay and quartz, that is rich in sesquioxides and poor in humus and is highly weathered. It occurs as red mottles in a platy, polygonal, or reticulate pattern. Repeated wetting and drying changes plinthite to ironstone hardpan or irregular aggregates (USDA, 1975). See also: *laterite.*

Pliocene An epoch of the Tertiary period, after the Miocene and before the Pleistocene; also, the corresponding worldwide series of rocks. It is considered to be a period when the Tertiary is designated as an era.

pliomagmatic zone *eugeosyncline.*

pliothermic Pertaining to a period in geologic history characterized by more than average warmth of climate. Cf: *miothermic.* Rarely used.

plocoid Said of a massive scleractinian corallum in which corallites have separated walls and are united by costae, dissepiments, or coenosteum.

plombierite A mineral: $Ca_5H_2Si_6O_{18} \cdot 6H_2O$(?).

plot To place survey data upon a map or plat; to draw to scale. The term was formerly used in noun form as a syn. of *plat.*

plowshare A wedge-shaped feature developed on a snow surface by deposition of windblown snow, and sometimes accentuated by ablation. Cf: *sastrugi; foam crust.*

plow sole A *pressure pan* representing a layer of soil compacted by repeated plowing to the same depth. Also spelled: *plowsole.* Syn: *hardpan.*

plucking [glac geol] The process of glacial erosion by which sizable rock fragments, such as blocks, are loosened, detached, and borne away from bedrock by the freezing of water along joints and stratification surfaces with resulting removal of rock as the ice advances. See also: *sapping [glac geol].* Syn: *quarrying [geomorph].*

plucking [streams] *hydraulic plucking.*

pluck side The downstream, or lee, side of a roche moutonnée, roughened and steepened by glacial plucking. Ant: *scour side.*

plug [drill] n. A watertight or gastight seal, such as an interval of cement or a *bridge plug,* installed in a borehole or well to prevent movement of fluids.—v. To set a plug in a borehole; to fill in or seal off fractures, cavities, or other porosity in the walls of a borehole.

plug [paleont] *umbilical plug.*

plug [pat grd] (a) A cohesive, commonly vertical column of gravelly material with considerable fines, representing the continuance at depth of a *sorted circle* in a gravel beach, as on Victoria Island, Canada (Washburn, 1956, p.844). (b) A similar columnlike feature occurring with a *mud circle* (Bird, 1967, p.194).

plug [sed] A mass of sediment filling the part of a stream channel abandoned by the formation of a cutoff; e.g. a *clay plug* or a *sand plug.* See also: *valley plug.*

plug [volc] (a) A vertical, pipelike body of magma that represents the conduit to a former volcanic *vent.* Cf: *neck [volc].* (b) A crater filling of lava, the surrounding material of which has been removed by erosion.

plug dome A volcanic dome characterized by an upheaved, consolidated conduit filling (Williams, 1932).

plugging The act or process of stopping the flow of water, oil, or gas in strata penetrated by a borehole or well, so that fluid from one stratum will not escape into another or to the surface; specif. the sealing of a well that is dry and is to be abandoned. It is usually accomplished by pumping several intervals of cement into the *mud column,* setting a surface *plug,* and capping the hole with a metal plate.

plugging back The act or process of cementing off a lower section of a *well bore,* or of blocking fluids below from rising to a higher section.

plug reef A small triangular reef that grows with its apex pointing seaward through openings between linear shelf-edge reefs (Maxwell, 1968, p.101). Its outline is analogous with that of a sand ridge formed in the lower reach of a large river. Plug reefs are found off the coast of Australia where high tide range results in strong currents.

plum A clast embedded in a matrix of a different kind; esp. a pebble in a conglomerate.

plumalsite A mineral: $Pb_4Al_2(SiO_3)_7$.

plumasite A coarsely xenomorphic-granular hypabyssal rock of variable composition, composed chiefly of corundum crystals enclosed in oligoclase grains. The name, given by Lawson in 1903, is for Plumas County, California. Not recommended usage.

plumbago A syn. of *graphite.* The term has also been applied to graphitic rock, to an impure graphite, and to graphitoid minerals such as molybdenite.

plumb bob A conical metal weight suspended by a cord, used to project a point vertically in space for short distances. Syn: *plumb.*

plumbic ocher *lead ocher.*

plumb line The line of force in the geopotential field; a continuous curve to which the direction of gravity is everywhere tangential.

plumboferrite A dark hexagonal mineral: $PbFe_4O_7$. Cf: *magnetoplumbite.*

plumbogummite (a) A mineral: $PbAl_3(PO_4)_2(OH)_5 \cdot H_2O$. (b) A group of isostructural minerals consisting of plumbogummite, gorceixite, goyazite, crandallite, florencite, and dussertite, and related to alunite and other sulfates isostructural with it.

plumbojarosite A mineral of the alunite group: $PbFe_6(SO_4)_4(OH)_{12}$.

plumbomicrolite A variety of microlite containing Pb.

plumbonacrite A mineral: $Pb_{10}(CO_3)_6(OH)_6O$ (?).

plumboniobite A dark-brown to black mineral of complex composition, consisting of a niobate of yttrium, uranium, lead, iron, and rare earths. It resembles samarskite, and may be a lead-bearing variety of samarskite.

plumbopalladinite A hexagonal mineral: Pd_3Pb_2.

plumbopyrochlore A mineral of the pyrochlore group: $(Pb,Y,U,Ca)_{2-x}Nb_2O_6(OH)$.

plum-cake rock A quarryman's term applied in northern England to a breccia or breccia-conglomerate having sporadic fragments embedded in a matrix like plums in a pudding.

plume A featherlike inclusion (flaw) in a gem, as in a "plume agate".

plume structure On the surface of a master joint, a ridgelike tracing in a plumelike pattern, usually oriented parallel to the upper and lower surfaces of the containing rock unit. Syn: *plumose structure.* Less-preferred syn: *feather fracture.*

plumicome A sponge spicule (hexaster) in which the terminal branches are S-shaped and arranged in several tiers, forming a plumelike structure.

plumose mica A feathery variety of muscovite mica.

plumose spiculofiber A spiculofiber of an axinellid or ectyonine sponge skeleton, in which some or all of the component spicules face obliquely outward.

plumose structure *plume structure.*

plumosite An antimony-sulfide mineral having a feathery form; e.g. jamesonite and boulangerite. Syn: *plumose ore.*

plum-pudding stone *puddingstone.*

plumule *epicotyl.*

plunge [struc geol] The inclination of a fold axis or other linear structure, measured in the vertical plane. It is mainly used for the geometry of folds. Cf: *dip.*

plunge [surv] v. (a) To set the horizontal cross wire of a theodolite in the direction of a grade when establishing a grade between two points of known level. (b) *transit.*

plunge basin A deep, relatively large hollow or cavity scoured in the bed of a stream at the foot of a waterfall or cataract by the force and eddying effect of the falling water. It is often called a *plunge pool.* Cf: *pothole [streams].*

plunge line *breaker line.*

plunge point The line along which a plunging wave curls over and collapses as it approaches the shore.

plunge pool (a) The water in a plunge basin. (b) A deep, circular lake occupying a plunge basin after the waterfall has ceased to exist or the stream has been diverted. Syn: *waterfall lake.* (c) A small, deep *plunge basin.*

plunging breaker A type of *breaker* whose crest curls over and collapses suddenly, with complete disintegration of the wave. Cf: *surging breaker; spilling breaker.*

plunging cliff A sea cliff bordering directly on deep water, having a base that lies well below water level.

plunging fold A fold of which the hinge line is inclined to the horizontal. Cf: *nonplunging fold; doubly plunging fold; pitching fold.* See also: *plunging inclined fold; plunging normal fold.*

plunging inclined fold A fold with a plunging axis and inclined axial plane (Turner & Weiss, 1963, p. 119).

plunging normal fold A fold with an inclined axis and vertical axial plane (Turner & Weiss, 1963, p. 119).

pluricolumnal Two or more crinoid columnals attached to one another.

plush copper ore *chalcotrichite.*

plus sight *backsight.*

plutology The study of the interior of the Earth.

pluton (a) An igneous *intrusion.* (b) A body of rock formed by metasomatic replacement.—The term originally signified only deep-seated or plutonic bodies of granitoid texture. See also: *plutonism.*

plutonic (a) Pertaining to igneous rocks formed at great depth. See also: *plutonic rock.* Cf: *hypabyssal.* (b) Pertaining to rocks formed by any process at great depth. —Syn: *abyssal; deep-seated; hypogene.*

plutonic cognate ejecta Pyrogenic fragments that were solidified at depth but were brought to the surface by eruption of the magma.

plutonic event A term proposed by Gilluly (1966, p. 97) for "concentrations of radiometric dates" in his criticism of some of the criteria used for naming orogenies.

plutonic metamorphism Deep-seated regional metamorphism at high temperatures and pressures, often accompanied by strong deformation; batholithic intrusion with accompanying metasomatism, infiltration, and injection (or, alternatively, differential fusion or anatexis) is characteristic. Cf: *injection metamorphism.*

plutonic nodule *nodule [ign.]*

plutonic rock A rock formed at considerable depth by crystallization of magma and/or by chemical alteration. It is characteristically medium- to coarse-grained, of granitoid texture.

plutonic water *Juvenile water* in, or derived from, magma at a considerable depth, probably several kilometers. Cf: *magmatic water; volcanic water.*

plutonism (a) A general term for the phenomena associated with the formation of *plutons.* (b) The concept of the formation of the Earth by solidification of a molten mass. The theory was promulgated by Hutton in the 18th Century.

plutonist A believer in the theory of plutonism as promulgated by Hutton. Ant: *neptunist.* Syn: *volcanist.*

pluvial [clim] Said of a climate characterized by relatively high precipitation, or of the time interval during which such a climate prevailed. Formerly equated with the glacial stage of the Quaternary glacial/interglacial sequence, pluvial intervals are now regarded more as transitional, or, in low latitudes, as typical of interglacials.

pluvial [geomorph] Said of a geologic episode, change, process, deposit, or feature resulting from the action or effects of rain; e.g. *pluvial* denudation, a landslide, or gully erosion and the consequent spreading-out of the eroded material below. The term sometimes includes the fluvial action of rainwater flowing in a stream channel, esp. in the channel of an ephemeral stream.

pluvial [meteorol] (a) Pertaining to rain, or more broadly, to precipitation. Syn: *pluvious.* (b) Characterized by or regularly receiving abundant rain. Syn: *pluviose.*—Cf: *hyetal.*

pluvial lake A lake formed in a period of exceptionally heavy rainfall; specif. a lake formed in the Pleistocene epoch during a time of glacial advance, and now either extinct or existing as a remnant. Example: Lake Bonneville, a prehistoric lake in present Utah, eastern Nevada, and southern Idaho.

pluviofluvial Pertaining to the combined action of rainwater and streams; e.g. *pluviofluvial* denudation.

pluviometer *rain gage.*

pneumatocyst A hollow inflated structure of a stipe that serves to keep some of the *brown algae* afloat.

pneumatogenic Said of a rock or mineral deposit formed by a gaseous agent. Cf: *hydatogenic; hydatopneumatogenic.* Syn: *pneumatolytic [ore dep].*

pneumatolysis Alteration of a rock or crystallization of minerals by gaseous emanations derived from solidifying magma. Adj: *pneumatolytic* [petrology].

pneumatolytic [ore dep] *pneumatogenic.*

pneumatolytic [petrology] (a) Formed by *pneumatolysis.* (b) Sometimes applied to the surface products of gaseous emanations near volcanoes. (c) Applied to the stage of magmatic differentiation between the pegmatitic and hydrothermal stages. (d) Said of the effects of contact metamorphism adjacent to deep-seated intrusions.

pneumatolytic differentiation Magmatic differentiation by the process of *gaseous transfer.* Syn: *gaseous transfer differentiation.*

pneumatolytic metamorphism Contact metamorphism accompanied by strong metasomatism resulting from the chemical action of magmatic gases on both country rock and intrusion.

pneumatolytic stage That stage in the cooling of a magma during which the solid and gaseous phases are in equilibrium.

pneumotectic Said of processes and products of magmatic consolidation affected to some degree by gaseous constituents of the magma.

pocket [coast] An enclosed or sheltered place along a coast, such as a reentrant between rocky cliffed headlands or a bight on a lee shore.

pocket [cryoped] The downfolded or sagging convex part of a layer in a *congeliturbate.* Ant: *festoon.*

pocket [eco geol] (a) A small, discontinuous occurrence or patch of ore, e.g. a mineralized cavity or crevice. (b) A localized enrichment of an ore deposit. Syn: *belly.*

pocket [geog] (a) A *water pocket* in the bed of an intermittent stream. (b) A hollow or glen in a mountain.

pocket [grd wat] A colloquial term for a small body of ground water.

pocket [ice] A rarely used syn. of *blind lead.*

pocket [speleo] A solutional concavity in the ceiling, walls, or floor of a cave. See also: *ceiling cavity.*

pocket beach A small, narrow beach formed in a *pocket [coast],* commonly crescentic in plan and concave toward the sea, and generally displaying well-sorted sand or gravel; a *bayhead beach.*

pocket lens *hand lens.*
pocket penitente A *nieve penitente* on the north side of which there is a water-filled depression in the ice, 30 to 60 cm deep, rounded or oval in outline, and with perpendicular walls; at the bottom is a thin layer of debris that absorbed solar heat and melted the ice (Workman, 1914, p.306). Such features are formed in late summer in the Himalayas.
pocket rock A term used in the SW U.S. for a desert boulder with a hard case of desert varnish.
pocket transit A small, compact surveyor's transit that fits in a pocket; specif. a *Brunton compass.*
pocket valley A valley whose head is enclosed by steep walls at the base of which underground water emerges as a spring. Syn: *blind valley.* See also: *steephead.*
pocosen *pocosin.*
pocosin A local term along the Atlantic coastal plain south of Virginia for a swamp or marsh on a flat upland, bordering on or near the sea, in many places enclosing knobs or hummocks. Etymol: American Indian (Delaware). Also spelled: *pocoson; pocosen.* Syn: *dismal.*
pocoson *pocosin.*
pod [eco geol] An orebody of *podiform* shape.
pod [geomorph] A term used in the steppes of southern Russia for a very shallow depression as much as 10 km in diameter, containing an intermittent lake or lakes; it may indicate uneven loess deposition, preloess topography, deflation, or solution.
pod [meta] A term formerly used to describe certain bodies that are long in one dimension and short in two dimensions and are enclosed in schist with the long axis parallel to the schistosity.
podial pore A pore admitting the passage of a tube foot between the ambulacral plates of an echinoderm (as of an asteroid). See also: *tentacle pore.*
podiform Said of an orebody, either diffuse or sharply demarcated, of an elongate lenticular or rodlike shape, e.g. chromite in alpine-type peridotite.
podite A limb segment of an arthropod; e.g. a joint of a biramous appendage of a trilobite.
podium The cylindrical outer part of a tube foot of an echinoderm. The term has also been applied to the tube foot itself. Pl: *podia.*
podoconus Internal cone within the central capsule of a nasselline radiolarian.
podocope Any ostracode belonging to the order Podocopida, characterized by a calcified shell with a curved dorsal margin or a straight dorsal margin shorter than the total length of the shell, and by a muscle scar pattern usually consisting of a few secondary scars. Range, Lower Ordovician to present.
podolite *carbonate-apatite.*
podomere An individual leg *segment* of an arthropod, connected by articulation with adjoining segments.
podophthalmite The distal segment of the eyestalk of a decapod crustacean, bearing the corneal surface of the eye. Cf: *basiophthalmite.*
podostyle A mass of cytoplasm that projects from the test aperture of monothalamous foraminifers and gives rise to pseudopodia (TIP, 1964, pt.C, p.62).
Podsol *Podzol.*
Podzol A great soil group in the 1938 classification system, a group of zonal soils having an organic mat and a very thin organic-mineral layer overlying a gray, leached A2 horizon and a dark brown, illuvial B horizon enriched in iron oxide, alumina, and organic matter. It develops under coniferous or mixed forests or under heath, in a cool to temperate moist climate (USDA, 1938). Also spelled: *Podsol.* Etymol: Russian *podsol,* "ash soil".
podzolization The process by which a soil becomes more acid owing to depletion of bases, and develops surface layers that are leached of clay and develop illuvial B horizons; the development of a *podzol.*
poëchore A climatic term for the part of the Earth's surface represented by steppes.
poecilitic The original spelling of *poikilitic.* Now obsolete in American usage, it is still the most accepted European spelling.
poeciloblast *poikiloblast.*
poeciloblastic *poikiloblastic.*
poikilitic Said of the texture of an igneous rock in which small grains of one mineral (e.g. plagioclase) are irregularly scattered without common orientation in a typically anhedral larger crystal of another mineral (e.g. pyroxene); also, said of the enclosing crystal, or *oikocryst.* In hand specimen, this texture produces lustrous patches (*luster mottling*) due to reflection from cleavage planes of the oikocrysts. Originally spelled *poecilitic.* Cf: *ophitic; endoblastic.* Nonrecommended syn: *semipegmatitic.*
poikilo- A prefix meaning "mottled", "spotted".
poikiloblast A large crystal formed by recrystallization during metamorphism and containing numerous inclusions of small grains. See also: *poikiloblastic.* Also spelled: *poeciloblast.*
poikiloblastic (a) Pertaining to a *poikiloblast.* (b) A metamorphic texture in which small grains of one constituent lie within larger metacrysts. Modern usage favors this meaning. Syn: *sieve texture.* (c) Said of a metamorphic texture due to the development, during recrystallization, of a new mineral around numerous relicts of the original minerals, thus simulating the poikilitic texture of igneous rocks. Cf: *helicitic.*—Also spelled: *poeciloblastic.*
poikilocrystallic A syn. of *poikilotopic.* The term was introduced by Phemister (1956, p.74).
poikilophitic Said of *ophitic* texture characterized by lath-shaped feldspar crystals completely included in large, anhedral pyroxene crystals; an intermediate texture between *ophitic* and *sporophitic.* Cf: *poikilitic.*
poikilotherm A poikilothermic organism; e.g. a frog. Syn: *heterotherm.*
poikilothermic Said of an organism having no internal mechanism for temperature regulation; having a body temperature that varies with the temperature of the environment; cold-blooded. Syn: *heterothermic.* Cf: *homoiothermic.*
poikilotope A large crystal enclosing smaller crystals of another mineral in a sedimentary rock showing poikilotopic fabric; e.g. a large calcite crystal enclosing smaller relics of incompletely replaced dolomite crystals in a dedolomitized rock, or a large gypsum crystal enclosing numerous grains of quartz and/or feldspar.
poikilotopic Said of the fabric of a recrystallized carbonate rock or a chemically precipitated sediment in which the constituent crystals are of more than one size and in which the larger crystals enclose smaller crystals of another mineral. The term was proposed by Friedman (1965, p.651). Cf: *porphyrotopic.* Syn: *poikilocrystallic.*
point [coast] A tapering tract of relatively low land, such as a small *cape,* projecting from the shore into a body of water; specif. the tip section or extremity of such a projection, or the sharp outer end of any land jutting out into a body of water. Syn: *tongue.*
point [gem] A unit of weight for diamonds and other gemstones, equal to 1/100 *carat,* or 2 mg. A stone weighing 32/100 carat is called a 32-point stone or a 32 pointer.
point [geomorph] A sharp projecting rocky prominence; esp. a peak of a mountain range.
point [surv] (a) One of the 32 precisely marked equidistant spots about the circumference of a circular card attached to a compass, which indicate the direction in which the various parts of the horizon lie; e.g. a *cardinal point.* The term is also applied to the angular distance of 11.25 degrees between two such successive points, and to the part of the horizon indicated precisely or approximately by a point of a compass card. (b) A *position* on a reference system determined by a survey and represented by a fix; e.g. a "point of observation".
point bar One of a series of low, arcuate ridges of sand and gravel developed on the inside of a growing meander by the slow addition of individual accretions accompanying migration of the channel toward the outer bank. Cf: *channel bar; meander scroll.* Syn: *meander bar.*
point-bar deposit A deposit consisting of a series of alternating *point bars* and intervening troughs.
point-counter analysis A statistical method involving the estimation of the frequency of occurrence of an object, such as a fossil or mineral species, in a sample, determined by counting the number of times it occurs at specified intervals throughout the sample. The analysis is commonly made with an automatic point counter attached to a microscope.
point defect A type of *crystal defect* occurring at a particular point and involving a particular atom in a lattice. Cf: *line defect; plane defect.* See also: *interstitial defect; Frenkel defect; Schottky defect.*
point diagram A *fabric diagram* in which poles representing lineations, normals to fabric planes, or crystallographic directions have been plotted. Syn: *scatter diagram.*
point group *class [cryst].*
point mass Any mass whose geophysical response is equivalent to the theoretical response of the same mass if concentrated at a

point. In gravity, a uniform sphere can be treated as if all its mass were concentrated at its center. At large distances, nonspherical masses can be approximated by point masses.

point maximum *maximum.*

point of origin *initial point.*

point sample A sample of the sediment contained at a single point in a body of water. It is obtained either by an instantaneous sampler or a time-integrating sampler.

point-source pollution Pollution resulting from any confined, discrete source, such as a pipe, ditch, tunnel, well, container, concentrated animal-feeding operation, or floating craft. Cf: *non-point-source pollution.*

poised stream (a) A stream that is neither eroding nor depositing sediment. Cf: *graded stream.* (b) A stream that possesses stability from an engineering viewpoint.

Poiseuille's law A statement in physics that the velocity of flow of a liquid through a capillary tube varies directly as the pressure and the fourth power of the diameter of the tube and inversely as the length of the tube and the coefficient of viscosity (Webster, 1967, p. 1751).

Poisson distribution A discrete *frequency distribution* similar in concept to the binomial distribution; but based on the assumptions that the events occur independently, the probability that an event will occur is proportional to the length of time preceding the event, and the probability that two events will occur at the same time is vanishingly small (Davis, 1973, p. 274).

Poisson's number The reciprocal of *Poisson's ratio.*

Poisson's ratio The ratio of the lateral unit strain to the longitudinal unit strain in a body that has been stressed longitudinally within its elastic limit. It is one of the *elastic constants.* Symbol: σ. See also: *Poisson's number; Poisson's relation.*

Poisson's relation In experimental structural geology, a model of elastic behavior that takes *Poisson's ratio* as equal to 0.25, i.e. approximating that of many solids.

poitevinite A mineral: $(Cu,Fe,Zn)SO_4 \cdot H_2O$.

pokelogan A term of Algonquian origin for a marshy cove or inlet of a stream or lake. Sometimes the term, used primarily in Wisconsin, is shortened to "logan". Syn: *bogan; peneloken.*

polar Relating or pertaining to the region of either or both of the two poles of the Earth.

polar air mass An *air mass* originating over oceans or continents between 50° and 60° north and 50° and 60° south of the Equator.

polar area The part of a pollen grain poleward from the ends of the colpi and their associated structures.

polar-area index The ratio between the polar area of a pollen grain and its diameter.

polar axis (a) The primary axis of direction or the fixed reference line from which the angle coordinate is measured in a system of polar coordinates; e.g. the axis of rotation of the Earth. (b) An axis of symmetry that has different crystal faces at opposite ends. (c) An imaginary line connecting the two poles of spores and pollen grains.

polar cap [astrogeol] An area at each pole of the planet Mars, covered with solid carbon dioxide and ice, and varying in extent with the Martian seasons.

polar cap [glaciol] (a) An ice sheet centered at the South Pole, i.e. Antarctica. (b) A term incorrectly applied to the sea ice of the Arctic Ocean.

polar circle The *Arctic Circle* or the *Antarctic Circle,* two parallels of latitude lying approx. 23°27′ from the poles of the Earth.

polar climate A type of climate of polar latitudes (above 66°33′), characterized by temperatures of 10°C and below. The two types of polar climate in Köppen's classification are *tundra climate* and *perpetual frost climate.* Cf: *frigid climate.*

polar convergence The line of *convergence* of polar and subpolar water masses in the ocean. It is indicated by a sharp change in water-surface temperatures.

polar coordinate (a) One of two coordinates that represent the distance from a central point of reference (the pole or origin) along a line to a point whose position is being defined, and the direction (angle) this line makes with a fixed line. (b) A coordinate used to define the position of a point in space with respect to an arbitrarily chosen origin by means of three polar coordinates (two directions or angles and one distance).

polar desert A high-latitude desert where the moisture present is frozen in ice sheets, and thus is unavailable for plant growth (Stone, 1967, p. 239). Syn: *arctic desert.*

polar firn Firn formed at low temperatures with no melting or liquid water present.

polar front The line of discontinuity, developed under favorable conditions between air originating in polar regions and air from low latitudes, on which the majority of the *depressions [meteorol]* of temperate latitudes develop.

polar glacier A glacier whose temperature is below freezing to considerable depth, or throughout, and on which there is no melting even in summer. See also: *high-polar glacier; temperate glacier; subpolar glacier.* Syn: *cold glacier.*

polar ice Any sea ice more than one year old and more than 3 m thick. It is heavily hummocked and usually the thickest form of sea ice. Syn: *arctic pack.*

polarimeter An instrument for measuring the amount of polarization of light or the proportion of polarized light in a partially polarized ray caused by some property of a substance. Sometimes called a *polariscope.*

polarimetry The art or process of using a *polarimeter.*

polariscope (a) Any of several optical instruments for observing the properties of polarized light or the effects produced on polarized light by various materials, e.g. the refraction of a crystal. It consists of a polarizer and an analyzer. (b) A *polarimeter.*

polarite A mineral: Pd(Bi,Pb).

polarity In *cladism,* the direction from *primitive characters* toward *derived characters* within a *morphocline.*

polarity-change horizon A surface or interface in the succession of rock strata marked by a change in magnetic polarity. Syn: *magnetostratigraphic polarity-change horizon.*

polarity-chronologic unit A division of time distinguished on the basis of the record of magnetopolarity as expressed by *polarity-chronostratigraphic units.* In order of decreasing magnitude, ranks are *polarity period, polarity epoch,* and *polarity event* (Oriel et al., 1976, p. 276-277).

polarity-chronostratigraphic unit A subdivision of rock considered solely as the magnetic polarity record of a specific interval of geologic time (Oriel et al., 1976, p. 276).

polarity epoch (a) A period of time during which the Earth's magnetic field was predominantly or entirely of one polarity. The chronology of the epochs and their events forms the *geomagnetic polarity time scale.* (b) Specif., the time during which rocks of the corresponding *polarity interval* formed; the *polarity-chronologic unit* of middle rank (Oriel et al., 1976, p. 277).—Syn: *geomagnetic polarity epoch.*

polarity event The shortest *polarity-chronologic unit* (Oriel et al., 1976, p. 277). Syn: *geomagnetic polarity event.*

polarity interval The fundamental unit of worldwide polarity-chronostratigraphic classification. The term is applied to rock, not time; it is used in a spatial sense (Oriel et al., 1976, p. 276).

polarity period The longest *polarity-chronologic unit* (Oriel et al., 1976, p. 277).

polarity reversal *geomagnetic reversal.*

polarity rock-stratigraphic unit *magnetopolarity unit.*

polarity zone The fundamental unit of polarity lithostratigraphic classification. A polarity zone is a unit of rock characterized by its polarity signature (Oriel et al., 1976, p. 276). Syn: *magnetostratigraphic polarity zone.*

polarization [elect] (a) The production of dipoles or higher-order multipoles in a medium. (b) The polarity or potential near an electrode.

polarization [magnet] A nonrecommended syn. of *magnetization.*

polarization [optics] The modification of light so that its vibrations are restricted to a single plane. Polarized light is used in the study of thin sections of minerals and rocks, by the polarizing microscope.

polarization brush *isogyre.*

polarized light Light that has been changed by passage through a prism or other polarizer so that its transverse vibrations occur in a single plane, or in a circular or elliptical pattern. (In general, the term polarized light is taken to mean plane-polarized light). It is used in the polarizing microscope for optical analysis of minerals or rocks in thin section.

polarizer An apparatus for polarizing light; in a polarizing microscope, it may be the lower *Nicol prism* or the *Polaroid.* Cf: *analyzer.* Partial syn: *nicol.*

polarizing angle The angle at which unpolarized light is incident upon a surface so that it acquires the maximum plane polarization. Syn: *Brewster angle.* See also: *Brewster's law.*

polarizing microscope A microscope that uses polarized light and a revolving stage for analysis of petrographic thin sections. Two prisms, one above and the other below the stage, polarize and analyze the light; the stage rotates about the line-of-sight axis. Syn: *petrographic microscope.*

polarizing prism *Nicol prism.*

polar lake A lake whose surface temperature never exceeds 4°C. Cf: *tropical lake.*

polar migration *polar wandering.*

polarograph An instrument for analyzing solutions by electrolysis using a cathode consisting of falling mercury drops. See also: *polarography.*

polarography An electrolytic technique for chemical analysis based on diffusion rates of ions to an electrode as a measure of the concentration of ions in the solution. A readily polarized electrode consisting of falling mercury drops is used to keep the ion concentration in the solution uniform. See also: *polarograph.*

Polaroid A trademark name for a *polarizer* that consists of a sheet of cellulose that is impregnated with crystals of quinine iodosulphate. The crystals are aligned so that their optical axes are parallel, and they polarize light in two directions at right angles: in one direction most visible light is absorbed, and in the other, essentially white light is transmitted. The term is often used in the noncapitalized form.

polar projection One of a group of projections that are centered on a pole of a sphere. Examples include any of several azimuthal map projections (polar stereographic projection, polar gnomonic projection, and polar orthographic projection).

polar space A four-sided region resulting from formation of *basal leaf cross* in an acantharian radiolarian.

polar spine (a) One of the modified coccoliths located at the ends of nonmotile fusiform coccolithophores (such as *Calciosolenia*). (b) A spine normal to the equator, and defining one axis, of the shell in a radiolarian; e.g. a radial spine disposed according to the *Müllerian law* and marking a zone in an acantharian comparable to the polar zone of the terrestrial globe.

polar stereographic projection A *stereographic projection* generally on an equatorial plane, having its center located at one of the poles of the sphere. It is suitable for maps of the polar regions on planes cutting the Earth north of 60°, and it serves as the base of the Universal Polar Stereographic (UPS) Military Grid System for latitudes between 80° and 90°. It is also widely used in optical mineralogy and in structural geology. See also: *Wulff net.*

polar symmetry A type of crystal symmetry in which the two ends of the central crystallographic axis are not symmetrical. Such a crystal is said to display *hemimorphism;* hemimorphite is the characteristic example.

polar tubule One of the external cylinders occurring at opposed poles in the main axis of an elliptic shell of a spumellarian radiolarian.

polar view The view of a spore or pollen grain from more or less directly above one of the poles. See also: *amb.*

polar wandering (a) Short-period movement of the Earth's poles, resulting from wobbling of its axis. (b) Long-period, more or less systematic displacement of the Earth's poles, which may have occurred during the passage of geologic time. Syn: *polar migration; Chandler motion.*

polder A generally fertile tract of flat, low-lying land (as in Netherlands and Belgium) reclaimed and protected from the sea, a lake, a river, or other body of water by the use of embankments, dikes, dams, or levees; e.g. a marsh that has been drained and brought under cultivation, or a lake that has been dried out by pumping. The term is usually reserved for coastal areas that are at or below sea level and that are constantly protected by an organized system of maintenance and defense. Etymol: old Flemish *poelen,* "to dig out". Syn: *polderland.*

polderization The creation of a polder or polders, esp. the draining and bringing under cultivation of a low-lying area reclaimed from the sea. See also: *empolder.* Syn: *empoldering.*

polderland *polder.*

pole [cryst] In crystallography, a line that is perpendicular to a crystal face and that passes through the center of the crystal. See also: *face pole.*

pole [geog] Either extremity of an axis of a sphere or spheroid, or one of the two points of intersection of its surface and its axis; specif. a *geographic pole* of the Earth.

pole [paleont] (a) An end of the axis of coiling in planispirally coiled shells or tests, as in the fusulinids. (b) An end of the theca in cystoids; e.g. "oral pole" containing the mouth, or "aboral pole" opposite the mouth and usually marking the end to which the column is attached.

pole [palyn] Either termination of the axis of a pollen grain or spore running from the center of the original tetrad to the center of the distal side of the grain, hence the center of both distal and proximal surfaces. The term is esp. useful for angiosperm pollen in which it is not apparent which is the proximal and which the distal surface.

pole [struc petrol] A point on a stereographic or equal-area projection that represents the projection of a lineation, the normal to a fabric plane, or a crystallographic direction. The term is also used for a point which represents the normal to any plane on a fabric diagram; thus the normal to a great-circle girdle would be called the *pole* of that distribution.

pole [surv] (a) A bar, staff, or rod used as a target in surveying; e.g. a *range rod.* (b) A unit of length measuring 16.5 ft or equivalent to a *perch* or *rod.* Also, a unit of area measuring 30.25 square yards or equal to a square perch or square rod. (c) The origin of a system of polar coordinates.

pole chain *Gunter's chain.*

pole coccolith A modified coccolith found at the flagellar and tail ends in flagellate coccolithophores exhibiting dimorphism (such as *Acanthoica*). Cf: *tail coccolith.*

pole-dipole array An electrode array in which one current electrode is placed at infinity while one current electrode and two potential electrodes in close proximity are moved across the structure to be investigated. The separation between the near current electrode and the closest potential electrode is an integral number times the spacing between the potential electrodes. The array is used in resistivity and induced-polarization surveys and in drill-hole logging.

pole-fleeing force A component of forces resulting from the Earth's rotation that carry the crust away from the poles, toward the equator. It was supposed by Wegener and Staub to have been sufficiently great to have displaced free-moving continents, an example cited being the closing-up of Tethys between the northern and southern continents of the Eastern Hemisphere. Jeffreys showed that the effectiveness of such a force is negligible. Etymol: German, Polflucht. Cf: *Coriolis force.*

pole of inaccessibility (a) A syn. of *ice pole.* (b) Point in Antarctica or Siberia most remote from the sea.

pole-pole array An electrode array, used in profiling or in logging, in which one current electrode and one potential electrode are removed to infinity while the other current electrode and the other potential electrode are kept in close proximity and traversed across the structure.

polianite A syn. of *pyrolusite,* esp. in well-formed tetragonal crystals.

polish An attribute of surface texture of a rock or particle, characterized by high luster and strong reflected light, produced by various agents; e.g. *desert polish, glacial polish,* or the coating formed on a gastrolith. Syn: *gloss.*

polished section A section of rock or mineral that has been highly polished. It is used for study of opaque minerals by plane or polarized reflected light. Cf: *polished thin section.* Partial syn: *microsection.*

polished surface *slickenside.*

polished thin section A thin section similar to that used in petrography but finished with a polished surface and not covered with a cover glass. It is useful for study by both transmitted and reflected light and by the electron microprobe. Cf: *polished section.* Partial syn: *microsection.*

polje A syn. of *interior valley.* Also spelled: *polya.* Etymol: Serbo-Croatian, "field".

pollen The several-celled microgametophyte of seed plants, enclosed in the microspore wall. Fossil pollen consists entirely of the microspore wall or exine, from which the microgametophyte itself was removed during or before lithification. The term "pollen" is a collective plural noun, and it is incorrect to say "a pollen". See also: *pollen grain.*

pollen analysis (a) A branch of palynology dealing with the study of Quaternary (esp. late Pleistocene and postglacial) sediments by employing pollen diagrams and isopollen maps to show the relative abundance of various pollen types in space and time; e.g. the identification and percentage determination of frequency of pollen grains of forest trees in peat bogs and lake beds as a means of dating fossil remains. It is used as a geochronologic and paleoecologic tool, often in collaboration with archaeology. (b) A term used prior to 1944 as a synonym of what is now known as *palynology*. —Syn: *pollen statistics.*

pollen diagram Any diagram of pollen abundance showing stratigraphic fluctuation; strictly, the graphical presentation of relative abundances of various genera of pollen and spores at successive levels of cores of Quaternary sediment studied in pollen analysis. Syn: *pollen profile.*

pollen grain One of the dustlike particles of which *pollen* is made up; a single unit of pollen. Syn: *grain [palyn].*

pollenite An olivine-bearing *phonolite,* also containing phenocrysts of sanidine, plagioclase, nepheline, clinopyroxene, amphibole, and biotite, in a glassy groundmass. The name, given by Lacroix in 1907, is from Vallone di Pollena, Monte Somma, Italy. Not recommended usage.

pollen mother cell A *mother cell* that is derived from the hypodermis of the pollen sac of a seed plant and that gives rise by meiosis to four cells, each of which develops into a pollen grain. See also: *spore mother cell.*

pollen profile A vertical section of an organic deposit (such as a peat bog) showing the sequence of buried or fossil pollen; a *pollen diagram.* Syn: *profile [palyn].*

pollen rain The total deposit of pollen and spores in a given area and period of time, as estimated by study of sediment samples and by pollen-trapping devices.

pollen sac One of the pouches in a seed plant that contain the pollen; e.g. the *anther* of an angiosperm or flowering plant.

pollen spectrum One of the characteristic horizontal lines in a pollen diagram, showing the relative abundances of the various sorts of pollen and spores diagrammed in a single sample analyzed from a single level.

pollen statistics *pollen analysis.*

pollen sum In pollen analysis, a portion of the total pollen count from which certain sorts of pollen are excluded by definition. The most usual pollen sum excludes all nonarborescent pollen and some arborescent pollen as well. Where pollen sums are used, pollen abundances are calculated as ratios of given sorts of pollen to the pollen sum, rather than to the raw total count.

pollen symbol An arbitrary sign used in Quaternary pollen diagrams, representing a genus or other group of plants, and serving as an internationally understood identification of a line in the pollen diagram.

pollen tube A more or less cylindric extension that develops from the wall of a pollen grain and that protrudes through one of its apertures when the grain germinates on contact with the stigmatic surface of flowering plants or the megasporangium of gymnosperms. The tube acts as a food-absorbing organ in lower seed plants such as cycads, but in flowering plants its primary function is to conduct the male nuclei to the vicinity of the female gametophyte to effect fertilization.

pollination The fertilization of a seed plant; specif. the transfer of pollen from a stamen or anther to an ovule or megasporangium.

pollinium A large, coherent mass of pollen, usually the contents of a whole locule of an anther, shed in the mature stage as a unit (as in the milkweed *Asclepias*). Pl: *pollinia.* Cf: *polyad; massula.*

pollucite A colorless, transparent zeolite mineral: $(Cs,Na)_2Al_2Si_4O_{12}\cdot H_2O$. It occurs massive or in cubes, and is used as a gemstone. Syn: *pollux.*

pollute (a) To make physically impure or unclean. (b) Often considered synonymous with "contaminate", but in some environmental regulations *pollute* represents a more advanced state of degradation. To pollute is to introduce any substance that alters some aspect of the environment in a manner to make it unfit for a particular use.

pollution *contamination [water].*

pollux Obsolete syn. of *pollucite.*

polster A stunted, perennial plant, usually isolated, that grows in a dense cushiony hummock capped with concentric layers of moss or lichen which incorporate sandy silt and small pebbles as they grow. It is esp. abundant on the snout of Matanuska Glacier, Alaska, and it resembles a *glacier mouse.* Etymol: German *Polster,* "cushion".

polya *polje.*

polyactine A sponge spicule having many rays diverging from a common center along more than four axes. Syn: *polyact.*

polyad A group of more than four mature pollen grains shed from the anther as a unit (as in *Acacia*). The grains within the polyad are usually in multiples of four. Cf: *pollinium; dyad; tetrad.*

polyannulate A long pore canal formed from a centrifugal extension of an enlarged annulus (Tschudy & Scott, 1969, p. 27).

polyargyrite A gray to black mineral: $Ag_{24}Sb_2S_{15}$ (?).

polyaxon A sponge spicule in which the rays grow along many axes of development emanating from a central point.

polybasite An iron-black to steel-gray metallic-looking mineral: $(Ag,Cu)_{16}Sb_2S_{11}$. Cf: *pearceite.*

polycentric Said of a corallite formed by a polyp retaining polystomodaeal condition permanently.

polychroism *pleochroism.*

polyclinal fold One of a group of adjacent folds, the axial surfaces of which have various orientations.

polyconic projection (a) A map projection (neither conformal nor equal-area) in which a series of right circular cones are each tangent to the Earth's surface at successive latitudes, each parallel thus constructed serving as if it were the chosen standard parallel for a simple *conic projection.* All parallels (developed from the bases of the cones) are arcs of nonconcentric circles with their centers on the straight line generally representing the central meridian, all other meridians being curved lines drawn through the true divisions of the parallels. The scale along each parallel and along the central meridian is true, but it increases on the meridians with increasing distance from the central meridian. The projection is suitable for maps of small areas and for areas of great longitudinal extent (such as Chile). (b) *modified polyconic projection.*

polycrase A black mineral: $(Y,Ca,Ce,U,Th)(Ti,Nb,Ta)_2O_6$. It is isomorphous with euxenite and occurs in granite pegmatites.

polycrystal An assemblage of crystal grains of a mineral, of unspecified number, shape, size, orientation, or bonding, that together form a solid body. Cf: *glomerocryst.*

polycyclic A term favored by many geomorphologists in place of the syns. *multicyclic* or *multicycle,* esp. for a stream whose course reflects base-leveling to more than one former sea level.

polydemic Said of an organism that is native to several regions.

polydymite A mineral of the linnaeite group: Ni_3S_4. Much so-called polydymite is really violarite.

polygene adj. (a) *polygenetic.* (b) A term applied specif. by Naumann (1850, p. 433) to an igneous rock composed of two or more minerals. Cf: *monogene; polymineralic; polymictic [sed].* Syn: *polymere; polygenic.*

polygenetic (a) Resulting from more than one process of formation, derived from more than one source, or originating or developing at various places and times; e.g. said of a mountain range resulting from several orogenic episodes. (b) Consisting of more than one type of material, or having a heterogeneous composition; e.g. said of a conglomerate composed of materials from several different sources. —Cf: *monogenetic.* Syn: *polygene; polygenic.*

polygenic (a) *polygenetic.* (b) *polygene.*

polygeosyncline A geosynclinal-geoanticlinal belt along the continental margin and receiving sediments from a borderland on its oceanic side (Schuchert, 1923). Cf: *monogeosynlcine.* See also: *sequent geosyncline.*

polygon A form of horizontal patterned ground whose mesh is tetragonal, pentagonal, or hexagonal. Its formation is favored by intensive frost action. See also: *sorted polygon; nonsorted polygon.*

polygonal ground A form of patterned ground marked by polygonal arrangements of rock, soil, and vegetation, produced on a level or gently sloping surface by frost action; esp. a ground surface consisting of a large-scale network of ice-wedge polygons. Syn: *polygon ground; polygonal soil; Polygonboden; polygonal markings; cellular soil.*

polygonal karst *cockpit karst.*

polygonal soil *polygonal ground.*

polygonization A process through which crystals reach an equilibrium number of dislocations per unit volume during deformation. Newly created dislocations that do not have the time to glide to a crystal boundary may climb into walls. Both dislocation density and wall ("subgrain boundaries") spacing are proportional to the applied differential stress.

polyhalite A mineral: $K_2MgCa_2(SO_4)_4\cdot 2H_2O$. It is often brick-red owing to iron oxide.

polyhedric projection A projection for large-scale topographic maps in which a small quadrangle on the sphere or spheroid is projected onto a plane trapezoid, the rectilinear parallels and meridians corresponding closely to arc distances on the sphere or spheroid.

polykinematic mélange A mélange that includes elements derived from an earlier mélange.

polylitharenite A lithic arenite with a diversity of sand-sized rock particles—volcanic, sedimentary, and metamorphic (Folk, 1968, p. 135).

polylithionite A mineral of the mica group: $KLi_2AlSi_4O_{10}(F,OH)_2$. It is related to lepidolite.

polymere An adjective applied by Rosenbusch (1898, p. 17) to an igneous rock composed of two or more minerals. Syn: *polygene; polymineralic.*

polymetallic Said of deposits that contain economically important quantities of three or more metals.

polymetamorphism Polyphase or multiple metamorphism, whereby two or more successive metamorphic events have left their imprint upon the same rocks. The superimposed metamorphism may be of a higher or lower grade than the earlier type. See also: *retrograde metamorphism; prograde metamorphism.* Syn: *superimposed metamorphism.*

polymict breccia A brecciated meteorite containing fragments of differing composition. Cf: *monomict breccia.*

polymictic [lake] Said of a lake that is continually mixing and has no persistent thermal stratification; e.g. a lake of great area, moderate or little depth, in a region of low humidity or at great altitude. Cf: *oligomictic [lake].*

polymictic [sed] (a) Said of a clastic sedimentary rock composed of many rock types, such as an arkose or graywacke, or a conglomerate with more than one variety of pebble; also, said of the clasts of such a rock. Polymictic rocks are characteristic of mobile (unstable) conditions such as are found in orogenic belts. Cf: *oligomictic.* (b) Said of a clastic sedimentary rock composed of more than one mineral species. Cf: *monomictic; polygene.*—Syn: *polymict.*

polymignyte A black mineral: $(Ca,Fe^{+2},Y,Zr,Th)(Nb,Ti,Ta,Fe^{+3})O_4$. Also spelled: *polymignite.*

polymineralic Said of a rock composed of two or more minerals; esp. said of an igneous rock consisting of more than one essential mineral. Cf: *monomineralic.* Syn: *polymere; polygene.*

polymodal distribution A frequency distribution characterized by two or more localized modes, each having a higher frequency of occurrence than other immediately adjacent individuals or classes. Cf: *bimodal distribution.*

polymodal sediment A sediment whose particle-size distribution shows one or more secondary maxima.

polymorph [cryst] A crystal form of a substance that displays *polymorphism.* Syn: *polymorphic modification; allomorph.*

polymorph [evol] An organism exhibiting *polymorphism [evol];* also, one of the forms of such an organism. Syn: *pleomorph.*

polymorphic Said of a chemical substance that displays *polymorphism,* and said of the different crystal forms so displayed. Syn: *polymorphous; pleomorphous; allomorphic; allomorphous.*

polymorphic modification *polymorph.*

polymorphism [cryst] The characteristic of a chemical substance to crystallize in more than one form, e.g. rhombic and monoclinic sulfur. Such forms are called *polymorphs.* Adj: *polymorphic.* See also: *dimorphism; trimorphism; tetramorphism; polytypism; allotropy.*

polymorphism [evol] The existence of a species in several forms independent of sexual variations; esp. referring to different types of individuals within a colony, as in bryozoans. Adj: *polymorphic.* See also: *polymorph [evol]; polytypy [evol].* Syn: *pleomorphism.*

polymorphous *polymorphic.*

polynigritite A type of *nigritite* that occurs finely dispersed in argillaceous rocks. Cf: *exinonigritite; humonigritite; keronigritite.*

polynya Any nonlinear opening enclosed in ice, esp. a large expanse of water, other than a *lead [ice],* surrounded by sea ice, but not large enough to be called *open water [ice];* commonly found off the mouth of a large river. Pl: *polynyas; polynyi.* Etymol: Russian *polyn'ya.* See also: *shore polynya; recurring polynya.* Syn: *pool [ice]; glade [ice]; ice clearing; clearing.*

polyp A typical coelenterate individual, with a hollow and tubular or columnar body terminating anteriorly in a central mouth surrounded by tentacles directed upward. It is posteriorly closed, and attached to the bottom (as in *Hydra*) or more or less directly continuous with other individuals of a compound animal (as in most corals).

polypary The common investing structure or tissue in which coral polyps are embedded; a coral colony as a whole. Syn: *polyparium.*

polyphyletic (a) Evolving from more than one ancestral stock. Cf: *monophyletic.* (b) In *cladism,* pertaining to a higher taxon that contains species descended from different ancestral stocks but incorrectly united on the basis of one or more misinterpreted *shared derived characters.*

polypide The parts of a bryozoan autozooid that undergo periodic replacement, viz. tentacles, tentacle sheath, alimentary canal, associated musculature, and nerve ganglion.

polyplacophoran A marine mollusk, considered a subclass of the *amphineurans,* characterized by a protective girdle consisting of overlapping calcareous valves or plates, usually eight. Range, Upper Cambrian to present. See also: *aplacophoran.* Syn: *chiton.*

polyplicate Said of pollen grains (such as those of *Ephedra*) with multiple longitudinal linear thinnings in the exine that resemble, but are not, true colpi.

polyquartz A group term for $Al(PO_4)$, $Al(AsO_4)$, $B(PO_4)$, etc.

polyschematic Said of mineral deposits having more than one textural element. Cf: *monoschematic.*

polysomatic chondrule A chondrule consisting of several crystals. Cf: *monosomatic chondrule.*

polystomodaeal budding A type of budding in scleractinian corals in which more than three stomodaea are developed within a common tentacular ring. Examples: *intramural budding; circummural budding; circumoral budding.*

polystromatic Said of a plant part, e.g. an algal thallus, composed of many layers of cells. Cf: *monostromatic; oligostromatic.*

polysynthetic twinning *Repeated twinning* of three or more individuals according to the same twin law and on parallel composition planes; e.g. albite twinning of plagioclase. It is often revealed megascopically by striated surfaces. Cf: *cyclic twinning.*

polythalamous Many-chambered; esp. said of a foraminifer or foraminiferal test composed of numerous chambers. See also: *multilocular.*

polytopism The independent origin of a species or other systematic group in more than one geographic area, presumably as a result of identical change in scattered individuals of its ancestor. Cf: *monotopism.*

polytype A type of *polymorph [cryst]* whose different possible forms result from different stackings of similar atomic structural units. For example, in metals, hexagonal close-packed (ABABAB) sequences and cubic close-packed (ABCABC) sequences would be called polytypes. Adj: *polytypic.*

polytypic (a) Said of a species consisting of subspecies that replace each other geographically. (b) Said of a taxon that contains two or more taxa of the next lower rank. Cf: *monotypic.* (c) Adj. of *polytype.*

polytypism The property of a mineral to crystallize in more than one form, owing to more than one possible mode of atomic packing; a form of one-dimensional *polymorphism [cryst].* Such a mineral is a *polytype.* Syn: *polytypy [cryst].*

polytypy [cryst] *polytypism.*

polytypy [evol] The existence of a species in several geographically separated forms; the existence of a species containing several geographic subspecies. Adj: *polytypic.* Cf: *polymorphism [evol].*

polyvalent Said of foraminiferal specimens or individuals forming a vegetative and accidental association (probably due to crowding) with two or more embryonic apparatuses always of the same generation (microspheric or megalospheric) and of approximately the same age (TIP, 1964, pt.C, p.62).

polyxene *platinum.*

polyzoan *bryozoan.*

polyzoic Said of a habitat that supports a wide variety of animals.

polzenite A group of *lamprophyres* characterized by the presence of olivine and melilite; also, any rock in that group, e.g. *modlibovite, luhite, vesecite.* The name, given by Scheumann in 1913, is for Polzen, Czechoslovakia. Not recommended usage.

pond (a) A natural body of standing fresh water occupying a small surface depression, usually smaller than a lake and larger than a pool. (b) A term frequently used interchangeably with *lake* and *pool* and applied indiscriminately to water bodies in various sections of the U.S. (c) A body of water formed in a stream by *ponding.* (d) A small, artificial body of water, used as a source of water. In Great Britain, the term usually refers only to a small body of standing water of artificial formation.

pondage land Land on which water is stored as dead water during flooding. It does not contribute to the downstream passage of flow. Syn: *flood fringe.*

ponding (a) The natural formation of a pond in a stream by an interruption of the normal streamflow, either by a transverse uplift whose rate of elevation exceeds that of the stream's erosion, or by a dam caused by landsliding, glacial deposition, volcanism, or strong flow of water from a side valley. (b) The artificial impoundment of stream water to form a pond.

pongo A term used in South America (esp. in Peru) for a canyon or gorge, esp. one cutting through a ridge or mountain range; also, a narrow and dangerous ford. Etymol: Quechua *puncu,* "door".

ponor A Yugoslavian term for a hole in karstic topography that acts as a *sinking stream.*

Pontian Eastern European stage: uppermost Miocene (above Pannonian, below Kimmerian). Also used loosely for Upper Miocene, esp. in continental strata. It is considered as Pliocene in some older literature, because it was mistakenly correlated with marine Pliocene of western Europe by Lyell in 1865. See also: *Turolian.*

pontic Pertaining to sediments or facies deposited in comparatively deep and motionless water, such as an association of black shales and dark limestones deposited in a stagnant basin. Etymol: Greek *pontos,* "sea". Cf: *euxinic.*

ponzaite A term proposed by Reinisch in 1912 embracing feldspathoidal and feldspathoid-free *trachytes.* Cf: *ponzite.* Not recommended usage.

ponzite A feldspathoid-free *trachyte* containing augite and pyroxenes that may be rimmed with acmite or acmite-augite. Little or no biotite or amphibole occurs as phenocrysts. Its name, given by Washington in 1913, is derived from the Ponza Islands, Italy. Cf: *ponzaite.* Not recommended usage.

pool [coast] (a) *tide pool.* (b) *beach pool.*

pool [ice] (a) *polynya.* (b) A large *puddle.*

pool [petroleum] A subsurface accumulation of oil and/or gas in porous and permeable rock. See also: *oil pool; gas pool.*

pool [water] (a) A small, natural body of standing water, usually fresh; e.g. a stagnant body of water in a marsh, or a transient puddle in a depression following a rain, or a still body of water within a cave. (b) A small, quiet, rather deep *reach* of a stream, as between rapids or where there is little current. See also: *plunge pool.* (c) A body of impounded water, artificially confined above a dam or the closed gates of a lock.

pool spring A spring fed from a deep source, sometimes related to a fault, and forming a pool. A pool spring may develop the shape of a jug, because a peripheral platform is developed over the water by vegetation and sediments blown in by the wind (Meinzer, 1923, p. 55). Cf: *mound spring.*

poop shot *weathering shot.*

poorly graded (a) A geologic term for *poorly sorted.* (b) An engineering term pertaining to a *nongraded* soil or unconsolidated sediment in which all the particles are of about the same size or in which a continuous distribution of particle sizes from the coarsest to the finest is lacking.—Ant: *well-graded.*

poorly sorted Said of a clastic sediment or of a cemented detrital rock that is not *sorted* or that consists of particles of many sizes mixed together in an unsystematic manner so that no one size class predominates and that has a sorting coefficient in the range of 3.5 to 4.5 and higher. Based on the phi values associated with the 84 and 16 percent lines, Folk (1954, p. 349) suggests *sigma phi* limits of 1.00-2.00 for poorly sorted material. Ant: *well-sorted.* Syn: *unsorted; assorted; nongraded; poorly graded.*

poort A term used in southern Africa for a mountain pass, esp. a water gap or a gorge cut by a river through a ridge or a range of hills or mountains. Etymol: Afrikaans, "gate".

popple rock An English term for *pebble bed.*

popular name *vernacular name.*

population [ecol] (a) All the individuals of the same species, or of a group of closely related species. (b) In former usage, all organisms occupying a certain area or environment.

population [stat] Any theoretical group of items or samples, all of which are capable of being measured statistically in one or more respects; all possible values of a variable, either finite or infinite, continuous or discrete. Syn: *universe.*

porate Said of pollen grains having a pore or pores in the exine.

porcelain clay A clay suitable for use in the manufacture of porcelain; specif. *kaolin.*

porcelainite (a) *porcellanite.* (b) *mullite.*

porcelain jasper A hard, naturally baked, impure clay or *porcellanite* which, because of its red color, was long considered a variety of jasper.

porcelain stone *china stone [ign].*

porcelaneous Resembling unglazed porcelain; e.g. said of a foraminiferal test having a calcareous wall with a dull white luster, or said of a rock consisting of chert and carbonate impurities or of clay and opaline silica. Also spelled: *porcellaneous; porcelanous; porcelainous.*

porcelaneous chert A hard, opaque to subtranslucent *smooth chert,* having a smooth fracture surface and a typically china-white appearance resembling chinaware or glazed porcelain (Ireland et al., 1947, p. 1485).

porcelanite *porcellanite.*

porcellanite A dense siliceous rock having the texture, dull luster, hardness, conchoidal fracture, and general appearance of unglazed porcelain; it is less hard, dense, and vitreous than chert. The term has been used for an impure chert, in part argillaceous (see also *siliceous shale*); for an indurated or baked clay or shale often found in the roof or floor of a burned-out coal seam (see also *porcelain jasper*); and for a fine-grained, acidic tuff compacted by secondary silica (see also *hälleflinta*). Etymol: Italian *porcellana,* "porcelain". Also spelled: *porcelanite; porcelainite.*—Syn: *thermuticle.*

pore [geol] A small to minute opening or passageway in a rock or soil; an *interstice.*

pore [paleont] (a) A small opening in an echinoderm; e.g. a *hydrospire pore* of a blastoid, or an opening from the exterior through the thecal plates of a cystoid. The term has also been used for a horizontal perforation (tube, canal, or slit) occupying parts of two adjoining thecal plates of a cystoid (TIP, 1967, pt.S, p.113). (b) One of numerous small openings from the exterior of a sponge; e.g. the terminus of a canal at any surface, an opening surrounded by a single cell, or a smaller-sized opening serving for inward flow of water. Cf: *osculum; ostium; skeletal pore.* (c) A hole through the wall, septum, or tabula in archaeocyathids; it may be round, oval, slitlike, rectangular, or polygonal (TIP, 1972, pt. E, p. 41).

pore [palyn] One of the external, more or less circular or slightly oval thinnings or openings in the exine of pollen grains. Pores may occur by themselves or in association with colpi. Cf: *colpus.* See also: *germ pore.*

pore canal (a) A minute tubular passageway extending through the shell of an ostracode; e.g. *normal pore canal* and *radial pore canal.* (b) A perforation in a thecal plate of an echinoderm. (c) A pore in relatively thick skeletal elements of archaeocyathids, in which the length of the opening is longer than its diameter. They may be straight or curved, circular or hexagonal in cross section, and normal or oblique to the wall surface. (d) A duct passing through the exoskeleton of a trilobite, ranging from about 1 to 75 μm in diameter. Some authors restrict the term to those ducts of about 1 to 2 μm diameter, using the term *setal duct* for the larger ones.

pore chamber A *dietella* of a cheilostome bryozoan.

pore diameter The diameter of a pore in a rock; it is measured as the diameter of the largest sphere that may be contained within the pore.

pore frame The raised edge around the area enclosing a minute opening in a radiolarian.

pore ice Ground ice that fills or partially fills pore spaces in the ground; it is formed by freezing of pore water in situ with no addition of water.

pore interconnection A constricted opening connecting pores in a pore system (Choquette & Pray, 1970, p. 214). Syn: *pore throat.*

pore pair An ambulacral pore of an echinoid, divided by a wall of stereome and through which a single tube foot passes.

pore plate (a) The flat, pore-bearing base of the podoconus in a nassellarian radiolarian. (b) In cheilostome bryozoans, a thin part of a calcareous wall forming the skeletal part of a septula. It is perforated by one or more pores. Syn: *rosette plate.*

pore plug Minute, single, organic, microporous plates lying at the base of external openings in certain foraminifers (TIP, 1964, pt.C, p.62).

pore pressure *neutral stress.*

pore rhomb One of the diamond-shaped structures on the surface of thecal plates of cystoids, consisting of a group of parallel, laterally directed perforations (tubes, grooves, slits) each end of which occupies parts of two adjacent plates (so that each plate of a pair bears one half of the rhomb). The ends may be exposed to the outside or covered by thin calcareous layers. Cf: *diplopore.* See also: *pectinirhomb.* Syn: *rhomb.*

pore space The open spaces in a rock or soil, considered collectively (Stokes & Varnes, 1955, p. 112). Syn: *pore volume.*

pore system All the openings in a rock or sediment, considered as a unit (Choquette & Pray, 1970, p. 214).

pore throat *pore interconnection.*

pore-tube In archaeocyathids, a commonly hexagonal thin-walled tube, formed by horizontal, oblique, or curved wall plates, or by scooplike bracts or peaks (TIP, 1972, pt. E, p. 41).

pore volume *pore space.*

pore water *interstitial water.*

pore-water pressure *neutral stress.*

poriferan *sponge.*

porocyte One of the large tubular cells that constitute the wall of the inhalant canals of some sponges, completely enclosing or surrounding a pore and capable of regulating its size by expansion or contraction.

porolith A coccolith in the form of a polygonal prism with an axial perforation; an axially perforated *prismatolith.* The term was introduced for the elements of the coccolithophorid *Thoracosphaera,* which electron-microscopic studies have shown to be both perforate and imperforate.

poros A coarse-grained limestone occurring in the Peloponnesus and extensively used as a building material by the ancient Greeks.

porosimeter An instrument that measures porosity.

porosity The percentage of the bulk volume of a rock or soil that is occupied by interstices, whether isolated or connected. Cf: *effective porosity.* Syn: *total porosity.* See also: *primary porosity; secondary porosity; porous.*

porosity log A generic term for *well log* curves whose measurements relate easily to formation porosity (*sonic log, density log, neutron log, epithermal-neutron log*).

porosity pod A potential reservoir for oil and gas in an area of monoclinal dip owing to local depositional variations in a lenticular sandstone; a local area of porosity within a sandstone lens (Forgotson & Forgotson, 1975, p. 1113-1114).

porosity trap A *trap [petroleum]* formed by lateral variation in porosity of the reservoir rock, e.g. as a result of cementation, the presence of clay minerals, or a decrease in grain size. Syn: *stratigraphic trap.*

porous Having numerous interstices, whether connected or isolated. "Porous" usually refers to openings of smaller size than those of a *cellular* rock. Cf: *cavernous.* See also: *porosity.*

porpezite A mineral consisting of a native alloy of gold and 5-10% of palladium. Syn: *palladium gold.*

porphyrin A large complex organic ring compound made up of, in addition to other rings, four substituted pyrrole rings. Chlorophyll is a porphyrin with magnesium coordinated in the center of the ring; heme (of hemoglobin) is a porphyrin with iron coordinated in the center. Porphyrins are found not only in plants, but also in carbonaceous shale, crude oil, and coal.

porphyrite An obsolete term synonymous with *porphyry.* The term was originally used to distinguish porphyries that contain plagioclase phenocrysts from those that contain alkali feldspar phenocrysts.

porphyritic (a) Said of the texture of an igneous rock in which larger crystals (*phenocrysts*) are set in a finer-grained *groundmass,* which may be crystalline or glassy or both. Also, said of a rock with such texture, or of the mineral forming the phenocrysts. (b) Pertaining to or resembling porphyry.

porphyro-aphanitic Said of the texture of a porphyritic igneous rock, esp. an extrusive rock, consisting of phenocrysts in an aphanitic groundmass. Also, said of a rock with such texture.

porphyroblast A pseudoporphyritic crystal in a rock produced by metamorphic recrystallization. Adj: *porphyroblastic.* Syn: *metacryst; pseudophenocryst.*

porphyroblastic Pertaining to the texture of a recrystallized metamorphic rock having large idioblasts of minerals possessing high *form energy* (e.g. garnet, andalusite) in a finer-grained crystalloblastic matrix. See also: *pseudoporphyroblastic.*

porphyroclast A paleoblast within a finer-grained recrystallized and/or pulverized matrix. See also: *porphyroclastic.*

porphyroclastic Said of a heterogranular metamorphic texture characterized by volumetrically significant amounts of both *porphyroclasts* and *neoblasts.* Also, said of a rock with such a texture. This follows "a longstanding descriptive usage in geology, and does not imply the earlier genetic and etymological connotation of breakage or fracture. Indeed, it is believed that porphyroclastic rocks are the product of plastic deformation and dynamic recrystallization rather than brittle deformation or cataclasis" (Harte, 1977). Semantically correct, though less used, syn: *blastogranular. Blastolaminar* applies to the most strongly *laminated* facies of porphyroclastic rocks.

porphyroclastic structure *mortar structure.*

porphyrocrystallic A syn. of *porphyrotopic.* The term was introduced by Phemister (1956, p.74).

porphyrogranulitic Nonrecommended term for an ophitic texture characterized by large phenocrysts of feldspar and augite or olivine in a groundmass of smaller lath-shaped feldspar crystals and irregular augite grains; a combination of porphyritic and intergranular textures. Also, said of a rock having such texture.

porphyroid n. A blastoporphyritic or sometimes porphyroblastic metamorphic rock of igneous origin, or a feldspathic metasedimentary rock having the appearance of a porphyry. It occurs in the lower grades of regional metamorphism.—adj. Said of or pertaining to such a rock.

porphyroid neomorphism A term introduced by Folk (1965, p. 22) for *aggrading neomorphism* in which small crystals are converted to large ones by growth of a few large crystals in and replacing a static matrix; e.g. the replacement of an aragonite shell by calcite mosaic. Cf: *coalescive neomorphism.*

porphyroskelic Said of an arrangement in a soil fabric whereby the plasma occurs as a dense matrix in which skeleton grains are set in the manner of phenocrysts in a porphyritic rock (Brewer, 1964, p.170). Cf: *agglomeroplasmic; intertextic.*

porphyrotope A large crystal enclosed in a finer-grained matrix in a sedimentary rock showing porphyrotopic fabric; e.g. a large dolomite crystal in finer-grained calcitic matrix.

porphyrotopic Said of the fabric of a recrystallized carbonate rock or a chemically precipitated sediment in which the constituent crystals are of more than one size and in which the larger crystals are enclosed in a finer-grained matrix. The term was proposed by Friedman (1965, p.649). Cf: *poikilotopic.* Syn: *porphyrocrystallic.*

porphyry An igneous rock of any composition that contains conspicuous *phenocrysts* in a fine-grained groundmass; a porphyritic igneous rock. The term (from a Greek word for a purple dye) was first applied to a purple-red rock quarried in Egypt and characterized by phenocrysts of alkali feldspar. The rock name descriptive of the groundmass composition usually precedes the term, e.g. diorite porphyry. Obsolete syn: *porphyrite.*

porphyry copper deposit A large body of rock, typically porphyry, that contains disseminated chalcopyrite and other sulfide minerals. Such deposits are mined in bulk on a large scale, generally in open pits, for copper and by-product molybdenum. Most deposits are 3 to 8 km across, and of low grade (less than 1% Cu). They are always associated with intermediate to felsic hypabyssal porphyritic intrusive rocks. Distribution of sulfide minerals changes outward from dissemination to veinlets and veins. *Supergene enrichment* has been very important at most deposits, as without it the grade would be too low to permit mining. Cf: *porphyry molybdenum deposit; porphyry tin deposit.*

porphyry molybdenum deposit A large, low-grade molybdenite deposit, in which sulfide minerals are distributed as stockworks and disseminated grains in and near siliceous porphyritic intrusive rocks with quartz-latite to rhyolite composition; associated hydrothermal alteration is characterized by potassium feldspathization, sericitization, argillization and silicification. These deposits are characteristically found in the western North American Cordillera (Clark, 1972), and appear more continentally affiliated than *porphyry copper deposits.* See also: *porphyry tin deposit.* Syn: *stockwork molybdenum deposit.*

porphyry tin deposit A center of tin mineralization in which a tin-bearing mineral, usually cassiterite, is distributed in porphyritic igneous stocks of intermediate composition as disseminated grains, stockworks, or intrusive-breccia fillings, with associated hydrothermal alteration, especially sericitization-silicification, of the wallrocks (Sillitoe et al., 1975). The term is considered to convey the same implications as *porphyry copper deposit* and *porphyry molybdenum deposit,* and is applied to many of the tin occurrences in Bolivia.

"porphyry" uranium deposit A uranium deposit in granitic rocks cut by aplite and pegmatite. The ore minerals, which include primary uraninite and uranothorite, are disseminated and also occur in joints and openings; they are microscopic to submicroscopic in size. Most of the uranium is contained in "felsic and varietal minerals" of the rock (Armstrong, 1974, p. 629). Examples are cited in Africa, Canada, Australia, and the U. S., but the designation "porphyry" remains provisional.

portal The mouth of an adit or tunnel.

Porterfield North American stage: Middle Ordovician (above Ashby, below Wilderness) (Cooper, 1956).

porticus A distinctly asymmetrical apertural flap in the tests of some planktonic foraminifers (such as *Ticinella* and *Praeglobo-*

truncana). It was originally defined as imperforate. Pl: *portici.*

portland cement A *cement [mater]* produced by fine-grinding a carefully proportioned mixture of limestone and shale (or equivalent raw materials); heating the mixture to incipient fusion in a rotary kiln; and fine-grinding the resulting clinker. It was developed in 1824 by Joseph Aspdin in England; the name is for a resemblance to Portland stone, a widely used British building stone. The U.S. industry dates from 1875.

Portlandian European stage: uppermost Jurassic (above Kimmeridgian, below Purbeckian-Tithonian). See also: *Bononian; LaCasitan; Lower Volgian.*

portlandite A mineral: $Ca(OH)_2$. It occurs as hexagonal plates in contact-metamorphic rocks and also in portland cement.

Portland stone (a) A yellowish-white, oolitic limestone from the Isle of Portland (a peninsula in southern England), widely used for building purposes. (b) A purplish-brown sandstone (*brownstone*) from Portland, Conn.

pošepnyte A light-green to red-brown resin with a high oxygen content (18%), found in plates and nodules in the Great Western mercury mine, Lake County, Calif. Also spelled: *pošepnyite; posepnyte.*

position (a) Data that define the location of a *point* with respect to a reference system in surveying. (b) The place occupied by a point on the surface of the Earth or in space. (c) The coordinates that define the location of a point on the geoid or spheroid. (d) A prescribed reading of the graduated horizontal circle of a direction instrument theodolite to be used for the observation on the initial station of a series of stations which are to be observed.

positive [optics] Said of anisotropic crystals: of a uniaxial crystal in which the ordinary index of refraction is greater than the extraordinary index; and of biaxial crystal in which the intermediate index of refraction β is closer in value to α, and in which Z is the acute bisectrix. Cf: *negative [optics].*

positive [photo] A photographic transparency or print having approximately the same or similar rendition of tones or colors as that of the original subject. Cf: *diapositive; negative [photo].*

positive [tect] n. An area of a craton that tends to stand higher than the surrounding area; a *positive element.*

positive area *positive element.*

positive birefringence *Birefringence* in which the velocity of the ordinary ray is greater than that of the extraordinary ray.

positive confining bed The upper confining bed of an aquifer whose head is above the upper surface of the zone of saturation, i.e. above the water table. Little used.

positive element A structural feature or area, characterized by conspicuous upward movement (uplift, emergence), by relative stability, or by subsidence that is less rapid or less frequent than that of adjacent *negative elements.* Syn: *positive area; archibole; positive.*

positive elongation In a section of an anisotropic crystal, a *sign of elongation* that is parallel to the slower of the two plane-polarized rays. Cf: *negative elongation.*

positive estuary An estuary in which there is a measurable dilution of seawater by land drainage. Ant: *inverse estuary.* See also: *freshwater estuary.*

positive landform An upstanding topographic form, such as a mountain, hill, or plateau, or a volcanic feature formed by an excess of material (as a cinder cone). Ant: *negative landform.*

positive movement (a) An upward movement of the Earth's surface relative to an adjacent part of the surface, such as produced by an uplift or by isostatic recovery; a positive movement of the land may result in a *negative movement* of sea level. (b) A relative rise of sea level with respect to the land, as produced by a *negative movement* of the Earth's crust or by an advance of the sea.

positive ore An orebody that has been exposed and developed on four sides, and for which tonnage and quality estimates have been made. Cf: *developed reserves; proved reserves.*

positive pole The north-seeking member of the *magnetic poles.* Cf: *negative pole; dipole field.*

positive shoreline *shoreline of submergence.*

posnjakite A mineral: $Cu_4(SO_4)(OH)_6 \cdot H_2O$.

possible ore A mineral deposit whose existence and extent is postulated on the basis of past geologic and mining experience. Syn: *future ore; geological ore.* Cf: *inferred ore; indicated ore; potential ore.* See also: *extension ore.*

post (a) An old English term, now largely obsolete, for a thick bed of sandstone or limestone. (b) A mass of slate traversed by so many joints as to be useless for building purposes.

postabdomen (a) The slender, attenuated posterior part of the abdomen of a scorpion, composed of five segments and a telson modified as a poison gland; the narrow posterior part of the abdomen of a merostome. (b) The *telson* of a crustacean. (c) A joint succeeding the third segment (abdomen) of the shell of a nassellarian radiolarian.

postadaptation More perfect adjustment to an adaptive zone after an organism has entered it.

postcervical groove One of the most conspicuous of the grooves on the carapace of decapods. It extends from about the middle of the dorsum of the carapace, or somewhat behind it, downward and slightly forward, its end meeting the hepatic groove (Holthuis, 1974, p. 733).

postcingular series The series of plates immediately behind the girdle of a dinoflagellate theca, usually fewer in number and often larger in size than those of the *precingular series.*

Post-Classic n. In New World archaeology, the final cultural stage before the arrival of the European colonists; it follows the Classic.—adj. Pertaining to the Post-Classic.

postcollarette A fine membrane prolonging the collar and appearing to close the pseudostome in some chitinozoans. It may have served for temporary closure or be a remnant of an attachment; it is commonly ragged or folded back and ornamented by a filamentous network of clear lines.

poster Part of the orifice in ascophoran cheilostomes (bryozoans) that is proximal to the condyles and leads to the ascus. Cf: *anter.*

posterior adj. Situated toward the back of an animal, or at or toward the hinder part of the body, as opposed to *anterior;* e.g. in a direction (in the plane of symmetry or parallel to it) toward the pedicle and away from the mantle cavity of a brachiopod, or (in the plane of bilateral symmetry) opposite the position of the head of a bivalve mollusk, or in a typically apical direction along the midline axis of a gastropod and opposite the head.—n. The hinder part or end of an animal, e.g. the part of a brachiopod shell occupied by the viscera and including the area nearest to the pedicle opening (the side defined by the position of the hinge line); the end defined by the position of the pallial sinus of a bivalve mollusk; or the end opposite the aperture of a gastropod.

posterior margin The posterior part of the junction between the edges of brachiopod valves. It may be a hinge line or a cardinal margin.

posterior oblique muscle One of a pair of muscles in discinacean brachiopods, originating posteriorly and laterally on the pedicle valve, and converging dorsally to insertions on the brachial valve between posterior adductor muscles (TIP, 1965, pt.H, p.150).

posterior side The back or rear end of a conodont; e.g. the concave side of the cusp (facing in the direction toward which the tip of the cusp points) in simple conodont elements, or the concave side of cusps and denticles in compound conodont elements, or the distal end of the plate in many platelike conodont elements (but orientation varies). Ant: *anterior side.*

postglacial Pertaining to the time interval since the total disappearance of continental glaciers in middle latitudes or esp. from a particular area; e.g. "postglacial rebound".

posthumous fold A kind of recurrent folding that occurs in younger sedimentary rocks overlying a buried fold belt. The term is little used.

posthumous movement Movement on a pre-existing structure.

posticum An obsolete term for the opening of an exhalant canal on the external or spongocoel surface of a sponge. Pl: *postica.* See also: *ostium.*

postkinematic *posttectonic.*

postmagmatic An indefinite term applied generally to reactions or events occurring after crystallization of the bulk of a magma, and usually including the hydrothermal stage. Cf: *deuteric.*

postmineral adj. In economic geology, said of a structural or other feature formed after mineralization. Cf: *premineral; intermineral.*

Postnormapolles A group of Cretaceous and Cenozoic porate pollen without the usual pore apparatus or other features of the *Normapolles* group, from which it presumably derived.

postobsequent stream A strike stream developed after the obsequent stream into which it flows (Varney, 1921, p. 198).

postorogenic Said of a geologic process or event occurring after a period of orogeny; or said of a rock or feature so formed. Cf: *posttectonic.*

postorogenic phase The final phase of an orogenic event, following the climactic orogeny. Cf: *preorogenic phase; orogenic phase;*

orogenic cycle.

postorogenic pluton An igneous intrusion emplaced after a period of orogenic activity.

postseptal passage An opening that connects all chamberlets of the same chamber of a foraminiferal test (as in the Alveolinidae), located between wall and septum at the rear of the chamber. Cf: *preseptal passage.*

post stone An English term for any fine-grained sandstone or limestone. Also spelled: *poststone.*

posttectonic Said of a geologic process or event occurring after any kind of tectonic activity; or said of a rock or feature so formed. Cf: *postorogenic.* Syn: *postkinematic.*

pot [coal] (a) *pot bottom.* (b) *caldron bottom.*

pot [geomorph] A general term for any hole, pit, or depression produced naturally in the ground (and often containing water) that suggests the shape or form of a pot or kettle; specif. any of various kinds of *pothole.*

pot [permafrost] A sedimentary deposit in the shape of a pot in distinctively different host materials. Typical pots of the Upland Gravels of Maryland and Virginia are 2 m to 8 m wide; they are composed of sandy-gravelly silt in gravel. A seasonal frost origin is proposed (Conant et al., 1976).

pot [seis] A colloquial syn. of *seismic detector.*

pot [speleo] *vertical cave.*

potable water Water that is safe and palatable for human use; *fresh water* in which any concentrations of pathogenic organisms and dissolved toxic constituents have been reduced to safe levels, and which is, or has been treated so as to be, tolerably low in objectionable taste, odor, color, or turbidity and of a temperature suitable for the intended use.

potamic Pertaining to rivers or river navigation; e.g. "potamic transport", or transportation of sediments by river currents.

potamoclastic rock *fluvioclastic rock.*

potamography The description of rivers. Etymol: Greek *potamos,* "river".

potamology The scientific study of rivers. Etymol: Greek *potamos,* "river".

potarite A silver-white tetragonal mineral: PdHg. It is a natural alloy of palladium and mercury. Syn: *palladium amalgam.*

potash (a) Potassium carbonate, K_2CO_3. (b) A term that is loosely used for potassium oxide, potassium hydroxide, or even for potassium, in such informal expressions as *potash spar.*

potash alum (a) *alum.* (b) *kalinite.*

potash bentonite *potassium bentonite.*

potash feldspar A misnomer for *potassium feldspar.*

potash kettle *kettle [glac geol].*

potash lake An *alkali lake* whose waters are rich in dissolved potassium salts. Examples occur in the western part of the sandhills region of north-central Nebraska.

potash mica A misnomer for potassium-rich mica; specif. *muscovite.*

potash spar An informal term for potassium feldspar, i.e. *orthoclase* or *microcline,* or for a feldspar mixture assaying at least 10% K_2O (Rogers & Neal, 1975, p. 638). Syn: *K-spar.* Cf: *soda spar.*

potassic Said of a rock or mineral containing a significant amount of potassium.

potassium alum *alum.*

potassium-argon age method Determination of the age of a mineral or rock in years, based on measurement of the ratio of radiogenic argon-40 to potassium-40 and the known radioactive decay rate of potassium-40 to argon-40. Cf: *argon-40/argon-39 age method.* Abbrev: K-Ar age method. Syn: *potassium-argon dating.*

potassium-argon dating *potassium-argon age method.*

potassium bentonite A potassium-bearing clay of the illite group, formed by alteration of volcanic ash; a *metabentonite* consisting of randomly interstratified layers of illite and montmorillonite with a ratio of 4 to 1 (potassium occupying about 80% of the exchangeable-cation positions of the mica portion). Syn: *K-bentonite; potash bentonite.*

potassium-calcium age method The determination of the age of a mineral or rock in years based on measurement of the ratio of radiogenic calcium-40 to potassium-40 and the known radioactive decay rate of potassium-40 to calcium-40. The method is not in common use, as initially there is apt to be a significant amount of calcium-40 present.

potassium feldspar An alkali feldspar containing the Or molecule ($KAlSi_3O_8$); e.g. orthoclase, microcline, sanidine, and adularia. See also: *potash spar.* Syn: *K-feldspar; K-spar.*

potassium-40 A radioactive isotope of potassium having a mass number of 40, a half-life of approximately 1.31 x 10^9 years, and an atomic abundance of 0.000122 grams per gram of potassium. Potassium-40 decays by beta emission to calcium-40 and by electron capture to argon-40. Potassium-40 and its decay product argon-40 are commonly used to date geologic materials (*potassium-argon age method*).

pot bottom A large boulder or concretion in the roof of a coal seam, having the rounded appearance of the bottom of an iron pot and easily detached. Cf: *caldron bottom; bell; kettle bottom; camel back; tortoise.* Syn: *pot [coal]; potstone [coal].*

potch An Australian term for an opal of inferior quality that does not exhibit play of color; it is found in association with precious opal.

pot clay (a) A refractory clay (fireclay) suitable for the manufacture of the melting pots in which glass is produced. (b) A clay bed associated with coal measures. (c) A kaolin-rich residual clay.

pot earth *potter's clay.*

potential Any of several different scalar quantities, each of which involves energy as a function of position or of condition; e.g. the *fluid potential* of ground water.

potential barrier The resistance to change from one energy state to another in a chemical system, which must be overcome by *activation energy.*

potential density The *density* of a unit of water after it is raised by an *adiabatic* process to the surface, i.e., determined from in-situ salinity and potential temperature. Cf: *in-situ density.*

potential difference The difference in electric potential between two points that represents the work involved or the energy released in the transfer of a unit amount of electricity between them.

potential disturbance *disturbing potential.*

potential drop The difference in pressure between two equipotential lines. Syn: *piestic interval.*

potential electrode One of two electrodes between which potential is measured.

potential gradient *hydraulic gradient.*

potential head *elevation head.*

potential of disturbing masses *disturbing potential.*

potential of random masses *disturbing potential.*

potential ore (a) As yet undiscovered mineral deposits. (b) A known mineral deposit for which recovery is not yet economically feasible.—Cf: *possible ore; inferred ore; indicated ore.* See also: *resources; reserves.*

potential temperature [meteorol] The temperature that a given unit of air would attain if it were reduced to a pressure of 1,000 millibars without any heat transfer to or from it.

potential temperature [oceanog] The temperature of a unit of water, measured after it is raised by an *adiabatic* process to the surface. Cf: *in-situ temperature.*

potential well In a field of force, a sharply defined area of minimum potential.

potentiometer An electrical instrument for the precise measurement of low-level direct-current voltages.

potentiometric map A map showing the elevation of a potentiometric surface of an aquifer by means of contour lines or other symbols. Syn: *pressure-surface map.*

potentiometric surface An imaginary surface representing the total head of ground water and defined by the level to which water will rise in a well. The water table is a particular potentiometric surface. Syn: *piezometric surface; pressure surface.*

pothole [coast] A small, rounded, steep-sided depression or pit in a coastal marsh, containing water at or below low-tide level (Veatch & Humphrys, 1966, p. 245). Syn: *rotten spot.*

pothole [geomorph] Any pot-shaped pit or hole.

pothole [glac geol] (a) *giant's kettle.* (b) A term applied in Michigan to a small pit depression (1-15 m deep), generally circular or elliptical, occurring in an outwash plain, a recessional moraine, or a till plain (Veatch & Humphrys, 1966, p. 244).—Cf: *kettle.*

pothole [glaciol] *moulin.*

pothole [lake] A shallow depression, generally less than 10 acres in area, occurring between dunes on a *prairie* (as in Minnesota and the Dakotas), often containing an intermittent pond or marsh and serving as a nesting place for waterfowl.

pothole [salt] A term used in Death Valley, Calif., for a circular opening, about a meter in diameter, filled with brine and lined with halite crystals.

pothole [speleo] *vertical cave.*

pothole [streams] A smooth, bowl-shaped or cylindrical hollow, generally deeper than wide, formed in the rocky bed of a stream by the grinding action of a stone or stones, or of coarse sediment (sand, gravel, pebbles, boulders), whirled around and kept in motion by eddies or the force of the stream current in a given spot, as at a strong rapid or the foot of a waterfall. Cf: *plunge basin.* Syn: *pot; kettle; eversion hollow; rock mill; churn hole; eddy mill; colk.*
pothole erosion *eversion.*
potholer *caver.*
potholing *caving [speleo].*
pot lead Graphite used on the bottoms of racing boats.
potrero An elongate, island-like beach ridge, surrounded by mud flats and separated from the coast by a lagoon and barrier island, and made up of a series of accretionary dune ridges (Fisk, 1959, p. 113); e.g. Potrero Lopeno, rising 10 m above the Laguna Madre Flats along the southern Texas coast.
Potsdam ellipsoid The geodetic reference ellipsoid used in most of Europe, Africa, the Middle East, and the Soviet Union.
Potsdam gravity *gravity standard.*
Potsdam system A system of gravity values based on the determination of absolute gravity at Potsdam, Germany in 1906. Recent determinations show that the Potsdam value is about 13 milligals too high.
potstone [coal] *pot bottom.*
potstone [mineral] (a) A dark-green or dark-brown impure steatite or massive talc, used in prehistoric times to make cooking pots and vessels. (b) A term used in Norfolk, England, for *paramoudra.* —Also spelled: *pot stone.*
potter's clay A plastic clay free from iron and devoid of fissility, suitable for modeling or making of pottery or adapted for use on a potter's wheel. It is white after burning. Syn: *potter's earth; pot earth; argil.*
Pottsvillian North American provincial stage: Lower Pennsylvanian (above Mauch Chunk, below Allegheny).
poughite A mineral: $Fe_2(TeO_3)_2(SO_4)\cdot 3H_2O$.
Poulter seismic method A type of *air shooting* in which the explosive is set on poles above the ground.
Pourtalès plan The arrangement of septa in some scleractinian corals (notably in the family Dendrophylliidae) characterized by much greater development of exosepta than of entosepta. Named after Louis F. de Pourtalès (1824-1880), Swiss naturalist.
pow A Scottish term for a pool.
powder *powder snow.*
powder avalanche *dry-snow avalanche.*
powder diffraction X-ray diffraction by a powdered, crystalline sample, commonly observed by the Debye-Scherrer camera method or by a recording diffractometer.
powder method *Debye-Scherrer method.*
powder pattern In the *powder method* of X-ray diffraction analysis, the display of lines made on film by the Debye-Scherrer method or on paper by a recording *diffractometer.* See also: *powder photograph.*
powder photograph The *powder pattern* made on film in the Debye-Scherrer method of X-ray diffraction analysis.
powder snow Dry snow of low density, composed of loose crystals that accumulated under conditions of low temperature and no wind and that have not been compacted. Cf: *sand snow.* Syn: *powder.*
powellite A tetragonal mineral: $CaMoO_4$. It is isomorphous with scheelite and is a minor ore of molybdenum.
power efficiency The probablity of rejecting a statistical hypothesis when it is false. Syn: *power.*
power of crystallization *form energy.*
power spectrum The sum of the squared Fourier coefficient values (Harbaugh & Merriam, 1968, p. 120).
Poynting's law A special case of the *Clapeyron equation,* in which the fluid is removed as fast as it forms, e.g. under metamorphic stress, so that its volume may be ignored.
pozzolan Siliceous material such as diatomaceous earth, opaline chert, and certain tuffs, which can be finely ground and combined with *portland cement* (in a proportion of 15 to 40 percent by weight). The pozzolan reacts with calcium hydroxide that is liberated as concrete hardens, forming compounds with cementitious properties. Pozzolans also counteract the adverse effects of certain undesirable *aggregates* that may have to be used in concrete. Portland-pozzolan cements are highly resistant to penetration and corrosion by salt water. The name comes from the town of Pozzuoli, Italy, near which occurs a leucite tuff that was used in cement in Roman times. Also spelled: *pozzolana; puzzolan; puzzuolana.*
PPI *plan-position indicator.*
pradolina A syn. of *urstromtal.* Etymol: Polish, "ancient valley". Pl: *pradoliny.*
Praetiglian North European climatostratigraphic and floral stage: Middle Pliocene. Equivalent to Lower Villafranchian.
praevestibulum A chamber within the ektexine of spores and pollen but not contiguous to the endexine (Tschudy & Scott, 1969, p. 27).
prairie (a) An extensive tract of level to rolling grassland, generally treeless, in the temperate latitudes of the interior of North America (esp. in the Mississippi Valley region), characterized by a deep, fertile soil and by a covering of tall, coarse grass and herbaceous plants. See also: *steppe; black prairie.* (b) One of a series of grassy *plains* into which the true prairies of the Mississippi Valley region merge on the west, whose treeless state is due to aridity. (c) A low, wet grass-grown tract or sink, often water-covered, in the pinewoods of Florida.— Etymol: French, "meadow, grassland".
Prairie soil A great soil group in the 1938 classification system, a group of zonal soils having a surface horizon that is dark or grayish brown, grading through brown soil into lighter-colored parent material. These soils are two to five feet thick, and develop under tall grass in a temperate and humid climate (USDA, 1938). Most are now classified as *Udolls* and *Xerolls.* Cf: *Reddish Prairie soil.* Syn: *Brunizem.*
prairillon A small prairie.
Prandtl number In fluid mechanics, a nondimensional parameter that is the ratio of kinematic viscosity to thermometric conductivity. It is named after Ludwig Prandtl, German physicist (d. 1953).
prase (a) A translucent dull green or yellow-green variety of chalcedony. (b) Crystalline quartz containing a multitude of green hairlike crystals of actinolite.—Syn: *mother-of-emerald.*
prasinite A greenschist in which the proportions of the hornblende-chlorite-epidote assemblage are more or less equal (Holmes, 1928, p.189).
prasopal A green variety of common opal containing chromium. Syn: *prase opal.*
pratincolous Said of an organism that lives in meadows or low grassy areas.
Pratt hypothesis A concept of the mechanism of *isostasy,* proposed by G.H. Pratt, that postulates an equilibrium of crustal blocks of varying density; thus the topographically higher mountains would be less dense than topographically lower units, and the depth of crustal material would be everywhere the same. Cf: *Airy hypothesis.*
praya A beach or waterfront. Etymol: Portuguese *praia.*
preabdomen The enlarged anterior part of the abdomen of a scorpion, composed of seven segments; the broad anterior part of the opisthosoma of a merostome.
preadaptation Appearance of nonadaptive or inadaptive characters that later prove to be adaptive in a different or changed environment.
prealpine facies A geosynclinal facies characteristic of neritic areas, displaying thick limestone deposits and coarse terrigenous material, and resembling epicontinental platform sediments. It is generally overlain by flysch, as in the Alpine region.
Preboreal n. A term used primarily in Europe for an interval of Holocene time (from about 10,000 to 9000 years ago) following the Younger Dryas and preceding the Boreal, during which the inferred climate was somewhat colder and wetter than during the Boreal; a unit of the *Blytt-Sernander climatic classification,* characterized by birch and pine vegetation. Also spelled: *Pre-Boreal.*— adj. Pertaining to the postglacial Preboreal interval and to its climate, deposits, biota, and events.
Precambrian All geologic time, and its corresponding rocks, before the beginning of the Paleozoic; it is equivalent to about 90% of geologic time. Precambrian time has been divided according to several different systems, all of which use the presence or absence of evidence of life as a criterion. See also: *Azoic; Proterozoic.*
Precambrian W That part of the U.S. Geological Survey's purely temporal, fourfold division of the Precambrian corresponding to the time span 2600 million years ago and older. Cf: *Precambrian Z; Precambrian Y; Precambrian X.*
Precambrian X That part of the U.S. Geological Survey's purely temporal, fourfold division of the Precambrian corresponding to the time span 1700-2600 million years ago. Cf: *Precambrian Z;*

Precambrian Y; Precambrian W.

Precambrian Y That part of the U.S. Geological Survey's purely temporal, fourfold division of the Precambrian corresponding to the time span 800-1700 million years ago. Cf: *Precambrian Z; Precambrian X; Precambrian W.*

Precambrian Z That part of the U.S. Geological Survey's purely temporal, fourfold division of the Precambrian corresponding to the time span 570-800 million years ago. Cf: *Precambrian Y; Precambrian X; Precambrian W.*

precession camera An X-ray camera used to register the diffraction from a single crystal showing individual layers of the reciprocal lattice without distortion. Cf: *Buerger precession method.*

precession method *Buerger precession method.*

precession of the equinoxes A consequence of the precession of the Earth's spin axis wherein the intersection of the ecliptic with the celestial equator advances along the equator. It produces approximately a twenty-second difference in length between the calendar year and the sidereal year.

precingular archeopyle An *archeopyle* formed in a dinoflagellate cyst by loss of the middorsal plate of the precingular series.

precingular series The series of plates between the apical series and the girdle in dinoflagellate theca. Cf: *postcingular series.*

precious Said of the finest variety of a gem or mineral; e.g. "precious jade" (true jadeite that is wholly or partly deep green) or "precious scapolite" (gem-quality scapolite). See also: *oriental; precious stone.*

precious garnet (a) An unusually purple and brilliant almandine. (b) An unusually red and brilliant pyrope.

precious metal A general term for gold, silver, or any of the minerals of the platinum group.

precious opal A gem variety of opal that exhibits a brilliant play of delicate colors; e.g. *white opal* and *black opal.* Cf: *common opal.*

precious serpentine A green massive translucent variety of the mineral serpentine.

precious stone (a) A gemstone that, owing to its beauty, rarity, durability, and hardness, has the highest commercial value and traditionally has enjoyed the highest esteem since antiquity; specif. diamond, ruby, sapphire, and emerald (and sometimes pearl, opal, topaz, and chrysoberyl). (b) Strictly, any genuine gem material.—Cf: *semiprecious stone.* See also: *precious.*

precipice A very steeply inclined, vertical, or overhanging wall or surface of rock; e.g. the high, steep face of a cliff. Syn: *sheer.*

precipitation Water that falls to the surface from the atmosphere as rain, snow, hail, or sleet. It is measured as a liquid-water equivalent regardless of the form in which it fell. Cf: *rainfall.*

precipitation excess The volume of water from precipitation that is available for direct runoff. Cf: *rainfall excess; abstraction [water]; effective precipitation.*

precipitation facies Facies characteristics that provide evidence of depositional conditions, as revealed mainly by sedimentary structures (such as cross-bedding and ripple marks) and by primary constituents (esp. fossils) (Sonder, 1956). Cf: *alimentation facies.*

precision (a) The degree of agreement or uniformity of repeated measurements of a quantity; the degree of refinement in the performance of an operation or in the statement of a result. It is exemplified by the number of decimal places to which a computation is carried and a result stated. Precision relates to the quality of the operation by which a result is obtained, as distinguished from *accuracy,* but it is of no significance unless accuracy is also obtained. (b) The deviation of a set of estimates or observations from their mean. (c) A term applied in surveying to the degree of perfection in the methods and instruments used when making measurements and obtaining results of a high order of accuracy.

precision depth recorder An *echo sounder* having an accuracy better than 1 in 3000. Abbrev: PDR.

Pre-Classic *Formative.*

preconsolidation pressure The greatest effective stress to which a soil or sediment has been subjected; the pressure exerted on unconsolidated sediment by present or former overlying material, or by desiccation of a silt or clay, resulting in compaction. Syn: *prestress.*

precoxa An occasionally occurring limb segment proximal to the coxa of a crustacean.

precurrent mark A structure produced on the surface of an unconsolidated sediment before the arrival of a turbidity current; e.g. an animal track.

predazzite A recrystallized limestone containing brucite and calcite, with calcite predominating.

prediagenesis A term used by Chilingar et al. (1967, p. 322) for that part of *syngenesis* responsible for "those parts that were introduced subsequently" to *syndeposition* but "before the principal processes of diagenesis began"; e.g. internal sedimentation of clastic material.

prediluvian *antediluvian.*

preferred orientation In structural geology, nonrandom orientation of planar or linear fabric elements, including crystallographic directions (*lattice-preferred orientation*) or elongation/flattening axes of crystals (*shape-preferred orientation*). See also: *foliation.*

pregeologic (a) Antedating reliable geologic data or theory. (b) Before the time when the surface of the Earth became generally similar to what it is today; e.g. "pregeologic time", or the part of geologic history that antedates the oldest rocks (about 3-4.5 b.y. ago).

preglabellar field That part of a trilobite cranidium between the glabella and the anterior border furrow.

preglacial (a) Pertaining to the time preceding a period of glaciation; specif. that immediately before the Pleistocene Epoch. (b) Said of material underlying glacial deposits; e.g. the loose sand and gravel lying beneath till in Iceland, where the term "preglacial drift" is (incorrectly) used.

prehistoric (a) Said of or pertaining to something in the past that is prior to the written records of man. (b) Pertaining to prehistory, i.e., the study of man during the time prior to his written records.

prehnite A pale-green, yellow-brown, or white orthorhombic mineral: $Ca_2Al_2Si_3O_{10}(OH)_2$. It usually occurs in crystalline aggregates having a botryoidal or mammillary and radiating structure, and is commonly associated with zeolites in geodes, druses, fissures, or joints in altered igneous rocks.

pre-Imbrian (a) Pertaining to the oldest lunar topographic features and lithologic map units, constituting a system of rocks that appear in the mountainous terrane and are well displayed in the southern part of the visible lunar surface and over much of the reverse side. (b) Said of the stratigraphic period during which the pre-Imbrian System was developed.

preliminary waves The body waves of an earthquake, including both P waves and S waves.

premagadiniform Said of the loop, or of the early stages in the development of the loop, of a terebratellid brachiopod, marked by growth and completion of descending branches from both the cardinalia and the median septum and by the appearance of a tiny hood developing into a ring on the septum (TIP, 1965, pt.H, p.151). Cf: *magadiniform.*

premineral adj. In economic geology, said of a structural or other feature extant before mineralization. Cf: *intermineral; postmineral.*

preobrazhenskite A mineral: $Mg_3B_{11}O_{15}(OH)_9$.

preoccupied name In taxonomy, a name that is a junior, i.e. a later, *homonym,* unavailable for use because given previously to a different taxon.

preoral cavity The depression above the gullet in a tintinnid.

preorogenic phase The initial phase of an *orogenic cycle,* prior to the climactic orogeny. This phase is the time of formation of geosynclines, most of which are clearly divisible into eugeosynclinal (internal) and miogeosynclinal (external) parts, the first characterized by abundant submarine volcanism, the second by little magmatism and by carbonate-quartzite sedimentation. Preorogenic plutonic rocks include ultramafic bodies and rare early granitic plutons. Cf: *orogenic phase; postorogenic phase; orogenic cycle.*

prepollen Functional pollen grains that have *haptotypic characters* like those of spores, usually a trilete mark. They usually have also a colpus and such other pollenlike features as vesicles. Prepollen grains are typical of extinct primitive gymnosperms (mostly Mississippian to Permian).

preseptal passage An opening that connects all chamberlets of the same chamber of a foraminiferal test (as in the Alveolinidae), located in the anterior part of the chamber. Cf: *postseptal passage.*

presque isle A promontory or peninsula extending into a lake, nearly or almost forming an island, its head or end section connected with the shore by a sag or low gap only slightly above water level, or by a strip of lake bottom exposed as a land surface by a drop in lake level (Veatch & Humphrys, 1966, p. 246). Type example: Presque Isle (Mich.), extending into Lake Huron. Etymol: "presque" is French for "almost".

pressed amber *amberoid.*

pressolution *pressure solution.*

pressolved Said of a sedimentary bed or rock in which the grains have undergone pressure solution; e.g. "pressolved quartzite" whose toughness and homogeneity is due to a tightly interlocked texture of quartz grains subjected to pressure solution. Term was introduced by Heald (1956, p. 22).

pressure (a) The force exerted across a real or imaginary surface divided by the area of that surface; the force per unit area exerted on a surface by the medium in contact with it. (b) A commonly used short form for *geostatic pressure.*

pressure altimeter *barometric altimeter.*

pressure arch A wavelike prominence, formed by pressure, on the surface of a glacier.

pressure bulb The zone in a loaded soil mass bounded by an arbitrarily selected isobar of stress (ASCE, 1958, term 277).

pressure burst *rock burst.*

pressure cone *shatter cone.*

pressure decay The decline, usually gradual, from a temporary, abnormal pressure condition toward a pressure that is more nearly in balance with permanent or steady-state environmental conditions.

pressure detector *hydrophone.*

pressure dome *tumulus.*

pressure drag *pressure resistance.*

pressure figure A pattern resembling a six-rayed star, produced by intersecting lines of parting due to gliding, when certain minerals, esp. mica, are compressed by a blunt point. The rays are similar in character, but not necessarily in position, to *percussion figures.*

pressure fringe *pressure shadow.*

pressure gradient (a) The rate of variation of pressure in a given direction in space at a fixed time. In the ocean, pressure gradients are caused by the vertical distribution of density (which depends on water temperature and salinity), by the slope of the sea surface with respect to the level surface, and by the difference at atmospheric pressure at the sea surface. (b) Loosely, the magnitude of the pressure gradient.—Cf: *hydraulic gradient.*

pressure head The height of a column of liquid supported, or capable of being supported, by pressure *p* at a point in the liquid. See also: *static head; total head.*

pressure ice A general term for ice, esp. sea ice, whose surface in places has been deformed by stresses generated by wind, currents, or waves; it includes pieces of ice squeezed against the shore or each other, or forced upwards or downwards. Pressure ice may be "rafted", "hummocked", or "tented". See also: *deformed ice; rough ice; screw ice.* Ant: *level ice.*

pressure maintenance The introduction of a fluid into an oil or gas reservoir to maintain the pressure and thereby improve *ultimate recovery.*

pressure melting Melting of ice in a place where its melting point is lowered by application of increased pressure; esp. the first part of the *regelation* process within a glacier, occurring at places where the overlying ice is especially thick.

pressure-melting temperature The temperature at which ice can melt at a given pressure. See also: *warm ice.*

pressure pan An induced soil *pan* having a higher bulk density and a lower total porosity than the soil directly above or below it, produced as a result of pressure applied by normal tillage operations or by other artificial means (SSSA, 1965, p.341). Cf: *genetic pan; plow sole.* Syn: *traffic pan.*

pressure penitente A *nieve penitente* composed of brilliantly white ice, shaped into a slender ridge by lateral pressure of converging morainal streams and by melting of the adjacent debris-covered ice (Workman, 1914, p. 316-317). Such features have broken upper surfaces and sharply inclined sides (resembling pointed cones, wedges, or pyramids), and usually occur on the lower parts of glaciers where morainal streams are strongly developed.

pressure plateau An uplifted area of a thick, ponded lava flow, measuring up to three or four square meters, uplift of which is due to intrusion of new lava from below that does not reach the surface. The lower part of such a flow may remain fluid for weeks.

pressure-release jointing *Exfoliation* that occurs in once deeply buried rock that erosion has brought nearer the surface, thus releasing its confining pressure. See also: *lift joint; sheeting.*

pressure resistance In fluid dynamics, a normal stress caused by acceleration of the fluid, which results in a decrease in pressure from the upstream to the downstream side of an object, and acting perpendicular to the boundary (Chow, 1957). Cf: *shear resistance.* Syn: *pressure drag.*

pressure ridge [glaciol] (a) A ridge of glacier ice, produced by horizontal pressure associated with glacier flow. (b) *ice-pushed ridge.*

pressure ridge [ice] A rugged, irregular wall of broken floating ice buckled upward by the lateral pressure of wind or current forcing or squeezing one floe against another; it may be fresh or weathered, and extend many kilometers in length and up to 30 m in height. Cf: *ice keel.* Syn: *ridge [ice].*

pressure ridge [seis] A seismic feature due to transverse pressure and shortening of the land surface; a *slice ridge.*

pressure ridge [volc] An elongate uplift of the congealing crust of a lava flow, probably due to the pressure of underlying still-flowing lava.

pressure shadow In structural petrology, aggregates of new grains growing on opposed sides of a host porphyroblast or detrital grain, thereby producing an elongate structure. This structure is generally aligned parallel to a foliation and may define a lineation (Hobbs, Means and Williams, 1976, p. 274). Syn: *pressure fringe; strain shadow; stress shadow.*

pressure solution Solution (in a sedimentary rock) occurring preferentially at the contact surfaces of grains (crystals) where the external pressure exceeds the hydraulic pressure of the interstitial fluid. It results in enlargement of the contact surfaces and thereby reduces pore space and tightly welds the rock. See also: *solution transfer.* Syn: *pressolution.*

pressure surface A less-preferred syn. of *potentiometric surface.*

pressure-surface map A less-preferred syn. of *potentiometric map.*

pressure tendency The net rise or fall of barometric pressure within a specified time (usually three hours) before a particular observation. Syn: *barometric tendency.*

pressure texture *cataclastic texture.*

pressure tube A deep cylindrical hole formed in a glacier by the sinking of an isolated stone that has absorbed more solar radiation than the surrounding ice (Mallet, 1838, p. 326-327).

pressure vessel *bomb [geochem].*

pressure wave *P wave.*

prestress *preconsolidation pressure.*

presuppression Holding the gain of a seismic amplifier low until after the appearance of the very strong *first arrivals.*

prevailing current The ocean current most frequently observed during a given period, as a month, season, or year.

prevailing westerlies The belt of westerly winds between 35° and 60° north and south of the Equator. These winds blow from the subtropical anticyclones toward the subarctic lows, from the southwest in the Northern Hemisphere and from the northwest in the Southern.

previtrain The woody lenses in lignite that are equivalent to vitrain in coal of higher rank (Schopf, 1960, p.30).

Priabonian European stage: Upper Eocene. It is believed to consist of Auversian and Bartonian (above Lutetian, below Rupelian).

priceite A snow-white earthy mineral: $Ca_4B_{10}O_{19}\cdot 7H_2O$ (?). Syn: *pandermite.*

priderite A red mineral: $(K,Ba)(Ti,Fe)_8O_{16}$.

prill An English term for a running stream.

primanal The proximal anal plate in the posterior interray of camerate crinoids.

primärrumpf A term proposed by W. Penck (1924) for a low, convex, rather featureless erosional landscape or plain produced by waxing uplift that proceeded so slowly with respect to the rate of denudation that there was no net rise of the surface or increase in its relief; an "expanding dome" that represents the universal and initial geomorphic unit. Etymol: German *Primärrumpf,* "primary torso". Cf: *endrumpf; old-from-birth peneplain.* Syn: *primary peneplain.*

Primary A term applied in the early 19th century as equivalent to *Primitive* or the period of time and associated rocks now referred to as Precambrian. It was later extended to include the lower Paleozoic, and still later restricted to the whole of the Paleozoic Era. The term was abandoned in the late 19th century in favor of *Paleozoic.* See also: *Secondary.*

primary [coast] Said of a youthful coast or shoreline where waves have not had time to produce notable effects and having features that are produced chiefly by nonmarine agencies (Shepard, 1937, p. 605); e.g. coasts shaped by diastrophism, volcanism, subaerial deposition, or land erosion. Cf: *secondary [coast].*

primary [eco geol] Said of a mineral deposit unaffected by supergene enrichment.
primary [metal] Said of a metal obtained from ore rather than from scrap. Ant: *secondary [metal].* Syn: *virgin.*
primary allochthony In coal formation, accumulation of plant remains in a region that does not correspond to that in which the plants grew. Cf: *secondary allochthony.*
primary ambulacral radius One of the three radii of edrioasteroids: anterior, right lateral, and left lateral. They extend out from the center of the oral area and are marked by the anterior and transverse oral midlines. The two lateral radii bifurcate to form the two lateral pairs of ambulacra, I-II and IV-V; the anterior radius continues as ambulacrum III (Bell, 1976).
primary arc Obsolete term for an *island arc* that is convex outward from a continent and overlies a deep-seated tectonic feature.
primary axial septulum A *primary septulum* in a foraminiferal test, representing an *axial septulum* observable in sagittal (equatorial) section (as in *Lepidolina* and *Yabeina*).
primary banding *phase layering.*
primary clay A clay found in the place where it was formed; a *residual clay.* Cf: *secondary clay.*
primary crater (a) An *impact crater* produced directly by the high-velocity impact of a meteorite or other projectile; e.g. any of the lunar craters formed by collision of the Moon with objects from space. Cf: *secondary crater.* (b) *true crater.*
primary creep Elastic deformation that is time-dependent and results from a constant differential stress acting over a long period of time. Cf: *secondary creep.* Syn: *transient creep.*
primary dip The slight dip of a bedded deposit assumed at its moment of deposition. Syn: *original dip; depositional dip.* Cf: *initial dip.*
primary dolomite A dense, finely textured dolomite rock, made up of grains less than 0.01 mm in diameter, formed in place by direct chemical or biochemical precipitation from seawater or lake water. Characteristically it is well-stratified, unfossiliferous, and interbedded with anhydrite, clay, and micritic limestone. Also, a similarly textured dolomite rock made up of clastic particles formed by direct accumulation. Some authors consider the rock to be syndiagenetic (Fairbridge, 1967, p. 66-67). See also: *dolomicrite.* Syn: *orthodolomite.*
primary fabric *apposition fabric.*
primary flat joint An approximately horizontal joint plane in igneous rocks. Syn: *L-joint.*
primary flowage Movement within an igneous rock that is still partly fluid (Cloos, 1946).
primary fumarole A *fumarole* formed over a volcanic fissure and fed directly from the main source of activity, thus giving a true index of internal conditions. Cf: *rootless fumarole.*
primary geosyncline Peyve & Sinitzyn's term for an *orthogeosyncline* (1950).
primary gneiss A rock that exhibits planar or linear structures characteristic of metamorphic rocks but lacks observable granulation or recrystallization and is therefore considered to be of igneous origin. Cf: *protoclastic.*
primary interstice *original interstice.*
primary lamella The first half-whorl of each brachiopod spiralium distal from its attachment to a crus (TIP, 1965, pt.H, p.151).
primary layer The outer shell layer immediately beneath the periostracum of a brachiopod, deposited extracellularly by columnar outer epithelium of the outer mantle lobe. It forms a well-defined calcareous layer, devoid of cytoplasmic strands, in most articulate brachiopods. Cf: *secondary layer.* Syn: *lamellar layer.*
primary layering *phase layering; compositional layering.*
primary ligament The part of a ligament of a bivalve mollusk representing the original condition of structure, consisting of periostracum and ostracum, but excluding secondary additions.
primary magma A magma originating below the Earth's crust. It is sometimes used as a syn. of *parental magma.*
primary mineral A mineral formed at the same time as the rock enclosing it, by igneous, hydrothermal, or pneumatolytic processes, and that retains its original composition and form. Cf: *secondary mineral.*
primary optic axis One of two *optic axes* in a crystal that are perpendicular to the circular sections of the *indicatrix* and along which all light rays travel with equal velocity. Cf: *secondary optic axis.*
primary oral In edrioasteroids, one of a number of large oral plates, with intrathecal extensions, that participate in formation of the underlying *oral frame* and have a fixed position within the theca relative to ambulacral and other oral plates (Bell, 1976).
primary orogeny Orogeny that is characteristic of the *internides* and that involves deformation, regional metamorphism, and granitization. Cf: *secondary orogeny.*
primary peneplain A syn. of *primärrumpf,* originally a German word with no adequate English equivalent. The term is unsatisfactory because the Davisian peneplain is developed by a different process and has different characteristics.
primary phase The only crystalline phase capable of existing in equilibrium with a given liquid; it is the first to appear on cooling from a liquid state, and the last to disappear on heating to the melting point.
primary phase region On a phase diagram, the locus of all compositions having a common primary phase.
primary porosity The *porosity* that developed during the final stages of sedimentation or that was present within sedimentary particles at the time of deposition. "Primary porosity includes all predepositional and depositional porosity of a particle, sediment, or rock" (Choquette & Pray, 1970, p. 249). Cf: *secondary porosity.*
primary precipitate A *precipitate* formed directly; e.g. an evaporite formed by evaporation of a saline solution, or a sediment formed as a result of a reaction between dissolved material and suspended clay or as a result of a change in acidity or a shift in the oxidation-reduction potential.
primary precipitate crystal *cumulus crystal.*
primary productivity In a body of water, the rate of photosynthetic carbon fixation by plants and bacteria forming the base of the food chain. See also: *productivity [biol]; production.*
primary rocks (a) Rocks of which the constituents are newly formed particles that have never been constituents of previously formed rocks and that are not the products of alteration or replacement, esp. igneous rocks formed directly by solidification from a magma. Cf: *secondary rocks.* (b) A "more appropriate" syn. of *primitive rocks* (Humble, 1843, p.210). The term in this usage is now obsolete.
primary sedimentary structure A syngenetic *sedimentary structure* determined by the conditions of deposition (mainly current velocity and rate of sedimentation) and developed before lithification of the rock in which it is found. It includes bedding in the broad sense (esp. the external form of the beds and their continuity and uniformity of thickness), bedding-plane markings such as ripple marks and sole marks, and those deformational structures produced by preconsolidation movement due to unequal loading or to downslope sliding or slumping. Syn: *primary structure [geol].*
primary septulum A major partition of a chamberlet in a foraminiferal test; e.g. *primary axial septulum* and *primary transverse septulum.* Cf: *secondary septulum.*
primary septum (a) In rugose and scleractinian corals, one of the six first-formed septa in the corallite. Syn: *protoseptum.* Cf: *alar septum; cardinal septum; counter septum; counter-lateral septum.* (b) The septum immediately following the proseptum in ammonoid conchs.
primary spine The first-formed and usually largest spine of a plate of the corona of an echinoid. It is situated over the growth center of the plate except on a *compound plate.* Cf: *secondary spine.*
primary stratification Stratification developed when the sediments were first deposited. Syn: *direct stratification.*
primary structure [geol] (a) A structure in an igneous rock that originated contemporaneously with the formation or emplacement of the rock, but before its final consolidation; e.g. pillow structure developed during the eruption of a lava, or layering developed during solidification of a magma. (b) *primary sedimentary structure,* e.g. bedding or ripple marks. (c) The structure pre-existing the deformation and re-equilibration associated with the emplacement at shallow depth of a metamorphic body of deep origin during an orogeny (e.g. Alpine lherzolitic massifs).—Cf: *secondary structure [geol].*
primary structure [paleont] Fine vacuoles or spaces in the wall of a tintinnid lorica. Cf: *secondary structure [paleont]; tertiary structure.*
primary suture The line of junction between the primary septum and the wall of an ammonoid conch.
primary tectogenesis Vertical uplift over a *geotumor* (Haarman, 1930). Obsolete.
primary tectonite A tectonite whose fabric is *depositional fabric.* Most tectonites, however, are *secondary tectonites.* See also: *fu-*

sion tectonite.

primary tissue Plant tissue derived directly by differentiation from an apical or intercalary meristem (Cronquist, 1961, p.880).

primary transverse septulum A *primary septulum* in a foraminiferal test that has a plane approximately normal to the axis of coiling and that is observable in axial section (as in *Lepidolina* and *Yabeina*).

primary type A specimen on which the description of a new *species* is based, wholly or in part; e.g. a *holotype, syntype,* or *lectotype* (Frizzell, 1933, p. 662).

primary-type coal *banded ingredients.*

primary wall The first wall proper formed in a developing plant cell. It is the only wall in many types of cells (Esau, 1965, p.36-37).

primary wave *P wave.*

primary zooid A single zooid (ancestrula), or one of the multiple, simultaneously formed zooids resulting from the metamorphosis of a single larva to found a cheilostome bryozoan colony (Boardman & Cheetham, 1973, p. 173).

primatology The study of fossil and living primates.

prime meridian An arbitrary meridian selected as a reference line having a longitude of zero degrees and used as the origin from which other longitudes are reckoned east and west to 180 degrees; specif. the *Greenwich meridian.* Local or national prime meridians are occasionally used. Syn: *zero meridian; initial meridian; first meridian.*

primeval Pertaining to the earliest ages of the Earth; e.g. said of lead that is associated with so little uranium (as in some meteorites) that the Pb-isotope composition has not changed appreciably in five billion years. See also: *primordial.*

primeval-fireball hypothesis *"big bang" hypothesis.*

primeval lead *primordial lead.*

primibrachial A plate of the proximal brachitaxis of a crinoid. It may or may not be an axillary, and it may or may not comprise part of theca. Syn: *primibrach.*

Primitive A name applied from the teachings of A.G. Werner in the 1790's to the group or series of rocks that were considered the first chemical precipitates derived from the ocean before emergence of land areas and that were believed to extend uninterruptedly around the world. The rocks included the larger intrusive igneous masses, all highly metamorphosed rocks, and roughly the rocks that later came to be known as Precambrian in age. See also: *Primary; Transition.*

primitive area Land belonging to the U.S. government that is to be preserved in its natural state. The only changes permitted are those for fire prevention.

primitive character In *cladism,* a character or character state possessed by an ancestral species. Cf: *homology (c).* Syn: *plesiomorphy.*

primitive circle That circle on a *stereographic projection* which is the intersection of the stereographic plane with the sphere of reflection; it is the sphere's equatorial circle.

primitive lattice A crystal lattice described in terms of a *unit cell,* i.e. one with lattice points only at its corners. Cf: *centered lattice.* Syn: *simple lattice.*

primitive rocks A term applied by Lehmann (1756) to crystalline rocks devoid of fossils and rock fragments, and believed to be of chemical origin, having formed prior to the advent of life; also, a term for the rocks believed to have been first formed, being irregularly crystallized and aggregated without cement. They include gneiss, schist, primary limestone, and plutonic rocks such as granite. The term is obsolete as many of these rocks are found in all ages and formations. Cf: *secondary rocks.* Syn: *primary rocks.*

primitive unit cell *unit cell.*

primitive water Water imprisoned in the interior, in either molecular or dissociated form, since the formation of the Earth (Meinzer, 1923, p. 31). Cf: *juvenile [water].*

primocryst A crystal in equilibrium with the magma where primary crystallization occurred. A primocryst becomes a *cumulus crystal* after settling out.

Primordial An obsolete term formerly applied to what is now called *Cambrian.* It was used by Joachim Barrande (1799-1883), French paleontologist, for the oldest or lowest fossiliferous strata as developed in Bohemia.

primordial Original, first in development, earliest, or existing from the beginning; e.g. "primordial ocean basin" or a "primordial magma". See also: *primeval.*

primordial lead Lead with isotopic ratios unaffected by addition of lead from radioactive decay of uranium and thorium since the formation of the earth (and solar system). It is often taken to be lead with an isotopic composition the same as that found in the troilite phase of meteorites. Syn: *primeval lead.*

primordial plate One of the first plates formed following metamorphosis in each plate system of an echinoid.

primordial valve A chitinous plate in certain cirripede crustaceans, having a distinctive honeycomb appearance, and developing at incipient umbones of the terga, scuta, and carina during metamorphosis (TIP, 1969, pt.R, p.100).

primordium The rudiment or earliest trace of development of any plant structure or organ (Swartz, 1971, p. 379).

principal axis [cryst] That crystallographic axis which is the most prominent. In the tetragonal and hexagonal systems, it is the vertical or *c* axis; in the orthorhombic, monoclinic, or triclinic systems, it is also usually the *c* axis, although in monoclinic minerals such as epidote it may be the *b* axis.

principal axis [exp struc geol] In experimental structural geology, a *principal axis of stress* or a *principal axis of strain.*

principal axis of strain One of the three mutually perpendicular axes corresponding to the three axes of the body that were also mutually perpendicular before deformation; also described as the axes of the strain ellipsoid. The longest or greatest is the axis of elongation, and the shortest or least is the axis of shortening. Syn: *strain axis; principal axis.*

principal axis of stress One of the three mutually perpendicular axes that are perpendicular to the *principal planes of stress.* Syn: *stress axis; principal axis.*

principal distance The perpendicular distance from the internal perspective center to the plane of a particular finished negative or print.

principal earthquake *main shock.*

principal focus The *focus* for a beam of incident rays parallel to the axis of a lens or optical system. Syn: *focal point.*

principalia The main parenchymal megascleres (spicules) in lyssacine hexactinellid sponges.

principal layer The main layer of a trilobite cuticle, commonly showing some internal lamination. Some authors recognize three distinct zones: outer, central, and inner. Cf: *prismatic layer.*

principal line [geochem] The *spectral line* that is most easily excited or observed.

principal line [photo] The trace of the principal plane upon a photograph; e.g. the line through the principal point and the nadir of a tilted photograph.

principal meridian A *central meridian* on which a rectangular grid is based; specif. one of a pair of coordinate axes (along with the *base line*) used in the U.S. Public Land Survey system to subdivide public lands in a given region. It consists of a line extending north and south along the astronomic meridian passing through the initial point and along which standard township, section, and quarter-section corners are established. The principal meridian is the line from which the survey of the township boundaries is initiated along the parallels.

principal plane (a) The vertical plane through the internal perspective center and containing the perpendicular from that center to the plane of a tilted photograph. (b) Any plane perpendicular to the axis of an optical system and passing through its principal points.

principal plane of stress One of three mutually perpendicular planes, upon each of which the resultant stress is normal, i.e. on which shear stress is zero. See also: *principal axis of stress.*

principal point (a) The foot of the perpendicular from the interior perspective center of a lens to the plane of the photograph; the geometric center of an aerial photograph, or the point where the optical axis of the lens meets the film plane in an aerial camera. Symbol: p. See also: *fiducial mark; photograph center.* Syn: *center point.* (b) Either of two points on the axis of a lens such that a ray from any point of the object directed toward one principal point will emerge from the lens in a parallel direction but directed through the other principal point. (c) The point at which a principal visual ray intersects a perspective plane.

principal ray (a) The one ray within a bundle of incident rays that, upon entering an optical instrument from any given point of the object, passes through the optical center of the lens. (b) *principal visual ray.*

principal section In a uniaxial *indicatrix,* any plane passing through the optic axis.

principal spine One of the large regularly placed spikes or needles in acantharian and spumellarian radiolarians.

principal stress A stress that is perpendicular to one of three mutually perpendicular planes that intersect at a point in a body on which the shearing stress is zero; a stress that is normal to a principal plane of stress. The three principal stresses are identified as least or minimum, intermediate, and greatest or maximum. See also: *mean stress.*

principal visual ray A perpendicular extending from a station point to a perspective plane and theoretically passing exactly along the visual axis of a viewing eye. Syn: *principal ray.*

principle of uniformity *uniformitarianism.*

print A photographic copy made from a negative or from a positive transparency by photographic means. Prints may be positive (the usual case) or negative. Cf: *positive [photo].*

prionodont Having a sawlike row of many simple and similar teeth; e.g. said of a hinge in a bivalve mollusk in which the teeth are developed in a direction transverse to the margin of the cardinal area, or an ostracode hinge resembling an adont hinge but distinguished by the presence of crenulations along elongate elevations and depressions. Cf: *taxodont.*

priorite A black mineral: (Y,Ca,Th) $(Ti,Nb)_2O_6$. It is isomorphous with aeschynite. Syn: *blomstrandine.*

priority State of being earlier or first; see *law of priority.*

prior river A term applied in Australia to a river system that is older than the present system but postdates the *ancestral river.*

Prior's rules A relationship among the chemical constituents of chondritic meteorites, first recognized by George T. Prior (1862-1936), an English mineralogist, and generally stated: "The less the amount of nickel-iron in a chondritic stone, the richer it is in nickel, and the richer in iron are the magnesium silicates" (Mueller and Saxena, 1976, p. 87-88).

prisere A *sere* that takes place in a barren area undisturbed by man's activities.

prism [cryst] A crystal form having three, four, six, eight, or twelve faces, with parallel intersection edges, and which is open only at the two ends of the axis parallel to the intersection edges of the faces.

prism [sed] (a) A long, narrow, wedge-shaped sedimentary body whose width/thickness ratio is greater than 5 to 1 but less than 50 to 1 (Krynine, 1948, p.146); e.g. an alluvial fan adjacent to an escarpment, or one of the great conglomerates of the geologic record. It is typical of orogenic sediments formed during periods of intense crustal deformation, such as the arkoses found in fault troughs. Cf: *tabular; shoestring.* Syn: *wedge.* (b) *geosynclinal prism.*

prismatic (a) Said of a sedimentary particle whose length is 1.5 to 3 times its width (Krynine, 1948, p.142). Cf: *tabular.* (b) Pertaining to a sedimentary *prism.* (c) Pertaining to a crystallographic *prism.* (d) Said of a crystal that shows one dimension markedly longer than the other two. (e) Said of a metamorphic texture in which a large proportion of grains are prismatic and have approximately parallel orientation, so that a lineated appearance is usually visible in hand specimen or thin section.

prismatic class That crystal class in the monoclinic system having symmetry $2/m$. Prisms of this system have four faces, are rhombic in cross section, and have as their axis either the c axis, the a axis, or any lattice row parallel to the b axis.

prismatic cleavage Mineral cleavage parallel to the faces of a prism, e.g. the (110) cleavage of amphibole.

prismatic compass A small *magnetic compass* held in the hand when in use and equipped with peep sights and a glass prism so arranged that the magnetic bearing or azimuth of a line can be read (through the prism) from a circular graduated scale at the same time that the line is sighted over.

prismatic jointing *columnar jointing.*

prismatic layer (a) The middle layer of the shell of a mollusk, consisting essentially of prisms of calcium carbonate (calcite or aragonite). (b) Part of the *secondary layer* in some articulate brachiopods, secreted extracellularly as prismatic calcite. (c) The most common type of outer layer of trilobite cuticle, generally about 20 to 30 μm thick. In some species the outer layer is pigmented or phosphatic rather than prismatic. Cf: *principal layer.*

prismatic structure *columnar jointing.*

prismatolith A coccolith constructed of polygonal prisms. It may be solid or axially perforated. See also: *porolith.*

prism crack A mud crack that develops in polygonal patterns on the surface of drying mud puddles and that breaks the sediment into prisms standing normal to bedding (Fischer, 1964, p.114).

prism level A type of *dumpy level* in which the level bubble can be viewed from the eyepiece end by means of an attached prism at the same time the rod is being read.

prismoid A solid body resembling a prism, having similar but unequal parallel polygonal ends. It is a textural term used for sedimentary particles. Adj: *prismoidal.*

prismoidal Adj. of *prismoid.* It is a term used in sedimentary petrology, and is not to be confused with *prismatic,* which is a crystallographic term.

prism twin law A rare, normal twin law in monoclinic or triclinic feldspars, having a twin plane of (110) or $(1\bar{1}0)$.

proancestrula The first-formed or basal part of the *ancestrula* of stenolaemate bryozoans. Syn: *basal disc; protoecium.*

probability A statistical measure (where zero is impossibility and one is certainty) of the likelihood of occurrence of an event.

probable ore (a) A syn. of *indicated ore.* (b) A mineral deposit adjacent to developed ore but not yet proven by development. Cf: *extension ore.*

probe n. Any measuring device that is placed in the environment to be measured, e.g. a potential electrode, a density probe in a drill hole, or oceanographic instruments that are lowered into the sea.

probertite A colorless monoclinic mineral: $NaCaB_5O_9 \cdot 5H_2O$. Syn: *kramerite.*

problematic fossil A natural object, structure, or marking in a rock, resembling a fossil but having a doubtful organic nature or origin. Cf: *pseudofossil.* Syn: *dubiofossil.*

problematicum A marking, object, structure, or other feature in a rock whose nature presents a problem, such as a doubtful "fossil" that may be of inorganic origin or whose organic nature is uncertain; esp. an undoubted organic remain (such as a trace fossil) with a more or less obscure nature. Pl: *problematica.*

proboscis A distal cylindrical tube extending from an astropyle of a phaeodarian radiolarian.

Procellarian (a) Pertaining to lunar lithologic map units and topographic forms constituting, or closely associated with, the maria. Such features were formerly mapped as the Procellarian System, but are now considered a unit of the Imbrian System. (b) Said of the time interval during which the Procellarum Group was developed.

procephalic Pertaining to, forming, or situated on or near the front of the head; e.g. the "procephalic lobe", or anterior preoral part, of a merostome embryo.

prochlorite *ripidolite.*

prochoanitic Said of a septal neck of a cephalopod directed forward (adorally, or toward the aperture). Ant: *retrochoanitic.*

prochronic Before time or creation. Ant: *diachronic.*

proclade A *clade* or bar in the ebridian skeleton that arises from the end of an upper actine and is directed toward the anterior or nuclear pole. Cf: *opisthoclade.*

procline Said of the ventral and anterior *inclination* of the cardinal area in the pedicle valve of a brachiopod, lying in the bottom right or second quadrant moving counterclockwise from the orthocline position (TIP, 1965, pt.H, p.60, fig.61).

Proctor Pertaining to or determined by a procedure designed by Ralph R. Proctor (1894-1962), U.S. civil engineer, to establish water content-density relationships of a remolded soil by application of compactive effort under standardized conditions; e.g. the "Proctor curve" (or compaction curve), the "Proctor compaction test", and the "Proctor density" of soil.

procumbent Said of a stem that trails or lies flat on the ground but does not take root.

prod cast The cast of a *prod mark,* consisting of a short ridge that rises downcurrent and ends abruptly. Originally defined by Dzulynski and Slaczka (1958, p. 232). Syn: *impact cast.*

prodelta The part of a delta that is below the effective depth of wave erosion, lying beyond the *delta front,* and sloping gently down to the floor of the basin into which the delta is advancing and where clastic river sediment ceases to be a significant part of the basin-floor deposits; it is entirely below the water level. Cf: *intradelta.*

prodelta clay The fine-grained river-borne material (very fine sand, silt, and clay) deposited as a broad fan on the floor of a sea or lake beyond the main body of a delta; the material in a *bottomset bed.*

prodissoconch The rudimentary or earliest-formed shell of a bivalve mollusk preserved at the tip of the beak of some adult shells.

prod mark (a) An indicator of slip direction on a slickensided fault surface, consisting of a groove made by a clast. (b) A short *tool mark* oriented parallel to the current of a stream and produced by an object that plowed into and was then raised above the bottom; its longitudinal profile is asymmetric. The mark deepens gradually downcurrent where it ends abruptly (unlike a flute). Cf: *bounce mark.* Syn: *impact mark.*

producer [ecol] An organism (e.g. most plants) that can form new organic matter from inorganic matter such as carbon dioxide, water, and soluble salts. Cf: *consumer.*

producer [petroleum] A well that produces oil or gas.

producing zone The rock stratum of an oil field that will produce petroleum or gas when penetrated by a well. Often incorrectly referred to as "producing horizon".

productid *productoid.*

production (a) The growth of organisms in a lake. (b) A time-rate unit of total amount of organisms grown. See also: *primary productivity; productivity [biol]; yield [lake]; carrying capacity.*

productivity (a) A general term for the organic fertility of a body of water. (b) The capacity of a lake to produce a particular organism. (c) *primary productivity.* (d) *production.*

productoid Any articulate brachiopod belonging to the suborder Productidina, characterized by a pseudopunctate shell having a flat or concave, rarely convex, brachial valve and a convex pedicle valve. This group includes the largest and most aberrant brachiopods yet known. Range, Lower Devonian to Upper Permian. Var: *productid.*

proepistome *interantennular septum.*

profile [geomorph] (a) The outline produced where the plane of a vertical section intersects the surface of the ground; e.g. the *longitudinal profile* of a stream, or the profile of a coast or hill. Syn: *topographic profile.* (b) *profile section.*

profile [geophys] A graph or drawing that shows the variation of one property such as elevation or gravity, usually as ordinate, with respect to another property, such as distance.

profile [palyn] *pollen profile.*

profile [seis] In seismic prospecting, the data recorded from one shot point by a number of groups of detectors. Syn: *seismic profile.*

profile [struc geol] Cross section of a region of cylindrical folds drawn perpendicular to the fold axes. Syn: *tectonic profile; right section.*

profile [water] A vertical section of a water table or other potentiometric surface, or of a body of surface water.

profile line The top line of a *profile section,* representing the intersection of a vertical plane with the surface of the ground. Cf: *line of section.*

profile method *two-dimensional method.*

profile of equilibrium [coast] The slightly concave slope of the floor of a sea or lake, taken in a vertical plane and extending away from and transverse to the shoreline, being steepest near the shore, and having a gradient such that the amount of sediment deposited by waves and currents is balanced by the amount removed by them; the transverse slope of a *graded shoreline.* The profile is easily disturbed by strong winds, large waves, and exceptional high tides. The concept is hypothetical; see, for example, Bradley & Griggs, 1976. Syn: *equilibrium profile; graded profile.*

profile of equilibrium [streams] The longitudinal profile of a graded stream or of one whose gradient at every point is just sufficient to enable the stream to transport the load of sediment made available to it. It has long been regarded as a smooth, parabolic curve, gently concave to the sky, practically flat at the mouth and steepening toward the source; but current thought is that it need not be smooth. Syn: *equilibrium profile; graded profile.*

profiler A marine seismic system that achieves only limited penetration into the sedimentary section. It usually employs a low-energy source, often a *sparker,* and only one or two recording channels. Often the only record produced is a single-channel plot on electrosensitive paper.

profile section A diagram or drawing that shows along a given line the configuration or slope of the surface of the ground as it would appear if intersected by a vertical plane. The vertical scale is often exaggerated. See also: *line of section; profile line.* Syn: *profile.*

profiling Exploration wherein sensors are moved along a line and a profile is developed. Cf: *sounding [elect].* Syn: *lateral search.*

profluent stream A stream that is flowing copiously or smoothly.

profundal adj. Pertaining to or existing in the deeper part of a lake, below the limit of well developed zones of vegetation.

progenitor In biology, an ancestor.

proglacial Immediately in front of or just beyond the outer limits of a glacier or ice sheet, generally at or near its lower end; said of lakes, streams, deposits, and other features produced by or derived from the glacier ice.

proglacial lake A lake formed just beyond the frontal margin of an advancing or retreating glacier, generally in direct contact with the ice. Cf: *glacier lake.* Syn: *ice-marginal lake.*

proglyph A hieroglyph consisting of a cast (Vassoevich, 1953, p.36); specif. a *groove cast.*

progradation The building forward or outward toward the sea of a shoreline or coastline (as of a beach, delta, or fan) by nearshore deposition of river-borne sediments or by continuous accumulation of beach material thrown up by waves or moved by longshore drifting. Ant: *retrogradation.* Cf: *advance.*

prograde metamorphism Metamorphic changes in response to a higher pressure or temperature than that to which the rock last adjusted itself. Cf: *polymetamorphism; retrograde metamorphism.*

prograde motion The predominant motion eastward of a body on the celestial sphere. Cf: *retrograde motion.*

prograding shoreline A shoreline that is being built forward or outward into a sea or lake by deposition and accumulation. Ant: *retrograding shoreline.*

program A plan for solution of a problem contained in a sequence of coded instructions for insertion into a device such as a computer.

progression *advance [coast].*

progressionism The special belief that accompanied *catastrophism,* that each successive creation consisted of animals of a higher order than the previous one.

progressive deformation A continuous sequence of configurations through which a body passes. Distinguished from a *deformation,* which involves just two discrete states of the body (commonly the initial and final states) (Flinn, 1962).

progressive fault *growth fault.*

progressive metamorphism Progressive change in the degree of metamorphism from lower to higher grade across a metamorphic terrane. The term may be applied to rocks in contact aureoles or to rocks traced through the different isograds or facies of regional metamorphism. It usually implies sequential passage of rocks up through all the lower grade zones.

progressive overlap A general term used by Grabau (1906, p.569) for a "regular progressive" onward movement or spreading of the "zones of deposition", and including what are now known as onlap, offlap, and continental transgression.

progressive sand wave A term used by Bucher (1919, p.168) for a sand wave that migrates downcurrent. Ant: *regressive sand wave.*

progressive sorting Sorting in the downcurrent direction, resulting in a systematic downcurrent decrease in the mean grain size of the sediment (Pettijohn, 1957, p. 541).

progressive wave A water wave, the wave form of which appears to move progressively. Cf: *standing wave.*

progressive-wave theory A former theory of the tides involving the formation of two tidal waves in the Southern Ocean, one following the Moon and the other on the opposite diameter of the Earth (Monkhouse, 1965, p. 249); it is replaced by *oscillatory-wave theory.*

projected profile A diagram that includes only those features of a series of profiles, usually drawn along several regularly spaced and parallel lines on a map, that are not obscured by higher intervening ground (Monkhouse & Wilkinson, 1952); it gives a panoramic effect with a distant skyline, a middleground, and a foreground, and it represents an outline landscape-drawing showing only summit detail. Cf: *superimposed profile; composite profile.*

projection (a) A systematic, diagrammatic representation on a plane (flat) surface of three-dimensional space relations, produced by passing lines from various points to their intersection with a plane; esp. a *map projection.* (b) Any orderly method by which a projection is made; the process or operation of transferring a point from one surface to a corresponding position on another surface by graphical or analytical means, so that each point of one corresponds to one and only one point of the other.

projection net *net [struc petrol].*

prokaryote One of a group of organisms comprising the bacteria and the blue-green algae, characterized by relatively simple protoplasmic structure, without a vesicular nucleus or membrane-bounded organelles. Cf: *eukaryote.*

prolapsed bedding A term used by Wood & Smith (1958, p.172) for bedding characterized by a series of flat folds with near-horizontal axial planes contained entirely within a bed having undisturbed boundaries.

prolate Extended or elongated in the direction of a line joining the poles; e.g. "prolate pollen" whose equatorial diameters are much shorter than the dimensions from pole to pole. Ant: *oblate.*

proloculus The initial or first-formed chamber of a foraminiferal test, typically at the small end of a series or at the center of a coil. Pl: *proloculi.* Syn: *proloculum.*

proloculus pore A single circular opening in a proloculus, leading to the next-formed chamber of a foraminiferal test (as in fusulinids).

proluvium A complex, friable, deltaic sediment accumulated at the foot of a slope as a result of an occasional torrential washing of fragmental material. Adj: *proluvial.*

promontory (a) A high, prominent projection or point of land, or cliff of rock, jutting out boldly into a body of water beyond the coastline; a *headland.* Syn: *beak; cobb; reach; ness; nook.* (b) A *cape,* either low-lying or of considerable height, with a bold termination. (c) A bluff or prominent hill overlooking or projecting into a lowland.

prong [geomorph] *spur [geomorph].*

prong [streams] A term applied in the southern Appalachian Mountains to a fork or branch of a stream or inlet.

prong reef A *wall reef* that has developed irregular buttresses normal to its axis in leeward and (to a smaller degree) seaward directions (Maxwell, 1968, p.99 & 101).

proostracum The anterior (adoral) horny or calcareous bladelike prolongation of the dorsal border of the phragmocone of belemnites and related cephalopods, forming a protecting shield over the visceral mass of the animal. Pl: *proostraca.*

propaedeutic stratigraphy *prostratigraphy.*

propagule The minimum number of individuals of a species required for the successful colonization of a habitable island (MacArthur & Wilson, 1967, p.190).

propane An inflammable gaseous hydrocarbon, formula C_3H_8, of the methane series. It occurs naturally in crude petroleum and natural gas. It is also produced by cracking and is used primarily as a fuel and in the making of chemicals.

proparea One of a pair of flattened subtriangular halves of the pseudointerarea of a brachiopod, divided medially by various structures (such as homeodeltidium, intertrough, or pedicle groove).

proparian adj. Of or concerning a trilobite whose facial sutures extend outward from the eyes to the lateral margin of the cephalon; e.g. a "proparian facial suture" that crosses the dorsal surface of the cephalon, passes along the medial edge of the eye, and intersects the lateral border of the cephalon in front of or at the genal angle. Cf: *opisthoparian.* Syn: *proparous.*—n. A proparian trilobite; specif. a trilobite of the order Proparia (now obsolete) in which the posterior branch of the facial suture cuts the lateral margin of the cephalon.

propeltidium An anterior sclerite of a segmented carapace of an arachnid, situated in front of the *mesopeltidium.*

proper cave A *cave [speleo]* large enough for a person to enter.

propodosoma A section of the body of an acarid arachnid, bearing the first and second pairs of legs. Cf: *metapodosoma.*

propodus The sixth or penultimate segment of the pereiopod of a malacostracan crustacean, bounded proximally by the carpus and distally by the dactylus. Pl: *propodi.* Syn: *propodite.*

proportional counter A radiation detector consisting of a gas-filled tube in which the amplitude of the discharge pulses is proportional to the energy of the ionizing particles.

proportional limit The highest value of stress that a material can undergo before it loses its linear relationship between stress and strain, i.e. before it ceases to behave according to Hooke's law.

propylite A hydrothermally altered *andesite* resembling a *greenstone* and containing calcite, chlorite, epidote, serpentine, quartz, pyrite, and iron oxides. The term was first used by Richtofen in 1868. See also: *propylitization.*

propylitization A deuteric or hydrothermal process involving the formation of a *propylite* by the introduction of, or replacement by, an assemblage of minerals including carbonates, epidote, quartz, and chlorite.

proration Restriction of oil and gas production by a regulatory commission, usually in anticipation of market demand. It is the basis on which *allowables* are assigned.

prorsiradiate Said of an ammonoid rib inclined forward (adorally) from the umbilical side toward the venter. Cf: *rursiradiate; rectiradiate.*

proseptum The septum closing the protoconch in an ammonoid shell.

prosicula The proximal, initially formed part of the *sicula* of a graptolite, secreted as a single conical unit. Cf: *metasicula.*

prosiphon A small, threadlike structure extending from the adapical part of the caecum to the wall of the protoconch of an ammonoid shell.

prosobranch Any gastropod belonging to the subclass Prosobranchia, characterized in most cases by the presence of a shell, commonly with an operculum, and by the anterior position of the auricle with respect to the ventricle. Range, Cambrian to present.

prosochete An *inhalant canal* of a sponge.

prosocline (a) Said of the hinge teeth (and, in some genera, of the body of the shell) of a bivalve mollusk, sloping from the lower end in the anterior or forward direction. (b) Said of the growth lines that incline forward relative to the growth direction of a gastropod shell.—Cf: *opisthocline.*

prosodus A small canal of uniform diameter in a sponge, leading from an inhalant canal to a prosopyle of approximately the same cross-sectional area. Pl: *prosodi.* Cf: *aphodus.*

prosogyrate Said of the umbones (of a bivalve mollusk) curved so that the beaks point in the anterior or forward direction. Ant: *opisthogyrate.* Syn: *prosogyral.*

prosoma (a) The anterior part of the body of various invertebrates; esp. the *cephalothorax* of an arachnid or merostome. See also: *opisthosoma.* (b) The *prosome* of a copepod.

prosome (a) The anterior region of the body of a copepod crustacean, commonly limited behind by major articulation. See also: *urosome.* Syn: *prosoma.* (b) A structure within the neck of the body of a chitinozoan, extended to various intermediate positions, or even projecting beyond the collar. Its upper surface may be flat, convex, conical, or truncate, and an upper flange may lie against the pseudostome; its top may be marked by dark radial fibers, and the tubular area commonly has many dark annular rings. Cf: *opisthosome.*

prosopite A colorless mineral: $CaAl_2(F,OH)_8$.

prosopon Sculpture on the external surface of a trilobite exoskeleton, with presumed but unknown functional significance. This term is preferred by some authors to *ornamentation.*

prosopore The entrance opening of an inhalant canal of a sponge. Cf: *apopore.*

prosopyle Any aperture through which water enters a flagellated chamber of a sponge. Cf: *apopyle.*

prospect n. (a) An area that is a potential site of mineral deposits, based on preliminary exploration. (b) Sometimes, an area that has been explored in a preliminary way but has not given evidence of economic value. (c) An area to be searched by some investigative technique, e.g. geophysical prospecting. (d) A geologic or geophysical anomaly, especially one recommended for additional exploration.—A prospect is distinct from a *mine* in that it is nonproducing. See also: *prospecting.*

prospect hole A general term for any shaft, pit, adit, drift, tunnel, or drill hole made for the purpose of prospecting mineral-bearing ground. More specific terms, such as *prospect shaft* and *prospect pit,* are generally used.

prospecting (a) Searching for economically valuable deposits of fuel or minerals. Cf: *exploration.* (b) *geophysical exploration.*

prospecting seismology *applied seismology.*

prospection *prospecting.*

prospective ecospace As proposed by Valentine (1969, p.687), "the total ecospace that an organism or other ecological unit may utilize if it is physically available". Cf: *realized ecospace.*

prospector An individual engaged in prospecting for valuable mineral deposits, generally working alone or in a small group, and on foot with simple tools or portable detectors. The term implies an individual searching on his own behalf, rather than an employee of a mining company.

prospect pit *prospect hole.*

prospect shaft *prospect hole.*

prostal A sponge spicule (megasclere) that protrudes from the surface of a hexactinellid sponge. Pl: *prostals* or *prostalia.*

prostratigraphy A term proposed by Schindewolf (1954) for "preliminary stratigraphy" including lithologic and paleontologic studies without consideration of the time factor; the "raw material" for stratigraphy, consisting of local observation, description, and arrangement of strata, but not yet methodically linked together by the concept of time. See also: *protostratigraphy; topostratigraphy.* Syn: *propaedeutic stratigraphy.*

prosuture The line of junction of a proseptum with the walls of an ammonoid shell.

protactinium-ionium age method Calculation of an age in years for deep-sea sediments formed during the last 150,000 years, based on the assumption that the initial protactinium-231 to ionium (thorium-230) ratio for newly formed sediments has remained constant for the total section of sediments under study. The age depends on the gradual change with time of the protactinium-231 to ionium ratio because of the difference in half-lives. See also: *uranium-series age method.* Syn: *protactinium-231/thorium-230 age method; thorium-230/protactinium-231 excess method.*

protactinium-231/thorium-230 age method *protactinium-ionium age method.*

protalus rampart An arcuate ridge of coarse, angular blocks of rock derived by single rockfalls from a cliff or steep rocky slope above, marking the downslope edge of an existing or melted snowbank. The blocks roll and slide across the snowbank but no fine material reaches its edge. After the snowbank melts, the rampart or ridge stands some distance beyond any talus near the base of the cliff (Bryan, 1934, p. 656). Cf: *winter protalus ridge.*

protaspis An early juvenile trilobite whose small, oval exoskeleton is not yet divisible into cephalon, thorax, and pygidium. Cf: *anaprotaspis; metaprotaspis.* Pl: *protaspides.*

protaxis An antique term for the central axis of a mountain chain, supposedly consisting of the oldest rocks and structures; for example, an "Archean protaxis".

protected thermometer A *reversing thermometer* that is protected against hydrostatic pressure by a glass shell. Cf: *unprotected thermometer.*

protectite A rock formed by the crystallization of a primary magma. See also: *anatexite, syntectite.*

protegulal node The apical, commonly raised portion of an adult brachiopod shell, representing the site of the protegulum and later growth up to the brephic stage (TIP, 1965, pt.H, p.151).

protegulum The smooth biconvex first-formed shell of organic material (chitin or protein) of a brachiopod, secreted simultaneously by both mantles.

proterobase A *diabase* in which the mafic mineral is primary hornblende. The term, originated by Gümbel in 1874, is obsolete.

proterogenesis *neoteny.*

proteroglacial Pertaining to the earlier great ice age (Hansen, 1894, p. 128). Cf: *deuteroglacial.*

Proterophytic *Archeophytic.*

proterosoma The anterior section of the body of an acarid arachnid, ending behind the second pair of legs.

Proterozoic (a) The more recent of two great divisions of the Precambrian. Cf: *Archeozoic.* Syn: *Algonkian; Agnotozoic.* (b) The entire *Precambrian.*

prothallus The *gametophyte* of a fern or other pteridophyte, usually a flattened thalluslike structure attached to the soil. Pl: *prothalli.* Syn: *prothallium.*

protheca (a) The proximal part of graptoloid theca before it is differentiated from the succeeding theca. It is considered equivalent to the *stolotheca* in those graptolites with more than one type of theca. (b) A primary element of the wall of a fusulinid, comprising diaphanotheca and tectum.

protist Any organism assigned to the kingdom Protista, which includes forms with both plant and animal affinities, e.g. protozoans, bacteria, and some algae, fungi, and viruses. No agreement exists on the limits of nomenclature of the Protista. Var: *protistan.*

protobitumen Any of the fats, oils, waxes, or resins present as unaltered or nearly unaltered plant and animal products from which fossil bitumens are formed (Tomkeieff, 1954). See also: *labile; stabile.*

Protochordata A phylum or subphylum of animals that possess a notochord during some part of their life history but lack a bony skeleton or spinal column. They occupy a position intermediate between the invertebrates and vertebrates; they are included by some in the Chordata.

protoclase Leith's term for a rock possessing what he considered to be primary cleavage, e.g. bedding planes in sedimentary rock, formed concurrently with the rock (1905, p. 12). Cf: *metaclase.*

protoclastic (a) Said of igneous rocks in which the earlier formed crystals have been broken or deformed due to differential flow of the magma before complete solidification. (b) Said of an igneous rock containing deformed *xenocrysts.* (c) Said of the texture characteristic of an early stage of *cataclasis,* with a very small amount of finite strain.

protoconch (a) The first portion of the embryonic shell of a cephalopod, its preservation in fossil and in living forms being uncertain. The term is sometimes applied to the first camera (chamber) of the shell, located at the apex of the phragmocone or at the center of the coil, and closed in an ammonoid by the proseptum. (b) The apical, usually smooth whorl of a gastropod shell, usually well-demarcated from the *teleoconch.* The term applies to the fully formed embryonic shell of a gastropod and should not be used synonymously with *nucleus,* although it has been restricted by some authors to the simple cap-shaped plate that constitutes the first shell rudiment (see Knight, 1941).

protocorallite The first-formed corallite of a colony.

protodolomite (a) A crystalline calcium-magnesium carbonate with a disordered structure in which the metallic ions occur in the same crystallographic layers instead of in alternate layers as in the dolomite mineral. (b) An imperfectly crystallized artificial material of composition near $CaMg(CO_3)_2$.

protoecium (a) *proancestrula.* (b) *ancestrula.*

protoenstatite An artificial, unstable modification of $MgSiO_3$, produced by decomposition of talc by heating and convertible to enstatite by grinding or by heating to a high temperature.

protoforamen The primary aperture of a foraminiferal test associated with a fully developed or rudimentary tooth plate. Cf: *deuteroforamen.*

protogene An old term for a primary rock. Adj: *protogenous.* Cf: *deuterogene.* Syn: *protogine.*

protogenesis Reproduction by budding.

protogenous Adj. of *protogene.*

protogine A granitic rock, occurring in the Alps, that has gneissic structure, contains sericite, chlorite, epidote, and garnet, and shows evidence of a composite origin or crystallization (or partial recrystallization) under stress after consolidation. Also spelled: *protogene.* The term, dating from 1806, is obsolete.

protogranular Said of the xenomorphic, granoblastic texture of a rock, characterized by sinuous mineral boundaries, intergrowths, and other features suggesting previous equilibrium with a melt (Mercier and Nicolas, 1975).

protointraclast A genetic term suggested by Bosellini (1966) for a limestone component that resulted from a premature attempt at resedimentation while still being in an unconsolidated and viscous or plastic state and that never existed as a free, clastic entity. Cf: *intraclast; plasticlast.*

protolith The unmetamorphosed rock from which a given metamorphic rock was formed by metamorphism. Syn: *parent rock.*

protomatrix A term introduced by Dickinson (1970, p. 702), for unrecrystallized clayey material in weakly consolidated graywackes and arkoses.

protomylonite (a) A mylonitic rock produced from contact-metamorphosed rock, with granulation and flowage being due to overthrusts following the contact surfaces between intrusion and country rock (Holmes, 1920). (b) A coherent crush breccia whose characteristically lenticular, megascopic particles faintly retain primary structures. It is a lower grade in the development of *mylonite* and *ultramylonite* (Waters & Campbell, 1935, p.479).

protonema The green filamentous gametophyte of mosses.

proton-precession magnetometer A type of *nuclear resonance magnetometer* that accurately measures total magnetic intensity by the use of the precession of protons in a hydrogen-rich liquid about the magnetic field direction. Precession frequency is proportional to field strength. See also: *proton-vector magnetometer.*

proton-vector magnetometer A type of *proton-precession magnetometer* with a system of auxiliary coils that permits measurement of horizontal intensity H or vertical intensity Z as well as total intensity F.

protonymph The first postembryonic stage in the arachnid order Acarida.

protophyte A unicellular organism or primitive plant (Swartz, 1971, p. 384). Cf: *metaphyte.*

protopod The proximal portion of a limb of a crustacean, consisting of coxa, basis, and sometimes precoxa, often fused to each other. Its distal edge generally bears the endopod and exopod. Syn: *protopodite; sympod.*

protopore A single fine opening or perforation in a foraminiferal test, rounded at least on the inner wall. Cf: *deuteropore.*

protopygidium The postcephalic portion of a trilobite *protaspis.* Pl: *protopygidia.*

protoquartzite A well-sorted, quartz-enriched sandstone that lacks the well-rounded grains of an *orthoquartzite;* specif. a lithic sandstone intermediate in composition between subgraywacke and orthoquartzite, containing 75-95% quartz and chert, less than 15% detrital clay matrix, and 5-25% unstable materials in which the rock fragments exceed the feldspar grains in abundance (Pettijohn, 1954, p.364). It commonly forms *shoestring sands.* Examples: Venango Formation (Upper Devonian) of New York and Pennsylvania, and Hartshorne Sandstone (Pennsylvanian) of Oklahoma and Arkansas. The term was used by Krynine (1951) for a "cleaned-up" graywacke (matrix washed out), intermediate in composition between quartzose graywacke and orthoquartzite. Syn: *quartzose subgraywacke.*

protore The rock below the sulfide zone of *supergene enrichment;* the primary, subeconomic material. See also: *oxidized zone; sulfide zone.*

protoscience Attempts to explain the natural world preceding or leading up to systematic studies.

protoseptum *primary septum.*

protostele A *stele* with a solid xylem core (Swartz, 1971, p. 384).

protostratigraphy A term proposed by Henningsmoen (1961) for preliminary or introductory stratigraphy, including lithostratigraphy and biostratigraphy; *prostratigraphy.*

prototheca The roughly conical or cup-shaped structure constituting the embryonic exoskeleton of a coral.

prototype An ancestral form; the most primitive form in a group of related organisms. Syn: *archetype.*

protoxylem The first-formed, primary xylem of a plant. Cf: *metaxylem; annular tracheid; spiral tracheid.*

protozoan A single-celled organism belonging to the *protist* phylum Protozoa, characterized by the absence of tissues and organs. Some members of the phylum have both plant and animal affinities (flagellates); other members are characterized by their development of calcareous and siliceous skeletons (foraminifers, radiolarians). Nonpreferred syn: *eozoan.*

Protozoic (a) That part of Precambrian time represented by rocks in which traces of life appear. Cf: *Azoic.* (b) The lower Paleozoic.—The term is obsolete.

protractor An instrument used in drawing and plotting, designed for laying out or measuring angles on a flat or curved surface, and consisting of a plate marked with units of circular measure.

protractor muscle A muscle that extends an organ or part; e.g. an *outside lateral muscle* or *middle lateral muscle* in some lingulid brachiopods, or a longitudinal fibril in connective tissue of the pedicle of some articulate brachiopods (TIP, 1965, pt.H, p.151). Cf: *retractor muscle.*

protriaene A sponge triaene in which the cladi curve away from the rhabdome, making an angle to the rhabdome that is noticeably greater than that of a normal tetraxon. Cf: *orthotriaene.*

protrusion A proposed term for a rock mass that has been tectonically intruded in the solid state; it is in contrast to an igneous intrusion (Lockwood, 1971). Adj: *protrusive.*

proudite A monoclinic mineral: $Cu_{0\text{-}1}Pb_{7.5}Bi_{9.5}(S,Se)_{22}$.

proustite A cochineal-red rhombohedral mineral: Ag_3AsS_3. It is isomorphous with pyrargyrite, and is a minor ore of silver. Cf: *xanthoconite.* Syn: *light ruby silver; light red silver ore.*

prove v. In economic geology, to establish, by drilling, trenching, underground openings, or other means, that a given deposit of a valuable substance exists, and that its grade and dimensions equal or exceed some specified amounts. See also: *proved reserves.*

proved ore *proved reserves.*

proved reserves Reserves of metallic and nonmetallic minerals, and of oil and gas, for which reliable quantity and quality estimates have been made. Cf: *developed reserves; positive ore.* Syn: *proved ore.*

provenance A place of origin; specif. the area from which the constituent materials of a sedimentary rock or facies are derived. Also, the rocks of which this area is composed. Cf: *distributive province.* Syn: *provenience; source area; sourceland.*

provenience *provenance.*

proven reserves Oil that has been discovered and determined to be recoverable but is still in the ground.

province [ecol] (a) Part of a *region,* isolated and defined by its climate and topography and characterized by a particular group of organisms. (b) A group of temporally and spatially associated plant or animal communities.

province [geog] Any large area or region considered as a whole, all parts of which are characterized by similar features or by a history differing significantly from that of adjacent areas; specif. a *geologic province* or a *physiographic province.*

provincial alternation The overlapping of sedimentary petrologic provinces, caused by oscillation of the boundary between two provinces during time (Pettijohn, 1957, p. 573-574).

provincial series A *series [stratig]* recognized only in a particular region and involving a major division of time within a period; e.g. the Wolfcampian Series within the Permian System in west Texas and New Mexico.

provincial succession A succession of sedimentary petrologic provinces, produced by changes in provenance leading to mineral associations that change with time (Pettijohn, 1957, p. 574).

provinculum A taxodont hinge composed of minute teeth developed in some bivalve mollusks before the permanent dentition.

provitrain *Vitrain* in which some plant structure is microscopically visible. Cf: *euvitrain.* See also: *periblain; suberain; xylain.* Syn: *telain.*

provitrinite A variety of the maceral *vitrinite* characteristic of provitrain and including the varieties periblinite, suberinite, and xylinite. Plant cell structure is visible under the microscope. The term *telinite* has been proposed as a preferable synonym. Cf: *euvitrinite.* Syn: *phyllovitrinite.*

prowersite An orthoclase- and biotite-rich *minette.* Its name, given by Rosenbusch in 1908, is derived from Prowers County, Colorado. Not recommended usage.

proximal [eco geol] Said of an ore deposit formed immediately adjacent to, perhaps part of, a volcanic hearth, pile, or fumarole to which it is genetically related and from which its constituents have been derived. Cf: *distal [eco geol].*

proximal [paleont] Next to or nearest the point of attachment or place of reference, a point conceived of as central, or the point of view. Examples in invertebrate morphology: "proximal direction" toward the dorsal pole or mouth of a crinoid; "proximal ray" of a sponge spicule, directed inward from a bounding surface of the sponge; and "proximal direction" toward the ancestrula or origin of growth of a bryozoan colony. Ant: *distal.*

proximal [palyn] Said of the parts of pollen grains or spores nearest or toward the center of the original tetrad; e.g. said of the side of a monocolpate pollen grain opposite the colpus, or said of the side of a trilete spore provided with contact areas. Ant: *distal [palyn].*

proximal [sed] Said of a sedimentary deposit consisting of coarse clastics and formed nearest the source area; e.g. a "proximal turbidite" consisting of thick sandy varves. Cf: *distal.*

proximale The noncirriferous topmost columnal of a crinoid, typically distinguished by enlargement and permanent attachment to the dorsal cup.

proximal onlap *Onlap* in the direction of the source of clastic supply (Mitchum, 1977, p. 208). Cf: *distal onlap.*

proximate admixture A term applied by Udden (1914) to an *admixture* (in a sediment of several size grades) whose particles are most similar in size to those of the dominant or maximum grade; material in one of the two classes adjacent to the maximum histogram class.

proximate analysis The determination of compounds contained in a mixture; for coal, the determination of moisture, volatile matter, ash, and fixed carbon (by difference) (ASTM). Cf: *ultimate analysis.*

proximate cyst A dinoflagellate cyst of nearly the same size as, and closely resembling, the motile theca. The ratio of the diameter of the main body to the total diameter of the cyst exceeds 0.8. The term refers to the supposed proximity of the main cyst wall to the theca at the time of encystment. See also: *chorate cyst; proximochorate cyst.*

proximochorate cyst A dinoflagellate cyst having sutured outgrowths that readily indicate the tabulate character. The ratio of the diameter of the main body to the total diameter of the cyst is between 0.6 and 0.8. See also: *chorate cyst; proximate cyst.*

proxistele The proximal region of a crinoid column near the theca, generally not clearly delimited from the mesistele. Cf: *dististele.*

proxy v. To substitute one ion or atom for another in a crystal structure.—adj. Said of such a substituted ion or atom.

proxying *ionic substitution.*

przhevalskite A bright greenish-yellow mineral: $Pb(UO_2)_2(PO_4)_2 \cdot 2H_2O$.

Psamment In U.S. Dept. of Agriculture soil taxonomy, a suborder of the soil order *Entisol*, characterized by well-sorted sands of sand dunes or of sandy natural levees or beaches. They have a texture of loamy fine sand or coarser, and a coarse-fragment content of less than 35%. Psamments range in age from very recent to Pliocene or older (USDA, 1975). Cf: *Aquent; Arent; Fluvent; Orthent.*

psammite (a) A clastic sediment or sedimentary rock composed of sand-size particles; a sandstone. The term is equivalent to the Latin-derived term, *arenite.* (b) A term formerly used in Europe for a fine-grained, fissile, clayey sandstone (as distinguished from a more siliceous and gritty one) in which "the component grains are scarcely distinguishable by the unassisted eye" (Oldham, 1879, p.44). (c) A term regarded by Tyrrell (1921, p.501-502) as the metamorphic derivative of arenite.—Etymol: Greek *psammos,* "sand". See also: *psephite; pelite.* Also spelled: *psammyte.*

psammitic (a) Pertaining to or characteristic of psammite; *arenaceous.* Cf: *sandy.* (b) Said of a metamorphic rock derived from a psammite; e.g. a "psammitic gneiss" or a "psammitic schist".

psammobiotic Said of an organism that lives in sand or sandy areas.

psammofauna The animals associated with sandy substrates.

psammogenic dune A dune "caused by the effect of sand surfaces in trapping more sand" (Schieferdecker, 1959, term 0148).

psammon The interstitial organisms found between sand grains.

psammophilic Said of an organism or of the fauna found in sand. Noun: *psammophile.*

psammophyte A plant preferring sand or very sandy soil for growth.

psammosere A *sere* that develops in a sandy environment.

psammyte *psammite.*

psephicity A term used by Mackie (1897, p.301) for the "coefficient of roundability" of a pebble- or sand-size mineral fragment, expressed as the ratio of specific gravity to hardness (as measured in air) or the quotient of specific gravity minus one divided by hardness (as measured in water).

psephite (a) A sediment or sedimentary rock composed of large fragments (coarser than sand) set in a matrix varying in kind and amount; e.g. rubble, talus, breccia, glacial till, tillite, shingle, gravel, and esp. conglomerate. The term is equivalent to the Latin-derived term, *rudite.* (b) A term regarded by Tyrrell (1921, p.501-502) as the metamorphic derivative of rudite.—Etymol: Greek *psephos,* "pebble". See also: *psammite; pelite.* Also spelled: *psephyte.*

psephitic (a) Pertaining to or characteristic of psephite. (b) Said of a metamorphic rock derived from a psephite.—Cf: *rudaceous; gravelly.*

psephonecrocoenosis A necrocoenosis of dwarf individuals.

psephyte [lake] A lake-bottom deposit consisting mainly of coarse, fibrous plant remains (Veatch & Humphrys, 1966, p.248). Cf: *pelphyte.*

psephyte [sed] *psephite.*

pseudatoll *pseudoatoll.*

pseudo- A prefix meaning "false" or "spurious".

pseudoactine A raylike arm or branch of a sponge spicule that contains no axial filament or axial canal.

pseudoallochem An object resembling an *allochem* but produced in place within a calcareous sediment by a secondary process such as recrystallization (Folk, 1959, p. 7).

pseudoaquatic Said of an organism living in moist or wet but not truly aquatic conditions.

pseudoatoll (a) An *atoll* that rises from the outer margin of a rimless shoal; a reef platform encircling a shallow pool of water, rising to low-tide level from the continental shelf. Syn: *bank atoll; shelf atoll; lagoon atoll.* (b) A ring-shaped island or reef composed of material other than true coral-reef limestone.—Syn: *pseudatoll.*

pseudoautunite A pale-yellow to white mineral: $(H_3O)_4Ca_2(UO_2)_2(PO_4)_4 \cdot 5H_2O$(?). It is not a member of the autunite group.

pseudobed A group of nearly parallel plane surfaces that dip upcurrent in climbing-ripple laminae and that are formed either by nondeposition or by erosion on the upcurrent sides of migrating superimposed ripple laminae: "between successive pseudobeds are sets of laminae that dip steeply in the opposite direction, formed by deposition on the lee side of each ripple crest and resembling, in general, the foresets of tabular planar cross-beds" (McKee in Middleton, 1965, p.250). See also: *cross-bedding.*

pseudobedding [petrology] *pseudostratification.*

pseudobedding [sed] Bedding developed by concentration or combining of ripple laminae representing the approach slopes of ripple deposits (McKee, 1939, p.72); i.e. bedding produced by the bounding surfaces between sets of cross-beds deposited by *climbing ripples.* See also: *false bedding; pseudo cross-bedding.*

pseudobivalved Said of rostroconch mollusks in which the larval shell is univalved and the adult shell bivalved. In this group there are always one or more shell layers continuous across the dorsal margin, and a dorsal commissure such as occurs in pelecypods is lacking; thus the two valves cannot be separated from one another.

pseudoboleite A mineral: $Pb_5Cu_4Cl_{10}(OH)_8 \cdot 2H_2O$. Also spelled: *pseudoboléite.*

pseudobomb *lava ball.*

pseudobreccia A partially and irregularly dolomitized limestone, characterized by a mottled appearance that gives the rock a texture mimicking that of a breccia, or by a weathered surface that appears deceptively fragmental. It is produced diagenetically by selective grain growth in which localized, patchy, and irregularly shaped recrystallized masses of coarse calcite (usually visible to the naked eye: 1-20 mm in diameter) are embedded in a lighter-colored and less-altered matrix of calcareous mud. The boundaries between the "clasts" and the matrix are indistinct or gradational. The term was introduced by Tiddeman in Strahan (1907, p. 10-15), and used by Dixon & Vaughan (1911, p. 507) and Wallace (1913). Cf: *pseudopsephite.* Syn: *recrystallization breccia.*

pseudobrookite A brown or black orthorhombic mineral: Fe_2TiO_5. It resembles brookite.

pseudocannel coal Cannel coal that contains much humic matter. Syn: *humic-cannel coal.*

pseudocarina A perforate, ridgelike thickening of the peripheral part of a chamber wall of a foraminiferal test, situated approximately in the plane of coiling.

pseudoceratite A Jurassic and Cretaceous ammonoid cephalopod having a suture similar to that of a ceratite. It is explained as a reversionary or atavistic modification of a normal ammonite.

pseudoceratitic suture A type of suture in ammonoids that approximates a *ceratitic suture* in form but is not related to ceratites; specif. a suture in pseudoceratites.

pseudochamber A partly subdivided cavity of a foraminiferal test (as in the family Tournayellidae), indicated by a slight protuberance or an incipient septum.

pseudochitin A resistant organic substance, the exact chemical structure of which is uncertain, though it apparently consists of compounds of C-H-O-N. The behavior of pseudochitin is similar to that of *chitin,* but by definition it does not yield a positive chitin staining reaction. Various fossils, including graptolites and chitinozoans, contain or consist mostly of this substance.

pseudochitinous Consisting of *pseudochitin.*

pseudochlorite (a) *swelling chlorite.* (b) *septechlorite.* (c) An artificial product obtained by adsorbing magnesium salts on montmorillonite and precipitating magnesium hydroxide between the layers of the mineral (Youell, 1960).

pseudochrysolite *moldavite [astron].*

pseudocirque A term used by Freeman (1925) and recommended by Charlesworth (1957, p. 244) for a feature that is similar but not homologous to a glacial cirque. See also: *cirque [geomorph].*

pseudocol A term proposed by Chamberlin (1894a) for a landform represented by a constriction of the valley of a stream diverted by glacial ponding, formed by the cutting through of a cover of drift and subsequent exposure of a former col; the feature occurs in regions of reversed drainage along the border of ancient glacial formations, as along several segments of the Ohio River valley.

pseudocolpus A colpuslike modification of the exine of pollen grains, differing from a true *colpus* in that it is not a site of pollen-tube emergence. Pl: *pseudocolpi.*

pseudoconcretion A subspherical sedimentary structure resembling a concretion but not formed by orderly precipitation of mineral matter in the pores of a sediment; e.g. an armored mud ball or certain algal structures.

pseudoconformity A term used by Fairbridge (1946, p.88) for a stratigraphic relationship that appears conformable but is characterized by nonaccumulation or deficiency of sediment, such as a slump gap in which an entire formation slipped away off the crest of a rising anticline or in which no trace of a hiatus is immediately apparent from the structure.

pseudoconglomerate A rock that resembles, or may easily be mistaken for, a normal sedimentary conglomerate; e.g. a *crush conglomerate* consisting of cemented fragments that have been rolled and rounded nearly in place by orogenic forces; a sandstone packed with many rounded concretions; or an aggregate of rounded boulders produced in place by *spheroidal weathering* and surrounded by clayey material. Term introduced by Van Hise (1896, p.679). Cf: *pseudopsephite.*

pseudocotunnite A mineral: K_2PbCl_4 (?).

pseudo cross-bedding (a) An inclined bedding produced by deposition in response to ripple-mark migration, and characterized by foreset beds that appear to dip into the current. See also: *pseudobedding.* (b) A structure resembling cross-bedding, caused by distortion-free slumping and sliding of a semiconsolidated mass of sediments.—Also spelled: *pseudocross-bedding.* Syn: *pseudo cross-stratification.*

pseudocruralium An excessive thickening of the secondary shell of a brachiopod, bearing dorsal adductor impressions, and elevated anteriorly above the floor of the valve.

pseudocrystal A substance that appears to be crystalline but does not give a diffraction pattern that confirms it as truly crystalline.

pseudodeltidium A single convex or flat plate affording variably complete cover of the delthyrium of a brachiopod, but always closing the apical angle when the pedicle foramen is supra-apical or absent, and dorsally enclosing the apical foramen (TIP, 1965, pt.H, p.151). Cf: *deltidium.*

pseudo-diffusion Mixing of thin layers of slowly accumulated marine sediments by the action of water motion and/or subsurface organisms. This phenomenon can lead to serious errors in determining the rate of sedimentation if the disturbed sediments are dated by carbon-14 or other radiometric methods (Bowen, 1966, p. 208).

pseudofault A term coined by Palmer (1920, p.851) for a faultlike feature resulting from weathering along joints, shrinkage cracks, or bedding planes.

pseudofibrous peat Peat that is fibrous in texture but that is plastic and incoherent. Cf: *fibrous peat; amorphous peat.*

pseudofossil A natural object, structure, or mineral of inorganic origin that may resemble or be mistaken for a fossil. Cf: *problematic fossil.*

pseudogalena *sphalerite.*

pseudogley A densely packed, silty soil that is alternately waterlogged and rapidly dried out (Kubiëna, 1953).

pseudogradational bedding A structure in metamorphosed sedimentary rock in which the original textural gradation (coarse at the base, finer at the top) appears to be reversed, due to the formation of porphyroblasts in the finer-grained part of the rock.

pseudo gravity The expected gravity field based on the observed magnetic field assuming Poisson's relation.

pseudogritty structure A type of *mortar structure* in which the larger relics are angular, due to fracture along cleavage planes. The term is not in common use.

pseudohexagonal Said of a crystal form, e.g. some orthorhombic forms, that simulate the hexagonal form.

pseudointerarea The somewhat flattened posterior sector of the shell of some inarticulate brachiopods "secreted by posterior sector of mantle not fused with that of opposite valve" (TIP, 1965, pt.H, p.151). Cf: *interarea.*

pseudokame *residual kame.*

pseudokarst A topography that resembles *karst* but was not formed by the dissolution of rock. Processes and forms involving *piping* and *thermokarst* are included (Otvos, 1976); some authors also include terrain characterized by lava tubes, sea caves, and blowouts. Pseudokarst has been applied to covered karst, and to karst produced by the dissolution of rocks that are relatively insoluble, such as quartzite and granite, but more general usage regards these as varieties of true karst. The term was first used by von Knebel in 1906.

pseudokettle *pingo remnant.*

pseudolagoon The shallow pool of water encircled by a *pseudoatoll.* Syn: *miniature lagoon.*

pseudolaueite A monoclinic mineral: $MnFe_2(PO_4)_2(OH)_2 \cdot 7\text{-}8H_2O$. Cf: *laueite.*

pseudoleucite A pseudomorph after leucite, consisting of a mixture of nepheline, orthoclase, and analcime, such as occur in certain syenites in Arkansas, Montana, and Brazil.

pseudomalachite A bright-green to blackish-green mineral: $Cu_5(PO_4)_2(OH)_4 \cdot H_2O$ (?). It resembles malachite and occurs in the oxidized zone of hydrothermal copper deposits. Syn: *dihydrite; phosphochalcite; phosphorochalcite; tagilite.*

pseudomatrix A term introduced by Dickinson (1970, p. 702) for "a discontinuous interstitial paste formed by the deformation of weak detrital grains" in graywackes and arkoses.

pseudomicrite A genetic term applied to calcareous micrite that has formed by secondary changes such as "degenerative" recrystallization (crystal diminution) of faunal and floral material (Chilingar et al., 1967, p. 319). Cf: *orthomicrite.*

pseudomicrosparite A genetic term applied to microsparite that has developed by recrystallization or by grain growth (Chilingar et al., 1967, p. 228). Cf: *orthomicrosparite.*

pseudomonoclinic Said of a triclinic crystal form, e.g. that of microcline, that simulates the monoclinic form.

pseudomorph A mineral whose outward crystal form is that of another mineral species; it has developed by alteration, substitution, incrustation, or paramorphism. A pseudomorph is described as being "after" the mineral whose outward form it has, e.g. quartz after fluorite. See also: *pseudomorphism; paramorph.* Adj: *pseudomorphous.* Syn: *false form; allomorph.*

pseudomorphism The process of becoming, and the condition of being, a *pseudomorph.*

pseudomorphous Adj. of *pseudomorph.*

pseudomountain A term used by Tarr (1902) for a mountain formed by differential erosion, as contrasted with one produced by uplift; e.g. a *plateau mountain.*

pseudo mud crack A term used by Ksiazkiewicz (1958, pl.16, fig.2) for a sedimentary structure now known as a *parting cast.* See also: *false mud crack.*

pseudonodule A primary sedimentary structure consisting of a ball-like mass of sandstone enclosed in shale or mudstone, characterized by a rounded base with upturned or inrolled edges, and resulting from the settling of sand into underlying clay or mud which welled up between isolated sand masses. The term was introduced by Macar (1948) who attributed the structure to horizontal displacement or vertical foundering. See also: *ball-and-pillow structure; flow roll.* Syn: *sand roll.*

pseudo-oolith A spherical or roundish pellet or particle (generally less than 1 mm in diameter) in a sedimentary rock, externally resembling an oolith in size or shape but of secondary origin and amorphous or crypto- or micro-crystalline, and lacking the radial or concentric internal structure of an oolith; e.g. a fecal pellet, a worn calcite grain, a shell fragment, a glauconite granule, or an oolith whose peripheral layers have been resorbed or replaced. Cf: *oolitoid; ooid.* Also spelled: *pseudoolith.* Syn: *false oolith.*

pseudo-ophitic Said of a texture of rock gypsum that is formed by a diagenetic rather than a metamorphic process, and that is characterized by large, platy selenite crystals enclosing small, well formed euhedra. The large crystals are probably of later origin than the matrix in which they are found (Pettijohn, 1957, p.479).

pseudo-orthorhombic Said of a monoclinic or triclinic crystal that approximates an orthorhombic crystal in lattice geometry or crystal form.

pseudophenocryst *porphyroblast.*

pseudophite A general name for compact, massive chlorite resembling serpentine, in part clinochlore and in part penninite (Hey, 1962, p. 569).

pseudoplankton *epiplankton.*

pseudopluton An igneous rock mass that resembles a pluton but lacks dike clusters and peripheral selvages, e.g. a *rheoignimbrite.* Rare.

pseudopodium A temporary or semipermanent projection or retractile process of the protoplasm of a cell (such as a unicellular organism) that serves for locomotion, attachment, and food gathering and that changes in shape, character, and position with the activity of the cell. It may be lobose, filamentous, bifurcating, or anastomosing. Examples: *axopodium; reticulopodium; rhizopodium.* Pl: *pseudopodia.* Syn: *pseudopod.*

pseudopore [paleont] (a) A tissue-filled space in the calcified layer of a body wall closed by an outer cuticle in many bryozoans. (b) A pore in the outer covering of various calcisponges, the covering being formed by outgrowth from the peripheral part of the inhalant canals.

pseudopore [palyn] An esp. thin area in the *leptoma* of certain coniferous pollen (as in the families Cupressaceae and Taxaceae).

pseudoporphyritic (a) Said of the texture of an igneous rock in which larger crystals have developed in a macrocrystalline groundmass, but were formed, at least in part, after the rock solidified (e.g. large potassium-feldspar crystals in a granite). The term was first used by Lasaulx in 1875 (Johannsen, 1939, p.230). (b) A syn. of *porphyroblastic.*

pseudoporphyroblastic Pertaining to a structure resembling *porphyroblastic* texture but due to processes other than growth, e.g. to differential granulation.

pseudopsephite The equivalent of *pseudobreccia* or *pseudoconglomerate* (Read, 1958).

pseudopuncta A conical deflection of the secondary shell of a brachiopod, with or without a *taleola,* pointing inwardly and commonly anteriorly so as to appear on the internal surface of the valve as a tubercle. It may weather out in fossil shells, leaving a tiny opening that may be mistaken for a *puncta* in punctate shells. Pl: *pseudopunctae.* Syn: *pseudopunctum.*

pseudopylome A prominent thickening of the wall at the antapical end of the vesicle in some acritarchs, resembling the rim of a pylome. In some species, such as *Axisphaeridium* and *Polyancistrodorus,* a central depression or notch seemingly continues as a canal.

pseudoraphe On the frustule of some pennate diatoms, a clear area on the valve between striae or costae.

pseudo ripple mark A term used by Kuenen (1948, p.372) for a bedding-plane feature resembling a ripple mark but attributed to lateral pressure caused by slumping (such as a mudflow structure imitating a ripple mark) or by local, small-scale tectonic deformation (such as a corrugation on the cleavage face of slate). See also: *crinkle mark; creep wrinkle.*

pseudorostrum The anterior part of the gnathothorax of a malacostracan crustacean, formed by a pair of anterior and lateral parts of the cephalic shield projecting forward and meeting in front of the true rostrum.

pseudorutile A mineral: $Fe_2Ti_3O_9$. It is an oxidation product of ilmenite, and is common in beach sands. Cf: *arizonite [mineral].*

pseudosaccus An ektexinous sac attached to a fossil spore (such as *Endosporites*), resembling the *saccus* of some pollen grains but typically not showing internal structure. The distinction between a pseudosaccus and a *vesicle [palyn]* is slight.

pseudoscience Procedures that use the appearance and language of science to obtain dubious and often fraudulent results.

pseudosecondary inclusion A fluid inclusion formed by healing of a fracture occurring during growth of the host crystal (Roedder, 1967, p. 522).

pseudo section A display of resistivity and induced-polarization data, obtained with the pole-dipole or dipole-dipole array, in which the observed data values are plotted in section at the intersections of lines drawn at 45 degrees from the mid-point of the current and potential electrode pairs (midpoint of potential pair and through the near current electrode for pole-dipole array); an artifice used to present all of the data from a sounding-profiling in one section. The vertical dimension of the pseudo section bears no simple relationship to the geologic section.

pseudoseptal pinnacle One of the sclerenchymal pillars constructed on the upper surface of a tabula in the zone between parent corallite and offset during the early stages of increase in corals (Fedorowski & Jull, 1976, p. 37). Cf: *septal pinnacle.*

pseudoseptum (a) A spinelike or toothlike skeletal projection in octocorals of the order Coenthecalia. Pseudosepta "bear no constant relationship with soft septa of polyps" (TIP, 1956, pt.F, p.174). (b) The plane of junction in a nautiloid conch between hyposeptal deposits of one septum and episeptal deposits on the preceding septum (TIP, 1964, pt.K, p. 58), probably representing remnants of the cameral mantle.

pseudoskeleton A sponge skeleton consisting of foreign bodies not secreted by the sponge. Cf: *autoskeleton.*

pseudosparite A limestone consisting of relatively large, clear calcite crystals that have developed by recrystallization (Folk, 1959, p. 33). Cf: *orthosparite.*

pseudosphärolith A spherulite consisting of two minerals, one with parallel and one with inclined extinction, growing from the same center (Johannsen, 1939, p. 193).

pseudospicule *spiculoid.*

pseudospondylium A cup-shaped chamber accommodating the ventral muscle field of a brachiopod and comprising an undercut callus contained between discrete dental plates. Cf: *spondylium.*

pseudostome (a) An aperture in a thecamoebian test from which pseudopodia protrude. It may be a simple opening or have definite structure. (b) An opening at the end of a chitinozoan neck. It may be simple, bordered by a small lip, or have a tubular collar.

pseudostratification [glac geol] A concentric structure, resembling stratification, that occurs in till deposits overridden by ice, formed partly by plastering-on of layers of debris and partly by shearing of the till due to pressure of superincumbent ice.

pseudostratification [ign] Apparent layering in some igneous rocks caused by structural features, esp. horizontal jointing, that have the appearance of stratification. Cf: *false stratification.* Syn: *pseudobedding.*

pseudostratification [struc geol] *sheeting.*

pseudosymmetry Apparent symmetry of a crystal, resembling that of another system; it is generally due to twinning.

pseudotachylyte (a) A dense rock produced in the compression and shear associated with intense fault movements, involving extreme mylonitization and/or partial melting. Similar rocks, such as the Sudbury breccias, contain shock-metamorphic effects and may be injection breccias emplaced in fractures formed during meteoric impact. Cf: *ultramylonite.* (b) A dark gray or black rock that externally resembles *tachylyte* and that typically occurs in irregularly branching veins. The material carries fragmental enclosures, and shows evidence of having been at high temperature. Miarolitic and spherulitic crystallization has sometimes taken place in the extremely dense base. Some pseudotachylyte has behaved like an intrusive and has no structures obviously related to local crushing.

pseudo telescope structure A term proposed by Blissenbach (1954, p.181) for an alluvial-fan structure resulting from slumping of unconsolidated deposits, such as that formed in a fan that has been cut by a series of small normal faults.

pseudotheca The false wall of a coral, formed by the thickening and fusion of the outer ends of septa.

pseudotill A nonglacial deposit resembling a glacial till.

pseudotillite A term proposed by Schwarzbach (1961) for a definitely nonglacial tillitelike rock, such as a *pebbly mudstone* formed on land by flow of nonglacial mud or deposited by a subaqueous turbidity flow; an indurated *pseudotill.* Harland et al. (1966, p.233) urge the use of "pseudotillite" as an "unambiguous" and "negative" term for tillitelike rocks found to be nonglacial: "a deposit so named is ... likely to show positive characters which will lead to different nomenclature". The term is equivalent to *tilloid* as used by Pettijohn (1957, p.265).

pseudoumbilicus A deep depression, either wide or narrow, between the inner umbilical chamber walls of a trochospirally enrolled foraminiferal test where the sharply angled umbilical shoulder occurs (as in *Globorotalites*).

pseudounconformity A term used by Fairbridge (1946, p.88) for a stratigraphic relationship that appears unconformable but is characterized by superabundance or excess accumulation of sediment, such as due to submarine slumping penecontemporaneous with sedimentation off the sides of a rising anticline or dome.

pseudoviscous flow *secondary creep.*

pseudovitrinite A maceral of coal that is superficially similar to *vitrinite* but is higher in reflectance and has slitted structure, remnant cellular structures, uncommon fracture patterns, higher relief, and paucity or absence of pyrite inclusions (Benedict et al., 1968, p.125).

pseudovitrinoid Pseudovitrinite that occurs in bituminous coals (Benedict et al., 1968, p. 126).

pseudovolcano A large crater or circular hollow believed not to be associated with volcanic activity; e.g. a crater that is possibly meteoritic in origin but may be the result of phreatic explosion or cauldron subsidence. Adj: *pseudovolcanic.*

pseudowavellite *crandallite.*

psi A negative logarithmic transformation (to the base 2) of settling velocity in cm/sec, analogous to the *phi* transformation of grain size. Proposed by Middleton (1967, p. 484).

psilate Said of the smooth walls of pollen and spores that lack sculpture. The term is usually applied to exines with pits less than one micron in diameter. Syn: *laevigate.*

psilomelane (a) A general field term for mixtures of manganese minerals, or for a botryoidal, colloform manganese oxide whose mineral composition is not specifically determined. Cf: *wad.* Syn: *manganomelane.* (b) A manganese-oxide mineral; specif. *romanechite.*

psilophyte An extinct *psilopsid;* a member of the Psilophytales, ranging from Silurian to (questionably) Carboniferous.

psilopsid Any primitive vascular plant referable to the division Psilopsida. It is generally without roots or leaves but has spore-bearing organs at the stem tips. Such plants range from the Silurian and occur chiefly during the Devonian. See also: *psilophyte.*

psittacinite *mottramite.*

Psychozoic n. A now obsolete term for the era in geologic time characterized and initiated by the appearance of man on Earth.

psychrophilic A syn. of *cryophilic.* Noun: *psychrophile.*

psychrophyte A plant adapted to arctic or alpine conditions.

psychrotolerant Said of an organism that lives at 0°C and tolerates temperatures above 20°C.

pterate chorate cyst A dinoflagellate *chorate cyst* characterized

by a pronounced equatorial outgrowth in the form of solid processes linked distally or in meshlike fashion, as in *Wanea.*
pteridophyte A fernlike, vascular plant that reproduces by spores. Members of this division, which appeared in the Devonian, include lycopods, horsetails or scouring rushes, and ferns. Cf: *spermatophyte; bryophyte; thallophyte.*
Pteridophytic *Paleophytic.*
pteridosperm A gymnosperm with fernlike foliage and true seeds borne on leaves, not in cones; a *seed fern.* It ranges from the Late Devonian to the Mesozoic.
pterocavate Said of a dinoflagellate cyst having a pronounced equatorial pericoel, as in *Stephodinium.*
pterodactyl (a) Strictly, any member of the Pterodactyloidea, the more advanced of two suborders into which the archosaurian order Pterosauria is divided, characterized by a reduced tail and a tendency toward loss of teeth and increase in size. Range, Middle Jurassic to Upper Cretaceous. (b) More loosely, any *pterosaur.*
pteropod Any *opisthobranch* belonging to the order Pteropoda, which includes pelagic forms sometimes with shells. The shells are generally conical and composed of aragonite. Range, Cretaceous to present.
pteropod ooze A *pelagic deposit* containing at least 30% calcareous remains of pteropods. It is a *calcareous ooze,* generally found at much shallower depth than other oozes.
pteropsid A vascular plant (of the class Pteropsida) that has leaf gaps, as in ferns and seed plants.
pterosaur A member of the order Pterosauria, archosaurian reptiles highly adapted to flight. They were characterized by extreme elongation of the fourth digit of the hand for support of a membranous wing, and by reduction of the hind limbs. Range, Upper Triassic to Upper Cretaceous. Partial syn: *pterodactyl.*
pterygostomial region The anterolateral part of the carapace on the ventral surface of some decapods, located on opposite sides of the buccal cavity (TIP, 1969, pt. R, p. 92).
ptilolite *mordenite.*
PTRM *partial thermoremanent magnetization.*
ptycholophe A brachiopod lophophore in which the brachia are folded into one or more lobes in addition to median indentation (TIP, 1965, pt.H, p.151).
ptychopariid Any trilobite belonging to the order Ptychopariida, characterized generally by opisthoparian sutures, more than three segments in the thorax, and a simple glabella. Range, Lower Cambrian to Middle Permian.
ptygma Granitic material within migmatite or gneiss, having the appearance of disharmonic folds (Dietrich, 1959, p. 358). The genesis of this type of "folding" remains controversial, with hypotheses favoring both primary and secondary origins (Dietrich, 1960a, p. 140). Syn: *ptygmatic fold.*
ptygmatic fold *ptygma.*
pubescent Said of a plant that is covered with soft, downy hairs.
public domain Land owned, controlled, or heretofore disposed of by the U.S. Federal government. It includes the land that was ceded to the government by the original thirteen States, together with certain subsequent additions acquired by cession, treaty, and purchase. At its greatest extent, the public domain occupied more than 1,820,000,000 acres. See also: *public land.*
public land Land owned by a government, esp. a national government; specif. the part of the U.S. *public domain* to which title is still vested in the Federal government and that is subject to appropriation, sale, or disposal under the general laws.
public-land survey A survey of public lands; specif. the U.S. Public Land Survey system (USPLS) by which much of the United States was surveyed and divided into a rectangular grid system using townships, sections, and fractions of sections.
pucherite A red-brown orthorhombic mineral: $BiVO_4$.
puddingstone (a) A popular name applied chiefly in Great Britain to a *conglomerate* consisting of well-rounded pebbles whose colors are in such marked contrast with the abundant fine-grained matrix or cement that the rock suggests an old-fashioned plum pudding. Example: the Hertfordshire Puddingstone (lower Eocene) in England, composed of black or brown flint pebbles cemented by white silica, with or without brown iron hydroxide. Syn: *plum-pudding stone.* (b) A siliceous rock cut into blocks for furnace linings.—Also spelled: *pudding stone.*
puddle A small accumulation of meltwater in a depression or hollow on the surface of any form of ice, produced mainly by the melting of snow and ice, and in most cases fresh and potable.
puddle-core dam An earth dam in which the impervious core is constructed of puddled clay.
puddle wall The impervious core of a dam, or an impervious cutoff wall in natural materials, made of puddled clay.
Puercan North American continental stage: lowermost Paleocene (above Cretaceous, below Dragonian).
puerto A term used in the SW U.S. for a pass over or through an escarpment or mountain range. Etymol: Spanish.
puff n. A high spot or elevation in *gilgai.* Cf: *crab hole.*
puffing hole *blowhole [coast].*
pug *gouge [ore dep].*
puglianite A coarse-grained *foid monzogabbro* composed of euhedral augite, leucite, anorthite, sanidine, hornblende, and biotite. It occurs as fragments in the lavas of Monte Somma, Italy. Lacroix proposed the term in 1917. Cf: *sebastianite.* Not recommended usage.
pulaskite A light-colored, feldspathoid-bearing, granular or trachytoid alkali *syenite* composed chiefly of alkali feldspar, sodic pyroxene, arfedsonite, and nepheline. The term has also been applied to quartz-bearing syenites. The name, given by Williams in 1891, is from Pulaski County, Arkansas. Not recommended usage.
pull-apart n. A precompaction sedimentary structure resembling *boudinage,* consisting of beds that have been stretched and torn apart into relatively short slabs, the intervening cracks being filled in from the top (or in some cases possibly from below) (Natland & Kuenen, 1951, p.89-90); e.g. stiff clay embedded in more mobile, water-soaked sand, or compact sandstone embedded in hydroplastic clayey rock.—adj. Said of a structure or bed characterized by pull-aparts.
pulmonate n. Any terrestrial or freshwater gastropod belonging to the subclass Pulmonata, characterized by the modification of the mantle cavity for air breathing and by the presence of a shell but rarely of an operculum (TIP, 1960, pt.I, p.153).
pulp cavity A cavity in the base of any vertebrate tooth, or one of several in the base of an unreduced scale, which in life contains vascular and nerve tissue. The term has been incorrectly used for the *basal cavity* of a conodont.
pulpit rock *chimney rock.*
pulps A term used by Allen & Day (1935, p. 65) for "a fine, mealy, opaline silica", much like sand.
pulpy peat *sedimentary peat.*
pulsating spring *geyser.*
pulsation [stratig] (a) A term used by Grabau (1936a) for a long rhythm, conceived to be nearly the length of a geologic-time period, representing a eustatic movement of sea level that resulted in simultaneous transgression and regression of widespread and semipermanent seas over whole continents. Cf: *oscillation [stratig].* (b) A distinct step or change in a series of rhythmical or regularly recurring movements.
pulsation [tect] A minor time of deformation, or a subdivision of a more prolonged epoch of orogeny. Cf: *event; disturbance.*
pulse In ecology, a sudden increase in the number of organisms or kinds of organisms, usually at regularly recurring intervals.
pulsed-neutron-capture log A specialized *radioactivity log* of *neutron-gamma log* type. A neutron generator produces short bursts of high-energy neutrons (about 14 Mev), which are slowed in the borehole and nearby rocks to thermal-energy level (about 0.025 ev); capture by nuclei (especially chlorine) results in gamma-ray emissions. The log is used in cased oil wells to determine changes in fluid saturation during oil production. It is marketed under several trade names, e.g. Neutron Lifetime Log, Thermal Decay Time Log.
pulverite A sedimentary rock composed of silt- or clay-sized aggregates of constructional (nonclastic) origin, simulating in texture a lutite of clastic origin; e.g. a rock formed of diatom frustules. The term was introduced by Grabau (1911, p. 1007). Syn: *pulveryte.*
pulverization *comminution.*
pulverulent Said of a mineral that is easily powdered.
pumice A light-colored vesicular glassy rock commonly having the composition of *rhyolite.* It is often sufficiently buoyant to float on water and is economically useful as a lightweight aggregate and as an abrasive. The adjectival form, *pumiceous,* is usually applied to pyroclastic ejecta. Cf: *scoria.* Syn: *foam; volcanic foam; pumicite; pumice stone.*
pumice fall The descent of pumice from an eruption cloud by *airfall deposition.* Cf: *ash fall.*
pumice flow A type of *pyroclastic flow* in which a large proportion of the fragments are of pumice. Cf: *ash flow.*
pumiceous Said of the texture of a pyroclastic rock, e.g. *pumice,*

characterized by numerous small cavities presenting a spongy, frothy appearance; finer than *scoriaceous.* Also, said of a rock exhibiting such texture.

pump A mechanical device for transferring either liquids or gases from one place to another, or for compressing or attenuating gases.

pumpage (a) The quantity of water or other liquid pumped, as of ground water. (b) The act of pumping.

pumpellyite A greenish, epidote-like mineral: $Ca_4Al_4(Al,Fe^{+2},Fe^{+3},Mg,Mn)_2Si_6O_{23}(OH)_3 \cdot 2H_2O$. It is probably related to clinozoisite. See also: *chlorastrolite.* Syn: *zonochlorite; lotrite.*

pumpellyite-prehnite facies The set of metamorphic mineral assemblages (facies) in which metagreywackes contain albite + quartz + prehnite + pumpellyite + chlorite + sphene (Coombs et al., 1959). It is generally believed to represent pressure-temperature conditions between those of the *zeolite facies* and the *greenschist facies.*

Pumpelly's rule The generalization that the axes and axial surfaces of minor folds of an area are congruent with those of the major fold structures of the same phase of deformation (Pumpelly, 1894, p. 158).

punch register To make punch holes in multiple pieces of cartographic copy which are in perfect registry for the insertion of pins or studs so that registry can be maintained in subsequent cartographic steps and in the preparation of negatives to be used in making color printing plates. Cf: *register punch.*

puncta (a) One of the minute, closely spaced pores, perforations, or tubules extending perpendicularly a variable distance from the inner or outer surface of a brachiopod shell. The term is also used as a plural of *punctum.* See also: *endopuncta; exopuncta; pseudopuncta.* Syn: *punctum.* (b) Any of various thin places arranged in characteristic pattern in the frustule of pennate diatoms, being smaller and simpler than an *areola;* specif. the smallest structure on a diatom valve, such as one of the pores having diameters as small as 0.037 μm (but commonly 0.5 to 1.0 μm), occurring either scattered or in rows, and sometimes having fine porous plates at their inner extremity. (c) A hole in the external wall of a foraminiferal chamber.—Pl: *punctae.*

punctate Minutely pitted, or having minute dots, punctae, or depressions, such as a "punctate leaf"; specif. said of a brachiopod or brachiopod shell possessing endopunctae. Cf: *impunctate.*

punctation The condition of being punctate.

punctuated equilibria That evolutionary pattern exhibited by successive species in an evolving *lineage,* in which rapid and substantial morphologic change is concentrated within geologically short moments, separating or "punctuating" much longer intervals before and after, when the ancestral and descendant species maintained relatively constant or "equilibrium" characters. Syn: *rectangular speciation.* Ant: *phyletic gradualism.*

punctum A small area marked off in any way from a surrounding surface; specif. a minute pit on the shell surface of a gastropod (not a tubule penetrating the shell substance), or a *puncta* in the shell of a brachiopod. Pl: *puncta.*

punky Said of a semi-indurated rock, such as a leached limestone; esp. said of a tuff that is weakly welded.

pup A term used in Alaska for a small tributary stream.

puppet *loess doll.*

Purbeckian European stage (Great Britain): uppermost Jurassic (above Tithonian, below Wealden).

pure coal An informal syn. of *vitrain.*

pure rotation *rotational strain.*

pure shear A particular example of *irrotational strain* in which the body is elongated in one direction and shortened at right angles to this. Cf: *simple shear.*

purga A violent arctic snowstorm. Etymol: Karelian, *purgu,* "snowstorm".

purgatory (a) A term used in New England for a long, deep, narrow, steep-sided cleft or ravine along a rugged coast, into which waves rush during a storm with great noise and violence; a rock chasm without a stream, often covered at the bottom with large, angular rocks, and difficult to traverse. (b) A swamp that is dangerous or difficult to traverse.

purl A swirling or eddying stream or rill, moving swiftly around obstructions; a stream making a soft, murmuring sound.

purple blende *kermesite.*

purple copper ore *bornite.*

purpurite A deep-red or reddish-purple mineral: $(Mn^{+3},Fe^{+3})PO_4$. It is isomorphous with heterosite.

push [hydraul] A force exerted directly by the wind upon the exposed sides of wave crests (Strahler, 1963, p. 308). Cf: *drag [hydraul].*

push [volc] *squeeze-up.*

push moraine A broad, smooth, arc-shaped morainal ridge consisting of material mechanically pushed or shoved along by an advancing glacier. Examples are common in the Netherlands and NW Germany. See also: *ice-pushed ridge.* Syn: *shoved moraine; push-ridge moraine; upsetted moraine; thrust moraine.*

push-pull wave *P wave.*

pustule A minute boss on an asterozoan ossicle, having a central depression in which a spine articulates.

pusule apparatus The sacklike vacuole in a dinoflagellate connected with the exterior by a slender canal opening into a flagellar pore.

Putnam anomaly *average level anomaly.*

putrefaction The decomposition of organic matter by slow distillation, in the presence of water, without air. Methane and other gaseous products (H_2, NH_3, H_2S) are formed. Cf: *disintegration [coal]; moldering; peat formation.*

puy A small, remnant volcanic cone; it is the French word for such structures in the Auvergne district of central France.

puzzolan *pozzolan.*

p-veatchite A mineral: $Sr_2B_{11}O_{16}(OH)_5 \cdot H_2O$. It is dimorphous with veatchite, and has a space group $P2_1/m$. Syn: *paraveatchite.*

P wave That type of seismic *body wave* that involves particle motion (alternating compression and expansion) in the direction of propagation. It is the fastest of the seismic waves, traveling 5.5-7.2 km/sec in the crust and 7.8-8.5 km/sec in the upper mantle. Sound waves are P-waves. The P stands for primary; it is so named because it is the *first arrival.* Syn: *longitudinal wave; irrotational wave; pressure wave; dilatational wave; primary wave; compressional wave; push-pull wave.* Cf: *S-wave; surface wave.*

pycnite A variety of topaz occurring in massive columnar aggregations.

pycnocline (a) A density gradient; esp. a vertical gradient marking a sharp change. Cf: *thermocline.* (b) A layer of water in the ocean, characterized by a rapid change of density with depth.

pycnogonid Any marine arthropod belonging to the subphylum Pycnogonida, resembling the *chelicerates* in having one pair of chelae but lacking a well-developed abdomen. Range, Lower Devonian to present.

pycnostromid An *algal biscuit* produced by *Pycnostroma.*

pycnotheca The dense, nonalveolate inner layer of the test wall of schwagerinid fusulinids, penetrated by septal pores, and wedged between tectum and keriotheca or antetheca.

pygidium A caudal structure or terminal body region of various invertebrates; esp. the posterior part or tail piece of an exoskeleton of a trilobite, consisting of several fused segments. Pl: *pygidia.* Adj: *pygidial.* Cf: *abdomen.*

pylome (a) A more or less circular opening in an acritarch, commonly closed by an operculum, and probably serving for emergence from a cyst. (b) A large opening in spumellarian radiolarians, commonly only in the outermost of concentric shells.

pyrabol *pyribole.*

pyrabole *pyribole.*

pyralmandite A garnet intermediate in chemical composition between pyrope and almandine.

pyralspite A group of garnets of formula: $M_3Al_2(SiO_4)_3$, where M = Mg, Fe^{+2}, or Mn^{+2}. It includes pyrope, almandine, and spessartine, and their intermediate forms.

pyramid [cryst] An open crystal form consisting of three, four, six, eight, or twelve nonparallel faces that meet at a point. Cf: *dipyramid.* Adj: *pyramidal.*

pyramid [paleont] A large beaklike or winglike element of Aristotle's lantern in the interambulacral position of an echinoid. See also: *demipyramid.*

pyramidal Having the symmetry of a *pyramid.*

pyramidal cleavage Mineral cleavage parallel to the faces of a pyramid, e.g. the (101) cleavage of scheelite.

pyramidal dune *star dune.*

pyramidal peak *horn [glac geol].*

pyramidal system *tetragonal system.*

pyramid pebble *dreikanter.*

pyranometer An *actinometer* that measures the combined intensity of incoming direct solar radiation and diffuse sky radiation.

pyrargyrite A dark-red, gray, or black rhombohedral mineral: Ag_3SbS_3. It is isomorphous with proustite and polymorphous with pyrostilpnite, and is an important ore of silver. Syn: *dark ruby* *silver; dark red silver ore.*

pyrene The small stone of a drupe or similar fruit.

Pyrenean orogeny One of the 30 or more short-lived orogenies during Phanerozoic time identified by Stille, in this case during the late Eocene, between the Bartonian and Ludian stages.

Pyrenean-type facies series Rocks produced in a type of dynamothermal regional metamorphism characteristic of the Pyrenees, in which the pressure range is 3500-5000 bars. The mineral sequence, in order of rising temperature, is staurolite - andalusite - sillimanite - cordierite (Hietanen, 1967, p.193). Cf: *Idahoan-type facies series.*

pyreneite *melanite.*

pyrgeometer An *actinometer* that measures the effective terrestrial radiation.

pyrheliometer An *actinometer* that measures the intensity of direct solar radiation.

pyribole A mnemonic term coined by Johannsen in 1911 in his classification of igneous rocks to indicate the presence of either or both a pyroxene and/or an amphibole. Also spelled: *pyrabole; pyrabol; pyrobol.* Etymol: *pyr*oxene + amph*ibole.*

pyric pond A pool of water that collects in a shallow hole or sink formed as a result of fires and subsequent subsidence in peat deposits, lignite, and coal beds.

pyrite A common, pale-bronze or brass-yellow, isometric mineral: FeS_2. It is dimorphous with marcasite, and often contains small amounts of other metals. Pyrite has a brilliant metallic luster and an absence of cleavage, and has been mistaken for gold (which is softer and heavier). It commonly crystallizes in cubes (whose faces are usually striated), octahedrons, or pyritohedrons, and it also occurs in shapeless grains and masses. Pyrite is the most widespread and abundant of the sulfide minerals and occurs in all kinds of rocks, such as in nodules in sedimentary rocks and coal seams or as a common vein material associated with many different minerals. Pyrite is an important ore of sulfur, less so of iron, and is burned in making sulfur dioxide and sulfuric acid; it is sometimes mined for the associated gold and copper. Cf: *pyrites.* Syn: *iron pyrites; fool's gold; mundic; common pyrites.*

pyrites (a) Any of various metallic-looking sulfides of which pyrite ("iron pyrites") is the commonest. The term is used with a qualifying term that indicates the component metal; e.g. "copper pyrites" (chalcopyrite), "tin pyrites" (stannite), "white iron pyrites" (marcasite), "arsenical pyrites" (arsenopyrite),"cobalt pyrites"(linnaeite), and "nickel pyrites" (millerite). When used popularly and without qualification, the term usually signifies *pyrite.* (b) An obsolete term for a stone that may be used for striking fire.

pyritization Introduction of, or replacement by, pyrite; e.g. the replacement of the original material of the hard parts of certain fossil animals and plants by pyrite. Pyritization is a common process of hydrothermal alteration and often involves the introduction of fine-grained pyrite disseminated as specks in rock adjacent to veins.

pyritohedral Adj. of *pyritohedron.*

pyritohedron A crystal form that is a *dodecahedron* consisting of 12 pentagonal faces that are not regular. Its symmetry is $2/m\bar{3}$ and its indices are {210}. It is named after pyrite, which characteristically has this crystal form. Adj: *pyritohedral.* Cf: *rhombic dodecahedron.* Syn: *pentagonal dodecahedron; regular dodecahedron; pyritoid.*

pyritoid As a noun, a syn. of *pyritohedron.*

pyroaurite A golden-yellow or brownish rhombohedral mineral: $Mg_6Fe_2(CO_3)(OH)_{16}\cdot 4H_2O$. It is dimorphous with sjögrenite, and may contain up to 5% MnO.

pyrobelonite A fire-red to deep brilliant-red orthorhombic mineral: $PbMn(VO_4)(OH)$.

pyrobiolite An organic rock containing organic remains that have been altered by volcanic action.

pyrobitumen Any of the dark-colored, fairly hard, nonvolatile native substances composed of hydrocarbon complexes, which may or may not contain oxygenated substances and are often associated with mineral matter. The nonmineral constituents are infusible, insoluble in water, and relatively insoluble in carbon disulfide. On heating, pyrobitumens generally yield bitumens, i.e. decompose rather than melt.

pyrobituminous Pertaining to substances which yield bitumens upon heating.

pyrobole *pyribole.*

pyrochlore (a) A pale-yellow, red, brown, or black isometric mineral: $(Na,Ca)_2(Nb,Ta)_2O_6(OH,F)$. It is isomorphous with microlite, with Nb greater than Ta, and it usually contains cerium and titanium. Pyrochlore occurs in pegmatites derived from alkalic igneous rocks and constitutes an ore of niobium. Syn: *pyrrhite.* (b) A group of minerals of the general formula: $A_2B_2O_6(O,OH,F)$, where A = Na, Ca, K, Fe^{+2}, U^{+4}, Sb^{+3}, Pb, Th, Ce, or Y, and B = Nb, Ta, Ti, Sn, Fe^{+3}, or W. It includes minerals such as pyrochlore, microlite, betafite, obruchevite, and pandaite.

pyrochroite A hexagonal mineral: $Mn(OH)_2$. It is white when fresh, but darkens on exposure; it is similar to brucite in appearance.

pyroclast An individual particle ejected during a volcanic eruption. It is usually classified according to size. Cf: *pyroclastics.*

pyroclastic Pertaining to clastic rock material formed by volcanic explosion or aerial expulsion from a volcanic vent; also, pertaining to rock texture of explosive origin. It is not synonymous with the adjective "volcanic".—In the plural, the term is used as a noun. See: *pyroclastics.*

pyroclastic breccia *explosion breccia.*

pyroclastic flow A syn. of *ash flow,* used in a more general, genetic sense. Cf: *base surge.*

pyroclastics A general term for a deposit of *pyroclasts.* Syn: *tephra.*

pyroelectricity The simultaneous development, in any crystal lacking a center of symmetry, of opposite electric charges at opposite ends of a crystal axis, due to certain changes in temperature.

pyrogenesis A broad term encompassing the intrusion and extrusion of magma and its derivatives. Adj: *pyrogenic.*

pyrogenetic mineral (a) An anhydrous mineral of an igneous rock, usually crystallized at high temperature in a magma containing relatively few volatile components. (b) Any mineral crystallized directly from a magma, as distinct from minerals formed by alteration or replacement.

pyrogenic Said of a process or of a deposit involving the intrusion and/or extrusion of magma. See also: *pyrogenesis.* Syn: *pyrogenetic; pyrogenous.*

pyrogenic rock *igneous rock.*

pyrogenous A syn. of *pyrogenic,* originally a syn. of *igneous.*

pyrogeology A synonym of *volcanology* that was proposed by Grabau (1924).

pyrognomic Said of a *metamict* mineral that easily becomes incandescent when heated. The term is little used.

pyrolite A model proposed by Ringwood (Green and Ringwood, 1963) for the material of the upper mantle, composed of one part basalt to three parts dunite and consisting mainly of olivine and pyroxenes. It is so designed that a partial melt will yield a basaltic magma.

pyrolith A term proposed by Grabau (1904) for igneous rock. Obsolete.

pyrolusite A soft iron-black or dark steel-gray tetragonal mineral: MnO_2. It is the most important ore of manganese and is dimorphous with ramsdellite. Pyrolusite is generally massive or reniform, sometimes with a fibrous or radiate structure. Syn: *polianite; gray manganese ore.*

pyromagma A highly mobile lava, oversaturated with gases, that exists at shallower depths than *hypomagma.* Cf: *epimagma.*

pyromelane *brookite.*

pyromeride A devitrified *rhyolite* characterized by spherulitic texture; a nodular rhyolite. An obsolete term dating from 1814.

pyrometamorphism Metamorphic changes taking place without the action of pressure or water vapor, at temperatures near the melting points of the component minerals; it is a local, intense type of *thermal metamorphism,* resulting from the unusually high temperatures at the contact of a rock with magma, e.g. in xenoliths (Turner, 1948). Cf: *igneous metamorphism.*

pyrometasomatic Formed by *metasomatic* changes in rocks, principally in limestone, at or near intrusive contacts, under influence of magmatic emanations and high temperature and pressure.

pyrometasomatism The formation of contact-metamorphic mineral deposits at high temperatures by emanations issuing from the intrusive and involving replacement of enclosing rock with addition or subtraction of materials; *skarn* formation. See also: *metasomatism.*

pyrometer An instrument that measures high temperature, e.g. of molten lavas, by electrical or optical means. See also: *optical pyrometer; pyrometry.*

pyrometric cone *Seger cone.*

pyrometry The measurement of high temperatures by electrical or optical means, using a *pyrometer.* Its geologic application is to incandescent lavas.

pyromorphite A green, yellow, brown, gray, or white mineral of the apatite group: $Pb_5(PO_4)_3Cl$. It is isomorphous with mimetite and vanadinite, and may contain arsenic or calcium. Pyromorphite is found in the oxidized zone of lead deposits, and is a minor ore of lead. Syn: *green lead ore.*

pyrope (a) The magnesium-aluminum end-member of the garnet group, characterized by a deep fiery-red color: $(Mg,Fe)_3Al_2(SiO_4)_3$. It rarely occurs in crystals, but is found in detrital deposits as rounded and angular fragments, or associated with olivine and serpentine in basic igneous rocks such as kimberlite. See also: *Cape ruby; Bohemian garnet.* Syn: *rock ruby.* (b) An obsolete name for a bright red gem, such as a ruby.

pyrophane (a) *fire opal.* (b) An opal (such as hydrophane) artificially impregnated with melted wax.

pyrophanite A blood-red mineral: $MnTiO_3$. It is isomorphous with ilmenite.

pyrophyllite A white, gray, or brown mineral: $AlSi_2O_5(OH)$. It resembles talc and occurs in a foliated form or in compact masses in quartz veins, granites, and esp. metamorphic rocks. Syn: *pencil stone.*

pyropissite An earthy, nonasphaltic pyrobitumen made up primarily of water, humic acid, wax (it is a source of *montan wax*), and silica. It is frequently found associated with *brown coal,* which is then called pyropissitic brown coal.

pyroretinite A type of retinite found in the brown coals of Aussig (Usti nad Labem), Bohemian Czechoslovakia.

pyroschist A schist or shale that has a sufficiently high carbon content to burn with a bright flame, or to yield volatile hydrocarbons, when heated.

pyrosmalite A colorless, pale-brown, gray, or grayish-green mineral: $(Fe^{+2},Mn)_8Si_6O_{15}(OH,Cl)_{10}$. Cf: *manganpyrosmalite.*

pyrosphere The zone of the Earth below the lithosphere, probably partly molten; it is equivalent to the *barysphere.* Syn: *magmosphere.*

pyrostibite *kermesite.*

pyrostilpnite A hyacinth-red monoclinic mineral: Ag_3SbS_3. It is polymorphous with pyrargyrite. Syn: *fireblende.*

pyroxene (a) A group of dark rock-forming silicate minerals, closely related in crystal form and composition and having the general formula: $ABSi_2O_6$, where A = Ca, Na, Mg, or Fe^{+2}, and B = Mg, Fe^{+2}, Fe^{+3}, Fe, Cr, Mn, or Al, with silicon sometimes replaced in part by aluminum. It is characterized by a single chain of tetrahedra with a silicon:oxygen ratio of 1:3; by short, stout prismatic crystals; and by good prismatic cleavage in two directions parallel to the crystal faces and intersecting at angles of about 87° and 93°. Colors range from white to dark green or black. Pyroxenes may crystallize in the orthorhombic or monoclinic systems; they constitute a common constituent of igneous rocks, and are similar in chemical composition to the amphiboles (except that the pyroxenes lack hydroxyls). (b) A mineral of the pyroxene group, such as enstatite, hypersthene, diopside, hedenbergite, acmite, jadeite, pigeonite, and esp. augite.—Etymol: Greek *pyros,* "fire", + *xenos,* "stranger", apparently so named from the mistaken belief that the pyroxenes "were only accidentally caught up in the lavas that contain them" (Challinor, 1978, p. 250). Pron: *pie*-rok-seen or *peer*-ok-seen.

pyroxene alkali syenite In Tobi's classification of the *charnockite series* (1971, p. 202), a member with less than 20% quartz and characterized by the presence of microperthite.

pyroxene-hornblende gabbronorite In the *IUGS classification,* a plutonic rock satisfying the definition of *gabbro* in which pl/(pl+hbl+px) is between 10 and 90, and px/(pl+hbl+px) and hbl/(pl+hbl+px) are greater than 5.

pyroxene-hornblende peridotite In the *IUGS classification,* a plutonic rock with M equal to or greater than 90, ol/(ol+hbl+px) between 40 and 90, px/(ol+hbl+px) greater than 5, and hbl/(ol+hbl+px) greater than 5.

pyroxene hornblendite In the *IUGS classification,* a plutonic rock with M equal to or greater than 90, ol/(ol+hbl+px) less than 5, and hbl/(px+hbl) between 50 and 90.

pyroxene-hornfels facies The set of metamorphic mineral assemblages (facies) in which basic rocks are represented by diopside + hypersthene + plagioclase, with amphibole typically absent. Pelitic assemblages exhibit the association of sillimanite (or andalusite) and cordierite with potassium feldspar; muscovite is absent and biotite usually small in amount. Marbles should ideally contain wollastonite and calcite + forsterite + periclase (Turner, 1968). The facies is typical of high-grade thermal metamorphism, as in the inner parts of contact aureoles. It corresponds to temperatures in excess of about 550°C, and to relatively low pressures.

pyroxene monzonite In Tobi's classification of the *charnockite series* (1971, p. 202), a quartz-poor member containing approximately equal amounts of microperthite and plagioclase; *mangerite.*

pyroxene peridotite In the *IUGS classification,* a plutonic rock with M equal to or greater than 90, ol/(ol+hbl+px) between 40 and 90, and hbl/(ol+hbl+px) less than 5.

pyroxene-perthite Lamellar intergrowths of any of several pyroxenes, as with the feldspars.

pyroxene syenite In Tobi's classification of the *charnockite series* (1971, p. 202), a quartz-poor member having more microperthite than plagioclase; a *mangerite-syenite.*

pyroxenide An informal term, used in the field, for any holocrystalline, medium- to coarse-grained igneous rock composed chiefly of pyroxene; e.g. a pyroxenite (Johannsen, 1931).

pyroxenite (a) In the *IUGS classification,* a plutonic rock with M equal to or greater than 90 and ol/(ol+opx+cpx) less than 40. (b) An ultramafic plutonic rock chiefly composed of pyroxene, with accessory hornblende, biotite, or olivine. Syn: *pyroxenolite.*

pyroxenoid Any mineral chemically analogous to pyroxene but with the SiO_4-tetrahedra connected in chains with a repeat unit of 3, 5, 7, or 9; e.g. wollastonite and rhodonite.

pyroxenolite A term proposed by Lacroix in 1894 as a synonym for *pyroxenite* of English-speaking petrologists, as the French usage of *pyroxenite* was confined to metamorphic rocks. Not recommended usage.

pyroxferroite A yellow mineral of the pyroxenoid group found in Apollo 11 lunar samples: $(Fe,Mn,Ca)SiO_3$. It is the iron analogue of pyroxmangite.

pyroxmangite A red or brown triclinic mineral of the pyroxenoid group: $(Mn,Fe,Ca,Mg)SiO_3$.

pyrrhite *pyrochlore.*

pyrrhotine *pyrrhotite.*

pyrrhotite A common red-brown to bronze pseudohexagonal mineral: $Fe_{1-x}S$. It has a defect structure in which some of the ferrous ions are lacking. Some pyrrhotite is magnetic. The mineral is darker and softer than pyrite; it is usually found massive and commonly associated with pentlandite, often containing as much as 5% nickel, in which case it is mined as an ore of nickel. Syn: *pyrrhotine; magnetic pyrites; dipyrite.*

pythmic Pertaining to the bottom of a lake (Klugh, 1923, p.372).

Q (a) A measure of the loss of energy by absorption. Q is 2π times the ratio of the peak energy in a cycle to the energy dissipated in that cycle. The larger the value of Q, the less the absorption. The dimensionless parameter Q is much used to describe the intrinsic attenuation of seismic waves. Syn: *attenuation coefficient.* (b) Abbrev. of *quartile.*

qanat A term used in Iran for an ancient, gently inclined, underground channel or conduit dug so as to conduct ground water by gravity from alluvial gravels and the foot of hills to an arid lowland; a horizontal well. Etymol: Arabic. Cf: *foggara; karez.* Syn: *kanat.*

Q-joint A partial syn. of *cross joint,* used for a cross joint that is perpendicular to flow structure.

Q-mode factor analysis *Factor analysis* concerned with relationships among samples or objects. Cf: *R-mode factor analysis.*

quad Shortened form of *quadrangle.*

quadrangle (a) A rectangular area bounded by parallels of latitude and meridians of longitude, used as a unit in systematic mapping. The dimensions of a quadrangle are not necessarily the same in both directions, and its size and the scale at which it is mapped are determined by the prime purpose of the map. (b) A sheet representing a quadrangle.—Syn: *quad.*

quadrangle map A rectangular map bounded by parallels of latitude and meridians of longitude and generally published in a series with prescribed scale.

quadrant [paleont] The space in the interior of a rugose corallite, bounded by the cardinal septum and an alar septum or by the counter septum and an alar septum.

quadrant [surv] (a) An instrument formerly used in surveying and astronomy for measuring angles and altitudes, consisting of a graduated arc of 90 degrees (180 degrees in range) equipped with a sighting device and a movable index or vernier and usually a plumb line or spirit level for fixing the vertical or horizontal direction. It is now largely superseded by the *sextant.* (b) A quarter of a circle, an arc of 90 degrees, or an arc subtending a right angle at the center. Also, the area bounded by a quadrant and two radii.

quadrat In ecology, a sample area (usually a square) chosen as the basis for studying a particular assemblage of organisms. Cf: *transect.*

quadratic elongation A measure of the change in the length of a line, specifically the square of the ratio of its final length to its initial length (Ramsay, 1967, p. 52). Cf: *stretch [exp struc geol].*

quadrature [astron] Either of two points in the Moon's orbit about the Earth when the Moon is in its first or third quarters, or when a line from Earth to Moon makes a right angle with respect to the line from Earth to Sun. Cf: *syzygy [astron].*

quadrature [geophys] A component of a vector that has a phase difference of one-quarter cycle as compared to the primary quantity.

quagmire (a) A soft marsh or bog that gives under pressure. (b) *quaking bog.*

quake n. A syn. of *earthquake;* also, a *seismic event* on another planetary body.

quake sheet A well-defined bed resembling a *slump sheet* but produced by an earthquake and resulting in load casting without horizontal slip (Kuenen, 1958, p.20).

quaking bog A peat bog that is either floating or is growing over water-saturated ground, so that it shakes or trembles when walked on. *Quagmire* is sometimes used as a synonym.

quantitative geomorphology The assignment of dimensions of mass, length, and time to all descriptive parameters of landform geometry and geomorphic processes, followed by the derivation of empirical mathematical relationships and formulation of rational mathematical models relating those parameters. In the study of stream networks, these dimensions are related to topological measures such as order, link distance, and diameter.

quantitative system *CIPW classification.*

quantum detector A semiconductor used in radiometers and scanner systems to count the quantity of photons striking a sensitive element. A photoconductive, photovoltaic, or photoelectromagnetic detector may be used. The photons striking the quantum detector interact with the crystal lattice, freeing electrons.

quantum evolution Rapid evolution from one established type of biologic adaptation to another completely different type under the influence of some strong selection pressure.

quaquaversal adj. Said of strata and structures that dip outward in all directions away from a central point. It is an old term. Ant: *centroclinal.* The term has also been used as a syn. of *periclinal.*—n. A geologic structure, such as a dome or ridge, having a quaquaversal dip. Cf: *pericline.* Ant: *centrocline.*

quaquaversal fold *dome [struc geol].*

quar A Welsh term for sandstone.

quarfeloid A term coined by Johannsen in 1911 in his classification of igneous rocks to indicate the mineral combinations of quartz and feldspars, and feldspars and feldspathoids (Johannsen, 1939, p. 194).

quar ice A term used in Labrador for ice formed during the spring by meltwater running off the land onto an icefoot or fast ice, where it refreezes (ADTIC, 1955, p. 64).

quarry Open workings, usually for the extraction of stone.

quarrying [geomorph] (a) *plucking [glac geol].* (b) *hydraulic plucking.*

quarrying [mining] The extraction of building stone or other valuable nonmetallic constituent from a surficial mine, or quarry. See also: *opencut mining.*

quarry sap *quarry water.*

quarry water Subsurface water retained in freshly quarried rock. Syn: *quarry sap.*

quarter post A post marking a corner of a *quarter section* of the U.S. Public Land Survey system. It is located midway between the controlling section corners, or 40 chains (0.5 mi) from the controlling section corner, depending on location within the township.

quarter-quarter section A sixteenth of a normal *section* of the U.S. Public Land Survey system, representing a piece of land normally a quarter mile square and containing 40 acres as nearly as possible; a quarter section divided into four parts. It is usually identified as the northeast quarter, northwest quarter, southeast quarter, or southwest quarter of a particular quarter section and section.

quarter section A fourth of a normal *section* of the U.S. Public Land Survey system, representing a piece of land normally a half mile square and containing 160 acres as nearly as possible. It is usually identified as the northeast quarter, northwest quarter, southeast quarter, or southwest quarter of a particular section.

quarter-wave plate *mica plate.*

quartile Any one of three particle-size values (diameters) dividing a frequency distribution into four classes, obtained graphically from a cumulative curve by following the 25, 50, or 75 percent line to its intersection with the curve and reading the value on the diameter scale directly below the intersection; e.g. the first quartile (the 25 percentile) is the size such that 25% of the particles are larger than itself and 75% smaller, this size being larger than the third quartile (the 75 percentile) which is the size such that 75% of the particles are larger than itself and 25% smaller. Abbrev: Q.

quartz (a) Crystalline silica, an important rock-forming mineral: SiO_2. It is, next to feldspar, the commonest mineral, occurring either in transparent hexagonal crystals (colorless, or colored by impurities) or in crystalline or cryptocrystalline masses. Quartz is the commonest gangue mineral of ore deposits, forms the major proportion of most sands, and has a widespread distribution in igneous (esp. granitic), metamorphic, and sedimentary rocks. It has a vitreous to greasy luster, a conchoidal fracture, an absence of cleavage, and a hardness of 7 on the Mohs scale (scratches glass easily, but cannot be scratched by a knife); it is composed exclusively of silicon-oxygen tetrahedra with all oxygens joined together in a three-dimensional network. It is polymorphous with cristobalite, tridymite, stishovite, coesite and keatite. Symbol: Q. Abbrev: qtz; qz. Etymol: German provincial *Quarz.* Cf: *tridymite; cristobalite; coesite; stishovite.* (b) A general term for a variety of noncrystalline or cryptocrystalline minerals having the same chemical composition as that of quartz, such as chalcedony, agate, and opal.

quartz andesite *dacite.*

quartz anorthosite In the *IUGS classification,* a plutonic rock with Q between 5 and 20, P/(A+P) greater than 90, and color index less than 10.

quartzarenite A sandstone that is composed primarily of quartz; specif. a sandstone containing more than 95% quartz framework grains (excluding detrital chert grains) and having a clay matrix and any sorting, rounding, texture, or hardness (Folk, 1968). McBride (1963, p.667), who included chert and quartzite in the 95% quartz content, coined the term as a contracted form of "quartz arenite", a term used by Williams, Turner and Gilbert (1954, p.294 & 316) for a mature sandstone containing more than 80% quartz, chert, and quartzite and less than 10% each of argillaceous matrix, feldspars, and unstable fine-grained rock fragments. The term is essentially equivalent to *orthoquartzite.*

quartz-banded ore A term used in Scandinavia for a metamorphosed *iron formation.*

quartz-bearing diorite A syn. of *quartz diorite,* although Streckeisen (1967, p. 157) suggests that the term be restricted to diorite in which quartz constitutes 5 to 20 percent of the light-colored components.

quartz-bearing monzonite As recommended by Streckeisen (1967, p. 157), any *monzonite* in which quartz constitutes from 5 to 20 percent of the light-colored components. In the most recent Soviet classification, it is a syn. of *quartz monzonite.*

quartz crystal Quartz that is transparent or nearly so, is usually colorless, and has a low refractive index resulting in low brilliancy. It is used for lenses, wedges, and prisms in optical instruments and for frequency control in electronics, or is fashioned into beads or other ornamental objects. It may or may not be in distinct crystals. Syn: *rock crystal; pebble [gem].*

quartz diorite (a) In the *IUGS classification,* a plutonic rock with Q between 5 and 20, P/(A+P) greater than 90, and plagioclase more sodic than An_{50}. (b) A group of plutonic rocks having the composition of *diorite* but with an appreciable amount of quartz, i.e. between 5 and 20 percent of the light-colored constituents, according to Streckeisen (1967, p. 157); also, any rock in that group; the approximate intrusive equivalent of *dacite.* Quartz diorite grades into *granodiorite* as the alkali feldspar content increases. Syn: *tonalite; quartz-bearing diorite.*

quartzfels *silexite [ign].*

quartz felsite *quartz porphyry.*

quartz-flooded limestone A limestone characterized by an abundance of quartz particles that were imported suddenly from a nearby source by wind or water currents, but that gradually die out upward and disappear within a few centimeters (Shrock, 1948, p. 87).

quartz-free wacke A wacke with more than 90% unstable materials (Fischer, 1934).

quartz gabbro In the *IUGS classification,* a plutonic rock with Q between 5 and 20, P/(A+P) greater than 90, and plagioclase more calcic than An_{50}.

quartz graywacke A term used by Williams, Turner and Gilbert (1954, p.294) for a graywacke containing abundant grains of quartz and chert and less than 10% each of feldspar and rock fragments. See also: *quartzose graywacke.*

quartzic *quartziferous.*

quartziferous Quartz-bearing. The term is applied to a rock (such as a limestone or syenite) that contains a minor proportion of quartz, to distinguish it from a variety (usually commoner) of the same rock that contains no quartz. Cf: *quartzose.* Syn: *quartzic.*

quartz index [petrology] A derived quantity (qz) in the Niggli system of rock classification, which may be either positive or negative, and is an indicator of a rock's degree of silica saturation.

quartz index [sed] A term used by Dapples et al. (1953, p. 294 & 304) to indicate the mineralogic maturity of a sandstone by measuring the percentage of detrital quartz. It is expressed as the ratio of quartz and chert to the combined percentage of sodic and potassic feldspar, rock fragments, and clay matrix. The index is used as a basis for evaluating the degree of weathering of the source rock and the degree to which the sediment has been transported. Values for sandstones range between 3 and 19.

quartzine Chalcedony characterized by fibers having a positive crystallographic elongation (parallel to the *c*-axis). Syn: *quartzin.*

quartzite [meta] A granoblastic metamorphic rock consisting mainly of quartz and formed by recrystallization of sandstone or chert by either regional or thermal metamorphism; *metaquartzite.* Cf: *orthoquartzite.*

quartzite [sed] A very hard but unmetamorphosed sandstone, consisting chiefly of quartz grains that have been so completely and solidly cemented with secondary silica that the rock breaks across or through the grains rather than around them; an *orthoquartzite.* The cement grows in optical and crystallographic continuity around each quartz grain, thereby tightly interlocking the grains as the original pore spaces are filled. Skolnick (1965) believes that most sedimentary quartzites are compacted sandstones developed by pressure solution of quartz grains.

quartzitic arkose *arkosite.*

quartzitic sandstone A term used by Krynine (1940, p.51) for a sandstone that contains 100% quartz grains cemented with silica. The term is essentially equivalent to *orthoquartzite.* Cf: *quartzose sandstone.*

quartz latite An obsolescent syn. of *rhyodacite.*

quartz mengwacke A wacke with 10-33% unstable materials (Fischer, 1934).

quartz mine A miner's term for a mine in which the valuable constituent, e.g. gold, is found in siliceous veins rather than in placers. It is so named because quartz is the chief accessory mineral.

quartz monzodiorite In the *IUGS classification,* a plutonic rock with Q between 5 and 20, P/(A+P) between 65 and 90, and plagioclase more sodic than An_{50}.

quartz monzogabbro In the *IUGS classification,* a plutonic rock with Q between 5 and 20, P/(A+P) between 65 and 90, and plagioclase more calcic than An_{50}.

quartz monzonite (a) In the *IUGS classification,* a plutonic rock with Q between 5 and 20 and P/(A+P) between 35 and 65. (b) In former U.S. usage, granitic rock in which quartz comprises 10-50% of the felsic constituents, and in which the alkali feldspar/total feldspar ratio is between 35% and 65%; the approximate intrusive equivalent of *rhyodacite.* With an increase in plagioclase and femic minerals, it grades into *granodiorite,* and with more alkali feldspar, into a *granite.* As introduced by Brögger in 1895, the term designated monzonite containing only small amounts of quartz, and it is still used in this sense by Soviet geologists. According to Tröger (1935, p.47), Lindgren, in 1900, changed the definition to apply to andesine-bearing granites. Now the term is applied by most British petrologists to granites with quartz constituting 20-60% of the light-colored components and with a plagioclase/total feldspar ratio of 35/65. Streckeisen (1967, p. 167) recommends replacing the term with *quartz-bearing monzonite.* Syn: *adamellite.*

quartz norite In Tobi's classification of the *charnockite series* (1971, p. 202), a member that contains plagioclase but no potassium feldspar.

quartzolite In the *IUGS classification,* a plutonic rock with Q greater than or equal to 90: preferred over *silexite [ign].*

quartzose Containing quartz as a principal constituent; esp. applied to sediments and sedimentary rocks (such as sands and sandstones) consisting chiefly of quartz. Cf: *quartziferous.* Syn: *quartzous; quartzy.*

quartzose arkose A term used by Hubert (1960, p.176-177) for a sandstone containing 50-85% quartz, chert, and metamorphic quartzite, 15-25% feldspars and feldspathic crystalline-rock fragments, and 0-25% micas and micaceous metamorphic-rock fragments. Cf: *quartzose graywacke.*

quartzose chert A vitreous, sparkly, shiny chert, which under high magnification shows a heterogeneous mixture of pyramids, prisms, and faces of quartz (Grohskopf & McCracken, 1949, pl.3), but also including chert in which the secondary quartz is largely anhedral. Also known as "drusy chert".

quartzose graywacke (a) A term used by Hubert (1960, p.176-177) for a sandstone containing 50-85% quartz, chert, and metamorphic quartzite, 15-25% micas and micaceous metamorphic-rock fragments, and 0-25% feldspars and feldspathic crystalline-rock fragments. Cf: *quartzose arkose.* (b) A term used by Krynine (1951) for a graywacke that has lost its micaceous constituents through abrasion and thus tends to approach an orthoquartzite. It is equivalent to *subgraywacke* of Folk (1954). See also: *quartz graywacke.*

quartzose sandstone (a) A well-sorted sandstone that contains (if pure) more than 95% clear quartz grains and 5% or less of matrix and cement (Krumbein & Sloss, 1963, p.169-170). (b) A sandstone that contains at least 95% quartz, but is not cemented with silica (Krynine, 1940, p.51). Cf: *quartzitic sandstone.* (c) A sandstone that contains 99% quartz and quartz cement (Shrock, 1948a). (d) A sandstone that contains 90% quartz grains (Dunbar & Rodgers, 1957).—See also: *orthoquartzite.* Syn: *quartz sandstone.*

quartzose shale A green or gray shale composed dominantly of rounded quartz grains of silt size. It is commonly associated with highly mature sandstones (orthoquartzites), and represents the reworking of residual clays as transgressive seas encroached on old land areas (marked by relatively stable conditions with gentle rates of subsidence).

quartzose subgraywacke *protoquartzite.*

quartz-pebble conglomerate A term used by Krumbein & Sloss (1963, p.163) for *orthoquartzitic conglomerate.*

quartz plate *quartz wedge.*

quartz porphyry A porphyritic extrusive or hypabyssal rock containing phenocrysts of quartz and alkali feldspar in a microcrystalline or cryptocrystalline groundmass; a *rhyolite.* European petrologists called pre-Tertiary and Tertiary extrusive equivalents of granite "quartz porphyry", and post-Tertiary equivalents "rhyolite" (Streckeisen, 1967, p. 189). Syn: *quartz felsite.* Cf: *granite porphyry.*

quartz-rich granitoid In the *IUGS classification,* a plutonic rock with Q between 60 and 90. Cf: *tarantulite.*

quartz sandstone *quartzose sandstone.*

quartz schist A schist whose foliation is due mainly to streaks and lenticles of nongranular quartz. Mica is present but in lesser quantities than in mica schist. A syn. of "schistose quartzite" of some petrologists.

quartz syenite In the *IUGS classification,* a plutonic rock with Q between 5 and 20 and P/(A+P) between 10 and 35.

quartz topaz A frequently used but incorrect syn. of *citrine.* Cf: *topaz quartz.*

quartz trachyte A fine-grained igneous rock consisting mostly of alkali feldspar, with normative quartz between 5 and 20 percent; the volcanic equivalent of *quartz syenite* (Williams et al., 1954, p. 100).

quartz wacke A gray to buff, moderately well-sorted, commonly fine-grained sandstone containing up to 90% quartz and chert, and with more than 10% argillaceous matrix (largely sericite and chlorite), less than 10% feldspar, and less than 10% rock fragments (bits of coal, shale, etc.) (Williams, Turner & Gilbert, 1954, p.292-293). Krumbein & Sloss (1963, p.172-173) give a lower limit of 15-20% matrix, and regard the rock as a "washed" graywacke such as one occurring under conditions of moderate subsidence in an unstable depositional area. The rock is equivalent to *subgraywacke* as originally defined by Pettijohn (1949, p.227 & 256). Term introduced by Fischer (1934) for a wacke with less than 10% unstable materials. Examples include many coal-measures sandstones of Pennsylvanian age (such as those of the Atokan Series). Also spelled: *quartzwacke.*

quartz wedge In an optical system such as a polarizing microscope, an elongate wedge of clear quartz that is used in analysis of a mineral's fast and slow vibration-plane traces, optical sign, and interference colors. Syn: *quartz plate.*

quartzy *quartzose.*

quasicratonic Said of semiconsolidated regions in which paratectonic deformation (*germanotype tectonics*) tends to dominate (Stille, 1940). Obsolete. Syn: *semicratonic.*

quasi-equilibrium The state of balance or grade in a stream's cross section, whereby "conditions of approximate equilibrium tend to be established in a reach of the stream as soon as a more or less smooth longitudinal profile has been established in that reach even though downcutting may continue" (Leopold & Maddock, 1953, p. 51).

quasi-geoid A nonequipotential surface in the vicinity of the geoid that is defined as the locus of points whose distances below the terrain are the normal heights.

quasi-instantaneous Geologically instantaneous; occurring within an interval of geologic time too small to be subdivided (Termier & Termier, 1956).

Quaternary The second period of the Cenozoic era, following the Tertiary; also, the corresponding system of rocks. It began two to three million years ago and extends to the present. It consists of two grossly unequal epochs: the Pleistocene, up to about 8,000 years ago, and the Holocene since that time. The Quaternary was originally designated an era rather than a period, with the epochs considered to be periods, and it is still sometimes used as such in the geologic literature. The Quaternary may also be incorporated into the Neogene, when the Neogene is designated as a period of the Tertiary era. See also: *age of man.*

quaternary sediment A sediment consisting of a mixture of four components or end members; e.g. a sediment with a clastic component (such as quartz), a secondary mineral (such as a clay mineral), a chemical component (such as calcite), and an organic residue.

quaternary system A chemical system having four principal components.

quay A *wharf* of solid construction, built roughly parallel to the shoreline and accommodating vessels on one side only. Pron: *key.* Cf: *pier.*

quebrada A term used in the SW U.S. for a ravine or gorge, esp. one that is usually dry but is filled by a torrent during a rain; a *barranco.* Also, a stream or brook. Etymol: Spanish.

queenstownite *Darwin glass.*

queluzite A rock composed chiefly of the mineral spessartine, occasionally with amphiboles, pyroxenes, or micas. Economically significant manganese ores are derived from this rock. The name, given by Derby in 1901, is from Queluz, Minas Gerais, Brazil. Not recommended usage.

quenching In experimental petrology, the very rapid cooling of a heated charge in order to preserve certain physical-chemical characteristics of the high-temperature state which would be changed by slow cooling.

quenselite A black monoclinic mineral: $PbMnO_2(OH)$ or $Pb_2Mn_2O_5 \cdot H_2O$.

quenstedtite A mineral: $Fe_2(SO_4)_3 \cdot 10H_2O$.

quernstone An English term for a millstone, and (in Norfolk) used as a syn. of *carstone.*

Querwellen wave *Love wave.*

quetzalcoatlite A hexagonal mineral: $Zn_8Cu_4(TeO_3)_3(OH)_{18}$.

quick [mineral] A local term used in the western U.S. for *quicksilver.*

quick [ore dep] Said of an economically valuable or productive mineral deposit, in contrast to a *dead* ground or area. An ore is said to be quickening as its mineral content increases. Syn: *alive.*

quick [sed] Said of a sediment that, when mixed with water, becomes extremely soft and incoherent, and is capable of flowing easily under load or by force of gravity; e.g. "quick clay" of glacial or marine origin, which, if disturbed, loses practically all its shear strength and flows plastically.

quick [soil] (a) Said of a condition of soil in which an increase in pore-water pressure decreases particle-to-particle attraction and reduces significantly the soil's bearing capacity. (b) Said of a highly porous soil that readily absorbs heat.

quick clay A clay that loses all or nearly all its shear strength after being disturbed; a clay that shows no appreciable regain in strength after remolding.

quicksand (a) A mass or bed of fine sand, as at the mouth of a river or along a seacoast, that consists of smooth rounded grains with little tendency to mutual adherence and that is usually thoroughly saturated with water flowing upward through the voids, forming a soft, shifting, semiliquid, highly mobile mass that yields easily to pressure and tends to suck down and readily swallow heavy objects resting on or touching its surface. Syn: *running sand.* (b) An area marked by the presence of one or more such beds. (c) The sand found in a bed of quicksand.

quicksilver A term applied to *mercury* where it occurs as a native mineral or has been mined but not yet used (as in "flasks of quicksilver").

quickstone A consolidated rock that had flowed under the influence of gravity before lithification; a *quick* sediment that has become lithified.

quickwater The part of a stream characterized by a strong current.

quiet reach *stillwater.*

quilted surface A land surface characterized by broad, rounded, uniformly convex hills separating valleys that are comparatively narrow "like the seams by which a quilt is furrowed" (Davis, 1918, p. 124).

quinary system A chemical system having five main components, e.g. Na_2O - CaO - K_2O - Al_2O_3 SiO_2.

quinqueloculine Said of a foraminiferal test having five externally visible chambers, as a result of growth in varying planes about an elongate axis; specif. pertaining to the foraminifer *Quinqueloculina.*

quinqueradiate Said of radial symmetry of certain echinoderms characterized by five rays extending from the mouth. Syn: *pentamerous.*

quisqueite A highly sulfurous asphaltum; a black, brittle, lustrous substance mostly composed of sulfur (37%) and carbon (43%) and accompanying the vanadium ores of Peru.

Q wave *Love wave.*

ra A Norwegian term, used esp. in southern Norway, for a morainal ridge covered with a surface layer of large stones (Stamp, 1961, p. 383); most of them are in or near the sea.

rabbittite A pale greenish-yellow secondary mineral: $Ca_3Mg_3(UO_2)_2(CO_3)_6(OH)_4 \cdot 18H_2O$.

rabdolith *rhabdolith.*

race [paleont] A group of organisms with similar characteristics but not sufficiently distinctive to be classified as a *species* or *subspecies.*

race [sed] Small calcium-carbonate concretions commonly found in brick clay; bits of chalk set in a clayey matrix. Syn: *rance.*

race [water] (a) A strong or rapid current of water flowing through a narrow channel or river, e.g. a *tide race.* (b) The constricted channel or river in which such a current flows. It may occur by the meeting of two tides, as near a headland separating two bays, or it may be artificial and used for an industrial purpose, such as conveying water to or away from the waterwheel of a mill. Syn: *water race.*

racemization A process in which an optically active stereoisomer is converted to a mixture of two isomers which possesses no optical activity (Cram & Hammond, 1959, p. 131). Cf: *racemization age method.*

racemization age method A method of geochronology based on the chemical *racemization* (or *epimerization*) of amino acids. Amino acids of living organisms consist virtually entirely of the L-enantiomer (or diastereomer). After death the L-enantiomer (or diastereomer) for each amino acid is slowly racemized and eventually forms an equilibrium mixture consisting of equal amounts of the D- and L-enantiomers (or diastereomers). The increase in D/L ratio can be used to obtain a measure of the time that has elapsed since the organism died. The method is sensitive to temperature, pH, and other environmental factors (Schroeder & Bada, 1976). See also: *racemization; epimerization.* Syn: *amino-acid racemization age method.*

rachis The axis of a large compound leaf, or of a flowering spike or raceme (Swartz, 1971, p. 395).

racial senescence *phylogerontism.*

radar (a) An electronic detection device or *active system* for locating or tracking a distant object by measuring elapsed circuit time of travel of ultrahigh-frequency radio waves of known propagation velocity emitted from a transmitter and reflected back by the object to or near the point of transmission in such a way that range, bearing, height, and other characteristics of the object may be determined. Radar operation is unaffected by darkness, but moisture in the form of fog, snow, rain, or heavy clouds may cause varying degrees of attenuation or reflection of the radio energy. The term is usually prefixed by a code letter indicating the frequency band for certain wavelength ranges; e.g. K-band (about 2 cm), X-band (about 4 cm), and P-band (about one meter). (b) A name applied to the method or technique of locating or tracking objects by means of radar, such as the observation and analysis of minute radio signals reflected from the objects and displayed in a radar system.—Etymol: *ra*dio *d*etecting *a*nd *r*anging. See also: *monostatic radar; bistatic radar.*

radar imagery Imagery provided by scanning devices using microwave radiation.

radar shadow A dark area of no return on a radar image that extends in the far-range direction from an object on the terrain that intercepts the radar beam.

radial [paleont] adj. (a) Belonging to or in the direction of a ray of an echinoderm. (b) A syn. of *ambulacral;* e.g. referring to the position of a line extending from the centrally placed mouth to the aboral end of any ambulacrum of a blastoid. (c) Directed outward from the center of the umbilicus of an ammonoid and at right angles to the axis of coiling and growth; transverse. (d) Said of the microstructure of hyaline calcareous foraminiferal tests consisting of calcite or aragonite crystals with *c*-axes perpendicular to the surface and exhibiting between crossed nicols a black cross with concentric rings of color mimicking a negative uniaxial interference figure (TIP, 1964, pt.C, p.63).—n. A *radial plate* together with all structures borne by it.

radial [palyn] Pertaining to trilete-spore features associated closely with the arms of the laesura. Cf: *interradial [palyn].*

radial [photo] A line or direction from the center (principal point, isocenter, nadir point, or substitute center) to any point on a photograph.

radial array *azimuthal survey.*

radial beam An internal rod usually connecting concentric lattice shells of spumellarian radiolarians.

radial canal (a) A tube extending radially from the ring canal beneath or along an ambulacrum of an echinoderm. It is closed at its outer end and bears rows of podia. The ambulacral system of echinoderms contains five radial canals. (b) One of the numerous minute canals, lined with choanocytes, radiating from the spongocoel in some sponges and ending just below the surface of the sponge.

radial coefficient A ratio expressing the number of radial elements divided by the diameter of the cup in archaeocyathids (TIP, 1972, pt. E, p. 13). Abbrev: RK.

radial drainage pattern A drainage pattern in which consequent streams radiate or diverge outward, like the spokes of a wheel, from a high central area; it is best developed on the slopes of a young, unbreached domal structure or of a volcanic cone. Cf: *centripetal drainage pattern.* Syn: *centrifugal drainage pattern.*

radiale (a) A *radial plate* of a crinoid. Pl: *radialia.* (b) A proximal bone on the radial (thumb) side of the wrist of primitive tetrapods.

radial facet A smooth or sculptured distal face of a radial plate of a crinoid, bearing marks of ligamentary or muscular articulation with the first primibrachial. It is lacking in radial plates that bear no arms.

radial fault One of a group of faults that radiate from a central point.

radial plate An echinoderm ray plate that, in crinoids, lies between the basal and brachial plates and bears an arm; in blastoids, rhombiferan cystoids, and parablastoids, lies between the basal and deltoid plates and bears part or all of an ambulacrum; in asteroids, is a prominent aboral plate at the base of each arm.

radial pore canal One of a series of tubules or *pore canals* in an ostracode leading from the line of concrescence through the area of the duplicature to the free margin of the valve, usually housing sensory setae protecting the gape of the open carapace. Cf: *normal pore canal.*

radial shield In many ophiuroids, one of a pair of relatively large ossicles adjacent to the base of an arm on the aboral surface of the disc.

radial spine A tangential rod or needle in the skeleton of an acantharian or phaeodarian radiolarian.

radial suture A suture in a heterococcolith corresponding to a radius in a circular coccolith, or to a straight line drawn through the nearest focus or connecting the foci of an elliptic coccolith.

radial symmetry The condition, property, or state of having similar parts of an organism regularly arranged about a common central axis (as in a starfish); e.g. a type of symmetry exemplified in a flower that can be separated into two approximate halves by a longitudinal cut in any plane passing through the center of the flower. Cf: *bilateral symmetry.*

radial triangulation A triangulation procedure in which direction lines from the centers of overlapping vertical or oblique photographs to control points imaged on the photographs are measured and used for horizontal-control extension by successive intersection and resection.

radial tube A centrifugal cylinder in an acantharian radiolarian.

radial zone The chamber portion of a foraminiferal test with essentially radial elements, situated between the marginal zone and the central complex (as in Orbitolinidae).

radianal Primitively the interradial of the C ray in inadunate crinoids; it is generally shifted up and to the left, into the posterior interray.

radiance Radiant flux per unit solid angle per unit area.

radiant n. An organism or group of organisms, such as a species, that has arrived at its present geographic location as the result of dispersal from its main place of origin. Cf: *radiation [evol]*. Obsolete.

radiante That point in the skeletal end of an anthaspidellid sponge from which the trabs radiate (Rigby & Bayer, 1971, p. 609).

radiant emittance Radiant flux emitted per unit area of a source.

radiant energy Energy transferred by electromagnetic waves, measured in joules or ergs.

radiant flux The rate of transfer of radiant energy, measured in watts or ergs per second. See also: *spectral radiant flux*. Syn: *radiant power*.

radiant intensity Radiant flux per unit solid angle.

radiant power *radiant flux*.

radiant-power peak The wavelength at which the maximum electromagnetic energy is radiated at a particular temperature.

radiate aperture A foraminiferal-test opening consisting of numerous diverging slits (as in the superfamily Nodosariacea).

radiated Said of an aggregate of acicular crystals that radiate from a central point. Cf: *spherulitic*.

radiate mud crack A mud crack that displays an incomplete radiate pattern and that lacks normal polygonal development (Kindle, 1926, p.73).

radiation [evol] The dispersal of a group of organisms into different environments accompanied by divergent change in the evolutionary structure. Cf: *convergence; radiant*.

radiation [surv] A method of surveying in which points are located by a knowledge of their distances and directions from a central point.

radiation damage The damage done to a crystal lattice (or glass) by passage of fission particles or alpha-particles from the nuclear decay of a radioactive element residing in the lattice. The damage paths (*alpha-particle recoil tracks* or *fission tracks*) can be enlarged to microscopic size by suitable etching techniques and used to determine an age for the material.

radiation detector *detector*.

radiation fog A type of *fog* that occurs at night when the air cools by radiation to a temperature below the *dew point*. Syn: *ground fog*.

radiation log *radioactivity log*.

radiaxial Radially axial; e.g. "radiaxial calcite" in sedimentary rocks, occurring as cavity linings composed of subparallel individual crystals elongated normal to the cavity wall (Fischer, 1964, p.148).

radiaxial calcite A type of calcite cement forming a crust of crystals elongated perpendicular to the encrusted surface; crystals have curved cleavage and twinning planes (concave outward), and optic axes of the crystals converge outward (Bathurst, 1959).

radiciform Said of a rootlike epithecal process (outgrowth) of a corallite wall, serving for fixation.

radicle [bot] The root-primordium of an embryo, which grows into the primary root of a seedling (Fuller & Tippo, 1954, p. 969).

radicle [paleont] (a) A bryozoan rootlike structure composed of one or more kenozooids. Syn: *rhizoid*. (b) An individual rootlike branch of a crinoid radix.

radii Plural of *radius*.

radioactivation analysis *activation analysis*.

radioactive Pertaining to or exhibiting *radioactivity*.

radioactive age determination *radiometric dating*.

radioactive chain *radioactive series*.

radioactive clock A *geochronometer* consisting of a radioactive isotope, e.g. carbon-14, rubidium-87, or potassium-40, whose decay constant is known and is low enough to be calibrated to time units, usually years. Radioactive clocks are the basis of absolute-age determinations and the specific element being used is sometimes designated as a clock, e.g. carbon clock. Syn: *atomic clock; nuclear clock*.

radioactive constant A less-preferred syn. of *decay constant*.

radioactive dating *radiometric dating*.

radioactive decay The spontaneous disintegration of the atoms of certain *nuclides* into new nuclides, which may be stable or undergo further decay until a stable nuclide is finally created. Radioactive decay involves the emission of *alpha particles, beta particles*, and other energetic particles, and usually is accompanied by emission of *gamma rays* and by atomic de-excitation phenomena. It always results in the generation of heat. Cf: *radioactivity*. Syn: *radioactive disintegration*.

radioactive disintegration *radioactive decay*.

radioactive equilibrium A relationship between a *parent* and one or more radioactive *daughters* in which the ratio of daughter activity to parent activity is constant.

radioactive heat Heat produced within a medium as a result of the decay of its constituent radioactive elements.

radioactive series A series or succession of *nuclides*, each of which becomes the next by radioactive decay, until a stable nuclide is formed. There are three important natural radioactive series, the *actinium series, thorium series*, and *uranium series*. See also: *parent; daughter; end product*. Syn: *radioactive chain*.

radioactive spring A spring whose water has a high and readily detectable radioactivity.

radioactive waste "Equipment and materials from nuclear operations that are radioactive and for which there is no further economic use. Waste is generally classified as high-level (having concentrations of radioactivity of hundreds to thousands of curies per gallon or cubic foot or cubic meter), low-level (averaging in the range of 1 microcurie per cubic foot or so), or intermediate-level (between these extremes). Waste is classified by its level of radioactivity at the time of burial. As radioactive decay proceeds, the level diminishes according to known and predictable mathematical expressions, depending on the radioactive nuclides present" (National Research Council Panel on Land Burial, 1976, p. 7).

radioactivity (a) The emission of energetic particles and/or radiation during *radioactive decay*. (b) A particular radiation component from a radioactive source, such as gamma radioactivity. (c) A *radionuclide*, such as a radioactivity produced in a bombardment. — Adj: *radioactive*.

radioactivity anomaly A deviation from expected results found when making a radioactivity survey. Such anomalies are important in mineral exploration.

radioactivity log The generic name for *well logs* whose curves derive from reactions of atomic nuclei involving the behavior of gamma rays and/or neutrons. Except for the natural *gamma-ray log* and the *spectral gamma-ray log*, they record the response of rocks very near the well bore to bombardment by gamma rays or neutrons from a source in the logging *sonde*. Most can be obtained in cased, empty, or fluid-filled well bores. See also: *density log; neutron log; neutron-activation log; epithermal-neutron log; pulsed-neutron-capture log*. Syn: *radiation log; nuclear log*.

radioassay An assay procedure involving the measurement of alpha, beta, or gamma radiation intensity of a sample.

radioautograph *autoradiograph*.

radiocarbon Radioactive carbon, esp. *carbon-14*, but also carbon-10 and carbon-11.

radiocarbon age *carbon-14 age*.

radiocarbon dating *carbon-14 dating*.

radiochemistry The chemical study of irradiated and naturally occurring radioactive materials and their behavior. It includes their use in tracer studies and other chemical problems.

radioecology The branch of ecology concerned with the relationship between natural communities and radioactive material.

radiogenic Said of a product of a radioactive process, e.g. heat, lead.

radiogenic age determination *radiometric dating*.

radiogenic argon (a) Argon-40 formed by the decay of potassium-40. (b) Argon-40 in a rock or mineral formed by the decay of potassium-40 in situ since the rock or mineral formed. Cf: *atmospheric argon; excess argon; inherited argon*.

radiogenic dating *radiometric dating*.

radiogenic isotope An isotope that was produced by the decay of a *radionuclide*, but which itself may or may not be radioactive. See also: *radioisotope*.

radiogenic lead (a) Lead formed as a result of radioactive decay of uranium or thorium. (b) Lead in a rock or mineral formed by radioactive decay of uranium or thorium in situ after formation of the rock or mineral. (c) Lead which has $^{207}Pb/^{204}Pb$ and $^{206}Pb/^{204}Pb$

ratios greater than they would be if the lead had evolved in a single stage, because it evolved in an integrated system with $^{238}U/^{204}Pb$ about 9. Cf: *common lead; primeval lead.*

radiogenic strontium (a) Strontium-87 formed from decay of rubidium-87. (b) Strontium-87 occurring in rocks and minerals that is the direct result of decay of rubidium-87 in situ since the rock or mineral formed. Cf: *common strontium.*

radiogeology A syn. of *isotope geology.* The term was introduced by the Russian geologist Vernadskiy.

radiograph A less-preferred syn. of *autoradiograph.*

radiohydrology The study of the hydrologic relationships of extraction, processing, and use (including use in hydrologic investigations) of radioactive materials and disposal of the associated waste products.

radioisotope A radioactive *isotope* of an element. Also used, though incorrectly, as a syn. of *radionuclide.* See also: *radiogenic isotope.* Syn: *unstable isotope.*

radiolarian Any actinopod belonging to the subclass Radiolaria, characterized mainly by a siliceous skeleton and a marine pelagic environment. Range, Cambrian to present. In some classifications the radiolarians are grouped with the rhizopods.

radiolarian chert A well-bedded, microcrystalline radiolarite that has a well-developed siliceous cement or groundmass.

radiolarian earth A *siliceous earth* composed predominantly of the remains (lattice-like skeletal framework) of Radiolaria; the unconsolidated equivalent of *radiolarite.*

radiolarian ooze A deep-sea *pelagic sediment* containing at least 30% opaline-silica tests of radiolarians. It is a *siliceous ooze.*

radiolarite [paleont] A fossil shell of the Radiolaria.

radiolarite [sed] (a) The comparatively hard fine-grained chert-like homogeneous consolidated equivalent of radiolarian earth. (b) Indurated radiolarian ooze. (c) A syn. of *radiolarian earth.*

radiolite A spherulite composed of radially arranged acicular crystals.

radiolitic [ign] Said of the texture of an igneous rock characterized by radial, fanlike groupings of acicular crystals, resembling sectors of *spherulites.*

radiolitic [sed struc] Said of limestones in which the components radiate from central points, with the cement comprising less than 50 percent of the total rock (Krumbein & Sloss, 1963, p.179).

radioluminescence Luminescence that is stimulated by the impact of radioactive particles.

radiolysis Chemical decomposition caused by radiation.

radiometer A nonimaging device for measuring radiant energy, especially thermal radiation.

radiometric Pertaining to the measurement of geologic time by the study of parent and/or daughter isotopic abundances and known disintegration rates of the radioactive parent isotopes, e.g. *radiometric dating.*

radiometric age An age expressed in years and calculated from the quantitative determination of radioactive elements and their decay products. A common syn. is *absolute age.* Syn: *isotopic age.*

radiometric age determination *radiometric dating.*

radiometric dating Calculating an age in years for geologic materials by measuring the presence of a short-life radioactive element, e.g. carbon-14, or by measuring the presence of a long-life radioactive element plus its decay product, e.g. potassium-40/argon-40. The term applies to all methods of age determination based on nuclear decay of naturally occurring radioactive isotopes. Syn: *isotopic age determination; radiometric age determination; radioactive age determination; radioactive dating; radiogenic age determination; radiogenic dating; nuclear age determination.*

radionuclide A radioactive *nuclide.* The term *radioisotope* is incorrectly used as a synonym. See also: *radioactivity.*

radiosonde An instrument package, carried aloft by a balloon, that is used for the gathering and radio transmission of meteorologic data.

radio-wave method Any electromagnetic exploration method wherein electromagnetic waves transmitted from radio broadcast stations are used as an energy source in determining the electrical properties of the Earth.

radiozone A *para-time-rock unit* representing a zone or succession of strata established on common radioactivity criteria (Wheeler et al., 1950, p.2364).

radius (a) A ray of an echinoderm, such as any of five radiating ossicles in Aristotle's lantern of an echinoid; esp. a *radial plate.* (b) The lateral part of a compartmental plate of a cirripede crustacean, overlapping the ala of an adjoining plate, and differentiated from the paries by a change in direction of the growth lines. (c) An imaginary radial line dividing the body of a radially symmetrical animal into similar parts. (d) The more medial of the two bones of the forearm of tetrapods.—Pl: *radii.*

radius of influence The radial distance from the center of a well bore to the edge of its area of influence.

radius ratio The radius of a cation divided by that of an ion. Relative ionic radii are pertinent to coordinations in ionic crystal structures.

radix Rootlike distal anchorage of a crinoid column.

radon-220 A radioactive, gaseous isotope of radon; it is a member of the *thorium series* and a daughter of radium-224. Syn: *thoron.*

radula A chitinous band or strip of horny material in nearly all univalve mollusks that bears numerous transverse rows of filelike or rasplike and usually very minute teeth on its dorsal surface, that can be protruded through the mouth from its position on the floor of the digestive canal, and that serves to gather and tear up food and draw it into the mouth. Pl: *radulae.*

radulifer Said of hook-shaped or rodlike brachiopod crura that arise on the ventral side of the hinge plates and project toward the pedicle valve.

rafaelite A nepheline-free orthoclase-bearing hypabyssal *syenite* that also contains analcime and calcic plagioclase. The name (Johannsen, 1938) is from the San Rafael Swell, Utah. Not recommended usage.

raft [coal] *float coal.*

raft [ign] A rock fragment caught up in a magma and drifting freely, more or less vertically. Cf: *xenolith.*

raft [streams] An accumulation or jam of floating logs, driftwood, dislodged trees, or other debris, formed naturally in a stream by caving of the banks, and acting as an impedance to navigation. See also: *raft lake.*

raft breccia A breccia having fragments that remained unworn during transportation, as by an iceberg or floating vegetation such as trees or seaweed (Norton, 1917, p.172).

rafted ice A form of pressure ice in which one floe overrides another.

rafting (a) The transporting of land-derived rocks, soil, and the like, by floating organic material, as seaweed or logs, or by floating ice (*ice-rafting*). (b) A form of deformation of floating ice wherein one piece overrides another, thus creating *rafted ice.* It is most common in new ice and young ice.

raft lake A relatively short-lived body of water impounded along a stream by a *raft;* examples are commonly found in the Red River of Louisiana during times of high water.

rag In British usage, any of various hard, coarse, rubbly or shelly rocks that weather with a rough irregular surface; e.g. a flaggy sandstone or limestone used as a building stone. The term appears in certain British stratigraphic names, as the Kentish Rag (a Cretaceous sandy limestone in East Kent). See also: *coral rag.* Syn: *ragstone.*

raggioni Large ray-crystals of former aragonite, now converted to calcite mosaic. They are typically 1-10 cm long, generally with squared-off ends and hexagonal cross-sections; they occur in some sabkha-type carbonates. Pron: rah-*joe*-nee. Plural form commonly used; sing: raggione. Etymol: Italian, "large rays".

raglanite A nepheline *diorite* composed of oligoclase, nepheline, and corundum with minor amounts of mica, calcite, magnetite, and apatite. Cf: *craigmontite.* The name, given by Adams and Barlow in 1910, is from Raglan Township, Ontario. Not recommended usage.

ragstone *rag.*

raguinite A mineral: $TlFeS_2$.

rain A falling of numerous particles, such as the long unending deposition of pelagic matter to the bottom of the ocean, or of micrometeorite dust from interplanetary space. Also, the falling particles themselves, such as *pollen rain.*

rainbeat *raindrop impact.*

rainbow Chromatic iridescence observed in drilling fluid that has been circulated in a well, indicating contamination or contact with fresh hydrocarbons.

rainbow quartz An *iris* quartz that exhibits the colors of the rainbow.

rain crust A type of *snow crust* formed by refreezing of surface snow that had been melted or wetted by liquid precipitation.

rain desert A desert in which rainfall is sufficient to maintain a sparse general vegetation. Cf: *runoff desert.*

raindrop impact The action of raindrops striking or falling upon

the surface of the ground. Syn: *rainbeat.*
raindrop impression *rain print.*
raindrop imprint *rain print.*
rainfall (a) The quantity of water that is precipitated out in the atmosphere as rain, in a given period of time. (b) The liquid product of precipitation in whatever form. In this sense the term is synonymous with *precipitation.*
rainfall excess The volume of water from rainfall that is available for direct runoff (Langbein & Iseri, 1960). Cf: *abstraction [water]; precipitation excess.* Syn: *excess water.*
rainfall penetration The depth below the soil surface to which water from a given rainfall is able to infiltrate.
rain forest A forest where the annual rainfall is at least 100 inches. The region is characterized by tall, lush evergreen trees.
rain gage A device used to measure precipitation (melted snow, sleet, or hail as well as rain). It consists of a receiving funnel, a collecting vessel, and a measuring cylinder. Syn: *pluviometer; hyetometer; snow gage.*
rain pillar A minor landform consisting of a column of soil or soft rock capped and protected by pebbles or concretions, produced by the differential erosion effected by the impact of falling rain (Stokes & Varnes, 1955, p. 118).
rain print A small, shallow craterlike pit surrounded by a slightly raised rim, formed in soft fine sand, silt, or clay, or in the mud of a tidal flat, by the impact of a falling raindrop, and sometimes preserved on the bedding planes of sedimentary rocks or as casts on the underside of overlying sandstone beds. See also: *hail imprint; spray print.* Syn: *raindrop imprint; raindrop impression.*
rain shadow A very dry region on the lee side of a topographic obstacle, usually a mountain range, where the rainfall is noticeably less than on the windward side. The White Mountains in east-central California are in the rain shadow of the Sierra Nevadas.
rain-shadow desert A desert occurring on the lee side of a mountain or mountain range that deflects moisture-laden air upward on the windward side. See also: *middle-latitude desert.* Syn: *orographic desert.*
rainwash (a) The washing-away of loose surface material by rainwater after it has reached the ground but before it has been concentrated into definite streams; specif. *sheet erosion.* Also, the movement downslope (under the action of gravity) of material loosened by rainwater. It occurs esp. in semiarid or scantily vegetated regions. Syn: *hillwash.* (b) The material that originates by the process of rainwash; material transported and accumulated, or washed away, by rainwater. (c) The rainwater involved in the process of rainwash.—Also spelled: *rain wash.*
rainwater Water that has fallen as rain and has not yet collected soluble matter from the soil, thus being quite soft.
rain-wave train *Overland flow* in the form of wave trains or series of uniformly spaced waves and involving nearly all the runoff. Rain-wave trains are usually associated with heavy rains, esp. cloudbursts (Horton, 1945, p. 313).
raise A mine shaft driven upward from a lower to a higher level.
raised bank *natural levee.*
raised beach An ancient beach occurring above the present shoreline and separated from the present beach, having been elevated above the high-water mark either by local crustal movements (uplift) or by lowering of sea level, and often bounded by inland cliffs. Cf: *marine terrace; elevated shoreline.* See also: *strandline.*
raised bog An area of acid, peaty soil, especially that developed from moss, in which the center is relatively higher than the margins. Syn: *Hochmoor.*
raised peat bog A *highmoor bog* with a thick accumulation of peat in the center, giving it a convex surface.
raised reef An *organic reef* elevated and standing above sea level.
raite An orthorhombic mineral: $Na_4Mn_3Si_8(O,OH)_{24} \cdot 9H_2O$ (?).
rake *pitch [struc geol].*
ralstonite A colorless, white, or yellowish mineral: $Na_xMg_xAl_{2-x}(F,OH)_6 \cdot H_2O$. It occurs in octahedral crystals.
ram An underwater ledge or projection from an ice wall, ice front, iceberg, or floe, caused by the more intensive melting and erosion of the unsubmerged part. Syn: *apron [ice]; spur [ice].*
Raman effect The resonant absorption and re-emission of radiant energy by specific interatomic valence bonds in a sample. When intense monochromatic radiation (as from a laser) is scattered by a sample, *Raman lines* appear in addition to the normal Rayleigh scattering of the exciting radiation. These lines are shifted relative to the exciting wavelength (Raman shifts) by amounts characteristic of the vibrational and rotational frequencies of the specific bonds causing the effect. Named in honor of C. V. Raman (1883-1970), Indian physicist.
Raman lines Shifted lines in the *Raman spectrum.* They are typically at longer wavelengths than the exciting radiation (Stokes lines), but shorter wavelengths (anti-Stokes lines) may occasionally appear.
Raman spectroscopy The observation of a *Raman spectrum* and all processes of recording and measuring that go with it.
Raman spectrum The characteristic spectrum observed when monochromatic light is scattered by a transparent substance. See also: *Raman effect.*
ramassis A local term in southern Louisiana for a mass of decomposed plant debris, dried plant remains, driftwood, and other flotsam occurring in coastal marshes or *flotants* (Russell, 1942, p. 96-97).
rambla A dry ravine, or the dry bed of an ephemeral stream. Etymol: Spanish, from Arabic *ramlah,* "sand".
ramdohrite A dark-gray mineral: $PbAgSb_3S_6$. It is closely related to andorite and fizelyite.
rameauite A monoclinic mineral: $K_2CaU_6O_{20} \cdot 9H_2O$.
ramentum A thin, chaffy scale on a leaf, e.g. on some ferns.
rami Plural of *ramus.*
rammell An English term for a rock containing a mixture of shale (or clay) and sand.
rammelsbergite A gray mineral: $NiAs_2$. It is dimorphous with pararammelsbergite and related to loellingite. Syn: *white nickel.*
ramose Consisting of or having branches; e.g. said of a bryozoan colony consisting of erect, round, or moderately flattened branches. Syn: *dendroid.*
ramosite A basic *scoria.* Obsolete.
ramp [paleont] The abapically sloping surface of a gastropod whorl next below a suture.
ramp [snow] A drift of snow that forms an inclined plane between land or land ice and sea or shelf ice.
ramp [struc geol] The steepening inclined segment of a thrust fault, esp. where a bedding thrust or décollement changes from a stratigraphically lower to a higher *lubricating layer.*
ramp anticline An anticline formed in a thrust sheet as a result of movement up a *ramp.*
rampart [geomorph] (a) A narrow, wall-like ridge, 1-2 m high, built up by waves along the seaward edge of a reef flat, and consisting of boulders, shingle, gravel, or reef rubble, commonly capped by dune sand. (b) A wall-like ridge of unconsolidated material formed along a beach by the action of strong waves and currents.
rampart [lake] *lake rampart.*
rampart [volc] A crescentic or ringlike deposit of pyroclastics around the top of a volcano.
rampart wall A *rimming wall* formed along the outer or seaward margin of a terrace, as on various "high limestone" Pacific islands (Flint et al., 1953, p. 1258).
ram penetrometer *ramsonde.*
ramp region The region between the base and top of a *ramp,* extending for its entire length (Morse, 1977; Serra, 1977).
ramp trough *ramp valley.*
ramp valley A valley that is bounded by high-angle thrust faults, or *ramps.* Syn: *ramp trough.*
ram resistance The resistance of a snow layer to penetration. A cone at the tip of a steel rod is driven vertically into the snow by repeated blows from a drop-hammer. The energy per unit of distance penetrated gives a ram resistance index in units of force.
ramsayite *lorenzenite.*
ramsdellite An orthorhombic mineral: MnO_2. It is dimorphous with pyrolusite.
ramsonde A cone-tipped metal rod that is driven downward into snow by repeated blows from a drop-weight in order to measure its penetration resistance. Syn: *ram penetrometer.* Also spelled: *ramsond; Rammsonde.*
ramule A minor branch of a crinoid arm, differing from a pinnule in less regular occurrence; in some crinoids it bears pinnules.
ramus (a) In the vertebrates, a branch or process, esp. of a bone, nerve, or blood vessel. (b) A projecting part or elongated process of an invertebrate; e.g. a branch of a crustacean limb, or the main branch of a crinoid arm. Pl: *rami.*
raña A Spanish term for a consolidated mudflow deposit containing angular blocks of rock of all sizes; e.g. a *fanglomerate.*
rance (a) A dull red marble, with blue-and-white markings, from Hainaut province in Belgium. Etymol: French. (b) *race [sed].*
Rancholabrean North American (California) continental stage:

Upper Pleistocene (above Irvingtonian).
rancieite A mineral: $(Ca,Mn^{+2})Mn_4{}^{+4}O_9 \cdot 3H_2O$.
rand (a) An English term for the low, marshy border of a lake or of a river overgrown with reeds. (b) A term used in South Africa for a long, low, rocky ridge or range of hills often covered with scrub; e.g. Witwatersrand (popularly contracted to "The Rand"), a ridge containing rich gold-bearing rocks, 100 km long, situated near Johannesburg.
randannite (a) A dark variety of diatomaceous earth containing humic material, occurring in the Puy-de-Dôme (Randan) region of France. (b) An earthy form of opal.—Also spelled: *randanite.*
randkluft A crevasse at the head of a *mountain glacier,* separating the moving ice and snow from the surrounding rock wall of the valley where no *ice apron* is present. It may be enlarged where heat radiating from the rock wall causes the ice to melt. Etymol: German *Randkluft,* "rim crevice". Cf: *bergschrund.*
random error Any *error* that is wholly due to chance and does not recur; an *accidental error.* Ant: *systematic error.*
random line (a) A trial surveying line that is directed as closely as circumstances permit toward a fixed terminal point that cannot be seen from the initial point. (b) *random traverse.*
random process *stochastic process.*
random sample A subset of a statistical population in which each item has an equal and independent chance of being chosen; e.g. a sample chosen to determine (within defined limits) the average characteristics of an orebody.
random traverse A survey traverse run from one survey station to another station which cannot be seen from the first station in order to determine their relative positions. Syn: *random line.*
random variable A real-valued mathematical function or *variate,* arising from a mathematical process, that is defined through a sample space.
rang One of the units of the *CIPW classification* of igneous rocks; a subdivision of an *order.*
range [eco geol] An area in which a mineral-bearing formation crops out, e.g. the "iron range" and "copper range" of the Lake Superior region; a mineral belt.
range [ecol] The geographic area over which an organism or group of organisms is distributed. Syn: *distribution [ecol].*
range [geomorph] (a) A *mountain range.* Also, a line of hills if the heights are comparatively low. (b) A term sometimes used in Australia for a single mountain. (c) Mountainous country; term usually used in the plural.
range [hydrog] An established or well-defined line or course whose position is known and along which soundings are taken in a hydrographic survey.
range [radioactivity] The distance that radiation penetrates a medium before its velocity becomes no longer detectable.
range [sed] A measure of the variability between the largest and smallest particle sizes of a sediment or sedimentary rock.
range [stat] The numerical difference between the highest and lowest values in any series.
range [stratig] *stratigraphic range.*
range [surv] (a) Any series of contiguous *townships* (of the U.S. Public Land Survey system) aligned north and south and numbered consecutively east and west from a principal meridian to which it is parallel (e.g. "range 3 east" indicates the third range or row of townships to the east from a principal meridian). Also, any series of contiguous sections similarly situated within a township. Abbrev (when citing specific location): R. Cf: *tier.* (b) The distance between any two points, usually an observation point and an object under observation; also, two points in line with the point of observation. Two or more objects in line are said to be "in range".
range chart A chart that records for a given area the local range zone (and often the peak zone) of each significant fossil taxon encountered in terms of genera and species.
range direction *look direction.*
range finder A *tachymeter* designed for finding the distance from a single point of observation to other points at which no instruments are placed. It uses the parallax principle, and is usually constructed to give a rapid mechanical solution of a triangle having the target at its apex and the range finder at one corner of its base. See also: *telemeter.*
range line One of the imaginary boundary lines running north and south at six-mile intervals and marking the relative east and west locations of townships in a U.S. public-land survey; a meridional township boundary line. Cf: *township line.*
range midpoint *midrange.*
range-overlap zone *concurrent-range zone.*
range pole A *range rod.* Syn: *ranging pole.*
range rod A wooden or metal *level rod,* rounded or octagonal in section, 6-8 ft long and one inch or less in diameter, fitted with a sharp-pointed metal shoe, usually painted in one-foot bands of alternate red and white, and used for sighting points and lines in surveying or for showing the position of a ground point. Syn: *range pole; lining pole; line rod; sight rod.*
range-zone A body of strata representing the total range of occurrence of any selected element of the total assemblage of fossil forms in a stratigraphic sequence. The word "range" implies extent in both horizontal and vertical directions (ISG, 1976, p. 53). The range-zone comprises the rocks that contain the taxon whose name it bears. Range-zones do not usually coincide with *assemblage-zones* named for the same fossil. The term is more or less synonymous with certain usages of *biozone* (Hedberg, 1958, p.1888; and Teichert, 1958a, p.114-115), and is most commonly used (when unmodified) in referring to the interregional range of an individual taxon. Cf: *acme-zone.* See also: *local range-zone; concurrent-range zone; multifossil range-zone.* Syn: *acrozone; zonite.*
rank [bot] (a) A vertical row of leaves. Cf: *decussate; distichous.* (b) The position of a category in the taxonomic hierarchy of plants.
rank [coal] Degree of metamorphism. It is the basis of *coal classification* into a natural series from lignite to anthracite. Cf: *type [coal]; grade [coal].*
rank [meta] *metamorphic rank.*
rankamaite A mineral: $(Na,K,Pb)_3(Ta,Nb,Al)_{11}(O,OH)_{30}$.
rankar A term used mostly in Europe and Asia for a soil whose humic layer lies directly on the parent rock, which is usually lime-deficient and siliceous.
rankinite A monoclinic mineral: $Ca_3Si_2O_7$. It is dimorphous with kilchoanite.
Ranney collector *collector well.*
ranquilite *haiweeite.*
ransomite A sky-blue mineral: $CuFe_2(SO_4)_4 \cdot 6H_2O$.
Raoult's law In its original sense, the statement that the partial vapor pressure of a solvent liquid is proportional to its mole fraction. It is now usually used, however, in a more general form to specify a model for the *ideal solution:* The statement that the activity of each component in a solution is equal to its mole fraction. It is obeyed by all solutions for the major component in sufficiently concentrated regions, and approximately by many solutions over large compositional ranges.
rapakivi n. In the U.S., *granite* or *quartz monzonite* that is characterized by alkali feldspar phenocrysts (commonly ellipsoidal) that are mantled with plagioclase (oligoclase, albite, or andesine). The term was introduced in 1694 by Urban Hjarne to denote crumbly stone in certain weathered outcrops in Finland. Etymol: Finnish, "rotten stone". Syn: *wiborgite.*—adj. Said of volcanic as well as plutonic rocks having alkali feldspar phenocrysts that are mantled with plagioclase.
rapakivi texture A texture observed in igneous and metamorphic rocks, in which rounded crystals of potassium feldspar, a few centimeters in diameter, are surrounded by a mantle or rim of sodium feldspar in a finer-grained matrix, usually composed of quartz and colored minerals. This texture was first described in Finnish granites.
raphe (a) That portion of the funiculus of an ovule that is adnate to the integument, usually represented by a ridge. It is present in most anatropous ovules (Lawrence, 1951, p.767). (b) A vertical, unsilicified groove or cleft in the valve of some pennate diatoms (Scagel et al., 1965, p.630).
raphide A very thin, hairlike sponge spicule (monaxon or oxea).
rapid flow Water flow whose velocity exceeds the velocity or propagation of a long surface wave in still water. Cf: *tranquil flow.* Syn: *shooting flow; supercritical flow.*
rapids (a) A part of a stream where the current is moving with a greater swiftness than usual and where the water surface is broken by obstructions but without a sufficient break in slope to form a waterfall, as where the water descends over a series of small steps. It commonly results from a sudden steepening of the stream gradient, from the presence of a restricted channel, or from the unequal resistance of the successive rocks traversed by the stream. The singular form "rapid" is rarely used. See also: *cascade; cataract.* (b) A swift, turbulent flow or current of water through a rapids.

raqqaite An extrusive rock having the composition of a *pyroxenite* (Streckeisen, 1967, p. 188). Cf: *komatiite.* Not recommended usage.

rare In the description of coal constituents, less than 5% of a particular constituent occurring in the coal (ICCP, 1963). Cf: *common; very common; abundant; dominant.*

rare earths Oxides of a series of fifteen metallic elements, from lanthanum (atomic number 57) to lutetium (71), and of three other elements: yttrium, thorium, and scandium. These elements are not especially rare in the Earth's crust, but concentrations are. The rare earths are constituents of certain minerals, esp. monazite, bastnaesite, and xenotime. Abbrev: REO.

rarefaction *kataseism.*

ras A cape or headland. Etymol: Arabic *ra's,* "head".

rash Very impure coal, so mixed with noncoal material (clay, shale, or other argillaceous substances from the top or bottom of the coal seam) as to be unsalable; a dark substance intermediate in character between coal and shale; a dirty coal. Not to be confused with *rashing.*

rashing A soft, friable, and flaky or scaly shale or clay immediately benenath a coal seam, often containing much carbonaceous material (numerous slickensided surfaces and streaks of coal), and readily mixed with the coal in mining. It may also overlie or be interstratified with the coal. The term often used in the plural. It is not to be confused with *rash.*

rasorite *kernite.*

raspberry spar (a) *rhodochrosite.* (b) Pink tourmaline.

raspite A yellow or brown monoclinic mineral: $PbWO_4$. It is dimorphous with stolzite.

Rassenkreis A *polytypic [paleont]* species. Etymol: German *Rasse,* "race", plus *Kreis,* "cycle". Plural: *Rassenkreise.*

rasskar A Norwegian term for a cirque which has served as "an old scree channel" and has been "carved upward by weathering" (Termier & Termier, 1963, p. 412).

rasvumite An orthorhombic mineral: $K_3Fe_9S_{14}$.

rate-of-change map A derived stratigraphic map that shows the rate of change of structure, thickness, or composition of a given stratigraphic unit (Krumbein & Sloss, 1963, p.484). It is based on analysis of the contour lines on an initial map (structure-contour map, isopach map, facies map, etc.).

rate of sedimentation The amount of sediment accumulated in an aquatic environment over a given period of time, usually expressed as thickness of accumulation per unit time. There appears to have been a progressive increase in the rate with decreasing geologic age of sediments, with an overall average of about 22 cm of thickness per 1000 years (Pettijohn, 1957, p. 688). Syn: *sedimentation rate.*

ratfish *Holocephali.*

rathite A dark-gray mineral: $(Pb,Tl)_3As_5S_{10}$.

rathite-II *liveingite.*

rating curve *stage-discharge curve.*

ratio map A *facies map* that depicts the ratio of thicknesses between rock types in a given stratigraphic unit; e.g. a "sand-shale ratio map" showing the ratio of sandstone thickness to shale thickness in a given unit.

ratiometer An instrument used to measure the ratio of two differences in potential.

rational face A crystal face naturally suggested by and peculiar to the internal molecular structure of the mineral species to which the crystal belongs. Such faces usually have low *Miller indices.*

rational formula In hydraulics, the expression of peak discharge (in cfs units) as equal to rainfall (in inches/hr) times drainage area (in acres) times a runoff coefficient depending on drainage-basin characteristics (Chow, 1957).

rational horizon (a) A *celestial horizon;* e.g. a great circle 90 degrees from the zenith and constituting the equator of the horizon system of coordinates. (b) *actual horizon.*

rattlesnake ore A gray, black, and yellow mottled ore of carnotite and vanoxite, its spotted appearance resembling that of a rattlesnake.

rattle stone A *concretion* composed of concentric laminae of different composition, in which the more soluble layers have been removed by solution, leaving the central part detached from the outer part, such as a concretion of iron oxide filled with loose sand that rattles on being shaken. Syn: *klapperstein.* Also spelled: *rattlestone; rattle-stone.*

rauenthalite A mineral: $Ca_3AsO_4)_2 \cdot 10H_2O$.

rauhaugite A *carbonatite* that contains ankerite or dolomite. The name, given by Brögger in 1921, is from Rauhaug in the Fen complex, Norway. Cf: *beforsite.*

rauk A Swedish term for a *stack.* Pl: *raukar.*

Rauracian Substage in Great Britain: Upper Jurassic (middle Lusitanian: above Argovian, below Sequanian).

rauvite A purplish-black to bluish-black mineral: $Ca(UO_2)_2V_{10}O_{28} \cdot 16H_2O$.

raveling ground A tunnelman's term for ground that begins to drop out of the roof or sides of a tunnel some time after being exposed. See also: *firm ground; flowing ground; running ground; squeezing ground; swelling ground.*

ravinated Said of a landform or landscape having or cut by ravines.

ravine (a) A small narrow deep depression, smaller than a *gorge* or a canyon but larger than a *gully,* usually carved by running water; esp. the narrow excavated channel of a mountain stream. (b) A stream with a slight fall between rapids.—Etymol: French, "mountain torrent".

ravinement [geomorph] (a) The formation of a ravine or ravines. (b) *gully erosion.*

ravinement [stratig] A term introduced by Stamp (1921, p.109) for "an irregular junction which marks a break in sedimentation", such as an erosion line occurring where shallow-water marine deposits have "scooped down into" (or "ravined") slightly eroded underlying beds; a small-scale disconformity caused by periodic invasions of the sea over a deltaic area. Etymol: French, "hollowing out (by waters), gullying".

raw Said of a mineral, fuel, or other material in its natural, unprocessed state, as mined.

raw humus *mor.*

ray [bot] (a) In a composite inflorescence, the corolla of a marginal flower. (b) A *vascular ray.* (c) In dasycladacean algae, a branch.

ray [lunar] One of the long, relatively bright, almost white streaks, loops, or lines observed on the Moon's surface and appearing to radiate from a large, well-formed lunar crater, in some examples extending for hundreds of kilometers. Rays are brightest at high Sun angles and nearly invisible at low Sun angles except for rough ground. They are believed to be formed in some way by fine-grained debris explosively ejected from craters either by impact or by volcanic activity.

ray [paleont] (a) Any of the radiating divisions of the body of an echinoderm together with all structures borne by it; e.g. a segment of an echinoderm body that includes one ambulacral axis, or a radial plate or an arm of a crinoid. Also, a radial direction established by the position of an *ambulacrum.* (b) One of the primary subdivisions of a sponge spicule containing an axial filament or an axial canal. (c) In Osteichthyes and Chondrichthyes, one of many fine rods, respectively bony or horny, that support a fin.

ray [phys] A vector normal to the wave surface, indicating the direction and sometimes the velocity of propagation.

ray crater A large, relatively young lunar crater with visible rays; e.g. Copernicus.

Rayleigh criterion The relationship between surface roughness, depression angle, and wavelength that determines whether a surface will respond in rough or smooth fashion to a radar pulse. Cf: *rough criterion; smooth criterion.*

Rayleigh scattering Selective scattering of electromagnetic radiation by particles in the atmosphere that are small relative to the wavelength. The scattering is inversely proportional to the fourth power of the wavelength. See also: *Mie theory.*

Rayleigh wave A type of *surface wave* having a retrograde, elliptical motion at the free surface. It is named after Lord Rayleigh, the English physicist who predicted its existence. See also: *Rg wave.* Syn: *R wave.*

ray parameter A function p that is constant along a given seismic ray when horizontal velocity is constant. It is defined as $p = v^{-1} \sin i$, where v is the velocity and i is the angle that the ray makes with the vertical.

raypath The imaginary line along which wave energy travels. A raypath is always perpendicular to the wave front in isotropic media. Syn: *path [seis]; trajectory [seis].*

razorback A sharp, narrow ridge, resembling the back of a razorback hog. "There is little or no implication as to geologic structure, hence the term is not quite so specific as *hogback*" (Stokes & Varnes, 1955, p. 119).

razor stone *novaculite.*

Rb-Sr age method *rubidium-strontium age method.*

reach [coast] (a) An arm of the sea extending up into the land; e.g.

an estuary or a bay. (b) *promontory.*

reach [geog] (a) A continuous and unbroken expanse or surface of water or land. Syn: *stretch.* (b) An unstated but specific distance; an interval.

reach [hydraul] (a) The length of a channel, uniform with respect to discharge, depth, area, and slope. (b) The length of a channel for which a single gage affords a satisfactory measure of the stage and discharge. (c) The length of a stream between two specified gaging stations.—See also: *test reach.*

reach [lake] (a) A relatively long, straight section of water along a lake shore; also, a narrow arm of a lake, reaching into the land. (b) A straight, narrow expanse of shore or land extending into a lake.

reach [streams] (a) A straight, continuous, or extended part of a stream, viewed without interruption (as between two bends) or chosen between two specified points; a straight section of a restricted waterway, much longer than a narrows. See also: *sea reach.* (b) The level expanse of water between locks in a canal.

reactance That part of the impedance of an alternating-current circuit that is due to capacitance and/or inductance. It is expressed in ohms.

reaction *nuclear reaction.*

reaction border *reaction rim.*

reaction boundary *reaction line.*

reaction curve *reaction line.*

reaction line In a ternary system, a special case of the *boundary line,* along which one of the two crystalline phases reacts with the liquid, as the temperature is decreased, to form the other crystalline phase. Syn: *reaction boundary; reaction curve.*

reaction pair Any two minerals, one of which is formed at the expense of the other by reaction with liquid; esp., any two adjacent minerals in a *reaction series.*

reaction point A usually isobarically invariant point on a liquidus diagram in which the composition of the liquid cannot be stated in terms of position quantities of all the solid phases in equilibrium at the point. In a binary system it is equivalent to an incongruent melting point, or *peritectic point.*

reaction principle The concept of a *reaction series.*

reaction rim A peripheral zone around a mineral; it is composed of another mineral species and represents the reaction of the earlier solidified mineral with the surrounding magma. Cf: *corrosion border; corona [petrology]; kelyphytic rim.* Syn: *reaction border.*

reaction series A series of minerals in which any early-formed mineral phase tends to react with the melt, later in the differentiation, to yield a new mineral further down in the series; e.g. early-formed crystals of olivine react with later liquids to form pyroxene crystals, and these in turn may further react with still later liquids to form amphiboles. There are two different series, a *continuous reaction series* and a *discontinuous reaction series.* This concept is frequently referred to as *Bowen's reaction series,* after N.L. Bowen, who first proposed it, or as the *reaction principle.* See also: *reaction pair.*

reactivation surface An inclined bedding surface, separating otherwise conformable cross-beds; it is formed by slight erosion of the lee side of a sand wave, megaripple, or bar, during a period when deposition is temporarily interrupted. Originally described by Collinson (1970) from braided river sands, but more abundant in sands deposited by the migration of megaripples in the tidal environment.

readvance (a) A new *advance* made by a glacier after receding from the position reached in an earlier advance. (b) A time interval during which a readvance occurred.

real-aperture radar A *SLAR* system in which azimuth resolution is determined by the length of the antenna and by the wavelength. The radar returns are recorded directly to produce images. Syn: *brute-force radar.*

realgar A bright-red to orange-red monoclinic mineral: AsS. It occurs as nodules in ore veins and as a massive or granular deposit from some hot springs, and it is frequently associated with orpiment. Syn: *red arsenic; sandarac; red orpiment.*

realized ecological hyperspace *biospace.*

realized ecospace That portion of the *ecospace* actually utilized by an organism (Valentine, 1969, p.687). Cf: *prospective ecospace.*

realm (a) A portion of the Earth consisting of several *regions [ecol].* (b) A large *region [ecol].*

real time Processing of data such that results are available immediately, allowing influence on further processing of data.

reamer A rotary-drilling tool with a special bit used for enlarging, smoothing, or straightening a drill hole, or making the hole circular when the drill has failed to do so.

recapitulation theory A theory in biology stating that an organism passes through successive stages resembling its ancestors so that the ontogeny of the individual is a recapitulation of the phylogeny of its group. See also: *palingenesis [paleont].* Syn: *Haeckel's law.*

recemented glacier *glacier remanié.*

Recent *Holocene.*

recentered projection A term preferred by the British National Committee for Geography (1966, p.33) to the synonym *interrupted projection.*

receptacle The apex of a pedicel (or peduncle) from which the organs of a flower grow out; also, the inflated tip of certain brown algae within which gametangia are borne (Fuller & Tippo, 1954, p. 969).

receptaculitid Any of a group of early and middle Paleozoic fossils (Ordovician to Devonian, possibly Carboniferous) of uncertain systematic position, belonging to the family Receptaculitidae, characterized by ovoid, globose, or discoidal calcareous hard parts. The receptaculitids have been variously classified as calcareous algae, foraminifers, sponges, echinoderms, and an independent phylum.

recess [fold] An area in which the axial traces of folds are concave toward the outer edge of the folded belt. Ant: *salient [fold].*

recess [geomorph] An indentation in a surface; e.g. a cleft in a steep rock bank. See also: *reentrant.*

recession [coast] A continuing landward movement of a shoreline or beach undergoing erosion. Also, a net landward movement of the shoreline or beach during a specified period of time. Ant: *advance.* Cf: *retrogradation.* Syn: *retrogression.*

recession [geomorph] (a) The backward movement or retreat of an eroded escarpment; e.g. the slow wasting-away of a cliff under the influence of weathering and erosion. (b) The moving-back of a slope from a former position without a change in its angle. (c) The gradual upstream retreat of a waterfall or nickpoint.

recession [glaciol] (a) A decrease in length of a glacier, resulting in a backward displacement of the terminus, caused when processes of ablation (usually melting and/or calving) exceed the speed of ice flow; normally measured in meters per year. Syn: *retreat; glacial retreat.* (b) An overall decrease in the volume of a glacier.—Syn: *glacial recession; glacier recession.*

recessional moraine An end or lateral moraine built during a temporary but significant pause in the final retreat of a glacier. Also, a moraine built during a slight or minor readvance of the ice front during a period of general recession. Syn: *peripheral moraine; retreatal moraine; stadial moraine.*

recession curve A hydrograph showing the decrease of the runoff rate after rainfall or the melting of snow. Direct runoff and base runoff are usually given separate curves as they recede at different rates. The use of the term *depletion curve* in reference to the base-runoff recession is considered incorrect (Langbein & Iseri, 1960).

recessive Said of a characteristic of an organism that must be inherited from both parents if it is to be exhibited by offspring. A recessive character can be passed on to offspring without being exhibited by a parent.

recharge The processes involved in the absorption and addition of water to the zone of saturation; also, the amount of water added. This does not include water reaching only the belt of soil water or the intermediate belt. Syn: *intake; replenishment [grd wat]; ground-water replenishment; ground-water recharge; ground-water increment; increment.*

recharge area An area in which water is absorbed that eventually reaches the zone of saturation in one or more aquifers. Cf: *catchment area [grd wat]; discharge area.* Syn: *intake area.*

recharge basin A basin constructed in sandy material to collect water, as from storm drains, for the purpose of replenishing ground-water supply.

recharge well A well used to inject water into one or more aquifers in the process of artificial recharge. Syn: *injection well.*

reciprocal bearing *back bearing.*

reciprocal lattice A lattice array of points formed by drawing perpendiculars to each plane (hkl) in a crystal lattice through a common point as origin. Points are located on each perpendicular at a distance from the origin (000) inversely proportional to spacing of the specific lattice planes (hkl). The axes of the reciprocal lattice are the *a^* axis,* the *b^* axis,* and the *c^* axis,* which are perpendicular, respectively, to (100), (010), and (001) of the crystal lattice.

The coordinates of each reciprocal lattice point are (*hkl*) or whole multiples (n*h*, n*k*, n*l*) in terms of the unit lengths a*, b*, and c*. Cf: *direct lattice.*

reciprocal leveling Trigonometric leveling in which vertical angles have been observed at both ends of the line in order to eliminate instrumental errors; e.g. leveling across a wide river by establishing a turning point on each bank of the river from one side and taking a backsight on each to determine the height of instrument on the other side. The mean of the differences in level represents the true difference.

reciprocal salt-pair diagram *Jänecke diagram.*

reciprocal strain ellipsoid In elastic theory, an ellipsoid of certain shape and orientation which under homogeneous strain is transformed into a sphere. Cf: *strain ellipsoid.*

reciprocity (a) The statement that the same seismic wave form would result if source and receiver were interchanged. (b) The statement in electrical studies that the potential at a point A due to a source at a point B is identical to the potential at point B due to a source at point A. (c) The statement that a transmitting coil at point A will produce an electric or magnetic field at point B equal to the field observed at A when the transmitting coil is at B.

reclined (a) Said of a graptoloid rhabdosome with stipes extending above the sicula and enclosing an angle less than 180 degrees between their dorsal sides. Cf: *reflexed; declined; deflexed.* (b) Said of a tabulate corallite growing and opening obliquely with respect to the surface of corallum (TIP, 1956, pt.F, p.250).

reclined fold A fold whose hingeline plunges parallel to the direction of dip of the axial surface (Turner & Weiss, 1963, p. 119).

recomposed granite (a) An arkose consisting of consolidated feldspathic residue (produced by surface weathering of an underlying granitic rock) that has been so little reworked and so little decomposed that upon cementation the rock looks very much like the granite itself. It has a faint bedding, an unusual range of particle sizes (unlike the even-grained or porphyritic texture of true granite), and a greater percentage of quartz than is normal for granite. Syn: *reconstructed granite.* (b) A conglomerate that has been recrystallized by strong metamorphism into a rock that simulates granite, as in the Lake Superior region.—Cf: *meta-arkose.*

recomposed rock A rock produced in place by the cementation of the fragmental products of surface weathering; e.g. a *recomposed granite.* The term has been applied to a rock of intermediate character straddling an unconformable surface between the breccia of the lower formation and the conglomeratic base of the upper formation (Leith, 1923).

reconnaissance (a) A general, exploratory examination or survey of the main features (or certain specific features) of a region, usually conducted as a preliminary to a more detailed survey; e.g. an engineering survey in preparing for triangulation of a region. It may be performed in the field or office, depending on the extent of information available. (b) A rapid geologic survey made to gain a broad, general knowledge of the geologic features of a region.

reconnaissance map A map based on the information obtained in a reconnaissance survey and on data obtained from other sources.

reconnaissance survey A preliminary survey, usually executed rapidly and at relatively low cost, prior to mapping in detail and with greater precision.

reconnoiter To make a reconnaissance of; esp. to make a preliminary survey of an area for military or geologic purposes.

reconsequent *resequent [streams].*

reconstructed glacier *glacier remanié.*

reconstructed granite *recomposed granite.*

reconstructed stone A gem material made by the fusing or sintering of small particles of the genuine stone; e.g. amberoid and reconstructed turquoise. Cf: *synthetic stone.*

reconstructive transformation A type of crystal *transformation* that involves the breaking of either first- or second-order coordination bonds. It is usually a slow transformation. An example is quartz-tridymite. Cf: *dilatational transformation; displacive transformation; rotational transformation; substitutional transformation.*

record (a) *geologic record.* (b) *stratigraphic record.* (c) *seismic record.*

record section A display of seismic *traces* side-by-side to show the continuity of events along a line of profile.

recovery [grd wat] (a) The rise in static water level in a well that occurs when discharge from that well or a nearby well is stopped. (b) *ground-water withdrawal.*

recovery [mining] In mining, the percentage of valuable constituent derived from an ore, or of coal from a coal seam; a measure of mining or extraction efficiency.

recovery [struc petrol] Any of the processes through which the number of grain dislocations (i.e. strain energy) produced during rock deformation can be reduced. It includes *polygonization,* by which dislocations guide and climb into walls or "subgrain boundaries", and *recrystallization,* by which new strain-free material is formed at the expense of *paleoblasts.*

recovery [surv] A visit to a survey station to identify its mark or monument as authentic and in its original location and to verify or revise its description.

recrystallization The formation, essentially in the solid state, of new crystalline mineral grains in a rock. The new grains are generally larger than the original grains, and may have the same or a different mineralogical composition. Specif., it is the way in which a deformed crystal aggregate releases stored strain energy due to deformation. It consists of several discrete phases: annihilation of crystal defects and their rearrangement into stable arrays, which may lead to the formation of polygonal subgrains; primary crystallization, involving the nucleation of new grains and their growth by the migration of high-angle boundaries, producing an aggregate of strain-free grains; and normal grain growth, in which the average grain size increases, driven by a tendency to decrease the interfacial energies of adjacent grains. Secondary recrystallization may follow, in which some grains grow very large by consuming neighboring grains. See also: *Riecke's principle.*

recrystallization breccia *pseudobreccia.*

recrystallization calcite Patchy mosaics of calcite crystals, interrupting or replacing a finer-grained calcite fabric in sedimentary rocks (Leighton & Pendexter, 1962, p. 60).

recrystallization flow *Flow [exp struc geol]* in which there is molecular rearrangement by solution and redeposition, solid diffusion, or local melting.

rectangular coordinates Two- or three-dimensional coordinates on any system in which the axes of reference intersect at right angles; *plane coordinates.* Also, a coordinate system using rectangular coordinates. Syn: *rectilinear coordinates.*

rectangular cross ripple mark An *oscillation cross ripple mark* consisting of two sets of ripples intersecting at right angles and enclosing a rectangular pit; it is formed by waves that oscillate at right angles to the direction of the original ripple mark.

rectangular drainage pattern A drainage pattern in which the main streams and their tributaries display many right-angle bends and exhibit sections of approximately the same length; it is indicative of streams following prominent fault or joint systems that break the rocks into rectangular blocks. It is more irregular than the *trellis drainage pattern,* as the side streams are not perfectly parallel and not necessarily as conspicuously elongated, and secondary tributaries need not be present (Zernitz, 1932, p. 503). Examples are well developed along the Norwegian coast and in parts of the Adirondack Mountains. See also: *angulate drainage pattern; joint valley.* Syn: *lattice drainage pattern.*

rectangular joint structure *mural joint structure.*

rectangular speciation *punctuated equilibria.*

rectification [coast] The simplification and straightening of the outline of an initially irregular and crenulate shoreline by marine erosion cutting back headlands and offshore islands, and by deposition of waste resulting from erosion or of sediment brought down by neighboring rivers.

rectification [eng] A new alignment to correct a deviation of a stream channel or bank.

rectification [photo] The process of projecting a tilted or oblique aerial photograph onto a horizontal reference plane, the angular relation between photography and plane being determined by ground reconnaissance. Cf: *transformation [photo].*

rectify

rectilinear coordinates *rectangular coordinates.*

rectilinear current *reversing current.*

rectilinear evolution Continued change of the same type and in the same direction within a line of descent over a considerable length of time. It is similar to *orthogenesis,* but without implying how the direction is determined and maintained.

rectilinear shoreline A long, relatively straight shoreline, or one with nearly right-angle bends, such as are caused by intersecting joints.

rectimarginate Said of a brachiopod having a plane (straight) anterior commissure; also, said of such a commissure.

rectiradiate Said of an ammonoid rib in straight radial position, bending neither forward nor backward. Cf: *prorsiradiate; rursiradiate.*

rectorite A white clay-mineral mixture with a regular interstratification of two mica layers (pyrophyllite and vermiculite) and one or more water layers. Syn: *allevardite.*

recumbent fold An overturned fold, the axial surface of which is horizontal or nearly so (Turner & Weiss, 1963, p. 119).

recurrence horizon In peat bogs, the demarcation between older, more decomposed peat and younger material; a parting or horizon marking an abrupt change in lithology of a peat bog, reflecting climatic change. Syn: *grenz.*

recurrence interval (a) The average time interval between occurrences of a hydrological event of a given or greater magnitude. (b) In an annual flood series, the average interval in which a flood of a given size recurs as an annual maximum. (c) In a partial duration series, the average interval between floods of a given size, regardless of their relationship to the year or any other period of time. This distinction holds even though for large floods recurrence intervals are nearly the same on both scales.—(ASCE, 1962).

recurrent Said of an organism or group of organisms that reappears in an area from which it had been previously expelled; e.g., a fossil present in two different rock units separated by a unit or units in which it is absent.

recurrent folding A type of folding that results from periodic deformation or subsidence and is characterized by thinning or disappearance of formations at the crest of uplifts. Cf: *supratenuous fold.* Syn: *revived folding.*

recurring polynya A *polynya* that is found in the same region every year.

recurve A feature produced by the successive landward extension of a spit.

recurved spit A spit whose outer end is turned landward by current deflection, by the opposing action of two or more currents, or by wave refraction. Syn: *hook; hooked spit.*

recycled grain A grain derived from a pre-existing sedimentary rock and incorporated into a new sediment or sedimentary rock; as contrasted to a grain derived directly from an igneous or metamorphic rock.

recycling The processes by which particles are weathered from an existing sedimentary rock, transported, deposited, and incorporated into a new sediment or sedimentary rock.

red algae A group of algae corresponding to the phylum Rhodophyta, which owes its reddish color to the presence of the pigment phycoerythrin. Its members may be filamentous, membranous, branched, or encrusting. Red algae have a worldwide distribution. Cf: *blue-green algae; brown algae; green algae; yellow-green algae.*

red antimony *kermesite.*

red arsenic *realgar.*

red beds Sedimentary strata composed largely of sandstone, siltstone, and shale, with locally thin units of conglomerate, limestone, or marl, that are predominantly red in color due to the presence of ferric oxide (hematite) usually coating individual grains; e.g. the Permian and Triassic sedimentary rocks of western U.S., and the Old Red Sandstone facies of the European Devonian. At least 60% of any given succession must be red before the term is appropriate, the interbedded strata being of any color (Hatch & Rastall, 1965, p. 371). Also spelled: *redbeds.* Syn: *red rock.*

red clay A pelagic deposit that is fine-grained and bright to reddish brown or chocolate-colored, formed by the slow accumulation of material a long distance from the continents and at depths generally greater than 3500 meters. It contains relatively large proportions of windblown particles, meteoric and volcanic dust, pumice, shark teeth, whale earbones, manganese nodules, and debris rafted by ice. The content of $CaCO_3$ ranges from 0 to 30%. Syn: *brown clay.*

red cobalt *erythrite.*

red copper ore *cuprite.*

Red Desert soil A great soil group in the 1938 classification system, a group of zonal soils having a light, friable, reddish brown surface over a heavy reddish brown or red horizon, underneath which is an accumulation of lime. It is developed in deserts of tropical to warm-temperate climate. Most of these soils are now classified as *Ustolls.* Cf: *Reddish Brown soil.*

reddingite A pink-white or yellow-white orthorhombic mineral: $(Mn,Fe)_3(PO_4)_2 \cdot 3H_2O$. It is isomorphous with phosphoferrite.

Reddish-Brown Lateritic soil A great soil group in the 1938 classification system, a group of zonal soils developed from a mottled red lateritic parent material and characterized by a reddish-brown surface horizon and an underlying red-clay B horizon. These soils are now classified as *Humults* and *Udults.*

Reddish Brown soil A great soil group in the 1938 classification system, a group of zonal soils having a reddish, light brown surface horizon overlying a heavier, more reddish horizon and a light-colored horizon with calcium-carbonate accumulation. These soils are developed in warm, temperate to tropical, semiarid climate under shrub and short-grass vegetation (USDA, 1938). Most are now classified as *Ustalfs, Orthids,* and *Argids.* Cf: *Red Desert soil.*

Reddish Chestnut soil A great soil group in the 1938 classification system, a group of zonal soils having a thick surface horizon that ranges from dark brown to reddish or pinkish, below which is a heavier, reddish-brown horizon and a carbonate accumulation. They are developed under mixed grasses with some shrubs, in a warm to temperate, semiarid climate (USDA, 1938). These soils are now classified as *Ustalfs* and *Ustolls.* Cf: *Chestnut soil.*

Reddish Prairie soil A great soil group in the 1938 classification system, a group of zonal soils having a surface horizon that is acidic and dark reddish brown, and that grades through a heavier reddish soil to the parent material. It is developed under tall grass in a warm to temperate, humid to subhumid climate (USDA, 1938). Most of these soils are now classified as *Ustolls.* Cf: *Prairie soil.*

reddle *red ocher.*

red earth A general term for the soil that is characteristic of a tropical climate; it is leached, red, deep, and clayey. Syn: *red loam.*

redeposition Formation in a new accumulation, such as the deposition of sedimentary material that has been picked up and moved (reworked) from the place of its original deposition, or the solution and reprecipitation of mineral matter. See also: *resedimentation.*

red hematite A syn. of *hematite.* Cf: *brown hematite; black hematite.*

redifferentiation Processes, such as partial melting, by which the initial continental crust of the Earth became vertically zoned, with granitic rocks in the upper layers and granulitic rocks in the lower layers (Lowman, 1976, p. 1).

redingtonite A pale-purple mineral: $(Fe,Mg,Ni)(Cr,Al)_2(SO_4)_4 \cdot 22H_2O$.

red iron ore *hematite.*

red lake A lake containing reddish water. The color may be due to iron-secreting bacteria, reddish plankton, ferrous-iron compounds in solution, or red clay held in suspension.

red lead *minium.*

red lead ore *crocoite.*

redledgeite A mineral: $Mg_4Cr_6Ti_{23}Si_2O_{61}(OH)_4$ (?). Syn: *chromrutile.*

red loam *red earth.*

red manganese A reddish manganese mineral; specif. rhodonite and rhodochrosite. Syn: *red manganese ore.*

Red Mediterranean soil An obsolete term for a reddish soil that forms in a Mediterranean climate. Cf: *Brown Mediterranean soil.*

red mud A type of *mud [marine geol]* that is terrigenous and contains as much as 25% calcium carbonate. Its color is due to the presence of ferric oxide.

red ocher A red clayey or earthy hematite used as a pigment. Syn: *reddle; ruddle.*

Redonian European stage: Upper Pliocene (above Plaisancian, below Calabrian). Equivalent to Astian.

red ore A red-colored ore mineral; specif. hematite or metahewettite.

red orpiment *realgar.*

red oxide of copper *cuprite.*

red oxide of zinc *zincite.*

redoxomorphic stage A term introduced by Dapples (1962) for the earliest geochemical stage of diagenesis, characterized by mineral changes primarily due to oxidation and reduction reactions (reversible reactions). It is typical of unlithified sediment and preceeds the *locomorphic stage.* See also: *phyllomorphic stage.*

Red Podzolic soil Formerly, a great soil group in the 1938 classification system; it was reclassified as *Red-Yellow Podzolic soil* in the 1949 revision (Thorp and Smith, 1949).

red rock (a) *red beds.* (b) A driller's term for any reddish sedimentary rock.

redruthite *chalcocite.*

red schorl (a) *rubellite.* (b) *rutile.*

red silver ore A red silver-sulfide mineral; specif. "dark red silver ore" (pyrargyrite) and "light red silver ore" (proustite). Syn: *red silver.*

red snow A general name for snow colored by the presence of various red or pink microscopic algae (such as *Sphaerella* and *Protococcus*) in the upper layers of snow in arctic and alpine regions. Cf: *green snow; yellow snow.* Syn: *pink snow.*

redstone A reddish sedimentary rock; specif. a deep-red, clayey sandstone or siltstone representing a flood-plain micaceous arkose, as in the Triassic deposits of Connecticut.

red tide A type of *water bloom* that is caused by dinoflagellates.

reduced latitude The angle at the center of a sphere tangent to a reference ellipsoid along the equator, between the plane of the equator and a radius to the point intersected on the sphere by a straight line perpendicular to the plane of the equator. Syn: *parametric latitude.*

reduced mud *black mud.*

reducing flame In blowpiping, the blue part of the flame, in which oxygen in the compound being tested is partly burned away. Cf: *oxidizing flame.*

reduction [geomorph] The lowering of a land surface by erosion.

reduction [geophys] *gravity reduction.*

reduction body A multicellular mass resulting from the disorganization of a sponge and capable of reorganizing into a sponge with a functional aquiferous system.

reduction index The rate of wear of a sedimentary particle subject to abrasion in the course of transportation, expressed as the difference between the mean weight of the particle before and after transport divided by the product of mean weight before transport and the distance traveled (Wentworth, 1931, p. 25). Abbrev: RI. Cf: *durability index.*

reduction sphere A white, leached spheroidal mass produced in a reddish or brownish sandstone by a localized reducing environment, commonly surrounding an organic nucleus or a pebble and ranging in size from a poorly defined speck to a large perfect sphere more than 25 cm in diameter (Hamblin, 1958, p.24-25); e.g. in the Jacobsville Sandstone of northern Michigan.

reduction to sea level The application of a correction to a measured distance or other quantity on the Earth's surface to reduce it to its corresponding value at sea level.

reduction to the pole A method of removing dependance on the angle of magnetic inclination. It converts data that have been recorded in the Earth's inclined magnetic field to what the data would have looked like if the magnetic field had been vertical.

reduzates Sediments accumulated under reducing conditions and thus characteristically rich in organic carbon and in iron sulphide; coal and black shale are principal examples. It is one of Goldschmidt's groupings of sediments or analogues of differentiation stages in rock analysis. Cf: *resistates; evaporates; hydrolyzates; oxidates.*

red vitriol *bieberite.*

Red-Yellow Podzolic soil A great soil group introduced in 1949 to include acidic, zonal soils having a leached, light-colored surface layer and a subsoil containing clay and oxides of aluminum and iron, ranging in color from red through yellowish red to bright yellowish brown. Its parent material is clayey but siliceous, and of variegated color. It is developed under forest vegetation in a warm, temperate, or tropical and humid climate (Thorp and Smith, 1949). Most of these soils are now classified as *Udults.* See also: *Yellow Podzolic soil; Red Podzolic soil.*

red zinc ore *zincite.*

reed cast A vertical and cylindrical cast of sand presumably representing the filling of a mold left by a reed.

reedmergnerite A colorless triclinic mineral of the feldspar group: $NaBSi_3O_8$. It is the boron analogue of albite.

reed peat *telmatic peat.*

reef (a) A ridgelike or moundlike structure, layered or massive, built by sedentary calcareous organisms, esp. corals, and consisting mostly of their remains; it is wave-resistant and stands above the surrounding contemporaneously deposited sediment. Also, such a structure built in the geologic past and now enclosed in rock, commonly of differing lithology. See also: *bank; bioherm; biostrome.* Syn: *organic reef.* (b) A mass or ridge of rocks, esp. coral and sometimes sand, gravel, or shells, rising above the surrounding sea or lake bottom to or nearly to the surface, and dangerous to navigation; specif. such a feature at 10 fathoms (formerly 6) or less. See also: *shoal.* (c) A provincial term for a metalliferous mineral deposit, esp. gold-bearing quartz (e.g. *saddle reef*).

reefal Pertaining to a reef and its integral parts. Not recommended usage.

reef apron The gently sloping surface of sediment accumulation behind or surrounding a reef (esp. a *reef flat*). Cf: *reef flank.*

reef breccia A rock formed by the consolidation of limestone fragments broken off from a reef by the action of waves and tides. Cf: *reef-rock breccia.*

reef buttress A long, narrow sloping ridge or vertical promontory, standing out or projecting seaward beyond the steep *fore-reef* slope or vertical *reef wall,* and flanked by shallow valleys or reentrants. Cf: *spur-and-groove structure.*

reef cap A deposit of fossil-reef material overlying or covering an island or mountain. Cf: *coral cap.*

reef cluster A group of reefs of wholly or partly contemporaneous growth, found within a circumscribed area or geologic province. See also: *hermatopelago.*

reef complex A solid reef and the heterogeneous and contiguous fragmentary material derived from it by abrasion; the aggregate of reef, fore-reef, back-reef, and interreef deposits, bounded on the seaward side by basin sediments and on the landward side by lagoonal sediments (Nelson et al., 1962, p.249). Term introduced by Henson (1950, p.215-216) to include the reef and "all genetically(?) associated sediments". Cf: *reef tract.*

reef conglomerate *reef talus.*

reef core Within an *organic reef,* the centrally located solid rock mass constructed in place by reef-building organisms; the solid reef proper. See also: *reef flank.*

reef crest The sharp break in slope at the seaward margin or edge of the *reef flat,* located at the top of the *reef front;* marked by dominance of a particular coral species (such as *Acropora palmata* throughout the Caribbean) or by an algal ridge and/or surge channels. Cf: *reef edge; reef front.*

reef debris *reef detritus.*

reef detritus Fragmental material derived from the erosion of an organic reef; some is produced by mechanical breakage by waves or surf, some by accumulation of shells, and some by boring animals detaching solid materials. The finer particles tend to be carried away by waves, while the coarser fragments often form a *talus apron* around the reef. See also: *reef talus.* Syn: *reef debris.*

reef edge The seaward margin of the reef flat, commonly marked by surge channels. Cf: *reef crest; reef front.*

reef flank The part of a reef that surrounds, interfingers with, and locally overlies the *reef core,* often indicated by beds of *reef detritus* dipping away from the core. It is the relatively narrow zone where the biologic forces of reef expansion contend with the physical and biologic forces of reef destruction. Cf: *reef apron.*

reef flat A stony platform of dead reef-rock, commonly strewn with coral fragments and coral sand, generally dry at low tide and formed as the summit of the reef above low tide. It may include shallow pools, irregular gullies, low islands of sand or rubble (often vegetated, esp. by palms), and scattered colonies of the more hardy species of coral.

reef frame *growth lattice.*

reef front The upper part of the outer or seaward slope of a reef, extending to the *reef edge* from above the depth limit of abundant living coral and coralline algae. Cf: *reef crest.*

reef-front terrace A shelflike or benchlike eroded surface, sometimes veneered with organic growth, often sloping gently seaward; may be developed at various depths, as at 6, 8, 20, and 35 m on reefs off northern Eleuthera. Syn: *reef terrace.*

reef knoll (a) A bioherm or fossil coral reef now represented by a small, prominent hill, up to 100 m high; specif. a small, pinnacle-like or conical mass of coralline limestone, more or less circular in ground plan and commonly surrounded by rock of different lithology, as in the type area of the Craven district in Yorkshire, England. (b) A present-day reef in the form of a knoll; a small *reef patch* developed locally and upward rather than outward or laterally. The term was first used by Tiddeman (1890, p.600) for a reef feature that originated as a knoll. Syn: *knoll reef.*

reef limestone A limestone consisting of the remains of active reef-building organisms, such as corals, sponges, and bryozoans, and of sediment-binding organic constituents, such as calcareous algae. See also: *coral-reef limestone.*

reef milk A very fine-grained matrix material of the back-reef facies, consisting of white, opaque microcrystalline calcite and/or aragonite derived from abrasion of the reef core and reef flank.

reefoid Resembling a reef; e.g. "reefoid rocks".

reef patch A growth of coral formed independently on a shelf of less than 70 m depth, often in the lagoon of a barrier reef or atoll, ranging from an expanse several kilometers across down to that of a single large colony. See also: *reef knoll; shoal reef.* Cf: *patch reef.*

reef pinnacle *pinnacle [reef].*

reef ring *atoll.*

reef rock A resistant massive unstratified rock consisting of the calcareous remains of reef-building organisms, often intermingled with carbonate sand and shingle, the whole cemented by calcium carbonate. Also spelled: *reefrock.* Cf: *biolithite; boundstone.* Syn: *hermatolith; hermatobiolith.*

reef-rock breccia A *coral rag* "in which masses of coral retain the attitude and position of growth, and to which the varied animal and vegetal life of the reef contributes" (Norton, 1917, p. 179). Cf: *reef breccia.*

reef segment A part of an organic reef lying between passes, gaps, or channels.

reef slope The face or flank of a reef, rising from the sea floor (Maxwell, 1968, p.106-107).

reef talus Massive or thick-bedded inclined strata consisting of coarse *reef detritus,* usually deposited along the seaward margin of an organic reef; it is one type of *reef flank* deposit. Syn: *reef conglomerate.*

reef terrace *reef-front terrace.*

reef tract An extensive offshore area in which many reefs are found, such as the Florida reef tract. Cf: *reef complex.*

reef tufa Drusy, prismatic, fibrous calcite deposited directly from supersaturated water upon the void-filling internal sediment of the calcite mudstone of a reef knoll (Bissell & Chilingar, 1967, p.165). See also: *stromatactis.*

reef wall (a) A wall-like upgrowth of living coral and the skeletal remains of dead coral and other reef-building organisms, reaching intertidal level where it acts as a partial barrier between adjacent environments (Henson, 1950, p.227); an elongate *reef core.* See also: *wall reef.* (b) The vertical submarine cliff, extending below most reef or coral growth, well down into fore-reef depths, as from 50 to 130 m off Jamaica and Belize. Syn: *deep fore-reef.*

reefy (a) Containing reefs, as a "reefy harbor". (b) Containing sedimentary material that resembles the material of an organic reef.

reentrant adj. Reentering or directed inward; e.g. a *reentrant* angle in a coastline or on a twinned crystal.—n. A prominent, generally angular indentation in a landform; e.g. an inlet between two promontories along a coastline, or a transverse valley extending into an escarpment. Also spelled: *re-entrant.* Ant: *salient [geomorph].* See also: *recess [geomorph].*

reentrant angle The angle between two plane surfaces on a solid, in which the external angle is less than 180°.

reevesite A mineral: $Ni_6Fe_2(OH)_{16}(CO_3)\cdot 4H_2O$.

reference axis *fabric axis.*

reference ellipsoid A theoretical figure whose dimensions closely approach the dimensions of the geoid, and whose exact dimensions are determined by various considerations of the section of the Earth's surface concerned. Cf: *ellipsoid.*

reference level A *datum plane*; e.g. a standard level (in the study of underwater sound) to which sound levels can be related.

reference line Any line that serves as a reference or base for the measurement of other quantities; e.g. a *datum line.*

reference locality A locality containing a reference section, established to supplement the *type locality.*

reference plane *datum plane.*

reference section A rock section, or group of sections, designated to supplement the *type section,* or sometimes to supplant it (as where the type section is no longer exposed), and to afford a standard for correlation for a certain part of the geologic column; e.g. an auxiliary section of particular regional or facies significance, established through correlation with the type section, and from which lateral extension of the boundary horizons may be made more readily than from the type section. See also: *standard section; hypostratotype.*

reference seismometer In seismic prospecting, a detector placed on the Earth's surface to record successive shots under similar conditions, to permit overall time comparisons. It is used in connection with the shooting of wells for velocity measurements.

reference source An electrically heated cavity in thermal infrared scanners and radiometers that is maintained at a known radiant temperature, used for calibrating the radiant temperature detected from a target.

reference station A place where tidal constants previously have been determined and which is used as a standard for the comparison of simultaneous observations at a second station. Also, a place where independent daily predictions are given in the tide and tidal-current tables, from which corresponding predictions are obtained for other stations by means of differences or factors (CERC, 1966, p. A26). Cf: *tide station.* British syn: *standard port.*

referencing The process of measuring the horizontal (or slope) distances and directions from a survey station to nearby landmarks, reference marks, and other permanent objects which can be used in the recovery or relocation of the station.

refikite A white, very soft mineral occurring in modern resins: $C_{20}H_{32}O_2$ (?). Also spelled: *reficite.*

reflectance The ratio of the energy reflected by a body to that incident upon it. See also: *spectral reflectance.*

reflected infrared Wavelengths from 0.7 to about 3 μm that are primarily reflected solar radiation. Syn: *near infrared; solar infrared.*

reflected wave An elastic wave that has been reflected at an interface between media with different elastic properties. It is indicated by such symbols as SS, SP, and PSS. Cf: *converted wave.*

reflecting goniometer *reflection goniometer.*

reflecting horizon In seismic profiling of the ocean floor, a major layer of *reflection*. It may be either sedimentary (chert) or igneous (basalt). Three layers are distinguished: *horizon A, horizon beta,* and *horizon B.*

reflection The return of a wave incident upon a surface to its original medium. See also: *law of reflection; total reflection.* Also, in seismic prospecting, the indication on a record of such reflected energy. Cf: *refraction; diffraction.*

reflection angle *Bragg angle.*

reflection coefficient The ratio of the amplitude of the reflected wave to that of the incident wave. The ratio of the reflected energy to the incident energy is the reflection coefficient squared. Syn: *reflectivity.*

reflection configuration *seismic reflection configuration.*

reflection goniometer A *goniometer* that measures the angles between crystal faces by reflection of a parallel beam of light from the successive crystal faces. Cf: *contact goniometer; two-circle goniometer.*

reflection plane *plane of mirror symmetry.*

reflection pleochroism *bireflectance.*

reflection profile A seismic recording from a number of individual seismometer groups arranged in a line and at relatively short distances from the shotpoint, in which the data displayed result from reflected seismic waves. Cf: *refraction profile.*

reflection shooting A type of seismic survey based on measurement of the travel times of waves that originate from an artificially produced disturbance and are reflected back at near-vertical incidence from subsurface boundaries separating media of different elastic-wave velocities. Cf: *refraction shooting.*

reflection spectrum The spectrum seen when incident waves are selectively altered by a reflecting substance. It is analyzed with a *spectroreflectometer.*

reflection twin A crystal twin whose symmetry is formed by apparent mirror image across a plane. Cf: *rotation twin.*

reflectometer An apparatus for measuring the reflectivity of a substance, using some form of radiant energy such as light.

reflector An interface between media of different elastic properties that reflects seismic waves.

reflexed Said of a graptoloid rhabdosome with stipes extending above the sicula, their initial parts enclosing an angle less than 180 degrees between their dorsal sides, and their distal parts tending to the horizontal. Cf: *reclined; deflexed; declined.*

reflux A return flow, especially the return flow of concentrated brine through the floor or across the barrier sill of an evaporite basin. Because such brines may be enriched in magnesium compared to seawater, reflux is believed to contribute to the dolomitization of carbonate rocks in some basinal sequences (Adams & Rhodes, 1960).

refoliation A foliation that is subsequent to and oriented differently from an earlier foliation.

refracted cleavage Cleavage that changes orientation from layer to layer where layers are of different rock types.

refracted wave *head wave.*

refraction The deflection of a ray of light or of an energy wave (such as a seismic wave) due to its passage from one medium to another of differing density, which changes its velocity. Cf: *reflection; diffraction.* See also: *single refraction; birefringence.*

refraction angle *angle of refraction.*

refraction profile A seismic profile obtained by designing the spread geometry in such a manner as to enhance the recording of energy that has traveled horizontally through a medium with large surface velocity. Cf: *reflection profile.*

refraction shooting A type of seismic survey based on the measurement of the travel times of seismic waves that have moved nearly parallel to the bedding in high-velocity layers, in order to map such layers. Cf: *reflection shooting.*

refractive index *index of refraction.*

refractive power *refractivity.*

refractivity The power of a substance to refract light. Such ability can be quantitatively expressed by the *index of refraction.* See also: *specific refractivity.* Syn: *refractive power; refringence.*

refractometer An apparatus for measuring the indices of refraction of a substance, either solid or liquid. Various types are designed for various substances; the chief type used for analysis of gems and minerals is the *Abbe refractometer.*

refractometry The measurement of indices of refraction, by means of a *refractometer.*

refractory adj. Said of an ore from which it is difficult or expensive to recover its valuable constituents.

refractory clay *fireclay.*

refringence *refractivity.*

Refugian North American provincial stage: Eocene and Oligocene (above Fresnian-Narizian, below Zemorrian).

refugium (a) An isolated area that underwent little environmental change, permitting a fauna or flora to persist locally long after it had been exterminated elsewhere. Syn: *asylum.* (b) A restricted area in which plants and animals persisted during a period of continental climatic change that made surrounding areas uninhabitable; esp. an ice-free or unglaciated area within or close to a continental ice sheet or upland ice cap, where hardy biotas eked out an existence during a glacial phase. It later served as a center of dispersal for the repopulation of surrounding areas after climatic readjustment. Pl: *refugia.* Syn: *glacial refuge.*

reg An extensive desert plain from which fine sand has been removed by the wind, leaving a sheet of coarse, smoothly angular, wind-polished gravel and small stones lying on an alluvial soil and strongly cemented by mineralized solutions to form a broad *desert pavement;* a *stony desert* of the plains, as in the Algerian Sahara and parts of American deserts. Etymol: Hamitic. Pl: *regs.* See also: *serir; hammada.* Syn: *gravel desert.*

regelation A two-fold process involving the melting of ice under excess pressure (*pressure melting*) and the refreezing of the derived meltwater upon release of that pressure. For the process to operate, heat must be able to flow from the region where water freezes to te region where ice melts. The term is sometimes restricted to the refreezing part of the process, but in some European literature it has been applied to often-repeated freezing and thawing (or *multigelation*).

regelation layer Ice at the bottom of a glacier or an ice sheet that has refrozen during the *regelation* process (Paterson, 1969, p. 129).

regenerated anhydrite Anhydrite produced by the dehydration of gypsum that was itself formed by the hydration of anhydrite (Goldman, 1961).

regenerated crystal A large crystal that has grown in a mass of crushed material, such as mylonite.

regenerated flow control Control of glacial drainage by modified morainal features, resulting from the readvance of a previously stagnant glacier (Gravenor & Kupsch, 1959, p. 56).

regenerated glacier (a) *glacier remanié.* (b) A glacier that becomes active after a period of stagnation.

regenerated rock A clastic rock. The term "regenerirte Gesteine" was used by Zirkel (1866, p.3).

regeneration The renewal, regrowth, or restoration of a body or of a part, tissue, or substance of a body, following injury or as a normal bodily process.

regime (a) A regular or systematic pattern of occurrence or action, or a condition or style having widespread influence, as a sedimentation regime. (b) The existence in a stream channel of a balance or grade between erosion and deposition over a period of years. (c) The condition of a stream with respect to the rate of its average flow as measured by the volume of water passing different cross sections in a specified period of time. In this sense the term is used incorrectly as a syn. of *regimen.* (d) In glaciology, a syn. of *balance.*

regimen [glaciol] *balance.*

regimen [lake] An analysis of the total quantity of water involved with a lake over a specified period of time (usually a year), including water losses (seepage, evaporation, transpiration, outflow, diversion) and gains (precipitation, inflow, ground-water migration, water pumped or drained into the lake basin) (Veatch & Humphrys, 1966, p.172). Syn: *hydrologic regimen.*

regimen [streams] The flow characteristics of a stream; specif. the habits of an individual stream (including low flows and floods) with respect to such quantities as velocity, volume, form of and changes in the channel, capacity to transport sediment, and amount of material supplied for transportation. Cf: *regime.*

regimen [water] The characteristic behavior and the total quantity of water involved in a drainage basin, determined by measuring such quantities as rainfall, surface and subsurface storage and flow, and evapotranspiration. Syn: *hydrologic regimen; water regimen.*

regime stream A stream with a mobile (erodible) boundary, making at least part of its boundary from its transported load and part of its transported load from its boundary, carrying out the process at different places and times in a balanced or alternating manner that prevents unlimited growth or removal of the boundary (Blench, 1957, p. 2). Cf: *graded stream.*

regime theory A theory of the formation of a channel in material carried and deposited by its stream.

region [ecol] A major division of the Earth having distinctive climatic and topographic features and floral and faunal provinces. Cf: *realm.*

region [geog] A very large expanse of land usually characterized or set apart by some aspect such as its being a political division or area of similar geography.

regional correlation Correlation of rock units, major structures, or other geologic features over or across wide areas of the Earth's surface.

regional dip The nearly uniform inclination of strata over a wide area, generally at a low angle, as in the Atlantic and Gulf coastal plains and parts of the Midcontinent region. Cf: *homocline.* Syn: *normal dip.*

regional geology The geology of any relatively large region, treated broadly and primarily from the viewpoint of the spatial distribution and position of stratigraphic units, structural features, and surface forms. Cf: *areal geology.*

regional-gravity map A gravity map showing only gradual changes of gravity.

regional metamorphism A general term for metamorphism affecting an extensive region, as opposed to *local metamorphism* that is effective only in a relatively restricted area. As introduced in the nineteenth century, the term covered only those changes due to deep burial metamorphism; today it is used almost synonymously with *dynamothermal metamorphism* (Holmes, 1920). Cf: *dynamic metamorphism.*

regional metasomatism Metasomatic processes affecting extensive areas. Cf: *contact metasomatism.*

regional overstep A term proposed by Swain (1949, p.634) for an *overstep* in which an unconformity occurs "widespread, but not universally, over very large parts of a craton (platform, shelf)".

regional snowline The level above which, averaged over a large area, snow accumulation exceeds ablation year after year. See also: *climatic snowline.* Cf: *snowline; orographic snowline.*

regional unconformity An unconformity that extends continuously throughout an extensive region. It may be nearly continent-wide and usually represents a relatively long period. Cf: *local unconformity.*

register mark A small figure, cross, circle, or other pattern at each corner of a map that is to be printed in more than one color. The accuracy of printing of each color is checked by synchronization of the register marks on each printing plate.

register punch The equipment used in *punch register* systems for making multiple copies of cartographic material. There are many designs, from a simple hand-operated punch to elaborate automated systems with material held down by vacuum during simultaneous punching of numerous holes. The punched holes may be round or slotted, and many sizes and spacing arrangements are possible. Cf: *punch register.*

registration The process of superposing two or more images or photographs so that equivalent points coincide.

regmagenesis The production of global strike-slip displacements (Sonder, 1956). Also spelled: *rhegmagenesis.* Adj: *regmatic.* Obsolete.

regmaglypt Any of various small, well-defined, characteristic indentations or pits on the surface of meteorites, frequently resembling the imprints of fingertips in soft clay. They are polygonal, round, almond-shaped, or elliptic; their diameters range from a few millimeters to many centimeters. Syn: *piezoglypt; pezograph.*

regmatic Adj. of *regmagenesis;* said of global strike-slip displacements related to a presumed simple stress pattern.

regolith A general term for the layer or mantle of fragmental and unconsolidated rock material, whether residual or transported and of highly varied character, that nearly everywhere forms the surface of the land and overlies or covers the bedrock. It includes rock debris of all kinds, volcanic ash, glacial drift, alluvium, loess and eolian deposits, vegetal accumulations, and soil. The term was originated by Merrill (1897, p. 299). Etymol: Greek *rhegos,* "blanket", + *lithos,* "stone". See also: *lunar regolith.* Syn: *mantle [geol]; soil [eng geol]; mantle rock; rock mantle; overburden.* Also spelled: *rhegolith.*

Regosol A great soil group introduced in 1949, an azonal group of soils that develop from deep, unconsolidated deposits and that have no definite genetic horizons (Thorp and Smith, 1949). Most of these soils are now classified as *Psamments, Orthents,* and *Andepts.*

regradation The formation by a stream of a new profile of equilibrium, as when the former profile, after *gradation,* became deformed by crustal movements, climatic change, or piracy.

regrading stream A stream that is simultaneously upbuilding (aggrading) and downcutting (degrading) along different parts of its profile.

regression [evol] (a) A hypothetical reversal in the direction of evolution that is sometimes used to explain certain paleontologic phenomena such as the extinction of the graptolites. (b) The trend exhibited by offspring, in respect to their inherited characteristics, away from advanced or specialized characters exhibited by their parents and toward a simpler state.

regression [stratig] The retreat or contraction of the sea from land areas, and the consequent evidence of such withdrawal (such as enlargement of the area of deltaic deposition). Also, any change (such as fall of sea level or uplift of land) that brings nearshore, typically shallow-water environments to areas formerly occupied by offshore, typically deep-water conditions, or that shifts the boundary between marine and nonmarine deposition (or between deposition and erosion) toward the center of a marine basin. Ant: *transgression [stratig].* Cf: *offlap.*

regression [streams] The name given to the theory that some rivers have their sources on the rainier sides of mountain ranges and gradually erode their heads backward until the ranges are cut through.

regression analysis A statistical technique applied to paired data to determine the degree or intensity of mutual association of a dependent variable with one or more independent variables.

regression coefficient A coefficient in a *regression equation;* the slope of the regression line.

regression conglomerate A coarse sedimentary deposit formed during a retreat of the sea.

regression curve The curve that is formed by the means of the distributions of the independent variable for different values of the dependent variable (Freund, 1960, p. 321).

regression equation An experimentally determinable equation of a *regression curve;* e.g. an approximate, generally linear relation connecting two or more quantities and derived from the *correlation coefficient.*

regression line A *regression curve* that is a straight line; the line or curve from a family of curves that best fits the empirical relation between a dependent variable and an independent variable.

regressive diagenesis *hypergenesis.*

regressive overlap *offlap.*

regressive reef One of a series of nearshore reefs or bioherms superimposed on basinal deposits during the rising of a landmass or the lowering of the sea level, and developed more or less parallel to the shore (Link, 1950). Cf: *transgressive reef.*

regressive ripple A term used by Jopling (1961) for an asymmetric ripple mark formed by a locally reversed current (e.g. in the lee of a sand wave or small delta) and therefore oriented in a direction opposite to the general movement of current flow.

regressive sand wave A term proposed by Bucher (1919, p.165) to replace *antidune* as used by Gilbert (1914, p.31). Most authors prefer the original term. Ant: *progressive sand wave.*

regressive sediments Sediments deposited during the retreat or withdrawal of water from a land area or during the emergence of the land, and characterized by an *offlap* arrangement.

regular In paleontology, pertaining to an echinoid of the Regularia division having a more or less globular symmetrical shell with 20 meridional rows of plates and displaying an *endocyclic* test in which the periproct is located within the oculogenital ring. Cf: *irregular.*

regular dissepimentarium A *dissepimentarium* in rugose corals in which the dissepiments are developed only in spaces between major septa and minor septa.

regular dodecahedron *pyritohedron.*

regulation Artificial manipulation of stream flow.

Regur A dark calcareous soil, high in montmorillonitic clay, formed from rocks low in silica. Regur soils occur on the Deccan plateau of India. They are now classified as *Vertisols.* Etymol: Hindi. Syn: *black cotton soil.*

reibungsbreccia A syn. of *fold breccia.* Etymol: German *Reibung,* "rubbing, friction."

Reichenbach's lamellae Thin platy inclusions of foreign minerals (usually troilite, schreibersite, or chromite) occurring in iron meteorites. Named after Karl von Reichenbach (1788-1869), German chemist.

Reid mechanism *elastic-rebound theory.*

reinerite A pale yellow-green mineral: $Zn_3(AsO_3)_2$. Not to be confused with renierite.

rejected recharge Water that infiltrates to the water table but then discharges because the aquifer is full and cannot accept it.

rejuvenated Said of a structural feature, e.g. a fault scarp, along which the original movement has been renewed. Syn: *revived.*

rejuvenated fault scarp A fault scarp freshened by renewed movement along an old fault line after the initial scarp had been partly dissected or eroded. Syn: *revived fault scarp.*

rejuvenated stream A stream that, after having developed to maturity or old age, has had its erosive ability renewed as a result of rejuvenation. It may be characterized by entrenched meanders, stream terraces, and meander cusps. Syn: *revived stream.*

rejuvenated water Water returned to the terrestrial water supply as a result of compaction and metamorphism. See also: *water of compaction.*

rejuvenation (a) The action of stimulating a stream to renewed erosive activity, as by uplift or by a drop of sea level; the renewal or restoration of youthful vigor in a stream that has attained maturity or old age. The causes of rejuvenation may be dynamic, eustatic, or static. (b) The development or restoration of youthful features of a landscape or landform in an area previously worn down nearly to base level, usually caused by regional uplift or eustatic movements, followed by renewed downcutting by streams; a change in conditions of erosion, leading to the initiation of a new cycle of erosion. (c) The renewal of any geologic process, such as the reactivation of a fissure.—Syn: *revival.*

rejuvenation head A *knickpoint* resulting from rejuvenation or from an uplift.

relative abundance The number of individuals of a taxon in comparison with the number of individuals of other taxa in a certain area or volume. See also: *abundance; absolute abundance.*

relative age The *geologic age* of a fossil organism, rock, geologic feature, or event, defined relative to other organisms, rocks, features, or events rather than in terms of years. Cf: *absolute age.*

relative aperture The diameter of the stop, diaphragm, or other physical element that limits the size of the bundle of rays traversing an optical instrument from a given point. It is expressed as a fraction of the focal length of the camera lens, with the symbol f being used instead of 1 as the numerator; e.g. a lens whose relative aperture is 1/4.5 of its focal length has a relative aperture of f/4.5 or f:4.5. See also: *f-number; speed.*

relative chronology *Geochronology* in which the time-order is based on superposition and/or fossil content rather than on an age expressed in years. Cf: *absolute chronology.*

relative consistency *consistency index.*

relative dating The proper chronological placement of a feature, object, or event in the *geologic time scale* without reference to its absolute age.

relative density The ratio of the difference between the void ratio of a cohesionless soil in the loosest state and any given void ratio to the difference between its void ratios in the loosest and in the densest states (ASCE, 1958, term 296).

relative dispersion *dispersive power.*

relative fugacity The ratio of the fugacity in a given state to the fugacity in a defined standard state.

relative-gravity instrument A device for measuring the difference in the gravity force or acceleration at two or more points. There are two principal types: a static type, or *gravimeter,* in which a linear or angular displacement is observed or nulled by an opposing force, and a dynamic type, in which the period of oscillation is a function of gravity and is the quantity directly observed. In another type, the Eötvös torsion balance, the gravity-field distortion is measured. Cf: *absolute-gravity instrument.*

relative humidity The ratio, expressed as a percentage, of the actual amount of water vapor in a given volume of air to the amount that would be present if the air were saturated at the same temperature. See also: *saturation [meteorol].* Cf: *absolute humidity; specific humidity; mixing ratio.*

relative index of refraction An *index of refraction* that is the ratio of the velocity of light in one crystal to that in another crystal. Syn: *relative refractive index.*

relative permeability The ratio between the *effective permeability* to a given fluid at a partial saturation and the permeability at 100% saturation (the *absolute permeability*). It ranges from zero at a low saturation to 1.0 at a saturation of 100% (Levorsen, 1967, p. 110).

relative refractive index *relative index of refraction.*

relative relief (a) *local relief.* (b) Within a drainage basin, the ratio of *basin relief* to *basin perimeter.* Symbol: R_{hp}. Cf: *relief ratio.*

relative time *Geologic time* determined by the placing of events in a chronologic order of occurrence; esp. time as determined by organic evolution or superposition. Cf: *absolute time; fossil time.*

relative time scale An uncalibrated *geologic time scale,* based on layered rock sequences and the paleontologic evidence contained therein, giving the relative order for a succession of events. Cf: *biologic time scale; atomic time scale.*

relative water content *liquidity index.*

relaxation [exp struc geol] In experimental structural geology, the release of applied stress with time, due to any of various creep processes.

relaxation [geophys] In an elastic medium, the decrease of elastic restoring force under applied stress, resulting in permanent deformation.

relaxation time The time required for a substance to return to its normal state after release of stress. See also: *rheidity.*

release adiabat A curve or locus of points that defines the succession of states through which a mass, shocked to a high-pressure state, passes while monotonically returning to zero pressure. The process operates over a short time-interval compared with the characteristic time for heat flow in the material.

released mineral A mineral formed during the crystallization of a magma as a consequence of an earlier phase failing to react with the liquid. Thus the failure of earlier formed olivine to react with the liquid portion of a magma to form pyroxene may result in the enrichment of the liquid in silica, which finally crystallizes as quartz, the "released mineral".

release fracture A fracture developed as a consequence of the relief of stress in one particular direction. The term is generally applied to a fracture formed when the maximum principal stress decreases sufficiently that it becomes the minimum principal stress; the fracture is an extension fracture oriented perpendicular to the then-minimum principal-stress direction.

release joint *sheeting structure.*

relic [geomorph] A landform that has survived decay or disintegration, such as an *erosion remnant;* or one that has been left behind after the disappearance of the greater part of its substance such as a *remnant island.* The term is sometimes used adjectively as a synonym of *relict,* but this usage is not recommended.

relic [meta] *relict [meta].*

relic [sed] A vestige of a particle in a sedimentary rock, such as a trace of skeletal material in a carbonate rock or an incompletely recrystallized mineral in a diagenetic rock.

relict [geomorph] adj. Said of a topographic feature that remains after other parts of the feature have been removed or have disappeared; e.g. a "relict beach ridge" or a "relict hill". Cf: *relic [geomorph]; residual [geomorph].* Syn: *relicted.*—n. A relict landform.

relict [meta] adj. Pertaining to a mineral, structure, or feature of an earlier rock that has persisted in a later rock in spite of processes tending to destroy it.—n. Such a mineral, structure, or other feature. See also: *stable relict; unstable relict.*—Also spelled: *relic.* Cf: *palimpsest [meta].*

relict [paleont] n. A remnant of an otherwise extinct flora, fauna, or kind of organism that has persisted since the extinction of the rest of the group.—Adj. Said of a remnant of an extinct group.

relict aperture One of the short radial slits around the umbilicus of a planktonic foraminiferal test that remain open when the umbilical parts of the equatorial aperture are not covered by succeeding chambers (as in Planomalinidae) or that, even when secondarily closed, allow the elevated apertural lips or flanges to remain visible around the umbilicus (as in *Planomalina* and *Hastigerinoides*) (TIP, 1964, pt.C, p.63).

relict dike In a granitized mass, a tabular body of crystalloblastic texture that represents a dike emplaced prior to granitization and relatively resistant to the granitization process (Goodspeed, 1955, p. 146).

relict glacier A remnant of an older and larger glacier.

reliction The slow and gradual withdrawal of the water in the sea, a lake, or a stream, leaving the former bottom as permanently exposed and uncovered dry land; it does not include seasonal fluctuations in water levels. Legally, the added land belongs to the owner of the adjacent land against which it abuts. Also, the land left uncovered by reliction. Cf: *dereliction; accretion.*

relict karst Karst formed in an earlier geologic epoch and never covered by later deposits. See also: *paleokarst.* Syn: *fossil karst.*

relict lake A lake that survives in an area formerly covered by the sea or a larger lake, or a lake that represents a remnant resulting from a partial extinction of an original body of water; a lake that has become separated from the sea by gentle uplift of the sea bottom.

relict mountain *mountain of circumdenudation.*

relict permafrost Permafrost that was formed in the past and persists in places where it could not form today (Hopkins et al., 1955).

relict sediment A sediment that had been deposited in equilibrium with its environment, but that is now unrelated to its present environment even though it remains unburied by later sediments; e.g. a land-laid or shallow-marine sediment occurring in deep water (as near the seaward edge of the continental shelf).

relict texture In mineral deposits, an original texture that remains after partial or total replacement.

relief [geomorph] (a) A term used loosely for the physical shape, configuration, or general unevenness of a part of the Earth's surface, considered with reference to variations of height and slope or to irregularities of the land surface; the elevations or differences in elevation, considered collectively, of a land surface. The term is frequently confused with *topography,* although the use of the two terms in the sense of surface configuration is "thoroughly established both in general speech and in technical geomorphological literature" in the U.S. (C.D. Harris, in Stamp, 1961, p. 454). Syn: *topographic relief.* (b) The vertical difference in elevation between the hilltops or mountain summits and the lowlands or valleys of a given region. A region showing a great variation in elevation has "high relief", and one showing little variation has "low relief". See also: *local relief; available relief.*

relief [geophys] The range of values over an anomaly or within an area, e.g. the "gravity relief" for the magnitude of a gravity anomaly.

relief [optics] An apparently rough surface of a crystal section under the microscope. High relief indicates a great difference in *index of refraction* between the crystal and its mounting medium. The relief is positive if the refractive index of the mineral is greater than that of the medium, and negative in the reverse case. Syn: *shagreen.*

relief displacement The geometric distortion on vertical aerial photographs. The tops of objects are located on the photograph radially outward from the base.

relief feature *landform.*

relief limonite *Indigenous limonite* that is porous and cavernous in texture, commonly botryoidal after chalcocite.

relief map A map that depicts the surface configuration or relief of an area by any method, such as by use of contour lines (contour map) and hachures, by hill shading (shaded-relief map), by photography (photorelief map), by layer tinting, by pictorial symbols (physiographic diagram), by molding plastic in three dimensions (plastic relief map), or by a combination of these methods. Cf: *relief model.*

relief model A three-dimensional representation of the physical features or relief of an area, in any size or medium, but not necessarily constructed to true scale (the vertical scale is generally exaggerated to accentuate the relief). Cf: *relief map.*

relief ratio Within a drainage basin, the ratio of *basin relief* to *basin length;* it is a measure of the overall steepness of the basin and the intensity of erosion on its slopes. Symbol: R_h. Cf: *relative relief.*

relief shading *hill shading.*

relief well A well used to relieve excess hydrostatic pressure, as to reduce waterlogging of soil or to prevent blowouts on the land side of levees or dams at times of high water. Cf: *drainage well.*

Relizian North American provincial stage: Miocene (above Saucesian, below Luisian).

remainder mountain *mountain of circumdenudation.*

remanent magnetization That component of a rock's *magnetization* that has a fixed direction relative to the rock and is independent of moderate, applied magnetic fields such as the Earth's magnetic field. Cf: *induced magnetization.* See also: *hysteresis; natural remanent magnetization.*

remanié adj. A French word meaning "reworked" or "rehandled", applied in geology to fragments or entities derived from older materials, esp. to fossils removed from or washed out of an older bed and redeposited in a new one (*remanié assemblage*). The term is also applied to boulders in a glacial till, country rock engulfed in a batholith, and glaciers that have been "reconstructed" (*glacier remanié*). *Cf: reworked; derived.* Anglicized sometimes as *remanie.*—n. A fragment of an older formation incorporated in a younger deposit.

remanié assemblage *mixed assemblage.*

remnant *erosion remnant.*

remobilized Having become mobile again, as a once-molten rock that was remelted.

remolded soil Soil that has had its natural internal structure modified or disturbed by manipulation so that it loses shear strength and gains compressibility.

remolding Disturbance of the naturally occurring internal structure of clay or cohesive soil.

remolding index The ratio of the modulus of deformation of a soil in the undisturbed state to that of a soil in the remolded state.

remolding sensitivity *sensitivity ratio.*

remolinite *atacamite.*

remote sensing The collection of information about an object by a recording device that is not in physical contact with it. The term is usually restricted to mean methods that record reflected or radiated electromagnetic energy, rather than methods that involve significant penetration into the Earth. The technique employs such devices as the camera, infrared detectors, microwave frequency receivers, and radar systems.

renardite A yellow mineral: $Pb(UO_2)_4(PO_4)_2(OH)_4 \cdot 7H_2O$.

Rendoll In U.S. Dept. of Agriculture soil taxonomy, a suborder of the soil order *Mollisol,* developed under humid-region forests in highly calcareous parent material. Rendolls lack argillic or calcic horizons but have 40% or more $CaCO_3$ equivalent within or just below a mollic epipedon (USDA, 1975). Cf: *Alboll; Aquoll; Boroll; Udoll; Ustoll; Xeroll.*

Rendzina soil A great soil group in the 1938 classification system, an intrazonal, calcimorphic group of soils having a brown or black, friable surface horizon and a light gray or yellow, soft, calcareous underlying horizon. They are developed from highly calcareous parent material under grasses or grasses with forest, in a humid to semiarid climate (USDA, 1938). Most of these soils are now classified as *Rendolls.*

renewed consequent stream *resequent stream.*

renierite A mineral: $Cu_3(Fe,Ge,Zn)(S,As)_4$. Not to be confused with reinerite.

reniform Kidney-shaped. Said of a crystal structure in which radiating crystals terminate in rounded masses; also said of mineral deposits having a surface of rounded, kidneylike shapes. Cf: *colloform; botryoidal.*

rensselaerite A compact fibrous talc, pseudomorphous after pyroxene, found in Canada and northern New York. It is harder than talc, takes a good polish, and is often made into ornamental articles.

REO Abbrev. of "rare-earth oxides", i.e. *rare earths.*

repeated twinning Crystal twinning that involves more than two simple crystals; it may be *cyclic twinning* or *polysynthetic twinning.* See also: *multiple twin.*

repetition The duplication of certain stratigraphic beds at the surface or in any specified section owing to disruption and displacement of the beds by faulting or intense folding. Ant: *omission.*

Repettian North American provincial stage: Lower Pliocene (above Delmontian, below Venturian).

repi A group term for "all lakes, ponds, or other standing water bodies related to sinks, or to subsidence, of land surface" (Veatch & Humphrys, 1966, p.264). Etymol: Greek.

replacement [meta] The process of practically simultaneous capillary solution and deposition, by which a new mineral of partly or wholly differing chemical composition may grow in the body of an old mineral or mineral aggregate.

replacement [paleont] A process of fossilization involving substitution of inorganic matter for the original organic constituents of an organism.

replacement [stratig] The gradual movement of the sea either toward or away from land areas, such as "marine replacement" (or transgression) and "continental replacement" (or regression).

replacement dike A dike formed by gradual transformation of wall rock by solutions along fractures or permeable zones (Goodspeed, 1955, p. 146).

replacing overlap A term, now obsolete, used by Grabau (1920, p.398) for a nonmarine overlap involving a receding shoreline, occurring where continental sediments are deposited and progressively "replace" the corresponding and all but contemporaneous marine sediments into which they grade (as described by Grabau, 1906, p.628-629). The misuse of *overlap* in this sense for a facies change from marine to continental sediments of the same age is confusing and "ungeological" (Lovely, 1948, p.2295).

replat (a) A French term for a horizontal surface (such as a bench, shelf, or shoulder), wider than a ledge, occurring along the steep side of a U-shaped valley (Stamp, 1961, p. 391). (b) A French term for a stretch of relatively horizontal land interrupting a slope.

replenishment [grd wat] *recharge.*

replenishment [speleo] The stage in development of a cave in which the presence of air in the passages allows the deposition of speleothems.

repose imprint A term used by Kuenen (1957, p.232) for a *sole mark* formed by an animal lying on or taking cover in bottom sediment.

repose period The interval of solfataric or fumarolic activity between volcanic outbursts.

representative fraction The *scale* of a map, expressed in the form of a numerical fraction that relates linear distances on the map to the corresponding actual distances on the ground, measured in the same unit (centimeters, inches, feet); e.g. "1/24,000" indicates that one unit on the map represents 24,000 equivalent units on the ground. Abbrev: RF. Syn: *natural scale.*

represo *charco.*

reproduction *synthetic stone.*

reptant (a) Creeping or prostrate; esp. said of a corallite with a creeping habit, growing attached along one side to some foreign body. Syn: *reptoid.* (b) Said of a bryozoan colony consisting of largely separate recumbent zooids attached to the substrate.

reptation A syn. of *surface creep* (Scheidegger, 1961, p. 290).

reptile Any vertebrate of the class Reptilia; cold-blooded tetrapods that are air-breathing at all stages of development. Range, Pennsylvanian to present.

reptilian age *age of reptiles.*

resaca A term applied in SW U.S. to a long, narrow, meandering lake occupying the bed of a former stream channel; a series of connected *bancos.* Also, the dry channel or the former marshy course of a stream, now containing a resaca. Etymol: American Spanish, from Spanish *resacar,* "to redraw".

resection (a) A method in surveying by which the horizontal position of an occupied point is determined by drawing lines from the point to two or more points of known position. The most usual problem in resection is the *three-point problem* when three known positions are observed to locate the occupied station. Cf: *intersection.* (b) A method of determining a plane-table position by orienting along a previously drawn foresight line and drawing one or more rays through the foresight from previously located stations.

resedimentation (a) Sedimentation of material derived from a pre-existing sedimentary rock; *redeposition* of sedimentary material. (b) Mechanical deposition of material in cavities of post-depositional age, such as the deposition of carbonate muds and silts by internal mechanical erosion or solution of a limestone. (c) The general process of subaqueous, downslope movement of sediment under the influence of gravity, such as the formation of a turbidity-current deposit.

resedimented rock (a) A rock consisting of reworked sediments.

(b) A turbidity-current deposit; e.g. a flysch or other similar graywacke, showing graded bedding, and alternating with shales in a thick sequence.

resequent [geomorph] Said of a geologic or topographic feature that resembles or agrees with a *consequent* feature but that developed from such a feature at a later date; esp. said of a block mountain that is similar in form to the original tilted block or horst but that is shaped by differential erosion after the original topography was destroyed and uplifted. Ant: *obsequent.*

resequent [streams] adj. Said of a stream, valley, or drainage system whose course or direction follows an earlier pattern but on a newer and lower surface, as in an area of ancient folding subjected to long-continued erosion. Etymol: *re* + con*sequent.* Syn: *reconsequent.*—n. *resequent stream.*

resequent fault-line scarp A *fault-line scarp* that faces in the same direction as the original fault scarp (i.e. facing the downthrown block), or in which the downthrown block is topographically lower than the upthrown block. Cf: *obsequent fault-line scarp.*

resequent stream A stream that flows down the dip of underlying strata in the same direction as an original consequent stream but developed later at a lower level than the initial slope (as on formerly buried resistant strata) and generally tributary to a subsequent stream; e.g. a stream flowing down the back slope of a cuesta. Syn: *resequent; renewed consequent stream.*

reserves *Identified resources* of mineral- or fuel-bearing rock from which the mineral or fuel can be extracted profitably with existing technology and under present economic conditions (Brobst & Pratt, 1973, p. 2). The concept can be used in global, regional, or local senses, or applied as a measure of the remaining effective life of an individual mine. See also: *resources.* Syn: *mineral reserves.*

reservoir [grd wat] *ground-water reservoir.*

reservoir [paleont] The enlarged posterior part of the gullet in some motile cells in protists such as Cryptophyceae and Euglenophyta.

reservoir [petroleum] (a) A subsurface volume of rock that has sufficient porosity and permeability to permit the accumulation of crude oil or natural gas under adequate trap conditions. (b) A *pool* of gas or oil.

reservoir [water] An artificial or natural storage place for water, such as a lake or pond, from which the water may be withdrawn as for irrigation, municipal water supply, or flood control.

reservoir energy The energy or "drive" within a petroleum *reservoir.* See: *dissolved-gas drive; gas-cap drive; water drive.*

reservoir gas-oil ratio The number of cubic feet of gas per barrel of oil in the reservoir. See also: *gas-oil ratio.*

reservoir pressure *bottom-hole pressure.*

reservoir rock In petroleum geology, any porous and permeable rock that yields oil or gas. Sandstone, limestone and dolomite are the most common reservoir rocks, but accumulation in fractured igneous and metamorphic rocks is not unknown.

residence time "The average amount of time a particular substance spends within a designated earth system. The residence time is inversely proportional to the rate of movement within the system and directly proportional to the size of the system" (Laporte, 1975, p. 175).

residual [geomorph] adj. Said of a topographic or geologic feature, such as a rock, hill, or plateau, that represents a small part or trace of a formerly greater mass or area, and that remains above the surrounding surface which has been reduced by erosion. Cf: *relict [geomorph].*—n. *erosion remnant.*

residual [geophys] n. In geophysics, that which is left after the regional has been subtracted; a field from which gross effects have been removed, in order to emphasize local anomalies. Cf: *regional [geophys].*—adj. Said of such an anomaly or gradient, e.g. residual gravity.

residual [ore dep] adj. Said of a mineral deposit formed by mechanical concentration, e.g. a placer, or by chemical alteration in the zone of weathering, e.g. kaolinite from feldspar.

residual [stat] *residual error.*

residual [weath] adj. Pertaining to or constituting a *residue;* esp. said of material left after the weathering of rock in place, such as a *residual deposit* or a *residual soil.* Syn: *eluvial.*

residual anticline In salt tectonics, a relative structural *high* that is created as the result of the depression of two adjacent *rim synclines.* Syn: *residual dome.*

residual arkose An arkose formed *in situ* by disintegration of a granite; an untransported arkose, commonly grading into the underlying granite (Barton, 1916, p. 447). See also: *grus.*

residual boulder *boulder of weathering.*

residual clay Clay material formed in place by the weathering of rock, derived either from the chemical decay of feldspar and other rock minerals or from the removal of nonclay-mineral constituents by solution from a clay-bearing rock (such as an argillaceous limestone); a soil or a product of the soil-forming processes. Cf: *secondary clay.* Syn: *primary clay.*

residual community *winnowed community.*

residual compaction The difference between the amount of compaction that will occur ultimately for a given increase in applied stress, and that which has occurred at a specified time (Poland et al., 1972).

residual deposit (a) The *residue* formed by weathering in place. (b) An ore deposit formed in clay by the conversion of metallic compounds (as of manganese, iron, lead, or zinc) into oxidized forms by weathering at or near the Earth's surface.

residual dome *residual anticline.*

residual error The difference between any measured value of a quantity in a series of observations (corrected for known systematic errors) and the computed value of the quantity obtained after the adjustment of that series. In practice, it is the residual errors that enter into a computation of probable error. Syn: *residual [stat].*

residual geosyncline *autogeosyncline.*

residualizing In applied geophysics, separating a curve or surface into its low-frequency parts (the "regional") and its high-frequency parts (the "residual"). Residualizing is an attempt to sort out of the total field those anomalies that result from local structure; that is, to find local anomalies by subtracting gross (regional) effects (Sheriff, 1973, p. 184). See also: *grid residual.*

residual kame A ridge or mound of sand or gravel formed by the denudation of glaciofluvial material that had been deposited in glacial lakes or on the flanks of hills of till (Gregory, 1912, p. 175). Syn: *pseudokame.*

residual liquid The still-molten part of a magma that remains in the magma chamber after some crystallization has taken place during a series of differentiations. Syn: *residual liquor; residual magma; rest magma; ichor.*

residual liquor *residual liquid.*

residual magma *residual liquid.*

residual map A stratigraphic map that displays the small-scale variations (such as local features in the sedimentary environment) of a given stratigraphic unit (Krumbein & Sloss, 1963, p.486). It is superimposed on the underlying pattern of a *trend map.*

residual material Unconsolidated or partly weathered *parent material* of a soil, presumed to have developed in place, by weathering, from the consolidated rock on which it lies; it is the material from which soils are formed. See also: *residual soil.* Cf: *transported soil material; cumulose.*

residual oil Oil that is left in the *reservoir rock* after the pool has been depleted.

residual sediment *resistate.*

residual soil A soil that developed from *residual material.* The term is obsolete; the adjective "residual" is more correctly applied to the parent material (USDA, 1957, p.766). Syn: *sedentary soil.*

residual strength *Ultimate strength* that develops along a surface in soil or a discontinuity in rock. For soil or rock not previously sheared, a marked decrease in strength usually occurs with increasing displacement until the residual value is reached.

residual swelling "The difference between the original prefreezing level of the ground and the level reached by the settling after the ground is completely thawed" (Muller, 1947, p.221).

residual valley A trough intervening between uplifted mountains, as in the Basin and Range Province of western U.S. (Gilbert, 1875, p. 63).

residue (a) An accumulation of rock debris formed by weathering and remaining essentially in place after all but the least soluble constituents have been removed, usually forming a comparatively thin surface layer concealing the unweathered or partly altered rock below. See also: *chemical residue; physical residue.* Syn: *residuum [weath]; residual deposit; eluvium [weath].* (b) *insoluble residue.*

residuite The translucent *residuum* that occurs in *clarain.*

residuo-aqueous sand A term used by Sherzer (1910, p.627) for a sand containing water-rounded particles that were subsequently subjected to weathering. Cf: *aqueo-residual sand.*

residuum [coal] The structureless groundmass of microscopically unresolvable constituents, consisting of particles of one to two microns or less, usually opaque, and of a dark color. It is the same as the lower range of fine *micrinite*. See also: *desmite; residuite.*

residuum [weath] *residue.*

resilience The ability of a material to store the energy of elastic strain. This ability is measured in terms of energy per unit volume.

resilifer A spoon-shaped recess or process on the hinge plate of some bivalve mollusks (as in *Mactra*) to which the resilium is attached or by which it is supported. See also: *chondrophore.* Syn: *resiliifer.*

resilium The internal ligament within the hinge line of a bivalve mollusk, compressed by the hinge plate when the valves are closed. Pl: *resilia.*

resin Any of various hard, brittle, transparent or translucent substances formed esp. in plant secretions and obtained as exudates of recent or fossil origin (as from pine or fir trees, or from certain tropical trees) by the condensation of fluids on the loss of volatile oils. Resins are yellowish to brown, with a characteristic luster; they are fusible and flammable, are soluble in ether and other organic solvents but not in water, and represent a complex mixture of terpenes, resin alcohols, and resin acids and their esters. See also: *amber; fossil resin; mineral resin.* Syn: *natural resin.*

resin canal *resin duct.*

resin duct A long, narrow intercellular canal in wood, surrounded by one or more layers of parenchyma cells of the epithelium (Record, 1934, p.72). See also: *resin rodlet.* Syn: *resin canal.*

resinite A maceral of coal within the *exinite* group, consisting of resinous compounds, often in elliptical or spindle-shaped bodies representing cell-filling matter or resin rodlets. Cf: *alginite; cutinite; sporinite.*

resin jack *rosin jack.*

resinoid A group term for the macerals in the resinite series.

resin opal A wax-yellow or honey-yellow variety of common opal with a resinous luster.

resinous coal Coal, usually younger coal, in which the attritus contains a large proportion of resinous material.

resinous luster The luster on the fractured surfaces of certain minerals (such as opal, sulfur, amber, and sphalerite) and rocks (such as pitchstone) that resembles the appearance of resin.

resin rodlet A fossil resinous secretion that may be isolated from coal. It was presumably deposited in a *resin duct* by a secretory epithelium.

resin tin *rosin tin.*

resistance The ratio of an applied constant electromotive force to the current it produces. Cf: *conductance.*

resistates Sediments composed of chemically resistant minerals, enriched in weathering residues; thus highly quartzose sediments characteristically rich also in zircon, ilmenite, rutile, and, more rarely, cassiterite, monazite, and gold. It is one of Goldschmidt's groupings of sediments or analogues of differentiation stages in rock analysis. Cf: *hydrolyzates; oxidates; reduzates; evaporates.*

resistivity (a) *electrical resistivity.* (b) *thermal resistivity.*

resistivity factor *formation factor.*

resistivity log A *well log* consisting of one or more resistivity curves, which may be of the *normal log, lateral log,* or *focused-current log* types (or their *microresistivity log* equivalents). As a spontaneous-potential curve is commonly also present, the term resistivity log is often used as a syn. of *electric log.*

resistivity method Any electrical exploration method in which current is introduced into the ground by two contact electrodes and potential differences are measured between two or more other electrodes.

resistivity profile A survey by the resistivity method in which an array of electrodes is moved along profiles to determine lateral variations in resistivity.

resistivity sounding *electrical-resistivity sounding.*

resistivity-thickness product The product of resistivity and thickness of a relatively high-resistivity layer present in a sequence of other, more conductive layers. It is often used to characterize the amplitude of response obtained from the resistive layer in direct-current resistivity soundings. Units are ohm-meters^2.

resolution A measure of the ability of individual components, and of remote-sensing systems, to define closely spaced targets. Syn: *resolving power.*

resolving power *resolution.*

resonance A buildup of amplitude in a physical system when the frequency of an applied oscillatory force is close to the natural frequency of the system. Cf: *nuclear resonance; electron spin resonance; nuclear magnetic resonance.*

resonant frequency The frequency at which maximum response of a system occurs.

resorbed reef A reef characterized by embayed margins and by numerous isolated patches of reef that are closely distributed about the main mass (Maxwell, 1968, p.106-107). Resorbed reefs frequently rise from larger, submerged platforms, and suggest restrictive growth or degeneration of the reef mass.

resorption (a) The act or process of reabsorption or readsorption; specif. the partial or complete re-fusion or solution, by and in a magma, of previously formed crystals or minerals with which the magma is not in equilibrium or, owing to changes of temperature, pressure (depth), or chemical composition, with which it has ceased to be in equilibrium. "The term is often wrongly applied to immature crystals, and to crystals which have decomposition borders through change of pressure or otherwise" (Holmes, 1928, p. 198). (b) The geologic capture of a mineral by another that is relatively fixed (Pryor, 1963).

resorption border A *corrosion border* representing partial resorption and recrystallization by a molten magma of previously crystallized minerals. Syn: *resorption rim.*

resorption rim A *resorption border* as seen in section.

resources *Reserves* plus all other mineral deposits that may eventually become available — either known deposits that are not economically or technologically recoverable at present, or unknown deposits, rich or lean, that may be inferred to exist but have not yet been discovered (Brobst & Pratt, 1973, p. 2). They represent the mineral endowment, global, regional, or local, ultimately available for man's use. See also: *identified subeconomic resources; hypothetical resources; speculative resources.* Syn: *mineral resources.*

responsivity In infrared detector terminology, the ratio of signal output to incident radiant flux, usually expressed as volts/watt (Bernard, 1970, p.56).

rest-hardening The increase of strength, with time, of a clay subsequent to its deposition, remolding, or modification by the application of shear stress. It appears to be an electrochemical bonding phenomenon.

resting spore A spore that remains dormant for a period before germination; e.g. a *chlamydospore,* or a desmid *zygospore* having thick cell walls and able to withstand adverse conditions such as heat, cold, or drying out. See also: *statospore; cyst [palyn].*

restite An essentially nongenetic designation for all immobile or less mobile parts of migmatites. Cf: *metaster; stereosome.* See also: *paleosome.*

rest magma *residual liquid.*

restricted basin A depression in the ocean floor characterized by topographically restricted water circulation, often resulting in oxygen depletion. Syn: *silled basin; barred basin.*

reststrahlen Narrow bands of enhanced reflectance that occur in transparent materials in which the refractive index is high or the absorption coefficient is large. Etymol: German, "residual rays".

resupinate Inverted or reversed in position; esp. referring to reversal in relative convexity of postbrephic brachiopod shells in which the convex pedicle valve becomes concave and the concave brachial valve becomes convex during successive adult stages of growth.

resurgence (a) The rising again of a stream from a cave, the water having entered the cave as a *sinking stream.* See also: *exsurgence.* (b) The point where any underground stream appears at the surface to become a surface stream. Syn: *rise; rising; emergence [streams]; debouchure.*

resurgent [petrology] Said of magmatic water or gases that were derived from sources on the Earth's surface, from the atmosphere, or from country rock adjacent to the magma. Cf: *juvenile [water].*

resurgent [pyroclast] In the classification of pyroclastics, the equivalent of *accessory.* Cf: *juvenile.*

resurgent cauldron A *cauldron [volc]* in which the cauldron block, following subsidence, has been uplifted, usually in the form of a structural dome (Smith & Bailey, 1968, p.613).

resurgent gas *resurgent vapor.*

resurgent vapor (a) Ground water volatilized by contact with hot rock. (b) Gas in magma, possibly derived from dissolved or assimilated country rock.—Syn: *resurgent gas.* Cf: *phreatic gas; juvenile [water].*

resurrected (a) Said of a surface, landscape, or feature (such as a mountain, peneplain, or fault scarp) that has been restored by *exhumation* to its previous status in the existing relief. Syn: *exhumed.* (b) Said of a stream that follows an earlier drainage system after a period of brief submergence had slightly masked the old course by a thin film of sediments. Syn: *palingenetic.*

resurrected karst A *paleokarst* that has been reactivated.

resurrected-peneplain shoreline A shoreline of submergence formed where the sea rests against an inclined resurrected peneplain (Johnson, 1925, p. 27); it may be remarkably straight for long distances.

retained water Water retained in a rock or soil after the *gravity ground water* has drained out. It is no longer ground water but has become vadose water. Most of it is held by molecular attraction, but part may be in isolated interstices or held by other, more or less obscure forces, and part remains as water vapor in interstices from which water has drained (Meinzer, 1923, p. 27-28).

retaining wall A wall designed to resist the lateral pressure of the material behind it, e.g. a bulkhead preventing an earth slide.

retard A permeable bank-protection structure situated at and parallel to the toe of a slope and projecting into a stream channel, designed to reduce stream velocity and induce silting or accretion.

retardation [cryst] In crystal optics, the amount by which the slow wave falls behind the fast wave during passage through an anisotropic crystal plate. Retardation depends on the plate's thickness and the difference in refractive indices of its two privileged directions.

retardation [tides] *lunitidal interval.*

retention The amount of water from precipitation that has not escaped as runoff or through evapotranspiration; "the difference between total precipitation and total runoff on a drainage area" (Nelson & Nelson, 1967).

reteporiform Said of a rigid, erect fenestrate colony in cheilostome bryozoans, firmly attached by a calcified base.

retgersite A tetragonal mineral: $NiSO_4 \cdot 6H_2O$. It is dimorphous with nickel-hexahydrite.

reticle A system of wires, cross hairs, threads, dots, or very fine etched lines, placed in the eyepiece of an optical instrument (such as a surveyor's telescope) perpendicular to its principal focus, to define the line of sight of the telescope or to permit a specific pointing to be made on a target or signal or a reading to be made on a rod or scale. Syn: *reticule.*

reticulate [ore dep] Said of a vein or lode with netlike texture, e.g. *stockwork.*

reticulate [paleont] (a) Said of a netted pattern of an invertebrate, or of one resembling a network, e.g. a "reticulate layer" consisting of ornamental ridges at the surface of a foraminiferal test, or a "reticulate ornamentation" on the exterior of a brachiopod shell, commonly involving the intersection of concentric rugae with radial costae or costellae. (b) Said of evolutionary change that involves repeated intercrossing between a number of lines; specif. a change involving the complex recombination of genes from varied strains of a diversified interbreeding population.

reticulate [palyn] Said of pollen and spores having sculpture consisting of a more or less regular network of ridges.

reticulate [petrology] Said of a rock texture in which crystals are partially altered to a secondary mineral, forming a network that encloses remnants of the original mineral. Cf: *mesh texture.* Syn: *reticulated; reticular.*

reticulated bar One of a group of slightly submerged sandbars in two sets, both of which are diagonal to the shoreline, forming a crisscross pattern (Shepard, 1952, p. 1909). Reticulated bars are observed in bays and lagoons on the inside of barrier islands.

reticule *reticle.*

reticulite *thread-lace scoria.*

reticulopodium A foraminiferal *pseudopodium* that bifurcates and anastomoses to form a network. Pl: *reticulopodia.*

retiform wall A wall in archaeocyathids in which the pores are so closely spaced that the entire wall appears netlike. The pores may be of various shapes (TIP, 1972, pt. E, p. 9).

retinalite A massive, honey-yellow or greenish serpentine mineral with a waxy or resinous luster; a variety of chrysotile.

retinasphalt A light-brown variety of retinite usually found with lignite.

retinite (a) A general term for a large group of fossil resins of variable composition (oxygen content generally 6-15%), characterized by the absence of succinic acid, and found in the younger (brown) coals or in peat. (b) Any fossil resin of the retinite group, such as glessite, krantzite, muckite, and ambrite. (c) A general name applied to fossil resins.

retinosite A microscopic constituent of torbanite, consisting of translucent orange-red discs (Dulhunty, 1939).

retractor muscle A muscle that draws in an organ or part, e.g. a "siphonal retractor muscle" serving to withdraw the siphon of a bivalve mollusk partly or wholly into the shell, a "pallial retractor muscle" withdrawing marginal parts of mantle into a bivalve-mollusk shell where there is no distinct line of muscle attachment, or an *anterior lateral muscle* in a lingulid brachiopod. Cf: *protractor muscle.*

retral Posterior, or situated at or toward the back; e.g. "retral processes" in foraminiferal tests, consisting of backward-pointing extensions of chamber cavity and enclosed protoplasm, located beneath external ridges on the chamber wall, and ending blindly at the chamber margins (as in *Elphidium*).

retreat A decrease in length of a glacier, resulting in a displacement upvalley or upslope of the position of the terminus, caused when processes of ablation (usually melting and/or calving) exceed the speed of ice flow: normally measured in meters per year. Cf: *recession [glaciol].*

retreatal moraine *recessional moraine.*

retrochoanitic Said of a septal neck of a cephalopod directed backward (adapically). Ant: *prochoanitic.*

retrodiagenesis *hypergenesis.*

retrogradation The backward (landward) movement or retreat of a shoreline or of a coastline by wave erosion; it produces a steepening of the beach profile at the breaker line. Ant: *progradation.* Cf: *recession.*

retrograde boiling The separation of a gas phase in a cooling magma as a result of its residual enrichment in the dissolved gaseous components by progressive crystallization of the magma.

retrograde condensation Appearance of a *condensate* in deep formations as the reservoir pressure is reduced through production of natural gas. The gas condenses to form a liquid, instead of the usual pattern of liquid changing to gas; hence the term retrograde.

retrograde diagenesis *hypergenesis.*

retrograde metamorphism A type of *polymetamorphism* by which metamorphic minerals of a lower grade are formed at the expense of minerals characteristic of a higher grade of metamorphism, a readjustment necessitated by a change in physical conditions, e.g. lowering of temperature. Cf: *prograde metamorphism.* Syn: *diaphthoresis; retrogressive metamorphism.*

retrograde motion A period when a planet moves westward on the celestial sphere. For example, when the Earth is overtaking Jupiter in terms of their mutual angular motion about the Sun, Jupiter will be in retrograde motion before and after opposition. Cf: *prograde motion.*

retrograding shoreline A shoreline that is being moved backward by wave attack. Ant: *prograding shoreline.* Syn: *abrasion shoreline.*

retrogression [coast] *recession [coast].*

retrogression [evol] The passage from a higher to a lower or from a more to a less specialized state or type of organization or structure during the development of an organism.

retrogressive erosion *headward erosion.*

retrogressive metamorphism *retrograde metamorphism.*

return In *SLAR,* a pulse of microwave energy reflected by the terrain and received at the radar antenna. The strength of a return is referred to as "return intensity".

return flow Irrigation water not consumed by evapotranspiration but returned to its source or to another body of ground or surface water. Water discharged from industrial plants is also considered return flow (Langbein & Iseri, 1960). Syn: *waste water; return water.*

returns Those surface waves on the record of a large earthquake that have traveled around the Earth's surface by the long arc between epicenter and station (greater than 180°), or have passed the station and returned after traveling the entire circumference of the earth.

return water *return flow.*

retusoid Said of spores, mostly Devonian, with prominent *contact areas* and *curvaturae* perfectae.

retzian A brown orthorhombic mineral: $Mn_2Y(AsO_4)(OH)_4$.

reverberation *singing.*

reversal *geomagnetic reversal.*

reverse Pertaining to the basal side of an incrusting or freely growing bryozoan colony.—Cf: *obverse.*

reverse bearing *back bearing.*

reverse branch A seismic event where right-to-left orientation on the event corresponds to the reverse orientation (left-to-right) on the reflector. Such an event could originate from a sharp syncline, which makes it appear to have anticlinal curvature on an unmigrated seismic section. For zero offset seismic data, this occurs where the radius of curvature for the concave-upward reflector is less than the reflector depth. See also: *buried focus.*

reversed *overturned.*

reversed chevron mark A *chevron mark* that points upstream (Dzulynski and Walton, 1965, p. 110).

reversed consequent stream A consequent stream whose direction of flow is contrary to that normally consistent with the geologic structure; e.g. the part of a captured consequent stream between the escarpment slope and the elbow of capture. See also: *obsequent stream.*

reversed fault *reverse fault.*

reversed gradient A local gradient opposite to the general gradient; esp. a valley gradient at the downstream side of a glacially overdeepened valley.

reversed magnetic field *reversed polarity.*

reversed magnetization A *natural remanent magnetization* that is opposite to the present ambient geomagnetic field.

reversed polarity (a) A *natural remanent magnetization* opposite to the present ambient geomagnetic-field direction. See also: *geomagnetic reversal.* Syn: *reversed magnetic field.* (b) A configuration of the Earth's magnetic field with the magnetic positive pole, where field lines leave the Earth, located near the geographic north pole.—Cf: *normal polarity.*

reversed stream A stream whose direction of flow has been reversed, as by glacial action, landsliding, gradual tilting of a region, or capture.

reversed tide An oceanic tide that is out of phase with the apparent motions of the tide-producing body, so that low tide is directly under the tide-producing body and is accompanied by a low tide on the opposite side of the Earth. Cf: *direct tide.* Syn: *inverted tide.*

reversed zoning *Zoning* in a plagioclase crystal in which the core is more sodic than the rim. Cf: *normal zoning.* Syn: *inverse zoning.*

reverse fault A fault that dips toward the block that has been relatively raised. Cf: *normal fault.* Partial syn: *thrust fault.* Syn: *reversed fault; reverse slip fault.*

reverse-flowage fold A fold in which flow from deformation has thickened the anticlinal crests and thinned the synclinal troughs, contrary to the normal flow pattern of a *flow fold.*

reverse grading A type of bedding that displays an increase in grain size with distance up from the base. Cf: *graded bedding.*

reverse saddle A mineral deposit associated with the trough of a synclinal fold and following the bedding plane. Syn: *trough reef.* Cf: *saddle reef.*

reverse scarplet An *earthquake scarplet* facing in toward the mountain slope and enclosing a trench, produced by reversal of earlier movement along a fault (Cotton, 1958, p. 165-166); examples are numerous in New Zealand. Syn: *earthquake rent.*

reverse similar fold A fold the strata of which are thickened on the limbs and thinned on the axes, contrary to the pattern of a *similar fold.*

reverse slip fault *reverse fault.*

reversible pendulum A pendulum that can swing around either of two knife edges placed in such a way that the period is the same for both; it is used in absolute-gravity determinations.

reversible process A thermodynamic process in which an infinitesimal change in the variables characterizing the state of the system can change the direction of the process.

reversing current A tidal current that flows in an alternating pattern of opposite directions for approximately equal lengths of time, with a slack period of no movement at each reversal. Reversing currents occur in estuaries, restricted channels, and inland bodies of water. Cf: *rotary current.* Syn: *rectilinear current.*

reversing dune A dune that tends to develop unusual height but migrates only a limited distance "because seasonal shifts in direction of dominant wind cause it to move alternately in nearly opposite directions" (McKee, 1966, p. 10). Its general shape may resemble that of a barchan or a transverse dune, but it differs in the complexity of its internal structural orientation due to reversals in direction of the slip face.

reversing thermometer A mercury-in-glass thermometer used to measure temperatures of the sea at depth. The temperature is recorded when the thermometer is inverted; and the recording is maintained until it is once again upright. A *protected thermometer* and an *unprotected thermometer* are usually used as a pair, attached to a *Nansen bottle.* See also: *thermometric depth.*

reversion A return toward an ancestral type or condition, such as the reappearance in an organism of an ancestral characteristic. Syn: *atavism.*

revet-crag A term proposed by Gilbert (1877, p. 26) for one of a series of narrow, pointed outliers or ridges of eroded strata inclined like a revetment against a mountain spur.

revetment A facing of stone, concrete, or other material, built to protect an embankment or a shore structure from wave erosion.

revier A term applied in SW Africa to a deeply cut river bed that usually remains dry.

revival *rejuvenation.*

revived *rejuvenated.*

revived fault scarp *rejuvenated fault scarp.*

revived folding *recurrent folding.*

revived stream *rejuvenated stream.*

revolution A term formerly popular among geologists for a time of profound orogeny and other crustal movements, on a continent-wide or even worldwide scale, the assumption being that such revolutions produced abrupt changes in geography, climate, and environment. Schuchert (1924) classed all orogenies at the close of geologic eras as revolutions, in contrast to *disturbances,* or orogenies within the eras. The basic premises of all these concepts are dubious, and the term revolution is little used today.

reworked Said of a sediment, fossil, rock fragment, or other geologic material that has been removed or displaced by natural agents from its place of origin and incorporated in recognizable form in a younger formation, such as a "reworked tuff" carried by flowing water and redeposited in another locality. Cf: *derived; remanié.*

reyerite A mineral: $(Na,K)_4Ca_{14}(Si,Al)_{24}O_{60}(OH)_5 \cdot 5H_2O$.

Reynolds critical velocity *critical velocity (b).*

Reynolds number A numerical quantity used as an index to characterize the type of flow in a hydraulic structure in which resistance to motion depends on the viscosity of the liquid in conjunction with the resisting force of inertia. It is the ratio of inertia forces to viscous forces, and is equal to the product of a characteristic velocity of the system (e.g. the mean, surface, or maximum velocity) and a characteristic linear dimension, such as diameter or depth, divided by the kinematic viscosity of the liquid; all expressed in consistent units in order that the combinations will be dimensionless. The number is chiefly applicable to closed systems of flow, such as pipes or conduits where there is free water surface, or to bodies fully immersed in the fluid so the free surface need not be considered (ASCE, 1962).

rezbanyite A gray mineral: $Pb_3Cu_2Bi_{10}S_{19}$.

RF *representative fraction.*

Rg wave A slow, short-period *Rayleigh wave* that travels only along a nonoceanic path. The "g" refers to the granitic layer. Cf: *Lg wave.*

rhabd (a) An obsolete term for *monaxon.* Syn: *rhabdus.* (b) *rhabdome.*

rhabdacanth A trabecula of a rugose coral in which the fibers are related to any number of separate, transient (shifting) centers of growth grouped around a main one (TIP, 1956, pt.F, p.235). Cf: *monacanth; rhipidacanth.*

rhabde The lower or axial branch in the triaene spicule of an ebridian skeleton.

rhabdite A syn. of *schreibersite,* esp. occurring in rods or needle-shaped crystals.

rhabdodiactine A seemingly monaxonic sponge spicule, formed by suppression of two of the axes of a hexactine which are preserved internally as an axial cross. See also: *diactine.*

rhabdoglyph A collective term used by Fuchs (1895) for a *trace fossil* consisting of a presumed worm trail appearing on the undersurface of flysch beds (sandstones) as a nearly straight bulge with little or no branching. Cf: *graphoglypt; vermiglyph.*

rhabdolith A calcareous, spinose, rodlike or clublike, supposedly perforate *coccolith,* averaging 3 μm in diameter, having a shield surmounted by a long stem. Rhabdoliths are found both at the surface and on the bottom of the ocean, and they have been classed as protozoans and as algae. Also spelled: *rabdolith.*

rhabdomancy A form of *dowsing* using a rod or twig. Cf: *pallomancy.*

rhabdome The long ray of a triaene sponge spicule. Syn: *rhabd.*

rhabdophane A brown, pinkish, or yellowish-white mineral: (Ce,

La)$PO_4 \cdot H_2O$. It contains yttrium and rare-earth elements. Syn: *rhabdophanite.*

rhabdosome The skeleton of a graptolithine colony.

Rhaetian European stage: uppermost Triassic (above Norian, below Hettangian of Jurassic). It is interpreted as lowermost Jurassic in some areas (as in France and Great Britain) or as a transitional stage between the Triassic and Jurassic. Syn: *Rhaetic.*

rhagon (a) The earliest developmental stage of a sponge with a functional aquiferous system having several flagellated chambers. Also, a sponge or sponge larva in such a stage. (b) A term used incorrectly as a syn. of *leucon.*—Adj: *rhagonoid.*

rhax A kidney-shaped sterraster (sponge spicule).

rhegmagenesis *regmagenesis.*

rhegolith *regolith.*

rheid A substance (below its melting point) which deforms by viscous flow during the time of applied stress at an order of magnitude at least three times that of the elastic deformation under similar conditions.

rheid fold A fold in which the strata have deformed by flow as if they were fluid. Cf: *flow fold.* Syn: *rheomorphic fold.*

rheidity *Relaxation time* of a substance, multiplied by 1000.

rhenium-osmium age method The determination of an age in years based on the known radioactive decay rate of rhenium-187 to osmium-187. The low crustal abundance of rhenium limits the application of this method.

rheoglyph A hieroglyph produced by syngenetic deformation, e.g. slumping (Vassoevich, 1953, p.55).

rheoignimbrite An ignimbrite, on the slope of a volcanic crater, that developed secondary flowage due to high temperatures.

rheologic settling Failure of a sediment under a stress load by plastic deformation or flow.

rheology The study of the deformation and flow of matter.

rheomorphic Said of a rock whose form and internal structure indicate a finite amount of flow in a plastic state; also, said of the phenomena causing such a rock. See also: *rheomorphism.*

rheomorphic fold *rheid fold.*

rheomorphic intrusion The injection of country rock that has become mobilized (rheomorphic) into the igneous intrusion that caused the rheomorphism. Such an intrusion usually resembles the metamorphosed country rock.

rheomorphism The process by which a rock becomes mobile as a result of at least partial fusion, commonly accompanied by, if not promoted by, addition of new material by diffusion. The term was created by H. G. Backlund in 1937. Cf: *mobilization.*

rheopexy The accelerated gelation of a thixotropic sol by agitating it in some manner, e.g. stirring.

rheophile adj. Said of an organism that lives in or prefers flowing water.

rheotaxis *Taxis [ecol]* resulting from mechanical stimulation by a stream of fluid, such as water. Cf: *rheotropism.*

rheotropism *Tropism* resulting from mechanical stimulation by a stream of fluid, such as water. Cf: *rheotaxis.*

rhexistasy The mechanical breaking-up and transport of old soils or other surface residual materials (Erhart, 1955). Etymol: Greek *rhexis,* "act of breaking", + *stasis,* "condition of standing". Adj: *rhexistatic.* See also: *biorhexistasy; biostasy.*

rhinestone (a) An inexpensive and lustrous imitation of diamond, consisting of glass that has been backed with a thin leaf of metallic foil to simulate the brilliancy of a diamond. (b) Originally, a syn. of *quartz crystal.* (c) Ocasionally, parti-colored, single-colored, or colorless glass.

rhinocanna *nasal tube.*

rhipidacanth A compound trabecula in Paleozoic rugose corals with spindle-shaped septa assigned to the Phillipsastraeidae. Each is composed of a central primary trabecula from which short secondary trabeculae radiate at varying angles perpendicular to the long axis of the septum. Cf: *monacanth; rhabdacanth.*

Rhipidistia A suborder of crossopterygian fish which remained primarily freshwater during its entire history; it includes the best candidates for tetrapod ancestry. Range, Lower Devonian to Permian.

rhizic water A syn. of *soil water* proposed by Meinzer (1939) as one of three classes of *krematic water.*

rhizoclone A monocrepid desma (of a sponge) consisting of a straight or curved body bearing branching outgrowths along its entire length. See also: *megarhizoclone.*

rhizoconcretion A small concretionlike structure in a sedimentary rock, cylindrical or conical, usually branching or forked, resembling a root of a tree. It may consist of material such as caliche or chert. Cf: *rhizocretion; pedotubule.* See also: *root sheath.* Syn: *rhizomorph; root cast.*

rhizocretion A term used by Kindle (1923, p.631) for a hollow concretionlike mass that had formed around the root of a living plant. Cf: *rhizoconcretion.*

rhizoid [bot] A unicellular or multicellular, rootlike filament that attaches some nonvascular plants and gametophytes of some vascular plants to the substrate (Scagel et al., 1965, p.630). Cf: *rhizome; rhizophore.*

rhizoid [paleont] adj. Resembling a root; e.g. "rhizoid spine" of a brachiopod, resembling a plant rootlet and serving for attachment either by entanglement or by extending along and cementing itself to a foreign surface.—n. A *radicle* of a bryozoan.

rhizome An underground stem that lies horizontally and that is often enlarged in order to store food. Not to be confused with *rhizoid.* Cf: *rhizophore.*

rhizomorph A term used by Northrop (1890) for a structure now known as a *rhizoconcretion.*

rhizophore A naked branch that grows down into the soil and develops roots from the apex (Swartz, 1971, p. 405). Cf: *rhizoid; rhizome.*

rhizophytous Said of a sponge that is fastened to the substrate by branching extensions of the body.

rhizopod A protozoan belonging to the class Sarcodina, subclass Rhizopoda, generally characterized by lobose pseudopodia and by zoned protoplasm in shelled forms and protoplasm differentiated into endo- and ectoplasm in nonshelled forms. Cf: *actinopod.* Range, Ordovician to present.

rhizopodial Said of a morphologic type or growth form in which the cell is somewhat amoeboid.

rhizopodium A bifurcating and anastomosing ectoplasmic *pseudopodium* that is typical of many foraminifers. Pl: *rhizopodia.* Syn: *rhizopod.*

rhizosphere The soil in the immediate vicinity of plant roots, in which the abundance or composition of the microbial population is affected by the presence of the roots. Syn: *root zone [soil].*

Rhodanian orogeny One of the 30 or more short-lived orogenies during Phanerozoic time identified by Stille, in this case at the end of the Miocene.

rhodesite A mineral: $(Ca,Na_2,K_2)_8Si_{16}O_{40} \cdot 11H_2O$. Cf: *mountainite.*

rhodite A mineral consisting of a native alloy of rhodium (about 40%) and gold.

rhodium A cubic mineral: (Rh,Pt).

rhodizite A mineral: $CsAl_4Be_4B_{11}O_{25}(OH)_4$.

rhodochrosite A rose-red or pink to gray rhombohedral mineral of the calcite group: $MnCO_3$. It is isomorphous with calcite and siderite, and commonly contains some calcium and iron; it is a minor ore of manganese. Syn: *dialogite; manganese spar; raspberry spar.*

rhodolite A pink, rose, or purple to violet garnet that is intermediate in chemical composition between pyrope and almandine, characterized by a lighter tone and a higher degree of transparency than either of the other two, and used as a gem.

rhodolith A nodule of red (coralline) algae, concentrically encrusted, often rolled by bottom currents (Bosellini & Ginsburg, 1971). Term proposed as rhodolite but this was already a mineral name. Cf: *oncolite.*

rhodonite A pale-red, rose-red, or flesh-pink to brownish-red or red-brown triclinic mineral of the pyroxenoid group: $MnSiO_3$. It sometimes contains calcium, iron, magnesium, and zinc, and is often marked by black streaks and veins of manganese oxide. Rhodonite is used as an ornamental stone, esp. in Russia. Syn: *manganese spar; manganolite.*

rhodostannite A hexagonal mineral: $Cu_2FeSn_3S_8$.

rhodusite A monoclinic mineral of the amphibole group: $Na_2(Fe^{+2},Mg)_3Fe_2{}^{+3}Si_8O_{22}(OH)_2$. It is near riebeckite in chemical composition.

rhohelos A stream-crossed, nonalluvial marsh typical of filled lake areas.

rhomb [cryst] An oblique, equilateral parallelogram; in crystallography, a *rhombohedron.*

rhomb [paleont] (a) *pore rhomb.* (b) A six-sided, roughly equidimensional crystal composing some heterococcoliths.

rhombic (a) Adj. of *rhomb.* (b) Adj. of *orthorhombic.*

rhombic-dipyramidal class That crystal class in the orthorhombic system having symmetry $2/m\ 2/m\ 2/m$.

rhombic-disphenoidal class That crystal class in the orthorhombic system having symmetry 222.

rhombic dodecahedron A crystal form in the cubic system that is a *dodecahedron,* the faces of which are equal rhombs. Cf: *pyritohedron.*

rhombic-pyramidal class That crystal class in the orthorhombic system having symmetry mm2.

rhombic section That plane in a triclinic feldspar containing both the b axis and the normal to b lying in (010).

rhombic system A syn. of *orthorhombic system.* It is an undesirable term because it may be confused with *rhombohedral.*

rhombochasm A term used by Carey (1958) for a parallel-sided gap in the sialic crust occupied by simatic crust, interpreted as due to spreading and separation. Cf: *sphenochasm.* See also: *rift [tect].*

rhomboclase A mineral: $HFe^{+3}(SO_4)_2 \cdot 4H_2O$. It occurs in colorless rhombic plates with basal cleavage.

rhombohedral class That crystal class in the rhombohedral division of the hexagonal system having symmetry $\bar{3}$.

rhombohedral cleavage Mineral cleavage parallel to the faces of the rhombohedron, e.g. the $(10\bar{1}1)$ cleavage of calcite.

rhombohedral iron ore (a) *hematite.* (b) *siderite [mineral].*

rhombohedral lattice A centered lattice of the trigonal system in which the unit cell is a rhombohedron. It may occur in crystal classes having one threefold axis. The unit cell contains three lattice points: one at the corners and two equally spaced along one long diagonal.

rhombohedral packing The "tightest" manner of systematic arrangement of uniform solid spheres in a clastic sediment or crystal lattice, characterized by a unit cell of six planes passed through eight sphere centers situated at the corners of a regular rhombohedron (Graton & Fraser, 1935). An aggregate with rhombohedral packing has the minimum porosity (25.95%) that can be produced without distortion of the grains. Cf: *cubic packing.* See also: *close packing.*

rhombohedral system A division of the *trigonal system* in which the unit cell is a rhombohedron.

rhombohedron A trigonal crystal form that is a parallelepiped whose six identical faces are rhombs. It is characteristic of the hexagonal system. Syn: *rhomb [cryst].*

rhomboidal lattice structure A common feature on sand beaches, resulting from two sets of small grooves that trend diagonally along the beach slope, and formed immediately after or during the last stages of backward flow. "The grooves are probably the surface expression of planar zones of high shear stress in which the grains become more closely packed than elsewhere in the beach sand" (Stauffer et al., 1976, p. 1667). Cf: *rhomboid ripple mark; rhomboid rill mark.*

rhomboid current ripple *rhomboid ripple mark.*

rhomboid rill mark A sandy-beach pattern consisting of shallow grooves, incised by breaker swash or backwash. The grooves form a fairly uniform rhomboid patterned network; the long diagonal of the rhombs parallels the direction of the creating current (Otvos, 1965, p. 271). They are distinguished from rhomboid ripple marks by lower relief and absence of slip faces; they are not a type of ripple, but a network of lineations. Cf: *rhomboid ripple marks; rhomboidal lattice structure.*

rhomboid ripple mark An aqueous *current ripple mark* characterized by diamond-shaped tongues of sand arranged in a reticular pattern resembling the scales of certain fish. Tongues range from 12 to 25 mm in width and 25 to 50 mm in length; each has one acute angle (formed by two steep sides) pointing downcurrent, and another acute angle (formed by the gentle side extending into the reentrant angle of the steep sides of two tongues of the following) pointing upcurrent; it is common on sand beaches where it forms during the final stages of backwash of each retreating wave. The sides are not more than 1 mm high. The term "rhomboid" applied to a ripple or ripple mark was introduced by Bucher (1919, p.153). Syn: *rhomboid current ripple; overhanging ripple.*

rhombolith *scapholith.*

rhomb-porphyry Porphyritic alkalic *syenite* or *trachyte* containing phenocrysts of augite, sparse olivine, and anorthoclase or potassium oligoclase with rhombohedron-shaped cross sections, in a groundmass composed chiefly of alkali feldspars. Var: *rhombenporphyry; rhombenporphyr.* Named by Buch in 1810. Usually applied to lavas and dikes of the Oslo district, Norway, but also used elsewhere as a textural term.

rhomb spar A *dolomite* mineral in rhombohedral crystals.

rhönite A triclinic mineral: $Ca_2(Mg,Ti,Al,Fe)_6(Si,Al)_6O_{20}$.

rhopaloid septum A rugose corallite septum characterized by distinctly thickened axial edge, appearing club-shaped in cross section (TIP, 1956, pt.F, p.250).

rhourd A pyramid-shaped sand dune, formed by the intersection of other dunes (Aufrere, 1931, p. 376). Cf: *khurd.*

rhumb line A curved line on the surface of the Earth that crosses successive meridians at a constant oblique angle and that spirals around and toward the poles in a constant true direction but theoretically never reaches them; a straight line on a Mercator projection, representing a line of constant bearing or compass direction. Syn: *loxodrome; loxodromic curve; Mercator track.*

rhyacolite *sanidine.*

Rhynchocephalia An order of lepidosaurian reptiles that includes large beaked animals with crushing dentitions, of Triassic age, together with the living lizardlike *Sphenodon* and its extinct relatives. Range, Lower Triassic to Recent.

rhyncholite A fossil beak or part of a cephalopod jaw; specif. the calcified tip of a jaw of a Triassic nautiloid.

rhynchonelloid Any articulate brachiopod belonging to the order Rhynchonellida, characterized by a rostrate shell, a functional pedicle, and a delthyrium partially closed by deltidial plates. Range, Middle Ordovician to present. Var: *rhynchonellid.*

rhyoandesite A term applied to the extrusive (aphanitic and aphanophyric) equivalent of *granodiorite.* Cf: *dacite.*

rhyocrystal One of a group of idiomorphs arranged in "streamlines" (Lane, 1902, p. 386).

rhyodacite A group of extrusive porphyritic igneous rocks intermediate in composition between dacite and rhyolite, with quartz, plagioclase, and biotite (or hornblende) as the main phenocryst minerals and a fine-grained to glassy groundmass composed of alkali feldspar and silica minerals; the extrusive equivalent of *granodiorite* or *quartz monzonite.* Also, any member of that group. Syn: *quartz latite; dellenite; toscanite.* The term was proposed by Winchell in 1913.

rhyodiabasic A nonrecommended term applied to the ophitic texture of an igneous rock in which the plagioclase phenocrysts are more or less parallel.

rhyolite A group of extrusive igneous rocks, typically porphyritic and commonly exhibiting flow texture, with phenocrysts of quartz and alkali feldspar in a glassy to cryptocrystalline groundmass; also, any rock in that group; the extrusive equivalent of *granite.* Rhyolite grades into *rhyodacite* with decreasing alkali feldspar content and into *trachyte* with a decrease in quartz. The term was coined in 1860 by Baron von Richtofen (grandfather of the World War I aviator). Etymol: Greek *rhyo-,* from *rhyax,* "stream of lava". Syn: *liparite.* Cf: *quartz porphyry.*

rhyotaxitic texture *flow texture.*

rhythmic crystallization A phenomenon, observed in igneous rocks, in which different minerals crystallize in concentric layers, giving rise to *orbicular* structure. See also: *cyclic crystallization.*

rhythmic layering Readily observable *layering* in an igneous intrusion, in which there is repetition of zones of varying composition having the pattern abc, abc, abc. It results from recurring crystallization of certain phases as the magma periodically reaches some critical composition. See also: *cyclic crystallization; cryptic layering; zebra layering.*

rhythmic sedimentation The consistent repetition, through a sedimentary succession, of a regular sequence of two or more rock units organized in a particular order and indicating a frequent and predictable recurrence or pattern of the same sequence of conditions. It may involve only two components (such as interbedded laminae of silt and clay), or broad changes in sediment character spanning whole systems (or longer intervals) and units up to hundreds of meters thick, or any sequence intermediate between these two extremes. See also: *cyclic sedimentation.*

rhythmic succession A succession of rock units showing continual and repeated changes of lithology. The term was used by Hudson (1924) for a continual repetition of a more or less complete suite comprising successive beds of certain kinds of sediments accompanied by an equally marked variation in the kind of fossils they contain.

rhythmic unit (a) *rhythmite.* (b) A layer or band of a rhythmically layered intrusive igneous rocks.

rhythmite An individual unit of a rhythmic succession or of beds developed by rhythmic sedimentation; e.g. a *cyclothem.* The term was used by Bramlette (1946, p.30) for the couplet of distinct types of sedimentary rock, or the graded sequence of sediments, that forms a unit in rhythmically bedded deposits. The term implies no limit as to thickness or complexity of bedding and it carries no time or seasonal connotation. See also: *laminite*. Syn: *rhythmic unit*.

rhytidome The technical term for *outer bark*.

RI *reduction index*.

ria (a) Any long, narrow, sometimes wedge-shaped inlet or arm of the sea (but excluding a fjord) whose depth and width gradually and uniformly diminish inland and which is produced by drowning due to submergence of the lower part of a narrow river valley or of an estuary; it is shorter and shallower than a *fjord*. Originally, the term was restricted to such an inlet produced where the trend of the coastal rock structure is at right angles to the coastline; it was later applied to any submerged land margin that is dissected transversely to the coastline. (b) Less restrictedly, any broad or estuarine river mouth, including a fjord, and not necessarily an embayment produced by partial submergence of an open valley.— See also: *estuary*. Etymol: Spanish *ría*, from *río*, "river".

ria coast A coast having several long parallel *rias* extending far inland and alternating with ridge-like promontories; e.g. the coasts of SW Ireland and NW Spain. It is especially developed where the trend of the coastal structures is transverse to that of the coastline.

ria shoreline A shoreline characterized by numerous *rias* and produced by drowning due to partial submergence of a land margin subaerially dissected by numerous river valleys (Johnson, 1919, p. 173).

rib [bot] In a leaf or similar organ, the primary *vein [bot]*.

rib [geomorph] A layer or dike of rock forming a small ridge on a steep mountainside.

rib [paleont] A radial or transverse fold upon a shell; e.g. any radial ornament on a brachiopod shell, a *costa* on the shell of a bivalve mollusk, or a raised radial ridge on the coiled conch of a nautiloid or ammonoid.

rib and furrow A term used by Stokes (1953, p.17-21) for the bedding-plane expression of *micro cross-bedding*. It consists of small transverse arcuate markings, convex upcurrent, occurring in sets and confined to relatively long grooves, 3-5 cm wide, oriented parallel to the current flow and separated from one another by narrow discontinuous ridges. It represents the eroded edges of upturned arcuate laminae. Syn: *rib-and-furrow structure*.

riband jasper *ribbon jasper*.

ribbed moraine One of a group of irregularly subparallel, locally branching, generally smoothly rounded and arcuate ridges that are convex in the downstream direction of a glacier but that curve upstream adjacent to eskers (J.A. Elson in Fairbridge, 1968, p. 1217). They are most common in the continental ice sheets, and are abundant in the Arctic.

ribble *ripple till*.

ribbon [ore dep] adj. Said of a vein having alternating streaks of ore with gangue or country rock, or simply of varicolored ore minerals. Cf: *banded; book structure*.

ribbon [petrology] One of a set of parallel bands or streaks in a mineral or rock, e.g. ribbon jasper; when the lines of contrast are on a larger scale, the term *banding* is used. When occurring in slate, the structure is known as *slate ribbon*. Syn: *stripe*.

ribbon banding A *banding* produced in the bedding of a sedimentary rock by thin strata of contrasting colors, giving the rock an appearance suggesting bands of ribbons.

ribbon bomb A type of volcanic bomb that is elongate and flattened, and derived from a rope of lava.

ribbon diagram A single, continuous geologic cross section drawn in perspective along a curved or sinuous line.

ribbon injection A tonguelike igneous intrusion along the cleavage planes of a foliated rock.

ribbon jasper Beautifully banded jasper with parallel, ribbon-like stripes of alternating colors or shades of color (as of red, green, and esp. brown). Syn: *riband jasper*.

ribbon reef A linear reef within the Great Barrier Reef off the NE coast of Australia, having inwardly curved extremities, and forming a festoon along the precipitous edge of the continental shelf. They are variable in length (3-24 km), less so in width (300-470 m).

ribbon rock A rock characterized by a succession of thin layers of differing composition or color; e.g. gray shale interspersed with thin seams of brown dolomite and lighter-colored limestone (Goldring, 1943), or a vein rock with narrow quartz bands separated by stripes of altered wall rock.

ribbon slate Slate produced by incomplete metamorphism, with still visible bedding that cuts across the cleavage surface; slate characterized by varicolored ribbons.

richellite A dubious mineral: $Ca_3Fe_{10}(PO_4)_8(OH,F)_{12} \cdot nH_2O$. It occurs in amorphous yellow masses.

richetite A mineral: an oxide of Pb and U.

Richmondian North American stage: Upper Ordovician (above Maysvillian, below Alexandrian of Silurian).

richterite (a) A brown, yellow, or rose-red monoclinic mineral of the amphibole group: $(Na,K)_2(Mg,Mn,Ca)_6Si_8O_{22}(OH)_2$. (b) An end-member of the amphibole group: $Na_2CaMg_5Si_8O_{22}(OH)_2$. Cf: *soda tremolite*.

Richter scale A numerical scale of *earthquake magnitude*, devised in 1935 by the seismologist C.F. Richter. Very small earthquakes, or microearthquakes, can have negative magnitude values. In theory there is no upper limit to the magnitude of an earthquake, but the strength of Earth materials produces an actual upper limit of slightly less than 9. Usually the scale refers to *local magnitude*, but for large earthquakes it often refers to *surface-wave magnitude*.

rickardite A deep-purple mineral: Cu_7Te_5.

ricolettaite A dark-colored, coarse-grained rock containing extremely calcic plagioclase, along with olivine, biotite, and augite. The name, given by Johannsen in 1920, is from Ricoletta in the Tyrolean Alps. Not recommended usage.

rideau A small ridge or mound of earth, or a slightly elevated piece of ground. Etymol: French.

ridge [beach] (a) *beach ridge*. (b) A low mound that is sometimes found above the water level on the foreshore of a sand beach during low tide. See also: *runnel*.

ridge [geomorph] (a) A general term for a long, narrow elevation of the Earth's surface, usually sharp-crested with steep sides, occurring either independently or as part of a larger mountain or hill; e.g. an extended upland between valleys. A ridge is generally less than 8 km long (Eardley, 1962, p.6). (b) A term occasionally applied to a range of hills or mountains. (c) The top or upper part of a hill; a narrow, elongated crest of a hill or mountain.

ridge [ice] *pressure ridge [ice]*.

ridge [marine geol] An elongate, steep-sided elevation of the ocean floor, having rough topography; Syn: *submarine ridge*.

ridge [paleont] An elevated body part of an animal, projecting from a surface; e.g. a relatively long narrow elevation of secondary shell of a brachiopod, or a *transverse ridge* on a crinoid. Also, an area separating adjacent pairs of ambulacral pores of a regular echinoid.

ridge-and-ravine topography Hack's term for landscapes in dynamic equilibrium, undergoing active regional erosion in a humid temperate climate, such as the *ridge-and-valley topography* of the Appalachians.

ridge-and-swale topography Long subparallel ridges and swales aligned obliquely across the regional trend of the contours and converging with the shoreline at an angle (Swift et al., 1972, p. 501); e.g. the inner-shelf topography of the Middle Atlantic Bight.

ridge-and-valley topography A land surface characterized by a close succession of parallel or nearly parallel ridges and valleys, and resulting from the differential erosion of highly folded strata of varying resistances. Type region: Ridge and Valley region in the Appalachian Mountains, lying to the west of the Blue Ridge. Cf: *ridge-and-ravine topography*.

ridged ice Sea ice having readily observed surface features in the form of one or more *pressure ridges;* it is usually found in first-year ice. See also: *ropak; ridging*.

ridge line A *slope line* linking *peaks* via *passes* or *pales*(Warntz, 1975, p. 211).

ridge-ridge transform fault A *transform fault* that offsets *mid-oceanic ridges*.

ridge-top trench A trench, occasionally found at or near the crest of high, steep-sided mountain ridges, formed by the creep displacement of a large slab of rock along shear surfaces more or less parallel with the side slope of the ridge. Trenches are usually parallel with the crest of the ridge. See also: *sackungen*.

ridging A form of deformation of floating ice, caused by lateral pressure, whereby ice is forced or piled haphazardly one piece over another to form *ridged ice*. Cf: *tenting*.

riebeckite A blue or black monoclinic mineral of the amphibole group: $Na_2(Fe,Mg)_5Si_8O_{22}(OH)_2$. It occurs as a primary constituent in some acid or sodium-rich igneous rocks. See also: *crocidolite*.

Riecke's principle The statement in thermodynamics that solution of a mineral tends to occur most readily at points where external pressure is greatest, and that crystallization occurs most readily at points where external pressure is least. It is applied to recrystallization in metamorphic rocks with attendant change in mineral shapes. It is named after the German physicist E. Riecke (1845-1915) although it was actually discovered and described by Sorby in 1863.

riedel shear A slip surface which develops during the early stage of shearing. Such shears are typically arranged en échelon, usually at inclinations of between 10° and 30° to the direction of relative movement (Riedel, 1929).

riedenite A *melteigite* composed of large tabular biotite crystals in a granular groundmass of nosean, biotite, pyroxene, and small amounts of sphene and apatite. The name, given by Brauns in 1922, is from Rieden, Laacher See district, Germany. Not recommended usage.

riegel A low, transverse ridge or barrier of bedrock on the floor of a glacial valley, esp. common in the Alps; it separates a rock basin from the gently sloping valley bottom farther downstream. See also: *rock step.* Etymol: German *Riegel,* "crossbar". Syn: *rock bar; threshold [glac geol]; verrou.*

Riel discontinuity A seismic-velocity discontinuity noted in Alberta that may be equivalent to the *Conrad discontinuity.*

riffle (a) A natural shallows or other expanse of shallow bottom extending across a stream bed over which the water flows swiftly and the water surface is broken into waves by obstructions wholly or partly submerged; a shallow rapids of comparatively little fall. See also: *rift [streams].* (b) An expanse of shallow water flowing over a riffle or at the head of a rapids. (c) A low bar or bedrock irregularity in a stream, resembling a riffle. (d) A wave of a riffle.—Syn: *ripple.*

riffle hollow A shallow depression in a stream bed, commonly 8 to 30 cm deep, produced by differential erosion of alternate layers of hard and soft rock (Bryan, 1920, p. 192).

riffler *sample splitter.*

rift [eco geol] A quarrymen's term for a direction of parting in a massive rock, e.g. granite, at approximately right angles to the *grain [eco geol].*

rift [geomorph] A narrow cleft, fissure, or other opening in rock (as in limestone), made by cracking or splitting.

rift [speleo] A narrow, high passage in a cave, the shape of which is controlled by a joint or by a bedding or fault plane.

rift [streams] A shallow or rocky place in a stream, forming either a ford or a rapids. The term is used in NE U.S. as a syn. of *riffle.*

rift [tect] (a) A long, narrow continental trough that is bounded by normal faults; a *graben* of regional extent. It marks a zone along which the entire thickness of the lithosphere has ruptured under extension. Cf: *paar.* (b) A belt of strike-slip faulting of regional extent.

rift fault A fault that bounds a rift valley. The term has been applied to normal faults and to large strike-slip faults.

rift lake *sag pond.*

rift trough *rift valley.*

rift valley (a) A valley that has developed along a *rift [tect].* Syn: *fault trough; rift trough.* (b) The deep central cleft in the crest of the *mid-oceanic ridge.* Syn: *central valley; median valley; mid-ocean rift.*

rift-valley lake *sag pond.*

rift zone (a) A system of crustal fractures; a *rift [tect].* (b) In Hawaii, a zone of volcanic features associated with underlying dike complexes. Syn: *volcanic rift zone.*

rig *drilling rig.*

right bank The bank of a stream situated to the right of an observer who is facing downstream.

right-handed [cryst] Said of an optically active crystal that rotates the plane of polarization of light to the right. Cf: *left-handed.* Syn: *dextrorotatory.*

right-handed [paleont] *dextral.*

right-handed separation *right-lateral separation.*

right-lateral fault A fault on which the displacement is *right-lateral separation.* Syn: *dextral fault; right-lateral slip fault; right-slip fault.*

right-lateral separation Displacement along a fault such that, in plan view, the side opposite the observer appears displaced to the right. Cf: *left-lateral separation.* Syn: *right-handed separation.*

right-lateral slip fault *right-lateral fault.*

right section *profile [struc petrol].*

right-slip fault *right-lateral fault.*

right valve The valve lying on the right-hand side of a bivalve mollusk when the shell is placed with the anterior end pointing away from the observer, the commissure being vertical and the hinge being uppermost. Abbrev: RV. Ant: *left valve.*

rigidity The property of a material to resist applied stress that would tend to distort it. A fluid has zero rigidity.

rigidity modulus *modulus of rigidity.*

rigolet A term applied in the Mississippi River valley to a small stream, creek, or rivulet. Etymol: French *rigole,* "trench, small ditch, channel".

rijkeboerite A mineral of the pyrochlore group: $Ba(Ta,Nb)_2(O,OH)_7$. It is the barium analogue of microlite.

rilandite A mineral: $(Cr,Al)_6SiO_{11} \cdot 5H_2O$ (?).

rill [beach] (a) A small, transient *runnel* carrying to the sea or a lake the water of a wave after it breaks on a beach, esp. one formed following an outgoing tide. It may be 2-10 mm wide, 0.5 m or more long, and about 1 mm deep. (b) The minute stream or thin sheet of water flowing in a rill.

rill [lunar] *rille.*

rill [stream] (a) A very small brook or trickling stream of water; a streamlet or rivulet. (b) A small channel eroded by a rill, esp. in soil; one of the first and smallest channels formed by runoff, such as a *shoestring rill.* Syn: *rill channel.*

rill cast A term used by Dzulynski & Slaczka (1958, p.230) for a sole mark that is a type of *flute cast.* See also: *rill mark.*

rille One of several trenchlike or cracklike valleys, up to several hundred km long and 1-2 km wide, commonly occurring on the Moon's surface. Rilles may be extremely irregular with meandering courses ("sinuous rilles"), or they may be relatively straight ("normal rilles"); they have relatively steep walls and usually flat bottoms. Rilles are essentially youthful features and apparently represent fracture systems originating in brittle material. Syn: *rill; rima.*

rillenkarren Downslope *solution grooves* about 1 cm wide, with sharp intergroove crests. Etymol: German, "rill tracks". See also: *karren; rinnenkarren; deckenkarren.*

rillenstein Tiny solution grooves of about one millimeter or less in width, formed on the surface of a soluble rock. Etymol: German, "rilled rock".

rill erosion The development of numerous minute closely spaced channels resulting from the uneven removal of surface soil by running water that is concentrated in streamlets of sufficient discharge and velocity to generate cutting power. It is an intermediate process between *sheet erosion* and *gully erosion.* Cf: *channel erosion.* Syn: *rill wash; rilling; rillwork.*

rill flow Surface runoff flowing in small, irregular channels too small to be considered rivulets.

rilling *rill erosion.*

rill mark (a) A small, dendritic channel, groove, or furrow formed on the surface of beach mud or sand by a wave-generated *rill* or by a retreating tide; esp. one formed on the lee side of a half-buried pebble, shell, or other obstruction, and usually branching upstream. (b) A dendritic channel formed by a small stream debouching on a sand flat or a mud flat; it shows a downslope bifurcation.—See also: *lobate rill mark.*

rill wash A syn. of *rill erosion.* Also spelled: *rillwash.*

rillwork *rill erosion.*

rim [geomorph] The border, margin, edge, or face of a landform, such as the curved brim surrounding the top part of a crater or caldera; specif. the *rimrock* of a plateau or canyon.

rim [glac geol] A ridge of morainal material, generally unbroken and of uniform height, surrounding a central depression (Gravenor & Kupsch, 1959, p. 52).

rim [ign] (a) *reaction rim.* (b) *kelyphytic rim.* (c) *corona [petrology].* (d) The outermost portion of a zoned crystal.

rim [paleont] (a) One of the paired bones of the axial skeleton that helps support the body wall in Osteichthyes and tetrapods. (b) One of the two flanges of a caneolith coccolith peripheral to the wall; e.g. the distal "upper rim" and the proximal "lower rim". (c) The outer, usually flangelike component of a *wheel* of a holothurian. It may be recurved, and the inner margin of its upper side is commonly denticulate or dentate. The rim is inclined to or within the plane of the wheel.

rima A long narrow aperture, cleft, or fissure; specif. a lunar *rille.* Pl: *rimae.*

rim cementation A term used by Bathurst (1958, p. 21) for *secondary enlargement* in detrital sediments; e.g. the chemical deposition of calcium carbonate forming a single, completely enveloping rim on a grain of the same composition, as in a crinoidal limestone where each grain (or crinoidal fragment) is a single crystal and is permeated by the calcite cement in lattice or optical continuity. Cf: *granular cementation.*

rime A deposit of rough ice crystals formed as a result of contact between supercooled droplets of fog and a solid object at a temperature below 0°C (McIntosh, 1963, p.216).

rim gypsum Gypsum in thin films between anhydrite crystals, believed to have been introduced in solution rather than produced by replacement (Goldman, 1952, p. 2).

rim height The maximum height of the rim of a crater above the original ground surface.

rimmed kettle A morainal depression with raised edges (Gravenor & Kupsch, 1959, p. 53); a kettle in stratified drift with raised edges.

rimming wall A steep, ridgelike erosional remnant of continuous layers of porous, permeable, poorly cemented, detrital limestones, believed to form under tropical or subtropical conditions (as on Okinawa and other Pacific islands) by surface-controlled secondary cementation on an original steep slope followed by differential erosion that brings the cemented zone into relief (Flint et al., 1953). See also: *rampart wall.*

rimpylite A group name for several green and brown hornblendes having high contents of $(Al,Fe)_2O_3$.

rim ridge A minor ridge of till defining the edge of a *moraine plateau* (Hoppe, 1952, p. 5).

rimrock [eco geol] The bedrock forming or rising above the margin of a placer or gravel deposit. Also spelled: *rim rock.*

rimrock [geomorph] (a) An outcrop of a horizontal layer of resistant rock, such as a lava flow, exposed at the edge of a plateau, butte, or mesa; it generally forms a cliff or ledge. (b) The edge or face of an outcrop of rimrock, esp. a cliff or a relatively vertical face of rock in the wall of a canyon. Syn: *rim [geomorph].*

rimrocking Prospecting for carnotite, specifically on the Colorado Plateau, where the favorable beds, more or less flat-lying, crop out in cliffs, or rims.

rimstone A thin crustlike deposit of calcite that forms a ring around an overflowing basin or pool of water.

rimstone barrier *rimstone dam.*

rimstone dam A deposit of *rimstone* that forms a pool or basin called a *rimstone pool.* Syn: *gour; rimstone barrier; travertine dam.*

rimstone pool The pool or basin of water that is formed of and bounded by a *rimstone dam.* Syn: *gour.*

rimstone shelf *shelfstone.*

rim syncline In salt tectonics, a local depression that develops as a border around a salt dome, as the salt in the underlying strata is displaced toward the dome. See also: *residual anticline.* Syn: *peripheral sink.*

rimule A *sinus* in the operculum-bearing orifice of an ascophoran cheilostome (bryozoan).

rincon (a) A term used in the SW U.S. for a square-cut recess or hollow in a cliff or a reentrant in the borders of a mesa or plateau. Cf: *cove [geomorph].* (b) A term used in the SW U.S. for a small, secluded valley, and for a bend in a stream.—Etymol: Spanish *rincón,* "inside corner, nook".

rindle An English syn. of *runnel.*

ring [geol] *ring structure.*

ring [paleont] The precursor to the ascending branches of the premagadiniform loop of terebratellid brachiopods, "consisting of thin circular ribbon, narrow ventrally and broadening dorsally to its attachment on median septum" (TIP, 1965, pt.H, p.152).

ring canal An internal tube encircling the oral region of an echinoderm, from which five radial canals branch.

ring complex An association of *ring dikes* and *cone sheets.*

ring depression The annular, structurally depressed area surrounding the central uplift of a *cryptoexplosion structure.* Faulting and folding may be involved in its formation. Syn: *ring syncline; peripheral depression.*

ring dike A dike that is arcuate or roughly circular in plan and is vertical or inclined away from the axis of the arc. Ring dikes are commonly associated with *cone sheets* to form a *ring complex.* Syn: *ring-fracture intrusion.*

ringer A thin bed of tough, tightly cemented, fine-grained sandstone that gives out a clear, resonant sound when struck with a hammer.

ring fault A steep-sided fault pattern that is cylindrical in outline and is associated with *cauldron subsidence.* Syn: *ring fracture.*

ring fissure A roughly circular desiccation crack formed on a playa around a point, generally a *phreatophyte.*

ring fracture *ring fault.*

ring-fracture intrusion *ring dike.*

ring-fracture stoping Large-scale *magmatic stoping* associated with *cauldron subsidence.* Cf: *ring dike; piecemeal stoping.*

ring hill An isolated, till-covered hill in Lapland, which remained above the marine limit and is surrounded by a very pronounced ring of bedrock washed clean of material (Stephens & Synge, 1966, p. 28).

ringing *singing.*

ringite An igneous rock formed by the mixing of silicate and carbonatite magmas. The name, given by Brögger in 1921, is from Ringe in the Fen complex, Norway. Not recommended usage.

ring mark A *skip mark* produced by a fish vertebra, consisting of a ring-like ridge whose higher side is upcurrent; often the ring is incomplete, forming a semicircle that is concave downcurrent.

ring moor A *string bog* with concentric ridges.

ring ore *cockade ore.*

ring plain A lunar crater of exceptionally large diameter and with a relatively smooth interior. See also: *walled plain.* An obsolete term.

ring-porous wood Wood in which the pores (vessels) of one part of an annual ring are distinctly different in size or number (or both) from those in the other part of the ring (Fuller & Tippo, 1954, p. 969).

ring reef *atoll.*

ring septum In some stenolaemate bryozoans, a perforated skeletal diaphragm that forms a washerlike transverse structure in living chambers of feeding zooecia.

ring silicate *cyclosilicate.*

ring structure A general term for an epigenetic structure with a ring-shaped trace in plan; e.g. a ring dike, or a lunar crater resulting from a meteorite impact. Syn: *ring.*

ring syncline *ring depression.*

ringwall A bordering wall that encircles a mare or crater on the surface of the Moon, formed in part by the mountains and lesser eminences of lunarite. An obsolete term.

ringwoodite A purple mineral of the spinel group: $(Mg,Fe)_2SiO_4$. It is a cubic dimorph of olivine.

rinkite *mosandrite.*

rinkolite *mosandrite.*

rinneite A colorless, pink, violet, or yellow rhombohedral mineral: NaK_3FeCl_6. It is isomorphous with chlormanganokalite.

rinnenkarren Downslope solution grooves about 0.5 m wide, with sharp intergroove crests. Etymol: German, "channel tracks". See also: *karren; rillenkarren.*

Rinnental A syn. of *tunnel valley.* Etymol: German, "channel valley".

rio A term used in SW U.S. for a river or stream, usually a permanent stream. Etymol: Spanish *río.*

rip (a) A turbulent agitation of water, generally caused in the sea by the meeting of water currents or the interaction of currents and wind, or in a river or a nearshore region by currents flowing rapidly over an irregular bottom; an *overfall.* See also: *tide rip; current rip.* (b) An abbreviated form of *ripple [currents],* often used in the plural.

ripa A legal term for the bank of a stream or lake (Veatch & Humphrys, 1966, p. 268).

riparian (a) Pertaining to or situated on the bank of a body of water, esp. of a watercourse such as a river; e.g. "riparian land" situated along or abutting upon a stream bank, or a "riparian owner" who lives or has property on a riverbank. Cf: *riverain.* Syn: *riparial; riparious.* (b) *littoral.*

riparian water loss Discharge of water through evapotranspiration by vegetation growing along a watercourse. Discharged water may be derived from the watercourse, adjacent ground water, and/or soil moisture.

rip channel A channel, often more than 2 m deep, carved on the shore by a rip current.

rip current A strong, narrow surface or near-surface current, of short duration (a few minutes to an hour or two) and high velocity (up to 2 knots), flowing seaward from the shore through the breaker zone at nearly right angles to the shoreline, appearing as a visible band of agitated water returning to the sea after being piled up on the shore by incoming waves and wind; it consists of a *feeder current,* a *neck,* and a *head.* Cf: *undertow.* Often miscalled a *rip tide.*

ripe Said of peat that is in an advanced state of decay. Cf: *unripe.*

ripe snow Snow that has reached the melting point and has its minimum capillary requirements satisfied, so that additional melting can produce meltwater runoff. See also: *ripe-snow area.*

ripe-snow area The area of a drainage basin where coarsely crystalline snow is in a condition to discharge meltwater upon the addition of heat (as by rain); expressed in percent of drainage basin or in square kilometers. Abbrev: RSA. See also: *ripe snow.*

Riphean The most recent era of Precambrian time, as defined by Russian geologists. Approximately equivalent terms are *Sinian, Beltian,* and *Eocambrian.*

ripidolite A mineral of the chlorite group, intermediate between clinochlore and chamosite in composition: $(Mg,Fe^{+2})_9Al_6Si_5O_{20}(OH)_{16}$. The name is sometimes applied to clinochlore. Syn: *prochlorite; aphrosiderite.*

ripple [currents] (a) A syn. of *capillary wave.* (b) The light ruffling of the surface of the water by a breeze.—Syn: *rip.*

ripple [sed struc] (a) A small ridge of sand resembling or suggesting a ripple of water and formed on the bedding surface of a sediment; specif. a *ripple mark,* or a small sand wave similar to a dune in shape. (b) A general term for all sand waves with shapes similar to small-scale ripples, no matter what the scale (e.g. Reineck and Singh, 1973, p. 14-47).—Syn: *sedimentary ripple.*

ripple [streams] (a) A shallow reach of running water in a stream, roughened or broken into small waves by a rocky or uneven bottom. (b) *riffle.*

ripple bedding (a) A bedding surface characterized by ripple marks. (b) A term preferred by Hills (1963, p.10-11) to *current bedding* when used for "the small-scale ripple-like bedding of rapidly deposited sand".

ripple biscuit A bedding structure produced by lenticular lamination of sand in a bay or lagoon (Moore, 1966, p.99).

ripple cross-lamination Small-scale cross-lamination formed by migrating current ripples developed during deposition, characterized by individual laminae whose thicknesses range between 0.08 cm (1/32 in.) and 0.3 cm (1/8 in.) (McKee, 1939, p.72). See also: *ripple lamina.* Syn: *rolling strata.*

ripple drift A term used by Sorby (1852, p. 232; see also 1857, p. 278) for small-scale cross-lamination formed by migrating ripples. See also: *drift bedding; climbing ripple.*

ripple height The vertical distance from crest to trough of a ripple on a ripple-marked surface. If the ripple is asymmetric, the height is measured from the trough adjacent to the steeper (downcurrent) slope. The term was used by Allen (1963, p.192). See also: *ripple-mark amplitude.*

ripple index The ratio of ripple-mark wavelength to ripple-mark amplitude. The ratio usually ranges from 6 to 22 for ripples produced by water currents or waves and from 20 to 50 for ripples produced by wind. Syn: *ripple-mark index; vertical form index.*

ripple lamina An internal sedimentary structure formed in sand or silt by currents or waves, as opposed to a ripple mark formed externally on a surface. The term, as commonly used in the plural, "includes sets of laminae in incomplete ripple profiles and isolated ripple lenses, as well as series of superposed rippled layers" (McKee, 1965, p.66). McKee proposed: "ripple laminae-in-rhythm", a general term for all ripple structures superimposed in an orderly sequence; and "ripple laminae-in-phase", a general term for ripple laminae in which the crests of vertically succeeding laminae (as seen in sections parallel to the direction of current or wave motion) are directly above one another. See also: *ripple cross-lamination.* Syn: *ripple lamination.*

ripple load cast A *load cast* of a ripple mark that shows signs of penecontemporaneous deformation (caused by unequal loading, settling, and compaction) in the accentuation of its trough and crest and in the oversteepening of component laminae (Kelling, 1958, p.120-121). Now generally termed a *load-casted ripple.*

ripple mark (a) An undulatory surface or surface sculpture consisting of alternating subparallel small-scale ridges and hollows formed at the interface between a fluid and incoherent sedimentary material (esp. loose sand). It is produced on land by wind action and subaqueously by currents or by the agitation of water in wave action, and generally trends at right angles or obliquely to the direction of flow of the moving fluid. It is no longer regarded as evidence solely of shallow water. Syn: *ripple.* (b) One of the small and fairly regular ridges, of various shapes and cross sections, produced on a ripple-marked surface; esp. a ripple preserved in consolidated rock as a structural feature of original deposition on a sedimentary surface and useful in determining the environment and order of deposition. The term was formerly restricted to *symmetric ripple mark,* but now includes *asymmetric ripple mark.* See also: *sand wave; wavemark.* Syn: *fossil ripple; ripple ridge.* (c) A corrugation on a snow surface, produced by wind.—The singular form may be used to denote general ripple structure (as well as a specific ripple), and the plural form to describe a particular example. Also spelled: *ripple-mark.*

ripple-mark amplitude The height of a ripple on a ripple-marked surface, measured as the vertical distance between the crest of the ripple and the adjacent trough; it is generally a centimeter or less. This use of the term "amplitude" is at variance with the convention used in physics and mathematics in which "amplitude" refers to displacement relative to a mean or equilibrium value. A better term is *ripple height.*

ripple-mark index *ripple index.*

ripple-mark wavelength *chord.*

ripple ridge *ripple mark.*

ripple scour A large shallow linear trough with superimposed transverse ripple mark (Potter & Glass, 1958, pl.5).

ripple spacing *chord.*

ripple symmetry index A term used by Tanner (1960, p.481) to express the degree of symmetry of a ripple mark, defined as the ratio of the horizontal length of the gentle (upcurrent) side to that of the steep (downcurrent) side; an asymmetric ripple mark has an index greater than 1. Abbrev: RSI. See also: *horizontal form index.*

ripple till A till sheet containing low, winding, smooth-topped ridges, 6-15 m high and 200-3000 m long, lying at right angles to the direction of ice movement, and grouped into narrow belts up to 80 km long that are generally parallel to the direction of ice movement (F.K. Hare in Stamp, 1961, p. 395); found in parts of northern Ontario. Syn: *ribble.*

ripple wavelength *chord.*

rippling A surface characterized by ripple mark; a collective term for a series of occurrences of ripples or ripple marks.

riprap (a) A layer of large, durable fragments of broken rock, specially selected and graded, thrown together irregularly (as offshore or on a soft bottom) or fitted together (as on the upstream face of a dam). Its purpose is to prevent erosion by waves or currents and thereby preserve the shape of a surface, slope, or underlying structure. It is used for irrigation channels, river-improvement works, spillways at dams, and sea walls for shore protection. (b) The stone used for riprap.

rip tide A popular, but improper, term used as a syn. of *rip current.* The usage is erroneous because a rip current has no relation to the tide. Also spelled: *riptide.*

rip-up (a) Said of a sedimentary structure formed by shale clasts (usually of flat shape) that have been "ripped up" by currents from a semiconsolidated mud deposit and transported to a new depositional site. (b) Said of a clast in a rip-up structure.

rischorrite A variety of *nepheline syenite* in which nepheline is poikilitically enclosed in microcline perthite. The name, given by Kupletsky in 1932, is for Rischorr in the Khibina complex, Kola Peninsula, U.S.S.R. (Sørensen, 1974, p. 572). Not recommended usage.

rise [geomorph] (a) An upward slope in the land. (b) The top part of a hill or other landform that is higher than the surrounding ground.

rise [marine geol] A broad, elongate, smooth elevation of the ocean floor. Syn: *swell [marine geol].*

rise [streams] *resurgence (b).*

rise pit A pit through which an underground stream rises to the surface.

riser The vertical or steeply sloping surface of one of a series of natural steplike landforms, as those of a glacial stairway or of successive stream terraces. Cf: *tread.*

rise time The time history of the fault dislocation at the focus of an earthquake.

rising *resurgence (b).*

rising dune *climbing dune.*

rising tide That part of a tide cycle between low water and the following high water, characterized by landward or advancing movement of water. Also, an inflowing tidal river. Ant: *falling tide.* Syn: *flood tide.*

Riss (a) European stage: Pleistocene (above Mindel, below Würm). (b) The third classical glacial stage of the Pleistocene Epoch in the Alps, after the Mindel-Riss interglacial stage. See also: *Illinoian; Saale.*—Etymol: Riss River, Germany. Adj: *Rissian.*

Riss-Würm A term applied in the Alps to the third classical interglacial stage of the Pleistocene Epoch, after the Riss glacial stage and before the Würm. See also: *Sangamon.*

rithe An English term for a small stream. Syn: *rive.*

Rittmann norm A norm calculated after the manner of the *Niggli molecular norm* but in which greater variability is introduced into the composition of the minerals in order to bring the resulting mineral percentages into closer agreement with the mode of the rock (Rittmann, 1973).

rivadavite A mineral: $Na_6MgB_{24}O_{40}\cdot22H_2O$.

rive *rithe.*

river [coast] A term used in place names for an estuary, lagoon, tidal river, inlet, or strait; e.g. York River, Va., and Indian River, Fla.

river [gem] A pure-white diamond of very high grade. See also: *extra river.* Cf: *water [gem].*

river [streams] (a) A general term for a natural freshwater surface *stream* of considerable volume and a permanent or seasonal flow, moving in a definite channel toward a sea, lake, or another river; any large stream, or one larger than a brook or a creek, such as the trunk stream and the larger branches of a drainage system. (b) A term applied in New England to a small watercourse which elswhere in the U.S. is known as a *creek.*

riverain Pertaining to a riverbank; situated on or near a river. The term has a wider meaning than *riparian.*

river bar A ridge-like accumulation of alluvium in the channel, along the banks, or at the mouth, of a river. It is commonly emergent at low water and constitutes a navigational obstruction.

river-bar placer *bench placer.*

river basin The entire area drained by a river and its tributaries. Cf: *drainage basin.*

river bed The channel containing or formerly containing the water of a river. Also spelled: *riverbed.*

river bluff A *bluff* or steep hillslope or line of slopes above a river bank. Cf: *river cliff.*

river bottom The low-lying alluvial land along a river.

river breathing Fluctuation of the water level of a river (ASCE, 1962). Syn: *breathing.*

river cliff The steep *cutbank* formed by the lateral erosion of a river. Cf: *river bluff.*

river-delta marsh A brackish or freshwater marsh bordering the mouth of a distributary stream.

river-deposition coast A deltaic coast characterized by lobate seaward bulges crossed by river distributaries and bordered by lowlands (Shepard, 1948, p. 72).

river drift Rock material deposited by a river in one place after having been moved from another.

river end The lowest point of a river with no outlet to the sea, situated where its water disappears by percolation or evaporation (Swayne, 1956, p. 121).

river engineering A branch of civil engineering that deals with the control of rivers, their improvement, training, and regulation; and with flood mitigation.

riveret An obsolete syn. of *rivulet.*

river flat An *alluvial flat* adjacent to a river; a bottom.

river forecasting Forecasting the river stage and discharge, by hydrology and meteorology, including research into forecasting methods. In some countries, the term hydrometeorology is used with this limited meaning (ASCE, 1962).

riverhead The source or beginning of a river.

river ice (a) Ice formed on a river. (b) Ice carried by a river.

river icing A common *icing* in stream courses. Syn: *flood icing; aufeis.*

riverine (a) Pertaining to or formed by a river; e.g. a "riverine lake" created by a dam across a river. (b) Situated along the banks of a river; e.g. a "riverine ore deposit".

river morphology The study of the *channel pattern* and the *channel geometry* at several points along a river channel, including the network of tributaries within the drainage basin. Syn: *channel morphology; fluviomorphology; stream morphology.*

river pattern *channel pattern.*

river-pebble phosphate A term used in Florida for a transported, dark variety of *pebble phosphate* obtained from bars and flood plains of rivers. Cf: *land-pebble phosphate.* Syn: *river pebble; river rock.*

river plain *alluvial plain.*

river profile The *longitudinal profile* of a river.

river rock A syn. used in South Carolina for *river-pebble phosphate.*

river-run gravel Natural gravel as found in deposits that have been subjected to the action of running water (Nelson, 1965, p.373).

rivershed The drainage basin of a river.

riversideite A white mineral: $Ca_5Si_6O_{16}(OH)_2\cdot2H_2O$.

river system A river and all of its tributaries. Syn: *water system.*

river terrace *stream terrace.*

river valley An elongate depression of the Earth's surface, carved by a river during the course of its development.

riverwash (a) Soil material that has been transported and deposited by rivers. (b) An alluvial deposit in a river bed or flood channel, subject to erosion and deposition during recurring flood periods.

riviera A resort coastline much frequented by tourists, usually having extensive sandy beaches and a mild climate. Type locality: the Riviera along the coast of the Mediterranean Sea between Marseilles, France, and La Spezia, Italy.

riving The splitting off, cracking, or fracturing of rock, esp. by frost action. See also: *congelifraction.*

rivotite A mixture of malachite and stibiconite.

rivulet (a) A small stream; a brook or a runnel. (b) A small river. (c) A streamlet developed by rills running down a steep slope.—Obsolete syn: *riveret.*

rizalite A *philippinite* tektite from Rizal.

rizzonite A local variety of *limburgite* occurring on Monte Rizzoni, Italy. Named by Doelter in 1903. Obsolete.

RK *radial coefficient.*

r-meter A device that measures X-ray or gamma-ray intensity. Syn: *roentgen meter.*

R-mode factor analysis *Factor analysis* concerned with relationships among variables. Cf: *Q-mode factor analysis.*

road [coast] A *roadstead.* Term is usually used in the plural; e.g. Hampton Roads, Va.

road [glac geol] One of a series of erosional terraces in a glacial valley, formed as the water level dropped in an ice-dammed lake. See also: *parallel roads.*

road log A descriptive record of the route taken on a field trip and of the geology observed along it. Syn: *guidebook.*

road metal Rock suitable for surfacing macadamized roads, and for foundations for asphalt and concrete roadways; also used without asphaltic binder as the traffic-bearing surface, generally on secondary roads.

roadstead An area of water near a shore, sheltered by a reef, sandbank, or island, or an open anchorage, usually a shallow indentation in the coast, where vessels may lie in relative safety from winds and heavy seas; it is often outside, and less sheltered than, a harbor. An "open roadstead" is unprotected from the weather. Syn: *road.*

roaring sand A *sounding sand,* found on a desert dune, that sets up a low roaring sound that sometimes can be heard for a distance of 400 m. See also: *booming sand.*

robbery *capture [streams].*

robertsite A monoclinic mineral: $Ca_3Mn_4{}^{+3}(PO_4)_4(OH)_6\cdot3H_2O$.

robinsonite A mineral: $Pb_4Sb_6S_{13}$.

rocdrumlin *rock drumlin.*

Roche limit The distance from the surface of a larger primary planetary body to a smaller body at which the tidal stress across the smaller body exceeds its tensile strength. Named for Edouard Roche, a French mathematician, who defined this limit based on a body with no tensile strength; therefore, a body technically has to be inside the Roche limit before it will disrupt.

roche moutonnée A small elongate protruding knob or hillock of bedrock, so sculptured by a large glacier as to have its long axis oriented in the direction of ice movement, an upstream (stoss or scour) side that is gently inclined, smoothly rounded, and striated, and a downstream (lee or pluck) side that is steep, rough, and hackly. It is usually a few meters in height, length, and breadth. The term is supposed to have originated with Saussure, but there is no record that he ever used it (Longwell, 1933). Saussure (1786, par. 1061, p. 512-513) did use *moutonnée,* a French adjective meaning "fleecy, ruffled, or curled", in describing an assemblage of rounded knobs in the Alps. *Roche moutonnée* came to connote a resemblance between a single knob of the character described and a grazing sheep. Much later, the term was applied to a single glaciated knob so roughened by plucking as to resemble a sheep's back in surface texture as well as in general form, but this kind of surface is not generally regarded as essential. Pl: *roches moutonnées.* Pron: rosh moo-to-nay. Syn: *sheepback rock; sheep rock; whaleback; embossed rock.*

roche rock Quartz-schorl rock, consisting entirely of quartz and tourmaline; a product of *tourmalinization.*

rochlederite The soluble resin extracted from *melanchym* by alcohol. See also: *melanellite.*

rock (a) An aggregate of one or more minerals, e.g. granite, shale, marble; or a body of undifferentiated mineral matter, e.g. obsidian, or of solid organic material, e.g. coal. (b) Any prominent peak, cliff, or promontory, usually bare, when considered as a mass, e.g. the Rock of Gibraltar. (c) A rocky mass lying at or near the surface of a body of water, or along a jagged coastline, esp. where dangerous to shipping. (d) A slang term for a gem or diamond.

rockallite A coarse-grained, mafic, alkalic *granite* composed of quartz, acmite, albite, and microcline. The name, given by Judd in 1897, is for Rockall Bank in the North Atlantic Ocean 350 km west of Scotland. Not recommended usage.

rock asphalt *asphalt rock.*

rock association A group of igneous rocks within a petrographic province that are related chemically and petrographically, generally in a systematic manner such that chemical data for the rocks plot as smooth curves on variation diagrams. See also: *tribe.* Syn: *rock kindred; kindred; association [petrology].*

rock avalanche The very rapid downslope flowage of rock fragments, during which the fragments may become further broken or pulverized. Rock avalanches typically result from large rockfalls and rockslides, and their patterns of displacement have led to the term *rock-fragment flow* (Varnes, 1958). Characteristic features include chaotic distribution of large blocks, flow morphology and internal structure, relative thinness in comparison to large areal extent, high porosity, angularity of fragments, and lobate form. Cf: *debris flow.* Preferred syn: *sturzstrom.*

rock baby An odd-shaped, protruding hill of sandy bedrock, produced by differential erosion in the desert region of the Henry Mountains, Utah (Hunt et al. 1953, p. 175).

rock bar *riegel.*

rock basin A depression in solid rock, sometimes of great extent; esp. one formed by local erosion of the uneven floor of a cirque or glacial valley in a mountainous region, and usually containing a lake. See also: *glacial basin.*

rock-basin lake A glacial lake occupying a rock basin; e.g. a *paternoster lake.*

rock bench (a) A narrow valley-side niche developed during backweathering of weaker beds in a section of stratified rocks; a *structural bench* cut in solid rock. (b) A *wave-cut bench* produced on a rock surface.

rock bind An English term for a sandy shale or a banded or nonbanded siltstone.

rock bit *drill bit.*

rock bolt A bar, usually of steel, used in *rock bolting.* It is generally at least one meter in length and about 2 cm in diameter, and it is provided with a device for expanding the leading end so that it may be anchored firmly in rock. Rock bolts are classified according to the means by which they are secured or anchored: expansion, wedge, grouted, and explosive. Also spelled: *rockbolt.* Cf: *roof bolt.*

rock bolting A method of securing or strengthening closely jointed or highly fissured rocks in mine workings, tunnels, or rock abutments by inserting and firmly anchoring *rock bolts* in predrilled holes that range in length from less than one meter to about 12 m.

rock borer Any of certain bivalve mollusks that live in cavities they have bored into soft rock, concrete, or other material, usually by rotating the shell. Cf: *saxicavous.*

rockbridgeite A mineral: $(Fe^{+2},Mn)Fe_4^{+3}(PO_4)_3(OH)_5$.

rock burst A sudden and often violent breaking of a mass of rock from the walls of a tunnel, mine, or deep quarry, caused by failure of highly stressed rock and the rapid or instantaneous release of accumulated strain energy. It may result in closure of a mine opening, or projection of broken rock into it, accompanied by ground tremors, rockfalls, and air concussions. See also: *burst [rock mech]; strain burst.* Syn: *pressure burst.* Also spelled: *rock-burst; rockburst.*

rock city An area of large blocks of rock formed in situ, usually on a hillcrest or summit, as a result of weathering along several joint systems. The blocks and open joint fissures resemble the buildings and streets of a city.

rock cork *mountain cork.*

rock crystal *quartz crystal.*

rock cycle A sequence of events involving the formation, alteration, destruction, and reformation of rocks as a result of such processes as magmatism, erosion, transportation, deposition, lithification, and metamorphism. A possible sequence involves the crystallization of magma to form igneous rocks that are then broken down to sediment as a result of weathering, the sediments later being lithified to form sedimentary rocks, which in turn are altered to metamorphic rocks.

rock-defended terrace (a) A river terrace protected from later undermining by a projecting ledge or outcrop of resistant rock at its base (or at successively lower levels of the river). (b) A marine terrace protected from wave erosion by a mass of resistant rock at the base of the wave-cut cliff formed in the overlying coastal-plain sediments.—Syn: *rock-perched terrace.*

rock desert An upland desert area in which bedrock has been exposed after the removal of sand and dust particles by wind, or in which bedrock is covered by a thin veneer of coarse rock fragments; e.g. a *hammada.* Cf: *stony desert.* Syn: *rocky desert.*

rock doughnut A raised annular ridge encircling a *weather pit,* such as occur on certain granite domes of central Texas (Blank, 1951). Syn: *doughnut.*

rock drift *creep [mass move].*

rock drill (a) A machine for boring or making holes in rock, such as by percussion (e.g. a jackhammer) or by abrasion (e.g. a rotary drill). (b) A conical bit for drilling hard rock.

rock drumlin A smooth, streamlined hill, having a core of bedrock usually veneered with a layer of till; it is modelled by glacial erosion, and its long axis is parallel to the direction of ice movement. It is similar in outline and form to a true *drumlin,* but is generally less symmetrical and less regularly shaped. Syn: *false drumlin; rocdrumlin; rock drum; drumlinoid.*

rock face An exposed surface of rock in a wall or cliff.

rock failure *failure.*

rockfall (a) The relatively free falling or precipitous movement of a newly detached segment of bedrock (usually massive, homogeneous, or jointed) of any size from a cliff or other very steep slope; it is the fastest form of mass movement and is most frequent in mountain areas and during spring when there is repeated freezing and thawing of water in cracks in the rock. Movement may be straight down, or in a series of leaps and bounds down the slope; it is not guided by an underlying slip surface. Syn: *sturzstrom.* (b) The mass of rock moving in or moved by a rockfall; a mass of fallen rocks.—Also spelled: *rock fall.*

rockfall avalanche A rockfall that has turned into a flow, occurring only when large rockfalls and rockslides, involving millions of metric tons, attain extremely rapid speeds; most common in a rugged mountainous region, as that which occurred in 1903 at Frank, Alberta (McConnell & Brock, 1904). Cf: *debris flow.* Preferred syn: *sturzstrom.*

rockfall talus An accumulation of coarse, angular rock fragments, derived by falling from a cliff or steep rocky slope above. Blocks may be derived by pressure release and freeze-thaw action in previously formed cracks; heavy rain also helps release blocks. The rocks may fall and shatter, roll, or bounce; large sizes collect at the bottom. The angle of slope is greater than 32° and may reach 42° (White, 1967, p. 237). Syn: *scree.*

rock fan An eroded, convex, fan-shaped bedrock surface having its apex at the point where a mountain stream debouches upon a piedmont slope, and occupying the zone where a pediment meets the mountain slope (assuming that the mountain front retreats as a result of lateral planation). According to Johnson (1932), a *pediment* evolves from a coalescence of rock fans, although the term "rock fan" is often considered an equivalent of "pediment".

rock-fill dam A dam composed primarily of large, broken, loosely placed or pervious rocks, with either an impervious core or an upstream facing.

rock-floor robbing A form of *sheetflood erosion* in which sheet-floods remove crumbling debris from rock surfaces in desert mountains (Cotton, 1958, p. 258).

rock flour Finely comminuted, chemically unweathered material, consisting of silt- and clay-sized angular particles of rock-forming minerals, chiefly quartz, formed when rock fragments are pulverized while being transported or are crushed by the weight of superincumbent material. The term is most commonly applied to the very fine powder that is formed when stones embedded in a glacier or ice sheet abrade the underlying rocks, and that is deposited as the matrix in till or in outwash deposits. Syn: *glacier meal; glacial meal; glacial flour; rock meal.*

rock flowage *flow [exp struc geol].*

rock-forming Said of those minerals that enter into the composition of rocks, and determine their classsification. The more important rock-forming minerals include quartz, feldspars, micas, amphiboles, pyroxenes, olivine, calcite, and dolomite.

rock-fragment flow *sturzstrom.*

rock generator *spinner magnetometer.*

rock glacier A mass of poorly sorted angular boulders and fine material, with interstitial ice a meter or so below the surface (ice-cemented) or containing a buried ice glacier (ice-cored). It occurs in high mountains in a permafrost area, and is derived from a cirque wall or other steep cliff. Rock glaciers have the general appearance and slow movement of small valley glaciers, ranging from a few hundred meters to several kilometers in length, and having a distal area marked by a series of transverse arcuate ridges. When active, they may be 50 m thick with a surface movement (resulting from the flow of interstitial ice) of 0.5-2 m/yr. Rock glaciers are classified in plan as lobate, tongue-shaped, or spatulate. Cf: *block stream; chrystocrene.*

rock-glacier creep The slow creep of tongues of rock glaciers.

rock gypsum A sedimentary rock composed chiefly of the mineral gypsum; it is generally massive, and ranges from coarsely crystalline to finely granular. It often shows disturbed bedding owing to expansion during hydration of the parent anhydrite. Syn: *gyprock.*

rock hole An Australian term for a *rock tank.*

rock hound (a) An amateur mineralogist. (b) A petroleum exploration geologist. —See also: *pebble pup.*

rocking stone A stone or boulder, often of great size, so finely poised on its foundation that it can be rocked or moved backward and forward with little force. It may be a glacial erratic or a rounded residual block formed in place by weathering. See also: *logan stone; balanced rock; elephant rock.* Syn: *roggan.*

rock island (a) *meander core.* (b) A bedrock hill surrounded by alluvium in an aggraded stream valley.

rock kindred *rock association.*

rock leather *mountain leather.*

rock magnetism The study of the origins and characteristics of magnetization in rocks and minerals.

rock mantle *regolith.*

rock meal (a) *moonmilk.* (b) *rock flour.*

rock mechanics The theoretical and applied science of the physical behavior of rocks, representing a "branch of mechanics concerned with the response of rock to the force fields of its physical environment" (NAS-NRC, 1966, p.3).

rock milk *moonmilk.*

rock mill A *pothole* in a stream bed.

rock pedestal *pedestal [geomorph].*

rock pediment A *pediment* developed on a bedrock surface.

rock pendant *pendant [speleo].*

rock-perched terrace *rock-defended terrace.*

rock phosphate *phosphate rock.*

rock pillar (a) A column of rock produced by differential weathering or erosion, as along a joint plane; a *hoodoo.* (b) In a cave, a columnlike structure that is residual bedrock rather than a speleothem.

rock plane A term used by Johnson (1932) as a syn. of *pediment;* the term is not recommended in this usage as there are many approximately planate rock surfaces that lack the areal extent and climatic restriction of a pediment.

rock platform (a) A *wave-cut platform* eroded on a rock surface. (b) A *high-water platform* eroded on a rock surface.

rock pool A *tide pool* formed along a rocky shoreline.

rock pressure (a) The pressure exerted by surrounding solids on the support system of underground openings, including that caused by the weight of the overlying material, residual unrelieved stresses, and pressures associated with swelling clays (Stokes & Varnes, 1955, p.125). (b) The compressive stress within the solid body of underground geologic material. (c) *geostatic pressure.*

rock ruby A fine red variety of garnet; specif. *pyrope.*

rock salt (a) Coarsely crystalline *halite* occurring as a massive, fibrous, or granular aggregate, and constituting a nearly pure sedimentary rock that may occur in domes or plugs or as extensive beds resulting from evaporation of saline water. It is frequently stained by iron or mixed with fine-grained sediments. (b) Artificially prepared salt in the form of large crystals or masses.

rock sea *block field.*

rock series *igneous-rock series.*

rockshelter A cave, commonly formed in nonsoluble rock, that extends only a short way underground, with a roof of overlying rock that usually extends beyond its sides. Syn: *shelter cave.* Partial syn: *sandstone cave.*

rock silk A silky variety of asbestos.

rockslide (a) A slide involving a downward and usually sudden and rapid movement of newly detached segments of bedrock sliding or slipping over an inclined surface of weakness, as a surface of bedding, jointing, or faulting, or other pre-existing structural feature. The moving mass is greatly deformed and usually breaks up into many small independent units. Rockslides frequently occur in high mountain ranges, as the Alps or Canadian Rockies. (b) The mass of rock moving in or moved by a rockslide.—Also spelled: *rock slide.* Syn: *rock slip.*

rockslide avalanche An obsolescent syn. of *rockslide.*

rock slip *rockslide.*

rock soap *mountain soap.*

rock stack *stack [coast].*

rock step (a) A *knickpoint* produced by the outcrop of a resistant rock. (b) One of a series of ledges or other irregularities of gradient in the upper reaches of a hanging valley; an abrupt descent in the floor of a glacial valley. See also: *riegel.*

rock-stratigraphic unit *lithostratigraphic unit.*

rock stratigraphy *lithostratigraphy.*

rock stream *block stream.*

rock stripe *stone stripe.*

rock tank A natural *tank* formed in rock by differential weathering or differential erosion. See also: *sand tank.* Syn: *rock hole.*

rock terrace A *stream terrace* produced on the side of a valley by erosion in horizontal beds of unequal resistance, composed of strong bedrock that is worn back less rapidly than the weaker beds above and below. Cf: *alluvial terrace.* Syn: *stream-cut terrace; cut terrace; erosion terrace.*

rock train A term suggested by Kendall & Wroot (1924, p. 448) for the rock material in "process of transport at the sides and in the middle of a glacier" and "subject to the dynamic forces of the glacier".

rock type [coal] *banded ingredients.*

rock type [petrology] (a) One of the three major groups of rocks: igneous, sedimentary, metamorphic. (b) A particular kind of rock having a specific set of characteristics. It may be a general classification, e.g. a basalt, or a specific classification, e.g. a basalt from a particular area and having a unique description.

rock unit *lithostratigraphic unit.*

rock waste *debris.*

rock weathering The chemical decomposition and mechanical disintegration of rocks in place, at or near the Earth's surface.

rock wood *mountain wood.*

rock wool *mineral wool.*

rock wreath *sorted circle.*

rocky desert *rock desert.*

Rocky Mountain orogeny A name proposed by W.H. White (1959) for a time of major folding and thrusting during Late Cretaceous and Paleocene time in the Rocky Mountains of eastern British Columbia and adjacent Alberta; it is broadly equivalent to the *Laramide orogeny* of western United States.

rod [paleont] (a) An elongate holothurian sclerite having a circular cross section, one or more axes, and an *eye* at its end. (b) A part of a heterococcolith having one dimension large and two much smaller. (c) A thin cylindrical or prismatic skeletal element in archaeocyathids, commmonly radially oriented.

rod [sed] A rod-like or prolate shape of a sedimentary particle, defined in *Zingg's classification* as having a width/length ratio less than 2/3 and a thickness/width ratio greater than 2/3. Syn: *roller [sed].*

rod [surv] (a) A bar or staff for measuring, such as a graduated pole used as a target in surveying; specif. a *level rod.* (b) A unit of length equal to 16.5 ft. Also called a *perch* or *pole.*

rodalquilarite A triclinic mineral: $H_3Fe_2{}^{+3}(TeO_3)_4Cl$.

rodding In metamorphic rocks, a linear structure in which the stronger parts, such as vein quartz or quartz pebbles, have been shaped into parallel rods. Whether the structure is formed parallel to the direction of transport or parallel to the fold axes has been debated. Cf: *mullion.*

roddon A term used in East Anglia, England, for a natural levee built of sediment carried upstream by the tide rather than downstream by a river.

rodingite A medium- to coarse-grained gabbroic rock, commonly enriched in calcium, barium, and strontium, and containing, as essential minerals, grossular and clinopyroxene. More highly altered varieties also contain prehnite or serpentine or both. The name was applied by Bell in 1911.

rodite An obsolete syn. of *diogenite.*

rod level A spirit level attached to a level rod or stadia rod to assure a vertical position of the rod prior to instrument reading.

rodman One who uses or carries a surveyor's level rod; a *chainman.*

roeblingite A white mineral: $Pb_2Ca_7Si_6O_{14}(OH)_{10}(SO_4)_2$.

roedderite A meteorite mineral of the osumilite group: $(Na,K)_2(Mg,Fe)_5Si_{12}O_{30}$.

roemerite A rust-brown to yellow mineral: $Fe^{+2}Fe_2^{+3}(SO_4)_4 \cdot 14H_2O$. Also spelled: *römerite.* Syn: *louderbackite.*

roentgenite *röntgenite.*

roentgen meter *r-meter.*

roepperite A variety of fayalite with Mn^{+2} and Zn replacing some Fe^{+2}.

roesslerite A monoclinic mineral: $MgH(AsO_4) \cdot 7H_2O$. It is isomorphous with phosphorroesslerite. Also spelled: *rösslerite.*

roestone *oolite.*

rofla A term used by E. Desor for an extremely narrow, tortuous gorge, formed by meltwater streams flowing from a glacier (Marr, 1900, p. 172 & 314); e.g. the gorge of the Trient, near Vernayaz, Switzerland. Pl: *roflas.*

rogenstein An oolite in which the ooliths are united by argillaceous cement. Etymol: German *Rogenstein,* "roestone". Also spelled: *roggenstein.*

roggan *rocking stone.*

roggianite A mineral: $NaCa_6Al_9Si_{13}O_{46} \cdot 20H_2O$.

rognon (a) A small rocky peak or ridge surrounded by glacier ice in a mountainous region. Also, a similar peak projecting above the bed of a former glacier (ADTIC, 1955, p.67). Cf: *nunatak.* (b) A rounded nunatak (Lliboutry, 1958, p.264); *nunakol.*

roil A small section of a stream, characterized by swiftly flowing, turbulent water.

roily (a) Said of muddy or sediment-filled water. Cf: *turbid.* (b) Said of turbulent, agitated, or swirling water.

roll [coal] (a) An elongate protrusion of shale, siltstone, or sandstone (locally limestone) from the roof into a coal seam, causing a thinning of the seam and sometimes replacing it almost entirely; cf: *cutout.* A roll is commonly overlain by a thin coal stringer. (b) An elongate upheaval of the floor material into a coal seam, causing thinning of the seam (Woolnough, 1910). Syn: *horseback [coal].* (c) Various minor deformations or dislocations of a coal seam, such as *washouts,* small monoclinal folds, or faults with little displacement.

roll [ore dep] *roll orebody.*

roll [sed] A primary sedimentary structure produced by deformation involving subaqueous slump or vertical foundering; e.g. a flow roll or a pseudonodule.

rolled garnet *rotated garnet.*

roller [sed] *rod [sed].*

roller [waves] A general term, usually meaning one of a series of massive, long-crested waves that roll in upon a coast (as after a storm), usually retaining its form until it reaches the beach or shoal. Cf: *comber.*

roll-front deposit *roll-front orebody.*

roll-front orebody A *roll orebody* of the Wyoming type, which is bounded on the concave side by oxidized altered rock typically containing hematite or limonite, and on the convex side by relatively reduced altered rock typically containing pyrite and organic matter. Cf: *bifacies.*

rolling beach The upper part of an accumulation at the base of a sea cliff of boulders and pebbles being ground to sand and finer particles (Shaler, 1895).

rolling prairie A term used in Texas for "a plain of undulating or rounded hilly relief" (Hill, 1900, p. 7).

rolling strata (a) *ripple cross-lamination.* (b) *wavy bedding.*

rolling topography Any land surface having a gradual succession of low, rounded hills or undulations that impart a wave effect to the surface; esp. a land surface much varied by many small hills and valleys.

roll mark One of a series of similar *tool marks* following each other in a line parallel to the direction of the current, produced by an object that was rolled along the bottom. Originally proposed (as "roll-spuren") by Krejci-Graf in 1932; first usage in English was by Dzulynski and Slaczka (1958, p. 234). Cf: *skip mark.*

roll-off The rate at which seismic attenuation changes with frequency in a filter.

roll orebody A uranium and/or vanadium orebody in a sandstone lens or layer, which cuts across bedding in sharply curving forms, commonly C-shaped or S-shaped in cross section. Two types can be distinguished: the Colorado Plateau type, named in 1956, and the Wyoming type, named in 1962. Roll orebodies of the Colorado Plateau type are of highly variable geometry, with their longest dimension in plan view parallel to the axes of buried sandstone lenses representing former stream channels, and surrounded by a wide halo of reduced (altered) rock. Orebodies of the Wyoming type are crescent-shaped in cross section and typically form in relatively thick, tabular, or elongate sandstone bodies, with the tips of the crescent thinning and becoming tangent to mudstone layers above and below. See: Bailey & Childers (1977). See also: *roll-front orebody.* Syn: *ore roll; roll-type orebody; roll [ore dep].*

rollover A feature of some Gulf Coast *growth faults,* in which the beds of the downthrown block dip toward the fault surface in an orientation opposite to that produced by drag. Syn: *dip reversal.*

roll-type orebody *roll orebody.*

roll-up structure *convolutional ball.*

romanechite An iron-black to steel-gray mineral: $BaMn^{+2}Mn_8^{+4}O_{16}(OH)_4$. Calcium, potassium, sodium, cobalt, and copper are sometimes present. Romanechite has a brownish-black streak, and commonly occurs massive, botryoidal, reniform, or stalactitic. It is an important ore of manganese. Syn: *psilomelane; black hematite.*

romanzovite A dark-brown variety of grossular garnet.

romarchite A tetragonal mineral: SnO.

romeite A honey-yellow to yellow-brown mineral occurring in minute octahedrons: $(Ca,Fe,Mn,Na)_2(Sb,Ti)_2O_6(O,OH,F)$. Also spelled: *roméite.*

römerite *roemerite.*

rond A British term for a narrow *washland* that separates a broad from the river; it is connected with the river by an artificial passageway.

rongstockite A medium- to fine-grained plutonic rock composed of zoned plagioclase, orthoclase, some cancrinite, augite, mica, hornblende, magnetite, sphene, and apatite. The rock resembles *essexite* but contains less nepheline and has sodic rather than calcic plagioclase. The name, given by Tröger in 1935, is from Rongstock (Roztoky), Czechoslovakia. Not recommended usage.

röntgenite A wax-yellow to brown mineral: $Ca_2(Ce,La)_3(CO_3)_5F_3$. Also spelled: *roentgenite.*

roof [intrus rocks] The country rock bordering the upper surface of an igneous intrusion. Cf: *floor [intrus rocks].*

roof [ore dep] The rock above an orebody; the *back.*

roof bolt A *rock bolt* used to support the roof of a mine or mine shaft.

roof collapse *roof foundering.*

roof control The engineering study of the behavior of unsupported rocks in mining operations, the systematic measurement of the movement of roof strata and the stresses involved, and the most effective measures to prevent or reduce roof movements (Nelson, 1965, p.378). Syn: *strata control.*

roofed mud crack *vaulted mud crack.*

roof foundering Collapse of rocks into an underlying reservoir or magma, usually following the evacuation of a large quantity of the magma. Less preferred syn: *roof collapse.*

roof pendant A downward projection of country rock into an igneous intrusion. Cf: *cupola.* Syn: *pendant.*

roof rock A shale or other impervious rock that acts as a barrier to the movement of oil or gas; it overlies a *reservoir rock* to form a *trap.*

roof thrust The upper boundary of a *duplex fault zone* (Dahlstrom, 1970, p. 418).

room An expanded part of a cave *passage.*

room-and-pillar (a) Said of a system of mining in which the ore is mined in rooms separated by pillars of undisturbed rock left for roof support. (b) Said of a coral-reef structure characterized by interconnected and roofed-over surge channels or caverns.

rooseveltite A white monoclinic mineral: $BiAsO_4$.

rooster tail A plumelike form of water and sometimes spray that occurs at the intersection of two crossing waves.

root [fold] The basal part of a fold nappe that was originally linked to its source, or *root zone.*

root [ore dep] (a) Syn. of *bottom [ore dep].* (b) The conduit leading up through the basement to an ore deposit in the superjacent rocks.

root [paleont] An expanded, branching, treelike extension at the distal end of an echinoderm.

root [tect] According to the *Airy hypothesis,* the downward extension of lower-density crustal material as isostatic compensation for its greater mass and high topographic elevation. Syn: *downward bulge.* Cf: *antiroot.*

root cap A thimblelike cellular tissue that fits over the growing tip of a rootlet and protects it.

root cast (a) A slender, nearly vertical, commonly downward-branching sedimentary structure, formed by the filling of a tubular opening left by a root. (b) *rhizoconcretion.*

root clay An *underclay* characterized by the occurrence of fossil roots of coal plants. See also: *rootlet bed.*

rootless fumarole A *fumarole* that derives its gases from the lava flow or ash flow on which it occurs, rather than from some deep source. Syn: *secondary fumarole.*

rootless vent A source of lava that is not directly connected to a volcanic vent or magma source; it may be an accumulation of overflow or an outflow from an otherwise solidified lava flow.

rootlet bed A stratum characterized by the occurrence of fossil rootlets of plants; e.g. a *root clay* beneath a coal bed.

root level The place within a sediment at which plant roots are found in the living position.

root-mean-square deviation *standard deviation.*

root scar *root zone.*

root sheath A hollow *rhizoconcretion.*

root-tuft A tuft of subparallel, elongate spicules protruding from the base of a sponge and serving to fix it in the substrate.

root zone [fault] That zone in the crust from which thrust faults emerge.

root zone [fold] The source or original attachment of the *root* of a nappe. Syn: *root scar.*

root zone [soil] *rhizosphere.*

ropak A pinnacle or slab of sea ice standing vertically, rising as high as 8 m above the surrounding ice, and representing an extreme formation of *ridged ice.* Etymol: Russian.

ropy lava *pahoehoe.*

roquesite A tetragonal mineral: $CuInS_2$.

rosasite A pale-green or sky-blue mineral: $(Cu,Zn)_2CO_3(OH)_2$. Cf: *glaukosphaerite.*

roscherite A dark-brown monoclinic mineral: $(Ca,Mn,Fe)_3Be_3(PO_4)_3(OH)_3 \cdot 2H_2O$.

roscoelite A vanadium-bearing mineral of the mica group, having the approximate formula: $K(V,Al,Mg)_2(AlSi_3)O_{10}(OH)_2$. It is tan to greenish brown and occurs in minute scales or flakes in the cement of certain sandstones and in certain gold-quartz deposits.

rose [gem] n. (a) *rose cut.* (b) *rose diamond.* (c) A diamond so small that it can be cut little if at all.—adj. Said of a gem having a rose, pink, or lilac color; e.g. "rose topaz".

rose [sed] *rosette [sed].*

rose [surv] *compass rose.*

rose cut An early style of cutting for a gemstone, now used primarily on small diamonds; it usually has a flat, unfaceted base and a somewhat dome-shaped top that is covered by a varied number of triangular facets and terminates in a point. Syn: *rose [gem]; rosette [gem].*

rose diagram A circular or semicircular star-shaped graph indicating values or quantities in several directions of bearing, consisting of radiating rays drawn proportional in length to the value or quantity; e.g. a current rose, a structural diagram for plotting strikes of planar features, or a "histogram" of orientation data.

rose diamond A rose-cut diamond. Syn: *rose [gem].*

roselite A rose-red monoclinic mineral: $Ca_2(Co,Mg)(AsO_4)_2 \cdot 2H_2O$. It is isomorphous with brandtite and dimorphous with beta-roselite.

rosenbuschite A mineral: $(Ca,Na)_3(Zr,Ti)Si_2O_8F$.

Rosenbusch's law A statement of the sequence and crystallization of minerals from magmas, proposed in 1882 by the German petrologist Harry Rosenbusch, to which there are many exceptions.

rosenhahnite A mineral: $Ca_3Si_3O_8(OH)_2$.

rose opal An opaque variety of common opal having a fine red color.

rose quartz A pink to rose-red and commonly massive variety of crystalline quartz often used as a gemstone or ornamental stone. The color is perhaps due to titanium, and is destroyed or becomes paler on exposure to strong sunlight. Syn: *Bohemian ruby.*

rosette [gem] *rose cut.*

rosette [paleont] (a) A delicate calcareous plate formed of metamorphosed basal plates, centrally located within the radial pentagon in some free-swimming crinoids. (b) The group of five petal-shaped ambulacra on certain echinoids. (c) A cluster of parts in circular form; e.g. a discoaster. (d) A flower-shaped button within a hexagonal pore frame in a radiolarian skeleton. (e) The cellular part of a *septula* in cheilostome and ctenostome bryozoans.

rosette [sed] A radially symmetric, sand-filled crystalline aggregate or cluster with a fancied resemblance to a rose, formed in sedimentary rocks by barite, marcasite, or pyrite. See also: *barite rosette.* Syn: *rose.*

rosette plate *pore plate.*

rosette texture A flowerlike or scalloped pattern of a mineral aggregate.

rosickyite A mineral: γ-S. It consists of native *sulfur* in the gamma crystal form. Syn: *gamma-sulfur.*

rosieresite A yellow to brown mineral consisting of a hydrous phosphate of lead, copper, and aluminum.

rosin jack A yellow variety of sphalerite. Syn: *resin jack.*

rosin tin A reddish or yellowish variety of cassiterite. Syn: *resin tin.*

Rosiwal analysis In petrography, a quantitative method of estimating the volume percentages of the minerals in a rock. Thin sections of a rock are examined with a microscope fitted with a micrometer which is used to measure the linear intercepts of each mineral along a particular set of lines. This method "is based on the assumption that the area of a mineral on an exposed surface is proportional to its volume in the rock mass" (Nelson & Nelson, 1967, p. 320).

rosolite *landerite.*

ross A promontory. Etymol: Celtic.

Rossi-Forel scale One of the earthquake *intensity scales,* devised by the Italian geologist Michele Stefano de Rossi and the Swiss naturalist Francois Alphonse Forel in 1883. It has a range of one to ten. It has been replaced by the *modified Mercalli scale.* Cf: *Richter scale.*

rossite A yellow mineral: $CaV_2O_6 \cdot 4H_2O$.

rösslerite *roesslerite.*

rostellum A low projection between the anterior adductor muscle scars of the pedicle valve of some craniacean brachiopods to which the internal oblique muscles are attached.

rosterite *vorobyevite.*

rosthornite A brown to garnet-red variety of retinite with a low oxygen content (4.5%), found in lenticular masses in coal.

rostral carina The ridge forming the lateral margin of the rostrum on some decapods, often passing into the orbital carina (Holthuis, 1974, p. 735).

rostral incisure The anterior opening in the carapace of many ostracodes having a rostrum, which persists when the valves are closed, serving for the permanent protrusion of the antennae; sometimes referred to as the "permanent opening" of the carapace.

rostral notch The indentation, as seen in lateral view, below the rostrum in certain marine ostracodes (such as *Cyridina*).

rostral plate (a) An anteriorly projecting movable median extension of the carapace of a malacostracan crustacean. (b) A small median ventral plate of the head region in a trilobite, immediately anterior to the hypostome. Syn: *epistome.*

rostral suture That portion of the facial suture of a trilobite along the join of the anterior of the rostral plate and the cranidium.

rostrate Having a rostrum; specif. said of a brachiopod with a prominent beak of the pedicle valve projecting over a narrow cardinal margin.

rostroconch Any benthic marine mollusk belonging to the class Rostroconchia, characterized by an uncoiled univalved larval shell that straddles the dorsal midline and a bivalved adult shell with one or more shell layers continuous across the dorsal margin so that a dorsal commissure is lacking (Pojeta & Runnegar, 1976).

rostrolateral One of a pair of plates in certain cirripede crustaceans lying between the rostrum and the lateral. Syn: *rostral latus.*

rostrum (a) A part of an arachnid suggesting a bird's bill, e.g. the tubelike "beak" in the order Solpugida, or the anterior spike of the carapace of the order Eophrynidae, similar to the rostrum of a lobster (TIP, 1955, pt.P, p.62). (b) The anteriorly projecting spine-like median extension of the carapace of a crustacean; e.g. an unpaired plate adjacent to the scuta of a cirripede. See also: *rostral plate.* (c) An elevation of the secondary shell on the inner surface of the brachial valve of some craniacean brachiopods, in front of the anterior adductor muscles, consisting of a pair of low club-shaped protuberances forming the seat of attachment for the brachial protractor muscles. Also, the beak of an articulate brachiopod. (d) The grooved extension of a gastropod shell protecting the siphon; the attenuated extremity of the last whorl other than the siphonal canal, as in *Tibia.* Also, the snout of a gastropod when nonretractile. (e) A pointed projection of peristome on the venter of an ammonoid. Also, the *guard* of a belemnite. (f) A tubular extension of the posterodorsal part of the shell of some rostroconchs. (g) In cheilostome bryozoans, the rounded or pointed skeletal rim around the palate on which the mandible occludes in an avicularium. Used as a syn. of *palate.* Syn: *beak.*—Pl: *rostra* or *rostrums.*

rotaliid Any trochospirally, rather than planispirally, coiled foraminifer, excluding trochospirally coiled planktonic genera.

rotary current A tidal current of variable velocity that flows continually (generally clockwise in the Northern Hemisphere) and changes direction progressively through 36° during a tide cycle, and that returns to its original direction after a period of 12.42 hours; occurs in open ocean and along the coast where flow is not restricted. Cf: *reversing current.*

rotary drilling The chief method of drilling deep wells, esp. for oil and gas. A hard-toothed *drill bit* at the bottom of a rotating *drill pipe* grinds a hole in the rock. Lubrication and cooling are provided by continuously circulating *drilling mud,* which also brings the *well cuttings* to the surface. Cf: *cable-tool drilling.*

rotary fault *rotational fault.*

rotary polarization *optical activity.*

rotary table In *rotary drilling,* a power-driven circular platform on the derrick floor that rotates the *kelly, drill pipe,* and *drill bit.* It is sometimes used as the zero-depth reference for downhole measurements. Cf: *kelly bushing.*

rotate Wheel-shaped, as in a flower whose parts are flat and radiating.

rotated garnet A garnet crystal that shows evidence that it has been rotated during metamorphism (Knopf and Ingerson, 1938). Syn: *rolled garnet; pinwheel garnet; spiral garnet; smowball garnet.*

rotating dipole A source that consists of two fixed dipoles, either magnetic or electric-current, oriented with their axes mutually perpendicular and energized sequentially so that the resultant magnetic and electric fields they produce rotate in space.

rotation (a) *internal rotation.* (b) *external rotation.*

rotational bomb A pyroclastic bomb whose shape is formed by spiral motion or rotation during flight; rotation produces such types as spheroidal, tear-shaped, and spindle-shaped bombs. See also: *fusiform bomb.*

rotational cylindroidal fold A cylindrical fold, the axial surface of which has been distorted by a subsequent or cross fold (Whitten, 1959).

rotational deflection The deflection of currents of air or water by rotation of the Earth, as stated in *Ferrel's law.*

rotational fault A fault on which rotational movement is exhibited; a partial syn. of *hinge fault.* Cf: *scissor fault.*

rotational flow Flow in which each fluid element rotates about its own mass center.

rotational landslide A slide in which shearing takes place on a well defined, curved shear surface, concave upward, producing a backward rotation in the displaced mass (Hutchinson, 1968). It may be single, successive (repeated up- and down-slope), or multiple (as the number of slide components increases). Syn: *rotational slump.* See also: *Toreva block.*

rotational movement Apparent fault-block displacement in which the blocks have rotated relative to one another, so that alignment of formerly parallel features is disturbed. Cf: *translational movement.* See also: *rotational fault.* Less-preferred syn: *rotatory movement.*

rotational slump *rotational landslide.*

rotational strain *Strain* in which the orientation of the principal axes of strain is different before and after deformation. Not to be confused with noncoaxial *progressive deformation.* Cf: *irrotational strain.* Syn: *pure rotation.*

rotational transformation A type of crystal *transformation* that is a change from an ordered phase to a partially disordered phase by rotation of groups of atoms. It is usually a rapid process. Cf: *dilatational transformation; displacive transformation; reconstructive transformation; substitutional transformation.*

rotational wave *S wave.*

rotation axis *symmetry axis.*

rotation method A method of X-ray diffraction analysis using a rotating single crystal, monochromatic radiation, and a cylindrical film coaxial with the rotation axis of the crystal.

rotation twin A crystal twin whose symmetry is formed by apparent axial rotation of 180°. Cf: *reflection twin.*

rotatory dispersion In crystal optics, the breaking-up of white light into colors by passing it through an optically active substance, such as quartz.

rotatory movement A less-preferred syn. of *rotational movement.*

rotatory reflection axis *rotoreflection axis.*

Rotliegende European series (esp. in Germany): Lower and Middle Permian (below Zechstein). It contains the Autunian and Saxonian stages. Obsolescent spelling: *Rothliegende.*

rotoinversion axis A type of crystal symmetry element that combines a rotation of 60°, 90°, 120°, or 180° with inversion across the center. Syn: *symmetry axis of rotoinversion; symmetry axis of rotary inversion.*

rotoreflection axis A type of symmetry element that combines a rotation of 60°, 90°, 120°, or 180° with reflection across the plane perpendicular to the axis. Syn: *rotatory reflection axis.*

rotten ice Ice in which the grains or crystals are loosened, one from the other, forming a *honeycomb structure [ice],* due to melting along grain boundaries. See also: *candle ice.*

rotten spot *pothole [coast].*

rottenstone Any highly decomposed but still coherent rock; specif. a soft, friable, lightweight, earthy residue consisting of fine-grained silica and resulting from the decomposition of siliceous limestone (or of a highly shelly sandstone) whose calcareous material has been removed by the dissolving action of water. Cf: *tripoli.* Syn: *terra cariosa.*

rotula One of the five massive radial elements in ambulacral position at the top of Aristotle's lantern of an echinoid. Pl: *rotulae.*

roubaultite A mineral: $Cu_2(UO_2)_3(OH)_{10} \cdot 5H_2O$.

rougemontite A coarse-grained *gabbro* composed of anorthite, titanaugite, and small amounts of olivine and iron oxides. Its name, given by O'Neill in 1914, is derived from Rougemont, Quebec. Not recommended usage.

rough n. An uncut gemstone.—adj. Pertaining to an uncut or unpolished gemstone; e.g. a *rough* diamond.

rough criterion In radar, the relationship between surface roughness, depression angle, and wavelength that determines whether a surface will scatter the incident radar pulse in rough or intermediate fashion. Cf: *smooth criterion; Rayleigh criterion.*

rough ice An expanse of ice having an uneven surface caused by formation of *pressure ice* or by *growlers* frozen in place (ADTIC, 1955, p. 68).

roughneck A general workman in a drilling crew; also, one who builds and repairs oil derricks. Cf: *roustabout.*

roughness coefficient A factor in formulas for computing the average velocity of flow of water in a conduit or channel which represents the effect of roughness of the confining material upon the energy losses in the flowing water.

rounded Round or curving in shape; specif. said of a sedimentary particle whose original edges and corners have been smoothed off to rather broad curves and whose original faces are almost completely removed by abrasion (although some comparatively flat surfaces may be present), such as a pebble with a roundness value between 0.40 and 0.60 (midpoint at 0.500) and few (0-5) and greatly subdued secondary corners that disappear at roundness 0.60 (Pettijohn, 1957, p. 59). The original shape is still readily apparent. Also, said of the *roundness class* containing rounded particles.

roundness The degree of abrasion of a clastic particle as shown by the sharpness of its edges and corners, expressed by Wadell (1932) as the ratio of the average radius of curvature of the several edges or corners of the particle to the radius of curvature of the maximum inscribed sphere (or to one-half the nominal diameter of the particle). The value is more conveniently computed from a plane figure (a projection or cross section); thus, roundness may be defined as the ratio of the average radius of curvature of the corners of the particle *image* to the radius of the maximum inscribed circle. A perfectly rounded particle (such as a sphere) has a roundness value of 1.0; less-rounded particles have values less than 1.0. The term has been used carelessly and should not be confused with *sphericity*: a nearly spherical particle may have sharp corners and be angular, while a flat pebble, far from spherical in shape, may be well-rounded. Cf: *flatness.* See also: *angularity; roundness* *class.*

roundness class An arbitrarily defined range of *roundness* values for the classification of sedimentary particles. Pettijohn (1957, p.58-59) recognizes five classes: *angular; subangular; subrounded; rounded; well-rounded.* Powers (1953) adds a sixth class: *very angular.* Syn: *roundness grade.*

roundness grade *roundness class.*

roundstone (a) A term proposed by Fernald (1929) for any naturally rounded rock fragment of any size larger than a sand grain (diameter greater than 2 mm), such as a boulder, cobble, pebble, or granule. Cf: *sharpstone.* (b) *cobblestone.*

roustabout A common laborer called upon to do any of the unskilled jobs in an oil field or refinery, or around a mine. Cf: *roughneck.*

routhierite A tetragonal mineral: $TlHgAsS_3$.

routine A sequence of computer instructions for accomplishing a specific, well-defined, or limited task; e.g. a subdivision of a computer *program.*

routivarite A fine-grained garnet-bearing *anorthosite,* named by Sjögren in 1893 for Routivara, Norrbotten, Sweden. Not recommended usage.

rouvillite A light-colored *theralite* composed predominantly of labradorite or bytownite and nepheline, with small amounts of titanaugite, hornblende, pyrite, and apatite. The name, given by O'Neill in 1914, is from Rouville County, Quebec. Not recommended usage.

roweite A light-brown mineral: $Ca_2Mn_2B_4O_7(OH)_6$.

rowlandite A massive dark-green mineral, approximately: $Y_3(SiO_4)_2(OH,F)$.

royal agate A mottled obsidian.

royalty The landowner's share of the value of minerals produced on a property. It is commonly a fractional share of the current market value (oil and gas) or a fixed amount per ton (mining).

rozenite A mineral: $FeSO_4 \cdot 4H_2O$.

rozhkovite An orthorhombic mineral: $(Cu,Pd)_{3-x}Au_2$.

RSA *ripe-snow area.*

r selection The evolutionary process favoring development of *r strategists* or *opportunistic species* (Wilson & Bossert, 1971, p. 110-111).

RSI *ripple symmetry index.*

r strategist *opportunistic species.*

RT *rotary table.*

rubasse (a) A crystalline variety of quartz, stained a ruby red by numerous small scales or flecks of hematite distributed within it. Syn: *Mont Blanc ruby.* (b) An imitation produced by artificially staining crackled quartz red.

rubble (a) A loose mass of angular rock fragments, commonly overlying outcropping rock; the unconsolidated equivalent of a *breccia.* Cf: *talus; volcanic rubble.* (b) Loose, irregular pieces of artificially broken stone as it comes from the quarry. (c) Fragments of floating or grounded *sea ice,* in hard, roughly spherical blocks measuring 0.5 to 1.5 m in diameter, resulting from the breakup of larger ice formations. When afloat, it is commonly called *brash ice.*

rubble beach A beach composed of angular rock fragments or rubble.

rubble breccia (a) A breccia in which no matching fragments are parted by initial planes of rupture; the fragments are close-set and in touch (Norton, 1917, p.161). (b) A tectonic breccia characterized by prominent relative displacement of fragments and by some rounding (Bateman, 1950, p.133). Cf: *shatter breccia.*

rubble drift (a) An English term for a rubbly deposit or *congeliturbate* formed by solifluction under periglacial conditions; e.g. *head [mass move]* and *coombe rock.* (b) A coarse mass of angular debris and large blocks set in an earthy matrix of glacial origin.

rubble island *debris island.*

rubble ore A term used in Brazil for iron ore found on the surface of *itabirite* and derived from it by "the breaking up of the thinner intercalated layers, more or less completely freed from the associated siliceous elements by rain and wind action" (Derby, 1910, p.818). Cf: *sandy ore; canga.*

rubblerock *breccia [geol].*

rubblestone (a) A graywacke (Humble, 1843, p.224). (b) *rubble.*

rubble tract The part of a reef flat immediately behind and lagoonward of the *reef front,* paved with cobbles, pebbles, blocks, and other coarse reef-rock fragments; when consolidated it forms a *reef breccia.*

rubellite A pale rose-red to deep ruby-red transparent lithian variety of tourmaline, used as a gemstone. Syn: *red schorl.*

rubicelle A yellow or orange-red gem variety of spinel. See also: *ruby spinel.*

rubidium-strontium age method Determination of an age for a mineral or rock in years based on the ratio of radiogenic strontium-87 to rubidium-87 and the known radioactive decay rate of rubidium-87. If ratios are measured for more than one phase of a single rock, or for a number of related rocks that differ in rubidium content, an *isochron* may be drawn. Syn: *rubidium-strontium dating; Rb-Sr age method.*

rubidium-strontium dating *rubidium-strontium age method.*

rubidium-vapor magnetometer A type of *optically pumped magnetometer* that uses magnetic atoms of rubidium. Cf: *cesium-vapor magnetometer.*

rubidium-87 The radioactive isotope of rubidium, which decays to strontium-87 with a half-life of 4.89×10^{10} years; ^{87}Rb constitutes 27.84% of total Rb.

rubinblende A name applied to the red silver-sulfide minerals pyrargyrite, proustite, and miargyrite. Syn: *ruby blende.*

Rubrozem A soil in which the A horizon is highly organic and low in bases, and the B horizon is reddish to brown and prismatic.

ruby The red variety of corundum, containing small amounts of chromium, used as a gemstone, and found esp. in the Orient (Burma, Ceylon, Thailand). Cf: *sapphire.*

ruby blende (a) A brownish-red or reddish-brown transparent variety of sphalerite. Syn: *ruby zinc.* (b) *rubinblende.*

ruby copper Cuprous oxide; specif. *cuprite.* Syn: *ruby copper ore.*

ruby sand A red-colored beach sand containing garnets, as at Nome, Alaska.

ruby silver A red silver-sulfide mineral; specif. "dark ruby silver" (pyrargyrite) and "light ruby silver" (proustite). Syn: *ruby silver ore.*

ruby spinel A clear-red gem variety of magnesian spinel, $MgAl_2O_4$, containing small amounts of chromium and having the color but none of the other attributes of true ruby. See also: *spinel ruby; balas ruby; rubicelle.*

ruby zinc A deep-red transparent zinc mineral; specif. *ruby blende* and *zincite.*

rudaceous Said of a sedimentary rock composed of a significant amount of fragments coarser than sand grains; pertaining to a *rudite.* The term implies no special size, shape, or roundness of fragments throughout the gravel range, and is broader than "pebbly", "cobbly", and "bouldery". Also said of the texture of such a rock. Term introduced by Grabau (1904, p.242). Cf: *psephitic.*

ruddle *red ocher.*

rudemark *rute-mark.*

rudimentary Vestigial; said of structures that are not as highly developed in the descendants as they were in the ancestors; very imperfectly developed.

rudist Any bivalve mollusk belonging to the superfamily Hippuritacea, characterized by an inequivalve shell, usually attached to a substrate, and either solitary or gregarious in reeflike masses. "Although the first rudists were only slightly inequivalve, their descendants very early became strongly so, with the two valves of individuals usually differing greatly from each other in size, shape, and shell wall structure" (TIP, 1969, pt.N, P.751). They are frequently found in association with corals. Range, Upper Jurassic to Upper Cretaceous, possibly Paleocene.

rudite A general name used for consolidated sedimentary rocks composed of rounded or angular fragments coarser than sand (granules, pebbles, cobbles, boulders, or gravel or rubble); e.g. conglomerate, breccia, and calcirudite. The term is equivalent to the Greek-derived term, *psephite,* and was introduced as *rudyte* by Grabau (1904, p.242) who used it with appropriate prefixes in classifying coarse-grained rocks (e.g. "autorudyte", "autosilicirudyte", "hydrorudyte", and "hydrocalcirudyte"). Etymol: Latin *rudus,* "crushed stone, rubbish, debris, rubble". See also: *lutite; arenite.*

rudyte Var. of *rudite.*

Rudzki anomaly A gravity anomaly calculated by replacing the surface topography by its mirror image within the geoid.

ruffle (a) A ripple mark produced by an eddy (Hobbs, 1917). Obsolete. (b) A roughness or disturbance of a surface, such as a ripple on a surface of water.

ruffled groove cast A *groove cast* with a feather pattern, consisting of a groove with lateral wrinkles that join the main cast in the downcurrent direction at an acute angle (Ten Haaf, 1959, p.32). See also: *vibration mark.*

ruga A visceral fold or wrinkle; e.g. a concentric or oblique wrinkling of the external surface of a brachiopod shell, or a *growth ruga* on the shell of a bivalve mollusk. Pl: *rugae.*

ruggedness number A dimensionless number formed by the product of maximum basin relief and drainage density within a given drainage basin; it expresses the essential geometric characteristics of the drainage system, and implicitly suggests steepness of slope (Strahler, 1958, p.289). Symbol: N_4.

rugose Coarsely wrinkled, uneven, rough (Swartz, 1971, p. 409).

rugose coral Any zoantharian belonging to the order Rugosa, characterized by calcareous corallites that may be solitary and cone-shaped or cylindrical, curved or erect, compound and branching or massive. Range, Ordovician to Permian.

rugulate Said of sculpture of pollen and spores consisting of wrinklelike ridges that irregularly anastomose.

ruin agate A brown variety of agate displaying on a polished surface markings that resemble the outlines of ruined buildings.

ruin marble A brecciated limestone that, when cut and polished, gives a mosaic effect suggesting the appearance of ruins or ruined buildings.

rule of tautonymy *tautonymy rule.*

rule of V's The outcrop of a formation that crosses a valley forms an acute angle (a V) that points in the direction in which the formation lies underneath the stream. The V points upstream where the outcrops of horizontal beds parallel the topographic contours, where the beds dip upstream, or where the beds dip downstream at a smaller angle than the stream gradient; the V points downstream where the beds dip downstream at a larger angle than the stream gradient.

rumänite A brittle, yellow-brown to red or black variety of amber containing 1-3% sulfur and found in Rumania. Also spelled: *rumanite.*

rumpffläche A plain extending across a region underlain by massive or undifferentiated rocks; a term used "purely to express relief, with no implication as to position in the cycle of erosion" (W. Penck, 1953, p. 420). The term has often been used to indicate a *peneplain.* Etymol: German *Rumpffläche,* "torso plain". Syn: *torso plain.*

run [intrus rocks] n. A branching or fingerlike extension of the feeder of an igneous intrusion. Runs typically spread laterally along several stratigraphic levels.

run [ore dep] n. A flat irregular ribbonlike orebody following the stratification of the host rock.

run [streams] n. A small, swiftly flowing watercourse; a brook or a small creek.

rundle Var. of *runnel.*

runic *graphic.*

runite A syn. of *graphic granite.* The term, first used by Pinkerton in 1811, was suggested by Johannsen (1939, p. 273).

runlet *runnel.*

runnel [beach] (a) A trough-like hollow, larger than that between ripple marks, formed landward of a *ridge* on the foreshore of a tidal sand beach by the action of tides or waves. It carries the water drainage off the beach as the tide retreats, and is flooded as the tide advances. Cf: *rill.* (b) *swale.* (c) *trough.*

runnel [streams] (a) A little brook; a rivulet or a streamlet. Syn: *runlet; rundle; rindle.* (b) The channel eroded by a runnel.

running ground (a) In mining, incoherent surface material; earth, soil, or rock that will not stand, esp. when wet, and that tends to flow into mine workings. See also: *mudrush.* (b) A tunnelman's term for soil that runs into a tunnel on removal of roof or side support; for example, dry cohesionless sand. See also: *firm ground; flowing ground; raveling ground; squeezing ground; swelling ground.*

running water Water that is flowing in a stream or that is not stagnant or brackish. Ant: *standing water.*

runoff [coal] The collapse of a pillar of coal in a steeply dipping coal seam; the pillar is said to have run off.

runoff [water] That part of precipitation appearing in surface streams. It is more restricted than *streamflow,* as it does not include stream channels affected by artificial diversions, storage, or other works of man. With respect to promptness of appearance after precipitation, it is divided into *direct runoff* and *base runoff;* with respect to source, into *surface runoff, storm seepage,* and *ground-water runoff.* It is the same as *total runoff* used by some workers (Langbein & Iseri, 1960). Syn: *virgin flow.* Cf: *water yield.* See also: *runoff cycle.*

runoff coefficient The percentage of precipitation that appears as runoff. The value of the coefficient is determined on the basis of climatic conditions and physiographic characteristics of the drainage area and is expressed as a constant between zero and one (Chow, 1964, p. 20-8, 21-37). Symbol: *C.*

runoff cycle That part of the hydrologic cycle involving water between the moment of its precipitation onto land and its subsequent evapotranspiration or discharge through stream channels. See also: *runoff [water].*

runoff desert An arid region in which local rain is insufficient to support any perennial vegetation except in drainage or runoff channels. Cf: *rain desert.*

runoff intensity The excess of rainfall intensity over infiltration capacity, usually expressed in inches in depth of rainfall per hour. Strictly, the volume of water derived from an area of land surface per hour. Symbol: Q. Syn: *runoff rate.*

runoff rate *runoff intensity.*

run-of-mine Said of ore in its natural, unprocessed state; pertaining to ore just as it is mined.

runout *water yield.*

runup *uprush.*

runway The channel of a stream.

Rupelian European stage: Middle Oligocene (above Tongrian, below Chattian). Syn: *Stampian.*

rupestral Said of an organism living among rocks or in rocky areas. Syn: *rupestrine; rupicolous; lithophilous; saxicolous; saxigenous.*

rupestrine *rupestral.*

rupicolous *rupestral.*

rupture *fracture [exp struc geol].*

rupture envelope *Mohr envelope.*

rupture strength The differential stress that a material sustains at the instant of rupture. The term is normally applied when deformation occurs at atmospheric confining pressure and room temperature.

rupture velocity The speed with which rupture propagates along the fault surface during an earthquake.

rupture zone The region immediately adjacent to the boundary of an explosion crater, characterized by excessive in-place crushing and fracturing where the stresses produced by the explosion exceeded the ultimate strength of the medium. Cf: *plastic zone.*

rursiradiate Said of an ammonoid rib inclined backward (adapically) from the umbilical area toward the venter. Cf: *prorsiradiate; rectiradiate.*

rusakovite A mineral: $(Fe,Al)_5(VO_4,PO_4)_2(OH)_9 \cdot 3H_2O$.

Ruscinian European stage: Lower Pliocene (above Turolian, below Villafranchian).

russellite A mineral: Bi_2WO_6.

rust An English term for a black shale discolored by ocher.

rustenburgite A cubic mineral: $(Pt,Pd)_3Sn$.

rustumite A mineral: $Ca_4Si_2O_7(OH)_2$.

rusty gold Native gold that has a thin coat of iron oxide or silica that prevents it from amalgamating readily.

rute-mark (a) A type of polygonal ground found in arctic regions, consisting of a polygon enclosed by a row of stones. Etymol: Norwegian *rutemark,* "route mark". Also spelled: *rutmark; rutemark; rudemark.* (b) A crack in soil or mud, similar in form to a rute-mark.

ruthenarsenite An orthorhombic mineral: (Ru,Ni)As.

rutheniridosmine A hexagonal mineral: (Os,Ir,Ru).

ruthenium A hexagonal mineral: Ru.

ruthenosmiridium A cubic mineral: (Ir,Os,Ru).

rutherfordine A yellow secondary mineral: $(UO_2)(CO_3)$.

rutilated quartz Sagenitic quartz characterized by the presence of enclosed needlelike crystals of rutile. See also: *sagenite.* Syn: *Venus hairstone.*

rutile A usually reddish-brown tetragonal mineral: TiO_2. It is trimorphous with anatase and brookite, and often contains a little iron. Rutile forms prismatic crystals in other minerals (esp. quartz); it occurs as a primary mineral in some acid igneous rocks (esp. those rich in hornblende), in metamorphic rocks, and as residual grains in sediments and beach sands. It is an ore of titanium. Syn: *red schorl.*

rutmark *rute-mark.*

rutterite A medium-grained equigranular dark-pink nepheline-bearing *syenite,* composed chiefly of microperthite, microcline, and albite, with small amounts of nepheline, biotite, amphibole, graphite, and magnetite. The name, given by Quirke in 1936, is from Rutter, Ontario. Not recommended usage.

ruware A term used in southern Africa for a low, flattish or gently domed, granitic pediment or outcrop of bare rock. It occurs where flat or gently dipping joint systems are prominent.

RV *right valve.*

R wave *Rayleigh wave.*

Ryazanian European stage: lowermost Cretaceous (above Volgian, below Valanginian).

Saale The term applied in northern Europe to the third classical glacial stage of the Pleistocene Epoch, after the Elster glacial stage and before the Weichsel; equivalent to the *Riss* and *Illinoian* glacial stages.

Saalian North European climatostratigraphic and floral stage: Upper Pleistocene (above Holsteinian, below Eemian). Equivalent in time to Riss glaciation.

Saalic orogeny One of the 30 or more short-lived orogenies during Phanerozoic time recognized by Stille, in this case early in the Permian between the Autunian and Saxonian stages.

sabach A term used in Egypt for a calcareous accumulation; specif. *caliche.* Syn: *sabath.*

sabellariid reef A small *organic reef* composed of the closely packed, sand-walled or arenaceous tubes constructed by sabellariid worms (Polychaeta; Annelida), as off east-central peninsular Florida. Cf: *serpulid reef.*

Sabinas North American (Gulf Coast) provincial series: Upper Jurassic (above older Jurassic, below Coahuilan) (Murray, 1961).

Sabinian North American (Gulf Coast) stage: Eocene (above Midwayan, below Claibornian). It includes strata most commonly grouped as *Wilcoxian.*

sabkha (a) A supratidal environment of sedimentation, formed under arid to semiarid conditions on restricted coastal plains just above normal high-tide level. It is gradational between the land surface and the intertidal environment. Sabkhas are characterized by evaporite-salt, tidal-flood, and eolian deposits, and are found on many modern coastlines, e.g. Persian Gulf, Gulf of California. (b) In the rock record, a sabkha facies may be indicated by evaporites, absence of fossils, thin flat-pebble conglomerates, stromatolitic laminae, desiccation features such as mud cracks, and diagenetic modifications, for example disrupted bedding, dissolution and replacement phenomena, and dolomitization. The sabkha environment may have been significant in the formation of certain petroleum and sulfide-mineral deposits (Kinsman, 1969; Renfro, 1974). Etymol: Arabic. Also spelled: *sebkha.*

sabkha process Shallow subsurface circulation, characteristic of brine of coastal *sabkhas,* suggested as a cause of transportation and deposition of metals in certain stratiform deposits (Renfro, 1974).

sabugalite A yellow secondary mineral: $HAl(UO_2)_4(PO_4)_4 \cdot 16H_2O$.

sabulous Sandy or gritty; *arenaceous.* Syn: *sabulose; sabuline.*

sac A pouch within an animal or plant; e.g. *pollen sac* or *air sac.*

saccate Like or having the form of a sac or pouch; e.g. the "saccate mantle canal" of a brachiopod, without terminal branches and not extending to the anterior and lateral periphery of the mantle, or "saccate pollen" containing vesicles. See also: *vesiculate.*

saccharoidal (a) Said of a granular or crystalline texture resembling that of loaf sugar; specif. said of the xenomorphic-granular texture typically developed in aplites, or said of the crystalline granular texture seen in some sandstones, evaporites, marbles, and dolomites. (b) Said of a white or nearly white equigranular rock having a saccharoidal texture.—See also: *aplitic.* Syn: *sucrosic; sugary.* Etymol: Latin *saccharum,* "sugar".

saccus A winglike extension or *vesicle [palyn]* of the exine in gymnospermous pollen and prepollen. Pl: *sacci.* Cf: *pseudosaccus.*

sackungen Deep-seated rock creep which has produced a *ridge-top trench* by gradual settlement of a slablike mass into an adjacent valley. The top of the settled slab is usually parallel to the crest line of the ridge (Zischinsky, 1969).

saddle [coal] A less preferred syn. of *baum pot.*

saddle [geomorph] (a) A low point in the crest line of a ridge, commonly on a divide between the heads of streams flowing in opposite directions. (b) A broad, flat gap or pass, sloping gently on both sides, and resembling a saddle in shape; a *col.*

saddle [ore dep] *saddle reef.*

saddle [paleont] An element or inflexion of a suture line in a cephalopod shell that forms an angle or curve whose convexity is directed forward or toward the aperture (or away from the apex). Ant: *lobe.*

saddle [struc geol] A low point, sag, or depression along the surface axis or axial trend of an anticline.

saddleback A hill or ridge having a concave outline along its crest.

saddle fold A type of fold that has an additional flexure near its crest, at right angles to that of the parent fold and much larger in radius.

saddle reef A mineral deposit associated with the crest of an anticlinal fold and following the bedding planes, usually found in vertical succession, esp. the gold-bearing quartz veins of Australia. Syn: *saddle; saddle vein.* Cf: *reverse saddle.*

saddle vein *saddle reef.*

safe yield A syn. of *economic yield* that is also applied to surface-water supplies. Use of the term is discouraged because the feasible rate of withdrawal depends on the location of wells in relation to aquifer boundaries and rarely can be estimated in advance of development.

safflorite A tin-white orthorhombic mineral: $CoAs_2$. It is isomorphous with loellingite and dimorphous with clinosafflorite. It usually contains considerable iron.

sag [geomorph] (a) A saddlelike pass or gap in a ridge or mountain range. (b) A shallow depression in an otherwise flat or gently sloping land surface; a small valley between ranges of low hills or between swells and ridges in an undulating terrain.

sag [sed] (a) A depression in a coal seam. (b) A *sag structure.*

sag [struc geol] (a) A basin or downwarp of regional extent; a broad, shallow structural basin with gently sloping sides, such as the Michigan and Illinois basins. (b) A depression produced by downwarping of beds on the downthrown side of a fault such that they dip toward the fault. (c) *fault sag.*

sag-and-swell topography An undulating surface characteristic of till sheets, as in the landscape of midwestern U.S.; it may include moraines, kames, kettles, and drumlins. Cf: *swell-and-swale topography.* Syn: *sag and swell.*

sag correction A *tape correction* applied to the apparent length of a level base line to counteract the sag in the measuring tape. It is the difference between the effective length of the tape (or part of it) when supported continuously throughout its length and when supported at a limited number of independent points. Base tapes usually are used with three or five points of support and hang in *catenaries* between them.

sagenite (a) An acicular variety of rutile that occurs in reticulated twin groups of needlelike crystals crossing at 60 degrees, and is often enclosed in quartz or other minerals. See also: *Venus hair.* (b) A crystal of sagenite. Also, a similar crystal of tourmaline, goethite, actinolite, or other minerals penetrating quartz. (c) Sagenitic quartz; esp. *rutilated quartz.*—Etymol: Latin *sagena,* "large fishing net".

sagenitic quartz Transparent quartz, colorless to nearly colorless, containing needle-shaped crystals of rutile, tourmaline, goethite, actinolite, or other minerals. See also: *rutilated quartz.*

sagger A coarse *fireclay,* often forming the floor of a coal seam, so called because it is used for making saggers or protective boxes in which delicate ceramic pieces are placed while being baked. Etymol: corruption(?) of "safeguard". Also spelled: *seggar; sagre.*

sagittal Pertaining to or situated in the median anterior-posterior plane of a body having bilateral symmetry, or in any plane parallel thereto, e.g. a "sagittal plane" dividing an edrioasteroid or trilobite into two similar halves; a "sagittal axis" of the frustule of a pennate diatom; a "sagittal section" of a foraminiferal test in an *equatorial* plane; a "sagittal ring" in certain radiolarians; or a "sagittal triradiate" of a sponge, having two mirror-imaged rays and a coplanar third ray pointing away from them along the axis of fourfold symmetry.

sagittate Like an arrowhead in form; triangular, with the basal lobes pointing downward or concavely toward the stalk.

sag pond A small body of water occupying an enclosed depression or *sag* formed where active or recent fault movement has impounded drainage; specif. one of many ponds and small lakes along the San Andreas Fault in California. Also spelled: *sagpond.* Cf: *swag.* Syn: *fault-trough lake; rift lake; rift-valley lake.*

sag structure A general term for load casts and related sedimentary structures.

sagvandite A carbonate rock with a high content of enstatite and magnesite (Johannsen, 1939, p.278).

sahamalite A mineral: $(Mg,Fe)Ce_2(CO_3)_4$.

sahlinite A sulfur-yellow mineral: $Pb_{14}(AsO_4)_2O_9Cl_4$.

sahlite *salite.*

sai A term used in central Asia for a gravelly talus, a river bed filled with stones, and a dry wash (Stone, 1967, p. 258), and also for a piedmont plain covered with pebbles showing desert varnish (Termier & Termier, 1963, p. 413).

saif *seif.*

sainfeldite A mineral: $H_2Ca_5(AsO_4)_4 \cdot 4H_2O$.

Saint Venant substance A material that demonstrates *elastico-plastic* behavior: it behaves elastically below a *yield stress,* but deforms continuously under a constant stress equal to yield stress.

sakhaite A mineral: $Ca_{12}Mg_4(CO_3)_4(BO_3)_7Cl(OH)_2 \cdot H_2O$.

sakharovaite A mineral: $(Pb,Fe)(Bi,Sb)_2S_4$.

Sakmarian European stage: lowermost Permian (above Stephanian of Carboniferous, below Artinskian).

sakuraiite A mineral: $(Cu,Zn,Fe)_3(In,Sn)S_4$.

sal *sial.*

salada A term used in SW U.S. for a salt-covered plain where a lake has evaporated. Etymol: Spanish, feminine of *salado,* "salted, salty". See also: *playa.*

sal ammoniac An isometric mineral: NH_4Cl. It is a white crystalline volatile salt that occurs esp. as an encrustation around volcanoes. Syn: *salmiac.*

salaquifer *saline aquifer.*

salar A term used in SW U.S. and in the Chilean nitrate fields for a salt flat or for a salt-encrusted depression that may represent the basin of a salt lake. Etymol: Spanish, "to salt". Pl: *salares; salars.* See also: *playa.* Pron: sah-*lar.*

salband The *selvage* of an igneous mass or of a mineral vein. Etymol: German *Salband* or *Sahlband.*

salcrete A term suggested by Yasso (1966) for a thin, hard crust of salt-cemented sand grains, occurring on a marine beach that is periodically saturated by saline water.

saléeite A lemon-yellow mineral of the autunite group: $Mg(UO_2)_2(PO_4)_2 \cdot 8H_2O$.

salesite A blue-green mineral: $Cu(IO_3)(OH)$.

salfemic One of five classes in the *CIPW classification* of igneous rocks, in which the ratio of salic to femic minerals is less than five to three but greater than three to five. Cf: *dosalic; dofemic.*

salic Said of certain light-colored silicon- or aluminum-rich minerals present in the norm of igneous rocks; e.g. quartz, feldspars, feldspathoids. Also, applied to rocks having one or more of these minerals as major components of the norm. Etymol: a mnemonic term derived from *s*ilicon + *al*uminum + *ic*. Cf: *femic; mafic; felsic.*

salic horizon A diagnostic subsurface soil horizon, at least 15cm thick, characterized by enrichment with soluble salts. It contains at least 2.0% salts.

salient [fold] An area in which the axial traces of folds are convex toward the outer edge of the folded belt. Ant: *recess [fold].*

salient [geomorph] adj. Projecting or jutting upward or outward; e.g. a *salient* point, or one formed by a conspicuous outward projection from the coast.—n. A landform that projects or extends outward or upward from its surroundings; e.g. a cape along a shoreline, or a spur from the side of a mountain. Ant: *reentrant.*

Salientia A superorder of lissamphibians that includes frogs and toads. Range, Lower Triassic to Recent.

saliferous Salt-bearing; esp. said of strata producing, containing, or impregnated with salt. See also: *saline.*

salina (a) A place where crystalline salt deposits are formed or found, such as a salt flat or pan, a salada, or a salt lick; esp. a salt-encrusted *playa* or a *wet playa.* (b) A body of saline water, such as a salt pond, lake, well, or spring, or a playa lake having a high concentration of salts. (c) *saltworks.* (d) *salt marsh.*—Etymol: Spanish, "salt pit, salt mine, saltworks". Anglicized equivalent: *saline.*

Salinan North American provincial stage: Upper Silurian (above Lockportian, below Tonolowayan).

salinastone A general term proposed by Shrock (1948a, p. 127) for a sedimentary rock composed dominantly of saline minerals (which are usually precipitated but may be fragmental); e.g. anhydrock and gyprock.

saline n. (a) A natural deposit of halite or of any other soluble salt; e.g. an evaporite. See also: *salines.* (b) An anglicized form of *salina.* In this usage, a "saline" may refer to various features such as a playa, a salt flat, a salt pan, a salt marsh, a salt lake, a salt pond, a salt well, or a saltworks. (c) *salt spring.* (d) A term used along the coast of Louisiana for a body of water behind a barrier island.—adj. (a) Salty; containing dissolved sodium chloride, e.g. seawater. (b) Having a salinity appreciably greater than that of seawater, e.g. a *brine.* (c) Containing dissolved salts at concentrations great enough to allow the precipitation of sodium chloride; *hypersaline.* (d) Said of a taste resembling that of common salt, esp. in describing the properties of a mineral.

saline-alkali soil *saline-sodic soil.*

saline aquifer An aquifer containing salty water. Syn: *salaquifer.*

saline deposit *evaporite.*

salinelle A *mud volcano* erupting saline mud.

saline residue *evaporite.*

salines (a) A general term for the naturally occurring soluble salts, such as common salt, sodium carbonate, sodium nitrate, potassium salts, and borax. (b) A general term for salt mines, salt springs, salt beds, salt rock, and salt lands.

saline sodic soil An older term referring to a *salt-affected soil* with a content of exchangeable sodium greater than 15% and sufficient soluble salts to give specific conductance greater than 4 mmhos/cm. The pH value is usually less than 9.5. Cf: *sodic soil; saline soil; nonsaline sodic soil.*

saline soil A nonsodic *salf-affected soil* having a content of soluble salts sufficient to give specific conductance greater than 4 mmhos/cm. Its exchangeable-sodium percentage is less than 15, and its pH value is below 8.5. Cf: *saline sodic soil; sodic soil.* See also: *Solonchak soil.*

saliniferous Said of a stratum that yields salt.

salinity The total quantity of dissolved salts in seawater, measured by weight in parts per thousand, with the following qualifications: all the carbonate has been converted to oxide, all the bromide and iodide to chloride, and all the organic matter has been completely oxidized. Salinity is usually computed from some other factor, such as chlorinity. It may also be defined in terms of electric conductivity relative to normal seawater.

salinity current A *density current* in the ocean, the flow of which is caused, controlled, or maintained by its relatively greater density due to excessive salinity.

salinity log *chlorine log.*

salinity meter *salinometer.*

salinization In a soil of an arid, poorly drained region, the accumulation of soluble salts by evaporation of the waters that bore them to the soil zone.

salinometer An instrument that is used to measure the salinity of seawater, e.g. by electrical conductivity. Syn: *salinity meter.*

salite A mineral of the clinopyroxene group: $Ca(Mg,Fe)Si_2O_6$. It is a gray-green to black variety of diopside with more magnesium than iron. Syn: *sahlite.*

salitral A term used in Patagonia for a swampy place where salts (esp. potassium nitrate) become encrusted in the dry season. Etymol: Spanish, "saltpeter bed".

salitrite A *lamprophyre* composed chiefly of sphene and clinopyroxene, with accessory apatite, microcline, and occasionally anorthoclase and baddeleyite; a sphene-rich *jacupirangite.* Named by Tröger in 1928 for the Salitre Mountains, Miras Gerais, Brazil. Not recommended usage.

salmiac *sal ammoniac.*

salmoite *tarbuttite.*

salmonsite A buff-colored mineral: $Mn_9Fe_2^{+3}(PO_4)_8 \cdot 14H_2O$ (?). Cf: *landesite.*

Salopian European stage: Upper Silurian (above Llandoverian, below Devonian). Includes Wenlockian and Ludlovian.

salpausselkä A Finnish term for a steep recessional moraine, usually interpreted as a series of end moraines, like the one trending east-west across Finland.

salpingiform Shaped like a trumpet; e.g. said of a cyrtolith coccolith with a trumpet-shaped central structure (as in *Discosphaera tubifer*).

salsima According to Van Bemmelen (1949), the theoretical layer of the Earth's crust beneath the sial and above the Mohorovičić discontinuity that is considered to be of basaltic composition. Also spelled: *sialsima.*. Cf: *sifema.*

salsuginous Said of a plant growing in soil or water with a high content of salts; i.e. a *halophyte.*

salt [geog] In geographic terminology, a *salt marsh,* esp. one flooded by the tide.

salt [sed] n. A general term for naturally occurring sodium chloride, NaCl. Syn: *halite; common salt; rock salt.* —adj. Containing salt, as *salt water,* or containing salt water, as *salt marsh.*

salt-affected soil A general term for a soil that is not suitable for the growth of crops because of an excess of salts, of exchangeable sodium, or both; e.g. *saline sodic soil, saline soil, sodic soil.*

salt-and-pepper Said of a sand or sandstone consisting of a mixture of light- and dark-colored particles, such as a strongly cherty graywacke (Krynine, 1948, p.152) or a lighter-colored and speckled subgraywacke (Pettijohn, 1957, p.319); e.g. the Bow Island Sandstone of Cretaceous age in Alberta.

salt anticline A diapiric or piercement structure, like a *salt dome* except that the salt core is linear rather than equidimensional, e.g. the salt anticlines in the Paradox basin of the central Colorado Plateau. Syn: *salt wall.*

saltation [evol] Sudden evolution of a new type of organism derived, in a single generation, from older ones without transitional intermediate forms. This process appears to be almost impossible genetically. See also: *saltatory evolution.*

saltation [sed] A mode of sediment transport in which the particles are moved progressively forward in a series of short intermittent leaps, jumps, hops, or bounces from a bottom surface; e.g. sand particles skipping downwind by impact and rebound along a desert surface, or bounding downstream under the influence of eddy currents that are not turbulent enough to retain the particles in suspension and thereby return them to the stream bed at some distance downstream. It is intermediate in character between suspension and the rolling or sliding of *traction.* Etymol: Latin *saltare,* "to jump, leap".

saltation load The part of the *bed load* that is bouncing along the stream bed or is moved, directly or indirectly, by the impact of bouncing particles.

saltation mark *skip mark.*

saltatory evolution The theory of evolution by *saltation.*

salt bottom A flat piece of relatively low-lying ground encrusted with salt.

salt burst Rock destruction caused by soluble salts that enter pores and crystallize from nearly saturated solutions. In deserts, salt bursts may be due to crystallization pressure, to the volumetric expansion of salts in capillaries, and to hydration pressures of the entrapped salts (Winkler & Wilhelm, 1970).

salt cake Commercial term for sodium sulfate, Na_2SO_4.

salt corrie A cirquelike hollow, resembling a crater or caldera, produced by the solution of salt.

salt crust A salt deposit formed on an ice surface by crystal growth forcing salt out of young sea ice and pushing it upward.

salt-crystal cast A *crystal cast* formed by solution of a soluble salt crystal, followed by filling with mud or sand or by crystallization of a pseudomorph (such as calcite after halite). See also: *hopper.*

salt-crystal growth The growth of salt crystals in openings in rock or soil, capable of exerting powerful stresses and producing granular disintegration in a dry climate. See also: *salt weathering.*

salt desert A desert with a saliferous soil; e.g. a *kavir.*

salt dome A *diapir* or piercement structure with a central, nearly equidimensional *salt plug,* generally one to two kilometers or more in diameter, which has risen through the enclosing sediments from a mother salt bed 5 km to more than 10 km beneath the top of the plug. Many salt plugs have a *cap rock* of less soluble evaporite minerals, esp. anhydrite. Most plugs have nearly vertical walls, but some overhang. The enclosing sediments are commonly turned up and complexly faulted next to a salt plug, and the more permeable beds serve as reservoirs for oil and gas. Salt domes are characteristic features of the Gulf Coastal Plain in North America and the North German Plain in Europe, but occur in many other regions. Cf: *salt anticline.* See also: *salt tectonics.*

salt-dome breccia A breccia found in deep shale sequences, occurring as a dome-shaped mass in a broad zone surrounding a salt plug. It is believed to be a result of differential pressure caused by diapiric intrusion of salt into shale (Kerr & Kopp, 1958).

saltern (a) A *saltworks* where salt is produced by boiling or evaporation of salt or brine. (b) *salt garden.*

saltfield An area overlying a salt deposit of economic value.

salt flat The level, salt-encrusted bottom of a lake or pond that is temporarily or permanently dried up; e.g. the Bonneville Salt Flats west of Salt Lake City, Utah. See also: *playa; alkali flat.*

salt flower An *ice flower* forming on surface sea ice around a salt-crystal nucleus.

salt garden A large, shallow basin or pond where seawater is evaporated by solar heat. Syn: *saltern.*

salt glacier A gravitational flow of salt down the slopes of an exposed salt plug, following the pre-existing structure. It can be compared with the *coulee* of a lava flow.

salt hill An abrupt hill of salt, with sinkholes and pinnacles at its summit (Thornbury, 1954, p. 521).

saltierra A deposit of salt left by evaporation of a shallow, inland lake. Etymol: Spanish, "salt earth".

salting (a) A British term for the slightly higher part of a *salt marsh,* flooded only by spring tides, containing little bare mud, and supporting grassy vegetation. Syn: *high marsh.* (b) A term used in parts of Great Britain for land regularly covered by the tide, as distinguished from a salt marsh.—The term is usually used in the plural.

salt lake An inland body of water situated in an arid or semiarid region, having no outlet to the sea, and containing a high concentration of dissolved salts (principally sodium chloride). Examples include the Great Salt Lake in Utah, and the Dead Sea in the Near East. See also: *alkali lake; bitter lake.* Syn: *brine lake.*

salt lick A place to which animals (e.g. deer, cattle, bison) go to lick up salt lying on the surface of the ground, as in an area surrounding a salt spring. The term has been used for the spring itself, but this usage is improper because a lick is dry. Syn: *lick.*

salt marsh Flat, poorly drained land that is subject to periodic or occasional overflow by salt water, containing water that is brackish to strongly saline, and usually covered with a thick mat of grassy halophytic plants; e.g. a coastal marsh periodically flooded by the sea, or an inland marsh (or *salina*) in an arid region and subject to intermittent overflow by water containing a high content of salt. Cf: *tidal marsh; marine marsh.* See also: *low marsh; salting; sea marsh; open-coast marsh; tidal-delta marsh; salt-marsh plain.* Syn: *salt.*

salt-marsh plain A *salt marsh* that has been raised above the level of the highest tide and has become dry land.

salt meadow A meadow subject to overflow by salt water.

salt pan (a) An undrained small shallow natural depression in which water accumulates and evaporates, leaving a salt deposit. Also, a shallow lake of brackish water occupying a salt pan. See also: *playa; pan [salt]; marsh pan.* (b) A large pan for recovering salt by evaporation.—Also spelled: *saltpan.*

saltpeter (a) Naturally occurring potassium nitrate; *niter.* Cf: *Chile saltpeter; Peru saltpeter; wall saltpeter.* (b) A speleologic term for earthy cave deposits of nitrate minerals.—Also spelled: *saltpetre.*

saltpeter earth A cave deposit containing nitrocalcite; it can be mined for conversion to saltpeter.

salt pillow An embryonic salt dome rising from its source bed, still at depth.

salt pit (a) A pit in which seawater is received and evaporated and from which salt is obtained. Syn: *vat; wich.* (b) A body of salt water occupying a salt pit.—Also spelled: *saltpit.*

salt plug The salt core of a *salt dome.* It is nearly equidimensional, about one to two kilometers in diameter, and has risen through the enclosing sediments from a mother salt bed 5 to 10 kilometers below.

salt polygon A structure of salt on a playa, having three to eight sides marked by ridges of material formed as a result of the expansive forces of crystallizing salt, and ranging in width from several centimeters to 30 m (Stone, 1967, p. 244).

salt pond (a) A large or small body of salt water in a marsh or swamp along the seacoast. (b) An artificial pond used for evaporation in the production of salt from seawater.

salt prairie *soda prairie.*

salt ribbon A ribbonlike growth of salt from a network of small cracks (Brown, 1946).

salt spring A *mineral spring* whose water contains a large quantity of common salt; a spring of salt water. See also: *salt lick.* Syn: *saline; brine spring.*
salt stock A general term for a diapiric salt body of whatever shape.
salt table The flat upper surface of a salt stock, along which ground-water solution leads to the formation of cap rock by freeing anhydrite (Goldman, 1952).
salt tectonics A general term for the study of the structure and mechanism of emplacement of *salt domes* and other salt-controlled structures. Syn: *halokinesis.*
salt wall *salt anticline.*
salt water A syn. of *seawater;* the ant. of fresh water in general.
salt-water encroachment Displacement of fresh surface or ground water by the advance of salt water due to its greater density, usually in coastal and estuarine areas but also by movement of brine from beneath a playa lake toward wells discharging fresh water. Encroachment occurs when the total head of the salt water exceeds that of adjacent fresh water. Syn: *sea-water intrusion; intrusion [grd wat]; salt-water intrusion; sea-water encroachment.* See also: *Ghyben-Herzberg ratio.*
salt-water front The interface between fresh and salty water in a coastal aquifer or in an estuary. Under certain conditions, a similar front may by found inland.
salt-water intrusion *salt-water encroachment.*
salt-water underrun A type of density current occurring in a tidal estuary, due to the greater salinity of the bottom water (ASCE, 1962).
salt-water wedge An intrusion, into an estuary or tidal river dominated by freshwater circulation, of salty ocean water in the form of a wedge that underlies the fresh water, slopes slightly downward in the upstream direction, and is characterized by a pronounced increase in salinity with depth.
salt weathering The granular disintegration or fragmentation of rock material effected by saline solutions or by *salt-crystal growth* (Wellman & Wilson, 1965). See also: *exsudation.*
salt wedge A wedge-shaped mass of salt water from an ocean or sea which intrudes the mouth and lower course of a river. The denser salt water underlies the fresher river water. Extent is regulated by river discharge and tides.
salt well A drilled or dug well from which brine is obtained. See also: *brine pit.*
saltworks Any installation where salt is produced commercially, as by extraction from seawater, from wells, or from the brine of salt springs. Syn: *salina; saltern.*
salty Pertaining to, containing, or resembling salt; *saline.*
samara A dry, indehiscent, usually one-seeded winged fruit, such as that of elm or maple.
samarium-neodymium age method A method of age determination based on the alpha decay of samarium-147 to neodymium-143 ($\lambda=6.54 \times 10^{-12}yr^{-1}$). The ratios $^{147}Sm/^{144}Nd$ and $^{143}Nd/^{144}Nd$ are measured and plotted on an isochron diagram. Lunar rocks 4.40 $\times$ 10^9 years old have been dated by this method.
samarskite A velvet-black to brown, commonly metamict mineral: $(Y,Ce,U,Ca,Fe,Pb,Th)(Nb,Ta,Ti,Sn)_2O_6$. It has a splendent vitreous or resinous luster, and is found in granite pegmatites. Syn: *ampangabeite; uranotantalite.*
samiresite A mineral that is possibly a variety of betafite containing lead, but perhaps an independent species.
sammelkristallization The action, depending on surface tension or on total free surface energy, by which smaller grains become unstable in relation to larger grains and will eventually be devoured by the larger grains (Barth, 1962, p.399). There is no English equivalent for this term. Etymol: German.
sample In statistics, the part or subset of a statistical population that if properly chosen may be used to estimate parameters.
sampleite A blue mineral: $NaCaCu_5(PO_4)_4Cl \cdot 5H_2O$.
sample log A *log* depicting the sequence of lithologic characteristics of the rocks penetrated in drilling a well, compiled by a geologist from microscopic examination of *well cuttings* and *cores.* The information is referred to depth of origin and is plotted on a *strip log* form. See also: *interpretive log; percentage log.* Syn: *lithologic log; graphic log.*
sample splitter A device for separating dry incoherent material (such as sediment) into truly representative samples of workable size for laboratory study. Syn: *riffler.*
sampling In economic geology, the gathering of specimens of ore or wall rock for appraisal of the orebody. As the average of many samples may be used, representative sampling is crucial. The term is usually modified to indicate the mode or locality, e.g. hand sampling, mine sampling.
samsonite A steel-black monoclinic mineral: $Ag_4MnSb_2S_6$.
samuelsonite A monoclinic mineral: $(Ca,Ba)Ca_8(Fe^{+2},Mn)_4Al_2(PO_4)_{10}(OH)_2$.
sanbornite A white triclinic mineral: $BaSi_2O_5$.
sancyite A light-colored *trachyandesite* containing tridymite (Sorensen, 1974, p. 572). The name, given by Lacroix in 1923, is for Puy Sancy, Auvergne region, France. Not recommended usage.
sand [drill] A driller's term applied loosely to any visibly granular sediment, or to any fluid-productive porous sedimentary unit or objective zone of a well. See also: *oil sand.*
sand [eng] Rounded rock or mineral grains with diameters between 0.074 mm (retained on U.S. standard sieve no.200) and 4.76 mm (passing U.S. standard sieve no.4). See also: *coarse sand; medium sand; fine sand.*
sand [geomorph] (a) A tract or region of sand, such as a sandy beach along the seashore, or a desert land. (b) A sandbank or a sandbar.—The term is usually used in the plural; e.g. "sea sands".
sand [sed] (a) A rock fragment or detrital particle smaller than a granule and larger than a coarse silt grain, having a diameter in the range of 1/16 to 2 mm (62-2000 microns, or 0.0025-0.08 in., or 4 to -1 phi units, or a size between that at the lower limit of visibility of an individual particle with the unaided eye and that of the head of a small wooden match), being somewhat rounded by abrasion in the course of transport. In Great Britain, the range of 0.1-1 mm has been used. See also: *very coarse sand; coarse sand; medium sand; fine sand; very fine sand.* (b) A loose aggregate of unlithified mineral or rock particles of sand size; an unconsolidated or moderately consolidated sedimentary deposit consisting essentially of medium-grained clastics. The material is most commonly composed of quartz resulting from rock disintegration, and when the term "sand" is used without qualification, a siliceous composition is implied; but the particles may be of any mineral composition or mixture of rock or mineral fragments, such as "coral sand" consisting of limestone fragments. Also, a mass of such material, esp. on a beach or a desert or in a stream bed. (c) *sandstone.*
sand [soil] (a) A term used in the U.S. for a rock or mineral particle in the soil, having a diameter in the range of 0.05-2 mm; prior to 1947, the range 1-2 mm was called "fine gravel". The diameter range recognized by the International Society of Soil Science is 0.02-2 mm. (b) A textural class of soil material containing 85% or more of sand, with the percentage of silt plus 1.5 times the percentage of clay not exceeding 15; specif. such material containing 25% or more of very coarse sand, coarse sand, and medium sand, and less than 50% of fine sand or very fine sand (SSSA, 1965, p.347). The term has also been used for a soil containing 90% or more of sand.—See also: *very coarse sand; coarse sand; medium sand; fine sand; very fine sand.*
sand apron A deposit of sand, mostly carbonate, along the shore of the lagoon of a reef.
sandar Plural of *sandur.*
sandarac A syn. of *realgar.* Also spelled: *sandarach.*
sand avalanche Movement of large masses of sand down a dune face when the angle of repose is exceeded, or when the dune is disturbed (Stone, 1967, p. 245). Cf: *sand run.*
sandbag In the roof of a coal seam, a deposit of glacial debris formed by scour and fill subsequent to coal formation.
sandbank (a) A submerged ridge of sand in the sea, a lake, or a river, usually exposed during low tide; a sandbar. (b) A large deposit of sand, esp. in a shallow area near the shore.
sandbar A *bar* or low ridge of sand that borders the shore and is built to, or nearly to, the water surface by currents in a river or by wave action along the shore of a lake or sea. Syn: *sand reef.*
sandblast (a) A stream of windblown sand driven against an exposed rock surface. (b) A gust of wind, carrying sand.
sandblasting A type of *blasting* in which the particles are hard mineral grains (usually quartz) of sand sizes. Syn: *sandblast action.*
sandblow A patch of coarse sandy soil denuded of vegetation by wind action.
sand boil A spring that bubbles through a river levee, with an ejection of sand and water, as a result of water in the river being forced through permeable sands and silts below the levee during flood stage. Syn: *blowout [grd wat].*
sand-calcite A calcite crystal containing a large percentage of

sand-grain inclusions; a *sand crystal* of calcite. See also: *Fontainebleau sandstone.*

sand cay A British syn. of *sand key.*

sand cone [geomorph] A conical deposit of sand, produced esp. in an alluvial cone.

sand cone [glaciol] A low *debris cone* whose protective veneer consists of sand.

sand crystal A large euhedral or subhedral crystal (as of barite, gypsum, and esp. calcite) loaded with detrital-sand inclusions (up to 60%), developed by growth in an incompletely cemented sandstone during cementation. See also: *sand-calcite; crystal sandstone.*

sand dike A sedimentary dike consisting of sand that has been squeezed or injected upward into a fissure. See also: *injection dike; neptunian dike.*

sand dome *dome [beach].*

sand drift (a) A general term for surface movement of wind-blown sand, occurring in deserts or along the shore. (b) An accumulation of sand formed in the lee of some fixed obstruction, such as a rock or a bush; usually smaller than a dune. See also: *sand shadow.*

sand drip A rounded or crescentic surface form on beach sand, resulting from the sudden absorption of overwash (Rosalsky, 1949, p. 12).

sand dune An accumulation of loose sand heaped up by the wind, commonly found along low-lying seashores above high-tide level, more rarely on the border of large lakes or river valleys, as well as in various desert regions, where there is abundant dry surface sand during some part of the year. See also: dune [geomorph]; sand hill.

sanderite A mineral: $MgSO_4 \cdot 2H_2O$.

sand fall An accumulation of sand swept over a cliff or escarpment (Stone, 1967, p. 245). It may occur in a submarine canyon as well as on land.

sandfall The *slip face* on the lee side of a dune. Syn: *sandfall face.*

sand flag Fine-grained sandstone that can be readily split into *flagstones.*

sand flat A sandy *tidal flat* barren of vegetation. Cf: *mud flat.*

sand flood A vast body of sand moving over or borne along on a desert floor, as in the Arabian deserts.

sand flow [marine geol] In a submarine canyon, a discontinuous movement of sand down the axis, in a series of slumps. See also: *sand fall.*

sand flow [mass move] A flow of wet sand, as along banks of noncohesive clean sand that is subject to scour and to repeated fluctuations in pore-water pressure due to rise and fall of the tide (Varnes, 1958, p. 41).

sand flow [pyroclast] (a) An obsolete syn. of *ash flow.* (b) A term applied by Fenner in 1923 to an unsorted rhyolitic tuff in the vicinity of Mt. Katmai, Alaska.

sand gall *sand pipe.*

sand glacier (a) An accumulation of sand that is blown up the side of a hill or mountain and through a pass or saddle, and then spread out on the opposite side to form a wide fan-shaped plain. (b) A horizontal plateau of sand terminated by a steep talus slope.

sand hill A ridge of sand; esp. a *sand dune* in a desert region. See also: *chop hill.*

sandhills A region of *sand hills,* as in north-central Nebraska.

sand hole A small pit, 7-8 mm in depth and a little less wide than deep, with a raised margin, formed on a beach by waves expelling air from a formerly saturated mass of sand; it resembles an impression made by a raindrop.

sand horn A pointed sand deposit extending from the shore into shallow water. Cf: *sand lobe.*

sandia An oblong, oval, or rounded mountain mass resembling a watermelon; e.g. the Sandia Mountains in New Mexico. Etymol: Spanish *sandía,* "watermelon".

sanding The accumulation or building-up of sand, as by the action of currents in filling a harbor.

sanding up Filling-in or choking with sand, as in a well that produces sand mixed with oil and gas.

sand key A small sandy island close to the shore. Syn: *sand cay.*

sand levee A *whaleback* in the desert.

sand line [drill] A *wire line* used in cable-tool drilling to raise and lower such tools as a bailer, or in rotary drilling to operate a *swab.* Also spelled: *sandline.*

sand line [glac geol] An "easily overlooked" mark made by glacier ice, about 5-10 cm long, fine as a hair and similar to one of the marks made by the "finest sandpaper" (Campbell, 1865, p. 4).

sand lobe A rounded sand deposit extending from the shore into shallow water. Cf: *sand horn.*

sand pavement A sandy surface derived from coarse-grained sand ripples, developed on the lower, windward slope of a dune or rolling sand area during a period of intermittent light, variable winds (E. Holm, 1957).

sand pipe A *pipe* formed in sedimentary rocks, filled with sand. Cf: *gravel pipe.* Syn: *sand gall.*

sand plain [geomorph] A sand-covered plain. The large sand plains in Western Australia have an uncertain origin: they may originate by deflation of sand dunes, the lower limit of erosion being governed by the ground-water level. Also spelled: *sandplain.*

sand plain [glac geol] (a) A small *outwash plain* consisting chiefly of sand deposited by meltwater streams flowing from a glacier. (b) A term used in New England for an *esker delta* or *delta kame.*

sand plateau (a) *esker delta.* (b) *delta kame.*

sand plug A mass of sand that fills the upper end of a stream channel abandoned by the formation of a chute cutoff.

sandr *sandur.*

sand reef *sandbar.*

sand ridge (a) A generic name for any low ridge of sand formed at some distance from the shore, and either submerged or emergent. Examples include a *longshore bar* and a *barrier beach.* (b) One of a series of long, wide, extremely low, parallel ridges believed to represent the eroded stumps of former longitudinal sand dunes, as in western Rhodesia. (c) A seaward-pointing landform found on a sandy beach; e.g. a *beach cusp.* (d) *sand wave.*

sand ripple A ripple composed of sand. See also: *ripple mark.*

sand river A river that deposits much of its sand load along its middle course, to be subsequently removed by the wind; e.g. the Red River in Texas. Cf: *sand stream.*

sandrock (a) A field term for a sandstone that is not firmly cemented (Tieje, 1921, p.655). (b) A term used in southern England for a sandstone that crumbles between the fingers. (c) *sandstone.*

sand roll *pseudonodule.*

sand run (a) A fluidlike motion of dry sand. (b) A mass of dry sand in motion. Cf: *sand avalanche.*

sand sea [desert] An extensive assemblage of sand dunes of several types in an area where a great supply of sand is present, characterized by an absence of travel lines or other directional indicators, and by a wavelike appearance of dunes separated by troughs, much as though storm sea waves were frozen into place. See also: *erg.*

sand sea [pyroclast] The flat, rain-smoothed plain of volcanic ash and other pyroclastics on the floor of a caldera.

sand shadow A lee-side accumulation of sand, such as a small turret-shaped dune, formed in the shelter of, and immediately behind, a fixed obstruction, like a clump of vegetation. See also: *sand drift.*

sandshale A sedimentary deposit consisting of thin alternating beds of sandstone and shale.

sand-shale ratio A term introduced by Sloss et al. (1949, p.100) for the ratio of the thickness or percentage of sandstone (and conglomerate) to that of shale in a stratigraphic section, disregarding the amount of nonclastic material; e.g. a ratio of 3.2 indicates that the section contains an average of 3.2 m of sandstone per meter of shale. Cf: *clastic ratio.*

sand sheet A thin accumulation of coarse sand or fine gravel formed of grains too large to be transported by saltation, characterized by an extremely flat or plainlike surface broken only by small sand ripples.

sand size A term used in sedimentology for a volume greater than that of a sphere with a diameter of 1/16 mm (0.0025 in.) and less than that of a sphere with a diameter of 2 mm (0.08 in.).

sand snow Cohesionless dry snow that has fallen at such cold temperatures (usually below -25°C) that intergranular adhesion is inhibited and surface friction is high. Its surface has the consistency of dry sand. Cf: *powder snow.*

sand spit A spit consisting chiefly of sand.

sand splay A *flood-plain splay* consisting of coarse sand particles.

sand stalagmite A stalagmite that is developed on sand and that is composed of sand cemented by calcite.

sandstone (a) A medium-grained clastic sedimentary rock composed of abundant rounded or angular fragments of sand size set in a fine-grained matrix (silt or clay) and more or less firmly united by a cementing material (commonly silica, iron oxide, or calcium carbonate); the consolidated equivalent of *sand,* intermediate in texture between conglomerate and shale. The sand particles usually consist of quartz, and the term "sandstone", when used without qualification, indicates a rock containing about 85-90% quartz (Krynine, 1940). The rock varies in color, may be deposited by water or wind, and contains numerous primary features (sedimentary structures and fossils). Sandstones may be classified according to composition of particles, mineralogic or textural maturity, fluidity index, diastrophism, primary structures, and type of cement (Klein, 1963). (b) A field term for any clastic rock containing individual particles that are visible to the unaided eye or slightly larger.—Syn: *sand; sandrock.*

sandstone-arenite A term used by Folk (1968, p.124) for a *sedarenite* composed chiefly of sandstone fragments.

sandstone cave A partial syn. of *rockshelter.*

sandstone cylinder *sandstone pipe.*

sandstone dike (a) A *clastic dike* composed of sandstone or lithified sand; a lithified *sand dike.* (b) *stone intrusion.*

sandstone pipe A *clastic pipe* consisting of sandstone. It may originate in various ways: gravitational foundering of sand into underlying water-saturated mud; filling of a spring vent; filling of a cavity caused by solution of underlying limestone or by volcanic explosion; or penecontemporaneous sag due to removal of support by flowage. Syn: *sandstone cylinder.*

sandstone sill A tabular mass of sandstone that has been emplaced by sedimentary injection parallel to the structure or bedding of pre-existing rock in the manner of an igneous sill, such as one injected at the mud-water interface by the underflow of a dense slurry.

sand streak A low, linear ridge formed at the interface of sand and air or water, oriented parallel to the direction of flow, and having a symmetric cross section.

sand stream A small sand delta spread out at the mouth of a gully, or a deposit of sand along the bed of a small creek, formed by a torrential rain (Stephenson & Veatch, 1915, p. 112). Cf: *sand river.*

sand stretch A striation worn in a rock surface by windblown sand.

sand strip A long, narrow ridge of sand extending for a long distance downwind from each horn of a dune.

sand tank A *rock tank* filled with sand.

sand trap A device designed to remove sand and other particles from flowing water; e.g. for separating heavy, coarse particles from the cuttings-laden drilling fluid overflowing a drill collar.

sand tube A tubular *fulgurite* formed in sand.

sand tuff (a) A tuffaceous sandstone. (b) A tuff whose component fragments are in the size range of sand. This usage is obsolete.

sand twig A small, twiglike aggregate of sand that stands more or less upright on the surface of a sand dune undergoing wind scour, apparently forming around a root or stem of a plant exposed on the dune surface (Carroll, 1939, p. 20-21).

sandur Icelandic term signifying "sand", but generally adopted for *outwash plain.* Pl: *sandar.* Also spelled: *sandr.*

sand volcano An accumulation of sand resembling a miniature volcano or low volcanic mound (maximum diameter 15 m), produced by expulsion of liquefied sand to the sediment surface. Examples are found on top of slump sheets or on the upper surface of highly contorted layers of laminated sediments.

sandwash A sandy or gravel stream bed, devoid of vegetation, containing water only during a sudden heavy rainstorm.

sand wave (a) A general term for a wavelike *bed form* in sand. (b) A generally large and asymmetrical bed form in sand, with a wavelike form but lacking the deep scour associated with *dunes* and *megaripples* (Boothroyd and Hubbard, 1974). Large sand waves generally have a high *ripple index,* and lee slopes inclined at less than the angle of repose of sand. (c) A general term to describe very large subaqueous *sand ripples.*

sand wedge A body of sand shaped like a vertical wedge with the apex downward. In areas of patterned ground, esp. Antarctica, it forms by infilling of debris in winter contraction cracks.

sand-wedge polygon A *nonsorted polygon* formed by infilling of sand and gravel in intersecting fissures resulting from thermal contraction. Surface diameter ranges from a few meters to tens of meters (Péwé 1959). Cf: *tesselation.*

sandy n. A low stream terrace whose upper surface rises upstream where aggradation keeps pace with the growth of a downstream dam, as along Big Sandy River in eastern Kentucky (Shaw, 1911, p. 489). Cf: *island hill; muddy [geomorph].*

sandy breccia A term used by Woodford (1925, p.183) for a breccia containing at least 80% rubble and 10% sand, and no more than 10% of other material.

sandy chert Chert with oolithlike structures, formed when silica replaces cement or fills pore spaces in sandy beds and incorporates large, rounded sand grains in a cherty body or matrix. The perimeters of the sand grains are commonly resorbed, giving gradational contacts between them and the secondary silica.

sandy clay (a) An unconsolidated sediment containing 10-50% sand and having a ratio of silt to clay less than 1:2 (Folk, 1954, p.349). (b) An unconsolidated sediment containing 40-75% clay, 12.5-50% sand, and 0-20% silt (Shepard, 1954). (c) A soil containing 35-55% clay, 45-65% sand, and 0-20% silt (SSSA, 1965, p.347).

sandy clay loam A soil containing 20-35% clay, 45-80% sand, and less than 28% silt. Cf: *clay loam.*

sandy conglomerate (a) A conglomerate containing 30-80% sand and having a ratio of sand to mud (silt + clay) greater than 9:1 (Folk, 1954, p.347); a consolidated sandy gravel. (b) A conglomerate containing more than 20% sand (Krynine, 1948, p.141).

sandy desert An area of sand accumulation in an arid region, usually having an undulating surface of dunes; an *erg* or *koum.*

sandy gravel (a) An unconsolidated sediment containing 30-80% gravel and having a ratio of sand to mud (silt + clay) greater than 9:1 (Folk, 1954, p.346); if the ratio is between 1:1 and 9:1, the sandy gravel is "muddy". (b) An unconsolidated sediment containing more particles of gravel size than of sand size, more than 10% sand, and less than 10% of all other finer sizes (Wentworth, 1922, p.390). (c) An unconsolidated sediment containing 50-75% sand and 25-50% pebbles (Willman et al., 1942, p.343-344). Cf: *pebbly sand.*

sandy loam A soil containing 43-85% sand, 0-50% silt, and 0-20% clay, or containing at least 52% sand and no more than 20% clay and having the percentage of silt plus twice the percentage of clay exceeding 30, or containing 43-52% sand, less than 50% silt, and less than 7% clay (SSSA, 1965, p.347); specif. such a soil containing at least 30% very coarse sand, coarse sand, and medium sand, and less than 25% very coarse sand and less than 30% fine sand or very fine sand. It is subdivided into coarse sandy loam, fine sandy loam, and very fine sandy loam. Sandy loam contains sufficient silt or clay to make the soil somewhat coherent. Cf: *loamy sand.*

sandy mud An unconsolidated sediment containing 10-50% sand and having a ratio of silt to clay between 1:2 and 2:1 (Folk, 1954, p.349).

sandy ore A term used in Brazil for iron ore found along the bottoms of valleys and derived from *rubble ore* "by the natural sluicing of the streams" (Derby, 1910, p.818).

sandy silt (a) An unconsolidated sediment containing 10-50% sand and having a ratio of silt to clay greater than 2:1 (Folk, 1954, p.349). (b) An unconsolidated sediment containing 40-75% silt, 12.5-50% sand, and 0-20% clay (Shepard, 1954). (c) An unconsolidated sediment containing more particles of silt size than of sand size, more than 10% silt, and less than 10% of all other sizes (Wentworth, 1922).

sandy siltstone (a) A consolidated *sandy silt.* (b) A siltstone containing more than 20% sand (Krynine, 1948, p.141).

Sangamon Pertaining to the third classical interglacial stage of the Pleistocene Epoch in North America, after the Illinoian glacial stage and before the Wisconsinan. Etymol: Sangamon County, Ill. See also: *Riss-Würm.* Syn: *Sangamonian.*

sanidaster A spinose rodlike monaxonic sponge spicule (streptaster).

sanidine A high-temperature mineral of the alkali feldspar group: $KAlSi_3O_8$. It is a highly disordered monoclinic form, occurring in clear, glassy, often tabular crystals embedded in unaltered acid volcanic rocks such as trachyte; it appears to be stable under equilibrium conditions above approximately 500°C. Sanidine forms a complete solid-solution series with high albite, and some sodium is always present. Syn: *glassy feldspar; ice spar; rhyacolite.*

sanidine nephelinite A nepheline-rich mafic *phonolite* with abundant magnetite, described by Nieland in 1931 from Katzenbuckel, Germany. Nosean is sometimes present instead of nepheline. The rock is not a *nephelinite,* and the term should not be applied at other localities.

sanidinite An igneous rock (*syenite* or *trachyte*) composed almost entirely of sanidine. The term has also been applied to rocks composed of other alkali feldspars. The name was proposed by Nose in 1808. Not recommended usage.

sanidinite facies The set of metamorphic mineral assemblages (facies) in which are found tridymite, mullite, monticellite, larnite, and sanidine. The term was introduced by Eskola (1915) to cover occurrences of metamorphism at maximum temperature and minimum pressure, e.g. xenoliths in basic lavas, fragments in tuffs, or narrow contact zones bordering shallow basic pipes. Many rocks show evidence of partial to complete fusion, e.g. *buchite.*

sanitary landfill A land site where municipal solid waste is buried in a manner engineered to minimize environmental degradation. Commonly the waste is compacted and periodically covered with soil or other earth material.

sanjuanite A mineral: $Al_2(PO_4)(SO_4(OH)\cdot 9H_2O$.

sanmartinite A mineral: $ZnWO_4$. The zinc is sometimes replaced by iron, manganese, or calcium.

sannaite A mafic *phonolite* containing phenocrysts of barkevikite, clinopyroxene, and biotite (in order of decreasing abundance) in a fine-grained groundmass of alkali feldspar, acmite, chlorite, calcite, and pseudomorphs of mica after nepheline. The name, given by Brögger in 1921, is from Sannavand, in the Fen complex, Norway. Not recommended usage.

Sannoisian European stage: Lower Oligocene (above Ludian, below Stampian). Syn: *Tongrian.*

sansicl An unconsolidated sediment consisting of a mixture of *san* d, *si* lt, and *cl* ay, in which no component forms 50% or more of the whole aggregate.

Sanson-Flamsteed projection A syn. of *sinusoidal projection.* Named after Nicolas Sanson (1600-1667), French geographer, and John Flamsteed (1646-1719), English astronomer.

santafeite A black orthorhombic mineral: $Na_2(Mn,Ca,Sr)_6Mn_3{}^{+4}(V,As)_6O_{28}\cdot 8H_2O$.

santanaite A hexagonal mineral: $Pb_{11}CrO_{16}$.

santite A mineral: $KB_5O_8\cdot 4H_2O$.

Santonian European stage: Upper Cretaceous (above Coniacian, below Campanian). See also: *Emscherian.*

santorinite (a) As defined by Washington in 1897, a light-colored extrusive rock containing approximately 60-65 percent silica, and calcic plagioclase (labradorite to anorthite) as the only feldspar. (b) As defined by Becke in 1899, a hypersthene andesite containing plagioclase crystals that have labradorite cores and sodic rims and a groundmass with microlites of sodic oligoclase. Its name is derived from the island of Santorini (or Thera), Greece. "Santorinite, with two meanings, should be abandoned" (Johannsen, 1937, p. 174).

sanukite An *andesite* characterized by orthopyroxene and garnet as the mafic minerals, andesine as the plagioclase, and a glassy groundmass. Cf: *orthoandesite.* The name, given by Weinschenk in 1890, is from Sanuki, Japan. Not recommended usage.

sanukitoid *orthoandesite.*

sapanthracite Sapropelic coal of anthracitic rank; it is the highest stage in the *sapropelic series.* Cf: *humanthracite.*

sapanthracon Sapropelic coal of Carboniferous age; it is the fifth stage in the *sapropelic series.* Cf: *humanthracon.*

saphir d'eau A syn. of *water sapphire* (variety of cordierite). Etymol: French, "water sapphire".

saponite A white or light-colored trioctahedral magnesium-rich clay mineral of the montmorillonite group: $(Ca/2,Na)_{0.33}(Mg,Fe)_3(Si_{3.67}Al_{0.33})O_{10}(OH)_2\cdot 4H_2O$. It represents an end member in which the replacement of aluminum by magnesium in the octahedral sheets is essentially complete. Saponite occurs in masses that fill veins and cavities in serpentine and basaltic rocks; it has an unctuous feel and is somewhat plastic, but does not adhere to the tongue. Syn: *bowlingite; mountain soap; piotine; soapstone.* Etymol: Greek *sapon,* "soap".

sappare (a) *sapphire.* (b) *kyanite.*

sapperite A natural pure white cellulose, $(C_6H_{10}O_5)_n$, which occurs in brown coal and fossil wood.

sapphire (a) Any pure, gem-quality corundum other than *ruby;* esp. the fine blue transparent variety of crystalline corundum of great value, containing small amounts of oxides of cobalt, chromium, and titanium, used as a gemstone, and found esp. in the Orient (Kashmir, Burma, Thailand, and Ceylon). Other colors, such as pink, purple, yellow, green, and orange, are included under *fancy sapphire.* Syn: *sappare.* (b) Any gem from a corundum crystal.

sapphire quartz (a) A rare, opaque, indigo-blue variety of quartz colored by included nonparallel fibers of silicified crocidolite. Cf: *hawk's-eye.* Syn: *azure quartz; blue quartz; siderite [mineral].* (b) A term used in the western U.S. for chalcedony having a light to pale sapphire-blue color.

sapphirine-1Tc A triclinic mineral: $(Mg,Al)_8(Al,Si)_6O_{20}$. It is related to aenigmatite.

sapphirine-2M (a) A green or pale-blue monoclinic mineral: $(Mg,Al)_8(Al,Si)_6O_{20}$. It is a principal constituent of certain high-grade silica-deficient metamorphic rocks and occurs usually in granular form. (b) A name applied to certain blue minerals such as hauyne and blue chalcedony.

sapping [geomorph] (a) The natural process of erosion along the base of a cliff by the wearing-away of softer layers, thus removing the support for the upper mass which breaks off into large blocks falling from the cliff face. See also: *landslide sapping.* Syn: *cliff erosion; undermining.* (b) *spring sapping.*

sapping [glac geol] (a) *basal sapping.* (b) Sometimes used as a syn. of *plucking [glac geol].*

Saprist In U.S. Dept. of Agriculture soil taxonomy, a suborder of the soil order *Histosol,* characterized by almost completely decomposed plant remains, black color, and a bulk density above 0.2. Unless artificially drained they are saturated more than half the year, but the water level fluctuates, permitting advanced decomposition. *Fibrists* and *Hemists* that are drained artificially may become Saprists in a few decades (USDA, 1975). Cf: *Fibrist; Folist; Hemist.*

saprobe *saprophyte.*

saprobic *saprophytic.*

saprocol Indurated sapropel; it is the second stage in the *sapropelic series.* Also spelled: *saprokol.* Cf: *humocoll.*

saprodil A sapropelic coal of Tertiary age; it is the third stage in the *sapropelic series.* Cf: *humodil.*

saprodite Sapropelic coal of brown-coal rank; it is the fourth stage in the *sapropelic series.* Cf: *humodite.*

saprogen An organism that lives on dead organic matter and can cause its decay. Adj: *saprogenic.*

saprogenic Said of an organism that produces decay or putrefaction. Cf: *saprophytic.* Syn: *saprogenous.* Noun: *saprogen.*

saprohumolith series Organic materials and coals intermediate between the *sapropelite series* and the *humolith series,* with sapropelic materials predominating. Cf: *humosapropelic series.*

saprokol *saprocol.*

saprolite A soft, earthy, typically clay-rich, thoroughly decomposed rock, formed in place by chemical weathering of igneous, sedimentary, and metamorphic rocks. It often forms a layer or cover as much as 100 m thick, esp. in humid and tropical or subtropical climates; the color is commonly some shade of red or brown, but it may be white or gray. Saprolite is characterized by preservation of structures that were present in the unweathered rock. The term was proposed by Becker (1895). Cf: *geest; laterite.* Also spelled: *saprolith; sathrolith.*

sapromyxite *boghead coal.*

sapront *saprophyte.*

sapropel An unconsolidated, jellylike ooze or sludge composed of plant remains, most often algae, macerating and putrefying in an anaerobic environment on the shallow bottoms of lakes and seas. It may be a source material for petroleum and natural gas. Cf: *dy; gyttja; humopel.*

sapropel-calc A sedimentary deposit in which the amount of calcareous-algae remains exceeds that of sapropel.

sapropel-clay A sedimentary deposit in which the amount of clay exceeds that of sapropel.

sapropelic Pertaining to or derived from *sapropel;* indicating a high sulfate or reducing environment.

sapropelic coal Coal that is derived from organic residues (finely divided plant material, spores, algae) in stagnant or standing bodies of water. Putrefaction under anaerobic conditions, rather than peatification, is the formative process. The main types of sapropelic coal are *cannel coal, boghead coal,* and *torbanite.* Sapropelic coals are high in volatiles, generally dull, massive, and relatively uncommon. Cf: *humic coal; nonbanded coal.* See also: *sapropelite series.* Syn: *sapropelite.*

sapropelite *sapropelic coal.*

sapropelite series The organic materials of *sapropelic coal* in metamorphic rank: *sapropel, saprocol, saprodil, saprodite, sapanthracon,* and *sapanthracite* (Heim & Potonié, 1932, p.146). Cf: *humolith series; humosapropelic series; saprohumolith series.*

sapropel-peat *peat-sapropel.*

saprophilous *saprophytic.*

saprophyte A plant that lives on decayed or decaying organic matter. Syn: *saprobe; sapront.* Adj: *saprophytic.*

saprophytic Said of a plant that receives its nourishment from the products of decaying organic matter; i.e. a saprophyte. Syn: *saprophilous; saprobic; saprozoic.* Cf: *saprogenic.*

sapropsammite *sandy sapropel.*

saprovitrinite Vitrinite in vitrain of sapropelic coal. Cf: *humo-*

vitrinite.

saprozoic *saprophytic.*

Saracen stone A syn. of *sarsen.* The term originally signified a pagan stone or monument. Syn: *Saracen's stone.*

sarcodictyum The outermost layer of cytoplasm in a radiolarian; a network of protoplasm on the surface of the calymma of a radiolarian.

Sarcodina A class of protozoans characterized mainly by their ability to form pseudopodia. Among the members of the class are rhizopods and actinopods. Range, Cambrian to present.

sarcolite A tetragonal mineral: $(Ca,Na)_4Al_3(Al,Si)_3Si_6O_{24}$(?).

sarcopside A mineral: $(Fe,Mn,Mg)_3(PO_4)_2$.

Sarcopterygii A subclass of extinct bony fish, characterized by fins with an axial fleshy lobe supported by a skeletal projection from the body wall, and presumably by a paired swim bladder that functioned as a lung. Syn: *lobefin.*

sarcotesta The fleshy outer layer of a seed coat, as in *Trigonocarpus* (Scott, 1923, p. 204). Cf: *sclerotesta.*

sard A translucent brown to deep orange-red variety of chalcedony, similar to *carnelian* but having less intense color; it is classed by some as a variety of carnelian. Syn: *sardius; sardine.*

Sardic orogeny One of the 30 or more short-lived orogenies during Phanerozoic time identified by Stille, in this case near the end of the Cambrian.

sardius Original name for *sard.*

sardonyx A gem variety of chalcedony that is like *onyx* in structure but includes straight parallel brownish-red or reddish-brown bands of sard alternating with white, brown, black, or other colored bands of another mineral. The name is applied incorrectly to carnelian and sard.

Sargasso Sea A warm region of the open North Atlantic Ocean to the east and south of the Gulf Stream, characterized by a large mass of floating vegetation that is mainly sargasso (gulfweed), a seaweed (brown alga) of the genus *Sargassum.*

sarkinite A flesh-red monoclinic mineral: $Mn_2(AsO_4)(OH)$.

Sarmatian Eastern European stage: Middle Miocene (above Bademian, below Pannonian). In continental usage (now obsolescent), above continental Vindobonian, below continental Pontian or Vallesian.

sarmientite A yellow mineral: $Fe_2(AsO_4)(SO_4)(OH)\cdot 5H_2O$. It is isomorphous with diadochite.

sarnaite A feldspathoid-bearing *syenite* containing cancrinite and clinopyroxene. Its name is derived from Särna, Sweden. Named by Brögger in 1890. Not recommended usage.

sarospatakite A micaceous clay mineral composed of mixed layers of illite and montmorillonite, found in Sárospatak, Hungary. Cf: *bravaisite.* Syn: *sarospatite.*

sarsden stone *sarsen.*

sarsen A large residual mass of stone left after the erosion of a once continuous bed of which it formed a part; specif. one of the large, rounded, gray blocks or fragments of silicified sandstone or conglomerate strewn over the surface of the English chalk downs (esp. in Wiltshire) and far from any similar beds, being the only remnants of the former Tertiary (Eocene?) cover. Etymol: alteration of *Saracen,* a Moslem or "outlandish stranger". Syn: *sarsen stone; Saracen stone; sarsden stone; greywether; druid stone.*

sartorite A dark-gray monoclinic mineral: $PbAs_2S_4$.

sarule A *sceptrule* with a single spinose terminal outgrowth that resembles a brush.

saryarkite A white tetragonal mineral: $(Ca,Y,Th)_2Al_4(SiO_4,PO_4)_4(OH)\cdot 9H_2O$.

sassolite A white or gray mineral consisting of native boric acid: $B(OH)_3$ or H_3BO_3. It usually occurs in small pearly scales as an incrustation or as tabular triclinic crystals around fumaroles or vents of sulfurous emanations. Syn: *sassoline.*

sastrugi Irregular ridges up to 5 cm high, formed in a level or nearly level snow surface by wind erosion, often aligned parallel to the wind direction, with steep, concave or overhanging ends facing the wind; or cut into snow dunes previously deposited by the wind. Syn: *zastrugi.* Cf: *erosion ridge; wind ridge; skavl.*

satelite Fibrous serpentine with a slight chatoyant effect, being pseudomorphous after asbestiform tremolite that has been silicified. It occurs in Tulare County, Calif.

satellite A secondary celestial body, natural or man-made, that revolves about another, primary body, e.g. the Moon about the Earth.

satellite geodesy Application of near-earth satellites for determining positions on the Earth's surface and the gravity field.

satellite navigation A method of determining location by using signals from a "transit" satellite in polar orbit above the Earth. Latitude and longitude can be determined by observing the Doppler frequency shift during the passage of such a satellite, and by utilizing the information that the satellite broadcasts about its own location. A "fix" can be obtained once per satellite pass, or about every 90 minutes. See also: *alert.*

satellitic crater *secondary crater.*

sathrolith *saprolite.*

satimolite A mineral: $KNa_2Al_4B_6O_{15}Cl_3\cdot 13H_2O$.

satin ice *acicular ice.*

satin spar (a) A white, translucent, fine fibrous variety of gypsum, characterized by chatoyancy or a silky luster. (b) A term used incorrectly for a fine fibrous or silky variety of calcite or aragonite. Syn: *satin stone.*

satpaevite A yellow mineral: $Al_{12}V_2{}^{+4}V_6{}^{+5}O_{37}\cdot 30H_2O$.

saturated [geol] (a) Said of a rock having quartz in its norm. (b) Said of a mineral that can form in the presence of free silica, i.e. one that contains the maximum amount of combined silica. (c) Said of an igneous rock composed chiefly of such minerals, or of its magma.—Cf: *oversaturated; undersaturated; unsaturated.*

saturated [water] Said of the condition in which the interstices of a material are filled with a liquid, usually water. It applies whether the liquid is under greater than or less than atmospheric pressure, so long as all connected interstices are full. See also: *zone of saturation.*

saturated permafrost Permafrost that contains no more ice than the ground could hold if the water were in the liquid state; permafrost in which all available pore spaces are filled with ice.

saturated pool An oil pool with an excess of gas, which forms a *gas cap* above the oil. Cf: *undersaturated pool.*

saturated surface *water table.*

saturated zone *zone of saturation.*

saturation The maximum possible content of water vapor in the Earth's atmosphere for a given temperature. See also: *dew point.*

saturation curve A curve showing the weight of solids per unit volume of a saturated earth mass as a function of its water content.

saturation line [glaciol] The boundary on a glacier between the *soaked zone* and the *percolation zone.*

saturation line [petrology] The line on a variation diagram of an igneous-rock series that represents saturation with respect to silica. Rocks to the right of the line are oversaturated and those to the left are undersaturated.

saturation magnetization The maximum possible magnetization of a material, i.e. alignment of all magnetic ions.

saturation pressure *bubble-point pressure.*

Saucesian North American stage: Oligocene and Miocene (above Zemorrian, below Relizian).

sauconite A trioctahedral, zinc-bearing clay mineral of the montmorillonite group: $Na_{0.33}Zn_3(Si_{3.47}Al_{0.53})O_{10}(OH)_2\cdot 4H_2O$. It represents an end member in which the replacement of aluminum by zinc in the octahedral sheets is essentially complete.

saukovite *metacinnabar.*

sault A waterfall or rapids in a stream. Etymol: Latin *saltus,* past participle of *salire,* "to leap". Pron: *sue.*

Saurischia One of the two orders of archosaurian reptiles commonly treated as *dinosaurs,* characterized by a triradiate pelvis. It includes bipedal predators (Carnosauria, Coelurosauria), quadrupedal long-necked herbivores (Sauropods), and presumptive ancestors of the latter (Prosauropoda). Range, Middle Triassic to Upper Cretaceous. Cf: *Ornithischia.*

sausage structure *boudinage.*

saussurite A tough, compact, and white, greenish, or grayish mineral aggregate consisting of a mixture of albite (or oligoclase) and zoisite or epidote, together with variable amounts of calcite, sericite, prehnite, and calcium-aluminum silicates other than those of the epidote group. It is produced by alteration of calcic plagioclase. Saussurite was originally thought to be a specific mineral.

saussuritization The replacement, esp. of plagioclase in basalts and gabbros, by a fine-grained aggregate of zoisite, epidote, albite, calcite, sericite, and zeolites. It is a metamorphic or deuteric process and is frequently accompanied by chloritization of the ferromagnesian minerals.

savanna (a) An open, grassy, essentially treeless plain, esp. as developed in tropical or subtropical regions. Usually there is a distinct wet and dry season; what trees and shrubs are found are drought-resistant. (b) The tall grass characteristic of a *savanna.* (c)

Along the southeastern Atlantic Coast of the U.S. the term (often spelled *savannah*) is used for marshy alluvial flats with occasional clumps of trees.

Savic orogeny One of the 30 or more short-lived orogenies during Phanerozoic time identified by Stille, in this case in the late Oligocene, between the Chattian and Aquitanian stages.

Savonius rotor current meter A sensor device used in oceanography for measuring the speed of a current. It utilizes an S-shaped rotor, or a pair of them with their axes perpendicular to each other.

saw-cut A large canyon that cuts across a terrace "with startling abruptness" so that it "cannot be seen until one has almost reached the edge" (Smith, 1898, p. 469).

saw-toothed (a) *serrate.* (b) Descriptive of *hacksaw structure.*

saxicavous Said of an organism that bores into rock, i.e., of a *rock borer.*

saxicolous *rupestral.*

saxifragous Said of a plant that grows in crevices of rock and promotes its splitting; i.e., of a *chasmophyte.*

saxigenous *rupestral.*

Saxonian European stage: Middle Permian (above Autunian, below Thuringian).

Saxonian-type facies series Rocks produced in a type of dynamothermal regional metamorphism for which the classical locality is the granulite area in Saxony. The pressure range of formation is extensive (2000-8000 bars) and the temperature range from 100° to 700°C, probably involving polymetamorphism. The rocks differ little mineralogically from those formed in the *Barrovian-type facies series,* except at the highest pressure and temperature values, at which kyanite- and orthoclase-bearing granulites are formed (Hietanen, 1967, p. 201).

saxonite A *peridotite* composed chiefly of olivine and orthopyroxene. It is considered by some petrologists as a syn. of *harzburgite* and by others as distinct from harzburgite owing to its lower percentage of opaque oxide minerals. The term was coined by Wadsworth in 1884, for Saxony, Germany. Not recommended usage.

sazhinite An orthorhombic mineral: $Na_3CeSi_6O_{15} \cdot 6H_2O$.

sborgite A triclinic mineral: $NaB_5O_8 \cdot 5H_2O$.

scabland An elevated area, underlain by flat-lying basalt flows, with a thin soil cover and sparse vegetation, and usually with deep, dry channels scoured into the surface. An example is the Columbia lava plateau of eastern Washington, which was widely and deeply eroded by glacial meltwaters. See also: *channeled scabland; scabrock.*

scabrate Rough or scaly; esp. said of sculpture of pollen and spores consisting of more or less isodiametric projections less than one micron in diameter.

scabrock (a) An outcropping of *scabland.* (b) Weathered material of a scabland surface.

scacchite (a) A mineral: $MnCl_2$. (b) A name applied to various minerals, including monticellite, a doubtful selenide of lead, and a brick-red powdery fluoride containing rare earths.

scaglia A dark, very fine-grained, more or less calcareous shale typically developed in the Upper Cretaceous and lower Tertiary of the northern Apennines. Etymol: Italian, "scale, chip". Pron: *skahl*-ya.

scalariform Pertaining to the ladderlike thickenings with intervening pits on the walls of certain xylem cells.

scale [cart] The ratio between linear distance on a map, chart, globe, model, or photograph and the corresponding distance on the surface being mapped. It may be expressed in the form of a direct or verbal statement using different units (e.g. "1 inch to 1 mile" or "1 inch = 1 mile"); a *representative fraction* or numerical ratio (e.g. "1/24,000" or "1:24,000", indicating that one unit on the map represents 24,000 identical units on the ground); or a graphic measure (such as a bar or line marked off in feet, miles, or kilometers). The scale of a photograph is usually taken as the ratio of the principal distance of the camera to the altitude of the camera station above mean ground elevation. See also: *small-scale map; large-scale map.* Syn: *map scale.*

scale [paleont] (a) A small platelike structure attached to septal grooves and interseptal ridges in some rugose corallites, as in *Tryplasma.* Also, a thin flat or nearly flat sclerite in an octocoral. (b) *scaphocerite.* (c) A flat to slightly curved projection from the lower lip of a pore, rising obliquely to the wall in archaeocyathids. (d) A small, more or less flattened, rigid, and definitely circumscribed plate forming part of the integumentary body covering of certain vertebrates.

scale factor A multiplier for reducing a distance obtained from a map by computation or scaling to the actual distance on the datum of the map.

scalenohedral Having the form or symmetry of a *scalenohedron.*

scalenohedron A closed crystal form whose faces are scalene triangles; the *hexagonal scalenohedron* has twelve faces, and the *tetragonal scalenohedron* has eight. Adj: *scalenohedral.*

scaler An electronic instrument that counts the pulses from a *nuclear radiation* detector.

scaling A type of *exfoliation* that produces thin flakes, laminae, or scales.

scallop [paleont] (a) Any of many marine bivalve mollusks of the family Pectinidae having a shell that is characteristically rather flat, radially ribbed, and marginally undulated, that has a single large adductor muscle, and that is able to swim by opening and closing the valves. (b) One of the valves of the shell of a scallop. Syn: *scallop shell.*

scallop [sed] *scalloping.*

scallop [speleo] One of a mosaic of small shallow intersecting hollows formed on the surface of soluble rock by turbulent dissolution. They are steeper on the upstream side, and smaller sizes are formed by faster-flowing water. Syn: *flute [speleo]; solution ripple.*

scalloped upland "The region near or at the divide of an upland into which glacial cirques have cut from opposite sides" (Stokes & Varnes, 1955, p. 129). Cf: *fretted upland.*

scalloping A term used by Gruner et al. (1941, p.1621-1622) for a sedimentary structure of uncertain origin, superficially resembling oscillation ripple mark, and having a concave side that is always oriented toward the top of the bed. It may have formed by differential expansion or shrinkage of adjoining layers of mud before complete consolidation. Syn: *scallop [sed].*

scalped anticline *breached anticline.*

scaly (a) Said of the texture of a mineral, esp. a mica, in which small plates break or flake off from the surface like scales. (b) Referring to the general appearance of a rock, as in *argille scagliose.*

scan n. A graphic or photographic depiction of the distribution of radioactivity of a substance. See also: *autoradiograph; skiagram.*

scandent Said of a graptoloid with stipes that grew erect along or enclosing the virgula.

scan line The narrow strip on the ground that is swept by the instantaneous field of view of a detector in a *scanner* system.

scanner An optical-mechanical imaging system in which a rotating or oscillating mirror sweeps the instantaneous field of view of a *detector* across the terrain. Scanners may be airborne or stationary. Nonrecommended syn: *line scanner.*

scanner distortion The geometric distortion that is characteristic of scanner images. The scale of the image is constant in the direction parallel with the flight direction of the aircraft or spacecraft. At right angles to this direction, however, the image scale becomes progressively smaller from the nadir line outward toward either margin of the image. Linear features, such as roads, that trend diagonally across a scanner image are distorted into S-shaped curves. Distortion becomes more pronounced with a larger angular field of view.

scanning electron microscope An *electron microscope* in which a finely focused beam of electrons is electrically or magnetically moved across the specimen to be examined, from point to point, again and again, and the reflected and emitted electron intensity measured and displayed, sequentially building up an image. The ultimate magnification and resolution is less than for the conventional electron microscope, but opaque objects can be examined, and great depth of field is obtained. Abbrev: SEM.

scanning transmission electron microscope A *transmission electron microscope* that has the capability of forming the electron beam into a fine probe (< 100Å in diameter) and scanning it across a thin specimen. The transmitted scanned beam is collected below the specimen by a solid-state detector and is reproduced electronically as an image on a cathode-ray tube. Image contrast is the same as in the conventional transmission electron microscope; the advantage of the scanning transmission electron microscope is the fine probe and the electronic manipulation of the detected transmitted beam. Electron diffraction patterns can be obtained from areas tens of angstroms in diameter, and, in conjunction with an X-ray detection system, compositional data can be obtained with spatial resolutions of 500-1000Å In some instruments, the electron-beam size can be reduced to several angstroms in diameter, resulting in high-resolution images of single large atoms, e.g. uranium. Abbrev: STEM.

scanoite An igneous rock similar to *ghizite* but containing normative nepheline and no biotite. The name, given by Lacroix in 1924, is from the Scano lava flow, Monte Ferru, Sardinia, Italy. Not recommended usage.

scan skew Distortion of scanner images caused by forward motion of the aircraft or satellite during the time required to complete a scan.

scaphocerite A flattened plate on the second joint of the antennae of many crustaceans; e.g. a scalelike exopod of an antenna of a eumalacostracan. Syn: *scale [paleont].*

scapholith An elongate, diamond- or boat-shaped heterococcolith with a central area of parallel laths. Syn: *rhombolith.*

scaphopod Any benthic marine univalve mollusk belonging to the class Scaphopoda, characterized by an elongate body completely surrounded by mantle and a tubular calcareous shell open at both ends. Range, Devonian to present.

scapolite (a) A group of minerals of general formula: $(Na,Ca,K)_4[Al_3(Al,Si)_3Si_6O_{24}](Cl,F,OH,CO_3,SO_4)$. It consists of generally white or gray-white minerals crystallizing in the dipyramidal class of the tetragonal system, and commonly forms an isomorphous series between marialite and meionite. Scapolite minerals characteristically occur in calcium-rich metamorphic rocks or in igneous rocks as the products of alteration of basic plagioclase feldspars. (b) A specific mineral of the scapolite group, intermediate in composition between marialite and meionite (Ma:Me from 2:1 to 1:3), containing 46-54% silica, and resembling feldspar when massive but having a fibrous appearance and a higher specific gravity. Syn: *wernerite.* (c) A member of the scapolite group, including scapolite, marialite, meionite, and mizzonite.

scapolitization Introduction of, or replacement by, scapolite. Plagioclase is commonly so replaced. The replacement may involve introduction of chlorine.

scar [geomorph] (a) A cliff, precipice, or other steep, rocky eminence or slope (as on the side of a mountain) where bare rock is well exposed to view; e.g. a limestone face in northern England. Originally, the term referred to a crack or breach; later, an isolated or protruding rock. Etymol: Old Norse *sker,* "skerry". Syn: *scaur; scaw.* (b) A rocky *shore platform.* (c) *landslide scar.* (d) *meander scar.*

scar [paleont] (a) *muscle scar.* (b) *cicatrix.*

scarbroite A white mineral: $Al_{14}(CO_3)_3(OH)_{36}$. Syn: *tucanite.*

scarp (a) A line of cliffs produced by faulting or by erosion. The term is an abbreviated form of *escarpment,* and the two terms commonly have the same meaning, although "scarp" is more often applied to cliffs formed by faulting. See also: *fault scarp; erosion scarp.* (b) A relatively straight, clifflike face or slope of considerable linear extent, breaking the general continuity of the land by separating surfaces lying at different levels, as along the margin of a plateau or mesa. A scarp may be of any height. The term should not be used for a slope of highly irregular outline. See also: *scarp slope.* (c) *beach scarp.* (d) A steep surface on the undisturbed ground at the edge of a landslide, caused by movement of slide material away from the undisturbed ground; also, a similar but smaller feature on the disturbed material, produced by differential movements within the sliding mass.

scarped plain An area marked by a succession of faintly inclined or gently folded strata, e.g. the eastern part of the Great Plains of the U.S. The inclination of strata has perceptible influence on even the smaller elements of the topography.

scarped ridge *cuesta.*

scarp face *scarp slope.*

scarp-foot spring A spring that flows onto the land surface at or near the foot of an escarpment.

scarpland A region marked by a succession of nearly parallel cuestas separated by lowlands.

scarplet (a) A miniature scarp, ranging in height from several centimeters to 6 m or more; specif. a *piedmont scarp.* Also, a small scarp formed on a wave-cut platform by the outcropping of resistant rocks. (b) *earthquake scarplet.*

scarp retreat The *slope retreat* of a scarp.

scarp slope (a) The relatively steeper face of a *cuesta,* facing in a direction opposite to the dip of the strata. Cf: *dip slope; back slope.* Syn: *scarp face; inface; front slope.* (b) A *scarp* or an *escarpment.*

scarp stream An *obsequent stream* flowing down a scarp, such as down the scarp slope of a cuesta.

scarred pebble *pitted pebble.*

scatter diagram [stat] A graphic representation of paired measurements, usually along Cartesian axes, that aids in visualizing the relationships between two or more variables. Syn: *distribution scatter.*

scatter diagram [struc petrol] *point diagram.*

scattered ice An obsolete term for sea-ice concentration of 1/10 to 5/10; now replaced generally by *open pack ice* and *very open pack ice.*

scatterometer A nonimaging radar device that records *back scatter* as a function of incidence angle.

scavenger An organism that feeds on dead animal matter, refuse, or matter detrimental to the health of humans.

scavenger well A well located between a well or wells that yield usable water and a source of potential contamination; it is pumped (or allowed to flow) as waste to prevent the contaminated water from reaching the good wells. The most common application is in coastal areas, where scavenger wells are used to prevent salt water from reaching water-supply wells.

scawtite A colorless monoclinic mineral: $Ca_7Si_6O_{18}(CO_3)\cdot 2H_2O$.

scene The area on the ground that is covered by an image or photograph.

sceptrule A hexactinellid-sponge spicule (microsclere) that consists of one long ray with one end containing an axial cross and usually bearing various anaxial outgrowths. See also: *sarule; scopule; clavule; lonchiole.*

schachnerite A hexagonal mineral: $Ag_{1.1}Hg_{0.9}$.

schafarzikite A red to brown tetragonal mineral: $FeSb_{2-x}(O,OH,H_2O)_4$.

schairerite A colorless rhombohedral mineral: $Na_{21}(SO_4)_7F_6Cl$. Cf: *galeite.*

schallerite A brown mineral: $(Mn,Fe)_8Si_6As(O,OH,Cl)_{26}$.

schalstein An altered tuff with shear structures; it is usually basic or calcareous (Holmes, 1928). Etymol: German. Cf: *adinole; spotted slate.*

schapbachite A syn. of *matildite.* The term has also been applied to an intimate intergrowth of matildite and galena and to a high-temperature polymorph of matildite.

schaurteite A mineral: $Ca_3Ge(SO_4)_2(OH)_6\cdot 3H_2O$.

scheelite A yellow-white or brown tetragonal mineral: $CaWO_4$. It is found in pneumatolytic veins associated with quartz, and fluoresces to show a blue color. Scheelite is isomorphous with powellite, and is an ore of tungsten.

schefferite A brown to black monoclinic mineral of the pyroxene group: $(Ca,Mn)(Mg,Fe,Mn)Si_2O_6$. It is a variety of diopside containing manganese and frequently much iron.

scheibeite *phoenicochroite.*

S-chert Stratigraphically controlled chert, occurring in beds (bedded chert), or in groups of nodules (nodular chert) distributed parallel to bedding (Dunbar & Rodgers, 1957, p. 248).

schertelite A mineral: $(NH_4)_2MgH_2(PO_4)_2\cdot 4H_2O$.

scheteligite A mineral: $(Ca,Y,Sb,Mn)_2(Ti,Ta,Nb,W)_2O_6(O,OH)$.

schiefer A general term referring to a rock's laminated or foliated structure, commonly used to describe rocks ranging from shale to schist: e.g. schieferton or argillaceous shale, tonschiefer or slate. Adj: *schiefrig.* Etymol: German.

schiefrig Adj. of *schiefer.*

schiller A syn. of *play of color.* Etymol: German. See also: *schillerization.*

schillerization The development of *schiller* or play of color in a mineral, due to the arrangement of minute inclusions in the crystal.

schiller spar A syn. of *bastite.* Also spelled: *schillerspar.*

schirmerite A mineral: $AgPb_2Bi_3S_7$.

schist A strongly foliated crystalline rock, formed by dynamic metamorphism, that can be readily split into thin flakes or slabs due to the well developed parallelism of more than 50% of the minerals present, particularly those of lamellar or elongate prismatic habit, e.g. mica and hornblende. The mineral composition is not an essential factor in its definition unless specifically included in the rock name, e.g. quartz-muscovite schist. Varieties may also be based on general composition, e.g. calc-silicate schist, amphibole schist; or on texture, e.g. spotted schist.

schist-arenite A light-colored sandstone containing more than 20% rock fragments derived from an area of regionally metamorphosed rocks (Krynine, 1940); specif. a lithic arenite having abundant fragments of schist. The term was applied by Krynine (1937, p.427) to the medium-grained clastic rocks of the Siwalik Series in northern India, averaging about 40% quartz, 15% feldpsar, 35-

40% schist and phyllite fragments, and 5-10% accessory materials.

schistic *schistose.*

schistoid adj. Resembling *schist.*

schistose Said of a rock displaying *schistosity.* Cf: *gneissic.* Syn: *schistic.*

schistosity The foliation in schist or other coarse-grained, crystalline rock due to the parallel, planar arrangement of mineral grains of the platy, prismatic, or ellipsoidal types, usually mica. It is considered by some to be a type of *cleavage.* Adj: *schistose.*

schizocarp A dry fruit that, at maturity, splits apart into several one-seeded, indehiscent carpels.

schizochroal eye A trilobite eye with large lenses, each with its own cornea, separated by scleral projections that ensure that each ommatidium receives light from only its own individual lens. Cf: *abathochroal eye; holochroal eye.*

schizodont (a) Said of the dentition of a bivalve mollusk with the median tooth of the left valve broad and divided into two equal parts (bifid), and characterized by coarse, variable, and amorphous teeth diverging sharply from beneath the beak. (b) Said of a subclass of amphidont hinges in ostracodes, having anterior tooth and socket of one valve both bifid, and a reverse arrangement of elevations and depressions in the opposed valve (TIP, 1961, pt.Q, p.54).

schizohaline Said of an environment characterized by extreme variation from hypersaline to brackish or fresh conditions, as in coastal lagoons subjected to seasonal or sporadic drought or storms (Folk & Siedlecka, 1974). Pron: skizo-haline. Etymol: Greek; *schizo,* "split", + haline, "salty".

schizolite A light-red variety of pectolite containing manganese.

schizolophe A brachiopod lophophore indented anteriorly and medially to define a pair of brachia that bear a row of paired filamentary appendages, at least distally (TIP, 1965, pt.H, p.152).

schizomycete An organism of the class Schizomycetes, a group of unicellular or noncellular organisms that are variously classified with fungi, with blue-green algae, or separately. See: Melchior & Werdermann, 1954, p. 42. Cf: *myxomycete; eumycete.*

schizoporellid Said of bryozoans characterized by a median sinus at the proximal margin of the orifice, as in the cheilostome family Schizoporellidae (TIP, 1953, pt. G, p. 14).

schizorhysis A *skeletal canal* in dictyonine hexactinellid sponges passing completely through the dictyonal framework as well as connecting laterally. It is covered by exopinacoderm. Pl: *schizorhyses.*

schlanite The soluble resin extracted from *anthracoxene* by ether. See also: *anthracoxenite.*

schlieren Tabular bodies, generally a few inches to tens of feet long, that occur in plutonic rocks. They have the same general mineralogy as the plutonic rocks, but because of differences in mineral ratios they are darker or lighter; the boundaries with the rock tend to be transitional. Some schlieren are modified inclusions, others may be segregations of minerals. Etymol: German for a flaw in glass due to a zone of abnormal composition. Singular *schliere.* Also spelled: *schliere.* Adj: *schlieric.* Cf: *flow layer.*

schlieren arch A term introduced by Balk (1937, p. 56) for an intrusive igneous body with flow layers along its borders but poorly developed or absent in its interior. Cf: *schlieren dome.* Not recommended usage.

schlieren dome A term introduced by Balk (1937, p. 56) for an intrusive igneous body more or less completely outlined by flow layers which culminate in one central area. Cf: *schlieren arch.* Not recommended usage.

Schlumberger An informal term designating an *electric log* or other well log. Named from the brothers Schlumberger, who first logged a borehole in 1927. Commonly pron: *slum*-bur-jay.

Schlumberger array An electrode array in which two closely spaced potential electrodes are placed midway between two current electrodes.

schmeiderite A mineral: $(Pb,Cu)_2SeO_4(OH)_2$(?).

Schmidt field balance A sensitive but now obsolete instrument for measuring variations in the horizontal or vertical magnetic field by means of a counterbalanced magnet. Cf: *torsion magnetometer.*

Schmidt net A coordinate system used to plot a *Schmidt projection,* used in crystallography for statistical analysis of data obtained esp. from universal-stage measurements, and in structural geology for plotting azimuths as angles measured clockwise from north and about a point directly beneath the observer.

Schmidt projection A term used in crystallography and structural geology for a *Lambert azimuthal equal-area projection* of the lower hemisphere of a sphere onto the plane of a meridian. Named after Walter Schmidt (1885-1945), Austrian petrologist and mineralogist, who first used the projection in structural geology (Schmidt, 1925, p.395-399). See also: *Schmidt net.*

schmitterite A mineral: $(UO_2)TeO_3$.

schneiderhöhnite A triclinic mineral: $Fe_8{}^{+2}As_{10}{}^{+3}O_{23}$.

schoderite An orange monoclinic mineral: $Al_2(PO_4)(VO_4){\cdot}8H_2O$.

schoenfliesite A cubic mineral: $MgSn(OH)_6$.

Schoenflies notation A system of describing crystal classes by means of symbols used esp. by physicists and chemists. Cf: *Hermann-Mauguin symbols.* Syn: *Schoenflies symbols.*

Schoenflies symbols *Schoenflies notation.*

schoenite A syn. of *picromerite.* Also spelled: *schönite.*

schoepite A yellow secondary mineral: $UO_3{\cdot}2H_2O$. See also: *metaschoepite; paraschoepite.* Syn: *epiianthinite.*

scholzite A colorless to white mineral: $CaZn_2(PO_4)_2{\cdot}2H_2O$.

schönfelsite A very dark-colored picritic *basalt* containing approximately 53% olivine and 16% augite as phenocrysts in a fine-grained groundmass of bytownite, glass, apatite, magnetite, augite, and bronzite. The name, given by Uhlemann in 1909, is from Altschönfels, Germany. Cf: *oceanite.* Not recommended usage.

schorenbergite A hypabyssal nepheline *leucitite* containing phenocrysts of nosean, or sometimes leucite, in a groundmass of leucite, nepheline, and acmite. Feldspar is absent. The name, given by Brauns in 1921, is from Schorenberg, Laacher See district, Germany. Not recommended usage.

schorl (a) A term commonly applied to *tourmaline,* esp. to the black, iron-rich, opaque variety. (b) An obsolete term for any of several dark minerals other than tourmaline; e.g. hornblende.—Also spelled: *shorl.* Syn: *schorlite.*

schorlomite A black mineral of the garnet group: $Ca_3(Fe,Ti)_2(Si,Ti)_3O_{12}$. Cf: *melanite.*

schorl rock A term used in Cornwall, England, for a granular rock composed essentially of aggregates of needle-like crystals of black tourmaline (schorl) associated with quartz, and resulting from the complete tourmalinization of granite.

schorre A Dutch term for that part of a sandy beach covered by the sea only during spring tides.

schott *shott.*

Schottky defect In a crystal structure, the absence of an atom; it is a type of point defect. Cf: *Frenkel defect; interstitial defect; defect lattice.*

schreibersite A silver-white to tin-white, highly magnetic, tetragonal meteorite mineral: $(Fe,Ni)_3P$. It contains small amounts of cobalt and traces of copper, and tarnishes to brass yellow or brown. Schreibersite occurs in tables or plates as oriented inclusions in iron meteorites. Syn: *rhabdite.*

Schreinemakers' analysis The method pioneered by H. A. Schreinemakers in the 1920's for determination of the topology of phase diagrams based on the number and composition of phases in the system.

schreyerite A monoclinic mineral: $V_2Ti_3O_9$.

schriesheimite A *hornblende peridotite* that contains diopside. Hornblende encloses olivine. Its name, given by Salomon and Nowomejsky in 1904, is derived from Schriesheim, Germany. Not recommended usage. Cf: *cortlandtite; scyelite.*

schroeckingerite A green-yellow secondary mineral: $NaCa_3(UO_2)(CO_3)_3(SO_4)F{\cdot}10H_2O$. It is an ore of uranium. Also spelled: *schröckingerite.* Syn: *dakeite.*

schrötterite An opaline variety of allophane rich in aluminum. Material from the type locality has been shown to be a mixture of glassy halloysite and earthy variscite.

schrund *bergschrund.*

schrund line A term introduced by Gilbert (1904, p. 582) for "the base of the bergschrund at a late stage in the excavation of the cirque basin". The line separates the steeper slope of the cirque wall from the gentler, usually scalable, slope below.

schubnelite A mineral: $Fe_{+3}VO_4{\cdot}H_2O$.

schuetteite A yellow mineral: $Hg_3(SO_4)O_2$.

schuilingite A blue mineral: $Pb_3Ca_6Cu_2(CO_3)_8(OH)_6{\cdot}6H_2O$.

schultenite A colorless mineral: $PbHAsO_4$.

Schulze's reagent An oxidizing mixture very commonly used in palynologic macerations, consisting of a saturated aqueous solution of $KClO_3$ and varying amounts of concentrated HNO_3 (Schulze, 1855). Named after Franz F. Schulze (1815-1873), German chemist. Syn: *Schulze's mixture; Schulze's solution.*

schungite *shungite.*

schuppen structure A syn. of *imbricate structure [tect].* Etymol: German *Schuppenstruktur.* The German word is sometimes used in English geologic literature as *schuppenstruktur.*

Schürmann series A list of metals so arranged that the sulfide of any one is precipitated at the expense of the sulfide of any lower metal in the series.

schwagerinid Any fusulinid belonging to the subfamily Schwagerininae.

schwartzembergite A mineral: $Pb_6(IO_3)_2Cl_4O_2(OH)_2$.

schwatzite A variety of tetrahedrite containing mercury.

sciaphilic Shade-loving; referring to marine organisms encrusting hard substrates that are sheltered or protected from direct sunlight, such as the undersides of corals and the ceilings of reef caverns. Cf: *coelobitic; cryptic.*

scientific hydrology Hydrologic study devoted principally to fundamental processes and relationships of the hydrologic cycle.

scientific method A general term for the lines of reasoning that scientists tend to follow in attempting to explain natural phenomena. It typically includes observation, analysis, synthesis, classification, and inductive inference, in order to arrive at a *hypothesis* that seems to explain the problem. Hypothesis becomes *theory* if it withstands repeated testing and application. Deductive use of the theory may then explain additional problems. Since the term actually covers several methods, it is often used in the plural. See also: *induction; deduction.*

scientific name The formal Latin name of a taxon. Cf: *vernacular name.*

scientific stone A synthetic, reconstructed, or imitation gemstone. Syn: *scientific gem.*

scintillation [gem] The flashing, twinkling, or sparkling of light, or the alternating display of reflections, from the polished facets of a gemstone.

scintillation [radioactivity] A small flash of light produced by an ionizing agent (such as radioactive particles) in a phosphor or *scintillator.* See also: *scintilloscope.*

scintillation counter An instrument that measures ionizing radiation by counting individual scintillations of a substance. It consists of a phosphor and a photomultiplier tube that registers the phosphor's flashes. It may be smaller and more efficient than a Geiger-Müller counter. It is used in spectrometry as well as prospecting. Syn: *scintillometer.*

scintillation spectrometer An instrument for measuring a mass or energy spectrum, therein similar to a *gamma-ray spectrometer,* and determining its frequency distribution by the use of a *scintillation counter.*

scintillator Any transparent material (crystalline, liquid, or organic) that emits small flashes of light when bombarded by an ionizing agent such as radioactive particles.

scintillometer *scintillation counter.*

scintilloscope An instrument that displays the *scintillation [radioactivity]* of a substance on a screen. Also spelled: *scintilliscope.*

sciophyte A plant preferring growth in light of low intensity.

scissor fault A fault on which there is increasing offset or separation along the strike from an initial point of no offset, with reverse offset in the opposite direction. The separation may be due to a scissorlike or pivotal movement on the fault, or it may be the result of uniform strike-slip movement along a fault across a synclinal or anticlinal fold. The terminology is not rigorous; *pivotal fault, hinge fault, rotary fault,* and *rotational fault* are similarly used. See also: *node; hinge line.* Syn: *differential fault.*

scleracoma A collective term for the hard skeletal parts of radiolarians.

scleractinian Any zoantharian coelenterate belonging to the order Scleractinia, characterized by solitary and colonial forms with calcareous exoskeletons secreted by the ectoderm. This order includes most post-Paleozoic and living corals. Range, Middle Triassic to present. Syn: *hexacoral.*

sclere A minute skeletal element; esp. a sponge spicule. See also: *megasclere; microsclere.*

sclerenchyma (a) Thick-walled strengthening tissue in a plant. It may consist of either elongate cells called fibers or shorter cells called stone cells. (b) *sclerenchyme.*

sclerenchyme (a) The calcareous tissue of rugose corallites, esp. the notably thickened parts of the skeleton (TIP, 1956, pt.F, p.250). (b) The vesicular skeletal structure between corallites in colonial coralla, such as the stony substance secreted by the coenenchyme of a scleractinian coral (Shrock & Twenhofel, 1953, p.133).—Cf: *mesenchyme; stereome.* Syn: *sclerenchyma; scleroderm.*

sclerite A hard chitinous or calcareous plate, piece, or spicule of an invertebrate, e.g. a hardened, chitinized cover forming part of the external skeleton of a merostome or arachnid; a calcareous ossicle (anchor, hook, rod, wheel, or disc) of a holothurian; a calcareous skeletal element of the mesogloea of an octocoral, irrespective of form; or a thickened line in the operculum, mandible, or frontal membrane of a bryozoan.

scleroblast (a) One of the cells of a sponge by which a spicule is formed; a mother cell of one or more sclerocytes. Also, a *sclerocyte.* (b) One of the ectodermal cells of octocorallian mesogloea that produce calcareous spicules; e.g. *axoblast.*

sclerocyte A cell that secretes all or part of a sponge spicule. Syn: *scleroblast; spiculoblast.*

scleroderm The hard *sclerenchyme* of the skeleton of a scleractinian coral.

sclerodermite (a) The center of calcification and surrounding cluster of calcareous (aragonitic) fibers making up a septum of a scleractinian coral. Sclerodermites are the apparent primary elements in septa and they are variously arranged in vertical series to make trabeculae. (b) A spine or plate of a holothurian. (c) The hard integument of an arthropod segment.

sclerometer An instrument used in mineral analysis to determine hardness by measuring the pressure required to scratch a polished surface of the material with a diamond point.

scleroseptum A calcareous radial septum of a coral.

sclerosome A continuous deposit in a calcareous sponge of nonspicular calcium carbonate that may form part or all of the skeleton.

sclerosponge A member of the sponge class Sclerospongiae, characterized by spherulitic tabular calcareous skeletons with embedded spicules.

sclerotesta The hard, bony coat of a seed, e.g. of a cycad. Cf: *sarcotesta.*

sclerotinite A maceral of coal within the *inertinite* group, consisting of the sclerotia of fungi or of fungal spores characterized by a round or oval form and varying size.

sclerotium In the eumycetes, a resting body composed of a hardened mass of hyphae, frequently rounded in shape; in the myxomycetes, a hard plasmodial resting stage.

sclerotized Said of the covering of an invertebrate (esp. an arthropod) hardened by substances other than chitin.

scolecite A zeolite mineral: $CaAl_2Si_3O_{10}\cdot3H_2O$. It usually occurs in delicate radiating groups of white fibrous or acicular crystals, and in some forms it shows a wormlike motion when heated.

scolecodont The fossil jaw, with denticles, of an *annelid.* It is composed of silica and chitin, the chitin being carbonized to a jet black during fossilization.

scolite *scolithus.*

scolithus Any of various tubular or vermiform trace-fossil structures found in Cambrian and Ordovician quartz-rich sandstones (and also in upper Precambrian rocks), consisting of narrow, vertical and usually straight tubes or tube fillings, about 0.2-1 cm in diameter, commonly but not always closely crowded, and generally flaring out into cuplike depressions at their tops. They are believed to be the fossil burrows of marine worms, possibly phoronids, and are assigned to the "genus" *Scolithus* (properly *Skolithos*). See also: *worm tube.* Syn: *scolite; pipe-rock burrow.*

scopule A *sceptrule* in which the terminal outgrowths are a pair or ring of spines whose ends may be clubbed and bear rings of recurved teeth.

scopulite A rodlike or stemlike crystallite that terminates in brushes or plumes.

score *scoring.*

scoria [coal] *clinker [coal].*

scoria [volc] A bomb-size pyroclast that is irregular in form and generally very vesicular (Macdonald, 1972, p. 126). In less restricted usage, a vesicular cindery crust on the surface of andesitic or basaltic lava, the vesicular nature of which is due to the escape of volcanic gases before solidification; it is usually heavier, darker, and more crystalline than *pumice.* The adjective form, *scoriaceous,* is usually applied to pyroclastic ejecta. *Cinder* is sometimes used synonymously. See also: *thread-lace scoria.*

scoriaceous [pyroclast] Said of the texture of a coarsely *vesicular* pyroclastic rock (e.g. *scoria*), usually of andesitic or basaltic composition, and coarser than a *pumiceous* rock. The walls of the vesicles may be either smooth or jagged. Also, said of a rock exhibiting such texture. Syn: *scoriform; scorious.*

scoriaceous [sed] Said of a sedimentary rock whose surface is pit-

ted and irregular like that of volcanic scoria; e.g. "scoriaceous limestone" produced by dissolution of the nodules of a nodular limestone.
scoria tuff A deposit of fragmented scoria in a fine-grained tuff matrix.
scoriform *scoriaceous [pyroclast].*
scorilite A volcanic glass (Hey, 1962, p. 593).
scoring (a) The formation of parallel scratches, lines, or grooves in a bedrock surface by the abrasive action of rock fragments transported by a moving glacier. (b) A scratch, line, or groove produced by scoring. Syn: *score.*
scorious *scoriaceous [pyroclast].*
scorodite A green or brown orthorhombic mineral: $FeAsO_4 \cdot 2H_2O$. It is isomorphous with mansfieldite and represents a lesser ore of arsenic.
scorzalite A blue mineral: $(Fe^{+2},Mg)Al_2(PO_4)_2(OH)_2$. It is isomorphous with lazulite.
Scotch pebble A rounded fragment of agate, carnelian, cairngorm, or other variety of quartz, found in the gravels of parts of Scotland, and used as a semiprecious stone.
Scotch topaz A yellow transparent variety of quartz resembling *topaz;* specif. *cairngorm.*
Scotch-type volcano A volcanic form characterized by concentric cuestas and produced by cauldron subsidence (Guilcher, 1950).
scour [eng] An artificial current or flow of water intended to remove mud or other granular material from a stream bed; also, the structure built to produce such a current.
scour [geomorph] (a) The powerful and concentrated clearing and digging action of flowing air, water, or ice, esp. the downward erosion by stream water in sweeping away mud and silt on the outside curve of a bend, or during time of flood. (b) A place in a stream bed swept (scoured) by running water, generally leaving a gravel bottom.
scour [tides] *tidal scour.*
scour and fill [geomorph] A process of alternate excavation and refilling of a channel, as by a stream or the tides; esp. such a process occurring in time of flood when the discharge and velocity of an aggrading stream are suddenly increased, causing the digging of new channels that become filled with sediment when the flood subsides. Cf: *cut and fill.*
scour and fill [sed struc] A sedimentary structure consisting of a small erosional channel, generally ellipsoidal, that is subsequently filled; a small-scale *washout.*
scour cast A sole mark consisting of a cast of a scour mark; specif. a *flute cast.*
scour channel A large groove-like erosional feature produced in sediments by scour.
scour depression A crescentic hollow produced in a stream bed near the outside of a bend by water that scours below the grade of the stream (Bryan, 1920, p. 191).
scouring The process of erosion by the action of flowing air, ice, or water, esp. *glacial scour.* See also: *scour [geomorph].*
scouring rush *sphenopsid.*
scouring velocity The velocity of water that is necessary to dislodge stranded solids from the stream bed.
scour lineation A smooth, low ridge, 2-5 cm wide, formed on a sedimentary surface and believed to result from the scouring action of a current of water. It is characterized by symmetrical ends so that the line of current movement, but not its direction, can be ascertained.
scour mark A *current mark* produced by the cutting or scouring action of a current of water flowing over the bottom; e.g. a flute. See also: *transverse scour mark.* Syn: *scour marking.*
scour side The upstream, or stoss, side of a roche moutonnée, smoothed, striated, and rounded by glacial abrasion. Ant: *pluck side.*
scourway A channel produced by a strong glacial stream near the margin of an ice sheet.
scrap mica Mica whose size, color, or quality is below specifications for sheet mica; e.g. flake mica, or the mica obtained as a by-product or waste from the preparation of sheet mica.
scratch *striation.*
scree A term commonly used in Great Britain as a loose equivalent of *talus* in each of its senses: broken rock fragments; a heap of such fragments; and the steep slope consisting of such fragments. Some authorities regard scree as the material that makes up the sloping land feature known as talus; others consider scree as a sheet of any loose, fragmental material lying on or mantling a slope (cf. *block field*), and talus as that material accumulating specif. at the base of, and obviously derived from, a cliff or other projecting mass (cf. *alluvial talus, avalanche talus, rockfall talus*).
scree creep The gradual and steady downhill movement of individual large blocks of rock on a slope that is often gentle; it is most noticeable where the rocks are massive or well-jointed. See also: *talus creep.*
screen [cart] To apply *screen tints* to copy for printing; a screen tint of the desired percent of full color is placed between the copy and the light-sensitive film or printing material, so that the copy will print in tints of the solid color. Cf: *screen tint.*
screen [eco geol] An apparatus used to separate material according to size of its particles or to allow the passage of smaller particles while preventing that of larger (as in grading coal, ore, rock, or aggregate); it usually consists of a perforated plate or sheet, or of meshed wire or woven cloth, with regularly spaced holes of uniform size, mounted in a suitable frame. Cf: *sieve.*
screen analysis Determination of the particle-size distribution of a soil, sediment, or ore by measuring the percentage of the particles that will pass through standard screens of various sizes.
screening The operation of passing loose materials (such as gravel or coal) through a screen so that constituent particles are separated into defined sizes.
screen tint A photomechanical impression on glass or plastic, consisting of evenly or randomly oriented squares or dots whose size increases in direct proportion to the intensity of tone to be printed. Screen tints may be negative (clear image and opaque background) or positive (opaque image and clear background). For printing, they are exposed on light-sensitized printing plates; screen tints may range from 4 to 95 percent of full color. See also: *screen [cart].*
screw axis A type of crystal symmetry element that is a combination of a rotation of $360°/n$ with a translation of a/mn where a is a lattice period (usually the a, b, or c crystal axis), n may be 1, 2, 3, 4, or 6, and m is an integer between 0 and n.
screw dislocation A type of *line defect* in a crystal: a row of atoms along which a crystallographic plane seems to spiral. See also: *spiral growth.*
screw ice (a) Small ice fragments in heaps or ridges, produced by the crushing together of ice cakes. (b) A small formation of *pressure ice.*
screwing A general term used in the older literature to describe the processes giving rise to *pressure ridges.*
scribing The process of preparing a map or other drawing for reproduction by cutting the detail to be shown into an opaque medium that coats a sheet of transparent plastic, using a scriber (an instrument holding one of a set of needles or blades of various diameters or cross-sectional shapes, sharpened to desired dimensions). The result of the process is a negative of the material to be reproduced.
scrobicule One of the smooth, shallow, depressed rings or trenches surrounding the bases of echinoid tubercles and serving for attachment of muscles of spines. See also: *areole.* Syn: *scrobicula; scrobiculus.*
scrobis septalis The inframarginal, asymmetrical, sometimes deep indentation or concave surface of the apertural face of a foraminiferal test (as in *Alabamina*) (TIP, 1964, pt.C, p.63). Etymol: Latin. See also: *infundibulum.* Syn: *inframarginal sulcus.*
scroll (a) One of a series of crescentic deposits built by a stream on the inner bank of a shifting channel; e.g. a *flood-plain scroll.* (b) *meander scroll.*
scroll meander A *forced-cut meander* in which the building of meander scrolls on the inner bank is the cause of erosion on the outer bank of the meander (Melton, 1936, p. 597). See also: *lacine meander.*
scrub Low-growing or stunted vegetation on poor soil or in semiarid regions, which sometimes forms inpenetrable masses.
sculpture [geomorph] (a) The carving-out of surficial features of the Earth's surface by erosive agents, such as rain, running water, waves, glaciers, and wind. The term has been loosely applied to include also the processes of deposition and earth movement. Syn: *earth sculpture; land sculpture; glyptogenesis.* (b) A landform resulting from a modification or sculpturing of an existing form.
sculpture [paleont] Strongly developed *ornamentation* of preserved hard parts of an animal; e.g. the relief pattern on the surface of a gastropod shell or on superficial dermal bones of a vertebrate.
sculpture [palyn] The external textural modifications (such as spines, warts, granules, pila, pits, streaks, and reticulations) of the

exine of pollen grains and spores. It is always a feature of ektexine. Cf: *structure [palyn].* Syn: *ornamentation [palyn].*

scum A film that floats on a liquid, such as a stagnant pool. The film may be composed of soap, of precipitated calcium carbonate, or of putrid matter.

scutum (a) One of a pair of opercular valves, adjacent to the rostrum in cirripede crustaceans, with adductor-muscle attachments. Cf: *tergum.* (b) A lateral *marginal spine,* generally broad and flat, overarching the frontal area in some anascan cheilostomes (bryozoans).—Pl: *scuta.*

scyelite A coarse-grained *hornblende peridotite* characterized by poikilitic texture resulting from the inclusion of olivine crystals in crystals of other minerals, esp. amphiboles. Mica and some magnetite are also present. Its name, given by Judd in 1885, is derived from Loch Scye, Scotland. Essentially the same rock as *cortlandtite* and *schriesheimite.* Not recommended usage.

scyphozoan Any marine coelenterate belonging to the class Scyphozoa, characterized by the predominance of medusoid forms. Range, Precambrian or Cambrian to present.

Scythian European stage: Lower Triassic (above Permian, below Anisian). Also spelled: *Skythian.* See also: *Werfenian.*

S-dolostone Stratigraphically controlled dolostone, occurring in extensive beds generally intertongued with limestone (Dunbar & Rodgers, 1957, p. 238). Cf: *T-dolostone; W-dolostone.*

se In structural petrology, a fabric defined by the preferred orientation of grains external to a porphyroblast. It may or may not be parallel to the preferred orientation of minerals within the porphyroblast. Cf: *si.*

sea [astrogeol] *mare.*

sea [oceanog] (a) An inland body of salt water. (b) A geographic division of an *ocean.* (c) An ocean area of wave generation.

sea [waves] A series of short-period asymmetric waves generated by wind, that lies within the area of generation. Such a wave is called a *sea wave* or a *wind wave;* when it leaves its area of generation, it becomes *swell [waves].*

sea arch An opening through a headland, formed by wave erosion or solution (as by the enlargement of a sea cave, or by the meeting of two sea caves from opposite sides) and leaving a bridge of rock over the water. Syn: *marine arch; marine bridge; sea bridge; natural arch; natural bridge.*

sea ball A spherical mass of somewhat fibrous living or dead vegetation (esp. algae), produced mechanically in shallow waters along a seashore by the compacting effect of wave movement. Cf: *lake ball.*

sea bank (a) *seashore.* (b) A *sandbank* adjacent to the sea. (c) *seawall.*

seabeach A beach lying along a sea or ocean.

seabeach placer *beach placer.*

seaboard (a) The strip of land bordering a seacoast. (b) *seacoast.*

sea bridge *sea arch.*

sea-captured stream A stream, flowing parallel to the seashore, that is cut in two as a result of marine erosion and that may enter the sea by way of a waterfall (Cleland, 1925).

sea cave A cleft or cavity in the base of a sea cliff, excavated where wave action has enlarged natural lines of weakness in easily weathered rock; it is usually at sea level and affected by the tides. Syn: *marine cave; sea chasm; cave.*

sea chasm A deep, narrow *sea cave.*

sea cliff A cliff or slope produced by wave erosion, situated at the seaward edge of the coast or the landward side of the wave-cut platform, and marking the inner limit of beach erosion. It may vary from an inconspicuous slope to a high, steep escarpment. Example: Gay Head on Marthas Vineyard, Mass. See also: *wave-cut cliff; shore cliff.* Also spelled: *seacliff.* Syn: *cliff; marine cliff.*

sea coal An old British syn. of *bituminous coal,* named after coal washed ashore and used for fuel; the name was extended to mined coal, as well.

seacoast The coast adjacent to a sea or ocean. Syn: *seaboard.*

sea cucumber A *holothuroid* echinoderm having a body shape resembling a cucumber, a flexible body wall, and the ability to creep along the sea floor.

sea-cut *marine-cut.*

sea fan *submarine fan.*

sea-floor spreading A hypothesis that the oceanic crust is increasing by convective upwelling of magma along the *mid-oceanic ridges* or *world rift system,* and by a moving-away of the new material at a rate of one to ten centimeters per year. This movement provides the source of dynamic thrust in the hypothesis of *plate tectonics.* See also: *expanding Earth.* Syn: *ocean-floor spreading; spreading concept; spreading-floor hypothesis.*

sea-floor trench *trench [marine geol].*

sea-foam *sepiolite.*

seafront The land, buildings, or section of a town along a seashore or bordering a sea.

sea gate (a) A restricted passage leading or giving access to the sea. (b) A gate that protects against the sea.

sea gully *slope gully.*

sea ice (a) Any form of ice originating from the freezing of seawater (thus excluding icebergs). Ant: *land ice.* See also: *field ice.* (b) A mariner's term for any ice that is floating in the sea or that has drifted to the sea.

sea-ice shelf Sea ice floating in the vicinity of its formation and separated from fast ice (of which it may have been a part) by a tide crack or a family of such cracks.

seaknoll *knoll [marine geol].*

sea level A popular syn. of *mean sea level.*

sea-level datum A determination of *mean sea level* that has been adopted as a standard datum for heights or elevations, based on tidal observations over many years at various tide stations along the coasts; e.g. the Sea-Level Datum of 1929 used by the National Geodetic Survey, the year referring to the last general adjustment of the level.

sealing-wax structure A term used by Fairbridge (1946, p.85 & 87) for a primary sedimentary flow structure produced by slumping, characterized by a lack of a sharply defined slip plane at the base or a contemporaneous erosion plane at the top, and occupying a zone of highly fluid contortion in otherwise normal sedimentary succession.

sea loch A *fjord* along the western Highland coast of Scotland.

seam [ore dep] A particular bed or vein in a series of beds; it is usually said of coal but may also pertain to metallic minerals.

seam [stratig] (a) A thin layer or stratum of rock separating two distinctive layers of different composition or greater magnitude. (b) Strictly, the line of separation between two different strata, resembling the seam between two parts of a garment.

seamanite A pale-yellow to wine-yellow orthorhombic mineral: $Mn_3(PO_4)(BO_3)\cdot 3H_2O$.

sea marsh A *salt marsh* periodically flooded by the sea. Syn: *sea meadow.*

sea mat A *bryozoan,* esp. an incrusting bryozoan.

sea meadow *sea marsh.*

sea meadows Those upper layers of the open ocean that have such an abundance of phytoplankton that they provide food for marine organisms. The term is usually used in the plural.

sea moat *moat [marine geol].*

seamount An elevation of the sea floor, 1000 m or higher, either flat-topped (called a *guyot*) or peaked (called a *seapeak*). Seamounts may be either discrete, arranged in a linear or random grouping, or connected at their bases and aligned along a ridge or rise.

sea mud Mud from the sea; specif. a rich, slimy deposit in a salt marsh or along a seashore, sometimes used as a manure. Syn: *sea ooze.*

sea ooze *sea mud.*

seapeak A type of *seamount* that has a pointed summit.

sea peat A rare type of peat, formed from seaweeds.

sea plain *plain of marine erosion.*

seapoose A term used along the shore of Long Island, N.Y., for a shallow inlet or tidal river. Etymol: Algonquian, akin to Delaware *sepus,* "small brook".

seaquake An earthquake that occurs beneath the ocean and that can be felt on board a ship in the vicinity of the epicenter. Syn: *submarine earthquake.*

sea reach The *reach* of the lower course of a stream where it approaches the sea.

sea rim The apparent horizon as actually observed at sea; the sea-level horizon.

searlesite A white mineral: $NaB(SiO_3)_2\cdot H_2O$.

seascarp A relatively long, high, rectilinear submarine cliff or wall.

seashore (a) The narrow strip of land adjacent to or bordering a sea or ocean. Syn: *seaside.* (b) A legal term for all the ground between the ordinary tide levels; the *foreshore.*—Syn: *seastrand.*

seashore lake A lake along the seashore, containing either fresh water or salt water, and isolated from the sea by a barrier of sediment built by waves or by a river on a delta.

seaside *seashore.*
sea slick A smooth area on the surface of an ocean or body of fresh water, caused by organic material, e.g. *water bloom.*
sea slope A slope of the land toward the sea.
sea snow The drifting descent of organic detritus in the ocean. Syn: *plankton snow; marine snow.*
seasonally frozen ground Ground that is frozen by low seasonal temperatures and remains frozen only during the winter; it corresponds to the *active layer* in permafrost regions. Syn: *frost zone.*
seasonal recovery Recharge to ground water during and after a wet season, with a rise in the level of the water table.
seasonal stream An *intermittent stream* that flows only during a certain climatic season; e.g. a winterbourne.
sea stack *stack [coast].*
sea state A description of the roughness of the ocean surface, either numerical or in words. Syn: *state of the sea.*
seastrand *seashore.*
seat clay *underclay.*
seat earth A British term for a bed of rock underlying a coal seam, representing an old soil that supported the vegetation from which the coal was formed; specif. *underclay.* A highly siliceous seat earth is known as *ganister.* Also spelled: *seatearth.* Syn: *seat rock; seat stone; spavin; coal seat.*
sea terrace *marine terrace.*
seat rock *seat earth.*
seat stone *seat earth.*
sea urchin An *echinoid* having a globular shape and a theca of calcareous plates, commonly with sharp movable spines.
sea wall (a) A long, steep-faced embankment of shingle or boulders, built by powerful storm waves along a seacoast at the high-water mark. (b) A man-made wall or embankment of stone, reinforced concrete, or other material along a shore to prevent wave erosion.—Also spelled: *seawall.*
seawater The water of the oceans, characterized by its salinity and distinguished from the fresh water of lakes, streams, and rain. *Salt water* is sometimes used synonymously. Also spelled: *sea water; sea-water.*
sea-water encroachment *salt-water encroachment.*
sea-water intrusion *salt-water encroachment.*
sea wave One of a series of waves collectively known as a *sea.* See also: *wind wave.*
seaworn Diminished or wasted away by the sea, as a *seaworn* shore.
sebastianite A *gabbro* composed of euhedral anorthite, biotite, augite, and apatite, but without feldspathoids or quartz. This rock has been found as fragments in extrusive rocks of Monte Somma, Italy. The name was given by Lacroix in 1917, for San Sebastian, Monte Somma. Cf: *puglianite.* Not recommended usage.
sebkha *sabkha.*
Secchi disc An instrument used to measure seawater *transparency* or clarity; a white disc of fixed diameter is lowered into the water, and an average is taken of the depth at which it disappears and at which it reappears when raised.
sechron The maximum interval of geologic time occupied by a given depositional sequence, defined at the points where the boundaries of the sequence change laterally from unconformities to conformities along which there is no significant hiatus (Mitchum, 1977, p. 210).
second (a) A unit of time equal to 1/60 of a minute or 1/3600 of an hour; specif. the cgs unit of time, originally equal to 1/86,400 part of the mean solar day but now defined as the duration of 9,-192,631,770 cycles of frequency associated with the transition between two hyperfine levels of the fundamental state of the atom of cesium-133. Abbrev: sec; s (in physical tables). (b) A unit of angular measure equal to 1/60 of a minute of arc or 1/3600 of a degree. Symbol: ″. (c) An informal oceanographic term used to describe distance or depth, equal to about 1463 m (4800 ft) or the distance that sound will travel through seawater during one second of time.
Secondary A term applied in the early 19th century as a syn. of *Floetz*. It was later applied to the extensive series of stratified rocks separating the older *Primary* and the younger *Tertiary* rocks, and ranging from the Silurian to the Cretaceous; still later, it was restricted to the whole of the Mesozoic Era. The term was abandoned in the late 19th century in favor of *Mesozoic.*
secondary [coast] Said of a mature coast or shoreline whose features are produced chiefly by present-day marine processes (Shepard, 1937, p. 605); e.g. coasts shaped by wave erosion, marine deposition, or marine organisms. Cf: *primary.*
secondary [eco geol] *supergene.*
secondary [metal] Said of metal obtained from scrap rather than from ore. Ant: *primary [metal].*
secondary allochthony In coal formation, accumulation of plant remains in a region characterized by erosion, transport, and resedimentation of coal masses previously deposited elsewhere than in place. Cf: *primary allochthony.*
secondary arc A mountain arc that lacks deep-seated connections. It is generally located behind the junction point of two primary arcs and is convex in the opposite direction. Obsolete.
secondary ash *extraneous ash.*
secondary axial septulum In a fusulinid foraminiferal test, an *axial septulum* located between primary axial septula. Cf: *secondary septulum.*
secondary clay A clay that has been transported from its place of formation and redeposited elsewhere. Cf: *residual clay; primary clay.*
secondary cleavage An old term for *cleavage,* used by Leith (1905, p. 11) to emphasize its development after consolidation of the rock, by deformation or metamorphism. See also: *metaclase.*
secondary consequent stream (a) A tributary of a subsequent stream, flowing parallel to or down the same slope as the original consequent stream; it is usually developed after the formation of a subsequent stream, but in a direction consistent with that of the original consequent stream. (b) A stream flowing down the flank of an anticline or syncline. Syn: *lateral consequent stream.*—Syn: *subconsequent stream.*
secondary consolidation Consolidation of sedimentary material, at essentially constant pressure, resulting from internal processes such as recrystallization.
secondary corner One of the minor convexities seen in the profile of a sedimentary particle. Fifteen to 30 may occur on angular particles, but secondary corners quickly disappear during abrasion and are absent at a roundness value of 0.60 (Pettijohn, 1957, p. 58-59).
secondary crater An *impact crater* produced by the relatively low-velocity impact of fragments ejected from a large *primary crater;* e.g. any of several lunar craters ("splash structures") formed by fragments thrown up from the Moon's surface as a result of violent primary impacts. Syn: *satellitic crater.*
secondary creep Deformation of a material under a constant differential stress, with the strain—time relationship as a constant. Cf: *primary creep.* Syn: *steady-state creep.*
secondary enlargement Deposition, around a clastic mineral grain, of material of the same composition as that grain and in optical and crystallographic continuity with it, often resulting in crystal faces characteristic of the original mineral (Pettijohn, 1957, p. 119); e.g. the addition of a quartz overgrowth around a silica grain in sandstone, or the growth of new material around detrital nuclei such as calcite, feldspar, and tourmaline. Cf: *rim cementation.* Syn: *secondary growth.*
secondary enrichment *supergene enrichment.*
secondary fumarole *rootless fumarole.*
secondary geosyncline (a) A geosyncline appearing at the culmination of or after geosynclinal orogeny, such as an *exogeosyncline, epieugeosyncline,* or *intradeep* (Peyve & Sinitzyn, 1950). (b) Haug's term for a *sequent geosyncline* (1900, p. 617-711).
secondary glacier A small valley glacier that joins a larger trunk glacier as a *tributary glacier.*
secondary growth *secondary enlargement.*
secondary inclusion A fluid inclusion formed by any process after crystallization of the host mineral is essentially complete. Most are formed during recrystallizational healing of fractures within a crystal (Roedder, 1967, p. 522).
secondary interstice An *interstice* that formed after the formation of the enclosing rock. Cf: *original interstice.*
secondary layer The inner shell layer of a brachiopod deposited by the outer epithelium median of the outer mantle lobes. It may be fibrous, laminar, or prismatic. Cf: *primary layer.*
secondary limestone Limestone deposited from solution in cracks and cavities of other rocks; esp. the limestone accompanying the salt and gypsum of the Gulf Coast salt domes.
secondary maximum A term used by Udden (1914) for a particle size of a sediment or rock, having greater frequency than other size ranges surrounding it but not greater than the modal diameter. It may be obtained graphically by locating the second highest peak of the frequency curve. A given sample may have more than

one secondary maximum. Syn: *secondary mode.*

secondary mineral A mineral formed later than the rock enclosing it, usually at the expense of an earlier-formed *primary mineral,* as a result of weathering, metamorphism, or solution.

secondary mode *secondary maximum.*

secondary optic axis One of two *optic axes* in a crystal along which all light rays travel with equal velocity. Secondary optic axes are close to but do not necessarily coincide with *primary optic axes.*

secondary oral A small to large oral plate in edrioasteroids, without intrathecal extensions and commonly with a fixed position relative to the primary orals; frequently only a surficial element (Bell, 1976).

secondary orifice The opening at the outer end of the peristome in some ascophoran cheilostomes (bryozoans). Syn: *peristomice.*

secondary porosity The *porosity* developed in a rock after its deposition or emplacement, through such processes as solution or fracturing. Cf: *primary porosity.*

secondary production The organic matter that *zooplankton* herbivores produce within a given marine area or volume in a given time. Cf: *net primary production.*

secondary recovery Production of oil or gas as a result of artificially augmenting the *reservoir energy,* as by injection of water or other fluid. Secondary-recovery techniques are generally applied after substantial depletion of the reservoir. See also: *water flooding.*

secondary reflection *multiple reflection.*

secondary rocks (a) Rocks composed of particles derived from the erosion or weathering of pre-existing rocks, such as residual, chemical, or organic rocks formed of detrital, precipitated, or organically accumulated materials; specif. clastic sedimentary rocks. Cf: *primary rocks.* (b) A term applied by Lehmann (1756) to fossiliferous and stratified rocks, containing material eroded from the older *primitive rocks.*

secondary septulum A minor partition of a chamberlet in a foraminiferal test, reaching downward (adaxially) from the spirotheca (as in Neoschwagerininae); e.g. *secondary axial septulum* and *secondary transverse septulum.* Cf: *primary septulum.*

secondary spine An intermediate-sized echinoid spine, appearing later than the *primary spine.*

secondary stratification Stratification developed when sediments already deposited are thrown into suspension and redeposited. Syn: *indirect stratification.*

secondary structure [geol] A structure that originated subsequently to the deposition or emplacement of the rock in which it is found, such as a fault, fold, or joint produced by tectonic movement; esp. an epigenetic *sedimentary structure*, such as a concretion or nodule produced by chemical action, or a sedimentary dike formed by infilling. Cf: *primary structure [geol].*

secondary structure [paleont] Coarse structure, commonly between distinct laminae, in the wall of a tintinnid lorica. Cf: *primary structure [paleont]; tertiary structure.*

secondary succession An association of plants that develops after the destruction of all or part of the original plant community.

secondary sulfide zone *sulfide zone.*

secondary tectogenesis Gravity-driven sliding off a *geotumor* and the resulting deformation (Haarman, 1930). Obsolete.

secondary tectonite A tectonite whose fabric is *deformation fabric.* Most tectonites are of this type. Cf: *primary tectonite.*

secondary tissue Plant tissue developed from a lateral or secondary meristem, e.g. cambium or cork cambium (Fuller and Tippo, 1954, p. 970).

secondary transverse septulum In a foraminiferal test, a *transverse septulum* whose plane is approximately normal to the axis of coiling. Cf: *secondary septulum.*

secondary twinning *deformation twinning.*

secondary vein In mining law, a vein discovered subsequent to the one on which the mining claim was based; an incidental vein. Cf: *discovery vein.*

secondary wall A cell-wall layer deposited internally on the primary wall layer in a plant cell. It is generally of different composition from the primary wall (Cronquist, 1961, p. 881).

secondary wave *S wave.*

second boiling point The development of a gas phase from a liquid phase upon cooling. During the cooling, crystallization of large quantities of compounds low in or lacking volatile materials, e.g. feldspar, results in a sufficient increase in the concentration of volatile materials, e.g. water, in the residual liquid that finally the vapor pressure of this liquid becomes greater than the confining pressure and a gas phase develops, i.e. the liquid boils.

second bottom The first terrace above the normal flood plain (or *first bottom*) of a river.

second-class ore An ore that needs preliminary treatment before it is of a sufficiently high grade to be acceptable for shipment or market. Cf: *first-class ore.* Syn: *milling ore.*

second-cycle conglomerate A conglomerate made of clasts that themselves show evidence of having been derived from a previous conglomerate.

second-derivative map A contour map of the second vertical derivative of a potential field such as the Earth's gravity or magnetic field. The values are normally derived by mathematical processing of the observed component.

second law of thermodynamics The statement concerning entropy as a function of the state of the system, which says that for all reversible processes, the change in entropy is equal to the heat which the system exchanges with the outside world divided by the absolute temperature. In irreversible processes, the change in entropy is greater than the quotient of heat and temperature.

second-order leveling Spirit leveling that has less stringent requirements than those of *first-order leveling,* in which lines between bench marks established by first-order leveling are run in only one direction using first-order instruments and methods (or other lines are divided into sections, over which forward and backward runnings are to be made) and in which the maximum allowable discrepancy is 8.4 mm times the square root of the length of the line (or section) in kilometers (or 0.035 ft times the square root of the distance in miles). Cf: *third-order leveling.*

second-order pinacoid In a monoclinic or triclinic crystal, any $\{h0l\}$ or $\{\bar{h}0l\}$ pinacoid. Cf: *first-order pinacoid; third-order pinacoid; fourth-order pinacoid.*

second-order prism A crystal form: in a tetragonal crystal, the $\{100\}$ prism; in a hexagonal crystal, the $\{11\bar{2}0\}$ prism; and in an orthorhombic crystal, any $\{h0l\}$ prism. Cf: *first-order prism; third-order prism; fourth-order prism.* See also: *macrodome.*

second water The quality or luster of a gemstone next below *first water,* such as that of a diamond that is almost perfect but contains slight flaws or turbid patches. Cf: *third water.*

second-year ice Sea ice that has survived only one summer's melt; it is thicker (2 meters or more) and less dense than first-year ice and therefore stands higher out of the water: any hummocks present show weathering. Syn: *two-year ice; young polar ice.*

secretion (a) The act or process by which animals and plants transform mineral material from solution into skeletal forms. (b) A secondary structure formed of material deposited from solution within a cavity in a rock, esp. a deposit formed on or parallel to the walls of the cavity; e.g. a mineral vein, an amygdule, or a geode. The space may be completely or only partly filled. Cf: *concretion.*

sectile Said of a mineral that can be cut with a knife; e.g. argentite.

section [geol] (a) An exposed surface or cut, either natural (such as a sea cliff or stream bank) or artificial (such as a quarry face or road cut), through a part of the Earth's crust; it may be vertical or inclined. (b) A description, or graphic representation drawn to scale, of the successive rock units or the geologic structure revealed by such an exposed surface, or as they would appear if cut through by any intersecting plane, such as a diagram of the geologic features or mine workings penetrated in a shaft or drilled well; esp. a *vertical section.* See also: *structure section.*—Syn: *geologic section.*

section [petrology] *thin section.*

section [stratig] (a) *columnar section.* (b) *geologic section.* (c) *type section.*

section [surv] (a) One of the 36 units of subdivision of a *township* of the U.S. Public Land Survey system, representing a piece of land normally one square mile in area (containing 640 acres as nearly as possible), with boundaries conforming to meridians and parallels within established limits. Sections within a normal township are numbered consecutively beginning with number one in the northeast section and progressing west and east alternately with progressive numbers in the tiers to number 36 in the southeast section. See also: *fractional section; half section; quarter section; quarter-quarter section.* (b) The part of a continuous series of measured differences of elevation that is recorded and abstracted as a unit. It always begins and ends on a bench mark.

section-gage log *caliper log.*

section line The boundary line of a section in surveying.

sector graben A *volcanic graben* on the slope of a volcanic cone.

secular Said of a process or event lasting or persisting for an indefinitely long period of time, e.g. secular variation; progressive or cumulative rather than *cyclic.*

secular equilibrium A relationship between a radioactive *parent* and one or more radioactive *daughters* in which the activity of each daughter (or each set of daughters if decay is branching) is equal to the activity of the parent.

secular movements Systematic, persistent movements of the Earth's crust, either upward or downward, that take place slowly and imperceptibly over long periods of geologic time.

secular variation Changes in the core-derived Earth's magnetic field, measured in centuries. See also: *westward drift.* Syn: *geomagnetic secular variation.*

secule A syn. of *moment.* Term suggested by Jukes-Browne (1903, p.37) for the duration of a biostratigraphic zone. Etymol: Latin *seculum,* "age".

secundine dike A dike which has been intruded into hot country rock. Pegmatites and aplites commonly occur in this mode. See also: *welded dike.*

sedarenite A term used by Folk (1968, p.124) for a *litharenite* composed chiefly of sedimentary rock fragments; e.g. *sandstone-arenite* and *shale-arenite.* It may have any clay content, sorting, or rounding.

sedentary Said of a sediment or soil that is formed in place, without transportation, by the disintegration of the underlying rock or by the accumulation of organic material.

sedentary soil *residual soil.*

sederholmite A mineral: beta-NiSe.

sedge peat *carex peat.*

sedifluction The subaqueous or subaerial movement of material in unconsolidated sediments, occurring in the primary stages of diagenesis (Richter, 1952).

sediment (a) Solid fragmental material that originates from weathering of rocks and is transported or deposited by air, water, or ice, or that accumulates by other natural agents, such as chemical precipitation from solution or secretion by organisms, and that forms in layers on the Earth's surface at ordinary temperatures in a loose, unconsolidated form; e.g. sand, gravel, silt, mud, till, loess, alluvium. (b) Strictly, solid material that has settled down from a state of suspension in a liquid.—In the singular, the term is usually applied to material held in suspension in water or recently deposited from suspension. In the plural, the term is applied to all kinds of deposits, and refers to essentially unconsolidated materials. Cf: *deposit.*

sedimental Formed of or from sediment.

sedimentary adj. (a) Pertaining to or containing sediment; e.g. a "sedimentary deposit" or a "sedimentary complex". (b) Formed by the deposition of sediment (e.g. a "sedimentary clay"), or pertaining to the process of sedimentation (e.g. "sedimentary volcanism").—n. A sedimentary rock or deposit.

sedimentary ash *extraneous ash.*

sedimentary breccia A *breccia* formed by sedimentary processes; e.g. a talus breccia. It is usually (but not necessarily) characterized by imperfect mechanical sorting and either by the predominance of material from one local source or by the presence of a marked variety of materials indiscriminately jumbled together (Wentworth, 1935, p.227). Syn: *sharpstone conglomerate.*

sedimentary-contact shoreline A shoreline formed by the partial submergence of the slope left by the removal of weak beds from one side of a straight sedimentary contact (Johnson, 1925, p. 19).

sedimentary cover *cover.*

sedimentary cycle *cycle of sedimentation.*

sedimentary differentiation The progressive separation (by erosion and transportation) of a well-defined rock mass into physically and chemically unlike products that are resorted and deposited as sediments over more or less separate areas; e.g. the segregation and dispersal of the components of an igneous rock into sandstones, shales, limestones, etc.

sedimentary dike A tabular mass of sedimentary material that cuts across the structure or bedding of preexisting rock in the manner of an igneous dike. It is formed by the filling of a crack or fissure from below, above, or laterally, by forcible injection or intrusion of sediments under abnormal pressure (as by gas pressure or by the weight of overlying rocks, or by earthquakes), or from above by simple infilling; esp. a *clastic dike.* See also: *sediment vein.*

sedimentary facies A term used by Moore (1949, p.32) for a stratigraphic *facies* representing any areally restricted part of a designated stratigraphic unit (or of any genetically related body of sedimentary deposits) which exhibits lithologic and paleontologic characters significantly different from those of another part or parts of the same unit. It comprises "one of any two or more different sorts of deposits which are partly or wholly equivalent in age and which occur side by side or in somewhat close neighborhood" (Moore, 1949, p.7).

sedimentary fault *growth fault.*

sedimentary injection *injection [sed].*

sedimentary insertion A term proposed by Challinor (1978, p. 273) for the emplacement of sedimentary material among deposits or rocks already formed, such as by infilling, injection or intrusion, or localized solution subsidence.

sedimentary intrusion *intrusion [sed].*

sedimentary laccolith A term introduced by Raaf (1945) for an intrusion of plastic sedimentary material (such as clayey salt breccia) forced up under high pressure and penetrating parallel or nearly parallel to the bedding planes of the invaded formation, and characterized by a very irregular thickness.

sedimentary lag Delay between the formation of potential sediment by weathering and its removal and deposition.

sedimentary mantle *cover.*

sedimentary marble *crystalline limestone.*

sedimentary ore A sedimentary rock of ore grade; an ore deposit formed by sedimentary processes, e.g. saline residues, phosphatic deposits, or iron ore of the *Clinton ore* type.

sedimentary peat Peat formed under water, usually lacustrine, and consisting mainly of algae and related forms. Syn: *lake peat; pulpy peat; dredge peat.*

sedimentary petrography The description and classification of sedimentary rocks. Syn: *sedimentography.*

sedimentary petrologic province An area underlain by sediments with a common provenance (Pettijohn, 1957, p. 573). Cf: *dispersal shadow.* Syn: *petrologic province.*

sedimentary petrology The study of the composition, characteristics, and origin of sediments and sedimentary rocks. Often miscalled "sedimentation".

sedimentary quartzite *orthoquartzite.*

sedimentary ripple *ripple [sed struc].*

sedimentary rock (a) A rock resulting from the consolidation of loose *sediment* that has accumulated in layers; e.g. a *clastic rock* (such as conglomerate or tillite) consisting of mechanically formed fragments of older rock transported from its source and deposited in water or from air or ice; or a *chemical rock* (such as rock salt or gypsum) formed by precipitation from solution; or an *organic rock* (such as certain limestones) consisting of the remains or secretions of plants and animals. The term is restricted by some authors to include only those rocks consisting of mechanically derived sediment; others extend it to embrace all rocks other than purely igneous and completely metamorphic rocks, thereby including pyroclastic rocks composed of fragments blown from volcanoes and deposited on land or in water. Syn: *stratified rock; derivative rock.* (b) Less restrictedly, a general term for any sedimentary material, unconsolidated or consolidated. This usage should be avoided, "notwithstanding the fact that geologists have agreed to call loose unconsolidated material 'rock' " (Challinor, 1978, p.273).

sedimentary structure A *structure* in a sedimentary rock, formed either contemporaneously with deposition (a *primary sedimentary structure*) or by sedimentary processes subsequent to deposition (a *secondary structure* [geol]).

sedimentary tectonics Folding and deformation in geosynclinal basins caused by geosynclinal subsidence and buckling of strata in the geosyncline. An example of a large anticline developed at depth in a geosyncline is the Cedar Creek anticline in the Williston Basin of North America (Gussow, 1970). Cf: *gravity orogenesis.*

sedimentary trap An area in which sedimentary material accumulates instead of being carried farther, as in an area between a high-energy and low-energy environment.

sedimentary tuff (a) A tuff containing a subordinate amount of nonvolcanic detrital material. (b) A deposit of reworked tuff and other detrital material.

sedimentary volcanism The expulsion or extrusion through overlying formations of a mixture of sediment (e.g. sand or clay), water, and gas, with gas under pressure furnishing the driving force. Also, the production of phenomena such as sand volcanoes and mud volcanoes. Sedimentary volcanism may result from diapiric intrusion, volcanism in a fumarolic stage, escaping hydrocarbons, oozing-out of material (during thaw) in a permafrost region, or orogenic pressure release as during an earthquake.

sedimentation (a) The act or process of forming or accumulating sediment in layers, including such processes as the separation of rock particles from the material from which the sediment is derived, the transportation of these particles to the site of deposition, the actual deposition or settling of the particles, the chemical and other (diagenetic) changes occurring in the sediment, and the ultimate consolidation of the sediment into solid rock. (b) Less broadly, the process of *deposition* of sediment. (c) Strictly, the act or process of depositing sediment by mechanical means from a state of suspension in air or water. (d) The accumulation of deposits of colluvium and alluvium derived from accelerated erosion of the soil. (e) *silting up.* (f) A term often used erroneously for "sedimentary petrology" and "sedimentology".

sedimentation analysis Determination of the particle-size distribution of a sediment by measuring settling velocities of different size fractions.

sedimentation balance An apparatus used to measure the settling rate of small particles dispersed in a liquid.

sedimentation curve An experimentally derived curve showing cumulatively the quantity of sediment deposited or removed from an originally uniform suspension in successive units of time (Krumbein & Pettijohn, 1938, p. 112-115).

sedimentation diameter A measure of particle size, equal to the diameter of a hypothetical sphere of the same specific gravity and the same settling velocity as those of a given sedimentary particle in the same fluid; twice the *sedimentation radius.* Cf: *equivalent radius; nominal diameter.*

sedimentation radius One half of the *sedimentation diameter.*

sedimentation rate *rate of sedimentation.*

sedimentation trend The direction in which sediments were laid down.

sedimentation unit A layer or deposit resulting from one distinct act of sedimentation, defined by Otto (1938, p.574) as "that thickness of sediment which was deposited under essentially constant physical conditions"; the deposit made during a time period when the prevailing current has a mean velocity and deposits some mean size, such as a cross-bedded layer of sand formed under conditions of essentially constant flow and sediment discharge. It is distinguished from like units by changes in particle size and/or fabric indicating changes in velocity and/or direction of flow.

sediment binder A sessile-benthonic organism that encrusts several adjacent uncemented particles, producing a single larger mass less likely to be moved by water currents.

sediment charge The ratio of the weight or volume of sediment in a stream to the weight or volume of water passing a given cross section per unit of time (ASCE, 1962).

sediment concentration The ratio of the dry weight of the sediment in a water-sediment mixture (obtained from a stream or other body of water) to the total weight of the mixture. It is usually expressed in percent for high concentration values, or in parts per million for low concentration values.

sediment-delivery ratio The ratio of sediment yield of a drainage basin to the total amount of sediment moved by sheet erosion and channel erosion; expressed in percent (Chow, 1964, p. 17-12).

sediment discharge The amount of sediment moved by a stream in a given time, measured by dry weight or by volume; the rate at which sediment passes a section of a stream. Syn: *sediment-transport rate.*

sediment-discharge ratio The ratio between the total discharge of a stream and the discharge of its load.

sediment feeder *deposit feeder.*

sediment load The solid material (*load*) that is transported by a natural agent, esp. by a stream. The total sediment load of a stream is equal to *bed-material load* plus *wash load,* and is expressed as the dry weight of all sediment that passes a given point in a given period of time.

sedimento-eustatism Worldwide changes in sea level produced by a change in the capacity of the ocean basins because of sediment accumulation. Cf: *diastrophic eustatism; glacio-eustatism.* See also: *eustasy.*

sedimentogenesis The formation of sediments.

sedimentography *sedimentary petrography.*

sedimentology The scientific study of sedimentary rocks and of the processes by which they were formed; the description, classification, origin, and interpretation of sediments. Sometimes miscalled "sedimentation".

sediment-production rate Sediment yield per unit of drainage area, derived by dividing the annual sediment yield by the area of the drainage basin; expressed as tons or acre-feet of sediment per square mile per year (Chow, 1964, p. 17-11).

sediment stabilizer A sessile-benthonic organism that grows out on unconsolidated sediment and thereby protects it from being removed or eroded.

sediment station A vertical cross-sectional plane of a stream, usually normal to the mean direction of flow, where samples of suspended load are collected on a systematic basis for determining concentration, particle-size distribution, and other characteristics.

sediment transport The movement and carrying-away of sediment by natural agents; esp. the conveyance of a stream load by suspension, saltation, solution, or traction.

sediment-transport rate *sediment discharge.*

sediment trapper A sessile-benthonic organism that projects up into sediment-laden water, slows its flow, and thereby causes some of the suspended sediment to settle out around the organism and accumulate on the sea floor.

sediment vein A *sedimentary dike* formed by the filling of a fissure from above with sedimentary material.

sediment yield The amount of material eroded from the land surface by runoff and delivered to a stream system.

sedovite A brown to red-brown mineral: $U(MoO_4)_2$.

seed [bot] The ripened ovule of a plant containing the *embryo.*

seed [cryst] *seed crystal.*

seed crystal A small, suitably oriented piece of crystal used in *crystal seeding.* Syn: *seed [cryst].*

seed fern *pteridosperm.*

seed leaf *cotyledon.*

seed plant *spermatophyte.*

Seelandian Northern European stage (Denmark): Upper Paleocene (essentially equivalent to Thanetian).

seeligerite A mineral: $Pb_3(IO_3)Cl_3O$.

seep n. An area, generally small, where water or oil percolates slowly to the land surface. For water, it may be considered as a syn. of *seepage spring,* but it is used by some for flows too small to be considered as springs. Cf: *oil seep.* Syn: *seepage [water].*—v. To move slowly through small openings of a porous material.

seepage [petroleum] *oil seep.*

seepage [water] (a) The act or process involving the slow movement of water or other fluid through a porous material such as soil. Cf: *influent seepage; effluent seepage.* (b) The amount of fluid that has been involved in seepage. (c) *seep.*

seepage face A belt along a slope, such as the bank of a stream, along which water emerges at atmospheric pressure and flows down the slope. See also: *seepage line.*

seepage lake (a) A *closed lake* that loses water mainly by seepage through the walls and floor of its basin. Cf: *drainage lake.* (b) A lake that receives its water mainly from seepage, as from irrigation waters in parts of western U.S.

seepage line The uppermost level at which flowing water emerges along a *seepage face;* an outcrop of the water table. Syn: *line of seepage; phreatic line.*

seepage loss Loss of water by influent seepage from a stream, canal, or other body of surface water.

seepage spring This term may be used as a syn. of *filtration spring,* but is often limited to springs of small discharge (Meinzer, 1923, p. 50). See also: *seep.* Syn: *weeping spring.*

seepage stress The force that is transferred from water flowing through a porous granular medium to the medium itself by means of viscous friction.

seepage velocity The rate at which seepage water is discharged through a porous medium per unit area of pore space perpendicular to the direction of flow.

segelerite An orthorhombic mineral: $CaMgFe^{+3}(PO_4)_2(OH) \cdot 4H_2O$.

Seger cone A small cone, made in the laboratory of a mixture of clay and salt, that softens at a definite, known temperature. It is used in the manufacture of refractories. It has also been used in volcanology to determine the approximate temperature of a molten lava. Syn: *pyrometric cone.*

seggar *sagger.*

segment (a) One of the constituent parts into which an invertebrate body is divided; esp. any of the succeeding or repeated body parts of an arthropod, many of which are likely to be similar in form and function (such as one of the components of the thorax of a trilobite, connected by articulation with adjoining segments), or the somite of a crustacean. See also: *podomere; article; joint [paleont].* (b) Any of the parts into which a heterococcolith naturally separates or divides. (c) In chambered cephalopod conchs, any part of the siphuncle between two successive septa.

segmentation The process by which a coastal lagoon is subdivided into smaller patches of water by the accumulation of *transverse bars* (Price, 1947).

segregated ice A syn. of *Taber ice.* Also known as *segregation ice.*

segregated vein A fissure whose mineral filling is derived from the country rock by the action of percolating water. Syn: *exudation vein.* Cf: *infiltration vein.* See also: *lateral secretion.*

segregation [petrology] (a) *magmatic segregation.* (b) A concentration of crystals of a particular mineral or minerals that accumulated during an early stage of consolidation as a result of magmatic segregation. (c) A "liquid segregation", e.g. from an early-formed immiscible liquid or a late-formed residual melt.

segregation [sed] A secondary feature formed as a result of chemical rearrangement of minor constituents within a sediment after its deposition; e.g. a nodule of iron sulfide, a concretion of calcium carbonate, or a geode.

segregation banding A compositional banding in gneisses that is not primary in origin, but rather is the result of segregation of material from an originally more nearly homogeneous rock (Billings, 1954). Cf: *cleavage banding.*

seiche (a) A free or standing-wave oscillation of the surface of water in an enclosed or semi-enclosed basin (as a lake, bay, or harbor) that varies in period, depending on the physical dimensions of the basin, from a few minutes to several hours, and in height from several centimeters to a few meters; that is initiated chiefly by local changes in atmospheric pressure, aided by winds, tidal currents, and small earthquakes; and that continues, pendulum fashion, for a time after cessation of the originating force. It usually occurs in the direction of longest diameter of the basin, but occasionally it is transverse. The term has also been applied to an oscillation superimposed on the tidal waves of the open ocean. (b) A term used in the Great Lakes area for any sudden rise (whether oscillatory or not) in the water of a harbor or lake.—Etymol: French, supposedly from Latin *siccus,* "dry"; a term used locally to describe the occasional rise and fall of water at the narrow end of Lake Geneva, Switzerland, where the phenomenon was first observed. Pron: *saysh.* Cf: *internal seiche.*

seidozerite A brown-red mineral: $(Na,Ca)_2(Zr,Ti,Mn)_2Si_2O_7(O,F)_2$.

seif A very large, sharp-crested, tapering *longitudinal dune* or chain of sand dunes, commonly found in the Sahara Desert; its crest in profile consists of a succession of peaks and cols, and it bears on one side a succession of curved slip faces produced by strong but infrequent cross winds that tend to increase its height and width. A seif dune may be as much as 200 m high, and from 400 m to more than 100 km long (300 km in Egypt). Etymol: Arabic *saif,* "sword"; the term originated in North Africa but is applied elsewhere to similar dunes of appreciably smaller size. Pron: *safe.* Syn: *seif; sif; saif; sword dune.*

seif dune *seif.*

seinäjokite An orthorhombic mineral: $(Fe,Ni)(Sb,As)_2$.

seis A colloquial syn. of *seismic detector.*

seism *earthquake.*

seismic Pertaining to an earthquake or Earth vibration, including those that are artificially induced.

seismic activity *seismicity.*

seismic area (a) An *earthquake zone.* (b) The region affected by a particular earthquake.

seismic belt An elongate *earthquake zone,* esp. a zone of subduction or sea-floor spreading.

seismic constant In building codes dealing with earthquake hazards, an amount of horizontal acceleration that a building must be designed to withstand.

seismic creep Relatively slow movement on a fault, as contrasted to the sudden movement associated with an earthquake.

seismic detector An instrument, e.g. a *seismometer* or *geophone,* that converts seismic impulses into electrical voltage or otherwise makes them evident. Colloquial syn: *pot [seis]; seis.*

seismic discontinuity *discontinuity [seis].*

seismic efficiency The percentage of earthquake-generated energy that goes into the production of elastic-wave energy.

seismic-electric effect The variation of resistivity because of elastic deformation of rocks.

seismic event *event [seis].*

seismic exploration *applied seismology.*

seismic facies analysis The description and geologic interpretation of seismic reflection patterns, based on reflection configuration, continuity, amplitude, frequency, and interval velocity (Mitchum, 1977, p. 210). See also: *seismic facies unit.*

seismic facies unit A mappable three-dimensional seismic unit composed of groups of reflections whose parameters, such as reflection configuration, continuity, amplitude, frequency, or interval velocity, differ from those of adjacent facies units. Once the interval reflection parameters, the external form, and the three-dimensional associations of the seismic facies are delineated, the unit can then be interpreted in terms of environmental setting, depositional processes, and estimates of lithology (Mitchum, 1977, p. 210). See also: *seismic facies analysis.*

seismic gap A segment of an active fault zone that has not experienced a *principal earthquake* during a time interval when most other segments of the zone have. Seismologists commonly consider seismic gaps to have a high future-earthquake potential.

seismic intensity The average rate of flow of seismic wave energy through a unit cross section perpendicular to the direction of propagation.

seismicity The phenomenon of Earth movements. Cf: *specific seismicity.* Syn: *seismic activity.*

seismic log The variation of acoustic impedance with arrival time or depth, determined from measurements of the amplitudes of successive reflection events on a *seismic record.* Generation of a seismic log from seismic data generally implies that amplitude variations result from reflectivity only, and that reflections occur at normal incidence. Seismic logs also often assume either that density does not vary or that it varies in a systematic way with velocity, so that the seismic log may be labeled as indicating velocity, i.e., as a pseudosonic log.

seismic map A contour map constructed from seismic data. Values may be in either time or depth; data may be plotted with respect to the observing station (producing an "unmigrated map") or with respect to the subsurface reflecting or refracting point locations (a "migrated map").

seismic moment A measure of the strength of an earthquake, particularly of the low-frequency wave motion. The seismic moment is equal to the product of the force and the moment arm of the double-couple system of forces that produces ground displacements equivalent to that produced by the actual earthquake dislocation. The seismic moment also is equal to the product of the rigidity modulus of the Earth material, the fault area, and the average dislocation along the fault surface.

seismic noise *microseism.*

seismic profile *profile [seis].*

seismic prospecting *applied seismology.*

seismic record (a) In seismic prospecting, a photographic or magnetic record of the energy received by a spread of geophone groups with time following the shot or energy release. (b) In earthquake seismology, a record of all seismic activity during a period of time, including background noise, body waves, and surface waves, from both natural and artificial events. Syn: *record [seis].*

seismic reflection *reflection.*

seismic reflection configuration The geometric patterns and relations of seismic reflections that are interpreted to represent configuration of strata generating the reflections (Mitchum, 1977, p. 209). Syn: *reflection configuration.*

seismic refraction *refraction.*

seismic sea wave *tsunami.*

seismic sequence analysis The seismic identification and interpretation of depositional sequences by subdividing the seismic section into packages of concordant reflections separated by surfaces of discontinuity, and interpreting them as depositional sequences (Mitchum, 1977, p. 210).

seismic shooting A method of geophysical prospecting in which elastic waves are produced in the Earth by the firing of explosives or by other means. See also: *reflection shooting; refraction shooting.*

seismic source (a) A device for generating seismic waves. (b) The point at which seismic waves are generated. (c) The point of origin of an earthquake.

seismic spectrum A curve showing the amplitude of the ground motion as a function of frequency or period, obtained by a Fourier analysis of the ground motion.

seismic spread *spread [seis].*

seismic stratigraphy The study of stratigraphy and depositional facies as interpreted from seismic data (Mitchum, 1977, p. 210).

seismic surge *tsunami.*

seismic surveying The gathering of seismic data from an area. Cf: *reflection shooting; refraction shooting.*

seismic velocity The rate of propagation of an elastic wave, usually measured in km/sec. The wave velocity depends on the type of wave, as well as the elastic properties and density of the Earth material through which it travels. Cf: *interval velocity; normal-moveout velocity; average velocity [seis].*

seismic wave (a) A general term for all elastic waves produced by earthquakes or generated artificially by explosions. It includes both *body waves* and *surface waves [seis].* Obsolete syn: *earth wave.* (b) A seismic sea wave, or tsunami.—Syn: *earthquake wave; elastic wave.*

seismite *Fault-graded beds* that are interpreted as an earthquake record, or "paleoseismogram" (Seilacher, 1969).

seismogram The record made by a *seismograph.* Syn: *earthquake record.*

seismograph An instrument that detects, magnifies, and records vibrations of the Earth, especially earthquakes. The resulting record is a *seismogram.* Cf: *seismometer; seismic detector; geophone.*

seismograph array A network of seismographs arranged in a spatial pattern so as to enable the signal from earthquake ground motion to be enhanced with respect to the noise of microseismic ground motion.

seismography An obsolete syn. of *seismology.*

seismologist One who applies the methods or principles of seismology, as in earthquake prediction or seismic exploration.

seismology The study of earthquakes, and of the structure of the Earth, by both natural and artificially generated seismic waves. Obsolete syn: *seismography.*

seismometer An instrument that detects Earth motions. Syn: *seismic detector; geophone; hydrophone.*

seismometer plant *plant [seis].*

seismometer spread *spread [seis].*

seismometry The instrumental aspects of seismology.

seismoscope An instrument that merely indicates the occurrence of an earthquake. It is considered by some, however, to be the equivalent of a *seismometer.*

sejunction water Capillary water bounded by menisci and in static equilibrium in the soil above the capillary fringe. This water may occur dispersed or as a coherent body (Schieferdecker, 1959, term 0243). Cf: *funicular water.*

sekaninaite A violet analogue of cordierite in which magnesium is largely replaced by ferrous iron: $(Fe,Mg)_2Al_4Si_5O_{18}$. Syn: *iron cordierite.*

Selachii An order, largely marine, of elasmobranch fishes characterized by a fusiform body and more or less predaceous habit; sharks. Range, Upper Devonian to Recent.

selagite A mica *trachyte* characterized by abundant tabular biotite crystals in a holocrystalline groundmass of alkali feldspar and clinopyroxene, and possibly olivine and secondary quartz. The name was derived by Haüy in 1822 from a Greek word meaning "to beam brightly", referring to the glistening mica. Obsolete.

selbergite A hypabyssal *phonolite* containing phenocrysts of leucite, nosean, sanidine, acmite-augite, and biotite in a fine-grained groundmass of nepheline, alkali feldspar, and acmite. The phenocrysts comprise a higher percentage of the volume of the rock than the groundmass. Its name, given by Brauns in 1922, is derived from Selberg, Laacher See district, Germany. The rock differs from *schorenbergite* in containing alkali feldspar. Not recommended usage.

selection *natural selection.*

selective fusion The fusion of only a portion of a mixture such as a rock. The liquid portion will generally contain a larger proportion of the more fusible components than the parent material did. Cf: *anatexis.*

selective weathering *differential weathering.*

selenate A mineral compound characterized by discrete $(SeO_4)^{-2}$ groups. An example is olsacherite, $Pb_2(SeO_4)(SO_4)$.

selenide A mineral compound that is a combination of selenium with a more positive element or radical. An example is eucairite, CuAgSe.

seleniferous plant A plant that absorbs and retains large quantities of selenium from the soil. Syn: *selenophile.*

selenite The clear, colorless variety of gypsum, occurring (esp. in clays) in distinct, transparent monoclinic crystals or in large crystalline masses that easily cleave into broad folia. Syn: *spectacle stone.*

selenite butte A small tabular mound, rising 1-3 m above a playa, composed of lake sediments capped with a veneer of selenite formed by deflation of the playa or by the effects of rising ground water (Stone, 1967, p. 246).

selenite plate *gypsum plate.*

selenium A trigonal mineral: Se.

selenizone A sharply defined spiral band of closely spaced crescentic growth lines or linear ridges generated by a narrow notch or slit in the outer lip of the aperture of a gastropod, generally at the periphery of a shell whorl. It typically marks the positions of the notch or slit during earlier stages of growth.

selenjoseite *laitakarite.*

selenochronology Chronology of the Moon.

selenodesy "Geodesy" of the Moon.

selenofault A term used by Fielder (1965, p. 172) for a large-scale fault on the Moon's surface.

selenographic chart A map representing the surface of the Moon, on which positions are measured in latitude from the Moon's equator and in longitude from a reference meridian.

selenography (a) The science of the physical features of the Moon; the observation and recording of lunar features. (b) The topography or physical geography of the Moon.

selenolite [mineral] A white mineral: SeO_2.

selenolite [sed] A sedimentary rock composed of gypsum or anhydrite.

selenology A branch of astronomy that deals with the Moon; the science of the Moon, including *lunar geology.*

selenomorphology "Geomorphology" of the Moon; the study of lunar landforms and their origin, evolution, and distribution.

selenophile *seleniferous plant.*

selenotectonics Tectonics of the Moon; the study of lunar structures and their movements as a result of the development of the Moon as a whole.

selen-tellurium A blackish-gray mineral: (Se,Te). The Te:Se ratio is nearly 3:2. Syn: *hondurasite.*

self-grown stream An *autogenetic* stream developed independently on an undisturbed land surface, diverging upstream in the manner of the branches of a tree (Willis, 1907, p. 8).

self-potential curve *spontaneous-potential curve.*

self-potential method An electrical exploration method in which one determines the spontaneous electrical potentials (*spontaneous polarization*) that are caused by electrochemical reactions associated with clay or metallic mineral deposits. Syn: *spontaneous-potential method.*

self-reading leveling rod A *level rod* with graduations designed to be read by the observer at the leveling instrument. Syn: *speaking rod.*

self-reversal Acquisition by a rock of a natural remanent magnetization opposite to the ambient magnetic field direction at the time the rock was formed.

self-rising ground Puffy irregular surface or near-surface zone of certain playas, formed by the effects of capillary rise of ground water, and consisting of a thin clay crust underlain by loose, friable, granular sediment (silt, clay, and salt) (Stone, 1967, p. 246).

seligmannite A lead-gray orthorhombic mineral: $PbCuAsS_3$.

sellaite A colorless tetragonal mineral: MgF_2.

selvage [fault] The altered, clayey material found along a fault zone; *fault gouge.* Also spelled: *selvedge.*

selvage [ign] A marginal zone of a rock mass, having some distinctive feature of fabric or composition; specif. the chilled border of an igneous mass (as of a dike or lava flow), usually characterized by a finer grain or sometimes a glassy texture, such as the glassy inner margins on the pillows in pillow lava. Syn: *selvedge; salband.*

selvage [ore dep] *gouge [ore dep].*

selvage [paleont] The principal ridge of the contact margin in an ostracode, serving to hermetically seal the carapace when it is closed.

selvedge *selvage.*

SEM *scanning electron microscope.*

semblance In seismic surveying, a measure of multichannel *coherence.* It is the energy of a summed trace divided by the energy of the components of the sum.

semenovite A tetragonal mineral: $(Ca,Ce,La)_{12}(Be,Si)_8Si_{12}O_{40}(O,OH,F)_8 \cdot H_2O$.

semianthracite Coal having a fixed-carbon content of 86% to

92%. It is between bituminous coal and anthracite in metamorphic rank, although its physical properties more closely resemble those of anthracite.

semiarid Said of a type of climate in which there is slightly more precipitation (25-50 cm) than in an arid climate, and in which sparse grasses are the characteristic vegetation. In Thornthwaite's classification, the moisture index is between -20 and -40. Syn: *subarid.*

semibituminous coal Coal that ranks between bituminous coal and semianthracite; it is harder and more brittle than bituminous coal. It has a high fuel ratio and burns without smoke. Syn: *smokeless coal.* Cf: *metabituminous coal.*

semibolson A wide desert basin or valley that is drained by an intermittent stream flowing through canyons at each end and reaching a surface outlet (such as another stream, a lower basin, or the sea); its central playa is absent or poorly developed. It may represent a *bolson* where the alluvial fill reached a level sufficient to permit occasional overflow across the lowest divide.

semibright coal A type of banded coal defined microscopically as consisting of between 80% and 61% of bright ingredients such as vitrain, clarain, and fusain, with clarodurain and durain composing the remainder. Cf: *semidull coal; bright coal; dull coal; intermediate coal.*

semicannel coal *lean cannel coal.*

semiconductor radiation detector A solid-state detector of ionizing radiation, fabricated from material such as germanium or silicon. Lithium is commonly diffused ("drifted") into the semiconductor to compensate for impurities. Energy of the detected radiation is determined by collecting the electrical charge produced along the ionizing path. Semiconductor detectors usually yield one to two orders of magnitude greater energy resolution than scintillation-type detectors. See also: *silicon detector; lithium-drifted germanium detector.*

semicratonic *quasicratonic.*

semicrystalline *hyalocrystalline.*

semidesert An area intermediate in character between a desert and a grassland and often located between them.

semidiurnal Said of a tide or tidal current having a period that is approximately equal to half a lunar day, or 12.42 solar hours.

semidull coal A type of banded coal defined microscopically as consisting mainly of clarodurain and durain, with from 40% to 21% bright ingredients such as vitrain, clarain, and fusain. Cf: *semibright coal; bright coal; dull coal; intermediate coal.*

semifusain A coal lithotype transitional between fusain and vitrain, but predominantly fusain. Cf: *fusovitrain.*

semifusinite A maceral of coal within the *inertinite* group whose optical properties are intermediate between those of vitrinite and those of fusinite. It has some woody structure preserved.

semifusinoid Fusinite, the optical properties of which are transitional between those of fusinoid and those of associated xylinoids, vitrinoids, and anthrinoids (ASTM, 1970, p.364).

semifusite A microlithotype of coal, consisting of 95% or more of *semifusinite.*

semilogarithmic Said of graph or plotting paper or of a chart made on such paper having a logarithmic scale on one axis and an arithmetic scale or uniform spacing on the other axis usually at right angles to the first. Syn: *semilog.*

semiopal A loosely used term for common opal, hydrophane, and any partly dehydrated or impure opal, as distinguished from precious opal and fire opal. Syn: *hemiopal.*

semipegmatitic A term suggested by Lacroix in 1900 for a type of pegmatitic texture in which the grains of one mineral, enclosed in larger crystals of another mineral, have variable extinction angles while the enclosing mineral has uniform extinction (Johannsen, 1939, p.232); a nonrecommended syn. of *poikilitic.*

semiperched ground water Unconfined ground water separated by a low-permeability, but saturated, bed from a body of confined water whose hydrostatic level is below the water table (Meinzer, 1923, p. 41). Because of the gradation between unconfined and confined ground water, the term has little significance and is not in common use.

semiprecious stone Any gemstone other than a *precious stone,* or any gemstone of lower commercial value than a precious stone; specif. a mineral that is less than 8 on the Mohs scale of hardness. A gemstone may also be regarded as semiprecious because of its comparative abundance, inferior brilliance, or unfamiliarity to the public, or owing to the whims of fashion. This arbitrary classification is misleading, as it does not recognize, for example, that a ruby of poor quality may be far less costly than a fine specimen of jadeite.

semisplint coal A type of banded coal that is intermediate in composition and character between *bright-banded coal* and *splint coal;* it corresponds to *duroclarain.* It is defined quantitatively as having 20-30% opaque attritus and more than 5% *anthraxylon.* Cf: *clarodurain.*

semitropical *subtropical.*

sempatic In the *CIPW classification* of igneous rocks, those rocks in which the ratio of groundmass to phenocrysts is less than five to three but greater than three to five. Rarely used. Cf: *dopatic; dosemic.*

semseyite A gray to black mineral: $Pb_9Sb_8S_{21}$.

senaite A black mineral: $Pb(Ti,Fe,Mn)_{21}O_{38}$. It occurs in rounded crystals and grains in diamond-bearing sands.

senarmontite A colorless, white, or gray isometric mineral: Sb_2O_3. It is polymorphous with valentinite.

Senecan North American provincial series: lower Upper Devonian (above Erian, below Chautauquan).

senegalite An orthorhombic mineral: $Al_2(PO_4)(OH)_3 \cdot H_2O$.

senescence [geomorph] The part of the developmental sequence of a landform, or the cycle of erosion of a landscape, when the landform or region enters upon the stage of *old age.* Cf: *senility.*

senescence [paleont] Old age; esp. the later stages in the life cycle of a species or other group. See also: *racial senescence.*

senescent (a) Pertaining to the stage in the developmental sequence of a landform, or in the cycle of erosion, when the processes of erosion become slow and ineffective in producing further topographic modification; esp. said of a landscape that is growing old or aging (such as one characterized by a pediplain). (b) Said of a lake that is nearing extinction, as from filling by the remains of aquatic vegetation.

senescent lake A lake that is approaching extinction as a result of sediment filling, erosion of the outlet, or other cause. See also: *extinct lake.*

senesland A term proposed by Davis (1932, p. 429) for a land surface that has "lost the full measure of relief that characterizes maturity"; a land surface intermediate between a *matureland* and a peneplain (Maxson, 1950, p. 101). Cf: *oldland.*

sengierite A yellow-green mineral: $Cu(UO_2)_2(VO_4)_2 \cdot 8\text{-}10H_2O$.

senile Pertaining to the stage of *senility* of the cycle of erosion; esp. said of a landscape or topography that is approaching a base-level plain or the end of the erosion cycle, or of a stream approaching an ultimate stage that is seldom fully reached, characterized by a sluggish current and a tendency to meander through a peneplain of slight relief only a little above base level.

senility The stage of the cycle of erosion in which erosion has reached a minimum and base level has been approached. Cf: *old age; senescence.*

senior In zoological nomenclature, said of the earlier published of two synonyms or homonyms.

Senonian European stage: Upper Cretaceous (above Turonian, below Danian). It includes: Coniacian, Santonian, Campanian, and Maestrichtian.

sensible horizon (a) The plane tangent to the Earth's surface at the observer's position; the *apparent horizon.* (b) *astronomic horizon.*

sensitivity [phys] (a) The least change in an observed quantity that can be perceived on the indicator of a given instrument. (b) The displacement of the indicator of a recording unit of an instrument per unit of change of a measurable quantity.

sensitivity [soil] The effect of *remolding* on the shear strength and consolidation characteristics of a clay or cohesive soil. A "sensitive" clay is one whose shear strength is decreased to a fraction of its former value on remolding at constant moisture content. See also: *sensitivity ratio.*

sensitivity ratio A measure of the degree of *sensitivity* of a clay or cohesive soil. It is the ratio between the unconfined compressive strength of an undisturbed specimen and that of the same specimen at the same moisture content but in a remolded state (Terzaghi & Peck, 1967, p. 31). The sensitivity ratio for most clays ranges between 2 and 4, although extrasensitive clays may have values greater than 8. Symbol: S_t. Syn: *remolding sensitivity.*

sensu lato Literally, "in a broad sense", referring to the wide application of the name of a *taxon.* Etymol: Latin. Abbrev: s.l. Cf: *sensu stricto.*

sensu stricto Literally, "in a narrow sense", referring to the narrow or restricted application of the name of a *taxon.* Etymol:

Latin. Abbrev: s.s. Cf: *sensu lato.*

sepal One of the separate parts of the calyx, the outermost part of the perianth of a flower; usually green and more or less leaflike in texture.

separate *soil separate.*

separation The distance between any two parts of an index plane (e.g. bed or vein) disrupted by a fault. See: *horizontal separation; vertical separation; stratigraphic separation.*

separation disc A breakage area in a filament of blue-green algae, formed by the death of a cell.

separation layer *abscission layer.*

sepiolite A chain-lattice clay mineral: $Mg_4(Si_2O_5)_3(OH)_2 \cdot 6H_2O$. It is a white to light-gray or light-yellow material, extremely lightweight, absorbent, and compact, that is found chiefly in Asia Minor and is used for making tobacco pipes, cigar and cigarette holders, and ornamental carvings. Sepiolite occurs in veins with calcite, and in alluvial deposits formed from weathering of serpentine masses. Syn: *meerschaum; sea-foam.*

septa Plural of *septum.*

septal angle The angle between tangents drawn from the apex of a planispiral nautiloid shell to two successive septa and measured on a section made along the plane of symmetry (TIP, 1964, pt.K, p.58).

septal cycle All septa belonging to a single stage in ontogeny of a scleractinian corallite, as determined by the order of appearance of septal groups (six protosepta comprising the first cycle, and later-formed exosepta and entosepta in constantly arranged succession being introduced in sextants) (TIP, 1956, pt.F, p.250).

septal flap The extension of each lamella in the tests of foraminifers of the superfamily Rotaliacea, formed on the inner side of the chamber over the distal face of the previous chamber, and resulting in secondarily doubled septa (TIP, 1964, pt.C, p.63).

septal fluting One of the folds, wrinkles, or corrugations of septa (and antetheca) in a fusulinid test transverse to the axis of coiling, generally strongest in the lower (adaxial) part of septa and toward the poles and decreasing in intensity toward the top.

septal foramen (a) An intercameral opening in the test of a foraminifer. It may be homologous with the aperture or be secondarily formed. Cf: *foramen.* (b) An opening in the septum at the siphuncle of a cephalopod, allowing passage of the siphuncular cord.

septal furrow (a) The narrow middorsal region of a nautiloid in which the mural part of the septum (attached to the wall of the conch) is lacking (TIP, 1964, pt.K, p.58). Syn: *dorsal furrow.* (b) An *external furrow* of a fusulinid.

septal groove A longitudinal groove on the outer surface of the wall of a corallite, corresponding in position to a septum on the inner surface of the wall. Cf: *interseptal ridge.*

septalial plate One of the *crural plates* forming the floor of the septalium of a brachiopod and united with the earlier-formed part of the median septum.

septalium A troughlike structure of the brachial valve of a brachiopod, situated between hinge plates (or homologues), consisting of septalial plates (or homologues) enveloping and buttressed by the median septum. It does not carry adductor muscles.

septal neck A funnel-like or tubelike forward or backward flexure or extension of a septum around the septal foramen of a cephalopod shell. Syn: *neck [paleont].*

septal pinnacle One of the upper tips of septa located in the zone between parent corallite and offset during the early stages of increase in Paleozoic corals. Cf: *pseudoseptal pinnacle* (Fedorowski & Jull, 1976, p. 41).

septal pore A small perforation in a septum (and antetheca) of the test of a fusulinid. Cf: *mural pore.*

septal tooth A small projection along the upper margin of a septum of a scleractinian coral, formed by extension of a trabecula beyond calcareous tissue connecting it with others.

septaria Plural of *septarium.*

septarian Said of the irregular polygonal pattern of internal cracks developed in septaria, closely resembling the desiccation structure of mud cracks; also said of the epigenetic mineral deposits that may occur as fillings of these cracks.

septarian boulder *septarium.*

septarian nodule *septarium.*

septarium (a) A large roughly spheroidal concretion, 8 to 90 cm in diameter, usually of an impure argillaceous carbonate such as clay ironstone. It is characterized internally by irregular polyhedral blocks formed by a series of radiating cracks that widen toward the center and that intersect a series of cracks concentric with the margins, the cracks invariably filled or partly filled by crystalline minerals (most commonly calcite) that cement the blocks together. Its origin involves the formation of an aluminous gel, case hardening of the exterior, shrinkage cracking due to dehydration of the colloidal mass in the interior, and vein filling. The veins sometimes weather in relief, thus producing a septate pattern. Cf: *melikaria.* Syn: *septarian nodule; septarian boulder; beetle stone; turtle stone.* (b) A crystal-lined crack or fissure in a septarium.—Pl: *septaria.*

septechlorite A name given to a group of minerals (amesite, cronstedtite, berthierine) having chloritelike formulas; they have basal spacings of 7 angstroms and therefore belong to the kaolinite-serpentine group. Syn: *pseudochlorite.*

septifer Said of brachiopod crura having the form of septa that descend directly from the dorsal side of the hinge plates to the floor of the brachial valve.

septotheca A wall of a scleractinian or rugose corallite, formed by thickened outer parts of septa along the axis of trabecular divergence. Cf: *synapticulotheca; paratheca.*

septula A communication organ in cheilostome and ctenostome bryozoans, consisting of a *pore plate* and a *rosette* of cells. Pl: *septulae.* Syn: *septulum; septule.*

septulum (a) *septula.* (b) A small septum; e.g. a ridge or small partition extending adaxially downward from the lower surface of the spirotheca in the test of a fusulinid foraminifer so as to partially subdivide the chambers of the test.—Pl: *septula.*

septum [bot] (a) In filamentous algae or fungi, a crosswall, generally perpendicular to the length of the trichome or hypha. (b) In a dinoflagellate cyst, a more or less membranous, linear projection that is perpendicular to the wall.

septum [paleont] (a) One of the transverse internal calcareous partitions dividing the shell of a cephalopod, such as a partition that divides the phragmocone of a nautiloid into camerae and that is attached to the inside of the wall of the conch. (b) One of several radially disposed longitudinal calcareous plates or partitions of a corallite, occurring between or within mesenterial pairs. It presumably supported the basal disk and lower wall of the polyp. Also, a thin radial noncalcareous partition, composed of soft tissue, dividing the gastrovascular cavity of an octocorallian polyp. (c) A partition or wall dividing a foraminiferal test interiorly into chambers, commonly formed by previous outer wall or apertural face. (d) A relatively long, narrow, commonly bladelike elevation of the secondary shell of a brachiopod; a median ridge in either valve of a brachiopod. It is indicated in articulates (within the underlying floor of a valve) by persistent high narrow deflections of fibrous calcite originating near the primary layer, and in inarticulates by comparable deflections of shell lamellae. (e) The wall separating the two rows of thecae in a biserial graptoloid. (f) A radial, longitudinal, normally perforate plate connecting the inner and outer walls of an archaeocyathid. Formerly known as *pariety.* (g) A platelike structure in an echinoid spine, radiating from the axial zone toward the anterior of the spine and seen in a cross section of the spine.—Pl: *septa.* Adj: *septal.*

Sequanian Substage in Great Britain: Upper Jurassic (upper Lusitanian; above Rauracian Substage, below Kimmeridgian Stage).

Sequanian river A river that rises in comparatively low ground and that decreases in volume in the summer owing to evaporation and vegetative absorption. Type river: the Seine in France.

sequence (a) A succession of geologic events, processes, or rocks, arranged in chronologic order to show their relative position and age with respect to geologic history as a whole. (b) A major informal lithostratigraphic unit of greater than group or supergroup rank, traceable over large areas of a continent, and bounded by unconformities of interregional scope, such as in the cratonic interior of North America (Sloss, 1963); a geographically discrete succession of major rock units that were deposited under related environmental conditions (Silberling & Roberts, 1962). See also: *sub-sequence.* Syn: *stratigraphic sequence.* (c) A term, now obsolete, used by Moore (1933, p.54) for the rocks formed during an era; an *erathem.* (d) A faunal succession.

sequence of crystallization *order of crystallization.*

sequent geosyncline The constituent geosynclines of a *polygeosyncline,* separated from one another by the development of geanticlines (Schuchert, 1923). Syn: *secondary geosyncline (b).*

sequential landform One of an orderly succession of smaller landforms that are developed by the erosion, weathering, and mass-wasting of larger *initial landforms;* it includes "erosional landforms" resulting from the progressive removal of earth materials, and "depositional landforms" resulting from the accumulation of the products of erosion. Cf: *ultimate landform; destructional.* Syn: *sequential form.*

serac A jagged pinnacle, sharp ridge, needlelike tower, or irregularly shaped block of ice on the surface of a glacier (commonly among intersecting crevasses, as on an icefall), formed where the glacier is periodically broken as it passes over a steep slope. Etymol: French *sérac,* a solid curdy white cheese made in the Alps and sold in quadrangular blocks, which glacial seracs resemble in appearance and shape. Both the French form and the anglicized version are in regular use. Pron: seh-*rahk.* Cf: *nieve penitente.*

seral The adj. of *sere.*

seral succession *sere.*

serandite A rose-red monoclinic mineral: $Na(Mn,Ca)_2Si_3O_8(OH)$. Cf: *pectolite.*

sere A sequence of ecologic communities that succeed one another in development from the *pioneer* stage to climax. Adj: *seral.* See also: *paleosere; succession [ecol].* Syn: *seral succession.*

serendibite An indigo-blue to grayish blue-green mineral: $Ca_2(Mg,Al)_6(Si,Al,B)_6O_{20}$.

serial homology The similarity that exists between members of a series of structures in an organism; e.g. the resemblance to each other of the vertebrae in the vertebral column.

serial sampling A method of gathering *samples* by a set pattern, such as a grid, to insure randomness.

seriate Said of the texture of an igneous rock, typically porphyritic, in which the sizes of the grains vary gradually or in a continuous series. Cf: *hiatal.*

sericite A white, fine-grained potassium mica occurring in small scales and flakes as an alteration product of various aluminosilicate minerals, having a silky luster, and found in various metamorphic rocks (esp. in schists and phyllites) or in the wall rocks, fault gouge, and vein fillings of many ore deposits. It is usually muscovite or very close to muscovite in composition, and may also include much illite.

sericitic sandstone A sandstone in which sericite (derived by decomposition of feldspar) intermingles with finely divided quartz and fills the voids between quartz grains.

sericitization A hydrothermal, deuteric, or metamorphic processes involving the introduction of, alteration to, or replacement by sericitic muscovite.

series [cart] *map series.*

series [geol] Any number of rocks, minerals, or fossils having characteristics, such as growth patterns, succession, composition, or occurrence, that make it possible to arrange them in a natural sequence.

series [ign] (a) *igneous-rock series.* (b) A term that is often misused for a sequence of rocks resulting from a succession of eruptions or intrusions, and that is usually preceded by an adjective such as "eruptive", "intrusive", or "volcanic" to indicate the origin of the rock. The term "group" should replace "series" in this usage (ACSN, 1961, art. 9f).

series [radioactivity] *radioactive series.*

series [soil] *soil series.*

series [stratig] (a) A chronostratigraphic unit generally classed next in rank below *system* and above *stage*, properly based on a clearly designated stratigraphic interval in a type area (although many series have come to be adopted quite generally without explicit indication of their limits) (ACSN, 1961, art.30); the rocks formed during an *epoch* of geologic time. Some series are recognized as worldwide units, others as *provincial series*. Most systems have been divided into series and most series into stages. The term is not restricted to stratified rocks but may be applied to intrusive rocks in the same time-stratigraphic sense. Formal series names are binomial, usually consisting of a geographic name (generally but not necessarily with the adjectival ending "-an" or "-ian") and the word "Series", the initial letter of both terms being capitalized. (b) A term that is often misused in a lithostratigraphic sense for a succession of continuously deposited sedimentary rocks; e.g. an assemblage of formations (a group) or an assemblage of formations and groups (a supergroup), esp. in the Precambrian. The term "should no longer be so used" (ACSN, 1961, art.9f). See also: *subsequence.* (c) A term used in England for a lithostratigraphic unit, generally for a large unit or esp. one throughout which the strata are conformable.

series circuit Elements in an electrical circuit so connected that there is a single continuous path through each element.

serir A desert plain strewn with rounded pebbles, older than the gravel-covered *reg;* a *stony desert* from which the sand has been blown away, as in the Sahara of Libya and Egypt. Etymol: Arabic, "dry". Pl: *serir.* See also: *pebble armor; hammada.* Syn: *shore [desert].*

Serozem *Sierozem.*

serpenticone An evolute, many-whorled cephalopod shell with the whorls hardly overlapping, resembling a coiled snake or rope.

serpentine (a) A group of common rock-forming minerals having the formula: $(Mg,Fe)_3Si_2O_5(OH)_4$. Serpentines have a greasy or silky luster, a slightly soapy feel, and a tough, conchoidal fracture; they are usually compact but may be granular or fibrous, and are commonly green, greenish yellow, or greenish gray and often veined or spotted with green, and white. Serpentines are always secondary minerals, derived by alteration of magnesium-rich silicate minerals (esp. olivines), and are found in both igneous and metamorphic rocks; they generally crystallize in the monoclinic system, but only as pseudomorphs. Translucent varieties are used for ornamental and decorative purposes, often as a substitute for jade. (b) A mineral of the serpentine group, such as chrysotile, antigorite, lizardite, parachrysotile, and orthochrysotile. (c) A term strictly applied in place of *chrysotile* as a species name, chrysotile becoming a variety and antigorite a separate species not included in the term "serpentine" (Hey, 1962, p. 595).—Etymol: Latin *serpentinus,* "resembling a serpent", from the mottled shades of green.

serpentine asbestos *chrysotile.*

serpentine jade A variety of the mineral serpentine resembling jade in appearance and used as an ornamental stone; specif. *bowenite [mineral].*

serpentine marble *verd antique.*

serpentine rock *serpentinite.*

serpentine spit A spit that is extended in more than one direction due to variable or periodically shifting currents. Syn: *serpent spit.*

serpentine-talc A material with the composition: $Mg_6Si_6O_{15}(OH)_6$. It is intermediate in composition and physical characteristics between serpentine and talc.

serpentinite A rock consisting almost wholly of serpentine-group minerals, e.g. antigorite and chrysotile or lizardite, derived from the alteration of ferromagnesian silicate minerals such as olivine and pyroxene. Accessory chlorite, talc, and magnetite may be present. Syn: *serpentine rock.*

serpentinization The process of hydrothermal alteration by which magnesium-rich silicate minerals (e.g. olivine, pyroxenes, and/or amphiboles in dunites, peridotites, and/or other ultrabasic rocks) are converted into or replaced by serpentine minerals.

serpent kame A term introduced by Shaler (1889, p. 549) for a sinuous *esker* and known in New England also as an *Indian ridge.*

serpent stone A stone, usually a highly absorbent aluminous gem, once believed to be formed by snakes and to be efficacious in drawing out poison; specif. *adder stone.*

serpierite A bluish-green mineral: $Ca(Cu,Zn)_4(SO_4)_2(OH)_6 \cdot 3H_2O$.

serpophite A metacolloidal variety of the mineral serpentine.

serpulid Any annelid that belongs to the family Serpulidae and that characteristically builds a contorted calcareous or leathery tube on a submerged surface. See also: *serpulid reef.*

serpulid reef A small reef patch, with a raised rim and cup-shaped central parts, constructed largely of calcareous tubes secreted by *serpulid* worms of the family Serpulidae (Polychaeta; Annelida). Cf: *sabellariid reef; vermetid reef; worm reef.*

serra (a) *sierra.* (b) A term used in Brazil for an elevated mountain zone supporting luxuriant vegetation.—Etymol: Latin, "saw".

serrate [bot] With teeth pointing toward the apex, as in some leaf margins.

serrate [geomorph] adj. Said of topographic features that are notched or toothed on the edge, or have a saw-edged profile; e.g. a *serrate* divide. Syn: *saw-toothed; serrated.*—n. A rocky mountain summit having a serrate profile (Stone, 1967, p. 246).

serrate [petrology] Said of saw-toothed contacts between minerals, usually resulting from replacement; e.g. the serrate texture of megacrysts in contact with plagioclase in igneous rocks.

Serravallian European stage: Middle Miocene (above Langhian, below Helvetian-Tortonian).

serule *microsere.*

sesquan A *cutan* consisting of sesquioxides or hydroxides (of aluminum or iron) (Brewer, 1964, p.214).

sessile Said of a plant or animal that is permanently attached to a substrate and is not free to move about. Cf: *vagile.*

sessile cruralium Cruralium united with the floor of the brachial valve of a brachiopod without intervention of the supporting median septum.

sessile spondylium Spondylium united with the floor of the pedicle valve of a brachiopod without intervention of the supporting median septum.

seston A general term for all suspended matter in water, both organic and inorganic.

set [currents] The compass direction toward which a current is flowing; a direction of flow. Syn: *current direction.*

set [exp struc geol] *permanent set.*

set [stratig] A term introduced by McKee & Weir (1953, p.382-383) for "a group of essentially conformable strata or cross-strata, separated from other sedimentary units by surfaces of erosion, nondeposition, or abrupt change in character"; it is composed of two or more consecutive beds of the same lithology, and is the smallest and most basic group unit. See also: *coset.*

seta [bot] (a) The sporophyte stalk in the mosses and liverworts. (b) A stiff, short hair on a plant.—Syn: *awn.*

seta [paleont] A slender, typically rigid or bristly and springy organ or part of an invertebrate, e.g. a movable whiplike part of a vibraculum in cheilostome bryozoans; a chitinous bristle arising from the invagination of the mantle groove of a brachiopod; or a hairlike process of the external membrane of a crustacean. Pl: *setae.*

setaceous Bristlelike, as of a plant part covered with stiff, short hairs.

setal duct A larger duct, up to 75 μm in diameter, through a trilobite exoskeleton. Cf: *pore canal.* Syn: *gland duct.*

seter A Norwegian term for a wave-cut rock terrace.

settled snow A loose term indicating snow that has become more or less compacted under gravity to a stable density. Cf: *old snow.*

settlement (a) The subsidence of a structure, caused by compression or movement of the soil below the foundation. See also: *differential settlement.* (b) The lowering of the overlying strata in a mine, owing to extraction of the mined material.

settling [mass move] The sag in outcrops of layered strata, caused by rock creep (Sharpe, 1938, p. 33). Syn: *outcrop curvature.*

settling [sed] (a) The deposition of sediment. (b) A sediment or precipitate.

settling [snow] Time-dependent compaction of snow under its own weight.

settling basin (a) An artificial basin or trap designed to collect the suspended sediment of a stream before it flows into a reservoir and thereby prevent the rapid siltation of the reservoir; e.g. a *desilting basin.* It is usually provided with means to draw off the clear water. (b) A sedimentation structure designed to remove pollutant materials from mill effluents; a tailings pond.

settlingite A hard, brittle, pale-yellow to deep-red hydrocarbon (H:C about 1.53) found in resinous drops on the walls of a lead mine at Settling Stones in Northumberland, England. Syn: *Settling Stones resin.*

settling reservoir A reservoir consisting of a series of shallow basins arranged in steps and connected by long conduits, allowing the removal of only the clear upper layer of water in each basin.

settling velocity The rate at which suspended solids subside and are deposited. Syn: *fall velocity.*

setup (a) The assembly and arrangement of the equipment and apparatus required for the performance of a surveying operation; specif. a surveying instrument (transit or level) placed in position and leveled, ready for taking measurements. (b) The actual physical placing of a leveling instrument over an instrument station. (c) *instrument station.* (d) The horizontal distance from the fiducial mark on the front end of a surveyor's tape (or the part of a tape which is in use at the time), measured in a forward direction to the point on the ground mark or monument to which the particular measure is being made.—Also spelled: *set-up.*

Sevier orogeny A name proposed by R.L. Armstrong (1968) for the well known deformations that occurred along the eastern edge of the Great Basin in Utah (eastern edge of Cordilleran miogeosyncline) during times intermediate between the Nevadan orogeny farther west and the Laramide orogeny farther east, culminating early in the Late Cretaceous. During the orogeny, the folding and eastward thrusting of the miogeosynclinal rocks over their foreland was largely completed.

sexine The outer division of the exine of pollen, more or less equivalent to *ektexine.* Cf: *nexine.*

sexiradiate A sponge spicule in the form of six equidistant coplanar rays arising from a common center.

sextant A double-reflecting, hand instrument used for measuring the angular distance between two objects in the plane defined by the two objects and the point of observation. It was originally characterized by having an arc of 60 degrees (and a range of 120 degrees), but the term is now applied to similar instruments regardless of range, esp. those used in navigation for measuring apparent altitudes of celestial bodies from a moving ship or airplane, or in hydrographic surveying for measuring horizontal angles at a point in a moving boat between shore objects. Cf: *quadrant; astrolabe.*

sexual dimorphism A condition in which the two sexes of an organism are markedly dissimilar in appearance.

seybertite *clintonite.*

seyrigite A variety of scheelite containing molybdenum.

sferic A rapid, transient variation in the Earth's electromagnetic field, usually caused by lightning. Etymol: A short form of *atmospheric.* Also spelled: *spheric.*

s-fold A fold whose profile has the form of the letter "s".

SF-tectonite A metamorphic rock with planar fabric in which the fabric elements were produced by fracture and/or shear along a pervasive set of subparallel surfaces. It is characterized by mesoscopic subparallel fractures, commonly slickensided, that are independent of the arrangement and/or orientation of mineral grains within the rock. A complete gradation probably exists between *S-tectonite* and SF-tectonite (Raymond, 1975, p. 8).

Sg *mean spring range.*

shabka A desert landscape formed by wind erosion of alluvial basins. Etymol: Arabic, "network, fiber".

shackanite An analcime *trachyte* containing rhombohedral phenocrysts of alkali feldspar in a groundmass of analcime, alkali feldspar, and glass. The name, given by Daly in 1912, is for Shackan railroad station, British Columbia. Not recommended usage.

shaded-relief map A map of an area whose relief is made to appear three-dimensional by the method of *hill shading.*

shading *hill shading.*

shadlunite A mineral of the pentlandite group: $(Fe,Cu)_8(Pb,Cd)S_8$.

shadow weathering Mechanical weathering in which disintegration of rock occurs along the margin of sunlight and shade; a kind of *insolation,* or the effect of sunlight on weathering of rock.

shadow zone [geomorph] *wind shadow.*

shadow zone [seis] (a) An area in which there is little penetration of acoustic waves. (b) A region 100°-140° from the epicenter of an earthquake in which, due to refraction from the *low-velocity zone* inside the core boundary, there is no direct penetration of seismic waves. Syn: *blind zone.*

shaft [mining] A vertical or inclined excavation through which a mine is worked.

shaft [paleont] (a) The main part of the spine of an echinoid. (b) The ridgelike or stalklike proximal part of the cardinal process of a brachiopod, supporting the myophore. (c) The rodlike part of a *rhyncholite* (TIP, 1964, pt. K, p. 22).

shaft [speleo] A passage in a cave that is vertical or nearly so.

shagreen *relief [optics].*

shakehole *sinkhole.*

shaking prairie A term used in Louisiana to describe delta land with a surface of matted vegetation resting on water, or on waterlogged peat or sand, which trembles when walked on. Syn: *trembling prairie.*

shale A fine-grained detrital sedimentary rock, formed by the consolidation (esp. by compression) of clay, silt, or mud. It is characterized by finely laminated structure, which imparts a fissility approximately parallel to the bedding, along which the rock breaks readily into thin layers and that is commonly most conspicuous on weathered surfaces, and by an appreciable content of clay minerals and detrital quartz; a thinly laminated or fissile claystone, siltstone, or mudstone. It normally contains at least 50% silt, with 35% "clay or fine mica fraction" and 15% chemical or authigenic materials (Krynine, 1948, p.154-155). Shale is generally soft but sufficiently indurated so that it will not fall apart on wetting; it is less firm than argillite and slate, commonly has a splintery fracture and a smooth feel, and is easily scratched. Its color may be red, brown, black, or gray. A review of the origin and use of the term "shale" is given by Tourtelot (1960), who notes that it originally meant a "laminated clayey rock" but historically has also been applied to the "general class of fine-grained rocks", and who states that the general trend prior to 1850 in the U.S. "seems to have been to use 'shale' for almost any clayey rock of Paleozoic age; afterwards the term came to be applied to many clayey rocks of all ages" (p.341). Etymol: Teutonic, probably Old English *scealu,* "shell, husk", akin to German *Schale,* "shell".

shale-arenite A term used by Folk (1968, p.124) for a *sedarenite* composed chiefly of shale fragments.

shale-ball A meteorite partly or wholly converted to iron oxides by weathering. Syn: *oxidite.*

shale baseline The line drawn by a *well log* analyst through the *spontaneous potential curve* corresponding to shales or their electrochemical equivalents. Where freshwater drilling muds and saline formation waters prevail, this line forms the right-hand reference from which calculation of formation-water resistivity in beds invaded by mud filtrate may be attempted. Syn: *shale line.*

shale break A thin layer or *parting* of shale between harder strata or within a bed of sandstone or limestone. Cf: *break [drill].*

shale crescent A term used by Shrock (1948, fig.86) for a crescent formed by the filling of a ripple-mark trough by shale. It is a syn. of *flaser structure.*

shale ice A mass of thin, brittle plates of river or lake ice, formed when sheets of skim ice break up into small pieces which are gathered into bunches.

shale line *shale baseline.*

shale oil A crude oil obtained from *oil shale* by submitting it to destructive distillation.

shale-out A *stratigraphic trap* formed by lateral variation or facies change of a porous sandstone or limestone, in which the clay content increases until porosity and permeability disappear and the bed grades into claystone or shale. Cf: *pinch-out.*

shale shaker In *rotary drilling,* an inclined vibrating sieve over which the *drilling mud* passes on its return to the surface and is screened to remove the *well cuttings* and be conditioned for recirculation.

shaley *shaly.*

shaliness The quality of being shaly; specif. the property of clay-rich rocks of splitting with concave or "shelly" surfaces roughly parallel to the bedding planes (Grabau, 1924, p.785).

shallower-pool test A well located within the known limits of an oil or gas pool and drilled with the object of searching for new producing zones above the producing zone of the pool (Lahee, 1962, p. 134).

shallow-focus earthquake An earthquake with a focus at a depth of less than 70 km. Most earthquakes are of this type. Cf: *intermediate-focus earthquake; deep-focus earthquake.*

shallow percolation Precipitation that moves downward and laterally toward streams. Syn: *storm seepage.* Cf: *deep percolation.*

shallow phreatic Said of cave formation near the top of the water-saturated zone. See also: *deep phreatic.*

shallows An indefinite term applied to a shallow place or area in a body of water, or to an expanse of shallow water; a *shoal.*

shallow scattering layer A stratified area of marine organisms over a continental shelf that scatter sound waves from an echo sounder. Cf: *deep scattering layer; surface scattering layer.*

shallow-water wave A wave on the surface of a body of water, the wave length of which is 25 or more times the water depth, and for which the water depth is an influence on the shape of the orbital and on the velocity. Cf: *deep-water wave; transitional-water wave.* Syn: *long wave [water].*

shallow well (a) A water well, generally dug by hand or by excavating machinery, or put down by driving or boring, that taps the shallowest aquifer in the vicinity. The water is generally unconfined ground water. (b) A well whose water level is shallow enough to permit use of a shallow-well (suction) pump, the practical lift of which is taken as 7 m.—Cf: *water-table well; deep well.*

shaly (a) Pertaining to, composed of, or having the character of shale; esp. readily split along closely spaced bedding planes, such as "shaly structure" or "shaly parting". Also, said of a fine-grained, thinly laminated sandstone having the characteristic fissility of shale owing to the presence of thin layers of shale; or said of a siltstone possessing bedding-plane fissility. (b) Said of bedding that consists of laminae ranging in thickness from 2 to 10 mm (Payne, 1942).—Cf: *argillaceous.* Also spelled: *shaley.* Syn: *shelly [sed].*

shandite A rhombohedral mineral: $Ni_3Pb_2S_2$.

Shand's classification A classification of igneous rocks based on crystallinity, degree of saturation with silica, degree of saturation with alumina, and color index. This system was developed in 1927 by S. J. Shand (Shand, 1947).

shank [fold] An obsolete syn. of *limb* of a fold.

shank [paleont] (a) The connection between the flukes and the stock of an *anchor* of a holothurian. (b) The part connecting the eye and spear of a *hook* of a holothurian.

shantung A monadnock in the process of burial by *huangho deposits* (Grabau, 1936, p. 266). Type locality: Shantung rocky mass of northern China.

Shantung soil An early name for *Noncalcic Brown soil.* Var: *Shantung Brown soil.*

shape *particle shape.*

shape class The general group of shapes (oblate, prolate, or intermediate) to which a pollen grain belongs in terms of the ratio between its equatorial diameter and pole-to-pole dimension.

shape-preferred orientation The preferred orientation of elongated or flattened axes of crystals, as a result of *crystal gliding, dynamic recrystallization* or magmatic settling or flow. Cf: *lattice-preferred orientation.*

shapometer A device for measuring the shapes of sedimentary particles (Tester & Bay, 1931).

shard (a) A vitric fragment in pyroclastics; some have a characteristically curved surface of fracture. Shards generally consist of bubble-wall fragments produced by disintegration of pumice during or after the eruption. (b) A syn. of *sherd.*

shared character A *character* held in common by two or more organisms or taxa; it may be either a *homology* or an *analogy.*

shared derived character In *cladism,* a *character* shared by two or more organisms or taxa and inherited from an immediately preceding or recent common ancestor; an advanced character held in common by two or more taxa. Cf: *homology (c); derived character; shared character.* Syn: *synapomorphy.*

shared primitive character In *cladism,* a *character* shared by two or more organisms or taxa and inherited from a remote or much earlier common ancestor. Cf: *homology (c); primitive character; shared character.* Syn: *symplesiomorphy.*

shark A member of the order *Selachii.*

sharkskin pahoehoe A type of *pahoehoe* whose surface displays innumerable tiny spicules or spines produced by escaping gas bubbles.

shark-tooth projection A structure formed by the pulling or tearing apart of plastic lava into fine, sharp points several centimeters in length. Such projections may occur along the edge of a flow or along a slump scarp.

sharpite A greenish-yellow mineral: $(UO_2)(CO_3)\cdot H_2O$ (?).

sharp sand A sand composed of angular grains, nearly or wholly free from foreign particles (as of clay), and used in making mortar.

sharpstone (a) A collective term proposed by Shrock (1948a) for any rock fragment larger than a sand grain (diameter greater than 2 mm) having angular edges and corners; a clastic constituent of rubble. Cf: *roundstone.* (b) A term used in Yorkshire, England, for a fine-grained, nonargillaceous sandstone that breaks into angular fragments. Also spelled: *sharp stone.*

sharpstone conglomerate *sedimentary breccia.*

Shasta Provincial series in California: Lower Cretaceous.

shastaite An obsolete term originally applied to a *dacite.* Iddings in 1913 derived the name from Mount Shasta, California.

shastalite An obsolete term, proposed by Wadsworth in 1884, originally applied to unaltered glassy andesites as distinct from *weiselbergite.*

shatter belt A less-preferred syn. of *fault zone.*

shatter breccia A tectonic breccia composed of angular fragments that show little rotation (Bateman, 1950, p.133). Cf: *rubble breccia.* Syn: *crackle breccia.*

shatter cone A distinctively striated conical fragment of rock, ranging in length from less than a centimeter to several meters, along which fracturing has occurred; generally found in nested or composite groups in the rocks of *cryptoexplosion structures;* and generally believed to have been formed by shock waves generated by meteorite impact (Dietz, 1959). Shatter cones superficially resemble *cone-in-cone structure* in sedimentary rocks. They are most common in fine-grained homogeneous rocks such as limestone and dolomite, but are also known in shale, sandstone, quartzite, and granite. The striated surfaces radiate outward from the apex in horsetail fashion; the apical angle varies but is close to 90 degrees. Syn: *shear cone; pressure cone.*

shatter-cone segment A part of an incompletely developed *shatter cone,* consisting of a single curved, striated surface, generally 10 to 45 degrees of cross section, of a cone whose apical angle may range from 90 to 120 degrees (Manton, 1965, p. 1021). Most shatter-coned rocks display only segments.

shatter coning A mode of rock failure characterized by the development of shatter cones (Manton, 1965, p. 1021).

shattering The breaking-up into angular blocks of a hard rock that has been subjected to severe stresses; the fractures may cut across mineral grains and structures in the rock.

shatter zone An area, often a belt or linear zone, of fissured or cracked rock that may be filled by mineral deposits, forming a network pattern of veins.

shattuckite A blue mineral: $Cu_5(SiO_3)_4(OH)_2$. Cf: *planchéite.*

shcherbakovite A monoclinic mineral: $(K,Na,Ba)_3(Ti,Nb)_2(Si_2O_7)_2$.

shcherbinaite An orthorhombic mineral: V_2O_5.

sheaf structure A bundled arrangement of crystals that is characteristic of certain fibrous minerals, e.g. stibnite.

shear A deformation resulting from stresses that cause or tend to cause contiguous parts of a body to slide relatively to each other in a direction parallel to their plane of contact. It is the mode of failure of a body or mass whereby the portion of the mass on one side of a plane or surface slides past the portion on the opposite side. In geological literature the term refers almost invariably to strain rather than to stress. It is also used to refer to surfaces and zones of failure by shear, and to surfaces along which differential movement has taken place.

shear cleavage A syn. of *slip cleavage;* it is a general term that refers to cleavage where there is displacement of pre-existing surfaces across the cleavage plane by movement parallel to it.

shear-cleavage fold *cleavage fold.*

shear cone *shatter cone.*

shear drag *shear resistance.*

shear fold A fold model of which the mechanism is shearing or slipping along closely spaced planes parallel to the fold's axial surface. The resultant structure is a *similar fold.* Syn: *slip fold; glide fold.* See also: *cleavage fold.*

shear fracture A fracture that results from stresses that tend to shear one part of a rock past the adjacent part. See also: *shear joint.* Cf: *tension fracture.*

shear joint A joint that formed as a *shear fracture.* Less-preferred syn: *slip joint.*

shear modulus *modulus of rigidity.*

shear moraine A debris-laden surface or zone along the margin of an ice sheet or ice cap, dipping in toward the center of the ice sheet but becoming parallel to the bed at the base. The name originated because it was thought that such zones were formed by shearing at the boundary between stagnant ice and active ice. This has been disputed and it has been suggested that shear moraines form through the refreezing of meltwater at the base of a glacier. Cf: *shear plane [glaciol].* Syn: *Thule-Baffin moraine.*

shear plane [exp struc geol] *shear surface.*

shear plane [glaciol] A planar surface in a glacier, usually laden with rock debris, attributed to discontinuous shearing or overthrusting. Apparent displacements on some shear planes may simply be due to differential ablation. Cf: *shear moraine.*

shear resistance In fluid dynamics, a tangential stress caused by the fluid viscosity, taking place along a boundary of the flow in the tangential direction of local motion (Chow, 1957). Cf: *pressure resistance.* Syn: *shear drag.*

shear slide A slide produced by shear failure, usually along a plane of weakness, as bedding or cleavage.

shear sorting Sorting of sediments in which the smaller grains tend to move toward the zone of greatest shear strain and the larger grains toward the zone of least shear. In sand dunes, it is characterized by a lamination of fine-grained dark minerals in the zone between moving and residual sand (Stone, 1967, p. 246).

shear strain A measure of the amount by which parallel lines have been sheared past one another by deformation, specif. the tangent of the change in angle between initially perpendicular lines (the ordinary or engineering shear strain); or half the tangent of the change in angle between initially perpendicualr lines (the tensor measure that appears in *infinitesimal-strain theory*).

shear strength The internal resistance of a body to shear stress, typically including a frictional part and a part independent of friction called *cohesion.*

shear stress That component of *stress* which acts tangential to a plane through any given point in a body; any of the tangential components of the stress tensor.

shear structure Any rock structure caused by shearing, e.g. crushing, crumpling, or cleavage.

shear surface A surface along which differential movement has taken place parallel to the surface. Syn: *shear plane.*

shear velocity The square root of the product of the acceleration due to gravity, the hydraulic mean depth of flow, and the slope of the energy grade line (ASCE, 1962).

shear wave *S wave.*

shear zone A tabular zone of rock that has been crushed and brecciated by many parallel fractures due to shear strain. Such an area is often mineralized by ore-forming solutions. See also: *sheeted-zone deposit.*

sheath [bot] (a) A tubular, enrolled part or organ of a plant, e.g. the lower part of a leaf in the grasses. (b) In blue-green algae, a mucilaginous covering external to the cell wall.

sheath [paleont] An investing cover or case of an animal body or body part; e.g. a tentacle sheath of a bryozoan, a receptacle or container in radiolarians, an expanded basal part of a bar, blade, or limb of a conodont element, or a finely perforate external layer in archaeocyathids.

shed A divide of land; e.g. a *watershed.*

sheen A subdued and often iridescent or metallic glitter that approaches but is just short of optical reflection and that modifies the surface luster of a mineral; e.g. the optical effect still visible in the body of a gem (such as tiger's-eye) after its silky surface appearance has been removed by polishing.

sheepback rock A term used as a syn. of *roche moutonnée* in reference to the fanciful resemblance of the landform to a sheep's back. Syn: *sheep rock; sheepback.*

sheep rock *roche moutonnée; sheepback rock.*

sheer A steep face of a cliff; a precipice.

sheet [intrus rocks] A general term for a tabular igneous intrusion, esp. those that are concordant or only slightly discordant. In this general sense, the term *dike* is used for a vertical or steeply dipping tabular body, and the term *sill* for a horizontal or gently dipping one. Cf: *intrusive vein; sole injection.*

sheet [ore dep] A term used in the Upper Mississippi lead-mining region of the U.S. for galena occurring in thin, continuous masses.

sheet [sed] *blanket.*

sheet [speleo] In a cave, a thin *flowstone* coating of calcite.

sheet [water] *sheetflood.*

sheet crack A planar crack attributed to shrinkage of sediment due to dewatering (Fischer, 1964, p.148). It is commonly parallel to the bedding and filled with sparry calcite or mud.

sheet deposit A mineral deposit that is generally stratiform, more or less horizontal, and areally extensive relative to its thickness.

sheet drift An evenly spread deposit of glacial drift that did not significantly alter the form of the underlying rock surface.

sheeted Said of an igneous rock such as a granite that has undergone *pressure-release jointing* or *exfoliation,* sometimes giving it the appearance of being stratified.

sheeted-zone deposit A mineral deposit consisting of veins or lodes filling a zone of shear faulting, or *shear zone.*

sheet erosion The removal of thin layers of surface material more or less evenly from an extensive area of gently sloping land, by broad continuous sheets of running water rather than by streams flowing in well-defined channels; e.g. erosion that occurs when rain washes away a thin layer of topsoil. Cf: *channel erosion; rill erosion; gully erosion.* Syn: *sheetflood erosion; sheetwash; unconcentrated wash; rainwash; slope wash; surface wash.*

sheet fabric One of three major types of materials recognized in electron-microscope study of graptolithine periderm as a fundamental structural element in the periderm. It is homogeneous and electron-dense; it commonly encloses or bounds *fusellar fabric* to form fusellar tissue (Urbanek & Towe, 1974, p. 5). See also: *cortical fabric.*

sheetflood A broad expanse of moving, storm-borne water that spreads as a thin, continuous, relatively uniform film over a large area in an arid region and that is not concentrated into well defined channels; its distance of flow is short and its duration is measured in minutes or hours. Sheetfloods usually occur before runoff is sufficient to promote channel flow, or after a period of sudden and heavy rainfall. Cf: *streamflood.* See also: *sheet flow [geomorph].* Also spelled: sheet flood. Syn: *sheetwash; sheet [water].*

sheetflood erosion A form of *sheet erosion* caused by a sheetflood. See also: *rock-floor robbing.*

sheet flow [geomorph] An *overland flow* or downslope movement of water taking the form of a thin, continuous film over relatively smooth soil or rock surfaces and not concentrated into channels larger than rills. Cf: *streamflow.* See also: *sheetflood.*

Also spelled: sheetflow.
sheet flow [hydraul] *laminar flow [hydraul].*
sheet ground A term used in the Joplin district of Missouri for extensive disseminated low-grade zinc-lead deposits.
sheet ice Ice formed in a smooth, relatively thin layer by the rapid freezing of the surface layer of a body of water. Not to be confused with *ice sheet.* Cf: *ice table.*
sheeting A type of jointing produced by pressure release, or *exfoliation.* Sheeting may separate large rock masses, e.g. of granite, into tabular bodies or lenses, roughly parallel with the rock surface, that become thicker, flatter, and more regular with depth. It is a useful characteristic of the rock in many quarries. Cf: *pressure-release jointing; lift joint.* Syn: *sheet structure [struc geol]; expansion joint; release joint; pseudostratification [struc geol]; bedding [mining].*
sheet jointing *exfoliation.*
sheet mica Mica that is relatively flat and sufficiently free from structural defects so that it can be punched or stamped into shapes for use by the electronic and electrical industries.
sheet mineral *phyllosilicate.*
sheet pile A *pile* with a generally flat cross section, which may be meshed or interlocked with adjacent similar members to form a diaphragm, wall, or bulkhead, and designed to resist lateral earth pressure or to reduce ground-water seepage. Syn: *sheeting pile.*
sheet sand *blanket sand.*
sheet silicate *phyllosilicate.*
sheet spar A sheet crack filled with spar (Fischer, 1964, p.148).
sheet structure [sed struc] *vertical sheet structure.*
sheet structure [struc geol] *sheeting.*
sheetwash (a) A *sheetflood* occurring in a humid region. (b) The material transported and deposited by the water of a sheetwash. (c) A term used as a syn. of *sheet flow* (a movement) and *sheet erosion* (a process).—Also spelled: *sheet wash.*
shelf [geomorph] (a) Bedrock or other solid rock beneath alluvial soil or deposits; a flat-surfaced layer or stratum. (b) A flat, projecting layer or ledge of rock, as on a slope.
shelf [marine geol] *continental shelf.*
shelf [paleont] The subhorizontal part of the whorl surface next to a suture in a gastropod shell, bordered on the side toward the periphery of the whorl by a sharp angulation or by a carina.
shelf [tect] A stable cratonic area that was periodically flooded by shallow marine waters and received a relatively thin, well-winnowed cover of sediments. Cf: *platform [tect].*
shelf atoll *pseudoatoll.*
shelf break An abrupt change in slope, marking the boundary between the continental shelf and the continental slope. Cf: *shelf edge.*
shelf channel A shallow, somewhat discontinuous valley along the continental shelf, e.g. the extension of the Hudson River Channel across the continental shelf.
shelf delta A delta-shaped feature standing above the surrounding shelf floor in the proximity of transverse shelf valleys, suggesting a deltaic origin during stillstands of a lower sea level (Swift et al., 1972, p. 506).
shelf edge The demarcation between the continental shelf and the continental slope. Cf: *shelf break.*
shelf-edge reef A reef located along the break in slope between a shallow flat shelf and the adjacent deeper basin (Walker, 1973). Cf: *bank reef.*
shelf facies A sedimentary facies that contains sediments produced in the neritic environment of the shelf seas marginal to a low-lying, stable land surface. It is also known as *shelly facies* in recognition of the importance of its characteristic carbonate rocks and fossil shells. Cf: *geosynclinal facies.* Syn: *platform facies; foreland facies.*
shelf ice (a) The ice of an ice shelf; a term introduced by Nordenskjöld in 1909 to describe the type of floating freshwater ice formed in, or broken away from, the feature then known as an "ice barrier" but now referred to as an *ice shelf.* Syn: *barrier ice.* (b) Nonrecommended syn. of *ice shelf.*
shelf lagoon A broad, shallowly submerged marine shelf or platform, such as the interior of the modern Great Bahama Bank or the sea of Permian time across Kansas.
shelf sea A shallow *marginal sea* situated on the continental shelf, rarely exceeding 150 fathoms (275 m) in depth, e.g. the North Sea.
shelfstone A speleothem formed as a horizontally projecting ledge at the edge of a cave pool. Syn: *crusted strand; folia; lily pad; rimstone shelf.*
shelf-valley complex A group of morphologic elements that occur along the paths of retreat of estuary mouths on autochthonous shelves, composed of deltas, shelf valleys, and *shoal-retreat massifs* (Stanley & Swift, 1976, p. 323).
shell [drill] A driller's term for a thin, hard layer of rock encountered in drilling a well. Cf: *shale break.* Syn: *shelly formation.*
shell [geol] (a) The crust of the Earth. Also, any of the continuous and distinctive concentric zones or layers composing the interior of the Earth (beneath the crust). The term was formerly used for what is now called the "mantle". Syn: *Earth shell.* (b) A thin and generally hard layer of rock; esp. such a stratum encountered in drilling a well.
shell [paleont] (a) The hard, rigid outer covering of an animal, commonly largely calcareous but in some cases chiefly or partly chitinous, siliceous, horny, or bony; e.g. the hard parts of an ammonoid (including the protoconch and the conch, but excluding the aptychus and the beaks or jaw structures), or of a cirripede crustacean (including compartmental plates, calcareous basis, and opercular valves). (b) A shell-bearing animal; esp. a shell-bearing mollusk.
shell [sed] A sedimentary deposit consisting primarily of animal shells.
shell bank A *bank [sed]* consisting largely of shells, esp. of brachiopods and pelecypods. Also spelled: shell-bank.
shell hash A sediment layer composed of coquina (Russell, 1968, p. 76).
shell ice Thin ice originally formed on a sheet of water and remaining as an unbroken shell after the underlying water has been withdrawn. Syn: *cat ice.*
shell marl (a) A sandy, clayey, or limy deposit, loose or weakly consolidated, containing abundant molluscan shells; a common term in the coastal plain of SE U.S. (b) A light-colored calcareous deposit formed on the bottoms of small freshwater lakes, composed largely of uncemented mollusk shells and precipitated calcium carbonate, along with the hard parts of minute organisms. See also: *falun.*
shell sand A marine sand containing up to 5% of shell fragments (Hatch & Rastall, 1965, p. 93).
shelly [paleont] (a) Pertaining to the shell of an animal; chitinous, siliceous, or testaceous. (b) Having a shell.
shelly [sed] (a) Said of a sediment or sedimentary rock containing the shells of animals; e.g. a "shelly sand" composed of a high proportion of loose shell fragments, or a "shelly limestone" composed chiefly of fossil shell fragments. (b) Said of land abounding in or covered with shells, such as a "shelly seashore". (c) Var. of *shaly.*
shelly facies A nongeosynclinal sedimentary facies that is commonly characterized by abundant calcareous fossil shells, dominant carbonate rocks (limestones and dolomites), mature orthoquartzitic sandstones, and paucity of shales. The term is frequently used in reference to lower Paleozoic strata, as in the upper Mississippi Valley and the Great Lakes area. The facies is also known as *shelf facies* in recognition of the presumed structural stability of the site of deposition.
shelly formation *shell [drill].*
shelly pahoehoe A type of *pahoehoe* whose surface contains large open tubes and blisters; its crust is 1-30 cm thick.
shelter cave *rockshelter.*
shelter porosity A type of primary interparticle porosity defined by Choquette & Pray (1970, p. 249) as the porosity "created by the sheltering effect of relatively large sedimentary particles which prevent the infilling of pore space beneath them by finer clastic particles".
sherd (a) A fragment or broken piece of pottery. (b) A syn. of *shard.*
shergottite An achondritic stony meteorite composed mainly of pigeonite and maskelynite.
sheridanite A talclike mineral of the chlorite group: $(Mg,Al)_6(Si,Al)_4O_{10}(OH)_8$. It is pale green to nearly colorless. It is a variety of clinochlore.
sherry topaz A variety of topaz resembling sherry wine in color. It is one of the more valuable and important varieties of topaz.
sherwoodite A blue-black tetragonal mineral: $Ca_5AlV_3{}^{+4}V_{11}{}^{+5}O_{40}\cdot 28H_2O$.
sheugh A Scottish term for a small ravine, esp. one containing a stream. Also spelled: *sheuch.*
Shidertinian European stage: Upper Cambrian (above Tuorian, below Tremadocian).
shield [eng] A framework or diaphragm of steel, iron, or wood,

used in tunneling and mining in unconsolidated materials. It is moved forward at the end of the tunnel or adit in process of excavation, and is used to support the ground ahead of the lining and to aid in its construction.

shield [paleont] (a) A protective cover or structure on an animal, likened to or resembling a shield; e.g. an ossicle of an ophiuroid, the carapace of a crustacean, or a large scale on the head of a lizard. (b) A flat or curved lateral outgrowth at one or more levels of a tangential rod or needle in the skeleton of an acantharian radiolarian, forming by fusion of the lattice shell. (c) One of the discoidal elements of the placolith coccolith.

shield [speleo] A speleothem composed of two parallel hemicircular plates that form a sandwich separated by a planar crack. Growth occurs at the rim, where water issues from the crack. Syn: *palette.*

shield [tect] A large area of exposed *basement rocks* in a craton, commonly with a very gently convex surface, surrounded by sediment-covered *platforms;* e.g. Canadian Shield, Baltic Shield. The rocks of virtually all shield areas are Precambrian. Syn: *continental shield; cratogene; continental nucleus.*

shield basalt A basaltic lava flow that erupted from numerous small closely spaced shield-volcano vents, and coalesced to form a single unit. It is generally of smaller extent than a *plateau basalt.*

shielding n. A grounded metallic enclosure intended to reduce noise, capacitively or inductively coupled into an electrical circuit.

shield volcano A volcano in the shape of a flattened dome, broad and low, built by flows of very fluid basaltic lava or by rhyolitic ash flows. Cf: *lava shield; ignimbrite shield.* Syn: *lava dome; basaltic dome.*

shift The relative displacement of the units affected by a fault but outside the fault zone itself; a partial syn. of *slip [struc geol].* See also: *strike shift; dip shift.*

shifting [coast] The fluctuation or oscillation of sea level; the change in position of a shoreline.

shifting [streams] The movement of the crest of a divide away from a more actively eroding stream (as on the steeper slope of an asymmetric ridge) toward a weaker stream on the gentler slope; the change in position of a divide (and of the stream channel) where one stream is captured by another. See also: *monoclinal shifting; leaping; creeping.* Syn: *migration.*

shinarump A colloquial term used in southwestern U.S. for *silicified wood.* Also spelled: *chinarump.*

shingle (a) Coarse loose well-rounded waterworn detritus or alluvial material of various sizes; esp. beach gravel, composed of smooth and spheroidal or flattened pebbles, cobbles, and sometimes small boulders, generally measuring 20-200 mm in diameter; it occurs typically on the higher parts of a beach. The term is more widely used in Great Britain than in the U.S. (b) A place strewn with shingle; e.g. a *shingle beach.*—Etymol: probably Scandinavian, akin to *singel,* "coarse gravel that sings or crunches when walked on".

shingle beach A narrow beach, usually the first to form on a coastline having resistant bedrock and cliffs, composed of *shingle,* and commonly having a steep slope on both its landward and seaward sides. Syn: *cobble beach; shingle.*

shingle-block structure An obsolete syn. of *imbricate structure [tect].*

shingle rampart A *rampart* or ridge, 1 or 2 m high, of *shingle* built up along the seaward edge of a reef.

shingle ridge A steeply sloping bank of *shingle* heaped up on and parallel with the shore.

shingle structure [eco geol] The arrangement of closely spaced veins overlapping in the manner of shingles on a roof.

shingle structure [sed] *imbricate structure.*

shingling *imbrication [sed].*

shingly Composed of or containing abundant *shingle;* e.g. a "shingly beach".

Shipek bottom sampler The commercial name of a popular type of *grab sampler;* it obtains a disturbed sample of the upper few centimeters of bottom sediment.

shipping channel The region of deeper water between the mainland and the principal offshore area of growth of a major barrier reef; as much as 100 feet deep and several miles wide, it is mostly free of patch reefs hazardous to navigation, so that large ocean-going ships can safely travel along it parallel to the coastline, as off northeastern Australia. Cf: *boat channel.*

shipping ore *first-class ore.*

shiver An old English term for soft and crumbly shale, or slate clay approaching shale. Etymol: Middle English *scifre.* Adj: *shivery.*

shiver spar A variety of calcite with slaty structure; specif. argentine. Syn: *slate spar.*

shoal adj. Having little depth; shallow.—n. (a) A relatively shallow place in a stream, lake, sea, or other body of water; a *shallows.* (b) A submerged ridge, bank, or bar consisting of or covered by sand or other unconsolidated material, rising from the bed of a body of water to near the surface so as to constitute a danger to navigation; specif. an elevation, or an area of such elevations, at a depth of 10 fathoms (formerly 6) or less, composed of material other than rock or coral. It may be exposed at low water. Cf: *reef.* (c) A rocky area on the sea floor within soundings. (d) A growth of vegetation on the bottom of a deep lake, occurring at any depth.—v. To become shallow gradually; to cause to become shallow; to fill up or block off with a shoal; to proceed from a greater to a lesser depth of water.

shoal breccia A submarine breccia formed by the action of waves and tides on a shoal, commonly of limestone (Norton, 1917, p.177).

shoaling A bottom effect that describes, in terms of wave heights, the alteration of a wave as it proceeds from deep water into shallow water; it is shown by an initial decrease in height of the incoming wave, followed by an increase in height as the wave arrives on the shore.

shoal reef (a) Any formation in which reef growth develops in irregular patches amidst submerged shoals of calcareous *reef detritus* derived from a large reef (Henson, 1950, p.227); it is smaller in area than a *bank reef.* See also: *reef patch.* (b) A term sometimes used as a syn. of *bank reef.*

shoal-retreat massif A broad, shelf-transverse sand ridge that marks the retreat path of a zone of *littoral drift* convergence (Stanley & Swift, 1976, p. 323).

shock n. *earthquake.*

shock breccia A fragmental rock formed by the action of shock waves; e.g. *suevite* formed by meteorite impact.

shock lamellae *planar features.*

shock lithification The conversion of originally loose fragmental materials into coherent aggregates, e.g. *instant rock,* by the action of shock waves, such as by those generated by explosions or meteorite impacts (Short, 1966). It apparently involves such mechanisms as fracturing, compaction, and intergranular melting.

shock loading The process of subjecting material to the action of high-pressure shock waves generated by artificial explosions or by meteorite impact.

shock melting Fusion of material as a result of the high temperatures produced by the action of high-pressure shock waves.

shock-metamorphic facies A discredited term for an association of mineralogic features (such as multiple parallel planes and thetomorphic glasses) formed by a particular degree of shock metamorphism. As it implies a near approach to equilibrium, the term is not recommended; the term "stages of shock metamorphism" is preferred.

shock metamorphism The totality of observed permanent physical, chemical, mineralogic, and morphologic changes produced in rocks and minerals by the passage of high-pressure shock waves acting over time intervals ranging from a few microseconds to a fraction of a minute (French, 1966). The only known natural mechanism for producing shock-metamorphic effects is the hypervelocity impact of large meteorites, but the term also includes identical effects produced by shock waves generated in small-scale laboratory experiments and in nuclear and chemical explosions. See also: *impact metamorphism.*

shock remanent magnetization Magnetization caused by impulsive stress of a material in the presence of a magnetic field.

shock wave (a) A compressional wave formed whenever the speed of a body relative to a medium exceeds that at which the medium can transmit sound, having an amplitude that exceeds the elastic limit of the medium in which it travels, and characterized by a disturbed region of small but finite thickness within which very abrupt changes occur in the pressure, temperature, density, and velocity of the medium; e.g. the wave sent out through the air by the discharge of the shot initiating an explosion. In rock, it travels at supersonic velocities and is capable of vaporizing, melting, mineralogically transforming, or strongly deforming rock materials. See also: *blast wave.* (b) *blast [geophys].*

shock zone A volume of rock in or around an impact or explosion crater in which a distinctive shock-metamorphic deformation or transformation effect is present.

shoe stone A sharp-grained sandstone used as a whetstone by shoemakers and other leatherworkers. Also, the whetstone so used.
shoestring A long, relatively narrow, and straight or curving sedimentary body whose width/thickness ratio is less than 5 to 1, and is usually on the order of 1 to 1 or even smaller (Krynine, 1948, p.147); e.g. a channel fill, a bar, a dune, or a beach deposit. Cf: *prism.*
shoestring rill One of several long narrow uniform channels, closely spaced and roughly parallel with one another, that merely score the homogeneous surface of a relatively steep slope of bare soil or weak clay-rich bedrock, and that develop wherever overland flow is intense.
shoestring sand A *shoestring* composed of sand or sandstone, usually buried in the midst of mud or shale; e.g. a buried sandbar or channel fill. Syn: *shoestring sandstone.*
shonkinite (a) In the *IUGS classification,* a plutonic rock in which F is between 10 and 60, P/(A+P) is 10 or less, and M is between 60 and 90. (b) A dark-colored syenite composed chiefly of augite and alkali feldspar, and possibly containing olivine, hornblende, biotite, and nepheline. Its name, given by Weed and Pirsson in 1895, is derived from Shonkin, the Indian name for the Highwood Mountains of Montana.
shoot [ore dep] A syn. of *ore shoot.* Also spelled: *chute.*
shoot [seis] In seismic prospecting, to explore an area by employing seismic techniques; to set off an explosion to generate seismic waves.
shoot [streams] (a) A place where a stream flows or descends swiftly. (b) A natural or artificial channel, passage, or trough through which water is moved to a lower level. (c) A rush of water down a steep place or a rapids.—Etymol: French *chute.* See also: *chute.*
shooting *seismic shooting.*
shooting flow (a) *jet flow.* (b) *rapid flow.*
shooting-flow cast A term proposed by Wood & Smith (1958, p.169) for one of a series of "strong parallel ridges", representing the cast of a groove up to 10 cm deep, 30 cm wide, and 2 m long.
shooting star A visual *meteor* (meteoroid) appearing as a thin, temporary streak or trace of light in the nighttime sky. It is not a true star. Syn: *falling star.*
shor A salt lake in Turkestan. Etymol: Russian.
shoran A precise electronic measuring system for indicating distance from an airborne or shipborne station to each of two fixed ground stations simultaneously by recording (by means of cathode-ray screens) the time required for round-trip travel of radar signals or high-frequency radio waves and thereby determining the position of the mobile station. Its range is effectively limited to line-of-sight distances (about 40 nautical miles). Shoran is used in control of aerial photography, airborne geophysical prospecting, offshore hydrographic surveys, and geodetic surveying for measuring long distances. Cf: *loran.* Etymol: *sho*rt-*ra*nge *n*avigation.
shore [coast] (a) The narrow strip of land immediately bordering any body of water, esp. a sea or a large lake; specif. the zone over which the ground is alternately exposed and covered by tides or waves, or the zone between high water and low water. The shore is the most seaward part of the *coast;* its upper boundary is the landward limit of effective wave action at the base of the cliffs and its seaward limit is the low-water line. Subdivided into a *foreshore* and a *backshore.* See also: *beach; strand.* (b) The term is commonly used in the sense of the *shoreline* and of the *foreshore.* (c) A nautical term for land as distinguished from the sea.
shore [desert] *serir.*
shore cliff A cliff at the edge of a body of water or extending along the shore. See also: *sea cliff.*
shore drift *littoral drift.*
shore dune A sand dune produced by wind action on beach sands along a shore.
shoreface (a) The narrow, rather steeply sloping zone seaward or lakeward from the low-water shoreline, permanently covered by water, over which beach sands and gravels actively oscillate with changing wave conditions (Johnson, 1919, p. 161). The zone lies between the seaward limit of the shore and the more nearly horizontal surface of the *offshore* zone. The term "shore face" was originally used by Barrell (1912, p. 385-386), in his study of deltas, for the relatively narrow slope developed by breaking waves and separating the subaerial plain from the subaqueous one below. Not to be confused with *beach face.* See also: *inshore.* (b) A relatively steep but short concave inner portion of the continental shelf (Price, 1954, p. 81).
shoreface terrace A wave-built terrace in the shoreface region, composed of gravel and coarse sand swept from the wave-cut bench into deeper water. See also: *offshore terrace.*
Shore hardness scale An empirical scale of hardness of rocks, metals, ceramics, or other materials as determined by a Shore scleroscope which utilizes the height of rebound of a small standard object (such as a diamond-tipped hammer) dropped from a fixed height onto the surface of a specimen. Named after Albert F. Shore, U.S. manufacturer, who proposed the technique in 1906.
shore ice (a) The basic form of *fast ice,* attached to the shore and, in shallow water, also grounded. (b) Sea ice that has been driven ashore and beached by wind, waves, currents, tides, or the pressure of adjacent ice. Cf: *stranded ice.* (c) Floating sea ice adjacent to the shore; it may or may not be attached to the shore.
shore-ice belt *icefoot.*
shoreland Land along a shore, or bordering a body of water.
shore lead A *lead* between pack ice and the shore, or between pack ice and an ice front. Formerly included what is now known as a *flaw lead.* Pron: shore leed.
shoreline (a) The intersection of a specified plane of water with the shore or beach; it migrates with changes of the tide or of the water level. The term is frequently used in the sense of "high-water shoreline" or the intersection of the plane of mean high water with the shore or beach, or the landward limit of the intermittently exposed shore. Syn: *waterline; shore; strandline.* (b) The general configuration or outline of the shore.—The terms *shoreline* and *coastline* are often used synonymously, but there is a tendency to regard "coastline" as a limit fixed in position for a relatively long time and "shoreline" as a limit constantly moving across the beach.
shoreline cycle The succession of changes through which coastal features normally pass during the development of a shoreline, from the time when the water first assumed its level and rested against the new shore to the time when the water can do no more work (either erosion or deposition).
shoreline-development ratio A ratio indicating the degree of irregularity of a lake shoreline, given as the length of the shoreline to the circumference of a circle whose area is equal to that of the lake (Veatch & Humphrys, 1966, p.289).
shoreline of depression A *shoreline of submergence* that implies an absolute subsidence of the land.
shoreline of elevation A *shoreline of emergence* that implies an absolute rise of the land. Not to be confused with *elevated shoreline.*
shoreline of emergence A shoreline resulting from the dominant relative *emergence* of the floor of an ocean or lake; the water surface comes to rest against the partially emerged land which is marked by marine-produced forms and structures. The shoreline is straight or gently curving, with no bays or promontories; it is simpler in outline than a *shoreline of submergence,* and is bordered by shallow water. The term carries no implication as to whether it is the land or the sea that has moved (Johnson, 1919, p. 173). See also: *shoreline of elevation.* Syn: *emerged shoreline; negative shoreline.*
shoreline of submergence A shoreline resulting from the dominant relative *submergence* of a landmass; the water surface comes to rest against the partially submerged land which is marked by subaerially produced forms and structures. The shoreline is characterized (in its youthful stage) by bays, promontories, offshore islands, spits, bars, cliffs, and other minor features; it is more irregular in outline than a *shoreline of emergence,* and is bordered by water of variable (and locally considerable) depth. The term carries no implication as to whether it is the land or the sea that has moved (Johnson, 1919, p. 173). See also: *shoreline of depression.* Syn: *submerged shoreline; positive shoreline.*
shore platform A descriptive term for the horizontal or gently sloping surface produced along a shore by wave erosion; specif. a *wave-cut bench.* Also, sometimes used as a purely descriptive term for *wave-cut platform.* Syn: *scar.*
shore polynya A *polynya* between pack ice and the coast, or between pack ice and an ice front, and formed by currents or by wind.
shore reef *fringing reef.*
shore terrace (a) A terrace produced by the action of waves and currents along the shore of a lake or sea; e.g. a *wave-built terrace.* (b) *marine terrace.*
shoreward Directed or moving toward the shore.
shorl *schorl.*
short eruption rate *age-specific eruption rate.*
shortite A mineral: $Na_2Ca_2(CO_3)_3$.

short period A period of seismic oscillation that is less than six seconds in duration. Cf: *long period.*

short-range order The state of a crystal structure in which the probability of every atom having the correct nearest or second-nearest neighbors is high. Cf: *long-range order.*

short shot *weathering shot.*

short wave *deep-water wave.*

shoshonite A *trachyandesite* composed of olivine and augite phenocrysts in a groundmass of labradorite with alkali feldspar rims, olivine, augite, a small amount of leucite, and some dark-colored glass. Shoshonite grades into *absarokite* with an increase in olivine and into *banakite* with more sanidine. Its name, given by Iddings in 1895, is derived from the Shoshone River, Wyoming.

shot [seis] The explosive charge or other energy source used in the shooting technique of seismic prospecting.

shot [soil] Hard, rounded particles, generally of sand size, that occur in soils. They may be aggregates or concretions. Syn: *buckshot.*

shot break In seismic prospecting, a record of the instant of generation of seismic waves, as by an explosion. Syn: *time break; shot instant.*

shot copper Small, rounded particles of native copper, molded by the shape of vesicles in basaltic host rock, and resembling shot in size and shape.

shotcrete *Gunite* that commonly includes coarse aggregate (up to 2 cm).

shot depth In seismic work, the vertical distance from the surface to an explosive charge.

shot elevation The elevation of the explosive charge in the shot hole. Not to be confused with *shothole elevation.*

shothole In seismic prospecting, a borehole in which an explosive is placed for generating seismic waves.

shothole bridge An obstruction in a shothole that prevents an explosive charge from going deeper. It may be accidental or intentional.

shothole drill A drill (esp. a rotary drill) used for making a shothole.

shothole elevation The elevation of the top of a shothole. Not to be confused with *shot elevation.*

shot instant *shot break.*

shotpoint In seismic shooting, the location of the explosive charge or other source of energy.

shotpoint gap A greater distance between geophone groups on each side of the shotpoint in an otherwise uniform spread, so that the groups nearest the shotpoint will be far enough from it to minimize adverse effects of hole noise.

shott (a) A shallow and brackish or saline lake or marsh in southern Tunisia or on the plateaus of northern Algeria, usually dry during the summer; a playa lake, often many tens of kilometers in diameter. (b) A closed basin containing a shott; esp. the dried bed existing after the water has disappeared, characterized by salt deposits and frequently by absence of vegetation.—Etymol: Arabic *shatt.* See also: *sabkha.* Also spelled: *chott; schott.*

shoulder [geomorph] (a) A short, rounded spur projecting laterally from the side of a mountain or hill. (b) The sloping part of a mountain or hill below the summit. (c) *valley shoulder.*

shoulder [glac geol] A bench on the side of a glaciated valley, occurring at the marked change of slope where the steep side of the inner, glaciated valley meets the much gentler slope above the level of glaciation. Cf: *alp.* Syn: *trimline.*

shoulder [paleont] (a) The girdle of the gnathostome anterior limb, together with its musculature. (b) The salient angulation of a gastropod-shell whorl, parallel to the coiling and forming the abaxial edge of a subsutural ramp. (c) The ventral and lateral blunt angle of a whorl of an ammonoid shell (TIP, 1959, pt.L, p.5).—See also: *umbilical shoulder.*

shoulder [struc geol] A structure formed on the face of a joint by the intersection of *plume structure* with *fringe joints.*

shoved moraine *push moraine.*

show (a) A trace of oil or gas detected in a core, cuttings, or circulated drilling fluid, or interpreted from the electrical or geophysical logs run in a well. Partial syn: *oil show.* (b) A small particle of gold found in panning a gravel deposit.

shrinkage The decrease in volume of clayey soil or sediment owing to reduction of void volume, principally by drying.

shrinkage crack A crack produced in fine-grained sediment or rock by the loss of contained water during drying or dehydration; e.g. a *desiccation crack* and a *syneresis crack.*

shrinkage factor The factor that is applied to convert a barrel of oil in the reservoir into an equivalent amount of gas-free oil in a stock tank at the surface. Shrinkage factors generally range from 0.63 to 0.88. Cf: *formation-volume factor.*

shrinkage index The numerical difference between the plastic limit of a material and its shrinkage limit.

shrinkage limit That moisture content of a soil below which a decrease in moisture content will not cause a decrease in volume, but above which an increase in moisture will cause an increase in volume.

shrinkage polygon *desiccation polygon.*

shrinkage pore A term used by Fischer (1964, p.116) for an irregular pore formed in muddy sediment by shrinkage (desiccation). It may become a bird's-eye (in a limestone) when filled with sparry calcite. Syn: *fenestra.*

shrinkage ratio The ratio of a volume change to the moisture-content change above the shrinkage limit.

shrub-coppice dune A small, streamlined dune that forms to the lee of bush-and-clump vegetation on a smooth surface of very shallow sand.

shuga An accumulation of spongy lumps of white sea ice, measuring a few centimeters across, and formed from grease ice or sludge, and sometimes from anchor ice rising to the surface (U.S. Naval Oceanographic Office, 1968, p. B36).

shungite A hard black amorphous material containing over 98% carbon, found interbedded among Precambrian schists. It is probably the metamorphic equivalent of bitumen, but it may represent merely impure graphite. Also spelled: *schungite.*

shut-in A narrow, steep-sided gorge along the course of an otherwise wide and shallow stream valley.

shut-in pressure *Reservoir pressure* as recorded at the wellhead when the valves are closed and the oil or gas well is shut in. Syn: *closed-in pressure.*

shutterridge A ridge formed by vertical, lateral, or oblique displacement on a fault traversing a ridge-and-valley topography, with the displaced part of a ridge "shutting in" the adjacent ravine or canyon (Buwalda, 1937).

SI *solidification index.*

si In structural petrology, a fabric defined by the preferred orientation of grains within or internal to a porphyroblast. It may or may not be parallel to the preferred orientation of grains outside the porphyroblast. Cf: *se.*

sial A petrologic name for the upper layer of the Earth's crust, composed of rocks that are rich in silica and alumina; it may be the source of granitic magma. It is characteristic of the upper *continental crust.* Etymol: an acronym for *si*lica + *al*umina. Adj: *sialic.* Cf: *sialma.* Syn: *sal; granitic layer.*

sialic Adj. of *sial.*

sialite *clay mineral.*

siallite (a) A group name for the kaolin clay minerals and allophane. (b) A rock composed of siallite minerals.

siallitic An old term used to describe weathered rock material consisting mainly of alumino-silicate clay minerals, highly leached of alkalies and alkaline earths (SSSA, 1970, p.14).

sialma A layer of the Earth's crust that is intermediate in both depth and composition between the *sial* and the *sima.* Etymol: an acronym for *si*lica + *al*umina + *ma*gnesia.

sialsima *salsima.*

siberite A violet-red or purple lithian variety of tourmaline; rubellite from Siberia.

sibirskite A mineral: $CaHBO_3$.

sibling species One of two or more species that are closely related, very similar morphologically, but reproductively isolated.

Sicilian European stage: Upper Pleistocene (above Emilian, below Milazzian).

sicklerite A dark-brown mineral: $Li(Mn^{+2},Fe^{+3})PO_4$. It is isomorphous with ferrisicklerite.

sickle trough A flat-bottomed, crescent-shaped rock basin sculptured by a glacier. Syn: *skärtråg.*

sicula The skeleton secreted by the initial zooid of a graptolite colony, divisible into a conical *prosicula* and a distal *metasicula.* Pl: *siculae.*

side (a) A slope of a mountain, hill, or bank; e.g. hill*side.* (b) A bank, shore, or other land bordering a body of water; e.g. sea*side.* (c) A geographic region; e.g. country*side.*

side canyon A ravine or other valley smaller than a canyon, through which a tributary flows into the main stream.

side-centered lattice A type of *centered lattice* that is centered on the side faces only.

sidelap The area common to two aerial or space photographs or images in adjacent parallel flight lines. Cf: *overlap.*

side-looking airborne radar An airborne radar system in which a long, narrow, stabilized antenna, aligned parallel to the motion of an aircraft or satellite, projects radiation at right angles to the flight path. It makes possible extremely fine-resolution photography and mapping of the ground surface. Abbrev: SLAR.

side moraine *lateral moraine.*

side muscle One of the lateral pedal and/or visceral muscles of primitive rostroconchs whose insertions form left and right linear connections between anterior and posterior median muscle insertions (Pojeta and Runnegar, 1976, p. 47).

side pinacoid In an orthorhombic, monoclinic, or triclinic crystal, the {010} pinacoid. Cf: *front pinacoid; basal pinacoid.* Syn: *brachypinacoid.*

side plate A small ambulacral plate in a blastoid; esp. one of two (rarely three) serially repeated wedge-shaped plates lying between the central lancet plate and the adjacent radial or deltoid. Each pair of side plates bears a single erect brachiole.

sideraerolite *stony-iron meteorite.*

siderazot A mineral: Fe_5N_2. Also spelled: *siderazote.*

sidereal Pertaining to the stars.

sidereal day The interval between two successive transits of a star over the meridian; the time required for the Earth to rotate once on its axis, or approx. 86,166 seconds. Cf: *solar day.*

sidereal month The mean time of the moon's revolution in its orbit. It is equal to 27 days 7 hours 43 minutes 11.47 seconds of mean *solar time.*

sidereal time Time based on the *sidereal day.* Cf: *solar time.*

siderite [meteorite] A general name for *iron meteorites,* composed almost wholly of iron alloyed with nickel. Obsolete syn: *aerosiderite.*

siderite [mineral] (a) A rhombohedral mineral of the calcite group: $FeCO_3$. It is isomorphous with magnesite and rhodochrosite, and commonly contains magnesium and manganese. Siderite is usually yellow-brown, brown-red, or brown-black, but is sometimes white or gray; it is often found in impure form in beds and nodules (of *clay ironstone*) in clays and shales, and as a directly precipitated deposit partly altered into iron oxides. Siderite is a valuable ore of iron. Syn: *chalybite; spathic iron; sparry iron; rhombohedral iron ore; iron spar; siderose; white iron ore.* (b) An obsolete syn. of *sapphire quartz.* (c) An obsolete term formerly applied to various minerals, such as hornblende, pharmacosiderite, and lazulite.

siderodot A variety of *siderite* containing calcium.

sideroferrite A variety of native iron occurring as grains in petrified wood.

siderogel A mineral consisting of truly amorphous FeO(OH) and occurring in some bog iron ores.

siderolite *stony-iron meteorite.*

sideromelane An approximate syn. of *tachylyte.* The term was introduced by Waltershausen in 1853, and revived by Peacock and Fuller in 1928 (Johannsen, 1937, p. 324).

sideronatrite An orange to straw-yellow mineral: $Na_2Fe^{+3}(SO_4)_2(OH)\cdot 3H_2O$.

siderophile (a) Said of an element concentrated in the metallic rather than in the silicate and sulfide phases of meteorites, and probably concentrated in the Earth's core relative to the mantle and crust (in Goldschmidt's scheme of element partition in the solid Earth). Cf: *chalcophile; lithophile.* (b) Said of an element with a weak affinity for oxygen and sulfur, and readily soluble in molten iron (Rankama & Sahama, 1950, p.88). Examples are: Fe, Ni, Co, P, Pt, Au.

siderophyllite A mineral of the mica group: $K(Fe_2^{+2}Al)(Al_2Si_2)O_{10}(F,OH)_2$.

siderophyre A *stony-iron meteorite* containing crystals of bronzite and tridymite in a network of nickel-iron. Syn: *siderophyry.*

siderose adj. Containing or resembling iron. The term was proposed to replace *ferruginous* when designating a form of iron other than iron oxide; e.g. "siderose cement" consisting of iron carbonate in a sandstone.—n. A syn. of *siderite.*—Also spelled: *sidérose.*

siderosphere A term used for the *inner core* of the Earth.

siderotil A mineral: $(Fe,Cu)SO_4\cdot 5H_2O$.

side shot A reading or measurement from a survey station to locate a point that is off the traverse or that is not intended to be used as a base for the extension of the survey. It is usually made to determine the position of some object that is to be shown on a map.

side stream A *tributary* that receives its water from a drainage area separate from that of the main stream into which it flows.

sidetracked hole n. A well "kicked-off" at a previously drilled depth in order to bypass an obstruction or to straighten the hole; or to redirect the deeper portion by redrilling to an alternate bottom-hole location. See also: *directional drilling; deviation.*

sidetracking The deliberate act or process of deflecting and redrilling the lower part of a borehole away from a previous course; e.g. drilling to the side of and beyond a piece of drilling equipment that is permanently lost in the hole, or drilling another oil well beside a nonproducing well, making use of the upper part of the nonproducing well. Cf: *directional drilling.*

sidewall core A core or rock sample extracted from the wall of a drill hole, either by shooting a retractable hollow projectile, or by mechanically removing a sample.

sidewall sampling The process of obtaining *sidewall cores,* usually by percussion (shooting hollow retractable cylindrical bullets into the walls). Syn: *sidewall coring.*

Siegenian European stage: Lower Devonian (above Gedinnian, below Emsian).

siegenite A mineral of the linnaeite group: $(Co,Ni)_3S_4$. It may contain copper or iron or both in appreciable amounts.

sienna Any of various brownish-yellow earthy limonitic pigments for oil stains and paints. It becomes orange red to reddish brown when burnt, and is generally darker and more transparent in oils than *ochers.* Named after Siena, a town in Tuscany, Italy. Cf: *umber.*

Sierozem A great soil group of the 1938 classification system, a group of zonal soils having a brownish-gray surface horizon and a light-colored subsurface horizon overlying a layer of carbonate accumulation and, sometimes, hardpan. It is developed under conditions of temperate to cool aridity, and under mixed shrub vegetation (USDA, 1938). Etymol: Russian *seroze,* "gray earth". These soils are now classified as *Orthids* and *Argids.* Also spelled: *Serozem; Cerzem.* Syn: *Gray Desert soil; gray earth.*

sierra (a) A high range of hills or mountains, esp. one having jagged or irregular peaks that when projected against the sky resemble the teeth of a saw; e.g. the Sierra Nevada in California. The term is often used in the plural, and is common in the SW U.S. and in Latin America. Syn: *serra.* (b) A mountainous region in a sierra.—Etymol: Spanish, "saw", from Latin *serra,* "saw".

sierranite A rock, consisting of onyx and chert, found in the Sierra Nevada of California.

sieve An apparatus used to separate soil or sedimentary material according to the size of its particles; it is usually made of brass, with a wire-mesh cloth having regularly spaced square holes of uniform diameter. Cf: *screen.*

sieve analysis Determination of the particle-size distribution in a soil, sediment, or rock by measuring the percentage of the particles that will pass through standard sieves of various sizes.

sieve deposition A term proposed by Hooke (1967, p. 454) for the formation of coarse-grained lobate masses on an alluvial fan whose material is sufficiently coarse and permeable to permit complete infiltration of water before it reaches the toe of the fan.

sieve diameter The size (diameter) of a sieve opening (mesh) through which a given particle will just pass.

sieve lobe A coarse-grained lobate mass produced by *sieve deposition* on an alluvial fan.

sieve membrane A sievelike, partly closing membrane in the areolae of a locular-walled diatom. It may occur at an outer or inner position of the wall.

sieve plate [bot] In phloem tissue, the perforated wall or wall portion between two sieve elements, e.g. in a sieve tube.

sieve plate [paleont] (a) A perforated diaphragm extending across the oscular end of the cloaca of a sponge. (b) A unilaminar and circular, subcircular, or polygonal perforate plate of a holothurian (TIP, 1966, pt.U, p.653). (c) A minute discoidal plate with numerous circular, triangular, and polygonal micropores arranged in concentric rows, contained in a pore canal of certain foraminifers. Also, a *trematophore.* (d) A flat, circular, porous plate in spumellarian radiolarians.

sieve texture A syn. of *poikiloblastic* texture.

sieve tissue *phloem.*

sieve tube A phloem tube formed from several sieve elements set end to end (Cronquist, 1961, p.881).

sieving The shaking of loose materials in a sieve so that the finer particles pass through the mesh. It is the most common method of measuring particle sizes of sediments, esp. in the range 1/16 mm (very fine sand) to about 30 mm (coarse pebbles).

sif A syn. of *seif.* Pl: *siuf.*

sifema According to Van Bemmelen (1949), the theoretical ultrabasic layer underlying the sima; it is the equivalent of the sima of some authors, and of the ultrasima of others. Cf: *salsima.*

siferna A term that has been used for the sima, in a scheme in which the ultrasima is referred to as the sima (Schieferdecker, 1959, term 4547).

sight (a) An observation (such as of the altitude of a celestial body) taken for determining direction or position. Also, the data obtained by such an observation; e.g. a bearing or angle taken with a compass or transit when making a survey. (b) A device with a small aperture through which objects are seen and by which their directions are determined; e.g. an "open sight" of an alidade.

sight rod *range rod.*

sigillarian n. An arborescent club moss of the genus *Sigillaria* that occurs in Carboniferous deposits.—adj. Pertaining to *Sigillaria.*—Cf: *lepidodendrid.*

sigloite A triclinic mineral: $(Fe_{+3},Fe_{+2})Al_2(PO_4)_2(O,OH)\cdot 8H_2O$.

sigma A C-shaped siliceous monaxonic sponge spicule (microsclere). Cf: *sigmaspire.* Pl: *sigmata* or *sigmas.*

sigma phi The verbalized expression for σ_ϕ or the standard deviation (sorting) of a particle-size distribution computed in terms of phi units of the sample. It is a measure of *degree of sorting.*

sigmaspire An S-shaped siliceous monaxonic sponge spicule (microsclere), smooth or spinose; a *sigma* twisted in the form of a spiral of about one revolution. Cf: *spinispire.*

sigma-t *density [oceanog].*

sigmoidal dune An S-shaped, steep-sided, sharp-crested sand dune formed under the influence of alternating and opposing winds of roughly equal velocities (E. Holm, 1957); a transitional dune between a crescentic form and some of the dune complexes, being up to 50 m high, 1-2 km long, and 50-200 m wide.

sigmoidal fold A recumbent fold, the axial surface of which is so curved as to resemble the Greek letter sigma.

signal [geophys] In geophysics, a desired indication on a reading. Cf: *noise.*

signal [surv] A natural or artificial object or structure located at or near a survey station and used as a sighting point or target for survey measurements; e.g. a flag on a pole, or a rigid structure erected over or close to a triangulation station.

signal correction In seismic analysis, a correction to eliminate the time differences between reflection times, resulting from changes in the outgoing signal from shot to shot.

signal-to-noise ratio The ratio of the amplitude of desired seismic energy (signal) to the amplitude of unwanted energy (noise). Abbrev: S/N.

signature (a) A characteristic or combination of characteristics by which a material or object may be identified, as on an image or photograph. (b) A wave form characteristic of an earthquake or other source of energy. (c) A graph of deflection versus time for points passed over by a wave.

significance level The probability that a stated statistical hypothesis will be rejected when in fact it is true. Syn: *alpha level.*

significant wave A statistical term for a fictitious wave whose height and period are equal to the average height and average period of the highest one-third of the actual waves that pass a given point.

sign of elongation In hexagonal and tetragonal crystals of prismatic habit, the sign of the long crystallographic direction; a negative sign of elongation indicates that the trace of the vibration plane of the fast component is parallel to the long axis; a positive sign indicates that it is the slow component that is parallel to that axis. A crystal with a negative sign is said to be "length fast", and a crystal with a positive sign is termed "length slow". See also: *negative elongation; positive elongation.* Syn: *elongation sign.*

sike (a) A British term for a small stream, esp. one that flows through flat or marshy ground and that is often dry in summer. (b) A British term for a gully, trench, drain, or hollow.

sikussak Very old, rough-surfaced sea ice trapped in a fjord, as along the north coast of Greenland; it resembles glacier ice because snow accumulation contributes to its formation and perpetuation. According to Koch (1926, p. 100), to be called sikussak "the ice must be at least 25 years old". Etymol: Eskimo, "very old ice". Cf: *fjord ice.*

sil *yellow ocher.*

silan A *cutan* consisting of silica in its various forms, esp. silt- or clay-size quartz and poorly crystalline chalcedony (Brewer, 1964, p.216).

silcrete (a) A term suggested by Lamplugh (1902) for a conglomerate consisting of surficial sand and gravel cemented into a hard mass by silica. Examples occur in post-Cretaceous strata of the U.S. (b) A siliceous *duricrust.*—Etymol: *sil*iceous + con*crete.* Cf: *calcrete; ferricrete.*

Silesian European stage: Middle and Upper Carboniferous (above Dinantian, below Permian). Includes Stephanian, Westphalian, and Namurian.

silex (a) The French term for *flint.* (b) Silica; esp. quartz, such as a pure or finely ground form for use as a filler. (c) An old term formerly applied to a hard, dense rock, such as basalt or compact limestone.—Etymol: Latin, "hard stone, flint, quartz". The term was used by Pliny for quartz.

silexite [ign] An igneous rock composed essentially of primary quartz (60-100 percent). The term was first used by Miller (1919, p. 30) to include a quartz dike, segregation mass, or inclusion inside or outside its parent rock. Syn: *igneous quartz; peracidite; quartzfels; quartzolite.* Cf: *tarantulite. Quartzolite* is the preferred term, because silexite in French means *chert.*

silexite [sed] The French term for *chert;* specif. chert occurring in calcareous beds (Cayeux, 1929, p. 554).

silhydrite An orthorhombic mineral: $3SiO_2\cdot H_2O$.

silica The chemically resistant dioxide of silicon: SiO_2. It occurs naturally in five crystalline polymorphs (the minerals quartz, tridymite, cristobalite, coesite, and stishovite); in cryptocrystalline form (chalcedony); in amorphous and hydrated forms (opal); in less pure forms (e.g. sand, diatomite, tripoli, chert, flint); and combined in silicates as an essential constituent of many minerals.

silica coefficient In Osann's chemical classification of igneous rocks, the number that expresses the ratio of the total silica in a rock to the silica in the feldspars and metasilicates (Johannsen, 1939, p. 61-82).

silica glass A glass or supercooled liquid consisting of pure or nearly pure silica, such as naturally occurring *lechatelierite* and artificially prepared *vitreous silica.* The term has been applied to impactites and to tektites.

silicalemma The three-layered organic membrane of a diatom cell in which silica is deposited and which probably forms the basis of the organic skin of the mature diatom wall.

silicalite A term used by Wadsworth (1893, p. 92) for any rock composed of silica, such as quartz, jasper, or diatomaceous earth.

silica rock An industrial term for certain sandstones and quartzites that contain at least 95% silica (quartz). It is used as a raw material of glass and other products. Cf: *silica sand.*

silica sand An industrial term for a sand or an easily disaggregated sandstone that has a very high percentage of silica (quartz). It is a source of silicon and a raw material of glass and other industrial products. Cf: *silica rock.*

silicastone A term suggested by Shrock (1948a, p. 125) for any sedimentary rock composed of siliceous minerals.

silicate A compound whose crystal structure contains SiO_4 tetrahedra, either isolated or joined through one or more of the oxygen atoms to form groups, chains, sheets, or three-dimensional structures with metallic elements. Silicates were once classified according to hypothetical oxyacids of silicon (see *metasilicate* and *orthosilicate*) but are now classified according to crystal structure (see *nesosilicate, sorosilicate, cyclosilicate, inosilicate, phyllosilicate, tectosilicate*).

silicated Said of a rock in which the process of *silication* has occurred.

silicate-facies iron formation An *iron formation* in which the principal iron minerals are silicates, such as greenalite, stilpnomelane, minnesotaite, and iron-rich chlorite (James, 1954, p.263-272).

silication The process of converting into or replacing by silicates, esp. in the formation of *skarn* minerals in carbonate rocks. Cf: *silicification.* Adj: *silicated.*

siliceous [ecol] *silicicolous.*

siliceous [petrology] Said of a rock containing abundant silica, esp. free silica rather than as silicates.

siliceous cyst A resting stage common in various yellow-green algae that is endogenous, flasklike or bottle-shaped, and from six to ten microns or rarely as large as 20 microns in size. It is composed of cellulose or pectin that is highly impregnated with silica, and it is closed by an organic plug.

siliceous earth A friable, porous, fine-grained sediment, usually white, consisting chiefly of siliceous (opaline) material, having a dry earthy feel and appearance, and generally derived from the remains of organisms; e.g. *diatomite* and *radiolarian earth.* Cf: *tripoli.*

siliceous fireclay A fireclay composed mainly of fine white clay mixed with clean sharp sand.

siliceous limestone (a) A dense, dark, commonly thin-bedded limestone representing an intimate admixture of calcium carbonate and chemically precipitated silica that are believed to have accumulated simultaneously. It is found in geosynclinal associations. (b) A silicified limestone, bearing evidence of replacement of calcite by silica.

siliceous ooze Any pelagic deep-sea sediment containing at least 30% siliceous skeletal remains, e.g. *radiolarian ooze, diatom ooze.*

siliceous residue An *insoluble residue* chiefly composed of siliceous material, such as quartz or chert.

siliceous sandstone A sandstone cemented with quartz or cryptocrystalline silica; e.g. an orthoquartzite.

siliceous sediment A sediment composed of siliceous materials that may be fragmental, concretionary, or precipitated, and of either organic or inorganic origin; e.g. chert, novaculite, geyserite, or diatomite. Siliceous sediments may be formed by primary deposition of silica or by secondary silicification and replacement.

siliceous shale A hard, fine-grained rock of shaly texture with an exceptional amount of silica (as much as 85%); it may have formed by silicification of normal shale (as by precipitation of silica derived from opal or devitrified volcanic ash) or by accumulation of organic material (such as diatom and radiolarian tests) at the same time the clay was deposited. Tarr (1938, p.20) prefers to describe such rock as *porcellanite* because it is not truly a shale.

siliceous sinter The lightweight porous opaline variety of silica, white or nearly white, deposited as an incrustation by precipitation from the waters of geysers and hot springs. The term has been applied loosely to any deposit made by a geyser or hot spring. Syn: *sinter; pearl sinter; geyserite; fiorite.*

siliceous sponge Any sponge having a skeleton composed of siliceous spicules.

silicic Said of a silica-rich igneous rock or magma. Although there is no firm agreement among petrologists, the amount of silica is usually said to constitute at least 65 percent or two-thirds of the rock. In addition to the combined silica in feldspars, silicic rocks generally contain free silica in the form of quartz. Granite and rhyolite are typical silicic rocks. The synonymous terms *acid* and *acidic* are used almost as frequently as "silicic". Syn: *oversaturated; persilicic.* Cf: *basic [geol]; intermediate; ultrabasic.*

siliciclastic Pertaining to clastic noncarbonate rocks "which are almost exclusively silicon-bearing, either as forms of quartz or as silicates" (Braunstein, 1961).

silicicolous Said of an organism living in siliceous soil. Syn: *siliceous [ecol].*

silicification [meta] The introduction of, or replacement by, silica, generally resulting in the formation of fine-grained quartz, chalcedony, or opal, which may fill pores and replace existing minerals. Cf: *silication.* Adj: *silicified.* Syn: *silification.*

silicification [paleont] A process of fossilization whereby the original organic components of an organism are replaced by silica, as quartz, chalcedony, or opal.

silicified Adj. of *silicification [meta].*

silicified wood A material formed by permineralization of wood by silica in such a manner that the original form and structure of the wood is preserved (Schopf, 1975, p. 29). The silica is generally in the form of opal or chalcedony. Syn: *petrified wood; woodstone; agatized wood; opalized wood; shinarump.*

silicilith (a) A term suggested by Grabau (1924, p.298) for a quartz (sedimentary) rock. Syn: *silicilyte.* (b) A sedimentary rock composed principally of the siliceous remains of organisms (Pettijohn, 1957, p. 429); e.g. a diatomite.

silicinate An adjective restricted by Allen (1936, p. 23) to designate the silica cement of a sedimentary rock.

siliciophite A mixture consisting of serpentine penetrated by opal.

silicoflagellate Any chrysomonad protozoan belonging to the family Silicoflagellidae and characterized by a skeleton composed of siliceous rings and spines. Range, Cretaceous to present.

silicomagnesiofluorite A mineral: $Ca_4Mg_3Si_2O_5(OH)_2F_{10}$.

silicon detector A *semiconductor radiation detector* of silicon rather than germanium.

silicon-oxygen tetrahedron A complex ion formed by four oxygen ions surrounding a silicon ion in a tetrahedral configuration, with a negative charge of 4 units. It is the basic unit of the silicates. It is commonly written as SiO_4.

silicon-32 age method A method of age determination based on measurement of the activity of silicon-32 (half-life approximately 350 years), a nuclide formed in the upper atmosphere. The method has been applied to rapidly deposited siliceous oozes.

silicotelic *telechemic.*

silification *silicification.*

silk Microscopically small, needlelike inclusions of rutile crystals in a natural gem, such as ruby, sapphire, or garnet, from which subsurface reflections produce a whitish sheen resembling that of silk fabric.

silky luster A type of mineral luster characteristic of certain fibrous minerals, e.g. chrysotile.

sill [intrus rocks] A tabular igneous intrusion that parallels the planar structure of the surrounding rock. Cf: *dike [intrus rocks]; sheet [intrus rocks]; sole injection.*

sill [marine geol] (a) A submarine ridge or rise at a relatively shallow depth, separating a basin from another basin or from an adjacent sea and causing the basin to be partly closed, e.g. in the Straits of Gibraltar. (b) A ridge of bedrock or earth material at a shallow depth near the mouth of a fjord, separating the deep water of the fjord from the deep ocean water outside. Syn: *threshold.*

sillar (a) The deposit from an ash cloud or nuée ardente that became indurated by recrystallization due to escaping gases rather than by welding, as is the case with *welded tuff;* it is a type of *ignimbrite.* (b) A nonwelded ash-flow tuff. Etymol: Peruvian.

sill depth Depth to the top of a sill responsible for partly closing a basin.

silled basin *restricted basin.*

sillenite A cubic mineral: Bi_2O_3. It occurs in greenish and earthy or waxy masses, and is polymorphous with bismite. Also spelled: *sillénite.*

sillimanite (a) A brown, gray, pale-green, or white orthorhombic mineral: Al_2SiO_5. It is trimorphous with kyanite and andalusite. Sillimanite occurs in long, slender, needlelike crystals often found in wisplike or fibrous aggregates in schists and gneisses; it forms at the highest temperatures and pressures of a regionally metamorphosed sequence and is characteristic of the innermost zone of contact-metamorphosed sediments. Syn: *fibrolite.* (b) A group of aluminum-silicate minerals including sillimanite, kyanite, andalusite, dumortierite, topaz, and mullite.

silt [eng] Nonplastic or slightly plastic material exhibiting little or no cohesive strength when air-dried, consisting mainly of particles having diameters less than 0.074 mm (passing U.S. standard sieve no.200). Cf: *clay [eng].*

silt [sed] (a) A rock fragment or detrital particle smaller than a very fine sand grain and larger than coarse clay, having a diameter in the range of 1/256 to 1/16 mm (4-62 microns, or 0.00016-0.0025 in., or 8 to 4 phi units; the upper size limit is approximately the smallest size that can be distinguished with the unaided eye), being somewhat rounded by abrasion in the course of transport. In Great Britain, the range of 0.01-0.1 mm has been used. See also: *coarse silt; medium silt; fine silt; very fine silt.* (b) A loose aggregate of unlithified mineral or rock particles of silt size; an unconsolidated or moderately consolidated sedimentary deposit consisting essentially of fine-grained clastics. It varies considerably in composition but commonly has a high content of clay minerals. The term is sometimes applied loosely to a sediment containing considerable sand- and clay-sized particles, and incorrectly to any clastic sediment (such as the muddy sediment carried or laid down by streams or by ocean currents in bays and harbors). (c) Sedimentary material (esp. of silt-sized particles) suspended in running or standing water; mud or fine earth in suspension.

silt [soil] (a) A rock or mineral particle in the soil, having a diameter in the range of 0.002-0.05 mm; prior to 1937, the range was 0.005-0.05 mm. The diameter range recognized by the International Society of Soil Science is 0.002-0.02 mm. (b) A soil containing more than 80% silt-size particles, less than 12% clay, and less than 20% sand.

siltage A mass of silt.

siltation *silting.*

silting The deposition or accumulation of silt that is suspended throughout a body of standing water or in some considerable portion of it; esp. the choking, filling, or covering with stream-deposited silt behind a dam or other place of retarded flow, or in a reservoir. The term often includes sedimentary particles ranging in size from colloidal clay to sand. Syn: *siltation.*

silting up The filling, or partial filling, with silt, as of a reservoir that receives fine-grained sediment brought in by streams and surface runoff. The term has been used synonymously with *sedimentation* without regard to any specific grain size.

siltite A term used by Kay (1951) for a *siltstone.*

silt load A *suspended load* consisting essentially of silt.

silt loam A soil containing 50-88% silt, 0-27% clay, and 0-50% sand; e.g. one with at least 50% silt and 12-27% clay, or one with 50-88% silt and less than 12% clay (SSSA, 1965, p.347).

silt shale A consolidated sediment consisting of no more than 10% sand and having a silt/clay ratio greater than 2:1 (Folk, 1954, p.350); a fissile siltstone.

silt size A term used in sedimentology for a volume greater than that of a sphere with a diameter of 1/256 mm (0.00016 in.) and less than that of a sphere with a diameter of 1/16 mm (0.0025 in.). See also: *dust size.*

siltstone An indurated silt having the texture and composition of shale but lacking its fine lamination or fissility; a massive *mudstone* in which the silt predominates over clay; a nonfissile silt shale. Pettijohn (1957, p.377) regards siltstone as a rock whose composition is intermediate between those of sandstone and shale and of which at least two-thirds is material of silt size; it tends to be flaggy, containing hard, durable, generally thin layers, and often showing various primary current structures. Syn: *siltite.*

silttil A friable, brownish to buff, open-textured silt containing a few small siliceous pebbles, representing a chemically decomposed and eluviated till that may originally have been clayey, and developed in an undulatory, well drained area, as the drift sheets in Illinois (Leighton & MacClintock, 1930, p. 41). Pronounced as if spelled "silt-till". Cf: *mesotil; gumbotil.*

silty breccia A term used by Woodford (1925, p.183) for a breccia containing at least 80% rubble and 10% silt, and no more than 10% of other material.

silty clay (a) An unconsolidated sediment containing 40-75% clay, 12.5-50% silt, and 0-20% sand (Shepard, 1954). (b) An unconsolidated sediment containing more particles of clay size than of silt size, more than 10% silt, and less than 10% of all other coarser sizes (Wentworth, 1922). (c) A soil containing 40-60% clay, 40-60% silt, and 0-20% sand (SSSA, 1965, p.347).

silty clay loam A soil containing 27-40% clay, 60-73% silt, and less than 20% sand. Cf: *clay loam.*

silty sand (a) An unconsolidated sediment containing 50-90% sand and having a ratio of silt to clay greater than 2:1 (Folk, 1954, p.349). (b) An unconsolidated sediment containing 40-75% sand, 12.5-50% silt, and 0-20% clay (Shepard, 1954). (c) An unconsolidated sediment containing more particles of sand size than of silt size, more than 10% silt, and less than 10% of all other sizes (Wentworth, 1922).

silty sandstone (a) A consolidated *silty sand.* (b) A sandstone containing more than 20% silt (Krynine, 1948, p.141).

Silurian A period of the Paleozoic, thought to have covered the span of time between 440 and 400 million years ago; also, the corresponding system of rocks. The Silurian follows the Ordovician and precedes the Devonian; in the older literature, it was sometimes considered to include the Ordovician. It is named after the Silures, a Celtic tribe. See also: *age of fishes.*

silvanite *sylvanite.*

silver A soft white isometric mineral, the native metallic element Ag. It occurs in strings and veins in volcanic and sedimentary rocks and in the upper parts of silver-sulfide lodes, and is often associated with small amounts of gold, mercury, copper, lead, tin, platinum, and other metals. Silver is ductile, malleable, and resistant to oxidation or corrosion, though it tarnishes brown; it has the highest thermal and electric conductivity of any substance. It is used for coinage, jewelry, and tableware, in photography, dentistry, and electroplating, and as a catalyst.

silver amalgam Naturally occurring *amalgam.*

silver Cape A *Cape diamond* having a very slight tint of yellow.

silver-copper glance *stromeyerite.*

silver glance *argentite.*

silver jamesonite *owyheeite.*

silver-lead ore Galena containing more than one percent silver; argentiferous galena.

silvicolous Said of an organism that lives in wooded areas.

sima A petrologic name for the lower layer of the Earth's crust, composed of rocks that are rich in silica and magnesia. It is equivalent to the *oceanic crust* and to the lower portion of the *continental crust,* underlying the *sial*. Etymol: an acronym for *si*lica + *ma*gnesia. Adj: *simatic*. Cf: *sialma.* Syn: *intermediate layer; basaltic layer.*

simatic Adj. of *sima.*

simetite A deep-red to light orange-yellow or brown variety of amber, having a high content of sulfur and oxygen and a low content of succinic acid, and occurring in the waters off Sicily.

similar fold A fold in which the orthogonal thickness of the folded strata is greater in the hinge than in the limbs, but the distance between any two folded surfaces is constant when measured parallel to the axial surface. Cf: *reverse similar fold; concentric fold.*

simonellite A hydrocarbon mineral: $C_{15}H_{20}$.

simple coral *solitary coral.*

simple crater A meteorite impact crater of relatively small diameter, characterized by a uniformly concave-upward shape and a maximum depth in the center, and lacking a central uplift and marginal slumping (Dence, 1968, p. 171); e.g. Barringer Crater (Meteor Crater) in Coconino County, Ariz. Cf: *complex crater.*

simple cross-bedding Cross-bedding in which the lower bounding surfaces are nonerosional (McKee & Weir, 1953, p.385); it is formed by deposition alone.

simple cuspate foreland A foreland in which beach ridges, swales, and other symmetrical lines of growth are oriented parallel with both shores of the cusp (Johnson, 1919, p. 325). Cf: *complex cuspate foreland.*

simple fold A single fold or flexure. Cf: *compound fold.*

simple lattice *primitive lattice.*

simple operculum A dinoflagellate operculum that is not at all, or only incompletely, divided by accessory archeopyle sutures. Cf: *compound operculum.*

simple ore An ore of a single metal. Cf: *complex ore.*

simple pit A *pit* [bot] in which there is no over-arching wall. Cf: *bordered pit.*

simple shear A particular type of constant-volume, plane-strain deformation characterized by fixed orientation of one of the circular sections of the *strain ellipsoid.* Simple shear can be closely approximated by shearing a deck of cards in one direction.

simple skeletal wall In stenolaemate bryozoans, a wall calcified by epidermis located on growing edges and one side (Boardman & Cheetham, 1969, p. 211). Cf: *compound skeletal wall.*

simple spit A spit, either straight or recurved, without the development of minor spits at its end or along its inner side. Cf: *compound spit; complex spit.*

simple stream A stream, generally small, whose drainage basin is "of practically one kind of structure and of one age" (Davis, 1889b, p. 218).

simple trabecula A *trabecula* of a scleractinian coral, composed of a series of single sclerodermites. Cf: *compound trabecula.*

simple twin A twinned crystal composed of only two individuals in twin relation.

simple valley A valley that maintains a constant relation to the general structure of the underlying strata; e.g. a longitudinal valley, or a transverse valley. Term introduced by Powell (1874, p. 50). Cf: *complex valley; compound valley.*

simplotite A dark-green monoclinic mineral: $CaV_4O_9 \cdot 5H_2O$.

simpsonite A hexagonal mineral: $Al_4(Ta,Nb)_3(O,OH,F)_{14}$.

simulated stone Any substance fashioned as a gemstone that imitates it in appearance: an *imitation.*

simulation The representation of a physical system by a device such as a computer or model that imitates the behavior of the system; a simplified version of a situation in the real world. Cf: *model.*

sincosite A green tetragonal mineral: $CaV_2^{+4}(PO_4)_2(OH)_4 \cdot 3H_2O$.

Sinemurian European stage: Lower Jurassic (above Hettangian, below Pliensbachian).

singing A seismic resonance phenomenon that is produced by short-path multiples in a water layer. Syn: *reverberation; ringing.*

singing sand *sounding sand.*

single cut A simplified *brilliant cut* consisting of 18 facets: a table, a culet, 8 bezel facets, and 8 pavilion facets. It is used mostly on small stones (*mêlée*) of low quality.

single-cycle mountain A fold mountain that has been destroyed without re-elevation of any important part (Hinds, 1943).

single-ended spread A type of seismic spread in which the shot point is located at one end of the arrangement of geophones.

single-grain structure A type of structure of a noncoherent soil in which there is no aggregation or orderly arrangement. It is characteristic of coarse-grained soils.

single-line stream A watercourse too narrow to depict (at the given scale on a map) by two lines representing the banks. Cf: *double-line stream; split stream.*
single refraction *Refraction* in an isotropic crystal, as opposed to the *birefringence* of an anisotropic crystal.
single tombolo A single, simple bar connecting an island with the mainland or with another island. Cf: *double tombolo.*
singular crystal form *fixed form.*
sinhalite A brown orthorhombic mineral: $MgAl(BO_4)$. It is structurally related to olivine.
Sinian An approximate equivalent of *Riphean.*
sinistral Pertaining, inclined, or spiraled to the left; specif. pertaining to the reversed or counterclockwise direction of coiling of some gastropod shells. A sinistral gastropod shell in apical view (apex toward the observer) has the whorls apparently turning from the right toward the left; when the shell is held so that the axis of coiling is vertical and the apex or spire is up, the aperture is open toward the observer to the left of the axis. Actually, the definition depends on features of soft anatomy: with genitalia on the left side of the head-foot mass, the soft parts and shell are arranged as in a mirror image of a dextral shell (TIP, 1960, pt.I, p.133). Ant: *dextral.* Syn: *left-handed [paleont].*
sinistral fault *left-lateral fault.*
sinistral fold An asymmetric fold with the asymmetry of a Z as opposed to that of an S when seen in profile. The long limb appears to be offset to the left. Cf: *dextral fold.*
sinistral imbrication The condition in a heterococcolith in which each segment overlaps the one to the left when viewed from the center of the cycle. Ant: *dextral imbrication.*
sink [desert] A depression containing a central playa or saline lake with no outlet, as where a desert stream comes to an end or disappears by evaporation; e.g. Carson Sink in Nevada.
sink [glac geol] An obsolete term for a depression in a terminal moraine.
sink [karst] *sinkhole.*
sink [volc] A circular or ellipsoidal depression on the flank of or near to a volcano, formed by collapse. It has no lava flows or rim surrounding it. Cf: *collapse caldera.* Syn: *pit crater; volcanic sink.*
sinkhole A circular depression in a karst area. Its drainage is subterranean, its size is measured in meters or tens of meters, and it is commonly funnel-shaped. Syn: *doline; sink [karst]; shakehole; leach hole; sotch.* Partial syn: *collapse sinkhole; solution sinkhole.* See also: *karst valley; sinking stream; cockpit.*
sinkhole karst The typical karst of temperate regions. See also: *cockpit karst; tower karst.*
sinking (a) *subsidence.* (b) The downward movement of surface water, generally caused by converging currents or by a water mass that becomes denser than the surrounding water. Ant: *upwelling.* Syn: *downwelling.*
sinking creek *lost stream.*
sinking stream A surface stream that disappears underground in a karst region. Syn: *insurgence; streamsink; swallet; swallow hole; sunken stream; disappearing stream; lost stream.*
sink lake *karst pond.*
sinnerite A mineral: $CU_6As_4S_9$. Cf: *luzonite.*
sinoite A meteorite mineral: Si_2N_2O.
sinopite A brick-red earthy ferruginous clay mineral used by the ancients as a red paint.
sinople A red or brownish-red variety of quartz containing inclusions of hematite. Also spelled: *sinopal; sinopel.*
sinter A chemical sedimentary rock deposited as a hard incrustation on rocks or on the ground by precipitation from hot or cold mineral waters of springs, lakes, or streams; specif. *siliceous sinter* and *calcareous sinter* (travertine). The term is indefinite and should be modified by the proper compositional adjective, although when used alone it usually signifies "siliceous" sinter. Etymol: German *Sinter,* "cinder". Cf. *tufa.*
sintering The process by which bonds develop when grains of solid material are brought into contact. In principle, progressive growth of bonds, or "necks", between grains can be brought about by inelastic deformation under stress; molecular diffusion through the vapor phase; surface diffusion of molecules; or volume diffusion of molecules. In the sintering of ice grains, it appears that vapor diffusion is the dominant mechanism.
sinuosity Ratio of the length of the channel or thalweg to the down-valley distance (Leopold & Wolman, 1957, p.53). Channels with sinuosities of 1.5 or more are called "meandering".
sinupalliate Said of a bivalve mollusk possessing a pallial line with a posterior embayment (pallial sinus). Cf: *integripalliate.*
sinus [bot] The space or recess between two lobes or divisions of a leaf or other expanded organ (Lawrence, 1951, p.770).
sinus [paleont] (a) A curved, moderately deep groove, indentation, or re-entrant in the outer lip of the aperture of a gastropod shell. It is progressively filled in as the shell grows and forms a distinct band, and is distinguished from the *slit* by nonparallel sides. (b) Any part of a transverse feature (apertural margin, rib, growth line) of a cephalopod, concave toward the aperture. (c) A major undulation or rounded depression along the commissure of a brachiopod (generally found on the pedicle valve), with the crest directed ventrally and commonly but not invariably associated with the ventral fold and the dorsal sulcus. The term is also used, irrespective of commissure, as a syn. of *sulcus.* (d) A slit or a rounded or V-shaped notch in the proximal lip of the orifice in some ascophoran cheilostomes (bryozoans), serving as an inlet to the ascus; syn: *rimule.* Also, loosely used for a similar slit or notch in the *secondary orifice.* Also, in many stenolaemate bryozoans, a groove on either side of the keel in endozonal walls of zooids growing from encrusting or median colony walls. (e) A V-shaped indentation of blastoid ambulacrum along the margins of deltoid plates and radial plates. (f) A *pallial sinus* of a bivalve mollusk.
sinusbed A term suggested by Engelund (1966) as a substitute for *antidune,* because, according to the definition adopted by hydraulic engineers, an antidune is a bed form in phase with surface water waves, and may move downstream, or be stationary, rather than move upstream.
sinusoidal projection An equal-area map projection representing the limiting form of the *Bonne projection,* using the equator as the standard parallel, and showing all parallels as equally spaced parallel straight lines drawn to scale. The meridians are sine curves, concaving toward the central meridian (a straight line, one half the length of, and at right angles to, the equator) along which the scale is true. The projection shows the entire globe but suffers from extreme distortion (shearing) in marginal zones at high latitudes; it is often used in atlases for the map of Africa. Syn: *Sanson-Flamsteed projection; Mercator equal-area projection.*
siphon [hydraul] A water conduit in the shape of an inverted U, in which the water is in hydrostatic equilibrium. Also spelled *syphon.* Cf: *inverted siphon.*
siphon [paleont] (a) Either of a pair of posterior tubelike extensions of the mantle in many bivalve mollusks, serving for the passage of water currents; e.g. an inhalant ventral tube that conducts water to the mouth and gills and confines the current flowing into the mantle cavity, and an exhalant dorsal tube that carries away waste water and confines the current flowing from the mantle cavity. (b) An anterior channel-shaped prolongation of the mantle in many gastropods, serving for the passage of water to the gills, and often being protected by a grooved extension of the margin of the shell. (c) The membranous siphuncle of cephalopods. (d) An internal tube extending inward from a foraminiferal aperture.
siphon [speleo] A part of a cave passage in which the ceiling dips below water level. Syn: *water trap; trap [speleo].* See also: *conduit [speleo]; inverted siphon; streamtube; sump [speleo]; periodic spring; torricellian chamber.*
siphonal canal The tabular or troughlike extension of the anterior (abapical) part of the apertural margin of a gastropod shell, serving for the shielding of the inhalant siphon.
siphonal deposit Calcareous structures within the siphuncle of a cephalopod, attaining considerable thickness in some nautiloids.
siphonal fasciole A band of abruptly curved growth lines near the foot of the columella of a gastropod, marking successive positions of the siphonal notch.
siphonal notch The narrow sinus of the apertural margin near the foot of the columella of a gastropod, serving for the protrusion of the inhalant siphon. It virtually separates the inner lip and the outer lip.
siphonoglyph A strongly ciliated groove extending down one side of the pharynx of a coral.
siphonostomatous Said of a gastropod shell with the apertural margin interrupted by a canal, spout, or notch for the protrusion of the siphon; e.g. said of various marine snails having the front edge of the aperture prolonged in the form of a channel for the protection of the siphon. Cf: *holostomatous.*
siphonozooid A degenerate octocorallian polyp with reduced or no tentacles and commonly with reduced septal filaments (thickened, convoluted edges of septa). It is usually much smaller than

an *autozooid* and is believed to regulate the water supply of the colony.

siphuncle (a) A long membranous tube extending through all the camerae and septa from the protoconch to the base of the body chamber of a cephalopod shell, and consisting of soft and shelly parts, including septal necks, connecting rings, calcareous deposits, and siphuncular cord. (b) The tubular or funnel-shaped shelly septal structures that ensheathe and support the siphuncle.

siphuncular cord The fleshy interior tissues of the siphuncle of a cephalopod.

sirloin-type ice A term used by Higashi (1958) and now regarded as a syn. of *Taber ice.*

siserskite *iridosmine.*

sismondine A magnesium-bearing chloritoid.

sister group Within a *cladogram,* any pair of taxa united at a single *node.*

sitaparite *bixbyite.*

site investigation The collection of basic facts about, and the testing of, surface and subsurface materials (including their physical properties, distribution, and geologic structure) at a site, for the purpose of preparing suitable designs for an engineering structure or other use.

sitting on a well Waiting at a well location for the well to be drilled into a producing formation. The geologist examines cuttings and cores, to ascertain what formations are penetrated and to look for signs of hydrocarbons.

six coordination The state of an atom when it is surrounded by six nearest neighbors. Cf: *octahedral coordination.*

sixth-power law A law asserting that the carrying power of a stream is proportional to the sixth power of its velocity; e.g. if the stream flows twice as rapidly, the size of the particles carried may be increased 64 times. The law postulates a complete transfer of kinetic energy from the water to the particle and makes no allowance for the effect of viscous drag.

size *particle size.*

size analysis *particle-size analysis.*

size distribution *particle-size distribution.*

size-frequency analysis *particle-size analysis.*

size-frequency distribution *particle-size distribution.*

sizing The arrangement, grading, or classification of particles according to size; e.g. the separation of mineral grains of a sediment into groups each of which has a certain range of size or maximum diameter, such as by sieving or screening.

sjögrenite A hexagonal mineral: $Mg_6Fe_2(CO_3)(OH)_{16}\cdot4H_2O$. It is dimorphous with pyroaurite. Also spelled: *sjogrenite.*

S-joint *longitudinal joint.*

Sk *skewness.*

skarn As used by Fennoscandian geologists, an old Swedish mining term for silicate gangue (amphibole, pyroxene, garnet, etc.) of certain iron-ore and sulfide deposits of Archean age, particularly those that have replaced limestone and dolomite. Its meaning has been generally expanded to include lime-bearing silicates, of any geologic age, derived from nearly pure limestone and dolomite with the introduction of large amounts of Si, Al, Fe and Mg (Holmes, 1920, p.211). In American usage the term is more or less synonymous with *tactite.* Cf: *endoskarn; exoskarn.*

skärtråg The Swedish term for *sickle trough.*

skauk A term introduced by Taylor (1951, p.620) for an extensive field of crevasses in a glacier.

skavl A Norwegian term for a large wind-eroded ridge of snow on a glacier. Pl: *skavler.* Skavler are generally equivalent to *sastrugi.*

skedophyre A porphyritic rock characterized by *skedophyric* texture. Obsolete.

skedophyric A term, now obsolete, applied by Cross et al. (1906, p.703) to porphyritic igneous rocks in which the phenocrysts are more or less uniformly scattered throughout the groundmass; of or pertaining to a *skedophyre.*

skeletal (a) Pertaining to material derived from organisms and consisting of the hard parts secreted by the organisms or of the hard material around or within organic tissue. (b) A term used by Nelson et al. (1962, p. 234) to refer to a limestone that consists of, or owes its characteristics to, virtually in-place accumulation of skeletal matter (as distinguished from a fragmental limestone formed by mechanical transport); but regarded by Leighton & Pendexter (1962) as synonymous with "bioclastic", indicating faunal or floral fragments, or whole components of organisms, that are not in their place of origin.

skeletal canal A canal-like cavity in a coherent skeletal framework of a sponge. It may or may not correspond to a canal of the aquiferous system. Examples: *amararhysis; diarhysis; schizorhysis; surface groove.*

skeletal crystal growth Microscopic development of the outline or framework of a crystal, with incomplete filling in of the crystal faces. Crystals formed in this way are called skeleton crystals.

skeletal duplicature The outer exoskeletal layers or molted skin of a branchiopod crustacean, shed during ecdysis. See also: *duplicature.*

skeletal fiber Any fiberlike structure of the sponge skeleton, such as a spiculofiber, a spicule tract, a sclerosomal trabecula, or a spongin fiber.

skeletal framework A coherent meshwork in a sponge, built of sclerosomal trabeculae, fused spicules, interlocking spicules, spongin-cemented spicules or sand grains, or spongin alone. Syn: *skeletal mesh; skeletal net.*

skeletal pore An opening between spicules or between skeletal fibers of the regular skeletal framework in a sponge, as distinct from larger openings (such as ostia or oscula) that interrupt the regular net and as distinguished from the true *pores* of the soft parts, with which the skeletal pores may not correspond.

skeletal residue An insoluble residue whose constituent material comprises less than 25% of the volume, and containing rhombohedral (dolomoldic) or spheroidal (oomoldic) openings (Ireland et al., 1947, p. 1482-1483). Cf: *lacy residue.*

skeletal soil *Lithosol.*

skeletan A *cutan* consisting of skeleton grains adhering to the surface (Brewer, 1964, p.217); e.g. bleached sand and silt grains high in quartz and low in feldspar.

skeletogenesis The process of forming the hard or skeletal parts of an organism.

skeleton The hard or bony structure that constitutes the framework supporting the softer parts of an animal and protecting or covering its internal organs; e.g. the mesh of spicules of a sponge, the shell of a brachiopod or mollusk, the calcareous layers of the body wall of a bryozoan, the chitinous covering of an arthropod, or the bones of a vertebrate. See also: *endoskeleton; exoskeleton.*

skeleton grain A relatively stable and not readily translocated grain of a soil material, concentrated or reorganized by soil-forming processes (Brewer & Sleeman, 1960); e.g. a mineral grain, or a resistant siliceous or organic body larger than colloidal size. Cf: *plasma [soil].*

skeleton layer The structure formed at the bottom of sea ice while freezing, consisting of vertically oriented platelets of ice separated by layers of brine; this layer has almost no mechanical strength.

skerry A small rugged island or reef; an isolated rock detached from the mainland, rising above sea level from a shallow strandflat, and covered by the sea during high tides or stormy weather. Examples occur along the coasts of Scotland and Scandinavia. Etymol: Scandinavian, akin to Old Norse *sker.* See also: *stack.*

skerry-guard A line, belt, or fringe of skerries, parallel to and extending along a coast for hundreds of kilometers, seemingly acting as a breakwater or "guard". The term is a common, but incorrect, translation of the Norwegian term *skjergaard,* "skerry enclosure", which properly refers to an area of calm water between a line of skerries and the mainland or enclosed by skerries.

sketch map An outline map drawn freehand from observation or from loose and uncontrolled surveys rather than from exact survey measurements, showing only the main features of an area. It preserves the essential space relationships, but does not truly preserve scale or orientation.

skewed projection Any standard projection, used in construction of maps or charts, that does not conform to a general north-south format with relation to the *neat lines* of the map or chart.

skewness (a) The quality, state, or condition of being distorted or lacking symmetry; esp. the quality or state of asymmetry shown by a *frequency distribution* that is bunched on one side of the average and tails out on the other side. It results from lack of coincidence of the mode, median, and arithmetic mean of the distribution. (b) A measure of asymmetry of a frequency distribution; specif. the quotient of the difference between the arithmetic mean and the mode divided by the standard deviation. The logarithm of this measure to the base ten has been used to express the asymmetry of the central half of the particle-size distribution of a sediment: $Q_1Q_3/(Md)^2$, where Q_1 and Q_3 are the particle diameters, respectively, at the 25% and 75% intersections on the cumulative frequency distribution, and Md is the median particle diameter.

Positive skewness is defined for the longer slope of the plotted distribution in the direction of increasing variate values (mean greater than mode, or coarser particles exceed finer particles in a particle-size distribution); negative skewness is defined for the longer slope of the plotted distribution in the direction of decreasing variate values (mode greater than mean, or finer particles exceed coarser particles in a particle-size distribution). Several coefficients of skewness have been devised in an attempt to assign genetic significance to sediment distributions. Abbrev: Sk.—Cf: *kurtosis.*

skiagite A hypothetical end-member of the garnet group: $Fe_3{}^{+2}Fe_2{}^{+3}(SiO_4)_3$.

skiagram An obsolete type of *scan* using X-ray shadows.

skialith A vague remnant of country rock in granite, obscured by the process of granitization. Cf: *schlieren; xenolith.*

skid boulder *sliding stone.*

Skiddavian *Arenigian.*

skim ice First formation of a thin layer of ice on the water surface. Syn: *skin.*

skimming (a) Diversion of water from a stream or conduit by shallow overflow in order to avoid diverting sand, silt, or other debris carried as bottom load (Langbein & Iseri, 1960). (b) Withdrawal of fresh ground water from a thin body or lens floating on salt water by means of shallow wells or infiltration galleries.

skin *skim ice.*

skin depth *depth of penetration [remote sensing].*

skin effect [drill] (a) The frequency-dependent reduction of *resistivity log* measurements in conductive formations due to inductive interaction between the current paths; the *induction logs* now operating at about 20 KHz are most affected. (b) The reduction of formation permeability in the vicinity of a well bore caused by drilling and completion operations.

skin effect [elect] The concentration of alternating current in a conductor towards its exterior boundary. See also: *depth of penetration [elect].*

skin friction [eng geol] (a) The frictional resistance developed between soil and an engineering structure. (b) The shearing resistance of the ground developed on the sides of a pile, pipe, or probing rod. See also: *negative skin friction.*

skin friction [hydraul] In hydraulics, the friction between a fluid and the surface of a solid moving through it (ASCE, 1962). Syn: *surface drag.*

skinnerite A monoclinic mineral: Cu_3SbS_3.

skiodrome A term used in optical mineralogy for an orthographic projection of curves of equal velocity as they would appear on a sphere, assuming that the light source is at the center of the sphere.

skiou A facetious term used by Davis (1912, p. 116) for a *morvan.*

skip cast The cast of a *skip mark.* Term originally proposed by Dzulynski et al. (1959, p. 1117).

skip mark One of a linear series of regularly spaced, crescent-shaped *tool marks* produced by an object that intermittently impinged on or skipped along the bottom of a stream. Cf: *ring mark; roll mark.* Syn: *saltation mark.*

skjergaard *skerry-guard.*

skleropelite A term proposed by Salomon (1915) for argillaceous and allied rocks that have been indurated by low-grade metamorphism. Skleropelite is more massive and dense than shale, and differs from slate by the absence of cleavage. Cf: *hornfels.*

sklodowskite A strongly radioactive, citron-yellow, orthorhombic secondary mineral: $Mg(UO_2)_2Si_2O_7 \cdot 6H_2O$. It is isostructural with uranophane and cuprosklodowskite. Also spelled: *sklodovskite.*

skolite A scaly, dark-green variety of glauconite rich in aluminum and calcium and deficient in ferric iron.

skomerite An altered *andesite* containing microscopic grains and crystals of augite and olivine, and phenocrysts of decomposed plagioclase (probably albite), in a groundmass of plagioclase that is thought to be more calcic than the phenocrysts (Johannsen, 1939, p. 280). The name, given by Thomas in 1911, is from Skomer Island, Pembrokeshire, Wales. Cf: *marloesite.* Not recommended usage.

skull The skeleton of the head in Osteichthyes and tetrapods, including both cartilage-replacement and dermal components. The term is not strictly applicable to the head skeleton of Agnatha, Placodermi, and Chondrichthyes because of the absence of dermal components in these groups.

skutterudite A tin-white to silver-gray isometric mineral: $(Co,Ni)As_3$. It may contain considerable iron, and it represents a minor ore of cobalt and nickel. See also: *smaltite.*

skylight (a) The component of light that is scattered by the atmosphere and consists predominantly of shorter wavelengths. (b) A submariner's term for a *polynya* or *lead [ice]* during the winter; it is covered by relatively thin ice (usually less than 1 m thick) and has a normally flat undersurface. Cf: *lake [ice].*

skystone *meteorite.*

Skythian *Scythian.*

s.l. *sensu lato.*

slab avalanche A snow avalanche that starts from a fracture line, in *wind slab* snow possessing a certain amount of cohesion. Cf: *loose-snow avalanche; wind-slab avalanche.*

slab jointing Jointing produced in rock by the formation of closely spaced parallel fissures dividing the rock into thin slabs.

slab pahoehoe A type of *pahoehoe* whose surface consists of a jumbled arrangement of plates or slabs of flow crust, presumably so arranged due to the draining-away of the underlying molten lava.

slabstone A rock that readily splits into slabs; *flagstone.*

slack [coast] A hollow or depression between lines of shore dunes or in a sandbank or mudbank on a shore.

slack [geomorph] (a) A British term for a hollow or dip in the ground; e.g. a pass between hills, a small shallow valley, or a depression in a hillside. (b) A British term for a soft, boggy piece of ground; a marsh or morass.

slack [water] *slack water.*

slack [weath] *grus.*

slack ice A syn. of *broken ice,* esp. if floating on slowly moving water.

slacking index *weathering index.*

slack tide *slack water.*

slack water (a) The condition of a tidal current or horizontal motion of water when its velocity is very weak (less than 0.1 knot) or zero, esp. at the turn of the tide when there is a reversal between ebb current and flood current. Also, the interval of time during which slack water occurs. Syn: *slack tide.* (b) A quiet part of, or a still body of water in, a stream; e.g. on the inside of a bend, where the current is slight. Syn: *slack [water].*

slag [mater] Material from the iron blast furnace, resulting from the fusion of *fluxstone* with coke ash and the siliceous and aluminous impurities remaining after separation of iron from the ore. Slag is also produced in steelmaking. Formerly a solid waste, slag is now utilized for various purposes, chiefly in construction.

slag [pyroclast] A scoriaceous or cindery pyroclastic rock.

slag [sed] A British term for a friable shale with many fossils.

slaking (a) The crumbling and disintegration of earth materials upon exposure to air or moisture; specif. the breaking-up of dried clay or soil when saturated with or immersed in water, or of clay-rich sedimentary rocks when exposed to air. (b) The disintegration of the walls of tunnels in swelling clay, owing to inward movement and circumferential compression (Stokes & Varnes, 1955, p.137). (c) The treating of lime with water to give hydrated (slaked) lime.

slant drilling *directional drilling.*

slant range In *SLAR,* the distance measured along a straight line between antenna and target.

slant-range image In *SLAR,* an image in which objects are located at positions corresponding to their *slant range* distances from the aircraft flight path. The scale in the range direction is compressed in the near-range region.

slant well *directional well.*

slap A British term for a pass, notch, or gap between hills.

SLAR *side-looking airborne radar.*

slash [beach] A term used in New Jersey for a wet or marshy *swale* between two parallel beach ridges.

slash [geog] (a) A local term in eastern U.S. for a marsh or a low swampy area overgrown with dense underbrush, and often covered by water. (b) An open or cutover tract in a forest, strewn with debris (logs, bark, branches, etc.), as from logging or fire. Also, the debris in such a tract. Syn: *slashing.*

slate (a) A compact, fine-grained metamorphic rock that possesses *slaty cleavage* and hence can be split into slabs and thin plates. Most slate was formed from shale. (b) A coal miner's term for any shale accompanying coal; also, sometimes the equivalent of *bone coal.*

slate clay A clay more or less transformed into slate; specif. a fireclay occurring in the English coal measures.

slate ground A term used in southern Wales for a dark fissile shale, resembling slate.

slate ribbon A relict *ribbon* structure on the cleavage surface of slate, consisting of varicolored and straight, wavy, or crumpled stripes. It is generally a trace of bedding.

slate spar *shiver spar.*

slatiness The quality of being slaty, such as a sedimentary rock splitting into thin layers or plates parallel to the bedding with essential regularity of surfaces similar to true slaty cleavage (Grabau, 1924, p.786).

slaty cleavage A pervasive, parallel foliation of fine-grained, platy minerals (mainly chlorite and sericite) in a direction perpendicular to the direction of maximum finite shortening, developed in slate or other homogeneous sedimentary rock by deformation and low-grade metamorphism. Most slaty cleavage is also *axial-plane cleavage.* Syn: *flow cleavage.*

slavikite A greenish-yellow rhombohedral mineral: $NaMg_2Fe_5(SO_4)_7(OH)_6 \cdot 33H_2O$.

slawsonite A mineral of the feldspar group: $(Sr,Ca)Al_2Si_2O_8$.

sleeve exploder A marine seismic-energy source in which propane or butane is exploded in a thick rubber bag (the sleeve), and from which the waste gases are vented to the air rather than into the water.

sleugh British var. of *slough* in the sense of a small marsh.

slew A syn. of *slough,* esp. a wet spot not large enough to be called a swamp or a marsh.

slice [stratig] An arbitrary informal division (either of uniform thickness or constituting some uniform vertical fraction) of an otherwise indivisible stratigraphic unit, distinguished for individual facies mapping or analysis.

slice [struc geol] Var. of *thrust slice.*

slice ridge A narrow linear ridge, a meter to 100 m high, representing a slice of rock squeezed up within a fault zone (esp. along a strike-slip fault). Syn: *fault-slice ridge; pressure ridge [seis].*

slick n. A glassy smooth elongate patch or weblike net on an otherwise rippled water surface, occurring in coastal and inland waters. It is caused by a monomolecular layer of organic material that has the effect of reducing surface tension.

slickens Extremely fine-grained material, such as finely pulverized tailings discharged from hydraulic mines or a thin layer of extremely fine silt deposited by a stream during flood.

slickenside A polished and smoothly striated surface that results from friction along a fault plane. A surface bearing slickenside is said to be "slickensided". Etymol: English dialect, *slicken,* "slick, smooth". Cf: *mullion structure; slip-scratch; striation.* Syn: *polished surface.*

slickolite A field term proposed by Bretz (1940, p. 338) for a vertically discontinuous striation produced by slippage and shearing and developed on strongly dipping bedding planes of limestone that forms the molding on the wall of a solution cavity. The structure resembles a slickenside but shows evidence of some solution and the development of incipient asymmetric stylolites oriented parallel to bedding.

slick spot On a soil, an area with an accumulation of sodic salts.

slide [fault] A term proposed by Fleuty (1964) for a fault formed in close connection with folding, and conformable with the fold limb or axial surface. It is accompanied by thinning and/or disappearance of the folded beds. Cf: *fold fault.*

slide [mass move] (a) A mass movement or descent resulting from failure of earth, snow, or rock under shear stress along one or several surfaces that are either visible or may reasonably be inferred; e.g. *landslide; snowslide; rockslide.* The moving mass may or may not be greatly deformed, and movement may be rotational or planar. A slide can result from lateral erosion, lateral pressure, weight of overlying material, accumulation of moisture, earthquakes, expansion owing to freeze-thaw of water in cracks, regional tilting, undermining, and human agencies. (b) The track of bare rock or furrowed earth left by a slide. See also: *landslide track.* (c) The mass of material moved in or deposited by a slide. (d) A shortened form of *landslide.*

slide cast The cast of a *slide mark,* commonly smoothly curved and less than a meter in length.

slide mark A scratch or *groove* left on a sedimentary surface by subaqueous sliding or slumping (Kuenen, 1957, p.251); it tends to be wider and shallower than a typical *drag mark,* and may be formed by sliding objects such as a mass of sediment, a plant mat, or a large soft-bodied animal. See also: *slide cast.* Syn: *olistoglyph.*

sliding *gravitational sliding.*

sliding stone An isolated angular block of stone resting on the floor of a *playa,* derived from an outcrop near the playa margin, and associated with a trail or mark indicating that the stone recently slid across the playa surface, in some instances as much as 300 m (Sharp & Dwight, 1976). Syn: *skid boulder.*

slieve An Irish term for a mountain.

slikke A Dutch term used also in France for a *tidal flat* or a *mud flat,* esp. one rich in decaying organic matter mixed with sand and crossed by tidal channels (Termier & Termier, 1963, p. 414).

slim hole (a) A rotary borehole having a diameter of 5 in. or less. (b) A drill hole of the smallest practicable size, often drilled with a truck-mounted rig, used primarily for mineral exploration or as a *stratigraphic test* or *structure test.*

slip [clay] A suspension of fine clay in water, having the consistency of cream, and used in the decoration of ceramic ware. See also: *slip clay.*

slip [coast] (a) An extension of navigable water into the space between adjacent structures (as piers) within which vessels can be berthed; a *dock.* (b) A sloping ramp extending into the water and serving as a landing place for vessels.

slip [cryst] *crystal gliding.*

slip [geomorph] A narrow mountain pass; a defile.

slip [struc geol] (a) The relative displacement of formerly adjacent points on opposite sides of a fault, measured in the fault surface. Partial syn: *shift.* Syn: *total displacement.* (b) A small fracture along which there has been some displacement.

slipband A planar region of microscopic size, within and parallel to which intracrystalline slip has occurred. Cf: *deformation lamella.*

slip bedding A term used by Ksiazkiewicz (1949) for *convolute bedding* supposedly produced by subaqueous sliding.

slip block A separate rock mass that has "slid away from its original position and come to rest some way down the slope without being much deformed" (Kuenen, 1948, p. 371).

slip clay An easily fusible clay containing a high percentage of fluxing impurities, used to produce a natural glaze on the surface of clayware. See also: *slip [clay].*

slip cleavage A type of cleavage that is superposed on slaty cleavage or schistosity, and is characterized by finite spacing of cleavage planes (*spaced cleavage*) between which occur thin, tabular bodies of rock displaying a crenulated cross-lamination. Syn: *crenulation cleavage; shear cleavage; strain-slip cleavage; close-joints cleavage.*

slip dike A dike that has been intruded along a fault plane.

slip face (a) The steeply sloping surface on the lee side of a dune, standing at or near the angle of repose of loose sand, and advancing downwind by a succession of slides wherever that angle is exceeded. Syn: *sandfall.* (b) The leeward surface of a sand wave, exhibiting foreset bedding. Syn: *slip slope.*—Also spelled: *slipface.*

slip fiber Veins of *fibrous* minerals, esp. asbestos, in which the fibers are more or less parallel to slickensided vein walls. Cf: *cross fiber.*

slip fold *shear fold.*

slip joint A less-preferred syn. of shear joint; it is descriptive and nongenetic.

slip-mark A term proposed by Challinor (1978) for the markings on a rock surface made by the movement over it of another rock mass, but of a type other than *slickenside.* Cf: *groove [fault]; striation; chattermark; mullion structure.* Syn: *slip-scratch.*

slip-off slope The long, low, relatively gentle slope on the inside of a stream meander, produced on the downstream face of the meander spur by the gradual outward migration of the meander as a whole; located opposite to the *cutbank.*

slip-off slope terrace A local terrace on the *slip-off slope* of a meander spur, formed by a brief halt during the irregular incision by a meandering stream.

slip plane [cryst] *glide plane.*

slip plane [mass move] A planar *slip surface.*

slip-scratch *slip-mark.*

slip sheet A stratum or rock unit on the limb of an anticline that, having become fractured at its base, has slid down and away from the anticline. It is a gravity-collapse structure.

slip slope The *slip face* of a sand wave.

slip surface A landslide displacement surface, often slickensided, striated, and subplanar. It is best exhibited in argillaceous materials and in those materials which are highly susceptible to clay alteration when granulated. See also: *slip plane.* Syn: *landslide shear surface; gliding surface.*

slip-vector analysis Determination of the movement on a fault from a study of the first motion from an earthquake. The first motion appears as a compression or a rarefaction, depending on the location of the observing station with respect to the epicenter and the direction of motion involved in the earthquake. See also: *fault-plane solution.*

slit A parallel-sided re-entrant in the outer lip of the aperture of a gastropod shell, ranging from a shallow incision to a deep fissure as much as half a whorl in extent (TIP, 1960, pt.I, p.133). Cf: *sinus [paleont].*

slither Loose rubble or talus; angular debris.

slob (a) A dense accumulation of heavy sludge of sea ice. Syn: *slob ice.* (b) A term used in Newfoundland for soft snow or mushy ice.

slob land (a) *low marsh.* (b) A term used in Ireland for muddy ground or soil, or for a level tract of muddy ground; esp. alluvial land that has been reclaimed.

sloot *sluit.*

slope [geomorph] (a) *gradient.* (b) The inclined surface of any part of the Earth's surface, as a *hillslope;* also, a broad part of a continent descending toward an ocean, as the Pacific *slope.*

slope [stream] *stream gradient.*

slope correction *grade correction.*

slope curvature The rate of change of angle of slope with distance.

slope-discharge curve A graphic presentation of the discharge at a given gaging station, taking into account the slope of the water surface as well as the gage height.

slope element A curved part of a slope profile; a smooth concave or convex area of a slope or portion of the slope profile. Cf: *slope segment.*

slope facet *slope segment.*

slope failure Gradual or rapid downslope movement of soil or rock under gravitational stress, often as a result of man-caused factors, e.g. removal of material from the base of a slope.

slope gully A small, discontinuous submarine valley, usually formed by slumping along a fault scarp or the slope of a river delta. Syn: *sea gully.*

slope length (a) The linear distance from the divide to the stream channel at the base of a valley-side slope, measured along the ground surface. (b) *length of overland flow.*

slope line An imaginary line on the ground surface indicating the direction of steepest gradient at a given point, and therefore intersecting the contour lines at right angles (Warntz, 1975, p. 211).

slope map A map that shows the distribution of the degree of surface inclination; e.g. *isotangent map* and *isosinal map.*

slope-over-wall structure A geologic structure produced where coastal talus deposits are cut back by marine denudation to form a truncated deposit overlying a cliff face.

slope retreat The progressive recession of a scarp or of the side of a hill or mountain; suggested causes include backwearing (*parallel retreat of slope*) and downwearing. Syn: *slope recession; scarp retreat.*

slope sector The part of a slope element on which the slope curvature remains constant.

slope segment A rectilinear part of a slope or slope profile. Cf: *slope element.* Syn: *slope facet.*

slope sequence The part of a hillside surface that consists, in succession, of a waxing slope, a maximum slope segment, and a waning slope.

slope stability The resistance of a natural or artificial slope or other inclined surface to failure by landsliding. See also: *bank stability.*

slope unit (a) The smallest feature of a slope profile, consisting either of a slope element or of a slope segment. (b) A system consisting of a base level of denudation and the correlated slope lying above it (Penck, 1953, p. 129). Slope units are separated by breaks of gradient. Etymol: English translation of German *Forsystem.*

slope wash (a) Soil and rock material that is or has been transported down a slope by mass wasting assisted by running water not confined to channels. Cf: *colluvium.* (b) The process by which slope-wash material is moved; specif. *sheet erosion.*—Also spelled: *slopewash.*

slough [drill] Fragmentary rock material that has crumbled and fallen away from the sides of a borehole or mine working. It may obstruct a borehole or be washed out during circulation of the drilling mud. Pron: sluff.

slough [geog] (a) A small marsh; esp. a marshy tract lying in a swale or other local shallow undrained depression on a piece of dry land, as on the prairies of the Midwest U.S. Also, a dry depression that becomes marshy or filled with water. Syn: *slew; slue; sleugh.* (b) A large wetland, as a swamp; e.g. in the Everglades of Florida. (c) A term used esp. in the Mississippi Valley for a creek or sluggish body of water in a tidal flat, bottomland, or coastal marshland. (d) A sluggish channel of water, such as a side channel of a river, in which water flows slowly through low, swampy ground, as along the Columbia River, or a section of an abandoned river channel, containing stagnant water and occurring in a flood plain or delta. Also, an indefinite term indicating a small lake, a marshy or reedy pool or inlet, a bayou, a pond, or a small and narrow backwater. Syn: *slew; slue.* (e) A small bay in eastern England. (f) A piece of soft, miry, muddy, or waterlogged ground; a place of deep mud, as a mudhole. The term "slough" is an obsolete syn. of "mud" or mire.—Pron: sloo.

slough ice Slushy ice or snow. Rarely used.

slow ray In crystal optics, that component of light in any birefringent crystal section that travels with the lesser velocity and has the higher index of refraction. Cf: *fast ray.*

slud [ice] A rarely used syn. of *young ice.*

slud [mass move] (a) The muddy material that has moved downslope by solifluction. (b) Ground that behaves as a viscous fluid, including material moved by solifluction as well as by mechanisms not limited to gravitational flow (Muller, 1947, p. 221).—Etymol: a provincial English word for a soft, wet, slippery mass, as mud or mire.

sludge [ice] A dense, soupy accumulation of new sea ice formed during an early stage of freezing, and consisting of incoherent floating frazil crystals that may or may not be slightly frozen together; it forms a thin gluey layer and gives the sea surface a steel-gray or leaden color. See also: *ice gruel.* Syn: *slush [ice]; cream ice; sludge ice.*

sludge [sed] (a) A soft, soupy or muddy bottom deposit, such as found on tideland or in a stream bed; specif. black ooze on the bottom of a lake (Twenhofel, 1937, p. 90). (b) A semifluid, slushy, murky mass of sediment resulting from treatment of water, sewage, or industrial and mining wastes, and often appearing as local bottom deposits in polluted bodies of water.

sludge cake An accumulation of hardened sludge (sea ice) strong enough to bear the weight of a person. See also: *sludge floe.*

sludge cast *furrow flute cast.*

sludge floe A large *sludge cake.*

sludge ice *sludge [ice].*

sludge lump An irregular mass of sludge (sea ice) shaped by strong wind action.

sludging *solifluction.*

slue *slough.*

slug flow Movement of an isolated body of water, such as *free water* moving downward in the zone of aeration. The term is based on slang for a small amount of liquid, such as a slug of whiskey.

sluggish Said of a stream in which the peaks of a flood form slowly because of the decrease in the slope as the age of the stream system advances, or as the flow is reduced or retarded by withdrawal or storage in upstream reaches.

sluice (a) A conduit or passage for carrying off surplus water, often at high velocity. It may be fitted with a valve or gate for stopping or regulating the flow. (b) A gate, such as a *floodgate.* (c) A body of water flowing through or stored behind a floodgate.

sluiceway *overflow channel.*

sluicing Concentrating heavy minerals, e.g., gold or cassiterite, by washing unconsolidated material through boxes (sluices) equipped with riffles that trap the heavier minerals on the floor of the box.

sluit An African term for a narrow, usually dry ditch, gully, or gulch, produced by the washing of heavy rains in a large natural fissure; it is shallower than a ravine. Also, a similar watercourse produced artificially for irrigation or drainage. Etymol: Afrikaans, from Dutch *sloot,* "ditch". Syn: *sloot.*

slump (a) A landslide characterized by a shearing and rotary movement of a generally independent mass of rock or earth along a curved slip surface (concave upward) and about an axis parallel to the slope from which it descends, and by backward tilting of the mass with respect to that slope so that the slump surface often exhibits a reversed slope facing uphill. Syn: *slumping.* (b) The sliding-down of a mass of sediment shortly after its deposition on an underwater slope; esp. the downslope flowage of soft, unconsolidated marine sediments, as at the head or along the side of a submarine canyon. This is the "commonest usage in geology in Britain", although "*subaqueous slump* would be more precise" (Challinor, 1978, p. 283). Syn: *subaqueous gliding.* (c) The mass of material slipped down during, or produced by, a slump. See also: *slump block.*

slump ball A relatively flattened mass of sandstone resembling a large concretion, measuring 2 cm to 3 m across, commonly thinly laminated with internal contortions and a smooth or lumpy external form, and formed by subaqueous slumping (Kuenen, 1948, p.369). Cf: *crumpled ball; spiral ball.* Syn: *snowball.*
slump basin A shallow basin near the base of a canyon wall and on a shale hill or ridge, formed by small, irregular slumps, and usually containing a short-lived lake (Worcester, 1939).
slump bedding A term applied loosely to any disturbed bedding; specif. deformed bedding produced by subaqueous slumping or lateral movement of newly deposited sediment.
slump block The mass of material torn away as a coherent unit during *slumping.* It may be 2 km long and as thick as 300 m.
slump breccia A contorted sedimentary bed produced by subaqueous slumping and exhibiting brecciation.
slump fold An intraformational fold produced by slumping of soft sediments.
slumping The downward movement, such as sliding or settling, of a slump. Syn: *slump.*
slump mark A mark made by sand (wet or dry) avalanching down the lee side of a sand wave or dune.
slump overfold A fold consisting of hook-shaped masses of sandstone produced during slumping (Crowell, 1957, p.998).
slump scarp A low cliff or rim of thin solidified lava occurring along the margins of a lava flow and against the valley walls, or around *steptoes,* after the central part of the lava crust collapsed due to outflow of still molten underlying layers; the inward-facing cliff may be several meters high. Term introduced by Finch (1933), but Sharpe (1938, p.70) prefers "lava subsidence scarp" or "lava slump scarp".
slump sheet A well-defined bed of limited thickness and wide horizontal extent, containing slump structures (Kuenen, 1948, p.373).
slump structure A generic term for any sedimentary structure produced by subaqueous slumping.
slurried bed A bed of muddy and sandy or pebbly material that is unsorted and ungraded except at the base, where there is commonly a thin graded layer (Wood & Smith, 1958, p. 173). It is thought to form by slumping and fragmentation of partially consolidated sediments, and may move downslope as a subaqueous debris flow. Syn: *slurry bedding.*
slurry (a) A highly fluid mixture of water and finely divided material, e.g. a naturally occurring muddy lake-bottom deposit, or a man-made mixture of pulverized coal and water, as for movement by pipeline. (b) *grout.*
slurry bedding *slurried bed.*
slurry slump A slump in which the incoherent sliding mass is mixed with water and disintegrates into a quasi-liquid slurry (Dzulynski & Slaczka, 1958, p. 217).
slush [geog] A soft mud; mire.
slush [ice] (a) A syn. of *sludge [ice].* (b) Snow that is saturated or mixed with water; it is found on land or ice, or floating in water after a heavy snowfall.
slush [snow] Snow saturated with water, occurring on land or ice surfaces, or as a viscous floating mass in water after a heavy snowfall. Syn: *snow slush.*
slush avalanche *wet-snow avalanche.*
slush field *snow swamp.*
slushflow (a) A mudflowlike outburst of water-saturated snow along a stream course, commonly occurring in the Arctic after intense thawing produces more meltwater than can drain through the snow, and having a width generally several times greater than that of the stream channel (Washburn & Goldthwait, 1958). (b) A flow of clear slush on a glacier, as in Greenland.
slush pit A surface excavation or diked area to impound water or *drilling mud* for use in drilling or to retain fluids discharged from a well. Also spelled: *slushpit.* Syn: *mud pit.*
slush pond A pool or lake containing slush, on the ablation surface of a glacier. It is especially common during summer.
slush pump *mud pump.*
small boulder A *boulder* having a diameter in the range of 256-512 mm (10-20 in., or -8 to -9 phi units).
small circle A curve formed on the surface of a sphere by the intersection of any plane that does not pass through the center of the sphere; specif. a circle on the Earth's surface, the plane of which does not pass through the center of the Earth, such as any parallel of latitude other than the equator. Cf: *great circle.*
small-circle girdle *cleft girdle.*
small cobble A geologic term for a *cobble* having a diameter in the range of 64-128 mm (2.5-5 in., or -6 to -7 phi units).
smaller foraminifera An informal term generally used to designate those foraminifers that are studied with the aid of thin sectioning. Cf: *larger foraminifera.*
small-scale map A map at a scale (in the U.S. Geological Survey, smaller than 1/250,000) such that a large area can be covered showing only generalized detail. The Defense Mapping Agency defines a small-scale map as one at 1/600,000 or smaller. Cf: *intermediate-scale map; large-scale map.*
small spore A term that is sometimes used as if synonymous with *microspore;* more accurately, a term that includes pollen and prepollen, as well as spores other than megaspores. The term is therefore nearly synonymous with *miospore* but lacks its precise size definition.
small watershed A drainage basin that is "so small that its sensitivities to high-intensity rainfalls of short durations and to land use are not suppressed by the channel-storage characteristics" and in which "the effect of overland flow rather than the effect of channel flow is a dominating factor affecting the peak runoff" (Chow, 1957, p. 379); its size may be a few acres to 1000 acres, or even up to 130 sq km (50 sq mi).
smaltite (a) A tin-white or pale-gray isometric mineral: $(Co,Ni)As_{3-x}$. It is an arsenic-deficient variety of *skutterudite.* Smaltite usually contains some iron, often occurs with cobaltite, and represents an ore of cobalt and nickel. Syn: *smaltine; tin-white cobalt; gray cobalt; white cobalt; speisscobalt.* (b) A term applied to undetermined, apparently isometric arsenides of cobalt or to a mixture of cobalt minerals.
smaragd *emerald.*
smaragdite A fibrous or thinly foliated green amphibole (near actinolite) pseudomorphous after pyroxene (such as omphacite) in rocks such as eclogite.
smectite (a) A name for the montmorillonite group of clay minerals. The term is in common use to designate dioctahedral (montmorillonite) and trioctahedral (saponite) clay minerals (and their chemical varieties) that possess swelling properties and high cation-exchange capacities. Syn: *montmorillonite (a).* (b) A term that was originally applied to fuller's earth and later to the mineral montmorillonite, and that has also been applied to certain clay deposits that are apparently bentonite and to a greenish variety of halloysite (Kerr & Hamilton, 1949, p. 59).
smirnovite *thorutite.*
Smithian European stage: Lower Triassic (above Dienerian, below Spathian).
smithite A red monoclinic mineral: $AgAsS_2$.
smithsonite (a) A white to yellow, gray, brown, or greenish mineral of the calcite group: $ZnCO_3$. It is a secondary mineral associated with sphalerite and often found as a replacement in limestone; it is commonly reniform, botryoidal, stalactitic, or granular, and is distinguished from hemimorphite by its effervescence with acids. Smithsonite is an ore of zinc. Syn: *dry-bone ore; calamine; zinc spar; szaskaite.* (b) A term sometimes used as a syn. of *hemimorphite.*
smokeless coal (a) *semibituminous coal.* (b) Any coal that burns without smoke, from semibituminous to superanthracitic.
smokestone *smoky quartz.*
smoking crest The crest of a dune, along which sand grains are being winnowed.
smoky quartz A smoky-yellow to brown-gray and often transparent variety of crystalline quartz sometimes used as a semiprecious gemstone. It may contain inclusions of both liquid and gaseous carbon dioxide. The color is probably due to exposure to high-energy radiation, such as produced by radioactive minerals. Syn: *cairngorm; smokestone.*
smoky topaz A trade name for smoky quartz used for jewelry.
smolianinovite An orthorhombic mineral: $(Co,Ni,Ca,Mg)_2(Fe^{+3},Al)(AsO_4)_2(OH)\cdot 5H_2O$.
smooth chert Hard, dense, homogeneous chert (as seen in insoluble residues), characterized by a conchoidal to even fracture surface that is devoid of roughness and by a lack of crystallinity, granularity, or other distinctive structure (Ireland et al., 1947, p. 1484). See also: *chalcedonic chert; ordinary chert; porcelaneous chert.* Cf: *granular chert; chalky chert.*
smooth criterion In radar, the relationship between surface roughness, depression angle, and wavelength that determines whether a surface will scatter the incident radar pulse in smooth or intermediate fashion. Cf: *rough criterion; Rayleigh criterion.*
smooth phase The part of stream traction wherein a mass of sedi-

ment travels as a sheet with gradually increasing density from the surface downward (Gilbert, 1914, p. 30-34). Cf: *dune phase; antidune phase.*

smothered bottom A term introduced by Shrock (1948, p.307-308) for a sedimentary surface on which complete, well-preserved, and commonly fragile and delicate fossils were saved by an influx of mud that buried them instantly. Such surfaces are common in sequences composed of alternating marine limestone and shale layers.

smythite A rhombohedral mineral: $(Fe,Ni)_9S_{11}$.

S/N *signal-to-noise ratio.*

snake hole A horizontal or nearly horizontal borehole used for blasting, drilled approximately on a level with the floor of a quarry or under a boulder to be broken up.

snaking stream A winding, sinuous stream; a *meandering stream.*

snapper *grab sampler.*

Snell's law *law of refraction.*

snout [geog] A protruding mass of rock; a promontory.

snout [glaciol] The protruding lower extremity, leading edge, or front of a glacier. Syn: *terminal face; terminus; glacier snout; ice front; front [glaciol].*

snout [paleont] The enlarged anterior part of the shell of some rostroconchs, separated from the body of the shell by a sulcus and by differences in sculpture (Pojeta and Runnegar, 1976, p.74).

snow (a) A form of ice composed of small white or translucent hexagonal crystals of frozen water, formed directly by sublimation of atmospheric water vapor around solid nuclei at a temperature below the freezing point. The crystals grow while floating or falling to the ground, and are often agglomerated into snowflakes. (b) A consolidated mass of fallen snow crystals. (c) *snowfall.* (d) A region covered with permanent snow. The term is usually used in the plural, as the high *snows.*

snow avalanche An *avalanche* consisting of relatively pure snow, although considerable earth and rock material may also be carried downward. Cf: *full-depth avalanche.* Syn: *snowslide.*

snowball A *slump ball* or a *spiral ball.* Term used by Hadding (1931, p.390) for a structure attributed to subaqueous sliding. The name is misleading as the ball did not roll down a slope picking up new layers of sediment (Kuenen, 1948, p.369).

snowball garnet *rotated garnet.*

snowbank A mound of snow, often the remnant of a snowdrift.

snowbank glacier *nivation glacier.*

snow banner A stream of snow blowing off a mountain top, streaming out several miles from its source.

snow barchan A small crescentic or horseshoe-shaped *snow dune* with the ends pointing downwind.

snow blanket A surface accumulation of snow.

snowblink A bright, white glare in the sky near the horizon or on the underside of a cloud layer, produced by light reflected from a snow-covered surface (as a snowfield); brighter than *iceblink [meteorol].* Also spelled: *snow blink.* Syn: *snow sky; snow sheen.*

snowbridge An arch or layer of snow that has drifted across a *crevasse* in a glacier, or a connecting splinter of ice or snow which allows a person or vehicle to cross a crevasse. Also spelled: *snow bridge.* Not to be confused with *bridge [snow].*

snow cap (a) A covering of snow on a mountain peak or ridge when no snow exists at lower elevations. (b) An accumulation of snow on the surface of a frozen lake.—Also spelled: snowcap.

snow cornice *cornice.*

snow course A line or a series of connecting lines of regularly spaced observation stations (usually 10 or more) at which snow samples are taken for measuring depth, density, and water equivalent for forecasting subsequent runoff. See also: *snow survey.*

snow cover (a) All snow that has accumulated on the ground, including that derived from snowfall, drifting or blowing snow, avalanches, frozen or unfrozen rain stored in the snow, rime, and frost. (b) The areal extent of ground partly or wholly covered with snow in a particular area, usually expressed as a percent of the total area. (c) The average depth of accumulated deposited snow on the ground in a particular area, usually expressed in centimeters.

snowcreep The slow internal deformation of a *snowpack* resulting from the stress of its own weight and metamorphism of snow crystals; it usually involves shear parallel to the slope and compaction perpendicular to the slope.

snow crust A firm or hard surface of snow overlying a layer of softer snow; it may be formed by the melting and refreezing of surface snow, by wind action, or by freezing of water on the surface. Cf: *ice crust [glaciol]; wind crust; sun crust; rain crust.*

snow crystal (a) A single ice crystal nucleated and grown in the atmosphere. (b) A single crystal of deposited snow. Cf: *snowflake.*

snow density The mass of snow per unit volume, usually given as mg/m^3. In these units it is numerically equal to specific gravity.

snowdrift Snow deposited by the wind in the lee of an obstacle or in other places where turbulent eddies are formed. Also spelled: *snow drift.*

snowdrift glacier *drift glacier.*

snow dune Snow deposited by the wind as a gently rounded hummock, usually with its long axis parallel with the wind direction. A dune is much larger than ripples, sastrugi, or snow barchans.

snowfall (a) The deposition, on the ground or other surface, of snow precipitated out in the atmosphere. (b) The rate of deposition of snow.

snowfield (a) A broad expanse of terrain covered with snow, relatively smooth and uniform in appearance, occurring usually at high latitudes or in mountainous regions above the snowline, and persisting throughout the year. (b) A region of permanent snow cover, as at the head of a glacier; the *accumulation area* of a glacier.

snowflake An aggregation of several single *snow crystals* that have collided and joined while falling through relatively still air.

snowflake obsidian An *obsidian* that contains white, gray, or reddish spherulites ranging in size from microscopic to a meter or more in diameter.

snow gage *rain gage.*

snow glide The slow sliding of a snowpack over an inclined surface. Cf: *snowcreep.*

snow grain A separate particle of snow, which may consist of one or more crystals. Cf: *grain [glaciol].*

snow ice Ice that has been formed when snow slush, a mixture of snow and water, has frozen. It has a whitish appearance if air bubbles are included.

snowline (a) A temporary line delimiting or defining the altitude of a snow-covered area; in a zone of patchy snow, the area or altitude of more than 50 percent snow cover. Cf: *climatic snowline.* Syn: *transient snowline.* (b) The line separating areas in which deposited snow disappears in summer from areas in which it remains throughout the year; on glaciers it is identical to the *firn line.* Cf: *regional snowline; equilibrium line.* (c) The ever-changing limit of the Earth's broad belt within which no snow falls. Its position depends on such physical conditions as altitude and nearness to the sea. The term is applied esp. to the winter snowline in the Northern Hemisphere.—Also spelled: *snow line.*

snowmelt The water resulting from the melting of snow. Also spelled: snow melt. Syn: *snow water.*

snow moisture *free-water content.*

snow niche *nivation hollow.*

snowpack (a) Any snow cover. (b) The amount of annual accumulation of snow at higher elevations in mountains, which provides water for hydroelectric power and irrigation.—Also spelled: *snow pack.*

snow patch An isolated mass of perennial snow and firn not large enough to be called a glacier.

snow-patch erosion *nivation.*

snow penitente A *nieve penitente* consisting of compacted snow.

snow pillow A device used to record the changing weight of the snow cover at a point. It consists of a fluid-filled bladder or metal container lying on the ground, the internal pressure of which measures the weight of overlying snow.

snowquake The sudden collapse of one or more layers of surface or subsurface snow, often accompanied by a sound that may resemble that of a distant explosion (ADTIC, 1955, p. 76). Syn: *snow tremor.*

snow roller A cylindrical mass of moist, cohesive snow, formed by rolling down a slope. It may be as large as 1 m in length and more than 2 m in circumference. Cf: *sun ball.*

snow sampler A snow-surveying instrument in the form of a tube, used to collect a core of deposited snow, which is subsequently weighed to determine density and water equivalent.

snowshed A drainage basin primarily supplied by snowmelt.

snow sheen *snowblink.*

snow sky *snowblink.*

snowslide A *snow avalanche.* The term has also been used for a smaller mass of downward-moving snow.

snow slush *slush [snow].*

snow survey The process of determining the depth, density, and water equivalent of snow that has fallen on a particular area by sampling representative points along a *snow course.* Snow surveys made in the spring are used for forecasting subsequent snowmelt runoff.

snow swamp An area of water-saturated snow having a soupy consistency, in which men and animals readily sink. If saturated to the surface, the snow is bluish; if not, it is similar in color to the surrounding snow. Syn: *slush field.*

snow tremor *snowquake.*

snow water *snowmelt.*

snub-scar A term used by Wentworth (1936) for a "push-off" end or edge (or lee-end pressure spall) characteristic of a glacial cobble.

soakaway A British term for a sink or depression in the Earth's surface, into which waters flow and drain away.

soaked zone The area on a glacier where considerable surface melting occurs in summer and meltwater percolates through the whole mass of the snow layer, bringing it to the melting temperature. However, melting is not sufficient to remove all the snow, and snow persists at the end of summer. The soaked zone may be bordered at higher altitudes by the *saturation line* and at lower altitudes by the *equilibrium line.* Cf: *percolation zone.*

soap clay *bentonite.*

soap earth Massive talc; *steatite.*

soap hole A term used in Wyoming for a hole resulting from the wetting of the outcrop surface of bentonite.

soaprock (a) *soapstone [mineral].* (b) *soapstone [rock].*

soapstone [mineral] (a) A mineral name applied to *steatite,* or massive talc. Syn: *soaprock.* (b) *saponite.* (c) A term loosely applied to much agalmatolite.

soapstone [rock] (a) A metamorphic rock of massive, schistose, or interlaced fibrous or flaky texture and soft, unctuous feel, composed essentially of talc with varying amounts of micas, chlorite, amphibole, pyroxenes, etc. and derived from the alteration of ferromagnesian silicate minerals. It may be sawed into laboratory-bench tops, switchboards, and other types of special-purpose dimension stone. (b) A miner's and driller's term for any soft, unctuous rock such as micaceous shale or sericitic schist.

soapy Said of a type of mineral texture that is slippery, smooth, and soft; e.g. that of talc. Syn: *unctuous.*

sobolevskite A hexagonal mineral: PdBi.

sobotkite A mineral of the montmorillonite group: $Mg_2Al(Si_3Al)O_{10}(OH)_2 \cdot 5H_2O$.

socket A recess or depression along the hinge line of a bivalve for the reception of a projecting hinge tooth from the opposite valve; esp. a *dental socket* of a brachiopod.

socket ridge A linear elevation of secondary shell extending laterally from the cardinal process of a brachiopod and bounding the margin of dental sockets.

soda Sodium carbonate, Na_2CO_3; especially the decahydrate, $Na_2CO_3 \cdot 10H_2O$. Loosely used for sodium oxide, sodium hydroxide, sodium bicarbonate, and even for sodium in informal expressions such as *soda spar.*

soda alum A cubic mineral of the alum group: $NaAl(SO_4)_2 \cdot 12H_2O$. Cf: *mendozite.* Syn: *sodium alum.*

soda ash Commercial term for sodium carbonate, Na_2CO_3.

sodaclase *albite.*

soda feldspar A misnomer for *sodium feldspar.*

soda hornblende *arfvedsonite.*

soda lake An *alkali lake* whose waters contain a high content of dissolved sodium salts, chiefly sodium carbonate accompanied by the chloride and the sulfate. Examples occur in Mexico and Nevada. Syn: *natron lake.*

soda leucite A hypothetical sodium-rich variety of leucite, postulated as the original material of some pseudoleucites.

soda-lime feldspar A misnomer for *sodium-calcium feldspar.*

sodalite (a) A mineral of the feldspathoid group: $Na_4Al_3Si_3O_{12}Cl$. It is usually blue or blue-violet, but may be white, greenish, gray, pink, or yellow, and it occurs in various sodium-rich igneous rocks. (b) A group of bluish feldspathoid minerals containing sodium silicate, including sodalite, hauyne, nosean, and lazurite.

sodalithite An igneous rock in which sodalite is the only light-colored mineral present. The term was introduced by Ussing in 1911, for layered cumulus rocks of the Ilimaussaq complex, SW Greenland. Johannsen (1938, p. 346) ascribed the layering in the type example to an extrusive origin, but extrusive or intrusive occurrence should not enter the definition. Cf: *naujaite.*

sodalitite An *urtite* composed chiefly of sodalite, with smaller amounts of acmite, eudialyte, and alkali feldspar.

soda mica *paragonite.*

soda microcline A variety of microcline in which sodium replaces potassium; specif. *anorthoclase.*

soda minette An alkalic *minette* named by Brögger in 1898, containing alkali feldspar (specif. cryptoperthite), dark-brown mica, acmite, apatite, and sphene. This dike rock is more accurately called mafic *trachyte.* Not recommended usage.

soda niter A white or colorless transparent hexagonal mineral: $NaNO_3$. It is a deliquescent, soluble crystalline salt that occurs esp. in *caliche* in Chile, and that is associated with halite and sandy and clayey material. Soda niter is a source of nitrates. Cf: *niter.* Syn: *nitratine; Chile saltpeter; Peru saltpeter.*

soda orthoclase *loxoclase.*

soda prairie An extensive level barren tract of land covered with a whitish efflorescence of sodium carbonate (natron), as in parts of SW U.S. and Mexico. Syn: *salt prairie.*

soda sanidine A mineral of the alkali feldspar group, containing 40-60% albite in solid solution.

soda spar An informal term for sodic feldspar, i.e. *albite,* or for a feldspar mixture assaying at least 7% Na_2O (Rogers & Neal, 1975, p. 638). Syn: *Na-spar.* Cf: *potash spar.*

soda straw *tubular stalactite.*

soda tremolite A monoclinic mineral of the amphibole group: $Na_2CaMg_5Si_8O_{22}(OH)_2$. It differs from tremolite in having sodium in place of half of the calcium. Cf: *richterite.*

soddyite A pale-yellow orthorhombic mineral: $(UO_2)_5Si_2O_9 \cdot 6H_2O$. Also spelled: *soddite.*

sodic soil A *salt-affected soil* having 15% or more exchangeable sodium. Cf: *saline soil; saline sodic soil.* Obsolete syn: *alkali soil.*

sodium alum *soda alum.*

sodium autunite A yellow mineral of the autunite group: $Na(UO_2)(PO_4) \cdot 4H_2O$.

sodium betpakdalite A monoclinic mineral: $(Na,Ca)_3Fe_2{}^{+3}(As_2O_4)(MoO_4)_6 \cdot 15H_2O$.

sodium boltwoodite An orthorhombic mineral: $(H_3O)(Na,K)(UO_2)(SiO_4) \cdot H_2O$.

sodium-calcium feldspar A syn. of *plagioclase.* See also: *soda-lime feldspar; lime-soda feldspar.*

sodium feldspar An alkali feldspar containing the Ab molecule ($NaAlSi_3O_8$); specif. *albite.* See also: *soda feldspar.*

sodium illite *brammallite.*

sodium uranospinite A yellow-green to lemon and straw-yellow secondary mineral: $(Na_2,Ca)(UO_2)_2(AsO_4)_2 \cdot 5H_2O$.

sodium-zippeite An orthorhombic mineral: $Na_4(UO_2)_6(SO_4)_3(OH)_{10} \cdot 4H_2O$.

soengei *sungei.*

sofar (a) A method of determining the location of an underwater explosion from points on shore. (b) A sound channel in the deep ocean that propagates acoustic waves (*channel waves*) for long distances with little attenuation.—Etymol: an acronym; *so*und *f*ixing *a*nd *r*anging. Also spelled: SOFAR.

soffione Steam-type fumaroles; the term was originally applied to boric-acid fumaroles of the Tuscany region of Italy. Etymol: Italian *soffio,* "puff" or "blast".

soft coal (a) A syn. of *bituminous coal.* Cf: *hard coal.* (b) Outside the U.S., the term is sometimes used for brown coal and/or lignite, or for any coal with a calorific value of 5700 kcal/kg (10,260 BTU/lb) or less on a moist, mineral-matter-free basis.

softening Reduction of the hardness of water by removing hardness-forming ions (chiefly calcium and magnesium) by precipitation or ion exchange, or sequestering them as by combining them with substances such as certain phosphates, that form soluble but nonionized salts.

soft ground [eng] (a) Ground that is too moist or yielding to support weight and allows an object to sink in. (b) Soil that does not stand well and requires heavy timbering, such as that about an underground opening.

soft ground [mining] That part of an ore deposit that can be mined without drilling and blasting. It is usually the upper, weathered portion of the deposit.

soft hail *graupel.*

soft magnetization Magnetization that is easily destroyed; specifically, remanent magnetization with a small coercive force. Cf: *hard magnetization.*

soft mineral A mineral that is softer than quartz, i.e. ranking less than seven on the Mohs scale. Cf: *hard mineral.*

soft ore A term used in the Lake Superior region for an earthy, incoherent iron ore mainly composed of hematite or limonite (goethite) and containing 45-60% iron. Cf: *hard ore.*

soft rock (a) A term used loosely for sedimentary rock, as distinguished from igneous or metamorphic rock. (b) A rock that is relatively nonresistant to erosion. (c) Rock that can be removed by air-generated hammers, but cannot be handled economically by pick. (d) A term used loosely by drillers for a post-Cretaceous sedimentary rock that is drilled relatively rapidly and that produces samples difficult to classify as to exact depth.—Cf: *hard rock.*

soft-rock geology A colloquial term for geology of sedimentary rocks, as opposed to *hard-rock geology.*

soft shore A shore composed of peat, muck, mud, or soft marl, or of marsh vegetation. Ant: *hard shore.*

software Any program or method to be used in controlling *hardware* or equipment, e.g. program, routine, subroutines, etc.

soft water Water that lathers readily with ordinary soap; water containing not more than 60 mg/l of hardness-forming constituents expressed as $CaCO_3$ equivalent. Cf: *hard water; hardness [water].*

softwood The wood of a gymnospermous tree, lacking vessels and wood fibers. Actually, such wood may be either soft or relatively hard; it tends to be more uniform than the wood of angiosperms. Cf: *hardwood.*

sogdianite A mineral of the osumilite group: $(K,Na)_2Li_2(Li,Fe,Al)_2ZrSi_{12}O_{30}$.

soggendalite A dark-colored dolerite containing abundant pyroxene. Named by Kolderup in 1896, for Soggendal, Norway. Not recommended usage. Cf: *mimosite.*

sogrenite A black organic material containing uranium.

söhngeite A mineral: $Ga(OH)_3$.

soil [eng geol] All unconsolidated materials above bedrock. This is the meaning of the term as used by early geologists and in some recent geologic reports, and has been vigorously advocated by Legget (1967, 1973). It is the common usage among engineering geologists (see, for example, *compaction; soil mechanics*). In recent years the approx. syn. *regolith* has come into wide geological use.

soil [lake] The bed or bottom of a lake.

soil [lunar] *lunar regolith.*

soil [soil] (a) The natural medium for growth of land plants. (b) A term used in soil classification for the collection of natural earthy materials on the Earth's surface, in places modified or even made by man, containing living matter, and supporting or capable of supporting plants out-of-doors. The lower limit is normally the lower limit of biologic activity, which generally coincides with the common rooting of native perennial plants.—Etymol: Latin *solum,* "ground".

soil association Two or more soils, occurring together in a characteristic pattern in a given geographic area, that are distinguishable among themselves but that, on all but detailed soil maps, are grouped together because of their intricate areal distribution. See also: *catena; soil complex.*

soil atmosphere The part of ground air that is in the soil and is similar to the air of the atmosphere but depleted or enriched in certain constituents, such as carbon dioxide. Cf: *subsurface air; ground air.*

soil binder A grass or other plant capable of preventing soil erosion by forming dense mats or roots; also, in an engineering sense, a fine material such as clay that gives cohesiveness to a coarse aggregate such as sand and gravel.

soil blister *frost mound.*

soil caliche *caliche.*

soil category A class of taxonomically related soils, defined at the same level of generalization that includes all soils. In the USDA soil taxonomy there are six categories: order, suborder, great group, subgroup, family, series.

soil circle A term used loosely for any circular form of patterned ground developed on a soil surface, either sorted or nonsorted, and with or without vegetation. Syn: *earth circle.*

soil climate The moisture and temperature of a soil.

soil colloids The inorganic and organic matter in soils having very small particle size and a correspondingly large surface area per unit of mass.

soil complex A mapping unit used in detailed soil surveys where two or more soils are so intermixed geographically that they cannot be separated at the scale being used. Cf: *soil association.*

soil-cover complex (a) A group of similar areas in which the soils, slopes, plant litter, and vegetation cover have comparable physical characteristics (Chow, 1964, p. 22-51). (b) The combination of a specific soil and a specific vegetation cover, used as a parameter in estimating the runoff in a drainage basin (Chow, 1964, p. 21-11).

soil creep The gradual, steady downhill movement of soil and loose rock material on a slope that may be very gentle but is usually steep. Syn: *surficial creep.*

soil discharge The release of water from the soil by evaporation and transpiration. The water may have been derived from the soil or from the zone of saturation by way of the capillary fringe. Syn: *soil evaporation.* See also: *vadose-water discharge.*

soil erosion Detachment and movement of topsoil, or soil material from the upper part of the profile, by the action of wind or running water, esp. as a result of changes brought about by human activity (such as unsuitable or mismanaged agricultural methods). It includes rill erosion, gully erosion, sheet erosion, and wind erosion.

soil evaporation *soil discharge.*

soilfall A *debris fall* involving soil material.

soil family *family [soil].*

soil fertility The status of a soil with respect to the amount and availability to plants of elements necessary for plant growth (SSSA, 1970, p.7). It is not synonymous with *soil productivity.*

soil flow *solifluction.*

soil fluction *solifluction.*

soil formation *soil genesis.*

soil-formation factors The natural conditions and substances that interact to produce a soil: parent material, climate, plants and other organisms, topography, and time.

soil genesis (a) The mode of origin of the soil, with special reference to the processes of soil-forming factors responsible for the development of the solum from the parent material. (b) A division of soil science concerned with soil genesis.—Syn: *soil formation; pedogenesis.*

soil horizon A layer of a soil that is distinguishable from adjacent layers by characteristic physical properties such as structure, color, or texture, or by chemical composition, including content of organic matter or degree of acidity or alkalinity. Soil horizons are generally designated by a capital letter, with or without a numerical annotation, e.g. A horizon, A2 horizon. Syn: *horizon [soil]; soil zone.*

soil ice Any ice formed in situ during freezing of unconsolidated materials, where the moisture source was the contained moisture or that brought in by capillary action during the freezing process. *Segregated ice* and *pore ice* are the most common forms (Mackay, 1966, p. 61).

soil map A map showing the distribution of kinds of soil in relation to prominent physical and cultural features of the Earth's surface. Kinds of soil are expressed in terms of soil taxonomic units, such as series, or as phases of series. Maps showing single soil characteristics or qualities, such as slope, texture, depth, fertility, or erodibility are not soil maps.

soil mechanics The application of the principles of mechanics and hydraulics to engineering problems dealing with the behavior and nature of soils, sediments, and other unconsolidated accumulations of solid particles; the detailed and systematic study of the physical properties and utilization of soils, esp. in relation to highway and foundation engineering and to the study of other problems relating to soil stability.

soil moisture *soil water.*

soil-moisture tension *moisture tension.*

soil-moisture weathering Accelerated weathering of granite below an old soil line, often causing the steepening of margins of granitic inselbergs (Stone, 1967, p.249).

soil order A group of soils in the broadest category. For example, in the 1938 classification system, the three soil orders were zonal soil, intrazonal soil, and azonal soil. In the present USDA classification there are 10 orders, differentiated by the presence or absence of diagnostic horizons: Alfisols, Aridisols, Entisols, Histosols, Inceptisols, Mollisols, Oxisols, Spodosols, Ultisols, and Vertisols Orders are divided into lower categories. Cf: *soil category.*

soil patterns Obsolete syn. of *patterned ground.*

soil phase A subdivision of any taxonomic unit in any category of the natural system of soil classification that is based on any characteristic or combination of characteristics significant to the use and management of the soils. Most phases are subdivisions of soil series.

soil physics The organized body of knowledge concerned with the physical characteristics of soils and with the methods and instruments used in determining these characteristics.

soil polygon A group term for forms of polygonal ground developed on a soil surface, frequently in permafrost areas but also in regions where contraction occurs (as in playa lakes and deserts), and occurring with or without a stone border. Diameter: a few millimeters to many tens of meters. The term is misleading because soil need not be present.

soil productivity The capacity of a soil, in situ, to produce a specified plant or sequence of plants under a specified system of management (SSSA, 1975, p. 13). The term is not synonymous with *soil fertility.*

soil profile A vertical section of a soil that displays all its horizons.

soil reaction The degree of a soil's acidity or alkalinity, expressed by its pH value.

soil science The study of the properties, occurrence, and management of soil as a natural resource. Generally it includes the chemistry, microbiology, physics, morphology, and mineralogy of soils, as well as their genesis and classification. See also: *pedography.* Syn: *pedology.* Obs syn: *agrology.*

soil separate A group of rock and mineral particles in the soil, obtained in separation (as in mechanical analysis), having equivalent diameters less than 2 mm, and ranging between specified size limits (from "very coarse sand" to "clay"). Cf: *coarse fragment.* Syn: *separate.*

soil series The lowest category in soil classification, more specific than a soil family; a group of soils having genetic horizons of similar characteristics and arrangement in the soil profile, except for texture of the surface soil, and developed from a particular type of parent material. See also: *family [soil].* Syn: *series [soil].*

soil slip *debris slide.*

soil solution Soil water considered as a solution of various salts, organic compounds, gases, etc., that are of significance to plant growth or to the consequences of flushing the soil solution to a body of ground or surface water.

soil stabilization Chemical or mechanical treatment designed to increase or maintain the stability of a soil mass or otherwise to improve its engineering properties (ASCE, 1958, term 337), as by increasing its shear strength, reducing its compressibility, or decreasing its tendency to absorb water. Stabilization methods include physical compaction and treatment with cement, lime, and bitumen.

soil-stratigraphic unit A soil with physical features and stratigraphic relations that permit its consistent recognition and mapping as a stratigraphic unit. It is distinct from both rock-stratigraphic and pedologic units (ACSN, 1961, p. 654).

soil strip *soil stripe.*

soil stripe A *sorted stripe* whose texture is considerably finer than that of a *stone stripe.* Syn: *soil strip; earth stripe.*

soil structure [pat grd] A term formerly used (usually in the plural) for *patterned ground* (Sharp 1942, p.275), but now discarded because humus and a soil profile may be absent in patterned ground. Syn: *structure soil.*

soil structure [soil] The combination or aggregation of primary soil particles into aggregates or clusters (peds), which are separated from adjoining peds by surfaces of weakness. Soil structure is classified on the basis of size, shape, and distinctness into classes, types, and grades.

soil survey A general term for the systematic examination of soils in the field and in the laboratory, their description and classification, the mapping of kinds of soil, and the interpretation of soils for many uses, including suitability for growing various crops, grasses, and trees, or for engineering uses, and predicting their behavior under different management systems.

soil type A phase or subdivision of a soil series based primarily on texture of the surface soil to a depth at least equal to plow depth (about 15 cm). In Europe, the term is roughly equivalent to the term *great soil group.*

soil ulmin *humus.*

soil water Water in the *belt of soil water.* Syn: *soil moisture; rhizic water.*

soil-water belt *belt of soil water.*

soil-water zone *belt of soil water.*

soil zone *soil horizon.*

sol (a) A homogeneous suspension or dispersion of colloidal matter in a fluid (liquid or gas). (b) A completely mobile mud.—A sol is in a more fluid form than a *gel.*

sola Plural of *solum.*

solar attachment An auxiliary instrument which may be attached to a surveyor's transit or compass for determining the true meridian directly from the Sun. Cf: *solar compass.*

solar compass A surveying instrument that permits the establishment and surveying of the astronomic meridian or parallel directly by observation on the Sun. It has been replaced by the *solar attachment* in combination with a transit.

solar constant The rate at which solar radiant energy is received outside the atmosphere on a surface normal to the incident radiation, at the Earth's mean distance from the Sun. The value of the mean solar constant is 1.94 gram calories per minute per square centimeter. Cf: *insolation [meteorol].*

solar day The interval between successive passages of the sun over the meridian. It is slightly longer than the *sidereal day* because of the orbital motion of the Earth. A solar day is approx. 86,400 seconds. Cf: *sidereal day.*

solar infrared *reflected infrared.*

solar lake A lake that has no connection to the sea and whose water temperature and salinity increase with depth.

solar nebula The cloud of gases and dispersed solids from which the Sun, and other objects in the solar system, condensed and/or accreted.

solar salt Crystalline salt obtained by evaporating seawater or other brine by the heat of the sun.

solar system All the matter and interstitial space within the gravitational retention of the Sun.

solar tide The part of the tide caused solely by the tide-producing force of the Sun. Cf: *lunar tide.*

solar time Time based on the *solar day.* Cf: *sidereal time.*

solar wind The motion of interplanetary plasma or ionized particles away from the Sun and towards the Earth, near which it interacts with the Earth's magnetic field (McIntosh, 1963, p. 235).

sole (a) The under surface of a rock body or vein, esp. the bottom of a sedimentary stratum. (b) The fault plane underlying a *thrust sheet.* (c) The middle and lower parts of the shear surface of a landslide. (d) The lower part or basal ice of a glacier, which often contains rock fragments, appears dirty, and is separated from clean ice by an abrupt boundary.

sole cast A *sole mark* preserved as a swelling or positive feature on the underside of a bed immediately overlying a finer-grained bed containing a primary sedimentary structure (depression).

soled boulder A stone with blunted corners and smoothed or flattened (and sometimes striated) sides, esp. one shaped by glacial grinding.

sole fault A low-angle thrust fault forming the base of a thrust nappe; also, the basal main fault of an imbrication. Syn: *décollement fault; detachment fault; detachment thrust; basal thrust plane.*

sole injection An igneous intrusion that was emplaced along a thrust-fault plane.

sole mark A general descriptive term applied to a directional structure or to a small, wavelike, mainly convex irregularity or penetration found on the underside of a bed of sandstone or siltstone along its contact with a finer-grained layer such as shale. The term usually refers to a filling of a primary sedimentary structure, e.g. a crack, track, groove, or other depression, formed on the surface of the underlying mud by such agents as currents, organisms, and unequal loading, and preserved as a *sole cast* after the underlying material had consolidated and weathered away. Examples: *load cast; flute cast; groove cast.* Syn: *sole marking.*

Solenhofen stone A *lithographic limestone* of Late Jurassic age found at Solenhofen (Solnhofen), a village in Bavaria, West Germany. It is evenly and thinly stratified and contains little clay.

solfatara A type of *fumarole,* the gases of which are characteristically sulfurous. Cf: *solfataric stage.* Etymol: the Solfatara volcano, Italy.

solfatara field A group of solfataras (Rittmann, 1962, p. 10). Cf: *fumarole field.*

solfataric stage A late or decadent type of volcanic activity, characterized by the emission of sulfurous gases from the vent. See also: *solfatara.* Cf: *fumarolic stage.*

solid Sedimentary material that is in solution or suspension but when freed of solvent or suspending medium has the form and properties of a solid. The term is usually used in the plural; e.g. *dissolved solids.*

solid diffusion Diffusion of chemical species through a rock that remains essentially solid. Cf: *metamorphic diffusion.*

solid earth *lithosphere.*

solid flow Flow in a solid by rearrangement among or within the constituent particles. Cf: *liquid flow; viscous flow.*

solid geology A British term for *bedrock* geology.

solidification The process of becoming solid or hard; esp. the change from the liquid to the solid state on the cooling of a magma. The term *lithification* is more generally applied in the case of sedimentary rocks. See also: *consolidation.*

solidification index A chemical parameter of igneous rocks, equal to 100 X $MgO/(MgO+FeO+Fe_2O_3+Na_2O+K_2O)$. Its usual range is from about 40 (basalt) to 100 (rhyolite). It reflects the steady decrease (common to all trends) of MgO relative to total iron and to the alkalies throughout the greater part of fractionation. It appears to decrease at nearly the same rate as the amount of residual liquid in a crystallizing magma decreases; hence its name (Kuno, 1957). Abbrev: SI.

solid map A British term for a geological map showing the extent of *solid rock,* on the assumption that all surficial deposits, other than alluvium, are absent or removed (Nelson & Nelson, 1967, p.352). Cf: *drift map.*

solid rock A British term for *bed rock.*

solid solution A single crystalline phase that may be varied in composition within finite limits without the appearance of an additional phase. Syn: *mix-crystal; mixed crystal.*

solid-solution series *isomorphous series.*

solid stage That stage in the cooling of a magma during which the magma becomes completely solid.

solidus On a temperature-composition diagram, the locus of points in a system at temperatures above which solid and liquid are in equilibrium and below which the system is completely solid. In binary systems without solid solutions, it is a straight line; in binary systems with solid solutions, it is a curved line or a combination of straight and curved lines; in ternary systems, it is a flat plane or a curved surface.

soliflual Said of debris resulting from solifluction (Baulig, 1957, p.927). Syn: *solifluidal.*

solifluction The slow viscous downslope flow of waterlogged soil and other unsorted and saturated surficial material, normally at 0.5-5.0 cm/yr; esp. the flow occurring at high elevations in regions underlain by frozen ground (not necessarily permafrost) that acts as a downward barrier to water percolation, initiated by frost action and augmented by meltwater resulting from alternate freezing and thawing of snow and ground ice. The term was proposed by Andersson (1906, p.95-96) as "the slow flowing from higher to lower ground of masses of waste saturated with water", but as he did not state explicitly that it referred to flow over frozen ground, the term has been extended to include similar movement in temperate and tropical regions; also, it has been used as a syn. of *soil creep,* although solifluction is generally more rapid. It is preferable to restrict the term to slow soil movement in periglacial areas. Also spelled: *solifluxion.* Syn: *soil flow; soil fluction; sludging.*

solifluction lobe An isolated, tongue-shaped feature, up to 25 m wide and 150 m long, formed by more rapid solifluction on certain sections of a slope showing variations in gradient. It commonly has a steep front (15°-25°) and a relatively smooth upper surface. Syn: *solifluction tongue.*

solifluction mantle The unsorted, water-saturated, locally derived material moved downslope by solifluction. Syn: *flow earth.*

solifluction sediment A sediment that has resulted from solifluction.

solifluction sheet A broad deposit of *solifluction mantle,* occurring evenly across a wide slope.

solifluction slope A smooth curvilinear slope of 2° to 30°, produced by solifluction or along which solifluction occurs.

solifluction step The flattish area at the front of a small solifluction lobe; the tread of a small, turf-banked terrace usually restricted to immediately above timberline.

solifluction stream A narrow, laterally confined streamlike deposit of *solifluction mantle.*

solifluction stripe A form of *striped ground* associated with solifluction. The term was used by Washburn (1947, p.94) as a syn. of *nonsorted stripe,* but solifluction may also be associated with sorted stripes.

solifluction terrace A low terrace or bench formed by solifluction at the foot of a slope; it may have a lobate margin reflecting uneven movement.

solifluction tongue *solifluction lobe.*

solifluidal *soliflual.*

solifluxion *solifluction.*

soligenous Said of peat deposits whose moisture content depends on both rainfall and surface water. Cf: *ombrogenous; topogenous.*

solimixtion A term introduced by Rosauer (1957, p.65) for "a relatively homogeneous blending in the vertical plane of two different materials due to frost action with little or no macro-optical structures", as at the contact of two different layers in a *loess* profile.

soliqueous A term proposed by Leet & Leet (1965, p. 620) to describe a state of matter, e.g. that of the materials in the Earth's mantle, that is neither solid, liquid, nor gaseous, but a mixture of all three; it is maintained and controlled by pressure, and it has neither a crystalline structure nor the chilled-liquid molecular arrangements of a glass.

solitaire A single diamond or sometimes other gem, set alone; a diamond *nonpareil.*

solitary coral A coral that does not form part of a colony; an individual corallite (of a polyp) that exists unattached to other corallites. Cf: *colonial.* Syn: *simple coral; cup coral; horn coral.*

Solod soil *Soloth soil.*

Solonchak soil A great soil group in the 1938 classification system, an intrazonal, halomorphic group of soils with a high content of soluble salts, usually having a light color but no characteristic structure; a *saline soil.* It is developed under salt-tolerant vegetation, in conditions of a semi-arid or desert climate and poor drainage (USDA, 1938). These soils are now classified as *Orthids* and *Aquepts.* Cf: *Solonetz soil; Soloth soil.*

Solonetz soil A great soil group in the 1938 classification system, an intrazonal group of soils that is characterized by a hard dark B horizon with columnar structure. It is usually highly alkaline (USDA, 1938). These soils would now be classified as *Mollisols, Alfisols,* and *Aridisols* that have a natric horizon. Cf: *Solonchak soil.*

solongoite A monoclinic mineral: $Ca_2B_3O_4(OH)_4Cl$.

Soloth soil A great soil group in the 1938 classification system, an intrazonal, halomorphic group of soils that has developed from saline material; a degraded, desalinized, decalcified *Solonetz soil.* It has a brown, friable surface layer, below which is a light-colored, leached horizon and an underlying dark horizon (USDA, 1938). Most of these soils would now be classified as *Aqualfs, Aquolls, Borolls,* and *Xeralfs.* Cf: *Solonchak soil.* Also spelled: *Solod soil.*

solstitial tide A tide occurring near the times of the solstices, when the Sun reaches a maximum north or south declination.

soluan A *cutan* consisting of crystalline salts, such as carbonates, sulfates, and chlorides of calcium, magnesium, and sodium (Brewer, 1964, p.216).

solubility The equilibrium concentration of a solute in a solution saturated with respect to that solute at a given temperature and pressure.

solubility product A syn. of *dissociation constant* that refers to a very slightly soluble compound.

solum The upper part of a soil profile, including the A and B horizons, in which soil-forming processes occur. Plural: *sola.* Syn: *true soil.*

solusphere That zone of the Earth in which water solutions affect geologic, chemical and life processes.

solution A process of chemical weathering by which mineral and rock material passes into solution; e.g. removal of the calcium carbonate in limestone or chalk by carbonic acid derived from rainwater containing carbon dioxide acquired during its passage through the atmosphere. Syn: *dissolution.*

solution banding In roll-front uranium deposits, "concentric color bands smoothly cross-cutting a sand body in arcuate patterns" (Bailey & Childers, 1977, p. 52). Color banding is in tints of red, yellow, and rusty brown; it probably reflects slight changes in the chemical composition of the ground water during mineralization.

solution basin A shallow surface depression, either man-made or natural, produced by solution of surface material, or resulting from the settlement of a surface through the removal in solution of underlying material (such as salt or gypsum); specif. a "solution depression" in a karstic region.

solution bench A low bench produced on limestone coasts by the action of fresh water (Wentworth, 1939). Presence of such benches on many desert coasts requires broadening of the definition. Preferred term: *solution platform.*

solution breccia A *collapse breccia* formed where soluble material has been partly or wholly removed by solution, thereby allowing the overlying rock to settle and become fragmented; e.g. a breccia consisting of chert fragments gathered from a limestone whose carbonate material has been dissolved away. See also: *evaporite-solution breccia.* Syn: *ablation breccia.*

solution cave A cave formed in a soluble rock. Cf: *lava tube; sea* *cave; rockshelter.*

solution channel Tubular or planar channel formed by solution in carbonate-rock terranes, usually along joints and bedding planes. It is the main water carrier in carbonate rocks. Cf: *solution opening.*

solution cleavage *Spaced cleavage* formed by rock dissolution, as in limestones of northern Italy. It is a "stylolitic cleavage that can represent tens of percent shortening parallel to bedding" (Alvarez et al., 1978, p. 263).

solution collapse Abrupt collapse of nonsoluble strata due to the dissolution of soluble underlying rock. See also: *solution subsidence.*

solution corridor A straight trench about 3 m wide in a karst area. Syn: *karst corridor; karst street; corridor [karst]; bogaz; zanjón.* See also: *solution fissure; karst valley.*

solution depression A general term for solution basin occurring in a karst region.

solution facet A nearly plane face, commonly bounded by a narrow rim or raised edge, developed on an exposed pebble or boulder of a soluble rock such as limestone by progressive dissolution by rainwater.

solution fissure One of a series of vertical open cracks about 0.5 m wide dissolved along joints, separating *limestone pavement* into *clints.* Syn: *grike; cutter; kluftkarren* (pl.). See also: *solution corridor.*

solution grooves *karren.*

solution lake (a) A syn. of *karst pond.* (b) A lake occupying a basin formed by surface solution of bedrock.

solution load *dissolved load.*

solution mining (a) The in-place dissolution of water-soluble mineral components of an ore deposit by permitting a leaching solution, usually aqueous, to trickle downward through the fractured ore to collection galleries at depth. It is a type of *chemical mining.* (b) The mining of soluble rock material, esp. salt, from underground deposits by pumping water down wells into contact with the deposit and removing the *artificial brine* thus created.

solution morel A surface texture resembling that of a morel mushroom, produced on impure limestone mainly by dissolution and to a lesser extent by local redeposition in arid or semi-arid climates (Scott, 1947). See also: *morel basin.*

solution nip A horizontal cavity formed in soluble rock at the edge of a water body. Syn: *nip [speleo].*

solution opening (a) An opening produced by direct solution by water penetrating pre-existing interstices. (b) An opening resulting from the decomposition of less soluble rocks by water penetrating pre-existing interstices, followed by solution and removal of the decomposition products. (c) *solution channel.*

solution pan A shallow flat-floored basin with overhanging sides, formed by solution. It ranges from a few centimeters to several meters in diameter, and from a centimeter to a meter in depth. Syn: *panhole; etched pothole; tinajita; kamenitza.* Partial syn: *opferkessel.* See also: *morel basin.*

solution pendant *pendant [speleo].*

solution pipe A vertical cylindrical hole, formed by solution and often without surface expression, that is filled with detrital matter.

solution pit An indentation up to about 1 mm in diameter formed on a rock surface by solution.

solution plane A plane containing lines of chemical weakness in a crystal, along which solution tends to occur under certain physical circumstances, e.g. great pressure.

solution platform An intertidal surface on carbonate rocks, nearly horizontal but not abraded, produced primarily by solution but with contributions by intertidal weathering and biological erosion and deposition (Bloom, 1978, p. 448). See also: *solution bench.* Syn: *low-tide platform.*

solution ripple *scallop [speleo].*

solution sinkhole The most common type of *sinkhole,* which grows when closely spaced fissures underneath it enlarge and coalesce. See also: *collapse sinkhole.*

solution subsidence Gradual subsidence of nonsoluble strata due to the solution of underlying rock. See also: *solution collapse.*

solution transfer The process of *pressure solution* of detrital grains at points of contact, followed by chemical redeposition of the dissolved material on the less-strained parts of the grain surfaces. See also: *Riecke's principle.*

solution valley *karst valley.*

Solvan European stage: Middle Cambrian (above Caerfaian, below Menevian).

solvate A chemical compound consisting of a dissolved substance and its solvent, e.g. hydrated calcium sulfate.

solvation The chemical union of a dissolved substance and its dissolving liquid.

solvsbergite A fine-grained, holocrystalline, rarely porphyritic hypabyssal *trachyte,* composed chiefly of sodic feldspar and a smaller amount of potassium feldspar, sodic pyroxene or amphibole, and little or no quartz. The name, given by Brögger in 1894, is for Sölvsberget, Norway. Not recommended usage. Cf: *lindoite.*

solvus The curved line in a binary system, or the surface in a ternary system, that separates a field of homogeneous solid solution from a field of two or more phases that may form from the homogeneous one by exsolution. Cf: *hypersolvus; subsolvus.*

somal unit A stratigraphic unit that intertongues laterally with its neighbor; e.g. lithosome or biosome.

somite One of the longitudinal series of body segments into which many animals (such as articulates and vertebrates) are more or less distinctly divided; esp. the basic embryologic unit of segmentation of the body of an arthropod, approximately equivalent to the part of the body covered by a single exoskeletal "body ring" and bearing no more than one pair of limbs.

somma n. A circular or crescentic ridge that is steep on its inner side and represents the rim of an ancient volcanic crater or caldera. Etymol: Mt. Somma, the ancient crater rim that surrounds Vesuvius. Syn: *somma ring.*—adj. Said of a volcanic crater with a central cone surrounded by a somma.

sommaite A *leucite monzonite* occurring as ejected blocks in lavas. Olivine and clinopyroxene form phenocrysts in a groundmass of leucite, alkali feldspar, and mafites. Its name, given by Lacroix in 1905, is derived from Monte Somma, Italy. Not recommended usage. Cf: *ottajanite.*

somma ring *somma.*

sonar An acronym of *so*und *na*vigation and *r*anging, a method used in oceanography to study the ocean floor.

sondage A deep trench, often of restricted area, dug to investigate the stratigraphy of an archaeological site (Bray & Trump, 1970, p. 215).

sondalite A metamorphic rock consisting of cordierite, quartz, garnet, tourmaline, and kyanite (Holmes, 1928, p.213).

sonde The elongate cylindrical tool assembly used in a borehole to acquire a *well log.* It is 6 to 40 feet in length and 2 to 6 inches in diameter, and contains various energy-input devices and/or response sensors. The sonde is lowered into the borehole by a multiconductor cable, or *wire line.* Syn: *logging tool.*

song of the desert The booming or roaring sound made by a sounding sand on a desert. Syn: *voices of the desert.*

sonic depth-finder *echo sounder.*

sonic log An *acoustic log* showing the interval-transit time of compressional seismic waves in rocks near the well bore of a liquid-filled borehole. First used for seismic-velocity information, it is now used chiefly for estimating porosity and lithology by the empirical *Wyllie time-average equation.* Syn: *velocity log; continuous-velocity log.*

sonic wave *acoustic wave.*

sonobuoy A buoy, generally free-floating, that contains a hydrophone and a radio transmitter that broadcasts information picked up by the hydrophone. It is used in seismic refraction surveying.

sonolite A mineral of the humite group: $Mn_9(SiO_4)_4(OH,F)_2$.

sonoprobe A type of echo sounder that generates sound waves and records their reflections. It is used in subbottom profiling.

sonoraite A monoclinic mineral: $Fe^{+3}Te^{+4}O_3(OH)\cdot H_2O$.

Sonstadt solution A solution of mercuric iodide in potassium iodide that is used as a *heavy liquid;* its specific gravity is 3.2. Cf: *bromoform; Klein solution; Clerici solution.* Syn: *Thoulet solution.*

sooty chalcocite *sooty ore.*

sooty ore A black, pulverulent variety of chalcocite, digenite, or djurleite, generally found coating pyrite and the supergene sulfides of porphyry copper deposits. Syn: *sooty chalcocite.*

sorbyite A mineral: $Pb_{17}(Sb,As)_{22}S_{50}$.

sordawalite An obsolete syn. of *tachylyte.*

soredium In lichens, a mass of algal cells surrounded by fungus hyphae, extruded through the outer or upper cortex of a thallus.

sorensenite A mineral: $Na_4Be_2SnSi_6O_{16}(OH)_4$.

sorkedalite A feldspathoid-free *monzonite* or *monzodiorite* containing olivine, clinopyroxene, and antiperthitic andesine, resembling *kjelsasite* but with high titanium, iron, and phosphorus content as reflected by high percentages of opaque minerals and apatite. Brögger in 1933 derived the name from Sörkedal, Oslo district, Norway. Not recommended usage.

sorosilicate A class or structural type of *silicate* characterized by the linkage of two SiO_4 tetrahedra by the sharing of one oxygen, with a Si:O ratio of 2:7. An example of a sorosilicate is hemimorphite, $Zn_4Si_2O_7(OH)_2 \cdot H_2O$. Cf: *nesosilicate; cyclosilicate; inosilicate; phyllosilicate; tectosilicate.*

sorption water *pellicular water.*

sorted [part size] Said of an unconsolidated sediment or of a cemented detrital rock consisting of particles of essentially uniform size or of particles lying within the limits of a single grade; *graded.* See also: *well-sorted; moderately sorted; poorly sorted.*

sorted [pat grd] Said of a nongenetic group of *patterned ground* features displaying a border of stones (including boulders) commonly surrounding or alternating with fines (including sand, silt, and clay). Ant: *nonsorted.*

sorted bedding A type of *graded bedding* in which only one particle size is present at each horizon within the bed, the size decreasing upward. Not in common usage.

sorted circle A form of *patterned ground* "whose mesh is dominantly circular and has a sorted appearance commonly due to a border of stones surrounding finer material" (Washburn, 1956, p.827); developed singly or in groups. Diameter: a few centimeters to more than 10 m; the stone border may be 35 cm high and 8-12 cm wide. See also: *debris island; stone pit; plug [pat grd].* Syn: *stone circle; stone ring; stone wreath; rock wreath; frost circle.*

sorted crack A form of nearly horizontal *patterned ground* consisting of a concentration of boulders along a straight line, as in the Swedish Caledonides. Cf: *nonsorted crack.*

sorted net A form of *patterned ground* "whose mesh is intermediate between that of a sorted circle and a sorted polygon and has a sorted appearance commonly due to a border of stones surrounding finer material" (Washburn, 1956, p.830). Diameter: a few centimeters to 3 m.

sorted polygon A form of *patterned ground* "whose mesh is dominantly polygonal and has a sorted appearance commonly due to a border of stones surrounding finer material" (Washburn, 1956, p.831); never developed singly. Diameter: a few centimeters to 10 m. Syn: *stone polygon; stone ring; stone net; stone mesh.*

sorted step A form of *patterned ground* "with a steplike form and a sorted appearance due to a downslope border of stones embanking an area of finer material upslope" (Washburn, 1956, p.833); formed in groups, rarely if ever singly. Dimensions: 1-3 m wide; up to 8 m long in downslope direction. See also: *nonsorted step; stone garland; stone-banked terrace.*

sorted stripe One of the alternating bands of finer and coarser material comprising a form of *patterned ground* characterized by "a striped pattern and a sorted appearance due to parallel lines of stones and intervening strips of dominantly finer material oriented down the steepest available slope" (Washburn, 1956, p.836). It never forms singly, usually occurring as one of many evenly spaced, sometimes sinuous, stripes that often exceed 100 m in length on slopes as steep as 30°. An individual stripe may be a few centimeters to 2 m wide, with the intervening area two to five times wider. See also: *block stripe; soil stripe; stone stripe; striped ground.*

sorting (a) The dynamic process by which sedimentary particles having some particular characteristic (such as similarity of size, shape, or specific gravity) are naturally selected and separated from associated but dissimilar particles by the agents of transportation (esp. by the action of running water). (b) The result of sorting; the degree of similarity of sedimentary particles in a sediment. (c) A measure of sorting, or of the spread or range of the particle-size distribution on either side of an average.—Cf: *gradation [part size].*

sorting coefficient A *sorting index* developed by Trask (1932), being a numerical expression of the geometric spread of the central half of the particle-size distribution of a sediment, and defined as the square root of the ratio of the larger quartile, Q_1 (the diameter having 25% of the cumulative size-frequency distribution larger than itself), to the smaller quartile, Q_3 (the diameter having 75% of the cumulative size-frequency distribution larger than itself). It is indicative of the range of conditions present in the transporting fluid (velocities, turbulence, etc.) and to some extent indicative of the distances of transportation. A perfectly sorted sediment has a coefficient of 1.0; less perfectly sorted sediments have higher coefficients. The Trask sorting coefficient is not considered a useful particle-size measure and is no longer used by sedimentologists, who now mainly use *phi standard deviation.*

sorting index A measure of the degree of sorting or of uniformity of particle size in a sediment, usually based on the statistical spread of the frequency curve of particle sizes; e.g. *sorting coefficient* and *grading factor.*

sorus A cluster of sporangia, as on the leaf of a fern.

sosie A seismic method that employs a pseudo-random series of seismic impulses to generate seismic waves. The recorded data have to be correlated with this series to produce an interpretable result. See Barbier & Viallix, 1973, p. 673-683.

sótano In Mexico, a deep vertical shaft in a karst area that may or may not lead to a cave. Etymol: Spanish, "cellar". Syn: *vertical cave.*

sotch *sinkhole.*

soufriere A common name for a volcanic crater or area of solfataric activity, used especially in the West Indies and other French-speaking regions. Also spelled: *soufrière.*

sound [coast] (a) A relatively long, narrow waterway connecting two larger bodies of water (as a sea or lake with the ocean or another sea) or two parts of the same body, or an arm of the sea forming a channel between a mainland and an island; it is generally wider and more extensive than a *strait [coast].* (b) A long, large, rather broad inlet of the ocean, generally extending parallel to the coast; e.g. Long Island Sound between New England and Long Island, N.Y. (c) A lagoon along the SE coast of the U.S.; e.g. Pamlico Sound, N.C. (d) A long bay or arm of a lake; a stretch of water between the mainland and a long island in a lake.

sound [geophys] Elastic waves in which the direction of particle motion is longitudinal, i.e. parallel with the direction of propagation. The term is sometimes restricted to such waves in gases, particularly air, and in liquids, particularly water, but it is also applied to wave motion in solids. It is the type of wave motion most often used in reflection-seismic exploration.

sound channel That region in a column of water in which the velocity of sound is a minimum, in which the sound waves are trapped by increasing temperature above and increasing pressure below. Syn: *sofar.*

sounding [elect] Mapping of (nominally) horizontal interfaces by resistivity, induced polarization, or electromagnetics. It usually involves variation of electrode interval for resistivity and induced polarization sounding, but may involve variation of either frequency or coil separation in electromagnetics. See also: *parametric sounding; geometric sounding.* Cf: *profiling.*

sounding [eng] Measuring the thickness of soil or depth to bedrock by driving a pointed steel rod into the ground or by using a penetrometer.

sounding [geophys] A determination of how some quantity varies with depth.

sounding [oceanog] The measurement of water depth taken from ship by either an *echo sounder* or a *lead line.*

sounding line *lead line.*

sounding sand Sand, usually clean and dry, that emits a musical, humming, or crunching sound when disturbed, such as a desert sand when sliding down the slip face of a dune or a beach sand when it is stirred or walked over. Examples: *musical sand; booming sand; roaring sand; whistling sand.* Syn: *singing sand.*

sound ranging The location of a source of seismic energy by acoustic trilateration, i.e. by recording signals on receivers at known positions.

sound wave *acoustic wave.*

sour Said of crude oil or natural gas containing significant fractions of sulfur compounds. Cf: *sweet.*

source [seis] *seismic source.*

source [streams] (a) The point of origin of a stream of water; the point at which a river rises or begins to flow. Syn: *fountain.* (b) A *headwater,* or one of the headwaters, of a stream; e.g. a fountainhead.

source area *provenance.*

source-bed concept The theory of sulfide ore genesis that postulates an original syngenetic deposition of sulfides, and their later migration and concentration, due, for example, to a rise in temperature of the rock.

source bias An effect in which azimuthally dependent departures from standard traveltimes in the upper mantle beneath the hypocenter result in consistent errors in the estimated epicenter (Herrin & Taggart, 1968).

source-bordering lee dune *umbrafon dune.*

sourceland *provenance.*

sourceline In computer modelling of hydrothermal systems, a line formed by connecting the starting points of all fluid packets whose pathlines pass through a given fixed point or small volume in an area of interest in a hydrothermal system (Norton, 1978, p. 23).

source link A magnitude-1 link that joins another magnitude-1 link at its downstream fork (Mock, 1971, p. 1558). Symbol: S. Cf: *link; magnitude.*

source-receiver product In seismic prospecting, the product of the number of detectors per trace and the number of sources used simultaneously.

source region That extensive area of the Earth's surface over which an *air mass* develops and acquires its distinctive characteristics.

sourceregion The region from which all fluids circulating through a pluton during some time interval were derived (Norton, 1978, p. 25). Cf: *sourceline.*

source rock [petroleum] Sedimentary rock in which organic material under pressure, heat, and time was transformed to liquid or gaseous hydrocarbons. Source rock is usually shale or limestone.

source rock [sed] *parent rock.*

source-rock index A term used by Dapples et al. (1953, p. 297 & 304) to indicate the extent of contribution to a sandstone of fragments from igneous and metamorphic rocks by measuring the degree of mixing of "arkose" and "graywacke" types. It is expressed as the ratio of sodic and potassic feldspar (arkose tendency) to the total of assorted rock fragments plus matrix of clay and micas (graywacke tendency). The index indicates source-rock types regardless of subsequent depositional history or diagenesis. Values greater than 3 indicate arkose; values less than 0.75 indicate graywacke.

South African ruby *Cape ruby.*

south geographic pole *south pole.*

southing A *latitude difference* measured toward the south from the last preceding point of reckoning; e.g. a linear distance southward from an east-west reference line.

south pole [astron] The south *celestial pole* representing the zenith of the heavens as viewed from the south geographic pole.

south pole [geog] The *geographic pole* in the southern hemisphere of the Earth at lat. 90°S, representing the southernmost point of the Earth or the southern extremity of its axis of rotation. Also spelled *South Pole.* Syn: *south geographic pole.*

souzalite A green mineral: $(Mg,Fe)_3(Al,Fe)_4(PO_4)_4(OH)_6 \cdot 2H_2O$.

sövite A *carbonatite* that contains calcite as a dominant phase. The name, given by Brögger in 1921, is for Söve, Fen complex, Norway.

sowback A long, low hill or ridge shaped like the back of a female pig; e.g. *hogback; horseback; drumlin.*

sowneck A narrow divide between two expanses of lowland, or a narrow boundary between two bodies of water, formed by a gentle rise of ground.

SP *spontaneous potential.*

sp. *species.*

spa (a) A *medicinal spring.* (b) A place where such springs occur, often a resort area or hotel.—The name is derived from that of a town in eastern Belgium where medicinal springs occur.

spaced cleavage In schist, the spacing or separation of cleavage planes from a few millimeters to a microscopic scale, e.g. *slip cleavage.* It is a *nonpenetrative* texture. Cf: *continuous cleavage.*

space geology *astrogeology.*

space group A set of symmetry operations of the indefinite repetition of motif in space; there are 230 of them. Cf: *plane group.*

space lattice *crystal lattice.*

space-time unit A stratigraphic unit whose lateral limits are determined by geographic coordinates and whose vertical extent is measured in terms of geologic time (Wheeler, 1958).

spad An iron, brass, or tin nail, up to 5 cm long, with a hook or eye at the head for suspending a plumb line, used to mark an underground survey station (as in a mine or tunnel).

spadaite A mineral: $MgSiO_2(OH)_2 \cdot H_2O$ (?).

spall (a) A chip or fragment removed from a rock surface by weathering; esp. a small, relatively thin curved piece of rock produced by exfoliation. (b) A similar rock fragment produced by chipping with a hammer; e.g. a piece of ore broken by spalling.

spallation The ejection of atomic particles from a nucleus following the collision of an atom and a high-energy particle (e.g. a cosmic ray), which results in the formation of a different isotope that is not a *fission [isotope]* product.

spalling (a) The chipping, fracturing, or fragmentation, and the upward and outward heaving, of rock caused by the interaction of a shock (compressional) wave at a free surface. (b) *exfoliation.*

spalmandite A garnet intermediate in chemical composition between spessartine and almandine; a variety of spessartine rich in iron.

span [sed struc] A term introduced by Allen (1968, p. 63) for the continuous length of the crest of a ripple mark, measured at right angles to the observed or inferred flow direction.

span [stratig] (a) The length of a time interval. (b) An informal designation for a local geologic-time unit.

spandite A garnet intermediate in chemical composition between spessartine and andradite; a variety of spessartine rich in calcium and iron.

spangolite A dark-green hexagonal mineral: $Cu_6Al(SO_4)(OH)_{12}Cl \cdot 3H_2O$.

Spanish chalk A variety of steatite from the Aragon region of Spain.

Spanish topaz (a) Any orange, orange-brown, or orange-red variety of quartz resembling the color of topaz; e.g. heat-treated amethyst. (b) A wine-colored or brownish-red citrine occurring in Spain.

spar [mineral] A term loosely applied to any transparent or translucent light-colored crystalline mineral, usually readily cleavable and somewhat lustrous, esp. one occurring as a valuable nonmetallic mineral; e.g. Iceland spar (calcite), fluorspar (fluorite), heavy spar (barite), or feldspar. Obsolete syn: *spath.*

spar [mining] A miner's term for a small *clay vein* in a coal seam.

sparagmite A collective term for the late Precambrian fragmental rocks of Scandinavia, esp. the feldspathic sandstones of the Swedish Jotnian, consisting mainly of coarse arkoses and subarkoses (characterized by high proportions of microcline), together with polygenetic conglomerates and graywackes. Etymol: Greek *sparagma,* "fragment, thing torn, piece".

sparite (a) A descriptive term for the crystalline transparent or translucent interstitial component of limestone, consisting of clean, relatively coarse-grained calcite or aragonite that either accumulated during deposition or was introduced later as a cement. It is more coarsely crystalline than *micrite,* the grains having diameters that exceed 10 microns (Folk, 1959) or 20 microns (Chilingar et al., 1967, p. 320). Syn: *sparry calcite; calcsparite.* (b) A limestone in which the sparite cement is more abundant than the micrite matrix. Syn: *sparry limestone.*

sparker A marine seismic-energy source employing a high-voltage electrical discharge underwater.

spark spectrum The spectrum of light emitted by a substance, usually a gas or vapor, when an electric spark is passed through it. The spectrum is representative of the ionized atoms. Cf: *arc spectrum.*

Sparnacian European stage: upper Upper Paleocene (above Thanetian, below Ypresian of Eocene or partially equivalent to the lower Ypresian).

sparry (a) Pertaining to, resembling, or consisting of *spar;* e.g. "sparry vein" or "sparry luster". (b) Pertaining to *sparite,* esp. in allusion to the relative clarity, both in thin section and hand specimen, of the calcite cement; abounding with sparite, such as a "sparry rock".

sparry calcite Clean, coarse-grained calcite crystal; *sparite.*

sparry iron *siderite.*

sparry limestone (a) *sparite.* (b) A coarsely crystalline *marble.*

sparse biomicrite A *biomicrite* in which the skeletal grains make up 10-50% of the rock. Cf: *packed biomicrite.*

spartalite *zincite.*

spasmodic turbidity current A single, rapidly developed turbidity current, such as one initiated by a submarine earthquake. Cf: *steady turbidity current.*

spastolith A deformed oolith; e.g. a chamositic oolith that has been closely twisted or misshapen due to its soft condition at the time of burial (Pettijohn, 1957, p. 97).

spate (a) A sudden flood on a river, caused by heavy rains or rapidly melting snow higher up the valley; a *freshet.* (b) A Scottish term for a flood.

spath An obsolete syn. of *spar [mineral].*

Spathian European stage: Lower Triassic (above Smithian, below Anisian).

spathic Resembling spar, esp. in regard to having good cleavage.

Syn: *spathose.*

spathic iron Ferrous-carbonate mineral with good rhombohedral cleavage; specif. *siderite.* Syn: *spathic iron ore; spathose iron.*

spathite A *tubular stalactite,* usually composed of aragonite, which consists of a vertical series of segments that flare downward.

spathization Widely distributed crystallization of sparry carbonates such as calcite and dolomite (Sander, 1951, p. 3); development of relatively large sparry crystals that have good cleavage.

spathose *spathic.*

spatial dendrite A type of snow crystal somewhat like a *stellar crystal* except that branched arms form an irregular three-dimensional structure instead of building a pattern of hexagonal symmetry in a single plane.

spatial frequency The number of wave cycles per unit of distance in a given direction, often the direction of the seismic *spread.* Syn: *wave number.*

spatial sediment concentration The sediment contained in a unit volume of flow used to measure *transport concentration.*

spatiography A science that deals with space beyond the Earth's atmosphere; esp. the description of the physical characteristics of the Moon and the planets (Webster, 1967, p. 2184). The term is obsolete.

spatium A localized widening of an axial canal of a crinoid columnal opposite interarticular sutures. Pl: *spatia.*

spatter [meteorite] Droplets on the surface of meteorites, often partly fused with the crust.

spatter [pyroclast] An accumulation of very fluid pyroclasts, coating the surface around a vent. Syn: *driblet.*

spatter cone A low, steep-sided cone of *spatter* built up on a fissure or vent; it is usually of basaltic material. Syn: *volcanello; agglutinate cone.*

spatulate Spoon-shaped; oblong with an attenuated base.

spavin An English term for a hard, unstratified, sandy clay or mudstone underlying a coal seam; *seat earth.*

SP curve *spontaneous-potential curve.*

speaking rod *self-reading leveling rod.*

spear The recurved part of a *hook* of a holothurian.

spear pyrites A form of *marcasite* in twin crystals showing reentrant angles that resemble the head of a spear. Cf: *cockscomb pyrites.*

special creation The theory, strongly supported before the theory of evolution was generally accepted, that each species of organism inhabiting the Earth was created fully formed and perfect by some divine process.

special erosion Erosion effected by agents (such as wind, waves, and glaciers) that are important only within strictly limited areas or that work with help from the agents of *normal erosion* (Cotton, 1958, p. 38). "The modern tendency ... is to regard any distinction between 'normal' and 'special' agencies as unreal" (Stamp, 1961, p. 340).

specialist species *equilibrium species.*

special-purpose map Any map designed primarily to meet specific requirements, and usually omitting or subordinating nonessential or less important information. Cf: *general-purpose map.*

speciation (a) The production of new species of organisms from pre-existing ones during evolution. (b) The sorting of a collection of many fossil or living specimens into groups, each of which represents one species.

species [mineral] A mineral distinguished from others by its unique chemical and physical properties; it may have *varieties.*

species [paleont] A group of organisms, either plant or animal, that may interbreed and produce fertile offspring having similar structure, habits, and functions. As a fundamental unit in the hierarchy of classification, species ranks next below *genus.* The name of a species is a *binomen;* e.g. *Nuculana diversa.* Adj: *specific.* Abbrev: *sp.* Plural: *species.* Cf: *subspecies; epithet.*

species gap zone A zone designated by the gap between the end of the range of one zonal marker species and the commencement of the range of a second zonal species (McLean, 1968, p. 10).

species index zone A zone designated by the total range of a designated species (McLean, 1968, p. 10).

species overlap zone "A zone designated by the interval of range overlaps of two or more species whose ranges overlap in a restricted interval" (McLean, 1968, p. 10).

species postlap zone "A zone designated by the part of the range of the marker species occurring *after* the cutoff of range of another marker species" (McLean, 1968, p. 10).

species prelap zone "A zone designated by that portion of a marker species' range prior to the commencement of a marker which ends the zone by definition" (McLean, 1968, p. 10).

species zone The *taxon-range-zone* of a species (ISG, 1976, p. 54).

specific (a) The adj. of *species.* (b) Precise or limiting.

specific absorption The capacity of water-bearing material to absorb liquid, after removal of free water; the ratio of the volume of water absorbed to the volume of the saturated material. It is equal to *specific yield* except when the water-bearing material has been compacted due to the weight of overlying rocks.

specific activity (a) The *activity* of a radioactive isotope, measured per unit weight of the element in the sample. (b) The activity per unit weight of a sample of radioactive material. (c) The activity per unit mass of a pure radionuclide.

specific capacity The rate of discharge of a water well per unit of drawdown, commonly expressed in gallons per minute per foot. It varies slowly with duration of discharge. If the specific capacity is constant except for the time variation, it is proportional to the hydraulic diffusivity of the aquifer.

specific character A particular characteristic that serves to distinguish one species from others.

specific conductivity With reference to the movement of water in soil, a factor expressing the volume of transported water per unit of time in a given area.

specific discharge *Discharge [hydraul]* per unit area. It is often used to define the magnitude of a flood.

specific energy The energy of water in a stream; it equals the mean depth plus velocity head of the mean velocity (ASCE, 1962).

specific-gravity liquid *heavy liquid.*

specific head The height of the *energy line* above the bed of a conduit (ASCE, 1962).

specific humidity The mass of water vapor per unit mass of moist air, usually expressed in g/g or g/kg. Cf: *absolute humidity; relative humidity.*

specific magnetization Magnetic moment per unit mass; magnetization divided by density.

specific mineral *essential mineral.*

specific name (a) The second term of a *binomen.* Cf: *epithet.* (b) A less preferred syn. of binomen.

specific permeability A factor expressing the permeability of a stream bed; it equals a constant times the square of representative pore diameter. Symbol: k. Syn: *intrinsic permeability.*

specific refractivity The *refractivity* of a substance divided by its density.

specific retention The ratio of the volume of water that a given body of rock or soil will hold against the pull of gravity to the volume of the body itself. It is usually expressed as a percentage. Cf: *field capacity.*

specific rotation The angle of rotation of plane-polarized light passing through a substance, measured in degrees per decimeter for liquids and solutions and in degrees per millimeter for solids.

specific seismicity The square root of the energy, per unit area per unit time, released by the earthquakes of a given region.

specific susceptibility Susceptibility divided by density; the ratio of specific induced magnetization to the strength H of the magnetic field causing the magnetization. Syn: *mass susceptibility.*

specific tenacity The ratio of a material's tensile strength to its density.

specific yield The ratio of the volume of water that a given mass of saturated rock or soil will yield by gravity to the volume of that mass. This ratio is stated as a percentage. Cf: *effective porosity; storage coefficient; specific absorption.*

specimen A sample, as of a fossil, rock, or ore; cf: *hand specimen.* Among miners, it is often restricted to selected or handsome samples, such as fine pieces of ore, crystals, or fragments of quartz showing visible gold.

specimen ore A particularly rich or well crystallized orebody.

speckstone An early name for talc or steatite. Etymol: German *Speckstein,* "bacon stone", alluding to its greasy feel. See also: *bacon stone.*

spectacle stone *selenite.*

spectra Pl. of *spectrum.*

spectral Pertaining to a *spectrum [phys],* e.g. *spectral line.*

spectral absorptance *Absorptance* measured at a specified wavelength. Cf: *total absorptance.*

spectral emittance *Emittance* measured at a specific wavelength.

spectral gamma-ray log The *radioactivity log* curves of the intensity of natural gamma radiation within discrete energy bands characteristic of specific radioactive series (uranium-radium, thorium) or isotopes (potassium-40). It is used in correlation where other criteria fail; also in uranium exploration where thorium or potassium minerals contribute significantly to total gamma radiation. See also: *gamma-ray log; neutron-activation log.*

spectral line One component line in the array of intensity values of a spectrum emitted by a source. See also: *principal line [geochem].*

spectral log *spectral gamma-ray log.*

spectral radiance Radiance per unit wavelength interval at a particular wavelength. Symbol: N_λ.

spectral radiant flux *Radiant flux* measured at a specific wavelength.

spectral reflectance *Reflectance* measured at a specific wavelength.

spectrochemical analysis Chemical analysis based on the spectral characteristics of substances. Syn: *spectrum analysis.*

spectrochemistry The branch of chemistry concerned with *spectrochemical analysis.*

spectrocolorimeter Essentially an absorption *spectrophotometer,* used to measure the absorptance of solutions over an entire spectrum, providing quantitative information about the composition of the solution. It is often used as a syn. of spectrophotometer. See also: *colorimeter.*

spectrocolorimetry The art or process of using the *spectrocolorimeter* for the quantitative study of color.

spectrogram A map, photograph, or other picture of a spectrum, usually produced by a *spectrograph.*

spectrograph A *spectroscope* designed to map or photograph a spectrum.

spectrography The art or process of using a *spectrograph* to photograph or map a spectrum.

spectrometer A device for measuring intensity of radiation as a function of wavelength.

spectrometry The art or process of using a *spectrometer.*

spectrophotometer A *photometer* for measuring and comparing the intensity of light in different parts of a spectrum as a function of wavelength. Common usage assigns this term mainly to analytical instruments that measure the characteristic absorption spectra of chemicals. See also: *flame photometer; spectrocolorimeter.*

spectrophotometry The art or process of using a *spectrophotometer* to measure the intensity of light in different parts of a spectrum as a function of wavelength.

spectroreflectometer An instrument for measuring and analyzing the *reflection spectrum* of a source.

spectroscope An instrument for producing and visually observing a spectrum.

spectroscopy The production and observation of a spectrum and all methods of recording and measuring, including the use of the *spectroscope,* that go with them.

spectrum [palyn] *pollen spectrum.*

spectrum [phys] n. (a) An array of visible light ordered according to its constituent wavelengths (colors) by being sent through a prism or diffraction grating. (b) An array of intensity values ordered according to any physical parameter, e.g. energy spectrum, mass spectrum, velocity spectrum. (c) Amplitude and phase response as a function of frequency for the components of a wavetrain, such as given by Fourier analysis, or as used to specify filter-response characteristics. Pl: *spectra.*—Adj: *spectral.*

spectrum analysis *spectrochemical analysis.*

specular coal *pitch coal.*

specular iron A syn. of *specularite.* Also called: *specular iron ore.*

specularite A black or gray variety of hematite with a splendent metallic luster, often showing iridescence. It occurs in micaceous or foliated masses, or in tabular or disklike crystals. Syn: *specular iron; gray hematite; iron glance.*

specular schist Metamorphosed *oxide facies iron formation* characterized by a high percentage of strongly aligned flakes of specular hematite.

specular surface A surface that is smooth with respect to the wavelength incident upon it.

speculative resources Undiscovered mineral resources that may occur either in known types of deposit in a favorable geologic setting where no discoveries have yet been made, or in as-yet-unknown types of deposit that remain to be recognized (Brobst & Pratt, 1974, p. 2). Cf: *hypothetical resources; identified resources.*

speed In photography, the response or sensitivity of a photographic film, plate, or paper to light; also, the light-gathering power of a lens or optical system, expressed as the *relative aperture* of the lens. See also: *f-number.*

speisscobalt *smaltite.*

spelean Said of or pertaining to a cave.

speleochronology The dating or chronology of a cave's formation, or of its mineral deposits or filling. The dating may be either relative or absolute.

speleogen In a cave, any surface that is formed by solution, such as a scallop, pendant, or domepit. Etymol: Greek,, "cave born".

speleogenesis The process of cave formation.

speleologist A scientist engaged in *speleology.* Nonrecommended syn: *speologist.* See also: *caver.*

speleology The exploration and scientific study of caves, both physical and biological, including geologic studies of their genesis, morphology, and mineralogy. The term was first published by Martel in 1896. See also: *caving [speleo]; speleologist.* Nonrecommended syn: *speology.*

speleothem Any secondary mineral deposit that is formed in a cave by the action of water. Syn: *cave formation; formation [speleo].* See also: *cave onyx; dripstone.* Etymol: Greek, "cave deposit".

spelunker *caver.*

spelunking *caving [speleo].*

spencerite (a) A pearly-white monoclinic mineral: $Zn_4(PO_4)_2(OH)_2 \cdot 3H_2O$. (b) An artificial substance: $(Fe,Mn)_3(C,Si)$.

spencite *tritomite-(Y).*

speologist A nonrecommended syn. of *speleologist.*

speology A nonrecommended syn. of *speleology,* derived from a less common French form *spéologie.*

spergenite A name proposed by Pettijohn (1949, p. 179 & 301) for a biocalcarenite that contains ooliths and fossil debris (such as bryozoan and foraminiferal fragments) and that has a quartz content not exceeding 10%. Type locality: Spergen Hill, situated a few miles SE of Salem, Ind., where the Salem Limestone (formerly the Spergen Limestone) is found. Syn: *Bedford limestone; Indiana limestone.*

spermatophyte A vascular plant that produces seeds, e.g. a gymnosperm or angiosperm; a *seed plant.* Such plants range from the Carboniferous. Cf: *pteridophyte; phanerogam.*

spermatozoid A motile male gamete characteristically produced in an antheridium by the gametophyte generation in pteridophytes. Fusion of the spermatozoid with an egg cell completes the fertilization process.

sperone A vesicular *leucitite* that contains small melanite crystals. Obsolete.

sperrylite A tin-white isometric mineral: $PtAs_2$.

spessartine The manganese-aluminum end-member of the garnet group: $Mn_3Al_2(SiO_4)_3$. It has a brown-red to yellow-brown color, and usually contains some iron, magnesium, and other elements in minor amounts. Spessartine is rather rare; it occurs in pegmatites and granites. Syn: *spessartite [mineral].*

spessartite [mineral] *spessartine.*

spessartite [rock] A *lamprophyre* composed of phenocrysts of green hornblende or clinopyroxene in a groundmass of sodic plagioclase, with accessory olivine, biotite, apatite, and opaque oxides. Rosenbusch in 1896 derived the name from Spessart, Germany.

spew frost *pipkrake.*

sphaeraster A many-rayed sponge spicule (euaster) in which the rays radiate from a prominent solid spherical centrum. Also spelled: *spheraster.*

sphaerite A light-gray or bluish mineral consisting of hydrous aluminum phosphate in globular concretions. It is perhaps the same as *variscite.* Also spelled: *spherite.*

sphaeroclone An *ennomoclone* in which six or more proximal arms radiate from one side of a frequently spherical centrum and terminate in cuplike zygomes.

sphaerocobaltite *spherocobaltite.*

sphaerocone A coiled, depressed, involute, globular cephalopod shell that has a small nearly occluded umbilicus and a round venter, and that commonly opens out suddenly along the last whorl (as in *Sphaeroceras*).

sphaerolite *spherulite [petrology].*

sphaerolitic *spherulitic.*

sphaeroplast A lorica-forming granule representing part of the shieldlike mass formed during reproduction in tintinnids when a single cell divides into two theoretically equal parts.

sphaerosiderite *spherosiderite.*

sphagnum atoll An *atoll moor* containing sphagnum.
sphagnum bog An acid freshwater bog containing abundant sphagnum, which may ultimately form a deposit of sphagnum peat (highmoor peat). See also: *balsam bog.*
sphagnum moss A moss of the genus *Sphagnum,* often forming peat; *peat moss.*
sphagnum peat *highmoor peat.*
sphalerite A yellow, brown, or black isometric mineral: (Zn,Fe)S. It is dimorphous with wurtzite, and often contains manganese, arsenic, cadmium, and other elements. Sphalerite has a highly perfect dodecahedral cleavage and a resinous to adamantine luster. It is a widely distributed ore of zinc, commonly associated with galena in veins and other deposits. Syn: *blende; zinc blende; jack; blackjack; steel jack; false galena; pseudogalena; mock ore; mock lead.*
Sphärokrystal A spherulite composed of a single mineral species.
Sphenacodontia A suborder of pelycosaurs characterized by adaptation to a predaceous habit. Range, Upper Pennsylvanian to ?Upper Permian.
sphene A usually yellow or brown mineral: $CaTiSiO_5$. It often contains other elements such as niobium, chromium, fluorine, sodium, iron, manganese, and yttrium. Sphene occurs in wedge-shaped or lozenge-shaped monoclinic crystals as an accessory mineral in granitic rocks and in calcium-rich metamorphic rocks. Syn: *titanite; grothite.*
sphenitite A sphene-rich *jacupirangite.* The term was introduced by Allen in 1914. Syn: *sphenite.* Not recommended usage.
sphenochasm A triangular gap of oceanic crust separating two continental blocks and converging to a point; it is interpreted by Carey (1958) as having originated by the rotation of one of the blocks with respect to the other. Cf: *rhombochasm.*
sphenoconformity A term used by Crosby (1912, p.297) for the relation between conformable strata that are thinner in one locality than in the other, though fully represented in both.
sphenoid An open crystal form having two nonparallel faces that are symmetrical to an axis of twofold symmetry. It occurs in monoclinic crystals of the sphenoidal class. Cf: *dome [cryst]; disphenoid.*
sphenoidal class That crystal class in the monoclinic system having symmetry 2.
sphenolith [intrus rocks] A wedgelike igneous intrusion, partly concordant and partly discordant.
sphenolith [paleont] A coccolith having a prismatic base formed by radial elements surmounted by a cone.
sphenopsid n. A pteridophyte of the class Articulatae (sphenopsida), characterized by distinctly jointed stems with whorled leaves and sporangia. Sphenopsids range from the Middle Devonian; the only living members of the group are assigned to the genus *Equisetum.* Syn: *horsetail; scouring rush.*
spheraster *sphaeraster.*
sphere A standard shape taken as a reference form in the analysis of sedimentary-particle shapes; the limiting shape assumed by many rock and mineral fragments upon prolonged abrasion, being a solid figure bounded by a uniformly curved surface, any point on which is equidistant from the center. It has the least surface area for a given volume and the greatest settling velocity of any possible shape (under conditions of low velocity and of constant volume and density). See also: *spheroid [sed].*
sphere ore *cockade ore.*
spheric *sferic.*
spherical bomb *spheroidal bomb.*
spherical cap In gravimetry, part of a spherical shell limited by a circular cone with the apex in the center of the sphere.
spherical coordinates (a) Three coordinates that represent a distance and two angles in space, consisting of two polar coordinates in a plane and the angle between this plane and a fixed plane containing the primary axis of direction (polar axis). The term includes coordinates on any spherical surface or on any surface approximating a sphere (such as the surface of the Earth). (b) A system of polar coordinates in which the origin is the center of a sphere and the points all lie on the surface of the sphere; a coordinate system used to define the position of a point with reference to two great circles that form a pair of axes (such as longitude and latitude) at right angles to each other or with reference to an origin and a great circle through the point.
spherical divergence The decrease in amplitude of seismic body waves with distance from the source. The amplitude varies inversely as the square of the distance. Cf: *cylindrical divergence.*
spherical triangle A *triangle* on the surface of a sphere, having sides that are arcs of three great circles.
spherical wave A seismic wave propagated from a point source whose front surfaces are concentric spheres (U.S. Naval Oceanographic Office, 1966, p.154).
spherical weathering *spheroidal weathering.*
sphericity The relation to each other of the various diameters (length, width, thickness) of a particle; specif. the degree to which the shape of a sedimentary particle approaches that of a sphere. True sphericity, as originally defined by Wadell (1932), is the ratio of the surface area of a sphere of the same volume as the particle to the actual surface area of the particle. Due to the difficulty of determining the actual surface area and volume of irregular solids, Wadell (1934) developed an operational definition expressed as the ratio of the true nominal diameter of the particle to the diameter of a circumscribing sphere (generally the longest diameter of the particle). A perfect sphere has a sphericity of 1.0; all other objects have values less than 1.0. Many other meanings of sphericity have been proposed. Not to be confused with *roundness* or *angularity.*
spherite [mineral] *sphaerite.*
spherite [sed] (a) A sedimentary rock composed of gravel-sized aggregates of constructional (nonclastic) origin, simulating in texture a rudite of clastic origin; e.g. a rock formed of volcanic bombs. The term was introduced by Grabau (1911, p. 1007). Syn: *spheryte.* (b) An individual spherical grain in a sedimentary rock, such as a concentric oolith in an oolite or a radial *spherulite* in a limestone.
spheroclast A rounded *phenoclast,* such as a pebble or cobble of a conglomerate. Cf: *anguclast.*
spherocobaltite A peachblossom-red mineral of the calcite group: $CoCO_3$. It occurs in spherical masses. Also spelled: *sphaerocobaltite.* Syn: *cobaltocalcite.*
spheroid [geodesy] Any figure that generally differs little from a sphere, specif. in geodesy the *spherop* whose potential is identical to that of the *geoid.* It is sometimes used as a syn. of *ellipsoid of revolution.* Cf: *ellipsoid.*
spheroid [sed] A spherical, or equant or equiaxial, shape of a sedimentary particle, defined in *Zingg's classification* as having width/length and thickness/width ratios greater than 2/3. See also: *sphere.*
spheroidal (a) Having the shape of a spheroid. (b) Composed of spherules. (c) Said of the texture of a rock composed of numerous spherules.
spheroidal bomb A rotational volcanic bomb in the shape of an oblate spheroid. Syn: *spherical bomb.*
spheroidal jointing *spheroidal parting.*
spheroidal parting A series of concentric and spheroidal or ellipsoidal cracks produced about compact nuclei in fine-grained, homogeneous rocks. Cf: *spheroidal weathering.* Syn: *spheroidal jointing.*
spheroidal symmetry *axial symmetry.*
spheroidal weathering A form of chemical weathering in which concentric or spherical shells of decayed rock (ranging in diameter from 2 cm to 2 m) are successively loosened and separated from a block of rock by water penetrating the bounding joints or other fractures and attacking the block from all sides. It commonly forms a rounded *boulder of decomposition.* It is similar to the larger-scale *exfoliation* produced usually by mechanical weathering. See also: *spheroidal parting.* Syn: *onion-skin weathering; concentric weathering; spherical weathering.*
spherop An equipotential surface in the normal gravity field of the Earth; a surface such that the spheropotential is constant and the normal gravity is perpendicular to it at every point. See also: *spheropotential number.* Syn: *spheropotential surface.*
spherophyre An igneous rock in which the phenocrysts are aggregations of crystals in the form of spherulites. Obsolete.
spheropotential The potential function of the normal gravity defined as either the external gravity potential of a level ellipsoid approximating the Earth, or as a function consisting of the first few terms of some expansion of the *geopotential.* See also: *disturbing potential.*
spheropotential number The spheropotential difference between the marigraph *spherop* and the spherop through an observation point. It is expressed in geopotential units. Syn: *normal geopotential number.*
spheropotential surface *spherop.*
spherosiderite A variety of siderite occurring in globular concretionary aggregates of bladelike crystals radiating from a center, generally in a clayey matrix (such as those in or below underclays

associated with coal measures). It appears to be the result of weathering of water-logged sediments in which iron, leached out of surface soil, is redeposited in a lower zone characterized by reducing conditions. Also spelled: *sphaerosiderite.*

spherule A little sphere or spherical body; e.g. a "magnetic spherule" in a deep-sea sediment, or an object that appears to be an amygdule or a spherulite.

spherulite [petrology] A rounded or spherical mass of acicular crystals, commonly of feldspar, radiating from a central point. Spherulites may range in size from microscopic to several centimeters in diameter (Stokes & Varnes, 1955, p. 140). Cf: *variole; spheruloid; orbicule.* Also spelled: *sphaerolite.*

spherulite [sed] (a) Any more or less spherical body or coarsely crystalline aggregate with a radial internal structure arranged around one or more centers, varying in size from microscopic grains to objects many centimeters in diameter, formed in a sedimentary rock in the place where it is now found; e.g. a minute particle of chalcedony in certain limestones, or a large carbonate concretion or nodule in shale. Cf: *spherite.* (b) A small (0.5-5 mm in diameter), spherical or spheroidal particle composed of a thin, dense calcareous outer layer with a sparry calcite core. It can originate by recrystallization or by biologic processes.

spherulitic Said of the texture of a rock composed of numerous spherulites; also, said of a rock containing spherulites. Cf: *variolitic; radiated.* Syn: *globular; sphaerolitic.*

spheruloid n. A spherule or nodule that lacks radial structure, as in *perlitic lava* (Nelson & Nelson, 1967, p.355). Cf: *spherulite [petrology].*

sphinctozoan Any calcisponge having a skeleton that consists of straight, curved, or branched series of hollow spheroidal bodies (TIP, 1955, pt.E, p.100).

spicularite A sediment or rock composed principally of the siliceous spicules of invertebrates; esp. a *spongolite* composed principally of sponge spicules. Syn: *spiculite; sponge-spicule rock.*

spiculation The formation, or the form and arrangement, of spicules. Also, a spicular component (as of a sponge).

spicule (a) One of the numerous minute calcareous or siliceous bodies, having highly varied and often characteristic forms, occurring in and serving to stiffen and support the tissues of various invertebrates, and frequently found in marine-sediment samples and in Paleozoic and Cretaceous cherts. Examples: a discrete skeletal element of a sponge, typically a needlelike rod or a fused cluster of such rods; a long sharp calcareous skeletal element of the mesogloea of an octocoral; a discrete elongate or needlelike skeletal element of many radiolarians; a scalelike calcareous object borne on the girdle of a primitive chiton; an irregular calcareous body secreted within the connective tissue of a brachiopod; and a minute cylindrical or radiate skeletal element of an asterozoan. (b) The empty siliceous shell of a diatom.

spicule tract A linear series or bundle of separate sponge spicules, usually held together by spongin.

spiculin The chemically undetermined protein substance that forms the axial filament of sponge spicules.

spiculite [petrology] A spindle-shaped belonite thought to have formed by the coalescence of globulites.

spiculite [sed] *spicularite.*

spiculoblast *sclerocyte.*

spiculofiber A fiberlike structure built of a bundle of sponge spicules held together by mutual interlocking or fusion, or by spongin or sclerosome.

spiculoid A discrete autochthonous element of a sponge skeleton, resembling a spicule but made of organic material only. Syn: *pseudospicule.*

spike The known amount of an isotope added to a sample to determine the unknown amount present in analysis by *isotope dilution.*

spilite An altered *basalt,* characteristically amygdaloidal or vesicular, in which the feldspar has been albitized and is typically accompanied by chlorite, calcite, epidote, chalcedony, prehnite, or other low-temperature hydrous crystallization products characteristic of a greenstone. Spilite often occurs as submarine lava flows and exhibits pillow structure. Adj: *spilitic.* The name, given by Brongniart in 1827, is widely used, in spite of Johannsen's suggestion (1937, p. 300) that it be dropped.

spilitic suite A group of altered extrusive and minor intrusive basaltic rocks that characteristically have a high albite content. The group is named for its type member, *spilite.*

spilitization Albitization of a basalt to form a spilite.

spill bank A term used in Great Britain and India for a *natural levee.*

spilling breaker A breaker whose crest collapses gradually over a nearly flat bottom for a relatively long distance, forming a foamy patch at the crest, the water spilling down continuously over the advancing wave front. See also: *comber.* Cf: *plunging breaker; surging breaker.*

spillpoint The point of maximum filling of a structural trap by oil or gas.

spill stream An *overflow stream* from a river.

spillway [eng] A passage or outlet through which surplus water flows from a dam or natural obstruction.

spillway [glac geol] *overflow channel.*

spilosite A rock representing an early stage in the formation of *adinole* or *spotted slate.*

spinach jade Dark-green nephrite.

spindle-shaped bomb A rotational volcanic bomb of fusiform shape, with earlike projections at its ends. See also: *fusiform bomb.* Syn: *almond-shaped bomb.*

spindle stage A single-axis *stage* of a microscope, consisting of a liquid-filled cell in which the crystal is immersed; it may be rotated 180°.

spine [bot] A thorn; a rigid, sharply pointed structure that may represent a modified leaf or leaf part, petiole, or stipule.

spine [paleont] A projection of the shell surface found on various invertebrates, e.g. a movable calcareous shaft mounted on, and articulating with, a tubercle on the test of an echinoid or asteroid; a cylindrical or elongated triangular projection from the external shell surface of a brachiopod; or a hollow tubular skeletal projection with contained body cavity (*lumen*) at or near the margin of the orifice of a cheilostome bryozoan.

spine [volc] A pointed mass or monolith of solidified lava that sometimes occurs over the throat of a volcano. It may be formed by slow, forced extrusion of viscous lava, or it may represent magma in the pipe that was exposed by weathering.

spinel (a) A mineral: $MgAl_2O_4$. The magnesium may be replaced in part by ferrous iron, and the aluminum by ferric iron. Spinel has great hardness, usually forms octahedral crystals (isometric system), varies widely in color (from colorless to purple-red, green, and yellow to black), and is used as a gemstone. It occurs typically as a product of contact metamorphism of impure dolomitic limestone, and less commonly as an accessory mineral of basic igneous rocks; it also occurs in alluvial deposits. (b) A group of minerals of general formula: AB_2O_4, where A represents magnesium, ferrous and ferric iron, zinc, or manganese, or any combination of them, and B represents aluminum, ferric and ferrous iron, or chromium; specif. an isomorphous series of oxides, $(Mg,Fe,Zn,Mn)Al_2O_4$, consisting of spinel, hercynite, gahnite, and galaxite. (c) A member of the spinel group or spinel series. (d) A substance (such as a sulfide) that has a similar formula and the same crystal structure as a spinel. (e) An artificial substance, similar to the mineral spinel, that is used as a gemstone, a refractory, or instrument bearings; e.g. ferrospinel.—Also spelled: *spinelle; spinell.* Syn: *spinelite.*

spinellid A mineral of the spinel group.

spinellide A name applied to the spinel group.

spinellite A medium- to coarse-grained, hypidiomorphic-granular, titaniferous, magnetite-rich igneous rock containing up to 20% spinel. "Spinel-magnetitite (or spinel-Kirunavaarite) is a better name" (Johannsen, 1938, p. 469).

spinel ruby A deep-red gem variety of spinel. The term is sometimes used inappropriately as a syn. of *ruby spinel.*

spinel twin law A twin law in crystals of the hexoctahedral, isometric system, e.g. spinel, having a twin axis of threefold symmetry with the twin plane parallel to one of the octahedron's faces.

spinifex texture Interpenetrating lacy elongate olivine crystals in *komatiite,* commonly considered to have been formed by quenching. Their disposition resembles the intermesh of an Australian grass for which the texture is named.

spinispire A siliceous monaxonic sponge spicule (microsclere) in the form of a spiral of more than one revolution. Cf: *sigmaspire.* Syn: *spiraster.*

spinneret A spinning organ, or an organ for producing threads of silk from the secretion of the silk glands of an arachnid; specif. an abdominal appendage of spiders, with spinning tubes at the end, or a special spinning organ on the movable finger (galea) of a chelicera of pseudoscorpions.

spinner magnetometer A laboratory instrument that continuously rotates the specimen whose remanent magnetization it is measuring, to produce an alternating voltage in a nearby coil by electromagnetic induction. Syn: *rock generator.*

spinose Spinelike, or full of or armed with spines; e.g. said of a foraminiferal test having fine elongate solid spines on its surface (as in *Hastigerinella*) with each spine comprising a single calcite crystal elongated along the *c* axis.

S-P interval In earthquake seismology, the time interval between the first arrivals of longitudinal and transverse waves, which is a measure of the distance from the earthquake source.

spinulus A closed-end process extending from the surface of a valve of a diatom frustule.

spiracle [paleont] (a) A large, generally rounded opening near the adoral tip of a deltoid plate of a blastoid and excavated within it. It opens into the space enclosed by a hydrospire. (b) A *stigma* of an arachnid.

spiracle [volc] A fumarolic vent in a lava flow, formed by a gaseous explosion in lava that is still fluid, probably due to generation of steam from underlying wet material. It is usually about one meter in diameter and up to five meters high, although in the NW U.S., where spiracles are common, they may be larger. Cf: *pipe vesicle.*

spiracular slit An elongate spiracle at the side of a blastoid ambulacrum, excavated in adjoining radial and deltoid plates.

spiraculate Having spiracles.

spiral angle The angle formed between two straight lines tangent to the periphery of two or more whorls on opposite sides of a gastropod shell. It is commonly determined by drawing tangents to the lowermost whorls of the spire. Syn: *spire angle.*

spiral ball A term used by Hadding (1931, p.389) for a sandstone body having a rolled-up, spiral structure due to lateral mass flowage of thin interbedded sands and shales. Cf: *slump ball.* Syn: *snowball.*

spiral canal The part of the canal system in the umbilical region of a foraminiferal test (as in *Elphidium*) parallel to and inside the lateral-chamber margins.

spiral garnet *rotated garnet.*

spiral growth Growth of a crystal along a *screw dislocation.* It may result in a *growth island* on the surface of the crystal.

spiralium A spiral brachidium; one of a pair of spirally ribbonlike calcareous supports for the deuterolophe or the spirolophe of certain brachiopods, composed of secondary shell. Pl: *spiralia.* Syn: *spire.*

spiral lamina The coiled or winding part of the lorica of a tintinnid.

spiral suture A line of contact between two whorls in the coiled test of a foraminifer.

spiral tracheid A *tracheid* in which secondary cell-wall material is deposited in a spiral configuration; common in *protoxylem.* Cf: *annular tracheid.*

spiramen A pore in the proximal wall of the peristome in some ascophoran cheilostome bryozoans, leading into the space enclosed by the peristome outward from the operculum-bearing orifice.

spiraster A spiral sponge spicule; e.g. a *streptosclere* or a *spinispire.*

spire (a) The adapical visible upper part of a spiral gastropod shell, including the whole series of whorls except the last or body whorl. (b) A *spiralium* of a brachiopod. (c) *spirillum.*

spiriferoid Any articulate brachiopod belonging to the order Spiriferida, characterized generally by a spiral brachidium and a biconvex, rarely planoconvex, shell. Range, Middle Ordovician to Jurassic. Var: *spiriferid; spirifer.*

spirilline Said of a foraminiferal test consisting of a planispiral nonseptate tube enrolled about a globular proloculus; specif. pertaining to the foraminiferal genus *Spirillina.*

spirillum (a) A helical or coiled morphologic form of a bacterial cell. Syn: *spire; spiril.* (b) Any bacterium of the genus *Spirillum,* now restricted to elongated forms having tufts of flagella at one or both ends.—Pl: *spirilla.*

spirit level (a) A sensitive device for finding a horizontal line or plane, consisting of a small closed glass tube or vial of circular cross section, nearly filled with a liquid of low viscosity (ether or alcohol) with enough free space being left for the formation of a bubble of air or gas that will always assume a position at the top of the tube. See also: *circular level.* Syn: *level [surv].* (b) An instrument using a spirit level to establish a horizontal line of sight.

spirit leveling A type of *leveling* using a spirit level to establish a horizontal line of sight.

spiroffite A red to purple monoclinic mineral: $(Mn,Zn)_2Te_3O_8$.

spirogyrate Said of the umbones of a bivalve mollusk, coiled outward from an anteriorly and posteriorly directed (sagittal) plane of symmetry.

spirolophe A brachiopod lophophore in which the brachia are spirally coiled and bear a single row of paired filamentary appendages. Cf: *deuterolophe; plectolophe.*

spirotheca The outer or upper spiral wall of the test in fusulinids.

spiroumbilical aperture An *interiomarginal aperture* in a foraminiferal test, extending from umbilicus to periphery and finally onto the spiral (dorsal) side where all whorls are visible.

spit (a) A small point or low tongue or narrow embankment of land, commonly consisting of sand or gravel deposited by longshore drifting and having one end attached to the mainland and the other terminating in open water, usually the sea; a fingerlike extension of the beach. (b) A relatively long, narrow shoal or reef extending from the shore into a body of water.

spitskop A term used in South Africa for a hill with a sharply pointed top. Etymol: Afrikaans. Cf: *tafelkop.*

spitz Nonrecommended syn. of *tip [paleont].*

spitzkarren Solution grooves about 0.5 m apart, separated by rows of sharp-crested pyramidal peaks. Etymol: German, "peak tracks". See also: *karren.*

***s* plane** In structural geology, a nongenetic term for any planar fabric element, e.g. *foliation* or *bedding.* Syn: *s* surface.

splash cup A concavity at the top of a stalagmite.

splash erosion The dislodgment and movement of soil particles under the impact of falling raindrops.

splash zone The shoreline area that is affected by the splashing of seawater from breaking waves. Cf: *spray zone.*

splay [fault] One of a series of minor faults at the extremities of a major fault; the fault pattern formed by *splaying out.* It is associated with rifts.

splay [geomorph] *flood-plain splay.*

splaying crevasse A *crevasse* in a valley glacier that is parallel to the direction of flow in the center of the glacier but curves toward the margin downstream. Cf: *longitudinal crevasse; marginal crevasse; transverse crevasse.*

splaying out (a) The breakup of a fault into a number of minor faults. (b) The dying-out of a fault by its dispersal into a number of minor faults. See also: *splay [fault].*

splendent luster A mineral luster of the highest intensity.

splent coal *splint coal.*

spline In *residualizing,* a smooth curve representing the regional gravity field.

splint *splint coal.*

splint coal A type of banded coal that is hard, dull, blocky, and greyish black, with rough, uneven fracture and granular texture. It is defined quantitatively as having more than 5% anthraxylon and more than 30% opaque attritus. Also spelled: *splent coal.* Cf: *semisplint coal; durain.* Syn: *splint.*

splintery fracture The property shown by certain minerals or rocks of breaking or fracturing into elongated fragments like splinters of wood.

split A coal seam that is separated from the main seam by a thick parting of other sedimentary rock. Syn: *split coal; coal split.*

split coal *split.*

split spread A type of seismic spread in which the shot point is at the center of the arrangement of geophones. It is commonly used for continuous profiling and for dip shooting. Syn: *straddle spread; symmetric spread.*

split stream (a) A stream shown on a map by a single line and containing an island that divides the stream into two channels. (b) A *single-line stream* that divides into branches in separate drainage areas.

splitting [paleont] In taxonomy, the practice of classifying species and genera on the basis of relatively minute differences. A taxonomist known for his preference for finely drawn distinctions is called a "splitter". Cf: *lumping.*

splitting [sed] (a) Abrasion of a rock fragment resulting in the production of two or three subequal parts or grains. (b) The property or tendency of a stratified rock of separating along a plane or surface of *parting.* (c) The sampling of a large mass of loose material (e.g. a sediment) by dividing it into two or more parts; e.g. *quartering.*

spodic horizon A soil horizon that is characterized by the illuvial accumulation of black or reddish amorphous materials that have a high cation-exchange capacity and consist of aluminum and organic carbon, sometimes with iron. The spodic horizon is moist or wet and has a loamy or sandy texture, a high pH-dependent exchange capacity, and few bases (USDA, 1975).

spodiosite A mineral whose status is doubtful: $Ca_2(PO_4)F$.

Spodosol In U.S. Dept. of Agriculture soil taxonomy, a soil order characterized by the presence of a spodic horizon. Usually spodosols have an O horizon and an albic horizon above the spodic horizon, and a few have a fragipan or an argillic horizon below it. They generally form from sandy parent materials. In cool, humid climates they usually form under coniferous forests. In hot, humid intertropical areas they form in quartz-rich sands that have fluctuating ground water near the surface. Spodosols are naturally infertile but may be highly responsive to good management (USDA, 1975). Suborders and great soil groups of this order have the suffix -od. See also: *Aquod; Ferrod; Humod; Orthod.*

spodumene A mineral of the clinopyroxene group: $LiAlSi_2O_6$. It occurs in white to green prismatic crystals, often of great size, esp. in granitic pegmatites. Spodumene is an ore of lithium. See also: *kunzite; hiddenite.* Syn: *triphane.*

spoil Overburden, nonore, or other waste material removed in mining, quarrying, dredging, or excavating.

spoil bank A bank, mound, or other accumulation composed of spoil; e.g. a submerged embankment of waste earth material dredged from a channel and dumped along it. See also: *spoil heap.*

spoil ground An area where dredged or excavated material is deposited or dumped.

spoil heap A pile of refuse material from an excavation or mining operation; e.g. a pile of dirt removed from, and stacked at the surface of, a mine in a conical heap or in layered deposits, such as a tip heap from a coal mine. See also: *spoil bank.*

spoke A radial and typically flat component of a *wheel* of a holothurian, connecting the central part and the rim.

spondylium A trough-shaped or spoonlike plate serving for muscle attachment and accommodating a ventral muscle field in the posterior part of the pedicle valve of a brachiopod, composed of dental plates in various stages of coalescence, usually with a median septum. Pl: *spondylia.* Cf: *pseudospondylium.*

spondylium duplex A spondylium formed by convergence of dental plates and supported by variably developed median septum arising from the floor of the pedicle valve.

spondylium simplex A spondylium formed by convergence of dental plates and supported ventrally by a variably developed simple median septum.

spong A term used in the Pine Barrens, N.J., for a *cripple* without a growth of cedar or having flowing water only after a rain; sometimes defined as any lowland area where highbush blueberries grow. Pron: spung.

sponge A many-celled aquatic invertebrate belonging to the phylum *Porifera* and characterized by an internal skeleton composed most frequently of opaline silica and less commonly of calcium carbonate. Range, Precambrian to present. Syn: *poriferan.*

sponge-spicule rock A lithified *spicularite.*

spongework An entangled net of irregular interconnecting cavities of various sizes produced by solution in the walls of limestone caves and separated by intricate perforated partitions and remnants of partitions. The relations are as complicated as those of the pores of a sponge. Partial syn: *anastomosis [speleo].*

spongework cave A cave consisting of irregular interconnected passages. See also: *anastomotic cave; network cave.*

spongin (a) A variety of collagen or scleroprotein (an insoluble fibrous protein) secreted by sponges and forming their skeletons. (b) A general term for any fibrous, organic skeletal material secreted by sponges. This usage is not recommended, although it is the sense in which the term has been used in all but the most recent literature.

spongioblast A cell that secretes spongin. Syn: *spongoblast; spongiocyte.*

spongocoel The *cloaca* of a sponge.

spongolite A sediment or rock composed principally of the remains of sponges; esp. a *spicularite.* Syn: *spongolith.*

spontaneous fission A rare mode of *radioactive decay,* in which the nucleus of a heavy atom produces two fission products and several neutrons. Significant for ^{238}U and ^{244}Pu.

spontaneous fission dating A method of determining the age in years of uranium minerals based on the known rate of spontaneous fission of uranium-238 to xenon.

spontaneous fission-track dating *fission-track dating.*

spontaneous generation An early concept in which living matter was thought to appear from dead material without the influence of outside or supernatural forces. Syn: *abiogenesis.*

spontaneous magnetization The magnetization within a domain in the absence of an applied magnetic field, due to spontaneous magnetic order caused by exchange forces.

spontaneous polarization Development of differences in static electrical potential between points in the Earth as a result of chemical reactions, differences in solution concentration, or the movement of fluids through porous media. See also: *self-potential method.*

spontaneous-potential curve The *electric log* curve that records changes in natural potential along an uncased borehole. Small voltages are developed between mud filtrate and formation water of an invaded bed, and also across the shale-to-mud interface. These electrochemical components are augmented by an electrokinetic potential (streaming potential) developed when mud filtrate moves toward a formation region of lower fluid pressure through the *mud cake.* Where formation waters are less resistive (more saline) than drilling-mud filtrate, the spontaneous-potential curve deflects to the left from the *shale baseline.* First used about 1932, the curve was added to the *resistivity log* to make up the basic electric log of well-logging practice. Syn: *SP curve; self-potential curve.*

spontaneous-potential method *self-potential method.*

sporadic permafrost A region of dominantly unfrozen ground containing scattered areas of permafrost (*permafrost islands*); it occurs along the southern limits of regions where summer frost conditions are usual, and in alpine areas. Cf: *discontinuous permafrost; continuous permafrost.*

sporadosiderite A stony meteorite containing disseminated iron.

sporae dispersae Pollen and spores obtained by maceration of rocks, in contrast with those that have been found within the sporangia that bore them.

sporal Pertaining to or having the special characteristics of a spore. The term is not in good usage in palynology.

sporangiospore A spore produced in a sporangium. The term is not in good usage in palynology.

sporangite A fossilized spore case of a plant. The term is not in good usage in palynology.

sporangium An organ within which spores are produced or borne; e.g. an organ in embryophytic plants in which spores are produced, such as a pollen sac of a gymnosperm or an anther of an angiosperm. Pl: *sporangia.* See also: *microsporangium; megasporangium.* Syn: *spore case.*

sporbo A term used in the San Joaquin Valley, Calif., for *oolite.* Pl: *sporbo.* Etymol: *s*mooth-*po*lished-*r*ound-*b*lack (*b*lue or *b* rown)-*o*bject (Galliher, 1931, p. 257).

spore Any of a wide variety of minute unicellular reproductive bodies or cells that are often adapted to survive unfavorable environmental conditions and that are capable of developing independently into new organisms, directly if asexual or after union with another spore if sexual; e.g. one of the haploid, dispersed reproductive bodies of embryophytic plants, having a very resistant outer wall, and frequently occurring as fossils from Silurian to the present.

spore case A less-preferred syn. of *sporangium.*

spore coat *sporoderm.*

spore mother cell The *mother cell* in the microsporangium of a spore-bearing plant (as in the anther of an angiosperm, or in the pollen sac of other seed plants), which, by reduction division, produces the tetrad of haploid microspores. See also: *pollen mother cell.* Syn: *sporocyte.*

sporinite A maceral of coal within the *exinite* group, consisting of spore exines that are generally compressed parallel to the stratification. Cf: *cutinite; alginite; resinite.*

sporite A coal microlithotype containing 95% or more of sporinite.

sporocyte *spore mother cell.*

sporoderm The wall of a spore or pollen grain, generally consisting of an inner layer (intine) and an outer layer (exine), and sometimes an extra third layer (perisporium) outside the exine. Syn: *spore coat.*

sporogenous Producing or adapted to the production of spores, or reproducing by spores; e.g. "sporogenous tissue" in sporangium from which spore mother cells originate.

sporologic Of or pertaining to palynology. The term is sometimes used as a synonym of the more current term "palynologic".

sporomorph A fossil pollen grain or spore. Cf: *palynomorph.*

sporophitic Said of the ophitic texture of an igneous rock, in which large pyroxene crystals enclose much smaller, widely separated plagioclase crystals (Walker, 1957, p.2); a form of poikilitic texture. Cf: *poikilophitic; hyalopilitic.*

sporophyll A spore-bearing leaf; commonly a much modified structure, as in seed plants.

sporophyte The individual or asexual generation of a plant that produces spores; e.g. the diploid generation of an embryophytic plant, produced by fusion of egg and spermatozoid in lower vascular plants or by fusion of egg nucleus and the sperm nucleus produced by the pollen of seed plants. Cf: *gametophyte.*

sporopollenin The very resistant and refractory organic substance of which the exine of spores and pollen is composed and which gives the sporomorph its extreme durability during geologic time, being readily destroyed only by oxidation. It is a high-molecular-weight polymer of C-H-O (primarily of monocarboxylic or dicarboxylic fatty acids), but its exact structural composition has not yet been resolved. Also spelled: *sporopollenine.*

sport An individual that exhibits a marked deviation well beyond the normal limits of individual variation and not anticipated or suggested by the characteristics of its parents; it therefore is usually regarded as the result of a *mutation.*

spot correlation In seismology, the correlation of reflections on isolated seismograms by noting similarities in character and interval.

spot elevation (a) An elevation shown on a topographic map at a critical point (such as a break in slope, a road intersection, or a point on a stream divide) to supplement the map information given by contour lines and bench marks from which contours can be correctly drawn. (b) A point on a map or chart whose elevation is noted; a *spot height.*

spot height A predominantly British term for a point, indicated on a map, whose elevation above a given datum has been correctly measured on the ground but, in contrast to a bench mark, is seldom indicated on the ground. See also: *spot elevation.*

spot medallion *frost scar.*

spotted *maculose.*

spotted schist *spotted slate.*

spotted slate A shaly, slaty, or schistose argillaceous rock the spotted appearance of which is the result of incipient growth of porphyroblasts in response to contact metamorphism of low to medium intensity. Cf: *desmosite; schalstein; spilosite; adinole.* See also: *fleckschiefer; fruchtschiefer; garbenschiefer; knotenschiefer.* Syn: *spotted schist; knotted schist; knotted slate.*

spotty Said of a mineral deposit or mineralized zone in which the valuable constituent occurs in scattered masses of high-grade material.

spout (a) A discharge or jet of water ejected with some violence, either continuously (e.g. a spring) or periodically (e.g. a geyser). (b) A rush of water to a lower level; e.g. a waterfall. (c) *waterspout.*

spouter *blowhole [coast].*

spouting horn A sea cave with a rearward or upward opening through which water spurts or sprays after waves enter the cave. Syn: *chimney; oven.*

spouty land Land so wet that it spouts water when walked on; e.g. a marshland.

spray ice Ice formed from ocean spray blown along the shore or upon floating ice.

spray print A small pit or depression similar to a *rain print* but formed by spray or windblown drops of water. Syn: *spray pit.*

spray ridge An ice formation on an icefoot, formed by the freezing of wind-blown ocean spray.

spray zone The area along a coast affected by the spray of breaking waves. Cf: *splash zone.*

spread [gem] The surface or width at the girdle in proportion to the depth of a cut stone, e.g. a diamond.

spread [seis] The layout of geophone groups from which data from a single shot are recorded simultaneously. Syn: *seismometer spread; seismic spread.*

spread [streams] A marsh or shallow body of water resulting from the expansion in width of a stream, as where a natural obstruction, aquatic vegetation, or sediment infilling chokes or impedes streamflow. Syn: *widespread.*

spread correction A correction for *normal moveout.*

spreading concept *sea-floor spreading.*

spreading-floor hypothesis *sea-floor spreading.*

spread stone A diamond brilliant cut with a large table and a thin crown.

spring A place where ground water flows naturally from a rock or the soil onto the land surface or into a body of surface water. Its occurrence depends on the nature and relationship of rocks, esp. permeable and impermeable strata, on the position of the water table, and on the topography.

spring alcove A term used along the Snake River Canyon in Oregon and Idaho for a short amphitheater or box-headed canyon formed by spring sapping along the edge of a basalt plateau.

spring dome A descriptive term suggested by Williamson (1961) for a nondiastrophic limestone structure consisting of a circular or elliptical mound, usually with a central hollow or crater, believed to result from the expulsion of water from an underlying source (such as from a semirigid sediment).

Springeran North American stage: Lower Pennsylvanian (above Mississippian, below Morrowan).

spring-fed lake *spring lake.*

springhead The *fountainhead* of a stream.

spring lake (a) A lake, usually of small size, that is created by the emergence of a spring or springs, esp. one having visibly flowing springs on its shore or springs rising from its bottom. (b) A lake that receives all or part of its waters directly from a spring.—Syn: *limnokrene; spring-fed lake.*

spring line A line of springs marking the intersection of the water table with the land surface, as at the foot of an escarpment or along the base of a permeable bed at its contact with an aquiclude.

spring mound A roughly circular mound of sand and silt, 5-6 m high and 10-12 m across, formed by a spring rising to the surface and depositing its load. See also: *mound spring.*

spring neck A long, narrow trench, commonly 60-90 cm wide and a few meters deep, formed by percolating water flowing toward the central level of a playa from a spring at its margin (Stone, 1967, p. 249).

spring pit A small crater formed on a sand beach by ascending water, characterized by coarse sand in the center and finer sand around the edge, and measuring 30-60 cm across and about 15 cm deep (Quirke, 1930).

spring pot A shallow depression formed on the margin of a pluvial lake bed or a modern playa by spring flow, measuring 90-120 cm across and 60-90 cm deep (Stone, 1967, p. 249).

spring range *mean spring range.*

spring sapping The erosion of a hillside around the fountainhead of a strongly flowing stream, causing small landslides and resulting in the retreat of the valley head. Syn: *springhead sapping.*

spring snow A coarse, granular wet snow developed during the spring, generally by high temperatures and *melt metamorphism.* Syn: *corn snow.*

spring tide A tide occurring twice each month at or near the times of new moon (conjunction) and full moon (opposition) when the gravitational pull of the Sun reinforces that of the Moon, and having an unusually large or increased tide range. Cf: *neap tide.* Syn: *syzygy tide.*

springwood A syn. of *early wood.* Also spelled: *spring wood.* Cf: *summerwood.*

spruit A term used in southern and eastern Africa for a small stream, esp. one that is usually dry but nourished by sudden floods. Etymol: Afrikaans, from Middle Dutch *spruten,* "to sprout".

spud v. To break ground with a drilling rig at the start of well-drilling operations. Syn: *spud-in.*

spudder (a) A small drilling rig, used primarily to start a new well. (b) A drill used for drilling seismic shotholes in hard rock or gravel. (c) The special drill bit used to begin a borehole.

spud in *spud.*

spumellarian Any radiolarian protozoan belonging to the suborder Spumellina, characterized by a thick-walled central capsule perforated by fine, evenly distributed pores.

spur [eng] An artificial obstruction, such as a pier dam, extending outward from the bank of a stream for the purpose of deflecting the current or of protecting the shore from erosion.

spur [geomorph] (a) A subordinate ridge or lesser elevation that projects sharply from the crest or side of a hill, mountain, or other high land surface; a small hill extending from a prominent range of hills or mountains. Syn: *prong [geomorph].* (b) *meander spur.*

spur [ice] *ram.*

spur [marine geol] (a) A ridge or other prolongation of a terrestrial mountain range, extending from the shore onto or across the continental shelf or insular shelf; e.g. the Bahama Spur in the Atlantic Ocean. (b) A subordinate ridge projecting outward from a larger submarine feature or elevation.

spur [ore dep] A small vein branching from a main one.

spur [paleont] A dependent projection of the basal margin of a tergum of a cirripede crustacean.

spur-and-groove structure A comb-tooth structure common to many reef fronts, best developed on the windward side, consisting of elongate channels or grooves a few meters wide and deep, separated by seaward-extending ridges or spurs (Maxwell, 1968, p. 110). Cf: *reef buttress.* Syn: *groove-and-spur structure.*
spur-end facet *triangular facet.*
spur furrow A groove on the outer surface of a tergum of a cirripede crustacean, extending along a spur to the apex. Syn: *spur fasciole.*
spurrite A light-gray mineral: $Ca_5(SiO_4)_2(CO_3)$.
squall (a) A strong, sudden wind, often accompanied by precipitation, thunder, and lightning. There are many local names for various types of squalls. (b) A severe local storm of short duration.
squall line A narrow band of powerful cumulonimbus clouds traveling in advance of a *cold front.* Tornadoes are often associated with the squall line.
Squamata An order of lepidosaurian reptiles that includes the lizards and snakes and their marine relatives the mosasaurs, of Mesozoic age. Range, Upper Triassic to Recent.
squamiform cast One of a group of "crowded, lobate casts overlapping downcurrent" (Ten Haaf, 1959, p. 46), resembling sagged flute casts but supposedly having an opposite orientation with regard to current direction. It appears to be a variety of loaded *flute, longitudinal furrow,* and *longitudinal ridge,* with a fleur-de-lis pattern.
squamose Scaly; said of a plant that is covered with small scales. Cf: *lepidote.*
squamula A small plate projecting subhorizontally in an eavelike manner from the wall of a tabulate corallite toward the axis, as in *Emmonsia.* Pl: *squamulae.*
square emerald cut An *emerald cut* with a square girdle outline modified by corner facets.
squeaking sand A term used by Humphries (1966, p.135) for *whistling sand.*
squeeze [drill] v. (a) To inject cement slurry into a well in order to recement a channeled area behind the casing or to close off perforations. See also: *squeeze job.* (b) To inject fluid under high pressure, as in *hydraulic fracturing.*—n. The plastic movement of a soft rock in the walls of a borehole. Cf: *heaving shale.*
squeeze [eng geol] (a) The rapid or gradual closing of a mine working by the displacement of weak floor strata from beneath supporting pillars into adjacent mine rooms. (b) A form of failure of bearing capacity, common in coal mines, caused by the mining-induced stress increase on supporting coal pillars. Bending of the roof strata is often reflected as subsidence at the ground surface. (c) An abnormal increase in load on tunnel or mine supports, often with some displacement of the floor around resisting roof-support members. (d) A mine area (such as a section in a coal seam) undergoing a squeeze.
squeeze job The forcing of cement slurry into a borehole. See also: *squeeze [drill].*
squeeze-up A small extrusion of viscous lava, or *toothpaste lava,* from a fracture or opening on the solidified surface of a flow, caused by pressure. It may take various forms, generally bulbous or linear, and may be from a few centimeters to almost a meter in height. It may be marked by vertical grooves. See also: *grooved lava.* Syn: *push [volc].*
squeezeway A cave passage that is traversable only with difficulty.
squeezing ground A tunnelman's term for soil or rock that creeps into a tunnel at constant volume. Clay, or rock containing much clay, constitutes typical squeezing ground. See also: *firm ground; flowing ground; raveling ground; running ground; swelling ground.*
squid magnetometer A sensitive magnetometer based on the use of a Josephson junction to detect changes in the magnetic field. Etymol: derived from the expression *S*uperconducting *Q*uantum *I*nterference *D*evice.
s.s. *sensu stricto.*
***s* surface** *s plane.*
stabile Resistant to chemical change, or decomposing with difficulty; e.g. "stabile protobitumen", a plant or animal product (such as wax, resin, spores, or leaf cuticle) that forms fossil carbonaceous deposits such as amber or cannel coal. Ant: *labile.*
stability [eng] The resistance of a structure, slope, or embankment to failure by sliding or collapsing under normal conditions for which it was designed; e.g. *bank stability* and *slope stability.*
stability [geochem] In thermodynamics, an equilibrium state to which a system will tend to move from any other state under the same external conditions. Since it is never possible to examine all alternative states, assertions about the stability of real systems must always contain, at least implicitly, reference to the alternative states relative to which stability is claimed.
stability field The range of conditions within which a mineral or mineral assemblage is stable.
stability series A grouping of minerals arranged according to their persistence in nature; i.e. to their resistance to alteration or destruction by weathering, abrasion during transportation, and postdepositional solution (Goldich, 1938); e.g. olivine (least stable), augite, hornblende, biotite (most stable). The most stable minerals are those that tend to be at equilibrium at the Earth's surface.
stability-time hypothesis The paleoecologic concept that long-term environmental stability, esp. of climate, tends to produce great faunal *diversity,* narrow ecologic *niches,* fast *speciation* rates, and *ecosystems* with complex tropic or feeding relationships. Syn: *time-stability hypothesis.*
stabilization [ecol] The characteristic of a *climax,* in which the greatest degree of adjustment between organisms and environment has been attained.
stabilization [eng] *soil stabilization.*
stabilized dune *anchored dune.*
stabilizing force The ordinary restoring force in an unstable gravimeter.
stable [radioactivity] Said of a substance that is not spontaneously radioactive. Cf: *unstable [radioactivity].*
stable [sed] (a) Said of a constituent of a sedimentary rock that effectively resists further mineralogic change and that represents an end product of sedimentation (often resulting from more than one cycle of erosion and deposition); e.g. quartz, quartzite, chert, and accessory minerals such as zircon, rutile, muscovite, and tourmaline. (b) Said of a mature sedimentary rock (such as orthoquartzite) consisting of stable particles that are rounded or subrounded, well-sorted, and composed essentially of silica.
stable [tect] Said of an area or part of the Earth's crust that shows neither uplift nor subsidence or that is not readily deformed; e.g. a "stable shoreline" that is neither advancing nor receding.
stable equilibrium A state of equilibrium of a body, such as a pendulum, when it tends to return to its original position after being displaced. Cf: *unstable equilibrium; dynamic equilibrium.*
stable gravimeter An instrument that uses a high order of optical and/or mechanical magnification so that an extremely small change in the position of a weight or associated property can be accurately measured.
stable isotope A nuclide that does not undergo *radioactive decay.* Cf: *unstable isotope.*
stable magnetization Remanent magnetization which does not change over geologic time, i.e., does not show magnetic viscosity. In practice it is nearly the same as *hard magnetization.*
stable relict A *relict [meta]* that was not only stable under the conditions of its formation but also under the newly imposed conditions of metamorphism. Cf: *unstable relict.*
stack [coast] An isolated pillar-like rocky island or mass near a cliffy shore, detached from a headland by wave erosion assisted by weathering; esp. one showing columnar structure and roughly horizontal stratification. Examples occur off the chalk cliffs of the Normandy coast. A stack may also form along the shore of a large lake. See also: *skerry; chimney; chimney rock.* Syn: *sea stack; marine stack; rauk.*
stack [geomorph] An upstanding, steep-sided mass of rock rising above its surroundings on all sides from a slope or hill.
stack [seis] The sum of several seismic traces that have been corrected for moveout and statics.
stack deposit A uranium deposit of irregular shape, associated with *trend deposits* and "frequently controlled in part by structure" (Bailey & Childers, 1977, p. 27).
stacking chart A diagram showing the interrelationships among the seismic traces from *common-depth-point shooting.* It is used to determine the proper traces for stacking, and for determining parameters for shifting traces.
stacking fault A type of *plane defect* in a crystal, caused by a mistake in the *stacking sequence.* Syn: *fault [cryst].*
stacking sequence The manner in which individual layers of a layered crystal structure are arranged in space.
stacking velocity *normal-moveout velocity.*
stade A substage of a glacial stage marked by a glacial readvance; "a climatic episode within a glaciation during which a secondary

advance of glaciers took place" (ACSN, 1961, art. 40). Syn: *stadial.*

stadia (a) A surveying technique or method using a stadia rod in which distances from an instrument to the rod are measured by observing through a telescope the intercept on the rod subtending a small known angle at the point of observation, the distance to the rod being proportional to the rod intercept. The angle is usually defined by two fixed lines in the reticle of the telescope. (b) *stadia rod.* (c) An instrument used in a stadia survey; esp. an instrument with stadia hairs.—Pl: *stadias.* The term is also used as an adjective in such expressions as "stadia surveying", "stadia distance", and "stadia station".

stadia constant (a) The ratio by which the intercept on the stadia rod must be multiplied to obtain the distance to the rod. On most surveying instruments, it is 100. (b) The ratio by which the sum of the stadia intervals of all sights of a continuous series of measured differences of elevation is converted to the length of the series in kilometers.

stadia hairs Horizontal cross hairs equidistant from the central horizontal cross hair; esp. two horizontal parallel lines or marks in the reticle of a transit telescope, arranged symmetrically above and below the line of sight, and used in the stadia method of surveying. Syn: *stadia wires.*

stadia interval The length of stadia rod subtended between the top and bottom cross hairs in the leveling instrument as these are projected against the face of the rod.

stadial adj. Pertaining to or formed during a *stade.*—n. *stade.*

stadial moraine *recessional moraine.*

stadia rod A graduated rod used with an instrument having stadia hairs to measure the distance from the observation point to the place where the rod is positioned. Syn: *stadia.*

stadia table A mathematical table from which may be found, with minimal computation, horizontal distance and the difference in elevation, knowing the stadia distance and the angle of sight.

stadia traverse A surveying traverse (such as a transit traverse or a traverse accomplished by planetable methods) in which distances are measured by the stadia method.

stadia wires *stadia hairs.*

stadimeter A surveying instrument for determining the distance to an object of known height by measuring the angle subtended at the observer by the object.

staffelite *francolite.*

staff gage A graduated scale or *gage* on a staff, wall, pier, or other vertical surface, used in gaging water-surface elevation. Cf: *chain gage.*

Staffordian European stage: middle Upper Carboniferous (above Yorkian, below Radstockian). It is equivalent to part of upper Westphalian.

stage [geochron] An obsolete term for a geologic-time unit of lesser duration than *age [geochron],* during which the rocks of a *formation [stratig]* were formed. See also: *substage [geochron].*

stage [geomorph] (a) A phase in the development of a cycle of erosion in which the features of the landscape have characteristic forms that distinguish them from similar features in other parts of the cycle, e.g. the *stages* of youth, maturity, and old age in the development of a stream or region; also, the interval of time during which such a phase persists. The stage is one of the factors that determines the development of landforms in the Davisian cycle of erosion, although Davis (1899) originally referred to this factor as "time" (changed by later writers to "stage"). (b) A particular phase in the historical development of a geologic feature; e.g. the Calumet *stage* of Lake Chicago.

stage [glac geol] A time term for a major subdivision of a glacial epoch; specif. a major Pleistocene subdivision equated with a rock unit of formation rank. It includes glacial stage and interglacial stage. The American Commission on Stratigraphic Nomenclature (1961, art. 31b) rejects this usage because of conflict with the definition of "stage" (as a time-stratigraphic unit) and the requirement that stages be extended geographically on the basis of time-equivalent criteria.

stage [hydraul] The height of a water surface above an arbitrarily established datum plane. Syn: *gage height.*

stage [optics] In a microscope or similar optical apparatus, the small platform on which the object to be studied is placed. See also: *universal stage; mechanical stage; spindle stage.*

stage [stratig] (a) A chronostratigraphic unit next in rank below *series* and above *substage,* commonly based on a succession of biostratigraphic zones that are considered to approximate closely time-equivalent deposits (ACSN, 1961, art.31); the rocks formed during an *age* of geologic time. The stage is the basic working unit of local chronostratigraphy because it is suited in scope and rank to the practical needs and purposes of interregional classification (ISG, 1976, p. 70-71). Most stage names are based on lithostratigraphic units, although preferably a stage should have a geographic name not previously used in stratigraphic nomenclature; the adjectival ending for the geographic name is most commonly "-an" or "-ian", although it is permissible to use the geographic name without any special ending, such as "Claiborne Stage". (b) An informal term used to indicate "any sort" of chronostratigraphic unit of approximate stage rank "which is not a part of the standard hierarchy" of named chronostratigraphic units (ISST, 1961, p.24-25). (c) A *para-time-rock unit* consisting of two or more zones (Wheeler et al., 1950, p.2364). (d) A term used in England for a rock-stratigraphic unit.

stage-capacity curve A graphic illustration of the relationship between the surface elevation of the water in a reservoir and the volume of water (Langbein & Iseri, 1960, p. 17). Cf: *stage-discharge curve.*

stage-discharge curve A graphic illustration of the relationship between gage height and volume of flowing water, expressed as volume per unit of time (Langbein & Iseri, 1960, p. 17-18). Cf: *stage-capacity curve.* Syn: *rating curve; discharge-rating curve.*

staghorn coral Any coral (esp. a scleractinian belonging to the genus *Acropora*) characterized by large branching colonies which resemble antlers.

stagnant basin An isolated or barred basin containing essentially motionless water, rich in organic accumulations and noxious substances, but deficient in oxygen and capable of supporting only anaerobic organisms.

stagnant glacier *dead glacier.*

stagnant ice *dead ice.*

stagnation The condition or quality of water unstirred by a current or wave, or of a glacier that has ceased to flow.

stagnation point On the surface of a solid immersed in a flowing fluid, that point at which the stream lines separate.

stagnation-zone retreat A concept of glacier retreat set forth by Currier (1941). Ice in a marginal zone, 3 to 10 miles wide, ceases to move and is separated from the active ice mass by rapid thinning and burial by glacial sediments. The glacier continues to move forward behind this outer zone, and as successive outer zones progressively stagnate and become covered with debris, a complex system of ice-contact deposits forms, chiefly over and around immobile ice masses.

stagnicolous Said of an organism that lives in or frequents stagnant water.

stagnum A small lake or pool of water lacking an outlet.

stained stone A gemstone whose color has been altered by the use of a coloring agent, such as a dye, or by impregnation with a substance, such as sugar, followed by either chemical or heat treatment, which usually produces a permanent color; e.g. green chalcedony. Cf: *heated stone; burnt stone.*

stainierite *heterogenite.*

staircase pond One of a sequent group of a dozen to a hundred ponds following the approximate axis of a poorly developed watercourse on a sloping, thinly soil-mantled flat in a high-altitude valley, and resulting from the "armoring" and binding of naturally created bars by dense and rapidly growing grass (Ives, 1941, p.287-290); e.g. in Albion Valley, Colo.

stairway (a) *glacial stairway.* (b) *cirque stairway.*

stalactite [speleo] A conical or cylindrical speleothem that hangs from the ceiling of a cave. It is deposited from drops of water and is usually composed of calcite but may be formed of other minerals. See also: *stalagmite [speleo].* Partial syn: *tubular stalactite.* Etymol: Greek, "dripping". The term was first used by Worm in 1642.

stalactite [volc] A conical formation of lava hanging from the roof or walls of a lava tunnel or other cavity and developed by the dripping of fluid lava. It generally measures about 15-30 cm in length. Cf: *stalagmite [volc].*

stalactostalagmite *column [speleo].*

stalagmite [speleo] A conical or cylindrical speleothem that is developed upward from the floor of a cave by the action of dripping water. It is usually composed of calcite but may be formed of other minerals. See also: *stalactite [speleo].* Etymol: Greek, "drop". The term was first used by Worm in 1642.

stalagmite [volc] A conical formation of lava that is built up from the floor of a cavity in a lava flow, and formed as a corresponding feature to a *stalactite* of lava. It generally measures up to 30 cm in height and up to 10 cm in diameter.

stalk That part of a plant by which a part is attached and supported, e.g. the petiole of a leaf, the stipe of an ovary, or the peduncle of a fruit.

stamen That organ of a flower which produces pollen.

staminate Said of a flower that has stamens but no pistil. Cf: *pistillate.*

Stampian *Rupelian.*

stamukha A hummock or ridge of grounded ice, typically on an isolated shoal, formed by heaping-up of blocks. A stranded floe may act as a nucleus. On broad shelves in the Arctic Ocean, they occur along the 20-meter isobath and mark the seaward boundary of the *fast ice.* Pl: *stamukhi.* Etymol: Russian.

stand [drill] Two or more lengths or connected joints of *drill pipe* or *casing* handled as a unit in rotary drilling.

stand [tides] *stand of tide.*

standard atmosphere A standard model of the atmosphere, which uses a temperature of 15°C, a pressure of 1,013.25 millibars measured at mean sea level, and a standard vertical gradient of temperature, pressure, and density. It is used as a representative model in various types of atmospheric analysis.

standard-cell method A method of studying the chemical relationships between rocks by calculating the number of various cations in the rock per 160 oxygen ions (Barth, 1948).

standard depth One of a series of depths (in meters) at which, by international agreement, physical measurements of seawater are to be taken.

standard deviation The square root of the average of the squares of deviations about the mean of a set of data. It is a statistical measure of *dispersion [stat].* Symbol: σ. Syn: *root-mean-square deviation.*

standard-deviation map A *vertical-variability map,* or *moment map,* that shows the degree of statistical dispersion of one lithologic type (in a given stratigraphic unit) about its mean position in the unit. Cf: *center-of-gravity map.*

standard Earth An Earth model in which each surface of P or S seismic velocity in the interior of the Earth is spherical, and encloses the same volume as the corresponding surface of equal velocity in the actual Earth (Runcorn et al., 1967, v. 2, p. 1437).

standard error A measure of the accuracy of a sample mean as an estimator for the population mean; the *standard deviation* of the sampling distribution of a statistical parameter, or the standard deviation of a sample mean; the standard deviation divided by the square root of the number of observations of a sampled variate.

standard meridian (a) The meridian used for determining standard time. (b) A meridian of a map projection, along which the scale is as stated.

standard mineral *normative mineral.*

standard parallel (a) Any parallel of latitude that is selected as a standard axis on which to base a grid system; specif. one of a set of parallels of latitude (other than the base line) of the U.S. Public Land Survey system, passing through a selected township corner on a principal meridian, and on which standard township, section, and quarter-section corners are established. Standard parallels are usually at intervals of 24 miles north or south of the base line, and they are used to limit the convergence of range lines that intersect them from the south so that nominally square sections and townships can be laid out. Syn: *correction line.* (b) A parallel of latitude that is used as a control line in the computation of a map projection; e.g. the parallel of a normal-aspect conical projection along which the principal scale is preserved. (c) A parallel of latitude on a map or chart along which the scale is as stated for that map or chart.

standard port British term for *reference station.*

standard project flood The high-water stage expected from the most severe combination of meteorologic and hydrologic conditions that are considered reasonably characteristic of the geographic region.

standard rig An archaic term for a *cable-tool drilling* rig.

standard seawater *normal water.*

standard section A *reference section* showing as completely as possible a sequence of all the strata in a certain area, in their correct order, thus affording a standard for correlation. It supplements (and sometimes supplants) the type section, esp. for time-stratigraphic units.

standard state A condition in the rocks in which the pressure is the same in all directions at any point, as a result of the weight of the overlying rocks. Cf: *load.*

standard zone A stratigraphic zone "based on a type section in which specified beds yield characteristic fauna. The base of the lowest bed yielding this fauna is defined as the base of the zone. The top of the zone is not explicitly defined". It is named for "one member of the characteristic fauna, preferably one of the most common, or most characteristic" (Callomon, 1965, p. 82).

standing crop *biomass.*

standing floe A separate floe standing vertically or inclined and enclosed by rather smooth ice.

standing level The water level in a well (or other excavation) penetrating the zone of saturation, from which water is not being withdrawn, whether or not it is affected by withdrawals from nearby wells or other ground-water sources. Cf: *static level.* Syn: *standing water level.*

standing oscillation *standing wave.*

standing water Surface water that has no perceptible flow and that remains essentially in place, such as the water of some lakes and ponds; stagnant water, such as that enclosed in marshes and swamps. Ant: *running water.*

standing water level *standing level.*

standing wave A water wave, the wave form of which oscillates vertically between two points or nodes, without progressive movement. Syn: *standing oscillation; stationary wave.*

stand of tide The time during which there is no appreciable change in the height of the tide; it occurs at high water and at low water, and its duration is generally shorter when the tide range is large and longer when the tide range is small. Syn: *tidal stand; stand [tides].*

standstill *stillstand.*

stanfieldite A meteorite mineral: $Ca_4(Mg,Fe,Mn)_5(PO_4)_6$.

stannic Relating to or containing tin in its tetravalent state. Cf: *stannous; stanniferous.*

stanniferous Relating to or containing tin, e.g. stanniferous ore. Cf: *stannic; stannous.*

stannite (a) A steel-gray or iron-black tetragonal mineral: Cu_2FeSnS_4. Zinc often replaces iron. Stannite has a metallic luster and usually occurs in granular masses in tin-bearing veins, associated with cassiterite. Syn: *tin pyrites; bell-metal ore.* (b) Impure cassiterite.—Syn: *stannine.*

stannoidite A mineral: $Cu_5(Fe,Zn)_2SnS_8$.

stannopalladinite A mineral: Pd_3Sn_2.

stannous Relating to or containing tin in its bivalent state. Cf: *stannic; stanniferous.*

stantienite A black variety of retinite having a very high oxygen content (23%). Syn: *black amber.*

stapes *columella (d).*

star n. (a) A rayed figure in a crystal, consisting of two or more intersecting bands of light radiating from a bright center, and observed best under strong illumination; an optical phenomenon caused by reflected light from inclusions or channels, and brought to sharp lines in gem materials by cabochon cutting. Stars usually have four, six, or twelve rays, but three-, five-, seven-, or nine-rayed stars occur, or are possible due to the absence of inclusions in a portion of the stone. See also: *asterism.* (b) A gemstone showing such a figure. (c) *star facet.* (d) *star cut.*—adj. Said of a mineral, crystal, or gemstone that exhibits asterism; e.g. "star agate". Syn: *asteriated.*

star cluster A system of gravitationally interacting stars, numbering from hundreds to millions, sharing a common evolution. Two fairly distinct types are recognized. The globar clusters possess a spherical, or halo, distribution with respect to the center of the galaxy; they show high central condensation, are very massive, and are relatively old (billions of years). The open or galactic clusters are distributed in the galactic plane, especially within the spiral arms; they are much less massive, show no central condensation, are sometimes as young as a few million years, and are relatively enriched in the heavy elements. Syn: *galactic cluster.*

star cut A general term that refers to any brilliant-cut stone whose pavilion facets present a star effect when viewed through the table. Syn: *star.*

star dune An isolated hill of sand, its base resembling in plan a star, and its sharp-crested ridges converging from the basal points to a central peak that may be as high as 100 m above the surrounding plain; it tends to remain fixed in place for centuries in an area where the wind blows from all directions. Syn: *pyramidal dune; heaped dune.*

star facet One of the eight triangular facets between the main *bezel facets* and bounding the table of a round brilliant-cut gem. Syn:

star.

staringite A mineral: $(Fe,Mn)_x(Nb,Ta)_{2x}(Sn,Ti)_{6-3x}O_{12}$.

starkeyite A mineral: $MgSO_4 \cdot 4H_2O$. Syn: *leonhardtite.*

star quartz A variety of quartz containing within the crystal whitish or colored starlike radiations along the diametral planes. The asterism is due to the inclusion of submicroscopic needles of some other mineral arranged in parallel orientation.

star ruby A semiopaque to semitransparent asteriated variety of ruby normally with six chatoyant rays.

star sapphire A semiopaque to semitransparent asteriated variety of sapphire normally with six rays resulting from the presence of microscopic crystals (such as rutile needles) in various orientations within the gemstone. Syn: *asteria.*

star stone (a) An *asteria;* esp. a star sapphire. (b) Less correctly, any asteriated stone, including even petrified wood containing small starlike figures in its more transparent parts.

starved basin A sedimentary basin in which the rate of subsidence is more rapid than the rate of sedimentation. Sediment thickness is greater at the margins than at the center (Adams et al., 1951).

starved ripple mark *incomplete ripple mark.*

stassfurtite A massive variety of boracite from Stassfurt, Germany, sometimes having a subcolumnar structure and resembling a fine-grained white marble or granular limestone.

Stassfurt salt Potassium salt from deposits in Stassfurt, Germany.

state-line fault A tongue-in-cheek term for the discontinuity of geologic structures appearing at the borders of geologic maps of adjacent geographic areas, owing to differences in interpretation.

state of the sea *sea state.*

static granitization Formation of a granitic rock by metasomatism in the absence of compressive force.

static head *static level (c).* See also: *head [hydraul]; hydrostatic head.*

static level (a) *hydrostatic level.* (b) *standing level.* (c) That water level of a well that is not being affected by withdrawal of ground water.

static metamorphism A variety of regional metamorphism brought about by the action of heat and solvents at high lithostatic pressures, not at pressures induced by orogenic deformation. See also: *load metamorphism.* Cf: *thermal metamorphism; geothermal metamorphism.*

static modulus A *modulus of elasticity* that is produced by a very slow application of load.

static pressure Pressure that is "standing" or stabilized because it has reached the maximum possible from its source, and is not being diminished by loss.

static rejuvenation A kind of *rejuvenation* resulting from a decrease in stream load, an increase in runoff (owing to increased rainfall), or an increase in stream discharge through acquisition of new drainage; it involves neither uplift of the land nor eustatic lowering of sea level.

statics Time corrections applied to seismic traces to eliminate delays caused by variations in elevation, weathering-layer thickness, and/or velocity.

static zone A term suggested for the zone below the lowest point of discharge, i.e. below the *zone of discharge,* supposedly where there is little or no water movement. This concept is inaccurate as there is substantial movement below this level in both surface- and ground-water bodies.

station [geophys] A position at which a geophysical observation is made.

station [surv] (a) A definite point on the Earth's surface whose position and location has been or will be determined by surveying methods; e.g. *triangulation station* and *instrument station.* It may or may not be marked on the ground. (b) A length of 100 ft, measured on a traverse along a given line, which may be straight, broken, or curved.

stationary field A physical field that does not vary with time, e.g. a magnetic field, either artificial or natural.

stationary flow *steady flow.*

stationary mass In some seismometers, a heavy weight, either suspended or supported, that tends to remain quiescent during an earthquake. Syn: *steady mass.*

stationary scanner A *scanner* designed for use from a fixed location. Coverage in the horizontal direction is provided by a faceted mirror that rotates about a vertical axis. Coverage in the vertical direction is provided by a planar mirror that tilts about a horizontal axis. Stationary-scanner images are individual frames, rather than continuous strips as acquired by airborne scanners.

stationary wave *standing wave.*

stationary-wave theory *oscillatory-wave theory.*

station error *deflection of the vertical.*

station pole A pole, rod, or staff for making stations in surveying; e.g. a *range rod* or a *level rod.* Also known as a "station rod" or "station staff".

statistical lithofacies A *lithofacies* that grades laterally into its neighbors and whose boundaries are vertical arbitrary-cutoff planes (Weller, 1958, p.633). Cf: *intertongued lithofacies.*

statistics (a) The pure and applied science of devising, applying, and developing techniques such that the uncertainty of numerical inferences may be calculated. (b) The art of reducing numerical data and their interrelationships to comprehensible summaries or parameters. (c) Numbers describing samples taken from any population.

statobiolith A term introduced by Sander (1967, p. 326) for a rock composed mainly of the remains of sessile reef- or shoal-building organisms in their positions of growth.

statoblast In the bryozoan class Phylactolaemata, a bud encapsulated in an envelope of chitin serving as an asexual reproductive body. It generally serves to preserve the species in winter, and develops into a new colony in the spring. See also: *hibernaculum.*

statospore A *resting spore,* e.g. a siliceous thick-walled resistant *cyst* formed within the frustules of various chiefly marine centric diatoms, and characterized by two overlapping convex valves and by absence of a girdle; or an intracellular cyst in various algae of the division Chrysophyta.

statute mile A measure of distance used on land, equal to 5280 ft, 1760 yd, 1609.35 m, 1.61 km, 880 fathoms, 80 chains, 320 rods, or 8000 links, and roughly equivalent to 0.87 international nautical mile. It is usually referred to as *mile.*

stauractine A spicule of hexactinellid sponges in the form of four coplanar rays at approximately right angles to one another.

staurodisc A hexactinellid-sponge spicule (microsclere) in the form of two interpenetrating amphidiscs at right angles to one another. Cf: *hexadisc.*

staurolite A brown to black orthorhombic mineral: $(Fe,Mg)_2Al_9Si_4O_{23}(OH)$. Twinned crystals often resemble a cross (six-sided prisms intersecting at 90° and 60°). It is a common constituent in rocks such as mica schists and gneisses that have undergone medium-grade metamorphism. Syn: *staurotide; cross-stone; grenatite; fairy stone.*

stauroscope A type of *polariscope* that is used to determine the position of vibration-plane traces on a crystal for the accurate measurement of angles of extinction.

staurotide *staurolite.*

stead A term used in south Wales for very thin bands of ironstone in coal measures.

steady flow In hydraulics, flow that remains constant in magnitude or in direction of the velocity vector. Cf: *unsteady flow.* Syn: *stationary flow.*

steady mass *stationary mass.*

steady-state creep *secondary creep.*

steady-state stream *graded stream.*

steady-state theory A model of the Universe as a stationary, expanding world that does not change in appearance through space and time. It has no beginning or end, and has a constant density of matter which is continuously and spontaneously created throughout space (Rogers, 1966, p. 15). Syn: *continuous-creation hypothesis.*

steady turbidity current A persistent turbidity current, such as one produced where a stream heavily laden with sediment flows into a body of deep standing water. Cf: *spasmodic turbidity current.*

steam quality In geothermal development, the quality of steam produced from underground as measured in terms of the weight of steam required to generate one kilowatt-hour of electrical energy. Measurement is in pounds or kilograms.

steam vent A type of hot spring from which superheated steam is rapidly and violently expelled.

steatite (a) A compact, massive, fine-grained, fairly homogeneous rock consisting chiefly of talc but usually containing much other material; an impure talc-rich rock. See also: *soapstone [rock].* (b) A term originally used as an alternative mineral name for *talc,* often restricted to gray-green or brown massive talc that can be easily carved into ornamental objects. Syn: *lardite; lard stone; soapstone [mineral]; soap earth.* (c) *steatite talc.*

steatite talc A relatively pure or high-grade variety of talc suitable for use in electronic insulators. It is the purest commercial form of talc. Syn: *steatite.*

steatitization Introduction of, alteration to, or replacement by, talc (steatite); esp. the act or process of hydrothermal alteration of ultrabasic rocks that results in the formation of a talcose rock (such as steatite, soapstone, or relatively pure concentrations of talc).

Stebinger drum A delicate vertical-angle adjustment for the *vernier* on the *explorer's alidade,* graduated in hundredths of a revolution. Cf: *tangent screw.*

S-tectonite A tectonite whose fabric is dominated by planar fabric elements caused by deformation, e.g. slate. Cf: *L-tectonite; B-tectonite.*

steel galena Galena having a fine-grained texture resulting from mechanical deformation or from incipient transformation to anglesite.

steel jack A colloquial syn. of *sphalerite.*

steel ore A name given to various iron ores (esp. to siderite) because they were readily used for making steel.

steenstrupine A dark-brown to black, hexagonal mineral: $(Ce,La,Na,Mn)_6(Si,P)_6O_{18}(OH)$.

steephead A nearly vertical, semicircular wall at the head of a *pocket valley,* at the base of which springs emerge (Sellards & Gunter, 1918, p. 27).

steepness ratio *wave steepness.*

steep-to Said of a shore, coast, bank, shoal, or other coastal feature that has a precipitous, almost vertical, slope.

steer To introduce a time shift into an ensemble of seismic traces so that energy approaching from a given direction appears on all traces at the same time. It is used in studying earthquakes with large arrays and in other applications. See also: *beam steering.*

Stefan-Boltzmann law The statement that the radiant flux of a black body is equal to the temperature to the fourth power times the Stefan-Boltzmann constant of $5.68 \cdot 10^{-12}$ watt cm^{-2} $°K^{-4}$.

stegidium The convex plate closing the gap between the delthyrial plate and the brachial valve of a spiriferoid brachiopod and consisting of a series of concentric layers deposited by the outer epithelium associated with the atrophying pedicle migrating dorsally (TIP, 1965, pt.H, p.153).

Stegocephalia An obsolescent term, which has been replaced by *Labyrinthodonta.* It formerly also included certain *Lepospondyli.*

steigerite A canary-yellow mineral: $AlVO_4 \cdot 3H_2O$.

steilwand A syn. of *gravity slope.* Etymol: German *Steilwand,* "steep wall" (Penck, 1924).

steinkern Rock material consisting of consolidated mud or sediment that filled the hollow interior of a fossil shell (such as a bivalve shell) or other organic structure. Also, the fossil thus formed after dissolution of the mold. Etymol: German *Steinkern,* "stone kernel". Syn: *internal cast; endocast.* See also: *internal mold.*

Steinmann trinity The association of serpentines, pillow lavas, and radiolarian cherts, which is characteristic of the axial belts of many geosynclines (Steinmann, 1926). The term was coined by Bailey in 1952.

stele The primary vascular structure of a plant stem or root, together with the tissues (such as the pith) which may be enclosed (Cronquist, 1961, p.873). Adj: *stelar.* Syn: *central cylinder.*

stell An English term for a brook.

stellar coal *stellarite.*

stellar crystal A common and beautiful type of snow crystal, having the shape of a flat hexagonal star, often with intricate branches. Cf: *spatial dendrite.*

stellarite A variety of *albertite* from Stellarton, Nova Scotia. Syn: *stellar coal.*

stellate Said of an aggregate of crystals in a starlike arrangement; e.g. wavellite.

stellate structure In receptaculitids, a distal, calcified portion of the merome, generally consisting of four ribs.

stellerite An orthorhombic zeolite mineral: $CaAl_2Si_7O_{18} \cdot 7H_2O$. It is a variety of stilbite.

STEM *scanning transmission electron microscope.*

stem [drill] *drill stem.*

stem [paleont] A narrow structure by which a sessile animal is made fast; e.g. the *column* of an echinoderm.

stemflow Water from precipitation that reaches the ground by running down the trunks of trees or plant stems. Cf: *throughfall.*

stemming (a) The material (sand, clay, limestone) that fills a shothole after the explosive charge has been inserted. It is packed between the charge and the outer end of the shothole and is used to prevent the explosive from "blowing" out along the hole. (b) The act of inserting the stemming into a shothole.—See also: *tamping.*

stem stream *trunk stream.*

stenecious Said of an organism that can adjust or survive in only a limited range of environments.

stengel gneiss A syn. of *pencil gneiss.* Etymol: German.

stenhuggarite A bright yellow tetragonal mineral: $Ca_2Fe^{+2}{}_2Sb^{+5}{}_2(As^{+3}O_3)_4O_3$.

steno- A prefix signifying "narrow", "close", "limited".

stenobathic Said of a marine organism that tolerates only a narrow range of depth. Cf: *eurybathic.*

stenobiontic Said of an organism that requires a stable and uniform environment.

stenohaline Said of a marine organism that tolerates only a narrow range of salinity. Cf: *euryhaline.*

stenolaemate Said of bryozoans belonging to the class Stenolaemata, characterized by colonies made up of long tubular members.

stenonite A monoclinic mineral: $(Sr,Ba,Na)_2Al(CO_3)F_5$.

stenoplastic Having limited capacity for modification and adaptation to new environmental conditions; incapable of major evolutionary differentiation. Cf: *euryplastic.*

stenopodium A slender, elongate limb of a crustacean, composed of rodlike segments. Pl: *stenopodia.* Cf: *phyllopodium.*

stenoproct Said of a sponge in which the cloaca is cylindrical.

stenosiphonate Said of nautiloids with relatively narrow siphuncles. Cf: *eurysiphonate.*

stenothermal Said of a marine organism that tolerates only a narrow range of temperature. Cf: *eurythermal.*

stenotopic (a) Said of an organism that is restricted to one habitat or to relatively few habitats. (b) Said of an organism that has little adaptability to changes in the environment.—Cf: *stenotropic.*

stenotropic Said of an organism that has a narrow range of tolerance for variations in particular environmental factors. Cf: *stenotopic.* Ant: *eurytropic.*

stentorg A well-defined felsenmeer, usually on the crest and flanks of an esker, and often striped by former shorelines and beach ridges (Stamp, 1961, p. 430). Etymol: Swedish, "stone marketplace". Rarely used.

step [coast] (a) The nearly horizontal section, not necessarily permanent, that roughly divides the beach from the shoreface, located just seaward of the low-tide shoreline, and in many cases marked by the presence of coarse sediment (such as gravel on a sand beach). Syn: *toe.* (b) An abrupt downward inflection composed of coarse sand or gravel that marks the breakpoint of relatively small waves and swells prevailing during the summer.

step [geomorph] (a) A *canyon bench* greatly broadened by erosion, such as those characteristic of the high plateaus of the western U.S.; a steplike landform on a hillside or valley slope that is otherwise smoothly rising. (b) *rock step.*

step [pat grd] A form of *patterned ground* characteristic of moderate slopes, and having a steplike form; it is transitional between a *polygon* (upslope) and a *stripe* (downslope). A step typically develops as a lobate solifluction terrace with the lower border convex downslope. See also: *sorted step; nonsorted step.* Obsolete syn: *terracette.* Syn: *lobate soil.*

step-and-platform topography A landscape developed in a region in which the rocks dip very gently in one direction over a large area and are composed of alternating layers of hard and soft material; characterized by a sequence of lowlands and cuestas (Marbut, 1896, p. 31-32).

stepanovite A mineral oxalate: $NaMgFe^{+3}(C_2O_4)_3 \cdot 8\text{-}9H_2O$. Cf: *zhemchuzhnikovite.*

step cline An irregular or broken *cline.*

step cut A style of cutting, widely used on colored gemstones, in which long, narrow four-sided facets form in a series or row parallel to the girdle and decrease in length as they recede above and below the girdle, giving the appearance of steps. The number of rows, or steps, may vary, although it is usually three on the crown and three on the pavilion. Different shapes of step cuts are described by their outline, such as rectangular step cut or square step cut. Cf: *emerald cut; brilliant cut; mixed cut.* See also: *baguette.* Syn: *trap cut.*

step delta One of a series of deltas built in a body of water whose level was alternately standing and falling, the upper delta being the oldest (Dryer, 1910, p. 259); examples are numerous in the Finger Lakes region of New York.

step fault (a) One of a set of parallel, closely spaced faults over which the total displacement is distributed. Cf: *fault zone.* Syn: *multiple fault; distributive fault.* (b) One of a series of low-angle thrust faults in which the fault planes step both down and laterally in the stratigraphic section to lower glide planes. Step faulting is due to variation in the competence of the beds in the stratigraphic section (Jones, P.B., 1971).

step fold An abrupt downward flexure of horizontal strata; it is a monoclinal structure.

Stephanian European stage: Upper Carboniferous (Upper Pennsylvanian; above Westphalian, below Asselian-Sakmarian of Permian).

stephanite An iron-black orthorhombic mineral: Ag_5SbS_4. It is an ore of silver. Syn: *brittle silver ore; black silver; goldschmidtine.*

stephanocolpate Said of pollen grains having more than three colpi, meridionally arranged.

stephanocolporate Said of pollen grains having more than three colpi, meridionally arranged and provided with pores.

stephanolith A crown- or star-shaped coccolith.

stephanoporate Said of pollen grains having more than three pores, disposed on the equator.

step lake A lake occupying one of a series of rock basins on a glacial stairway; e.g. a *paternoster lake.*

step-out A well drilled at a distance from a producing oil or gas well in an effort to extend the productive limits of a field. Cf: *extension well.*

stepout *moveout.*

stepout correction A correction for *normal moveout.*

steppe An extensive, treeless grassland area in the semiarid midlatitudes of southeastern Europe and Asia. It is generally considered drier than the *prairie* which develops in the subhumid midlatitudes of the U.S.

stepped crescent One of a succession of crescentic scarps (several centimeters to a meter high at the center) along the course of an arroyo, resembling together a stair with broad steps and shallow risers (Sharpe, 1938, p. 25).

stepped plain A plain that has a succession of levels, "like the steps of a staircase belonging to one flight of stairs" (Schwarz, 1912, p. 95). Syn: *klimakotopedion.*

stepping method A rapid method of determining differences in elevations by the use of stadia hairs. The method is now replaced by one using the Beaman stadia arc.

stepping stone An island used by a species that spread (or is spreading) from one region to another (MacArthur & Wilson, 1967, p.191).

step tablet *step wedge.*

step terrace (a) A man-made terrace having several steplike levels on which crops are cultivated. (b) A terrace with a vertical drop on the downhill side.

steptoe An isolated hill or mountain of older rock surrounded by a lava flow. Syn: *dagala.*

step vein A vein that alternately conforms with and cuts through the country-rock bedding.

step wedge An *optical wedge* whose transparency diminishes in discrete, graduated adjacent steps from one end to the other. Cf: *gray scale.* Syn: *step tablet.*

step zone A zone along a shoreline, located slightly below mean sea level and characterized by sediments that are coarser than those above (on the beach) or below (on the shoreface).

stercorite A white mineral: $HNa(NH_4)(PO_4)\cdot 4H_2O$. Syn: *microcosmic salt.*

stereocomparator A stereoscope for accurately measuring the three space coordinates of the image of a point on a photograph; it is used in making topographic measurements by the comparison of stereoscopic photographs.

stereogenic The adj. form of *stereosome.* Also spelled: *stereogenetic.* Little used.

stereogram [geol] A graphic diagram on a plane surface, giving a three-dimensional representation, such as projecting a set of angular relations; e.g. a block diagram of geologic structure, or a *stereographic projection* of a crystal.

stereogram [photo] A *stereoscopic pair* of photographs correctly oriented and mounted for viewing with a stereoscope. Syn: *stereograph.*

stereograph (a) A stereometer with a pencil attachment, used to plot topographic detail from a stereogram. (b) *stereogram [photo].*

stereographic net *net [struc petrol].*

stereographic projection (a) A perspective, conformal, azimuthal map projection in which meridians and parallels are projected onto a tangent plane, with the point of projection on the surface of the sphere diametrically opposite to the point of tangency of the projecting plane. Any point of tangency may be selected (at a pole, on the equator, or a point in between). It is the only azimuthal projection that is conformal. Stereographic projections are much used for maps of a hemisphere and are useful in showing geophysical relations (such as patterns of island arcs, mountain arcs, and their associated earthquake epicenters). (b) A similar projection used in optical mineralogy and structural geology, made on an equatorial plane (passing through the center of the sphere) with the point of projection at the south pole. Syn: *stereogram [geol].*—See also: *polar stereographic projection.*

stereome (a) The calcareous tissue in the mesodermal endoskeleton of a living echinoderm. Cf: *stroma.* (b) The more or less dense calcareous skeletal deposit generally covering and thickening various parts of a scleractinian or rugose corallite. Cf: *sclerenchyme.*—Also spelled: *stereom.*

stereometer A device used to measure heights of Earth features on a stereoscopic pair of aerial photographs, containing a micrometer movement by means of which the separation of two index marks can be changed to measure parallax dfference. Syn: *parallax bar.*

stereo model A three-dimensional mental impression produced by viewing the left and right images of an overlapping pair of photographs or images with the left and right eye respectively.

stereomodel *stereoscopic image.*

stereonet A term used in structural geology for a *Wulff net.*

stereo pair Two overlapping images or photographs that may be viewed stereoscopically.

stereopair *stereoscopic pair.*

stereophytic Said of a sedimentary rock of organic origin (such as coral rocks and some algal limestones) that has been "built up directly, from the beginning, as a quite solid material" (Tyrrell, 1926, p. 234).

stereoplasm Gelated protoplasm; specif. the relatively solid axis of granular-reticulose pseudopodia in foraminifers, surrounded by a granular fluid outer portion (rheoplasm). It is noted in foraminifers *Peneroplis* and *Elphidium,* but is not visible in most agglutinated types (TIP, 1964, pt.C, p.64).

stereoscope A binocular optical instrument for assisting the observer to view two properly oriented photographs or diagrams to obtain the mental impression of a three-dimensional model (ASP, 1975, p. 2105).

stereoscopic fusion The mental process by which two perspective views or images (one for each eye) are combined and brought to a focus on the retina of each eye to give the impression of a three-dimensional model. Syn: *fusion [photo].*

stereoscopic image The mental impression of a three-dimensional model that results from stereoscopic fusion. Syn: *stereoscopic model; stereomodel.*

stereoscopic model *stereoscopic image.*

stereoscopic pair An *overlapping pair* of photographs that, when properly oriented and used with a stereoscope, gives a three-dimensional view of the area of overlap. See also: *stereogram.* Syn: *stereopair.*

stereoscopic principle The formation of a single, three-dimensional image by simultaneous vision with both eyes of two photographic images of the same terrain taken from different camera stations.

stereoscopic radius The limiting distance at which an object can be seen in stereoscopic relief; it is about 450 m with unaided human vision.

stereoscopic vision Simultaneous vision with both eyes in which the mental impression of depth and distance is obtained, usually by means of two different perspectives of an object (such as two photographs of the same area taken from different camera stations); the viewing of an object in three dimensions. Syn: *stereoscopy; stereovision.*

stereoscopy (a) The science and art that deals with the use of simultaneous vision with both eyes for observation of a pair of overlapping photographs or other perspective views, and with the methods by which such viewing is performed and such effects are produced. (b) *stereoscopic vision.*

stereosome The part of a chorismite that remained solid at all times during the process of formation. Adj: *stereogenic.* Cf: *metaster; restite; chymogenic.* See also: *paleosome.* Little used.

stereosphere A term that was originally proposed for the innermost shell of the Earth's mantle, but is also used as equivalent to the lithosphere. Cf: *chalcosphere.*

stereostatic pressure *geostatic pressure.*

stereotheca The inner layer of a thecal plate of a cystoid. It is thicker than *epitheca.*

stereotriangulation A triangulation procedure that uses a stereoscopic plotting instrument to establish horizontal and/or vertical control by means of successive orientation of stereoscopic pairs of photographs in a continuous strip.

stereovision *stereoscopic vision.*

stereozone An area of dense skeletal deposits in a corallite, generally peripheral or subperipheral in position.

steric Pertaining to phenomena involving molecular dimensions or arrangements in space; e.g. a steric sea level resulting from variations in the density (as by changes in water temperature or salinity) of a column of seawater without change in mass.

sterigma A minute, spiculelike stalk at the apex of a besidium, on which the spore is developed in the basidiomycete fungi (Swartz, 1971, p. 447).

sternal pore A pore in radiolarian skeletons of the subfamily Trissocyclinae, directly below the vertical spine and partly framed by the tertiary-lateral bars. It is not present on specimens having a vertical pore.

sternbergite A dark-brown or black mineral: $AgFe_2S_3$. It occurs in tabular crystals or soft flexible laminae. Syn: *flexible silver ore.*

Sternberg's law The decline in size of a clastic particle transported downstream is proportional to the weight of the particle in water and to the distance it has traveled, or to the work done against friction along the bed: $W = W_0e^{-as}$, where W is the weight at any distance s, W_0 is the initial weight of the particle, and a is the coefficient of size reduction (Pettijohn, 1957, p. 530). This relation was observed by H. Sternberg in 1875.

sternite The ventral part of a somite of an arthropod; e.g. the chitinous plate forming the ventral cover of an abdominal or thoracic segment of an insect, the sclerotized plate forming the ventral cover of a somite of an arachnid or merostome, or the sclerotized ventral surface of a single somite of a crustacean.

sternum (a) In birds, mammals, and their closest extinct relatives, the ventral midsagittal bone with which the distal ends of most anterior ribs articulate. (b) The ventral surface of the body, of a single tagma, or of a somite of an arthropod; the whole ventral wall of the thorax of an arthropod (as of an arachnid). Pl: *sterna.*

steroid Any one of several complex hydrocarbons occurring naturally in living organisms and having a polycyclic structure the same as found in the sterols. The fatty acids found in bile and vitamin D are examples of steroids.

sterraster A sponge spicule (euaster) of globular or kidney shape with a granular surface, formed ontogenetically by expansion of a centrum to engulf all but the tips of the rays.

sterrettite A syn. of *kolbeckite.* It was formerly described as a hydrous phosphate of aluminum: $Al_6(PO_4)_4(OH)_6 \cdot 5H_2O$.

sterryite A mineral: $Pb_{12}(Sb,As)_{10}S_{27}$.

stetefeldtite A mineral of the stibiconite group: $Ag_2Sb_2(O,OH)_7$(?).

stevensite A mineral: $Mg_3Si_4O_{10}(OH)_2$. Syn: *aphrodite.*

stewartite (a) A brownish-yellow mineral: $Mn^{+2}Fe_2{}^{+3}(PO_4)_2(OH)_2 \cdot 8H_2O$. It usually occurs in minute crystals and tufts of fibers in pegmatites. (b) A steel-gray, ash-rich, fibrous variety of bort containing iron, having magnetic properties, and found in the diamond mines of Kimberley, South Africa.

stibarsen A hexagonal mineral: SbAs.

stibianite *stibiconite.*

stibiconite A mineral: $Sb_3O_6(OH)$. It is usually yellow to chalky white, and occurs as an alteration product of stibnite. Syn: *stibianite.*

stibiocolumbite An orthorhombic mineral: $Sb(Nb,Ta)O_4$. It is isomorphous with stibiotantalite.

stibiopalladinite A silver-white to steel-gray isometric mineral: Pd_5Sb_2.

stibiotantalite A brown or yellow orthorhombic mineral: $Sb(Ta,Nb)O_4$. It is isomorphous with stibiocolumbite.

stibium An ancient name for *stibnite* used (as in Egypt) as a cosmetic for painting the eyes.

stibnite A lead-gray mineral: Sb_2S_3. It has a brilliant metallic luster, differs from galena by ease of fusion, and often contains gold and silver. Stibnite occurs in massive forms and in prismatic orthorhombic crystals that show highly perfect cleavage and are striated vertically. It is the principal ore of antimony. Syn: *antimonite; antimony glance; gray antimony; stibium.*

stichtite A lilac-colored rhombohedral mineral: $Mg_6Cr_2(CO_3)(OH)_{16} \cdot 4H_2O$. It is dimorphous with barbertonite and may contain some iron.

stick-slip A jerky, sliding motion associated with fault movement in laboratory experiments. It may be a mechanism in shallow-focus earthquakes.

sticky limit The lowest water content at which a soil will adhere to a metal blade drawn across the surface of the soil mass (ASCE, 1958, term 353). Cf: *sticky point.*

sticky point A condition of consistency at which a soil material barely fails to adhere to a foreign object; specif. the moisture content of a well-mixed, kneaded soil material that barely fails to adhere to a polished nickel or stainless-steel surface when the shearing speed is 5 cm/sec (SSSA, 1965, p.349). Cf: *sticky limit.*

stictolith Chorismite with spotted appearance (Dietrich & Mehnert, 1961). Little used.

stiff clay Clay of low plasticity.

stiffening limit The water content (expressed as percent by weight of dry soil) at which a thoroughly stirred thixotropic soil still flows under its own weight in a test tube of 11 mm diameter after exactly one minute of rest (Mielenz & King, 1955, p.223).

stigma [bot] That part of a pistil (usually at the tip of the style) which receives pollen grains and on which they germinate.

stigma [paleont] An opening of an arachnid; esp. an opening of an air-conveying tube forming the respiratory system or an opening into a saccular breathing organ (book lung). Pl: *stigmata.* Syn: *spiracle [paleont].*

Stigmaria A genus of Permian and Carboniferous plants, identified by features of their rhizophores and roots. *Stigmaria* is regarded as a form genus; it is approximately correlated with *Sigillaria, Lepidodendron, Bothrodendron,* and others.

stilbite (a) A zeolite mineral: $NaCa_2Al_5Si_{13}O_{36} \cdot 14H_2O$. It occurs in sheaflike aggregates of crystals and also in radiated masses. Syn: *desmine; epidesmine.* (b) A term used by German mineralogists as a syn. of *heulandite.*

stilleite A mineral: ZnSe.

stillstand (a) Stability of an area of land, as a continent or island, with reference to the Earth's interior or mean sea level, as might be reflected, for example, by a relatively unvarying base level of erosion between periods of crustal movement. (b) A period of time during which there is a stillstand.—Syn: *standstill; stand.*

stillwater (a) A reach of a stream that is so nearly level as to have no visible current or other motion. Syn: *quiet reach.* (b) A sluggish stream, the water of which appears to be marked by little or no agitation.—Also spelled: *still water.*

stillwaterite A hexagonal mineral: Pd_8As_3.

stillwellite A rhombohedral mineral: $(Ce,La,Ca)BSiO_5$.

stilpnomelane A black or green-black mineral, approximately: $K(Fe^{+2},Fe^{+3},Al)_{10}Si_{12}O_{30}(OH)_{12}$. It occurs in micalike plates, fibrous forms, and velvety bronze-colored incrustations. Syn: *chalcodite.*

stinkquartz A variety of quartz that emits a fetid odor when struck.

stinkstein A syn. of *stinkstone.* Etymol: German *Stinkstein.*

stinkstone A stone that emits an odor on being struck or rubbed; specif. a *bituminous limestone* (or brown dolomite) that gives off a fetid smell (owing to decomposition of organic matter) when rubbed or broken. It may emit a "sweet-and-sour" smell if the carbonate rock is rich in organic-phosphatic material. See also: *anthraconite.* Syn: *stinkstein.*

stipe [bot] The stalk of a pistil or other small organ when axile in origin; also, the petiole of a fern leaf (Lawrence, 1951, p.771). Adj: *stipitate.*

stipe [paleont] One branch of a branched graptolite colony (rhabdosome), made up of a series of overlapping tubes (thecae). In an unbranched rhabdosome, a stipe is the entire graptolithine colony.

stipitate Said of a plant having a *stipe* or special stalk.

stipoverite *stishovite.*

stipule A basal appendage of a petiole in some dicotyledenous plants. There are usually two stipules on a complete leaf.

stirrup-pore In archaeocyathids, a pore excavated from the axial edges of a septum where it joins the inner wall (TIP, 1972, pt. E, p. 19).

stishovite A tetragonal mineral: SiO_2. It is a high-pressure, extremely dense (4.35 g/cm^3) polymorph of quartz, produced under static conditions at pressures above about 100 kb and found naturally associated with coesite and only in shock-metamorphosed quartz-bearing rocks such as those from Barringer Crater (Meteor

Crater), Ariz., and the Ries basin, Germany. Its occurrence provides a criterion for meteorite impact. Stishovite has a closely packed rutile type of structure in which the silicon has a coordination number of 6 (instead of 4 as in quartz and coesite); it forms at higher pressures than coesite and is apparently less stable at lower pressures after formation. Syn: *stipoverite.*

stistaite A mineral: SnSb.

stochastic hydrology That branch of hydrology involving the "manipulation of statistical characteristics of hydrologic variables with the aim of solving hydrologic problems, using the stochastic properites of the events" (Hofmann, 1965, p. 120). Cf: *parametric hydrology; synthetic hydrology.*

stochastic process A process in which the *dependent variable* is random (so that prediction of its value depends on a set of underlying probabilities) and the outcome at any instant is not known with certainty. Ant: *deterministic process.* Syn: *random process.*

stock [intrus rocks] An igneous intrusion that is less than 40 sq mi (100 sq km) in surface exposure, is usually but not always discordant, and resembles a batholith except in size. Cf: *boss [ign].*

stock [ore dep] A rarely used term for a chimneylike orebody; a syn. of *pipe [ore dep].*

stock [paleont] The terminal bar of an *anchor* of a holothurian, perpendicular to the shank and of varying shape.

stockade Piling that serves as a breakwater.

stockwork A mineral deposit consisting of a three-dimensional network of planar to irregular veinlets closely enough spaced that the whole mass can be mined. Cf: *chambered; reticulate.* Syn: *network deposit; stringer lode.*

stockwork molybdenum deposit *porphyry molybdenum deposit.*

stoichiometric With reference to a compound or a phase, pertaining to the exact proportions of its constituents specified by its chemical formula. It is generally implied that a stoichiometric phase does not deviate measurably from its ideal composition.

stoichiometric coefficient One of the numerical coefficients in an equation that specify the combining proportions of the reactants and products of the reaction described.

Stokes' formula A formula first published in 1849 for computing the undulations of *compensated geoids* from gravity anomalies.

stokesite A colorless orthorhombic mineral: $CaSnSi_3O_9 \cdot 2H_2O$.

Stokes' law A formula that expresses the rates of settling of spherical particles in a fluid: $V = Cr^2$, where V is velocity (in cm/sec), r is the particles' radius (in cm), and C is a constant relating relative densities of fluid and particle, acceleration due to gravity, and the viscosity of the fluid. It is named after Sir George Stokes, British mathematician and physicist. Cf: *impact law.*

stolidium The thin marginal extension of one or both valves of certain brachiopods (as in the superfamily Stenoscismatacea) forming a frill protruding at a distinct angle to the main contour of the shell (TIP, 1965, pt.H, p.153).

stolon (a) A slender internal threadlike tubule from which graptolithine thecae appear to originate, lying within the common canal. (b) A creeping and ribbonlike or membranous basal expansion from which certain octocorallian polyps arise (e.g. those of Stolonifera and Telestacea). (c) A slender tube of kenozooids bearing autozooids of ctenostomate bryozoans along its length. (d) A small calcareous tubelike projection serving as a connection between chambers in the test of an orbitoid foraminifer.

stolotheca A type of graptolithine theca (tube) that encloses the main stolon and proximal parts of three new thecae (*autotheca* and *bitheca,* and daughter stolotheca). It is probably secreted by an immature autothecal zooid, constituting in effect the proximal part of the autotheca. Stolotheca is equivalent to *protheca* of graptoloids.

stolzite A tetragonal mineral: $PbWO_4$. It is isomorphous with wulfenite and dimorphous with raspite.

stoma A pore flanked by guard cells in the epidermis of a leaf, through which gases within the leaf are exchanged. Pl: *stomata.*

stomach stone *gastrolith.*

stomatal coccolith One of the modified coccoliths surrounding the flagellar field in flagellate coccolithophores exhibiting dimorphism.

stomodaeum An esophaguslike tubular passageway or pharynx leading from the mouth of a coral polyp to the gastrovascular cavity (TIP, 1956, pt.F, p.251). Pl: *stomodaea. Adj: stomodaeal.*

stomostyle A thickened outer membrane invaginated in cytoplasm of the apertural region of some foraminifers, from which the mass of cytoplasm emerges giving rise to pseudopodia.

stone (a) A general term for rock that is used in construction, either crushed for use as *aggregate* or cut into shaped blocks as *dimension stone.* (b) One of the larger fragments in a variable matrix of a sedimentary rock; a *phenoclast* or *megaclast.* (c) A *stony meteorite.* (d) A cut and polished natural gemstone; a gem or precious stone. The term is used incorrectly for an artificial reproduction of, or a substitute for, a gem.

Stone Age In archaeology, a cultural level that was originally the first division of the *three-age system,* and was subsequently divided into the *Paleolithic, Mesolithic, and Neolithic.* It is characterized by the use of materials other than metal, e.g. stone, wood, or bone, for technical purposes. Correlation of relative cultural levels with actual age (and, therefore, with the time-stratigraphic units of geology) varies from region to region; e.g. this oldest cultural level has been discovered to exist in recent times.

stone band *dirt band [coal].*

stone-banked terrace A *sorted step* whose steep slope is bordered by stones. The term should be reserved for a terracelike feature that lacks a regular pattern and is not a well-defined form of patterned ground (Washburn, 1956, p.833). Cf: *stone garland.*

stone bind An English term for interbedded layers of sandstone and shale, or for a rock (such as siltstone) intermediate between a sandstone and a mudstone (Arkell & Tomkeieff, 1953, p.115).

stone-bordered strip *stone stripe.*

stone bubble *lithophysa.*

stone canal A calcified, typically short tube or *canal* leading from the madreporite to the ring canal in the water-vascular system of an echinoderm.

stone cells Plant cells that have a strongly thickened wall, are usually lignified, commonly occur isolated in pulpy tissue, and are either isodiametric or highly irregular in form. Syn: *sclerid.*

stone circle *sorted circle.*

stone clunch A term used in England for a very hard underclay (*clunch*) with interbeds of sand.

stone coal *anthracite.*

stone eye *stone intrusion.*

stone field *block field.*

stone gall A clay concretion found in certain sandstones. Also spelled: *stonegall.*

stone garland A *sorted step* consisting of a tongue-shaped mass of fine material enclosed on the downslope side by a crescentic stony embankment; similar to but smaller than a *stone-banked terrace.* Syn: *stone semicircle; garland.*

stone guano A secondary deposit formed by leaching of guano and consequent enrichment in insoluble phosphates.

stone intrusion An irregular, bulbous, sometimes much distorted *sandstone dike* occurring within a coal seam or penetrating it (frequently from top to bottom), and always connected with a similar sandstone in the roof or in higher strata. The British usage of the term *intrusion* for such a mass of sedimentary rock in a coal seam is not recommended (BSI, 1964, p. 10). Cf: *drop [coal].* Syn: *stone eye.*

stone lace *stone lattice.*

stone land Legally, an area that is economically valuable for some variety of stone, such as granite or sandstone, that can be quarried. Cf: *mineral lands.*

stone lattice A *honeycomb structure [weath]* produced on a rock face in a desert by sandblast that "pecks away at the softer places and leaves the harder ones in relief" (Hobbs, 1912, p. 205). Syn: *stone lace.*

Stoneley wave A type of *guided wave* that is propagated along an internal surface of discontinuity; an interface wave. Cf: *channel wave.*

stone line A broken line of angular and subangular rock fragments, paralleling a sloping topographic surface and lying just above the parent material of a soil at a depth of a few meters below that surface (Sharpe, 1938, p. 24); it outcrops in natural and artificial cuts. Syn: *carpedolith.*

stone mesh *sorted polygon.*

stone net *sorted polygon.*

stone packing "A frost structure exclusively occurring on pebble beaches in arctic areas, consisting of a large, flat-lying boulder surrounded by a cluster of edgewise-lying flat stones, in an arrangement resembling the petals of a rose" (Schieferdecker, 1959, term 2403).

stone pavement An accumulation of rock fragments, esp. pebbles and boulders, in which the surface stones lie with a flat side up and are fitted together like a mosaic (Washburn, 1973, p. 149). Cf:

subnival boulder pavement.

stone peat The dark, compacted peat at the bottom of a bog.

stone pit A shallow *sorted circle,* less than a meter in diameter, consisting of a floor of isolated and dominantly circular stones (without finer material) surrounded by thick vegetation; term introduced by Lundqvist (1949, p.336).

stone pitch *Pitch* that is as hard as stone.

stone polygon *sorted polygon.*

stone reef A longshore bar whose upper 3-4 m has been solidly cemented by calcium carbonate derived from organic material. Examples occur off the coast of Brazil near Recife.

stone ring A syn. of *sorted circle* and *sorted polygon;* the term refers to the circular or polygonal border of stones surrounding a central area of finer material.

stone river A term used in the Falkland Islands for a *rock stream* formed by solifluction. Cf: *block stream.*

stone semicircle *stone garland.*

stone stream *block stream.*

stone stripe A *sorted stripe* consisting of coarse rock debris, and occurring between wider stripes of finer material. Cf: *block stripe; soil stripe.* Syn: *stone-bordered strip; rock stripe.*

stone wall *hogback.*

stoneware clay A clay suitable for manufacture of stoneware (ceramic ware fired to a hard, dense condition and with an absorption of less than 5%), used for items such as crocks, jugs, and jars. It possesses good plasticity, fusible minerals, and a long firing range.

stonewort *charophyte.*

stone wreath *sorted circle.*

stony desert A desert area whose surface has been deflated, leaving a concentration of coarse fragments after the removal of sand and dust particles, as a gravel-strewn *reg* or a pebble-strewn *serir;* a desert surface covered with *desert armor.* Cf: *rock desert.*

stony-iron meteorite A general name for relatively rare meteorites containing large (at least 25%) and approximately equal amounts (by weight) of both nickel-iron and heavy basic silicates (such as pyroxene and olivine); e.g. *pallasite* and *mesosiderite.* Syn: *stony-iron; siderolite; iron-stony meteorite; lithosiderite; sideraerolite.* Obsolete syn: *syssiderite; aerosiderolite.*

stony meteorite A general name for meteorites consisting largely or entirely of silicate minerals (chiefly olivine, pyroxene, and plagioclase); e.g. *chondrite* and *achondrite.* Stony meteorites resemble ultramafic rocks in composition, and they constitute more than 90% of all meteorites seen to fall. Syn: *meteoric stone; asiderite.* Obsolete syn: *aerolite; brontolith; meteorolite.*

stop A dam or weir.

stopbank An Australian term for a levee.

stope An underground excavation formed by the extraction of ore. Cf: *stoping.*

stoping [intrus rocks] *magmatic stoping.*

stoping [mining] Extraction of ore in an underground mine by working laterally in a series of levels or steps in the plane of a vein. It is generally done from lower to upper levels, so that the whole vein is ultimately removed. The process is distinct from working in a shaft or tunnel or in a room in a horizontal drift, although the term is used in a general sense to mean the extraction of ore. See also: *stope.*

storage (a) Artificially impounded water, in surface or subsurface reservoirs, for future use. Also, the amount of water so impounded. (b) Water naturally detained in a drainage basin, e.g. ground water, depression storage, and channel storage.

storage coefficient (a) For surface waters such as a reservoir, a coefficient expressing the relationship of the surface area to the mean annual flow that supplies it. (b) For an aquifer, the volume of water released from storage in a vertical column of 1.0 sq ft when the water table or other piezometric surface declines 1.0 ft. In an unconfined aquifer, it is approximately equal to the specific yield (Theis, 1938, p. 894). Syn: *coefficient of storage.*

storage curve *capacity curve.*

storage ratio The net available storage of a body of water divided by the mean annual flow (Langbein & Iseri, 1960, p. 18).

storage-required frequency curve A graphic illustration of the frequency with which storage equal to or greater than selected amounts will be required to maintain selected rates of regulated flow (Langbein & Iseri, 1960, p. 18).

storied Said of a landform or landscape characterized by two or more adjacent levels; e.g. a *storied* peak plain with summit levels at different altitudes. See also: *two-story cliff.*

storis A floating mass of closely crowded icebergs and floes, esp. the remnants of the thickest polar ice drifting from the Arctic Ocean into the North Atlantic between Spitsbergen and Greenland. Pl: *storis.* Etymol: Danish, "large ice".

storm A temporary, substantial disturbance of a geophysical field, e.g. a magnetic storm.

storm beach (a) A low, rounded ridge of coarse gravel, cobbles, and boulders, piled up by powerful storm waves behind or at the inner margin of a beach, above the level reached by normal high spring tides or by ordinary waves. Syn: *storm terrace.* (b) A beach as it appears immediately after an exceptionally violent storm, characterized either by removal or deposition of beach materials.

storm berm A low ridge along a beach, marking the limit of wave action during storms. See also: *winter berm.*

storm cusp A transient *cusp,* developed during a period of relatively heavy seas. Distance between storm cusps is 70 to 120 m. Term introduced by Evans (1938).

storm delta *washover.*

stormflow *direct runoff.*

storm icefoot An icefoot produced by the breaking of a heavy sea and consequent freezing of the wind-driven spray.

storm microseism A long-period microseism (25+ sec) caused by ocean waves.

storm roller A term used by Chadwick (1931) for a wave-formed sedimentary structure "mismentioned" as a concretion. The feature is now regarded as a *ball-and-pillow structure* or *flow roll* formed by load deformation.

storm runoff *direct runoff.*

storm seepage The *runoff [water]* infiltrating the surface soil and moving toward streams as ephemeral shallow perched ground water above the main ground-water level. It is usually considered part of direct runoff (Langbein & Iseri, 1960). Syn: *subsurface storm flow; subsurface runoff; subsurface flow; shallow percolation; interflow [water].* Cf: *surface runoff; ground-water runoff.*

storm surge An abnormal, sudden rise of sea level along an open coast during a storm, caused primarily by onshore-wind stresses, or less frequently by atmospheric pressure reduction, resulting in water piled up against the coast. It is most severe when accompanied by a high tide. Erroneous syn: *tidal wave; storm tide.* Syn: *surge [waves]; storm wave; hurricane surge; hurricane wave.*

storm terrace *storm beach.*

storm tide An erroneous syn. of *storm surge.*

storm water *direct runoff.*

storm wave *storm surge.*

stoss Said of the side of a hill or knob that faces the direction from which an advancing glacier or ice sheet moved; facing the upstream side of a glacier, and most exposed to its abrasive action. Etymol: German *stossen,* "to push, thrust". Ant: *lee.*

stoss-and-lee topography An arrangement, in a strongly glaciated area, of small hills or prominent rocks having gentle slopes on the stoss side and somewhat steeper, plucked slopes on the lee side; this arrangement is the reverse of *crag and tail.* Syn: *onset-and-lee topography.*

stottite A brown tetragonal mineral: $FeGe(OH)_6$.

straat A term used in the Kalahari region of southern Africa for the trough between dunes, often floored with clayey sand, and generally 100-150 m wide. Etymol: Afrikaans, "street". Pl: *straate.* See also: *street.*

straddle spread *split spread.*

straight-line evolution *orthogenesis.*

straight suture An externally visible line of articular contact perpendicular to the longitudinal axis of adjoined crinoid ossicles.

strain Change in the shape or volume of a body as a result of *stress;* a change in relative configuration of the particles of a substance. Syn: *deformation.*

strain ellipse An ellipse in the deformed state that is derived from a circle in the undeformed state.

strain ellipsoid An ellipsoid in the deformed state that is derived from a sphere in the undeformed state. The sphere is considered to have unit radius, and the ellipsoid accordingly has principal semi-axes equal in length to the principal strains expressed as *stretches.* Syn: *deformation ellipsoid.*

strain field The array of states of strain at each point within a volume of material.

strain gage A general term for a device with which mechanical strain can be measured, commonly by an electrical signal, e.g. a wire strain gage.

strain hardening The behavior of a material whereby each additional increment of strain requires an additional increment of differential stress.

strain recrystallization Recrystallization in which a deformed mineral changes to a mosaic of undeformed crystals of the same mineral, e.g. strained to unstrained calcite (Folk, 1965, p.15).

strain seismometer A seismometer that is designed to detect deformation of the ground by measuring relative displacement of two points.

strain shadow (a) *undulatory extinction.* (b) *pressure shadow.*

strain-slip cleavage *slip cleavage.*

strait [coast] A relatively narrow waterway connecting two larger bodies of water, as the Strait of Gibraltar linking the Atlantic Ocean with the Mediterranean sea; a small *channel.* The term is often used in the plural. Cf: *sound.*

strait [geog] (a) A neck of land. (b) An obsolete term for a gorge.

strand [coast] (a) A syn. of *shore* and *beach;* the land bordering any large body of water, esp. the beach of a sea or an arm of the ocean, or the bank of a large river. (b) An Anglo-Saxon term for the narrow strip of land lying between high water and low water, being alternately exposed and covered by the tide.

strand [streams] A British term for a stream or current, and for a channel.

strand crack A fissure at the junction of a sheet of inland ice, an ice piedmont, or an ice rise with an ice shelf, the latter being subject to the rise and fall of the tide (Armstrong & Roberts, 1958, p.96).

stranded ice Floating ice that is deposited on the shore by retreating high water. Cf: *grounded ice; shore ice.*

strandflat (a) Any *wave-cut platform;* esp. a low, flat platform up to 65 km wide, extending for many hundreds of kilometers along the rocky coast of western Norway, either partly submerged or standing slightly above the present sea level as a result of isostasy, and supporting thousands of stacks, skerries, and other small islands. (b) A discontinuous shelf of land inside a fjord, reaching to about 30 m in height, having a rounded and dissected form.—Sometimes spelled: *strand flat.*

strandline (a) The ephemeral line or level at which a body of standing water, e.g. the sea, meets the land; the *shoreline,* esp. a former shoreline now elevated above the present water level. (b) A beach, esp. one raised above the present sea level.—Also spelled: *strand line.* See also: *raised beach.*

strand mark Any inorganic sole mark on a sedimentary surface along the shore (Clarke, 1918). Cf: *undertow mark.* Syn: *strand marking.*

strand plain A prograded shore built seaward by waves and currents, and continuous for some distance along the coast (Cotton, 1958, p. 431). Syn: *foreland.*

stranskiite A blue triclinic mineral: $Zn_2Cu(AsO_4)_2$.

strashimirite A mineral: $Cu_8(AsO_4)_4(OH)_4 \cdot 5H_2O$.

strata Plural of *stratum.*

strata-bound Said of a mineral deposit confined to a single stratigraphic unit. The term can refer to a *stratiform* deposit, to variously oriented orebodies contained within the unit, or to a deposit containing veinlets and alteration zones that may or may not be strictly conformable with bedding. See: Wolf, 1976. Cf: *bedded [ore dep].*

strata control *roof control.*

stratal Pertaining to a stratum or strata; e.g. "stratal dip" or "stratal unit".

strata time *Geologic time* estimated from the thickness of strata and the rate of deposition (Kobayashi, 1944a, p. 476).

strategic materials Materials that are vital to the security of a nation, but that must be procured entirely or in large part from foreign sources because the available domestic production will not meet the nation's requirements in time of war; e.g. *strategic minerals.*

strategic minerals Minerals that are considered to be *strategic materials;* e.g. chromium- and tin-bearing minerals, quartz crystal, and sheet mica were some of the "strategic minerals" during World War II.

strath [geomorph] (a) An extensive undissected terracelike remnant of a broad, flat valley floor that has undergone dissection following uplift; e.g. a continuous river terrace along a valley wall, interrupted in its development during the mature stage of a former erosion cycle. Bucher (1932) prefers the term *strath terrace* for this feature. Bascom (1931) proposed that "strath" be replaced by *berm.* (b) A broad, flat valley bottom formed in bedrock and resulting from degradation, "first by lateral stream cutting and later by whatever additional processes of degradation may be involved" (Bucher, 1932, p. 131); a level valley floor representing a local base level. It is usually covered by a veneer of alluvium, and is wider and flatter than a *glen.* Etymol: Scottish, from the Gaelic *strath,* "valley". Syn: *incipient peneplain.*

strath [marine geol] An elongate, broad, steep-sided depression on the continental shelf, usually glacial in origin. It is often deeper on its nearshore side.

strath stage The stage in the peneplanation of a region when the main streams have carved broad, flat-floored valleys that are graded to the same regional base level.

strath terrace (a) A term used by Bucher (1932) for an extensive remnant of a *strath* (i.e. a flat valley bottom) that belonged to a former erosion cycle and that has undergone dissection by a rejuvenated stream following uplift. The term is synonymous with *strath* and *berm* as used by other writers. Cf: *fillstrath terrace.* (b) A *strath* (i.e. a remnant) of greater extension than that of a narrow ribbon along one valley (Von Engeln, 1942, p. 222).

strath valley (a) A stream valley characterized by the development of a flat valley bottom (strath) resulting from degradation (Bucher, 1932, p.131). (b) A valley abandoned by a stream whose course was dislocated (Von Engeln, 1942, p. 224).

stratic Pertaining to or designating the order or sequence of strata; stratigraphic, or pertaining to stratigraphy. Grabau (1924, p.821) referred to a disconformity as a "stratic unconformable relation".

straticulate Characterized by numerous very thin parallel layers, whether separable or not, either of sedimentary deposition (as a bed of clay) or of deposition from solution (as in a stalagmite or banded agate).

straticulation The formation of a straticule or straticules.

straticule A French term for a thin sedimentary layer, or *lamina.*

stratification [lake] The state of a body of water consisting of two or more horizontal layers of differing characteristics; esp. the arrangement of the waters of a lake in layers of differing densities. See also: *density stratification; thermal stratification.*

stratification [sed] (a) The formation, accumulation, or deposition of material in layers; specif. the arrangement or disposition of sedimentary rocks in strata. See also: *bedding [stratig].* (b) A structure produced by deposition of sediments in strata; a stratified formation, or stratum. It may be due to differences of texture, hardness, cohesion or cementation, color, internal structure, and mineralogic or lithologic composition. (c) The state of being stratified; a term describing a layered or bedded sequence, or signifying the existence of strata.

stratification [snow] Layering in a mass of snow, firn, or ice; it is caused by discontinuous deposition, by sedimentation of different kinds of snow, by accumulation of rock dust during summer periods, or by the development of layers of *depth hoar* at times of rapid changes in temperature.

stratification index A measure of the "beddedness" of a stratigraphic unit, expressed as the number of beds in the unit per 100 feet of section (Kelley, 1956, p.295). It is determined by multiplying the number of beds times 100, and dividing by the unit's thickness in feet. See also: *isostratification map; phyllofacies.* Syn: *beddedness index; bedding index.*

stratification plane *bedding plane.*

stratified Formed, arranged, or laid down in layers or strata; esp. said of any layered sedimentary rock or deposit. See also: *bedded [stratig].*

stratified cone A less-preferred syn. of *stratovolcano.*

stratified drift Glaciofluvial, glaciolacustrine, or glaciomarine drift, consisting of sorted and layered material deposited by a meltwater stream or settled from suspension in a body of quiet water adjoining the glacier. Cf: *till.* Obsolete syn: *washed drift; modified drift.*

stratified estuary An estuary in which salinity increases with depth as well as along its length. An estuary is "highly stratified" if there is a density discontinuity separating surface river flow and bottom seawater; it is "slightly stratified" if the amount of increase in salinity with depth is not significant. Ant: *vertically mixed estuary.*

stratified lake A lake exhibiting *stratification* of its waters.

stratified rock A rock displaying stratification. The term is virtually synonymous with *sedimentary rock,* although some sedimentary rocks (such as tillite) are without internal stratification. The term is sometimes, but not generally, applied to layered igneous rocks.

stratiform [ore dep] Said of a special type of *strata-bound* deposit in which the desired rock or ore constitutes, or is strictly coextensive with, one or more sedimentary, metamorphic, or igneous layers; e.g. beds of salt or iron oxide, or layers rich in chromite or platinum in a layered igneous complex. See: Wolf, 1976. Cf: *bedded [ore dep].*

stratiform [sed struc] Having the form of a layer, bed, or stratum; consisting of roughly parallel bands or sheets, such as a "stratiform intrusion". Incorrect spellings: "strataform", "stratoform".

stratiform intrusion *layered intrusion.*

stratify To lay down or arrange in strata.

stratigrapher A geologist who studies or specializes in stratigraphy.

stratigraphic break *break [stratig].*

stratigraphic classification The arbitrary but systematic arrangement, zonation, or partitioning of the sequence of rock strata of the Earth's crust into units with reference to any or all of the many different characters, properties, or attributes which the strata may possess (Hedberg, 1958, p.1881-1882).

stratigraphic code A usefully comprehensive, yet concisely stated, formulation of generally accepted views on stratigraphic principles, procedures, and practices, designed to obtain the greatest possible uniformity in applying such principles, etc.; specif. "a systematic collection of rules of formal stratigraphic classification and nomenclature" (ACSN, 1961, art.3). It is applicable to all kinds of rocks (sedimentary, igneous, metamorphic).

stratigraphic column *geologic column.*

stratigraphic condensation The mingling or intimate association, in a very thin stratigraphic interval or even a single bed, of fossils representative of different ages and environments (ISG, 1976, p. 48).

stratigraphic control [ore dep] The influence of stratigraphic features on ore deposition, e.g. ore minerals selectively replacing calcareous beds. Cf: *structural control.*

stratigraphic control [stratig] The degree of understanding of the stratigraphy of an area; the body of knowledge that can be used to interpret its stratigraphy or geologic history.

stratigraphic correlation The process by which stratigraphic units in two or more separated areas are demonstrated or determined to be laterally similar in character or mutually correspondent in stratigraphic position, as based on geologic age (time of formation), lithologic characteristics, fossil content, or any other property; *correlation [geol]* in the usual or narrowest sense. Unless otherwise stated, the term usually means *time-correlation.* See also: *lithologic correlation.*

stratigraphic cutoff *cutoff [stratig].*

stratigraphic facies Facies distinguished primarily on the basis of form, nature of boundaries, and mutual relations, to which appearance and composition are subordinated (Weller, 1958, p.627). These facies are all stratigraphic bodies of one kind or another; they may occur in vertical succession and have boundaries that are more or less horizontal stratigraphic planes (e.g. systems, formations, biostratigraphic zones, and lithostromes), or they may be laterally intergrading parts of some kind of a stratigraphic unit and separated at more or less arbitrary vertical cutoff boundaries (e.g. lithofacies), or they may bear both lateral and vertical relations to each other and have irregular boundaries (e.g. the magnafacies of Caster, 1934). See also: *facies.* Cf: *petrographic facies.*

stratigraphic geology *stratigraphy.*

stratigraphic guide In mineral exploration, a rock unit known to be associated with an ore. Cf: *lithologic guide.* See also: *ore guide.*

stratigraphic heave An obsolete syn. of *gap [fault]* and of *overlap [fault].*

stratigraphic interval The body of strata between two stratigraphic *markers* (ISG, 1976, p. 14). Syn: *interval [stratig].*

stratigraphic leak The deposition of sediments and/or fossils of a younger age within or under rocks of an older age; such a deposit may be said to be laid down in intraposition (Foster, 1966). It frequently involves microfossils, such as conodonts, which have descended through crevices or solution channels to lodge in a lower stratum where they become associated with fossils of greater age. Var: *stratigraphic leakage.* See also: *intrapositional deposit.*

stratigraphic map A map that shows the areal distribution, configuration, or aspect of a stratigraphic unit or surface. It involves a span of geologic time. Examples include isopach map, structure-contour map, facies map, and vertical-variability map.

stratigraphic nomenclature The proper names given to specific stratigraphic units; for example, *Trenton* Formation, *Jurassic* System, *Dibunophyllum* Range-zone (ISG, 1976, p. 13).

stratigraphic overlap An obsolete syn. of *overlap [fault].*

stratigraphic paleontology The study of fossils and of their distribution in various geologic formations, emphasizing the stratigraphic relations (time and sequence) of the sedimentary rocks in which they are contained. Cf: *biostratigraphy.*

stratigraphic range The distribution or spread of any given species, genus, or other taxonomic group of organisms through geologic time, as indicated by its distribution in strata whose geologic age is known. Also, the persistence of a fossil organism through the stratigraphic sequence. Syn: *range [stratig]; geologic range; time-rock span.*

stratigraphic record The *geologic record* based on or derived from a study of the stratigraphic sequence; the rocks arranged chronologically as in a geologic column. Syn: *record.*

stratigraphic reef A thick, laterally restricted mass of carbonate rock, without genetic connotations (Dunham, 1970, p. 1931). Cf: *ecologic reef.*

stratigraphic section *geologic section.*

stratigraphic separation The thickness of the strata that originally separated two beds brought into contact at a fault. Syn: *stratigraphic throw.*

stratigraphic sequence A chronologic succession of sedimentary rocks from older below to younger above, essentially without interruption; e.g. a *sequence* of bedded rocks of interregional scope, bounded by unconformities.

stratigraphic terminology The unit terms used in stratigraphic classification, such as *formation, stage, biozone* (ISG, 1976, p. 13).

stratigraphic test A hole drilled to obtain information on the thickness, lithology, sequence, porosity, and permeability of the rock penetrated, or to locate the position of a key bed. It is frequently drilled to evaluate a potentially productive oil or gas zone. Cf: *structure test.* Syn: *strat test.*

stratigraphic throw *stratigraphic separation.*

stratigraphic trap A *trap* for oil or gas that is the result of lithologic changes rather than structural deformation. See also: *shale-out; pinch-out.* Cf: *structural trap; combination trap.* Syn: *porosity trap.*

stratigraphic unconformity *disconformity.*

stratigraphic unit A stratum or body of adjacent strata recognized as a unit in the classification of a rock sequence with respect to any of the many characters, properties, or attributes that rocks may possess (ISG, 1976, p. 13), for any purpose such as description, mapping, and correlation. Rocks may be classified stratigraphically on the basis of lithology (lithostratigraphic units), fossil content (biostratigraphic units), age (chronostratigraphic units), or properties (such as mineral content, radioactivity, seismic velocity, electric-log character, chemical composition) in categories for which formal nomenclature is lacking. A geologic-time unit is not a stratigraphic unit.

stratigraphy (a) The science of rock strata. It is concerned not only with the original succession and age relations of rock strata but also with their form, distribution, lithologic composition, fossil content, geophysical and geochemical properties — indeed, with all characters and attributes of rocks *as strata;* and their interpretation in terms of environment or mode of origin, and geologic history. All classes of rocks, consolidated or unconsolidated, fall within the general scope of stratigraphy. Some nonstratiform rock bodies are considered because of their association with or close relation to rock strata (ISG, 1976, p. 12). Syn: *stratigraphic geology.* (b) The arrangement of strata, esp. as to geographic position and chronologic order of sequence. (c) The sum of the characteristics studied in stratigraphy; the part of the geology of an area or district pertaining to the character of its stratified rocks. (d) A term sometimes used to signify the study of *historical geology.*

stratlingite A trigonal mineral: $Ca_2Al_2SiO_7 \cdot 8H_2O$.

stratofabric The arrangement of strata in any body of stratified rock, "from the dimensions of a thin section to those of a sedimentary basin" (Fischer, 1964, p.148).

stratomere Any segment of a rock sequence, irrespective of magnitude (Challinor, 1978, p. 295).

stratose Arranged in strata.

stratosphere The outer layer of the atmosphere, overlying the *troposphere.*

stratotectonic A term used, mainly in Australia, to describe tectonic evolution in relation to stratigraphy.

stratotype The original, or subsequently designated, type representative of a named stratigraphic unit, or of a stratigraphic boundary identified as a point in a specific sequence of rock strata. It constitutes the standard for the definition and recognition of that stratigraphic unit or boundary (ISG, 1976, p. 14). See also: *component-stratotype; composite-stratotype; boundary stratotype.* According to the International Stratigraphic Guide (1976), stratotype is a syn. of *type section.*

stratous Composed of strata.

stratovolcano A volcano that is constructed of alternating layers of lava and pyroclastic deposits, along with abundant dikes and sills. Viscous, acidic lava may flow from fissures radiating from a central vent, from which pyroclastics are ejected. Syn: *composite volcano; composite cone.* Less-preferred syn: *bedded volcano; stratified cone.*

strat test *stratigraphic test.*

stratum A tabular or sheetlike body or *layer [stratig]* of sedimentary rock, visually separable from other layers above and below; a *bed [stratig].* It has been defined as a stratigraphic unit that may be composed of a number of beds (Dana, 1895, p.91), as a layer greater than 1 cm in thickness and constituting a part of a bed (Payne, 1942, p.1724), and as a general term that includes both "bed" and "lamination" (McKee & Weir, 1953, p.382). The term is more frequently used in its plural form, *strata.* Cf: *lamina.*

stratum plain A plain having a *stripped structural surface.* Examples in the Colorado Plateau region of the U.S. commonly form isolated buttes and mesas, or benchlike or terracelike areas along valley sides, but some have considerable areal extent. See also: *dip plain; cut plain.* Syn: *structural plain; stripped plain.*

Straumanis camera method In X-ray diffraction analysis, a method of mounting film in a cylindrical X-ray camera to allow for recording of both front and back reflections on both sides of the exit and entry ports, enabling the determination of film diameter from the measurements. Cf: *Wilson technique.*

straw stalactite *tubular stalactite.*

stray (a) A lenticular or discontinuous rock unit encountered unexpectedly in drilling *development wells.* (b) A thin, local rock unit separated by a short interval from a thicker, more persistent formation of similar lithology; e.g. *stray sand.*

stray current An electric current that is introduced into the Earth by leakage of currents from cultural installations.

stray sand A *stray* consisting of sandstone.

streak [geog] A long, narrow, irregular strip or stretch of land or water.

streak [mineral] The color of a mineral in its powdered form, usually obtained by rubbing the mineral on a *streak plate* and observing the mark it leaves. Streak is an important characteristic in mineral identification; it is sometimes different from the color of the sample, and is generally constant for the same mineral.

streak [sed] (a) A comparatively small and flattish or elongate sedimentary body, visibly differing from the adjacent rock, but without the sharp boundaries typical of a lens or layer (Stokes & Varnes, 1955, p.143). (b) A long, narrow body of sand, perhaps representing an old shoreline; a shoestring. (c) The outcropping edge of a coal bed.

streaked-out ripples A term used to describe sedimentary *flame structure.*

streaking *mineral streaking.*

streak plate In mineral identification, a piece of unglazed porcelain used for rubbing a sample to obtain its powder color, or *streak.* It has a hardness of about seven.

stream [glaciol] (a) *ice stream [glaciol].* (b) A stream of meltwater.

stream [streams] (a) Any body of running water that moves under gravity to progressively lower levels, in a relatively narrow but clearly defined channel on the surface of the ground, in a subterranean cavern, or beneath or in a glacier. It is a mixture of water and dissolved, suspended, or entrained matter. Cf: *river.* (b) A term popularly applied to a *brook* (as in Maine) or to a small river. (c) The water flowing in a stream. (d) A term used in quantitative geomorphology interchangeably with *channel.*

stream action *fluviation.*

stream azimuth Orientation of the mean line of a stream from head to mouth, measured in degrees from some arbitrary direction, generally north (Strahler, 1954, p.346). Symbol: α.

stream bed The channel containing or formerly containing the water of a stream. Also spelled: *streambed.*

stream-built terrace *alluvial terrace.*

stream capture *capture [streams].*

stream channel The bed where a natural stream of water runs or may run; the long narrow depression shaped by the concentrated flow of a stream and covered continuously or periodically by water. Syn: *streamway.*

stream current (a) A relatively narrow, deep ocean current, well-defined and fast-moving, e.g. the Gulf Stream; a *drift current* deflected by an obstruction, such as a shoal or land. Syn: *stream.* (b) A steady current in a terrestrial stream or river.

stream-cut terrace *rock terrace.*

stream deposition The accumulation of any transported rock particles on the bed or adjoining flood plain of a stream, or on the floor of a body of standing water into which a stream empties.

stream-entrance angle *axil angle.*

streamer A marine seismic recording cable containing hydrophones and designed for recording seismic signals while under continuous tow. Reflection surveys use streamers up to 3 km in length. Cf: *bay cable.*

stream erosion The progressive removal, by a stream, of bedrock, overburden, soil, or other exposed matter from the surface of its channel, as by hydraulic action, corrasion, and corrosion.

streamflood A flood of water in an arid region, characterized by the "spasmodic and impetuous flow" of a *sheetflood* but confined to a definite, shallow channel that is normally dry (Davis, 1938).

streamflow A type of *channel flow,* applied to that part of surface runoff traveling in a stream whether or not it is affected by diversion or regulation. Also spelled: *stream flow.* Cf: *sheet flow [geomorph]; overland flow.*

streamflow depletion The amount of water that flows into a given land area minus the amount that flows out of that area.

streamflow wave A traveling wave caused by a sudden increase of water flow (ASCE, 1962).

stream frequency Ratio of the number of streams of all orders within a drainage basin to the area of that basin; a measure of topographic texture. Symbol: F. Syn: *channel frequency.*

stream gaging Measurement of the velocity of a stream of water in a channel or open conduit and of the cross-sectional area of the water, in order to determine discharge. See also: *chemical gaging.* Syn: *gaging.*

stream gold Gold occurring in alluvial placers.

stream gradient The angle between the water surface (of a large stream) or the channel floor (of a small stream) and the horizontal, measured in the direction of flow; the "slope" of the stream. Symbol: S. See also: *law of stream gradients.* Syn: *stream slope; slope [stream].*

stream-gradient ratio Ratio of the gradient of a stream channel of a given order to that of a stream of the next higher order in the same drainage basin. Symbol: R_s. Syn: *channel-gradient ratio.*

streamhead The source or beginning of a stream.

streaming flow [glaciol] Glacier flow in which the ice moves without cracking or breaking into blocks, as where the walls and bottom are relatively smooth for a long distance.

streaming flow [hydraul] *tranquil flow.*

streaming lineation *parting lineation.*

streaming potential *electrofiltration potential.*

stream length The length of a stream segment of a given order u. Symbol: L_u. See also: *law of stream lengths.* Syn: *channel length.*

stream-length ratio Ratio of the mean stream length of a given order to the mean stream length of the next lower order within a specified drainage basin (Horton, 1945, p.296). Symbol: R_L.

streamlet A small stream.

stream line An imaginary line connecting a series of fluid particles in a moving fluid so that, at a given instant, the velocity vector of every particle on that line is tangent to it.

streamline flow *laminar flow [hydraul].*

stream load (a) The solid material that is transported by a stream, either as visible sediment (carried in suspension, or moved along the stream bed by saltation and traction) or in chemical or colloidal solution. (b) The quantity or amount of such material at any given time or passing a point in a given period of time, and expressed as a weight or volume per unit time.—Material in solution is sometimes excluded in the usage of the term. See also: *suspended load; bed load; dissolved load.*

stream morphology *river morphology.*

stream number A syn. of *number of streams;* see also: *law of stream numbers.*

stream order A classification of the relative position of streams in a channel network, assigning each *link* an integer order number determined by the pattern of confluences in the tributary network headward of the given link. The most widely employed ordering method was developed by Strahler (1952b): all exterior links are order 1; proceeding downstream the confluence of two links of order u generates a resultant stream of order u + 1; the confluence of tributary orders u and v, where u > v, leads to a resultant stream of order u. The Horton (1945) ordering method modifies the Strahler stream orders by the headward extension of order numbers greater than 1: at each junction of equal-order tributary streams, the order of the resultant downstream link is extended headward along the tributary most nearly parallel to the downstream link; if both tributaries enter at the same angle, the order of the downstream link is assigned to the longer tributary. See also: *basin order; stream segment; link.* Syn: *channel order.*

stream piracy *capture [streams].*

stream placer *alluvial placer.*

stream profile The *longitudinal profile* of a stream.

stream robbery *capture [streams].*

stream segment A *link,* or sequence of links, along a stream channel, extending from the fork where the stream achieves a given *stream order* to the downstream fork where it joins a stream of equal or higher order. Syn: *channel segment.*

streamsink *sinking stream.*

stream slope *stream gradient.*

stream terrace One of a series of level surfaces in a stream valley, flanking and more or less parallel to the stream channel, originally occurring at or below, but now above, the level of the stream, and representing the dissected remnants of an abandoned flood plain, stream bed, or valley floor produced during a former stage of erosion or deposition. See also: *alluvial terrace; rock terrace; meander terrace; inset terrace.* Syn: *terrace [geomorph]; river terrace.*

stream tin Cassiterite occurring in the form of waterworn pebbles in alluvial or placer deposits or on bedrock along valleys or streams, such as that resulting from the wearing away of pneumatolytic veins associated with acid rocks. Cf: *lode tin.* Syn: *alluvial tin.*

stream transportation The movement by a stream of weathered or eroded rock material in chemical solution, in turbulent suspension, or by rolling, dragging, or bouncing along the stream bed.

streamtube A passage in a cave that is or has been completely filled with turbulent water and that has a surface containing *scallops [speleo].* See also: *conduit [speleo]; siphon [speleo]; tube [speleo].*

stream underflow Percolating water in the permeable bed of a stream and flowing parallel to the stream (ASCE, 1962).

stream valley An elongate depression on the Earth's surface, carved by a stream during the course of its development.

stream velocity The rate of flow, measured by distance per time unit, e.g. ft/sec.

streamway (a) The current of a stream. (b) *stream channel.*

street A part of a bare desert floor that forms a gap separating chains of sand dunes. See also: *straat.*

strelkinite An orthorhombic mineral: $Na_2(UO_2)_2(VO_4)_2 \cdot 6H_2O$.

strengite A pale-red orthorhombic mineral: $FePO_4 \cdot 2H_2O$. It is isomorphous with variscite and dimorphous with phosphosiderite, and may contain some manganese.

strength A term used in experimental structural geology that is meaningful only when all the environmental conditions of the experiment are specified; in general, the ability to withstand differential stress, measured in units of stress.

streptaster A sponge spicule (microsclere) having the form of a modified aster in which the rays do not arise from a common center but radiate from an axis; e.g. a streptosclere, a sanidaster, or a discorhabd. Cf: *euaster.*

streptosclere A siliceous sponge spicule (streptaster) in which long raylike spines are given off in spiral succession about a central axis, and intergrade with simple euasters. Syn: *spiraster.*

streptospiral Said of a foraminiferal test coiled like a ball of twine.

stress (a) In a solid, the force per unit area, acting on any surface within it, and variously expressed as pounds or tons per square inch, or dynes or kilograms per square centimenter; also, by extension, the external pressure which creates the internal force. The stress at any point is mathematically defined by nine values: three to specify the normal components and six to specify the shear components, relative to three mutually perpendicular reference axes. Cf: *strain.* See also: *normal stress; shear stress.* (b) A commonly used short form for *differential stress.*

stress axis *principal axis of stress.*

stress difference The difference between the greatest and least of the three principal stresses.

stress ellipsoid A geometric representation of the state of stress at a point that is defined by three mutually perpendicular principal stresses and their intensities.

stress field The state of stress, either homogeneous or varying from point to point, in a given domain.

stress mineral A term suggested by Harker (1918) for minerals such as chlorite, chloritoid, talc, albite, epidote, amphiboles, and kyanite, whose formation in metamorphosed rocks he believed was favored by shearing stress. The term has become largely obsolete. Cf: *antistress mineral.*

stress pillars *vertical sheet structure.*

stress shadow *pressure shadow.*

stress-strain curve The plot of conventional strain in percent shortening or elongation, as the abscissa, and true longitudinal differential stress, i.e. the difference between the greatest and least principal stresses, as the ordinate. Syn: *stress-strain diagram.*

stress-strain diagram *stress-strain curve.*

stress tensor A description of the state of stress at a point, which involves nine components, referred to three orthogonal coordinate axes. Three components are normal stresses, acting perpendicular to the coordinate planes. The remaining six components are shear stresses, acting parallel to the coordinate planes.

stretch [exp struc geol] A measure of the change in length of a line, specifically the ratio of the final length to the initial length of the line (Malvern, 1969, p. 164). Cf: *quadratic elongation; extension [exp struc geol].*

stretch [geog] A *reach* of water or land.

stretched Said of a structure or texture produced by dynamic metamorphism in which the constituents are stretched and commonly broken in the same direction, e.g. stretch-pebble conglomerate. A stretched condition should not be confused with *lineation.*

stretch fault *stretch thrust.*

stretching In metamorphic rocks, the elongation of mineral grains, gas bubbles, or other features; a type of lineation. Cf: *mineral streaking.*

stretch modulus *Young's modulus.*

stretch thrust A little-used term for a reverse fault formed by shear in the inverted limb of an overturned fold. Syn: *stretch fault.*

strewn field (a) A restricted geographic area within which a specific group of tektites is found. Examples include western Czechoslovakia, the southern half of Australia, the Ivory Coast, and southern U.S. (Texas and Georgia). (b) *dispersion ellipse.*—Syn: *tektite field.*

stria (a) One of a series of parallel straight lines on the surface of a crystal, as in pyrite, indicative of an oscillation between two crystal forms; also, one of a series of such lines on the cleavage planes of a mineral, as of plagioclase, calcite, or corundum, indicative of *polysynthetic twinning.* Syn: *striation.* (b) One of a series of fine grooved lines or threads on the surface of some shells, esp. on otherwise smooth shells, e.g. one of the parallel minute grooves on nautiloid and ammonoid conchs, separated by *lirae* and not easily discernible with the unaided eye.—Adj: *striate.* Pl: *striae.*

striae Plural of *stria.*

striate (a) Adj. of *stria.* (b) Said of spores and pollen having a streaked sculpture characterized by multiple, usually parallel, grooves and ribs in the exine; specif. referring to the Striatiti. (c) *striated.*

striated (a) Adj. of *striation.* (b) *striate.*

striated ground *striped ground.*

striation (a) *stria.* (b) One of multiple scratches or minute lines, generally parallel, inscribed on a rock surface by a geologic agent, i.e. glaciers (*glacial striation*), streams (cf. *drag mark*), or faulting (cf. *slickenside*). Syn: *scratch.* (c) The condition of being striated; the disposition of striations.—Adj: *striated; striate.*

striation cast The cast of a striation produced on a sedimentary surface; it is usually found on the underside of a thin siltstone or fine sandstone bed interlayered with mudstone. Cf: *groove cast.* Syn: *microgroove cast.*

Striatiti Abundant upper Paleozoic and lower Mesozoic pollen with characteristic striate sculpture in the exine of the body of the pollen grain, the grooves and ribs usually oriented perpendicular to the axes of the vesicles (if these are present). They are presumably pollen of conifers and gnetaleans.

stricture A contraction between successive shell joints of the skeleton of a nasselline radiolarian.

striding level (a) A spirit level so mounted that it can be placed above and parallel with the horizontal axis of a surveying instrument and so supported that it can be used for precise leveling of the horizontal axis of the instrument or for measuring any remaining inclination of the horizontal axis. (b) A demountable spirit level that can be attached to the telescope tube to level the line of sight.

strigovite (a) A dark-green mineral of the chlorite(?) group: $Fe_3^{+2}(Al,Fe^{+3})_3Si_3O_{11}(OH)_7$. (b) A hypothetical end-member of the chlorite group: $(Mg,Fe)_2Al_2Si_2O_7(OH)_4$.

strike [eco geol] n. The discovery of a mineral deposit, esp. if sudden or unexpected.—v. To discover or reach a mineral deposit suddenly or unexpectedly, e.g. to "strike" oil.

strike [struc geol] n. The direction or trend taken by a structural surface, e.g. a bedding or fault plane, as it intersects the horizontal. See also: *attitude.* Cf: *trend; trace.* Syn: *line of strike.*—v. To be aligned or to trend in a direction at right angles to the *line of dip.*

strike fault A fault that strikes parallel with the strike of the strata involved. Cf: *dip fault; oblique fault.*

strike fold *longitudinal fold.*

strike joint A joint that strikes parallel to the strike or lineation of the enclosing rock. Cf: *dip joint.*

strike-overlap A term proposed by Melton (1947, p.1870) for truncation of sedimentary rocks below unconformities, esp. for a slow, extremely low-angle regional truncation of contrasting depositional strike below a regional unconformity. The term is essentially synonymous with *overstep* if it is assumed that "in most bodies of marine, or interfingering marine and nonmarine rock, angular unconformities eventually pass downdip into disconformities, which in turn disappear farther out in the basin" (Swain, 1949, p.634).

strike separation In a fault, the distance of *separation* of two formerly adjacent beds on either side of the fault surface, measured parallel to the strike of the fault. Cf: *dip separation; strike slip.*

strike-separation fault *lateral fault.*

strike shift In a fault, the *shift* or relative displacement of the rock units parallel to the strike of the fault, but outside the fault zone itself; a partial syn. of *strike slip.*

strike-shift fault *strike-slip fault.*

strike slip In a fault, the component of the movement or slip that is parallel to the strike of the fault. Cf: *dip slip; strike separation; oblique slip.* Syn: *horizontal displacement; horizontal separation.* Partial syn: *strike shift.*

strike-slip fault A fault on which the movement is parallel to the fault's strike. Cf: *dip-slip fault; lateral fault.* See also: *transcurrent fault.* Syn: *strike-shift fault.*

strike stream A *subsequent stream* that follows the strike of the underlying strata.

strike valley A *subsequent valley* eroded in, and developed parallel to the strike of, underlying weak strata; a valley containing a *strike stream.*

string (a) A syn. of *drill string.* (b) The casing, tubing, or pipe, of one size, used in a well.

string bog A linear periglacial muskeg or moor with an undulating surface, occurring in the boreal needle-tree forest zone of the northern hemisphere (esp. western Siberia and the Hudson Bay area), and characterized by shallow water-filled depressions and festoons of lenticular ridges (up to 2 m high) consisting of floating fen or moss vegetation. Its origin is controversial, but it seems to be related to frost action and gravity movements in bog areas. See also: *ring moor.*

stringer [ore dep] A mineral veinlet or filament, usually one of a number, occurring in a discontinuous subparallel pattern in host rock. See also: *stringer lode.*

stringer [seis] A thin *high-speed layer,* usually with limited lateral continuity.

stringer [stratig] A thin sedimentary bed.

stringer lode A zone of shattered host rock containing a network of *stringers;* a *stockwork.*

stringhamite A monoclinic mineral: $CaCuSiO_4 \cdot 2H_2O$.

strip [ice] A long narrow area of pack ice about 1 km or less in width, usually composed of small fragments detached from the main mass of ice, and run together under the influence of wind, swell, or current; it is more limited than a *belt.*

strip [photo] *flight strip.*

stripe [meta] *ribbon.*

stripe [pat grd] One of the alternating bands of fine and coarse surficial material, or of rock or soil and vegetation-covered ground, comprising a form of patterned ground characteristic of slopes steeper than those of *steps.* It is usually straight, but may be sinuous or branching, and is probably the result of solifluction acting in conjunction with other processes, such as rillwork. See also: *sorted stripe; nonsorted stripe; contraction stripe.*

striped ground A form of *patterned ground* marked by alternating *stripes* produced on a sloping surface by frost action. See also: *sorted stripe.* Syn: *striped soil; striated ground.*

striped soil *striped ground.*

stripe hummock A *nonsorted stripe,* probably closely related to an earth hummock, but formed on sloping ground.

strip log A graphic record of a drilled well, plotted to scale on a *log strip,* e.g. *drilling-time log; sample log.*

strip mining *opencut mining.*

stripped bedding plane The exposed top surface of a resistant stratum that forms a *stripped structural surface* when extended over a considerable area.

stripped illite *degraded illite.*

stripped peneplain *fossil peneplain.*

stripped structural surface An *erosion surface* developed during the mature stage of an erosion cycle in an area underlain by horizontal or gently inclined strata of unequal resistance, the overlying softer beds having been removed by erosion so as to expose the more or less smooth surface of a resistant stratum that has served as a local base level and thereby controlled the depth of erosion; specif. the surfaces produced on a *structural plateau,* a *stratum plain,* and a *structural terrace.* Syn: *stripped surface.*

stripped surface *stripped structural surface.*

stripping *layer stripping.*

strip thrust An obsolete syn. of *décollement.*

strobilus An axis bearing a conelike aggregation of sporophylls, as in club mosses and many conifers. Cf: *cone [bot].*

stroma (a) The supporting framework of an animal organ, such as organic tissue in the mesodermal endoskeleton of a living echinoderm. Cf: *stereome.* (b) A compact mass of fungous cells, or of mixed host and fungous cells, in or on which spores or sporocarps are formed.—Pl: *stromata.*

stromatactis A sedimentary structure characterized by a horizontal or nearly flat bottom, up to about 10 cm in diameter, and an irregular or convex-upward upper surface, consisting of sparry-calcite cement, usually in the central part of a *reef core* (Chilingar et al., 1967, p. 321); sometimes called *reef tufa.* They have been interpreted as fillings of original cavities caused by the burial and decay of soft-bodied but rigid frame-building organisms (known as *Stromatactis* according to Lowenstam, 1950), although they may represent syngenetic voids in calcareous sediments; some examples represent recrystallized sheetlike bryozoan colonies.

stromatite Chorismite with two or more textural elements arranged in essentially parallel layers. Little used.

stromatolite An organosedimentary structure produced by sediment trapping, binding, and/or precipitation as a result of the growth and metabolic activity of micro-organisms, principally cyanophytes (blue-green algae) (Walter, 1976, p. 1). It has a variety of gross forms, from nearly horizontal to markedly columnar, domal, or subspherical. The term was introduced by Kalkowsky in 1908 as *stromatolith.* Cf: *oncolite.* Syn: *algal stromatolite.*

stromatolith [intrus rocks] A complex, sill-like igneous intrusion that is interfingered with sedimentary strata (Foye, 1916).

stromatolith [sed] *stromatolite.*

stromatology A term, now obsolete, proposed to embrace "the history of the formation of the stratified rocks" (Page, 1859, p.340).

stromatoporoid A general name for any of a group of extinct sessile benthic marine organisms of uncertain biologic affinities (probably phylum Porifera, possibly Coelenterata or Cyanophyta). They secreted a calcareous skeleton, generally a few tens of centimeters across, of tabular, encrusting, dendroidal, domal, or bulbous form; internal structural elements were arranged in subhorizontal or concentric laminae or lines of dissepiments separated by small radial pillars, or in an irregular open network. Stromatoporoids were especially abundant in Ordovician-Devonian reefs. Range, Cambrian(?) to Cretaceous.

Strombolian-type bomb A general type of volcanic bomb produced from lava that is less fluid than that of a *Hawaiian-type bomb.* It is usually large and pear-shaped.

Strombolian-type eruption A type of volcanic eruption characterized by jetting of clots or "fountains" of fluid, basaltic lava from a central crater. Etymol: Stromboli, Lipari Islands of Italy. Cf:

Hawaiian-type eruption; Peléean-type eruption; Vulcanian-type eruption.

stromeyerite A dark steel-gray orthorhombic mineral with a blue tarnish: CuAgS. Syn: *silver-copper glance.*

stromoconolith A layered igneous intrusion that is either conical or funnel-shaped (Tomkeieff, 1961).

strong Said of large or important mineral veins or faults.

strong motion *Ground motion* that is sufficiently strong to be of interest in engineering seismology.

strongyle A rodshaped sponge spicule (monaxon) with both ends blunt. Also spelled: *strongyl.*

strongylote Said of a sponge spicule having one end rounded.

strontianite A pale-green, white, gray, or yellowish orthorhombic mineral of the aragonite group: $SrCO_3$.

strontioborite A mineral: $SrB_8O_{13}\cdot 2H_2O$ (?).

strontioginorite A mineral: $(Sr,Ca)_2B_{14}O_{23}\cdot 8H_2O$. It is a strontian variety of ginorite.

strontiohilgardite A variety of hilgardite containing Sr.

strontium-apatite A pale-green to yellowish-green mineral of the apatite group: $(Sr,Ca)_5(PO_4)_3(OH,F)$.

strontium-87 A stable isotope of strontium, which is produced by beta decay of naturally occurring radioactive ^{87}Rb. Its abundance in sea water is 6.98% of the strontium present.

strophic Said of a brachiopod shell in which the true hinge line is parallel to the hinge axis. Cf: *nonstrophic.*

strophomenid Any articulate brachiopod belonging to the order Strophomenida, characterized chiefly by a plano- to concavo-convex shell that may be resupinate or geniculate. Range, Lower Ordovician to Lower Jurassic.

strophotaxis *Taxis [ecol]* in which an organism tends to turn in response to some external stimulus. Cf: *phobotaxis; thigmotaxis.*

structural Of or pertaining to rock deformation or to features that result from it.

structural adjustment A term proposed by Salisbury (1904, p. 710) for the rearrangement of the drainage of an area so as to conform to the geologic structure; esp. the flowing of streams along the strike of the strata. Cf: *topographic adjustment.*

structural analysis This term is meant to be an English translation of Sander's Gefügekunde. It includes the analysis of structural features on all scales from thin section to discontinuous outcrop. It also involves placing interpretations on the movements and stress fields that were responsible for deformation. Cf: *structural petrology; structural geology.*

structural basin *basin [struc geol].*

structural bench A bench representing the resistant edge of a *structural terrace* that is being reduced by erosion (Cotton, 1958, p. 94-95). Syn: *rock bench.*

structural closure *closure [struc geol].*

structural contour *structure contour.*

structural control The influence of structural features on ore deposition, e.g. ore minerals filling fractures. Cf: *stratigraphic control.*

structural crystallography Study of the internal arrangement and spacing of atoms and molecules composing crystalline solids.

structural datum *datum horizon.*

structural depression A topographically low area resulting from structural deformation of the Earth's crust.

structural diagram A figure illustrating the spatial array of orientation of objects.

structural dome *dome [struc geol].*

structural feature A feature produced by deformation or displacement of the rocks, such as a fold or fault. For such features the more colloquial term *structure* (used as a specific noun) is now generally accepted.

structural formula A chemical formula giving the relative number of atoms present in a particular mineral. Syn: *chemical composition [mineral]; composition [mineral].*

structural geology The branch of geology that deals with the form, arrangement, and internal structure of the rocks, and especially with the description, representation, and analysis of *structures,* chiefly on a moderate to small scale. The subject is similar to *tectonics,* but the latter is generally used for the broader regional or historical phases.

structural high *high.*

structural lake *tectonic lake.*

structural landform A landform developed by erosion and controlled by the structure of the rocks. Cf: *tectonic landform.*

structural low *low [struc geol].*

structural map *structure-contour map.*

structural nose *nose.*

structural petrology The analysis of fabric on the thin-section or micro scale. It includes the study of grain shapes and relationships (microstructure) and the study of crystallographic preferred orientations. Recently the transmission electron microscope has also been employed to examine the substructures of deformed crystals. Cf: *structural analysis.* Syn: *fabric analysis; petrofabric analysis; petrofabrics; microtectonics.*

structural plain *stratum plain.*

structural plateau A plateaulike landform with a *stripped structural surface.* Syn: *stripped structural plateau.*

structural province A region whose geologic structure differs significantly from that of adjacent regions. It is generally coextensive with a *physiographic province.*

structural relief (a) The vertical distance between stratigraphically equivalent points at the crest of an anticline and in the trough of an adjacent syncline. (b) More generally, the difference in elevation between the highest and lowest points of a bed or stratigraphic horizon in a given region.

structural terrace (a) A local shelf or steplike flattening in otherwise uniformly dipping strata, composed of a synclinal bend above and an anticlinal bend at a lower level. (b) A terracelike landform controlled by the structure of the underlying rocks; esp. a terrace produced by the more rapid erosion of weaker strata lying on more resistant rocks in a formation with horizontal bedding. Cf: *structural bench.*

structural trap A *trap* for oil or gas that is the result of folding, faulting, or other deformation. Cf: *stratigraphic trap; combination trap.*

structural unconformity *angular unconformity.*

structural valley A valley that owes its origin or form to the underlying geologic structure. Cf: *tectonic valley.*

structure [geomorph] A comprehensive term for the assemblage of rocks upon which erosive agents are, and have been, at work; the terrane underlying a landscape. The term indicates the product of all constructional agencies, and includes the arrangement and disposition of the rocks, their nature and mode of aggregation, and even their initial forms prior to erosion.

structure [mineral] (a) The form assumed by a mineral; e.g. bladed, columnar, tabular, or fibrous. (b) *crystal structure.*

structure [palyn] (a) The internal complexity in the makeup of the ektexine of pollen grains and spores, usually consisting of *columellae* that may be branched and more or less fused laterally. Cf: *sculpture.* (b) A term that is less desirably used to describe major morphologic characteristics of spores, esp. those of the Paleozoic.

structure [petrology] A megascopic feature of a rock mass or rock unit, generally seen best in the outcrop rather than in hand specimen or thin section, such as columnar structure, blocky fracture, platy parting, or foliation. The term is also applied to the appearance, or to a smaller-scale feature, of a rock in which the texture or composition is different in neighboring parts; e.g. banded structure, orbicular structure. The term *texture* is generally used for the smaller features of a rock or for the particles composing it; although the two terms are often used interchangeably, they should not be considered synonymous, even though some textures may parallel major structural features. See also: *sedimentary structure.*

structure [struc geol] (a) The general disposition, attitude, arrangement, or relative positions of the rock masses of a region or area; the sum total of the *structural features* of an area, consequent upon such deformational processes as faulting, folding, and igneous intrusion. (b) A term used in petroleum geology for any physical arrangement of rocks (such as an anticline or reef) that may hold an accumulation of oil or gas.

structure contour A *contour* that portrays a structural surface such as a formation boundary or a fault. Syn: *subsurface contour.* See also: *structure-contour map.*

structure-contour map A map that portrays subsurface configuration by means of *structure contour* lines. See also: *contour map; tectonic map.* Syn: *structural map; structure map.*

structure ground A term used by Antevs (1932, p.48) but now replaced by its syn. *patterned ground.*

structure map *structure-contour map.*

structure-process-stage The name given to the Davisian principle (Davis, 1899) that the development of all landforms in the cycle of erosion is a function of three basic factors: geologic structure, geomorphic process, and stage of development. Davis originally referred to "structure, process, and time", but later writers have changed this to "structure-process-stage".

structure section A *vertical section* that shows the observed geologic structure on a vertical or nearly vertical surface, or, more commonly, one that shows the inferred geologic structure as it would appear on a vertical plane cutting through a part of the Earth's crust. The vertical scale is often exaggerated.

structure soil *soil structure [pat grd].*

structure test A generally shallow hole drilled primarily to obtain information on geologic structure, although other types of information may be acquired during drilling. It is frequently drilled to a structural datum that is normally short of a known or expected oil-producing zone or zones. Cf: *stratigraphic test.*

structure type A group of crystals having the same atomic structure, i.e. having the constituent atoms arranged in a geometrically analogous way. An example is the NaCl structure type, in which equal numbers of cations and anions occur in six coordination; it includes sylvite, periclase, and galena.

struggle for existence The natural process by which members of a population compete automatically for a limited supply of vital necessities, thus resulting in *natural selection.*

Strukturboden A term formerly used for what is now known as *patterned ground.* Etymol: German, "structure ground" or "structure soil".

Strunian European stage: uppermost Devonian, transitional into Carboniferous (below Tournaisian). See also: *Etroeungtian.*

strunzite A straw-yellow monoclinic mineral: $MnFe_2(PO_4)_2(OH)_2 \cdot 8H_2O$. It is polymorphous with laueite.

strut thrust An obsolete term for a fault initiated by the shearing of a strut, or competent bed.

strüverite A tetragonal mineral: $(Ti,Ta,Fe)_3O_6$. It forms a series with ilmenorutile.

struvite A colorless to yellow or pale-brown orthorhombic mineral: $Mg(NH_4)(PO_4) \cdot 6H_2O$.

stubachite An altered *peridotite* characterized by the presence of tremolite, talc, serpentine, magnetite, pyrite, and magnesite. Obsolete.

Student's t test A statistical test used to determine whether two samples could have been drawn from the same *population.* Named after Student, pseudonym of William S. Gosset, 20th century Irish statistician. Syn: *t test.*

studtite A monoclinic mineral: $UO_4 \cdot 4H_2O$.

stuffed mineral A mineral having extra ions of a foreign element within the large interstices of its structure; e.g. garnet with an extra cation.

stumpflite A hexagonal mineral: Pt(Sb,Bi).

stunted fauna *dwarf fauna*

sturtite A black mineral, a non-crystalline hydrous manganese silicate.

sturzstrom A huge mass of rapidly moving rock debris and dust, derived from the collapse of a rock cliff or mountainside, flowing down steep slopes and across gentle to flat ground, often for several kilometers. A sturzstrom may include compressed air, water, or mud between the rock fragments, but ordinarily it is considered to be a mass of cohesionless blocks of rock dispersed in a cloud of dust, flowing at speeds of more than 100 km/hr. Sturzstroms are the most catastrophic of all forms of mass movement. Several have been identified on the Moon. Term first used by A. Heim, 1881; see Hsu, 1975. Cf: *debris flow; rockfall.* Syn: *rock avalanche; rockfall avalanche; rock-fragment flow.*

stützite A mineral: $Ag_{5-x}Te_3$. It was formerly regarded as identical with empressite.

stylaster Any one of a group of hydrozoan coelenterates belonging to the order Stylasterina, characterized by a calcareous skeleton and by sexual individuals that remain attached to the colony. Cf: *hydroid; millepore.* Range, Cretaceous to present.

style [bot] The usually attenuated part of the pistil that connects the stigma to the ovary.

style [paleont] (a) A sponge spicule (monaxon) with one blunt end and one pointed end; e.g. *tylostyle.* (b) A tubule that arises from the galea in phaeodarian radiolarians. (c) A central calcareous process in certain pores of a stylaster coral. (d) The *telson* of a crustacean.

style [tect] *tectonic style.*

styliform columella A solidly fused and longitudinally projecting coral *columella.* It is fused to scleractinian entosepta by secondary stereome.

styliform cyrtolith A cyrtolith coccolith with a long spinose central structure; e.g. a *pole coccolith.*

stylocerite A rounded or spiniform process on the outer part of the antennular peduncle in some decapod crustaceans.

stylolite (a) A surface or contact, usually occurring in homogeneous carbonate rocks and more rarely in sandstones and quartzites, that is marked by an irregular and interlocking penetration of the two sides: the columns, pits, and teeth-like projections on one side fit into their counterparts on the other. As usually seen in cross section, it resembles a suture or the tracing of a stylus. The seam is characterized by a concentration of insoluble constituents of the rock, e.g. clay, carbon, or iron oxides, and is commonly parallel to the bedding. Stylolites are supposedly formed diagenetically by differential vertical movement under pressure, accompanied by solution. See also: *microstylolite; suture joint.* Syn: *stylolite seam.* (b) A straight, vertically grooved column, of the same material as the rock in which it occurs, commonly less than a centimeter in length, fitting into a corresponding socket in a stylolitic seam and being highly inclined or at right angles to the bedding plane. It often results from the slipping under vertical pressure of a part capped by a fossil shell through adjacent parts not so capped.—Term introduced by Klöden (1828, p.28). Etymol: Greek *stylos,* "pillar", + *lithos,* "stone". Obsolete syn: *epsomite; crystallite; toenail; devil's toenail; crowfoot; lignilite.*

stylolitic Pertaining to a stylolite, as a "stylolitic seam" or "stylolitic column".

stylotypite A syn. of *tetrahedrite.* The name has been applied esp. to tetrahedrite containing considerable silver.

Styrian orogeny One of the 30 or more short-lived orogenies during Phanerozoic time identified by Stille, in this case in the Miocene, between the Burdigalian and Aquitanian stages.

suanite A mineral: $Mg_2B_2O_5$. Syn: *magnioborite.*

subactive volcano *dormant volcano.*

subaerial Said of conditions and processes, such as erosion, that exist or operate in the open air on or immediately adjacent to the land surface; or of features and materials, such as eolian deposits, that are formed or situated on the land surface. The term is sometimes considered to include *fluvial.* Cf: *subaqueous; subterranean.* See also: *surficial.*

subaerial bench A term used by Lawson (1915) for a nonalluviated, concave-upward pediment.

subaerialism The doctrine that the landscape and its landforms are formed chiefly by subaerial agents (esp. rainwash) and processes.

subage A seldom-used term for a geologic-time unit shorter than an *age,* during which the rocks of the corresponding *substage* were formed. It is usually characterized by the occurrence of some specific phenomenon, such as the deposition of loess. Syn: *episode (b); time (d); phase [geochron] (a).*

subalkalic (a) A group term applied to rocks of the tholeiitic and calc-alkaline series. (b) Said of an igneous rock that contains no alkali minerals other than feldspars. (c) Used to describe an igneous rock of the Pacific suite.

subalkaline basalt As proposed by Chayes (1964), a replacement for the terms *tholeiitic basalt* and *tholeiite.* Basalts in which neither nepheline nor acmite appear in the CIPW norm fall in the subalkaline category.

suballuvial bench A term used by Lawson (1915, p. 34) for the outward or basinward extension of a pediment, covered by alluvium as the basin slowly filled (the thickness increasing basinward to several hundred meters), and exhibiting a convex-upward longitudinal profile. Cf: *concealed pediment.*

subalpine *montane.*

subalpine peat *hill peat.*

subaluminous Said of an igneous rock in which there is little or no excess of aluminum oxide over that required to form feldspars or feldspathoids; one of Shand's (1947) groups of igneous rocks, classified on the basis of the degree of aluminum-oxide saturation. Cf: *peralkaline; peraluminous; metaluminous.*

subangular Somewhat angular, free from sharp angles but not smoothly rounded; specif. said of a sedimentary particle showing definite effects of slight abrasion, retaining its original general form, and having faces that are virtually untouched and edges and corners that are rounded off to some extent, such as a glacial boulder with numerous (10-20) secondary corners and a roundness value between 0.15 and 0.25 (midpoint at 0.200) (Pettijohn, 1957, p.59), or one with one-third of its edges smooth (Krynine, 1948, p.142). Also, said of the *roundness class* containing subangular particles.

subaquatic plant A hydrophyte that is not a *submerged aquatic plant.*

subaqueous Said of conditions and processes, or of features and deposits, that exist or are situated in or under water, esp. fresh water, as in a lake or stream. Cf: *subaerial.*

subaqueous gliding *slump [mass move].*

subaqueous sand dune *dune [stream].*

subaqueous till Berg till; also, till deposited from a glacier terminating in water.

subarctic Pertaining or relating to the regions directly adjacent to the Arctic Circle, or to areas that have climate, vegetation, and animals similar to those of arctic regions.

subarkose A sandstone that does not have enough feldspar to be classed as an *arkose,* or a sandstone that is intermediate in composition between arkose and pure quartz sandstone. A quantitative definition: an arkosic sandstone containing 75-95% quartz and chert, less than 15% detrital clay matrix, and 5-25% unstable materials in which the feldspar grains exceed the rock fragments in abundance (Pettijohn, 1954, p.364). Other definitions have been given by Folk, (1954, p.354; 1968, p.124), and McBride (1963, p.667). Pettijohn (1957, p.322) later used 10-25% unstable fragments, so that a subarkose might have as little as 5% feldspar. The rock is roughly equivalent to *feldspathic arenite* of Williams, Turner and Gilbert (1954). Syn: *feldspathic quartzite; feldspathic sandstone.*

subarkosic wacke Essentially a wacke (over 15% matrix) with 5 to 25% feldspar; a species of feldspathic graywacke (Pettijohn, Potter & Siever, 1973, p. 171). Syn: *feldspathic wacke; subfeldspathic lithic wacke.*

subartesian Said of confined ground water that is under sufficient pressure to rise above the water table but not to the land surface; i.e. *mesopiestic water.*

Subatlantic n. A term used primarily in Europe for a period of Holocene time (approximately the last 2500 years, or from 500 B.C. to the present) following the Subboreal, during which the inferred climate became generally milder and wetter; a unit of the *Blytt-Sernander climatic classification,* characterized by beech and linden vegetation. Also spelled: *Sub-Atlantic.*—adj. Pertaining to the postglacial Subatlantic interval and to its climate, deposits, biota, and events.

subautomorphic (a) Said of the holocrystalline texture of an igneous or metamorphic rock characterized by crystals bounded in part by their own *rational faces* and in part by faces resulting from growth interference with surrounding crystals. Also said of a rock with such a texture. *Hypidiomorphic* (Rosenbusch, 1887, p. 11) and *hypautomorphic* (Rohrbach, 1885, p. 17-18) are considered as syns., though originally used to describe individual mineral crystals (*subhedral* crystals) in igneous rocks. (b) A syn. of *subhedral* in European usage, now obsolete in American usage. Syn: *subidiomorphic.*—Cf: *automorphic; xenomorphic.*

subbase A base or supporting material placed below that which ordinarily forms the base; specif. a layer of earth or rock placed between the *base course* and the *subgrade,* designed to give additional support, to distribute the load, or to form a pervious layer; e.g. the first layer of large-diameter crushed stone laid down in constructing a road, airstrip, or other graded surface.

subbentonite *metabentonite.*

subbiozone A subdivision of a *biozone* "to express finer biostratigraphic detail" (ISG, 1976, p. 48-49).

subbituminous A coal A type of *subbituminous coal,* characteristically nonagglomerating, having 10,500 to 13,000 BTU/lb (moist, mineral-matter-free). It is differentiated from high-volatile C bituminous by agglomerating characteristics.

subbituminous B coal A type of *subbituminous coal* having 9,500 to 10,500 BTU/lb (moist, mineral-matter-free).

subbituminous C coal A type of *subbituminous coal* having 8,300 to 9,500 BTU/lb (moist, mineral-matter-free).

subbituminous coal A black coal, intermediate in rank between lignite and bituminous coals; or, in some classifications, the equivalent of *black lignite.* It is distinguished from lignite by higher carbon and lower moisture content. Further classification of subbituminous coal is made on the basis of calorific value. See also: *subbituminous A coal; subbituminous B coal; subbituminous C coal; lignite.* Cf: *gloss coal; metalignitous coal.*

Subboreal n. A term used primarily in Europe for an interval of Holocene time (from about 4500 to 2500 years ago) following the Atlantic and preceding the Subatlantic, during which the inferred climate became generally cooler and drier; a unit of the *Blytt-Sernander climatic classification,* characterized by oak, ash, and linden vegetation. Also spelled: *Sub-Boreal.*—adj. Pertaining to the postglacial Subboreal interval and to its climate, deposits, biota, and events.

subboreal (a) Said of a climate that is very cold or approaching frigidity. (b) Pertaining to a biogeographic zone that approaches a boreal climatic condition. (c) Pertaining to the Subboreal postglacial period, and to the climate of such a period. Also spelled: *sub-Boreal.*

subcannel coal Cannel coal of brown-coal to subbituminous rank. Cf: *metacannel coal; lean cannel coal.*

subcapillary interstice An *interstice* sufficiently smaller than a *capillary interstice* that the molecular attraction of its walls reaches across the entire opening. Water held in it by adhesive forces is immovable except by forces in excess of pressures commonly found in subsurface water. "The conditions existing in these interstices are, however, only very imperfectly understood and are largely a matter of speculation" (Meinzer, 1923, p. 19). Cf: *supercapillary interstice.*

subchela The distal prehensile or grasping part of a crustacean limb formed by folding dactylus against propodus or dactylus and propodus against carpus. Pl: *subchelae.*

subclass In the hierarchy of classification of plants and animals, a subcategory of *class.*

subconsequent stream (a) *secondary consequent stream.* (b) An obsolete syn. of *subsequent stream.*

subcontinent (a) A division or part of a continent having characteristics that distinguish it from the rest of the continent, e.g. the Indian subcontinent. This subdivision is typically based on geologic or geomorphic characteristics. (b) A large land mass, such as Greenland or Antarctica, that is smaller than any of the seven recognized continents.

subcortical crypt An inhalant aquiferous cavity lying beneath a cortex in a sponge and differentiated from a canal by virtue of its larger size and distinctive shape.

subcrevasse channel A shallow channel eroded in subglacial material by a stream flowing along the bottom of a crevasse that completely penetrated a glacier (Leighton, 1959, p. 340).

subcritical flow *tranquil flow.*

subcrop (a) An occurrence of strata in contact with the undersurface of an inclusive stratigraphic unit that succeeds an important unconformity on which *overstep* is conspicuous; a "subsurface outcrop" that describes the areal limits of a truncated rock unit at a buried surface of unconformity. (b) An area within which a formation occurs directly beneath an unconformity.—The term, in common use in petroleum geology, appears to have been used first by Swesnick (1950, p.401) at the suggestion of Thom H. Green.

subcrop map A geologic map that shows the distribution of formations that have been preserved and remain covered beneath a given stratigraphic unit or immediately underlying an unconformity; properly, a map of an area where the overlapping formation is still present. The term "may be considered a generalization" of the term *paleogeologic map* (Krumbein & Sloss, 1963, p.448). Cf: *supercrop map.*

subcrustal Said of a material or region beneath the Earth's crust.

subdelta A small delta, forming a part of a larger delta or complex of deltas.

subdeltoid An anal deltoid plate of a blastoid, typically horseshoe-shaped, located on the adoral and lateral margins of the anal opening and on the aboral border of the superdeltoid (TIP, 1967, pt. S, p. 350). See also: *cryptodeltoid.*

subdermal space A *vestibule* of a sponge.

subdiabasic Said of an igneous-rock texture that is similar to ophitic texture except that the augite of the groundmass is not optically continuous but is divided into granular aggregates.

subdivide A drainage divide between the tributaries of a main stream; a subordinate divide.

subdivision A category in the hierarchy of botanical classification intermediate between *division* and *class.*

subdorsal carina One of two longitudinal ridges on some decapods, usually granulose or spinulose, extending backward from the submedian region of the rostrum and usually diverging.

subdrainage Drainage from beneath, either natural or artificial.

subdrift topography Topography of a bedrock surface underlying unconsolidated glacial drift.

subduction The process of one lithospheric *plate* descending beneath another. A related concept was originally used by Alpine geologists. See also: *subduction zone.*

subduction zone A long, narrow belt in which *subduction* takes place, e.g. along the Peru-Chile trench, where the Pacific plate descends beneath the South American plate.

subdued Said of a landform or landscape that is marked by a broadly rounded form and by moderate height, as if produced by long-continued weathering and erosion; esp. said of a mountain in the stage of senescence in a cycle of erosion, sufficiently worn down to have lost its peaks and cliffs, and having its moderately steep slopes covered with its own detritus. Cf: *feral.*

subepoch A term proposed by Sutton (1940, p. 1402) for a geologic-time unit representing the first division of an *epoch.* It is applied only to a few portions of geologic time. Cf: *subseries.*

subera A little-used term referring to a "portion of an *era* comprised of two or more periods" (Sutton, 1940, p. 1410).

suberain A kind of *provitrain* in which the cellular structure is derived from corky material. Cf: *periblain; xylain.*

suberin An organic compound similar to *cutin* that occurs in corkified cell walls of bark and on roots, stems, and fruits, as a protection against desiccation. Cf: *suberinite.*

suberinite (a) A variety of provitrinite characteristic of suberain and consisting of corky tissue. (b) A maceral of brown coal and lignite derived from the *suberin* layer in corkified cell walls of some Mesozoic and younger plants.—Cf: *periblinite; xylinite; telinite.*

subfabric The array of only one kind of structural element in a rock. See also: *fabric.*

subface The basal or lower surface of a stratigraphic unit.

subfacies A subdivision of a facies, as of a broadly defined sedimentary facies, or of a metamorphic facies based on compositional differences rather than pressure-temperature relations.

subfamily In the hierarchy of classification of plants and animals, a subcategory of *family.* In zoology, the name of a subfamily characteristically has the ending -inae; e.g. Cytheredeinae.

subfeldspathic (a) Said of a mature lithic wacke (or lithic graywacke) in which quartz grains and fragments of siliceous and argillaceous rocks predominate, and feldspars make up less than 10% of the rock and may be altogether lacking (Williams, Turner & Gilbert, 1954, p.302-303). Such rocks have also been called *subgraywackes.* (b)Said of a mature lithic arenite containing abundant quartz grains and fragments of the more stable rocks (such as cherts), and less than 10% feldspar grains (Williams, Turner & Gilbert, 1954, p.304 & 307).

subfeldspathic lithic arenite An arenite with 10% or less of feldspar and a larger quantity of rock fragments (Williams, Turner & Gilbert, 1954, p.304).

subfeldspathic lithic wacke A lithic wacke containing less than 10% feldspar; a species of feldspathic graywacke (Williams, Turner & Gilbert, 1954, p.302). Syn: *subarkosic wacke.*

subfluvial Situated or formed at the bottom of a river; e.g. a "subfluvial deposit".

subfossil n. A fossil that is younger than what would be considered typical fossil age (i.e., preserved since about 6,000 years ago, by common convention) but not strictly recent or present-day.—adj. Applied to an organism that would be considered a "subfossil". Cf: *fossil.*

subgelisol The zone of unfrozen ground (*talik*) beneath permafrost.

subgenus In the hierarchy of classification of plants and animals, a subcategory of *genus.* The name of a subgenus is placed in parentheses after the genus name and is followed by the name of the species, e.g. *Palaeoneilo (Koenenia) emarginata.*

subglacial (a) Formed or accumulated in or by the bottom parts of a glacier or ice sheet; said of meltwater streams, till, moraine, etc. Syn: *infraglacial.* (b) Pertaining to the area immediately beneath a glacier, as *subglacial* eruption or *subglacial* drainage.

subgrade A layer, stratum, or surface immediately beneath some principal surface; specif. a layer of earth or rock that is graded to receive the foundation of an engineering structure. Often it is the soil or natural ground that is prepared and compacted to support, and that lies directly below, a road, pavement, building, airfield, or railway. Cf: *subbase.*

subgraphite *meta-anthracite.*

subgraywacke (a) A term introduced by Pettijohn (1949, p.227 & 255-256) for a sedimentary rock that has less feldspar and more and better-rounded quartz grains than *graywacke;* specif. a sandstone containing 15-85% quartz and chert, 15-75% detrital clay matrix (chiefly sericite and chlorite), less than 10-15% feldspar, and an appreciable quantity (5%) of rock fragments. This rock, as originally defined, is equivalent to *quartz wacke* of Krumbein & Sloss (1963), to *low-rank graywacke* of Krynine (1948), to *lithic graywacke* of Pettijohn (1954), and to the *subfeldspathic* wackes of Williams, Turner and Gilbert (1954). (b) A term redefined by Pettijohn (1957, p.316-320) as the most common type of sandstone, intermediate in composition between graywacke and orthoquartzite, containing less than 75% quartz and chert (commonly 30-65%), less than 15% detrital clay matrix, and an abundance (more than 25%) of unstable materials (feldspar grains and rock fragments) in which the rock fragments (at least 15%) exceed the feldspars; and having voids and/or mineral cement (esp. carbonates) exceeding the amount of clay matrix. The rock is lighter-colored and better-sorted, and has less matrix, than graywacke, and commonly forms great lenticular bodies as a result of paralic sedimentation from normal subaqueous currents. Example: the Oswego Sandstone (Upper Ordovician) of central Pennsylvania. (c) A term used by Folk (1954, p.354) for a sedimentary rock that does not have enough rock fragments to be classed as a graywacke; specif. a sandstone with 5-25% micas and metamorphic-rock fragments and less than 10% feldspars and igneous-rock fragments, and having any clay content or sorting. This rock is equivalent to *quartzose graywacke* of Krynine (1951).

subgroup A formally differentiated assemblage of formations within a *group* (ACSN, 1961, art.9d).

subhedral (a) Said of a mineral grain that is bounded partly by its own *rational faces* and partly by surfaces formed against preexisting grains as a result of either crystallization or recrystallization. (b) Said of the shape of such a crystal.—Intermediate between *euhedral* and *anhedral.* The term was proposed, originally in reference to igneous-rock components, by Cross et al. (1906, p.698) in preference to the synonymous terms *hypidiomorphic* and *hypautomorphic* (as these were originally defined).

subhedron A geometrical term for a solid figure partly limited by plane surfaces. The term was introduced by Cross et al. (1906) for an igneous-rock component (crystal) only partly bounded by its own *rational faces.* Pl: *subhedrons; subhedra.*

subhepatic region A part of the carapace of some decapods extending below the edge of the hepatic region.

Subhercynian orogeny One of the 30 or more short-lived orogenies during Phanerozoic time recognized by Stille, in this case in the Late Cretaceous, between the Turonian and Senonian stages.

subhumid Said of a climate type that is transitional between humid and semiarid types according to quantity and distribution of precipitation. In Thornthwaite's classification, the moisture index is between zero and -20.

subhydrous (a) Said of coal containing less than 5% hydrogen, analyzed on a dry, ash-free basis. (b) Said of a maceral of low hydrogen content, e.g. fusinite.—Cf: *orthohydrous; perhydrous.*

subida A rock-floored belt produced by wind scour and "potentially reaching to the base of a mountain range" (Stone, 1967, p. 250). Etymol: Spanish, "ascent, acclivity".

subidioblast *hypidioblast.*

subidiomorphic *hypidiomorphic; subautomorphic.*

subimposed Said of a subterranean stream that becomes a surface stream, as when the roof of a limestone cavern falls in. An obsolete term, originally proposed by Russell (1898b, p. 246).

subirrigation Irrigation of plants with water delivered to the roots from underneath, either naturally or artificially.

subjacent [geomorph] Being lower, but not necessarily lying directly below; e.g. "hills and subjacent valleys".

subjacent [intrus rocks] Said of an igneous intrusion, generally discordant and without a known floor, that presumably enlarges downward to an unknown depth.

subjacent [stratig] Said of a stratum situated immediately under a particular higher stratum or below an unconformity. Ant: *superjacent.* Syn: *underlying.*

subjacent karst *interstratal karst.*

subjective synonym In taxonomy, one of two or more competing names for the same taxon which are based on different *types.* Cf: *objective synonym.*

subjoint A minor joint associated with a major joint, either divergent or parallel.

subkingdom In the hierarchy of classification of plants and animals, a subcategory of *kingdom.* It is sometimes considered synonymous with *phylum* and sometimes ranked above it.

sublacustrine Existing or formed beneath the waters, or on the bottom, of a lake; e.g. a "sublacustrine channel" eroded in the lake bed by a surface stream before the lake was there or by a strong current within the lake.

sublimate A solid that has been deposited by a gas or vapor; in volcanology it refers to such a deposit made by a volcanic gas, e.g. metals around the mouth of a fumarole.

sublimation [chem] The process by which a solid substance vaporizes without passing through a liquid stage. Cf: *evaporation.*

sublimation [ore dep] The process of ore deposition, as of sulfur or mercury, by vapors; the volatilization and transportation of minerals followed by their deposition at reduced temperatures and pressures. Sublimation deposits are generally associated with fumarolic activity.

sublimation ice Hoarfrost crystals in open or closed cavities in permafrost, produced by condensation of water vapor.

sublimation loss Loss of water through the direct evaporation of ice and snow on lakes or from any body of ice or snow.

sublitharenite (a) A term introduced by McBride (1963, p.667) for a sandstone that does not have enough rock fragments to be classed as a *litharenite,* or a sandstone that is intermediate in composition between litharenite and pure quartz sandstone; specif. a sandstone with 5-25% fine-grained rock fragments, 65-95% quartz, quartzite, and chert, and less than 10% feldspar. (b) A term used by Folk (1968, p.124) for a sandstone, regardless of clay content or texture, with 75-95% quartz and metamorphic quartzite and a content (5-25%) of fine-grained volcanic, metamorphic, and sedimentary rock fragments (including chert) that exceeds that of feldspar and fragments of gneiss and granite.

sublithistid Said of a sponge containing desmoids.

sublithographic Said of a limestone whose texture approaches the exceedingly fine grain of lithographic limestone. Also, said of the texture of such a rock.

sublithwacke A wacke with 5-25% detrital rock particles; a sublitharenite with over 15% matrix (Pettijohn, Potter & Siever, 1973).

sublittoral Said of that part of the *littoral* zone that is between low tide and a depth of about 100 m. Syn: *neritic.*

submarginal ambulacral suture The zone of contact between the ambulacral coverplates and floorplates in the Edrioasteroidea (Bell, 1976).

submarginal channel A channel formed by a meltwater stream flowing near the ice margin but also cutting across spurs or "behind small outlying hills" (Rich, 1908, p. 528).

submarginal moraine *lodge moraine.*

submarginal resources Low-grade *resources* that are recoverable at prices more than 1.5 times those prevailing now, i.e. are of lower grade than *paramarginal resources* (Brobst & Pratt, 1973, p. 3).

submarginal ring A prominent circlet of thick plates exposed on both oral and aboral sides of the theca of cyclocystoids and representing their "most conspicuous and best-preserved feature" (TIP, 1966, pt.U, p.201). Cf: *marginal ring.*

submarginal suture line In the Edrioasteroidea, the intersection of the submarginal ambulacral suture with the thecal surface. It extends along the line of contact of the adradial ends of the coverplates and the externally exposed parts of the floorplates (Bell, 1976).

submarine bank (a) A general term for limestone deposits that are locally abnormally thick and that appear to have formed over submerged shallow areas that rose above the general level of the surrounding sea floor (Harbaugh, 1962, p. 13). Submarine banks lack the hard, wave-resistant character of organic reefs. See also: *bank [sed]; carbonate buildup.* Syn: *marine bank.* (b) *bank [oceanog].*

submarine bar A *longshore bar* that is always submerged, even at low tide.

submarine barchan A large-scale lunate asymmetric ripple mark on the sea floor, ranging in length from 1 m to 100 m or more. Examples occur in the shallow-water areas of the Bahamas.

submarine canyon (a) A steep-sided, V-profile trench or valley winding along the continental shelf or continental slope, having tributaries and resembling an unglaciated, river-cut land canyon. (b) A general term for all valleys of the deep-sea floor. Syn: *submarine valley.*

submarine cone *submarine fan.*

submarine delta *submarine fan.*

submarine earthquake *seaquake.*

submarine fan A terrigenous, cone- or fan-shaped deposit located seaward of large rivers and submarine canyons. Syn: *submarine cone; abyssal cone; abyssal fan; subsea apron; deep-sea fan; submarine delta; sea fan; fan [marine geol]; cone [marine geol].*

submarine geology *geological oceanography.*

submarine geomorphology That aspect of geological oceanography which deals with the relief features of the ocean floor and with the forces that modify them.

submarine meadow A grassland consisting of marine plants such as turtle grass.

submarine plain A syn. of *plain of marine erosion.* Term is not recommended because some of these plains have been uplifted. Syn: *submarine platform.*

submarine plateau *plateau [marine geol].*

submarine platform *submarine plain.*

submarine ridge *ridge [marine geol].*

submarine spring A large offshore emergence of fresh water, usually associated with a coastal karst area but sometimes with lava tubes.

submarine valley *submarine canyon.*

submarine volcano A volcano on the ocean floor, commonly of tholeiitic basalt. See also: *volcanic island.*

submarine weathering *halmyrolysis.*

submask geology The geology of the surface underlying a cover of alluvium, glacial drift, windblown sand, low-angle overthrust sheets, or water (as under shallow lakes and bays) (Kupsch, 1956). Rarely used.

submature [geomorph] Said of a topographic feature that has passed through the stage of youth but is not completely mature; e.g. a *submature* shoreline characterized by the cutting-back of headlands and by the near closing of baymouths by bars, thus simplifying an earlier intricately embayed shoreline (Cotton, 1958, p. 456).

submature [sed] Pertaining to the second stage of textural maturity (Folk, 1951); said of a clastic sediment intermediate in character between an *immature* and a *mature* sediment, characterized by little or no clayey material and by poorly sorted and angular grains. Example: a clean "submature sandstone" containing less than 5% clay and commonly occurring in stream channels. Cf: *supermature.*

submeander A small meander contained within the banks of the main channel (Langbein & Iseri, 1960, p. 19); it is associated with relatively low discharges.

submerged aquatic plant A *hydrophyte* the main part of which grows below the surface of the water. Cf: *subaquatic plant.*

submerged coastal plain The continental shelf representing the seaward continuation of a coastal plain on the land. Syn: *coast shelf.*

submerged contour A contour on the bed of a lake or reservoir, joining points of equal elevation where the elevation is related to a datum (usually mean sea level) used for mapping adjacent land (BNCG, 1966, p.14). Cf: *isobath [oceanog].*

submerged forest Forest remains, e.g. stumps still rooted in peaty soil, seen at low tide or found below sea level, indicating a rise in sea level or a subsidence of the coast.

submerged land A legal term for the land at the bottom of a lake, or the land covered by water when the lake is at its mean high-water level or at a level set by court decree (Veatch & Humphrys, 1966, p.324).

submerged rib A rib generated by the commissural denticles on the inside of the outer shell layers of some rostroconchs, and covered by the growth of inner shell layers (Pojeta and Runnegar, 1976, p. 47).

submerged shoreline *shoreline of submergence.*

submerged valley A *drowned valley,* such as a ria.

submergence A rise of the water level in relation to the land, so that areas formerly dry land become inundated; it results either from a sinking of the land or from a rise of the water level. Ant: *emergence.*

submersible A small self-propelled underwater vehicle for direct sea-floor observation and sampling.

submesothyridid Said of a brachiopod pedicle foramen located mainly in the delthyrium and partly in the ventral umbo (TIP, 1965, pt.H, p.153). Cf: *permesothyridid.*

submetallic luster A mineral luster between *metallic* and *nonmetallic.* Chromite, for example, has a metallic to submetallic luster.

subnival *periglacial.*

subnival boulder pavement A pavement of tightly packed blocks and boulders, with flat faces upward simulating a paved Roman highway, on a mountain valley floor in or near a stream channel and beneath a transient snowbank. Boulder pavements form downwind from sharp breaks in a valley floor and usually above treeline. They may be the result of the raising of blocks by *congeliturbation,* the rotation and flattening of the blocks by their own weight in fluid mud and by the weight of wind-packed snow and possible creep of the snowbank, and by meltwater saturation and later removal of fine sediment between blocks (White, 1972, p. 195). Cf: *stone pavement.*

subnormal-pressure surface A potentiometric surface that is below the upper surface of the zone of saturation (Meinzer, 1923, p. 39). The term is not in general use among hydrogeologists. Cf: *normal-pressure surface; artesian-pressure surface.*

subophitic Said of the texture of an igneous rock in which the feldspar crystals are approximately the same size as the pyroxene and are only partially included by them. The term *ophitic* generally includes such textures.

suborder In the hierarchy of classification of plants and animals, a subcategory of *order.* It is sometimes considered equivalent to *superfamily* and sometimes as the next higher rank.

suboutcrop *blind apex.*

subperiod A geologic-time unit that is a portion of a *period [geochron],* but longer than an *epoch [geochron],* during which the rocks of the corresponding *subsystem* were formed.

subpermafrost water Ground water in the unfrozen ground beneath permafrost.

subphyllarenite A *phyllarenite* containing 3-25% rock fragments.

subphylum In the hierarchy of zoological classification, a category intermediate between *phylum* and *class.*

subpolar glacier A glacier on which there is some surface melting during the summer but which is below the freezing temperature throughout most of its mass. Cf: *polar glacier; high-polar glacier.*

subrosion Subsurface erosion, apparently from solution of salts and subsequent adjustment/collapse.

subrounded Partially rounded; specif. said of a sedimentary particle showing considerable but incomplete abrasion and an original general form that is still discernible, and having many of its edges and corners noticeably rounded off to smooth curves, such as a cobble with a reduced number (5-10) of secondary corners, a considerably reduced area of the original faces, and a roundness value between 0.25 and 0.40 (midpoint at 0.315) (Pettijohn, 1957, p. 59), or one with two-thirds of its edges smooth (Krynine, 1948, p.142). Also, said of the *roundness class* containing subrounded particles.

subsea apron *submarine fan.*

subseptate Having imperfect or partial septa; e.g. having slight protuberances or incipient septa that form pseudochambers in a foraminiferal test (as in the family Tournayellidae).

sub-sequence A term applied by R.C. Moore (1958, p.80) to a Precambrian rock division (often called a *series* in Canada) that "cannot be correlated from one region to another". The term requires a hyphen in order to distinguish it from the noun "subsequence". See also: *sequence.*

subsequent [geomorph] Said of a post-*consequent* geologic or topographic feature whose development is controlled by differences in the erosional resistance of the underlying rocks; e.g. a subsequent ridge formed by differential erosion of a consequent ridge, a subsequent waterfall produced where a downcutting stream encounters rock of exceptional hardness, or a subsequent valley developed along the strike of a weakly resistant homoclinal bed.

subsequent [streams] adj. Said of a stream, valley, or drainage system that is developed independently of, and subsequent to, the original relief of a land area, as by shifting of divides, stream capture, or adjustment to rock structure. The concept was originally discussed by Jukes (1862, p. 393-395).—n. *subsequent stream.*

subsequent fold *cross fold.*

subsequent stream A tributary that has developed its valley (mainly by headward erosion) along a belt of underlying weak rock and is therefore adjusted to the regional structure; esp. a stream that flows approximately in the direction of the strike of the underlying strata and that is subsequent to the formation of a consequent stream of which it is a tributary. Obsolete syn: *subconsequent stream.* Syn: *subsequent; strike stream; longitudinal stream.*

subsere (a) A secondary ecologic *succession [ecol]* that arises on a denuded area following an ecologic climax. (b) A seral *community* that is prevented from reaching ecological climax by a temporary interference, e.g. by human activity.

subseries A term proposed by Sutton (1940, p.1402) for the first division of a series, representing the rocks formed during a *subepoch.*

subsidence (a) The sudden sinking or gradual downward settling of the Earth's surface with little or no horizontal motion. The movement is not restricted in rate, magnitude, or area involved. Subsidence may be caused by natural geologic processes, such as solution, thawing, compaction, slow crustal warping, or withdrawal of fluid lava from beneath a solid crust; or by man's activity, such as subsurface mining or the pumping of oil or ground water. See also: *cauldron subsidence.* Syn: *land subsidence; bottom subsidence.* (b) A sinking or downwarping of a large part of the Earth's crust relative to its surrounding parts, such as the formation of a rift valley or the lowering of a coast due to tectonic movements.—Syn: *sinking.*

subsidence caldera *collapse caldera.*

subsidence/head-decline ratio The ratio between land subsidence and the hydraulic head decline in the coarse-grained beds of the compacting aquifer system (Poland et al., 1972).

subsidence theory A theory of coral-atoll and barrier-reef formation according to which upward reef growth kept pace uninterruptedly over a long period with slow subsidence of a volcanic island, forming first a fringing reef that became a barrier reef and later an atoll when the island was completely submerged; it accounts satisfactorily for many Pacific reefs. Theory was proposed by Charles Darwin in 1842. Cf: *glacial-control theory; antecedent-platform theory.*

subsidiary fold *parasitic fold.*

subsilicic A term proposed by Clarke (1908, p. 357) to replace *basic.* Cf: *persilicic; mediosilicic.*

subsoil (a) A syn. of *B horizon,* in a soil profile having distinct horizons. (b) The soil below the *surface soil;* this is an older meaning.—Cf: *topsoil.*

subsoil ice *ground ice [permafrost].*

subsoil weathering A term used by Davis (1938) for the chemical decomposition that produces spheroidal boulders beneath the regolith in granitic areas of the desert by percolation of water along joints followed by exposure and exfoliation.

subsolidus A chemical system that is below its melting point, and in which reactions may occur in the solid state.

subsolvus Said of those granites, syenites, and nepheline syenites that are characterized by both potassium feldspar and plagioclase (Bowen and Tuttle, 1958, p. 129). Cf: *solvus; hypersolvus.*

subspeciation Division into or formation of subspecies.

subspecies In the hierarchy of classification of plants and animals, a subcategory of *species.* Groups within a species that are geographically isolated from one another are geographic subspecies; groups separated in geologic time are chronologic subspecies. The name of a subspecies is a trinomen; e.g. *Bollia americana zygocornis.* Cf: *variety [taxon].*

substage [geochron] An obsolete term for a geologic-time unit of lesser duration than *stage [geochron],* during which the rocks of a *member* were formed.

substage [glac geol] A time term for a subdivision of a glacial stage during which there was a secondary fluctuation of glacial advance and retreat; specif. a Pleistocene subdivision equated with a rock unit of member rank, such as the "Woodfordian substage" of the Wisconsinan. stage.

substage [optics] In a microscope, an attachment for holding polarizers or other attachments below the stage.

substage [stratig] A subdivision of a *stage [stratig];* the rocks formed during a *subage* of geologic time. The frequently used synonym *zone* is not recommended (ISST, 1961, p. 13).

substitute Any substance represented to be, or used to imitate, a gemstone; e.g. plastic, glass, doublet, synthetic ruby, or natural spinel all could be substitutes for natural ruby.

substitution *ionic substitution.*

substitutional transformation A type of crystal *transformation* of a disordered phase (a substitutional solid solution) to an ordered phase. It is usually a slow transformation. Cf: *dilatational transformation; displacive transformation; reconstructive transformation; rotational transformation; order-disorder transformation.*

substitution solid solution A crystal in which a particular atomic site can be occupied by any of two or more elements. Cf: *omission solid solution.*

substrate [cryst] The surface on which *epitaxy* occurs.

substrate [ecol] The substance, base, or nutrient on which, or the medium in which, an organism lives and grows, or the surface to which a fixed organism is attached; e.g. soil, rocks, water, and leaf tissues, or perhaps a gel for the accumulation and preservation of prebiologic organic matter. Syn: *substratum [ecol].*

substratum [ecol] *substrate.*

substratum [soil] Any layer lying beneath the solum. It is applied to both parent materials and to other layers unlike the parent material, below the B horizon or subsoil.

subsurface n. (a) The zone below the surface, whose geologic features, principally stratigraphic and structural, are interpreted on the basis of drill records and various kinds of geophysical evidence. (b) Rock and soil materials lying beneath the Earth's surface.—adj. Formed or occurring beneath a surface, esp. beneath the Earth's surface. Cf: *surficial.* See also: *subterranean.*

subsurface air Gas in interstices in the zone of aeration that open directly or indirectly to the surface and that is therefore at or near atmospheric pressure. Its composition is generally similar though not identical to that of the atmosphere (Meinzer, 1923, p. 21). Cf: *soil atmosphere; included gas; ground air.*

subsurface contour A syn. of *structure contour,* used to distinguish it from a surface or topographic contour.

subsurface drainage The removal of surplus water from within the soil by natural or artificial means, such as by drains placed below the surface to lower the water table below the root zone.

subsurface flow *storm seepage.*

subsurface geology Geology and correlation of rock formations, structures, and other features beneath the land or sea-floor surface as revealed or inferred by exploratory drilling, underground workings, and geophysical methods. Ant: *surface geology.*

subsurface ice *ground ice [permafrost].*

subsurface map A map depicting geologic data or features below the Earth's surface; esp. a plan of mine workings, or a structure-contour map of a petroleum reservoir or an underground ore deposit, coal seam, or key bed.

subsurface perched stream Vadose water flowing toward the water table in fractures in solution openings. There may be solid rock instead of a true zone of aeration beneath such a stream, and it could be considered simply as gravity ground water on its way to the water table by the easiest route. According to Meinzer (1923, p. 22), "Water in transit from the surface to a zone of saturation may for convenience be regarded as passing through a zone of aeration or may be regarded as an irregular and perhaps temporary projection of the zone of saturation".

subsurface runoff *storm seepage.*

subsurface storm flow *storm seepage.*

subsurface water Water in the lithosphere in solid, liquid, or gaseous form. It includes all water beneath the land surface and beneath bodies of surface water. Syn: *subterranean water; underground water; ground water.*

subsystem [chem] Any part of a system that may be treated as an independent system.

subsystem [stratig] A subdivision of a *system [stratig].* "Special circumstances have suggested the occasional need" for subsystems, e.g. the Mississippian Subsystem of the Carboniferous System (ISG, 1976, p. 73-74).

subtalus buttress The convex-upward rock surface developed under a rising talus slope as the cliff above it weathers back (Howard, 1942, p. 27).

subterrane n. The bedrock beneath a surficial deposit or below a given geologic formation. Syn: *subterrain.* —adj. *subterranean.*

subterranean adj. Formed or occurring beneath the Earth's surface, or situated within the Earth. Cf: *subaerial.* See also: *subsurface.* Syn: *subterrestrial; subterrane.*

subterranean cutoff Diversion of a surface stream by the development of a shorter underground course across a meander neck.

subterranean ice *ground ice [permafrost].*

subterranean stream A body of subsurface water flowing through a cave or a group of communicating caves, as in a karstic region. See also: *underground stream.*

subterranean stream piracy Capture of a surface stream perched on soluble rocks, and its diversion through an underground channel to an adjacent entrenched stream.

subterranean water (a) A syn. of *subsurface water.* (b) A syn. of *ground water* in less preferred usage.

subterrestrial *subterranean.*

subtropical Said of the climate of the subtropics, which borders that of the tropics and is intermediate in character between tropical and temperate, though more like the former than the latter. Syn: *semitropical.*

subulate Awl-shaped; said of a leaf that tapers from its base to a sharp point.

subvective system A system for gathering and transporting food particles to the mouth of an echinoderm. It cannot be separated morphologically from the *ambulacral system* (TIP, 1966, pt.U, p.155).

subvolcanic *hypabyssal.*

subvolcano A small intrusion not far from the surface.

subweathering velocity The seismic velocity of the layer immediately underlying the near-surface *low-velocity layer.* This velocity is distinctly greater than that in the weathered zone. Cf: *weathering velocity.*

subzone (a) A subdivision of a *biozone,* next lower in rank than the zone itself (ACSN, 1961, art.20e). It is a biostratigraphic unit that is usually recognizable across a continent, and it may include any number of *zonules.* (b) A subdivision of an informal stratigraphic zone of any kind (ISST, 1961, p.29).

succession [ecol] The progressive change in a biologic community as a result of the response of the member species to the environment. See also: *sere; faunal succession.* Syn: *ecologic succession.*

succession [stratig] (a) A number of rock units or a mass of strata that succeed one another in chronologic order; e.g. an inclusive stratigraphic sequence involving any number of stages, series, systems, or parts thereof, as shown graphically in a geologic column or seen in an exposed section. (b) The chronologic order of rock units.

successional speciation The gradual evolution from one species to another, eventually leading to replacement by the latter.

succinic acid A crystalline dicarboxylic acid, formula $HOOCCH_2CH_2COOH$, a constituent of wood bark and occurring in the amber group (but not in the retinite group) of fossil resins.

succinite (a) An old name for *amber,* esp. that mined in East Prussia or recovered from the Baltic Sea. (b) A light-yellow, amber-colored variety of grossular garnet.

sucrosic A syn. of *saccharoidal.* The term is commonly applied to idiotopic dolomite rock. Syn: *sucrose.*

sudburite An augite-bearing hypersthene *basalt* characterized by pillow structure and containing bytownite and magnetite. "It differs from normal basalts in containing neither glass nor olivine and in having an equigranular texture" (Johannsen, 1937, p. 305). Its name, given by Coleman in 1914, is derived from the Sudbury district, Ontario. Not recommended usage.

sudburyite A hexagonal mineral: (Pd,Ni)Sb.

Sudetic orogeny One of the 30 or more short-lived orogenies during Phanerozoic time identified by Stille, in this case between the Early and Late Carboniferous.

sudoite A dioctahedral aluminum-rich mineral of the chlorite group: $(Al,Fe,Mg)_{4-5}(Si,Al)_4O_{10}(OH)_8$.

Suess effect A decrease in the concentration of carbon-14 in atmospheric carbon dioxide due to dilution by nonradioactive carbon from the burning of fossil fuels.

Suess torsion balance A double-variometer torsion balance with visual observation, which replaced the single balance.

suevite A grayish or yellowish *breccia* that is associated with meteorite impact craters and that contains both shock-metamorphosed rock fragments and glassy inclusions that occur typically as aerodynamically shaped bombs. It closely resembles a tuff breccia or pumiceous tuff but is of nonvolcanic origin and can be distinguished by the presence of shock-metamorphic effects. The term was originally applied to material from the Ries basin, Germany, but is now used to designate similar brecciated material (*impactites*) found at other meteorite impact structures.

suffosion The bursting-out on the surface in little eruptions of highly mobile or water-saturated material; esp. a destructive process operating under periglacial conditions, in which underground water, resulting from partial melting of ground ice, exerts upward pressure and bursts through a hard dried upper skin to deposit a mound of mud, clay, sand, and/or boulders. Suffosional forms due to corrasion by underground water include dimpling, pits, blind valleys, shafts, and cavern entrances.

suffosion knob A *frost mound* (Muller, 1947, p.222).

sugar iceberg An iceberg consisting of porous glacier ice that is formed at very low temperatures.

sugarloaf A conical hill or mountain comparatively bare of timber, resembling the shape of a loaf of sugar; e.g. Sugarloaf Mountain, Maine.

sugar sand A sandstone that breaks up into granules that resemble sugar.

sugar stone Compact, white to pink datolite from the Michigan copper district (Pough, 1967, p. 140).

sugary *saccharoidal.*

sugilite A mineral of the osumilite group: $(K,Na)(Na,Fe^{+3})_2(Li_2Fe^{+3})Si_{12}O_{30}$.

suicidal stream A stream that rises in desert mountains and that loses its small amount of water by evaporation and infiltration soon after reaching the desert plain below (Stone, 1967, p. 250).

suite [ign] (a) A set of apparently comagmatic igneous rocks. (b) A collection of rock specimens from a single area, generally representing related igneous rocks. (c) A collection of rock specimens of a single kind, e.g. granites from all over the world.

suite [stratig] (a) A term used by Caster (1934, p.18) for a body of rocks intermediate between monothem (formation) and member, consisting of several intimately related members bracketed together; esp. a repeated sequence of such closely associated strata. (b) A local stratigraphic subdivision used in the USSR, corresponding approximately to a formation. (c) A term proposed for rock masses of diverse types that are either of igneous or high-grade metamorphic origin, e.g. Tuolumne Intrusive Suite (Sohl, 1977, p. 249).

sukulaite A mineral of the pyrochlore group: $Sn_2Ta_2O_7$.

sulcal notch A ventral indentation in the margin of some apical archeopyles in a dinoflagellate cyst, corresponding to the *sulcal tongue.*

sulcal plate One of the plates of the ventral furrow region in dinoflagellates possessing a theca. The plates are subdivided as to left or right, and anterior or posterior position.

sulcal tongue An extension of the operculum of a dinoflagellate cyst, in the position of the first apical plate of the theca. It is normally bordered by those parts of the archeopyle suture corresponding to thecal structures separating the first apical plate from the first precingular plate, the last precingular plate, and the intervening anterior sulcal plate. Cf: *sulcal notch.*

sulcate Pertaining to a sulcus, or scored with furrows, grooves, or channels, esp. lengthwise; e.g. said of a form of alternate folding in brachiopods, with the brachial valve bearing a median sulcus and an anterior commissure median sinus (TIP, 1965, pt.H, p.153). Ant: *uniplicate.* Syn: *sulcated.*

sulcus [paleont] (a) A major longitudinal depression in the surface of either valve of a brachiopod, externally concave in transverse profile and radial from the umbo, and usually median in position. It is typically associated with the *fold [paleont].* See also: *sinus.* (b) An elongate shallow depression in the lateral surface of an ostracode valve, extending from the dorsal region toward the venter. (c) A radial depression of the surface of the shell of a bivalve mollusk. (d) A longitudinal groove on the venter of an ammonoid shell.—Pl: *sulci.*

sulcus [palyn] (a) A relatively broad longitudinal furrow in the exine of pollen grains; a *colpus.* The term is usually applied only to the distal furrow of monocolpate pollen grains. (b) A longitudinal posterior groove lying on the surface of dinoflagellate thecae and containing the trailing flagellum.

suldenite A *hornblende andesite* differing from *ortlerite* in having an andesitic rather than a microdioritic groundmass (Holmes, 1920, p. 219). Obsolete.

sulfate A mineral compound characterized by the sulfate radical SO_4. Anhydrous sulfates, such as barite, $BaSO_4$, have divalent cations linked to the sulfate radical; hydrous and basic sulfates, such as gypsum, $CaSO_4 \cdot 2H_2O$, contain water molecules. Cf: *chromate.*

sulfate-reducing bacteria *sulfur bacteria.*

sulfide A mineral compound characterized by the linkage of sulfur with a metal or semimetal, such as galena, PbS, or pyrite, FeS_2. See also: *sulfosalt.*

sulfide enrichment Enrichment of a deposit by replacement of one sulfide by another of higher value, as pyrite by chalcocite.

sulfide-facies iron formation An *iron formation* consisting essentially of pyritic carbonaceous slate. It was formed in the deeper, reducing parts of seas or basins.

sulfide zone (a) That part of a sulfide deposit that has not been oxidized by near-surface waters. Cf: *oxidized zone; protore.* (b) A generally manto-shaped deposit in which secondary sulfide enrichment has occurred as a part of ore-deposit oxidation. Syn: *secondary sulfide zone.*

sulfidization *sulfurization.*

sulfoborite A mineral: $Mg_3B_2(SO_4)(OH)_{10}$. Also spelled: *sulphoborite.*

sulfohalite *sulphohalite.*

sulfosalt A type of *sulfide* in which both a metal and a semimetal are present, forming a double sulfide, e.g. enargite, Cu_3AsS_4.

sulfur (a) An orthorhombic mineral, the native nonmetallic element S. It occurs in yellow crystals at hot springs and fumaroles, and in masses or layers associated with limestone, gypsum, and anhydrite, esp. in salt-dome caprock and bedded deposits. Sulfur exists in several allotropic forms, including the ordinary yellow orthorhombic alpha form stable below 95.5°C and the pale-yellow monoclinic crystalline beta form. See also: *rosickyite.* Syn: *brimstone.* (b) A mining term used for iron sulfide (pyrite) occurring in coal seams and with zinc ores in Wisconsin and Missouri. Also spelled: *sulphur.*

sulfur bacteria Anaerobic bacteria that obtain the oxygen needed in metabolism by reducing sulfate ions to hydrogen sulfide or elemental sulfur. Accumulations of sulfur in this way are *bacteriogenic* ore deposits. Cf: *iron bacteria.*

sulfur ball [coal] A pyritic impurity in coal, occurring as a spheroidal or irregular mass. Cf: *coal ball.*

sulfur ball [pyroclast] A sulfurous mud skin that forms on a bubble of hot volcanic gas and becomes firm on contact with the air.

sulfurization "Reaction between sulfur from an external source and cations such as iron, nickel, and copper in solid solution in common rock-forming minerals or in igneous magma," considered as an ore-forming process (Naldrett & Kullerud, 1965). Syn: *sulfidization.*

sulfur-mud pool *mud pot.*

sulfur spring A spring containing sulfur water.

sulfur water Generally, water containing enough hydrogen sulfide to smell and taste. Except for the hydrogen sulfide, it may not differ in mineral content from ordinary potable water, or it may qualify as *saline* water. In either case, it is usually considered a mineral water.

sullage Mud and silt deposited by flowing water.

sulphatite Free sulfuric acid (H_2SO_4) found in some waters.

sulphoborite *sulfoborite.*

sulphohalite A mineral: $Na_6(SO_4)_2FCl$. Also spelled: *sulfohalite.*

sulphophile Said of an element, e.g. fluorine or chlorine, that occurs preferentially in a mineral that is oxygen-free, i.e. as a sulfide, selenide, telluride, arsenide, or antimonide. The term incorporates the chalcophile elements and some of the siderophile elements of Goldschmidt's classification. The term is rarely used.

sulphur *sulfur.*

sulphur ore A mining term used for both pyrite and native sulfur.

sulvanite A bronze-yellow isometric mineral: Cu_3VS_4. Not to be confused with *sylvanite.*

sumacoite A magnetite-rich extrusive *trachyandesite* containing abundant phenocrysts of plagioclase and augite and rare olivine in a groundmass of andesine, alkali feldspar, nepheline, and hauyne. Its name (Johannsen, 1938) is derived from Sumaco crater, Ecuador. Not recommended usage.

summation method In seismology, a method of correcting the arrival times of reflected waves for the time they spend in the low-velocity zone.

summer balance The change in mass of a glacier from the maximum value in a certain year to the following minimum value of that year; sometimes called *apparent ablation* or (erroneously) *net ablation.* Cf: *winter balance.*

summer berm A berm built on the backshore by the uprush of waves during the summer. Cf: *winter berm.*

summer season In glaciology, that period of a year when the *balance* of a glacier decreases from a maximum value to a minimum value for the year. This is a period when, on the average, ablation exceeds accumulation. Cf: *winter season.* Syn: *ablation season.*

summer surface An observable or measurable horizon (e.g. *dirt band*) in a glacier marking the time of minimum mass of the glacier in a year's time. See also: *balance year; net balance.*

summerwood A syn. of *late wood.* Also spelled: *summer wood.* Cf: *springwood.*

summit (a) The top, or the highest point or level, of an undulating land feature, as of a hill, mountain, volcano, or rolling plain; a peak. See also: *crest [geomorph].* (b) Loosely, a divide or pass; e.g. Donner Summit, Calif.

summit concordance Equal or nearly equal elevation of ridgetops or mountain summits over a region. The concordance is commonly thought to indicate the existence of an ancient erosion plain of which only scattered patches are preserved. See also: *accordant summit level; even-crested ridge.* Syn: *accordance of summit lev-*

els; concordance of summit levels.

summit eruption An eruption at the top of a volcanic mountain. Cf: *flank eruption.*

summit graben A *volcanic graben* on the summit of a volcanic cone, more or less rectangular or triangular; also, a graben crossing the crest of a *resurgent cauldron* (Smith & Bailey, 1968). Cf: *sector graben.*

summit level [eng] The highest point of a road, railroad, or canal; the highest of a series of elevations over which a road or canal is carried.

summit level [geomorph] The elevation of a summit plane. See also: *accordant summit level.*

summit plain *peak plain.*

summit plane The plane passing through a series of accordant summits. See also: *gipfelflur.*

summitpoint The point of maximum elevation on the cross section of a ripple mark. Term introduced by Allen (1968). Cf: *brinkpoint.*

sump [geog] (a) An excavation in which the drainage water of an area is collected for subsequent use in irrigation or wild-fowl conservation. (b) A dialect term for a swamp or morass, and for a stagnant pool or puddle of dirty water. (c) An English term for a cove or a muddy inlet.

sump [speleo] (a) A pool of water in a cave, the outlet of which lies beneath its surface. See also: *siphon [speleo].*

sun ball A snowball formed by a lump of snow falling onto a slope of moist snow and rolling downhill. Cf: *snow roller.*

sunburn The collective effects produced in lunar rocks and soils by the long-term bombardment by high-energy particles from the sun or other cosmic sources. These effects include particle tracks, nuclear reactions, and the implantation of atoms from the solar wind. Syn: *suntan.*

sun crack A crack in sediment or rock, formed by the drying action of the Sun's heat; esp. a *mud crack.*

sun crust A type of *snow crust* formed by refreezing of surface liquid that had been melted by the sun; it is usually thin and has a smooth surface.

sun cup A cuspate hollow or depression in a snow surface, formed during sunny weather by complex ablation processes. In some environments, a sun cup may grow into a *nieve penitente.*

sundtite *andorite.*

sungei A marine channel cutting across the Aroe Islands north of Australia, representing a Quaternary antecedent stream that was invaded by the Flandrian Transgression (Fairbridge, 1951). Etymol: Malay, "river, large stream". Also spelled: *soengei.*

sunken caldera *collapse caldera.*

sunken island A high-relief feature of a lake basin, such as a basin divide or the crest of a knob, covered by a shallow depth of water; it was "never originally above water level" and is therefore "not due to subsidence" (Veatch & Humphrys, 1966, p.317). Cf: *blind island.*

sunken stream *sinking stream.*

sun opal *fire opal.*

sun spike *nieve penitente.*

sunspot A relatively dark area on the Sun's surface, representing lower temperature and consisting of a dark central umbra surrounded by a penumbra which is intermediate in brightness between the umbra and the surrounding surface of the photosphere (NASA, 1966, p. 47).

sunstone An *aventurine feldspar,* usually a brilliant, translucent variety of oligoclase that emits a reddish or golden billowy reflection from minute scales or flakes of hematite spangled throughout and arranged parallel to planes of repeated twinning. Cf: *moonstone.* Syn: *heliolite.*

sun-synchronous Said of an Earth-satellite orbit in which the orbit plane is near polar and the altitude such that the satellite passes over all places on Earth having the same latitude twice daily at the same local sun time.

suolunite A mineral: $Ca_2Si_2O_5(OH)_2 \cdot H_2O$.

superanthracite A syn. of *meta-anthracite.* Not listed by ASTM as a rank classification.

superbiozone A unit consisting of "several biozones with common biostratigraphic features" (ISG, 1976, p. 48).

supercapillary interstice An *interstice* sufficiently larger than a *capillary interstice* that surface tension will not hold water far above a free water surface. Water moving in it, as by *supercapillary percolation,* may develop currents and eddies (Meinzer, 1923, p. 18). Cf: *subcapillary interstice.*

supercapillary percolation Percolation through *supercapillary interstices.*

supercell The unit cell of a *superlattice.*

supercooling The process of lowering the temperature of a phase or assemblage below the point or range at which a phase change should occur at equilibrium, i.e. making the system metastable by lowering the temperature. It generally refers to a liquid taken below its liquidus temperature. Cf: *superheating.* Syn: *undercooling.*

supercritical Said of a system that is at a temperature higher than its critical temperature; also, said of the temperature itself.

supercritical flow *rapid flow.*

supercrop map A geologic map that shows the distribution of stratigraphic units lying immediately above a given rock body or a surface at a given time. Cf: *subcrop map.* Syn: *worm's-eye map.*

supercycle A group of cycles of eustatic or relative change of sea level in which a cumulative rise to a higher position is followed by a cumulative fall to a lower position (Mitchum, 1977, p. 211). Commonly, one or two major falls rather than a succession of them are evident at the end of a supercycle. See also: *cycle of relative change of sea level; paracycle of relative change of sea level.*

superdeltoid An anal deltoid plate on the border of the mouth opening of a blastoid, associated either with the subdeltoid or with a pair of cryptodeltoids abutting its aboral margin and in some genera bordering the anal opening (TIP, 1967, pt. S, p. 350).

superface The top or upper surface of a stratigraphic unit.

superfacies A large-scale stratigraphic facies, generally consisting of two or more subordinate facies; e.g. a laterally equivalent and contrasting part of a formation, within which smaller-scale, laterally equivalent, and contrasting parts are recognized.

superfamily In the hierarchy of classification of plants and animals, a category next above *family.* It may be considered as equivalent to *suborder,* or as intermediate between suborder and family.

superficial Pertaining to, or lying on or in, a surface or surface layer; e.g. "superficial weathering" of a rock, or a "superficial structure" formed in a sediment by surface creep. The term is used esp. in Great Britain; the syn. *surficial* is more generally applied in the U.S.

superficial deposit *surficial deposit.*

superficial fold *décollement fold.*

superficial moraine *surficial moraine.*

superficial oolith An oolith with an incomplete or single layer; specif. one in which the thickness of the accretionary coating is less than the radius of the nucleus (Beales, 1958, p. 1863). Cf: *oopellet.*

superfluent lava flow A flow of lava issuing from a summit crater and streaming down the flanks of the volcano (Dana, 1890); an obsolete term. Cf: *effluent lava flow; interfluent lava flow.*

supergene Said of a mineral deposit or *enrichment* formed near the surface, commonly by descending solutions; also, said of those solutions and of that environment. Cf: *hypogene; mesogene.* Syn: *hypergene; secondary [eco geol].*

supergene enrichment The *supergene* processes of mineral deposition. Near-surface oxidation produces acidic solutions that leach metals, cary them downward, and reprecipitate them, thus enriching sulfide minerals already present. Supergene enrichment has been important in upgrading porphyry copper deposits to the status of ore. Syn: *enrichment; secondary enrichment.* See also: *oxidized zone.*

superglacial Carried upon, deposited from, or pertaining to the top surface of a glacier or ice sheet; said of meltwater streams, till, drift, etc.

supergroup An assemblage of related *groups,* or of formations and groups, having significant lithologic features in common (ISG, 1976, p. 34). Cf: *megagroup.*

superheating (a) The addition of more heat than necessary to complete a given phase change. (b) In a magma, the addition of more heat than is necessary to cause complete melting. The temperature increase above liquidus is called the superheat. (c) The process of increasing heat beyond that point at which a phase or assemblage changes at equilibrium, i.e. to a metastable state in the sense analogous to *supercooling.*

superimposed [stratig] Said of rocks that are layered or stratified.

superimposed [streams] Said of a stream or drainage system let down from above by erosion, through the formations on which it was developed, onto rocks of different structure lying unconformably beneath. The term was first applied by Powell (1875, p. 165-166) to the valley thus formed, although Maw (1866, p. 443-444) earlier discussed the concept. Syn: *superposed; inherited; epigenetic.*

superimposed drainage Drainage by *superimposed streams.*

superimposed fan A term proposed by Blissenbach (1954, p. 180-181) for a newly deposited alluvial fan that has a steeper gradient than the older fan upon which it is developed; it results from tectonic movements that initiate a new stage of deposition.

superimposed fold A syn. of *cross fold.* Var: *superposed fold.*

superimposed ice Ice formed when meltwater percolates down through a snowpack on a glacier and refreezes at the base of the snowpack, or as it is trapped on a lower horizon of reduced permeability such as a firn-ice boundary. This ice appears at the surface of a glacier in the *superimposed-ice zone.*

superimposed ice stream Ice from a tributary glacier that rests on the surface of a larger glacier but does not sink down into it, such as where a tributary glacier flows onto the surface of a trunk glacier. Cf: *inset ice stream.*

superimposed-ice zone On a glacier or ice sheet where surface melting occurs, much of the meltwater is refrozen at the base of the snowpack as *superimposed ice,* and the snowpack is removed, exposing superimposed ice at the surface. The zone is bordered at higher altitudes by the *firn line,* and at lower altitudes by the *equilibrium line.*

superimposed metamorphism *polymetamorphism.*

superimposed profile A diagram on which a series of profiles drawn along several regularly spaced and parallel lines on a map are placed one on top of the other (Monkhouse & Wilkinson, 1952); it may emphasize such features as accordant summit levels and erosion platforms. Cf: *composite profile; projected profile.*

superimposed stream A stream that was established on a new surface and that maintained its course despite different lithologies and structures encountered as it eroded downward into the underlying rocks. Syn: *superinduced stream.*

superimposition [stratig] *superposition.*

superimposition [streams] The establishment, originally on a cover of rocks now removed by erosion, of a stream or drainage system on existing underlying rocks independently of their structure. Gilbert (1877, p. 144) recognized superimposition from an unconformable cover of sediments, from a surface of alluvium, and from a surface produced by planation. Syn: *epigenesis.*

superincumbent Said of a *superjacent* layer, esp. one that is situated so as to exert pressure.

superindividual An aggregate of grains that behaves as a unit in the fabric of a deformed rock.

superinduced stream *superimposed stream.*

superjacent Said of a stratum situated immediately upon or over a lower stratum or an unconformity. Ant: *subjacent.* See also: *superincumbent.*

superlattice The crystal lattice of a substance in which, because of chemical substitutions or atomic ordering, one or more translational periodicities are multiples of those in the related unsubstituted or ordered form. Syn: *superstructure [cryst].*

supermature Pertaining to the fourth and last stage of textural maturity (Folk, 1951); said of a *mature* clastic sediment whose well-sorted grains have become subrounded to well-rounded, such as a clay-free "supermature sandstone" whose sand-size quartz grains have an average roundness that exceeds 0.35 and that is presumed to form mainly as dune sands. Cf: *immature; submature.*

superparamagnetism The *paramagnetic* behavior of an assembly of extremely small particles of ferromagnetic or ferrimagnetic minerals.

superperiodicity A lattice constant in an ordered superlattice which is a simple multiple of a corresponding direction in the disordered sublattice; the periodicity of a *superlattice.*

superplasticity The plastic behavior of very fine-grained material that yields a total strain larger than what can be expected from the internal deformation of the individual crystals; it implies grain-boundary sliding. This behavior is experimentally shown by ceramics and may explain the *fluidal texture* and inferred rheology of the deepest peridotite xenoliths in kimberlites.

superposed A term introduced by McGee (1888) as a shortened form of *superimposed.*

superposed fold Var. of *superimposed fold.*

superposition (a) The order in which rocks are placed or accumulated in beds one above the other, the highest bed being the youngest. (b) The process by which successively younger sedimentary layers are deposited on lower and older layers; also, the state of being superposed.—See also: *law of superposition.* Syn: *superimposition [stratig]; supraposition.*

superprint *overprint.*

supersaline *hypersaline.*

supersaturated permafrost Permafrost that contains more ice than the ground could possibly hold if the water were in the liquid state.

supersaturated solution A solution which contains more of the solute than is normally present when equilibrium is established between the saturated solution and undissolved solute.

superseries An infrequently used term for a group of two or more *series [stratig]* (ISG, 1976, p.72).

superstage A chronostratigraphic unit consisting of several adjacent *stages [stratig]* (ISG, 1976, p. 72).

superstructure [cryst] *superlattice.*

superstructure [tect] The upper structural layer in an orogenic belt, subjected to relatively shallow or near-surface deformational processes, in contrast to an underlying and more complexly deformed and metamorphosed *infrastructure.* Also spelled: *suprastructure.*

supersystem A chronostratigraphic unit next higher in rank than a *system [stratig].* "Special circumstances have suggested the occasional need" for supersystems, e.g. the Karroo Supersystem (ISG, 1976, p. 73-74). See also: *erathem.*

superterranean Occurring on or above the Earth's surface. Syn: *superterrene; superterrestrial.* Rarely used.

superzone An assemblage of two or more stratigraphic zones of any kind. Syn: *megazone.*

supplementary contour A contour line, generally dashed or screened, drawn at less than the regular interval in order to increase the topographic expression of an area, such as in an area of extremely low relief. Syn: *auxiliary contour.*

suppressed Marked or affected by suppression, or being vestigial; e.g. said of aborted conodont denticles that could not develop into mature structures owing to crowded conditions along the growing edge of the structure.

supra-anal plate The dorsal, posteriorly produced part of the telson of a branchiopod crustacean.

supracrustal (a) Said of rocks that overlie the *basement.* (b) At the surface of the Earth. Cf: *infracrustal.*

supraembryonic area Circular apical area over the megalospheric proloculus in some foraminifers of the family Orbitolinidae.

suprafan An upbulging zone at the downslope end of a *fan valley,* confined to fans composed of relatively coarse sediment (Kelling & Stanley, 1976, p. 387).

supragelisol *suprapermafrost layer.*

suprageneric name The name of any taxon above the level of *genus.*

supraglacial *superglacial.*

supralithion Animals that swim above rock but are dependent on it as their source of food.

supralittoral Pertaining to the shore area marginal to the *littoral* zone, just above high-tide level. Syn: *supratidal.*

supraorbital carina The ridge extending backward from the supraorbital spine on some decapods (Holthuis, 1974, p. 735).

suprapelos Animals that swim above soft mud but are dependent on it as their source of food.

suprapermafrost layer The layer of ground (or soil) above the permafrost, consisting of the active layer and, wherever present, taliks and the pereletok. Syn: *supragelisol.*

suprapermafrost water Ground water existing above the impervious permafrost table (Muller, 1947, p.222).

supraposition *superposition.*

suprapsammon Animals that swim above sand but are dependent on it as their source of food.

suprastructure *superstructure.*

supratenuous fold A pattern of fold in which there is thickening at the synclinal troughs and thinning at the anticlinal crests. It is formed by differential compaction on an uneven basement surface. See also: *compaction fold.*

supratidal *supralittoral.*

suranal plate One of the first-formed and largest plates of the periproctal system of an echinoid, often filling the central and anterior parts of the periproct; but not recognizable in many echinoids. Syn: *suranal.*

surf (a) The wave activity in the *surf zone.* (b) A collective term for *breakers.*

surface (a) The exterior or outside part of the solid Earth or ocean; the top of the ground or the exposed part of a rock formation. (b) A two-dimensional boundary between geologic features such as formations or structures, including *bedding surface* and *fault surface,* or an imaginary surface such as the *axial surface* of a fold; usually an internal boundary, rather than one occurring on the outside of a feature (Challinor, 1978, p. 302). It need not be flat. Cf: *plane.*

surface anomaly A geophysical anomaly caused by an irregularity at or near the Earth's surface that interferes with geophysical measurements; *surface interference.*

surface axis *axial trace.*

surface cauldron subsidence *Cauldron subsidence* in which the ring faults extend to the surface. Cf: *underground cauldron subsidence.*

surface conductivity Conduction along the surfaces of certain mineral grains, due to excess ions in the diffuse layer of adsorbed cations.

surface correction A correction of a geophysical measurement to remove the influence of varying surface elevation and of surface anomalies.

surface creep The slow downwind advance of large sand grains along a surface by impact of smaller grains in saltation, as in the shifting or movement of a sand dune. See also: *surficial creep.* Syn: *reptation.*

surface curve *surface profile.*

surface density (a) The density of the surface material within the range of the elevation differences of the gravitational surface. Both the Bouguer correction and the terrain correction depend on the density of the surface material. (b) A quantity (as mass and electricity) per unit area distributed over a surface.

surface detention *detention.*

surface drag *skin friction [hydraul].*

surface drainage The removal of unwanted water from the surface of the ground, or the prevention of its entry into the soil, by natural or artificial means, such as by grading and smoothing the land to remove barriers and to fill in depressions, by terracing or digging ditches, or by diverting runoff from adjacent areas to natural waterways.

surface factor *fineness factor.*

surface forces Any of the forces acting over the surface of a body of material. Cf: *body forces.*

surface geology (a) Geology and correlation of rock formations, structures, and other features as seen at the Earth's surface. Ant: *subsurface geology.* (b) *surficial geology.*

surface groove A *skeletal canal* in the form of a groove on the surface of the skeletal framework of a sponge and generally corresponding to an exhalant canal of the soft parts.

surface hoar A type of frost consisting of leaf- or plate-shaped ice crystals formed directly on a snow surface by condensation from vapor. Cf: *depth hoar; frost flower.*

surface interference *surface anomaly.*

surface moraine *surficial moraine.*

surface of concentric shearing In flexural folding, the bedding plane along which slip occurs. Var: *concentric shearing surface.*

surface of no strain A surface along which the original configuration of an array of points remains unchanged after deformation of the body in which it occurs. In two-dimensional models sometimes referred to as the *neutral axis.* Syn: *neutral surface.*

surface of rupture (a) In a *slide [mass move],* the projection or extension of the major scarp surface under the disturbed material. (b) The surface of rock from which the material of a landslide or slump was removed.

surface of unconformity The surface of contact between two groups of rocks displaying an unconformable relationship, such as a buried surface of erosion or of nondeposition separating younger strata from underlying older rocks. Syn: *unconformity.*

surface phase In metamorphism, a thin layer of material having properties that may differ from those of the volume phases on either side (Turner and Verhoogen, 1960, p.461). Phase in this context is used in the geochemical sense. Syn: *volume phase.*

surface pipe The uppermost length of *casing* set in a well, used to prevent unconsolidated sediments from entering the hole and to shut off and protect shallow freshwater sands from contamination by drilling mud and deeper saline waters. It also serves as a conductor to return the drilling mud through loose, near-surface formations during deep open-hole drilling. Syn: *surface string.*

surface profile The longitudinal profile assumed by the surface of a stream of water in an open channel. See also: *backwater curve.* Syn: *surface curve.*

surface runoff The *runoff [water]* that travels over the soil surface to the nearest surface stream; runoff of a drainage basin that has not passed beneath the surface since precipitation. The term is misused when applied in the sense of *direct runoff* (Langbein & Iseri, 1960). Cf: *storm seepage; ground-water runoff.*

surface scattering layer An area of marine organisms near the ocean surface that scatter sound waves from an echo sounder. Cf: *shallow scattering layer; deep scattering layer.*

surface-ship gravimeter An instrument designed to produce useful gravity observations aboard a surface ship underway. Due to large rotations and accelerations encountered, a complex system is required to stabilize the gravity sensor and by filtering to distinguish between inertial forces and the gravity value sought. Accuracies attained range from 1 to 5 milligals depending largely on quality of the navigation and state of the sea.

surface slope The inclination of the water surface expressed as change of elevation per unit of slope length; the sine of the angle which the water surface makes with the horizontal. The tangent of that angle is ordinarily used; no appreciable error results except for the steeper slopes (ASCE, 1962).

surface soil The *A horizon.* The term is sometimes applied to the upper 12 to 20 cm of a soil that is tilled. Cf: *subsoil.* Partial syn: *topsoil.*

surface texture The aggregate of the surface features of sedimentary particles, independent of size, shape, or roundness; e.g. polish, frosting, or striations.

surface wash *sheet erosion.*

surface water [oceanog] A *water mass* of varying salinity and temperature, occurring at the ocean surface and having a thickness of 300 meters or less. Cf: *intermediate water; deep water; bottom water.*

surface water [water] All waters on the surface of the Earth, including fresh and salt water, ice and snow.

surface wave [seis] A *seismic wave* that travels along the surface of the Earth, or along a subsurface interface. Surface waves include the *Rayleigh wave,* the *Love wave,* and the *coupled wave.* See also: *hydrodynamic wave.* Cf: *body wave; guided wave; P wave.* Obsolete syn: *circumferential wave; long wave [seis]; large wave.* Syn: *L wave.*

surface wave [water] (a) A progressive gravity wave in which the particle movement is confined to the upper limits of a body of water; strictly, a gravity wave whose celerity is a function of wavelength only. (b) *deep-water wave.*

surface-wave magnitude An *earthquake magnitude* determined at teleseismic distances, using the logarithm of the amplitude of 20-sec-period *surface waves.*

surf base The depth at which a wave begins to peak and, under storm conditions, to break; it is usually 10-20 meters. Cf: *wave base.* Syn: *surge base.*

surf beat A wind-generated ocean wave, traveling shoreward, caused by interference of two different wind-wave trains, and usually associated with the onset of heavy surf. It has a long wave period (1-10 min), and a beat frequency and amplitude of a few centimeters.

surficial Pertaining to, or occurring on, a surface, esp. the surface of the Earth. Cf: *subsurface.* See also: *subaerial.* Syn: *superficial.*

surficial creep A syn. of *soil creep.* See also: *surface creep.*

surficial deposit Unconsolidated and residual, alluvial, or glacial deposits lying on bedrock or occurring on or near the Earth's surface; it is generally unstratified and represents the most recent of geologic deposits. Syn: *superficial deposit.*

surficial geology Geology of *surficial deposits,* including soils; the term is sometimes applied to the study of bed rock at or near the Earth's surface. See also: *surface geology.*

surficial moraine A moraine, such as a lateral or medial moraine, in transit on the surface of a glacier. Syn: *superficial moraine; surface moraine.*

surf ripple A general term proposed by Kuenen (1950, p.292) for a ripple mark formed on a sandy beach by wave-generated currents in the surf zone.

surfusion An obsolete term proposed by Fournet (1844) for a condition under which the fusing points of substances are lowered to temperatures much below the points at which they usually solidify (see Zittel, 1901, p.342).

surf zone The area bounded by the landward limit of wave uprush and the farthest seaward breaker. Syn: *surf; breaker zone.*

surge [glaciol] The period of very rapid flow of a *surging glacier;* also, the displacement or advance of ice resulting from the very rapid flow. Syn: *glacier surge; catastrophic advance.*

surge [hydraul] In fluid flow, long-interval variations in velocity and pressure that are not necessarily periodic and may even be transient.

surge [waves] (a) *storm surge.* (b) Horizontal oscillation of water with a comparatively short period, accompanying a seiche. (c) Water transported up a beach by breaking waves.

surge base *surf base.*

surge channel A transverse channel that cuts across the outer edge of an organic reef, especially cutting through an algal ridge, in which the water level rises and falls as the result of wave and tidal action.

surging breaker A type of *breaker* that peaks and then surges onto the beach face without spilling or plunging. It forms over a very steep bottom gradient. Cf: *plunging breaker; spilling breaker.*

surging glacier A glacier that alternates periodically between brief periods (usually one to four years) of very rapid flow, called surges, and longer periods (usually 10 to 100 years) of near stagnation. During a *surge,* a large volume of ice from an ice-reservoir area is displaced downstream at speeds up to several meters per hour into an ice-receiving area, and the affected portion of the glacier is chaotically crevassed. Only in exceptional cases does the glacier advance beyond its former limit. In the interval between surges, the ice reservoir is slowly replenished by accumulation and normal ice flow, and the ice in the receiving area is greatly reduced by ablation.

surinamite A mineral: $(Al,Mg,Fe)_3(Si,Al)_2(O,OH)_8$.

sursassite A mineral: $Mn_5Al_4Si_5O_{21} \cdot 3H_2O$ (?).

survey v. To determine and delineate the form, extent, position, boundaries, value, or nature of a tract of land, coast, harbor, or the like, esp. by means of linear and angular measurements and by the application of the principles of geometry and trigonometry.—n. (a) The orderly and exacting process of examining, determining, finding, and delineating the physical or chemical characteristics of the Earth's surface, subsurface, or internal constitution by topographic, geologic, geophysical, or geochemical measurements; esp. the act or operation of making detailed measurements for determining the relative positions of points on, above, or beneath the Earth's surface. (b) The associated data or results obtained in a survey; a map or description of an area obtained by surveying. (c) An organization engaged in making surveys; e.g. a government agency such as the U.S. Geological Survey.

surveying (a) The art of making a survey; specif. the applied science that teaches the art of making such measurements as are necessary to determine the area of any part of the Earth's surface, the lengths and directions of the boundary lines, and the contour of the surface, and of accurately delineating the whole on paper. (b) The act of making a survey; the occupation of one that surveys.

surveyor's compass A surveying instrument used for measuring horizontal angles; specif. one designed for determining a magnetic bearing of a line of sight by the use of a sighting device, a graduated horizontal circle, and a pivoted magnetic needle. The surveyor's compass used on the early land surveys in U.S. had a pair of peep sights to define the line of sight and was usually mounted on a *Jacob's staff.* See also: *circumferentor.* Syn: *land compass.*

surveyor's level A *leveling instrument* consisting of a telescope (with cross hairs) and a spirit level mounted on a tripod, revolving on a vertical axis, and having leveling screws that are used to adjust the instrument to the horizontal.

surveyor's measure A system of measurement used in land surveying, having the surveyor's chain (one chain = 4 rods = 66 feet = 100 links = 1/80 mile) as a unit. Cf: *Gunter's chain.*

surveyor's rod *level rod.*

survival of the fittest The tendency for the environmentally better-adapted members of a population to survive to reproductive age and thus to contribute more strongly to the genetic composition of the next generation.

susannite A rhombohedral mineral: $Pb_4(SO_4)(CO_3)_2(OH)_2$. It is dimorphous with leadhillite.

susceptibility [elect] The ratio of the electric polarization to the electric intensity in a polarized dielectric.

susceptibility [magnet] The ratio of *induced magnetization* to the strength H of the magnetic field causing the magnetization. See also: *susceptibility anisotropy.* Syn: *magnetic susceptibility; volume susceptibility.*

susceptibility anisotropy In minerals of low crystal symmetry or in rocks with planar or linear fabric, magnetic *susceptibility* which is not perfectly parallel with the inducing magnetic field, because it depends on direction and induced magnetization. Syn: *magnetic anisotropy.*

suspended current A *turbidity current,* in a body of standing water, that is not in contact with the bottom (as on a slope or where the current overrides denser underlying water) and that continues in suspension (Dzulynski & Radomski, 1955).

suspended load (a) The part of the total *stream load* that is carried for a considerable period of time in suspension, free from contact with the stream bed; it consists mainly of clay, silt, and sand. (b) The material collected in, or computed from samples collected with, a suspended-load sampler.—Syn: *suspension load; suspensate; silt load; wash load.*

suspended-load sampler A device that collects a sample of water with its sediment load without separating the sediment from the water.

suspended water *vadose water.*

suspensate *suspended load.*

suspension (a) A mode of sediment transport in which the upward currents in eddies of turbulent flow are capable of supporting the weight of sediment particles and keeping them indefinitely held in the surrounding fluid (such as silt in water or dust in air). Cf: *traction.* (b) The state of a substance in such a mode of transport; also, the substance itself.

suspension current *turbidity current.*

suspension feeder An animal that feeds by selecting microorganisms and detritus suspended in the surrounding water. Cf: *filter feeder.*

suspension flow Flow of a water-sediment mixture in which the sediment is maintained in suspension by the combination of water turbulence and relatively low settling velocities of particles. It commonly occurs in a turbidity current.

suspension load *suspended load.*

sussexite [mineral] A white mineral: $(Mn,Mg)BO_2(OH)$. It is isomorphous with szaibelyite.

sussexite [rock] A porphyritic tinguaitic rock composed chiefly of nepheline and acmite, with subordinate feldspar; a feldspar-bearing *urtite.* Its name, given by Brögger in 1894, is derived from Sussex County, New Jersey. Not recommended usage. Cf: *muniongite.*

sustained runoff *base runoff.*

sutural Pertaining to a suture, or corresponding to sutures in position; e.g. "sutural supplementary apertures" in a foraminiferal test.

sutural element One of the major undulations of a suture of a cephalopod shell, such as a lobe or a saddle.

sutural pore (a) One of the pores bordered by sutures along the meeting branches of two or more adjacent radial spines in an acantharian radiolarian. Cf: *parmal pore.* (b) A pore lying along any suture between plates of an echinoderm theca, commonly restricted to nonambulacral areas.

suture [bot] On a fruit, the line of dehiscence or splitting; a groove marking a natural division or union. Syn: *commissure.*

suture [paleont] (a) The line of junction of a septum of a cephalopod's shell with the inner surface of the shell wall. It is commonly more or less undulated or plicated, and is visible only when the shell wall is removed. (b) The line of contact between two whorls of a gastropod shell. It is typically a spiral on the outer surface, and also on the inner surface around the axis of coiling where the umbilicus is not closed. (c) A plane of junction between adjacent plates of an echinoderm. Also, the boundary line on the surface marking the area of contact between the adjacent plates. (d) A line of demarcation between two fused or partly fused limb segments or body somites of a crustacean; e.g. a line or seam at the juncture of two compartmental plates of a cirripede, or a weak or uncalcified narrow seam between parts of a trilobite exoskeleton that separate at time of molting. (e) A line of contact between two chambers or two whorls of a foraminiferal test. It may be reflected on the outer wall by a groovelike or ridgelike feature. (f) The boundary between segments of heterococcoliths. (g) In vertebrates, an articulation in which bones of different embryonic centers of origin are united by fibrous connective tissue, and function as a solid or nearly solid unit in the adult.

suture [palyn] The line along which the laesura of an embryophytic spore opens on germination; loosely, the *laesura.* See also:

commissure.

sutured Said of the texture found in rocks in which mineral grains or irregularly shaped crystals interfere with their neighbors, producing interlocking, irregular contacts without interstitial spaces, resembling the sutural structures in bones. Also, said of the crystal contacts in rock with this texture. Syn: *consertal.*

suture joint A very small *stylolite.*

svabite A colorless, yellowish-white, or gray mineral of the apatite group: $Ca_5(AsO_4)_3F$. It may contain phophorus, lead, magnesium, or manganese.

svanbergite A colorless to yellow, rose, or reddish-brown rhombohedral mineral: $SrAl_3(PO_4)(SO_4)(OH)_6$. It is isomorphous with corkite, hinsdalite, and woodhouseite.

svetlozarite A mineral of the zeolite group: $(Ca,K_2,Na_2)(Al_2Si_{12})O_{28}\cdot 6H_2O$.

sviatonossite An andradite-bearing *syenite* in which the pyroxene is acmite-augite and the plagioclase is oligoclase. The name, given by Eskola in 1922, is for Sviatoy Noss, Transbaikalia, U.S.S.R. Not recommended usage.

swab n. A pistonlike device equipped with an upward-opening check valve and provided with flexible circular suction cups, lowered into a cased well bore by means of a wire line for the purpose of cleaning out drilling mud or inducing flow of formation fluids.—v. (a) To draw out oil from a well by means of a swab; to clean out with a swab. (b) In vernacular usage, to obtain confidential or proprietary information by adroit or devious questioning.

swag (a) A shallow pocket or closed depression in flat or gently rolling terrain, often filled with water (as in the bottomlands of the lower Mississippi Valley). (b) A shallow water-filled hollow produced by subsidence resulting from underground mining.—Cf: *sag pond.*

swale (a) A slight depression, sometimes swampy, in the midst of generally level land. (b) A shallow depression in an undulating ground moraine due to uneven glacial deposition. (c) A long, narrow, generally shallow, trough-like depression between two *beach ridges,* and aligned roughly parallel to the coastline. Syn: *low; furrow; slash; runnel.*

swallet *sinking stream.*

swallow hole (a) A closed depression or *doline* into which all or part of a stream disappears underground. Partial syn: *swallet; ponor.* (b) An underwater outlet of a lake.

swamp An area intermittently or permanently covered with water, having shrubs and trees but essentially without the accumulation of peat. Cf: *marsh; bog.*

swamp ore *bog iron ore.*

swamp theory *in-situ theory.*

swarm *earthquake swarm.*

swartzite A green monoclinic mineral: $CaMg(UO_2)(CO_3)_3\cdot 12H_2O$.

swash bar (a) A small, transitory bar built above the stillwater level by wave uprush, and forming a tiny lagoon on its landward side (King & Williams, 1949, p. 81). (b) A bar over which the sea washes. Syn: *swash.*

swash channel (a) A narrow sound or secondary channel of water lying within a sandbank or between a sandbank and the shore, or passing through or shoreward of an inlet or river bar. Syn: *swash; swashway; swatch.* (b) A channel cut on an open shore by flowing water in its return to the parent body.

swash mark A thin delicate wavy or arcuate line or very small ridge on a beach, marking the farthest advance of wave uprush. It is convex landward and consists of fine sand, mica flakes, bits of seaweed, and other debris. Syn: *wave line; wavemark; debris line.*

swash pool A shallow pool of water formed behind a swash bar.

swashway *swash channel.*

swash zone The sloping part of the beach that is alternately covered and uncovered by the uprush of waves, and where longshore movement of water occurs in a zigzag (upslope-downslope) manner.

swatch A British syn. of *swash channel.* Syn: *swatchway.*

S wave A seismic *body wave* propagated by a shearing motion that involves oscillation perpendicular to the direction of propagation. It does not travel through liquids, or through the outer core of the Earth. Its speed is 3.0-4.0 km/sec in the crust and 4.4-4.6km/sec in the upper mantle. The *S* stands for secondary; it is so named because it arrives later than the *P wave* (primary wave). Syn: *shear wave; secondary wave; rotational wave; tangential wave; equivoluminal wave; distortional wave; transverse wave.*

swedenborgite A mineral: $NaBe_4SbO_7$.

Swedish mining compass A compass in which a magnetic needle is suspended on a jewel and a stirrup so that it can rotate about both a horizontal and a vertical axis.

sweeping The progressive down-valley movement or shift of a system of meanders. See also: *wandering.* Syn: *sweep.*

sweepstakes route A biogeographic dispersal path that constitutes a formidable obtacle to the migration of plants and animals but is occasionally conquered. The term implies that the odds against crossing the barrier are as great as those against someone winning a lottery, i.e., great but not impossible.

sweet Said of crude oil or natural gas that contains few or no sulfur compounds. Cf: *sour.*

sweet water *fresh water.*

swell [eng geol] The increase in volume, or dilatancy, exhibited by soil or rock on removal of in-situ stress, esp. as a result of absorption of water.

swell [marine geol] *rise [marine geol].*

swell [ore dep] An enlarged place in, or part of, an orebody, as opposed to a *pinch.*

swell [struc geol] A general, imprecise term for *dome* and *arch.*

swell [waves] (a) One of a series of regular, long-period, somewhat flat-crested waves that has traveled out of its generating area. See also: *ground swell.* (b) The slow and regular undulation of the surface of the open ocean; a series of unbroken waves.—Ant: *sea.*

swell-and-swale topography A low-relief, undulating landscape characteristic of the ground moraine of a continental glacier, exhibiting gentle slopes and well-rounded hills interspersed with shallow depressions. Cf: *sag-and-swell topography.*

swelling chlorite A chloritelike mineral, found in clays, that behaves like a chlorite on heating but has its basal spacing expand on glycerol treatment. It contains incomplete hydroxide (brucite or gibbsite) layers, and might be regarded as a special interlayering of chlorite with smectite or vermiculite (Martin Vivaldi & MacEwan, 1960). See also: *corrensite.* Syn: *pseudochlorite.*

swelling clay Clay that is capable of absorbing large quantities of water, thus increasing greatly in volume; e.g. bentonite. It shrinks and cracks on drying.

swelling ground A tunnelman's term for rock or soil that increases in volume on excavation; it usually contains clay minerals with high swelling capacity. See also: *firm ground; flowing ground; raveling ground; running ground; squeezing ground.*

swelling pressure The pressure exerted by a clay or shale when it absorbs water in a confined space.

swept ripple mark Ripple mark that is sharply skewed relative to the general flow. The most common type, swept *catenary ripple mark,* appears to be generally formed in two stages: first by formation of a ripple mark transverse to the flow, and then by modification of the original crest to a catenary form as the flow changes direction. Term introduced by Allen (1968, p. 66).

swimming leg The hindmost prosomal appendage of a merostome, serving as a swimming organ.

swimming stone *floatstone [mineral].*

swinefordite A monoclinic mineral: $(Li,Ca,Na)(Al,Li,Mg)_4(Si,Al)_8O_{20}(OH,F)_4$.

swinestone *anthraconite.*

swinging The steady lateral movement of a meander belt from one side of the valley floor to the other. See also: *wandering.* Syn: *swing.*

swinging dip *migrating dip.*

swing mark A circular or semicircular sedimentary structure formed by wind action on an anchored plant stem or root that is swept to and fro, or round and round, across a sandy surface.

swither A colloquial term used in the Wisconsin lead-mining region for an offshoot or branch of a main lode.

switzerite A mineral: $(Mn,Fe)_3(PO_4)_2\cdot 4H_2O$.

sword dune *seif.*

sychnodymite *carrollite.*

sycon A sponge or sponge larva in which separate flagellated chambers open directly into a central spongocoel lined with pinacoderm. Cf: *ascon; leucon.* Adj: *syconoid.*

syenide An informal term, for field use, applied to any holocrystalline, medium- to coarse-grained igneous rock containing one or more feldspars and, generally, biotite or hornblende. The dark minerals comprise less than half of the rock. The term includes syenites and light-colored diorites.

syenite (a) In the *IUGS classification,* a plutonic rock with Q between 0 and 5, and P/(A+P) between 10 and 35. (b) A group of plutonic rocks containing alkali feldspar (usually orthoclase, microcline, or perthite), a small amount of plagioclase (less than in *monzonite*), one or more mafic minerals (esp. hornblende), and quartz, if present, only as an accessory; also, any rock in that group; the intrusive equivalent of *trachyte.* With an increase in the quartz content, syenite grades into *granite.* Its name is derived from Syene, Egypt. A. G. Werner in 1788 applied the name in its present meaning to rock at Plauenscher Grund, Dresden, Germany; the Egyptian rock is a granite containing much more quartz.

syenitoid In the *IUGS classification,* a preliminary term (for field use) for a plutonic rock with Q less than 20 or F less than 10, and P/(A+P) less than 65.

syenodiorite A group of plutonic rocks intermediate in composition between *syenite* and *diorite,* containing both alkali feldspar (usually orthoclase) and plagioclase feldspar, commonly more of the former; also, any rock in that group. Generally considered a syn. of *monzonite,* but may also include both monzonite and rocks intermediate between monzonite and diorite (Streckeisen, 1967, p. 170).

syenogabbro A plutonic rock differing in composition from *gabbro* by the presence of alkali feldspar.

sylvanite A steel-gray, silver-white, or brass-yellow monoclinic mineral: $(Ag,Au)Te_4$. It often occurs in implanted crystals resembling written characters. Not to be confused with *sulvanite* or *sylvinite.* Also spelled: *silvanite.* Syn: *graphic tellurium; white tellurium; yellow tellurium; goldschmidtite.*

sylvine *sylvite.*

sylvinite A mixture of halite and sylvite, mined as a potash ore; a rock that contains chiefly impure potassium chloride. Not to be confused with *sylvanite.*

sylvite A white or colorless isometric mineral: KCl. It is the principal ore mineral of potassium compounds. Sylvite occurs in beds as a saline residue with halite and other evaporites. It has a sharper taste than that of halite. Syn: *sylvine; leopoldite.*

symbiosis The relationship that exists between two different organisms that live in close association, with at least one being helped without either being harmed. Cf: *mutualism; commensalism; parasitism.* Adj: *symbiotic.*

symbiotic The adj. of *symbiosis.*

symbol A diagram, design, letter, color hue, abbreviation, or other graphic device placed on maps, charts, and diagrams, which by convention, usage, or reference to a legend is understood to represent a specific characteristic, feature, or object, such as structural data, rock outcrops, or mine openings.

symmetrical fold (a) A fold whose axial surface is perpendicular to the *enveloping surface.* (b) A fold whose limbs have the same angle of dip relative to the axial surface. Cf: *asymmetric fold.* Syn: *normal fold.*

symmetric bedding Bedding characterized by lithologic types or facies that follow each other in a "retracing" arrangement illustrated by the sequence 1-2-3-2-1-2-3-2-1. Cf: *asymmetric bedding.*

symmetric ripple mark A *ripple mark* having a symmetric profile in cross section, being similarly shaped on both sides of the crest, which in plan view is predominantly straight; specif. *oscillation ripple mark.* Ant: *asymmetric ripple mark.*

symmetric spread *split spread.*

symmetry [cryst] The repeat pattern of similar crystal faces that indicates the ordered internal arrangement of a crystalline substance. More precisely, that aspect of a crystal that renders its appearance, structure, and directional properties invariant to certain operations, such as rotation, reflection, inversion, and their combinations.

symmetry [paleont] See *bilateral symmetry; radial symmetry.*

symmetry [struc petrol] The symmetry of a fabric is the combined symmetry of all the elements making up the fabric. There are five possible symmetries: (1) spherical, for fabrics having the symmetry of a sphere; (2) axial, for fabrics having the symmetry of a spheroid; (3) orthorhombic, for fabrics having the symmetry of a triaxial ellipsoid; (4) monoclinic, for fabrics having only one unique plane of symmetry; and (5) triclinic, for fabrics having no planes of symmetry (Turner and Weiss, 1963, p. 43-44).

symmetry axis In a crystal, an imaginary line about which the crystal may be rotated, during which there may be two, three, four, or six repetitions of its appearance (lines, angles, or faces); it is one of the *symmetry elements.* Syn: *axis of symmetry; rotation axis; symmetry axis of rotation.*

symmetry axis of rotary inversion *rotoinversion axis.*

symmetry axis of rotation *symmetry axis.*

symmetry center *center of symmetry.*

symmetry elements The axes, plane, and center of symmetry, by which crystal symmetry can be described. There are 32 possible arrangements of the elements of symmetry; each arrangement is a crystal *class.* Syn: *elements of symmetry.*

symmetry operations Various movements of a crystal that leave the crystal and its directional properties invariant; these are rotation about an axis, reflection across a plane, inversion, and rotary inversion.

symmetry plane *plane of mirror symmetry.*

symmetry principle A statement of the way in which the symmetry of an observed fabric is related to the symmetry of the factors that produced it. "The symmetry of any physical system must include those symmetry elements that are common to all the independent factors (physical fields and physical properties of the medium) that contribute to the system, and it may include additional symmetry elements; however, any symmetry elements absent in the system must also be absent in at least one of the independent contributing factors" (Paterson and Weiss, 1961, p. 859).

symmict Said of a structureless sedimentation unit, as in a varved clay or a graded series, composed of material in which coarse- and fine-grained particles are mixed to a greater extent in the lower part due to rapid flocculation. Also, said of the sedimentary structure so formed. Also spelled: *symminct.*

symmictite [ign] A term used by Sederholm (1924, p. 148) for a homogenized eruptive breccia composed of a mixture of country rock and intrusive rock.

symmictite [sed] *diamictite.*

symmicton *diamicton.*

symminct *symmict.*

symmixis Flocculation induced in sediments by certain electrolytes (esp. sodium chloride), resulting in the mixing of silt and clay particles and the formation of a nearly homogeneous or nonlaminated clay.

symon fault A syn. of *horseback [coal],* named after such a structure in the Coalbrookdale coalfield of England that was originally thought to be a large fault.

sympatric Said of populations occupying the same geographic area without losing their identity as a result of interbreeding. Noun: *sympatry.* Cf: *allopatric.*

sympatric species Related species whose geographic distributions overlap or coincide.

symphytium A single plate formed in certain brachiopods by fusion of deltidial plates dorsally or anteriorly from the pedicle foramen and lacking a median line of junction.

symplectic Said of a rock texture produced by the intimate intergrowth of two different minerals; sometimes the term is restricted to such textures of secondary origin. One of the minerals may assume a vermicular habit. Also, said of a rock exhibiting such texture, or of the intergrowth itself, i.e. *symplectite.* Also spelled: *symplektic; symplectitic; symplektitic.* Cf: *dactylitic.*

symplectite An intimate intergrowth of two different minerals, sometimes restricted to those of secondary origin; also, a rock (igneous or thermally metamorphosed) characterized by *symplectic* texture. Also spelled: *symplektite.* Cf: *pegmatite.*

symplesiomorphy *shared primitive character.*

symplesite A blue, blue-green, or pale-indigo triclinic mineral: $Fe_3(AsO_4)_2 \cdot 8H_2O$. Cf: *parasymplesite; ferrisymplesite.*

sympod The *protopod* of a crustacean. Syn: *sympodite.*

sympodium A plant stem made up of a series of branches that grow on each other. It gives the effect of a simple stem.

symptomatic mineral *diagnostic mineral.*

syn- A prefix meaning "together", "with".

synadelphite A black mineral: $(Mn,Mg,Ca,Pb)_4(AsO_4)(OH)_5$.

synaeresis *syneresis.*

synangium An aggregated, exannulate fern sporangium forming a series of locules, as in *Marattia.*

synantectic Said of a primary mineral formed by the reaction of two other minerals, as in the formation of a reaction rim. See also: *deuteric.*

synantexis *Deuteric* alteration.

synapomorphy *shared derived character.*

Synapsida A subclass of reptiles characterized by a single lower temporal opening; mammal-like reptiles *sensu lato.* Range, Lower Pennsylvanian to Upper Triassic. Cf: *Therapsida.*

synapticula (a) One of numerous transverse calcareous rods or bars that extend between and connect the opposed faces of adjacent septa of some corals and that perforate the mesenteries between them. A "compound synapticula" consists of a broad bar formed by fusion of opposed ridges on adjacent septa. The term is also used as a plural of *synapticulum.* (b) An anaxial bar of secondarily secreted silica connecting adjacent spicules in hexactinellid sponges. (c) A rodlike structure extending between septa of an archaeocyathid.—Pl: *synapticulae.*

synapticulotheca The porous outer wall of a scleractinian corallite, formed by union of one or more rings of synapticulae along the axis of trabecular divergence. Cf: *septotheca; paratheca.*

synapticulum A coral synapticula. Pl: *synapticula.*

syncarpous Having carpels united, as in a plant ovary; also, said of pistils partially united within a flower (Lawrence, 1951, p. 772). Cf: *apocarpous.*

synchisite *synchysite.*

synchronal *synchronous.*

synchrone (a) A zone representing equal time. (b) A stratigraphic surface on which every point has the same geologic age; a *time plane.*

synchroneity The state of being *synchronous* or simultaneous; coincident existence, formation, or occurrence of geologic events or features in time, such as "glacial synchroneity". Syn: *synchronism.*

synchronism *synchroneity.*

synchronogenic Formed in the same part of geologic time (R.C. Moore, 1958, p.21); e.g. said of rocks possessing identical or nearly identical geologic ages. Cf: *syntopogenic.*

synchronous Occurring, existing, or formed at the same time; contemporary or simultaneous. The term is applied to rock surfaces on which every point has the same geologic age, such as the boundary between two ideal time-stratigraphic units in continuous and unbroken succession. It is also applied to growth (or depositional) faults and to plutons emplaced contemporaneously with orogenies. Cf: *isochronous; diachronic.* Syn: *synchronal; synchronic.*

synchysite A mineral: $(Ce,La)Ca(CO_3)_2F$. Also spelled: *synchisite.*

synchysite-(Y) A mineral: $(Y,Ce)Ca(CO_3)_2F$. It is related to parisite. Syn: *doverite.*

synclinal n. An obsolete form of *syncline.*—adj. Pertaining to a syncline.

synclinal axis Axis of a syncline.

syncline A fold of which the core contains the stratigraphically younger rocks; it is generally concave upward. Ant: *anticline.* See also: *synform; synclinal.*

synclinorium A composite synclinal structure of regional extent composed of lesser folds. Cf: *anticlinorium.* See also: *geosyncline.* Pl: *synclinoria.*

syncolpate Said of pollen grains in which the colpi join, normally near the pole.

syndeposition A term used by Chilingar et al. (1967, p. 322) for that part of *syngenesis* comprising "processes responsible for the formation of the sedimentary framework". Cf: *prediagenesis.*

syndepositional fold A fold structure that forms contemporaneously with sedimentation. It is a feature associated with sedimentary tectonics.

syndiagenesis A term used by Bissell (1959) for the sedimentational, prediastrophic phase of diagenesis, including alterations occurring during transportation (halmyrolysis) and during deposition of sediments, and continuing through the early stages of compaction and cementation but ending before that of deep burial (less than 100 m). It is characterized by the presence of large amounts of interstitial or connate water that is expelled only very slowly and by extreme variations in pH and Eh. It is equivalent to *early diagenesis.* See also: *epidiagenesis; anadiagenesis.* Cf: *syngenesis.* Adj: *syndiagenetic.*

syndromous load cast A term used by Ten Haaf (1959, p.48) for an elongate, shallow load cast having sharp creases that combine to form a dendritic pattern, the junctures always occurring downcurrent.

syneclise A negative or depressed structure of the continental platform; it is of broad, regional extent (tens to hundreds of thousands of square kilometers) and is produced by slow crustal downwarp during the course of several geologic periods. The term is used mainly in the Russian literature, e.g. the Caspian syneclise. Ant: *anteclise.*

synecology The study of the relationships between communities and their environments. Cf: *autecology.*

syneresis The spontaneous separation or throwing-off of a liquid from or by a gel or flocculated colloidal suspension during aging, resulting in shrinkage and in the formation of cracks, pits, mounds, cones, or craters. Also spelled: *synaeresis.*

syneresis crack A *shrinkage crack* formed by the spontaneous throwing-off of water by a gel during aging.

syneresis vug A vug formed by syneresis, esp. in a sedimentary carbonate rock.

synform A fold whose limbs close downward in strata for which the stratigraphic sequence is unknown. Cf: *syncline.* Ant: *antiform.*

synformal anticline An *anticline* the limbs of which close downward as in a *synform* (Turner & Weiss, 1963, p. 106).

syngenesis (a) A term introduced by Fersman (1922) for the formation, or the stage of accumulation, of unconsolidated sediments in place, including the changes affecting detrital particles still in movement in the waters of the depositional basin. The term is in dispute by Russian geologists (Dunoyer de Segonzac, 1968, p. 170): some would apply it to initial diagenesis (designating exchange phenomena between fresh sediment and the sedimentary environment), others would extend it to all the transformations undergone by a sediment before its compaction. The term is equivalent to *early diagenesis* as used in the U.S. Cf: *syndiagenesis.* (b) A term used by Chilingar et al. (1967, p. 322) for the "processes by which sedimentary rock components are formed simultaneously and penecontemporaneously"; it includes *syndeposition* and *prediagenesis.*

syngenetic [ore dep] Said of a mineral deposit formed contemporaneously with, and by essentially the same processes as, the enclosing rocks. Cf: *epigenetic [ore dep]; diplogenetic.* Syn: *idiogenous.*

syngenetic [sed] (a) Said of a primary sedimentary structure, such as a ripple mark, formed contemporaneously with the deposition of the sediment. (b) Pertaining to sedimentary syngenesis.—Cf: *epigenetic [sed].*

syngenetic karst Karst that has developed simultaneously with the lithification of eolian calcarenite or reef limestone.

syngenite A colorless or white monoclinic mineral: $K_2Ca(SO_4)_2 \cdot H_2O$.

synglyph A *hieroglyph* formed contemporaneously with sedimentation (Vassoevich, 1953, p.33).

synkinematic *syntectonic.*

synneusis A mechanism by which small plagioclase crystals float into growing phenocrysts of potassium feldspar. Also, said of the texture of a rock showing such crystals. Etymol: Greek, "to swim together". Cf: *gregaritic.*

synnyrite A *nepheline syenite* containing fine-grained intergrowths of alkali feldspar with kalsilite (Sørensen, 1974, p. 573). The name, given by Zhidkov in 1962, is for Synnyr in the northern Baikal region, U.S.S.R. Not recommended usage.

synodic month The time between one new moon and the next; that is, the time necessary for the moon to complete one revolution about the earth. It is equal to 29 days 12 hours 44 minutes 2.78 seconds. It is longer than a *sidereal month,* because of the orbital motion of the earth.

synonym One of two or more names applied to the same *taxon* (Jeffrey, 1973, p. 67). See also: *synonymy.*

synonymy (a) The relationship between two or more different names that have been applied to the same *taxon.* (b) A list of *synonyms* that have been applied to a particular taxon.

synoptic Pertaining to simultaneously existing meteorologic conditions that together give a description of the weather; also, said of a weather map or chart that shows such conditions.

synoptic oceanography Continuous gathering and reporting of simultaneous oceanographic data. It has become more important and feasible with the development of satellites. Syn: *hydropsis.*

synorogenic Said of a geologic process or event occurring during a period of orogenic activity; or said of a rock or feature so formed. Cf: *syntectonic.*

synorogenic pluton An igneous intrusion emplaced during a period of orogenic activity. Cf: *syntectonic pluton.*

synrhabdosome An association of graptoloid rhabdosomes, commonly biserial, attached distally by their nemas around a common center (TIP, 1970, pt. V, p. 11).

synsedimentary Accompanying deposition; specif., said of a sedimentary ore deposit in which the ore minerals formed contemporaneously with the enclosing rock.

synsedimentary fault *growth fault.*

syntactic Recommended adj. of *syntaxy.*

syntactic growth *syntaxy.*

syntaphral Descriptive of a type of tectonics involving gravity sliding of unconsolidated sediments toward the axis of a geosyncline (Carey, 1963, p.A6). Cf: *diataphral; apotaphral.*

syntaxial Adj. of *syntaxy.*

syntaxial rim An optically oriented crystal overgrowth of a detrital grain, developed during diagenesis.

syntaxic Adj. of *syntaxy.*

syntaxis A sharp bend in an orogenic belt, accompanied by a fraying into several strands. The term is Sollas's translation of the term *Schaarung,* first used by Suess in 1901. Staub in 1928 and Bucher in 1933 used the term as an ant. of *virgation.* Current usage follows Suess's original definition.

syntaxy Similar crystallographic orientation in a mineral grain and its overgrowth. Adj: *syntactic; syntaxic; syntaxial.* Cf: *topotaxy; epitaxy.* Syn: *syntactic growth.*

syntectic The adj. of *syntexis.*

syntectite A rock formed by syntexis. See also: *anatexite; protectite.*

syntectonic Said of a geologic process or event occurring during any kind of tectonic activity; or of a rock or feature so formed. Cf: *synorogenic.* Syn: *synkinematic.*

syntectonic pluton An igneous intrusion emplaced during an interval of tectonic activity. Cf: *synorogenic pluton.*

syntexis (a) The formation of magma by melting of two or more rock types and assimilation of country rock; *anatexis* of two or more rock types. (b) Modification of the composition of a magma by *assimilation.* (c) Any kind of reaction between a rising body of magma and the crustal rocks with which it comes into contact.—Adj: *syntectic.*

synthem An unconformity-bounded stratigraphic unit (ISG, 1976, p. 92).

synthetic [fault] Pertaining to minor normal faults that are of the same orientation as the major fault with which they are associated. Ant: *antithetic.* Syn: *homothetic [fault].*

synthetic [gem] adj. Said of a substance produced artificially, such as a gemstone (ruby, sapphire) made by the *Verneuil process,* or a diamond produced by subjecting a carbonaceous material to extremely high temperature and pressure.—n. Something produced by synthesis; a *synthetic stone.*

synthetic-aperture radar A *SLAR* system in which high-azimuth resolution is achieved by storing and processing the return information to give the effect of a very long antenna.

synthetic group A lithostratigraphic unit consisting of two or more formations that are associated because of similarities or close relationships between their fossils or lithologic characters (Weller, 1960, p.434). Cf: *analytic group.*

synthetic hydrology A catchall term for new techniques in hydrologic analysis involving the generation of hydrologic information or sequences of hydrologic events by means other than direct measurement or observation. It has been suggested that the term be abandoned and replaced by the terms *parametric hydrology* and *stochastic hydrology* (Hofmann, 1965, p. 119).

synthetic ore Material that is the equivalent of, or better than, natural ore, that can be put to the same uses, and is produced by means other than ordinary concentration, calcining, sintering, or nodulizing (U.S. Bureau of Mines, 1968, p.1113-1114).

synthetic seismogram An artificial seismic reflection record, manufactured from velocity-log data by convolving the reflectivity function with a waveform that includes the effects of filtering by the Earth and the recording system. It is compared with an actual seismogram to aid in identifying events or in predicting how stratigraphic variation might affect a seismic record.

synthetic stone A man-made stone that has the same physical, optical, and chemical properties, and the same chemical composition, as the genuine or natural stone that it reproduces. Many gem materials have been made synthetically as a scientific experiment, but only corundum, spinel, emerald, rutile, garnet, quartz, chrysoberyl (alexandrite), opal, and turquoise have been made commercially and cut as gemstones for the jewelry trade. Cf: *imitation; reconstructed stone.* Syn: *synthetic [gem]; reproduction.*

syntopogenic Formed in the same or similar place, or denoting similar conditions of origin (R.C. Moore, 1958, p.21); e.g. said of sedimentary rocks deposited in identical or nearly identical environments in marine waters or on land. Cf: *synchronogenic.*

syntype Any of the specimens on which the description of a species or subspecies is based when no *holotype* has been designated. Nonpreferred syn: *cotype.*

synusia A subdivision of an ecologic community or habitat having a characteristic life pattern or uniform conditions. Plural: *synusiae.* Adj: *synusial.*

syphon *siphon.*

syrinx A tube of secondary shell located medially on the ventral side of the delthyrial plate of certain brachiopods (as in *Syringothyris*) and split along its ventral and anterior surface. Pl: *syringes* or *syrinxes.*

sysertskite A variety of iridosmine containing 50-80% osmium (or 20-50% iridium). Syn: *siserskite.*

syssiderite An obsolete syn. of *stony-iron meteorite.*

system [chem] (a) Any portion of the material universe that can be isolated completely and arbitrarily from the rest for consideration of the changes that may occur within it under varied conditions. (b) A conceptual range of compositions defined by a set of components in terms of which all compositions in the system can be expressed, e.g. the system CaO - MgO - SiO_2. Ricci has proposed to distinguish this sense from the ordinary thermodynamic one by capitalizing the word.

system [cryst] *crystal system.*

system [geol] A group of related natural features, objects, or forces; e.g. a *drainage system* or a *mountain system.*

system [stratig] (a) A major chronostratigraphic unit of worldwide significance, representing the fundamental unit of chronostratigraphic classification of Phanerozoic rocks, extended from a type area or region and correlated mainly by its fossil content (ACSN, 1961, art.29); the rocks formed during a *period* of geologic time. It is next in rank above *series [stratig]* and below *erathem.* The systems commonly recognized in the U.S. (in order of increasing age): Quaternary, Tertiary, Cretaceous, Jurassic, Triassic, Permian, Pennsylvanian, Mississippian, Devonian, Silurian, Ordovician, and Cambrian. Internationally, there is considerable divergence of opinion with respect to classification and nomenclature, and to boundaries for almost all systems. Most of the systems in current use lack adequate type or reference sections and were not established according to any systematic plan for geochronologic division of the Earth's strata as a whole (ISST, 1961, p.27). Systems in the Precambrian have only local significance, have not been placed in widely accepted orderly successions, and do not serve as the fundamental units of chronostratigraphic classification. (b) An informal term sometimes used locally for a large lithostratigraphic unit that does not coincide with, and is somewhat larger in scope than, the formal or standard system of chronostratigraphic classification; e.g. "Karroo System" of Africa and "Hokonui System" of New Zealand. The informal term "sequence" has been suggested for this usage (ISST, 1965, p.1698).

systematic error Any *error* that persists and cannot be considered as due entirely to chance, or an error that follows some definite mathematical or physical law or pattern and that can be compensated, at least partly, by the determination and application of a correction; e.g. an error whose magnitude changes in proportion to known changes in observational conditions, such as an error caused by the effects of temperature or pressure on a measuring instrument or on the object to be measured. Ant: *random error.* See also: *constant error; instrument error.*

systematic joints Joints that occur in sets or patterns. They cross other joints, their surfaces are plane or only broadly curved, they are oriented perpendicular to the boundaries of the enclosing rock unit, and the structures on their faces are oriented. Cf: *nonsystematic joints.*

systematic paleontology That branch of paleontology involving the classification and treatment of fossil forms according to their taxonomic position or order. See also: *systematics; taxonomy.*

systematics The study of the types and diversity of organisms and their relationships. Cf: *taxonomy; classification.*

syzygial (a) Pertaining to syzygy; e.g. a "syzygial pair" consisting of two crinoid ossicles joined by syzygy. (b) Pertaining to zygosis in sponges.

syzygy [astron] Either of two points in the Moon's orbit about the Earth when the Moon is new (in conjunction) or full (in opposition), or when the Sun, Moon, and Earth are in a nearly straight-line configuration. Etymol: Greek *syzygos,* "yoked together". Cf: *quadrature [astron].*

syzygy [paleont] (a) Ligamentary articulation of crinoid plates with opposed joint faces bearing numerous fine culmina that radiate from the axial canal, the culmina meeting one another instead of fitting into crenellae. It forms a single segment and allows very slight mobility of joined ossicles in all directions. (b) The segment formed by syzygy.

syzygy tide *spring tide.*

szaibelyite A white to yellowish acicular mineral occurring in nodular masses: $MgBO_2(OH)$. It is isomorphous with sussexite. Szaibelyite is the principal Russian source of boron compounds. Also spelled: *szájbelyite.* Syn: *ascharite.*

szaskaite *smithsonite.*

szmikite A monoclinic mineral: $MnSO_4 \cdot H_2O$.

szomolnokite A yellow or brown monoclinic mineral: $FeSO_4 \cdot H_2O$.

taaffeite A lilac mineral: $BeMgAl_4O_8$. It resembles mauve-colored spinel.

tabasheer Translucent to opaque and white to bluish-white opaline silica of organic origin (deposited within the joints of the bamboo), valued in the East Indies as a medicine and used in native jewelry. Var: *tabaschir; tabashir.*

tabbyite A variety of solid *asphalt* found in veins in Tabby Canyon, Utah.

tabella One of several small subhorizontal plates in the central part of a corallite, forming part of an *incomplete tabula.* Pl: *tabellae.*

Taber ice Ice films, seams, lenses, pods, or layers, generally 1-150 mm thick, that grow in the ground by drawing-in water as the ground freezes (Péwé, 1976, p. 37). Named for Stephen Taber (1882-1963), American geologist. See also: *epigenetic ice.* Syn: *segregated ice; sirloin-type ice.*

tabetification The process of forming a *talik* (Bryan, 1946, p.640).

tabetisol *talik.*

Tabianian European stage: Lower Pliocene (above Messinian, below Plaisancian). See also: *Zanclean.*

table [gem] (a) The large facet that caps the crown of a faceted gemstone. In the standard round brilliant, it is octagonal in shape and is bounded by eight star facets. (b) *table diamond.*

table [geomorph] (a) The flat summit of a mountain, such as one capped with horizontal sheets of basalt. (b) A term used in the western U.S. for *tableland.*

table cut (a) An early style of fashioning diamonds, in which opposite points of an octahedron were ground down to squares to form a large culet and a larger table, and the remaining parts of the eight octahedral faces were polished. (b) A term sometimes used loosely to describe any one of the variations of the *bevel cut,* provided it has the usual large table of that cut.

table diamond A relatively flat, table-cut diamond. Syn: *table [gem].*

table iceberg *tabular iceberg.*

tableknoll *guyot.*

tableland (a) A general term for a broad, elevated region with a nearly level or undulating surface of considerable extent; e.g. South Africa. (b) A *plateau* bordered by abrupt clifflike edges rising sharply from the surrounding lowland; a mesa.

tablemount *guyot.*

table mountain A mountain having a comparatively flat summit and one or more precipitous sides. See also: *mesa.*

table reef A small, isolated, flat-topped organic reef, with or without islands, that does not enclose a lagoon. Cf: *patch reef; platform reef.*

tablet (a) A tabular crystal. (b) A table-cut gem.

tabula (a) One of the transverse and nearly plane or upwardly convex or concave partitions (septa) of a corallite, extending to the outer walls or occupying only the central part of the corallite. See also: *complete tabula; incomplete tabula.* (b) A subhorizontal perforate plate in the intervallum of an archaeocyathid, extending from one septum to another or in some genera supplanting the septa. (c) A six-sided crystalline element of a heterococcolith, with two dimensions equal and the third smaller.—Pl: *tabulae.*

tabular (a) Said of a feature having two dimensions that are much larger or longer than the third, such as an igneous dike, or of a geomorphic feature having a flat surface, such as a plateau. (b) Said of the shape of a sedimentary body whose width/thickness ratio is greater than 50 to 1, but less than 1000 to 1 (Krynine, 1948, p.146); e.g. a graywacke formation in a geosynclinal deposit. Cf: *blanket; prism.* (c) Said of a sedimentary particle whose length is 1.5-3 times its thickness (Krynine, 1948, p.142). Cf: *prismatic.* (d) Said of a crystal form that shows one dimension markedly smaller than the other two. (e) Said of a metamorphic texture in which a large proportion of grains are tabular and have approximately parallel orientation (Harte, 1977).

tabula-rasa theory A theory according to which during the Pleistocene Epoch the entire Scandinavian peninsula became covered with ice and its fauna and flora were completely destroyed, and subsequent immigration from central Europe, England, and Siberia produced an entirely new biota (Dahl, 1955, p. 1500). Etymol: Latin, "a blank tablet".

tabular berg *tabular iceberg.*

tabular cross-bedding Cross-bedding in which the cross-bedded units, or *sets,* are bounded by planar, essentially parallel surfaces, forming a tabular body; e.g. *torrential cross-bedding.*

tabular dissepiment A nearly flat plate extending across an entire scleractinian corallite or confined to its axial part.

tabular iceberg A flat-topped iceberg that may be very large (up to 160 km long and more than 500 m thick), with clifflike sides. Tabular icebergs are usually detached from an ice shelf and are numerous in the Antarctic. See also: *ice island.* Syn: *tabular berg; table iceberg.*

tabularium The axial part of the interior of a corallite in which tabulae are developed. Cf: *marginarium.*

tabular spar *wollastonite.*

tabular structure The structure of a mineral or rock that makes it tend to separate into plates or laminae.

tabulate adj. (a) Having tabulae; specif. said of a coral characterized by prominent tabulae. (b) Having plates; e.g. said of phaeodarian radiolarians having smooth plates, or of dinoflagellate thecae having armored plates.—n. Any zoantharian belonging to the order Tabulata. Their stratigraphic range is Middle Ordovician to Permian, possibly also Triassic to Eocene.

tacharanite A mineral: $Ca_{12}Al_2Si_{18}O_{51} \cdot 18H_2O$.

tacheometer *tachymeter.*

tachygenesis The extreme crowding and eventual loss of those primitive phylogenetic stages that are represented early in the life of an individual. Cf: *acceleration.*

tachygraphometer A tachymeter with alidade for surveying.

tachyhydrite A yellowish mineral: $CaMg_2Cl_6 \cdot 12H_2O$. Syn: *tachydrite; tachhydrite.*

tachylyte A volcanic glass that may be black, green, or brown because of abundant crystallites. It is formed from basaltic magma and is commonly found as chilled margins of dikes, sills, or flows. Syn: *tachylite; hyalobasalt; basalt glass; jaspoid; basalt obsidian; sordawalite; sideromelane; wichtisite.* Cf: *hyalomelane; hydrotachylyte.* See also: *pseudotachylyte.*

tachymeter A surveying instrument designed for use in the rapid determination from a single observation of the distance, direction, and elevation difference of a distant object; esp. a transit or theodolite with stadia hairs, or an instrument in which the base line for distance measurements is an integral part of the instrument. Tachymeters include *range finders* with self-contained bases, although range finders do not usually afford the means for the determination of elevation. Syn: *tacheometer.*

tachymetry A method of rapid surveying using the tachymeter; e.g. the stadia method of surveying used in U.S.

tachytely A phylogenetic phenomenon characterized by a rapid temporary spurt in evolution that occurs as populations shift from one major zone of adaptation to another; episodic evolution. Cf: *bradytely; horotely; lipogenesis.*

Taconian orogeny *Taconic orogeny.*

Taconic orogeny An orogeny in the latter part of the Ordovician period, named for the Taconic Range of eastern New York State, well developed through most of the northern Appalachians in the U.S. and Canada. In places it can be strictly defined as Late Ordovician by limiting fossiliferous strata, or it can be extended to include many pulsations that occurred from place to place from early in the Ordovician to early in the Silurian. Syn: *Taconian orogeny.*

taconite (a) A local term used in the Lake Superior iron-bearing district of Minnesota for any bedded ferruginous chert or variously tinted jaspery rock, esp. one that enclosed the Mesabi iron ores (granular hematite); an unleached *iron formation* containing magnetite, hematite, siderite, and hydrous iron silicates (greenalite, minnesotaite, and stilpnomelane). The term is specifically applied to this rock when the iron content, either banded or disseminated, is at least 25%, so that natural leaching can convert it into a low-grade iron ore, with 50 to 60% iron. (b) Since World War II, a low-grade iron formation suitable for concentration of magnetite and hematite by fine grinding and magnetic treatment, from which pellets containing 62 to 65% iron can be manufactured.

tactite A rock of complex mineralogical composition, formed by contact metamorphism and metasomatism of carbonate rocks. It is typically coarse-grained and rich in garnet, iron-rich pyroxene, epidote, wollastonite, and scapolite. Approx. syn: *skarn.*

tactoid A spindle-shaped body, e.g. in vanadium pentoxide sol, visible under a polarizing microscope.

tactosol A sol that contains tactoids in a spontaneous, parallel arrangement.

tadjerite A black, semiglassy, chondritic stony meteorite composed of bronzite and olivine.

tadpole nest A small, irregular *cross ripple mark* characterized by a polygonal or cell-like pattern, formed by the intersection of two sets of ripples approximately at right angles and once believed to have been made by tadpoles. The height of the ripple mark is considerably greater than in the equivalent form associated with transverse ripple mark. Term introduced by Hitchcock (1858, p.121-123).

tadzhikite A monoclinic mineral: $Ca_3(Ce,Y)_2(Ti,Al,Fe)B_4Si_4O_{22}$.

taele (a) Older form of the Norwegian term *tele.* (b) Anglicized version of the Swedish term *tjäle;* see *tjaele.*

taenia An irregularly bent small plate in the intervallum of an archaeocyathid (TIP, 1955, pt.E, p.7).

taeniolite A white or colorless mineral of the mica group: $KLiMg_2Si_4O_{10}F_2$. Also spelled: *tainiolite.*

taenite A meteorite mineral consisting of the face-centered cubic gamma-phase of a *nickel-iron* alloy, with a nickel content ranging from about 27% up to 65%. It occurs in iron meteorites as lamellae or strips flanking bands of kamacite.

tafelberg A term used in South Africa for a mesa or a table mountain; a large *tafelkop.* Etymol: Afrikaans.

tafelkop A term used in South Africa for an isolated hill with a flat top; a butte. Etymol: Afrikaans. Cf: *spitskop.*

tafone (a) A Corsican dialect term for one of the natural cavities in a *honeycomb structure,* formed by *cavernous weathering* on the face of a cliff in a dry region or along the seashore. The hole or recess may reach a depth of 10 cm, and is explained as due to solution of free salts in crystalline rock (granite, gneiss) following heating by insolation. (b) A granitic or gneissic block or boulder hollowed out by cavernous weathering.—Pl: *tafoni.*

tafrogenesis *taphrogenesis.*

Tagg's method A method of interpretation of resistivity sounding data obtained with a *Wenner array* over a two-layered Earth.

Taghanican North American stage: uppermost Middle Devonian (above Tioughniogan, below Fingerlakesian).

tagilite *pseudomalachite.*

tagma A major division of the body of an arthropod, each composed of several somites; e.g. cephalon, thorax, and abdomen. Pl: *tagmata.*

tahitite A feldspathoid-bearing *trachyandesite* that contains hauyne phenocrysts and generally more abundant sodic plagioclase than potassium feldspar. Its name, given by Lacroix in 1917, is derived from Tahiti. Not recommended usage.

tahoma (a) A generic term indicating a high snowy or glacier-clad mountain in the NW U.S., like Mount Hood. (b) The V-shaped residual ridge between two cirques on a glacially carved volcanic cone in the NW U.S., e.g. one of several on Mount Rainier (Russell, 1898a, p. 382-383).—Etymol: An approximation of one of several Indian names for Mount Rainier; presumably the higher the mountain, the longer the second syllable was drawn out, as "tahoooom-ah".

taiga A swampy area of coniferous forest sometimes occurring between *tundra* and *steppe* regions. Etymol: Russian.

tail [coast] (a) A bar or barrier formed behind a small island or a skerry. Syn: *trailing spit; banner bank.* (b) The outermost part of a projecting bar.

tail [glac geol] A small ridge of rock, or an accumulation or streak of till, tapering down-glacier from the lee side of a knob of resistant rock. See also: *crag and tail.*

tail [paleont] Any of various backwardly directed and usually posterior parts on the body of an invertebrate, esp. if attenuated.

tail [sed] The rear part of a turbidity current, which is less dense than the *nose* and moves more slowly.

tail [streams] (a) The downstream section of a pool or stream. (b) The comparatively calm water after a current or a reach of rough water.

tail coccolith A modified coccolith found at the end opposite the flagellar field in flagellate coccolithophores exhibiting dimorphism (such as *Calciopappus*). Cf: *pole coccolith.*

tail dune A dune that accumulates on the leeward side of an obstacle and that tapers away gradually for a distance of up to 1 km. Cf: *head dune.*

tail fan *caudal fan.*

tailing adj: *leggy.*

tailings Those portions of washed or milled ore that are regarded as too poor to be treated further, as distinguished from the *concentrates,* or material of value.

tail-land A term proposed by Dryer (1899, p. 273) for a meander lobe that slopes regularly from an elevated mainland or meander neck to a low point.

tail water The water downstream from a structure, as below a dam.

taimyrite A nosean-bearing *phonolite.* It was named after the Taimyr River, Siberia. Named by Chrustschoff in 1894. Not recommended usage.

Taimyr polygon An *ice-wedge polygon,* so-called from its occurrence in northern Siberia.

tainiolite *taeniolite.*

takanelite A hexagonal mineral: $(Mn^{+2},Ca)Mn_4{}^{+4}O_9 \cdot H_2O$.

takir A clay-silt *playa* (Cooke & Warren, 1973, p. 215). Etymol: Russian. Also spelled: *takyr.*

takovite A mineral: $Ni_5Al_4O_2(OH)_{18} \cdot 6H_2O$.

takyr *takir.*

tala A term introduced by Berkey & Morris (1924, p. 105) for a broad structural basin of internal drainage, formed in the Gobi Desert by subsidence or warping and bounded by inconspicuous divides or mountain ranges. Cf: *gobi.* Etymol: Mongolian, "open steppe-country".

talc (a) An extremely soft, light green or gray monoclinic mineral: $Mg_3Si_4O_{10}(OH)_2$. It has a characteristic soapy or greasy feel and a hardness of 1 on the Mohs scale, and it is easily cut with a knife. Talc is a common secondary mineral derived by alteration (hydration) of nonaluminous magnesium silicates (such as olivine, enstatite, and tremolite) in basic igneous rocks, or by metamorphism of dolomite rocks, and it usually occurs in foliated, granular, or fibrous masses. Talc is used as a filler, coating, and dusting agent, in ceramics, rubber, plastics, lubricants, and talcum powder. Originally spelled: *talck.* See also: *steatite.* (b) In commercial usage, a talcose rock; a rock consisting of talc, tremolite, chlorite, anthophyllite, and related minerals. (c) A thin sheet of muscovite mica.

talcite (a) A massive variety of talc. (b) *damourite.*

talcoid n. A mineral: $Mg_3Si_5O_{12}(OH)_2$. It is possibly a mixture of talc and quartz.—adj. Resembling talc; e.g. "talcoid schist".

talcose (a) Pertaining to or containing talc; e.g. "talcose schist". (b) Resembling talc; e.g. a "talcose rock" that is soft and soapy to the touch.

talc schist A schist in which talc is the dominant schistose mineral. Common associates are mica and quartz.

talc slate An impure, hard, slaty variety of talc, a little harder than French chalk; "indurated talc".

taleola A cylinder or rod of granular, nonfibrous calcite in the axial region of some *pseudopunctae* of brachiopods. Pl: *taleolae.*

talet A term used in the High Atlas of Morocco for a dried-out torrential gully (Termier & Termier, 1963, p. 415). Etymol: Berber.

talik A Russian term for a layer of unfrozen ground above, within, or beneath permafrost; occurs in regions of *discontinuous permafrost.* It may be permanent or temporary. See also: *subgelisol; suprapermafrost layer.* Syn: *tabetisol.*

talmessite *arsenate-belovite.*

talnakhite A mineral: $Cu_9(Fe,Ni)_8S_{16}$.

talpatate (a) A surficial rock formed by the cementing action of calcium carbonate on sand, soil, or volcanic ash, and partly equivalent to caliche. See also: *tepetate.* (b) A poor, thin soil consisting of partly decomposed volcanic ash that is more or less consolidated.—Etymol: American Spanish, from Nahuatl (Aztec) *tepetatl,* "stone matting". Syn: *talpetate.*

talus [geol] Rock fragments of any size or shape (usually coarse and angular) derived from and lying at the base of a cliff or very steep, rocky slope. Also, the outward sloping and accumulated heap or mass of such loose broken rock, considered as a unit, and formed chiefly by gravitational falling, rolling, or sliding. Cf: *alluvial talus; avalanche talus; rockfall talus.* See also: *scree.* Syn: *rubble.*—Etymol: French *talu,* later *talus,* "a slope", originally in the military sense of fortification for the outside of a rampart or sloping wall whose thickness diminishes with height; from Latin *talutium,* a gossan zone or slope indicative of gold (probably of Iberian origin). Pl: *taluses.*

talus [reef] *reef talus.*

talus apron A poorly sorted but well-bedded accumulation of coarse *reef detritus,* usually much larger in volume than the parent reef, with a surface inclination up to 40°.

talus breccia A breccia formed by the accumulation and consolidation of talus.

talus cave A cave formed accidentally by a fall of talus.

talus cone A small cone-shaped or apron-like landform at the base of a cliff, consisting of poorly sorted rock debris that has accumulated episodically by rockfall or alluvial wash. Also, a similar feature of fluvial origin, tapering up into a gully. Cf: *alluvial talus.*

talus creep The slow downslope movement of talus, either individual rock fragments or the mass as a whole. Cf: *scree creep.*

talus fan *alluvial fan.*

talus slope A steep, concave slope formed by an accumulation of loose rock fragments; esp. such a slope at the base of a cliff, formed by the coalescence of several *rockfall taluses* or *alluvial taluses;* the surface profile of an accumulation of talus. Etymologically, "talus slope" is a tautology, as the term "talus" originally signified a slope, although incorrectly used for the material composing the slope. See also: *scree.*

talus spring A spring occurring at the base of a talus slope and originating from water falling upon or seeping into the slope.

taluvium A term introduced by Wentworth (1943) for a detrital cover consisting of talus and colluvium; the fragments range from large blocks to silt. Obsolete.

talweg *thalweg.*

tamaraite A dark-colored hypabyssal *lamprophyre* containing augite, hornblende, biotite, nepheline, plagioclase, alkali feldspar, minor accessories, and secondary cancrinite and analcime; a nepheline-rich *camptonite.* The name, given by Lacroix in 1918, is for Tamara Island, Los Archipelago, Guinea. Not recommended usage.

tamarugite A colorless mineral: $NaAl(SO_4)_2 \cdot 6H_2O$. It was originally named *lapparentite.*

tamping (a) The act or an instance of filling up a drill hole above a blasting charge with moist, loose material (mud, clay, earth, sand) in order to confine the force of the explosion to the lower part of the hole. (b) The material used in tamping.—See also: *stemming.*

tang A Scottish term for a low, narrow cape.

tangeite A syn. of *calciovolborthite.* Also spelled: *tangueite; tangeïte.*

tangent n. (a) A straight line that touches, but does not transect, a given curve or surface at one and only one point; a line that touches a circle and is perpendicular to its radius at the point of contact. (b) The part of a traverse included between the point of tangency (the point in a line survey where a circular curve ends and a tangent begins) of one curve and the point of curvature (the point in a line survey where a tangent ends and a circular curve begins) of the next curve. (c) A great-circle line that is tangent to a parallel of latitude at a township corner in the U.S. Public Land Surveys system. (d) A term sometimes applied to a long straight line of a traverse whether or not the termini of the line are points of curve. (e) The ratio of the length of the leg opposite an acute angle in a right-angled triangle to the length of the leg adjacent to the angle.—adj. Said of a line or surface that meets a curve or surface at only one point.

tangential cross-bedding Cross-bedding in which the foreset beds appear in section as smooth arcs meeting the underlying surface at low angles; large-scale tangential cross-bedding is commonly believed to imply deposition by wind. Cf: *angular cross-bedding.*

tangential ray A sponge-spicule ray that lies approximately parallel to the surface of the sponge.

tangential section (a) A slice through part of a foraminiferal test parallel to the axis of coiling or growth but not through the proloculus. (b) A section at a right angle to the zooids in the exozone of stenolaemate bryozoans, generally parallel to and just under the surface of a colony. (c) A section of a cylindrical organ (such as a stem) cut lengthwise and at right angles to the radius of the organ.

tangential stress *shear stress.*

tangential wave *S wave.*

tangent screw A very fine, slow-motion screw giving a tangential movement for making the final setting to a precision surveying instrument (such as for completing the alignment of sight on a theodolite or transit by gentle rotation of the reading circle about its axis).

tangi A term used in Baluchistan for a narrow, transverse gorge or cleft through which a stream penetrates a longitudinal ridge. Etymol: Persian *tang,* "narrow".

tangiwai A term used by the Maoris of New Zealand for *bowenite.* Syn: *tangiwaite; tangawaite.*

tangle sheet Mica with intergrowths of crystals or laminae resulting in books that split well in some places but tear to produce a large proportion of partial films (Skow, 1962, p. 170).

tangue Calcareous mud occurring in the shallow bays along the coast of Brittany (NW France), consisting in part of a fluviolacustrine silt reworked by recent marine transgressions and in part of finely powdered molluscan shell material transported and deposited by the tides. It contains 25-60% calcium carbonate, is coherent (even when dry), and has a permeability similar to that of fine sand. Pron: *tong.*

tank (a) A term applied in SW U.S. to a natural depression or cavity in impervious rocks (usually crystalline) in which rainwater, floodwater, snowmelt, and seepage are collected and preserved during the greater part of the year. See also: *rock tank; charco.* (b) A natural or artificial pool, pond, or small lake occupying a tank; esp. an artificial reservoir for supplying water for livestock. (c) A term used in Ceylon and the drier parts of peninsular India for an artificial pond, pool, or lake formed by building a mud wall across the valley of a small stream to retain the monsoon rainwater.

tannbuschite A rarely used term applied to dark-colored *olivine nephelinite* or *ankaratrite.* Johannsen (1938) coined the name after Tannbusch, Czechoslovakia. Not recommended usage.

tantalaeschynite-(Y) An orthorhombic mineral: $(Y,Ce,Ca)(Ta,Ti,Nb)_2O_6$. It forms a series with aeschynite-(Y).

tantalite A black mineral: $(Fe,Mn)(Ta,Nb)_2O_6$. It is isomorphous with *columbite* and dimorphous (orthorhombic phase) with tapiolite; it occurs in pegmatites, and is the principal ore of tantalum.

tanteuxenite A black or brown mineral: $(Y,Ce,Ca)(Ta,Nb,Ti)_2(O,OH)_6$. It is a variety of euxenite with Ta largely or almost entirely substituting for Nb. Syn: *delorenzite; eschwegeite.*

tanzanite A sapphire-blue gem variety of zoisite that exhibits strong pleochroism.

tape A continuous ribbon or strip of steel, invar, dimensionally stable alloys, specially made cloth, or other suitable material, having a constant cross section and marked with linear graduations, used by surveyors in place of a *chain* for the measurement of lengths or distances.

tape correction A quantity applied to a taped distance to eliminate or reduce errors due to the physical condition of the tape and the the manner in which it is used; e.g. a correction based on the length, temperature, tension, or alignment of the tape. See also: *sag correction; slope correction.*

tapeman *chainman.*

tapetum Tissue of nutritive cells in the sporangium of embryophytic plants, digested during development of the spores. In angiosperms, it is the inner wall of the anther locules and provides nutritive substances for the developing pollen. Pl: *tapeta.*

taphocoenose *thanatocoenosis.*

taphocoenosis *thanatocoenosis.*

taphoglyph A *hieroglyph* representing the impression of the body of a dead animal (Vassoevich, 1953, p.72).

taphonomy The branch of paleoecology concerned with the manner of burial and the origin of plant and animal remains. It includes *Fossildiagenese* and *biostratonomy.* Syn: *para-ecology.*

taphrogenesis *taphrogeny. Also spelled: tafrogenesis.*

taphrogenic Adj. of *taphrogeny.*

taphrogeny A general term for the formation of rift phenomena, characterized by high-angle normal faulting and associated subsidence (Krenkel, 1922). Etymol: Greek, *taphre,* "trench". Also spelled: *tafrogeny.* Adj: *taphrogenic.* Syn: *taphrogenesis.*

taphrogeosyncline A geosyncline developed as a rift or trough between faults (Kay, 1945). Cf: *aulacogen.*

taphrolith A trough-shaped lava flow, poured out along the bounding faults into a trough or graben. The term is little used.

taping The operation of measuring distances on the ground by means of a surveyor's tape. Cf: *chaining.*

tapiolite A mineral: $Fe(Ta,Nb)_2O_6$. It is isomorphous with mossite and dimorphous (tetragonal phase) with tantalite; it occurs in pegmatites or detrital deposits, and is an ore of tantalum.

tapoon A subsurface dam built in a dry wash, either to increase recharge to nearby wells or to impound water for direct use. In the latter case, a pipe is run from the dam to the point of use.

tar A thick brown to black viscous organic liquid, free of water, which is obtained by condensing the volatile products of destructive distillation of coal, wood, oil, etc. It has a variable composition, depending on the temperature and material used to obtain it.

taramellite A reddish-brown mineral: $Ba_4(Fe,Mg)Fe_2{}^{+3}TiSi_8O_{24}(OH)_2$.

taramite A black monoclinic mineral of the amphibole group, approximately: $(Ca,Na,K)_3Fe_5(Si,Al)_8O_{22}(OH)_2$.

taranakite A yellowish-white clayey mineral: $KAl_3(PO_4)_3(OH)\cdot 9H_2O$.

tarantulite A plutonic rock containing more than 50% quartz, less alkali feldspar, over half of which is potassium feldspar and the rest albite, and up to 5% dark minerals. The term was proposed by Johannsen (1920, p. 54) as a substitute for *alaskite-quartz.* The rock is transitional between *alaskite* and *silexite [ign].* Its name is derived from Tarantula Spring, Nevada. Cf: *quartz-rich granitoid* for preferred usage.

tarapacaite A yellow mineral: K_2CrO_4.

tarasovite A mixed-layer clay mineral: $(Ca,Na)_{0.4}KNa(H_3O)Al_8(Si,Al)_{16}O_{40}(OH)_8\cdot 2H_2O$.

taraspite A mottled, compact dolomite rock from Tarasp, Switzerland, used as decorative stone.

tarbuttite A triclinic mineral: $Zn_2(PO_4)(OH)$. It is isomorphous with paradamite. Syn: *salmoite.*

tar coal Resinous, bitumen-rich brown coal.

tare A discontinuity in data, indicating an error in measurement or computation rather than a sudden change in the quantity being measured. Sometimes spelled: *tear.*

target [photo] (a) The distinctive marking or instrumentation of a ground point to aid in its identification on a photograph. It is a material marking so arranged and placed on the ground as to form a distinctive pattern over a geodetic or other control-point marker, on a property corner or line, or at the position of an identifying point above an underground facility or feature (ASP, 1975, p. 2106). (b) The image pattern on an aerial photograph of the actual mark or target placed on the ground prior to photography.

target [surv] The vane or sliding sight on a surveyor's *level rod;* a device, object, or point upon which sights are made.

target rod A *level rod* with an adjustable target that is moved into position by the rodman in accordance with signals given by the man at the leveling instrument and that is read and recorded by the rodman when it is bisected by the line of collimation. Cf: *self-reading leveling rod.*

tarn (a) A landlocked pool or small lake such as those that occur within tracts of swamp, marsh, bog, or muskeg in Michigan (Davis, 1907, p. 116), and on moors in northern England. In unrestricted usage, "any small lake may be called a tarn" (Veatch & Humphrys, 1966, p. 326). (b) A relatively small and deep, steep-banked lake or pool amid high mountains, esp. one occupying an ice-gouged rock basin amid glaciated mountains. (c) *cirque lake.*—Etymol: Icelandic.

tarnish A thin film that forms on the surface of certain minerals, esp. those containing copper. Its color and luster are different from those of the fresh mineral.

tar pit An area in which an accumulation of natural bitumen is exposed at the land surface, forming a trap into which animals (esp. vertebrates) fall and sink, their hard parts being preserved in the bitumen.

tar sand A type of *oil sand* from which the lighter fractions of crude oil have escaped, leaving a residual asphalt to fill the interstices.

tarsus (a) The ankle of a tetrapod. (b) The distal part of the limb of an arthropod; e.g. the last segment (sometimes subsegmented) of a leg of an arachnid, or a joint of the distal part of a prosomal appendage of a merostome. Pl: *tarsi.*

Tartarian *Tatarian.*

tasco A fireclay from which melting pots are made. Also spelled: *tasko.*

tasmanite [coal] An impure coal, transitional between cannel coal and oil shale. Syn: *combustible shale; yellow coal; Mersey yellow coal; white coal.*

tasmanite [ign] An intrusive rock similar in composition to *ijolite* but containing melilite, and zeolites in place of nepheline. Among the zeolites are natrolite, thomsonite, and phillipsite. The rock is named after Tasmania (Johannsen, 1938). Not recommended usage.

tasmanites An informal term for members of the genus *Tasmanites,* a large spherical palynomorph with a thick perforate wall and probably representing the resting body of certain green algae. Its fossils (ranging from Ordovician to Cenozoic) are usually classed with the acritarchs; certain organic-rich shales (also known as *tasmanite,* as in Australia) contain enormous numbers of these fossils.

Tatarian A syn. of *Chideruan.* Also spelled: *Tartarian.*

tatarskite A mineral: $Ca_6Mg_2(SO_4)_2(CO_3)_2Cl_4(OH)_4\cdot 7H_2O$.

tauactine A spicule of hexactinellid sponges, having three coplanar rays of which two lie along the same axis.

taurite A sodic *rhyolite* containing acmite and differing from *comendite* in having a spherulitic or granophyric groundmass. Obsolete.

tautirite A nepheline *trachyandesite* or *hawaiite* composed of potassium feldspar, andesine, nepheline, and amphibole, with abundant accessory sphene. Cf: *pollenite.* The name, given by Iddings in 1918, is for Tautira beach, Tahiti. Not recommended usage.

tautonym A *binomen* or *trinomen* in which the term designating the genus is the same as that for the species or subspecies; e.g. *Troglodytes troglodytes.* See also: *tautonymy rule.*

tautonymy rule A rule of binomial nomenclature by which a species having a name identical to the name of the genus to which it belongs automatically designates the type species of that genus. See also: *tautonym.*

tavistockite *carbonate-apatite.*

tavolatite A *phonolite,* occurring as blocks in volcanic breccia, containing large phenocrysts of leucite, hauyne, acmite-augite, and garnet in a microphyric groundmass of the same minerals with interstitial leucite, alkali feldspar, hauyne, labradorite, acmite-augite, biotite, garnet, and nepheline. The name, given by Washington in 1906, is for Osteria di Tavolato near Rome. Not recommended usage.

tavorite A yellow mineral: $LiFe^{+3}(PO_4)(OH)$.

tawite A plutonic rock, similar in composition to *ijolite,* containing 30 to 60 percent mafic minerals, but with sodalite in place of nepheline as the dominant sodium feldspathoid. The name, given by Ramsay in 1894, is from Tawojak, Kola Peninsula, U.S.S.R. A preferred name would be "sodalite ijolite".

tawmawite A yellow or green to dark-green variety of epidote containing chromium and found in Tawmaw, upper Burma.

taxa The plural of *taxon.*

taxichnic Said of a dolomite rock in which the original texture or structure of a limestone is preserved or distinguishable. Term introduced by Phemister (1956, p. 74).

taxis [ecol] A movement or orientation of an organism with respect to a source of stimulation. Cf: *tropism.*

taxis [paleont] A definite linear series of plates in any part of the crown of a crinoid; e.g. anitaxis and brachitaxis. Pl: *taxes.*

taxite A general term for a volcanic rock that appears to be clastic because of its mixture of materials of varying texture and structure from the same flow. See also: *ataxite; eutaxite.*

taxodont adj. Said of the dentition of a bivalve mollusk characterized by numerous short subequal hinge teeth forming a continuous row, with some or all of the teeth transverse to the margin of the hinge. The term is virtually synonymous with *prionodont.*—n. A taxodont mollusk; specif. a bivalve mollusk of the order Taxodonta, having numerous unspecialized hinge teeth and equally developed adductor muscles. Cf: *heterodont.*

taxon A named group of organisms of any rank, such as a particular species, family, or class; also, the name applied to that unit. A taxon may be designated by a formal Latin name or by a letter, number, or other symbol. The term was proposed for general systematic use by H. J. Lam in 1948 (Lam, 1957, p. 213; Morton, 1957, p. 155; Rickett, 1958, p. 37). Plural: *taxa; taxons.* See also: *paratax-on.*

taxonomy The theory and practice of classifying plants and animals. The terms taxonomy and *systematics* are usually distinguished, the latter having broader connotation, but they may also be used more or less synonymously. Cf: *classification.*

taxon-range-zone The body of strata representing the total range of occurrence (horizontal and vertical) of specimens of a particular taxon (species, genus, family, etc.) (ISG, 1976, p. 54).

Tayloran North American (Gulf Coast) stage: Upper Cretaceous (above Austinian, below Navarroan).

taylorite [clay] An obsolete name first applied by Knight (1897) to the rock subsequently designated *bentonite.* Named after William Taylor, who made the first commercial shipments of the clay from the Rock Creek district, Wyo.

taylorite [mineral] A white mineral: $(K,NH_4)_2SO_4$. It is a variety of arcanite containing ammonia, and occurs in compact lumps in the guano beds on certain offshore Peruvian islands.

tazheranite A cubic mineral: $(Zr,Ca,Ti)O_2$.

TBM *temporary bench mark.*

tcheremkhite An algal sapropelic deposit found in the vicinity of Cheremkhovo, USSR, which has been interpreted as an aggregation of peaty matter washed from other deposits (Twenhofel, 1939, p.434).

T-chert Tectonically controlled chert, occurring in irregular masses related to fractures and orebodies (Dunbar & Rodgers, 1957, p. 248).

Tchornozem *Chernozem.*

TD *total depth.*

T.D. curve *time-distance curve.*

***t* direction** A term used in crystal plasticity to denote the direction of slip in a crystallographic slip plane. See also: *f* axis; *T plane.*

T-dolostone Tectonically controlled dolostone, occurring in irregular masses related to fracture systems (Dunbar & Rodgers, 1957, p. 239). Cf: *S-dolostone; W-dolostone.*

TΔT analysis A method of measuring the velocity of seismic waves from *normal moveout* and *arrival time* measurements.

teallite A black or black-gray mineral: $PbSnS_2$.

tear *tare.*

tear fault A steep to vertical fault associated with a low-angle overthrust fault and occurring in the hanging wall. It strikes perpendicular to the strike of the overthrust; displacement may be horizontal, and there may be a scissor effect. It is considered by some to be a type of strike-slip fault.

tear-shaped bomb A rotational volcanic bomb shaped like a teardrop and having an ear at its constricted end; it ranges in size from 1 mm to more than 1 cm in length. Cf: *Pele's tears; spindle-shaped bomb.*

technical scale A standard of fifteen minerals by which the *hardness* of a mineral may be rated. The scale includes, from softest to hardest and numbered one to fifteen: talc; gypsum; calcite; fluorspar; apatite; orthoclase; pure silica glass; quartz; topaz; garnet; fused zircon; corundum; silicon carbide; boron carbide; and diamond. Cf: *Mohs scale.*

tecnomorph In dimorphic ostracodes, the adult form that has the general shape of the juvenile instars. It is generally presumed to be the male of the species. Cf: *heteromorph.*

tectate Said of a pollen grain whose ektexine has an outer surface supported by more or less complicated inner structure usually consisting of columellae that support the tectum.

tectine An albuminoid (proteinaceous) organic substance in the wall of some foraminifers, having the appearance of, but chemically distinct from, chitin (TIP, 1964, pt.C, p.64).

tectite *tektite.*

tectocline *geotectocline.*

tectofacies A lithofacies that is interpreted tectonically. The term was introduced by Sloss et al. (1949, p.96) for "a group of strata of different tectonic aspect from laterally equivalent strata", and was defined by Krumbein & Sloss (1951, p.383) as "laterally varying tectonic aspects of a stratigraphic unit". The term appears to be of "very limited practical value" because generally the nature of a tectofacies is noted "only after the area of the tectofacies has been outlined on the basis of other considerations" (Weller, 1958, p.635). Not to be confused with *tectonic facies.* See also: *facies.*

tectogene (a) A long, relatively narrow unit of downfolding of sialic crust considered to be related to mountain-building processes. The term was proposed by Haarmann (1926, p. 107) as a substitute for *orogene.* Syn: *geotectogene.* (b) The downfolded portion of an orogene (Hess, 1938). Syn: *downbuckle.*

tectogenesis *orogeny.*

tectonic Said of or pertaining to the forces involved in, or the resulting structures or features of, *tectonics.* Syn: *geotectonic.*

tectonic accretion *continental accretion.*

tectonic axis *fabric axis.*

tectonic block A mass of rock that has been transported with respect to adjacent rock masses through the operation of tectonic processes (Berkland et al., 1972, p. 2296).

tectonic breccia A *breccia* formed as a result of crustal movements, usually developed from brittle rocks. Cf: *fault breccia; fold breccia; crush breccia.*

tectonic conglomerate *crush conglomerate.*

tectonic creep Slow, apparently continuous movement on a fault.

tectonic cycle The cycle that relates the larger structural features of the Earth's crust to gross crustal movements and to the kinds of rocks that form in the various stages of development of these features; *orogenic cycle.*

tectonic denudation The stripping of an underbody, such as basement or other competent rock, by the movement of an upper stratified layer over it. During movement of rootless masses of the upper rocks by gravity sliding, the surface of the underbody is laid bare in places. Cf: *décollement.*

tectonic earthquake An earthquake due to faulting rather than to volcanic activity. The term is little used. Cf: *volcanic earthquake.*

tectonic enclave A body of rock that has become detached or isolated from its source by tectonic forces. Cf: *tectonic inclusion.*

tectonic fabric *deformation fabric.*

tectonic facies A collective term for rocks that owe their present characteristics mainly to tectonic activity; e.g. mylonites and some phyllites. This concept was introduced by Sander (1912). Not to be confused with *tectofacies.*

tectonic flow *tectonic transport.*

tectonic framework The combination or relationship in space and time of subsiding, stable, and rising tectonic elements in sedimentary provenance and depositional areas. Var: *framework [tect].*

tectonic gap *lag fault.*

tectonic inclusion A body of rock that has become detached or isolated from its source by tectonic disruption, and that is enclosed or included in the surrounding rock, e.g. a boudin. The term was used by Rast (1956, p.401) to replace *boudin,* a term that is often misused for any such inclusion. Cf: *tectonic enclave.*

tectonic lake A lake occupying a basin produced mainly by crustal movements; e.g. caused by the impoundment of a drainage system as a result of upwarping, or occupying a graben (such as Lake Baikal in Russia). Syn: *structural lake.*

tectonic land Linear fold ridges and volcanic islands that had an ephemeral existence in the internal parts of an orogenic belt during the early or geosynclinal phase. Kay (1951) compares them with modern island arcs, and proposes that their existence probably accounts for many of the features formerly ascribed to *borderlands.*

tectonic landform A landform produced by earth movements. Cf: *structural landform.*

tectonic lens A body of rock similar to a *boudin* but interpreted as having been formed by distortion of a continuous incompetent layer enclosed between competent layers, rather than vice versa as for a boudin.

tectonic line A major, extensive fault, the displacement on which is both lateral and vertical, that cuts across or delineates an orogenic belt.

tectonic map A map that portrays the architecture of the outer part of the Earth. It is similar to a *structure-contour* map, which primarily shows dipping strata, folds, faults, and the like, but the tectonic map also presents some indication of the ages and kinds of rocks from which the structures were made, as well as their historical development. Cf: *paleotectonic map.*

tectonic mélange A *mélange* produced by tectonic processes (Berkland et al., 1972, p. 2296).

tectonic moraine An aggregation of boulders incorporated in the base or sole of an overthrust mass. It is often mistaken for a conglomerate as it has a local concordant relation to the associated strata (Pettijohn, 1957, p. 281).

tectonic overpressure The pressure exceeding the load pressure, at times during metamorphism, by amounts that depend on the strength of the rocks (1000-2000 bars).

tectonic profile *profile [struc petrol].*

tectonics A branch of geology dealing with the broad architecture of the outer part of the Earth, that is, the regional assembling of structural or deformational features, a study of their mutual relations, origin, and historical evolution. It is closely related to *structural geology,* with which the distinctions are blurred, but tectonics generally deals with larger features. Adj: *tectonic.* Syn: *geotectonics.*

tectonic style The total character of a group of related structures that distinguishes them from other groups of structures, in the same way that the style of a building or an art object distinguishes it from others of different periods or influences. Syn: *style [tect].*

tectonic transport A kinematic term used strictly only for a *deformation plan* that possesses monoclinic symmetry. The direction of tectonic transport is the kinematic a-direction or direction of maximum displacement in the unique symmetry plane. Syn: *tectonic flow.*

tectonic valley A valley that is produced mainly by crustal movements, such as by faulting or folding. Cf: *structural valley.*

tectonism *diastrophism.*

tectonite Any rock whose fabric reflects the history of its deformation; a rock whose fabric clearly displays coordinated geometric features that indicate continuous solid flow during formation (Turner and Weiss, 1963, p.39). Also spelled: *tektonite.*

tectonization A term sometimes used as a generalized synonym of orogenesis, diastrophism, etc. Thus, deformed rocks in orogenic belts are said to have been "tectonized". This usage is not recommended.

tectono-eustatism *diastrophic eustatism.*

tectonophysics A branch of geophysics that deals with the forces responsible for movements in, and deformation of, the Earth's crust.

tectonosphere The zone or layer of the Earth above the level of isostatic equilibrium, in which crustal or tectonic movements originate. Cf: *crust [interior Earth].*

tectono-stratigraphic unit A mixture of lithostratigraphic units resulting from tectonic deformation; e.g. a *mélange.*

tectorium The internal lining of a foraminiferal chamber (as in fusulinids), composed of dense calcite formed at or near the same time as that in which the tunnel in the test was excavated. It may include the *lower tectorium* and the *upper tectorium.* Pl: *tectoria.*

tectosequent Said of a surface feature that reflects the underlying geologic structure. Ant: *morphosequent.*

tectosilicate A class or structural type of *silicate* characterized by the sharing of all four oxygens of the SiO_4 tetrahedra with neighboring tetrahedra, and with a Si:O ratio of 1:2. Quartz, SiO_2, is an example. Cf: *nesosilicate; sorosilicate; cyclosilicate; inosilicate; phyllosilicate.* Syn: *framework silicate.* Also spelled: *tektosilicate.*

tectosome A term used by Sloss (in Weller, 1958, p.625) for a "body of strata indicative of uniform tectonic conditions"; the sedimentary rock record of a uniform tectonic environment or of a tectotope. The term replaces *tectotope* as that term was originally defined.

tectosphere A layer or shell of the Earth that has been variously equated with the lithosphere, the asthenosphere, and the tectonosphere.

tectostratigraphic Pertaining to facies aspects determined by tectonic conditions and influences; said of an interpretative (rather than an objective) stratigraphic facies characterized lithologically "in whatever way is considered to be most significant tectonically" (Weller, 1958, p.630).

tectotope An area of uniform tectonic environment. The term was originally defined by Sloss et al. (1949, p.96) as a "stratum or succession of strata with characteristics indicating accumulation in a common tectonic environment", but was used by Krumbein & Sloss (1951, p.381) to designate a tectonic environment. Sloss (in Weller, 1958, p.616) later regarded the term as referring to an area, rather than to a stratigraphic body or an environment, and notes that it "is an almost pure abstraction dependent upon interpretation" of a *tectosome.* Weller (1958, p.636) considers the term "superfluous" because tectonic areas are "generalized and extensive" and "not subject to the differentiation possible in the consideration or description of sedimentary environments".

tectum [paleont] (a) The thin, dense outermost layer of the spirotheca in fusulinids. Cf: *diaphanotheca.* (b) Marginal prolongation of a chamber in trochospirally coiled foraminiferal tests, making sutures of the dorsal (spiral) side more inclined than those of the ventral (umbilical) side. This usage is not recommended because of prior adoption of the term for fusulinids.—Pl: *tecta.*

tectum [palyn] (a) The surface of tectate pollen grains. (b) A term sometimes inaccurately used to designate a projecting flap of exine associated with the laesura on a fossil spore.

teepee butte *tepee butte.*

teepleite A mineral: $Na_2BO_2Cl \cdot 2H_2O$.

teggoglyph *load cast.*

tegillum An umbilical covering in a planktonic foraminiferal test (as in *Globotruncana* and *Rugoglobigerina*), comprising an extension from a chamber comparable to a highly developed apertural lip but extending across the umbilicus, thus completely covering the primary aperture (main opening of the test) and attached at its farther margin or at the tegillum of an earlier chamber. It may have small openings along its margin or be pierced centrally. Pl: *tegilla.*

tegmen The oral surface of an echinoderm body; strictly the calcareous adoral part of a crinoid calyx roofing the dorsal cup, situated at the origin of the free arms and occupying the space between them. It may include calcareous ambulacral and interambulacral plates or be composed entirely of soft tissue (Beerbower, 1968, p.400).

teilchron A term proposed by Arkell (1933) for the locally recognizable time span of a taxonomic entity. It is synonymous with *teilzone* as defined by Pompeckj (1914).

teilzone (a) A time term introduced in the German literature as "Teilzone" by Pompeckj (1914) to designate the local duration of existence of a species. Syn: *teilchron.* (b) A syn. of *local range-zone.* —Etymol: German *Teilzone,* "part zone". Also spelled: *teil-zone.*

teineite A blue mineral: $CuTeO_3 \cdot 2H_2O$.

tejon A term used in the SW U.S. for a solitary, disk-shaped eminence separated by erosion from the mass of which it was originally a part. Etymol: Spanish *tejón,* "round gold ingot". Pron: tay-*hone.* Cf: *huerfano.*

tektite A rounded pitted jet-black to greenish or yellowish body of silicate glass of nonvolcanic origin, usually walnut-sized, found in groups in several widely separated areas of the Earth's surface and generally bearing no relation to the associated geologic formations. Most tektites are high in silica (68-82%) and very low in water content (average 0.005%): their composition is unlike that of obsidian and more like that of shale. Some, especially from Australia, have shapes strongly suggesting aerodynamic *ablation* during hypersonic flight. Tektites average a few grams in weight (largest weights 3.2 kg). They are believed to be of extraterrestrial origin (e.g. moon splash, formed as ejecta following large lunar impacts), or alternatively the product of large hypervelocity meteorite impacts on terrestrial rocks. Term proposed by Suess (1900) who believed they were meteorites which at one time underwent thorough melting. Etymol: Greek tektos, "molten". Syn: *tectite; obsidianite.*

tektite field *strewn field.*

tektonite *tectonite.*

tektosilicate Var. of *tectosilicate.*

telain A syn. of *provitrain.* It is used in the names of transitional coal lithotypes, e.g. clarotelain.

telargpalite A cubic mineral: $(Pd,Ag)_{16}(Pb,Bi)(Te,Se)_4$(?).

tele A Norwegian term for *frozen ground,* often used erroneously as a syn. of *permafrost.* Also spelled: *taele.*

telechemic Said of the earliest minerals to crystallize during solidification of a magma; e.g. zircon, apatite, corundum. Syn: *silicotelic.*

teleconnection Identification and correlation of a series of varves, esp. over great distances or even worldwide, in order to construct a uniform time scale for a part of the Pleistocene Epoch.

telemagmatic Said of a hydrothermal mineral deposit located far from its magmatic source. Cf: *apomagmatic; perimagmatic; cryptomagmatic.*

telemeter n. A surveying instrument for measuring the distance of an object from an observer; e.g. a telescope with stadia hairs in which the angle subtended by a short base of known length is measured. See also: *range finder.* —v. To transmit data by radio or microwave links.

teleoconch The entire gastropod shell exclusive of the *protoconch.*

teleodont Said of the dentition of a bivalve mollusk (e.g. *Venus*) having differentiated cardinal teeth and lateral teeth, similar to *diagenodont* dentition but with additional elements giving rise to a more complicated hinge.

Teleostei An infraclass of ray-finned/bony fish that includes the great bulk of living fishes, both marine and freshwater, in 9 superorders encompassing some 29 orders. Range, Upper Triassic to Recent.

telescoped Said of ore deposits in which the normal upward range of high- to low-temperature mineral assemblages is vertically compressed.

telescope structure A term proposed by Blissenbach (1954, p.180) for an alluvial-fan structure "characterized by younger fans with flatter gradients spreading out from between fan mesas of older fans with steeper gradients"; e.g. in the Santa Catalina Mountains, Ariz.

telescopic alidade An *alidade* used with a planetable, consisting of a telescope mounted on a straightedge ruler, fitted with level bubble, scale, and vernier to measure angles, and calibrated to measure distances.

teleseism An earthquake that is distant from the recording station.

telethermal Said of a hydrothermal mineral deposit formed at shallow depth and relatively low temperatures, with little or no wall-rock alteration, presumably far from the source of hydrothermal solutions. Also, said of that environment. See also: *telemagmatic.* Cf: *hypothermal; mesothermal; epithermal; xenothermal; leptothermal.*

teleutospore A *fungal spore* developed in the final stage of the life cycle of rust fungi, whose thick walls may be composed of chitin. Such spores may occur as microfossils in palynologic preparations. Cf: *urediospore.* Syn: *teliospore.*

telinite (a) A maceral of coal within the *vitrinite* group, characteristic of vitrain and consisting of cell-wall material. Cf: *suberinite; xylinite.* (b) A preferred syn. of *provitrinite.*

teliospore *teleutospore.*

telite A microlithotype of coal, a variety of *vitrite,* consisting of 95% or more of telinite.

tell An Arabic term for archaeological *mound.*

tellurantimony A trigonal mineral: Sb_2Te_3.

tellurate A salt of telluric acid; a compound containing the radical $(TeO_4)^{-2}$. An example is montanite, $Bi_2TeO_6 \cdot 2H_2O$.

tellurbismuth *tellurobismuthite.*

telluric Pertaining to the Earth, esp. the depths of the Earth, e.g. as applied to natural electric fields or currents.

telluric bismuth *tetradymite.*

telluric current Natural electric current that flows in large sheets on or near the Earth's surface. Syn: *earth current; ground current.*

telluric method An electrical exploration method in which the Earth's natural electric field is measured at two or more stations simultaneously and a quantitative estimate of the electric properties of the section is obtained thereby.

telluric ocher *tellurite.*

telluric water Water formed by the combination of hydrogen with the oxygen of the atmosphere at high temperature and pressure (Swayne, 1956, p. 137). Cf: *juvenile water.*

telluride A mineral compound that is a combination of tellurium with a metal. An example is hessite, Ag_2Te.

tellurite A white or yellowish orthorhombic mineral: TeO_2. It is dimorphous with paratellurite. Syn: *telluric ocher.*

tellurium A silvery-white to brownish-black mineral, the native semimetallic element Te. It is occasionally found in pyrites and sulfur, or in the fine dust of gold-telluride ores.

tellurium glance *nagyagite.*

tellurium-130/xenon-130 age method A method of age determination applicable to tellurium-rich minerals in which the $^{130}Te/^{130}Xe$ ratio is measured and combined with the decay rate of ^{130}Te to ^{130}Xe by double beta emission (half life approximately 2.83×10^{21} years). It has been applied to samples ranging from 7 to 57 million years in age.

tellurobismuthite A pinkish mineral: Bi_2Te_3. It is often intergrown with tetradymite. Syn: *tellurbismuth.*

telluroid A surface near the terrain that is the locus of points in which the spheropotential coincides with the geopotential of corresponding points on the terrain.

Tellurometer Trade name of a rugged portable electronic device that measures ground distances precisely by determining the velocity of a phase-modulated, continuous, microwave radio signal transmitted between two instruments operating alternately as master station and remote station. It has a range up to 65 km (35-40 miles). Cf: *geodimeter.*

telmaro A term proposed by Veatch & Humphrys (1966, p. 326) for a river traversing a peat marsh or peat swamp. Etymol: Greek *telma,* "marsh".

telmatic peat A *lowmoor peat* developed in very shallow water and consisting mainly of reeds. Syn: *reed peat.*

telmatology The study of wetlands, marshy areas and swamps.

teloclarain A transitional lithotype of coal characterized by the presence of telinite, though in lesser amounts than other macerals. Cf: *clarotelain.* Syn: *teloclarite.*

teloclarite *teloclarain.*

telodurain A coal lithotype transitional between durain and telain, but predominantly durain. Cf: *durotelain.*

telofusain A coal lithotype transitional between fusain and telain, but predominantly fusain. Cf: *fusotelain.*

telogenetic A term proposed by Choquette & Pray (1970, p. 220-221) for the period during which long-buried carbonate rocks are affected significantly by processes related to weathering and subaerial and subaqueous erosion. Also applied to the porosity that develops during the telogenetic stage. Cf: *eogenetic; mesogenetic.*

telome Idealized concept of a single terminal segment of a branching axis. See also: *telome theory.*

telome theory A theory that plants evolved through formation of vascular tissue, stomata, and terminal sporangia from an undifferentiated axis (Swartz, 1971, p. 467). See also: *telome.*

telson (a) The last somite of the body of a crustacean, bearing the anus and commonly the caudal furca. Syn: *postabdomen; style [paleont].* (b) A dorsal, postanal extension of the body of an arachnid, articulated to the last abdominal segment; a postanal spine or plate of a merostome. (c) A terminal or anal segment of a trilobite. The term is sometimes incorrectly used for a spine mounted on the terminal or one of the near-terminal segments of a trilobite and directed posteriorly along the midline, such as the first macrospine on the posterior part of the thorax in certain Olenellidae (TIP, 1959, pt.O, p.126).

telum A guardlike structure in aulacocerid cephalopods, consisting of concentric layers of mostly organic material (conchiolin?) with a few interspersed calcitic layers. See: *rostrum.*

temagamite An orthorhombic mineral: Pd_3HgTe_3.

temblor A syn. of *earthquake.* Etymol: Spanish, a "trembling".

temiskamite *maucherite.*

Temnospondyli An order of labyrinthodont amphibians characterized by reduction of the centrum and emphasis on the intercentrum in each vertebra. Range, Upper Mississippian to Upper Triassic.

temperate Said of a temperature that is moderate or mild. The term is also used to describe temperatures of the middle latitudes, whether moderate or not.

temperate glacier A glacier characteristic of the temperate zone, in which at the end of the ablation season the firn and ice of which it consists are near the melting point (Ahlmann, 1933). Its temperature is in fact close to the melting point of ice throughout, except during the winter, when its upper part is frozen to a depth of several meters. Examples: almost all the glaciers in Scandinavia and the Alps, and in the U.S. outside of northern Alaska. Cf: *polar glacier.* Syn: *warm glacier.*

temperature A basic property of a system in thermal equilibrium, measured by various scales based on changes in volume, electrical resistance, thermal electromotive force, or length. Systems in thermal equilibrium with each other have the same temperature.

temperature gradient *thermal gradient.*

temperature-gradient metamorphism A process of modification of ice crystals in deposited snow, brought about by vapor transfer under strong gradients of temperature and vapor pressure. It produces grain growth, the most extreme products being complex-shaped crystals, usually with stepped or layered surfaces, termed *depth hoar.* Cf: *equitemperature metamorphism.* Archaic syn: *constructive metamorphism.*

temperature log A *well log,* usually run in a cased hole, that depicts temperature variation with depth. It is useful to identify the top of cement that is curing behind casing, or points of gas entry. Syn: *thermal log.*

temperature-salinity diagram A plotting of temperature and salinity in a column of water, thus identifying the water masses within it, its stability, and sigma-t value. Syn: *T-S diagram.*

temperature survey Measurement of temperature in drill holes. An absolute accuracy of about 0.05°C and a precision of about 0.005°C can be obtained. Maps of isotherm surfaces can be constructed that help to detect anomalies in geologic structure or subsurface ground-water conditions.
temperature zone A general term for a region characterized by a relatively uniform temperature or temperature range. It may refer to a region of a particular *climatic classification* (a latitudinal division) or to a particular temperature belt on a mountainside (a vertical or altitudinal division). Cf: *climatic zone.*
tempestite A storm deposit, showing evidence of violent disturbance of pre-existing sediments followed by their rapid redeposition, all in a shallow-water environment (Ager, 1974, p. 86).
templet In photogrammetry, a transparent celluloid overlay made on an aerial photograph and showing the center of the photograph and all radial lines through images of control points, as well as azimuth lines connecting the center with images of points that show on the photograph and are themselves the centers of other photographs. The term is a var. of "template".
temporal transgression A name given by Wheeler & Beesley (1948, p.75) to the principle that rock units and unconformities vary in age from place to place.
temporary base level Any *base level,* other than sea level, below which a land area cannot be reduced, for the time being, by erosion; e.g. a level locally controlled by a resistant stratum in a stream bed, the surface of a lake, or the level of the main stream into which a tributary flows. Cf: *ultimate base level.* Syn: *local base level.*
temporary bench mark A supplementary *bench mark* of less enduring character than a *permanent bench mark,* intended to serve for only a comparatively short period of time, such as a few years; e.g. an intermediate bench mark established at a junction of sections of a continuous series of measured differences of elevation for the purpose of holding the end of a completed section and serving as an initial point from which the next section is run. Its elevation determination may not be precise. Examples include a chiseled square or cross on masonry, a nail and washer in the root of a tree, or a bolt on a bridge. Abbrev: TBM.
temporary extinction The *extinction* of a lake by the loss of water only (such as due to climatic changes), the lake basin remaining intact to receive water at some future time.
temporary hardness *carbonate hardness.*
temporary lake *intermittent lake.*
temporary plankton *meroplankton.*
temporary stream *intermittent stream.*
temporary wilting A degree of wilting from which a plant can recover by decreasing its rate of transpiration and without adding water to the soil. Cf: *permanent wilting; wilting point.*
tenacity The property of the particles or molecules of a substance to resist separation; *tensile strength.*
tenebrescence In optics, the absorption of light by a crystal under the influence of radiation.
tengerite A mineral: $CaY_3(CO_3)_4(OH)_3 \cdot 3H_2O$ (?). The name was applied originally to a supposed beryllium-yttrium carbonate.
tennantite A blackish lead-gray isometric mineral: $(Cu,Fe)_{12}As_4S_{13}$. It is isomorphous with tetrahedrite, and sometimes contains zinc, silver, or cobalt replacing part of the copper. It is an important ore of copper. Syn: *fahlore; gray copper ore.*
tenor *grade [ore dep].*
tenorite A triclinic mineral: CuO. It occurs in minute shining steel-gray or iron-gray scales, in black powder, or in black earthy masses, generally in the oxidized (weathered) zones or gossans of copper deposits. Tenorite is an ore of copper. Syn: *melaconite; black copper.*
tensile strength The maximum applied tensile stress that a body can withstand before failure occurs. Syn: *tenacity.*
tensile stress A *normal stress* that tends to cause separation across the plane on which it acts. Cf: *compressive stress.*
tension A state of stress in which tensile stresses predominate; stress that tends to pull a body apart.
tension crack A fracture caused by tensile stress. Cf: *shear crack.*
tension fault A genetic term for any fault caused by tension. Cf: *extension fault.*
tension fracture A fracture that is the result of tensional stress in a rock. Cf: *shear fracture.* See also: *extension fracture.*
tension gash A short tension fracture along which the walls have been pulled apart. Tension gashes may be open or filled, and commonly have an en echelon pattern. They may occur diagonally in fault zones.
tension joint A joint that is a *tension fracture.*
tension zone *ecotone.*
tensor resistivity Representation of the resistivity of a material by a matrix of values, three of which represent resistance to current flow in the direction of the applied electric field and six of which represent resistance to current flow at right angles to the applied electric field. It is also used in two dimensions, with two principal values and two cross-coupling values.
tentacle Any of various elongate and flexible processes of differing functions that are borne by invertebrates, e.g. a movable tubular extension of soft integument rising from the oral disk of a coral polyp, closed at the tip, and serving primarily for food getting; a tube foot of an asterozoan or an arm of a crinoid; a short, slender armlike sensory appendage extending in front of the head of a cephalopod or gastropod; or one of the numerous small ciliated processes borne on the arms of a brachiopod or the lophophore of a bryozoan and used primarily for food getting.
tentacle pore A term commonly used for *podial pore* of ophiuroids (TIP, 1966, pt.U, p.30).
tentacle sheath A thin, delicate membranous part of the body wall of a bryozoan, which introverts to enclose the tentacles when the polypide is retracted.
tentaculitid A marine invertebrate animal characterized by radial symmetry, a small conical shell with transverse rings of variable size and spacing, longitudinal striae, an embryonic chamber with a bluntly pointed apex, and small pores in the shell wall. The tentaculitids are referred to the order Tentaculitida and questionably assigned to the mollusks. Range, Lower Ordovician to Upper Devonian, the oldest representatives belonging to the genus *Tentaculites.* Var: *tentaculite.*
tented ice Sea ice deformed by *tenting.*
tent hill A term used in Australia for a butte or flat-topped hill resembling a canvas tent; it often is capped by resistant rock from a former plateau surface. Cf: *tepee butte.*
tenting The vertical displacement upward of sea ice under lateral pressure to form a flat-sided arch over a cavity between the raised ice and the water beneath; a type of *ridging.*
tenuitas A thin area in the exine of pollen and spores, usually germinal in function, as the annular tenuitas of *Classopollis.* A tenuitas is a less distinct feature than a colpus or pore.
tepee butte A conical hill or knoll resembling an American Indian tepee; esp. an isolated, residual hill formed by a capping of resistant rock that protects the underlying softer material from erosion, e.g. one of the partly exhumed *bioherms* in the Pierre Shale of Colorado, or one of the sandstone-capped hills in the Painted Desert, Ariz. Cf: *tent hill; klint [reef].* Also spelled: *teepee butte.*
tepee structure A disharmonic sedimentary structure consisting of a fold which in cross section resembles a chevron, an "inverted depressed V", or the profile of a peaked dwelling of the North American Indians. It is believed to be an early diagenetic structure formed at the margins of large polygons produced by expansion of surface sediments (Assereto & Kendall, 1977). The term was introduced by Adams & Frenzel (1950). See also: *enterolithic.*
tepetate (a) An evaporite consisting of a calcareous crust coating solid rocks on or just beneath the surface of an arid or semiarid region; a deposit of *caliche.* (b) A term used in Mexico for a volcanic tuff, or a secondary volcanic or chemical nonmarine deposit, very commonly calcareous (Brown & Runner, 1939, p. 338).—Etymol: Mexican Spanish, from Nahuatl (Aztec) *tepetatl,* "stone matting". See also: *talpatate.*
tephra A general term for all *pyroclastics* of a volcano.
tephrite A group of extrusive rocks, of basaltic character, primarily composed of calcic plagioclase, augite, and nepheline or leucite as the main feldspathoids, with accessory alkali feldspar; also, any member of that group; the extrusive equivalent of *theralite.* With the addition of olivine, the rock would be called a *basanite.*
tephritoid n. A term proposed by Bücking in 1881, but never adopted, for a group of rocks intermediate in composition between *basalt* and *tephrite* (Johannsen, 1939, p. 283), i.e. having the chemical composition of a tephrite but with a soda-rich glassy groundmass in place of nepheline.—adj. Said of a rock resembling tephrite.
tephrochronology The collection, preparation, petrographic description, and approximate dating of *tephra.* It is a relatively new field; "the study of tephra today remains an inexact discipline" (Steen-McIntyre, 1977).
tephroite A mineral of the olivine group: Mn_2SiO_4. It occurs with

zinc and manganese minerals.

teppe A Persian term for archaeological *mound.*

terebrataliiform Said of the loop, or of the growth stage in the development of the loop, of a dallinid brachiopod (as in *Terebratalia*), consisting of long descending branches with connecting bands to the median septum, then recurving into ascending branches that meet in transverse band (TIP, 1965, pt.H, p.154). The terebrataliiform loop is morphologically similar to the *terebratelliform* loop.

terebratellacean n. Any terebratuloid belonging to the superfamily Terebratellacea, characterized by a long brachial loop. Range, Upper Cretaceous to present.—adj. Said of a terebratuloid having a long brachial loop, or of its shell. Not to be confused with *Terebratulacean.*

terebratellid Any terebratuloid belonging to the suborder Terebratellidina, characterized by a loop that develops in connection with both cardinalia and median septum. Range, Lower Devonian to present.

terebratelliform Said of the loop, or of the growth stage in the development of the loop, of a terebratellid brachiopod (as in the subfamily Terebratellinae), consisting of long descending branches with connecting bands to the median septum, then recurving into ascending branches that meet in transverse band (TIP, 1965, pt.H, p.154). The terebratelliform loop is morphologically similar to the *terebrataliiform* loop.

terebratulacean Any terebratuloid belonging to the superfamily Terebratulacea, characterized by the development of the cardinal process and outer hinge plates and by the absence of inner hinge plates. Range, Upper Triassic to present. Not to be confused with *terebratellacean.*

terebratuliform Said of a short, typically U- or W-shaped loop found in most terebratulacean brachiopods.

terebratuliniform Said of a short brachiopod loop in which crural processes are fused medially to complete a ringlike or boxlike apparatus (TIP, 1965, pt.H, p.154).

terebratuloid Any articulate brachiopod belonging to the order Terebratulida, characterized chiefly by a punctate shell with a teardrop-shaped outline, pointed at the posterior end. Range, Lower Devonian to present. Var: *terebratulid.*

tergal fold *epimere.*

tergite The dorsal plate or dorsal part of the covering of a segment of an articulate animal; e.g. the sclerotized dorsal surface of a single crustacean somite, a hardened chitinous plate on the dorsal surface of an arachnid body segment, or a plate forming the dorsal cover of a merostome somite.

tergopore A polymorph on the reverse side of some stenolaemate bryozoan colonies, with a cross section comparable in size to that of associated feeding zooids.

tergum (a) One of a pair of opercular valves, adjacent to the carina in cirripede crustaceans, lacking adductor-muscle attachments. Cf: *scutum.* (b) The back or dorsal surface of the body of an animal.—Pl: *terga.*

terlinguaite A yellow monoclinic mineral: Hg_2ClO.

terminal curvature A sharp, local change in the dip of strata or cleavage near a fault. Not commonly used in the U.S.

terminal diaphragm A membranous or skeletal partition in stenolaemate bryozoans, extending across the aperture of a zooid to seal off the living chamber from the environment.

terminal face The lower extremity, or *snout,* of a glacier.

terminal moraine (a) The end moraine, extending across a glacial valley as an arcuate or crescentic ridge, that marks the farthest advance or maximum extent of a glacier; the outermost end moraine of a glacier or ice sheet. It is formed at or near a more-or-less stationary edge, or at a place marking the cessation of an important glacial advance. Obsolete syn: *marginal moraine.* (b) A term sometimes used as a syn. of *end moraine.*

terminal plate A single plate at the end of an arm of an asteroid, appearing very early in ontogeny. Syn: *terminal.*

terminal tentacle A terminal podium of a radial vessel of the water-vascular system of an echinoid, extending through an *ocular pore.*

terminal velocity The limiting velocity reached asymptotically by a particle falling under the action of gravity in a still fluid (ASCE, 1962).

terminator The line separating the illuminated and the unilluminated parts of a celestial body (a planet, the Moon, etc.); the dividing line between day and night as observed from a distance.

terminus The lower margin or extremity of a glacier; the *snout.* Syn: *glacier snout.*

termitarium A mound of mud built by termites, ranging up to 4 m in height, and commonly found in the lateritic soil belts of the tropics and subtropics. Pl: *termitaria.* Syn: *termite mound; ant-hill.*

ternary diagram A triangular diagram that graphically depicts the composition of a three-component mixture or ternary system.

ternary feldspar Any feldspar containing more than 5% of a third component; e.g. anorthoclase, soda sanidine, and potassian oligoclase or potassian andesine.

ternary sediment A sediment consisting of a mixture of three components or end members; e.g. a sediment with one clastic component (such as feldspar) and two chemical components (such as calcite and quartz), or an aggregate containing sand, silt, and clay.

ternary system A system having three components, e.g. $CaO - Al_2O_3 - SiO_2$.

ternovskite A monoclinic mineral of the amphibole group: $Na_2(Mg,Fe^{+2})_3Fe_2{}^{+3}Si_8O_{22}(OH)_2$. It is near riebeckite in chemical composition.

terra A bright upland or mountainous region on the surface of the Moon, characterized by a lighter color than that of a *mare,* by relatively high albedo, and by a rough texture formed by large intersecting or overlapping craters. It probably represents a remnant of an ancient lunar surface, sculptured largely by impact of meteorites; it may also be attributed to igneous and volcanic activity from within the Moon (Lowman, 1976). Etymol: Latin, "earth". Pron: *ter*-uh. Pl: *terrae.* Syn: *continent.*

terra cariosa A syn. of *rottenstone.* Etymol: Latin, "rotten earth".

terrace [coast] (a) A narrow, gently sloping constructional coastal strip extending seaward or lakeward, and veneered by a sedimentary deposit; esp. a *wave-built terrace.* See also: *marine terrace.* (b) Loosely, a stripped wave-cut platform that has been exposed by uplift or by lowering of the water level; an elevated wave-cut bench.

terrace [geomorph] (a) Any long, narrow, relatively level or gently inclined surface, generally less broad than a plain, bounded along one edge by a steeper descending slope and along the other by a steeper ascending slope; a large bench or steplike ledge breaking the continuity of a slope. The term is usually applied to both the lower or front slope (the riser) and the flattish surface (the tread), and it commonly denotes a valley-contained, aggradational form composed of unconsolidated material as contrasted with a *bench* eroded in solid rock. A terrace commonly occurs along the margin and above the level of a body of water, marking a former water level; e.g. a *stream terrace.* (b) A term commonly but incorrectly applied to the deposit underlying the tread and riser of a terrace, esp. the alluvium of a stream terrace; "this deposit ... should more properly be referred to as a fill, alluvial fill, or alluvial deposit, in order to differentiate it from the topographic form" (Leopold et al., 1964, p. 460). (c) *structural terrace.*

terrace [marine geol] A bench-like structure on the ocean floor.

terrace [soil] A horizontal or gently sloping ridge or embankment of earth built along the contours of a hillside for the purpose of conserving moisture, reducing erosion, or controlling runoff.

terrace cusp *meander cusp.*

terraced flowstone A series of *rimstone dams.*

terraced flute cast A flute cast with external sculpturing resembling differentially weathered laminae, but "in reality, a cast of differentially eroded laminations in the underlying shale and unrelated to internal structure of the cast" (Pettijohn & Potter, 1964, p.347).

terraced pool One of the shallow, rimmed pools on a reef surface, circular around the point where the water reaches the surface, produced by the growth of lime- and silica-secreting algae, and "arranged in successively lower terraces as sections of a circle around the overflow from the higher level" (Kuenen, 1950, p.431-432).

Terrace epoch An obsolete term formerly applied informally to the earlier part of the Holocene Epoch characterized by the formation of stream terraces in drift-filled valleys of the regions glaciated during the preceding Pleistocene Epoch. Syn: *Terracian.*

terrace flight A series of terraces resembling a series of stairs, formed by the swinging meanders of a degrading stream that continuously excavates its valley. Syn: *flight.*

terrace line A fine raised ridge on the surface of a trilobite exoskeleton, particularly on the ventral surface of the doublure.

terrace meander A meander formed by the incision of a free meander associated with a former valley floor whose remnants now form a terrace (Schieferdecker, 1959, term 1492).

terrace placer *bench placer.*

terrace plain A well-developed stream terrace that represents a narrow but "true" plain (Tarr, 1902, p. 88).

terrace slope The scarp or bluff below the outer edge of a terrace; the front or face of a terrace.

terracette [mass move] A small ledge, bench, or steplike form, or a series of such forms, produced on the surface of a slumped soil mass along a steep grassy slope or hillside, ranging from several centimeters to 1.5 m in height and averaging a meter in width, and developed as a result of small landslides and subsequent backward tilting of the soil surface. See also: *catstep.*

terracette [pat grd] An obsolete syn. of *step [pat grd].*

Terracian *Terrace epoch.*

terracing (a) The formation of terraces, as by the shrinkage of a glacier. (b) A terraced structure, feature, or contour.

terra-cotta A fired or kiln-burnt clay of a peculiar brownish-red or yellowish-red color, used for statuettes, figurines, and vases, and for ornamental work on the exterior of buildings. Also, an object made of terra-cotta. Etymol: Italian, "baked earth".

terra-cotta clay A term applied loosely to any fine-textured, fairly plastic clay that acquires a natural vitreous skin in burning and that is used in the manufacture of terra-cotta. It is characterized by low shrinkage, freedom from warping, strong bonding, and absence of soluble salts.

terradynamics The study of projectile penetration of natural earth materials (Colp, 1967, p. 38).

terrae Plural of *terra.*

terrain (a) A tract or region of the Earth's surface considered as a physical feature, an ecologic environment, or a site of some planned activity of man, e.g. an engineering location; or in terms of military science, as in *terrain analysis* (Fairbridge, 1968, p. 1145). (b) An obsolescent syn. of *terrane.*

terrain analysis A study of the natural and man-made features of an area as they may be expected to affect military operations.

terrain correction A correction applied to observed values obtained in geophysical surveys in order to remove the effect of variations due to the topography. Syn: *topographic correction [geophys].*

terrain profile recorder *airborne profile recorder.*

terra Lemnia (a) *Lemnian bole.* (b) A clay, perhaps cimolite (Dana, 1892, p. 689).

terra miraculosa *bole.*

terrane An obsolescent term applied to a rock or group of rocks and to the area in which they crop out. The term is used in a general sense and does not imply a specific rock unit. Obsolete syn: *terrain.*

terraqueous zone That part of the lithosphere that is penetrated by water.

terra rosa *terra rossa.*

terra rossa A reddish-brown residual soil found as a mantle over limestone bedrock, typically in the karst areas around the Adriatic Sea, under conditions of Mediterranean-type climate. Also spelled: *terra rosa.* Etymol: Italian, "red earth".

terra verde A syn. of *green earth.* Etymol: Italian.

terreplein An embankment of earth with a broad level top.

terrestrial (a) Pertaining to the Earth. Cf: *planetary; extraterrestrial.* (b) Pertaining to the Earth's dry land. Cf: *continental.*

terrestrial deposit (a) A sedimentary deposit laid down on land above tidal reach, as opposed to a marine deposit, and including sediments resulting from the activity of glaciers, wind, rainwash, or streams; e.g. a lake deposit, or a *continental deposit.* (b) Strictly, a sedimentary deposit laid down on land, as opposed to one resulting from the action of water; e.g. a glacial or eolian deposit. (c) A sedimentary deposit formed by springs or by underground water in rock cavities.—Cf: *terrigenous deposit.*

terrestrial equator The *equator* on the Earth's surface. Cf: *astronomic equator; geodetic equator.*

terrestrial latitude Latitude on the Earth's surface; angular distance from the equator.

terrestrial longitude Longitude on the Earth's surface.

terrestrial magnetism *geomagnetism.*

terrestrial meridian A *meridian* on the Earth's surface; specif. an *astronomic meridian.*

terrestrial peat Peat that is developed above the water table.

terrestrial planet One of the four inner planets of the solar system: Mercury, Venus, Earth, and Mars. The Moon is sometimes considered a terrestrial planet.

terrestrial pole *geographic pole.*

terrestrial radiation The total infrared radiation emitted from the Earth's surface. Cf: *counterradiation.* See also: *effective terrestrial radiation.* Syn: *Earth radiation.*

terre verte A syn. of *green earth.* Etymol: French.

terrigenous Derived from the land or continent.

terrigenous deposit Shallow marine sediment consisting of material eroded from the land surface. Cf: *hemipelagic deposit; pelagic deposit.*

territorial sea The coastal waters (and the accompanying seabed) under the jurisdiction of a maritime nation or state, usually measured from the mean low-water mark or from the seaward limit of a bay or river mouth and, as originally defined under international law, extending 3 nautical miles (about 5.6 km) outward to the *high seas,* although the U.S. officially accepts a 6-mile limit and attempts have been made to extend it to as much as 15 miles (28 km). See also: *marginal sea.* Syn: *territorial waters.*

territorial waters (a) The surface waters under the jurisdiction of a nation or state, including *inland waters* and *marginal sea.* (b) *territorial sea.*

tersia A syn. of *tersioid outgrowth.* Pl: *tersiae.*

tersioid outgrowth In archaeocyathids, a fingerlike or roughly cylindrical holdfast, composed of close subparallel to curved plates connected by dissepiments, generally an expansion of the cup from the intervallum (TIP, 1972, pt. E, p. 42). Syn: *tersia; tersioid process.*

Tertiary The first period of the Cenozoic era (after the Cretaceous of the Mesozoic era and before the Quaternary), thought to have covered the span of time between 65 and three to two million years ago. It is divided into five epochs: the Paleocene, Eocene, Oligocene, Miocene, and Pliocene. It was originally designated an era rather than a period; in this sense, it may be considered to have either five periods (Paleocene, Eocene, Oligocene, Miocene, Pliocene) or two (Paleogene and Neogene), with the Pleistocene and Holocene included in the Neogene.

tertiary septum A third order of septum in scleractinian and rugose corals, normally shorter than a *major septum* or a *minor septum* (Fedorowski & Jull, 1976, p. 62).

tertiary structure Very coarsely irregular shell material in the lorica of a tintinnid. Cf: *primary structure; secondary structure.*

tertschite A mineral: $Ca_4B_{10}O_{19}\cdot 20H_2O$.

teruggite A monoclinic mineral: $Ca_4MgAs_2B_{12}O_{22}(OH)_{12}\cdot 12H_2O$.

teschemacherite A yellowish to white mineral: $(NH_4)HCO_3$.

teschenite A granular hypabyssal rock containing calcic plagioclase, augite, sometimes hornblende, and a small amount of biotite, with interstitial analcime. It is of darker color than *bogusite* and is distinguished from *theralite* by the presence of analcime in place of nepheline. Named by Hohenegger in 1861, for Teschen, Czechoslovakia. A preferred name would be "analcime theralite".

tesla A unit of *magnetic-field intensity,* equal to 10^4 gauss or 10^9 gammas. Cf: *nanotesla.*

tesselation A geomorphic feature resembling a mosaic pattern, for example a surface of *sand-wedge polygons* or of the salty crust of certain dry lakes in Australian deserts. Cf: *mud crack.* Also spelled: *tessellation.*

tessellate Said of a plating arrangement in echinoderms, in which sutures between contiguous elements are vertical or nearly so and adjacent plates abut one another to form a mosaic, like tiles in a floor (Bell, 1976).

tesserae A syn. of *felder.* Its singular form is *tessera.*

test [paleont] (a) The external shell, secreted exoskeleton, mesodermal endoskeleton, or other hard or firm covering or supporting structure of many invertebrates, such as the plates of the coronal, apical, periproctal, and peristomial systems of an echinoid; esp. a gelatinous, calcareous, or siliceous foraminiferal shell composed of secreted platelets or solid wall, agglutinated foreign particles, or a combination of these. A test may be enclosed within an outer layer of living tissue, such as a protozoan shell enclosed in cytoplasm. (b) The *theca* of a dinoflagellate.

test [petroleum] (a) Any procedure for sampling the content of an oil or gas reservoir; e.g. a *drill-stem test* or a *wire-line test.* (b) An informal syn. of *test well.*

testa The seed coat of a gymnosperm or a flowering plant developed from the *integument* of the ovule.

testaceous Having or consisting of a shell; specif. pertaining to the test of an invertebrate.

tester A service-company representative who supervises *drill-stem test* operations.

testibiopalladite A cubic mineral: Pb(Sb,Bi)Te.

test reach A *reach [hydraul],* esp. one between two gaging stations, that is long enough to be used in the determination of slope.

test well [petroleum] An *exploratory well* drilled to determine the presence and commercial value of oil or gas in an unproven area. Informal syn: *test.*

test well [water] A well dug or drilled in search for water; e.g. a well drilled adjacent to a lake to determine the relationship between the ground-water level and the lake level.

tetartohedral Said of that crystal class in a system, the general form of which has only one fourth the number of equivalent faces of the corresponding form in the *holohedral* class of the same system. Cf: *merohedral.*

tetartohedron Any crystal form in the tetartohedral class of a crystal system.

tetartoid An isometric, closed crystal form having 12 faces that correspond to one fourth of the faces of a hexoctahedron. A tetartoid may be right-handed or left-handed.

tetartoidal class That crystal class of the isometric system having symmetry 23.

Tethys An ocean that occupied the general position of the Alpine-Himalayan orogenic belt between the Hercynian and Alpine orogenies. It was largely obliterated by the Alpine-Himalayan continental collision.

tetraclone A four-armed desma of a sponge, built about a tetraxial crepis; e.g. a *trider.*

tetracoral A coral with fourfold symmetry.

tetracrepid Said of a desma (of a sponge) with a tetraxial crepis.

tetractine A sponge spicule having four rays. Syn: *tetract.*

tetrad A symmetric grouping of four embryophytic spores (or pollen grains) that result from meiotic division of one mother cell. A number of pollen types regularly remain in united tetrads as mature pollen when shed by the pollen sacs (as in the fossil *Classopollis* or in the living *Rhododendron*). Cf: *dyad; polyad.*

tetrad scar *laesura.*

tetradymite A pale steel-gray mineral: Bi_2Te_2S. It occurs usually in foliated masses in auriferous veins, often with tellurobismuthite. Syn: *telluric bismuth.*

tetraene A sponge spicule with one long ray and four short rays at one end.

tetraferroplatinum A tetragonal mineral: Pt,Fe.

tetragonal dipyramid A crystal form of eight faces consisting of two tetragonal pyramids repeated across a mirror plane of symmetry. A cross section perpendicular to the unique fourfold axis is ideally square. Its indices are $\{h0l\}$ or $\{hhl\}$ in most tetragonal crystals, also $\{hkl\}$ or $\{khl\}$ in class $4/m$.

tetragonal dipyramidal class That crystal class in the tetragonal system having symmetry $4/m$.

tetragonal disphenoid A crystal form consisting of four faces, ideally isosceles triangles, in which the unique $\bar{4}$-axis joins those two edges that are at right angles. Its indices are $\{hhl\}$, $\{hhl\}$ in $\bar{4}2m$, or $\{h0l\}$, $\{h\bar{h}l\}$, or $\{hkl\}$ in $\bar{4}$. Cf: *tetrahedron; orthorhombic disphenoid.*

tetragonal-disphenoidal class That crystal class in the tetragonal system having symmetry $\bar{4}$.

tetragonal prism A crystal form of four equivalent faces parallel to the symmetry axis that are, ideally, square in cross section. Its indices are {100} or {110} with symmetry $4/m$ $2/m$ $2/m$, or $\{hk0\}$ in $4/m$.

tetragonal pyramid A crystal form consisting of four equivalent faces, ideally isosceles triangles, in a pyramid that is square in cross section. Its indices are $\{h0l\}$ and $\{hhl\}$ in symmetry 4mm, also $\{hkl\}$ in symmetry 4.

tetragonal-pyramidal class That crystal class in the tetragonal system having symmetry 4.

tetragonal-scalenohedral class That crystal class in the tetragonal system having symmetry $\bar{4}2m$.

tetragonal scalenohedron A *scalenohedron* of eight faces, with symmetry $\bar{4}2m$ and indices $\{hkl\}$. It resembles a disphenoid. Cf: *hexagonal scalenohedron.*

tetragonal system One of the six *crystal systems,* characterized by three mutually perpendicular axes, the vertical one of which is a fourfold rotation or symmetry axis; it is longer or shorter than the two horizontal axes, which are of equal length. Cf: *isometric system; hexagonal system; orthorhombic system; monoclinic system; triclinic system.* Syn: *pyramidal system.*

tetragonal-trapezohedral class That crystal class in the tetragonal system having symmetry 422.

tetragonal trapezohedron A crystal form of eight faces, each of which is a trapezium. Its indices are $\{hkl\}$ in symmetry 422, and it may be right- or left-handed.

tetragonal trisoctahedron *trapezohedron.*

tetragonal tristetrahedron *deltohedron.*

tetrahedral Having the symmetry or form of a *tetrahedron.*

tetrahedral coordination An atomic arrangement in which an ion is surrounded by four ions of opposite sign, whose centers form the points of a tetrahedron around it. It is typified by SiO_4.

tetrahedral radius The radius of a cation when in tetrahedral coordination.

tetrahedrite A steel-gray to iron-black isometric mineral: $(Cu,Fe)_{12}Sb_4S_{13}$. It is isomorphous with tennantite, and often contains zinc, lead, mercury, cobalt, nickel, or silver replacing part of the copper. Tetrahedrite commonly occurs in characteristic tetrahedral crystals associated with copper ores. It is an important ore of copper and sometimes a valuable ore of silver. Syn: *fahlore; gray copper ore; panabase; stylotypite.*

tetrahedron A crystal form in cubic crystals having symmetry $\bar{4}3m$ or 23. It is a four-faced polyhedron, each face of which is a triangle. Adj: *tetrahedral.*

tetrahedron hypothesis An obsolete hypothesis that the Earth, because of shrinking, tends to assume the form of a tetrahedron. For discussion see Holmes, 1965, p. 32. Cf: *contracting Earth.*

tetrahexahedron An isometric crystal form having 24 faces that are isosceles triangles and that are arranged four to each side of a cube. Its indices are $\{hk0\}$ and its symmetry is $4/m\ \bar{3}\ 2/m$. Syn: *tetrakishexadron.*

tetrakalsilite A mineral: $(K,Na)AlSiO_4$. It is a polymorph of kalsilite with an a-axis of about 20 angstroms. Cf: *trikalsilite.*

tetrakishexadron *tetrahexahedron.*

tetramorph One of four crystal forms displaying *tetramorphism.*

tetramorphism That type of *polymorphism [cryst]* in which there are four crystal forms, known as *tetramorphs.* Adj: *tetramorphous.* Cf: *dimorphism; trimorphism.*

tetramorphous Adj. of *tetramorphism.*

tetranatrolite A tetragonal mineral of the zeolite family: $Na_2(Al_2Si_3)O_{10}\cdot 2H_2O$.

tetrapod n. An animal with four limbs; an informal term to distinguish amphibians, reptiles, birds, and mammals from aquatic classes in which paired limbs are absent or are fins instead of legs.—adj. Four-legged.

tetratabular archeopyle An *apical archeopyle* formed in a dinoflagellate cyst by the loss of four plates.

tetrawickmanite A tetragonal mineral: $MnSn(OH)_6$. It is dimorphous with wickmanite.

tetraxon A sponge spicule in which the rays grow along four axes arranged like the diagonals of a tetrahedron.

texas tower A radar-equipped offshore platform, mounted on the continental shelf or a shoal, and designed in part to provide oceanographic and meteorological data. Etymol: *texas,* a structure on the deck of a steamboat.

textulariid Any agglutinated foraminifer belonging to the family Textulariidae. Range, Carboniferous to present.

textural maturity A type of sedimentary *maturity* in which a sand approaches the textural end product to which it is driven by the formative processes that operate upon it. It is defined in terms of uniformity of particle size and perfection of rounding and depends upon the stability of the depositional site and the input of modifying wave and current energy; it is independent of mineral composition (Folk, 1951). Cf: *compositional maturity; mineralogic maturity.*

texture [geomorph] *topographic texture.*

texture [petrology] The general physical appearance or character of a rock, including the geometric aspects of, and the mutual relations among, its component particles or crystals; e.g. the size, shape, and arrangement of the constituent elements of a sedimentary rock, or the crystallinity, granularity, and fabric of the constituent elements of an igneous rock. The term is applied to the smaller (megascopic or microscopic) features as seen on a smooth surface of a homogeneous rock or mineral aggregate. The term *structure* is generally used for the larger features of a rock. The two terms should not be used synonymously, although certain textural features may parallel major structural features. Confusion may arise because in some languages, e.g. French, the usage of *texture* and *structure* are the reverse of the English usage.

texture [remote sensing] The frequency of change and arrangement of tones on an image.

texture [soil] The physical nature of the soil according to the relative proportions of sand, clay, and silt. Cf: *loam; silt loam; clay.*
texture ratio Ratio of the greatest number of streams crossed by a contour line within a drainage basin to the length of the upper basin perimeter intercept (Smith, 1950, p.656); a measure of *topographic texture.* Symbol: T.
tey A muddy islet near the mouth of the Rhône River (Reclus, 1872, p. 358).
TGA *thermogravimetric analysis.*
TGS *transcontinental geophysical survey.*
thalassic (a) Pertaining to the deep ocean. (b) Pertaining to seas and gulfs.—Not commonly used.
thalassocratic (a) Adj. of *thalassocraton.* (b) Said of a period of high sea level in the geologic past. Cf: *epeirocratic.*
thalassocraton A *craton* that is part of the oceanic crust; the English version of *Tiefkraton* (Stille, 1940). The concept of cratonic areas in oceanic crust is now outmoded. Cf: *hedreocraton; epeirocraton.* Adj: *thalassocratic.*
thalassogenesis A Russian term synonymous with *basification.*
thalassoid A lunar *mare basin* not filled, or only partly filled, with mare material; e.g. the Nectaris Basin.
thalcusite A tetragonal mineral: $Cu_{3-x}Tl_2Fe_{1+x}S_4$.
thalenite A flesh-red or pink mineral: $Y_2Si_2O_7$. Cf: *thortveitite; yttrialite.*
thallite A yellowish-green *epidote.*
thallogen *thallophyte.*
thallophyte A nonvascular plant without differentiated roots, stems, leaves, flowers, or seeds. Algae and fungi are thallophytes. Cf: *bryophyte; pteridophyte.* Syn: *thallogen.*
thallus A plant body without differentiation into root, stem, and leaf, as in *thallophytes.*
thalweg [coast] *midway.*
thalweg [geomorph] The line of continuous maximum descent from any point on a land surface; e.g. the line of greatest slope along a valley floor, or the line crossing all contour lines at right angles, or the line connecting the lowest points along the bed of a stream. Etymol: German *Talweg,* "valley way". Also spelled: *talweg.*
thalweg [grd wat] A subsurface ground-water stream percolating beneath and generally in the same direction as the course of a surface stream or valley.
thalweg [streams] (a) The line connecting the lowest or deepest points along a stream bed or valley, whether under water or not; the *longitudinal profile* of a stream or valley; the line of maximum depth. Syn: *valley line.* (b) The median line of a stream; the *valley axis.* (c) *channel line.*—Etymol: German *Thalweg* (later *Talweg*), "valley way". Pron: *tal*-veg. Also spelled: *talweg.*
thamnasterioid Said of a massive corallum characterized by absence of corallite walls and by confluent septa that join neighboring corallites, with a pattern of septa resembling lines of force in a magnetic field.
thanatocenosis *thanatocoenosis.*
thanatocoenosis (a) A set of fossils brought together after death by sedimentary processes, rather than by virtue of having originally lived there collectively. Cf: *biocoenosis.* Syn: *death assemblage.* (b) A group of fossils that may represent the *biocoenosis* of an area or the biocoenosis plus the thanatocoenosis of another environment; all the fossils present at a particular place in a sediment.—The term was introduced by the German hydrobiologist Wasmund in 1926. Var: *thanatocenosis; thanatocoenose; thanatocenose.* Plural: *thanatocoenoses.* Etymol: Greek *thanatos,* "death" + *koinos,* "general, common". Syn: *taphocoenose; taphocoenosis.*
thanatotope The total area in which the dead specimens of a taxon or taxa are deposited.
Thanetian European stage: uppermost Paleocene (above Montian, below Ypresian of Eocene). Essentially equivalent to Seelandian.
thaumasite A white mineral: $Ca_3Si(OH)_6(CO_3)(SO_4)\cdot 12H_2O$.
thaw v. To go from a frozen state, as ice, to a liquid state; to melt.—n. The end of a frost, when the temperature rises above the freezing point, and ice or snow undergo melting. Also, the transformation of ice or snow to water.
thaw depression A hollow in the ground resulting from subsidence following the local melting of ground ice in a permafrost region. See also: *cave-in lake.* Syn: *thermokarst depression.*
thaw hole A vertical hole in sea ice, formed where a surface puddle melts through to the underlying water.
thaw lake [glaciol] A pool formed on the surface of a large glacier by accumulation of meltwater.
thaw lake [permafrost] *cave-in lake.*
thaw sink A closed *thaw depression* with subterranean drainage, believed to have originated as a cave-in lake (Hopkins, 1949).
THDM *translucent humic degradation matter.*
theca (a) An echinoderm skeleton consisting of calcareous plates and enclosing the body and internal organs; e.g. the dorsal cup of the calyx of a crinoid. The term is generally applied to all fossilized parts, and includes ambulacra but excludes the column or stem and appendages such as free arms and brachioles. (b) An individual tube or cup (other than the sicula) that housed a zooid of a graptolite colony; the term is commonly used for the autotheca. (c) The external skeletal deposit of a coelenterate, such as the calcareous *wall* enclosing a corallite and presumably the sides of a coral polyp. (d) The sometimes resistant-walled coat or external covering, formed of numerous plates, of the nonencysted or actively swimming stage of the life cycle of some dinoflagellates. Syn: *test [paleont].*—Pl: *thecae.*
thecal plate Any of numerous calcareous plates that form an element in the theca of an echinoderm. It is usually distinguished from a plate of an ambulacrum or arm.
thecamoebian Any one of a group of usually freshwater testaceous protozoans, the fossil members of which belong to the orders Arcellinida and Gromida and to part of the suborder Allogromiina of the foraminifers. Range, Eocene to present.
thecodont adj. Pertaining to vertebrate teeth set in sockets, or to the implantation of such teeth.
Thecodontia An order of primitive archosaurian reptiles of varied body forms, some foreshadowing dinosaurs and some crocodilians. Range, Upper Permian to Upper Triassic.
Theis curve Log-log plot of drawdown or recovery of head against time used in an aquifer test based on the *Theis equation.*
Theis equation An equation relating drawdown or recovery of ground-water head to rate of withdrawal or addition of water and to the hydraulic characteristics of the aquifer (Theis, 1935, p. 520). See also: *Theis curve.*
thenardite A white or brownish orthorhombic mineral: Na_2SO_4. It occurs in masses or crusts often in connection with salt lakes.
theodolite A precision surveying instrument that can be rotated on a horizontal base so as to be sighted first upon one point and then upon another and that is used for measuring angular distances in both vertical and horizontal planes. It consists of a telescope so mounted on a tripod as to swivel vertically in supports secured to a revolvable table carrying a vernier for reading horizontal angles (azimuths), and usually includes a compass, a spirit level, and an accurately graduated circle for determining vertical angles (altitudes). See also: *transit.*
theory A *hypothesis* that is supported to some extent by experimental or factual evidence but that has not been so conclusively proved as to be generally accepted as a law; e.g. the "theory of continental drift."
theralite (a) In the *IUGS classification,* a plutonic rock with F between 10 and 60, and P/(A+P) greater than 90. Cf: *foid diorite.* Syn: *foid gabbro.* (b) A group of mafic plutonic rocks composed of calcic plagioclase, feldspathoids, and augite, with lesser amounts of sodic sanidine and sodic amphiboles and accessory olivine; also, any rock in that group; the intrusive equivalent of *tephrite.* Theralite grades into nepheline monzonite with an increase in the alkali feldspar content, into gabbro as the feldspathoid content diminishes, and into diorite with both fewer feldspathoids and increasingly sodic plagioclase. The term, defined by Rosenbusch in 1887, is derived from the Greek word for "eagerly looked for", and not from the island of Thera (Santorini).
Therapsida An order of synapsid reptiles of nearly mammalian grade in limb structure, tail length, and general pattern of skull; mammal-like reptiles *sensu stricto.* Range, Upper Permian to Upper Triassic. Cf: *Synapsida.*
Theria A subclass that includes all living mammals except the egg-laying Monotremata. Range, Upper Jurassic to Recent.
Theriodontia Suborder of therapsid-synapsid reptiles characterized by a tendency toward predaceous habit and a mammalian morphological and functional grade. Range, Upper Permian to Upper Triassic.
thermal [glac geol] n. A syn. of *interglacial stage.*
thermal [meteorol] n. A vertically moving current of air that is caused by differential heating of the ground below it.
thermal [phys] Pertaining to or caused by heat. Syn: *thermic.*
thermal analysis The study of chemical and/or physical changes in materials as a function of temperature, i.e. the heat evolved or

absorbed during such changes. See also: *differential thermal analysis.* Syn: *thermoanalysis.*
thermal aureole *aureole.*
thermal band *thermal infrared.*
thermal bar A boundary separating regions of a large lake having considerable differences in surface temperature; commonly occurring in the spring and autumn in large lakes in temperate climates. In the spring, water shoreward of the thermal bar is commonly warmer than the main body of the lake, and in the autumn the shoreward water is colder.
thermal capacity *heat capacity.*
thermal conduction *heat conduction.*
thermal conductivity (a) The time rate of transfer of heat by conduction, through unit thickness, across unit area for unit difference of temperature. (b) A measure of the ability of a material to conduct heat. Typical values of thermal conductivity for rocks range from 3 to 15 millicalories/cm-sec-°C.—Syn: *heat conductivity.*
thermal contraction crack *frost crack.*
thermal demagnetization A technique of partial *demagnetization* by heating the specimen to a temperature T then cooling to room temperature in a nonmagnetic space; this destroys a *partial thermoremanent magnetization* for that temperature interval but leaves unaffected a partial thermoremanent magnetization for temperature intervals above T. Cf: *alternating-field demagnetization; chemical demagnetization.*
thermal diffusivity *Thermal conductivity* of a substance divided by the product of its density and specific heat capacity. In rock, the common range of values is from 0.005 to 0.025 cm^2/sec.
thermal energy yield In volcanology, the thermal energy equivalent of a volcanic eruption, based on the thermal characteristics and volume of the volcanic products (Yokoyama, 1956-57, p.75-97).
thermal equator *oceanographic equator.*
thermal exfoliation A hypothetical type of *exfoliation* caused by the heating of rock during the day and its rapid cooling at night.
thermal fracture The disintegration of, or the formation of a fracture or crack in, a rock as a result of sudden temperature changes. It is thought to occur where rock-forming minerals have varying coefficients of expansion, where a cliff face receives and loses radiation rapidly, or where there is a quick drop of temperature after sundown. Also, the result of such a process.
thermal gradient The rate of change of temperature with distance. When applied to the Earth, the term *geothermal gradient* may be used. Syn: *temperature gradient.*
thermal infrared Referring to that portion of the infrared spectral region that ranges in wavelength from 3.0 to 15 μm. The thermal-infrared band includes the *radiant-power peak* of the Earth. Abbrev: *thermal IR.* Syn: *thermal band.*
thermal IR *thermal infrared.*
thermal log *temperature log.*
thermal maximum A term proposed by Flint & Deevey (1951) as a substitute for *climatic optimum* and *Altithermal.*
thermal metamorphism A type of metamorphism resulting in chemical reconstitution controlled by a temperature increase, and influenced to a lesser extent by the confining pressure; there is no requirement of simultaneous deformation. See also: *pyrometamorphism.* Cf: *geothermal metamorphism; static metamorphism; load metamorphism.* Syn: *thermometamorphism.*
thermal-neutron log A *radioactivity log* of the neutron-neutron type, in which the detector discriminates for slowed neutrons of thermal-energy level (about 0.025 ev). Cf: *neutron-gamma log; epithermal neutron log.* Syn: *n-tn log.*
thermal niche A niche produced by rapid thaw and removal of permafrost at waterlines.
thermal prospecting *geothermal prospecting.*
thermal resistivity The reciprocal of thermal conductivity.
thermal shock Failure of a material, esp. a brittle material, due to the thermal stress of rapidly rising temperature.
thermal spring A spring whose water temperature is appreciably higher than the local mean annual atmospheric temperature. A thermal spring may be a *hot spring* or a *warm spring* (Meinzer, 1923, p. 54).
thermal stratification The *stratification* of a lake produced by changes in temperature at different depths and resulting in horizontal layers of differing densities. See also: *density stratification.*
thermal stress Stress in a body caused by a local temperature gradient within the body.
thermal structure An arrangement of zones of increasing metamorphic grade in some distinct structural pattern, for example a thermal anticline or a thermal dome. Such features are associated with orogenesis and are produced by a localized heat source, possibly accompanied by anatexis (Winkler, 1967).
thermal water Water, generally of a spring or geyser, whose temperature is appreciably above the local mean annual air temperature.
thermic *thermal.*
thermic temperature regime A soil temperature regime in which the mean annual temperature at 50cm depth is at least 15°C but less than 22°C, with a variation of more than 5°C between summer and winter measurements (USDA, 1975). Cf: *isothermic temperature regime.*
thermistor chain A chain that is towed astern, carrying instruments to measure seawater temperatures.
thermite An old name for any fossil substance that is combustible.
thermoanalysis *thermal analysis.*
thermobarometer *hypsometer.*
thermocline [lake] (a) The horizontal plane in a thermally stratified lake located at the depth where temperature decreases most rapidly with depth. (b) The horizontal layer of water characterized by a rapid decrease of temperature and increase of density with depth; sometimes arbitrarily defined as the layer in which the rate of temperature decrease with depth is equal to at least 1°C per meter. This is an older and less preferred definition, and one that is often used in engineering literature. Syn: *metalimnion.*
thermocline [oceanog] A vertical, negative gradient of temperature that is characteristic of the layer of ocean water under the mixed layer; also, the layer in which this gradient occurs. A thermocline may be either seasonal or permanent. Cf: *halocline.* See also: *discontinuity layer.*
thermodynamic equilibrium constant *equilibrium constant.*
thermodynamic potential Any thermodynamic function of state, an extremum of which is a necessary and sufficient criterion of equilibrium for a system under specified conditions. Thus, for example, the *Gibbs free energy* is the thermodynamic potential for a system at constant pressure and temperature, the *Helmholtz free energy* is the thermodynamic potential for a system at constant temperature and volume. See also: *free energy.*
thermodynamic process A change in any macroscopic property of a thermodynamic system.
thermodynamics The mathematical treatment of the relation of heat to mechanical and other forms of energy.
thermoelastic effect A fall in temperature under tension or a rise in temperature under compression during elastic deformation.
thermoerosional niche A niche resulting from undercutting produced by bank erosion in north-flowing Arctic rivers during the limited summer flow, esp. in Alaska and Siberia where niches as wide as 8 m have been formed in one year (Hamelin & Cook, 1967, p. 123).
thermogene Pertaining to the formation of minerals primarily under the influence of temperature (Kostov, 1961). Cf: *piezogene.*
thermogenesis A rise in temperature in a body from reactions within it, as by oxidation or the decay of radioactive elements.
thermogram An image acquired by a stationary scanner operating in the thermal infrared spectral region. The term is sometimes used erroneously as a syn. of a *thermal infrared* image.
thermography A term suggested (R. Williams, 1972) as a replacement for the phrase "thermal infrared imagery", analogous to the term "photography". No confusion would then exist between "photograph" (a record of reflected solar energy) and "thermograph" (a record of emitted thermal energy), such as now exists between infrared photography and infrared imagery.
thermogravimetric analysis A syn. of *thermogravimetry.* Abbrev: TGA.
thermogravimetry A method of analysis that measures the loss or gain of weight by a substance as the temperature of the substance is raised or lowered at a constant rate. Syn: *thermogravimetric analysis.*
thermohaline circulation Vertical movement of seawater, generated by density differences; these are caused by variations in temperature and salinity, which induce convective overturning and consequent mixing. Syn: *thermohaline convection.*
thermohaline convection *thermohaline circulation.*
thermokarst (a) Karstlike topographic features produced in a permafrost region by the local melting of ground ice and the subsequent settling of the ground. Cf: *glaciokarst.* (b) A region marked by *thermokarst topography.* (c) The process of formation of a ther-

mokarst topography.—Syn: *cryokarst.*

thermokarst depression *thaw depression.*

thermokarst lake *cave-in lake.*

thermokarst mound A residual polygonal hummock, bordered by depressions that were formed by the melting of ground ice in a permafrost region.

thermokarst topography An irregular land surface containing cave-in lakes, bogs, caverns, pits, and other small depressions, formed in a permafrost region by the melting of ground ice; in exterior appearance, it resembles the uneven *karst topography* formed by the solution of limestone.

thermolabile Said of a material that is decomposable by heat.

thermoluminescence The property possessed by many substances of emitting light when heated. It results from release of energy stored as electron displacements in the crystal lattice.

thermoluminescent dating A method of dating applicable to materials that have once been heated (e.g. pottery, lava flows). A fraction of the energy released by decay of long-lived radioactive nuclides is stored as trapped electrons, and this energy is released as light upon heating. The age of a sample can be determined if the natural thermoluminescence is measured, the thermoluminescence induced by a known radiation dose is measured, and the radiation dose received by the sample per unit time in the past is measured.

thermomer A relatively warm period within the Pleistocene Epoch, such as an *interstade* (Lüttig, 1965, p. 582). Ant: *kryomer.*

thermometamorphism *thermal metamorphism.*

thermometric depth The depth, in meters, at which a pair of *reversing thermometers* are inverted.

thermometric leveling A type of indirect leveling in which elevations above sea level are determined from observed values of the boiling point of water. Cf: *barometric leveling.*

thermonatrite A white mineral: $Na_2CO_3 \cdot H_2O$. It is found in some lakes and alkali soils, and as a saline residue.

thermo-osmosis Osmosis occurring under the influence of a temperature difference between fluids on either side of a semipermeable membrane, with movement from the warmer to the cooler side, as when water flows in small openings from the warmer to the cooler parts of a soil mass.

thermophilic Said of an organism that prefers high temperatures, esp. bacteria that thrive in temperatures between 45 and 80°C. Noun: *thermophile.*

thermophyte A plant preferring high temperatures for growth.

thermoremanence *thermoremanent magnetization.*

thermoremanent magnetization Remanent magnetization acquired as a rock cools in a magnetic field from above the *Curie point* down to room temperature. It is very stable and is exactly parallel to the ambient field at the time of cooling. Abbrev: TRM. Syn: *thermoremanence.*

thermuticle *porcellanite.*

thesocyte An amoebocyte in a sponge, filled with inclusions of stored by-products (metabolites) of protoplasmic activity.

Thetis hairstone A variety of *hairstone* containing or penetrated by tangled or ball-like inclusions of green fibrous crystals of hornblende, asbestos, and esp. actinolite.

thetomorph A glass or glassy phase, commonly of quartz or feldspar composition, produced by solid-state alteration of an originally crystalline mineral by the action of shock waves, and retaining the form and original textures (such as fractures, twin lamellae, and grain boundary shapes) of the pre-existing mineral or grain. Cf: *maskelynite.*

thetomorphic Pertaining to a thetomorph; e.g. "thetomorphic silica glass" having a refractive index of 1.46 and retaining the morphology of the host quartz from which it was formed by shock. Term introduced by Chao (1967a, p. 211-212). Etymol: Greek *thetos,* "adopted", + *morphe,* "form". Cf: *diaplectic.*

thick bands In banded coal, vitrain bands from 5.0 to 50.0 mm thick (Schopf, 1960, p.39). Cf: *thin bands; medium bands; very thick bands.*

thick-bedded A relative term applied to sedimentary beds variously defined as more than 6.4 cm (2.5 in.) to more than 100 cm (40 in.) in thickness; specif. said of a bed whose thickness is in the range of 60-120 cm (2-4 ft), a bed greater than 120 cm being "very thick-bedded" (McKee & Weir, 1953, p.383). Cf: *thin-bedded; medium-bedded.* See also: *stratification index.*

thicket *coral thicket.*

thicket reef A reef consisting of closely interfingered delicate branching corals. Cf: *pillar reef; coral thicket.*

thickness [geol] The extent of a tabular unit from its bottom boundary surface to its top surface, usually measured normal to these surfaces; e.g. the distance at right angles between the hanging wall and the footwall of a lode, or the dimension of a stratigraphic unit measured at right angles to the bedding surface. See also: *true thickness; apparent thickness.*

thickness [paleont] (a) The greatest distance between the two valves of a brachiopod shell at right angles to the length and width. It is equal to the *height* in biconvex, plano-convex, and convexo-plane shells. (b) The distance between the inner and outer surfaces of the wall of a bivalve-mollusk shell. Also, a term used to denote the shell measurement of a bivalve mollusk commonly called *inflation.*

thickness contour *isopach.*

thickness line *isopach.*

thickness map *isopach map.*

thief formation A rock unit responsible for excessive fluid loss during drilling operations.

thigmotaxis *Taxis [ecol]* in response to mechanical or tactile stimuli. Cf: *strophotaxis; phobotaxis.*

thill A British term for the floor of a coal mine or coal seam; hence, *underclay.* The term is also used for a thin stratum of fireclay.

thin bands In banded coal, vitrain bands from 0.5 to 2.0 mm thick (Schopf, 1960, p.39). Cf: *medium bands; thick bands; very thick bands.*

thin-bedded A relative term applied to sedimentary beds variously defined as less than 30 cm (1 ft) to less than 1 cm (0.4 in.) in thickness; specif. said of a bed whose thickness is in the range of 5-60 cm (2 in. to 2 ft.), a bed less than 5 cm but more than 1 cm thick being "very thin-bedded" (McKee & Weir, 1953, p.383). Cf: *thick-bedded; medium-bedded.* See also: *stratification index.*

thinic Pertaining to a sand dune (Klugh, 1923, p. 374).

thin-layer chromatography An essentially adsorptive chromatographic technique for separating components of a sample by moving it in a mixture or solution through a uniformly thin deposit of an adsorbent on rigid supporting plates in such a way that the different components have different mobilities and thus become separated (May & Cuttitta, 1967, p.116). Abbrev: TLC. See also: *chromatography.*

thinolite (a) A pale-yellow to light-brown variety of calcite, often terminated at both ends by pyramids. It may be pseudomorphous after gaylussite. (b) *thinolitic tufa.*

thinolitic tufa A tufa deposit consisting in part of layers of delicate prismatic skeletal crystals of *thinolite,* up to 20 cm long and 1 cm thick; it occurs in the desert basins of NW Nevada, as in domelike masses along the shore of extinct Lake Lahontan, where it overlies *lithoid tufa* and underlies *dendroid tufa.*

thin out To grow progressively thinner in one direction until extinction. The term is applied to a stratum, vein, or other body of rock that decreases gradually in thickness so that its upper and lower surfaces eventually meet and the layer of rock disappears. The thinning may be original or due to truncation beneath an unconformity. Syn: *pinch out; wedge out.*

thin section A fragment of rock or mineral mechanically ground to a thickness of approximately 0.03 mm, and mounted between glasses as a microscope slide. This reduction renders most rocks and minerals transparent or translucent, thus making it possible to study their optical properties. Syn: *section [petrology].*

thin-skinned structure A term used by Rodgers (1963) for an interpretation that folds and faults of miogeosynclinal and foreland rocks in an orogenic belt involve only the upper strata, and lie on a *décollement* beneath which the structure differs; it is also called the *no-basement interpretation.* Proposed examples are in the Valley and Ridge and Plateau provinces of the Appalachian belt, and in the Jura Mountains. "Thick-skinned structure" is a contrasting interpretation.

thiospinel A general term for minerals with the spinel structure having the general formula: AR_2S_4.

third-law entropy The difference in entropy between a substance at some finite temperature and at absolute zero, as defined by the *third law of thermodynamics.*

third law of thermodynamics The statement that the entropy of any perfect crystalline substance becomes zero at the absolute zero of temperature. See also: *third-law entropy.*

third-order leveling Spirit leveling that does not attain the quality of *second-order leveling,* in which lines are not extended more than 30 miles (48.3 km) from lines of first- or second-order leveling and must close upon lines of equal or higher order of accuracy and in which the maximum allowable discrepancy is 12 mm times the square root of the length of the line in kilometers (or 0.05 ft times the square root of the distance in miles). It is used for subdividing loops of first- and second-order leveling and in providing local vertical control for detailed surveys. Cf: *first-order leveling.*

third-order pinacoid In a triclinic crystal, any $\{hk0\}$ or $\{\bar{h}k0\}$ pinacoid, with symmetry $\bar{1}$. Cf: *first-order pinacoid; second-order pinacoid; fourth-order pinacoid.*

third-order prism A crystal form: in a tetragonal crystal, a $\{hk0\}$ prism, with symmetry $4m$, $\bar{4}$, or 4; in a hexagonal crystal, a $\{hki0\}$ prism, with symmetry $6/m$, 6, or $\bar{3}$; and in orthorhombic and monoclinic crystals, any $\{hk0\}$ prism. Cf: *first-order prism; second-order prism; fourth-order prism.*

third water The quality or luster of a gemstone next below *second water,* such as that of a diamond that is clearly imperfect and contains flaws. Cf: *first water.*

thixotropic clay A clay that displays *thixotropy,* i.e. weakens when disturbed and strengthens when left undisturbed. Syn: *false body.*

thixotropy The property of certain colloidal substances, e.g. a *thixotropic clay,* to weaken or change from a gel to a sol when shaken but to increase in strength upon standing.

tholeiite A silica-oversaturated (quartz-normative) *basalt,* characterized by the presence of low-calcium pyroxenes (orthopyroxene and/or pigeonite) in addition to clinopyroxene and calcic plagioclase. Olivine may be present in the mode, but neither olivine nor nepheline appear in the norm. The term, first used in 1840 by Steininger, was derived from Tholey, Saarland, Germany, and was applied to a sill of altered *andesite.* The term was given its present meaning by Kennedy in 1933. Chayes (1966) reviewed the history of the term and recommended its replacement by *subalkaline basalt.* Cf: *basalt.* Syn: *tholeiitic basalt; subalkaline basalt.*

tholeiitic basalt *tholeiite.*

tholeiitic rock series Those series of comagmatic silica-oversaturated volcanic rocks that are characterized mineralogically by the presence of groundmass augite (in basalts) or pigeonite (in andesites and dacites) and chemically by an early stage of strong enrichment in iron relative to magnesium (Kuno, 1959). See also: *calc-alkali rock series.*

tholoid *volcanic dome.*

tholus A term established by the International Astronomical Union for an isolated domical mountain or hill on Mars. Most are considered volcanic. Generally used as part of a formal name for a Martian landform, such as Ceraunius Tholus (Mutch et al., 1976, p. 57). Etymol: Latin *tholus,* dome.

Thompson diagram *AFM projection.*

thomsenolite A white monoclinic mineral: $NaCaAlF_6 \cdot H_2O$.

thomsonite A zeolite mineral: $NaCa_2Al_5Si_5O_{20} \cdot 6H_2O$. It has considerable replacement of CaAl by NaSi, and sometimes contains no sodium. It usually occurs in masses of radiating crystals. Syn: *ozarkite.*

thoraceton The opisthosoma or abdomen of a merostome.

thoracic Pertaining to, located within, or involving the thorax; e.g. a "thoracic limb" attached to any somite of the thorax of a crustacean. Syn: *thorasic.*

thoracomere A somite of the thorax of a crustacean.

thoracopod A limb of any thoracic somite of a crustacean; a thoracic limb, such as a maxilliped or a pereiopod. Syn: *thoracopodite.*

thorax (a) The skeleton of the anterior trunk of a mammal, including ribs, sternum, and usually about 12 distinctive dorsal vertebrae. The thorax is not definable in other tetrapod groups. (b) The central tagma of the body of an arthropod, consisting of several generally movable segments, e.g. the nearly always limb-bearing tagma between cephalon and abdomen of a crustacean; the middle part of an exoskeleton of a trilobite, extending between cephalon and pygidium and consisting of several freely articulated segments; or the middle of the three chief divisions of the body of an insect. See also: *cephalothorax.* Syn: *trunk [paleont].* (c) The second joint of the shell of a nasselline radiolarian.—Pl: *thoraxes* or *thoraces.*

thorbastnaesite A brown mineral: $Th(Ca,Ce)(CO_3)_2F_2 \cdot 3H_2O$. It occurs as an accessory mineral in iron-rich albitites and in selvages of veinlets and stockworks.

thoreaulite A brown mineral: $SnTa_2O_7$.

thorianite A mineral: ThO_2. It is isomorphous with uraninite, often contains rare-earth metals and uranium, and is strongly radioactive.

thorite A brown, black, or sometimes orange-yellow tetragonal mineral: $ThSiO_4$. It is isostructural with thorogummite, strongly radioactive, and usually metamict, and may contain as much as 10% uranium. Thorite resembles zircon and occurs as a minor accessory mineral of granites, syenites, and pegmatites. It is dimorphous with huttonite.

thorium-lead age method Calculation of an age in years for geologic material based on the known radioactive rate of thorium-232 to lead-208. It is part of the more inclusive *uranium-thorium-lead age method* in which the parent-daughter pairs are considered simultaneously.

thorium series The radioactive series beginning with thorium-232.

thorium-230/protactinium-231 deficiency method Calculation of an age in years for fossil coral, shell, or bone 10,000 to 250,000 years old, based on the growth of uranium daughter products from uranium isotopes that enter the carbonate or phosphate material shortly after its formation or burial. The age depends on measurement of the thorium-230/uranium-234 and protactinium-231/uranium-235 activity ratios, which vary with time. See also: *uranium-series age method; ionium-deficiency method.*

thorium-230/protactinium-231 excess method *protactinium-ionium age method.*

thorium-230/thorium-232 age method *ionium-thorium age method.*

thorn A short, sharply pointed triangular or conical surface extension in the skeleton of a spumellarian radiolarian.

Thornthwaite's classification A *climate classification,* formulated by the U.S. agricultural climatologist Warren Thornthwaite, that is based on ratios of precipitation to evaporation. Five humidity provinces are distinguished: perhumid, humid, subhumid, semiarid, and arid. Cf: *Köppen's classification.*

thorogummite A secondary mineral: $Th(SiO_4)_{1-x}(OH)_{4x}$. It is isostructural with thorite and may contain as much as 31.4% uranium. Syn: *mackintoshite; maitlandite.*

thoron *radon-220.*

thorosteenstrupine A dark-brown to nearly black mineral: $(Ca,Th,Mn)_3Si_4O_{11}F \cdot 6H_2O$.

thorotungstite *yttrotungstite.*

thoroughfare (a) A tidal channel or creek providing an entrance to a bay or lagoon behind a barrier or spit. Syn: *thorofare.* (b) A navigable waterway, as a river or strait; esp. one connecting two bodies of water.

thortveitite A gray-green mineral: $(Sc,Y)_2Si_2O_7$. It is a source of scandium. Cf: *thalenite.*

thorutite A black mineral: $(Th,U,Ca)Ti_2(O,OH)_6$. Syn: *smirnovite.*

Thoulet solution *Sonstadt solution.*

thread (a) A thin stream of water. (b) The middle of a stream. (c) The line along the surface of a stream connecting points of maximum current velocity. Cf: *channel line.*

thread-lace scoria A *scoria* in which the vesicle walls have burst and are represented only by an extremely delicate three-dimensional network of glass threads. Syn: *reticulite.*

three-age system In archaeology, the original classification scheme of relative prehistoric time, including *Stone Age, Bronze Age,* and *Iron age.* Since its formulation in the early nineteenth century, it has been expanded by division of the Stone Age into the *Paleolithic, Mesolithic,* and *Neolithic,* and by addition of the *Copper Age* between the Neolithic and Bronze ages (Bray and Trump, 1970, p.231).

three array An electrode array used in profiling, in which one current electrode is placed at infinity while one current electrode and two potential electrodes are in close proximity and are moved across the structure to be investigated. When the separation between the potential electrodes equals the separation between the near current electrode and the closest potential electrode, the three array is reduced to the pole-dipole array. It is used in resistivity and induced polarization surveys and in drill-hole logging.

threefold coordination *triangular coordination.*

three-layer structure A type of *layer structure* having three unit layers to the full repeat unit; e.g. some phlogopites, which have one octahedral and two tetrahedral layers per unit along the c axis. Such micas are usually hexagonal. Cf: *two-layer structure.*

threeling *trilling.*

three-mile limit The seaward limit of the *territorial sea* or *marginal sea* of 3 nautical miles (about 5.6 km), the one-time range of a cannon shot.

three-phase inclusion An inclusion in a gemstone consisting of tiny crystals, gas, and liquid. Cf: *two-phase inclusion.*

three-point method A geometric method of calculating the dip and strike of a structural surface from three points of varying elevation along the surface.

three-point problem (a) The problem of locating the horizontal geographic position of a point of observation from data comprising two observed horizontal angles subtended by three known sides of a triangle (or situated between three points of known position). It is solved analytically by trigonometric calculation in triangulation, mechanically by means of a three-arm protractor, or graphically by trial-and-error change in the orientation of the board in planetable surveying. Cf: *two-point problem.* (b) A name applied to the method of solving the three-point problem in planetable surveying, commonly by taking backsights on three previously located stations.—See also: *resection.*

three-swing cusp A *meander cusp* formed by three successive swings of a meander belt, the scar produced by the third swing meeting the point of the cusp formed by the first two swings (Lobeck, 1939, p. 240-241); it may be seen on the edge of a rock-defended terrace.

threshold [geochem] The entrance, boundary, or beginning of a new domain; the lowest detectable value; the point at which a process or effect commences.

threshold [glac geol] *riegel.*

threshold [marine geol] *sill [marine geol].*

threshold pressure *yield stress.*

threshold velocity The minimum velocity at which wind or water, in a given place and under specified conditions, will begin to move particles of soil, sand, or other material.

throat plane A plane passed through the centers of the spheres in a layer of rhombohedral packing.

thrombolite A cryptalgal structure like a *stromatolite* but with an obscurely clotted, rather than laminated, internal structure. Cf: *rhodolith.*

throughfall Water from precipitation that falls through the plant cover directly onto the ground or that drips onto the ground from branches and leaves. Cf: *interception; stemflow.*

through glacier A double-ended glacier, consisting of two valley glaciers situated in a single depression, from which they flow in opposite directions. A "through-glacier system" is a body of glacier ice consisting of interconnected through glaciers that may lie in two or more drainage systems. Cf: *transection glacier.*

through valley A flat-floored depression or channel eroded across a divide by glacier ice or by meltwater streams; a valley excavated by a through glacier.

throw (a) On a fault, the amount of vertical displacement. Cf: *heave.* See also: *upthrow; downthrow.* (b) The vertical component of the net slip.

throwing clay Clay plastic enough to be shaped on a potter's wheel.

throwout Fragmental material ejected from an impact or explosion crater during formation and redeposited on or outside the crater lip. Cf: *fallout [crater]; fallback.*

thrust (a) An overriding movement of one crustal unit over another, as in thrust faulting. (b) *thrust fault.*

thrust block *thrust sheet.*

thrust fault A fault with a dip of 45° or less over much of its extent, on which the hanging wall appears to have moved upward relative to the footwall. Horizontal compression rather than vertical displacement is its characteristic feature. Cf: *normal fault.* Partial syn: *reverse fault.* Syn: *thrust; overthrust.*

thrust moraine (a) A moraine produced by the overriding and pushing forward, by a regenerated glacier, of dead ice and its deposits (Gravenor & Kupsch, 1959). (b) *push moraine.*

thrust nappe *thrust sheet.*

thrust outlier *klippe.*

thrust plane The *thrust surface* of a thrust fault, when the surface is planar.

thrust plate *thrust sheet.*

thrust pond A small, shallow, roughly circular pond on the floor of a slightly inclined mountain valley, and bordered by a raised rim of thick soil resulting from *ice push* and supporting a very dense growth of coarse alpine grass (Ives, 1941, p.290); e.g. in the high parts of the Rocky Mountains in Colorado.

thrust scarp The scarp along the forward edge of a thrust fault or nappe.

thrust sheet The body of rock above a large-scale thrust fault whose surface is horizontal or very gently dipping. Syn: *thrust block; thrust nappe; thrust plate.*

thrust slice A relatively thin body of rock bounded above and below by thrust faults within a zone of thrusting. Var: *slice.*

thrust surface The surface, usually a plane, along which thrust faulting occurs. See also: *thrust plane.*

thucholite A brittle, jet-black mixture or complex of organic matter (hydrocarbons) and uraninite, with some sulfides, occurring esp. in gold conglomerates (as in the Witwatersrand of South Africa) or in pegmatites (as in Canada). It may contain up to 48% thorium in the ash.

thufa An Icelandic term for *earth hummock.* Pl: *thufur.*

Thulean province A region of Tertiary volcanic activity (basalt flood) including Iceland and most of Britain and Greenland. Etymol: Greek *Thule,* "the northernmost part of the habitable world."

Thule-Baffin moraine *shear moraine.*

thulite A pink, rose-red, or purplish-red variety of zoisite containing manganese and used as an ornamental stone.

Thumper Trade name for a device for generating seismic waves by dropping a heavy weight (often a 3-megagram weight dropped from a 3-meter elevation).

thunder egg A popular term for a small, geodelike body of chalcedony, opal, or agate that has weathered out of the welded tuffs of central Oregon.

Thuringian European stage: Upper Permian (above Saxonian, below Triassic).

thuringite An olive-green or pistachio-green mineral of the chlorite group, a ferrian variety of chamosite. It is isomorphous with pennantite.

thurm A term used in Nova Scotia for a ragged and rocky headland swept by the sea. Syn: *thurm cap.*

Thyssen gravimeter An early gravity meter of the unstable equilibrium type.

tibia (a) The more medial of the two elements of the tetrapod leg. (b) The fifth segment of a typical leg or pedipalpus of an arachnid, following upon the patella which may be completely fused with it (TIP, 1955, pt.P, p.63). (c) A joint of the distal part of a prosomal appendage of a merostome.—Pl: *tibiae.*

tickle (a) Any narrow passage connecting larger bodies of water. (b) A term used in the Gulf of St. Lawrence region for an inlet of the sea into a lagoon.

tidal basin A dock or basin in a tidal region, in which water is maintained at a desired level by means of a gate; it is filled at high tide by water that is retained and then released at low tide.

tidal bedding Sedimentary bedding caused by tides in a tidal channel, a tidal flat, or a marsh; esp. bedding produced where currents of high tides are stronger than those of low tides flowing in the opposite direction, as where a layer of coarse sediments deposited by a high tide is not destroyed by the low tide.

tidal bench mark A durable bench mark fixed rigidly in stable ground and set to reference a tide staff at a tide station.

tidal bore *bore [tide].*

tidal bulge The tidal effect on the side of the Earth nearest to the Moon, where lunar attraction is greatest. Cf: *antipodal bulge.*

tidal channel (a) A major channel followed by the tidal currents, extending from offshore into a tidal marsh or a tidal flat. (b) *tidal inlet.*

tidal compartment The part of a stream that "intervenes between the area of unimpeded tidal action and that in which there is a complete cessation or absence of tidal action" (Carey & Oliver, 1918, p. 8).

tidal constant A parameter of a tide that, for a given locality, usually remains constant. *Tide amplitude* and *tidal epoch* are harmonic tidal constants; *tide range* is a disharmonic constant.

tidal correction A correction to gravity measurements, applied to compensate for the effects of the Sun and Moon (i.e., for Earth tides). It is sometimes included in the *drift correction* and may be determined by a series of observations at a fixed base station.

tidal creek A relatively small tidal inlet or estuary. Syn: *creek.*

tidal current The periodic horizontal movement of ocean water associated with the vertical rise and fall of the tides and resulting from the gravitational attraction of the Moon and Sun upon the Earth. In the open ocean, its direction rotates 360° on a diurnal or semidiurnal basis; in coastal areas, however, topography influences its direction. Incorrect syn: *tide.* British syn: *tidal stream.* Syn: *periodic current.*

tidal cycle *tide cycle.*

tidal datum A *chart datum* based on a phase of the tide.

tidal day The interval between two consecutive high waters of the tide at a place, averaging 24 hours and 51 minutes. Cf: *lunar day.*
tidal delta A delta formed at the mouth of a tidal inlet on both the seaward and lagoon sides of a barrier island or baymouth bar by changing tidal currents that sweep sand in and out of the inlet.
tidal-delta marsh A *salt marsh* found around distributary patterns of tidal rivers inside a tidal inlet.
tidal divide A divide between two adjacent tidal channels.
tidal efficiency The ratio of the fluctuation of water level in a well to the tidal fluctuation causing it, expressed in the same units such as feet of water. Symbol: C. Cf: *barometric efficiency.*
tidal epoch A *tidal constant* representing the phase difference or time elapsing between the Moon's transit over a fixed meridian and the ensuing high tide. Syn: *epoch [tides]; phase lag.*
tidal flat An extensive, nearly horizontal, marshy or barren tract of land that is alternately covered and uncovered by the tide, and consisting of unconsolidated sediment (mostly mud and sand). It may form the top surface of a deltaic deposit. See also: *tidal marsh; mud flat.* Syn: *tide flat.*
tidal flushing The removal of sediment, as from an estuary, by the ebb and flow of tidal currents that are stronger, more constant, or more prevalent than the incoming river flow.
tidal friction The frictional effect of the tides, resulting in dissipation of energy, esp. along the bottoms of shallow seas where the tidal epoch is lengthened and the Earth's rotational velocity is retarded, thereby slowly increasing the length of the day and theoretically causing the Moon to recede and accelerate over the course of geologic time.
tidal glacier *tidewater glacier.*
tidal inlet Any *inlet* through which water flows alternately landward with the rising tide and seaward with the falling tide; specif. a natural inlet maintained by tidal currents. Syn: *tidal outlet; tidal channel.*
tidalite A sediment deposited by tidal tractive currents, by an alternation of tidal tractive currents and tidal suspension deposition, or by tidal slack-water suspension sedimentation. Tidalites occur both in the intertidal zone and in shallow, subtidal, tide-dominated environments. See also: *intertidalite.*
tidal marsh A marsh bordering a coast (as in a shallow lagoon or sheltered bay), formed of mud and of the resistant mat of roots of salt-tolerant plants, and regularly inundated during high tides; a marshy *tidal flat.* Cf: *salt marsh.*
tidal outlet *tidal inlet.*
tidal pool *tide pool.*
tidal prism The volume of water that flows in or out of a harbor or estuary with the movement of the tide, and excluding any freshwater flow; it is computed as the product of the tide range and the area of the basin at midtide level, or as the difference in volume at mean high water and at mean low water.
tidal range *tide range.*
tidal resonance theory George Darwin's postulation of the cause of continental drift; the Moon separated from the Earth, leaving the Pacific Ocean as a scar, and the consequent shortage of continental crust caused a global tension, resulting in drifting (Fairbridge, 1966, p.83).
tidal river A river whose lower part for a considerable distance is influenced by the tide of the body of water into which it flows; the movement of water in and out of an estuary or other inlet as a result of the alternating rise and fall of the tide. Syn: *tidal stream.*
tidal scour The downward and sideward erosion of the sea floor by powerful tidal currents resulting in the removal of inshore sediments and the formation of deep channels and holes. Syn: *scour.*
tidal stand *stand of tide.*
tidal stream (a) *tidal river.* (b) A British syn. of *tidal current.*
tidal swamp A swamp partly covered during high tide by the backing-up of a river (Stephenson & Veatch, 1915, p. 37). Cf: *upland swamp.*
tidal water *tidewater.*
tidal wave An erroneous syn. of both *storm surge* and *tsunami.*
tidal wedge A tidal channel that is narrower and shallower at the downstream end.
tide (a) The rhythmic, alternate rise and fall of the surface (or water level) of the ocean, and of bodies of water connected with the ocean such as estuaries and gulfs, occurring twice a day over most of the Earth, and resulting from the gravitational attraction of the Moon (and, in lesser degree, of the Sun) acting unequally on different parts of the rotating Earth. (b) An incorrect syn. of *tidal current.* (c) *earth tide.* (d) *atmospheric tide.*
tide amplitude A *tidal constant* representing one-half of the *tide range;* the elevation of tidal high water above mean sea level.
tide crack A *crack [ice],* usually parallel to the shore, at the junction line between an immovable icefoot or ice wall and fast ice, and caused by the rise and fall of the tide which moves the fast ice upward and downward.
tide curve A graphic record of the height (rise and fall) of the tide, with time as abscissa and tide height as ordinate. Syn: *marigram.*
tide cycle A period that includes a complete set of tide conditions or characteristics, such as a tidal day or a lunar month. Syn: *tidal cycle.*
tide flat A syn. of *tidal flat.* Also spelled: *tideflat.*
tide gage A device for measuring the height (rise and fall) of the tide; esp. an instrument automatically making a continuous graphic record of tide height versus time. See also: *tide staff; marigraph.*
tideland (a) The coastal area that is alternately covered and uncovered by the ordinary daily tides; land that is covered by *tidewater* during a flood tide. (b) Land that underlies the ocean beyond the low-water mark, but within the territorial waters of a nation. The term is often used in the plural.
tide pole *tide staff.*
tide pool A pool of water, as in rock basin, left on a beach or reef by an ebbing tide. Syn: *tidal pool.*
tide race A type of *race [current]* caused by a greater tide range at one end of the channel than at the other end.
tide range A *tidal constant* representing the difference in height between consecutive high water and low water at a given place; it is twice the *tide amplitude.* Cf: *mean range.* Syn: *tidal range.*
tide rip A *rip* produced by the meeting of opposing tides (as where tidal currents are converging and sinking), or by a tidal current suddenly entering shallow water.
tide staff A *tide gage,* either fixed or portable, consisting of a long, vertical, graduated rod or board, from which the height of the tide can be read directly at any time. Syn: *tide pole.*
tide station A place where tide observations are obtained. Cf: *reference station.*
tidewater (a) Water that overflows the land during a flood tide; water that covers the *tideland.* Also, stream water that is affected by the rise and fall of the tide. Syn: *tidal water.* (b) A broad term for the seacoast, or low-lying coastal land traversed by tidewater streams.
tidewater glacier A glacier that terminates in the sea, where it usually ends in an ice cliff from which icebergs are discharged. Syn: *tidal glacier.*
tie (a) A survey connection from a point of known position to a point whose position is desired. (b) A survey connection to close a survey on a previously determined point.
tie bar *tombolo.*
tied island An island connected with the mainland or with another island by a tombolo. Syn: *tombolo island.*
tie-in In geophysics, the relating of a new station or value to ones already established.
tie line [chem] A line at constant temperature that connects any two phases that are in equilibrium at the temperature of the tie line. See also: *conjugation line.* Syn: *conode.*
tie line [surv] A line measured on the ground to connect some object to a survey; e.g. a line joining opposite corners of a four-sided figure, thereby enabling its area to be checked by triangulation.
tiemannite A dark-gray or nearly black mineral: HgSe.
tienshanite A mineral: $Na_2BaMnTiB_2Si_6O_{20}$.
tie point (a) A point to which a tie is made; esp. a point of closure of a survey either on itself or on another survey. (b) An image point identified on oblique aerial photographs in the overlap area between two or more adjacent strips of photography. They tie individual sets of photographs into a single flight unit and adjacent flights into a common network.
tier Any series of contiguous *townships* (of the U.S. Public Land Survey system) aligned east and west and numbered consecutively north and south from a base line. Also, any series of contiguous sections similarly situated within a township. Cf: *range.*
tierra blanca A Spanish term for "white ground" or "white earth", and applied to white calcareous deposits such as tufa, caliche, and chalky limestone.
tiff A *sparry* mineral. The term is applied to calcite in SW Missouri and to barite in SE Missouri.
Tiffanian North American land-mammal age: Middle Paleocene

(above Torrejonian, below Clarkforkian).

tiger's-eye A chatoyant, translucent to semitranslucent, yellowish-brown or brownish-yellow gem and ornamental variety of quartz, pseudomorphous after crocidolite, whose fibers (penetrating the quartz) are changed to iron oxide (limonite); silicified crocidolite stained yellow or brown by iron oxide. Upon heating, the limonite turns to hematite and produces a red to brownish-red sheen. Cf: *hawk's-eye; cat's-eye [mineral].* Also spelled: *tiger-eye; tigereye.* Syn: *tigerite.*

tight fold A fold with an inter-limb angle between 0° and 30° (Fleuty, 1964, p. 470).

tight hole A drilling or completed well about which information is kept secret by the operator.

tight sand A sand whose interstices are filled with fine grains or with matrix material, thus effectively destroying porosity and permeability. The term is used in petroleum geology.

Tiglian North European climatostratigraphic and floral stage: Upper Pliocene (above Pretiglian, below Eburonian).

tikhonenkovite A monoclinic mineral: $SrAlF_4(OH) \cdot H_2O$.

tilaite A mafic *gabbro* having crystalline-granular texture and containing abundant green clinopyroxene and olivine, and a little orthopyroxene with about 20 percent calcic plagioclase. The name, given by Duparc and Pamfil in 1910, is for Tilai Kamen, northern Urals, U.S.S.R. Not recommended usage.

tilasite A violet-gray monoclinic mineral: $CaMg(AsO_4)F$. It is isomorphous with isokite.

tile ore A red or brownish earthy variety of cuprite often mixed with red iron oxide.

tilestone An English term for a flagstone (flaggy sandstone) used for roofing.

till Dominantly unsorted and unstratified drift, generally unconsolidated, deposited directly by and underneath a glacier without subsequent reworking by meltwater, and consisting of a heterogeneous mixture of clay, silt, sand, gravel, and boulders ranging widely in size and shape. Cf: *stratified drift.* See also: *moraine.* Syn: *boulder clay; glacial till; ice-laid drift.*

till ball An *armored mud ball* whose core is made of till, occurring in certain Pleistocene glacial deposits.

till billow An undulating or swelling accumulation of glacial drift irregularly disposed with regard to the direction of movement of the ice (Chamberlin, 1894b, p. 523).

till crevasse filling A ridge of unstratified morainal material deposited in a crack of a wasting glacier and left standing after the ice melted; term introduced by Gravenor (1956, p. 10). Cf: *crevasse filling.*

tilleyite A white mineral: $Ca_5(Si_2O_7)(CO_3)_2$.

tillite A consolidated or indurated sedimentary rock formed by lithification of glacial till, esp. pre-Pleistocene till (such as the Late Carboniferous tillites in South Africa and India).

tilloid A term introduced by Blackwelder (1931a, p.903) for a till-like deposit of "doubtful origin", but redefined by Pettijohn (1957, p.265) as a nonglacial *conglomeratic mudstone* (also known as *geröllton*) varying from "a chaotic unassorted assemblage of coarse materials set in a mudstone matrix to a mudstone with sparsely distributed cobbles", such as a sedimentary deposit resulting from extensive slides or flows of mud on the margin of a geosyncline. Harland et al. (1966, p.251) urge the use of "tilloid" as a nongenetic term for a rock resembling tillite in appearance but whose origin is in doubt or unknown. Cf: *pebbly mudstone; pseudotillite.*

till plain An extensive area, with a flat to undulating surface, underlain by till with subordinate end moraines; such plains occupy parts of Indiana, Illinois, and Iowa. See: *ground moraine.*

till-shadow hill A glacial hill, without a core of resistant rock, that has a gentle south slope on which till thickens but does not form a well-developed tail (Coates, 1966); examples occur in central New York State.

till sheet A sheet, layer, or bed of till, without reference to its topographic expression. It may form ground moraine.

tillstone A boulder or other stone in a deposit of glacial till. Also spelled: *till stone.*

till tumulus A low, stony mound representing the immature nucleus of a drumlin (Chamberlin, 1894b, p. 522-523).

till wall A ridge consisting of morainal material squeezed upward into a crevasse by the pressure of overlying ice (Gravenor & Kupsch, 1959, p. 58).

tilly Composed or having the nature of glacial till; e.g. *tilly* land.

tilt (a) The angle at the perspective center between the plumb line and the perpendicular from the interior perspective center to the plane of the photograph. (b) The lack of parallelism (or the angle) between the plane of the photograph from a downward-pointing aerial camera and the horizontal plane (normal to the plumb line) of the ground.

tilt block A *fault block* that has become tilted, perhaps by rotation on a hinge line. Syn: *tilted fault block.*

tilted fault block *tilt block.*

tilted iceberg A tabular iceberg that has become unbalanced due to melting or calving, so that its flat top is inclined.

tilted photograph An aerial photograph taken with a camera whose plane of film is not parallel (horizontal) with the plane of the ground at the time of exposure.

tilth The physical condition of a soil relative to its fitness for the growth of a specified plant or sequence of plants.

tiltmeter An instrument that measures slight changes in the tilt of the Earth's surface, usually in relation to a liquid-level surface or to the rest position of a pendulum. It is used in volcanology and in earthquake seismology.

timberline The elevation (as on a mountain) or the latitudinal limits (on a regional basis) at which tree growth stops. Syn: *tree line.*

time (a) Measured or measurable duration; a nonmaterial dimension of the universe, representing a period or interval during which an action, process or condition exists or continues. See also: *geologic time.* (b) A reference point from which duration is measured; e.g. the instant at which a seismic event occurs relative to a chosen reference time such as a shot instant. (c) A reckoning of time, or a system of reckoning duration. (d) An informal term proposed by the International Subcommission on Stratigraphic Terminology (1961, p. 13 & 25) for the geologic-time unit next in order of magnitude below *age [geochron] (a),* during which the rocks of a *substage* (or of any time-stratigraphic unit of lesser rank than a *stage*) were formed. It is a syn. of *subage; episode (b); phase [geochron] (a).* (e) Any division of geologic chronology, such as "Paleozoic time" or "Miocene time".

time break *shot break.*

time-correlation The demonstration or determination of age equivalence or mutual time relations (such as contemporaneity of origin) of stratigraphic units in two or more separated areas. It is the oldest kind of *stratigraphic correlation,* and is established mainly by fossil content.

time-depth chart A graphical expression of the relation between velocity and arrival time of vertically travelling seismic reflections. It permits the time increments to be converted to the corresponding depths. Syn: *time-depth curve.*

time-depth curve *time-depth chart.*

time-distance curve A syn. of *traveltime curve.* Abbrev: T. D. curve.

time domain (a) Measurements as a function of time, or operations in which time is the variable, in contrast to the *frequency domain.* (b) Transmission of a single or repetitive pulse of electromagnetic energy, frequently of square wave form, and reception of electromagnetic energy, as a function of time, in the time interval after the transmitted wave form has been turned off. It is used with induced electrical polarization and electromagnetic methods.

time-domain electromagnetic method An exploration method based on measurements of the variation of electromagnetic quantities with time.

time lag A delay in the arrival time of seismic energy from the time expected. Time lags may be produced by an abnormal low-velocity layer, phase shifts in filtering, or other factors.

time lead The arrival of seismic energy earlier than expected, indicating that part of the travel path involved high velocity. It is an indication of a salt dome in *fan shooting.* Syn: *lead [seis].*

time line (a) A line indicating equal age in a geologic cross section or correlation diagram; e.g. a line separating two time-stratigraphic units. (b) A rock unit represented by a time line; e.g. an intraformational conglomerate formed by subaqueous slump and turbidity flows of but a few hours' duration.

time of concentration *concentration time.*

time-parallel (a) Said of a surface that is parallel to or that closely approximates a synchronous surface and that involves a geologically insignificant amount of time, such as the surface of a rapidly transgressed unconformity. (b) Said of a stratum bounded by time-parallel surfaces.

time plane A stratigraphic horizon identifying an *instant* in geologic time. Syn: *synchrone.*

time-rock span *stratigraphic range.*
time-rock unit "An undesirable term for chronostratigraphic unit" (ISSC, Rept. 6, 1971, p. 6, footnote).
time scale *geologic time scale.*
time section A graphical representation of reflection seismic records shot along a line, in which the vertical scale is two-way seismic-travel time and the horizontal scale is surface distance.
time series A series of statistical data collected at regular intervals of time; a frequency distribution in which the independent variable is time.
time-stability hypothesis *stability-time hypothesis.*
time standard Any category of physical or biologic phenomena or processes by which segments of time can be measured or subdivided; e.g. radioactive decay of elements, orderly evolution of forms of life, rotation of the Earth on its axis, revolution of the Earth around the Sun, and human artifacts. All such time standards are "partial" with respect to time in the abstract (Jeletzky, 1956, p. 681).
time-stratigraphic facies A stratigraphic facies that is recognized on the basis of the amounts of geologic time during which sedimentary deposition and nondeposition occurred; a facies that is a laterally segregated, statistical variant of a stratigraphic interval and whose boundaries (vertical surfaces or arbitrary cutoffs) extend from the bottom to the top of the interval (Wheeler, 1958, p.1060).
time-stratigraphic unit *chronostratigraphic unit.*
time-stratigraphy *chronostratigraphy.*
time tie The identification of seismic events on different records by their arrival times, when they possess common raypaths.
time-transgressive *diachronous.*
time-transitional Said of a rock unit including within itself an important geologic time plane and thus consisting of strata that belong to two adjacent chronostratigraphic units (such as systems).
time unit *geologic-time unit.*
time value The interval of geologic time represented by or involved in producing a stratigraphic unit, an unconformity, the range of a fossil, or any geologic feature or event. See also: *hiatus (b).*
timing line One of a series of marks or lines placed on seismic records at precisely determined intervals of time (usually 0.01 or 0.005 sec) for the purpose of measuring the arrival time of recorded events.
Timiskamian A division of the Archeozoic of the Canadian Shield. Also spelled: *Timiskaming.*
Timiskaming Var. of *Timiskamian.*
tin (a) A bluish-white mineral, the native metallic element Sn. (b) A term used loosely to designate cassiterite and concentrates containing cassiterite with minor amounts of other minerals.
tinaja (a) A term used in SW U.S. for a *water pocket* developed below a waterfall, esp. when partly filled with water. (b) A term used loosely in New Mexico for a temporary pool, and for a spring too feeble to form a stream.—Etymol: Spanish, "large earthen jar". Pron: te-*na*-ha.
tinajita *solution pan.* Etymol: Spanish, "little water jar".
tinaksite A mineral: $K_2NaCa_2TiSi_7O_{19}(OH)$.
tincal An old name for crude *borax* formerly obtained from Tibetan-lake shores and deposits and once the chief source of boric compounds.
tincalconite A colorless to dull-white rhombohedral mineral: $Na_2B_4O_7 \cdot 5H_2O$. Syn: *mohavite; octahedral borax.*
tind A Norwegian term for a glacial *horn* that is detached from the main mountain range by the lateral recession of cirques cutting through an upland spur between two glacial troughs (Thornbury, 1954, p. 373). Syn: *monument.*
tinder ore An impure variety of jamesonite. Syn: *pilite.*
tinguaite A textural variety of *phonolite,* typically found in dikes, and characterized by conspicuous acicular crystals of acmite arranged in radial or criss-cross patterns in the groundmass. The phenocrysts are of equigranular alkali feldspar and nepheline. The name, given by Rosenbusch in 1887, is derived from the Tingua Mountains (Serra de Tingua) near Rio de Janeiro, Brazil. Adj: *tinguaitic.* Cf: *muniongite.* Not recommended usage.
tin ore *cassiterite.*
tin pyrites *stannite.*
tinsel A flagellum having a central axis from which extend many fine short hairs (mastigonemes) in one or two rows along its length.
tinstone *cassiterite.*
tint *hypsometric tint.*
tinticite A creamy-white mineral: $Fe_6(PO_4)_4(OH)_6 \cdot 7H_2O$.
tintinaite A mineral: $Pb_5(Sb,Bi)_8S_{17}$.
tintinnid A ciliate protozoan belonging to the family Tintinnidae and characterized by a lorica that is almost always inflated in the oral region. Range, Jurassic to present.
tin-white cobalt *smaltite.*
tinzenite A yellow monoclinic mineral: $(Ca,Mn,Fe)_3Al_2(BO_3)(Si_4O_{12})(OH)$. It is a variety of axinite rich in manganese.
Tioughniogan North American provincial stage: Middle Devonian (above Cazenovian, below Taghanican).
tip [mass move] The point on the *toe [mass move]* farthest from the top of a slide (Varnes, 1958, pl. 1).
tip [paleont] The inversely conical initial part of the cup in archaeocyathids (TIP, 1972, pt. E, p. 42). Nonrecommended syn: *spitz.*
tiphic Pertaining to a pond or ponds. Cf: *ombrotiphic.*
tiphon *diapirism.*
tipper Correlation matrix between the vertical and horizontal magnetic fields measured in a survey by the *magnetotelluric method.*
tirodite A monoclinic mineral of the amphibole group: $(Mg,Mn)_7Si_8O_{22}(OH)_2$.
tissue An aggregate of similar cells into a structural unit that performs a particular function.
titanaugite A variety of augite rich in titanium and occurring in basaltic rocks: $Ca(Mg,Fe,Ti)(Si,Al)_2O_6$.
titanic iron ore A syn. of *ilmenite.* Also called: *titaniferous iron ore.*
titanite *sphene.*
titanochromite A lunar mineral which is a solid solution of the components $FeCr_2O_4$, $TiFe_2O_4$, $FeAl_2O_4$, and MgO.
titanomaghemite A general term applied to an abnormal titanium-bearing magnetite with varying cation vacancies in an oxygen framework of spinel structure.
titanomagnetite (a) A titaniferous variety of magnetite: $Fe(Fe^{+2},Fe^{+3},Ti)_2O_4$. It is strictly a homogeneous cubic solid solution of ilmenite in magnetite. (b) A term loosely used for mixtures of magnetite, ilmenite, and ulvöspinel.
titanorhabdophane *tundrite.*
Tithonian European (Great Britain) stage: Lower Cretaceous (above Portlandian, below Berriasian). Southern European equivalent of Portlandian.
title box *cartouche.*
tjaele A syn. of *frozen ground.* The term has been used erroneously as a syn. of *permafrost* (Bryan, 1951). Etymol: Swedish *tjäle,* "frozen ground". Also spelled: *tjäle; taele.*
tjosite A dark-colored porphyritic nepheline-bearing *phonolite,* tending toward the composition of *jacupirangite* or *lamprophyre,* forming dikes in *larvikite.* Its name, given by Brögger in 1906, is derived from Tjose, Larvik district, Norway. Not recommended usage.
tlalocite A mineral: $(Cu,Zn)_{16}(TeO_3)(TeO_4)_2Cl(OH)_{25} \cdot 27H_2O$.
TLC *thin-layer chromatography.*
toadback marl A term used in Lancashire, England, for unlaminated marl with lumpy fracture. Cf: *beechleaf marl.*
toad's-eye tin A reddish or brownish variety of cassiterite occurring in botryoidal or reniform shapes that display an internal concentric and fibrous structure. Syn: *toad's-eye.*
toadstone A fossilized object, such as a fish tooth or palatal bone, that was thought to have formed within a toad and was frequently worn as a charm or an antidote to poison.
toadstool rock *mushroom rock.*
Toarcian European stage: Lower Jurassic (above Pliensbachian, below Bajocian).
tobacco jack A miner's term for *wolframite.*
tobacco rock A term used in SW U.S. for a favorable host rock for uranium, characterized by light yellow or gray color and by brown limonite stains.
tobermorite A mineral: $Ca_5Si_6O_{16}(OH)_2 \cdot 4H_2O$.
tochilinite A triclinic mineral: $6Fe_{0.9}S \cdot 5(Mg,Fe)(OH)_2$.
tocornalite A mineral consisting of silver mercury iodide.
toddite A mixture of columbite and samarskite.
todorokite A mineral: $(Mn,Ca,Mg)Mn_3{}^{+4}O_7 \cdot H_2O$. It may contain some barium and zinc. Syn: *delatorreite.*
toe [coast] *step.*
toe [drill] The bottom of a drill hole (esp. one used for blasting), as distinguished from its open end.

toe [fault] The leading edge of a thrust sheet.
toe [mass move] The lower, usually curved, margin of the disturbed material of a landslide pushed over onto the undisturbed slope; it is most distant from the place of origin. Cf: *foot; tip [mass move].*
toe [slopes] The lowest part of a slope or cliff; the downslope end of an alluvial fan.
toe [volc] *lava toe.*
toellite *tollite.*
toenail (a) A curved joint intersecting a sheet structure, usually along the strike or sometimes differing from it by 45° or more. (b) Obsolete syn. of *stylolite.*
toernebohmite *törnebohmite.*
toeset The forward part of a tangential foreset bed.
toe-tap flood plain The outer end of a meander lobe, built by a stream as it meanders down-valley.
toft A British term for an isolated hill, knoll, or other eminence in a flat region, esp. one suitable for a homesite.
toise An old French unit of length used in early geodetic surveys and equal to 6 French feet, 6.396 U.S. feet, or 1.949 meters.
tokeite A dark-colored picritic *basalt* similar in composition to *schönfelsite,* in which the plagioclase is labradorite. Cf: *oceanite.* The name, given by Duparc and Molly in 1928, is for Arête de Toké, Gouder Valley, Ethiopia. Not recommended usage.
Tolkowsky theoretical brilliant cut A style of diamond cutting having those proportions and facet angles calculated by Marcel Tolkowsky in 1919 to produce the maximum brilliancy with a high degree of fire in a round brilliant cut diamond. The "ideal" proportions are a table diameter of 53% of the girdle diameter; thickness above the girdle, 16.2%; thickness below the girdle 43.1%, girdle thickness 0.7-1.7%. Syn: *American cut; ideal cut.* Cf: *brilliant cut.*
tollite A hypabyssal rock containing phenocrysts of hornblende, andesine, garnet, and some biotite, in a groundmass of white mica, quartz, and alkali feldspar. No dark-colored minerals occur in the granophyric groundmass. The name was given by Pichler in 1875, from Töll in the Tyrolean Alps. Obsolete. Syn: *toellite.*
tolt A term used in Newfoundland for an isolated peak rising abruptly from a plain.
tombarthite A mineral: $Y_4(Si,H_4)_4O_{12-n}(OH)_{4+2n}$.
tombolo A sand or gravel bar or barrier that connects an island with the mainland or with another island. Etymol: Italian, "sand dune"; from Latin *tumulus,* "mound". Pl: *tombolos.* Syn: *connecting bar; tie bar; tying bar.*
tombolo cluster *complex tombolo.*
tombolo island *tied island.*
tombolo series *complex tombolo.*
tomite *boghead coal.*
tonalite In the *IUGS classification,* a plutonic rock with Q between 20 and 60, and P/(A+P) greater than 90. The name, given by Rath in 1864, is derived from Tonale Pass, northern Italy. See also: *adamellite.*
Tonawandan Stage in New York State: middle Middle Silurian.
Tongrian European stage: Lower Oligocene (above Ludian of Eocene, below Rupelian). Syn: *Sannoisian.*
tongue [coast] (a) A *point,* or long low strip of land projecting from the mainland into the sea or other body of water. (b) *inlet.*
tongue [glaciol] A long narrow extension of the lower part of a glacier, either on land or afloat. Cf: *glacier tongue afloat; glacial lobe.* Syn: *glacier tongue.* Nonpreferred syn: *ice tongue.*
tongue [ice] A projection of the *ice edge* up to several kilometers in length, caused by wind and current.
tongue [intrus rocks] A branch or offshoot of a larger intrusive body. See also: *epiphysis.* Syn: *apophysis.*
tongue [oceanog] An extension of one type of water into water of differing salinity or temperature, e.g. salt water into the mouth of a river.
tongue [stratig] n. A minor lithostratigraphic unit of limited geographic extent, being a subdivision of a formation or *member,* and disappearing laterally (usually by facies change) in one direction; "a member that extends outward beyond the main body of a formation" (ACSN, 1961, art. 7). Cf: *lentil.*—v. To thin laterally to disappearance.
tongue [streams] *meander lobe.*
tongue [volc] (a) A lava flow that is an offshoot from a larger flow; it may be as much as several kilometers in length. (b) A syn. of *coulee [volc].*
tonnage-volume factor In economic geology, the number of cubic feet in a ton of ore.
Tonolowayan North American provincial stage: Upper Silurian (above Salinan, below Keyseran).
tonsbergite An altered red igneous rock that is sometimes porphyritic and resembles *larvikite,* the feldspar being represented by orthoclase and andesine. Brögger in 1898 derived the name from Tönsberg, Norway. Obsolete.
tonstein A compact argillaceous rock containing the clay mineral kaolinite in a variety of forms together with occasional detrital and carbonaceous material, commonly occurring as a thin band in a Carboniferous coal seam (or locally in the roof of a seam), and often used as an aid in correlating European strata of Westphalian age. Etymol: German *Tonstein,* "claystone".
tool mark A *current mark* produced by the impact against a muddy bottom of a solid object swept along by the current, and generally preserved as a cast on the underside of the overlying bed. The mark may be produced by an object in continuous contact with the bottom (e.g. a groove or a striation), in intermittent contact with the bottom (e.g. a skip mark or a prod mark), or rolling along the bottom (e.g. a roll mark). The engraving "tools" include shell fragments, sand grains, pebbles, fish bones, seaweed, and wood chips. Originally defined by Dzulynski & Sanders (1962, p. 72). Syn: *tool marking.*
toolpusher The general supervisor of operations on a *drilling rig.*
tooth (a) In gnathostomes, any very hard element of the same mineral composition as bone but of divergent histologic structure and derived from ectodermal as well as mesodermal embryonic tissue, borne primarily on one of the bones of the jaws and specialized for seizing and/or processing food. (b) Any of various horny, chitinous, or calcareous projections of an invertebrate that function like or resemble the vertebrate tooth, as on the radula of a gastropod. (c) A toothlike process on the margin of a bivalve shell; specif. *hinge tooth.* (d) A calcareous rod located in the pyramid of Aristotle's lantern of an echinoid. Its upper end is uncalcified. (e) A projection in the aperture of a foraminiferal test. It may be simple or complex, and single or multiple.
toothpaste lava Viscous lava that is extruded as a *squeeze-up.*
tooth plate An internal, apertural modification of a foraminiferal test, commonly consisting of a contorted plate that extends from the aperture through the chamber to the previous septal foramen. One side may be attached to the chamber wall or to the proximal border of the foramen, the opposite side being free and folded (TIP, 1964, pt.C, p.64).
top [gem] *crown [gem].*
top [mass move] The highest point of contact between the disturbed material of a landslide and the scarp face along which it moved. Cf: *tip [mass move].*
top [ore dep] A quarrymen's syn. of *overburden.*
top [stratig] A term used in petroleum geology for the uppermost surface of a formation where it is encountered during drilling, usually characterized by the first appearance of a distinctive feature (such as a marked change in lithology or the occurrence of a guide fossil). It is often determined by a distinctive configuration on an electric log, and it is widely used in correlation and structure-contour mapping.
topaz (a) A white or lightly colored orthorhombic mineral: $Al_2SiO_4(F,OH)_2$. It occurs as a minor constituent in highly siliceous igneous rocks and tin-bearing veins as translucent or transparent prismatic crystals and masses, and as rounded waterworn pebbles. Topaz has a hardness of 8 on the Mohs scale. (b) A transparent topaz used as a gemstone. (c) A yellow quartz that resembles topaz in appearance, such as smoky quartz turned yellow by heating; specif. *false topaz* and *Scotch topaz.* See also: *Spanish topaz.* (d) A term used for a green-yellow to orange-yellow mineral resembling topaz in appearance, such as "oriental topaz" (a yellow corundum).
topazfels *topazite.*
topazite A hypabyssal rock composed almost entirely of quartz and topaz (Johannsen, 1920). Syn: *topazfels; topazoseme; topazogene.*
topazogene *topazite.*
topazolite A green-yellow to yellow-brown variety of andradite garnet, having the color and transparency of topaz.
topazoseme *topazite.*
topaz quartz Topaz-colored quartz; specif. *citrine.* Cf: *quartz topaz.*
top conglomerate A conglomerate formed of gravels lying at the top of a stratum, and not separated from it by an erosional surface (Twenhofel, 1939, p.203-204).

top-discordance A term used in seismic stratigraphy to refer to a lack of parallelism between a sequence of strata and its upper boundary, owing to either *truncation [stratig]* or *toplap* (Mitchum, 1977, p. 206). Cf: *base-discordance.*
tophus A syn. of *tufa.* Etymol: Latin. Pl: *tophi.*
toplap Termination of strata against an overlying surface, mainly as a result of nondeposition (sedimentary bypassing) with perhaps minor erosion. Each unit of strata laps out in a landward direction at the top of the unit, but the successive terminations lie progressively seaward (Mitchum, 1977, p. 211).
topocentric horizon *apparent horizon.*
topocline A *cline* related to a geographic zone and usually unrelated to any ecologic condition.
topogenous Said of peat deposits whose moisture content is dependent on surface water. Cf: *ombrogenous; soligenous.*
topographic (a) Pertaining to *topography.* (b) Surveying or representing the topography of a region; e.g. a "topographic survey" or a "topographic map".—Syn: *topographical.*
topographic adjustment The condition existing where the gradient of a tributary is harmonious with that of the main stream. Cf: *structural adjustment.*
topographic adolescence *adolescence.*
topographic-bathymetric map A map with both the relief of the land and the relief of the offshore areas shown by contours. Datum for land contours is mean sea level and for bathymetric contours is mean low water on the east coast and mean lower low water on the west coast of the U.S. Contour intervals on land and on sea bottom may not be the same.
topographic contour *contour.*
topographic correction [cart] Correction of errors in a topographic map.
topographic correction [geophys] *terrain correction.*
topographic deflection of the vertical An expression used to indicate that the deflection of the vertical has been computed from the topography. This method may be used when it is not possible to compare astronomic and geodetic positions directly.
topographic depression *closed depression.*
topographic desert A desert of low rainfall because of its location in the middle of a continent, far from the ocean, or on the lee side of high mountains, cut off from prevailing winds.
topographic divide A drainage *divide.*
topographic expression The effect achieved by shaping and spacing contour lines on a map so that topographic features can be interpreted with the greatest ease and fidelity.
topographic feature A prominent or conspicuous *topographic form* or noticeable part thereof (Mitchell, 1948, p. 80). Cf: *physiographic feature.*
topographic form A *landform* considered without regard to its origin, cause, or history (Mitchell, 1948, p. 80). Cf: *physiographic form.*
topographic grain The alignment and direction of the topographic-relief features of a region.
topographic infancy *infancy.*
topographic license The freedom to adjust, add, or omit contour lines, within allowable limits, in order to attain the best topographic expression; it does not permit the adjustment of contours by amounts that significantly impair their accuracy.
topographic map A map showing the topographic features of a land surface, commonly by means of contour lines. It is generally on a sufficiently large scale to show in detail selected man-made and natural features, including relief and such physical and cultural features as vegetation, roads, and drainage. Cf: *planimetric map.*
topographic maturity *maturity [topog].*
topographic old age *old age [topog].*
topographic profile *profile [geomorph].*
topographic relief *relief [geomorph].*
topographic survey A survey that determines the configuration (relief) of the Earth's ground surface and the location of natural and artificial features thereon. Also, an organization making such a survey.
topographic texture Disposition, grouping, or average size of the topographic units composing a given topography; usually restricted to a description of the relative spacing of drainage lines in stream-dissected regions. See also: *coarse topography; fine topography; texture ratio.* Syn: *texture [geomorph].*
topographic unconformity (a) The relationship between two parts of a landscape or two kinds of topography that are out of adjustment with one another, due to an interruption in the ordinary course of the erosion cycle of a region; e.g. a lack of harmony between the topographic forms of the upper and lower parts of a valley, due to rejuvenation. (b) A land surface exhibiting topographic unconformity.
topographic youth *youth [topog].*
topography (a) The general configuration of a land surface or any part of the Earth's surface, including its relief and the position of its natural and man-made features. See also: *geomorphy.* Cf: *relief.* Syn: *lay of the land.* (b) The natural or physical surface features of a region, considered collectively as to form; the features revealed by the contour lines of a map. In nongeologic usage, the term includes man-made features (such as are shown on a topographic map). (c) The art or practice of accurately and graphically delineating in detail, as on a map or chart or by a model, selected natural and man-made surface features of a region. Also, the description, study, or representation of such features. Cf: *chorography.* (d) Originally, the term referred to the detailed description of a particular place or locality (such as a city, parish, or tract of land) as distinguished from the general geography of a country or other large part of the world, and also to the science or practice of such a description; this usage is practically obsolete.—Etymol: Greek *topos,* "place", + *graphein,* "to write".
topologically distinct Said of *channel nets* of a given magnitude "whose schematic map projections cannot be continuously deformed and rotated in the plane of projection so as to become congruent" (Shreve, 1966, p. 27).
topologic path length *link distance.*
topology (a) The analytical, detailed study of minor landforms, requiring fairly large scales of mapping (Matthes, 1912, p. 336). Cf: *topometry.* (b) The topographic study of a particular place; specif. the history of a region as indicated by its topography (Webster, 1967, p. 2411).
topometry The art, process, or science of making large-scale, high-precision maps (1:20,000 or larger) upon which geomorphic features are "measured in with mathematical accuracy, practically nothing being 'sketched in' by eye" (Matthes, 1912, p. 338). Cf: *topology.*
toposequence A sequence of kinds of soil in relation to position on a slope. See also: *catena.*
topostratigraphic unit A term proposed by Jaanusson (1960, p.218) for a "convenient regional stratigraphic unit" consisting of a combined rock unit and biostratigraphic unit.
topostratigraphy Preliminary or introductory stratigraphy, including lithostratigraphy and biostratigraphy; *prostratigraphy.*
topotactic Adj. of *topotaxy.*
topotaxial Adj. of *topotaxy.*
topotaxy Strong preferred orientation of a crystal aggregate, produced by *transformation* of a polymorph. It occurs in displacive transformations with no breaking of primary interatomic bonds, as in low- to high-quartz transformation. The degree of topotaxy is low in reconstructive transformations. Adj: *topotactic; topotaxial.* Cf: *epitaxy; syntaxy.*
topotype A specimen of a particular species that comes from the same locality as the *type specimen* of that species (Frizzell, 1933, p. 665).
topozone A syn. of *local range-zone.* The term was proposed by Moore (1957) for a paleontologically defined horizon or zone recognizable in a single locality.
topset A *topset bed.*
topset bed One of the nearly horizontal layers of sediments deposited on the top surface of an advancing delta and continuous with the landward alluvial plain; it truncates or covers the edges of the seaward-lying *foreset beds.* See also: *bottomset bed.* Also spelled: *top-set bed.* Syn: *topset.*
topsoil (a) A presumably fertile soil used to cover areas of special planting. (b) A partial syn. of *surface soil.* (c) A syn. of *A horizon.* (d) The dark-colored upper portion of a soil, varying in depth according to soil type.—Cf: *subsoil; loam.*
tor A high, isolated crag, pinnacle, or rocky peak; or a pile of rocks, much-jointed and usually granitic, exposed to intense weathering, and often assuming peculiar or fantastic shapes, e.g. the granite rocks standing as prominent masses on the moors of Devon and Cornwall, England. Linton (1955) suggests that a tor is a residual mass of bedrock resulting from subsurface rotting through the action of acidic ground water penetrating along joint systems, followed by mechanical stripping of loose material. Periglacial processes may also be important in the formation of tors. Etymol:

Celtic(?). See also: *core-stone.*

torbanite Essentially synonymous with *boghead coal,* but often considered as a highly carbonaceous oil shale. It is named from its type locality, Torbane Hill, in Scotland. Cf: *cannel coal; wollongongite.* Syn: *kerosine shale; bitumenite.*

torbernite A green radioactive tetragonal mineral: $Cu(UO_2)_2(PO_4)_2 \cdot 8\text{-}12H_2O$. It is isomorphous with autunite. Torbernite is commonly a secondary mineral and occurs in tabular crystals or in foliated form. Syn: *chalcolite; copper uranite; cuprouranite; uran-mica.*

torch peat A waxy, resinous peat derived mainly from pollen.

tordrillite A light-colored *rhyolite* that lacks mafic minerals and has the same chemical composition as *alaskite.* It was named by Spurr in 1900 after the Tordrillo Mountains, Alaska. Not recommended usage.

Toreva block A slump block consisting essentially of a single large mass of unjostled material which, during descent, has undergone a backward rotation toward the parent cliff about a horizontal axis that roughly parallels it (Reiche, 1937). See also: *rotational landslide.*

tornado A small-scale *cyclone,* generally less than 500 m in diameter and with very strong winds. Tornadoes commonly occur as dark funnel-like features suspended from low-lying cumulonimbus clouds. See also: *squall line.*

törnebohmite A green mineral: $Ce_3Si_2O_8(OH)$. Also spelled: *toernebohmite.*

tornote A monaxonic sponge spicule having abruptly pointed ends. Cf: *oxea.*

toroid A cast, commonly consisting of sand, of a circular scour pit made in firm, shallow-water sediments (such as hard mud) by an eddy or whirlpool in flowing water. It has a characteristic swirled shape like a folded bun, but with a homogeneous internal structure and texture.

torose load cast One of a group of elongate load casts that pinch and swell along their trends and that may terminate downcurrent in bulbous, tear-drop, or spiral forms (Crowell, 1955, p.1360).

torque The effectiveness of a force that tends to rotate a body; the product of the force and the perpendicular distance from its line of action to its axis.

Torrejonian North American continental stage: Lower Eocene (above Dragonian, below Tiffanian).

torrent (a) A violent and rushing stream of water; e.g. a flooded river, or a rapidly flowing stream in a mountain ravine, or a stream suddenly raised by heavy rainfall or rapid snowmelt and descending a steep slope. Also, any similar stream, as of lava. (b) A mountain channel that is intermittently filled with rushing water at certain times or seasons.—Adj: *torrential.*

torrential cross-bedding A variety of *angular cross-bedding* in which the beds make a nearly uniform but relatively large angle with the layers that enclose them (Hobbs, 1906, p. 291). It is essentially planar *tabular cross-bedding.* It is thought, probably erroneously, to result from rapid deposition.

torrential plain An early term used by McGee (1897) for a feature now known as a *pediment.*

torrent tract *mountain tract.*

Torrert In U.S. Dept. of Agriculture soil taxonomy, a suborder of the soil order *Vertisol,* characterized by formation in an arid region. Because there is infrequent wetting, deep cracks may stay open throughout the year, and there may be no *gilgai* relief (USDA, 1975). Cf: *Udert; Ustert; Xerert.*

torreyite A mineral: $(Mg,Mn,Zn)_7(SO_4)(OH)_{12} \cdot 4H_2O$. Cf: *mooreite.* Syn: *delta-mooreite.*

torricellian chamber An air-filled cave room that is sealed by water and has a pressure less than atmospheric pressure and an air-water surface higher than adjacent free air-water surfaces. See also: *periodic spring; siphon [speleo].*

Torridonian A division of the upper Precambrian in Scotland.

Torrox In U.S. Dept. of Agriculture soil taxonomy, a suborder of the soil order *Oxisol,* characterized by formation in an aridic soil moisture regime. Torrox soils apparently formed during a former period of higher rainfall. They are dominantly red, and are low in organic matter; have relatively high base saturation; and with irrigation can be highly productive (USDA, 1975). Cf: *Aquox; Humox; Orthox; Ustox.*

torsion The state of stress produced by two force couples of opposite moment acting in different but parallel planes about a common axis.

torsion balance A geophysical prospecting instrument that is used to determine distortions in the gravitational field. It consists of a pair of masses suspended by a sensitive torsion fiber and so supported that they are displaced both horizontally and vertically from each other. A measurement is made of the rotation of the suspended system about the fiber; the rotation is caused by slight differences in the gravitational attraction on the two masses. Syn: *Eötvös torsion balance.*

torsion coefficient The resistance of a material to torsional stress, measured as the work necessary to overcome it, in c.g.s. units.

torsion fault *wrench fault.*

torsion magnetometer An instrument that is both a *horizontal-field balance* and a *vertical-field balance* and consists of a suspended permanent magnet. Cf: *Schmidt field balance.*

torsion modulus *modulus of rigidity.*

torsion period The natural period of oscillation of the suspended system in a torsion balance.

torso mountain A mountain rising above a peneplain; a *monadnock.*

torso plain *rumpffläche.*

torticone A cephalopod shell coiled in a three-dimensional spiral with progressive twisting of the conch, like most gastropods, as distinguished from one coiled in a plane spiral. See also: *helicoid.* Syn: *trochoceroid.*

tortoise *camel back.*

Tortonian European stage: Upper Miocene (above Serravallian, below Messinian). See also: *Diestian.*

tortuosity [elect] The inverse ratio of the length of a rock specimen to the length of the equivalent path of electrolyte within it.

tortuosity [hydraul] The ratio of the actual length of a river channel, measured along the middle of the main channel, to the axial length of the river (ASCE, 1962).

tortuous flow *turbulent flow.*

torus (a) An invagination or protuberance of exine more or less paralleling the laesura of a spore. Cf: *kyrtome.* (b) The thickening of the closing membrane in a bordered pit.

toryhillite A plutonic rock (albite-rich *urtite*) containing albite, nepheline, clinopyroxene, garnet, iron oxides, apatite, and calcite. Sodalite may be present; potassium feldspar is absent. Johannsen in 1920 derived the name from Toryhill, Monmouth Township, Ontario. Not recommended usage.

tosca (a) A term used in Patagonia for a white deposit of calcium carbonate occurring in the loess of the pampas. (b) A term used in Mexico for various rocks, such as clayey vein matter, a talc seam, and soft, decomposed porphyry. (c) A soft coral limestone, used in Puerto Rico for masonry, road surfacing, and as fertilizer.—Etymol: Spanish feminine of *tosco,* "rough, coarse, unpolished".

toscanite A *rhyodacite* containing sanidine, plagioclase (intermediate between calcic and sodic), and accessory hypersthene, biotite, apatite, and opaque oxides, in a silica-rich glassy groundmass. The name, given by Washington in 1897 for Tuscany, Italy, is synonymous with *dellenite* and *quartz latite* (both obsolescent) and with *rhyodacite.* Not recommended usage.

tosudite A dioctahedral mineral consisting of interlayered chlorite and montmorillonite.

total absorptance *Absorptance* measured over all wavelengths of incident energy. Cf: *spectral absorptance.*

total depth The greatest depth reached by a well bore, measured along its axis; not necessarily a vertical depth.

total displacement A syn. of *slip [struc geol].* Cf: *normal displacement.*

total field The vector sum of all components of a field such as a magnetic field. Syn: *total intensity [geophys].*

total field resistivity Using the bipole-dipole array, computation of *apparent resistivity,* ignoring information about direction of the received electric field and using only its magnitude.

total-fusion age An argon-40/argon-39 age in which the sample is fused and all of the argon analyzed in a single experiment. It is comparable to a conventional potassium-argon age.

total hardness *hardness [water].*

total head The sum of the *elevation head, pressure head,* and *velocity head* of a liquid. For ground water, the velocity-head component is generally negligible.

total intensity [geophys] *total field.*

total intensity [magnet] The magnitude of the intensity of the magnetic field, symbolized by F; it is one of the *magnetic elements.* Syn: *total magnetic intensity.*

total magnetic intensity *total intensity [magnet].*

total passing Transportation of all sediment across an area without any being deposited. Cf: *bypassing.*

total porosity *porosity.*

total reflection *Reflection* in which all of the incident wave is returned.

total-rock *whole-rock.*

total runoff *runoff [water].*

total slip *net slip.*

total strain The strain relating the configuration of a body at some point in its strain history to its initial, unstrained configuration. Also referred to as "finite strain" or "total finite strain".

total time correction The sum of all corrections applied to reflection traveltime in seismic prospecting, to express times as those that would have been obtained if source and detectors were located on a selected datum plane, in the absence of a low-velocity layer or variations in elevation.

tot structure In crystallography, an acronym for a sheet-silicate structure consisting of units containing a sandwich of *t*etrahedral-*o*ctrahedral-*t*etrahedral layers.

touchstone A black, flinty stone, such as a silicified shale or slate, or a variety of quartz allied to chert or jasper, whose smoothed surface was formerly used to test the purity or fineness of alloys of gold and silver by comparing the streak left on the stone when rubbed by the metal with that made by an alloy of predetermined composition. Syn: *Lydian stone; basanite; flinty slate.*

tour The work shift of a *rotary drilling* crew. Pron: tower.

tourmaline (a) A group of minerals of general formula: $(Na,Ca)(Mg,Fe^{+2},Fe^{+3},Al,Li)_3Al_6(BO_3)_3Si_6O_{18}(OH)_4$. It sometimes contains fluorine in small amounts. (b) Any of the minerals of the tourmaline group, such as buergerite, elbaite, and dravite.—Tourmaline occurs in 3-, 6-, or 9-sided prisms, usually vertically striated, or in compact or columnar masses; it is commonly found as an accessory mineral in granitic pegmatites, and is widely distributed in acid igneous rocks and in metamorphic rocks. Its color varies greatly and gives a basis for naming the varieties; when transparent and flawless, it may be cut into gems. See also: *schorl.* Also spelled: *turmaline.*

tourmalinization Introduction of, or replacement by, tourmaline.

tourmalite A rock composed almost entirely of tourmaline and quartz, with a mottled appearance and a texture ranging from dense to granular to schistose. It is of secondary origin, resulting from metasomatic and pneumatolytic effects along the margins of igneous intrusions (Johannsen, 1939, p.22). See also: *schorl rock; luxullianite.*

Tournaisian European stage: Lower Carboniferous (Lower Mississippian; above Famennian of Devonian, below Viséan).

tour report A record, filled out on a tabulated form by the chief of the crew that drills an oil or gas well, showing drill progress (such as number of feet drilled each day), bit descriptions, drilling tools used, size of the hole, rock encountered (including its character, color, and description), personnel present during the tour ("tower") of duty, and other pertinent facts having to do with the drilling (including any unusual event).

towan A coastal sand dune in Cornwall, England.

tower A very high rock formation or peak marked by precipitous sides; e.g. Devils Tower, Wyo.

tower karst A type of tropical karst that is characterized by isolated, steep-sided limestone hills that may be flat-topped and that are surrounded by a flat plain usually underlain by alluvium. Syn: *turmkarst.* See also: *cockpit karst; sinkhole karst.*

towhead A low alluvial island or shoal in a river; esp. a sandbar covered with a growth of cottonwoods or young willows.

township The unit of survey of the U.S. Public Land Survey system, representing a piece of land that is bounded on the east and west by meridians approximately six miles apart (exactly six miles at its south border) and on the north and south by parallels six miles apart, and that is normally subdivided into 36 *sections.* Townships are located with reference to the initial point of a principal meridian and base line, and are normally numbered consecutively north and south from a base line (e.g. "township 14 north" indicates a township in the 14th *tier* north of a base line). The term "township" is used in conjunction with the appropriate *range* to indicate the coordinates of a particular township in reference to the initial point (e.g. "township 3 south, range 4 west" indicates the particular township which is the 3rd township south of the base line and the 4th township west of the principal meridian controlling the surveys in that area). Abbrev (when citing specific location): T. See also: *fractional township.*

township line One of the imaginary boundary lines running east and west at six-mile intervals and marking the relative north and south locations of townships in a U.S. public-land survey. Cf: *range line.*

toxa A siliceous, monaxonic sponge spicule (microsclere) with a central arch and broadly recurved extremities that taper to a point, the whole resembling an archer's bow. Pl: *toxae* or *toxas.*

TP (a) *tree pollen.* (b) *turning point.*

T phase A seismic phase with a period of 1 sec or less, which travels through the ocean with the speed of sound in water. It is occasionally identified on the records of those earthquakes in which a large part of the path from epicenter to station is across the deep ocean.

T plane A term used in crystal plasticity to denote the crystallographic slip plane. See also: *f* axis; *t* direction. Syn: *glide plane; gliding plane; slip plane; translation plane.*

trab A complex beam or rod in anthaspidellid sponges, formed by union of the ray tips of dendroclones. The trabs appear like siderails in ladders, with individual spicules for the rungs (Rigby & Bayer, 1971, p. 609).

trabecula [bot] (a) In the wood of a gymnosperm, a small bar extending across the lumina of the ordinary tracheid from one tangential wall to another. It has no known function. (b) In a moss, a transverse bar of the teeth of the peristome. (c) In some dinoflagellate cysts, a narrow, solid rod connecting distal processes. (d) In the lycopod order Isoetales, a plate forming a partial septum in the microsporangium. (e) In the lycopod *Selaginella,* the lacunar tissue between the cortex and the central vascular strand.—Pl: *trabeculae.*

trabecula [paleont] (a) A rod or pillar of radiating calcareous fibers forming a skeletal element in the structure of the septum and related components of a coral. See also: *simple trabecula; compound trabecula.* (b) A branch separating the fenestrae in reteporiform cheilostome bryozoans. (c) One of the individual anastomosing filaments of a hexactinellid sponge, which form a web in which the flagellated chambers are suspended and constitute the pinacoderm. Also, any rodlike or beamlike skeletal element of a sponge other than a ray or branch of a single spicule; esp. such a structure of sclerosome. (d) A tiny rodlike structure, smaller and less regular than a *pillar,* connecting layers of sclerite in holothurians.—Pl: *trabeculae.*

trabecular columella A spongy *columella* in scleractinian corals, formed of trabeculae loosely joined with synapticulae or paliform lobes.

trabecular linkage The connection between corallite centers in scleractinian corals, reflecting in the hard parts the *indirect linkage* of stomodaea.

trabecular network (a) A network of sponge trabeculae made of sclerosome. (b) The cellular web of a hexactinellid sponge.

trabeculate chorate cyst A dinoflagellate *chorate cyst* possessing trabeculae (e.g. *Cannosphaeropsis*).

trace [geochem] A concentration of a substance that is detectable but too minute for accurate quantitative determination.

trace [meteorol] A quantity of precipitation that is insufficient to be measured by a gage.

trace [paleont] A sign, evidence, or indication of a former presence; specif. a mark left behind by an extinct animal, such as a *trace fossil.*

trace [seis] The record of the output of one geophone group with time after the shot, displayed on paper, film, or magnetic tape.

trace [struc geol] The intersection of a geological surface with another surface, e.g. the trace of bedding on a fault surface, or the trace of a fault or outcrop on the ground. Cf: *trend; strike.*

trace element (a) An element that is not essential in a mineral but that is found in small quantities in its structure or adsorbed on its surfaces. Although not quantitatively defined, it is conventionally assumed to constitute significantly less than 1.0% of the mineral. Syn: *accessory element; guest element.* (b) An element that occurs in minute quantities in plant or animal tissue and that is essential physiologically. Syn: *minor element; microelement.*

trace fossil A sedimentary structure consisting of a fossilized track, trail, burrow, tube, boring, or tunnel resulting from the life activities (other than growth) of an animal, such as a mark made by an invertebrate moving, creeping, feeding, hiding, browsing, running, or resting on or in soft sediment. It is often preserved as a raised or depressed form in sedimentary rock. Many trace fossils were formerly assumed to be bodily preserved plants or animals. Syn: *ichnofossil; trace [paleont]; vestigiofossil; lebensspur; bioglyph.*

tracer Any substance that is used in a process to trace its course, specif. radioactive material introduced into a chemical, biological, or physical reaction.

trace slip In a fault, that component of the net slip which is parallel to the trace of an index plane, such as bedding, on the fault plane. See also: *trace-slip fault.*

trace-slip fault A fault on which the net slip is *trace slip,* or slip parallel to the trace of the bedding or other index plane.

tracheid A pitted elongate cylindrical cell in xylem, with tapered ends and no cross walls, that serves for both strength and water conduction. *Tracheids* especially characterize the wood of *conifers.*

tracheophyte A *vascular plant;* includes all plants that have a vascular system, i.e. xylem and phloem (Swartz, 1971, p. 477).

trachographic map A map using perspective symbols to show local relief and average slope of the Earth's surface, after the style of Erwin Raisz (1959). Syn: *physiographic pictorial map.*

trachyandesite An extrusive rock, intermediate in composition between *trachyte* and *andesite,* with sodic plagioclase, alkali feldspar, and one or more mafic minerals (biotite, amphibole, or pyroxene). Although "trachyte" has been considered a syn. of *latite,* Streckeisen (1967, p. 185) suggests that they can be distinguished by considering trachyandesites as including latites plus latite-andesites. Approximately synonymous with *mugearite.*

trachybasalt An extrusive rock intermediate in composition between *trachyte* and *basalt,* characterized by the presence of both calcic plagioclase and alkali feldspar, along with clinopyroxene, olivine, and possibly minor analcime or leucite. According to Streckeisen (1967, p. 185), the term has been variously defined as synonymous with *latite,* intermediate between latite and basalt, and intermediate between trachyte and basalt, but he recommends applying it to latite and latite-basalt. The term was also used, by Rosenbusch, as a syn. of *trachydolerite* (Johannsen, 1939, p. 284). Approximately synonymous with *hawaiite.*

trachydiscontinuity A term proposed by Sanders (1957, p.293) for an unconformity characterized by an irregular surface. Cf: *leurodiscontinuity.* Etymol: Greek *trachys,* "rough", + discontinuity.

trachydolerite (a) An alkalic *basalt* composed of orthoclase or anorthoclase, along with labradorite and a small amount of feldspathoids. (b) A rock intermediate in composition between *trachyte* and *basalt,* and, in this sense, a syn. of *trachybasalt*, as used by Rosenbusch. Not recommended usage.

trachyophitic Said of the ophitic texture of an igneous rock in which the feldspar grains enclosed by pyroxene have a parallel or subparallel alignment; it also applies to *nesophitic* textures with such a microlitic fabric (Walker, 1957, p.2).

trachyostracous Thick-shelled; esp. said of a thick-shelled gastropod.

trachyte A group of fine-grained, generally porphyritic, extrusive rocks having alkali feldspar and minor mafic minerals (biotite, hornblende, or pyroxene) as the main components, and possibly a small amount of sodic plagioclase; also, any member of that group; the extrusive equivalent of *syenite.* Trachyte grades into *latite* as the alkali feldspar content decreases, and into *rhyolite* with an increase in quartz. Etymol: Greek *trachys,* "rough", in reference to the fact that rocks of this group are commonly rough to the touch.

trachytic (a) A textural term applied to volcanic rocks in which feldspar microlites of the groundmass have a subparallel arrangement corresponding to the flow lines of the lava from which they were formed. Cf: *trachytoid texture; pilotaxitic; orthophyric.* (b) Pertaining to or composed of *trachyte.*

trachytoid A textural term originally applied to phaneritic igneous rocks by analogy with the *trachytic* texture of some lava flows. In such rocks (e.g. many nepheline syenites), the feldspars have a parallel or subparallel disposition; *trachytoid* is now used for all similar textures, regardless of the composition of the rock in which they occur.

track [paleont] (a) A fossil structure consisting of a mark left in soft material by the foot of a bird, reptile, mammal, or other animal. Cf: *trail [paleont].* Syn: *footprint.* (b) *muscle track.*

track [photo] The path of an aircraft, spacecraft, or satellite over the Earth's surface.

trackway A continuous series of tracks left by a single organism.

tract [geog] A region or area of land that may be precisely or indefinitely defined.

tract [streams] A part of a stream, such as a *mountain tract* or a *valley tract.*

traction [exp struc geol] The stress vector acting across a particular plane in a body.

traction [sed] A mode of sediment transport in which the particles are swept along (on, near, or immediately above) and parallel to a bottom surface by rolling, sliding, dragging, pushing, or *saltation;* e.g. boulders tumbling along a stream bed, or sand carried by the wind over a desert surface or moved by waves and currents on a beach. The term was introduced into geology by Gilbert (1914, p. 15) for the entire complex process of carrying material along a stream bed. Cf: *suspension.*

tractionite A deposit of well-bedded, winnowed clastic sediments of sand size or larger, made by moving water or wind (Natland, 1976).

traction load *bed load.*

tractive current A current, in standing water, that transports sediment along and in contact with the bottom, as in a stream (Passega, 1957, p. 1961). Cf: *turbidity current.* Syn: *traction current.*

tractive force In hydraulics, drag or shear developed on the wetted area of the stream bed, acting in the direction of flow. As measured per unit wetted area, unit tractive force equals the specific weight of water times hydraulic radius times slope of the channel bed (Chow, 1957). See also: *critical tractive force.*

trade-wind desert *tropical desert.*

trade winds A major system of tropical winds moving from the subtropical highs to the equatorial low-pressure belt. It is northeasterly in the Northern Hemisphere and southeasterly in the Southern Hemisphere. Cf: *antitrades.*

trafficability The quality or suitability of the soil or a terrain to permit passage, such as the cross-country movement of military troops; esp. the capacity of the soil to support moving vehicles.

traffic pan *pressure pan.*

trail [glac geol] "A line or belt of rock fragments picked up by glacial ice at some localized outcrop and left scattered along a more or less well defined tract during the movement and melting of a glacier" (Stokes & Varnes, 1955, p. 154). See also: *train [glac geol].*

trail [mass move] *congeliturbate.*

trail [paleont] (a) A fossil structure consisting of a trace or sign of the passing of one or many animals; esp. a more or less continuous marking left by an organism moving over the bottom, such as a *worm trail.* Cf: *track [paleont].* (b) The extension of either valve of a brachiopod shell anterior to the geniculation (or anterior to the "visceral disc" or the part of the shell posterior to the geniculation) (TIP, 1965, pt.H, p.154 & 155).

trailing edge The rear edge of a thrust sheet with respect to a given stratigraphic unit where it terminates against another thrust fault (Dahlstrom, 1970). Syn: *footwall cutoff.* Cf: *leading edge.*

trailing spit *tail [coast].*

trail of a fault Crushed material along the fault surface that is used as an indication of the direction of displacement. Such material can be a source of mineral deposits (*drag ore*).

train [glac geol] A narrow glacial deposit extending for a long distance, such as a *valley train* or a *boulder train.* See also: *trail [glac geol]; rock train.*

train [seis] A series of oscillations on a seismograph record.

trajectory [exp struc geol] The curve that a moving body in a field, or a characteristic of a field, describes in space.

trajectory [seis] The path of a seismic wave. Syn: *raypath.*

tranquil flow Water flow whose velocity is less than that of a long surface wave in still water. Cf: *rapid flow.* Syn: *subcritical flow; streaming flow [hydraul].*

tranquillityite A hexagonal mineral: $Fe_8(Zr,Y)_2Ti_3Si_3O_{24}$.

transceiver An instrument that both transmits and receives radio-frequency or other waves.

transcontinental geophysical survey A comprehensive geological and geophysical study from coast to coast across a continent; specif. the study of a band 4° wide (about 440 km) centered on lat. 37°N, extending across the U.S. and offshore into the Atlantic and Pacific oceans. Abbrev: TGS.

transcurrent fault A large-scale *strike-slip fault* in which the fault surface is steeply inclined. Syn: *transverse thrust.*

transect n. In ecology, a sample area (usually elongate or linear) chosen as the basis for studying a particular assemblage of organisms. Cf: *quadrat.*

transection glacier A glacier that fills an entire valley system, concealing the divides between the valleys. Cf: *through glacier.*

transfer A single process occurring continuously in space-time in which erosion is followed by transportation and deposition of sediment (Wilson, 1959).

transfer impedance The complex ratio of voltage at one pair of terminals to the current at another pair in a four-terminal network.

transfer percentage For any element, the ratio of the amount present in sea water to the amount supplied to sea water during geologic time by weathering and erosion, multiplied by 100.

transfluence The flowing of glacier ice through a breach made by the headward growth of cirques on both sides of a mountain ridge.

transformation [cryst] The change from one crystal polymorph to another, by one of several processes. See: *dilatational transformation; displacive transformation; reconstructive transformation; rotational transformation; substitutional transformation; order-disorder transformation.* Syn: *inversion [cryst].*

transformation [isotope] *transmutation [isotope].*

transformation [petrology] *granitization.*

transformation [photo] The process of projecting a photograph mathematically, graphically, or photographically from its plane onto another plane by translation, rotation, and/or scale change. See also: *rectification [photo].*

trans-formational breccia A term used by Landes (1945) for a breccia occurring in a vertical body and cutting through a stratigraphic section, and believed to have been produced by collapse, such as above a dissolved salt bed.

transformation series *morphocline.*

transformation twin A crystal twin developed by a transformation from a higher to a lower symmetry; e.g. Dauphiné twinning in the transformation from high quartz to low.

transformed wave *converted wave.*

transform fault (a) A special variety of *strike-slip fault,* along which the displacement suddenly stops or changes form. Many transform faults are associated with *mid-oceanic ridges,* where the actual slip is opposite from the apparent displacement across the fault (Wilson, 1965). (b) A *plate boundary* that ideally shows pure strike-slip displacement (Dennis & Atwater, 1974, p. 1034).

transformism A theory that explains the origin of granite as a result of *granitization;* opposed to *magmatism.* A proponent of this theory is called a *transformist.*

transformist A proponent of the theory of *transformism.* Syn: *granitizer; antimagmatist.*

transfusion The entry and exit of any gaseous or hydrothermal fluid in solid rock to produce such rocks as granite. Cf: *granitization.*

transgression (a) The spread or extension of the sea over land areas, and the consequent evidence of such advance (such as strata deposited unconformably on older rocks, esp. where the new marine deposits are spread far and wide over the former land surface). Also, any change (such as rise of sea level or subsidence of land) that brings offshore, typically deep-water environments to areas formerly occupied by nearshore, typically shallow-water conditions, or that shifts the boundary between marine and nonmarine deposition (or between deposition and erosion) outward from the center of a marine basin. Ant: *regression [stratig].* Cf: *continental transgression; onlap.* Syn: *invasion [stratig]; marine transgression.* (b) A term used mostly in Europe for discrepancy in the boundary lines of continuous strata; i.e. *unconformity.*

transgressive Said of a minor igneous intrusion, typically tabular, that cuts across the bedding or foliation of the country rock rather than confining itself to a single horizon.

transgressive overlap *onlap.*

transgressive reef One of a series of nearshore reefs or bioherms superimposed on back-reef deposits of older reefs during the sinking of a landmass or a rise of the sea level, and developed more or less parallel to the shore (Link, 1950). Cf: *regressive reef.*

transgressive sediments Sediments deposited during the advance or encroachment of water over a land area or during the subsidence of the land, and characterized by an *onlap* arrangement.

transient [elect] n. A pulse that is of short time duration.

transient [evol] n. A term occasionally used to denote a subdivision of a species whose members varied with time; comparable to a subspecies for a subdivision of a species in space; a chronologic subspecies.

transient beach A beach whose sand is removed by storm waves but is quickly restored by longshore currents.

transient creep *primary creep.*

transient method An electrical method of geophysical exploration that depends either on the introduction into the ground of a sharp current pulse, such as may be produced by suddenly closing or opening an electrical circuit connected to grounded electrodes, or on impressing an electric current of a certain wave form on the ground. Measurements are made of the form of the current or more commonly of the form of the resulting potential.

transient snowline *snowline.*

transient strain A less precise, loosely used syn. of *creep recovery.*

transit n. (a) A *theodolite* in which the telescope can be reversed (turned end for end) in its supports without being lifted from them, by rotating it 180 degrees or more about its horizontal transverse axis. Syn: *transit theodolite.* (b) The act of reversing the direction of a telescope (of a transit) by rotation about its horizontal axis.—v. To reverse the direction of a telescope (of a transit) by rotating it 180 degrees about its horizontal axis. Syn: *plunge.*

Transition A name, now obsolete, applied by Jameson (1808) from the teachings of A.G. Werner in the 1790's to the group or series of rocks occurring between the older and more crystalline *Primitive* rocks and the younger and better stratified *Floetz,* and roughly corresponding to the upper Precambrian and to the lower Paleozoic strata now assigned to the Cambrian, Ordovician, and Silurian. The rocks, consisting of dikes and sills, thick graywackes, and thoroughly indurated limestones, were considered to have been the first orderly deposits formed from the ocean during the passage (transition) of the Earth from its chaotic state to its habitable state; they were laid down with original steep dip and contained the first traces of organic remains, and were believed to extend uninterruptedly around the world.

Transitional *Mesolithic.*

transitional fossils Fossils that collectively form sequences (*morphologic series*) showing gradual and continuous changes in morphologic characters from geologically earlier to later forms.

transitional series A series of *passage beds;* esp. a large and widespread series of beds that are transitional in the character of their fossils.

transitional-water wave A wave that is moving from deep water to shallow water, i.e. is transitional between a *deep-water wave* and a *shallow-water wave;* its wavelength is more than twice but less than 25 times the depth of the water, and the wave orbitals are beginning to be influenced by the bottom. Syn: *intermediate wave.*

transition point *inversion point.*

transition temperature *inversion point.*

transition zone (a) A region within the *upper mantle* bordering the lower mantle, at a depth of 410-1000 km, characterized by a rapid increase in density of about 20% and an increase in seismic-wave velocities; it is equivalent to the *C layer.* (b) A region within the *outer core,* transitional to the inner core; the *F layer.*—The zone may consist of several zones of rapid increase in seismic velocity corresponding to phase or chemical changes.

transit line Any line of a traverse that is projected, either with or without measurement, by the use of a transit or other device; an imaginary straight line between two transit stations.

transitory frozen ground Ground that is frozen by a sudden drop of temperature and that remains frozen for a short period, usually hours or days (Muller, 1947, p.223).

transitory pygidium The posterior fused portion of a meraspis trilobite; it will become the true pygidium in the adult form, together with a varying number of segments released into the thorax during the meraspid stage of development.

transit theodolite *transit.*

transit traverse A surveying traverse in which the angles are measured with a transit or theodolite and the lengths with a metal tape. It is usually executed for the control of local surveys.

translation A shift in position without rotation. When applied to plastic deformation, it refers to the movement of one block of atoms past another.

translational Pertaining to or said of a uniform movement in one direction, without rotation.

translational fault A fault in which there has been translational movement and no rotational component of movement; dip in the two walls remains the same. It can be strictly applied only to segments of faults (Dennis, 1967). Syn: *translatory fault.*

translational movement Apparent fault-block displacement in which the blocks have not rotated relative to one another, so that features that were parallel before movement remain so afterwards. Cf: *rotational movement.* See also: *translational fault.* Less-preferred syn: *translatory movement.*

translational slide A major group in the classification of landslides, involving the downslope displacement of soil-rock material on a surface that is roughly parallel to the general ground surface, in contrast to rockfalls and rotational landslides. The term includes such diverse landslide types as rock slides, block glides, slab or flake slides, and debris slides.

translation gliding *crystal gliding.*

translation lattice *crystal lattice.*

translation plane *glide plane.*

translation vector In tectonics, a term used by Bhattacharji (1958, p.626) for the vector representing the direction and the net displacement of material from a reference point; the sum of the vectors for compression and for flow.

translatory fault *translational fault.*

translatory movement A less-preferred syn. of *translational movement.*

trans link A *link* in a trunk stream channel bounded by tributaries that enter from opposite sides (James & Krumbein, 1969). The trunk channel is traced upstream by following the link of greater magnitude at each fork. Cf: *cis link; cis-trans link.*

translucent Said of a mineral that is capable of transmitting light, but is not *transparent.* Cf: *opaque; transopaque.*

translucent attritus Attritus consisting mostly of translucent humic degradation matter, mainly exinite, with minor quantities of opaque materials. Cf: *opaque attritus.* Syn: *humodurite.*

translucent humic degradation matter *Humic degradation matter* that is translucent and of the same deep red color as anthraxylon; humic degradation matter that is less than 14 microns in width, measured perpendicular to the bedding. Abbrev: THDM.

translunar Pertaining to phenomena, or to the space, beyond the Moon's orbit about the Earth. Cf: *cislunar.*

transmedian muscle One of a pair of muscles in some lingulid brachiopods, anterior to the umbonal muscle. One muscle originates on the left side of the pedicle valve and rises dorsally to be inserted on the right side of the brachial valve, whereas the other muscle originates on the right side of the pedicle valve and is inserted on the left side of the brachial valve (TIP, 1965, pt.H, p.154).

transmissibility coefficient *transmissivity [hydraul].*

transmission capacity In a column of soil of unit cross section, the volume of water that flows per unit of time, with a hydraulic gradient unity or with a hydraulic head equal to the length of the soil column (Horton, 1945, p. 308).

transmission constant An expression of the ability of a permeable medium to transmit a fluid under pressure. As applied to ground water, the discharge in cubic feet per minute through each square foot of cross-sectional area under a 100-percent hydraulic gradient (Tolman, 1937, p. 564).

transmission electron microscope An electron-optical microscope that utilizes an assembly of magnetic lenses and a beam of high-energy electrons (80 keV to 3 MeV) that are transmitted through a thin specimen. The main advantage is high resolution, $\approx$2Å, which results from the very small wavelengths of electrons (0.037Å at 100kV). Information is obtained on a fluorescent screen or photographic plate from images formed by diffraction-contract mechanisms using *bright field* or *dark field* modes, and from electron-diffraction patterns of selected areas. Abbrev: TEM.

transmission window *atmospheric window.*

transmissivity The rate at which water of the prevailing kinematic viscosity is transmitted through a unit width of the aquifer under a unit hydraulic gradient. It replaces *coefficient of transmissibility* because by convention it is considered a property of the aquifer, which is transmissive, whereas the contained liquid is transmissible. However, though spoken of as a property of the aquifer, it embodies also the saturated thickness and the properties of the contained liquid (Lohman et al., 1970, p. 41).

transmissometer An instrument that measures the capability of a fluid to transmit light; esp. one that measures the turbidity of water by determining the percent transmission of a light beam. See also: *turbidimeter.*

transmutation [evol] The change from one species to another.

transmutation [isotope] The transformation of one element into another. Radioactive decay is an example. Transmutation can also be accomplished by bombardment of atoms with high-speed particles. Syn: *transformation [isotope].*

transopaque Said of a mineral that is *transparent* in one part of the visible spectrum and *opaque* in another; e.g. goethite, hematite (Salisbury & Hunt, 1968).

transparency [oceanog] The ability of seawater to transmit light; the depth to which water is transparent may be measured by the use of a *Secchi disc.*

transparency [photo] A positive image, either black and white or in color, on a clear base (glass or film), intended to be viewed by transmitted light; a *diapositive.*

transparent Said of a mineral that is capable of transmitting light, and through which an object may be seen. Cf: *translucent; opaque; transopaque.*

transpiration The process by which water absorbed by plants, usually through the roots, is evaporated into the atmosphere from the plant surface. Cf: *guttation.*

transport [sed] A syn. of *transportation.* The term is favored in British usage (Stamp, 1961, p. 458), and often occurs in combined terms such as *sediment transport* and *mass transport.*

transport [struc petrol] *tectonic transport.*

transportation A phase of *sedimentation* that includes the movement by natural agents (such as flowing water, ice, wind, or gravity) of sediment or of any loose material, either as solid particles or in solution, from one place to another on or near the Earth's surface; e.g. the drifting of sand along a seashore under the influence of currents, the creeping movement of rocks on a glacier, or the conveyance of silt, clay, and dissolved salts by a stream. Syn: *transport [sed].*

transportation velocity *nonsilting velocity.*

transport concentration In a stream, the rate of flow of sediment passing through a given cross-sectional area perpendicular to the flow, compared with the rate of flow of the suspension of water and sediment passing through the same area (ASCE, 1962). See also: *spatial sediment concentration.*

transported Said of material that has been carried by natural agents from its former site to another place on or near the Earth's surface.

transported assemblage An *assemblage [paleoecol]* in which the specimens have been transported significant distances, thereby intermingling forms which did not originally live together. Cf: *fossil community; winnowed community; disturbed-neighborhood assemblage; mixed assemblage.* See also: *thanatocoenosis.*

transported soil material *Parent material* that has been moved and redeposited from the site of its parent rock. The adjective "transported" is also applied to the soil formed from such a parent material. Cf: *residual material.*

transporting erosive velocity The velocity of water in a channel that both maintains silt in movement and scours the bed. Cf: *noneroding velocity.*

transposed hinge A hinge in bivalve mollusks in which certain hinge teeth present in one valve occupy positions of teeth usually found in the other valve.

transposition structure A primary sedimentary structure resulting from hydroplastic or fluid flow of sediment after deposition and sometimes after partial consolidation of the sediment (Hills, 1963, p.30).

transpression In crustal deformation, an intermediate stage between compression and strike-slip motion; it occurs in zones with oblique compression. The tectonic style in Caledonian Spitzbergen provides evidence for a transpression regime (Harland, 1971).

transtension Crustal deformation in oblique zones of ocean spreading that consist of stepped *transform faults.* It combines the two elements of extension and strike-slip motion (Harland, 1971).

transverse [geomorph] Said of an entity that is extended in a crosswise direction; esp. of a topographic feature that is oriented at right angles to the *grain* or general strike of a region. Ant: *longitudinal.*

transverse [ore dep] Said of a vein or lode that is oriented across the bedding of the host rock or across any important planar feature.

transverse band The connecting lamella joining the posterior ends of the ascending branches of a brachiopod loop.

transverse bar (a) A slightly submerged sand ridge that extends more or less at right angles to the shoreline. It has been described as a "giant sand wave" and a "plateau-like sandbar". (b) In shallow rivers, a flat-topped sand or gravel body that grows by downcurrent additions to the slip-face margins.

transverse basin *exogeosyncline.*

transverse coastline *discordant coastline.*

transverse crevasse A *crevasse* developed across a glacier, roughly perpendicular to the direction of ice movement, and commonly convex on the downstream side. Cf: *marginal crevasse; splaying crevasse.*

transverse diameter The thecal diameter in edrioasteroids perpendicular to the *axial diameter* (Bell, 1976).

transverse dune A strongly asymmetrical sand dune elongated perpendicularly to the direction of the prevailing winds, having a gentle windward slope and a steep leeward slope standing at or near the angle of repose of sand; it generally forms in areas of sparse vegetation.

transverse fault A fault that strikes obliquely or perpendicular to the general structural trend of the region.

transverse flagellum A flagellum, often ribbonlike in form, encircling the body of a dinoflagellate in a nearly transverse plane, usually lodged in a deep encircling groove (girdle), and arising from the anterior pole near the proximal end of the girdle.

transverse fold *cross fold.*

transverse furrow An equatorial thinning in the exine of a dicotyledonous pollen grain, usually occurring at the equator, and always running perpendicular to a meridional colpus. Syn: *colpus transversalis.*

transverse joint *cross joint.*

transverse lamination Lamination transverse to bedding. Syn: *oblique lamination.*

transverse Mercator projection A cylindrical conformal map projection, equivalent to the regular *Mercator projection* turned (transversed) 90° in azimuth, so that the cylinder is tangent along a given meridian (or any pair of opposing meridians) rather than along the equator. The central meridian is represented by a straight line and is divided truly; all other meridians and all parallels (except the equator, if shown) are curved lines intersecting at right angles. Lines of constant direction (rhumb lines) are also curved lines. The projection is designed to minimize scale error or variation along a narrow zone by using a great cirle, centrally located to the area to be mapped, as the "theoretical equator". It is used for maps of areas extending short distances from the central meridian, for charts of polar regions, and as a worldwide standard for plotting military maps. A special case of the projection is used as the basis of the *universal transverse Mercator (UTM) projection.* See also: *Gauss projection.*

transverse oral midline A perradial line in edrioasteroids that extends across the exterior of the oral region between opposing anterior and posterior orals, perpendicular to the anterior primary ambulacral radius. It extends from the proximal tip of one lateral ambulacral bifurcation plate to the other (Bell, 1976).

transverse profile *cross profile.*

transverse projection A projection in which the projection axis is rotated 90° in azimuth; e.g. "transverse Mercator projection" or "transverse polyconic projection". Syn: *inverse projection.*

transverse resistance The product of average resistivity and thickness of a layer of rock. It is used in interpretation of direct-current resistivity soundings. Units are ohm-meters^2.

transverse resistivity Resistivity of rock measured across the direction of bedding. Cf: *longitudinal resistivity.*

transverse rib One of a series of ridges of pebbles, cobbles, or boulders oriented transverse to stream flow (McDonald & Banerjee, 1971, p. 1290). Rib spacing is regular and generally ranges from about 0.2 to 1.5 m. Transverse ribs are commonly formed by upper-regime, shallow flows, for example on braided outwash plains, either by a hydraulic jump that moves progressively upstream, or by antidunes.

transverse ridge A generally denticulate fulcral elevation on an articular face of an ossicle of a crinoid ray, disposed perpendicularly or slightly oblique to the axis extending from the dorsal toward the ventral side.

transverse ripple mark A ripple mark formed approximately at right angles to the direction of the current. Its profile may be asymmetric or symmetric.

transverse scour mark A *scour mark* whose long axis is transverse to the main direction of the current. The regular spacing of such marks may lead to confusion with ordinary transverse ripple marks. The term was apparently first used by Dzulynski and Sanders (1962, p. 68). See also: *current-ripple cast.*

transverse section *cross section.*

transverse septulum A minor partition of a chamber in a foraminiferal test, oriented transverse to the axis of coiling and observable in sagittal and parallel sections. See also: *primary transverse septulum; secondary transverse septulum.*

transverse septum One of a series of septa dividing the parietal tubes of a cirripede crustacean into a series of cells, oriented normal to a longitudinal septum and parallel to the basis (TIP, 1969, pt.R, p.103).

transverse thrust *transcurrent fault.*

transverse valley (a) A valley having a direction at right angles to the general strike of the underlying strata; a *dip valley.* (b) A valley that cuts across a ridge, range, or chain of mountains or hills at right angles (Conybeare & Phillips, 1822, p. xxiv).—Cf: *longitudinal valley.* Syn: *cross valley.*

transverse wall One of a pair of oppositely placed *vertical walls* bounding the ends of a cheilostome bryozoan zooid and commonly developed largely as an *interior wall.*

transverse wave *S wave.*

trap [cryst] An imperfection in a crystal structure that may trap a mobile electron, usually temporarily.

trap [eng] A device for separating suspended sediment from flowing water; e.g. a *sand trap.*

trap [ign] Any dark-colored fine-grained nongranitic hypabyssal or extrusive rock, such as a *basalt, peridotite, diabase,* or fine-grained *gabbro;* also, applied to any such rock used as crushed stone. Etymol: Swedish *trappa,* "stair, step", in reference to the stairstep appearance created by the abrupt termination of successive flows. Syn: *trapp; trap rock; trappide.* Cf: *whinstone.*

trap [petroleum] Any barrier to the upward movement of oil or gas, allowing either or both to accumulate. A trap includes a *reservoir rock* and an overlying or updip impermeable *roof rock;* the contact between these is concave as viewed from below. See also: *stratigraphic trap; structural trap; combination trap.*

trap [speleo] *siphon [speleo].*

trap cut *step cut.*

trap-door fault A curved fault bounding a block that is hinged along one edge; it is an *intrusion displacement* structure in the Little Rocky Mountains of Montana.

trap efficiency The ability of a storage reservoir to trap and retain sediment, expressed as the percent of incoming sediment (sediment yield) retained in the basin.

trapezohedral Said of those crystal classes in the tetragonal and hexagonal systems in which the general form is a trapezohedron.

trapezohedron (a) An isometric crystal form of 24 faces, each face of which is ideally a four-sided figure having no two sides parallel, or a trapezium. Syn: *tetragonal trisoctahedron; leucitohedron; icositetrahedron.* (b) A crystal form consisting of six, eight, or twelve faces, half of which above are offset from the other half below. Each face is, ideally, a trapezium. The tetragonal and hexagonal forms may be right- or left-handed.

trapezoidal projection A map projection in which equally spaced straight parallels and straight converging meridians divide the area into trapezoids.

trapp *trap [ign].*

trappide *trap [ign].*

trap rock *trap [ign].*

trapshotten gneiss A gneiss injected with flinty pseudotachylyte. The term was originated by King and Foote in 1864.

trash ice Broken or crumbled ice mixed with water. Syn: *trash.*

trash line A line on a beach, marking the farthest advance of high tide, and consisting of debris (Pettijohn & Potter, 1964, p. 350). Cf: *debris line.*

traskite A mineral: $Ba_9Fe_2Ti_2Si_{12}O_{36}(OH,Cl,F)_6 \cdot 6H_2O$.

trass A pumiceous tuff resembling pozzolan that is used in the production of hydraulic cement.

traveled Removed from the place of origin, as by streams or wind, and esp. by glacier ice, as a *traveled* stone; a syn. of *erratic.* Also spelled: *travelled.*

traveling beach A beach that is continually moving in one general direction under the influence of floods.

traveling dune *wandering dune.*

traveling wave *kinematic wave.*

traveltime The time required for a seismic wave train to travel from its source to a point of observation. Cf: *traveltime curve.*

traveltime curve In seismology, a plot of wave-train *traveltime* against corresponding distance along the Earth's surface from the source to the point of observation. Syn: *time-distance curve.* Obsolete syn: *hodochrone; hodograph [seis].*

traverse [geol] (a) A line across a thin section or other sample along which grains of various minerals are counted or measured. (b) A vein or fissure in a rock, running obliquely and in a transverse direction.

traverse [surv] n. (a) A sequence or system of measured lengths and directions of straight lines connecting a series of surveyed points (or *stations*) on the Earth's surface, obtained by or from field measurements, and used in determining the relative positions of the points (or stations). (b) *traverse survey.* (c) A line surveyed across a plot of ground.—v. To make a traverse; to carry out a traverse survey.

traverse map A map made from a traverse survey.

traverse survey A survey in which a series of lines joined end to end are completely determined as to length and direction, these lines being often used as a basis for triangulation. It is used esp. for long narrow strips of land (such as for railroads) and for underground surveys. Syn: *traverse.*

traverse table A mathematical table listing the lengths of the two sides opposite the oblique angles for each of a series of right-angled plane triangles as functions of every degree of angle (azimuth or bearing) and of all lengths of the hypotenuse from 1 to 100. Traverse tables are used in computing latitudes and departures in surveying and courses in navigation.

travertine (a) A dense, finely crystalline massive or concretionary limestone, of white, tan, or cream color, often having a fibrous or concentric structure and splintery fracture, formed by rapid chemical precipitation of calcium carbonate from solution in surface and ground waters, as by agitation of stream water or by evaporation around the mouth or in the conduit of a spring, esp. a hot spring. It also occurs in limestone caves, where it forms stalactites, stalagmites, and other deposits; and as a vein filling, along faults, and in soil crusts. The spongy or less compact variety is *tufa.* The term is also regarded as a mineral name for a variety of calcite or aragonite. See also: *onyx marble.* Syn: *calcareous sinter; calc-sinter.* (b) A term sometimes applied to any cave deposit of calcium carbonate. (c) A term used inappropriately as a syn. of *kankar.*—Etymol: Italian *tivertino,* from the old Roman name of Tivoli, a town near Rome where travertine forms an extensive deposit. Syn: *travertin.*

travertine dam *rimstone dam.*

travertine terrace A series of *rimstone dams.*

tread The flat or gently sloping surface of one of a series of natural steplike landforms, such as those of a glacial stairway or of successive stream terraces; a bench level. Cf: *riser.*

treanorite *allanite.*

treated stone A gemstone that has been heated, stained, oiled, or coated, or one that has been treated by various types of irradiation, in order to improve or otherwise alter its color; or laser-drilled to make flaws inconspicuous. Also, a stone that has been preserved from dehydration, such as an opal whose cracks have been filled with oil or other liquid; or one in which special effects have been produced, e.g. amber with "spangles" (tension cracks).

trechmannite A red rhombohedral mineral: $AgAsS_2$.

tree agate A moss agate whose dendritic markings resemble trees.

tree line *timberline.*

Tree of Life A figurative reference to the branching pattern of evolution, often represented by a fanciful picture of a tree.

tree ore A high-grade uranium ore consisting of buried *carbon trash* that has been replaced or enriched with uranium-bearing solutions.

tree pollen A syn. of *arborescent pollen.* Abbrev: TP.

tree ring *growth ring [geochron].*

tree-ring chronology *dendrochronology.*

trellis drainage pattern A drainage pattern characterized by parallel main streams intersected at or nearly at right angles by their tributaries, which in turn are fed by elongated secondary tributaries parallel to the main streams, resembling in plan the stems of a vine on a trellis. It is commonly developed where the beveled edges of alternating hard and soft rocks outcrop in parallel belts, as in a rejuvenated folded-mountain region or in a maturely dissected belted coastal plain of tilted strata; it is indicative of marked structural control emphasized by subsequent and secondary consequent streams. Examples are well displayed in the Appalachian Mountains region. Cf: *fault-trellis drainage pattern; rectangular drainage pattern.* Syn: *trellised drainage pattern; grapevine drainage pattern; espalier drainage pattern.*

trema An orifice, occurring singly or in series, in the outer wall of some gastropod shells for excretory functions. Pl: *tremata.*

Tremadocian European stage: Lower Ordovician (above Dolgellian of Cambrian, below Arenigian). In Great Britain, the stage has been traditionally classed as uppermost Cambrian, although its fauna has greater affinities with the Ordovician.

tremalith A centrally and minutely perforate coccolith, such as a placolith or a rhabdolith. The term is sometimes restricted to *placolith* only. Cf: *discolith.* Syn: *trematolith.*

trematolith *tremalith.*

trematophore A perforated plate over the aperture of the test in some miliolid foraminifers. Syn: *sieve plate.*

trembling prairie *shaking prairie.*

tremocyst A term used in the older literature of bryozoans for a calcareous layer with pseudopores, thought to form a secondary or tertiary deposit over an olocyst or a pleurocyst in the frontal shield of certain ascophoran cheilostomes.

tremolite A white to dark-gray monoclinic mineral of the amphibole group: $Ca_2Mg_5Si_8O_{22}(OH)_2$. It has varying amounts of iron, and may contain manganese and chromium. Tremolite occurs in long blade-shaped or short stout prismatic crystals and also in columnar, fibrous, or granular masses or compact aggregates, generally in metamorphic rocks such as crystalline dolomitic limestones and talc schists. It is a constituent of much commercial talc. Cf: *actinolite.*

tremopore A pseudopore in the tremocyst of an ascophoran cheilostome (bryozoan).

tremor A minor earthquake, esp. a foreshock or an aftershock. Syn: *earth tremor; earthquake tremor.*

tremor tract In coal mining, an area of complex folding, faulting, and gliding of coal seams and associated rocks. It may be formed by seismic shocks during the deposit's semicompacted state (Nelson & Nelson, 1967, p.387).

Trempealeauan North American stage: uppermost Cambrian (above Franconian, below Canadian).

trench [geomorph] (a) A long, straight, commonly U-shaped valley or depression between two mountain ranges, often occupied by two streams that drain in opposite directions. Syn: *trough [geomorph].* (b) A narrow, steep-sided canyon, gully, or other depression eroded by a stream. (c) Any long, narrow cut or excavation produced naturally in the Earth's surface by erosion or tectonic movements. Also, a similar feature produced artificially, such as a ditch.

trench [marine geol] A narrow, elongate depression of the deep-sea floor, with steep sides, oriented parallel to the trend of the continent and between the continental margin and the abyssal hills. Such a trench is about 2 km deeper than the surrounding ocean floor, and may be thousands of kilometers long. Cf: *foredeep; trough.* Syn: *oceanic trench; marginal trench; sea-floor trench.*

trend [geophys] That component in a geophysical anomaly map which is relatively smooth, generally produced by regional geologic features.

trend [paleont] In evolutionary paleontology, the evolution of a specific structure or morphologic characteristic within a group, esp. in taking an overall view of a large group, such as an order or a class; e.g., the evolution of the form of the septal suture, from simple to complex, in tracing the ammonoids as a group from the Devonian to the Triassic.

trend [stat] (a) The direction or rate of increase or decrease in magnitude of the individual members of a time series of data when random fluctuations of individual members are disregarded; the general movement through a sufficiently long period of time of some statistical progressive change. (b) *trend line [stat].*

trend [struc geol] A general term for the direction or bearing of the outcrop of a geological feature of any dimension, such as a layer, vein, ore body, fold, or orogenic belt. Cf: *strike; trace.* Syn: *direction.*

trend deposit A uranium orebody in sandstone along a mineralized belt or trend. Trend deposits are generally tabular and are subparallel to the gross stratification (Bailey & Childers, 1977, p. 27).

trend line A straight line or other statistical curve expressing the best empirical relationship between two variables (commonly a regression line) or showing the tendency of some function to increase or decrease through a period of time. Syn: *trend [stat].*

trend map A stratigraphic map that displays the relatively systematic, large-scale features of a given stratigraphic unit, such as those indicating broad postdepositional structural and erosional changes or those controlled by regional deposition (Krumbein & Sloss, 1963, p.485-488). Cf: *residual map.*

trend-surface analysis A method for fitting and evaluating the degree of fit of a set of data (usually contoured) to a calculated mathematical surface of linear, quadratic, or higher degree.

Trentonian North American provincial stage: Middle Ordovician (above Wilderness, below Edenian); it is equivalent to the Barneveld of upper Mohawkian. It has also been regarded as a substage (upper Mohawkian Stage), above the Blackriverian.

trepostome Any ectoproct bryozoan belonging to the order Trepostomata, characterized by tubular zooecia with distinct endozone and exozone and a terminal aperture. Adj: *trepostomatous.* Range, Ordovician to Permian.

treppen Nacreous tablets in the mollusks arranged in a steplike pattern (Wise, 1970).

treppen concept The concept that, on a surface reduced to old age by streams and then uplifted, the rejuvenated streams develop second-cycle valleys first near their mouths and that these young valleys are extended headward to form *piedmont steps.* Etymol: German *Treppen,* "stair steps".

treptomorphic Adj. of *treptomorphism;* isochemically metamorphosed.

treptomorphism *isochemical metamorphism.*

trevalganite A tourmaline *granite* containing large phenocrysts of pink feldspar and/or quartz in a groundmass of *schorl rock.* Obsolete.

trevor A dark-brown granitic rock from Wales, used to make curling stones. Local usage only.

trevorite A black or brownish-black mineral of the magnetite series in the spinel group: $NiFe_2O_4$.

triactine A sponge spicule having three rays. Syn: *triact.*

triad Said of a symmetry axis that requires a rotation of 120° to repeat the crystal's appearance. Cf: *diad.*

triaene (a) A tetraxon in which three similar rays differ from the fourth; specif. an elongated sponge spicule with one long ray (the rhabdome) and three similar short rays (the cladi), which are sometimes branched or otherwise modified, divergent at one end. See also: *dichotriaene; phyllotriaene; discotriaene.* (b) The initial four-rayed or tridentlike spicule of the ebridian skeleton.

triakisoctahedron *trisoctahedron.*

triakistetrahedron *trigonal tristetrahedron.*

triangle An ordinarily plane figure bounded by three straight-line sides and having three internal angles. Cf: *spherical triangle.*

triangle closure The amount by which the sum of the three measured angles of a triangle fails to equal exactly 180 degrees plus the spherical excess (the amount by which the sum of the three angles of a triangle on a sphere exceeds 180 degrees); the *error of closure* of triangle. Also known as "closure of triangle".

triangular coordination An atomic arrangement in which an ion is surrounded by three ions of opposite sign. Syn: *threefold coordination.*

triangular diagram A method of plotting compositions in terms of the relative amounts of three materials or components, involving an equilateral triangle in which each apex represents a pure component. The perpendicular distances of a point from each of the three sides will then represent the relative amounts of each of the three materials represented by the apexes opposite those sides.

triangular facet A physiographic feature, having a broad base and an apex pointing upward; specif. the face on the end of a *faceted spur,* usually a remnant of a fault plane at the base of a block mountain. A triangular facet may also form by wave erosion of a mountain front or by glacial truncation of a spur. Syn: *spur-end facet.*

triangular organelle A small sensory structure near the peristome of a tintinnid.

triangular texture In mineral deposits, texture produced when exsolved or replacement mineral crystals are arranged in a triangular pattern, following the crystallographic directions of the host mineral.

triangulate To divide into triangles; esp. to use, or to survey, map, or determine by, triangulation. Etymol: back-formation from *triangulation.*

triangulation (a) A trigonometric operation for finding the directions and distances to and the coordinates of a point by means of bearings from two fixed points a known distance apart; specif. a method of surveying in which the stations are points on the ground at the vertices of a chain or network of triangles, whose angles are measured instrumentally, and whose sides are derived by computation from selected sides or base lines the lengths of which are obtained by direct measurement on the ground or by computation from other triangles. Triangulation is generally used where the area surveyed is large and requires the use of geodetic methods. Cf: *trilateration.* (b) The network or system of triangles into which any part of the Earth's surface is divided in a trigonometric survey.

triangulation net A *net* or series of adjoining triangles covering an area in such a manner that the lengths and relative directions of all lines forming the triangles can be computed successively from a single base line; arcs of triangulation connected together to form a system of loops or circuits extending over an area. See also: *base net.*

triangulation station A surveying *station* whose position is determined by triangulation. It is usually a permanently marked point that has been occupied (such as one identified by a bench mark), as distinguished from secondary points such as church spires, chimneys, water tanks, and prominent summits located by intersection. See also: *trigonometric point.*

triangulation tower An engineering structure used to elevate the line of sight above intervening obstacles (such as trees and topographic features), usually consisting of two separate towers built one within the other, the central one supporting the theodolite and the outer one supporting the observing platform; e.g. the Bilby steel tower consisting of two steel tripods that are demountable and easily erected.

triangulum In certain receptaculitids, a triangular plate which, in conjunction with an interpositum, is associated with an increase in the number of meromes per whorl.

Trias *Triassic.*

Triassic The first period of the Mesozoic era (after the Permian of the Paleozoic era, and before the Jurassic), thought to have covered the span of time between 225 and 190 million years ago; also, the corresponding system of rocks. The Triassic is so named because of its threefold division in the rocks of Germany. Syn: *Trias.*

triaxial compression test A test in which a cylindrical specimen of rock encased in an impervious membrane is subjected to a confining pressure and then loaded axially to failure. See also: *unconfined compression test. Cf: extension test.* Syn: *compression test.*

triaxial extension test A test in which a cylindrical specimen of rock encased in an impervious membrane is subjected to a confining pressure and the axial load is decreased to failure. Cf: *triaxial compression test.* Syn: *extension test.*

triaxial state of stress A stress system in which none of the principal stresses is zero.

triaxon A siliceous sponge spicule in which six rays or their rudiments grow along three mutually perpendicular axes.

tribe A subdivision of the *rock association* or *kindred.* A tribe is made up of *clans [petrology].*

tributary n. (a) A stream feeding, joining, or flowing into a larger stream or into a lake. Ant: *distributary.* Syn: *tributary stream; affluent; feeder; side stream; contributory.* (b) A valley containing a tributary stream. (c) *tributary glacier.*—adj. Serving as a tributary.

tributary bifurcating link A link of magnitude μ that is formed at its upstream fork by the confluence of two links, each of a magnitude of $1/2\mu$, and that flows at its downstream fork into a link of a magnitude greater than or equal to 2μ (Mock, 1971, p. 1559). Symbol: TB. Cf: *link; magnitude.*

tributary glacier A glacier that flows into a larger glacier. See also: *secondary glacier.*

tributary link A link of magnitude μ that is formed at its upstream fork by the confluence of two links of unequal magnitude and that flows at its downstream fork into a link of a magnitude greater than or equal to 2μ (Mock, 1971, p. 1559). Symbol: T. Cf: *link; magnitude.*

tributary source link A magnitude-1 link that joins a link of magnitude greater than 1 at its downstream fork (Mock, 1971, p. 1558). Symbol: TS. Cf: *magnitude; link.*

tricentric Said of a corallite formed by a polyp retaining tristomodaeal condition permanently.

trichalcite *tyrolite.*

trichite [paleont] A hairlike siliceous sponge spicule occurring in fascicles.

trichite [petrology] A straight or curved hairlike crystallite, usually black. Trichites occur singly or radially arranged in clusters and are found in glassy igneous rocks.

trichobothrium A sensory hair arising from the center of a disklike membrane on the legs or pedipalpi of an arachnid and serving for perception of air currents (TIP, 1955, pt.P, p.63). Also, a sensory organ consisting of one or more such hairs together with supporting structures. Pl: *trichobothria.*

trichotomocolpate Said of monocolpate pollen grains in which the colpus is triangular and may simulate a trilete laesura. Syn:

trichotomosulcate.

trichotomosulcate *trichotomocolpate.*

trichroic Said of a mineral that displays *trichroism.*

trichroism *Pleochroism* of a crystal that is indicated by three colors. A mineral showing trichroism is said to be *trichroic.* Cf: *dichroism.*

triclinic system One of the six *crystal systems,* characterized by a onefold axis of symmetry, and having three axes whose lengths and angles of intersection are unconstrained.

tricolpate Said of pollen grains having three meridionally arranged colpi that are not provided with pores. Tricolpate pollen are typical of dicotyledonous plants, and they first appeared in the fossil record in the Aptian-Albian interval (upper part of Lower Cretaceous).

tricolporate Said of pollen grains having three colpi that are provided with pores or with other, usually equatorial, modifications.

Triconodonta One of two orders of mammals (the other being Docodonta) of primitive structure and uncertain subclass assignment, mostly of Triassic and Jurassic age.

tricranoclone An *ennomoclone* with three proximal rays.

trider A *tetraclone* consisting of three similar arms differing from the fourth.

tridymite A mineral: SiO_2. It is a high-temperature polymorph of quartz, and usually occurs as minute tabular white or colorless crystals or scales, in cavities in acidic volcanic rocks such as trachyte and rhyolite. Tridymite is stable between 870° and 1470°C; it has an orthorhombic structure (alpha-tridymite) at low temperatures and a hexagonal structure (beta-tridymite) at higher temperatures. Cf: *cristobalite.* Syn: *christensenite.*

trifilar suspension A construction used in some gravity meters in order to increase their sensitivity. It consists of a disk supported by a helical spring at its center and three equally spaced wires at its circumference (Heiland, 1940, p.131).

trigonal dipyramid A crystal form of six faces, ideally isosceles triangles, consisting of two trigonal pyramids repeated across a mirror plane of symmetry. It is trigonal in cross section, and its indices are $\{h0l\}$ in symmetry $\bar{6}m2$, also $\{hhl\}$ in $\bar{6}$ and 32, and $\{hkl\}$ in $\bar{6}$.

trigonal dipyramidal class That crystal class in the hexagonal system having symmetry $\bar{6}$.

trigonal prism A crystal form of three faces that are parallel to a threefold symmetry axis. Its indices are $\{100\}$ in symmetry $\bar{6}m2$, $\bar{6}$, $3m$, and 3; or $\{110\}$ in symmetry $\bar{6}$, 32, and 3; or $\{hk0\}$ in $\bar{6}$ and 3.

trigonal pyramid A crystal form consisting of three faces, ideally isosceles triangles, in a pyramid with a triangular cross section. Its indices are $\{h0l\}$ in symmetry $3m$, and $\{h0l\}$, $\{hhl\}$, and $\{hkl\}$ in symmetry 3.

trigonal-pyramidal class That class in the rhombohedral division of the hexagonal system having symmetry 3.

trigonal-scalenohedral class *hexagonal-scalenohedral class.*

trigonal system A crystal system of threefold symmetry that is often considered as part of the ***hexagonal system*** since the lattice may be either hexagonal or rhombohedral. See also: *rhombohedral system.*

trigonal-trapezohedral class That crystal class in the rhombohedral division of the hexagonal system having symmetry 32.

trigonal trapezohedron A crystal form of six faces, having a threefold axis and three twofold axes, but neither mirror planes nor a center of symmetry. The top and bottom trigonal pyramids are rotated less than 30° about c with respect to each other. It may be either right-handed or left-handed. Its indices are $\{hkl\}$.

trigonal trisoctahedron *trisoctahedron.*

trigonal tristetrahedron A *tristetrahedron* whose faces are triangular rather than quadrilateral, as in the *deltohedron.* Syn: *triakistetrahedron.*

trigoniid Any bivalve mollusk belonging to the family Trigoniidae, characterized by a variably shaped and ornamented shell, generally with opisthogyrate umbones and with the ornamentation of the posterior portion of the valves differing from that of the flank.

trigonite A yellow to brown monoclinic mineral: $MnPb_3H(AsO_3)_3$. It occurs in triangular wedge-shaped crystals.

trigonododecahedron An obsolete syn. of *deltohedron.*

trigonometric leveling A type of indirect leveling in which differences of elevation are determined by means of observed vertical angles combined with measured or computed horizontal or inclined distances. Syn: *vertical angulation.*

trigonometric point A fixed point determined with great accuracy in the triangulation method of surveying; a *triangulation station* being the vertex of the triangle. Shortened form: *trig point.*

trigonometric survey A survey accomplished by triangulation and by trigonometric calculation of the elevations of points of observation. It is generally preliminary to a topographic survey, and is performed after careful measurement of a base line and of the angles made with this line by the lines toward points of observation.

trihedron A geometric form composed of three planes that meet at a central point, e.g. the trigonal pyramid crystal form.

trikalsilite A hexagonal mineral: $(K,Na)AlSiO_4$. It is a polymorph of kalsilite with an *a*-axis of 15 angstroms. Cf: *tetrakalsilite.*

trilateration A method of surveying in which the lengths of the three sides of a series of touching or overlapping triangles are measured (usually by electronic methods) and the angles are computed from the measured lengths. Cf: *triangulation.*

trilete adj. Said of an embryophytic spore and some pollen having a laesura consisting of a three-pronged mark somewhat resembling an upper-case "Y". Cf: *monolete.*—n. A trilete spore. The usage of this term as a noun is improper.

trill *trilling.*

trilling A cyclic crystal twin consisting of three individuals. Cf: *twoling; fourling; fiveling; eightling.* Syn: *threeling; trill.*

trilobite Any marine arthropod belonging to the class Trilobita, characterized by a three-lobed, ovoid to subelliptical exoskeleton divisible longitudinally into axial and side regions and transversely into cephalon (anterior), thorax (middle), and pygidium (posterior). Range, Lower Cambrian to Permian.

triloculine Having three chambers; specif. said of a foraminiferal test having three externally visible chambers, resembling *Triloculina* in form and plan.

trimaceral Said of a coal microlithotype consisting of three macerals. Cf: *monomaceral; bimaceral.*

trimacerite A coal microlithotype consisting of macerals from the three groups, vitrinite, exinite and inertinite, each more than 5% in abundance (ICCP, 1971). Cf: *duroclarite; clarodurite; vitrinertoliptite.*

trimerite A salmon-colored mineral: $Be(Ca,Mn)(SiO_4)$.

trimerous Said of the radial symmetry of certain echinoderms, such as edrioasteroids, characterized by three primary rays extending from the mouth, each of two lateral rays giving off two branches.

trimetal detector A device for detecting thermal infrared radiation. A common trimetal detector is an alloy of mercury, cadmium, and tellurium.

trimetric projection A projection based on representation of a spherical triangle by a plane triangle whose sides are lines of zero distortion, and in which the three spatial axes are represented as unequally inclined to the plane of projection (equal distances along the axes are drawn unequal).

trimetrogon A system of three cameras arranged systematically at fixed angles and simultaneously taking photographs (one vertical photograph, and two oblique right and left photographs along the flight line at 60 degrees from the vertical) at regular intervals over the area being mapped. Etymol: originally equipped with wide-angle Metrogon lenses.

trimline A sharp boundary line delimiting the maximum upper level of the margins of a glacier that has receded from an area. It usually coincides with a break in slope and the upper limit of unweathered rock on a valley wall or a nunatak; but the trimline of a long-extinct glacier may be marked by a sharp change in the age, constitution, or density of vegetation. Syn: *shoulder [glac geol].*

trimming The elimination of spurs that jut across a widening stream valley, effected by lateral erosion where a stream flows against and undercuts the sides of the spurs in going around meanders.

trimorph One of three crystal forms displaying *trimorphism.*

trimorphism That type of *polymorphism [cryst]* in which there are three crystal forms, known as *trimorphs.* Adj: *trimorphous.* Cf: *dimorphism; tetramorphism.*

trimorphous Adj. of *trimorphism.*

trinacrite A tuff having the composition of palagonite (Hey, 1962, p. 629).

Trinitian North American (Gulf Coast) stage: Lower Cretaceous (above Nuevoleonian, below Fredericksburgian).

trinomen A name of an animal that consists of three words, the first designating the genus, the second the species, and the third the subspecies; e.g. *Odontochile micrurus clarkei.* See: ICZN, 1964, p. 153. Syn: *trinomial.*

trinomial n. A syn. of *trinomen.*

trioctahedral Pertaining to a layered-mineral structure in which all possible octahedral positions are occupied. Cf: *dioctahedral.*

triode (a) The initial triradial spicule (in which one ray is atrophied) of the ebridian skeleton. (b) A *triradiate* (sponge spicule). Also spelled: *triod.*

trip The operation in *rotary drilling* of "pulling out" (trip out) and "running in" (trip in) the drill string, as required to replace a worn bit, extract a core, or recover a fish. Syn: *round trip.*

tripartite method A method of determining the apparent surface velocity and direction of propagation of microseisms or earthquake waves by determining the times at which a given wave passes three separated points.

tripestone (a) A concretionary variety of anhydrite composed of contorted plates suggesting pieces of tripe. (b) A stalactite resembling intestines. (c) A variety of barite.—Also spelled: *tripe stone.*

triphane *spodumene.*

triphylite A gray-green or blue-gray orthorhombic mineral: $Li(Fe^{+2},Mn^{+2})PO_4$. It is isomorphous with lithiophilite.

triple junction A point where three lithospheric *plates* meet.

triple point An invariant point at which three phases coexist in a unary system. When not otherwise specified, it usually refers to the coexistence of solid, liquid, and vapor of a pure substance.

triple stomodaeal budding A type of budding in scleractinian corals similar to *tristomodaeal budding* in which the three stomodaea invariably form a triangle and only one interstomodaeal couple of mesenteries occurs between each pair of neighboring stomodaea.

triplet An *assembled stone* of two main parts of gem materials bonded by a layer of cement or other thin substance (the third part of the triplet), which gives the stone color. Cf: *doublet.*

triplite A dark-brown mineral: $(Mn,Fe,Mg,Ca)_2(PO_4)(F,OH)$. Syn: *pitchy iron ore.*

triploblastic Said of the structure of animals having three layers (ectoderm, mesoderm, and endoderm).

triploidite A yellowish to reddish-brown mineral: $(Mn,Fe)_2(PO_4)(OH)$. It is isomorphous with wolfeite.

tripod (a) A sponge spicule having three equal rays that radiate as if from the apex of a pyramid. (b) A stool-shaped shell formed from divergent rods united at a common center in nasselline radiolarians.

tripoli (a) A light-colored porous friable siliceous (largely chalcedonic) sedimentary rock, which occurs in powdery or earthy masses that result from the weathering of chert or of siliceous limestone (Tarr, 1938, p. 27). It has a harsh, rough feel, and is used for the polishing of metals and stones. (b) An incompletely silicified limestone from which the carbonate has been leached; *rottenstone.* (c) A term that was originally, but is now incorrectly, applied to a *siliceous earth* that closely resembles tripoli; specif. *diatomite,* such as the type deposit of northern Africa. See also: *tripolite.*

tripoli-powder An obsolete syn. of *diatomite.*

tripolite A term that has been applied as a syn. of *diatomite,* in reference to the material from the north African location of Tripoli. It has also been used, less correctly, as a syn. of *tripoli.*

triporate Said of pollen grains having three pores, usually disposed at 120° from each other in the equator.

trippkeite A blue-green mineral: $CuAs_2O_4$. It has excellent prismatic cleavage that permits crystals to be broken into flexible fibers.

tripuhyite A greenish-yellow to dark-brown mineral: $FeSb_2O_6$. Syn: *flajolotite.*

triradiate n. A sponge spicule in the form of three coplanar rays radiating from a common center. Syn: *triode.*—adj. Descriptive of trilete spores.

triradiate crest In a trilete spore, the three-rayed raised figure on the proximal surface caused by the intersection of the contact areas. Syn: *triradiate ridge.*

triserial Arranged in, characterized by, or consisting of three rows or series; specif. said of the chambers of a foraminiferal test arranged in three columns or in a series of three parallel or alternating rows, such as a trochospiral test with three chambers in each whorl.

trisoctahedron An isometric crystal form of 24 faces, each of which is an isosceles triangle. Its indices are {*hhk*}. Syn: *triakisoctahedron; trigonal trisoctahedron.*

tristanite Silica-saturated to undersaturated igneous rocks intermediate between *trachyandesite* and *trachyte,* with a differentiation index between 65 and 75 and K_2O:Na_2O greater than 1:2. The name, given by Tilley and Muir in 1964, is for the island of Tristan da Cunha.

tristetrahedron An isometric crystal form having 12 faces that are either triangular (*trigonal tristetrahedron*) or quadrilateral (*deltohedron*). Its indices are {*hkk*} and its symmetry is $\bar{4}3m$ or 23.

tristomodaeal budding A type of budding in scleractinian corals in which three stomodaea are developed within a common tentacular ring and either occur in series or form a triangle, and two interstomodaeal couples of mesenteries are located between the original and each new stomodaeum. See also: *triple stomodaeal budding.*

tritium A radioactive isotope of hydrogen having two neutrons and one proton in the nucleus.

tritium dating Calculation of an age in years by measuring the concentration of radioactive hydrogen-3 (tritium) in a substance, usually water. Maximum possible age limit is about 30 years. The method also provides a means of tracing subsurface movements of water and of determining its velocities.

tritomite A dark-brown mineral: $(Ce,La,Y,Th,Zr)_5(Si,B)_3(O,OH,F)_{13}$ (?).

tritomite-(Y) A mineral: $(Y,Ca,La,Fe)_5(Si,B,Al)_3(O,OH,F)_{13}$(?).

tritonymph The third developmental stage in the arachnid order Acarida.

trittkarren Crescentic solution pockets about 30 cm in diameter on limestone surfaces. Etymol: German, "step tracks". See also: *karren.*

trituration *comminution.*

trivariant Pertaining to a chemical system having three degrees of freedom, i.e. having a variance of three.

trivium (a) The three anterior ambulacra of an echinoid. (b) The part of an asterozoan containing three rays, excluding the bivium. The term is not recommended as applied to asterozoans (TIP, 1966, pt.U, p.30).—Pl: *trivia.* Cf: *bivium.*

TRM *thermoremanent magnetization.*

trochanter (a) The second segment of a pedipalpus or leg of an arachnid, so articulated to coxa and femur as to permit motion of the entire leg in any direction (TIP, 1955, pt.P, p.63). It corresponds physiologically to vertebrate hip articulation. (b) A joint of the proximal part of a prosomal appendage of a merostome. (c) One of up to five processes near the proximal end of the tetrapod femur.

trochiform Shaped like a top; e.g. said of a gastropod shell (e.g. *Trochus*) with a flat base, evenly conical sides, and a not highly acute spire.

trochite A wheel-shaped joint of the stem of a fossil crinoid.

trochoceroid A syn. (esp. in the older literature) of *torticone.*

trochoid adj. (a) Said of a horn-shaped solitary corallite with sides regularly expanding from the apex at an angle of about 40 degrees. Cf: *turbinate; patellate.* (b) Said of a foraminiferal test with spirally or helically coiled chambers, evolute on one side of the test and involute on the opposite side. Syn: *trochospiral.*—n. A trochoid corallite or foraminiferal test.

trochoidal fault A type of *pivotal fault,* the pivotal point of which has also moved or slipped along the fault surface (Nelson & Nelson, 1967).

trocholophe A brachiopod lophophore disposed as a ring surrounding the mouth, bearing a single row of unpaired (or more rarely, a double row of paired) filamentary appendages (TIP, 1965, pt.H, p.154).

trochospiral Said of a foraminiferal test with spirally coiled chambers; *trochoid.*

troctolite (a) In the *IUGS classification,* a plutonic rock satisfying the definition of *gabbro,* in which pl/(pl+px+ol) is between 10 and 90 and px/(pl+px+ol) is less than 5. (b) A *gabbro* that is composed chiefly of calcic plagioclase (e.g. labradorite) and olivine with little or no pyroxene.—Syn: *forellenstein; troutstone.* Such rocks commonly are speckled like trout, hence the three synonyms, derived from Greek, German, and English.

troegerite A lemon-yellow mineral: $(UO_2)_3(AsO_4)_2 \cdot 12H_2O$. Also spelled: *trögerite.*

Tröger's classification A quantitative mineralogic classification of igneous rocks proposed by E. Tröger in 1935.

troglobiont A *troglodyte,* esp. one living in the lightless waters of caves.

troglobite An organism that must live its entire life underground. See also: *troglodyte; troglophile; trogloxene.*

troglodyte Any organism that lives in a cave or rock shelter. Adj: *troglodytic.* Cf: *troglobite; troglophile; trogloxene.* Syn: *troglobiont.*

troglophile Any organism that completes its life cycle in a cave but that also occurs in certain environments outside the cave. Cf: *troglodyte; trogloxene; troglobite.*

trogloxene Any organism that regularly or accidentally enters a cave but that must return to the surface to maintain its existence. Cf: *troglophile; troglodyte; troglobite.*

trogschluss A syn. of *trough end.* Etymol: German *Trogschluss.*

trogtalite An isometric mineral: $CoSe_2$. It is dimorphous with hastite.

troilite A hexagonal mineral present in small amounts in almost all meteorites. FeS. It is a variety of pyrrhotite with almost no ferrous-iron deficiency.

trolleite A mineral: $Al_4(PO_4)_3(OH)_3$.

trona A white or yellow-white monoclinic mineral: $Na_3(CO_3)(HCO_3)\cdot 2H_2O$. It occurs in fibrous or columnar layers and thick beds in saline residues. Trona is a major source of sodium compounds.

trondhjemite A light-colored plutonic rock primarily composed of sodic plagioclase (esp. oligoclase), quartz, sparse biotite, and little or no alkali feldspar; a *quartz diorite* with oligoclase as the sole feldspar. Its name, given by Goldschmidt in 1916, is derived from Trondhjem, Norway. Also spelled: *trondjemite; trondheimite.*

troostite A mineral: $(Zn,Mn)_2SiO_4$. It is a reddish variety of willemite containing manganese and occurring in large crystals.

Tropept In U.S. Dept. of Agriculture soil taxonomy, a suborder of the soil order *Inceptisol,* characterized by formation in an isomesic or warmer temperature regime on moderate to steep slopes in the humid tropics. Most tropepts have either an ochric or an umbric epipedon and a cambic horizon (USDA, 1975). Cf: *Andept; Aquept; Ochrept; Plaggept; Umbrept.*

trophic Of or pertaining to nutrition.

trophic level A stage of nourishment representing one of the segments of the food chain, and characterized by organisms that all obtain food and energy in basically similar fashion.

trophism Nutrition involving metabolic exchanges in the tissues.

trophocyte An amoebocyte of a sponge that serves as a nourishing cell of an oocyte, embryo, or gemmule.

trophogenic Said of the upper or illuminated zone of a lake in which photosynthesis converts inorganic matter to organic matter. Cf: *tropholytic.* See also: *euphotic zone.*

tropholytic Said of the deeper part of a lake in which organic matter tends to be dissimilated. Cf: *trophogenic.* See also: *profundal.*

tropic n. The area of the Earth falling between the tropics of Cancer and Capricorn. Generally used in the plural.—adj. Pertaining to features, climate, vegetation and animals characteristic of the region between the tropics of Cancer and Capricorn.

tropical (a) Said of a climate characterized by high temperature and humidity and by abundant rainfall. An area of tropical climate borders that of *equatorial* climate. (b) Pertaining to the tropic regions.

tropical cyclone An intense *cyclone* that forms over the tropical oceans and ranges from 100 to 1000 km in diameter. It moves first westward, then northeastward in the Northern Hemisphere and southeastward in the Southern Hemisphere; its wind velocity is moderate at its fringe area but may increase to as much as 200 knots (230 mph) near its center. The center itself, or *eye,* is often calm and clear. Cf: *hurricane; typhoon.*

tropical desert A hot, dry desert lying between latitude 15° and 30° north or south of the Equator, more specifically near the tropics of Cancer or Capricorn, where subtropical high-pressure air masses prevail, producing conditions of very low, sporadic rainfall. See also: *west-coast desert.* Syn: *low-latitude desert; trade-wind desert.*

tropical lake A lake whose surface temperature is constantly above 4°C. Cf: *polar lake.*

tropical spine A radial spine disposed according to the Müllerian law and marking a zone in an acantharian radiolarian comparable to the tropical zone of the terrestrial globe.

tropic current A twice-monthly type of tidal current that occurs in conjunction with the maximum declinations of the Moon.

Tropic of Cancer The parallel of latitude approx. 23 1/2° north of the Equator, which indicates the northernmost latitude reached by the Sun's vertical rays.

Tropic of Capricorn The parallel of latitude approx. 23 1/2° south of the Equator, which indicates the southernmost latitude reach by the Sun's vertical rays.

tropic tide A tide occurring twice monthly when the Moon is at its maximum declination north or south of the Earth's equator (when the Moon is nearly above the tropics of Cancer and Capricorn), and displaying the greatest diurnal inequality. Cf: *equatorial tide.*

tropism An involuntary orientational movement or growth in which an organism or one of its parts turns or curves as a positive or negative response to a stimulus. It may be indistinguishable from *taxis [ecol]* in motile organisms.

tropophilous Said of an organism that has adapted physiologically to periodic changes in the environment, e.g. seasonal changes.

tropophyte A plant adapted to seasonal changes in moisture and temperature; e.g. the deciduous trees of temperate and tropical regions.

troposphere That portion of the atmosphere next to the Earth's surface, in which temperature generally decreases rapidly with altitude, clouds form, and convection is active. In middle latitudes the troposphere generally includes the first 10 to 12 km above the Earth's surface. Cf: *stratosphere.*

tropotaxis *Taxis [ecol]* resulting from the simultaneous comparison of stimuli of different intensity that are acting on separate ends of the organism.

trottoir A narrow, organic, intertidal reef construction, composed of either a solid mass or a simple crust covering a rocky substratum, separating the shoreline from the sea in the same manner that a sidewalk separates the street from the adjoining houses. Etymol: French, "sidewalk".

trough [beach] A small linear depression formed just offshore on the bottom of a sea or lake and on the landward side of a *longshore bar.* It is generally parallel to the shoreline, and is always under water. A trough may be excavated by the extreme turbulence of wave and current action in the zone where breakers collapse. Inappropriate syn: *runnel; low.*

trough [fault] *graben.*

trough [fold] The lowest point of a given stratum in any profile through a fold. Cf: *trough line.*

trough [geomorph] (a) Any long, narrow depression in the Earth's surface, such as one between hills or with no surface outlet for drainage; esp. a broad, elongate U-shaped valley, such as a *glacial trough* or a *trench.* (b) The channel in which a stream flows.

trough [marine geol] An elongate depression of the sea floor that is wider and shallower than a *trench*, with less steeply dipping sides. Troughs and trenches are gradational forms; a trough may develop from a trench by becoming filled with sediment.

trough [paleont] (a) The furrow on the posterior part of the pedicle valve of an inarticulate brachiopod beneath the apex which provides space for the pedicle (Moore et al., 1952, p.207). (b) *hinge trough.*

trough [sed] *geosynclinal trough.*

trough banding Rhythmic layering or alignment of minerals in an igneous rock, confined to troughlike depressions.

trough cross-bedding Cross-bedding in which the lower bounding surfaces are curved surfaces of erosion (McKee & Weir, 1953, p.385); it results from local scour and subsequent deposition. See also: *festoon cross-bedding.* Syn: *crescent-type cross-bedding.*

trough end The steep, semicircular rock wall forming the abrupt head or end of a glacial trough. See also: *oversteepened wall.* Syn: *trough wall; trogschluss.*

trough fault A fault, generally a normal fault, which bounds a graben or other structural depression. Cf: *ridge fault.*

trough-in-trough Said of a cross profile depicting two or more glaciations, each of which shaped its own troughlike valley, esp. where a steep-sided inner trough lies within a wider trough with a flatter bottom.

trough line The line joining the *trough* points of a given stratum. Cf: *trough.*

trough reef *reverse saddle.*

trough surface A surface that connects the *trough lines* of the beds of a syncline.

trough valley *U-shaped valley.*

trough wall *trough end.*

troutstone *troctolite.*

trowlesworthite A coarse-grained plutonic rock characterized by the presence of red orthoclase, tourmaline, fluorite, and quartz. According to Johannsen (1939, p. 285), it may represent a vein in granite that has undergone pneumatolysis. The name, given by Worth in 1887, is for Trowlesworthy, Devonshire, England. Obsolete.

trudellite A mixture of chloraluminite and natroalunite.

true azimuth The *azimuth [surv]* measured clockwise from true north through 360 degrees.

true bearing The *bearing* expressed as a horizontal angle between a geographic meridian and a line on the Earth; esp. a horizontal angle measured clockwise from true north. Cf: *magnetic bearing.*

true cleavage A quarrymen's term for the dominant cleavage in a rock, e.g. slaty cleavage, to distinguish it from minor or *false cleavage.* Geologically, the term is misleading and should be avoided.

true crater A primary depression, formed by impact or explosion, before modification by slumping or by deposition of ejected material; the crater prior to *fallback* of debris. "The true crater is defined as the boundary between the loose, broken fallback material and the underlying material that has been crushed and fractured but has not experienced significant vertical displacement" (Nordyke, 1961, p. 3447). Cf: *apparent crater.* Syn: *primary crater.*

true dip A syn. of *dip,* used in comparison with *apparent dip.* Syn: *full dip.*

true folding Folding due to lateral compression. Ant: *false folding.* Syn: *buckle folding.*

true granite A syn. of *two-mica granite,* used by Rosenbusch. Obsolete.

true homology *homology.*

true horizon (a) A *celestial horizon.* Also, the horizon at sea. (b) A horizontal plane passing through a point of vision or a perspective center. The *apparent horizon* approximates the true horizon only when the point of vision is very close to sea level.

true north The direction from any point on the Earth's surface toward the geographic north pole; the northerly direction of any geographic meridian or of the meridian through the point of observation. It is the universal zero-degree (or 360-degree) mapping reference. True north differs from *magnetic north* by the amount of magnetic declination at the given point. Syn: *geographic north.*

true resistivity The resistivity of a locally homogeneous medium. Cf: *apparent resistivity.*

true soil *solum.*

true thickness The *thickness* of a stratigraphic unit or other tabular body, measured at right angles to the direction of extension of the unit or body. See also: *vertical [stratig].* Cf: *apparent thickness.*

truffite Nodular masses of woody lignite occurring within Cretaceous lignite of France. It is so named because of its trufflelike odor.

trug A term used in Devon, England, for red limestone.

Truman gravimeter One of the earliest practical field gravity meters of the unstable equilibrium type (Nettleton, 1940, p.32).

trumpet log *Microlaterolog.*

trumpet valley A narrow valley or gorge that cuts through the central part of a lobe (in the morainal landscape of a former piedmont glacier) and opens out into a broad funnel as it reaches the glaciofluvial sand and gravel cone or fan of the lower piedmont. Examples are numerous along the northern Alpine foothills in Bavaria.

truncate v. In crystal structure, to replace the corner of a crystal form with a plane. Such a crystal form is said to be truncated.

truncated [geomorph] Said of a landform (such as a headland or mountain), or of a geologic structure, that has been abbreviated by *truncation;* esp. said of a conical eminence (such as a volcano) whose apical part has been replaced by a plane section parallel to the land or ocean surface.

truncated [soil] Said of a soil profile in which upper horizons are missing.

truncated spur A spur that projected into a preglacial valley and was partially worn away or beveled by a moving glacier as it widened and straightened the valley. See also: *faceted spur.*

truncated wave-ripple lamina One of a set of parallel laminae that were deposited conformably with the original wave-rippled surface but were partly eroded before deposition of the next set of wave-ripple laminae. Ripple wavelengths range from 10 cm to 10 m (Campbell, 1966, p. 825). Cf: *hummocky cross-stratification.*

truncation [geomorph] An act or instance of cutting or breaking-off the top or end of a geologic structure or landform, as by erosion. Cf: *beveling.*

truncation [paleont] The natural loss, in life, of the apical portion of a nautiloid shell.

trunk [paleont] (a) The *thorax* of an arthropod; esp. the postcephalic part of the body of a crustacean. (b) The axial skeleton of a vertebrate, esp. a tetrapod, less neck and tail and with or without associated soft parts.

trunk [streams] The principal channel of a system of tributaries; the channel of a trunk stream.

trunk glacier A central or main valley glacier formed by the union of several tributary glaciers. See also: *dendritic glacier.*

trunk stream A *main stream* having an axial or central position in a drainage system. Syn: *stem stream.*

truscottite A mineral: $(Ca,Mn)_2Si_4O_9(OH)_2$. It is perhaps identical with reyerite.

trüstedtite A mineral: Ni_3Se_4.

tschermakite (a) A mineral of the amphibole group: $Ca_2Mg_3(Al,Fe^{+3})_2(Al_2Si_6)O_{22}(OH,F)_2$. Not to be confused with *Ca-Tschermak molecule.* (b) A gray-white feldspar (albite?) containing some magnesium but no calcium, from Bamble, Norway. (c) A plagioclase feldspar (oligoclase or albite) with composition ranging from $Ab_{95}An_5$ to $Ab_{80}An_{20}$.

Tschermak molecule *Ca-Tschermak molecule.*

tschermigite A mineral of the alum group: $(NH_4)Al(SO_4)_2 \cdot 12H_2O$. Syn: *ammonia alum; ammonium alum.*

Tschernosem *Chernozem.*

Tschernosiom *Chernozem.*

T-S diagram *temperature-salinity diagram.*

tsilaisite Manganese-rich variety of tourmaline.

tsingtauite A porphyritic *granite* that contains only feldspars as phenocrysts. The feldspars are microperthite and sodic plagioclase. The name, given by Rinne in 1904, is for Tsingtau, Korea. Not recommended usage.

tsumcorite A monoclinic mineral: $PbZnFe^{+2}(AsO_4)_2 \cdot H_2O$.

tsumebite An emerald-green mineral: $Pb_2Cu(PO_4)(SO_4)(OH)$.

tsunami A gravitational sea wave produced by any large-scale, short-duration disturbance of the ocean floor, principally by a shallow submarine earthquake, but also by submarine earth movement, subsidence, or volcanic eruption. It is characterized by great speed of propagation (up to 950 km/hr), long wavelength (up to 200 km), long period (varying from 5 min to a few hours, generally 10-60 min), and low observable amplitude on the open sea, although it may pile up to heights of 30 m or more and cause much damage on entering shallow water along an exposed coast (often thousands of kilometers from the source). Etymol: Japanese, "harbor wave". Pron: (t)soo-*nah*-me. Pl: *tsunamis; tsunami.* Adj: *tsunamic.* Erroneous syn: *tidal wave.* Syn: *seismic sea wave; seismic surge; earthquake sea wave; tunami.*

t test *Student's t test.*

tube [paleont] (a) The central cylinder connecting the two shields of a placolith coccolith. (b) One of the siphons of a bivalve mollusk.

tube [speleo] A smooth-sided cave *passage* that is elliptical to circular in cross section. See also: *half-tube; streamtube.*

tube foot One of numerous small tentacular flexible organs of echinoderms, being the end of a branch of the water-vascular system and serving for grasping, adhesion, locomotion, respiration, feeding, or combination of these.

tubercle (a) One of the small rounded knoblike structures on the outer surface of the test plates of an echinoid, bearing a movable spine. (b) A low rounded prominence of intermediate size on the surface of an ostracode valve, commonly along the free margin (TIP, 1961, pt.Q, p.55). See also: *eye tubercle.* (c) Any fine low rounded protuberance on either surface of a brachiopod valve, irrespective of origin (TIP, 1965, pt.H, p.154). (d) A small, rounded, moderately prominent elevation on the surface of a gastropod shell (TIP, 1960, pt.I, p.133). (e) Any small process of a bone.

tubercle texture In mineral deposits, a texture in which gangue is replaced by automorphic minerals. Cf: *atoll texture.*

tuberose Said of a mineral whose form is that of irregular, rootlike shapes or branches.

tube-wave A seismic surface wave traveling along the walls of a borehole; it is involved in acoustic logging.

tube well (a) *driven well.* (b) *tubular well.*—Also spelled: *tube-well.*

tubing (a) A small-diameter removable pipe, suspended and immobilized in a well inside a large-diameter *casing* and opening at a producing zone, through which fluids are produced (brought to the surface). (b) The act or process of placing tubing in a well.

tubular spring A gravity or artesian spring whose water issues from rounded openings, such as lava tubes or solution channels.

tubular stalactite A thin hollow stalactite that maintains the diameter of a drop of water. Syn: *straw stalactite; soda straw.*

tubular well General term for a drilled, driven, or bored well of circular cross section, and of a depth that is relatively great compared to diameter. Syn: *tube well.*

tubule A hollow twiglike calcareous concretion, characteristic of loess deposits.

tubulospine A foraminiferal chamber produced radially into a long hollow extension (as in *Schackoina*).

tubulus In the archaeocyathids, a radiating porous-walled prismatic tubule in the intervallum of the Syringocnemidida (TIP, 1972, pt. E, p. 17). Pl: *tubuli.*

tucanite *scarbroite.*

tufa A chemical sedimentary rock composed of calcium carbonate, formed by evaporation as a thin, surficial, soft, spongy, cellular or porous, semifriable incrustation around the mouth of a hot or cold calcareous spring or seep, or along a stream carrying calcium carbonate in solution, and exceptionally as a thick, bulbous, concretionary or compact deposit in a lake or along its shore. It may also be precipitated by algae or bacteria. The hard, dense variety is *travertine.* The term is rarely applied to a similar deposit consisting of silica. It is not to be confused with *tuff.* Etymol: Italian *tufo.* Cf: *sinter.* Syn: *calcareous tufa; calc-tufa; tuft; petrified moss; tophus.*

tufaceous Pertaining to or like tufa. Not to be confused with *tuffaceous.*

tuff A general term for all consolidated pyroclastic rocks. Not to be confused with *tufa.* Adj: *tuffaceous.*

tuffaceous Said of sediments containing up to 50% *tuff.*

tuff ball An individual unit in a mass of *accretionary lapilli.*

tuff breccia A pyroclastic rock consisting of more or less equal amounts of ash, lapilli, and larger fragments.

tuffeau A term used in France for tufa, micaceous chalk, and soft, very porous, extremely coarse limestone made up of bryozoan fragments.

tuffisite A term proposed by Cloos (1941) for fragmented country rock in pipes located in Swabia, southwest Germany. Syn: *intrusive tuff.*

tuffite A tuff containing both pyroclastic and detrital material, but predominantly pyroclasts.

tufflava An extrusive rock containing both pyroclastic and lava-flow characteristics, so that it is considered to be an intermediate form between a lava flow and a *welded-tuff* type of *ignimbrite.* Whether or not it is actually a genetically distinct type of rock is a matter of debate. Also spelled: *tuffolava; tuff lava; tuflava.* Cf: *ignispumite.*

tuffolava *tufflava.*

tuff ring A wide, low-rimmed, well-bedded accumulation of hyaloclastic debris built around a volcanic vent located in a lake, coastal zone, marsh, or area of abundant ground water. Syn: *maar.*

tuffstone A sandstone that contains pyroclasts of sand-grain size.

tuft A term used in England for any porous or soft stone, such as the sandstone in the Alston district of Cumberland, and the rock now known as *tufa* (Arkell & Tomkeieff, 1953, p. 121).

tugtupite A mineral: $Na_4BeAlSi_4O_{12}Cl$. Syn: *beryllosodalite.*

tuhualite A mineral: $(Na,K)_2(Fe^{+2},Fe^{+3},Al)_3Si_7O_{18}(OH)_2$.

tulameenite A tetragonal mineral: Pt_2FeCu.

tulare A syn. of *tule land.* Etymol: Spanish, "tule field".

tule *tule land.*

tule land A local term given in the Sacramento River valley, California, to a large tract of overflowed land (or *flood basin*) on which the tule, a variety of bulrush, is the dominant or characteristic native plant. The land is often referred to as *tule* or "the tules". Etymol: Spanish. Syn: *tulare.*

tumescence The swelling of a volcanic edifice due to accumulation of magma in the reservoir. It may or may not be followed by eruption. Syn: *bulge; inflation.*

tump (a) A mound, hummock, hillock, or other small rise of ground. (b) A clump of vegetation, such as trees, shrubs, or grass; esp. one forming a small island in a marsh or swamp.

tumuli Plural of *tumulus.*

tumulus [paleont] (a) A secondary deposit on the chamber floor of a foraminiferal test, appearing in cross section as more or less a symmetric node with a rounded summit. (b) A minor bulge in the wall of the cup in archaeocyathids; it may be perforated by one or many pores.—Pl: *tumuli.*

tumulus [volc] A doming or small mound on the crust of a lava flow, caused by pressure due to the difference in rate of flow between the cooler crust and the more fluid lava below. Unlike a *blister [volc],* it is a solid structure. Pl: *tumuli.* Syn: *pressure dome.*

tunami *tsunami.*

tundra A treeless, level or gently undulating plain characteristic of arctic and subarctic regions. It usually has a marshy surface, which supports a growth of mosses, lichens, and numerous low shrubs and is underlain by a dark, mucky soil and permafrost.

tundra climate A type of *polar climate* having a mean temperature in the warmest month of between 0° and 10°C. Cf: *perpetual frost climate.*

tundra crater A circular or shapeless "island" of silt (without vegetation) found in tundra and formed during the period of thawing by the forced rise of silt to the surface and its pouring-out like lava onto the surface.

tundra ostiole A *mud circle* in the tundra soil of northern Quebec. Syn: *ostiole [geomorph].*

tundra peat Peat occurring in subarctic areas and derived from mosses, heaths, and birch and willow trees.

tundra polygon *ice-wedge polygon.*

Tundra soil A great soil group in the 1938 classification system, a group of zonal soils having dark brown, highly organic upper horizons and grayish lower horizons. It is developed over a permafrost substratum in the tundra, under conditions of cold, humidity, and poor drainage (USDA, 1938). Most of these soils are now classified as *Aquepts.*

tundrite A mineral: $Na_3\,(Ce,La)_4\,(Ti,Nb)_2\,(SiO_4)_2\,(CO_3)_3\,O_4\,(OH)\cdot 2H_2O$. Syn: *titanorhabdophane.*

tundrite-(Nd) A triclinic mineral: $Na_3(Nd,La)_4(Ti,Nb)_2(SiO_4)_2(CO_3)_3O_4(OH)\cdot 2H_2O$.

tunellite A colorless monoclinic mineral: $SrB_6O_{10}\cdot 4H_2O$.

tungomelane A variety of psilomelane containing tungsten.

tungstate A mineral compound characterized by the radical WO_4, in which the hexavalent tungsten ion and the four oxygens form a flattened square rather than a tetrahedron. An example of a tungstate is wolframite, $(Fe,Mn)WO_4$. Tungsten and molybdenum may substitute for each other. Cf: *molybdate.*

tungsten (a) An obsolete term formerly applied to tungsten minerals such as scheelite and wolframite. (b) A metallic element with atomic number 74. Syn: *wolfram.*

tungstenite A mineral: WS_2. It occurs in small lead-gray folia or scales. The term is "sometimes erroneously given as a translation from German or Russian for *wolframite*" (Fleischer, 1966, p. 1317).

tungstenite-3R A trigonal mineral: WS_2.

tungstic ocher (a) *tungstite.* (b) *ferritungstite.*—Syn: *wolfram ocher.*

tungstite An earthy mineral: $WO_3\cdot H_2O$. It occurs in yellow or yellowish-green pulverulent masses. Syn: *tungstic ocher; wolframine.*

tungusite A mineral: $Ca_4Fe_2Si_6O_{15}(OH)_6$.

tuning-fork spicule A triradiate sponge spicule in which two of the rays are subparallel and at approximately 180 degrees to the third ray. They are found in the class Calcarea.

tunisite A tetragonal mineral: $NaHCa_2Al_4(CO_3)_4(OH)_{10}$.

tunnel [mining] Strictly speaking, a passage in a mine that is open to the surface at both ends. It is often used loosely as a synonym for *adit* or *drift.*

tunnel [paleont] A low slitlike opening representing a resorbed area at the base of septa in the central part of the test in many fusulinids, serving to facilitate communication between adjacent chambers.

tunnel [speleo] *natural tunnel.*

tunnel cave *natural tunnel.*

tunneldal A syn. of *tunnel valley.* Etymol: Danish, "tunnel valley".

tunnel erosion *piping.*

tunneling (a) The operation of excavating, driving, and lining tunnels. (b) A form of failure occurring in earth dams and embankments, in which a tunnel is created when cracks, developed in the structure under dry conditions, collapse internally when brought into sudden contact with water. It starts at the wet face or upstream side and proceeds downstream.

tunnel lake A glacial lake occupying a *tunnel valley.*

tunnel valley A shallow trench cut in drift and other loose material, or in bedrock, by a subglacial stream not loaded with coarse sediment. See also: *ice-walled channel.* Syn: *tunneldal; Rinnen-*

tal.

tunnerite A mineral that may be identical with woodruffite.

Tuorian European stage: Upper Cambrian (above Mayan, below Shidertinian).

turanite An olive-green mineral: $Cu_5(VO_4)_2(OH)_4$ (?).

turbation Churning, stirring, mixing, or other modifications of a sediment or soil by agents not determined. The term is generally preceded by a prefix denoting agent, if known; e.g. *congeli-,* frost, or *bio-,* organisms.

turbid Stirred up or disturbed, such as by sediment; not clear or translucent, being opaque with suspended matter, such as of a sediment-laden stream flowing into a lake; cloudy or muddy in physical appearance, such as of a feldspar containing minute inclusions. Cf: *roily.*

turbidimeter An instrument for measuring or comparing the turbidity of liquids in terms of the reduction in intensity of a light beam passing through the medium. See also: *transmissometer.*

turbidimetry The measurement of the amount of suspended or slow-settling matter in a liquid; the measurement of the decrease in intensity of a light beam passed through a medium. Cf: *nephelometry.*

turbidite A sediment or rock deposited from, or inferred to have been deposited from, a turbidity current. It is characterized by graded bedding, moderate sorting, and well-developed primry structures in the sequence noted in the *Bouma cycle.*

turbidity (a) The state, condition, or quality of opaqueness or reduced clarity of a fluid, due to the presence of suspended matter. (b) A measure of the ability of suspended material to disturb or diminish the penetration of light through a fluid.

turbidity current A *density current* in water, air, or other fluid, caused by different amounts of matter in suspension, such as a dry-snow avalanche or a descending cloud of volcanic dust; specif. a bottom-flowing current laden with suspended sediment, moving swiftly (under the influence of gravity) down a subaqueous slope and spreading horizontally on the floor of the body of water, having been set and/or maintained in motion by locally churned- or stirred-up sediment that gives the water a density greater than that of the surrounding or overlying clear water. Such currents are known to occur in lakes, and are believed to have produced the submarine canyons notching the continental slope. They appear to originate in various ways, such as by storm waves, tsunamis, earthquake-induced sliding, tectonic movement, over-supply of sediment, and heavily charged rivers in spate with densities exceeding that of sea water. The term was introduced by Johnson (1939, p. 27), and is applied to a current due to turbidity, not to one showing that property. See also: *turbidity flow; suspended current.* Cf: *tractive current.* Syn: *suspension current.*

turbidity fan A local fan-shaped area of turbid water at the mouth of a stream flowing into a lake or adjacent to an eroding bank of a lake (Veatch & Humphrys, 1966, p.334).

turbidity flow A tongue-like flow of dense, muddy water moving down a slope; the flow of a *turbidity current.*

turbidity limestone A limestone indicating resedimentation by turbidity currents (Bissell & Chilingar, 1967, p. 168).

turbidity size analysis A kind of particle-size analysis based upon the amount of material in turbid suspension, the turbidity decreasing as the particles settle.

turbinate (a) Said of a horn-shaped solitary corallite with sides expanding from apex at an angle of about 70 degrees. Cf: *trochoid; patellate.* (b) Shaped like a top; e.g. said of a spiral gastropod shell with a generally rounded base, a broadly conical spire, and whorls decreasing rapidly from base to apex, or said of a protist shaped like a cone with the point down. (c) Pertaining to trabeculate, coiled sheets of bone springing from the inner walls of the nasal passage in mammals, which support a large surface of mucous membrane in the air passage.

turbodrill In *rotary drilling,* a drill bit that is directly rotated by a turbine attached to the drill pipe at the bottom of the hole and driven by drilling mud pumped under high pressure. It was developed in the U.S.S.R. for drilling deep oil wells.

turboglyph A current-produced hieroglyph (Vassoevich, 1953, p.36 & 65); specif. a *flute cast.*

turbulence *turbulent flow.*

turbulence spectrum *eddy spectrum.*

turbulent diffusion *eddy diffusion.*

turbulent flow Water flow in which the flow lines are confused and heterogeneously mixed. It is typical of flow in surface-water bodies. Cf: *laminar flow [hydraul]; mixed flow.* Syn: *turbulence; tortuous flow.*

turbulent flux *eddy flux.*

turbulent velocity That velocity of water in a stream above which the flow is turbulent, and below which it may be either laminar or turbulent. Cf: *laminar velocity.*

turf Peat that has been dried for use as a fuel.

turf-banked terrace A *nonsorted step* whose riser is covered by vegetation and whose tread is composed of fine soil. The term should be reserved for an irregular, terracelike feature that is not a clearly defined form of patterned ground (Washburn, 1956, p.835). Syn: *turf garland.*

turf garland *turf-banked terrace.*

turf hummock A hummock of turf rather than soil. Cf: *earth hummock.*

turgite A red fibrous mineral: $Fe_2O_3 \cdot nH_2O$. It is equivalent to hematite with adsorbed water or to an iron ore intermediate between hematite and goethite (hematite being in excess). It occurs as a ferruginous cement in sandstones. Syn: *hydrohematite.*

turjaite A dark-colored plutonic *melilitolite* containing nepheline and 60 to 90 percent mafic minerals. The presence of melilite distinguishes turjaite from *melteigite.* The name, given by Ramsay in 1921, is derived from Turja, on the Kola Peninsula, U.S.S.R. Cf: *okaite.* Not recommended usage.

turjite An *alnoite* containing calcite, melanite, and analcime. The name, given by Belijankin in 1924, is for Turij Mis (Turja), Kola Peninsula, U.S.S.R. Not recommended usage.

turkey-fat ore A popular name used in Arkanasas and Missouri for smithsonite colored yellow by greenockite. Syn: *turkey ore.*

turkey shoot A comparison of the results of recording with two or more sets of seismic instruments simultaneously under the same field conditions.

Turkey stone (a) A very fine-grained siliceous rock, containing up to 25% calcite, quarried in central Turkey and used as a whetstone; *novaculite.* Syn: *Turkey slate.* (b) *turquoise.*

turlough An Irish term for a winter lake that is dry or marshy in summer. Also, the ground or hollow periodically flooded to form a turlough. Etymol: Gaelic *turloch,* "dry lake".

turma An artificial suprageneric grouping of form genera of fossil spores and pollen (mostly pre-Cenozoic) based on morphology. It is subdivided into other groups such as "subturmae" and "infraturmae". The system is not governed by the International Code of Botanical Nomenclature. Pl: *turmae.* See also: *ante-turma.*

turmaline Var. of *tourmaline.* The mineral name was originally spelled: "turmalin".

turmkarst A syn. of *tower karst.* Etymol: German.

turning point (a) A surveying point on which a level rod is held, after a foresight has been made on it, and before the differential-leveling instrument is moved to another station so that a backsight may be made on it to determine the height of instrument after the resetting; a point of intersection between survey lines, such as the intervening point between two bench marks upon which rod readings are taken. It is established for the purpose of allowing the leveling instrument to be moved forward (alternately leapfrogging with the rod) along the line of survey without a break in the series of measured differences of elevation. Abbrev: TP. (b) A physical object representing a turning point, such as a steel pin or stake driven into the ground.

turnover [ecol] (a) The process by which some species become extinct in a region and are replaced by other species (MacArthur and Wilson, 1967, p.191). (b) The number of animal generations that replace each other during a given length of time (Thorson, 1957, p.491).

turnover [lake] A period (usually in the fall or spring) of uniform vertical temperature when convective circulation occurs in a lake; the time of an *overturn.* See also: *circulation [lake].*

Turolian European land-mammal age: Late Miocene (above Vallesian, below Ruscinian). See also: *Pontian; Pannonian.*

Turonian European stage: Upper Cretaceous, or Middle Cretaceous of some authors (above Cenomanian, below Coniacian).

turquoise A triclinic mineral: $CuAl_6(PO_4)_4(OH)_8 \cdot 5H_2O$. It is isomorphous with chalcosiderite. Turquoise is blue, blue-green, or yellowish green; when sky blue it is valued as the most important of the nontransparent gem materials. It usually occurs as reniform masses with a botryoidal surface, in the zone of alteration of aluminum-rich igneous rocks (such as trachytes). Syn: *turquois; Turkey stone; calaite.*

turrelite An asphaltic shale found in Texas.

turriculate Turreted, or furnished with or as if with turrets; esp.

said of a gastropod shell (e.g. of *Turritella*) with an acutely or highly conical spire composed of numerous flat-sided whorls.

turtleback An extensive smooth curved topographic surface, apparently unique to the Death Valley (Calif.) region, that resembles the carapace of a turtle or a large, elongate dome with an amplitude up to a few thousand meters. Turtlebacks were first mapped, described, and named by Curry (1938).

turtle stone A large, flattened oval *septarium* released from its matrix and so weathered and eroded that the interior vein-filled system of cracks may be seen. Its form has a rough resemblance to that of a turtle and its polished surface bears a fancied resemblance to a turtle's back. Such concretions are abundant in the Devonian shales of eastern North America. Also spelled: *turtlestone.* Formerly called: *beetle stone.*

tusculite A melilite *leucitite* that contains small quantities of pyroxene, ilmenite, and feldspar. Its name, given by Cordier in 1868, is derived from Tusculum, Italy. Not recommended usage.

tussock A dense tuft of grass or grasslike plants usually forming one of many firm hummocks in the midst of a marshy or boggy area.

tussock-birch-heath polygon A *vegetation polygon* characterized by the assemblage indicated (Hopkins & Sigafoos, 1951, p.52-53, 87-92); permafrost appears essential for its formation. Diameter: 2-5 m.

tussock ring A *nonsorted circle* consisting of tussocks surrounding a patch of bare soil.

Tuttle lamellae Planes of inclusions in quartz, oriented randomly rather than with reference to the enclosing crystal. Cf: *Boehm lamellae.*

tutvetite A light-reddish trachytoid rock composed chiefly of alkali feldspars, a decomposed mafic mineral (possibly acmite), and accessory pyrite and possibly anatase and nordenskiöldine; an altered *trachyte.* Its name (Johannsen, 1938) is derived from Tutvet, Norway. Not recommended usage.

tuxtlite A clinopyroxene intermediate in composition between jadeite and diopside: $NaCaMgAlSi_4O_{12}$. It is a pea-green variety of jadeite containing magnesium and calcium and found in Tuxtla, SE Mexico. It may be identical with omphacite. Syn: *diopside-jadeite.*

tuya A flat-topped and steep-sided volcano that erupted into a lake thawed in a glacier by the volcano's heat. Examples of tuyas occur in northern British Columbia.

tvalchrelidzeite A monoclinic mineral: $Hg_{12}(Sb,As)_8S_{15}$.

tveitåsite A dark-colored, medium- to fine-grained contact igneous rock composed chiefly of a clinopyroxene (acmite-diopside) and alkali feldspar (orthoclase, cryptoperthite, microperthite, albite), with accessory sphene, apatite, pyroxene, and possibly nepheline and calcite. The rock is probably a hybrid (Johannsen, 1939, p. 285). The name, given by Brögger in 1921, is from Tveitåsen, Fen complex, Norway. Cf: *fenite.* Not recommended usage.

T wave A short-period (0.5 sec) acoustic wave in the sea.

twenty-degree discontinuity The break in the traveltime curve of seismic *P* waves, originally defined as occurring at an angular distance of about 20° and at a depth of 413 km. Later studies determined the angular distance to be 15°. The term is also written as "20° discontinuity".

twig *divining rod.*

twilight zone *disphotic zone.*

twin A rational intergrowth of two or more single crystals of the same mineral in a mathematically describable manner, so that some lattices are parallel whereas others are in reversed position. The symmetry of the two parts may be reflected about a common plane, axis, or center. See also: *twinning.* Syn: *twin crystal; twinned crystal.*

twin axis The crystal axis about which one individual of a twin crystal may be rotated (usually 180°) to bring it into coincidence with the other individual. It cannot be coincident with the axes of twofold, fourfold, or sixfold symmetry. Cf: *twin plane; twin center.* Syn: *twinning axis.*

twin center The crystal point about which the individuals of a twin may be symmetrically arranged. Cf: *twin plane; twin axis.*

twin crystal *twin.*

twin gliding *Crystal gliding* that results in the formation of crystal twins.

twin law A definition of a twin relationship in a given mineral or mineral group, specifying the twin axis, center, or plane, defining the composition surface or plane if possible, and giving the type of twin.

twinned crater A lunar surface feature consisting of two craters with overlapping rims.

twinned crystal *twin.*

twinning The development of a twin crystal by growth, transformation, or gliding.

twinning axis *twin axis.*

twinning displacement Displacement in a crystal due to twin gliding.

twinning plane *twin plane.*

twinnite A mineral: $Pb(Sb,As)_2S_4$.

twin plane The common plane across which the individual components of a crystal twin are symmetrically arranged or reflected. It is parallel to a possible crystal face but not to a plane of symmetry of a single crystal. It is usually identical with the *composition surface.* Cf: *twin axis; twin center.* Syn: *twinning plane.*

twin shell Shell of a spumellarian radiolarian with a median transverse constriction.

two-circle goniometer A *goniometer* that measures the azimuth and polar angles for the pole of each face on a crystal by reflection of a parallel beam of light from a cross slit. The two circle angles are readily plotted in stereographic or gnomonic projection for indexing in any crystal system. Cf: *contact goniometer; reflection goniometer.*

two-cycle coast A coast characterized by *two-story cliffs* (Cotton, 1922, p. 426).

two-cycle valley A valley produced by rejuvenation, as by headward erosion or by differential earth movements, and characterized by a *valley-in-valley* cross profile. Syn: *two-story valley.*

two-dimensional method A simplified method for calculating the effect on gravity of geological structures in section, in which these structures are assumed to be infinitely long at right angles to the section. Syn: *profile method.*

two-layer structure A type of *layer structure* having two layers to the full repeat unit; e.g. kaolinite, which has one octahedral and one tetrahedral layer per unit along the c axis. Cf: *three-layer structure.*

twoling A crystal twin consisting of two individuals. Cf: *trilling; fourling; fiveling; eightling.*

two-mica granite A *granite* containing both dark mica (biotite) and light mica (muscovite). This granite was called *true granite* by Rosenbusch and *binary granite* by Keyes. Cf: *aplogranite; granitelle.*

two-phase inclusion An angular cavity in a gemstone consisting of a gas bubble in a liquid; the cavity may or may not coincide with a possible crystal form of the host mineral. Examples occur in corundum. Cf: *three-phase inclusion.*

two-point problem A problem in planetable surveying of determining the position of a point with the known factor being the length of one line that does not include the point to be located. Cf: *three-point problem [surv].*

two-story cliff A sea cliff consisting of an ancient, uplifted cliff of a former shoreline cycle separated from a lower-lying cliff of a later cycle by a narrow wave-cut bench. Syn: *two-storied cliff.*

two-story valley *two-cycle valley.*

two-sweep cusp A *meander cusp* formed by the sweep of two successive meanders migrating downstream while the stream remains on the same side of the flood plain (Lobeck, 1939, p. 241).

two-swing cusp A *meander cusp* formed by two successive swings of a meander belt, the scar produced by the first swing intersecting the scar of the second swing in such a way that a Y-shaped feature results, with the handle of the Y pointing either upstream or downstream (Lobeck, 1939, p. 240-241).

two-year ice *second-year ice.*

tychite A white isometric mineral: $Na_6Mg_2(CO_3)_4(SO_4)$.

tychopotamic Said of an aquatic organism adapted to living chiefly in still, fresh water. Cf: *autopotamic; eupotamic.*

tying bar *tombolo.*

tylaster A small tylote aster (sponge spicule).

Tyler standard grade scale A *grade scale* for the particle-size classification of sediments and soils, devised by the W.S. Tyler Company of Cleveland, Ohio; it is based on the square root of 2, with the midpoint values of each size class being simple whole numbers or common fractions. It is used as specifications for sieve mesh.

tylosis A proliferation of a wood parenchyma cell through a pit-pair into the lumen of an adjacent inactive tracheary element, where it may or may not divide (Record, 1934, p.68) Cf: *tylosoid.*

tylosoid A tylosislike intrusion of a parenchyma cell into an intercellular space. It differs from a *tylosis* in that it does not pass through the cavity of a pit (Record, 1934, p.70).

tylostyle A *style* (sponge spicule) in which the blunt end is swollen or knobbed. Syn: *tylostylus*.

tylote n. A slender, elongate sponge spicule (monaxon) with a knob at each end.—adj. Said of a sponge spicule with the ends of the rays knobbed or swollen.

tylotoxea A rodlike sponge spicule tapering to a sharp point at one end and to a knob at the other.

tympanoid Said of a squat, drum-shaped scleractinian corallite.

Tyndall figure A water- and vapor-filled cavity in the shape of a snowflake, oriented parallel to the basal plane within an ice crystal, and formed through melting by radiation absorbed at points of defect in the ice lattice. It is called a "vapor figure" if the water in the cavity is refrozen. Named after John Tyndall (1820-1893), British physicist. Syn: *Tyndall flower; Tyndall star; negative snowflake*.

type [coal] A *coal classification* based on the constituent plant materials. Cf: *rank [coal]; grade [coal]*.

type [petrology] *rock type*.

type [taxon] The standard reference for determining the application of a scientific name. Unless otherwise qualified, a type usually refers to the *holotype* of a *species*. The type of a species is a single specimen and the species includes all organisms regarded as conspecific with the *type specimen*. The nomenclatural type of a genus is a species and the genus includes all species regarded as congeneric with its species type.

type area The geographic territory surrounding the *type locality* of a stratigraphic unit or boundary. Syn: *type region*.

type-boundary section *boundary stratotype*.

type concept A principle for stabilizing the application of scientific nomenclature by recognizing a permanent association of a *taxon* with one of its constituent elements, designated as its nomenclatural *type*, which serves as a point of reference. The nomenclatural type is not necessarily the most typical or representative of a taxon; it is that element with which the name is permanently associated (ICBN, 1972, p. 18).

type curve *master curve*.

type fossil A term occasionally used as a syn. of *index fossil*.

type genus The *genus* that serves as a permanent nomenclatural reference for application of the name of a *family*, and the ranks of super- and sub-taxa that have the same common point of nomenclatural reference.

type locality (a) The place at which a *stratotype* is situated and from which it ordinarily derives its name. It contains the type section, and is contained within the type area. Cf: *reference locality*. (b) The place where a geologic feature (such as an ore occurrence, a particular kind of igneous rock, or the type specimen of a fossil species or subspecies) was first recognized and described.

type material All the specimens upon which the description of a new species is based. Cf: *hypodigm*.

type region *type area*.

type section (a) The originally described sequence of strata that constitute a stratigraphic unit. It serves as an objective standard with which spatially separated parts of the unit may be compared, and it is preferably in an area where the unit shows maximum thickness and is completely exposed (or at least shows top and bottom). Type sections for lithostratigraphic units can never be changed (ACSN, 1961, art.13h): there is only one type section, although there may be more than one typical section. Cf: *reference section*. (b) According to the International Stratigraphic Guide (1976), a syn. of *stratotype*, thus constituting not only the type representative of a stratigraphic unit but also that of a stratigraphic boundary or horizon.

type species That *species* on which the original description of a genus or subgenus is largely or entirely based; the *type* of a genus or subgenus. Syn: *genotype*.

type specimen The single specimen on which the original description of a particular species is based, which serves as a permanent point of nomenclatural reference for application of the name of that species. The type specimen may be a *holotype*, a *neotype*, or a *lectotype*.

typhoon A *tropical cyclone*, esp. in the western Pacific. Etymol: Chinese, "great wind".

typomorphic mineral A mineral that is typically developed in only a narrow range of temperature and pressure. The term was originated by Becke. Cf: *critical mineral; index mineral*.

tyretskite A mineral: $Ca_3B_8O_{13}(OH)_4$ (?).

tyrolite A mineral: $Cu_5Ca(AsO_4)_2(CO_3)(OH)_4 \cdot 6H_2O$. Syn: *trichalcite*.

tyrrellite A mineral: $(Cu,Co,Ni)_3Se_4$.

Tyrrhenian European stage: uppermost Pleistocene (above Milazzian, below Versilian).

tysonite *fluocerite*.

tyuyamunite An orthorhombic mineral: $Ca(UO_2)_2(VO_4)_2 \cdot 5\text{-}8H_2O$. It is an ore of uranium, and occurs in yellow incrustations as a secondary mineral. Syn: *calciocarnotite*.

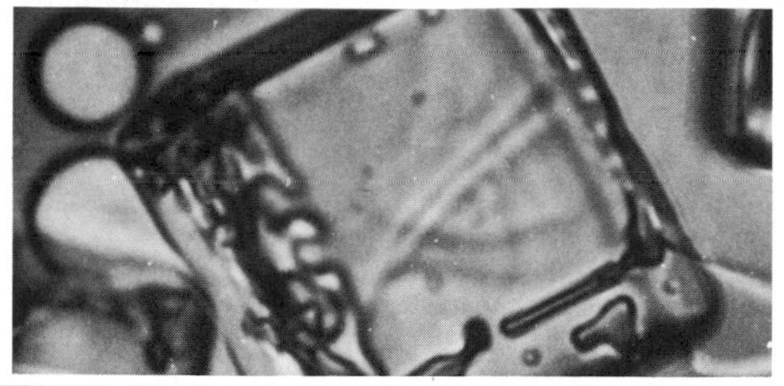

ubac A mountain slope so oriented as to receive the minimum available amount of light and warmth from the Sun; esp. a northward-facing slope of the Alps. Etymol: French dialect, "shady side". Cf: *adret.*

ubehebe A low volcanic cone composed of pyroclasts which are mostly accidental. Rarely used.

Udalf In U.S. Dept. of Agriculture soil taxonomy, a suborder of the soil order *Alfisol,* characterized by formation in a udic moisture regime and in a mesic or warmer temperature regime. Undisturbed soils have a thin A1 horizon, may have a relatively thin A2 horizon, and have a brownish argillic horizon. They have been intensively farmed, so that in many places much of the A horizon has been eroded away, and the plow layer consists of material most of which was once part of the argillic horizon (USDA, 1975). Cf: *Aqualf; Boralf; Ustalf; Xeralf.*

Udden grade scale A logarithmic *grade scale* devised by Johan A. Udden (1859-1932), U.S. geologist; it uses 1 mm as the reference point and progresses by the fixed ratio of 1/2 in the direction of decreasing size and of 2 in the direction of increasing size, such as 0.25, 0.5, 1, 2, 4 (Udden, 1898). See also: *Wentworth grade scale.*

Udert In U.S. Dept. of Agriculture soil taxonomy, a suborder of the soil order *Vertisol*, characterized by formation in a humid region. Cracks in the soil open and close one or more times in most years, but they do not remain open for 60 consecutive days in the summer, and in some years they may not open at all. Most Uderts have a dark gray to black surface horizon over a gray to brownish clay subsoil (USDA, 1975). Cf: *Torrert; Ustert; Xerert.*

udic moisture regime A soil moisture regime common to soils of humid climates that have well-distributed rainfall. In most years the soil is not dry for as many as 90 cumulative days (USDA, 1975).

Udoll In U.S. Dept. of Agriculture soil taxonomy, a suborder of the soil order *Mollisol,* characterized by formation in a udic moisture regime in humid continental climates of the midlatitudes. Most formed under a tall grass prairie and have either a cambic or an argillic horizon below the mollic epipedon (USDA, 1975). Corn and soybeans are very common on these soils. Cf: *Alboll; Aquoll; Boroll; Rendoll; Ustoll; Xeroll.*

Udult In U.S. Dept. of Agriculture soil taxonomy, a suborder of the soil order *Ultisol*, characterized by a low to moderate organic-carbon content, a light-colored A horizon, an argillic horizon that is reddish or yellowish, and by formation in a udic moisture regime and a mesic or warmer temperature regime. Some have a fragipan or plinthite in or below the argillic horizon. Many Udults are cultivated, either by using fertilizers or in a system of shifting cultivation (USDA, 1975). Cf: *Aquult; Humult; Ustult; Xerult.*

ufertite *davidite.*

ugandite An extrusive rock containing leucite, clinopyroxene, and abundant olivine, in a soda-rich glassy groundmass; *olivine leucitite.* The name, given by Holmes and Harwood in 1937, is derived from Uganda. Not recommended usage.

ugrandite A group name for the calcium garnet minerals uvarovite, grossular, goldmanite, and andradite.

uhligite (a) A black isometric mineral consisting of a titanate and zirconate of calcium and aluminum. (b) An amorphous variscite or fischerite.

uintahite A black, shiny asphaltite, with a brown streak and conchoidal fracture, which is soluble in turpentine; it occurs primarily in veins in the Uinta Basin, Utah. See also: *wurtzilite.* Also spelled: *uintaite.* Syn: *gilsonite.*

uintaite *uintahite.*

Uintan North American continental Stage: Upper Eocene (above Bridgerian, below Duchesnean).

Uinta structure A basement diapir or doming or upwarp in the form of a regional, flattened anticlinal flexure on which denudation has exposed the basement-rock core. It is named after the Uinta Mountains of northeastern Utah.

uklonskovite A mineral: $NaMg(SO_4)(OH)\cdot 2H_2O$.

Ulatisian North American stage: Middle Eocene (above Penutian, below Narizian).

ulexite A white triclinic mineral: $NaCaB_5O_9\cdot 8H_2O$. It forms rounded reniform masses of extremely fine acicular crystals and is usually associated with borax in saline crusts on alkali flats in arid regions. Syn: *boronatrocalcite; natroborocalcite; cotton ball.*

ulinginous Said of an organism living in wet or swampy ground.

ullmannite A steel-gray to black mineral: NiSbS. It usually contains a little arsenic. Syn: *nickel-antimony glance.*

ulmain A kind of *euvitrain* that consists completely of ulmin but that is not precipitated from solution. Cf: *collain.*

ulmic acid *ulmin.*

ulmification The process of peat formation. See also: *ulmin.* Syn: *paludification.*

ulmin Vegetable-degradation material occurring in coal as an amorphous, brown to black substance or gel and that is insoluble in alkaline solution. It is abundant in peat and lignite, and forms vitrinite. Syn: *ulmic acid; humin; carbohumin; humogelite; fundamental jelly; jelly; gélose; fundamental substance; vegetable jelly.*

ulminite (a) A maceral of brown coal within the *huminite* group, consisting of gelified plant-cell walls (ICCP, 1971). (b) A variety of euvitrinite characteristic of ulmain and consisting of gelified but not precipitated plant material. Cf: *collinite.*

ulrichite [mineral] A syn. of *uraninite;* specif. the original unoxidized UO_2.

ulrichite [rock] A dark-colored hypabyssal rock composed of large phenocrysts of nepheline, alkali feldspar, sodic pyroxene, and amphibole, and smaller accessory olivine phenocrysts in a groundmass of feldspar, analcime, pyroxene, and amphibole; an olivine-bearing *phonolite.* The name, given by Marshall in 1906, is in honor of G. H. F. Ulrich, discoverer of alkaline rocks at Dunedin, New Zealand. Not recommended usage.

ULSEL *ultra-long-spaced electric log.*

Ulsterian North American provincial series: Lower Devonian (above Cayugan of Silurian, below Erian).

ultimate analysis The determination of the elements in a compound; for coal, the determination of carbon, hydrogen, sulfur, nitrogen, ash, and oxygen. Cf: *proximate analysis.*

ultimate base level The lowest possible *base level;* for a stream, it is sea level, projected inland as an imaginary surface beneath the stream. Cf: *temporary base level.* Syn: *general base level.*

ultimate bearing capacity The average load per unit of area required to produce shear failure by rupture of a supporting soil mass. See also: *bearing capacity.*

ultimate landform The theoretical landform produced near the end of a cycle of erosion. Cf: *initial landform; sequential landform.* Syn: *ultimate form.*

ultimate recovery The quantity of oil or gas that a well, pool, field, or property will produce. It is the total obtained or to be obtained from the beginning of production to final abandonment.

ultimate shear strength The maximum shearing stress (i.e., half the differential stress) corresponding with the *ultimate strength.*

ultimate strength The maximum differential stress that a material can sustain under the conditions of deformation. Beyond this point, rock *failure* occurs. See also: *ultimate shear strength.*

Ultisol In U.S. Dept. of Agriculture soil taxonomy, a soil order characterized by the presence of an argillic horizon, a low supply of bases, particularly at depth, and a mesic or warmer temperature regime. They form at mid to low latitudes in a warm humid climate with enough excess seasonal precipitation to remove bases released by weathering. Bases are highest in the surface soil because of plant recycling. Ultisols have low native fertility, but, because they are warm and moist, they can be highly productive with fertilizers (USDA, 1975). See also: *Aquult; Humult; Udult; Ustult; Xerult.*

ultrabasic Said of an igneous rock having a silica content lower than that of a basic rock. Percentage limitations are arbitrary and vary with different petrologists, although the upper limit was originally set at 44%. The term is frequently used interchangeably with *ultramafic.* Although most ultrabasic rocks are also ultramafic, there are some exceptions; e.g. monomineralic rocks composed of pyroxenes are ultramafic but are not ultrabasic because of their high SiO_2 content. A monomineralic rock composed of anorthite would be considered ultrabasic ($SiO_2 = 43.2$ percent) but not ultramafic. "Ultrabasic" is one subdivision of a widely used system for classifying igneous rocks on the basis of silica content; the other subdivisions are *acidic, basic,* and *intermediate.* Cf: *silicic.*

ultrabasite *diaphorite.*

ultra-long-spaced electric log A *resistivity log* curve derived from 2-electrode spacing of as much as 1000 feet. See also: *normal log.* Abbrev: ULSEL.

ultramafic Said of an igneous rock composed chiefly of mafic minerals, e.g. monomineralic rocks composed of hypersthene, augite, or olivine. Cf: *ultrabasic.*

ultramafic rock In the *IUGS classification,* a general name for plutonic rock with color index M greater than or equal to 90, including, among others, dunite, peridotite, and pyroxenite.

ultramarine A syn. of *lazurite.* The term is also applied to artificial lazurite and to compounds allied to it; e.g. the brilliant blue pigment formerly made by powdering lapis lazuli and characterized by the durability of its color.

ultrametamorphism Metamorphic processes at the extreme upper range of temperatures and pressures, at which partial to complete fusion of the affected rocks takes place and magma is produced. The term was originated by Holmquist in 1909.

ultramicroearthquake An earthquake having a body-wave magnitude of zero or less on the Richter scale. Such a limit is arbitrary, and may vary according to the user. Cf: *microearthquake; major earthquake.*

ultramylonite An ultra-crushed variety of *mylonite,* in which primary structures and porphyroclasts have been obliterated so that the rock becomes homogeneous and dense, with little if any parallel structure (Quensel, 1916). Cf: *protomylonite; pseudotachylyte.* Syn: *flinty crush rock.*

ultraplankton The smallest plankton; they are five microns and smaller. Cf: *nannoplankton; microplankton; macroplankton; megaloplankton.*

ultrasima The supposedly ultrabasic layer of the Earth below the sima, immediately below the Mohorovičić discontinuity.

ultrastructure The internal structure and character of plant and animal tissues, esp. skeletal tissues, as revealed by the electron microscope. Cf: *microstructure.*

ultraviolet absorption spectroscopy The observation of an *absorption spectrum* in the ultraviolet frequency region and all processes of recording and measuring that go with it.

ultravulcanian eruption A type of volcanic eruption characterized by violent, gaseous explosions of lithic dust and blocks, with little if any incandescent scoria. It is commonly observed during the opening or reopening of a volcanic vent. Its type occurrence is the explosion of Krakatoa in 1883.

ulvite *ulvöspinel.*

ulvöspinel A mineral of the spinel group: Fe_2TiO_4. It usually occurs as fine exsolution lamellae, intergrown with magnetite. Syn: *ulvite.*

umangite A dark-red mineral: Cu_3Se_2.

umbel (a) An umbrellalike structure consisting of multiple recurved teeth attached to the tip of a ray or the pseudoactine of a sponge spicule (such as of an amphidisc). (b) A sponge spicule consisting of a single shaft with an umbel at one end; e.g. a paraclavule or one type of a clavule.

umbelliferous (a) Said of a tabulate corallum having corallites arranged like ribs of an umbrella, growing outward in whorls. (b) Producing umbels.

umber A naturally occurring brown earth that is darker than ocher and sienna and that consists of manganese oxides as well as hydrated ferric oxide, silica, alumina, and lime. It is highly valued as a permanent paint pigment, and is used either in the greenish-brown natural state ("raw umber") or in the dark-brown or reddish-brown calcined state ("burnt umber").

umbilical area The inner part or surface of a whorl of a cephalopod conch, between the umbilical shoulder and the umbilical seam. It is called an "umbilical wall" if it rises steeply from the spiral plane and "umbilical slope" if it rises gently (TIP, 1957, pt. L, p. 6).

umbilical lobe The large primary lobe of a suture of an ammonoid, centered on or near the umbilical seam, and forming part of both external and internal sutures.

umbilical perforation The vacant space or opening around the axis of coiling of a cephalopod shell, connecting the umbilici on opposite sides.

umbilical plug (a) The deposit of secondary skeletal or shell material in the axis or umbilical region of certain coiled foraminiferal tests (e.g. in *Rotalia*). (b) The calcareous deposit filling the umbilicus of a cephalopod.—Syn: *plug [paleont].*

umbilical seam The helical line of junction or overlap of adjacent whorls of a coiled cephalopod conch. Syn: *umbilical suture.*

umbilical shoulder (a) The part of a cephalopod shell bordering the umbilicus and forming its outer margin; e.g. the strongly bent part of a whorl of a nautiloid shell between the flank and the inner part of the umbilical area. (b) The angulation of whorls at the margin of and within the umbilicus of a gastropod shell (Moore et al., 1952, p.289). (c) The part of a foraminiferal test bordering the umbilicus.—See also: *shoulder [paleont].*

umbilical suture (a) A continuous line separating successive whorls as seen in the umbilicus of phaneromphalous gastropod shells. (b) An *umbilical seam* of a cephalopod.

umbilical tooth One of the projections forming a triangular modification of the apertural lip of a foraminiferal test, with those of successive chambers in forms (e.g. *Globoquadrina*) with umbilical aperture giving a characteristic serrate border to the umbilicus.

umbilicus (a) A cavity or depression in the center of the base of a spiral shell of a univalve mollusk; e.g. the conical opening formed around the central axis of a spiral gastropod shell between faces of adaxial walls of whorls where these do not coalesce, or an external depression centered around the axis of coiling of a cephalopod shell and formed by the diminishing width of whorls toward the axis. (b) A circular depression or pit in the axis of a coiled foraminiferal test; e.g. the closed shallow depressed area formed by curvature of overlapping chamber walls in involute forms, or the space formed between inner margins of the walls of chambers belonging to the same whorl of the test.—Pl: *umbilici.*

umbo (a) The "humped" part of the shell of a bivalve mollusk, or the elevated and relatively convex part of a valve surrounding the point of maximum curvature of the longitudinal dorsal profile and extending to the beak when not coinciding with it. The term is often used synonymously with *beak,* but with most shells two distinct terms are needed. (b) The relatively convex, apical part of either valve of a brachiopod, just anterior to or containing the beak. (c) The apical part of either valve of the bivalved carapace of a crustacean; e.g. the point on the plate from which successive growth increments extend in a cirripede. (d) A prominence on the frontal shield in some cheilostome bryozoans, proximal to the orifice or on the ovicell. (e) A central round elevated structure in discoidal foraminiferal tests. It is commonly due to lamellar thickening and may occur on one or both sides of the test. (f) A central projection on the thecal plate of an echinoderm, representing part of its ornamentation.—Pl: *umbones* or *umbos.* Syn: *umbone.*

umbonal angle (a) The approximate angle of divergence of the posterior/dorsal and anterior/dorsal parts of the longitudinal profile of bivalve-mollusk shells; specif. the angle of divergence of umbonal folds in pectinoid shells. (b) The angle subtended at the umbo of a brachiopod by the region of the shell surface adjacent to the umbo.

umbonal chamber One of a pair of posteriorly and laterally located cavities in either valve of a brachiopod, bounded in the pedicle valve by dental plates and shell walls and limited medially in the brachial valve by crural plates (or homologues) and shell walls.

umbonal fold The ridge originating at the umbo of a pectinoid shell and setting off the body of the shell from the auricle.

umbonal muscle A single muscle occurring in some lingulid brachiopods, thought to be homologous with the posterior adductor muscles, and consisting of two bundles of fibers, posteriorly and slightly asymmetrically placed (TIP, 1965, pt.H, p.154).

umbonate Having or forming an umbo; e.g. having an umbo on one or both sides of an enrolled foraminiferal test. Also, said of a foraminifer bearing a convex elevation in the center.

umbone A syn. of *umbo* [paleont]. The term "umbones" is the usual plural for "umbo".

umbonuloid adj. Pertaining to an ascophoran cheilostome (bryozoan) characterized by a *frontal shield* formed by calcification of the inner wall of a double-walled fold overarching the frontal membrane; esp. said of a frontal shield formed in that way. —n. An ascophoran cheilostome having such a structure.

umbozerite A noncrystalline mineral: $Na_3Sr_4ThSi_8(O,OH)_{24}$.

umbra (a) The completely shadowed region of an eclipse. (b) The inner, darker region of a sunspot.—Cf: *penumbra.*

umbracer dune A *lee dune* tapering to a point downwind, formed under constant wind direction commonly behind a clump of bushes or a prominent bedrock obstacle (Melton, 1940, p. 120). See also: *wind-shadow dune.*

umbrafon dune A *lee dune* developed to the leeward of a source or area of loose sand where the sand supply is constantly replenished (Melton, 1940, p. 122); e.g. a dune on the lee side of a stream flood plain or landward from a sandy beach. Syn: *lee-source dune; source-bordering lee dune.*

Umbrept In U.S. Dept. of Agriculture soil taxonomy, a suborder of the soil order *Inceptisol,* characterized by formation in hilly to mountainous regions of mid to high latitudes. The climate is temperate to cold with high precipitation. Umbrepts are acid, dark or brown, and high in organic matter. Most have an umbric epipedon and a cambic horizon (USDA, 1975). Cf: *Andept; Aquept; Ochrept; Plaggept; Tropept.*

umbric epipedon A diagnostic surface horizon that is similar to a *mollic epipedon* except for having a base saturation of less than 50%, measured at a pH of 7 (USDA, 1975). Cf: *ochric epipedon.*

umohoite A black to bluish-black mineral: $(UO_2)MoO_4 \cdot 4H_2O$.

umptekite A *syenite* composed chiefly of microperthite and sodic amphibole, with accessory sphene, apatite, and opaque oxides, and occasionally small amounts of interstitial nepheline; a sodic syenite resembling *pulaskite.* Its name, given by Ramsay in 1884, is derived from Umptek (now Khibine), Kola Peninsula, U.S.S.R. Not recommended usage; "the name umptekite seems quite unnecessary since the composition of pulaskite is very similar" (Johannsen, 1938, p. 9).

unaka (a) A term proposed by Hayes (1899, p. 22) for a large residual mass rising above a peneplain that is less advanced than one having a *monadnock,* and sometimes displaying on its surface the remnants of a peneplain older than the one above which it rises; an erosion remnant of greater size and height than a monadnock. (b) A group or sprawling mass of monadnocks, often occurring near the headwaters of stream systems where erosion has not yet reduced the area to the level of a peneplain (Lobeck, 1939, p. 633). locality: Unaka Mountains of eastern Tennessee and western North Carolina.

unakite An epidote-rich *granite,* which besides epidote contains pink orthoclase, quartz, and minor opaque oxides, apatite, and zircon. The name is derived from the type locality, the Unaka Range, Great Smoky Mountains, eastern Tennessee.

unarmored Said of naked dinoflagellates, such as those of the order Gymnodiniales, that lack a plate-constructed theca or cell wall and are enclosed by a thin, structureless pellicle. Ant: *armored.*

unary system A chemical system that has only one component. Syn: *unicomponent system.*

unavailable moisture *unavailable water.*

unavailable water Water that cannot be utilized by plants because it is held in the soil by adsorption or other forces; water in the soil in an amount below the *wilting point.* Syn: *unavailable moisture.*

unbalanced force A force that is not opposed by another force acting along the same line in the opposite sense of direction; an unbalanced force causes translation of a body.

uncinate A hexactinellid-sponge spicule (diactinal monaxon) covered on all sides with short thornlike spines directed toward one end. Cf: *cleme.*

uncompahgrite A plutonic rock composed chiefly of melilite, along with pyroxene, opaque oxides, perovskite, apatite, calcite, anatase, melanite, and occasionally phlogopite; a member of the *melilitolite* group. The name, given by Larsen and Hunter in 1914, is for Mount Uncompahgre, Colorado. Not recommended usage.

unconcentrated flow *overland flow.*

unconcentrated wash *sheet erosion.*

unconfined aquifer An aquifer having a water table; an aquifer containing unconfined ground water. Syn: *water-table aquifer.*

unconfined compression test A special condition of a *triaxial compression test* in which no confining pressure is applied. Syn: *crushing test.*

unconfined ground water Ground water that has a free water table, i.e. water not confined under pressure beneath relatively impermeable rocks. Ant: *confined ground water.* Syn: *phreatic water; nonartesian ground water; free ground water; unconfined water.*

unconfined water *unconfined ground water.*

unconformability The quality, state, or condition of being unconformable, such as the relationship of unconformable strata; *unconformity.*

unconformable Said of strata or stratification exhibiting the relation of unconformity to the older underlying rocks; not succeeding the underlying rocks in immediate order of age or not fitting together with them as parts of a continuous whole. In the strict sense, the term is applied to younger strata that do not "conform" in position or that do not have the same dip and strike as those of the immediately underlying rocks. Also, said of the contact between unconformable rocks. Cf: *conformable [stratig].* Syn: *discordant [stratig].*

unconformity (a) A substantial break or gap in the geologic record where a rock unit is overlain by another that is not next in stratigraphic succession, such as an interruption in the continuity of a depositional sequence of sedimentary rocks or a break between eroded igneous rocks and younger sedimentary strata. It results from a change that caused deposition to cease for a considerable span of time, and it normally implies uplift and erosion with loss of the previously formed record. An unconformity is of longer duration than a *diastem.* (b) The structural relationship between rock strata in contact, characterized by a lack of continuity in deposition, and corresponding to a period of nondeposition, weathering, or esp. erosion (either subaerial or subaqueous) prior to the deposition of the younger beds, and often (but not always) marked by absence of parallelism between the strata; strictly, the relationship where the younger overlying stratum does not "conform" to the dip and strike of the older underlying rocks, as shown specif. by an angular unconformity. Cf: *conformity.* Syn: *unconformability; transgression [stratig].* (c) *surface of unconformity.*—Common types of unconformities recognized in U.S.: *nonconformity; angular unconformity; disconformity; paraconformity.* Since the essential feature of an unconformity, as understood in Great Britain, is structural discordance rather than a time gap, the British do not recognize disconformity and paraconformity as unconformities. For an historical study of unconformities, see Tomkeieff (1962).

unconformity iceberg An iceberg consisting of two or more distinct layers or lenses that differ in composition and are separated by surface discontinuities; e.g. an iceberg in which crevassed glacier ice is overlain successively by a layer of silt and a layer of firn.

unconformity trap A *trap* for oil or gas associated with an unconformity.

unconsolidated material (a) A sediment that is loosely arranged or unstratifid, or whose particles are not cemented together, occurring either at the surface or at depth. (b) Soil material that is in a loosely aggregated form.

uncontrolled mosaic An aerial *mosaic [photo]* formed solely by matching detail of overlapping photographs without spatial or directional adjustments to control points.

uncovers *dries.*

unctuous *soapy.*

unda adj. A term applied by Rich (1951, p. 2) to the environment of sedimentation that lies in the zone of wave action. It may be used alone or as a combining form. See also: *undaform; undathem.* Cf: *clino; fondo.* Etymol: Latin *unda,* "wave".

undaform The subaqueous land form produced by the erosive and constructive action of the waves during the development of the subaqueous profile of equilibrium (Rich, 1951, p. 2). It is the site of the *unda* environment of deposition. Cf: *clinoform; fondoform.*

undathem Rock units formed in the *unda* environment of deposition (Rich, 1951, p. 2). Cf: *clinothem; fondothem.*

undation theory A theory proposed by Van Bemmelen (1933) that explains the structural and tectonic features of the Earth's crust by vertical upward and downward movements caused by waves that are generated by deep-seated magma. Cf: *blister hypothesis.*

undaturbidite A term proposed by Rizzini & Passega (1964, p. 71) for a sediment formed from a suspension produced by violent storms; a deposit intermediate between an ordinary wave deposit and a turbidite. Cf: *fluxoturbidite.*

underclay A layer of fine-grained detrital material, usually clay, lying immediately beneath a coal bed or forming the floor of a coal seam. It represents the old soil in which the plants (from which the coal was formed) were rooted, and it commonly contains fossil roots (esp. of the genus *Stigmaria*). It is often a *fireclay,* and some underclays are commercial sources of fireclay. Syn: *underearth; seat earth; seat clay; root clay; thill; warrant; coal clay.*

underclay limestone A thin, dense, nodular, relatively unfossiliferous *freshwater limestone* underlying coal deposits, so named because it is closely related to underclay.

undercliff [geomorph] A terrace or subordinate cliff along a coast, formed of material fallen from the cliff above; the lower part of a cliff whose upper part underwent landsliding.

undercliff [sed] A term used in southern Wales for a shale forming the floor of a coal seam.

underconsolidation Consolidation (of sedimentary material) less than that normal for the existing overburden; e.g. consolidation resulting from deposition that is too rapid to give time for complete settling. Ant: *overconsolidation.*

undercooling *supercooling.*

undercurrent A current of water flowing beneath a surface current at a different speed or in a different direction; e.g. the Mediterranean Undercurrent off Gibraltar. See also: *equatorial undercurrent.*

undercut A reentrant in the face of a cliff, produced by undercutting.

undercutting The removal of material at the base of a steep slope or cliff or other exposed rock by the erosive action of falling or running water (such as a meandering stream), of sand-laden wind in the desert, or of waves along the coast.

underearth (a) A hard fireclay forming the floor of a coal seam; *underclay.* (b) The soil beneath the Earth's surface. (c) The depths of the Earth.

underfit stream A *misfit stream* that appears to be too small to have eroded the valley in which it flows; a stream whose volume is greatly reduced or whose meanders show a pronounced shrinkage in radius. It is a common result of drainage changes effected by capture, by glaciers, or by climatic variations.

underflow (a) The movement of ground water in an *underflow conduit;* the flow of water through the soil or a subsurface stratum, or under a structure. (b) The rate of discharge of ground water through an underflow conduit. (c) The water flowing beneath the bed or alluvial plain of a surface stream, generally in the same direction as, but at a much slower rate than, the surface drainage; esp. the water flowing under a dry stream channel in an arid region.

underflow conduit A permeable deposit that underlies a surface stream channel, that is more or less definitely limited at its bottom and sides by rocks of relatively low permeability, and that contains ground water moving in the same general direction as the stream above it (Meinzer, 1923, p. 43). See also *underflow.*

underground cauldron subsidence *Cauldron subsidence* in which the ring faults do not extend to the surface (Billings, 1972, p. 359). Cf: *surface cauldron subsidence.*

underground ice *ground ice [permafrost].*

underground stream A body of water flowing as a definite current in a distinct channel below the surface of the ground, usually in an area characterized by joints or fissures; legally, such a stream discoverable by men without scientific instruments. Application of the term to ordinary aquifers is incorrect. Cf: *subterranean stream; percolating water.*

underground water (a) A syn. of *ground water.* (b) A syn. of *subsurface water* in less preferred usage.

underhand stoping The working of a block of ore from an upper to a lower level.

underlay In mining, the extension of a vein or ore deposit beneath the surface; also, the inclination of a vein or ore deposit from the vertical, that is, *hade.* Syn: *underlie.*

underlie [mining] *underlay.*

underlie [stratig] v. To lie or be situated under, to occupy a lower position than, or to pass beneath. The term is usually applied to certain rocks over which younger rocks (usually sedimentary or volcanic) are spread out. Ant: *overlie.*

underloaded stream A stream that carries less than a full load of sediment and that erodes its bed.

underlying *subjacent.*

undermass Harder rock material (or the basement) lying beneath the *cover mass,* characterized by a more complex or intensely deformed structure; the material below the surface of an angular unconformity. See also: *compound structure.*

undermelting The melting from below of any floating ice (Huschke, 1959, p. 601). Rarely used.

undermining The action of wearing away supporting material, as the *undermining* of a cliff by stream erosion; *sapping.*

underplight A substratum, once consisting of soft mud, that preserves the form of an overlying thin layer of sand or gravel that has been contorted by alternate freezing and thawing (Spurrell, 1887).

undersaturated (a) Said of an igneous rock consisting of *unsaturated* minerals, e.g. feldspathoids and olivine. (b) Said of a rock whose norm contains feldspathoids and olivine, or olivine and hypersthene. Cf: *critically undersaturated; oversaturated; saturated.*

undersaturated permafrost Permafrost that contains less ice than the ground could hold if the water were in the liquid state.

undersaturated pool A pool in which all the gas present is dissolved in the oil. Cf: *saturated pool.*

underthrust fault A type of thrust fault in which it is the lower rock mass that has been actively moved under the upper, passive rock mass. An underthrust may be difficult to distinguish from an *overthrust.*

undertow The seaward return flow, near the bottom of a sloping beach, of water that was carried onto the shore by waves. Cf: *rip current.*

undertow mark A channeled structure on a sedimentary surface, believed to have been made by currents dragging heavy objects in very shallow water adjacent to a beach (Clarke, 1918). Cf: *strand mark.* Syn: *undertow marking.*

underwater gravimeter An instrument capable of measuring gravity when lowered to the sea bottom from a stationary surface vessel; it is leveled and read in a few minutes by remote control and has an accuracy of about 0.1 milligal.

underwater ice Ice formed below the surface of a body of water; e.g. *anchor ice.*

undisturbed Said of a soil sample that has not been subjected to disturbance by boring, sampling, or handling, and that thus closely represents the in-situ characteristics of the material. Most sampling procedures produce, at best, only "relatively undisturbed samples".

undivided Said of a surface, landscape, or area that has no noticeable feature separating the drainage of neighboring streams.

undulate Wavy, as of the margin of a leaf or petal.

undulating fold A minor fold with rounded apexes; a fold whose beds are bent into alternate elevations and depressions.

undulation [geodesy] The separation or height of the geoid above or below the *reference ellipsoid.*

undulation [geomorph] (a) A landform having a wavy outline or form; e.g. a desert sand deposit similar to a *whaleback* but shorter and lacking the definite form of the whaleback (Stone, 1967, p. 252). (b) A rippling or scalloped land surface, having a wavy outline or appearance, or resembling waves in form; e.g. a rolling prairie.

undulatory extinction A type of *extinction* that occurs successively in adjacent areas, as the microscope's stage is turned. Cf: *parallel extinction; inclined extinction.* Syn: *strain shadow; oscillatory extinction; wavy extinction.*

uneven fracture A general type of mineral fracture that is rough and irregular.

ungaite A general term suggested by Iddings in 1913 for oligoclase-bearing *dacite.* Its name is derived from Unga Island, Kamchatka, U.S.S.R. Obsolete.

ungemachite A colorless to yellowish rhombohedral mineral: $Na_8K_3Fe(SO_4)_6(OH)_2 \cdot 10H_2O$.

unglaciated Said of a land surface that has not been modified by the action of a glacier or an ice sheet; "never-glaciated". Cf: *deglaciation.*

uniaxial Said of a crystal having only one optic axis, e.g. a tetragonal or hexagonal crystal. Cf: *biaxial.*

uniclinal shifting *monoclinal shifting.*

unicline An obsolete syn. of *monocline.*

unicomponent system *unary system.*

uniform channel In hydraulics, a channel having a uniform cross section and a constant roughness and slope (ASCE, 1962).

uniform development The production of a landscape where the rate of uplift is equal to the rate of downward erosion, characterized by constant relief and straight slopes. Cf: *accelerated development; declining development.*

uniform flow Flow of a current of water in which there is neither convergence nor divergence.

uniformitarian n. A believer in the doctrine of uniformitarianism.—adj. Pertaining to the doctrine of uniformitarianism.

uniformitarianism (a) The fundamental principle or doctrine that geologic processes and natural laws now operating to modify the Earth's crust have acted in the same regular manner and with essentially the same intensity throughout geologic time, and that past geologic events can be explained by phenomena and forces observable today; the classical concept that "the present is the key to the past". The doctrine does not imply that all change is at a uniform rate, and does not exclude minor local catastrophes. The term was originated by William Whewell to describe the basic approach to geology of Charles Lyell. Cf: *catastrophism.* Syn: *actualism; principle of uniformity.* (b) The logic and method by which geologists attempt to reconstruct the past using the principle of uniformitarianism.

uniformity coefficient A numerical expression of the variety in particle sizes in mixed natural soils, defined as the ratio of the sieve size through which 60% (by weight) of the material passes to the sieve size that allows 10% of the material to pass. It is unity for a material whose particles are all of the same size, and it increases with variety in size (as high as 30 for heterogeneous sand).

uniform strain *homogeneous strain.*

unilaminate Said of an encrusting or erect bryozoan colony consisting of a single layer of zooids with or without extrazooidal parts.

unilateral Said of a stream or drainage system in which all tributaries come in from one side, while the other walls of the main valleys are practically unbroken (Rich, 1915, p. 145).

unilobite A descriptive term for a *trace fossil* consisting of a one-lobed (unilobate) trail. About 80 percent of all invertebrate tracks are unilobites. The term is seldom used.

unilocular Containing a single chamber or cavity; e.g. said of a single-chambered foraminifer. Syn: *monothalamous.*

unimodal sediment A sediment whose particle-size distribution shows no secondary maxima; e.g. a modern beach gravel.

uninverted relief A topographic configuration that reflects the underlying geologic structure, as where mountains mark the sites of anticlines and valleys mark the sites of synclines. Ant: *inverted relief.*

uniplicate Said of a form of alternate folding in brachiopods with the pedicle valve bearing a median sulcus and an anterior commissure median plica (TIP, 1965, pt.H, p.155). Ant: *sulcate.*

uniserial Arranged in, characterized by, or consisting of a single row or series, e.g. a "uniserial arm" of a crinoid composed of brachial plates arranged in a single row; a "uniserial test" of a foraminifer whose chambers are arranged in a single linear or curved series; or a "uniserial rhabdosome" of a graptoloid consisting of a single row of thecae. Cf: *biserial.*

unit bar In a braided stream, a sand or gravel bar that is relatively unmodified, with a morphology "determined by mainly depositional processes" (Smith, 1974, p. 210).

unit cell The smallest volume or parallelepiped within the three-dimensional repetitive pattern of a crystal that contains a complete sample of the atomic or molecular groups that compose this pattern. Crystal structure can be described in terms of the translatory repetition of this unit in space in accordance with one of the space lattices. Syn: *primitive unit cell.*

unit character A natural characteristic that is dependent on the presence or absence of a single gene.

unit circle In a gnomonic projection, the circle that is the projection of the equatorial plane of the sphere of projection. Its radius gives the scale used in plotting the projection.

unit coal Pure coal, free of moisture and noncoal mineral matter, calculated from analysis. Unit coal is expressed by the equation: unit coal$=1.00-(W+1.08A+0.55S)$, in which W=water, A=ash, and S=sulfur.

unit dry weight *dry unit weight.*

unit form A crystal form in a system other than the cubic, having intercepts on the chosen crystal axes that define the axial ratio. Unit forms have Miller indices {111}, {110}, {011}, {101}.

unit stratotype The type section of strata serving as the standard for the definition and recognition of a stratigraphic unit. The upper and lower limits of a unit stratotype are its *boundary stratotypes* (ISG, 1976, p. 24).

unit value The monetary value of a mineral or rock product per ton or other unit of measurement.

unit weight A term applied esp. in soil mechanics to the weight per unit of volume, such as grams per cubic centimeter; the density of a material. Symbol: γ. See also: *dry unit weight; effective unit weight; wet unit weight.*

univalve adj. Having or consisting of one valve only. Cf: *bivalve.* Syn: *univalved.*—n. (a) A univalve animal; specif. a mollusk with a univalve shell, such as a gastropod, a cephalopod, or a scaphopod. (b) A shell of a univalve animal; specif. a mollusk shell consisting of one piece.

univariant Said of a chemical system having one degree of freedom; said of an equilibrium system in which the arbitrary variation of more than one physical condition will result in the disappearance of one of the phases.

universal stage A *stage [optics]* of three, four, or five axes, attached to the rotating stage of a polarizing microscope, that enables the thin section under study to be tilted about two horizontal axes at right angles. It is used for optical study of low-symmetry minerals or for determining the orientation of any mineral relative to the section surface and edge directions. Syn: *U-stage; Fedorov stage.*

universal time Time defined by the rotation of the Earth and determined from its apparent diurnal motions. Because of variations in the rate of rotation, universal time is not rigorously uniform. Also called Greenwich mean time. Abbrev: UT.

universal transverse Mercator projection *transverse Mercator projection*

universe In statistics, a syn. of *population.*

unloading The removal by denudation of overlying material.

unmatched terrace *unpaired terrace.*

unmixing [chem] A syn. of *exsolution* that is also applied to the separation of immiscible liquids.

unmixing [sed] Segregation and concentration of sedimentary material during diagenesis.

unoriented [geol] Said of a rock or other geologic specimen whose original position in space, when collected, is unknown or not definitely ascertained.

unoriented [surv] Said of a map or surveying instrument whose internal coordinates are not coincident with corresponding directions in space.

unpaired terrace A *stream terrace* with no corresponding terrace on the opposite side of the valley, usually produced by a meandering stream swinging back and forth across a valley. See also: *meander terrace.* Cf: *paired terrace.* Syn: *unmatched terrace.*

unprotected thermometer A *reversing thermometer* that is not protected against hydrostatic pressure. Cf: *protected thermometer.*

unradiogenic lead Lead which has $^{207}Pb/^{204}Pb$ and $^{206}Pb/^{204}Pb$ ratios less than they would be in a single-stage development because of an integrated $^{238}U/^{204}Pb$ less than about 9.

unripe Said of peat that is in an early stage of decay, and in which original plant structures are visible. Cf: *ripe.*

unroofed anticline *breached anticline.*

unsaturated Said of a mineral that does not form in the presence of free silica; e.g. nepheline, leucite, olivine, feldspathoids. Cf: *undersaturated; saturated; oversaturated.*

unsaturated flow The flow of water in an undersaturated soil by capillary action and gravity.

unsaturated zone *zone of aeration.*

unsorted *poorly sorted.*

unstable [radioactivity] Said of a radioactive substance. Cf: *stable [radioactivity].*

unstable [sed] (a) Said of a constituent of a sedimentary rock that does not effectively resist further mineralogic change and that represents a product of rapid erosion and deposition (as in a region of tectonic activity and high relief); e.g. feldspar, pyroxene, hornblende, and various fine-grained rock fragments. (b) Said of an immature sedimentary rock (such as graywacke) consisting of unstable particles that are angular to subrounded, poorly to moderately sorted, and composed of feldspar grains or rock fragments.—Cf: *labile [geol].*

unstable equilibrium A state of equilibrium from which a chemical system, or a body (such as a pendulum), will depart in response to the slightest perturbation. Cf: *stable equilibrium.*

unstable gravimeter *astatic gravimeter.*

unstable isotope A syn. of *radioisotope.* Cf: *stable isotope.*
unstable relict A *relict [meta]* that is unstable under the newly imposed conditions of metamorphism but persists in a perhaps altered but still recognizable form owing to the low velocity of transformation. A preferable term would be *metastable relict.* Cf: *stable relict.* See also: *armored relict.*
unstable remanent magnetization *viscous magnetization.*
unsteady flow In hydraulics, flow that changes in magnitude or direction with time. Cf: *steady flow.* Syn: *nonsteady flow.*
unstratified Not formed or deposited in strata; specif. said of *massive* rocks or sediments with an absence of layering, such as granite or glacial till.
unweathered *fresh [weath].*
Unwin's critical velocity *critical velocity (e).*
upbank thaw A thaw or marked rise of temperature occurring at hill or mountain level while the frost is unbroken in the valley below.
upbuilding The building-up of a sedimentary deposit, as by a stream or in the ocean. Cf: *aggradation.*
upconcavity The persistent downstream decrease in gradient as seen on the channel profiles of most streams.
upconing *coning.*
updating A change, commonly a decrease, in the radiometric age of a rock, caused by a complete or partial disturbance of the isolated radioactive system by thermal, igneous, or tectonic activities. This results in a loss, usually of daughter products, from the system of radiogenic isotopes (only rarely in a gain of radioactive isotopes). See also: *hybrid age; mixed ages; overprint [geochron].*
updip A direction that is upwards and parallel to the dip of a structure or surface. Cf: *downdip.*
updip block The rocks on the *upthrown* side of a fault. Cf: *downdip block.*
updrift The direction opposite that of the predominant movement of littoral materials.
upfaulted Said of the rocks on the *upthrown* side of a fault, or the *updip block.* Cf: *downfaulted.*
upgrading *aggradation.*
uphole Said of any location in a well bore that is above a given depth. Cf: *downhole.*
uphole shooting In seismic exploration, the setting-off of successive shots in a shothole at varying depths in order to determine velocities and velocity variation of the materials forming the walls of the hole.
uphole time In seismic exploration, the time required for the seismic impulse to travel from a given depth in a shothole to the surface.
upland (a) A general term for high land or an extensive region of high land, esp. far from the coast or in the interior of a country. Sometimes used synonymously with *fastland.* (b) The higher ground of a region, in contrast with a valley, plain, or other low-lying land; a plateau. (c) The elevated land above the low areas along a stream or between hills; any elevated region from which rivers gather drainage. Also, an area of land above flood level, or not reached by storm tides.—Ant: *lowland.*
upland plain A relatively level area of land lying at a considerable altitude; esp. a high-lying erosion surface.
upland swamp A swamp that "probably" occupies the site of a former shallow sound or coastal lagoon which has become land "through uplift and retreat of the sea" (Stephenson & Veatch, 1915, p. 37). Cf: *tidal swamp.*
uplift [eng] Any force that tends to raise an engineering structure and its foundation relative to its surroundings. It may be caused by pressure of subjacent ground, surface water, expansive soil under the base of the structure, or lateral forces such as wind.
uplift [tect] A structurally high area in the crust, produced by positive movements that raise or upthrust the rocks, as in a dome or arch. Cf: *depression.*
uplimb thrust fault A contraction fault developed on the limb of an anticline in which the direction of tectonic transport is uplimb and the fault dips initially in the same direction as the limb but at a steeper angle. Continued rotation of the limb toward the vertical results in associated limb-contraction faults that dip toward the fold axis (Perry & DeWitt, 1977). Syn: *flexural-slip thrust fault.*
upper Pertaining to rocks or strata that are normally above those of earlier formations of the same subdivision of rocks. The adjective is applied to the name of a chronostratigraphic unit (system, series, stage) to indicate position in the geologic column and corresponds to *late* as applied to the name of the equivalent geologic-time unit; e.g. rocks of the Upper Jurassic System were formed during the Late Jurassic Period. The initial letter of the term is capitalized to indicate a formal subdivision (e.g. "Upper Devonian") and is lowercased to indicate an informal subdivision (e.g. "upper Miocene"). The informal term may be used where there is no formal subdivision of a system or series. Cf: *lower [stratig]; middle [stratig].*
upper break *head [struc geol].*
Upper Carboniferous In European usage, the approximate equivalent of the *Pennsylvanian.* Cf: *Lower Carboniferous.*
upper keriotheca The abaxial (upper) part of *keriotheca* in the wall of a fusulinid, characterized by fine alveolar structure (as in *Schwagerina*). Cf: *lower keriotheca.*
upper mantle That part of the *mantle* which lies above a depth of about 1000 km and has a density of 3.40 g/cm^3, in which *P*-wave velocity increases from about 8 to 11 km/sec with depth and *S*-wave velocity increases from about 4.5 to 6 km/sec with depth. It is presumed to be peridotitic in composition. It is sometimes referred to as the *asthenosphere,* and includes the *transition zone;* it is equivalent to the B and C layers. Syn: *outer mantle; peridotite shell.*
upper Paleolithic n. The third and most recent division of the *Paleolithic,* characterized by *Homo sapiens* and the appearance of man in Australia and the Americas. Cf: *lower Paleolithic; middle Paleolithic.*—adj. Pertaining to the upper Paleolithic.
upper plate The hanging wall of a fault. Cf: *lower plate.*
upper tectorium The abaxial secondary layer of spirotheca in the wall of a fusulinid, next above the tectum (as in *Profusulinella*). Cf: *tectorium; lower tectorium.*
Upper Volgian European stage: Upper Jurassic (above Lower Volgian, below Cretaceous-Berriasian). Equivalent to upper Tithonian.
upright fold A fold having an essentially vertical axial surface; a *vertical fold.*
uprush The advance of water up the foreshore of a beach or structure, following the breaking of a wave. Cf: *backwash.* Syn: *runup.*
upsetted moraine *push moraine.*
upside-down channel *ceiling channel.*
upsiloidal dune A general term for a U-shaped or V-shaped dune whose form is concave toward the wind; e.g. a *parabolic dune.*
upslope n. A slope that lies upward; uphill.—adj. In an upward or uphill direction, or ascending; e.g. an *upslope* ripple that climbed a sloping surface.
upstream In the direction from which a stream or glacier is flowing. Similarly, *upriver.* Ant: *downstream.*
upthrow n. (a) The upthrown side of a fault. (b) The amount of upward vertical displacement of a fault.—Cf: *downthrow; heave.*
upthrown Said of that side of a fault that appears to have moved upward, compared with the other side. Cf: *downthrown.*
upthrown block *upthrow.*
up-to-basin fault A term used in petroleum geology for a fault whose upthrown side is toward the adjacent basin.
upward continuation Calculation of the potential field at an elevation higher than that at which the field is known. The *continuation* involves the application of Green's theorem and is rigorous if the field is completely known over the lower surface and if no sources are present between the surfaces (as is usually true for gravity and magnetic fields). Upward continuation is used to smooth out near-surface effects and to tie aeromagnetic surveys flown at different heights. Cf: *downward continuation.*
upwarping The uplift of a regional area of the Earth's crust, usually as a result of the release of isostatic pressure, e.g. melting of an ice sheet. Cf: *downwarping.*
upwelling [currents] The rising of cold, heavy subsurface water toward the surface, esp. along the western coasts of continents (as along the coast of southern California); the displaced surface water is transported away from the coast by the action of winds parallel to it or by diverging currents. Upwelling may also occur in the open ocean where cyclonic circulation is relatively permanent, or where southern trade winds cross the equator. Ant: *sinking [currents].*
upwelling [volc] The relatively quiet eruption of lava and volcanic gases, without much force.
uraconite A name that has been used for various uranium sulfates, but that "lacks specific meaning and should be abandoned" (Frondel et al., 1967, p. 44).
uralborite A mineral: $CaB_2O_4 \cdot 2H_2O$.

Uralian Stage in Russia: uppermost Carboniferous (above Gzhelian, below Sakmarian of Permian). Equivalent to Orenburgian-Gzhelian.

Uralian emerald (a) Emerald from near Sverdlovsk in the Ural Mountains, U.S.S.R. (b) *demantoid.*

uralite A green, generally fibrous or acicular variety of secondary amphibole (hornblende or actinolite) occurring in altered rocks and pseudomorphous after pyroxene (such as augite).

uralite diabase *uralitite.*

uralitite A term suggested for a *diabase* that contains augite altered to uralite. Syn: *uralite diabase.* Not recommended usage.

uralitization The development of amphibole from pyroxene; specif. a late-magmatic or metamorphic process of replacement whereby uralitic amphibole results from alteration of primary pyroxene. Also, the alteration of an igneous rock in which pyroxene is changed to amphibole; e.g. the alteration of gabbro to greenstone by pressure metamorphism.

uralolite A mineral: $CaBe_3(PO_4)_2(OH)_2 \cdot 4H_2O$.

Ural-type glacier *drift glacier.*

uramphite A bottle-green to pale-green mineral: $(NH_4)(UO_2)(PO_4) \cdot 3H_2O$.

uraninite A black, brown, or steel-gray octahedral or cubic mineral, essentially UO_2, but usually partly oxidized. It is strongly radioactive, and is the chief ore of uranium. It is isomorphous with thorianite. Uraninite often contains impurities such as thorium, radium, the cerium and yttrium metals, and lead; when heated, it yields a gas consisting chiefly of helium. It occurs in veins of lead, tin, and copper minerals and in sandstone deposits, and is a primary constituent of granites and pegmatites. See also: *pitchblende.* Syn: *ulrichite; coracite.*

uranite A general term for a mineral group consisting of uranyl phosphates and arsenates of the autunite, meta-autunite, and torbernite type.

uranium-isotope age *uranium-uranium age.*

uranium-lead age method Calculation of an age in years for geologic material based on the known radioactive decay rate of uranium-238 to lead-206 and uranium-235 to lead-207. It is part of the more inclusive *uranium-thorium-lead age method* in which the parent-daughter pairs are considered simultaneously. Syn: *lead-uranium age method.*

uranium ocher *gummite.*

uranium series The radioactive series beginning with uranium-238.

uranium-series age method Calculation of an age in years for Quaternary materials based on the general finding that the decay products uranium-234, thorium-230, and protactinium-231 in natural materials are commonly in disequilibrium with their parent isotopes, uranium-238 and uranium-235, either deficient or in excess. The age is determined from the measured activity ratios of these isotopes. See also: *ionium-thorium age method; ionium-excess method; ionium-deficiency method; thorium-230/protactinium-231 deficiency method; protactinium-ionium age method; uranium-234 age method.*

uranium-thorium-lead age method Calculation of an age in years for geologic material, often zircon, based on the known radioactive decay rate of uranium-238 to lead-206, uranium-235 to lead-207, and thorium-232 to lead-208, whose ratios give three independent ages for the same sample. The determined lead-207/lead-206 ratio can be used to compute a fourth age (*lead-lead age*). The method is most applicable to minerals that are Precambrian in age. Whether all four possible dates are concordant or discordant is useful for evaluating the results of this method, used alone or in comparison with other methods, and in determining whether the initially closed system has been disturbed. Partial syn: *uranium-lead age method; thorium-lead age method.* Syn: *uranium-thorium-lead dating.*

uranium-thorium-lead dating *uranium-thorium-lead age method.*

uranium-uranium age An age in years calculated from the ratio of uranium-235 to uranium-238; a by-product of the *uranium-thorium-lead age method.* Syn: *uranium-isotope age.*

uranium-234 age method The calculation of an age in years for fossil coral or shell (limited to those formed during the last million years), based on the assumption that the initial uranium-234/uranium-238 ratio is known for the fossil. The change in this ratio is directly related to passage of time as the two isotopes have very different half-lives. See also: *uranium-series age method.* Syn: *uranium-234 excess method; uranium-234/uranium-238 age method; uranium-238/uranium-234 disequilibrium method.*

uranium-234 excess method *uranium-234 age method.*

uranium-234/uranium-238 age method *uranium-234 age method.*

uranium-238/uranium-234 disequilibrium method *uranium-234 age method.*

uran-mica A uranite, esp. torbernite.

uranocher A general name used chiefly for uranium sulfates (such as uranopilite) and for some uranium oxides. Also spelled: *uranochre.*

uranocircite A yellow-green mineral of the autunite group: $Ba(UO_2)_2(PO_4)_2 \cdot 8H_2O$.

uranophane A strongly radioactive yellow orthorhombic secondary mineral: $Ca(UO_2)_2Si_2O_7 \cdot 6H_2O$. It is isostructural with sklodowskite and cuprosklodowskite, and dimorphous with beta-uranophane. Syn: *uranotile.*

uranopilite A yellow secondary mineral: $(UO_2)_6(SO_4)(OH)_{10} \cdot 12H_2O$.

uranosphaerite An orange-red or brick-red secondary mineral: $Bi_2U_2O_9 \cdot 3H_2O$. Also spelled: *uranospherite.*

uranospinite A green to yellow secondary mineral of the autunite group: $Ca(UO_2)_2(AsO_4)_2 \cdot 10H_2O$. It is isomorphous with zeunerite.

uranotantalite *samarskite.*

uranothallite *liebigite.*

uranothorianite A variety of thorianite containing uranium; an intermediate member in the uraninite-thorianite isomorphous series.

uranothorite A variety of thorite containing uranium.

uranotile A syn. of *uranophane.* Also spelled: *uranotil.*

urantsevite A hexagonal mineral: $Pd(Bi,Pb)_2$.

urbainite An *ilmenitite* that contains 10-20 percent rutile and 3-5 percent sapphirine. Its name, given by Warren in 1912, is derived from St. Urbain, Quebec. Not recommended usage.

urban geology The application of geologic knowledge and principles to the planning and management of cities and their surroundings. It includes geologic studies for physical planning, waste disposal, land use, water-resources management, and extraction of usable raw materials. See also: *environmental geology.*

urea A tetragonal mineral: $CO(NH_2)_2$.

urediospore A yellow, orange, or reddish *fungal spore* of brief vitality, whose thin walls may be composed of chitin. Such spores may occur as microfossils in palynologic preparations. Cf: *teleutospore.* Also spelled: *uredospore.*

ureilite An achondritic stony meteorite composed essentially of olivine and clinobronzite, with accessory amounts of nickel-iron, troilite, diamond, and graphite. It is the only achondrite with an appreciable amount of nickel-iron.

ureyite A meteorite mineral of the pyroxene group: $NaCrSi_2O_6$. Syn: *kosmochlor; cosmochlore.*

uricite A monoclinic mineral: $C_5H_4N_4O_3$ (2,6,8-trihydroxy purine).

Uriconian A division of the Precambrian in Great Britain.

Urodela An order of caudate lissamphibians that includes salamanders and newts. Range, Upper Jurassic to present.

uropod Either of the flattened leaflike appendages of the last abdominal segment of various crustaceans, which with the telson form the caudal fan; e.g. limb of the sixth abdominal somite of a eumalacostracan, or one of the last three abdominal appendages of an amphipod. The term is sometimes applied to any abdominal appendage of a crustacean. Syn: *uropodite.*

urosome The part of the body of a copepod crustacean behind the major articulation that marks the posterior boundary of a *prosome.* Syn: *urosoma.*

ursilite A lemon-yellow mineral: $(Ca,Mg)_2(UO_2)_2Si_5O_{14} \cdot 9\text{-}10H_2O$.

urstromtal A wide, shallow, trenchlike valley or depression excavated by a temporary meltwater stream flowing parallel to the front margin of a continental ice sheet, esp. one of the east-west depressions across northern Germany; a large-scale overflow channel. Etymol: German *Urstromtal,* "ancient river valley". Pl: *urstromtäler.* Syn: *pradolina.*

urtite (a) In the *IUGS classification,* a plutonic rock in which F is between 60 and 100, M is 30 or less, and sodium exceeds potassium. Cf: *italite.* (b) A light-colored member of the *ijolite* series that is composed chiefly of nepheline and 0-30% mafic minerals, esp. acmite and apatite. Cf: *melteigite.*—The name, given by Ramsay in 1896, is for Lujavr-Urt (now Lovozero), Kola Peninsula, U.S.S.R.

usamerite A term proposed by Boswell (1960, p.157) for a rock comparable to the type *graywacke* and characterized by size grades ranging from gravel to sand, by poor sorting with a "substantial" quantity of matrix, and by variable rock and mineral fragments that are predominantly angular to subangular. Etymol: *U*nited *S*tates of *Amer*ica + *ite*.

U-shaped dune A dune having the form of the letter "U", its open end facing upwind.

U-shaped valley A valley having a pronounced parabolic cross profile suggesting the form of a broad letter "U", with steep walls and a broad, nearly flat floor; specif. a valley carved by glacial erosion, such as a *glacial trough*. Cf: *V-shaped valley*. Syn: *U-valley; trough valley*.

usovite A mineral: $Ba_2MgAl_2F_{12}$.

Ussherian Pertaining to the biblical chronology compiled by James Ussher (d.1656), Irish archbishop, who calculated from studies of the Scriptures that the Earth was formed in 4004 B.C.

ussingite A reddish-violet mineral: $Na_2AlSi_3O_8(OH)$.

U-stage *universal stage*.

Ustalf In U.S. Dept. of Agriculture soil taxonomy, a suborder of the soil order *Alfisol*, characterized by formation in an ustic moisture regime and in a temperature regime that is usually thermic or warmer. Most Ustalfs are reddish and occur in warm subhumid to semiarid regions. Moisture penetrates to deeper soil layers only in occasional years, and droughts are common. Native vegetation is often savanna, and where cropped, sorghum and cotton are common (USDA, 1975). Cf: *Aqualf; Boralf; Udalf; Xeralf.*

ustarasite A gray mineral: $Pb(Bi,Sb)_6S_{10}$.

Ustert In U.S. Dept. of Agriculture soil taxonomy, a suborder of the soil order *Vertisol*, characterized by formation in an isohyperthermic temperature regime in areas of monsoon climates, tropical and subtropical areas having two rainy and two dry seasons, and temperate regions having low summer rainfall. Cracks close once or twice each year for extended periods but are open more than 90 cumulative days each year. Usterts are extensive in Texas, Australia, Africa, and India (USDA, 1975). Cf: *Torrert; Udert; Xerert.*

ustic moisture regime A soil moisture regime that is intermediate between the acidic and udic moisture regimes. It is one of limited moisture, but the moisture is present when conditions are suitable for plant growth (USDA, 1975).

Ustoll In U.S. Dept. of Agriculture soil taxonomy, a suborder of the soil order *Mollisol*, characterized by formation in an ustic moisture regime and in a mesic or warmer temperature regime. They are soils of mid to low latitudes with subhumid to semiarid climates. Because summer rainfall is erratic, drought is frequent and often severe. Productivity without irrigation is limited. Most Ustolls have a cambic, argillic, or natric horizon and a horizon of lime accumulation (USDA, 1975). Cf: *Alboll; Aquoll; Boroll; Rendoll; Udoll; Xeroll.*

Ustox In U.S. Dept. of Agriculture soil taxonomy, a suborder of the soil order *Oxisol*, characterized by formation in an ustic soil moisture regime and by a mean annual soil temperature of 15°C or more. Ustox soils are generally red and are low in organic matter. They tend to occur near the tropics of Cancer and Capricorn (USDA, 1975). Cf: *Aquox; Humox; Orthox; Torrox.*

Ustult In U.S. Dept. of Agriculture soil taxonomy, a suborder of the soil order *Ultisol*, characterized by formation in warm regions with high rainfall but a pronounced dry season. Ustults have a low organic-carbon content and an ochric epipedon that rests directly on a reddish argillic horizon. They develop in an ustic moisture regime and a thermic or warmer temperature regime (USDA, 1975). Cf: *Aquult; Humult; Udult; Xerult.*

UT *universal time*.

utahite (a) *jarosite*. (b) *natrojarosite*.

utahlite A syn. of *variscite*, esp. that found in compact, nodular masses in Utah.

UTM projection *universal transverse Mercator projection*.

utricle Any bladder-shaped plant appendage.

uvala A syn. of *karst valley*. Etymol: Serbo-Croatian.

U-valley *U-shaped valley*.

uvanite A brownish-yellow mineral: $U_2V_6O_{21} \cdot 15H_2O$ (?).

uvarovite The calcium-chromium end-member of the garnet group, characterized by an emerald-green color: $Ca_3Cr_2(SiO_4)_3$. It may have considerable alumina. Also spelled: *uwarowite; ouvarovite*.

uvite A mineral of the tourmaline group: $CaMg_3(Al_5Mg)(BO_3)_3Si_6O_{18}(OH)_4$.

uwarowite *uvarovite*.

uzbekite *volborthite*.

V

vacancy A vacant site in a crystal structure, due to the absence of an atom or ion from its normal structural position. Syn: *hole [cryst].*

vacuity *degradation vacuity.*

vacuole [paleont] A cavity in the cytoplasm of a cell of a plant or protozoan, often containing a watery solution enclosed by a membrane, and performing various functions such as digestion (food vacuole) and hydrostatic relation (contractile vacuole); e.g. one of the irregularly shaped *alveoles* in a foraminiferal-test wall. Also, the globular fluid inclusion or droplet enclosed in a vacuole.

vacuole [petrology] A syn. of *vesicle;* such usage is usually French.

vadose solution Solution action by vadose water above the level of the water table. Cf: *phreatic solution.*

vadose water Water of the zone of aeration. Syn: *kremastic water; suspended water; wandering water.*

vadose-water discharge The release, by evaporation, of water not originating in the zone of saturation. It may be in the form of *vegetal discharge* or *soil discharge.*

vadose zone *zone of aeration.*

vaesite An isometric mineral with pyrite structure: NiS_2.

vagile Said of a plant or animal that is free to move about. Cf: *sessile.*

vake The French term for *wacke* or soft, compact, mixed claylike material with a flat, even fracture, commonly associated with basaltic rocks.

vakite A rock composed predominantly of vake. The term is not recommended.

val A longitudinal, synclinal valley in the folded Jura Mountains of the European Alps. Etymol: French, "narrow valley". Pl: *vaux.* Cf: *cluse.* See also: *combe.* Syn: *vallon.*

Valanginian European stage: Lower Cretaceous (above Berriasian, below Hauterivian).

valbellite A fine-grained black hypabyssal rock containing bronzite, olivine, hornblende, and magnetite; a hornblende *harzburgite.* The name, given by Schaefer in 1898, is for Valbella, Piedmont, Italy. Obsolete.

vale (a) A lowland, usually containing a stream; e.g. the depression between two parallel cuestas. It often forms the wider and flatter part of a valley. (b) A rift valley or tectonic valley; e.g. the Vale of Arabia. (c) A poetic var. of *valley,* esp. one that is relatively broad and flat.

valencianite A variety of adularia from Guanajuato, Mexico.

Valentian *Llandoverian.*

valentinite A white orthorhombic mineral: Sb_2O_3. It is polymorphous with senarmontite. Syn: *antimony bloom; white antimony.*

valid (a) Said of a name that must, under the rules of zoological nomenclature, be adopted for a *taxon* with a particular rank, position, and description. (b) Said of publication in accordance with articles 32-45 of the International Code of Botanical Nomenclature (ICBN, 1972), which cover effective publication, form of name, and descriptive and illustrative requirements.

valleriite A mineral: $4(Fe,Cu)\cdot3(Mg,Al)(OH)_2$.

Vallesian European stage: Upper Miocene (above Maremmian, below Turolian).

valleuse A French term for a *hanging valley,* as on the chalk cliffs of France.

vallevarite A light-colored monzonitic igneous rock composed chiefly of andesine, microcline, and antiperthite, with small quantities of clinopyroxene, biotite, and apatite. The name, given by Gavelin in 1915, is for Vallevara, Sweden. Obsolete.

valley [geomorph] (a) Any low-lying land bordered by higher ground; esp. an elongate, relatively large, gently sloping depression of the Earth's surface, commonly situated between two mountains or between ranges of hills or mountains, and often containing a stream with an outlet. It is usually developed by stream erosion, but may be formed by faulting. (b) A broad area of generally flat land extending inland for a considerable distance, drained or watered by a large river and its tributaries; a river basin. Example: the Mississippi Valley.—Etymol: Latin *vallis.* Syn: *vale; dale.*

valley [marine geol] A wide, low-relief depression of the ocean floor with gently sloping sides, as opposed to a submarine canyon.

valley axis A term used by Woodford (1951, p. 803a) to replace *thalweg,* signifying "the surface profile along the center line of the valley".

valley bottom *valley floor.*

valley braid *anabranch.*

valley drift Outwash material constituting a *valley train.*

valley fill The unconsolidated sediment deposited by any agent so as to fill or partly fill a valley.

valley flat (a) The low or nearly level ground lying between valley walls and bordering a stream channel; esp. the small plain at the bottom of a narrow, steep-sided valley. Howard (1959, p. 239) recommends that the term be applied noncommittally to a flat surface that cannot be identified with certainty as a flood plain or terrace. Syn: *flat.* (b) A bedrock surface produced by lateral erosion, commonly veneered with the alluvium of a *flood plain* (Thornbury, 1954, p. 130).

valley floor The comparatively broad, flat bottom of a valley; it may be excavated and represent the level of a former erosion cycle, or it may be buried under a thin cover of alluvium. Syn: *valley bottom; flood plain.*

valley-floor basement The gently sloping, degraded bedrock underlying the *valley-floor side strip* and the valley floor proper, developed in a humid climate by lateral extension of the valley floor at the expense of the enclosing slopes, and covered with slowly creeping soil and flood-plain deposits (Davis, 1930).

valley-floor divide A divide in a valley; a dividing height located between two parts of the same valley, each part draining to a different river basin.

valley-floor increment The loose material coming to and lying upon the valley floor (Malott, 1928b, p. 12).

valley-floor side strip The narrow, level to slightly concave surface between the wash slope and the valley floor proper (the flood plain), produced by degradation and recession of the valley-side slope. See also: *valley-floor basement.*

valley glacier A glacier flowing between the walls of a mountain valley in all or part of its length; an *alpine glacier.* Nonrecommended syn: *ice stream [glaciol].*

valley head The upper part of a valley.

valley-head cirque A cirque formed at the head of a valley. Cf: *hanging cirque.*

valley iceberg An iceberg eroded in such a manner that a large U-shaped slot, which may be awash, extends through the ice, separating pinnacles or slabs of ice. Syn: *drydock iceberg.*

valley-in-valley (a) Said of the condition, structure, or cross profile of a valley form whose side is marked by a *valley shoulder* separating a steep-sided, youthful valley below from a more widely opened, older valley above. (b) Pertaining to a *two-cycle valley.*

valley line *thalweg [streams].*

valley-loop moraine *loop moraine.*

valley meander One of a series of curves of a *meandering valley.*

valley-moraine lake A glacial lake formed in a valley by the damming action of a recessional moraine produced by a mountain glacier. Cf: *drift-barrier lake.*

valley of elevation A syn. of *anticlinal valley.* The term was introduced in 1825 by Buckland (1829, p. 123).

valley of subsidence A syn. of *synclinal valley.* The term was used by Hitchcock (1841, p. 178).

valley plain (a) A continuous flood plain (Cotton, 1940). (b) *valley floor.*

valley-plain terrace A term used by Cotton (1940, p. 28-29) for the remnant of a formerly continuous flood plain or valley floor; it would include the features now known as a *strath terrace* and a *fillstrath terrace.*

valley plug A local constriction in a stream channel, which may be formed by any of several types of channel obstructions and may cause rapid deposition. See also: *plug [sed].*

valley profile The *longitudinal profile* of a valley.

valley shoulder A bedrock surface made in a *valley-in-valley* form, representing the sharp angle or break in slope between the side or floor of the upper, older valley and the side of the lower, newer valley. It is a remnant of the valley floor formed during a previous erosion cycle, marking the former base level of erosion, and extending across rocks of varying lithology. Syn: *shoulder [geomorph].*

valley-side moraine *lateral moraine.*

valley-side slope (a) A measure, generally expressed in degrees, of the steepest inclination of the side of a valley in stream-eroded topography. Maximum slope is measured at intervals along the valley walls on the steepest parts of the contour orthogonals running from divides to adjacent stream channels. Symbol: θ. Syn: *ground slope.* (b) The surface between a drainage divide and the valley floor. Syn: *valley wall.*

valley sink *karst valley.*

valley spring A type of *depression spring* issuing from the side of a valley at the outcrop of the water table.

valley storage (a) The volume of water in a body of water below the water-surface profile. (b) The natural storage capacity or volume of water of a stream in flood that has overflowed its banks; it includes both the water within the channel and the water that has overflowed.—(ASCE, 1962).

valley system A valley and all of its tributary valleys.

valley tract The middle part of a stream course, characterized by a moderate gradient and a fairly wide valley. Cf: *mountain tract; plain tract.*

valley train A long, narrow body of outwash, deposited by meltwater streams far beyond the terminal moraine or the margin of an active glacier and confined within the walls of a valley below the glacier; it may or may not emerge from the mouth of the valley to join an *outwash plain.* See also: *gravel train; valley drift.* Syn: *outwash train.*

valley wall *valley-side slope.*

valley wind A daytime *anabatic wind* moving up a valley or mountain slope. Cf: *mountain wind.*

vallis A valley on another planet or on the Moon; generally used as a proper name, e.g. Vallis Shröteri. A system of valleys is labeled by the plural form, e.g. Valles Marineris. No particular origin is implied by the term (Mutch, 1970, p. 224). Etymol: Latin.

vallon A syn. of *val.* Etymol: French, "small valley".

Valmeyeran Provincial series in Illinois: Lower and Upper Mississippian (equivalent to Osagian and Meramecian elsewhere).

value In economic geology, (a) the valuable constituents of an ore; (b) their percentage in an orebody, or *assay grade;* (c) their quantity in an orebody, or *assay value.* See also: *unit value.*

valve (a) One of the distinct and usually articulated pieces that make up the shell of certain invertebrates, e.g. one of the two curved calcareous plates that constitute the shell of a bivalve mollusk; one of the two halves of the carapace of a crustacean, divided by articulation along the mid-dorsal line; or one of the two curved plates that form the shell of a brachiopod. (b) One of the two silicified pieces or encasing membranes forming the top or bottom surface of a diatom frustule; e.g. epivalve and hypovalve.

valverdite A rounded or lenticular glass object containing crystalline inclusions, found near Del Rio in Val Verde County, Texas. It is weathered obsidian.

valvular Resembling or having the function of a valve in the body of an invertebrate; e.g. "valvular pyramid" of a cystoid or edrioasteroid, composed of several more or less triangular plates covering the anus or a gonopore.

van A term used in the French Alps for *cirque [glac geol]* (Schieferdecker, 1959, term 1667).

vanadate A mineral compound characterized by pentavalent vanadium and oxygen in the anion. An example is vanadinite, $Pb_5(VO_4)_3Cl$. Cf: *arsenate; phosphate.*

vanadinite A red, yellow, or brown mineral of the apatite group: $Pb_5(VO_4)_3Cl$. It is isomorphous with pyromorphite, and commonly contains arsenic or phosphorus. Vanadinite often forms globular masses encrusting other minerals in lead mines, and is an ore of vanadium and lead.

vanado-magnetite *coulsonite.*

vanalite A bright-yellow mineral: $NaAl_8V_{10}O_{38}\cdot 30H_2O$.

Van Allen belt A zone of charged particles (protons and electrons) surrounding the Earth, beginning at about 1,000 km altitude; produced largely by geomagnetic trapping of solar and cosmic particulate radiation. Radiation belts are also known around Jupiter.

vandenbrandeite A blackish-green mineral: $CuUO_4\cdot 2H_2O$.

vandendriesscheite A yellow or amber-orange mineral: $PbU_7O_{22}\cdot 12H_2O$.

van der Kolk method A test used in refractometry to determine the index of refraction of a mineral relative to that of the liquid medium in which it is immersed. When an obstacle blocks the light rays used for illumination, its shadow appears on the same side as itself when the mineral grain has a relatively higher refractive index, and on the opposite side when the mineral's refractive index is relatively lower than that of the medium.

Vandyke brown n. *black earth [coal].* Etymol: its use by the 17th-Century Flemish painter Van Dyck.

vane (a) The target on a *level rod.* (b) One of the sights of a compass or quadrant.

vane test An in-place test to measure the shear strength of fine-grained cohesive soils and other soft deposits. A rod with four flat radial blades, or vanes, projecting at 90-degree intervals is forced into the soil and rotated; the torque required to rotate the rod is a measure of the material's shear strength.

vanoxite A black mineral: $V_4^{+4}V_2^{+5}O_{13}\cdot 8H_2O$ (?).

van't Hoff equation An equation giving the temperature dependence of the *equilibrium constant* of a reaction: $d \ln K/dT = \Delta H°/RT^2$, where K=equilibrium constant, T=absolute temperature, $\Delta H°$=enthalpy change for the hypothetical reaction with all substances in their standard states, and R=gas constant.

vanthoffite A colorless mineral: $Na_6Mg(SO_4)_4$.

van't Hoff law The statement in phase studies that, when a system is in equilibrium, of the two opposed interactions, the endothermic one is promoted by raising the temperature, and the exothermic one by lowering it.

vanuralite A citron-yellow mineral: $Al(UO_2)_2(VO_4)_2(OH)\cdot 11H_2O$.

vanuranylite A bright-yellow mineral: $[(H_3O),Ba,Ca,K]_{1.6}(UO_2)_2(VO_4)_2\cdot 4H_2O$(?).

vanuxemite A mixture of sauconite and hemimorphite.

Van Veen grab sampler A type of *grab sampler* that encloses ocean-bottom material in two hemicylindrical buckets that rotate shut on a hinge when the sampler strikes the bottom. Syn: *Peterson grab.*

Vaporchoc A trade name for a marine seismic energy source in which a quantity of superheated steam under high pressure is injected into the water. Subsequent condensation of the steam attenuates bubble oscillation. Also called "steam gun".

vaporization *evaporation.*

vara Any of various old Spanish units of length used in Latin America and SW U.S., equal in different localities to between 31 and 34 inches; e.g. a unit equal to 33.3333 inches in Texas, to 33.372 inches in California, to 33.00 inches in Arizona and New Mexico, and to 32.9931 inches and 32.9682 inches (among others) in Mexico. For other values, see ASCE (1954, p. 169-170).

variability [grd wat] The ratio of the difference between maximum and minimum discharge of a spring to its average discharge, expressed as a percentage.

variability [paleont] The quality or attribute of an organism that causes it to exhibit variation.

variable (a) Any measurable or changeable statistical quality or quantity; e.g. *independent variable* and *dependent variable.* See also: *attribute; variate; parameter [stat].* (b) A quantity that can assume any of a given set of values at different stages in a computer program.

variable-amplitude trace *wiggle trace.*

variable-area recording A method of displaying seismic data in which the height of a blackened area on film or paper is proportional to the signal amplitude.

variable-density recording A method of displaying seismic data in which the darkness of the image is proportional to the signal amplitude.

variance [chem] *degrees of freedom.*

variance [stat] The square of the standard deviation. Symbol: σ^2.

variant An individual exhibiting *variation.*

variate A quantitative *variable;* e.g. a *random variable.*

variation Divergence in the structural or functional characteristics of an organism from those that are considered typical of the group to which it belongs. See also: *variant.*

variation diagram A binary or ternary diagram that shows the relations among various chemical parameters (e.g., oxide percentages, Niggli numbers, differentiation indexes) of the igneous rocks in a suite. Syn: *Harker diagram.*

variegated Said of a sediment or sedimentary rock, such as red beds or sandstone, showing variations of color in irregular spots, streaks, blotches, stripes, or reticulate patterns. Cf: *mottled.*

variegated copper ore *bornite.*

varietal mineral A mineral that is either present in considerable amounts in a rock or characteristic of the rock; a mineral which distinguishes one variety of rock from another. Syn: *distinctive mineral.*

variety [mineral] In gemology, a mineral that is a type of the mineral *species,* distinguished by color or other optical phenomenon or characteristic: e.g. emerald and aquamarine are varieties of beryl.

variety [taxon] A category in the hierarchy of botanical classification subordinate in rank to *subspecies.* Such infraspecific taxa are excluded from the present Code of Zoological Nomenclature.

varigradation A term used by McGee (1891, p. 261-267) for the process by which all streams of progressively increasing discharge tend constantly, in a degree varying inversely with the discharge, to depart slightly from the normal gradients.

variole A pea-size spherule, usually composed of radiating crystals of plagioclase or pyroxene. This term is generally applied only to such spherical bodies in basic igneous rock, e.g. variolite. Cf: *spherulite.*

variolite A general term, coined by Aldrovande in 1648 (Johannsen, 1938, p. 300), for aphanitic and fine-grained igneous rocks containing *varioles.* The term is derived from the Latin for "smallpox", in allusion to the spotted appearance of weathered surfaces. Not recommended usage.

variolitic Said of the texture of a rock, esp. a basic igneous rock, composed of pea-size spherical bodies (varioles) in a finer-grained groundmass. Cf: *spherulitic.*

variometer An instrument for measuring the variation of a magnetic element, using the torque on a permanent magnet in a uniform magnetic field.

Variscan orogeny The late Paleozoic orogenic era of Europe, extending through the Carboniferous and Permian. By current usage, it is synonymous with the *Hercynian orogeny.* Cf: *Armorican orogeny; Altaides.*

Variscides Hercynian mountain chains in Europe. Approx. syn: *Hercynides.*

variscite A green orthorhombic mineral: $AlPO_4 \cdot 2H_2O$. It is isomorphous with strengite and dimorphous with metavariscite. Variscite is a popular material for *cabochons* and various kinds of carved objects, and is often used as a substitute for turquoise. See also: *sphaerite.* Syn: *lucinite; utahlite.*

varix (a) One of the transverse elevations of the surface of a gastropod shell that is more prominent than a costa and that represents a halt in growth during which a thickened outer lip was developed (TIP, 1960, pt.I, p.134). (b) A thickening of an ammonoid shell marked on an internal mold by a transverse groove (Moore et al., 1952, p.366).—Pl: *varices.*

varlamoffite A mineral: $(Sn,Fe)(O,OH)_2$. It is perhaps fine-grained cassiterite.

varnish *desert varnish.*

varnsingite A coarse-grained light-colored hypabyssal rock containing albite (over 50 percent), pyroxene, sphene, magnetite, apatite, and secondary epidote, prehnite, chlorite, amphibole, and muscovite. The name, given by Sobral in 1913, is for Våstra Värnsingen, Sweden. Not recommended usage.

varulite A dull olive-green mineral: $(Na_2,Ca)(Mn^{+2},Fe^{+2})_2(PO_4)_2$. It is isomorphous with hühnerkobelite.

varve (a) A sedimentary bed or lamina or sequence of laminae deposited in a body of still water within one year's time; specif. a thin pair of graded glaciolacustrine layers seasonally deposited, usually by meltwater streams, in a glacial lake or other body of still water in front of a glacier. A glacial varve normally includes a lower "summer" layer consisting of relatively coarse-grained, light-colored sediment (usually sand or silt) produced by rapid melting of ice in the warmer months, which grades upward into a thinner "winter" layer, consisting of very fine-grained (clayey), often organic, dark sediment slowly deposited from suspension in quiet water while the streams were ice-bound. Counting and correlation of varves have been used to measure the ages of Pleistocene glacial deposits. (b) Any cyclic sedimentary *couplet,* as in certain shales and evaporites.—Etymol: Swedish *varv,* "layer" or "periodical iteration of layers" (DeGeer, 1912, p. 242).

varved clay A distinctly laminated lacustrine sediment consisting of clay-rich varves; also the upper, fine-grained, "winter" layer of a glacial varve. Syn: *varve clay.*

varvite An indurated rock consisting of ancient varves.

varvity The property of being varved; the seasonal and alternating lamination in varves.

varzea A term used in Brazil and Portugal for an alluvial flood plain or the bank of a river; also, a field or a level tract of land, esp. one that is sowed and cultivated. Etymol: Portuguese *várzea.*

vascular bundle In a vascular plant, conductive tissue composed of a strand of xylem and a strand of phloem, commonly separated by cambium and containing sclerenchymatous supporting tissue (Swartz, 1971, p. 492).

vascular plant A plant with a well-developed conductive system and structural differentiation; a *tracheophyte.* The majority of visible terrestrial plants are vascular.

vascular ray A ribbonlike aggregate of cells extending radially in stems through xylem and, often, phloem (Fuller & Tippo, 1954, p. 974). Syn: *ray [bot].*

vascular tissue The conducting tissue of vascular plants, composed of xylem and phloem.

vase Freshwater silt deposited in estuaries along the Atlantic coast of Europe and Africa, consisting of a mixture of sandy and pulverulent grains of quartz, calcite, clay minerals, and diatom shells, with a binder of *algon* (Bourcart, 1941). Etymol: French, "slime, mud". Pron: *vahz.*

vashegyite A white, yellow, or rust-brown mineral: $2Al_4(PO_4)_3(OH)_3 \cdot 27H_2O$ (?).

vastitas A term established by the International Astronomical Union for an extensive plain on Mars. Used as part of a formal name for a Martian landform, such as Vastitas Borealis (Mutch, et al., 1976, p. 57). Etymol: Latin *vastus.* empty, immense.

vat (a) *salt pit.* (b) A term used in SW U.S. for a dried and encrusted margin around a water hole.

vaterite A rare hexagonal mineral: $CaCO_3$. It is trimorphous with calcite and aragonite, and is a relatively unstable form of calcium carbonate.

vaterite-A Artificial calcite.

vaterite-B Artificial vaterite.

vauclusian spring A fountaining spring in a karst region, generally of exceptionally large discharge. Etymol: from Fontaine de Vaucluse in southern France. Syn: *gushing spring.* See also: *boiling spring.*

vaughanite A term suggested by Kindle (1923a, p. 370) for a pure, dense, homogeneous, dove-colored, fine-textured limestone that breaks with a smooth and more or less pronounced conchoidal fracture, that contains relatively few fossils, and that typically has a white, chalky appearance on weathered surfaces. Named after T. Wayland Vaughan (1870-1952), U.S. paleontologist.

vaugnerite A dark-colored, coarse-grained hypabyssal *quartz diorite* containing abundant biotite, along with green hornblende, white plagioclase, and quartz, with accessory alkali feldspar, apatite, magnetite, pyrite, and sphene. The name was given by Fournet in 1836 for Vaugneray, France. Not recommended usage.

vault [geomorph] A structure in the Earth's crust, resembling or suggesting a vault or an arched room; e.g. a cavern or a volcanic crater.

vault [paleont] (a) The part of a blastoid theca above the dorsal region (from aboral tips of ambulacra to dorsal pole). (b) An arched covering of calcareous plates between crinoid arms.

vaulted mud crack A raised mud crack on a playa, shaped like an inverted "V", and formed by salts and clayey material rising by capillary action through the narrow mud crack (Stone, 1967, p. 252). Syn: *roofed mud crack.*

vauquelinite A green to brownish-black mineral: $Pb_2Cu(CrO_4)(PO_4)(OH)$. It is isomorphous with fornacite.

vaux Pl. of *val.*

vauxite A blue triclinic mineral: $Fe^{+2}Al_2(PO_4)_2(OH)_2 \cdot 6H_2O$. It has less water than metavauxite and paravauxite.

väyrynenite A mineral: $MnBe(PO_4)(OH,F)$.

V-bar A *cuspate bar* whose seaward angle is fairly sharp, as where a secondary spit trails abruptly back toward the shore from the point of a primary spit.

V-coal Microscopic coal particles that are predominantly vitrain and clarain, as found in miners' lungs. Cf: *F-coal; D-coal.*

VE *vertical exaggeration.*

veatchite A white mineral: $Sr_2B_{11}O_{16}(OH)_5 \cdot H_2O$. It is dimorphous with p-veatchite, and has a space group A2/a.

vector structure *directional structure.*

veenite A mineral: $Pb_2(Sb,As)_2S_5$.

vegasite A mineral that may be identical with plumbojarosite.

vegetable jelly *ulmin.*

vegetal discharge The release, through the transpiration of plants, of water derived either from the zone of areation or from the zone of saturation by way of the capillary fringe. See also: *vadose-water discharge.*

vegetation anomaly As seen on aerial or space photographs or images, a deviation from the normal distribution or properties of vegetation. It may be caused by conditions along faults, trace elements in the soil, or other factors.

vegetation arabesque *vegetation polygon.*

vegetation coast A coast that is being extended seaward by the growth of vegetation, such as the mangrove trees in the Everglades of Florida.

vegetation polygon A small *nonsorted polygon* whose fissured borders are emphasized by thick vegetation (usually moss, lichen, or willow) and whose center consists of fine-textured material or a mixture of fines and stones. Diameter: about 1 m. See also: *lichen polygon; tussock-birch-heath polygon.* Syn: *vegetation arabesque.*

vegetation stripe (a) A form of *nonsorted stripe.* (b) A *sorted stripe* emphasized by vegetation (Sigafoos, 1951, p.289).

vegetative reproduction Plant reproduction by vegetable parts such as buds and gemmae; nonsexual reproduction (Swartz, 1971, p. 494).

veil [cryst] An aggregate of minute bubbles creating a whitish or cloudlike appearance in quartz.

veil [paleont] A variously formed weblike or netlike film in a radiolarian; e.g. *patagium.*

vein [bot] One of the vascular bundles of a leaf. See also: *venation.*

vein [ice] (a) A narrow water channel within land ice; also, the stream of water flowing through such a channel. (b) A narrow lead or lane in pack ice.

vein [intrus rocks] A thin, sheetlike igneous intrusion into a fissure. Not recommended usage.

vein [ore dep] An epigenetic mineral filling of a fault or other fracture in a host rock, in tabular or sheetlike form, often with associated replacement of the host rock; a mineral deposit of this form and origin. Cf: *lode [ore dep].*

vein [streams] (a) A narrow waterway or channel in rock or earth. Also, a stream of water flowing in such a channel. (b) An archaic term for the flow or current of a stream.

vein bitumen Any one of the black or dark-brown bitumens that give off a pitchy odor, burn readily with a smoky flame, and occupy fissures in rocks or less frequently form basin-shaped deposits on the surface (Nelson & Nelson, 1967).

vein-dike A pegmatitic intrusion that has the characteristics of both a vein and a dike. Also spelled: *veindike; vein dike.*

veined gneiss A composite gneiss with irregular layering. The term is generally used in the field and has no genetic implications (Dietrich, 1960, p. 50). Cf: *venite; arterite; phlebite; composite gneiss.*

vein quartz A rock composed chiefly of sutured quartz crystals of pegmatitic or hydrothermal origin and commonly of variable size.

vein system An assemblage of veins of a particular area, age, or fracture system, usually inclusive of more than one *lode [ore dep].*

velar Having the form of a veil, frill, or curtain.

velardenite *gehlenite.*

velar dimorphism In certain ostracodes, a kind of dimorphism in which the velar structures of the two sexes differ in size and/or curvature. Less-preferred syn: *velate dimorphism.*

velar ridge A velar structure developed as a low, usually rounded ridge in an ostracode valve. Less-preferred syn: *velate ridge.*

velar structure Any elongate elevated structure parallel to the free edge of the valve in straight-backed ostracodes, commonly developed as a *frill* (velum) or a *velar ridge;* if a carina is present, the velar structure lies ventral to it. Less-preferred syn: *velate structure.*

Vela Uniform A research program, sponsored by the Advanced Research Projects Agency of the U.S. Dept. of Defense, which had the objective of improving the capability of detecting underground nuclear explosions and of distinguishing them from earthquakes.

veld An open grassland area of South Africa. The number of terms incorporating "veld", e.g. *bushveld,* denotes the diversity of the area regarding elevation, vegetation, soil, etc. Also spelled: *veldt.*

velocimeter An instrument for measuring the velocity of sound in water, sometimes used to correct Doppler-sonar data for salinity and temperature variations.

velocity analysis Calculation of velocity distribution of seismic signals using normal-moveout times.

velocity coefficient A numerical factor, always less than unity, that expresses the ratio between the actual velocity of water issuing from an orifice or other hydraulic structure or device and the theoretical velocity which would exist if there were no friction losses due to the orifice. The square of the velocity coefficient is a measure of the efficiency of a structure as a waterway. It is a dimensionless number (ASCE, 1962).

velocity discontinuity *discontinuity [seis].*

velocity distribution The relationship between seismic velocity and location, including both horizontal and vertical distribution.

velocity gradient The rate of change of velocity with respect to distance normal to the direction of flow (ASCE, 1962).

velocity head Ground-water *head* resulting from kinetic energy of the water. It is negligible in most geological contexts. See also: *total head.*

velocity-head coefficient A correction factor applied to the velocity head of the mean velocity to correct for nonuniformity of velocity in a cross section. The factor is 1.0 where velocities are identical across a section and greater than 1.0 where velocities vary across a section (ASCE, 1962).

velocity/height ratio In airborne surveying, the ground speed of an aircraft divided by the aircraft's altitude above the terrain. Abbrev: V/H ratio.

velocity log *sonic log.*

velocity profile A seismic arrangement used to record reflections over a large range of shot-to-geophone distances, which is used to determine seismic velocity from the time-distance relationship.

velu A term used in the Maldive Islands of the Indian Ocean for the lagoon of a *faro.*

velum (a) A sail- or frill-like structure along the distal part of the valve of an ostracode, typically developed as the double-walled outfold of the carapace. (b) A thin membrane in larval mollusks.—Pl: *vela.*

velvet copper ore *cyanotrichite.*

venanzite A holocrystalline porphyritic extrusive *leucitite* or *melilitite* composed of olivine and phlogopite phenocrysts in a fine-grained groundmass of these minerals plus melilite, leucite, and magnetite. Syn: *euktolite.* The name, given by Sabatini in 1898, is for San Venanzo, Umbria, Italy. Not recommended usage.

venation The arrangement of the vascular bundles (*veins*) in a leaf. See also: *parallel venation; net venation.*

veneer (a) A thin but extensive layer of sediments covering an older geologic formation or surface; e.g. a *veneer* of alluvium covering a pediment. (b) A weathered or otherwise altered coating on a rock surface; e.g. desert varnish.

Vening Meinesz zone A belt of negative gravity anomalies, generally related to island arcs and/or oceanic deeps. Syn: *negative strip.*

venite Migmatite of which the mobile portion(s) were formed by exudation (secretion) from the rock itself (Dietrich & Mehnert, 1961). Originally proposed, along with the term *arterite,* to replace *veined gneiss* with terms of genetic connotation (Mehnert, 1968, p. 17). Cf: *phlebite; composite gneiss; diadysite.* Not widely used.

vent The opening at the Earth's surface through which volcanic materials are extruded; also, the channel or conduit through which they pass. Cf: *neck [volc]; pipe [volc].* See also: *feeder [volc]; chimney [volc].*

venter [bot] In the female gametophyte of certain plants, the enlarged base of an archegonium that contains the egg.

venter [paleont] (a) The outer and convex part of the shell of a curved or coiled cephalopod or gastropod, or the peripheral wall of a cephalopod whorl comprising the part of the shell radially farthest from the protoconch; the underside of a nautiloid and of its conch, distinguished generally by the hyponomic sinus and often by a conchal furrow (TIP, 1964, pt.K, p.59). (b) The median region of the shell of a productid brachiopod, situated between the valve surfaces on either side of the median sector of the shell (TIP, 1965, pt.H, p.155). (c) The belly region or the lower part of the carapace of an ostracode.—Cf: *dorsum.*

ventifact A general term introduced by Evans (1911) for any stone or pebble shaped, worn, faceted, cut, or polished by the abrasive or sandblast action of windblown sand, generally under desert conditions; e.g. a *dreikanter.* See also: *windkanter.* Syn: *glyptolith; wind-worn stone; wind-cut stone; wind-polished stone; wind-grooved stone; wind-scoured stone; wind-shaped stone.*

ventral (a) Pertaining or belonging to the abdominal or lower surface of an animal or of one of its parts that is opposite the back, e.g. in the direction toward the pedicle valve from the brachial valve of a brachiopod; pertaining to the region of the shell of a bivalve mollusk opposite the hinge, where the valves open most widely; or pertaining to the side of the stipe on which the thecal apertures of a graptoloid are situated, or to the inferior (commonly umbilical or apertural) side of a foraminiferal test. (b) Referring to the direction or side of an echinoderm toward or containing the mouth, normally upward; adoral or oral.—Ant: *dorsal.*

ventrallite A plagioclase-bearing *phonolite* containing more potassium feldspar than calcic plagioclase and containing nepheline as the essential feldspathoid. The name (Johannsen, 1939) is for Ventralla, Monte Vico, Italy; however, Johannsen (1938) spells the rock *vetrallite* and the locality Vetralla. Not recommended usage.

ventral lobe The main adapical lobe or inflection of a suture on the venter of a cephalopod shell. See also: *external lobe.* Cf: *dorsal lobe.*

ventral process A medially located excessive thickening of secondary shell of a brachiopod underlying the pseudodeltidium and projecting dorsally to fit between lobes of the cardinal process.

ventral shield An ossicle of secondary origin on the oral side of an arm in an ophiuroid. Cf: *dorsal shield.*

ventral valve The *pedicle valve* of a brachiopod.

ventromyarian Said of a nautiloid in which the retractor muscles of the head-foot mass are attached to the shell along the interior areas of the body chamber adjacent to, or coincident with, its ventral midline (TIP, 1964, pt.K, p.59). Cf: *dorsomyarian; pleuromyarian.*

Venturian North American stage: Middle Pliocene (above Repettian, below Wheelerian).

Venus hair Needlelike crystals of reddish-brown or yellow rutile, forming tangled swarms of inclusions in quartz. See also: *sagenite.*

Venus hairstone *rutilated quartz.*

verd antique A dark green massive serpentine, commonly with veinlets of calcium carbonate and magnesium carbonate. It is capable of being polished and is commercially considered a marble. Also spelled: *verde antique.* See also: *ophicalcite.* Syn: *serpentine marble.*

verdelite A green variety of tourmaline.

verdite A green rock, consisting chiefly of impure fuchsite and clayey matter, used as an ornamental stone.

vergence The direction of overturning or of inclination of a fold. The term is a translation of the German *Vergenz,* "overturn", coined by Stille (1930, p. 379) for the direction in which a geologic structure or family of structures is facing. Cf: *facing.*

verglas A thin film or layer of clear, hard, smooth ice on a rock surface, formed as a result of a frost following rain or snowmelt or by rime. Etymol: French. Cf: *black ice.*

verite A black extrusive *trachyte* containing phlogopite or biotite, clinopyroxene, and olivine crystals in a glassy groundmass; the glassy periphery of *fortunite.* Its name, given by Osann in 1889, is derived from the town of Vera, Cabo de Gata, Spain. Not recommended usage.

vermeil (a) An orange-red garnet. Syn: *vermilion.* (b) A reddish-brown to orange-red gem corundum; ruby. (c) An orange-red spinel.—Syn: *vermeille.*

vermetid reef A small *organic reef* composed of the irregularly entwined tubelike calcareous shells of vermetid (wormlike) gastropods, as in Bermuda. Cf: *serpulid reef; worm reef.*

vermicular quartz Quartz occurring in wormlike forms intergrown with or penetrating feldspar, as in myrmekite.

vermiculated Said of a stone, carbonate sediment, or any corroded geologic feature that has the appearance of having been eaten into by worms.

vermiculite (a) A group of platy or micaceous clay minerals closely related to chlorite and montmorillonite, and having the general formula: $(Mg,Fe,Al)_3(Al,Si)_4O_{10}(OH)_2 \cdot 4H_2O$. The minerals are derived generally from the alteration of micas (chiefly biotite and phlogopite) in the zone of weathering, and they vary widely in chemical composition. They are characterized by marked exfoliation when heated at 800° to 1100°C; granules expand 6 to 20 times at right angles to the cleavage as the contained water is converted into steam. The result is elongated wormlike particles that entrap air and produce a lightweight material that is used as an insulator and as an aggregate in concrete and plaster. Vermiculites differ from montmorillonites in that the characteristic exchangeable cation is Mg^{+2}, the lattice expands only to a limited degree (hydration and dehydration is limited to two layers of water), and they have higher layer charges per formula unit (0.6-0.9) and higher cation-exchange capacities. Vermiculites are found mostly with basic rocks such as dunite and pyroxenite. (b) Any mineral of the vermiculite group, such as maconite, jefferisite, and lennilite.

vermiform Wormlike or having the form of a worm; e.g. "vermiform problematica", consisting of long thin more or less cylindrical tubes.

vermiglyph A collective term used by Fuchs (1895) for a trace fossil consisting of a presumed worm trail appearing on the undersurface of flysch beds (mostly sandstones) as a threadlike, unbranched, and irregular relief form a few millimeters wide with a straight or variously winding course. Cf: *graphoglypt; rhabdoglyph.*

vermilion (a) *cinnabar.* (b) An orange-red garnet; *vermeil.*—Syn: *vermillion.*

vernacular name In biologic nomenclature, the common name of a plant or animal as opposed to its formal Latin name; e.g. sugar maple is the vernacular name of *Acer saccharum.* Syn: *popular name.* Cf: *scientific name.*

vernadite A name used in the U.S.S.R. for a mineral of supposed composition $MnO_2 \cdot nH_2O$.

vernadskite *antlerite.*

Verneuil process A method developed by Auguste V.L. Verneuil (1856-1913), French mineralogist and chemist, for the manufacture of large synthetic crystals of corundum and spinel, in which an alumina powder of the desired composition is melted in an oxyhydrogen flame, producing a series of drops that build up the *boules* of the synthetic gems.

vernier A short, uniformly divided, auxiliary scale that slides along the primary scale of a measuring device and that is used to measure accurately fractional parts of the smallest divisions of the primary scale or to obtain one more significant figure of a particular measurement. It is graduated such that the total length of a given number of divisions on a vernier is equal to the total length of one more or one less than the same number of divisions on the primary scale; parts of a division are determined by observing what line on the vernier coincides with a line on the measuring device. Named after Pierre Vernier (1580-1637), French mathematician.

vernier compass A surveyor's compass with a vernier, used for measuring angles without the use of the magnetic needle by means of a compensating adjustment made for magnetic variation.

Verona earth A naturally occurring iron silicate from Verona, Italy; specif. *celadonite.* Syn: *veronite.*

verplanckite A mineral: $Ba_2(Mn,Fe,Ti)Si_2O_6(O,OH,Cl,F)_2 \cdot 3H_2O$.

verrou A syn. of *riegel.* Etymol: French, "bolt".

verrucate Warty, or covered with wartlike knobs or elevations; e.g. said of spores and pollen having sculpture consisting of such projections. Syn: *verrucose.*

versant (a) The slope or side of a mountain or mountain chain. (b) The general slope of a region; e.g. the Pacific versant of the U.S. — An obsolete term.

Versilian European stage: Recent (above Tyrrhenian).

verst A Russian unit of distance equal to 0.6629 mile or 1.067 km.

vertebra (a) One of the bony or cartilaginous elements that together make up the spinal column of a vertebrate. (b) One of a fused pair of opposite ambulacral plates of an asterozoan, articulating with neighboring vertebrae by ball-and-socket joints.—Pl: *vertebrae.*

Vertebrata A subphylum of the Chordata characterized by an internal skeleton of cartilage or bone, and by specialized organization of the anterior end of the animal; the front of the body is a head that bears organs of sight, smell, taste, and hearing, and the front of the central nervous system is a brain.

vertebrate paleontology The branch of paleontology dealing with fossil vertebrates. Syn: *vertebrate paleozoology.*

vertebrate paleozoology *vertebrate paleontology.*

vertex The culmination or high point of a feature. The term was used by Sollas as the English version of Suess's term "Scheitel", the apex or nucleus of a continental or other large structure; example, the *Angara Shield,* supposed to have been the vertex of Asia.

vertical [geophys] adj. Said of a direction that is perpendicular to a *horizontal* plane.

vertical [stratig] Said of the direction at right angles to the direction of extension of the strata, as in the measurement of *true thickness.* The term has also been used to indicate a direction at right angles to the surface of the land, as in the measurement of *apparent thickness.* Cf: *lateral [stratig].*

vertical accretion Upward growth of a sedimentary deposit; e.g. settling of sediment from suspension in a stream subject to overflow. Cf: *lateral accretion.*

vertical-accretion deposit *flood-plain deposit.*

vertical angle An angle in a vertical plane; the angle between the horizontal and an inclined line of sight, measured on a vertical circle either upward or downward from the horizon. One of the directions that form a vertical angle in surveying is usually either the direction of the vertical (the angle being termed the *zenith distance*) or the line of intersection of the vertical plane in which the angle lies with the plane of the horizon (the angle being termed the altitude). Cf: *horizontal angle.*

vertical angulation *trigonometric leveling.*

vertical axis The line through the center of a theodolite or transit about which the alidade rotates. Cf: *horizontal axis.*

vertical cave A cave containing vertical or nearly vertical shafts, commonly but not necessarily at the entrance. Syn: *pothole [speleo]; pit [speleo]; pot [speleo]; sótano; gouffre.* Partial syn: *cenote; aven.*

vertical circle Any great circle of the celestial sphere passing through the zenith.

vertical collimator A telescope so mounted that its collimation axis may be made to coincide with the vertical (or direction of the plumb line). It may be used for centering a theodolite on a high tower exactly over a station mark on the ground. See also: *collimator.*

vertical control A series of measurements taken by surveying methods for the determination of elevation with respect to an imaginary level surface (usually mean sea level) and used as fixed references in positioning and correlating map features.

vertical corrasion Corrasion of the stream bed, causing a deepening of the channel.

vertical dip slip *vertical slip.*

vertical erosion *downcutting.*

vertical exaggeration [cart] (a) A deliberate increase in the vertical scale of a relief model, plastic relief map, block diagram, or cross section, while retaining the horizontal scale, in order to make the model, map, diagram, or section more clearly perceptible. (b) The ratio expressing vertical exaggeration; e.g. if the horizontal scale is one inch to one mile and the vertical scale is one inch to 2000 ft, the vertical exaggeration is 2.64. Abbrev: VE.

vertical exaggeration [photo] The apparent increase in the relief as seen in a stereoscopic image.

vertical fault A fault with a dip of 90 degrees. Cf: *horizontal fault.*

vertical-field balance An instrument that measures the vertical component of the magnetic field by means of the torque that the field component exerts on a horizontal permanent magnet. The two most common types are the *Schmidt field balance* and the *torsion magnetometer.* Cf: *horizontal-field balance.*

vertical fold A fold having a vertical axis.

vertical form index A term used by Bucher (1919, p.154) for a ratio also known as *ripple index.* Cf: *horizontal form index.*

vertical gradiometer An instrument for measuring the vertical gradient of gravity.

vertical intensity The vertical component of the *magnetic-field intensity;* it is one of the *magnetic elements,* and is symbolized by Z. It is usually considered positive if downward, negative if upward. Cf: *horizontal intensity.*

vertical interval The difference in vertical height between two points on a land surface; specif. *contour interval.* Abbrev: VI. Cf: *horizontal equivalent.* Syn: *vertical distance.*

vertical limb A graduated arc attached to a surveying instrument and used to measure vertical angles.

vertical-loop method An inductive electromagnetic method of prospecting in which the transacting coil has its plane vertical (i.e., axis horizontal).

vertically mixed estuary An estuary in which there is no measurable variation in salinity with depth, although salinity may increase laterally from the head to the mouth; occurs where tidal currents are very strong relative to river flow. Ant: *stratified estuary.* Syn: *vertically homogeneous estuary.*

vertical photograph An aerial photograph made with the camera axis vertical (camera pointing straight down) or as nearly vertical as possible in an aircraft. Cf: *oblique photograph.*

vertical pore A pore surrounding the vertical spine in the back of the lattice shell in the radiolarian skeletons of the subfamily Trissocyclinae.

vertical profiling A seismic survey in a deep borehole, in which a cable containing a number of geophones is lowered into the hole. The energy source is usually on the surface but sometimes it is placed in the borehole.

vertical section (a) A natural or artificial vertical exposure of rocks or soil, as in a sea cliff or canyon wall. (b) A *section* or diagram representing a vertical segment of the Earth's crust either actually exposed or as it would appear if cut through by any intersecting vertical plane; e.g. a *columnar section* or a *structure section.*

vertical separation In a fault, the distance measured vertically between two parts of a displaced marker such as a bed. Cf: *horizontal separation.*

vertical shaft *domepit.*

vertical sheet structure Slightly sinuous, subparallel, mainly vertical streaks in a sedimentary rock, distinguished from the surrounding rock by their lighter color and greater resistance to weathering. In plan view they show a straight or slightly wavy pattern of subparallel lines...and thus have a sheetlike three-dimensional form (Laird, 1970, p. 428). The structure is thought to be formed after deposition of a sand bed by escape of fluid, which washes out some of the finer particles. Lowe (1975) called the same type of structures *stress pillars.* Cf: *pillar structure.*

vertical shift In a fault, the vertical component of the shift.

vertical slip In a fault, the vertical component of the net slip; it equals the vertical component of the dip slip. Cf: *horizontal slip.* Syn: *vertical dip slip.*

vertical stack A composite or mix of the records from several shots made in nearly the same location without correcting for offset differences. It is used especially with surface sources, in which the records from several successive weight drops or vibrations are added together without making static or dynamic corrections beforehand.

vertical tectonics The tectonics of Pratt-type isostasy, in which vertical movements of crustal units of varying specific gravity cause topographic relief.

vertical-variability map A stratigraphic map that depicts the relative vertical positions, thicknesses, and number of occurrences of specific rock types in a sequence of strata or within a designated stratigraphic unit; e.g. a "number-of-sandstones map" indicating the number of discrete sandstone units in a given stratigraphic body, or a "limestone mean-thickness map" indicating the average thickness of limestone units in a given stratigraphic body. The map gives information about the internal geometry of the stratigraphic unit in terms of a designated rock component or property, or it may show the degree of differentiation of the unit into subunits of different lithologic types. Cf: *facies map.* See also: *center-of-gravity map; standard-deviation map; multipartite map; interval-entropy map.*

vertical-velocity curve A graphic presentation of the relationship between depth and velocity, at a given point along a vertical line, of water flowing in an open channel or conduit (ASCE, 1962). Syn: *mean velocity curve; depth-velocity curve.*

vertical wall One of the walls of a bryozoan zooid which are entirely or in part at high angles to the basal wall and the wall bearing the orifice.

verticil A *whorl* or circular arrangement of similar parts around an axis, as leaves of arthrophytes. Adj: *verticillate.*

verticillate Arranged in or having *verticils;* e.g. having successive whorls of branches arranged like the spokes of a wheel. Syn: *whorled.*

vertikalschichtung Nacreous tablets in a mollusk shell, arranged in vertical columns (Wise, 1970).

Vertisol In U.S. Dept. of Agriculture soil technology, a soil order that contains at least 30% clay, usually smectite. It is found in regions where soil moisture changes rapidly from wet to dry, and is characterized by pronounced changes in volume with changes in moisture, and by deep, wide cracks and high bulk density between the cracks when dry. Gilgai microrelief and slickensides are further evidence of the churning and self-mulching of these shrink-swell soils (USDA, 1975). Suborders and great soil groups of this soil order have the suffix -ert. See also: *Torrert; Udert; Ustert; Xerert.* Obsolete syn: *Grumusol.*

very angular A term used by Powers (1953) to describe a sedimentary particle with a roundness value between 0.12 and 0.17 (midpoint at 0.14). Also, said of the *roundness class* containing very angular particles. Cf: *angular.*

very close pack ice Pack ice in which the concentration approaches 10/10; the floes are tightly packed with very little, if any, seawater visible. See also: *close ice.*

very coarsely crystalline Descriptive of an interlocking texture of a carbonate sedimentary rock having crystals whose diameters are in the range of 1-4 mm (Folk, 1959).

very coarse pebble A geologic term for a *pebble* having a diameter in the range of 32-64 mm (1.3-2.5 in., or -5 to -6 phi units) (AGI, 1958).

very coarse sand (a) A geologic term for a *sand* particle having a diameter in the range of 1-2 mm (1000-2000 microns, or zero to -1 phi units). Also, a loose aggregate of sand consisting of very coarse sand particles. (b) A soil term used in the U.S. for a *sand* particle having a diameter in the range of 1-2 mm. Obsolete syn: *fine gravel.*

very common In the description of coal constituents, 10-30% of a particular constituent occurring in the coal (ICCP, 1963). Cf: *rare; common; abundant; dominant.*

very fine clay A geologic term for a *clay* particle having a diameter in the range of 1/4096 to 1/2048 mm (0.24-0.5 microns, or 12 to 11 phi units). Also, a loose aggregate of clay consisting of very fine clay particles.

very finely crystalline Descriptive of an interlocking texture of a carbonate sedimentary rock having crystals whose diameters are in the range of 0.004-0.016 mm (Folk, 1959).

very fine pebble A term used by AGI (1958) as a syn. of *granule [sed].*

very fine sand (a) A geologic term for a *sand* particle having a diameter in the range of 0.062-0.125 mm (62-125 microns, or 4 to 3 phi units). Also, a loose aggregate of sand consisting of very fine sand particles. Syn: *flour sand.* (b) A soil term used in the U.S. for a *sand* particle having a diameter in the range of 0.05-0.10 mm. (c) Soil material containing 85% or more of sand-size particles (percentage of silt plus 1.5 times the percentage of clay not exceeding 15) and 50% or more of very fine sand (SSSA, 1965, p. 347).

very fine silt A geologic term for a *silt* particle having a diameter in the range of 1/256 to 1/128 mm (4-8 microns, or 8 to 7 phi units). Also, a loose aggregate of silt consisting of very fine silt particles.

very large boulder A *boulder* having a diameter in the range of 2048-4096 mm (80-160 in., or -11 to -12 phi units).

very open pack ice Pack ice in which the concentration is 1/10 through 3/10 with water predominating over ice; the floes are loose and widely spaced. See also: *scattered ice.*

very thick bands In banded coal, vitrain bands exceeding 50.0 mm in thickness (Schopf, 1960, p.39). Cf: *thin bands; medium bands; thick bands.*

vesbite A melilite- and leucite-rich *fergusite* or *italite* containing clinopyroxene and accessory apatite and opaque oxides. Washington in 1920 derived the name from Vesbius (Latin for Vesuvius), Italy, where the rock occurs as ejected blocks. Not recommended usage.

vesecite A monticellite-bearing *melilitite* that contains olivine, melilite, phlogopite, and nepheline, but no clinopyroxene. Cf: *modlibovite; polzenite.* The name, given by Scheumann in 1922, is for Vesec Svetla, Czechoslovakia. Not recommended usage.

vesicle [ign petrol] A cavity of variable shape in a lava, formed by the entrapment of a gas bubble during solidification of the lava. Syn: *vacuole.*

vesicle [paleont] (a) A plant or animal structure having the general form of a membranous cavity; e.g. the space enclosed in the interior of a corallite. (b) A term incorrectly applied to a *dissepiment* in a corallite.

vesicle [palyn] An expanded, bladdery projection of ektexine extending beyond the main body of a pollen grain and typically showing more or less complex internal structure. One or more vesicles are characteristic of many gymnospermous (especially coniferous) genera, in which they tend to reduce the specific gravity of, and give buoyancy to, the pollen grains. Cf: *pseudosaccus.* Syn: *wing [palyn]; bladder; air sac; saccus.*

vesicle cylinder A cylindrical zone in a lava, in which there are abundant vesicles, probably formed by the generation of steam from underlying wet material. This feature occurs in the lavas of the northwestern U.S.

vesicular [paleont] (a) Pertaining to or containing vesicles in a plant or animal. (b) Having dissepiments in corals. This usage is not recommended.

vesicular [petrology] Said of the texture of a rock, esp. a lava, characterized by abundant vesicles formed as a result of the expansion of gases during the fluid stage of the lava. Cf: *cellular; scoriaceous.*

vesiculate Possessing *vesicles [palyn];* e.g. said of *saccate* pollen.

vesiculation (a) The process of forming vesicles. (b) The arrangement of vesicles in a rock.

vesignieite A greenish mineral: $BaCu_3(VO_4)_2(OH)_2$. Also spelled: *vesigniéite.*

vessel A xylem tube formed from several vessel segments (modified tracheids with imperfect or no end walls) set end to end (Cronquist, 1961, p.883).

vestibulate Having or resembling a vestibule or vestibulum.

vestibule (a) An inhalant cavity, other than a canal, of the aquiferous system of a sponge, located close to the surface and receiving water from one or more ostia. Syn: *subdermal space.* (b) The distal (near-surface) part of the zooecium in a cryptostome bryozoan, often delimited basally by hemisepta. Syn: *vestibulum.* (c) In the bryozoans, a space surrounded by membranous wall below the orifice through which the lophophore and tentacles pass in protruding. (d) The space between the duplicature and the calcareous wall (outer lamella) composing the externally visible shell of an ostracode. (e) A subcylindrical prolongation of the pedicle valve dorsal of the brachial valve of a brachiopod (TIP, 1965, pt.H, p.155).

vestibulum (a) The space between the external opening (*exopore*) in the ektexine and the internal opening (*endopore*) in the endexine of a pollen grain with a complex porate structure. The openings are about the same size. Cf: *atrium [palyn].* (b) The *vestibule* in a cryptostome bryozoan.—Pl: *vestibula.*

vestige A small and imperfectly developed or degenerate bodily part or organ that is a remnant of one more fully developed in an earlier stage in the life cycle of the individual, in a past generation, or in closely related forms. Adj: *vestigial.* Syn: *vestigial structure.*

vestigial Of, pertaining to, or being a *vestige.*

vestigial structure *vestige.*

vestigiofossil *trace fossil.*

vestured pit In certain plants, a *bordered pit* with a cavity wholly or partly lined with projections from a secondary wall (Swartz, 1971, p. 496).

Vesulian Stage in Great Britain: Middle Jurassic (above Bajocian, below Bathonian).

vesuvian (a) *vesuvianite.* (b) A mixture of calcite and hydromagnesite. (c) *leucite.*

Vesuvian garnet An early name for leucite whose crystal form resembles that of garnet.

vesuvianite A mineral: $Ca_{10}Mg_2Al_4(SiO_4)_5(Si_2O_7)_2(OH)_4$. It is usually brown, yellow, or green, sometimes contains iron and fluorine, and is commonly found in contact-metamorphosed limestones. Syn: *idocrase; vesuvian.*

Vesuvian-type eruption *Vulcanian-type eruption.*

vesuvite A *tephrite* containing abundant leucite. The name, given by Lacroix in 1917, is for Vesuvius, Italy. Not recommended usage.

veszelyite A greenish-blue mineral: $(Cu,Zn)_3PO_4(OH)_3 \cdot 2H_2O$. Syn: *arakawaite.*

VHA basalt Acronym for a lunar rock type, very high alumina basalt. The term is based on chemical composition, and the material may be a mixture of other rock types (Hubbard, 1973, p. 339).

V/H ratio *velocity/height ratio.*

VI *vertical interval.*

vibertite *bassanite.*

vibetoite A coarse-grained biotite-hornblende *pyroxenite* containing much calcite and apatite. It is either an altered alkalic ultramafic rock or a hybrid product of *carbonatite* with ultramafic silicate-rock affinities. The name, given by Brögger in 1921, is for Vibeto farm, Fen complex, Norway. Not recommended usage.

vibex A line, presumably marking a cuticular insertion, in the skeletal material on the basal side of a branch in the generally reteporiform colonies of certain ascophoran cheilostomes (bryozoans) (Lagaaij, 1952, p. 16). Pl: *vibices.*

Vibracorer Coring device used to obtain a continuous sample from coarse-grained or semilithified sediments. The core barrel is supported by a tripod which rests on the sediment surface, and the barrel is driven into the sediment by a vibrator.

vibraculum A heterozooid in cheilostome bryozoans, having the equivalent of the operculum or the mandible setae in the form of a bristle or whip. Pl: *vibracula.*

vibration gravimeter A device that affords a measurement of gravity by observation of the period of transverse vibration of a thin wire tensioned by the weight of a known mass. It is useful for observations at sea.

vibration magnetometer A type of magnetometer for individual rock specimens that uses the alternating voltage generated by relative vibration of the specimen and a coil. Syn: *Foner magnetometer.*

vibration mark A term used by Dzulynski & Slaczka (1958, p.234) for a sedimentary structure representing a modified groove consisting of crescentic depressions (concave upcurrent) presumed to result from the unsteady inscribing action of a solid object moved by the current. Cf: *chevron mark; chattermark.* See also: *ruffled groove cast.*

vibration plane In optics, a plane of polarized light, including the directions of propagation and vibration. Syn: *plane of polarization; plane of vibration.*

Vibroseis Trade name for a seismic method in which a vibrator is used as an energy source to generate a wave train of controlled frequencies.

vicarious (a) Said of an *avicularium* of a cheilostome bryozoan, which is intercalated in a linear series in a space as large as or larger than those occupied by autozooids. (b) Pertaining to closely related kinds of organisms that occur in similar environments or as fossils in corresponding strata but in distinct and often widely separated areas. Also, characterized by the presence of, or consisting of, such organisms.

vicinal face A crystal face that modifies a normal crystal face, which it closely approximates in angle.

Vickers hardness test The testing of metals by indenting with a pyramidal diamond point and measuring the area of indentation.

Vicksburgian North American (Gulf Coast) stage: Oligocene (above Jacksonian, below Frio).

vicoite An extrusive rock composed of leucite, sodic sanidine, calcic plagioclase, and augite; a feldspathic *leucitite.* The name, applied by Washington in 1906, is for Vico volcano, Italy. Not recommended usage.

vidicon A storage-type electronically scanned photoconductive television camera tube, which often has a response to radiations beyond the limits of the visible region.

Villafranchian European stage: continental Upper Pliocene, essentially equivalent to Astian. Villafranchian in the broad sense spans Upper Pliocene to Lower Pleistocene; its middle and upper part is equivalent to the Calabrian.

villamaninite A black isometric mineral of the pyrite group: $(Cu,Ni,Co,Fe)(S,Se)_2$.

villiaumite A carmine to colorless isometric mineral: NaF.

vimsite A monoclinic mineral: $CaB_2O_2(OH)_4$.

vincentite A mineral: $(Pd,Pt)_3(As,Sb,Te)$.

vincularian *vinculariiform.*

vinculariiform Said of a rigid, ramose, erect colony in cheilostome bryozoans, firmly attached by a calcified base and having subcylindrical branches (Lagaaij & Gautier, 1965, p. 51).

vinculum Calcareous portion of connecting ring of siphuncle segment attached to adapical surface of septum in discosorid and, more rarely, in oncocerid conchs in which it also may be present on the adoral side of septum (TIP, 1964, pt. K, p. 59).

Vindobonian European stage: Lower Miocene (above Burdigalian, below Sarmatian). Obsolescent.

vinogradovite A white to colorless mineral: $(Na,Ca,K)_4Ti_4AlSi_6O_{23}\cdot 2H_2O$.

vintlite A porphyritic hypabyssal rock composed of phenocrysts of labradorite or bytownite and hornblende in a fine-grained groundmass of feldspar, hornblende, and quartz; a hornblende *diorite.* The name, given by Pichler in 1875, is for Vintl, Tyrolean Alps. Obsolescent.

violaite A highly pleochroic mineral of the clinopyroxene group: $Ca(Mg,Fe)(SiO_3)_2$.

violan A translucent massive blue or blue-violet variety of diopside, containing MnO and Mn_2O_3. Syn: *violane.*

violarite A violet-gray mineral of the linnaeite group: Ni_2FeS_4.

virgal One of an articulated series of more or less rod-shaped or cylindrical ossicles forming a structure that extends outward from an ambulacral plate of certain asterozoans. Pl: *virgals* or *virgalia.*

virgation [fault] A divergent, branchlike pattern of fault distribution. The term is used in the Russian literature.

virgation [fold] A fold pattern in which the axial surfaces diverge or fan out from a central bundle.

virgation [geomorph] A sheaflike pattern, as shown on a map, of mountain ranges diverging from a common center. Ant: *syntaxis.*

virgella A spine developed during formation of the metasicula of a graptolithine. It is embedded in the wall of the metasicula and projects freely from the apertural margin.

virgence The converging or diverging of fans of *axial-plane cleavage.*

Virgilian North American provincial series: uppermost Pennsylvanian (above Missourian, below Wolfcampian of Permian).

virgin *primary [metal].*

virgin clay Fresh clay, as distinguished from that which has been fired.

virgin flow Streamflow unaffected by artificial obstructions, storage, or other works of man in the stream channels or drainage basin; a syn. of *runoff [water].*

Virglorian *Anisian.*

virgula A tubular prolongation from the apex of the prosicula of scandent graptolites. It is homologous with *nema,* and the term is used where the prolongation is enclosed within a scandent rhabdosome (biserial rhabdosomes) or incorporated in the dorsal wall (as in monograptids).

viridine A grass-green variety of andalusite containing manganese. Syn: *manganandalusite.*

viridite (a) A general term formerly applied to the indeterminable or obscure green alteration products (such as chlorite and serpentine) occurring in scales and threads in the groundmasses of porphyritic rocks. Cf: *opacite; ferrite.* (b) An iron-rich chlorite containing considerable ferric iron.

virtual geomagnetic pole A conventional form of expressing a measured remanent magnetization; the pole location of the dipole magnetic field for which the field direction at the rock's location is parallel to its measured remanent magnetization. Syn: *paleomagnetic pole.*

visceral Pertaining to or located on or among the internal organs of a body; e.g. the "visceral hump" consisting of the part of the body of a mollusk behind the head and above the foot, in which the digestive and reproductive organs are concentrated.

visceral skeleton In the vertebrates, a system of serially arranged bars ("gill bars") of bone or cartilage, jointed and movable in gnathostomes but fixed in Agnatha, which supports the gills behind and below the head. The visceral skeleton also includes the jaws and hyoid apparatus of gnathostomes.

viscoelastic Said of a material for which there is a limiting strain at a constant stress, but the strain is approached asymptotically rather than instantaneously (as for an ideal *elastic* material).

viscometer An instrument used to measure viscosity. Syn: *viscosimeter.*

viscometry The measurement of viscosity. Syn: *viscosimetry.*

viscosimeter *viscometer.*

viscosimetry *viscometry.*

viscosity The property of a substance to offer internal resistance to flow; its *internal friction.* Specifically, the ratio of the shear stress to the rate of shear strain. This ratio is known as the *coefficient of viscosity.* See also: *Newtonian liquid.*

viscosity coefficient A numerical factor that measures the internal resistance of a fluid to flow; it equals the shearing force in dynes/sq cm transmitted from one fluid plane to another that is 1 cm away, and generated by the difference in fluid velocities of 1 cm/sec in the two planes. The greater the resistance to flow, the larger the coefficient. Syn: *absolute viscosity; dynamic viscosity.*

viscous creep Inelastic, time-dependent strain in which the rate of strain is constant at constant differential stress.

viscous flow A syn. of *Newtonian flow.* Cf: *liquid flow; solid flow.*

viscous magnetization A component of magnetization that behaves as *remanent magnetization* during the time needed for laboratory measurement and like *induced magnetization* during geologic time, thus showing *magnetic viscosity.* Syn: *viscous remanent magnetization; unstable remanent magnetization.*

viscous remanent magnetization *viscous magnetization.* Abbrev: VRM.

viscous stress The resistive force of water. It is proportional to the speed of the current, but acts opposite to its direction of flow (U.S.

Naval Oceanographic Office, 1966, p. 175).

Viséan European stage: Lower Carboniferous (lowermost Upper Mississippian; above Tournaisian, below lower Namurian). Also spelled: *Visean.*

viséite A mineral: $NaCa_5Al_{10}(SiO_4)_3(PO_4)_5(OH)_{14}\cdot 16H_2O$(?). It is regarded as a zeolite with structure analogous to analcime but with some vacant lattice positions in the $(Al,Si,P)O_2$ net (Hey, 1962, 17.6.8). Also spelled: *viseite.*

vishnevite A sulfate-bearing variety of cancrinite: $(Na,K,Ca)_{6\text{-}8}(Al_6Si_6O_{24})(SO_4,CO_3)\cdot nH_2O$, where n varies between 1 and 5.

Vishnu A division of the Archeozoic in the Grand Canyon region.

visible horizon *apparent horizon.*

visor The more or less inclined, overhanging surface directly above the wave-cut notch in a sea cliff, most commonly found in limestones in a tropical region.

vitalism The theory that some internal force or driving energy of organisms exerts a directional effect that more or less determines how variation and evolution will proceed. Cf: *holism.*

viterbite [mineral] A mixture of allophane and wavellite(?).

viterbite [rock] An extrusive rock composed chiefly of sodic sanidine and large phenocrysts of leucite, with smaller quantities of calcic plagioclase, augite, biotite, apatite, and opaque oxides; a leucite *phonolite.* Its name, given by Washington in 1906, is derived from Viterbo, Italy. Not recommended usage.

vitr- A prefix meaning "glass". Cf: *hyalo-.*

vitrain A coal *lithotype* characterized macroscopically by brilliant, vitreous luster, black color, and cubic cleavage with conchoidal fracture. Vitrain bands or lenticles are amorphous, usually 3-5 mm thick, and their characteristic microlithotype is *vitrite.* Cf: *clarain; durain; fusain.* See also: *euvitrain; provitrain.* Syn: *pure coal.*

vitreous (a) *glassy.* (b) Said of a hyaline foraminifer having the appearance and luster of glass.

vitreous copper *chalcocite.*

vitreous luster A type of luster resembling that of glass. Quartz, for example, has a vitreous or *glassy luster.*

vitreous silver *argentite.*

vitric Said of pyroclastic material that is characteristically glassy, i.e. contains more than 75% glass.

vitric tuff A tuff that consists predominantly of volcanic glass fragments. Cf: *crystal-vitric tuff.*

vitrification Formation of a glass. Syn: *vitrifaction.*

vitrinertite A coal microlithotype that contains a combination of vitrinite and inertinite totalling at least 95%, and containing more of each than of exinite. It generally occurs in high-ranking bituminous coals.

vitrinertoliptite A microlithotype of coal, a variety of *trimacerite* consisting of exinite in greater abundance than inertinite and vitrinite (ICCP, 1971). Cf: *duroclarite; clarodurite.*

vitrinite (a) A coal maceral group that is characteristic of vitrain and is composed of humic material. It includes *provitrinite* and *euvitrinite* and their varieties. Cf: *inertinite; exinite.* (b) A coal maceral group distinguished by a middle level of reflectance, higher than exinite and lower than inertinite in the same coal. It includes *telinite, collinite,* and *vitrodetrinite* macerals (ICCP, 1971). Some coal petrologists distinguish "vitrinite A" and "vitrinite B", and others *pseudovitrinite.*

vitrinization A process of *coalification* in which vitrain is formed. Cf: *incorporation; fusinization.*

vitrinoid Vitrinite that occurs in bituminous caking coals and that has a reflectance of 0.5-2.0% (ASTM, 1970, p.466). Cf: *xylinoid; anthrinoid.*

vitriol peat Peat that contains abundant iron sulfate.

vitriphyric A term, now obsolete, applied by Cross et al. (1906, p.703) to the texture of a porphyritic igneous rock in which the groundmass is microscopically glassy. Cf: *vitrophyric.*

vitrite A coal microlithotype group that contains vitrinite macerals totalling at least 95%. Cf: *inertite; liptite; vitrain.*

vitroclarain A transitional lithotype of coal characterized by the presence of vitrinite, but in lesser amounts than other macerals. Cf: *clarovitrain.* Syn: *vitroclarite.*

vitroclarite *vitroclarain.*

vitroclastic Pertaining to a pyroclastic rock structure characterized by fragmented bits of glass; also, said of a rock having such a structure.

vitrodetrinite A maceral of coal within the vitrinite group, consisting of small fragments with more or less angular and variable outlines due to extensive comminution during transport and deposition (ICCP, 1971). Particles are usually less than 10 microns across.

vitrodurain A coal lithotype transitional between durain and vitrain, but predominantly durain. Cf: *durovitrain.*

vitrophyre Any porphyritic igneous rock having a glassy groundmass. Adj: *vitrophyric.* Cf: *felsophyre; granophyre.* Syn: *glass porphyry.*

vitrophyric (a) Said of a porphyritic igneous rock having large phenocrysts in a glassy groundmass. Syn: *vitroporphyric.* (b) A term used by Vogelsang in 1872 and applied by Cross et al. (1906, p.703) to the texture of porphyritic igneous rocks in which the groundmass is megascopically glassy. Cf: *vitriphyric.*

vitrophyride A porphyritic volcanic glass; suggested for field use only (Johannsen, 1939, p. 287).

vitroporphyric *vitrophyric.*

vitta One of a pair of oppositely placed, membrane-covered linear furrows near the lateral margins of the frontal shield in certain ascophoran cheilostomes (bryozoans). A series of pores in each furrow communicates with the zooidal body cavity. Pl: *vittae.*

vivianite A mineral: $Fe_3(PO_4)_2\cdot 8H_2O$. It is colorless, blue, or green when unaltered, but grows darker on exposure; it occurs as monoclinic crystals, fibrous masses, or earthy forms in copper, tin, and iron ores and in clays, peat, and bog iron ore. It is dimorphous with metavivianite. Syn: *blue iron earth; blue ocher.*

vladimirite A mineral: $Ca_5H_2(AsO_4)_4\cdot 5H_2O$.

vlasovite A colorless monoclinic mineral: $Na_2ZrSi_4O_{11}$.

vlei A Dutch word used in the Middle Atlantic states and in southern Africa for a shallow lake or marshy area, esp. one developed in the poorly drained valley of an intermittent stream. Also spelled: *vley; vly.*

vley *vlei.*

vloer A term used in South Africa for a flat surface of caked mud with a high salt content and generally destitute of vegetation; a *playa.* It has a more irregular shape, greater area, and shallower depth than a *pan [salt],* and usually has an outlet. Etymol: Afrikaans, "floor".

vltavite *moldavite [astron].*

vly *vlei.*

vogesite A *lamprophyre* composed of hornblende phenocrysts in a groundmass of alkali feldspar and hornblende. Clinopyroxene, olivine, and plagioclase feldspar also may be present. Vogesite contains less biotite than *minette.* The name, given by Rosenbusch in 1887, is for the Vosges Mountains, France.

voglite An emerald-green to bright grass-green mineral: $Ca_2Cu(UO_2)(CO_3)_4\cdot 6H_2O$ (?).

voices of the desert *song of the desert.*

void *interstice.*

voidal concretion A large tubelike iron-oxide concretion with a central hollow or cavity and a hard, dense limonitic rim, commonly found in sands and sandstones and in some clays. It appears to be a product of weathering (oxidation) of a sideritic concretion.

void ratio The ratio of the volume of void space to the volume of solid substance in any material consisting of voids and solid material, such as a soil sample, sediment, or sedimentary rock. Symbol: e. Syn: *voids ratio.*

volatile (a) adj. Readily vaporizable. (b) n. A syn. of *volatile component.*

volatile combustible *volatile matter.*

volatile component A material in a magma, such as water or carbon dioxide, whose vapor pressures are sufficiently high for them to be concentrated in any gaseous phase. Syn: *fugitive constituent; volatile; volatile flux.*

volatile flux *volatile component.*

volatile matter In coal, those substances, other than moisture, that are given off as gas and vapor during combustion. Standardized laboratory methods are used in analysis. Syn: *volatiles; volatile combustibles.*

volatiles *volatile matter.*

volatile transfer *gaseous transfer.*

volborthite A green or yellow mineral: $Cu_3(VO_4)_2\cdot 3H_2O$. It may contain some calcium and barium, and it represents a principal ore of vanadium. Syn: *uzbekite.*

volcan (a) The component of the Earth's crust made up of volcanoes and various hypabyssal rocks (Makiyama, 1954, p.146). (b) *volcano.*—A little-used term.

volcanello (a) A small, active cone within the central crater of a volcano, e.g. Mount Nuevo in Vesuvius. (b) A *spatter cone.*—Etymol: Italian, "small volcano".

volcanic (a) Pertaining to the activities, structures, or rock types of a volcano. (b) A syn. of *extrusive.*

volcanic accident A departure from the *normal cycle* of erosion, caused by the outbreak of volcanic activity.

volcanic arc *island arc.*

volcanic arenite A term used by Williams, Turner and Gilbert (1954, p.308) for a lithic arenite composed chiefly of volcanic detritus and having a low quartz content. It is a common rock among Tertiary and Mesozoic sediments around the Pacific Basin. Folk (1968, p.124) used "volcanic-arenite" for a litharenite composed chiefly of volcanic rock fragments, and having any clay content, sorting, or rounding; for more detailed specification, terms such as "basalt-arenite" and "andesite-arenite" may be used.

volcanic ash *ash [volc].*

volcanic ball *lava ball.*

volcanic blowpiping *gas fluxing.*

volcanic bomb *bomb [pyroclast].*

volcanic breccia (a) A pyroclastic rock that consists of angular volcanic fragments that are larger than 64 mm in diameter and that may or may not have a matrix. Cf: *agglomerate.* (b) A rock that is composed of accidental or nonvolcanic fragments in a volcanic matrix. Syn: *alloclastic breccia; lava breccia.*—See: Parsons, 1969.

volcanic butte An isolated hill or mountain resulting from the differential weathering or erosion and consequent exposure of a volcanic neck, or of a narrow, vertical igneous intrusion into overlying weaker rock; e.g. Ship Rock, N. Mex. Cf: *mesa-butte.*

volcanic center A site at which volcanic activity is occurring or has occurred in the past.

volcanic chain Linear arrangement of a number of volcanoes, apparently associated with a controlling geologic feature such as a fault.

volcanic clay *bentonite.*

volcanic cloud *eruption cloud.*

volcanic cluster A group of volcanic vents in apparent random arrangement.

volcanic conduit The channelway that brings volcanic material up from depth. Cf: *vent.*

volcanic cone A conical hill of lava and/or pyroclastics that is built up around a volcanic vent. Syn: *cone [volc].*

volcanic conglomerate A water-deposited conglomerate containing over 50% volcanic material, esp. coarse pyroclastics.

volcanic cycle A regular sequence of changes in the behavior of a volcano.

volcanic debris *volcanic rubble.*

volcanic dome A steep-sided, rounded extrusion of highly viscous lava squeezed out from a volcano, and forming a dome-shaped or bulbous mass of congealed lava above and around the volcanic vent. Portions of older lavas may be elevated by the pressure of the new lava rising from below. The structure generally develops inside a volcanic crater or on the flank of a large volcano, and is usually much fissured and brecciated (Williams, 1932). Cf: *lava dome.* Syn: *bulbous dome; tholoid; dome volcano; dome [volc]; cumulo-dome; cumulo-volcano.*

volcanic earthquake An earthquake associated with volcanic rather than tectonic forces. Cf: *tectonic earthquake.*

volcanic field A more or less well-defined area that is covered with volcanic rocks, e.g. the San Francisco volcanic field, Arizona. Cf: *ash field; lava field.*

volcanic flow drain *lava tube.*

volcanic foam *pumice.*

volcanic focus The subterranean seat or center of volcanism of a region, or of a volcano.

volcanic gases Volatile matter, released during a volcanic eruption, that was previously dissolved in the magma. Water vapor forms about 90% of the gases; other constituents include carbon gases, esp. carbon dioxide; sulfur gases, esp. sulfur dioxide at high temperatures and hydrogen sulfide at low temperatures; hydrogen chloride; nitrogen as a free element, and others (Krauskopf, 1967). See also: *gas phase.*

volcanic glass A natural glass produced by the cooling of molten lava, or a liquid fraction of it, too rapidly to permit crystallization. Examples are obsidian, pitchstone, tachylyte, and the glassy *mesostasis* of many extrusive rocks.

volcanic graben A straight-walled collapse structure on the summit or flanks of a volcanic cone. See also: *summit graben; sector graben.*

volcanic gravel A pyroclastic deposit, the individual clasts of which are in the size range 0.5 cm or smaller, but still distinguishable by the unaided eye. The finer fractions are also called *volcanic sand.* Cf: *block [volc]; cinder; lapilli.*

volcanic graywacke *volcanic wacke.*

volcanic harbor A natural harbor formed by the sea breaking through a gap in the rim of a volcanic crater; e.g. that of St. Paul Rocks in the southern Atlantic Ocean.

volcanic island A *submarine volcano* that has been sufficiently built up to be exposed above sea level. Syn: *island volcano.*

volcanicity *volcanism.*

volcaniclastic Pertaining to a clastic rock containing volcanic material in whatever proportion, and without regard to its origin or environment.

volcanic mud A mixture of water and volcanic ash, either just erupted and hot or already cooled. The mixture may form a *mudflow* down the slope of the volcano.

volcanic pisolite *pisolite [volc].*

volcanic plain Surface formed by extensive lava or ash flows that cover topographic irregularities.

volcanic province A petrographic province in which the visible igneous rocks are largely volcanic.

volcanic rain *eruption rain.*

volcanic rent A great volcanic depression, bordered by fissures that are usually concentric in plan, caused by the pressure of magmatic injection or by the overloading of cone material on a weak substratum.

volcanic rift zone *rift zone.*

volcanic rock (a) A generally finely crystalline or glassy igneous rock resulting from volcanic action at or near the Earth's surface, either ejected explosively or extruded as lava; e.g. basalt. The term includes near-surface intrusions that form a part of the volcanic structure. See also: *volcanics.* Cf: *plutonic rock.* Syn: *volcanite [petrology].* (b) A general term proposed by Read (1944) to include the effusive rocks and associated high-level intrusive rocks; they are dominantly basic. Cf: *neptunic rock; plutonic rock.*

volcanic rubble The unconsolidated equivalent of volcanic breccia. Syn: *volcanic debris.*

volcanics Those igneous rocks that have reached or nearly reached the Earth's surface before solidifying. The common use of the term for *volcanic rocks* should be avoided (USGS, 1958, p.86).

volcanic sand A pyroclastic deposit, the individual clasts of which are in the size range 2-5 mm; the finer fractions of *volcanic gravel.*

volcanic sandstone An indurated deposit of rounded, water-worn pyroclastic fragments and a subordinate amount of nonvolcanic detritus.

volcanic shield *shield volcano.*

volcanic sink *sink [volc].*

volcanic wacke A term used by Williams, Turner and Gilbert (1954, p.303) for a lithic wacke composed chiefly of detritus derived from intermediate (andesitic) and basic (basaltic) volcanic rocks and having a low quartz content. It is a common rock among Tertiary and Mesozoic sediments deposited in orogenic belts bordering the Pacific Ocean. Syn: *volcanic graywacke.*

volcanic water Water in, or derived from, magma at the Earth's surface or at a relatively shallow level; *juvenile* water of volcanic origin. Cf: *plutonic water.*

volcanism The processes by which magma and its associated gases rise into the crust and are extruded onto the Earth's surface and into the atmosphere. Also spelled: *vulcanism.* Syn: *volcanicity.*

volcanist An obsolete syn. of *volcanologist.* Also spelled: *vulcanist.* Syn: *plutonist.*

volcanite [mineral] An old mineral name suggested as a synonym of pyroxene and of a variety of sulfur containing less than one percent selenium.

volcanite [rock] (a) An extrusive rock composed chiefly of anorthoclase, andesine, and augite phenocrysts in a glassy groundmass containing feldspar and augite microlites. The name, given by Hobbs in 1893, is derived from Volcano in the Lipari Islands. Obsolete. (b) A syn. of *volcanic rock.*—Also spelled: *vulcanite.*

volcano (a) A vent in the surface of the Earth through which magma and associated gases and ash erupt; also, the form or structure, usually conical, that is produced by the ejected material. (b) Any eruption of material, e.g. mud, that resembles a magmatic volcano.—Obsolete var: *vulcano.* Pl: *volcanoes.* Etymol: the Roman deity of fire, Vulcan.

volcanogenic Formed by processes directly connected with volcanism; specif., said of mineral deposits (massive sulfides, exhalites, banded iron formations) considered to have been produced through volcanic agencies and demonstrably associated with volcanic phenomena. Also spelled: *volcanigenic.*

volcano-karst A terrain of freshly erupted volcanic materials (esp. pyroclastics such as certain tuffs and agglomerates that contain unstable minerals) that have been corroded by rainwater to develop microrelief forms resembling limestone karst. Term introduced by Naum et al. (1962) as "vulcanokarstul".

volcanologist One who studies *volcanology.* Obsolete syn: *volcanist.*

volcanology The branch of geology that deals with volcanism, its causes and phenomena. Also spelled: *vulcanology.* See also: *volcanologist.* Less-preferred syn: *pyrogeology.*

volcano shoreline A roughly circular, steeply-sloping shoreline formed where fragmental volcanic materials or flows of lava occur in a coastal location, or where an active volcano projects above the water surface, building a cone upward and outward by continued addition of ejected materials.

volcano-tectonic depression A large-scale depression, usually linear, that is controlled by both tectonic and volcanic processes (van Bemmelen, 1933). An example is the Toba trough in northern Sumatra.

volhynite A quartz-bearing *kersantite* composed of phenocrysts of plagioclase, hornblende, and sometimes biotite in a groundmass of quartz, feldspar, and abundant chlorite. Its name, given by Ossovski in 1885, is derived from Volhynia, U.S.S.R. Obsolete.

volkonskoite A bluish-green, chromium-bearing clay mineral of the montmorillonite group; specif. a variety of nontronite containing chromium. Also spelled: *volchonskoite; wolchonskoite.*

volkovskite A mineral: $(Ca,Sr)B_6O_{10} \cdot 3H_2O$.

voltage gradient *electric-field intensity.*

voltaite A dull oil-green to brown or black mineral: $K_2Fe_5^{+2}Fe_4^{+3}(SO_4)_{12} \cdot 18H_2O(?)$.

volt-second *weber.*

voltzite A yellowish or reddish material consisting of wurtzite plus an organometallic compound of zinc.

volume control *gain control.*

volume elasticity *bulk modulus.*

volume law *Lindgren's volume law.*

volume magnetization *magnetization.*

volume phase *surface phase.*

volume susceptibility *susceptibility [magnet].*

volumetric analysis Quantitative chemical analysis where the amount of a substance in a solution is determined by adding a fixed volume of a standard solution to the prepared sample until a reaction occurs. The amount of standard solution needed to produce the desired reaction indicates the amount of substance in the original sample.

volumetric shrinkage The decrease in volume, expressed as a percentage of the soil mass when dried, of a soil mass when the water content is reduced from a given percentage to the shrinkage limit (ASCE, 1958, term 411).

volution A *whorl* of a spiral shell. Syn: *volute.*

volynskite A mineral: $AgBiTe_2$.

von Baer's law The principle stated by Karl E. von Baer (1792-1876), Estonian embryologist and geologist, according to which the rotation of the Earth causes an asymmetrical, lateral erosion of stream beds.

von Kármán constant A dimensionless number which relates the mixing length to the flow condition in turbulent flow (Middleton, 1965, p. 252). It is symbolized by k in formulas for velocity and sediment distribution in turbulent flow.

von Schmidt wave *head wave.*

vonsenite A coal-black orthorhombic mineral: $(Fe^{+2},Mg)_2Fe^{+3}BO_5$. It is isomorphous with ludwigite.

Von Sterneck-Askania pendulum A device for measuring the vertical component of gravity, characterized by four pendulums in a single case (Jakosky, 1950, p. 277).

von Wolff's classification A quantitative chemical-mineralogic classification of igneous rocks proposed in 1922 by F. von Wolff.

vorobievite A rose-red, purplish-red, or pinkish gem variety of beryl containing cesium. Appreciable amounts of sodium and other alkalies may be present. Also spelled: *worobieffite; vorobyevite.* Syn: *morganite; rosterite.*

vortex [hydraul] A fluid flow that has a revolving motion in which the stream lines are concentric circles, and in which the total head for each stream line is the same.

vortex [struc geol] A vertical, cylindrical fold formed in incompetent rock by late-stage deformation during deep-zone orogeny (Wynne-Edwards, 1957, p. 643).

vortex cast *flute cast.*

Vraconian European stage: uppermost Lower Cretaceous or lowermost Upper Cretaceous.

vrbaite A gray-black to dark-red orthorhombic mineral: $Tl_4Hg_3Sb_2As_8S_{20}$.

vredenburgite (a) *beta-vredenburgite.* (b) *alpha-vredenburgite.*

VRM *viscous remanent magnetization.*

V-shaped valley A valley having a pronounced cross profile suggesting the form of the letter "V", characterized by steep sides and short tributaries; specif. a young, narrow valley resulting from downcutting by a stream. The "V" becomes broader as the amount of mass wasting increases. Cf: *U-shaped valley.* Syn: *V-valley.*

V's, rule of *rule of V's.*

V-terrace A triangular terrace, commonly formed in a long, narrow arm of an old lake, having one side built against an even coast and the opposite angle pointing toward open water (Gilbert, 1890, p. 58).

vuagnatite An orthorhombic mineral: $CaAl(SiO_4)(OH)$.

vug A small cavity in a vein or in rock, usually lined with crystals of a different mineral composition from the enclosing rock. Etymol: Cornish *vooga,* "underground chamber, cavern, cavity". Adj: *vuggy.* Cf: *druse; geode.* Syn: *bug hole.* Var: *vugh.*

vuggy Pertaining to a *vug* or having numerous vugs.

vuggy porosity In petroleum geology, porosity resulting from the presence of openings ("vugs") from the size of a small pea upwards; it is usually used with reference to limestones.

vugh (a) A relatively large and usually irregular void in a soil material, not normally interconnected with other voids of comparable size (Brewer, 1964, p.189). (b) Var. of *vug.*

Vulcanian-type eruption A type of volcanic eruption characterized by the explosive ejection of fragments of new lava, commonly incandescent when they leave the vent but either solid or too viscous to assume any appreciable degree of rounding during their flight through the air. With these there are often *breadcrust bombs* or *blocks,* and generally large porportions of ash (Macdonald, 1972, p. 223). Also spelled: *Vulcano-type eruption.* Syn: *Vesuvian-type eruption; paroxysmal eruption.*

vulcanism *volcanism.*

vulcanite [mineral] An orthorhombic mineral: CuTe.

vulcanite [petrology] *volcanite [petrology].*

vulcanorium A structure transitional between a mid-oceanic ridge and an island arc, e.g. the Arctic vulcanorium between the Nansen and Amundsen deeps in the Arctic Ocean (Runcorn et al., 1967, p.161).

Vulcano-type eruption *Vulcanian-type eruption.*

vulpinite A scaly, granular, grayish-white variety of anhydrite. It sometimes has an admixture of silica.

vulsinite An extrusive *trachyte,* composed chiefly of alkali feldspar and also containing calcic plagioclase and augite with or without feldspathoids. The name, given by Washington in 1896, is from the Vulsinii, an Etruscan tribe in the Bolsena region of Italy. Not recommended usage.

vuonnemite A triclinic mineral: $Na_4TiNb_2Si_4O_{17} \cdot 2Na_3PO_4$.

vysotskite A tetragonal mineral: $(Pd,Ni)S$.

Waalian North European climatostratigraphic and floral stage: Lower Pleistocene (above Eburonian, below Menapian). Equivalent in time to Donau/Günz interglacial.

wacke (a) A "dirty" sandstone that consists of a mixed variety of angular and unsorted or poorly sorted mineral and rock fragments and of an abundant matrix of clay and fine silt; specif. an impure sandstone containing more than 10% argillaceous matrix (Williams, Turner & Gilbert, 1954, p.290). The term is used for a major category of sandstone, as distinguished from *arenite.* (b) A term used by Fischer (1934) for a clastic sedimentary rock in which the grains are almost evenly distributed among the several size grades; e.g. a sandstone consisting of sediment "poured in" to a basin of deposition at a comparatively rapid rate without appreciable selection or reworking by currents after deposition, or a mixed sediment of sand, silt, and clay in which no component forms more than 50% of the whole aggregate. (c) A term commonly used as a shortened form of *graywacke.* This usage is not recommended. (d) Originally, a term applied to a soft and earthy variety of basalt, or to the grayish-green to brownish-black clay-like residue resulting from the partial chemical decomposition in place of basalts, basaltic tuffs, and related igneous rocks. Syn: *vake.*—Etymol: German *Wacke,* an old provincial mining term signifying a large stone or "stoniness" in general. Pron: wacky.

wackestone A term used by Dunham (1962) for a mud-supported carbonate sedimentary rock containing more than 10% grains (particles with diameters greater than 20 microns); e.g. a calcarenite. The use of this term is discouraged because it is apt to lead to confusion with terrigenous rocks of the "wacke type". Cf: *mudstone; packstone.*

wad [coast] A Dutch term for *tidal flat.* Pl: *wadden.* The spelling "wadd" is incorrect.

wad [mineral] (a) An earthy, dark-brown or black mineral substance consisting chiefly of an impure mixture of manganese oxides and other oxides, with varying amounts of other compounds (such as copper, cobalt, and silica) and 10-20% water. It is commonly very soft, soiling the hand, but is sometimes hard and compact, and it has a low apparent specific gravity. Wad generally occurs in damp, marshy areas as a result of decomposition of manganese minerals. Cf: *psilomelane.* Syn: *bog manganese; black ocher; earthy manganese.* (b) A general term applied to hydrated oxides of manganese (or of manganese and other metals) whose true nature is unknown or which have variable and uncertain compositions and at least some of which may be amorphous. (c) An English dialectal term for graphite.

wadden Plural of the Dutch term *wad,* meaning a "tidal flat". The form "waddens" is incorrect.

waddy *wadi.*

wadeite A mineral: $K_2CaZr(SiO_3)_4$.

wadi (a) A term used in the desert regions of SW Asia and northern Africa for a stream bed or channel, or a steep-sided and bouldery ravine, gully, or valley, or a dry wash, that is usually dry except during the rainy season, and that often forms an oasis. (b) The intermittent and torrential stream that flows through a wadi and ends in a closed basin. (c) A shallow, usually sharply defined, closed basin in which a wadi terminates.—Etymol: Arabic. Variant plurals: *wadis; wadies; wadian; widan.* See also: *arroyo; nullah.* Also spelled: *wady; waddy.* Syn: *oued; widiyan.*

wagnerite A yellow, red, or greenish monoclinic mineral: $Mg_2(PO_4)F$. Ferrous iron and calcium may be present.

wairakite A zeolite mineral: $CaAl_2Si_4O_{12}\cdot 2H_2O$. It is isostructural with analcime.

wairauite A mineral: CoFe.

wakabayashilite A monoclinic mineral: $(As,Sb)_{11}S_{18}$.

wake dune A sand dune occurring on the leeward side of a larger dune, and trailing away in the direction of the wind.

wakefieldite A mineral: YVO_4.

walchowite A honey-yellow variety of retinite containing a little nitrogen, found in brown coal at Walchow in Moravia, Czechoslovakia.

walker's earth A syn. of *fuller's earth.* Etymol: German *Walkererde.*

walking beam A rigid, oscillating lever balanced on a fulcrum, used to activate the cable in cable-tool drilling, or to pump oil from a well, by alternating up-and-down motion.

walking leg (a) A prosomal appendage of a merostome, serving for walking. (b) A *pereiopod* of a malacostracan crustacean. (c) The inner branch of the biramous appendage of a trilobite.

walking out A simple method of correlation by which stratigraphic units are traced from place to place along continuous outcrops.

wall [fault] The rock mass on one side of a fault, e.g. hanging wall, footwall. See also: *wall rock [fault].* Syn: *fault wall.*

wall [mining] The side of a lode, or of mine workings. Cf: *footwall; hanging wall.*

wall [paleont] (a) An external layer surrounding internal parts of an invertebrate, e.g. a skeletal deposit, formed in various corallites, that encloses the column of a scleractinian polyp and unites the outer edges of the septa, such as septotheca, paratheca, and synapticulotheca; or the part of a cephalopod conch comprising the external shell. (b) The raised margin in a caneolith. (c) The limiting skeletal element of the cup in archaeocyathids.

wall [speleo] (a) The side of a cave passage. (b) A series of *columns* along a joint crack that have fused together into a solid mass. See also: *partition.*

Wallace's line The hypothetical boundary that separates the distinctly different floras and faunas of Asia and Australia. It is usually drawn between the islands of Bali and Lombok, through the Strait of Macassar, between Borneo and Celebes, and south of the Philippines. Named after Alfred Russel Wallace (1822-1913), English naturalist. Cf: *Weber's line.*

Wallachian orogeny One of the 30 or more short-lived orogenies during Phanerozoic time identified by Stille, in this case at the end of the Pliocene.

wall-contact log A *well log* curve produced by closely spaced electrodes or source-to-detector components of the *sonde* placed in direct contact with the walls of the borehole. Contact resistivity devices produce the various *microresistivity logs* and the *dipmeter.* Radioactivity logs of contact design include the density log and certain neutron logs. Caliper logs require wall contact. Syn: *contact log.*

walled lake A lake bordered along its shore by *lake ramparts.*

walled plain A large lunar crater characterized by a broad, nearly level floor and filled by dark material similar to that which forms the floors of the maria. It is not as deep as a cup-shaped crater. See also: *ring plain.* Syn: *cirque [lunar].*

wallisite A mineral: $PbTl(Cu,Ag)As_2S_5$.

wall niche *meander niche.*

wallongite *wollongongite.*

wallow A depression or area, often filled with water or mud, that suggests a place where animals have wallowed.

wallpaper effect The swelling and blocking action (upon the introduction of water) shown by interstitial clay minerals of the expanding-lattice type.

wall reef An elongate steep-sided coral reef, away from shore, usually without islands on its reef flat. Wall reefs range up to several km long but are generally under 1 km wide. See also: *reef wall.*

wall rock [eco geol] The rock forming the walls of a vein, lode, or disseminated deposit; the *country rock* of an ore deposit.

wall rock [fault] The rock mass comprising the *wall* of a fault.

wall rock [intrus rocks] *country rock.*

wall-rock alteration Alteration of country rock adjacent to hydrothermal ore deposits by the fluids responsible for or derived during the formation of the deposits; also, the alteration products themselves.

wall saltpeter Naturally occurring calcium nitrate; nitrocalcite occurring on walls of limestone caves. Cf: *saltpeter.*

wall-sided glacier A glacier that is on a steep slope and not laterally confined by valley walls.

walpurgite A yellow to yellow-orange mineral: $(BiO)_4(UO_2)(AsO_4)_2 \cdot 3H_2O$. Syn: *waltherite.*

walstromite A mineral: $BaCa_2Si_3O_9$.

waltherite A mineral formerly believed to be a poorly defined carbonate of bismuth, but now known to be identical to *walpurgite.*

Waltonian British climatostratigraphic stage: Lower Pleistocene (above Pliocene Icenian, below Ludhamian). Probably correlative with Eburonian. Equivalent to Merksemian of lowlands.

wandering The slow, winding compound movement of a stream, consisting of the *sweeping* of meanders and the *swinging* of a meander belt.

wandering dune A sand dune, such as a barchan, that is slowly shifted more or less as a unit in the leeward direction of the prevailing winds, and that is characterized by insufficient vegetation to anchor it. Cf: *anchored dune.* Syn: *migrating dune; traveling dune.*

wandering water *vadose water.*

waning development *declining development.*

waning slope The lower part of a hillside surface, tending to become concave below the *constant slope,* having an angle that decreases continuously downslope as the hillside stretches to the valley floor or other local base level (Wood, 1942). Cf: *wash slope.* Ant: *waxing slope.* Syn: *concave slope.*

want A zone in which the coal of a coal seam is missing, owing to a low-angle normal fault or a *washout, squeeze,* or *roll.* Cf: *nip; pinch.* Syn: *cutout.*

Warburg impedance The ion-diffusion impedance that arises at the interface between an electronic conductor, such as a sulfide grain, and an electrolytic conductor (Sumner, 1976, p. 255).

warden A term used in south Wales for a strong massive sandstone associated with coal.

wardite A light blue-green mineral: $NaAl_3(PO_4)_2(OH)_4 \cdot 2H_2O$.

wardsmithite A mineral: $Ca_5MgB_{24}O_{42} \cdot 30H_2O$.

warm front The sloping *front* between an advancing warm air mass and a colder air mass over which it moves. Its passage is usually preceded by considerable precipitation, whereas there is little or no precipitation afterwards. Cf: *cold front.*

warm glacier *temperate glacier*

warm ice Ice at the *pressure-melting temperature.*

warm loess Continental loess composed of desert dust, such as that formed in the inland basins and steppes encircling the modern deserts of central Asia between lat. 52°N and 56°N. Cf: *cold loess.*

warm sector The warm tongue of air in a developing mid-latitude *cyclone* lying between the warm front and the trailing cold front.

warm spring A *thermal spring* whose temperature is appreciably above the local mean annual atmospheric temperature, but below that of the human body (Meinzer, 1923, p. 54). Cf: *hot spring.*

warp [mass move] *congeliturbate.*

warp [sed] (a) An English provincial term for the fine mud and silt held in suspension in waters artificially introduced over low-lying land. (b) A general term for a bed or layer of sediment deposited by water; e.g. an estuarine clay, or the alluvium laid down by a tidal river.—See also: *warping [sed].*

warp [tect] n. A slight flexure or bend of the Earth's crust, either upward or downward, and usually on a broad or regional scale. See also: *warping [tect].*

warped fault A fault, usually a thrust fault, that has been slightly folded. Cf: *folded fault.*

warping [sed] (a) The intentional flooding at high tides of low-lying land near an estuary or tidal river by water loaded with fine mud and silt (warp), the water remaining until the warp is deposited and then being allowed to run off clear during low tides; a means of fertilizing or of raising the general level of large, low tracts, e.g. the conversion of a lagoon or tidal flat into a marsh. Cf: *colmatage.* (b) The filling up of hollows or the choking of a channel with warp.

warping [tect] The slight flexing or bending of the Earth's crust on a broad or regional scale, either upwards (*upwarping*) or downwards (*downwarping*); the formation of a *warp* [tect].

warpland Low-lying land that has been built up or fertilized by the process of *warping.* Also spelled: *warp land.*

warrant A term used in England for a particularly "hard and tough" *underclay* (Nelson, 1965, p. 503). The term "possibly has the sense of a token or guarantee of the presence of coal" (Arkell & Tomkeieff, 1953, p. 124).

warrenite (a) A general term for gaseous and liquid bitumens consisting mainly of a mixture of paraffins and isoparaffins: a variety of petroleum rich in paraffins. (b) A pink variety of smithsonite containing cobalt. (c) A name applied to a mineral that may be owyheeite or jamesonite.

warwickite A dark-brown to dull-black orthorhombic mineral: $(Mg,Ti,Fe^{+3},Al)_2(BO_3)O$.

Wasatchian North American continental stage: Lower Eocene (above Clarkforkian, below Bridgerian).

wash [coast] (a) A piece of land that is washed, or alternately covered and uncovered, by a sea or river; e.g. a sandbank or mudbank, or an area of such banks, alternately submerged and exposed by the tide. (b) The shallowest part of a river, estuary, or arm of the sea. (c) A bog, fen, or marsh.

wash [eco geol] An alluvial placer.

wash [geomorph] (a) Erosion effected by wave action. (b) The wearing away of soil by runoff water, as in gullying or sheet erosion; *rainwash.*

wash [sed] (a) Loose or eroded surface material (such as gravel, sand, silt) collected, transported, and deposited by running water, as on the lower slopes of a mountain range; esp. coarse alluvium. (b) A fan-shaped deposit, as an *alluvial fan* or an *alluvial cone,* or a mound of detritus below a cliff opening. (c) *downwash.*

wash [speleo] *fill [speleo].*

wash [streams] (a) A term applied in the western U.S. (esp. in the arid and semiarid regions of the SW) to the broad, gravelly, normally dry bed of an intermittent stream, often situated at the bottom of a canyon; it is occasionally filled by a torrent of water. Syn: *dry wash; washout.* (b) A shallow body of water; esp. a shallow creek.

washboard moraine (a) One of several small, parallel, regularly spaced ridges that are oriented transverse to the ice movement in a general sense and that collectively resemble a washboard. They are abundant in north-central U.S. and in the western plains of Canada. (b) A subglacial disintegration feature formed at the base of a thrust plane in glacier ice by the periodic recession and readvance of a glacier which pushes previously deposited ground moraine into a ridge (Gravenor & Kupsch, 1959, p. 54). Examples are common on the swell-and-swale topography of Alberta and Saskatchewan.

wash cone *outwash cone.*

washing (a) Erosion or wearing-away by the action of waves or running water. (b) The selective sorting, and removal, of fine-grained sediment by water currents. Cf: *winnowing.*

washings Material abraded or transported by the action of water.

Washitan North American (Gulf Coast) stage: Lower and Upper Cretaceous (above Fredericksburgian, below Woodbinian).

Washita stone A porous, uniformly textured *novaculite* found in the Ouachita ("Washita") Mountains, used esp. for sharpening woodworking tools. Syn: *Ouachita stone.*

washland An embanked, low-lying land bordering a river or estuary, usually part of the natural flood plain, over which floodwaters are allowed to flow periodically in order to control high water levels in the river. See also: *rond.*

wash load That part of a stream's sediment load that consists of grain sizes finer than those of the bed (Einstein, 1950, p. 4). It is that part of the *suspended load* that is not derived from the bed but is supplied to the stream by bank erosion, sheetwash, and mass wasting. Cf: *bed-material load.*

washout [geomorph] (a) The washing-out or away of earth materials as a result of a flood or a sudden and concentrated downpour, often causing extensive scouring and bank caving. (b) A place where part of a road or railway has been washed away by the waters of a freshet or local flood.

washout [mining] In a coal seam, a mass of shale, siltstone, or sandstone filling a channel that was cut into the coal swamp during the time of deposition. Cf: *cutout; horseback [coal].*

washout [sed struc] A channel or channel-like feature produced in a sedimentary deposit by the scouring action of flowing water and later filled with the sediment of a younger deposit. Cf: *channel cast.* Syn: *scour and fill; cut and fill.*

washout [streams] A narrow channel or gully cut by a swiftly flowing stream during and after a heavy rainfall; a *wash.*
washover (a) Material deposited by the action of overwash; specif. a small delta built on the landward side of a bar or barrier, separating a lagoon from the open sea, produced by storm waves breaking over low parts of the bar or barrier and depositing sediment in the lagoon. Cf: *blowover.* Syn: *wave delta; storm delta.* (b) The process by which a washover is formed.
washover crescent A term used by Tanner (1960, p. 484) for a type of shallow-water ripple mark that resemble tiny barchans "but actually are coalesced flat-topped ripple marks having crescentic depressions between them".
washover fan A fan-like deposit consisting of sand washed onto the shore during a storm. Syn: *washover apron.*
wash plain (a) *outwash plain.* (b) An *alluvial plain* composed of coarse alluvium.
wash slope The lower, gentle slope of a hillside, lying at the foot of an escarpment or steep rock face and usually covered by an accumulation of talus; it is less steep than the *gravity slope* above and often consists of alluvial fans or pediments. Term introduced by Meyerhoff (1940). Cf: *waning slope.* See also: *foot slope.* Syn: *haldenhang; basal slope.*
washy Easily washing out or eroding, such as a *washy* hillside.
wastage [geomorph] A general term for the denudation of the Earth's surface. See also: *mass wasting.*
wastage [glaciol] *ablation [glaciol].*
waste Loose material resulting from weathering by mechanical and chemical means, and moved down sloping surfaces or carried short distances by streams to the sea; esp. *rock waste.*
waste bank A bank or other accumulation composed of waste material; e.g. a bank where excess earth excavated during the digging of a ditch is dumped parallel to it.
waste-disposal well A less-preferred syn. of *waste-injection well.*
waste-injection well A well used for the injection of waste water or other fluids into the subsurface. Because the wastes rarely degrade to an innocuous condition but are simply stored, they are not truly disposed of; hence, this term is preferred to *waste-disposal well.*
waste mantle Disintegrated and decomposed rock material that overlies bedrock.
waste plain (a) *alluvial plain* (b) *bajada.*
waste rock In mining, rock that must be broken and disposed of in order to gain access to and excavate the ore; valueless rock that must be removed or set aside in mining. Syn: *muck; mullock.*
waste stream The loose rock debris in transit to the sea or to rock-rimmed desert basins, consisting wholly of debris or of debris and water in varying proportions (Grabau, 1924, p. 541).
waste water (a) *return flow.* (b) Seepage of water from a ditch or reservoir.
wasting The gradual destruction or wearing away of a landform or surface by natural processes, including removal by wind, gravity, and rill wash, but excluding stream erosion; e.g. of a glacier by melting, or of rocks by weathering. See also: *mass wasting; backwasting.* Cf: *wearing.*
water [gem] A rarely used term referring to the color and clarity of a precious stone or pearl, and esp. of a diamond. Cf: *river [gem].*
water [geog] (a) A British term for lake, pond, pool, or other body of standing fresh water. (b) A Scottish term for a stream and for a stream bank or the land abutting a stream.
water agate *enhydros.*
water balance *hydrologic budget.*
water bed A term used in the upper Mississippi Valley for a bed of coarse gravel or pebbles occurring in the lower part of the upper till.
water biscuit *algal biscuit.*
water bloom An aquatic growth of algae in such concentration as to cause discoloration of the water. See also: *red tide.* Syn: *plankton bloom; bloom [oceanog].*
water-break (a) A place in a stream where the surface of the water is broken by bottom irregularities. (b) *breakwater.*
water budget *hydrologic budget.*
water capacity The maximum amount of water that a rock or soil can hold.
water color The apparent color of the surface waters of the ocean. Detrital, organic, or dissolved material in the water may affect its color.
water column *depth.*
water content [sed] Amount of water contained in a porous sediment or sedimentary rock, generally expressed as the ratio of the weight of the water in the sediment to that of the sediment when dried, multiplied by 100. See also: *moisture content [soil].*
water content [snow] This term is not recommended because it has been used for both *water equivalent* and *free-water content,* two different concepts.
watercourse (a) A natural, well-defined channel produced wholly or in part by a definite flow of water, continuous or intermittent. Also, a ditch, canal, aqueduct, or other artificial channel for the conveyance of water, as for the draining of a swamp. (b) Legally, a natural stream arising in a given drainage basin but not wholly dependent for its flow on surface drainage in its immediate area, flowing in a channel with a well-defined bed between visible banks or through a definite depression in the land, having a definite and permanent or periodic supply of water, and usually, but not necessarily, having a perceptible current in a particular direction and discharging at a fixed point into another body of water. (c) A legal right permitting the use of the flow of a stream, esp. of one flowing through one's land, or the receipt of water discharged upon land belonging to another.
water creep The movement of water under or around a structure, such as a dam, built on a semipermeable foundation. See also: *piping.*
water crop *water yield.*
water cupola A vaulted uprising of the surface of the ocean above a submarine volcanic explosion; the initial effect of a submarine explosion on the water surface. It immediately precedes the eruption of volcanic gases and ejectamenta. Syn: *water fountain.*
water cushion Water pumped into the drill pipe during a *drill-stem test* to retard fill-up and prevent collapse of the pipe under sudden pressure changes.
water cycle *hydrologic cycle.*
water drive Energy within an oil or gas pool, resulting from hydrostatic or hydrodynamic pressure transmitted from the surrounding aquifer. Cf: *dissolved-gas drive; gas-cap drive.*
water equivalent The amount (or depth) of water that would result from the complete melting of a sample of deposited snow. Not to be confused with *free-water content.* See also: *water content [snow].*
water escape structure *fluid escape structure.*
water eye A small, shallow depression formed as a result of chemical weathering in crystalline rock (Russell, 1968, p.94).
water-faceted stone *aquafact.*
waterfall (a) A perpendicular or steep descent of the water of a stream, as where it crosses an outcrop of resistant rock overhanging softer rock that has been eroded, or flows over the edge of a plateau or cliffed coast. See also: *cascade; cataract.* Syn: *fall.* (b) An obsolete term for a riffle or rapids in a swift stream. (c) A falling away of the ground such that water may be drained off.
waterfall lake *plunge pool.*
waterfinder (a) One who seeks sources of water supply, esp. a *dowser.* (b) Any instrument purported to indicate the presence of water (e.g. *divining rod*).
water-fit A Scottish term for the mouth of a river. Syn: *water-foot.*
waterflood Informal term for a *water flooding* operation.
water flooding A *secondary recovery* operation in which water is injected into a petroleum reservoir to force additional oil out of the reservoir rock and into producing wells. Syn: *waterflood.*
water fountain *water cupola.*
water gap A deep pass in a mountain ridge, through which a stream flows; esp. a narrow gorge or ravine cut through resistant rocks by an antecedent stream. Example: Delaware Water Gap, Penna. Cf: *wind gap.*
water gate A Scottish term for a natural watercourse.
waterhead The *headwater* of a stream.
water hemisphere That half of the Earth containing the bulk (about six-sevenths) of the ocean surface; it is mostly south of the equator, with New Zealand at its approx. center. Cf: *land hemisphere.*
water-holding capacity The smallest value to which the water content of a soil can be reduced by gravity drainage (ASCE, 1958, term 414).
waterhole (a) A natural hole, hollow, or small depression that contains water; esp. in an arid or semiarid region. (b) A spring in the desert. (c) A natural or artificial pool, pond, or small lake.
water horizon *aquifer.*
water humus Humus formed from organic deposits, allochthonous and autochthonous and including both plant and animal

matter, in rivers, lakes, and seas.

water-laid Deposited in or by water; sedimentary.

water level [grd wat] *water table.*

water level [petroleum] The surface below which the rock pores are virtually saturated with water and above which there is an exploitable concentration of hydrocarbons. Syn: *oil-water contact; gas-water contact.*

water level [surv] An instrument that shows the level by means of the surface of water in a trough or in the legs of a U-tube.

water leveling A type of leveling in which relative elevations are obtained by observing heights with respect to the surface of a body of still water (as of a lake).

water-level mark (a) A small horizontal wave-cut "terrace" on an inclined surface of unconsolidated sediment, marking a former water level. (b) *watermark.*

water-level weathering In coastal areas, a lateral widening of a pool of water, due to the alternate wetting and drying in and around it that causes the banks to retreat. By this process, beaches are created that are unrelated to the stands of the sea, esp. in porous or readily eroded rock (Russell, 1968, p.94).

waterlime *hydraulic limestone.*

waterline [coast] (a) The migrating line of contact between land and sea; the *shoreline.* (b) The actual line of contact, at a given time, between the standing water of a lake or sea and the bordering land. (c) The *limit of backrush* where waves are present on a beach.—Also spelled: *water line.*

waterline [grd wat] *water table.*

waterlogged Said of an area in which water stands near, at, or above the land surface, so that the roots of all plants except hydrophytes are drowned and the plants die.

water mass (a) A body of seawater having a characteristic temperature and salinity range defined by a temperature-salinity curve. (b) A mixture of two or more water types. See also: *intermediate water; deep water; bottom water; surface water [oceanog].*

water of capillarity *capillary water.*

water of compaction *Rejuvenated water* originating from the destruction of interstices by compaction of sediments.

water of crystallization Water in a crystal structure that is chemically combined but may be driven off by heat; molecular water, e.g. in gypsum: $CaSO_4 \cdot 2H_2O$.

water of dehydration Water that has been set free from its chemically combined state. Cf: *water of crystallization.*

water of dilation *water of supersaturation.*

water of hydration Water that is chemically combined in a crystalline substance to form a hydrate, but that may be driven off by heat.

water of imbibition (a) The amount of water a rock can contain above the water table. (b) *water of saturation.*

water of retention That part of the interstitial water in a sedimentary rock that remains in the interstices under capillary pressure and under conditions of unhindered flow; usually (though incorrectly) called connate water.

water of saturation The amount of water that can be absorbed by water-bearing material without dilation of that material. Syn: *water of imbibition.*

water of supersaturation Water in excess of that required for saturation; water in sedimentary materials that are inflated or dilated, such as plastic clay or flowing mud whose particles are not in contact and are separated by water. Syn: *water of dilation.*

water opal (a) *hyalite.* b) Any transparent precious opal.

water opening Any break in sea ice which reveals the water; e.g. a *lead.*

water parting A term suggested by Huxley (1877, p. 18) to replace *watershed* in the original meaning of that term (i.e. a drainage *divide*).

water-plasticity ratio *liquidity index.*

water pocket A small, bowl-shaped depression on a bedrock surface, where water may gather; esp. a *water hole* in the bed of an intermittent stream, formed at the foot of a cliff by the action of falling water when the stream is in the flood stage. Syn: *tinaja.*

waterpower The power of moving or falling water, once used to drive machinery directly, as by a water wheel, but now more commonly used to generate electricity by means of a power generator coupled to a turbine through which the water passes. Cf: *hydroelectric power; hydropower; white coal.*

water quality The fitness of water for use, being affected by physical, chemical, and biological factors.

water race A *race* or watercourse.

water regimen *regimen [water].*

water reserve (a) An area of land set aside for feeding streams that are used for water supply. (b) A general term for a quantity or source of water regarded as a supplemental or reserve supply.

water resources A general term referring to the occurrence, replenishment, movement, discharge, quantity, quality, and availability of water.

water-rolled Said of round, smooth sedimentary particles that have been rolled about by water.

waters The marine *territorial waters* of a nation or state.

water sand A porous sand with high or total water content. Cf: *oil sand.*

water sapphire (a) A light-colored blue sapphire. (b) An intense-blue variety of cordierite occurring in waterworn masses in certain river gravels (as in Ceylon) and sometimes used as a gemstone. Syn: *saphir d'eau.* (c) A term applied to waterworn pebbles of topaz, quartz, and other minerals from Ceylon.

water saturation *free-water content.*

watershed (a) A term used in Great Britain for a drainage *divide.* (b) A *drainage basin.*—Etymol: probably German *Wasserscheide,* "water parting". The original and "correct" meaning of the term "watershed" signifies a "water parting" or the line, ridge, or summit of high ground separating two drainage basins. However, the usage of the term, esp. in the U.S. and by several international agencies, has been changed to signify the region drained by, or contributing water to, a stream, lake, or other body of water. The term, when used alone, is ambiguous, and unless the context happens to suffice without aid from the word itself, "the uncertainty of meaning entailed by this double usage makes the term undesirable" (Meinzer, 1923, p. 16).

watershed area The total area of the watershed above the discharge-measuring points. Symbol: A. Cf: *basin area.*

watershed leakage Seepage or flowage of water underground from one drainage basin to an outlet in a neighboring drainage basin or directly to the sea.

watershed line A drainage *divide.*

watershed management Administration and regulation of the aggregate resources of a drainage basin for the production of water and the control of erosion, streamflow, and floods. Also includes the operational functions.

water sky Dark or gray streaks or patches in the sky near the horizon or on the underside of low clouds, indicating the small amount of light reflected from water features in the vicinity of sea ice; darker than *land sky.* See also: *blink.*

watersmeet A meeting place of two streams.

waterspace The ecologic and social interplay among land, water, and social institutions (Padfield & Smith, 1968).

watersplash A shallow ford in a stream.

water spreading Artificial recharge of ground water by spreading water over an absorptive area. Generally broadened to include all methods of artificial recharge involving surficial structures or shallow furrows, pits, or basins, as opposed to injection of water through wells or deep pits or shafts.

waterstead An English term for a stream bed.

water stone A mineral name that has been applied to moonstone, hyalite, enhydros, and jade.

waterstone An English term applied to a stratum whose surface has a watery appearance (like watered silk) and generally understood to express the water-bearing quality of the rock (Woodward, 1887, p. 227); specif. the Waterstones, certain flaggy micaceous sandstones and marls in the Keuper of the English Midlands, from which some water is available. The term should not be used in place of "aquifer" (Stamp, 1966, p. 485).

water supply A source or volume of water available for use; also, the system of reservoirs, wells, conduits, and treatment facilities required to make the water available and usable. Syn: *water system.*

water system (a) *river system.* (b) *water supply.*

water table The surface between the *zone of saturation* and the *zone of aeration;* that surface of a body of unconfined ground water at which the pressure is equal to that of the atmosphere. Syn: *waterline [grd wat]; water level [grd wat]; ground-water table; ground-water surface; plane of saturation; saturated surface; level of saturation; phreatic surface; ground-water level; free-water elevation; free-water surface.*

water-table aquifer *unconfined aquifer.*

water-table cement *ground-water cement.*

water-table divide *divide [grd wat].*

water-table map A map that shows the upper surface of the zone of saturation by means of contour lines.

water-table mound *ground-water mound.*

water-table rock Rock cemented at or near the level of the water table; a specific type of *hardpan* (Russell, 1968).

water-table stream Concentrated flow of ground water at the level of the water table in a structure or mass of rock having high permeability.

water-table well A well tapping unconfined ground water. Its water level may, but does not necessarily, lie at the level of the water table. Cf: *artesian well; nonflowing well; shallow well.*

water tagging The introduction of foreign substances (tracers) into water to detect its movement by measurement of the subsequent location and distribution of the tracers.

water trap *siphon [speleo].*

water type A body of seawater having a characteristic combination of temperature and salinity.

water-vascular system A fluid-filled system of tubular vessels or canals peculiar to echinoderms, used to control the movement of tube feet and perhaps also functioning in excretion and respiration. Primary components include a stone canal, a ring canal, radial canals, and tube feet. See also: *ambulacral system.*

water vein (a) Ground water in a crevice or fissure in dense rock. (b) A term popularly applied to any body of ground water, in part because dowsers commonly describe water as occurring in veins. Hence, the term is little used among hydrologists.

waterway (a) A way or channel, either natural (as a river) or artificial (as a canal), for conducting the flow of water. (b) A navigable body or stretch of water available for passage; a watercourse.

water well (a) A *well* that extracts water from the zone of saturation or that yields useful supplies of water. (b) A well that obtains ground-water information or that replenishes ground water. (c) A well drilled for oil but yielding only water.

water witch (a) A device for determining the presence of water, usually electrically. Cf: *divining rod.* (b) *dowser.*—Nonpreferred syn: *witch.*

water witching *dowsing.*

waterwork A tank, dock, canal lock, levee, sea wall, or other engineering structure built in, for, or as a protection against, water.

water yield The *runoff* from a drainage basin; precipitation minus evapotranspiration (Langbein & Iseri, 1960). Syn: *water crop; runout.*

wath A dialectal term for a *ford* in a stream.

watt A syn. of *tidal flat.* Pl: *watten.* Etymol: German *Watt.*

wattenschlick Tidal or intertidal mud. Etymol: German *Wattenschlick,* "tidal-flats mud".

wattevillite A colorless mineral: $Na_2Ca(SO_4)_2 \cdot 4H_2O$ (?). It occurs in hairlike monoclinic crystals. Also spelled: *wattevilleite.*

Waucoban North American provincial series: Lower Cambrian (above Precambrian, below Albertan). Also spelled: *Waucobian.* Syn: *Georgian.*

wave [seis] A *seismic wave.*

wave [water] An oscillatory movement of water manifested by an alternate rise and fall of a surface in or on the water.

wave age The state of development of a wind-generated, water-surface wave, expressed as the ratio of wave velocity to wind velocity (measured at about 8 m above stillwater level).

wave base The depth at which wave action no longer stirs the sediments; it is usually about 10 meters. Cf: *surf base.* Syn: *wave depth.*

wave-built Constructed or built up by the action of lake or sea waves, assisted by their currents. The term is widely used in regard to *marine-built* features. Cf: *wave-cut.*

wave-built platform A syn. of *wave-built terrace.* The term is inconsistent because a platform is usually regarded as an erosional feature.

wave-built terrace A gently sloping coastal feature at the seaward or lakeward edge of a wave-cut platform, constructed by sediment brought by rivers or drifted along the shore or across the platform and deposited in the deeper water beyond. See also: *marine terrace; beach plain; Syn: wave-built platform; built terrace.*

wave cross ripple mark *oscillation cross ripple mark.*

wave-current ripple mark A longitudinal *compound ripple mark* in which the material forming the crest is believed to have accumulated by the oscillation produced by wave action on a preexisting transverse (current) ripple mark (Straaten, 1953a; and Kelling, 1958, p.124).

wave-cut Carved or cut away by the action of lake or sea waves, assisted by their currents. The term is widely used in regard to *marine-cut* features. Cf: *wave-built.*

wave-cut bench A level to gently sloping narrow surface or platform produced by wave erosion, extending outward from above the base of the wave-cut cliff and occupying all of the shore zone and part or all of the shoreface (Johnson, 1919, p. 162); it is developed mainly above water level by the spray and splash of storm waves aided by subaerial weathering and rainwash. The bench may be bare, freshly worn rock or it may be temporarily covered by a beach; it may end abruptly or grade into the *abrasion platform.* See also: *wave-cut platform.* Syn: *shore platform; beach platform; high-water platform.*

wave-cut cliff A cliff, esp. a *sea cliff,* produced by the breaking-away of rock fragments after horizontal and landward undercutting by waves.

wave-cut notch A *notch* produced along the base of a sea cliff by wave erosion.

wave-cut pediment A *wave-cut platform* formed by erosion of a fault-scarp shoreline (Hinds, 1943, p. 792). The term is not recommended.

wave-cut plain *wave-cut platform.*

wave-cut platform (a) A gently sloping surface produced by wave erosion, extending far into the sea or lake from the base of the wave-cut cliff. It represents both the wave-cut bench and the abrasion platform. Syn: *wave-cut terrace; cut platform; erosion platform; wave platform; shore platform; wave-cut plain; strandflat.* (b) A term sometimes used more restrictedly as a syn. of *abrasion platform.*

wave-cut terrace A syn. of *wave-cut platform.* The term is inconsistent because a terrace is usually regarded as a constructional feature.

wave delta *washover.*

wave depth *wave base.*

wave drift The net translation of water in the direction of wave movement, caused by the open orbital motion of water particles with the passage of each surface wave.

wave energy The capacity of waves to do work. The energy of a wave system is theoretically proportional to the square of the wave height, and the actual height of the waves (being a relatively easily measured parameter) is a useful index to wave energy: a high-energy coast is characterized by breaker heights greater than 50 cm and a low-energy coast is characterized by breaker heights less than 10 cm. Most of the wave energy along equilibrium beaches is used in shoaling and in sand movement. See also: *coastal energy.*

wave erosion *marine abrasion.*

wave-etched shoreline A relatively straight shoreline made irregular by differential wave erosion acting on coastal materials of varying resistance.

wave forecasting The theoretical determination of future wave characteristics, usually from observed or predicted meteorological phenomena such as wind velocities, duration, and fetch. Cf: *wave hindcasting.*

wave front [optics] In optics, the locus of all the points reached by light that is sent outward in all directions from a point. In an isotropic medium, the wave front is a sphere; if the light is constrained to a beam, the wave front will be a plane surface. Cf: *wave normal.* Syn: *wave surface.*

wave front [seis] A surface representing the position of a traveling seismic disturbance at a particular time. Also spelled: *wavefront.*

wave-front chart A diagram showing the position of a traveling seismic disturbance at successive times. It usually shows *raypaths* also.

wave generation The creation and growth of waves by natural or mechanical means, as by a wind blowing over a water surface for a certain period of time.

wave guide A region, usually a layer, in the atmosphere, ocean, or solid Earth that tends to channel seismic energy.

wave hindcasting The calculation of ocean-wave characteristics that probably occurred for some past situation at a given time and place, based on historic synoptic wind charts giving direction, velocity, and duration of winds. Cf: *wave forecasting.*

wave interference ripple mark *oscillation cross ripple mark.*

wavelength (a) The distance between successive wave crests, or other equivalent points, in a series of harmonic waves. (b) In symmetrical, periodic fold systems, the distance between adjacent antiformal or synformal hinges. For asymmetrical and nonperiodic systems, various definitions have been proposed; see Fleuty

(1964a).

wavelet processing Processing of seismic data with the objective of shortening and simplifying the effective wave shape, making it constant along the entire line, or achieving a specific equivalent wave shape, often a zero-phase one.

wave line *swash mark.*

wavellite A white to yellow, green, or black orthorhombic mineral: $Al_3(PO_4)_2(OH)_3 \cdot 5H_2O$. It occurs usually in small hemispherical aggregates exhibiting a strongly developed internal radiating structure. See also: *fischerite.*

wavemark [coast] *swash mark.*

wavemark [sed] A *ripple mark* produced by wave action during the period of deposition.

wave meter An instrument for measuring and recording wave heights.

wave normal In optics, the line at a given point perpendicular to a plane that is tangent to the surface of a light wave at that point. Cf: *wave front [optics].*

wave number *spatial frequency.*

wave of oscillation *oscillatory wave.*

wave of translation A water wave in which the individual particles of water are significantly displaced in the direction of wave travel. Cf: *oscillatory wave.*

wave ogive A curved undulation in the surface of a glacier, forming an arc convex downslope, usually repeated periodically downstream and often merging into a *dirt-band ogive* formed at the base of certain icefalls. Cf: *Forbes band; dirt band [glaciol]; ogive.* Syn: *glacier wave.* Pron: wave + *o*-jive.

wave platform *wave-cut platform.*

wave pole A device for measuring the heights and periods of water-surface waves, consisting of a graduated, weighted vertical pole below which a disk is suspended at a depth sufficiently deep for the wave motion associated with deep-water waves to be negligible. Syn: *wave staff.*

wave ray *orthogonal.*

wave refraction (a) The process by which a water wave, moving in shallow water as it approaches the shore at an angle, tends to be turned from its original direction. The part of the wave advancing in shallower water moves more slowly than the part still advancing in deeper water, causing the wave crests to bend toward parallel alignment with the shoreline. (b) The bending of wave crests by currents.

wave ripple mark *oscillation ripple mark.*

wave spectrum (a) The description of wave energy with respect to frequency by mathematical function. The square of the wave height is related to the potential energy of the surface of the sea. (b) A graph or table that shows, for a region of the ocean, the distribution of wave height with respect to frequency.

wave staff *wave pole.*

wave steepness The ratio of the height of a water wave to its length. A wave with a ratio of 1/25 to 1/7 has "great" steepness; one with a ratio of less than 1/100 has "low" steepness. Syn: *steepness ratio.*

wave surface *wave front [optics].*

wave wash The erosion of shores or embankments by the lapping or breaking of waves; esp. the erosion of levees during floods.

wavy bedding (a) Bedding characterized by undulatory bounding surfaces. Syn: *rolling strata.* (b) A form of interbedded mud and ripple-cross-laminated sand, in which "the mud layers overlie ripple crests and more or less fill the ripple troughs, so that the surface of the mud layer only slightly follows the concave or convex curvature of the underlying ripples" (Reineck and Wunderlich, 1968, p. 101). Cf: *flaser structure; lenticular bedding.*

wavy extinction *undulatory extinction.*

wax A solid, noncrystalline hydrocarbon of mineral origin such as ozocerite and paraffin wax, composed of the fatty acid esters of the higher hydrocarbons.

waxing development *accelerated development.*

waxing slope The upper part of a hillside surface, tending to become convex by being rounded off just above an escarpment, having an angle that increases continuously downslope as the hillside is worn back (Wood, 1942). Ant: *waning slope.* Syn: *convex slope.*

wax opal Yellow opal with a waxy luster.

waxy luster A type of mineral luster that resembles the luster of wax, e.g. in chalcedony.

waylandite A white mineral: $(Bi,Ca)Al_3(PO_4,SiO_4)_2(OH)_6$.

W-chert Chert nodules formed by weathering (Dunbar & Rodgers, 1957, p. 249).

W-dolostone Dolostone produced by weathering (Dunbar & Rodgers, 1957, p. 239). Cf: *S-dolostone; T-dolostone.*

weak ferromagnetism *Antiferromagnetism* in which the opposing atomic magnetic moments do not cancel perfectly, so that there is a weak, spontaneous macroscopic magnetization. An example of a mineral displaying weak ferromagnetism is α Fe_2O_3, hematite. Syn: *parasitic ferromagnetism.*

weal A descriptive field term for one of the crisscrossing raised bands, 5-7.5 cm wide, occurring on a more or less evenly patterned sedimentary surface (Donaldson & Simpson, 1962, p.74). The bands in cross section are almost semicircular.

Wealden European (Great Britain) stage: Lower Cretaceous (above Purbeckian, below Gault).

wear The reduction in size or the change in shape of clastic fragments by one or more of the mechanical processes of abrasion, impact, or grinding (Wentworth, 1931, p. 24-25). See also: *wearing.*

wearing The gradual destruction of a landform or surface by friction or attrition. Cf: *wasting.* See also: *backwearing; downwearing; wear.*

weather [geol] v. To undergo changes, such as discoloration, softening, crumbling, or pitting of rock surfaces, brought about by exposure to the atmosphere and its agents. See also: *weathering.*

weather [meteorol] n. The condition of the Earth's atmosphere, specif. its temperature, barometric pressure, wind velocity, humidity, clouds, and precipitation.—adj. A syn. of *windward.*

weather chart *weather map.*

weather coal Brown coal that has been weathered and displays bright colors.

weathered ice Sea ice that has undergone a gradual elimination of surface irregularities by thermal and mechanical processes of removal and addition of material; ice whose hummocks and pressure ridges are smoothed and rounded.

weathered iceberg An iceberg that has undergone prolonged ablation, which generally gives it a very irregular but rounded shape.

weathered layer In seismology, that zone of the Earth that is immediately below the surface, characterized by low seismic-wave velocities.

weathering The destructive process or group of processes by which earthy and rocky materials on exposure to atmospheric agents at or near the Earth's surface are changed in color, texture, composition, firmness, or form, with little or no transport of the loosened or altered material; specif. the physical disintegration and chemical decomposition of rock that produce an in-situ mantle of waste and prepare sediments for transportation. Most weathering occurs at the surface, but it may take place at considerable depths, as in well-jointed rocks that permit easy penetration of atmospheric oxygen and circulating surface waters. Some authors restrict weathering to the destructive processes of surface waters occurring below 100°C and 1 kb; others broaden the term to include biologic changes and the corrasive action of wind, water, and ice. Syn: *demorphism; clastation.*

weathering boulder *boulder of weathering.*

weathering correction In seismic exploration, a correction applied to reflection and refraction data for variations in traveltime produced by irregularities in a low-velocity or weathered layer near the surface. Syn: *low-velocity-layer correction.*

weathering crust A term in use by European geologists to designate a regionally widespread and usually deep zone of weathered materials, formed over a geologically long interval by relatively uniform chemical weathering. The weathered materials typically consist of clay minerals, laterite (including bauxite), or both (Keller, 1977).

weathering escarpment An escarpment developed where gently dipping sedimentary rocks of varying resistance are subjected to degradation; the term is not appropriate because mass-wasting, sheetwash, and stream erosion are fully as important as weathering (Thornbury, 1954, p. 71-72).

weathering front The interface of fresh and weathered rock; a term proposed by Mabbutt (1961) to replace *basal surface.*

weathering index A measure of the weathering characteristics of coal, according to a standard laboratory procedure. Syn: *slacking index.*

weathering out The exposing of relatively resistant rock as the surrounding softer rock is reduced by weathering.

weathering-potential index A measure of the degree of susceptibility to weathering of a rock or mineral, computed from a chemical analysis, and expressed as the mol-percentage ratio of the sum of the alkalies and alkaline earths (less combined water) to the total mols present exclusive of water (Reiche, 1943, p. 66).

weathering rind An outer crust or layer on a pebble, boulder, or other rock fragment, formed by weathering.

weathering shot In seismic exploration, the detonation of a small explosive charge in the weathering or *low-velocity layer* to determine its velocity characteristics and thickness. Syn: *short shot; poop shot.*

weathering velocity That velocity with which a seismic *P wave* travels through the near-surface *low-velocity layer.* Cf: *subweathering velocity.*

weather map A chart that is used to show temperature, pressure, precipitation, wind direction and velocity, air masses, and fronts of a given area. Syn: *weather chart.*

weather pit A shallow depression on the flat or gently sloping summit of large exposures of granite or granitic rocks (as in the Sierra Nevada, Calif.), attributed to strongly localized solvent action of impounded water (Matthes, 1930, p.64); diameter is 30-45 cm, and depth ranges up to 15 cm. Cf: *rock doughnut; oven.*

weather shore A shore lying to the windward or in the direction from which the wind is blowing, and thereby exposed to strong wave action. Ant: *lee shore.*

weber The SI unit of magnetic flux: 10^8 *maxwell*. Syn: *volt-second.*

weberite A pale-gray mineral: Na_2MgAlF_7.

Weber number The relationship of the forces of inertia to those of surface energy, expressed as the product of density times velocity of flow squared times length divided by surface energy. It is important in the movement of water in porous media and capillaries (Chow, 1964, p. 7-5).

Weber's line A hypothetical boundary between the Asian and Australasian biogeographic regions. It generally coincides with the Australian-Papuan shelf and is sometimes used in preference to *Wallace's line.* Named after Max Weber (d. 1937), German zoologist.

websterite [mineral] *aluminite.*

websterite [rock] (a) In the *IUGS classification,* a plutonic rock with M equal to or greater than 90, ol/(ol+opx+cpx) less than 5, and both opx/(ol+opx+cpx) and cpx/(ol+opx+cpx) less than 90. (b) A pyroxenite composed chiefly of ortho- and clinopyroxene.—The name, given by G. H. Williams in 1890, is from Webster, North Carolina.

weddellite A mineral (calcium oxalate): $CaC_2O_4 \cdot 2H_2O$. It is found as small isolated crystals in urinary calculi and in mud at the bottom of Weddell Sea, Antarctica. Cf: *whewellite.*

wedge [optics] (a) *optical wedge.* (b) *quartz wedge.*

wedge [paleont] A five-sided crystalline element of a heterococcolith, having two dimensions subequal and the third dimension small at one edge and approaching zero at the other.

wedge [stratig] (a) The shape of a stratum, vein, or intrusive body that thins out; specif. a wedge-shaped sedimentary body, or *prism [sed].* (b) *sand wedge.*

wedge ice *foliated ground ice.*

wedge out v. To become progressively thinner or narrower to the point of disappearance; to *thin out.*

wedge-out n. The edge or line of *pinch-out* of a lensing or truncated rock formation.

wedge theory A corollary of the obsolete *contracting Earth* theory; it supposes that shrinking of the crust breaks it into wedge-shaped blocks, which are uplifted and laterally compressed along their margins, resulting in two-sided orogenic structures.

wedgework The action of rock disintegration by the wedgelike insertion of agents such as roots and esp. ice. See also: *frost wedging.* Also, the results of wedgework action.

wedging The splitting, breaking, or forcing apart of a rock as if by a wedge, such as by the growth of salt or mineral crystals in interstices; specif. *frost wedging.*

weedia A type of stromatolite consisting of algal crusts that are nearly flat or essentially parallel to the bedding and appear in cross section as a branching network of bedding planes that join in an irregular manner (Pettijohn, 1957, p.222 & 399).

weeksite A yellow orthorhombic mineral: $K_2(UO_2)_2(Si_2O_5)_3 \cdot 4H_2O$.

weeping rock A porous rock from which water oozes.

weeping spring A spring of small yield; a syn. of *seepage spring.*

Wegener hypothesis *continental displacement.*

wegscheiderite A triclinic mineral: $Na_5(CO_3)(HCO_3)_3$.

wehrlite [mineral] A mineral: BiTe. It is a native alloy of bismuth and tellurium, and was earlier formulated Bi_2Te_3. Syn: *mirror glance.*

wehrlite [rock] (a) In the *IUGS classification,* a plutonic rock with M equal to or greater than 90, ol/(ol+opx+cpx) between 40 and 90, and opx/(ol+opx+cpx) less than 5. (b) A peridotite composed chiefly of olivine and clinopyroxene with common accessory opaque oxides. Kobell in 1834 named the rock, which he thought to be a mineral, after Wehrle, who had analyzed it.

weibullite A mineral: $Pb_5Bi_8(S,Se)_{17}$.

Weichsel The term applied in northern Europe to the classical fourth and last glacial stage of the Pleistocene Epoch, after the Saale glacial stage; equivalent to the *Würm* and *Wisconsinan.* Adj: *Weichselian.*

weigelith An amphibole- and enstatite-bearing *peridotite,* named by Kretschmer in 1918 for Weigelsberg, Czechoslovakia. Obsolescent.

weight dropping A method used in seismic prospecting, in which a heavy weight is dropped to create seismic waves. See also: *Thumper.*

weighting A statistical method for expressing the relative importance of different measurements for a set of data; the purposeful addition of statistical bias.

weilerite A mineral: $BaAl_3(AsO_4)(SO_4)(OH)_6$ (?).

weilite A mineral: $CaHAsO_4$.

weinschenkite (a) A white mineral: $YPO_4 \cdot 2H_2O$. Syn: *churchite.* (b) A dark-brown variety of hornblende low in ferrous iron and high in ferric iron, aluminum, and water.

weir (a) A small dam in a stream, designed to raise the water level or to divert its flow through a desired channel; e.g. a "leaping weir". (b) A notch in a levee, dam, embankment, or other barrier across or bordering a stream, through which the flow of water is regulated; e.g. a "wasteweir".

weisbachite A variety of anglesite containing barium.

weiselbergite An altered glassy *basalt* or *andesite* characterized by labradorite, augite, and iron-oxide phenocrysts in a groundmass of plagioclase and augite microlites and interstitial glass. The name was proposed by Rosenbusch in 1877, after Weiselberg, Germany, for augite andesite. Wadsworth in 1884 applied the term to altered glassy andesite. Obsolete. Cf: *shastalite.*

Weisenboden A great soil group in the 1938 classification system, which was reclassified as *Humic Gley soil* in the 1949 revision (Thorp and Smith, 1949). Syn: *meadow soil.*

Weissenberg pattern The pattern of X-ray diffraction spots obtained from a single crystal using a Weissenberg camera and monochromatic radiation, by a moving-film method which enables unambiguous indexing of all diffractions for any properly oriented single crystal.

weissite A blue-black mineral: Cu_5Te_3.

welded contact Any intimate, closely fitting contact between two bodies of rock that have not been disrupted tectonically; e.g. a contact between two parallel limestone beds separated by a paraconformity. The term does not imply a preliminary softening by heat.

welded dike A *secundine dike* whose boundaries have become obscured by continued mineral growth of the granitic country rock into the intrusion.

welded flow breccia The lower part of the fragmented crust of aa and block-lava flows, where the fragments are thoroughly welded together rather than being loose as in the upper part of the crust.

welded texture A texture of pyroclastic rocks, especially those derived from ash flows and nuées ardentes, that is formed by the heat and pressure of still-plastic particles as they are deposited.

welded tuff A glass-rich pyroclastic rock that has been indurated by the welding together of its glass shards under the combined action of the heat retained by particles, the weight of overlying material, and hot gases. It is generally composed of silicic pyroclasts and appears banded or streaky. Cf: *sillar.* Syn: *tufflava.*

welding (a) Consolidation of sediments (esp. of clays) by pressure resulting from the weight of superincumbent material or from earth movement, characterized by cohering particles brought within the limits of mutual molecular attraction as water is squeezed out of the sediments (Tyrrell, 1926, p. 196). (b) The diagenetic process whereby discrete crystals and/or grains become attached to each other during compaction, often involving pressure solution and solution transfer (Chilingar et al., 1967, p. 322).

welinite A mineral: $(Mn^{+4},W)_{1-x}(Mn^{+2},W,Mg)_{3-y}Si(O,OH)_7$.

well [eng] A hollow cylinder of reinforced concrete, steel, timber, or masonry constructed in a pit or hole in the ground that reaches to hardpan or bedrock and used as a support for a bridge or building. Also, the pit or hole in which the well is built.

well [gem] The small dark nonreflecting area in the center of a fashioned stone, esp. in that of a colorless diamond cut too thick.

well [petroleum] A borehole or shaft sunk into the ground for the purpose of obtaining oil and/or gas from an underground source, or of introducing water or gas under pressure into an underground formation. See also: *oil well; gas well.*

well [water] (a) An artificial excavation (pit, hole, tunnel), generally cylindrical in form and often walled in, sunk (drilled, dug, driven, bored, or jetted) into the ground to such a depth as to penetrate water-yielding rock or soil and to allow the water to flow or to be pumped to the surface; a *water well.* (b) A term originally applied to a natural spring or to a pool formed by or fed from a spring; esp. a mineral spring. (c) A term used chiefly in the plural form for the name of a place where mineral springs are located or of a health resort featuring marine or freshwater activities; a spa.

well-bedded Said of a *bedded* rock whose beds are numerous and clearly defined.

well bore *borehole.*

well casing *casing.*

well cuttings Rock chips cut by a bit in the process of well drilling, and removed from the hole in the drilling mud in rotary drilling or by the bailer in cable-tool drilling. Well cuttings collected at closely spaced intervals provide a record of the strata penetrated. Syn: *cuttings; drill cuttings; well samples.*

well-data system A system designed for computer storage and retrieval of well data, including programs necessary to update the file.

well-graded (a) A geologic term for *well-sorted.* (b) An engineering term pertaining to a *graded* soil or unconsolidated sediment with a continuous distribution of particle sizes from the coarsest to the finest, in such proportions that the successively smaller particles almost completely fill the spaces between the larger particles.—Ant: *poorly graded.*

wellhead The source from which a stream flows; the place in the ground where a spring emerges.

well log A graphic record of the measured or computed physical characteristics of the rock section encountered in a well, plotted as a continuous function of depth. Measurements are made by a *sonde* as it is withdrawn from the borehole by a *wire line.* Several measurements are usually made simultaneously, and the resulting curves are displayed side by side on the common depth scale. Both the full display and the individual curves are called *logs.* Well logs are commonly referred to by generic type, e.g. *resistivity log, radioactivity log,* or by specific curve type, e.g. *sonic log, gamma-ray log.* See also: *sample log; driller's log; mud log.* Syn: *borehole survey; borehole log; geophysical log; wire-line log.*

well point A hollow vertical tube, rod, or pipe terminating in a perforated pointed shoe and fitted with a fine-mesh wire screen, connected with others in parallel to a drainage pump, and driven into or beside an excavation to remove underground water, to lower the water level and thereby minimize flooding during construction, or to improve stability.

well record A concise statement of the available data regarding a well; a full history or day-by-day account, from the day the location was surveyed to the day production ceased.

well-rounded Said of a sedimentary particle whose original faces, edges, and corners have been destroyed by abrasion and whose entire surface consists of broad curves without any flat areas; specif. said of a particle with no secondary corners and a roundness value between 0.60 and 1.00 (Pettijohn, 1957, p. 59). The original shape is suggested by the present form of the particle. Also, said of the *roundness class* containing well-rounded particles.

well samples *well cuttings.*

well shooting In seismic prospecting, a method of determining velocity as a function of depth by lowering a geophone into a borehole and recording energy from shots fired from surface shotholes. See also: *check shot.*

well site *location [drill].*

wellsite A zeolite mineral: $(Ba,Ca,K_2)Al_2Si_3O_{10} \cdot 3H_2O$.

well-sorted Said of a *sorted* sediment that consists of particles all having approximately the same size and that has a sorting coefficient less than 2.5. Based on the phi values associated with the 84 and 16 percent lines, Folk (1954, p. 349) suggests sigma phi limits of 0.35-0.50 for well-sorted material. Ant: *poorly sorted.* Syn: *well-graded.*

wellspring The *fountainhead* of a stream.

wellstrand A Scottish term for a stream flowing from a spring.

well ties The comparison of seismic measurements with geologic datum points at well locations.

well water Water obtained from a well; water from the zone of saturation or from a perched aquifer; *phreatic water.*

weloganite A mineral: $Sr_3Na_2Zr(CO_3)_6 \cdot 3H_2O$.

welt A nongenetic term used by Bucher (1933) for a raised part of the Earth's crust of any size with a distinct linear development. Cf: *furrow.*

Wemmelian European stage: Upper Middle Eocene (above Ledian, below Auversian).

wenkite A mineral: $(Ba,K)_4(Ca,Na)_6(Si,Al)_{20}O_{41}(OH)_2(SO_4)_3 \cdot H_2O$.

Wenlockian European stage: Middle Silurian (above Llandoverian, below Ludlovian).

wennebergite A quartz-bearing porphyry containing phenocrysts of orthoclase, biotite, and quartz in a microlitic, chloritic groundmass with abundant apatite and sphene. The name, given by Schuster in 1905, is from Wenneberg, Germany. Obsolete.

Wenner array An electrode array in which the four electrodes are in-line and equally spaced, and in which the outer pair is used to inject current into the ground while the inner pair is used to measure differences in potential.

Wentworth grade scale An extended version of the *Udden grade scale*, adopted by Chester K. Wentworth (1891-1969), U.S. geologist, who modified the size limits for the common grade terms but retained the geometric interval or constant ratio of 1/2 (Wentworth, 1922). The scale ranges from clay particles (diameter less than 1/256 mm) to boulders (diameter greater than 256 mm). It is the grade scale generally used by North American sedimentologists. See also: *phi grade scale.*

Werfenian European stage: lowermost Triassic (above Thuringian, below Virglorian). See also: *Scythian.*

wermlandite A mineral: $Ca_2Mg_{14}(Al,Fe)_4(CO_3)(OH)_{42} \cdot 29H_2O$.

Wernerian adj. Of or relating to Abraham G. Werner (1749-1817), German geologist and mineralogist, who classified minerals according to their external characters, advocated the theory of *neptunism,* and postulated a worldwide age sequence of rocks based on their lithology. Also, said of one who is a great, but dogmatic, teacher of geology.—n. An adherent of Wernerian beliefs; a *neptunist.*

wernerite A syn. of common *scapolite,* a specific mineral of the scapolite group intermediate between meionite and marialite.

wesselite A hypabyssal *nephelinite* containing biotite, barkevikite, titanaugite, hauyne, and nepheline. The name, given by Scheumann in 1922, is for Wesseln, Czechoslovakia. Not recommended usage.

west-coast desert A *coastal desert* found on the western edge of continents and occurring in the *tropical-desert* latitude, i.e. near the tropics of Cancer or Capricorn. The fluctuation, both annually and daily, is much lower than for inland tropical deserts (Strahler, 1963, p. 335).

westerveldite An orthorhombic mineral: (Fe,Ni,Co)As.

westerwaldite An extrusive *trachybasalt* composed of phenocrysts of serpentinized olivine, some with augite rims, in a groundmass of labradorite, sanidine, augite, and biotite, with interstitial nepheline. The name (Johannsen, 1938) is from Westerwald, Germany. Not recommended usage.

westgrenite A mineral of the pyrochlore group: $(Bi,Ca)(Ta,Nb)_2O_6(OH)$.

westing A *departure* (difference in longitude) measured to the west from the last preceding point of reckoning; e.g. a linear distance westward from the north-south (vertical) grid line that passes through the origin of the grid.

Westphal balance In mineral analysis, a balance used to determine specific gravity of a *heavy liquid.* Syn: *beam balance.*

Westphalian European stage: Upper Carboniferous (Middle Pennsylvanian; above upper Namurian, below Stephanian).

westward drift A component of the *secular variation* of the Earth's magnetic field; the movement is about 0.2° per year of the broad-scale departures of the actual geomagnetic field from an ideal dipole field.

wet analysis A method of estimating the effective diameters of particles smaller than 0.06 mm by mixing the sample in a measured volume of water and checking its density at intervals with a sensitive hydrometer (Nelson, 1965, p. 512).

wet assay Any type of assay procedure that involves liquid, generally aqueous, as a means of separation. Cf: *dry assay.*

wet avalanche *wet-snow avalanche.*

wet blasting Abrasion or attrition effected by the impact of water against an exposed surface; e.g. the formation of an *aquafact* by wave action.

wet-bulb temperature The lowest temperature to which air can be cooled at constant pressure by the evaporation of water into it. During the process, the heat required for the evaporation is supplied by the air itself, which is thereby cooled. Cf: *dew point.*

wet chemical analysis Any of the methods for chemical determinations using water or other liquids as part of the process. Typically it refers to *gravimetric analysis* and *volumetric analysis.*

wet gas A natural gas containing liquid hydrocarbons. Cf: *dry gas; condensate.*

wetlands A general term for a group of wet habitats, in common use by specialists in wildlife management. It includes areas that are permanently wet and/or intermittently water-covered, esp. coastal marshes, tidal swamps and flats, and associated pools, sloughs, and bayous.

wet playa A playa that is soft under foot, having a thin and puffy surface that is coated with white efflorescent salts indicating the active discharge of near-surface ground water by evaporation (Thompson, 1929); a *salina.* It is underlain by loose granular silt, salt crystals, and moist clay. Cf: *dry playa.* Syn: *moist playa.*

wet snow Deposited snow that contains liquid water. Cf: *dry snow.*

wet-snow avalanche An avalanche caused by a sudden spring thaw, which releases downslope a single blanket-like mass of heavy wet snow. Due to friction, it is the slowest-moving of the snow avalanches. Syn: *slush avalanche; wet avalanche; full-depth avalanche.*

wettability The ability of a liquid to form a coherent film on a surface, owing to the dominance of molecular attraction between the liquid and the surface over the cohesive force of the liquid itself.

wetted perimeter (a) The length of the wetted contact between a stream of flowing water and its containing conduit or channel, measured in a plane at right angles to the direction of flow. (b) The length of the perimeter of a conduit below the water surface. (c) The entire perimeter of a conduit flowing full. The wetted perimeter is used when computing the *hydraulic radius* (ASCE, 1962).

wetting front *pellicular front.*

wet unit weight The *unit weight* of soil solids plus water per unit of total volume of soil mass, irrespective of the degree of saturation. Syn: *mass unit weight.*

whaleback (a) A large mound or hill having the general shape of a whale's back, esp. a smooth elongated ridge of desert sand having a rounded crest and ranging widely in size (about 300 km long, 1-3 km wide, and perhaps 50 m high). It forms a coarse-grained platform or pedestal built up and left behind by a succession of longitudinal (seif) dunes along the same path. Syn: *sand levee.* (b) A rounded, elongated rock mass, commonly granite, found in tropical areas associated with tors. (c) A *roche moutonnée,* often of granitic composition, such as those in Canada and Finland.

wharf Any structure (such as a *pier* or *quay*) built out from the shore and serving as a berthing place for vessels.

wheel A holothurian sclerite in the form of a wheel, consisting of a *hub,* a *rim,* and *spokes.*

Wheelerian North American stage: Upper Pliocene (above Venturian, below Hallian).

wheelerite A yellowish variety of retinite that is soluble in ether and that fills fissures in, or is thinly interbedded with, lignite beds in northern New Mexico.

wheel ore The mineral bournonite, esp. when occurring in wheel-shaped twin crystals.

wherryite A pale-green mineral: $Pb_4Cu(CO_3)(SO_4)_2(Cl,OH)_2O$ (?).

whetstone Any hard fine-grained naturally occurring rock, usually siliceous, that is suitable for sharpening implements such as razors, knives, and mechanics' tools; e.g. *novaculite.*

whewellite A white or colorless monoclinic mineral (calcium oxalate): $CaC_2O_4 \cdot H_2O$. It occurs as a warty and somewhat opaline incrustation on marble. Cf: *weddellite.*

whinstone A colloquial British term for *dolerite, basalt,* and other dark fine-grained igneous rocks. The term is derived from the Whin Sill in northern England. Cf: *trap [ign].*

whiplash A smooth-surfaced flagellum (without mastigonemes) of some algae and protozoans, having a long rigid basal part and a short thinner distal region.

whipstock n. A long wedge-shaped steel device with a concave groove along its inclined face, placed in an oil well and used during drilling to deflect and guide the drill bit toward the direction in which the inclined grooved surface is facing.—v. To use a whipstock to drill a *directional well.*

whirl ball A spindle-shaped, tubular, ellipsoidal, or spherical mass of fine sandstone embedded in silt, its long axis being vertical or steeply inclined. It is attributed to vortices in mudflows (Dzulynski et al., 1957).

whirlpool A body of water moving rapidly in a circular path of relatively limited radius. It may be produced by a current's passage through an irregular channel or by the meeting of two opposing currents. Cf: *eddy; maelstrom.*

whirl zone A zone of transition between a slump sheet and the overlying strata.

whistling sand A *sounding sand,* often found on a beach, that gives rise to a high-pitched note when stepped on or struck with the hand, the sound apparently resulting from the translation of grain over grain. Syn: *squeaking sand; musical sand.*

Whitbian Stage in Great Britain: upper Lower Jurassic (above Domerian, below Yeovilian).

Whitcliffian European stage: Upper Silurian (above Leintwardian, below Devonian).

white agate A term sometimes applied to white or whitish chalcedony.

white alkali An older term for accumulation of salts with high levels of sodium that may develop as a crust. Cf: *black alkali.*

white antimony *valentinite.*

white band A layer in a glacier consisting of ice that is white and opaque because it contains numerous air bubbles. Cf: *blue band [glaciol].*

white-bedded phosphate A term used in Tennessee for a phosphatic limestone characterized by partial replacement of calcite by calcium phosphate, and by a matrix consisting of cryptocrystalline quartz. It occurs in regular bands alternating with thinner beds of chert. Cf: *brown rock; hard-rock phosphate.*

whitecap The white froth on the crest of a wave; it is caused by wind blowing the crest forward and over.

white chert A light-colored *chert,* or chert proper, as distinguished from the dark variety or *black chert.*

white clay *kaolin.*

white coal (a) A fanciful term for *waterpower.* (b) *tasmanite [coal].*

white cobalt (a) *cobaltite.* (b) *smaltite.*

white copperas (a) *goslarite.* (b) *coquimbite.*

whitedamp A term for carbon monoxide in coal mines. Cf: *blackdamp; afterdamp; firedamp.*

white earth A siliceous earthy material that is used as a pigment in paint.

white feldspar *albite.*

white garnet (a) A translucent variety of grossular, sometimes resembling white jade in appearance. (b) *leucite.*

white gold A pale alloy of gold that resembles silver or platinum; esp. gold alloyed with a high proportion of nickel or palladium to give it a white color, with or without other alloying metals (such as tin, zinc, or copper).

white ice (a) Sea ice of not more than one winter's growth, and a thickness of 30-70 cm, also known as "thin" *first-year ice.* (b) Coarsely granular, porous glacier ice formed by compaction of snow and appearing white. Cf: *black ice; blue ice.* (c) Snow ice that forms on top of river and lake ice. Whiteness results from scattering of light by entrapped air bubbles.

white iron ore A syn. of *siderite.*

white iron pyrites *marcasite [mineral].*

white lead ore *cerussite.*

white mica A light-colored mica; specif. *muscovite.*

white mundic *arsenopyrite.*

white nickel (a) *nickel-skutterudite.* (b) *rammelsbergite.*

white olivine *forsterite.*

white opal A form of precious opal of any light color, as distinguished from *black opal;* e.g. a pale blue-white gem variety of opal.

whiteout The diffusion of daylight by multiple reflection between fallen snow and overcast clouds, so that the horizon and surface features are impossible to discern.

white pyrite *marcasite [mineral].*

white pyrites (a) *arsenopyrite.* (b) *marcasite [mineral].*

Whiterock North American (California, Nevada, Oklahoma) stage: lowermost Middle Ordovician (above Lower Ordovician, below Marmor) (Cooper, 1956).

white sand (a) Quartzitic sand, pure enough to resist heat, used in steel furnaces. (b) Gypsum sand, as at White Sands National

Monument, New Mexico.

white sapphire The colorless or clear pure variety of crystallized corundum.

white schorl *albite.*

white stone A clear, colorless imitation gem, resembling the diamond.

white tellurium (a) *sylvanite.* (b) *krennerite.*

white top In coal mining, light gray shale that occurs above a coal seam, between the coal and the overlying dark-shale roof. It may be arenaceous and is usually unlaminated.

white trap A term used in the Midland Valley of Scotland for an intrusive igneous rock, usually of basic composition, that has been bleached at the contact with coal or other carbonaceous rock. It is created where gaseous hydrocarbons and carbon dioxide, resulting from the local breakdown of the sedimentary organic matter, invade the igneous body at a late stage during its cooling and convert the ferromagnesian and feldspar minerals into a mixture of carbonates and clay minerals.

white vitriol *goslarite.*

whiting A mass of muddy water in which abundant carbonate material is suspended, producing a white color. Whitings typically occur over shallow carbonate platforms and are elongated by wind or tidal currents. Most of them consist of stirred-up bottom sediment (Blatt et al., 1972, p. 426).

whitleyite An achondritic stony meteorite of the *aubrite* class, containing fragments of black chondrite.

whitlockite A mineral: $Ca_9(Mg,Fe)H(PO_4)_7$. Syn: *merrillite.*

whitmoreite A mineral: $Fe^{+2}Fe_2^{+3}(PO_4)_2(OH)_2 \cdot 4H_2O$.

Whitneyan North American continental stage: Upper Oligocene (above Orellan, below Arikareean).

whole-rock Adj. used in analytical geology to indicate that a portion of rock rather than individual minerals was examined. In the rubidium-strontium age method the rock may have remained a closed system for rubidium and strontium isotopes whereas the constituent minerals did not. Thus, a calculated age for the whole rock would give the apparent age of formation whereas the individual minerals might give discordant ages. This whole-rock, closed-system feature does not hold true for all isotopic systems. Syn: *total-rock.*

whorl (a) One of the turns of a spiral or coiled shell; specif. a single complete turn through 360 degrees of a gastropod shell, a cephalopod conch, or a foraminiferal test. See also: *body whorl.* Syn: *volution.* (b) An arrangement of two or more anatomical parts or organs of one kind in a circle around the same point on an axis; e.g. a circle of equally spaced branches around the stem of a plant, arranged like the spokes of a wheel. Syn: *verticil.*

whorl coccolith One of the modified coccoliths forming a whorl about the naked pole in nonmotile coccolithophores exhibiting dimorphism (such as *Ophiaster*).

whorled Having a circular arrangement of appendages, e.g. leaves, at nodes on an axis. Cf: *alternate; opposite; acyclic.* Syn: *verticillate.*

whorl height The height of a nautiloid or ammonoid whorl measured at right angles to the maximum *width,* comprising the distance from the middle of the venter to the middle of the dorsum plus the depth of the impressed area (TIP, 1959, pt.L, p.6). In practice, the "oblique whorl height" is commonly used, consisting of the distance from umbilical seam to the middle of the venter.

whorl section A transverse section of a cephalopod whorl.

whorl side The *flank* of a cephalopod conch; esp. the lateral wall of an ammonoid whorl between umbilical seam and ventral area.

wiborgite *rapakivi.*

wich A term used in England for a damp meadow or a marshy place, esp. where salt is found or has been worked; also, a *salt pit.* Also spelled: *wych.*

Wichita orogeny A name used by van der Gracht (1931) for the first major phase of deformation in the Wichita and Ouachita orogenic belts of southern Oklahoma. In the Wichita belt it is dated by adjacent strata as early Pennsylvanian (Morrow); in the Ouachita belt, it includes a Mississippian phase that produced the great flysch body of the Stanley and Jackfork formations.

wichtisite An obsolete synonym of *tachylyte.*

wickenburgite A mineral: $Pb_3Al_2CaSi_{10}O_{24}(OH)_6$.

wickmanite A mineral: $MnSn(OH)_6$.

widenmannite A mineral: $Pb_2(UO_2)(CO_3)_3$.

widespread *spread [streams].*

wide water A local term applied in northern Michigan to a wide, shallow expanse of water backed up behind a natural dam or produced by a widening in the course of a stream.

widiyan Var. of *wadi* used in the north African deserts (Stone, 1967, p. 264).

Widmanstätten structure A triangular pattern observed on polished and etched surfaces of iron meteorites (octahedrites), composed of parallel bands or plates of kamacite bordered by taenite and intersecting one another in two, three, or four directions. The kamacite bands, arranged parallel to the octahedral planes in the host taenite, are produced by exsolution from an originally homogeneous taenite crystal. As the bands become finer (thinner), the nickel content increases. Named after Aloys B. Widmanstätten (1753?-1849), Austrian mineralogist who discovered the structure in 1808. Also spelled: *Widmanstatten structure.* Syn: *Widmanstätten figure; Widmanstätten pattern.*

width (a) On a brachiopod, the maximum dimension measured perpendicular to the plane of symmetry (at right angles to the length and thickness or height). (b) In cephalopod conchs, the horizontal distance between points located between ribs or spines on opposite whorl sides; the maximum dimension perpendicular to the height.

Wiechert-Gutenberg discontinuity *Gutenberg discontinuity.*

Wien's displacement law The statement that when the temperature of a radiating blackbody increases, the wavelength corresponding to maximum radiance decreases in such a way that the product of the absolute temperature and wavelength is constant (Wien, 1894, p. 132-165). See also: *color temperature.*

wiggle stick *divining rod.*

wiggle trace A graph of amplitude against time, as on a conventional seismic recording with mirror galvanometers. Also called "squiggle recording". Syn: *variable-amplitude trace.*

wightmanite A colorless triclinic mineral: $Mg_5(BO_3)O(OH)_5 \cdot 2H_2O$.

wiikite A poorly defined mineral high in niobium, tantalum, titanium, and yttrium, found to be a variable mixture of euxenite and obruchevite.

Wilcoxian *Sabinian.*

wild Having very large and often unpredictable amplitudes, e.g. a noisy seismic channel at high *gain.*

wildcat *wildcat well.*

wildcat well An *exploratory well* drilled for oil or gas on a geologic feature not yet proven to be productive, or in an unproven territory, or to a zone that has never produced or is not known to be productive in the general area. Cf: *outpost well; deeper-pool test; shallower-pool test.* Syn: *wildcat.*

Wilderness North American stage: Middle Ordovician (above Porterfield, below Barneveld; it includes uppermost Black River and Rockland rocks) (Cooper, 1956). See also: *Blackriverian.*

wilderness An area or tract of land that is uncultivated and uninhabited by man.

wilderness area An area set aside by government for preservation of natural conditions for scientific or recreational purposes. See also: *natural area.*

wildflysch A type of *flysch* facies representing a mappable stratigraphic unit displaying large and irregularly sorted blocks and boulders resulting from tectonic fragmentation, and twisted, contorted, and confused beds resulting from slumping or sliding under the influence of gravity. The term was first applied by Kaufmann (1886) in the Alps.

wild land Uncultivated land, or land that is unfit for cultivation; e.g. a wasteland or a desert.

wild river (a) A river whose shores and waters remain essentially in a virgin condition, unmodified by man. (b) A torrential river.

wilkeite A rose-red or yellow mineral of the apatite group, containing hydroxyl, in which the phosphate is partly replaced by carbonate, sulfate, or silicate: $Ca_5(SiO_4,PO_4,SO_4)_3(O,OH,F)$.

wilkmanite A mineral: Ni_3Se_4.

willemite A rhombohedral mineral: Zn_2SiO_4. It is a minor ore of zinc and commonly contains manganese. Willemite varies in color from white or greenish yellow to green, reddish, and brown; it exhibits an intense bright-yellow fluorescence in ultraviolet light.

willemseite A talclike mineral: $(Ni,Mg)_3Si_4O_{10}(OH)_2$.

williamsite A massive yellow to green, impure variety of antigorite resembling jade in appearance and used for decorative purposes. It usually contains specks of chromite.

Williston's law The theory that evolution tends to reduce the number of similar parts in organisms and render them progressively more different from each other.

Willow Lake layering A type of *rhythmic layering* in which the layers are composed of elongate mineral grains oriented with their long axes roughly perpendicular to the layering (Taubeneck & Poldervaart, 1960, p. 1295).

willyamite A pseudocubic mineral: (Co,NI)SbS, with Co greater than Ni.

Wilson cycle The opening and closing of ocean basins through plate-tectonic processes. Named for J.T. Wilson. Cf: *geosyncline.*

wilsonite [mineral] A purplish-red material consisting of an aluminosilicate of magnesium and potassium, and representing an altered scapolite.

wilsonite [pyroclast] A tuff composed of fragments of pumice and andesite in a matrix of vitric and granular material (Holmes, 1928, p. 240).

Wilson technique In X-ray diffraction analysis, a method of mounting film in a cylindrical X-ray powder camera which enlarges the area for recording back-reflection diffractions on both sides of the entry port. Cf: *Straumanis camera method.*

wilting coefficient *wilting point.*

wilting percentage *wilting point.*

wilting point The point at which the water content of the soil becomes too low to prevent the *permanent wilting* of plants. As originally introduced, and to a certain extent today, the point at which a soil-water deficiency produces any degree of wilting is the wilting point. Syn: *wilting coefficient; wilting percentage.* Cf: *temporary wilting.*

wiluite (a) A green variety of grossular garnet. (b) A greenish variety of vesuvianite.

wind (a) Naturally moving air, of any direction or velocity. (b) More specifically, a meteorologic term for that component of air that moves parallel to the Earth's surface. Its direction and velocity can be measured.

wind abrasion A process of erosion in which windblown particles of rock material or snow scour and wear away exposed surfaces of any kind. Syn: *wind corrasion.*

wind corrasion *wind abrasion.*

wind crust A type of *snow crust* formed by the packing of previously fallen snow into a hard layer by wind action. Cf: *wind slab.*

wind current *drift [oceanog].*

wind-cut stone *ventifact.*

wind-deposition coast A coast built out into the sea by sand dunes advancing in the direction of the prevailing winds; generally found on the lee side of a sandy neck of land.

wind direction *direction of the wind.*

wind drift (a) *drift* [oceanog]. (b) "That portion of the total vector drift of sea ice from which the effects of the current have been subtracted" (Baker et al., 1966, p. 183). (c) The average direction of the wind over a period of time.

wind-driven current *drift [oceanog].*

wind erosion Detachment, transportation, and deposition of loose topsoil by wind action, esp. in duststorms in arid or semiarid regions or where a protective mat of vegetation is inadequate or has been removed. See also: *deflation.*

wind-faceted stone *windkanter.*

wind gap (a) A shallow notch in the crest or upper part of a mountain ridge, usually at a higher level than a water gap. (b) A former *water gap,* now abandoned (as by piracy) by the stream that formed it; a pass that is not occupied by a stream. Syn: *dry gap; air gap; wind valley.*

wind-grooved stone *ventifact.*

windkanter A *ventifact,* usually highly polished, bounded by one or more smooth faces or facets, curved or nearly flat, that intersect in one or more sharp edges or angles. The faces may be cut at different times, as when the wind changes seasonally or the pebble is undermined and turned over on its flattened face. Etymol: German *Windkanter,* "one having wind edges". See also: *einkanter; zweikanter; dreikanter; parallelkanter.* Syn: *faceted pebble; wind-faceted stone.*

wind noise (a) Random noise attributed principally to ground unrest caused by the wind moving plants and trees and shaking their roots. (b) Seismic background noise (in the absence of a shot) regardless of the source.

window [geomorph] The opening under a natural bridge (Gregory, 1917, p. 134).

window [paleont] An opening in the skeleton of an ebridian, such as a "lower window" between the opisthoclades, a "middle window" between the mesoclades and the actines, and an "upper window" between the proclades.

window [river] A part of a river surrounded by river ice, remaining unfrozen during all or part of the winter, and caused by local inflow of warm water or by turbulence or a strong current.

window [tect] An eroded area of a thrust sheet that displays the rocks beneath the thrust sheet. Syn: *fenêtre; fenster.*

wind packing The compaction of snow by wind action.

wind polish *desert polish.*

wind-polished stone A *ventifact* having a *desert polish.*

wind ridge A ridge of snow formed by the deposition of blowing snow at right angles to the direction of the prevailing wind. Its lee side is the steeper. Cf: *sastrugi.*

wind-rift dune A sand dune produced in a shrub-covered area by a strong wind of constant direction, typically marked by a gap or "rift" at the very tip or downwind end of a hairpin-shaped sand rim (the hairpin or elongated chevron is opened toward the wind), and extending up to 2 km in length and about 100 m in width (Melton, 1940, p. 129-130); it is usually found along a seashore, rarely on a desert. The term is also applied to the "doublet" of parallel sand ridges resulting from the formation of the "rift". The spelling "windrift" is not recommended owing to possible confusion with "wind-drift".

wind ripple [sed] One of many wavelike, asymmetric undulations produced on a sand surface by the saltatory movement of particles by wind and occasionally found in eolian rocks; it is generally longer and of smaller height than an aqueous ripple mark, but is similar in having a steep lee side (facing downcurrent) and a gentle windward side (facing upcurrent). See also: *sand ripple; granule ripple.* Cf: *antiripple.*

wind ripple [snow] One of a series of wavelike formations on a snow surface, lying at right angles to the wind direction, and formed as snow grains are moved along by the wind.

windrow (a) A low bank, heap, or other accumulation of material, formed naturally by the wind (as a snowdrift) or the tide (as a pile of beach shells), or artificially (as a ridge of construction material along a road). (b) Part of a *slick* that has broken up into a narrower and shorter band or bands at wind speeds greater than about 7 knots, its long axis always oriented along the wind direction.

windrow ridge A term used by Tanner (1960, p.482) for a shallow-water ripple mark that is parallel with and directly beneath a windrow on a water surface, that consists of a straight, tapered ridge becoming narrower and shorter in the downwind direction, and that cuts regularly across a pre-existing ripple mark.

wind scoop A saucerlike depression in the snow near an obstruction (such as a tree or rock), caused by the eddying action of the deflected wind.

wind-scoured basin *deflation basin.*

wind-scoured stone *ventifact.*

wind set-up (a) The vertical rise of the still-water level on the leeward side of a body of water, caused by the force of wind on the surface of the water; the difference between the leeward and windward sides of the form. The term is usually reserved for use on smaller bodies of water. It is a type of *meteorologic tide.* Syn: *wind tide.*

wind shadow The area in the lee of an obstacle, where air motion is not capable of moving material (such as sand in saltation) and thus traps it when it falls; the zone that is gradually filled with sand drift during the formation of a dune, and determines the shape of the dune. Syn: *shadow zone.*

wind-shadow dune A longitudinal *umbracer dune.*

wind-shaped stone *ventifact.*

wind slab A layer of snow that is packed tightly by the wind while being deposited. Cf: *wind crust.* Also spelled: *windslab.*

wind-slab avalanche An avalanche started by the dislodging or slipping of a rigid *wind slab* from the underlying snow. Cf: *slab avalanche; loose-snow avalanche.*

windsorite A light-colored quartz-monzonite *aplite* with a minor amount of biotite. It was named by Daly in 1903 after Windsor, Vermont. Not recommended usage.

wind stress The force per unit area of the wind acting on a water surface to produce waves and currents; its magnitude depends on wind speed, air density, and roughness of water surface.

wind sweep The trough-shaped part of the windward slope of an advancing dune, up which the main wind currents pass.

wind tide *wind set-up.*

wind valley *wind gap.*

windward adj. (a) Said of the side (as of a shore or reef) located toward the direction from which the wind is blowing; facing the wind, such as the "windward slope" of a dune, up which sand moves by saltation. Syn: *weather [meteorol].* (b) Said of a tide moving toward the direction from which the wind is blowing.—n. The part or side (as of a hill or shore) from which the wind is blowing; the side facing the wind. Also, the direction from which the wind is blowing, or the direction opposite to that toward which the wind is blowing. Ant: *leeward.*

windward reef In a *reef complex* or *reef tract,* a reef on the side from which the prevailing winds blow. Windward reefs tend to be better developed than *leeward reefs,* as on Eniwetok Atoll. See also: *outer reef; inner reef.*

wind wave A wind-generated wave; a *sea wave,* or part of *sea.*

wind-worn stone *ventifact.*

wineglass valley A valley resembling in plan view a goblet or champagne glass. It flares broadly open at its upper end, where it has a cup-shaped or funnel-shaped head; narrows sharply to form a constricted lower section; and flares open again on a spreading alluvial fan. The valley commonly forms at right angles to a fault scarp in an arid region. Syn: *goblet valley; hourglass valley.*

wing [geomorph] The forward extending, outer end of a dune; a *horn.*

wing [paleont] A solid or fenestrated extension from the side wall of the shell of a nasselline radiolarian.

wing [palyn] *vesicle [palyn].*

wing bar A sandbar that partly crosses the entrance to a bay or the mouth of a river.

wing dam *pier dam.*

winged headland A headland having spits extending from both sides in opposite directions. It may be produced by waves that are unable to move material to the bayhead. Syn: *winged beheadland.*

Winkler method In oceanography, a chemical method of determining the amount of dissolved oxygen in seawater.

winnowed community An *assemblage [paleoecol]* in which the organisms lived in the same place where their fossils are found, but from which many small-sized forms (both immatures of large species and adults of small species) have been removed by gentle water movements. Cf: *fossil community; disturbed-neighborhood assemblage; transported assemblage; mixed assemblage.* See also: *biocoenosis.* Syn: *residual community.*

winnowing The selective sorting, or removal, of fine particles by wind action, leaving the coarser grains behind. The term is often applied to removal by or sorting in water, but the term *washing* is more appropriate for such a process.

winter balance The change in mass of a glacier from the minimum value at the beginning of a balance year to the following maximum value; sometimes called *apparent accumulation,* or (erroneously) *net accumulation.* Cf: *summer balance.*

winter berm A berm built on the backshore by the uprush of large storm waves during the winter; it is landward of, and somewhat higher than, the *summer berm.* See also: *storm berm.*

winterbourne A regular *bourne* that breaks out every year at the same spot in the floor of a dry valley; specif. one that flows only or chiefly in winter, when the water table rises above the valley floor, as in the chalk regions of southern England.

winter ice A term formerly used for sea ice of less than one winter's growth, and a thickness ranging from 15 cm to 3.7 m (12 ft); it is being replaced by the term *first-year ice.*

winter moraine A minor end moraine formed during glacier readvance in a single winter, either on land or under water (Gravenor & Kupsch, 1959, p. 54).

winter protalus ridge A wall-like *protalus rampart* formed of blocks and boulders derived from cliffs above a snowbank-occupied cirque before the summer heat melts the snow across which the blocks roll.

winter season In glaciology, that period of a year when the *balance* of a glacier increases to the maximum for the year. This is the part of the year when, on the average, accumulation exceeds ablation. Cf: *summer season.* Syn: *accumulation season.*

winze A steeply inclined shaft driven to connect one mine level with a lower level.

wire adj. A syn. of *capillary [mineral];* said of native metals, e.g. wire silver.

wire line A general term for any flexible steel line or *cable [drill]* connecting a surface winch to a tool assembly lowered in a well bore. See also: *sand line [drill].* Also spelled: *wireline.*

wire-line coring Cutting and removing of a core sample (of soft sandstone or shale) with the drill bit still in place and without withdrawing and dismantling the drill pipes, as by raising the core in a retractable core barrel and lowering the same or an alternate barrel into place inside the drill pipe. Obsolescent.

wire-line log *well log.*

wire-line test A procedure for measuring the potential productivity of an oil reservoir by means of a tool lowered into a borehole by a *wire line,* in which a sample of fluid and the formation pressure are obtained. It is faster than a *drill-stem test.*

wire strain gage An instrument consisting of a fine wire used to indicate minute changes in strain by detecting corresponding changes in electrical resistance via elongation of the wire.

Wisconsin *Wisconsinan.*

Wisconsinan Pertaining to the classical fourth glacial stage (and the last definitely ascertained, although there appear to be others) of the Pleistocene Epoch in North America, following the Sangamonian interglacial stage and preceding the Holocene. See also: *Würm.* Also spelled: *Wisconsinian; Wisconsin.*

wiserite A mineral: $Mn_4B_2O_5(OH,Cl)_4$.

witch A nonpreferred syn. of *water witch.*

witching stick *divining rod.*

withamite A red to yellow variety of epidote containing a little manganese and occurring in andesites in Glencoe, Scotland. Cf: *piemontite.*

withdrawal The act of removing water from a source for use; also, the amount removed.

witherite A yellowish-white or grayish-white orthorhombic mineral of the aragonite group: $BaCO_3$.

witness butte *butte témoin.*

witness corner A monumented survey point near a *corner* and usually on a line of the survey, established as a reference mark where the true corner is inaccessible or cannot be monumented or occupied; e.g. a post set near the corner of a mining claim, with the distance and direction of the true corner indicated thereon.

witness mark A physical structure (such as a post, rock, stake, or tree) placed at a known distance and direction from a property corner, instrument, or other survey station, to aid in its recovery and identification; e.g. a blazed tree on the bank of a river, indicating the corner which is at the intersection of some survey line with the center line of the river and therefore cannot be marked directly.

witness point A monumented *station* on a line of survey, used to perpetuate an important location more or less remote from, and without special relation to, any regular corner.

witness rock *zeuge.*

witness tree *bearing tree.*

wittichenite A steel-gray to tin-white mineral: Cu_3BiS_3.

wittite A lead-gray mineral: $Pb_9Bi_{12}(S,Se)_{27}$.

wobbling of the pole An expression sometimes used to describe the period polar motion identified by a Chandler term (approximately a 14-month period) and an annual term. See also: *Chandler wobble.*

wodanite A variety of biotite containing titanium.

wodginite A black mineral: $(Ta,Nb,Sn,Mn,Fe)_{16}O_{32}$.

woebourne A *bourne* that is regarded in some English localities as appearing only when some disaster is about to happen.

woehlerite *wöhlerite.*

wöhlerite (a) A yellow or brown mineral: $NaCa_2(Zr,Nb)Si_2O_8(O,OH,F)$. Cf: *lavenite.* (b) A name for organic matter in carbonaceous chondrites.—Also spelled: *woehlerite.*

wolchonskoite *volkonskoite.*

wold A range of hills produced by differential erosion from inclined sedimentary rocks; a *cuesta.*

Wolfcampian North American series: lowermost Permian (above Virgilian of Pennsylvanian, below Leonardian).

wolfeite A mineral: $(Fe,Mn)_2(PO_4)(OH)$. It is isomorphous with triploidite.

wolfram (a) *wolframite.* (b) The metallic element *tungsten.*

wolframine (a) *tungstite.* (b) *wolframite.*

wolframite (a) A brownish or grayish-black mineral: $(Fe,Mn)WO_4$. It is isomorphous with and intermediate between huebnerite and ferberite, and occurs in monoclinic crystals (commonly twinned so as to imitate orthorhombic tabular forms) and in granular masses or columnar aggregates (as in pneumatolytic veins near granite masses and associated with tin ores). Wolframite is the principal ore of tungsten. See also: *tungstenite.* Syn: *wolfram; wolframine; tobacco jack.* (b) A name applied to an isomorphous mineral series consisting of the end-members huebnerite and ferberite and of wolframite.

wolfram ocher *tungstic ocher.*

wolframoixiolite A mineral: $(Nb,W,Ta,Fe,Mn)_3O_6$.

wolfsbergite *chalcostibite.*

wolgidite A *leucitite* that contains leucite, magnophorite, diopside, and minor amounts of olivine and phlogopite. The name, given by Wade and Prider in 1940, is from the Wolgidee Hills, Western Australia. Not recommended usage.

wollastonite A triclinic mineral of the pyroxenoid group: $CaSiO_3$. It is dimorphous with parawollastonite. Wollastonite is found in contact-metamorphosed limestones, and occurs usually in cleavable masses or sometimes in tabular twinned crystals; it may be white, gray, brown, red, or yellow. It is not a pyroxene. Symbol: Wo. Syn: *tabular spar.*

Wollaston prism In an optical system, a double-image prism consisting of two right-angled calcite prisms that produce two perpendicular beams of plane-polarized light.

wollongite *wollongongite.*

wollongongite A coal-like shale similar to *torbanite.* It is named from its type locality, Wollongong, New South Wales, Australia. Also spelled: *wollongite; wallongite.*

wölsendorfite A red or orange-red orthorhombic mineral: $(Pb,Ca)U_2O_7 \cdot 2H_2O$.

wood Technically, *xylem;* more popularly, the hard, fibrous xylem of trees and shrubs (Fuller & Tippo, 1954, p. 975).

wood agate A term used for *agatized wood,* esp. agate formed by siliceous permineralization of wood.

Woodbinian North American (Gulf Coast) stage: Upper Cretaceous (above Washitan, below Eaglefordian).

wood coal *woody lignite.*

wood copper A fibrous variety of olivenite.

woodendite An extrusive *mugearite* or *trachyandesite* containing olivine and clinopyroxene in an alkalic brown glassy groundmass. The name, given by Skeats and Summers in 1912, is for Woodend, Victoria, Australia. Cf: *macedonite.* Not recommended usage.

woodfordite *ettringite.*

wood hematite A finely radiated variety of hematite exhibiting alternate bands of brown or yellow or varied tints.

woodhouseite A colorless rhombohedral mineral: $CaAl_3(PO_4)(SO_4)(OH)_6$. It is isomorphous with svanbergite, corkite, and hinsdalite.

wood iron ore A fibrous variety of limonite from Cornwall, England.

wood opal A variety of common opal that has filled the cavities in, and replaced the organic matter of, wood and that often preserves the original features of the wood. See also: *opalized wood.* Syn: *xylopal; lithoxyl.*

woodruffite A mineral: $(Zn,Mn^{+2})Mn_3^{+4}O_7 \cdot 1\text{-}2H_2O$.

woodstone *silicified wood.*

wood tin A nodular or reniform brownish variety of cassiterite, having a concentric structure of radiating fibers resembling dry wood in appearance. Syn: *dneprovskite.*

woodwardite A bluish mineral: $Cu_4Al_2(SO_4)(OH)_{12} \cdot 2\text{-}4H_2O$ (?).

woodyard *forest bed.*

woody lignite Lignite that shows the fibrous structures of wood. Cf: *earthy lignite.* Syn: *xyloid lignite; xyloid coal; wood coal; board coal; bituminous wood.*

woody peat *fibrous peat.*

woody plant A vascular, usually perennial, plant with a large development of xylem.

wool An English term for a sandy shale or shaly flagstone with irregular curly bands or bedding.

Worden gravimeter A compact temperature-compensated gravity meter, in which a system is held in unstable equilibrium about an axis, so that an increase in the gravitational pull on a mass at the end of a weight arm causes a rotation opposed by a sensitive spring. The meter weighs 5 pounds and has a sensitivity of less than 0.1 mgal.

work v. To undergo gradual movement, such as heaving, sliding, or sinking; said of rock materials.

workings The system of openings or excavations made in mining or quarrying; esp. the area where the ore is actually mined.

World Data Centers Centers for the collection, exchange, and general availability of data from various geophysical disciplines, e.g. solid-earth geophysics, solar-terrestrial geophysics, oceanography, glaciology, meteorology, tsunamis. They were originally established for the International Geophysical Year, but are being continued under the auspices of the International Council of Scientific Unions (ICSU).

world geodetic system Any system that connects the major continental geodetic datums and land masses into a unified Earth-centered network.

world point A term proposed by Kobayashi (1944, p.745) for a single restricted outcrop regarded as representative of a geologic province or part of the world. Rarely used.

world rift system A major tectonic element of the Earth, consisting of midoceanic ridges and their associated rift valleys, such as those along the Mid-Atlantic Ridge. It is believed to be the locus of extensional splitting and upwelling of magma that has resulted in *sea-floor spreading.* Cf: *rift [tect].*

world time *Geologic time* as indicated by the life range of a single cosmopolitan fossil species (Kobayashi, 1944, p. 745).

worm boring *worm tube.*

worm cast (a) A sinuous fossil trail of a worm, preserved as a sand cast on the bedding plane of an arenaceous rock. (b) *worm casting.* —Also spelled: *wormcast.*

worm casting A cylindrical mass of earth or mud excreted by an earthworm. Syn: *erpoglyph; worm cast.*

worm reef A small *organic reef* built by wormlike organisms. Not recommended usage. Cf: *sabellariid reef; serpulid reef; vermetid reef.*

worm's-eye map (a) A term applied to what is more formally known as a *supercrop map* or a *lap-out map,* in reference to the pattern of formations that would be visible to an observer looking upward at the bottom of the rocks overlying a given surface. (b) A map showing overlap of sediments, or of progressive transgressions of a sea over a given surface.

worm trail A marking in a fossiliferous rock, formed by the passage of an ancient worm or worms.

worm tube (a) A fossilized tubular structure built by a marine worm and preserved in the top of a bed that was exposed for some time as the sea floor; e.g. a *scolithus.* Syn: *worm boring.* (b) A membranous tube, usually of calcium carbonate or particles of mud or sand, built on a submerged surface by a marine worm.

worobieffite *vorobievite.*

wrench fault A *lateral fault* in which the fault surface is more or less vertical. Syn: *basculating fault; torsion fault.*

Wright biquartz wedge *biquartz plate.*

wrinkle layer Pattern of irregularly transverse, very fine ridges or wrinkles on the inside of the shell, especially the body chamber, of many cephalopod conchs. Syn: *wrinkled layer; runzelschicht.*

wrinkle mark An irregular small ripplelike feature, composed of ridges 0.5 to 1 mm thick and a few millimeters long, parallel to each other or in a curved or honeycomb pattern. Described (as *Runzelmarken*) by Häntzschel and Reineck in 1968 and named wrinkle mark by Teichert (1970).

wrinkle ridge A sinuous, irregular, segmented, apparently smooth elevation occurring within the borders of a mare region of the Moon's surface and characterized by dikelike outcrops, crest-top craters, and longitudinal rifts. Wrinkle ridges are up to 35 km wide and 100 m high, and may extend for hundreds of kilometers. They probably originated in fissure eruptions or from volcanic activity along fractures. Syn: *mare ridge.*

wrist *carpus.*

wroewolfeite A monoclinic mineral: $Cu_4(SO_4)(OH)_6 \cdot 2H_2O$.

W-shaped valley A valley having an inverted and faintly pan-shaped cross profile suggesting the form of the letter "W", such as the valley of a river having the highest parts of its flood plain immediately near both banks (Lane, 1923).

wulfenite A yellow, orange, or red (sometimes grayish or green) tetragonal mineral: $PbMoO_4$. It is isomorphous with stolzite. Wulfenite occurs in tabular crystals and in granular masses, and is an ore mineral of molybdenum. Syn: *yellow lead ore.*

Wulff net (a) A coordinate system used in crystallography to plot a *polar stereographic projection* with conservation of equal angles, such as for plotting angular relations obtained from universal-stage measurements. (b) *stereonet.* —Named after Georgi Viktorovich Vulf (1862-1925), Russian crystallographer who introduced the net and whose name was transliterated into German.

Würm (a) European stage: uppermost Pleistocene (above Riss, below Holocene). (b) The classical fourth glacial stage of the Pleistocene Epoch in the Alps, after the Riss-Würm interglacial stage. See also: *Wisconsinan; Weichsel.* —Etymol: Würm, a lake in Germany. Adj: *Würmian.*

wurtzilite A black, massive asphaltic pyrobitumen, sectile and infusible, that is closely related to *uintahite,* but insoluble in turpentine, and derived from the metamorphism of petroleum. It is found in veins in Uinta County in Utah.

wurtzite A brownish-black hexagonal mineral: $(Zn,Fe)S$. It is dimorphous with sphalerite. Wurtzite occurs in hemimorphic pyramidal crystals, or in radiating needles and bundles within lamellar sphalerite. Many polymorphs with slight variants on the wurtzite structure are known, and separate names proposed for some of these are "superfluous and not generally accepted" (Hey, 1962, 3.4.3).

wüstite A mineral: FeO. Artificially prepared specimens are characteristically deficient in iron. Also spelled: *wustite.* Syn: *iozite.*

W wave An archaic term for a surface wave that returns through the antipode of the epicenter.

wyartite A violet-black secondary mineral: $Ca_3U^{+4}(UO_2)_6(CO_3)_2(OH)_{18} \cdot 3\text{-}5H_2O$. It was erroneously called *ianthinite.*

wych *wich.*

wye level A leveling instrument having a removable telescope, with attached spirit level, supported in Y-shaped rests, in which it may be rotated about its longitudinal (or collimation) axis, and from which it may be lifted and reversed, end for end, for testing and adjustment. Cf: *dumpy level.* Syn: *Y level.*

wyllieite A monoclinic mineral: $(Na,Ca,Mn)_2(Fe^{+2},Mg)_2Al(PO_4)_3$.

Wyllie time-average equation The empirical formula proposed by Wyllie (1957) that linearly relates interval transit time (reciprocal velocity) from a *sonic log* (Δt) to porosity (ϕ) of granular rocks in terms of the travel time of the solid rock (Δt_m) and of the interstitial liquid (Δt_f): $\Delta t = \phi\ \Delta t_f + \Delta t_m\ (1\text{-}\phi)$.

wyomingite A hypabyssal *lamproite* containing phlogopite phenocrysts in a fine-grained groundmass of leucite and diopside; a phlogopite-leucite *phonolite.* Its name, given by Cross in 1897, is derived from the state of Wyoming, U.S.A. Not recommended usage.

X In seismic prospecting, the distance from the shot point to the center of the spread, or to any particular geophone.

x-acline B twin law A complex twin law in feldspars with twin axis parallel to (010) and composition plane (100).

xalostocite *landerite.*

xanthiosite A yellow mineral: $Ni_3(AsO_4)_2$.

xanthite A yellowish to yellowish-brown variety of vesuvianite.

xanthochroite *greenockite.*

xanthoconite A brilliant-red, orange-yellow, or brown monoclinic mineral: Ag_3AsS_3. Cf: *proustite.*

xanthophyllite *clintonite.*

xanthosiderite *goethite.*

xanthoxenite A wax-yellow monoclinic mineral: $Ca_4Fe_2(PO_4)_4(OH)_2 \cdot 2H_2O$.

x-Carlsbad twin law A complex twin law in feldspar, having a twin axis at right angles to [001] and a composition plane (100). It is supposedly a true interpretation of the *acline-B twin law.*

xenoblast A mineral of low *form energy* that has grown during metamorphism without development of its characteristic crystal faces. It is a type of *crystalloblast.* The term was originated by Becke (1903). Cf: *idioblast; hypidioblast.* Syn: *allotrioblast.*

xenocryst A crystal that resembles a phenocryst in igneous rock but is foreign to the body of rock in which it occurs. See also: *disomatic.* Syn: *chadacryst.*

xenogenous A little-used syn. of *epigenetic [ore dep].* Cf: *idiogenous; hysterogenous.*

xenoikic A term, now obsolete, suggested by Cross et al. (1906, p. 704) for a poikilitic texture in which the ratio between enclosing crystals (*oikocrysts*) and enclosed crystals (*chadacrysts*) is less than 5 to 3 but greater than 3 to 5.

xenolith A foreign inclusion in an igneous rock. Cf: *autolith.* Syn: *exogenous inclusion; accidental inclusion.*

xenology The dating of early events in the chronology of the planetary system on the basis of excess xenon-129 in meteorites. Xenon-129 is a decay product of radioactive iodine-129, which has a half-life of approximately 17 million years.

xenomorphic (a) Said of the holocrystalline texture of an igneous or metamorphic rock, characterized by crystals not bounded by their own *rational faces* but with their forms impressed upon them by adjacent mineral grains. Also said of a rock with such a texture. The term *xenomorphisch* was proposed by Rohrbach (1885, p. 17-18) originally to describe in an igneous rock the individual mineral crystals (now known as *anhedral* crystals) whose mutual growths have prevented the assumption of *rational faces.* Current American usage tends to apply *xenomorphic* to a texture characterized by such crystals. (b) A syn. of *anhedral* in European usage.—Syn: *allotriomorphic; anidiomorphic; leptomorphic.* Cf: *automorphic; subautomorphic.*

xenomorphic-granular *xenomorphic.*

xenomorphism The state or condition of special sculpture at the umbonal region of the unattached valve of a bivalve mollusk, resembling the configuration of the substratum onto which the attached valve is or was originally fixed. It is known on the left valves in the Anomiidae and on the right valves in the Gryphaeidae and Ostreidae. Erroneous syn: *allomorphism.*

xenon-xenon age method A method of dating uranium-bearing minerals by means of mass spectrometric determination of xenon isotopic ratios. It is based on the compositional differences between xenon produced by spontaneous fission of uranium-238 in nature and xenon produced from uranium-235 fission induced by thermal neutrons in a nuclear reactor. Xenon is extracted incrementally, and an age spectrum produced. The method appears to be useful for rocks older than about 100 million years.

xenothermal Said of a hydrothermal mineral deposit formed at high temperature but shallow depth; also, said of that environment. Cf: *telethermal; epithermal; mesothermal; hypothermal; leptothermal.*

xenotime A brown, yellow, or reddish tetragonal mineral: YPO_4. It is isostructural with zircon, and often contains erbium, cerium, and other rare earths, as well as thorium, uranium, aluminum, calcium, beryllium, zirconium, or other elements. Xenotime occurs as an accessory mineral in granites and pegmatites.

xenotopic Said of the fabric of a crystalline sedimentary rock in which the majority of the constituent crystals are anhedral. Also, said of an evaporite, a chemically deposited cement, or a recrystallized limestone or dolomite with such a fabric. The term was proposed by Friedman (1965, p.648). Cf: *idiotopic; hypidiotopic.*

Xeralf In U.S. Dept. of Agriculture soil taxonomy, a suborder of the soil order *Alfisol,* characterized by formation in a xeric moisture regime and a mesic or thermic temperature regime. Agriculture is limited without irrigation, but with irrigation a wide variety of crops can be grown (USDA, 1975). Cf: *Aqualf; Boralf; Udalf; Ustalf.*

xerarch adj. Said of an ecologic succession (i.e. a *sere*) that develops under *xeric* conditions. Cf: *mesarch; hydrarch.* See also: *xerosere.*

Xerert In U.S. Dept. of Agriculture soil taxonomy, a suborder of the soil order *Vertisol,* characterized by formation in a xeric moisture regime and by a once-yearly opening and closing of its wide, open cracks. In most years the cracks are open more than 60 days in the summer and closed more than 60 days in the winter (USDA, 1975). Cf: *Torrert; Udert; Ustert.*

xeric Said of a habitat characterized by a low or inadequate supply of moisture; also, said of an organism or group of organisms existing in such a habitat. Cf: *mesic; hydric.* See also: *xerarch.*

xeric moisture regime A soil moisture regime that is characteristic of the cool, moist winter and warm, dry summer of a Mediterranean climate. The soil is dry for at least 45 consecutive days in the summer and is moist for at least 45 consecutive days in the winter. Winter rainfall often penetrates deeply into the soil and is particularly effective for leaching (USDA, 1975).

xerochore A climatic term for the part of the Earth's surface represented by waterless deserts.

xerocole *xerophilous.*

Xeroll In U.S. Dept. of Agriculture soil taxonomy, a suborder of the soil order *Mollisol,* characterized by formation in a xeric soil moisture regime. They develop under a savanna of annual or perennial grasses or under bunch grasses and shrubs. Most Xerolls have a relatively thick mollic epipedon, a cambic or argillic horizon, and a subsoil accumulation of lime. Some also have a duripan. Without irrigation they are used for wheat and grazing; with irrigation they support a wide variety of crops (USDA, 1975). Cf: *Alboll; Aquoll; Boroll; Rendoll; Udoll; Ustoll.*

xeromorphic (a) Said of a plant characterized by the morphology of a *xerophyte.* (b) Said of conditions favorable for the growth of *xerophilous* organisms.

xerophile n. *xerophyte.* adj. *xerophilous.*

xerophilous Said of an organism adapted to dry conditions. Syn: *xerophile; xerocole.* Cf: *xerophobous.*

xerophobous Said of a plant that cannot tolerate dry conditions. Cf: *xerophilous.*

xerophyte A plant adapted to dry conditions; a desert plant. Cf: *xerotherm; hydrophyte; mesophyte.* Syn: *eremophyte; xerophile.*

xerophytization Adaptation, esp. in the development of a species, to conditions of low moisture supply, i.e. to xeric conditions.

xerosere A *sere* that develops under extremely dry (i.e. xeric) conditions; a *xerarch* sere. Cf: *hydrosere; mesosere.*

xerotherm A plant adapted to hot dry conditions. Cf: *xerophyte.*

Xerothermic n. A term used to designate a postglacial interval of warmer and drier climate. See also: *Altithermal; Hypsithermal.*

xerothermic Said of a hot, dry climate; also, pertaining to the climate of the Xerothermic postglacial interval.

Xerult In U.S. Dept. of Agriculture soil taxonomy, a suborder of the soil order *Ultisol,* characterized by formation in a xeric soil moisture regime and a mesic or thermic temperature regime. Xerults have low to moderate organic-carbon content and an ochric epipedon that rests directly on a reddish or brownish argillic horizon. Most Xerults in the U.S. occur in mountainous areas of the Pacific states (USDA, 1975). Cf: *Aquult; Humult; Udult; Ustult.*

xingzhongite A cubic mineral: (Ir,Cu,Rh)S.

xiphosuran Any *merostome* belonging to the subclass Xiphosura, characterized by a trilobate dorsal shield. Horseshoe crabs are included in this group. Cf: *eurypterid.* Range, Cambrian to present.

xocomecatlite An orthorhombic mineral: $Cu_3(TeO_4)(OH)_4$.

xonotlite A pale-pink, white, or gray mineral: $Ca_6Si_6O_{17}(OH)_2$.

x-pericline twin law A complex twin law in feldspar, now considered to be equivalent to the *Carlsbad B twin law.*

X-ray Non-nuclear electromagnetic radiation of very short wavelength, in the interval of 0.1-100 angstroms (10^{-11}-10^{-8}m), i.e. between that of gamma rays and ultraviolet radiation. Also spelled: *x-ray.*

X-ray diffraction The *diffraction* of a beam of X-rays, usually by the three-dimensional periodic array of atoms in a crystal that has periodic repeat distances (lattice dimensions) of the same order of magnitude as the wavelength of the X-rays. See also: *electron diffraction.*

X-ray diffraction pattern The characteristic interference pattern obtained when X-rays are diffracted by a crystalline substance. The geometry of the pattern is a function of the repeat distances (lattice dimensions) of the periodic array of atoms in the crystal; the intensities of the diffracted beams give information about the atomic arrangement and unit-cell dimensions. See also: *electron diffraction pattern.*

X-ray emission spectroscopy The qualitative study of a substance by exciting its characteristic X-ray spectrum and measuring the wavelengths present. See also: *X-ray fluorescence spectroscopy.*

X-ray fluorescence spectroscopy A type of *X-ray emission spectroscopy* in which the characteristic X-ray spectrum of a substance is produced by using X-rays of short wavelength to induce the substance to emit X-rays of a longer wavelength. Abbrev: XRF.

X-ray photo-electron spectrometry The study of the photo-electron spectra produced by characteristic X-ray excitation. The energy carried by the emitted electrons is the difference between the energy of the exciting photon and the binding energy of the bound electron.

X-ray powder diffraction Diffraction of a beam of X-rays by a powdered crystalline sample. The random orientation of the tiny crystals comprising the powder enables one to record all possible diffracted beams simultaneously without rotating the sample.

X-ray powder pattern The pattern obtained in *X-ray powder diffraction.* See also: *X-ray diffraction pattern.*

X-ray scattering The phenomenon of changes in direction of X-ray transmission by interaction of the waves with objects or with the transmitting medium due to reflection, refraction, or diffraction.

X-ray spectrograph An instrument for producing, recording, and analyzing an *X-ray spectrum* by reflecting X-rays from a given sample, measuring the angle of diffraction, and thence determining the wavelengths of the X-rays. Sometimes called an *X-ray spectrometer.*

X-ray spectrometer *X-ray spectrograph.*

X-ray spectroscopy The observation of an *X-ray spectrum* and all processes of recording and measuring that go with it.

X-ray spectrum The spectrum of X-rays emitted when a substance is bombarded with energetic particles or radiation. It consists of a "characteristic spectrum", resulting from specific electronic transitions within the atoms of the substance, superimposed on a "continuous spectrum" resulting from inelastic collisions of particles or incoherent scattering of the exciting radiation. The X-ray spectrum may be excited by electrons (cathode rays) as in standard X-ray tubes and *electron microprobes,* by other particles such as protons as in particle-induced X-ray emission, or by higher energy X-rays or gamma rays as in X-ray fluorescence.

XRF *X-ray fluorescence spectroscopy.*

x twin law A normal twin law in feldspar.

xylain A kind of *provitrain* in which the cellular structure is derived from woody material. Cf: *periblain; suberain.*

xylem In vascular plants, a complex tissue that is involved in water conduction, food storage, and strengthening. Types of cells commonly found in xylem include tracheids (for conduction and strengthening), vessels (for conduction), fibers (for strengthening), and parenchyma (for storage). Syn: *wood.*

xylinite A variety of *provitrinite* characteristic of xylain and consisting of xylem or lignified tissue. Cf: *suberinite; periblinite; telinite.*

xylinoid Vitrinite that occurs in noncaking subbituminous coals and lignite and that has a reflectance of less than 0.5% (ASTM, 1970, p.466). Cf: *vitrinoid; anthrinoid.*

xylith A type of lignite that is composed almost entirely of anthraxylon (Parks, 1951, p.30).

xyloid coal *woody lignite.*

xyloid lignite *woody lignite.*

xylopal *wood opal.*

xylotile A delicately fibrous mineral, approximately: $(Mg,Fe^{+2})_3 Fe_2^{+3}Si_7O_{20} \cdot 10H_2O$. It is a serpentine mineral derived from alteration of asbestos or chrysolite.

xylovitrain *euvitrain.*

X^2 - T^2 analysis A method of determining seismic velocity from *normal moveout* measurements.

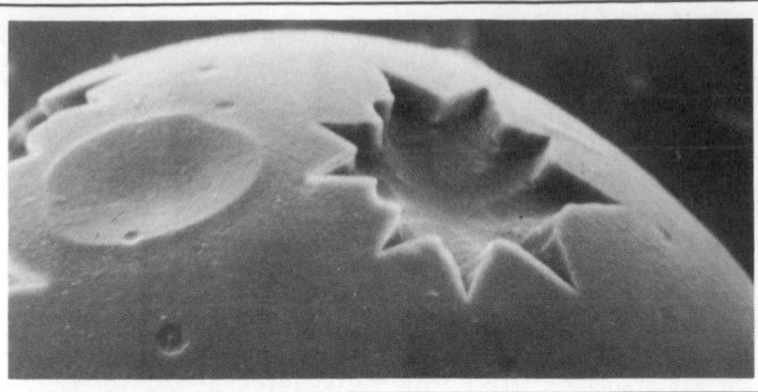

yagiite A mineral of the osumilite group: $(Na,K)_3Mg_4(Al,Mg)_6(Si,Al)_{24}O_{60}$.

yaila A term used in central Kurdistan (of eastern Turkey) for a small, grassy upland plain.

yakatagite A name proposed by Miller (1953, p.26) for a "conglomeratic sandy mudstone" from Yakataga, SE Alaska. It is a poorly indurated tillitelike glaciomarine sedimentary rock containing angular gravel-sized fragments.

yama A steep or perpendicular shaft that leads into a cave (Cvijić, 1924). See also: *lapiés*.

yamaskite A medium- to fine-grained *pyroxenite* containing hornblende, titanaugite, a small amount of anorthite, and accessory biotite and iron oxides; an amphibole-bearing *jacupirangite*. The name, given by Young in 1906, is from Mount Yamaska, Quebec. Not recommended usage.

yamatoite Hypothetical end-member of the garnet group: $Mn_3V_2(SiO_4)_3$.

yardang (a) A long, irregular, sharp-crested, undercut ridge between two round-bottomed troughs, carved on a plateau or unsheltered plain in a desert region by wind erosion, and consisting of soft but coherent deposits (such as clayey sand); it lies in the direction of the dominant wind, and may be up to 6 m high and 40 m wide. Syn: *yarding; jardang*. (b) A landscape form produced in a region of limestone or sandstone by infrequent rains combined with wind action, and characterized by "a surface bristling with a fine and compact lacework of sharp ridges pitted by corrosion" (Stone, 1967, p. 254).—Etymol: ablative of Turki *yar*, "steep bank".

yardang trough A long, shallow, round-bottomed groove, furrow, trough, or corridor excavated in the desert floor by wind abrasion, and separating two yardangs.

yardarm carina One of the oppositely placed *carinae* of a rugose coral that give cross sections of a septum the appearance of yardarms along a mast. Cf: *zigzag carina*.

yarding *yardang*.

Yarmouth Pertaining to the classical second interglacial stage of the Pleistocene Epoch in North America, after the Kansan glacial stage and before the Illinoian. Etymol: Yarmouth, a town in Iowa. See also: *Mindel-Riss*. Syn: *Yarmouthian*.

yaroslavite A mineral: $Ca_3Al_2F_{10}(OH)_2 \cdot H_2O$.

yatalite A pegmatitic rock composed chiefly of amphibole (replacing clinopyroxene), albite, magnetite, sphene, and some quartz. Benson in 1909 proposed the name, derived from Yatala, South Australia. Not recommended usage.

Yavapai A provincial series of the Precambrian in Arizona.

yavapaiite A monoclinic mineral: $KFe(SO_4)_2$.

yazoo (a) *yazoo stream*. (b) *deferred junction*.

yazoo stream A tributary that flows parallel to the main stream for a considerable distance before joining it at a *deferred junction;* esp. such a stream forced to flow along the base of a natural levee formed by the main stream. Type example: Yazoo River in western Mississippi, joining the Mississippi River at Vicksburg. Also spelled: *Yazoo stream*. Syn: *yazoo; Yazoo-type tributary; deferred tributary*.

yeatmanite A brown mineral: $(Mn,Zn)_{16}Sb_2Si_4O_{29}$.

yedlinite A trigonal mineral: $Pb_6CrCl_6(O,OH)_8$.

yellow arsenic *orpiment*.

yellow coal *tasmanite [coal]*.

yellow copperas *copiapite*.

yellow copper ore *chalcopyrite*.

yellow earth [mineral] Impure *yellow ocher*.

yellow earth [sed] Loess of northern China.

yellow-green algae A group of algae corresponding to the division Chrysophyta, that owes its yellowish green to golden brown color to chromatophores of that range of pigmentation. Such algae usually have a cell wall composed of overlapping segments. They are both unicellular and filamentous in habit and are most common in fresh water. Cf: *blue-green algae; green algae; brown algae; red algae*.

yellow ground Oxidized *kimberlite* of yellowish color found at the surface of diamond pipes (e.g. South Africa), above the zone of *blue ground*.

yellow lead ore *wulfenite*.

yellow ocher (a) A mixture of limonite usually with clay and silica, used as a pigment. See also: *yellow earth [mineral]*. Syn: *sil*. (b) A soft, earthy, yellow variety of limonite or of goethite.

yellow ore A yellow-colored ore mineral; specif. carnotite and chalcopyrite.

Yellow Podzolic soil A great soil group in the 1938 classification that was reclassified as *Red-Yellow Podzolic soil* in the 1949 revision (Thorp and Smith, 1949).

yellow pyrites *chalcopyrite*.

yellow quartz *citrine*.

yellow snow Snow tinted yellow by algae such as *Raphidonema* and *Scotiella*. Cf: *green snow; red snow*.

yellow substance A portion of the dissolved organic matter in seawater; commonly carbohydrate-humic acids.

yellow tellurium *sylvanite*.

yenite *ilvaite*.

yentnite A coarse-grained granitic rock originally thought to contain scapolite, plagioclase, and biotite. The "scapolite" was later discovered to be quartz, and the name was withdrawn (Johannsen, 1939, p. 288).

Yeovilian Stage in Great Britain: uppermost Lower Jurassic (above Whitbian, below Aalenian).

yield [exp struc geol] v. To undergo permanent deformation as a result of applied stress.

yield [lake] n. (a) The amount of water that can be taken continuously from a lake for any economic purpose. (b) The amount of organic matter (plant and animal) produced by a lake, either naturally or under management. See also: *production*.

yield point *yield stress*.

yield strength A syn. of *yield stress;* the stress at which a material begins to undergo permanent deformation.

yield stress The differential stress at which permanent deformation first occurs in a material. Syn: *yield point; yield strength; threshold pressure*.

yixunite A cubic mineral: PtIn.

Y level *wye level*.

Y-mark A trilete *laesura* on embryophytic spores and some pollen, consisting of a three-pronged mark somewhat resembling an upper-case "Y". It is commonly also a commissure or suture along which the spore germinates. The term is also applied to analogous marks, which are not laesurae, on pollen grains.

Ynezian North American provincial stage: Lower Paleocene (above Upper Cretaceous, below Bulitian).

yoderite A purple mineral: $(Mg,Al)_8Si_4(O,OH)_{20}$.

yofortierite A monoclinic mineral: $(Mn,Mg)_5Si_8O_{20}(OH)_2 \cdot 8\text{-}9H_2O$.

yogoite An obsolete term originally applied to a *monzonite* that contains approximately equal amounts of alkali feldspar, plagioclase, and clinopyroxene. Weed and Pirsson in 1895 derived the name from Yogo Peak, Little Belt Mountains, Montana.

yoked basin *zeugogeosyncline*.

yoke-pass *joch*.

yosemite A portion of a glacial valley, esp. in the Sierra Nevada of California, that is deeply U-shaped, with sheer walls, hanging troughs, and a wide almost level floor, and hence resembles the Yosemite Valley, Calif.

yoshimuraite An orange-brown mineral: $(Ba,Sr)_2TiMn_2(SiO_4)_2(PO_4,SO_4)(OH,Cl)$.

young [geomorph] Pertaining to the stage of *youth* of the cycle of erosion; esp. said of a stream that has not developed a profile of equilibrium, and of its valley. Syn: *youthful.*

young [struc geol] v. To *face,* in the sense "to present the younger aspect" of one formation toward another; e.g. if formation A "youngs" toward formation B, then B is younger than A unless some fold, fault, unconformity, or intrusion intervenes. Term coined by Bailey (1934, p. 469) and used "as a record of observation and not of stratigraphic deduction".

young coastal ice Sea ice in the initial stage of fast-ice formation, consisting of *nilas* or young ice of local origin, and having a width ranging from a few meters to 100-200 m from the shoreline.

Younger Dryas n. A term used primarily in Europe for an interval of late-glacial time (centered about 10,500 years ago) following the Allerød and preceding the Preboreal, during which the climate, as inferred from stratigraphic and pollen data (Iversen, 1954), deteriorated favoring either expansion or retarded retreat of the waning continental and alpine glaciers.—adj. Pertaining to the late-glacial Younger Dryas interval and to its climate, deposits, biota, and events.

young ice Newly formed, floating, flat sea ice in the transition stage between *nilas* and *first-year ice;* it is 10-30 cm thick. Includes: *gray ice; gray-white ice.* Syn: *slud [ice]; fresh ice.*

younging A colloquial syn. of *facing [struc geol].*

young lake A lake developed during the stage of *youth.* See also: *aging.*

youngland The land surface, with its plateaus and valleys, of the youthful stage of the cycle of erosion (Maxson & Anderson, 1935, p. 90).

young mountain A mountain that was formed during the Tertiary or Quaternary periods, esp. a *fold mountain* produced during the last great period of folding (i.e. the Alpine orogeny). Ant: *old mountain.*

young polar ice *second-year ice.*

Young's modulus A *modulus of elasticity* in tension or compression, involving a change of length. See also: *elastic compliance.* Syn: *stretch modulus.*

young stream A stream developed during the stage of *youth.*

youth [coast] A stage in the development of a shore, shoreline, or coast characterized by an ungraded profile of equilibrium. For a shoreline of submergence: an irregular or crenulate outline, vigorous wave action, formation of sea cliffs and associated erosional forms, a steep offshore profile, and the presence of bays, promontories, offshore islands, spits, bars, and other minor irregularities. For a shoreline of emergence: a usually straight and simple outline, larger waves breaking well offshore, smaller waves coming to land to produce a nip or low cliff, and the formation of barrier beaches, lagoons, and marshes. See also: *primary [coast].*

youth [streams] The first stage in the development of a stream, at which it has just entered upon its work of erosion and is increasing in vigor and efficiency, being able everywhere to erode its channel and having not reached a graded condition. It is characterized by: an ability to carry a load greater than the load it is actually carrying; active and rapid downcutting, forming a deep, narrow V-shaped valley (gorge or canyon) with a steep and irregular gradient and rocky outcrops; numerous waterfalls, rapids, and lakes; a swift current and clear water; a few short, straight tributaries; an absence of flood plains as the stream occupies all or nearly all of the valley floor; and an ungraded bed.

youth [topog] The first stage of the cycle of erosion in the topographic development of a landscape or region, in which the original surface or structure is still the dominant feature of the relief and the landforms are being accentuated or are tending toward complexity. It is characterized by a few small widely spaced young streams; broad, flat-topped interstream divides and upland surfaces, little modified by erosion; poorly integrated drainage systems, with numerous swamps and shallow lakes; and rapid and progressive increase of local relief, with sharp landforms, steep and irregular slopes, and a surface well above sea level. Cf: *infancy.* Syn: *topographic youth.*

youthful Pertaining to the stage of *youth* of the cycle of erosion; esp. said of a topography or region, and of its landforms (such as a plain or plateau), having undergone little erosion or being in an early stage of development. Cf: *infantile.* Syn: *young; juvenile [geomorph].*

Ypresian European stage: lowermost Eocene (above Thanetian of Paleocene, below Lutetian).

Y-shaped valley A valley having a cross profile suggesting the form of the letter "Y", such as a rejuvenated valley in which the grade of the river has recently been increased by uplift of the headwaters (Lane, 1923).

Y-tombolo A tombolo consisting of two embankments that extend shoreward from an island or seaward from the mainland and that unite "to form a single ridge before the connection is completed" (Johnson, 1919, p. 315); there is a body of water between the prongs of the "Y".

yttrialite An olive-green mineral: $(Y,Th)_2Si_2O_7$. Cf: *thalenite.*

yttrocerite A violet-blue variety of yttrofluorite containing cerium.

yttrocolumbite A mineral: $(Y,U,Fe)(Nb,Ta)O_4$. Cf: *yttrotantalite.*

yttrocrasite A black mineral: $(Y,Th,U,Ca)Ti_2(O,OH)_6$.

yttrofluorite A mineral: $(Ca,Y)F_{2\text{-}3}$. It is a variety of fluorite containing yttrium.

yttrotantalite A black or brown mineral: $(Y,U,Fe)(Ta,Nb)O_4$. Cf: *yttrocolumbite.*

yttrotungstite A mineral: $YW_2O_6(OH)_3$ (?). It may contain a little thorium. Syn: *thorotungstite.*

yugawaralite A zeolite mineral: $CaAl_2Si_6O_{16}\cdot4H_2O$.

yukonite (a) A noncrystalline mineral: $Ca_3Fe_7{}^{+3}(AsO_4)_6(OH)_9\cdot18H_2O$(?). (b) An obsolete term originally assigned to an igneous rock intermediate in composition between a *tonalite* and an *aplite.* It is named after the Yukon River, Alaska.

yuksporite A mineral: $(Na,K)_2(Ca,Sr,Ba)_2Ti_2Si_2O_8F\cdot2H_2O$ (?).

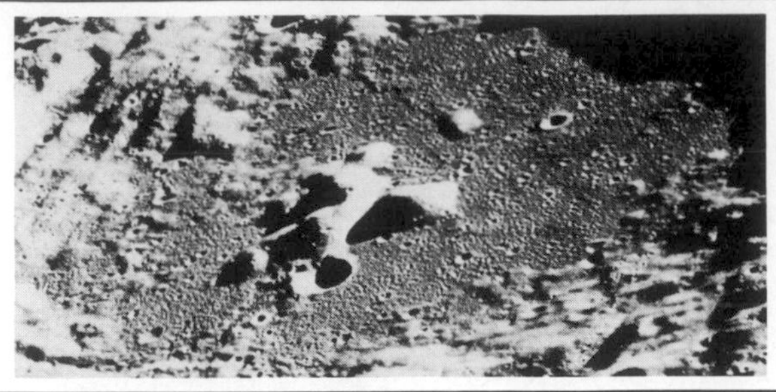

zairite A trigonal mineral of the crandallite group: $Bi(Fe,Al)_3(PO_4)_2(OH)_6$.

Zanclean European stage: Middle Pliocene (above Messinian, below Plaisancian). See also: *Tabianian.*

zanjón A Puerto Rican term for a *solution corridor.* Etymol: Spanish, "deep ditch".

zanoga A glacial *cirque* (von Engeln, 1942, p. 447).

zapatalite A tetragonal mineral: $Cu_3Al_4(PO_4)_3(OH)_9 \cdot 4H_2O$.

zap crater An informal syn. of *micrometeorite crater.*

zaphrentid A solitary coral with marked pinnate septal arrangement, lacking dissepiments or axial structures.

zaratite An emerald-green mineral: $Ni_3(CO_3)(OH)_4 \cdot 4H_2O$. It occurs in secondary incrustations or compact masses. Syn: *emerald nickel.*

zastrugi *sastrugi.*

zavaritskite A mineral: BiOF.

zawn An English term for a sandy cove in a cliff (Robson & Nance, 1959, p. 39) or a little inlet of the sea (Stamp, 1966, p. 495).

zeasite An opal, formerly an old name for "fire opal" but now applied to "wood opal".

zebra dolomite A term used in the Leadville district of Colorado for an altered dolomite rock that shows conspicuous banding (generally parallel to bedding) consisting of light-gray coarsely textured layers alternating with darker finely textured layers. See also: *zebra rock.*

zebraic chalcedony Chalcedony that under crossed nicols shows alternating black and white bands irregularly athwart the fibers, giving a zebralike appearance. The effect is produced by helical rotation of component quartz crystallites (McBride and Folk, 1977).

zebra layering *Rhythmic layering* in which dark and light bands alternate, generally reflecting differing amounts of pyroxene and plagioclase.

zebra limestone A term used by Fischer (1964, p. 135) for a limestone banded by parallel sheet cracks filled with calcite.

zebra rock (a) A term used in the Colville district of NE Washington State for a dolomite that shows narrow banding consisting of black layers (indicative of organic matter) alternating with white, slightly coarse-grained, and somewhat vuggy layers. See also: *zebra dolomite.* (b) A term used in Western Australia for a banded quartzose rock of Cambrian age.

Zechstein European series (esp. in Germany): Upper Permian (above Rotliegende). It contains the Thuringian Stage.

zeilleriid Said of a long brachiopod loop (as in the superfamily Zeilleriacea) not attached to the dorsal septum in adults.

zektzerite An orthorhombic mineral: $LiNaZrSi_6O_{15}$.

zellerite A lemon-yellow secondary mineral: $Ca(UO_2)(CO_3)_2 \cdot 5H_2O$.

zemannite A mineral: $(Zn,Fe)_2(TeO_3)_3Na_xH_{2-x} \cdot yH_2O$.

Zemorrian North American stage: Oligocene and Miocene (above Refugian, below Saucesian).

zenith The point on the celestial sphere that is directly above the observer and directly opposite to the *nadir [geodesy].* In a more general sense, the term denotes the stretch of sky overhead.

zenithal projection *azimuthal projection.*

zeolite (a) A generic term for a large group of white or colorless (sometimes red or yellow) hydrous aluminosilicates that are analogous in composition to the feldspars, with sodium, calcium, and potassium (rarely barium or strontium) as their chief metals; have a ratio of (Al + Si) to nonhydrous oxygen of 1:2; and are characterized by their easy and reversible loss of water of hydration and by their ready fusion and swelling when strongly heated under the blowpipe. Zeolites have long been known to occur as well-formed crystals in cavities in basalt. Of more significance is their occurrence as authigenic minerals in the sediments of saline lakes and the deep sea, and esp. in beds of tuff. They form "during and after burial, generally by reaction of pore waters with solid aluminosilicate materials (e.g. volcanic glass, feldspar, biogenic silica, and clay minerals)" (Hay, 1978, p. 135). (b) Any of the minerals of the zeolite group, including natrolite, heulandite, analcime, chabazite, stilbite, mesolite, scolecite, phillipsite, laumontite, mordenite, clinoptilolite, erionite, harmotome, and others less important, as well as minerals not yet classified. (c) Any of various silicates that are processed natural materials (such as glauconite) or artificial granular sodium aluminosilicates used in the base-exchange method of water softening and as gas adsorbents or drying agents. The term now includes such diverse groups of compounds as sulfonated organics or basic resins, which act in a similar manner to effect either cation or anion exchange.—Etymol: Greek *zein,* "to boil".

zeolite facies The set of metamorphic mineral assemblages (facies) that includes the zeolites analcime, heulandite, stilbite, laumontite, and wairakite (Coombs, 1960). Developed best in metagreywackes and metabasalts, it is the lowest grade of metamorphism, transitional between diagenesis (or unmetamorphosed rock) and the *prehnite-pumpellyite facies* or the *greenschist facies.* Various zeolite assemblages can be correlated with depth of burial (Miyashiro and Shido, 1970).

zeolitic ore deposits Ore deposits, particularly of native copper in basalts, which have zeolites as distinctive, though not necessarily abundant, gangue minerals.

zeolitization Introduction of, alteration to, or replacement by, a mineral or minerals of the zeolite group. This process occurs chiefly in rocks containing calcic feldspars or feldspathoids, and is sometimes associated with copper mineralization.

zeophyllite A white rhombohedral mineral: $Ca_4Si_3O_7(OH)_4F_2$. It sometimes contains iron.

zerdeb A term used in Algeria for an interdune hollow floored with calcareous tufa (Capot-Rey, 1945, p. 397).

zero curtain The zone immediately above the permafrost table, where zero temperature (0°C) lasts a considerable period of time during freezing and thawing of the overlying ground (Brown & Kupsch, 1974, p. 42).

zero-energy coast A coast characterized by average breaker heights of 3 cm or less. Cf: *low-energy coast.*

zero-length spring A special type of gravimeter spring so constructed that the total length is proportional to the applied force. It is used in astatic, or unstable, instruments.

zero-length-spring gravimeter *LaCoste-Romberg gravimeter.*

zero meridian *prime meridian.*

zeta potential Drop in potential across the diffuse layer in an electrolyte.

zeuge A tabular mass of resistant rock left standing on a pedestal of softer rocks, resulting from differential erosion by the scouring effect of windblown sand in a desert region; it may be 2 m to 50 m high. Etymol: German *Zeuge,* "witness". Pl: *zeugen.* Pron: *tzoy -geh.* See also: *mushroom rock.* Syn: *witness rock.*

zeugenberg A syn. of *butte témoin.* Etymol: German *Zeugenberg,* "witness hill".

zeugogeosyncline A *parageosyncline* with an adjoining uplifted area also in the craton, receiving clastic sediments; an intracratonic trough (Kay, 1945). Syn: *yoked basin.* Cf: *autogeosyncline.*

zeunerite A green secondary mineral of the autunite group: $Cu(UO_2)_2(AsO_4)_2 \cdot 10\text{-}16H_2O$. It is isomorphous with uranospinite.

zeylanite *ceylonite.*

zhemchuzhnikovite A green mineral oxalate: $NaMg(Al,Fe^{+3})(C_2O_4)_3 \cdot 8H_2O$. Cf: *stepanovite.*

zietrisikite Incorrect spelling of *pietricikite,* a variety of ozocerite.

zigzag carina One of the not quite oppositely placed *carinae* of a rugose coral on the two sides of a septum. Cf: *yardarm carina.*

zigzag cross-bedding *chevron cross-bedding.*

zigzag fold A *kink fold,* the limbs of which are of unequal length. Cf: *chevron fold.* See also: *knee fold.*

zigzag ridge A continuous ridge that trends first in one direction, then in another. Zigzag ridges are produced in folded mountains, e.g. the Appalachians, by truncation of plunging folds.

zigzag watershed A drainage divide through which rivers have broken by headward erosion, the divide retaining its original position between the drainage basins.

zinalsite A clay mineral: $Zn_7Al_4(SiO_4)_6(OH)_2 \cdot 9H_2O$ (?).

zinc A blue-white mineral, the native metallic element Zn. The occurrence of native zinc is unconfirmed, although it has been reported to have been found in basalt and also in auriferous sands in Victoria, Australia.

zincaluminite A light-blue mineral: $Zn_6Al_6(SO_4)_2(OH)_{26} \cdot 5H_2O$.

zinc blende *sphalerite.*

zinc bloom *hydrozincite.*

zincite A red to yellow brittle mineral: (Zn,Mn)O. It is an ore of zinc, as in New Jersey where it is associated with franklinite and willemite. Syn: *red zinc ore; red oxide of zinc; ruby zinc; spartalite.*

zinckenite *zinkenite.*

zinc-melanterite A monoclinic mineral: $(Zn,Cu,Fe)SO_4 \cdot 7H_2O$.

zincobotryogen A mineral: $(Zn,Mg,Mn)Fe(SO_4)_2(OH) \cdot 7H_2O$.

zincocopiapite A mineral: $ZnFe_4(SO_4)_6(OH)_2 \cdot 18H_2O$.

zincrosasite A mineral: $(Zn,Cu)_2(CO_3)(OH)_2$. It is a variety of rosasite with zinc greater than copper.

zincsilite A mineral: $Zn_3Si_4O_{10}(OH)_2 \cdot nH_2O$. It is the aluminum-free end-member of the montmorillonite-sauconite series.

zinc spar *smithsonite.*

zinc spinel *gahnite.*

zinc vitriol *goslarite.*

zinc-zippeite An orthorhombic mineral: $Zn_2(UO_2)_6(SO_4)_3(OH)_{10} \cdot 16H_2O$.

Zingg's classification A classification of pebble shapes, devised by Theodor Zingg (1905-), Swiss meteorologist and engineer, based on the graphical representation of the diameter ratio of intermediate (width) to maximum (length) plotted against the diameter ratio of minimum (thickness) to intermediate (Zingg, 1935). The classification distinguishes four shape classes: *spheroid; disk [sed]; blade; rod.*

zinkenite A steel-gray hexagonal mineral: $Pb_6Sb_{14}S_{27}$. Also spelled: *zinckenite.* Syn: *keeleyite.*

zinnwaldite A mineral of the mica group: $K_2(Li,Fe,Al)_6(Si,Al)_8O_{20}(OH,F)_4$. It is a pale-violet, yellowish, brown, or dark-gray variety of lepidolite containing iron, and is the characteristic mica of greisens.

zippeite An orthorhombic mineral: $K_4(UO_2)_6(SO_4)_3(OH)_{10} \cdot 4H_2O$.

zircon A mineral: $ZrSiO_4$. It occurs in tetragonal prisms, has various colors and is a common accessory mineral in siliceous igneous rocks, crystalline limestones, schists, and gneisses, in sedimentary rocks derived therefrom, and in beach and river placer deposits. It is the chief ore of zirconium, and is used as a refractory; when cut and polished, the colorless varieties provide exceptionally brilliant gemstones. Syn: *zirconite; hyacinth; jacinth.*

zirconite Gray or brownish zircon.

zirconolite *zirkelite [mineral].*

zircophyllite A mineral of the astrophyllite group: $(K,Na,Ca)_3(Mn,Fe_2)_7(Zr,Nb)_2Si_8O_{27}(OH,F)_4$. Also spelled: *zirkophyllite.*

zircosulfate A mineral: $Zr(SO_4)_2 \cdot 4H_2O$.

zirkelite [mineral] A mineral of the pyrochlore group: $(Ca,Th,Ce)Zr(Ti,Nb)_2O_7$. Syn: *zirconolite.*

zirkelite [rock] An altered basaltic glass. Obsolete.

zirklerite A mineral: $(Fe,Mg,Ca)_9Al_4Cl_{18}(OH)_{12} \cdot 14H_2O$ (?).

zirkophyllite *zircophyllite.*

zirsinalite A trigonal mineral: $Na_6(Ca,Mn,Fe)ZrSi_6O_{18}$.

zittavite A type of lustrous black lignite. It is harder and more brittle than *dopplerite.*

zoantharian Any anthozoan belonging to the subclass Zoantharia, characterized by paired mesenteries. They may or may not have a calcareous exoskeleton. Range, Ordovician to present.

zoarium (a) The collective skeletal parts (zooecia and extrazooidal skeleton) of an entire bryozoan colony. (b) A colony of bryozoans.—Pl: *zoaria.* Also spelled: *zooarium.*

ZoBell bottle A sterilized bottle used for collection of sea water for bacteriological analysis.

zobtenite A gabbro-gneiss characterized by *augen* of diallage surrounded by uralite and embedded in granular epidote and plagioclase (saussurite) (Holmes, 1928, p.242). Cf: *flaser gabbro.*

zodiacal dust *cosmic dust.*

zoecium *zooecium.*

zoichnic Said of a dolomite or recrystallized limestone in which animal fossils, though partly destroyed by recrystallization, are still recognizable in outline or by traces of internal structure (Phemister, 1956, p. 74). Cf: *zoophasmic.*

zoid *zooid.*

zoisite An orthorhombic mineral of the epidote group: $Ca_2Al_3Si_3O_{12}(OH)$. It often contains appreciable ferric iron, and is white, gray, brown, green, or rose red in color. Zoisite occurs in metamorphic rocks (esp. schists formed from calcium-rich igneous rocks), and in altered igneous rocks, and is an essential constituent of saussurite. Cf: *clinozoisite.*

zona *zone [palyn].*

zonal axis *zone axis.*

zonal equation The statement that if a given crystal face (hkl) belongs to a zone with the axis $[uvw]$, then $hu + kv + lw = 0$.

zonal guide fossil A *guide fossil* that makes possible the identification of a specific biostratigraphic zone and that gives its name to the zone. It need not necessarily be either restricted to the zone or found throughout every part of it. Syn: *zone fossil.*

zonal profile *composite profile.*

zonal soil In early U.S. classification systems, one of the *soil orders* that embraces soils with well-developed characteristics that presumably reflect the influence of the agents of soil genesis, esp. climate and plants; also, any soil belonging to the zonal order. Cf: *intrazonal soil; azonal soil.* Syn: *mature soil.*

zonal structure *zoning.*

zonal theory A theory of hypogene mineral-deposit formation, and the spatial distribution patterns of mineral sequences to be expected from change in a mineral-bearing fluid as it migrates away from a magmatic source (Park & MacDiarmid, 1970, p. 165). It also deals with thermal-chemical gradients associated with the genesis of ore deposits, whether of direct magmatic origin or not, and with metallogenic zoning on a regional scale. See also: *zoning of ore deposits.*

zonate Said of spores possessing a zone.

zonation The condition of being arranged or formed in zones; e.g. the distribution of distinctive fossils, more or less parallel to the bedding, in biostratigraphic zones.

zone [cryst] *crystal zone.*

zone [ecol] (a) Part of a biogeographic region characterized by uniform climatic conditions, fauna, and flora and usually by a sloping, band-formed area. (b) An area characterized by the dominance of a particular organism, such as the *Acropora palmata* zone on Caribbean reefs.

zone [geog] A term used generally, even vaguely, for an area or region of latitudinal character, more or less set off from surrounding areas by some special or distinctive characteristics; e.g. any of the five great belts or encircling regions into which the Earth's surface is divided with respect to latitude and temperature, viz. the torrid zone, the two temperate zones, and the two frigid zones.

zone [geol] A belt, band, or strip of earth materials, however disposed, characterized as distinct from surrounding parts by some particular property or content; e.g. the *zone of saturation,* the *zone of fracture* or a *fault zone.*

zone [meta] *aureole.*

zone [palyn] An annular, more or less equatorial, extension of a spore, having varying equatorial width and being as thick as or thinner than the spore wall. It is much thinner than a *cingulum.* The term is also used in a general sense for any equatorial extension of the spore wall. Cf: *flange; corona [palyn]; auricula; crassitude.* Syn: *zona.*

zone [stratig] A minor interval in any category of stratigraphic classification. There are many kinds of zones, depending on the characters under consideration — biozones, lithozones, chronozones, mineral zones, metamorphic zones, zones of reversed magnetic polarity, and so on (ISG, 1976, p. 14). The term should always be preceded by a modifier indicating the kind of zone to which reference is made.

zone axis That line or crystallographic direction through the center of a crystal which is parallel to the intersection edges of the crystal faces defining the *crystal zone.* Syn: *zonal axis.*

zone-breaking species A fossil species that is confined to a biostratigraphic zone in certain areas but transgresses the boundaries of that zone at other places (Arkell, 1933, p.32).

zone fossil A fossil characteristic of a zone; a *zonal guide fossil.*
zone of ablation *ablation area.*
zone of accumulation [snow] (a) *accumulation area.* (b) In respect to an avalanche, a syn. of *accumulation zone.*
zone of accumulation [soil] *B horizon.*
zone of aeration A subsurface zone containing water under pressure less than that of the atmosphere, including water held by capillarity; and containing air or gases generally under atmospheric pressure. This zone is limited above by the land surface and below by the surface of the *zone of saturation,* i.e. the *water table.* The zone is subdivided into the *belt of soil water,* the *intermediate belt,* and the *capillary fringe.* Syn: *vadose zone; unsaturated zone; zone of suspended water.*
zone of astogenetic change A part of a bryozoan colony in which zooids show morphologic differences from generation to generation in a more or less uniform progression distally, ending with a pattern capable of endless repetition of one or more kinds of zooids.
zone of astogenetic repetition A part of a bryozoan colony in which zooids show one or more repeated morphologies from generation to generation in a pattern capable of endless repetition.
zone of capillarity *capillary fringe.*
zone of cementation The layer of the Earth's crust below the *zone of weathering,* in which percolating waters cement unconsolidated deposits by the deposition of dissolved minerals from above.
zone of deposition "The area in which continental glaciers deposit materials derived from the *zone of erosion.* It is usually covered with drift and has the general aspect of a plain" (Stokes & Varnes, 1955, p. 164).
zone of discharge A term suggested for that part of the zone of saturation having a means of horizontal escape. Cf: *static zone.*
zone of erosion "The area from which continental glaciers have removed material by erosion. It is mostly a bare rock surface" (Stokes & Varnes, 1955, p. 164). Ant: *zone of deposition.*
zone of flow [glaciol] The inner, mobile main body or mass of a glacier, in which most of the ice flows without fracture. Cf: *zone of fracture [glaciol].*
zone of flow [interior Earth] *zone of plastic flow.*
zone of flowage *zone of plastic flow.*
zone of fracture [glaciol] The outer, rigid part of a glacier, in which the ice is much fractured. Cf: *zone of flow [glaciol].*
zone of fracture [interior Earth] The upper, brittle part of the Earth's crust, in which deformation is by fracture rather than by plastic flow; that region of the crust in which fissures can exist. Cf: *zone of plastic flow; zone of fracture and plastic flow.* Syn: *zone of rock fracture.*
zone of fracture and plastic flow That region of the Earth's crust intermediate in depth and pressure between the *zone of fracture* and the *zone of plastic flow,* in which deformation of the weaker rocks is by plastic flow, and of the stronger rocks by fracture.
zone of illuviation *B horizon.*
zone of intermittent saturation A term applied by Monkhouse (1965) to the temporary zone of saturation formed in the soil by infiltration from rainfall or snowmelt at a rate in excess of that at which the water can move downward to the main water table.
zone of mobility *asthenosphere.*
zone of plastic flow That part of the Earth's crust that is under sufficient pressure to prevent fracturing, i.e. is ductile, so that deformation is by flow. Cf: *zone of fracture; zone of fracture and plastic flow.* Syn: *zone of flow [interior Earth]; zone of rock flowage; zone of flowage.*
zone of rock flowage *zone of plastic flow.*
zone of rock fracture *zone of fracture [interior Earth].*
zone of saturation A subsurface zone in which all the interstices are filled with water under pressure greater than that of the atmosphere. Although the zone may contain gas-filled interstices or interstices filled with fluids other than water, it is still considered *saturated.* This zone is separated from the *zone of aeration* (above) by the *water table.* Syn: *saturated zone; phreatic zone.*
zone of soil water *belt of soil water.*
zone of suspended water *zone of aeration.*
zone of weathering The superficial layer of the Earth's crust above the water table that is subjected to the destructive agents of the atmosphere, and in which soils develop. Cf: *zone of cementation.*
zone symbol The symbol of the zone axis of a crystal in terms of the crystal lattice, e.g. the symbols for the zone axis of a series of ($hk0$) faces would be [001]. Cf: *indices of lattice row.*
zone time A syn. of *moment.* The term was suggested by Kobayashi (1944, p.742) for the average duration in years of a biostratigraphic zone in any given geologic system, ranging between about 300,000 and 5 million years. Cf: instant.
zoning [cryst] A variation in the composition of a crystal from core to margin, due to a separation of the crystal phases during its growth by loss of equilibrium in a continuous reaction series. The higher-temperature phases of the isomorphic series form the core, with the lower-temperature phases toward the margin. Cf: *armoring.* See also: *normal zoning.* Syn: *zonal structure.*
zoning [meta] The development of areas of metamorphosed rocks that may exhibit zones in which a particular mineral or suite of minerals is predominant or characteristic, reflecting the original rock composition, the pressure and temperature of formation, the duration of metamorphism, and whether or not material was added.
zoning of ore deposits Spatial distribution patterns of elements, minerals, or mineral assemblages; paragenetic sequences, either syngenetic or epigenetic (Park & MacDiarmid, 1970, p. 165). Zoning is especially well developed in the mineralization-alteration assemblages about subvolcanic occurrences such as porphyry base-metal deposits. See also: *zonal theory.* Syn: *mineral zoning.*
zonite A term proposed by Henningsmoen (1961) to replace *range-zone.*
zonochlorite A syn. of *pumpellyite* occurring in green pebbles of banded structure (as in the Lake Superior region). It was previously thought to be an impure prehnite.
zonolimnetic Pertaining to a definite depth zone, in a body of fresh water, inhabited by planktonic animals.
zonotrilete Said of a trilete spore characterized by an equatorial zone or other thickening.
zonule The smallest recognized subdivision of a *biozone,* generally consisting of a single stratum or small thickness of strata (ACSN, 1961, art.20f); a locally recognizable biostratigraphic unit, such as in a sedimentary basin or similar restricted area of sedimentation. It may be a subdivision of a *subzone,* or it may be distinguished as a minor component of a zone without subdividing the zone into subzones. The term was defined by Fenton & Fenton (1928, p.20-22) as the strata or stratum that contains a *faunule* or *florule,* its thickness and area being limited by the vertical and horizontal range of that faunule or florule.
zooarium *zoarium.*
zoobenthos Animal forms of the benthos.
zoochore A plant whose seeds or spores are distributed by living animals.
zooeciule A minute polymorph in some cheilostome bryozoans, having a polypide and operculum-bearing orifice but apparently unable to feed or produce gametes.
zooecium The skeleton of a bryozoan zooid. Pl: *zooecia.* Adj: *zooecial.* Also spelled: *zoecium.*
zooecology The branch of ecology concerned with the relationships between animals and their environment. Cf: *phytoecology.*
zoogenic rock A *biogenic rock* produced by animals or directly attributable to the presence or activities of animals; e.g. shell limestone, coral reefs, guano, and lithified calcareous ooze. Cf: *zoolith.* Syn: *zoogenous rock.*
zoogenous rock *zoogenic rock.*
zoogeography The branch of *biogeography* dealing with the geographic distribution of animals. Cf: *phytogeography.*
zooid (a) A more or less independent animal produced by other than direct sexual methods and therefore having an equivocal individuality; any individual of a colony, irrespective of its morphologic specifications, such as an octocorallian polyp or a soft-bodied graptolite individual inhabiting a theca. (b) One of the physically connected, asexually replicated morphologic units which, together with any extrazooidal parts present, comprise a bryozoan colony. It performs major functions with a system of organs comparable to that of a solitary animal.—Also spelled: *zoid.*
zoolite An animal fossil. Syn: *zoolith.*
zoolith [paleont] *zoolite.*
zoolith [sed] A *biolith* formed by animal activity or composed of animal remains; specif. *zoogenic rock.*
zoophasmic Said of a dolomite or recrystallized limestone that contains vague but unmistakable traces of the former presence of animal fossils (Phemister, 1956, p. 74). Cf: *zoichnic.*
zoophyte (a) Any member of the phylum Bryozoa. (b) In a broad sense, any nonmobile plantlike animal, e.g. sponges, sea anemones, hydroids, or bryozoans. Syn: *phytozoan.*

zooplankton The animal forms of *plankton*, e.g. jellyfish. They consume the *phytoplankton*.
zootrophic *heterotrophic*.
zooxanthella An algal cell living symbiotically in the cells of certain invertebrate animals, e.g. in the endoderm of hermatypic coral polyps.
zorite An orthorhombic mineral: $Na_2Ti(Si,Al)_3O_9 \cdot nH_2O$.
Z phenomenon A possible time lag (a few seconds or less) between the issuance of *P waves* and *S waves* from an earthquake focus (Runcorn et al., 1967).
Zuloagan North American (Gulf Coast) stage: Upper Jurassic (above older Jurassic, below LaCasitan; it is equivalent to European Oxfordian) (Murray, 1961).
zunyite A mineral: $Al_{13}Si_5O_{20}(OH,F)_{18}Cl$. It occurs in minute transparent terahedral crystals.
zussmanite A mineral: $K(Fe,Mg,Mn)_{13}(Si,Al)_{18}O_{42}(OH)_{14}$.
zvyagintsevite A mineral: $(Pd,Pt)_3(Pb,Sn)$.
zweikanter A *windkanter* or stone having two faces intersecting in two sharp edges. Etymol: German *Zweikanter*, "one having two edges". Pl: *zweikanters; zweikanter*.
zwieselite A monoclinic mineral: $(Fe^{+2},Mn)_2(PO_4)F$. It forms a series with triplite.
Zwischengebirge A term proposed by Kober in 1921 to designate an undeformed or little deformed area between the two marginal chains of a symmetrical *orogenic belt*. In Kober's original sense an orogen consists of two marginal chains (*Randketten*) and an intervening Zwischengebirge. The term has been translated into English as *intramontane space* (Longwell, 1923), *betwixt mountains* (Collet, 1927), and *median mass* (Bucher, 1933). Staub (1928) has used the term for any undeformed area between deformed belts of similar age, e.g. the Swiss plain.
zygal ridge The ridge uniting the median lobe (L2) and the proximal posterior lobe (L3) in many Paleozoic ostracodes.
zygolith A coccolith in the form of an elliptic ring with a crossbar arching slightly or strongly upward (e.g. distally), and bearing a knob or short spine.
zygolophe A brachiopod lophophore in which each brachium consists of a straight or crescentic side arm bearing two rows of paired filamentary appendages (TIP, 1965, pt.H, p.155).
zygome An articulatory structure of a desma in a sponge.
zygomorphic Said of an organism or organ that is bilaterally symmetrical or capable of division into essentially symmetric halves by only one longitudinal plane passing through the axis. Cf: *actinomorphic*.
zygosis The interlocking of sponge desmas, without fusion, by means of zygomes.
zygospore A *resting spore* of various nonvascular plants (such as desmids), produced by sexual fusion of two protoplasts. It often has a thick, resistant wall and can therefore occur as a palynomorph.
zygous basal plate One of the two large plates of the basalia of a blastoid, located in the right posterior (BD) or left anterior (DA) position and formed by fusion of a pair of antecedent small basal plates comparable to *azygous basal plate* in the AB interray (TIP, 1967, pt.S, p.350).

References Cited

A

Abraham, Herbert (1960) *Asphalts and allied substances,* v.1, 6th ed. Princeton, N.J.: Van Nostrand. 370p.

Ackermann, Ernst (1951) *Geröllton!* Geologische Rundschau, v.39, p.237-239.

Ackermann, Ernst (1962) *Büssersteine—Zeugen vorzeitlicher Grundwasserschwankungen.* Zeitschrift für Geomorphologie, n.F., Bd.6, p.148-182.

ACSN: American Commission on Stratigraphic Nomenclature (1957) *Nature, usage, and nomenclature of biostratigraphic units.* American Association of Petroleum Geologists. Bulletin, v.41, p.1877-1889. (Its Report 5).

ACSN: American Commission on Stratigraphic Nomenclature (1959) *Application of stratigraphic classification and nomenclature to the Quaternary.* American Association of Petroleum Geologists. Bulletin, v.43, p.663-675. (Its Report 6).

ACSN: American Commission on Stratigraphic Nomenclature (1961) *Code of stratigraphic nomenclature.* American Association of Petroleum Geologists. Bulletin, v.45, p.645-660.

ACSN: American Commission on Stratigraphic Nomenclature (1965) *Records of the Stratigraphic Commission for 1963-1964.* American Association of Petroleum Geologists. Bulletin, v.49, p.296-300. (Its Note 31).

ACSN: American Commission on Stratigraphic Nomenclature (1967) *Records of the Stratigraphic Commission for 1964-1966.* American Association of Petroleum Geologists. Bulletin, v.51, p.1862-1868. (Its note 34).

Adams, J.A.S., and Weaver, C.E. (1958) *Thorium-to-uranium ratios as indicators of sedimentary processes—example of concept of geochemical facies.* American Association of Petroleum Geologists. Bulletin, v.42, p.387-430.

Adams, J.E., and Frenzel, H.N. (1950) *Capitan barrier reef, Texas and New Mexico.* Journal of Geology, v.58, p.289-312.

Adams, J.E., and others (1951) *Starved Pennsylvanian Midland Basin.* American Association of Petroleum Geologists. Bulletin, v.35, p.2600-2607.

Adams, J.E., and Rhodes, M.L. (1960) *Dolomitization by seepage refluxion.* American Association of Petroleum Geologists. Bulletin, v.44, p.1912-1920.

ADTIC: U.S. Arctic, Desert, Tropic Information Center (1955) *Glossary of arctic and subarctic terms.* Maxwell Air Force Base, Ala.: Air University, Research Studies Institute. 90p. (ADTIC Publication A-105).

Agassiz, Louis (1866) *Geological sketches.* Boston: Ticknor & Fields. 311p.

Ager, D.V. (1963) *Principles of paleoecology: an introduction to the study of how and where animals and plants lived in the past.* New York: McGraw-Hill, Inc.

Ager, D.V. (1974) *Storm deposits in the Jurassic of the Moroccan High.* Palaeogeography, palaeoclimatology, palaeoecology, v.15, p.83-93.

AGI: American Geological Institute. Data Sheet Committee (1958) *Wentworth grade scale.* GeoTimes, v.3, no.1, p.16. (Its Data sheet 7).

Ahlmann, H.W. (1933) *Scientific results of the Swedish-Norwegian Arctic Expedition in the summer of 1931, Pt.8.* Geografiska Annaler, v.15, p.161-216, 261-295.

AIME: American Institute of Mining, Metallurgical, and Petroleum Engineers (1960) *Industrial minerals and rocks (nonmetallics other than fuels).* 3rd ed. New York: American Institute of Mining, Metallurgical, and Petroleum Engineers. 934p.

Aitken, J.D. (1967) *Classification and environmental significance of cryptalgal limestones and dolomites, with illustrations from the Cambrian and Ordovician of southwestern Alberta.* Journal of Sedimentary Petrology, v.37, p.1163-1178.

Alden, W.C. (1928) *Landslide and flood at Gros Ventre, Wyoming.* American Institute of Mining and Metallurgical Engineers. Transactions, v.76, p.347-361. (Technical publication no.140).

Allan, R.S. (1948) *Geological correlation and paleoecology.* Geological Society of America. Bulletin, v.59, p.1-10.

Allen, E.T., and Day, A.L. (1935) *Hot springs of the Yellowstone National Park.* Microscopic examinations by H. E. Merwin. Carnegie Institution of Washington. Publication 466. 525p.

Allen, J.R.L. (1960) *Cornstone.* Geological Magazine, v.97, p.43-48.

Allen, J.R.L. (1963) *Asymmetrical ripple marks and the origin of water-laid cosets of cross-strata.* Liverpool and Manchester Geological Journal, v.3, p.187-236.

Allen, J.R.L. (1968) *Current ripples: their relation to patterns of water and sediment motion.* Amsterdam: North Holland Publishing Co. 433p.

Allen, V.T. (1936) *Terminology of medium-grained sediments. With notes by P.G.H. Boswell.* National Research Council. Division of Geology and Geography. Annual report for 1935-1936, appendix I, exhibit B. 23p. (Its Committee on Sedimentation. Report, exhibit B).

Allen, V.T., and Nichols, R.L. (1945) *Clay-pellet conglomerates at Hobart Butte, Lane County, Oregon.* Journal of Sedimentary Petrology, v.15, p.25-33.

Alling, H.L. (1943) *A metric grade scale for sedimentary rocks.* Journal of Geology, v.51, p.259-269.

Alling, H.L. (1945) *Use of microlithologies as illustrated by some New York sedimentary rocks.* Geological Society of America. Bulletin, v.56, p.737-755.

Allum, J.A.E. (1966) *Photogeology and regional mapping.* New York: Pergamon. 107p.

Alvarez, Walter, Engelder, Terry, and Geiser, P.A. (1978) *Classification of solution cleavage in pelagic limestones.* Geology, v.6, p.263-266.

Amin, B.S., Lal, D., and Somayajulu, B.L.K. (1975) *Chronology of marine sediments using the ^{10}Be method: intercomparison with other methods.* Geochimica et Cosmochimica Acta, v.39, p.1187-1191.

Amiran, D.H.K. (1950-1951) *Geomorphology of the central Negev highlands.* Israel Exploration Journal, v.1, p.107-120.

Anderson, D.M. and others (1969) *Bentonite debris flows in northern Alaska.* Science, v.164, p.173-174.

Andersson, J.G. (1906) *Solifluction, a component of subaerial denudation.* Journal of Geology, v.14, p.91-112.

Andreev, P.F., and others (1968) *Transformation of petroleum in nature.* Translated from the Russian edition by R.B. Gaul and B.C. Metzner. Translation editors: E. Barghoorn and S. Silverman. New York: Pergamon. 468p. (*International series of monographs in earth sciences,* v.29).

Andresen, M.J. (1962) *Paleodrainage patterns: their mapping from subsurface data, and their paleogeographic value.* American Association of Petroleum Geologists. Bulletin, v.46, p.398-405.

Antevs, E.V. (1932) *Alpine zone of Mt. Washington Range.* Auburn, Me.: Merrill & Weber. 118p.

Antevs, E.V. (1948) *Climatic changes and pre-white man.* University of Utah. Bulletin, v.38, no.20, p.168-191.(*The Great Basin, with emphasis on glacial and postglacial times, 3.*).

Antevs, E.V. (1953) *Geochronology of the Deglacial and Neothermal ages.* Journal of Geology, v.61, p.195-230.

Apfel, E.T. (1938) *Phase sampling of sediments.* Journal of Sedimentary Petrology, v.8, p.67-68.

Archie, G.E. (1942) *The electrical resistivity log as an aid in determining some reservoir characteristics.* American Institute of Mining, Metallurgical, and Petroleum Engineers. Transactions, v.146, p.54-60.

Arkell, W.J. (1933) *The Jurassic system in Great Britain.* Oxford: Clarendon Press. 681p.

Arkell, W.J. (1956) *Jurassic geology of the world.* New York: Hafner. 806p.

Arkell, W.J. and Tomkeieff, S.I. (1953) *English rock terms, chiefly as used by miners and quarrymen.* London: Oxford University Press. 139p.

Arkley, R.J. and Brown, H.C. (1954) *The origin of Mima mound (hogwallow) microrelief in the far western states.* Soil Science Society of America. Proceedings, v.18, p.195-199.

Armstrong, F.C. (1974) *Uranium resources of the future—"porphyry" uranium deposits.* In: *Formation of uranium ore deposits.* Proceedings of a symposium, Athens, 6-10 May, 1974, p.625-635.

Armstrong, R.L. (1958) *Sevier orogenic belt in Nevada and Utah.* Geological Society of America. Bulletin, v.79, p.429-458.

Armstrong, Terence, and others (1966) *Illustrated glossary of snow and ice.* Cambridge, England: Scott Polar Research Institute. 60p.

Armstrong, Terence, and Roberts, Brian (1958) *Illustrated ice glossary.* Pt. 1. Polar Record, v.8, no.52, p.4-12.

Armstrong, Terence, and Roberts, Brian (1958) *Illustrated ice glossary.* Pt. 2. Polar Record, v.9, no.59, p.90-96.

Armstrong, Terence, Roberts, Brian, and Swithinbank, Charles (1977) *Proposed new terms and definitions for snow and ice features.* Polar Record, v.18, p.501-502.

Arndt, N.T., Naldrett, A.J., and Pyke, D.R. (1977) *Komatiitic and iron-rich tholeiitic lavas of Munro Township, northeast Ontario.* Journal of Petrology, v.18, p.319-369.

Arnold, C.A. (1947) *An introduction to paleobotany.* New York: McGraw-Hill. 433p.

Arrhenius, G. (1963) *Pelagic sediments.* In: *The sea,* v.3, ch.25, p.655-727. New York: Interscience.

ASCE: American Society of Civil Engineers. Hydraulics Division. Committee on Hydraulic Structures (1962) *Nomenclature for hydraulics.* Fred W. Blaisdell and others, eds. New York: American Society of Civil Engineers. 501p. (*Manuals and reports on engineering practice,* no.43.)

ASCE: American Society of Civil Engineers. Soil Mechanics and Foundations Division. Committee on Glossary of Terms and Definitions in Soil Mechanics (1958) *Glossary of terms and definitions in soil mechanics; report of the committee.* R.E. Fadum, chairman. Its Journal, 1958, no.SM4, pt.1. 43p. (American Society of Civil Engineers. Proceedings, v.84, paper 1826).

ASCE: American Society of Civil Engineers. Surveying and Mapping Division. Committee on Definitions of Surveying Terms (1954) *Definitions of surveying, mapping, and related terms.* Charles B. Breed, chairman. New York: American Society of Civil Engineers. 202p. (*Manuals of engineering practice,* no.34).

ASP: American Society of Photogrammetry (1975) *Manual of remote sensing.* R.G. Reeves, ed. Falls Church, Va.: American Society of Photogrammetry. Vol.1, 867p.; vol.2, 2144p.

ASP: American Society of Photogrammetry. Committee on Nomenclature (1966) *Definitions of terms and symbols used in photogrammetry.* Editor: Robert D. Turpin. In: American Society of Photogrammetry. *Manual of photogrammetry,* v.2, p.1125-1161. 3rd ed. Falls Church, Va.: American Society of Photogrammetry. 1199p.

Assereto, Riccardo, and Kendall, C.G. (1977) *Nature, origin, and classification of peritidal tepee structures and related breccias.* Sedimentology, v.24, p.153-210.

Association of Engineering Geologists (1969) *Definition of engineering geology.* Its Newsletter, v.12, no.4, p.3.

ASTM: American Society for Testing and Materials (1970) *Annual book of ASTM standards, part 33: glossary of ASTM definitions and index to ASTM standards.* Philadelphia: American Society for Testing and Materials. 706p.

Atterberg, Albert (1905) *Die rationelle Klassifikation der Sande und Kiese.* Chemiker-Zeitung, Jahrg.29, p.195-198.

Aubouin, Jean (1965) *Geosynclines.* Amsterdam: Elsevier. 335p.

Aufrère, L. (1931) *Le cycle morphologique des dunes.* Annales de Géographie, v.40, no.226, p.362-385.

B

Back, William (1966) *Hydrochemical facies and ground-water flow patterns in northern part of Atlantic Coastal Plain.* U.S. Geological Survey. Professional Paper 498-A. 42p.

Bagnold, R.A. (1941) *The physics of blown sand and desert dunes.* London: Methuen. 265p.

Bagnold, R.A. (1956) *The flow of cohesionless grains in fluids.* Royal Society of London. Philosophical Transactions, ser.A, v.249, p.235-297.

Bailey, E.B. (1934) *West Highland tectonics: Loch Leven to Glen Roy.* Geological Society of London. Quarterly Journal, v.90, p.462-525.

Bailey, R.V., and Childers, M.O. (1977) *Applied mineral exploration with special reference to uranium.* Boulder, Colo.: Westview Press. 542p.

Baker, A.A. (1959) *Imprisoned rocks: a process of rock abrasion.* Victorian Naturalist, v.76, p.206-207.

Baker, B.B., and others (1966) *Glossary of oceanographic terms.* 2nd ed. Washington, D.C.: U.S. Naval Oceanographic Office. 204p. (U.S. Naval Oceanographic Office. Special Publication 35).

Baker, H.A. (1920) *On the investigation of the mechanical constitution of loose arenaceous sediments by the method of elutriation, with special reference to the Thanet beds of the southern side of the London Basin.* Geological Magazine, v.57, p.321-332, 363-370, 411-420, 463-467.

Baker, P.E., Buckley, F., and Holland, J.G. (1974) *Petrology and geochemistry of Easter Island.* Contributions to Mineralogy and Petrology, v.44, p.85-100.

Balk, Robert (1937) *Structural behavior of igneous rocks (with special reference to interpretations by H. Cloos and collaborators).* Geological Society of America. Memoir 5. 177p.

Ballard, T.J., and Conklin, Q.E. (1955) *The uranium prospector's guide.* New York: Harper. 251p.

Banta, W.C. (1970) *The body wall of cheilostome Bryozoa: III, the frontal wall of* Watersipora arcuata *Banta, with a revision of the Cryptocystidea.* Journal of Morphology, v.131, p.37-56.

Barbier, M.G., and Viallix, J.R. (1973) *Sosie: a new tool for marine seismology.* Geophysics, v.38, p.673-683.

Barker, D.S. (1970) *Compositions of granophyre, myrmekite, and graphic granite.* Geological Society of America. Bulletin, v.81, p.3339-3350.

Barrell, Joseph (1907) *Geology of the Marysville mining district, Montana: a study of igneous intrusion and contact metamorphism.* U.S. Geological Survey. Professional Paper 57. 178p.

Barrell, Joseph (1912) *Criteria for the recognition of ancient delta deposits.* Geological Society of America. Bulletin, v.23, p.377-446.

Barrell, Joseph (1913) *The Upper Devonian delta of the Appalachian geosyncline.* American Journal of Science, ser.4, v.36, p.429-472.

Barrell, Joseph (1917) *Rhythms and the measurements of geologic time.* Geological Society of America. Bulletin, v.28, p.745-904.

Barth, T.F.W. (1948) *Oxygen in rocks: a basis for petrographic calculations.* Journal of Geology, v.56, p.50-60.

Barth, T.F.W. (1959) *Principles of classification and norm calculations of metamorphic rocks.* Journal of Geology, v.67, p.135-152.

Barth, T.F.W. (1962) *Theoretical petrology.* 2d ed. New York: Wiley. 416p.

Barton, D.C. (1929) *The Eötvös torsion balance method of mapping geologic structure.* American Institute of Mining, Metallurgical, and Petroleum Engineers. [Transactions, v.18], Geophysical prospecting, p.416-479.

Barton, P.B., jr., and Toulmin, Priestley, III (1964a) *Experimental data from the system Cu-Fe-S and their bearing on exsolution textures in ores.* Economic Geology, v.59, p.1241-1269.

Barton, P.B., jr., and Toulmin, Priestley, III (1964b) *The electron-tarnish method for the determination of the fugacity of sulfur in laboratory sulfide systems.* Geochimica et Cosmochimica Acta, v.28, p.619-640.

Bascom, Florence (1931) *Geomorphic nomenclature.* Science, v.74, p.172-173.

Bastin, E.S. (1909) *Chemical composition as a criterion in identifying metamorphosed sediments.* Journal of Geology, v.17, p.445-472.

Bateman, A.M. (1959) *Economic mineral deposits.* 2nd ed. New York: Wiley. 916p.

Bates, R.L. (1938) *Occurrence and origin of certain limonite concretions.* Journal of Sedimentary Petrology, v.8, p.91-99.

Båth, Markus (1966) *Earthquake seismology.* Earth-Science Reviews, v.1, p.69-86.

Bathurst, R.G.C. (1958) *Diagenetic fabrics in some British Dinantian limestones.* Liverpool and Manchester Geological Journal, v.2, pt.1, p.11-36.

Bathurst, R.G.C. (1959) *The cavernous structure of some Mississippian* (Stromatactis) *reefs in Lancashire, England.* Journal of Geology, v.67, p.506-521.

Bathurst, R.G.C. (1966) *Boring algae, micrite envelopes and lithification of molluscan biosparites.* Geological Journal, v.5, p.15-32.

Bathurst, R.G.C. (1971) *Carbonate sediments and their diagenesis.* New York: Elsevier. 620p. (*Developments in sedimentology,* 12.)

Batten, Roger (1966) *The lower Carboniferous gastropod fauna from the Hotwells limestone of Compton Martin, Somerset.* Part I. Palaeontographical Society. Monographs, v.119, p.1-52.

Baulig, Henri (1956) *Vocabulaire franco-anglo-allemand de géomorphologie.* Paris: Soc. Ed. Belles Lettres. 229p.

Baulig, Henri (1957) *Peneplains and pediplains.* Translated from the French by C.A. Cotton. Geological Society of America. Bulletin, v.68, p.913-929.

Baumhoff, M.A., and Heizer, R.F. (1965) *Postglacial climate and archaeology in the Desert West.* In: Wright, H.E., jr., and Frey, D.G., eds. *The Quaternary of the United States,* p.697-707. Princeton, N.J.: Princeton Univ. Press. 922p.

Bayly, Brian (1968) *Introduction to petrology.* Englewood Cliffs, N.J.: Prentice-Hall. 371p.

Beales, F.W. (1958) *Ancient sediments of Bahaman type.* American Association of Petroleum Geologists. Bulletin, v.42, p.1845-1880.

Beasley, H.C. (1914) *Some fossils from the Keuper Sandstone of Alton, Staffordshire.* Liverpool Geological Society. Proceedings, v.12, p.35-39.

Becke, F. (1903) *Über Mineralbestand und Struktur der Kristallinischen Schiefer.* Comptes Rendus, Congrès Géologique International, 9th, Vienna.

Becke, F. (1913) *Ueber Minerallesband und Struktur der krystallinischen Schiefer.* Akademie der Wissenschaften in Wien. Denkschriften, v.75, p.1-53.

Becker, G.F. (1895) *Reconnaissance of the gold fields of the southern Appalachians.* U.S. Geological Survey. Annual Report, 16th, pt.3, p.251-331.

Becker, Hans (1932) *Report on some work on sediments done in Germany in 1931.* National Research Council. Committee on Sedimentation. Report, 1930-1932, p.82-89. (National Research Council. Bulletin, no.89).

Beerbower, J.R. (1964) *Cyclothems and cyclic depositional mechanisms in alluvial plain sedimentation.* Kansas. State Geological Survey. Bulletin 169, v.1, p.31-42.

Beerbower, J.R. (1968) *Search for the past; an introduction to paleontology.* 2nd ed. Englewood Cliffs, N.J.: Prentice-Hall. 512p.

Bell, B.M. (1976) *A study of North American Edrioasteroidea.* New York State Museum and Science Service. Memoir 21, 447p.

Bell, Robert (1894) *Pre-Paleozoic decay of crystalline rocks north of Lake Huron.* Geological Society of America. Bulletin, v.5, p.357-366.

Benedict, L.G., and others (1968) *Pseudovitrinite in Appalachian coking coals.* Fuel, v.47, p.135-143.

Beneo, E. (1955) *Les résultats des études pour la recherche pétrolifère en Sicile.* World Petroleum Congress. 4th, Rome, 1955. Proceedings, sec.1, p.109-124.

Berger, W.H. (1974) *Deep-sea sedimentation.* In: Burk, C.A., and others, eds. *The geology of continental margins,* p.213-241. New York: Springer-Verlag.

Berkey, C.P., and Morris, F.K. (1924) *Basin structures in Mongolia.* American Museum of Natural History. Bulletin, v.51, p.103-127.

Berkland, J.O., and others (1972) *What is Franciscan?* American Association of Petroleum Geologists. Bulletin, v.56, p.2295-2302.

Bernard, B. (1970) *ABC's of infrared.* Indianapolis: Howard W. Sames & Co.

Berry, L.G., and Mason, Brian (1959) *Mineralogy: concepts, descriptions, determinations.* San Francisco: Freeman. 630p.

Berry, W.B.N. (1966) *Zones and zones—with exemplification from the Ordovician.* American Association of Petroleum Geologists. Bulletin, v.50, p.1487-1500.

Berthelsen, Asger (1970) *Globulith: a new type of intrusive structure, exemplified by metabasic bodies in the Moss area, SE Norway.* Norges Geologiske Undersokelse, [Publikasjoner]. No.266 (Årbok 1969).

Beus, A.A. (1978) *Lithogeochemistry.* Journal of Geochemical Exploration, v.9, p.110-111.

Beus, A.A., and others (1962) *Albitized and greisenized granite (apogranite).* (In Russian.) Moscow: Izd. Akad. Nauk SSSR. 195p.

Bhattacharji, Somdev (1958) *Theoretical and experimental investigations on crossfolding.* Journal of Geology, v.66, p.625-667.

Billings, B.H. (1963) *Optics.* In: *American Institute of Physics Handbook.* 2d ed.

Billings, M.P. (1954) *Structural geology.* 2nd ed. Englewood Cliffs, N.J.: Prentice-Hall. 514p.

Billings, M.P. (1972) *Structural geology.* 3rd ed. Englewood Cliffs, N.J.: Prentice-Hall. 606p.

Birch, A.F. (1952) *Elasticity and constitution of the Earth's interior.* Journal of Geophysical Research, v.57, p.227-286.

Bird, J.B. (1957) *The physiography of Arctic Canada, with special reference to the area south of Parry Channel.* Baltimore: Johns Hopkins Press. 336p.

Birkenmajer, Krzysztof (1958) *Oriented flowage casts and marks in the Carpathian flysch and their relation to flute and groove casts.* Acta geologica Polonica, v.8, p.139-148.

Birkenmajer, Krzyysztof (1959) *Classification of bedding in flysch and similar graded deposits.* Studia geologica Polonica, v.3, p.81-133.

Bissell, H.J. (1959) *Silica in sediments of the Upper Paleozoic of the Cordilleran area.* In: Ireland, H.A., ed. *Silica in sediments—a symposium.* Tulsa: Society of Economic Paleontologists and Mineralogists. 185p. (Its Special Publication, no.7, p.150-185).

Bissell, H.J. (1964) *Ely, Arcturus, and Park City groups (Pennsylvanian-Permian) in eastern Nevada and western Utah.* American Association of Petroleum Geologists. Bulletin, v.48, p.565-636.

Bissell, H.J. (1964a) *Patterns of sedimentation in Pennsylvanian and Permian strata of part of the eastern Great Basin.* Kansas. State Geological Survey. Bulletin 169, v.1, p.43-56.

Bissell, H.J., and Chillingar, G.V. (1967) *Classification of sedimentary carbonate rocks.* In: Chillingar, G.V., and others, eds. *Carbonate rocks; origin, occurrence and classification,* p.87-168. Amsterdam: Elsevier. (Developments in Sedimentology 9A, ch.4).

Black, R.F. (1954) *Permafrost—a review.* Geological Society of America. Bulletin, v.65, p.839-856.

Black, R.F. (1966) *Comments on periglacial terminology.* Biuletyn peryglacjalny, no.15, p.329-333.

Blackwelder, Eliot (1931) *Desert plains.* Journal of Geology, v.39, p.133-140.

Blackwelder, Eliot (1931a) *Pleistocene glaciation in the Sierra Nevada and Basin Ranges.* Geological Society of America. Bulletin, v.42, p.865-922.

Blackwelder, R.E. (1967) *Taxonomy.* New York: Wiley. 698p.

Blank, H.R. (1951) *"Rock doughnuts", a product of granite weathering.* American Journal of Science, v.249, p.822-829.

Blatt, Harvey, Middleton, Gerald, and Murray, Raymond (1972) *Origin of sedimentary rocks.* Englewood Cliffs, N.J.: Prentice-Hall. 634p.

Blench, Thomas (1957) *Regime behaviour of canals and rivers.* London: Butterworths. 138p.

Blissenbach, Erich (1954) *Geology of alluvial fans in semiarid regions.* Geological Society of America. Bulletin, v.65, p.175-189.

Bloom, A.L. (1978) *Geomorphology: A systematic analysis of Late Cenozoic landforms.* Englewood Cliffs, N.J.: Prentice-Hall. 510p.

BNCG: British National Committee for Geography (1966) *Glossary of technical terms in cartography.* London: The Royal Society. 84p.

Boardman, R.S. (1971) *Mode of growth and functional morphology of autozooids in some recent and Paleozoic tubular bryozoa.* Smithsonian Contributions to Paleobiology, no.8. 51p.

Boardman, R.S., and Cheetham, A.H. (1969) *Skeletal growth, intracolony variation, and evolution in bryozoa—a review.* Journal of Paleontology, v.43, no.2, p.205-233.

Boardman, R.S., and Cheetham, A.H. (1973) *Degrees of colony dominance in stenolaemate and gymnolaemate Bryozoa.* In: Boardman, R.S., and others, eds. *Animal colonies,* p.121-200. Stroudsburg, Pa.: Dowden, Hutchinson & Ross.

Bold, H.C. (1967) *Morphology of plants.* 2nd ed. New York: Harper & Row.

Bolli, H.M. (1969) *Report of working group for a biostratigraphic zonation of the Cretaceous and Cenozoic as a basis for correlation in marine geology.* International Union of Geological Sciences.

Geology Newsletter, v.1969, no.3, p.199-207.

Bonney, T.G. (1886) *The anniversary address of the President. "Metamorphic" rocks.* Geological Society of London. Quarterly Journal, v.42, (Proc.), p.38-115.

Boothroyd, J.C., and Hubbard, D.K. (1974) *Bedform development and distribution pattern, Parker and Essex estuaries, Massachusetts.* U.S. Army Corps of Engineers. Coastal Engineering Research Center. Miscellaneous Paper 1-74.

Bosellini, A. (1966) *Protointraclasts: texture of some Werfenian (Lower Triassic) limestones of the Dolomites (northeastern Italy).* Sedimentology, v.6, p.333-337.

Bosellini, Alfonso, and Ginsburg, R.N. (1971) *Form and internal structure of recent algal nodules (rhodolites) from Bermuda.* Journal of Geology, v.79, p.669-682.

Bostick, N.H. (1970) *Measured alteration of organic particles (phytoclasts) as an indicator of contact and burial metamorphism in sedimentary rocks.* Geological Society of America. Abstracts with Programs, v.2, no.2, p.74.

Boswell, P.G.H. (1960) *The term graywacke.* Journal of Sedimentary Petrology, v.30, p.154-157.

Boulton, G.S. (1970) *On the deposition of subglacial and melt-out tills at the margins of certain Svalbard glaciers.* Journal of Glaciology, v.9, p.231-245.

Bouma, A.H. (1962) *Sedimentology of some flysch deposits; a graphic approach to facies interpretation.* Amsterdam: Elsevier. 168p.

Bourcart, Jacques (1939) *Essai d'une définition de la vase des estuaires.* Académie des Sciences, Paris. Comptes rendus hebdomadaires des séances, v.209, no. 14, p.542-544.

Bourcart, Jacques (1941) *Essai de définition des vases des eaux douces.* Académie des Sciences, Paris. Comptes rendus hebdomadaires des séances, v.212, no.11, p.448-450.

Bowen, N.L. (1941) *Certain singular points on crystallization curves of solid solutions.* National Academy of Science. Proceedings, v.27, p.301-309.

Bowen, R. (1966) *Oxygen isotopes as climatic indicators.* Earth-science Reviews, v.2, p.199-224.

Boydell, H.C. (1926) *A discussion of metasomatism and the linear "force of growing crystals".* Economic Geology, v.21, p.1-55.

Boyer, S.E. (1976) *Formation of the Grandfather Mountain window duplex thrusting* (abstract). Geological Society of America, Annual Meeting, Denver. Abstracts with Programs, p.788-789.

Bradley, W.C. (1958) *Submarine abrasion and wave-cut platforms.* Geological Society of America. Bulletin, v.69, p.967-974.

Bradley, W.C. (1963) *Large-scale exfoliation in massive sandstones of the Colorado Plateau.* Geological Society of America. Bulletin, v.74, p.519-527.

Bradley, W.C., and Griggs, G.B. (1976) *Form, genesis, and deformation of central California wave-cut platforms.* Geological Society of America. Bulletin, v.87, p.433-449.

Bradley, W.H. (1930) *The behavior of certain mud-crack casts during compaction.* American Journal of Science, ser.5, v.20, p.136-144.

Bradley, W.H. (1931) *Origin and microfossils of the oil shale of the Green River Formation of Colorado and Utah.* U.S. Geological Survey. Professional Paper 168. 58 p.

Brady, J.B. (1975) *Chemical components and diffusion.* American Journal of Science, v.275, p.1073-1088.

Bramlette, M.N. (1946) *The Monterey Formation of California and the origin of its siliceous rocks.* U.S. Geological Survey. Professional Paper 212. 57p.

Branco, W., and Fraas, E. (1905) *Das kryptovulcanische Becken von Steinheim.* Akademie der Wissenschaften, Berlin. Physikalische Abhandlungen, Jahrg. 1905, Abh.1. 64p.

Braunstein, Jules (1961) *Calciclastic and siliciclastic.* American Association of Petroleum Geologists. Bulletin, v.45, p.2017.

Bray, Warwick, and Trump, David (1970) *The American Heritage guide to archaeology.* New York: American Heritage Press. 269p.

Breithaupt, August (1847) *Handbuch der mineralogie. Vol. 3.* Dresden & Leipzig: Arnoldische Buchhandlung. 496p.

Bretz, J H. (1929) *Valley deposits immediately east of the channeled scabland of Washington.* Journal of Geology, v.37, p.393-427, 505-541.

Bretz, J H. (1940) *Solution cavities in the Joliet Limestone of northeastern Illinois.* Journal of Geology, v.50, p.675-811.

Bretz, J H. (1942) *Vadose and phreatic features of limestone caverns.* Journal of Geology, v.50, p.675-811.

Brewer, Roy (1964) *Fabric and mineral analysis of soils.* New York: Wiley. 470p.

Brewer, Roy, and Sleeman, J.R. (1960) *Soil structure and fabric: their definition and description.* Journal of Soil Science, v.11, p.172-185.

Brigham, A.P. (1901) *A text-book of geology.* New York: Appleton. 477p.

Brobst, D.A., and Pratt, W.P. (1973) *United States mineral resources.* U.S. Geological Survey. Professional Paper 820. 722p.

Brongniart, Alexandre (1823) *Macigno.* Dictionaire des sciences naturelles, v.27, p.297-504.

Brook, G.A., and Ford, D.C. (1978) *The origin of labyrinth and tower karst and the climatic conditions necessary for their development.* Nature, v.275, p.493-496.

Brooks, H.K. (1954) *The rock and stone terms limestone and marble.* American Journal of Science, v.252, p.755-760.

Brown, J.S. (1943) *Suggested use of the word microfacies.* Economic Geology, v.38, p.325.

Brown, R.J.E., and Kupsch, W.O. (1974) *Permafrost terminology.* National Research Council of Canada. Technical memorandum no.111. 62p. (NRCC 14274.)

Brown, R.W. (1946) *Salt ribbons and ice ribbons.* Washington Academy of Sciences. Journal, v.36, p.14-16.

Brown, V.J., and Runner, D.G. (1939) *Engineering terminology; definitions of technical words and phrases.* 2nd ed. Chicago: Gillette. 439p.

Bryan, A.B. (1937) *Gravimeter design and operation.* Geophysics, v.2, no.4, p.301-308.

Bryan, Kirk (1920) *Origin of rock tanks and charcos.* American Journal of Science, 4th ser., v.50, p.186-206.

Bryan, Kirk (1923a) *Erosion and sedimentation in the Papago country, Arizona, with a sketch of the geology.* U.S. Geological Survey. Bulletin 730-B, p.19-90.

Bryan, Kirk (1923b) *Geology and ground-water resources of Sacramento Valley, California.* U.S. Geological Survey. Water-supply Paper 495. 285p.

Bryan, Kirk (1934) *Geomorphic processes at high altitude.* Geographical Review, v.24, p.655-656.

Bryan, Kirk (1940) *Gully gravure, a method of slope retreat.* Journal of Geomorphology, v.3, p.89-107.

Bryan, Kirk (1946) *Cryopedology, the study of frozen ground and intensive frost-action, with suggestions on nomenclature.* American Journal of Science, v.244, p.622-642.

Bryan, Kirk (1951) *The erroneous use of* tjaele *as the equivalent of perennially frozen ground.* Journal of Geology, v.59, p.69-71.

Bryant, D.G. (1968) *Intrusive breccias associated with ore, Warren (Bisbee) mining district, Arizona.* Economic Geology, v.63, p.1-12.

BSI: British Standards Institution (1964) *Glossary of mining terms; section 5: Geology.* London: British Standards Institution. 18p. (B.S. 3618).

Bucher, W.H. (1919) *On ripples and related sedimentary surface forms and their paleogeographic interpretation.* American Journal of Science, ser.4, v.47. p.149-210, 241-269.

Bucher, W.H. (1932) *"Strath" as a geomorphic term.* Science, v.75, p.130-131.

Bucher, W.H. (1933) *The deformation of the Earth's crust.* Princeton: Princeton University Press. 518p.

Bucher, W.H. (1952) *Geologic structure and orogenic history of Venezuela; text to accompany the author's geologic tectonic map of Venezuela.* Geological Society of America. Memoir 49. 113p.

Bucher, W.H. (1955) *Deformation in orogenic belts.* Geological Society of America. Special Paper 62, p.343-368.

Bucher, W.H. (1963) *Cryptoexplosion structures caused from without or from within the Earth? ("astroblemes" or "geoblemes"?).* American Journal of Science, v.261, p.597-649.

Bucher, W.H. (1965) *Role of gravity in orogenesis.* Geological Society of America. Bulletin, v.67, p.1295-1318.

Buckland, William (1817) *Description of the Paramoudra, a singular fossil body that is found in the Chalk of the North of Ireland; with some general observations upon flints in chalk, tending to illustrate the history of their formation.* Geological Society of London. Transactions, v.4, p.413-423.

Buckland, William (1829) *On the formation of the Valley of Kingsclere and other valleys by the elevation of the strata that enclose them; and on the evidences of the original continuity of the basins of London and Hampshire.* Geological Society of London. Transactions, ser.2, v.2, p.119-130.

Buckland, William (1829a) *On the discovery of coprolites, or fossil* *faeces, in the Lias at Lyme Regis, and in other formations.* Geological Society of London. Transactions, ser.2, v.3, p.223-236.

Buckman, S.S. (1893) *The Bajocian of the Sherborne district: its relation to subjacent and superjacent strata.* Geological Society of London. Quarterly Journal, v.49, p.479-522.

Buckman, S.S. (1902) *The term "hemera".* Geological Magazine, dec.4, v.9, p.554-557.

Bull, W.B. (1975) *Allometric change of landforms.* Geological Society of America. Bulletin, v.86, p.1489-1498.

Burt, D.M. (1974) *Concepts of acidity and basicity in petrology—the exchange operator approach* (abstract). Geological Society of America. Abstracts, v.6, p.674-676.

Burt, F.A. (1928) *Melikaria: vein complexes resembling septaria veins in form.* Journal of Geology, v.36, p.539-544.

Buwalda, J.P. (1937) *Shutterridges, characteristic physiographic features of active faults.* Geological Society of America. Proceedings, 1936, p.307.

C

Cady, G.H. (1921) *Coal resources of District IV.* Illinois State Geological Survey. Cooperative Mining Series, Bulletin 26. 247p.

Cairnes, D.D. (1912) *Some suggested new physiographic terms.* American Journal of Science, v.34, p.75-87.

Callomon, J.H. (1963) *Sexual dimorphism in Jurassic ammonides.* Leicester Literary and Philosophical Society. Transactions, v.57, p.21-56.

Callomon, J.H. (1965) *Notes on Jurassic stratigraphical nomenclature.* Carpatho-Balkan Geological Association. 7th Congress, Sofia, Sept. 1965. Reports, pt.2, v.1, p.81-85.

Cameron, W.S. (1972) *Comparative analyses of observations of lunar transient phenomena.* Icarus, v.16, p.339-387.

Campbell, C.V. (1966) *Truncated wave-ripple laminae.* Journal of Sedimentary Petrology, v.36, p.825-828.

Campbell, J.F. (1865) *Frost and fire; natural engines, tool-mark & chips with sketches taken at home and abroad by a traveller.* Vol.2. Philadelphia: Lippincott. 519p.

Campbell, K.M., and others (1977) *Langmuir circulation as a factor in the formation of depositional beach cusps.* In: Tanner, W.F., ed. *Coastal sedimentology,* p.245-262. Tallahassee: Department of Geology, Florida State University.

Campbell, M.R. (1896) *Drainage modifications and their interpretation.* Journal of Geology, v.4, p.567-581, 657-678.

Capot-Rey, R. (1945) *Dry and humid morphology in the Western Erg.* Geographical Review, v.35, p.391-407.

Carey, A.E., and Oliver, F.W. (1918) *Tidal lands; a study of shore problems.* London: Blackie. 284p.

Carey, S.W. (1958) *A tectonic approach to continental drift.* In: Carey, S.W., convener. *Continental drift: a symposium.* Hobart: University of Tasmania, Geology Dept. p.177-355.

Carey, S.W., convener (1963) *Syntaphral tectonics and diagenesis; a symposium.* Hobart: University of Tasmania, Geology Dept. 190p.

Carmichael, I.S.E., Turner, F.J., and Verhoogen, J. (1974) *Igneous petrology.* New York: McGraw-Hill. 739p.

Carozzi, A.V. (1957) *Micro-mechanisms of sedimentation in epicontinental environment.* Geological Society of America. Bulletin, v.68, p.1706-1707.

Carozzi, A.V., and Textoris, D.A. (1967) *Paleozoic carbonate microfacies of the eastern stable interior (U.S.A.).* Leiden: Brill. 41p. (*International sedimentary petrographical series, v.11.*)

Carroll, Dorothy (1939) *Movement of sand by wind.* Geological Magazine, v.76, p.6-23.

Casertano, L., Oliveri del Castillo, A., and Quagliariello, M.T. (1976) *Hydrodynamics and geodynamics in the Phlegraean Fields area of Italy.* Nature, v.264, p.161-164.

Cassidy, W.A. (1968) *Meteorite impact structures at Campo del Cielo, Argentina.* In: French, B.M., and Short, N.M., eds. *Shock metamorphism of natural materials,* p.117-128. Baltimore: Mono Book Corp. 644p.

Caster, K.E. (1934) *The stratigraphy and paleontology of northwestern Pennsylvania.* Pt. 1. Bulletins of American Paleontology, v.21, no.71. 185p.

Cayeux, Lucien (1929) *Les roches sedimentaires de France; roches siliceuses.* Paris: Imprimerie Nationale. 774p.

CERC: U.S. Army. Coastal Engineering Research Center (1966) *Shore protection, planning and design.* 3rd ed. Washington, D.C.: Government Printing Office. 580p. (Its Technical report no.4).

Chadwick, G.H. (1931) *Storm rollers.* Geological Society of America. Bulletin, v.42, p.242.

Chadwick, G.H. (1939) *Geology of Mount Desert Island, Maine.* American Journal of Science, v.237, p.355-363.

Chadwick, G.H. (1948) *Ordovician "dinosaur-leather" markings (exhibit).* Geological Society of America. Bulletin, v.59, p.1315.

Challinor, J., and Williams, K.E. (1926) *On some curious marks on a rock surface.* Geological Magazine, v.63, no.746, p.341-343.

Challinor, John (1962) *A dictionary of geology.* 1st ed. New York: Oxford University Press. 235p.

Challinor, John (1967) *A dictionary of geology.* 3rd ed. New York: Oxford University Press. 298p.

Challinor, John (1978) *A dictionary of geology.* 5th ed. New York: Oxford University Press. 365p.

Chamberlin, T.C. (1879) *Annual report of the Wisconsin Geological Survey for the year 1878.* Madison: Wisconsin Geological Survey. 52p.

Chamberlin, T.C. (1883) *Terminal moraine of the second glacial epoch.* U.S. Geological Survey. Annual Report, 3rd, p.291-402.

Chamberlin, T.C. (1888) *The rock-scorings of the great ice invasions.* U.S. Geological Survey. Annual Report, 7th, v.3, p.147-248.

Chamberlin, T.C. (1893) *The horizon of drumlin, osar and kame formation.* Journal of Geology, v.1, p.255-267.

Chamberlin, T.C. (1894a) *Pseudo-cols.* Journal of Geology, v.2, p.205-206.

Chamberlin, T.C. (1894b) *Proposed genetic classification of Pleistocene glacial formations.* Journal of Geology, v.2, p.517-538.

Chamberlin, T.C. (1897) *The method of multiple working hypotheses.* Journal of Geology, v.5, p.837-848.

Chao, E.C.-T. (1967) *Shock effects in certain rock-forming minerals.* Science, v.156, p.192-202.

Chao, E.C.-T (1967a) *Impact metamorphism.* In: Abelson, P.H., ed. *Researches in geochemistry,* v.2, p.204-233. New York: Wiley. 663p.

Charlesworth, J.K. (1957) *The Quaternary era, with special reference to its glaciation.* London: Arnold. 2 vols., 1700p.

Chayes, F. (1956) *Petrographic modal analysis.* New York: Wiley. 113p.

Chayes, Felix (1964) *A petrographic distinction between Cenozoic volcanics in and around the open oceans.* Journal of Geophysical Research, v.69, p.1573-1588.

Chayes, Felix (1966) *Alkaline and subalkaline basalts.* American Journal of Science, v.264, p.128-145.

Chebotarev, I.I. (1955) *Metamorphism of natural waters in the crust of weathering.* [Pt.] 1-3. Geochimica et cosmochimica acta, v.8, p.22-48, 137-170, 198-212.

Chilingar, G.V. (1957) *Classification of limestones and dolomites on basis of Ca/Mg ratio.* Journal of Sedimentary Petrology, v.27, p.187-189.

Chilingar, G.V., and others (1967) *Diagenesis in carbonate rocks.* In: Larsen, Gunnar, and Chilingar, G.V., eds. *Diagenesis in sediments, p.179-322.* Amsterdam: Elsevier. 551p. (*Developments in sedimentology* 8.)

Chilingarian, G.V., and Wolf, K.H., eds. (1975, 1976) *Compaction of coarse-grained sediments.* Amsterdam: Elsevier. Vol.1, 548p.; vol.2, 808p. (*Developments in sedimentology,* vols. 18A and 18B.)

Chinner, G.A. (1966) *The significance of the aluminum silicates in metamorphism.* Earth Science Reviews, v.2.

Choquette, P.W. (1955) *A petrographic study of the "State College" siliceous oölite.* Journal of Geology, v.63, p.337-347.

Choquette, P.W., and Pray, L.C. (1970) *Geologic nomenclature and classification of porosity in sedimentary carbonates.* American Association of Petroleum Geologists. Bulletin, v.54, p.207-250.

Chow, Van Te, chairman (1957) *Report of the Committee on Runoff, 1955-1956.* American Geophysical Union. Transactions, v.38, p.379-384.

Chow, Van Te, ed. (1964) *Handbook of applied hydrology; a compendium of water-resources technology.* New York: McGraw-Hill. 1418p.

Church, W.R. (1968) *Eclogites.* In: Hess, H.H., and Poldervaart, Arie, eds. *Basalts—the Poldervaart treatise on rocks of basaltic composition,* v.2, p.755-798. New York: Interscience.

Clapp, C.H. (1913) *Contraposed shorelines.* Journal of Geology, v.21, p.537-540.

Clark, K.F. (1972) *Stockwork molybdenum deposits in the western Cordillera of North America.* Economic Geology, v.67, p.731-758.

Clarke, J.I. (1966) *Morphometry from maps.* In: Dury, G.H., ed. *Essays in geomorphology,* p.235-274. New York: American Elsevier. 404p.

Clarke, J.M. (1911) *Observations on the Magdalen Islands.* New York State Museum. Bulletin 149, p.134-155.

Clarke, J.M. (1918) *Strand and undertow markings of Upper Devonian time as indications of the prevailing climate.* New York State Museum. Bulletin 196, p.199-238.

Clarke, J.W. (1958) *The bedrock geology of the Danbury quadrangle.* Connecticut State Geological and Natural History Survey. Quadrangle report no.7. 47p.

Claus, George, and Nagy, Bartholomew (1961) *A microbiological examination of some carbonaceous chondrites.* Nature, v.192, p.594-596.

Cleland, H.F. (1910) *North American natural bridges, with a discussion of their origin.* Geological Society of America. Bulletin, v.21, p.313-338.

Cleland, H.F. (1929) *Geology; physical and historical.* New York: American Book Co. 718p.

Cloos, Ernst (1946) *Lineation, a critical review and annotated bibliography.* Geological Society of America. Memoir 18. 122p.

Cloos, Hans (1939) *Hebung, Spaltung, Vulkanismus; Elemente einer geometrischen Analyse indischer Grossformen.* Geologische Rundschau, v.30, no.4a, p.405-527.

Cloos, Hans (1941) *Bau und Tätigkeit von Tuffschloten: Untersuchungen an dem schwäbischen Vulkan.* Geologische Rundschau, v.32, p.709-800.

Close, U., and McCormick, E. (1922) *Where the mountains walked.* National Geographic Magazine, v.41, p.445-464.

Cloud, P.E., jr. (1957) *Nature and origin of atolls; a progress report.* Pacific Science Congress. 8th, Philippines, 1953. Proceedings, v.3A, p.1009-1036.

Cloud, P.E., jr. (1974) *Rubey conference on crustal evolution.* Science, v.183, p.878-881.

Cloud, P.E., jr., and Barnes, V.E. (1957) *Early Ordovician sea in central Texas.* Geological Society of America. Memoir 67, v.2, p.163-214.

Cloud, P.E., jr., and others (1943) *Stratigraphy of the Ellenburger Group in central Texas—a progress report.* University of Texas. Publication 4301, p.133-161.

Clough, C.T., and others (1909) *The cauldron-subsidence of Glen Coe, and the associated igneous phenomena.* Geological Society of London. Quarterly Journal, v.65, p.611-678.

Coates, D.R. (1966) *Glaciated Appalachian Plateau: till shadows on hills.* Science, v.152, p.1617-1619.

Coats, R.R. (1968) *The Circle Creek Rhyolite, a volcanic complex in northern Elko County, Nevada.* Geological Society of America. Memoir 116, p.69-106.

Coats, R.R. (1968) *Basaltic andesites.* In: Hess, H.H., and Poldervaart, A., eds. *Basalts: the Poldervaart treatise on rocks of basaltic composition,* v.2, p.689-736. New York: Interscience.

Coffey, G.N. (1909) *Clay dunes.* Journal of Geology, v.17, p.754-755.

Cohee, G.V. (1962) *Stratigraphic nomenclature in reports of the U.S. Geological Survey.* Washington: U.S. Geological Survey. 35p.

Cole, J.P., and King, C.A.M. (1968) *Quantitative geography.* London: Wiley. 692p.

Coleman, Alice (1952) *Selenomorphology.* Journal of Geology, v.60, p.451-460.

Coleman, R.G. (1971) *Plate tectonic emplacement of upper mantle peridotites along continental edges.* Journal of Geophysical Research, v.76, p.1212-1222.

Coleman, R.G. (1977) *Ophiolites.* New York: Springer-Verlag. 229p.

Coleman, R.G., and others (1965) *Eclogites and eclogites—their differences and similarities.* Geological Society of America. Bulletin, v.76, p.483-508.

Collet, L.W. (1927) *The structure of the Alps.* London: Edward Arnold. 289p.

Collie, G.L. (1901) *Wisconsin shore of Lake Superior.* Geological Society of America. Bulletin, v.12, p.197-216.

Collinson, J.C. (1970) *Bedforms of the Tana River, Norway.* Geografiska Annaler, v.52, p.31-56.

Colp, J.L. (1967) *Terradynamics: a study of projectile penetration of natural earth materials.* Geological Society of America. Special Paper 115, p.38.

Compton, R.R. (1962) *Manual of field geology.* New York: Wiley. 378p.

Comstock, T.B. (1878) *An outline of general geology, with copious references.* Ithaca: University Press. 82p.

Conant, L.C., Black, R.F., and Hosterman, J.W. (1976) *Sediment-filled pots in upland gravel of Maryland and Virginia.* U.S. Geological Survey. Journal of Research, v.4, p.353-358.

Coney, P.J. (1970) *The geotectonic cycle and the new global tectonics.* Geological Society of America. Bulletin, v.81, p.739-747.

Conkin, J.E., and Conkin, B.M. (1975) *The Devonian-Mississippian and Kinderhookian-Osagean boundaries in the east-central United States are paracontinuities.* University of Louisville. Studies in Paleontology and Stratigraphy, no.4. 52p.

Conybeare, C.E.B., and Crook, K.A.W. (1968) *Manual of sedimentary structures.* Australia. Bureau of Mineral Resources, Geology and Geophysics. Bulletin 102. 327p.

Conybeare, W.D., and Phillips, William (1822) *Outlines of the geology of England and Wales, with an introductory compendium of the general principles of that science, and comparative views of the structure of foreign countries.* Part 1. London: William Phillips. 470p.

Cook, J.H. (1946) *Kame complexes and perforation deposits.* American Journal of Science, v.244, p.573-583.

Cook, P.L. (1965) *Notes on Cupuladriidae (Polyzoa, Anasca).* British Museum (Natural History). Bulletin, Zoology, v.13, p.159-160.

Cooke, R.U., and Warren, Andrew (1973) *Geomorphology in deserts.* London: Batsford. 374p.

Coombs, D.S. (1960) *Lower grade mineral facies in New Zealand.* International Geological Congress. 21st, Copenhagen, 1960. Report, part 13, p.339-351.

Coombs, D.S. (1961) *Some recent work on the lower grades of metamorphism.* The Australian Journal of Science, v.24, no.5, p.203-215.

Coombs, D.S., and others (1959) *The zeolite facies.* Geochimica et Cosmochimica Acta, v.17, p.53-107.

Cooper, B.N. (1945) *Industrial limestones and dolomites in Virginia; Clinch Valley district.* Virginia Geological Survey. Bulletin 66. 259p.

Cooper, B.N., and Cooper, G.A. (1946) *Lower Middle Ordovician stratigraphy of the Shenandoah Valley, Virginia.* Geological Society of America. Bulletin, v.57, p.35-113.

Cooper, G.A. (1956) *Chazyan and related brachiopods.* Smithsonian Miscellaneous Collections, v.127, pt.1-2. 1245p.

Cooper, J.R. (1943) *Flow structure in the Berea Sandstone and Bedford Shale of central Ohio.* Journal of Geology, v.51, p.190-203.

Cooper, W.S. (1958) *Terminology of post-Valders time.* Geological Society of America. Bulletin, v.69, p.941-945.

Cornish, Vaughan (1898) *On sea-beaches and sandbanks.* Geographical Journal, v.11, p.628-651.

Cornish, Vaughan (1899) *On kumatology.* Geographical Journal, v.13, p.624-628.

Correns, C.W. (1950) *Zur Geochemie der Diagenese; I: Das Verhalten von $CaCo_3$ und SiO_2.* Geochimica et Cosmochimica Acta, v.1, p.49-54.

Cottingham, Kenneth (1951) *The geologist's vocabulary.* Scientific Monthly, v.72, p.154-163.

Cotton, C.A. (1916) *Fault coasts in New Zealand.* Geographical Review, v.1, p.20-47.

Cotton, C.A. (1922) *Geomorphology of New Zealand.* Pt.1. Wellington: Dominion Museum. 462p. (New Zealand Board of Science and Art. Manual no.3).

Cotton, C.A. (1940) *Classification and correlation of river terraces.* Journal of Geomorphology, v.3, p.27-37.

Cotton, C.A. (1942) *Climatic accidents in landscape-making.* Christchurch N.Z.: Whitcombe & Tombs. 354p.

Cotton, C.A. (1948) *Landscape as developed by the processes of normal erosion.* 2nd ed. New York: Wiley. 509p.

Cotton, C.A. (1950) *Tectonic scarps and fault valleys.* Geological Society of America. Bulletin, v.61, p.717-757.

Cotton, C.A. (1958) *Geomorphology; an introduction to the study of landforms.* 7th ed., rev. Christchurch , N.Z.: Whitcombe and Tombs. 505p.

Cowan, S.T. (1968) *A dictionary of microbial taxonomic usage.* Edinburgh, Oliver & Boyd. 118p.

Cram, D.J., and Hammond, G.S. (1959) *Organic chemistry.* New York: McGraw-Hill. 712p.

Crickmay, C.H. (1933) *The later stages of the cycle of erosion; some weaknesses in the theory of the cycle of erosion.* Geological Magazine,v.70, p.337-347.

Crickmay, C.H. (1967) *A note on the term bocanne.* American Journal of Science, v.265, p.626-627.

Criswell, D.R., Lindsay, J.F., and Reasoner, D.L. (1975) *Seismic* *and acoustic emissions of a booming dune.* Journal of Geophysical Research, v.80, p.4963-4974.

Cronquist, Arthur (1961) *Introductory botany.* New York: Harper & Row. 892p.

Crosby, W.O. (1912) *Dynamic relations and terminology of stratigraphic conformity and unconformity.* Journal of Geology, v.20, p.289-299.

Cross, Whitman, and Howe, Ernest (1905) *Description of the Silverton quadrangle.* U.S. Geological Survey. Geologic Atlas, folio no. 120. 34p.

Cross, Whitman, and others (1902) *A quantitative chemico-mineralogical classification and nomenclature of igneous rocks.* Journal of Geology, v.10, p.555-690.

Cross, Whitman, and others (1906) *The texture of igneous rocks.* Journal of Geology, v.14, p.692-707.

Crowell, J.C. (1955) *Directional-current structures from the Prealpine Flysch, Switzerland.* Geological Society of America. Bulletin, v.66, p.1351-1384.

Crowell, J.C. (1957) *Origin of pebbly mudstones.* Geological Society of America. Bulletin, v.68, p.993-1009.

Crowell, J.C. (1964) *Climatic significance of sedimentary deposits containing dispersed megaclasts.* In: Nairn, A.E.M., ed. *Problems in palaeoclimatology,* p.86-99. New York: Interscience. 705p.

Cullison, J.S. (1938) *Origin of composite and incomplete internal moulds and their possible use as criteria of structure.* Geological Society of America. Bulletin, v.49, p.981-988.

Cumings, E.R. (1930) *List of species from the New Corydon, Kokomo, and Kenneth formations of Indiana, and from reefs in the Mississinewa and Liston Creek formations.* Indiana Academy of Science. Proceedings, v.39, p.204-211.

Cumings, E.R. (1932) *Reefs or bioherms?* Geological Society of America. Bulletin, v.43, p.331-352.

Cumings, E.R., and Shrock, R.R. (1928) *Niagaran coral reefs of Indiana and adjacent states and their stratigraphic relations.* Geological Society of America. Bulletin, v.39, p.579-619.

Curl, R.L. (1964) *On the definition of a cave.* National Speleological Society. Bulletin, v.26, p.1-6.

Curray, J.R. (1966) *Continental terrace.* In: *Oceanography,* p.207-214. New York: Reinhold. 1021p.

Currier, L.W. (1941) *Disappearance of the last ice sheet in Massachusetts by stagnation zone retreat* (abstract). Geological Society of America. Bulletin, v.52, p.1895.

Curry, H.D. (1938) *"Turtleback" fault surfaces in Death Valley, Calif.* Geological Society of America. Bulletin, v.49, p.1875.

Curry, H.D. (1954) *Turtlebacks in the central Black Mountains, Death Valley, California.* California Division of Mines. Bulletin 170, ch.4, [Pt.] 7, p.53-59.

Cuvillier, Jean (1951) *Corrélations stratigraphiques par microfaciès en Aquitaine occidentale.* Avec la collaboration de V. Sacal. Leiden: E.J. Brill. 23p. 90 plates.

Cvijic, Jovan (1924) *The evolution of lapiés, a study in karst topography.* Geographical Review, v.14, p.26-49.

Cys, J.M. (1963) *A new definition of the penesaline environment.* Compass, v.40, p.161-163.

D

Dahl, Eilif (1955) *Biogeographic and geologic indications of unglaciated areas in Scandinavia during the glacial ages.* Geological Society of America. Bulletin, v.66, p.1499-1519.

Dahlstrom, C.D.A. (1970) *Structural geology in the eastern margin of the Canadian Rocky Mountains.* Bulletin of Canadian Petroleum Geology, v.18, p.332-406. Reprinted, 1977, in Joint Wyoming-Montana-Utah Geological Associations Guidebook, p.407-439.

Dale, T.N. (1894) *Mount Greylock, its areal and structural geology.* U.S. Geological Survey. Monograph 23, p.119-203.

Dale, T.N. (1923) *The commercial granites of New England.* U.S. Geological Survey. Bulletin 738. 488p.

Dalrymple, G.B., Silver, E.A., and Jackson, E.D. (1974) *Origin of the Hawaiian Islands.* In: Greeley, Ronald, ed., *Guidebook to the Hawaiian Planetology Conference,* ch.3, p.23-36. Moffett Field, Calif.: NASA/Ames Research Center.

Daly, R.A. (1902) *The geology of the northeast coast of Labrador.* Harvard College. Museum of Comparative Zoology. Bulletin 38, Geological Series 5, p.205-270.

Daly, R.A. (1917) *Metamorphism and its phases.* Geological Society of America. Bulletin, v.28, p.375-418.

Damon, P.E. (1968) *Potassium-argon dating of igneous and metamorphic rocks with applications to the Basin Ranges of Arizona and Sonora.* In: Hamilton, E.I., and Farquhar, R.M., eds. *Radiometric dating for geologists,* p.1-71. New York: Interscience. 506p.

Dana, E.S. (1892) *The system of mineralogy of J. D. Dana, 1837-1868; descriptive mineralogy.* 6th ed. New York: Wiley. 1134p.

Dana, J.D. (1873) *On some results of the Earth's contraction from cooling including a discussion of the origin of mountains and the nature of the Earth's interior.* American Journal of Science, v.5, p.423-443.

Dana, J.D. (1874) *A text-book of geology.* 2nd ed. New York: Ivison, Blakeman, Taylor. 358p.

Dana, J.D. (1890) *Characteristics of volcanoes. . .* New York: Dodd, Mead. 399p.

Dana, J.D. (1895) *Manual of geology; treating of the principles of the science with special reference to American geological history.* 4th ed. New York: American Book Co. 1088p.

Dapples, E.C. (1962) *Stages of diagenesis in the development of sandstones.* Geological Society of America. Bulletin, v.73, p.913-933.

Dapples, E.C., and others (1953) *Petrographic and lithologic attributes of sandstones.* Journal of Geology, v.61, p.291-317.

Darling, F.F., and Boyd, J.M. (1964) *The Highlands and Islands.* London: Collins. 336p.

Darrah, W.C. (1939) *Principles of paleobotany.* Leiden, Holland: Chronica Botanica Co. 239p.

Daubrée, G.A. (1879) *Études synthétiques de géologie expérimentale.* Paris: Dunod. 828p.

Davies, H.G. (1965) *Convolute lamination and other structures from the Lower Coal Measures of Yorkshire.* Sedimentology, v.5, p.305-325.

Davis, C.A. (1907) *Peat, essays on its origin, uses, and distribution in Michigan.* Michigan Geological Survey. Annual Report 1906, p.93-395.

Davis, J.C. (1973) *Statistics and data analysis in geology.* New York: Wiley. 550p.

Davis, W.M. (1885) *Geographic classification, illustrated by a study of plains, plateaus, and their derivatives.* American Association for the Advancement of Science. Proceedings, v.33, p.428-432.

Davis, W.M. (1889a) *Topographic development of the Triassic formation of the Connecticut Valley.* American Journal of Science, 3rd ser., v.37, p.423-434.

Davis, W.M. (1889b) *The rivers and valleys of Pennsylvania.* National Geographic Magazine, v.1, p.183-253.

Davis, W.M. (1890) *The rivers of northern New Jersey, with notes on the classification of rivers in general.* National Geographic Magazine, v.2, p.81-110.

Davis, W.M. (1894) *Physical geography in the university.* Journal of Geology, v.2, p.66-100.

Davis, W.M. (1895) *The development of certain English rivers.* Geographical Journal, v.5, p.127-146.

Davis, W.M. (1897) *Current notes on physiography.* Science, n.s., v.6, p.22-24.

Davis, W.M. (1899) *The geographical cycle.* Geographical Journal, v.14, p.481-504.

Davis, W.M. (1902) *Base level, grade, and peneplain.* Journal of Geology, v.10, p.77-111.

Davis, W.M. (1909) *Geographical essays.* Boston: Ginn. 777p.

Davis, W.M. (1911) *The Colorado Front Range; a study in physiographic presentation.* Association of American Geographers. Annals, v.1, p.21-83.

Davis, W.M. (1912) *Relation of geography to geology.* Geological Society of America. Bulletin, v.23, p.93-124.

Davis, W.M. (1918) *A handbook of northern France.* Cambridge: Harvard Univ. Press. 174p.

Davis, W.M. (1922) *Peneplains and the geographical cycle.* Geological Society of America. Bulletin, v.33, p.587-598.

Davis, W.M. (1925a) *The undertow myth.* Science, v.61, p.206-208.

Davis, W.M. (1925b) *A Roxen lake in Canada.* Scottish Geographical Magazine, v.41, p.65-74.

Davis, W.M. (1927) *The rifts of southern California.* American Journal of Science, 5th ser., v.13, p.57-72.

Davis, W.M. (1930) *Rock floors in arid and in humid climates.* Journal of Geology, v.38, p.1-27,136-158.

Davis, W.M. (1932) *Piedmont benchlands and Primärrümpfe.* Geological Society of America. Bulletin, v.43, p.399-440.

Davis, W.M. (1933) *Granite domes of the Mojave Desert, California.* San Diego Society of Natural History. Transactions, v.7, no.20, p.211-258.

Davis, W.M. (1938) *Sheetfloods and streamfloods.* Geological Society of America. Bulletin, v.49, p.1337-1416.

Dean, W.E., Davies, G.R., and Anderson, R.Y. (1975) *Sedimentological significance of nodular and laminated anhydrite.* Geology, v.3, p.367-372.

Deane, R.E. (1950) *Pleistocene geology of the Lake Simcoe district, Ontario.* Canada. Geological Survey. Memoir 256. 108p.

Deevey, E.S., jr., and Flint, R.F. (1957) *Postglacial hypsithermal interval.* Science, v.125, p.182-184.

DeFord, R.K. (1946) *Grain size in carbonate rock.* American Association of Petroleum Geologists. Bulletin, v.30, p.1921-1928.

De Geer, G. (1912) *A geochronology of the last 12,000 years.* International Geological Congress. 11th, Stockholm, 1910. Report, v.1, p.241-253.

De la Beche, H.T. (1832) *Geological manual.* Philadelphia: Carey & Lea. 535p.

Dellwig, L.F. (1955) *Origin of the Salina salt of Michigan.* Journal of Sedimentary Petrology, v.25, p.83-110.

Dence, M.R. (1968) *Shock zoning at Canadian craters: petrography and structural implications.* In: French, B.M., and Short, N.M., eds. *Shock metamorphism of natural materials,* p.169-184. Baltimore: Mono Book Corp. 644p.

Dennis, J.G., and Atwater, T.M. (1974) *Terminology of geodynamics.* American Association of Petroleum Geologists. Bulletin, v.58, p.1030-1036.

Dennis, J.G., ed. (1967) *International tectonic dictionary; English terminology.* American Association of Petroleum Geologists. Memoir 7. 196p.

Derby, O.A. (1910) *The iron ores of Brazil.* In: *The iron ore resources of the world,* v.2, p.813-822. International Geological Congress. 11th, Stockholm, 1910. Stockholm Generalstabens Litografiska Anstalt. 2v. 1068p.

DeSitter, L.U. (1954) *Gravitational sliding tectonics—an essay on comparative structural geology.* American Journal of Science, v.252, p.321-344.

DeVoe, J.R., and Spijkerman, J.J. (1966) *Mössbauer spectrometry.* Analytical Chemistry, v.38, no.5, p.382R-393R.

Dewey, J.F. (1969) *Continental margins—a model for conversion of Atlantic type to Andean type.* Earth and Planetary Science Letters, v.6, p.189-197.

DeWiest, R.J.M. (1965) *Geohydrology.* New York: Wiley. 366p.

Dewolf, Y. (1970) *Les argiles à siles: paléosols ou pédolithes.* Association francaise pour l'étude du Quaternaire. Bulletin, no.2-3, p.117-119.

Dickinson, W.R. (1965) *Geology of the Suplee-Izee area, Crook, Grant, and Harney counties, Oregon.* Oregon Department of Geology and Mineral Industries. Bulletin 58. 109p.

Dickinson, W.R. (1966) *Structural relationships of San Andreas fault system, Cholame Valley and Castle Mountain Range, California.* Geological Society of America. Bulletin, v.66, p.707-725.

Dickinson, W.R. (1970) *Interpreting detrital modes of graywacke and arkose.* Journal of Sedimentary Petrology, v.40, p.695-707.

Dickinson, W.R., and Vigrass, L.W. (1965) *Geology of the Suplee-Izee area, Crook, Grant, and Harney counties, Oregon.* Oregon Department of Geology and Mineral Resources. Bulletin, no.58, 110p.

Dietrich, R.V. (1959) *Development of ptygmatic features within a passive host during partial anatexis.* Beiträge zur Mineralogie und Petrologie, v.6, no.6, p.357-365.

Dietrich, R.V. (1960) *Nomenclature of migmatitic and associated rocks.* GeoTimes, v.4, no.5, p.36-37, 50-51.

Dietrich, R.V. (1960a) *Genesis of ptygmatic features.* International Geological Congress, 21st, Copenhagen, 1960. Report, Norden. Proceedings, sec.14, p.138-148.

Dietrich, R.V. (1963) *Banded gneisses of eight localities.* Norsk Geologisk Tidsskrift, v.43, no.1, p.89-119.

Dietrich, R.V. (1969) *Hybrid rocks.* Science, v.163, p.557.

Dietrich, R.V., and Mehnert, K.R. (1961) *Proposal for the nomenclature of migmatites and associated rocks.* International Geological Congress, 21st, Copenhagen, 1960. Report, pt.26, sec.14, p.56-67.

Dietz, R.S. (1959) *Shatter cones in cryptoexplosion structures (meteorite impact?).* Journal of Geology, v.67, p.496-505.

Dietz, R.S. (1960) *Meteorite impact suggested by shatter cones in rock.* Science, v.131, p.1781-1784.

Dietz, R.S. (1961) *Astroblemes.* Scientific American, v.205, p.50-58.

Dietz, R.S. (1963) *Wave-base, marine profile of equilibrium, and wave-built terraces: a critical appraisal.* Geological Society of America. Bulletin, v.74, p.971-990.

Dietz, R.S., and Holden, J.C. (1966) *Miogeoclines (miogeosynclines) in space and time.* Journal of Geology, v.74, no.5, pt.1, p.566-583.

Dionne, J.-C. (1973) *Monroes: a type of so-called mud volcanoes in tidal flats.* Journal of Sedimentary Petrology, v.43, p.848-856.

Dixon, E.E.L., and Vaughan, Arthur (1911) *The Carboniferous succession in Gower (Glamorganshire), with notes on its fauna and conditions of deposition.* Geological Society of London. Quarterly Journal, v.67, p.477-571.

Dobkins, J.E., jr., and Folk, R.L. (1970) *Shape development on Tahiti-Nui.* Journal of Sedimentary Petrology, v.40, p.1167-1203.

Dobrin, M.B. (1952) *Introduction to geophysical prospecting.* New York: McGraw-Hill. 435p.

DOD: U.S. Department of Defense. Army Topographic Command (1969) *Glossary of mapping, charting, and geodetic terms.* 2nd ed. Washington, D.C.: U.S. Department of Defense. 281p.

DOD: U.S. Department of Defense. Army Topographic Command (1973) *Glossary of mapping, charting and geodetic terms.* 3rd ed. Washington.: U.S. Department of Defense. 272p.

Doe, B.R. (1970) *Lead isotopes.* New York: Springer-Verlag. 137p.

Donaldson, Douglas, and Simpson, Scott (1962) Chomatichnus, *a new ichnogenus, and other trace-fossils of Wegber Quarry.* Liverpool and Manchester Geological Journal, v.3, p.73-81.

Donnay, Gabrielle, and Donnay, J.D.H. (1953) *The crystallography of bastnaesite, parisite, roentgenite, and synchisite.* American Mineralogist, v.38, nos. 11&12, pp.932-963.

Dorr, J.V.N., II, and Barbosa, A.L. de Miranda (1963) *Geology and ore deposits of the Itabira district, Minas Gerais, Brazil.* U.S. Geological Survey. Professional Paper 341-C. 110p.

Dott, R.H., jr. (1964) *Wacke, graywacke and matrix—what approach to immature sandstone classification?* Journal of Sedimentary Petrology, v.34, p.625-632.

Douglas, R.J.W. (1950) *Callum Creek, Langford Creek, and Gap map areas, Alberta.* Geological Survey of Canada. Memoir 255. 124p.

Draper, J.W. (1847) *On the production of light by heat.* Philosophical Magazine, ser.3, v.30, p.345-360.

Drew, G.H. (1911) *The action of some denitrifying bacteria in tropical and temperate seas, and the bacterial precipitation of calcium carbonate in the sea.* Marine Biological Association of the United Kingdom. Journal, v.9, p.142-155.

Drewes, Harold (1959) *Turtleback faults of Death Valley, California: a reinterpretation.* Geological Society of America. Bulletin, v.70, p.1497-1508.

Dryer, C.R. (1899) *The meanders of the Muscatatuck at Vernon, Ind.* Indiana Academy of Science. Proceedings, 1898, p.270-273.

Dryer, C.R. (1901) *Certain peculiar eskers and esker lakes of northeastern Indiana.* Journal of Geology, v.9, p.123-129.

Dryer, C.R. (1910) *Some features of delta formation.* Indiana Academy of Science. Proceedings, 1909, p.255-261.

Duff, P.M.D., and Walton, E.K. (1962) *Statistical basis for cyclothems: a quantitative study of the sedimentary succession in the East Pennine Coalfield.* Sedimentology, v.1, p.235-255.

Dulhunty, J.A. (1939) *The torbanites of New South Wales, Pt.1.* Royal Society of New South Wales. Journal and Proceedings, 1938, v.72, p.179-198.

Dunbar, C.O., and Rodgers, John (1957) *Principles of stratigraphy.* New York: Wiley. 356p.

Dunham, R.J. (1962) *Classification of carbonate rocks according to depositional texture.* American Association of Petroleum Geologists. Memoir 1, p.108-121.

Dunham, R.J. (1970) *Stratigraphic reefs versus ecologic reefs.* American Association of Petroleum Geologists. Bulletin, v.54, p.1931-1932.

Dunn, J.A. (1942) *Granite and magmation and metamorphism.* Economic Geology, v.37, p.231-238.

Dunoyer de Segonzac, G. (1968) *The birth and development of the concept of diagenesis (1866-1966).* Earth-Science Reviews, v.4, p.153-201.

DuToit, A. (1920) *The Karroo dolerites of South Africa.* Geological Society of South Africa. Transactions, v.23, p.1-42.

DuToit, A. (1937) *Our wandering continents.* New York: Hafner. 379p.

Dzulynski, Stanislaw, and Kotlarczyk, J. (1962) *On load-casted ripples.* Société Géologique Pologne. Annales, v.32, p.148-159.

Dzulynski, Stanislaw, and others (1959) *Turbidites in flysch of the Polish Carpathian Mountains.* Geological Society of America.

Bulletin, v.70, p.1089-1118.

Dzulynski, Stanislaw, and Radomski, Andrzej (1955) *Pochodzenie śladów wleczenia na tle teorii pradów zawiesinowych.* Acta geologica Polonica, v.5, p.47-66.

Dzulynski, Stanislaw, Radomski, A., and Slaczka, A. (1957) *Sandstone whirl-balls in the silts of the Carpathian flysch.* Société Géologique Pologne. Annales, v.26, p.107-126.

Dzulynski, Stanislaw, and Sanders, J.E. (1962) *Current marks on firm mud bottoms.* Connecticut Academy of Arts and Sciences. Transactions, v.42, p.57-96.

Dzulynski, Stanislaw, and Slaczka, Andrzej (1958) *Directional structures and sedimentation of the Krosno beds (Carpathian flysch).* Polskie Towarzystwo Geologiczne. Rocznik, t.28, p.205-260.

Dzulynski, Stanislaw, and Walton, E.K. (1963) *Experimental production of sole markings.* Edinburgh Geological Society. Transactions, v.19, p.279-305.

Dzulynski, Stanislaw, and Walton, E.K. (1965) *Sedimentary features of flysch and greywackes.* New York: Elsevier. 274p.

E

Eakin, H.M. (1916) *The Yukon-Koyukuk region, Alaska.* U.S. Geological Survey. Bulletin 631. 88p.

Eardley, A.J. (1962) *Structural geology of North America.* 2nd ed. New York: Harper & Row. 743p.

Eaton, J.E. (1929) *The by-passing and discontinuous deposition of sedimentary materials.* American Association of Petroleum Geologists. Bulletin, v.13, p.713-761.

Eaton, J.E. (1951) *Inadvisability of restricting terms that designate the relative.* Geological Society of America. Bulletin, v.62, p.77-80.

Eckis, Rollin (1928) *Alluvial fans of the Cucamonga district, southern California.* Journal of Geology, v.36, p.225-247.

Ehrenberg, C.G. (1854) *Mikrogeologie; das Erde und Felsen schaffende wirken des unsichtbar kleinen selbstständigen Lebens auf der Erde.* Leipzig: L. Voss. 347p.

Einstein, H.A. (1950) *The bed-load function for sediment transportation in open channel flows.* U.S. Department of Agriculture. Technical Bulletin 1026. 70p.

Eisenack, Alfred (1931) *Neue Mikrofossilien des baltischen Silurs.* Pt.1. Palaeontologische Zeitschrift, Bd.13, p.74-118.

El-Etr, H.A. (1976) *Proposed terminology for natural linear features.* Utah Geological Association. Publication no.5, p.480-489.

Elias, M.K. (1965) *Cycleology—a discipline concomitant to paleoecology* (abstract). American Association of Petroleum Geologists. Bulletin, v.49, p.339.

Elliott, R.E. (1965) *A classification of subaqueous sedimentary structures based on rheological and kinematical parameters.* Sedimentology, v.5, p.193-209.

Elton, C.S. (1927) *The nature and origin of soil-polygons in Spitsbergen.* Geological Society of London. Quarterly Journal, v.83, p.163-194.

Emery, K.O. (1948) *Submarine geology of Bikini atoll.* Geological Society of America. Bulletin, v.59, p.855-859.

Emiliani, Cesare (1955) *Pleistocene temperatures.* Journal of Geology, v.63, p.538-578.

Emmons, W.H. (1933) *On the mechanism of the deposition of certain metalliferous lode systems associated with granitic batholiths.* In: The Committee on the Lindgren Volume, ed. *Ore deposits of the western states.* New York: American Institute of Mining and Metallurgical Engineers. 797p.

Engel, A.E.J., Engel, C.G., and Havens, R.G. (1965) *Chemical characteristics of oceanic basalts and the upper mantle.* Geological Society of America. Bulletin, v.76, p.719-734.

Engelhardt, Wolf von, and Stoffler, D. (1968) *Stages of shock metamorphism in crystalline rocks of the Ries basin, Germany.* In: French, B.M., and Short, N.M., eds. *Shock metamorphism of natural materials,* p.159-168. Baltimore: Mono Book Corp. 644p.

Engelund, Frank (1966) *On the possibility of formulating a universal spectrum function for dunes.* Technical University of Denmark. Hydraulics Laboratory. Basic Research Progress Rept. 18, p.1-4.

Enos, Paul (1974) *Reefs, platforms, and basins of Middle Cretaceous in northeast Mexico.* American Association of Petroleum Geologists. Bulletin, v.58, p.800-809.

Erhart, Henri (1955) *"Biostasie" et "rhexistasie", esquisse d'une théorie sur le rôle de la pédogenèse en tant que phénomène géologique.* Académie des Sciences, Paris. Comptes rendus hebdomadaires des séances, t.241, no.18, p.1218-1220.

Erhart, Henri (1956) *La genèse des sols; esquisse d'une théorie géologique et géochimique: biostasie et rhexistasis.* Paris: Masson. 90p.

Esau, Katherine (1965) *Plant anatomy.* 2d ed. New York: Wiley.

Eskola, Pentti (1915) *Om sambandet mellan kemisk och mineralogisk sammansättning hos Orijävitraktens metamorfa bergarter.* With an English summary of the contents. Finland, Commission Géologique. Bulletin, v.8, no.44, 145p.

Eskola, Pentti (1938) *On the esboitic crystallization of orbicular rocks.* Journal of Geology, v.46, p.448-485.

Eskola, Pentti (1939) *Die metamorphen Gesteine.* In: Barth, T.F.W., Correns, Carl W., and Eskola, Pentti. *Die Entstehung der Gesteine, ein Lehrbuch der Petrogenese.* Berlin: Julius Springer. 422p.

Eskola, Pentti (1948) *The problem of mantled gneiss domes.* Geological Society of London. Quarterly Journal, v.104, p.461.

Eskola, Pentti (1961) *Granitentschung bei Orogenese und Epeirogenese.* Geologische Rundschau, v.50, p.105-113.

ESSA: U.S. Environmental Science Services Administration (1968) *ESSA.* Washington: U.S. Government Printing Office. [8]p.

Evans, B.W. (1977) *Metamorphism of alpine peridotite and serpentinite.* Annual Review of Earth and Planetary Science, v.5, p.397-447.

Evans, J.W. (1911) *Dreikanter.* Geological Magazine, v.8, p.334-335.

Evans, O.F. (1938) *The classification and origin of beach cusps.* Journal of Geology, v.46, p.615-627.

Evitt, W.R. (1963) *A discussion and proposals concerning fossil dinoflagellates, hystrichospheres, and acritarchs.* Pt.2. National Academy of Sciences, U.S.A. Proceedings, v.49, p.298-302.

Eythórsson, Jón (1951) *Jökla-mys.* Journal of Glaciology, v.1, p.503.

F

Fairbairn, H.W. (1943) *Packing in ionic minerals.* Geological Society of America. Bulletin, v.54, p.1305-1374.

Fairbanks, H.W. (1906) *Practical physiography.* Boston: Allyn & Bacon. 542p.

Fairbridge, R.W. (1946) *Submarine slumping and location of oil bodies.* American Association of Petroleum Geologists. Bulletin, v.30, p.84-92.

Fairbridge, R.W. (1951) *The Aroe Islands and the continental shelf north of Australia.* Scope, v.1, no.6, p.24-29.

Fairbridge, R.W. (1954) *Stratigraphic correlation by microfacies.* American Journal of Science, v.252, p.683-694.

Fairbridge, R.W. (1958) *What is a consanguineous association?* Journal of Geology, v.66, p.319-324.

Fairbridge, R.W., ed. (1966) *The encyclopedia of oceanography.* New York: Reinhold. 1021p. (*Encyclopedia of earth sciences series,* v.1).

Fairbridge, R.W. (1967) *Phases of diagenesis and authigenesis.* In: Larsen, Gunnar, and Chilingar, G.V., eds. *Diagenesis in sediments.* Amsterdam: Elsevier. 551p. (*Developments in sedimentology* 8, p.19-89).

Fairbridge, R.W., ed. (1968) *The encyclopedia of geomorphology.* New York: Reinhold. 1295p. (*Encyclopedia of earth sciences series,* v.3).

Fairbridge, R.W. (1972) *Climatology of a glacial cycle.* Quaternary Research (Washington University), v.2, p.283-302.

Fairchild, H.L. (1904) *Geology under the planetesimal hypothesis of Earth origin.* Geological Society of America. Bulletin, v.15, p.243-266.

Fairchild, H.L. (1913) *Pleistocene geology of New York State.* Geological Society of America. Bulletin, v.24, p.133-162.

Fairchild, H.L. (1932) *New York moraines.* Geological Society of America. Bulletin, v.43, p.627-662.

Farmin, Rollin (1934) *"Pebble dikes" and associated mineralization at Tintic, Utah.* Economic Geology, v.29, p.356-370.

Fay, A.H. (1918) *A glossary of the mining and mineral industry.* Washington: Government Printing Office. 754p. (U.S. Bureau of Mines. Bulletin 95.)

Fedorowski, J., and Jull, R.K. (1976) *Review of blastogeny in Paleozoic corals and description of lateral increase in some Upper Ordovician rugose corals.* Acta Palaeontologica Polonica, v.21, no.1, p.37-78.

Fenneman, N.M. (1909) *Physiography of the St. Louis area.* Illinois State Geological Survey. Bulletin, v.12, 83p.

Fenneman, N.M. (1914) *Physiographic boundaries within the United States.* Association of American Geographers. Annals, v.4, p.84-134.

Fenner, C.N. (1948) *Incandescent tuff flows in southern Peru.* Geological Society of America. Bulletin, v.59, p.879-893.

Fenton, C.L., and Fenton, M.A. (1928) *Faunule and zonule.* American Midland Naturalist, v.11, p.1-23. (*Ecologic interpretations of some biostratigraphic terms,* 1).

Fenton, C.L., and Fenton, M.A. (1930) *Zone, subzone, facies, phase.* American Midland Naturalist, v.12, p.145-153. (*Ecologic interpretations of some biostratigraphic terms,* 2).

Fernald, F.A. (1929) *Roundstone, a new geologic term.* Science, n.s., v.70, p.240.

Fernald, M.L. (1950) *Gray's manual of botany.* 8th ed. New York: American Book Company.

Fersman, A.E. (1922) *Geokhimiia Rossii.* St. Petersburg: Nauchnoe Khimichesko-Tekhnicheskoe izdatel'stvo. Approx. 210p.

Fielder, Gilbert (1965) *Lunar geology.* Chester Springs, Penna.: Dufour. 184p.

Finch, R.H. (1933) *Slump scarps.* Journal of Geology, v.41, p.647-649.

Finsterwalder, Richard (1950) *Some comments on glacier flow.* Journal of Glaciology, v.1, no.7, p.383-388.

Fischer, A.G. (1964) *The Lofer cyclothem of the Alpine Triassic.* Kansas State Geological Survey. Bulletin 169, v.1, p.107-149.

Fischer, A.G. (1969) *Geological time-distance rates: the Bubnoff unit.* Geological Society of America. Bulletin, v.80, p.549-551.

Fischer, A.G., and others (1954) *Arbitrary cut-off in stratigraphy.* American Association of Petroleum Geologists. Bulletin, v.38, p.926-931.

Fischer, A.G., and Teichert, Curt (1969) *Cameral deposits in cephalopod shells.* University of Kansas. Paleontological Contributions, Paper 37. 30p.

Fischer, Georg (1934) *Die Petrographie der Grauwacken.* Preussische Geologische Landesanstalt. Jahrbuch, 1933, Bd.54, p.320-343.

Fisher, O. (1866) *On the warp (of Mr. Trimmer)—Its age and probable connexion with the last geological events.* Geological Society of London. Quarterly Journal, v.22. p.553-565.

Fisher, W.B. (1950) *The Middle East; a physical, social, and regional geography.* London: Methuen. 514p.

Fisk, H.N. (1959) *Padre Island and the Laguna Madre Flats, coastal south Texas.* Coastal Geography Conference, 2nd, Baton Rouge, La., 1959. [Proceedings], p.103-151.

Flanders, P.L., and Sauer, F.M. (1960) *A glossary of geoplosics: the systematic study of explosion effects in the Earth.* Kirtland Air Force Base, N.Mex.: U.S. Air Force Special Weapons Center. 34p. (Technical note 60-20).

Flawn, P.T. (1953) *Petrographic classification of argillaceous sedimentary and low-grade metamorphic rocks in subsurface.* American Association of Petroleum Geologists. Bulletin, v.37, p.560-565.

Fleischer, Michael (1975) *1975 Glossary of mineral species.* Bowie, Md.: Mineralogical Record. 145p. Additions and corrections: Mineralogical Record, v.7, no.2, March-April 1976; v.8, no.5, Sept.-Oct. 1977.

Fleuty, M.J. (1964) *Tectonic slides.* Geological Magazine, v.101, p.452-456.

Fleuty, M.J. (1964a) *The description of folds.* Geologists' Association. Proceedings, v.75, p.461-492.

Flinn, D. (1962) *On folding during three-dimensional progressive deformation.* Geological Society of London. Quarterly Journal, v.118, p.385-433.

Flint, D.E., and others (1953) *Limestone walls of Okinawa.* Geological Society of America. Bulletin, v.64, p.1247-1260.

Flint, R.F. (1928) *Eskers and crevasse fillings.* American Journal of Science, v.15, p.410-416.

Flint, R.F. (1955) *Pleistocene geology of eastern South Dakota.* U.S. Geological Survey. Professional Paper 262. 173p.

Flint, R.F. (1957) *Glacial and Pleistocene geology.* New York: Wiley. 553p.

Flint, R.F. (1971) *Glacial and Quaternary geology.* New York: Wiley. 892p.

Flint, R.F., and Deevey, E.S., jr. (1951) *Radiocarbon dating of late-Pleistocene events.* American Journal of Science, v.249, p.257-300.

Flint, R.F., and others (1960a) *Symmictite: a name for nonsorted terrigenous sedimentary rocks that contain a wide range of particle sizes.* Geological Society of America. Bulletin, v.71, p.507-509.

Flint, R.F., and others (1960b) *Diamictite, a substitute term for symmictite.* Geological Society of America. Bulletin, v.71, p.1809.

Fogg, G.E., and others (1973) *The blue-green algae.* New York: Academic Press. 459p.

Folk, R.L. (1951) *Stages of textural maturity in sedimentary rocks.* Journal of Sedimentary Petrology, v.21, p.127-130.

Folk, R.L. (1954) *The distinction between grain size and mineral composition in sedimentary-rock nomenclature.* Journal of Geology, v.62, p.344-359.

Folk, R.L. (1959) *Practical petrographic classification of limestones.* American Association of Petroleum Geologists. Bulletin, v.43, p.1-38.

Folk, R.L. (1962) *Spectral subdivision of limestone types.* American Association of Petroleum Geologists. Memoir 1, p.62-84.

Folk, R.L. (1965) *Some aspects of recrystallization in ancient limestones.* Society of Economic Paleontologists and Mineralogists. Special Publication no.13, p.14-48.

Folk, R.L. (1968) *Petrology of sedimentary rocks.* Austin, Tex.: Hemphill's Book Store. 170p.

Folk, R.L. (1973) *Evidence for peritidal deposition of Devonian Caballos Novaculite, Marathon Basin, Texas.* American Association of Petroleum Geologists. Bulletin, v.57, p.702-725.

Folk, R.L. (1974) *The natural history of crystalline calcium carbonate: effects of magnesium content and salinity.* Journal of Sedimentary Petrology, v.44, p.40-53.

Folk, R.L., and Assereto, Riccardo (1974) *Giant aragonite rays and baroque white dolomite in tepee-fillings, Triassic of Lombardy, Italy* (abstract). American Association of Petroleum Geologists. Annual meeting. Abstracts, v.1, p.34-35.

Folk, R.L., and Assereto, Riccardo (1976) *Comparative fabrics of length-slow and length-fast calcite and calcite-aragonite in a Holocene speleothem, Carlsbad Caverns, New Mexico.* Journal of Sedimentary Petrology, v.46, p.486-496.

Folk, R.L., and Siedlecka, Anna (1974) *The schizohaline environment: its sedimentary and diagenetic fabric as exemplified by late Paleozoic rocks of Bear Island, Svalbard.* Sedimentary Geology, v.11, p.1-15.

Folk, R.L., Roberts, H.H., and Moore, C.H. (1973) *Black phytokarst from Hell, Cayman Islands, British West Indies.* Geological Society of America. Bulletin, v.84, p.2351-2360.

Forgotson, J.M., jr. (1954) *Regional stratigraphic analysis of the Cotton Valley Group of Upper Gulf Coastal Plain.* American Association of Petroleum Geologists. Bulletin, v.38, p.2476-2499.

Forgotson, J.M., jr. (1957) *Nature, usage, and definition of marker-defined vertically segregated rock units.* American Association of Petroleum Geologists. Bulletin, v.41, p.2108-2113.

Forgotson, J.M., jr. (1960) *Review and classification of quantitative mapping techniques.* American Association of Petroleum Geologists. Bulletin, v.44, p.83-100.

Forgotson, J.M., jr., and Forgotson, J.M. (1975) *"Porosity pod": concept for successful stratigraphic exploration of fine-grained sandstones.* American Association of Petroleum Geologists. Bulletin, v.59, p.1113-1125.

Fortescue, J. (1965) *Exploration architecture.* Geological Survey of Canada. Paper 65-6, p.4-14.

Foster, A.S., and Gifford, E.M., jr. (1974) *Comparative morphology of vascular plants.* San Francisco: Freeman. 751p.

Foster, N.H. (1966) *Stratigraphic leak.* American Association of Petroleum Geologists. Bulletin, v.50, p.2604-2611.

Fournet, J. (1844) *Sur l'état de surfusion du quartz dans les roches éruptives et dans les filons métallifères.* Académie des Sciences, Paris. Comptes rendus, v.18, p.1050-1057.

Fournet, J. (1847) *Résultats d'une exploration des Vosges.* Société Géologique de France. Bulletin, series 2, v.4, p.220-254.

Fowler, G.M., and Lyden, J.P. (1932) *The ore deposits of the Tri-State district (Missouri-Kansas-Oklahoma).* American Institute of Mining and Metallurgical Engineers. Transactions, v.102, p.206-251.

Fowler, H.W. (1937) *A dictionary of modern English usage.* 1st ed, reprinted with corrections. London: Oxford University Press. 742p.

Fox, D.L. (1957) *Particulate organic detritus.* Geological Society of America. Memoir 67, v.1, p.383-389.

Foye, J.C. (1916) *Are the "batholiths" of the Halliburton-Bancroft area, Ont., correctly named?* Journal of Geology, v.24, p.783-791.

Freeburg, J.H. (1966) *Terrestrial impact structures—a bibliogra-*

phy. U.S. Geological Survey. Bulletin 1220. 91p.

Freeman, O.W. (1925) *The origin of Swimming Woman Canyon, Big Snowy Mountains, Montana, an example of a pseudo-cirque formed by landslide sapping.* Journal of Geology, v.33, p.75-79.

French, B.M. (1966) *Shock metamorphism of natural materials.* Science, v.153, p.903-906.

Freund, J.E. (1960) *Modern elementary statistics,* 2nd ed. Englewood Cliffs, N.J.: Prentice-Hall.

Friedman, G.M. (1965) *Terminology of crystallization textures and fabrics in sedimentary rocks.* Journal of Sedimentary Petrology, v.35, p.643-655.

Friedman, J.D. (1970) *The airborne infrared scanner as a geophysical research tool.* Optical Spectra, June, p.35-44.

Friedman, M., and others (1976) *Experimental folding of rocks under confining pressure: Part III. Faulted drape folds in multilithologic layered specimens.* Geological Society of America. Bulletin, v.87, p.1049-1066.

Fritsch, F.E. (1961) *The structure and reproduction of the algae.* v.1. Cambridge, England: Cambridge University Press. 791p.

Frizzell, D.L. (1933) *Terminology of types.* American Midland Naturalist, v.14, p.637-668.

Frondel, J.W., and others (1967) *Glossary of uranium- and thorium-bearing minerals.* 4th ed. U.S. Geological Survey. Bulletin 1250. 69p.

Frye, J.C., and Willman, H.B. (1960) *Classification of the Wisconsinan Stage in the Lake Michigan glacial lobe.* Illinois State Geological Survey. Circular 285. 16p.

Fuchs, Theodor (1895) *Studien über Fucoiden und Hieroglyphen.* Akademie der Wissenschaften, Vienna. Mathematisch-Naturwissen-schaftliche Classe. Denkschriften, v.62, p.369-448.

Fuller, H.J., and Tippo, Oswald (1954) *College botany.* Revised edition. New York: Holt, Rinehart & Winston.

Fuller, M.L. (1914) *The geology of Long Island, N.Y.* U.S. Geological Survey. Professional Paper 82. 231p.

Fyfe, W.S., and Turner, F.J. (1966) *Reappraisal of the metamorphic facies concept.* Contributions to Mineralogy and Petrology, v.12, p.354-364.

Fyfe, W.S., Turner, F.J., & Verhoogen, J. (1958) *Metamorphic reactions and metamorphic facies.* Geological Society of America. Memoir 73. 259p.

G

Galkiewicz, T. (1968) *Geocosmology.* International Geological Congress. 23rd, Prague, 1968. Report, sec.13, p.145-149.

Galliher, E.W. (1931) *Collophane from Miocene brown shales of California.* American Association of Petroleum Geologists. Bulletin, v.15, p.257-269.

Galloway, J.J. (1922) *Value of the physical characters of sand grains in the interpretation of the origin of sandstones.* Geological Society of America. Bulletin, v.33, p.104-105.

Galton, Francis (1889) *On the principle and methods of assigning marks for bodily efficiency.* Nature, v.40, p.649-653.

Gammon, J.B. (1966) *Fahlbands in the Precambrian of southern Norway.* Economic Geology, v.61, p.174-188.

Garland, G.D. (1971) *Introduction to geophysics: mantle, core, and crust.* Philadelphia: W.B. Saunders. 420p.

Gault, D.E., and others (1968) *Impact cratering mechanics and structures.* In: French, B.M., and Short, N.M., eds. *Shock metamorphism of natural materials,* p.87-100. Baltimore: Mono Book Corp. 644p.

Geikie, James (1898) *Earth sculpture, or the origin of land-forms.* New York: Putnam. 397p.

Geological Society of London (1967) *Report of the Stratigraphical Code Sub-Committee.* Geological Society of London. Proceedings, no.1638, p.75-87.

Gilbert, G.K. (1875) *Report on the geology of portions of Nevada, Utah, California, and Arizona, examined in the years 1871 and 1872.* U.S. Geographical Surveys West of the One Hundredth Meridian. Report, v.3, pt.1, p.17-187.

Gilbert, G.K. (1876) *The Colorado Plateau province as a field for geological study.* American Journal of Science, ser.3, v.12, p.16-24, 85-103.

Gilbert, G.K. (1877) *Report on the geology of the Henry Mountains.* Washington: U.S. Geographical and Geological Survey of the Rocky Mountain Region. 160p.

Gilbert, G.K. (1885) *The topographic features of lake shores.* U.S. Geological Survey. Annual Report, 5th, p.69-123.

Gilbert, G.K. (1890) *Lake Bonneville.* U.S. Geological Survey. Monograph 1. 438p.

Gilbert, G.K. (1898) *A proposed addition to physiographic nomenclature.* Science, n.s., v.7, p.94-95.

Gilbert, G.K. (1899) *Ripple-marks and cross-bedding.* Geological Society of America. Bulletin, v.10, p.135-140.

Gilbert, G.K. (1904) *Systematic asymmetry of crest lines in the high Sierra of California.* Journal of Geology, v.12, p.579-588.

Gilbert, G.K. (1914) *The transportation of débris by running water.* U.S. Geological Survey. Professional Paper 86. 263p.

Gilbert, G.K. (1928) *Studies of Basin Range structure.* U.S. Geological Survey. Professional Papper 153. 92p.

Gilbert, G.K., and Brigham, A.P. (1902) *An introduction to physical geography.* New York: Appleton. 380p.

Giles, A.W. (1918) *Eskers in the vicinity of Rochester, N.Y.* Rochester Academy of Science. Proceedings, v.5, p.161-240.

Gilluly, James (1966) *Orogeny and geochronology.* American Journal of Science, v.264, p.97-111.

Glaessner, M.F. (1945) *Principles of micropaleontology.* Melbourne: Melbourne Univ. Press. 296p.

Glaessner, M.F., and Teichert, C. (1947) *Geosynclines: a fundamental concept in geology.* American Journal of Science, v.245, p.465-482, 571-591.

Glen, J.W. (1955) *The creep of polycrystalline ice.* Royal Society of London. Proceedings, ser.A, v.228, p.519-538.

Glock, W.S. (1928) *An analysis of erosional forms.* American Journal of Science, ser.5, v.15, p.471-483.

Glock, W.S. (1932) *Available relief as a factor of control in the profile of a landform.* Journal of Geology, v.40, p.74-83.

Glynn, P.W. (1974) *Rolling stones among the Scleractinia: mobile coralliths in the Gulf of Panama.* Second International Coral Reef Symposium. Proceedings, v.2, p.183-198. Brisbane: The Great Barrier Reef Committee. 753p.

Godwin, Harry (1962) *Radiocarbon dating; fifth international conference.* Nature, v.195, p.943-945.

Goldich, S.S. (1938) *A study in rock weathering.* Journal of Geology, v.46, p.17-58.

Goldman, M.I. (1921) *Lithologic subsurface correlation in the "Bend" series of north-central Texas.* U.S. Geological Survey. Professional Paper 129, p.1-22.

Goldman, M.I. (1933) *Origin of the anhydrite cap rock of American salt domes.* U.S. Geological Survey. Professional Paper 175-D, p.83-114.

Goldman, M.I. (1952) *Deformation, metamorphism, and mineralization in gypsum-anhydrite cap rock, Sulphur salt dome, Louisiana.* Geological Society of America. Memoir 50. 169p.

Goldman, M.I. (1961) *Regenerated anhydrite redefined.* Journal of Sedimentary Petrology, v.31, p.611.

Goldring, Winifred (1943) *Geology of the Coxsackie quadrangle, New York.* New York State Museum. Bulletin, no.332. 374p.

Goldschmidt, V.M. (1937) *The principles of distribution of chemical elements in minerals and rocks.* Chemical Society (of London). Journal, v.1937, p.655-673.

Goldschmidt, V.M. (1954) *Geochemistry.* Oxford: Clarendon Press.

Goldsmith, Richard (1959) *Granofels: a new metamorphic rock name.* Journal of Geology, v.67, p.109-110.

Goldthwait, J.W. (1908) *Intercision, a peculiar kind of modification of drainage.* School Science and Mathematics, v.8, p.129-139.

Goode, H.D. (1969) *Geoevolutionism: a step beyond catastrophism and uniformitarianism.* Geological Society of America. Abstracts with programs for 1969, pt.5, p.29.

Goodspeed, G.E. (1947) *Xenoliths and skialiths.* American Journal of Science, v.246, p.515-525.

Goodspeed, G.E. (1948) *Origin of granites.* In: Gilluly, J. *Origin of granite.* Geological Society of America. Memoir 28. p.55-78.

Goodspeed, G.E. (1955) *Relict dikes and relict pseudodikes.* American Journal of Science, v.253, no.3, p.146-161.

Govett, G.J.S. (1978) *Lithogeochemistry.* Journal of Geochemical Exploration, v.9, p.109-110.

Grabau, A.W. (1903) *Paleozoic coral reefs.* Geological Society of America. Bulletin, v.14, p.337-352.

Grabau, A.W. (1904) *On the classification of sedimentary rocks.* American Geologist, v.33, p.228-247.

Grabau, A.W. (1905) *Physical characters and history of some New York formations.* Science, v.22, p.528-535.

Grabau, A.W. (1906) *Types of sedimentary overlap.* Geological Society of America. Bulletin, v.17, p.567-636.

Grabau, A.W. (1911) *On the classification of sand grains.* Science, n.s., v.33, p.1005-1007.

Grabau, A.W. (1913) *Principles of stratigraphy.* Reprinted, 1960, by Dover Publishing Co., New York.

Grabau, A.W. (1920) *Principles of salt deposition.* 1st ed. New York: McGraw-Hill. 435p. (*Geology of the non-metallic mineral deposits other than silicates,* v.1.)

Grabau, A.W. (1920a) *General geology.* Boston: Heath. 864p. (*A textbook of geology.* pt.1.)

Grabau, A.W. (1924) *Principles of stratigraphy.* 2nd ed. New York: A.G. Seiler. 1185p.

Grabau, A.W. (1932) *Principles of stratigraphy.* 3d ed. New York: A.G. Seiler. 1185p.

Grabau, A.W. (1936) *The Great Huangho Plain of China.* Association of Chinese and American Engineers. Journal, v.17, p.247-266.

Grabau, A.W. (1936a) *Oscillation or pulsation.* International Geological Congress. 16th, Washington, 1933. Report, v.1, p.539-553.

Grabau, A.W. (1940) *The rhythm of the ages; Earth history in the light of the pulsation and polar control theory.* Peking: Henri Vetch. 561p.

Graton, L.C., and Fraser, H.J. (1935) *Systematic packing of spheres, with particular relation to porosity and permeability.* Journal of Geology, v.43, p.785-909.

Gravenor, C.P. (1956) *Air photographs of the plains region of Alberta.* Research Council of Alberta. Preliminary Report 56-5. 35p.

Gravenor, C.P., and Kupsch, W.O. (1959) *Ice-disintegration features in western Canada.* Journal of Geology, v.67, p.48-64.

Gray, H.H. (1955) *Stratigraphic nomenclature in coal-bearing rocks.* Geological Society of America. Bulletin, v.66, p.1567-1568.

Green, D.H., and Ringwood, A.E. (1963) *Mineral assemblages in a model mantle composition.* Journal of Geophysical Research, v.68, p.937-945.

Greensmith, J.T. (1957) *The status and nomenclature of stratified evaporites.* American Journal of Science, v.255, p.593-595.

Gregory, H.E. (1917) *Geology of the Navajo country; a reconnaissance of parts of Arizona, New Mexico, and Utah.* U.S. Geological Survey. Professional Paper 93. 161p.

Gregory, J.W. (1912) *The relations of kames and eskers.* Geographical Journal, v.40, p.169-175

Gressly, Amanz (1838) *Observations géologiques sur le Jura Soleurois.* [Pt.1]. Allgemeine Schweizerische Gesellschaft für die gesammten Naturwissenschaften. Neue Denkschriften, v.2, [pt.6]. 112p.

Griffiths, J.C., and Rosenfeld, M.A. (1954) *Operator variation in experimental research.* Journal of Geology, v.62, p.74-91.

Grim, R.E. (1953) *Clay mineralogy.* New York: McGraw-Hill. 384p.

Grim, R.E. (1968) *Clay mineralogy.* 2nd ed. New York: McGraw-Hill. 596p.

Grim, R.E., and Güven, Necip (1978) *Bentonites.* Amsterdam: Elsevier. 256p.

Grim, R.E., and others (1937) *The mica in argillaceous sediments.* American Mineralogist, v.22, p.813-829.

Grohskopf, J.G., and McCracken, Earl (1949) *Insoluble residues of some Paleozoic formations of Missouri, their preparation, characteristics and application.* Missouri Division of Geological Survey and Water Resources. Report of Investigations, no.10 39p.

Grossman, W.L. (1944) *Stratigraphy of the Genesee group of New York.* Geological Society of America. Bulletin, v.55, p.41-75.

Grout, F.F. (1932) *Petrography and petrology, a texbook.* New York: McGraw-Hill. 522p.

Grubenmann, Ulrich (1904) *Die Kristallinen Schiefer.* Berlin: Gebrüder Borntraeger. 2 vol.

Grubenmann, Ulrich, and Niggli, P. (1924) *Die Gesteinsmetamorphose: I. Allgemeiner Teil.* Berlin: Gebrüder Borntraeger. 539p.

Gruner, J.W. (1950) *An attempt to arrange silicates in the order of reaction energies at relatively low temperatures.* American Mineralogist, v.35, p.137-148.

Gruner, J.W., and others (1941) *Structural geology of the Knife Lake area of northeastern Minnesota.* Geological Society of America. Bulletin, v.52, p.1577-1642.

Guilcher, André(1950) *Définition d'un type de volcan "écossais".* Association de Géographes Français. Bulletin, no.206-207, p.2-11.

Gulliver, F.P. (1896) *Cuspate forelands.* Geological Society of America. Bulletin, v.7, p.399-422.

Gulliver, F.P. (1899) *Shoreline topography.* American Academy of Arts and Sciences. Proceedings, v.34, p.149-258.

Gümbel, C.W. von (1868) *Geognostische Beschreibung des ostbayerischen Grenzgebirges oder des bayerischen und oberpfälzer Waldgebirges.* Gotha: Justus Perthes. 968p. (Bavaria. Bayerisches Oberbergamt Geognostische Abteilung. *Geognostische Beschreibung des Koenigreichs Bayern,* v.2).

Gümbel, C.W. von (1874) *Die paläolithischen Eruptivgesteine des Fichtelgebirges (als vorläufige Mittheilung).* Munich. 50p.

Gussow, W.C. (1954) *Differential entrapment of oil and gas: a fundamental principle.* American Association of Petroleum Geologists. Bulletin, v.38, p.816-853.

Gussow, W.C. (1958) *Metastasy or crustal shift.* Alberta Society of Petroleum Geologists. Journal, v.6, p.253-257.

H

Haarmann, E. (1926) *Tektogenese oder Gefugebildung statt Orogenese oder Gebirgsbildung.* Deutsche Geologische Gesellschaft. Zeitschrift, v.78, p.105-107.

Haarmann, E. (1930) *Die Oszillationstheorie; eine erklarung der krustenbewegungen von erder und mond.* Stuttgart: F. Enke, 260p.

Hack, J.T. (1960) *Interpretation of erosional topography in humid temperate regions.* American Journal of Science, v.258a, p.80-97.

Hadding, Assar (1931) *On subaqueous slides.* Geologiska Föreningen, Stockholm. Förhandlingar, Bd.53, p.377-393.

Hadding, Assar (1933) *On the organic remains of the limestones; a short review of the limestone forming organisms.* Acta universitatis Lundensis, n.s., Bd.29, pt.2, no.4. 93p. (*The pre-Quaternary sedimentary rocks of Sweden,* 5).

Haeckel, Ernst (1862) *Die Radiolarien (Rhizopoda radiaria); eine Monographie.* Berlin: Georg Reimer. 572p.

Hall, James (1843) *Geology of New York; part IV, comprising the survey of the fourth geological district.* Albany: Carroll and Cook. 683p.

Hall, James (1859) *Description and figures of the organic remains of the lower Helderberg Group and the Oriskany Sandstone.* New York Geological Survey. Paleontology, v.3. 532p.

Hamblin, W.K. (1958) *The Cambrian sandstones of northern Michigan.* Michigan Geological Survey. Publication 51. 149p.

Hamelin, Louis-Edmond (1961) *Périglaciaire du Canada: idées nouvelles et perspectives globales.* Cahiers de géographie de Québec, v.5, no.10, p.141-203.

Hamelin, Louis-Edmond, and Clibbon, Peter (1962) *Vocabulaire périglaciaire bilingue (français et anglais).* Cahiers de géographie de Québec, v.6, no.12, p.201-226.

Hamelin, Louis-Edmond, and Cook, F.A. (1967) *Le périglaciaire par l'image; illustrated glossary of periglacial phenomena.* Quebec: Presses de l'Université Laval. 237p. (Centre d'Etudes Nordiques. Travaux et documents 4.)

Hamilton, E.J. (1965) *Applied geochronology.* New York: Academic Press. 267p.

Hammer, W. (1914) *Das Gebiet der Bündnerschiefer im tirolischen Oberinntal.* Kaiserliche Köningliche Geologische Reichsanstalt (Austria). Jahrbuch, v.64, p.443-567.

Hammond, A.L. (1974) *Bright spot: better seismological indicators of oil and gas.* Science, v.185, p.515-517.

Hansen, A.M. (1894) *The glacial succession in Norway.* Journal of Geology, v.2, p.123-143.

Hapke, B.W. (1966) *Optical properties of the Moon's surface.* In: Hess, W.N., and others, eds. *The nature of the lunar surface,* p.141-154. Baltimore: Johns Hopkins Press. 320p.

Harbaugh, J.W. (1962) *Geologic guide to Pennsylvanian marine banks, southeast Kansas.* Kansas Geological Society. Guidebook; field conference, 27th, p.13-67.

Harbaugh, J.W., and Bonham-Carter, G.F. (1970) *Computer simulation in geology.* New York: Wiley. 575p.

Harbaugh, J.W., and Merriam, D.F. (1968) *Computer applications in stratigraphic analysis.* New York: Wiley. 282p.

Hardy, C.T., and Williams, J.S. (1959) *Columnar contemporaneous deformation.* Journal of Sedimentary Petrology, v.29, p.281-283.

Harker, Alfred (1904) *Tertiary igneous rocks of Skye.* Geological Survey of the United Kingdom. Memoir. Glasgow: H.M. Stationery Office. 481p.

Harker, Alfred (1909) *The natural history of igneous rocks.* New York: Macmillan. 377p.

Harker, Alfred (1918) *The anniversary address of the President.* Geological Society of London. Quarterly Journal, v.74, p.i-lxxx.

Harker, Alfred (1939) *Metamorphism.* 2d ed. London: Methuen.

362p.

Harland, W.B. (1971) *Tectonic transpression in Caledonian Spitzbergen.* Geological Magazine, v.108, p.27-42.

Harland, W.B., and others (1966) *The definition and identification of tills and tillites.* Earth Science Reviews, v.2, p.255-256.

Harland, W.B., Smith, A.S., and Wilcock, B., eds. (1964) *The Phanerozoic time-scale: a symposium.* Geological Society of London. Quarterly Journal, v.120, Supplement. 458p.

Harms, J.C., and others (1975) *Depositional environments as interpreted from primary sedimentary structures and stratification sequences.* Society of Economic Paleontologists and Mineralogists. Short course, lecture notes 2. 161p.

Harper, W.G. (1957) *Morphology and genesis of calcisols.* Soil Science Society of America. Proceedings, v.21, p.420-424.

Harrassowitz, H. (1927) *Anchimetamorphose, das Gebiet zwischen Oberflächen- und Tiefenumwandlung der Erdrinde.* Oberhessischen Gesellschaft Natur- und Heilkunde zu für Giessen. Naturwissenschaftliche Abteilung. Bericht, 12, p.9-15.

Harrington, H.J. (1946) *Las corrientes de barro ("mud-flows") de "El Volcán", quebrada de Humahuaca, Jujuy.* Sociedad Geológica Argentina. Revista, t.1, p.149-165.

Harris, S.E., jr. (1943) *Friction cracks and the direction of glacial movement.* Journal of Geology, v.51, p.244-258.

Harte, Ben (1977) *Rock nomenclature with particular relation to deformation and recrystallization textures in olivine-bearing xenoliths.* Journal of Geology, v.85, p.279-288.

Hartshorn, J.H. (1958) *Flowtill in southeastern Massachusetts.* Geological Society of America. Bulletin, v.69, p.477-482.

Harvey, R.D. (1931) *Glacial chutes from the Peruvian Cordillera.* American Journal of Science, 5th ser., v.31, p.220-231.

Harvie-Brown, J.A. (1910) *"Caledonia rediviva".* Scottish Geographical Magazine, v.26, p.93-94.

Hass, W.H. (1962) *Conodonts.* In: Moore, R.C., ed. *Treatise on Invertebrate Paleontology.* Part W, p.3-69.

Hatch, F.H., and Rastall, R.H. (1913) *The petrology of the sedimentary rocks; a description of the sediments and their metamorphic derivatives.* London: George Allen. 425p.

Hatch, F.H., and Rastall, R.H. (1965) *Petrology of the sedimentary rocks.* 4th ed. revised by J. Trevor Greensmith. London: Thomas Murby. 408p. (*Textbook of petrology,* v.2.)

Hatch, F.H., Wells, A.K., and Wells, M.K. (1961) *Petrology of the igneous rocks,* 12th ed. London: Thomas Murby. 515p.

Haug, Emile (1900) *Les géosynclinaux et les aires continentales; contribution à l'étude des transgressions et des regressions marines.* Société Géologique de France. Bulletin, ser.3, v.28, p.617-711.

Haug, Emile (1907) *Les phénomènes géologiques.* Paris: Librairie Armand Colin. 538p. (His *Traité de géologie,* 1).

Haupt, A.W. (1953) *Plant morphology.* New York: McGraw-Hill.

Haupt, L.M. (1883) *The topographer, his instruments and his methods.* Philadelphia: H.C. Baird. 247p.

Haupt, L.M. (1906) *Changes along the New Jersey coast.* New Jersey Geological Survey. Annual report of the State Geologist, 1905, p.27-95.

Haüy, R.J. (1801) *Traité de minéralogie.* T. 2. Paris: Chez Louis. 444p.

Hawkes, H.E. (1958) *Principles of geochemical prospecting.* U.S. Geological Survey. Bulletin 1000, p.225-353.

Hawkes, H.E., and Webb, J.S. (1962) *Geochemistry in mineral exploration.* New York: Harper & Row. 415p.

Hay, R.L. (1978) *Geologic occurrence of zeolites.* In: Sand, L.B., and Mumpton, F.A., eds. *Natural zeolites,* p.135-143. New York: Pergamon. 546p.

Hayes, C.W. (1899) *Physiography of the Chattanooga district, in Tennessee, Georgia, and Alabama.* U.S. Geological Survey. Annual Report, 19th, pt.2, p.1-58.

Haynes, Vance (1973) *The Calico site: artifacts or geofacts?* Science, v.181, p.305.

Heald, M.T. (1956) *Cementation of Simpson and St. Peter sandstones in parts of Oklahoma, Arkansas, and Missouri.* Journal of Geology, v.64, p.16-30.

Hedberg, H.D. (1958) *Stratigraphic classification and terminology.* American Association of Petroleum Geologists. Bulletin, v.42, p.1881-1896.

Hedberg, H.D. (1961) *The stratigraphic panorama (an inquiry into the bases for age determination and age classification of the Earth's rock strata).* Geological Society of America. Bulletin, v.72, p.499-517.

Heezen, B.C., and Menard, H.W. (1963) *Topography of the deep-sea floor.* In: *The Sea,* v.3, ch.12, p.233-280. New York: Interscience.

Heiken, G.H. (1971) *Tuff rings: examples from the Fort Rock—Christmas Lake valley basin, south-central Oregon.* Journal of Geophysical Research, v.76, p.5615-5626.

Heiland, C.A. (1940) *Geophysical exploration.* New York: Prentice-Hall. 1013p.

Heim, Arnold (1908) *Über rezente und fossile subaquatische Rutschungen und deren lithologische Bedeutung.* Neues Jahrbuch für Mineralogie, Geologie und Paläontologie, 1908, Bd.2, p.136-157.

Heim, Arnold (1934) *Stratigraphische Kondensation.* Eclogae Geologicae Helvetiae, v.27, p.372-383.

Heim, Arnold, and Potonié, Robert (1932) *Beobachtungen über die Entstehung der tertiären Kohlen (Humolithe und Saprohumolith) in Zentralsumatra.* Geologische Rundschau, v.23, no.3,4, p.145-172.

Heinrich, E.W. (1956) *Microscopic petrography.* New York: McGraw-Hill. 296p.

Heinrich, E.W. (1966) *The geology of carbonatites.* Chicago: Rand McNally. 555p.

Hemley, J.J., and Jones, W.R. (1964) *Chemical aspects of hydrothermal alteration with emphasis on hydrogen metasomatism.* Economic Geology, v.59, p.538-569.

Henbest, L.G. (1952) *Significance of evolutionary explosions for diastrophic division of Earth history.* Journal of Paleontology, v.26, p.299-318.

Henbest, L.G. (1968) *Diagenesis in oolitic limestones of Morrow (Early Pennsylvanian) age in northwestern Arkansas and adjacent Oklahoma.* U.S. Geological Survey. Professional Paper 594-H. 22p.

Hendricks, C.L. (1952) *Correlations between surface and subsurface sections of the Ellenburger group of Texas.* Texas Bureau of Economic Geology. Report of Investigations, no.11. 44p.

Henningsmoen, Gunnar (1961) *Remarks on stratigraphical classification.* Norges Geologiske Undersokelse, no.213, p.62-92.

Henson, F.R.S. (1950) *Cretaceous and Tertiary reef formations and associated sediments in Middle East.* American Association of Petroleum Geologists. Bulletin, v.34, p.215-238.

Herrick, C.L. (1904) *The clinoplains of the Rio Grande.* American Geologist, v.33, p.376-381.

Herrin, Eugene, and Taggart, James (1968) *Source bias in epicenter determinations.* Seismological Society of America. Bulletin, v.58, p.1791-1796.

Hess, H.H. (1938) *Gravity anomalies and island arc structures, with particular reference to the West Indies.* American Philosophical Society. Proceedings, v.79, p.71-96.

Hess, H.H. (1960) *Stillwater igneous complex, Montana.* Geological Society of America. Memoir 80. 230p.

Hess, W.N., Menard, D.H., and O'Keefe, J.A., eds. (1966) *The nature of the lunar surface.* Baltimore: Johns Hopkins Press. 320p.

Hesse, Richard (1924) *Tiergeographie auf ökologischer Grundlage.* Jena: G. Fischer. 613p.

Hesse, Richard, and others (1937) *Ecological animal geography.* New York: Wiley. 597p.

Hey, M.H. (1962) *An index of mineral species & varieties arranged chemically, with an alphabetical index of accepted mineral names and synonyms.* 2nd rev. ed., reprinted with corrections. London: British Museum (Natural History). 728p.

Hey, M.H. (1963) *Appendix to the second edition of An index of mineral species and varieties arranged chemically.* London: British Museum (Natural History). 135p.

Hietanan, Anna (1967) *On the facies series in various types of metamorphism.* Journal of Geology, v.75, p.187-214.

Higashi, Akira (1958) *Experimental study of frost heaving.* U.S. Army. Snow, Ice & Permafrost Research Establishment. Research Report 45. 20p.

Hill, R.T. (1891) *The Comanche series of the Texas-Arkansas region.* Geological Society of America. Bulletin, v.2, p.503-528.

Hill, R.T. (1900) *Physical geography of the Texas region.* U.S. Geological Survey. Topographic Atlas, folio 3. 12p.

Hills, E.S. (1940) *The lunette, a new land form of aeolian origin.* Australian Geographer, v.3, no.7, p.15-21.

Hills, E.S. (1963) *Elements of structural geology.* New York: Wiley. 483p.

Himus, G.W. (1954) *A dictionary of geology.* Baltimore: Penguin

Books. 153p.
Hind, Wheelton, and Howe, J.A. (1901) *The geological succession and palaeontology of the beds between the Millstone Grit and the limestone-massif at Pendle Hill and their equivalents in certain other parts of Britain.* Geological Society of London. Quarterly Journal, v.57, p.347-404.
Hinds, N.E.A. (1943) *Geomorphology, the evolution of landscape.* New York: Prentice-Hall. 894p.
Hitchcock, Edward (1841) *Elementary geology.* 2nd ed. New York: Dayton & Saxton. 346p.
Hitchcock, Edward (1843) *The phenomena of drift, or glacioaqueous action in North America between the Tertiary and alluvial periods.* Association of American Geologists and Naturalists. Reports, 1st-3rd mtngs, p.164-221.
Hitchcock, Edward (1844) *Report on ichnolithology or fossil footmarks, with description of several new species and the coprolites of birds, and of a supposed footmark from the valley of Hudson River.* American Journal of Science, v.47, p.292-322.
Hitchcock, Edward (1858) *Ichnology of New England; a report on the sandstone of the Connecticut Valley, especially its footmarks.* Boston: William White. 220p.
Hobbs, B.E., Means, W.D., and Williams, P.F. (1976) *An outline of structural geology.* New York: Wiley. 571p.
Hobbs, W.H. (1901) *The Newark system of the Pomperaug Valley, Connecticut.* U.S. Geological Survey. Annual Report, 21st, pt.3, p.7-160.
Hobbs, W.H. (1906) *Guadix formation of Granada, Spain.* Geological Society of America. Bulletin, v.17, p.285-294.
Hobbs, W.H. (1907) *On some principles of seismic geology.* Beiträge zur Geophysik, v.7, p.219-292.
Hobbs, W.H. (1911a) *Characteristics of existing glaciers.* New York: Macmillan. 301p.
Hobbs, W.H. (1911b) *Repeating patterns in the relief and in the structure of the land.* Geological Society of America. Bulletin, v.22, p.123-176.
Hobbs, W.H. (1912) *Earth features and their meaning; an introduction to geology for the student and the general reader.* New York: Macmillan. 506p.
Hobbs, W.H. (1917) *The erosional and degradational processes of deserts, with especial reference to the origin of desert depressions.* Association of American Geographers. Annals, v.7, p.25-60.
Hobbs, W.H. (1921) *Studies of the cycle of glaciation.* Journal of Geology, v.29, p.370-386.
Hoffmeister, J.E., and Ladd, H.S. (1944) *The antecedent-platform theory.* Journal of Geology, v.52, p.388-402.
Hofmann, H.J. (1972) *Precambrian remains in Canada: fossils, dubiofossils, and pseudofossils.* International Geological Congress, 24th, Montreal. Section 1, p.20-30.
Hofmann, Walter, chairman (1965) *Parametric hydrology and stochastic hydrology; report of the Committee on Surface Water Hydrology.* American Society of Civil Engineers. Proceedings, Hydraulics Division. Journal, v.91, no.HY6, p.119-122.
Hollingworth, S.E., Taylor, J.H., & Kellaway, G.A. (1950) *Large-scale superficial structures in the Northampton Ironstone Field.* Geological Society of London. Quarterly Journal, v.100, p.1-44.
Hollister, C.D., and Heezen, B.C. (1972) *Geologic effects of ocean bottom currents: western North Atlantic.* In: Gordon, A.L., ed. *Studies in physical oceanography,* v.2, p.37-66. New York: Gordon and Breach.
Holm, D.A. (1957) *Sigmoidal dunes; a transitional form.* Geological Society of America. Bulletin, v.68, p.1746.
Holm, E.A. (1957) *Sand pavements in the Rub' al Khali.* Geological Society of America. Bulletin, v.68, p.1746.
Holmes, Arthur (1920) *The nomenclature of petrology.* 1st ed. London: Thomas Murby. 284p.
Holmes, Arthur (1928) *The nomenclature of petrology.* 2nd ed. London: Thomas Murby. 284p.
Holmes, Arthur (1959) *A revised geological timescale.* Edinburgh Geological Society. Transactions, v.17, p.183-216.
Holmes, Arthur (1965) *Principles of physical geology.* 2nd ed. New York: Ronald Press. 1288p.
Holthuis, L.B. (1974) *The lobsters of the superfamily Nephropidea of the Atlantic Ocean.* Bulletin of Marine Science, v.24, no.4.
Hooke, R.L. (1967) *Processes on arid-region alluvial fans.* Journal of Geology, v.75, p.438-460.
Hopkins, D.M. (1949) *Thaw lakes and thaw sinks in the Imuruk Lake Area, Seward Peninsula, Alaska.* Journal of Geology, v.57, p.119-131.
Hopkins, D.M., and others (1955) *Permafrost and ground water in Alaska.* U.S. Geological Survey. Professional Paper 264-F, p.113-146.
Hopkins, D.M., and Sigafoos, R.S. (1951) *Frost action and vegetation patterns on Seward Peninsula, Alaska.* U.S. Geological Survey. Bulletin 974-C, p.51-100.
Hoppe, Gunnar (1952) *Hummocky moraine regions, with special reference to the interior of Norbotton.* Geografiska Annaler, v.34, p.1-71.
Horberg, Leland (1954) *Rocky Mountain and continental Pleistocene deposits in the Waterton region, Alberta, Canada.* Geological Society of America. Bulletin, v.65, p.1093-1150.
Hornibrook, N. de B. (1965) *A viewpoint on stages and zones.* New Zealand Journal of Geology and Geophysics, v.8, p.1195-1212.
Horton, R.E. (1932) *Drainage basin characteristics.* American Geophysical Union. Transactions, v.13, p.350-361.
Horton, R.E. (1945) *Erosional development of streams and their drainage basins; hydrophysical approach to quantitative morphology.* Geological Society of America. Bulletin, v.56, p.275-370.
Hörz, F., Hartung, J.B., and Gault, D.E. (1971) *Micrometeorite craters on lunar rock surfaces.* Journal of Geophysical Research, v.76, p.5770-5798.
Howard, A.D. (1942) *Pediment passes and pediment problem.* Journal of Geomorphology, v.5, p.1-31, 95-136.
Howard, A.D. (1959) *Numerical systems of terrace nomenclature; a critique.* Journal of Geology, v.67, p.239-243.
Howard, W.V., and David, M.W. (1936) *Development of porosity in limestones.* American Association of Petroleum Geologists. Bulletin, v.20, p.1389-1412.
Howell, J.V. (1922) *Notes on the pre-Permian Paleozoics of the Wichita Mountain area.* American Association of Petroleum Geologists. Bulletin, v.6, p.413-425.
Hsü, K.J. (1955) *Monometamorphism, polymetamorphism and retrograde metamorphism.* American Journal of Science, v.253, p.237-239.
Hsü, K.J. (1968) *Principles of mélanges and their bearing on the Franciscan-Knoxville paradox.* Geological Society of America. Bulletin, v.79, p.1063-1074.
Hsü, K.J. (1975) *Catastrophic debris streams (Sturzstroms) generated by rockfalls.* Geological Society of America. Bulletin, v.86, p.129-140.
Hsu, T.C. (1966) *The characteristics of coaxial and non-coaxial strain paths.* Journal of Strain Analysis, v.1, p.216-222.
Huang, W.H., and Walker, R.M. (1967) *Fossil alpha-particle recoil tracks: a new method of age determination.* Science, v.155, p.1103-1106.
Huang, W.T. (1962) *Petrology.* New York: McGraw-Hill. 480p.
Hubbard, N.J. (1973) *Chemistry of lunar basalts with very high alumina contents.* Science, v.181, p.339-342.
Hubert, J.F. (1960) *Petrology of the Fountain and Lyons formations, Front Range, Colorado.* Colorado School of Mines. Quarterly, v.55, no.1, p.1-242.
Hudson, G.H. (1909) *Some items concerning a new and an old coast line of Lake Champlain.* New York State Museum. Bulletin 133, p.159-163.
Hudson, G.H. (1910) *Joint caves of Valcour Island—their age and their origin.* New York State Museum. Bulletin 140, p.161-196.
Hudson, R.G. (1924) *On the rhythmic succession of the Yoredale Series in Wensleydale.* Yorkshire Geological Society. Proceedings, n.s., v.20, p.125-135.
Hulings, N.C., and Gray, J.S. (1971) *A manual for the study of meiofauna.* Smithsonian Contributions to Zoology, no.78. Washington: Smithsonian Institution. 84p.
Humble, William (1843) *Dictionary of geology and mineralogy, comprising such terms in botany, chemistry, comparative anatomy, conchology, entomology, palaeontology, zoology, and other branches of natural history, as are connected with the study of geology.* 2nd ed., with additions. London: Henry Washbourne. 294p.
Humphries, D.W. (1966) *The booming sand of Korizo, Sahara, and the squeaking sand of Gower, S. Wales: a comparison of the fundamental characteristics of two musical sands.* Sedimentology, v.6, p.135-152.
Hunt, C.B., and others (1953) *Geology and geography of the Henry Mountains region, Utah.* U.S. Geological Survey. Professional Paper 228. 234p.
Huschke, R.E., ed. (1959) *Glossary of meteorology.* Boston: American Meteorological Society. 638p.

Hutchinson, G.E. (1957) *A treatise on limnology.* Vol.1. New York: Wiley. 1015p.

Hutchinson, J.N. (1967) *The free degradation of London Clay cliffs.* Geotechnical conference on shear strength properties of natural soils and rocks, Oslo. Proceedings, 1, p.113-118.

Hutchinson, J.N. (1968) *Mass movement.* In: Fairbridge, R.W., ed. *Encyclopedia of geomorphology,* p.688-696. New York: Reinhold. 1295p.

Hutchinson, J.N., & Bhandari, R.K. (1971) *Undrained loading, a fundamental mechanism of mudflows and other mass movements.* Géotechnique, v.21, no.4, p.353-358.

Hutton, James (1788) *Theory of the Earth; or an investigation of the laws observable in the composition, dissolution, and restoration of land upon the globe.* Royal Society of Edinburgh. Transactions, v.1, p.209-304.

Huxley, T.H. (1862) *The anniversary address.* Geological Society of London. Quarterly Journal, v.18, p.xl-liv.

Huxley, T.H. (1877) *Physiography: an introduction to the study of nature.* London: Macmillan. 384p.

Hyde, H.A., and Williams, D.A. (1944) *The right word.* Pollen Analysis Circular, no.8, p.6.

I

ICBN: Stafleu, F.A., ed. (1972) *International code of botanical nomenclature.* International Botanical Congress. 11th, Seattle, 1969. Utrecht, The Netherlands: Oosthoek. 426p.

ICCP: International Committee for Coal Petrology (1963) *Handbook of coal petrology.* 2d ed. Paris: Centre National de la Recherche Scientifique.

ICCP: International Committee for Coal Petrology (1971) *Handbook of coal petrology.* 2nd ed., supplement. Paris: Centre National de la Recherche Scientifique.

ICZN: International Commission on Zoological Nomenclature (1964) *International code of zoological nomenclature, adopted by the XV International Congress of Zoology.* N.R. Stoll, Editorial Committee, chairman. 2nd ed. London: International Trust for Zoological Nomenclature. 176p.

Illing, L.V. (1954) *Bahaman calcareous sands.* American Association of Petroleum Geologists. Bulletin, v.38, p.1-95.

Imbrie, John, and Buchanan, Hugh (1965) *Sedimentary structures in modern carbonate sands of the Bahamas.* Society of Economic Paleontologists and Mineralogists. Special Publication, no.12, p.149-172.

Imbt, W.C., and Ellison, S.P., jr. (1947) *Porosity in limestone and dolomite petroleum reservoirs.* Drilling and Production Practice, 1946, p.364-372.

Ingle, J.C. (1966) *The movement of beach sand; an analysis using fluorescent grains.* New York: Elsevier. 221p. (*Developments in sedimentology,* 5).

Ireland, H.A., and others (1947) *Terminology for insoluble residues.* American Association of Petroleum Geologist. Bulletin, v.31, p.1479-1490.

Ireland, H.O. (1969) *Foundations for heavy structures.* Reviews in Engineering Geology, v.2, p.1-15.

Irvine, T.R., and Baragar, W.R.A. (1971) *A guide to the chemical classification of the common volcanic rocks.* Canadian Journal of Earth Science, v.8, p.523-548.

Irwin, J.S. (1926) *Faulting in the Rocky Mountain region.* American Association of Petroleum Geologists. Bulletin, v.10, p.105-129.

Isacks, Bryan, and others (1968) *Seismology and the new global tectonics.* Journal of Geophysical Research, v.73, p.5855-5899.

ISG: Hedberg, H.D., ed. (1976) *International stratigraphic guide. A guide to stratigraphic classification, terminology, and procedure.* International Union of Geological Sciences, Commission on Stratigraphy, International Subcommission on Stratigraphic Classification. New York: Wiley. 200p.

ISSC: International Subcommission on Stratigraphic Classification (1970) *Preliminary report on lithostratigraphic units.* H.D. Hedberg, ed. International Union of Geological Sciences, ISSC, Rept. 3. 30p.

ISSC: International Subcommission on Stratigraphic Classification (1971) *Preliminary report on biostratigraphic units.* H.D. Hedberg, ed. International Geological Congress. 24th, Montreal, 1972. Report 5. 50p.

ISST: International Subcommission on Stratigraphic Terminology (1961) *Stratigraphic classification and terminology.* Edited by H.D. Hedberg. International Geological Congress. 21st, Copenhagen, 1960. Report, pt.25. 38p.

ISST: International Subcommission on Stratigraphic Terminology (1965) *Definition of geologic systems.* American Association of Petroleum Geologists. Bulletin, v.49, p.1694-1703. (American Commission on Stratigraphic Nomenclature. Note 32).

IUGS: International Union of Geological Sciences, Subcommission on the Systematics of Igneous Rocks (1973) *Plutonic rocks, classification and nomenclature.* Geotimes, v.18, no.10, p.26-30.

Iversen, Johs. (1954) *The late-glacial flora of Denmark and its relation to climate and soil.* Denmark. Geologiske Undersogelse. Danmarks geologiske undersogelse, ser.2, no.80, p.87-119.

Ives, R.L. (1941) *Tundra ponds.* Journal of Geomorphology, v.4, p.285-296.

J

Jaanusson, Valdar (1960) *The Viruan (Middle Ordovician) of Öland.* Uppsala. University. Geological Institutions. Bulletin, v.38, p.207-288.

Jacks, G.V., and others, eds. (1960) *Multilingual vocabulary of soil science.* 2nd ed., revised. New York(?): United Nations, Food and Agriculture Organization, Land & Water Development Division. 430p.

Jackson, B.D. (1928) *A glossary of botanic terms with their derivation and accent.* 4th ed. Reprinted 1949, 1953. New York: Hafner. 481p.

Jackson, J.R. (1834) *Hints on the subject of geographical arrangement and nomenclature.* Royal Geographical Society. Journal, v.4, p.72-88.

Jaeger, J.C. (1965) *Application of the theory of heat conduction to geothermal measurements.* In: *Terrestrial heat flow,* ch.2. American Geophysical Union. Geophysical Monograph No.8.

Jahns, R.H. (1953) *Surficial geology of the Ayer quadrangle, Massachusetts.* U.S. Geological Survey. Map GQ-21.

Jahns, R.H. (1967) *Serpentinites of the Roxbury district, Vermont.* In: Wyllie, P.J., ed. *Ultramafic and related rocks,* p.137-160. New York: Wiley. 464p.

Jakosky, J.J. (1950) *Exploration geophysics.* 2nd ed. Los Angeles: Trija.

James, H.L. (1954) *Sedimentary facies of iron-formation.* Economic Geology, v.49, p.235-293.

James, N.P. (1977) *The deep and the past.* Third International Coral Reef Symposium. Proceedings, v.2, p.xxv-xxvii. Miami: University of Miami, Rosenstiel School of Marine and Atmospheric Science. 628p.

James, W.R., and Krumbein, W.C. (1969) *Frequency distributions of stream link lengths.* Journal of Geology, v.77, p.544-565.

Jameson, Robert (1808) *Elements of geognosy.* Edinburgh: William Blackwood. 368p. (His *System of mineralogy,* v.3).

Jamieson, J.A., and others (1963) *Infrared physics and engineering.* New York: McGraw-Hill.

Jarvis, R.S. (1972) *New measure of the topologic structure of dendritic drainage networks.* Water Resources Research, v.8, p.1265-1271.

Jeffrey, Charles (1973) *Biological nomenclature.* New York: Crane, Russak. 69p.

Jeletzky, J.A. (1956) *Paleontology, basis of practical geochronology.* American Association of Petroleum Geologists. Bulletin, v.40, p.679-706.

Jeletzky, J.A. (1966) *Comparative morphology, phylogeny, and classification of fossil Coleoidea.* University of Kansas. Paleontological Contributions, art.7. 162p.

Jennings, J.D. (1968) *Prehistory of North America.* New York: McGraw-Hill. 391p.

Johannsen, Albert (1917) *Suggestions for a quantitative mineralogical classification of igneous rocks.* Journal of Geology, v.25, p.63-97.

Johannsen, Albert (1920) *A quantitative mineralogical classification of igneous rocks—revised.* Journal of Geology, v.28, p.38-60, 158-177, 210-232.

Johannsen, Albert (1931) *Introduction, textures, classifications and glossary.* 1st ed. Chicago: University of Chicago Press. 267p. (*A descriptive petrography of the igneous rocks,* v.1.)

Johannsen, Albert (1931a) *The quartz-bearing rocks.* 1st ed. Chicago: University of Chicago Press. (*A descriptive petrography of the igneous rocks,* v.2.)

Johannsen, Albert (1937) *A descriptive petrography of the igne-*

ous rocks, v.3. Chicago: University of Chicago Press, 360p.

Johannsen, Albert (1938) *The feldspathoid rocks; the peridotites and perknites.* Chicago: University of Chicago Press. 523p. (*A descriptive petrography of the igneous rocks,* v.4.)

Johannsen, Albert (1939) *Introduction, textures, classifications and glossary.* 2nd ed. Chicago: University of Chicago Press. 318p. (*A descriptive petrography of the igneous rocks,* v.1.)

Johnson, D.W. (1916) *Plains, planes, and peneplanes.* Geographical Review, v.1, p.443-447.

Johnson, D.W. (1919) *Shore processes and shoreline development.* New York: Wiley. 584p.

Johnson, D.W. (1925) *The New England-Acadian shore line.* New York: Wiley. 608p.

Johnson, D.W. (1932) *Rock fans of arid regions.* American Journal of Science, 5th ser., v.23, p.389-416.

Johnson, D.W. (1939) *The origin of submarine canyons; a critical review of hypotheses.* New York: Columbia University Press. 126p.

Johnson, P.H., and Bhappu, R.B. (1969) *Chemical mining—a study of leaching agents.* New Mexico Bureau of Mines and Mineral Resources. Circular 99. 10p.

Jones, O.T. (1937) *On the sliding or slumping of submarine sediments in Denbighshire, North Wales, during the Ludlow period.* Geological Society of London. Quarterly Journal, v.93, p.241-283.

Jones, P.B. (1971) *Folded faults and sequence of thrusting in Alberta foothills.* American Association of Petroleum Geologists. Bulletin, v.55, p.292-306.

Jones, P.H. (1969) *Hydrology of Neogene deposits in the northern Gulf of Mexico basin.* Louisiana Water Resources Research Institute. Bulletin GT-2, 105p.

Joplin, G.A. (1968) *A petrography of Australian metamorphic rocks.* New York: American Elsevier. 262p.

Jopling, A.V. (1961) *Origin of regressive ripples explained in terms of fluid-mechanic processes.* U.S. Geological Survey. Professional Paper 424-D, art.299, p.15-17.

Joreskog, K.G., Klovan, J.E., and Reyment, R.A. (1976) *Geological factor analysis.* New York: Elsevier. 178p. (*Methods in geomathematics,* 1.)

Judson, S.S., jr. (1953) *Geology of the San Jon site, eastern New Mexico.* Smithsonian Miscellaneous Collections, v.121, no.1. 70p.

Jukes, J.B. (1862) *On the mode of formation of some of the river-valleys in the south of Ireland.* Geological Society of London. Quarterly Journal, v.18, p.378-403.

Jukes-Browne, A.J. (1903) *The term "hemera".* Geological Magazine, dec.4, v.10, p.36-38.

Jull, R.K. (1967) *The hystero-ontogeny of* Lonsdaleia *McCoy and* Thysanophyllum orientale *Thomson.* Palaeontology, v.10, p.617-628.

Jurine, Louis (1806) *Réflexions sur la nécessité d'une nouvelle nomenclature en géologie, et l'exposé de celle qu'il propose.* Journal des Mines, v.19, p.367-378.

K

Kahn, J.S. (1956) *The analysis and distribution of the properties of packing in sand-size sediments.* Journal of Geology, v.64, p.385-395, 578-606.

Kalkowsky, Ernst (1880) *Ueber die Erforschung der archäischen Formationen.* Neues Jahrbuch für Mineralogie, Geologie und Palaeontologie, Jahrg. 1880, Bd.1, p.1-28.

Kalkowsky, Ernst (1908) *Oolith und Stromatolith im norddeutschen Buntsandstein.* Deutsche Geologische Gesellschaft. Zeitschrift, Bd.60, p.68-125.

Kamen-Kaye, Maurice (1967) *Basement subsidence and hypersubsidence.* American Association of Petroleum Geologists. Bulletin, v.51, p.1833-1842.

Karcz, Iaakov (1967) *Harrow marks, current-aligned sedimentary structures.* Journal of Geology, v.75, p.113-121.

Karlstrom, T.N.V. (1956) *The problem of the Cochrane in late Pleistocene chronology.* U.S. Geological Survey. Bulletin 1021-J, p.303-331.

Karlstrom, T.N.V. (1961) *The glacial history of Alaska: its bearing on paleoclimatic theory.* New York Academy of Sciences. Annals, v.95, art.1, p.290-340.

Karlstrom, T.N.V. (1966) *Quaternary glacial record of the North Pacific region and worldwide climatic changes.* In: Blumenstock, D. I., ed. *Pleistocene and post-Pleistocene climatic variations in the Pacific area,* p.153-182. Honolulu: Bishop Museum Press. 182p.

Kaufmann, F.J. (1886) *Emmen- und Schlierengegenden nebst Umgebungen bis zur Brünigstrasse und Linie Lungern-Grafenort geologisch aufgenommen und dargestellt.* Beiträge zur geologischen Karte der Schweiz, v.24, 608p. pt.1. with atlas of 30 plates.

Kay, G.F. (1916) *Gumbotil, a new term in Pleistocene geology.* Science, v.44, p.637-638.

Kay, G.M. (1937) *Stratigraphy of the Trenton Group.* Geological Society of America. Bulletin, v.48, p.233-302.

Kay, G.M. (1945) *North American geosynclines: their classification* [abstract]. Geological Society of America. Bulletin, v.56, p.1172.

Kay, G.M. (1945a) *Paleogeographic and palinspastic maps.* American Association of Petroleum Geologists. Bulletin, v.29, p.426-450.

Kay, G.M. (1947) *Geosynclinal nomenclature and the craton.* American Association of Petroleum Geologists. Bulletin, v.31, no.7, p.1287-1293.

Kay, G.M. (1951) *North American geosynclines.* Geological Society of America. Memoir 48. 143p.

Keeton, W.T. (1967) *Biological science.* New York: Norton.

Keith, Arthur (1895) *Description of the Knoxville sheet.* U.S. Geological Survey. Geologic Atlas, folio 16. 6p.

Keith, M.L., and Degens, E.T. (1959) *Geochemical indicators of marine and freshwater sediments.* In: Abelson, P.H., ed. *Researches in geochemistry,* p.38-61. New York: Wiley. 511p.

Keller, B.M., and others (1968) *The main features of the Late Proterozoic paleogeography of the USSR.* International Geological Congress. 23rd, Prague. Proceedings, sec.4, p.189-202.

Keller, W.D. (1958) *Argillation and direct bauxitization in terms of concentrations of hydrolyzing aluminum silicates.* American Association of Petroleum Geologists. Bulletin, v.42, p.233-245.

Keller, W.D. (1963) *Diagenesis in clay minerals—a review.* National Conference on Clays and Clay Minerals. 11th, Ottawa, 1962. Proceedings, p.136-157. (*International series of monographs on earth sciences,* v.13.)

Keller, W.D. (1977) *Scan electron micrographs of kaolins collected from diverse environments of origin—IV, Georgia kaolin and kaolinizing source rocks.* Clays and Clay Minerals, v.25, p.311-345.

Keller, W.D., Viele, G.W., and Johnson, C.H. (1977) *Texture of Arkansas novaculite indicates thermally induced metamorphism.* Journal of Sedimentary Petrology, v.47, p.834-843.

Keller, W.D., Westcott, J.F., and Bledsoe, A.O. (1954) *The origin of Missouri fireclays.* National Academy of Sciences—National Research Council. Publication 327, p.7-46.

Kelley, V.C. (1956) *Thickness of strata.* Journal of Sedimentary Petrology, v.26, p.289-300.

Kelling, Gilbert (1958) *Ripple-mark in the Rhinns of Galloway.* Edinburgh Geological Society. Transactions, v.17, p.117-132.

Kelling, Gilbert, and Stanley, D.J. (1976) *Sedimentation in canyon, slope, and base of slope environments.* In: Stanley, D.J., and Swift, D.J.P., eds. *Marine sediment transport and environmental management.* Ch.17.

Kellogg, C.E. (1950) *Tropical soils.* 4th International Congress on Soil Science. Transactions, v.1, p.266-276.

Kemp, J.F. (1896) *A handbook of rocks, for use without the microscope.* New York: J.F. Kemp. 176p.

Kemp, J.F. (1900) *Handbook of rocks for use without the microscope.* New York: Van Nostrand. 185p.

Kemp, J.F. (1934) *A handbook of rocks for use without the microscope, with a glossary of the names of rocks and other lithological terms.* 5th ed. New York: Van Nostrand. 300p.

Kemp, J.F. (1940) *A handbook of rocks (for use without the petrographic microscope).* 6th ed., completely revised & edited by F.F. Grout. New York: Van Nostrand.

Kendall, P.F. (1902) *A system of glacier-lakes in the Cleveland Hills.* Geological Society of London. Quarterly Journal, v.58, p.471-571.

Kendall, P.F., and Bailey, E.B. (1908) *The glaciation of East Lothian south of the Garleton Hills.* Royal Society of Edinburgh. Transactions, v.46, p.1-31.

Kendall, P.F., and Wroot, H.E. (1924) *Geology of Yorkshire; an illustration of the evolution of northern England.* Vol. 1. Vienna: Hollinek Brothers. 660p.

Kennedy, J.F. (1963) *The mechanics of dunes and antidunes in erodible-bed channels.* Journal of Fluid Mechanics, v.16, p.521-544.

Kennedy, W.Q. (1933) *Trends of differentiation in basaltic magmas.* American Journal of Science, 5th series, v.25, p.239-256.

Keroher, G.C., and others (1966) *Lexicon of geologic names of the United States for 1936-1960.* Parts 1-3. U.S. Geological Survey. Bulletin 1200. 4341p.

Kerr, P.F., and Hamilton, P.K. (1949) *Glossary of clay mineral names.* New York: Columbia University. 66p. (American Petroleum Institute. Project 49: Clay Mineral Standards. Preliminary Report no.1).

Kerr, P.F., and Kopp, O.C. (1958) *Salt-dome breccia.* American Association of Petroleum Geologists. Bulletin, v.42, p.548-560.

Kerr, R.C., and Nigra, J.O. (1952) *Eolian sand control.* American Association of Petroleum Geologists. Bulletin, v.36, p.1541-1573.

Kerr, W.C. (1881) *On the action of frost in the arrangement of superficial earthy material.* American Journal of Science, v.21, p.345-358.

Kesling, R.V. (1951) *Terminology of ostracod carapaces.* Ann Arbor: University of Michigan. Contributions from the Museum of Paleontology, v.9, no.4, p.93-171.

Kesseli, J.E. (1941) *The concept of the graded river.* Journal of Geology, v.49, p.561-588.

Kessler, Paul (1922) *Über Lochverwitterung und ihre Beziehungen zur Metharmose (Umbildung) der Gesteine.* Geologische Rundschau, v.12, p.237-270.

Keyes, C.R. (1910) *Deflation and the relative efficiencies of erosional processes under conditions of aridity.* Geological Society of America. Bulletin, v.21, p.565-598.

Keyes, C.R. (1913) *Antigravitational gradation.* Science, n.s., v.38, p.206.

Kieffer, S.W. (1971) *Shock metamorphism of the Coconino sandstone at Meteor Crater, Arizona.* Journal of Geophysical Research, v.76, p.5449-5473.

Kinahan, G.H. (1878) *Manual of the geology of Ireland.* London: C.K. Paul. 444p.

Kindle, E.M. (1916) *Small pit and mound structures developed during sedimentation.* Geological Magazine, dec.6, v.3, p.542-547.

Kindle, E.M. (1917) *Recent and fossil ripple-mark.* Geological Survey of Canada. Museum Bulletin no.25. 121p. (Canada Department of Mines. Geological Series, no.34.)

Kindle, E.M. (1923) *Range and distribution of certain types of Canadian Pleistocene concretions.* Geological Society of America. Bulletin, v.34, p.609-648.

Kindle, E.M. (1923a) *Nomenclature and genetic relations of certain calcareous rocks.* Pan-American Geologist, v.39, p.365-372.

Kindle, E.M. (1926) *Contrasted types of mud cracks.* Royal Society of Canada. Proceedings and Transactions, ser.3, v.20, sec.4, p.71-75.

King, C.A.M., and Williams, W.W. (1949) *The formation and movement of sand bars by wave action.* Geographical Journal, v.113, p.70-85.

King, Clarence (1878) *Systematic geology.* Washington: U.S. Government Printing Office. 803p. (U.S. Geological Exploration of the Fortieth Parallel. Report, v.1).

King, L.C. (1948) *A theory of bornhardts.* Geographical Journal, v.112, p.83-87.

King, L.C. (1959) *Denudational and tectonic relief in south-eastern Australia.* Geological Society of South Africa. Transactions, v.62, p.113-138.

King, L.C. (1962) *The morphology of the Earth; a study and synthesis of world scenery.* New York: Hafner. 699p.

King, Philip (1959) *The evolution of North America.* Princeton: Princeton University Press. 190p.

Kinsman, D.J.J. (1969) *Modes of formation, sedimentary associations, and diagnostic features of shallow-water and supratidal evaporites.* American Association of Petroleum Geologists. Bulletin, v.53, p.830-840.

Kirkham, R.V. (1971) *Intermineral intrusions and their bearing on the origin of porphyry copper and molybdenum deposits.* Economic Geology, v.66, p.1244-1249.

Kirwan, Richard (1794) *Elements of mineralogy.* Vol.1. 2nd ed. London: J. Nichols. 510p.

Klein, G. de V. (1963) *Analysis and review of sandstone classifications in the North American geological literature, 1940-1960.* Geological Society of America. Bulletin, v.74, p.555-575.

Klöden, K.F. von (1828) *Beiträge zur mineralogischen und geognostischen Kenntniss der Mark Brandenburg.* [Pt.1]. Berlin: W. Dieterici. 108p.

Klotz, I.M. (1972) *Chemical thermodynamics,* 3rd ed. Menlo Park, Calif.: W.A. Benjamin. 444p.

Klugh, A.B. (1923) *A common system of classification in plant and animal ecology.* Ecology, v.4, p.366-377.

Knight, C.L. (1957) *Ore genesis—the source bed concept.* Economic Geology, v.53, p.808-817.

Knight, J.B. (1941) *Paleozoic gastropod genotypes.* Geological Society of America. Special Paper 32. 510p.

Knight, S.H. (1929) *The Fountain and the Casper formations of the Laramie Basin.* University of Wyoming. Publications in Science. Geology, v.1, p.1-82.

Knight, W.C. (1897) *"Mineral soap".* Engineering and Mining Journal, v.63, p.600-601.

Knight, W.C. (1898) *Bentonite.* Engineering and Mining Journal, v.66, p.491.

Knopf, E.B., and Ingerson, Earl (1938) *Structural petrology.* Geological Society of America. Memoir 6. 270p.

Knox, Alexander (1904) *Glossary of geographical and topographical terms.* London: Edward Stanford. 432p. (*Stanford's compendium of geography and travel;* suppl. vol.)

Knutson, R.M. (1958) *Structural sections and the third dimension.* Economic Geology, v.53, p.270-286.

Kobayashi, Teiichi (1944) *An instant in the Phanerozoic Eon and its bearings on geology and biology.* Imperial Academy, Tokyo. Proceedings, v.20, p.742-750. (*Concept of time in geology,* 3).

Kobayashi, Teiichi (1944a) *On the major classification of the geological age.* Imperial Academy, Tokyo. Proceedings, v.20, p.475-478. (*Concept of time in geology,* 1).

Koch, Lauge (1926) *Ice cap and sea ice in north Greenland.* Geographical Review, v.16, p.98-107.

Koppejan, A.W., van Wanelen, B.M., and Weinberg, L.J.H. (1948) *Coastal flow slides in the Dutch province of Zeeland.* International Conference on Soil Mechanics and Foundation Engineering. 2nd, Rotterdam, 1948? Proceedings, 5, p.89-96.

Kornicker, L.S., and Boyd, D.W. (1962) *Shallow-water geology and environments of Alacran reef complex, Campeche Bank, Mexico.* American Association of Petroleum Geologists. Bulletin, v.46, p.640-673.

Kostov, Ivan (1961) *Genesis of kyanite in quartz veins.* International Geology Review, v.3, p.645-651.

Kovalenko, V.I., and others (1971) *Topaz-bearing quartz keratophyre (ongonite), a new variety of subvolcanic igneous vein rock.* Doklady Akad. Nauk SSSR, v.199, p.132-135.

Krasheninnikov, G.F. (1964) *Facies, genetic types, and formations.* International Geology Review, v.6, p.1242-1248.

Krauskopf, K.B. (1967) *Introduction to geochemistry.* New York: McGraw-Hill. 721p.

Kremp, G.O.W. (1965) *Morphologic encyclopedia of palynology.* University of Arizona, Program in Geochronology. Contribution 100. Tucson: University of Arizona Press. 185p.

Krishtofovich, A.N. (1945) *The mode of preservation of plant fossils and their bearing upon the problem of coal formation.* Akademiya Nauk SSSP. Izvestiya, Seriya Geologicheskaya, no.2, p.136-150.

Krivenko, V., and Lapchik, T. (1934) *On the petrography of the crystalline rocks of the rapids of the Dnipro.* Ukrainskii Naukovo-Doslidchii Geologichnii Institute. Trudy, v.5, no.2, p.67-77.

Kruger, F.C. (1946) *Structure and metamorphism of the Bellows Falls quadrangle of New Hampshire and Vermont.* Geological Society of America. Bulletin, v.57, p.161-205.

Krumbein, W.C. (1934) *Size frequency distributions of sediments.* Journal of Sedimentary Petrology, v.4, p.65-77.

Krumbein, W.C. (1955) *Composite end members in facies mapping.* Journal of Sedimentary Petrology, v.25, p.115-122.

Krumbein, W.C., and Graybill, F.A. (1965) *An introduction to statistical models in geology.* New York: McGraw-Hill. 475p.

Krumbein, W.C., and Libby, W.G. (1957) *Application of moments to vertical variability maps of stratigraphic units.* American Association of Petroleum Geologists. Bulletin, v.41, p.197-211.

Krumbein, W.C., and Pettijohn, F.J. (1938) *Manual of sedimentary petrography.* New York: Appleton. 549p.

Krumbein, W.C., and Sloss, L.L. (1951) *Stratigraphy and sedimentation.* San Francisco: Freeman. 497p.

Krumbein, W.C., and Sloss, L.L. (1963) *Stratigraphy and sedimentation.* 2nd ed. San Francisco: Freeman. 660p.

Krynine, P.D. (1937) *Petrography and genesis of the Siwalik Series.* American Journal of Science, ser.5, v.34, no.204, p.422-446.

Krynine, P.D. (1940) *Petrology and genesis of the Third Bradford Sand.* Pennsylvania State College. Mineral Industries Experiment Station. Bulletin 29. 134p.

Krynine, P.D. (1941) *Differentiation of sediments during the life* *history of a landmass.* Geological Society of America. Bulletin, v.52, p.1915.

Krynine, P.D. (1945) *Sediments and the search for oil.* Producers Monthly, v.9, p.12-22.

Krynine, P.D. (1948) *The megascopic study and field classification of sedimentary rocks.* Journal of Geology, v.56, p.130-165.

Krynine, P.D. (1951) *Reservoir petrography of sandstones.* In: Payne, T.G., and others. *Geology of the Arctic Slope of Alaska,* sheet 2. U.S. Geological Survey. Oil and Gas Investigations, map OM 126. 3 sheets.

Ksiazkiewicz, Marian (1949) *Slip-bedding in the Carpathian flysch.* Société Géologique Pologne. Annales, v.19, p.493-501.

Ksiazkiewicz, Marian (1958) *Submarine slumping in the Carpathian Flysch.* Polskie Towarzystwo Geologicne. Rocznik, t.28, p.123-151.

Kubiëna, W.L. (1953) *The soils of Europe.* London: Thomas Murby. 317p.

Kuenen, P.H. (1943) *Pitted pebbles.* Leidsche geologische mededeelingen, v.13, p.189-201.

Kuenen, P.H. (1947) *Water-faceted boulders.* American Journal of Science, v.245, p.779-783.

Kuenen, P.H. (1948) *Slumping in the Carboniferous rocks of Pembrokeshire.* Geological Society of London. Quarterly Journal, v.104, p.365-385.

Kuenen, P.H. (1950) *Marine geology.* New York: Wiley. 568p.

Kuenen, P.H. (1953) *Significant features of graded bedding.* American Association of Petroleum Geologists. Bulletin, v.37, p.1044-1066.

Kuenen, P.H. (1957) *Sole markings of graded graywacke beds.* Journal of Geology, v.65, p.231-258.

Kuenen, P.H. (1958) *Experiments in geology.* Geological Society of Glasgow. Transactions, v.23, p.1-28.

Kulp, J.L. (1961) *Geologic time scale.* Science, v.133, p.1105-1114.

Kümmel, H.B. (1893) *Some rivers of Connecticut.* Journal of Geology, v.1, p.371-393.

Kuno, Hisachi (1957) *Differentiation of Hawaiian magmas.* Japanese Journal of Geology and Geography, v.28, p.193-194.

Kuno, Hisachi (1959) *Origin of Cenozoic petrographic provinces of Japan and surrounding areas.* Volcanological Society of Japan. Bulletin, series 2, tome 20, p.37-76.

Kuno, Hisachi (1960) *High-alumina basalt.* Journal of Petrology, v.1, p.121-145.

Kupsch, W.O. (1956) *Submask geology in Saskatchewan.* International Williston Basin Symposium. 1st, Bismarck (N.D.), 1956. Williston Basin Symposium, p.66-75.

L

Lacroix, A. (1922) *Minéralogie de Madagascar. v.2: minéralogie appliquée; lithologie.* Paris: Augustin Challamel. 694p.

Lagaaij, R. (1952) *The Pliocene Bryozoa of the Low Countries and their bearing on the marine stratigraphy of the North Sea region.* Netherlands, Geologische Stichtung, Mededeelingen, serie C-V, no.5, 233p.

Lagaaij, R. and Gautier, Y.V. (1965) *Bryozoan assemblages from marine sediments of the Rhône delta, France.* Micropaleontology, v.11, p.39-58.

Lahee, F.H. (1923) *Field geology.* 2nd ed. New York: McGraw-Hill. 649p.

Lahee, F.H. (1961) *Field geology.* 6th ed. New York: McGraw-Hill. 926p.

Lahee, F.H. (1962) *Statistics of exploratory drilling in the United States 1945-1960.* Tulsa, American Association of Petroleum Geologists. 135p.

Laird, M.G. (1970) *Vertical sheet structure—a new indicator of sedimentary fabric.* Journal of Sedimentary Petrology, v.40, p.428-434.

Lam, H.J. (1957) *What is a taxon?* Taxonomy, v.6, p.213-215.

Lamar, J.E. (1928) *Geology and economic resources of the St. Peter Sandstone of Illinois.* Illinois State Geological Survey. Bulletin no.53. 175p.

Lamb, H.H. (1970) *Volcanic dust in the atmosphere.* Royal Society of London, Philosophical Transactions, Series A, Mathematics and Physical Sciences, v.266, p.425-533.

Lambe, T.W. (1953) *The structure of inorganic soil.* American Society of Civil Engineers. Proceedings, v.79, separate no.315. 49p.

Lamont, Archie (1957) *Slow anti-dunes and flow marks.* Geological Magazine, v.94, p.472-480.

Lamplugh, G.W. (1895) *The crush-conglomerates of the Isle of Man.* With a petrographical appendix by W.W. Watts. Geological Society of London. Quarterly Journal, v.51, p.563-599.

Lamplugh, G.W. (1902) *"Calcrete".* Geological Magazine, n.s., dec.4, v.9, no.462, p.575.

Landes, K.K. (1945) *The Mackinac Breccia.* Michigan Geological Survey. Publication 44, Geological Series 37, ch.3, p.123-153.

Landes, K.K. (1957) *Chemical unconformities* (abstract). Geological Society of America. Bulletin, v.68, p.1759.

Landes, K.K. (1967) *Eometamorphism, and oil and gas in time and space.* American Association of Petroleum Geologists. Bulletin, v.51, p.828-841.

Lane, A.C. (1903) *Porphyritic appearance of rocks.* Geological Society of America. Bulletin, v.14, p.369-379.

Lane, A.C. (1923) *Communication.* Journal of Geology , v.31, p.348.

Lane, A.C. (1928) *Isontic?* Science, v.68, p.37.

Lang, W.T.B. (1937) *The Permian formations of the Pecos Valley of New Mexico and Texas.* American Association of Petroleum Geologists. Bulletin, v.21, p.833-898.

Langbein, W.B., and Iseri, K.T. (1960) *General introduction and hydrologic definitions.* U.S. Geological Survey. Water-supply Paper 1541-A. 29p. (*Manual of hydrology:* pt.1, p.1-29).

Laporte, L.F. (1975) *Encounter with the Earth.* San Francisco: Canfield Press. 538p.

Larsen, E.S. (1938) *Some new variation diagrams for groups of igneous rocks.* Journal of Geology, v.46, p.505-520.

Larsen, Gunnar, and Chilingar, G.V., eds. (1967) *Diagenesis in sediments.* Amsterdam: Elsevier. 551p. (*Developments in sedimentology,* 8).

Larsen, Ole, and Sorensen, H. (1960) *Principles of classification and norm calculations of metamorphic rocks: a discussion.* Journal of Geology, v.68, p.681-683.

Lasius, G.S.O. (1789) *Beobachtungen über die Harzgebirge, nebst einem Profilnisse, als ein Beytrag zur mineralogischen Naturkunde.* Hannover: Helwingischen Hofbuchhandlung. 559p. in 2 v.

Lasky, S.G. (1950) *How tonnage and grade relations help predict ore reserves.* Engineering and Mining Journal, v.151, no.4, p.81-85.

Lawrence, G.H.M. (1951) *Taxonomy of vascular plants.* New York: Macmillan. 823p.

Lawson, A.C. (1885) *Report on the geology of the Lake of the Woods region, with special reference to the Keewatin (Huronian?) belt of the Archean rocks.* Geological Survey of Canada. Annual Report, 1:CC. 151p.

Lawson, A.C. (1894) *The geomorphogeny of the coast of northern California.* University of California. Department of Geology. Bulletin, v.1, no.8, p.241-271.

Lawson, A.C. (1904) *The geomorphogeny of the upper Kern Basin.* University of California. Department of Geology. Bulletin, v.3, p.291-376.

Lawson, A.C. (1913) *The petrographic designation of alluvial-fan formations.* University of California. Department of Geology. Bulletin, v.7, p.325-334.

Lawson, A.C. (1915) *The epigene profiles of the desert.* University of California. Department of Geology. Bulletin, v.9, p.23-48.

Lawson, A.C., and others (1908) *The California earthquake of April 18, 1906; report of the State Earthquake Commission.* Vol.1. Carnegie Institution of Washington. Publication no.87. 451p.

Lee, C.A. (1840) *The elements of geology, for popular use; containing a description of the geological formations and mineral resources of the United States.* New York: Harper. 375p.

Lee, W.T. (1900) *The origin of the débris-covered mesas of Boulder, Colo.* Journal of Geology, v.8, p.504-511.

Lee, W.T. (1903) *The canyons of northeastern New Mexico.* Journal of Geography, v.2, p.63-82.

Lees, G.M. (1952) *Foreland folding.* Geological Society of London. Quarterly Journal, v.108, p.1-34.

Leet, L.D., and Leet, F.J. (1965) *The Earth's mantle.* Seismological Society of America. Bulletin, v.55, p.619-625.

Legget, R.F. (1962) *Geology and engineering.* 2nd ed. New York: McGraw-Hill. 884p.

Legget, R.F. (1967) *Soil: its geology and use.* Geological Society of America. Bulletin, v.78, p.1433-1460.

Legget, R.F. (1973) *Soil.* Geotimes, v.18, no.9, p.38-39.

Legrand, H.E. (1965) *Patterns of contaminated zones of water in the ground.* Water Resources Research, v.1, no.1, p.83-95.

Lehmann, J.G. (1756) *Versuch einer Geschichte von Flötz-Gebürgen, betreffend deren Entstehung, Lage, darinne befindliche Me-*

tallen, Mineralien und Fossilien, gröstentheils aus eigenen Wahrnehmungen, chymischen und physicalischen Versuchen, und aus denen Grundsätzen der Natur-Lehre hergeleitet, und mit nöthigen Kupfern versehen. Berlin:F.A. Lange. 329p.

Leighton, M.M. (1959) *Stagnancy of the Illinoian glacial lake east of the Illinois and Mississippi rivers.* Journal of Geology, v.67, p.337-344.

Leighton, M.M., and MacClintock, Paul (1930) *Weathered zones of the drift-sheets of Illinois.* Journal of Geology, v.38, p.28-53.

Leighton, M.W., and Pendexter, C. (1962) *Carbonate rock types.* American Association of Petroleum Geologists. Memoir 1, p.33-61.

Leighton, R.B., and others (1969) *Mariner 6 and 7 television pictures: preliminary analysis.* Science, v.166, no.3901, p.49-67.

Leith, C.K. (1905) *Rock cleavage.* U.S. Geological Survey. Bulletin 239. 216p.

Leith, C.K. (1923) *Structural geology.* Revised ed. New York: Holt. 390p.

Leith, C.K., and Mead, W.J. (1915) *Metamorphic geology; a textbook.* New York: Holt. 337p.

Leopold, L.B., and Langbein, W.B. (1962) *The concept of entropy in landscape evolution.* U.S. Geological Survey. Professional Paper 500-A, p.1-20.

Leopold, L.B., and Maddock, Thomas, jr. (1953) *The hydraulic geometry of stream channels and some physiographic implications.* U.S. Geological Survey. Professional Paper 252. 57p.

Leopold, L.B., and others (1964) *Fluvial processes in geomorphology.* San Francisco: Freeman. 522p.

Leopold, L.B., and others (1966) *Channel and hillslope processes in a semiarid area, New Mexico.* U.S. Geological Survey. Professional Paper 352-G, p.193-253.

Leopold, L.B., and Wolman, M.G. (1957) *River channel patterns—braided, meandering, and straight.* U.S. Geological Survey. Professional Paper 282-B, p.39-85.

Lesevich, V.V. (1877) *Opyt kriticheskogo izsledovaniia osnovonachal pozitivnoi filosofii.* St. Petersburg: M. Stasiulevich. 295p.—Also in:Lesevich, V.V. (1915) *Sobranie sochinenii,* v.1, p.189-454. Moscow: IU.V. Leontovich.

Leverett, Frank (1903) *Glacial features of lower Michigan.* Journal of Geology, v.11, p.117-118.

Levin, E.M., and others (1964) *General discussion of phase diagrams for ceramists.* M.K. Reser, ed. Columbus, Ohio: American Ceramic Society. 601p.

Levinson, A.A. (1974) *Introduction to exploration geochemistry.* Calgary: Applied Publishing Co. 612p.

Levorsen, A.I. (1933) *Studies in paleogeology.* American Association of Petroleum Geologists. Bulletin, v.17, p.1107-1132.

Levorsen, A.I. (1943) *Discovery thinking.* American Association of Petroleum Geologists. Bulletin, v.27, p.887-928.

Levorsen, A.I. (1960) *Paleogeologic maps.* San Francisco: Freeman. 174p.

Levorsen, A.I. (1967) *Geology of petroleum,* 2nd ed. San Francisco: Freeman. 724p.

Lewan, M.D. (1978) *Laboratory classification of very fine-grained sedimentary rocks.* Geology, v.6, p.745-748.

Lewis, D.W. (1964) *"Perigenic"; a new term.* Journal of Sedimentary Petrology, v.34, p.875-876.

Lilly, H.D. (1966) *Late Precambrian and Appalachian tectonics in the light of submarine exploration on the Great Bank of Newfoundland and in the Gulf of St. Lawrence; preliminary views.* American Journal of Science, v.264, p.569-574.

Lindgren, Waldemar (1912) *The nature of replacement.* Economic Geology, v.7, p.521-535.

Lindgren, Waldemar (1928) *Mineral deposits.* 3d ed. New York: McGraw-Hill. 1049p.

Lindgren, Waldemar (1933) *Mineral deposits.* 4th ed. New York: McGraw-Hill. 930p.

Lindsey, A.A., and others (1969) *Natural areas in Indiana and their preservation.* Lafayette: Indiana Natural Areas Survey, Department of Biological Sciences, Purdue University. 594p.

Link, T.A. (1950) *Theory of transgressive and regressive reef (bioherm) development and origin of oil.* American Association of Petroleum Geologists. Bulletin, v.34, p.263-294.

Linton, D.L.(1955) *The problem of tors.* Geographical Journal, v.121, p.470-487.

Lliboutry, Louis (1958) *Studies of the shrinkage after a sudden advance, blue bands and wave ogives on Glaciar Universidad (central Chilean Andes).* Journal of Glaciology, v.3, p.261-272.

Lobeck, A.K. (1921) *A physiographic diagram of the United States.* Chicago: Nystrom. Map. 1:3,000,000.

Lobeck, A.K. (1924) *Block diagrams and other graphic methods used in geology and geography.* New York: Wiley. 206p.

Lobeck, A.K. (1939) *Geomorphology; an introduction to the study of landscapes.* New York: McGraw-Hill. 731p.

Lockwood, J.P. (1971) *Sedimentary and gravity-slide emplacement of serpentinite.* Geological Society of America. Bulletin, v.82, p.919-936.

Loewinson-Lessing, F.J. (1899) *Note sur la classification et la nomenclature des roches éruptives.* International Geological Congress. 7th, St. Petersburg, 1897. Compte rendu, p.53-71.

Logan, W.E. (1863) *Report on the geology of Canada.* Geological Survey of Canada. Report of Progress to 1863.

Lohest, M.J.M., and others (1909) *Compte rendu de la session extraordinaire de la Société Géologique de Belgique tenue à Eupen et à Bastogne les 29, 30 et 31 août et 1,2 et 3 septembre 1908.* Société Géologique de Belgique. Annales, t.35, p.351-414.

Lohman, S.W., and others (1970) *Definitions of selected groundwater terms.* U.S. Geological Survey. Open-File Report, Aug. 1970.

Lombard, Augustin (1963) *Laminites: a structure of flysch-type sediments.* Journal of Sedimentary Petrology, v.33, p.14-22.

Long, A.E. (1960) *A glossary of the diamond-drilling industry.* U.S. Bureau of Mines. Bulletin 583. 98p.

Longwell, C.R. (1923) *Kober's theory of orogeny.* Geological Society of America. Bulletin, v.34, p.231-241.

Longwell, C.R. (1933) *Meaning of the term "roches moutonnées".* American Journal of Science, v.225, p.503-504.

Longwell, C.R. (1951) *Megabreccia developed downslope from large faults.* American Journal of Science, v.249, p.343-355.

Longwell, C.R., Knopf, Adolph, and Flint, R.F. (1969) *Physical geology.* New York: Wiley. 685p.

Lougee, R.J. and Lougee, C.R. (1976) *Late-glacial chronology.* New York: Vantage Press. 553p.

Lovely, H.R. (1948) *Onlap and strike-overlap.* American Association of Petroleum Geologists. Bulletin, v.32, p.2295-2297.

Lovén, S.L. (1874) *Etudes sur les Echinoïdées.* K. Svenska Vetenskaps-Akademien, Stockholm. Handlingar, ser.4, Bd.11, no.7, p.1-91. 53 plates.

Lovering, T.S. (1963) *Epigenetic, diplogenetic, syngenetic, and lithogene deposits.* Economic Geology , v.58, p.315-331.

Lowe, C.H.,jr. (1961) *Biotic communities in the Sub-Mogollon region of the inland Southwest.* Arizona Academy of Science. Journal, v.2, p.40-49.

Lowe, D.R. (1975) *Water escape structures in coarse-grained sediments.* Sedimentology, v.22, p.157-204.

Lowe, D.R., and LoPiccolo, R.D. (1974) *The characteristics and origins of dish and pillar structures.* Journal of Sedimentary Petrology, v.44, p.484-501.

Lowenstam, H.A. (1950) *Niagaran reefs of the Great Lakes area.* Journal of Geology, v.58, p.430-487.

Lowman, P.D., jr. (1976) *Crustal evolution in silicate planets: implications for the origin of continents.* Journal of Geology, v.84, p.1-26.

Lozinski, W.V. (1909) *Das Sandomierz-Opalower Lössplateau.* Globus, v.96, p.330-334.

Lucia, F.J. (1962) *Diagenesis of a crinoidal sediment.* Journal of Sedimentary Petrology, v.32, p.848-865.

Lundqvist, G. (1949) *The orientation of the block material in certain species of flow earth.* Geografiska Annaler, v.31, p.335-347.

Lüttig, Gerd (1965) *Interglacial and interstadial periods.* Journal of Geology, v.73, p.579-591.

Lyell, Charles (1830) *Principles of geology.* Vol.1. London: John Murray. 511p.

Lyell, Charles (1833) *Principles of geology.* Vol.3. London: John Murray. 398p. plus 109p. of appendix.

Lyell, Charles (1838) *Elements of geology.* London:John Murray. 543p.

Lyell, Charles (1839) *Elements of geology.* 1st American ed., from the 1st London ed. Philadelphia: J. Kay. 316p.

Lyell, Charles (1840) *Principles of geology.* 6th ed. London: John Murray. 3v.

Lyell, Charles (1854) *Principles of geology; or, the modern changes of the Earth and its inhabitants considered as illustrative of geology.* 9th rev. ed. New York: Appleton. 834p.

Lynch, E.J. (1962) *Formation evaluation.* New York: Harper & Row. 422p.

M

Mabbutt, J.A. (1961) *"Basal surface" or "weathering front".* Geologists' Association. Proceedings, v.72, p.357-358.

Macar, P. (1948) *Les pseudo-nodules du Famennien et leur origine.* Société Géologique de Belgique. Annales, v.72, p.47-74.

MacArthur, R.H. (1957) *On the relative abundance of bird species.* National Academy of Science. Proceedings, v.43, p.293-295.

MacArthur, R.H., and Wilson, E.O. (1967) *The theory of island biogeography.* Princeton, N.J.: Princeton University Press. 203p. (Monographs in population biology).

MacBride, T.H. (1910) *Geology of Hamilton and Wright cos.* Iowa Geological Survey. Volume 20, p.97-149.

MacClintock, Copeland (1967) *Shell structure of patelloid and bellerophontoid gastropods.* Yale University, Peabody Museum of Natural History. Bulletin 22. 140p.

Macdonald, G.A. (1953) *Pahoehoe, aa, and block lava.* American Journal of Science, v.251, p.169-191.

Macdonald, G.A. (1960) *Dissimilarity of continental and oceanic rock types.* Journal of Petrology, v.1, p.172-177.

Macdonald, G.A. (1972) *Volcanoes.* Englewood Cliffs, N.J.: Prentice-Hall. 510p.

Macdonald, Ray, and Bailey, D.K. (1973) *The chemistry of the peralkaline oversaturated obsidians.* U.S. Geological Survey. Professional Paper 440 (*The data of geochemistry,* 6th edition), ch.N, part 1. 37p.

MacEwan, D.M.C. (1947) *The nomenclature of the halloysite minerals.* Mineralogical Magazine, v.28, p.36-44.

Mackay, J.R. (1966) *Segregated epigenetic ice and slumps in permafrost, Mackenzie Delta area, NWT.* Canada Dept. Mines & Technical Surveys, Geographical Branch. Geographical Bulletin, v.8, p.59-80.

Mackay, J.R. (1972) *The world of underground ice.* Association of Americal Geographers. Annals, v.62, p.1-22.

Mackie, William (1897) *On the laws that govern the rounding of particles of sand.* Edinburgh Geological Society. Transactions, v.7, p.298-311.

Mackin, J.H. (1937) *Erosional history of the Big Horn Basin, Wyoming.* Geological Society of America. Bulletin, v.48, p.813-893.

Mackin, J.H. (1948) *Concept of the graded river.* Geological Society of America. Bulletin, v.59, p.463-511.

Mackin, J.H. (1950) *The down-structure method of viewing geologic maps.* Journal of Geology, v.58, p.55-72.

Mackinder, H.J. (1919) *Democratic ideals and reality; a study in the politics of reconstruction.* London: Constable. 272p.

MacNeil, F.S. (1954) *Organic reefs and banks and associated detrital sediments.* American Journal of Science, v.252, p.385-401.

Makiyama, Jiro (1954) *Syntectonic construction of geosynclinal neptons.* Kyoto University. College of Science. Memoirs, ser.B, v.21, p.115-149.

Mallet, Robert (1838) *On the mechanism of glaciers, being an attempt to ascertain the causes and effects of their peculiar, and in part, unobserved, motions.* Geological Society of Dublin. Journal, v.1, p.317-335.

Malott, C.A. (1928a) *An analysis of erosion.* Indiana Academy of Science. Proceedings, v.37, p.153-163.

Malott, C.A. (1928b) *The valley form and its development.* Indiana University Studies, v.15, no.81, p.3-34.

Malvern, L.E. (1969) *Introduction to the mechanics of a continuous medium.* Englewood Cliffs, N.J.: Prentice-Hall. 713p.

Mann, C.J. (1970) *Isochronous, synchronous, and coetaneous.* Journal of Geology, v.78, p.749-750.

Manton, W.I. (1965) *The orientation and origin of shatter cones in the Vredefort Ring.* New York Academy of Sciences. Annals, v.123, p.1017-1049.

Marbut, C.F. (1896) *Physical features of Missouri.* Missouri Geological Survey. Volume 10, p.11-109.

Marks, R.W. (1969) *The new dictionary and handbook of aerospace.* New York: Praeger. 531p.

Marr, J.E. (1900) *The scientific study of scenery.* London: Methuen. 368p.

Marr, J.E. (1901) *The origin of moels, and their subsequent dissection.* Geographical Journal, v.17, p.63-69.

Marr, J.E. (1905) *The anniversary address of the President.* Geological Society of London. Quarterly Journal, v.61, p.xlvii-lxxxvi.

Marschner, Hannelore (1969) *Hydrocalcite ($CaCO_3.H_2O$) and nesquehonite ($MgCO_3.3H_2O$) in carbonate scales.* Science, v.165, p.1119-1121.

Martel, E.A. (1896) *Speleology.* International Geographical Congress. 6th, London. Report, p.717-722.

Martin, H.G. (1931) *Insoluble residue studies of Mississippian limestones in Indiana.* Indiana Department of Conservation. Publication no.101. 37p.

Martin Vivaldi, J.L., and MacEwan, D.M.C. (1960) *Corrensite and swelling chlorite.* Clay Minerals Bulletin, v.4, no.24, p.173-181.

Mason, Brian (1958) *Principles of geochemistry.* New York: Wiley.

Matsukuma, Toshinori, and Horikoshi, Ei (1970) *Kuroko deposits in Japan, a review.* In: Tatsumi, T., ed. *Volcanism and ore genesis,* p.153-179. Tokyo: University of Tokyo Press.

Matthes, F.É. (1912) *Topology, topography and topometry.* American Geographical Society. Bulletin, v.44, p.334-339.

Matthes, F.É. (1930) *Geologic history of the Yosemite Valley.* U.S. Geological Survey. Professional Paper 160. 137p.

Matthes, F.É. (1939) *Report of Committee on Glaciers, April 1939.* American Geophysical Union. 20th annual meeting (1939). Transactions, pt.4, p.518-523.

Maw, George (1866) *On subaerial and marine denudation.* Geological Magazine, v.3, p.439-451.

Mawe, John (1818) *A new descriptive catalogue of minerals, consisting of more varieties than heretofore published, and intended for the use of students, with which they may arrange the specimens they collect.* 3rd ed. London: Longman, Hurst, Rees, Orme, and Brown. 96p.

Maxey, G.B. (1964) *Hydrostratigraphic units.* Journal of Hydrology, v.2, p.124-129.

Maxson, J.H. (1940) *Fluting and faceting of rock fragments.* Journal of Geology, v.48, p.717-751.

Maxson, J.H. (1940a) *Gas pits in non-marine sediments.* Journal of Sedimentary Petrology, v.10, p.142-145.

Maxson, J.H. (1950) *Physiographic features of the Panamint Range, California.* Geological Society of America. Bulletin, v.61, p.99-114.

Maxson, J.H., and Anderson, G.H. (1935) *Terminology of surface forms of the erosion cycle.* Journal of Geology, v.43, p.88-96.

Maxson, J.H., and Campbell, Ian (1935) *Stream fluting and stream erosion.* Journal of Geology, v.43, p.729-744.

Maxwell, W.G.H. (1968) *Atlas of the Great Barrier Reef.* New York: Elsevier. 258p.

May, Irving, and Cuttitta, Frank (1967) *New instrumental techniques in geochemical analysis.* In: Abelson, P.H., ed. *Researches in geochemistry,* p.112-142. New York: Wiley. 663p.

McBirney, A.R., and Williams, H. (1969) *Geology and petrology of the Galapagos Islands.* Geological Society of America. Memoir 118. 197p.

McBride, E.F. (1962) *Flysch and associated beds of the Martinsburg Formation (Ordovician), central Applachians.* Journal of Sedimentary Petrology, v.32, p.39-91.

McBride, E.F. (1962a) *The term graywacke.* Journal of Sedimentary Petrology, v.32, p.614-615.

McBride, E.F. (1963) *A classification of common sandstones.* Journal of Sedimentary Petrology, v.33, p.664-669.

McBride, E.F., and Folk, R.L. (1977) *Caballos Novaculite revisited, part II: Chert and shale members.* Journal of Sedimentary Petrology, v.47, p.1261-1286.

McBride, E.F., and Yeakel, L.S. (1963) *Relationship between parting lineation and rock fabric.* Journal of Sedimentary Petrology, v.33, p.779-782.

McCammon, R.B. (1968) *The dendrograph: a new tool for correlation.* Geological Society of America. Bulletin, v.79, p.1663-1670.

McConnell, Duncan (1973) *Biomineralogy of phosphates and physiological mineralization.* In: Griffith, E.J., and others, eds. *Environmental phosphorus handbook.* Ch.25, p.425-442. New York: Wiley.

McConnell, R.G., and Brock, R.W. (1904) *Report on the great landslide at Frank, Alberta.* Canada Department of the Interior. Annual report 1902-1903, pt.8, appendix. 17p.

McCullough, L.A. (1977) *Early diagenetic calcareous coal balls and roof shale concretions from the Pennsylvanian (Allegheny Series).* Ohio Journal of Science, v.77, p.125-134.

McElroy, C.T. (1954) *The use of the term "greywacke" in rock nomenclature in N.S.W.* Australian Journal of Science, v.16, p.150-151.

McGee, W J (1888) *The geology of the head of Chesapeake Bay.* U.S. Geological Survey. Annual Report, 7th, p.537-646.

McGee, W J (1891) *The Pleistocene history of northeastern Iowa.* U.S. Geological Survey. Annual Report, 11th, pt.1,p.189-577.

McGee, W J (1897) *Sheetflood erosion.* Geological Society of America. Bulletin, v.8, p.87-112.

McGee, W J (1908) *Outlines of hydrology.* Geological Society of America. Bulletin, v.19, p.193-200.

McIntosh, D.H., ed. (1963) *Meteorological glossary.* London: Her Majesty's Stationery Office. 288p.

McIver, N.L. (1961) *Upper Devonian marine sedimentation in the central Appalachians.* PhD thesis, Johns Hopkins University. 347p.

McKee, E.D. (1938) *The environment and history of the Toroweap and Kaibab formations of northern Arizona and southern Utah.* Carnegie Institution of Washington. Publication 492. 268p.

McKee, E.D. (1939) *Some types of bedding in the Colorado River delta.* Journal of Geology, v.47, p.64-81.

McKee, E.D. (1954) *Stratigraphy and history of the Moenkopi Formation of Triassic age.* Geological Society of America. Memoir 61. 133p.

McKee, E.D. (1965) *Experiments on ripple lamination.* Society of Economic Paleontologists and Mineralogists. Special Publication no.12, p.66-83.

McKee, E.D. (1966) *Structures of dunes at White Sands National Monument, New Mexico (and a comparison with structures of dunes from other selected areas.)* Sedimentology, v.7, p.1-69.

McKee, E.D., and Weir, G.W. (1953) *Terminology for stratification and cross-stratification in sedimentary rocks.* Geological Society of America. Bulletin, v.64, p.381-389.

McKinstry, H.E. (1948) *Mining geology; including a glossary of mining and geological terms.* New York: Prentice-Hall. 680p.

McLean, J.D. (1968) *Foraminiferal zones and zone charts—an analysis and a compilation.* Manual of Micropaleontological Techniques, v.7. Alexandria, Va.: McLean Paleontological Laboratory.

McManus, D.A. (1962) *A criticism of certain usage of the phi-notation.* Journal of Sedimentary Petrology, v.13, p.670-674.

McNitt, J.R. (1965) *Review of geothermal resources,* ch.9 in: *Terrestrial heat flow.* American Geophysical Union. Geophysical Monograph No.8, p.240-266.

McVaugh, R., Ross, R., and Stafleu, F.A. (1968) *An annotated glossary of botanical nomenclature.* Utrecht: International Bureau for Plant Taxonomy and Nomenclature. 31p. (Regnum Vegetabile, v.56.)

McWolfinvan, C., ed. (1971) *Orismologic progeny.* Bethesda (Md.)/ Washington, D.C.: Rebecca Elizabeth Wolf and Matthew Joseph Sullivan.

Mead, D.W. (1919) *Hydrology, the fundamental basis of hydraulic engineering.* 1st ed. New York: McGraw-Hill. 647p.

Means, W.D. (1976) *Stress and strain.* New York: Springer-Verlag. 339p.

Mechtly, E.A. (1964) *International system of units: physical constants and conversion factors.* National Aeronautics and Space Administration: George C. Marshall Space Flight Center. 19p.

Medlicott, H.B., and Blanford, W.T. (1879) *A manual of the geology of India.* Pt.1-2. 2v. Calcutta: Geological Survey of India. 817p.

Mehnert, K.R. (1968) *Migmatites and the origin of granitic rocks.* Amsterdam: Elsevier. 393p.

Meier, M.F. (1961) *Mass budget of South Cascade Glacier, 1957-1960.* U.S. Geological Survey. Professional Paper 424-B, p.206-211.

Meinzer, O.E. (1923) *Outline of ground-water hydrology, with definitions.* U.S. Geological Survey. Water-supply Paper 494. 71p.

Meinzer, O.E. (1939) *Discussion of question No.2 of the International Commission on Subterranean Water: definitions of the different kinds of subterranean water.* American Geophysical Union. Transactions, pt.4, p.674-677.

Meinzer, O.E. (1942) *Hydrology.* New York: Dover. 712p.

Meischner, K.D. (1964) *Allodapische Kalke, Turbidite in Riff-nahen Sedimentations-Becken.* In: Bouma, A.H., and Brouwer, A., eds. *Turbidites,* p.156-191. New York: Elsevier. (*Developments in Sedimentation,* v.3.)

Melchior, Hans, ed. (1954) *A. Engler's Syllabus der Pflanzenfamilien,* v.2. Berlin: Gebrüder Borntraeger. 666p.

Melchior, Hans, and Werdermann, Erich (1954) *A. Engler's syllabus der Pflanzenfamilien,* v.1. Berlin: Gebrüder Borntraeger. 367p.

Mellett, J.S. (1974) *Scatological origin of microvertebrate fossil accumulations.* Science, v.185, p.349-350.

Mellor, J.W. (1908) *A note on the nomenclature of clays.* English Ceramic Society. Transactions, v.8, p.23-30.

Melton, F.A. (1936) *An empirical classification of flood-plain streams.* Geographical Review, v.26, p.593-609.

Melton, F.A. (1940) *A tentative classification of sand dunes; its application to dune history in the southern High Plains.* Journal of Geology, v.48, p.113-173.

Melton, F.A. (1947) *Onlap and strike-overlap.* American Association of Petroleum Geologists. Bulletin, v.31, p.1868-1878.

Mercier, J.-C., and Nicolas, A. (1975) *Textures and fabrics of upper-mantle peridotites as illustrated by xenoliths from basalts.* Journal of Petrology, v.16, p.454-487.

Merriam, C.H. (1890) *Results of a biological survey of the San Francisco mountain region and desert of the Little Colorado in Arizona.* U.S. Department of Agriculture. Division of Ornithology and Mammalogy. North American fauna, no.3. 136p.

Merriam, D.F. (1962) *Late Paleozoic limestone "buildups" in Kansas.* Kansas Geological Society. Guidebook; field conference, 27th, p.73-81.

Merriam, D.F. (1963) *The geologic history of Kansas.* Kansas State Geological Survey. Bulletin 162. 317p.

Merriam, D.F. (1976) *CAI in geology.* Computers and Geoscience, v.2, p.3-7.

Merrill, G.P. (1897) *A treatise on rocks, rock-weathering and soils.* New York: Macmillan. 411p.

Meyerhoff, H.A. (1940) *Migration of erosional surfaces.* Association of American Geographers. Annals, v.30, p.247-254.

Middleton, G.V. (1967) *Experiments on density and turbidity currents. III. Deposition of sediment.* Canadian Journal of Earth Science, v.4, p.475-505.

Middleton, G.V., ed. (1965) *Primary sedimentary structures and their hydrodynamic interpretation.* Society of Economic Paleontologists and Mineralogists. Special Publication no. 12, 265p.

Mielenz, R.C., and King, M.E. (1955) *Physical-chemical properties and engineering performance of clays.* California Division of Mines. Bulletin 169, p.196-254. (National Conference on Clays and Clay Technology. 1st, Berkeley, Calif., 1952. Proceedings.)

Milankovitch, Milutin (1920) *Théorie mathématique des phénomènes thermiques produits par la radiation solaire.* Paris: Gauthier-Villars. 338p.

Milankovitch, Milutin (1941) *Kanon der Erdbestrahlung und seine Anwendung auf das Eiszeitenproblem.* Belgrade: Académie Royale serbe. 633p.

Miller, D.J. (1953) *Late Cenozoic marine glacial sediments and marine terrces of Middleton Island, Alaska.* Journal of Geology, v.61, p.17-40.

Miller, Hugh (1841) *The Old Red Sandstone; or, new walks in an old field.* Edinburgh: J.Johnstone. 275p.

Miller, Hugh (1883) *River-terracing: its methods and their results.* Royal Physical Society of Edinburgh. Proceedings, v.7, p.263-306.

Miller, J.A. (1972) *Dating Pliocene and Pleistocene strata using the potassium-argon and argon-40/argon-39 methods.* In: Bishop, W.W., and Miller, J.A., eds. *Calibration of hominoid evolution,* p.63-76. Edinburgh: Scottish Academic Press.

Miller, W.J. (1919) *Pegmatite, silexite, and aplite of northern New York.* Journal of Geology, v.27, p.28-54.

Milner, H.B. (1922) *The nature and origin of the Pliocene deposits of the county of Cornwall and their bearing on the Pliocene geography of the south-west of England.* Geological Society of London. Quarterly Journal, v.78, p.348-377.

Milner, H.B. (1940) *Sedimentary petrography, with special reference to petrographic methods of correlation of strata, petroleum technology and other economic applications of geology.* 3rd ed. London: Thomas Murby. 666p.

Milton, D.J. (1969) *Astrogeology in the 19th century.* GeoTimes, v.14, no.6, p.22.

Mitchell, H.C. (1948) *Definitions of terms used in geodetic and other surveys.* Washington: U.S. Government Printing Office. 87p. (U.S. Coast and Geodetic Survey. Special Publication no.242).

Mitchum, R.M., jr. (1977) *Glossary of terms used in seismic stratigraphy.* In: Payton, C.E., ed. *Seismic stratigraphy—applications to hydrocarbon exploration,* p.205-212. American Association of Petroleum Geologists. Memoir 26.

Miyashiro, Akiho (1961) *Evolution of metamorphic belts.* Journal of Petrology, v.2, p.227-311.

Miyashiro, Akiho (1968) *Metamorphism of mafic rocks.* In: Hess, H.H., and Poldervaart, Arie, eds. *Basalts; the Poldervaart treatise on rocks of basaltic composition,* v.2, p.799-834. New York: Interscience.

Miyashiro, Akiho (1973) *Metamorphism and metamorphic belts.*

New York: Wiley. 492p.
Miyashiro, Akiho, and Shido, F. (1970) *Progressive metamorphism in zeolite assemblages.* Lithos, v.3, p.251-260.
Mock, S.J. (1971) *A classification of channel links in stream networks.* Water Resources Research, v.7, p.1558-1566.
Monkhouse, F.J. (1965) *A dictionary of geography.* Chicago: Aldine. 344p.
Monkhouse, F.J., and Wilkinson, H.R. (1952) *Maps and diagrams; their compilation and construction.* London: Methuen. 330p.
Monroe, W.H. (1970) *A glossary of karst terminology.* U.S. Geological Survey. Water-Supply Paper 1899-K.
Moore, Derek (1966) *Deltaic sedimentation.* Earth-Science Reviews, v.1, p.87-104.
Moore, D.G., and Scruton, P.C. (1957) *Minor internal structures of some recent unconsolidated sediments.* American Association of Petroleum Geologists. Bulletin, v.41, p.2723-2751.
Moore, G.W., and Kennedy, M.P. (1975) *Quaternary faults at San Diego Bay, California.* U.S. Geological Survey. Journal of Research, v.3, p.589-595.
Moore, G.W., and Sullivan, G.N. (1978) *Speleology—the study of caves.* Teaneck, N.J.: Zephyrus Press. 150p.
Moore, P.F.(1958) *Nature, usage, and definition of marker-defined vertically segregated rock units.* American Association of Petroleum Geologists. Bulletin, v.42, p.447-450.
Moore, R.C. (1926) *Origin of inclosed meanders on streams of the Colorado Plateau.* Journal of Geology, v.34, p.29-57.
Moore, R.C. (1933) *Historical geology.* New York: McGraw-Hill. 673p.
Moore, R.C. (1936) *Stratigraphic classification of the Pennsylvanian rocks of Kansas.* Kansas State Geological Survey. Bulletin 22. 256p.
Moore, R.C. (1949) *Meaning of facies.* Geological Society of America. Memoir 39, p.1-34.
Moore, R.C. (1957) *Minority report.* In: American Commission on Stratigraphic Nomenclature. *Nature, usage, and nomenclature of biostratigraphic units,* p.1888. American Association of Petroleum Geologists. Bulletin, v.41, p.1877-1889.
Moore, R.C. (1957a) *Modern methods of paleoecology.* American Association of Petroleum Geologists. Bulletin, v.41, p.1775-1801.
Moore, R.C. (1958) *Introduction to historical geology.* 2nd ed. New York: McGraw-Hill. 656p.
Moore, R.C., and others (1952) *Invertebrate fossils.* 1st ed. New York: McGraw-Hill. 766p.
Moore, W.G. (1967) *A dictionary of geography; definitions and explanations of terms used in physical geography.* 3rd ed. New York: Praeger. 246p.
Morlot, Adolphe von (1847) *Ueber Dolomit und seine künstliche Darstellung aus Kalkstein.* Naturwissenschaftliche Abhandlungen, gesammelt und durch Subscription hrsg. von Wilhelm Haidinger, v.1, p.305-315.
Morse, James (1977) *Deformation in ramp regions of overthrust faults: experiments with small-scale rock models.* Joint Wyoming-Montana-Utah Geological Associations Guidebook, *Rocky Mountain thrust belt geology and resources,* p.457-470.
Morton, C.V. (1957) *The misuse of the term taxon.* Taxonomy, v.6, p.155. Also in: Rhodora, 1957, p.43-44.
Moss, J.H. (1951) *Early man in the Eden valley.* Philadelphia: University of Pennsylvania. University Museum Monograph. 92p.
Mueller, I.I., and Rockie, J.D. (1966) *Gravimetric and celestial geodesy; a glossary of terms.* New York: Ungar. 129p.
Mueller, R.F., and Saxena, S.K. (1976) *Chemical petrology.* New York: Springer-Verlag. 394p.
Muir, I.D., and Tilley, C.E. (1961) *Mugearites and their place in alkali igneous rock series.* Journal of Geology, v.69, p.186-203.
Muir-Wood, H.M., and Cooper, G.A. (1960) *Morphology, classification and life habits of the Productoidea (Brachiopoda).* Geological Society of America. Memoir 81. 447p.
Müller, German (1967) *Diagenesis in argillaceous sediments.* In: Larsen, Gunnar, and Chilingar, G.V., eds. *Diagenesis in sediments,* p.127-177. Amsterdam: Elsevier. 551p. (*Developments in sedimentology 8.*)
Müller, Johannes (1858) *Ueber die Thalassicollen, Polycystinen und Acanthometren des Mittelmeeres.* Akademie der Wissenschaften, Berlin. Abhandlungen (physikalische), 1858, p.1-62.
Muller, P.M., and Sjogren, W.L. (1968) *Mascons: lunar mass concentrations.* Science, v.161, p.680-684.
Muller, S.W. (1965) *Superposition of strata: part 1. Physical criteria for determining top and bottom of beds.* American Geological Institute: Data Sheet no. 10.
Muller, S.W., compiler (1947) *Permafrost or permanently frozen ground and related engineering problems.* Ann Arbor, Mich.: Edwards. 231p.
Murchison, R.I. (1839) *The Silurian System.* Pt.1. London: John Murray. 576p.
Murray, A.N. (1930) *Limestone oil reservoirs of the northeastern United States and Ontario, Canada.* Economic Geology, v.25, p.452-469.
Murray, G.E. (1961) *Geology of the Atlantic and Gulf Coast Province of North America.* New York: Harper. 692p.
Murray, G.E. (1965) *Indigenous Precambrian petroleum?* American Association of Petroleum Geologists. Bulletin, v.49, p.3-21.
Murray, R.C. (1960) *Origin of porosity in carbonate rocks.* Journal of Sedimentary Petrology, v.30, p.59-84.
Murray, R.C., and Tedrow, J.C.F. (1975) *Forensic geology.* New Brunswick, N.J.: Rutgers University Press. 217p.
Murzaevs, E., and Murzaevs, V. (1959) *Slovar' mestnykh geograficheskikh terminov.* Moscow: Geografizdat. 303p.
Mutch, T.A. (1970) *Geology of the moon.* Princeton, N.J.: Princeton University Press. 324p.
Mutch, T.A., and others (1976) *The geology of Mars.* Princeton, N.J.: Princeton University Press. 400p.

N

Nabholz, W.K. (1951) *Beziehungen zwischen Fazies und Zeit.* Eclogae geologicae Helvetiae, v.44, p.131-158.
Naldrett, A.J. (1973) *Nickel sulfide deposits—their classification and genesis, with special emphasis on deposits of volcanic association.* Canadian Institute of Mining and Metallurgy. Transactions, v.76, p.183-201.
Naldrett, A.J., and Kullerud, G. (1965) *Two examples of sulfurization in nature* (abstract). Economic Geology, v.60, p.1563.
NASA: National Aeronautics and Space Administration (1966) *Short glossary of space terms.* Its publication: SP-1.
NASA: National Aeronautics and Space Administration, Goddard Space Flight Center (1971) *NASA Earth resources data users' handbook.* (Document 71SD4249.)
NAS-NRC: National Academy of Sciences - National Research Council. Committee on Rock Mechanics (1966) *Rock-mechanics research; a survey of United States research to 1965, with a partial survey of Canadian universities.* Washington, D.C. 82p. (Its Publication 1466).
National Research Council Panel on Land Burial (1976) *The shallow land burial of low-level radioactively contaminated solid waste.* Washington: National Academy of Sciences. 150p.
Natland, M.L. (1976) *Classification of clastic sediments* (abstract). American Association of Petroleum Geologists. Bulletin, v.60, p.702.
Natland, M.L., and Kuenen, P.H. (1951) *Sedimentary history of the Ventura Basin, California, and the action of turbidity currents.* Society of Economic Paleontologists and Mineralogists. Special Publication no.2, p.76-107.
Naum, T., and others (1962) *Vulcanokarstul din Masivul Calimanului (Carpatii Orientali).* Bucharest. Universitatea. Analele; seria stiintele naturii: geologie-geografie, anul 11, no.32, p.143-179.
Naumann, C.F. (1850) *Lehrbuch der Geognosie.* Bd. 1. Leipzig: Wilhelm Engelmann. 1000p.
Naumann, C.F. (1858) *Lehrbuch der Geognosie.* Bd. 1. 2nd ed. Leipzig: Wilhelm Engelmann. 960p.
Nayak, V.K. (1970) *Geoplanetology: a new term for geology of the planets including the Moon.* Geological Society of America. Bulletin, v.81, p.1279.
Nelson, A. (1965) *Dictionary of mining.* New York: Philosophical Library. 523p.
Nelson, A., and Nelson, K.D. (1967) *Dictionary of applied geology, mining and civil engineering.* New York: Philosophical Library. 421p.
Nelson, H.F., and others (1962) *Skeletal limestone classification.* American Association of Petroleum Geologists. Memoir 1, p.224-252.
Nettleton, L.L. (1940) *Geophysical prospecting for oil.* 1st ed. New York: McGraw-Hill. 444p.
Neumann, A.C. (1966) *Observations on coastal erosion in Bermuda and measurements of the boring rate of the sponge,* Cliona lampa.

Limnology and Oceanography, v.11, p.92-108.

Nevin, C.M. (1949) *Principles of structural geology.* 4th ed. New York: Wiley. 410p.

Newell, N.D., and others (1953) *The Permian reef complex of the Guadalupe Mountains region, Texas and New Mexico - a study in paleoecology.* San Francisco: Freeman. 236p.

Nichols, R.A.H. (1967) *The ". . .sparite" complex: eosparite vs. neosparite.* Journal of Sedimentary Petrology, v.37, p.1247-1248.

Nichols, R.L. (1936) *Flow-units in basalt.* Journal of Geology, v.44, p.617-630.

Nichols, R.L. (1938) *Grooved lava.* Journal of Geology, v.46, p.601-614.

Nichols, R.L. (1946) *McCartys basalt flow, Valencia County, New Mexico.* Geological Society of America. Bulletin, v.57, p.1049-1086.

Nicodemus, F.E. (1971) *A proposed military standard on infrared terms and definitions,* Part I, Appendix C of the Philco-Ford Corp. final report, Contract No.DAAH01-71-C-0433. (Available as AD-758341 from the National Technical Information Service of the U. S. Department of Commerce, Springfield, Va.) 256p.

Niggli, Paul (1954) *Rocks and mineral deposits.* English translation by Robert L. Parker. San Francisco: Freeman. 559p.

Nikolaev, V.A. (1959) *Some structural characteristics of mobile tectonic belts.* International Geology Review, v.1, p.50-64.

Noble, L.F. (1941) *Structural features of the Virgin Spring area, Death Valley, California.* Geological Society of America. Bulletin, v.52, p.941-999.

Nockolds, S.R. (1934) *The production of normal rock types by contamination and their bearing on petrogenesis.* Geological Magazine, v.71, p.31-39.

Nordenskjöld, Otto (1909) *Einige Beobachtungen über Eisformen und Vergletscherung der antarktischen Gebiete.* Zeitschrift für Gletscherkunde, Bd.3, p.321-334.

Nordyke, M.D. (1961) *Nuclear craters and preliminary theory of the mechancis of explosive crater formation.* Journal of Geophysical Research, v.66, p.3439-3459.

Norris, D.K. (1958) *Structural conditions in Canadian coal mines.* Geological Survey of Canada. Bulletin 44. 54p.

Norris, D.K. (1964) *Microtectonics of the Kootenay Formation near Fernie, British Columbia.* Bulletin of Canadian Petroleum Geology, v.12, p.383-398.

Northrop, J.I. (1890) *Notes on the geology of the Bahamas.* New York Academy of Sciences. Transactions, v.10, p.4-22.

Norton, Denis (1978) *Sourcelines, sourceregions, and pathlines for fluids in hydrothermal systems related to cooling plutons.* Economic Geology, v.73, p.21-28.

Norton, W.H. (1917) *A classification of breccias.* Journal of Geology, v.25, p.160-194.

Nye, J.F. (1952) *The mechanics of glacier flow.* Journal of Glaciology, v.2, p.82-93.

Nye, J.F. (1958) *Surges in glaciers.* Nature, v.181, p.1450-1451.

O

Oakes, H., and Thorp, J. (1951) *Dark clay soils of warm regions variously called rendzina, black cotton soils, regur, and tirs.* Soil Science Society of America. Proceedings, v.15, p.347-354.

Ogilvie, I.H. (1905) *The high-altitude conoplain; a topographic form illustrated in the Ortiz Mountains.* American Geologist, v.36, p.27-34.

Ohio, Legislative Service Commission (1969) *Preservation of natural areas, and Report of Committee to Study Natural Areas.* Project officer: John P. Bay. Columbus: Ohio Legislative Service Commission. 20p. (Its Staff Research Report no.89).

Oldham, Thomas (1879) *Geological glossary for the use of students.* Edited by R.D. Oldham. London: Stanford. 62p.

O'Leary, D.W., Friedman, J.D., and Pohn, H.A. (1976) *Lineament, linear, lineation: some proposed new standards for old terms.* Geological Society of America. Bulletin, v.87, p.1463-1469.

Oliver, W.A., jr. (1968) *Some aspects of colony development in corals.* Journal of Paleontology, v.42, no.5, pt.2, p.16-34.

Oppel, Albert (1856-1858) *Die Juraformation Englands, Frankreichs und des südwestlichen Deutschlands; nach ihren einzelnen Gliedern einigetheilt und verglichen.* Stuttgart: Ebner & Seubert. 857p. (Württembergische Naturwissenschaftliche Jahreshefte. Jahrg.12, p.121-132, 313-556; 13, p.141-396; 14, p.128-291).

Oriel, S.S. (1949) *Definitions of arkose.* American Journal of Science, v.247, p.824-829.

Oriel, S.S., and others (1976) *Stratigraphic Commission Note 44—Application for addition to code concerning magnetostratigraphic units.* American Association of Petroleum Geologists. Bulletin, v.60, p.273-277.

Osborne, R.H. (1969) *Undergraduate instruction in geomathematics.* Journal of Geological Education, v.17, p.120-124.

Otterman, Joseph, and Bronner, F.E. (1966) *Martian wave of darkening: a frost phenomenon?* Science, v.153, p.56-60.

Otto, G.H. (1938) *The sedimentation unit and its use in field sampling.* Journal of Geology, v.46, p.569-582.

Otvos, E.G., jr. (1976) *"Pseudokarst" and "pseudokarst terrains": problems of terminology.* Geological Society of America. Bulletin, v.87, p.1021-1027.

Otvos, E.G., jr., and Price, W.A. (1979) *Problems of chenier genesis and terminology—an overview.* Marine Geology, v.31, p.251-263.

P

Packer, R.W. (1965) *Lag mound features on a dolostone pavement.* Canadian Geographer, v.9, p.138-143.

Packham, G.H. (1954) *Sedimentary structures as an important factor in the classification of sandstones.* American Journal of Science, v.252, p.466-476.

Packham, G.H., and Crook, K.A.W. (1960) *The principle of diagenetic facies and some of its implications.* Journal of Geology, v.68, p.392-407.

Padfield, H.I., and Smith, C.L. (1968) *Water and culture: new decision rules for old institutions.* Rocky Mountain Social Science Journal, v.5, no.2, p.23-32.

Page, David (1859) *Handbook of geological terms and geology.* Edinburgh: William Blackwood. 416p.

Palmer, A.R. (1965) *Biomere - a new kind of biostratigraphic unit.* Journal of Paleontology, v.39, p.149-153.

Palmer, L.A. (1920) *Desert prospecting.* Engineering and Mining Journal, v.110, p.850-853.

Park, C.F., jr. (1959) *The origin of hard hematite in itabirite.* Economic Geology, v.54, p.573-587.

Park, C.F.,jr., and MacDiarmid, R.A. (1970) *Ore deposits.* 2d ed. San Francisco: Freeman. 522p.

Park, James (1914) *A text-book of geology; for use in mining schools, colleges, and secondary schools.* London: Charles Griffin. 598p.

Parks, B.C. (1951) *Petrography of American lignites.* Economic Geology, v.46, no.1, p.23-50.

Parsons, W.H. (1969) *Criteria for the recognition of volcanic breccias: a review.* Geological Society of America. Memoir 115, p.263-304.

Passega, R. (1957) *Texture as characteristic of clastic properties.* American Association of Petroleum Geologists. Bulletin, v.41, p.1952-1984.

Paterson, M.S., and Weiss, L.E. (1961) *Symmetry concepts in the structural analysis of deformed rocks.* Geological Society of America. Bulletin, v.72, p.841-882.

Paterson, W.S.B. (1969) *The physics of glaciers.* New York: Pergamon. 250p.

Paul, C.R.C. (1968) *Morphology and function of dichoporite pore structures in cystoids.* Palaeontology, v.11, p.697-730.

Paul, C.R.C. (1972) *Morphology and function of exothecal pore-structures in cystoids.* Palaeontology, v.15, p.1-28.

Pavlov, A.P. (1889) *Geneticheskie tipy materikovykh obrazovanii lednikovoi i posle-lednikovoi epokhi.* Geologicheskogo Komiteta (St. Petersburg). Izvestiia, v.7, p.243-261.

Payne, T.G. (1942) *Stratigraphical analysis and environmental reconstruction.* American Association of Petroleum Geologists. Bulletin, v.26, p.1697-1770.

Peabody, F.E. (1947) *Current crescents in the Triassic Moenkopi Formation.* Journal of Sedimentary Petrology, v.17, p.73-76.

Peacock, M.A. (1931) *Classification of igneous rock series.* Journal of Geology, v.39, p.54-67.

Pearn, W.C. (1964) *Finding the ideal cyclothem.* Kansas State Geological Survey. Bulletin 169, v.2, p.399-413.

Peltier, L.C. (1950) *The geographic cycle in periglacial regions as it is related to climatic geomorphology.* Association of American Geographers. Annals, v.40, p.214-236.

Pelto, C.R. (1954) *Mapping of multicomponent systems.* Journal of Geology, v.62, p.501-511.

Penck, Albrecht (1900) *Geomorphologische Studien aus der Her-*

zegowina. Deutscher und Oesterreichischer Alpenverein. Zeitschrift, Bd. 31, p.25-41.

Penck, Albrecht (1919) *Die Gipfelflur der Alpen.* Akademie der Wissenschaften, Berlin. Sitzungsberichte, Jahrg. 1919, p.256-268.

Penck, Walther (1924) *Die morphologische Analyse; ein Kapitel der physikalischen Geologie.* Stuttgart: J. Engelhorns. 283p. (Geographische Abhandlungen, 2 Reihe, Hft.2).

Penck, Walther (1953) *Morphological analysis of land forms; a contribution to physical geology.* Translated by Hella Czech and K.C. Boswell. New York: St. Martin's Press. 429p.

Pennak, R.W. (1964) *Collegiate dictionary of zoology.* New York: Ronald. 583p.

Pepper, J.F., and others (1954) *Geology of the Bedford Shale and Berea Sandstone in the Appalachian basin.* U.S. Geological Survey. Professional Paper 259. 111p.

Pepper, T.B. (1941) *The Gulf underwater gravimeter.* Geophysics, v.6, p.34-44.

Perry, W.J., jr., and DeWitt, Wallace (1977) *A field guide to thin-skinned tectonics in the central Appalachians.* American Association of Petroleum Geologists. Annual Convention, Washington, D.C. Field Trip 4, Guidebook. 54p.

Peters, L.J. (1949) *The direct approach to magnetic interpretation and its practical application.* Geophysics, v.14, p.290-320.

Pettijohn, F.J. (1949) *Sedimentary rocks.* New York: Harper. 526p.

Pettijohn, F.J. (1954) *Classification of sandstones.* Journal of Geology, v.62, p.360-365.

Pettijohn, F.J. (1957) *Sedimentary rocks.* 2nd ed. New York: Harper. 718p.

Pettijohn, F.J., and Potter, P.E. (1964) *Atlas and glossary of primary sedimentary structures.* New York: Springer-Verlag. 370p.

Pettijohn, F.J., Potter, P.E., and Siever, Raymond (1972) *Sand and sandstone.* New York: Springer-Verlag. 600p.

Péwé, T.L. (1959) *Sand-wedge polygons (tesselations) in the McMurdo Sound region, Antarctica—a progress report.* American Journal of Science, v.257, p.545-552.

Péwé, T.L. (1976) *Permafrost.* McGraw-Hill Yearbook of Science and Technology, p.30-42. New York: McGraw-Hill.

Peyve, A.V., and Sinitzyn, V.M. (1950) *Certains problèmes fondamentaux de la doctrine des géosynclinaux.* Akademiya Nauk SSSR. Izvestiya, Seriya Geologicheskaya, v.4, p.28-52.

Phemister, James (1956) *Petrography.* In: Great Britain. Geological Survey. *The limestones of Scotland; chemical analyses and petrography.* Edinburgh: H.M.S.O. 150p. (Its Memoirs; special reports on the mineral resources of Great Britain, v.37, p.66-74).

Philipsborn, H. von (1930) *Zur chemisch-analytischen Erfassung der isomorphen Variation gesteinbildender Minerale. Die Mineralkomponenten des Pyroxengranulits von Hartmannsdorf (Sa.).* Chemie der Erde, v.5, p.233-253.

Phillips, F.C. (1954) *The use of stereographic projection in structural geology.* London: Edward Arnold. 86p.

Phillips, William (1818) *A selection of facts from the best authorities, arranged so as to form an outline of the geology of England and Wales.* London: William Phillips. 250p.

Picard, M.D. (1953) *Marlstone - a misnomer as used in Uinta Basin, Utah.* American Association of Petroleum Geologists. Bulletin, v.37, p.1075-1077.

Pike, R.J., and Wilson, S.J. (1971) *Elevation-relief ratio, hypsometric integral, and geomorphic area-altitude analysis.* Geological Society of America. Bulletin, v.82, p.1079-1084.

Pinkerton, John (1811) *Petralogy; a treatise on rocks.* 2 v. London: White, Cochrane. 599p. and 656p.

Pirsson, L.V. (1896) *A needed term in petrography.* Geological Society of America. Bulletin, v.7, p.492-493.

Pirsson, L.V. (1915) *Physical geology.* New York: Wiley. 404p. (A text-book of geology, pt.1).

Pittock, A.B., and others, eds. (1978) *Climatic change and variability: a southern perspective.* New York, Cambridge University Press. 455p.

Playfair, John (1802) *Illustrations of the Huttonian theory of the Earth.* Edinburgh: Wm. Creech. 528p.

Plumley, W.J., and others (1962) *Energy index for limestone interpretation and classification.* American Association of Petroleum Geologists. Memoir 1, p.85-107.

Pojeta, John, jr., and Runnegar, Bruce (1976) *The paleontology of the rostroconch mollusks and the early history of the phylum Mollusca.* U.S. Geological Survey. Professional Paper 968. 88p.

Poland, J.F., Lofgren, B.E., and Riley, F.S. (1972) *Glossary of selected terms useful in studies of the mechanics of aquifer systems and land subsidence due to fluid withdrawal.* U.S. Geological Survey. Water-supply Paper 2025. 9p.

Poldervaart, Arie, and Parker, A.B. (1964) *The crystallization index as a parameter of igneous differentiation in binary variation diagrams.* American Journal of Science, v.262, p.281-289.

Pompeckj, J.F. (1914) *Die Bedeutung des schwäbischen Jura für die Erdgeschichte.* Stuttgart: Schweizerbartsche. 64p.

Porsild, A.E. (1938) *Earth mounds in unglaciated Arctic northwestern America.* Geographical Review, v.28, p.46-58.

Post, Lennart von (1924) *Ur de sydsvenska skogarnas regionala historia under postarktist tid.* Geologiska Föreningen, Stockholm. Förhandlingar, Bd.46, p.83-128.

Potter, P.E., and Glass, H.D. (1958) *Petrology and sedimentation of the Pennsylvanian sediments in southern Illinois - a vertical profile.* Illinois State Geological Survey. Report of Investigations 204. 60p.

Pough, F.H. (1967) *The story of gems and semiprecious stones.* Irvington-on-Hudson: Harvey House. 142p.

Powell, J.W. (1873) *Some remarks on the geological structure of a district of country lying to the north of the Grand Cañon of the Colorado.* American Journal of Science, 3rd ser., v.5, p.456-465.

Powell, J.W. (1874) *Remarks on the structural geology of the Valley of the Colorado of the West.* Philosophical Society of Washington. Bulletin, v.1, p.48-51.

Powell, J.W. (1875) *Exploration of the Colorado River of the West and its tributaries.* Washington: Government Printing Office. 291p.

Powell, J.W. (1895) *Physiographic features.* National Geographic Monographs, v.1, no.2, p.33-64.

Power, F.D. (1895) *A glossary of terms used in mining geology.* Adelaide: Australasian Institute of Mining Engineers. 69p.

Powers, M.C. (1953) *A new roundness scale for sedimentary particles.* Journal of Sedimentary Petrology, v.23, p.117-119.

Powers, R.W. (1962) *Arabian Upper Jurassic carbonate reservoir rocks.* American Association of Petroleum Geologists. Memoir 1, p.122-192.

Pratt, W.P., and Brobst, D.A. (1974) *Mineral resources: potentials and problems.* U.S. Geological Survey. Circular 698. 20p.

Prentice, J.E. (1956) *The interpretation of flow-markings and load-casts.* Geological Magazine, v.93, p.393-400.

Presnall, D.C. (1969) *The geometrical analysis of partial fusion.* American Journal of Science, v.267, p.1178-1194.

Price, R.A. (1965) *Flathead map-area, British Columbia and Alberta.* Geological Survey of Canada. Memoir 336. 221p.

Price, W.A. (1947) *Equilibrium of form and forces in tidal basins of coast of Texas and Louisiana.* American Association of Petroleum Geologists. Bulletin, v.31, p.1619-1663.

Price, W.A. (1954) *Dynamic environments - reconnaissance mapping, geologic and geomorphic, of continental shelf of Gulf of Mexico.* Gulf Coast Association of Geological Societies. Transactions, v.4, p.75-107.

Prior, G.T. (1920) *The classification of meteorites.* Mineralogical Magazine, v.19, no.90, p.51-63.

Pryor, E.J. (1963) *Dictionary of mineral technology.* London: Mining Publications. 437p.

Pumpelly, Raphael (1894) *Geology of the Green Mountains in Massachusetts.* U.S. Geological Survey. Monograph 23, p.5-34.

Pumpelly, Raphael, Wolf, J.E., and Dale, T.N. (1894) *Geology of the Green Mountains in Massachusetts; part III, Mount Greylock: its areal and structural geology,* by T. Nelson Dale. U.S. Geological Survey. Monograph 23.

Purdy, R.C. (1908) *Qualities of clays suitable for making paving brick.* Illinois State Geological Survey. Bulletin no.9, p.133-278.

Purser, B.H., and Loreau, J.-P. (1973) *Aragonitic supratidal encrustations on the Trucial Coast, Persian Gulf.* In: Purser, B.H., ed. *The Persian Gulf.* New York: Springer-Verlag. 471p.

Pustovalov, L.V. (1933) *Geokhimicheskie fatsii, ikh znachenie v obshchei i prikladnoi geologii.* Problemy Sovetskoi geologii, 1933, [ser.1], t.1, no.1, p.57-80.

Q

Quensel, P. (1916) *Zur Kenntnis der Mylonitbildung, erläutert an Material aus dem Kebnekaisegebiet.* Uppsala. Universitet. Geologiska Institut. Bulletin, v.15, p.91-116.

Quirke, T.T. (1930) *Spring pits, sedimentation phenomena.* Journal of Geology, v.38, p.88-91.

R

Raaf, J.F.M. de (1945) *Notes on the geology of the southern Rumanian oil district with special reference to the occurrence of a sedimentary laccolith.* Geological Society of London. Quarterly Journal, v.101, p.111-134.

Raistrick, Arthur, and Marshall, C.E. (1939) *The nature and origin of coal and coal seams.* London: English Universities Press Ltd. 282p.

Raisz, E.J. (1959) *Landform maps - a method of preparation.* U.S. Office of Naval Research, Geography Branch. Part one of Final Report, Contract Nonr 2339(00). 23p.

Ramberg, Hans (1955) *Natural and experimental boudinage and pinch-and-swell structures.* Journal of Geology, v.63, p.512-526.

Ramsay, A.C. (1846) *On the denudation of South Wales and the adjacent counties of England.* Great Britain. Geological Survey. Memoirs, v.1, ch.2, p.297-335.

Ramsay, J.G. (1967) *Folding and fracturing of rocks.* New York: McGraw-Hill. 568p.

Rankama, Kalervo (1962) *Planetology and geology.* Geological Society of America. Bulletin, v.73, p.519-520.

Rankama, Kalervo (1967) *Megayear and gigayear: two units of geological time.* Nature, v.214, p.634.

Rankama, Kalervo, and Sahama, Th.G. (1950) *Geochemistry.* Chicago: University of Chicago Press. 912p.

Ransome, F.L., and Calkins, F.C. (1908) *The geology and ore deposits of the Coeur d'Alene district, Idaho.* U.S. Geological Survey. Professional Paper 62. 203p.

Rapp, Anders (1959) *Avalanche boulder tongues in Lappland.* Geografiska Annaler, v.41, p.34-48.

Rast, Nicholas (1956) *The origin and significance of boudinage.* Geological Magazine, v.93, p.401-408.

Raymond, L.A. (1975) *Tectonite and mélange—a distinction.* Geology, v.3, p.7-9.

Raymond, L.A. (1978) *A classification of mélanges and broken formations* (abstract). Geological Society of America. Abstracts with Programs, v.10, p.143.

Read, H.H. (1931) *The geology of central Sutherland (east-central Sutherland and south-western Caithness). (Explanation of sheets 108 and 109).* Geological Survey of Scotland. Memoir. 238p.

Read, H.H. (1944) *Meditation on granite, pt.2.* Geologists' Association. Proceedings, v.55, p.45-93.

Read, H.H. (1958) *A centenary lecture; Stratigraphy in metamorphism.* Geologists' Association. Proceedings, v.69, pt.2, p.83-102.

Rechard, P.A., and McQuisten, Richard, compilers (1968) *Glossary of selected hydrologic terms.* University of Wyoming. Water Resources Research Institute. Water Resources Series, no.1, 54p.

Reclus, Elisée (1872) *The Earth: a descriptive history of the phenomena of the life of the globe.* Translated by B.B. Woodward. Edited by Henry Woodward. New York: Harper. 573p.

Record, S.J. (1934) *Identification of the timbers of temperate North America, including anatomy and certain physical properties of wood.* New York: Wiley.

Reiche, Parry (1937) *The Toreva-block, a distinctive landslide type.* Journal of Geology, v.45, p.538-548.

Reiche, Parry (1938) *An analysis of cross-lamination; the Coconino Sandstone.* Journal of Geology, v.46, p.905-932.

Reiche, Parry (1943) *Graphic representation of chemical weathering.* Journal of Sedimentary Petrology, v.13, p.58-68.

Reiche, Parry (1945) *A survey of weathering processes and products.* Albuquerque: University of New Mexico Press. 87p. (University of New Mexico. Publications in Geology, no.1).

Reid, H.F. (1924) *Antarctic glaciers.* Geographical Review, v.14, p.603-614.

Reimnitz, Erk, Barnes, P.W., and Alpha, T.R. (1973) *Bottom features and processes related to drifting ice on the arctic shelf, Alaska.* U.S. Geological Survey. Map MF-532.

Reimnitz, Erk, and others (1977) *Ice gouge recurrence and rates of sediment reworking, Beaufort Sea, Alaska.* Geology, v.5, p.405-408.

Reineck, H.-E., and Singh, I.B. (1973) *Depositional sedimentary environments.* New York: Springer-Verlag. 439p.

Reineck, H.-E., and Wunderlich, F. (1968) *Classification and origin of flaser and lenticular bedding.* Sedimentology, v.11, p.99-104.

Renevier, E., and others (1882) *Rapport du comité suisse sur l'unification de la nomenclature.* International Geological Congress. 2nd, Bologna, 1881. Compte rendu, p.535-548.

Renfrew, J.M. (1973) *Palaeoethnobotany.* New York: Columbia University Press. 248p.

Renfro, A.R. (1974) *Genesis of evaporite-associated stratiform metalliferous deposits—a sabkha process.* Economic Geology, v.69, p.33-45.

Reusch, Hans (1894) *The Norwegian Coast Plain.* Journal of Geology, v.2, p.347-349.

Revelle, Roger, and Fairbridge, R.W. (1957) *Carbonates and carbon dioxide.* Geological Society of America. Memoir 67, v.1, p.239-295.

Rice, C.M. (1945) *Dictionary of geological terms.* 1st ed. Ann Arbor (Mich.): Edwards Brothers. 461p.

Rice, C.M. (1954) *Dictionary of geological terms.* 2nd ed. Ann Arbor (Mich.): Edwards Brothers.

Rich, J.L. (1908) *Marginal glacial drainage features in the Finger Lake region.* Journal of Geology, v.16, p.527-548.

Rich, J.L. (1914) *Certain types of stream valleys and their meaning.* Journal of Geology, v.22, p.469-497.

Rich, J.L. (1915) *Notes on the physiography and glacial geology of the northern Catskill Mountains.* American Journal of Science, 4th ser., v.39, p.137-166.

Rich, J.L. (1938) *A mechanism for the initiation of geosynclines and geo-basins.* Geological Society of America. Proceedings, p.106-107.

Rich, J.L. (1950) *Flow markings, groovings, and intra-stratal crumplings as criteria for recognition of slope deposits, with illustrations from Silurian rocks of Wales.* American Association of Petroleum Geologists. Bulletin, v.34, p.717-741.

Rich, J.L. (1951) *Geomorphology as a tool for the interpretation of geology and Earth history.* New York Academy of Sciences. Transactions, ser.2, v.13, p.188-192.

Rich, J.L. (1951) *Three critical environments of deposition and criteria for recognition of rocks deposited in each of them.* Geological Society of America. Bulletin, v.62, p.1-20.

Richter, C.F. (1958) *Elementary seismology.* San Francisco: Freeman. 768p.

Richter, Rudolf (1952) *Fluidal-Textur in Sediment- Gesteinen und über Sedifluktion überhaupt.* Hessisches Landesamt für Bodenforschung, Wiesbaden. Notizblatt, F.6, H.3, p.67-81.

Richtofen, Ferdinand von (1868) *The natural system of volcanic rocks.* California Academy of Sciences. Memoirs, v.1, pt.2, 94p.

Rickett, H.W. (1958) *So what is a taxon?* Taxonomy, v.7, p.37-38.

Riecke, Eduard (1894) *Über das Gleichgewicht zwischen einem festen, homogen deformirten Körper und einer flüssigen Phase, insbesondere über die Depression des Schmelzpunctes durch einseitige Spannung.* Gesellschaft der Wissenschaften zu Göttingen. Mathematisch-Physikalische Klasse. Nachrichten, 1894, no.4, p.278-284.

Riedel, W. (1929) *Zur Mechanik geologischer Brucherscheinungen.* Zentralblatt für Mineralogie, Geologie, und Palaeontologie B, p.354-368.

Rieke, H.H., III, and Chilingarian, G.V. (1974) *Compaction of argillaceous sediments.* Amsterdam: Elsevier. 424p. (*Developments in sedimentology,* 16.)

Ries, Heinrich (1937) *Economic geology.* 7th ed. New York: Wiley. 720p.

Rigby, J.K. (1959) *Possible eddy markings in the Shinarump Conglomerate of northeastern Utah.* Journal of Sedimentary Petrology, v.29, p.283-284.

Rigby, J.K., and Bayer, T.N. (1971) *Sponges of the Ordovician Maquoketa Formation in Minnesota and Iowa.* Journal of Paleontology, v.45, p.608-627.

Rittenhouse, Gordon (1943) *The transportation and deposition of heavy minerals.* Geological Society of America. Bulletin, v.54, p.1725-1780.

Rittmann, Alfred (1962) *Volcanoes and their activity.* New York: Interscience. 305p.

Rittmann, Alfred (1973) *Stable mineral assemblages of igneous rocks—a method of calculation.* Minerals, Rocks and Inorganic Materials, v.7. New York: Springer-Verlag.

Rivière, André (1952) *Expression analytique générale de la granulométrie des sédiments meubles.* Société Géologique de France. Bulletin, ser.6, v.2, p.155-167.

Rizzini, A., and Passega, R. (1964) *Évolution de la sédimentation et orogenèse, vallée du Santerno, Apennin septentrional.* In: Bouma, A.H., and Brouwer, A., eds. *Turbidites,* p.65-74. Amsterdam: Elsevier. 264p. (*Developments in sedimentology* 3).

Roberts, George (1839) *An etymological and explanatory diction-*

ary of the terms and language of geology; designed for the early student, and those who have not made great progress in that science. London: Longman, Orme, Brown, Green, & Longmans, 183p.

Roberts, R.J. (1951) *Geology of the Antler Peak quadrangle, Nevada.* U.S. Geological Survey. Map GQ10, scale 1:62,500, with text.

Robinove, C.J. (1963) *Photography and imagery - a clarifiction of terms.* Photogrammetric Engineering, v.29, p.880-881.

Robson, J., and Nance, R.M. (1959) *Geological terms used in S.W. England.* Royal Geological Society of Cornwall. Transactions, 1955-1956, v.19, pt.1, p.33-41.

Rodgers, John (1963) *Mechanics of Appalachian foreland folding in Pennsylvania and West Virginia.* American Association of Petroleum Geologists. Bulletin, v.47, p.1527-1536.

Rodgers, John (1967) *Chronology of tectonic movements in the Appalachian region of North America.* American Journal of Science, v.265, p.408-427.

Roedder, Edwin (1967) *Fluid inclusions as samples of ore fluids.* In: Barnes, H.L., ed. *Geochemistry of hydrothermal ore deposits,* p.515-574. New York: Holt, Rinehart and Winston.

Rogers, C.P., jr., and Neal, J.P. (1975) *Feldspar and aplite.* In: Lefond, S.J., ed. *Industrial Minerals and Rocks,* 4th ed., p.637-651. New York: American Institute of Mining, Metallurgical, and Petroleum Engineers.

Rogers, H.D. (1860) *On the distribution and probable origin of the petroleum, or rock oil, of western Pennsylvania, New York, and Ohio.* Royal Philosophical Society of Glasgow. Proceedings, v.4, p.355-359.

Rogers, H.H., compiler (1966) *Glossary of terms frequently used in cosmology.* New York: American Institute of Physics. 16p.

Rohrbach, C.E.M. (1885) *Ueber die Eruptivgesteine im Gebiete der schlesisch-mährischen Kreideformation.* Tschermak's mineralogische und petrographische Mitteilungen, N.F., Bd.7, p.1-63.

Roques, M. (1961) *Nomenclature de J. Jung et M. Roques pour certains types de migmatites.* In: Sorensen, H., ed., *Symposium on migmatite nomenclature,* p.68. International Geological Congress, 21st, Copenhagen, 1960. Report, pt.26, sec.14.

Rosalsky, M.B. (1949) *A study of minor beach features at Fire Island, Long Island, New York.* New York Academy of Sciences. Transactions, ser.2, v.12, p.9-16.

Rosauer, E.A. (1957) *Climatic conditions involved in glacial loess formation.* Doctoral dissertation, University of Bonn. 105p.

Rosenbusch, Harry (1887) *Mikroskopische Physiographie der massigen Gesteine.* 2nd ed. Stuttgart: Schweizerbart'sche Verlagshandlung (E.Koch). 877p. (His *Mikroskopische Physiographie der Mineralien und Gesteine,* Bd.2).

Rosenbusch, Harry (1888) *Microscopical physiography of the rock-making minerals: an aid to the microscopical study of rocks.* Translated and abridged by Joseph P. Iddings. New York: Wiley. 333p.

Rosenbusch, Harry (1898) *Elemente der Gesteinlehre.* Stuttgart: Schweizerbart'sche. 546p.

Ross, C.S., and Shannon, E.V. (1926) *The minerals of bentonite and related clays and their physical properties.* American Ceramic Society. Journal, v.9, p.77-96.

Ross, C.S., and Smith, R.L. (1961) *Ash-flow tuffs: their origin, geologic relations, and identification.* U.S. Geological Survey. Professional Paper 366. 81p.

Rousseau, Jacques (1949) *Modifications de la surface de la toundra sous l'action d'agents climatiques.* Revue Canadienne Géographie, v.3, p.43-51.

Royse, Frank, jr., Warner, M.A., and Reese, D.L. (1975) *Thrust belt structural geometry and related stratigraphic problems, Wyoming-Idaho-northern Utah.* In: Bolyard, D.W., ed. *Symposium on deep drilling in the central Rocky Mountains.* Denver: Rocky Mountain Association of Geologists.

Rubey, W.W., and Hubbert, M.K. (1959) *Role of fluid pressure in mechanics of overthrust faulting.* Geological Society of America. Bulletin, v.70, p.167-206.

Ruhe, R.V. (1969) *Quaternary landscapes in Iowa.* Ames: Iowa State University Press. 255p.

Runcorn, S.K., and others, eds. (1967) *International dictionary of geophysics.* New York: Pergamon. 2 volumes. 1728p.

Runnegar, B., and others (1975) *Biology of the Hyolitha.* Lethaia, v.8, p.181-191.

Russell, I.C. (1885) *Existing glaciers of the U.S.* U.S. Geological Survey. Annual Report, 5th, p.303-355.

Russell, I.C. (1898a) *Glaciers of Mount Rainier.* U.S. Geological Survey. Annual Report, 18th, pt.2, p.349-415.

Russell, I.C. (1898b) *Rivers of North America.* New York: Putnam. 327p.

Russell, R.J. (1942) *Flotant.* Geographical Review, v.32, p.74-98.

Russell, R.J., ed. (1968) *Glossary of terms used in fluvial, deltaic, and coastal morphology and processes.* Coastal Studies Institute, Louisiana State University. Technical Report No.63.

Rust, B.R. (1972) *Structure and process in a braided river.* Sedimentology, v.18, p.221-245.

Ruxton, B.P., and Berry, Leonard (1959) *The basal rock surface on weathered granitic rocks.* Geologists' Association. Proceedings, v.70, p.285-290.

Ryland, J.S. (1970) *Bryozoans.* London: Hutchinson University Library. 175p.

S

Sabins, F.F. (1978) *Remote sensing—principles and interpretation.* San Francisco: Freeman. 426p.

Sackin, M.J., and Merriam, D.F. (1969) *Autoassociation, a new geological tool.* International Association for Mathematical Geology. Journal, v.1, p.7-16.

Salisbury, J.W., and Hunt, G.R. (1968) *Martian surface materials: effect of particle size on spectral behavior.* Science, v.161, p.365-366.

Salisbury, R.D. (1904) *Three new physiographic terms.* Journal of Geology, v.12, p.707-715.

Salisbury, R.D, and others (1902) *The glacial geology of New Jersey.* New Jersey Geological Survey. Final Report 5. 802p.

Salomon, W. (1915) *Die Definitionen von Grauwacke, Arkose und Ton.* Geologische Rundschau, Leipzig; VI, p.398-404.

Sander, Bruno (1911) *Über Zusammenhänge zwischen Teilbewegung und Gefüge in Gesteinen.* Tschermaks Mineralogische und Petrographische Mitteilungen, v.30, p.281-314.

Sander, Bruno (1912) *Über tektonische Gesteinsfazies.* Austria. Geologische Reichsanstalt. Verhandlungen, 1912, no.10, p.249-257.

Sander, Bruno (1930) *Gefügekunde der Gesteine mit besonderer Berücksichtigung der Tektonite.* Vienna: Julius Springer. 352p.

Sander, Bruno (1936) *Beiträge zur Kenntnis der Anlagerungsgefüge (rhythmische Kalke und Dolomite aus der Trias).* [Pt.] 1-2. Mineralogische und petrographische Mitteilungen, Bd. 48, p.27-139, 141-209.

Sander, Bruno (1951) *Contributions to the study of depositional fabric.* Translated by E.B. Knopf. Tulsa: American Association of Petroleum Geologists. 207p.

Sander, Bruno (1970) *An introduction to the study of fabrics of geological bodies.* Translated by F.C. Phillips and G. Windsor. New York: Pergamon. 641p.

Sander, N.J. (1967) *Classification of carbonate rocks of marine origin.* American Association of Petroleum Geologists. Bulletin, v.51, p.325-336.

Sanders, J.E. (1957) *Discontinuities in the stratigraphic record.* New York Academy of Sciences. Transactions, ser.2, v.19, p.287-297.

Sanders, J.E. (1963) *Concepts of fluid mechanics provided by primary sedimentary structures.* Journal of Sedimentary Petrology, v.33, p.173-179.

Sangster, D.F. (1972) *Precambrian volcanogenic massive sulphide deposits in Canada: a review.* Geological Survey of Canada. Paper 72-22. 44p.

Sauer, C.O. (1930) *Basin and Range forms in the Chiricahua area.* University of California, Berkeley. Publications in Geography, v.3, p.339-414.

Saussure, H.-B. de (1786) *Voyages dans les Alpes, précédés d'un essai sur l'histoire naturelle des environs de Genève.* Vol.2. Genève: Barde, Manget. 641p.

Scagel, R.F., and others (1965) *An evolutionary survey of the plant kingdom.* Belmont, Calif.: Wadsworth.

Scheidegger, A.E. (1961) *Theoretical geomorphology.* Berlin: Springer-Verlag. 333p.

Scheidegger, A.E. (1963) *Principles of geodynamics.* 2nd ed. New York: Academic Press. 362p.

Schenck, H.G., and Muller, S.W. (1941) *Stratigraphic terminology.* Geological Society of America. Bulletin, v.52, p.1419-1426.

Schermerhorn, L.J.G. (1966) *Terminology of mixed coarse-fine sediments.* Journal of Sedimentary Petrology, v.36, p.831-835.

Schieferdecker, A.A.G., ed. (1959) *Geological nomenclature.* Gorinchem: Royal Geological and Mining Society of the Netherlands. 523p.

Schiffman, A. (1965) *Energy measurements in the swash-surf zone.* Limnology and Oceanography, v.10, p.255-260.

Schindewolf, O.H. (1950) *Grundlagen und Methoden der paläontologischen Chronologie.* 3rd ed. Berlin: Gebrüder Borntraeger. 152p.

Schindewolf, O.H. (1954) *Über einige stratigraphische Grundbegriffe.* Roemeriana, H.1, p.23-38.

Schindewolf, O.H. (1957) *Comments on some stratigraphic terms.* American Journal of Science, v.255, p.394-399.

Schindewolf, O.H. (1959) *On certain stratigraphic fundamentals.* International Geology Review, v.1, no.7, p.62-70.

Schipman, Henry, jr. (1968) *Buffalo rings.* National Parks Magazine, v.42, no.254, p.14-15.

Schlanger, S.O. (1957) *Dolomite growth in coralline algae.* Journal of Sedimentary Petrology, v.27, p.181-186.

Schmidt, Volkmar (1965) *Facies, diagenesis, and related reservoir properties in the Gigas Beds (Upper Jurassic), northwestern Germany.* Society of Economic Paleontologists and Mineralogists. Special Publication no.13, p.124-168.

Schmidt, Walter (1925) *Gefügestatistik.* Tschermak's mineralogische und petrographische Mitteilungen, N.F., Bd.38, p.392-423.

Schmoll, H.R., and Bennett, R.H. (1961) *Axiometer - a mechanical device for locating and measuring pebble and cobble axes for macrofabric studies.* Journal of Sedimentary Petrology, v.31, p.617-622.

Schofield, W. (1920) *Dumb-bell islands and peninsulas on the coast of South China.* Liverpool Geological Society. Proceedings, v.13, p.45-51.

Schopf, J.M. (1960) *Field description and sampling of coal beds.* U.S. Geological Survey. Bulletin 1111-B, p.25-70.

Schopf, J.M. (1975) *Modes of fossil preservation.* Review of Palaeobotany and Palynology, v.20, p.27-53.

Schroeder, R.A., and Bada, J.L. (1976) *A review of the geochemical applications of the amino acid racemization reaction.* Earth-Science Reviews, v.12, p.347-391.

Schuchert, Charles (1910) *Paleogeography of North America.* Geological Society of America. Bulletin, v.20, p.427-606.

Schuchert, Charles (1923) *Sites and natures of the North American geosynclines.* Geological Society of America. Bulletin, v.34, p.151-260.

Schuchert, Charles (1924) *Historical geology.* 2nd ed. New York: Wiley. 724p. (*A textbook of geology,* pt.2.)

Schulze, Franz (1855) *Bemerkungen über das Vorkommen wohlerhaltener Cellulose in Braunkohle und Steinkohle.* Akademie der Wissenschaften, Berlin. Bericht über die zur Bekanntmachung geeigneten Verhandlungen, 5 Nov 1855, p.676-678.

Schumm, S.A.(1956) *Evolution of drainage systems and slopes in badlands at Perth Amboy, New Jersey.* Geological Society of America. Bulletin, v.67, p.597-646.

Schwarz, E.H.L. (1912) *South African geology.* Glasgow: Blackie. 200p.

Schwarzbach, Martin (1961) *Das Klima der Vorzeit; eine Einführung in die Paläoklimatologie.* 2nd ed. Stuttgart: Ferdinand Enke. 275p.

Schytt, Valter (1963) *Fluted moraine surfaces.* Journal of Glaciology, v.4, p.825-827.

Scott, D.H. (1923) *Studies in fossil botany,* 3rd ed, v.2. London: A. & C. Black. 446p.

Scott, H.W. (1947) *Solution sculpturing in limestone pebbles.* Geological Society of America. Bulletin, v.58, p.141-152.

Scott, W.B. (1922) *Physiography, the science of the abode of man.* New York: Collier. 384p.

Scrivenor, J.B. (1921) *The physical geography of the southern part of the Malay peninsula.* Geographical Review, v.11, p.351-371.

SCSA: Soil Conservation Society of America (1976) *Resource conservation glossary.* Journal of Soil and Water Conservation, v.31, no.4.

Searle, A.B. (1912) *The natural history of clay.* New York: Putnam. 176p.

Searle, A.B. (1923) *Sands and crushed rocks.* Vol.1. London: Henry Frowde and Hodder & Stoughton. 475p. (Oxford technical publications).

Sears, F.W. (1958) *Mechanics, wave motion and heat.* Reading, Mass.: Addison-Wesley. 664p.

Sederholm, J.J. (1891) *Studien über Archäis die Eruptivgesteine aus dem sudwestlichen Finnland.* Tschermaks Mineralogische und Petrographische Mitteilungen, v.12, p.97-142.

Sederholm, J.J. (1907) *Om granit och gneis deras uppkomst, uppträdande och utbredning inom urberget i Fennoskandia.* Finland. Commission Géologique. Bulletin, no.23, 110p.

Sederholm, J.J. (1924) *Granit-gneisproblemen belysta genom iakttagelser i Åbo-Ålands skärgard.* Geologiska Föreningen, Stockholm. Förhandlingar, v.46, p.129-153.

Seglund, J.A. (1974) *Collapse-fault systems of Louisiana Gulf Coast.* American Association of Petroleum Geologists. Bulletin, v.58, p.2389-2397.

Seilacher, Adolf (1969) *Fault-graded beds interpreted as seismites.* Sedimentology, v.13, p.155-159.

Sellards, E.H., and Gunter, Herman (1918) *Geology between the Apalachicola and Ocklocknee rivers in Florida.* Florida Geological Survey. Annual Report, 10th-11th, p.9-56.

Serra, Sandro (1977) *Styles of deformation in the ramp regions of overthrust faults.* Joint Wyoming-Montana-Utah Geological Associations Guidebook, *Rocky Mountain thrust belt geology and resources,* p.487-498.

Shackleton, R.M. (1958) *Downward-facing structures of the Highland Border.* Geological Society of London. Quarterly Journal, v.113, p.361-392.

Shaler, N.S. (1889) *The geology of Cape Ann, Mass.* U.S. Geological Survey. Annual Report, 9th, p.529-611.

Shaler, N.S. (1890) *General account of the fresh-water morasses of the United States, with a description of the Dismal Swamp district of Virginia and North Carolina.* U.S. Geological Survey. Annual Report, 10th, pt.1, p.255-339.

Shaler, N.S. (1895) *Beaches and tidal marshes of the Atlantic coast.* National Geographic Society. National Geographic Monographs, v.1, no.5, p.137-168.

Shand, S.J. (1916) *The pseudotachylyte of Parijs (Orange Free State), and its relation to "trapshotten-gneiss" and "flint-crush-rock".* Geological Society of London. Quarterly Journal, v.72, p.198-220.

Shand, S.J. (1947) *Eruptive rocks, their genesis, composition, classification, and their relation to ore deposits, with a chapter on meteorites.* 3rd ed. London: Thomas Murby. 488p.

Sharp, J.V.A., and Nobles, L.H. (1953) *Mudflow of 1941 at Wrightwood, Southern California.* Geological Society of America. Bulletin, v.64, no.5, p.547-560.

Sharp, R.P. (1942) *Soil structures in the St. Elias Range, Yukon Territory.* Journal of Geomorphology, v.5, p.274-301.

Sharp, R.P. (1954) *Physiographic features of faulting in southern California.* California Division of Mines. Bulletin 170, ch.5, [pt.]3, p.21-28.

Sharp, R.P., and Carey, D.L. (1976) *Sliding stones, Racetrack Playa, California.* Geological Society of America. Bulletin, v.87, p.1704-1717.

Sharpe, C.F.S. (1938) *Landslides and related phenomena; a study of mass-movements of soil and rock.* New York: Columbia University Press. 136p.

Shatsky, N.S. (1945) *Outlines of the tectonics of Volga-Urals petroleum region and adjacent parts of the west slope of the southern Urals.* In: Mather, K.F., ed. *Source book in geology 1900-1950,* p.263-267. Cambridge, Mass.: Harvard University Press.

Shatsky, N.S. (1955) *O proiskhozhdenii Pachelmskogo progiba.* Moskovskoe Obshchestvo Ispytatelei Prirody. Biulleten'; otdel geologicheskii, n.s., v.30, no.5, p.3-26.

Shatsky, N.S., and Bogdanoff, A.A. (1957) *Explanatory note on the tectonic map of the U.S.S.R. and adjoining countries.* Moscow: State Scientific and Technical Publishing House. (English translation in International Geology Review, 1959, v.1, p.1-49).

Shatsky, N.S., and Bogdanoff, A.A. (1960) *La carte tectonique internationale de l'Europe au 2 500 000.* Akademiya Nauk SSSR, Izvestiya, Seriya Geologicheskaya, no.4.

Shaw, E.W. (1911) *Preliminary statement concerning a new system of Quaternary lakes in the Mississippi basin.* Journal of Geology, v.19, p.481-491.

Shawe, D.R. and Granger, H.C. (1965) *Uranium ore rolls—an analysis.* Economic Geology, v.60, p.240-250.

Shepard, F.P. (1937) *Revised classification of marine shorelines.* Journal of Geology, v.45, p.602-624.

Shepard, F.P. (1948) *Submarine geology.* New York: Harper. 348p.

Shepard, F.P. (1952) *Revised nomenclature for depositional coastal features.* American Association of Petroleum Geologists. Bulletin, v.36, p.1902-1912.

Shepard, F.P. (1954) *Nomenclature based on sand-silt-clay ratios.* Journal of Sedimentary Petrology, v.24, p.151-158.

Shepard, F.P. (1967) *The earth beneath the sea.* Revised ed. Baltimore: Johns Hopkins Press. 242p.

Shepard, F.P., and Cohee, G.V. (1936) *Continental shelf sediments off the mid-Atlantic States.* Geological Society of America. Bulletin, v.47, p.441-457.

Shepard, F.P., and Dill, R.F. (1966) *Submarine canyons and other sea valleys.* Chicago: Rand McNally. 381p.

Shepard, F.P., and Young, Ruth (1961) *Distinguishing between beach and dune sands.* Journal of Sedimentary Petrology, v.31, p.196-214.

Sheriff, R.E. (1968) *Glossary of terms used in geophysical exploration.* Geophysics, v.33, p.181-228.

Sheriff, R.E. (1973) *Encyclopedic dictionary of exploration geophysics.* Tulsa: Society of Exploration Geophysicists. 266p.

Sherzer, W.H. (1910) *Criteria for the recognition of the various types of sand grains.* Geological Society of America. Bulletin, v.21, p.625-662, 775-776.

Shipley, R.M. (1951) *Dictionary of gems and gemology including ornamental, decorative and curio stones,* 5th ed. Los Angeles: Gemological Institute of America. 261p.

Shipley, R.M. (1974) *Dictionary of gems and gemology,* 6th ed. Los Angeles: Gemological Institute of America.

Short, N.M. (1966) *Shock-lithification of unconsolidated rock materials.* Science, v.154, p.382-384.

Shreve, R.L. (1966) *Statistical law of stream numbers.* Journal of Geology, v.74, p.17-37.

Shreve, R.L. (1967) *Infinite topologically random channel networks.* Journal of Geology, v.75, p.178-186.

Shreve, R.L. (1969) *Stream lengths and basin areas in topologically random channel networks.* Journal of Geology, v.77, p.397-414.

Shrock, R.R. (1947) *Loiponic deposits.* Geological Society of America. Bulletin, v.58, p.1228.

Shrock, R.R. (1948) *Sequence in layered rocks, a study of features and structures useful for determining top and bottom or order of succession in bedded and tabular rock bodies.* New York: McGraw-Hill. 507p.

Shrock, R.R. (1948a) *A classification of sedimentary rocks.* Journal of Geology, v.56, p.118-129.

Shrock, R.R., and Twenhofel, W.H. (1953) *Principles of invertebrate paleontology.* 2nd ed. New York: McGraw-Hill. 816p.

Shvetsov, M.S. (1960) *K voprosu o diageneze.* In: Natsional'nyi komitet geologov Sovetskogo Soiuza. *Voprosy sedimentologii; doklady sovetskikh geologv k VI Mezhdunarodnomu kongressu po sedimentologii,* p.153-161. Moscow: Gos. nauchno-tekh. izd'vo lit'ry po geologii i okhrane nedr. 215p.

Siegel, R., and Howell, J.R. (1968) *Thermal radiation heat transfer,* v.1. National Aeronautics and Space Administration. SP-164.

Sigafoos, R.S. (1951) *Soil instability in tundra vegetation.* Ohio Journal of Science, v.51, p.281-298.

Sigal, Jacques (1964) *Une thérapeutique homéopathique en chronostratigraphie: les parastratotypes (ou prétendus tels).* France. Bureau de Recherches Géologiques et Minières. Dépt. d'Information Géologique. Bulletin trimestriel, année 16, no.64, p.1-8.

Silberling, N.J., and Roberts, R.J. (1962) *Pre-Tertiary stratigraphy and structure of northwestern Nevada.* Geological Society of America. Special Paper 72. 58p.

Sillitoe, R.H., Halls, C., and Grant, J.N. (1975) *Porphyry tin deposits in Bolivia.* Economic Geology, v.70, p.913-927.

Simons, D.B., and others (1961) *Flume studies using medium sand (0.45mm).* U.S.Geological Survey. Professional Paper 1498-A. 76p.

Simons, D.B., and Richardson, E.V. (1961) *Forms of bed roughness in alluvial channels.* American Society of Civil Engineers. Proceedings, v.87, no.HY3, p.87-105.

Simpson, E.S.W. (1954) *On the graphical representation of differentiation trends in igneous rocks.* Geological Magazine, v.91, p.238-244.

Simpson, G.G. (1940) *Types in modern taxonomy.* American Journal of Science, v.238, p.413-431.

Sjogren, W.L. (1974) *Lunar gravity via the Apollo 15 and 16 subsatellites.* The Moon, v.9, p.115-128.

Skolnick, Herbert (1965) *The quartzite problem.* Journal of Sedimentary Petrology, v.35, p.12-21.

Skow, M.L. (1962) *Mica; a materials survey.* U.S. Bureau of Mines. Information Circular 8125. 241p.

Sloane, R.L., and Kell, T.R. (1966) *The fabric of mechanically compacted kaolin.* National Conference on Clays and Clay Minerals. 14th, Berkeley, Calif., 1965. Proceedings, p.289-296. New York: Pergamon Press. 443p. (International series of monographs on earth sciences, v.26).

Sloss, L.L. (1953) *The significance of evaporites.* Journal of Sedimentary Petrology, v.23, p.143-161.

Sloss, L.L. (1963) *Sequences in the cratonic interior of North America.* Geological Society of America. Bulletin, v.74, p.93-113.

Sloss, L.L., and Laird, W.M. (1947) *Devonian System in central and northwestern Montana.* American Association of Petroleum Geologists. Bulletin, v.31, p.1404-1430.

Sloss, L.L., and others (1949) *Integrated facies analysis.* Geological Society of America. Memoir 39, p.91-123.

Smart, J.S. (1969) *Topological properties of channel networks.* Geological Society of America. Bulletin, v.80, p.1757-1774.

Smirnov, V.I. (1968) *The sources of the ore-forming fluid.* Economic Geology, v.63, p.380-389.

Smit, D.E., and Swett, Keene (1969) *Devaluation of "dedolomitization".* Journal of Sedimentary Petrology, v.39, p.379-380.

Smith, J.T., jr. (1968) *Glossary of aerial photographic terms.* In: Smith, J.T., jr., and Anson, Abraham, eds. *Manual of color aerial photography,* p.489-509. Falls Church, Va.: American Society of Photogrammetry. 550p.

Smith, K.G. (1950) *Standards for grading texture of erosional topography.* American Journal of Science, v.248, p.655-668.

Smith, N.D. (1974) *Sedimentology and bar formation in the upper Kicking Horse River, a braided outwash stream.* Journal of Geology, v.82, p.205-223.

Smith, R.A., and others (1968) *The detection and measurement of infra-red radiation.* 2nd ed. Oxford: Clarendon Press.

Smith, R.E. (1968) *Redistribution of major elements in the alteration of some basic lavas during burial metamorphism.* Journal of Petrology, v.9, p.191-219.

Smith, R.L. (1960) *Zones and zonal variations in welded ash flows.* U.S. Geological Survey. Professional Paper 354-F.

Smith, R.L., and Bailey, R.A.(1968) *Resurgent cauldrons.* Geological Society of America. Memoir 116, p.613-662.

Smith, W.O.(1961) *Mechanism of gravity drainage and its relation to specific yield of uniform sands.* U.S. Geological Survey. Professional Paper 402-A. 12p.

Smith, W.S.T. (1898) *A geological sketch of San Clemente Island.* U.S. Geological Survey. Annual Report, 18th, pt.2, p.459-496.

Smyth, H.L.(1891) *Structural geology of Steep Rock Lake, Ont.* American Journal of Science, ser.3, v.42, p.317-331.

Sneed, E.D., and Folk, R.L. (1958) *Pebbles in the lower Colorado River, Texas. A study in particle morphogenesis.* Journal of Geology, v.66, p.114-140.

Snoke, A.W., and Calk, L.C. (1978) *Jackstraw-textured talc-olivine rocks, Preston Peak area, Klamath Mountains, California.* Geological Society of America. Bulletin, v.89, p.223-230.

Sohl, N.F. (1977a) *Application for amendment of articles 8 and 10 of Code, concerning smallest formal rock-stratigraphic unit.* American Association of Petroleum Geologists. Bulletin, v.61, p.252.

Sohl, N.F. (1977b) *Stratigraphic Commission Note 45—Application for amendment concerning terminology for igneous and high-grade metamorphic rocks.* American Association of Petroleum Geologists. Bulletin, v.61, p.248-251.

Sonder, R.A. (1956) *Mechanik der Erde; Elemente und Studien zur tektonischen Erdgeschichte.* Stuttgart: Schweizerbartsche. 291p.

Sorby, H.C. (1852) *On the oscillation of the currents drifting the sandstone beds of the south-east of Northumberland, and on their general direction in the coalfield in the neighborhood of Edinburgh.* West Yorkshire Geological Society. Proceedings, v.3, p.232-240.

Sorby, H.C. (1857) *On the physical geography of the Tertiary estuary of the Isle of Wight.* Edinburgh New Philosophical Journal, n.s., v.5, p.275-298.

Sorby, H.C. (1863) *Über Kalkstein-Geschiebe mit Eindrücken.* Neues Jahrbuch für Mineralogie, Geologie, und Palaeontologie, Jahrg. 1863, p.801-807.

Sorby, H.C. (1879) *The anniversary address of the President.* Geological Society of London. Proceedings, session 1878-1879, p.39-95. (Geological Society of London. Quarterly Journal, 1879, v.35).

Southard, J.B. (1971) *Representation of bed configurations in depth-velocity-size diagram.* Journal of Sedimentary Petrology, v.41, p.903-915.

Sparks, R.S.J., and Walker, G.P.L. (1973) *The ground surge deposit: a third type of pyroclastic rock.* Nature, v.241, p.62-64.

Spate, O.H.K. (1954) *India and Pakistan; a general and regional geography.* With a chapter on Ceylon by B.H. Farmer. London: Methuen. 827p.

Speers, E.C. (1957) *The age relation and origin of common Sudbury breccia.* Journal of Geology, v.65, p.497-514.

Spencer, A.C. (1917) *The geology and ore deposits of Ely, Nevada.* U.S. Geological Survey. Professional Paper 96. 189p.

Spencer, E.W. (1969) *Introduction to the structure of the Earth.* New York: McGraw-Hill. 597p.

Spieker, E.M. (1956) *Mountain-building chronology and nature of geologic time scale.* American Association of Petroleum Geologists. Bulletin, v.40, no.8, p.1769-1815.

Spotts, J.H., and Weser, O.E. (1964) *Directional properties of a Miocene turbidite, California.* In: Bouma, A.H., and Brouwer, A., eds. *Turbidites,* p.199-221. Amsterdam: Elsevier. 264p. (*Developments in sedimentology* 3.)

Sprinkle, James (1973) *Morphology and evolution of blastozoan echinoderms.* Harvard University Museum of Comparative Zoology. Special Publication. 283p.

Sproule, J.C. (1939) *The Pleistocene geology of the Cree Lake region, Saskatchewan.* Royal Society of Canada. Transactions, ser.3, v.33, sec.4, p.101-109.

Spurr, J.E. (1923) *The filling of fissure veins.* Engineering & Mining Journal, v.116, Aug.25, p.329-330.

Spurr, J.E. (1944) *The Imbrian Plain region of the Moon.* Lancaster, Penna.: Science Press. 112p. (*Geology applied to selenology,* v.1.)

Spurrell, F.C.J. (1887) *A sketch of the history of the rivers and denudation of West Kent, etc.* Geological Magazine, dec.3, v.4, no.273, p.121-122.

Sørensen, H., ed. (1974) *The alkaline rocks.* New York: Wiley. 622p.

SSSA: Soil Science Society of America (1965) *Glossary of soil science terms.* Its Proceedings, v.29, p.330-351. Revised (1970), 27p.; (1975), 34p. Madison, Wis.: Its publication.

Stach, E. (1968) *Basic principles of coal petrology; macerals, microlithotypes and some effects on coalification.* In: Murchison, Duncan, and Westoll, T.S., eds. *Coal and coal-bearing strata,* p.3-17. New York: Elsevier. 418p.

Stamp, L.D. (1921) *On cycles of sedimentation in the Eocene strata of the Anglo-Franco-Belgian basin.* Geological Magazine, v.58, p.108-114, 146-157, 194-200.

Stamp, L.D., ed. (1961) *A glossary of geographical terms.* New York: Wiley. 539p.

Stamp, L.D., ed. (1966) *A glossary of geographical terms.* 2nd ed. London: Longmans, Green. 539p.

Staub, Rudolf (1928) *Der bewegungsmechanismus der erde dargelegt am bau der irdischen gebirgssysteme.* Berlin: Gebrüder Borntraeger. 270p.

Stauffer, M.R., Hajnal, Z., and Gendzwill, D.J. (1976) *Rhomboidal lattice structure: a common feature on sandy beaches.* Canadian Journal of Earth Science, v.13, p.1667-1677.

Stearns, D.W. (1978) *Faulting and forced folding in the Rocky Mountains foreland.* Geological Society of America. Memoir 151, p.1-37.

Stearns, H.T., and Macdonald, G.A. (1942) *Geology and groundwater resources of the island of Maui, Hawaii.* Hawaii Division of Hydrography. Bulletin, no.7. 344p.

Steen-McIntyre, Virginia (1977) *A manual for tephrochronology—collection, preparation, petrographic description, and approximate dating of tephra (volcanic ash).* Idaho Springs, Colo.: published by the author. 167p.

Stefan, Joseph (1879) *Über die Beziehung zwischen der Wärmestrahlung und der Temperatur.* Akademie der Wissenschaften in Wien. Sitzungsberichte, v.79, pt.2, p.391-428.

Steinmann, G. (1926) *Die ophiolitischen Zonen in den mediterranean Kettengebirgen.* International Geological Congress. 14th, Madrid, 1926. Session 14, Report, p.637.

Steno, Nicolaus (1669) *Nicolai Stenonis de solido intra solidum naturaliter contento dissertationis prodromus.* Florence. 79p. - See also: Steno, Nicolaus (1916) *The prodromus of Nicolaus Steno's dissertation concerning a solid body enclosed by process of nature within a solid.* English version with introduction & explanatory notes by J.G. Winter. Foreword by W.H. Hobbs. New York: Macmillan. 119p. (University of Michigan Studies; Humanistic Series, v.11, pt.2, p.165-283).

Stephens, N., and Synge, F.M. (1966) *Pleistocene shorelines.* In: Dury, G.H., ed. *Essays in geomorphology.* p.1-51. New York: Elsevier. 404p.

Stephenson, L.W., and Veatch, J.O. (1915) *Underground waters of the Coastal Plain of Georgia.* U.S. Geological Survey. Water-supply Paper 341. 539p.

Sternberg, H. (1875) *Untersuchungen über Längen- und Querprofil geschiebeführender Flüsse.* Zeitschrift für Bauwesen, v.25, p.483-506.

Stevens, R.E., and Carron, M.K. (1948) *Simple field test for distinguishing minerals by abrasion pH.* American Mineralogist, v.33, p.31-50.

Stewart, D.B. (1975) *Apollonian metamorphic rocks—the products of prolonged subsolidus equilibration* (abstract). Abstracts of papers submitted to the Sixth Lunar Science Conference, March 17-21, 1975. Lunar Science VI, Part II, p.774-776. Houston: Lunar Science Institute.

Stewart, H.B.,jr. (1956) *Contorted sediments in modern coastal lagoon explained by laboratory experiments.* American Association of Petroleum Geologists. Bulletin, v.40,p.153-161.

Stewart, J.H., and Poole, F.G. (1974) *Lower Paleozoic and uppermost Precambrian Cordilleran miogeocline, Great Basin, western United States.* Society of Economic Paleontologists and Mineralogists. Special Publication 22, p.28-57.

Stille, Hans (1930) *Über Einseitigkeiten in der germanotypen Tektonik Nordspaniens und Deutschlands.* Gesellschaft der Wissenschaften, Göttingen. Mathematisch-Physikalische Klasse. Nachrichten, 1930, H.3, p.379-397.

Stille, Hans (1935) *Der derzeitige tektonische Erdzustand.* Preussische Academie der Wissenschaften, Physikalisch-mathematische Klasse. Sitzungsberichte, v.13, p.179-219.

Stille, Hans (1936) *Present tectonic state of the Earth.* American Association of Petroleum Geologists. Bulletin, v.20, p.849-880.

Stille, H.W. (1940) *Einführung in den Bau Amerikas.* Berlin: Gebrüder Borntraeger. 717p.

Stillwell, F.L. (1918) *The metamorphic rocks of Adelie Land.* Australasian Antarctic Expedition, 1911-1914, under the leadership of Sir Douglas Mawson. Scientific reports, Series A, v.3, pt.1, p.7-230.

Stoces, Bohuslav, and White, C.H. (1935) *Structural geology, with special reference to economic deposits.* London: Macmillan. 460p.

Stockdale, P.B. (1939) *Lower Mississippian rocks of the east-central interior.* Geological Society of America. Special Paper 22. 248p.

Stockwell, C.H. (1964) *Fourth report on structural provinces, orogenics, and time-classification of rocks of the Canadian Precambrian Shield.* In: *Age determinations and geological studies,* part 2. Geological Survey of Canada. Geological studies. Paper 64-17.

Stokes, W.L. (1947) *Primary lineation in fluvial sandstones, a criterion of current direction.* Journal of Geology, v.55, p.52-54.

Stokes, W.L. (1953) *Primary sedimentary trend indicators as applied to ore finding in the Carrizo Mountains, Arizona and New Mexico.* Pt.1. Salt Lake City: University of Utah. 48p. (U.S. Atomic Energy Commission. Report RME-3043, pt.1).

Stokes, W.L., and Judson, S.S. (1968) *Introduction to geology, physical and historical.* Englewood Cliffs, N.J.: Prentice-Hall. 530p.

Stokes, W.L., and Varnes, D.J. (1955) *Glossary of selected geologic terms, with special reference to their use in engineering.* Denver: Colorado Scientific Society. 165p. (Colorado Scientific Society. Proceedings, v.16).

Stone, G.H. (1899) *The glacial gravels of Maine and their associated deposits.* U.S. Geological Survey. Monograph 34. 499p.

Stone, R.O. (1967) *A desert glossary.* Earth-science Reviews, v.3, p.211-268.

Storey, T.P., and Patterson, J.R. (1959) *Stratigraphy—traditional and modern concepts.* American Journal of Science, v.257, p.707-721.

Størmer, Leif (1966) *Concepts of stratigraphical classification and terminology.* Earth-Science Reviews, v.1, p.5-28.

Strahan, Aubrey (1907) *The country around Swansea; being an account of the region comprised in sheet 247 of the map.* London:H.M.S.O. 170p. (Great Britian. Geological Survey. Memoirs; England and Wales: the geology of the South Wales coal-field, pt.8).

Strahler, A.N. (1946) *Geomorphic terminology and classification of land masses.* Journal of Geology, v.54, no.1, p.32-42.

Strahler, A.N. (1952a) *Dynamic basis of geomorphology.* Geological Society of America. Bulletin, v.63, p.923-938.

Strahler, A.N. (1952b) *Hypsometric (area-altitude) analysis of erosional topography.* Geological Society of America. Bulletin, v.63,

p.1117-1142.
Strahler, A.N. (1954) *Quantitative geomorphology of erosional landscapes.* International Geological Congress. 19th, Algiers, 1952. Comptes rendus, sec.13, pt.3, fasc.15, p.341-354.
Strahler, A.N. (1956) *The nature of induced erosion and aggradation.* In: Thomas, W.L., jr., ed. *Man's role in changing the face of the Earth,* p.621-638. Chicago: University of Chicago Press. 1193p.
Strahler, A.N. (1958) *Dimensional analysis applied to fluvially eroded landforms.* Geological Society of America. Bulletin, v.69, p.279-299.
Strahler, A.N. (1963) *The earth sciences.* New York: Harper & Row. 681p.
Strahler, A.N. (1964) *Quantitative geomorphology of drainage basins and channel networks.* In: Chow, Ven Te, ed. *Handbook of applied hydrology; a compendium of water-resources technology,* sec.4, p.39-76. New York: McGraw-Hill. 29 sections.
Streckeisen, A.L. (1967) *Classification and nomenclature of igneous rocks (final report of an inquiry).* Neues Jahrbuch für Mineralogie. Abhandlungen, v.107, no.2, p.144-214.
Strickland, Cyril (1940) *Deltaic formation with special reference to the hydrographic processes of the Ganges and the Brahmaputra.* New York: Longmans, Green. 157p.
Stringfield, V.T. (1966) *Hydrogeology - definition and appliction.* Ground Water, v.4, no.4, p.2-4.
Strøm, K. (1966) *Geophysiography.* Atlas, v.2, no.1, p.8-9.
Stutzer, Otto, and Noé, A.C. (1940) *Geology of coal.* Chicago: University of Chicago Press. 461p.
Suess, F.E. (1900) *Die Herkunft der Moldavite und verwandter Gläser.* Austria. Geologische Reichsanstalt. Jahrbuch, Bd.50, H.2, p.193-382.
Suess, H.E., and Urey, H.C. (1956) *Abundances of the elements.* Reviews of Modern Physics, v.28, p.53-74.
Suggate, R.P. (1965) *The definition of "interglacial".* Journal of Geology, v.73, p.619-626.
Sullwold, H.H.,jr. (1959) *Nomenclature of load deformation in turbidites.* Geological Society of America. Bulletin, v.70, p.1247-1248.
Sullwold, H.H.,jr. (1960) *Load-cast terminology and origin of convolute bedding: further comments.* Geological Society of America. Bulletin, v.71, p.635-636.
Sumner, J.S. (1976) *Principles of induced polarization for geophysical prospecting.* New York: Elsevier. 277p.
Sutton, A.H. (1940) *Time and stratigraphic terminology.* Geological Society of America. Bulletin, v.51,p.1397-1412.
Svensson, Nils-Bertil (1968) *Lake Lappajärvi, central Finland: a possible meteorite impact structure.* Nature, v.217, p.438.
Sverdrup, H.U., Johnson, M.W., and Fleming, R.H. (1942) *The oceans, their physics, chemistry, and general biology.* Englewood Cliffs, N.J.: Prentice Hall. 1087p.
Swain, F.M. (1949) *Onlap, offlap, overstep, and overlap.* American Association of Petroleum Geologists. Bulletin, v.33, p.634-636.
Swain, F.M. (1958) *Organic materials of early Middle Devonian, Mt. Union area, Pennsylvania.* American Association of Petroleum Geologists. Bulletin, v.42, p.2858-2891.
Swain, F.M. (1963) *Geochemistry of humus.* In: Breger, I.A., ed. *Organic geochemistry,* p.87-147. New York: Macmillan. 658p. (Earth science series 16.)
Swain, F.M., and Prokopovich, N. (1954) *Stratigraphic distribution of lipoid substances in Cedar Creek Bog, Minnesota.* Geological Society of America. Bulletin, v.65, p.1183-1198.
Swann, D.H., and Willman, H.B. (1961) *Megagroups in Illinois.* American Association of Petroleum Geologists, Bulletin, v.45, p.471-483.
Swartz, Delbert (1971) *Collegiate dictionary of botany.* New York: Ronald. 520p.
Swayne, J.C. (1956) *A concise glossary of geographical terms.* London: George Philip & Son. 164p.
Swenson, H.N., and Woods, J.E. (1971) *Physical science for liberal arts students.* 2d ed. New York: Wiley.
Swesnick, R.M. (1950) *Goldren Trend of south-central Oklahoma.* American Association of Petroleum Geologists. Bulletin, v.34, p.386-422.
Swift, D.J.P., and others (1972) *Holocene evolution of the shelf surface, central and southern Atlantic shelf of North America.* In: Swift, D.J.P., Duane, D.B., and Pilkey, O.H., eds. *Shelf sediment transport: process and pattern,* ch.23, p.499-574. Stroudsburg, Pa.: Dowden, Hutchinson, & Ross.

T

Taber, Stephen (1943) *Perennially frozen ground in Alaska: its origin and history.* Geological Society of America. Bulletin, v.54, p.1433-1548.
Tanner, W.F. (1960) *Shallow water ripple mark varieties.* Journal of Sedimentary Petrology, v.30, p.481-485.
Tanner, W.F. (1975) *Beach processes, Berrien County, Michigan.* International Association for Great Lakes Research. Journal of Great Lakes Research, v.1, p.171-178.
Tanner, W.F. (1976) *Tectonically significant pebble types: sheared, pocked, and second-cycle examples.* Sedimentary Geology, v.16, p.69-83.
Tanton, T.L. (1944) *Conchilites.* Royal Society of Canada. Transactions, ser.3, v.38, sec.4, p.97-104.
Tarr, R.S. (1902) *The physical geography of New York.* New York: Macmillan. 397p.
Tarr, R.S. (1914) *College physiography.* New York: Macmillan. 837p.
Tarr, R.S., and Martin, Lawrence (1914) *Alaskan glacier studies of the National Geographic Society in the Yakutat Bay, Prince William Sound and lower Copper River regions.* Washington: The National Geographic Society. 498p.
Tarr, R.S., and Von Engeln, O.D. (1926) *New physical geography.* New York: Macmillan. 689p.
Tarr, W.A. (1938) *Terminology of the chemical siliceous sediments.* National Research Council. Division of Geology and Geography. Annual report for 1937-1938, appendix A, exhibit A, p.8-27. (Its Committee on Sedimentation. Report, exhibit A).
Tator, B.A. (1949) *Valley widening processes in the Colorado Rockies.* Geological Society of America. Bulletin, v.60, p.1771-1783.
Tator, B.A. (1953) *Pediment characteristics and terminology; part II.* Association of American Geographers. Annals, v.43, p.47-53.
Taubeneck, W.H., and Poldervaart, A. (1960) *Geology of the Elkhorn Mountains, northern Oregon: Part 2. Willow Lake intrusion.* Geological Society of America. Bulletin, v.71, p.1295-1322.
Taylor, G.L., and Reno, D.H. (1948) *Magnetic properties of "granite" wash and unweathered "granite".* Geophysics, v.13, p.163-181.
Taylor, S.R. (1975) *Lunar science: a post-Apollo view.* New York: Pergamon. 372p.
Taylor, T.G., ed. (1951) *Geography in the twentieth century.* New York: Philosophical Library. 630p.
Teall, J.J.H. (1887) *On the origin of certain banded gneisses.* Geological Magazine, v.4, no.11, p.484-493.
Teall, J.J.H. (1903) *On dedolomitisation.* Geological Magazine, n.s., dec.4, v.10, p.513-514.
Tebbutt, G.E., and others (1965) *Lithogenesis of a distinctive carbonate rock fabric.* University of Wyoming. Contributions to Geology, v.4, no.1, p.1-13.
Teichert, Curt (1958) *Concept of facies.* American Association of Petroleum Geologists. Bulletin, v.42, p.2718-2744.
Teichert, Curt (1958a) *Some biostratigraphic concepts.* Geological Society of America. Bulletin, v.69, p.99-119.
Teichert, Curt (1970) *Runzelmarken (wrinkle marks).* Journal of Sedimentary Petrology, v.40, p.1056-1057.
Ten Haaf, Ernst (1956) *Significance of convolution lamination.* Geologie en Mijnbouw, v.18, p.188-194.
Ten Haaf, Ernst (1959) *Graded beds of the northern Apennines.* PhD thesis, Rijks University of Groningen. 102p.
Termier, Henri, and Termier, Geneviève (1956) *La notion de migration en paléontologie.* Geologische Rundschau, v.45, p.26-42.
Termier, Henri, and Termier, Geneviève (1963) *Erosion and sedimentation.* Translated by D.W. Humphries and E.E. Humphries. New York: Van Nostrand. 433p.
Terzaghi, Karl, and Peck, R.B. (1967) *Soil mechanics in engineering practice,* 2nd ed. New York: Wiley. 729p.
Terzaghi, K.C. von (1943) *Theoretical soil mechanics.* New York: Wiley. 510p.
Tester, A.C., and Bay, H.X. (1931) *The shapometer; a device for measuring the shapes of pebbles.* Science, n.s., v.73, p.565-566.
Theis, C.V. (1935) *The relation between the lowering of the piezometric surface and the rate and duration of discharge of a well using ground-water storage.* American Geophysical Union. Transactions, 1935, p.519-524.
Theis, C.V. (1938) *The significance and nature of the cone of depression in ground-water bodies.* Economic Geology, v.33, p.889-902.

Thomas, G.E. (1962) *Grouping of carbonate rocks into textural and porosity units for mapping purposes.* American Association of Petroleum Geologists. Memoir 1, p.193-223.

Thomas, H.D. (1960) *Misuse of "bioclastic limestone".* American Association of Petroleum Geologists. Bulletin, v.44, p.1833-1834.

Thomas, H.D., and Larwood, G.P. (1956) *Some "uniserial" membraniporine polyzoan genera and a new American Albian species.* Geological Magazine, v.93, p.369-376.

Thompson, D.G. (1929) *The Mohave Desert region, California; a geographic, geologic, and hydrologic reconnaissance.* U.S. Geological Survey. Water-supply Paper 578. 759p.

Thompson, J.B., jr. (1957) *The graphical analysis of mineral assemblages in pelitic schists.* American Mineralogist, v.42, p.842-858.

Thorn, C.E. (1976) *Quantitative evaluation of nivation in the Colorado Front Range.* Geological Society of America. Bulletin, v.87, p.1169-1178.

Thornbury, W.D. (1954) *Principles of geomorphology.* New York: Wiley. 618p.

Thornton, C.P., and Tuttle, O.F. (1960) *Chemistry of igneous rocks—pt.1, differentiation index.* American Journal of Science, v.258, p.664-668.

Thorp, J., and Smith, G.D. (1949) *Higher categories of soil classification.* Soil Science, v.67, p.117-126.

Thorson, Gunnar (1957) *Bottom communities (sublittoral or shallow shelf).* Geological Society of America. Memoir 67, v.1, p.461-534.

Thrush, P.W., compiler (1968) *A dictionary of mining, mineral, and related terms.* Compiled and edited by P.W. Thrush and the Staff of the Bureau of Mines. Washington: U.S. Bureau of Mines. 1269p.

Thwaites, F.T. (1926) *The origin and significance of pitted outwash.* Journal of Geology, v.34, p.308-319.

Tiddeman, R.H. (1890) *On concurrent faulting and deposit in Carboniferous times in Craven, Yorkshire, with a note on Carboniferous reefs.* British Association for the Advancement of Science. Report, 1889, 59th, p.600-603.

Tieje, A.J. (1921) *Suggestions as to the description and naming of sedimentary rocks.* Journal of Geology, v.29, p.650-666.

Tilley, C.E. (1925) *Metamorphic zones in the southern highlands of Scotland.* Geological Society of London. Quarterly Journal, v.81, p.100-112.

Tilley, C.E., and Muir, I.D. (1964) *Intermediate members of the oceanic basalt-trachyte association.* Geologiska Föreningen i Stockholm. Förhandlingar, v.85, p.436-444.

Tilsley, J.E. (1977) *Placosols: another problem in exploration geochemistry.* Journal of Geochemical Exploration, v.7, p.21-30.

TIP: Moore, R.C., editor 1953-1974 (subsequent editors, Curt Teichert and R.A. Robison), *Treatise on invertebrate paleontology.* Prepared under the guidance of the Joint Committee on Invertebrate Paleontology. Lawrence, Kan.: University of Kansas Press and the Geological Society of America. Issued in separate parts identified by letter.

Tippo, Oswald (1942) *A modern classification of the plant kingdom.* Chronica Botanica, v.7, p.203-206.

Titkov, Nikolai, and others (1965) *Mineral formation and structure in the electrochemical induration of weak rocks.* New York: Consultants Bureau. 74p.

Tobi, A.C. (1971) *The nomenclature of the charnockitic rock suite.* Neues Jahrbuch für Mineralogie, Monatshefte, no.5, p.193-205.

Todd, J.E. (1902) *Hydrographic history of South Dakota.* Geological Society of America. Bulletin, v.13, p.27-40.

Todd, J.E. (1903) *Concretions and their geological effects.* Geological Society of America. Bulletin, v.14, p.353-368.

Tolman, C.F. (1937) *Ground water.* New York: McGraw-Hill. 593p.

Tomkeieff, S.I. (1943) *Megatectonics and microtectonics.* Nature, v.152, p.347-349.

Tomkeieff, S.I. (1946) *James Hutton's "Theory of the Earth", 1795.* Geologists' Association. Proceedings, v.57, p.322-328.

Tomkeieff, S.I. (1954) *Coals and bitumens and related fossil carbonaceous substances; nomenclature and classification.* London: Pergamon Press. 122p.

Tomkeieff, S.I. (1961) *Alkalic ultrabasic rocks and carbonatites of the U.S.S.R.* International Geology Review, v.3, p.739-758.

Tomkeieff, S.I. (1962) *Unconformity—an historical study.* Geologists' Association. Proceedings, v.73, p.383-417.

Törnebohm, A.E. (1880-1881) *Nagra ord om granit och gneis.* Geologiska Föreningen, Stockholm. Förhandlingar, v.5, p.233-248.

Tourtelot, H.A. (1960) *Origin and use of the word "shale".* American Journal of Science, v.258-A (Bradley volume), p.335-343.

Tower, W.S. (1904) *The development of cutoff meanders.* American Geographical Society. Bulletin, v.36, p.589-599.

Trask, P.D. (1932) *Origin and environment of source sediments of petroleum.* Houston: American Petroleum Institute. 323p.

Treiber, Erich (1957) *Die Chemie der Pflanzenzellwand.* Berlin: Springer-Verlag. 511p.

Trevisan, Livio (1950) *Genèse des terrasses fluviatiles en relation avec les cycles climatiques.* International Geographical Congress. 16th, Lisbon. Comptes Rendus, tome 2, p.511-528.

Tröger, W.E. (1935) *Spezielle Petrographie der Eruptivgesteine: Ein Nomenklatur-Kompendium.* Berlin: Verlag der Deutschen Mineralogischen Gesellschaft e.V. 360p.

Troll, Carl (1944) *Strukturböden, Solifluktion und Frostklimate der Erde.* Geologische Rundschau, Bd.34, p.545-694.

Trowbridge, A.C. (1921) *The erosional history of the Driftless Area.* State University of Iowa. Studies in Natural History, v.9, no.3. 127p. (State University of Iowa. Studies, 1st ser., no.40).

Trueman, A.E. (1923) *Some theoretical aspects of correlation.* Geologists' Association. Proceedings, v.34, p.193-206.

Trümpy, Rudolf (1955) *Wechselbeziehungen zwischen Palaogeographie und Deckenbau.* Naturforschende Gesellschaft in Zürich, Vierteljahrschrift, pt.C, p.217-231.

Tschudy, R.H., and Scott, R.D. (1969) *Aspects of palynology.* New York: Interscience. 510p.

Turner, F.J. (1948) *Mineralogical and structural evolution of the metamorphic rocks.* Geological Society of America. Memoir 30. 342p.

Turner, F.J. (1968) *Metamorphic petrology.* New York: McGraw-Hill. 403p.

Turner, F.J., and Verhoogen, Jean (1951) *Igneous and metamorphic petrology.* 1st ed. New York: McGraw-Hill. 602p.

Turner, F.J., and Verhoogen, Jean (1960) *Igneous and metamorphic petrology.* 2d ed. New York: McGraw-Hill. 694p.

Turner, F.J., and Weiss, L.E. (1963) *Structural analysis of metamorphic tectonites.* New York: McGraw-Hill. 545p.

Tuttle, O.F., and Bowen, N.L. (1958) *Origin of granite in the light of experimental studies in the system $NaAlSi_3O_8$-$KAlSi_3O_8$-SiO_2-H_2O.* Geological Society of America. Memoir 74. 153p.

Tuttle, O.F., and Gittins, J., eds. (1966) *Carbonatites.* New York: Interscience. 591p.

Twenhofel, W.H. (1937) *Terminology of the fine-grained mechanical sediments.* National Research Council. Division of Geology and Geography. Annual report for 1936-1937, appendix I, exhibit F, p.81-104. (Its Committee on Sedimentation. Report, exhibit F).

Twenhofel, W.H. (1939) *Principles of sedimentation.* New York: McGraw-Hill. 610p.

Twenhofel, W.H. (1950) *Principles of sedimentation.* 2nd ed. New York: McGraw-Hill. 673p.

Tyrrell, G.W. (1921) *Some points in petrographic nomenclature.* Geological Magazine, v.58, p.494-502.

Tyrrell, G.W. (1926) *The principles of petrology; an introduction to the science of rocks.* London: Methuen. 349p.

Tyrrell, G.W. (1950) *The principles of petrology; an introduction to the science of rocks.* London: Methuen; New York: E.P. Dutton.

Tyrrell, J.B. (1904) *Crystosphenes or buried sheets of ice in the tundra of northern America.* Journal of Geology, v.12, p.232-236.

Tyrrell, J.B. (1910) *"Rock glaciers" or chrystocrenes.* Journal of Geology, v.18, p.549-553.

Tyrrell, J.B., and Dowling, D.B. (1896) *Report on the country between Athabasca Lake and Churchill River with notes on two routes travelled between the Churchill and Saskatchewan rivers.* Geological Survey of Canada. Annual report 1895, n.s., v.8, report D. 120p.

U

Udden, J.A. (1898) *The mechanical composition of wind deposits.* Rock Island, Ill.: Augustana Library Publications, no.1. 69p.

Udden, J.A. (1914) *Mechanical composition of clastic sediments.* Geological Society of America. Bulletin, v.25, p.655-744.

Ulrich, E.O. (1911) *Revision of the Paleozoic systems.* Geological Society of America. Bulletin, v.22, p.281-680.

Umbgrove, J.H.F. (1933) *Verschillende Typen van tertiaire Geosynclinalen in den Indischen Archipel.* Leidsche Geologische Mededeelingen, v.5, no.1, p.33-43.

Underwood, L.B. (1957) *Rebound problem in the Pierre Shale at*

Oahe Dam, Pierre, South Dakota, Pt.1 (abstract). Geological Society of America. Bulletin, v.68, no.12, pt.2, p.1807-1808.

Urbanek, Adam, and Towe, K.M. (1974) *Ultrastructural studies on graptolites, 1: The periderm and its derivatives in the Dendroidea and in* Mastigograptus. Smithsonian Contributions to Paleobiology, no.20. 48p.

U.S. Federal Committee on Research Natural Areas (1968) *A directory of research natural areas on Federal lands of the United States of America.* Washington: U.S. Government Printing Office. 129p.

USDA: U.S. Department of Agriculture (1938) *Soils and men.* Washington: Government Printing Office. 1232p. (Its *Yearbook of agriculture,* 1938.)

USDA: U.S. Department of Agriculture (1957) *Soil.* Washington: Government Printing Office. 784p. (Its Yearbook of Agriculture, 1957).

USDA: U.S. Department of Agriculture (1975) *Soil taxonomy: a basic system of soil classification for making and interpreting soil surveys.* Agriculture Handbook 436. 754p.

USGS: U.S. Geological Survey (1958) *Suggestions to authors of the reports of the United States Geological Survey.* 5th ed. Washington, D.C.: U.S. Government Printing Office. 255p.

USGS: U.S. Geological Survey (1970) *Geologic time—The age of the Earth.* Washington, D.C.: U.S. Government Printing Office. 20p.

U.S. Naval Oceanographic Office (1968) *Ice observations.* 2nd ed. Washington: U.S. Naval Oceanographic Office. 42p. (H.O. Publication no.606-d).

V

Valentine, J.W. (1969) *Patterns of taxonomic and ecological structure of the shelf benthos during Phanerozoic time.* Palaeontology, v.12, pt.4, p.684-709.

Van Bemmelen, R.W. (1932) *De undatie theorie.* Natuurkundig Tijdschrift voor Nederlandsch-Indië, v.92, p.85-242, 373-402.

Van Bemmelen, R.W. (1933) *The undation theory of the development of the Earth's crust.* International Geological Congress. 16th, Washington, D.C. Proceedings, v.2, p.965-982, [1935].

Van Bemmelen, R.W. (1949) *The geology of Indonesia,* vol.1A. The Hague: Government Printing Office. 732p.

Van der Gracht, W.A.J.M. van Waterschoot (1931) *The Permo-Carboniferous orogeny in the south-central United States.* Koninklijke Akademie van Wetenschappen, Amsterdam. Afdeeling natuurkunde Verhandlingen, deel 27, no.3.

Van der Hammen, T., Wijmstra, T.A., and Zagwijn, W.H. (1971) *The floral record of the Late Cenozoic of Europe.* In: Turekian, K.K., ed. *Late Cenozoic glacial ages,* p.329-424. New Haven: Yale University Press.

Van Hise, C.R. (1896) *Principles of North American pre-Cambrian geology.* U.S. Geological Survey. Annual Report, 16th, pt.1, p.571-843.

Van Hise, C.R. (1904) *A treatise on metamorphism.* U.S. Geological Survey. Monograph 47. 1286p.

Van Hise, C.R., and Leith, C.K. (1911) *The geology of the Lake Superior region.* U.S. Geological Survey. Monograph 52. 641p.

Van Riper, J.E. (1962) *Man's physical world.* New York: McGraw-Hill. 637p.

Vanserg, Nicholas (1952) *How to write geologese.* Economic Geology, v.52, p.220-223.

Van Straaten, L.M.J.U. (1951) *Longitudinal ripple marks in mud and sand.* Journal of Sedimentary Petrology, v.21, p.47-54.

Van Straaten, L.M.J.U. (1953) *Megaripples in the Dutch Wadden Sea and in the basin of Arcachon (France).* Geologie en Mijnbouw, n.s., v.15, p.1-11.

Van Straaten, L.M.J.U. (1953a) *Rhythmic patterns on Dutch North Sea beaches.* Geologie en Mijnbouw, n.s., v.15, p.31-43.

Van Tuyl, F.M. (1916) *The origin of dolomite.* Iowa Geological Survey. Volume 25, p.251-421.

Varnes, D.J. (1958) *Landslide types and processes.* National Research Council. Highway Research Board. Special Report 29, p.20-47.

Varney, W.D. (1921) *The geological history of the Pewsey Vale.* Geologist' Association. Proceedings, v.32, p.189-205.

Vassoevich, N.B. (1948) *Evolyutsiya predstavlenii o geologicheskikh fatsiyakh.* Leningrad: Gostoptekhizdat. (Vsesouiznyi Neftianoi Nauchno-Issledovatel'skiy Geologo-Rezvedochnyi Institut. Litologicheskii sbornik, no.1).

Vassoevich, N.B. (1953) *O nekotorykh flishevykh teksturakh (znakakh).* L'vovskoe Geologicheskoe Obshchestvo. Trudy; geologicheskaia seriia, no.3, p.17-85.

Vassoevich, N.B. (1965) *Vernadskiy's views on the origin of oil.* International Geology Review, v.7, p.507-517.

Veatch, J.O., and Humphrys, C.R. (1966) *Water and water use terminology.* Kaukauna, Wisc.: Thomas Printing & Publishing Co. 381p.

Vella, P. (1964) *Biostratigraphic units.* New Zealand Journal of Geology and Geophysics, v.7, p.615-625.

Vening Meinesz, F.A. (1955) *Plastic buckling of the Earth's crust: the origin of geosynclines.* Geological Society of America. Special Paper 62, p.319-330.

Vernon, R.H. (1974) *Controls of mylonitic compositional layering during noncataclastic ductile deformation.* Geological Magazine, v.111, p.121-123.

Vernon, R.H. (1975) *Metamorphic processes, reactions and microstructure development.* New York: Wiley. 247p.

Viljoen, M.J., and Viljoen, R.P. (1969) *The geology and geochemistry of the lower ultramafic unit of the Onverwacht Group and a proposed new class of igneous rock.* Geological Society of South Africa. Special publication, *Upper Mantle Project,* v.2, p.221-244.

Vistelius, A.B. (1967) *Studies in mathematical geology.* New York: Consultants Bureau. 294p.

Vitaliano, D.B. (1968) *Geomythology: the impact of geologic events on history and legend, with special reference to Atlantis.* Indiana University. Journal of the Folklore Institute, v.5, p.5-30.

Vitaliano, D.B. (1973) *Legends of the earth: their geologic origins.* Bloomington: Indiana University Press. 305p.

Vogt, J.H.L. (1905) *Über anchi-eutektische und anchi-monomineralische Eruptivgesteine.* Norsk geologisk tidsskrift, v.1, no.1, paper 2. 33p.

Von Bernewitz, M.W. (1931) *Handbook for prospectors.* 2d ed. New York: McGraw-Hill.

Von Engeln, O.D. (1942) *Geomorphology; systematic and regional.* New York: Macmillan. 655p.

W

Wadell, H.A. (1932) *Volume, shape, and roundness of rock particles.* Journal of Geology, v.40, p.443-451.

Wadell, H.A. (1934) *Shape determinations of large sedimental rock fragments.* Pan-American Geologist, v.61, p.187-220.

Wadsworth, M.E. (1893) *A sketch of the geology of the iron, gold, and copper districts of Michigan.* Michigan Geological Survey. Report of the State Geologist for 1891-1892, p.75-174.

Wager, L.R. (1968) *Rhythmic and cryptic layering in mafic and ultramafic plutons.* In: Hess, H.H., and Poldervaart, Arie, eds. *Basalts: the Poldervaart treatise on rocks of basaltic composition,* v.2, p.483-862. New York: Interscience.

Wager, L.R., and Brown, G.M. (1967) *Layered igneous rocks.* San Francisco: Freeman. 588p.

Wager, L.R., and Deer, W.A. (1939) *The petrology of the Skaergaard intrusion, Kangerdlugssuaq, east Greenland.* Meddelelser om Grönland, v.105, no.4. 352p. (*Geological investigations in east Greenland,* pt.3.)

Wahlstrom, E.E. (1948) *Optical crystallography.* New York: Wiley. 206p.

Wahlstrom, E.E. (1969) *Optical crystallography,* 4th ed. New York: Wiley. 489p.

Walcott, R.H. (1898) *The occurrence of so-called obsidian bombs in Australia.* Royal Society of Victoria. Proceedings, n.s., v.11, pt.1, p.23-53.

Walker, Frederick (1957) *Ophitic texture and basaltic crystallization.* Journal of Geology, v.65, no.1, p.1-14.

Walker, K.R. (1973) *Major Middle Ordovician reef tract in east Tennessee.* American Journal of Science, v.273-A (Cooper volume), p.294-325.

Walker, T.L. (1902) *The geology of Kalahandi state, central provinces.* India, Geological Survey. Memoirs, v.33, part 3.

Wallace, R.C. (1913) *Pseudobrecciation in Ordovician limestones in Manitoba.* Journal of Geology, v.21, p.402-421.

Walter, M.R. (1976) *Stromatolites.* Amsterdam: Elsevier. 790p. (*Developments in sedimentology,* 20.)

Walther, Johannes (1893-1894) *Einleitung in die Geologie als historische Wissenschaft; Beobachtungen über die Bildung der Gesteine und ihrer organischen Einschlusse.* Jena: G. Fischer. 1055p.

Walton, E.K. (1956) *Limitations of graded bedding: and alternative* *criteria of upward sequence in the rocks of the Southern Uplands.* Edinburgh Geological Society. Transactions, v.16, p.262-271.

Walton, John (1940) *An introduction to the study of fossil plants.* London: Adam & Charles Black. 188p.

Wanless, H.R., and Weller, J.M. (1932) *Correlation and extent of Pennsylvanian cyclothems.* Geological Society of America. Bulletin, v.43, p.1003-1016.

Ward, F.K. (1923) *From the Yangtze to the Irrawaddy.* Geographical Journal, v.62, p.6-20.

Ward, W.H. (1953) *Glacier bands; conference on terminology.* Journal of Glaciology, v.2, no.13, p.229-232.

Warntz, William (1975) *Stream ordering and contour mapping.* Journal of Hydrology, v.25, p.209-227.

Washburn, A.L. (1947) *Reconnaissance geology of portions of Victoria Island and adjacent regions, arctic Canada.* Geological Society of America. Memoir 22. 142p.

Washburn, A.L. (1950) *Patterned ground.* Revue Canadienne Géographie, v.4, no.3-4, p.5-59.

Washburn, A.L. (1956) *Classification of patterned ground and review of suggested origins.* Geological Society of America. Bulletin, v.67, p.823-865.

Washburn, A.L. (1973) *Periglacial processes and environments.* London: Edward Arnold. 320p.

Washburn, A.L., and Goldthwait, R.P. (1958) *Slushflows* (abstract). Geological Society of America. Bulletin, v.69, p.1657-1658.

Washburne, C.W. (1943) *Some wrong words.* Journal of Geology, v.51, p.495-497.

Waters, A.C., and Campbell, C.D. (1935) *Mylonites from the San Andreas fault zone.* American Journal of Science, 5th ser., v.29, no.174, p.473-503.

Waterston, C.D. (1975) *Gill structures in the Lower Devonian eurypterid* Tarsopterella scotica. Fossils and Strata, no.4, p.241-254.

Wayland, E.J. (1920) *Some facts and theories relating to the geology of Uganda.* Uganda Geological Dept. Pamphlet no.1. 52p.

Wayland, E.J. (1934) *Peneplains and some other erosional platforms.* Uganda Geological Survey Dept. Annual Report and Bulletin, 1933, Bull.no.1, p.77-79.

Weast, R.C., ed. (1970) *Handbook of chemistry and physics.* 51st ed. Cleveland: The Chemical Rubber Co.

Weatherley, A.H., ed. (1967) *Australian inland waters and their fauna; eleven studies.* Canberra: Australian National University Press. 287p.

Weaver, J.E., and Clements, F.E. (1938) *Plant ecology.* 2nd ed. New York: McGraw-Hill. 601p.

Webb, R.W. (1936) *Kern Canyon Fault, southern Sierra Nevada.* Journal of Geology, v.44, p.631-638.

***Webster's third new international dictionary of the English language; unabridged* (1967)** P.B. Gove, ed. Springfield, Mass.: G. & C. Merriam. 2662p.

Weeks, L.G. (1952) *Factors of sedimentary basin development that control oil occurrence.* American Association of Petroleum Geologists. Bulletin, v.36, p.2071-2124.

Weeks, L.G. (1959) *Geologic architecture of circum-Pacific.* American Association of Petroleum Geologists. Bulletin, v.43, p.350-380.

Wegener, Alfred (1912) *The origin of the continents and oceans.* English translation, 1924. New York: Dutton.

Welch, P.S. (1952) *Limnology.* 2nd ed. New York: McGraw-Hill. 538p.

Weld, L.D., ed. (1937) *Glossary of physics.* 1st ed. New York: McGraw-Hill. 255p.

Weller, J.M. (1930) *Cyclical sedimentation of the Pennsylvanian Period and its significance.* Journal of Geology, v.38, p.97-135.

Weller, J.M. (1956) *Argument for diastrophic control of late Paleozoic cyclothems.* American Association of Petroleum Geologists. Bulletin, v.40, p.17-50.

Weller, J.M. (1958) *Stratigraphic facies differentiation and nomenclature.* American Association of Petroleum Geologists. Bulletin, v.42, p.609-639.

Weller, J.M. (1958a) *Cyclothems and larger sedimentary cycles of the Pennsylvanian.* Journal of Geology, v.66, p.195-207.

Weller, J.M. (1960) *Stratigraphic principles and practice.* New York: Harper. 725p.

Weller, J.M., and others (1942) *Stratigraphy of the fusuline-bearing beds of Illinois.* Illinois State Geological Survey. Bulletin 67, p.9-34.

Wellman, H.W., and Wilson, A.T. (1965) *Salt weathering, a neglected geological erosive agent in coastal and arid environments.* Nature, v.205, p.1097-1098.

Wells, F.G. (1949) *Ensimatic and ensialic geosynclines* (abstract). Geological Society of America. Bulletin, v.60, p.1927.

Wells, J.W. (1944) *Middle Devonian bone beds of Ohio.* Geological Society of America. Bulletin, v.55, p.273-302.

Wells, J.W. (1947) *Provisional paleoecological analysis of the Devonian rocks of the Columbus region.* Ohio Journal of Science, v.47, p.119-126.

Wentworth, C.K. (1922) *A scale of grade and class terms for clastic sediments.* Journal of Geology, v.30, p.377-392.

Wentworth, C.K. (1922a) *A method of measuring and plotting the shapes of pebbles.* U.S. Geological Survey. Bulletin 730-C, p.91-114.

Wentworth, C.K. (1922b) *The shapes of beach pebbles.* U.S. Geological Survey. Professional Paper 131-C, p.75-83.

Wentworth, C.K. (1925) *Chink-faceting; a new process of pebble-shaping.* Journal of Geology, v.33, p.260-267.

Wentworth, C.K. (1931) *Pebble wear on the Jarvis Island beach.* Washington University Studies, n.s., Science and Technology, no.5, p.11-37.

Wentworth, C.K. (1935) *The terminology of coarse sediments.* With notes by P.G.H. Boswell. National Research Council. Division of Geology and Geography. Committee on Sedimentation. Report for 1932-1934, p.225-246. (National Research Council. Bulletin, no.98).

Wentworth, C.K. (1936) *An analysis of the shapes of glacial cobbles.* Journal of Sedimentary Petrology, v.6, p.85-96.

Wentworth, C.K. (1939) *Marine bench-forming processes: II, Solution benching.* Journal of Geomorphology, v.2, p.3-25.

Wentworth, C.K. (1943) *Soil avalanches on Oahu, Hawaii.* Geological Society of America. Bulletin, v.54, p.53-64.

Wentworth, C.K., and Williams, Howel (1932) *The classification and terminology of the pyroclastic rocks.* National Research Council. Division of Geology and Geography. Committee on Sedimentation. Report for 1930-1932, p.19-53. (National Research Council. Bulletin, no.89)

Wentworth, C.M. (1967) *Dish structure, a primary sedimentary structure in coarse turbidites.* American Association of Petroleum Geologists. Bulletin, v.51, p.485.

Wheeler, H.E. (1958) *Time-stratigraphy.* American Association of Petroleum Geologists. Bulletin, v.42, p.1047-1063.

Wheeler, H.E. (1958a) *Primary factors in biostratigraphy.* American Association of Petroleum Geologists. Bulletin, v.42, p.640-655.

Wheeler, H.E. (1964) *Baselevel, lithosphere surface, and time-stratigraphy.* Geological Society of America. Bulletin, v.75, p.599-609.

Wheeler, H.E., and Beesley, E.M. (1948) *Critique of the time-stratigraphic concept.* Geological Society of America. Bulletin, v.59, p.75-85.

Wheeler, H.E., and Mallory, V.S. (1953) *Designation of stratigraphic units.* American Association of Petroleum Geologists. Bulletin, v.37, p.2407-2421.

Wheeler, H.E., and Mallory, V.S. (1956) *Factors in lithostratigraphy.* American Association of Petroleum Geologists. Bulletin, v.40, p.2711-2723.

Wheeler, H.E., and others (1950) *Stratigraphic classification.* American Association of Petroleum Geologists. Bulletin, v.34, p.2361-2365.

Whitaker, J.H.McD. (1973) *"Gutter casts", a new name for scour-and-fill structures: with examples from Llandoverian and Rjngerike and Malmöya, southern Norway.* Norsk Geologisk Tidsskrift, v.53, p.403-417.

White, C.A. (1870) *Report on the geological survey of the State of Iowa.* Vol.1. Des Moines: Mills & Co. 391p.

White, David (1915) *Some relations in origin between coal and petroleum.* Washington Academy of Sciences. Journal, v.5, no.6, p.189-212.

White, D.E. (1957) *Thermal waters of volcanic origin.* Geological Society of America. Bulletin, v.68, p.1637-1657.

White, S.E. (1967) *Rockfall, alluvial, and avalanche talus in the Colorado Front Range* (abstract). Geological Society of America. Special Paper 115, p.237.

White, S.E. (1972) *Alpine subnival boulder pavements in Colorado Front Range.* Geological Society of America. Bulletin, v.83, p.195-200.

White, S.E. (1976) *Rock glaciers and block fields, review and new data.* Quaternary Research, v.6, p.77-97.

White, W.H. (1959) *Cordilleran tectonics in British Columbia.* American Association of Petroleum Geologists. Bulletin, v.43, p.60-100.

Whitney, J.D. (1888) *Names and places; studies in geographical and topographical nomenclature.* Cambridge, England: Cambridge University Press. 239p.

Whitten, D.G.A., with Brooks, J.R.V. (1972) *The Penguin dictionary of geology.* Harmondsworth, England: Penguin Books. 495p.

Whitten, E.H.T. (1959) *A study of two directions of folding: the structural geology of the Monadhliath and mid-Strathspey.* Journal of Geology, v.67, p.14-47.

Whitten, E.H.T. (1966) *Structural geology of folded rocks.* Chicago: Rand McNally. 663p.

Whittington, H.B., and Rickards, R.B. (1968) *New tuboid graptolite from the Ordovician of Ontario.* Journal of Paleontology, v.42, p.61-69.

Wickman, F.E. (1966) *Repose period patterns of volcanoes. Part 1: Volcanic eruptions regarded as random phenomena.* Arkiv for Mineralogi och Geologi, v.4, no.7, p.291-301.

Wiegel, R.L. (1953) *Waves, tides, currents and beaches: glossary of terms and list of standard symbols.* Berkeley, Calif.: Council on Wave Research, The Engineering Foundation. 113p.

Wien, Willy (1894) *Temperatur und Entropie der Strahlung.* Annalen der Physik, ser.2, v.52, p.132-165.

Wilkinson, J.F.G. (1968) *The petrography of basaltic rocks.* In: Hess, H.H., and Poldervaart, A., eds. *Basalts: the Poldervaart treatise on rocks of basaltic composition,* v.1, p.163-214. New York: Interscience.

Willard, Bradford (1930) *Conglomerite, a new rock term.* Science, v.71, p.438.

Williams, Alwyn (1970) *Origin of laminar-shelled articulate brachiopods.* Lethaia, v.3, p.329-342.

Williams, B.J., and Prentice, J.E. (1957) *Slump-structures in the Ludlovian rocks of North Herefordshire.* Geologists' Association. Proceedings, v.68, p.286-293.

Williams, Howel (1932) *The history and character of volcanic domes.* University of California, Department of Geological Sciences. Bulletin, v.21, no.5, p.51-146.

Williams, Howel (1941) *Calderas and their origin.* University of California, Department of Geological Sciences. Bulletin, v.25, no.6, p.239-346.

Williams, Howel, Turner, F.J., and Gilbert, C.M. (1954) *Petrography—an introduction to the study of rocks in thin sections.* San Francisco: Freeman. 406p.

Williams, H.R., and Meyers, C.J. (1964) *Oil and gas terms; annotated manual of legal, engineering, tax words and phrases.* San Francisco: Matthew Bender. 449p.

Williams, H.S. (1893) *The elements of the geological time-scale.* Journal of Geology, v.1, p.283-295.

Williams, H.S. (1895) *Geological biology; an introduction to the geological history of organisms.* New York: Henry Holt. 395p.

Williams, H.S. (1901) *The discrimination of time-values in geology.* Journal of Geology, v.9, p.570-585.

Williams, M.Y. (1936) *Frost circles.* Royal Society of Canada. Transactions, ser.3, v.30, p.129-132.

Williams, R.S., jr. (1972) *Thermography.* Photogrammetric Engineering, v.38, p.881-883.

Williamson, I.A. (1961) *Spring domes developed in limestone.* Journal of Sedimentary Petrology, v.31, p.288-291.

Willis, Bailey (1893) *The mechanics of Appalachian structure.* U.S. Geological Survey. Annual Report, 13th, pt.2, p.211-281.

Willis, Bailey (1903) *Physiographgy and deformation of the Wenatchee-Chelan district, Cascade Range.* U.S. Geological Survey. Professional Paper 19, p.41-97.

Willis, Bailey (1907) *Geographic history of Potomac River.* U.S. Geological Survey. Water-Supply Paper 192, p.7-22.

Willis, Bailey (1928) *Dead Sea problem: rift valley or ramp valley?* Geological Society of America. Bulletin, v.39, p.490-542.

Willis, Bailey (1938) *Asthenolith (melting spot) theory.* Geological Society of America. Bulletin, v.49, p.603-614.

Willman, H.B., and others (1942) *Geology and mineral resources of the Marseilles, Ottawa, and Streator quadrangles.* Illinois State Geological Survey. Bulletin 66. 388p.

Wills, L.J. (1956) *Concealed coalfields; a palaeogeographical study of the stratigraphy and tectonics of mid-England in relation to coal reserves.* London: Blackie. 208p.

Wilmarth, M.G. (1938) *Lexicon of geologic names of the United States (including Alaska).* Parts 1-2. U.S. Geological Survey. Bulletin 896. 2396p.

Wilson, E.D., and Bossert, W.H. (1971) *A primer of population biology.* Stamford, Conn.: Sinauer Associates. 192p.

Wilson, Gilbert (1953) *Mullion and rodding structures in the Moine series of Scotland.* Geologists' Association. Proceedings, v.64, p.118-151.

Wilson, Gilbert (1961) *The tectonic significance of small scale structures and their importance to the geologist in the field.* Société Géologique de Belge. Annales, tome 84, p.423-548.

Wilson, J.A. (1959) *Transfer, a synthesis of stratigraphic processes.* American Association of Petroleum Geologists. Bulletin, v.43, p.2861-2862.

Wilson, J.A. (1971) *Note 39—records of the Stratigraphic Commission for 1968-1970.* American Association of Petroleum Geologists. Bulletin, v.55, p.1866-1872.

Wilson, J.T. (1950) *An analysis of the pattern and possible cause of young mountain ranges and island arcs.* Geological Association of Canada. Proceedings, v.3, p.141-166.

Wilson, J.T. (1965) *A new class of faults and their bearing on continental drift.* Nature, v.207, p.343-347.

Wilson, J.T. (1968) *Static or mobile earth.* In: *Gondwanaland revisited: New evidence for continental drift.* American Philosophical Society. Proceedings, v.112, p.309-320.

Wilson, L. (1976) *Explosive volcanic eruptions; III. Plinian eruption columns.* Royal Astronomical Society. Geophysical Journal, v.45, p.543-556.

Winchell, N.H., and Winchell, H.V. (1891) *The iron ores of Minnesota, their geology, discovery, development, qualities and origin, and comparison with those of other iron districts.* Minnesota Geological and Natural History Survey. Bulletin no.6. 430p.

Winkler, E.M., and Wilhelm, E.J. (1970) *Salt burst by hydration pressures in architectural stone in urban atmosphere.* Geological Society of America. Bulletin, v.81, p.567-572.

Winkler, H.G.F. (1967) *Petrogenesis of metamorphic rocks.* 2d ed. New York: Springer-Verlag. 237p.

Winsauer, W.O., and others (1952) *Resistivity of brine saturated sands in relation to pore geometry.* American Association of Petroleum Geologists. Bulletin, v.36, p.253-277.

Wise, Sherwood, jr. (1970) *Microarchitecture and mode of formation of nacre (mother-of-pearl) in pelecypods, gastropods, and cephalopods.* Eclogae Geologicae Helvetiae, v.63, p.775-797.

Woldenberg, M.J. (1966) *Horton's laws justified in terms of allometric growth and steady state in open systems.* Geological Society of America. Bulletin, v.77, p.431-434.

Wolf, K.H. (1960) *Simplified limestone classification.* American Association of Petroleum Geologists. Bulletin, v.44, p.1414-1416.

Wolf, K.H. (1965) *Littoral environment indicated by open-space structures in algal limestones.* Palaeogeography, Palaeoclimatology, Palaeoecology, v.1, p.183-223.

Wolf, K.H., ed. (1976) *Handbook of strata-bound and stratiform ore deposits,* v.1-7. Amsterdam: Elsevier. 2863p.

Wolfe, C.W., and others (1966) *Earth and space science.* Boston: Heath. 630p.

Wood, Alan (1935) *The origin of the structure known as guilielmites.* Geological Magazine, v.72, p.241-245.

Wood, Alan (1941) *"Algal dust" and the finer-grained varieties of Carboniferous limestone.* Geological Magazine, v.78, p.192-200.

Wood, Alan (1942) *The development of hillside slopes.* Geologists' Association. Proceedings, v.53, p.128-138.

Wood, Alan, and Smith, A.J. (1958) *The sedimentation and sedimentary history of the Aberystwyth Grits (upper Llandoverian).* Geological Society of London. Quarterly Journal, v.114, p.163-195.

Wood, W.F., and Snell, J.B. (1960) *A quantitative system for classifying land forms.* Natick, Mass.: U.S. Army Natick Laboratory. Technical Report EP-124. 20p.

Woodford, A.O. (1925) *The San Onofre Breccia; its nature and origin.* University of California. Department of Geological Sciences: Bulletin, v.15, no.7, p.159-280.

Woodford, A.O. (1951) *Stream gradients and Monterey sea valley.* Geological Society of America. Bulletin, v.62, p.799-851.

Woodward, H.B. (1887) *Geology of England and Wales: with notes on the physical features of the country.* 2d ed. London: G. Philip. 670p.

Woodward, H.B. (1894) *The Jurassic rocks of Britain.* Vol.4. London: Her Majesty's Stationery Office. 628p. (Great Britain. Geological Survey. Memoir).

Woodworth, J.B. (1894a) *Postglacial eolian action in southern New England.* American Journal of Science, ser.3, v.47, p.63-71.

Woodworth, J.B. (1894b) *Some typical eskers of southern New England.* Boston Society of Natural History. Proceedings, v.26, p.197-220.

Woodworth, J.B. (1901) *Pleistocene geology of portions of Nassau County and Borough of Queens.* New York State Museum. Bulletin, v.48, p.618-670.

Woodworth, J.B. (1912) *Geological expedition to Brazil and Chile, 1908-1909.* Harvard College. Museum of Comparative Zoology. Bulletin, v.56, no.1. 137p.

Woollacott, R.M., and Zimmer, R.L. (1972) *Origin and structure of the brood chamber in* Bugula meritina *(Bryozoa).* Marine Biology, v.16, p.165-170.

Woolnough, W.G. (1910) *Stone rolls in the Bulli coal seam of New South Wales.* Royal Society of New South Wales. Proceedings, v.44, p.334-340.

Woolnough, W.G. (1927) *The duricrust of Australia.* Royal Society of New South Wales. Journal and Proceedings, v.61, p.24-53.

Worcester, P.G. (1939) *A texbook of geomorphology.* New York: Van Nostrand. 565p.

Workman, W.H. (1914) *Nieve penitente and allied formations in Himalaya, or surface-forms of névé and ice created or modelled by melting.* Zeitschrift für Gletscherkunde, Bd. 8, p.289-330.

Wright, J.K. (1944) *The terminology of certain map symbols.* Geographical Review, v.34, p.653-654.

Wright, J.K. (1947) *Terrae incognitae: the place of the imagination in geography.* Association of American Geographers. Annals, v.37, p.1-15.

Wright, W.B. (1914) *The Quaternary ice age.* London: Macmillan. 464p.

Wright, W.B. (1926) *Stratigraphical diachronism in the Millstone Grit of Lancashire.* British Association for the Advancement of Science. Report, 94th, p.354-355.

Wulff, G.V. von (1902) *Untersuchungen im Gebiete der Optischen Eigenschaften isomorpher Krystalle.* Zeitschrift für Krystallographie und Mineralogie, Bd.36, p.1-28.

Wyckoff, R.D. (1941) *The Gulf gravimeter.* Geophysics, v.6, p.13-33.

Wyllie, M.R.J. (1957) *The fundamentals of well log interpretation,* 2nd ed. New York: Academic Press. 176p.

Wyllie, M.R.J., Gregory, A.R., and Gardner, L.W. (1956) *Elastic wave velocities in heterogeneous and porous media.* Geophysics, v.21, p.41-70.

Wyllie, P.J. (1966) *kExperimental petrology: an indoor approach to an outdoor subject.* Journal of Geological Education, v.14, p.93-97.

Wynne-Edwards, H.R. (1957) *Structure of the Westport concordant pluton in the Grenville, Ontario.* Journal of Geology, v.65, p.639-649.

Y

Yaalon, D.H. (1965) *Microminerals and micromineralogy.* Clay Minerals, v.6, p.71.

Yasso, W.E. (1966) *Heavy mineral concentration and sastrugi-like deflation furrows in a beach salcrete at Rockaway Point, New York.* Journal of Sedimentary Petrology, v.36, p.836-838.

Yoder, H.S., jr., and Tilley, C.E. (1962) *Origin of basalt magmas: an experimental study of natural and synthetic rock systems.* Journal of Petrology, v.3, p.342-532.

Yokoyama, Izumi (1956-57) *Energetics and active volcanoes.* Tokyo University. Earthquake Research Institute. Bulletin, v.35, pt.1, p.75-97.

Youell, R.F. (1960) *An electrolytic method for producing chlorite-like substances from montmorillonite.* Clay Minerals Bulletin, v.4, p.191-195.

Young, A.P. (1910) *On the glaciation of the Navis Valley in North Tirol.* Geological Magazine, v.7, p.244-258.

Z

Zankl, H., and Multer, H.G. (1977) *Origin of some internal fabrics in Holocene reef rocks, St. Croix, U.S. Virgin Islands.* Third International Coral Reef Symposium. Proceedings, v.2, p.127-133. Miami: University of Miami, Rosenstiel School of Marine and Atmospheric Science. 628p.

Zemansky, M.W. (1957) *Heat and thermodynamics.* New York: McGraw-Hill. 484p.

Zen, E-an (1966) *Construction of pressure-temperature diagrams for multicomponent systems after the method of Schreinemakers—a geometric approach.* U.S. Geological Survey. Bulletin 1225. 56p.

Zernitz, E.R. (1932) *Drainage patterns and their significance.* Journal of Geology, v.40, p.498-521.

Zingg, Theodor (1935) *Beitrag zur Schotteranalyse; die Schotteranalyse und ihre Anwendung auf die Glattalschotter.* Schweizerische mineralogische und petrographische Mitteilungen, Bd.15, p.39-140.

Zirkel, Ferdinand (1866) *Lehrbuch der Petrographie.* Bd.1. Bonn: Adolph Marcus. 607p.

Zirkel, Ferdinand (1876) *Microscopical petrography.* U.S. Geological Exploration of the Fortieth Parallel. Report, v.6, 297p. (U.S. Army. Engineer Dept. Professional Paper, no.18).

Zirkel, Ferdinand (1893) *Lehrbuch der Petrographie.* Bd.1. 2nd ed. Leipzig: Wilhelm Engelmann. 845p.

Zischinsky, Ulf (1969) *Über Sackungen.* Rock Mechanics, v.1, p.30-52.

Zittel, K.A. von (1901) *History of geology and palaeontology to the end of the nineteenth century.* Translated by M.M. Ogilvie-Gordon. London: Walter Scott. 562p.

Addendum

Barton, D. C. (1916) *The geological significance and classification of arkose deposits.* Journal of Geology, v. 24, p. 417–449.

Bertrand, Marcel (1892) *Les récents progrès de nos connaissances orogéniques.* Revue Générale des Sciences. Tome 3, no. 1, p. 5–12.

Clarke, F. W. (1908) *The data of geochemistry.* U.S. Geological Survey Bulletin 330. 716 p.

Cleland, H. F. (1925) *Geology, physical and historical.* New York: American Book Co. 718 p.

Fleischer, Michael (1966) *Index of new mineral names, discredited minerals, and changes of mineralogical nomenclature in volumes 1–50 of* The American Mineralogist. American Mineralogist, v. 51, p. 1248–1357.

Friedman, G. M., and Sanders, J. E. (1978) *Principles of sedimentology.* New York: Wiley. 792 p.

Gussow, W. C. (1962) *Energy source of intrusive masses.* Royal Society of Canada. Transactions, v. 56, series 3, sec. 3, p. 1–19.

Johannsen, Albert (1932) *A descriptive petrography of the igneous rocks,* v. 2. Chicago: University of Chicago Press. 428 p.

Krenkel, Erich (1922) *Die Bruchzonen Ostafrikas.* Berlin: Gebrüder Borntraeger.

McDonald, B. C., and Banerjee, Indranil (1971) *Sediments and bed forms on a braided outwash plain.* Canadian Journal of Earth Sciences, v. 8, p. 1282–1301.

Odum, E. P. (1959) *Fundamentals of ecology.* Philadelphia: W. B. Saunders.

Osborne, F. F. (1956) *The Grenville region of Quebec.* Royal Society of Canada. Special Publication 1, *The Grenville Problem,* p. 3–13. Toronto: University of Toronto Press. 119 p.

Otvos, E. G., jr. (1965) *Types of rhomboid beach surface patterns.* American Journal of Science, v. 263, p. 271–276.

Schardt, Hans (1893) *Sur l'origine des Alpes du Chablais et du Stockhorn, en Savoie et en Suisse.* Comptes Rendus des Séances de l'Academie des Sciences, Tome 117, no. 21, p. 707–709.

Stanley, D. J., and Swift, D. J. P., eds. (1976) *Marine sediment transport and environmental management.* New York: Wiley. 602 p.

Stearns, N. D. (1927) *Report on the geology and ground-water hydrology of the experimental area of the United States Public Health Service at Fort Caswell, North Carolina.* U.S. Hygienic Laboratory Bulletin 147, p. 137–168.

Stille, Hans (1936) *Wege und Ergebnisse der geologisch-tektonischen Forschung.* Berlin: 25 Jahre Kaiser Wilhelm-Gesellschaft zur Förderung der Wissenschaft, v. 2, p. 77–97.

U.S. Naval Oceanographic Office (1966) *Glossary of oceanographic terms.* 2nd ed. U.S. Naval Oceanographic Office, Special Publication 35. 204 p.

Glossary of Geology Second Edition

Composed in Century Schoolbook by
International Computaprint Corporation.

Printed offset on Perkins & Squier Offset
by Haddon Craftsmen, Book Manufacturing.

Bound in Holliston Roxite by
Haddon Craftsmen, Book Manufacturing.

Library of Congress Cataloging in Publication Data

Bates, Robert Latimer, 1912–
Glossary of geology.

First ed. (c1972) by M. Gary, R. McAfee, Jr., and C. L. Wolf.
Bibliography: 35 p.
1. Geology—Dictionaries. I. Jackson, Julia A., 1939– joint author. II. Gary, Margaret. Glossary of geology. III. Title.
QE5.G37 1980 550'.3 79–57360
ISBN 0–913312–15–0